P9-ELV-388

BIOCHEMISTRY

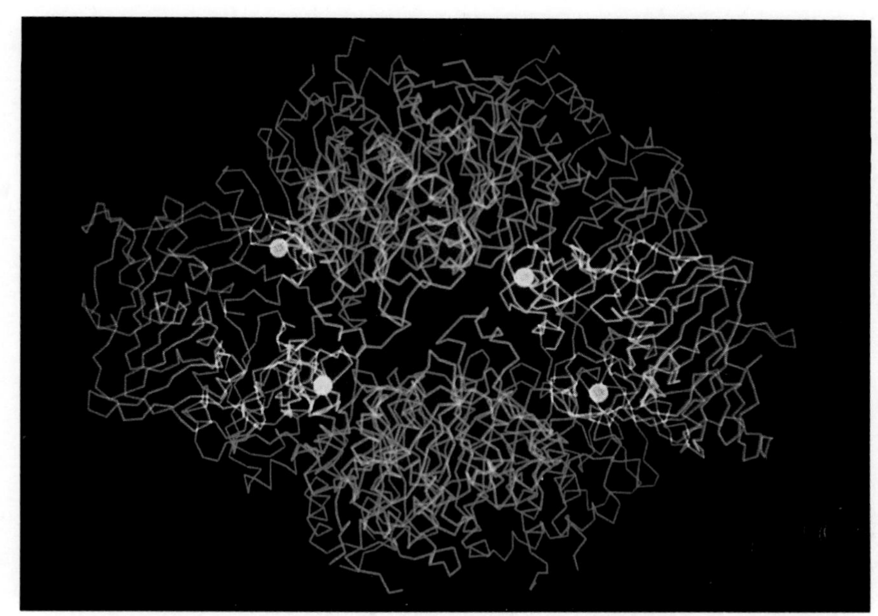

OF RELATED INTEREST FROM THE BENJAMIN/CUMMINGS SERIES IN LIFE SCIENCES AND CHEMISTRY

W. M. Becker, and D. W. Deamer
The World of the Cell, second edition (1991)

R. F. Boyer, and E. L. Uhlenhopp
Modern Experimental Biochemistry, second edition (1991)

N. A. Campbell
Biology, second edition (1990)

I. D. Campbell, and R. A. Dwek
Biological Spectroscopy (1984)

J. Chinnici, J. Frankel, and A. Sherald
Genetics (1991)

R. E. Dickerson, and I. Geis
Hemoglobin: Structure, Function, Evolution, and Pathology (1983)

D. S. Eisenberg, and D. M. Crothers
Physical Chemistry with Applications to the Life Sciences (1979)

P. B. Hackett, J. A. Fuchs, and J. W. Messing
An Introduction to Recombinant DNA Techniques: Basic Experiments in Gene Manipulation, second edition (1988)

L. E. Hood, I. L. Weissman, W. B. Wood, and J. H. Wilson
Immunology, second edition (1984)

W. Heidcamp
Experimental Cell Biology (1991)

G. M. Loudon
Organic Chemistry, second edition (1988)

R. Schleif
Genetics and Molecular Biology (1986)

J. D. Watson, N. H. Hopkins, J. W. Roberts, J. A. Steitz, and A. M. Weiner
Molecular Biology of the Gene, fourth edition, Volumes I and II (1987)

W. B. Wood, J. H. Wilson, R. M. Benbow, and L. E. Hood
Biochemistry: A Problems Approach, second edition (1981)

BIOCHEMISTRY

Christopher K. Mathews
Oregon State University

K. E. van Holde
Oregon State University

Illustration concepts by Audre W. Newman
with art contributions from Irving Geis

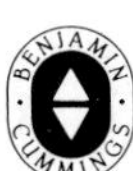

The Benjamin/Cummings Publishing Company, Inc.
Redwood City, California • Fort Collins, Colorado • Menlo Park, California
Reading, Massachusetts • New York • Don Mills, Ontario • Wokingham, U.K.
Amsterdam • Bonn • Sydney • Singapore • Tokyo • Madrid • San Juan

Cover

Dimer of *trp* repressor protein, with bound tryptophan (in blue). The protein binds to DNA and regulates expression of the *trp* genes that control tryptophan biosynthesis. Crystal structure by Paul Sigler et al.; image by Jane and David Richardson.

Frontispiece

Figure 11.15a The T state of aspartate transcarbamoylase, as determined by x-ray diffraction.

Sponsoring Editor: Diane Bowen
Developmental Editor: Susan Weisberg
Production Supervisor: Anne Friedman
Art Coordinator: Bruce Lundquist
Interior and Cover Designer: Gary Head
Dummy Artist: Wendy Calmenson
Copyeditor and Indexer: Mary Prescott
Principal Illustrator: Georg Klatt
Illustrators: Ken Miller, Elizabeth Morales-Denney, and Irene Imfeld
Proofreaders: Sheila Kennedy and Kathy Lee
Photo Research and Permissions: Rachel Menzi
Typesetter: York Graphic Services, Inc.
Film Preparation: Color Response, Inc.

Credits for photographs appear on pages xi–xiii

Library of Congress Cataloging-in-Publication Data

Mathews, Christopher, K., 1937–
Biochemistry/Christopher K. Mathews, K. E. van Holde; illustration concepts by Audre Newman with art contributions from Irving Geis.
p. cm.
Includes bibliographical references.
ISBN 0-8053-5015-2
1. Biochemistry. I. Van Holde, K. E. (Kensal Edward), 1928–
II. Title.
[DNLM: 1. Biochemistry. QU 4 M4294b]
QP514.2.M384 1990
574.192—dc20
DNLM/DLC
for Library of Congress 89-17922
CIP

ISBN: 0-8053-5015-2

BCDEFGHIJ-VH- 93210

The Benjamin/Cummings Publishing Company
390 Bridge Parkway
Redwood City, CA 94065

To our students, who taught us biochemistry
by showing us that the best way to learn a subject is to teach it.

About the Authors

Christopher K. Mathews is Professor and Chairman of the Department of Biochemistry and Biophysics at Oregon State University. He holds a B.A. from Reed College (1958) and a Ph.D. from The University of Washington (1962). Dr. Mathews was an Eleanor Roosevelt International Fellow at the Karolinska Institute in Stockholm in 1984–85. Dr. Mathews has published over 100 scientific papers dealing with molecular virology, metabolic regulation, nucleotide enzymology, and biochemical genetics. He is the author of *Bacteriophage Biochemistry* (1971) and coeditor of *Bacteriophage T4* (1983). His teaching experience includes undergraduate, graduate, and medical school biochemistry courses.

K. E. van Holde is Professor of Biochemistry and Biophysics at Oregon State University. He earned his degrees from the University of Wisconsin (B.S. 1949, Ph.D. 1952). Dr. van Holde's major research interest is the structure of chromatin; his work resulted in the award of an American Cancer Society Research Professorship in 1977. He has been at Oregon State University since 1967, and was named Distinguished Professor in 1988. He is a member of the National Academy of Sciences, and has received Guggenheim, NSF, and EMBO fellowships. He is the author of over 150 scientific papers; the author of two books, *Physical Biochemistry* (1971, 1985) and *Chromatin* (1988); and the coeditor of *The Origins of Life and Evolution* (1981). His teaching experience includes undergraduate and graduate chemistry, biochemistry, and biophysics, and also the physiology and molecular biology course at The Marine Biological Laboratory at Woods Hole.

Preface

The idea for this book was born in late spring 1984, after our department had completed its annual selection of textbooks for our three introductory biochemistry courses. Excellent textbooks by distinguished authors have been available in the past, but nothing seemed ideal for today's teaching situation. Our experience showed us that a contemporary biochemistry text had to meet several important criteria:

1. an overall organization that puts structure and catalysis earliest, followed by intermediary metabolism, and then by genetic biochemistry, yet *introduces* molecular genetics at an early point;

2. an effective balance of the chemistry of biochemistry with its biological context, with examples drawn from medicine, biology, nutrition, and agricultural and environmental science;

3. illustrations that are carefully developed and closely linked to the text so that the two work interdependently to teach biochemistry;

4. emphasis on the essential chemical concepts that underlie biochemistry, even if this necessitates repetition of some material from general, organic, and physical chemistry;

5. presentation of the historical and experimental basis for our current understanding, rather than a recitation of facts and reactions;

6. a style that kindles the student's interest, by emphasizing the unknown and identifying research opportunities for young scientists, as well as recounting and organizing the vast body of current information.

We aimed the book at students taking biochemistry as either advanced undergraduates or beginning graduate students, including those in the health professional schools. Regardless of the course, students using this book should have completed two years of college-level chemistry, one year of physics, and mathematics through calculus. Some background in physical chemistry is desirable as well, but essential concepts such as entropy, free energy, and chemical kinetics are presented here, in a form that the uninitiated should be able to grasp.

We have organized the field along rather traditional lines, having found from our own experience that this organization is the most practical pedagogically. Chemical concepts come first (Chapters 2–3), followed by macromolecular structure (Chapters 4–9), catalysis and design of metabolic pathways (Chapters 10–12) and intermediary metabolism (Chapters 13–23). The book concludes with informational metabolism—nucleic acid and protein synthesis, and the processing of information in multicellular organisms (Chapters 24–29).

Within these broad outlines, however, there are some quite nontraditional features. We include thermodynamic concepts and bioenergetics early, in Chapter 3, because one general theme of the book is to emphasize how energy relationships underlie all biochemical processes. We introduce nucleic acid structure in Chapter 4, before protein structure (Chapters 5–7), to more clearly illustrate how protein chemistry is shaped by the genetic information that directs the synthesis of proteins. Chapters 4, 5, and 7 include also a brief review of nucleic acid and protein synthesis, to aid that small fraction of students who will not have been exposed to this material in previous biology or chemistry courses. This section gives all students a common base of understanding, and makes possible a more cogent discussion of protein evolution than most books can achieve at this point. Further, it allows presentation of intermediary metabolism in a contemporary context—for example, in describing molecular genetic approaches to weed control (Chapter 21) or drug development (Chapter 22). In contrast to some texts, which present separate chapters on biosynthesis and catabolism, we have tried to integrate these processes, to emphasize that metabolism is a continuum and that distinction of synthetic from degradative processes is often artificial. The one significant exception is found in our treatment of carbohydrate metabolism, because the large amount of information that must be assimilated does require a more compartmentalized approach. Our integrative perspective extends to the concept of regulation, which we introduce as it is relevant to specific biochemical processes. In addition, we devote Chapter 23 to a treatment of integration and control of metabolic processes.

A major challenge to biochemistry teachers is keeping pace with the dizzying speed at which our science is advancing. This challenge is especially great for textbook authors, given the five-year gestation period for a book. Fortunately, our editors have been lenient in allowing us to insert new material long after a particular section or chapter had reached a state of "finality." Numerous colleagues, both in our department and elsewhere, have helped us to keep abreast of late-breaking developments in fields peripheral to our own.

One important point is our use of illustrations not simply to repeat information in the text, but as a means to extend the text and to present new information. As authors, we have been blessed with the opportunity to collaborate with a creative and biochemically sophisticated artist—Audre W. Newman, Ph.D. She has functioned essentially as a coauthor in working with us to design novel ways to introduce biochemical concepts. Examples include the recurring metabolic "theme" diagram shown first as Figure 12.1; the operation of "swinging arms" to show the operation of multienzyme complexes in metabolic pathways (Figures 14.9 and 17.23); the processing of glycoprotein carbohydrate chains in different cell compartments (Figure 16.17); the effects of transcription on DNA topology (Figure 26.11); and the anatomy and function of ribosomes (Figure 27.28). We were fortunate also to benefit from collaboration with the distinguished biochemical artist Irving Geis, whose excellent illustrations can be found primarily in the earlier chapters of the book.

Another challenge is ensuring that essential concepts of physical science are thoroughly understood, while at the same time whetting and sustaining the interest of students who are drawn to biochemistry because of the spectacular advances in the field and the opportunities these have created in the biological and medical sciences. We have designed Chapters 2 and 3 to review principles of strong and weak chemical bonding, ionization of weak acids and bases, and bioenergetics, so that these topics are well in hand before presentations of the structures and properties of biomolecules. Because some of that material is review of topics presented in chemistry courses, well-prepared classes can move rapidly over this material.

Because biochemistry is much more an experimental than a theoretical science, comprehension of the major concepts demands understanding of the experimental evidence on which those concepts are based. We have given this principle special emphasis by including the *Tools of Biochemistry* sections. Each of these essays is located at the end of the chapter in which a technique is first introduced and is cross-referenced throughout the book. The Tools describe the background and concept behind each of the major experimental methodologies in biochemistry, and may compensate in some small measure for the fact that many students today do not take a laboratory with their first biochemistry course. We have also included enough experimental-historical treatment in the main text that some or all of the Tools sections can be bypassed if the instructor so desires.

Creation of a textbook by a two-author team is something of a rarity in recent biochemistry texts. We have found the two-author relationship to be particularly rewarding. As colleagues in the same department, with offices just a few steps apart, we have been in close communication at all stages of the book's development. Equally important, we bring to the project complementary orientations and backgrounds—van Holde in structural biochemistry, developmental biology, proteins, physical biochemistry, and eukaryotic gene expression, and Mathews in enzymology, nucleotides and coenzymes, metabolic regulation, virology, genetics, and cell biology. It is our hope that our backgrounds and cooperation have led to a comprehensive, authoritative, and cohesive textbook.

We are grateful to our wives, Kate and Barbara, who shared with us over a very long time the omnipresence of this project. But for their patience and support, this book would never have been born.

Central to our creation of this book have been our relationships with dozens of reviewers, particularly G. Barrie Kitto, who reviewed the entire manuscript; with our editors, Diane Bowen and Susan Weisberg, who invested much of themselves in this project and earned our respect and gratitude over its several-year lifespan; with a talented production staff including production supervisor Anne Friedman, art coordinator Bruce Lundquist, and chief illustrator Georg Klatt; and with three young biochemists, who detected numerous errors that we had missed. The latter are our graduate students Walter Lang and Patrick Young, who read galley proofs, and Dr. Lawrence Mathews, who worked every problem and checked our answers. All five groups—artists, reviewers, editors, production staff, and checkers—shaped the book in innumerable ways as it progressed from nearly unreadable chapter drafts to this completed volume. Still, as authors we bear final responsibility for the quality of this contribution, which must be ultimately judged by its value in training the next generation of biochemists.

Christopher K. Mathews *K. E. van Holde*

Reviewers

Hugh Akers, *Lamar University*
David F. Albertini, *Tufts University*
Mark Alper, *University of California, Berkeley*
Dean R. Appling, *The University of Texas at Austin*
Thomas O. Baldwin, *Texas A&M University*
Clinton E. Ballou, *University of California, Berkeley*
Wayne M. Becker, *University of Wisconsin, Madison*
Helen M. Berman, *Fox Chase Cancer Center*
Loran L. Bieber, *Michigan State University*
Robert Blankenship, *Arizona State University*
John W. Bodnar, *Northeastern University*
Rodney F. Boyer, *Hope College*
Robert B. Buchanan, *University of California, Berkeley*
Neil A. Campbell, *San Bernardino Valley College*
W. Scott Champney, *East Tennessee State University*
John Coffin, *Tufts University*
Jeffrey A. Cohlberg, *California State University, Long Beach*
Anne Dell, *Imperial College, London*
John Elam, *Florida State University*
Donald M. Engleman, *Yale University*
David E. Fahrney, *Colorado State University*
Richard E. Fine, *Boston University School of Medicine*
William H. Fuchsman, *Oberlin College*
Reginald H. Garrett, *University of Virginia*
Arthur M. Geller, *University of Tennessee, Memphis*
John H. Golbeck, *Portland State University*
Lowell P. Hager, *University of Illinois, Urbana-Champaign*
Gerald W. Hart, *Johns Hopkins University*
Standish C. Hartman, *Boston University*
Glenn A. Herrick, *University of Utah*
John W. B. Hershey, *University of California, Davis*
C. H. W. Hirs, *The University of Colorado Health Sciences Center*
Laura L. M. Hoopes, *Occidental College*
Joyce E. Jentoft, *Case Western Reserve University*
Howard M. Jernigan, Jr., *University of Tennessee, Memphis*
Kenneth A. Johnson, *The Pennsylvania State University*
G. Barrie Kitto, *University of Texas*
Gunter B. Kohlhaw, *Purdue University*
Sydney R. Kushner, *University of Georgia*
Tomas M. Laue, *The University of New Hampshire*
Timothy M. Lohman, *Texas A&M University*
Kenneth J. Longmuir, *University of California, Irvine*
Ponzy Lu, *University of Pennsylvania*
Joan Lusk, *Brown University*
Judith K. Marquis, *Boston University School of Medicine*
Rowena G. Matthews, *University of Michigan, Ann Arbor*
William R. McClure, *Carnegie Mellon University*
David B. McKay, *University of Colorado, Boulder*
David Mount, *University of Arizona*
Burton L. Nesset, *Pacific Lutheran University*
Merle S. Olson, *University of Texas Health Science Center at San Antonio*
Stanley Parsons, *University of California, Santa Barbara*
Mulchand S. Patel, *Case Western Reserve University*
David M. Prescott, *University of Colorado, Boulder*
David G. Priest, *Medical University of South Carolina*
Norbert O. Reich, *University of California, Santa Barbara*
Thomas Schleich, *University of California, Santa Cruz*
Earl Shrago, *University of Wisconsin, Madison*
Elizabeth R. Simons, *Boston University School of Medicine*
Gerald R. Smith, *Fred Hutchinson Cancer Research Institute*
Thomas W. Sneider, *Colorado State University*
Lewis Stevens, *Stirling University*
Phyllis R. Strauss, *Northeastern Unviersity*
Charles C. Sweeley, *Michigan State University, East Lansing*
Robert L. Switzer, *University of Illinois at Urbana-Champaign*
Buddy Ullman, *Oregon Health Sciences University*
Dennis E. Vance, *University of Alberta*
Andrew H.-J. Wang, *Massachusetts Institute of Technology*
Peter J. Wejksnora, *University of Wisconsin, Milwaukee*
Beulah M. Woodfin, *University of New Mexico*

Photograph and Illustration Credits

Chapter 1 1.1: Science Source © Biophoto Associates/Photo Researchers, Inc. 1.5: J. Paulson and U. K. Laemmli, *Cell* 12(1977):817; © Cell Press. 1.6: Computer graphics modeling and photography, Arthur J. Olson, Ph.D. Research Institute of Scripps Clinic, © 1986. 1.11b: S. C. Holt, University of Texas Health Science Center, San Antonio/BPS. 1.12b: R. Rodewald, University of Virginia/BPS. 1.13b: Micrograph by W. P. Wergin; courtesy of E. H. Newcomb, University of Wisconsin–Madison. 1.14a: © Ed Reschke. 1.14b: Illustration provided by Dr. Dorothea Zucker-Franklin, New York University School of Medicine. 1.14c: Judith Croxdale, University of Wisconsin–Madison. 1.16a,b: Frederick A. Murphy, Centers for Disease Control. 1.16c: Courtesy of Dr. R. C. Williams, University of California–Berkeley. 1.2, 1.10, 1.12a, 1.13a: From W. M. Becker, *The World of the Cell* (Menlo Park, CA: Benjamin/Cummings, 1986).

Chapter 2 2.6: Courtesy of J. M. Burridge, IBM UK Scientific Centre.

Chapter 4 4.1: Courtesy of Lawrence Livermore National Laboratory. 4.9: Reprinted by permission from *Nature* 171(1953):740; © 1953 Macmillan Magazines Ltd. 4.10: Dreiding Stereo-Models, Büchi Laboratoriums-Technik Ltd. 4.15: Richard E. Dickerson, *Sci. Am.*, December 1983, pp. 100–104; © 1983 Dickerson and Geis. 4.16: N. L. Max, University of California/BPS. 4.18a: Drs. K. Dressler and K. Koths. 4.18b: R. Kavenoff, Designergenes Ltd./BPS. 4.18c: P. Wellauer and I. David, *Cold Spring Harbor Symp. Quant. Biol.* 38(1974):525. 4.20: Courtesy of J. C. Wang. 4.21: Courtesy of J. C. Wang. 4.24: Adapted from J. Marmur and P. Doty, *J. Mol. Biol.* 5(1962):120. Table 4.1: Courtesy of CRC Press.

Chapter 5 5.1: John C. Kendrew, *Sci. Am.*, December 1961, p. 98. 5.6: D. Wetlaufer, *Adv. Protein Chem.* 17(1962): 303–390. 5.18: Reprinted by permission from Blanchetot, *Nature* 301(1983):732–734; © 1983 Macmillan Magazines Ltd. Table 5.4: Data excerpted from *Enzyme Nomenclature,* Nomenclature Committee of the International Union of Biochemistry (Academic Press, 1984).

Chapter 6 6.8a: Courtesy of Richard J. Feldman, National Institutes of Health. 6.13d,e: Dr. Alan Hodges. 6.15a–c: Courtesy of Dr. Jane Richardson. 6.20: A. Ginsburg and W. R. Carroll, *Biochemistry* 4(1965):2159–2174. 6.25: From W. M. Becker, *The World of the Cell* (Menlo Park, CA: Benjamin/Cummings, 1986) p. 86. 6.26h: Bernard Roizman, University of Chicago/BPS. 6.27a: T. Pollard, reproduced from *J. Cell Biol.* 91(1981):156, by copyright permission of Rockefeller University Press. 6.27b: Courtesy of Dr. R. C. Williams, University of California–Berkeley. 6.28: C. C. F. Blake et al., *J. Mol. Biol.* 88(1974):1–12.

Chapter 7 7.7a: © T. Eisner (Cornell University). 7.7b: © Ed Reschke. 7.22: J. V. Kilmartin, *Br. Med. Bul.* 32(1976):209–212. 7.25: W. G. Wood, *Br. Med. Bul.* 32(1976):282–287. 7.30: Courtesy of T. Williams and R. Josephs. 7.31: Courtesy of B. Carragher, D. Bluemke, M. Potell, and R. Josephs. 7.33b: R. Tizard, *Immunology: An Introduction* (Philadelphia: Saunders, 1984). 7.34: Adapted with permission from *Molecular Cell Biology* by James Darnell et al.; © 1986 Scientific American Books, Inc. 7.37: A. G. Amit et al., *Science* 233(1986):747–753; © 1986 by the AAAS. 7.39: J. D. Capra and A. B. Edmundson, *Sci. Am.* 236(1)(1977):50–59.

Chapter 8 8.19a: Science Source © Biophoto Associates/Photo Researchers, Inc. 8.19b: © Dr. L. M. Beidler. 8.19c: Micrograph by Don Fawcett, M.D. 8.24: From W. M. Becker, *The World of the Cell* (Menlo Park, CA: Benjamin/Cummings), p. 103. 8.26a: J. A. Buckwalter and L. Rosenberg, *Collagen Relat. Res.* 3(1983):489–504. 8.26b: L. Rosenberg, in *Dynamics of Connective Tissue Macromolecules,* M. Burleigh and R. Poole, ed. (Amsterdam: North Holland, 1975). 8.31b: Courtesy of Dr. Susumu Ito.

Chapter 9 9.3: P. Julien, J.-P. Despres, and A Angel, *J. Lipid Res.* 30(1989):293–299. Used by permission. 9.16: J. Yu, A. Fischman, and T. L. Speck, *J. Supramol. Struct.* 1(1973):233. 9.17a,b: D. Branton et al., *Cell* 24(1981):24–32; © Cell Press. 9.10, 9.15, 9.25, 9.26, 9.27: From W. M. Becker, *The World of the Cell* (Menlo Park, CA: Benjamin/Cummings, 1986).

Chapter 10 10.11a: Courtesy of Dr. J. S. Richardson. 10.11b: T. C. Alber et al., *Cold Spring Harbor Symp. Quant. Biol.* 52(1987):603. 10.27: P. G. Schultz, *Science* 240(1988):425–432; © 1988 by the AAAS.

Chapter 11 11.18: S. Sprang, E. Goldsmith, and P. Fletterick, *Science* 237(1987):1012–1019; © 1987 by the AAAS. 11.21a: © Lennart Nilsson/Boehringer Ingelheim InterNational. 11.21b: Courtesy of Dr. Henry Slayter.

Chapter 12 12.6: From W. M. Becker, *The World of the Cell* (Menlo Park, CA: Benjamin/Cummings, 1986), p. 27. 12.7: Reproduced from J. J. Wolosewick and K. R. Porter, *J. Cell Biol.* 82(1979):114–139, by copyright permission of Rockefeller University Press. 12.8: G. K. Radda, *Science* 233(1988):640–645; © 1988 by the AAAS.

Chapter 13 13.8: C. L. Markert, *Science* 140(1963):1329; © 1963 by the AAAS. 13.9: Adapted with permission from A. Betz and B. Chance, *Arch. Biochem. Biophys.* 109(1964):586. 13.20a,b: Reprinted by permission of the publisher from R. H. Kretsinger, S. E. Rudwick, and L. J. Weissman, *J. Inorg. Chem.* 28(1986):293, © 1986 by Elsevier Science Publishing Co., Inc.

Chapter 14 14.5: R. M. Oliver and L. J. Reed, in *Electron Microscopy of Proteins*, Vol. 2, J. R. Harris, ed. (Orlando, Fla.: Academic Press, 1982). 14.11: Reproduced, with permission, from G. Wiegand and S. J. Remington, *Annu. Rev. Biophys. Biophys. Chem.*, Vol. 15; © 1986 by Annual Reviews, Inc.

Chapter 15 15.2, 15.5a–c, 15.9, 15.14a,b, 15.16a,b: A. Tzagoloff, *Mitochondria* (Plenum, 1982).

Chapter 17 17.2: R. N. Trelease, P. J. Gruber, W. M. Becker, and E. H. Newcomb, *Plant Physiology* 48(1971):461. 17.18: Reprinted with permission from M. D. Lane, J. Moss, and S. E. Polakis, *Curr. Top. Cell. Regul.* 8(1974):144–145. 17.22: S. J. Wakil, J. K. Stoops, and V. C. Joshi, reproduced, with permission, from *Annu. Rev. Biochem.* 52(1983):537–580; © 1983 by Annual Reviews Inc. 17.24: Adapted from S. J. Wakil, J. K. Stoops, and V. C. Joshi, *Annu. Rev. Biochem.* 52(1983):537–580; © 1983 by Annual Reviews Inc.

Chapter 18 18.13: Courtesy of Dr. Cedric Raine. 18.23a,b: R. G. W. Anderson, M. S. Brown, and J. L. Goldstein, *Cell* 10(1977):351–364; © 1977 by Cell Press. 18.24: Micrograph produced by John Heuser, Washington University School of Medicine.

Chapter 19 19.4a: Micrograph by M. W. Steer; photo provided by E. H. Newcomb. 19.4b: Micrograph by K. P. Wergin; photo provided by E. H. Newcomb. 19.5: N. Lang, University of California, Davis/BPS. 19.18: H. Michel and J. Deisenhofer, Max Planck Institute für Biochemie, Munich. 19.25: From N. A. Campbell, *Biology* (Menlo Park, CA: Benjamin/Cummings, 1987), p. 222.

Chapter 20 20.3: The Nitragin Company. 20.5a,b: Reprinted by permission from D. Eisenberg et al., *Nature* vol. 323(1986):304–309; © 1986 Macmillan Magazines Ltd. 20.13a,b: S. J. Benkovic, C. A. Fierke, and A. M. Naylor, *Science* 239(1988):1105–1110; © 1988 by the AAAS.

Chapter 21 21.12: C. C. Hyde et al., *J. Biol. Chem.* 263(1988):17857–17871; © American Society for Biochemistry and Molecular Biology.

Chapter 22 22.13: Adapted with permission, from L. Thelander and P. Reichard. *Annu. Rev. Biochem.* 48(1979):133–158. © 1979 by Annual Reviews, Inc. 22.14: B-M. Sjöberg, *EMBO J.* vols. 3 and 5(1986):2038. 22.20, 22.21a,b: L. W. Hardy, J. Finer-Moore, D. V. Santi, and R. Stroud, *Science* 235(1987):448–455; © 1987 by the AAAS.

Chapter 23 23.6: From E. N. Marieb, *Human Anatomy and Physiology* (Menlo Park, CA: Benjamin/Cummings, 1989), p. 8. 23.10: J. R. Marx, *Science* 238(1987):615–616; © 1987 by the AAAS provided by R. Lefkowitz. 23.14: A. M. DeVos et al., *Science* 239(1988):888–895; © 1988 by the AAAS provided by S-H. Kim. 23.16: Reproduced by permission, J. Ramachandran and A. Ullrich, *Trends Pharmacol. Sci.* 8(1987):28–31; © 1987 by Elsevier Science Publishing Co., Inc.

Chapter 24 24.1: L. S. Beese and T. A. Steitz, *Nucleic Acid and Molecular Biology*, vol. 3, F. Eckstein and D. M. J. Lilley eds. © 1989 Springer-Verlag. 24.4: J. Cairns, *Cold Spring Harbor Symp. Quant. Biol.* 28(1963):44. 24.5b: M. Schnös and R. B. Inman, *J. Mol. Biol.* 51(1970):61–73. 24.6: R. L. Rodriguez, M. S. Dalbey, and C. I. Davern, *J. Mol. Biol.* 74(1973):599–604. 24.7: G. C. Fareed et al., *J. Virol.* 10(1972):484. 24.9: E. B. Gyurasits and R. G. Wake, *J. Mol. Biol.* 68(1973):501–509. 24.13: P. S. Freemont, J. M. Friedman, L. S. Beese, M. R. Sanderson, and T. A. Steitz, *Proc. Natl. Acad. Sci. USA* 85(1988):8924–8928. 24.15: L. Simon and T. F. Anderson, 1967, Academic Press, Inc. 24.16: C. K. Mathews, in *Compr. Virol.* 7(1977):179–299, H. Fraenkel-Conrat and R. R. Wagner, eds. With permission from Plenum Publishing Corp. 24.26: A. Kornberg, *J. Biol. Chem.* 1988 minireview, © American Society for Biochemistry and Molecular Biology, Inc. 24.28: P. O. Brown and N. R. Cozzarelli, *Science* 206(1989):1081–1083; © 1989 by the AAAS. 24.30: T. Kirchhausen, J. C. Wang, and S. C. Harrison, *Cell* 41(1985):933–943; © 1985 by Cell Press. 24.36: H. Echols, *Science* 233(1986):1050–1056; © 1986 by the AAAS. 24.37: C. W. Fuller and C. C. Richardson, *J. Biol. Chem.* 260(1985):3185–3196.

Chapter 25 25.1: J. A. McClarin, et al., *Science* 234 (1986):1528–1541; © 1986 by the AAAS provided by J. M. Rosenberg. 25.3: G. R. Smith et al., *Cold Spring Harbor Symp. Quant. Biol.* 49(1984):485–495. 25.5: Courtesy of Catherine Z. Mathews. 25.17: H. Potter and D. Dressler, *Proc. Natl. Acad. Sci. USA* 74(1977):4168–4172. 25.21: L. D. Harris and J. Griffith, *J. Biol. Chem.* 262(1987):9285–9292. 25.22: M. Amrein et al., *Science* 240(1988):514–516; © 1988 by the AAAS provided by G. Travaglini. 25.23: G. R. Smith et al., Cold Spring Harbor Symp. Quant. Biol. 49(1984): 485–495, provided by A. F. Taylor.

Chapter 26 26.1: S. R. Jordan and C. O. Pabo, *Science* 242(1988):893–907; © 1988 by the AAAS. 26.8: D. Hawley and W. R. McClure, reproduced, with permission, from the *Annu. Rev. Biochem.* 54(1985):171–204; © 1985 by Annual Reviews Inc. 26.9: D. S. Oppenheim, G. N. Bennett, and C. Yanofsky, *J. Mol. Biol.* 144(1980):133–142. 26.10: U. Siebenlist and W. Gilbert, *Proc. Natl. Acad. Sci. USA* 77(1980):125. 26.18: T. Steitz, *Cold Spring Harbor Symp. Quart. Biol.* 47(1983):419. 26.20: From *A Genetic Switch*, M. Ptashne, Cell Press and BSI, 1986. 26.21: A. D. Johnson, B. J. Meyer, and M. Ptashne, *Proc. Natl. Acad. Sci. USA* 76(1979):5062. 26.24: From *A Genetic Switch*, M. Ptashne, Cell Press and BSI, 1986. 26.25: Courtesy of T. A. Steitz and

I. T. Weber. 26.27: From *A Genetic Switch,* M. Ptashne, Cell Press and BSI, 1986. 26.31: P. B. Sigler, *Nature* 334(1988):321–329; © 1988 Macmillan Magazines Ltd. 26.12, 26.29: From J. D. Watson et al., *Molecular Biology of the Gene* 4th ed. (Menlo Park, CA: Benjamin/Cummings, 1987).

Chapter 27 Figure 27.2a: Adapted from H. M. Dintzis, *Proc. Nat. Acad. Sci.* 47(1961):247. 27.9a,b: S. H. Kim et al., *Science* 185(1974):435; © 1974 by the AAAS. 27.15: Courtesy of R. Traut. 27.16: Courtesy of R. Gutell and H. Noller. 27.17a,b: J. A. Lake *Sci. Am.* 245(1981):86; © 1981 by Scientific American, Inc. All rights reserved. 27.18: Adapted from A. Spirin, *Ribosome Structure and Protein Synthesis* (Menlo Park, CA: Benjamin/Cummings, 1986). 27.25a: Courtesy of Barbara Hamkalo. 27.26a: Courtesy of P. Moore and V. Ramakrishnan. 27.26b: M. Capel et al., *Science* 238(1987):1403–1406; reprinted by permission of the AAAS. 27.26c: S. Stern, T. Powers, L.-M. Changchien, and H. Noller, *Science* 244(1989):783–790; © 1989 by the AAAS. 27.27: C. Bernabeu and J. H. Lake, *Proc. Natl. Acad. Sci. USA* 79(1982):3111. 27.30: D. M. Engelman and T. A. Steitz, *Cell* 23(1981):411; © 1981 Cell Press. 27.7a,b; 27.16; 27.25b: J. Watson et al., *Molecular Biology of the Gene,* 4th ed. (Menlo Park: Benjamin/Cummings, 1987).

Chapter 28 28.3b: P. Chambon, *Sci. Am.* 245(1981):60–70; © 1981 by Scientific American, Inc. All rights reserved. 28.4b: R. Rodwald, University of Virginia/BPS. 28.4c: Reprinted by permission of J. P. Stafstrom and L. A. Staehelin, *J. Cell Biol.* 98(1984):699; by copyright permission of the Rockefeller University Press. 28.5: G. F. Bahr, Armed Forces Institute of Pathology. 28.6: Courtesy of K. van Holde. 28.7: Micrograph by C. F. L. Woodcock, University of Massachusetts/Amherst. 28.8: Courtesy of A. Klug. 28.11: D. A. Jackson, S. J. McCready, P. R. Cook, *J. Cell Sci.,* suppl. 4(1984):59–79. 28.18a: Courtesy of Dr. A. Anunziato. 28.19b: Courtesy of O. L. Miller and B. Beatty. 28.21: From J. D. Watson, *Molecular Biology of the Cell,* 4th ed. (Menlo Park, CA: Benjamin/Cummings, 1987), p. 713. 28.22a: Adapted from J. Berg, *Proc. Natl. Acad. Sci. USA* 85(1988):99–102. 28.22b: J. Y. Tso, D. J. van den Berg, and L. J. Korn, *Nucleic Acid Res.* 14(1986):2187–2200. 28.26: G. Conway, J. Wooley, T. Bibring, and W. M. LeStourgeon, *Mol. Cell Biol.* 8(1988):2884; © 1988 by American Society of Microbiology. 28.37: J. W. Fristrom, R. Raikow, W. Petri, D. Stewert: *Problems in Biology: RNA in Development,* E. W. Hanly, ed. Permission granted by University of Utah Press. 28.38: E. B. Lewis, in *Embryonic Development Part A: Genetic Aspects* (New York: Alan Liss, 1982). 28.4a, 28.10, 28.14: From W. M. Becker, *The World of the Cell* (Menlo Park, CA: Benjamin/Cummings, 1986).

Chapter 29 29.6: Adapted from James Darnell et al., *Molecular Cell Biology.* © 1986 by Scientific American Books, Inc. Used by permission. 29.10: Manfred Kage and Peter Arnold. 29.12: *Principles of Neural Science,* E. R. Kandel and J. H. Schwartz, eds. © 1981 Elsevier Publishing Co. 29.17: R. G. Kessel and R. H. Kardon, *Tissues and Organs: A Text Atlas of Scanning Electron Microscopy,* © 1979 by W. H. Freeman and Co. 29.18: *Principles of Neural Science,* E. R. Kandel and J. H. Schwartz, eds. © 1981 Elsevier Publishing Co. 29.19: Courtesy of Dr. Wolfgang Kabsch. 29.20a, 29.22: Courtesy of T. Pollard. 29.24b: Courtesy of H. E. Huxley. 29.25: Courtesy of J. Heuser. 29.32: D. P. Kiehart, I. Mabuchi, and S. Inoue, *J. Cell Biol.* 94(1982):165, by copyright permission of the Rockefeller University Press. 29.34a: Courtesy of U. W. Goodenough, *J. Cell Biol.,* 96(1983):1610, by copyright permission of the Rockefeller University Press. 29.34b: Courtesy of C. J. Brokaw, California Institute of Technology. 29.35a,b: Photo Researchers. 29.35c: W. L. Dentler, University of Kansas/BPS. 29.39: B. J. Schnapp, R. D. Vale, M. P. Sheetz, and T. S. Reese, *Cell* 40(1985):455; © 1985 by Cell Press. 29.40a: Courtesy of J. Adler. 29.40b: J. Adler in *Cell Motility,* R. Goldman, T. Pollard, and J. Rosenbaum, eds. © 1976 by Cold Spring Harbor Press. 29.41: R. M. Macnab and M. K. Ornston, *J. Mol. Biol.* 112(1977):1–30; © 1977 by Academic Press, Inc. 29.2, 29.7, 29.23, 29.26a, 29.27, 29.29, 29.30, 29.36, 29.40b,c, 29.42: From W. M. Becker, *The World of the Cell* (Menlo Park, CA: Benjamin/Cummings, 1986). 29.17a, 29.35b, 29.38: From N. A. Campbell, *Biology* (Menlo Park, CA: Benjamin/Cummings, 1987).

Tools T1.1, T1.3: W. M. Becker, *The World of the Cell* (Menlo Park, CA: Benjamin/Cummings, 1986). T.1.2: T. Pollard and P. Maupiu in *Electron Microscopy in Biology,* vol. II, J. D. Griffith, ed.; © 1982 by John Wiley and Sons, Inc. T.1.4: © Lennart Nilsson, Boehringer Ingelheim International. T.5.5: Courtesy of Christin A. Frederick. T.5.6: Courtesy of Reinhard Gessner. T.5.7: K. D. Watenpaugh, L. K. Sieker, and L. H. Jensen, *J. Mol. Biol.* 138(1980):615–633; © 1980 by Academic Press, Inc. T.6.2: K. E. van Holde, *Physical Biochemistry,* 2nd ed., © 1985. Reprinted by permission of Prentice Hall, Inc. T.7.1: Used by permission of Beckman Instruments, Inc. T.10.6b: K. Wüthrich and G. Wagner, *J. Mol. Biol.* 130(1979):1–18; © 1979 by Academic Press, Inc. T.10.8: Reprinted with permission from A. Allerhand and E. Oldfield, *Biochemistry* 12(1973):3428–3433; © 1973 by American Chemical Society. T.12.3a,b: Y. Engström et al., *EMBO J.* 7(1988):1617. T12.4: G. J. Tortora, B. R. Funke, and C. L. Case, *Microbiology: An Introduction,* 3rd ed. (Menlo Park, CA: Benjamin/Cummings, 1989) p. 423. T.16.3: Courtesy of C. K. Mathews. T.16.4: D. M. Prescott, *Progr. Nucleic Acid Res.* vol. 3, 35(1964). T.25.4: K. E. van Holde, *Physical Biochemistry* 2nd ed.; © 1985. Adapted by permission of Prentice Hall, Inc., Englewood Cliffs, New Jersey. T21.1: J. D. Watson et al., *Molecular Biology of the Gene,* 4th ed. (Menlo Park, CA: Benjamin/Cummings, 1987), p. 609.

Brief Contents

Detailed Contents

3 The Energetics of Life 59

PART II Molecular Architecture of Living Matter 89

4 Nucleic Acids 91

5 Introduction to Proteins: The Primary Level of Protein Structure 133

PART IV Dynamics of Life: Energy, Biosynthesis, and Utilization of Precursors 431

13 Carbohydrate Metabolism I: Anaerobic Processes in Generating Metabolic Energy 433

14 Oxidative Processes: Citric Acid Cycle and Pentose Phosphate Pathway 467

23 Integration and Control of Metabolic Processes 779

PART V Information 815

24 Information Copying: Replication 817

Tools of Biochemistry

PART I

The Realm of Biochemistry

CHAPTER 1

The Scope of Biochemistry

The biological sciences are undergoing a revolution, and biochemistry is at the heart of that revolution. Nowhere is this better illustrated than in the award of the 1988 Nobel Prize for Medicine or Physiology, shared by Gertrude Elion and George Hitchings of the United States and Sir James Black of Great Britain. All were honored for their leadership in inventing new drugs—Elion and Hitchings for developing chemical analogs of nucleic acids and vitamins now used to treat leukemia, bacterial infections, malaria, herpes viral infections, and AIDS; and Black for developing beta blockers used to reduce the risk of heart attack. These discoveries were made possible by several decades of accumulated knowledge in central areas of biochemistry—protein structure and function, nucleic acid synthesis, enzyme mechanisms, receptors and metabolic control, vitamins and coenzymes, and comparative biochemistry (the study of chemical differences among organisms). The impact of biochemistry is felt in all of our lives through discoveries such as these and in the fuel these discoveries provide for the growth of all the life sciences.

Two main factors contribute to both the excitement of today's biochemistry and its impact on other life sciences. First, it is now well established that living matter obeys the same fundamental physical laws that govern all matter. Therefore, the full power of modern chemical and physical theory can be brought to bear on biological problems. Second, a wealth of powerful new research techniques is permitting scientists to ask penetrating questions about the basic processes of life—questions that could not have been imagined even a few years ago.

Living matter, exemplified by the cell depicted in Figure 1.1, is composed of chemical substances, and any biological function can be described in terms of the structures of those substances and the reactions that they undergo. To understand any life process, therefore, demands understanding of the chemistry of that process. This is true even for biological processes—such as evolution, differentiation, or behavior—that until recently were not thought to be amenable to dissection at the molecular level. Thus, biochemistry is penetrating into all disciplines of biology.

Aiding this penetration is the development of incredibly powerful research techniques, which have led, for example, to our present ability to positively identify an individual by analyzing the genetic material extracted from a single human hair or to introduce novel genes into living organisms. These techniques have opened windows into living cells and have led to

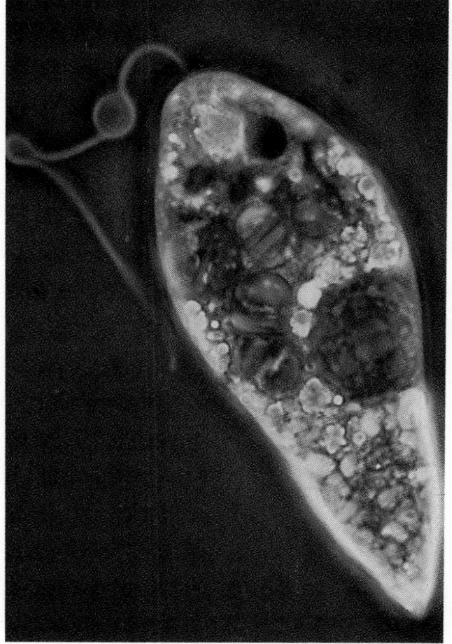

Figure 1.1
A one-celled organism, the photosynthetic protozoan Euglena. This colored micrograph reveals the complexity of structure within a single cell. The nucleus is the brown area at right center. To the left and above are the green chloroplasts, the photosynthetic organelles. The organism is propelled by the whiplike flagellum on the top left.

practical applications of the resulting knowledge, with progress being made much faster than was predicted by anyone even as recently as ten years ago. The student beginning the study of biochemistry today faces the exciting challenge of entering into a field that is developing at an unprecedented rate and is affecting virtually every area of human endeavor.

What Is Biochemistry?

The Goals of Biochemistry

Biochemistry seeks to describe the structure, organization, and functions of living matter in molecular terms. What are the chemical structures of the components of living matter? How do their interactions lead to the assembly of organized supramolecular structures, cells, and multicellular tissues and organisms? How does living matter extract energy from its surroundings in order to remain living? How does an organism store and transmit the information it needs to grow and to reproduce itself? What chemical changes accompany the reproduction, aging, and death of cells or organisms? How are chemical reactions controlled inside living cells? These are the major questions being approached by biochemists.

Biochemistry can be subdivided into three principal areas: (1) the **structural chemistry** of the components of living matter and the relationship of biological function to chemical structure, (2) the study of **metabolism,** the totality of chemical reactions that occur in living matter, and (3) the chemistry of processes and substances that store and transmit biological information; this latter area is also the province of **molecular genetics,** a field that seeks to understand heredity and the expression of genetic information in molecular terms.

The Roots of Biochemistry

Biochemistry has its origins as a distinct field of study in the early nineteenth century, with the pioneering studies of Friedrich Wöhler (Figure 1.2). Prior to Wöhler's time it was generally believed that living matter was composed of substances qualitatively different from those found in nonliving matter, substances that did not behave according to the known laws of physics and chemistry. In 1828 Wöhler showed that urea, a substance of biological origin, could be synthesized in the laboratory from the inorganic compound ammonium cyanate. As Wöhler phrased it in a letter to a colleague, "I must tell you that I can prepare urea without requiring a kidney or an animal, either man or dog."

$$\overset{+}{N}H_4NCO^- \longrightarrow H_2N-\overset{\overset{\displaystyle O}{\|}}{C}-NH_2$$

Ammonium cyanate **Urea**

Even after this demonstration that organic compounds could be derived from inorganic materials, there remained a persuasive viewpoint, called **vitalism,** which held that if not the compounds, at least the reactions that took place in living matter could occur only in living cells, through the agency of a mysterious "life force." This dogma was shattered in 1897, when two German brothers, Eduard and Hans Buchner, found that extracts from broken—and thoroughly dead—yeast cells could carry out the entire process of fermentation of sugar into ethanol. This discovery opened the door to analysis of biochemical reactions and processes **in vitro** (from Latin, "in glass," meaning in a test tube, rather than in intact living matter, **in vivo**). In succeeding decades other metabolic reactions and reaction pathways were reproduced in vitro, allowing identification of the reactants and products in these pathways, as well as of the **enzymes,** or biological catalysts, that promoted each biochemical reaction.

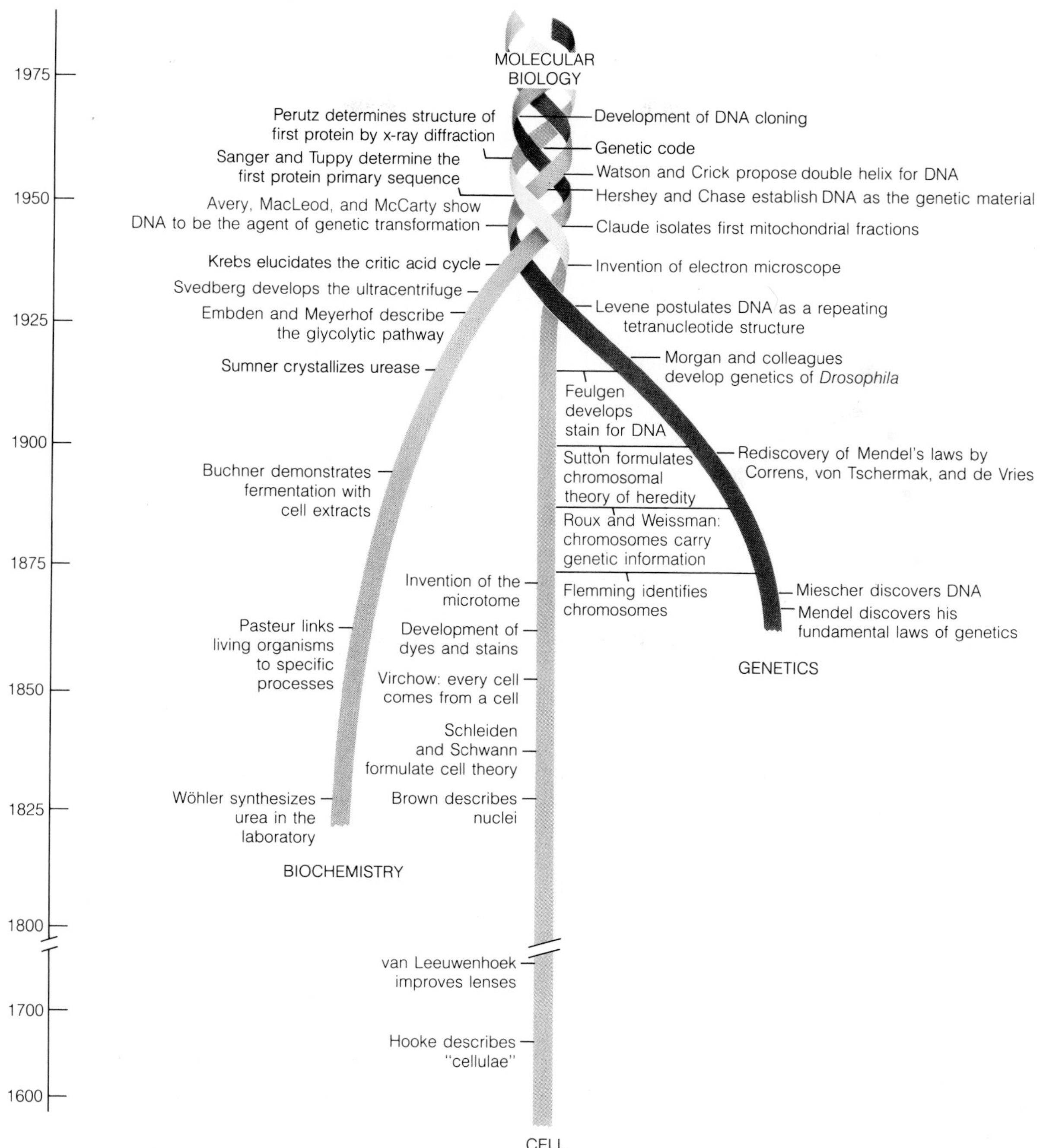

Figure 1.2
Interweaving of the historical tradition of biochemistry, cell biology, and genetics. These three disciplines, which originally were considered to be quite separate, have become intertwined to yield a true molecular biology, the subject matter of present-day biochemistry.

The nature of biological catalysis remained the last refuge of the vitalists, who held that the structures of enzymes (or "ferments") were too complex to be described in chemical terms. But in 1926 J. B. Sumner showed that the protein urease, an enzyme from jack beans, could be crystallized, like any organic compound. Although proteins have large and complex structures, they are just organic compounds and their structures are amenable to determination by the methods of chemistry.

In parallel with developments in biochemistry, **cell biologists** had been continually refining knowledge of cellular structure. Beginning with Robert Hooke's first observation of cells in the seventeenth century, steady improvements in microscopic technique led to the understanding that the cell was a complex, compartmental structure. Chromosomes were discovered in 1875 by Walter Flemming and identified as genetic elements by 1902.

The development of the electron microscope, between about 1932 and 1950, provided a whole new level of insight into cellular structure. Now subcellular organelles, like mitochondria and chloroplasts, could be studied, and it was realized that specific biochemical processes were localized in these subcellular particles.

Although developments in the first half of this century revealed in broad outline the chemical structures of biological materials, identified the reactions in many metabolic pathways, and localized these within the cell, biochemistry remained an incomplete science. We knew that the uniqueness of an organism is determined by the totality of its chemical reactions. However, we had little understanding of how those reactions are controlled in living tissue or of how the information that regulates those reactions is stored, transmitted when cells divide, or processed when cells differentiate.

The idea of the **gene,** a unit of hereditary information, arose in the mid-nineteenth century from the work of Gregor Mendel, and by about 1900 it was known that genes are found in chromosomes. However, until the mid-twentieth century no one had isolated a gene or determined its chemical structure.

Chromosomes are composed of proteins and nucleic acids. Nucleic acids had been isolated as early as 1869 (by Friedrich Miescher), but their chemical structures were poorly understood. Throughout the early 1900s nucleic acids were regarded as rather simple substances, fit only for structural roles in the cell. Most biochemists believed that only the proteins were structurally complex enough to carry genetic information.

This was dead wrong, as shown by experiments in the 1940s and early 1950s proving conclusively that **deoxyribonucleic acid (DNA)** is the bearer of genetic information. One of the most important scientific advances of this or any century occurred in 1953, when James Watson and Francis Crick described the double-helical structure of DNA. This immediately suggested ways in which information could be encoded in the structure of molecules and transmitted intact from one generation to the next.

At this point the strands of scientific development shown in Figure 1.2—biochemistry, cell biology, and genetics—became inextricably interwoven. A new science, **molecular biology,** emerged. The distinction between molecular biology and biochemistry is not always clear, since both disciplines take as their province the complete definition of life in molecular terms. However, the term molecular biology is often used in a narrower sense to denote the study of nucleic acid structure and function and the genetic aspects of biochemistry. Molecular biology and biochemistry are perhaps distinguished more readily by the orientations of their practitioners than by the research problems being addressed. It can be said that biochemists think like chemists, and molecular biologists think like biologists. Even this distinction is somewhat artificial, since successful scientists in either field must use the approaches of all relevant disciplines, including chemistry, biology—and physics. Two of the most powerful research techniques used by biochemists were developed by physicists: **electron microscopy,** which has revealed remarkable details of cellular structure (Tools of Biochemistry 1), and **x-ray diffraction,** which has revealed the precise three-dimensional structures of huge biological molecules (see Tools of Biochemistry 5 following Chapter 4).

Biochemistry as a Discipline and an Interdisciplinary Science

Biochemistry draws its major themes from many disciplines—from organic chemistry, which describes the properties of biomolecules; from medical research, which increasingly seeks to understand diseased states in molecular terms; from nutrition, which has illuminated metabolism by describing the dietary requirements for maintenance of health; from microbiology, which showed that single-celled organisms and viruses were ideally suited for the elucidation of many metabolic pathways and regulatory mechanisms; from physiology, which provided understanding of life processes at the cell and tissue levels and thereby opened doors for molecular investigation; from cell biology, which describes the biochemical division of labor within a cell; from biophysics, which uses the techniques of physics to help understand life at the molecular level; and from genetics, which describes mechanisms that give a particular cell or organism its biochemical identity. Biochemistry draws strength from all of these disciplines, and it nourishes them in return; it is truly an interdisciplinary science.

However, biochemistry is also a distinct discipline, with its own identity. It is distinctive in its emphasis on the structures of biomolecules and the reactions that they undergo, on enzymes and the nature of biological catalysis, on the elucidation of metabolic pathways and their control, and on the principle that life processes can be understood through the laws of chemistry. As you read this book, keep in mind both the uniqueness of biochemistry as a separate discipline and its absolute interdependence with other physical and life sciences.

Biochemistry as a Chemical Science

Though we often describe biochemistry as a life science and relate its developments to the history of biology, it remains first and foremost a chemical science. In order to understand the impact of biochemistry on biology, you must understand the chemical elements of living matter and the complete structures of hundreds of biological compounds, their functional groups, and their behavior during metabolic reactions. You will need to know the stoichiometry and mechanisms of many reactions, including those that produce the very large molecules (**biopolymers**) from low-molecular-weight intermediates, the forces that shape these large molecules in three dimensions and determine their interactions with other molecules, and their roles in life processes.

All forms of life, from the simplest and smallest to the largest and most complex, are constructed from the same chemical elements, which in turn are formed into the same types of molecules. In short, the chemistry of living matter is similar throughout the biological world. Undoubtedly, this continuity in biochemical processes reflects the common evolutionary ancestry of all cells and organisms. Let us begin a preliminary examination of the composition of living matter, starting with the chemical elements.

The Chemical Elements of Living Matter

Life is a phenomenon of the second generation of stars. This rather strange-sounding statement is based on the fact that life, as we conceive it, can come into being only when certain elements—C, H, O, N, and P—are abundant. The very early universe was made almost entirely of hydrogen and helium for these were the only elements produced in the condensation of matter following the primeval explosion, or "big bang." The first generation of stars contained no heavier elements from which to form planets. As these early stars matured over billions of years, they burned their hydrogen and helium in thermonuclear reactions. These reactions produced heavier elements—first carbon, nitrogen, and oxygen and later all the other members of the periodic table. In their maturation, large stars became unstable and exploded as novae and supernovae. The explosions of these early stars spread the heavier elements through the cosmic surroundings. This matter condensed again to form second-generation stars, complete with planetary systems rich in the heavier elements. Our present universe, which is rich in second-generation stars, has the approximate elemental composition shown in Figure 1.3a. In such a universe there is the possibility of life, and we know that on at least one planet, of one solar system, it does exist.

Why are elements heavier than H and He essential for life? The answer is that life, as we can imagine it, requires large and complex molecular structures. These can be formed only from certain elements and can be stable only under restricted environmental conditions. A universe of hydrogen and helium has *no* chemistry. Nor is chemistry possible in the heat of stars, where all compounds are broken into their elements. In environments as cold as the moon or space, a slow, simple chemistry may occur, but one cannot envision the formation of molecules as complex as proteins or nucleic acids. Only in the temperate environment of an appropriate planet, enriched with elements capable of forming complicated compounds, can life arise.

Living creatures on the earth are composed *mainly* of a very few elements, principally carbon, hydrogen, oxygen, and nitrogen (CHON). Comparing Figure 1.3a and c, you will find that these are also (with the addition of helium) the most abundant elements in the universe. Helium, being an inert and light gas, is not equipped for a role in life processes. It does not form stable compounds, and it is readily lost from planetary atmospheres.

A partial explanation for the abundance of oxygen and hydrogen in organisms lies in the major role played by water in life on Earth. We live in a highly aqueous world, and, as we shall see in Chapter 2, the solvent properties of water are indispensable in biochemical processes. The human body, in fact, is about 70% water.

The elements C, H, O, and N are important to life because of their strong tendencies to form covalent bonds. In particular, the stability of carbon–carbon bonds, the tetrahedral nature of the carbon bonding, and the possibility of forming single, double, or triple bonds give carbon the versatility to form an enormous diversity of chemical compounds.

But life is not built on these four elements alone. Many other elements are necessary for terrestrial organisms, as you can see from Table 1.1. A "second tier" of essential elements includes the covalent-bond-forming sulfur and phosphorus and the ions Na^+, K^+, Mg^{2+}, Ca^{2+}, and Cl^-. Sulfur is an important constituent of proteins, and phosphorus plays essential roles in energy metabolism and the structure of nucleic acids.

It is interesting to consider why some elements that are abundant on the earth's surface have *not* found their way into living organisms to an appreciable extent (see Figure 1.3b). The most conspicuous example is silicon, especially since silicon has, like carbon, the ability to form large molecules, with both Si—Si and Si—C bonds. Why is silicon not widely used in life? A likely reason can be seen by examining the scale of covalent bond energies shown in Figure 1.4. Silicon forms *extremely* strong bonds with oxygen, much stronger than any of the carbon bonds shown and stronger

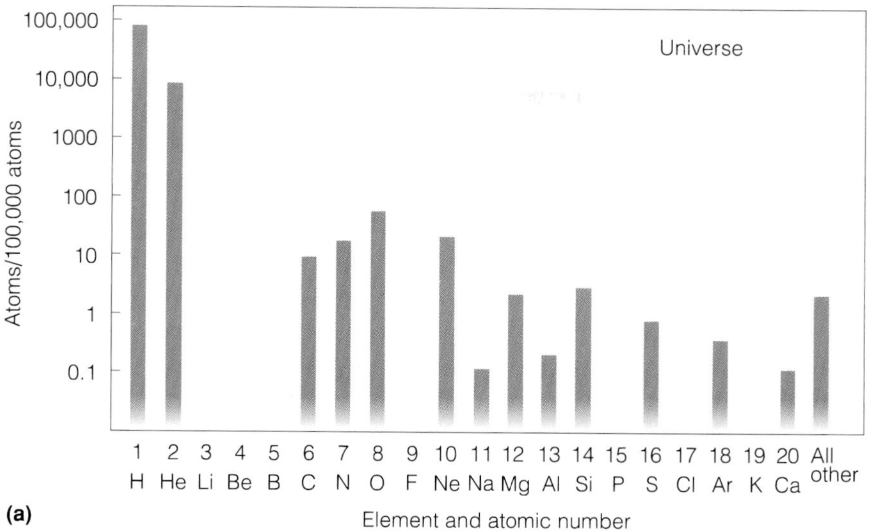

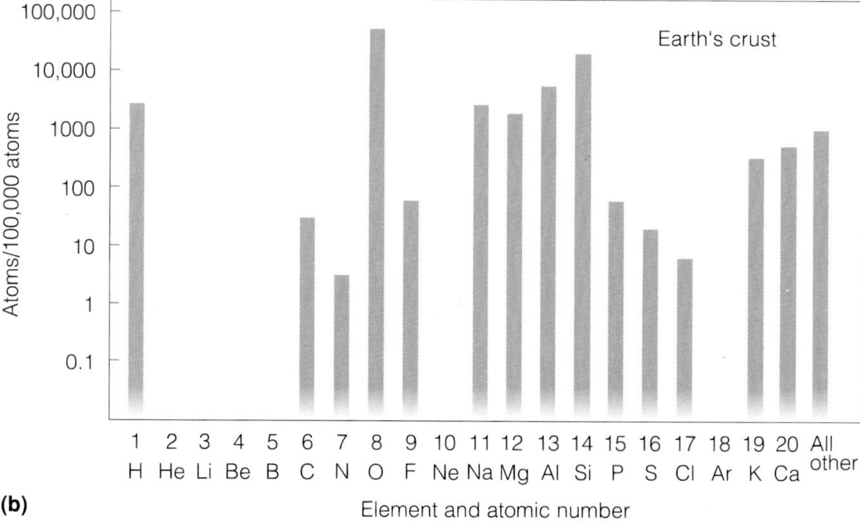

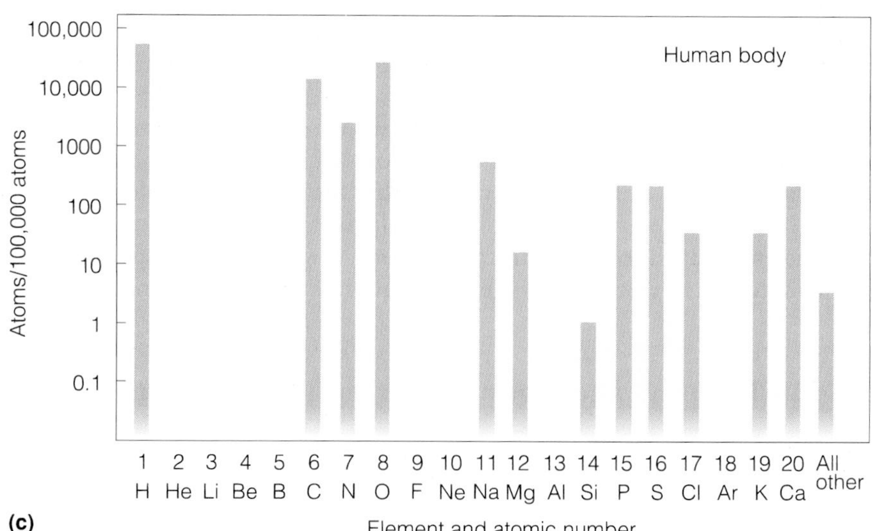

Figure 1.3
Composition of the universe (**a**), the earth's crust (**b**), and the human body (**c**). Amounts are expressed as number of atoms of each element per 100,000 atoms. Note that the scale is logarithmic. On a linear scale, H and He would greatly dominate in (**a**); O and Si in (**b**); and H, C, N, and O in (**c**). Remarkably, the human body resembles the whole universe in composition more than it does the earth's crust.

Table 1.1
Elements found in organisms

Element	Present in All Organisms	Present in Some Organisms	Comment
Carbon (C)	x		"First tier": most abundant in all organisms
Hydrogen (H)	x		
Nitrogen (N)	x		
Oxygen (O)	x		
Calcium (Ca)	x		"Second tier": much less abundant but found in all organisms
Chlorine (Cl)	x		
Magnesium (Mg)	x		
Phosphorus (P)	x		
Potassium (K)	x		
Sodium (Na)	x		
Sulfur (S)	x		
Cobalt (Co)	x		"Third tier": metals present in small amounts but essential to life
Copper (Cu)	x		
Iron (Fe)	x		
Manganese (Mn)	x		
Zinc (Zn)	x		
Aluminum (Al)		x	"Fourth tier": found in or required by *some* organisms in trace amounts
Arsenic (As)		x	
Boron (B)		x	
Bromine (Br)		x	
Chromium (Cr)		x	
Fluorine (F)		x	
Gallium (Ga)		x	
Iodine (I)		x	
Molybdenum (Mo)		x	
Selenium (Se)		x	
Silicon (Si)		x	
Vanadium (V)		x	

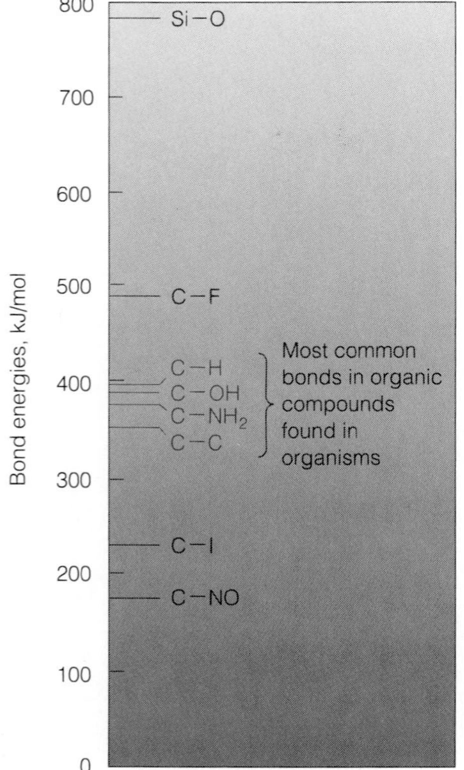

Figure 1.4
Scale of bond energies for some covalent bonds. Most of those found in biological compounds lie near 400 kJ/mol.

than Si—Si or Si—C bonds. Very strong bonds like Si—O lead to static, immutable structures, at least under conditions similar to those prevailing on the earth. Consequently, virtually all the silicon on earth is locked up in silicate minerals. It is probably not coincidental that the strengths of the covalent bonds important to life processes tend to cluster in a small range, as Figure 1.4 shows. Life is in a constant state of flux. The bonding between atoms in biomolecules must be sufficiently strong to ensure some structural stability but must not be *too* strong.

Beyond the first two tiers of elements (which correspond roughly to the most abundant elements of the first two rows of the periodic table), we come to those that play quantitatively minor, but often indispensable, roles. As Table 1.1 shows, most of these third-tier elements are metals, and some serve as aids to catalysis of biochemical reactions. In succeeding chapters we shall encounter many examples of the importance of these trace elements to life.

Biological Polymers and Their Monomeric Components

The complexity of life processes requires that many of the molecules that participate in these processes be of enormous size. A good example is DNA. Consider, for instance, the DNA molecules released from one human chro-

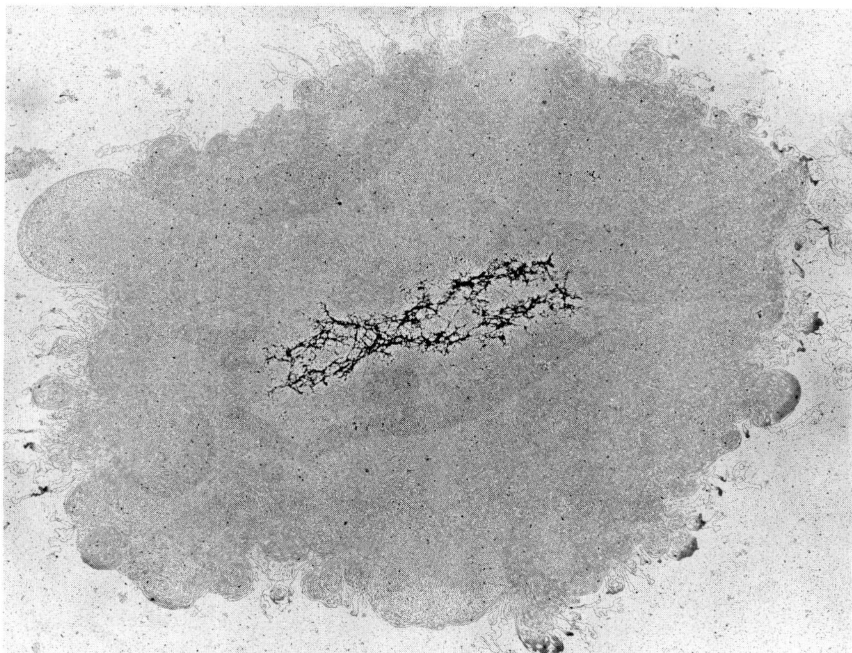

Figure 1.5
The DNA from a single human chromosome. The chromosome contains just two DNA molecules. When the chromosomal proteins are removed, as here, the DNA spills out as enormously long, twisted threads. If extended, each thread would be approximately 4 cm long.

mosome, as shown in Figure 1.5. The long thread you see corresponds to just two enormous molecules, each with a molecular weight of about 10^{12}—about 1 trillion. Even a simple organism such as the single-celled bacterium *Escherichia coli* contains a DNA molecule with a molecular weight of about 2 billion. Protein molecules are generally smaller but are still large, a typical protein having a mass of 30,000 daltons. To give an idea of the complexity of such a molecule, Figure 1.6 shows the three-dimensional structure of one protein molecule as revealed by x-ray crystallography.

These giant molecules, or **macromolecules,** constitute a large fraction of the mass of any cell. As we shall see in detail in later chapters, there are very good reasons for some biological materials to be so large. DNA molecules, for example, can be thought of as "tapes" from which genetic information is read out in a linear fashion. Since the amount of information needed to specify the structure of a multicellular organism is very great, these tapes must be extremely long. In fact, the DNA molecules in a single human cell, if stretched end to end, would reach a length of about 2 meters.

The synthesis of such very large molecules poses an interesting challenge to the cell. If the cell functioned like an organic chemist carrying out a complex laboratory synthesis bit by bit, millions of *different* types of reactions would be involved, and thousands of intermediates would accumulate. Instead, cells use a modular approach for constructing large molecules. All such structures are **polymers,** made by joining together prefabricated units, or **monomers,** which are of limited diversity and are linked together, or **polymerized,** by identical mechanisms. A simple example is the carbohydrate **cellulose,** a major constituent of the cell walls of plants. As you can see in Figure 1.7, cellulose is a polymer made by joining together thousands of molecules of glucose, a simple sugar; in this polymer all of the chemical linkages between glucose molecules are identical. Since cellulose is a polymer of a simple sugar, or **saccharide,** it is called a **polysaccharide.** Since this polymer is constructed from identical monomeric units, it is called a **homopolymer.** In contrast, nucleic acids and proteins (Figure 1.8) are **heteropolymers,** for each is constructed from a number of different kinds of monomer

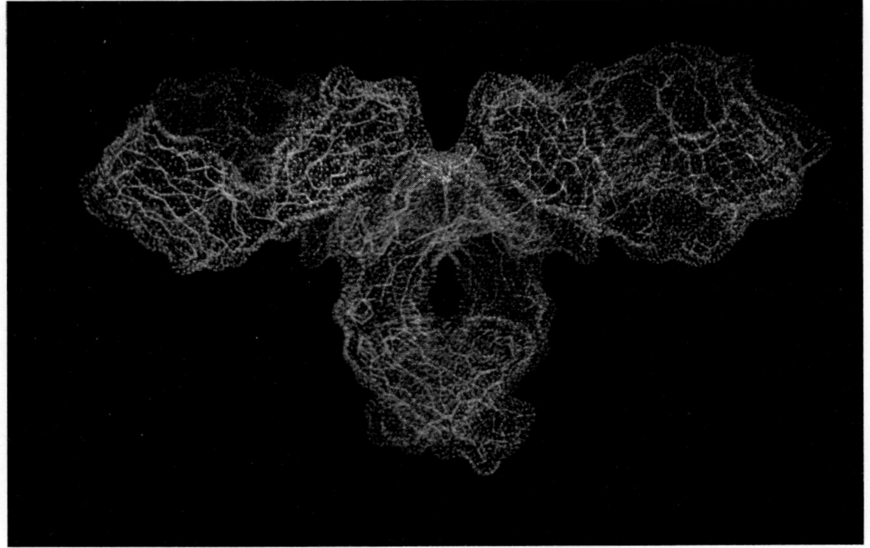

Figure 1.6
A protein molecule. Shown here is a model of an immunoglobulin molecule—a molecule that serves as an antibody in the immune reaction. It has a molecular weight of approximately 150,000.

units. Nucleic acids are polymers of four different **nucleotides,** so nucleic acids are also called **polynucleotides.** In a similar way, proteins are assembled from 20 different **amino acids.** Proteins are **polypeptides,** a term derived from the **peptide bond** that joins two amino acids together.

Such polymers form much of the structural and functional machinery of the cell. Polysaccharides serve both as structural components, such as cellulose, and as reserves of biological energy, such as **starch,** another type of glucose polymer found in plants. Nucleic acids participate in information storage, transmission, and expression. Deoxyribonucleic acid, or DNA, serves principally as a storehouse of genetic information, while the chemically similar **ribonucleic acid,** or **RNA,** is involved in the readout of information stored in DNA. Proteins, which have far more structural diversity than polysaccharides or nucleic acids, perform a more diverse set of biological functions; they play structural roles, as in keratin of hair or skin and the collagen of connective tissue; act as transport substances, as does hemoglobin, the oxygen-carrying protein of blood; transmit information between distant parts of an organism, as do dozens of protein **hormones** and the cell

Figure 1.7
Cellulose is a polymer of the monomer β-D-glucose. Covalent links between glucose units are formed by removing a water between two joining molecules; the portion of a glucose molecule remaining in the chain is called a glucose **residue.**

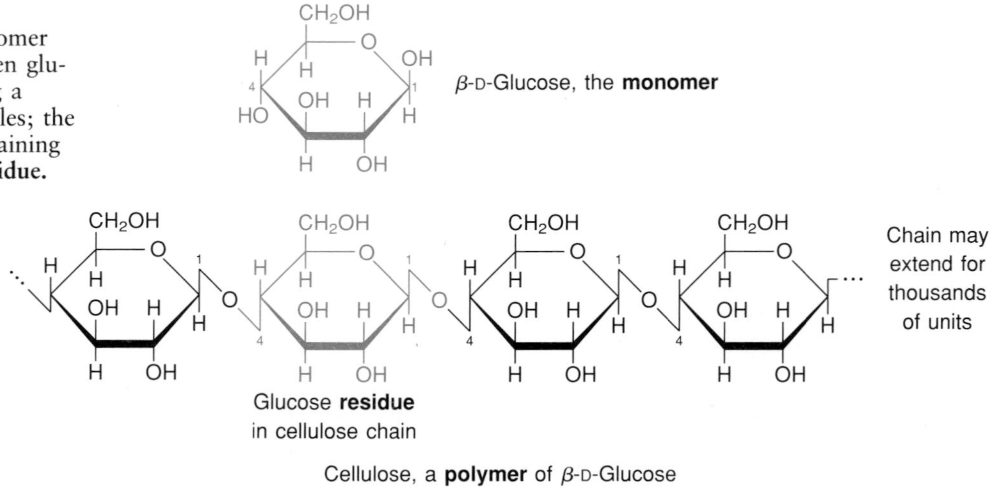

β-D-Glucose, the **monomer**

Glucose **residue** in cellulose chain

Chain may extend for thousands of units

Cellulose, a **polymer** of β-D-Glucose

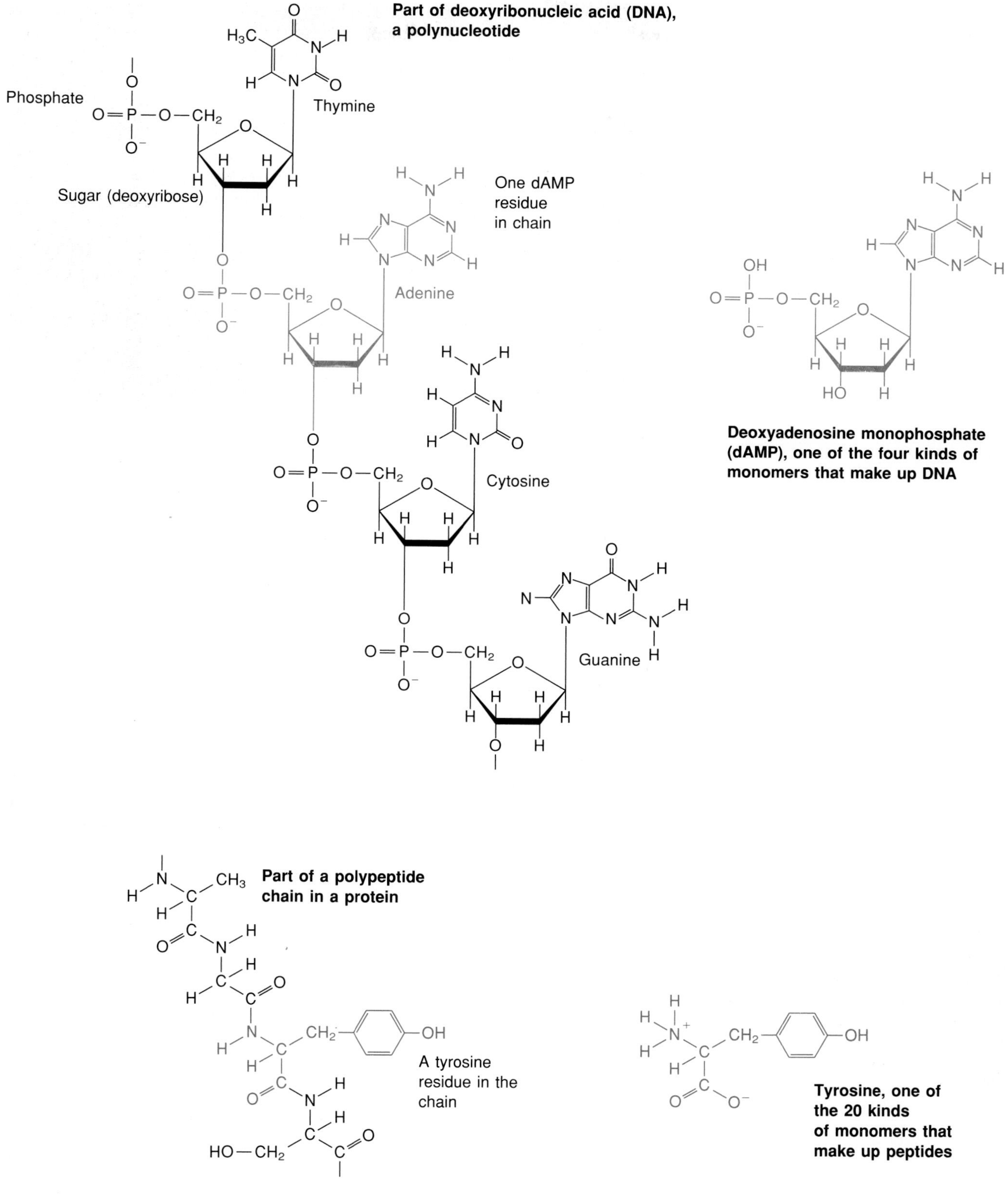

Figure 1.8
Polynucleotides and polypeptides as polymers. DNA, a polynucleotide, and one of its monomers, dAMP, as well as part of a protein, and one of its monomers, tyrosine, are shown.

surface **receptors** that receive the signals; and defend an organism against infection, as do the **antibodies.** Most important of all, proteins function as enzymes, catalyzing the thousands of chemical reactions that occur within an individual cell.

In addition to these macromolecules and the many small molecules involved in metabolism and as monomers in macromolecular synthesis, there is one other extremely important class of cellular constituents. These are the **lipids,** a chemically diverse group of compounds that are classified together because of their apolar structures, which give them very low solubility in the aqueous environment of the cell. This low solubility equips lipids for one of their most important functions—to serve as the major structural element of the membranes that surround and partition cells.

Biochemistry as a Biological Science

We must never lose sight of the fact that it is the chemistry of *life* that concerns us here. The complex chemical substances and reactions that we introduced above have their significance as parts of living matter and life processes. To see biochemistry from this perspective we should begin by asking: What is life?

Distinguishing Characteristics of Living Matter

What distinguishes living from nonliving matter? The first attribute that may come to mind is the sheer *complexity* of even the simplest living creature. But complexity is not enough; many things in the universe are complex but nonliving. A dead mouse is, for a little while at least, almost as complicated as a live one. Nor can we distinguish living from nonliving matter simply on the basis of *motility,* for many completely nonmotile things (like mushrooms) are very much alive. The major quality that distinguishes life is the constant *renewal* of a highly ordered structure, often accompanied by an increase in the complexity of that structure. Organisms create an elegant molecular order within themselves and pass a pattern of that order on to descendent organisms. This creation and perpetuation of order, out of often chaotic surroundings, is unique to life, and it seems to fly in the face of one of the fundamental laws of the universe, the **second law of thermodynamics.** One way of stating the law is this: the **entropy** (read: disorder) of the universe continually increases. But the second law does not require that disorder increase everywhere, at all times; living creatures are like little whirlpools of order in the stream of the universe. The laws of thermodynamics do require, however, that this local creation of order and complexity in matter be paid for by the continual expenditure of energy. This is why living organisms must forever take energy from their surroundings—either from sunlight, as plants do, or from foodstuffs, as animals do. To accomplish this, any organism must interact with its surroundings, and in many cases it must act on them, as we humans do. No living organism can be isolated from its surroundings.

Finally, and in a sense most important of all, life is *self-replicating.* It has been suggested that the most primitive "organisms," at the dawn of life, were nothing more than giant molecules that could direct their own replication. The reproductive process has become exceedingly complex through evolution, but its basis remains the same: *information* describing the structure of an organism is passed from one generation to the next. Thus, although every individual creature must lose its own battle with chaos, life itself continues.

The Unit of Biological Organization: The Cell

One of the first major discoveries in biology was Robert Hooke's observation (1665) that plant tissues (in this case, cork) were divided into tiny compartments, which he called *cellulae,* or **cells.** By 1840 improved observations on many tissues led Theodor Schwann to propose that all organisms exist as either single cells or aggregates of cells. More than a century of careful study by cytologists has confirmed this hypothesis.

Furthermore, cells, from whatever organism, are of quite similar size. Most bacterial cells are about 1–2 μm in diameter and most cells of higher organisms are only about 5–10 times larger. There are, to be sure, exceptions: there are very small bacteria (0.2 μm) and there are unusual cells like those in the nervous systems of vertebrates, some of which may be over 1 m long. But compared to the overall range of sizes of natural objects (Figure 1.9), all cells are much alike.

In both plants and animals, moreover, the size of the cells bears no relationship to the size of the organism. An elephant and a flea have cells of about the same size; the elephant just has more. Why is such uniformity in cell size maintained? A clue can be found in the fact that the surface/volume ratio for an object depends on its size (Figure 1.10). The complex chemical processes in a cell and the large molecules that participate in them require a significant volume. Yet the cell must effectively communicate with its surroundings, which in turn requires an appreciable surface area. Too large a cell will not have enough surface through which to exchange substances with its surroundings and support the active metabolism within. Cells of higher organisms can be very large only if they are highly elongated—such as the nerve cells mentioned above—which will increase the surface/volume ratio. The simplest microbes, on the other hand, can be very small because their metabolism is simple. Viruses, which are even smaller, do not enjoy their own metabolism but live as parasites on the cells they invade.

Since cells are the universal units of life, let us examine them more closely. Major differences between cell structures define the two great classes of organisms—prokaryotic and eukaryotic. The **prokaryotes,** which are always unicellular, include the **bacteria,** some primitive algae, and an ancient class called **archaebacteria.** A typical prokaryotic organism is shown schematically in Figure 1.11. Prokaryotic cells are surrounded by a membrane and usually a rigid cell wall as well. Within the membrane is contained the **cytosol,** a semiliquid concentrated solution or suspension. In prokaryotes the cytosol is not divided into compartments, and the genetic information is in the form of one or more DNA molecules (**nucleoids**) that exist free in the cytosol. Also suspended in the cytosol are the **ribosomes,** which constitute the molecular machinery for protein synthesis. The surface of a prokaryotic cell may carry **pili,** which aid in attaching the organism to other cells or surfaces, and **flagella,** which enable it to swim.

All other organisms are called **eukaryotes.** These include not only the multicellular plants, animals, and fungi but also the protozoans and some other unicellular organisms like yeasts and the true algae. Some of the many differences between eukaryotes and prokaryotes are summarized in Table 1.2. Schematic views of typical animal and plant cells are shown in Figures 1.12 and 1.13. Most eukaryotic cells are larger (by about tenfold) than

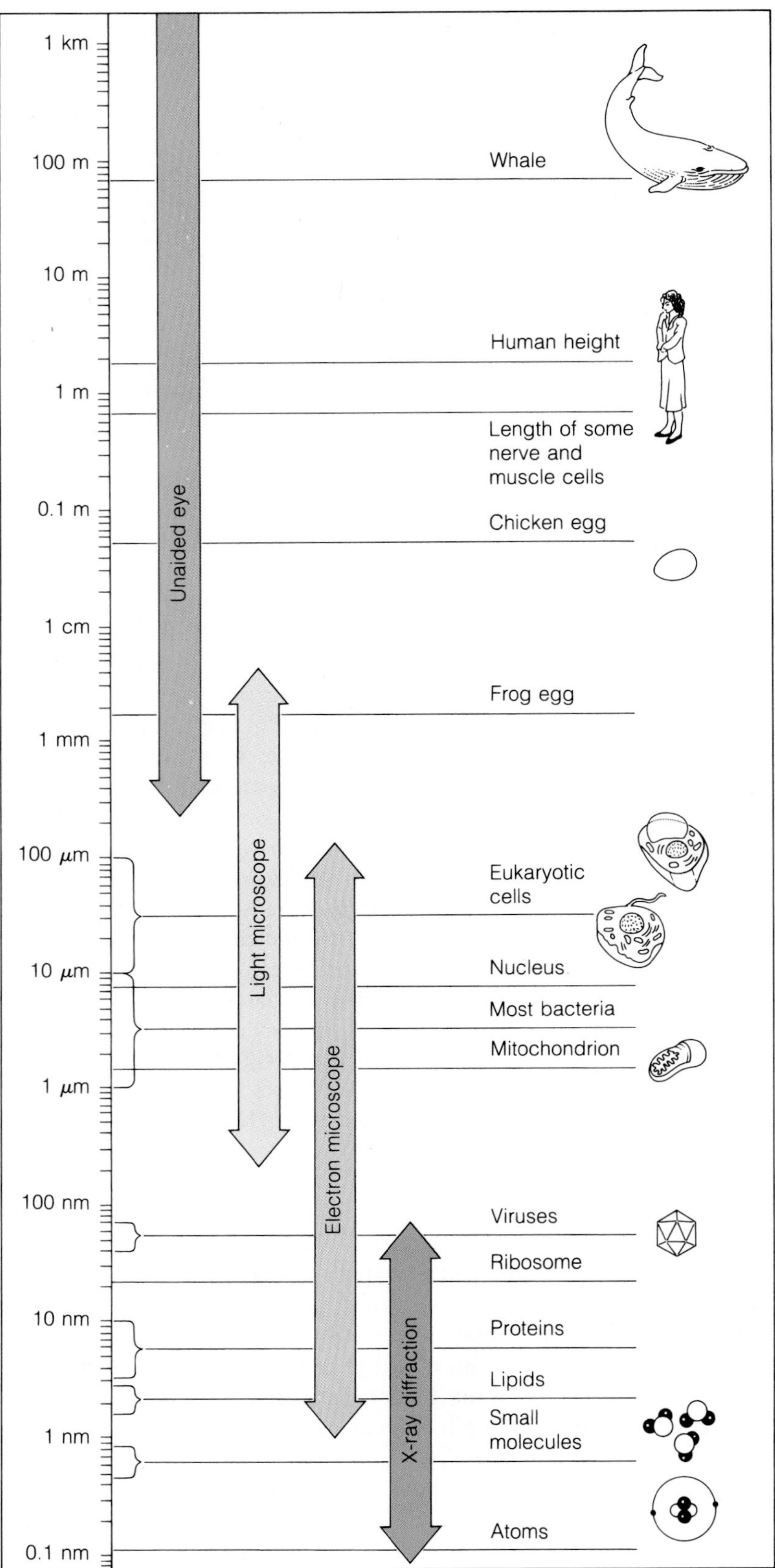

Figure 1.9
Range of sizes of objects studied by biochemists and biologists. The approximate ranges of different methods for observing structure are indicated by the arrows.

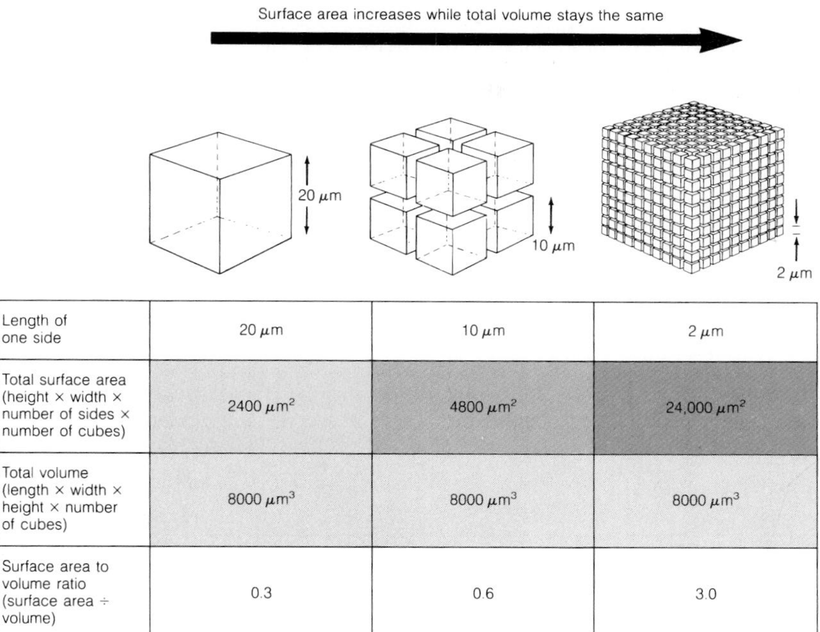

Length of one side	20 μm	10 μm	2 μm
Total surface area (height × width × number of sides × number of cubes)	2400 μm^2	4800 μm^2	24,000 μm^2
Total volume (length × width × height × number of cubes)	8000 μm^3	8000 μm^3	8000 μm^3
Surface area to volume ratio (surface area ÷ volume)	0.3	0.6	3.0

Figure 1.10
How the surface/volume ratio depends on size. If we divide a given volume into smaller and smaller elements, the ratio of surface to volume changes dramatically.

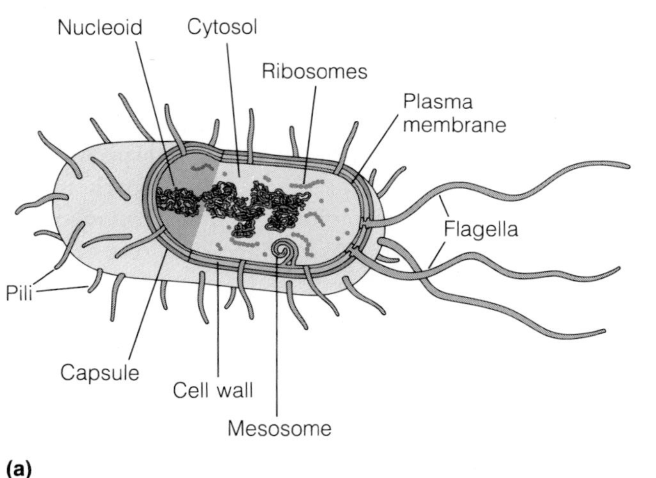

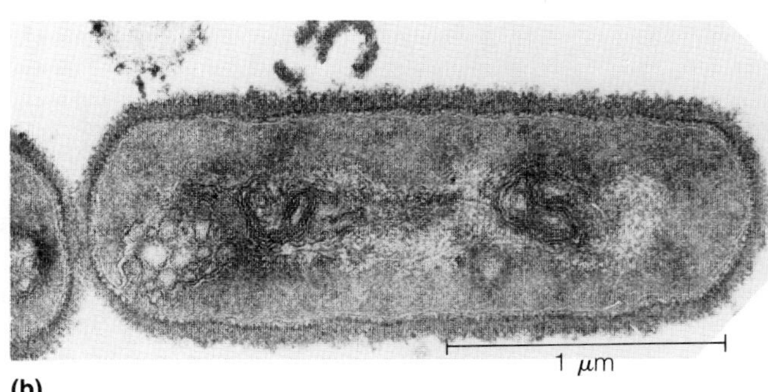

Figure 1.11
Prokaryotic cells. (**a**) Schematic view of a typical prokaryotic (bacterial) cell. The DNA molecule that is the genetic material is coiled up in a region called the nucleoid, which shares the fluid interior of the cell (the cytosol) with ribosomes (which synthesize proteins), other particles, and a large variety of dissolved molecules. The cell is bounded by a plasma membrane, which in some prokaryotes invaginates in places to form structures called mesosomes. Outside the plasma membrane are a fairly rigid cell wall and, often, an outer capsule. Projecting from the surface may be pili, which function to attach the cell to other cells or surfaces, and one or more flagella, which enable the cell to swim through a liquid environment.
(**b**) Electron micrograph of a thin section of a dividing cell of the bacterium *Bacillus coagulans*. The light areas represent the two nucleoids, and the dark granules are ribosomes.

prokaryotic cells, but they compensate for their large size by being *compartmentalized*. Specialized functions are carried out in **organelles**—membrane-surrounded structures lying within the surrounding cytosol. Major organelles common to all eukaryotic cells are the **mitochondria,** which specialize in oxidative metabolism, the **endoplasmic reticulum,** a folded membrane structure rich in ribosomes, the **Golgi complex,** membrane-bounded chambers that function in secretion and the maturation of carbohydrate-linked proteins, and the **nucleus.** The nuclei of eukaryotic cells contain the genetic information, encoded in DNA packaged into **chromosomes.** A portion of this DNA is subpackaged into a dense region within the nucleus, called the **nucleolus.** Surrounding the nucleus is a **nuclear envelope,** pierced by pores through which nucleus and cytosol communicate. **Basal bodies** act as anchors for cilia or flagella in cells that have those appendages.

Table 1.2
Comparison of some properties of prokaryotic and eukaryotic cells

	Prokaryotic Cells	Eukaryotic Cells
Size	0.2–5 μm in diameter	Most are 10–50 μm in diameter
Containment of DNA	Free in cytoplasm as nucleoid	In nucleus, condensed with proteins into multiple chromosomes
Ploidy[a]	Usually haploid	Almost always diploid or polyploid
Mechanism of cell replication	Simple division following DNA replication	Mitosis in somatic cells, meiosis in gametes[b]
Internal compartmentation	No	Yes, with several different kinds of organelles

[a] The term *ploidy* refers to the number of copies of the genetic information carried by each cell. Haploid cells have one copy, diploid cells two, polyploid cells more than two copies.

[b] In mitosis the diploid state is retained by chromosome duplication. This occurs in most *somatic,* or "body," cells of organisms. In the gametes (cells that lead to sperm or ova) there is a somewhat different process called meiosis, which leads to a haploid state. See Chapter 28 for details.

There are also organelles specific to plant and animal cells, as can be seen by comparing Figures 1.12 and 1.13. For example, animal cells contain digestive bodies called **lysosomes,** whereas plant cells have **chloroplasts,** the sites of photosynthesis, and **peroxisomes,** which aid in metabolism. Another common feature of most plant cells is the large, water-filled **vacuole** shown in Figure 1.13. Furthermore, whereas most animal cells are surrounded by a single **plasma membrane,** plant cells often have a tough cellulosic **cell wall** outside the membrane.

It is often useful to think of the cell as being like a factory, an analogy we shall frequently use in later chapters. Membranes enclose the structure and separate different organelles, which can be thought of as departments with specialized functions. The nucleus, for example, is the central administration. It contains in its DNA a library of information for cellular structures and processes. From it, instructions are issued for proper regulation of the business of the cell. The chloroplasts and mitochondria are power generators (the former being solar, the latter fuel-burning). The cytosol can be thought of as the general work area, where protein machinery (enzymes) carries out the formation of new molecules from imported raw materials. There are special molecular channels in the membranes between compartments and between the cell and its surroundings that monitor the flow of molecules in appropriate directions. Like factories, cells tend to specialize in function; for example, many of the cells of higher organisms are largely devoted to the production and export of one or a few molecular products.

The cells depicted in Figures 1.11, 1.12, and 1.13 are idealized representatives of their classes. Figure 1.14, which presents a gallery of cell types, gives a better idea of the diversity of forms that have evolved to suit various organismal needs. Diverse as these cells may be, they are all composed of the same kinds of proteins, lipids, polysaccharides, and nucleic acids as we discussed in the preceding section. Figure 1.15 gives an idea of how these types of molecules are distributed in a cell.

(Text continues on p. 22)

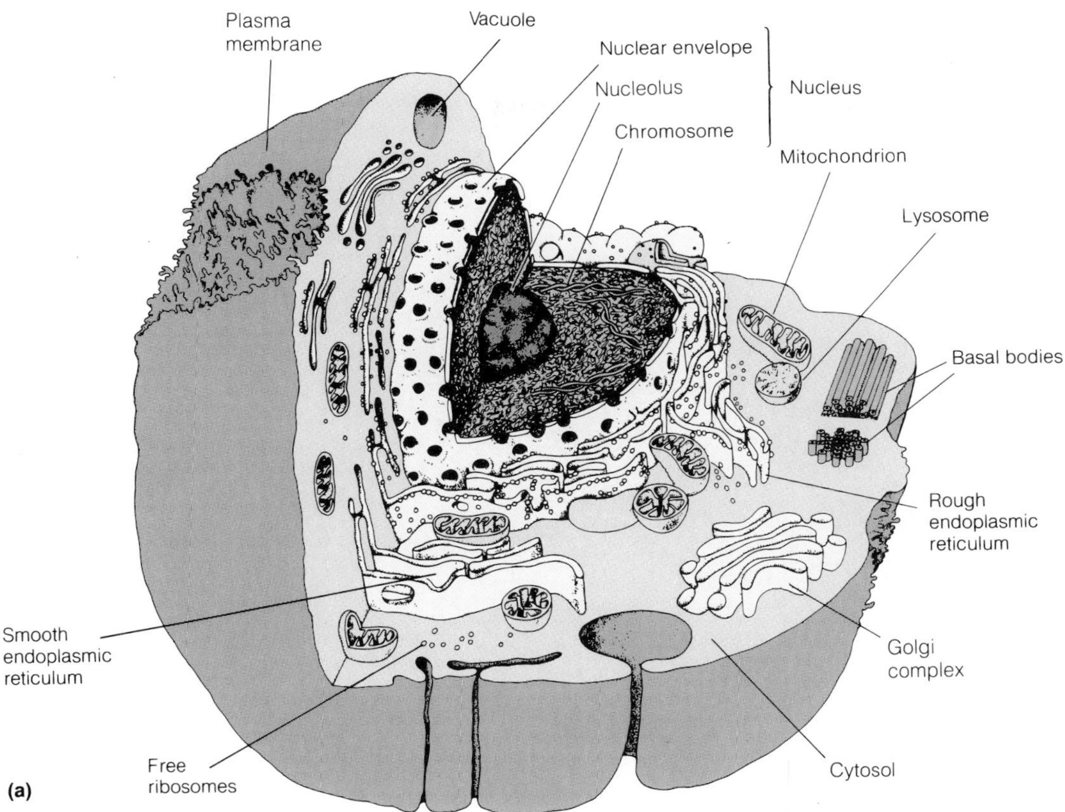

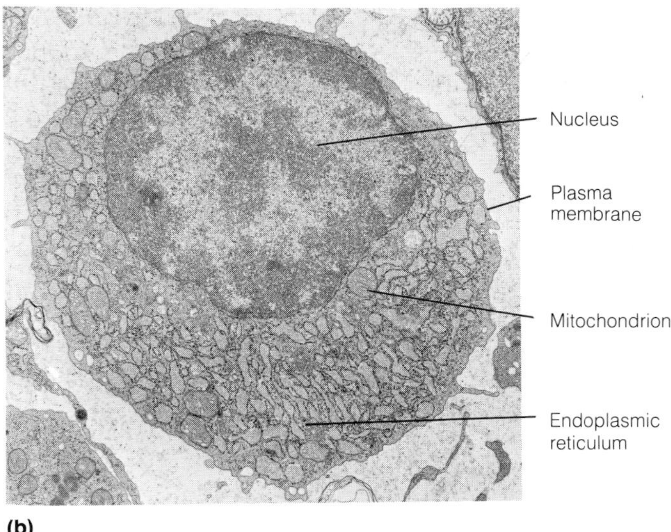

Figure 1.12
Animal cells. (**a**) Schematic view of a typical cell. (**b**) Electron micrograph of a thin section of a white blood cell.

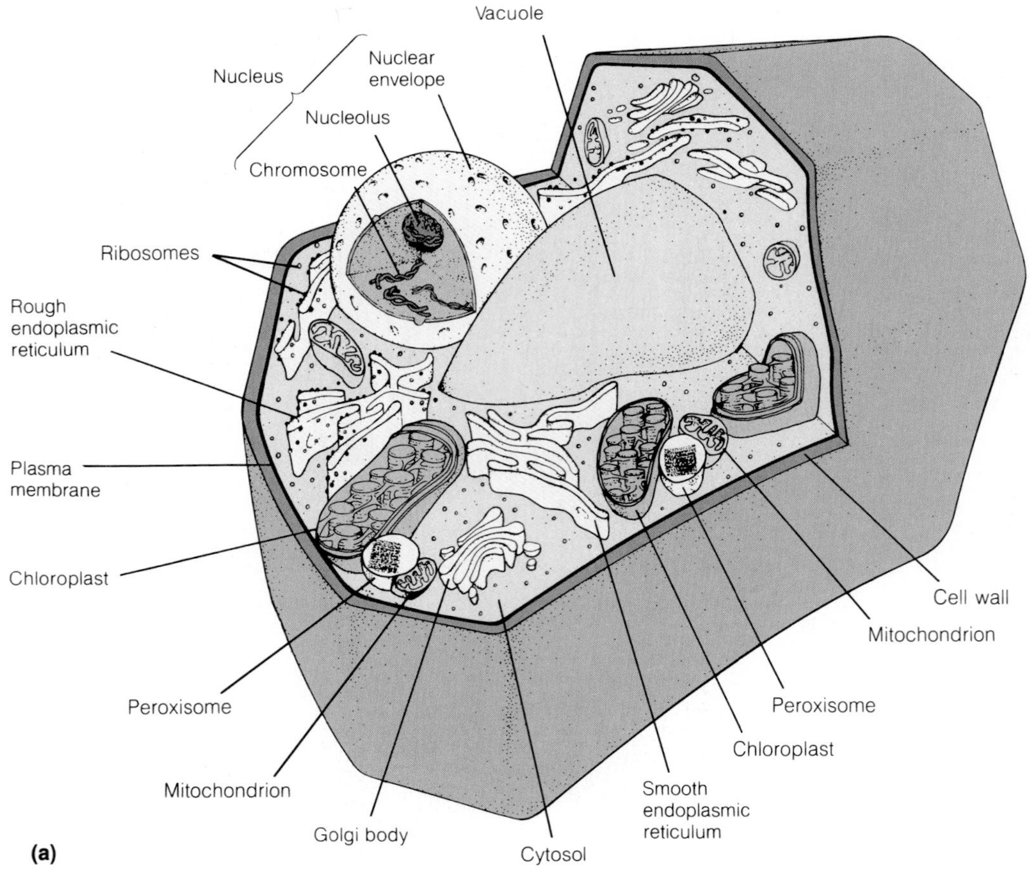

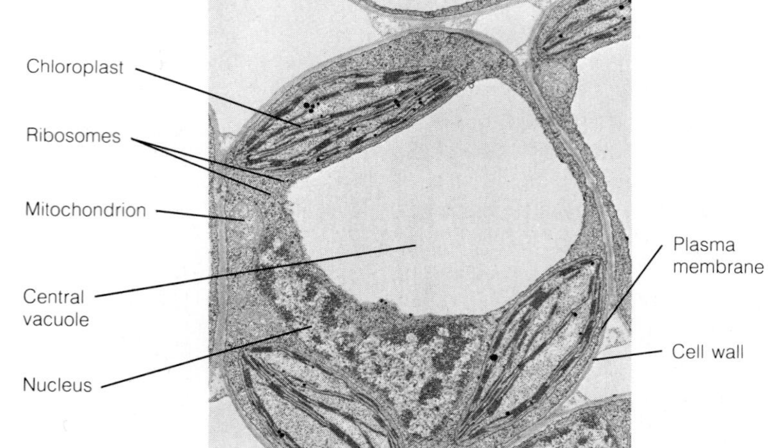

Figure 1.13
Plant cells. (**a**) Schematic drawing of a typical plant cell. Note the chloroplasts, large vacuole, and rigid cell wall, which distinguish this cell from the animal cell in Figure 1.12. (**b**) Electron micrograph of a thin section of a cell from a *Coleus* leaf.

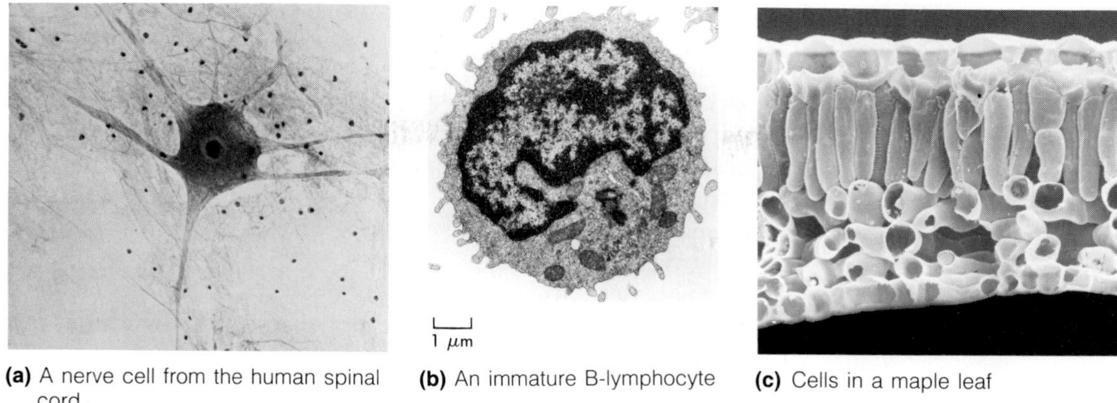

(a) A nerve cell from the human spinal cord **(b)** An immature B-lymphocyte **(c)** Cells in a maple leaf

Figure 1.14
A gallery of cells. Panel **a** is by light microscopy; panel **b** is by transmission electron microscopy; panel **c** is by scanning electron microscopy.

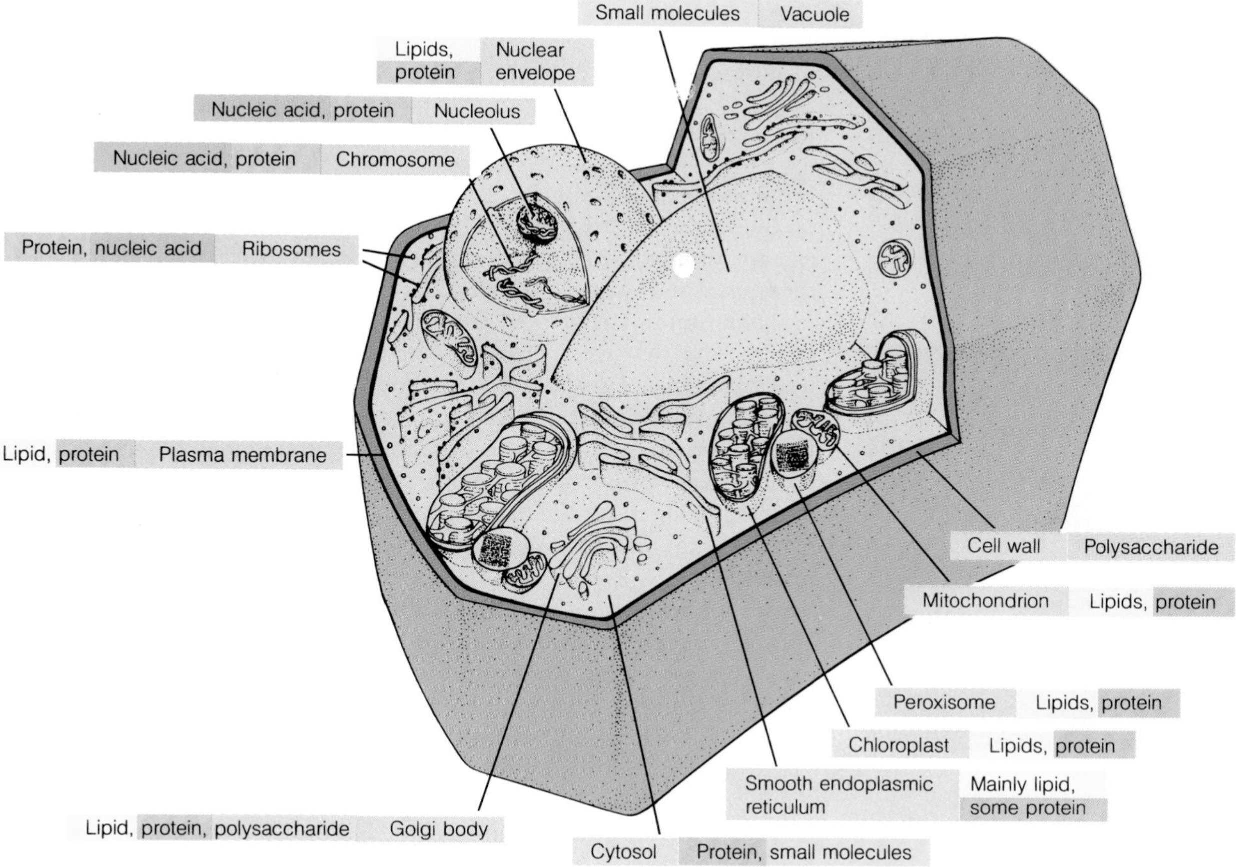

Figure 1.15
Distribution of biomolecules in a cell. The plant cell depicted in Figure 1.13(a) is shown here with the predominant molecular composition of different structural elements identified.

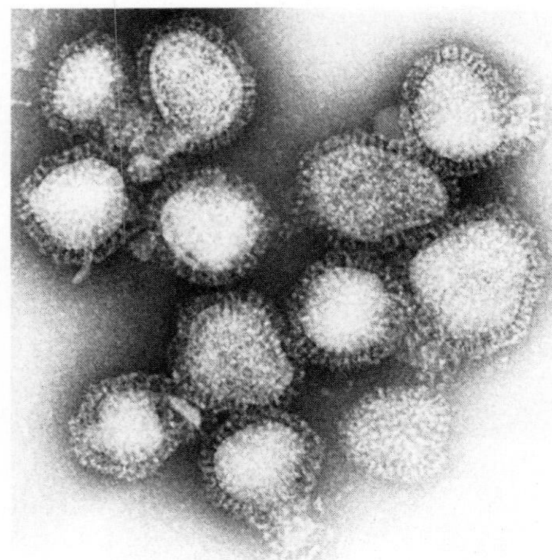

(a) Influenza virus

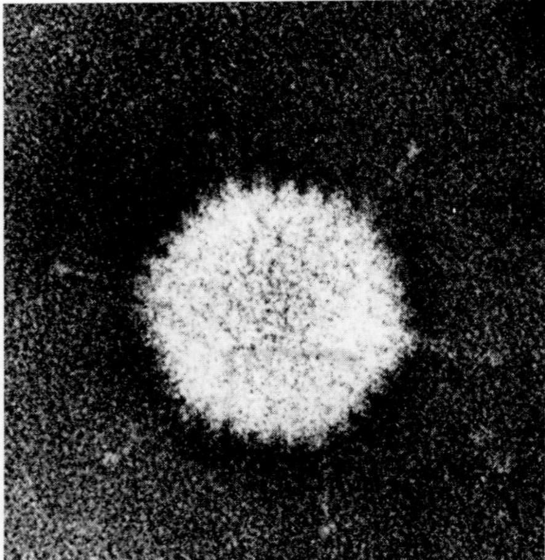

(b) Adenovirus

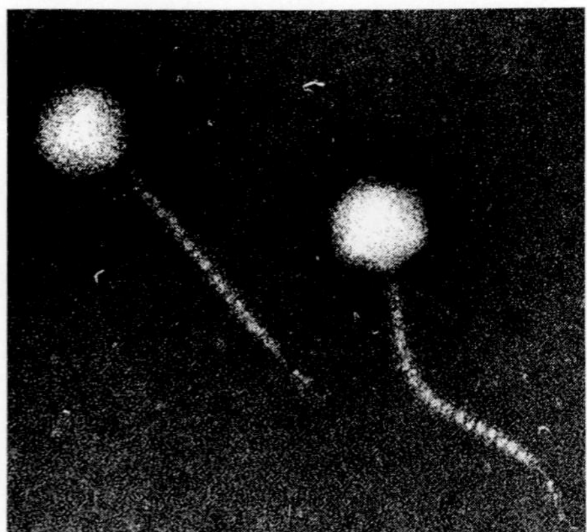

(c) Bacteriophage lamba

Figure 1.16
A gallery of viruses.

Windows on Cellular Function: The Viruses

In dissecting metabolism, biochemists have been aided immeasurably by the **viruses.** Viruses are intracellular parasites, biological entities that can grow only by invading cells. Viruses usually consist of one molecule of nucleic acid wrapped in an envelope made largely or completely of protein. The envelope is specialized to allow the virus particle to enter particular plant, animal, or bacterial cells. Figure 1.16 shows the structures of several representative viruses.

Because viruses contain no metabolic machinery of their own, they must use that of the host cell in order to reproduce. They thus provide useful windows onto the cellular functions that are co-opted during infection. For example, the smallest DNA-containing viruses replicate that DNA using host cell enzymes exclusively. Because of their small size, these viral DNA molecules can be isolated and characterized much more easily than the giant DNA molecules in cellular chromosomes. Many important aspects of cellular DNA replication have been revealed through analysis of intermediates in viral DNA replication. Larger viruses stimulate the formation of new enzymes after infection, with viral genes specifying the structures of these enzymes. Studies of the synthesis of these virus-coded enzymes have allowed insights into mechanisms controlling the expression of genes.

New Methodologies in the Biological Revolution

From its very beginnings, biochemistry has been much more an experimental than a theoretical science. Knowledge has developed more from attempts to explain laboratory observations than from experiments designed to test theories. The Buchner brothers, for example, did not set out to prove that fermentation of sugar to ethanol could occur outside the cell. Rather, their observation that this occurred spurred the developments that led to our understanding of carbohydrate metabolism.

The experimental nature of biochemistry has two important corollaries. First, in order to understand biochemistry, you must understand the critical experiments and experimental techniques on which our common understanding rests. That is why this book includes separate descriptions of experimental methodologies—the Tools of Biochemistry sections following many chapters. Second, the spectacular growth of biochemical knowledge over the past four decades is due directly to the development of many exceedingly powerful research methodologies—the equipment of the biological revolution. Some of this development is chronicled in Figure 1.17. Probably the example most familiar to you is gene cloning, which dates from about 1973. The ability to isolate any desired gene, to read its information, and to manipulate it so that any gene product can be formed in essentially unlimited amounts has transformed all of biology in the most profound ways. However, gene cloning is but one of a long series of experimental advances that have transformed biochemistry and biology in equally fundamental ways. For example, the introduction of radioisotopes was essential to the elucidation of most metabolic pathways, x-ray crystallography permitted the structural localization of every atom within a protein structure, and two-dimensional gel electrophoresis permitted the visualization and quantitation of virtually all of the thousands of protein molecules present in a cell. The truly vast amount of information that can be generated by the new methodologies would be almost unusable were it not for parallel advances in computer compilation and processing of data. For example, models like that shown in Figure 1.6 are generated entirely by computer graphics, using the x-ray diffraction data.

The Uses of Biochemistry

Biochemistry is a research discipline, but the results of biochemical research are used extensively in the world outside the laboratory.

Biochemistry finds its major applications in agriculture and in medical sciences. Areas relevant to both of these fields are human and animal nutrition, where biochemistry relates the dietary requirements of an animal to the metabolic utilization and fates of a nutrient. Why, for example, does fat have a higher caloric content than carbohydrate? Why are vitamins essential to health? In clinical chemistry, biochemical measurements on people reveal clues to diagnoses of illnesses or allow monitoring responses to treatment. For example, the detection of certain enzymes in blood serum is often a clue to internal damage to a tissue, which released that enzyme from its cells.

Pharmacology and toxicology are concerned with the effects of external chemical substances on metabolism. Drugs and poisons usually act by interfering with specific metabolic pathways. A good example is the antibiotic penicillin, which kills bacteria by inhibiting an enzyme that synthesizes an essential polysaccharide of the bacterial cell wall. Since animal cells do not synthesize these polysaccharides, they are not harmed by this inhibitor and so it can be used therapeutically. A particularly exciting prospect in contemporary biochemistry is that of creating so-called designer drugs. If the target site for action of a drug is a protein enzyme or receptor, then determination of the detailed molecular structure of that target should show us how to design inhibitors that bind to the target with great selectivity. In the mid-1980s the detailed molecular structures of viruses began to be determined, opening the possibility of rational design of antiviral agents as well.

Herbicides and pesticides, in many instances, act in similar ways—by blocking enzymes or receptors in the target organism. Again, biochemistry is involved in understanding the actions of these agents, in trying to increase their selectivity, and in understanding and dealing with mechanisms by which the target organisms become resistant to the agents. This is especially important because the first generations of pesticides and herbicides were so nonspecific in their effects that severe damage to the environment resulted from their indiscriminate use. Organisms other than the target populations were often affected, with unforeseen consequences. Thus, biochemistry has become an important component of environmental science.

As a final example, current studies on enzyme catalysis are opening doors to the use of biochemistry in many manufacturing industries. The precise dissection of the structures of enzymes is revealing general mechanisms by which proteins catalyze reactions, speeding them up enormously, under mild reaction conditions. The knowledge gained can be applied to the development of wholly new synthetic or semisynthetic catalysts, either protein or nonprotein in nature. Thus, reactions unlike any that occur in metabolism may be controlled by "designer enzymes," leading to a whole new era in manufacturing technology.

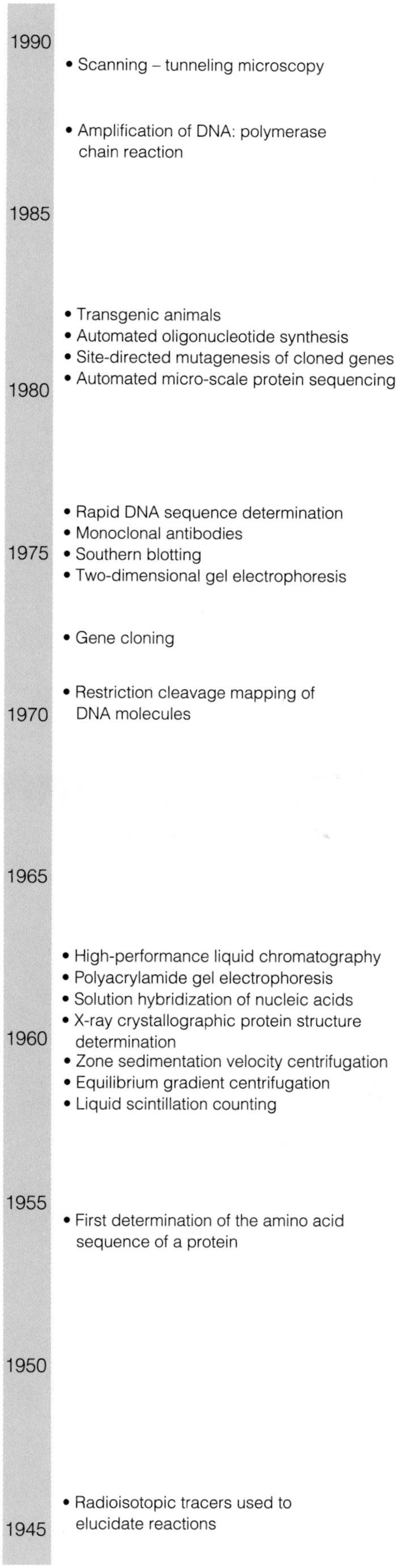

Figure 1.17
A chronicle of recent developments in experimental techniques in biochemistry.

1

TOOLS OF BIOCHEMISTRY 1

The Electron Microscope

Transmission Electron Microscopy

Much of our current knowledge concerning the fine structure of cells, viruses, and even very large molecules comes from electron microscopy. The most widely used instrument, the **transmission electron microscope (TEM)**, bears certain similarities to the familiar light microscope. Figure T1.1 compares them, with the light microscope inverted from its usual orientation to make the correspondence of parts more obvious. The major difference is in the nature of the radiation used—visible light or occasionally ultraviolet light in the light microscope, an electron beam emitted by a tungsten filament in the electron microscope. In the light microscope rays are focused by glass or quartz lenses, whereas in the electron microscope the magnetic fields produced by electromagnetic lenses focus a beam of charged electrons on a fluorescent screen or photographic film.

To understand why the electron microscope can be used to study such finer detail, we must consider a quantity, r, called the **resolution** of a microscope. The resolution is quantitatively the minimum distance between two objects that can just be distinguished as separate. It is given by the equation

$$r = \frac{0.61\lambda}{n \sin \alpha} \qquad \text{(T1.1)}$$

Here λ is the wavelength of the radiation used, n is the refractive index of the medium between the sample and the objective lens, and α is the **angular aperture** of the objective lens (see Figure T1.1). The quantity $\sin \alpha$ is basically a measure of the radiation-gathering power of the lens system. But the major difference between the light microscope and any electron microscope lies in the wavelength, λ, of the illumination used. Resolution depends primarily on wavelength, because the objects must be comparable in size to the wavelength in order to perturb the waves sufficiently to convey information. The angular apertures of the best light microscopes are about 70°, so if light of wavelength 450 nm is used and the medium between the sample and the objective lens is air ($n = 1$), we get

$$r = \frac{0.61 \times 450}{1.0 \times \sin 70°} \cong 300 \text{ nm} = 0.3\ \mu\text{m} \qquad \text{(T1.2)}$$

The wavelength associated with electrons is given by the relation

$$\lambda = \frac{h}{E} \qquad \text{(T1.3)}$$

where E is the energy of the electrons and h is Planck's constant (6.625×10^{-27} erg s = 6.625×10^{-34} J s). When electrons are accelerated by 50–100 thousand volts between the cathode and the anode, their wavelengths are very much shorter than that of visible light—in fact, less than 1 nm. This would predict a resolution of better than 1 nm for a transmission electron microscope. Practical considerations, however, give an operational limit of about 2 nm for most instruments. Still, this is about 100 times finer than even the best optical microscope can accomplish. There is no sense in magnifying an image beyond the point where the resolution, as defined above, becomes what the human eye can resolve. In practice, this means that the best light microscopes have a useful maximum magnifying power of about 1000–2000, since magnifying 0.3 μm by 2000 gives 0.6 mm. Further mag-

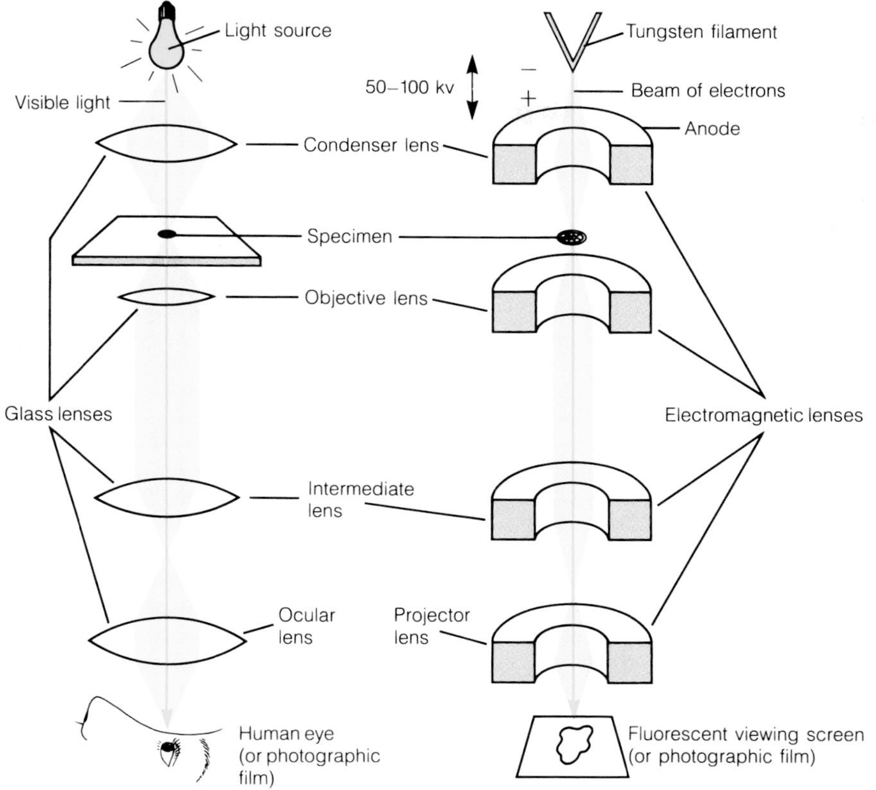

Figure T1.1
Structure of the transmission electron microscope compared to the optical microscope.

nification of the image does not help—the fuzziness just gets bigger. However, a good transmission electron microscope can usefully magnify to over 100,000 times.

Clear as this advantage may be, there are disadvantages to transmission electron microscopy. The electron beam requires that a high vacuum be maintained throughout the instrument, including the sample chamber. This, in turn, means that only completely dried samples can be examined. Despite the fact that many methods for very careful fixation and drying have been devised, there is always the possibility of inducing changes in samples. Living structures, of course, cannot be examined. Furthermore, the electron energies in most transmission microscopes do not allow penetration of thick samples (>100 nm). Thus, sections of cells must be fixed and sliced very thin, using an **ultramicrotome** (Figure T1.2a). Particles like viruses and large molecules can be deposited directly on a thin film supported by a copper grid. But in this case the contrast between particle and background will not be sufficient, so the sample is usually **negatively stained** (Figure T1.2b) or **shadowed** (Figure T1.2c). Representative electron micrographs obtained by each of the techniques described above are shown in the figure. Other special techniques such as *freeze fracturing* and *freeze etching* are discussed in later chapters.

Scanning Electron Microscopy

A quite different kind of technique is called **scanning electron microscopy** (SEM). A schematic diagram of a scanning microscope is shown in Figure T1.3. Here, the electron beam is scanned back and forth across the sample, in a pattern generated by the scan generator and beam deflector, and secondary electrons emitted from the

(Text continues on p. 28)

(a) Sectioning and staining with OsO_4

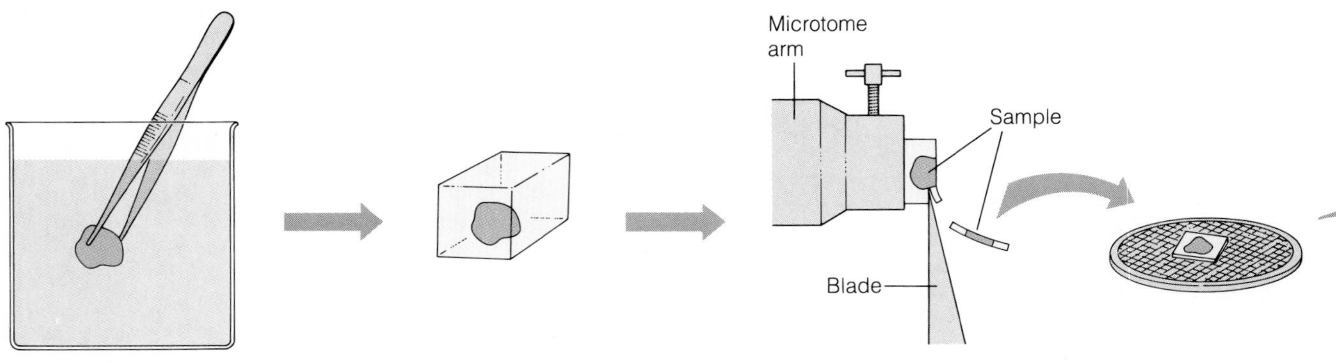

1. The sample is fixed in aldehyde and stained with OsO_4 to enhance contrast.
2. The sample is embedded in a block of plastic.
3. Thin sections are cut on an ultramicrotome.
4. Sections are laid on copper grids for examination.

(b) Negative staining

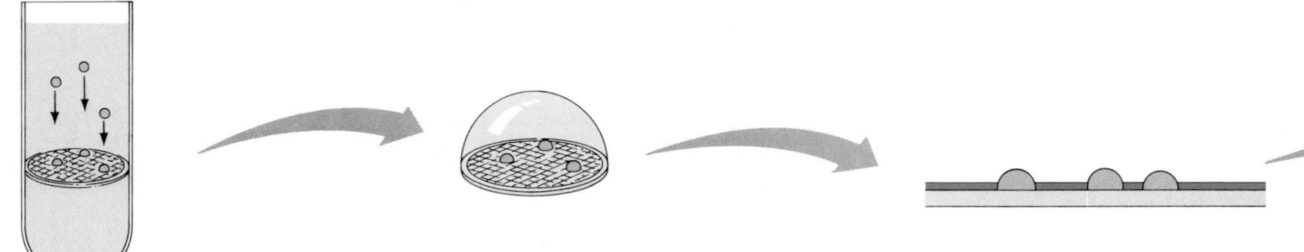

1. Particles are collected on copper grid covered with a thin plastic film.
2. A drop of heavy metal staining solution is placed on the grid.
3. The heavy metal forms a layer around the particle, which then appears more transparent than the background.

(c) Shadowing

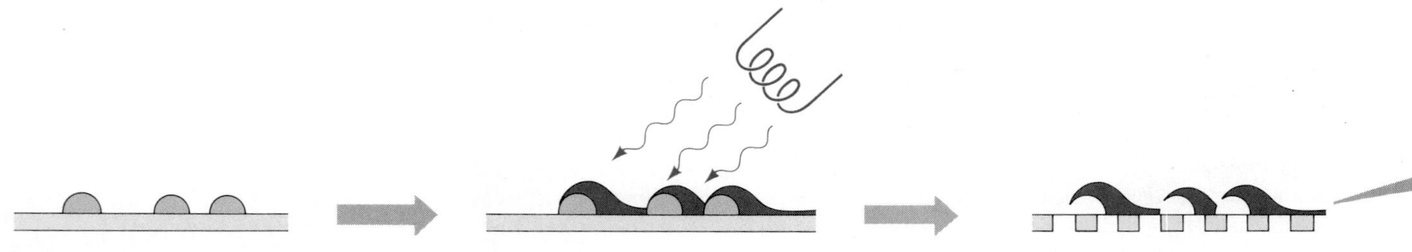

1. Particles are deposited on a mica plate.
2. Metal is deposited from an angle while in a vacuum, forming a replica of the specimen.
3. The specimen is dissolved away, and the metal replica is placed on a grid for examination.

Figure T1.2
Three methods of preparing samples for electron microscopy.

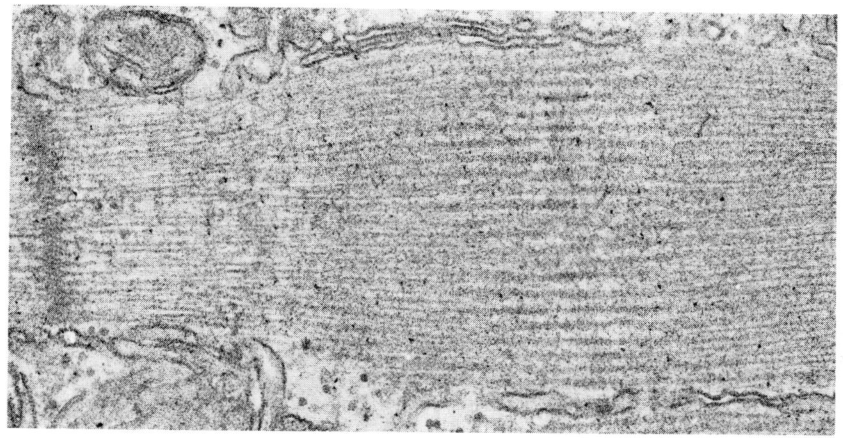

5. Stained skeletal muscle

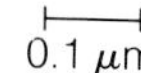

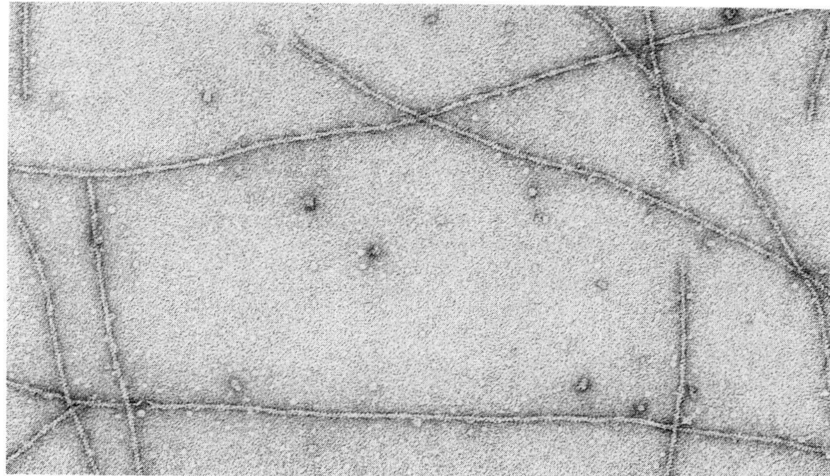

4. Negatively stained muscle protein fibers

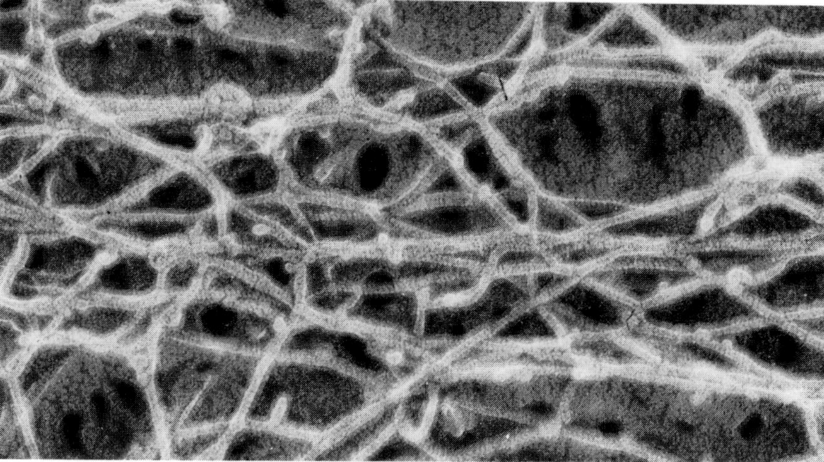

4. Shadowed muscle protein fibers

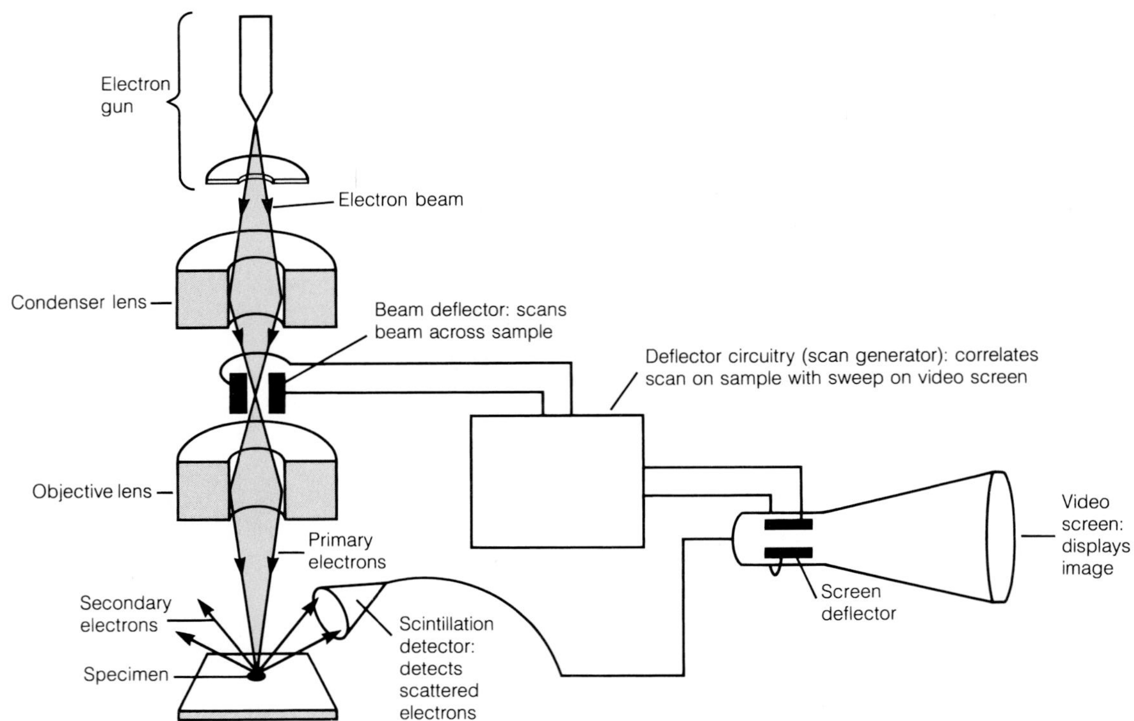

Figure T1.3
The principle of the scanning electron microscope.

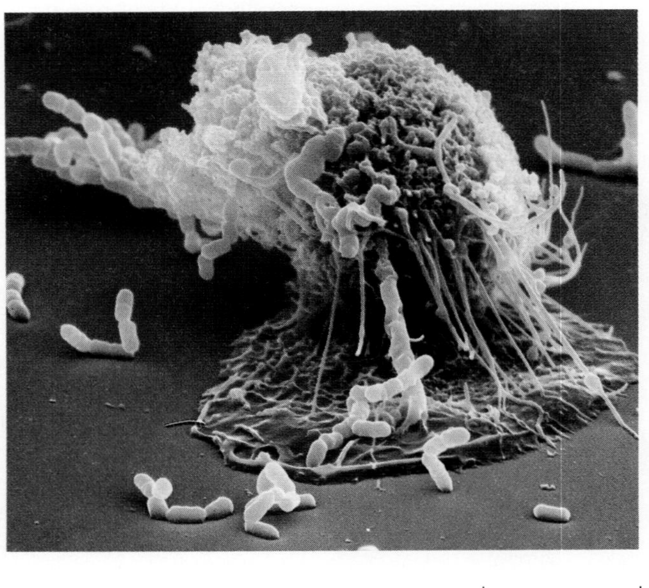

Figure T1.4
A scanning electron micrograph showing phagocytosis, magnified 4300×. A macrophage is engulfing several sausage-shaped *E. coli.*

point at which the beam impinges on the sample surface are picked up by a scintillation detector. The image is then displayed on a video screen, whose surface is scanned in register with the scanning of the sample. The SEM method does not have the resolution of TEM, but it is excellent for obtaining extremely clear views of the surfaces of minute objects (Figure T1.4). Preparation for SEM studies does not require sectioning, but the specimen must be fixed and dried to be stable in the high vacuum and is usually coated with a thin layer of gold to aid in emission of secondary electrons.

Another technique should be mentioned, although it is as yet employed in only a limited number of laboratories. This is **scanning transmission electron microscopy (STEM)**. In this method the beam is scanned over the specimen, as in SEM, but it is detected in transmission. The method has the advantage that unstained, unfixed specimens can sometimes be employed. Furthermore, analysis of the absorption of electrons of different energies can be used to obtain some information about the composition of different portions of the sample. Finally, in early 1989 a new technique, *scanning tunneling microscopy*, was applied to DNA, with exciting results (see Figure 4.1 in Chapter 4).

REFERENCE

Watt, I. M. (1985) *The Principles and Practice of Electron Microscopy*. Cambridge Univ. Press, Cambridge, England.

CHAPTER 2

The Matrix of Life: Weak Interactions in an Aqueous Environment

The macromolecules that participate in the structural and functional matrix of life are immense structures held together by strong, covalent bonds. Yet covalent bonding alone cannot begin to describe the complexity of molecular structure in biology. Most of the elegant cellular architecture visible in the electron micrographs of Chapter 1 is held together by much weaker forces—the **noncovalent interactions** between molecules and between parts of molecules. Consider the macromolecules we discussed in Chapter 1. DNA has a specific three-dimensional structure, maintained by noncovalent interactions. Each kind of protein is folded into a specific molecular conformation. Proteins interact with other protein molecules or DNA to form complex structures. All of this is accounted for by a myriad of noncovalent interactions within and between macromolecules. Moving up a step in the organization of life, we note that the cytosol of a cell is itself a highly organized structure, also held together for the most part by noncovalent interactions.

Thus, if we are to understand life, we must know something about these noncovalent forces. Furthermore, we must know how they will act in an aqueous environment, for every cell in every organism on earth is bathed in and permeated by water. This is as true for creatures living in the most arid deserts as for those in the depths of the sea. Water is, in fact, the major constituent of organisms, 70% or more of the total weight in most cases.

This chapter first describes the nature of noncovalent, or weak, interactions and then focuses on the properties of water that make it so important to life. We shall see that the forces between molecules and ions are greatly modified by the abundance of water in all cells and tissues.

The Nature of Noncovalent Interactions

Molecules and ions can interact with one another in a number of different ways, as summarized in Figure 2.1. All of these noncovalent forces are fundamentally electrostatic in nature, although this may not be immediately

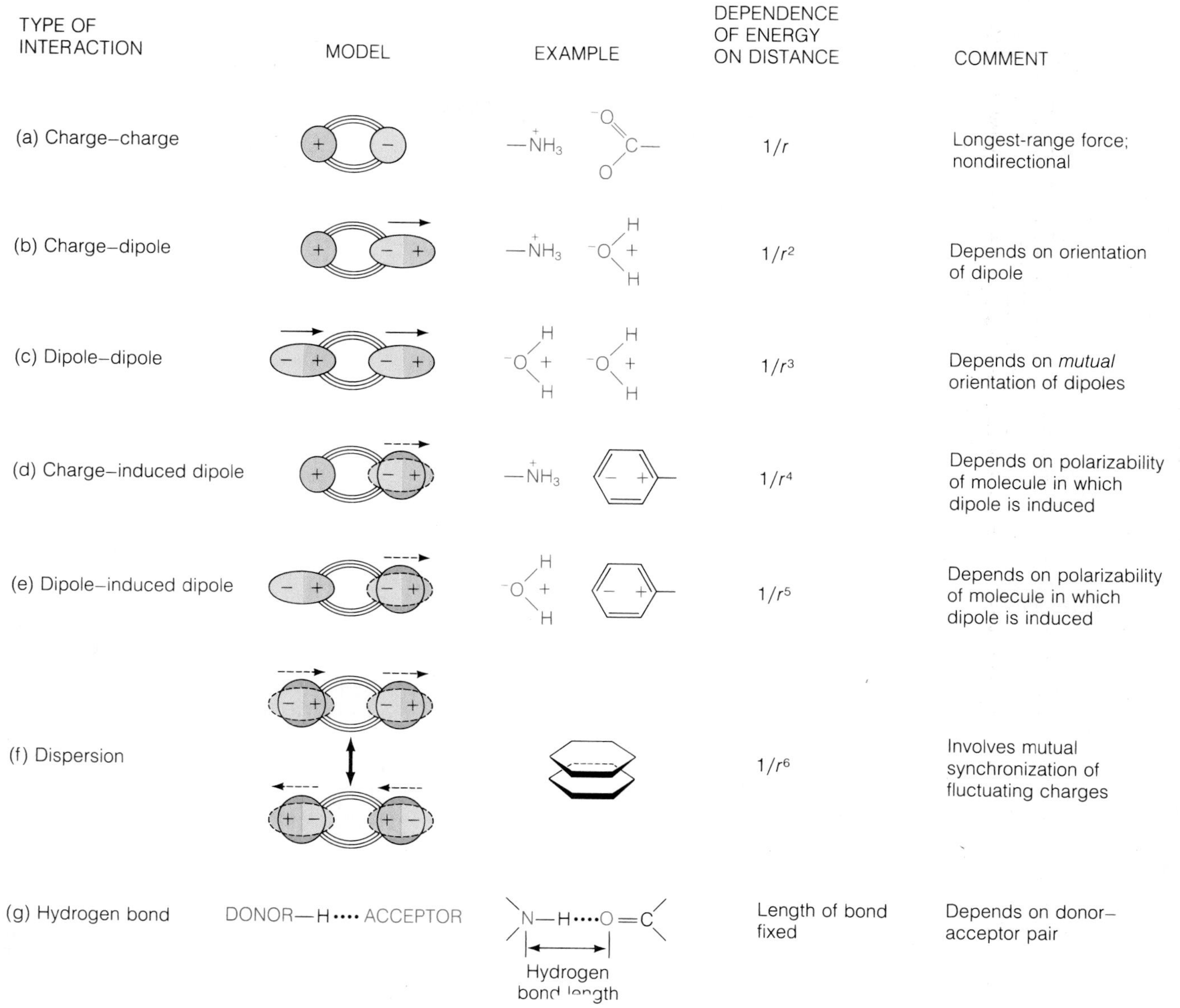

TYPE OF INTERACTION	MODEL	EXAMPLE	DEPENDENCE OF ENERGY ON DISTANCE	COMMENT
(a) Charge–charge		$-\overset{+}{N}H_3$ ^{-}O–C(=O)–	$1/r$	Longest-range force; nondirectional
(b) Charge–dipole		$-\overset{+}{N}H_3$ ^{-}O(H)(H)	$1/r^2$	Depends on orientation of dipole
(c) Dipole–dipole		^{-}O(H)(H) ^{-}O(H)(H)	$1/r^3$	Depends on *mutual* orientation of dipoles
(d) Charge–induced dipole		$-\overset{+}{N}H_3$	$1/r^4$	Depends on polarizability of molecule in which dipole is induced
(e) Dipole–induced dipole		^{-}O(H)(H)	$1/r^5$	Depends on polarizability of molecule in which dipole is induced
(f) Dispersion			$1/r^6$	Involves mutual synchronization of fluctuating charges
(g) Hydrogen bond	DONOR—H···· ACCEPTOR	N—H····O=C (Hydrogen bond length)	Length of bond fixed	Depends on donor–acceptor pair

Figure 2.1
Types of noncovalent interactions. Note that the induced dipole (**d, e**) and the dispersion forces (**f**) depend on a distortion (dashed lines) of the electron distribution in a nonpolar atom or molecule.

obvious in some cases. We shall begin with the simplest kind—the purely electrostatic interaction between a pair of charged particles.

Charge–Charge Interactions

Many of the molecules present in cells, including macromolecules like DNA or proteins, carry a net electrical charge. At the same time, the cell contains an abundance of small ions, both cations like Na^+, K^+, and Mg^{2+} and anions like Cl^- and $HOPO_3{}^{2-}$. All of these charged entities can exert forces on one another (see Figure 2.1a). The force between a pair of charges, q_1 and q_2, separated in a vacuum by a distance r is given by **Coulomb's law:**

$$F = k\,\frac{q_1 q_2}{r^2} \tag{2.1}$$

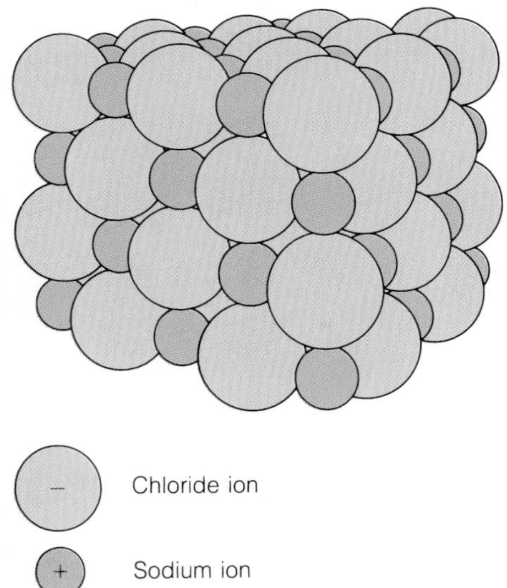

Figure 2.2
A crystal of sodium chloride. Ionic crystals like sodium chloride are held together by the charge–charge interactions between positive ions and negative ions. In sodium chloride, each positive sodium ion is surrounded by six negative chloride ions, while each negative chloride ion is surrounded by six positive sodium ions.

where k is a constant whose value depends on the units used.* If q_1 and q_2 have the same sign, the force is positive, which corresponds to repulsion. If one charge is + and the other −, the force is negative, signifying attraction. It is such interactions that stabilize a crystal of a salt, like that shown in Figure 2.2.

A fundamental problem with equation (2.1) as it stands is that the biological environment is never a vacuum. The charges are always separated by water or by other molecules or parts of molecules. The existence of such a **dielectric medium** between charges has the effect of screening them from one another, so that the actual force is always less than that given by equation (2.1). This screening effect is expressed by inserting a dimensionless number, the **dielectric constant** (ϵ), in equation (2.1):

$$F = k\frac{q_1 q_2}{\epsilon r^2} \tag{2.2}$$

The dielectric constant of water is very high, approximately 80. Organic substances usually have much lower values, in the range 1–10. We shall see presently the reason for this high value in water, but its major consequence is obvious. Charged particles like ions interact rather weakly in an aqueous environment unless they are very close to one another.

Since we will be concerned with the energy changes in biological processes, we will be interested in the **energy of interaction** (***U***). This is the energy required to separate two charged particles from a distance r to an infinite distance—in other words, to pull them apart. The energy is closely related to the force and is given by an equation very similar to (2.2):

$$U = k\frac{q_1 q_2}{\epsilon r} \tag{2.3}$$

The energy of an oppositely charged pair becomes more negative the closer the two particles approach one another. The zero value for the energy is taken to be that for complete separation of the particles. Two major characteristics of charge–charge interactions emerge from equation (2.3). First, the force is wholly *nondirectional,* depending only on the *distance* of separation, and second, the energy varies quite gradually with that distance (inverse first power of r). These characteristics are to be contrasted with those of some of the other noncovalent interactions described in Figure 2.1.

Interactions Involving Permanent Dipoles

Even molecules that carry no net charge may have an asymmetric internal distribution of charge. Consider the water molecule, as shown in Figure 2.3. Although the molecule as a whole is uncharged, the electron distribution within it is such that the end toward the oxygen atoms is slightly negative and the end toward the two hydrogen atoms is slightly positive. Such a molecule is called a **dipole** and is said to have a **permanent dipole moment**

---

* In the c.g.s. (centimeter-gram-second) system, with charges in electrostatic units, k is unity. In this book we use the SI, or international, system of units. Here q_1 and q_2 are in coulombs (C), r is in meters (m), and $k = 1/(4\pi\epsilon_0)$. The quantity ϵ_0 is the *permittivity of a vacuum* and has the value $8.85 \times 10^{-12}\ \mathrm{J^{-1}\ C^2\ m^{-1}}$, where J is the energy unit, joules.

($\boldsymbol{\mu}$). If a molecule has fractional charges $+q$ and $-q$, separated by a distance x, the dipole moment is a *vector,* whose magnitude is

$$\mu = qx \tag{2.4}$$

Some dipole moment values are given in Table 2.1. Note the large value for the amino acid *glycine,* which exists at neutral pH as the **zwitterion** form ($^+H_3NCH_2COO^-$), which has both + and − charges. In this case, whole electron charges are separated by the length of the molecule. In glycylglycine, which is made by covalently linking two glycine molecules, the moment is nearly twice as big.

The importance of permanent dipoles is that they can be attracted by a nearby ion or by one another, as shown in Figure 2.1b and c. Unlike the simple charge–charge interaction, dipolar interactions, which are vectorial, depend on the orientation of the dipoles. Furthermore, they are "shorter-range" interactions, since the energy depends on $1/r^2$ and $1/r^3$ (see Figure 2.1). Thus, a pair of permanent dipoles must be quite close together before the interaction becomes strong.

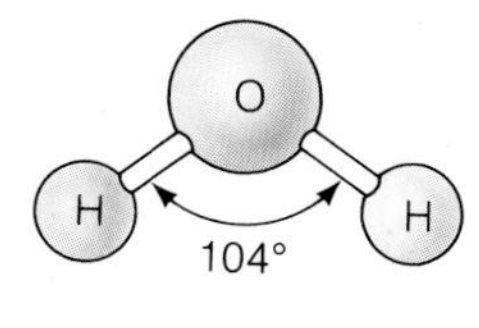

(a)

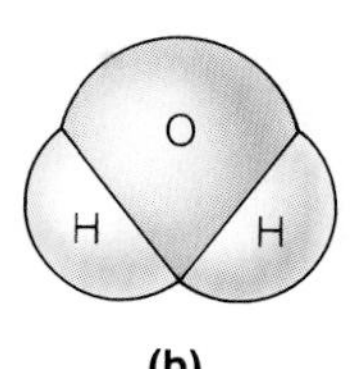

(b)

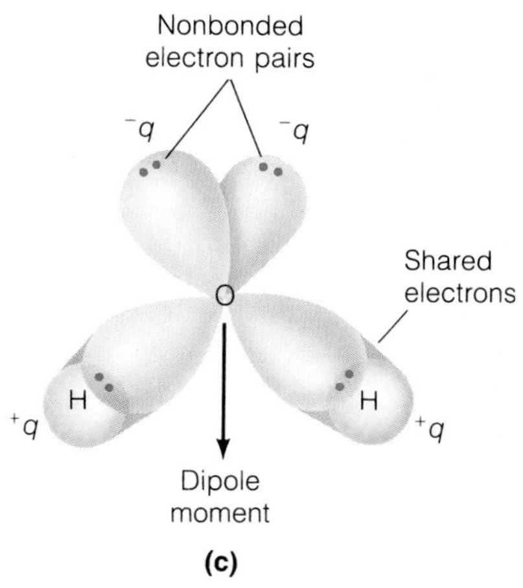

(c)

Figure 2.3
The water molecule as a dipole. (**a**) Ball-and-stick geometry of water, showing asymmetric structure. (**b**) Space-filling model with correct relative atomic dimensions. (**c**) Molecular orbitals of water: four sp^3 orbitals are directed toward the corners of a tetrahedron. Two of the electron pairs of oxygen are nonbonded and the other two electrons are shared with two hydrogens, creating a concentration of negative charge at one end of the molecule and a concentration of positive charge at the other. Because of this, the water molecule has an appreciable dipole moment, which can be thought of as a vector bisecting the angle between the O—H bonds.

Table 2.1
Dipole moments of some molecules

Molecule	Formula	Dipole Moment (D)[a]
Carbon dioxide	O=C=O	0
Ammonia	NH_3 (N bonded to three H)	1.48
Water	H_2O (O bonded to two H, bent)	1.83
Hydrogen sulfide	H_2S (S bonded to two H, bent)	1.02
Methane	CH_4 (H—C—H with H above and below)	0
ortho-Dichlorobenzene	benzene ring with Cl on adjacent carbons	2.59
para-Dichlorobenzene	Cl—C_6H_4—Cl (Cl on opposite carbons)	0
Urea	$O{=}C(NH_2)_2$	4.56
Glycine	$H_3\overset{+}{N}—CH_2—COO^-$	16.7
Glycylglycine	$H_3\overset{+}{N}—CH_2—C({=}O)—N(H)—CH_2COO^-$	28.6

[a]The common units of dipole moment are *debyes;* 1 debye (D) equals 3.34×10^{-30} C m.

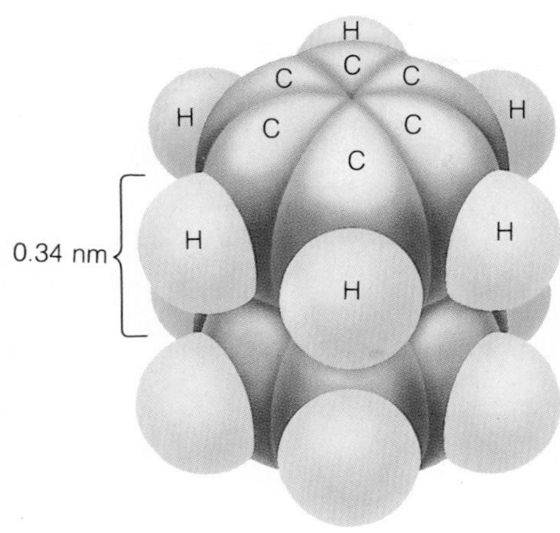

Figure 2.4
Stacking of planar molecules. Molecules like benzene have a strong tendency to stack in the fashion shown here. Fluctuations in the electron clouds in the rings interact with one another, an example of a *dispersion* force. Benzene has neither net charge nor a permanent dipole moment.

Induced Dipole Interactions

Even molecules that do not have permanent dipole moments can become dipolar in the presence of an electrical field. The field may be externally imposed, as in a laboratory instrument, or it may be produced by a neighboring charge or a permanent dipole in the molecular environment. A molecule in which a dipole can be so induced (Figure 2.1d and e) is said to be **polarizable.** Aromatic rings, for example, are very polarizable because the electrons can easily be displaced in the plane of the ring. As the figure shows, induced dipole interactions are even shorter range than permanent dipole interactions.

Even two molecules that have neither net charge nor a permanent dipole moment can attract one another if they are close enough (Figure 2.1f). The electronic charge in a molecule is never static, but fluctuates. When two molecules approach very closely, they synchronize their charge fluctuations so as to give a net attractive force. Such forces are called **dispersion forces,** and their attractive energy varies as the inverse sixth power of the distance. Dispersion forces can be thought of as a mutual dipole induction, significant only at very short range. They can become particularly strong when two planar molecules can stack on one another, as shown in Figure 2.4.

Molecular Repulsion at Extremely Close Approach: The van der Waals Radius

When molecules or atoms come so close together that their outer electron orbitals begin to overlap, there is a mutual repulsion between them. This repulsion increases *very* rapidly as the distance between their centers (r) decreases; it can be approximated as proportional to r^{-12}. If we combine this repulsive energy with one or more of the kinds of attraction between molecules described above, we see that the energy of a pair of molecules, atoms, or ions will vary with distance of separation in the manner depicted in Figure 2.5. Two points should be noted. First, there is a *minimum* in the energy curve at position r_0. This minimum corresponds to the most stable distance between the centers of the two particles. Second, the repulsive potential rises so steeply at shorter distances that it acts as a "wall," effectively barring approach closer than the distance r_v. This distance defines the so-called **van der Waals radius, *R*,** the effective radius for closest molecular packing. For a pair of identical spherical molecules, $r_v = 2R$; for molecules with van der Waals radii R_1 and R_2, $r_v = R_1 + R_2$.

Real molecules, of course, are not spherical objects like those depicted in Figure 2.5. The large molecules so common in biochemistry all have complicated shapes. So it is useful to extend the concept of van der Waals radius to atoms or groups of atoms. Some values are given in Table 2.2; they represent the distance of closest approach for another atom or group. This is the concept behind the construction of space-filling models, such as that shown in Figure 2.6. In this protein molecule each atom or group is represented by a sphere with the appropriate van der Waals radius. Noncovalent attractive forces within the molecule condense it into the compact shape shown, but each atom has its own inviolate space.

Table 2.2
Van der Waals radii of some atoms and groups of atoms

	R (nm)
Atoms	
H	0.12
O	0.14
N	0.15
C	0.17
S	0.18
P	0.19
Groups	
—OH	0.14
$—NH_2$	0.15
$—CH_2—$	0.20
$—CH_3$	0.20
Half-thickness of aromatic ring	0.17

Hydrogen Bonds

One particular kind of interaction is of the greatest importance in biochemistry: the **hydrogen bond** (Figure 2.1g). It is formed between a covalently

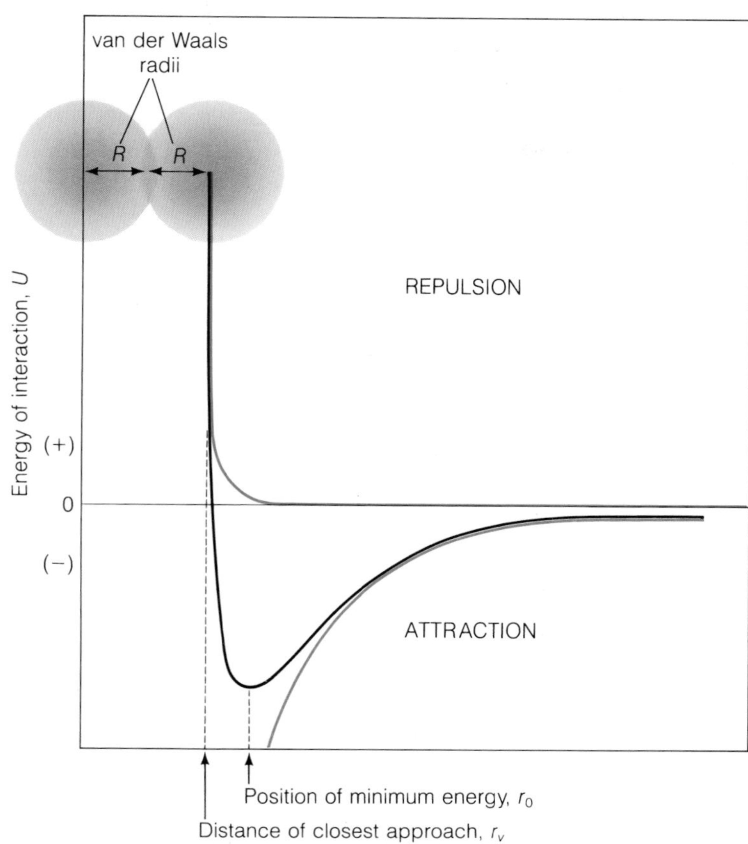

Figure 2.5
Noncovalent interaction energy between two approaching atoms or molecules. The energy (U) required to separate the particles when they are a distance r apart is graphed versus r (solid black line). This energy is the sum of two curves; the red line represents the attractive force, the blue line the repulsive force. The latter changes so rapidly with r that it acts effectively as a barrier, defining the distance of closest approach (r_v) and the van der Waals radii (R). The position of minimum energy (r_0) is usually very close to r_v.

bonded hydrogen atom on a *donor* group (like —OH or >N—H) and a pair of nonbonding electrons on an *acceptor* group (like O=C— or N≦). The strength of a donor depends very much on its electronegativity—that is, how much negative charge has been withdrawn from the hydrogen atom by the donor atom. Among the atoms encountered in biological compounds, only O and N have appropriate electronegativities to serve as donors. Thus, ⋗C—H groups do not form strong hydrogen bonds, but —OH groups do. In this book hydrogen bonds are represented by dotted lines. Examples of particularly strong hydrogen bonds of biological importance are listed in Table 2.3, and an example of hydrogen bonding in a biologically important structure is shown in Figure 2.7.

The hydrogen bond has characteristics of both noncovalent and covalent interactions. On the one hand, there is a major electrostatic contribution to the interaction; the partial positive charge on the H is attracted by the negative charge concentrated on the unpaired electrons of the acceptor. The covalent character of the bond is indicated by the fact that the distances between atoms in hydrogen bonds are considerably closer than would be expected from van der Waals radii. For example, the H···O distance in the bond (>N—H···O=C<) is only about 0.19 nm, whereas we would predict about 0.26 nm from the sum of the van der Waals radii given in Table 2.2. On the other hand, a *covalent* H—O bond has a length of only 0.10 nm. This partial covalent character is also reflected in two other ways. First, the energy of H bonds is considerably higher than that of most other noncovalent interactions. Second, hydrogen bonds are like covalent bonds

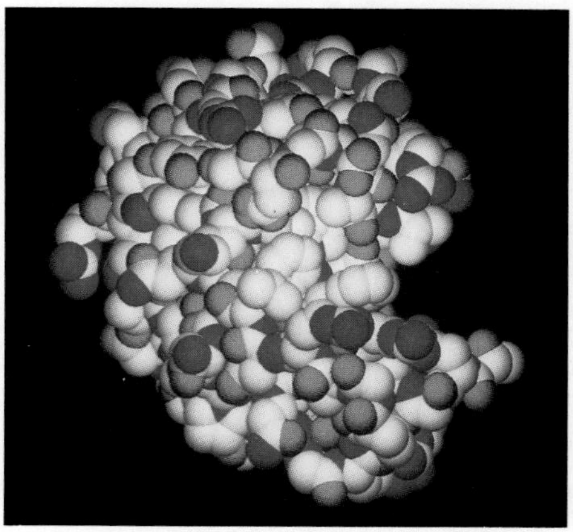

Figure 2.6
Space-filling model of a protein (lysozyme). All of the atoms in this giant molecule are represented as spheres with appropriate van der Waals radii. Note that the atoms pack very tightly against one another.

Table 2.3
Major types of hydrogen bonds found in biologically important molecules

Donor···Acceptor	Bond Length[a] (nm)	Comment
—OH···O(H)<	0.28 ± 0.01	H bond formed in water
—OH···O=C<	0.28 ± 0.01	Bonding of water to other molecules often involves these
>N—H···O(H)<	0.29 ± 0.01	
>N—H···O=C<	0.29 ± 0.01	Very important in protein and nucleic acid structures
>N—H···N<	0.31 ± 0.02	
>N—H···S<	0.37	Relatively rare; weaker than above

[a] Defined as distance from center of donor atom to center of acceptor atom. For example, in the N—H···O=C— bond it is the N—O distance.

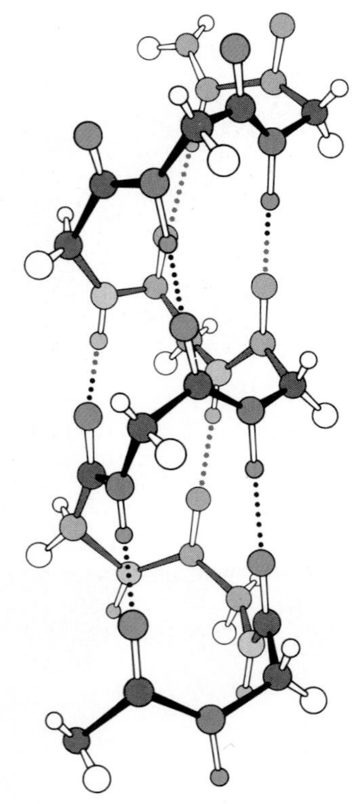

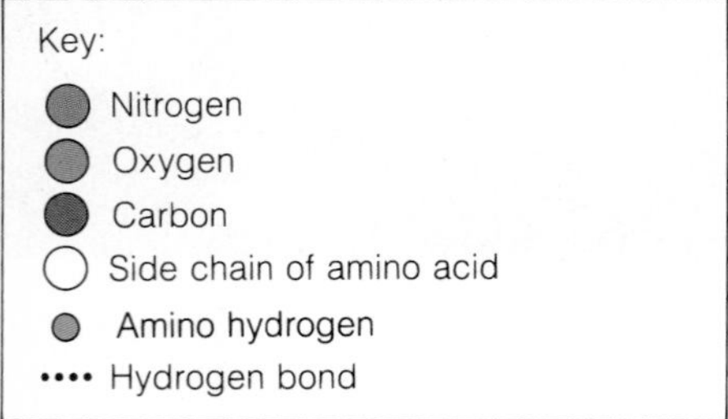

Figure 2.7
Example of hydrogen bonding in biological structure. Shown is a portion of a protein in an α-helix arrangement. The α helix, a common structural element in proteins, is maintained by >N—H···O=C< hydrogen bonds (dotted lines) between groups in the protein chain.

in being highly directional: the donor H bond tends to point directly at the acceptor electron pair. The importance of this directionality is seen in the role hydrogen bonds play in organizing a regular biochemical structure—the α helix in proteins (see Figure 2.7).

The key to the importance of noncovalent interactions is seen in Figure 2.8. These interaction energies are an order of magnitude smaller than covalent bond strengths. Why are weak interactions so important? The answer is that such interactions can continually be broken and reformed under physiological conditions. The dynamics of life depends on rapid changes in intra- and intermolecular interactions, and these could not occur if the forces were too strong.

The Role of Water in Biological Processes

The chemical and physical processes of life require that molecules be able to move about, encounter one another, and change partners frequently in the complicated processes of metabolism and synthesis. A fluid environment allows molecular mobility, and water, the most abundant fluid on earth, is admirably suited to this purpose. To see why, we must examine the properties of water in some detail.

The Structure and Properties of Water

Although we tend to take its properties for granted, water is really a most curious substance. Table 2.4, which contrasts H_2O (molecular weight 18) with other compounds of comparable molecular weight, reveals a remarkable fact. Most such low-molecular-weight compounds are gases at room temperature, having very much lower boiling points than does water. Why is water unique?

The answer lies mainly in the strong tendency of water to form hydrogen bonds. The structure of the water molecule is shown in Figure 2.3. Of the six electrons in the outer orbital of the oxygen atom, two are involved in covalent bonds to the hydrogens; the other four exist in nonbonded pairs. The nonbonded pairs are excellent hydrogen bond acceptors, and the hydrogens from other water molecules can act as donors (Figure 2.9). Thus water is *simultaneously* a hydrogen bond donor and a hydrogen bond acceptor. As a consequence of the strong hydrogen bonding, vaporization of water requires an unusual amount of energy for a molecule of its size. Thus, both the heat of vaporization and the boiling point of water are abnormally high (Table 2.4). Its tendency to such intermolecular interaction enables water to remain in the liquid state at temperatures characteristic of much of the earth's surface.

The hydrogen bonding between water molecules becomes most regular and clearly defined when water is frozen into ice and the regular lattice of hydrogen bonds shown in Figure 2.10 is established. In the ice lattice each water molecule interacts with four neighbors; it acts as a hydrogen bond *donor* for two of these and a hydrogen bond *acceptor* for two others. This rather open structure is only partially dismantled when ice melts, and some long-range order persists even at higher temperatures (Figure 2.11). The structure of liquid water has been described as "flickering clusters" of hydrogen bonds, with remnants of the ice lattice continually breaking and reforming as the molecules move about.

The partial collapse of the relatively open ice lattice in melting produces a rather unusual effect—liquid water is *more* dense than its solid form. This seemingly trivial fact is of the utmost importance for life on earth. If water behaved like most substances and became more dense when frozen, the ice formed on lake and ocean surfaces each winter would sink to the bottom. There, insulated by the overlying layers, it would have accumulated over the ages, and most of the water on earth would by now have become locked up in ice. It seems doubtful that life could have evolved if water at 0°C were even 9% less dense than it is, for then ice would sink.

Other unusual properties of water, as listed in Table 2.5, are also readily explained in terms of its molecular structure. Compared to most organic liquids, water has a relatively high *viscosity*—a characteristic that is also a consequence of the interlocked, hydrogen-bonded structure. This cohesiveness also accounts for the high *surface tension* of water. The high dielectric constant of water, already alluded to, results from its dipolar character. An electric field in an aqueous environment causes extensive orientation of water dipoles and a significant amount of induced polarization. These oriented dipoles contribute to a counterfield, reducing the effective field intensity between, for example, a pair of ions.

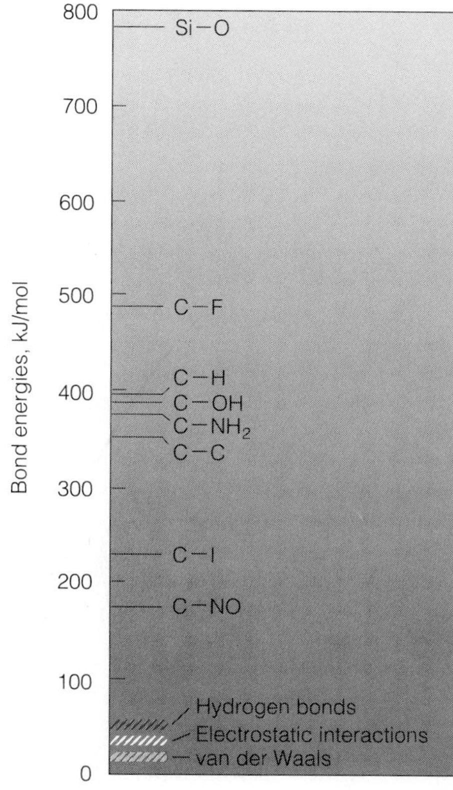

Figure 2.8
Comparison of noncovalent and covalent bond energies. Noncovalent bond energies (blue, white, red) are about one order of magnitude weaker than those in covalent bonds, which is why they are called weak interactions.

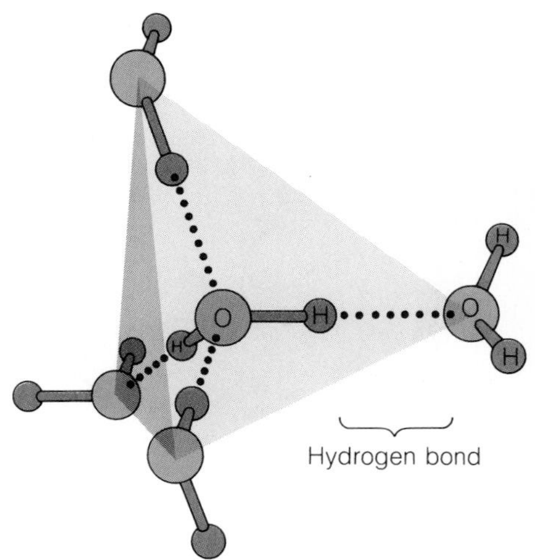

Figure 2.9
Hydrogen bonding in water. Each water molecule is capable of making four hydrogen bonds. It acts as a donor of hydrogen to two water molecules and as an acceptor of hydrogen from two more. Thus, each water molecule sits at the center of a tetrahedron of water molecules.

Table 2.4
Properties of water compared to some other low-molecular-weight compounds

Compound	Molecular Weight	Melting Point (°C)	Boiling Point (°C)	Heat of Vaporization (kJ/mol)
CH_4	16.04	−182	−162	8.16
NH_3	17.03	−78	−33	23.26
H_2O	18.02	0	+100	40.71
H_2S	34.08	−86	−61	18.66

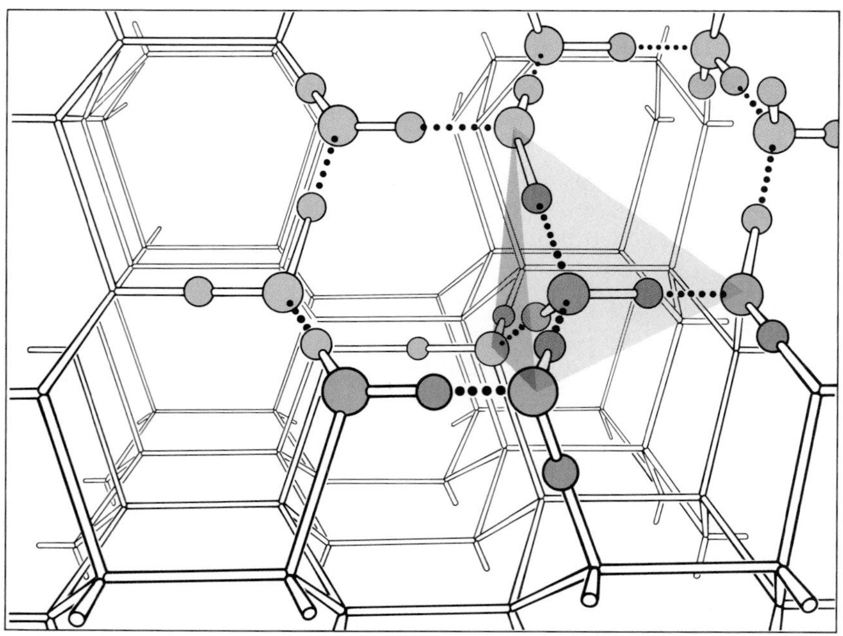

Figure 2.10
Structure of ice. The structure of ice can be considered an indefinite repetition of the tetrahedral hydrogen-bonding pattern shown for a single water molecule in Figure 2.9. Because of the length of the hydrogen bonds, the structure is a relatively open one, which accounts for the low density of ice.

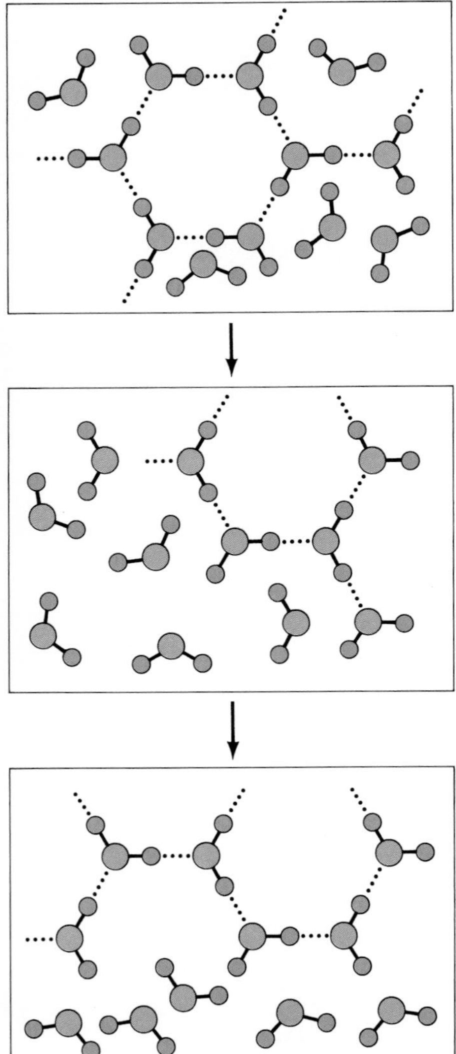

Figure 2.11
Structure of liquid water. When ice melts to water, the regular tetrahedral lattice is broken. However, substantial remnants of it remain, especially at low temperatures. The structure of liquid water can be best thought of as flickering clusters of molecules held together by hydrogen bonds that are continually breaking and reforming. In this schematic "motion picture," successive frames represent changes occurring in picoseconds (10^{-12} s).

Water as a Solvent

The processes of life require solubilization of a wide variety of ions and of molecules large and small. Water's remarkable solvent capabilities enable it to serve as the universal intracellular and extracellular medium. The solvent abilities of water arise primarily from the two properties we have been discussing: its tendency to form hydrogen bonds and its dipolar character. Any molecules that carry groups capable of forming hydrogen bonds can do so with water; such groups will tend to make these molecules **hydrophilic** ("water loving"). Thus, as Table 2.3 implies, water readily dissolves hydroxyl compounds, amines, sulfhydryl compounds, esters, ketones, and a wide variety of other organic compounds. When molecules that contain internal hydrogen bonds (such as the α helix shown in Figure 2.7) dissolve in water, some or all of their internal H bonds may be exchanged for H bonds to H_2O (Figure 2.12). Aliphatic and aromatic hydrocarbons and

Table 2.5
Important properties of liquid water compared to those of a nonpolar, non-hydrogen-bonding liquid[a]

Property	Water	*n*-Pentane
Molecular weight (g/mol)	18.02	72.15
Density (g/cm^3)	0.997	0.626
Boiling point (°C)	100	36.1
Dielectric constant ϵ	78.54	1.84
Viscosity (g/cm s)	0.890×10^{-2}	0.228×10^{-2}
Surface tension (dyne/cm)	71.97	17

[a] All data are for 25°C.

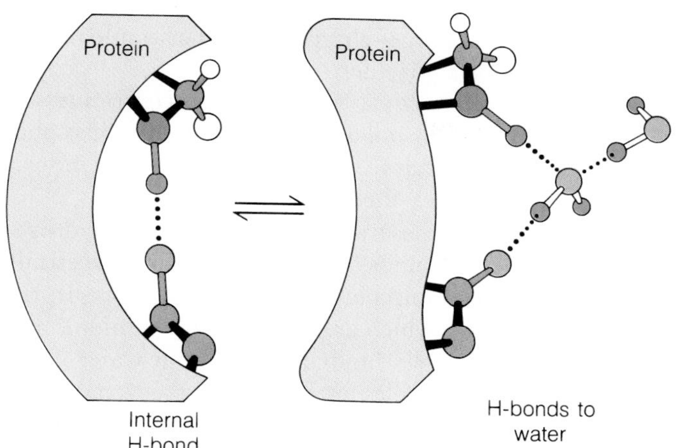

Figure 2.12
Competition of water hydrogen bonding with intramolecular hydrogen bonding. The section of a protein molecule shown in Figure 2.7 is depicted here making hydrogen bonds to solvent water instead of internal hydrogen bonds. Because of the competition with water, the internal hydrogen bonds contribute much less energy to protein structural stability than would be expected from their absolute strength.

many of their derivatives, which cannot form hydrogen bonds, are not water soluble.

But it is not only hydrogen bond acceptors or donors that dissolve well in water. In contrast to most organic liquids, water is an excellent solvent for ionic compounds. Indeed, substances like sodium chloride, which exist in the solid state as very stable lattices of ions, dissolve readily in water (Figure 2.13). The explanation lies in the dipolar nature of the water molecule. Dipoles interact with ions, so cations and anions in aqueous solution are **hydrated,** that is, surrounded by shells of water molecules, called **hydration shells.** The propensity of many ionic compounds like NaCl to dissolve in water can be accounted for largely by two factors. First, the formation of

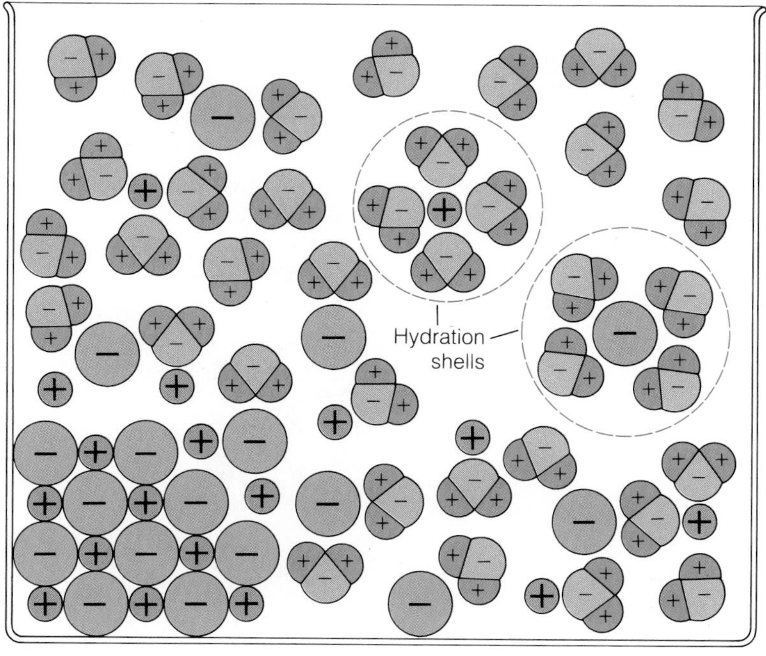

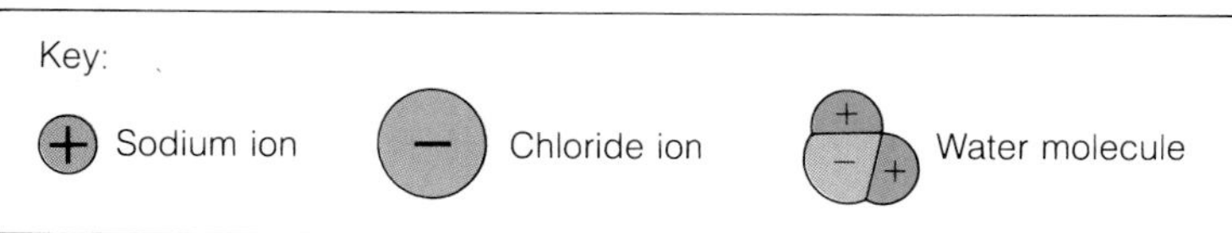

Figure 2.13
Salt crystal dissolving in water. The noncovalent interaction between the sodium and chloride ions and the dipolar, polarizable water molecules causes the formation of a hydration shell about each ion. The energy released in this interaction helps overcome the charge–charge interactions stabilizing the salt crystal.

hydration shells is energetically favorable. Second, the high dielectric constant of water decreases the charge–charge forces that would otherwise pull oppositely charged ions back together.

The dipolar nature of the water molecule also contributes to water's ability to dissolve nonionic, but polar, organic molecules like phenols, esters, and amides. These molecules often have large dipole moments, and dipole–dipole interaction favors their dissolution in water.

The solubility of all of these kinds of substances—hydrogen bond-formers, ions, and polar compounds—depends on their energetically favorable interaction with water molecules. It is therefore not surprising that substances like hydrocarbons, which are nonpolar and nonionic and cannot form hydrogen bonds, show only limited solubility in water. Such **hydrophobic** ("water fearing") molecules interact with the water solvent in a different way. They do not form hydration shells as hydrophilic substances do. Instead, the regular water lattice forms "cages" of icelike **clathrate** structures about nonpolar molecules (Figure 2.14). The ordering of water molecules corresponds to a decrease in the randomness of the mixture—an entropy decrease (see Chapter 3). This makes the solubility even less favorable. We shall return to consideration of these various kinds of water–solute interactions when we discuss the behavior of protein molecules in solution (Chapter 5).

A most interesting and important class of molecules exhibit both hydrophilic and hydrophobic properties simultaneously. Such **amphipathic** molecules have a head group that is strongly hydrophilic, coupled to a hydrophobic tail—usually a hydrocarbon (Figure 2.15). When one attempts to dissolve them in water, amphipathic substances form peculiar structures (Figure 2.16). They may form a **monolayer** on the water surface, with only the head groups immersed. Alternatively, if the mixture is vigorously stirred, **micelles** (spherical structures formed by a single layer of molecules) or **bilayer vesicles** may form. In such cases the hydrocarbon tails of the molecules tend to lie in roughly parallel arrays, which allows them to interact via van der Waals forces. The polar or ionic head groups are strongly hydrated by the water around them. Most important to biochemistry is the fact that amphipathic molecules form the basis of the biological membrane bilayers that surround cells and form the partitions between cellular compartments. We shall have much more to say about such membranes in Chapter 9.

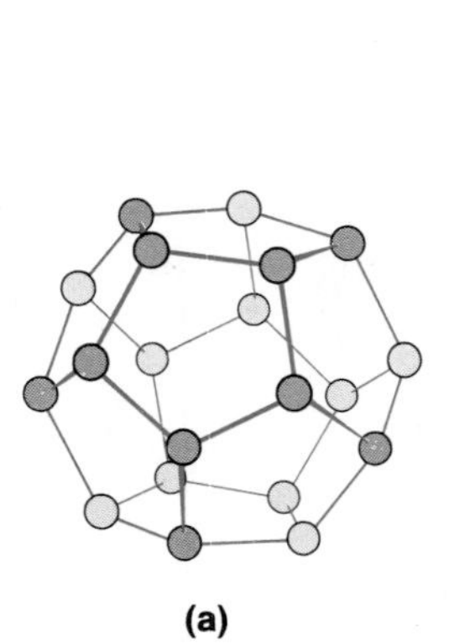

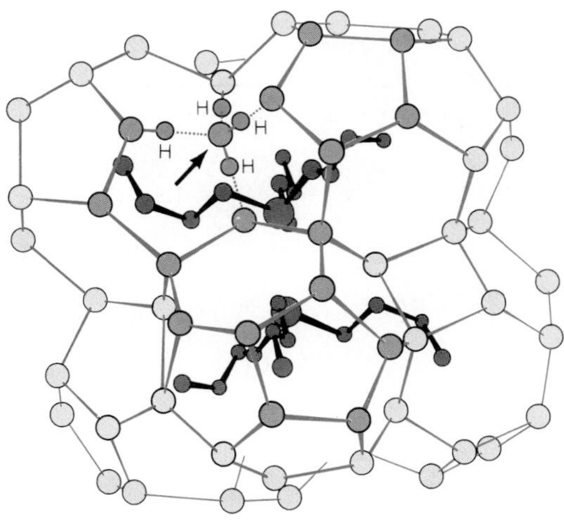

Figure 2.14
A clathrate structure. Molecules like hydrocarbons are only slightly soluble in water. When they do dissolve, clathrate structures (ordered cages of water molecules around hydrocarbon chains) are formed. (**a**) A dodecahedral cage of water molecules, a common building block in clathrates. (**b**) A portion of the cage structure of $(n\text{-}C_4H_9)_3S^+F^-\cdot 23\ H_2O$. This hydrocarbon (shown in black) nests within the hydrogen-bonded framework of the water molecules. In the intact framework, each oxygen is tetrahedrally coordinated to four others. One such oxygen atom and its associated hydrogens are shown by the arrow.

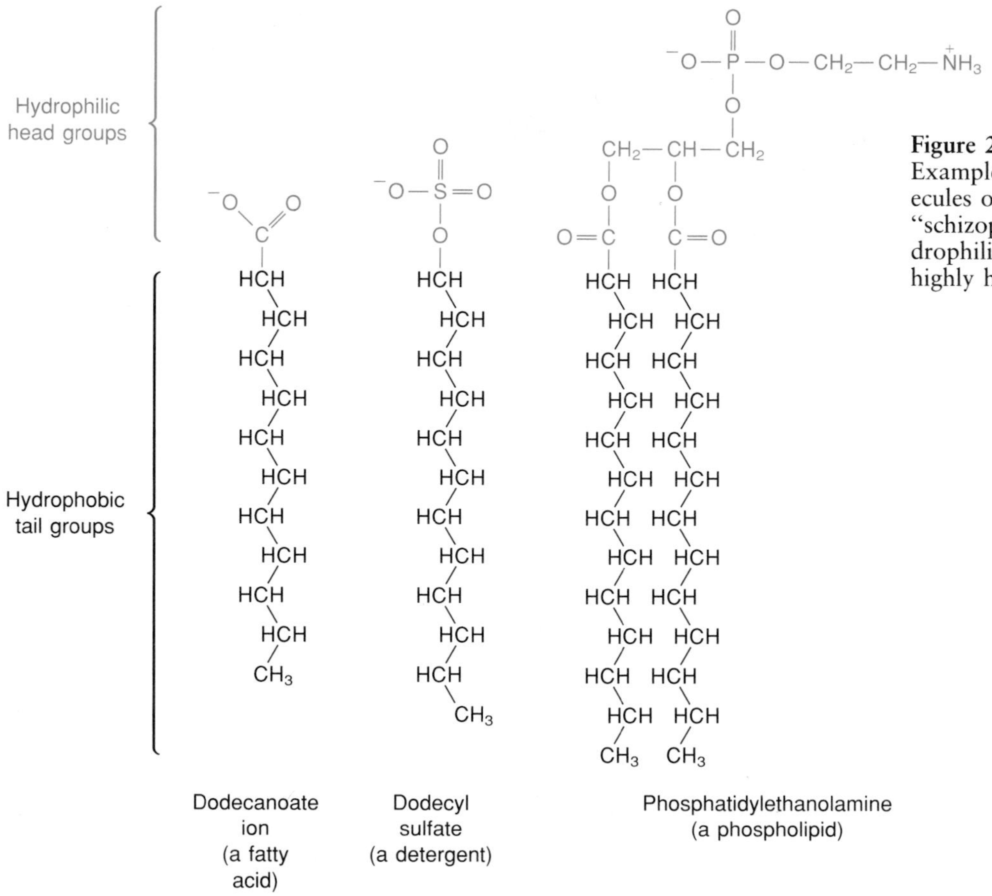

Hydrophilic head groups
Hydrophobic tail groups
Dodecanoate ion (a fatty acid)
Dodecyl sulfate (a detergent)
Phosphatidylethanolamine (a phospholipid)

Figure 2.15
Examples of amphipathic molecules. Molecules of the kind shown here have a "schizophrenic" structure in which a hydrophilic head group is attached to a highly hydrophobic tail.

Ionic Equilibria

Virtually all biochemical reactions occur in an aqueous environment; the exceptions are the few that occur within the hydrophobic interiors of membrane bilayers. The many substances dissolved in the aqueous cytosol and extracellular bodily fluids include not only free ions like K^+, Cl^-, and Mg^{2+} but also both small molecules and macromolecules carrying ionizable groups. To take one example, the very large protein molecules contain on their surfaces both acidic (e.g., carboxylate) and basic (e.g., amino) groups. The behavior of all these molecules in biochemical processes depends strongly on their state of ionization. Thus, it is important that we review briefly some aspects of ionic equilibrium.

SURFACE OF LIQUID
Monolayer
Amphipathic molecules
Micelle
AQUEOUS PHASE
Bilayer vesicle

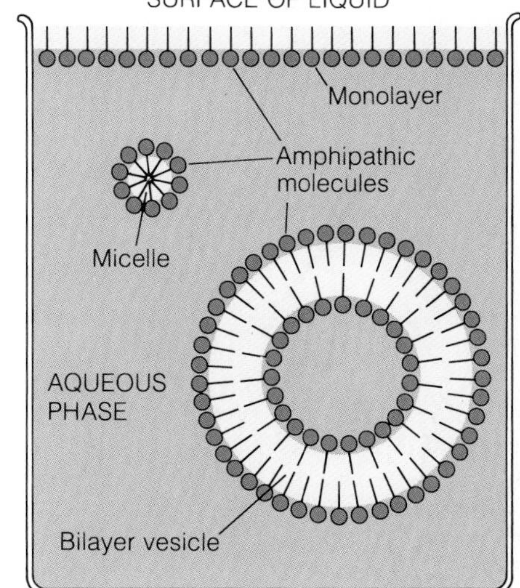

Figure 2.16
Ways in which amphipathic molecules interact with water. In each case, the hydrophilic head groups (blue) are in contact with the aqueous phase, whereas the hydrophobic tails associate with one another. Possible structures are a *monolayer* on the water surface, a *micelle,* and a bilayer *vesicle,* with water both inside and out.

Ionization of Water and the pH Scale

Although it is essentially a neutral molecule, water does have a slight tendency to ionize. The most accurate way to understand the ionization reaction is to note that water can act both as an acid and as a base: one water molecule can transfer a proton to another to yield a hydronium ion (H_3O^+) and a hydroxyl ion (OH^-),

$$H_2O + H_2O \rightleftharpoons H_3O^+ + OH^-$$

Even this is an oversimplification, for the transferred proton may be associated with different clusters of water molecules to yield species like $H_5O_2^+$

and $H_7O_3^+$. A proton in aqueous solution is very mobile, hopping from one water molecule to another with a period of about 10^{-15} second.

For practical purposes, it almost always suffices to write the above equation in the much simpler form

$$H_2O \rightleftharpoons H^+ + OH^-$$

as long as we remember that the proton *never* exists in aqueous solution as a free ion; it is always hydrated by being associated with one or more water molecules. Whenever a reaction is written in which H^+ is involved, keep in mind that it is really a *hydrated* proton.

The equilibrium described above can be expressed in terms of the **ion product**

$$[H^+][OH^-] = K_w = 1 \times 10^{-14}\ \text{M}^2 \qquad (2.5)$$

Since the ion product is a constant, $[H^+]$ and $[OH^-]$ cannot vary independently of one another. If we change either $[H^+]$ or $[OH^-]$ by adding acidic or basic substances to water, the other must change accordingly to satisfy equation (2.5).

If we take pure water at 25°C to which no acidic or basic substances have been added, all the H^+ and OH^- ions must come from the dissociation of the water itself. Under these circumstances the concentrations of H^+ and OH^- must be equal, and therefore

$$[H^+] = [OH^-] = 1 \times 10^{-7}\ \text{M} \qquad (2.6a)$$

and the solution is said to be *neutral.* It should be noted that since the ion product depends on temperature, a neutral solution does not have, under all circumstances, $[H^+]$ and $[OH^-]$ concentrations of 10^{-7} M. For example, at human body temperature (37°C) a neutral solution has

$$[H^+] = [OH^-] = 1.6 \times 10^{-7}\ \text{M} \qquad (2.6b)$$

Hydrogen ion concentration is almost always expressed as pH. In elementary chemistry you learned that the pH of a solution is defined* as

$$\text{pH} = -\log[H^+] \qquad (2.7)$$

A diagrammatic scale of pH values is shown in Figure 2.17, with the pH values for some well-known solutions indicated. Note that body fluids tend to have pH values in the range 6.5–8.0, which is often referred to as the **physiological pH range.** Most biochemistry occurs in this region of the scale.

Because of the sensitivity of biochemical processes to even small pH changes, monitoring pH is essential in most biochemical experiments. Although indicator dyes, often impregnated in "pH paper," can be used to

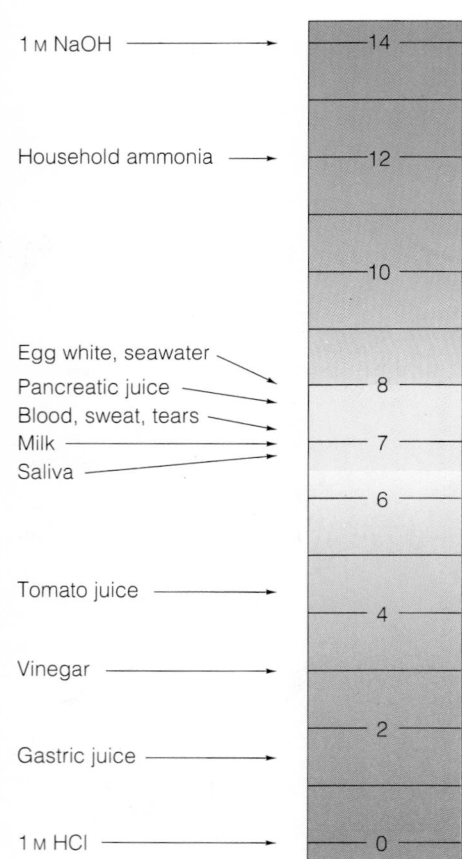

Figure 2.17
The pH scale. The pH values of some common substances and body fluids are listed. The area shaded in blue is basic; that shaded in red is acidic. The neutral pH range between 6.5 and 8.0 (yellow) is where most, but not all, physiological processes occur. Gastric juice, to cite one exception, has a pH between 1 and 2.

* More exactly, we should substitute the hydrogen ion *activity* for the concentration, but the distinction between molar concentrations and activities (activities being effective concentrations, corrected for solution nonideality) is almost always neglected in biochemistry. This is appropriate in most biochemical *experiments,* which are usually conducted in dilute solutions, but it may not be correct *in vivo,* where solutions are concentrated.

Table 2.6
Some weak acids and their conjugate bases

Acid (proton donor)		Conjugate Base (proton acceptor)		pK_a	K_a (mol/liter)
HCOOH formic acid	$\rightleftharpoons$	$HCOO^-$ formate ion	$+ H^+$	3.75	1.78×10^{-4}
CH_3COOH acetic acid	$\rightleftharpoons$	CH_3COO^- acetate ion	$+ H^+$	4.76	1.74×10^{-5}
$CH_3CH_2CH_2COOH$ butyric acid	$\rightleftharpoons$	$CH_3CH_2CH_2COO^-$ butyrate ion	$+ H^+$	4.81	1.54×10^{-5}
$CH_3CH(OH){-}COOH$ lactic acid	$\rightleftharpoons$	$CH_3CH(OH){-}COO^-$ lactate ion	$+ H^+$	3.86	1.38×10^{-4}
H_3PO_4 phosphoric acid	$\rightleftharpoons$	$H_2PO_4^-$ dihydrogen phosphate ion	$+ H^+$	2.14	7.24×10^{-3}
$H_2PO_4^-$ dihydrogen phosphate ion	$\rightleftharpoons$	HPO_4^{2-} monohydrogen phosphate ion	$+ H^+$	6.86	1.38×10^{-7}
HPO_4^{2-} monohydrogen phosphate ion	$\rightleftharpoons$	PO_4^{3-} phosphate ion	$+ H^+$	12.4	3.98×10^{-13}
H_2CO_3 carbonic acid	$\rightleftharpoons$	HCO_3^- bicarbonate ion	$+ H^+$	6.37	4.27×10^{-7}
HCO_3^- bicarbonate ion	$\rightleftharpoons$	CO_3^{2-} carbonate ion	$+ H^+$	10.25	5.62×10^{-11}
C_6H_5OH phenol	$\rightleftharpoons$	$C_6H_5O^-$ phenolate ion	$+ H^+$	9.89	1.29×10^{-10}
$\overset{+}{N}H_4$ ammonium ion	$\rightleftharpoons$	NH_3 ammonia	$+ H^+$	9.25	5.62×10^{-10}
$CH_3CH_2\overset{+}{N}H_3$ ethylamine ion	$\rightleftharpoons$	$CH_3CH_2NH_2$ ethylamine	$+ H^+$	10.8	1.58×10^{-11}

judge pH roughly, almost all measurements are now made with glass electrode pH meters. The electrode generates an electrical potential, which is converted by the instrument into a pH reading.

Acids and Bases

Many biologically important compounds contain weakly acidic and basic groups. The response of such groups to pH changes in or near the physiological range is often of considerable importance to their function. For example, the catalytic efficiency of many enzymes depends critically on the ionization state of certain groups. For such reasons, it is important that we review the dissociation equilibria of weak acids and bases.

You will recall from elementary chemistry that acids are proton donors, and bases are proton acceptors. A **strong acid,** such as HCl, dissociates completely to yield (in this case) H^+ and Cl^-. The H^+ concentration in a solution of HCl in water is almost exactly equal to the molar concentration of HCl added. Similarly, NaOH is called a **strong base** because it ionizes entirely to give OH^- ions, powerful proton acceptors. Most of the acidic and basic substances encountered in biochemistry are **weak acids** or **weak bases.** The dissociation of a weak acid is only partial; there is a measurable equilibrium between the weak acid and its corresponding, or *conjugate,* weak base. Examples are shown in Table 2.6.

In each case the reaction may be written as the dissociation of an acid:

Acid		*Conjugate base*	
HA^+	$\rightleftharpoons$	A	$+ H^+$
HA	$\rightleftharpoons$	A^-	$+ H^+$
HA^-	$\rightleftharpoons$	A^{2-}	$+ H^+$

Note that in some cases the conjugate base has a negative charge and in other cases not. The important point is that in *all* cases it has one less positive charge than the acid. We will, for convenience, always write the reaction as $HA \rightleftharpoons A^- + H^+$. The equilibrium constant for the dissociation of a weak acid is then

$$K_a = \frac{[H^+][A^-]}{[HA]} \tag{2.8}$$

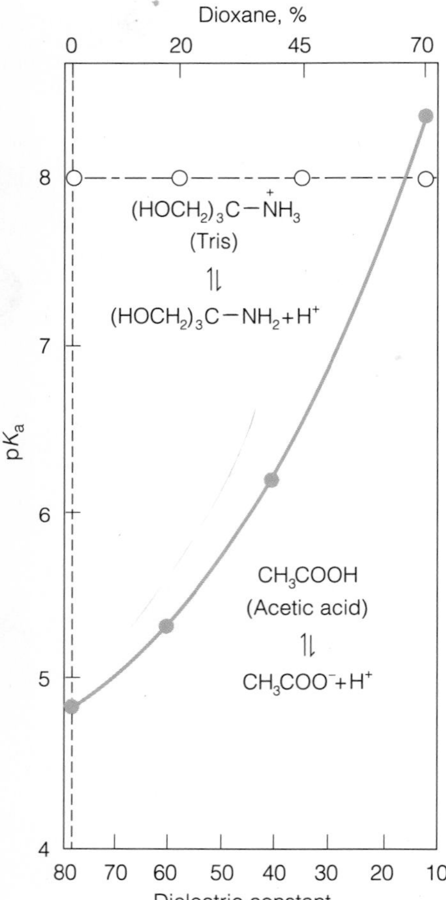

Figure 2.18
Variation of pK_a with dielectric constant of the medium. If dioxane is mixed with water, the dielectric constant is lowered. In such mixtures the pK_a of acetic acid (green) rises with increased dioxane content. Acetic acid becomes weaker because the electrostatic force between the acetate ion and the proton becomes greater. The ionization of tris(hydroxymethyl)aminomethane (Tris), a common buffer used in the laboratory, is uninfluenced by the dielectric constant change because the conjugate base is uncharged.

The larger K_a is, the greater is the tendency to dissociate, and therefore the stronger the acid. More often, the strength of weak acids is expressed in terms of the pK_a of the acid:

$$pK_a = -\log K_a \tag{2.9}$$

Since pK_a is the negative logarithm of K_a, a numerically small value of pK_a corresponds to a strong acid and a numerically large value to a weak acid.

A number of aspects of the behavior of pK_a values can be understood in light of our earlier discussion of the properties of water. Favoring the dissociation of an acid is the hydration of the proton and, in most cases, the conjugate base as well. Exceptions are positively charged acids like $\overset{+}{N}H_4$, which give an uncharged conjugate base. Hydration of $\overset{+}{N}H_4$ stabilizes it and is one reason it is such a weak acid. Opposing ionization is the electrostatic attraction between the proton and a negatively charged conjugate base. This effect can be seen by comparing the successive pK_a values for phosphoric acid dissociation in Table 2.6. As the charge on the conjugate base rises in going from $H_2PO_4^-$ to PO_4^{3-}, the pK_a rises as well. HPO_4^{2-} is a *very* weak acid. Another example of this effect can be seen in Figure 2.18, which shows the pK_a values for acetic acid in water–dioxane mixtures. As the dioxane concentration is increased, the dielectric constant is lowered and the attraction between H^+ and CH_3COO^- increases. Therefore acetic acid becomes a weaker acid as dioxane is added. Note, in comparison, the behavior of Tris in this figure. Since the conjugate base is electrically neutral in this case, no charge separation is involved in dissociation of the acid, and consequently the pK_a is independent of dielectric constant.

These examples are important, for they show how environmental effects can influence pK_a values. When we investigate proteins we will find that the pK_a values for supposedly identical groups can vary widely, depending on the local molecular environment.

Titration of Weak Acids: The Henderson–Hasselbalch Equation

A very useful equation can be obtained by taking the negative logarithm of both sides of equation (2.8):

$$-\log K_a = -\log\left\{\frac{[H^+][A^-]}{[HA]}\right\} \quad (2.10a)$$

or

$$pK_a = -\log\left\{\frac{[H^+][A^-]}{[HA]}\right\} \quad (2.10b)$$

Using the rules for operating with logarithms, this becomes

$$pK_a = -\log[H^+] - \log\left\{\frac{[A^-]}{[HA]}\right\}$$

$$= pH - \log\left\{\frac{[A^-]}{[HA]}\right\} \quad (2.11)$$

Simply by rearranging this, we obtain the *Henderson–Hasselbalch* equation:

$$pH = pK_a + \log\left\{\frac{[A^-]}{[HA]}\right\} \quad (2.12a)$$

or

$$pH = pK_a + \log\left\{\frac{[\text{conjugate base}]}{[\text{acid}]}\right\} \quad (2.12b)$$

Here it must be remembered that [HA] is the concentration of *undissociated* acid present in the solution and $[A^-]$ the concentration of its conjugate base. For example, in the case of formic acid the Henderson–Hasselbalch equation becomes

$$pH = 3.75 + \log\left\{\frac{[HCOO^-]}{[HCOOH]}\right\} \quad (2.13a)$$

and with ammonium ion

$$pH = 9.25 + \log\left\{\frac{[NH_3]}{[\overset{+}{N}H_4]}\right\} \quad (2.13b)$$

The usefulness of this equation comes from the fact that it relates pH to the [conjugate base]/[acid] ratio. Knowing one of these and the pK_a, we can calculate the other. Even more important, the Henderson–Hasselbalch equation allows us to describe the course of titration of a weak acid. Consider formic acid as an example, starting with a 1 M solution and adding strong base to titrate. First, we must ask: What is the initial pH? If 1 mol of formic acid is dissolved in 1 liter of water, some dissociation will occur. To a good approximation (which we will check later) we can say that this dissociation is so slight that the concentration of undissociated formic acid ([HCOOH]) is still about 1 M. Since the pK_a of formic acid is 3.75, $K_a = 10^{-3.75} = 1.78 \times 10^{-4}$. From the equilibrium relation

$$K_a = 1.78 \times 10^{-4} = \frac{[H^+][HCOO^-]}{[HCOOH]} \quad (2.14)$$

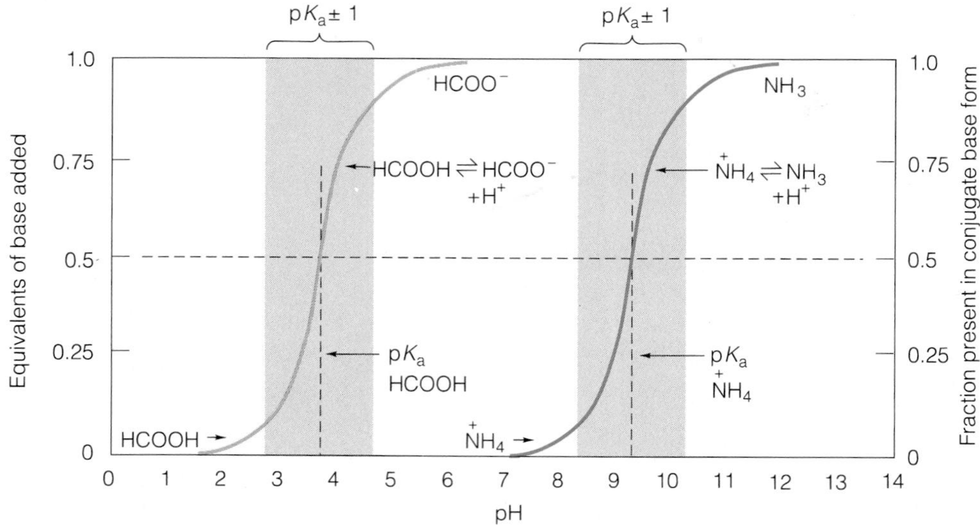

Figure 2.19
Titration curves for formic acid (HCOOH) and ammonium ion ($\overset{+}{N}H_4$). Note that the pK_a corresponds to the half-titration point, and most of the titration occurs in a range of one pH unit on either side of the pK_a (shaded areas). At pH = pK_a the change in pH with equivalents of base added is minimal; thus, this is the best buffering range.

we get $[HCOO^-] = [H^+] = \sqrt{1.78 \times 10^{-4}\,\text{M} \times 1\,\text{M}} = 1.33 \times 10^{-2}\,\text{M}$. The amount of formic acid dissociation is quite small (only about 1%) so our assumption is fairly good. If we wished to be more exact, we could do a series of successive approximations, next setting [HCOOH] = (1.0 − 0.0133) M, and repeating the calculation until no further significant change was observed.*

From the calculated hydrogen ion concentration, we know that the pH of the initial solution is pH = $-\log(1.33 \times 10^{-2}) = 1.88$. Now suppose we begin to titrate this solution by adding a solution of NaOH. As each small amount of NaOH is added, the hydroxyl ions react with hydrogen ions to keep the ion product $[H^+][OH^-] = 10^{-14}$. But as H^+ is removed by hydroxyl ions, more of the formic acid dissociates to maintain the equilibrium expressed by equation (2.14). Therefore, the ratio $[HCOO^-]/[HCOOH]$ increases continually and, as equation (2.13a) shows, the pH rises as the titration continues. The titration curve is shown in Figure 2.19, along with a corresponding curve for the ammonium ion. A point of special importance should be noted in Figure 2.19: At the midpoint of the titration curve, where exactly half of the acid has been titrated, pH = pK_a. This can also be deduced from the Henderson–Hasselbalch equation. When the titration is half complete, $[HA] = [A^-]$; half the substance is in the acid form, half in the form of the conjugate base. Equation (2.12) then becomes

$$\text{pH} = \text{p}K_a + \log 1 = \text{p}K_a \tag{2.15}$$

Thus, *the* pK_a *of a weak acid corresponds to the midpoint of its titration curve.* Furthermore, as examination of Figure 2.19 shows, most of the abrupt change in $[A^-]/[HA]$ occurs within ±1 pH unit of the pK_a. If we know the pK_a of a weak acid, we know the pH range in which it will titrate and precisely where the midpoint of the titration will occur.

It should be emphasized that such titration curves are reversible. If we were to take the final solution, at high pH, and begin adding a strong acid like HCl, the same curve would be retraced in the opposite direction.

---

* An alternative but less informative way to do the problem is to solve a quadratic equation. If $x = [HCOO^-] = [H^+]$, then $K_a = x^2/(1 - x)$.

Buffer Solutions

If we look at Figure 2.19 in a different way, another important point emerges. In the pH range near the pK_a, the pH changes only a little with each increment of base or acid added. In fact, the pH is least changed per increment of acid or base just at the pK_a. This is the principle behind **buffering** of solutions by the use of weak acid–base mixtures, a technique used in virtually every biochemical experiment.

Suppose a biochemist wishes to study a reaction at pH 4.00. The reaction may be one that generates or consumes protons. To prevent the pH from drifting during the reaction, the experimenter should use a buffer solution consisting of a predetermined mixture of a weak acid and its conjugate base. In this example a formic acid–formate buffer would be a good choice, because the pK_a of formic acid (3.75) is close to the pH value required. An acetic acid–acetate mixture would not be so satisfactory, since the pK_a of acetic acid (4.76) is nearly 1 pH unit away. The ratio of formate ion to formic acid required can be calculated from the Henderson–Hasselbalch equation:

$$4.00 = 3.75 + \log\left\{\frac{[HCOO^-]}{[HCOOH]}\right\} \tag{2.16}$$

which can be rewritten as

$$\frac{[HCOO^-]}{[HCOOH]} = 10^{0.25} = 1.78 \tag{2.17}$$

Such a mixture could be made, for example, by using 0.1 M formic acid and 0.178 M sodium formate. Alternatively, a solution of formic acid could be titrated to pH 4.00 with sodium hydroxide.

Since it is desirable to study many biochemical reactions near physiological pH, there is a particular need for mixtures that buffer the pH in the pH range 6.5–8.0. Of the acid–base pairs listed in Table 2.6, only mixtures of dihydrogen phosphate ion and monohydrogen phosphate ion, or possibly carbonic acid and bicarbonate ion, would be satisfactory. Phosphate buffers are often used, but they cannot serve under all circumstances. In some biochemical reactions, phosphate is consumed or produced. Furthermore, both phosphate-containing and carbonate-containing solutions precipitate some ions (Ca^{2+}, for example) that may be needed in the reaction. Therefore, a number of other compounds (some of which are naturally occurring, others synthetic) are employed as buffers in this range. Examples are given in Table 2.7.

Organisms themselves must maintain the pH inside cells and in most bodily fluids in the narrow pH range between about 6.5 and 8.0. Two buffer

Table 2.7
Some buffers commonly employed for biochemical studies

Buffer Substance (acid form)	Acronym	pK_a
Cacodylic acid	—	6.2
Bis(2-hydroxyethyl)-imino-tris(hydroxymethyl)methane	BISTRIS	6.5
Piperazine *N,N'*bis-(2 ethanesulfonic acid)	PIPES	6.8
Imidazole	—	7.0
N'-2-hydroxyethylpiperazine-*N'*-ethanesulfonic acid	HEPES	7.6
Tris(hydroxymethyl)aminomethane	Tris	8.3

systems that are of greatest importance within organisms are included in Table 2.6. The phosphate system, with a pK_a of 6.86, plays a major role in controlling intracellular pH, for phosphate is an abundant material in cells. In blood, which contains dissolved CO_2 as a waste product of metabolism, the carbonic acid–bicarbonate system, which has a pK_a value of 6.37, provides considerable buffering capacity. But a major role in the control of pH within organisms is played by proteins. As we shall see in Chapter 5, proteins contain many weakly acidic or basic groups, and some of these have pK_a values near 7.0. Since proteins are abundant in both cells and body fluids like blood and lymph, pH buffering is very strong.

Ampholytes, Polyampholytes, and Polyelectrolytes

So far, we have considered only molecules containing a single weakly acidic or basic group. But many molecules contain multiple ionizing groups and display more complex behavior on titration. A molecule that contains groups with both acidic and basic pK_a values is called an **ampholyte.** Consider, for example, the molecule glycine:

$$H_2N—CH_2—COOH$$

Glycine is an α-amino acid, one of a group of important amino acids that we shall encounter in Chapter 5 as constituents of proteins. The pK_a values of the carboxylate and amino groups on glycine are 2.34 and 9.6, respectively. If we dissolved glycine in a very acidic solution (say, pH 1.0), both the amino group and the carboxylate group would be protonated. If the pH were increased (by adding NaOH, for example) the following series of steps would occur:

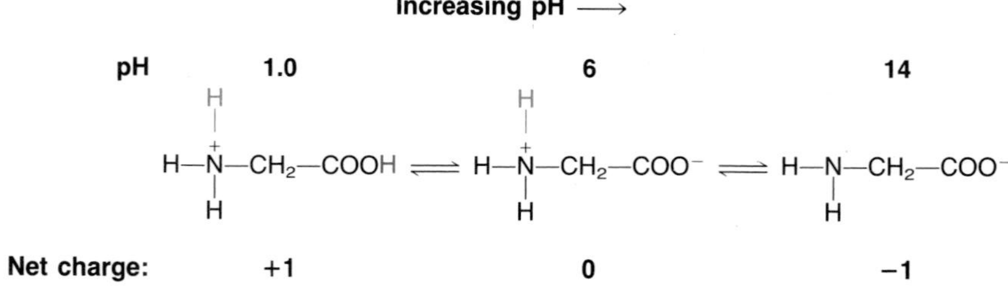

The actual course of the titration of glycine is shown in Figure 2.20. The titration occurs in two steps as the more acidic carboxylate and the less acidic amino groups successively lose their protons. Note that glycine can therefore serve as a good buffer in two quite different pH ranges.

The situation near neutral pH is an interesting one. In this region, the principal form is $H_3\overset{+}{N}—CH_2—COO^-$. Although this molecule has both positive and negative charges, its *net* charge is zero. An ampholyte in this state is called a *zwitterion.* At one particular point in this pH region, the average charge on glycine is zero. This is called the **isoelectric point** (pI). At the isoelectric point, most of the glycine molecules are in the zwitterion form, and the small amounts of residual $H_3\overset{+}{N}—CH_2—COOH$ and $H_2N—CH_2—COO^-$ are equal. We can calculate the isoelectric point by a simple application of the Henderson–Hasselbalch equation. The equation must be satisfied for both of the ionizing groups; that is, when pH = pI,

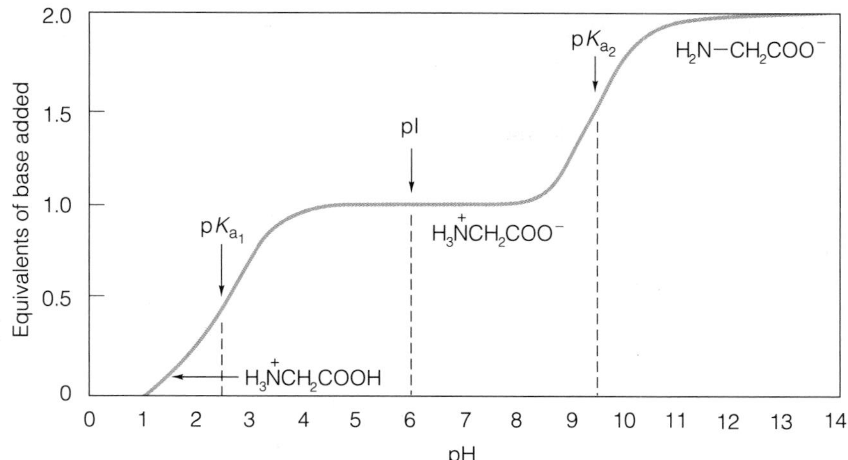

Figure 2.20
Titration of glycine. Since two groups, with quite different pK_a values, can be titrated, this is clearly a two-step titration curve. The calculated isoelectric point, pI, is shown.

$$\mathrm{pI} = \mathrm{p}K_{\mathrm{COOH}} + \log\left\{\frac{[\mathrm{H_3\overset{+}{N}CH_2COO^-}]}{[\mathrm{H_3\overset{+}{N}CH_2COOH}]}\right\} \tag{2.18a}$$

and

$$\mathrm{pI} = \mathrm{p}K_{\overset{+}{\mathrm{N}}\mathrm{H_3}} + \log\left\{\frac{[\mathrm{H_2NCH_2COO^-}]}{[\mathrm{H_3\overset{+}{N}CH_2COO^-}]}\right\} \tag{2.18b}$$

Note that in each case the quantity to the right is [conjugate base]/[acid].

Adding equations (2.18a) and (2.18b) and remembering that the sum of the logarithms of two quantities is the logarithm of their product, we get

$$2\ \mathrm{pI} = \mathrm{p}K_{\mathrm{COOH}} + \mathrm{p}K_{\overset{+}{\mathrm{N}}\mathrm{H_3}} + \log\left\{\frac{[\mathrm{H_2NCH_2COO^-}]}{[\mathrm{H_3\overset{+}{N}CH_2COOH}]}\right\} \tag{2.19}$$

Recalling that at pI we must have $[\mathrm{H_2NCH_2COO^-}] = [\mathrm{H_3\overset{+}{N}CH_2COOH}]$, we note that the right-hand term = log 1 = 0, so

$$2\ \mathrm{pI} = \mathrm{p}K_{\mathrm{COOH}} + \mathrm{p}K_{\overset{+}{\mathrm{N}}\mathrm{H_3}} \tag{2.20a}$$

or

$$\mathrm{pI} = \tfrac{1}{2}(\mathrm{p}K_{\mathrm{COOH}} + \mathrm{p}K_{\overset{+}{\mathrm{N}}\mathrm{H_3}}) \tag{2.20b}$$

The result is simple in this case: the pI is simply the average of the two pK_a's. If there are several positively and negatively charged groups, the calculation becomes more complicated (see Edsall and Wyman, *Biophysical Chemistry,* Chapter 9).

Large molecules such as proteins can have *many* acidic and basic groups. Such molecules are called **polyampholytes.** As long as both positively and negatively charged groups are present, the same general principle as above applies: there will always be an isoelectric point, at which the net average charge is zero. If acidic groups predominate, pI will be low, whereas if basic groups are in the majority, pI will be high. In Chapter 5 we shall find that this is an important consideration in dealing with solutions of proteins.

It is possible to determine the pI of an ampholyte or a polyampholyte by simple methods. In **electrophoresis** (see Tools of Biochemistry 2) an electric field is applied to a solution. Positively charged molecules migrate toward the negative cathode and negatively charged ones toward the positive anode. At its isoelectric point, an ampholyte moves in *neither* direction, since it has zero net charge. An even more precise method for determining pI is called **isoelectric focusing** (see Tools of Biochemistry 2).

There also exist macromolecules that carry multiples of only one kind of charge, cationic or anionic. A general term for these is **polyelectrolyte.** There are strong polyelectrolytes, like the nucleic acids (see Chapter 4), which are ionized over a very wide pH range. In addition, there are weak polyelectrolytes, like polylysine, a polymer of the amino acid lysine:

```
     +NH3        +NH3        +NH3
      |           |           |
    (CH2)4      (CH2)4      (CH2)4
      |           |           |
—N——C——C——N——C——C——N——C——C—
 |   |   ||   |   |   ||   |   |   ||
 H   H   O    H   H   O    H   H   O
```

Polylysine

When a number of weakly ionizing groups are carried on the same molecule, the pK_a of each group is influenced by the state of ionization of the others.

In a molecule like polylysine, the first protons are more easily removed than the last, since the strong positive charge on the fully protonated molecule helps drive protons away. Conversely, a molecule that develops a strong negative charge as protons are removed parts with the last ones only with difficulty—their pK_a values become anomalously high.

Interactions Between Macroions in Solution

Large polyelectrolytes such as the nucleic acids, or polyampholytes such as proteins, are classed together as **macroions.** Depending on the solution pH, they may carry a substantial net charge. The electrostatic forces of attraction or repulsion between such charged particles play a major role in determining their behavior in solution, as shown in Figure 2.21.

Particles of like net charge repel one another; thus, nucleic acid molecules tend to remain apart from each other in solution (see Figure 2.21a) and proteins tend to be soluble at pH values above or below their isoelectric points. On the other hand, if positively and negatively charged macromolecules are mixed, the electrostatic attraction between them makes them tend to associate with one another. A striking example of this is found in the chromosomes of higher organisms, where the negatively charged DNA is strongly associated with positively charged proteins termed *histones* (see Figure 2.21b) to form the complex called chromatin (see Chapter 28).

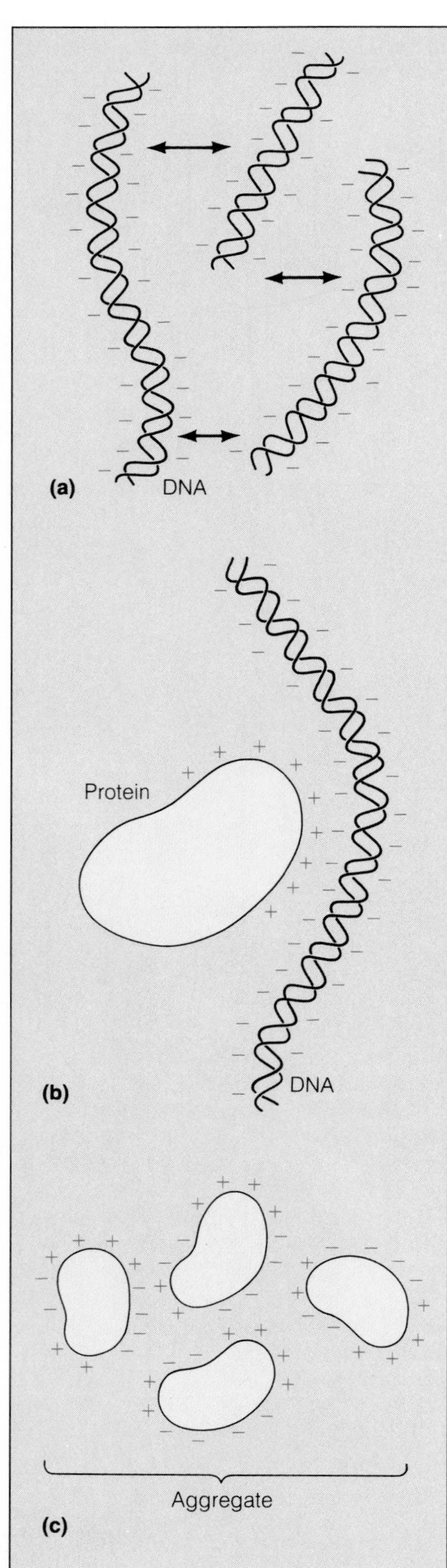

Figure 2.21
Electrostatic interactions between macroions. (**a**) DNA molecules, with many negative charges, repel one another strongly in solution. (**b**) If DNA is mixed with a positively charged protein, there is a strong tendency for these to associate. (**c**) In the isoelectric region, where polyampholytes have no *net* charge, they may still retain regions of + and − charge on their surfaces. These can lead to aggregation and precipitation.

Some electrostatic interactions are more subtle. For example, though a protein at the isoelectric point carries no *net* charge, negatively charged regions of one molecule are attracted to positively charged portions of the surface of another. Thus the molecules of a particular protein may tend to self-associate at the isoelectric pH (see Figure 2.21c).

This is only part of the story, however, for such interactions are strongly modified by the presence of small ions, such as those from salts, in the same solution. Each macroion will collect about it a **counterion atmosphere** enriched in oppositely charged small ions (Figure 2.22), and such a cloud of ions will tend to screen the molecules from one another. It is as if one macroion, in "looking" at another, "sees" not the other macroion itself but the macroion plus its oppositely charged surrounding cloud. Obviously, the larger the concentration of small ions present, the more effective this electrostatic screening will be. The quantitative expression of this effect for spherical macroions was analyzed by P. Debye and E. Hückel. The Debye–Hückel theory can be expressed in terms of an effective radius (r) of the counterion atmosphere. This radius may be taken as a measure of the distance at which two macroions "sense" one another's presence. According to the Debye–Hückel theory

$$r = \frac{K}{I^{1/2}} \tag{2.21}$$

where K is a constant that depends on the dielectric constant of the medium and the temperature, and I is a function of the concentration called the **ionic**

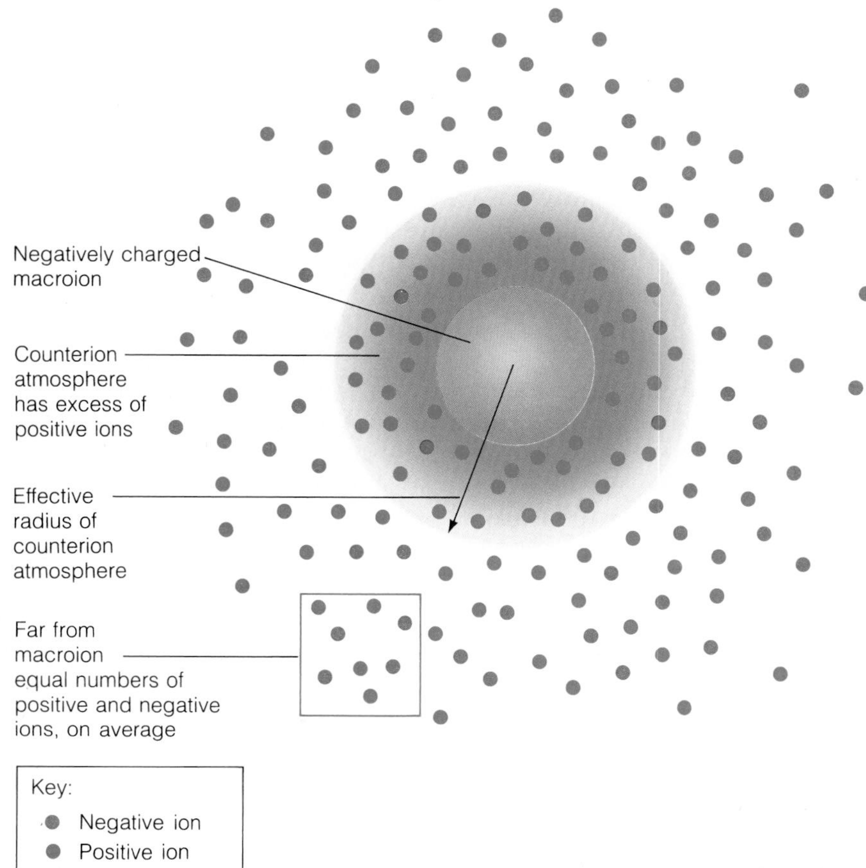

Figure 2.22
The Debye–Hückel ion atmosphere. When a macroion (in this example negatively charged) is placed in an aqueous salt solution, ions of opposite sign tend to cluster about it, forming a counterion atmosphere. Near this macroanion there are more cations than anions. Far away, their average concentrations are equal.

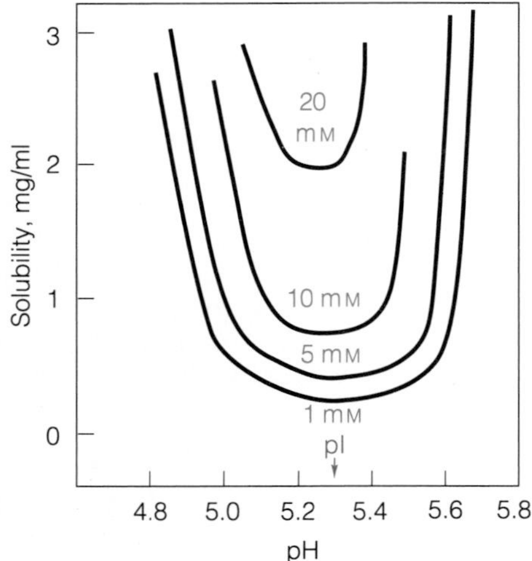

Figure 2.23
Dependence of protein solubility on both pH and ionic strength. The dependence of solubility on pH for the milk protein β-lactoglobulin is shown at four different ionic strengths of NaCl (blue). At all salt concentrations, solubility is lowest at the pI of the protein, as measured by mg/ml.

strength. The ionic strength is defined as

$$I = \tfrac{1}{2} \sum_i \text{M}_i Z_i^2 \qquad (2.22)$$

where the sum is taken over all small ions in the solution. For each ion type, M_i is its molarity and Z_i is its stoichiometric charge. For a 1:1 electrolyte like NaCl we have $Z_{\text{Na}^+} = +1$, $Z_{\text{Cl}^-} = -1$, and since $\text{M}_{\text{Na}^+} = \text{M}_{\text{Cl}^-} = \text{M}_{\text{NaCl}}$, we find that $I_{\text{NaCl}} = \text{M}_{\text{NaCl}}$. Thus, for 1:1 electrolytes, ionic strength equals salt molarity. This is obviously not true if divalent, trivalent (and so on), ions are involved. Such multivalent ions make greater individual contributions to the ion atmosphere than do monovalent ions, as reflected in the fact that the *square* of the ion charge is included in calculating the ionic strength. For such electrolytes $I > \text{M}$.

The effects of ionic strength of the medium on the interaction between charged macroions can be summarized as follows. At very low ionic strength, the counterion atmosphere is highly expanded and diffuse, and screening is ineffective. Like-charged particles repel strongly; unlike-charged particles attract one another strongly. As the ionic strength increases, the counterion atmosphere shrinks and becomes concentrated about the macroion. Screening becomes effective.

The effects of charge and ionic strength on the solubility of polyampholytes such as proteins can be explained in terms of these electrostatic interactions. Consider the behavior of the common milk protein β-lactoglobulin (Figure 2.23). The isoelectric point of this polyampholyte is about 5.3. Above or below this pH, the molecules all have either negative or positive charges and repel one another, so the protein is very soluble at either acidic or basic pH. At the isoelectric point, the net charge is zero but each molecule still carries surface patches of both positive charge and negative charge. The ionic interactions between them, together with other kinds of intermolecular interactions such as van der Waals forces, make the molecules tend to clump together and precipitate. Therefore, solubility is minimal at the isoelectric point. If the ionic strength is increased, however, the counterion atmosphere shrinks about the charged regions, and the attractive interactions between positive and negative groups are effectively screened. Hence, solubility increases, even at the isoelectric point. This effect of putting proteins into solution by increasing the salt concentration is called **salting in.**

Raising the salt concentration to very high levels (several molar, for example) introduces another effect. In very concentrated salt solutions, much of the water that would normally solvate the protein molecule is bound up in the hydration shells of the numerous salt ions. Thus, at very high salt concentration, the solubility of a protein again decreases, an effect called **salting out.** We shall see later how salting in and salting out are often used to purify proteins.

All these effects of ionic interactions on the behavior of biological macromolecules mean that the biochemist must pay close attention not only to the pH but also to the ionic strength of the solutions used in studying biochemical reactions. Usually, in addition to a buffer to control the pH, some amount of neutral salt (like NaCl or KCl) is added to yield the desired ionic strength. In doing so, the biochemist will often attempt to mimic the ionic strengths of cell and body fluids. Although these vary, a value of 0.1 to 0.2 M is often thought appropriate.

The meaningful study of the processes of life requires more than simply analyzing the behavior of biological molecules in a medium approximating that encountered in vivo. Cells and organisms are dynamic structures, whose processes demand a continual expenditure of energy. To understand life we must find how it is structured and fueled. To do that, we must examine how energy is gained and channeled by living structures; this is the topic we address in Chapter 3.

REFERENCES

Noncovalent Interactions

Eisenberg, D., and D. Crothers (1979) *Physical Chemistry with Applications to the Life Sciences,* Chapter 11. Benjamin/Cummings, Menlo Park, Calif. In addition to a thorough description of covalent bonding, this chapter contains an excellent discussion of dipole moments, polarizability, and noncovalent interactions.

Pauling, L. (1960) *The Nature of the Chemical Bond,* 3rd ed. Cornell Univ. Press, Ithaca, N.Y. The classic work on chemical bonds, with special emphasis on the hydrogen bond.

Watson, J. D., N. H. Hopkins, J. W. Roberts, J. A. Steitz, and A. M. Weiner (1987) *Molecular Biology of the Gene,* 4th ed. Benjamin/Cummings, Menlo Park, Calif. Chapter 5 contains a clear nonmathematical discussion of weak interactions in biology.

Water

Eigen, M., and L. DeMaeyer (1959) Hydrogen bond structure, proton hydration, and proton transfer in aqueous solutions. In: *The Structure of Electrolyte Solutions,* edited by W. J. Hamer, pp. 64–85. Wiley, New York. Although not recent, this remains an excellent, interesting review.

Hagler, A. T., and J. Moult (1978) Computer simulation of solvent structure around biological macromolecules. *Nature* 272:222.

Kamb, B. (1968) Ice polymorphism and the structure of water. In: *Structural Chemistry and Molecular Biology,* edited by A. Rich and N. Davidson, pp. 507–542. Freeman, San Francisco. A review of theories of water structure.

Tanford, C. (1973) *The Hydrophobic Effect.* Wiley-Interscience, New York.

Ionic Equilibria

Edsall, J. T., and J. Wyman (1958) *Biophysical Chemistry,* Vol. I. Academic Press, New York. An excellent in-depth treatment.

Williams, V. R., W. L. Mattice, and H. B. Williams (1978) *Basic Physical Chemistry for the Life Sciences.* Freeman, New York. Chapter 5 has a good discussion of pH measurement and buffers.

PROBLEMS

1. A chloride ion and a sodium ion are separated by a distance of 0.5 nm. Calculate the interaction energy (the energy required to pull them infinitely far apart) if the medium between them is (a) water and (b) *n*-pentane (see Table 2.5). Express the energy in joules per mole of ion pairs. [Note: the charge on an electron is 1.602×10^{-19} C.]

2. Explain why carbon dioxide has a dipole moment of zero. Would you expect the dipole moment for carbon monoxide also to be zero?

3. The accompanying graph depicts the interaction energy between two water molecules with parallel dipole moments. Sketch an approximate curve for the interaction between two water molecules oriented with *antiparallel* dipole moments.

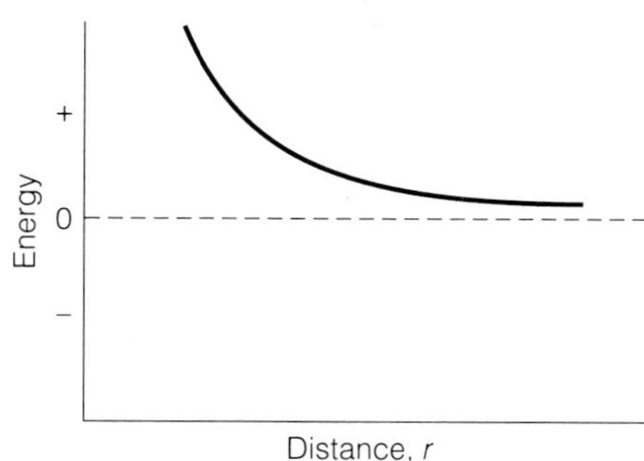

4. What is the pH of each of the following solutions?
   (a) 0.35 M hydrochloric acid
   (b) 0.35 M acetic acid
   (c) 0.035 M acetic acid

5. What is the pH of the following buffer mixtures?
   (a) 1 M acetic acid plus 0.5 M sodium acetate.
   (b) 0.3 M phosphoric acid plus 0.8 M KH_2PO_4

6. (a) Suppose you wanted to make a buffer of exactly pH 7.00 using KH_2PO_4 and Na_2HPO_4. If you had a solution of 0.1 M KH_2PO_4, what concentration of Na_2HPO_4 would you need?

(b) Now assume you wish to make a buffer at the same pH, using the same substances, but want the total phosphate molarity ($[HPO_4^{2-}] + [H_2PO_4^-]$) to equal 0.3. What concentrations of the KH_2PO_4 and Na_2HPO_4 would you use?

7. The amino acid *arginine* ionizes according to the following scheme:

I: H_3N^+–CH(COOH)–$(CH_2)_3$–NH–C(=NH_2^+)–NH_2 $\underset{+H^+}{\overset{pK_a = 1.82,\ -H^+}{\rightleftharpoons}}$ II: H_3N^+–CH(COO^-)–$(CH_2)_3$–NH–C(=NH_2^+)–NH_2 $\underset{+H^+}{\overset{pK_a = 8.99,\ -H^+}{\rightleftharpoons}}$ III: H_2N–CH(COO^-)–$(CH_2)_3$–NH–C(=NH_2^+)–NH_2 $\underset{+H^+}{\overset{pK_a = 13.2,\ -H^+}{\rightleftharpoons}}$ IV: H_2N–CH(COO^-)–$(CH_2)_3$–NH–C(=N–H)–NH_2

Calculate the isoelectric point of arginine. You can neglect contributions from form I. Why?

8. A student is carrying out a biological preparation that requires 1 M NaCl to maintain an ionic strength of 1.0. The student chooses to use 1.0 M ammonium sulfate instead. Why is this a serious error?

TOOLS OF BIOCHEMISTRY 2

Electrophoresis and Isoelectric Focusing

General Principles

A charged molecule, when placed in the electric field that exists between two electrodes immersed in a solution, migrates toward one electrode or the other. Molecules with a net positive charge migrate toward the negative cathode, whereas those with a net negative charge move toward the positive anode. This migration is called **electrophoresis.** The velocity of motion depends on two factors. Driving the motion is the force exerted by the electric field on the particle, $q\mathscr{E}$, where q is the molecule's charge (in coulombs) and $\mathscr{E}$ is the electrical field strength (in volts per meter). Resisting the motion is the frictional force exerted on the particle as it moves through the medium. This is given by fv, where v is the velocity of the particle and f is a quantity called the **frictional coefficient,** which depends on the size and shape of the molecules. Large or asymmetric molecules encounter more frictional resistance than small or compact ones and consequently have larger frictional coefficients.

When the electric field is turned on, the molecule quickly accelerates to a velocity at which these forces balance and then moves steadily at this rate. The basic equation for electrophoresis expresses this balance of driving and resisting forces:

$$fv = q\mathscr{E} \qquad \text{(T2.1)}$$

The equation can be rewritten to express the rate of motion per unit of field strength, which is called the **electrophoretic mobility** (μ) of the molecule:

$$\mu = \frac{v}{\mathscr{E}} = \frac{q}{f} = \frac{Ze}{f} \qquad \text{(T2.2)}$$

On the right-hand side of this equation we have expressed the charge on the molecule as the product of the unit of electron (or proton) charge (e) times the number of unit charges. Z will be a positive or negative integer. Equation (T2.2) tells us that the mobility of a molecule depends on its charge and on the molecular dimensions.*

* Equation (T2.2) is actually an approximation, for it neglects effects of the ion atmosphere. See the references at the end of this Tools section for more detail.

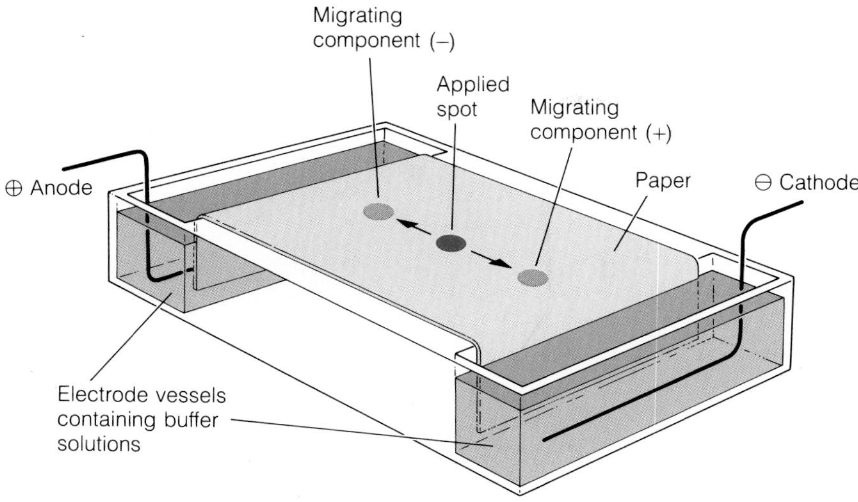

Figure T2.1
Paper electrophoresis.

Since the various ions and macroions studied by biochemists differ in both respects, their behavior in an electric field provides a powerful way of separating them. Electrophoretic separation is one of the most widely used methods in biochemistry.

Paper Electrophoresis and Gel Electrophoresis

Although electrophoresis can be carried out free in solution, it is more convenient to use some kind of *supporting medium.* Two kinds most often used are shown in Figures T2.1 and T2.2. Figure T2.1 depicts **paper electrophoresis,** often used for separating mixtures of small charged molecules. A piece of filter paper, wetted with a buffer solution to control the pH, is stretched between two electrode vessels. A drop of the mixture to be analyzed is placed at one point on the paper, and the electric current is turned on. After the molecules have migrated for a sufficient time, the paper is removed, dried, and stained with a dye that colors the substances to be examined. Each kind of charged molecule in the mixture will have migrated a certain distance toward either the anode or the cathode, depending on its particular charge and dimensions, and will show up as a stained spot on the paper at the new position. Usually, the spots can be identified by comparison with a set of standards, run on the same paper. If the unknowns are radioactive, spots can be cut out and their radioactivity quantitated by scintillation counting (see Tools of Biochemistry 16).

Figure T2.2 depicts **gel electrophoresis,** a technique much used with proteins and nucleic acids. A gel containing the appropriate buffer solution is cast as a thin slab between glass plates. Common gel-forming materials are polyacrylamide, a water-soluble, cross-linked polymer, and agarose, a polysaccharide. The slab is placed between electrode compartments as shown, and a small amount of a solution of the sample is carefully pipetted into precast notches on top of the gel. Usually glycerol and a water-soluble cationic or anionic "tracking" dye are added to the sample. The glycerol makes the sample solution dense, so that it does not mix into the buffer solution in the upper electrode chamber. The dye migrates faster than most macroions, so the experimenter is easily able to follow the progress of the experiment. The current is turned on and run until the tracking dye band is near the bottom of the slab. The gel is then removed from between the glass plates and is usually stained with a dye that binds to proteins or nucleic acids. At this point, a photograph is taken of the gel for a permanent record. Since the protein or nucleic acid mixture was applied as a narrow band at the top of the gel, components migrating with different mobilities appear as narrow bands on the gel, although the bands may be broadened somewhat by diffusion. Certain techniques (see Tools references) make it possible to sharpen the bands even further, so that individual types of macroions appear as narrow lines on the gel; Figure T2.3 shows an example of separa-

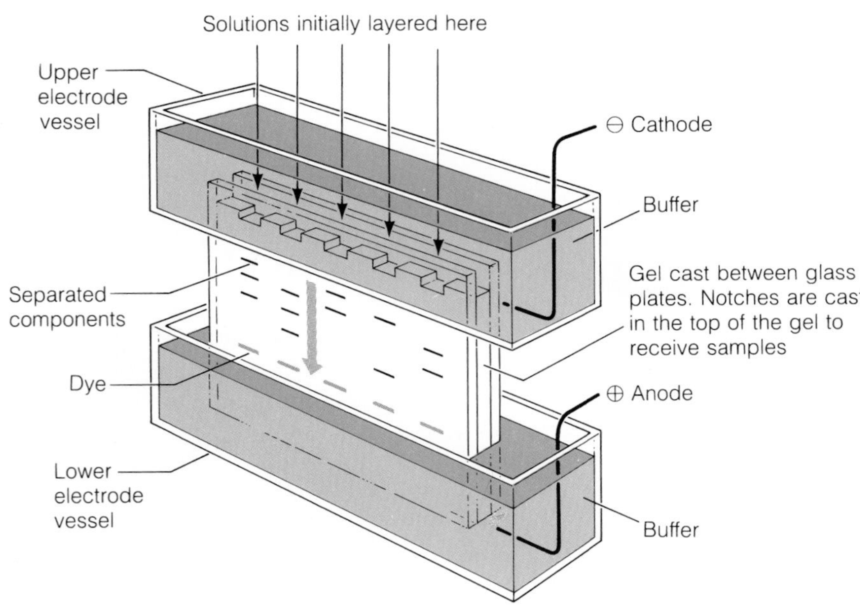

Figure T2.2
Gel electrophoresis.

tion of DNA fragments by this method. The *relative mobility* of each component is calculated from the distance it has moved relative to the tracking dye.

Principles of Separation in Gel Electrophoresis

When electrophoresis is carried out in a gel or other supporting medium, the mobility is lower than would be expected from equation (T2.2) because the gel or other matrix exhibits a molecular sieving effect. This can be seen by graphing mobility as a function of the concentration of the gel (Figure T2.4a). A graph of log μ versus % gel is usually linear; this is called a **Ferguson plot.** The limiting mobility approached as % gel approaches zero is called the **free mobility;** it should obey equation (T2.2). The steepness of the Ferguson plot depends on the size and shape of the macroion, for it reflects the difficulty the macroion experiences in passing through the molecular mesh of the gel.

As a result of these several factors, different kinds of molecules can exhibit widely different behaviors in gel electrophoresis (see Figure T2.4b). However, certain simple cases are of great importance. Polyelectrolytes like DNA or the polylysine molecule shown on p. 50 will have one unit charge on each residue. Thus, in a mixture of such molecules, each will have a charge (Ze) proportional to its molecular length. But the frictional factor also increases with molecular length. To a first approximation, therefore, macroions whose charge is proportional to length will have a free mobility almost independent of molecular size. The molecular sieving effect will determine the relative mobilities of such molecules at any given gel concentration (Figure T2.4c). This means that we can neatly separate molecules of this kind on the basis of *size alone* by gel electrophoresis, as shown in Figure T2.3. For extended molecules like nucleic acids, the relative mobility is often approximately a linear function of the logarithm of the molecular weight (Figure T2.4d). Usually, standards of known molecular weight are electrophoresed in one or more lanes on the gel. The molecular weight is then read from a graph like that in Figure T2.4d prepared for the standards.

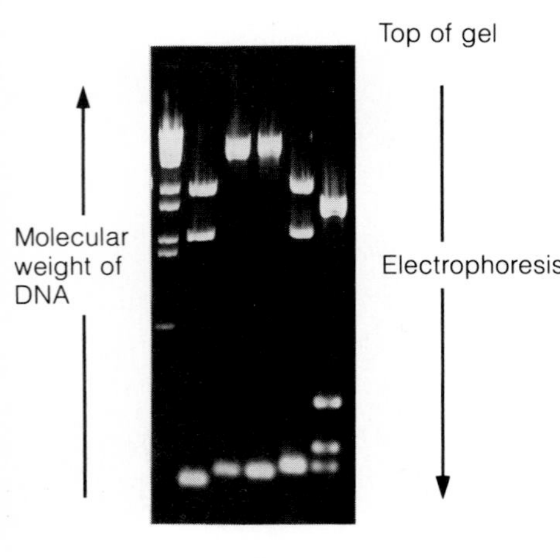

Figure T2.3
Gel showing separation of DNA fragments.

Isoelectric Focusing

Yet another technique allows separation of molecules purely on the basis of their charge characteristics. We have noted that every polyampholyte has an isoelectric point, at which its net charge is zero. It is possible to set up, in a gel, a stable pH gradient covering a pH range that includes the isoelectric points of the various polyampholytes in a mixture. This is done by using a mixture of low-molecular-

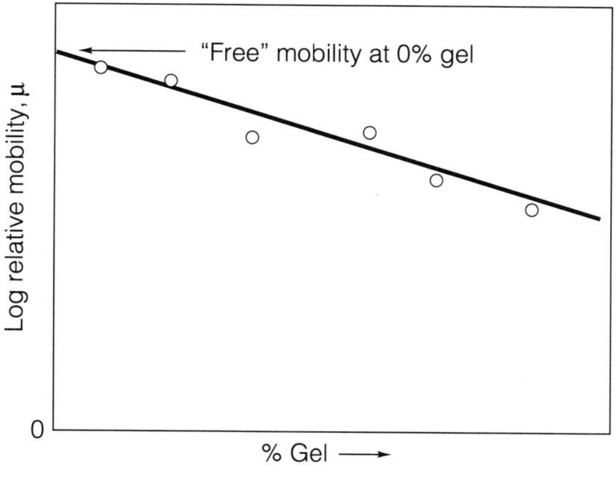

(a) Ferguson plot

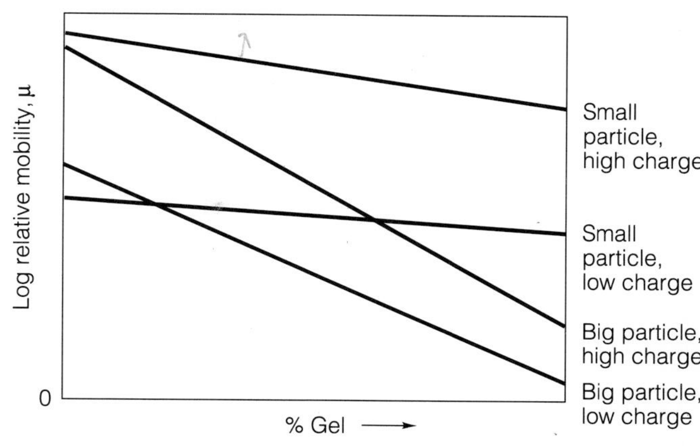

(b) Representative Ferguson plot for different kinds of molecules

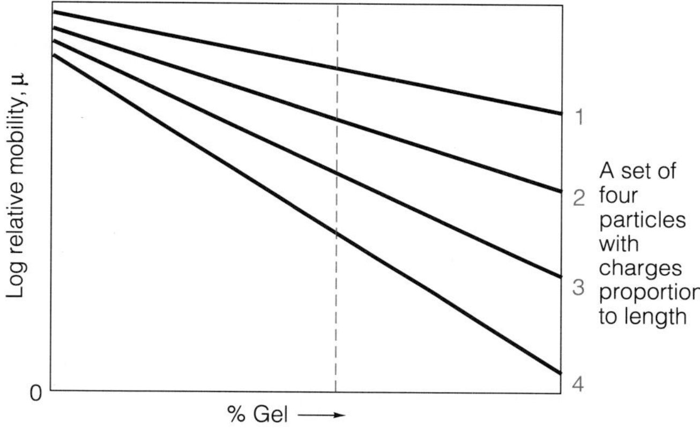

(c) Ferguson plot observed when charge is proportional to length

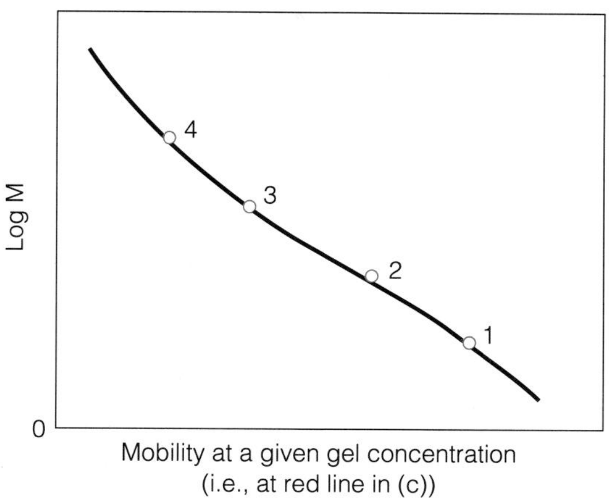

(d) Relationship between molecular weight (M) and mobility, at a given gel concentration for molecules like those shown in (c).

Figure T2.4
Separation in gel electrophoresis.

weight ampholytes as the gel buffer. These produce a gradient of pH between the electrodes. In such a gradient, each polyampholyte molecule migrates to the position of its isoelectric point. At that position its net charge becomes zero, so it remains there; bands of different polyampholytes accumulate at their respective isoelectric positions. As Figure T2.5 shows, the method is capable of resolving molecules with very small differences in isoelectric point.

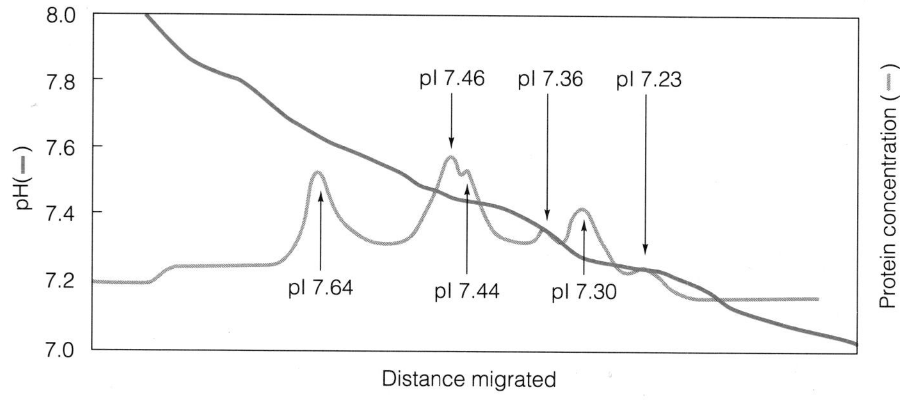

Figure T2.5
Isoelectric focusing of variants of a single protein, hemoglobin.

Other Methods

We discuss other electrophoretic techniques in later chapters and Tools sections. For example, in Tools of Biochemistry 11, we describe the use of SDS-gel electrophoresis for protein separation and molecular weight determination. What we have presented here is only a brief overview of a widely applied technique. For further information, consult the following list of references.

REFERENCES

Hames, B. D., and D. Rickwood, eds. (1981) *Gel Electrophoresis of Proteins.* IRL Press, Oxford, Washington, D.C., and Rickwood, D., and B. D. Hames, eds. (1982) *Gel Electrophoresis of Nucleic Acids.* IRL Press, Oxford, Washington, D.C. These two volumes are extremely useful laboratory manuals for gel electrophoresis techniques.

Osterman, L. A. (1984) *Methods of Protein and Nucleic Acids Research,* Vol. 1, Parts I and II. Springer-Verlag, New York. An up-to-date summary of electrophoresis and isoelectric focusing.

van Holde, K. E. (1985) *Physical Biochemistry,* 2nd ed., Chapter 6. Prentice-Hall, Englewood Cliffs, N.J. A somewhat more detailed discussion than given here.

CHAPTER 3

The Energetics of Life

A living cell is a dynamic structure. It grows, it moves, it synthesizes complex macromolecules, and it selectively shuttles substances in and out and between compartments. All of this activity requires energy, so every cell must obtain energy from its surroundings and expend it as efficiently as possible. Plants gather most of their energy from sunlight; animals use the energy stored in plants or other animals that they consume. The processing of this energy, tapping as much of it as possible to do the things a cell or organism needs to do, is what much of biochemistry is about. Most of the elegant molecular structure that exists in every cell is dedicated to this task.

Because of the central role of energy in life, it is appropriate that we begin a study of biochemistry with an introduction to **bioenergetics**—the quantitative analysis of how organisms gain and utilize energy. Bioenergetics may be regarded as a special part of the general science of energy transformations, which is called **thermodynamics.** In this chapter we shall explore just a bit of that field, choosing only concepts that are important to the biochemist or biologist.

Energy, Heat, and Work

A word we shall often use in our discussion is **system.** In this context, a system is any part of the universe that we choose for study. It can be a single bacterial cell, a Petri dish containing nutrient and millions of cells, the whole laboratory in which this dish rests, the earth, or the entire universe. A system must have defined boundaries, but otherwise there are few restrictions. The system may be *isolated,* or closed to exchange of either energy or matter (or both) with its surroundings, or it may be *open,* so that energy and matter can pass in and out.

Internal Energy and the State of a System

Any system will contain a certain amount of **internal energy,** which we denote by E. It is important to be specific as to what this internal energy includes. The system's atoms and molecules have kinetic energy of motion and energy of vibration and rotation. In addition, we should include all of the energy stored in chemical bonding between the atoms and the energy of noncovalent interaction between molecules. We should include, in fact, *any* kind of energy that might be changed by chemical processes. We need not include energy stored in the atomic nucleus, for this is unchanged in any chemical or biochemical reaction. The internal energy is a function of the **state** of a system. The thermodynamic state is defined by giving any two of three variables—the temperature (T), the pressure on the system (P), and the volume of the system (V)—plus the amounts of all substances present. It is essentially a recipe for producing the system in a defined way. For example, a system composed of 1 mol of O_2 gas in 1 liter at 273°K will have a defined state and therefore a definite internal energy value. This value is independent of any past history of the system.

Unless a system is isolated, it can in principle exchange energy with its surroundings and thereby change its internal energy. This exchange can happen in only two ways. First, **heat** may be transferred to or from the system. Second, the system may do **work** on its surroundings or have work done on it. Work can take many forms. It may include expansion of the system against an external pressure such as expansion of the lungs, electrical work such as that done by a battery or the pumping of ions across a membrane, expansion of a surface against surface tension, flexing of a flagellum to propel a protozoan, or lifting of a weight by contraction of a muscle. In all of these examples, a force is exerted against a resistance to produce a displacement, and work is done. Note that heat and work are not *properties* of the system. They may be thought of as "energy in transit" between the system and its surroundings. Certain conventions have been adopted to describe these ways of exchanging energy:

1. We denote heat by the symbol q. A positive value of q indicates that heat is absorbed by the system from its surroundings. A negative value means that heat flows from the system to its surroundings.

2. We denote work by the symbol w. A positive value of w indicates that work is done by the system on its surroundings. A negative value means that the surroundings do work on the system.

The First Law of Thermodynamics

Since the internal energy of a system can change only by heat or work exchanges, the change in internal energy, which we call ΔE, must be given by

$$\Delta E = q - w \tag{3.1}$$

This equation, which holds for all processes, expresses the **first law of thermodynamics.** It is simply a statement of the conservation of energy. Consider, for example, some process in which a certain amount of heat is absorbed by a system, while the system does an exactly equivalent amount of work on its surroundings. In this case, $q = w$ and $\Delta E = 0$. This is in agreement with common sense; if so much energy went in, and an equal quantity came out, then the energy within the system must be unchanged.

Changes in functions of state, like the internal energy, depend only on the initial and final states of a system and are independent of the path. But

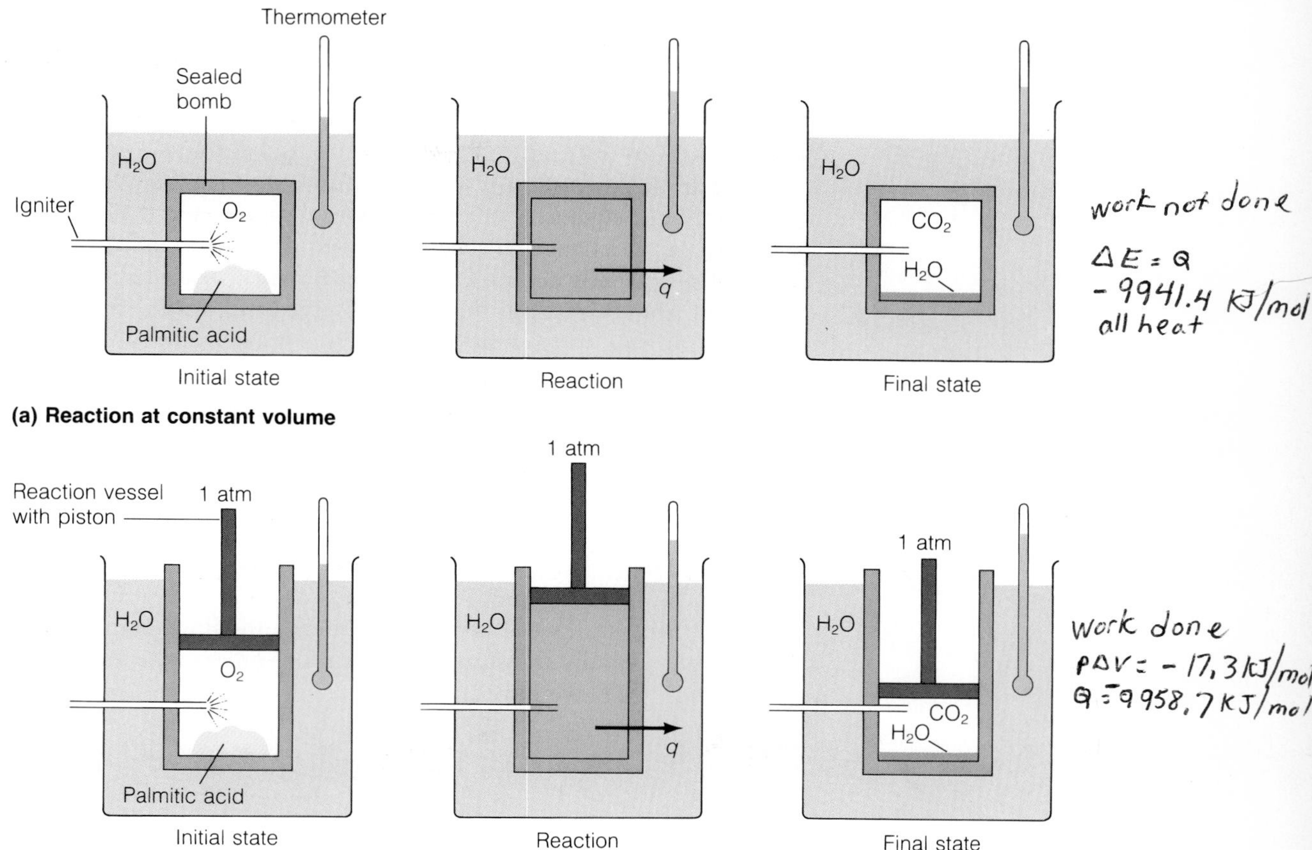

Figure 3.1
Two ways of studying the heat evolved in combustion. In (**a**) the reaction occurs in a sealed vessel, or "bomb," at constant volume. During the reaction, heat (q) is transferred to the surrounding water bath, which experiences a small increase in temperature. No work is done because the system is at constant volume. In (**b**) the reaction vessel is fitted with a piston held at 1 atm pressure. During the reaction the temporary heating of the gas in the vessel causes the piston to be pushed up, but the contraction in total gas volume in the reaction has the result that after the vessel and gas have cooled to the water temperature the final volume is actually smaller than the initial volume. Thus, net work is done *on* the system, and the total amount of heat delivered to the bath is slightly different from that in (**a**).

the amounts of heat and work exchanged in any process depend very much on the conditions we impose. To make this concrete, let us consider a specific chemical reaction—the complete oxidation of a fatty acid, palmitic acid:

$$CH_3(CH_2)_{14}COOH \text{ (solid)} + 23O_2 \text{ (gas)} \longrightarrow 16CO_2 \text{ (gas)} + 16H_2O \text{ (liquid)}$$

This is, in fact, an important biochemical reaction, which takes place in a much more indirect way in our bodies when we metabolize fats. We shall consider running this reaction in two ways, as shown in Figure 3.1. In Figure 3.1a the reaction is carried out by igniting the mixture in a sealed vessel immersed in a water bath. The container has heat-conducting walls, so heat can be transmitted to the surrounding bath. We can measure the heat passed from the reaction vessel to the water bath by the temperature change in the bath, knowing the mass of water and the heat capacity (per gram) of water. This instrument is a kind of **calorimeter,** of a type often used to measure the **heat of combustion.** During the combustion reaction, both the temperature and the pressure within the reaction vessel become very high, but the pressure falls again as the vessel cools by transferring heat to the surrounding water. In fact, it falls below the initial value, for the total amount of gaseous material has decreased by 7 mol per mole of palmitic acid (see reaction on p. 110). However, since the reaction vessel has a fixed volume, no work has been done against the surroundings or by the surroundings. Therefore, from equation (3.1),

$$\Delta E = q \tag{3.2}$$

The total heat that is transferred from the reaction vessel to the surroundings just equals the change in internal energy, and that energy change results from the changes in chemical bonding that occurred during the reaction. The presently accepted unit for heat, work, and energy is the **joule (J)**.* For the above reaction, the value observed for ΔE is -9941.4 kJ/mol.

Now suppose the same reaction is carried out at a constant pressure of 1 atmosphere as shown in Figure 3.1b. Since the system ultimately contracts by an amount proportional to the decrease in the number of moles of gas (23 to 16 mol), a certain amount of work will be done by the surroundings on the system. This can be calculated in the following way.

When volume (V) is changed against a constant pressure (P),

$$w = P\,\Delta V \tag{3.3}$$

In this case, if we assume that the initial and final temperatures of the system are essentially the same** (say 25°C) and that the gases are ideal, so that $PV = nRT$, we have

$$\Delta V = \Delta n \frac{RT}{P} \tag{3.4}$$

where R is the gas constant, T the absolute temperature in degrees Kelvin, and Δn the change in number of moles of gas per mole of palmitic acid oxidized. Then, inserting (3.4) in (3.3) we obtain

$$w = \Delta n\, RT \tag{3.5}$$

Since we would like w in joules per mole, we use $R = 8.314$ J/°mol, which gives, using equation (3.5), $w = -17.3$ kJ/mol palmitate.

The heat evolved in this constant-pressure combustion will then be

$$\begin{aligned} q &= \Delta E + w = \Delta E + P\,\Delta V = \Delta E + \Delta n\, RT \\ &= -9941.4 \frac{\text{kJ}}{\text{mol}} - 17.3 \frac{\text{kJ}}{\text{mol}} = -9958.7 \frac{\text{kJ}}{\text{mol}} \end{aligned} \tag{3.6}$$

A slightly greater amount of heat is released to the surroundings under these *constant-pressure* conditions than under the *constant-volume* conditions of Figure 3.1a. The reason for this should be clear: Under constant-pressure conditions the surroundings do work on the system, and this work appears as extra heat released.

We emphasize this difference because, in fact, most chemical reactions in the laboratory and virtually all biochemical processes occur under condi-

---

* In the past, biochemists tended to express energy, heat, and work in *calories* or *kilocalories*. However, the SI units *joules* and *kilojoules* are now replacing these. Conversion is easy: 1 cal = 4.184 J. Similarly, 1 kcal (kilocalorie or 10^3 calories) = 4.184 kJ (kilojoules).

** This may seem inconsistent with the observation that the reaction is heating the calorimeter. In practice, the volume of water used is so large that the total temperature change is small and can be neglected for this purpose.

tions more nearly approximating constant pressure than constant volume. If we are interested in the heat obtainable by oxidizing palmitic acid in an animal, then the heat evolved at constant pressure is what we want to know. However, this is not exactly ΔE, because of the PV work done.

Enthalpy

What we need is a function of state, the changes in which reflect the heat evolution or absorption in a constant-pressure reaction. Therefore, we define a new quantity, the **enthalpy**, which we give the symbol ***H***.

$$H = E + PV \tag{3.7}$$

Since E is a function of state, as is the product PV, H is also a function of state. The change, ΔH, depends only on the initial and final states of the process for which it is calculated. For reactions at constant pressure, ΔH becomes

$$\Delta H = \Delta E + P\,\Delta V \tag{3.8}$$

This is the same as the quantity calculated in equation (3.6). In other words, *when the heat of a reaction is measured at constant pressure, it is ΔH that is determined.*

The energy changes you will find tabulated throughout this book and other books on biochemistry will almost always be given as ΔH values. That is most appropriate, for in vivo these reactions occur under very nearly constant-pressure conditions. If a nutritionist wishes to know the energy available from the oxidation of palmitic acid in the body, for example, ΔH is the appropriate quantity.

Although we have pointed out the distinction between ΔE and ΔH, we should emphasize that for most biochemical reactions the quantitative difference between them is infinitesimal. Most of these reactions occur in solution and do not involve the consumption or formation of gases. The volume changes are thus exceedingly small, and $P\,\Delta V$ is a tiny quantity compared to ΔE or ΔH. For this reason, we are justified in most cases in thinking of ΔH as a direct measure of the energy change in a process. Note that even for the example given, the oxidation of palmitic acid, the difference between ΔH and ΔE is only 0.2%. In most cases, biochemists simply disregard the difference.

Reversible and Irreversible Processes

We have emphasized that changes in functions of state, like ΔH and ΔE, depend only on the initial and final states and are independent of the pathway between them. But q and w, which are *not* properties of the system, are definitely path dependent.

A very simple experiment will demonstrate this. It is not a very "biochemical" example, but it is a clear one and it teaches something important about processes relevant to biochemistry.

Suppose we have 1 mol of an ideal gas in a volume of 22.4 liters at 0°C under 1 atm pressure. We intend to expand it to 44.8 liters by dropping the pressure to $\frac{1}{2}$ atm. We keep the temperature constant by immersing the cylinder of gas in a constant-temperature bath. Note that in the apparatus shown in Figure 3.2 the pressure is produced by having an appropriate weight sitting on a weightless piston.

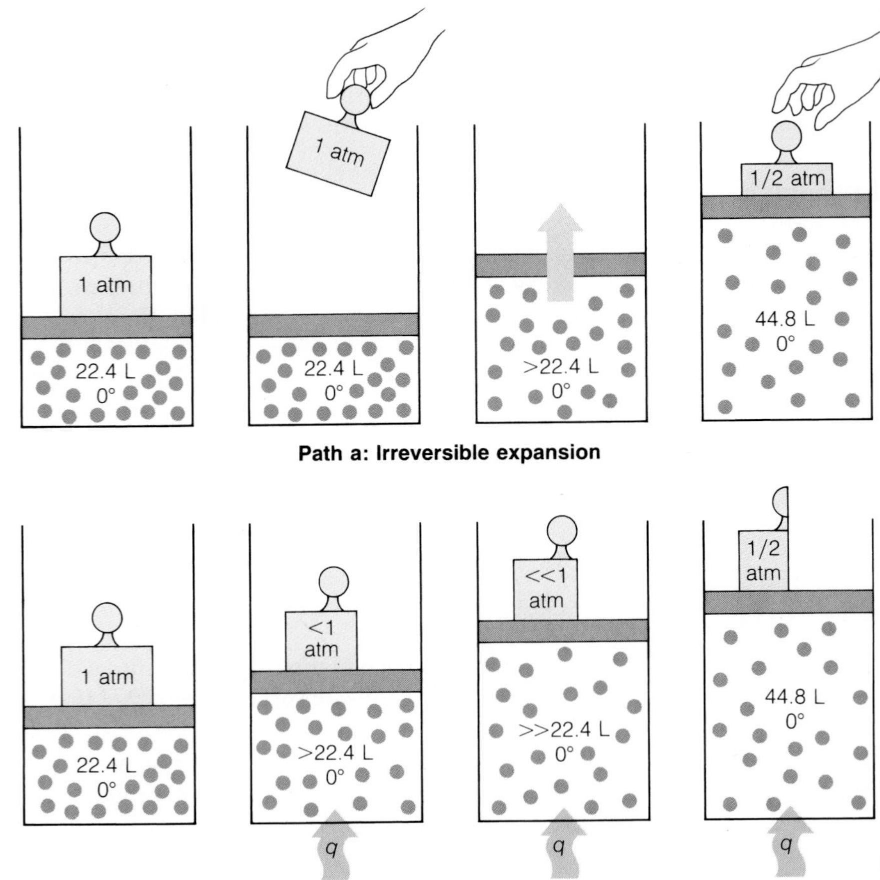

Figure 3.2
Two ways of expanding 1 mol of a gas from 22.4 to 44.8 L. Path (**a**) is an *irreversible* expansion: the 1-atm weight is simply snatched off the piston and replaced by a $\frac{1}{2}$-atm weight only after the piston has popped up to the 44.8-L mark. No work is done and no heat is absorbed. Path (**b**) approximates a *reversible* expansion. The weight is reduced bit by bit so that the system is always close to equilibrium. A maximum amount of weight is lifted in this way, and a maximum amount of work is done. Consequently, heat (q) must be absorbed from the surroundings to replace the energy expended as work.

There are many ways in which the expansion could be carried out. Following path *a* shown in the figure we could simply remove the weight, let the piston pop up against no resistance, and stop it when $V = 44.8$ L by putting on a weight corresponding to $\frac{1}{2}$ atm. Because the expansion is against zero pressure, $P\,\Delta V = 0 \cdot \Delta V = 0$, so *no* work is done at all. Furthermore, if this is an ideal gas, ΔE will be zero for the expansion. This is because the internal energy of an ideal gas is entirely in the form of kinetic energy of the molecules, which depends only on T. Ideal gas molecules do not interact with one another, so there is no dependence of E on the volume to which they are confined. For path *a* we have

$$\Delta E = 0 \tag{3.9a}$$

$$w = 0 \tag{3.9b}$$

and, since $\Delta E = q - w$,

$$q = 0 \tag{3.9c}$$

so by path *a* no heat is exchanged and no work is done.

Path *b* in Figure 3.2 is a path by which we can do the *maximum* amount of work. Suppose we reduce the weight bit by bit (by slowly filing it away, for example). In this case, at each point in the process, the expanding gas is lifting the most that it can. Calculating w is a little more complicated now, since P is continually changing. We must integrate over the path from V_1 to V_2:

$$w = \int_{V_1}^{V_2} P\,dV \tag{3.10}$$

The pressure is being decreased steadily, and the volume is increasing to keep pace with this decrease, so at every point in the expansion $P = RT/V$. Inserting this in (3.10), we get

$$w = RT\int_{V_1}^{V_2} \frac{dV}{V} = RT \ln\left\{\frac{V_2}{V_1}\right\} \tag{3.11}$$

Calculation yields

$$w = 8.314 \frac{\text{J}}{°\text{mol}} \times 273° \times \ln 2 = 1573 \frac{\text{J}}{\text{mol}} \tag{3.12}$$

Since E is a function of state, ΔE will be the same for this path as for path a. Therefore, $\Delta E = 0$, as before, but now $w = 1573$ J/mol, so we must have $q = w = 1573$ J/mol. The heat q would have to be absorbed from the surrounding constant-temperature bath to keep T constant. By path b, some heat has been converted into work.

The contrast between path a and path b illustrates some important physical concepts. In path b the system was kept very, very close to equilibrium at every point during the expansion. By adding a few grains of weight back onto the piston, we could have reversed the direction of the process at any point. Such a path is said to be **reversible.** It proceeds through a series of essentially equilibrium states. A reversible path *always* produces the maximum work from any process. It is easy to see why the maximum work is done in this example, for at each point in the expansion the gas was lifting the greatest possible weight. Path a, on the other hand, is an example of an **irreversible** path. The system is far from equilibrium at all times, so that at no point during the expansion could we reverse the direction of the process by an infinitesimal change.

An irreversible process is also often called a **spontaneous** process, but we prefer the word **favorable.** The word *spontaneous* tends to connote falsely that the process is rapid. Thermodynamics has nothing to say about how fast processes will be, but it can indicate which direction is favored. Here, the result accords with intuition; you would not expect the gas to contract, or even to keep the same volume, when the pressure was reduced. Either would be most definitely an unfavorable process. The distinction between reversible, favorable, and unfavorable processes is vital to bioenergetics. It can be expressed most succinctly by what is called the **second law of thermodynamics,** which tells us which processes are thermodynamically favorable.

The Direction of Processes: Entropy, Free Energy, and the Second Law

The first law of thermodynamics is just a bookkeeping rule. When a physical process or a chemical reaction has occurred, we can tote up the incomes and expenditures of energy, and the books must balance. Energy can be gained and spent in different ways, but, at least in chemical processes, it cannot be created or destroyed.

Randomness and Entropy

Useful as the first law may be, it cannot tell us the favored directions of processes. Consider the following examples:

> We place an ice cube in a glass of water at room temperature. It melts. Why doesn't the rest of the water freeze instead?
>
> We place an ice cube in a jar of carefully supercooled water. The whole freezes.
>
> We touch a match to a piece of paper. It burns to carbon dioxide and water. But we can mix carbon dioxide and water ad infinitum and they will never form paper or any similar substance.

The common theme in these examples is this: Favorable processes proceed spontaneously in certain directions. Why?

A first guess at an explanation might be that systems always go toward a lowest-energy state. Water runs downhill; objects spontaneously fall in the earth's gravitational field; the oxidation of palmitic acid releases a great deal of energy. But such an explanation cannot account for the melting of ice at 25°C; in fact, energy is *absorbed* in that process. And if we consider some other favorable processes, it becomes clear that another, very different factor must be at work. We can imagine another simple experiment that gives a clear indication of what this factor may be. Suppose we very carefully layer pure water (without stirring) on a sucrose solution. Over a period of time, we will observe that the whole system becomes more and more uniform; eventually the sucrose molecules will be evenly distributed throughout the container. Though there is practically no energy change, the process is clearly a favorable one. Certainly, the opposite process (self-segregation of the sucrose molecules into a portion of the solution volume) will never occur. What is clearly important here is the fact that *systems of molecules have a natural tendency to randomization.*

The degree of randomness of a system is measured by a function of state called the **entropy** (*S*). There are a number of ways of defining entropy, but the most useful for our applications is the following: If a given thermodynamic state of a system corresponds to a number (W) of substates of equal energy, the entropy is defined as

$$S = k \ln W \tag{3.13}$$

where k is the **Boltzmann constant,** the gas constant R divided by Avogadro's number. There will always be many fewer ways of putting a large number of molecules into an orderly structure than into a disorderly one. Therefore, the entropy of an ordered state is lower than that of a disordered state of the same system. In fact, the minimal value of entropy (zero) is found *only* for a perfect crystal at the absolute zero of temperature (0°K or −273°C).

To demonstrate the entropy change in a simple process, consider the diffusing sucrose molecules described above. In Figure 3.3, we imagine dividing the entire final volume of the solution into N_F cells, each large enough to hold a sucrose molecule. In the initial state, the dissolved molecules were placed in N_I cells. For convenience, we assume that we have

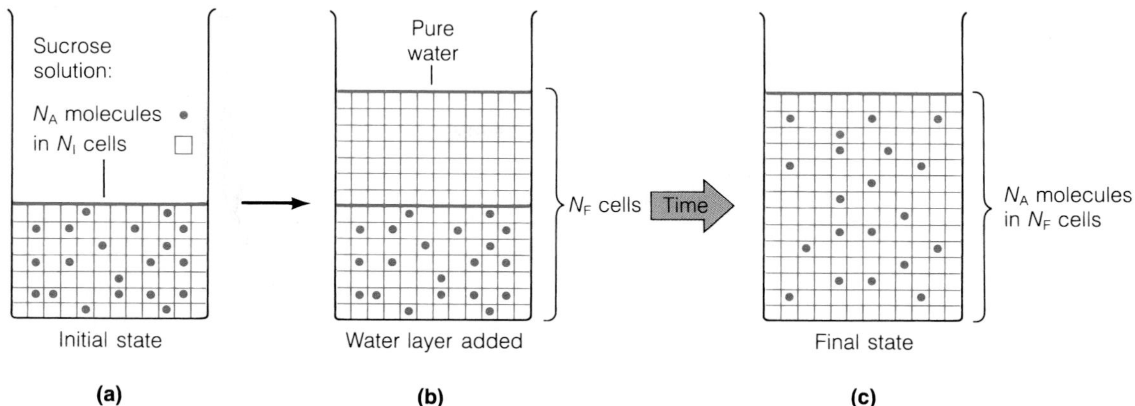

Figure 3.3
Diffusion as an entropy-driven process. Initially (**a**) we have a dilute solution of a solute (sucrose, as an example) in water. (**b**) We layer some pure water on top, being very careful not to mix. (**c**) If we watch for a long time, we find that ultimately the sucrose molecules are distributed uniformly through the solution.

The situation pictured in (**b**) is not in equilibrium, but (**c**) represents the final equilibrium state. The drive toward equilibrium is a consequence of the higher entropy, or randomness, in state (**c**).

We can calculate this entropy change, as shown in the text, by imagining the solution volume to be made up of cells, each big enough to hold one sucrose molecule. The equilibrium state (**c**) is more random than (**b**) in the sense that any cell in the whole volume in state (**c**) has an equal chance to be occupied. In (**b**) the cells in the upper half are not occupied. A system would never go spontaneously from state (**c**) to state (**b**). Exactly the same analysis could have been used for the expansion of a gas as shown in Figure 3.1.

1 mol (Avogadro's number, N_A) of sucrose molecules and that the initial and final volumes are so large that the solutions are quite dilute.

To calculate ΔS, we simply take

$$\Delta S = S_F - S_I = k \ln W_F - k \ln W_I = k \ln(W_F/W_I) \tag{3.14}$$

where W_F and W_I are the numbers of ways of arranging N_A sucrose molecules in the N_F and N_I cells in the final and initial states, respectively.

To a very good approximation, we can calculate W_F and W_I as follows. Consider W_I: there are N_I ways to put one molecule into N_I cells and *almost* N_I ways for the second and each following molecule. In other words, if N_I is a very big number and N_A is much less, we can make the approximations

$$W_I = N_I^{N_A} \tag{3.15a}$$

and

$$W_F = N_F^{N_A} \tag{3.15b}$$

so we find

$$\begin{aligned}\Delta S &= k \ln(W_F/W_I) = k \ln(N_F/N_I)^{N_A} \\ &= N_A k \ln(N_F/N_I)\end{aligned} \tag{3.16}$$

Since $N_A k = R$, by definition, and $N_F/N_I = V_F/V_I$, where V_F and V_I are the final and initial volumes, we find

$$\Delta S = R \ln(V_F/V_I) \tag{3.17}$$

Note what has happened here. Although the energy of the system was unchanged, a function of state, the entropy, increased during the process. In this particular case, it is the increase in randomness, or the entropy change, that can be thought of as the thermodynamic driving force. The process of diffusion evens out concentrations simply because there are more ways to distribute molecules over a large volume than over a small one. What we have expressed is one form of the **second law of thermodynamics:** *The entropy of an isolated system will tend to increase to a maximum value.* For sucrose to "dediffuse" into a corner of the container would involve an entropy decrease and a violation of the second law.

Free Energy

The form of the second law stated above is not very useful to biologists or biochemists because we never deal with isolated systems. Every biological system (e.g., cell, organism, population) is open to its environment. Since living systems are open to exchange energy with their surroundings, both energy *and* entropy changes will be of importance in determining the direction of thermodynamically favorable processes. We need, for such systems, a function of state that will include both factors. There are several such, but the one of importance in biochemistry is the **Gibbs free energy (G)** or, as we shall call it, the **free energy.** This combines an enthalpy term, which measures the energy change at constant pressure, and an entropy term, which takes into account the randomization factor. It is defined as

$$G = H - TS \tag{3.18}$$

where T is the absolute temperature. For a free energy change ΔG in a system at constant temperature and pressure we can write

$$\Delta G = \Delta H - T\,\Delta S \tag{3.19}$$

We can gain an insight into the importance of free energy by considering the factors that experience shows to make processes favorable. We said that a decrease in energy (ΔH is negative) and/or an increase in entropy (ΔS is positive) are typical of favorable processes. Either of these conditions will tend to make ΔG negative. In fact, another way to state the second law of thermodynamics is this: *The criterion for a favorable process in a nonisolated system, at constant temperature and pressure, is that* ΔG *be negative.* Conversely, a positive ΔG means that a process is *not* favorable, but the *reverse of that process is.* Processes accompanied by negative free energy changes are said to be **exergonic;** those for which ΔG is positive are **endergonic.** You will see these terms frequently throughout this book.

Now suppose that the ΔH and $T\,\Delta S$ terms in the free energy equation just balance one another. In this case $\Delta G = 0$, and the process is not favored to go either forward or backward. In fact, the system is in equilibrium. Under these conditions, the process is reversible; that is, it can be displaced in either direction by an infinitesimal push in one way or the other.

The Interplay of Enthalpy and Entropy: An Example

To make these ideas more concrete, let us consider in detail a process we mentioned before, the transition between liquid water and ice. This familiar example demonstrates the interplay of enthalpy and entropy in determining the state of a system. In an ice crystal there is a maximum number of hydrogen bonds between the water molecules (see Chapter 2). When ice melts, some of these bonds must be broken. The enthalpy difference between ice and water corresponds almost entirely to the energy required to break hydrogen bonds. As Figure 3.4 shows, the enthalpy change for the transition ice → water is positive and nearly constant over a considerable temperature range.

The entropy change in melting arises primarily from the fact that liquid water is more random in structure than ice. In an ice crystal, each water molecule has a fixed place in the lattice and binds to its neighbor in the same way as every other water molecule. On the other hand, molecules in liquid water are continually moving, exchanging hydrogen-bond partners as they

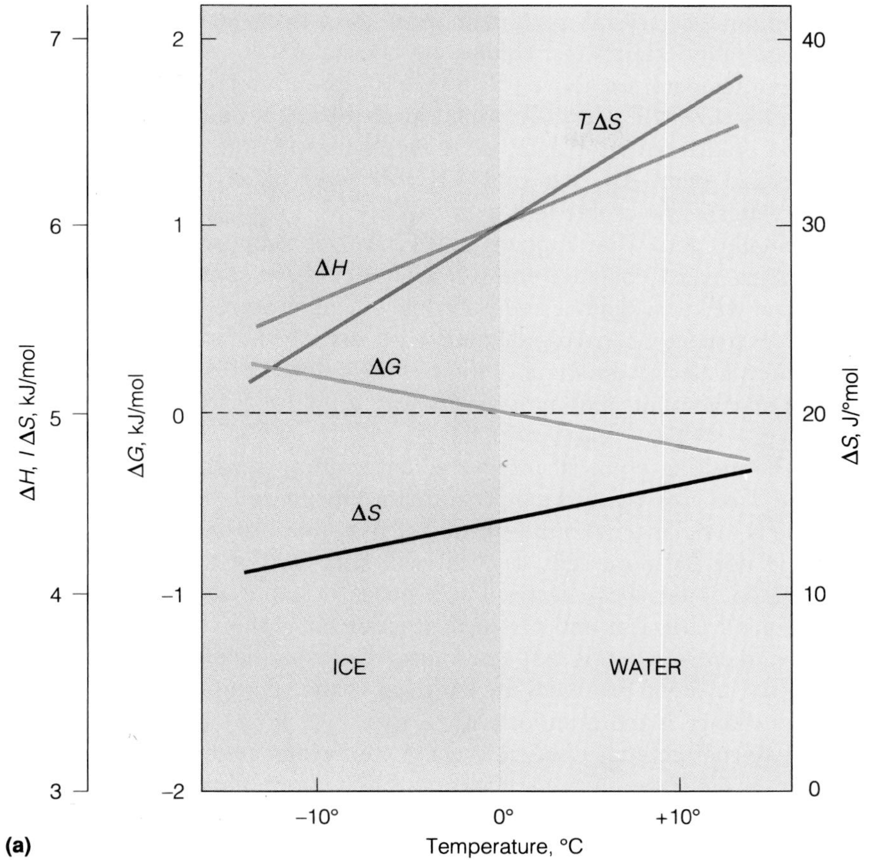

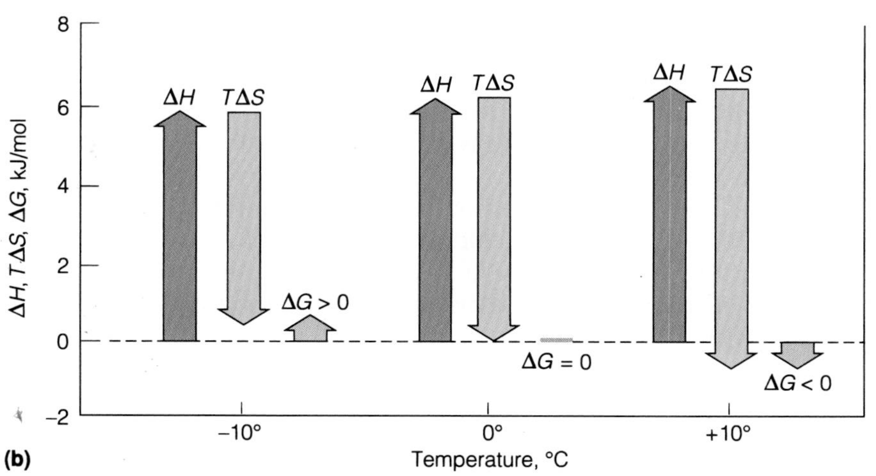

Figure 3.4
Thermodynamic functions for the ice–water transition. (**a**) Values are given for 1 mol of H_2O. For this transition, ΔH and ΔS are both nearly constant with temperature and both are positive. The ΔH and $T\,\Delta S$ curves cross at 0°C, at which point ΔG becomes zero. (**b**) The arrows graphically depict the competition between the ΔH and $-T\,\Delta S$ terms at three different temperatures.

do (compare Figures 2.10 and 2.11). Figure 3.4 shows that the entropy difference between water and ice is also nearly constant over a wide range. If we calculate the free energy change ($\Delta G = \Delta H - T\,\Delta S$) for the ice → water transition, we find the following: At low temperatures, ΔG is positive. For example, at −10°C we find $\Delta G = +213$ J/mol. This means that the transition ice → water is *not* favorable under these conditions. The opposite transition (water → ice) *is* favorable and therefore irreversible, as we can see from the behavior of supercooled water. If we disturb supercooled water, or

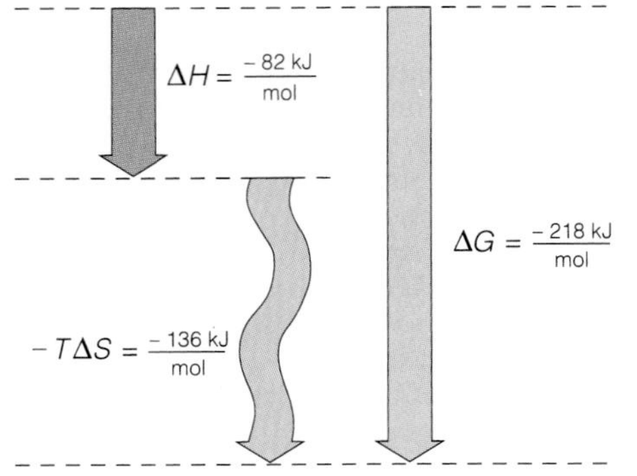

(a) Fermentation of glucose to ethanol

$C_6H_{12}O_6(s) \longrightarrow 2C_2H_5OH(l) + 2CO_2(g)$

Both enthalpy and entropy changes favor the reaction.

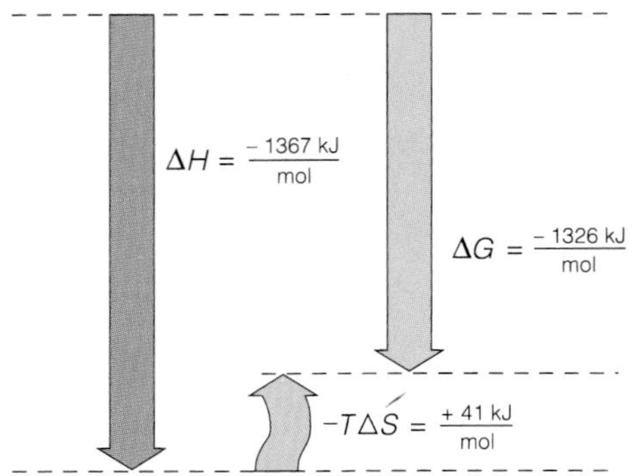

(b) Combustion of ethanol

$C_2H_5OH(l) + 3O_2(g) \longrightarrow 2CO_2(g) + 3H_2O(l)$

Enthalpy favors this reaction, but entropy opposes it. We could call this an "enthalpy-driven" reaction. If water *vapor* were the product, an entropy increase would favor the reaction as well.

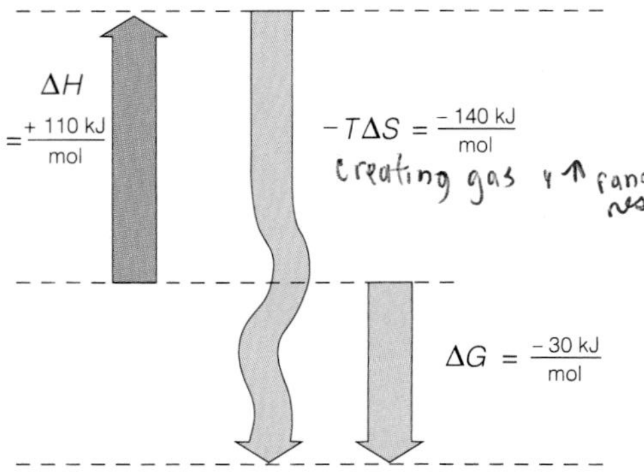

(c) Decomposition of nitrogen pentoxide

$N_2O_5(s) \longrightarrow 2NO_2(g) + 1/2\ O_2(g)$

This is a somewhat unusual chemical reaction in that it is "entropy-driven." The reaction actually absorbs heat but is favored by the large entropy increase resulting from the formation of gaseous products.

add a minute ice crystal to start the process of freezing, the entire sample will freeze. No infinitesimal change we can make will reverse this process.

At a temperature above 0°C, say +10°, the situation is entirely different. An ice cube will irreversibly melt under these conditions. Again, this is what we predict from the data, for if we calculate ΔG at +10°C, we obtain $\Delta G = -225$ J/mol. The sign of ΔG is now negative; at +10°C the process (ice → water) is favorable and irreversible.

Finally, note what happens at 0°C. At this temperature, the ΔH and $T\,\Delta S$ terms exactly balance and $\Delta G = 0$. *This is the condition for equilibrium,* and we know that ice and water are in equilibrium at 0°C. The change is now reversible; when ice and liquid water are together at 0°C, we can melt a bit more water by adding an infinitesimal amount of heat. Alternatively, we can take a minute amount of heat away from the system and freeze a bit more ice.

The melting point is simply the temperature at which the curves for ΔH and $T\,\Delta S$ cross; at this temperature the energetically favored process of freezing is in balance with the entropically favored process of melting. Neither ΔH nor ΔS alone tells us what will happen, but their combination, $\Delta H - T\,\Delta S$, prescribes exactly which phase is stable at any temperature.

For all chemical and physical processes it is the competition of enthalpy and entropy terms that determines the favorable direction. As Figure 3.5 shows, in some processes the enthalpy change dominates; in others the entropy change is more important.

Two matters that frequently cause confusion should be cleared up at this point. First, we must emphasize that the favorability of a process has nothing to do with its rate. Students frequently associate favorable processes with rapid ones, but this is not necessarily so. A reaction may have a very large negative free energy change but still proceed at a very low rate because of kinetic barriers. A surprising example of this is the simple reaction C (diamond) → C (graphite). The free energy change for this transformation, at room temperature, is −2.88 kJ/mol. Thus, diamond is unstable. Yet the reaction is imperceptibly slow. Diamonds are *not* forever—just almost. A catalyst may increase the rate for some reactions, but the favored direction is always dictated by ΔG and is independent of whether the reaction is catalyzed or not.

Second, there is no reason why the entropy of an *open* system cannot decrease. This happens, for example, whenever water freezes. More important to us, it happens all the time in living organisms. An organism takes in foodstuffs, often in the form of disorganized small molecules, and from these builds enormous, complex, highly ordered macromolecules like proteins and nucleic acids. From these it constructs elegantly structured cells, tissues, and organs. All of this involves a tremendous entropy decrease. The implication of equation (3.19) is that entropy *can* decrease in a favored process, but only if there is also a large energy decrease. *Energy must be expended to pay the price of organization.* This really is what life is all about. Living organisms spend energy to destroy entropy. For these processes to proceed, the *overall* free energy changes in the organism must be negative. Life is an irreversible process. An organism that comes to equilibrium with its surroundings is dead.

There is an even deeper philosophical implication of the thermodynamics of life. The universe as a whole is an isolated system. The entropy of

Figure 3.5
Competition of enthalpy and entropy for several processes. Each of these processes has a negative free energy change, but the change is accomplished in different ways. (Not to scale.)

the whole universe must be increasing. Thus each of us, as a living organism that locally and temporarily decreases entropy, must produce, somewhere in the world around us, an even greater increase. In a very real sense, we buy our lives through the entropic death of the universe.

Free Energy and Concentration

The sign of the free energy change in a process tells us whether that process or its reverse is thermodynamically favorable. The magnitude of ΔG is an indication of how far the process is from equilibrium. Clearly, ΔG is a quantity of fundamental importance in determining which processes will or will not occur in a cell. But to express these ideas quantitatively, in terms of changes in the concentrations of substances, we need to know how the free energy of a system depends on the amounts of various components in the mixture.

Chemical Potential

The answer is simple. If we have a mixture containing N_A moles of component A, N_B moles of component B, and so on, we may write

$$G = N_A\overline{G}_A + N_B\overline{G}_B + N_C\overline{G}_C + \cdots \tag{3.20}$$

The quantities $\overline{G}_A$, $\overline{G}_B$, and so forth, are called the **partial molar free energies** or **chemical potentials** of the various components. Each represents the contribution, per mole, of one component to the total free energy of the system. (In some texts, chemical potential is given the symbol μ.) We shall take the approximation, which is usually valid for dilute solutions, that each of the chemical potentials depends only on the concentration of the appropriate substance itself. For dilute solutions $\overline{G}_A$, $\overline{G}_B$, etc. turn out to be simple functions of the corresponding concentrations:

$$\overline{G}_A = G_A{}^\circ + RT\ln[\text{A}] \tag{3.21a}$$
$$\overline{G}_B = G_B{}^\circ + RT\ln[\text{B}] \tag{3.21b}$$
$$\text{etc.}$$

where [A], [B], etc. are the molar concentrations of the components. Note what happens when [A], for example, equals 1 M. The logarithmic term then vanishes ($\ln 1 = 0$) and $\overline{G}_A = \overline{G}_A{}^\circ$. This shows what $G_A{}^\circ$ and $G_B{}^\circ$, mean. They are reference, or standard state, values of the chemical potential. We always express the chemical potentials with respect to a **standard state,** which in this book will always represent a 1 M solution of the compound. In equations (3.21), T is the absolute temperature and R is the gas constant. In all calculations we will do based on equation (3.19) $R = 8.314$ J/°mol.

Use of the Chemical Potential: Application to Transport Across Membranes

The importance of equations (3.21) is that they allow us to apply general thermodynamic principles to practical problems. In particular, they allow us to predict the favored directions for real processes. One biochemical process to which these principles are particularly relevant is transport across membranes. To take a specific example, suppose we have two solutions of a

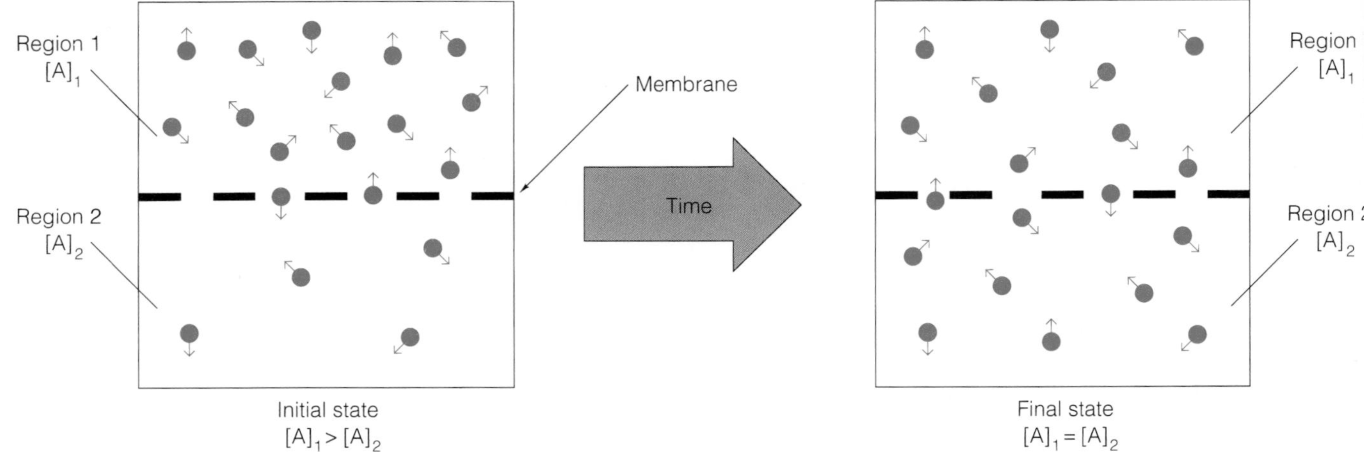

Figure 3.6
Equilibration across a membrane. Two solutions of A, of concentrations $[A]_1$ and $[A]_2$, are separated by a membrane through which A can pass in either direction. If the initial concentration of A is higher in region 1 (left panel), the chemical potential of A will also be higher in that region, and net transport will occur from 1 to 2 until equal concentrations (and chemical potentials) are obtained (right panel).

substance A separated by a membrane through which A can pass (Figure 3.6). Assume that in region 1 the concentration is initially made $[A]_1$ and in region 2 the concentration is $[A]_2$. To determine in which direction transport of A will be favored, imagine transferring a very small amount (ΔN moles) of A from region 1 to region 2. We make ΔN so small that the concentrations are not changed appreciably by the transfer. Now we calculate the free energy changes in regions 1 and 2 by using equations (3.20) and (3.21). Recall that we are taking ΔN moles of A out of region 1, where its concentration is $[A]_1$, and putting it into a region where its concentration is $[A]_2$.

$$\text{In region 1} \quad \Delta G_1 = -\Delta N\, G_A{}^\circ - \Delta N\, RT \ln[A]_1 \tag{3.22a}$$
$$\text{In region 2} \quad \Delta G_2 = +\Delta N\, G_A{}^\circ + \Delta N\, RT \ln[A]_2 \tag{3.22b}$$

The *total* free energy change is

$$\Delta G = \Delta G_1 + \Delta G_2 = \Delta N\, RT\{\ln[A]_2 - \ln[A]_1\} \tag{3.23}$$

Note that the ΔG° terms have canceled. Equation (3.23) can be rewritten in a more compact form as

$$\Delta G = \Delta N\, RT \ln([A]_2/[A]_1) \tag{3.24}$$

To obtain ΔG per *mole* of A transferred, we divide by ΔN

$$\Delta G \text{ (per mole)} = RT \ln([A]_2/[A]_1) \tag{3.25}$$

We will draw on this equation throughout our discussions of the transfer of substances across cell membranes. It is important to note two conditions that have an impact on such transfer:

1. If $[A]_2$ is less than $[A]_1$, ΔG is negative; transfer from region 1 to region 2 is favorable.

2. If $[A]_2$ is greater than $[A]_1$, ΔG is positive. Transfer in this direction is not favorable, but transfer in the opposite direction is.

Thus, we have the following conclusion: If a substance can be transported across a membrane, the direction of favorable transfer will always be from the region of high concentration to the region of low concentration. Furthermore, suppose $[A]_1 = [A]_2$. In this case, equation (3.23) states that $\Delta G = 0$. This means that the process is reversible; the system is at equilibrium. Thus, the equilibrium state will be that in which the concentrations of A on the two sides of the membrane have become equal. From whatever distribution of A we may start with, this equilibrium state of equal concentration will be approached.

We shall encounter situations in living cells that seem, at first glance, to violate this rule. There are cases in which substances pass readily from regions of low concentration to regions of high concentration. But we shall find that, in such circumstances, the necessary free energy price is paid by *coupling* the transport process to thermodynamically favorable chemical reactions. The cell may seem to avoid the laws of thermodynamics, but it never really does so.

Free Energy and Chemical Reactions: Chemical Equilibrium

The Free Energy Change and the Equilibrium Constant

We can likewise use the knowledge of how the chemical potentials of substances depend on their concentrations to describe quantitatively the free energy changes in chemical reactions. We are thereby able to predict the favorable directions for reactions. Suppose we have a reaction like

$$aA + bB \rightleftharpoons cC + dD$$

Even though the reaction can proceed in either direction, we have written it with C and D on the right, so we call these *products* and A and B *reactants*. We wish to calculate the change in free energy that occurs when a moles of A and b moles of B form c moles of C and d moles of D each at some given concentration.* For the reverse reaction, we need only change the sign of the calculated quantity.

The free energy change must be the free energy of the products minus that of the reactants.

$$\Delta G = G \text{ (products)} - G \text{ (reactants)} \qquad (3.26)$$

We can write these free energies in terms of the chemical potentials of the substances, each multiplied by the number of moles involved. For our example

$$\Delta G = c\overline{G}_C + d\overline{G}_D - a\overline{G}_A - b\overline{G}_B \qquad (3.27)$$

Now we insert the appropriate expression for $\overline{G}_c$ and so forth in terms of concentrations and obtain

$$\begin{aligned}\Delta G = {} & cG_C^\circ + cRT\ln[C] + dG_D^\circ + dRT\ln[D] \\ & - aG_A^\circ - aRT\ln[A] - bG_B^\circ - bRT\ln[B] \qquad (3.28a)\end{aligned}$$

*How *does* one carry out a finite amount of reaction while keeping concentrations of both reactants and products constant? Two ways could be imagined. First, the total amounts of reactants and products could be so enormous that a finite reaction would not appreciably change concentrations. Alternatively, we could imagine hypothetical processes that would remove products and add reactants so as to keep concentrations unchanged.

$$= cG_C^\circ + dG_D^\circ - aG_A^\circ - bG_B^\circ + RT\ln[C]^c + RT\ln[D]^d - RT\ln[A]^a - RT\ln[B]^b \qquad (3.28b)$$

$$= \Delta G^\circ + RT(\ln[C]^c + \ln[D]^d - \ln[A]^a - \ln[B]^b) \qquad (3.28c)$$

In going from (3.28a) to (3.28c) we have done two things: grouped the G° terms together and made use of the fact that $aRT\ln[A] = RT\ln[A]^a$. The group of G° terms (ΔG°) has a simple meaning: since G° is the free energy per mole of a substance in the standard state (1 M), ΔG° represents the **standard state free energy change** in the reaction. It is the free energy change that would be observed if a moles of A and b moles of B, each at 1 M concentration, formed c moles of C and d moles of D, each at 1 M. The terms containing logarithms can be combined, since each is multiplied by the same factor, RT. Therefore

$$\Delta G = \Delta G^\circ + RT\ln\left\{\frac{[C]^c[D]^d}{[A]^a[B]^b}\right\} \qquad (3.29)$$

The quantity ΔG represents the free energy change when a moles of A (at concentration [A]) and b moles of B (at concentration [B]) make c moles of C (at concentration [C]) and d moles of D (at concentration [D]). These concentrations may be anything we want them to be. When all are 1 M, equation (3.29) reduces to $\Delta G = \Delta G^\circ$, as we would expect. The importance of equation (3.29) is that it allows us to calculate ΔG under any conditions we wish.

Suppose, now, that the reaction has come to equilibrium. In that case, two things must be true. First, the concentrations in the factor in braces in equation (3.29) must be equilibrium concentrations, and therefore the factor in braces is now the *equilibrium constant* K for the reaction:

$$K = \left\{\frac{[C]^c[D]^d}{[A]^a[B]^b}\right\}_{eq} \qquad (3.30)$$

Second, if we carry out any amount of a reaction reversibly—that is, keeping the system very close to equilibrium—ΔG must equal zero. So

$$0 = \Delta G^\circ + RT\ln\left\{\frac{[C]^c[D]^d}{[A]^a[B]^b}\right\}_{eq} \qquad (3.31a)$$

or

$$-\Delta G^\circ = RT\ln K \qquad (3.31b)$$

or

$$K = e^{-\Delta G^\circ/RT} \qquad (3.31c)$$

Equations (3.31) establish the important relationship between ΔG° and the equilibrium constant. Because ΔG° is related to the logarithm of K, even quite small values of ΔG° lead to large values of K.

Equation (3.29) may best be thought of in the following way: ΔG° represents a reference value for the free energy change, whereby the intrinsic free energy changes in different reactions can be compared under equivalent circumstances (1 M concentrations). The magnitude of this term tells us the equilibrium constant. The second (concentration-dependent) term represents the extra free energy change (+ or −) involved if we were to carry out the reaction at some other, arbitrary set of concentrations. In applying these equations to biochemical problems, we must always keep in mind that it is ΔG, as determined by the actual concentrations in the cell, rather than $\Delta G°$ that determines whether or not a reaction is favored in vivo.

Application of Free Energy Calculations to a Biochemical Reaction

To make the application of these somewhat abstract ideas a bit clearer, let us consider an example—a very simple but important biochemical reaction, the isomerization of glucose-6-phosphate into fructose-6-phosphate:

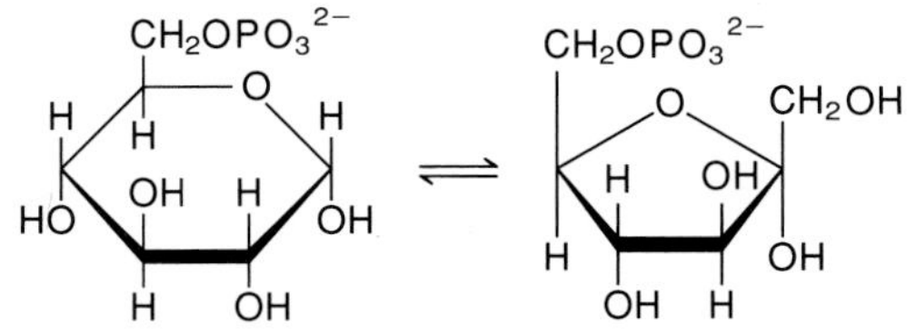

Glucose-6-phosphate $\rightleftharpoons$ **Fructose-6-phosphate**

or:

$$\text{glucose-6-phosphate} \rightleftharpoons \text{fructose-6-phosphate}, \qquad \Delta G° = +1.7 \text{ kJ/mol}$$
$$\text{G6P} \rightleftharpoons \text{F6P}$$

This is the second step in the glycolytic pathway, which is discussed in Chapter 13. The reaction is clearly endergonic under standard conditions. This means that the system is not at equilibrium when G6P and F6P are both at 1 M. Rather, equilibrium lies to the left, with a higher concentration of G6P than F6P. We can state this quantitatively by calculating the equilibrium constant from equation (3.31c).

$$K = e^{-\Delta G°/RT} = e^{-1700\,(\text{J/mol})/8.314\,(\text{J/°mol})\times 298°} \tag{3.32a}$$

$$K = 0.504 = \left\{\frac{[\text{F6P}]}{[\text{G6P}]}\right\}_{\text{eq}} \tag{3.32b}$$

where $([\text{F6P}]/[\text{G6P}])_{\text{eq}}$ is the ratio of the equilibrium concentration of fructose-6-phosphate to that of glucose-6-phosphate. The fact that $K < 1$ is another way of saying that the equilibrium lies to the left. But we can do better than this and show *where* the equilibrium lies. The total substance in the reaction is distributed between F6P and G6P. We can write

$$K = \frac{[\text{F6P}]_{\text{eq}}}{[\text{G6P}]_{\text{eq}}} = \frac{[\text{F6P}]_{\text{eq}}/\text{total concentration}}{[\text{G6P}]_{\text{eq}}/\text{total concentration}} = \frac{(f_{\text{F6P}})_{\text{eq}}}{(f_{\text{G6P}})_{\text{eq}}} \tag{3.33}$$

where f_{F6P} and f_{G6P} represent the fractions of the total material in each form. Since

$$f_{\text{G6P}} = 1 - f_{\text{F6P}} \tag{3.34}$$

we have

$$K = \frac{(f_{\text{F6P}})_{\text{eq}}}{1 - (f_{\text{F6P}})_{\text{eq}}} \tag{3.35}$$

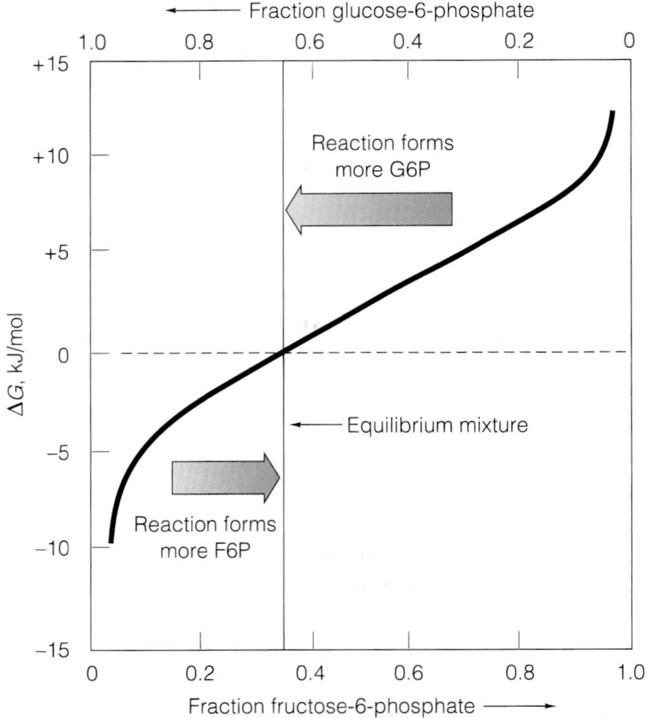

Figure 3.7
Free energy changes for the reaction glucose-6-phosphate ⇌ fructose-6-phosphate as a function of the fraction of either component. If the composition is initially set to the left of the equilibrium point, the reaction will proceed to the right. If the mixture is initially set at the right of the equilibrium point, the reaction will proceed to the left. In either case equilibrium will be approached.

Using the value of $K = 0.510$, we can solve this equation to yield $(f_{F6P})_{eq} = 0.335$. In other words, at equilibrium 33.5% of the sugar will be in the form of fructose-6-phosphate and the remaining 66.5% in the form of glucose-6-phosphate.

A further insight into the power of such analysis is provided by the application of equation (3.29) to this problem:

$$\Delta G = \Delta G^\circ + RT \ln\left\{\frac{[\mathrm{F6P}]}{[\mathrm{G6P}]}\right\} \tag{3.36a}$$

$$= \Delta G^\circ + RT \ln\left\{\frac{(f_{F6P})}{1 - (f_{F6P})}\right\} \tag{3.36b}$$

What we are doing now is calculating the ΔG value when f_{F6P} has any value we wish. We can pick a value and ask what the free energy change for converting 1 mol of glucose-6-phosphate to 1 mol of fructose-6-phosphate would be if the fraction f_{G6P} were set at that value. The results are graphed in Figure 3.7; their significance can be seen from the following examples.

1. Suppose we made a mixture with $f_{F6P} = 0.2$. Under these conditions, equation (3.36b) shows that ΔG is negative (-1.76 kJ/mol). The reaction to the right is thermodynamically favored. Fructose-6-phosphate will be formed at the expense of glucose-6-phosphate until f_{F6P} increases to the equilibrium value, 0.335. At this point, the process will have come to equilibrium.

2. If the initial mixture were prepared instead with $f_{F6P} = 0.8$, the initial value for ΔG would be $+5.11$ kJ/mol. The reverse reaction would now be favored, and fructose-6-phosphate

would be converted into glucose-6-phosphate until the equilibrium state was attained.

This example shows how thermodynamically favored (irreversible) processes are related to equilibrium. Whenever a system is displaced from equilibrium, it will spontaneously proceed in whichever direction is necessary to move toward that equilibrium state.

High-Energy Phosphate Compounds: Free Energy Sources in Biological Systems

The central role of free energy changes in determining the favorable directions for chemical reactions is of the greatest importance in biochemistry, for every metabolic pathway must, overall, be a thermodynamically favored process. However, individual reactions in a pathway often turn out to have positive, and therefore unfavorable, $\Delta G°$ values. How, then, can the reactions occur efficiently? In some in vivo cases, the clue is found in the fact that reactant and product concentrations are maintained far from equilibrium values. If reactants are present in high concentration, and products are continually being siphoned off in further reactions, the value of ΔG (as opposed to $\Delta G°$) may be quite negative. Such is the case, for example, in the conversion of glucose-6-phosphate to fructose-6-phosphate discussed earlier. At the steady-state concentrations of these substances maintained in the cell, $\Delta G \cong -2.5$ kJ/mol, as contrasted to the $\Delta G°$ value (+1.7 kJ/mol). Thus, the reaction continues to proceed to the right.

Coupled Reactions

Although many biological reactions can proceed favorably because their products are removed efficiently, there are also many cases where such assistance is not sufficient. There is another and very important way in which intrinsically unfavorable reactions can be driven. Suppose, for example, we have a reaction that is part of an essential pathway but is endergonic and therefore intrinsically unfavored:

$$\text{A} \rightleftharpoons \text{B} \qquad \Delta G° = +10 \text{ kJ/mol}$$

At the same time, suppose there is another process that is highly exergonic:

$$\text{C} \rightleftharpoons \text{D} \qquad \Delta G° = -30 \text{ kJ/mol}$$

If the cell can manage to *couple* these two reactions, the $\Delta G°$ for the *overall* process will be the algebraic sum of the individual values:

$$\text{A} + \text{C} \rightleftharpoons \text{B} + \text{D} \qquad \Delta G° = +10\frac{\text{kJ}}{\text{mol}} - 30\frac{\text{kJ}}{\text{mol}} = -20\frac{\text{kJ}}{\text{mol}}$$

In the overall process equilibrium now lies far to the right; the consequence is that B is efficiently produced from A.

Coupling of endergonic reactions to exergonic reactions is one of the most important principles in biochemistry. Such linkage of highly favored to unfavored processes is used not only to drive innumerable reactions but also to transport materials across membranes, transmit nerve impulses, and contract muscles.

PHOSPHATE COMPOUND		HYDROLYSIS PRODUCTS	ΔG°′ (kJ/mol)
Phosphoenolpyruvate (PEP)	+ H_2O	Pyruvate + $HO{-}PO_3^{2-}$	−62
1,3-Bis-phosphoglycerate	+ H_2O	3-Phosphoglycerate (3PG) + H^+ + $HO{-}PO_3^{2-}$	−49
Creatine phosphate (CP)	+ H_2O	Creatine + $HO{-}PO_3^{2-}$	−43
Pyrophosphate (PP_i)	+ H_2O	2 $HO{-}PO_3^{2-}$	−33
Adenosine triphosphate (ATP)	+ H_2O	Adenosine diphosphate (ADP) + H^+ + $HO{-}PO_3^{2-}$	−31
		Adenosine monophosphate (AMP) + H^+ + pyrophosphate	−31
Adenosine monophosphate (AMP)	+ H_2O	Adenosine + $HO{-}PO_3^{2-}$	−14
Glycerol-1-phosphate (G1P)	+ H_2O	Glycerol + $HO{-}PO_3^{2-}$	−10

Transfer potential: 60, 50, 40, 30, 20, 10

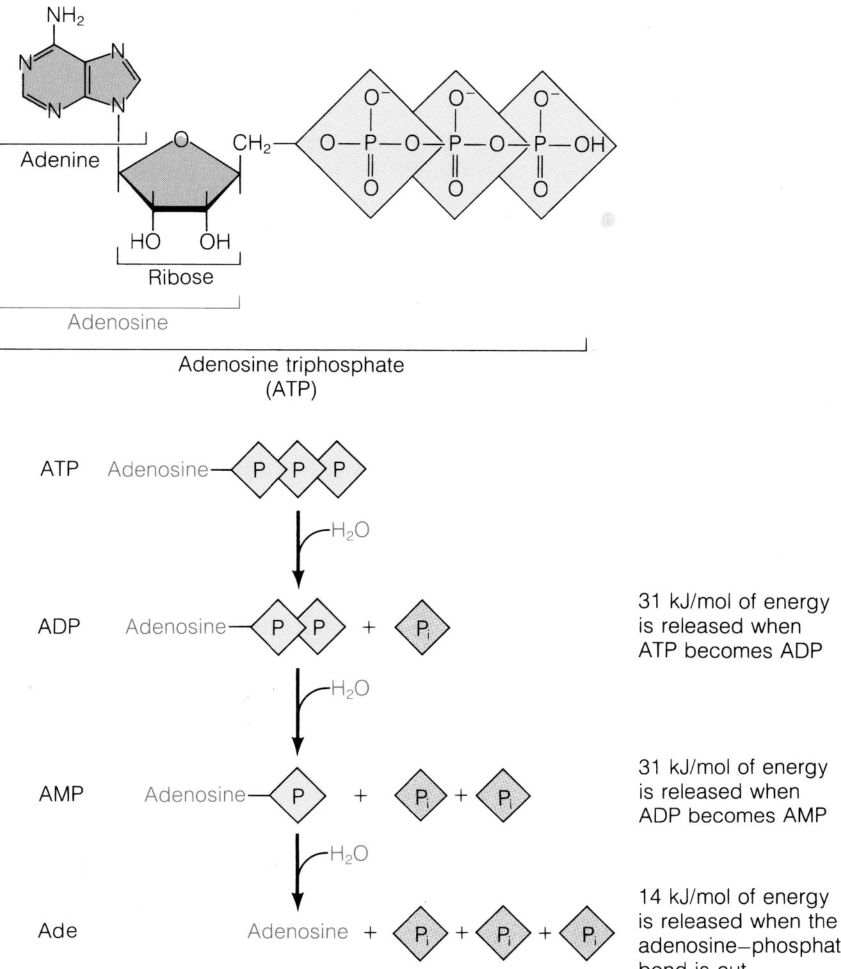

Figure 3.9
The ATP molecule and its hydrolysis reactions. Throughout this book ⓟ represents the tetrahedral phosphate group.

High-Energy Phosphate Compounds as Energy Stores

The fact that driving processes by coupling is so common must mean that there exist in the cell a number of compounds (like the hypothetical C above) that can undergo reactions with large negative free energy changes. Such substances can be thought of as stores of free energy in the cell. The most important of these stores are certain phosphate compounds, which can undergo hydrolytic release of their phosphate groups in aqueous solution. A number of such compounds and their hydrolysis reactions are shown in Figure 3.8. Some of them, like **phosphoenolpyruvate (PEP), creatine phosphate (CP),** and **adenosine triphosphate (ATP),** have very negative standard state free energies of hydrolysis. Perhaps the most important of these compounds, and the one you will encounter most often in this book, is ATP, the structure and hydrolysis reactions of which are shown in Figure 3.9. Hydrolysis of ATP is highly exergonic, with $\Delta G^{\circ\prime} = -31$ kJ/mol. A ΔG° of

---

Figure 3.8
Hydrolysis reactions for a number of biochemically important phosphate compounds. The labile phosphate group of each compound is shown in color. The more stable reaction product, P_i, is in gray. A scale of phosphate transfer potential is shown to the right.

−31 kJ/mol corresponds to an equilibrium constant greater than 10^5. Such an equilibrium lies so far to the right that the reaction can be considered essentially irreversible.

Figure 3.8 also shows that whereas some of these phosphate hydrolysis reactions are truly high-energy processes, others are not. At one time it was conjectured that compounds like PEP, CP, and ATP contained a special kind of "high-energy phosphate bond." However, careful investigation has revealed that there is nothing special about the bonds. Rather, the explanation for the high free energies of hydrolysis of some of these compounds lies in the special properties of both the reactants and products in the reactions. A variety of factors contribute to make these free energy changes large. Those that seem to be most important are described below.

RESONANCE STABILIZATION OF THE PHOSPHATE PRODUCTS. The **orthophosphate** ion, which is often abbreviated as $\mathbf{P_i}$ (inorganic phosphate), is capable of a wide variety of resonance forms. Both the bound proton and the oxygen bonding should be thought of as delocalized, so a more appropriate way to write the structure is as shown in Figure 3.10. The multiple forms, of equal energy, contribute to the high entropy of such a resonance structure (see equation (3.13)). Not all of these forms are possible when the phosphate is bound in an ester. Consequently, release of the phosphate is favored. Resonance stabilization applies in *all* of the phosphate hydrolysis reactions described in Figure 3.8.

ADDITIONAL HYDRATION OF THE HYDROLYSIS PRODUCTS. Release of the phosphate residue from its bonded state allows greater opportunities for hydration. This is especially true when both products are charged.

ELECTROSTATIC REPULSION BETWEEN CHARGED PRODUCTS. In the hydrolysis of phosphoenolpyruvate, bisphosphoglycerate, and adenosine triphosphate, both products of hydrolysis carry a negative charge. The repulsion between these products strongly favors the hydrolysis reaction.

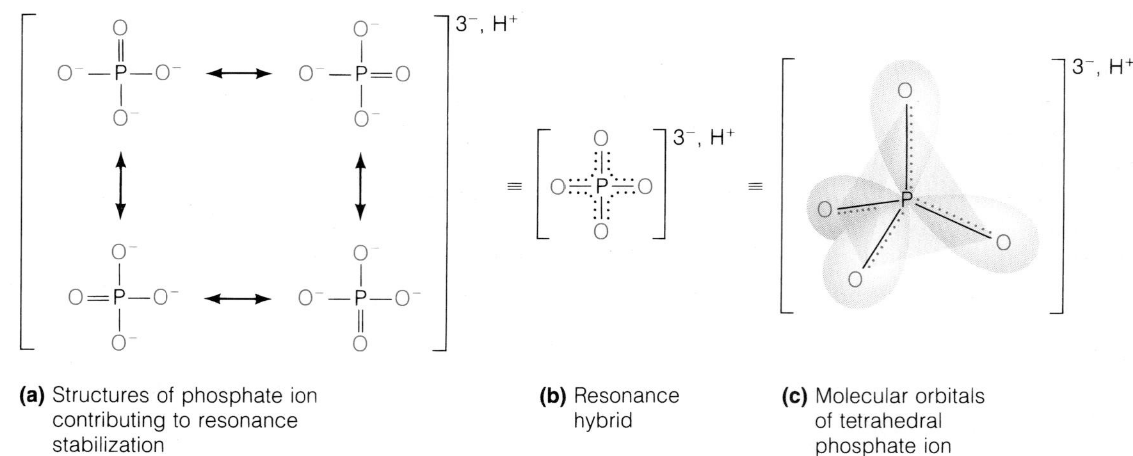

(a) Structures of phosphate ion contributing to resonance stabilization

(b) Resonance hybrid

(c) Molecular orbitals of tetrahedral phosphate ion

Figure 3.10
Resonance structures for orthophosphate HPO_3^{2-} (P_i). In (**a**) resonance is depicted as an equilibrium between four forms, with the H^+ not assigned permanently to any one of the four oxygens. This can be abbreviated as in (**b**), where the dotted lines represent partial bonds. That such partial bonds have physical significance is pointed out in (**c**), which represents the phosphate ion as a tetrahedral structure with four equivalent bonds.

ENHANCED RESONANCE STABILIZATION OF PRODUCT MOLECULES. Hydrolysis is sometimes favored because, in addition to the resonance stabilization of the phosphate, the other product can adopt more molecular forms. This is true for pyruvate, which can tautomerize:

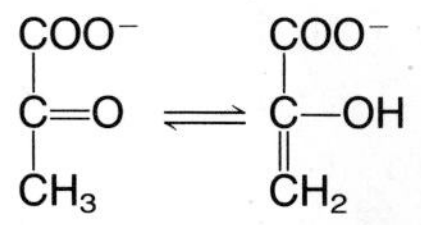

$$\begin{matrix} COO^- \\ | \\ C{=}O \\ | \\ CH_3 \end{matrix} \rightleftharpoons \begin{matrix} COO^- \\ | \\ C{-}OH \\ \| \\ CH_2 \end{matrix}$$

and for creatine, which has three resonance forms:

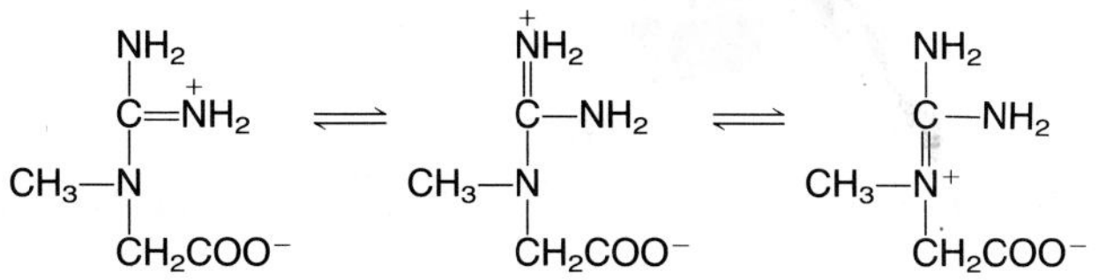

$$\begin{matrix} NH_2 \\ | \\ C{=}\overset{+}{N}H_2 \\ | \\ CH_3{-}N \\ | \\ CH_2COO^- \end{matrix} \rightleftharpoons \begin{matrix} \overset{+}{N}H_2 \\ \| \\ C{-}NH_2 \\ | \\ CH_3{-}N \\ | \\ CH_2COO^- \end{matrix} \rightleftharpoons \begin{matrix} NH_2 \\ | \\ C{-}NH_2 \\ \| \\ CH_3{-}N^+ \\ | \\ CH_2COO^- \end{matrix}$$

RELEASE OF A PROTON IN BUFFERED SOLUTIONS. In some of the reactions listed in Figure 3.8 a proton is released. This means that the hydrogen ion concentration (i.e., the pH) will influence the equilibrium. Note that we have designated the standard state free energies of hydrolysis for these reactions as $\Delta G^{\circ\prime}$. The superscript prime signifies that the standard state is taken at pH 7, near the pH value maintained for almost all physiological processes (see Chapter 2). Consider how we would write the true ΔG for a reaction like the hydrolysis of ATP to ADP, under circumstances where both H_2O and H^+ concentrations could be varied:

$$ATP^{4-} + H_2O \rightleftharpoons ADP^{3-} + HPO_4^{2-} + H^+$$

Then, by equation (3.29), we should have

$$\Delta G = \Delta G^\circ + RT\ln\left\{\frac{[ADP^{3-}][HPO_4^{2-}][H^+]}{[ATP^{4-}][H_2O]}\right\} \tag{3.37}$$

But since the hydrolysis reactions in vivo always occur in very dilute aqueous solution, near pH 7, it is inappropriate to retain the concentrations of H_2O and H^+ as variables.

The molar concentration of water in dilute aqueous solutions is always close to 55.6 M and will not change appreciably if a tiny bit is used to hydrolyze a bit of ATP. Because living cells are usually buffered at a pH very close to 7.0, we may consider the hydrogen ion concentration to be maintained at about 10^{-7} M. Since both $[H_2O]$ and $[H^+]$ are nearly constant, equation (3.37) may be conveniently rewritten as

$$\Delta G = \Delta G^\circ + RT\left\{\frac{[ADP][P_i]}{[ATP]}\right\} + RT\ln\left\{\frac{[H^+]}{[H_2O]}\right\} \tag{3.38}$$

The third term on the right is essentially constant. We may insert it into a redefined ΔG°, which we shall call $\Delta G^{\circ\prime}$:

$$\Delta G = \Delta G^{\circ\prime} + RT\ln\left\{\frac{[ADP][P_i]}{[ATP]}\right\} \tag{3.39}$$

where

$$\Delta G^{\circ\prime} = \Delta G^{\circ} + RT \ln\left\{\frac{[H^+]}{[H_2O]}\right\} \tag{3.40}$$

This is the true meaning of the $\Delta G^{\circ\prime}$ values given in Figure 3.8. The fact that the hydrolysis of ATP normally occurs in the presence of a vast excess of H_2O, and under conditions where the hydrogen ion concentration is kept very low, contributes enormously to the favorability of the reaction. The former condition is true of hydrolysis reactions in general; the latter is important whenever protons are released.

Phosphate Transfer Potential and the Central Role of ATP as a Free Energy Currency

Phosphate Transfer Potential

There is another, very useful way in which we can think about the $\Delta G^{\circ\prime}$ values for various high-energy phosphate compounds. As Figure 3.8 shows, they form a scale of phosphate **transfer potentials.** The potential is simply defined as $-\Delta G^{\circ\prime}$. Each compound is capable of driving the phosphorylation of compounds lower on the scale, provided that a suitable coupling mechanism is available. Consider, for example, the following reactions, which are written in the somewhat abbreviated form we will frequently employ:

(1) Hydrolysis of phosphoenolpyruvate	PEP $\rightleftharpoons$ pyruvate + P_i	$\Delta G^{\circ\prime} = -62\ \frac{kJ}{mol}$
(2) Phosphorylation of adenosine diphosphate	ADP + P_i $\rightleftharpoons$ ATP	$\Delta G^{\circ\prime} = +31\ \frac{kJ}{mol}$
(1) + (2): Coupled phosphorylation of ADP by PEP	PEP + ADP $\rightleftharpoons$ pyruvate + ATP	$\Delta G^{\circ\prime} = -31\ \frac{kJ}{mol}$

Thus, phosphoenolpyruvate, having the very high phosphate transfer potential of 62 kJ/mol, is capable of adding a phosphate group to ADP in a thermodynamically favored process. On the other hand, ATP can pass this phosphate on to glucose, since the phosphate transfer potential of glucose-6-phosphate lies still further down the scale:

(1) Hydrolysis of ATP	ATP $\rightleftharpoons$ ADP + P_i	$\Delta G^{\circ\prime} = -31\ \frac{kJ}{mol}$
(2) Phosphorylation of glucose	glucose + P_i $\rightleftharpoons$ glucose-6-phosphate	$\Delta G^{\circ\prime} = +14\ \frac{kJ}{mol}$
(1) + (2): Coupled phosphorylation of glucose by ATP	ATP + glucose $\rightleftharpoons$ ADP + glucose-6-phosphate	$\Delta G^{\circ\prime} = -17\ \frac{kJ}{mol}$

This emphasizes how ATP can act as a versatile phosphate transfer agent through coupled reactions. In each case, the coupling is accomplished by having the reactions take place on the surface of a large protein molecule, an

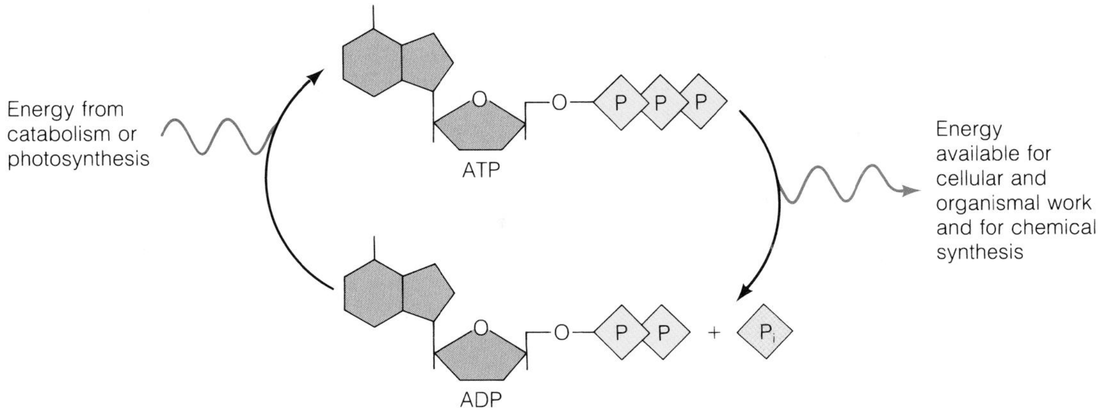

Figure 3.11
The central role of ATP in metabolism. Energy obtained from photosynthesis or breakdown of foodstuffs is stored in ATP and used, via coupled hydrolysis, to drive life processes.

enzyme. We shall study enzymes in detail in Chapters 10 and 11 and find that they can both facilitate such coupling and accelerate the reactions.

ATP as a Free Energy Currency

Adenosine triphosphate lies about midway on the scale of phosphate transfer potential. This is a strategic position, for ATP serves as the general "free energy currency" for virtually all cellular processes. Hydrolysis of ATP is used to drive innumerable biochemical reactions, including many that are not phosphorylations. It is a source of energy for cell motility, muscle contraction, and the specific transport of substances across membranes. The processes of photosynthesis and metabolism of foodstuffs are used mainly to produce ATP. It is probably no exaggeration to call ATP the single most important substance in biochemistry (Figure 3.11).

ATP Hydrolysis Under Cellular Conditions

Because of its central role in biochemical reactions, it is important to consider the hydrolysis of ATP in some detail. As illustrated in Figures 3.8 and 3.9, the hydrolysis can proceed in either of two ways. Either the terminal phosphate can be cleaved off to produce **adenosine diphosphate** (**ADP**) and phosphate (P_i), or cleavage may occur at the second phosphodiester bond to yield **adenosine monophosphate** (**AMP**) and **pyrophosphate** (**PP_i**). The former reaction is the more common one in vivo, but the free energy yield is about the same for both. Note that AMP is not a "high-energy" phosphate compound. On hydrolysis, it yields the uncharged adenosine molecule. Part of the driving force for ATP hydrolysis—namely, the repulsion between charged products and their strong hydration—is lacking in this case.

It must be emphasized that the $\Delta G^{\circ\prime}$ values for ATP hydrolysis do *not* represent the actual ΔG values to be expected under likely biological circumstances. There are many reasons for this:

1. ΔG depends on temperature, and different organisms live at different temperatures.

2. Even though $\Delta G^{\circ\prime}$ is defined at pH 7.0, the actual pH may vary from about 6.5 to 8.0 in different cells, tissues, and organisms. Because the reactants and products in the reaction all have

pK_a's in the neighborhood of neutrality, their degree of ionization varies markedly with pH. This in turn modifies ΔG. For example, one of the steps in ionization of ADP has a pK_a of 6.7:

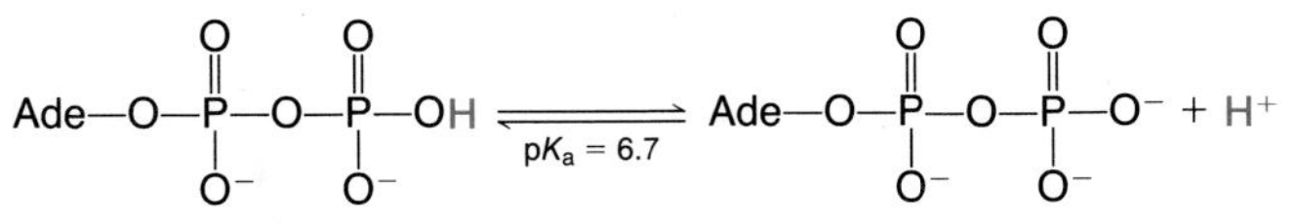

$$\text{Ade—O—}\overset{\text{O}}{\underset{\text{O}^-}{\overset{\|}{\underset{|}{\text{P}}}}}\text{—O—}\overset{\text{O}}{\underset{\text{O}^-}{\overset{\|}{\underset{|}{\text{P}}}}}\text{—OH} \underset{\text{p}K_a = 6.7}{\rightleftharpoons} \text{Ade—O—}\overset{\text{O}}{\underset{\text{O}^-}{\overset{\|}{\underset{|}{\text{P}}}}}\text{—O—}\overset{\text{O}}{\underset{\text{O}^-}{\overset{\|}{\underset{|}{\text{P}}}}}\text{—O}^- + \text{H}^+$$

which we can write more compactly as

$$ADP^{2-} \rightleftharpoons ADP^{3-} + H^+$$

At pH 7.0, ADP is about 30% in the ADP^{2-} form and 70% in the ADP^{3-} form. An increase in pH favors the latter form, which has a stronger electrostatic repulsion to the phosphate product. This in turn makes the ΔG for the hydrolysis of ATP to ADP even more negative in the alkaline pH range.

3. Divalent ions, like Mg^{2+}, are present in appreciable quantities in all cells. The reactants and products in the hydrolysis of ATP have varying affinities for these ions. Most ATP in cells, for example, exists in the form

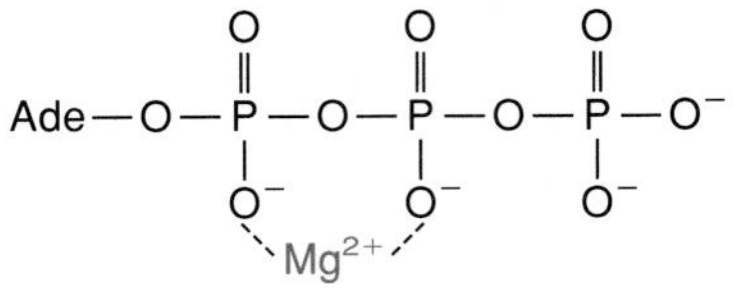

$$\text{Ade—O—}\overset{\text{O}}{\underset{\text{O}^-}{\overset{\|}{\underset{|}{\text{P}}}}}\text{—O—}\overset{\text{O}}{\underset{\text{O}^-}{\overset{\|}{\underset{|}{\text{P}}}}}\text{—O—}\overset{\text{O}}{\underset{\text{O}^-}{\overset{\|}{\underset{|}{\text{P}}}}}\text{—O}^- \quad (\text{first two O}^- \cdots \text{Mg}^{2+})$$

Varying levels of magnesium will change ΔG in complicated ways, depending on relative affinities of reactants and products for the magnesium ion.

4. By far the most important reason, however, is that the actual concentrations of ATP, ADP, and P_i in the cell are very different from the 1 M values of the standard state. Let us do a more realistic calculation of ΔG for ATP hydrolysis in vivo. In a typical bacterial cell growing at 37°C, the concentrations of ATP, ADP, and P_i are maintained at about 8, 1, and 8 mM, respectively. These values are also in the range found in many other prokaryotic and eukaryotic cells. If we insert these concentrations in equation (3.39) we find

$$\begin{aligned}\Delta G &= -31\,\frac{\text{kJ}}{\text{mol}} + 8.314 \times 10^{-3}\,\frac{\text{kJ}}{{}^\circ\text{mol}} \times 310^\circ \times \ln\left\{\frac{[1\times10^{-3}][8\times10^{-3}]}{[8\times10^{-3}]}\right\} \\ &= -31\,\frac{\text{kJ}}{\text{mol}} - 18\,\frac{\text{kJ}}{\text{mol}} = -49\,\frac{\text{kJ}}{\text{mol}} \qquad (3.41)\end{aligned}$$

As the calculation shows, the effective value for ΔG in a cell is likely to be in the vicinity of -50 kJ/mol. This very large negative free energy change explains why ATP hydrolysis can be so effective in driving a wide variety of cell processes.

The same distinction between standard state free energy changes and actual ΔG values under cellular conditions applies to all reactions. In general, we will find that ΔG at cellular concentrations is considerably more negative than ΔG°. Therefore, the phosphate transfer potentials given in Figure 3.8 are only a rough guide to what may happen in the cell. Cellular concentrations of reactants and products can vary greatly, depending on the state of the cell's metabolism.

The Energy Charge

Obviously, the capacity of a cell to carry out ATP-driven reactions must depend on the relative concentrations of ATP and its various hydrolysis products. One way to describe this capacity is to use the inverse of the concentration factor in equation (3.39), that is, $[\text{ATP}]/[\text{ADP}][\text{P}_i]$. This quantity gives a measure of the driving force for the ATP hydrolysis reaction. However, such a formulation overlooks the fact that hydrolysis of ADP is also a high-energy reaction, which can be used in some cellular processes. For this reason the energy state of the cell might better be described by a quantity called the **energy charge:**

$$\text{Energy charge} = \frac{[\text{ATP}] + \frac{1}{2}[\text{ADP}]}{[\text{ATP}] + [\text{ADP}] + [\text{AMP}]} \qquad (3.42)$$

In the numerator are concentrations of the forms that can act as free energy sources, whereas the denominator includes both ATP and all of its degradation products. The value of the energy charge can range between 0 (all present as AMP) and 1 (all present as ATP). Most healthy cells function at energy charge values of about 0.9.

The Importance of Metastability

One further aspect of ATP hydrolysis is essential if ATP is to play the role of free energy currency in the cell: The currency must not be squandered. ATP would be useless to the cell if its hydrolysis proceeded randomly, rather than being linked to specific processes. At this point the concept of **metastability** becomes most significant.

A **metastable** compound is one that is thermodynamically unstable but, in the absence of a catalyst, can only slowly break down. Although the hydrolysis of ATP under physiological conditions is highly favored *thermodynamically,* it is normally a very slow process *kinetically*. For example, if ATP is placed in an aqueous solution buffered at pH 7, the half-time for its hydrolysis is many hours. A catalyst is needed to promote ATP hydrolysis at any physiologically useful rate. In the cell, specific enzymes act as such catalysts, and they function to couple the hydrolysis of ATP to specific physiologically useful reactions. The elegance of the whole ATP system should now be apparent. Reactions exist whereby ATP can be made. Its hydrolysis produces a free energy change sufficiently negative to drive many different processes. But the compound is metastable; it undergoes useless hydrolysis only very slowly and is most often used instead for processes the cell "needs."

We shall find that the property of metastability is a quite general feature of the molecules of life. The processes of life can be directed so specifically because there is not only a driving force—free energy—but also *barriers* to the release of that energy. The energy is channeled into the paths that allow the organism to survive and perpetuate its kind. In the chapters that follow we shall see how those paths are organized and the structures that are necessary for this organization.

REFERENCES

This chapter has presented a very abbreviated treatment of thermodynamics. For the student who wishes more background in this field and more information about its applications to biochemistry, we recommend the following books.

Eisenberg, D., and D. Crothers (1979) *Physical Chemistry with Applications to the Life Sciences.* Benjamin/Cummings, Menlo Park, Calif. A very fine physical chemistry text, written by two physical biochemists. Strongly recommended, for it contains many biochemical applications of physical-chemical principles.

Klotz, I. (1957) *Energetics in Biochemical Reactions.* Academic Press, New York. A brief nonmathematical introduction to thermodynamics for biochemists. Some excellent examples and explanations.

Morowitz, H. J. (1970) *Entropy for Biologists.* Academic Press, New York. Broader than its title suggests, this is a good, concise thermodynamics text.

van Holde, K. E. (1985) *Physical Biochemistry,* 2nd ed. Prentice-Hall, Englewood Cliffs, N.J. The first three chapters contain a somewhat more extended treatment of thermodynamics, from much the same viewpoint as adopted here.

PROBLEMS

1. The enthalpy change (heat of fusion, ΔH_f) for the transition

$$\text{ice} \longrightarrow \text{water}$$

at 0°C and 1 atm pressure is +6.01 kJ/mol. The change in volume when 1 mol of ice is melted is -1.625 cm^3/mol = -1.625×10^{-6} m^3/mol. Calculate the difference between ΔH_f and ΔE_f for this process, and express is as a percentage of ΔE_f. [Note: 1 atm = 1.013×10^5 N/m^2 in SI units.]

2. Given the following reactions and their enthalpies:

	ΔH (kJ/mol)
$H_2(g) \longrightarrow 2H(g)$	+436
$O_2(g) \longrightarrow 2O(g)$	+495
$H_2(g) + \frac{1}{2}O_2(g) \longrightarrow H_2O(g)$	−242

(a) Devise a way to calculate ΔH for the reaction

$$H_2O(g) \longrightarrow 2H(g) + O(g)$$

(b) From this, estimate the H—O bond energy.

3. The decomposition of crystalline N_2O_5

$$N_2O_5(s) \longrightarrow 2NO_2(g) + \tfrac{1}{2}O_2(g)$$

is an example of a reaction that is thermodynamically favored even though it absorbs heat. At 25°C we have the following values for the standard state enthalpy and free energy changes of the reaction:

$$\Delta H^\circ = +109.6 \text{ kJ/mol}$$
$$\Delta G^\circ = -30.5 \text{ kJ/mol}$$

(a) Calculate ΔS° at 25°C.

(b) Why is the entropy change so favorable for this reaction?

(c) Calculate ΔE° for this reaction at 25°C.

(d) Why is ΔH° greater than ΔE°?

4. The combustion of glucose to CO_2 and water is a major source of energy in aerobic organisms. It is a reaction favored mainly by a large negative enthalpy change.

$$C_6H_{12}O_6(s) + 6O_2 \longrightarrow 6CO_2(g) + 6H_2O(l)$$

$$\Delta H^\circ = -2816 \text{ kJ/mol}, \quad \Delta S^\circ = +181 \text{ J/°mol}$$

(a) At 37°C, what is the value for ΔG°?

(b) In the overall reaction of aerobic metabolism of glucose, 38 mol of ATP is produced from ADP for every mole of glucose oxidized. Calculate the standard state free energy change for the *overall* reaction when glucose oxidation is coupled to the formation of ATP.

(c) What is the *efficiency* of the process in terms of the percentage of the available free energy change captured in ATP?

5. The first reaction in glycolysis is the phosphorylation of glucose:

$$P_i + \text{glucose} \rightleftharpoons \text{glucose-6-phosphate}$$

This is a thermodynamically unfavorable process, with $\Delta G^{\circ\prime} = +14$ kJ/mol.

(a) In a liver cell at 37°C the concentrations of both phosphate and glucose are normally maintained at about 5 mM each. What would the *equilibrium* concentration of glucose-6-phosphate be, according to the above data?

(b) This very low concentration of the desired product would be very unfavorable for glycolysis. In fact, the reaction is coupled to ATP hydrolysis to give the overall reaction

$$\text{ATP} + \text{glucose} \rightleftharpoons \text{glucose-6-phosphate} + \text{ADP}$$

What is $\Delta G^{\circ\prime}$ for the reaction now?

(c) If, in addition to the constraints on glucose concentration listed above, we also have in the liver cell [ATP] = 3 mM and [ADP] = 1 mM, what is the equilibrium concentration of glucose-6-phosphate? This is an absurdly high value for the cell and in fact is never approached in reality. Explain why.

6. In a liver cell like that described in Problem 5, with [ATP] = 3 mM, [ADP] = 1 mM, and an energy charge of 0.85, what is the concentration of AMP?

7. In another key reaction in glycolysis, dihydroxyacetone phosphate (DHAP) is isomerized into glyceraldehyde-3-phosphate (G3P):

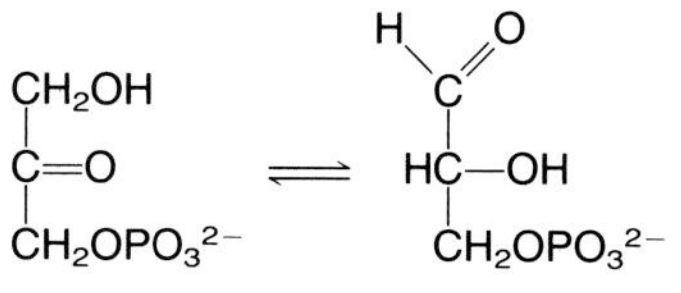

$$\underset{\textbf{DHAP}}{\begin{matrix} CH_2OH \\ | \\ C{=}O \\ | \\ CH_2OPO_3^{2-} \end{matrix}} \rightleftharpoons \underset{\textbf{G3P}}{\begin{matrix} H\diagdown\;\diagup\!\!\!\!/O \\ C \\ | \\ HC{-}OH \\ | \\ CH_2OPO_3^{2-} \end{matrix}} \qquad \Delta G^{\circ\prime} = +7.5\,\frac{kJ}{mol}$$

Obviously, equilibrium lies to the left.

(a) Calculate the equilibrium constant, and the equilibrium fraction of G3P from the above, at 37°C.

(b) In the cell, depletion of G3P makes the reaction proceed. What will ΔG be if the concentration of G3P is always kept at 1/100 of the concentration of DHAP?

8. A protein molecule, in its folded native state, has *one* favored conformation. But when it is denatured it becomes a random coil, with very many possible conformations.

(a) What must be the sign of ΔS for the change native → denatured?

(b) Will the contribution of ΔS to the free energy change be + or −? What requirement does this impose if proteins are to be stable structures?

9. Suppose a reaction has $\Delta H°$ and $\Delta S°$ values independent of temperature. Show from this, and equations given in this chapter, that

$$\ln K = \frac{-\Delta H°}{RT} + \frac{\Delta S°}{R}$$

where K is the equilibrium constant. How could you use values of K determined at different temperatures to determine $\Delta H°$ for the reaction?

10. The following data give the ion product (K_w; equation 2.5) for water at various temperatures:

T (°C)	K_w (M^2)
0	1.14×10^{-15}
25	1.00×10^{-14}
30	1.47×10^{-14}
37	2.56×10^{-14}

(a) Using the results from Problem 9, calculate $\Delta H°$ for the ionization of water.

(b) Use these data, and the ion product at 25°C, to calculate $\Delta S°$ for water ionization. [Caution: Note that K_w is not a true equilibrium constant. What must you do to calculate K from K_w?]

11. The phosphate transfer potentials for glucose-1-phosphate and glucose-6-phosphate are 21 kJ/mol and 14 kJ/mol, respectively.

(a) What is the equilibrium constant for this reaction at 25°C?

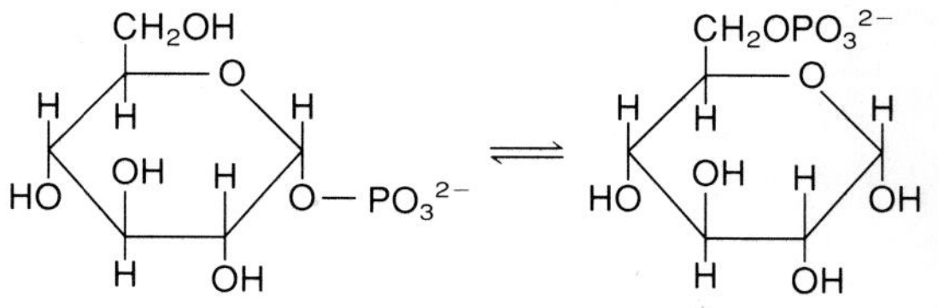

Glucose-1-phosphate ⇌ Glucose-6-phosphate

(b) If a mixture were prepared containing 1 M glucose-6-phosphate and 1×10^{-3} M glucose-1-phosphate, what would be the thermodynamically favored direction for the reaction?

PART II

Molecular Architecture of Living Matter

CHAPTER 4

Nucleic Acids

In the next several chapters we discuss three major classes of substances: nucleic acids, proteins, and carbohydrates. Together, they make up a large part of all living matter. As we saw in Chapter 1, these substances exist as macromolecules—some of them of giant size. We shall find all of these macromolecules to be polymers; each type is made by the linking together of a limited number of kinds of monomer units.

The Nature of Nucleic Acids

It is appropriate to begin this section with the **nucleic acids,** for in a certain sense they are the most fundamental and important constituents of the living cell. It seems probable that life itself began its evolution with nucleic acids, for only they, of all biological substances, carry the potential for self-duplication. Today, nucleic acids are the repositories and transmitters of genetic information for every cell, tissue, and organism. The blueprint for an organism has been encoded in its nucleic acid, in gigantic molecules like that shown in Figure 1.5 (Chapter 1). Physical development throughout any organism's life is preprogrammed in these remarkable molecules. The proteins that its cells will make and the functions that they will perform are all recorded on this molecular tape.

Recently, we have learned much about nucleic acids and their function in life processes. New microscopic techniques enable us to visualize these structures in almost atomic detail (Figure 4.1), and modern molecular biological methods allow us to manipulate them so as to change organisms. For the first time, we have the potential to direct the future development of life itself.

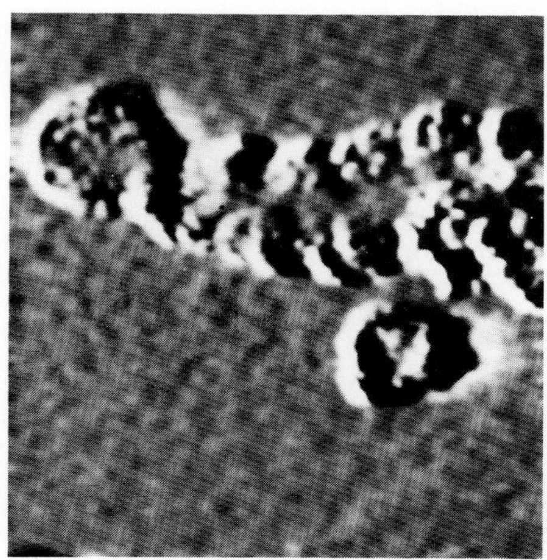

Figure 4.1
A nucleic acid molecule as seen by scanning-tunneling microscopy. In this new microscopic technique, an extremely fine wire electrode is passed just above the surface of molecules deposited on graphite. Electrons which tunnel through the molecule to the electrode produce an image of the structure. Here, DNA is shown at a detail that allows us to visualize its double-helical structure, magnified about 3,000,000 times.

In this chapter, and in the several that follow, we provide a brief introduction to the ways in which nucleic acids preserve and transmit genetic information. The details of these processes will be covered in Part V of this book, but it is important that we consider, at the beginning, the role that nucleic acids play in the formation of proteins and cellular structure.

The Two Types of Nucleic Acid: DNA and RNA

There are two types of nucleic acid, **ribonucleic acid (RNA)** and **deoxyribonucleic acid (DNA).** As Figure 4.2 shows, each is a **polymer** chain, with

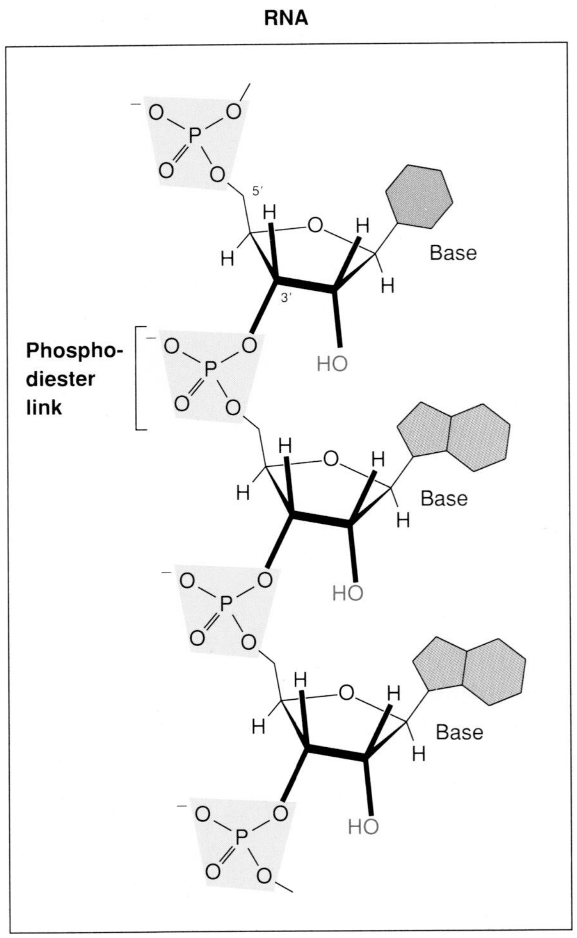

Figure 4.2
Chemical structures of ribonucleic acid (RNA) and deoxyribonucleic acid (DNA). The ribose–phosphate (or deoxyribose–phosphate) backbone of each chain is shown in detail. The bases shown schematically here are detailed in Figure 4.3.

similar **monomer** units connected by covalent bonds. The monomer units are shown below:

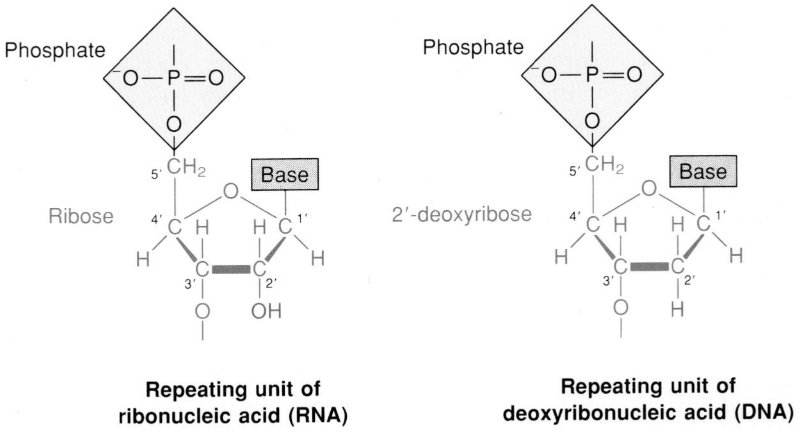

Repeating unit of ribonucleic acid (RNA)

Repeating unit of deoxyribonucleic acid (DNA)

In each case the monomer unit contains a five-carbon sugar (**ribose** in RNA, **2′-deoxyribose** in DNA), shown in blue in the structures above. The difference between the two sugars lies solely in the 2′ hydroxyl group on ribose. The connection between successive monomer units in nucleic acids is through a phosphate residue attached to the hydroxyl on the 5′ carbon of one unit and the 3′ hydroxyl of the next one. This forms a **phosphodiester**

Figure 4.3
Purine and pyrimidine bases found in DNA and RNA. A major difference between the two types of nucleic acids is that RNA has uracil (U) instead of thymine (T).

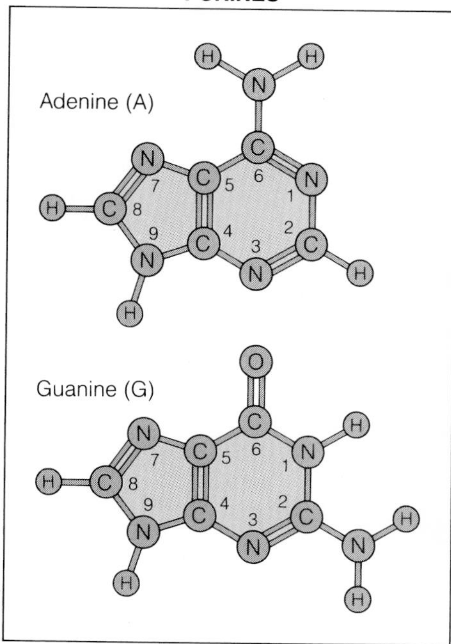

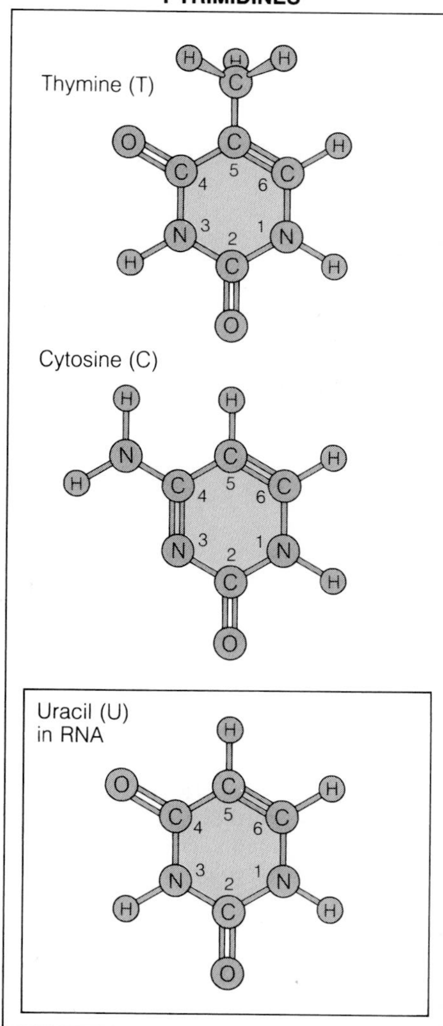

link between two residues (see Figure 4.2). In this way very long nucleic acid chains, sometimes containing billions of units, are built up. The phosphate group is a strong acid, with a pK_a of about 1; this is why DNA and RNA are called nucleic *acids.*

The phosphodiester-linked sugar residues form the backbone of the nucleic acid molecule. By itself, the backbone is a repetitious structure, incapable of encoding information. The importance of the nucleic acids in information storage and transmission derives from the fact that they are **heteropolymers.** Each monomer in the chain carries a *basic group,* always attached to the 1′ carbon of the sugar (see left and Figure 4.2). There are two types of these basic substances, called **purines** and **pyrimidines.**

The structures of the major bases* found in the nucleic acids are shown in Figure 4.3. Note that DNA has the purines **adenine (A)** and **guanine (G)** and the pyrimidines **cytosine (C)** and **thymine (T).** In RNA the bases are the same except that **uracil (U)** replaces thymine. The bases in Figure 4.3 are depicted in their major tautomeric forms; tautomeric shift to the alternate forms shown in Figure 4.4 is possible. Although each base spends only a minute fraction of its time in these alternate forms, we shall find later that the existence of such tautomers is important in the mutation of biological species.

Since DNA and RNA each contain four kinds of bases, each can be regarded as a polymer made from four kinds of monomers. The monomers are phosphorylated ribose or deoxyribose molecules with purine or pyrimidine bases attached to their 1′ carbons. In purines the attachment is through nitrogen 9, in pyrimidines through nitrogen 1. These monomers are called **nucleotides.** Each nucleotide can be considered the 5′ monophosphorylated derivative of a sugar-base adduct called a **nucleoside** (Figure 4.5). Thus, the nucleotides could also be called *nucleoside 5′-monophosphates.* You have already encountered one of these molecules—adenosine 5′-monophosphate, or AMP—in Chapter 3.

Since all of the nucleic acids may be regarded as polymers of nucleotides, they are often referred to by the generic name **polynucleotides.** Small polymers, containing only a few residues, are called **oligonucleotides.**

Stability and Formation of the Phosphodiester Linkage

If we compare the structures of the nucleotides shown in Figure 4.5 with the polynucleotide chains depicted in Figure 4.2, we see that, in principle, a polynucleotide could be generated from its nucleotide monomers by elimination of a water molecule between each pair of monomers. That is, we might imagine adding another nucleotide residue to a polynucleotide chain by the dehydration reaction shown in Figure 4.6. However, the free energy change in this dehydration reaction is quite positive, about +25 kJ/mol; therefore equilibrium lies far to the side of hydrolysis in an aqueous environment, such as that existing in vivo.

* RNA and, to a lesser extent, DNA also contain a small fraction of chemically modified bases. We discuss these in later sections.

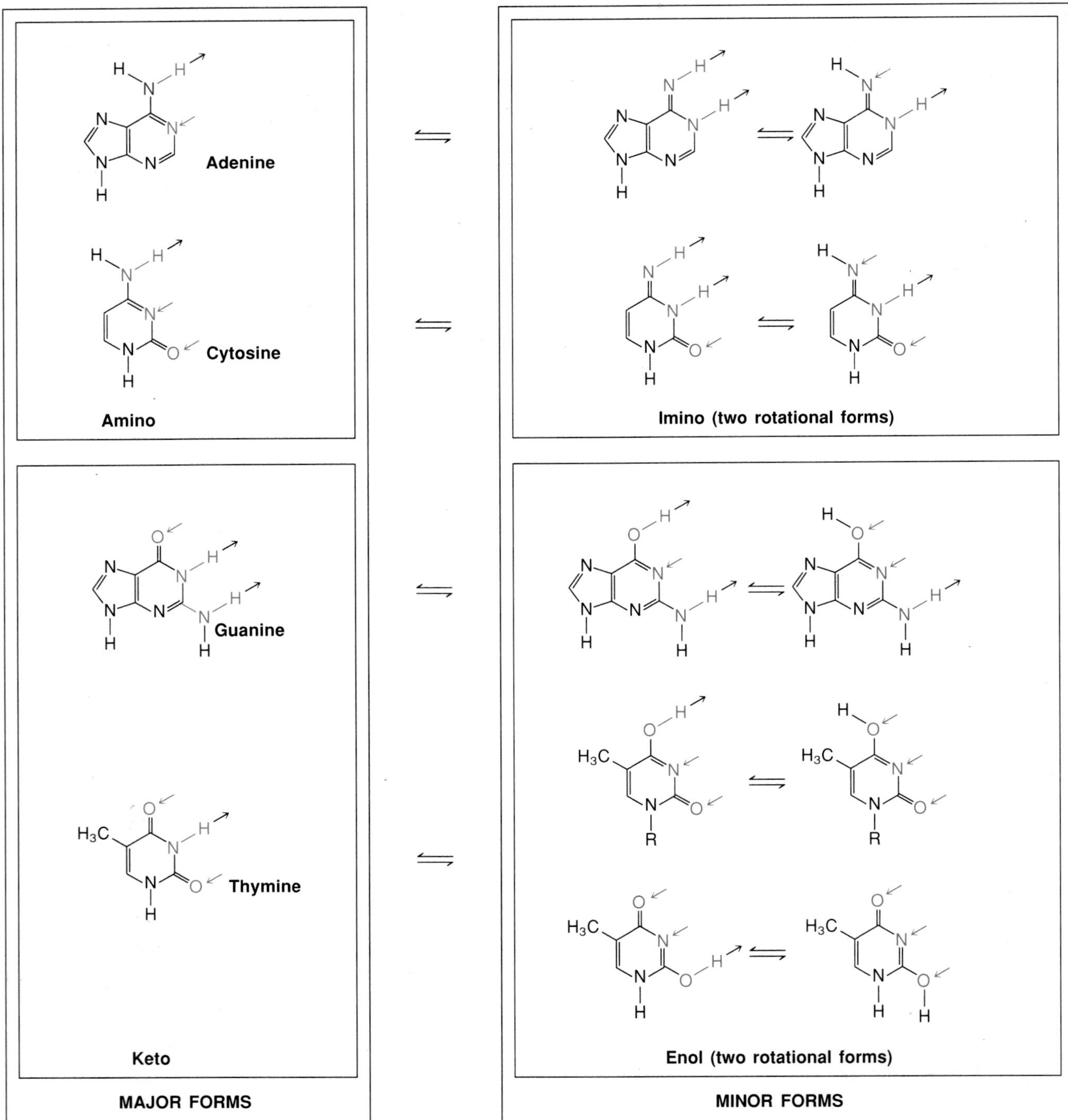

Figure 4.4
Tautomeric equilibria of the DNA bases. The tautomeric forms shown to the left are the major ones. The imino and enol forms shown to the right are present in very small amounts, even in the free bases. Atoms in color may participate in hydrogen bond formation. Hydrogen bond donors (black arrows) and acceptors (red arrows) are shown. Note that the hydrogen-bonding possibilities differ in different tautomers.

This is the first of many examples we shall see of the metastability (see Chapter 3) of biologically important compounds. Although polynucleotides are thermodynamically unstable in vivo, their hydrolysis is exceedingly slow unless catalyzed. This is of the greatest importance, for it ensures that the DNA in cells is sufficiently stable to serve as a useful repository of genetic information. Indeed, DNA is so stable that it has even been possible to recover fragments of DNA molecules from some not-too-ancient fossils. When catalysts *are* present, however, hydrolysis can be exceedingly rapid. You are able to break down polynucleotides in the foodstuffs you consume

NUCLEOSIDES

NUCLEOTIDES

Adenosine

Adenosine 5′-monophosphate (AMP)

Guanosine

Guanosine 5′-monophosphate (GMP)

Cytidine

Cytidine 5′-monophosphate (CMP)

Uridine

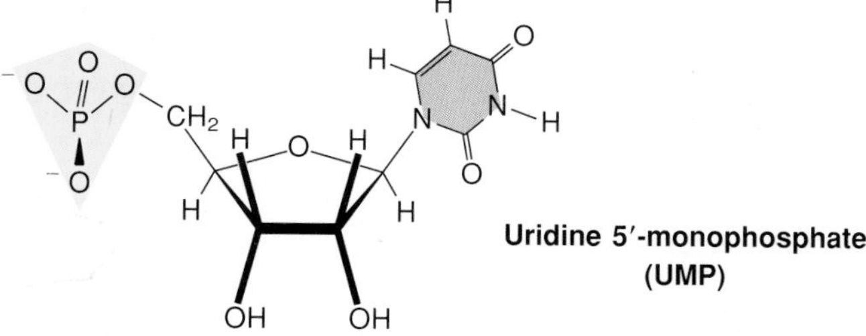

Uridine 5′-monophosphate (UMP)

Figure 4.5
Nucleosides and nucleotides. All bases are shown in their major tautomeric forms. Each nucleoside here is formed by coupling ribose to a base. A comparable set of deoxyribonucleosides and deoxyribonucleotides exists. The nucleotides, which can be considered to be the monomer units of the nucleic acids, are the 5′-monophosphates of the nucleosides.

because your digestive system contains enzymes called *nucleases* that catalyze this process. Examples of such enzymes are described in some detail in Chapter 11.

The unfavorable thermodynamics of the hypothetical reaction shown in Figure 4.6 leads us to ask: If polynucleotides cannot be synthesized in vivo by the direct elimination of water, how are they actually made? The answer is that their synthesis involves the high-energy nucleoside *triphosphates*. Although the process as it occurs in cells is quite complex, the basic reaction is simple. Instead of the dehydration reaction of Figure 4.6 what happens in living cells is shown in Figure 4.7. The residue being added to the growing chain is presented as a nucleoside triphosphate, like ATP or deoxy ATP (dATP). We can calculate the free energy change for this reaction by noting that it can be considered the sum of two reactions: hydrolysis of a nucleoside triphosphate and formation of a phosphodiester link by elimination of water:

				$\Delta G^{\circ\prime}$
	Nucleoside triphosphate + H_2O	$\rightleftharpoons$	nucleoside monophosphate + pyrophosphate (PP_i)	$-30\ \frac{kJ}{mol}$
	(Polynucleotide chain)$_N$ + nucleoside monophosphate	$\rightleftharpoons$	(polynucleotide chain)$_{N+1}$ + H_2O	$+25\ \frac{kJ}{mol}$
Sum:	(Polynucleotide chain)$_N$ + nucleoside triphosphate	$\rightleftharpoons$	(polynucleotide chain)$_{N+1}$ + pyrophosphate (PP_i)	$-5\ \frac{kJ}{mol}$

The coupled reaction is favorable because the *net* $\Delta G^{\circ\prime}$ is negative. The reaction is further favored by the fact that the hydrolysis of the pyrophosphate product (PP_i) to phosphate (P_i) also has a $\Delta G^{\circ\prime} = -30$ kJ/mol. Thus, the pyrophosphate is readily removed, driving the synthesis reaction even further to the right. This is an example of a principle we emphasized in Chapter 3—the use of favorable reactions to drive thermodynamically unfavorable ones.

It is important to appreciate how the energetics of such processes fit into the overall scheme of life. An organism obtains energy—either from photosynthesis, if it is a plant, or from metabolism of foodstuffs—and stores part of this energy by generating ATP, GTP, dATP, dGTP, and so forth. It uses these compounds in turn to synthesize macromolecules like DNA and RNA. This is a pattern that you will see again and again throughout this book.

Primary Structure of Nucleic Acids

The Meaning and Significance of Primary Structure

A closer examination of Figure 4.2 reveals two important features of all polynucleotides:

1. A polynucleotide chain has a sense of *direction;* the phosphodiester linkage between monomer units is always between the 3′ carbon of one monomer and the 5′ carbon of the next. Thus, the

Polynucleotide with *N* nucleoside residues

+

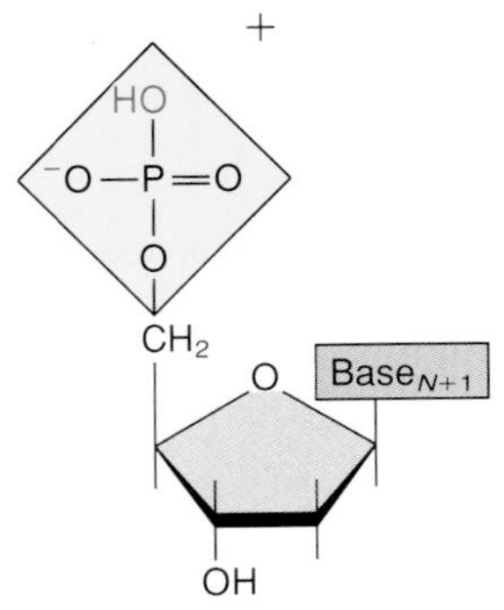

Deoxynucleoside monophosphate

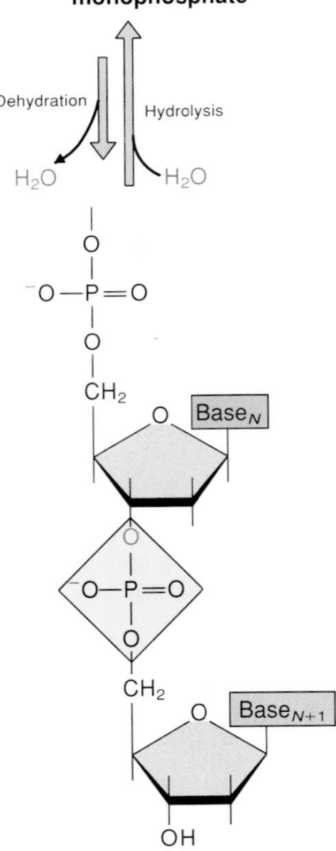

Polynucleotide with *N* + 1 nucleoside residues

Figure 4.6
A hypothetical dehydration reaction to add a nucleotide residue to a nucleic acid chain. This is thermodynamically unfavorable. The reverse reaction, hydrolysis, is favored.

two ends of a linear polynucleotide chain will be distinguishable. One end will normally carry an unreacted 5′ phosphate, the other an unreacted 3′ hydroxyl group.

2. A polynucleotide chain has *individuality*, determined by the sequence of its bases—that is, the *nucleotide sequence*. This sequence is called the **primary structure** of that particular nucleic acid. *It is in the primary structure of DNA that genetic information is stored.* A **gene** is nothing more than a particular DNA sequence, encoding information in a four-letter language in which each letter represents one of the bases.

If we want to describe a particular polynucleotide sequence (either DNA or RNA), it is exceedingly awkward to draw the molecule in its entirety as in Figure 4.2. Accordingly, some compact nomenclatures have been devised. If we state that we are describing a DNA molecule, or an RNA molecule, then most of the structure is understood. We can then abbreviate a small DNA as follows:

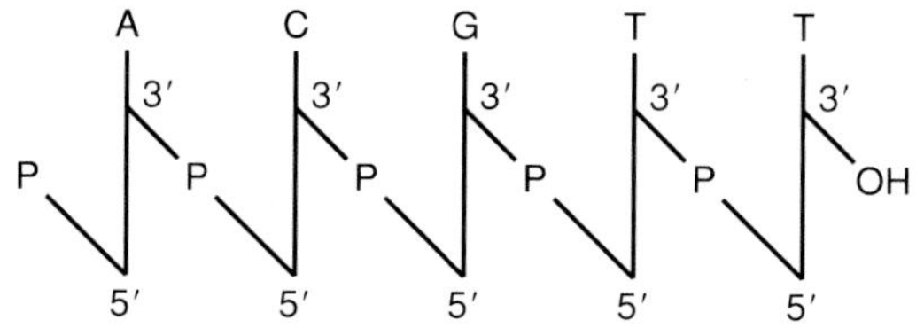

A C G T T
3′ 3′ 3′ 3′ 3′
P P P P P OH
5′ 5′ 5′ 5′ 5′

This notation shows (1) the sequence of nucleotides, by their letter abbreviations (A, C, G, T); (2) that all phosphodiester links are 3′ → 5′; and (3) that this particular molecule has a phosphate group at its 5′ end and an unreacted 3′ hydroxyl at its 3′end.

If all of the phosphodiester links can be assumed to be 3′ → 5′ (as is usually the case), a more compact notation is possible for the same molecule:

pApCpGpTpT

The 3′—OH group is understood to be present and unreacted. Were there a phosphate on the 3′ end and an unreacted hydroxyl on the 5′ end, we would write

ApCpGpTpTp

Finally, if we are concerned *only* with the sequence of bases in the molecule, as will often be the case, we can write it still more compactly as

ACGTT

Note that the sequences are always written, by convention, with the 5′ end to the left and the 3′ end to the right.

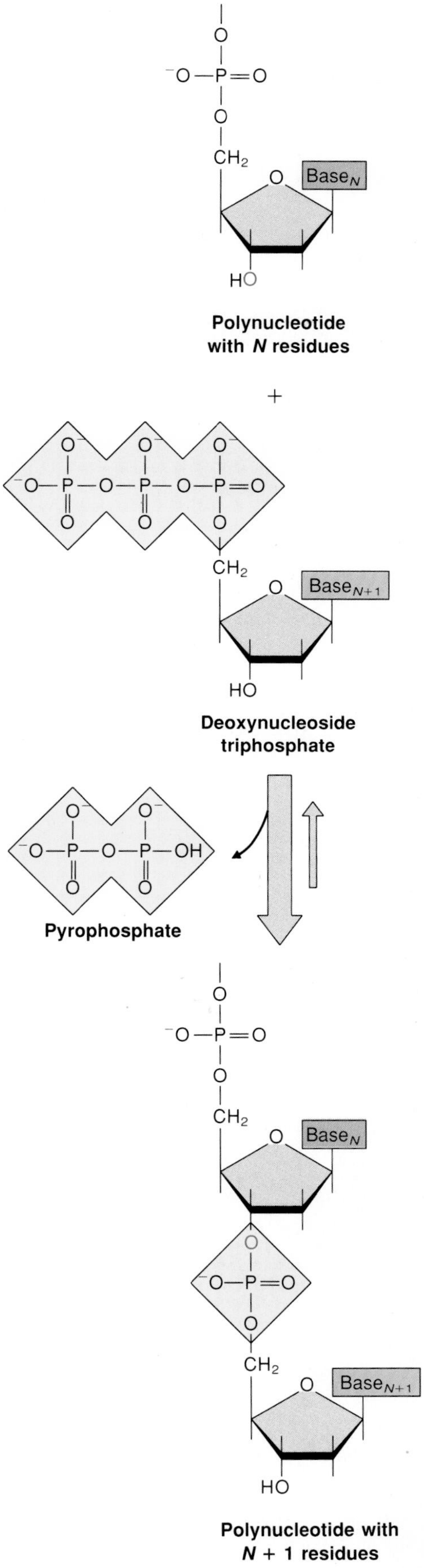

Figure 4.7
The way nucleotide residues from a nucleoside triphosphate are added to a nucleic acid chain. Because a nucleoside triphosphate is cleaved, this reaction is thermodynamically favorable.

Polynucleotide with *N* residues

+

Deoxynucleoside triphosphate

Pyrophosphate

Polynucleotide with *N* + 1 residues

Determination of Primary Structure: DNA Sequencing

We have emphasized that the basic information for biological structure and function is encoded in sequences of the DNA molecules of cells. Many of the remarkable advances in biology and medicine in recent years have come about because we are now able to read that information. Several methods for determining the sequences of natural DNA and RNA molecules have been developed that are extremely accurate and fast. Some of these "sequencing" techniques depend on sophisticated aspects of molecular biology and can only be discussed later (see Chapters 24–28). The DNA sequencing method described in Tools of Biochemistry 3 uses only simple reactions plus gel electrophoresis, which you have already encountered in Tools of Biochemistry 2.

Most naturally occurring DNA molecules are so immense that to determine the sequence of the whole molecule in one operation would be unthinkable. Consider, for example, the single DNA molecule that constitutes the entire genetic content of *Escherichia coli (E. coli),* an intestinal bacterium widely used in biochemical research. There are about 4×10^6 bases in each strand of *E. coli* DNA. It is difficult even to isolate such a molecule in intact form. Two techniques were essential before the sequencing of large DNA molecules could be contemplated.

First, a method to cleave the molecule in a defined fashion was required. This can be accomplished by catalysis with enzymes called **restriction endonucleases.** An **endonuclease** is a protein that catalyzes hydrolytic cleavages within a nucleic acid chain; the special class called restriction endonucleases cleave only at specific sequences. For example, an enzyme called *Hae*III cleaves in the middle of the sequence . . .GG↓CC. . . and nowhere else. Whenever this series of four bases occurs, this enzyme will catalyze cleavage. By using different enzymes of this kind (hundreds are known), researchers can obtain defined pieces of a large DNA and isolate them by gel electrophoresis.

To work with these pieces, a significant quantity of each fragment is necessary. This is a second major problem. Each *E. coli* cell contains only one giant DNA molecule and hence only one copy of each unique fragment. A method for multiplying specific sequences is required. This has been accomplished by **molecular cloning,** in which pieces of DNA are inserted into bacterial plasmids, small DNA molecules maintained in multiple copies in some bacteria. Alternatively, the desired DNA is sometimes inserted into the genome of a bacteriophage, which can replicate itself many times in a bacterial host. These **recombinant DNA techniques,** which are discussed in detail in Chapter 25, have allowed the preparation of substantial amounts of many specific DNA sequences.

With such pure samples of defined DNA molecules, the method described in Tools of Biochemistry 3 could be used to determine the nucleotide sequence. The technique is simple, and a skilled researcher can easily determine a sequence several hundred residues long in a few days.

By sequencing many overlapping fragments of larger DNA molecules, the entire sequences of very long DNA molecules can now be determined. For example, the complete sequence of the DNA of the Epstein–Barr virus, 172,282 nucleotides in length, has been obtained. To simply write this sequence in the most compact nomenclature would fill approximately 100 pages of this book. Such data are usually stored and analyzed by computer methods. A vast library of DNA primary structure information is being accumulated in computer banks and made available to scientists.

The very fact that this kind of analysis can be carried out at all proves that natural DNA molecules have very accurately reproducible, individually unique primary structures. The same is true for naturally occurring RNA molecules, for which comparable sequencing techniques are available.

In the manipulations of nucleic acids that are becoming so common in molecular biology, it is often necessary to synthesize oligonucleotides in the laboratory. In this way we can, for example, make new and modified genes and test their effects in organisms. The techniques for **oligonucleotide synthesis** have become sophisticated and highly automated. One method is described in Tools of Biochemistry 4.

Secondary Structure of Nucleic Acids

One of the most momentous discoveries in the history of science occurred in 1953. In that year James Watson and Francis Crick proposed a model for the three-dimensional structure of deoxyribonucleic acid. Their insight suddenly opened whole new directions in biology and biochemistry. Genetics had hitherto been a rather abstract science, concerned with entities called "genes" whose nature was unknown. Now it became clear that a gene was a DNA sequence. Genetics finally found its true roots in molecular structure. Thus, what we now think of as molecular biology had its inception in Watson and Crick's work.

The Watson–Crick Model: History and Development

Like most great scientific advances, Watson and Crick's discovery did not occur in a vacuum. For many decades, investigators had been seeking the "genetic substance," the material of which genes are made. Indeed, in the late 1800s, shortly after the German biochemist Friedrich Miescher had first isolated DNA from salmon sperm, some scientists suspected that DNA might be the genetic material. But subsequent studies showing that DNA contained only four kinds of monomers seemed to deny it such a complicated role. Early researchers thought it more likely that genes were made of proteins, for these were beginning to be recognized as much more complex molecules. For most of the first half of the twentieth century, nucleic acids were considered to be merely some kind of structural material in the cell nucleus.

Between 1944 and 1952 a series of crucial experiments clearly pointed to DNA as the genetic material. In 1944 Oswald Avery, Colin MacLeod, and Maclyn McCarty found that the DNA from pathogenic strains of the bacterium *Pneumococcus* could be transferred into nonpathogenic strains, making them pathogenic (Figure 4.8a). The transformation was genetically stable; succeeding generations of bacteria retained the new characteristics. However, it was an elegant experiment by Alfred Hershey and Martha Chase that finally convinced many scientists. Hershey and Chase studied the infection of the bacterium *E. coli* by a bacterial virus, the bacteriophage T2. Making use of the fact that the bacteriophage proteins contain sulfur but little phosphorus and that the DNA contains phosphorus but no sulfur, they labeled T2 bacteriophage with the radioisotopes ^{35}S and ^{32}P (Figure 4.8b). They then showed that when the bacteriophage attached to *E. coli,* it was mainly the ^{32}P (and hence the DNA) that was transferred to the bacteria. Even if the residual protein part of the bacteriophage was shaken off the

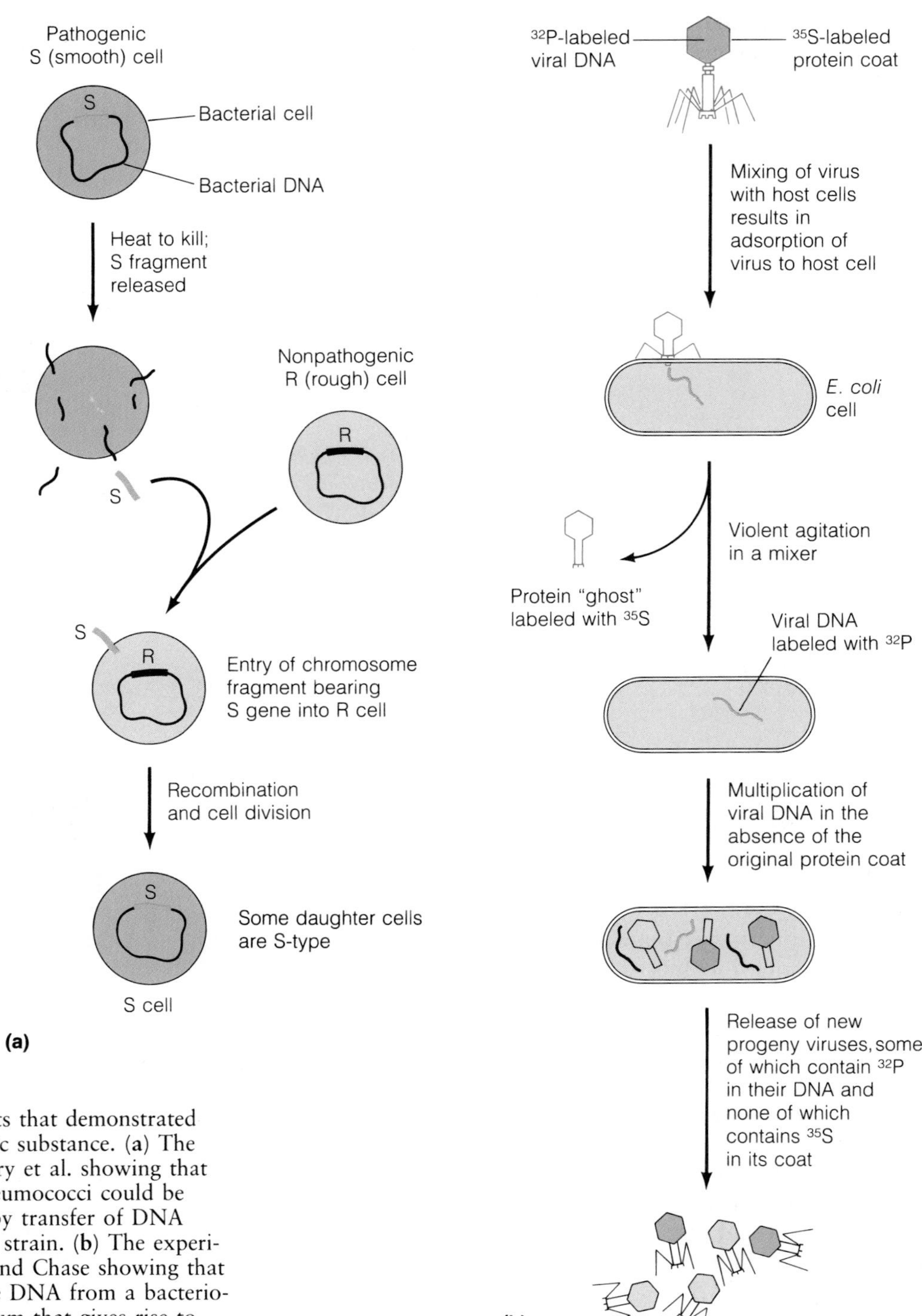

Figure 4.8
Crucial experiments that demonstrated DNA as the genetic substance. (**a**) The experiment of Avery et al. showing that nonpathogenic pneumococci could be made pathogenic by transfer of DNA from a pathogenic strain. (**b**) The experiment of Hershey and Chase showing that it is transfer of the DNA from a bacteriophage to a bacterium that gives rise to new bacteriophages.

bacteria, the inserted DNA was sufficient to direct the formation of new bacteriophage.

Through these and similar experiments it was generally recognized by 1952 that DNA must be the genetic substance. But how could it carry the enormous amount of information that a cell needed, how could it transmit this to the cell, and, above all, how could it be accurately replicated in cell division?

Watson and Crick sought the answers to these questions in the three-dimensional structure of DNA. For some time, a number of laboratories had been investigating fibers drawn from concentrated DNA solutions using the technique of **x-ray diffraction** (see Tools of Biochemistry 5). Watson and Crick, working at Cambridge University in England, had access to DNA diffraction patterns photographed by Rosalind Franklin, a researcher in the laboratory of Maurice Wilkins at King's College, London. These were for the B form of DNA (see below) and were some of the best patterns from wet DNA fibers that had yet been obtained. They clearly showed that the DNA in the fibers must have some kind of regular three-dimensional structure. We refer to such regular folding in polymers as **secondary structure,** as distinguished from the primary structure, which is simply the sequence of residues.

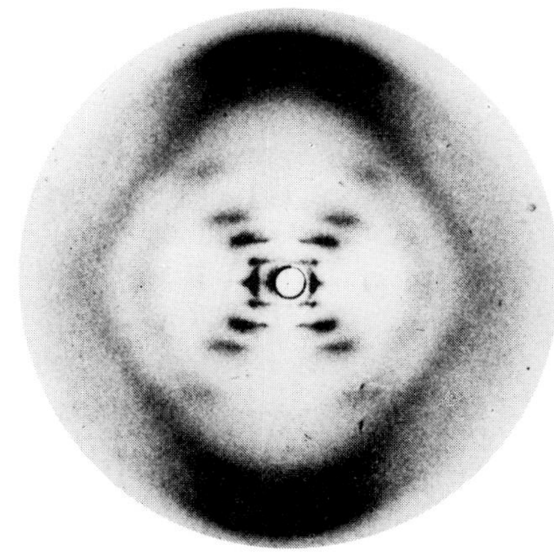

Figure 4.9
The x-ray diffraction pattern produced by wet DNA fibers. This photograph, obtained by Rosalind Franklin, played a key role in the elucidation of the DNA structure. A helical structure is indicated by the "cross" pattern. The strong spots at top and bottom correspond to the helical rise of 0.34 nm. The fact that the layer line spacing is one-tenth of the distance from the center to either of these spots indicates that there are 10 base pairs per repeat.

Watson and Crick quickly recognized that the DNA fiber diffraction exhibited a cross pattern typical of a helical secondary structure (Figure 4.9), and they noted that since the layer line spacing is one-tenth of the pattern repeat, there must be 10 residues per turn (see Tools of Biochemistry 5). Data on the density of the fibers suggested that there must be *two* DNA strands in each helical molecule. So far, this was only direct scientific deduction from the data. The great leap of intuition was the realization that a two-strand helix could be stabilized by hydrogen bonding between bases on opposite strands if the bases were paired in one particular way. A molecular model of the kind of structure Watson and Crick proposed is shown in Figure 4.10. As Figure 4.11 shows, A could neatly form hydrogen bonds with T, and G with C. Note that this particular pairing is possible only if the bases have the major tautomeric forms shown in Figure 4.3, rather than the minor forms in Figure 4.4. At the time the Watson–Crick model was developed, there was initial confusion because of uncertainty as to the correct forms. (See Watson, 1968, in References.) With the kind of base pairing proposed by Watson and Crick, the distances between the 1′ carbons on the deoxyribose moieties of A + T or G + C were the same—about 1.1 nm in each case. This meant that the double helix could be regular in diameter, an impossibility if purines paired with purines or pyrimidines paired with pyrimidines.

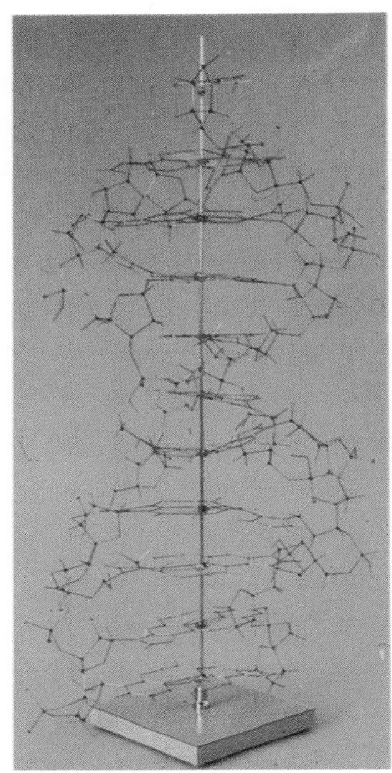

Figure 4.10
The structure originally proposed by Watson and Crick for DNA. Base planes are perpendicular to the helix axis and chains are antiparallel. The structure closely, but not exactly, corresponds to present views of the structure of B-form DNA.

As is often the case with a good theory or model, the Watson–Crick structure also explained other data that had not been understood until then. The biochemist Erwin Chargaff, who had measured the relative amounts of A, T, G, and C in DNAs from many organisms, had noted the perplexing fact that A and T were always present in equal or nearly equal quantities, as were G and C (Table 4.1). If most DNA in cells was double stranded, with the Watson–Crick base pairing, Chargaff's rule followed as a natural consequence.

In the Watson–Crick model the base pairs are stacked on one another with their planes nearly perpendicular to the helix axis. This stacking, as shown in Figure 4.12a, allows strong van der Waals interaction between the bases. The hydrophilic phosphate–deoxyribose backbones are on the outside, in contact with the aqueous environment. Each base pair is rotated by 36° with respect to the next to accommodate 10 base pairs in each turn of the helix. Since the diffraction pattern showed the repeat distance to be about 3.4 nm, the helix *rise,* that is, the distance for each base pair in the helix direction, had to be about 0.34 nm (Figure 4.12b). Note that this is just twice the van der Waals thickness of a planar ring (Table 2.2). Thus, the bases are closely packed within the helix.

Building molecular models of two-stranded DNA structures soon convinced Watson and Crick that the DNA strands must run in opposite direc-

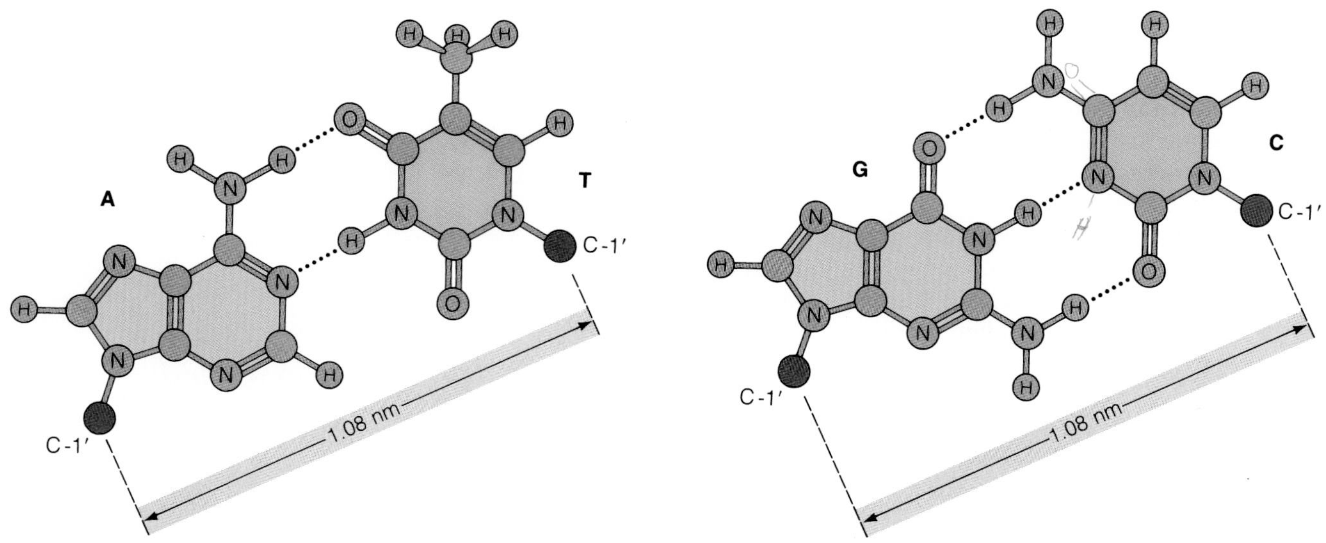

Figure 4.11
Base pairing in the DNA helix. The allowed base pairs for the Watson–Crick structures are shown. Note that such pairing allows the C-1′ carbons on the two strands to be exactly the same distance apart.

Table 4.1
Base compositions of DNAs from various organisms[a]

Source	Mol % of Bases					Ratios		
	Adenine (A)	Guanine (G)	Cytosine (C)	5-Methyl cytosine (5MeC)	Thymine (T)	$\frac{A}{T}$	$\frac{G}{C + 5MeC}$	% (G + C + 5MeC)
Bacteriophage ϕX-174	24.0	23.3	21.5	—[b]	31.2	0.77[c]	1.08[c]	44.8
Bacteriophage T7	26.0	23.8	23.6	—	26.6	0.98	1.01	47.4
Escherichia coli B	23.8	26.8	26.3	—	23.1	1.03	1.02	53.2
Neurospora	23.0	27.1	26.6	—	23.3	0.99	1.02	53.8
Corn (maize)	26.8	22.8	17.0	6.2	27.2	0.99	0.98	46.1
Tetrahymena	35.4	14.5	14.7	—	35.4	1.00	0.99	29.2
Octopus	33.2	17.6	17.6	—	31.6	1.05	1.00	35.2
Drosophila	30.7	19.6	20.2	—	29.5	1.03	0.97	39.8
Starfish	29.8	20.7	19.7	1.0	28.8	1.03	1.00	41.3
Salmon	28.0	22.0	20.0	1.8	27.8	1.01	1.01	44.1
Frog	26.3	23.5	21.8	2.0	26.8	1.00	0.99	47.4
Chicken	28.0	22.0	21.6	—	28.4	0.99	1.02	43.7
Rat	28.6	21.4	20.5	1.1	28.4	1.01	1.00	42.9
Calf	27.3	22.5	21.2	1.3	27.7	0.99	1.00	45.0
Human	29.3	20.7	20.0	—	30.0	0.98	1.04	40.7

[a]Data taken from *Handbook of Biochemistry,* 2nd ed., edited by H. E. Sober. Chemical Rubber Publishing Co., 1970. Values for higher organisms vary slightly from one tissue to another, probably as a result of experimental error.

[b]Dashes indicate absent or not detected.

[c]This bacteriophage has a single-strand DNA, which need not follow Chargaff's rule.

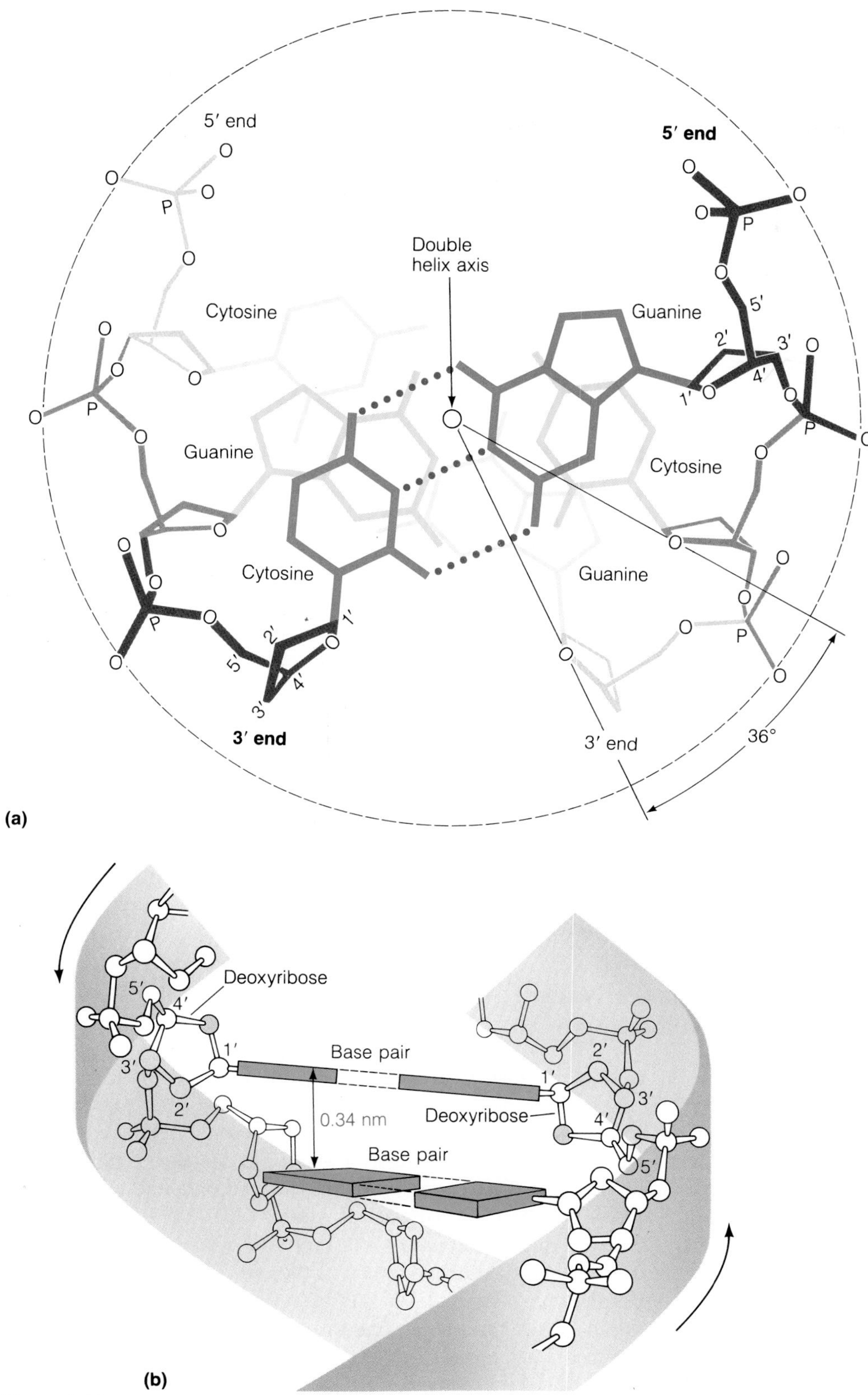

Figure 4.12
Fine structure in B-DNA. (**a**) A view *down* the helix axis, showing how the base pairs stack on one another. The helix axis and 36° rotation of one base pair with respect to the next are indicated. (**b**) A view from the side, illustrating the symmetry of orientation of the antiparallel strands. The 0.34-nm rise in the helix is indicated.

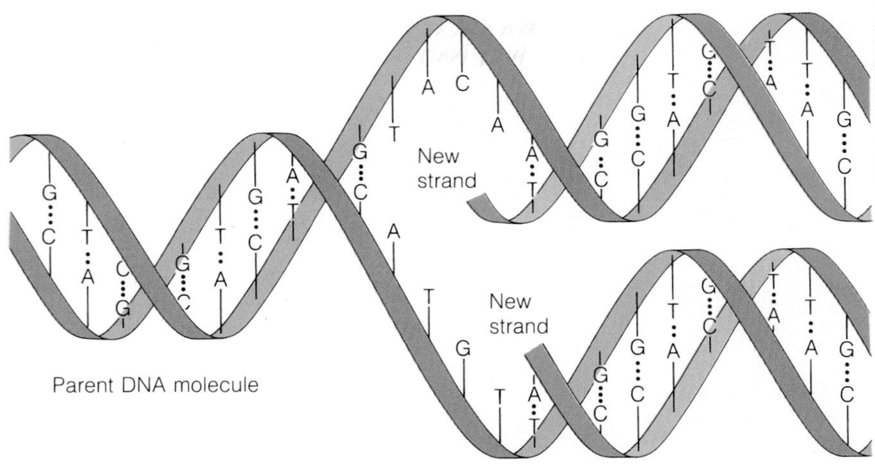

Figure 4.13
A model for DNA replication. Each strand acts as a template for a new complementary strand. Thus, when copying is complete, there will be two double-strand daughter DNA molecules, each identical in sequence to the parent molecule.

tions. This can be seen clearly in Figure 4.12b. The model presented was for a right-hand helix, although at this time evidence for the sense of the helix was weak.

The Watson–Crick model does not only explain the structure of DNA. It carries implications that go much further, into the very heart of biology. Note that since A always pairs with T, and G always with C, the two strands are *complementary*. If the strands could be separated and new DNA synthesized along each, following the same base-pairing rule, two double-strand DNA molecules would be obtained, each an *exact* copy of the original (Figure 4.13). This is precisely the property that the genetic material must have: when a cell divides, two complete copies of the genetic information carried in the original cell must be produced. In their 1953 paper (see References) announcing the model Watson and Crick express this in what may be the most understated scientific prediction ever made: "It has not escaped our notice that the specific pairing we have postulated immediately suggests a possible copying mechanism for the genetic material."

Alternative Nucleic Acid Structures: The Plasticity of the Double Helix

As is the case with most great discoveries, the Watson–Crick model for DNA was neither complete nor exactly correct. In the first place, several other possible structures for two-strand polynucleotide helices exist. Watson and Crick realized that this must be so, for at the time they proposed their model two quite different x-ray diffraction patterns had already been obtained. Those they had studied were of what is called the **B form;** it exists in fibers prepared in high humidity and therefore was expected to be the form found in the highly aqueous milieu of the cell. The most recent model for B-DNA is shown in Figure 4.14a and b. In this refined model, based on better data than Watson and Crick had available, the bases are not exactly perpendicular to the helix axis (they are about 6° off) and the sugar conformation is slightly different from that proposed by Watson and Crick. Fibers subjected to conditions of low humidity give an entirely different structure, the so-called **A form,** shown in Figure 4.14c and d. Double-strand RNA molecules always form the A structure, even under humid conditions. The same is true for **DNA–RNA hybrid molecules,** which are formed by the pairing of one DNA strand with one RNA strand.

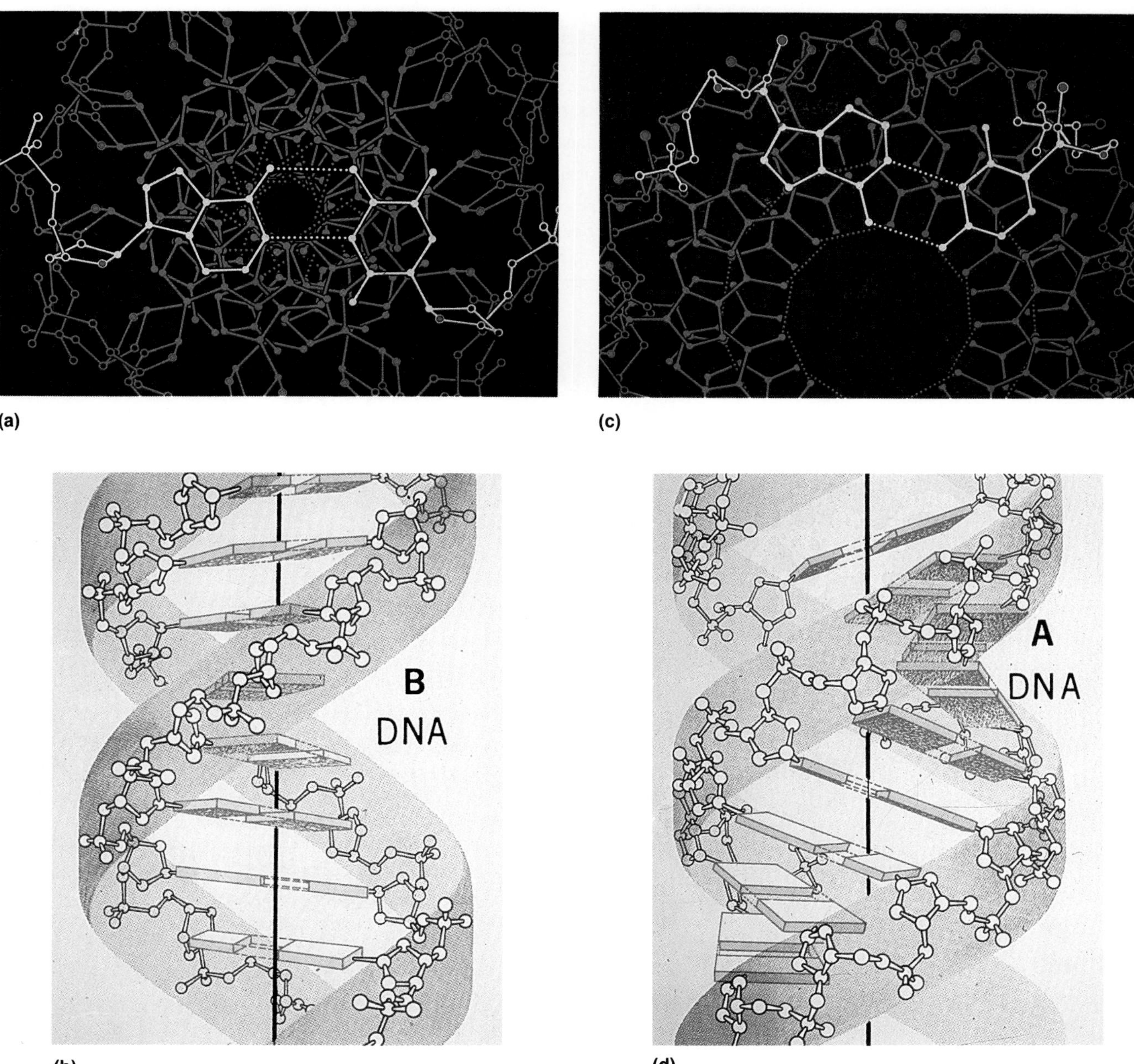

Figure 4.14
Comparison of B-DNA and A-DNA. Shown here, in both end-on and side views, are the structures of B-DNA (**a, b**) and A-DNA (**c, d**) as deduced from recent fiber diffraction studies. The ribose-phosphate backbones are shown in detail, and the bases are schematic.

As Figure 4.14 and Table 4.2 show, the A and B helices are very different. In the B helix the bases lie close to the helix axis, and this axis passes between the hydrogen bonds (note the end-on views of the helices in Figure 4.14a and c). In the A helix the bases lie further to the outside and are strongly tilted with respect to the helix axis. The surfaces of the helices are also very different; in the B helix there are two quite different grooves, called the *major* and *minor* grooves. In the A helix the two grooves are more nearly equal in depth. The grooves are important for interactions of DNA with other molecules, such as proteins, since they allow access to the bases themselves from outside the helix.

All fiber diffraction studies, including those that provided the information described above, suffer from a major limitation. In analyzing fiber patterns, researchers do not directly determine the details of a nucleic acid

Table 4.2
Parameters of polynucleotide helices

	A form	B form	Z form
Direction of helix rotation	Right	Right	Left
Number of residues per turn (n)	11	10	12 (6 dimers)
Rotation per residue ($= 360°/n$)	33°	36°	−60° per dimer; ~−30° per residue
Rise in helix per residue (h)	0.255 nm	0.34 nm	0.37 nm
Pitch of helix ($= nh$)	2.8 nm	3.4 nm	4.5 nm

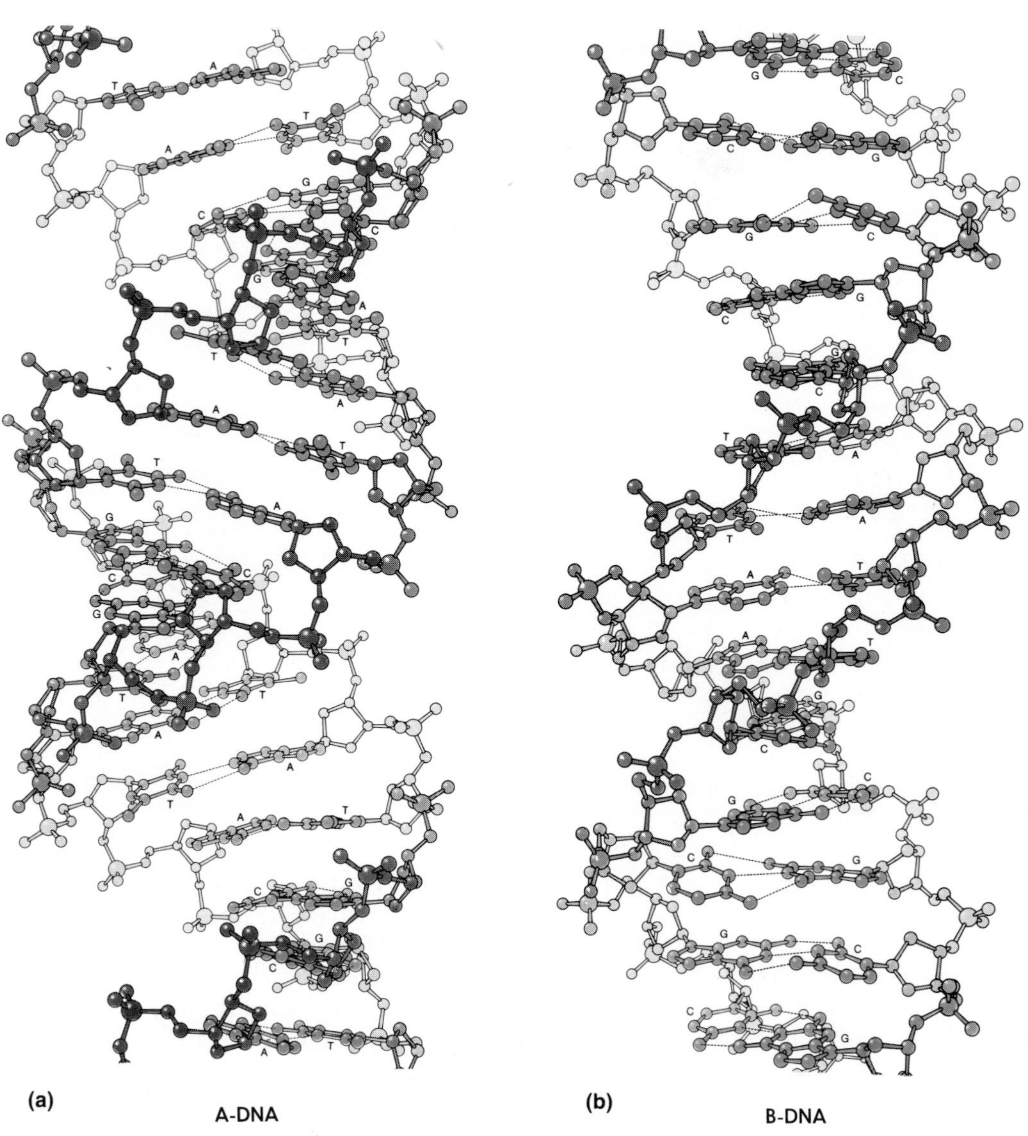

Figure 4.15
DNA structures from studies of molecular crystals. Small nucleic acid oligomers have been crystallized in the A, B, and Z forms. The structures found are shown in panels **a**, **b**, and **c**, respectively.

secondary structure. Instead, they propose models that best account for the positions and intensities of the spots on the diffraction pattern (see Tools of Biochemistry 5). This approach is necessary because fibers are never perfect crystals, and there is always some ambiguity in the interpretation of their diffraction patterns. Therefore, a major advance was made when R. E. Dickerson and his colleagues succeeded in *crystallizing* a small double-strand DNA fragment, which had the sequence

5′CGCGAATTCGCG3′
3′GCGCTTAAGCGC5′

Molecular crystallography of this fragment and other small DNA fragments has given us much more detailed information concerning secondary structure of polynucleotides. The results of such studies are summarized in Figure 4.15. (The figure also includes a third form, which is discussed below.) Here we can show DNA molecules with the position of every atom clearly specified. Although such pictures reveal fine details of the structures, they do not convey correctly the atomic packing. Therefore, we include Figure 4.16 to show how the B-DNA structure depicted in Figure 4.15b appears when atoms have their full van der Waals radii.

A first major point that emerges from the molecular crystal studies is that the models drawn from the fiber patterns represent oversimplifications of the structures. The real structure of B-DNA involves local variations in the angle of rotation between base pairs, the sugar conformation, the tilt of the bases, and even the rise distance. Nucleic acid secondary structure is not homogeneous. It varies in response to the local sequence and is probably easily changed by interaction with other molecules. The parameters given for various forms of DNA in Table 4.2 should therefore be thought of as *average* values, from which considerable local deviation is possible.

The molecular crystallographic studies also provide a possible explanation of why B-DNA is favored in an aqueous environment. The B form of DNA, but not A-DNA, can accommodate a spine of water molecules lying in the minor groove. The hydrogen bonding between these water molecules and the DNA may confer stability to the B form. According to this hypothesis, when this water is removed (as in fibers at low humidity) the A form becomes the more stable structure.

Why, then, do double-strand RNA and DNA–RNA hybrids always adopt the A form? The answer probably lies in the extra hydroxyl group on the ribose in RNA. This hydroxyl interferes sterically in the B form by lying too close to the phosphate and carbon 8 on the adjacent base. Therefore, ribonucleic acid *cannot* adopt the B form, even under conditions where hydration might favor it. In deoxyribonucleic acid, where the hydroxyl is replaced by hydrogen, such steric hindrance does not exist.

Z-DNA: A Left Helix

Since both the A and B forms of polynucleotide helices are right-handed, the discovery of a left-hand form in 1979 was a considerable surprise. Alexander Rich and his colleagues carried out x-ray diffraction studies of crystals of the small deoxyoligonucleotide

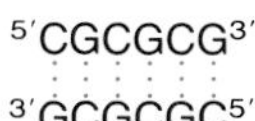

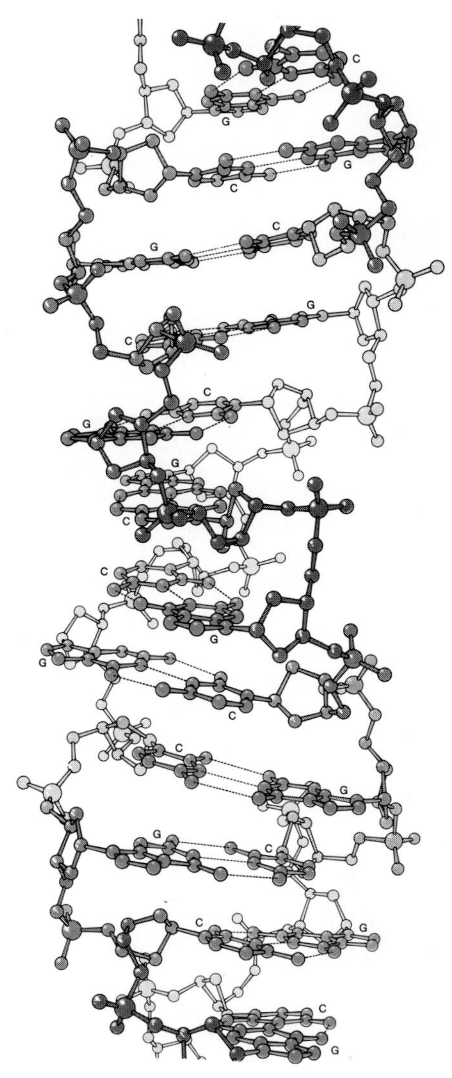

Figure 4.15 *Continued*

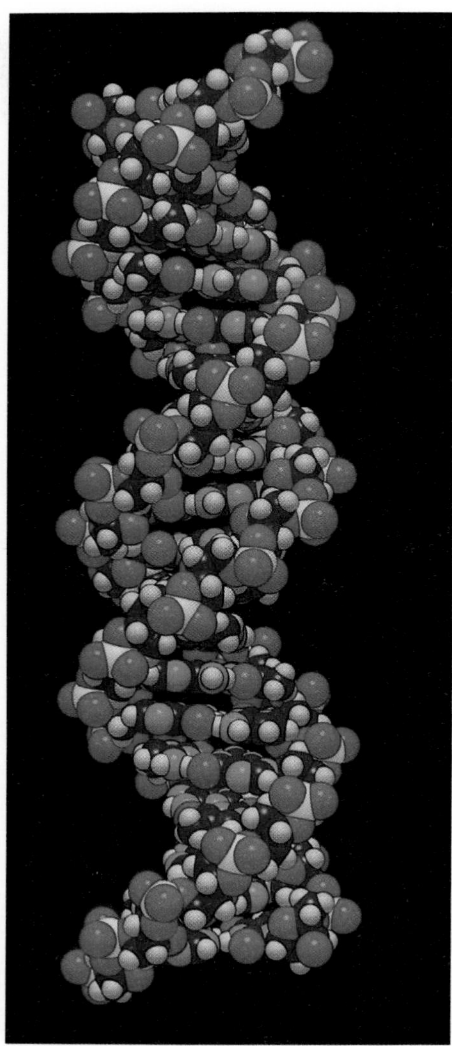

Figure 4.16
A space-filling model of B-DNA. DNA is shown here with each atom given its van der Waals radius. This shows more clearly how closely the bases are packed within the helix.

This was found to exist as a double-strand helix, with G-C base pairing, as expected. However, the data were only consistent with a peculiar *left-hand* structure called **Z-DNA.** A model for a long DNA molecule in the Z conformation is shown in Figure 4.15c.

In addition to the reverse sense of the helix, Z-DNA exhibits other structural peculiarities. There are, in polynucleotides, two most stable orientations of the bases with respect to their deoxyribose rings; these are called *syn* and *anti:*

***Syn* deoxy A** ***Anti* deoxy A**

In both A- and B-form polynucleotides, all bases are in the *anti* orientation. In Z-DNA, however, the *pyrimidines* are always *anti,* but the *purines* are always *syn.* Since Z-DNA is most often found in polynucleotides with alternating purines and pyrimidines in each strand,

···CGCGCG···
···GCGCGC···

for example, this means that the base orientations will alternate. Parameters for Z-DNA (Table 4.2) reflect this, in that the repeating unit is not one base pair but two base pairs. Furthermore, this alternation gives a zigzag pattern to the phosphates, hence the name Z-DNA (see Figure 4.15c).

Polynucleotide Structures in Vivo

The relative importance of the various polynucleotide forms in vivo can be judged from the conditions under which they occur. Since cells contain much water, we would expect most DNA to be in the B form or something very like it. There is evidence that B-DNA dissolved in solution is a little different in conformation from the B-form fiber, having about 10.5 base pairs per turn instead of the expected 10.0. Double-strand RNA and DNA–RNA hybrids, which we discuss below, adopt the A conformation. The existence of Z-DNA or Z-RNA in vivo is more problematic. Normally, the Z conformation is adopted only by regions of polynucleotides having alternating purine–pyrimidine sequences and then only at salt concentrations much higher ($\geq$3 M) than that found in vivo. However, the Z form can be adopted at physiological ionic strength *if* a substantial fraction of the C residues are methylated to form 5-methylcytosine:

5-Methylcytosine (5MeC)

Such methylation actually occurs as a base modification to a significant extent in natural DNA (see Table 4.1). In addition, the presence of divalent cations like Mg^{2+} or small, positively charged molecules like spermine and spermidine which are found in most cells, favors the Z form.

$$H_3\overset{+}{N}-CH_2-CH_2-CH_2-CH_2-\overset{+}{N}H_2-CH_2-CH_2-CH_2-\overset{+}{N}H_3$$

Spermidine

$$H_3\overset{+}{N}-CH_2-CH_2-CH_2-\overset{+}{N}H_2-CH_2-CH_2-CH_2-CH_2-\overset{+}{N}H_2-CH_2-CH_2-CH_2-\overset{+}{N}H_3$$

Spermine

Therefore, the presence in vivo of left-hand helices in some CG-rich regions is certainly possible.

Researchers are beginning to realize that unusual DNA structures may exist in vivo and play important roles. Recent evidence, for example, suggests that some alternating purine–pyrimidine sequences may adopt unusual, multi-strand conformations that may be important in genetic recombination (see Wells et al, 1988, references). Nevertheless, most parts of DNA molecules in cells are undoubtedly B-form, with the two strands being complementary copies. However, some DNA viruses carry single-strand DNA molecules. Examples of some naturally occurring DNAs are listed in Table 4.3. Some of them, like those in the chromosomes of eukaryotes, are truly immense molecules. The DNA from the *Drosophila* (fruit fly) chromosome listed would be 2 cm long if fully extended! Another way to gain an impres-

Table 4.3
Properties of DNA molecules found in various organisms

Source	Single Strand (SS) or Double Strand (DS)	Circular or Linear	Number of Base Pairs (bp) or Bases (b)	Molecular Weight	Length[a]	% (G + C)
Simian virus 40 (genome)[b]	DS	Circular	5243 bp	3.293×10^6	1.78 μm	40.80
ϕX-174 virus (genome)[b]	SS	Circular	5386 b	1.664×10^6	[c]	44.76
Bacteriophage M13 (genome)[b]	SS	Circular	6407 b	1.977×10^6	[c]	40.75
Cauliflower mosaic virus (genome)[b]	DS	Circular	8031 bp	4.962×10^6	2.73 μm	40.19
Adenovirus AD-2 (genome)[b]	DS	Linear	35,937 bp	2.221×10^7	12.2 μm	55.20
Epstein–Barr virus (genome)[b]	DS	Circular	172,282 bp	1.065×10^8	58.6 μm	59.94
Bacteriophage T2 (genome)	DS	Linear	$\sim 1.7 \times 10^5$ bp	$\sim 1.1 \times 10^8$	~58 μm	~52
Bacterium *E. coli* (genome)	DS	Circular	$\sim 4 \times 10^6$ bp	$\sim 2.6 \times 10^9$	~1.4 mm	~53
Fruitfly (*Drosophila melanogaster*) (one chromosome)	DS	Probably linear	$\sim 6.5 \times 10^7$ bp	$\sim 4.3 \times 10^{10}$	~2 cm	~40

[a] Calculated for double-strand DNA of known sequence: 0.34 nm × the number of base pairs (assumes B form). Values for nonsequenced DNAs are from electron microscopy or other methods.

[b] These have been completely sequenced, so numbers of base pairs, molecular weights, and % (G + C) can be given exactly.

[c] The lengths of single-strand DNAs are not well defined; they depend very much on solvent conditions.

Table 4.4
Properties of some naturally occurring RNA molecules

Source (Organism)	Designation	Function	Size (b or bp)
tRNA (transfer RNA)			
E. coli	$tRNA^{leu}$	Transfers leucine in protein synthesis	87 b
Yeast	$tRNA^{phe}$	Transfers phenylalanine in protein synthesis	76 b
Rat	$tRNA^{ser}$	Transfers serine in protein synthesis	85 b
rRNA (ribosomal RNA)			
E. coli	5S RNA	Part of ribosome structure	120 b
E. coli	16S RNA	Part of ribosome structure	1542 b
E. coli	23S RNA	Part of ribosome structure	2904 b
mRNA (messenger RNA)			
Chicken	$mRNA_{LYS}$	Messenger RNA for protein lysozyme	584 b
Rat	$mRNA_{SA}$	Messenger RNA for protein serum albumin	~2030 b
vRNA (viral RNA)			
Polio virus	Polio RNA	Genome of the virus	7440 b
Cytoplasmic polyhedrosis virus of tussock moth	CPV RNA	Genome of the virus	Ten double-strand molecules, ~890 to ~5150 bp

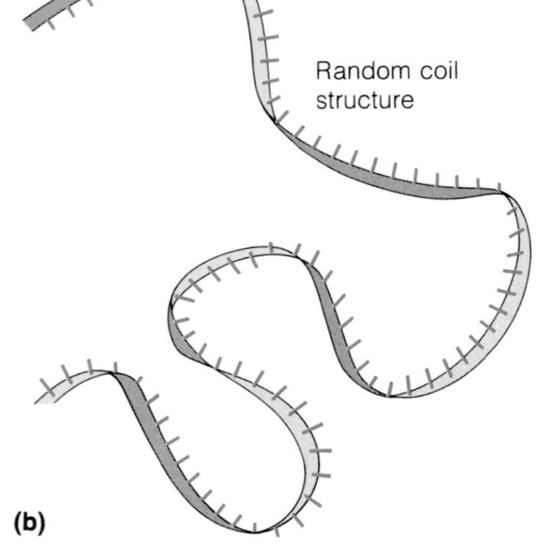

Figure 4.17
Conformations of single-strand nucleic acids. (**a**) Under conditions where hydrogen bonding and base stacking are favored, single-strand nucleic acids will exhibit some degree of secondary structure, either by base pairing in self-complementary regions or by transient stacking. (**b**) Under denaturing conditions all vestiges of secondary structure are lost, and the molecule adopts a random coil conformation.

sion of the size of such molecules is from electron micrographs like that in Figure 4.18, which shows the DNA spilling from a single *E. coli* bacterial cell.

Most RNA molecules are found in cells in single-strand form, although again viruses contain exceptions—some have double-strand RNA (Table 4.4).

It should not be assumed that single-strand DNA or RNA molecules will be entirely devoid of three-dimensional structure. Even single-strand polynucleotides show a propensity to local base stacking, although such structure is much less stable than that observed in double helices. Furthermore, segments exhibiting base complementarity may form local double-strand regions, even in single-strand molecules (see Figure 4.17a). We shall see striking examples of this in later discussions of RNA.

Under extreme conditions, such as high temperature and/or strongly hydrogen-bonding solvents like formamide, all vestiges of double-strand structure and base stacking may be lost. Under such circumstances, DNA and RNA molecules exist in solution as **random coils** (Figure 4.17b). When polynucleotide structure is not constrained (as it *is* in a helix) there is great freedom of rotation about the bonds in the chain backbone. This means that the molecule can freely twist and gyrate in solution, rapidly passing through many, many different conformations. Such conformations are often highly extended because of the *polyelectrolyte* character of all polynucleotides; the negative charges spaced regularly along the chain repel one another. When the molecules are returned to physiological conditions of temperature and solvent, hydrogen bonding and stacking reform and double-strand structures are regained where possible. Details of this helix $\rightleftharpoons$ random coil transition are described below.

Tertiary Structure of Nucleic Acids

Circular DNA and Supercoiling

Many naturally occurring DNA molecules are actually circular, with no free 5′ or 3′ end. For example, the DNA of the simian tumor virus 40 (see Table 4.3) is a circle of 5243 base pairs (bp). The single chromosome of the common bacterium *E. coli* is a giant circle of about 4×10^6 bp (Figure 4.18b). Such circular DNA molecules have a very interesting and inportant property—they are often **supercoiled.** To understand what supercoiling means, consider the following thought experiment (Figure 4.19). Suppose we took a long linear DNA molecule, laid it flat on a surface, and brought the strand ends together 5′ to 3′. If the DNA molecule contained just enough base pairs to make exactly L complete turns of the helix, the 5′ and 3′ ends of each strand would be in position to meet one another, and we could covalently join them (Figure 4.19a). (Certain enzymes can do this in the cell.)

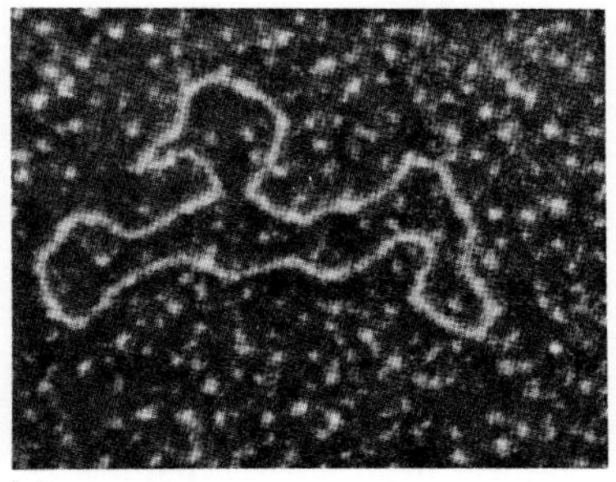

(a)

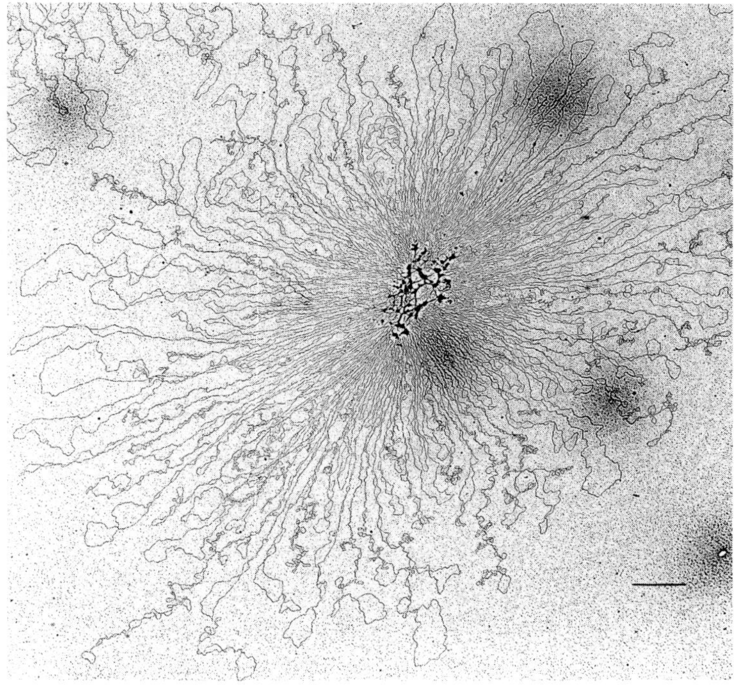

(b)

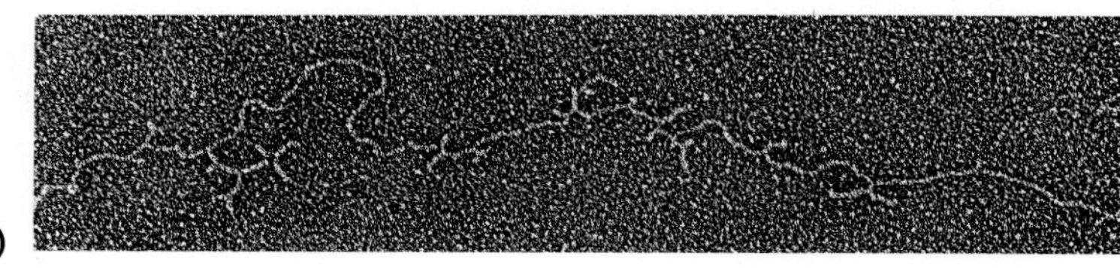

(c)

Figure 4.18
A gallery of nucleic acid molecules as seen by the electron microscope. (**a**) The circular, single-strand DNA of the small virus ϕX-174. (**b**) The large, double-strand circle of *E. coli* DNA. (**c**) The large (45S) precursor of eukaryotic ribosomal RNA. Although this is a single RNA strand, it has many local regions of self-complementarity, which form the double-strand loops.

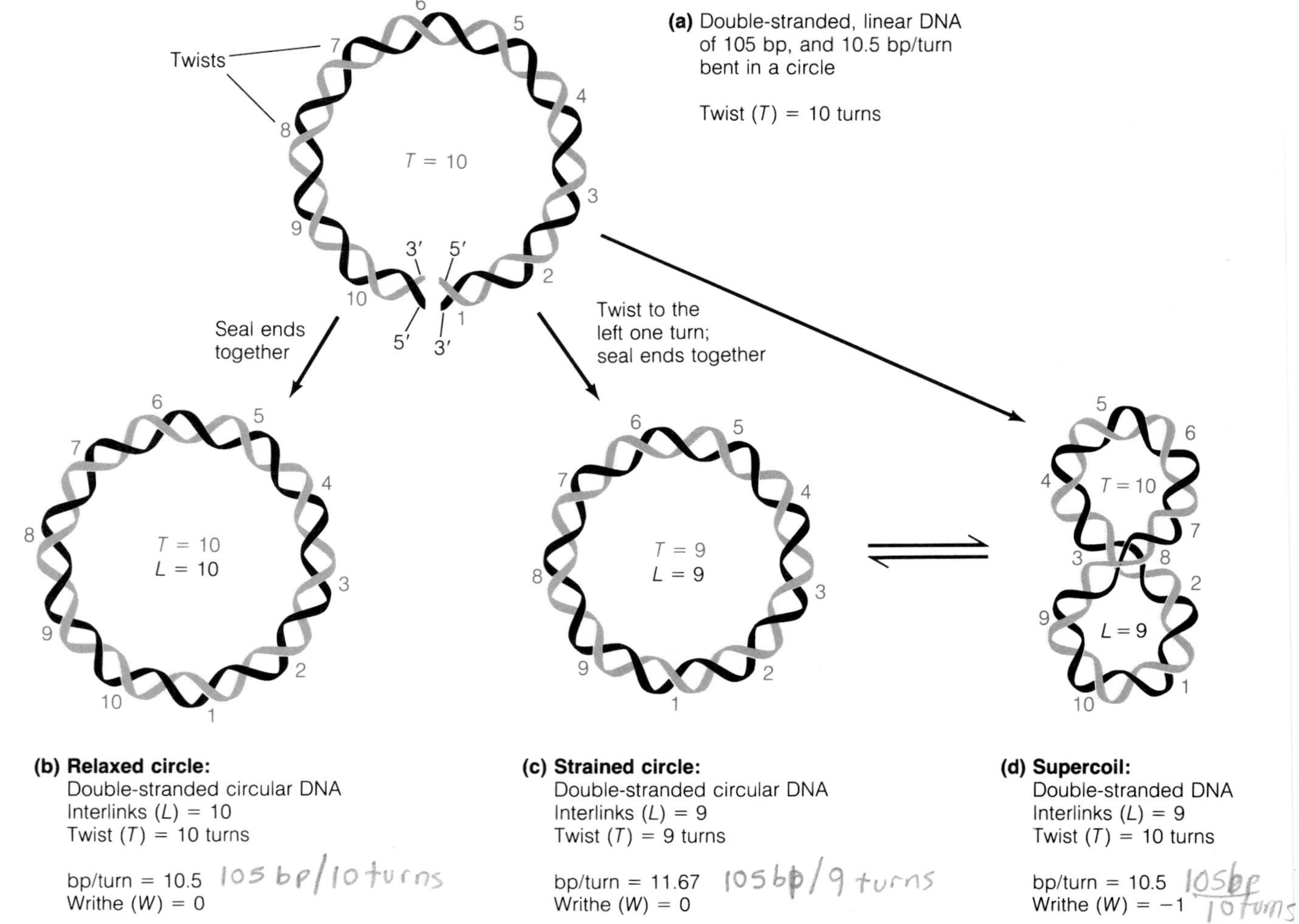

Figure 4.19
DNA supercoiling. (**a**) A linear DNA containing an integral number of turns of the DNA helix is laid on a surface. It is assumed that the DNA has 10.5 bp/turn, as is usual for DNA in solution. The DNA is assumed to have 105 bp, so there will be 10 turns; the twist $T = 10$. (**b**) Since there is an integral number of DNA turns, we can link the 5′ and 3′ ends together without twisting, to make a *relaxed circle*. For this circle the linking number $L = 10$ and the writhe $W = 0$. (**c**) Now suppose we reduce the number of turns in the DNA before joining, by unwinding the DNA by one turn. The closed circle now has $L = 9$. If we still require the DNA molecule to lie flat on a surface, the twist of the helix must change. We will now have $T = 9$ and $W = 0$. With $T = 9$ in 105 bp, we have the DNA forced into a conformation with $105/9 = 11.67$ bp/turn. (**d**) Rather than absorbing the imposed strain in a change in twist, the DNA molecule may writhe, or *supercoil*. We now have $L = 9$ and $T = 10$, so $W = -1$. Imposing supercoils is actually easier than changing twist in most cases and explains the contorted structures seen for the closed-circular, double-strand DNAs shown in Figures 4.18b and 4.20c.

This would make a relaxed circle, which could lie flat on the surface (Figure 4.19b). Since each chain is seen to cross the other L times, we would say that this closed circular molecule has a **linking number** of value L. But suppose *before* joining the ends we made some number (ΔL) of extra turns in one end. This would place a strain on the DNA, and it could respond in two different ways:

1. The molecule could remain flat on the surface. If it did so, however, the **twist** (*T*) of the DNA strands would have to change (Figure 4.19c). The number of base pairs per turn of the helix would now have to become greater or smaller, depending on whether we had twisted the molecule to left or right before joining.

2. The molecule could retain its original twist but could wrap about itself in some number of **superhelical turns** (see Figure 4.19d). This superhelical coiling is called **writhe** (***W***). The molecule can no longer lie flat; it crosses over itself. The molecule now possesses a higher order of three-dimensional structure. We call this the **tertiary structure** level.

Actually, the strain imposed by putting ΔL extra turns in the DNA helix distributes itself between a change in twist and writhing. The simple formula describing this is

$$\Delta L = \Delta T + \Delta W \tag{4.1}$$

where ΔL is positive or negative depending on whether the extra turns are right or left handed. The same sign conventions apply to ΔT and ΔW; more right-hand twisting, or a right-hand writhing of the molecule about itself, is taken as positive.

The superhelicity of DNA molecules is often expressed in terms of the **superhelix density** $\sigma = \Delta L/L_0$, where L_0 is the linking number for the DNA in the relaxed state. Many naturally occurring DNA molecules have superhelix densities of about -0.06. To get an idea of what this means, consider a hypothetical DNA molecule of 10,000 bp, which is in the "classical" B form, with 10.0 bp/turn. Then L_0 is 10,000 bp/(10.0 bp/turn) = 1000 turns. Each DNA strand crosses the other 1000 times in the relaxed circle. If the molecule had a superhelical density of -0.06, then $\Delta L = -0.06L_0$, or $\Delta L = -60$. This could be accommodated, for example, by the helix axis writhing about itself 60 times in a left-hand sense, which would correspond to $\Delta W = -60$, $\Delta T = 0$; the molecule would have 60 left-hand superhelical turns. Alternatively, the twist of the molecule could change so that it had 940 turns in 10,000 bp ($T = 940$) or 10,000/940 = 10.64 bp/turn. This would correspond to $\Delta W = 0$, $\Delta T = -60$. Actually, although any combination of ΔT and ΔW that sums to -60 could occur, the real molecule would probably release the strain mainly by writhing into superhelical turns, because it is easier to bend long DNA than to untwist it. A comparison of a DNA in relaxed circle and supercoil forms is shown in Figure 4.20.

Such differences in supercoiling can be detected by gel electrophoresis. As described in Tools of Biochemistry 2, the rate at which a molecule can be electrophoresed through a gel matrix depends on its dimensions and, hence, on its tertiary structure. Figure 4.21 shows the electrophoretic pattern of a circular DNA that has been relaxed by making a single-strand nick and then resealed. Multiple bands appear because the DNA molecules undergoing molecular motion at the moment of resealing have been caught with various values of *L*, the linking number. Such DNA molecules, identical except for different numbers of superhelical turns, are called **topoisomers.*** The upper

---

* Enzymes called **topoisomerases** are capable of changing the linking number of DNA. We shall have more to say about these in later chapters.

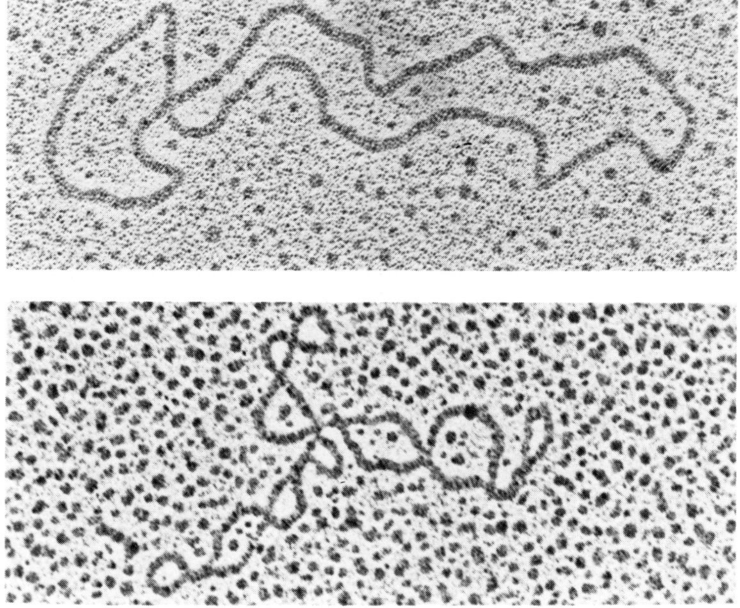

Figure 4.20
The DNA of the bacteriophage PM2 in two topological forms: (**a**) relaxed circle and (**b**) supercoiled. The latter is the native form.

band represents the relaxed, least strained molecule. All other bands represent topoisomers with positive or negative supercoiling. They migrate faster in the gel because they are more compact.

Supercoiling is more than a laboratory curiosity; the naturally occurring *E. coli* DNA shown in Figure 4.18 is supercoiled. As we shall see in later chapters, the strain placed on DNA by supercoiling probably plays an important role in the information-reading processes of the cell. Indeed, it is possible that the B → Z transition can be triggered by increased negative supercoiling. This is because switching of a portion of the DNA from a right to a left helix releases part of the strain imposed by the negative supercoils, since the twist in this portion of the DNA has been reversed.

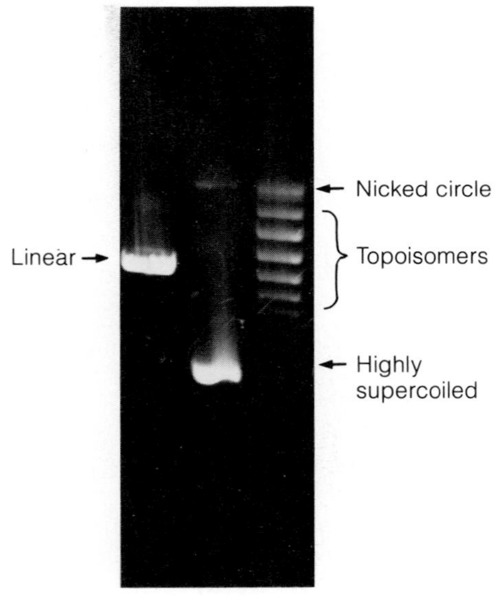

Figure 4.21
Gel electrophoresis of supercoiled DNA. The top band is a relaxed circle obtained by nicking one strand. The multiple bands seen below this correspond to different **topoisomers,** supercoiled DNA molecules differing only in linking number.

Tertiary Structure in RNA Molecules

The RNA molecules found in nature are usually much smaller than DNA molecules and in most cases exist as single strands (see Table 4.4). Even RNA molecules that are produced as single strands may have some double-strand regions, as a consequence of self-complementarity. The existence of such self-complementary double-strand regions often folds RNA molecules into complicated three-dimensional structures such as that shown in Figure 4.18c or 4.22. Like the twisting of circular DNA into supercoils, such a higher order of folding (above the secondary structure of the local double-strand regions) is referred to as the *tertiary structure* of the molecule.

The molecule shown in Figure 4.22 is one of a class of important ribonucleic acids called **transfer RNAs,** or **tRNAs.** These play a role in assembling amino acids into proteins. We discuss their properties in considerable detail in Chapter 27. The structure of these molecules is known in such detail because the folded molecules can crystallize, and x-ray diffraction can be carried out with the crystals. Many ribonucleic acid molecules other than tRNAs also show specific secondary and tertiary structures. In each case, the overall tertiary structure and the local regions of secondary structure within it are dictated by the primary structure of the molecule.

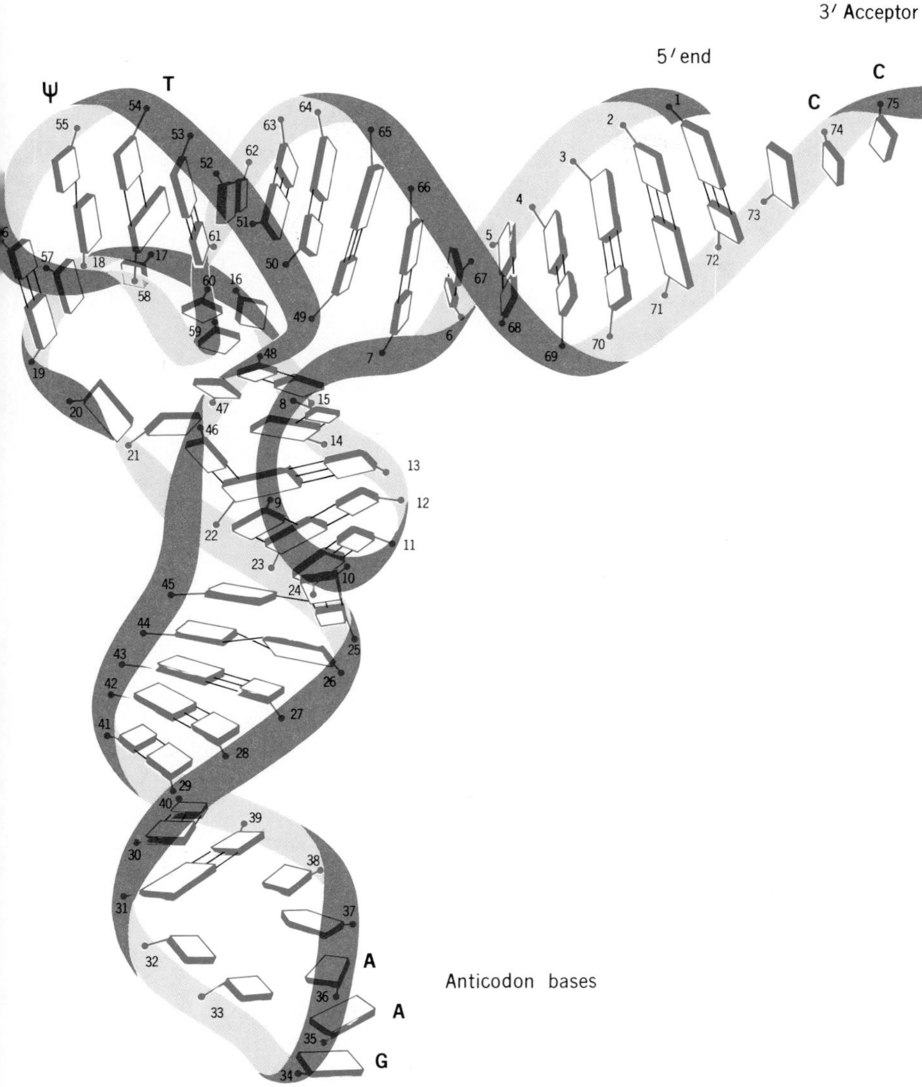

Figure 4.22
The tertiary structure of a transfer RNA. Shown here is the tRNA that transfers the amino acid phenylalanine into proteins synthesized in yeast cells. The ribose-phosphate backbone is shown as a red ribbon, the paired bases in blue.

Stability of Secondary and Tertiary Structure

The Helix-to-Random-Coil Transition: Nucleic Acid Denaturation

Secondary and tertiary polynucleotide structures are relatively stable under physiological conditions. Yet they must not be *too* stable, for many biochemical processes—DNA replication, for example—require that the double-helix structure be opened up. This loss of secondary structure is called **denaturation** (see Figure 4.17). Competing factors create a balance between structured and unstructured forms of nucleic acids.

Two major factors favor dissociation of double helices into randomly coiled single chains. The first is the electrostatic repulsion between the chains. At physiological pH, every residue on a DNA or RNA molecule carries a negative charge on the phosphate group. Even though this charge is partially neutralized by small counterions (like K^+, Na^+, and Mg^{2+}) present in the medium, a substantial net negative charge remains on each chain in the helix and tends to drive the two chains apart.

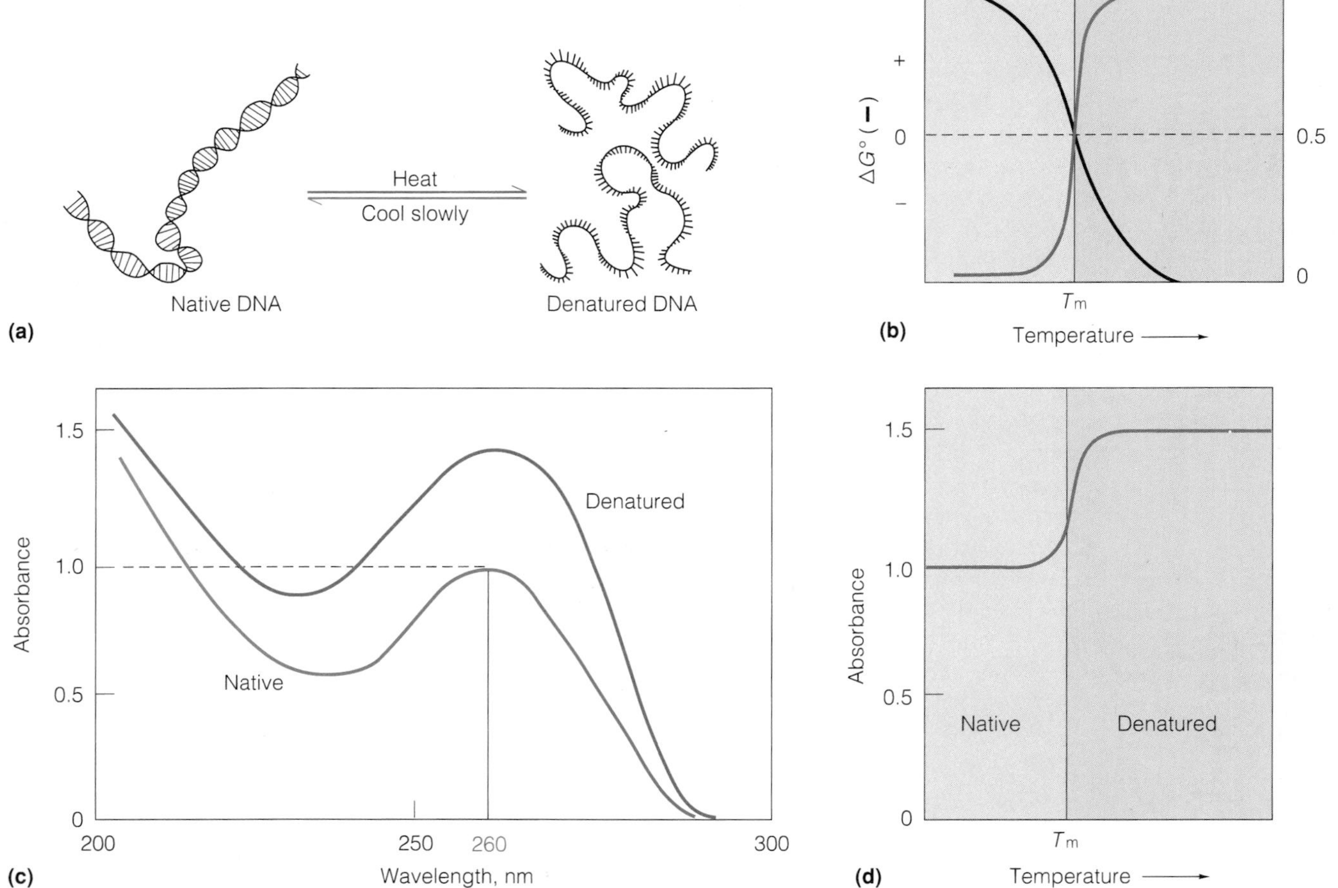

Figure 4.23
Denaturation of DNA. (**a**) A double strand DNA, if heated above its melting temperature, will separate into two random coil single strands. (**b**) The change in ΔG and fraction denatured with temperature. At low T, ΔG is positive and the process is not favored. But as T increases, the $-T\,\Delta S$ term overcomes the ΔH term, and ΔG becomes negative; denaturation is now thermodynamically favorable. The midpoint of the curve marks the "melting" temperature, T_m. (**c**) The hypochromicity of double-strand nucleic acids, which can be used to follow denaturation. Absorption spectra for DNA are shown in the native (double-strand) form (blue line) and the denatured (single-strand) form (red line). The maximum difference in absorbance occurs at a wavelength of 260 nm. (**d**) How absorbance changes as DNA is heated.

A more subtle factor favoring denaturation is the fact that the random coil structure has a higher entropy. This is because of the greater randomness of the denatured form, with its many possible configurations. Consider equation (3.13) (from Chapter 3): if a rigid double helix separates into two flexible random coils, the number of configurations accessible to the molecule greatly increases (Figure 4.23a). The free energy change in going from a regular two-strand polynucleotide secondary structure (such as B-form DNA) to individual random coil strands is given by the usual formula:

$$\Delta G = \Delta H - T\,\Delta S \qquad \text{(helix} \rightleftharpoons \text{random coil)} \quad (4.2)$$

Since ΔS is positive, the term $(-T\,\Delta S)$ makes a negative contribution to the free energy change.

Thus, two factors favor the helix → coil transition: the higher randomness of the random coil ($\Delta S > 0$) and the electrostatic repulsion between chains ($\Delta H_{el} < 0$). If the double-strand helical structure is to be stable under any conditions, ΔG for the unfolding reaction must be positive. We must look for a large positive contribution to ΔH to compensate for the factors mentioned above. The sources of such a positive ΔH are the hydrogen bonds between the base pairs and van der Waals interactions between stacked bases. Much energy must be put in to break these, and hence the total ΔH is positive.

Since ΔH and ΔS in equation (4.2) are both positive, the sign of ΔG will change as T is increased. At low temperature, the term $T\,\Delta S$ will be less than ΔH; ΔG will be >0, and the helix will be stable. But as the temperature is increased, ΔG will pass through zero and become negative; at higher temperatures the double-strand structure is unstable and will fall apart (see Figure 4.23b). The process is somewhat analogous to the process of ice melting into water as described in Chapter 3.

It is possible to follow this melting process by observing the absorbance of ultraviolet light of wavelength about 260 nm in a DNA solution. All nucleotides and nucleic acids absorb light strongly in this wavelength region because of the aromaticity of the purine and pyrimidine bases. When the nucleotides are polymerized into a polynucleotide and this polynucleotide is packed into a helical structure, the absorption of light is reduced (Figure 4.23c). This phenomenon, called **hypochromism,** results from close interaction of the light-absorbing groups, the purine and pyrimidine rings. If the secondary structure is lost, the absorbance increases and becomes closer to that of a mixture of the free nucleotides. Therefore, raising the temperature of a DNA solution, with accompanying breakdown of the secondary structure, will result in an absorbance change like that shown in Figure 4.23d.

The remarkable thing about this helix-to-random-coil transition is that it is so sharp. It occurs over a very small temperature range, almost like the melting of a crystalline solid. Therefore, it is sometimes referred to as a *melting* of the polynucleotide double helix, even though the term is not technically correct. We shall encounter such abrupt changes in configuration in other macromolecules later. They are always characteristic of what are called **cooperative transitions.** What this means in the case of DNA or RNA is that a double helix cannot melt bit by bit. If you examine the kinds of structure shown in Figures 4.15 and 4.22, you will see that it would be very difficult for a single base to "flip out" of the stacked, hydrogen-bonded structure. Rather, the whole structure holds together until it is at the verge of instability and then denatures over a very narrow temperature range. The "melting temperature" at which this happens is characteristic of the (G + C)/(A + T) ratio of the DNA. Since each G-C base pair forms three hydrogen bonds and each A-T pair only two, ΔH is greater for the melting of GC-rich polynucleotides. The value of T_m corresponds to the temperature at which $\Delta G = 0$ (see Figure 4.23). Thus,

$$0 = \Delta H - T_m \Delta S \tag{4.3}$$

or

$$T_m = \frac{\Delta H}{\Delta S}$$

On a per-base-pair basis ΔS is about the same for all polynucleotides, but ΔH depends on base composition, as described above. This is why T_m increases with increasing G + C content. Figure 4.24 shows a graph of T_m versus percent (G + C) for a number of naturally occurring DNAs.

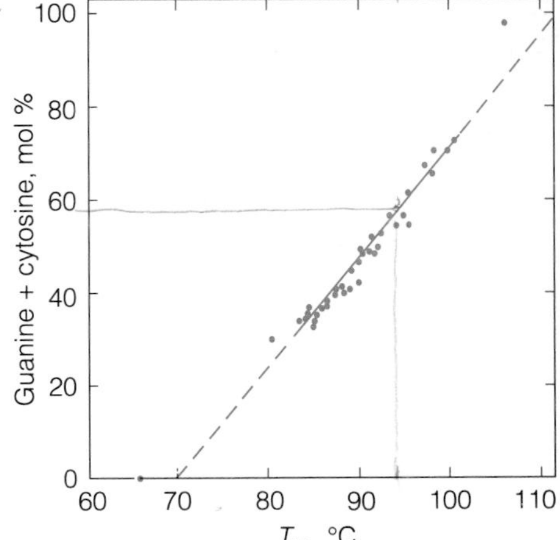

Figure 4.24
Relationship between melting point and % (G + C) in DNAs.

Renaturation: C_0t Analysis

If a solution of a double-strand nucleic acid that has been denatured by heating above T_m is slowly cooled to just below T_m, the double-strand structure will eventually reform. However, this **renaturation** process is difficult, for the complementary strands must find one another and then align properly to form the correct base pairings. If a mixture of DNA molecules con-

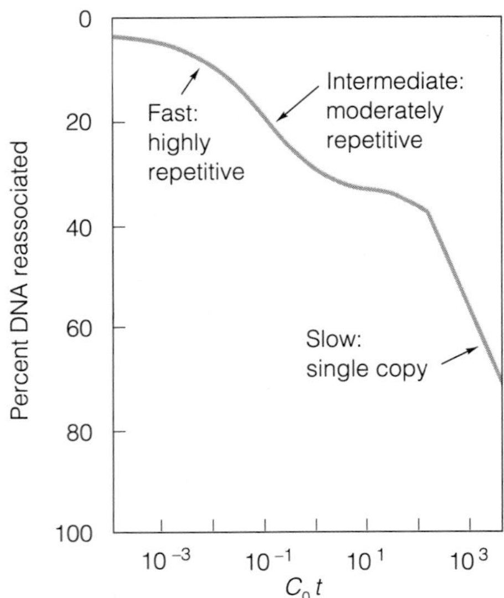

Figure 4.25
Analysis of DNA renaturation. The fraction of DNA remaining denatured is plotted against the product C_0t, where C_0 is the DNA concentration (moles of base pairs per liter) and t is time of renaturation. Data are for human DNA. Three classes of DNA are distinguished. While single-copy sequences are unique, the distinction between moderately and highly repetitive is somewhat arbitrary. For such experiments the genomic DNA is usually sheared into pieces about 300 bp in length.

tains some sequences that are more common than others, the frequently occurring ones will renature first, for they can find complementary partners more easily. This is the basis of the C_0t plot method for determining the relative frequency of different sequences. For example, the renaturation analysis in Figure 4.25 reveals three broad classes of sequences in the human genome. A substantial portion of the DNA is present as sequences that occur only once, or a very few times each, in the whole genome. These are the slowest to reassociate. Then there is a fraction of sequences that exist in hundreds or thousands of copies each, which reassociate more rapidly. Finally, a small fraction of the DNA exists as a highly reiterated form (up to millions of copies), which reassociates most rapidly of all. As we shall see in Chapter 28, each of these classes of DNA sequences plays a different role in the eukaryotic genome.

Many studies of the denaturation and renaturation of nucleic acid structures have consistently pointed to the same conclusion: Complex as the secondary and tertiary structures may be, they represent the stable conformations of these molecules under physiological conditions. The structures may be quite easily perturbed, but when native conditions are reestablished, they spontaneously reform. The information for the structure of a molecule like tRNA is built into the sequence; no more is needed.

The Biological Functions of Nucleic Acids: A Preview

We have emphasized that the fundamental role of the nucleic acids is the storage and transmission of genetic information, and we will continue to develop this theme throughout the book. Details of how information is passed from parent cell to daughter cell or from an organism to its descendents and of how it is expressed in directing biochemical processes and the formation of other complex molecules like proteins are elaborated in Part V. In this and the following chapter we present a preliminary overview of these nucleic acid functions. From this, you will gain some appreciation of the relationships between nucleic acid and protein structures, how evolution occurs at the molecular level, and how we can modify microbes, plants, and animals through genetic engineering.

A defined polynucleotide sequence is a coded string of information. The DNA contained in each cell constitutes the **genome,** or total genetic information content, of the organism. A substantial fraction of this DNA is capable of being transcribed, or "read," to allow the expression of this information in directing synthesis of RNA and protein molecules. The segments that can be transcribed are referred to as **genes.** The DNA in each cell of every organism contains at least one copy (and sometimes several) of the gene carrying the information to make each protein that the organism requires. Much of the "low-copy-number" (slowly reassociating) DNA referred to in Figure 4.25 consists of these protein-coding genes. In addition, there are genes (often reiterated) for the many specific RNA molecules, such as the tRNA shown in Figure 4.22, that are essential for cellular function.

Expression of the genetic information always involves as a first step the **transcription** of genes into complementary RNA molecules. This production of specific RNA molecules is easy to visualize. Just as a DNA strand can direct **replication**—the synthesis of a new DNA strand complementary to itself, as shown in Figure 4.13—it can equally well direct the formation of a complementary RNA strand (Figure 4.26). Of course, different monomers are required in transcription than in replication. Instead of the deoxy-

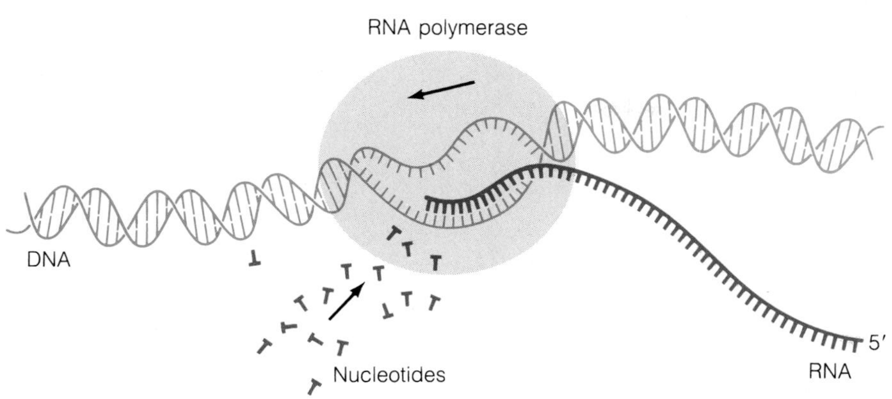

Figure 4.26
The basic principle of transcription. An enzyme *(RNA polymerase)* travels along DNA, making an RNA transcript by adding nucleotides one at a time, copying the oligonucleotide sequence from one of the DNA strands.

ribonucleoside triphosphates, the ribonucleoside triphosphates ATP, GTP, CTP, and UTP are needed to make RNA. (Note that U in the new RNA pairs with A in the DNA template.) Each of these processes, DNA replication and transcription, requires a special set of enzyme catalysts known as **polymerases.** We discuss these, and the details of the processes they catalyze, in Chapters 24, 26, and 28.

Transcription alone is sufficient for the production of the many functional RNA molecules of the cell, such as the tRNAs or ribosomal RNAs in Table 4.4. However, the synthesis of specific proteins, as dictated by their genes, is a more complex matter. The problem is that, as we shall see in Chapter 5, proteins are polymers made from 20 different kinds of amino acid monomers. Since there are only four different nucleotide monomer types in DNA, there could not be a one-to-one relationship between the sequence of nucleotides in a DNA molecule and the sequence of amino acids in a protein. In order to specify each of the 20 amino acids, the protein-coding information is "read" by the cell in blocks of three nucleotide residues or **codons,** each corresponding to a separate amino acid. The set of rules that specifies which nucleic acid codons correspond to which amino acids is known as the **genetic code.** We describe it in the next chapter.

The production of proteins does not proceed directly from the DNA. For the information to be translated from the DNA sequences of the genes into amino acid sequences of proteins, a special class of RNA molecules are used as intermediates. Complementary copies of the genes to be expressed are transcribed from the DNA in the form of **messenger RNA (mRNA)** molecules (examples are listed in Table 4.4). The mRNAs are read by the protein-synthesizing machinery of the cell to make the appropriate proteins. This process, which takes place on subcellular particles called **ribosomes,** is referred to as **translation** (Figure 4.27). We outline its main features in Chapter 5 and describe it in detail in Chapters 27 and 28. The flow of genetic information in the cell can be summarized by the simple schematic diagram shown in Figure 4.28.

As we shall see in the following chapters, proteins are the major structural and functional molecules in most cells. What a cell is like and what it can do depend largely on the proteins it contains. These, in turn, are dictated by the information stored in the cell's DNA, transcribed into mRNA, and expressed by the protein-synthesizing machinery. We are now learning how to modify and manipulate this stored information and its expression by **recombinant DNA** techniques.

These methods have completely revolutionized biology and biochemistry. To discuss them in detail at this point would be premature, for they

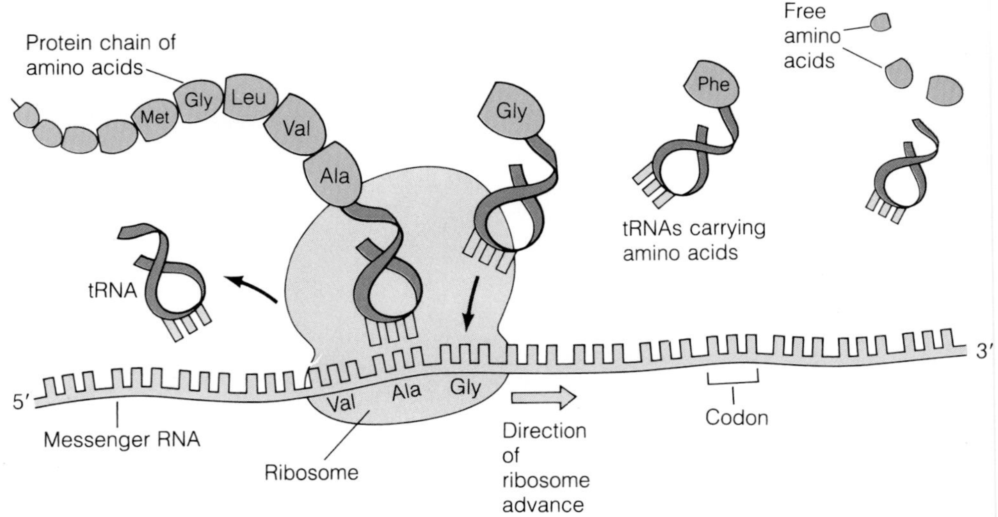

Figure 4.27
Translation. A messenger RNA is bound to a particle called a *ribosome*. Individual amino acids are brought to the ribosome, one at a time, by tRNA molecules. Each tRNA identifies the appropriate *codon* on the messenger RNA and adds this amino acid to the growing protein chain. The ribosome travels along the mRNA, so that the whole genetic message can be read and *translated* into protein.

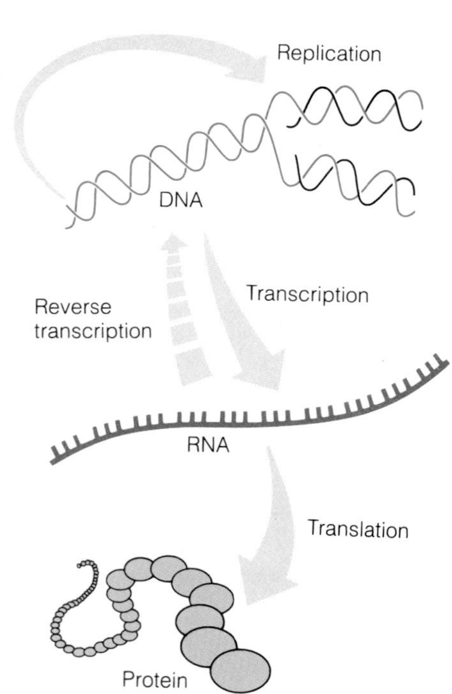

Figure 4.28
The flow of genetic information. DNA can either replicate or be transcribed into RNA. Messenger RNAs are then translated into protein sequences. Under some special circumstances, RNA can be reverse transcribed to produce DNA.

depend on complicated enzymatic processes that can be understood fully only at a later point. You will find them detailed in Chapter 25. But it is appropriate to list here briefly a few of the kinds of things that can now be accomplished:

1. Individual genes or other desired DNA sequences can be retrieved from the whole genome of a higher organism and **cloned** (reproduced identically) in bacteria such as *E. coli.* This allows the large-scale production of particular DNA regions, to be used, for example, in sequencing.

2. Genes of higher organisms can be expressed in bacteria, allowing the production of large amounts of some normally rare eukaryotic proteins.

3. The cloned genes can be modified in desired ways and their modified products studied.

4. In some cases, the modified genes can be reinserted into the organism from which they came, to allow study of the physiological and biochemical effects of the modification.

Such methods will clearly lead to enormous increases in our understanding of biochemical and biological processes. Even more significant in the long term may be their potential for human manipulation of the genetic information in bacteria, plants, and animals. Like most great technical advances, genetic engineering may be used for good or ill. The ethical questions involved will be debated for a long time. However these debates are resolved, the knowledge now exists and humanity—in fact, life itself—has passed a new threshold. For the first time in the long history of evolution, that process itself can now be consciously controlled.

REFERENCES

General References

Dickerson, R. E. (1983) The DNA helix and how it is read, *Sci. Am.* 249:(6)94–111. A short, popularly written, but very lucid description of DNA structure.

Saenger, W. (1984) *Principles of Nucleic Acid Structure.* Springer-Verlag, New York.

Sarma, R. H. (ed.) (1981) *Biomolecular Stereodynamics*, Vols. 1 and 2. Adenine Press, Guilderland, N.Y. The Saenger and Sarma references provide much greater detail concerning nucleic acid structure than is given in this book.

Historical

Avery, O. T., C. M. MacLeod, and M. McCarty (1944) Studies on the chemical transformation of pneumococcal types, *J. Exp. Med.* 79:137–158. The pioneering study that lent credence to the idea that DNA is the genetic substance.

Hershey, A. D., and M. Chase (1952) Independent function of viral protein and nucleic acid on growth of bacteriophage, *J. Gen. Physiol.* 36:39–56. The convincing evidence that DNA is the genetic material.

Judson, H. (1979) *The Eighth Day of Creation.* Simon & Schuster, New York. A detailed (and fascinating) account of the development of modern ideas about nucleic acids.

Watson, J. D. (1968) *The Double Helix.* Atheneum, New York (trade and paperback editions); New American Library, New York (paperback). A delightful (and illuminating) recounting of the elucidation of the DNA structure.

Watson, J. D., and F. H. C. Crick (1953) Molecular structure of nucleic acids. A structure for deoxyribose nucleic acid, *Nature* 171:737–738. Two pages that shook the world.

Specialized Papers of Importance

Bauer, W. R., F. H. C. Crick, and J. H. White (1980) Supercoiled DNA, *Sci. Am.* 243:(1)118–133. A readable description of supercoiling.

Levene, S. D., and D. M. Crothers (1986) Topological distribution and the torsional rigidity of DNA. A Monte-Carlo study of DNA circles, *J. Mol. Biol.* 189:73–83.

Maniatis, T. R., C. Hardison, E. Lacey, J. Lauer, C. O'Connell, D. Quon, G. K. Sim, and A. Efstratiadis (1978) The isolation of structural genes from libraries of eukaryotic DNA, *Cell* 15:687–701.

Pohl, F. M., and T. M. Jovin (1972). Salt-induced cooperative conformational change of a synthetic DNA: Equilibrium and kinetic studies with poly(dG-dC), *J. Mol. Biol.* 67:375–396. The first hint of Z-DNA.

Saenger, W., W. N. Hunter, and O. Kennard (1986) DNA conformation is determined by economics in the hydration of phosphate groups, *Nature* 324:385–388.

Wang, A. H.-J., G. J. Quigley, F. J. Kolpak, J. L. Crawford, J. H. van Boom, G. van der Marel, and A. Rich (1979) Molecular structure of a left-handed DNA fragment at atomic resolution, *Nature* 282:680–686. Discovery of Z-DNA.

Wells, R., D. A. Collier, J. C. Hanvey, M. Shimizu, and F. Wohlrab (1988) The chemistry and biology of unusual DNA structures adopted by oligopurine-oligopyrimidine sequences. *FASEB J.* 2:2939–2949.

Wing, R. M., H. R. Drew, T. Takano, C. Brodka, S. Tanaka, K. Itakura, and R. E. Dickerson (1980) Crystal structure analysis of a complete turn of B-DNA, *Nature* 287:755–758. First crystallographic study of a B-DNA structure.

PROBLEMS

1. A viral DNA is analyzed and found to have the following base composition, in mole percent: A = 32, G = 16, T = 40, C = 12. What can you immediately conclude about this DNA?

2. Given the following sequence for one strand of a double-strand oligonucleotide, write the sequence of the complementary strand:

   5′ ACCGTAAGGCTTTAG 3′

3. Some naturally occurring polynucleotide sequences are *palindromic*. This means that they are self-complementary about an axis of symmetry. Such a sequence is

   . . .TC↓AAGTCCATGGACTT↓GG
   AG↑TTCAGGTACCTGAA↑CC

   Show how such a structure might form a "double hairpin" conformation in the region between the arrows.

4. The gel pattern in the accompanying illustration was obtained in Maxam–Gilbert sequencing (described in Tools of Biochemistry 3) of a DNA. Read the sequence.

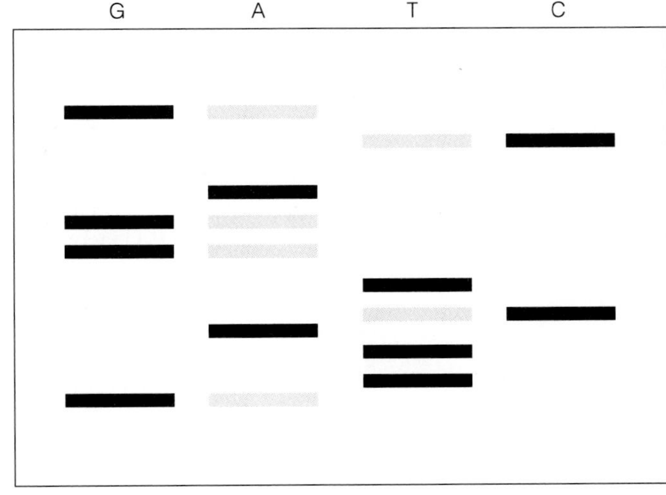

5. A variant of the B-DNA structure, called C-DNA, exists in the presence of Li^+ ions. It has 9.33 base pairs per turn.
   (a) How many base pairs are contained in one repeat of this structure? How many turns are in one repeat?
   (b) Is this DNA twisted more tightly or less tightly than normal B-DNA?

6. A circular, double-strand DNA contains 2100 base pairs. The solution conditions are such that DNA has 10.5 bp/turn.

   (a) What is L_0 for this DNA?

   (b) The DNA is found to have 12 left-hand superhelical turns. What is the superhelix density σ?

7. In a supercoiled DNA, a stretch of about 10 base pairs changes from the B form to the Z form. What is the change in (a) T, (b) L, and (c) W?

8. Calculate the length of the DNA molecule of Epstein–Barr virus (172,282 base pairs) if it were extended in a linear A structure.

9. Suppose a particular sequence A appears on the average every X base pairs in a genome.

   (a) The rate of reassociation of A will be given by

$$\frac{d[\mathrm{A}]}{dt} = -k[\mathrm{A}]^2$$

   where k is the second-order rate constant. Integrate this and show that f, the fraction of A sequences associated at time t, is given by

$$f = \frac{1}{1 + 2A_0kt}$$

   where $f = [\mathrm{A}]/A_0$ and A_0 is the total molar concentration of A. Noting that $A_0 = C_0/X$, where C_0 is the total molar concentration of *base pairs* in the solution, show that

$$(C_0t)_{1/2} = \frac{X}{2k}$$

   where $(C_0t)_{1/2}$ is the value of C_0t when half of the sequence A is reassociated.

   (b) Under given renaturation conditions, the synthetic polynucleotide

   . . .AAAAAA. . .
   . . .TTTTTT. . .

   renatures with $(C_0t)_{1/2} = 0.01$ (mol/L) sec. Under the same conditions, a particular class of sequences in the human genome renatures with $(C_0t)_{1/2} = 100$ (mol/L) sec. If the human genome size is 10^{10} bp, how many copies of this sequence exist in the human genome? [Hint: $X = 1$ for the synthetic polynucleotide. Why?]

10. A particular double-strand DNA has, under the conditions used in Figure 4.24, a melting point of 94°. Estimate the base composition (in mole percent) of this DNA.

3 TOOLS OF BIOCHEMISTRY 3

Determining Nucleic Acid Sequences

The development in the late 1970s of rapid methods for sequencing DNA revolutionized biochemistry. We shall describe here the technique first described by Allan Maxam and Walter Gilbert in 1977.

The Maxam–Gilbert sequencing technique depends on the existence of chemical reactions that can specifically cleave a polynucleotide chain wherever a particular kind of residue (A, T, G, or C) appears. Two basic reactions are used:

1. Dimethyl sulfate methylates purine residues, as shown in Figure T3.1. The methylated derivatives are susceptible to **depurination,** or removal of the purine moeity. The polynucleotide chain can then be cleaved by hydrolysis in hot 0.1 N alkali, which removes the depurinated sugar. Selectivity is obtained by the fact that N_7-methylguanine depurinates preferentially at neutral pH, whereas acid conditions lead to preferential depurination of N_3-methyladenine.

2. The cleavage at pyrimidines starts with **hydrazinolysis.** Again, conditions can be chosen to favor reaction of one base or another; the reactions used are depicted in Figure T3.2. The hydrazone derivatives, when treated with piperidine, will hydrolyze at the 3′ phosphodiester bond.

We have, then, four different reaction conditions, each of which can lead to the preferential cleavage of the DNA chain adjacent to specific residue types.

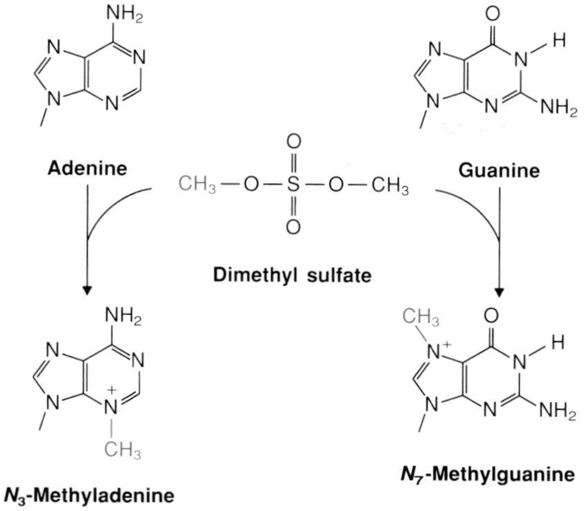

Figure T3.1
Reaction of purines with dimethyl sulfate.

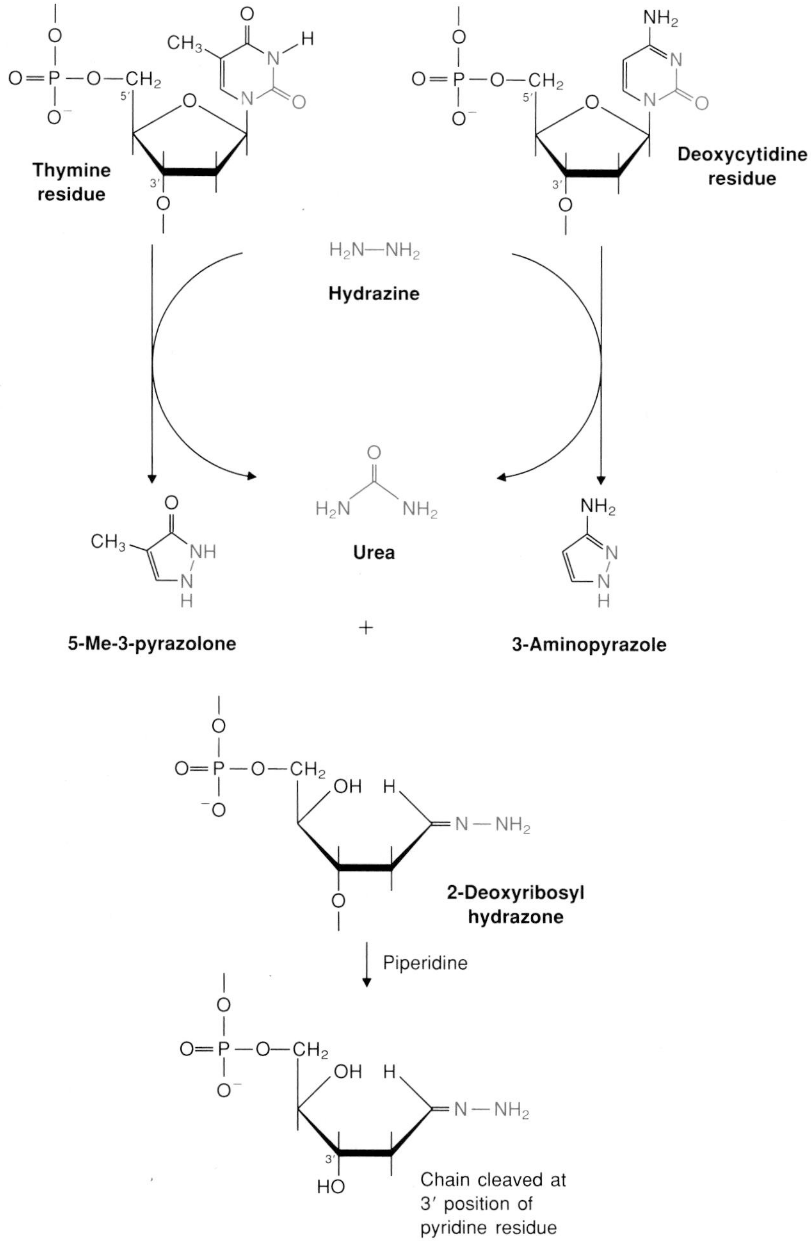

Figure T3.2
Hydrazinolysis of pyrimidines.

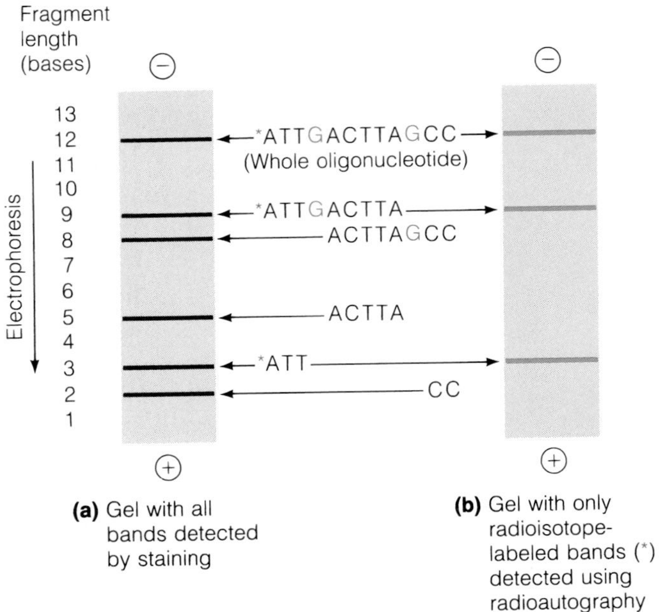

Figure T3.3
Gel electrophoresis of fragments produced by cleavage of an oligonucleotide at G residues.

(a) Gel with all bands detected by staining

(b) Gel with only radioisotope-labeled bands (*) detected using radioautography

Suppose we carried out the guanine-specific cleavage on the sequence ATTGACTTAGCC. We should expect the following possibilities:

ATTGACTTAGCC, or:	ATTGACTTAGCC, or:	ATTGACTTAGCC
↓	↓	↓
ATT + ACTTAGCC	ATTGACTTA + CC	ATT + ACTTA + CC

The entire mix of products shown above would be obtained. These could be separated (for example, by gel electrophoresis), but the mix would be very difficult to interpret (Figure T3.3a). But suppose the 5′ end of the DNA had been radioactively labeled, by attaching a ^{32}P-labeled phosphate to the 5′ hydroxyl group, and we detected *only* radiolabeled oligonucleotides on the gel. Of the whole collection above, we would then see only *ATT and *ATTGACTTA, and residual whole oligonucleotides (see Figure T3.3b). The identification of labeled oligonucleotides 3 and 9 residues in length after the G reaction tells us immediately that there are G residues at positions 4 and 10 in the chain.

This is the fundamental idea of the method. To "sequence" a DNA molecule, then, the following steps are necessary, as diagrammed in Figure T3.4.

1. A homogeneous sample of the DNA chain to be sequenced must be prepared. If the DNA (like many natural DNAs) occurs in a double-strand form, the two strands must be separated and the desired strand isolated. This can usually be done by high-resolution electrophoresis.

2. The 5′ end of the DNA is labeled with a radioactive phosphate group on the 5′ hydroxyl residue. Enzymes called **polynucleotide kinases** will catalyze this reaction. If the DNA already has a nonradioactive phosphate in this position, this must first be removed using catalysis by a **phosphatase.**

3. The labeled DNA sample is divided into four portions and the four site-specific reactions are carried out, one on each portion. Conditions are chosen so that each molecule will be cleaved, on the average, about once. Since cleavage at each kind of site is random, this will give a good set of fragments. (If the reaction were carried out too extensively, only the shortest labeled fragments would show up.)

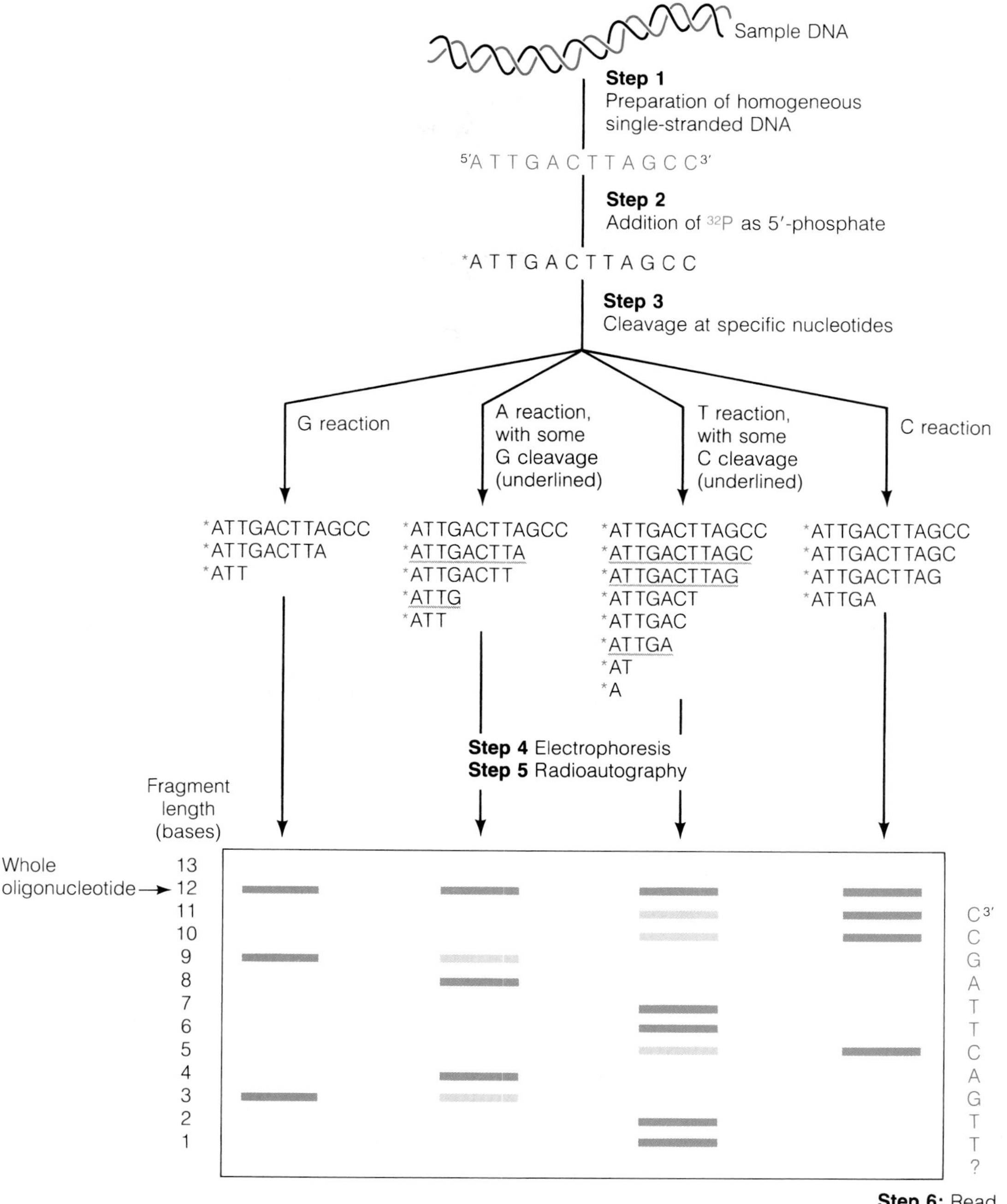

Figure T3.4
Sequencing an oligonucleotide by the Maxam-Gilbert method.

4. The products of the four reactions are electrophoresed in adjacent lanes on a polyacrylamide gel under denaturing conditions. A gel capable of resolving nucleotides differing by only one residue is used. The smallest fragments will run most rapidly on the gel.

5. The gel is dried and laid against a photographic film for **autoradiography.** Only the bands that are radioactive will expose the film, marking their positions.

6. The investigator then simply starts at the bottom of the film and reads the sequence as far as bands can be resolved—often extending to over 200 residues. You can see how this is done for our model oligonu-

cleotide in Figure T3.4. Note that the first residue (at the 5′ end) will never be seen, for this is destroyed in the appropriate reaction. Sometimes it is difficult to deduce the next few as well, for very short oligonucleotides are produced by the cleavages near the 5′ end. In practice, this is rarely a problem, for most DNAs occur in nature as double-strand molecules with oppositely oriented complementary sequences. The sequence near the 5′ end of one strand can be deduced from the 3′ end of the complementary strand. Furthermore, when long DNA molecules are being sequenced, the experimenter will always break them, by using different restriction nucleases, into a series of overlapping fragments. Thus, residues lying near the end of one fragment will lie in the interiors of other fragments. Also note that there is some "crossover" in the reactions. The A reaction gives some G cleavage, and the T reaction gives some C cleavage. However, the extra bands are usually faint and will not cause confusion.

Simple as the Maxam–Gilbert method may be, it is rapidly being supplanted by other methods. Most important of these is the dideoxy method, which is described in Chapter 25. We will not go into this technique here, for in its most useful form it involves rather sophisticated molecular biological methods. Suffice it to say that the dideoxy method is readily adaptable to automation and is so efficient that a skilled practitioner can sequence several thousand bases a week.

REFERENCE

Maxam, A. M., and W. Gilbert (1977) *Proc. Natl. Acad. Sci. USA* 74:560–564.

4

TOOLS OF BIOCHEMISTRY 4

Chemical Synthesis of Oligonucleotides

Modern molecular biology has been greatly stimulated by the new ability to synthesize, in the laboratory, oligonucleotides of defined sequence. These have been used in x-ray studies of DNA structure, in gene modification, and as "primers" in DNA sequence determination (see Chapter 25; Dideoxy Method). Several methods for accomplishing such syntheses now exist, some capable of preparing pure polynucleotides over 100 bases in length. We shall describe one such technique, which can be employed in automated **DNA synthesizers**, or "gene machines."

First, several requirements must be met to obtain a pure product in high yields:

1. The residues must be added one at a time, and the reactions must be very efficient. If even a few growing chains in the reaction mixture fail to add the expected residue at each step, an impure product will result.

2. Conditions must be found to make the reactions thermodynamically favorable. This can be accomplished by using nonaqueous solvents and "activated," highly reactive forms of the nucleotides. The reactions can be facilitated by the use of condensing agents to remove water from the reaction.

3. To avoid unwanted side reactions, all potentially reactive groups on the molecules, except for those where reaction is desired, must be chemically blocked.

Suppose we wish to synthesize a deoxyribonucleic acid with a specific sequence, for example,

ATTGACTTAGCC

Figure T4.1 illustrates one method. The monomers used are *phosphoramidites*, highly reactive molecules with *trivalent* phosphorus.

DMTr—O—CH_2 Base*

Base, with blocked NH_2 groups on A, C, G

Dimethoxytrityl group blocks 5′ position

P—OCH_3

$(CH_3)_2CH—\overset{+}{N}H—CH(CH_3)_2$

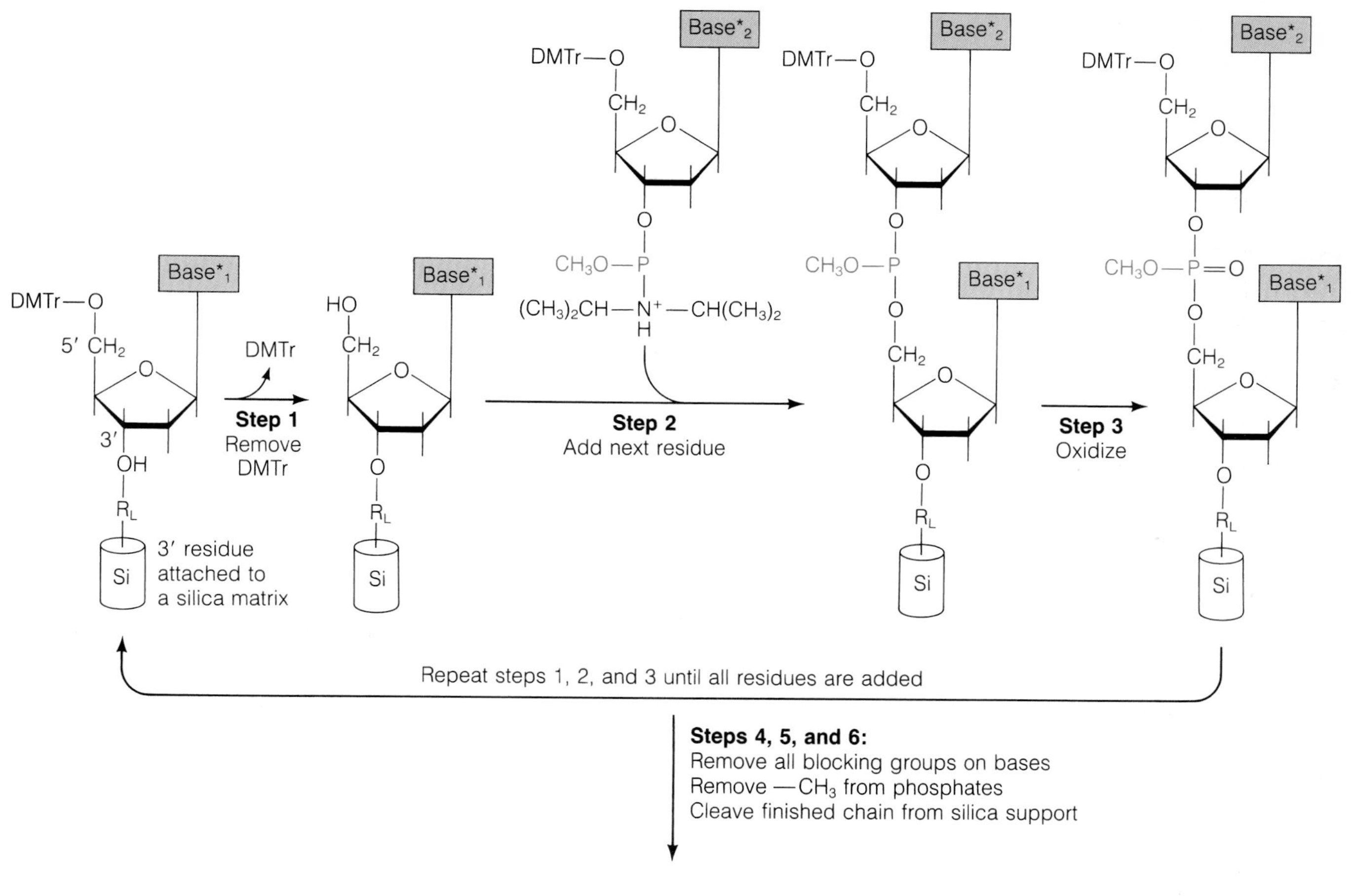

5′ 3′

Oligonucleotide chain

*Reactive groups on all bases are blocked by chemical reagents

$R_L = —\underset{\|\atop O}{C}—(CH_2)_2—\underset{\|\atop O}{C}—NH—(CH_2)_3—O—Si$

Figure T4.1
Solid-phase synthesis of oligonucleotides by the phosphoramidite method.

Note that potentially reactive groups are blocked so that side reactions will not occur. The structures of commonly used blocking groups are

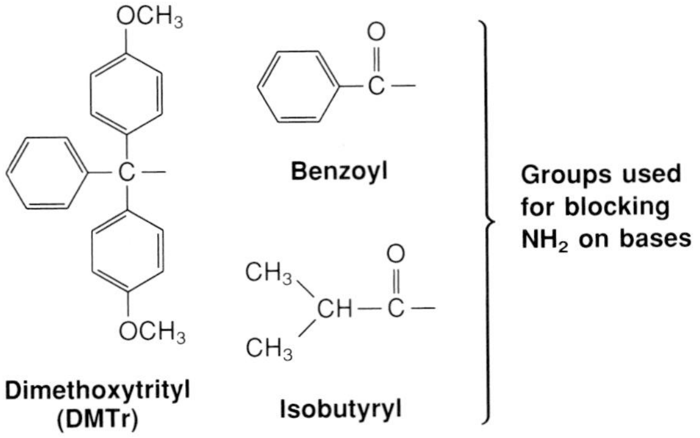

OCH_3

C—

OCH_3

Dimethoxytrityl (DMTr)

Benzoyl

CH_3 CH—C— CH_3

Isobutyryl

Groups used for blocking NH_2 on bases

As shown in Figure T4.1, the nucleotide selected to be at the 3′ end of the chain is coupled through its 3′ hydroxyl to a solid silica support. In the first step of the synthesis the DMTr group is cleaved off and the next monomer added in a protonated form. To stabilize the product, the trivalent phosphorus is oxidized to the pentavalent state, forming a *phosphotriester*. This process is repeated, in a wholly automated fashion, until the entire sequence is completed. Finally, the blocking groups are removed and the chain is hydrolyzed from the support.

The necessity for high reaction yields at each step can be seen from the following calculation. In the example above, 11 addition steps are required. If the yield was 90% at each step, the final yield of the desired product would be only 31%. In practice, stepwise yields of about 98% can be routinely accomplished, which would give an 80% final yield.

REFERENCES

Beaucage, S. L., and M. H. Caruthers (1981) *Tetrahedron Lett.* 22:1859–1862.

Mattencci, M. D., and M. H. Caruthers (1981) *J. Am. Chem. Soc.* 103:3185–3191.

5 TOOLS OF BIOCHEMISTRY 5

An Introduction to X-Ray Diffraction

Only a few decades ago, virtually nothing was known concerning the three-dimensional structures of nucleic acids, proteins, and polysaccharides. Today, largely as a result of the technique of x-ray diffraction, many of these molecules are understood at a level of detail that would have astounded the biochemists of 1950. The method is complicated and it is only possible to give here a brief introduction, describing what is measured and what can be obtained.

When radiation of any kind passes through a regular, repeating structure, diffraction is observed. This simply means that radiation scattered by the repeating elements in the structure shows positive reinforcement of the scattered waves in certain specific directions. A simple example is given in Figure T5.1, which shows radiation being scattered from a row of equally spaced atoms. Only in certain directions will the scattered waves be in phase and therefore reinforce one another. In all other directions they will be out of phase and destructively interfere with one another. Thus, a **diffraction pattern** is generated. For the diffraction pattern to be sharp, it is essential that the wavelength of the radiation used be somewhat shorter than the regular spacing between the elements of the structure. This is why x-rays are

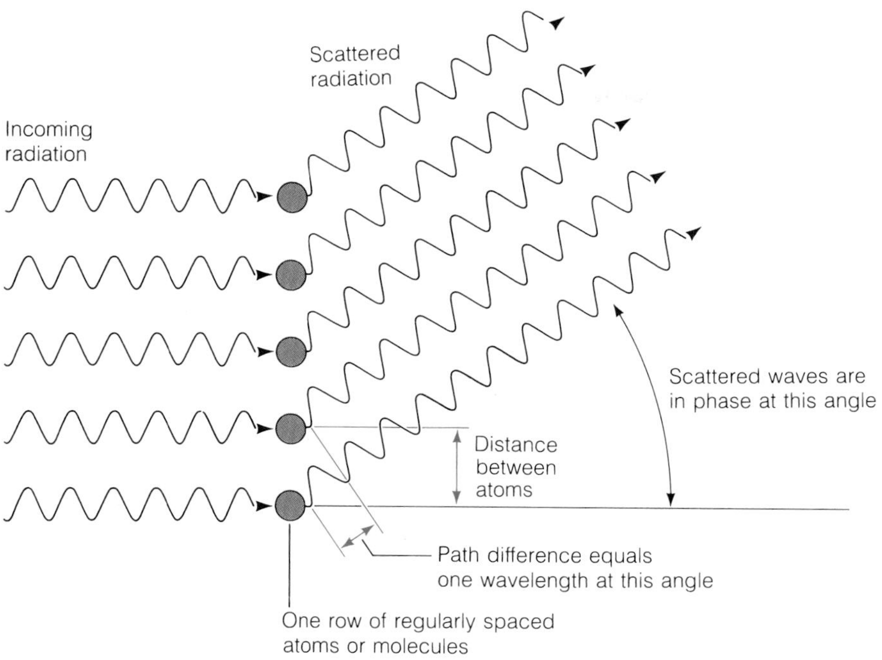

Figure T5.1
Diffraction from a very simple structure—a row of atoms or molecules.

used in studying molecules, for x-rays typically have a wavelength of only a few tenths of a nanometer. If the regular spacing in the object being studied is large (as in a window screen), we can observe exactly the same phenomenon with visible light, which has a wavelength thousands of times longer than x-rays. We will find that a point source, seen through a screen, gives a rectangular pattern of spots.

The rule relating the periodic spacings in object and diffraction pattern is simple: short spacings in the periodic structure correspond to large spacings in the diffraction pattern, and vice versa. In addition, by determining the relative intensities of different spots, we can tell how matter is distributed within each repeat of the structure.

Fiber Diffraction

We consider first the diffraction from helical molecules, aligned approximately parallel to the axis of a stretched fiber. A helical molecule, like the one shown schematically in Figure T5.2, is characterized by certain parameters:

> The *repeat* (c) of the helix is the distance parallel to the axis in which the structure exactly repeats itself. The repeat will contain some integral number (m) of polymer residues. In Figure T5.2, $m = 4$.
>
> The *pitch* (p) of the helix is the distance parallel to the helix axis in which the helix makes one turn. If there is an integral number of residues per turn (as here), the pitch and repeat will be equal.
>
> The *rise* (h) of the helix is the distance parallel to the axis from one residue to the next. Obviously, $h = c/m$.

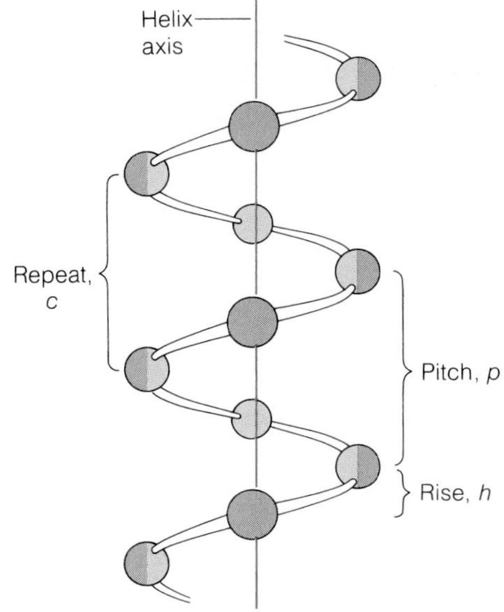

Figure T5.2
A simple helical molecule.

Suppose we wish to investigate a polymer with the helical structure shown in Figure T5.2. A fiber is pulled from a concentrated solution of the polymer. Stretching the fiber further will produce approximate alignment of the long helical molecules with the fiber axis. The fiber is then placed in an x-ray beam and a photographic film positioned behind it, as shown in Figure T5.3a. The diffraction pattern, which consists of spots or short arcs, will look like that in Figure T5.3b. It can be

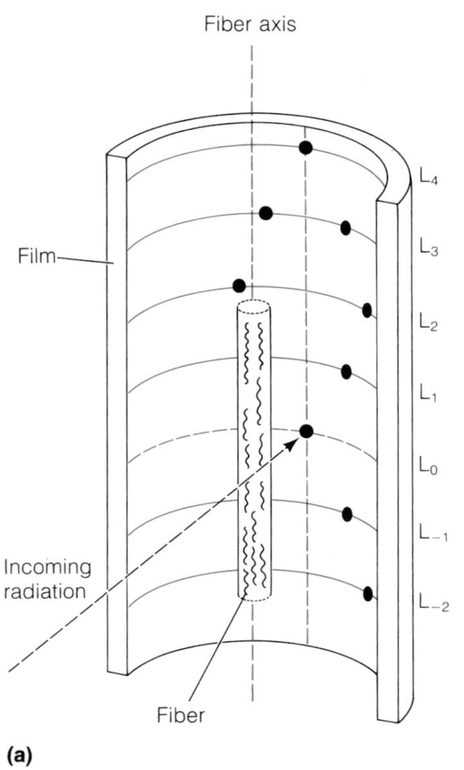

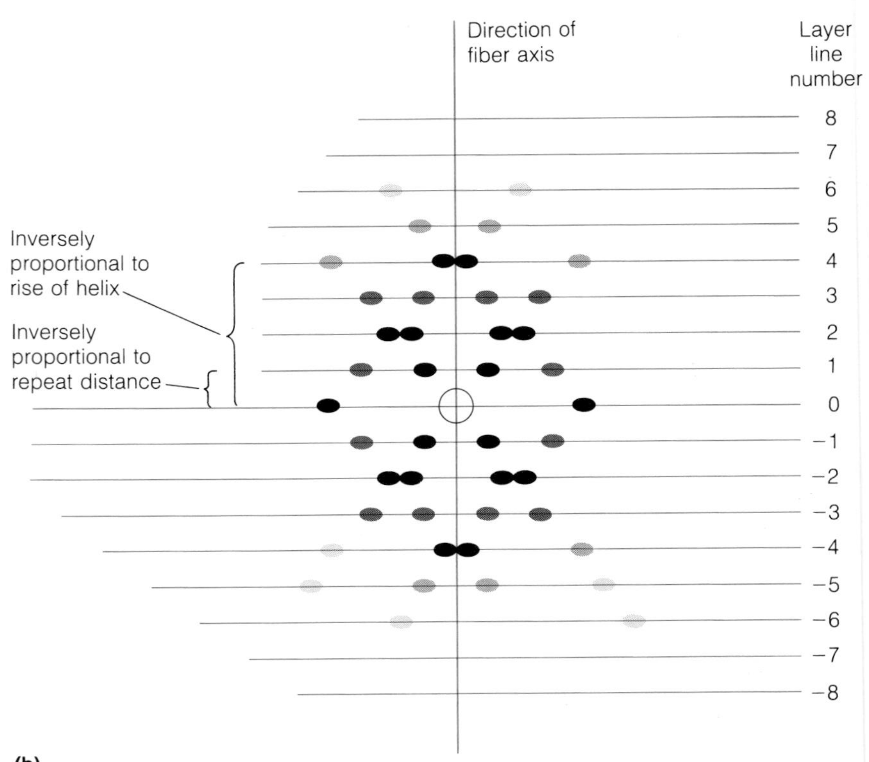

Figure T5.3
Diffraction from fibers: (**a**) a fiber in an x-ray beam; (**b**) the diffraction pattern.

read as follows: according to the mathematics of diffraction theory, a helix always gives rise to this kind of cross-shaped pattern. Therefore, we know we are dealing with a helical structure. The spots all lie on lines perpendicular to the fiber axis; these are called **layer lines.** The spacing between these lines is inversely proportional to the repeat of the helix, c, which in this case equals the pitch. Note that the cross pattern repeats itself on every fourth layer line. This tells us that there are exactly 4.0 residues per turn in the helix. Thus, the rise in the helix is $c/4$.

The information above is given directly by the pattern. To find out exactly how all of the atoms in each residue are arranged in each repeat, a more detailed analysis is necessary. Usually, a model is made using the correct repeat, pitch, and rise. Model making is simplified because we know approximate bond lengths and the angles between many chemical bonds. The model must also be inspected to see that no two atoms approach closer than their van der Waals radii. From such a model, the intensities of the various spots can be predicted. These are compared with the observed intensities, and the model is readjusted until a best fit is obtained. The initial determination of the structure of DNA was done in just this way. As you can see from Figure 4.9, real fiber diffraction patterns are not as neat as the idealized example, mainly because of incomplete alignment of the molecules.

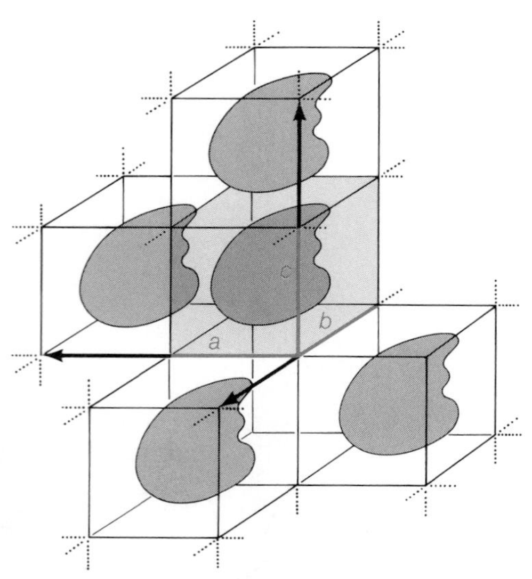

Figure T5.4
Schematic drawing of a molecular crystal.

Crystal Diffraction

To study molecular crystals, as are formed by small oligonucleotides, molecules like tRNA, and globular proteins, the experimenter faces a rather different problem and proceeds in a quite different way. A schematic drawing of such a crystal is shown in Figure T5.4. The repeating unit is now the **unit cell,** which may contain one, two, or more molecules. The unit cell may be thought of as the basic building block of the crystal. Repetition of the unit cell in three dimensions (marked by arrows on the figure) creates the whole crystal. A simple two-dimensional analog of the crystal unit cell is the repeating pattern in wallpaper. No matter how random a wallpaper pattern may seem, if you stare at it you can always find a unit that, by repetition, fills the entire wall.

Just as in fiber diffraction, passing an x-ray beam through a molecular crystal produces a diffraction pattern. The pattern shown in Figure T5.5 was obtained from a crystal of a small DNA. Again, the spacing of the spots allows one to determine the repeating distances in the periodic structure—in this case the x, y, and z dimensions of the unit cell labeled a, b, and c in Figure T5.4. But the important information in crystal diffraction studies is just how the atoms are arranged *within* each unit cell, for that arrangement describes the molecule. Again, this information is contained in the relative intensities of the diffraction spots in a pattern like that shown in Figure T5.5. But in this case more exact information can be extracted than from a fiber diffraction pattern, because the corresponding molecules in each unit cell are of the same shape and are oriented in the same way. In fiber diffraction, the helical molecules may all have their long axes pointed in the same direction, but they are rotated randomly about these axes. This difference in exactness of arrangement can be appreciated by comparing the sharpness of the crystal diffraction pattern shown in Figure T5.5 with the fiber pattern depicted in Figure 4.9.

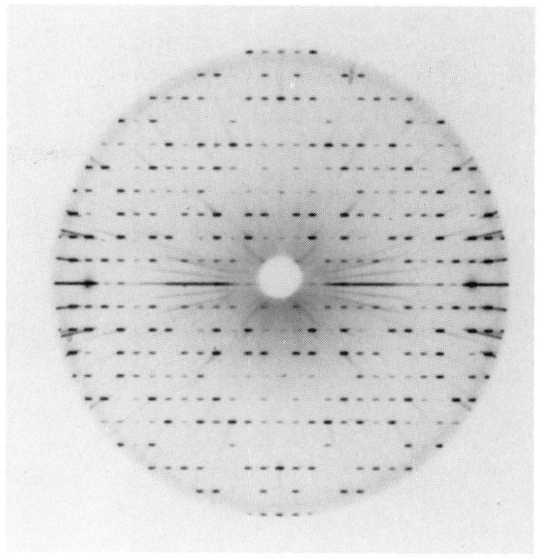

Figure T5.5
The kind of diffraction pattern produced by a molecular crystal of a small DNA.

After obtaining the diffraction pattern from a molecular crystal, the experimenter measures the intensities of a large number of the spots. If the molecule being studied is a small one, it is possible to proceed in much the same manner as with fiber patterns. A structure is guessed at, and expected intensities are calculated and compared with the observed intensities. The structure is refined until the relative intensities of all spots are correctly predicted. However, such a procedure won't work with a molecule as complex as the tRNA shown in Figure 4.22—there is simply no way to guess such a structure.

Why not proceed directly from spot intensities to the structure? The difficulty is that some of the information contained in the spot intensities is hidden. To greatly simplify a complex problem, we may say that it is as if the quantities the experimenter needed to deduce the structure (which are called **structure factors**) were the square roots of the intensities.* If the intensity has a value of, say, 25, the investigator knows that the number needed is +5 or −5. But which? This is the essence of the *phase problem,* which prevented progress in large-molecule crystallography for many years. One way of solving the problem was discovered in the early 1950s. Suppose a heavy metal atom can be introduced into some point in the molecule in such a way that the molecule and crystal are otherwise unchanged. This is called an **isomorphous replacement.** Now suppose the heavy metal contributes a value of +2 to the structure factor for the spot we were discussing above. If the original value was +5, its new value is +7 and its square is 49. If the original value was −5, the value now becomes −3 and its square is 9. The investigator takes a diffraction photograph of the crystal with the heavy metal inserted. If the new crystal has an intensity for this spot of 9, the original structure factor must have been −5, not +5. Although an oversimplification, this example gives the essence of the problem. Usually multiple isomorphous replacements are necessary to determine the phases of the structure factors.

Given structure factors for all of the spots, the investigator can calculate the positions of all atoms in the unit cell. What is actually calculated is an **electron density** distribution (Figure T5.6), but this amounts to the same thing, for regions of high electron density are where the atoms are. In this particular view, we are looking at a "slice" through the three-dimensional electron density distribution.

It is now appropriate to review the steps that must be taken to determine the three-dimensional structure of a macromolecule from crystal diffraction studies:

1. Obtain satisfactory crystals. This is often the hardest part of the procedure, for the crystals must be of good quality and about 1 mm in minimum dimension. Crystals that are too small will not give sharp diffraction patterns. Getting macromolecules to crystallize well is an art.

2. Record the diffraction pattern from the crystal, and measure the intensities of many of the spots.

3. Find some way to make isomorphous replacements in the molecule; usually two or more are required.

* For the mathematically more sophisticated reader, we note that the structure factors are usually complex numbers and can thus be represented as vectors in the complex plane. What is determined from the intensities are their *amplitudes,* but what is not known are their *phases.*

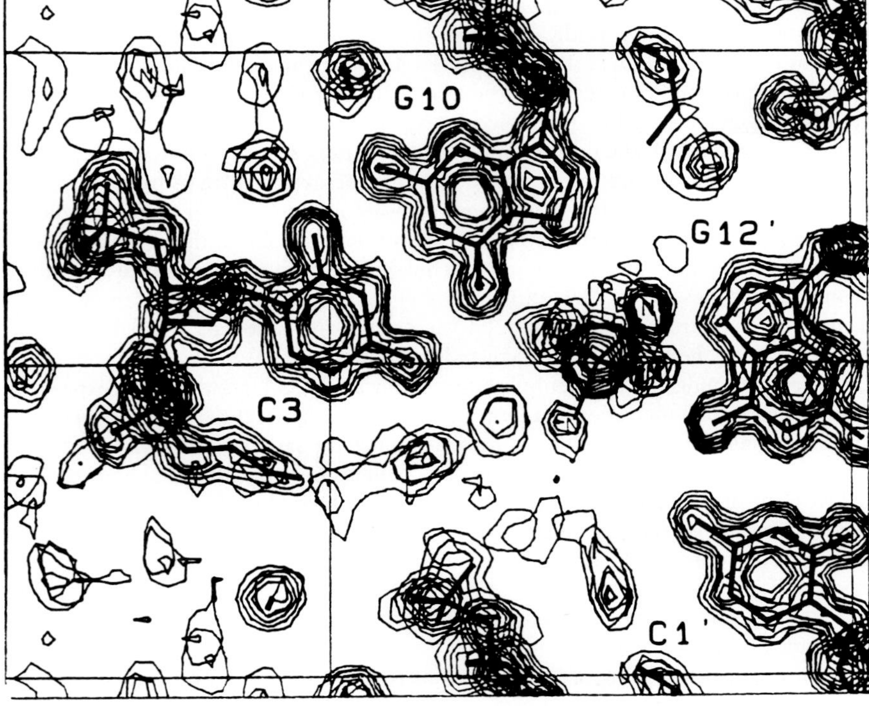

Figure T5.6
Part of electron density map for the DNA in Figure T5.5.

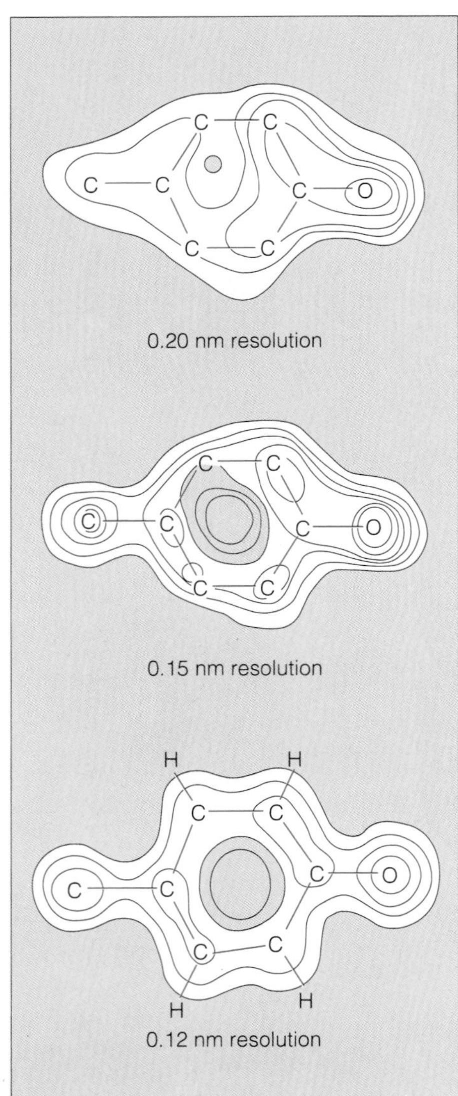

Figure T5.7
Effect of increased resolution on molecular detail observed by x-ray diffraction.

4. Repeat steps 1 and 2 for each isomorphous derivative.

5. Calculate structure factors and the electron density distribution. This is usually done on a large computer.

In most cases, the investigator may first carry through this analysis with a relatively small number of spots. This will give a *low-resolution* structure. If all is going well, more spots will be measured and the calculations refined to give higher resolution. With the best crystals it is now possible to obtain resolutions of about 0.1 nm. This is sufficient to identify individual groups and even some atoms and to show how they interact with one another. The detail in the phenolic ring of a protein side chain revealed at different resolutions is shown in Figure T5.7. The shaded areas show electron density below adjacent contours. Thus, the hole in the center of the ring becomes apparent only at a resolution of about 0.15 nm.

Most of the detailed three-dimensional structures of biological macromolecules shown in this book have been determined by x-ray diffraction studies of crystals. At present, several hundred such structures are known. This has represented an enormous amount of labor in many laboratories, but the results allow us to understand macromolecular function at a level that would have been unbelievable only a short time ago.

REFERENCE

van Holde, K. E. (1985) *Physical Biochemistry*, 2nd ed., Chapter 11. Prentice-Hall, Englewood Cliffs, N.J. A more detailed but not overly sophisticated treatment of x-ray diffraction of biopolymers. This chapter also gives references to more advanced treatments.

CHAPTER 5

Introduction to Proteins: The Primary Level of Protein Structure

We have seen that one class of biopolymers, the nucleic acids, store and transmit the genetic information of the cell. Much of that information is expressed in another class of biopolymers, the **proteins.** Proteins play an enormous variety of roles: Some carry out the transport and storage of small molecules; others make up a large part of the structural framework of cells and tissues. Antibodies are proteins, and so are the blood clotting factors. Perhaps the most important of all classes of proteins are the **enzymes**—the catalysts that promote the enormous variety of reactions that channel metabolism into essential pathways. Altogether, each type of cell has several thousand kinds of proteins.

In keeping with the multiplicity of their functions, proteins are extremely complex molecules. This you can see by a glance at Figure 5.1, which depicts the molecular structure of myoglobin, a protein used to store oxygen in animal tissues. In this and the following two chapters we will analyze in detail the structure of a number of proteins, including myoglobin. We will see that each protein has a logical, functional structure of its own, as well as certain features in common with all other proteins. We begin with a description of the units of chemical structure common to all proteins, the amino acids.

Amino Acids

Structure of the α-Amino Acids

All proteins are polymers, and the monomers that combine to make them are ***α*-amino acids.** A representative α-amino acid, valine, is shown in Figure 5.2a. Note that the amino group is attached to the α-carbon, the carbon next to the carboxyl group. This is why valine is called an α-amino acid. To the α-carbon of every amino acid are also attached a *hydrogen atom* and a *side chain.* Different α-amino acids are distinguished by their different side chains. Thus we can write the general structure for an α-amino acid in the fashion shown to the left in Figure 5.2b, with R representing the side chain. This representation, although chemically correct, ignores the conditions

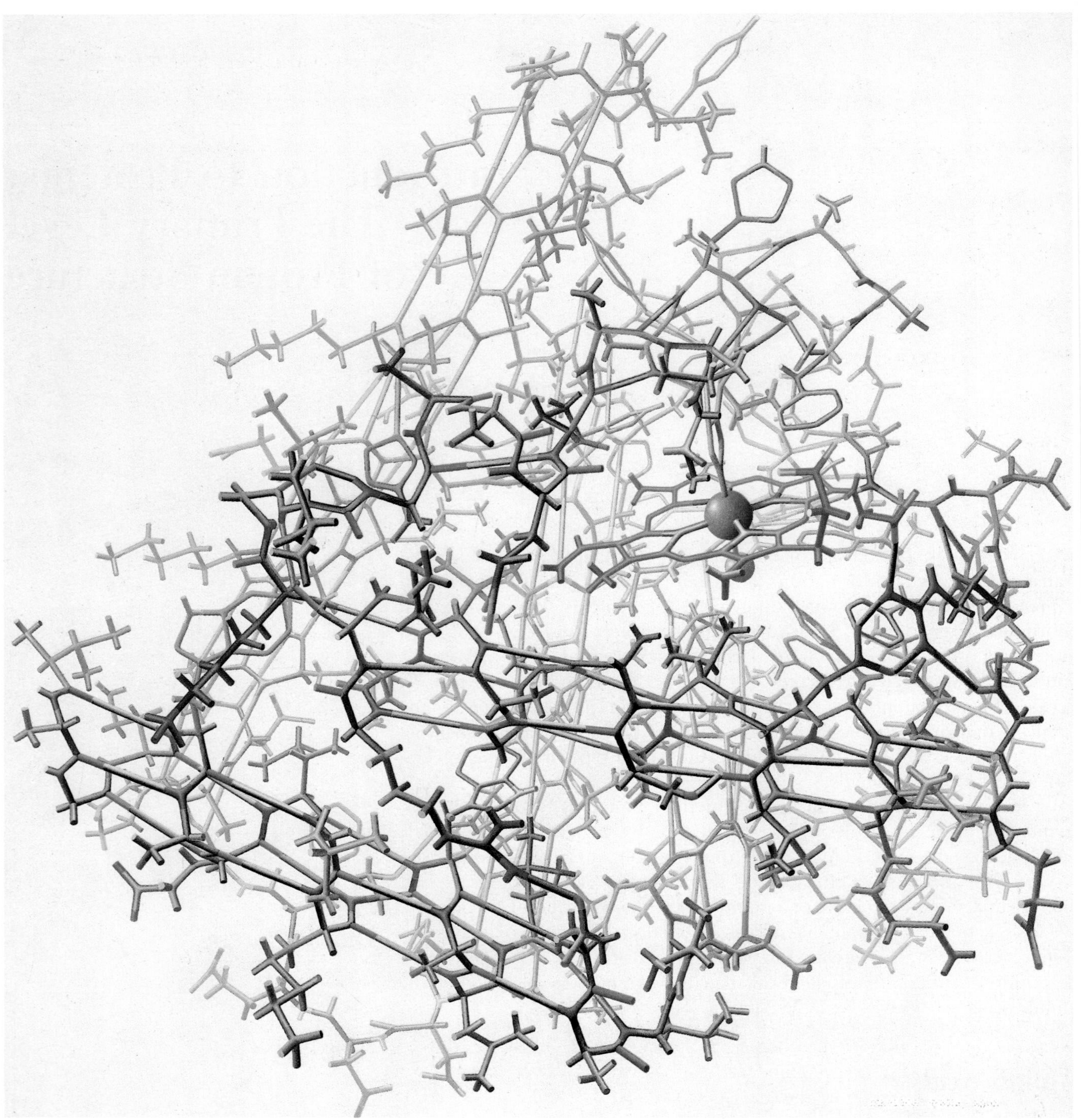

Figure 5.1
The three-dimensional structure of a protein, myoglobin. This painting, by Irving Geis in collaboration with John Kendrew, depicts as a stick model the first protein whose structure was deduced by x-ray diffraction—sperm whale myoglobin. The figure emphasizes both the complexity and specificity of protein structure. From John C. Kendrew, "The Three-Dimensional Structure of a Protein Molecule."

in vivo. As pointed out in Chapter 2, most biochemistry occurs in the physiological pH range near neutrality. The pK_a's of the carboxyl and amino groups of the α-amino acids are about 2 and 10, respectively. Therefore, in the vicinity of neutral pH the carboxylate group will have lost a proton, and the amino group will have picked up a proton, to yield the *zwitterion* form shown to the right in Figure 5.2b. This is the form in which we will customarily write amino acid structures.

Twenty different kinds of amino acids are coded for in the genes and incorporated into proteins. The complete structures of these amino acids are

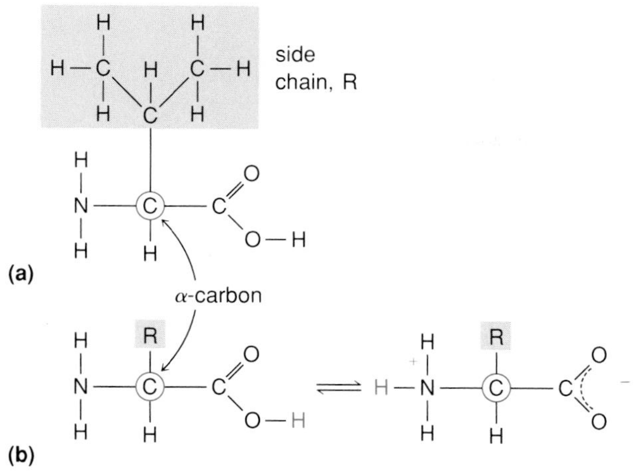

Figure 5.2
α-Amino acid structure. (**a**) Valine, a representative amino acid, has both a carboxyl group and an amino group on the α-carbon, as well as a side chain (R) that gives it its unique properties. (**b**) General structure of an α-amino acid. Amino acids exist most commonly as zwitterions in which the carboxyl group has lost a proton and the amino group has gained one. Note that the negative charge on the carboxyl is delocalized between the two oxygen atoms.

shown in Figure 5.3; other important data are given in Table 5.1. The 20 amino acids contain, in their 20 different side chains, a remarkable collection of diverse chemical groups; these form the vocabulary that allows proteins to exhibit such a great variety of structures and properties.

Note that proline has been set apart in Figure 5.3. Proline is a cyclic amino acid, since the side chain is bonded back to the nitrogen atom, forming a ring. It is sometimes incorrectly referred to by biochemists as an imino acid.

Stereochemistry of the α-Amino Acids

Planar formulas like those shown in Figure 5.3 fail to reveal some important features of amino acid structure. Bonding about the α-carbon is tetrahedral, as you would expect (Figure 5.4a). Therefore, representation of an amino acid in this three-dimensional manner is more realistic. The same feature can be represented more efficiently in the way shown in Figure 5.4b.

Whenever a carbon atom has four different substituents attached to it, forming an asymmetric molecule, the carbon is said to be **chiral,** or a **center of chirality,** or a **stereocenter.** In such cases, two distinguishable **stereoisomers** exist; these are mirror images of one another, as shown in Figure 5.5. The forms of alanine shown in that figure are called the L and D **enantiomers,** respectively.* The L and D enantiomers can be distinguished from one another experimentally by the fact that their solutions rotate the plane of polarized light in opposite directions. Thus, they are sometimes called **optical isomers.** All amino acids except glycine can exist in D and L forms, since in each case the α-carbon is chiral. Glycine is the exception because two of the groups on the α-carbon are the same (H).

The important fact for our purposes is that *all of the amino acids incorporated by organisms into proteins are of the L form.* It is not clear why this should be so, for L-amino acids have no obvious inherent superiority

---

* Those who are familiar with modern organic chemistry will know that there are two systems for distinguishing stereoisomers—the older D, L system and the newer, more comprehensive *R*, *S* system. Both are discussed in more detail in Chapter 8. For now, it should suffice to note that the D- and L-amino acid conformations are named with respect to the absolute configurations of D- and L-glyceraldehyde.

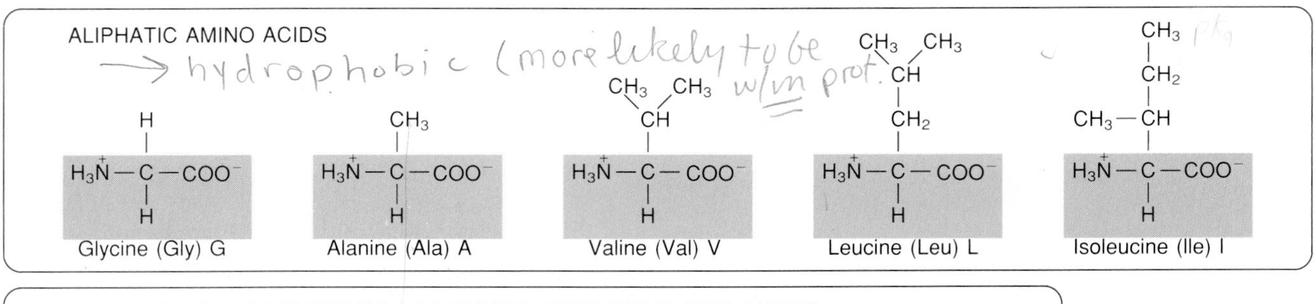

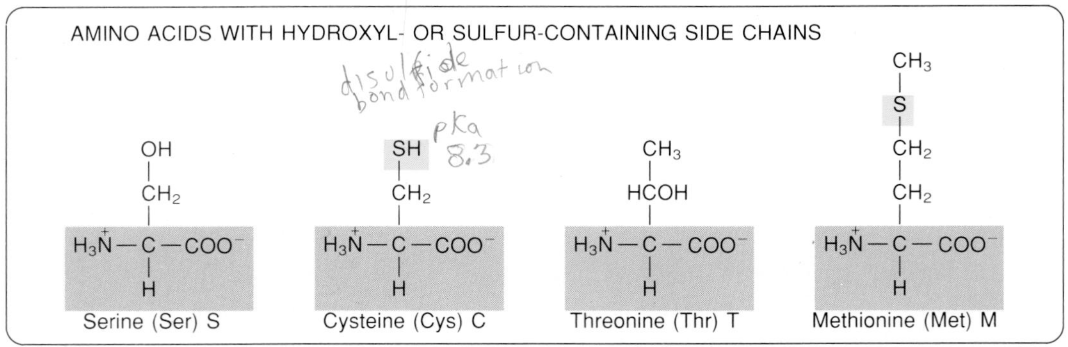

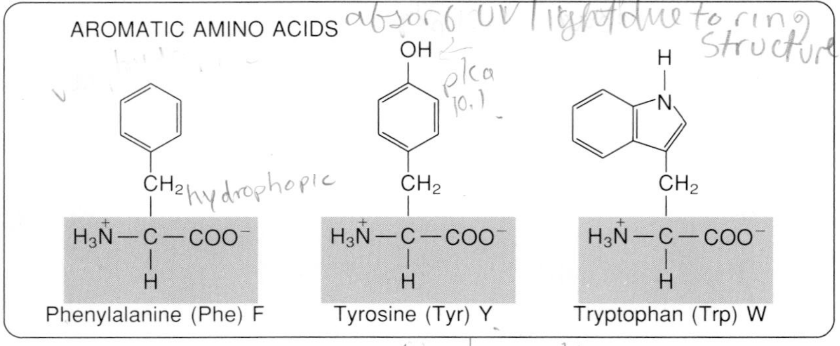

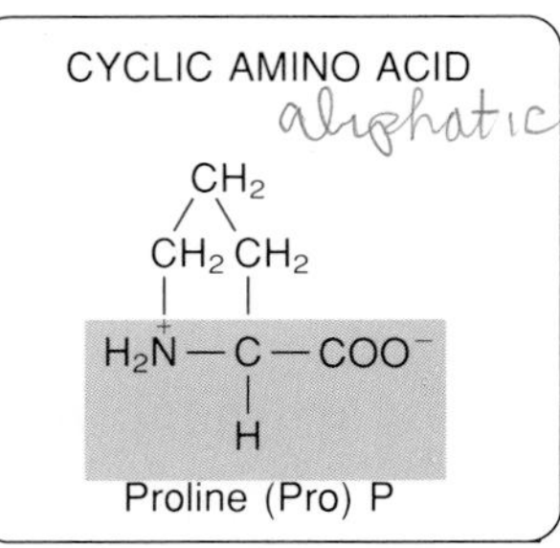

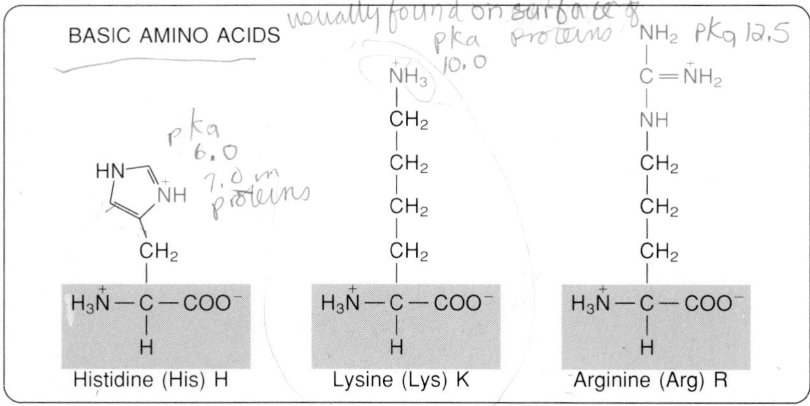

ACIDIC AMINO ACIDS AND THEIR AMIDES

Aspartic acid (Asp) D

Glutamic acid (Glu) E

Asparagine (Asn) N

Glutamine (Gln) Q

Figure 5.3
The 20 amino acids that are incorporated into proteins. These are arranged in the order discussed in the text. Below each, along with its name, is given a three-letter abbreviation (e.g., Gly) and a one-letter abbreviation (G) often used in describing amino acid sequences in proteins.

Table 5.1
Properties of the amino acids found in proteins

Name	pK_a of α-Carboxyl Group	pK_a of α-Amino Group	pK_a of Ionizing Side Chain[a]	Residue[b] Mass (daltons)	Occurrence[c] in Proteins (mol %)
Alanine	2.3	9.7	—	71.08	9.0
Arginine	2.2	9.0	12.5	156.20	4.7
Asparagine	2.0	8.8	—	114.11	4.4
Aspartic acid	2.1	9.8	3.9	115.09	5.5
Cysteine	1.8	10.8	8.3	103.14	2.8
Glutamine	2.2	9.1	—	128.14	3.9
Glutamic acid	2.2	9.7	4.2	129.12	6.2
Glycine	2.3	9.6	—	57.06	7.5
Histidine	1.8	9.2	6.0	137.15	2.1
Isoleucine	2.4	9.7	—	113.17	4.6
Leucine	2.4	9.6	—	113.17	7.5
Lysine	2.2	9.0	10.0	128.18	7.0
Methionine	2.3	9.2	—	131.21	1.7
Phenylalanine	1.8	9.1	—	147.18	3.5
Proline	2.0	10.6	—	97.12	4.6
Serine	2.2	9.2	—	87.08	7.1
Threonine	2.6	10.4	—	101.11	6.0
Tryptophan	2.4	9.4	—	186.21	1.1
Tyrosine	2.2	9.1	10.1	163.18	3.5
Valine	2.3	9.6	—	99.14	6.9

[a] Approximate values found for residues in the *free* amino acids. The α-carboxyl and α-amino groups have pK_as of about 2 and 9–10, respectively, in the free acids.

[b] To obtain the mass of the amino acid itself, add the mass of a mole of water, 18.02 g.

[c] Average for a large number of proteins.

over their D isomers for biological function. Indeed, D-amino acids do exist in nature, and some play important biochemical roles (some examples are given in Table 5.2), but they are never found in proteins. It has been hypothesized that very early in the evolution of life the L isomers were "chosen" by chance, or had some small advantage, and that the machinery for synthesizing and utilizing proteins became fixated on these forms. But we really do not know the answer to this puzzle.

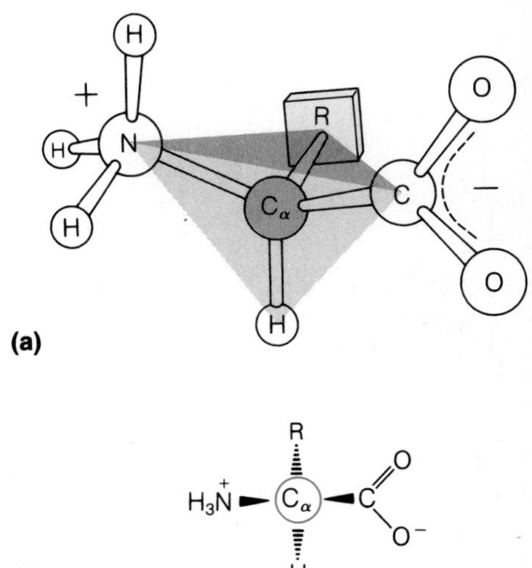

Figure 5.4
Three-dimensional representations of amino acids. Since the α-carbon is chiral, with tetrahedral bonding, an α-amino acid should actually be represented as shown in (**a**). A somewhat more compact rendition is shown in (**b**).

Properties of Amino Acid Side Chains: Classes of α-Amino Acids

These, then, are the units that make up proteins: 20 L-α-amino acids bearing different side chains. If we examine Figure 5.3, it becomes evident that there are several different classes of side chains, distinguished by their dominant chemical features. These features include hydrophobic or hydrophilic character, polar or nonpolar nature, and presence or absence of ionizable groups. Many ways have been proposed to group the amino acids into

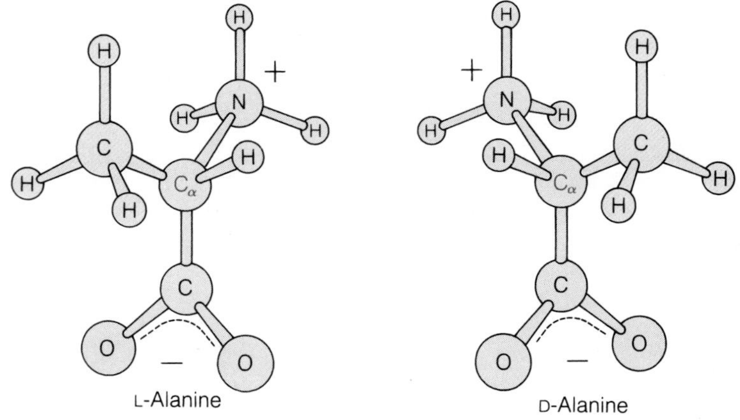

Figure 5.5
Stereoisomers of α-amino acids. L-Alanine and its mirror image isomer D-alanine are shown as ball-and-stick models. The $—CH_3$ side chain is indicated in orange.

Table 5.2
Some biologically important amino acids not found in proteins

Name	Formula	Biochemical Source, Function
β-Alanine	$H_3\overset{+}{N}—CH_2—CH_2—COO^-$	Found in the vitamin pantothenic acid and in some important natural peptides
D-Alanine	$H_3\overset{+}{N}—C(H)(CH_3)—COO^-$	In polypeptides in some bacterial cell walls
γ-Aminobutyric acid GABA	$H_3\overset{+}{N}—CH_2—CH_2—CH_2—COO^-$	Brain, other animal tissues; functions as neurotransmitter
D-Glutamic acid	$H_3\overset{+}{N}—C(H)(CH_2—CH_2—COO^-)—COO^-$	In polypeptides in some bacterial cell walls
Homoserine	$H_3\overset{+}{N}—C(H)(CH_2—CH_2OH)—COO^-$	Many tissues; an intermediate in amino acid metabolism
Ornithine	$H_3\overset{+}{N}—C(H)(CH_2—CH_2—CH_2\overset{+}{N}H_3)—COO^-$	Many tissues; an intermediate in arginine synthesis
Sarcosine	$CH_3—N(H)—CH_2—COO^-$	Many tissues; intermediate in amino acid synthesis
Thyroxine	$H_3\overset{+}{N}—C(H)(COO^-)—CH_2—C_6H_2I_2—O—C_6H_2I_2—OH$	Thyroid gland; is thyroid hormone

classes, but none are wholly satisfactory. We shall discuss them in an order that makes them easy to remember: the arrangement shown in Figure 5.3, which proceeds from the simplest to the more complex.

AMINO ACIDS WITH ALIPHATIC SIDE CHAINS. **Glycine, alanine, valine, leucine,** and **isoleucine** have aliphatic side chains. As we progress from left to right along the top row of Figure 5.3, the R group becomes more extended and more hydrophobic. Isoleucine, for example, has a much greater preference to transfer from water to a hydrocarbon solvent than does glycine. The more hydrophobic amino acids prefer an environment *within* a protein molecule, where they are shielded from water. **Proline,** which is difficult to fit into any category, shares many properties with the amino acids listed above. Although it is a cyclic amino acid, its side chain has primarily aliphatic character.

AMINO ACIDS WITH HYDROXYL- OR SULFUR-CONTAINING SIDE CHAINS. In this category we can place **serine, cysteine, threonine,** and **methionine.** These amino acids, because of their weakly polar side chains, are generally more hydrophilic than their aliphatic analogs, although methionine is fairly hydrophobic. Among the group, cysteine is noteworthy in two respects. First, the side chain can ionize at high pH:

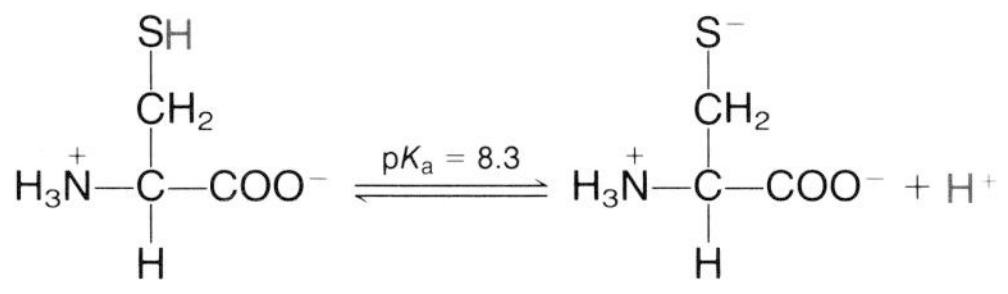

$$H_3\overset{+}{N}-\underset{H}{\overset{CH_2SH}{C}}-COO^- \xrightleftharpoons{pK_a = 8.3} H_3\overset{+}{N}-\underset{H}{\overset{CH_2S^-}{C}}-COO^- + H^+$$

Second, oxidation can occur between pairs of cysteine side chains to form a **disulfide bond.**

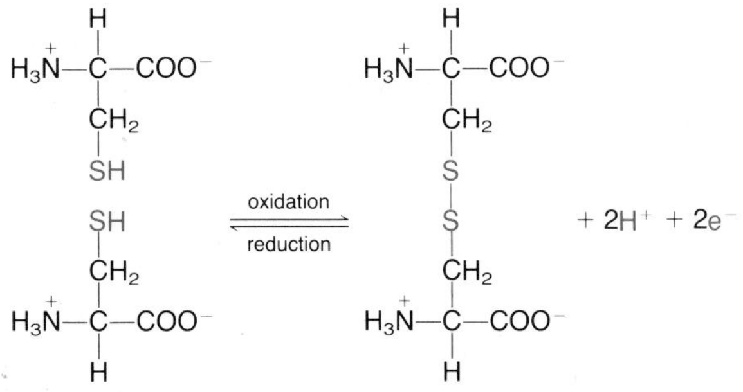

$$2\ H_3\overset{+}{N}-\underset{H}{\overset{}{C}}(CH_2SH)-COO^- \underset{reduction}{\overset{oxidation}{\rightleftharpoons}} H_3\overset{+}{N}-CH(COO^-)-CH_2-S-S-CH_2-CH(COO^-)-\overset{+}{N}H_3 + 2H^+ + 2e^-$$

Cysteine **Cystine**

The product of this oxidation is given the name **cystine.** We do not list it among the 20 amino acids because cystine is always formed by oxidation of two cysteine side chains. As we shall see, such disulfide bonds can form between cysteine residues in proteins and often play an important structural role.

AROMATIC AMINO ACIDS. Three amino acids, **phenylalanine, tyrosine,** and **tryptophan,** carry aromatic side chains. Phenylalanine, together with valine, leucine, and isoleucine, is one of the most hydrophobic amino acids. Tyrosine and tryptophan have some hydrophobic character as well, but it is

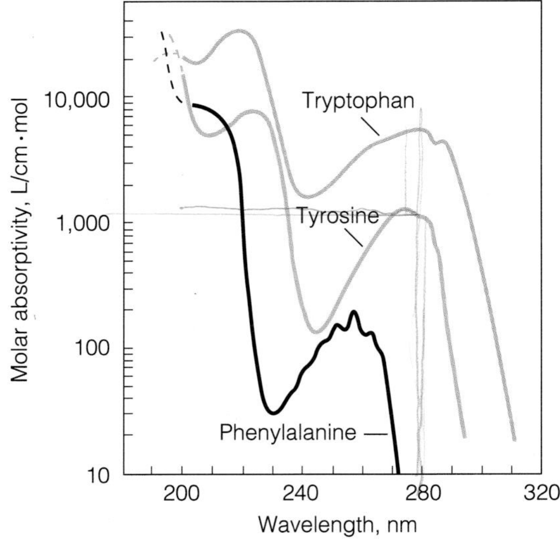

Figure 5.6
Absorption spectra of the aromatic amino acids in the near-ultraviolet region. Note that the absorbance scale is logarithmic. Tryptophan and tyrosine produce most of the UV absorbance by proteins in the region around 280 nm.

tempered by the polar groups in their side chains. In addition, tyrosine can ionize at high pH:

H–O–(C₆H₄)–CH_2–C(H)(H_2N)–COO^- $\underset{}{\overset{pK_a = 10.1}{\rightleftharpoons}}$ O^-–(C₆H₄)–CH_2–C(H)(H_2N)–COO^- + H^+

The aromatic amino acids, like most compounds carrying conjugated rings, exhibit strong absorption of light in the near-ultraviolet region of the spectrum (Figure 5.6). This absorption is frequently used for the analytical detection of proteins.

Table 5.3
Typical ranges observed for pK_a values of groups in proteins

Group Type	Typical pK_a Range[a]
α-Carboxyl	3.5–4.0
Side chain carboxyls of aspartic and glutamic acids	4.0–4.8
Imidazole (histidine)	6.5–7.4
Cysteine (—SH)	8.5–9.0
Phenolic (tyrosine)	9.5–10.5
α-Amino	8.0–9.0
Side chain amino (lysine)	9.8–10.4
Guanidinyl (arginine)	~12

[a] Values outside these ranges are observed. For example, side chain carboxyls have been reported with pK_a values as high as 7.3.

BASIC AMINO ACIDS. **Histidine, lysine,** and **arginine** carry basic groups in their side chains. They are represented in Figure 5.3 in the form that exists at pH values slightly lower than neutrality. As the titration curve for histidine (Figure 5.7a) shows, the imidazole ring in the side chain of the free amino acid loses its proton at about pH 6 (pK_a values for the side chains of free amino acids are given in Table 5.1). When histidine is incorporated into proteins, the pK_a is raised to about 7 (Table 5.3). Because the histidine side chain can exchange protons near physiological pH, it often plays a role in enzymatic catalysis involving proton transfer.

Lysine and arginine are more basic amino acids and, as their pK_a values indicate (Tables 5.1 and 5.3), their side chains are positively charged under physiological conditions.

The basic amino acids are strongly polar and, as a consequence, are usually found on the exterior surfaces of proteins, where they can be hydrated by the surrounding aqueous environment.

ACIDIC AMINO ACIDS AND THEIR AMIDES. **Aspartic acid** and **glutamic acid** are the only amino acids that carry negative charges at pH 7; they are

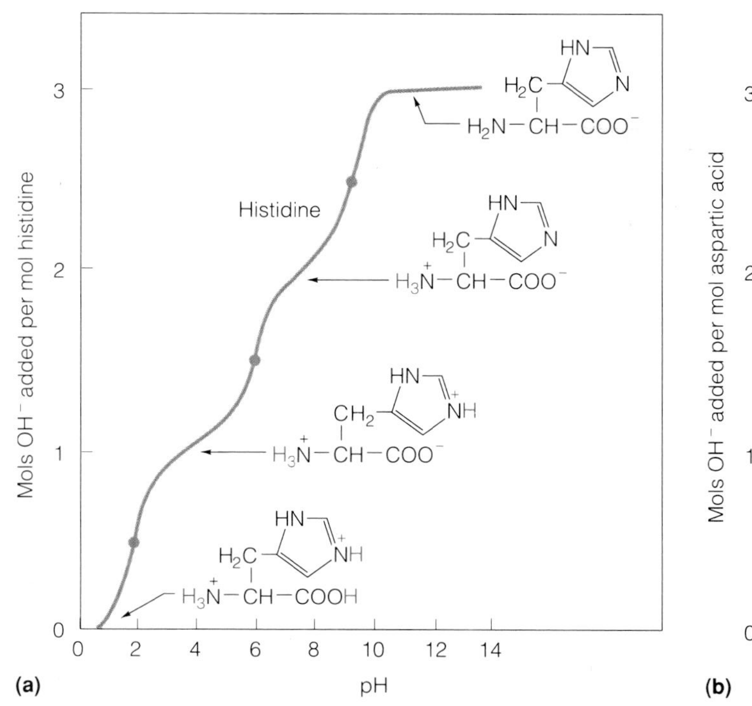

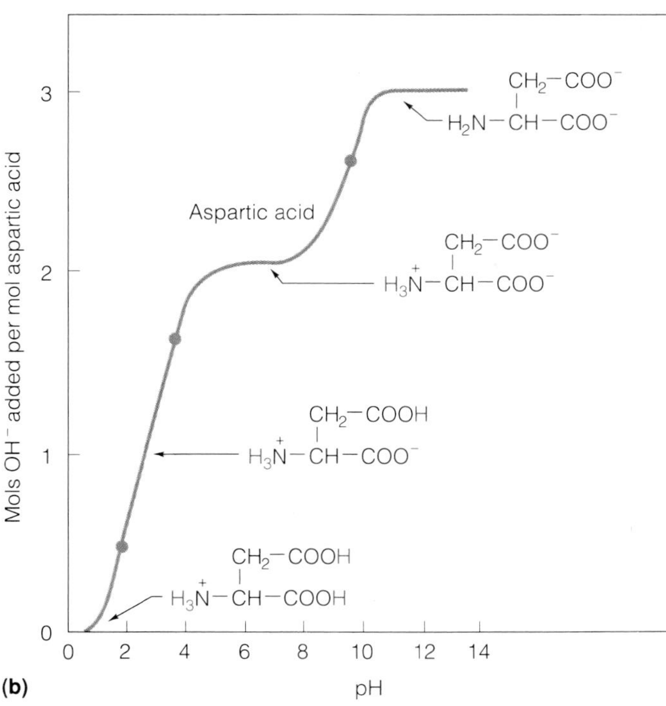

Figure 5.7
Titration curves of amino acids with ionizing side chains. (**a**) Histidine. (**b**) Aspartic acid. Dots correspond to pK_a values, and forms predominating at different pH values are indicated. Labile hydrogens are shown in red.

depicted in the anionic forms in Figure 5.3. The negative charge will, of course, be retained at higher pH; only by decreasing the pH below the pK_a will the carboxyl groups be neutralized. A titration curve is shown in Figure 5.7b, and pK_a values are given in Table 5.1. These pK_a values are so low, even when the amino acids are incorporated into proteins, that the negative charge will be retained under physiological conditions. Hence, these are often referred to as **aspartate** and **glutamate.**

Companions to aspartic and glutamic acids are their amides, **asparagine** and **glutamine.** Unlike their acidic analogs, asparagine and glutamine have uncharged side chains, though they are decidedly polar. Like the basic and acidic amino acids, they are definitely hydrophilic and tend to be on the surface of a protein molecule, in contact with the surrounding water.

Modified Amino Acids

We have now considered all the amino acids that are coded for in DNA and incorporated directly into proteins. The repertoire of amino acids in proteins is embellished, however, by the fact that certain amino acids may occasionally become chemically modified after they are assembled into proteins. Several such *modified amino acids* are depicted below:

o-Phosphoserine **4-Hydroxyproline** **δ-Hydroxylysine** **γ-Carboxyglutamic acid**

We shall not consider these again until we encounter specific proteins in which such modification has occurred.

The amino acids found in proteins are by no means the only ones to occur in living organisms. Many other amino acids play important roles in metabolism. A partial list is given in Table 5.2; note that not all of them are α-amino acids. We shall encounter all of these amino acids again in later chapters.

Peptides and the Peptide Bond

Peptides

Amino acids can be covalently linked together by formation of an **amide bond** between α-amino and α-carboxyl groups. The products formed by such a linkage are called **peptides.** An example of such a reaction is shown in Figure 5.8. The product in this case is a **dipeptide,** since two amino acids have been combined. The reaction can be viewed as simple elimination of a water molecule between the carboxyl group of one amino acid and the amino group of the other. Note that the reaction in Figure 5.8 still leaves an $H_3\overset{+}{N}$— group available on one end of the dipeptide and an unreacted carboxyl group on the other. Thus, the reaction could in principle be continued by adding glutamic acid to one end and lysine to the other to yield the **tetrapeptide** shown in Figure 5.9. As each amino acid is added to the chain,

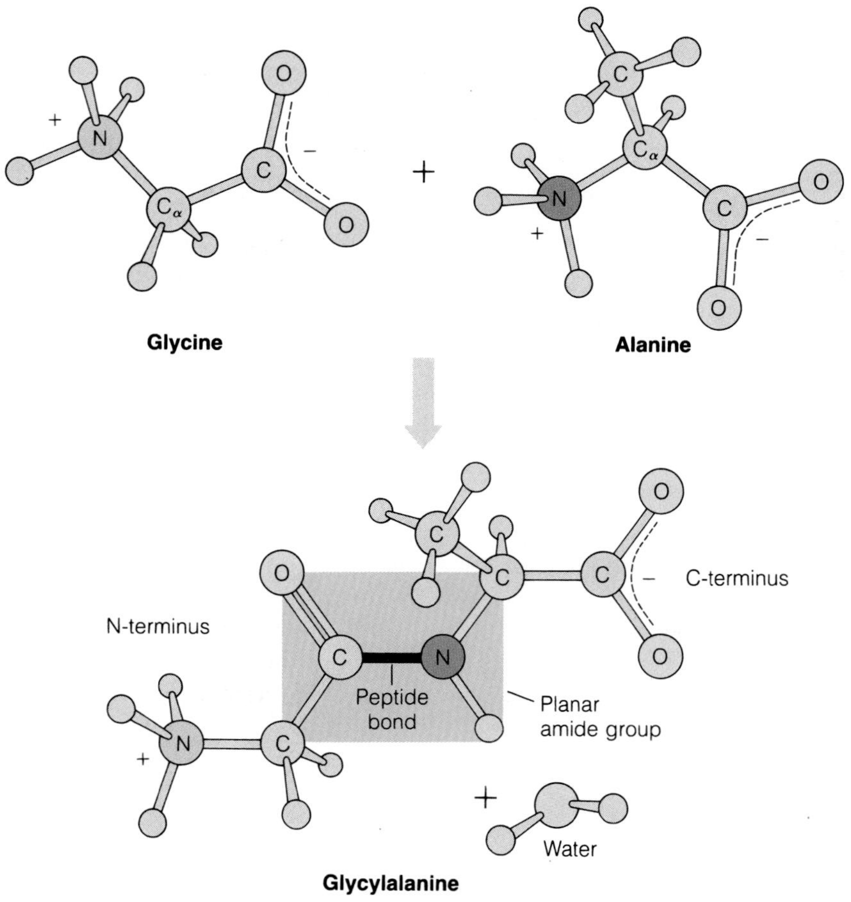

Figure 5.8
Formation of a peptide. When two amino acids join, a **peptide bond** forms. Here the dipeptide glycylalanine (Gly–Ala) is formed by removal of a water molecule when glycine is linked to alanine. The resulting planar amide group is shown in green.

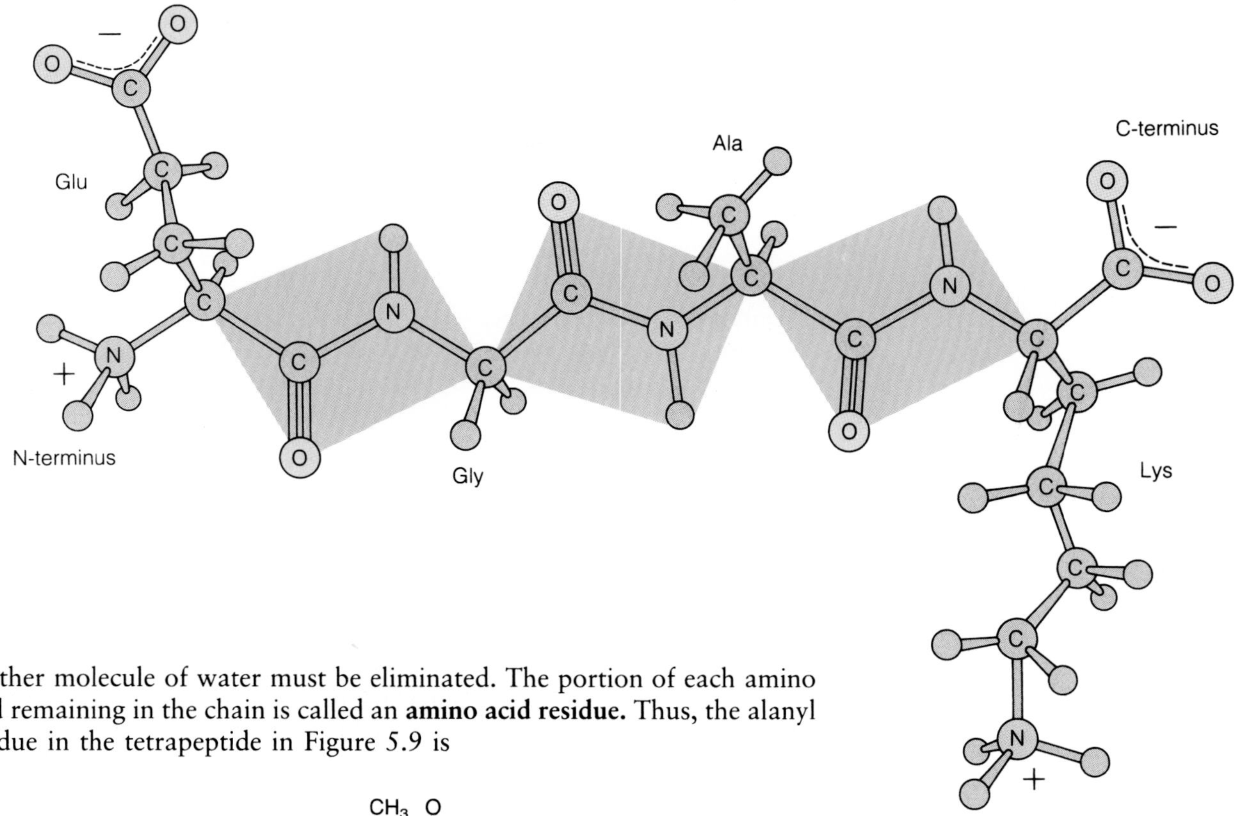

another molecule of water must be eliminated. The portion of each amino acid remaining in the chain is called an **amino acid residue.** Thus, the alanyl residue in the tetrapeptide in Figure 5.9 is

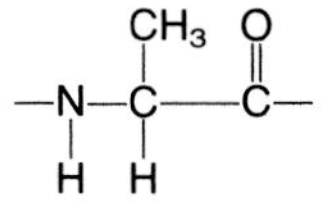

$$-\underset{|}{\underset{\text{H}}{\text{N}}}-\overset{\text{CH}_3}{\underset{\text{H}}{\text{C}}}-\overset{\text{O}}{\overset{\|}{\text{C}}}-$$

Chains containing only a few amino acid residues (like a tetrapeptide) are collectively referred to as **oligopeptides.** If the chain is very long, it is called a **polypeptide.** Most oligopeptides and polypeptides still retain an unreacted amino group at one end (called the **amino terminus** or **N-terminus**) and an unreacted carboxyl at the other end (the **carboxyl terminus** or **C-terminus**). Exceptions are certain small **cyclic oligopeptides,** in which the N- and C-termini have been linked. In addition, many proteins have N-termini blocked by *N*-formyl or *N*-acetyl groups, and a few have C-terminal carboxyls that have been modified to amides (Figure 5.10).

All proteins are polypeptides. This is why understanding the nature of polypeptides and the peptide bond is so important a part of biochemistry.

Polypeptides as Polyampholytes

In addition to the free amino group at the N-terminus and the free carboxyl group at the C-terminus, polypeptides usually contain some amino acids that have ionizable groups on their side chains. These various groups have a wide range of pK_a values, as shown in Table 5.3, but are all weakly acidic or basic groups. Thus, polypeptides are excellent examples of the polyampholytes described in Chapter 2.

The kind of behavior that is seen as one titrates a polypeptide is exemplified by the tetrapeptide (Glu–Gly–Ala–Lys) in Figure 5.9. We can imagine starting with the tetrapeptide in a very acidic solution, say pH 0. At this pH, which is below the pK_a of any of the groups present, all of the ionizable residues will be in their protonated forms:

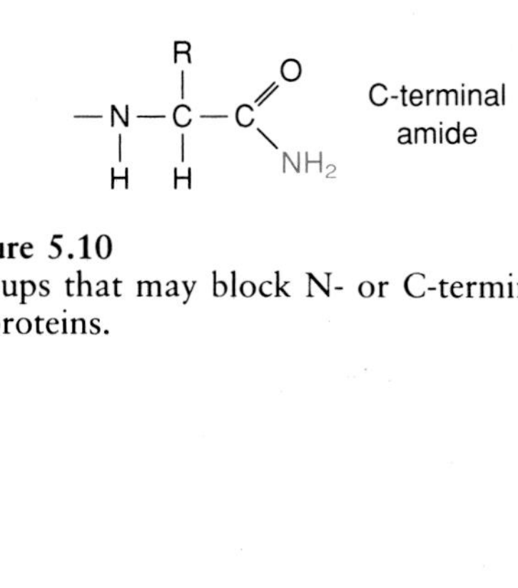

Figure 5.9
A tetrapeptide, Glu–Gly–Ala–Lys. We have imagined that glutamic acid and lysine are added to the amino and carboxyl ends, respectively, of the dipeptide (Gly–Ala) shown in Figure 5.8.

Figure 5.10
Groups that may block N- or C-termini in proteins.

$H_3\overset{+}{N}$–CH(CH_2CH_2COOH)–CO–NH–CH_2–CO–NH–CH(CH_3)–CO–NH–CH(COOH)–$H_2C-CH_2-CH_2-CH_2-\overset{+}{N}H_3$

Glu : Gly : Ala : Lys

All amino groups are positively charged, and each carboxyl has zero charge. Therefore, the whole molecule has a charge of +2 at this pH. If we now imagine removing protons from the solution (by titrating with NaOH, for example), the various groups will lose protons at pH values in the vicinity of their pK_a values.

The progress of the titration of the tetrapeptide is shown in Figure 5.11. As protons are removed, raising the pH, more groups become deprotonated. The positive charge decreases, passes through zero, and the molecule becomes negatively charged, ultimately reaching a net charge of −2 at very high pH.

As pointed out in Chapter 2, there is one pH at which the net charge on an ampholyte is zero: the *isoelectric point*. Every amino acid, peptide, and protein has an isoelectric point; the pH at which it lies depends on the relative numbers and kinds of acidic and basic groups in the molecule. Note that the molecule still possesses some charged groups at the isoelectric point; it is only the *net* charge that is zero.

We will find these effects of changing pH to be of importance in biochemistry in general and protein chemistry in particular. Sometimes even a small shift in pH will significantly alter the constellation of charges with which a protein molecule faces its environment, and thereby significantly modify its behavior. Solubility of many proteins is minimal at the isoelectric

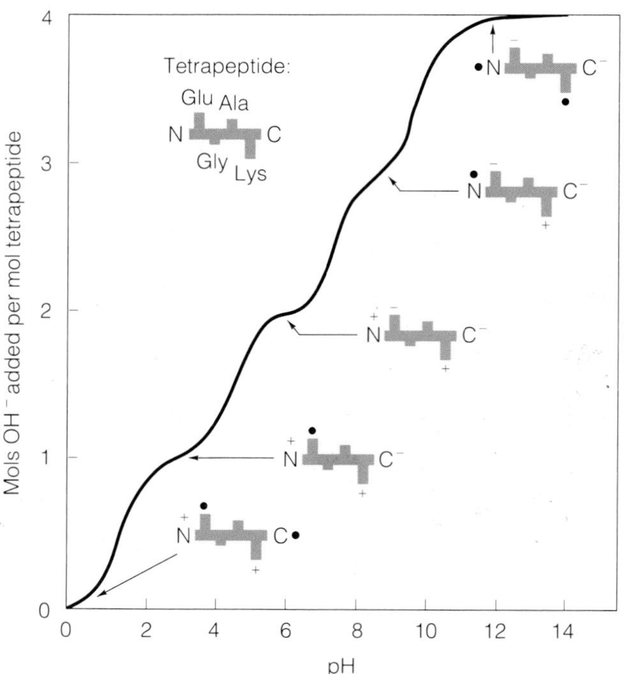

Figure 5.11
Titration of the tetrapeptide shown in Figure 5.9. The major forms present are shown at several pH values (arrows). Groups that may be charged are indicated by +, −, or •, depending on whether they carry a positive, negative, or zero charge, respectively.

point, since the molecules no longer repel one another when their net charge is zero (see Figure 2.23).

The fact that different proteins and oligopeptides have different net charges at a given pH is often used to advantage in their separation, either by electrophoresis (Tools of Biochemistry 2) or by ion-exchange chromatography (Tools of Biochemistry 6).

The Structure of the Peptide Bond

Now let us examine the nature of the bond that has been formed in linking amino acids together. In the dipeptide of Figure 5.8 (Gly–Ala), the shaded portion contains what is called the **peptide bond.** This amide bond, which is found between every pair of residues in a protein, has some properties very important to protein structure. For example, almost invariably, the —C=O and —N—H bonds are parallel, and there is very little twisting about the C—N bond. This is because the peptide bond has a substantial fraction of double-bond character; it can be considered a resonance hybrid of two forms:

$$\mathrm{O{=}C{-}N{-}H} \longleftrightarrow \mathrm{{}^{-}O{-}C{=}\overset{+}{N}{-}H}$$

or

$$\mathrm{O^{\delta-}{\cdots}C{\cdots}N^{\delta+}{-}H}$$

A schematic depiction of the electron density about the peptide bond is shown in Figure 5.12a and bond lengths and angles are given in Figure 5.12b.

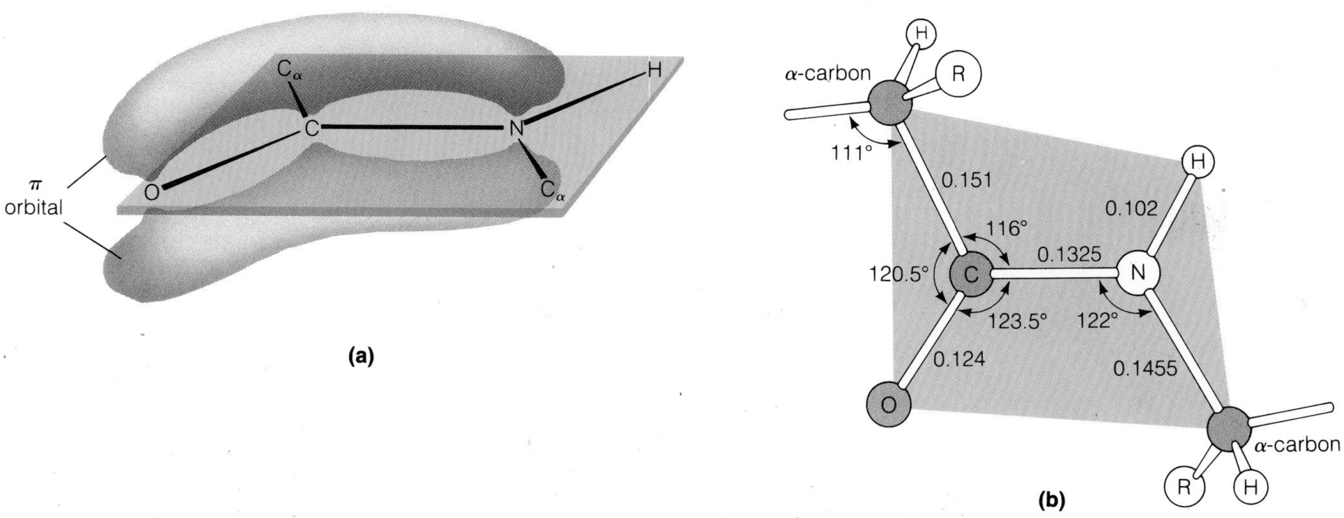

Figure 5.12
Structure of the peptide bond. (**a**) Delocalization of the π-electron orbitals over O—C—N, which accounts for the partial double bond character of the C—N bond. (**b**) Presently accepted values for bond angles and lengths. Bond lengths are in nanometers (nm).

Even though it is planar, the group of atoms could exist about the peptide bond in two possible conformations, trans and cis:

Trans Cis

In fact, the trans form is usually favored, for in the cis conformation the bulky R groups on adjacent α-carbons usually interfere. The major exception is the bond in the sequence X–Pro, where X is any other amino acid. In this the cis configuration is favored.

Stability and Formation of the Peptide Bond

Figure 5.8 and our discussion of it implied that a peptide bond could be formed by the elimination of a water molecule between two amino acids. In fact, in an aqueous environment this is not a favored process. The free energy change for this reaction at room temperature in aqueous solution is

Table 5.4
Some proteolytic enzymes[a]

N-terminal ···—NH—C(R_1)H—CO—▲—NH—C(R_2)H—CO—··· C-terminal

Enzyme	Preferred Site	Source
Trypsin	R_1 = Lys, Arg	From digestive systems of animals, many other sources
Chymotrypsin	R_1 = Tyr, Phe, Leu, Ileu, Val, Trp, and His at high pH	Same as trypsin
Pepsin	R_1 = Phe, Leu, many others	Same as trypsin but confined to stomach, where pH is low
Thrombin	R_1 = Arg	From blood; involved in coagulation
Papain	R_1 = Arg, Lys, Phe-X (CO side of residue next to Phe)	From papaya latex
Bromelain	R_1 = Lys, Ala, Tyr, Gly	From pineapple
Thermolysin	R_2 = same residues as chymotrypsin	From *Bacillus thermoproteolyticus*
Subtilisin	Very little specificity	From various bacilli
Carboxypeptidase A	R_2 = C-terminal amino acid	From digestive systems of animals

[a]Data excerpted from *Enzyme Nomenclature,* Nomenclature Committee of the International Union of Biochemistry, Academic Press, Orlando, Fla., 1984.

about +10 kJ/mol. Therefore, the thermodynamically favored reaction under these conditions is the *hydrolysis* of the peptide bond:

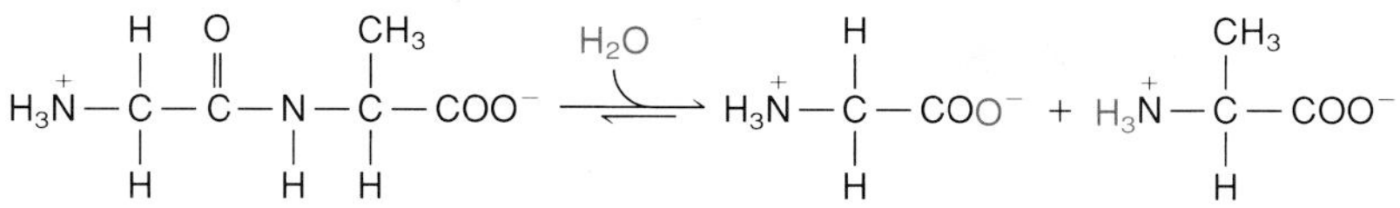

$$H_3\overset{+}{N}-CH_2-\overset{O}{\overset{\|}{C}}-NH-\underset{H}{\overset{CH_3}{\overset{|}{C}}}-COO^- \xrightarrow[\longleftarrow]{H_2O} H_3\overset{+}{N}-CH_2-COO^- + H_3\overset{+}{N}-\underset{H}{\overset{CH_3}{\overset{|}{C}}}-COO^-$$

with equilibrium lying well to the right. Like polynucleotides, polypeptides are metastable, easily hydrolyzing when catalysts are present.

Peptide hydrolysis can be catalyzed in several ways. A general method, which cleaves all peptide bonds, is heating in strong mineral acid (usually 6 M HCl). More specific catalysis is provided by **proteolytic enzymes** or **proteases.** A number of such enzymes, together with their cleavage specificities, are listed in Table 5.4. Some of these enzymes are secreted into the digestive tracts of animals, where they break down proteins for further digestion. Others, such as papain, are found in certain plant tissues. The existence of a battery of such enzymes, with specific cutting sites, is of great utility to the biochemist, for they allow the cleavage of polypeptides in well-defined ways. Another, nonenzymatic, reaction that cleaves a specific peptide bond uses the reagent cyanogen bromide (BrC≡N). Cyanogen bromide cleaves specifically to the carboxyl side of methionine residues (Figure 5.13).

As with polynucleotides, the thermodynamic instability of polypeptides raises the question of how they can be synthesized in the aqueous medium of the cell. You may already have guessed the answer—that coupling of the synthetic reaction to ATP hydrolysis is required. This is correct, but the details are somewhat more complex than in the synthesis of polynucleotides from nucleoside triphosphates. We shall give a brief outline of the

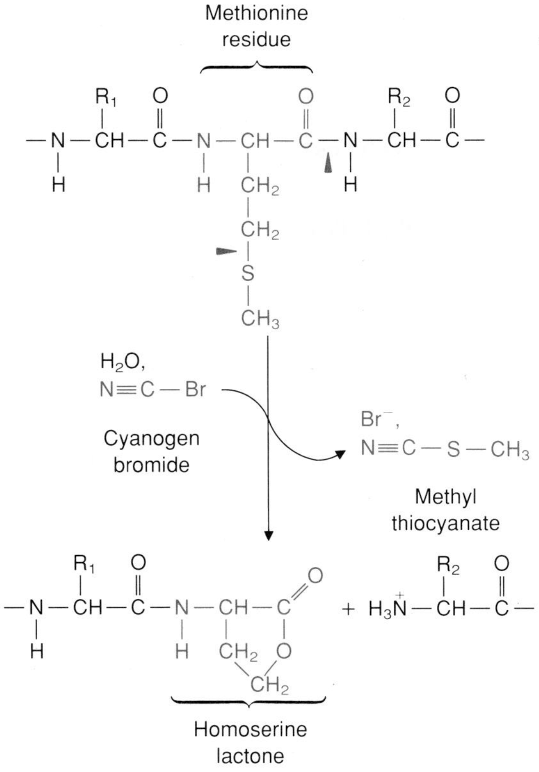

Figure 5.13
The cyanogen bromide reaction. This reaction specifically cleaves the peptide bond to the carboxyl side of methionine in any polypeptide and converts Met to homoserine lactone. The cleavage sites are indicated by ▲.

	NA1	NA2	A1	A2	A3	A4	A5	A6	A7	A8	A9	A10	A11	A12	A13
Number	1	2	3	4	5	6	7	8	9	10	11	12	13	14	15
Human	G	L	S	D	G	E	W	Q	L	V	L	N	V	W	G
Whale	V	L	S	E	G	E	W	Q	L	V	L	H	V	W	A

	B12	B13	B14	B15	B16 *	C1	C2	C3	C4	C5	C6	C7	CD1	CD2	CD3
Number	31	32	33	34	35	36	37	38	39	40	41	42	43	44	45
Human	R	L	F	K	G	H	P	E	T	L	E	K	F	D	K
Whale	R	L	F	K	S	H	P	E	T	L	E	K	F	D	R

	E4	E5	E6	E7	E8	E9	E10	E11	E12	E13	E14	E15	E16	E17	E18
Number	61	62	63	64	65	66	67	68	69	70	71	72	73	74	75
Human	L	K	K	H	G	A	T	V	L	T	A	L	G	G	I
Whale	L	K	K	H	G	V	T	V	L	T	A	L	G	A	I

	F6	F7	F8	F9	FG1	FG2	FG3	FG4	FG5	G1	G2	G3	G4	G5	G6
Number	91	92	93	94	95	96	97	98	99	100	101	102	103	104	105
Human	Q	S	H	A	T	K	H	K	I	P	V	K	Y	L	E
Whale	Q	S	H	A	T	K	H	K	I	P	I	K	Y	L	E

	GH3	GH4	GH5	H1	H2	H3	H4	H5	H6	H7	H8	H9	H10	H11	H12
Number	121	122	123	124	125	126	127	128	129	130	131	132	133	134	135
Human	G	D	F	G	A	D	A	Q	G	A	M	N	K	A	L
Whale	G	N	F	G	A	D	A	Q	G	A	M	N	K	A	L

Figure 5.14
The amino acid sequences of sperm whale myoglobin and human myoglobin. The amino acids are given by their single-letter abbreviations. Identical amino acids are in blue; conservative substitutions (e.g., isoleucine for leucine) are overlaid in green; nonconservative substitutions are in yellow. Of the 153 amino acid residues, 128 or 84% are identical in humans and whales. If we include the 16 conservative substitutions, the two proteins are 94% homologous. Numbers are counting from the N-terminus. The number–letter combinations above the residues indicate regions of secondary structure; they will be discussed in Chapters 6 and 7.

process later in this chapter. First it is appropriate to describe the most important class of polypeptides—the proteins.

Proteins: Polypeptides of Defined Sequence

Proteins are not just polypeptides. They are polypeptides of defined sequence. Every protein has a defined order of amino acid residues. As with the nucleic acids, this sequence is referred to as the *primary structure* of the protein.

Figure 5.14 shows the amino acid sequence of sperm whale **myoglobin,** the protein whose structure we saw in Figure 5.1. Above it is listed the sequence of human myoglobin, the protein that serves the same oxygen storage function in humans. Two points are immediately obvious from examination of these sequences.

First, proteins are *long* polypeptides. Sperm whale myoglobin contains 153 amino acids; so does human myoglobin. Yet these are among the smaller proteins; some proteins have sequences extending for many hundreds or even thousands of amino acid residues.

Second, although the two myoglobin sequences are similar, they are not identical. Their similarity is sufficient for each to serve the same biochemical purpose; therefore we call each a *myoglobin.* But they are not quite the same, for evolution has proceeded for many millions of years since sperm whales and humans had a common ancestor. Proteins evolve, and they evolve by changes in their amino acid sequences.

A14	A15	A16	AB1	B1	B2	B3	B4	B5	B6	B7	B8	B9	B10	B11
16	17	18	19	20	21	22	23	24	25	26	27	28	29	30
K	V	E	A	D	I	P	G	H	G	Q	E	V	L	I
K	V	E	A	D	V	A	G	H	G	Q	D	I	L	I

CD4	CD5	CD6	CD7	CD8	D1	D2	D3	D4	D5	D6	D7	E1	E2	E3
46	47	48	49	50	51	52	53	54	55	56	57	58	59	60
F	K	H	L	K	S	E	D	E	M	K	A	S	E	D
F	K	H	L	K	T	E	A	E	M	K	A	S	E	D

E19	E20	EF1	EF2	EF3	EF4	EF5	EF6	EF7	EF8	F1	F2	F3	F4	F5
76	77	78	79	80	81	82	83	84	85	86	87	88	89	90
L	K	K	K	G	H	H	E	A	E	I	K	P	L	A
L	K	K	K	G	H	H	E	A	E	L	K	P	L	A

G7	G8	G9	G10	G11	G12	G13	G14	G15	G16	G17	G18	G19	GH1	GH2
106	107	108	109	110	111	112	113	114	115	116	117	118	119	120
F	I	S	E	C	I	I	Q	V	L	Q	S	K	H	P
F	I	S	E	A	I	I	H	V	L	H	S	R	H	P

H13	H14	H15	H16	H17	H18	H19	H20	H21	H22	H23	H24	H25	H26	HC1	HC2	HC3	HC4
136	137	138	139	140	141	142	143	144	145	146	147	148	149	150	151	152	153
E	L	F	R	K	D	M	A	S	N	Y	K	E	L	G	F	Q	G
E	L	F	R	K	D	I	A	A	K	Y	K	E	L	G	Y	Q	G

We must also emphasize the uniqueness of a given protein in a particular species of organism. Every sample of sperm whale myoglobin, taken from any sperm whale, will have the same amino acid sequence (unless, by rare chance, a sample is taken from a mutant whale).

Biochemists come to understand such a complex structure as the myoglobin molecule bit by bit, by analyzing the molecule at successively higher levels of complexity. To begin any such study of a protein, it is necessary to prepare the protein in a pure form, free of contamination by other proteins or other cellular substances. Methods for doing this are described in Tools of Biochemistry 6. Having purified a protein, a researcher might first wish to know its amino acid composition: What are the relative amounts of the different amino acids in the protein? Table 5.1 gives average results for a large number of proteins; individual proteins differ widely in composition. The amino acid composition is determined by first hydrolyzing the polypeptide into its constituent amino acids by treatment with concentrated mineral acid at high temperature. The mixture of amino acids can then be separated by ion exchange chromatography and the amount of each measured. Just how this is done is described in Tools of Biochemistry 7.

Composition determination gives limited information about a protein. What is much more important to the biochemist or biologist is the sequence of amino acids. As a first step in sequence analysis, it is often desirable to find out what the N-terminal and C-terminal amino acids are. Procedures for such analysis are described in Tools of Biochemistry 8. This may seem an unnecessary step, for techniques exist for sequencing the entire chain, but there is a good reason for beginning in this way. Some proteins contain

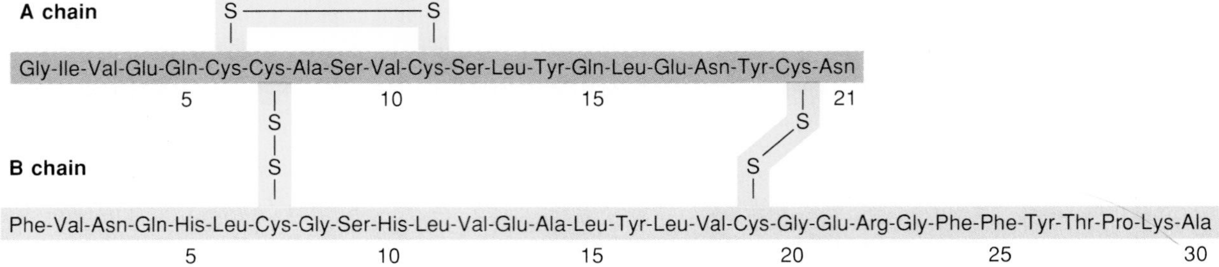

Figure 5.15
The primary structure of bovine insulin. This protein is composed of two polypeptide chains (A and B) joined by disulfide bonds. The A chain also contains an internal disulfide bond.

several polypeptide chains in strong (or even covalent) association, and one can usually find this out by noting the presence of two or more different N- or C-terminal residues.

Multichain proteins are common. For example, **hemoglobin,** which is closely related to myoglobin, has four polypeptide chains, two each of two different but similar sequences (see Chapter 7). These are held together by noncovalent forces and can be separated from one another quite easily. An example of a protein with covalently connected polypeptide chains is the hormone **insulin** (Figure 5.15). The two polypeptide chains, called A and B, are held together by disulfide bonds. An analysis of the terminal residues of insulin will immediately reveal the multichain character by the presence of two different N-terminal groups, glycine and phenylalanine, and two different C-terminal residues, asparagine and alanine.

The complete sequence of amino acids in a protein can be determined in two quite different ways. The most direct way is to carry out the sequencing study on the polypeptide chain (or chains) of the protein itself. This procedure, described in Tools of Biochemistry 9, is usually done automatically by instruments known as protein **sequenators.**

Another technique, rapidly gaining favor, is to determine the protein sequence from the nucleotide sequence of the gene that codes it (see Tools of Biochemistry 3). As we explained in Chapter 4, the primary structure of every protein is dictated by a particular gene. Since we now know the code that relates DNA sequence to protein sequence, determination of the nucleotide sequence of a gene allows us to read the corresponding protein sequence. In fact, since determining DNA sequences has become very easy, determining a protein primary structure in this way is often easier and quicker than by direct sequencing of the polypeptide chain, and it will probably be the method of preference in the future. So much DNA sequencing has already been accomplished that there exist many known polypeptide sequences for which the corresponding protein functions are still unrecognized.

The problems of how protein-defining sequences in the genome are identified and how the genes are retrieved and cloned are technical aspects that will be discussed in Part V of this text.

From Gene to Protein

The Genetic Code

In Chapter 4 we stated that the DNA sequences of genes are transcribed into messenger RNA molecules, which are in turn translated into proteins. But there are only 4 kinds of nucleotides in DNA, each of which transcribes to a particular nucleotide in RNA, and 20 kinds of amino acids. Obviously, a 1:1 correspondence between nucleotide and amino acid is impossible. By

SECOND POSITION

FIRST POSITION (5′ end)	U	C	A	G	THIRD POSITION (3′ end)
U	Phe	Ser	Tyr	Cys	U
	Phe	Ser	Tyr	Cys	C
	Leu	Ser	Stop	Stop	A
	Leu	Ser	Stop	Trp	G
C	Leu	Pro	His	Arg	U
	Leu	Pro	His	Arg	C
	Leu	Pro	Gln	Arg	A
	Leu	Pro	Gln	Arg	G
A	Ile	Thr	Asn	Ser	U
	Ile	Thr	Asn	Ser	C
	Ile	Thr	Lys	Arg	A
	Met	Thr	Lys	Arg	G
G	Val	Ala	Asp	Gly	U
	Val	Ala	Asp	Gly	C
	Val	Ala	Glu	Gly	A
	Val	Ala	Glu	Gly	G

Figure 5.16
The genetic code. The table is arranged so that it is possible to quickly find any amino acid from the three letters (written in 5′ → 3′ direction) of the codon. For example, the amino acid Ile corresponding to 5′ AUA 3′ is found by looking in the first position A row, the second position U column, and the third position A space. Codons are expressed in terms of bases in mRNA instead of DNA (i.e., U instead of T).

using triplets of nucleotides (codons) for each amino acid, $4^3 = 64$ different combinations are possible. This is more than enough to code for 20 amino acids, so most amino acids have multiple codons. The genetic code is virtually universal—that is, all organisms use the same codons to translate their genomes into proteins. The only exceptions are slight differences in codon usage in certain protozoans and in mitochondria. We discuss details of the genetic code and how it was deduced in Chapter 27.

Figure 5.16 depicts the genetic code in terms of the mRNA triplets that correspond to the different amino acid residues. Three triplets—UAA, UAG, and UGA—do not code for any amino acids but serve as "stop" signals to end translation at the C-terminus of the chain. The codon AUG, which normally codes for methionine, also serves as a "start" signal. When starting a polypeptide chain, AUG directs the placement of *N*-formylmethionine (in prokaryotes) or methionine (in eukaryotes) at the N-terminal position (Figure 5.17). The implication is that prokaryotic proteins should start with *N*-formylmethionine and eukaryotic proteins with methionine. Generally this is true, though in many cases the N-terminal residue or several residues are cleaved off in the cell by specific proteases immediately after translation. Figure 5.18 shows the relationship between DNA, mRNA, and polypeptide sequences for the N-terminal portion of seal myoglobin. In this case the N-terminal methionine is removed.

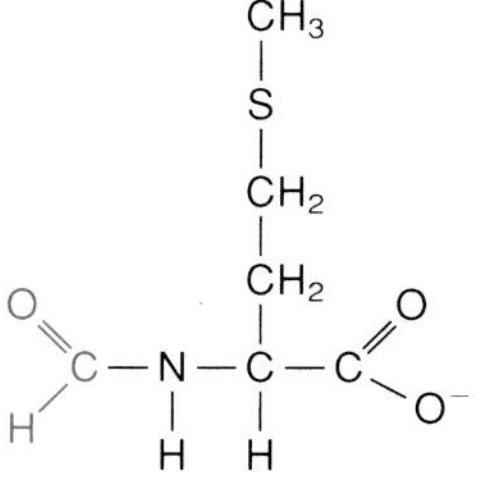

Figure 5.17
N-Formylmethionine, the initiating amino acid in prokaryotic translation.

STOP: UAA
UAG
UGA
START: AUG

Translation

We are now in a position to discuss the process of translation of mRNA into protein in a bit more detail. We noted earlier in this chapter that the thermodynamics of peptide bond formation requires that amino acids be activated before they can be added to a polypeptide chain. This activation is accomplished by coupling each amino acid to the 3′ end of an appropriate **transfer**

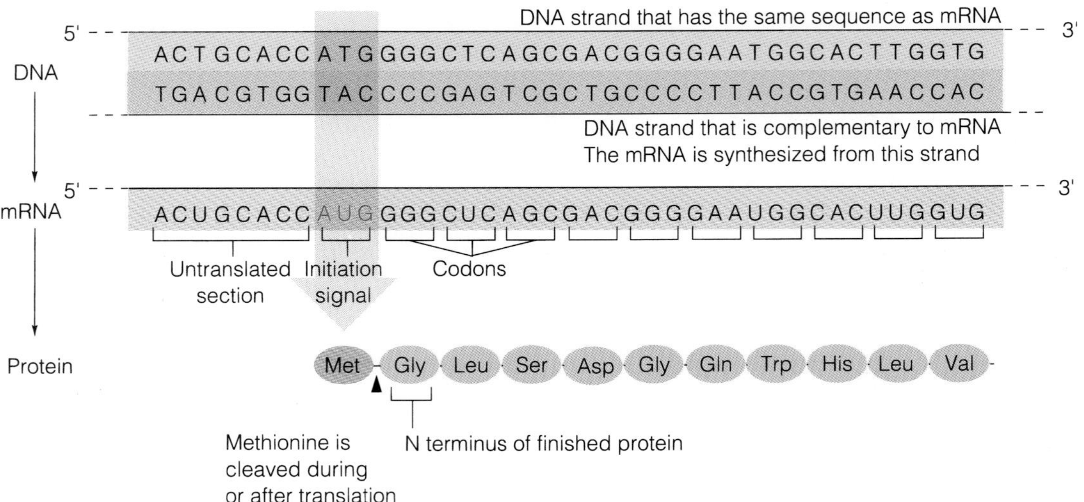

Figure 5.18
Relationship between DNA, mRNA, and polypeptide chain for the first 10 residues of gray seal myoglobin. Note that the DNA strand that is transcribed from is complementary to the mRNA.

RNA (Figure 5.19). The process requires the hydrolysis of ATP to AMP and pyrophosphate, which constitutes the free energy source in the reaction. The coupling is catalyzed by specific enzymes called **aminoacyl-tRNA synthetases.** Each such enzyme recognizes both a particular amino acid and its appropriate tRNA. The tRNA contains, in a region known as the **anticodon loop,** a nucleotide sequence (the **anticodon**) that is complementary to the appropriate codon in the message. The messenger RNA has been bound to a particle called a **ribosome;** ribosomes are the sites of protein synthesis in the cell. The aminoacyl tRNAs also bind here, matching their anticodons to the codons on the message (Figure 5.20). The amino acid carried by each tRNA is transferred to the growing peptide chain, and the tRNA is then released. After each amino acid is added, the ribosome moves one codon length along the message, allowing the next tRNA to come into place, carrying *its* amino acid. In every cell, this remarkable machinery translates the information coded in thousands of different genes into thousands of different proteins. In Chapters 27 and 28 we discuss details of how the process starts, stops, and is energized.

Posttranslational Processing of Proteins

As the ribosome moves along the messenger RNA, it eventually encounters a "stop" codon. At this point, the polypeptide chain is released. But it is not

Figure 5.19
Activation of amino acids for incorporation into proteins. The enzyme aminoacyl-tRNA synthetase recognizes both a particular amino acid and the appropriate tRNA, carrying the anticodon corresponding to that amino acid. It catalyzes the formation of an aminoacyl tRNA with accompanying hydrolysis of one ATP.

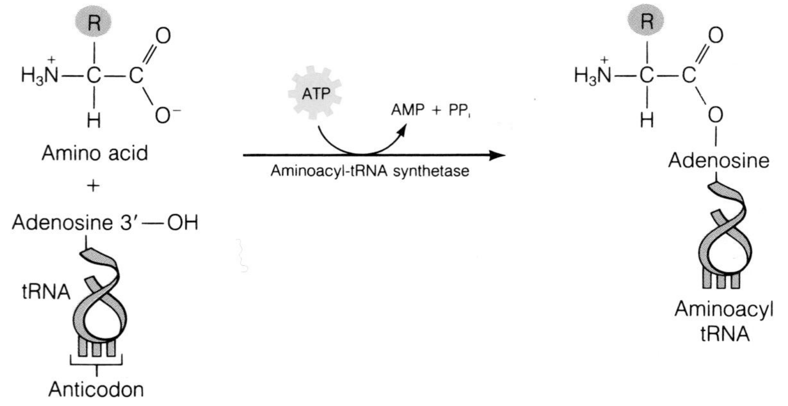

Amino acid + Adenosine 3′—OH (tRNA, Anticodon) —ATP → AMP + PP_i (Aminoacyl-tRNA synthetase)→ Aminoacyl tRNA

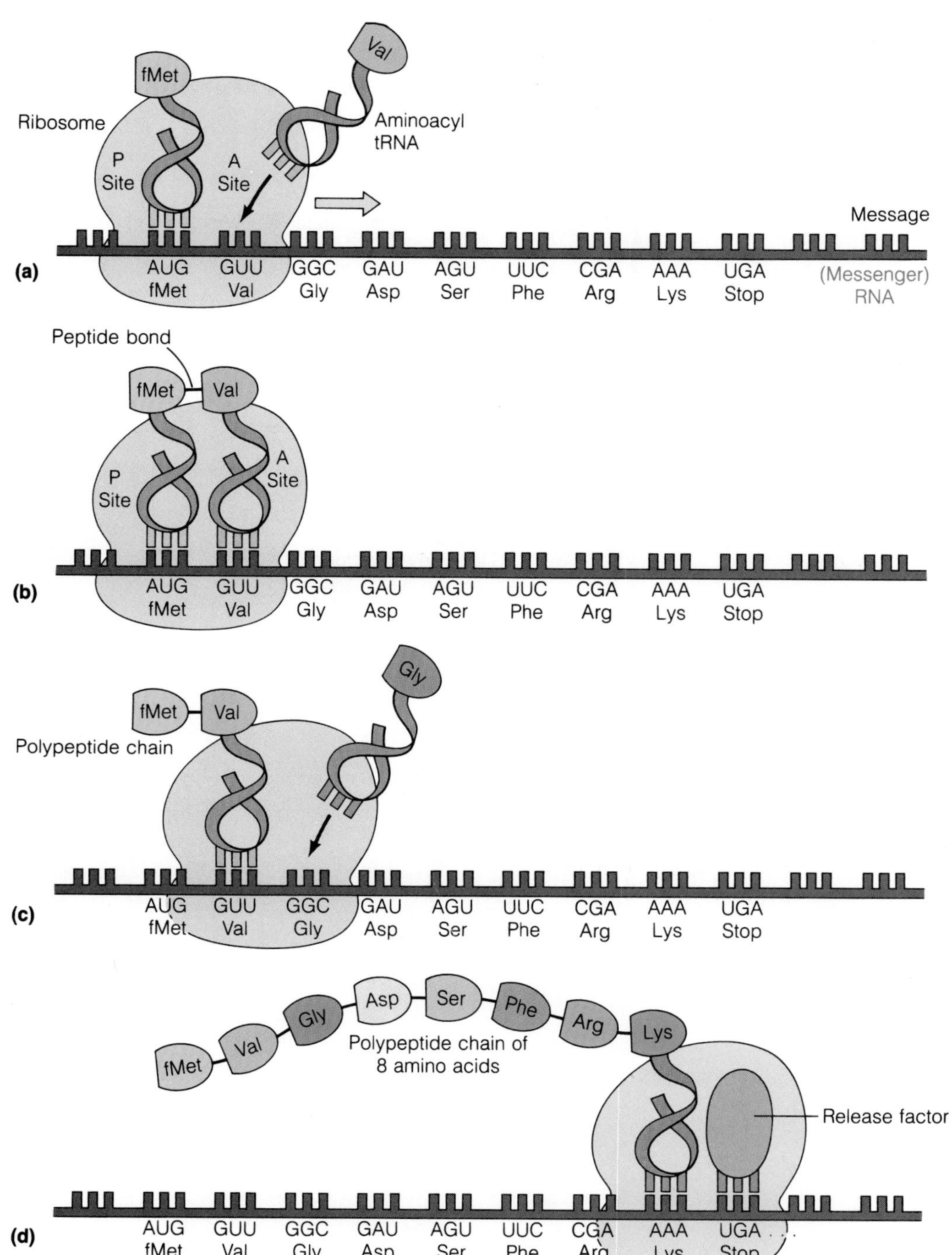

Figure 5.20
Translation of an RNA message into a protein. As the ribosome moves along the message, it successively accepts aminoacyl tRNAs, selecting them by matching the anticodon on the tRNA to the codon on the message (**a**). The amino acid is transferred to the growing polypeptide chain (**b**), and the ribosome moves on to the next codon (**c**) to repeat the process. This continues until a *stop* signal is read (**d**), whereupon a protein release factor causes both polypeptide and mRNA to be released. The example shown here is an unrealistically short polypeptide, to illustrate both starting and stopping. Many details of the process have been omitted in this introduction to translation (see Chapters 27, 28).

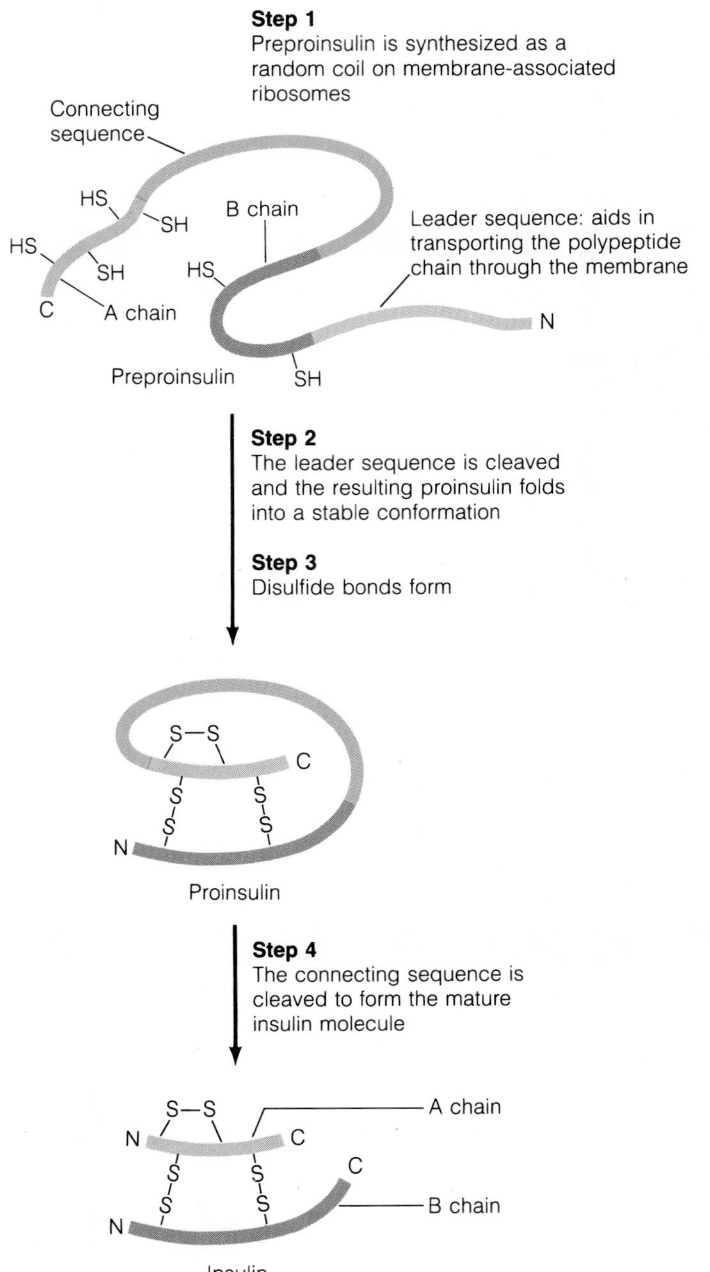
Step 1
Preproinsulin is synthesized as a random coil on membrane-associated ribosomes
Connecting sequence
HS
SH
B chain
Leader sequence: aids in transporting the polypeptide chain through the membrane
HS
SH
HS
C
A chain
N
Preproinsulin
SH
Step 2
The leader sequence is cleaved and the resulting proinsulin folds into a stable conformation
Step 3
Disulfide bonds form
S—S
C
N
Proinsulin
Step 4
The connecting sequence is cleaved to form the mature insulin molecule
S—S
A chain
N
C
C
B chain
N
Insulin

Figure 5.21
Structure of preproinsulin and its conversion to insulin.

necessarily finished. It must fold into its correct three-dimensional structure, and in some cases disulfide bonds must form. Certain amino acids may be modified by enzymes in the cell to produce, for example, the kinds of modifications shown on page 141.

Many proteins are further modified by specific proteolytic cleavage to shorten the chain length. A remarkable example is found in the synthesis of insulin. We have encountered insulin as a two-chain protein held together by disulfide bonds. But insulin is actually synthesized as a single, much longer polypeptide chain, called **preproinsulin** (Figure 5.21). The residues at the beginning of the molecule (the exact number varies with the species) serve as a "signal peptide" (also called a leader sequence) to help the preproinsulin molecule be transported through the very hydrophobic cell membranes. The signal peptide is then cut off, leaving **proinsulin.** Proinsulin folds into a specific three-dimensional structure, which helps it to form the correct disulfide bonds. The intervening sequence between the A chain and the B chain is then cut out by protease action, yielding the "finished" insulin molecule.

The primary structure of a protein molecule is a sequence of *information*. The 20 kinds of amino acid side chains can be thought of as words in a long sentence. These words have been translated from another language, the language of nucleic acid sequences stored in the genes. After translation, the sentence has been edited, with certain words modified and others deleted in the posttranslational processing. In the next chapter we shall see that the information contained in the "sentence" of a protein sequence dictates how that protein folds in three dimensions. This folding, in turn, prescribes the function of the protein—how it interacts with small molecules and ions, with other proteins, and with substances like nucleic acids, carbohydrates, and lipids. The information expressed in protein sequences plays a primary role in determining how cells and organisms function.

REFERENCES

General References

A number of excellent books provide more detailed or supplementary information on protein structure. We particularly recommend the following for Chapters 5, 6, and 7.

Creighton, T. E. (1983) *Proteins: Structure and Molecular Properties.* Freeman, San Francisco. An elegant, thorough, contemporary exposition of all aspects of protein chemistry.

Dickerson, R. E., and I. Geis (1962) *The Structure and Action of Proteins.* Benjamin/Cummings, Menlo Park, Calif. Concise, well written and illustrated.

Hirs, C. H. W., and S. N. Timasheff (eds.) (1983) *Enzyme Structure,* Part I, in *Methods in Enzymology,* Vol. 91. Academic Press, New York. Contains a series of useful chapters on protein analytical methods.

Neurath, H., and R. L. Hill (eds.) (1979) *The Proteins,* 3rd ed., Vols. 1–3. Academic Press, New York. The latest edition of a classic, comprehensive treatise. Volumes 1–3 contain a series of useful chapters on protein separation and analysis.

Reviews and Papers on Special Topics

Blake, C. C. F., and L. N. Johnson (1984) Protein structure. *Trends Biochem. Sci.* 9:147–151.

Brown, J. R., and B. S. Hartley (1966) Location of disulfide bridges by diagonal paper electrophoresis. *Biochem. J.* 101:214–228.

Dayhoff, M. O. (1972) *Atlas of Protein Sequence and Structure.* National Biomedical Research Foundation, Washington, D.C. (See also supplements to this volume published in subsequent years.)

Doolittle, R. (1985) Proteins. *Sci. Am.* 253(4):88–96.

Greenstein, J. P., and M. Winitz (1961) *Chemistry of the Amino Acids.* Wiley, New York.

Rose, G. D., A. R. Geselowitz, G. J. Lesser, R. H. Lee, and M. H. Zehfus (1985) Hydrophobicity of amino acid residues in globular proteins. *Science* 229:834–838.

Wilbur, P. J., and A. Allerhand (1977) Titration behavior and tautomeric states of individual histidine residues of myoglobin. *J. Biol. Chem.* 252:4968–4975.

PROBLEMS

1. The melanocyte-stimulating peptide hormone *α-melanotropin* has the following sequence:

   Ser Tyr Ser Met Glu His Phe Arg Trp Gly Lys Pro Val

   (a) Write the sequence using the one-letter abbreviations.

   (b) Calculate the molecular weight of α-melanotropin.

2. (a) Sketch the titration curve you would expect for α-melanotropin (Problem 1). (Assume pK_a values in the middle of ranges given in Table 5.3.)

   (b) Approximately what charge would you expect at pH values of 11, 5, and 1?

   (c) Estimate the isoelectric point of α-melanotropin.

3. What peptides are expected to be produced when α-melanotropin (Problem 1) is cleaved by (a) trypsin, (b) cyanogen bromide, and (c) thermolysin?

4. There is another melanocyte-stimulating hormone called *β-melanotropin.* Cleavage of β-melanotropin with trypsin produces the following peptides plus free aspartic acid.

   WGSPPK
   DSGPYK
   MEHFR

(a) If you assume maximum sequence homology between α-melanotropin and β-melanotropin, what must be the sequence of the latter?

(b) What *single* additional cleavage, plus sequencing of the fragments, would confirm the answer to (a)?

5. *Apamine* is a small protein toxin present in the venom of the honeybee. It has the sequence

CNCKAPETALCARRCQQH

(a) It is known that apamine forms disulfide bonds and does not react with iodoacetate. How many disulfide bonds are present?

(b) Devise a strategy for locating the disulfide bond(s) in apamine.

6. (a) Write a possible sequence for an mRNA segment coding for apamine.

(b) Do you think apamine has been synthesized in the form shown in Problem 5, or is it more likely a product of proteolytic cleavage of a larger peptide? Explain.

7. Assume the following portion of an mRNA. Find a start signal, and write the amino acid sequence that is coded for.

5′···GCCAUGUUUCCGAGUUAUCCCAAAGAUAAAAAAGAG···3′

6

TOOLS OF BIOCHEMISTRY 6

Ways to Isolate and Purify Proteins and Other Macromolecules

The biochemist, in attempting to understand the chemical processes in cells and organisms, is faced with a formidable task. Even the simplest cell is a complex structure, made up of thousands of different kinds of compounds. Before their interactions and changes can be understood, it is necessary to *isolate* these various cellular constituents, identify them, and study their structures and properties.

To separate molecules, the biochemist seizes on differences between them that can be exploited. Such differences may be in size, mass, electrical charge, or affinity for other molecules. Electrophoresis, which uses electrical charge differences in separation, has been discussed in Tools of Biochemistry 2. But electrophoresis is primarily an *analytical* tool. Though it is sometimes used to purify small quantities of proteins and nucleic acids, its capacity is limited. To prepare the quantities of substances needed for characterization and study, other methods are required. Usually, to go from cell or tissue to a purified material, a number of different techniques are used sequentially. We will describe some of these here.

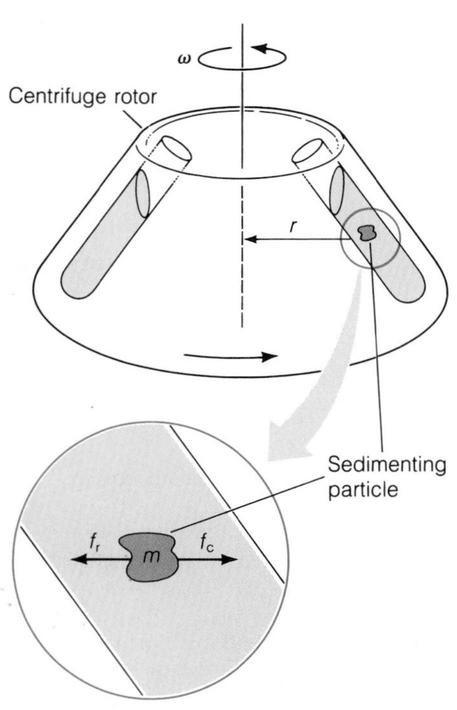

Figure T6.1
Forces on a particle in a centrifugal field in a fixed-angle rotor.

Centrifugation

Any molecule, or particle, that is subjected to a centrifugal field by being spun in a centrifuge rotor can be said to be subject to a centrifugal force (Figure T6.1). For a particle of mass m this force is given by

$$f_c = m(1 - \bar{v}\rho)\omega^2 r \qquad \text{(T6.1)}$$

Here r is the distance of the particle from the center of the rotor, which is spinning at an angular velocity of ω radian/s ($\omega = (2\pi/60) \times$ RPM, where RPM is the number of revolutions per minute). The factor $(1 - \bar{v}\rho)$ is the **buoyancy factor.** It contains the solution density ρ (g/ml) and the specific volume of the particle $\bar{v}$ (ml/g). This factor takes into account the fact that the particle is buoyed up by the surrounding solution. The specific volume $\bar{v}$ can be considered equal to $1/\rho_P$, where ρ_P is the effective density of the particle. So the factor $1 - \bar{v}\rho = 1 - \rho/\rho_P$. Obviously, if $\rho_P = \rho$, there is no net force on the particle, for it displaces its own weight of solution. As in electrophoresis, the particle's motion is resisted by a force $f_r = fv$, where v is the velocity and f is the **frictional coefficient.** A steady velocity of sedimentation is established, where $f_r = f_c$: Therefore

$$fv = m(1 - \bar{v}\rho)\omega^2 r \tag{T6.2}$$

As in electrophoresis, we divide velocity by field strength, in this case the centrifugal field strength $\omega^2 r$, and obtain

$$S = \frac{v}{\omega^2 r} = \frac{m(1 - \bar{v}\rho)}{f} = \frac{M(1 - \bar{v}\rho)}{Nf} \tag{T6.3}$$

where M is molecular weight and N is Avogadro's number ($M = Nm$). The quantity S is the **sedimentation coefficient.** Note that equation (T6.3) is analogous to equation (T2.2) for the electrophoretic mobility. In each case the quantity (S or μ) is defined as velocity/field strength and is expressed in molecular terms (on the right) as the ratio of a factor describing the driving force to a factor accounting for the frictional resistance.

The sedimentation coefficient has the units of seconds. Typical values for molecules are in the neighborhood of 10^{-13} s, so this quantity has been designated one **Svedberg unit** (S). For example, the sedimentation coefficient of hemoglobin is about 4×10^{-13} s, or 4S. Often cellular particles are referred to by their S value, for example, 70S ribosomes.

Clearly S increases with particle mass, but the relationship is neither linear nor simple because the frictional coefficient, f, increases with the particle size and also depends on the particle shape. Although we can use sedimentation as a rough measure of molecular weight, the relationship is only approximate.

Since different-sized particles or molecules differ in S, and hence differ in sedimentation velocity, sedimentation is a useful tool for separation. A number of techniques are used, depending on what one wants to separate. If one wants simply to remove large particles or aggregates from a solution of molecules, a low-speed centrifuge with a rotor carrying fixed-angle tubes (Figure T6.1) may suffice. For example, a first step in isolating proteins often involves breaking cells and separating the soluble proteins of the cytosol from cell nuclei, cell wall fragments, and other heavy debris. Sedimentation in a fixed-angle rotor at a few thousand RPM for 10–20 min usually accomplishes this. For more difficult separations, in which several kinds of large molecules or particles must be separated, the **sucrose gradient** technique shown in Figure T6.2 may be employed. It could be used, for example, to obtain individual proteins from the supernatant obtained from a low-speed centrifugation of broken cells. For such separations of protein molecules, much higher rotor speeds (often up to 60,000 RPM) are required.

It is also possible to use sedimentation to analyze, rather than preparatively separate, mixtures. In such cases, an **analytical ultracentrifuge** is used, in which the progress of sedimentation of individual components can be followed during the experiment. By such experiments, the sedimentation coefficients of macromolecules can be determined with accuracy, using equation (T6.3). The use of sedimentation experiments for the actual determination of molecular weights will be described in Tools of Biochemistry 11.

Chromatography

Much of modern biochemistry depends on the use of **column chromatographic** methods to separate molecules. There are many such techniques, but the principle common to most is illustrated in Figure T6.3. A column is packed with some material that can selectively adsorb molecules on the basis of some difference in their chemical structure. The column initially is wetted with the appropriate buffer solution. The mixture of the molecules to be separated is then placed on top of the column and slowly washed through the column with buffer. Fractions are taken, using an automatic fraction collector, as this process of **elution** of the column continues. Some kinds of molecules are adsorbed only weakly or not at all, and these are eluted first. The most strongly adsorbed are eluted last. Sometimes the composition

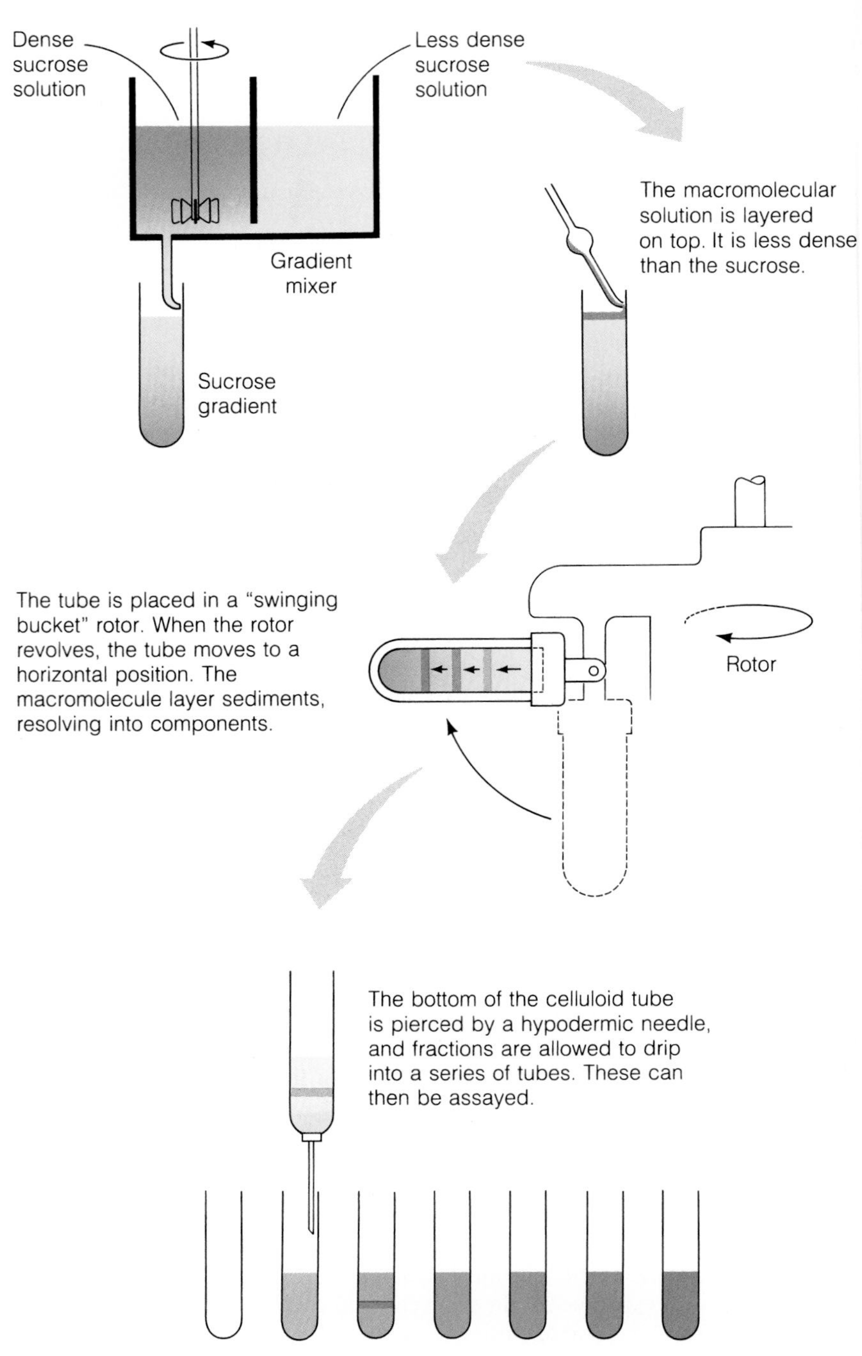

Dense sucrose solution
Less dense sucrose solution
Gradient mixer
Sucrose gradient
The macromolecular solution is layered on top. It is less dense than the sucrose.
The tube is placed in a "swinging bucket" rotor. When the rotor revolves, the tube moves to a horizontal position. The macromolecule layer sediments, resolving into components.
Rotor
The bottom of the celluloid tube is pierced by a hypodermic needle, and fractions are allowed to drip into a series of tubes. These can then be assayed.

Figure T6.2
Preparative centrifugation, using the sucrose gradient method.

of the buffer solution must be changed during the elution to remove the more tightly bound molecules. The column material used and the method of elution depend on the basis of separation desired. We describe some important methods below.

Ion-Exchange Chromatography

This method is used to separate molecules on the basis of their electrical charge. **Ion-exchange resins** are used, which are either *polyanions* or *polycations*. Suppose one wishes to separate three kinds of molecules, one of which is negatively charged, one weakly positively charged, and one strongly positively charged. An *anionic* resin, carrying negatively charged groups, would be used for the column material. The negatively charged molecules in the mixture would pass through without adsorption and be found in early fractions. The two kinds of positively charged mole-

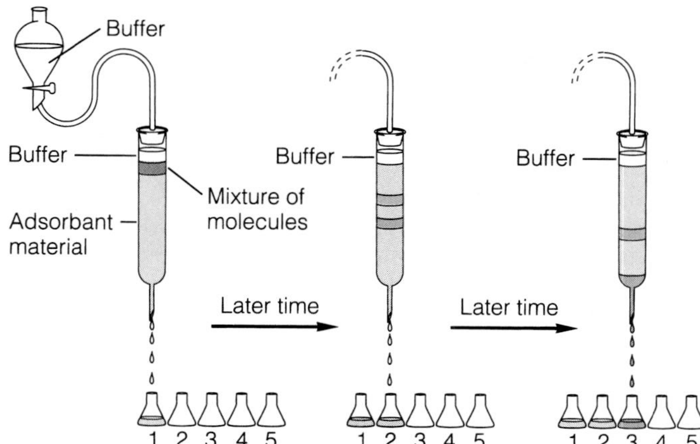

Figure T6.3
The principle of chromatography.

cules would both be bound by the resin, but the weakly positive ones would be bound less tightly. Since increased salt concentrations tend to disrupt electrostatic interactions of this kind, a *salt gradient* might be used after the first fractions were collected. That is, the eluting buffer would be made gradually more concentrated in salt (by using NaCl, for example). The weakly bound cations would be eluted next, and the tightly bound, strongly positive ones would be eluted only at a higher salt concentration.

Affinity Chromatography

This is a more specific kind of chromatography, ideal for isolating one or a few proteins from a complex mixture. Many proteins exhibit quite strong interactions with other molecules. Examples are the interactions of enzymes with analogs of their substrates or with cofactors (see Chapter 10). The appropriate molecules, covalently coupled to an inert matrix material, will act as molecular "fishhooks" to pick up the desired protein. All remaining proteins will simply pass through the column. The captured protein molecules can then be released by eluting the column with a buffer solution containing free copies of the particular molecules, or some other reagent that can break the interaction. In a variant of this method, *antibodies* (see Chapter 7) to a particular substance are coupled to the resin and provide very specific retention of the desired material.

High-Performance Liquid Chromatography (HPLC)

Chromatography is often a slow process. In the usual procedures, only a small hydrostatic pressure is applied to force the fluids through the column, and elution takes many hours. This is not only time-consuming but also sometimes deleterious to sensitive materials. Furthermore, the sample tends to spread out, because of diffusion, as it moves down the column. The longer the experiment takes, the more serious the spreading will be, and resolution of components will suffer. Because of all these factors, the development of high-performance liquid chromatographic methods has been welcomed by biochemical researchers. In these techniques, pressures of 5,000–10,000 psi are used to force the solutions rapidly through the resin. This has required the development of noncompressible resin materials and strong metal columns in which to carry out the process. In this way, separations that formerly required hours can now be done in minutes, yet with higher resolution.

Gel Filtration

This is often a convenient method for separating different-sized macromolecules or for removing low-molecular-weight contaminants from solutions of large molecules. The apparatus used is very similar to that described for column chromatography. A column is packed with porous gel beads, usually a cross-linked polysaccharide material. The porosity of the gel is chosen so that the smaller molecules in the mixture can penetrate the beads, whereas the larger ones cannot. The sample is applied to the top of the column, as in chromatography, and eluted with a buffer. As the sample moves down the column the larger molecules move faster, for they cannot enter the gel beads and can only flow through the interstices between them. The small molecules can wander into the gel beads, and they loiter behind (Figure T6.4). If fractions are collected as buffer is eluted through the column, the earlier fractions will contain the larger molecules. By choosing the bead porosity properly, mixtures in different size ranges can be separated. The column can even be calibrated with known substances to allow a rough measure of molecular size from the point at which a particular component emerges. In a sense, gel filtration is a form of chromatography in which the basis of separation is molecular size rather than chemical properties. Thus, it often provides an alternative to sedimentation.

Dialysis and Ultrafiltration

It is possible to obtain **semipermeable membranes** that have pores large enough to permit small molecules to pass freely but present a barrier to proteins and other macromolecules. Such membranes are used in both purification and concentration of biopolymers.

Dialysis is used routinely to remove small-molecule contaminants or to change buffer conditions gently. A solution of a protein, for example, is placed in a closed bag made from such a membrane and immersed in a much larger volume of buffer solution. Over many hours, the low-molecular-weight contaminants leak out and the original environment of the protein molecules is replaced by the outside buffer solution. (Often the outside solution is replaced several times during the process.) **Ultrafiltration** is sometimes used to concentrate solutions of macromolecules. A similar semipermeable membrane is used, but pressure is applied to force solvent and small molecules through the membrane.

Fractional Precipitation

A traditional step in protein purification, which is still employed, is fractional precipitation by salts like ammonium sulfate. In concentrated salt solutions, some proteins are more soluble than others, and a crude separation of protein fractions can be made in this way.

An Example: How to Prepare Myoglobin

Having described a number of different separation and purification methods, it seems appropriate to give an example of how some of these might be used in an actual protein purification. The protein chosen is myoglobin, both because the pro-

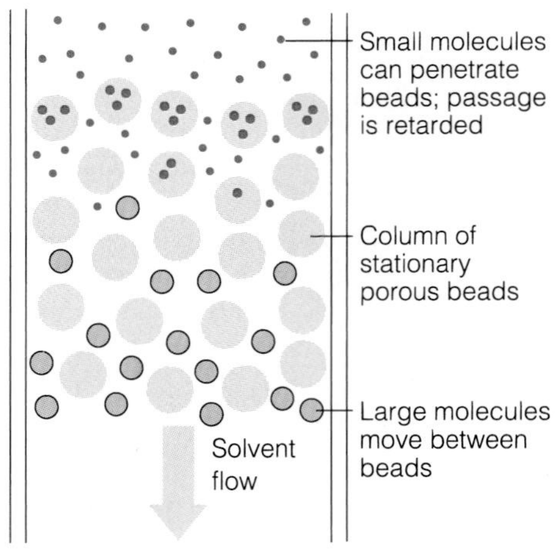
Small molecules can penetrate beads; passage is retarded
Column of stationary porous beads
Large molecules move between beads
Solvent flow

Figure T6.4
The principle of gel filtration.

Table T6.1
How to make myoglobin[a]

Steps	Rationale
1. Begin with frozen red muscle tissue ↓	A source rich in the protein desired should be used. It has been kept frozen to avoid degradation of the material.
2. Cut up, homogenize in Waring blender with 65% saturated ammonium sulfate solution. Keep at 4°C. ↓	The homogenization breaks the muscle cells, releasing the myoglobin and other cytosolic proteins. The ammonium sulfate precipitates many of the soluble proteins, but *not* myoglobin.
3. Centrifuge at 8000 RPM in fixed-angle rotor for 30 min. Collect supernatant; discard pellet at bottom of tube. ↓	This removes nuclei, subcellular particles, broken cell walls, muscle fibers, etc., together with proteins precipitated by the $(NH_4)_2SO_4$. The supernatant is a crude preparation of myoglobin.
4. Dialyze at 4°C for 24 h against a buffer at pH 8.70, changing several times. ↓	This removes small molecules that were present in the cytosol and leaves the protein dissolved in the desired buffer. pH is chosen as one at which myoglobin is especially stable.
5. Chromatograph dialyzed solution over a DEAE[b]-cellulose ion-exchange resin. The main myoglobin component is the first heme-containing (red) fraction to emerge from the column.	This provides a final separation from other proteins, including minor myoglobin variants present in the mixture. The product should be a quite pure preparation of myoglobin.

[a] Adapted from T. E. Hugli and F. R. N. Gurd, *J. Biol. Chem.* 245:1930–1938, 1970.
[b] DEAE = diethylaminoethyl.

cedure is relatively short and simple and because we will use myoglobin extensively as an example of a typical protein. It is appropriate to know how it might be made. The overall scheme is outlined in Table T6.1, along with reasons for each step.

The purity of the material obtained at the end of step 5 could be checked by gel electrophoresis or isoelectric focusing. A single band should be observed on the gel.

It should be emphasized that this is a particularly simple protein preparation. Myoglobin is very abundant in the kind of tissue used, and relatively few steps are required. Most purifications are much more difficult than this one.

REFERENCES

Cantor, C. R., and P. R. Schimmel (1980) *Biophysical Chemistry*, Vol. 2. Freeman, New York. Detailed coverage of a number of techniques is given in this volume.

Freifelder, D. (1982) *Physical Biochemistry: Applications to Biochemistry and Molecular Biology*. Freeman, New York. More emphasis is placed on practical separation techniques than in Cantor and Schimmel.

Scopes, R. (1982) *Protein Purification: Principles and Practice*. Springer-Verlag, New York.

van Holde, K. E. (1985) *Physical Biochemistry*, 2nd ed. Prentice-Hall, Englewood Cliffs, N.J.

TOOLS OF BIOCHEMISTRY 7

Amino Acid Analysis of Proteins

Contemporary methods for determination of the amino acid composition of proteins involve three basic steps:

1. *Hydrolysis* of the protein to its constituent amino acids.
2. *Separation* of the amino acids in the mixture.
3. *Quantitation* of the individual amino acids.

What is described here is a technique used, with minor variations, in many laboratories. A small sample of the protein is first purified, using, for example, some combination of the methods described in Tools of Biochemistry 6. The protein is dissolved in 6 M HCl, and the solution is sealed in an evacuated ampoule. It is then heated at 105–110°C for about 24 hours. Under these conditions, the metastable peptide bonds between the residues will be completely hydrolyzed.

The hydrolyzed sample is then separated into the constituent amino acids on a cation-exchange column. The kinds of resin typically used are sulfonated polystyrenes:

$$-CH_2-CH-CH_2-CH-CH_2-CH-CH_2-CH-$$

SO_3^- SO_3^-

Such a resin separates amino acids in two ways. First, since it is negatively charged, it tends to pass acidic amino acids first and retain basic ones. The pH of the eluting buffer is increased during elution to facilitate this separation. Second, the hydrophobic nature of the polystyrene itself tends to hold up the more hydrophobic amino acids such as leucine and phenylalanine. An example of such an analysis is shown in Figure T7.1. Note the order of appearance of the amino acids, proceeding from the more acidic to the more basic. Modern amino acid analyzers are completely preprogrammed and carry out both the chromatographic separation of the amino acids and their quantitation. Increasingly, researchers are turning to HPLC (see Tools of Biochemistry 6) for separation of the amino acids. The advantages are rapidity and even better resolution.

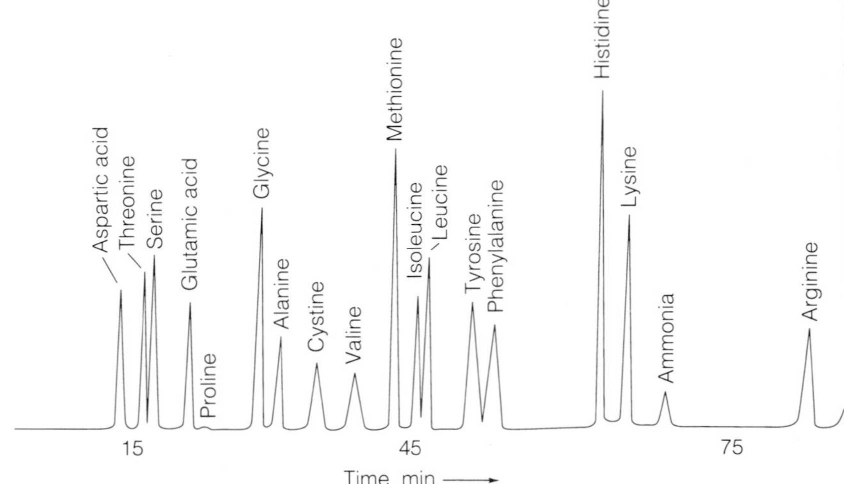

Figure T7.1
Analysis of a protein hydrolysate on a single-column amino acid analyzer.

There are many methods for detection of the amino acids eluting from the column. Older techniques employed the **ninhydrin** reaction (Figure T7.2). Ninhydrin was mixed with and reacted at 100°C with the eluting amino acids in the analyzer; the absorbance of each peak (which is proportional to the amino acid concentration) was then detected spectrophotometrically in the same instrument. This technique can allow detection of a few nanomoles (10^{-9} mol) of an amino acid.

More recently, the sensitivity of analysis has been greatly improved by the use of fluorescent reagents and fluorescence detection. For example, the amino acids may be reacted with *o*-phthalaldehyde to yield a fluorescent complex:

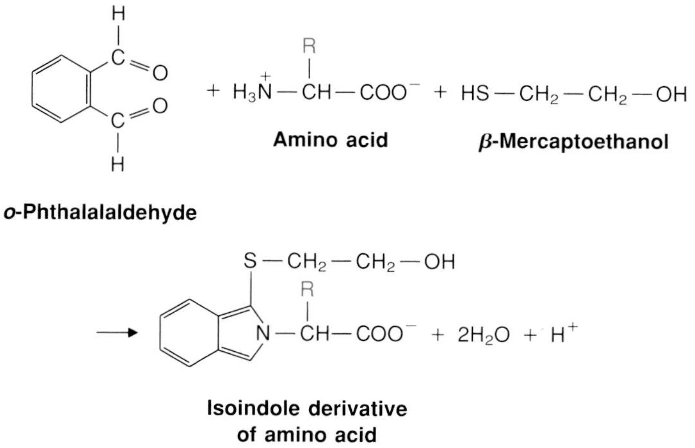

o-Phthalalaldehyde + $H_3\overset{+}{N}-CH(R)-COO^-$ **Amino acid** + $HS-CH_2-CH_2-OH$ **β-Mercaptoethanol**

→ **Isoindole derivative of amino acid** + $2H_2O$ + H^+

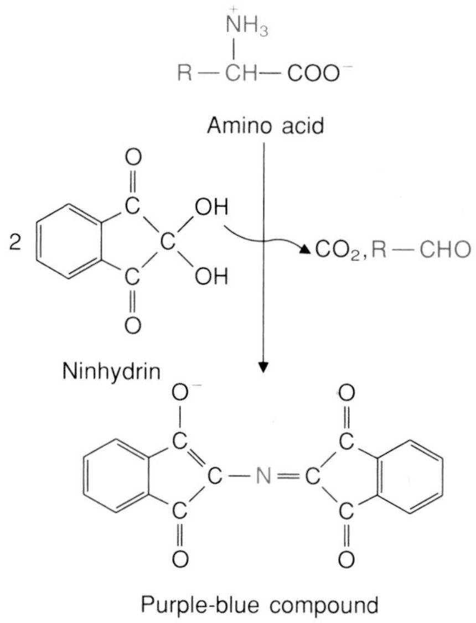

Figure T7.2
The ninhydrin reaction.

Such techniques easily give sensitivity to the picomole (pmol; 10^{-12} mol) range. Recent advances using a new micro electrophoresis system and fluorescence detection have extended this to the *attomole* (1 amol; 10^{-18} mol) range. This is an amount corresponding to only a few thousand molecules. Indeed, amino acid analysis techniques have proceeded to the point where the amount of protein contained in one spot in two-dimensional gel electrophoresis can easily be analyzed.

Of course, the whole procedure is not as simple and trouble free as the foregoing discussion might imply. Some amino acids give problems in reaction with the compounds used for detection; proline in particular, since it is cyclic amino acid, often reacts differently or not at all. Furthermore, some amino acids tend to be partially destroyed during the severe hydrolysis. Tryptophan is troublesome in this respect and must be determined by a separate analysis, usually based on its strong ultraviolet absorbance (see Figure 5.6). Serine, threonine, and tyrosine also tend to be degraded during long hydrolysis.

To a considerable extent these difficulties can be circumvented either by carrying out protective reactions first or by measuring the apparent content of the amino acid at different hydrolysis times and extrapolating to zero hydrolysis time. Finally, asparagine and glutamine are invariably hydrolyzed to aspartic and glutamic acids, so that the total content of these acids observed includes the amides. This, as well as the other degradation reactions mentioned above, can be avoided by using an enzymatic hydrolysis, with a mixture of proteolytic enzymes, in place of the acid hydrolysis. However, this method also has its drawbacks, since it is sometimes difficult to achieve complete hydrolysis and the enzymes themselves must be removed before analysis.

Despite such minor complications, amino acid analysis, using automated analyzers, has become a routine operation in many laboratories. One of the first analyses of any newly discovered protein will invariably be for amino acid composition.

REFERENCES

Chang, J.-Y., R. Knecht, and D. Braun (1981) Amino acid analysis at the picomole level. *Biochem. J.* 199:547–555.

Cheng, Y.-F., and N. Dovichi (1988) Subattomole amino acid analysis by capillary zone electrophoresis and laser-induced fluorescence. *Science* 242:562–564.

Hill, R. L. (1965) Hydrolysis of proteins. *Adv. Protein Chem.* 20:37–107.

Liu, T.-Y. (1972) Determination of tryptophan. *Methods Enzymol.* 25:44–55.

TOOLS OF BIOCHEMISTRY 8

Determination of the N-Terminal and C-Terminal Residues of a Protein

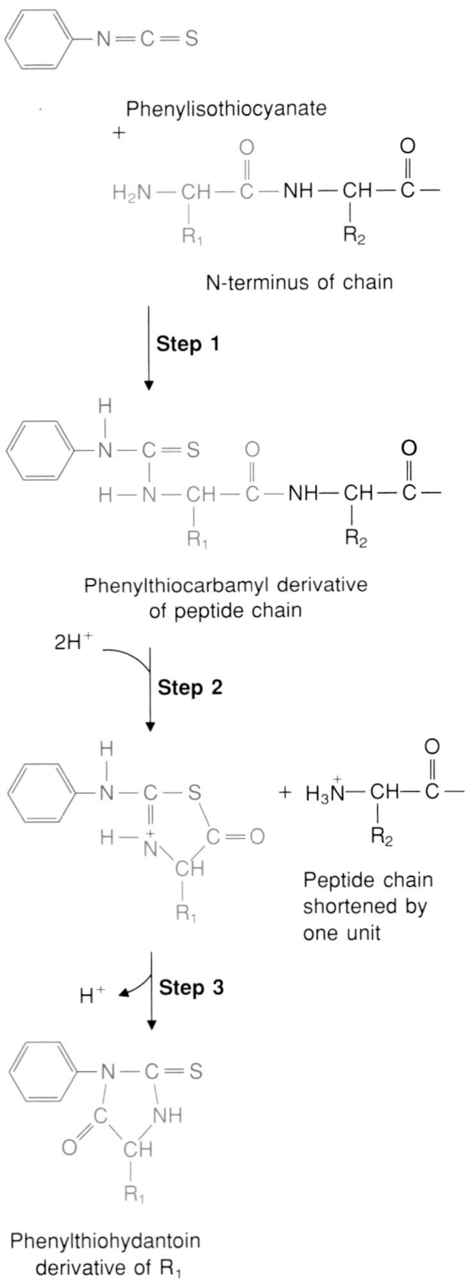

Phenylisothiocyanate

+

$H_2N-CH(R_1)-CO-NH-CH(R_2)-CO-$

N-terminus of chain

Step 1

Phenylthiocarbamyl derivative of peptide chain

$2H^+$ Step 2

$+ \; H_3\overset{+}{N}-CH(R_2)-CO-$

Peptide chain shortened by one unit

H^+ Step 3

Phenylthiohydantoin derivative of R_1

Figure T8.1
The Edman degradation.

N-Termini

Several methods exist for the quick identification of the N-terminal residues of a polypeptide.

Sanger's Reagent

The compound 2,4-dinitrofluorobenzene (DNF) reacts with the N-terminal amino group of a polypeptide in alkaline solution:

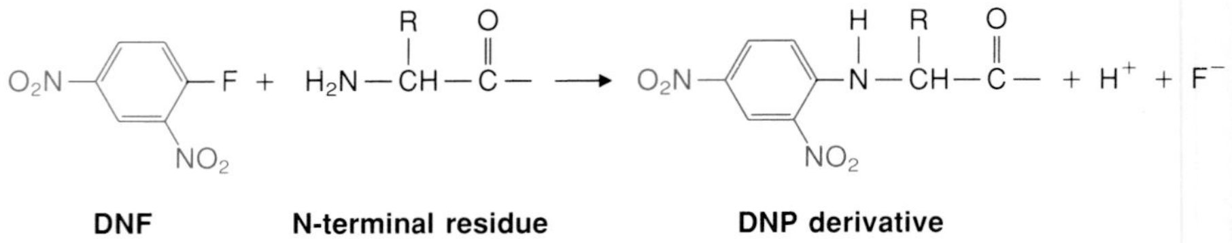

$$O_2N-C_6H_3(NO_2)-F + H_2N-CH(R)-CO- \longrightarrow O_2N-C_6H_3(NO_2)-NH-CH(R)-CO- + H^+ + F^-$$

DNF **N-terminal residue** **DNP derivative**

If the polypeptide is now hydrolyzed, the N-terminal amino acid will be labeled with the dinitrobenzene group, which imparts a bright yellow color. The amino acids can be separated by paper electrophoresis or paper chromatography (Tools of Biochemistry 2 and 6), and the N-terminal residue spot, which is yellow, can be identified by comparison with the migration of known DNP (dinitrophenyl) derivatives.

Dansyl Chloride

This compound also reacts with N-terminal amino groups at high pH:

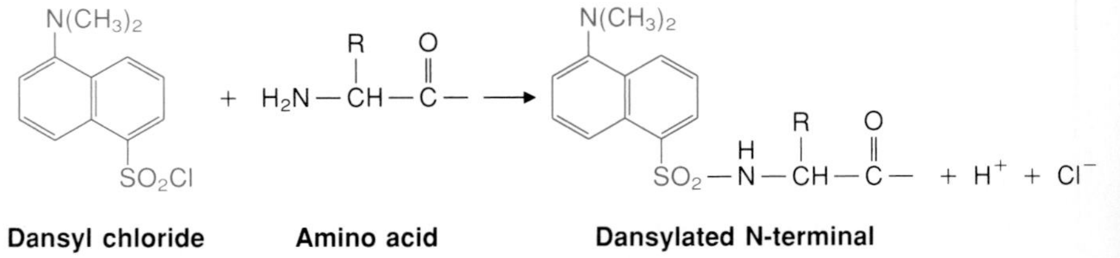

$$(CH_3)_2N-C_{10}H_6-SO_2Cl + H_2N-CH(R)-CO- \longrightarrow (CH_3)_2N-C_{10}H_6-SO_2-NH-CH(R)-CO- + H^+ + Cl^-$$

Dansyl chloride **Amino acid** **Dansylated N-terminal**

The reaction and analysis are very similar to the Sanger method, but dansyl chloride has the advantage that it is intensely fluorescent. Using this reagent, a concentration of residue as low as 1 nM (10^{-9} mol) can be detected.

The Edman Reagent

The sequence of reactions discovered by Pehr Edman is of great importance, for it not only identifies N-termini but also, when used repeatedly, provides a method for the *complete* sequencing of long polypeptides (see Tools of Biochemistry 9). The compound phenylisothiocyanate is reacted in alkali with the terminal amino group to yield a phenylthiocarbamyl derivative of the peptide (step 1, Figure T8.1). This derivative is then treated with a strong anhydrous acid, which results in cleavage of the peptide bond between residues 1 and 2 (step 2).

The derivative of the N-terminal residue then rearranges to yield a phenylthiohydantoin (PTH) derivative of the amino acid (step 3). Two important things have been accomplished. First, the N-terminal residue has been marked with an identifiable label, as in the Sanger and dansyl chloride methods. But in the Edman reaction, the rest of the polypeptide has not been destroyed; it has simply been shortened by one residue. The whole sequence of reactions can now be repeated and the second residue determined. By continued repetition a long polypeptide can be "read," starting from the N-terminal end. This is the basis of one important method for sequencing polypeptide chains.

C-Termini

Not so many good methods exist for determining carboxyl-terminal residues of polypeptides. The major chemical method uses **hydrazinolysis:**

$$\underbrace{H_3\overset{+}{N}-\overset{R_1}{\overset{|}{CH}}-\overset{O}{\overset{\|}{C}}\text{------}NH-\overset{R_{N-1}}{\overset{|}{CH}}-\overset{O}{\overset{\|}{C}}}_{\text{Polypeptide of } N \text{ amino acids}}-\underbrace{NH-\overset{R_N}{\overset{|}{CH}}-COO^-}_{\text{C-terminal residue}}$$

$$\Big\downarrow\; H_2N-NH_2 \text{ Hydrazine} + H^+$$

$$H_3\overset{+}{N}-\overset{R_1}{\overset{|}{CH}}-\overset{O}{\overset{\|}{C}}-NHNH_2 \;+\; \cdots \;+\; H_3\overset{+}{N}-\overset{R_{N-1}}{\overset{|}{CH}}-\overset{O}{\overset{\|}{C}}-NHNH_2 \qquad + \; H_3\overset{+}{N}-\overset{R_N}{\overset{|}{CH}}-COO^-$$

Hydrazine derivatives of all amino acids except the C-terminal residue; C-terminal residue

This reaction leads to cleavage, with modification of every residue except the C-terminal one, which can then easily be distinguished. However, the existence of a number of complicating reactions has led to another technique gaining favor.

The other technique depends on the use of a proteolytic enzyme called carboxypeptidase. There are several such enzymes; each specifically catalyzes the hydrolysis of the last (most C-terminal) peptide bond in a protein:

$$-NH-\overset{R_{N-2}}{\overset{|}{CH}}-\overset{O}{\overset{\|}{C}}-NH-\overset{R_{N-1}}{\overset{|}{CH}}-\overset{O}{\overset{\|}{C}}-\underbrace{NH-\overset{R_N}{\overset{|}{CH}}-COO^-}_{\text{C-terminal residue}}$$

$$\Big\downarrow\; H_2O,\ \text{Carboxypeptidase}$$

$$\underbrace{-NH-\overset{R_{N-2}}{\overset{|}{CH}}-\overset{O}{\overset{\|}{C}}-\underset{H}{\underset{|}{N}}-\overset{R_{N-1}}{\overset{|}{CH}}-C\genfrac{}{}{0pt}{}{\nearrow O}{\searrow O^-}}_{\text{Peptide chain shortened by one residue}} \;+\; \underbrace{H_3\overset{+}{N}-\overset{R_N}{\overset{|}{CH}}-COO^-}_{\text{C-terminal residue}}$$

Earlier studies employed carboxypeptidase A, but carboxypeptidase Y is now favored, for it will remove *any* C-terminal residue, even proline. The reaction will, of course, continue, chewing away residues one by one from the C-terminal end. If the experimenter is skillful (and lucky) it will be possible, by following the production of free amino acids, to tell which comes off first, which second, and so forth. In this way, several residues may be identified before the results become too confusing.

REFERENCES

Gray, W. R. (1972) End group analysis using dansyl chloride. *Methods Enzymol.* 25:121–138.

Schroeder, W. A. (1972) Hydrazinolysis. *Methods Enzymol.* 25:138–143. This volume of *Methods in Enzymology* also contains a number of other chapters on end-group analysis.

9

TOOLS OF BIOCHEMISTRY 9

How to Sequence a Protein

Determination of the primary structure of a protein, like similar analysis of nucleic acids, is commonly referred to as sequencing. Today, virtually all direct sequencing is via the Edman degradation (Tools of Biochemistry 8) and is done almost entirely automatically in instruments known as sequenators. Such a device is able to carry out the entire set of reactions shown in Figure T8.1 over and over again. The sequenator will accumulate in a separate tube the phenylthiohydantoin derivative of each amino acid residue in the polypeptide, starting with the N-terminal residue and proceeding for as many cycles as the operator desires or precision allows. The PTH derivatives are usually identified by high-performance liquid chromatography.

The protein we shall use as an example is bovine insulin. This is appropriate, for it was the first protein ever sequenced, by Frederick Sanger and his co-workers in the early 1950s. (Of course, the methods we describe are much more sophisticated than those available to Sanger in his pioneering work.)

The researcher intending to sequence a protein must first make sure that the material is pure. The protein can be separated from other proteins by some combination of the methods described in Tools of Biochemistry 6. To check for purity, electrophoresis and/or isoelectric focusing may be used. Next, it must be determined whether the material contains more than one polypeptide chain, for in some cases disulfide bridges will covalently bond chains together. End-group determination (Tools of Biochemistry 8) and sodium dodecyl sulfate (SDS) gel electrophoresis in the presence and absence of reducing agents can answer this question. In the insulin example, the investigator would find that there are two chains, A and B (Figure T9.1). These chains must be separated and sequenced individually. To break disulfide bonds and thus separate the chains, several reactions are available. The two most important are described below.

Performic acid oxidation. As shown in step 1 of Figure T9.1, the strong oxidizing agent performic acid will yield cysteic acid:

$$\begin{matrix} -NH-CH(-CH_2-S-)-C(=O)- \\ | \\ -NH-CH(-CH_2-S-)-C(=O)- \end{matrix} \xrightarrow[\text{Performic acid}]{H-C(=O)-OOH} \begin{matrix} -NH-CH(-CH_2-SO_3^-)-C(=O)- \\ \\ -NH-CH(-CH_2-SO_3^-)-C(=O)- \end{matrix}$$

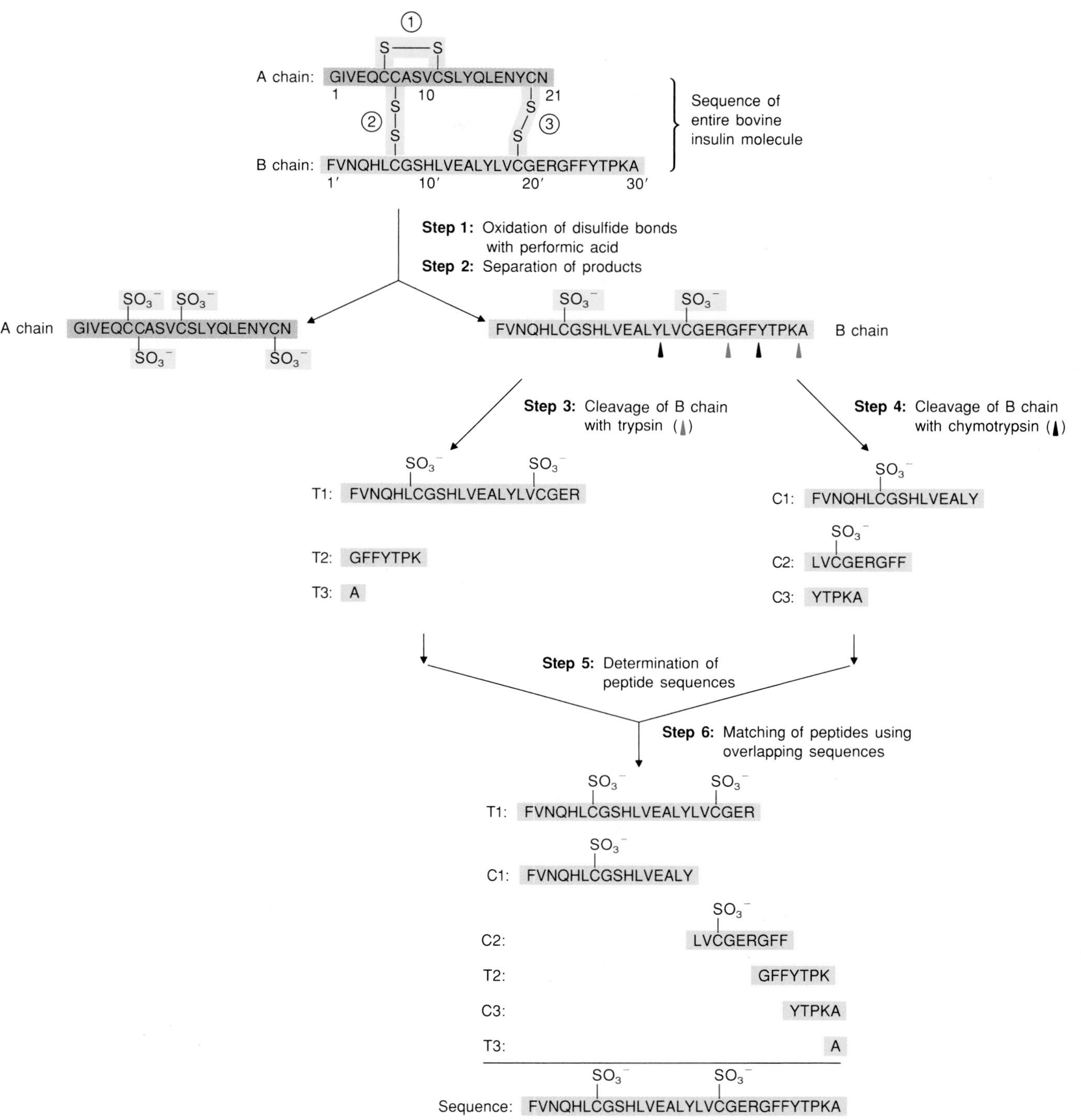

Figure T9.1
Sequencing the B chain of insulin.

Reduction with β-mercaptoethanol:

$$\begin{array}{c}\text{—NH—CH—C(=O)—} \\ | \\ CH_2 \\ | \\ S \\ | \\ S \\ | \\ CH_2 \\ | \\ \text{—NH—CH—C(=O)—}\end{array} + 2(HS\text{—}CH_2CH_2OH) \longrightarrow \begin{array}{c}\text{—NH—CH—C(=O)—} \\ | \\ CH_2 \\ | \\ SH \\ \\ SH \\ | \\ CH_2 \\ | \\ \text{—NH—CH—C(=O)—}\end{array} + \begin{array}{l} S\text{—}CH_2CH_2OH \\ | \\ S\text{—}CH_2CH_2OH \end{array}$$

β-Mercaptoethanol

This reaction leaves free sulfhydryl groups, often positioned so that reoxidation to reform the disulfide bond is likely. Therefore, the sulfhydryls are usually blocked to prevent this. A common blocking reagent is iodoacetate:

$$\begin{array}{c}\text{—NH—CH—C(=O)—} \\ | \\ CH_2 \\ | \\ SH \end{array} + \begin{array}{c} I \\ | \\ CH_2 \\ | \\ COO^- \end{array} \longrightarrow \begin{array}{c}\text{—NH—CH—C(=O)—} \\ | \\ CH_2 \\ | \\ S \\ | \\ CH_2 \\ | \\ COO^- \end{array} + H^+ + I^-$$

Cysteine residue **Iodoacetate** **δ-Carboxymethyl-cysteine residue**

If one of these methods (for example, performic acid oxidation) were carried out with insulin, the whole protein would be cleaved into A and B chains as shown in step 2 of Figure T9.1. These could then be separated by chromatographic methods and the individual chains would be ready for sequence determination.

Before sequencing is started, the amino acid composition is usually determined (Tools of Biochemistry 7). This may point to unusual compositions and thus warn the operator of potential problems. Furthermore, composition data will serve as a check on the sequencing results, since the sequence determined must be consistent with the composition.

Now sequencing can begin. The researcher must recognize that even the best Edman sequencing techniques cannot give reliable results beyond about 40–60 residues. Beyond this point, accumulated impurities and incompleted reactions make results ambiguous. If SDS gel electrophoresis has indicated that the chain is this long or longer, it will be necessary first to cleave it into fragments of manageable size. In bovine insulin, the A and B chains are so short that modern techniques could sequence either in one sequenator run. But to demonstrate the methods needed for larger proteins, we will assume that the investigator must cleave the insulin chains into small polypeptides. This was indeed the case in Sanger's pioneering sequencing studies on insulin. Suppose the insulin B chain is to be sequenced. A first step would be to cleave separate aliquots of the chain with two or more of the specific cleavage reagents described previously. Trypsin and chymotrypsin, for instance, would yield the sets of peptides shown in steps 3 and 4 of Figure T9.1. The individual peptides would then be isolated from the mixture, using, for example, ion-exchange chromatography.

A number of instrumental techniques are currently used to sequence peptides via the Edman technique. We describe two of them below.

Spinning-Cup Sequenators (Figure T9.2a). In this method, the reaction vessel is a spinning cup. The polypeptide is precipitated on the wall of the cup, and reagents are added and products and excess reagents removed from this vessel. All reactions take place in the cup and the phenylthiohydantoin derivatives are collected in a

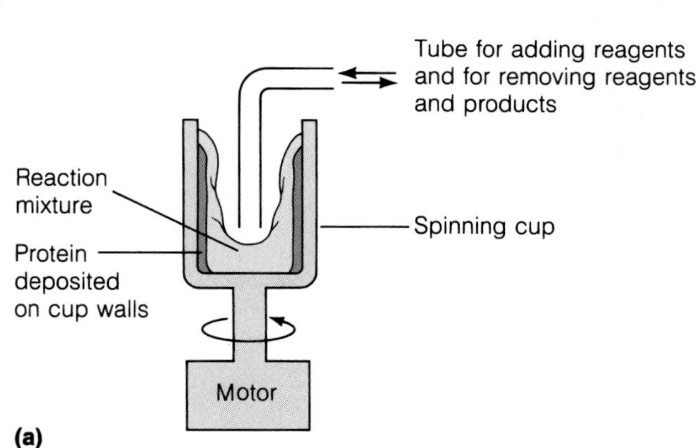

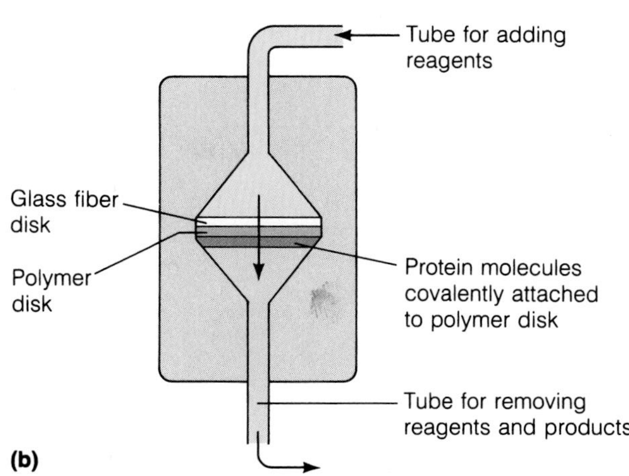

Figure T9.2
Reaction vessels in the two major types of automated sequenators. (**a**) Spinning cup. (**b**) Solid phase.

fraction collector. The spinning cup technique is the original automated method. It requires fairly large quantities of protein and has the disadvantage that it may be difficult to keep insoluble the short polypeptides corresponding to the last few amino acids.

Solid-Phase Sequenators. In such instruments, the polypeptide is covalently coupled, through its C-terminus, to a solid supporting matrix. This is usually on a thin polymer membrane, which is held in the reaction vessel (Figure T9.2b). Again, reagents are added, reactions occur, and products are removed automatically. In a recent variant of this technique, **gas-phase sequencing,** many of the reactants and products are gaseous, allowing for easier handling and less contamination. These solid-phase methods are capable of sequencing very small amounts of polypeptides; as little as 1 μg (10^{-3} mg) sometimes suffices.

Now suppose the peptides shown in Figure T9.1 have each been sequenced (step 5). Although the tryptic peptides alone cover the whole sequence, they are not sufficient to allow us to write down the sequence of the insulin B chain because we do not know the order in which they appear in the chain. However, we also have the chymotryptic peptides, which overlap the tryptic peptides; therefore, all ambiguity is removed. Only one arrangement of the whole chain is consistent with the sequences of these two sets of peptides, as can be seen by matching overlapping sequences (step 6).

Finally, a complete characterization of the covalent structure of a protein requires that the positions of any disulfide bonds be located. In preparation for sequencing, these bonds would have been destroyed, but the positions of all cysteines, some of which *might* have been involved in bonding, would have been determined. The question is, *which* cysteines are so linked in the native protein, and which is connected to which?

To determine the arrangement of disulfide bonds, the experimenter again starts with the native protein—insulin in the example shown in Figure T9.3. Reaction with radioactive labeled iodoacetate will mark any free cysteine residues and fragmentation of the protein into the same peptides used in sequencing will allow the positions of these nonbonded cysteines to be identified (step 1). Then the intact protein is cut with the same specific reagents as used before (step 2), but now without first cleaving disulfide bonds. Some peptides, which are connected by these bonds, will now be attached to one another. These can then be isolated and their disulfide bonds cleaved to find out which of the original set of peptides each contains.

We have described how the entire amino acid sequence, or primary structure, of a protein can be determined. Such analyses have been carried out on several thousand different proteins in the years since Sanger first determined the sequence of insulin. The very fact that such sequencing can be done is convincing proof that each protein has a unique primary structure.

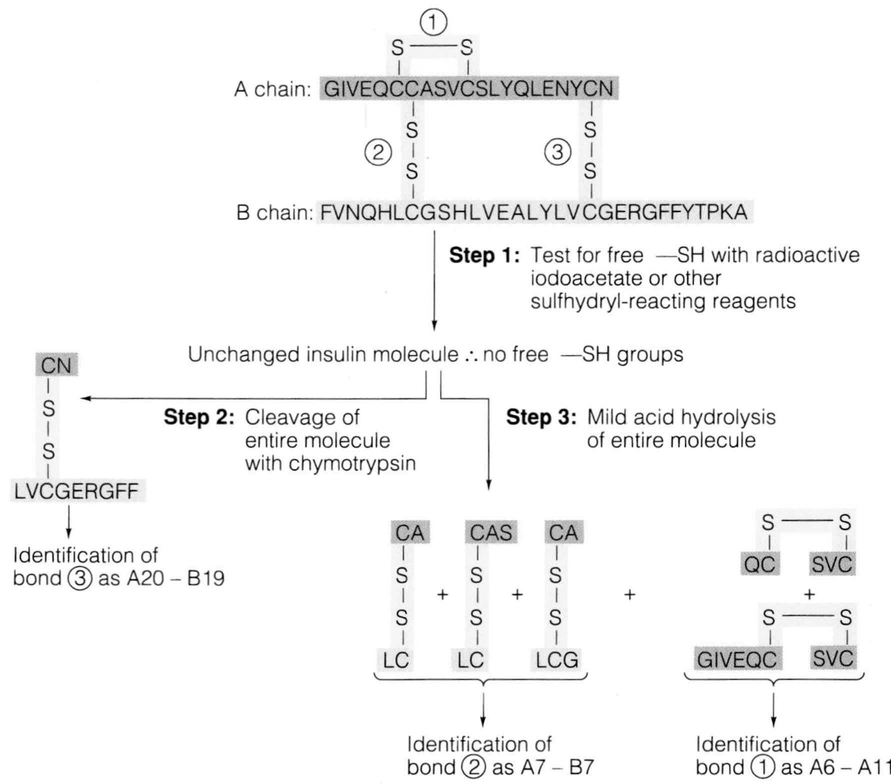

Figure T9.3
Locating the disulfide bonds in insulin.

REFERENCES

Edman, P., and G. Begg (1967) A protein sequenator. *Eur. J. Biochem.* 1:80–91. The first automated method.

Hunkapiller, M. W., J. E. Strickler, and K. J. Wilson (1984) Contemporary methodology for protein structure determination. *Science* 226:304–311.

Sanger, F. (1952) The arrangement of amino acids in proteins. *Adv. Protein Chem.* 7:1–67. A review of the earlier work.

Stein, S., and S. Udenfreund (1984) A picomole protein and peptide chemistry: Some applications to the opioid peptides. *Anal. Biochem.* 136:7–23.

Walsh, K. A., L. H. Ericsson, D. C. Parmelee, and K. Titani (1981) Advances in protein sequencing. *Annu. Rev. Biochem.* 50:261–284.

CHAPTER 6

The Three-Dimensional Structure of Proteins

In Chapter 5 we introduced the concept of protein *primary* structure. We emphasized that this first level of organization, the amino acid sequence, is dictated by the DNA sequence in the gene for each protein. Most proteins exhibit higher levels of structural organization. It is the specific three-dimensional structure of each protein that allows it to function in its particular biological role.

Figure 6.1 depicts the three-dimensional conformation of the myoglobin molecule. If you examine the figure closely, you can see that the polypeptide chain of myoglobin exhibits two distinguishable levels of three-dimensional folding. First, the chain appears to be locally wrapped into regions of helical structure. Such local *regular* folding is called the **secondary structure** of the molecule. The helically coiled regions are in turn folded into a specific compact structure for the entire polypeptide chain. We call this the **tertiary structure** of the molecule. Later in this chapter we shall find that some proteins consist of several polypeptide chains, arranged in a regular manner. This arrangement we designate the **quaternary** level of organization.

This chapter is devoted to an examination of the several levels of protein structure—their geometry, how they are stabilized, and their importance in protein function.

Secondary Structure: Regular Ways to Fold the Polypeptide Chain

The Discovery of Regular Polypeptide Structures

In the early 1950s, Linus Pauling and his collaborators began a systematic analysis of the possible regular conformations of the polypeptide chain. They postulated several principles that any such structure must obey:

1. The bond lengths and bond angles should be distorted as little as possible from those found in small peptides, as shown in Figure 5.12b (Chapter 5).

2. No two atoms should approach one another more closely than is allowed by their van der Waals radii.

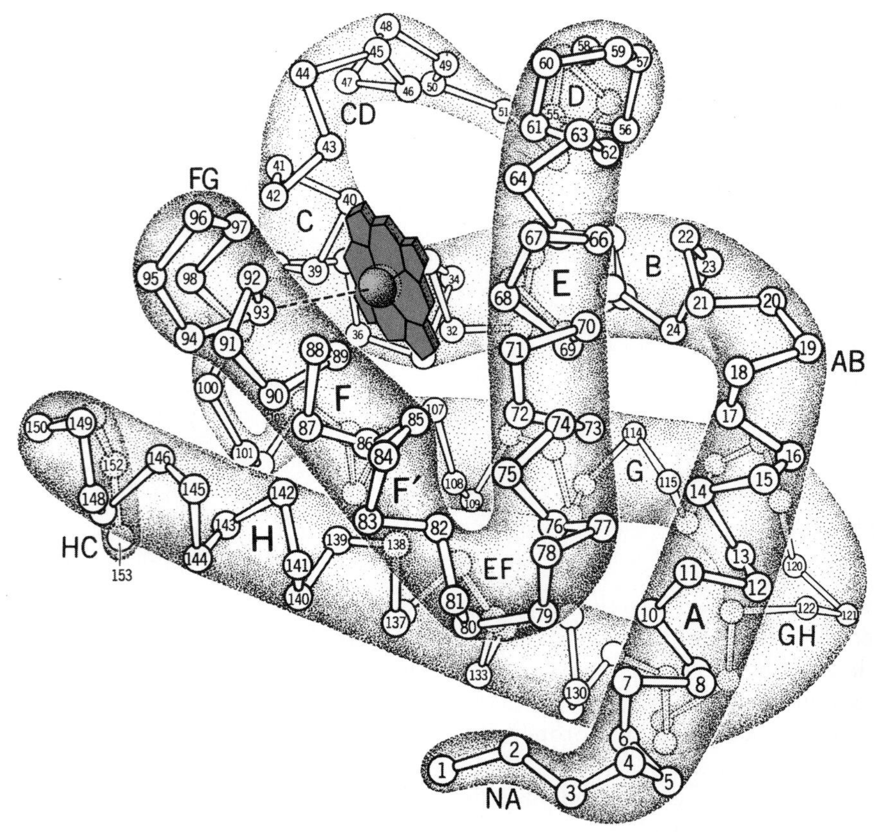

Figure 6.1
Three-dimensional folding of the protein myoglobin. Each amino acid is indicated by a circle corresponding to its α-carbon atom. Side chains are not shown, in order to emphasize the way in which the polypeptide backbone is wrapped into helices and folded. **N** and **C** refer to the N-terminus and C-terminus, respectively. Individual α-helical regions are labeled A–H, with turn regions designated by two letters (e.g., GH). This protein folds about a *heme* group, a planar heterocyclic structure that chelates iron and serves as the oxygen binding site.

3. The amide group must remain planar and in the trans configuration as shown in Figure 5.8b. (This feature had been recognized in earlier x-ray diffraction studies of small peptides.) Consequently, rotation is possible only about the two bonds adjacent to the α-carbon in each amino acid residue, as shown in Figure 6.2.

4. Some kind of noncovalent bonding is necessary to stabilize a regular folding. The most obvious possibility is hydrogen bonding between amide protons and carbonyl oxygens:

$$\rangle N{-}H \cdots O{=}C\langle$$

Thus, the preferred conformations must be those that allow a maximum amount of hydrogen bonding, yet satisfy criteria 1–3.

Working mainly with molecular models, Pauling and his associates were able to arrive at a small number of regular conformations that satisfied all of these criteria. Some were helical structures, and some were sheetlike. Figure 6.3a and b show two major structures proposed by Pauling and Corey—the α *helix* and β *sheet.* Figure 6.4 shows two other polypeptide helices that have since been defined. The 3_{10} helix is observed in some proteins but is not as common as the α helix. The π helix, while sterically possible, has not been observed, possibly because it has a hole down the middle that reduces van der Waals interactions but is too small to admit water molecules. All of the structures shown in Figures 6.3 and 6.4 satisfy

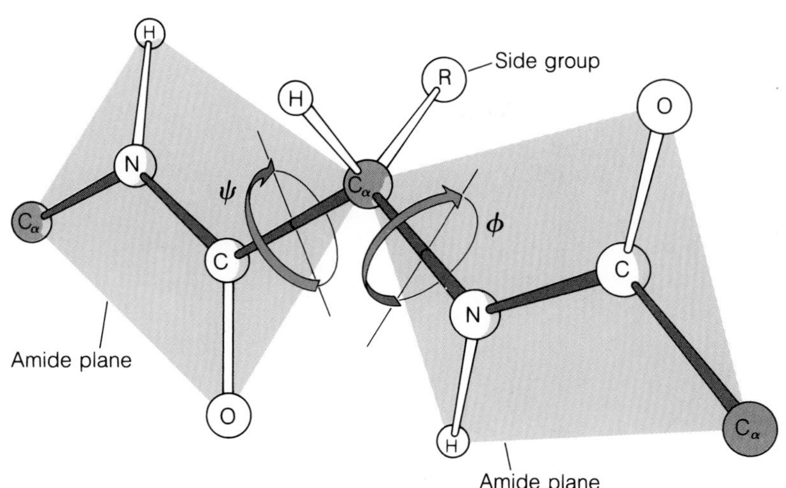

Figure 6.2
Rotation around bonds in a polypeptide chain. Two adjacent amide planes are shown in green. Rotation is allowed only about the N—C_α and C_α—$\overset{\overset{\displaystyle O}{\|}}{C}$— bonds. The angles of rotation about these bonds are defined as ϕ and ψ, with directions defined as positive rotation as shown by the arrows. The extended conformation of the chain shown here corresponds to $\phi = 180°$, $\psi = 180°$.

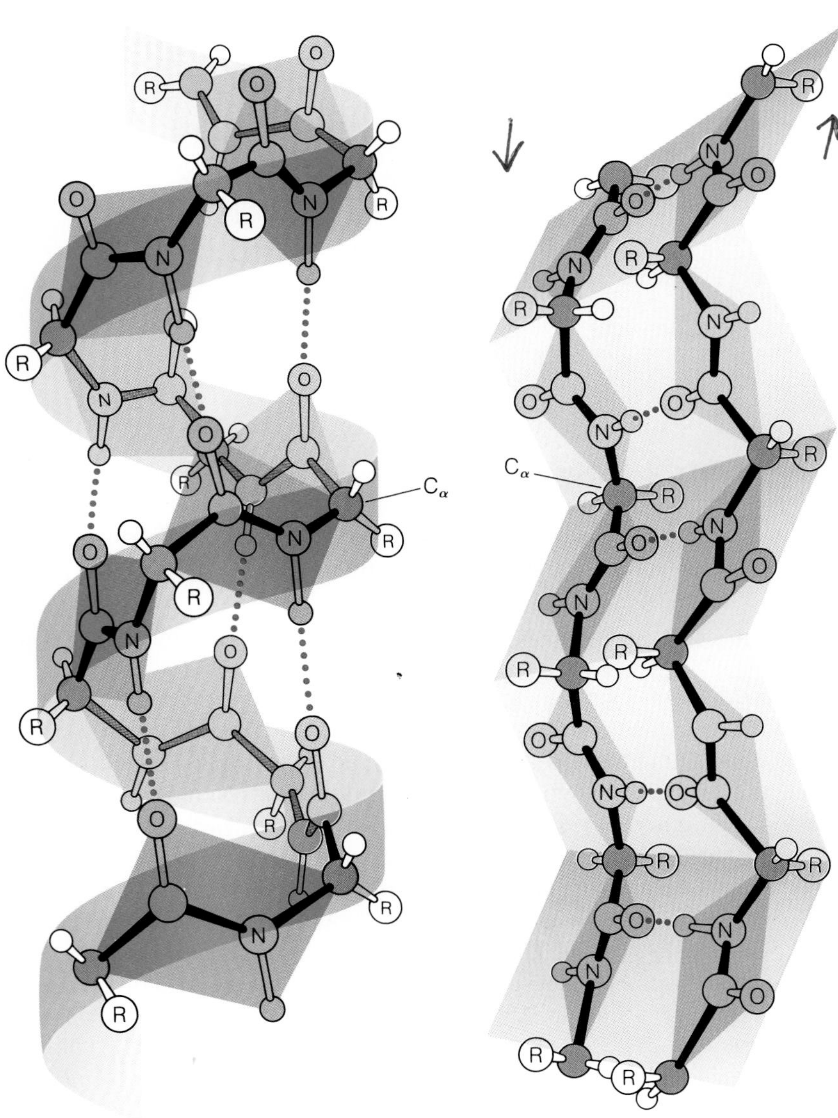

Figure 6.3
The α helix and β sheet. These are the two most important regular secondary structures of proteins. In the α helix (**a**) the hydrogen bonds (red) are within a single chain, whereas in the β sheet (**b**) hydrogen bonds are *between* chains that run side by side.

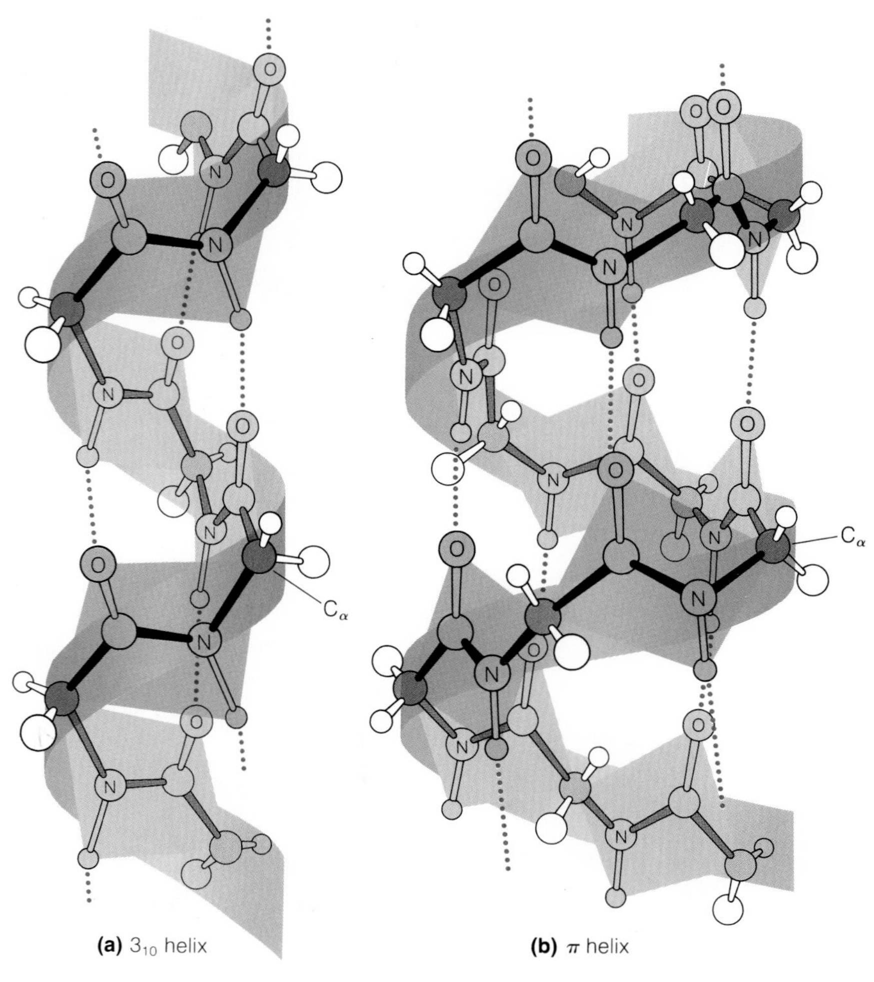

Figure 6.4
Two other possible helical structures of polypeptide chains: (**a**) The 3_{10} helix and (**b**) the π (or 4.4_{16}) helix. Hydrogen bonds are shown in red.

the criteria listed above. In particular, in each structure the peptide group is planar, and every amide proton and every carbonyl oxygen (except for a few near the ends of helices) are involved in hydrogen bonds. Each constitutes a possible kind of secondary structure in proteins.

Describing the Structures

In Chapter 4 (Tools of Biochemistry 5) we listed the distances that define a molecular helix: the crystallographic repeat (c), the pitch (p), and the rise (h). We also pointed out that helices may be either right-handed or left-handed and may contain either an integral number of residues per turn or a nonintegral number. We call the number of residues per turn n and the number of residues per repeat m. The latter number must always be an integer, since it defines an exact repeat of the structure. If there is an integral number of residues per turn, the pitch and the repeat will be equal, and $n = m$. Some helices of this kind are illustrated schematically in Figure 6.5. Note that as the number of residues per turn increases, the structure changes progressively from a flat ribbon to a broad helix and eventually to a closed ring with $p = 0$.

One of Pauling's major insights was to recognize that helices do not

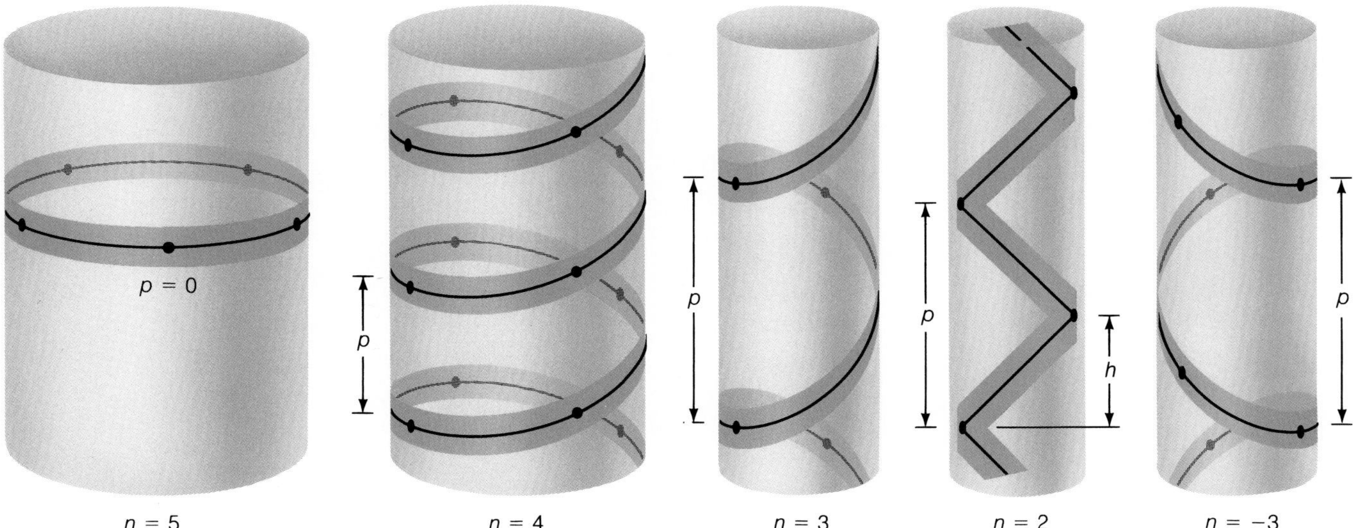

Figure 6.5
Idealized polypeptide helices with differing numbers of residues per turn (n). In each case the pitch (p) is indicated, and for $n = 2$ the rise (h) is also shown. Polypeptides can form helices ranging from a closed ring ($n = 5$, $p = 0$) to a twofold helix ($n = 2$), with each residue rotated by 180° with respect to the preceding one. The $n = 4$ and $n = 3$ helices are right handed, the $n = 5$ and $n = 2$ structures have no handedness, and the $n = -3$ helix shown is left handed. The right α helix (not shown here), with $n = 3.6$, is intermediate between the $n = 3$ and $n = 4$ structures.

have to have an integral number of residues per turn. For example, the α helix repeats after exactly 18 residues, which amounts to 5 turns. It has, therefore, 3.6 residues per turn. Since the pitch of a helix is given by $p = nh$, we have for the α helix, with a rise of 0.15 nm/residue, $p = 3.6$ (res/turn) × 0.15 (nm/res) = 0.54 nm/turn. Parameters for the other helices shown in Figure 6.4 are listed in Table 6.1.

The parameters defined above describe most molecular helices. For polypeptide helices, which involve hydrogen bonding, there is an additional important quantity. If you examine the model for the α helix, you will note that each carbonyl oxygen is hydrogen bonded to the amino protein on the *fourth* residue up the helix. Thus, if we include the hydrogen bond, a loop of 13 atoms is formed, as shown in Figure 6.6. Each of the helices shown in Figures 6.3 and 6.4 has a different number of atoms in such hydrogen-bonded loops. We shall call this number N. A quick way to describe a polypeptide helix, then, is by the shorthand n_N, where n is the number of residues per turn. The 3_{10} helix fits this description; there are exactly 3.0 residues per turn and a 10-member loop. The α helix could also be called a 3.6_{13} helix, and the π helix a 4.4_{16} helix.

Figure 6.6
Hydrogen bonding patterns for four different helices. The structures are represented in a diagrammatic way to simplify counting the atoms in each H-bonded loop. For example, there are 13 atoms in the H loop corresponding to the α (3.6_{13}) helix.

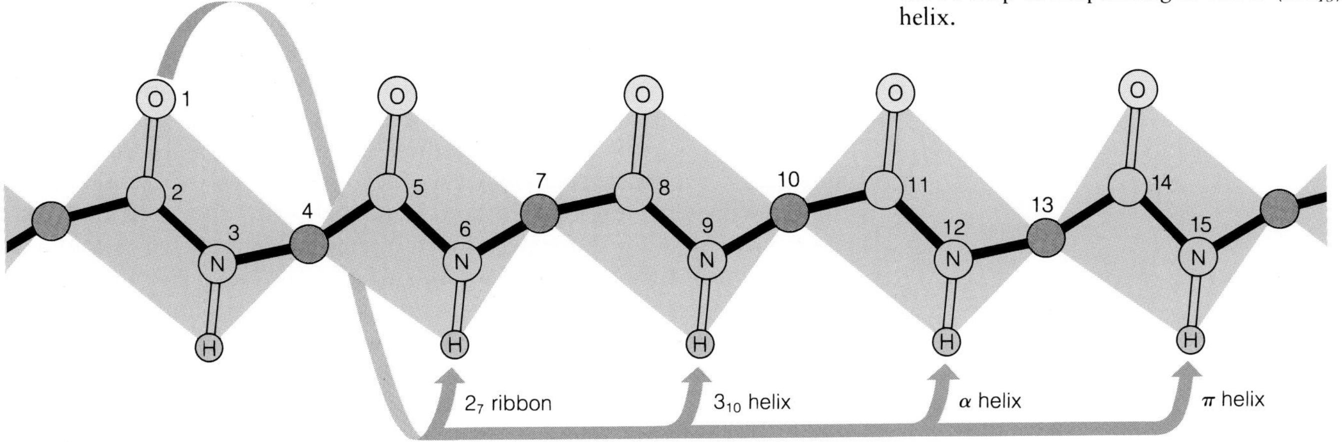

Table 6.1
Parameters of some polypeptide secondary structures

Structure Type	Residues/ Turn	Rise (nm)	Number of Atoms in H-Bonded Ring	ϕ (°)	ψ (°)
Antiparallel β sheet	2.0	0.34	[a]	−139	+135
Parallel β sheet	2.0	0.32	[a]	−119	+113
3_{10} helix	3.0	0.20	10	−49	−26
α helix (3.6_{13})	3.6	0.15	13	−57	−47
π helix (4.4_{16})	4.4	0.12	16	−57	−70

[a]Bonding is between polypeptide chains.

Because hydrogen bonds tend to be linear, the atoms —N—H···O= in polypeptide helices should lie on a straight line. If you examine Figures 6.3 and 6.4 you will see that this requirement is at least approximately satisfied for the 3_{10}, α, and π helices. However, it is very difficult to make helices with small n and linear hydrogen bonds. Therefore, the only $n = 2$ structure that is found is the *β-pleated sheet* structure shown in Figure 6.3b. Here, each residue is rotated by 180° with respect to the preceding one, which makes each chain an $n = 2$ helix. If the chains are also folded in an accordionlike fashion, linear hydrogen bonds can occur *between* adjacent chains. Actually, there are two ways in which this can be done (Figure 6.7). The chains can have their N → C directions running parallel, to make the *parallel β sheet,* or they can be antiparallel.

Thus, there are two general classes of possible secondary protein structures: various helices and at least two types of pleated sheet. But just because a structure can be drawn does not mean that it necessarily exists. Many kinds of helices can be imagined that are sterically impossible because atoms in the backbone and/or side chains would overlap. These steric restrictions can be fully appreciated only when one examines space-filling models, as shown in Figure 6.8. The α-helix, for example, is fully packed. To examine this question in a systematic way, we need a general procedure for describing polypeptide conformations.

Ramachandran Plots

As was shown in Figure 6.2, each residue in a polypeptide chain has only two bonds about which rotation is permitted. The angles of rotation about these bonds, defined as ϕ and ψ, may be adjusted to form different structures. Thus we could describe the backbone configuration of any particular residue in any protein by specifying these two angles. To make the definition meaningful, we must specify what we mean by a positive direction of rotation and the zero angle conformation of each. The directions of positive rotation about ϕ and ψ are given by the arrows in Figure 6.2. The conformation shown in that figure corresponds to $\phi = 180°$ and $\psi = 180°$.

With these conventions, the backbone conformation of any particular residue in a protein can be described by a point on a map (Figure 6.9) with coordinates ϕ and ψ. If a protein had all of its residues in a particular secondary structure (an α helix, for example), the points for all residues would superimpose. Thus, a single point on such a map also describes a

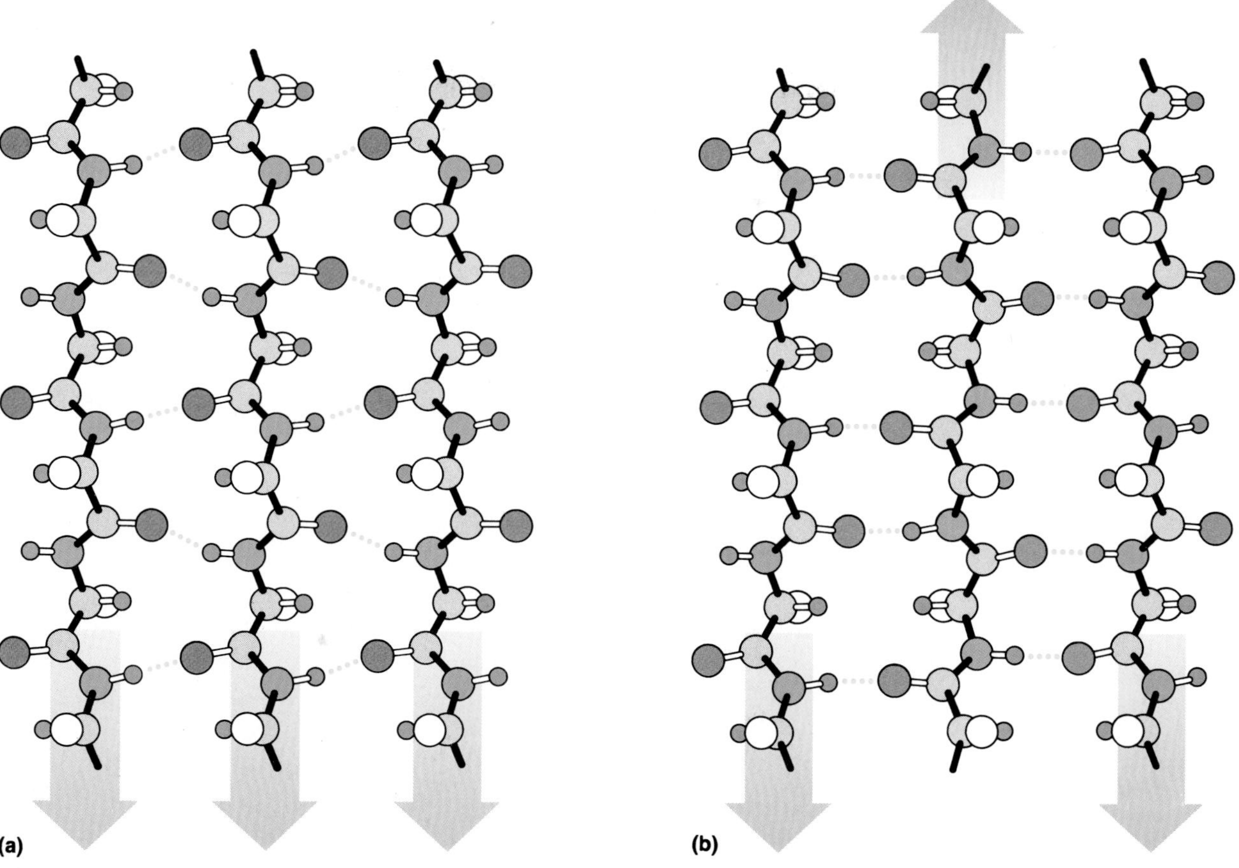

Figure 6.7
Parallel β sheet (**a**) and antiparallel β sheet (**b**). The arrows point in N → C directions in each chain. Gray = carbon, red = oxygen, blue = hydrogen; purple = nitrogen; dotted lines = hydrogen bonds.

given secondary structure. Such maps are called **Ramachandran plots,** after the biochemist who first made extensive use of them.

The Ramachandran plot in Figure 6.9 illustrates the positions of the various regular secondary structures discussed above, as well as some additional conceivable ones. Note that all conformations with a given number of residues per turn lie on one of the set of lines drawn across the map. Particularly important are the lines corresponding to the flat ribbon (n = 2) and the points corresponding to closed rings (n = 5). As these are passed, the handedness of the helix changes. This can easily be understood from the following thought experiment. Imagine unwinding a right-hand helix until it approaches a flat ribbon. The value of n will decrease until, at n = 2, the flat ribbon structure is reached. If one continues to wind in the same direction, a left-hand helix begins to form. Similarly, if one progresses to broader and broader helices a ring structure is eventually reached. With polypeptide helices this is the five-membered ring indicated on the diagram. If one continued to modify the structure in the same manner, a broad helix of opposite sign would be generated.

One of the most useful features of Ramachandran maps is that they allow us to describe very simply which structures are sterically possible and which are not. For many pairs of ϕ, ψ values atoms in the chain would

Figure 6.8
Space-filling model of a segment of α helix. The segment illustrated is from the E helix in sperm whale myoglobin.

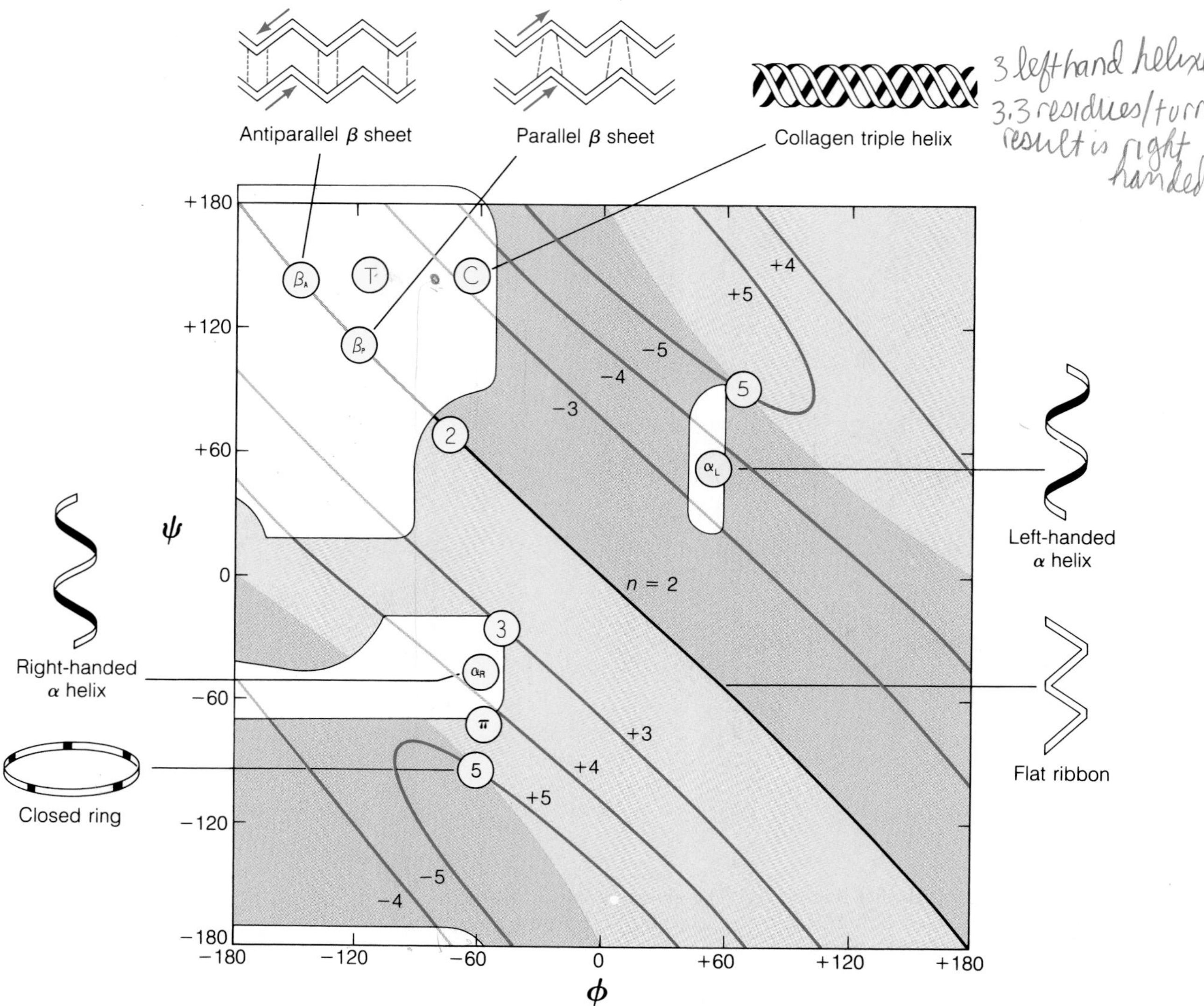

Figure 6.9
The Ramachandran diagram. The coordinates are ϕ, ψ angles defined as in Figure 6.2. The white areas correspond to those allowed when side chains are alanines. Circles with symbols correspond to important secondary structures:

- α_R, α_L: right and left α helices
- β_P, β_A: parallel and antiparallel β sheets
- T: twisted β sheet—parallel or antiparallel
- 3: the 3_{10} helix
- π: the 4.4_{16} helix
- 5: five-membered ring
- 2: the twofold ribbon
- C: the collagen helix

The dark contour lines running across the graph correspond to various values of n. Where values of n are positive, the helix is right handed; where negative, it is left handed. The contour line bisecting the drawing corresponds to $n = 2$. Areas for right helices are in red; those for left helices in blue.

approach closer than allowed by their van der Waals radii. Such conformations are sterically excluded. Ramachandran and other researchers have examined the entire map surface, using models and computers to determine which conformations are allowed. These are shown as the uncolored areas in Figure 6.9. Clearly, only a relatively small fraction of the conceivable conformations is actually possible. All of the regular structures we have discussed fall into or very close to these regions.

Although Figure 6.9 shows the left-hand α helix lying on the edge of an allowed region, it is, in fact, not nearly as favored as the right-hand form. This is a consequence of the fact that all amino acids in proteins are of the L form. With L-amino acids, steric interference between the side chains and the backbone of the helix is less with a right-hand helix. Such side-chain effects depend, of course, on the bulkiness of the side chain. The map shown was drawn with the assumption that all side chains are alanine (—CH_3) groups. If bulkier side chains were considered, the "allowed" region would be shrunk even more. Conversely, glycine allows more conformations than are shown here.

The foregoing analysis of protein structure is quite abstract. How does it square with reality? Quite well, as shown in Figure 6.10, which illustrates the individual ϕ, ψ angles for amino acid residues in a small protein. These

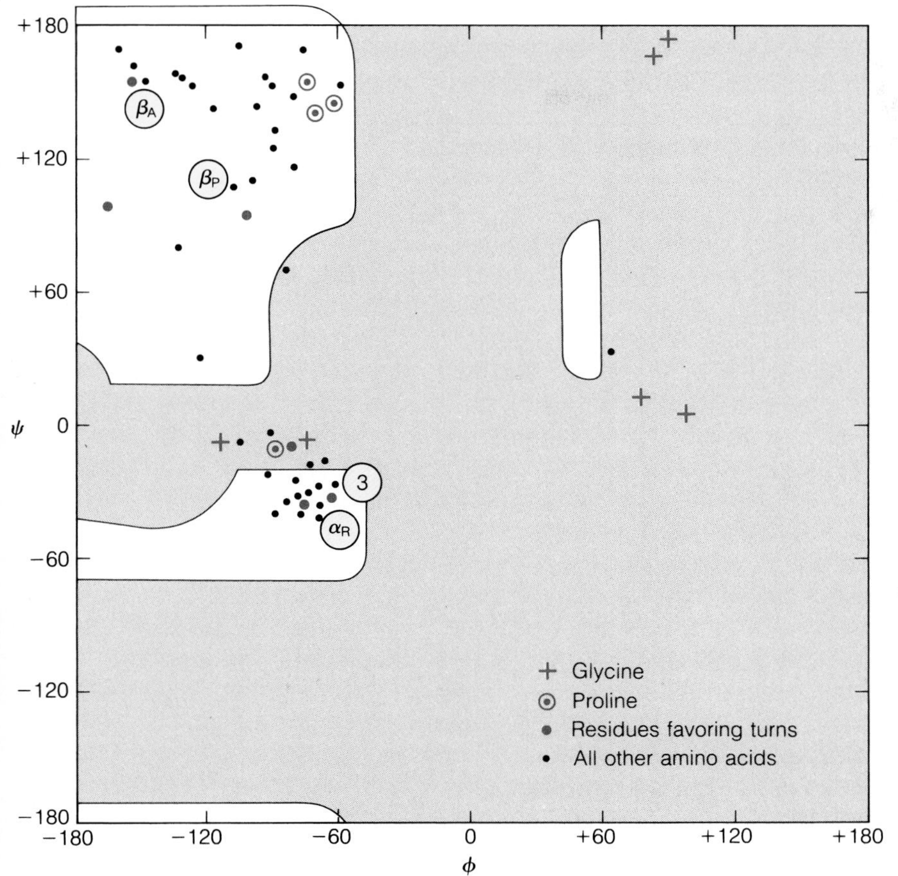

Figure 6.10
A Ramachandran plot of the amino acid residues in bovine trypsin inhibitor. The positions of major structures are indicated by yellow circles (see Figure 6.9). Note that most ϕ, ψ pairs fall within or close to the allowed regions (white). Many of the points in nonallowed regions correspond to glycines.

quantities have been calculated from a detailed examination of the actual molecular structure, as determined by x-ray diffraction. Most of the points in this map fall very close to the α_R position, or the β-sheet positions. But they do not correspond exactly to these points, testifying to the existence of distortions of these structures in a real protein. Although most of the points fall in "allowed" regions, a few lie in "nonallowed" regions. These are mainly glycines, for which a much wider range of ϕ, ψ angles is allowed because the side chain (—H) is so small.

Our discussion so far provides a background for understanding the basics of protein structure; it is now time to consider some specific cases. We begin with the observation that there exist two major classes of proteins, *fibrous* and *globular*, distinguished by major structural differences. Let us first consider the fibrous proteins.

Fibrous Proteins: Structural Materials of Cells and Tissues

Fibrous proteins, in most cases, play structural roles in animal cells and tissues; they hold things together. They include the major proteins of skin and connective tissue and of animal fibers like hair, wool, and silk. The amino acid sequence of each of these proteins favors a particular kind of secondary structure, which in turn confers a particular set of appropriate mechanical properties on the substance. Table 6.2 lists the amino acid composition of four specific examples, which we discuss below.

Table 6.2
Amino acid compositions of some fibrous proteins. The three most abundant amino acids in each protein are indicated in italics; values given are in mole percent

Amino Acid	Fibroin (silk)	α-Keratin (wool)	Collagen (bovine tendon)	Elastin (pig aorta)
Gly	*44.6*	8.1	*32.7*	*32.3*
Ala	*29.4*	5.0	*12.0*	*23.0*
Ser	*12.2*	*10.2*	3.4	1.3
Glu + Gln	1.0	*12.1*	7.7	2.1
Cys	0	*11.2*	0	[e]
Pro	0.3	7.5	*22.1*[a]	10.7[c]
Arg	0.5	7.2	5.0	0.6
Leu	0.5	6.9	2.1	5.1
Thr	0.9	6.5	1.6	1.6
Asp + Asn	1.3	6.0	4.5	0.9
Val	2.2	5.1	1.8	*12.1*
Tyr	5.2	4.2	0.4	1.7
Ileu	0.7	2.8	0.9	1.9
Phe	0.5	2.5	1.2	3.2
Lys	0.3	2.3	3.7[b]	3.6[d]
Trp	0.2	1.2	0	[e]
His	0.2	0.7	0.3	[e]
Met	0	0.5	0.7	[e]

[a] About 39% of this is hydroxyproline.
[b] About 14% of this is hydroxylysine.
[c] About 13% of this is hydroxyproline.
[d] Most (about 80%) is involved in cross-links.
[e] Essentially absent.

Fibroin

The fibers spun by the silkworm are made in large part of a single protein, *fibroin.* Its structure contains long regions of antiparallel β sheet, with the polypeptide chains running parallel to the fiber axis. The β-sheet regions comprise almost exclusively multiple repetitions of the sequence

[Gly–Ala–Gly–Ala–Gly–Ser–Gly–Ala–Ala–Gly–(Ser–Gly–Ala–Gly–Ala–Gly)$_8$]

Examining this sequence, you will note that almost every other residue is Gly and that between these lie either Ala or Ser residues. This alternation allows the sheets to fit together and pack on top of one another in the manner shown in Figure 6.11. The arrangement results in a fiber that is strong and relatively inextensible, because the covalently bonded chains are stretched to nearly their maximum possible length. Yet the fibers are very flexible, for bonding between the sheets involves only the weak van der Waals interactions between the side chains. Not all of the fibroin protein is in β sheets. As the amino acid composition in Table 6.2 shows, there are small amounts of other, bulky amino acids like valine and tyrosine, which would not fit into the structure shown. These are carried in folded regions that periodically interrupt the β-sheet segments, and they probably account for the amount of stretchiness that silk fibers have. In fact, different species of silkworms produce fibroins with different extents of such non-β-sheet structure and corresponding differences in stretchiness. The overall fibroin

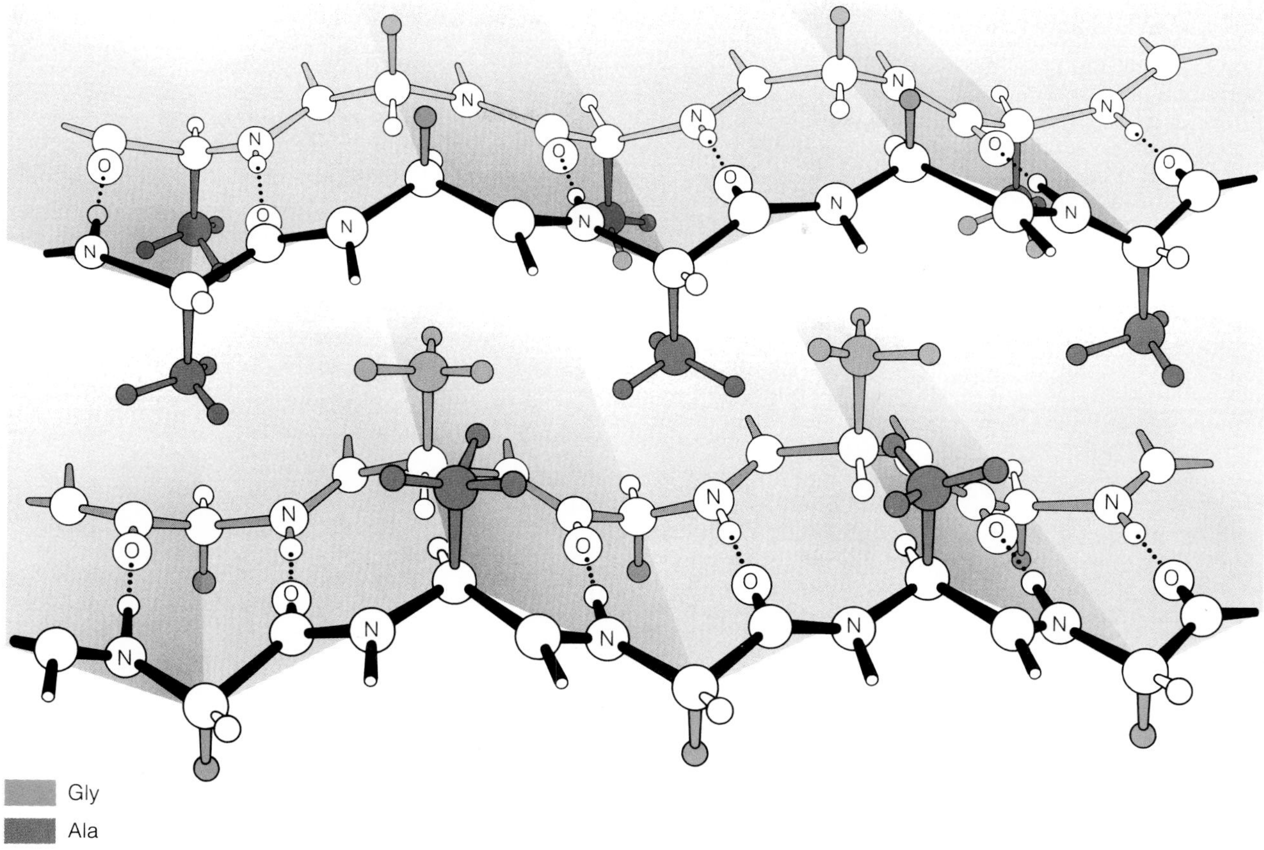

Gly
Ala

(a)

structure is a beautiful example of a protein molecule that has evolved to perform a particular function—to provide a tough, yet flexible fiber for the silkworm's cocoon.

The Keratins

There are two classes of proteins that have similar amino acid sequences and biological function, called α- and β-keratins. The **α-keratins** are the major proteins of hair and wool. These molecules are constructed primarily from α helices. The individual keratin molecules are very long and lie with their helix axes nearly parallel to the fiber axis. The individual α helices in a hair are wound together in a manner reminiscent of a rope or cable (Figure 6.12). First, three helices are wrapped about one another to form a **protofibril.** Then, in a 9 + 2 arrangement, the protofibrils are combined to form a **microfibril.** Numbers of these microfibrils are in turn packed together to form **macrofibrils.** As the figure shows, these are the principal structural elements of a hair.

Because of the helical structure of the polypeptide chains, hair fibers are quite extensible. If fibers are heated and stretched to about twice their normal length, the hydrogen bonds along the α helices are broken, and the protein is converted to a β structure. This elongation is not permanent, however, for the chains are held together by a network of disulfide bonds

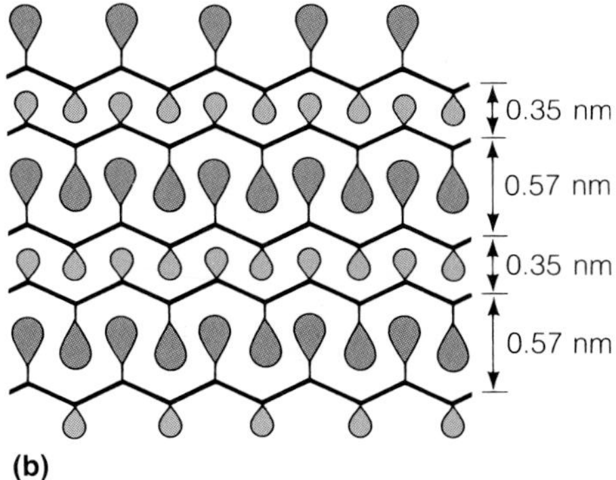

(b)

Figure 6.11
The structure of silk fibroin. (**a**) A three-dimensional view of stacked β sheets. (**b**) Interdigitation of alanine (or serine) side chains (purple) and glycine (hydrogen) side chains (orange) in alternating layers.

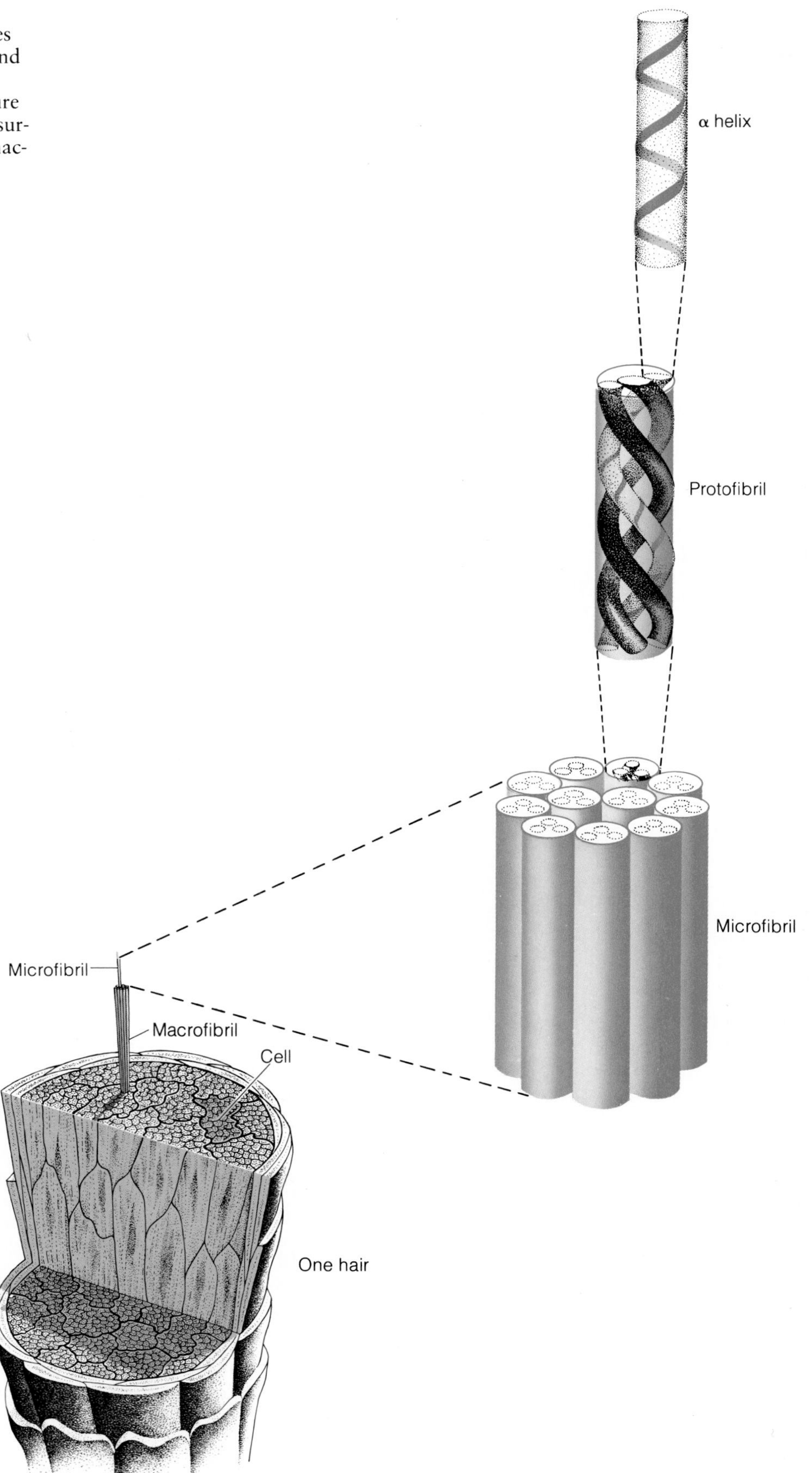

Figure 6.12
The structure of hair. α-Keratin helices wrap into protofibrils, which then bond together into microfibrils and macrofibrils. The microfibrils have a structure in which two central protofibrils are surrounded by nine others. Bundles of macrofibrils pass through and around the cells in a hair.

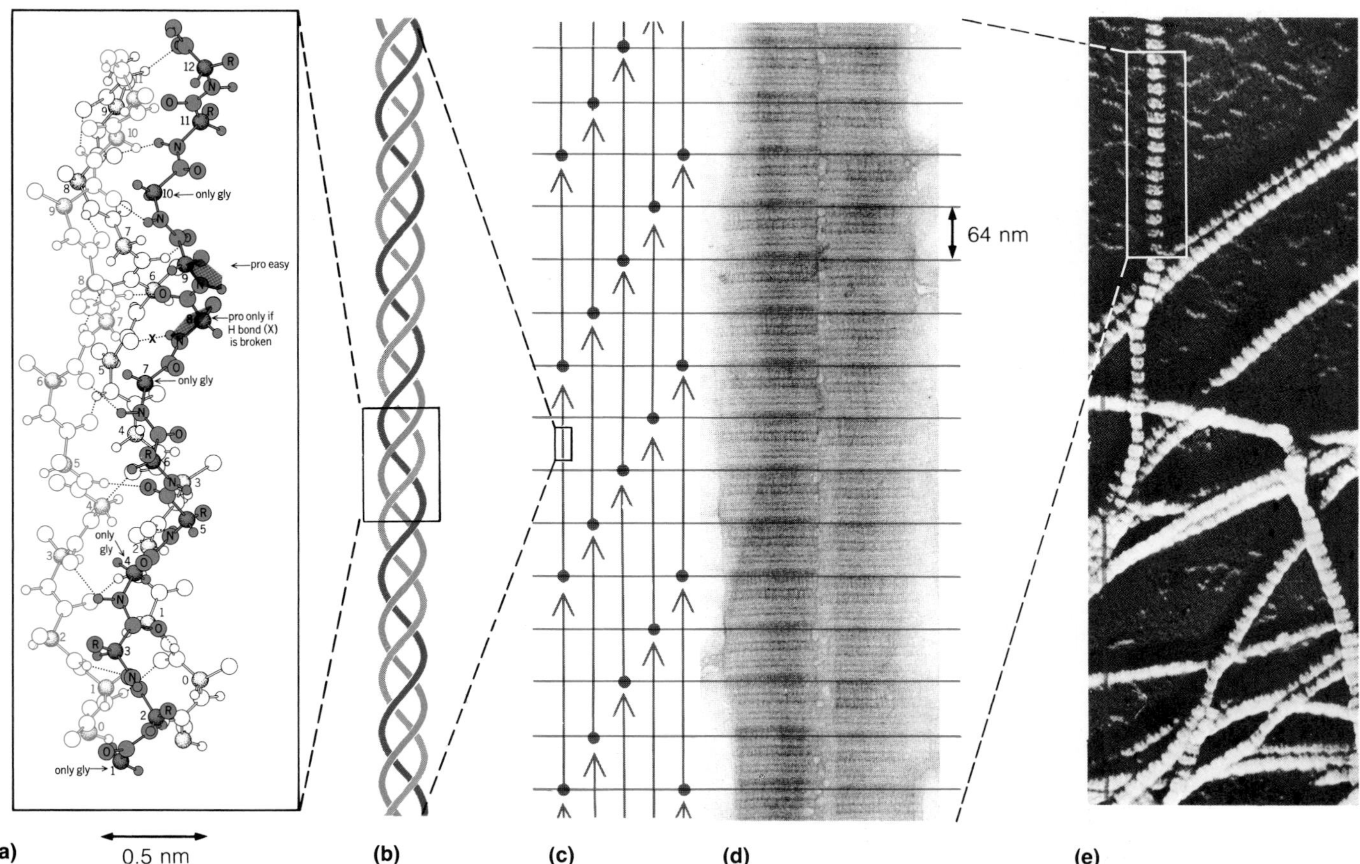

Figure 6.13
The structure of collagen fibers. The tropocollagen molecule is a triple helix (**a** and **b**). These molecules align side by side in a staggered fashion (**c**) to form the fiber. The periodic pattern can be seen in detail in electron micrographs (**d**) and (**e**).

(keratin is particularly rich in cysteine). These links cause the original structure to recover when tension is released. When human hair is given a "permanent," the disulfide bonds are broken purposely by reduction and allowed to reform after the hair has been shaped.

With some other forms of keratin, a β-sheet structure is the normal conformation. These **β-keratins** are found in feathers and in the skin, claws, beaks, and scales of most reptiles and birds. The structure is somewhat similar to that of silk fibroin but exhibits many modifications to provide the exact physical properties needed in different tissues.

Collagen

Because it performs such a wide variety of functions, **collagen** is the most abundant single protein in most vertebrates. In large animals, it may make up a third of the total protein mass. Collagen is the *matrix*, or cement, material in bone, on which the mineral constituents precipitate; it forms the major portion of tendons; and it is an important constituent of skin.

The basic unit of collagen is the **tropocollagen** molecule, a triple helix of three polypeptide chains, each about 1000 residues in length. This threefold helical structure, shown schematically in Figure 6.13a, is unique to collagen. The individual chains are left-hand helices, with about 3.3 residues/turn. Three of these wrap around one another in a right-hand sense, with hydrogen bonds extending *between* the chains. Examination of the model reveals that every third residue, which must lie near the center of the triple helix, can *only* be glycine (see Figure 6.13a). Any side chain other than

H— would be too bulky. Formation of the individual helices also requires that the preceding residue be proline or hydroxyproline. Thus, a repetitive theme in the sequence must be (Gly–X–Pro) or (Gly–X–Hypro), where X is some other amino acid. A glance at the partial sequence shown in Figure 6.13a confirms that this is indeed so. Again, it is clear how a particular kind of repetitive sequence dictates a particular structure.

Collagen is also unusual in the widespread modification of proline to hydroxyproline. Most of the hydrogen bonds between chains in the triple helix are from amide protons to carbonyl oxygens, but the —OH groups of hydroxyproline also seem to participate in stabilizing the structure. Hydroxylation of lysine residues in collagen also occurs but is much less frequent. It plays a different role, serving to form attachment sites for polysaccharides.

These hydroxylation reactions involve **vitamin C,** ascorbic acid (see Chapter 21). A symptom of extreme vitamin C deficiency, called **scurvy,** is the weakening of collagen fibers as hydroxyproline is not replaced. Consequences are as might be expected: skin and gums develop lesions, and blood vessels are weakened. The condition is quickly improved by administration of vitamin C.

The individual tropocollagen molecules pack together in a collagen fiber in a very specific way. Each is about 300 nm long, and each overlaps its neighbor by about 64 nm, producing a characteristic banded appearance of the fibers (Figure 6.13c, d, e). This structure clearly contributes remarkable strength: collagen fibers in tendons have a strength comparable to that of hard-drawn copper wire.

Part of the toughness of collagen is accounted for by the fact that tropocollagen molecules are cross-linked to one another via a reaction involving lysine side chains. Some of the lysine side chains are oxidized to aldehyde derivatives, which can then react with either a lysine residue or one another to produce an aldol cross-link:

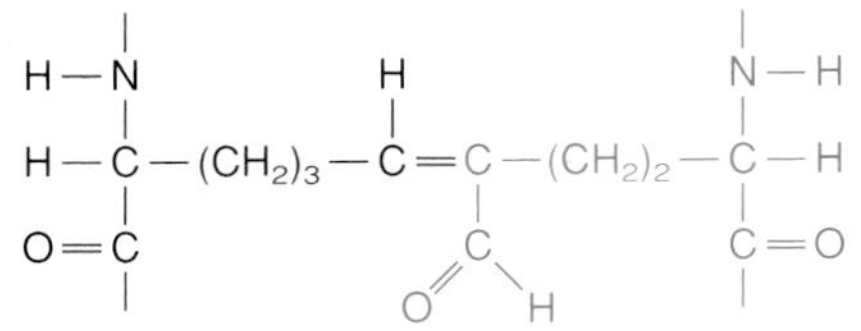

```
   |                              |
 H—N              H               N—H
   |              |               |
 H—C—(CH2)3—C=C—(CH2)2—C—H
   |                |             |
 O=C                C             C=O
   |              //  \           |
                 O     H
```

This process continues through life, and the accumulating cross-links make the collagen become steadily less elastic and more brittle. As a result, bones and tendons in older individuals are more easily snapped, and the skin loses much of its elasticity. Many of the signs we associate with aging are consequences of this simple cross-linking process.

Elastin

Whereas collagen is the protein used in tissues where strength or toughness is required, sometimes a highly elastic fiber is needed. Ligaments and arterial blood vessels are two such cases. Such tissues contain large amounts of the fibrous protein **elastin.**

The polypeptide chain of elastin is rich in glycine and alanine and is very flexible and easily extended. However, the sequence also contains frequent lysine side chains, which can be involved in cross-links. These cross-links prevent the elastin fibers from extending indefinitely and allow them to "snap back" on removal of tension. The cross-links in elastin are rather different from those in collagen, for they are designed to hold several chains together. Four lysine side chains can be combined so as to yield a **desmosine** cross-link:

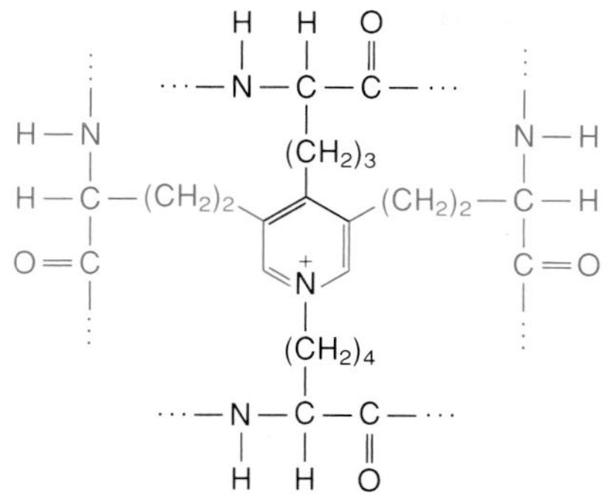

```
                      H   H   O
      ⋮               |   |   ‖               ⋮
      |          ···—N—C—C—···               |
 H—N                  |                      N—H
      |              (CH2)3                   |
 H—C—(CH2)2                  (CH2)2—C—H
      |                                       |
 O=C                  N+                     C=O
      |               |                       |
      ⋮              (CH2)4                   ⋮
                      |
                 ···—N—C—C—···
                      |   |   ‖
                      H   H   O
```

Since the α-carbons of *four* separate chains are connected, only a small amount of such cross-linking is needed to convert an elastin fiber into a highly interconnected, rubbery network.

This brief overview of a few of the structural proteins brings out several points. First, proteins can evolve to serve an almost infinite diversity of functions. Second, the structural fibrous proteins do so by taking advantage of the propensities of particular repetitive sequences of amino acid residues to favor one kind of secondary structure or another. Finally, posttranslational modification of proteins, such as cross-linking, is an important adjunct in tailoring a protein to its function. We will say more about such modification in Chapters 27 and 28 when we consider the whole process of protein synthesis.

These few examples do not exhaust the list of structural proteins. There are other, very important ones, such as **actin** and **myosin** of muscle and **tubulin** in microtubules. But these are constructed in a rather different way and will be discussed in later chapters (see especially Chapter 29).

Globular Proteins: Tertiary Structure Allows Functional Diversity

Different Folding for Different Functions

Abundant and essential as the structural proteins may be in any organism, they constitute only a small fraction of the kinds of proteins it will possess. Most of the chemical work of the cell, the synthesizing, transporting, and metabolizing, is carried out with the aid of an enormous class of **globular proteins.** These are given this name because their polypeptide chains are folded into compact structures very unlike the extended, filamentous forms of the fibrous proteins. Myoglobin is a globular protein. A glance at its three-dimensional structure, as compared to that of, say, collagen, immediately reveals this qualitative difference.

We now know a great deal about the structural details of very many globular proteins, largely through the use of x-ray diffraction methods. Often the resolution (the fineness of detail that can be discriminated) extends to less than 0.2 nm. This is sufficient to identify individual amino acid residues, so the polypeptide chain can be traced through the molecule.

As it passes through the molecule, the polypeptide chain is often locally folded into one or another of the kinds of secondary structure (α helix, β sheet, etc.) that we have already discussed. But to make the structure globular and compact, these regions must themselves be folded on one another. This folding is referred to as the **tertiary structure** of the protein. The distinction between secondary and tertiary structures can be clearly seen, for example, in the structure of myoglobin, as shown in Figure 6.1. Myoglobin is a protein with much α helix (about 70%), and that helix is bent and folded to form a compact molecule. Within that folding is formed a pocket to hold a **prosthetic group,** the **heme.** Many globular proteins carry prosthetic groups, small molecules that may be noncovalently or covalently bonded to the protein, to enable them to fulfill special functions. In this case, the noncovalently bound heme group carries the oxygen binding site of myoglobin (see Chapter 7).

Figure 6.14 depicts another globular protein, **ribonuclease.** Ribonuclease is an enzyme; as its name implies, it catalyzes the hydrolysis of ribonucleic acid. The secondary and tertiary structures of ribonuclease are entirely different from those of myoglobin. Ribonuclease has little α helix but a considerable amount of antiparallel β sheet. This β sheet has been somewhat twisted and deformed from the idealized structure shown in Figure 6.7b. The tertiary structure of ribonuclease wraps the molecule about a cleft in which the ribonucleic acid can be held while it is being cleaved.

The point we wish to emphasize is this: Every globular protein has a *unique* tertiary structure, made up of secondary structure elements (helices, β sheets, nonregular regions) folded in a specific way. As we examine proteins, we will find that each such conformation is suited to the particular functional role that protein plays.

Varieties of Protein Structure: Patterns of Protein Folding

At first glance, it might seem that there would be an almost infinite number of ways in which globular proteins could fold. In terms of details, this is true. Yet when a large number of the known structures are examined, certain common motifs and principles emerge. The first principle is that many proteins are made up of a number of domains. A **domain** is a compact, locally folded region of tertiary structure. Domains are connected by the polypeptide strand that runs through the whole molecule. Multiple domains are especially common in the larger globular proteins, whereas small proteins like myoglobin and ribonuclease tend to be single folded domains. As we shall see in later sections, such domains often perform different func-

Figure 6.14
The three-dimensional structure of the enzyme ribonuclease. (**a**) A ball-and-stick model. The only side chains shown are His 12, Lys 41, and His 119 (purple), which are important in the catalysis by this enzyme. SH bonds are shown in orange. (**b**) A more diagrammatic representation in which helices are represented as spirals (green) and β sheet by ribbons (gray) with arrows pointing in the C-terminal direction. Note that the β sheet is antiparallel but highly twisted. SH bonds have been omitted for clarity. The molecule that was crystallized to allow this structure to be determined is actually *ribonuclease S,* in which the peptide bond between residues 20 and 21 has been cleaved (dashed line). This form crystallizes better than native ribonuclease but has probably undergone some small rearrangement.

N-terminus
40
95
110
26
84
65
72
58
C-terminus

(a)

N-terminus
B
95
A
Lys 41
His 119
His 12
C
C-terminus

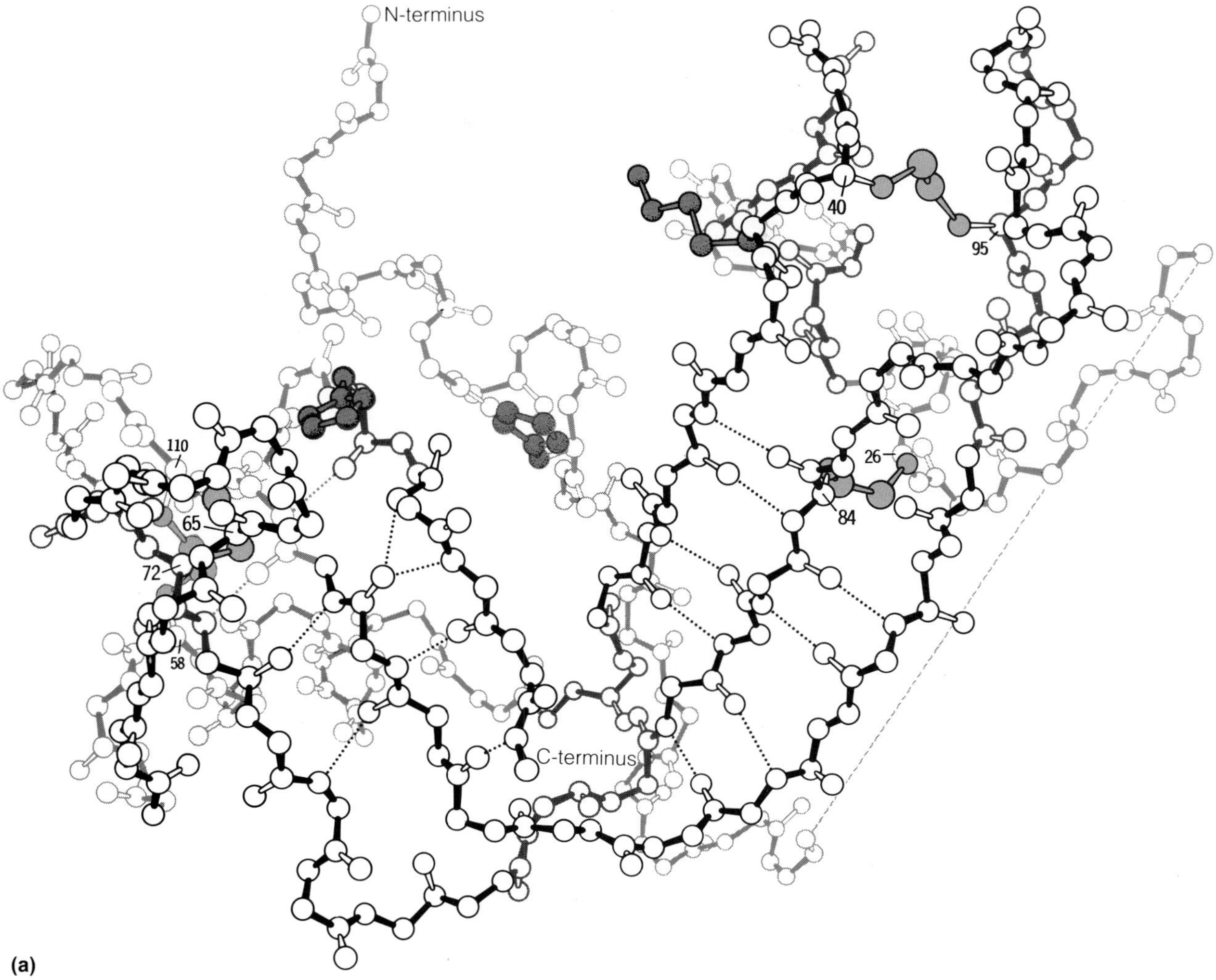

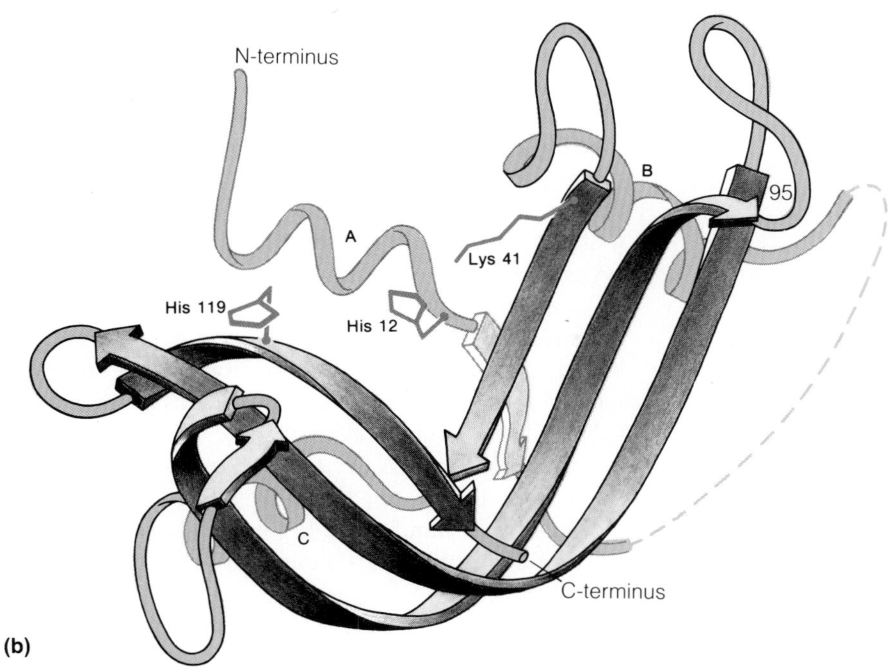

(b)

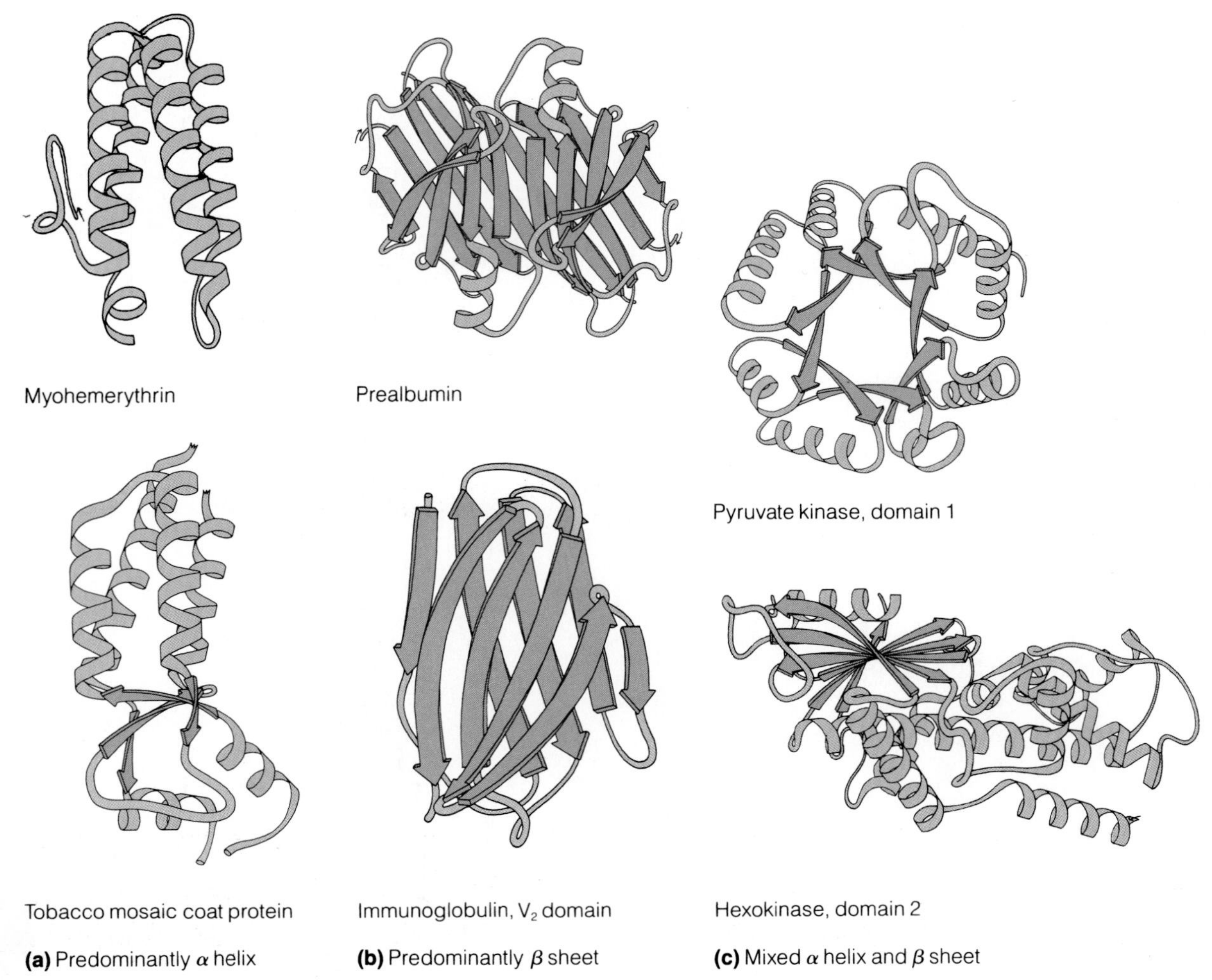

Figure 6.15
A gallery of type structures of globular proteins. (**a**) Predominantly α-helical (helix bundles). Top: Myohemerythrin. Bottom: Tobacco mosaic virus coat protein. (**b**) Predominantly β sheet. Top: Prealbumin. Bottom: Immunoglobulin; V_2 domain. These are both examples of antiparallel β barrels. (**c**) Mixed α helix and β sheet. Top: Pyruvate kinase, domain 1. This contains an excellent example of a parallel β barrel, seen here from the top. Bottom: Hexokinase, domain 2. This contains a very clear example of a twisted β sheet.

tions, and a given domain type can sometimes be recognized in several different proteins.

Among the domain varieties, several distinct classes have been recognized by Jane Richardson (see References). The major folding patterns are of two kinds: those that are built about a packing of α helices and those that are constructed on a framework of β-sheet structures. A gallery of examples is shown in Figure 6.15. As we examine these as representatives of globular protein structure, certain general rules become evident.

All globular proteins are constructed so as to have a defined inside and outside. When amino acid sequences are compared with the three-dimensional structure, it is invariably found that the hydrophobic residues are packed mostly on the inside, with hydrophilic residues in contact with the solvent. This is made evident for cytochrome *c* in Figures 6.16 and 6.17.

β-sheets are usually twisted, or wrapped into barrel structures. Clear examples can be seen in Figure 6.15. In most cases, such arrangements exhibit a right-hand twist. It has been argued that this tendency of β sheets to right-hand twisting is a consequence of the L configuration of the amino acid residues.

There are a number of ways in which the polypeptide chain can turn corners, so as to go from one β segment or α helix to the next. The most

1 10 20 30 40 50

G C V E K G K K I F V Q K C A Q C H T V E K G G K H K T G P N L H G L F G R K T G Q A P G F T Y T D

A N K N K G I T W K E E T L M E Y L E N P K K Y I P G T K M I F A G I K K K T E R E D L I A Y L K K A T N E

60 70 80 90 100

Figure 6.16
The sequence of horse heart cytochrome *c*. Hydrophobic residues are indicated in red, hydrophilic residues are in green, and ambivalent ones are left white. Note that each type appears to be scattered throughout the sequence.

compact kind of turn is called a β turn (Figure 6.18). There are several varieties of β turn, each able to accomplish a complete reversal of the polypeptide chain in only four residues. β turns are frequently found in antiparallel β-sheet structures. Bends and turns in α helices, on the other hand, frequently involve proline residues, which do not fit well within the helices themselves. Bends and turns most often occur at the surface of proteins.

Not all parts of globular proteins can be conveniently classified as helix, β sheet, or β turns. Examination of Figure 6.15, for example, reveals many strangely contorted loops and folds in the chains. These have sometimes been referred to as "random coil" regions, but this is a misnomer, for such sections of the chain are not flexible in the same way as a true random coil (see Chapter 4). Rather, each such region has its own particular folding, exactly the same in each example of the particular protein molecule. We know this because x-ray diffraction studies show the same arrangement in all molecules in the crystal. We might call these *irregularly structured regions.*

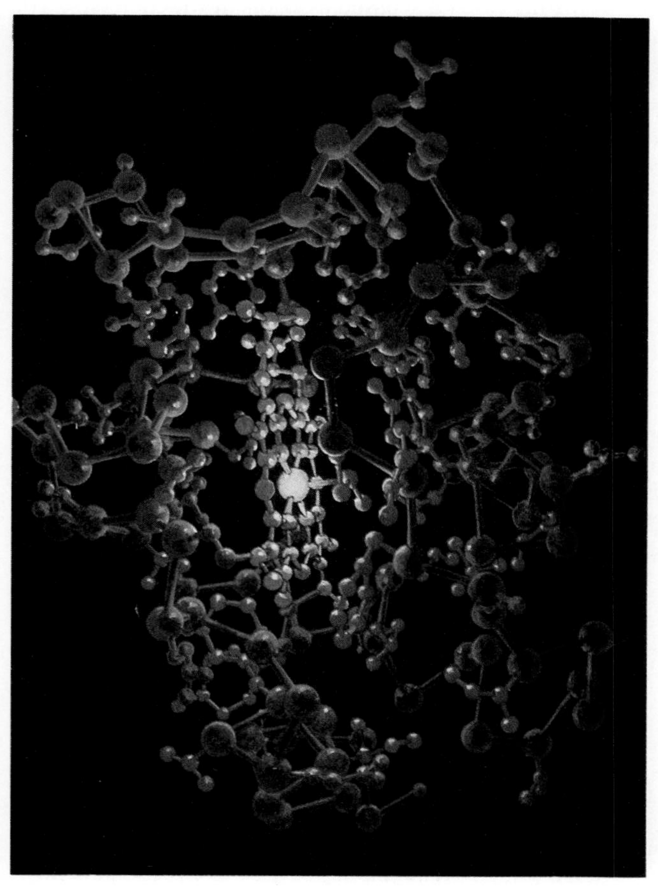

(a)

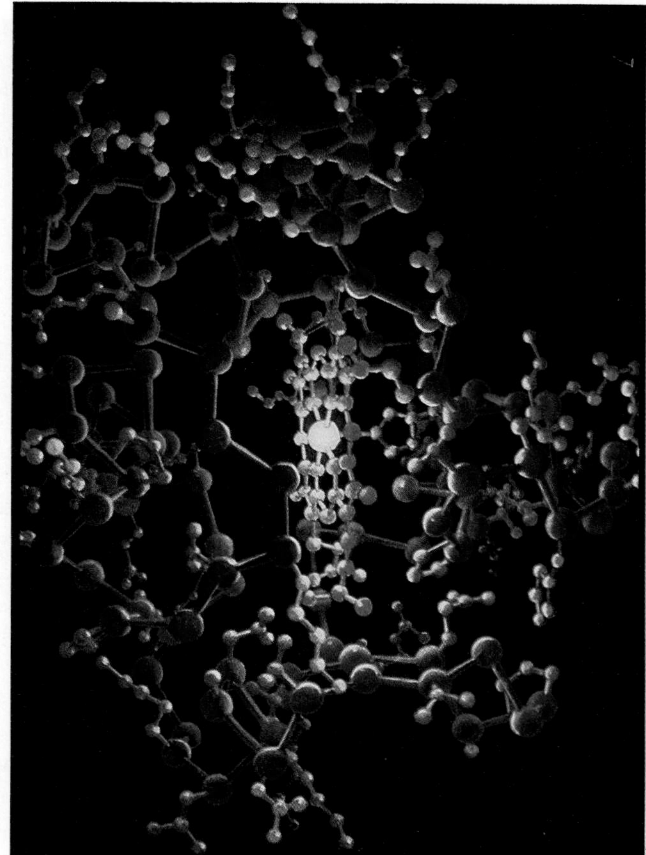

(b)

Figure 6.17
The structure of horse heart cytochrome *c*. In (**a**) the hydrophobic amino acids are shown in red. Note how in the three dimensional structure they cluster about the heme and on the inside of the molecule. In (**b**) the hydrophilic residues are shown in green. Note how they tend to lie on the surface of the molecule.

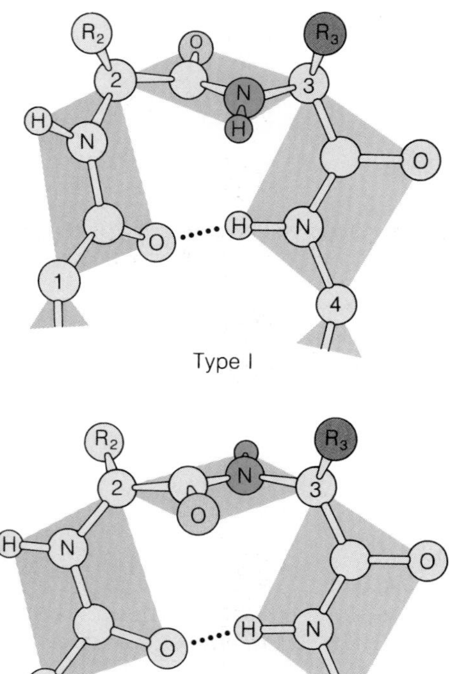

Figure 6.18
Examples of β turns. Two types (I and II) are depicted, either of which allows a very abrupt change in polypeptide chain direction. In the type II turn residue 3 is usually glycine, presumably because a bulky R group would clash with the carbonyl oxygen of residue 2.

In some proteins the folding is dominated by a prosthetic group. Myoglobin is an example. Although myoglobin might be roughly described as an α-helix bundle, the tertiary structure of this protein has been distorted to form a hydrophobic cage about the bulky heme group.

Factors Determining Secondary and Tertiary Structure

The Information for Protein Folding

What ultimately determines the complex mixture of secondary and tertiary folding that characterizes each globular protein? *The information for determining the three-dimensional structure of a protein is carried entirely in the amino acid sequence of that protein.* This can be demonstrated by experiments in which the "native," or natural, three-dimensional structure is broken up by changing the environmental conditions. If we raise the temperature sufficiently, or make the pH extremely acid or alkaline, or add to the solvent certain kinds of organic molecules, such as alcohols or urea, the protein structure will unfold (Figure 6.19). As with nucleic acids, this process is called *denaturation,* for the natural structure of the protein has been lost, along with many of its specific properties. The unfolded chain is a *random coil,* with freedom of rotation about bonds in both the polypeptide backbone and the side chains. It no longer has a single compact conformation but continually fluctuates between a very large number of extended conformations. Differences between the native and random coil conformations can be detected by one of the methods discussed in Tools of Biochemistry 10.

In the example shown in Figure 6.19 the enzyme ribonuclease has been denatured by raising the temperature. The tertiary and secondary structure have been lost, and ribonuclease will no longer act as a catalyst for the cleavage of ribonucleic acid. The process can be followed by various physical measurements, as shown in Figure 6.20. Remarkably, this total "scram-

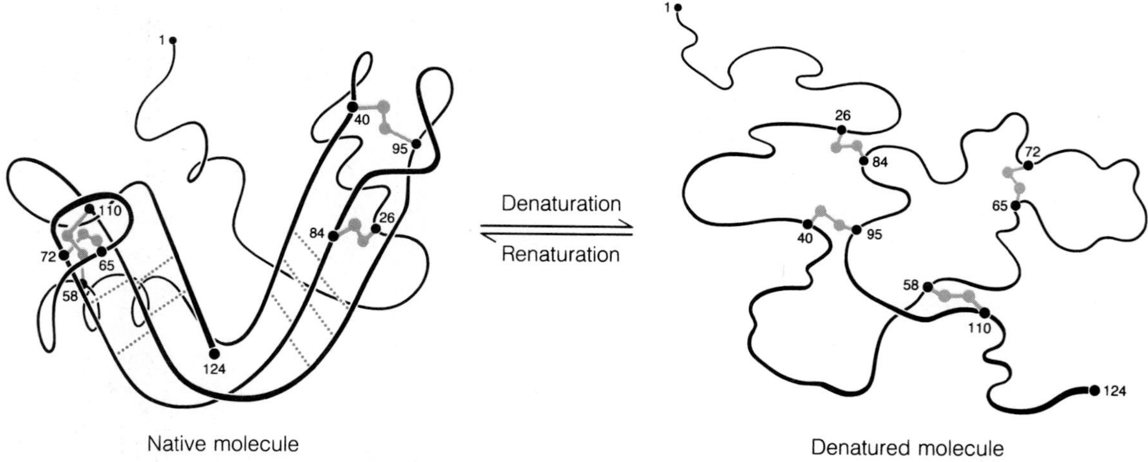

Figure 6.19
The thermal denaturation of ribonuclease. This schematic drawing depicts the conformational change to a disordered structure when ribonuclease is heated above its thermal denaturation temperature. Note that the disulfide bonds (orange) are still intact, providing some residual limitation to conformational flexibility.

bling" of protein structures is in most cases fully reversible. If ribonuclease is restored to physiological conditions of temperature and solvent it will spontaneously refold into its original structure. Furthermore, it will once again be a functional enzyme. Thus, the protein "knows" its own favored conformation; it needs no information to guide it other than that contained in its sequence. In the cell, a newly synthesized polypeptide chain will spontaneously fold into the proper conformation, ready to take up its role.

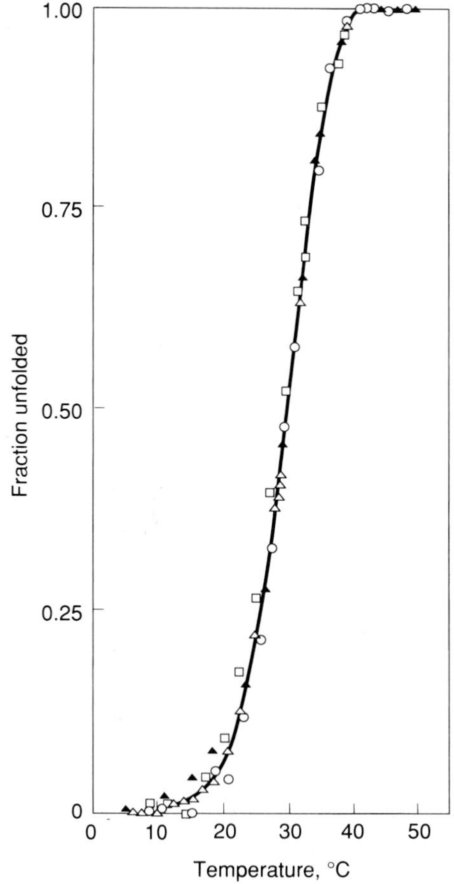

Figure 6.20
Denaturation of ribonuclease as shown by various physical methods. The fraction unfolded is measured by the increase in solution viscosity (□), change in optical rotation at 365 nm (○), and change in UV absorbance at 287 nm (△). The filled triangles (▲) represent a second denaturation after cooling. Note that all three techniques indicate the same fraction unfolding. The experiments were conducted at pH 2.1, ionic strength 0.019 M. Under physiological conditions ribonuclease is much more stable, not denaturing until about 70–80°C.

The Thermodynamics of Folding

The folding of a globular protein is clearly a thermodynamically favored process under physiological conditions. This tells us that the overall free energy change on folding must be negative. But it is clear that the folding process, which involves going from a multitude of random coil conformations to a *single* folded structure, must involve a decrease in randomness and thus a decrease in entropy. The free energy equation, $\Delta G = \Delta H - T\,\Delta S$, shows that this negative ΔS will yield a *positive* contribution to ΔG. The conformational entropy change must work *against* folding. Therefore, to seek the explanation for an overall negative ΔG in protein folding, we must seek features of the process that will yield either a large negative ΔH or an additional source of entropy *increase* on folding. Both can be found.

The major source for a negative ΔH is energetically favorable interactions between groups within the folded molecule. These include many of the noncovalent interactions described in Chapter 2.

ELECTROSTATIC INTERACTIONS. Such interactions can occur between positively and negatively charged side chain groups. For example, a lysine side chain amino group may be placed close to the γ-carboxyl group of some glutamic acid residue. Since at neutral pH one group will be charged positively and the other negatively, there is an electrostatic force between them. We could say that such a pair form a kind of "salt" within the protein molecule; consequently, such interactions are sometimes called **salt bridges.** Such bonds will be broken if the protein is taken to pH values high enough or low enough that either partner loses its charge. This loss of salt bridges is a partial explanation for acid or base denaturation of proteins. The mutual repulsion between the numerous similarly charged groups that are present in very acidic or basic solutions contributes further to the instability of the folded structure under these conditions.

INTERNAL HYDROGEN BONDS. Many of the amino acid side chains carry groups that are either good hydrogen bond donors or good acceptors. Examples are the hydroxyls of serine or threonine, the amino groups and carbonyl oxygens of asparagine or glutamine, and the ring nitrogens in histidine. Furthermore, if amide protons or carbonyls in the polypeptide backbone are not involved in secondary structure formation, they are potential candidates for interaction with side chain groups. A network of several types of internal hydrogen bonds is seen in the portion of the molecule of the enzyme lysozyme shown in Figure 6.21.

VAN DER WAALS INTERACTIONS. The weak interactions between uncharged molecular groups can also make significant contributions to protein stability. As space-filling models of proteins demonstrate, the interior is tightly packed, allowing maximum contact between side chain atoms.

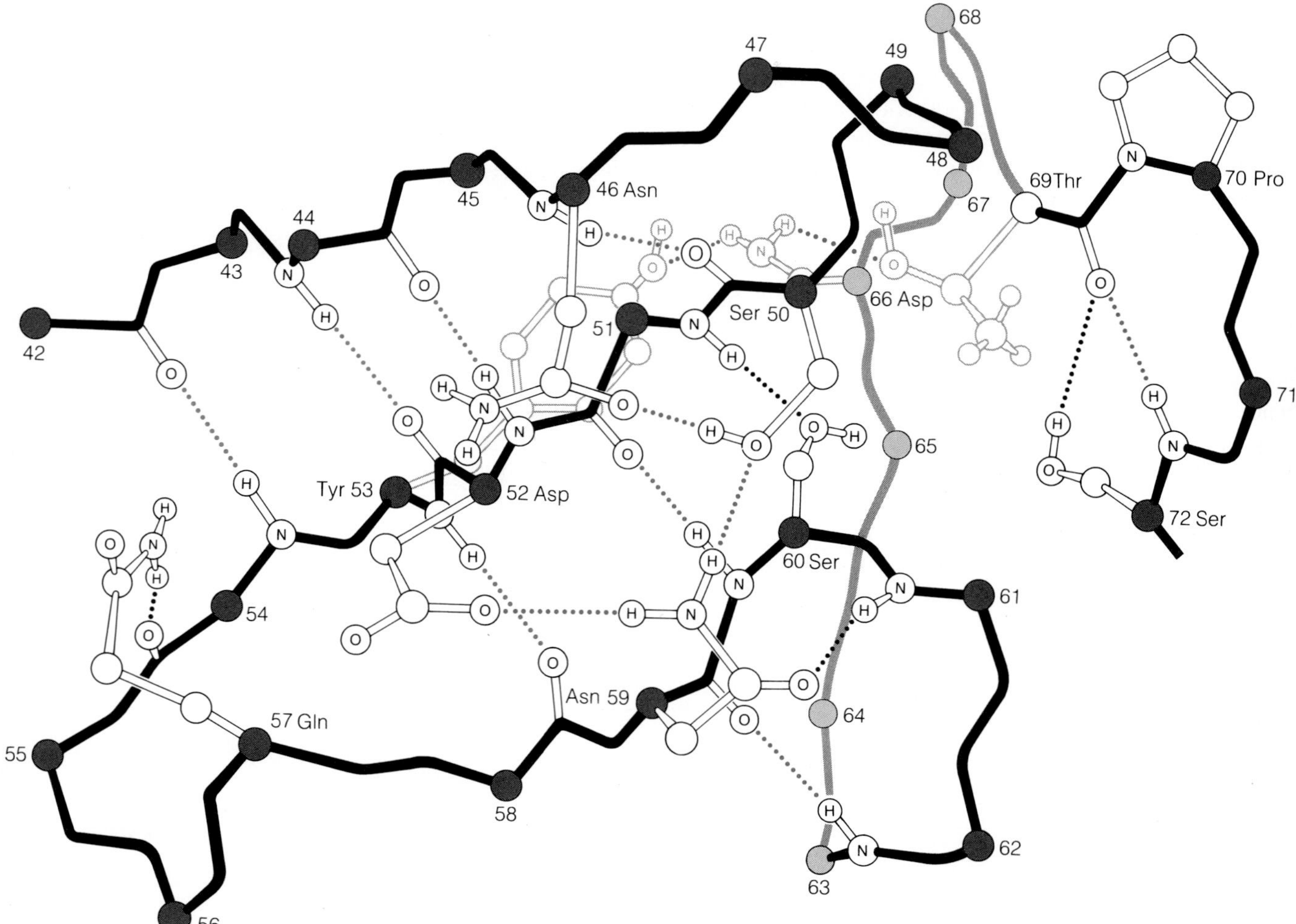

Figure 6.21
Details of intramolecular interactions in proteins. The figure shows a network of three kinds of hydrogen bonds within the enzyme *lysozyme*—those between side chain groups (···), those between side chain groups and backbone amide hydrogens or carbonyl oxygen (·:·), and those directly between backbone groups (···).

The contribution of any of these interactions to the enthalpy of folding is diminished by the fact that when a protein molecule folds from an expanded random coil, it gives up some favorable interactions with water. As we pointed out in Chapter 2, an unfolded α helix forms hydrogen bonds with water instead of forming internal bonds. It is the overall *difference* in interaction energy that is important.

Each individual interaction of the kinds we have listed can contribute only a small amount (at most only a few kilojoules) to the overall negative enthalpy of interaction. But the sum of many of them can yield powerful stabilization to the folded structure. Examples of the total enthalpy changes for folding are given for some representative proteins in Table 6.3. The unfavorable entropy of folding can be more than compensated for by a favorable energy contribution from the sum of intramolecular interactions.

The Hydrophobic Effect

There is another factor that makes a major contribution to the thermodynamic stability of many globular proteins. Recall from Chapter 2 that hydrophobic substances in contact with water cause the water molecules to form cagelike structures around them. This ordering corresponds to a loss of randomness in the system; the entropy is decreased. Suppose a protein contains, in its amino acid sequence, a substantial number of residues with hydrophobic side chains (for example, leucine, isoleucine, and phenylalanine). When the polypeptide chain is in an unfolded form, these residues will be in contact with water and cause ordering of the surrounding water structure. But when the chain folds into a globular structure, these hydrophobic residues can be buried within the molecule (see Figure 6.17a). The water molecules that they have ordered will then be released and gain freedom of motion. Thus, internalizing hydrophobic groups can *increase* the randomness of the whole system (protein plus water) and therefore yield an entropy *increase* on folding. This will make a negative contribution to the free energy of folding and increase the stability of the protein structure. The term *hydrophobic bonding* has sometimes been used to describe this effect, but this is really a misnomer. It is not that bonds are formed between hydrophobic groups (although van der Waals interactions surely do occur) but rather that the overall structure, with buried hydrophobic groups, is stabilized by an entropy effect. The most appropriate term to describe this source of protein stabilization is the **hydrophobic effect.**

Overall, then, the stability of the folded structure of a globular protein depends on the interplay of three factors:

1. The unfavorable conformational entropy terms.
2. The favorable enthalpy term arising from intramolecular side group interactions.
3. The favorable entropy term arising from burial of hydrophobic groups within the molecule.

A schematic picture of the way these terms contribute to the free energy of folding is shown in Figure 6.22. In different proteins, the apportionment of stabilization between side chain interactions and the hydrophobic effect differs, but the overall consequence is the same: some particular folded structure corresponds to a free energy minimum for the polypeptide under physiological conditions. This is why the chains fold spontaneously.

Examination of the data in Table 6.3 reveals an important aspect of protein stabilization. The relatively small ΔG corresponding to the folding reaction is usually the difference between large ΔH and $T\,\Delta S$ terms. These large values arise because a protein tends to fold in a *cooperative* manner; in a partially folded mixture there are mainly wholly unfolded and wholly folded molecules, with few intermediate structures. A consequence of this cooperativity is that the thermal denaturation of a protein usually occurs over a narrow temperature range, as is shown in Figure 6.20.

Table 6.3
Thermodynamic data for the folding of several proteins at 25°C

Protein	Conditions	$\Delta G_{\text{folding}}$ $\left(\frac{\text{kJ}}{\text{mol}}\right)$	$\Delta H_{\text{folding}}$ $\left(\frac{\text{kJ}}{\text{mol}}\right)$	$\Delta S_{\text{folding}}$ $\left(\frac{\text{J}}{^\circ\ \text{mol}}\right)$
Ribonuclease	pH 2.5	−7.3	−238	−774
Chymotrypsinogen	pH 3	−32	−163	−439
Myoglobin	pH 9	−57	−175	−397
β-Lactoglobulin	In 5 M urea	+1.7	+88	+301

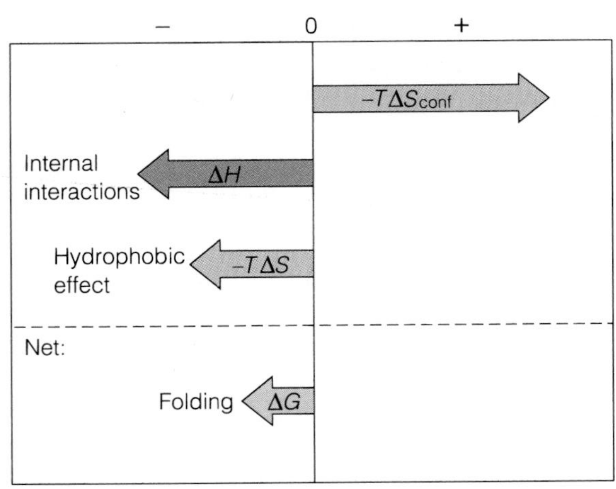

Figure 6.22
Factors contributing to the free energy of folding of globular proteins. The conformational entropy change (ΔS_{conf}) works against folding, but the enthalpy of internal interactions and the hydrophobic entropy change favor folding. Therefore the overall $\Delta G_{folding}$ is negative, and the folded structure is stable.

The Role of Disulfide Bonds

Once folding has occurred, the three-dimensional structure is in some cases further stabilized by the formation of disulfide bonds between cysteine residues. These bonds are very specific, and which residues are so cross-linked depends on the already established folding. A classic example is the disulfide bonding in the enzyme ribonuclease, depicted schematically in Figure 6.14. There are four disulfide bonds in this protein, between residues 26 and 84, 40 and 95, 58 and 110, and 65 and 72. There is very little chance that these bonds could form accidentally in this pattern, as can be shown by a simple calculation. Suppose the protein had been reduced and could reform disulfide bonds *at random*. The first —SH group to pick a partner would have 7 choices, the second group would have 5, the third 3, and the last only 1. There would then be $7 \times 5 \times 3 \times 1 = 105$ equally probable combinations. Thus, by *chance*, one would expect only about 1% of reduced ribonuclease to refold successfully. But in a classic study of protein folding, Christian Anfinsen observed that reduced, denatured ribonuclease, when returned to conditions where the native structure is stable, slowly reformed its correct disulfide bonds and regained enzymatic activity (Figure 6.23). This can be explained only if the protein refolds *first* (as depicted in the figure), placing the —SH groups in position for correct pairing. This means that the —S—S— bridges are not essential for correct refolding. They do, however, contribute to the stability of the folded structure. A molecule containing —S—S— bridges exhibits a smaller conformational entropy change (since the number of conformations available to the unfolded form is less) and is therefore stabilized.

This raises a question: why do not *most* proteins have disulfide bonds? In fact, such bonds are relatively rare and are found primarily in proteins that are exported from cells, such as ribonuclease and insulin. The answer seems to be that the environment inside most cells is quite reducing and tends to keep sulfhydryl groups in the reduced state.

The fact that folding precedes disulfide bond formation can also explain the necessity for the proinsulin structure we described in the last chapter. Attempts to reform insulin from A and B chains in vitro yield molecules with various incorrect —SH pairings as the major products. But proinsulin folds into a conformation that favors the correct pairing. The intervening sequence, which is later cut out, seems to be essential for correct folding of the molecule.

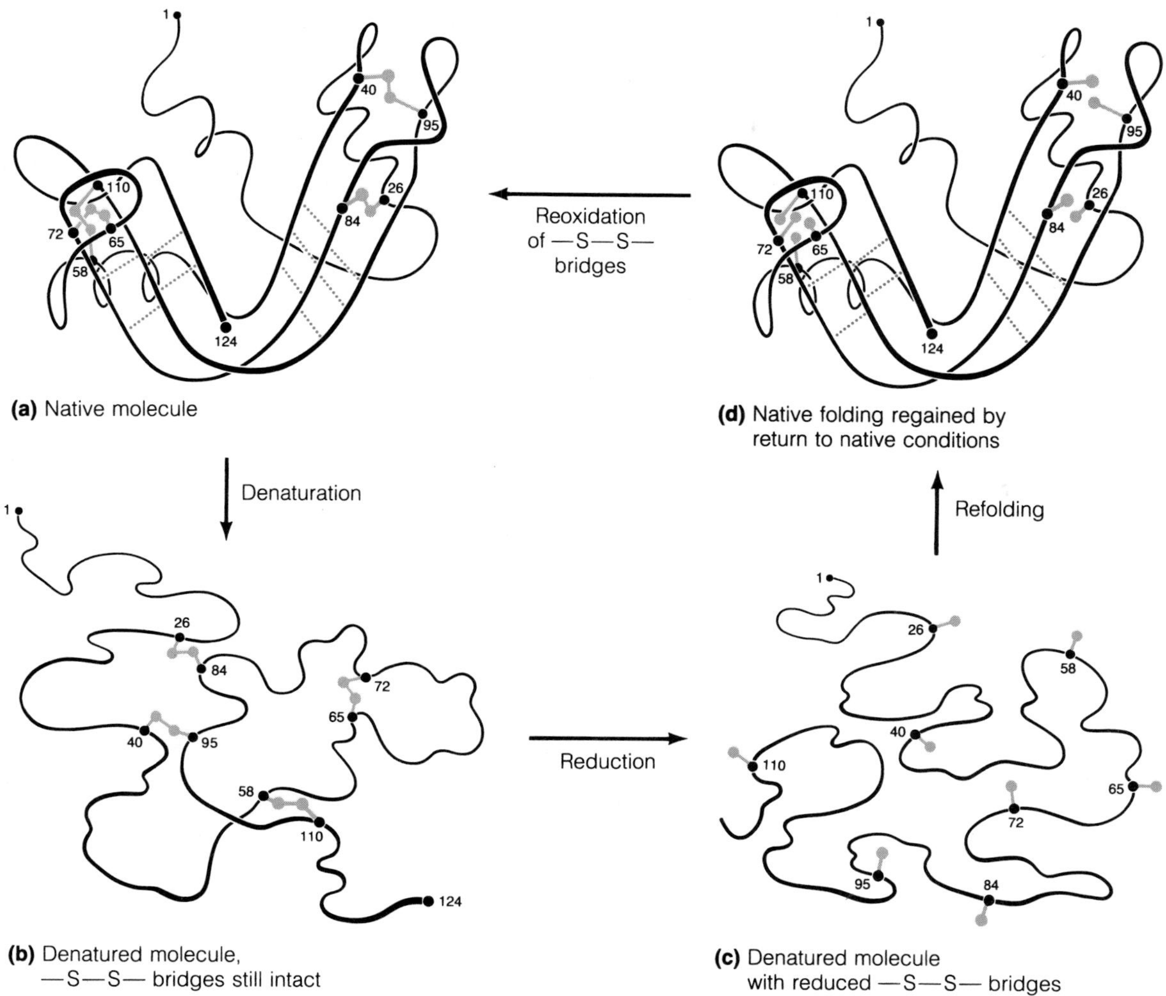

Figure 6.23
Ribonuclease refolds correctly even if disulfide bonds have been broken. (**a**) A schematic structure of the intact ribonuclease molecule. It is denatured by heat or a denaturing solvent (**b**) and then has its disulfide bonds (orange) reduced (**c**). Note that the molecule can now expand more and exhibit greater conformational freedom than the denatured ribonuclease shown in (**b**) or in Figure 6.19. The denatured molecule is returned to "native" environmental conditions (**d**), and its disulfide bonds are allowed to reform by oxidation of the sulfhydryls. Since it refolds correctly, even without —S—S— bridges, the correct sulfhydryl pairing occurs and the native structure is regained.

A powerful new method for studying the sources of protein structural stabilization has been developed from recombinant DNA techniques. Now that the genes for proteins can be cloned and expressed in bacterial carriers, it is possible to make specific changes at desired positions. This method of **site-directed mutagenesis** (see Tools of Biochemistry 22) allows us to test the effect of changing one or more amino acid residues or adding or removing disulfide bonds.

Dynamics of Globular Protein Structure

The structural and thermodynamic descriptions we have developed so far tend to give too static a picture of globular proteins. The x-ray diffraction results, which present precise pictures, are apt to give the impression of a rigid, inflexible structure. Likewise, the thermodynamic analysis of folding concentrates on the initial (unfolded) and final (folded) states. But it is now recognized that globular proteins fold via complex *kinetic* pathways and that even the folded structure, once attained, is far from static. This section explores some of these dynamic aspects of globular proteins.

Kinetics of Protein Folding

By the use of rapid kinetic methods (see Tools of Biochemistry 15), it has become possible to analyze some of the intermediate states that occur as a protein molecule folds. Once again, we take the enzyme ribonuclease as an example. The folding of ribonuclease has been studied by a number of physical techniques, including fluorescence, absorbance, and hydrogen exchange. The last technique works by measuring the rate at which certain protons in the protein can exchange with those in the surrounding solvent. The amide protons and those involved in side chain amino and carboxyl groups, for example, are potentially capable of exchanging. In the random coil these exchange very rapidly. However, in a folded protein many of these protons are involved in stable hydrogen bonds and may require many minutes, or even hours, to exchange. At present, it is known that the folding of ribonuclease, which is essentially complete in less than a minute, involves at *least* three steps, with distinguishable protein conformations designated by U_S, I_I, I_N and N:

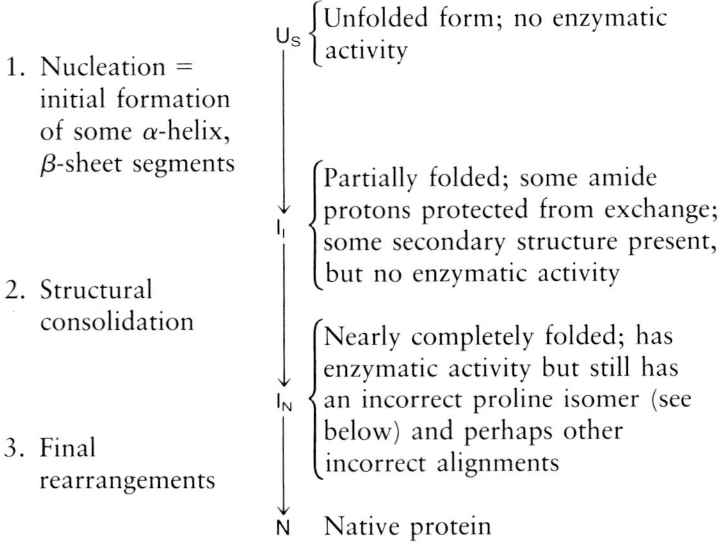
Us
Unfolded form; no enzymatic activity
1. Nucleation = initial formation of some α-helix, β-sheet segments
I_I
Partially folded; some amide protons protected from exchange; some secondary structure present, but no enzymatic activity
2. Structural consolidation
I_N
Nearly completely folded; has enzymatic activity but still has an incorrect proline isomer (see below) and perhaps other incorrect alignments
3. Final rearrangements
N
Native protein

From recent studies it appears that the antiparallel β-sheet fold in ribonuclease (see Figure 6.14) is formed early in the process, followed by slower readjustments. One of these slow processes is the cis–trans isomerization of the amide bond adjacent to a proline residue:

Cis ⇌ **Trans**

Unlike other peptide bonds in proteins, for which the trans isomer is highly favored, proline residues favor the trans form in the preceding peptide bond by only about 4:1. Therefore, when a protein folds there is a significant chance that the "wrong" isomer will form first. Conversion to the correct configuration may involve a significant chain rearrangement and hence be slow. There is now evidence that in the cell there are enzymes to catalyze this cis–trans isomerization.

Although we still know only a little about the kinetics of protein folding, we are certain of one thing: folding does not proceed by anything approaching a random search through all of the conformations possible to the unfolded form. A rough estimate indicates that there are on the order of 10^{100} possible conformations for a chain like ribonuclease, which has 124 residues. Even taking into account the rapidity of atomic motions, it would take the molecule about 10^{50} years to search them all! Therefore, it seems likely that the search for the final conformation begins with nucleation of elements of secondary structure, like β sheet or α helix, and that these then direct the further folding.

Motions Within Globular Protein Molecules

There is much evidence that various kinds of motions are continually occurring within folded protein molecules. These motions can be roughly grouped into several classes, as shown in Table 6.4. Motions in class 1 occur even within protein molecules in crystals and account, at least in part, for the limits of resolution obtainable in x-ray diffraction studies. The larger, slower motions in classes 2 and 3 are more likely to occur in solution. Some of them, like the opening and closing of clefts in molecules, are thought to be involved in the catalytic functions of enzymes. As we shall see in later chapters, how long it takes for a protein to bind or release a small molecule may depend on the time required for the protein to open or close a cleft. Similarly, the protein "gates" that pass molecules and ions through membranes open and close in a sporadic, irregular manner. Future research may show that the dynamic behavior of proteins is at least as important in their function as the static details of their structure.

Prediction of Protein Structure

Can protein structure be predicted? In one sense, the answer to this question must surely be *yes*. We know that the molecular information necessary to determine the secondary and tertiary structures is carried in the amino acid sequence itself; thus, the gene "predicts" the structure. The implication of this is that if we, as scientists, fully understood the rules of folding, we could

Table 6.4
Motions in protein molecules

Class	Type of Motion	Approximate Range of	
		Amplitude (nm)	Time (s)
1	Vibrations and oscillations of individual atoms and groups	0.2	10^{-15}–10^{-12}
2	Concerted motions of structural elements, like α helices, groups of residues	0.2–1	10^{-12}–10^{-8}
3	Motions of whole domains; opening and closing of clefts	1–10	$\geq 10^{-8}$

Table 6.5
Relative probabilities of amino acid residue occurrence in different globular protein secondary structures[a,b]

Amino Acid	α Helix (P_α)	β Sheet (P_β)	Turn (P_t)	
Ala	1.29	0.90	0.78	favor α helices
Cys	1.11	0.74	0.80	
Leu	1.30	1.02	0.59	
Met	1.47	0.97	0.39	
Glu	1.44	0.75	1.00	
Gln	1.27	0.80	0.97	
His	1.22	1.08	0.69	
Lys	1.23	0.77	0.96	
Val	0.91	1.49	0.47	favor β sheets
Ile	0.97	1.45	0.51	
Phe	1.07	1.32	0.58	
Tyr	0.72	1.25	1.05	
Trp	0.99	1.14	0.75	
Thr	0.82	1.21	1.03	
Gly	0.56	0.92	1.64	favor turns
Ser	0.82	0.95	1.33	
Asp	1.04	0.72	1.41	
Asn	0.90	0.76	1.23	
Pro	0.52	0.64	1.91	
Arg	0.96	0.99	0.88	

[a] Data adapted from M. Levitt (1978) *Biochemistry* 17:4277–4285.

[b] Rules for prediction from P. Y. Chou and G. D. Fasman (1974) *Biochemistry* 13:222–245. The symbols $\langle P_\alpha \rangle$, $\langle P_\beta \rangle$, and $\langle P_t \rangle$ denote *average* values of these quantities in a region of the sequence.

1. Any segment of six residues or more, with $\langle P_\alpha \rangle \geq 1.03$, as well as $\langle P_\alpha \rangle > \langle P_\beta \rangle$, and not including Pro, is predicted to be α helix.
2. Any segment of five residues or more, with $\langle P_\beta \rangle \geq 1.05$, and $\langle P_\beta \rangle > \langle P_\alpha \rangle$, is predicted to be β sheet.
3. Examine the sequence for tetrapeptides with $\langle P_\alpha \rangle < 0.9$, $\langle P_t \rangle > \langle P_\beta \rangle$. These have a good chance of being turns. The actual rules for predicting β turns are more complex, but this will work in most cases.

describe the entire three-dimensional conformation of any protein, starting with nothing more than a knowledge of its sequence. This cannot be done at the present time. We can predict secondary structure fairly well, but most tertiary folding is still too complicated.

Prediction of Secondary Structure

The successful approach to the prediction of secondary structure has been entirely empirical. From analysis of the *known* structures of a number of proteins, tables have been compiled to show the relative frequency (P_α, P_β, P_t) with which a particular kind of amino acid residue lies in α helices, β sheets, or turns. An example is given in Table 6.5. From these data certain clear distinctions can be made. For example, Leu, Met, and Glu are all strong "helix formers"; Gly and Pro are "helix breakers." Similarly, Ile, Val, and Phe are strong β-sheet formers, whereas Pro does not fit well into β sheets. Glycine is frequently found in β turns, whereas valine is not. We have already discussed why proline tends to lie in turns. Why other residues are generally found in one structure or another is not so clear. Various rules have been proposed to use these values to predict structures; those developed by P. Y. Chou and G. D. Fasman are listed in Table 6.5. In Figure 6.24 are shown the results of the application of such calculations to the small protein bovine pancreatic trypsin inhibitor. In most cases, the results compare quite favorably, although not exactly, with secondary structures experimentally determined by x-ray diffraction.

Tertiary Structure: Computer Simulation of Folding

Attempts to predict tertiary structure have been much less successful, probably because the higher-order folding depends so critically on specific side chain interactions, often between residues far removed from one another in the sequence. About all that can be done at present is to assert that if two proteins are similar in sequence, they will probably also be similar in tertiary structure.

In a few cases, attempts have been made to predict the overall three-dimensional structures of small globular proteins in a more direct way. The method depends on the fact that, in their spontaneous folding, proteins are seeking an energy minimum. A random coil chain is allowed, in computer simulation, to undergo a very large number of small permutations in its conformation, through rotation about individual bonds. The program keeps track of the total energy, in terms of possible interactions, and seeks an energy minimum. So far, such attempts have been only moderately successful. Furthermore, the amount of computer time required for such a search is enormous, even on the largest, fastest computers and with the smallest, simplest proteins. This approach is still in its infancy, yet it gives the hope that we may ultimately be able to predict not only how proteins are folded but also the pathways that lead to the final structures.

Quaternary Structure of Proteins

So far, we have explored increasingly complex levels of protein structure, from primary to secondary to tertiary. Functional protein organization can reach at least one more level, which is called **quaternary structure** (Figure 6.25). Many proteins exist in the cell (and in solution, under physiological conditions) as specific aggregates of two or more folded polypeptide chains, or *subunits*. Methods for determining whether or not a protein is composed of multiple subunits are described in Tools of Biochemistry 11. Such **multisubunit proteins** include many of the most important enzymes and transport proteins.

The interactions between the folded polypeptide chains in multisubunit proteins are of the same kinds that stabilize tertiary structure—salt bridges, hydrogen bonding, van der Waals forces, hydrophobic interactions, and sometimes disulfide bonding. These provide the energy to stabilize the multisubunit structure.

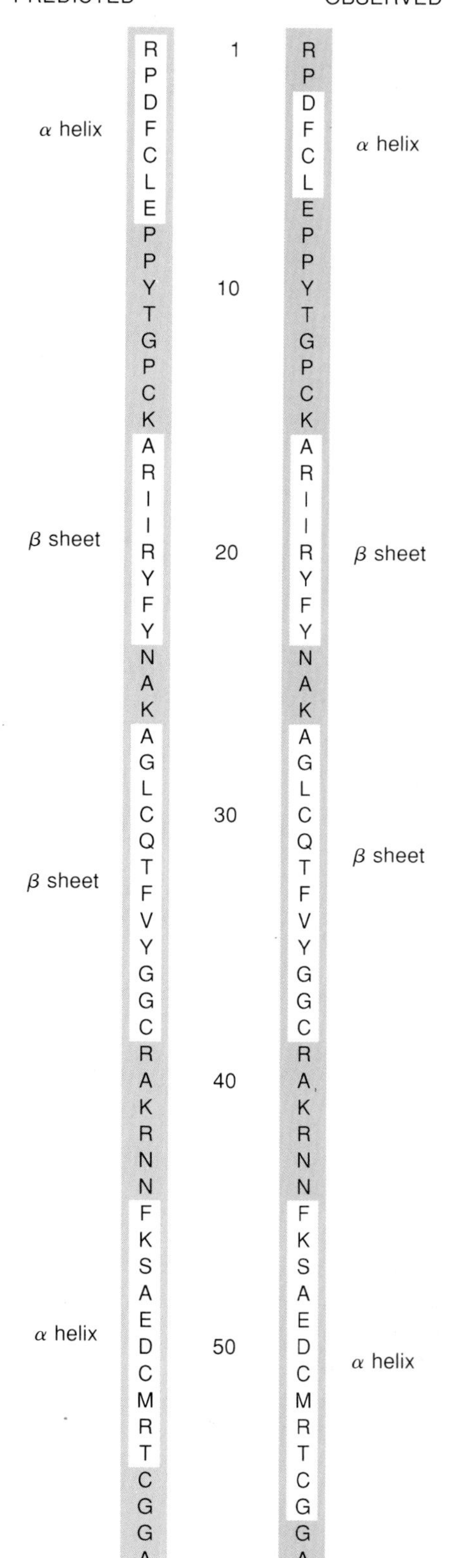

Figure 6.24
Prediction of the secondary structure of a protein. Bovine pancreatic trypsin inhibitor (BPTI) is a small globular protein that specifically inhibits the proteolytic action of the enzyme trypsin. To the left are shown the secondary structural elements as predicted by P. Y. Chou and G. D. Fasman. To the right, for comparison, are the results of x-ray diffraction studies of the same protein. The agreement between the predicted and observed structures is somewhat better here than for most proteins.

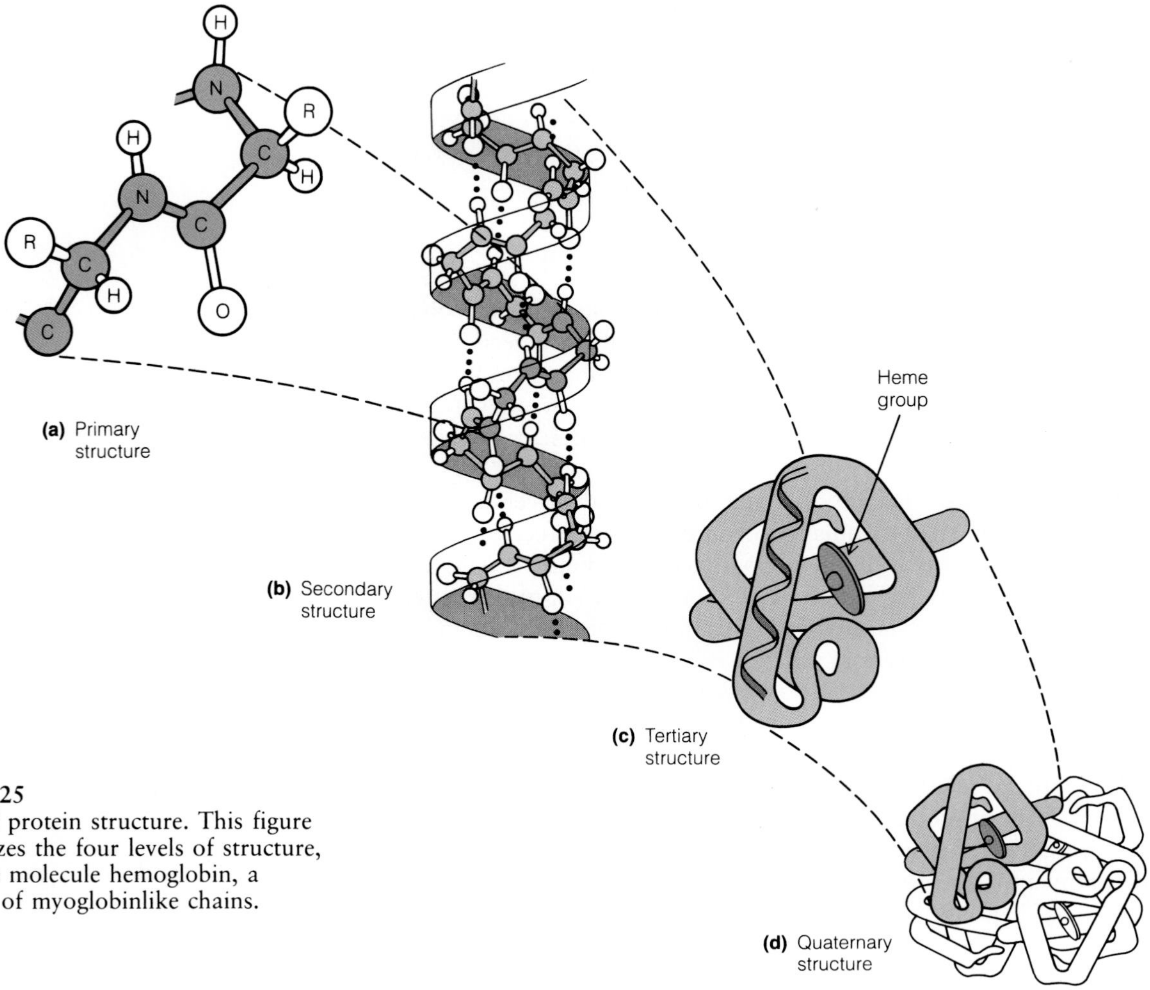

Figure 6.25
Levels of protein structure. This figure summarizes the four levels of structure, using the molecule hemoglobin, a tetramer of myoglobinlike chains.

Each polypeptide chain is an asymmetric unit in the aggregate, but the overall quaternary structure may exhibit a wide variety of symmetries, depending on the geometry of the interactions. For purposes of illustration, we shall use an asymmetric object familiar to everyone—a right shoe. Think of this as a polypeptide chain folded into a compact three-dimensional form. There are many ways in which we can stick shoes together. If the interacting surfaces (A and B) were at toe and heel, a linear aggregate could form:

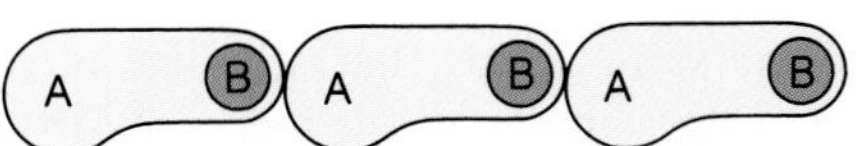

Since the two interacting groups lie on entirely different surfaces, we call this a **heterologous** interaction. Heterologous interactions must be very specially oriented to give a truly linear aggregate. More often, the interaction will be such that each unit is twisted through some angle with respect to the preceding one. This will give rise to a helical structure. Figure 6.26a shows an arrangement of shoes that forms a right helix with n units per turn. The top of the toe of each shoe is attached to the sole of the toe of the next, with a rotation of $360/n$ degrees. Two biological examples, both of **helical symmetry,** are shown in Figure 6.27: the helical coat of tobacco mosaic virus and the double helix of the muscle protein actin. Note that both linear and

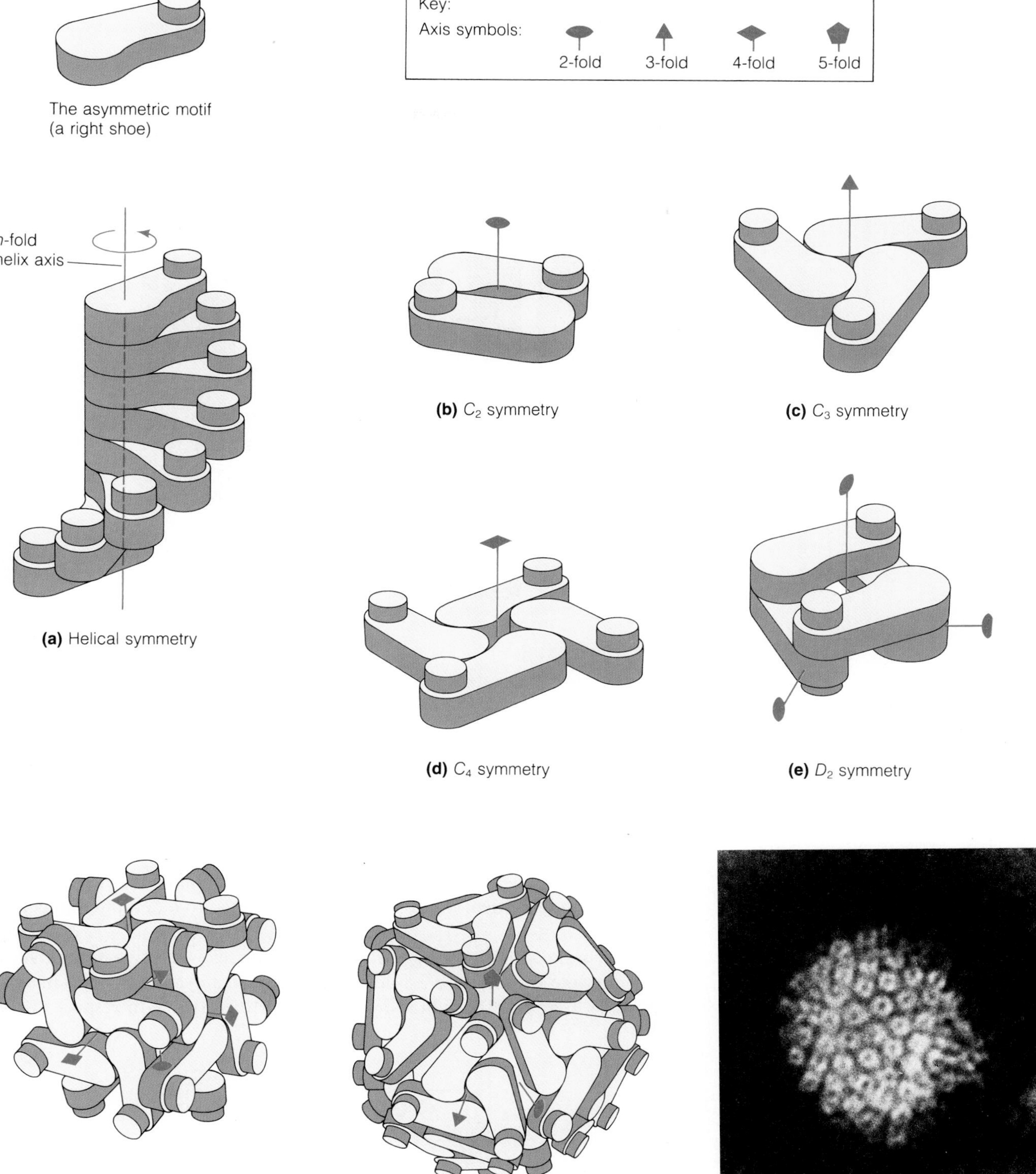

Figure 6.26
Examples of symmetry patterns adopted by proteins in forming quaternary structures. As an asymmetric motif, we use a right shoe. (**a**) A helix formed by rotation of each unit by $360/n$ degrees with respect to the preceding one. This will produce an n-unit-per-turn helix of indefinite length. (**b**) A dimer with C_2 symmetry: one 2-fold axis. (**c**) A trimer with C_3 symmetry: one 3-fold axis. (**d**) A tetramer with C_4 symmetry: one 4-fold axis. (**e**) A tetramer with D_2 symmetry: three 2-fold axes. (**f**) A 24-mer, exhibiting cubic symmetry. This structure has 4-fold, 3-fold, and 2-fold axes. (**g**) An icosahedral 60-mer, the kind of structure found in the protein coat of a number of viruses. (**h**) An electron micrograph of a virus (Herpes) with icosahedral symmetry.

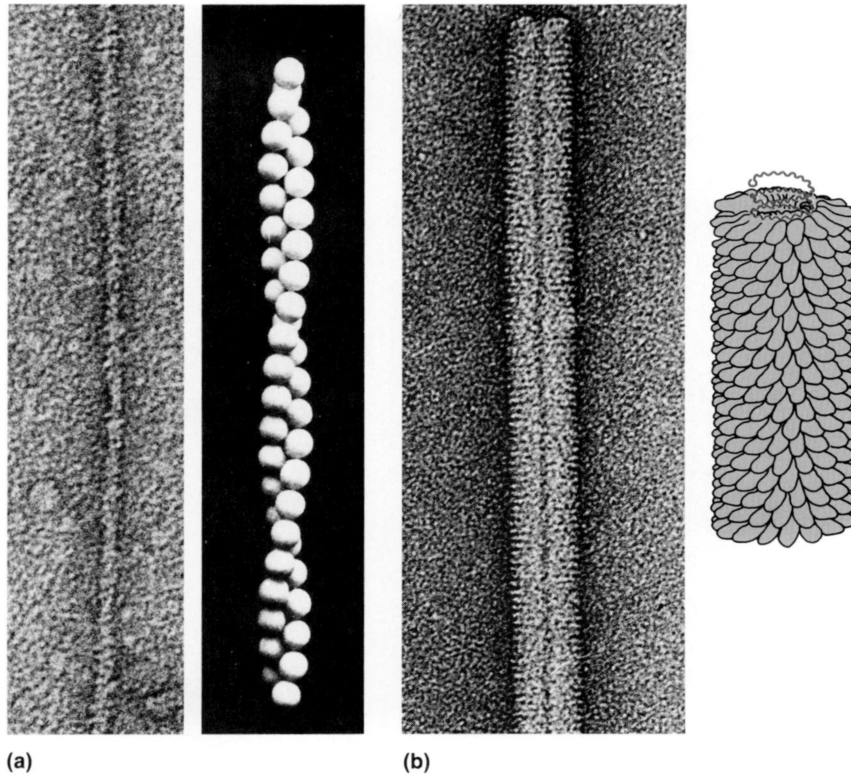

Figure 6.27
Two proteins forming helical aggregates. (**a**) Actin. (**b**) Tobacco mosaic virus coat protein. Beside each electron micrograph is shown a diagrammatic representation of the structure.

helical arrays are potentially capable of indefinite growth by addition of more subunits. Actin filaments can be thousands of units in length.

A third possible consequence of heterologous interaction is the formation of a closed ring structure with **cyclic symmetry.** The symmetry axis will be 3-fold, 4-fold, or in general n-fold (Figure 6.26c and d). An n-fold axis involves rotation of each subunit by $360/n$ degrees with respect to its neighbor. As for linear aggregates, the formation of higher cyclic symmetries requires very special orientation between the interacting groups. If these are not quite right, the ring will not close, and a helix will be generated. But if a closed ring is formed, the structure will be limited to a definite number of subunits.

A special situation of great importance arises when two subunits are related to one another by a 2-fold axis (also called a **dyad** axis). This means, of course, that each is rotated by 180° about this axis with respect to the other:

(This is visualized in three dimensions in Figure 6.26b). Imagine that there are interacting groups at A and B. For example, A could be a hydrogen bond donor, B an acceptor. Note that in this case the 2-fold symmetry means that *two* interactions occur, symmetrically placed about the dyad axis. Such a symmetric interaction is called **isologous;** it can occur whenever 2-fold axes are present. Dimers are the most common of all quaternary structures, and

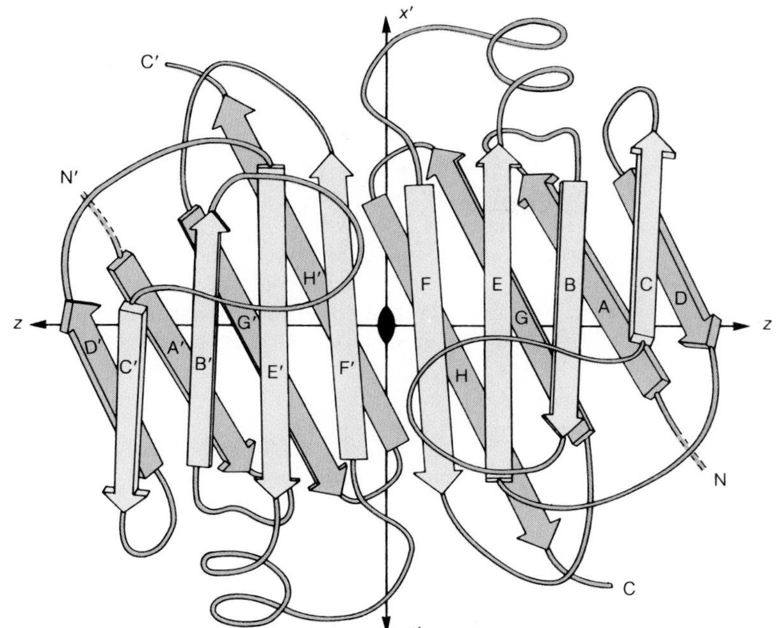

Figure 6.28
The dimer of prealbumin. The monomeric form of this same protein is shown in Figure 6.15b. The dimer has 2-fold symmetry about an axis perpendicular to the paper. The isologous interactions here are mainly hydrogen bonds between the β-sheet strands F and F′ and H and H′. The two monomers combine to form a complete β barrel. Prealbumin can also form a tetramer from two of these dimers by further isologous interactions.

they are almost always bound together in this way. An example is shown in Figure 6.28. Further isologous interaction can easily give rise to more complex quaternary structures of higher symmetry.

The example shown in Figure 6.26e is an example of **dihedral** symmetry, the most common structure for tetrameric proteins. It has three mutually perpendicular 2-fold axes. Other, more complex symmetries exist in some protein quaternary structures; examples are shown in Figure 6.26f and g. Note that the more complex structures involve both 2-fold axes and axes with $n > 2$. Such molecules will exhibit both isologous and heterologous interactions. Most important is the fact that molecules exhibiting any of these symmetries are always constrained to a definite number of subunits. Most multisubunit proteins exhibit this kind of association geometry, rather than the linear or helical aggregation that can lead to indefinite growth.

Two exceptions to this simple characterization of possible structures must be noted. First, although most dimers have 2-fold symmetry and isologous binding, it is possible to construct dimers in which the binding is heterologous, but indefinite association is still sterically blocked. Such dimers do not have 2-fold symmetry (Figure 6.29). A second complication is encountered whenever more than one type of polypeptide chain is incorporated into a specific multisubunit structure. In such cases the symmetry is reduced from the level you might expect from the total number of polypeptide chains, for two or more different chains will now form one asymmetric unit. We will consider one such example, *hemoglobin*, in the next chapter.

Note that each level of protein structure is built on the lower levels, as summarized in Figure 6.25. Tertiary structure can be thought of as a folding of elements of secondary structure, and quaternary structure is established by combining folded subunits. It is *all* dictated by primary structure, and ultimately by the gene.

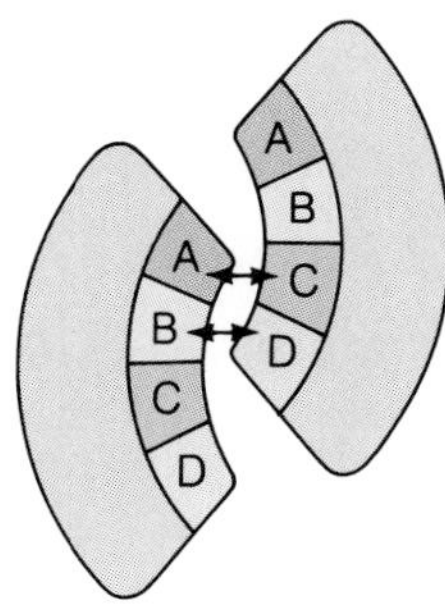

Figure 6.29
A dimer without symmetry. A schematic picture of how two subunits can associate by heterologous interactions and still not yield an indefinite chain. The interaction sites (A, B, C, D) are blocked from further reaction by the close fit of the surfaces.

REFERENCES

General

Creighton, T. E. (1983) *Proteins: Structures and Molecular Properties.* Freeman, San Francisco. A detailed, authoritative treatment of protein structure and function.

Dickerson, R. E., and I. Geis (1969) *The Structure and Action of Proteins.* Benjamin/Cummings, Menlo Park, Calif. Concise and well illustrated.

Fletterick, R. J., T. Schoer, and R. J. Matela (1985) *Molecular Structure: Macromolecules in Three Dimensions.* Blackwell, Oxford.

Oxender, D. L., and C. F. Fox (1987) *Protein Engineering.* Alan R. Liss, New York. An excellent collection of contemporary papers.

Schultz, G., and R. H. Schirmer (1979) *Principles of Protein Structure.* Springer-Verlag, New York.

Walton, A. G. (1981) *Polypeptides and Protein Structure.* Elsevier, North-Holland, New York.

Secondary Structure

Cantor, C. R., and P. R. Schimmel (1980) *Biophysical Chemistry.* Freeman, San Francisco.

Pauling, L., et al. (1951) The structure of proteins: Two hydrogen bonded helical conformations of the polypeptide chain. *Proc. Natl. Acad. Sci. USA* 37:205–211. The first description of the α helix.

Ramachandran, G. N., and V. Sassiekharan (1968) Conformation of polypeptides and proteins. *Adv. Protein Chem.* 28:283–437. Introduction of Ramachandran plots.

Fibrous Proteins

Cheah, K. S. E. (1985) Collagen genes and inherited connective tissue disease. *Biochem. J.* 222:287–303.

Fraser, R. D. B., and T. P. MacRae (1972) *Conformation in Fibrous Proteins and Related Synthetic Polypeptides.* Academic Press, New York.

Piez, K. A., and A. H. Reddi (1984) *Extracellular Matrix Biochemistry.* Elsevier Science Publishing, New York.

Globular Proteins: Secondary and Tertiary Structure

Matthews, B. W. (1987) Structural basis of protein stability and DNA–protein interactions. *The Harvey Lectures,* Ser. 81, pp. 33–51. Alan R. Liss, New York. Describes use of site-directed mutagenesis to analyze protein stability.

Richardson, J. S. (1981) The anatomy and taxonomy of protein structure. *Adv. Protein Chem.* 34:167–339.

Rose, G. (1979) Heirarchic organization of domains in globular proteins. *J. Mol. Biol.* 134:447–470.

Wetzel, R. (1988) Harnessing disulfide bonds using protein engineering. *Trends Biochem. Sci.* 12:478–482. A brief discussion of the importance of —S—S— bonds in protein function and stability.

Protein Folding

Anfinsen, C. B. (1973) Principles that govern the folding of protein chains. *Science* 181:223–230. Describes some of the important earlier experiments.

Harrison, S. C., and R. Durbin (1985) Is there a single pathway for the folding of a polypeptide chain? *Proc. Natl. Acad. Sci. USA* 82:4028–4030.

Kim, P. S., and R. L. Baldwin (1980) Structural intermediates trapped during the folding of ribonuclease A by amide exchange. *Biochemistry* 19:6124–6129.

Levitt, M. (1982) Protein conformation dynamics and folding by computer simulation. *Annu. Rev. Biophys. Bioeng.* 11:251–271.

Rossman, M. G., and P. Argos (1981) Protein folding. *Annu. Rev. Biochem.* 50:497–532.

Prediction of Secondary Structure

Chou, P. Y., and G. D. Fasman (1978) Empirical predictions of protein structure. *Annu. Rev. Biochem.* 47:251–276.

Kabsch, W., and C. Sander (1983) How good are predictions of protein secondary structure. *FEBS (Fed. Eur. Biochem. Soc.) Lett.* 155:179–182.

Presta, L. C., and G. D. Rose (1988) Helix signals in proteins. *Science* 240:1632–1641.

Protein Dynamics

Karplus, M., and J. A. McCannon (1986) The dynamics of proteins. *Sci. Am.* 254:(4)42–51. An excellent and readable introduction.

Ringe, D., and G. R. Petsko (1985) Mapping protein dynamics by x-ray diffraction. *Prog. Biophys. Mol. Biol.* 45:197–235.

Quaternary Structure

Caspar, D. L. P., and A. Klug (1962) Physical principles in the construction of regular viruses. *Cold Spring Harbor Symp. Quant. Biol.* 27:1–24.

Klotz, I. M., et al. (1970) Quaternary structure of proteins. *Annu. Rev. Biochem.* 39:25.

Matthews, B. W., and S. A. Bernhard (1973) Structure and symmetry of oligomeric enzymes. *Annu. Rev. Biophys. Bioeng.* 2:257–317.

Morgan, R. S., et al. (1979) The symmetry of self-complementary surfaces. *J. Mol. Biol.* 127:31–39.

PROBLEMS

1. Polyglycine, a simple polypeptide, can form a helix with $\phi = -80°$, $\psi = +150°$. From the Ramachandran plot, describe this helix with respect to (a) handedness and (b) number of residues per turn.

2. In the protein *adenylate kinase,* the C-terminal region is α-helical, with the sequence

   Val–Asp–Asp–**Val**–**Phe**–Ser–Gln–**Val**–Cys–
   Thr–His–**Leu**–Asp–Thr–**Leu**–Lys–

   The hydrophobic residues in this sequence are presented in boldface type. Suggest a possible reason for the periodicity in their spacing.

3. Although the bond energy for the hydrogen bond in vacuum is estimated to be about 20 kJ/mol, we find that each

hydrogen bond in a folded protein contributes much less—probably less than 5 kJ/mol—to the enthalpy of protein stabilization. Suggest an explanation for this difference.

4. Consider a small protein containing 101 amino acid residues. The protein will have 200 bonds about which rotation can occur. Assume that three orientations are possible about each of these bonds.

   (a) Based on these assumptions, about how many *random coil* conformations will be possible for this protein?

   (b) The estimate obtained in (a) is surely too large. Give one reason why.

5. (a) Based on a more conservative answer to Problem 4 (2.7×10^{92} conformations), estimate the configurational entropy change on folding a mole of this protein into a structure with only one conformation.

   (b) If the protein folds *entirely* into α helix with H bonds as the only source of enthalpy of stabilization, and each mole of H bonds contributes -5 kJ/mol to the enthalpy, estimate $\Delta H_{\text{folding}}$.

   (c) From your answers to (a) and (b), estimate $\Delta G_{\text{folding}}$ for this protein at 25°C. Is the folded form of the protein stable at 25°C?

6. The following sequence is part of a globular protein. Using Table 6.5 and the Chou–Fasman rules, predict the secondary structure in this region.

   . . . RRPVVLMAACLRPVVFITYGDGGTYYHWYH . . .

7. (a) A protein is found to be a tetramer of identical subunits. Name two symmetries possible for such a molecule. What kinds of interactions (isologous or heterologous) would stabilize each?

   (b) Suppose a tetramer, like hemoglobin, consists of two each of two types of chains, α and β. What is the highest symmetry now possible?

8. The peptide hormone *vasopressin* is used in the regulation of saltwater balance in many vertebrates. Porcine (pig) vasopressin has the sequence

   Asp–Tyr–Phe–Glu–Asn–Cys–Pro–Lys–Gly

   (a) Using the data in Figure 5.6 and data in Table 5.1, estimate the extinction coefficient ϵ, in units of cm^2/mg for vasopressin, using radiation with $\lambda = 280$ nm.

   (b) A solution of vasopressin is placed in a 0.5-cm-thick cuvette. Its absorbance at 280 nm is found to be 1.3. What is the concentration of vasopressin, in mg/cm^3?

   (c) What fraction of the incident light is passed through the cuvette in (b)?

TOOLS OF BIOCHEMISTRY 10

10

Spectroscopic Methods for Studying Macromolecular Conformation in Solution

X-ray diffraction (Tools of Biochemistry 5) is the method *par excellence* for determining the details of the three-dimensional structure of globular proteins and other biopolymers. Yet this technique has the fundamental limitation that it can be employed only when the molecules are crystallized, and this is not always easy or even possible. Nor can it easily be used to study conformational changes in response to changes in the molecules' environment. Therefore, investigators also use a variety of other methods that allow the study of molecules in the dissolved state. A number of these can be grouped in the category of **spectroscopic techniques.**

Absorption Spectroscopy

Proteins, carbohydrates, and nucleic acids are complex molecules and can absorb radiation over a wide spectral range. Yet the basic principles of their absorption can be explained in terms of the simplest kind of molecule, a diatomic molecule.

When two atoms interact to form a molecule, the potential energy curve for the lowest-energy electronic state (the **ground state**) will look like the lower curve in Figure T10.1a. **Excited electronic states** will have similar curves for energy versus interatomic distance, but these will lie at higher energies. For each electronic state of

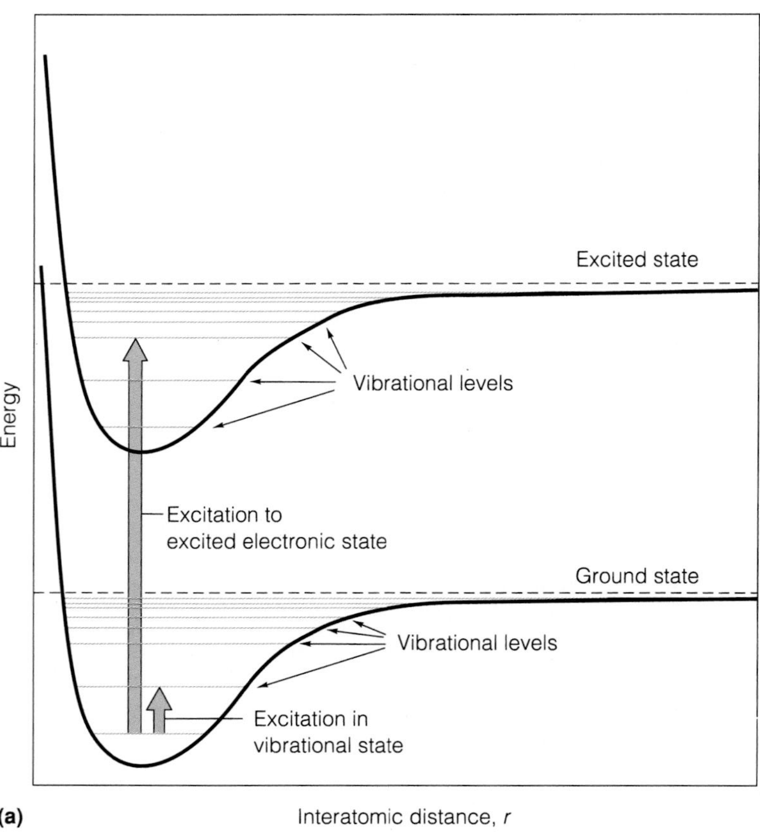

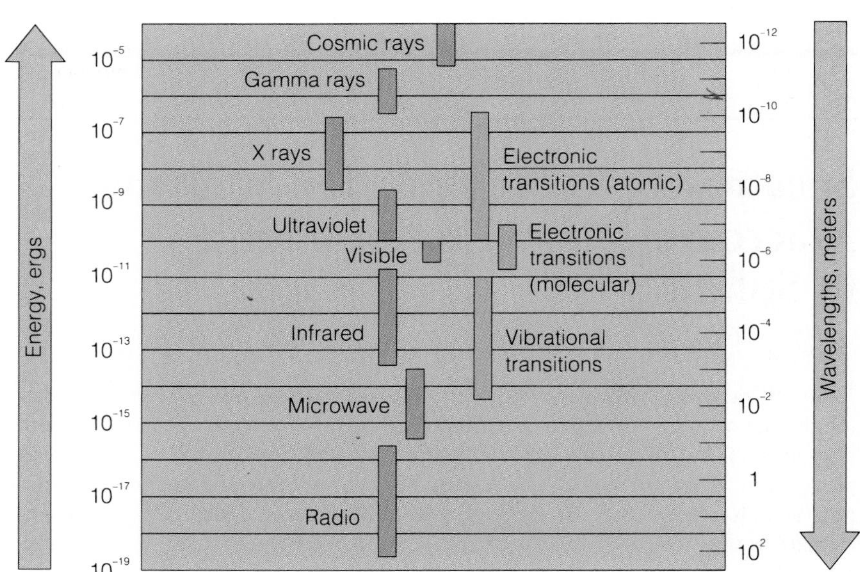

Figure T10.1
The principles of absorption spectroscopy. (**a**) Energy versus interatomic distance for ground state and an excited state of a diatomic molecule. (**b**) The electromagnetic spectrum.

the molecule, there will be a series of allowed **vibrational states,** indicated by horizontal lines in the figure. The basics of molecular spectroscopy can be understood by two simple rules: (1) transitions are possible only between allowed energy states of the molecule (energy levels are **quantized**), and (2) the energy (ΔE) that has to be absorbed in any transition determines the wavelength (λ) of the radiation that is absorbed to accomplish that transition. The energy in a **quantum of radiation** is inversely proportional to λ:

$$\Delta E = hc/\lambda; \qquad \Delta E = E_{\text{final state}} - E_{\text{initial state}} \tag{T10.1}$$

Here h is Planck's constant (6.627×10^{-34} J s), and c is the velocity of light (3×10^8 m/s). According to equation (T10.1) small energy differences correspond to long wavelengths and large difference to short wavelengths. This is in accord with Figure T10.1b, which indicates that the high-energy transitions between electronic states lead to absorption in the visible or ultraviolet region of the spectrum, whereas the low-energy transitions between different vibrational energy levels correspond to absorption of infrared energy.

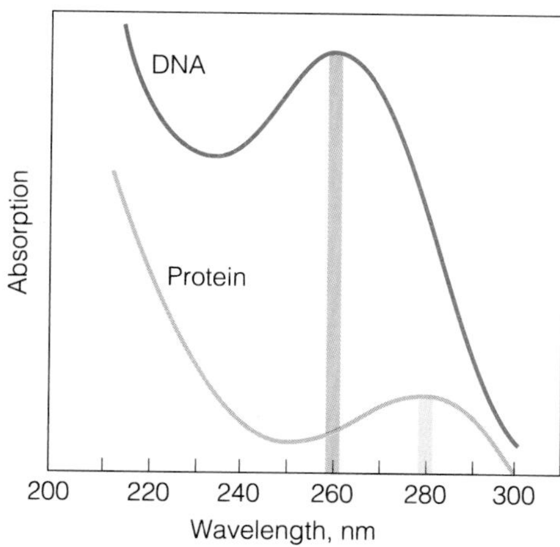

Figure T10.2
Comparison of the ultraviolet absorption spectra that would be expected for solutions of DNA and a typical protein.

Complex biopolymers like proteins and nucleic acids can undergo a multitude of kinds of molecular vibrations and oscillations. **Infrared spectroscopy** can provide direct information concerning macromolecular structure. For example, the exact positions of infrared bands corresponding to vibrations in the polypeptide backbone are sensitive to the conformational state (α helix, β sheet, etc.) of the chain. Thus, studies in this region of the spectrum are often used to investigate the conformations of protein molecules.

Most biopolymers do not significantly absorb visible light. There are some colored proteins, but these invariably contain prosthetic groups, such as the heme in myoglobin, or metal ions, such as copper, that confer the absorption. Blood and red meat owe their color to the heme groups carried by hemoglobin, myoglobin, and other heme proteins. Such absorption can often be exploited to investigate changes in the molecular environment of the prosthetic group. An example is the use of absorption spectroscopy in the visible spectrum to follow the oxygenation of myoglobin or hemoglobin.

By far the most common uses of spectroscopic techniques in biochemistry employ the ultraviolet region of the spectrum. In this region both proteins and nucleic acids absorb strongly (Figure T10.2). The protein absorption is found in two wavelength regions.

In the neighborhood of 270–290 nm, we see absorption by the aromatic side chains of phenylalanine, tyrosine, and tryptophan (see also Figure 5.6). Therefore, absorption at 280 nm is used routinely to measure protein concentrations. A **spectrophotometer** is used, in which a cuvette of thickness l is placed in a beam of radiation of intensity I_0 (Figure T10.3). The emerging intensity will be decreased to a value I because of absorption of part of the radiation. The **absorbance** at wavelength λ is defined as $A_\lambda = \log(I_0/I)$ and is related to l and the concentration c by **Beer's law:**

$$A_\lambda = \epsilon_\lambda lc \tag{T10.2}$$

Here ϵ_λ is the **extinction coefficient** at wavelength λ for the particular substance being studied. Its dimensions depend on the concentration units employed. If protein concentration is measured in mg/cm^3 and l in cm, then ϵ_λ must have the dimensions cm^2/mg, since A is a dimensionless quantity. Once the extinction coefficient for a particular protein has been determined (for example, by measuring the absorbance of a solution containing a known weight of the protein) the concentration of any other solution of that protein can be calculated from a simple absorbance measurement, using equation (T10.2). The same method is routinely used with nucleic acids, but in that case a wavelength of 260 nm is usually employed, for nucleic acids absorb most strongly in this spectral region.

A second region of strong absorption in the protein spectrum lies in the range 180–220 nm. Absorption at such wavelengths arises from electronic transitions in

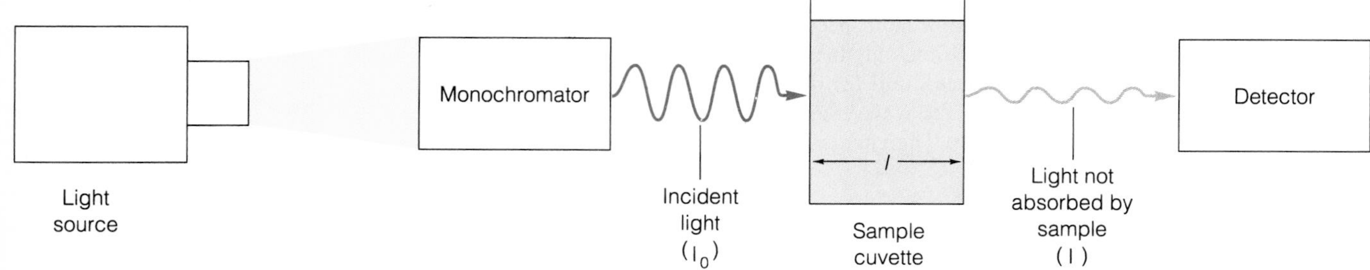

Figure T10.3
Measurement of light absorption with a spectrophotometer.

the polypeptide backbone itself and is therefore sensitive to the backbone conformation.

Actually, both regions of the protein absorption spectrum are somewhat affected by conformational state. Extinction coefficients change slightly when a molecule goes from a globular to a random coil form, because the local environment of all groups, including aromatic side chains, is modified. Since spectroscopy can be performed with great accuracy, even small changes can be measured and used to follow protein denaturation. An example is shown in Figure 6.20.

Fluorescence

In most cases, molecules raised to an excited electronic state by absorption of radiant energy return to the ground state by **radiationless transfer** of the excitation energy to surrounding molecules. In short, the energy reappears as heat. But occasionally, as shown in Figure T10.4a, a molecule will lose only part of its energy of excitation by transfer (yellow arrow) and will reradiate the larger part (green arrow). This gives rise to the phenomenon called **fluorescence.** Since, as the figure shows, the quantum of energy re-emitted as fluorescence is always smaller than the quantum that was initially absorbed (red arrow), the wavelength of the fluorescent light will be longer than the wavelength of the exciting light. The **fluorescence emission spectrum** of tyrosine is contrasted with the absorption spectrum in Figure T10.4b. In proteins, tyrosine and tryptophan are the major fluorescent groups. The environment of these residues can greatly modify the intensity of their fluorescence, and this technique can then be used to monitor changes in protein conformation. Further, excitation of fluorescence by plane polarized light (see below) provides a way of studying the dynamics of protein structure. If the excited residues are able to move or rotate appreciably before the fluorescent light is re-emitted, the fluorescence will be depolarized to some extent.

Excited state
Energy
Energy transfer
Fluorescence
Excitation
Ground state
(a) Interatomic distance, *r*

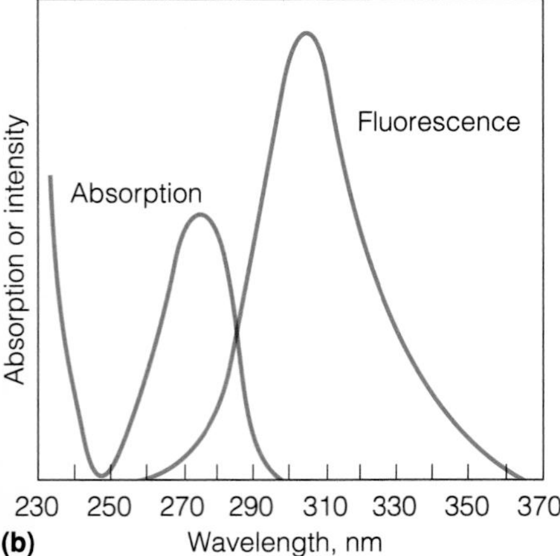

Figure T10.4
The principles of fluorescence.

Circular Dichroism

Although absorption spectroscopy and fluorescence can be very helpful in following molecular changes, such measurements are difficult to interpret directly in terms of changes of secondary structure. For this purpose, techniques involving the use of polarized light have become important.

There are various ways in which light can be polarized. Most familiar is **plane polarization,** in which the varying electric field of the radiation has a fixed orientation (Figure T10.5a). Less familiar, but equally important, is **circular polarization,** in which the electric field rotates with the frequency of the radiation (Figure T10.5b). If you observe a circularly polarized beam, the electric field will rotate in either a clockwise or counterclockwise direction. The former is called *right circularly polarized light,* the latter *left circularly polarized light.* Most of the molecules studied by biochemists are asymmetric—for example, L- and D-amino acids, right- and left-hand protein helices, and right- and left-hand nucleic acid helices. When such molecules absorb light, they exhibit a preference for the absorption of left or right circularly polarized light. For example, a right circularly polarized beam interacts differently with a right-hand α helix than a left circularly polarized beam. This

Figure T10.5
Polarized light and circular dichroism.

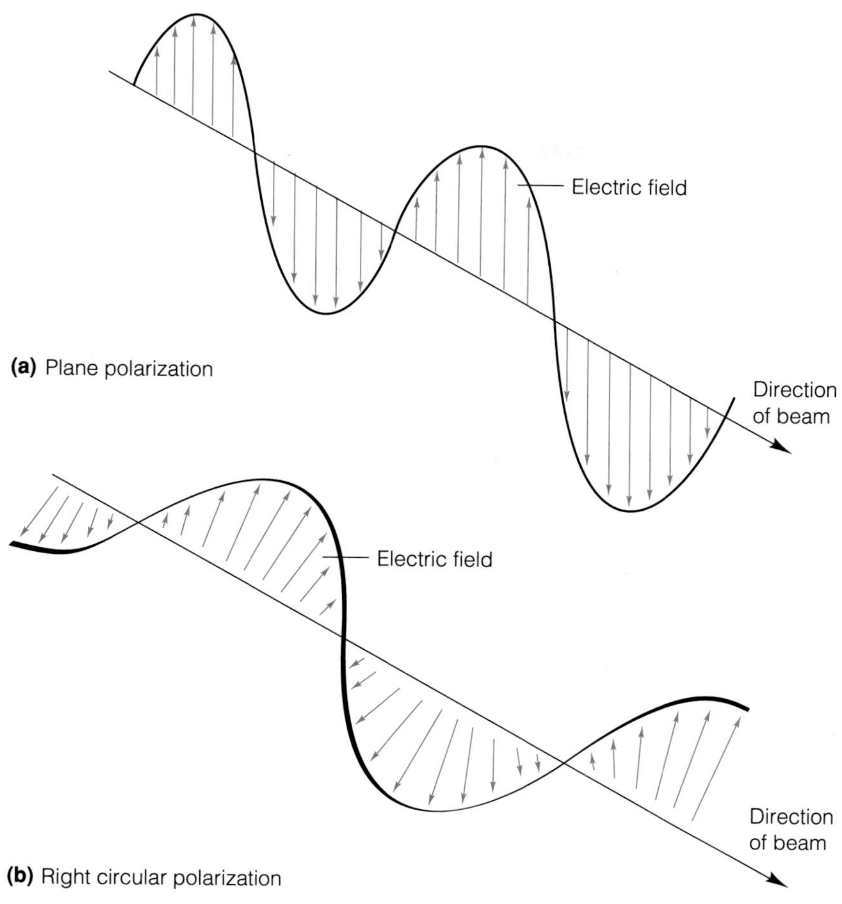

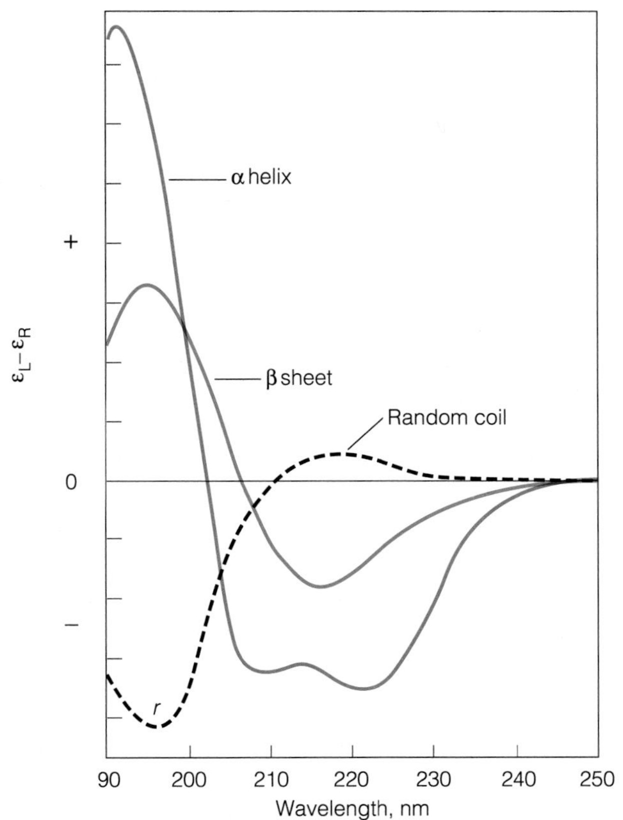

(c) Circular dichroism spectra for polypeptides in various conformations

difference in absorption, called **circular dichroism,** is defined as

$$\Delta A = \frac{A_L - A_R}{A} \quad \text{(T10.3)}$$

where A_L is the absorbance for left circularly polarized light, A_R is the corresponding quantity for right circular polarization, and A is the absorbance for unpolarized light. Obviously, ΔA can be either positive or negative, so a **circular dichroism spectrum** (or **CD spectrum**) is unlike a normal absorption spectrum in that both + and − values are allowed.

Figure T10.5c shows CD spectra for polypeptides in the α-helical, β-sheet, and random coil conformations. It should be obvious from the figure that circular dichroism can be a powerful tool for following conformational changes in proteins. For example, if a protein is denatured so that its native structure, containing α-helical and β-sheet regions, is transformed into an unfolded, random coil structure, this will be reflected in a dramatic change in its CD spectrum.

Circular dichroism can be used in another way, to estimate the content of α helix and β sheet in native proteins. Since the contributions of these different secondary structures to the circular dichroism at different wavelengths are known, we may attempt to match an observed spectrum for a protein by a combination of such contributions. This kind of analysis frequently turns out to agree with the secondary structure composition as judged by x-ray studies, and it has given support to the idea that the structures of globular proteins observed in crystals are preserved when these crystals are dissolved in buffer solutions at physiological pH.

Although it is an extremely useful technique, circular dichroism is not a very discriminating one. That is, it cannot, at present, tell us what is happening at a particular *point* in a protein molecule. A method that has the potential to do so is nuclear magnetic resonance.

Nuclear Magnetic Resonance

The nuclei of certain isotopes of some atoms have a property referred to as **spin,** which makes these nuclei behave like minute magnets. Only a limited number of isotopes have this property; some of use to biochemists are listed in Table T10.1. If an external magnetic field is applied to a sample containing such nuclei, different orientations of the nuclear spin will have different energies. Microwave radiation can flip these nuclei from one energy state to another. This phenomenon is called **nuclear magnetic resonance (NMR).** NMR is really a kind of spectroscopy. As Figure T10.6a shows, at a given magnetic field, a particular energy corresponding to a particular wavelength in the microwave region of the spectrum will correspond exactly to the energy difference between the spin states. So the experimenter could use a fixed magnetic field and change the microwave wavelength until "resonance" was obtained. More often, the experiment is done the other way around: the wavelength of the radiation is held fixed, and the magnetic field is varied to achieve resonance.

The energy levels of a spinning nucleus in a magnetic field are very sensitive to the environment surrounding the atom in question. Different hydrogens in a compound, for example, will reach resonance at different magnetic field strengths. These differences are usually expressed in terms of *chemical shifts* (δ) defined with respect to a reference material added in the sample:

$$\delta = \frac{H_{ref} - H}{H_{ref}} \times 10^6 \quad \text{(T10.4)}$$

Here H is the field strength for resonance for the nucleus in question, and H_{ref} is that for a reference nucleus.

With modern, large NMR instruments, it is possible to resolve resonances for most or all of the protons in even a large molecule; an NMR spectrum for the protein bovine pancreatic trypsin inhibitor, for example, is shown in Figure T10.6b. If these resonances can be identified, a task of some difficulty, it becomes possible to

Table T10.1
Nuclei most often used in biochemical NMR experiments

Isotope	Spin	Natural[a] Abundance (%)	Relative[b] Sensitivity	Applications
^{1}H	1/2	99.98	(1.000)	Almost every kind of biochemical study
^{2}H	1	0.02	0.0096	Studies of selectively deuterated compounds
^{13}C	1/2	1.11	0.0159	Studies of specific carbon-containing groups
^{19}F	1/2	100.00	0.834	Fluorine is substituted for H as a "probe" of local structure
^{31}P	1/2	100.00	0.0664	Studies of nucleic acids, phosphorylated compounds

[a]The number represents the percentage of this isotope in the naturally occurring mix of isotopes of each element. Those that are close to 100% can be studied directly in the naturally occurring biopolymers. Rare isotopes, such as ^{2}H (deuterium) and ^{13}C, usually have to be artificially enriched in substances to be studied.

[b]Indicates the sensitivity (relative to ^{1}H) of conventional NMR instruments to each isotope. Low values mean that the experiment will be more difficult or time-consuming.

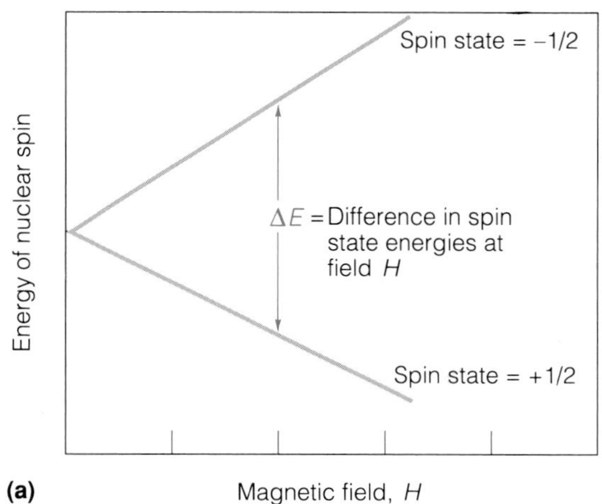

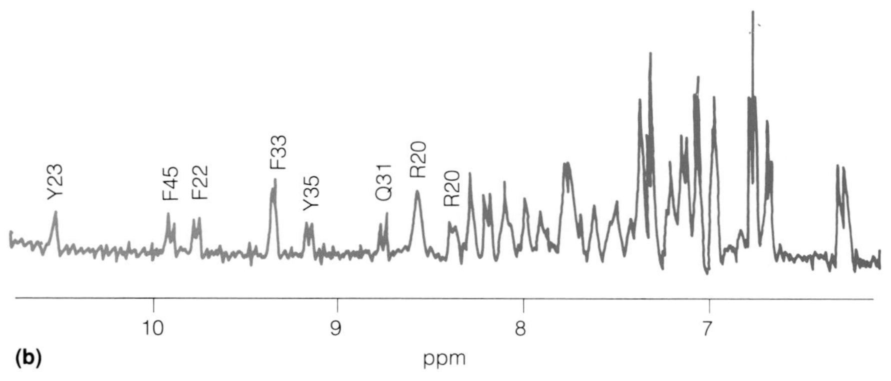

Figure T10.6
Nuclear magnetic resonance. (**a**) The principle of the method. (**b**) A portion of the ^{1}H NMR spectrum of bovine pancreatic trypsin inhibitor. The axis is δ, in parts per million (ppm). Letters refer to resonances assigned to specific groups (see Figure 6.24).

Figure T10.7
Titration of individual histidine residues in ribonuclease by NMR.

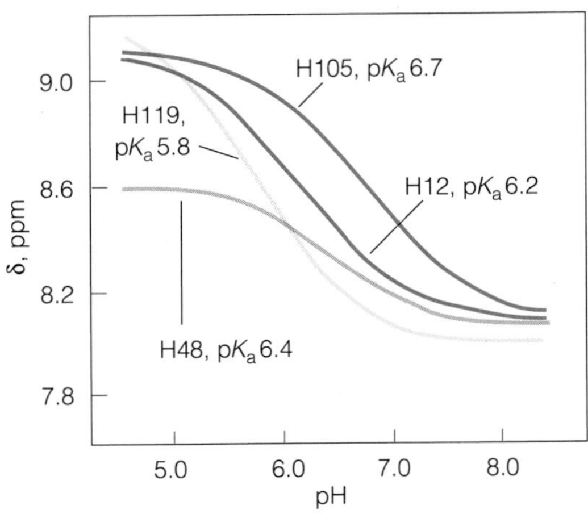

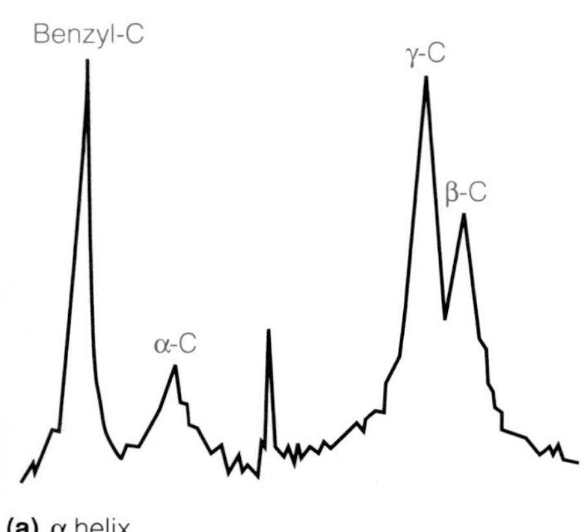

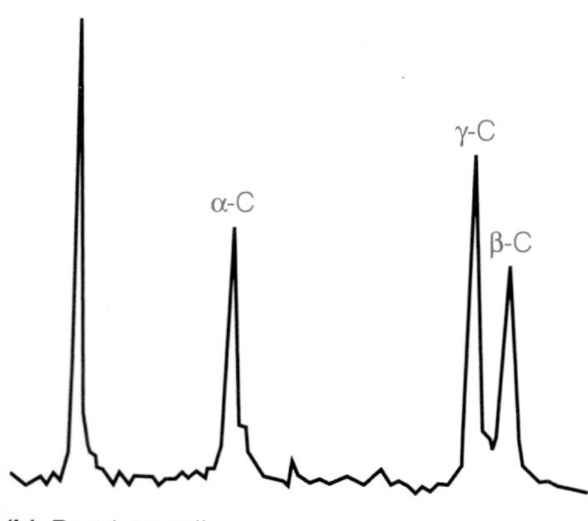

Figure T10.8
The ^{13}C NMR spectrum of the polypeptide poly(γ-benzyl)-L-glutamate in the α-helical (**a**) and random coil (**b**) conformations. Individual peaks for the α, β, γ, and benzyl carbons of the side chains are clearly resolved.

ask what is happening to a particular group or atom in a molecule as large as a protein. An example of such discrimination is shown in Figure T10.7, which traces the titration curves of individual histidine residues in the protein ribonuclease. The figure also graphically demonstrates a principle alluded to in Chapter 5: individual groups of a given type can show quite different pK_a values because of their different environments within the complex protein molecule.

Examples of the discriminatory power of NMR are legion. As another example, we can use ^{31}P NMR to ask what the phosphate groups in a nucleic acid molecule are doing. The ^{13}C NMR spectra in Figure T10.8 resolve the individual types of carbon atoms in the synthetic polypeptide poly-γ-benzl-L-glutamate. This figure demonstrates another important application of NMR. When atoms can move only slowly in solution, NMR lines are broadened. Note the contrast in sharpness of the lines between the helix and random coil forms of this polypeptide. In the random coil, the backbone and side chains are free to swing about in random motion, but in the helix each is locked in place and can move only with the sluggish motion of the whole large rodlike molecule. Thus, in this case NMR allows us to study in a most direct manner the helix → random coil transition.

Many other, even more sophisticated applications of NMR to biochemical problems are constantly being developed. Most important is **two-dimensional NMR,** which allows the study of interactions between specific groups in protein molecules. For details of this, and other new methods, consult the appropriate references.

REFERENCES

Bax, A., and L. Lerner (1986) Two-dimensional nuclear magnetic resonance spectroscopy. *Science* 232:960–967. An overview of some of the newer techniques that are proving especially important in protein structure analysis.

Campbell, I. D., and R. A. Dwek (1984) *Biological Spectroscopy.* Benjamin/Cummings, Menlo Park, Calif. An excellent practical description of many techniques. Numerous fine examples and problems.

Cantor, C. R., and P. R. Schimmel (1980) *Biophysical Chemistry,* Vol. 2. Freeman, San Francisco. Contains more detailed and rigorous treatments of these techniques than are given here.

Kline, A. D., W. Braun, and K. Wüthrich (1986) Studies by ^{1}H nuclear magnetic resonance and distance geometry of the solution conformation of the α-amylase inhibitor tendamistat. *J. Mol. Biol.* 189:377–382. An application of modern NMR methodology to a small protein.

van Holde, K. E. (1985) *Physical Biochemistry,* 2nd ed. Prentice-Hall, Englewood Cliffs, N.J. More detailed than the presentation given here, but still at a relatively elementary level.

TOOLS OF BIOCHEMISTRY 11

Determining Molecular Weights and the Number of Subunits in a Protein Molecule

When a new protein has been identified and purified, there are two immediate questions:

1. Does this protein exist under physiological conditions as a single polypeptide chain, or is it made up of multiple subunits?
2. If there is more than one subunit in the functional protein, are the subunits identical, or are there several kinds?

The answer to these questions can usually be obtained by first determining the molecular weight of the native protein and then subjecting it to conditions under which dissociation into subunits should occur. If subunits are held together by noncovalent interactions, changing the solvent environment will often effect dissociation. For example, the pH might be raised or lowered well outside the physiological range. Alternatively, denaturing solvents like concentrated solutions of urea or guanidine hydrochloride might be used. These compounds, which are excellent hydrogen bonders, destroy the regular water structure. For this reason they are sometimes called **chaotropic** ("chaos-forming") agents. Destruction of the water structure decreases the hydrophobic effect and thereby promotes the unfolding and dissociation of protein molecules. Even more effective are detergents like sodium dodecyl sulfate (SDS), which form micelles about individual polypeptide chains. By determining the molecular weights of the dissociated products in such a solvent and comparing them with the "native" molecular weight, we can tell how many subunits are involved.

H_2N—C(=O)—NH_2

Urea

$(H_2N)_2C{=}NH_2^+ + Cl^-$

Guanidine hydrochloride (guanidinium chloride)

Determining the Molecular Weight of the Native Structure

To determine the molecular weights of proteins in their physiological states, several techniques are available. Recall from Tools of Biochemistry 6 that the *sedimentation coefficient, S,* is related to the molecular weight:

$$S = \frac{M(1 - \bar{v}\rho)}{Nf} \quad \text{(T11.1)}$$

This equation contains the frictional coefficient, f, which also depends on molecular size. To eliminate f, one may make a separate determination of the **diffusion coefficient, *D*,** of the protein. This quantity measures how fast molecules diffuse in solution because of thermal Brownian movement. The diffusion coefficient depends on the temperature, T, and the frictional coefficient:

$$D = \frac{RT}{Nf} \quad \text{(T11.2)}$$

where R is the gas constant, 8.31×10^7 erg/° mol. The diffusion coefficient can be measured by determining how rapidly a boundary between a solution of the molecule and pure solvent is blurred, or by laser light-scattering techniques (see van

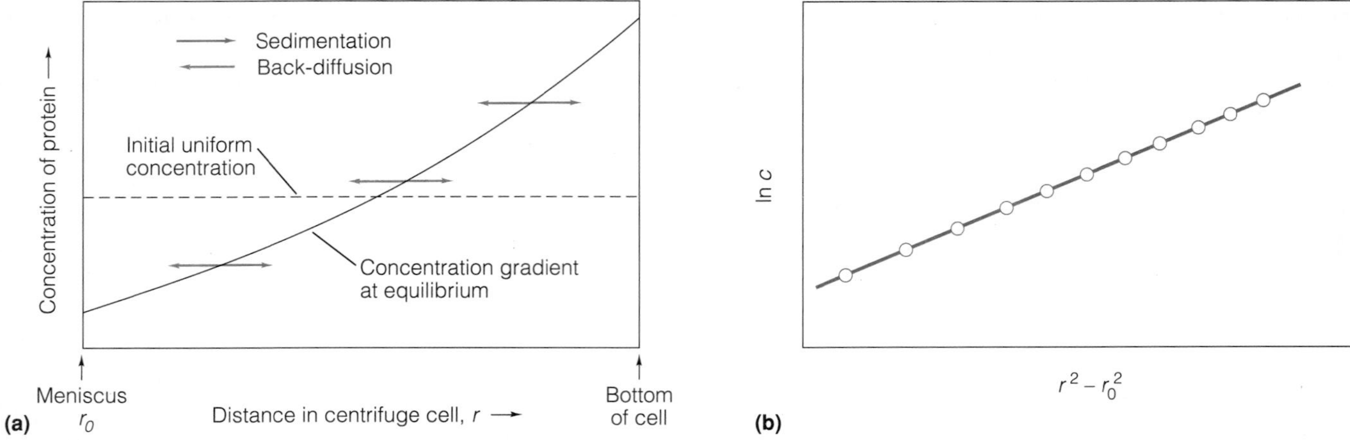

Figure T11.1
Sedimentation equilibrium as a method for molecular weight measurement. (**a**) Sedimentation equilibrium is attained when the outward flow of solute caused by its sedimentation is balanced by back-flow due to diffusion. (**b**) At the equilibrium state, the concentration gradient is such that a graph of ln c versus $r^2 - r_0^2$ is a straight line.

Holde, Chapter 4). Combining equation (T11.1) with (T11.2) we obtain

$$M = \frac{RTS}{D(1 - \bar{v}\rho)} \qquad \text{(T11.3)}$$

A more direct way to measure M uses the technique of **sedimentation equilibrium.** If a protein solution is sedimented for many hours at low rotor speed, an equilibrium will be established between the tendency of the molecules to sediment and their tendency to diffuse back. A steady concentration gradient will be set up (Figure T11.1a). It can be demonstrated that a homogeneous protein will give a gradient described by the equation

$$\frac{c(r)}{c(r_0)} = e^{M(1-\bar{v}\rho)\omega^2(r^2-r_0^2)/2RT} \qquad \text{(T11.4)}$$

where $c(r)$ is the concentration at a distance r from the center of rotation, $c(r_0)$ is the concentration at the meniscus position (r_0), and other quantities are as defined in Tools of Biochemistry 6. According to this equation, if ln $c(r)$ is graphed against $r^2 - r_0^2$ a straight line should be obtained (Figure T11.1b)

$$\ln c(r) = \ln c(r_0) + \frac{M(1 - \bar{v}\rho)\omega^2(r^2 - r_0^2)}{2RT} \qquad \text{(T11.5)}$$

The slope of the line is $M(1 - \bar{v}\rho)\omega^2/2RT$, allowing the determination of M. An analytical ultracentrifuge allows quite exact measurement of M by this technique. Details of sedimentation equilibrium and the S/D method, along with other physical techniques that can be used to determine the molecular weights of native proteins, are given in van Holde (see References).

Determination of Number and Weights of Subunits: SDS Gel Electrophoresis

Once the native molecular weight has been determined, the easiest way to find the molecular weight(s) of the subunits is to use gel electrophoresis in the presence of SDS. Under these conditions quaternary, tertiary, and secondary structures of proteins are all broken down. The chain is unfolded and surrounded by SDS molecules to form a *micelle* (Figure T11.2). The numerous negative charges carried by the many SDS molecules bound to the protein make the charge carried by the protein insignificant. The polypeptide chain is therefore transformed into an elongated micelle, the length and charge of which are each proportional to the length (and hence molecular weight) of the chain. As pointed out in Tools of Biochemistry 2, such particles will migrate in gel electrophoresis with relative mobilities depending only on their lengths. This is demonstrated by the graph shown in Figure T11.3. If electrophoresis of an unknown protein chain is carried out on the same gel as such a set of "standards," the molecular weight of the unknown can be measured by interpolation in a graph like Figure T11.3.

In investigating the subunits of a protein by this technique, it is advisable to do two experiments: one in the presence of a disulfide bond–reducing agent like β-mercaptoethanol (HS—CH_2CH_2OH) and one in its absence. This will distinguish between subunits that are held to one another by —S—S— bridges and those that are held together only by noncovalent forces. If a single band is found on each of these SDS gels, corresponding in molecular weight to that of the native protein, we may conclude that the protein exists under physiological conditions as a single polypeptide chain. If the band or bands observed are of much lower molecular weight, a multisubunit structure is indicated. Even though the molecular weight values obtained from gel electrophoresis may be approximate, they should be sufficiently accurate for a good guess as to the number of subunits.

Assuming that multiple subunits are indicated, is there only one kind or are there several? More than one band on the SDS gel is a clear indication of multiple types of subunits. But finding only one band does not prove that subunits are identical. There may, in fact, be several kinds of subunits with distinct amino acid sequences but nearly identical molecular weights; these usually cannot be resolved on SDS gels. To be satisfied that only one type of chain is present, the researcher must turn to other methods. If a way can be found to dissociate the protein without the use of detergents or chaotropic agents, isoelectric focusing can be a very sensitive technique (see Tools of Biochemistry 2).

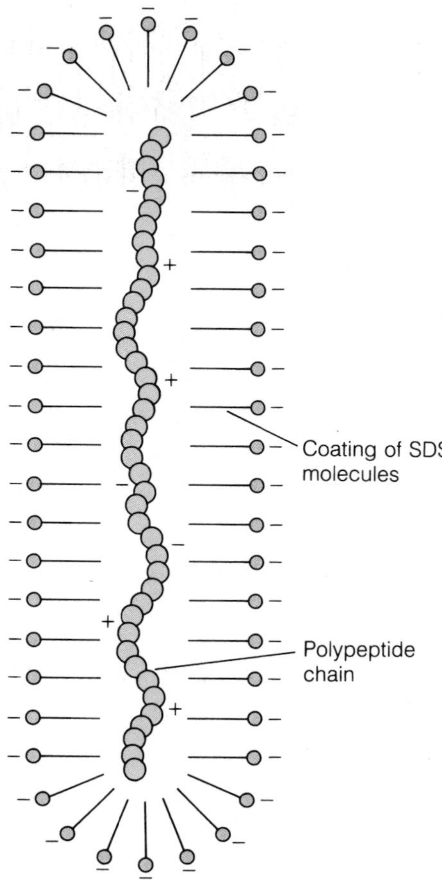

Figure T11.2
Protein molecules, when dissolved in an SDS solution, form the cores of elongated detergent micelles.

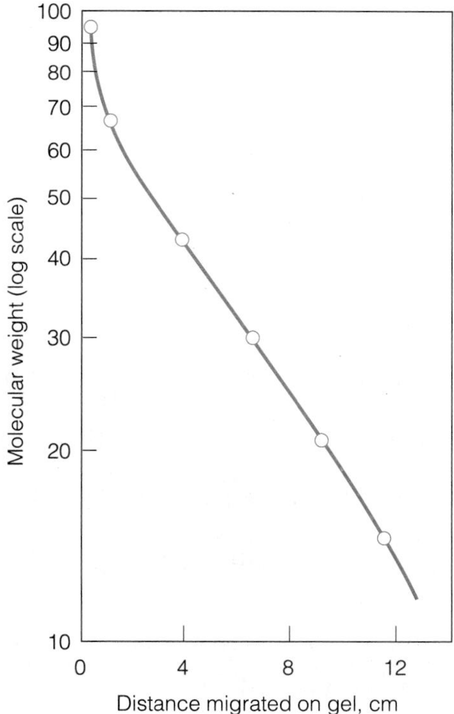

Figure T11.3
A graph of log *M* versus relative electrophoretic mobility for a series of proteins dissolved in a solution containing the detergent SDS.

REFERENCES

Hames, B. D., and D. Rickwood (eds.) (1981) *Gel Electrophoresis of Proteins.* IRL Press, Oxford, Washington, D.C.

van Holde, K. E. (1985) *Physical Biochemistry,* 2nd ed., Chapters 5 and 6. Prentice-Hall, Englewood Cliffs, N.J.

CHAPTER 7

Protein Function and Evolution

Now that we have developed an understanding of the complex, folded structures of globular proteins, we can look more closely at how such structures are related to the molecules' function. We will take as our examples in this chapter two groups of proteins whose main function is binding other molecules.

To begin with, we will examine **myoglobin** (abbreviated **Mb**) and its molecular relative **hemoglobin** (**Hb**). There are a number of reasons for this choice. First, the hemoglobins and myoglobins play vital roles in one of the most important aspects of animal metabolism—the acquisition and utilization of oxygen. The most efficient energy-generating mechanisms in animal cells require molecular oxygen for the oxidation of foodstuffs. Therefore, proteins that can deliver oxygen to cells and store it until needed are essential for any higher organism. In addition, hemoglobin plays a second role in removing CO_2 from tissues. Oxygen delivery and CO_2 uptake are carefully regulated to meet tissue needs. Thus, these examples can teach much about regulation of protein function. Finally, the relationship between hemoglobin and myoglobin gives some important insights into how protein function evolves.

We shall then turn to a group of proteins that perform a very different, but equally important, function—the **immunoglobulins**, or antibody molecules. Whereas the hemoglobins and myoglobins are devoted primarily to the binding of a single kind of molecule (oxygen), the immunoglobulins represent protein structures that can be produced in a multitude of forms to bind many different kinds of molecules in the immune response.

Oxygen Storage: Myoglobin

The Physiological Role of Myoglobin

Myoglobin is found in virtually all higher animals, especially in tissues such as muscle, which require large oxygen reserves for periods when energy demands are high. But myoglobinlike proteins are more widely distributed than this would imply. Even some one-cell organisms (*Paramecium,* for example) have such proteins, probably to store oxygen for periods of oxygen deficit.

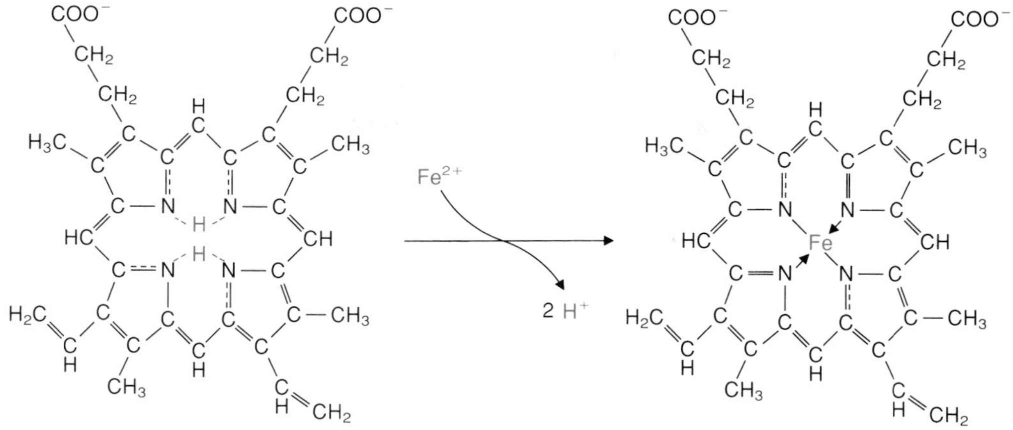

Fe^{2+} → $2\ H^+$

(a) Protoporphyrin IX

(b) Heme (Fe-protoporphyrin IX)

Figure 7.1
The structures of protoporphyrin IX (**a**) and heme (**b**). Since there is resonance delocalization of the electrons in the porphyrin ring, all bonds to the iron atom in heme are actually equivalent.

The key to the functional requirements for an oxygen-storing molecule is found in the word *storage*. Any such molecule must be able to bind O_2, not allow it to oxidize anything (which would reduce the O_2), and then give it up on demand.

How can oxygen be bound by a protein molecule? The organic structure of a protein is wholly unsuited to *direct* binding of oxygen. But certain transition metals, in their lower oxidation states (particularly Fe^{2+} and Cu^{+}), have a strong tendency to bind oxygen. In the evolution of the hemoglobin–myoglobin family of proteins, Fe^{2+} has been utilized in the O_2 binding site.*

There are a number of possible ways in which various iron-containing proteins hold Fe^{2+}. Throughout the myoglobin–hemoglobin family, the iron is chelated by a tetrapyrrole ring system called **protoporphyrin IX** (Figure 7.1a). Protoporphyrin IX is one of a large class of **porphyrin** compounds of similar structure. We will encounter others in chlorophyll, the cytochrome proteins, and some natural pigments. Like most compounds with large conjugated ring systems, the porphyrins are strongly colored. The iron-porphyrin in hemoglobin accounts for the red color of blood, while the magnesium-porphyrin in chlorophyll is responsible for the green of plants.

The complex of protoporphyrin IX with Fe^{2+} is called **heme** (Figure 7.1b). This prosthetic group is noncovalently bonded in a hydrophobic crevice in the myoglobin molecule (Figure 7.2). Ferrous iron is normally octahedrally coordinated, which means it should have six **ligands**, or binding groups, attached to it, in the geometry shown in Figure 7.3a. The nitrogen atoms of the porphyrin ring account for only four of these ligands. There are two remaining coordination sites available, and these lie along an axis perpendicular to the plane of the ring. In myoglobin one of these sites is occupied by the nitrogen of histidine residue number 93, which is a part of the F

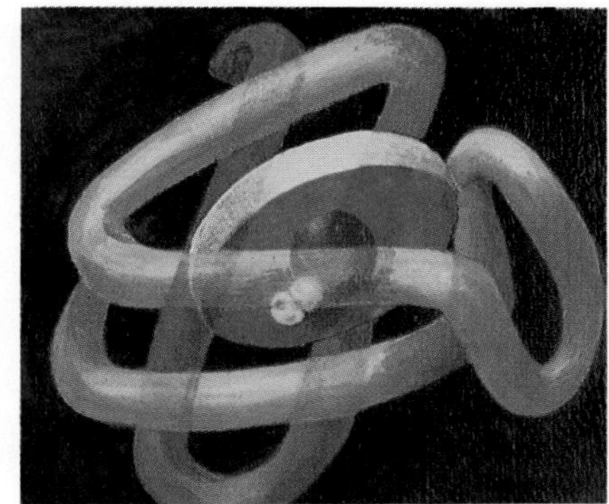

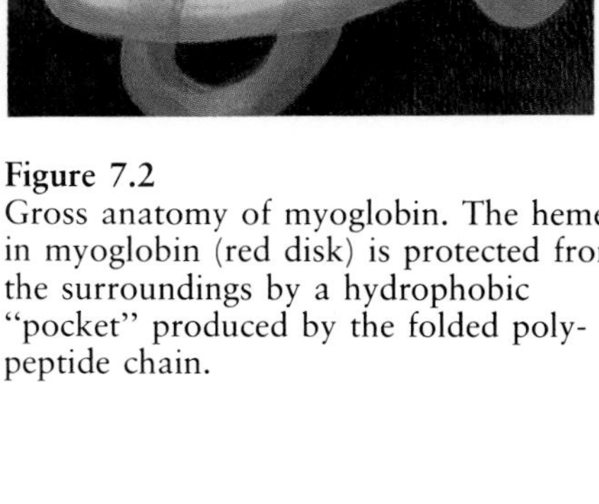

Figure 7.2
Gross anatomy of myoglobin. The heme in myoglobin (red disk) is protected from the surroundings by a hydrophobic "pocket" produced by the folded polypeptide chain.

*This is not, however, the *only* method of oxygen binding used. Most mollusks and some arthropods have a quite different oxygen-binding protein, hemocyanin, which contains copper; and still other invertebrates use a wholly unrelated iron-containing protein called hemerythrin. This shows that the same function can often be arrived at by quite different evolutionary routes.

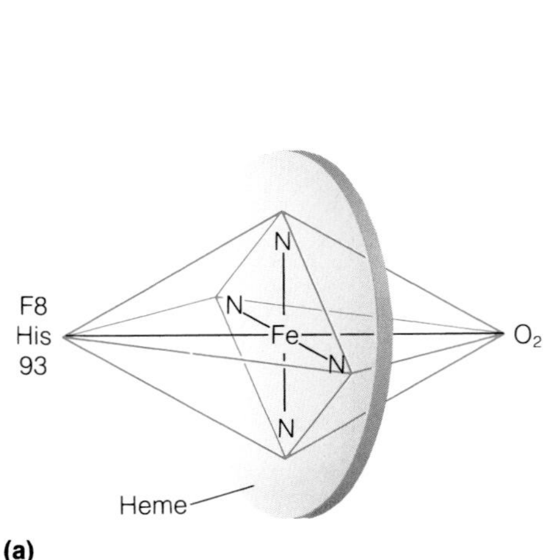

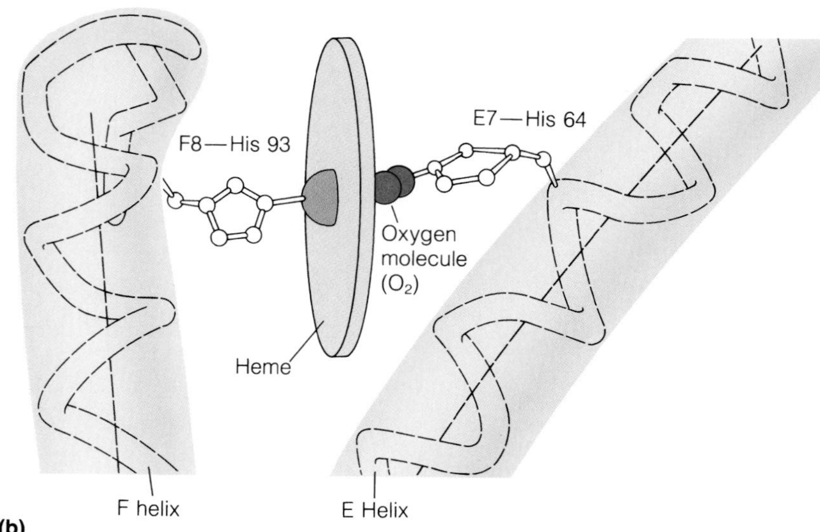

Figure 7.3
The geometry of iron coordination in oxymyoglobin. (**a**) The octahedral coordination of the iron atom, with the four nitrogens from the porphyrin and the Fe^{2+} lying (nearly) in a plane, and histidine F8 and O_2 at the axial positions. (**b**) Close-up of the heme pocket, with the proximal (His 93) and distal (His 64) histidine side chains.

helix of the protein. In the nomenclature often employed, this is called histidine F8 or the **proximal** histidine (see Figure 7.3b). In **deoxymyoglobin** the second coordination site, on the other side of the iron, is vacant. When oxygen is bound, to make **oxymyoglobin,** the O_2 molecule occupies this site. On one side oxygen is coordinated to the iron atom of the heme. On the other side of the O_2 lies another histidine, number 64, or E7, the **distal** histidine. Thus, the bound O_2 molecule is sandwiched between the ring nitrogen of the distal histidine and the iron atom. Although the site is ideally adapted to hold an O_2 molecule, it will accept some other small molecules. The most physiologically important of these is carbon monoxide, which is approximately the same size as O_2. However, CO is bound with much greater affinity to myoglobin and hemoglobin than is O_2, and the binding is not readily reversible. This is why CO is such a toxic gas; it irreversibly ties up oxygen binding sites.

Normally, an oxygen molecule in such close contact with a ferrous ion would oxidize the latter to the ferric state. The heme alone does not afford protection, for heme dissolved free in solution is readily oxidized by O_2. But in the hydrophobic, protected environment provided by the interior of the myoglobin molecule, the iron does not easily become oxidized. The oxygen is bound, and a temporary electron rearrangement occurs. When the oxygen is released, the iron remains in the ferrous state, able to bind another O_2. It is found that when myoglobin is stored in air, outside the cellular environment, the iron does slowly become oxidized, to form what is called **metmyoglobin.** When this happens, the binding site is inactivated. Metmyoglobin will not bind O_2; a water molecule occupies the O_2 site instead.

This protection from irreversible oxidation is the functional reason for the existence of myoglobin and hemoglobin. The two molecules provide environments in which the first step of an oxidation reaction (the binding of oxygen) is permitted, but the final step (oxidation) is blocked.

Analysis of Oxygen Binding by Myoglobin

The binding of oxygen by myoglobin must meet certain physiological requirements. Figure 7.4 shows a schematic of the oxygen delivery system of animals with lungs or gills. Myoglobin in tissues accepts oxygen from hemoglobin in the circulating arterial blood. It then delivers the oxygen to the oxygen-consuming organelles of the cells (the mitochondria) when their oxygen needs are sufficiently great. To understand these functions on a quantitative basis, we must examine how the binding of a ligand like oxygen depends on its concentration in the surroundings.

First, we need a way to measure the concentration of dissolved oxygen. We know from Henry's law that the concentration of any gas dissolved in a fluid is proportional to the *partial pressure* of that gas above the fluid. Therefore, we can conveniently regulate (and measure) the concentration of dissolved O_2 by regulating the partial O_2 pressure above the myoglobin solution being studied. In fact, we will express oxygen concentration *as* this partial pressure—P_{O_2}.

A second requirement for the study of binding is that we have a way of measuring the fraction of myoglobin molecules carrying oxygen. When myoglobin is oxygenated, there is a change in the absorption spectrum as a consequence of electron displacements in the porphyrin ring. This allows a spectrophotometric determination of the fraction of myoglobin molecules that are oxygenated. The results of such analysis, using myoglobin in solution at neutral pH, are shown in Figure 7.5. Such a graph is called a **binding curve,** for it describes how the fraction of the myoglobin sites that have oxygen bound to them (θ) depends on the concentration (partial pressure) of free oxygen.

The hyperbolic shape of the myoglobin binding curve can be explained very simply. Since the binding equilibrium is described by the reaction

$$\text{Mb} + \text{O}_2 \rightleftharpoons \text{Mb} \cdot \text{O}_2$$

we must have

$$K = \frac{[\text{MbO}_2]}{[\text{Mb}][\text{O}_2]} \tag{7.1}$$

where the equilibrium constant K is called an **affinity constant,** and the quantities in brackets denote molar concentrations of oxygenated myoglobin ($[MbO_2]$), nonoxygenated myoglobin ($[Mb]$), and free oxygen ($[O_2]$). The fraction of myoglobin sites occupied is defined as

$$\theta = \frac{\text{sites occupied}}{\text{total available sites}} \tag{7.2}$$

Since there is only one site per myoglobin molecule, the total number of available sites is proportional to the total concentration of myoglobin, $[MbO_2] + [Mb]$. Therefore

$$\theta = \frac{[\text{MbO}_2]}{[\text{MbO}_2] + [\text{Mb}]} \tag{7.3}$$

If we rewrite equation (7.1) as $[MbO_2] = K[Mb][O_2]$ and substitute this into equation (7.3), we obtain

$$\theta = \frac{K[\text{Mb}][\text{O}_2]}{K[\text{Mb}][\text{O}_2] + [\text{Mb}]} \tag{7.4}$$

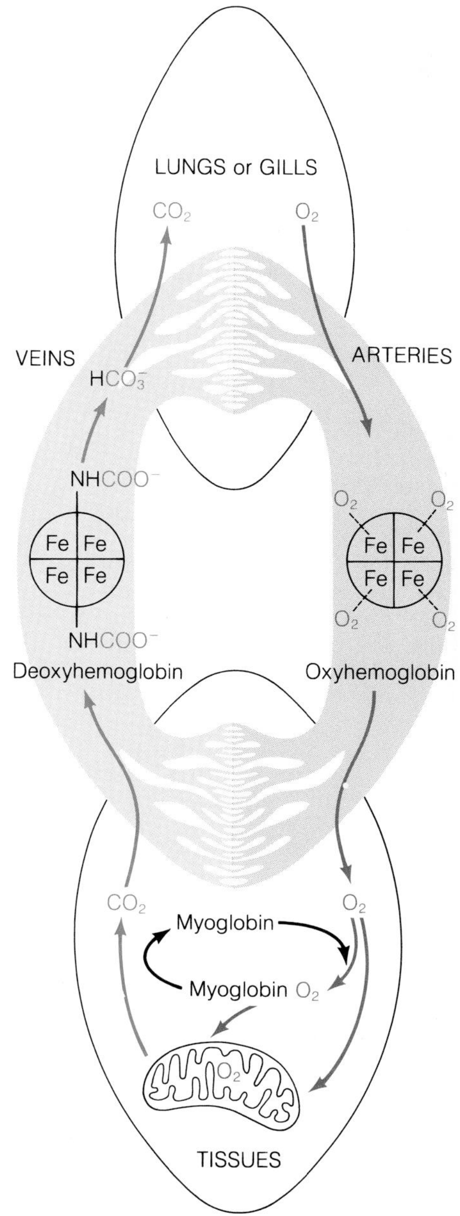

Figure 7.4
Schematic representation of respiratory transport in animals. Oxygen is transported by hemoglobin from the lungs or gills to the tissues. There, part of it may be utilized directly for metabolism. Another part may be stored by binding to myoglobin and therefore available when oxygen demand is heavy. CO_2 released by oxidative processes in tissues is carried back to the lungs or gills and expired.

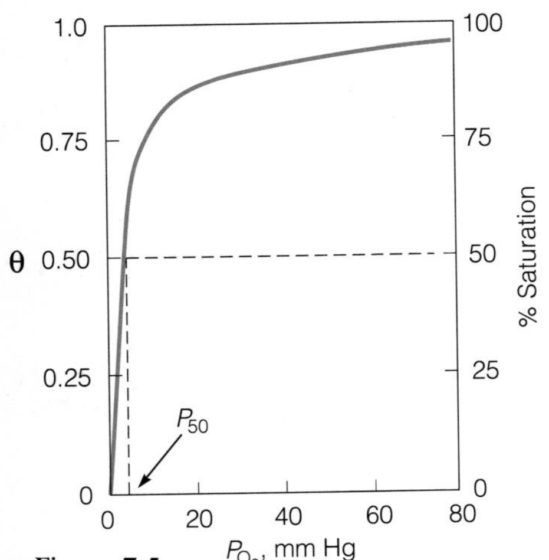

Figure 7.5
Oxygen binding curve for myoglobin. The free oxygen concentration is expressed as P_{O_2} and the fraction of sites saturated as θ. As P_{O_2} becomes very large, saturation is approached asymptotically. The value of P_{50} is indicated on the graph. This curve obeys equation (7.7).

The concentration of unliganded myoglobin ([Mb]) can be factored out of the top and bottom to give

$$\theta = \frac{K[O_2]}{K[O_2] + 1} = \frac{[O_2]}{[O_2] + 1/K} \tag{7.5}$$

or

$$\theta = \frac{[O_2]}{[O_2] + [O_2]_{1/2}} \tag{7.6}$$

where we have made use of the fact that $1/K = [O_2]_{1/2}$, the oxygen concentration when half of the molecules are saturated. It is easy to check this by setting ($\theta = 1/2$) in equation (7.5). Since oxygen concentration is proportional to oxygen partial pressure, equation (7.6) can equally well be written as

$$\theta = \frac{P_{O_2}}{P_{O_2} + P_{50}} \tag{7.7}$$

where P_{50} is the oxygen partial pressure for half-saturation.

Equation (7.7) describes the kind of *hyperbolic* binding curve shown in Figure 7.5. The P_{50} for myoglobin is very low (about 2 mm Hg), corresponding to a high oxygen affinity. This is appropriate for a protein that must extract oxygen from the blood. At the oxygen concentration existing in the capillaries (about 30 mm Hg) the myoglobin in adjacent tissues will be nearly saturated. When cells are metabolically active their internal P_{O_2} falls to much lower levels. Under these conditions myoglobin will deliver its oxygen.

Thus, the myoglobin molecule must not only provide an environment for the reversible binding of oxygen but also ensure that the affinity constant K (or P_{50}) is of just the right magnitude. We can gain some insight into how this is accomplished if we recall that the affinity constant is an equilibrium constant and therefore must be the ratio of two rate constants, k_1 for the *binding reaction* and k_{-1} for the *release reaction*. That is,

$$K = \frac{k_1}{k_{-1}} \tag{7.8}$$

We do not yet know much about the binding reaction, but rapid kinetic studies have revealed some details of the way in which oxygen is released. It appears that the rate-limiting process in oxygen release is the opening of a pathway for the O_2 molecule to escape from the heme pocket (Figure 7.6). In fact, the oxygen may spend some time rattling in its cage before the tertiary structure of the myoglobin shifts enough to let it escape. This is an explicit example of a principle set forth in the preceding chapter—the dynamic internal motions of globular protein molecules play important roles in regulating the processes they mediate. The key to just how *strongly* myoglobin binds oxygen may lie in the motional flexibility of the myoglobin molecule.

In summary, what we observe in myoglobin is a complicated molecular structure that has evolved to produce exactly the right environment to allow the binding and release of oxygen under the appropriate conditions.

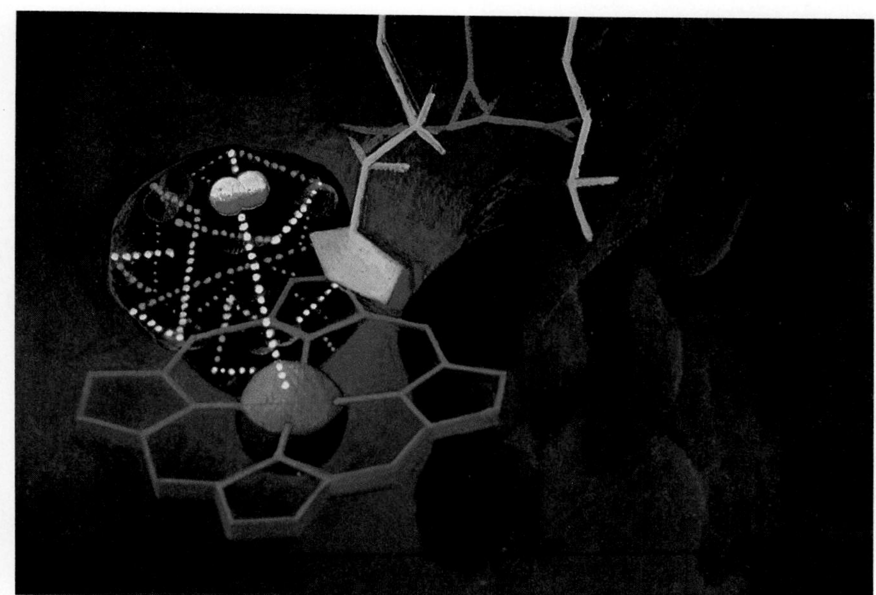

(a)

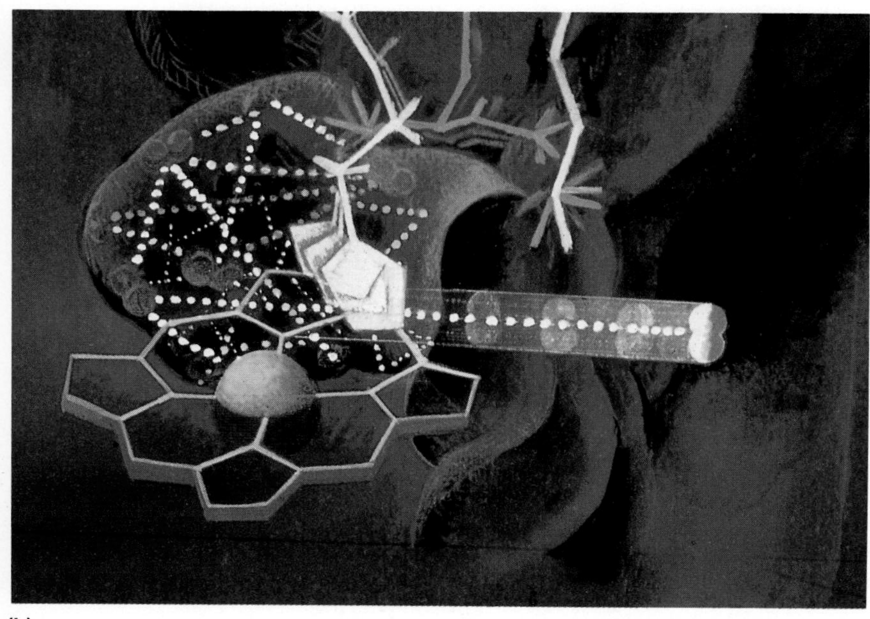

(b)

Figure 7.6
Dynamics of oxygen release by myoglobin. In (**a**) the O_2 molecule (orange and white double ball) has been released from the binding site, but bounces around inside a pocket in the protein molecule. It may rebind or, as shown in (**b**), escape if fluctuations in the protein structures open a pathway to the outside.

Oxygen Transport: Hemoglobin

Why a Transport Protein Is Needed

Any animal larger than a few millimeters in diameter faces a serious problem in carrying out **aerobic** (oxygen-requiring) metabolism. It must ensure a steady supply of oxygen to its body cells and remove metabolic waste products such as carbon dioxide. These gases will diffuse through tissues, but transport by diffusion becomes very slow if appreciable distances must be crossed. Insects solve the problem by having **tracheae,** invaginations in their body walls that reach down into the tissues (Figure 7.7a). They have, in

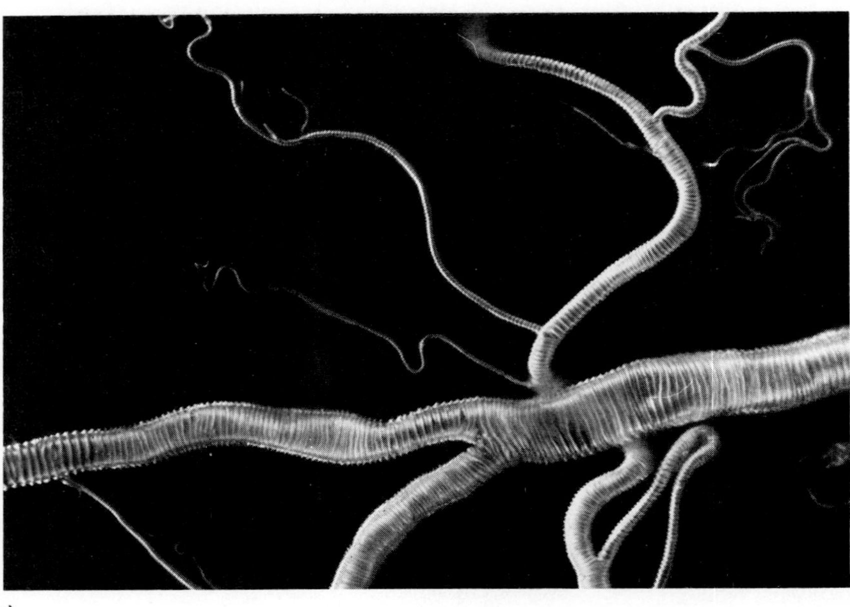
(a)

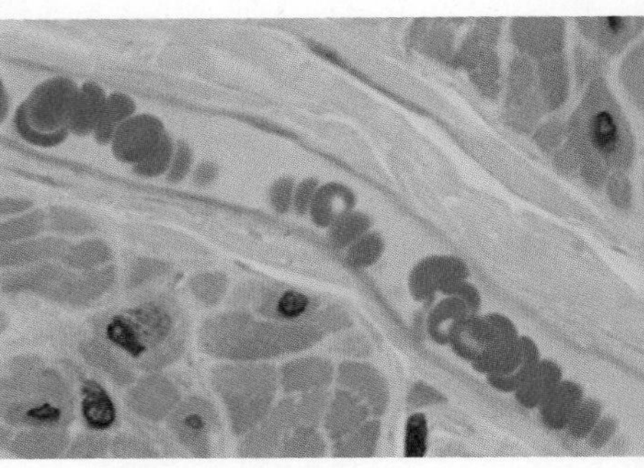
(b)

Figure 7.7
Two mechanisms of oxygen transport. (a) A small part of the tracheal system of a cockroach. In most insects, such branched tubes carry oxygen from the surface directly to the tissues. (b) Human red blood cells (erythrocytes) moving in a capillary. Each erythrocyte contains about 300 million hemoglobin molecules.

effect, increased their body surface area to the extent that diffusion is practicable. This works because insects are small. (Alternatively, we might say that insects are small because they rely on this mechanism for obtaining oxygen.*)

Almost all other animals pick up oxygen in lungs or gills and pump it in the blood through arteries to the tissues (see Figure 7.4). Carbon dioxide is returned in the venous blood and released in the lung or gill. In some primitive organisms, the gases are simply dissolved in the blood, but this is very inefficient, especially for oxygen, which has low solubility. Much blood must be pumped, at great metabolic expense, to deliver even a little oxygen in this way. Evolution of all higher organisms has been accompanied by the development of **oxygen transport proteins,** which allow the blood to carry a much greater load of oxygen than would be permitted by solubility alone. Oxygen transport proteins may be either dissolved in the blood (as in some invertebrates) or concentrated in specialized cells, like those shown in Figure 7.7b.

Cooperative Binding

Consider the peculiar demands placed on an oxygen transport protein. To be useful, it must accept oxygen efficiently at the partial pressure found in lungs or gills (approximately 100 mm Hg) and then deliver an appreciable fraction of it to tissues, at a pressure of about 30 mm Hg. In other words, an ideal oxygen transport protein would be nearly saturated at 100 mm Hg and unsaturated at about 20–40 mm Hg. In that way, each transport protein molecule could deliver a significant fraction of its oxygen load. If the transport protein had a hyperbolic binding curve like that of myoglobin, such behavior would be impossible to achieve. Figure 7.8a and b show that,

* This hypothesis raises an intriguing problem: How do we explain the giant dragonflies of the Carboniferous period, with wingspreads up to 2 feet? Was the oxygen content of the atmosphere higher at the time, or did they use some other mechanism for oxygen uptake?

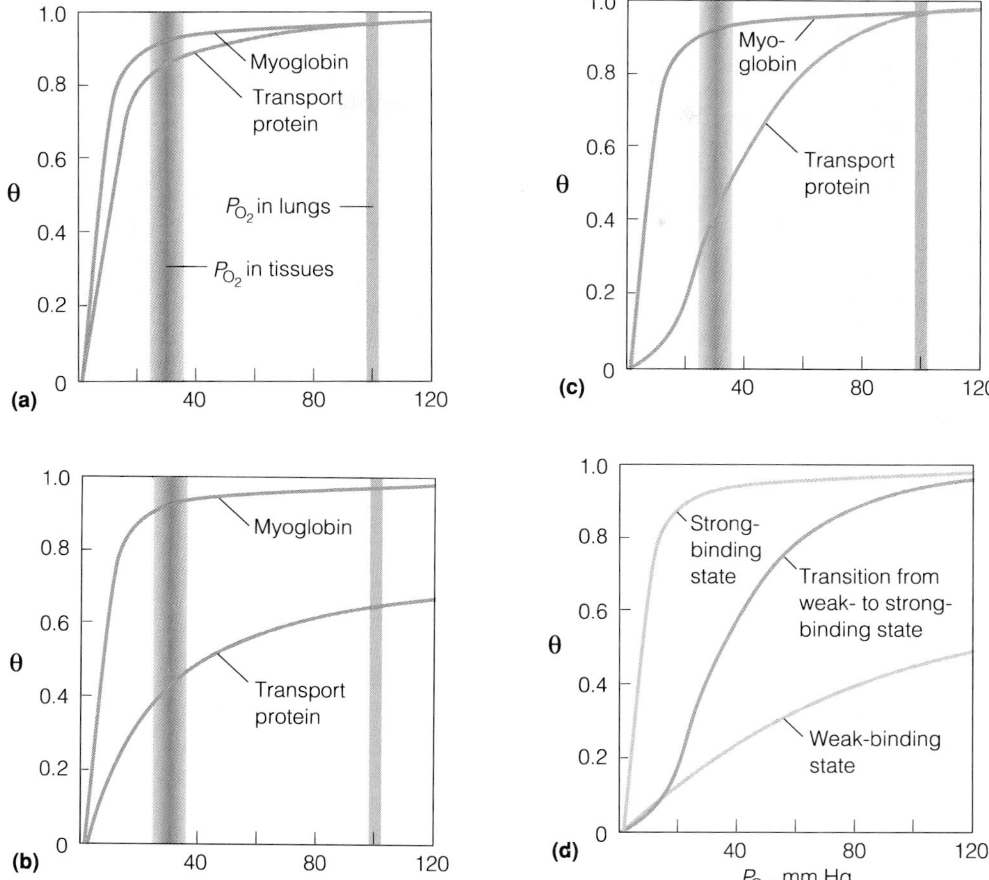

Figure 7.8
Why a sigmoidal binding curve is needed for an oxygen transport protein. (**a**) and (**b**) show the consequences that would result if the transport protein had a hyperbolic curve like that of myoglobin. If O_2 saturation were assured in the lungs (**a**), unloading of O_2 to myoglobin would be inefficient. Alternatively, if the O_2 affinity of the transport protein were lower (**b**) it could unload to myoglobin, but not saturate in the lungs. (**c**) shows how a sigmoidal curve can satisfy both criteria. (**d**) shows that such a sigmoidal curve can be visualized as one in which a switch occurs from a weak-binding state to a strong-binding state.

with such a binding curve, the protein would be inefficient either in uptake or in delivery.

The problem has been solved through the evolution of oxygen transport proteins that have the kind of binding curve shown in Figure 7.8c. Such a *sigmoidal* curve is very efficient for it allows nearly full saturation of the protein in the lungs or gills, yet quite efficient release in the capillaries. You can understand how such a curve can be possible by examining Figure 7.8d. At low oxygen pressures the protein acts as if it were binding oxygen very weakly, yet as more oxygen is bound the affinity becomes greater. This must mean that there is *cooperative interaction* between oxygen binding sites in the protein molecule. Filling of the first ones somehow increases the affinity of the remaining sites. This can be so only if there is some kind of mutual communication between binding sites. A single-site protein, such as myoglobin, cannot accomplish this, for one myoglobin molecule is completely ignorant of the state of another. They are independent entities.

By contrast, such communication is possible between the subunits of a multisubunit protein. This is exactly the evolutionary strategem that has been adapted; virtually all oxygen transport proteins are multisubunit structures, exhibiting cooperative interaction between their binding sites.

In the evolutionary line that led to the vertebrates, the protein used for oxygen transport is hemoglobin, which is highly concentrated in circulating blood cells called **erythrocytes** (see Figure 7.7b). Hemoglobin has evolved from myoglobin so as to form the kind of *tetrameric structure* shown in

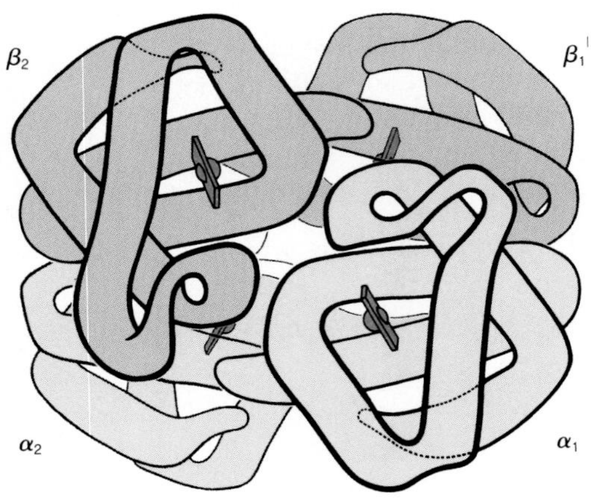

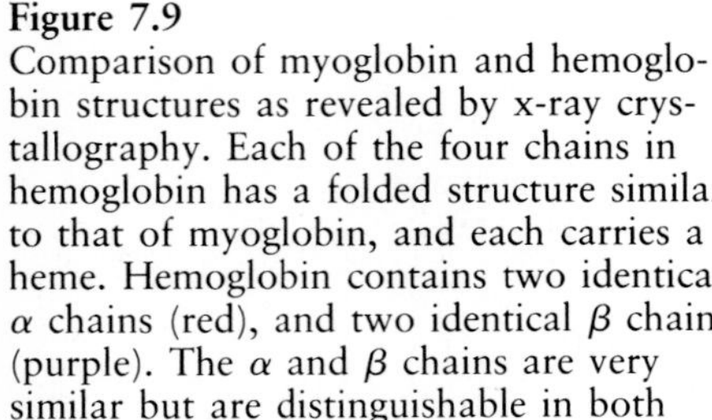

Figure 7.9
Comparison of myoglobin and hemoglobin structures as revealed by x-ray crystallography. Each of the four chains in hemoglobin has a folded structure similar to that of myoglobin, and each carries a heme. Hemoglobin contains two identical α chains (red), and two identical β chains (purple). The α and β chains are very similar but are distinguishable in both primary structures and folding.

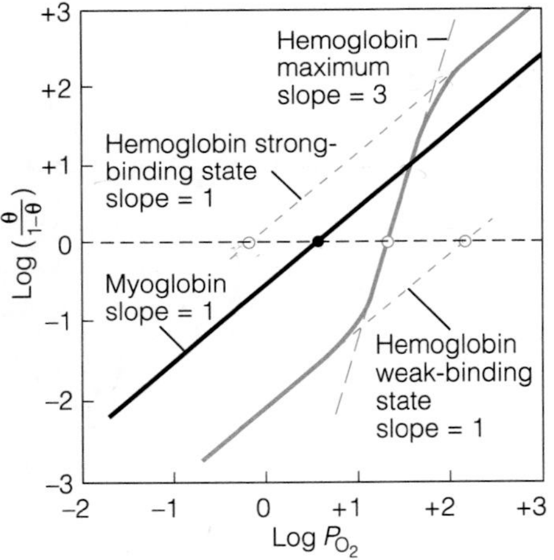

Figure 7.10
Hill plots of oxygen binding for myoglobin and hemoglobin. Myoglobin (black), which binds oxygen noncooperatively, gives a straight line with slope of unity. Hemoglobin (red) exhibits a switch from a weak-binding state to a strong-binding state. The maximum slope (line with long dashes) gives a Hill coefficient of about 3. These data are under near-physiological conditions. Dots indicate P_{50} values for the different proteins and conformations.

Figure 7.9. Each of the subunits has primary, secondary, and tertiary structures rather like those of myoglobin, but the amino acid side chains in hemoglobin also provide the necessary interactions—salt bridges, hydrogen bonds, and hydrophobic interactions—to stabilize a particular quaternary structure.

Because it is a tetramer, each hemoglobin molecule can bind four oxygens, in four myoglobinlike sites. The functional difference from myoglobin lies in the fact that these sites now exhibit cooperativity in binding.

We have emphasized that this cooperativity arises from a conformational "switch" from a weak binding state to a strong binding state. This switching is not easily visualized when binding curves are represented as in Figure 7.8c or d, nor do such curves give an easy way to measure the degree of cooperativity. A rearrangement of equation (7.7) is helpful. If we calculate the quantity $\theta/(1-\theta)$, we obtain

$$\frac{\theta}{1-\theta} = \frac{P_{O_2}}{P_{50}} \tag{7.9}$$

Or, taking logarithms of both sides

$$\log\left\{\frac{\theta}{1-\theta}\right\} = \log P_{O_2} - \log P_{50} \tag{7.10}$$

Graphing $\log(\theta/1-\theta)$ versus $\log P_{O_2}$ produces what is called a **Hill plot.** The Hill plot for noncooperative binding will, according to equation (7.10), be a straight line, with slope = 1. The abscissa value corresponding to $\log\{\theta/(1-\theta)\} = 0$ will equal $\log P_{50}$ (Figure 7.10). This may seem a roundabout way to handle an equation like (7.7), but consider the kind of Hill plot a cooperatively binding protein such as hemoglobin exhibits. As Figure 7.10 shows, hemoglobin begins binding (at low P_{O_2}) with a line of slope ≅ 1, corresponding to the weak binding state (high P_{50}). As binding progresses, the curve switches over to approach another, and parallel, line, which describes the strong binding state (small P_{50}). On a Hill plot, the transition between binding states is clear, and the behavior is unmistakably different for cooperative and noncooperative systems. Furthermore, the Hill plot gives a direct numerical measure of the degree of cooperativity from its maximum slope, n_H, which is called the **Hill coefficient.** Three cases may be considered, for a molecule with n binding sites:

1. $n_H = 1$: The molecule binds noncooperatively. This may happen even with a multisite protein if the sites do not communicate with one another.

2. $1 < n_H < n$: This is the usual situation for a cooperatively binding protein, as depicted in Figure 7.10.

3. $n_H = n$: This is a hypothetical situation in which the molecule is *wholly* cooperative. In such a situation, one molecule would fill up its sites before any others had taken oxygen, so that only wholly unliganded and wholly liganded molecules would be present at any point in the binding process. This case is never seen in reality. For example, the Hill coefficient of hemoglobin never exceeds about 3.5.

Figure 7.10 presents Hill plots for myoglobin and hemoglobin under near-physiological conditions. Under these conditions, the Hill coefficient for hemoglobin is about 3.0. Although lower than the maximum hypothetical value of 4, it still indicates considerable cooperativity.

Allostery and Mechanisms of Cooperative Binding

The cooperative binding of oxygen by hemoglobin is one example of what are referred to as **allosteric** effects. In allosteric binding, the uptake of one ligand influences the affinities of remaining unfilled binding sites. The ligands may be of the same kind, as in this case, or they may be different. As we shall see in Chapter 11, the modulation of binding by different ligands is an important mechanism for regulating the activity of enzymes. In particular, allostery allows one kind of small molecule to regulate the action of a protein on another kind of molecule. The fact that they allow allosteric regulation may be one of the reasons multisubunit proteins are so common.

Changes in Hemoglobin Structure Accompanying Binding of Oxygen

To understand the allosteric behavior of hemoglobin, it is necessary to examine the protein in more detail. The hemoglobin of higher vertebrates is made up of two types of chains, referred to as α and β. Their primary structures are compared to that of myoglobin in Figure 7.11. As you can see, the α and β sequences have considerable similarity to one another and some similarity to the sequence of myoglobin. Essential residues, like the proximal and distal histidines (F8 and E7, respectively), are conserved, and apparently those critical to the tertiary structure are as well, for the hemoglobin chains and myoglobin all have very similar tertiary structure. In the hemoglobin molecule there are two of each kind of chain, so the whole molecule can be called an $\alpha_2\beta_2$ tetramer. The chains are placed in a roughly tetrahedral arrangement as shown schematically in Figure 7.12. The way in which hemoglobin dissociates in urea solutions suggests that the closest and strongest contacts are between α and β chains, rather than α–α or β–β. In other words, the molecule could be thought of as a dimer of $\alpha\beta$ dimers. The

	Mb	Hbβ	Hbα
1	V	V	V
	—	H	—
	L	L	L
	S	T	S
	E	P	P
	G	E	A
	E	E	D
	W	K	K
	Q	S	T
	L	A	N
	V	V	V
	L	T	K
	H	A	A
	V	L	A
	W	W	W
	A	G	G
	K	K	K
	V	V	V
	E	—	G
	A	—	A
	D	N	H
	V	V	A
	A	D	G
	G	E	E
	H	V	Y
	G	G	G
	Q	G	A
	D	E	E
	I	A	A
	L	L	L
	I	G	G
	R	R	R
	L	L	M
	F	L	F
	K	V	L
	S	V	S
	H	Y	F
	P	P	P
	E	W	T
	T	T	T
40	L	Q	K
	E	R	T
	K	F	Y
	F	F	F
	D	E	P
	R	S	H
	F	F	F
	K	G	—
	H	D	D
	L	L	L
	K	S	S
	T	T	H

	Mb	Hbβ	Hbα	
	E	P	—	
	A	D	—	
	E	A	—	
	M	V	—	
	K	M	—	
	A	G	G	
	S	N	S	
	E	P	A	
	D	K	Q	
	L	V	V	
	K	K	K	
	K	A	G	
E7	H	H	H	
	G	G	G	
	V	K	K	
	T	K	K	
	V	V	V	
	L	L	A	
	T	G	D	
	A	A	A	
	L	F	L	
	G	S	T	
	A	D	N	
	I	G	A	
	L	L	V	
	K	A	A	
	K	H	H	
	K	L	V	
80	G	D	D	
	H	N	D	
	H	L	M	
	E	K	P	
	A	G	N	
	E	T	A	
	L	F	L	
	K	A	S	
	P	T	A	
	L	L	L	
	A	S	S	
	Q	E	D	
	S	L	L	
F8	H	H	H	
	A	C	A	
FG corner	T	D	H	
FG corner	K	K	K	
FG corner	H	L	L	
FG corner	K	H	R	
FG corner	I	V	V	FG5
	P	D	D	
	I	P	P	
	K	E	V	

	Mb	Hbβ	Hbα
	Y	N	N
	L	F	F
	E	R	K
	F	L	L
	I	L	L
	S	G	S
	E	N	H
	A	V	C
	I	L	L
	I	V	L
	H	C	V
	V	V	T
	L	L	L
	H	A	A
	S	H	A
	R	H	H
	H	F	L
120	P	G	P
	G	K	A
	D	E	E
	F	F	F
	G	T	T
	A	P	P
	D	P	A
	A	V	V
	Q	Q	H
	G	A	A
	A	A	S
	M	Y	L
	N	Q	D
	K	K	K
	A	V	F
	L	V	L
	E	A	A
	L	G	S
	F	V	V
	R	A	S
	K	N	T
	D	A	V
	I	L	L
	A	A	T
	A	H	S
	K	K	K
	Y	Y	Y
	K	H	R
	E		
	L		
	G		
	Y		
	Q		
153	G		

Figure 7.11
An alignment of amino acid sequences of whale myoglobin with the β and α chains of human hemoglobin. Gaps have been inserted where necessary to provide maximum alignment. Brown indicates residues common to myoglobin and hemoglobin chains, and purple those common only to both hemoglobin chains. The positions of certain critical residues are indicated. Numbers are for the myoglobin sequence.

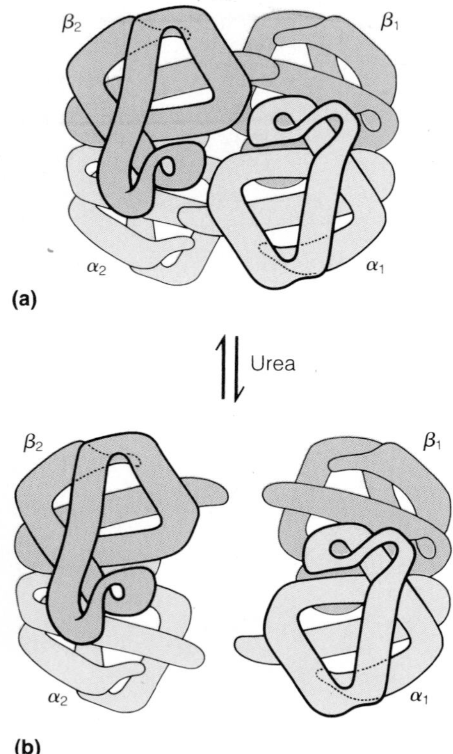

Figure 7.12
Dissociation of hemoglobin tetramers into $\alpha\beta$ dimers. When exposed to solutions containing urea, the hemoglobin molecule breaks into α–β dimers, suggesting that α–β interactions are stronger than α–α or β–β interactions.

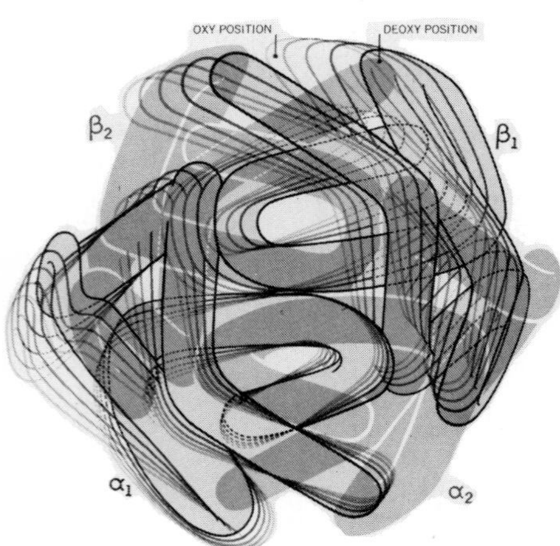

Figure 7.13
Motion of subunits during shifts from deoxy- to oxyhemoglobin. This somewhat fanciful "stroboscopic" side view shows how one $\alpha\beta$ dimer rotates and slides with respect to another in transition from the deoxy to the oxy state.

heme groups, with their O_2 binding sites, are all close to the surface but *not* close to one another. Therefore, we cannot seek the source of cooperative binding in anything so unsubtle as direct heme–heme interaction.

A key to what is actually happening during oxygenation can be seen in Figure 7.15a and b (pp. 228–229), which compare the structures of deoxyhemoglobin and the fully oxygenated molecule carrying four oxygens. What has occurred is a small change in the quaternary structure. Although the secondary and tertiary structures of the individual chains change hardly at all, one α/β pair rotates and slides with respect to the other so as to bring the β chains closer together (Figure 7.13). This has the effect of narrowing a central cavity in the molecule that is particularly noticeable in the deoxy form (see Figure 7.15a). To a first approximation, then, we can regard the hemoglobin molecule as having two states of quaternary structure, one characteristic of the deoxy form and the other favored by the oxygenated form. The oxy structure has the higher affinity for O_2, and it is the switch to this state that accounts for the cooperativity in binding.

We can now interpret the Hill plot shown in Figure 7.10 in terms of such an allosteric shift between two molecular conformations. Wholly deoxygenated hemoglobin molecules will be in the deoxy conformation. Therefore, as we add oxygen to a solution of such molecules, the binding initially occurs along the line corresponding to the weak-binding state. But partial oxygenation favors transition to the strong-binding oxy state. As oxygen is bound, more and more of the remaining available sites are in hemoglobin molecules that have this conformation. Therefore, the binding curve passes over to that for the strong-binding state. As the last few sites are filled, all the molecules have adopted the strong-binding form.

Models for the Allosteric Change

We know more about hemoglobin than most other allosteric proteins, yet the extent of our factual knowledge is largely restricted to the extreme (deoxy and oxy) states. The question of *how* the changes occur is still a matter for debate, for it is very difficult to obtain detailed structural information at intermediate states of oxygenation.

A number of theories have been developed to describe allosteric transitions. These may be generally grouped into three classes:

1. *Sequential models:* The prototype for such models is that of Koshland, Nemethy, and Filmer (Figure 7.14a). The KNF model assumes that the subunits can change their conformation one at a time and that the presence of some subunits carrying oxygen favors the strong-binding oxy form in adjacent subunits whose sites are not yet filled.

2. *Concerted models:* At the opposite extreme lies the theory of Monod, Wyman, and Changeux (Figure 7.14b). According to the MWC model, the entire hemoglobin tetramer exists in an equilibrium between two forms. In the deoxy (T) state, all subunits in each molecule are in the weak-binding conformation, and in the oxy (R) state, all are in the strong-binding form. (The symbols T and R stand for "tense" and "relaxed"; the significance of this will be seen below.) An equilibrium between these states is presumed to exist, and partial oxygenation shifts that equilibrium toward the R state. The shift is a *concerted* one, so that molecules with some subunits in the R state and some in the T state are specifically excluded.

1. No oxygen (•) bound. Almost all subunits in all molecules are in weak-binding state, ⌐. Only a few happen to be in the strong-binding state, ⌒.

2. Some oxygen bound (red). As each oxygen is bound it favors the transition of adjacent subunits to the strong-binding state, and promotes their binding of oxygen.

3. More oxygen is bound. More and more subunits sitting next to oxygen-occupied sites are being switched to the strong-binding state.

4. Approaching saturation. Almost all sites are filled, and almost all subunits are now in the strong-binding state.

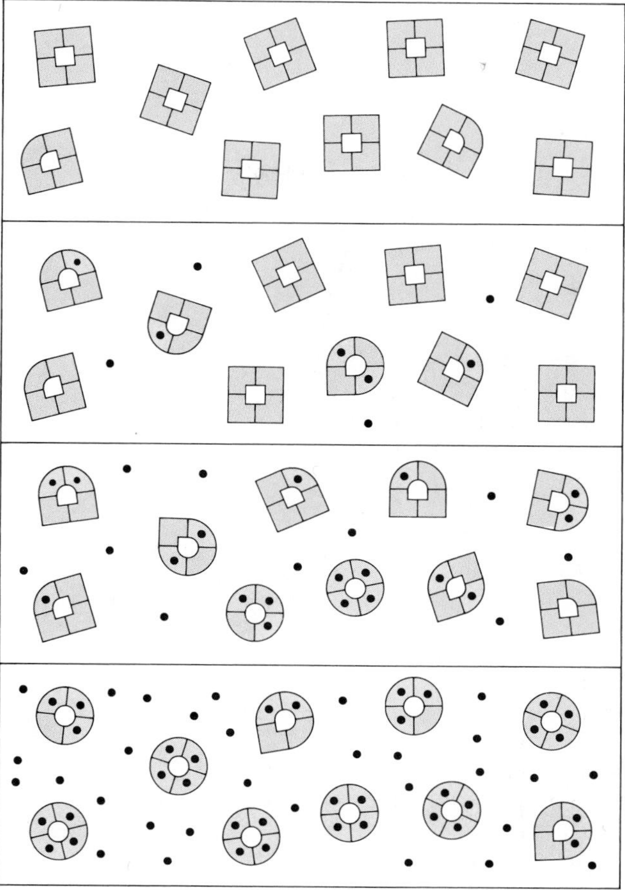

(a)

1. No oxygen (•) bound. Most tetramers are in the T state, ⊞, with only a few in the R state, ⊕.

2. Some oxygen bound (red). Preference is for binding to molecules in R state so T ⇌ R equilibrium is shifted toward R.

3. More oxygen bound. Now most molecules are in R state. Note that T binds oxygen also, but more weakly.

4. Approaching saturation. Almost all molecules have been shifted to R state. Almost all sites are filled.

(b)

Figure 7.14
Two theoretical models for allosteric binding to a tetramer. (**a**) The Koshland, Nemethy, and Filmer (KNF) model. Each *subunit,* in binding a ligand, changes its conformational state and thereby promotes a similar change in an adjacent subunit. (**b**) The Monod, Wyman, Changeux (MWC) model. In this model the *entire molecule* has two different states—tense (T) and relaxed (R)—which are in equilibrium. Binding of ligands favors the strong-binding (R) state, and therefore binding shifts the equilibrium to this side. Both models can explain cooperative binding.

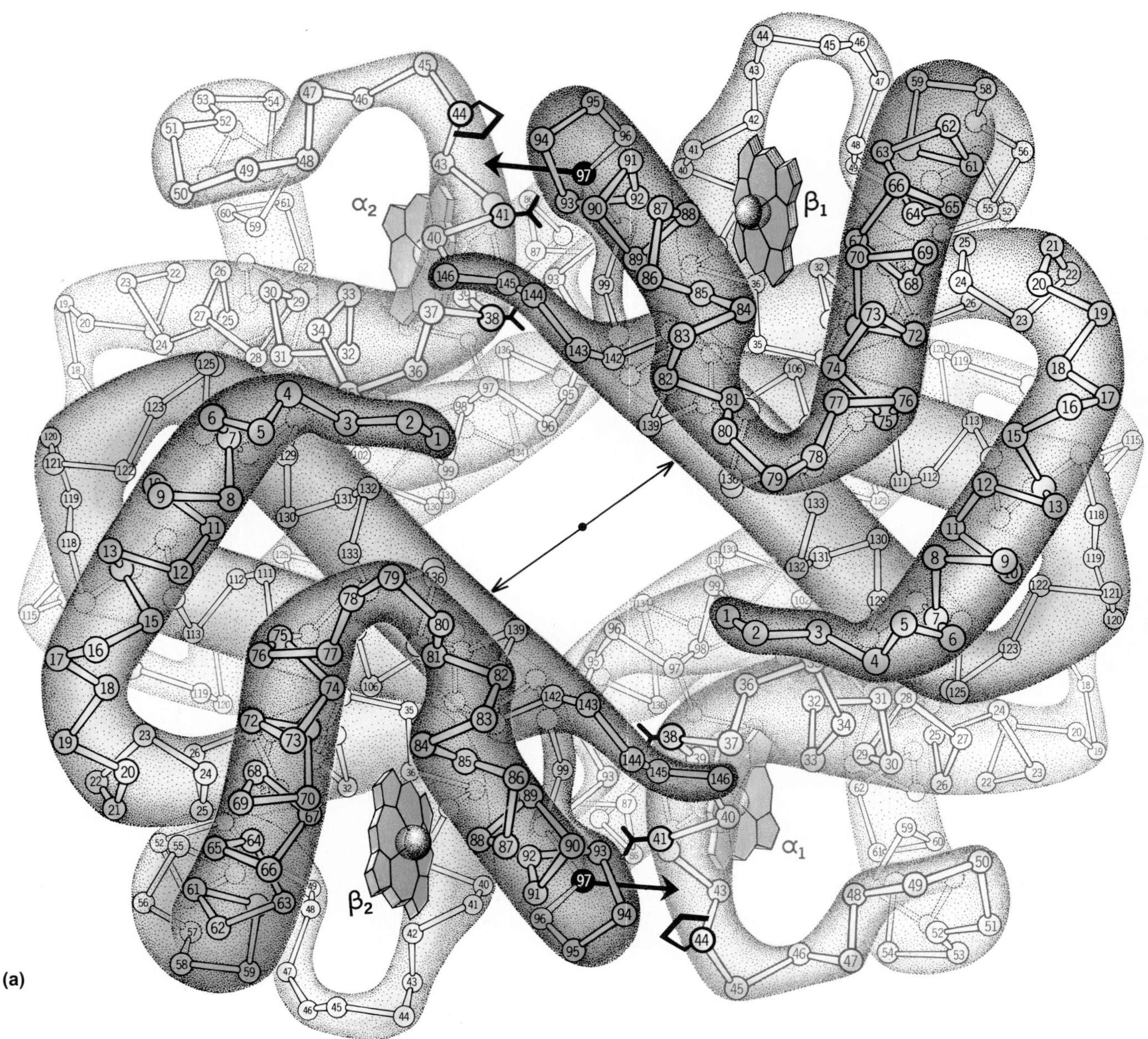

Figure 7.15
Top view of horse hemoglobin, looking down the 2-fold symmetry axis. (**a**) Deoxyhemoglobin. The two β subunits are in the foreground; α subunits are in the background. Note the breadth of the central cavity and the fact that residue 97 on one β chain lies between residues 41 and 44 on the adjacent α chain. (**b**) Oxyhemoglobin. The shift in going from the deoxy to the oxy state can be noted by the shrinkage of the central cavity and the shift in the contact of residue β97 with the α chain.

3. *Multistate models:* It has become clear in recent years that neither the KNF nor the MWC theory can *exactly* explain the allosteric behavior of proteins, including hemoglobin. Consequently, more complex models have been devised. Most such models retain the MWC concept of a concerted switch in conformation but involve more than two states for the entire molecule.

Mechanism of the Allosteric Change in Hemoglobin

Since it has been possible to determine by x-ray diffraction details of both the deoxy and oxy states of hemoglobin, it is possible to provide a quite complete description of the overall change and to speculate concerning its mechanism.

The transition from the deoxy to the oxy conformation involves major changes in the details of subunit–subunit interaction. Some idea of this may be gained from a closer study of Figure 7.15a. Note the region, to the lower

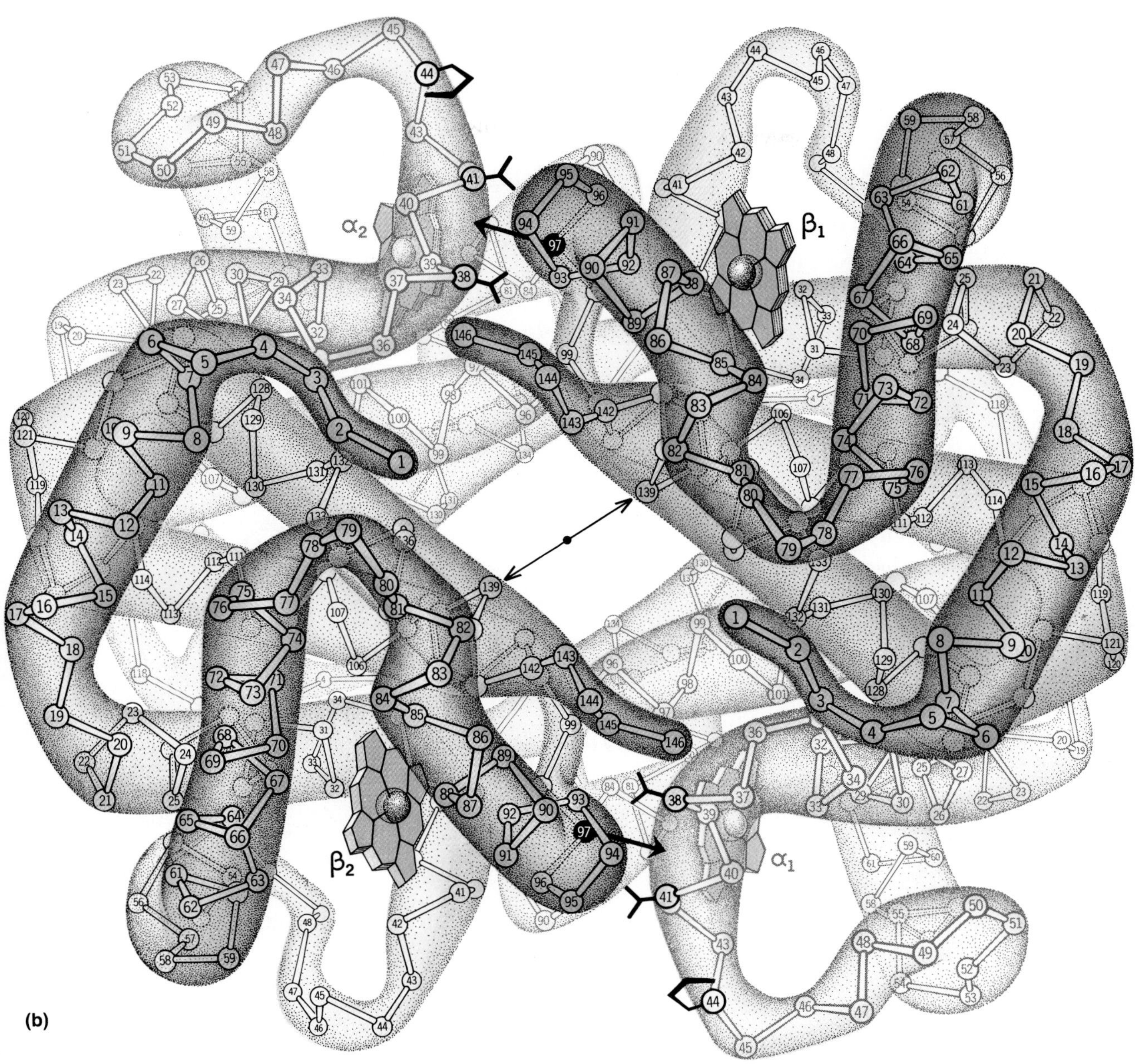

right, where the β_2 subunit interacts with the α_1 chain. In the deoxy form, the C-terminus of β_2 (residue 146) lies atop the C helix of α_1 (residues 36–42) and is held in this position by a network of hydrogen bonds and salt bridges. (These are shown, from a different point of view, in Figure 7.16.) His 97 in the FG corner of β_2 is pushed against the CD corner of α_1, between Thr 41 and Pro 44. In the oxy form, rotation and sliding of the subunits have pulled the C-termini of β chains away from α contacts (Figure 7.15b). The salt bridges and hydrogen bonds holding the C-terminus have been broken, and His 97 of β_2 now lies between Thr 38 and Thr 41 of α_1. Because of the symmetry of the structure, an exactly equivalent set of changes occurs in the $\alpha_2\beta_1$ interface. The molecule has, as it were, "switched" and clicked into a new set of interactions. In the process a number of strong interactions (those involving the C-termini in particular) have been broken. The energy price for this change is paid for by the binding of O_2 to the molecule. Once the O_2 has departed, the molecule will naturally

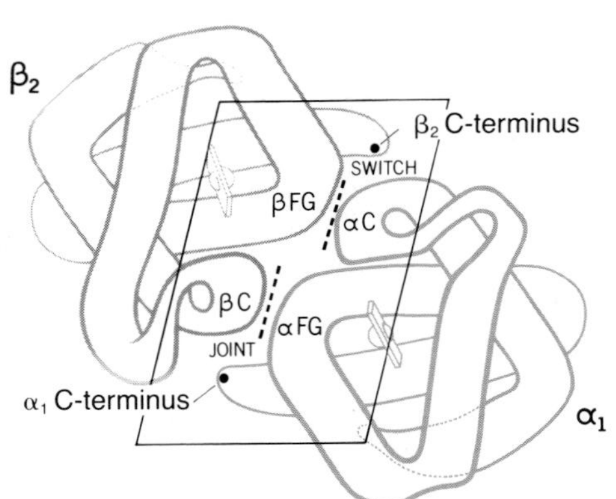

Figure 7.16
Details of some of the $\alpha\beta$ subunit interactions in deoxy- and oxyhemoglobin. The contacts between FG corners and C helices in adjacent subunits are shown. The molecule is shown in the deoxy conformation. Salt bridges and hydrogen bonds present in the deoxy state are shown in black; those formed in the oxy state are in red.

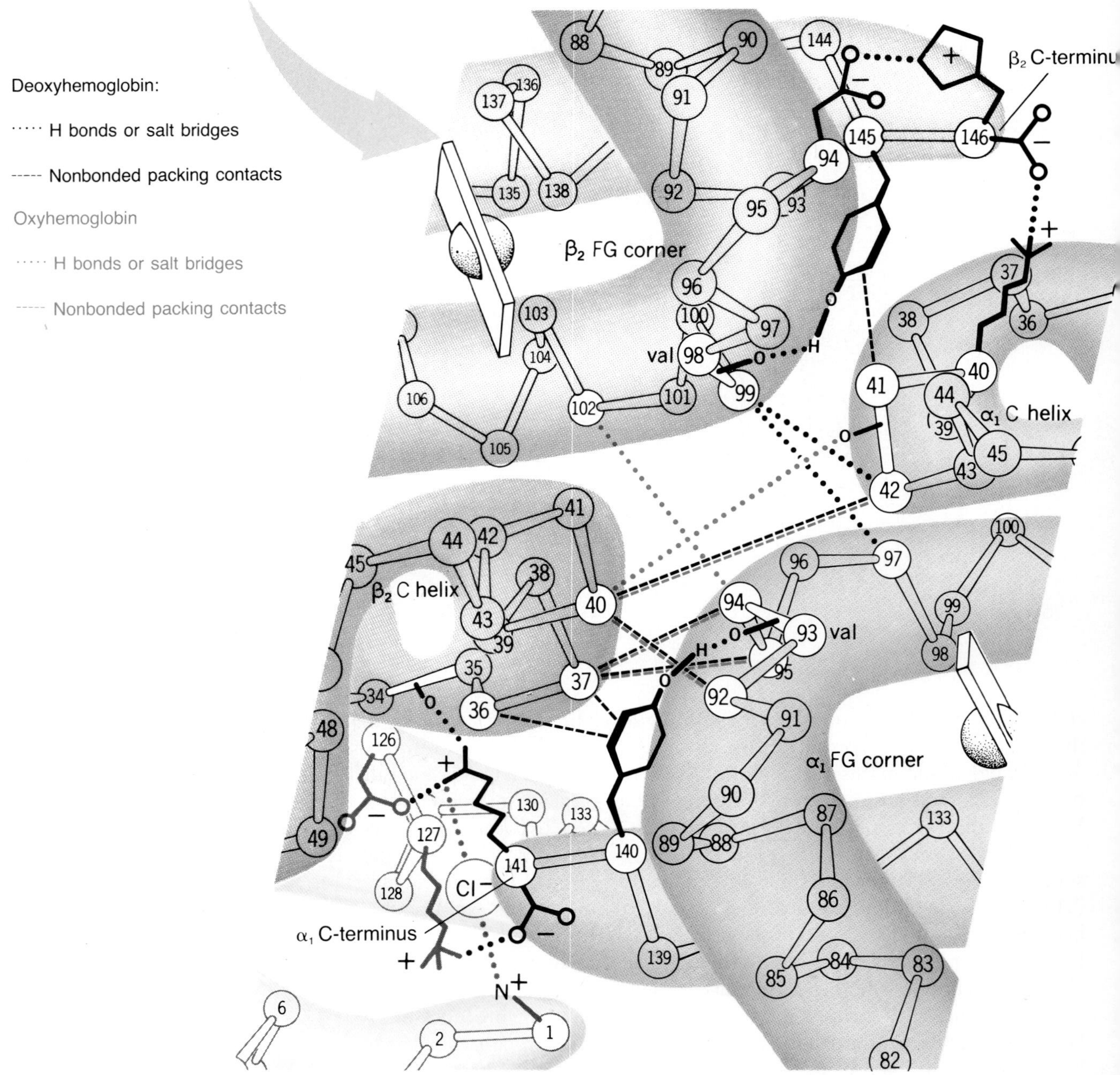

Deoxyhemoglobin:

····· H bonds or salt bridges

----- Nonbonded packing contacts

Oxyhemoglobin

····· H bonds or salt bridges

----- Nonbonded packing contacts

fall back into its lower energy deoxy conformation. This tighter conformation is called tense (T).

Exactly how is the energy of O_2 binding communicated to effect this molecular switching? Again, the details are complicated, but a partial idea can be gained from Figures 7.17 and 7.18. Figure 7.17a is a schematic picture of the relationship of histidine F8 and the neighboring Val (FG5) to the heme in deoxyhemoglobin. The figure illustrates an important fact not mentioned previously: Not only is the Fe^{2+} a bit above the mean heme plane but also the heme itself is not quite flat; it is "domed." Furthermore, in deoxymyoglobin or deoxyhemoglobin, F8 does not point directly perpendicular to the heme but is tilted by about 8°. A ligand like oxygen, binding to the other side, tends to pull the Fe^{2+} a very short distance down into the heme and flattens the heme (Figure 7.17b, c). This cannot happen without molecular rearrangement, for such motion would bring both the ϵ-hydrogen of histidine F8 and the side chain of Val FG5 too close to the heme. What happens is that the histidine changes its orientation toward the perpendicular, shifting the F helix and the FG corner as it does so. As Figure 7.18 shows, this breaks the H bond between Val 98 (FG5) and Tyr 145 (HC2). This in turn distorts and weakens the whole complex of H bonds and salt bridges that connect FG corners of one subunit with C helices of another (see Figure 7.16). Consequently, the rearrangement shown in Figures 7.13 and 7.14 occurs.

In the simplest terms, what has happened is that the binding of O_2, by pulling the Fe a fraction of a nanometer into the heme, has produced a much larger shift in the surrounding structure, particularly at the critical $\alpha\beta$ interfaces.

Effects of Other Ligands on the Allosteric Behavior of Hemoglobin

Cooperative binding and transport of oxygen is only part of the allosteric behavior of hemoglobin. The realities of animal physiology impose further demands. First, as oxygen is utilized in tissues, carbon dioxide is produced and must be transported back to the lungs or gills. Accumulation of CO_2 also lowers the pH in tissues and capillaries through the reaction

$$CO_2 + H_2O \rightleftharpoons HCO_3^- + H^+$$

At the same time, the high demand for oxygen, especially in muscle involved in vigorous activity, can result in oxygen deficit. As shown in Chapter 23, a consequence of this deficit is the production of lactic acid, which also lowers the pH. The falling pH in tissue and venous blood signals a demand for more oxygen delivery.

Hemoglobin functions efficiently to meet these requirements. It does so through its allosteric transition between structurally different high-affinity and low-affinity states. Carbon dioxide, protons, and other substances promote these changes and are called **allosteric effectors.**

Response to pH Changes: The Bohr Effect

A pH drop in the capillaries has the effect of lowering the oxygen affinity of hemoglobin, allowing even more efficient release of the last traces of oxy-

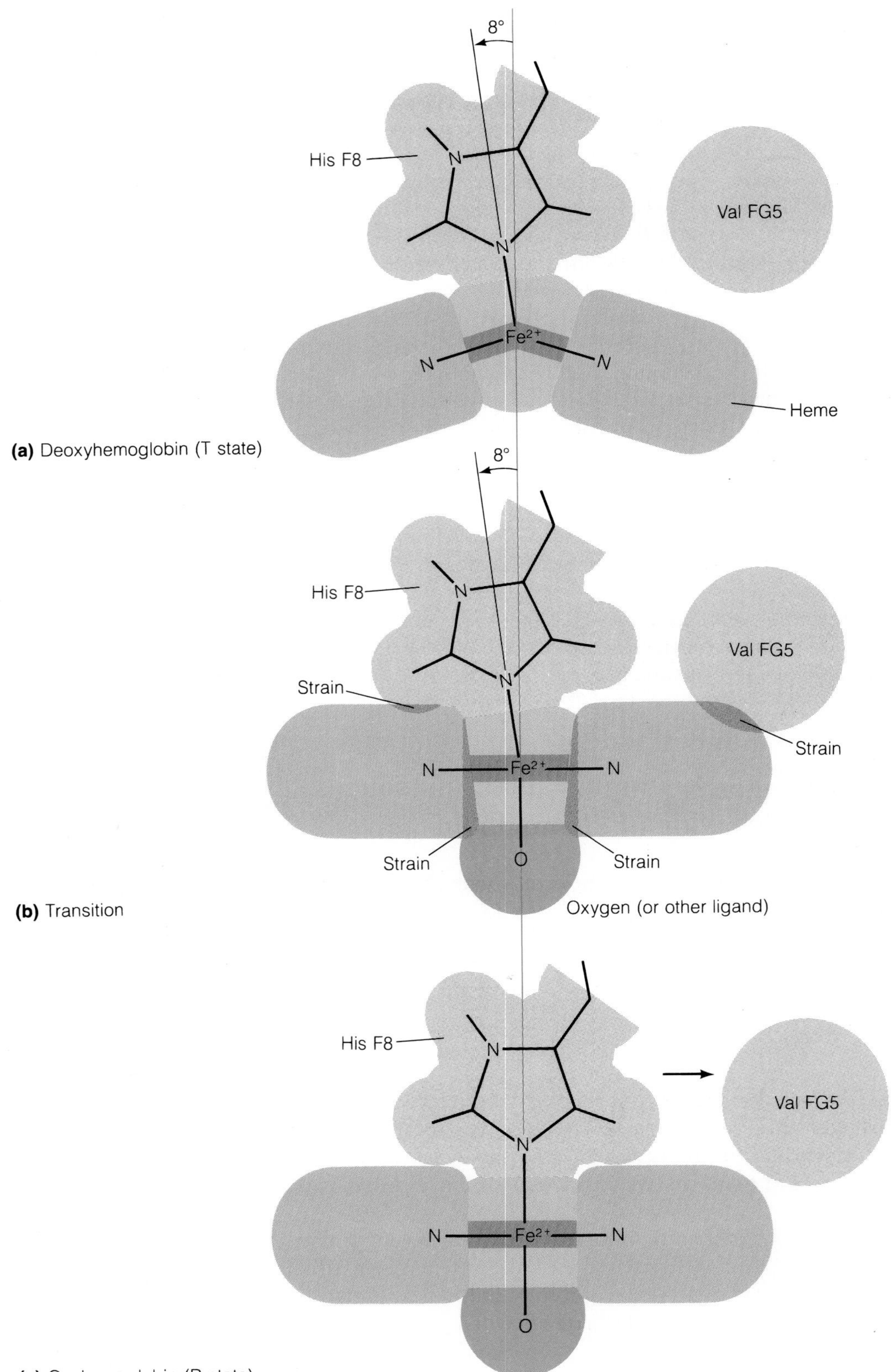

Figure 7.17
Mechanism of the T–R transition in hemoglobin: changes at the heme. Binding of the O_2 ligand pulls the Fe into the heme plane, causing strain as indicated. A shift in the orientation of His F8 relieves this, partly because Val FG5 is pushed to the right.

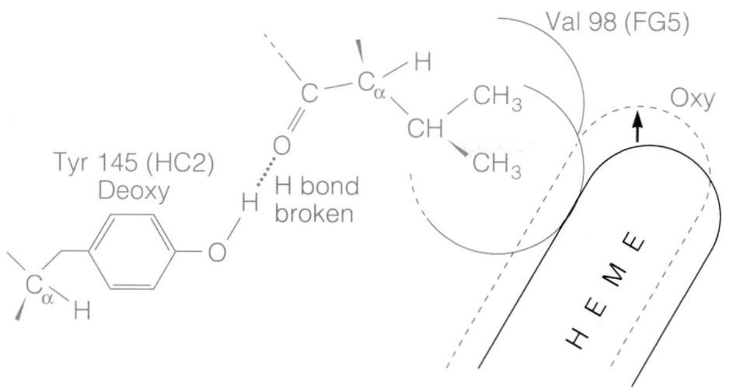

Figure 7.18
Mechanism of the T–R transition in hemoglobin: communicating changes at the heme to the $\alpha\beta$ interface. When F8 histidine tilts and Val FG5 changes position as shown in Figure 7.17, a consequence is the weakening of a hydrogen bond to Tyr 145 (HC2).

gen. The response of hemoglobin to pH change is called the **Bohr effect.** The overall reaction may be written

$$\mathrm{Hb \cdot 4O_2} + n\mathrm{H^+} \rightleftharpoons \mathrm{Hb} \cdot n\mathrm{H^+} + \mathrm{4O_2}$$

where n has a value somewhat greater than 2. Physiologically, this reaction has two consequences. First, in the capillaries, hydrogen ions promote the release of O_2 by driving the reaction to the right. Then, when the venous blood recirculates to the lungs or gills, the oxygenation has the effect of releasing the H^+ by shifting the equilibrium to the left. This, in turn, tends to release CO_2 from bicarbonate dissolved in the blood by the reversal of the bicarbonate reaction (above). The free CO_2 can then be expired.

The mechanism of the Bohr effect can be explained by the model that was used to explain cooperative binding of O_2. There are certain proton binding sites in hemoglobin that are of higher affinity in the deoxy form than in the oxy form. While a number of residues participate, a major contribution comes from histidine residue 146 at the C-termini of the two β chains. In the deoxy form, this residue can make a salt bridge with Asp 94 in the same chain, *if the histidine is protonated* (see Figure 7.16). As a consequence, this histidine residue has an abnormally high pK_a. The salt bridge stabilizes the proton against dissociation. But in the oxy form, this salt bridge simply cannot be formed, so the pK_a falls to its normal value of about 6.5. Consequently, at blood pH (~7.4), His 146 is largely unprotonated in *oxy* hemoglobin. The consequence is that a high concentration of protons, which favors protonation, also favors the deoxy form and thus promotes the release of oxygen. As mentioned above, other residues are also involved, including the N-terminal amino groups of the α chains, but the basic mechanism is the same; protons are allosteric effectors favoring the deoxy conformation. The effect of lowering pH on the oxygen affinity of hemoglobin is illustrated in Figure 7.19. Note that a decrease in pH of only 0.8 unit shifts the P_{50} from about 20 to over 40 mm Hg, nearly doubling the oxygen unloaded to myoglobin.

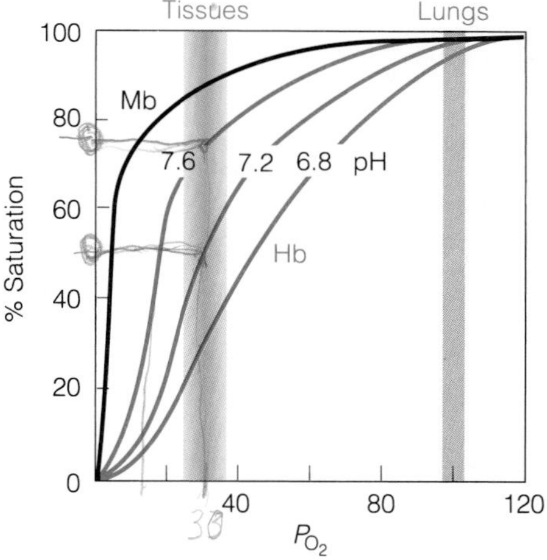

Figure 7.19
The Bohr effect in hemoglobin. Oxygen binding curves are shown at pH 7.6, 7.2, and 6.8. Note that the efficiency of unloading oxygen from hemoglobin (Hb; red lines), as measured by the difference between the curves at P_{O_2} = 30 mm Hg, increases greatly as the pH drops. Myoglobin (Mb, black line) has very little Bohr effect, so the curve shown is approximately correct at all three pH values.

Carbon Dioxide Transport

Release of carbon dioxide from respiring tissues lowers oxygen affinity in two ways. First, as mentioned above, some of the carbon dioxide becomes bicarbonate, releasing protons that contribute to the Bohr effect. Second, carbon dioxide itself reacts directly with hemoglobin, binding to the N-terminal amino groups of the chains to form **carbamates:**

$$-\overset{+}{N}H_3 + CO_2 \rightleftharpoons -\overset{H}{\overset{|}{N}}-COO^- + 2H^+$$

This carbamation reaction allows hemoglobin to transport CO_2 from tissues to lungs or gills. It has two additional effects. First, the protons released contribute to the Bohr effect. Second, a negatively charged group is introduced at the N-terminus of the chains, stabilizing salt bridge formation between α and β chains, which is characteristic of the deoxy state. Both the latter effect and the lower pH effect promote oxygen release when CO_2 is abundant. The reverse reaction that occurs in lungs or gills is equally important. Here, the high O_2 concentration favors oxygenation and hence the oxy form of the molecule. When this switch occurs, stabilization of the carbamated N-termini is decreased, and CO_2 is expelled and expired.

We may summarize the effects of H^+ and CO_2 in terms of the respiratory cycle shown in Figure 7.4: In the lungs or gills of an animal O_2 is abundant. Oxygenation favors the oxy conformation of hemoglobin, which stimulates the release of CO_2. As the blood then travels via arteries into the tissue capillaries, the lower pH and high CO_2 content favor the deoxy form, promoting O_2 release and binding of CO_2. Carbon dioxide, both in forming bicarbonate and in reacting with hemoglobin, releases more protons, further stimulating O_2 release and CO_2 binding. A consequence of the *lack* of CO_2 stimulation of O_2 release is seen in hyperventilation. If a person breathes too rapidly, CO_2 is effectively purged from the tissues, and subsequent release of oxygen is impaired. This leads to dizziness and, in extreme cases, unconsciousness.

Bisphosphoglycerate

Although H^+ and CO_2 are effectors that function rapidly to facilitate the exchange of O_2 and CO_2 in the respiratory cycle, there is one other major effector that operates over longer periods to permit organisms like humans to adapt to gradual changes in oxygen availability. It is a common observation that individuals who move to high altitudes at first experience some distress, but gradually acclimate to the lower oxygen pressure. In part, this is a consequence of increased synthesis of hemoglobin, but another effect is due to an allosteric effector called **2,3-bisphosphoglycerate (BPG)**, or glycerate-2,3-bisphosphate (Figure 7.20). (Note that bisphosphoglycerate was

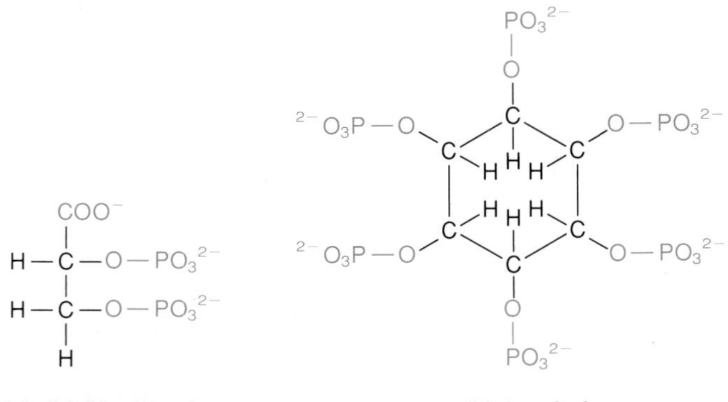

Figure 7.20
Two anionic compounds that bind to deoxyhemoglobin: (**a**) 2,3-bisphosphoglycerate (BPG) found in mammals; (**b**) inositol hexaphosphate (IHP) found in birds.

formerly known as diphosphoglycerate (DPG); you will sometimes find the older term in the literature.)

Like H^+ and CO_2, BPG acts to lower the oxygen affinity of hemoglobin. At first glance this seems like a strange way to adapt to lower O_2 pressure, but in fact the more efficient unloading of oxygen in the tissues more than compensates for the slight decrease in loading efficiency in the lungs. The action of BPG is easily explained. As Figure 7.21 shows, BPG binds in the cavity between the β chains, making electrostatic interactions with positively charged groups surrounding this opening. Comparison of Figures 7.13a and b will show you that this opening is much narrower in oxyhemoglobin than in deoxyhemoglobin. In fact, BPG cannot be accommodated in the oxy form. The higher the BPG content in red blood cells, the more stable the deoxy structure will be. Once again, a decrease in O_2 affinity is explained by stabilization of the deoxy structure. Increased BPG levels are also found in the blood of smokers, who, because of the carbon monoxide in smoke, also suffer from limitation in oxygen supply.

BPG plays one other subtle, but very important, role in the respiration of humans and other mammals. Consider the problem faced by a fetus, which must obtain oxygen from the mother's blood by exchange through the placenta. For this exchange to work well, fetal blood must have a higher O_2 affinity than the mother's blood. In fact, the human fetus has a hemoglobin different from the adult form. Whereas adult hemoglobin (HbA) has two α and two β chains ($\alpha_2\beta_2$), in the fetus the β chains have been replaced by similar, but distinctly different, polypeptides. These are called the γ chains, so fetal hemoglobin (HbF) is written as an $\alpha_2\gamma_2$ structure. The intrinsic oxygen affinity of HbF is very similar to that of HbA, but HbF has a much lower affinity for BPG than does HbA. This difference is largely accounted for by the fact that in fetal hemoglobin histidine 143 in the β chain has been replaced by a serine in the γ chain. As Figure 7.21 shows, the positively charged His 143 helps to bind the negative BPG molecule. Since the concentration of the small, diffusible molecule BPG is about the same in the circulatory systems of mother and fetus, the fetal oxygen affinity is higher, since it does not bind the available BPG so well.

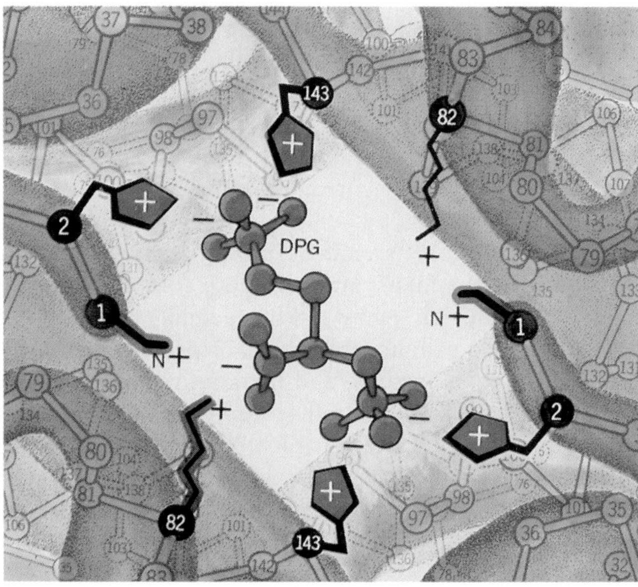

Figure 7.21
Binding of 2,3-bisphosphoglycerate to deoxyhemoglobin. The binding site, in the central cavity of the hemoglobin tetramer, is lined with eight positively charged groups that help bind the negatively charged BPG molecule. (Note that in this illustration the abbreviation DPG appears; this stands for the previously used term diphosphoglycerate.) Note the histidine residues (β143) that are replaced by serine in fetal hemoglobin.

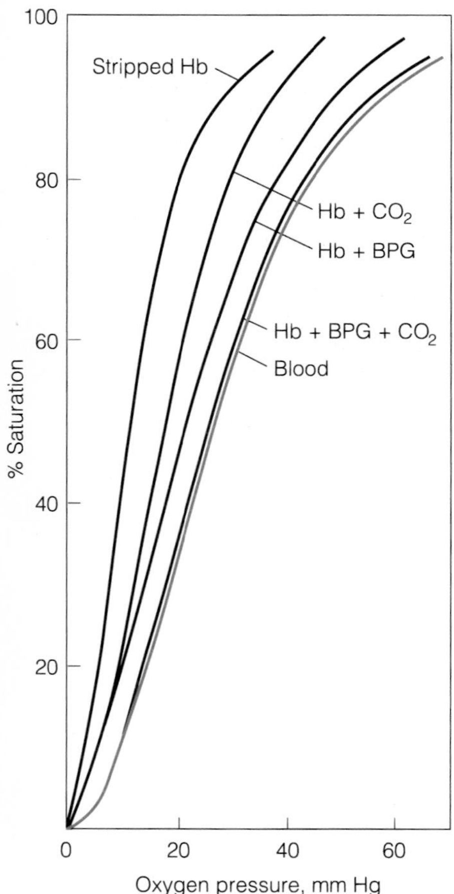

Figure 7.22
Combined effects of CO_2 and BPG on the oxygen binding curve of hemoglobin. Hemoglobin that has been stripped of both CO_2 and BPG has a very high oxygen affinity. Adding both of these at the levels found in blood emerging from the capillaries reproduces almost exactly the binding curve observed for whole blood.

The use of such effectors is not restricted to mammals. The blood of birds contains **inositol hexaphosphate** (see Figure 7.20b), and fishes use ATP for a similar purpose. All of these are molecules with strong negative charge, binding in the deoxy cleft.

In summary, it should be emphasized that all of these allosteric effectors, H^+, CO_2, and BPG, act in the same manner—by biasing the conformational equilibrium in hemoglobin toward the deoxy form. However, they interact at distinctly different sites, and therefore their effects can be additive, as illustrated in Figure 7.22.

Hemoglobin, as it has developed in the higher vertebrates, represents a sophisticated molecular machine, finely tuned for its function. In the following section we explore how this structure might have evolved.

Protein Evolution: Myoglobin and Hemoglobin as Examples

We have emphasized that for each polypeptide chain an organism produces there exists a corresponding gene. The nucleotide sequence in that gene dictates, via the genetic code, the amino acid sequence of the protein. Evolution of proteins occurs through accumulated changes in that nucleotide sequence. In this section we explore this process, using as an example the evolutionary development of the myoglobin–hemoglobin family of proteins. First we must examine in a bit more detail the structure of genes in eukaryotic organisms and the mechanisms through which mutation can occur.

The Structure of Eukaryotic Genes: Exons and Introns

In previous chapters we have implied that there is a direct correspondence between the nucleotide sequence in a gene, the mRNA that is produced, and the polypeptide chain that is translated from it. For most genes in prokaryotic organisms this simple rule is a correct summary. But as investigation of the eukaryotic genome began to yield detailed information, a surprising result was obtained: Within most eukaryotic genes there exist regions of DNA sequence that are never expressed in the polypeptide chain. In fact, these noncoding regions, called **introns,** alternate with regions that *are* expressed in the polypeptide sequence, called **exons.** The situation is exemplified by the structure of the β-globin gene shown in Figure 7.23. Only the parts of the gene shown in color correspond to portions of the polypeptide chains.

The explanation for this seemingly puzzling situation lies in the fact that mRNA production in eukaryotes is a more complex process than had been assumed. As Figure 7.23 shows, what actually happens is that transcription first produces a primary transcript, or **pre-mRNA,** corresponding to the whole gene—exons, introns, and flanking regions. The pre-mRNA, while still in the cell nucleus, is cut and spliced to remove the regions corresponding to introns and produce an mRNA that codes correctly for the polypeptide chain. We shall describe the details of this remarkable process in Chapter 28; for now, keep in mind that there are extensive regions within most eukaryotic genes that do not correspond to *any* part of the protein sequence.

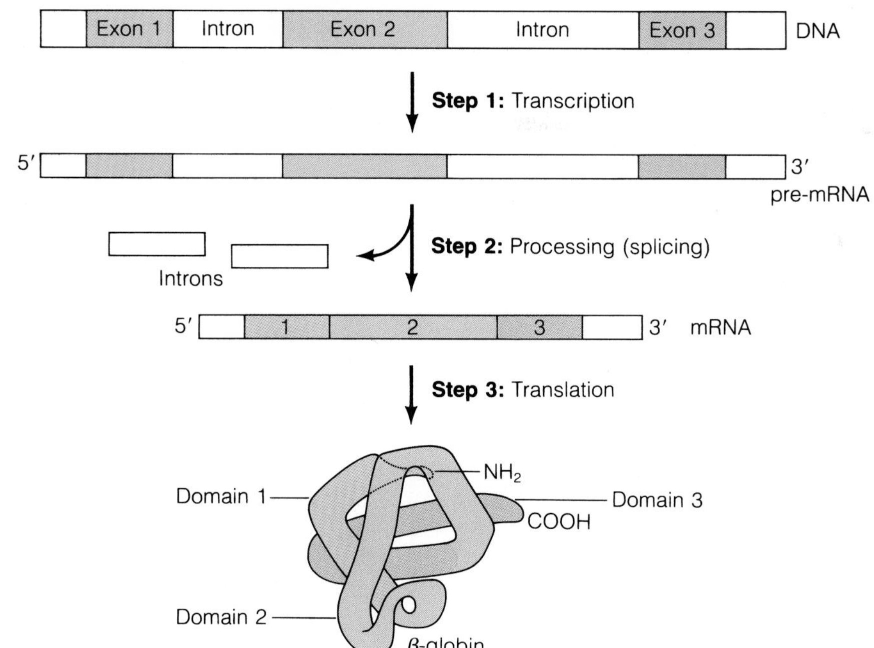

Figure 7.23
Exon–intron structure of globin genes. Shown at the top are the positions of exons and introns in the gene for human β hemoglobin. From this gene is produced a primary transcript, or pre-mRNA (step 1), which is then spliced (step 2) to yield the final mRNA. This mRNA is translated into the β chain, which adopts its favored three-dimensional structure. In the figure, colors allow you to follow each exon into the final folded chain. Note that the whole heme-binding region is coded for by one exon (exon 2), while the C-terminal region so important for allosteric interaction is coded by exon 3.

Mechanisms of Protein Mutation

As organisms reproduce, they copy their DNA. The copying is not always exact; occasionally mistakes are made. These may occur as random errors in copying, or they may be results of damage the DNA has suffered from radiation or chemical **mutagens** (substances that produce mutations). In any event, these alterations will appear as **mutations** in the DNA of the next and subsequent generations. There are two basic kinds of changes in the DNA sequence that may give rise to mutations in proteins.

Replacement of One DNA Base by Another. Replacement of bases can have several possible consequences. First, the base change may not affect the protein sequence at all. The change may occur in an intron. But even if it is in a protein-coding region (exon), the replacement may make no difference in sequence because the new codon codes for the same amino acid as the original one. The redundancy of the genetic code (see Chapter 5) is such that this happens fairly frequently. On the other hand, an amino acid residue in the original protein may be replaced by a different one in the mutated protein; this is called a **missense** mutation (Figure 7.24a). Occasionally, the codon for an amino acid residue within the original protein will be changed to a *stop* codon. We call this a **nonsense** mutation, for the protein will be terminated prematurely and usually become nonfunctional (Figure 7.24b). Sometimes the opposite happens—a stop codon mutates into a codon for an amino acid residue. In this case translation continues, elongating the chain.

Deletions or Insertions of Bases in the Gene. Deletions or insertions may be large or small. Large insertions or deletions in coding regions almost invariably prevent the production of useful protein; sometimes even whole protein genes may be deleted. The effect of short deletions or insertions depends on whether or not they involve multiples of three bases. If one, two,

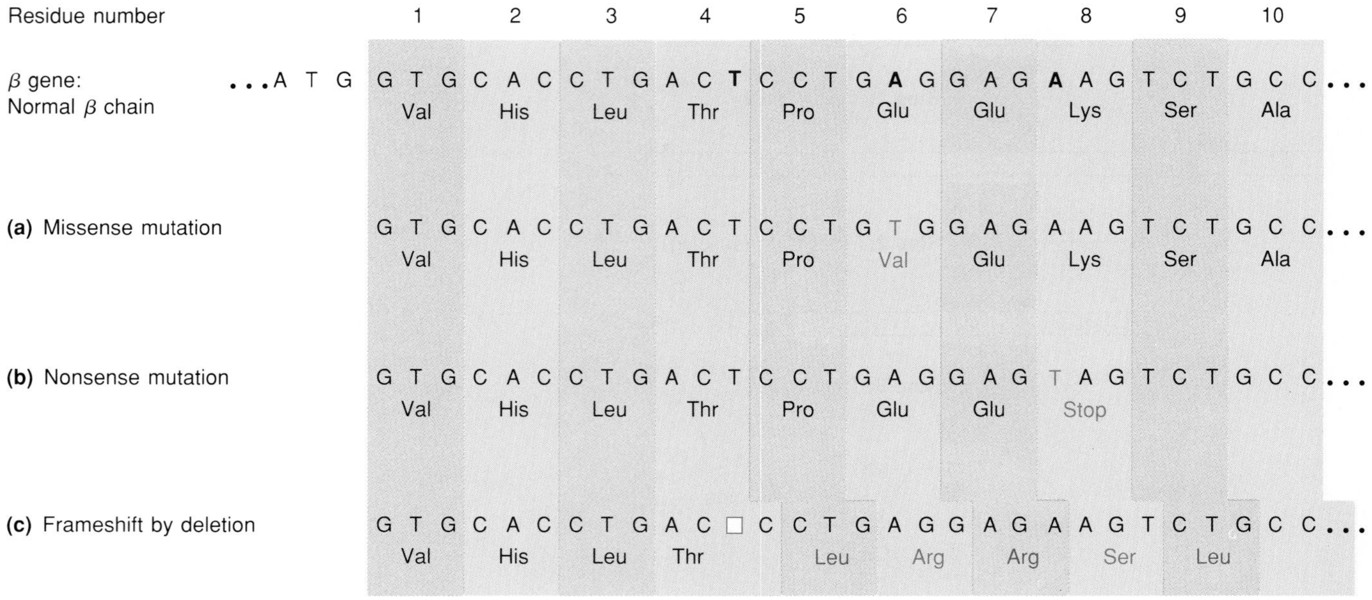

Residue number		1	2	3	4	5	6	7	8	9	10	
β gene:	...A T G	G T G	C A C	C T G	A C **T**	C C T	G **A** G	G A G	**A** A G	T C T	G C C	...
Normal β chain		Val	His	Leu	Thr	Pro	Glu	Glu	Lys	Ser	Ala	
(a) Missense mutation		G T G	C A C	C T G	A C T	C C T	G T G	G A G	A A G	T C T	G C C	...
		Val	His	Leu	Thr	Pro	Val	Glu	Lys	Ser	Ala	
(b) Nonsense mutation		G T G	C A C	C T G	A C T	C C T	G A G	G A G	T A G	T C T	G C C	...
		Val	His	Leu	Thr	Pro	Glu	Glu	Stop			

(c) Frameshift by deletion

G T G C A C C T G A C □ C C T G A G G A G A A G T C T G C C ...

Val His Leu Thr Leu Arg Arg Ser Leu

Figure 7.24
Some ways in which mutations could occur in the β hemoglobin chain. The first 10 residues of the normal human β chain, together with its DNA code, are shown at top. (**a**) A missense mutation has occurred in residue 6. (This is, in fact, the sickle-cell mutation.) (**b**) A nonsense mutation has introduced a *stop* signal after residue 7. This terminates the chain. (**c**) A frameshift mutation has occurred by deletion of a single T residue. The rest of the chain has completely altered sequence and will continue until a stop signal is encountered in this frame. Both (**b**) and (**c**) would result in β-thalassemia (see p. 248).

or more *whole codons* are removed or added, the consequence is the deletion or addition of a corresponding number of amino acid residues. However, a deletion or insertion in a coding region of any number of bases *other* than a multiple of three has a much more profound effect. It causes a shift in the reading frame during translation. Such **frameshift mutations** result in a complete change in the amino acid sequence in the C-terminal direction from the point of mutation (Figure 7.24c).

The effects of these kinds of mutations on the functionality of the protein product, and therefore on the organism itself, can be quite varied. Base substitutions may, in some cases, be neutral in effect, either not changing the amino acid coded for or changing it to another that functions equally well at that position in the protein. More often, the result is deleterious. Occasionally, such mutations increase the efficiency of a protein, and the mutated organisms may be selected for in future generations. Nonsense mutations and frameshift mutations, by contrast, almost always result in destruction of the protein function. If the protein is important to the life of the organism, such mutations are strongly selected against in the course of evolution.

Gene Duplications and Rearrangements

By accumulating small mutational changes over eons of time, proteins gradually evolve. The diversity of functions that they can perform is increased by two other phenomena: **gene duplication** and **exon recombination.**

Very occasionally, replication of the genome occurs in such a way that some DNA sequence, containing a particular gene, is copied twice. Initially, the only result of such duplication is that the descendents of this organism have two copies of the same gene. This may be advantageous, if the protein is one that is needed in large amounts, for the efficiency of its production will be increased. In such cases, there will be selective pressures to maintain two or even more copies of the same gene. Alternatively, the two copies may evolve independently. One copy may continue to express the protein fulfilling the original function, but the other may evolve into an entirely different protein with a new function. Another way in which the diversity of proteins may increase is through the *fusion* of two or more initially independent genes. This will lead to the production of multidomain proteins exhibiting new combinations of functions.

The intervening sequences in eukaryotic genes (introns) offer a further possibility for diversification of protein structure and function. Since these regions are not used for coding, they represent positions where genes can be cut and recombined in the process of **genetic recombination.** The mechanisms of recombination will be described in Chapter 25; at this point we are concerned only with the consequences. Suppose that an exon from one gene, which codes for a protein region with physiological function B, is inserted into an intron region in a gene for a protein carrying function A. The new hybrid protein is now capable of both functions A and B and may serve a new physiological function.

Through the combined effects of mutations, gene duplication, and genetic recombination, organisms are able to develop new abilities, adapt to new environments, and become new species. The process of organismal evolution, which we see exhibited in the fossil record and the incredible variety of existing plants, animals, and microorganisms, is largely a consequence of this molecular evolution of proteins.

Evolution of the Myoglobin–Hemoglobin Family of Proteins

We have already seen an example of the process of protein evolution. If we compare the sequence of sperm whale and human myoglobin (see Figure 5.14, Chapter 5), we find 25 amino acid changes. Since it is believed that the evolutionary lines that led to sperm whales and humans diverged from a common mammalian ancestor about 100 million years ago, we can gain an idea of the rate of this process. If the rate was uniform, there has been one replacement about every 4 million years.

If we compare human myoglobin with that of the shark, we find about 88 differences. Since these evolutionary lines diverged about 500 million years ago, the accumulated differences are about what we would expect from the example above. In other words, the number of amino acid substitutions in two related proteins is roughly proportional to the evolutionary time that has elapsed since the proteins (and the species) had a common ancestor. Using this principle, we can compare the sequences of both hemoglobins and myoglobins and attempt to construct a "family tree" of globin proteins. The tree is complicated by the fact that higher eukaryotes, including humans, carry genes for both myoglobin and several *different* hemoglobin chains. These different genes are expressed at different times in human development (Figure 7.25). The α and β chains, as mentioned earlier, are normally present in adults. But in the early embryo, the hemoglobin genes expressed are those for the embryonic chains, ζ and ϵ. As the fetus develops, these are replaced by α and γ chains (Figure 7.25), and finally, at about the time of birth, the γ chains are replaced by β chains. In addition, after birth a small amount of a δ chain is produced. These developmental types of hemoglobin chain are slightly different, and each is coded for by a separate gene in the human genome.

Comparison of the sequences of many hemoglobins from many different species yields the evolutionary tree shown in Figure 7.26. According to these results, very primitive animals had only a myoglobinlike, single-chain ancestral globin for oxygen storage. Most of these animals, like protozoans

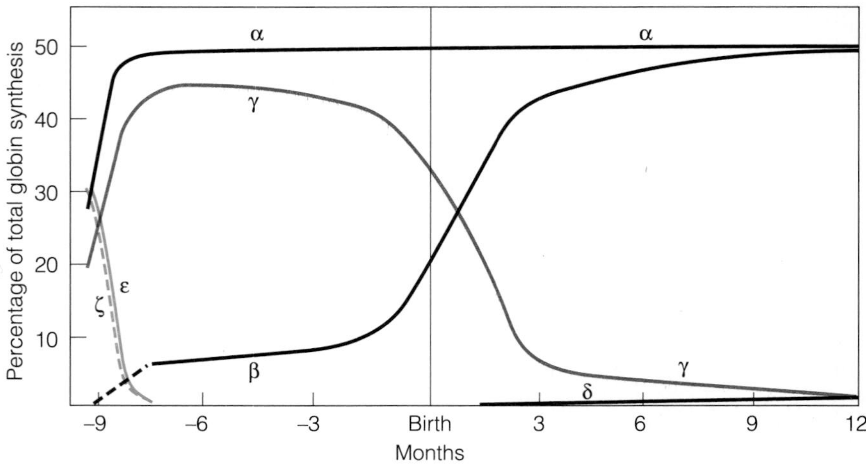

Figure 7.25
Expression of different human globin genes at different stages in development. The ζ and ϵ genes make the hemoglobin ($\zeta_2\epsilon_2$) used in the very early embryo. These are soon supplanted by the $\alpha_2\gamma_2$ hemoglobin of the fetus. At about the time of birth, transcription of the γ gene ceases and the β gene begins to be transcribed. By six months, the infant will have almost completely adult type ($\alpha_2\beta_2$; black lines) hemoglobin. The δ gene is never transcribed at high rates. There are two copies (α_1, α_2) of the α gene. Both contribute to the production of α chains.

and flatworms, were so small that they did not require a transport protein.

More than 800 million years ago, an important event occurred: the ancestral myoglobin gene was duplicated. One of the copies became the ancestor of the myoglobin genes of all higher organisms; the other evolved into the gene for an oxygen transport protein and gave rise to the hemoglobins. Along the evolutionary line leading to vertebrates and mammals, the most primitive animals to possess such a transport protein are the lampreys. Lamprey hemoglobin can form dimers but not tetramers and is only weakly cooperative. But subsequently a *second* gene duplication occurred, giving rise to the ancestors of the present-day α and β hemoglobin chain families. Reconstruction, from sequence comparison, indicates that this must have happened about 500 million years ago, at about the time of divergence of the sharks and bony fish. The evolutionary line of the latter led to the reptiles, and eventually mammals, all carrying both α- and β-globin genes and capable of forming tetrameric $\alpha_2\beta_2$ hemoglobins. Further gene duplications have occurred in the hemoglobin line, leading to the embryonic forms ζ and ϵ and the fetal γ. As Figure 7.26 shows, the duplications that led to a distinction between β and α gene subtypes coincide fairly well with the development of placental mammals. This is functionally appropriate, for in these mammals the later stages of embryo development occur within the mother, and a special hemoglobin, adapted to promote oxygen transfer through the placenta from the mother to the fetus, is essential.

During the long evolution of the hemoglobin–myoglobin family of proteins, only a few amino acid residues have remained invariant. These may mark the truly essential positions in the molecule. As Figure 7.11 shows, these include the histidines proximal and distal to the heme iron (F8 and E7; see Figure 7.3b). Interestingly, Val FG5, which has been implicated in the deoxy–oxy conformation change as described earlier, is invariant in hemoglobins, replacing the isoleucine found at this position in most myoglobins. Other regions highly conserved are those near the $\alpha_1\beta_2$ (and $\alpha_2\beta_1$) contacts. As shown in Figure 7.16, this is the part of the molecule most directly involved in the allosteric conformational change.

Despite the major changes that have occurred in the primary structure of the myoglobin–hemoglobin family over eons of evolution, the secondary and tertiary structures of these proteins have remained surprisingly unchanged. Figure 7.27 shows the backbone structure of members of this family ranging from insect to horse. All are recognizable as the same basic structure, and the similarity is particularly strong in the region that binds

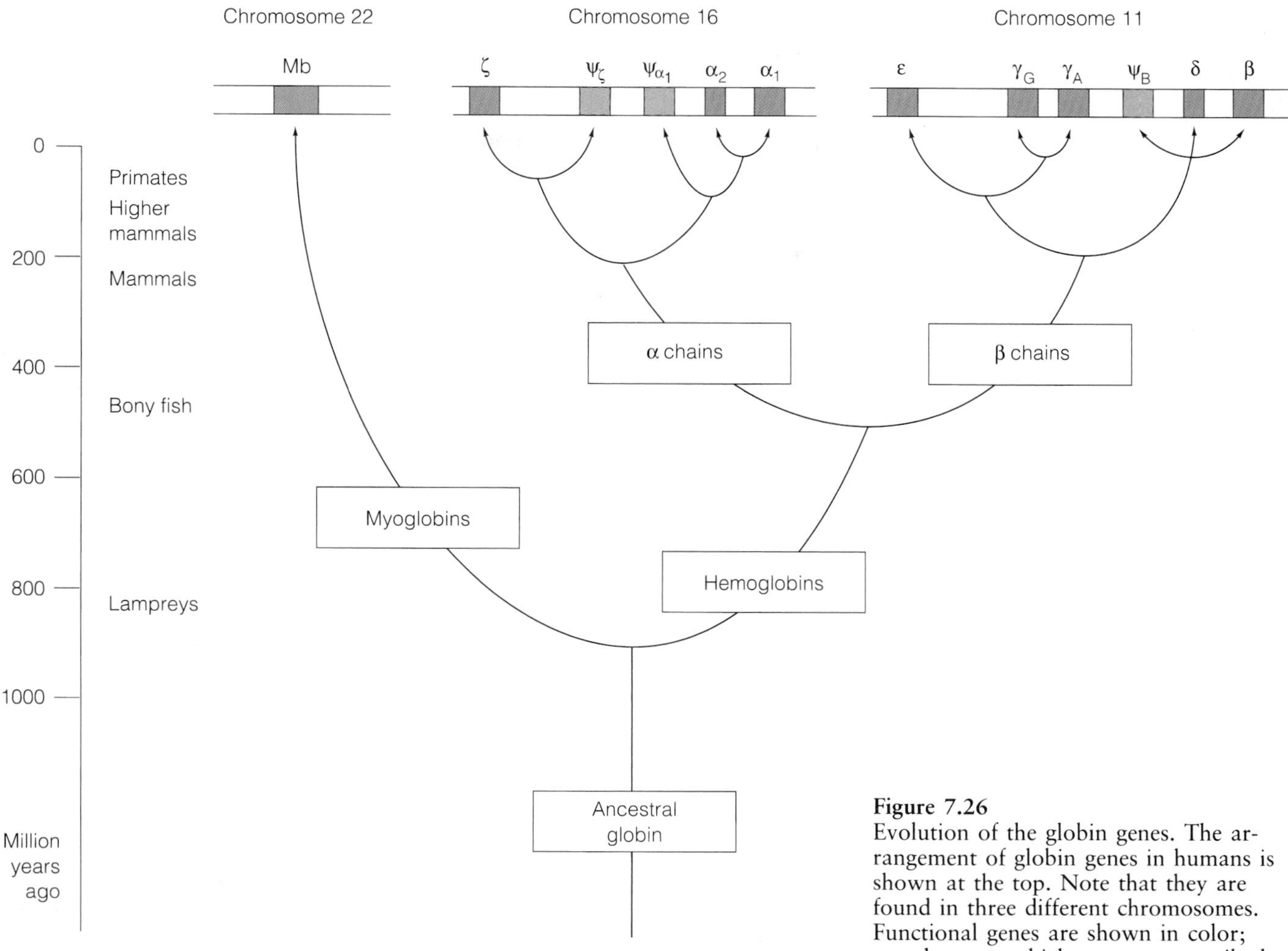

Figure 7.26
Evolution of the globin genes. The arrangement of globin genes in humans is shown at the top. Note that they are found in three different chromosomes. Functional genes are shown in color; pseudogenes, which are nontranscribed variants of a gene, are in gray. The diagram below shows the probable evolution of the globin gene family, based on sequence differences between the various globin genes in humans and in other animals. The times at which gene duplications occurred are inferred from a combination of sequence and fossil evidence and are only approximate. The two α genes and the two γ genes are too similar in sequence to allow us to judge the time of their divergence. We only know that it must have been quite recently in evolution.

the heme. At first glance this seems inconsistent with our earlier statements that primary structure determines secondary and tertiary structure. However, careful examination of many sequences shows that many of the replacements have been *conservative*—that is, an amino acid has been replaced by another of the same general class. Obviously, evolution of these proteins has proceeded not at random, but under the constraint of maintaining a physiologically functional structure. Survival of mutant proteins in the globin family has been restricted to those that maintain the basic "globin fold."

Hemoglobin Variants: Evolution in Progress

Variants and Their Inheritance

Evolution of hemoglobin genes is still going on. We can see this in the existence of hemoglobin variants or, as they are often called, abnormal hemoglobins. There exist, at present, several hundred recognized mutant hemoglobins within the human population. A "map" of positions of mutation on the tetramer is shown in Figure 7.28. It is probable that most pro-

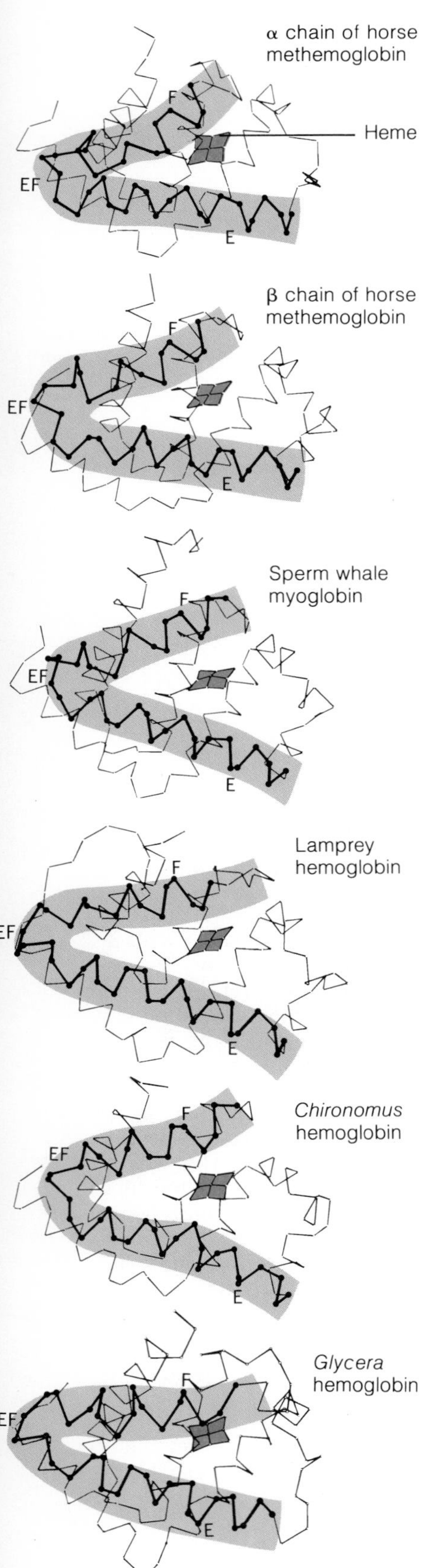

teins in existing plants and animals show comparable diversity. It is only because human hemoglobins have been so thoroughly studied that we can recognize such variety in this case. Each of these mutant forms exists in only a small fraction of the total human population; some have been recognized in only a few individuals. Many are deleterious and give rise to recognized pathologies; under conditions of natural selection these would eventually disappear. Others are, as far as we can tell, harmless, and these are often referred to as "neutral" mutations. A very few may have as yet unrecognized advantages and may come, in time, to dominate the population.

We shall consider only a few of these abnormal hemoglobins. In preface, it is necessary to say a bit about genetics. All human cells, except for the germ cells (sperm and ova), are **diploid;** that is, they carry two copies of each chromosome. Therefore, they carry two copies of each gene, one on each of the paired chromosomes. Suppose we consider a gene such as the adult β-hemoglobin gene, which can exist in two forms—the "normal" type, β, and a mutant "variant" type, β^*. An individual can have three possible combinations of these genes in his or her paired chromosomes:

A. $\beta + \beta$: **homozygous** (same genes) in the normal type.

B. $\beta + \beta^*$: **heterozygous** (mixed genes).

C. $\beta^* + \beta^*$: homozygous in the variant type.

Having genes for only the normal β-hemoglobin, individual A will produce only normal β-hemoglobin chains. Individual C, who has genes for only the variant type, will produce only variant hemoglobin chains. Individual B, with genes for both types, will produce both. If the mutation is deleterious, C will be in serious trouble. B, on the other hand, may do fairly well, since he or she will make normal protein chains along with the variant ones.

Consider now what happens when two individuals produce offspring, as illustrated in Figure 7.29. Each parent will donate to a child *one* of his or her copies of the β-hemoglobin gene, the selection of which will be random. If both parents carry only the normal gene, the child must receive the same. If both carry only the variant gene, the child must be homozygous for it as well. If both are heterozygous for the gene, the child has one chance in four of being homozygous normal, one in four of being homozygous in the variant gene, and two in four of being heterozygous. Since the variant hemoglobin genes are generally rare in the human population, only occasionally do we find an individual homozygous for the variant type. All of these factors must be taken into account to understand the effects of hemoglobin mutations on individuals.

Figure 7.27
Evolutionary conservation of the globin folding pattern. The drawings emphasize that the overall tertiary structure of myoglobin and hemoglobin chains has remained nearly constant despite extensive changes in the primary structure. The shaded regions delineate the E and F helices, which surround the heme. Note that they are almost invariant, and changes tend to be concentrated near the ends of the chains. The most primitive proteins shown are the single chain "hemoglobins" of the marine worm *Glycera* and the fly *Chironomus* (one of the few insects to have a hemoglobinlike protein). The lamprey is the most primitive creature to have distinct myoglobins and hemoglobin. The lamprey hemoglobin forms dimers and exhibits some cooperativity in binding. The α and β chains of horse hemoglobin are almost identical to those of all other mammals.

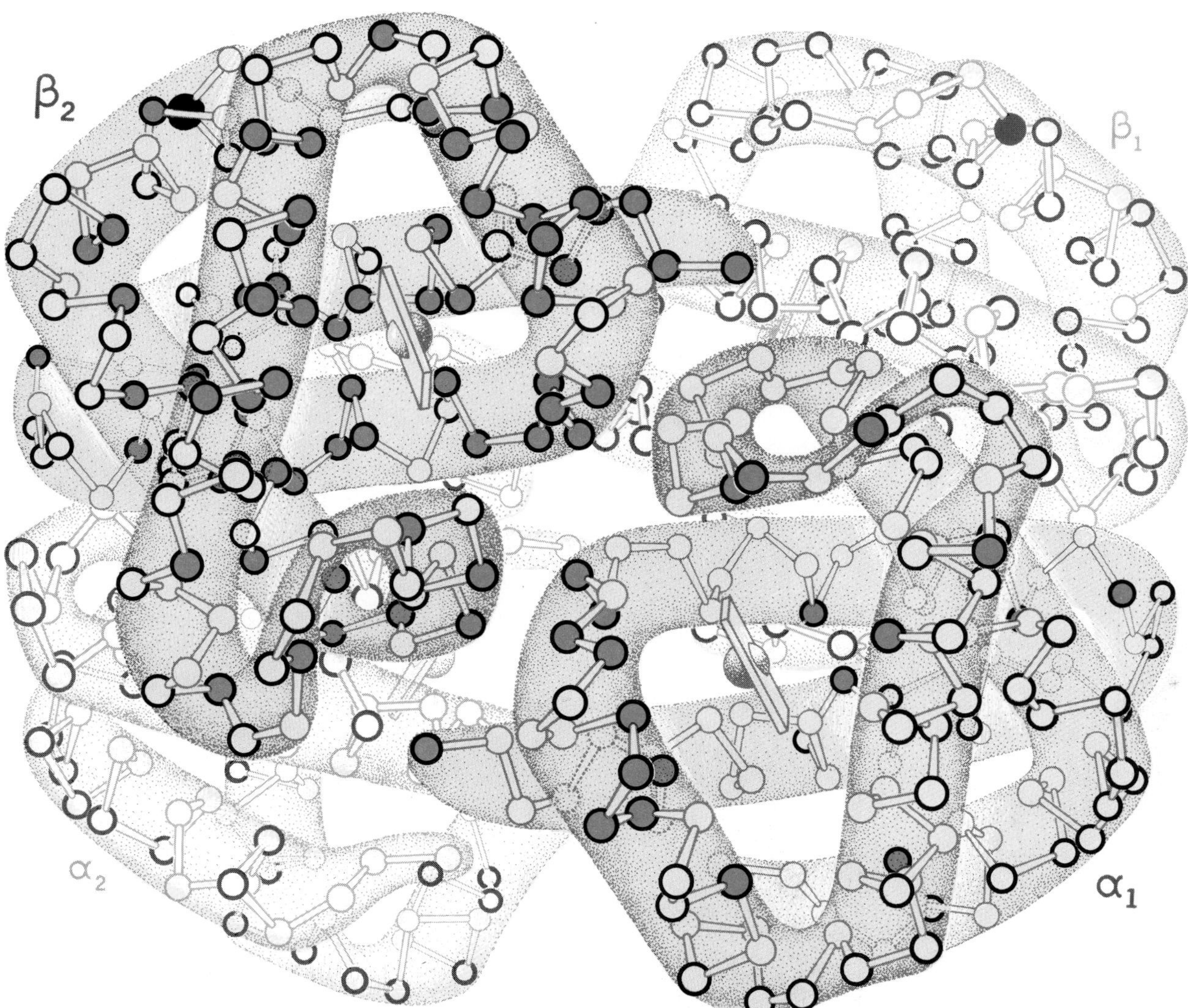

Figure 7.28
Positions of mutations in human hemoglobins. The circles with heavy outlines represent *all* positions at which amino acid substitutions have been found. In the α_1 and β_2 chains, those that have known pathological effects are shown in red. At many of these positions more than one different substitution has been observed. The $\beta6$ position, at which the sickle-cell mutation occurs, is shown in black.

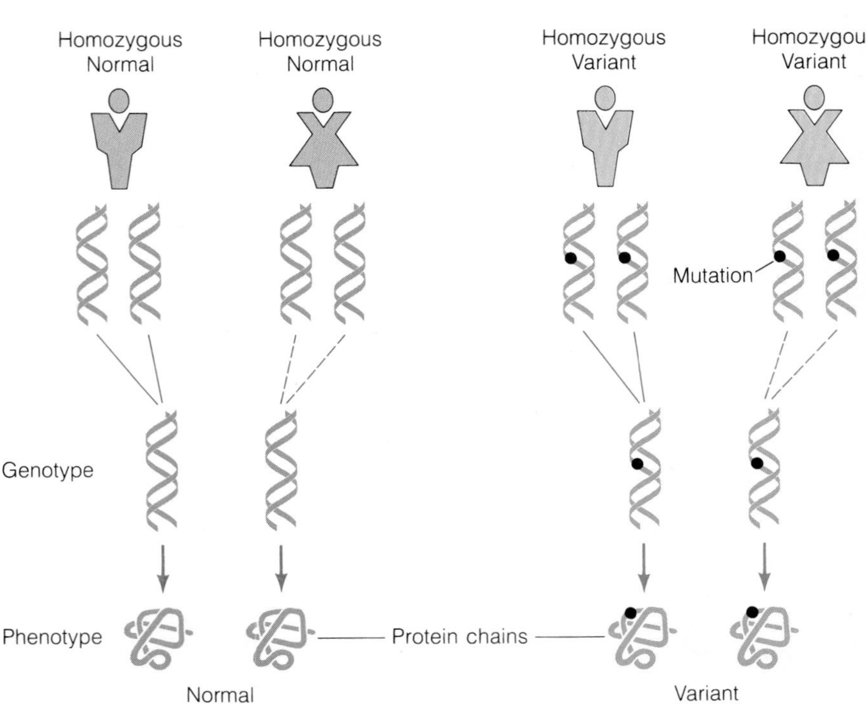

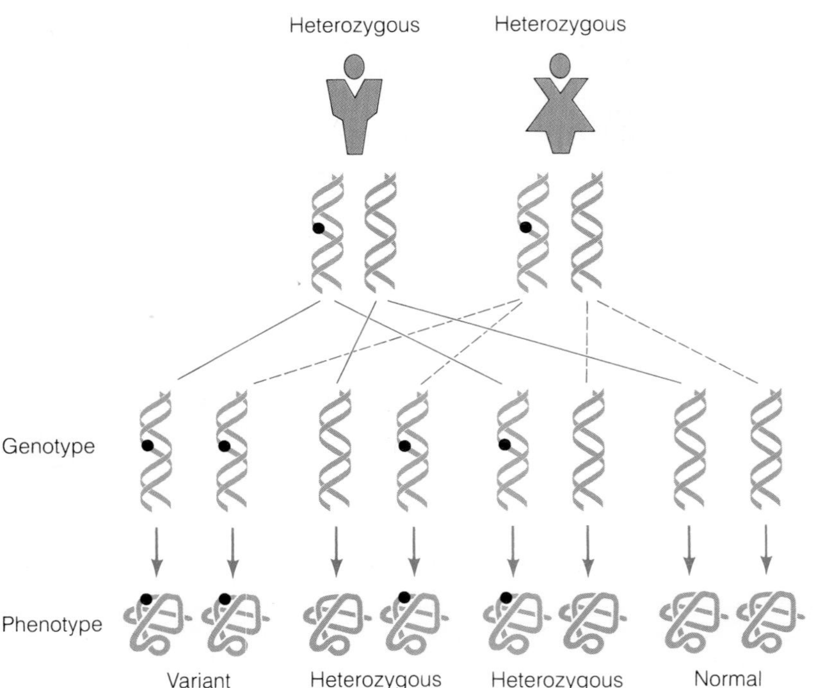

Figure 7.29
Inheritance of normal and variant (mutant) proteins. Diploid organisms can exist as one of three types with respect to any gene: homozygous normal (red), homozygous variant (blue), or heterozygous (lavender). The offspring of a homozygous normal pair will all be homozygous normal, and the offspring of a homozygous variant pair will all be homozygous variant. The offspring of a heterozygous pair may be of three types: homozygous normal, heterozygous, or homozygous variant, in a probability ratio of 1:2:1. We leave it as an exercise for the student to work out other possibilities—for instance, the offspring of a pair in which one is homozygous normal and the other homozygous variant.

Pathological Effects of Variant Hemoglobins

Of the large number of hemoglobin mutations, a significant fraction have deleterious effects. As Figure 7.28 shows, these are mostly clustered about the heme pockets and in the vicinity of the $\alpha\beta$ contact region that is so important in the allosteric transition. A few of the well-studied pathological

Table 7.1
Selected list of missense mutations in human hemoglobins

Effect	Residue Changed	Change	Name	Consequences of Mutation	Explanation
Sickling	β6 (A3)	Glu → Val	S	Sickling	Val fits into EF pocket in chain of another hemoglobin molecule
	β6 (A3)	Glu → Ala	G Makassar	Not significant	Ala probably does not fit the pocket as well
	β121 (GH4)	Glu → Lys	O Arab, Egypt	Enhances sickling in S/O heterozygote	β121 lies close to residue β6; Lys increases interaction between molecules
Change in O_2 affinity	α87 (F8)	His → Tyr	M Iwate	Forms methemoglobin, decreased O_2 affinity	The histidine normally ligated to Fe has been replaced by Tyr
	α141 (HC3)	Arg → His	Suresnes	Increased O_2 affinity by favoring R state	Replacement eliminates bond between Arg 141 and Asn 126 in deoxy state
	β74(E18)	Gly → Asp	Shepherds Bush	Increased O_2 affinity by decrease in BPG binding	The negative charge at this point decreases BPG binding
	β146 (HC3)	His → Asp	Hiroshima	Increased O_2 affinity, reduced Bohr effect	Disrupts salt bridge in deoxy state and removes His, which provides Bohr proton
Heme loss	β92 (F8)	His → Gln	St. Etienne	Loss of heme	The normal bond from F8 to Fe is lost, and the polar glutamine tends to open the heme pocket
	β42 (CD1)	Phe → Ser	Hammersmith	Unstable, loses heme	Replacement of hydrophobic Phe with Ser attracts water into heme pocket
Dissociation of tetramer	α95 (G2)	Pro → Arg	St. Lukes	Dissociation	Chain geometry altered in subunit contact region
	α136 (H19)	Leu → Pro	Bibba	Dissociation	Pro interrupts helix H

mutations are listed in Table 7.1. For example, a class of variants known as *hemoglobins M* tend to be readily oxidized to methemoglobin, which cannot bind O_2. Many of these involve replacement of either the proximal or distal histidine by other residues. Since the individual's hemoglobin is less effective in transporting oxygen, the symptom is anemia. Other variants that involve changes at the subunit interfaces have two kinds of effects. Some, like *hemoglobin St. Lukes,* destabilize the hemoglobin tetramer, whereas others (e.g., *hemoglobin Suresnes*) tend to stabilize either the oxy or deoxy conformation, inhibiting the allosteric switch. Finally, there are those like *hemoglobin Hammersmith* in which the tertiary structure of the molecule is unstable; some of these cannot hold the heme effectively.

The most infamous of all variant hemoglobins, *sickle-cell hemoglobin,* is a source of misery and early death to many humans. The variant has gained its name because it causes red blood cells to adopt an elongated,

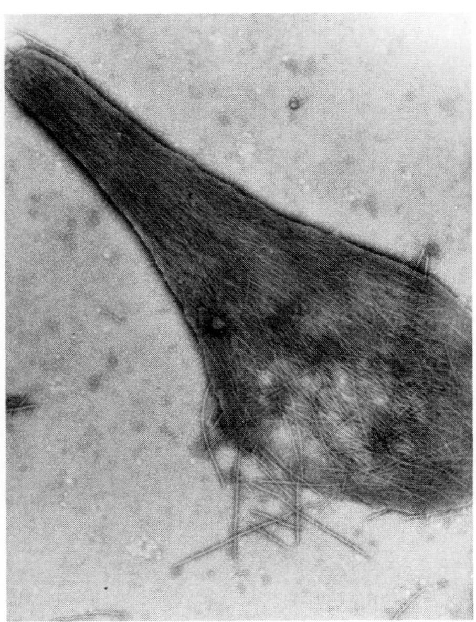

Figure 7.30
Scanning electron micrograph of a sickled erythrocyte. The hemoglobin-S fibers can be seen more or less aligned within the distorted cell. The cell has ruptured and hemoglobin fibers are spilling out.

sickle shape at low oxygen concentrations (Figure 7.30). This "sickling" is a consequence of the tendency of the sickle-cell mutant hemoglobin, in its deoxygenated state, to aggregate into long rodlike structures (Figure 7.31). The elongated cells tend to block capillaries, causing inflammation and considerable pain. Even more serious is the fact that the sickled cells are fragile; their breakdown leads to an anemia that leaves the victim susceptible to infections and diseases. Individuals who are homozygous for the sickle-cell mutation often do not survive into adulthood, and those who do are seriously debilitated. Heterozygous individuals, who can still produce some normal hemoglobin, frequently suffer little or no distress except under conditions of severe oxygen deprivation.

Remarkably, sickling stems from what we might expect to be an innocuous mutation in a part of the molecule far from the "critical" regions mentioned above (see Figure 7.28). The glutamic acid residue normally found at position 6 in β chains has been replaced by a valine. This hydrophobic valine can fit into a pocket at the EF corner of a β chain in *another* hemoglobin molecule and thus, as shown in Figure 7.32, adjacent hemoglobin molecules can fit together into a long rodlike helical fiber. Why sickling occurs with deoxyhemoglobin, but not with the oxygenated form, is simply explained: in the oxy form, the rearrangement of subunits has made the pocket inaccessible.

Sickle-cell anemia is a genetic disease confined largely to populations arising in tropical areas of the world. At first glance, this seems unexpected: why should a *genetic* disease be restricted geographically? The answer tells us something about the persistence of what would seem to be unfavorable traits. In fact, the high incidence of sickle-cell disease generally coincides with a high incidence of malaria, a tropical disease. Individuals *heterozygous* in sickle-cell hemoglobin have a higher resistance to malaria than those who do not carry the sickle-cell mutation. The malarial parasite spends a portion of its life cycle in red cells, and the increased fragility of the sickled cells, even in heterozygous individuals, tends to interrupt this cycle. Thus, such individuals have a higher survival rate (and therefore have a better chance of procreation) in malaria-infested regions. However, their high incidence in the populace leads to the birth of many people who are homozygous for the mutant trait.

Sickle-cell anemia is the kind of genetic disease that many scientists hope will eventually be alleviated by recombinant gene therapy. If a way could be found to introduce functional β-globin genes into an individual homozygous for the sickle-cell defect, he or she would be rendered effectively heterozygous, with greatly increased chances for an extended and productive life.

Thalassemias: Effects of Nonfunctional Hemoglobin Genes

The human hemoglobin variants we have mentioned so far are consequences of missense mutations. Because of a base substitution in a gene coding for one of the chains, one amino acid has been substituted for another. There are, however, other genetic defects involving hemoglobin in which one or more of the chains are simply not produced. The pathological condition that arises is called **thalassemia.** The condition of thalassemia can arise in several ways:

1. One or more of the genes coding for hemoglobin chains (Figure 7.26) may have been deleted.

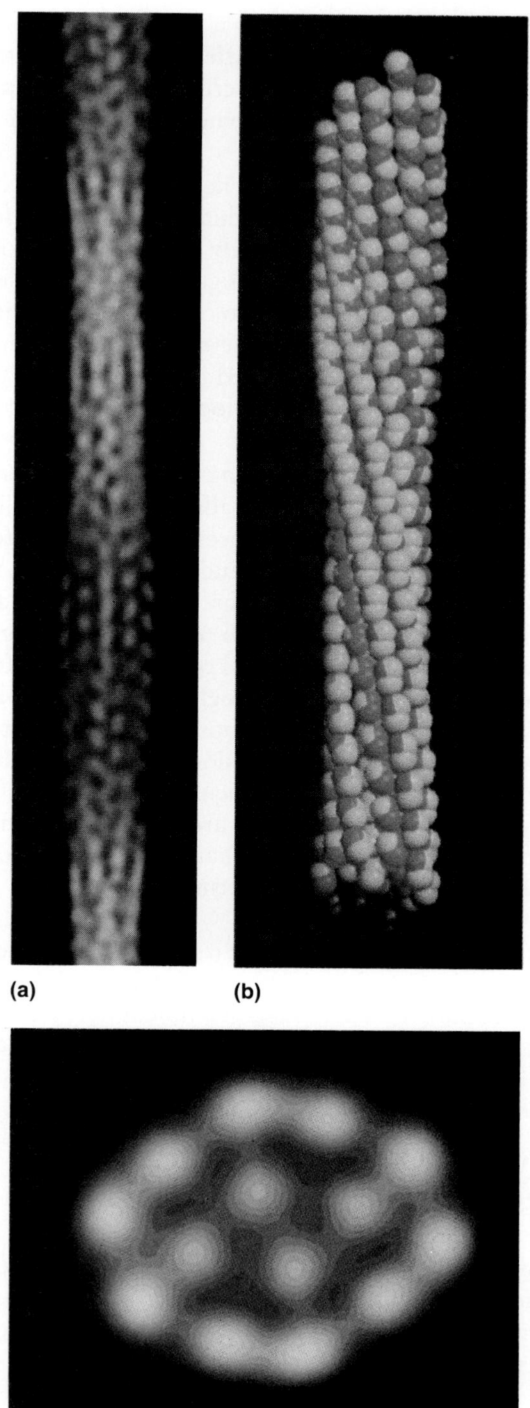

Figure 7.31
A fiber of sickle-cell hemoglobin. (**a**) shows an electron micrograph of one fiber and (**b**) a computer-graphic model of the same. In (**c**) a cross-section is shown depicting the 14 strands (in 7 pairs) in the fiber.

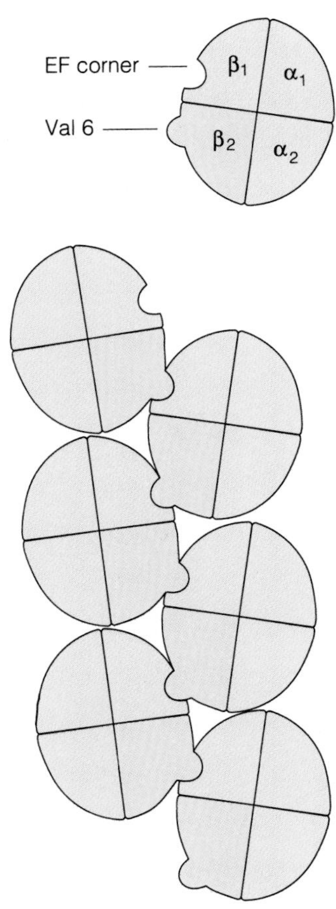

Figure 7.32
The molecular basis for strand formation in sickle-cell disease. Two strands of hemoglobin are locked together because Val 6 in the β chain of one hemoglobin molecule fits into a pocket in an adjacent molecule. Interaction of these, in turn, with other strands produces the multistrand structures shown in Figure 7.31.

2. All genes may be present, but one or more may have undergone a nonsense mutation so as to produce a shortened chain, or a frameshift mutation.

3. All genes may be present, but a mutation may have occurred outside the coding regions, leading to either a block in transcription or improper processing of the pre-mRNA to produce mRNA.

In case 1 or 2, no functional protein will be produced. In case 3 there may be limited transcription and translation of the correct polypeptide sequence.

Since the human genome contains a number of globin genes, corresponding to the protein chains used at different developmental stages, there are many varieties of thalassemia. We describe here only two major classes—those involving loss or misfunction of genes for the adult β and α chains.

β-THALASSEMIA. If the β hemoglobin gene is lost or cannot be expressed, a most serious condition arises in individuals homozygous for this defect. They can make *no* β chains and must rely on continued production of the fetal γ chains to make a functional hemoglobin, $\alpha_2\gamma_2$. While γ production may continue in such individuals well into childhood, most die before reaching maturity. Much less serious is the heterozygous state, in which one β gene is still functioning. In these so-called β^--thalassemias, the production of β-globin is limited but not entirely blocked.

α-THALASSEMIAS. Thalassemias involving the α chain present a more complicated situation, because there are two adjacent copies of the gene (α_1 and α_2) on the human chromosome. Their protein products differ by only one amino acid, and both are functional. An individual can have, therefore, 4, 3, 2, 1, or 0 copies of an α gene. Only if three or more genes are missing are serious effects observed. Individuals with only one α gene are anemic. The low level of α hemoglobin leads to partial compensation by formation of β_4 hemoglobin *(hemoglobin H)* and γ_4 hemoglobin *(hemoglobin Bart's)*. These can bind and carry oxygen, but they do not exhibit the allosteric transition, remaining always in the R state. Nor do they exhibit a Bohr effect. So unloading of oxygen to tissues is inefficient. In the condition known as *hydrops fetalis,* all four α gene copies are missing. Such individuals are inevitably stillborn, for the supply of γ chains, which falls near birth, is not in itself sufficient to support the near-term fetus.

Because there are two copies of the α gene but only one of the β gene, most of the deleterious mutations in mammalian hemoglobins occur in the β chains (see Figure 7.28). This may suggest a functional role for gene duplication: if two or more copies of a gene are present, the species is somewhat protected from the effects of harmful mutations.

Immunoglobulins: Variability in Structure Yields Versatility in Binding

In the remainder of this chapter, we describe a group of proteins whose primary function, like that of myoglobin and hemoglobin, is binding of other substances. These are the immunoglobulins, the antibodies involved in the immune response.

The Immune Response

When a foreign substance—a virus, a bacterium, or even a foreign protein—invades the tissues of a higher vertebrate (like a human) the organism defends itself by what is called the **immune response.** There are two facets to this defense. In the **humoral** immune response, a class of lymphatic cells called **B lymphocytes** synthesize specific immunoglobulin molecules that bind to the invading substance and either precipitate it or mark it for destruction by cells called **macrophages.** In the **cellular** immune response, another class of lymphatic cells (T lymphocytes), bearing immunoglobinlike molecules on their surfaces, recognize and kill foreign cells. In this section we shall be mainly concerned with the humoral immune response.

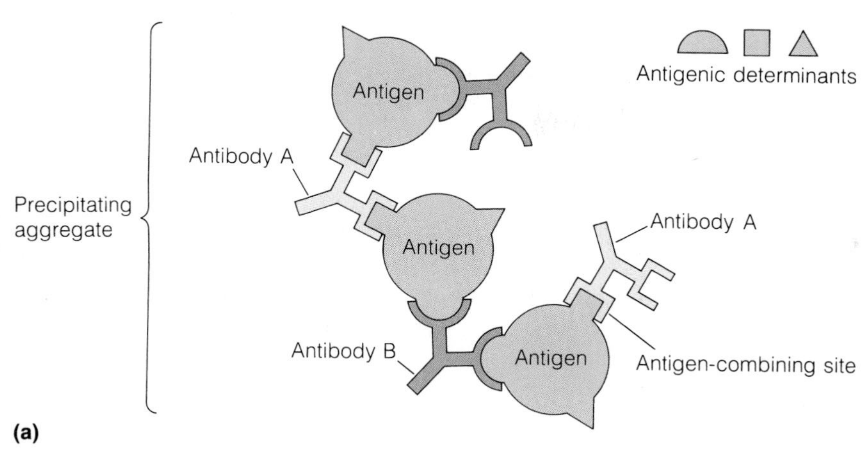

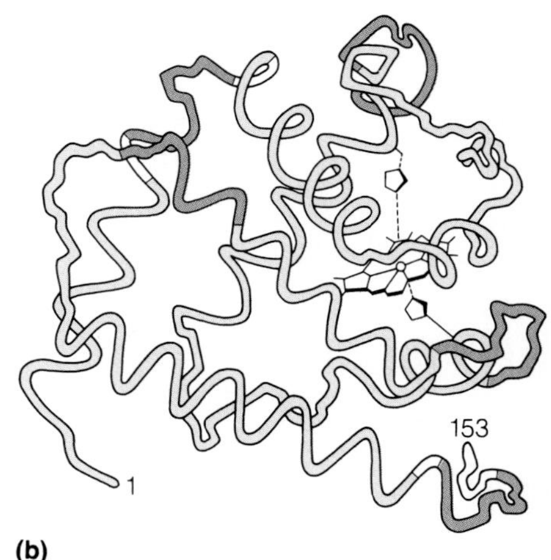

Figure 7.33
Antigenic determinants. (**a**) A foreign object such as a virus, bacterial cell, or even a foreign protein molecule may lead to the production of antibodies to several different antigenic determinants on its surface. When the foreign substance is mixed with this collection of antibodies, precipitation will occur. (**b**) The antigenic determinants on sperm whale myoglobin. The purple portion represents segments of the polypeptide chain that act as complete determinants. The white portions form part of the antigenic determinant with some antibodies.

The substance that elicits an immune response is called the **antigen,** and a specific immunoglobulin that binds to this substance is the **antibody** (Figure 7.33a). If the invasive particle is large, like a cell, a virus, or a protein, many different antibodies may be elicited, each type binding specifically to a given **antigenic determinant** (or **epitope**) on the surface of the particle. Such antigenic determinants may be, for example, groups of amino acids on a protein surface or groups of sugar residues in a carbohydrate; an example is shown in Figure 7.33b.

The immune response is our first line of defense against infection and probably against cancer cells as well. It is the crippling of the immune system that makes **AIDS** (**acquired immune deficiency syndrome**) a disease that has so far proved to be almost inevitably fatal. Victims of AIDS do not die of the disease itself—they perish from infectious diseases or cancers that their immune system is no longer able to defend against.

The immune response has some remarkable features. First, it is incredibly versatile, being able to respond to an enormous number of different foreign substances. These range from cells of another individual of the same species (the basis of tissue graft or organ transplant rejection) to synthetic molecules that could never have been encountered in nature. Second, the immune response has a kind of memory; after an initial exposure to a given antigen, a second exposure at a later date will result in rapid and much more massive production of the specific antibodies.

Scientists were long perplexed by the immune response. In particular, it seemed difficult to explain how specific antibodies would be generated against millions of different antigens, some of which no organism had ever encountered before. Early theories, called **instructive theories,** suggested that an antigenic determinant could somehow induce an antibody molecule to take up a particular tertiary folding, which would then serve as a binding site for that determinant. However, models of this kind violate the principle we discussed in Chapter 6: higher-order protein structure is dictated entirely by the primary structure. Evidence that the same principle held for antibody binding sites came from experiments in which an antibody to a specific determinant was denatured and then allowed to renature; the capacity to bind the antigen, which had been lost on denaturation, was regained even if refolding took place in the absence of antigen.

Thus, the evidence suggested that an immense diversity of antibodies with different amino acid sequences, able to bind an enormous range of antigens, somehow preexisted in the human body. How such a system can work is described in what is called the **clonal selection theory,** which is now

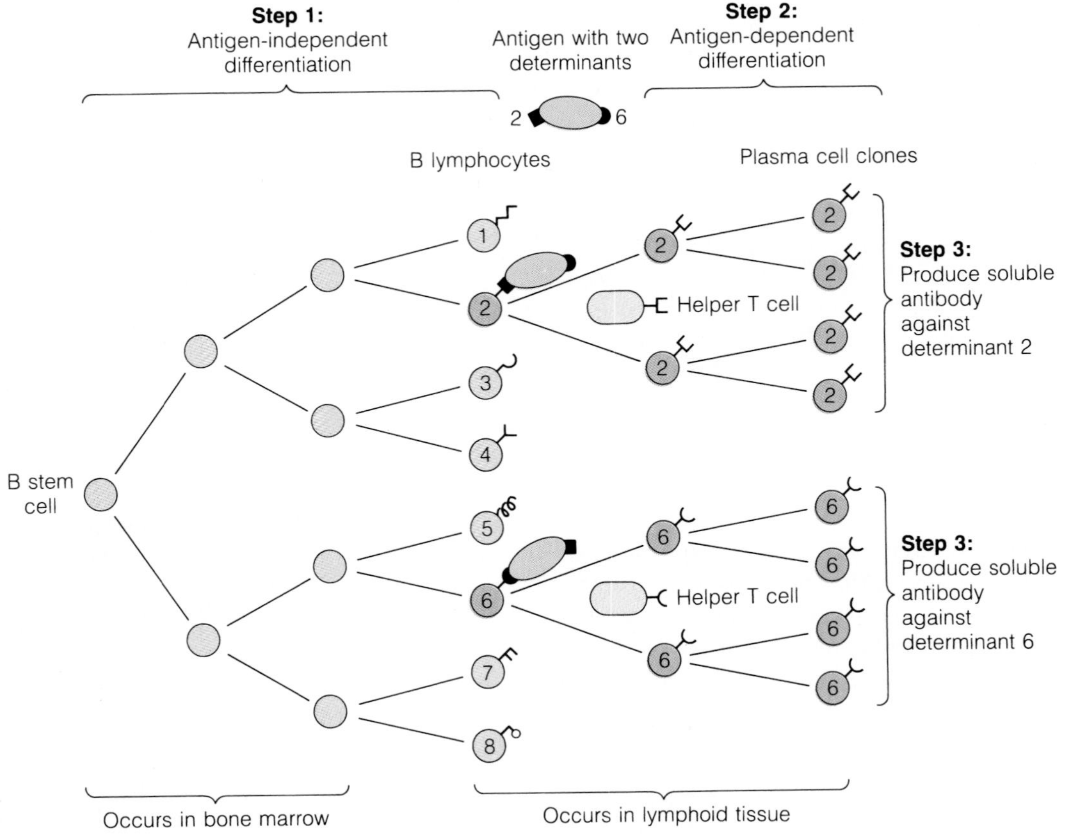

Figure 7.34
The clonal selection theory of the immune response. Stem cells in the bone marrow (left) differentiate and migrate to the lymphoid tissue. Individual cells each synthesize a specific antibody. These antibodies are present on the cell surfaces. When antigens are encountered, the cells carrying antibodies to the antigenic determinants are stimulated to multiply, and some to produce soluble antibodies.

supported by overwhelming evidence. The basic postulates of the clonal selection theory, illustrated in Figure 7.34, are:

1. **B stem cells** in the bone marrow differentiate to become B lymphocytes, each producing a slightly different immunoglobulin molecule, with a binding site that will recognize specific molecular shapes. These antibodies are exposed on the outer surfaces of the B lymphocytes.

2. Binding of an antigen to one of these antibodies stimulates the cell carrying it to replicate, generating a clone of plasma cells. (A **clone** is a collection of cells with identical genetic information.) This process is aided by a special class of T cells called "helper T cells." If these can also recognize the same antigen, they bind to B cells and provide them with a hormone, B cell growth factor. The AIDS virus attacks a class of helper T cells and thereby weakens the entire immune response.

3. Certain cells—**effector cells**—of a particular clone will now produce *soluble* antibodies, which are excreted into the circulatory system. These have the same antigen binding sites as the surface antibodies of the B cell from which they arose, but they lack the hydrophobic tail that bound the surface antibodies to the membranes. Another class of cells in the clone—**memory cells**—will persist for some time, even after antigen is no longer present. This persistence allows the rapid response to a second stimulation by the same antigen characteristic of the immune "memory" (Figure 7.35).

The clonal selection theory explains many features of the immune response and does so in accord with what we know about the determinants of protein folding. But a critical question may have occurred to you—why do we not find clones producing antibodies against *our own* proteins and tissues? The answer is a fascinating one that tells much about how biochemical "self" is established. When immature B cells in the fetus encounter substances that bind to their surface antibodies, they are *not* stimulated to replicate. Rather, they are destroyed. Thus, cells producing antibodies against all of the potential "self" antigens to which we might react are eliminated before birth. The only B cells that can mature are the ones that produce antibodies against substances that will henceforth be regarded as foreign.

To see how clonal selection actually works at the molecular level, we must explore the structure of the immunoglobulin molecules that constitute the antibody arsenal.

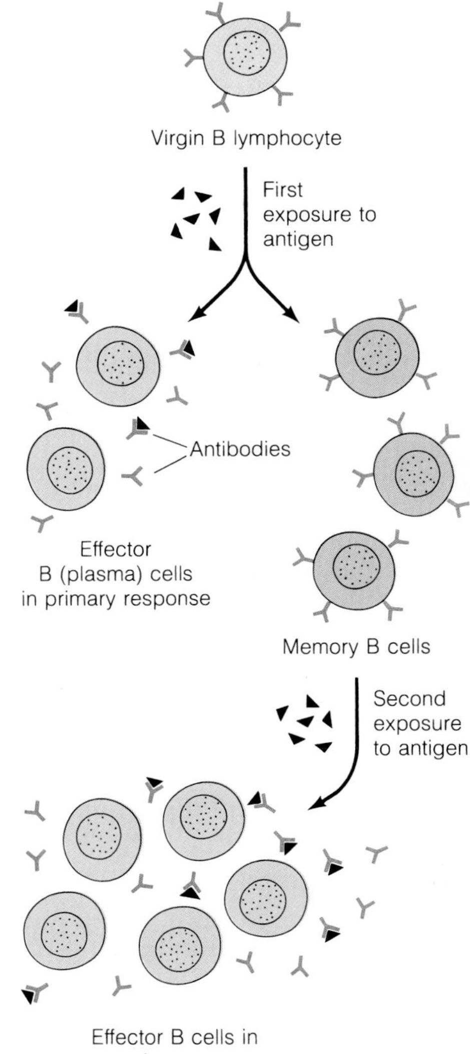

Figure 7.35
Two developmental paths for stimulated B lymphocytes. Exposure to antigen causes two kinds of cells to develop from B lymphocytes. Cells of one type (effector plasma cells) synthesize soluble antibody. Cells of the second class (memory cells) carry membrane-bound antibody to allow a rapid response to a second exposure.

The Structure of Antibodies

There are five classes of antibody molecules, which carry out various functions in the whole immune system (Table 7.2). However, all are built from the same basic immunoglobulin pattern, which is shown in schematic molecular detail in Figure 7.36. Different kinds of antibodies may contain from one to five immunoglobulin molecules; when more than one is present, they are linked by a second type of molecule, called a J chain (see Table 7.2). Each immunoglobulin molecule consists of four chains, two heavy chains ($M = 53{,}000$ each) and two light chains ($M = 23{,}000$ each), held together by disulfide bonds. In each chain there are **constant domains** (identical in all antibodies of a given class) and a **variable domain.** It is variations in the amino acid sequence (and therefore the tertiary structure) of the variable domains of the light and heavy chains that confers the multitudinous specificities of antigens to different determinants.

Note that the variable domains are carried at the ends of the Y-like fork of the molecule. The presence of two binding sites allows each antibody molecule to bind to two antigenic determinants. A structure like a large protein, virus, or bacterial cell will have on its surface many different potential antigenic determinants. Antibodies may be generated to several of them and thereby precipitate the antigen (see Figure 7.33a). If the antigen is so small as to have only one determinant, binding will occur but precipitation will not. Precipitation also requires that the antibody is *bivalent* (has two binding sites). By careful proteolysis, it is possible to cleave antibodies at the "hinge" region (see Figure 7.36) so as to produce **Fab fragments** that have only one binding site each. Such fragments will bind, but not precipitate, antigen.

The antigen binding site lies at the extreme end of the variable domains and involves amino acid residues from the variable regions of both heavy and light chains. Different sequences in these variable regions give rise to different local secondary and tertiary structure and can thereby define binding sites to fit different antigens. Figure 7.37 shows the results of x-ray diffraction studies of the interaction of an Fab fragment with a protein antigen, lysozyme. Close contacts are made between 16 lysozyme residues and 10 heavy chain and 7 light chain residues. The antigen and antibody surfaces fit together in a highly complementary fashion.

The constant domains of the heavy chains in the base of the Y-shaped molecule serve not only to hold the chains together. These regions also function as effectors, to signal macrophages in the circulatory system to

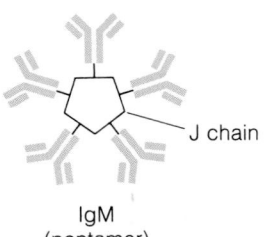

Table 7.2
The five classes of immunoglobulins

J chain IgM (pentamer)	IgM is produced during the early response to an invading microorganism. It is the largest immunoglobulin, containing five Y-shaped units of two light and two heavy chains each. The units are held together by a component called a J chain. The relatively large size of IgM restricts it to the bloodstream.
IgG (monomer)	IgG molecules, also known as *γ-globulin,* are the most abundant of circulating antibodies. A variant is attached to B-cell surfaces. IgG molecules consist of a single Y-shaped unit and can traverse blood vessel walls rather readily; they also cross the placenta to carry some of the mother's immune protection to the developing fetus. IgG also triggers an important mechanism for foreign cell destruction, called the complement system.
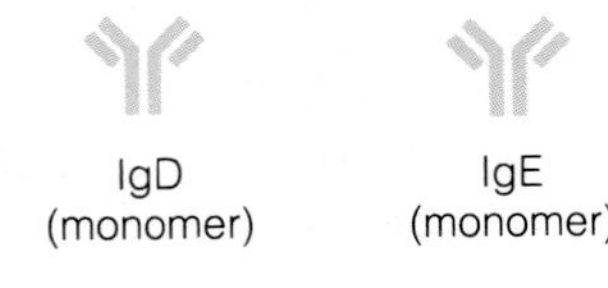 IgA (monomer or dimer)	IgA is found in body secretions, including saliva, sweat, and tears, and along the walls of the intestines. It is the major antibody of colostrum, the initial secretion from a mother's breasts after birth, and of milk. IgA occurs as a monomer or as double-unit aggregates of the Y-shaped protein molecule. IgA molecules tend to be arranged along the surface of body cells and to combine there with antigens, such as those on a bacterium, thus preventing the foreign substance from directly attaching to the body cell. The invading substance can then be swept out of the body together with the IgA molecule.
IgD (monomer) IgE (monomer)	Less is known about the IgD and IgE immunoglobulins. IgD molecules are found on the surface of B cells, though little is known about their function. IgE is associated with some of the body's allergic responses, and its levels are elevated in individuals who have allergies. The constant regions of IgE molecules can bind tightly to mast cells, a type of connective tissue cell that releases histamines as part of the allergic response. Both IgD and IgE consist of single Y-shaped units.

SOURCE: N. A. Campbell (1987) *Biology.* Benjamin/Cummings, Menlo Park, Calif.

attack particles or cells that have been labeled by antibody binding. Macrophages are large white blood cells that are specially adapted to engulf and digest foreign particles. In addition, differences in heavy chains identify immunoglobulin types for delivery to different tissues or for secretion (see Table 7.2).

Generation of Antibody Diversity

A logical question is how the enormous diversity of immunoglobulin molecules is generated so as to provide antibodies to an almost unlimited range of antigens. The human genome simply does not have enough room to encode, individually, the genes for each of the millions of different immunoglobulin molecules required. Instead, two special processes occur in the cells that produce antibodies.

The major source of antibody diversity is *recombination of exons.* The genomes of higher vertebrates contain "libraries" of exons corresponding to different portions of the immunoglobulin molecule and mechanisms for rearranging these exons to create different combinations in both the heavy

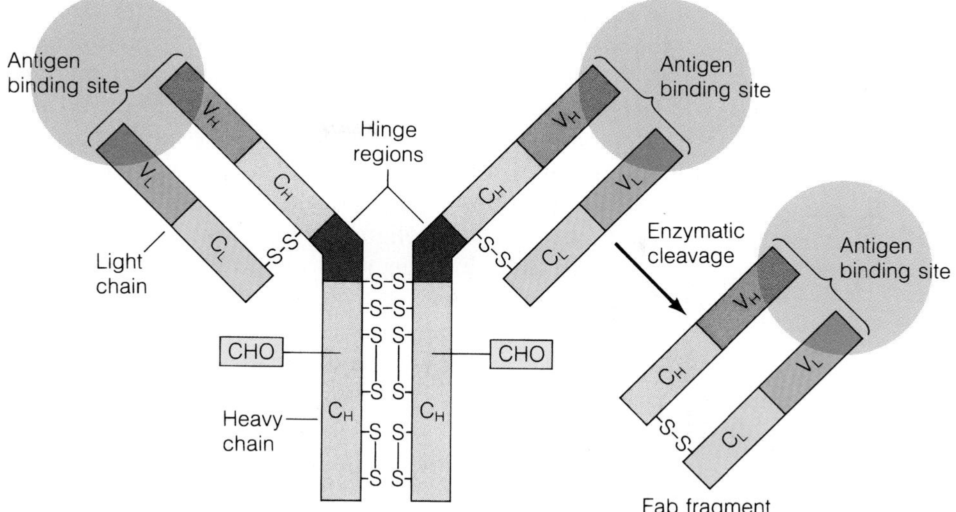

Figure 7.36
Schematic model of an antibody molecule. The molecule is made from two heavy chains and two light chains, all held together by disulfide bonds. Each chain contains both *constant* domains (C), which are the same in all antibody molecules of a given class (see Table 7.2), and *variable* domains (V), which convey the specificity to a given antigenic determinant. Cleavage by proteolytic enzymes at the *hinge region* allows production of monovalent *Fab fragments*. The carbohydrate (CHO) attached to the heavy chains aids in determining the destinations of antibodies in the tissues and in stimulating secondary responses such as phagocytosis. A molecular model of an immunoglobulin molecule, as derived from x-ray diffraction studies, is shown in Figure 1.6.

and light chains. We have already referred to such rearrangements, occurring in the germ-line cells which produce sperm and ova, as playing a role in protein evolution. The same process, occurring now in certain **somatic** (body) cells, creates new immunoglobulins in individual cells. The details and mechanism of this process will be described in Chapter 25; it suffices for now to note that it has been calculated that about 10 million combinations can be made from the library of immunoglobulin gene fragments available in the human genome.

An additional source of antibody diversity is *somatic mutations,* mutations that are not inherited, since they occur in somatic cells. In the cells that generate antibodies, certain portions of the variable regions in the immunoglobulin genes mutate at an unusually high rate. The reason for these localized high rates of mutation is still obscure, but this process, together with recombination of gene fragments, can account for the generation of an immense diversity of immunoglobulin molecules.

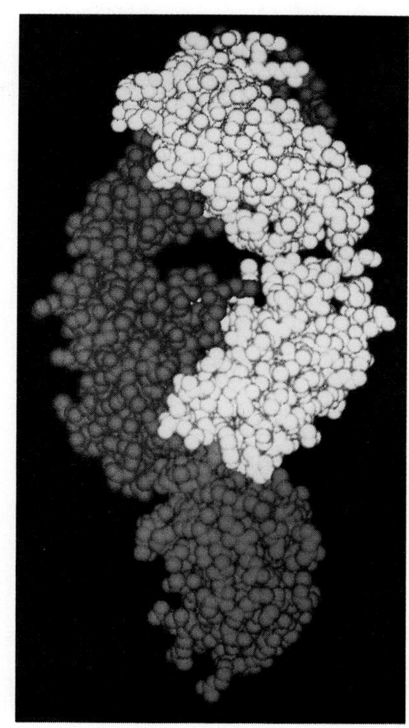

Figure 7.37
Binding of an Fab fragment to a specific antigen, the protein lysozyme. The antibody light chain is shown in yellow, the heavy chain fragment in blue, and the lysozyme molecule in green. The red dot corresponds to Glu 121 of lysozyme, which fits snugly into a cleft between the light and heavy chains of the antibody. The surfaces of the protein and the binding site of the antibody fit closely together, and a number of specific hydrogen bonds are made across this surface.

T Cells and the Cellular Response

Whereas the humoral immune response is based on precipitation—frequently followed by digestion by macrophages—the cellular response involves a quite different mechanism for killing of foreign cells. The cellular response plays a major role in tissue rejection and in destroying virus-infected cells. Although the mechanisms of the two processes are quite dif-

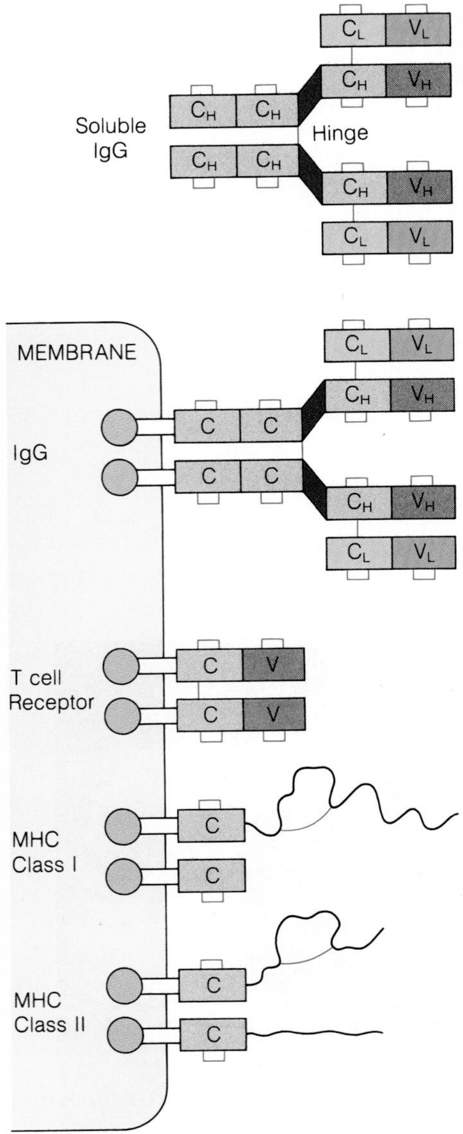

Figure 7.38
Similarity in structure between members of the immunoglobulin superfamily of proteins. C and V indicate constant and variable domains, each with a structure similar to that shown in Figure 7.39. MHC indicates major histocompatibility complex. Connecting disulfide bridges are shown in blue. Circles represent membrane-binding domains.

ferent, similar immunoglobulin molecules are involved in both cases, pointing to a common evolutionary origin for the humoral and cellular responses.

The major participants in the cellular immune response are **cytotoxic T cells,** also referred to as **killer T cells.** These cells carry on their surfaces receptor molecules that have considerable similarity to the Fab fragments of antibody molecules (Figure 7.38). Like antibodies, these fragments have a wide range of binding specificities, mostly directed toward short oligopeptide sequences. Such oligopeptides might be produced, for example, by a virus-infected cell when it partially digests virus particles within it. The T-cell receptor does not recognize *free* oligopeptides—they must be presented on the surface of the cell, bound to another class of immunoglobulin-like molecules, proteins of the **major histocompatibility complex (MHC** proteins; see Figure 7.38). When a killer T cell identifies (via its receptor) an appropriate antigen carried on the surface of another cell by an MHC protein, it releases a protein called **perforin.** This protein forms pores in the plasma membrane of the cell being attacked, allowing critical ions to diffuse out and killing the cell.

As Figure 7.38 emphasizes, there are elements of common structure in both the antibodies of the humoral response and the molecules involved in the cellular response. The similarity is even closer than the figure suggests. The domains in these molecules are built on a common motif—the "immunoglobulin fold" (Figure 7.39). This β-sheet structure probably represents the primitive structural element in the evolution of the immune response. Indeed, the immunoglobulin fold is also found in a number of other proteins that are involved in cell recognition.

It is instructive to compare the immunoglobulin family of proteins with the myoglobin–hemoglobin family. In both cases the primary function of the proteins is binding. In the myoglobin–hemoglobin family we see evidence for the progressive evolution of more and more sophisticated methods for regulating the binding of a particular molecule—oxygen—and coupling oxygen binding to that of CO_2. In the immunoglobulin family, evolution has led to an enormous diversification in binding function. A mechanism has evolved that allows the production of an immense range of molecules with different binding capacities. Different tasks require different tools.

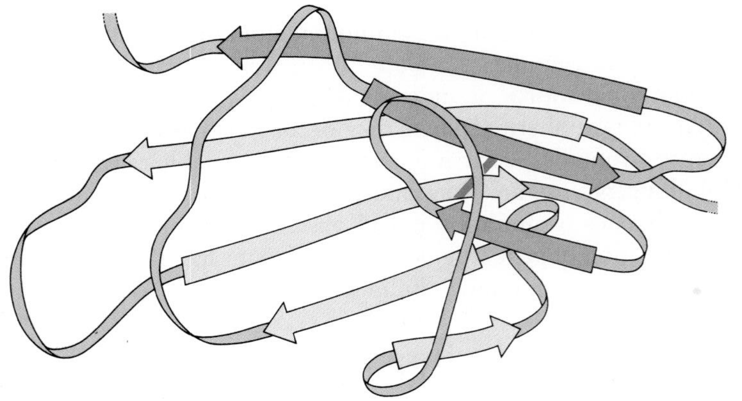

Figure 7.39
The immunoglobulin fold. This is a common structure in domains of many proteins in the immunoglobulin superfamily. Two layers of antiparallel β sheet are stacked face to face. The disulfide bridge is shown in blue.

REFERENCES

General

Dickerson, R. E., and I. Geis (1983) *Hemoglobin: Structure, Function, Evolution, and Pathology*. Benjamin/Cummings, Menlo Park, Calif. A wealth of detail on myoglobin and hemoglobin, including much on hemoglobin evolution and variants.

King, J. (1989) Deciphering the rules of protein folding. *Chem. Engr. News* 67:32–54. A well-written and illustrated review describing genetic and biophysical approaches to the folding problem.

van Holde, K. E. (1985) *Physical Biochemistry*, 2nd ed. Prentice-Hall, Englewood Cliffs, N.J. Chapter 3 contains a more detailed discussion of binding equilibrium than is presented here.

Allosteric Models

Imai, K. (1983) The Monod–Wyman–Changeux model describes haemoglobin oxygenation with only one adjustable parameter. *J. Mol. Biol.* 167:741–749. A careful application of the MWC model.

Koshland, D. E., G. Nemethy, and D. Filmer (1966) Comparison of experimental binding data and theoretical models in proteins containing subunits. *Biochemistry* 5:365–385.

Monod, J., J. Wyman, and J.-P. Changeux (1965) On the nature of allosteric transitions: A plausible model. *J. Mol. Biol.* 12:88–118.

The papers by Koshland et al. and Monod et al. presented the major classes of models for allostery.

Dynamics of Oxygen Binding and Release

Case, D. A., and M. Karplus (1979) Dynamics of ligand binding to heme proteins. *J. Mol. Biol.* 132:343–368. A pioneering paper on protein dynamics.

Antibodies

Darnell, J., H. Lodish, and D. Baltimore (1986) *Molecular Cell Biology*, Chapter 24. Scientific American Publishing Company, New York. An elegant, detailed presentation.

Tizard, I. (1984) *Immunology, an Introduction*. Saunders College Publishers, Philadelphia.

Tonegawa, S. (1985) The molecules of the immune system. *Sci. Am.* 253:(4)122–130. Very readable and concise description.

PROBLEMS

1. The following data describe the binding of oxygen to human myoglobin at 37°C.

P_{O_2} (mm Hg)	θ	P_{O_2} (mm Hg)	θ
0.5	0.161	6	0.697
1	0.277	8	0.754
2	0.434	12	0.821
3	0.535	20	0.885
4	0.605		

From these data estimate (a) P_{50}, (b) the fraction saturation of myoglobin at 30 mm, the partial pressure of O_2 in venous blood.

2. Measurements of oxygen binding by whole human blood, at 37°C, pH 7.4, and in the presence of 40 mm of CO_2 and normal physiological levels of BPG (5 mmol/liter of cells), give the following.

P_{O_2} (mm Hg)	% saturation (= 100 × θ)
10.6	10
19.5	30
27.4	50
37.5	70
50.4	85
77.3	96
92.3	98

(a) From these data, construct a binding curve, and estimate the percent oxygen saturation of blood at
   1. 100 mm Hg, the approximate partial pressure of O_2 in the lungs
   2. 30 mm Hg, the approximate partial pressure of O_2 in venous blood

(b) Under these conditions, what percentage of the oxygen bound in the lungs is delivered to the tissues?

(c) Using the data in Figure 7.19, repeat the calculation of part (b) if the pH drops to 6.8 in capillaries but goes back to 7.4 as CO_2 is unloaded in the lungs.

3. It is observed that chloride ion acts as a negative allosteric effector for hemoglobin. On the basis of Figure 7.16, explain why this should be so.

4. Precise data have been obtained for the oxygen binding of stripped human hemoglobin at 25°C.

P_{O_2} (mm Hg)	% saturation (= 100 θ)	P_{O_2} (mm Hg)	% saturation
0.10	0.315	5.75	76.0
0.350	0.990	7.94	90.9
0.794	3.06	12.88	96.9
1.748	9.09	29.51	99.0
2.884	24.0	67.60	99.7
4.467	50.0		

Use a Hill plot to determine (a) P_{50}, (b) n_H (maximum slope), and (c) the P_{50} values corresponding to the T and R states.

5. G. Ackers, M. L. Johnson, F. C. Mills, et al., (*Biochemistry* 14:5128–5134 (1975)) have observed that the P_{50} of purified hemoglobin decreases as the concentration of hemoglobin in solution is decreased. Suggest an explanation.

6. Oxygen binding by the hemocyanin of the shrimp *Callienassa* has been measured. Using the data below, prepare a Hill plot and determine (a) P_{50}; (b) n_H, the Hill coefficient; and (c) the *minimum* number of oxygen binding sites on the protein molecule.

P_{O_2} (mm Hg)	θ	P_{O_2} (mm Hg)	θ
1.1	0.003	136.7	0.557
7.7	0.019	166.8	0.673
10.7	0.035	203.2	0.734
31.7	0.084	262.2	0.794
71.9	0.190	327.0	0.834
100.5	0.329	452.0	0.875
123.3	0.487	736.7	0.913

7. Suggest probable consequences of the following real or possible hemoglobin mutations. [Note: consult Figures 7.16, 7.18, and 7.21.]
   (a) at β 146 (HC3) His → Asp
   (b) at β 92 (F8) His → Leu
   (c) at α 141 (HC3) Arg → Leu
   (d) at β 2 (NA2) His → Asp

8. The binding of antigens to antibodies can be thought of in ways similar to those we have used for discussing oxygen binding. Assuming we have *monovalent* antigens and n-*valent* antibodies, we can write for the number (r) of antigen molecules bound to an antibody at a concentration of free antigen (c) the equation:

$$r = \frac{nKc}{1 + Kc}$$

where K is the equilibrium constant for the binding (the affinity constant).

(a) Show that this equation can be rearranged to give

$$\frac{r}{c} = Kn - Kr$$

This is called the *Scatchard equation.* It predicts that a graph of r/c versus r will be a straight line.

(b) Use the following data and a graph according to the Scatchard equation to obtain n and K for an antibody–antigen reaction.

c (M)	r
1.43×10^{-5}	0.50
2.57×10^{-5}	0.77
6.00×10^{-5}	1.20
1.68×10^{-4}	1.68
3.70×10^{-4}	1.85

12

TOOLS OF BIOCHEMISTRY 12

Immunological Methods

Because of the ease with which antibodies against biological materials can be prepared, and because of their great specificity, antibodies form the core of many important analytical and preparative biochemical procedures. We shall discuss here some of the methods most important to biochemists. Other immunological techniques are described in Tools of Biochemistry 17 (radioimmunoassay).

Experiments in which an antigen is injected into a rabbit show that one antigen can elicit formation of several different antibodies, each of which recognizes one particular portion of the antigen molecule, called an **antigenic determinant,** or **epitope.** Figure 7.33b shows the epitopes that have been identified in sperm whale myoglobin. Each is a region encompassing five or six residues in the myoglobin sequence. Thus, an **antiserum** against myoglobin, or serum from an animal immunized against myoglobin, carries at least five different antimyoglobin antibodies, each directed against one of the five epitopes shown in the figure.

Generally speaking, immunogenic substances (those that elicit an immune reaction, or synthesis of antibodies) are macromolecules. However, some of the most useful immunological reagents are antibodies against low-molecular-weight substances. For example, radioimmunoassay is used to test for levels of steroid hormones or drugs, compounds that do not themselves stimulate antibody synthesis. However, when such a compound is coupled to an antigenic protein, the resultant conjugate is immunogenic, and some of the antibodies are directed against the low-molecular-weight, nonprotein constituent, which is called a **hapten.**

Most antibodies that are useful in biochemistry are of the IgG type (see Table 7.2). These are Y-shaped molecules, each with two antigen-combining sites. In an antigen–antibody reaction each site usually binds to a different antigen molecule if sufficient antigen is present. If the antigen has two or more antibody-combining sites, as most do, then a giant insoluble network of linked antigen and antibody molecules, called a **precipitin reaction,** is formed (see Figure 7.33a). The insoluble nature of this complex is at the heart of some of the older analytical techniques, such as immunodiffusion and immunoelectrophoresis.

In **immunodiffusion,** wells in an agar plate are filled with antigen and antibody. The antigen and antibody diffuse toward each other in the agar, and where they meet a precipitin reaction forms an insoluble precipitate, which can be visualized either directly or after staining with Coomassie blue or similar protein stain (Figure T12.1). This technique can be used to compare different antigens. For example, suppose you wanted to ask whether the protein cytochrome *c* from yeast was related to that from horse heart. You could place yeast cytochrome *c* in one well, the horse protein in another, and antibody to the horse protein in a third. If the proteins are closely related, the precipitin lines will intersect, as shown in part a of the figure, usually with a "spur" pointing toward the less reactive antigen (in this case the yeast protein). If the two antigens are unrelated, but each reacts with an antibody in the antiserum that was placed in the well, two precipitin lines are formed that simply cross each other (part b). If the proteins were immunologically identical, a single, unbroken precipitin line would be found (part c). Finally, if the antibody binds to one but not at all to the other, only a short line is seen (part d).

Immunodiffusion can be used to characterize mutant organisms that may be defective in synthesis of a biologically active protein. The presence of a cross-reacting antigen, such as seen in the pattern of Figure T12.1a, suggests that a complete protein is made but that one of the residues essential for biological activity has been changed by the mutation.

Immunoelectrophoresis allows analysis of more complex mixtures of antigens and antibodies. Here proteins in a complex mixture, such as serum, are subjected to electrophoresis, after which antibody is applied in a trough, as shown in Figure T12.2. This method provides many different precipitin reactions in one experiment, and with relatively simple variations it can be adapted to quantitate antigens in a mixture.

Much more widely used now to quantitate antigen–antibody reactions is the **enzyme-linked immunosorbent assay (ELISA).** Here one uses a cross-linking reagent to couple covalently an enzyme that can be easily assayed spectrophotometrically to an antibody that is being used as an assay for a particular substance. Although this method has many variations, the principle is to assay for bound antibody simply by analyzing for the activity of the conjugated enzyme. This technique forms the basis of many clinical diagnostic tests, such as the most widely used current test for AIDS infection. An enzyme-linked antibody to one of the surface proteins of the human immunodeficiency virus is used to detect the presence of the viral antigen in human blood samples.

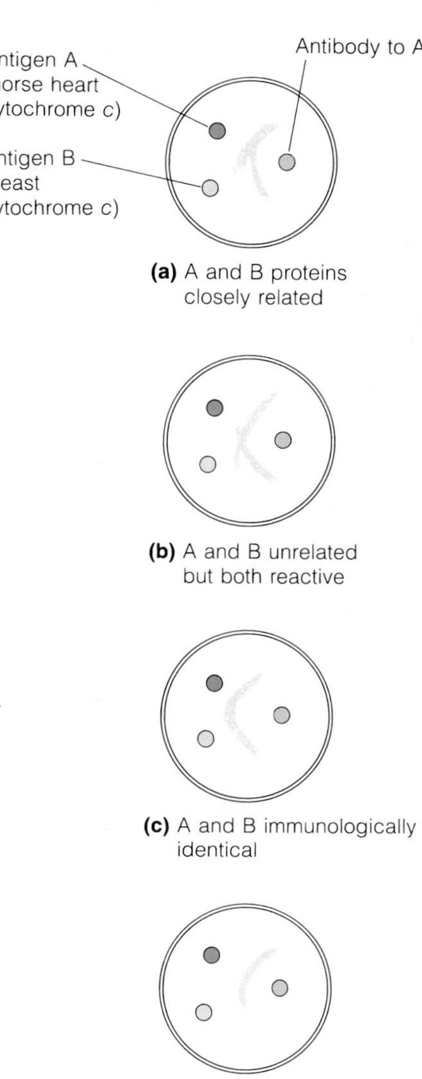

Figure T12.1
Gel diffusion precipitin test.

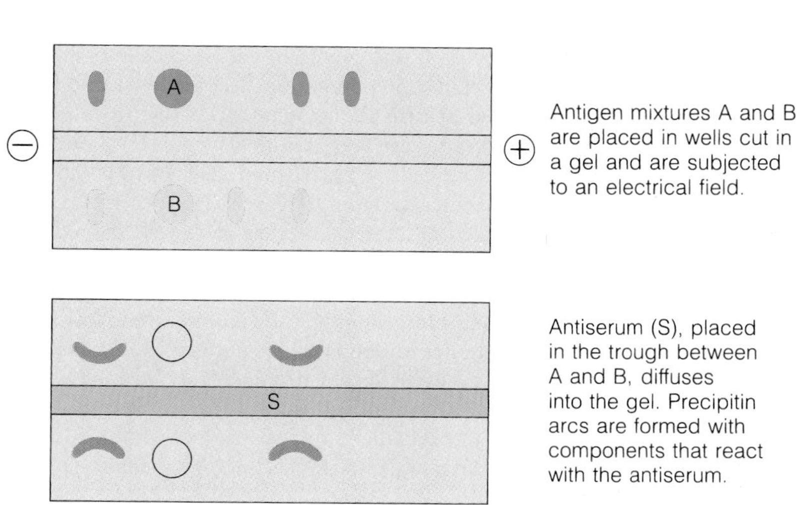

A
B
S
Antigen mixtures A and B are placed in wells cut in a gel and are subjected to an electrical field.
Antiserum (S), placed in the trough between A and B, diffuses into the gel. Precipitin arcs are formed with components that react with the antiserum.

Figure T12.2
Immunoelectrophoresis.

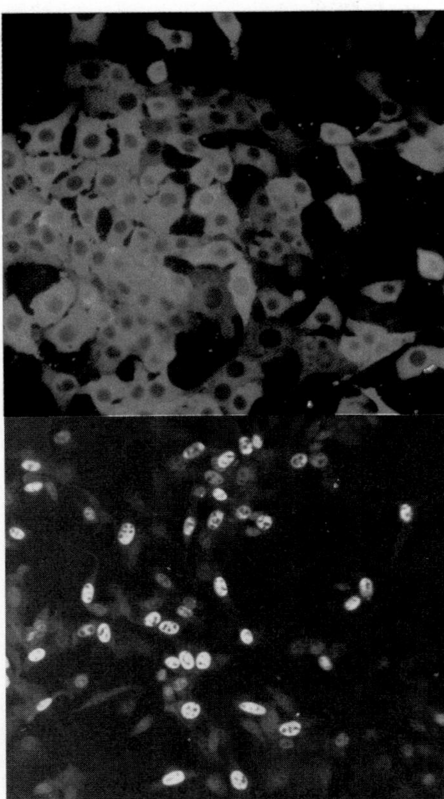

Figure T12.3
Use of immunofluorescence microscopy to differentiate between cytosolic (top) and nuclear (bottom) proteins in cultured 3T6 mouse cells.

Another recently introduced analytical technique is **Western blotting,** so called because of its superficial resemblance to a nucleic acid analytical technique called Southern blotting (described in Chapter 25). Western blotting is used to detect, in a mixture of proteins or fragments of proteins, those that react with the same antibody. It can be used, for example, in studying the posttranslational cleavage of proteins that is known to occur often as part of protein maturation. Here one analyzes the antibody-reactive proteins in a mixture by first resolving the proteins in that mixture by one-dimensional denaturing gel electrophoresis. After electrophoresis, the gel is placed in contact with a sheet of nitrocellulose and the proteins are transferred (or "blotted") to the nitrocellulose by an electric current. The proteins are bound irreversibly to the nitrocellulose sheet, so one can visualize antigen–antibody reactions after treatment of the sheet with antibody. Visualization involves either the use of antibody radiolabeled with iodine-125 followed by radioautography, or a form of the ELISA technique.

Because of their high specificity in protein binding, antibodies can also be used to purify proteins. In this technique, called **immunoaffinity chromatography,** one couples the antibody to a chromatographic support and uses a column of this material to adsorb selectively the protein being purified. The protein is then desorbed, usually by a pH adjustment in the eluting solution and often in a state very close to homogeneity.

Biochemists, like cell biologists, must be concerned with intracellular organization and with the location of enzymes that catalyze reactions of interest. An array of techniques with the generic term **immunocytochemistry** uses antibodies to help localize particular antigens in cytological preparations. In the simplest form, an antibody is conjugated with a fluorescent dye such as fluorescein. A thin section of cell or tissue is then immersed in a solution of the fluorescent antibody. After the excess is washed off, the bound antibody can be visualized by fluorescence microscopy (Figure T12.3). In the upper panel of this figure, a fluorescent antibody has been made against a cytoplasmic protein; in the lower panel a nuclear protein is probed. Alternatively, the antibody can be linked to the iron-binding protein ferritin, and the bound iron can be visualized from its high electron density in the electron microscope. Another technique involves the binding of antibody linked to horseradish peroxidase, an enzyme that can be visualized by either light- or electron-microscopic techniques.

As noted earlier, an antiserum prepared against a pure antigen, such as a protein, usually contains several different antibodies against that antigen. Moreover, the serum also contains all other antibodies that the animal carried in its bloodstream at the time that it was immunized against the protein of interest. Clearly the specificity, sensitivity, and reproducibility of all of the foregoing techniques would be increased greatly if they could be carried out with *pure* antibodies. Because of the chemical similarities among different IgGs, their purification by standard protein fractionation techniques is virtually impossible. Fortunately, there is an alternative way to attain the same goal, namely by use of **monoclonal antibodies.**

Each antibody-forming B lymphocyte is specialized for the synthesis and secretion of one and only one antibody. If one could grow these cells in tissue culture, one could in principle isolate a clone, or population derived from one cell, that synthesized one antibody directed against the protein of interest. Although these cells do not grow in culture, Georges Kohler and Cesar Milstein in 1975 discovered a way to propagate antibody-forming cells with cultured cells from a mouse with multiple myeloma, a neoplastic proliferation of white cells (Figure T12.4). One first immunizes a mouse with the antigen of interest and then fuses cells from the spleen of this mouse with the myeloma cells. This fusion gives rise to cell lines called **hybridomas**—cells that proliferate endlessly in culture, like cancer cells, but synthesize only one antibody. By screening a large number of clones resulting from a fusion and searching for those that synthesize antibody to the antigen of interest, one can usually isolate several hybridomas, each of which makes and secretes a different antibody reactive against the antigen. Antibodies can easily be purified from cultures of the appropriate hybridomas, and they have many uses in addition to the techniques described above. Another advantage of this technology is that one need not

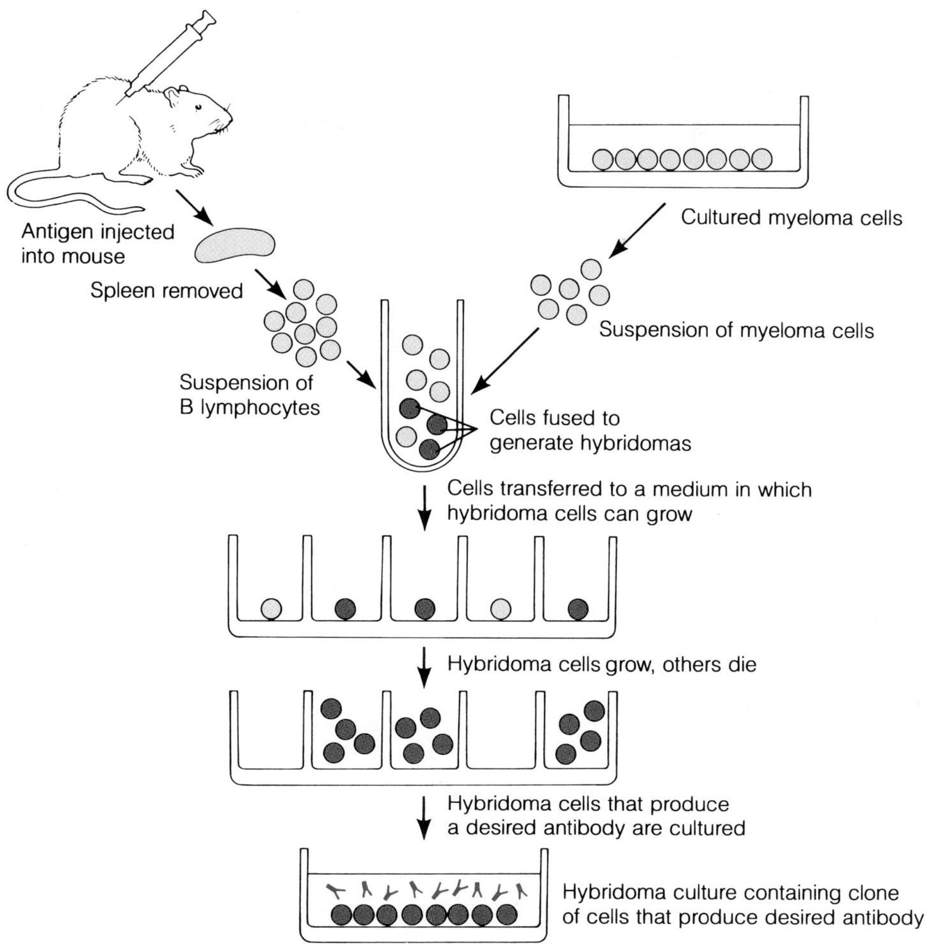
Antigen injected
into mouse
Spleen removed
Suspension of
B lymphocytes
Cultured myeloma cells
Suspension of myeloma cells
Cells fused to
generate hybridomas
Cells transferred to a medium in which
hybridoma cells can grow
Hybridoma cells grow, others die
Hybridoma cells that produce
a desired antibody are cultured
Hybridoma culture containing clone
of cells that produce desired antibody

Figure T12.4
Production of monoclonal antibodies.

use a homogeneous antigen. With sufficient patience to examine many hybridomas, one can immunize the animal at the outset with a crude preparation of the antigen of interest. Later, after a desired monoclonal antibody has been obtained, it can be used as the basis of an immunoaffinity purification of the antigen.

REFERENCE

Harlow, E., and D. Lane (1988) *Antibodies, a Laboratory Manual.* Cold Spring Harbor Laboratory, Cold Spring Harbor, New York.

Weir, D. M. (ed.) (1986) *Handbook of Experimental Immunology.* Oxford University Press, London/New York. In addition to the books listed in the references in Chapter 7, this provides a good description of current techniques.

CHAPTER 8

Carbohydrates

We turn now to the third great class of biological molecules, the **carbohydrates**, or **saccharides.** The simplest carbohydrates are small, monomeric molecules—the **monosaccharides,** or simple sugars such as glucose (Figure 8.1a). Many of the most important carbohydrates are formed by linking such monosaccharides together. If only a few monomer units are involved, we call the molecule an **oligosaccharide.** An example is **maltose** (Figure 8.1b), a **disaccharide** made by linking two glucose molecules together. Long polymers of the monosaccharides, like the starch **amylose** (Figure 8.1c), are called **polysaccharides.** Since there are many kinds of monosaccharides, polysaccharides may be very complex polymers.

Saccharides are often called by the more common name **carbohydrates** because many of them can be represented by the simple stoichiometric formula $(CH_2O)_n$. This formula is an oversimplification, however, for many saccharides also contain amino, sulfate, and phosphate groups.

The saccharides play an enormous variety of roles in living organisms. Indeed, much of the basic energy cycle on earth is built on carbohydrate metabolism. Before we turn to carbohydrate structure, let us look briefly at this cycle.

In **photosynthesis** plants take up CO_2 from the atmosphere and "fix" it into carbohydrates. The basic reaction can be described (in a vastly simplified way) as the reduction of CO_2, driven by the energy of sunlight:

$$nCO_2 + nH_2O \xrightarrow{\text{Light energy}} \underset{\textbf{Carbohydrate}}{(CH_2O)_n} + nO_2$$

Much of this carbohydrate is stored in the plants as starch or cellulose. Animals eat plants or other plant eaters and thus obtain the carbohydrates. Thus plant-generated carbohydrates ultimately become the principal sources of carbon in all animal tissues. Both plants and animals carry out, via oxidative metabolism, a reaction that is the reverse of photosynthesis:

$$nO_2 + (CH_2O)_n \longrightarrow nCO_2 + nH_2O$$

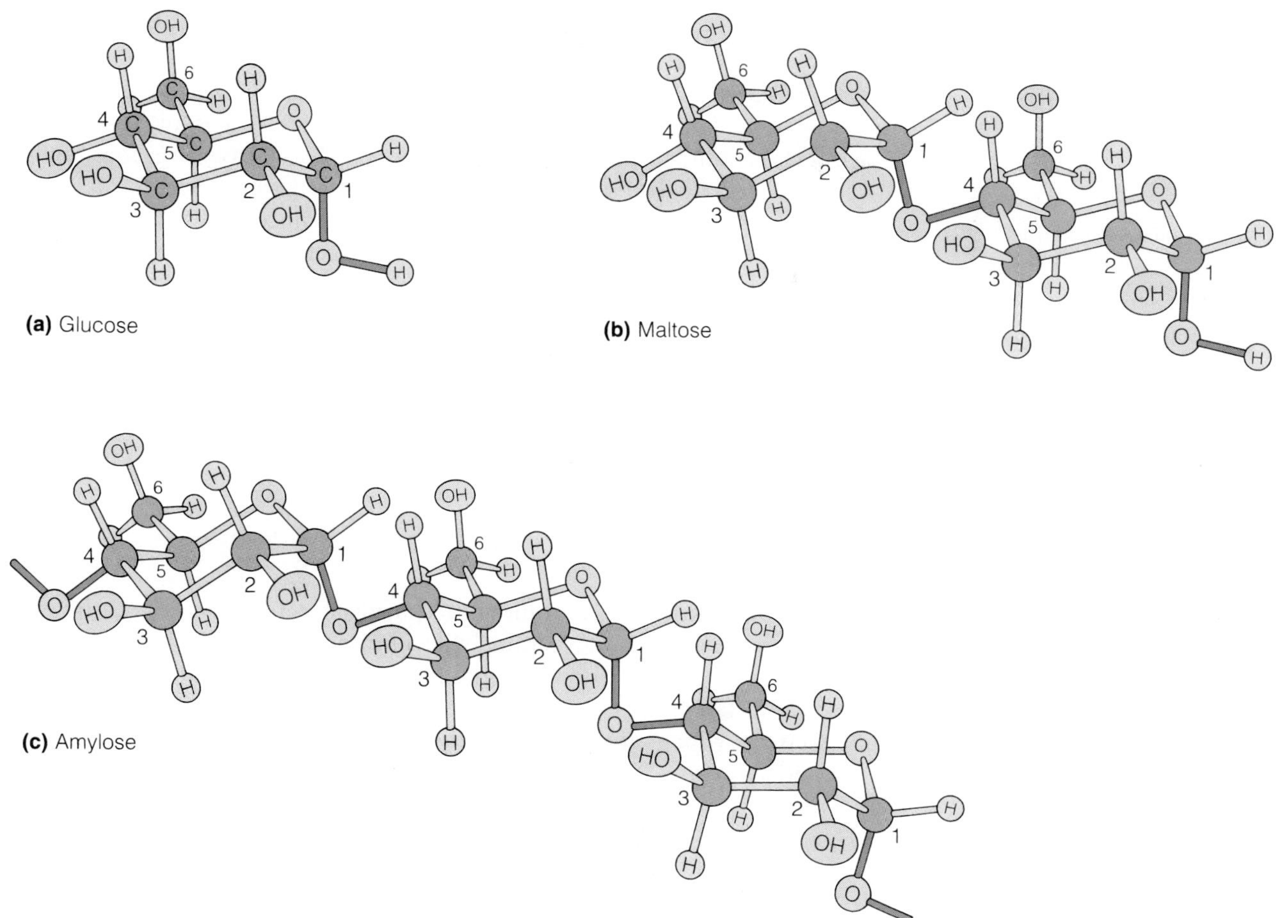

Figure 8.1
Representative carbohydrates. (**a**) A monosaccharide, glucose; (**b**) a disaccharide, maltose, containing glucose units; and (**c**) a portion of a polysaccharide, amylose. Amylose, a form of starch, is a polymer of glucose.

This oxidation of carbohydrates is the primary energy-generating process in metabolism. Of course, there is much, much more to metabolism than is indicated here, as we shall see in later chapters. But the central role of carbohydrates is obvious when one considers that a basic foodstuff of humans is the starch in grains such as rice or wheat. Even the meats we consume ultimately can be traced in large part to the carbohydrates eaten by grazing animals.

As critical as energy storage and generation are, they are not the only functions of carbohydrates. Many biological structural materials are polysaccharide in whole or in part; important examples are the cellulose of woody plants, the cell walls of bacteria, and the exoskeletons of insects. Polysaccharides on cell surfaces or attached to proteins play a different role: they aid in molecular recognition. Thus, like proteins, carbohydrates are extremely versatile molecules, essential to every kind of living organism.

Monosaccharides

We begin with the simple, monomeric sugars, the monosaccharides. The simplest compound with the empirical formula of the class $(CH_2O)_n$ is found when $n = 1$. However, **formaldehyde,** $H_2C{=}O$, has little in common with our usual concept of sugars; indeed, it is a noxious, poisonous gas. The smallest molecules usually regarded as monosaccharides are those with $n = 3$: **glyceraldehyde** and **dihydroxyacetone** (Figure 8.2). These molecules, simple as they are, already exhibit certain features that we shall encounter again and again in discussing sugars. Glyceraldehyde is an aldehyde, one of a class of monosaccharides called **aldoses.** (The suffix *ose* is commonly used to designate compounds as saccharides.) Dihydroxyacetone is a ketone; such monosaccharides are termed **ketoses.** Since each contains three carbons, both glyceraldehyde and dihydroxyacetone are called **trioses.** You will note that glyceraldehyde and dihydroxyacetone have the same atomic composition. They are **tautomers** and can interconvert via an unstable **enediol** intermediate, as shown in Figure 8.3. Such interconversions occur to a certain extent between all such pairs of aldose and ketose monosaccharides, but the reactions are usually very slow unless catalyzed. Thus, glyceraldehyde and dihydroxyacetone, although interconvertible, can each exist as a quite stable compound.

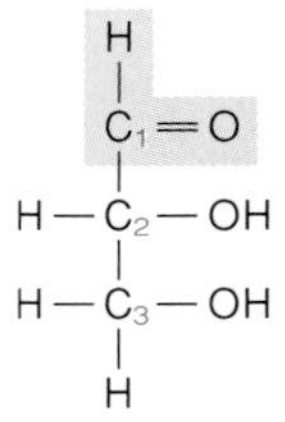

D-Glyceraldehyde (an aldose)

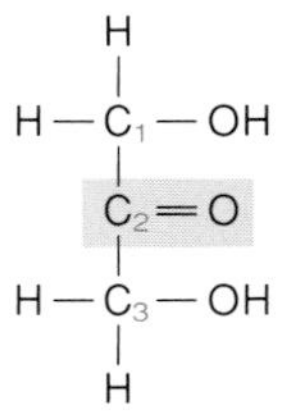

Dihydroxyacetone (a ketose)

Figure 8.2
The simplest monosaccharides. The figure illustrates the difference between aldose and ketose forms. These two molecules are tautomers. Carbon numbering begins in all aldoses with aldehyde, and in ketoses with the end carbon closest to the ketone group.

Stereoisomerism

An important feature of monosaccharide structure can be seen by examining the formula for glyceraldehyde a bit more carefully. The second carbon atom is a **chiral** carbon, like the α-carbon in most α-amino acids. It carries four different substituents. Therefore, there are two **enantiomers** of glyceraldehyde, which are mirror images of one another and are designated as the D and L forms. Three-dimensional drawings of D- and L-glyceraldehyde are shown in Figure 8.4. A more compact way to represent these is by what is called a **Fischer projection.** In a Fischer projection the hydroxyl is drawn to the right of the carbon for the D form, to the left for the L form. Thus, for D- and L-glyceraldehyde we have

D-Glyceraldehyde	**L-Glyceraldehyde**	Carbon number:
CHO	CHO	1
H—C—OH	HO—C—H	2
CH_2OH	CH_2OH	3

Note that we do not need to draw any special orientation about carbons 1 or 3 because they are not chiral centers.

Originally, the terms D and L were meant to indicate the direction of rotation of the plane of polarization of polarized light: D for right (dextro), L for left (laevo). It is true that a solution of D-glyceraldehyde does rotate the plane of polarization to the right, as do many other D-monosaccharides, but

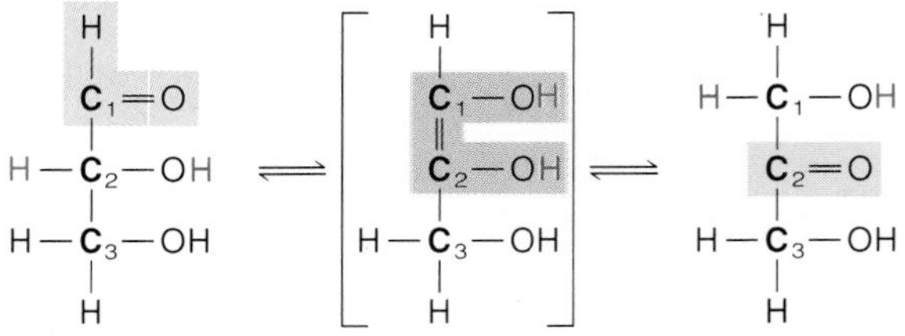

D-Glyceraldehyde (aldotriose) ⇌ Enediol intermediate ⇌ Dihydroxyacetone (ketotriose)

Figure 8.3
Aldose–ketose interconversion via an enediol intermediate. The intermediate is unstable and cannot be isolated.

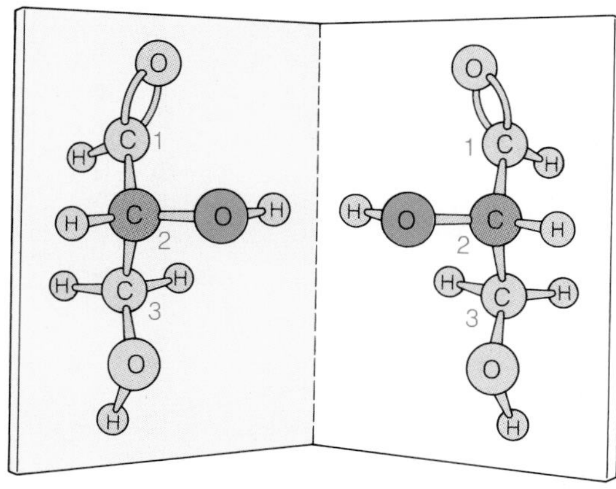

D-Glyceraldehyde L-Glyceraldehyde

Figure 8.4
The enantiomers of glyceraldehyde. The configuration of groups around the chiral carbon 2 (shown in black) distinguishes D from L.

this correspondence does not always hold. The magnitude and even the direction of optical rotation are a complicated function of the electronic structure surrounding the chiral center. For example, the common enantiomer of the sugar fructose (p. 269) is termed D because of the stereochemical orientation about the appropriate carbon. But this enantiomer actually rotates the plane of polarization to the left, and its mirror image, L-fructose, rotates the plane of polarization to the right. Another disadvantage of the D–L nomenclature is that it is not absolute; designation is always with respect to some reference compound. Accordingly, an *absolute* convention has been developed that allows us to assign a stereochemical designation to any compound from examination of its three-dimensional structure. An example of how this so-called R–S convention works is shown in Figure 8.5. Despite its limitations, the D–L terminology is still commonly used by biochemists.

Diastereoisomers

When we consider the structures of the **tetroses**—monosaccharides with the empirical formula $(CH_2O)_4$—a further structural complication appears. There are *two* chiral carbons in the aldose forms. Therefore, there will be *four* stereoisomers of an aldotetrose, as shown in Figure 8.6. In general, a molecule with n chiral centers will have 2^n stereoisomers, because there are two choices at each chiral center. The following convention attempts to give a rational method for naming and distinguishing them: The prefix D or L

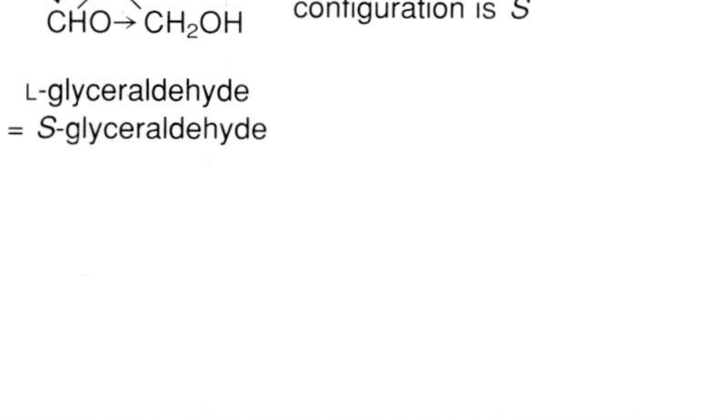

H OH
C
$HOCH_2 \leftarrow CHO$

D-glyceraldehyde
= *R*-glyceraldehyde

Rotate molecule so group of lowest priority (H) faces away

If priority of remaining groups decreases in clockwise direction, configuration is *R*

HO H
C
$CHO \rightarrow CH_2OH$

L-glyceraldehyde
= *S*-glyceraldehyde

Priority decreases in counterclockwise direction, configuration is *S*

Figure 8.5
Definition of absolute stereochemical configuration using the RS system. Each type of group attached to a chiral carbon (gray) is given a *priority,* according to a set of defined rules. Priorities for groups common in carbohydrate chemistry are —OR > —OH > NH_2 > CO_2H > CHO > CH_2OH > CH_3 > H. We then view the molecule with the group of lowest priority away from us. In the case here (glyceraldehyde), this will be H. If the priority of the remaining three groups *decreases* in a *clockwise* order, the absolute configuration is called *R* (from Latin *rectus* = right). If priority decreases in a *counterclockwise* order, the configuration is *S* (Latin *sinister* = left). Thus, in this notation D-glyceraldehyde is *R*-glyceraldehyde and L-glyceraldehyde is *S*-glyceraldehyde.

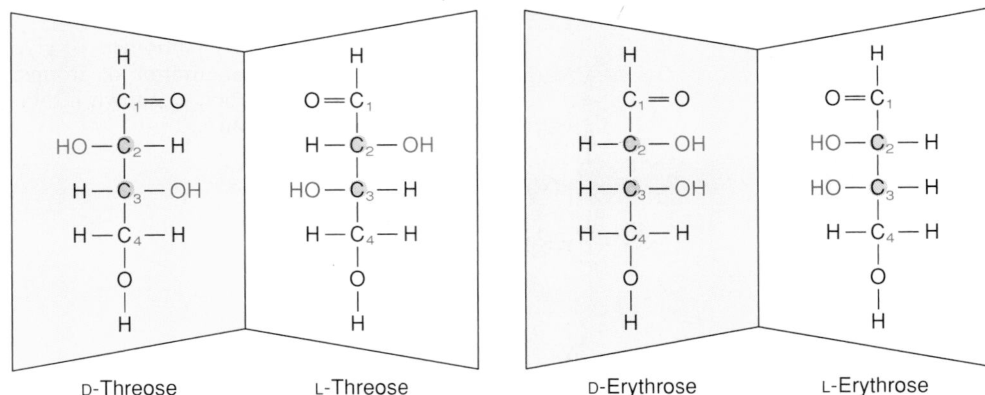

Figure 8.6
Enantiomers of aldotetroses. These molecules have two chiral carbons (2 and 3) and thus have two diastereomeric forms—threose and erythrose, each with two enantiomers. Note that the threose enantiomers have *opposite* configuration about carbons 2 and 3, whereas erythose enantiomers have the *same* configuration about these two carbons.

refers specifically to the orientation about the chiral carbon *farthest* from the carbonyl group—carbon number 3 in this case. Molecules with different orientations about the carbons preceding this are given separate names. Thus, **threose** and **erythrose** are two aldotetroses with opposite orientations about carbon 2. Such multiple forms are called **diastereoisomers.** Erythrose and threose are diastereoisomers, and each has two enantiomers (D and L) that are mirror images. Unfortunately, there is no logical rule for the specific names (such as erythrose and threose); they must simply be learned, like the names of the amino acids.

The four-carbon ketose, which is called **erythrulose,** has only one pair of enantiomers, because there is only one chiral carbon in this monosaccharide (Figure 8.7). Another naming convention appears at this point: Usually the ketose name is derived from the corresponding aldose name by insertion of the letters *ul*; thus *erythrose* becomes *erythrulose.* As with glyceraldehyde (and other monosaccharides), the ketose and aldose forms are interconvertible via tautomerization in dilute alkali. The aldose–ketose conversion also provides a route for interconversion of aldose diastereoisomers, using the ketose as an intermediate.

Adding one more carbon, we obtain the **pentoses.** The **aldopentoses** have three chiral centers; therefore we expect $2^3 = 8$ stereoisomers, in four pairs of enantiomers. The D forms are shown in Figure 8.8. Note that each of these has the D orientation about carbon 4 and that all possible combinations of orientations about carbons 2 and 3 are included. From here on in our illustration of carbohydrate structure we will show only the D forms; you can easily draw the L forms from the rules given above.

Ketopentoses will have two chiral carbons, and thus four isomers (two pairs of enantiomers) must exist. The diastereoisomers are called **D-ribulose** and **D-xylulose.** We leave it as an exercise to deduce what their Fischer projections must be; you should be able to do this on the basis of the discussion above.

Just as in the case of amino acids, one enantiomeric form of monosaccharides dominates in living organisms. In proteins it was the L-amino acids; in carbohydrates it is the D-monosaccharides. Again, there is no obvious reason why this preference was established in nature; but once fixed in early evolution it has persisted, for most of the cellular machinery has become geared to operate with D-sugars. However, just as D-amino acids are sometimes found in living organisms, so are L-monosaccharides. Like the "abnormal" amino acids, they play relatively special roles (Table 8.1).

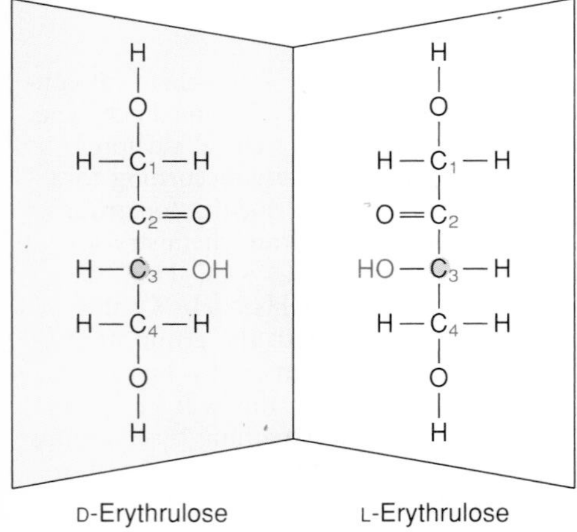

Figure 8.7
The two enantiomers of erythrulose. Unlike the four-carbon aldoses (Figure 8.6) the ketose has only one chiral carbon and only one enantiomer pair.

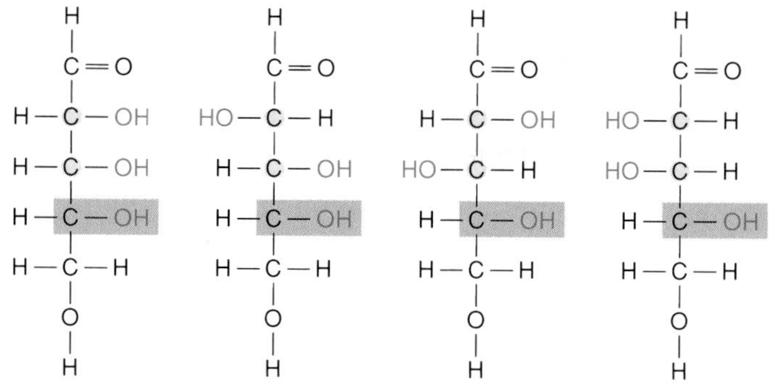

D-Ribose D-Arabinose D-Xylose D-Lyxose

Figure 8.8
D-enantiomers of the aldopentoses. These are all defined as D by the configuration about carbon 4, shaded in brown.

Ring Structures

With the aldopentoses, a new feature of monosaccharide chemistry becomes of major importance. There are now five carbons in the chain, so there is the potential for the formation of two types of very stable ring structures.* A molecule like D-ribose (see Figure 8.8) has a very strong tendency to cyclize via internal **hemiacetal** formation. Two reactions are possible, as shown in Figure 8.9. One involves formation of five-membered ring structures called **ribofuranoses;** the name reflects their structural similarity to the heterocyclic compound furan.

Alternatively, a six-membered ring can be obtained, if the reaction occurs with the C-5 hydroxyl. These six-membered rings are designated by the suffix **pyranose,** to indicate their relation to the heterocyclic compound pyran.

Both of the reactions shown in Figure 8.9 have equilibria highly in favor of the cyclic structures. Under physiological conditions, in solution, monosaccharides with five or more carbons are typically more than 99% in the ring forms. The distribution between pyranose and furanose forms, however, depends on the particular sugar structure, the pH, the solvent composition, and the temperature. Representative data are shown in Table 8.2. When the monomers are incorporated into polysaccharides, the structure of the polymer may influence the ring form chosen. For example, as Table 8.2 shows, D-ribose exists in solution, at equilibrium, as a mixture of the two ring forms. But in biological polysaccharides, specific forms are stabilized. Ribonucleic acid, for example, contains exclusively ribofuranose, whereas some plant cell wall polysaccharides have pentoses entirely in the pyranose form.

It is instructive to look a bit more closely at the ring structures shown in Figure 8.9. Cyclization has created a new asymmetric center at carbon 1. That is why we have had to draw two distinguishable forms of D-ribofuranose, referred to as α-D-ribofuranose and β-D-ribofuranose. There is a corresponding pair of ribopyranoses. Like other kinds of stereoisomers, these α and β forms rotate the plane of polarized light differently and can be distinguished in this way. Such isomers, differing in configuration at carbon 1, are called **anomers,** and carbon 1 is often referred to as the *anomeric*

*The bond angles characteristic of carbon and oxygen bonding are such that rings containing fewer than five atoms are relatively strained, whereas five- or six-membered rings are easily formed. It should be noted that aldotetroses can also form five-membered ring structures.

Table 8.1
Examples of occurrence and biochemical roles of monosaccharides

Monosaccharides	Natural Occurrence	Physiological Roles
Trioses		
Glyceraldehyde	Widespread (as phosphate)	The 3-phosphate is an important intermediate in glycolysis
Dihydroxyacetone	Widespread (as phosphate)	The 1-phosphate is an important intermediate in glycolysis
Tetroses		
D-Erythrose	Widespread	The 4-phosphate is an intermediate in carbohydrate metabolism
Pentoses		
D-Arabinose	Some plants, tuberculosis bacilli	Plant glycosides, cell walls
L-Arabinose	Widely distributed in plants, bacterial cell walls	Important constituent of cell walls, plant glycoproteins
D-Ribose	Widespread, in all organisms	Constituent of ribonucleic acid
2-Deoxyribose	Widespread, in all organisms	Constituent of deoxyribonucleic acid
D-Xylose	Woody materials	Constituent of plant polysaccharides
D-Erythropentulose	Many plants, animal tissues	Intermediate in photosynthesis and in metabolism
Hexoses		
D-Galactose	Widespread	Milk (as part of lactose); structural polysaccharides
L-Galactose	Agar-agar, other polysaccharides	Polysaccharide structures
D-Glucose	Widespread	A major energy source for animal metabolism; structural role in cellulose
D-Mannose	Plant polysaccharides, animal glycoproteins	Polysaccharide structures
D-Fructose	A major plant sugar; part of sucrose	Intermediate in glycolysis (phosphate esters)
Heptoses		
D-Sedoheptulose	Many plants	Intermediate in Calvin cycle in photosynthesis
Octoses		
D-Glycero-D-manno-octulose	Avocados	Unknown

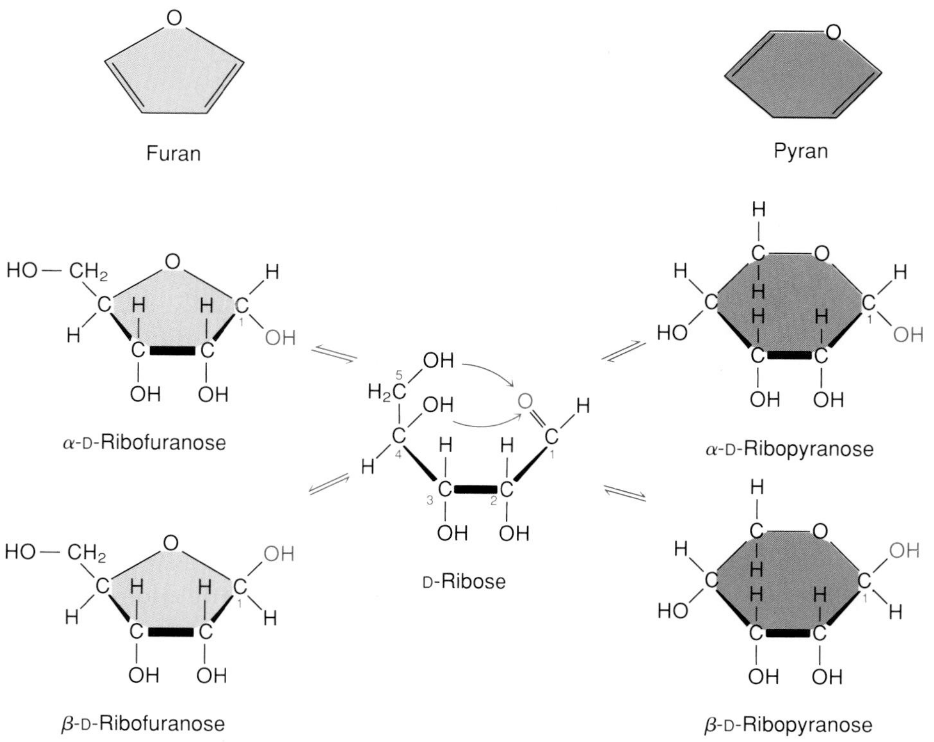

Furan

Pyran

α-D-Ribofuranose

D-Ribose

α-D-Ribopyranose

β-D-Ribofuranose

β-D-Ribopyranose

Figure 8.9
Formation of ring structures by pentoses. The example here is D-ribose which can form either a five-membered furanose ring (orange) or a six-membered pyranose ring (green). In each case, two anomeric forms (α and β) are possible. The reactions involve formation of hemiacetals from the aldehyde group.

carbon atom. The monosaccharides can undergo interconversion between the α and β forms, using the open-chain structure as an intermediate. This process is referred to as **mutarotation.** A purified anomer, dissolved in aqueous solution, will approach the equilibrium mixture, with an accompanying change in the optical rotation of the solution. Enzymes called **mutarotases** catalyze this process in vivo.

The representation of a cyclic sugar structure we have used in Figure 8.9 is called a **Haworth projection.** You are to imagine that you are seeing the ring in perspective, and the groups attached to the ring carbons (H, OH, CH_2OH) are pictured as being above or below the ring. The relationship between hydroxyl orientations in a Haworth and a Fischer projection is straightforward. Those represented to the right of the chain in a Fischer

Table 8.2
Relative amounts of tautomeric forms for some monosaccharide sugars at equilibrium in water at 40°C[a]

	Relative Amount (%)				
Monosaccharide	α-Pyranose	β-Pyranose	α-Furanose	β-Furanose	Total Furanose
Ribose	20	56	6	18	24
Lyxose	71	29	[b]	[b]	<1
Altrose	27	40	20	13	33
Glucose	36	64	[b]	[b]	<1
Mannose	67	33	[b]	[b]	<1
Fructose	3	57	9	31	40

[a] In all cases, the open-chain form is much less than 1%. For data on other sugars, see S. J. Angyal, The composition and conformation of sugars in solution, *Angew. Chem.* 8:157–226(1969).

[b] Much less than 1%.

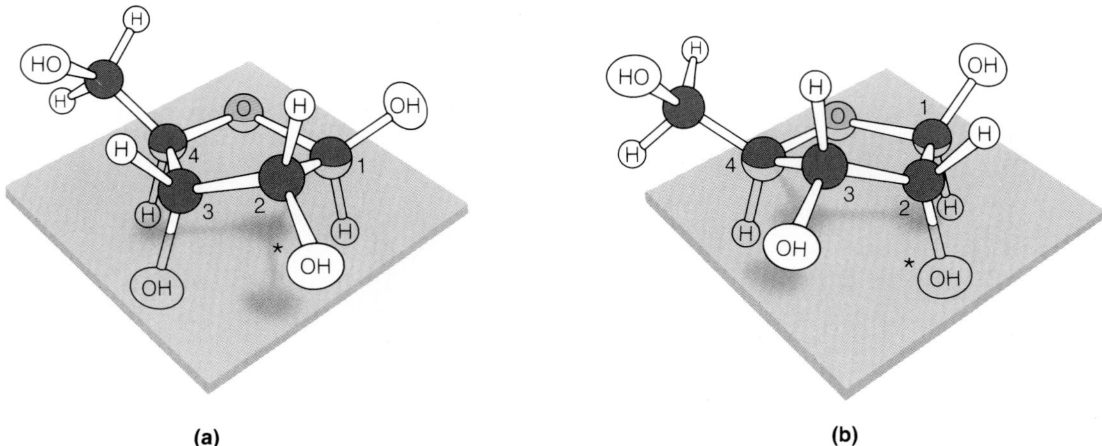

Figure 8.10
Two of the possible ring conformations for β-D-ribofuranose. In both drawings, C-1, O, and C-4 define a plane. In the C-2 *endo* conformation (**a**), C-2 is above this plane; in C-3 *endo* (**b**), C-3 is above this plane. These are the two most common conformations for ribose and deoxyribose in nucleic acids. (Note that in DNA the hydroxyl at carbon 2, indicated here by ∗, is replaced by hydrogen.) A C-3-*exo* conformation would look like the right figure, but C3 would be *below* the plane. (Carbon atoms are shown in gray.)

projection are shown below the ring in a Haworth. For example, Fischer projections of α-D-ribofuranose and β-D-ribofuranose would look like this:

H—C—OH
H—C—OH
H—C—OH
HO—CH_2—C—H
(ring O bridging C-1 and C-4)

α-D-Ribofuranose

HO—C—H
H—C—OH
H—C—OH
HO—CH_2—C—H
(ring O bridging C-1 and C-4)

β-D-Ribofuranose

Of course, even Haworth projections do not accurately depict the three-dimensional structure of molecules like ribofuranose or ribopyranose. Saturated five- and six-membered rings cannot be planar, for the C—C—C bond angles are about 109° and the C—O—C angle about 118°. More realistic representations of β-D-ribofuranose are shown in Figure 8.10.

The furanose and pyranose rings are not planar, and there are many different ways in which the ring can "pucker." Two of the major forms of β-D-ribofuranose (which we call **conformational isomers**) are shown in Figure 8.10.

We have already encountered β-D-ribofuranose (and its close relative β-D-2-deoxyribofuranose)* in Chapter 4. These sugars play a major role in biochemistry, for they are part of the backbone structure of the nucleic acids. The β anomers exclusively are involved in nucleic acid structure, and the 2-endo and 3-endo conformations shown in Figure 8.10 are favored, although there is some variation in ring conformation, even locally, along DNA and RNA chains.

This flexibility points up a fundamental difference between *conformation* and *configuration*. Conformational isomers can interchange by a sim-

* Note that 2-deoxyribose is the first exception we have encountered to the general rule that carbohydrates have the empirical formula $(CH_2O)_n$. There is no question, however, that deoxyribose should be regarded as a monosaccharide; it is, in fact, derived from ribose in nucleotide synthesis (Chapter 22). We shall encounter other compounds that depart from the simple stoichiometric formula, but in every case we may regard them as carbohydrate derivatives.

ple deformation of the molecule. But *configurational* isomers, such as the various kinds of stereoisomers described above, can interconvert only through the breaking and reformation of covalent bonds.

Like the aldopentoses, the ketopentoses exist almost entirely in the ring form. However, in this case, only the furanose form is possible. An example is α-D-ribulose, which we shall find to be a major intermediate in the carbon fixation processes in photosynthesis.

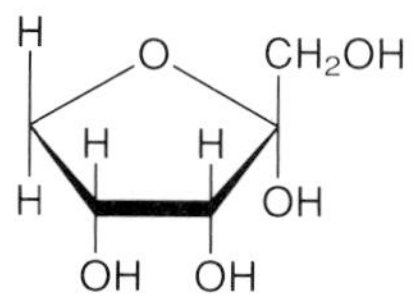

α-D-Ribulose

Monosaccharides containing six carbon atoms are called **hexoses.** As you might imagine, there are a large number of possible hexoses. To keep their structures in mind it is useful to relate them to the simpler pentoses, tetroses, and trioses. Figure 8.11 provides a compact summary of these relationships. The hexoses we will encounter again and again are **glucose** and **fructose. Mannose** and **galactose** are also widespread in nature. In fact, almost all of the hexoses play some significant biological role (see Table 8.1).

Like the pentoses, the hexoses exist primarily in the ring forms under physiological conditions. Again, two kinds of rings are found—five-membered furanose structures and six-membered pyranose rings. In each case, α and β anomers are possible; examples, illustrated as Haworth projections, are shown below.

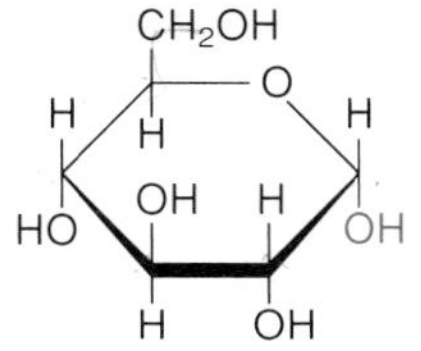

α-D-Glucopyranose

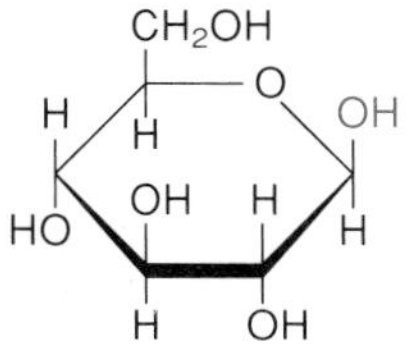

β-D-Glucopyranose

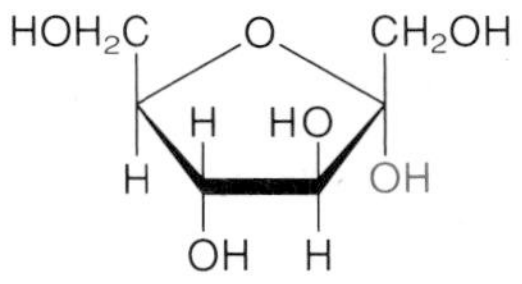

α-D-Fructofuranose

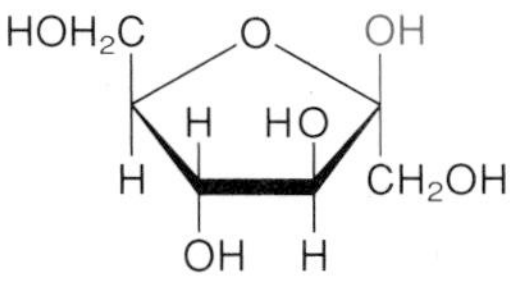

β-D-Fructofuranose

Table 8.2 gives the fractions of the different forms found at equilibrium for a number of hexoses. Clearly, which forms are favored depends very much on the structure of the particular sugar, although one can make the generalization that hexoses prefer pyranose rings. Environmental effects may also influence the preference for one anomeric form or another. For example, if glucose is crystallized from water, the crystals contain exclusively the α anomer, whereas crystallization from pyridine yields β-D-glucose crystals. Each mutarotates back to the equilibrium mixture if redissolved in water. The elucidation of the distribution of anomeric and tautomeric forms of the sugars existing in solutions has been greatly facilitated by the techniques of nuclear magnetic resonance (see Tools of Biochemistry 10). Only this technique has the exquisite sensitivity to structure in solution to allow such analyses. Figure 8.12 shows the structures of the four most common hexoses.

As in the case of furanose rings, Haworth projections of the pyranoses do not depict the actual three-dimensional structure in the correct way. Two major classes of pyranose conformations exist—the more stable "chair"

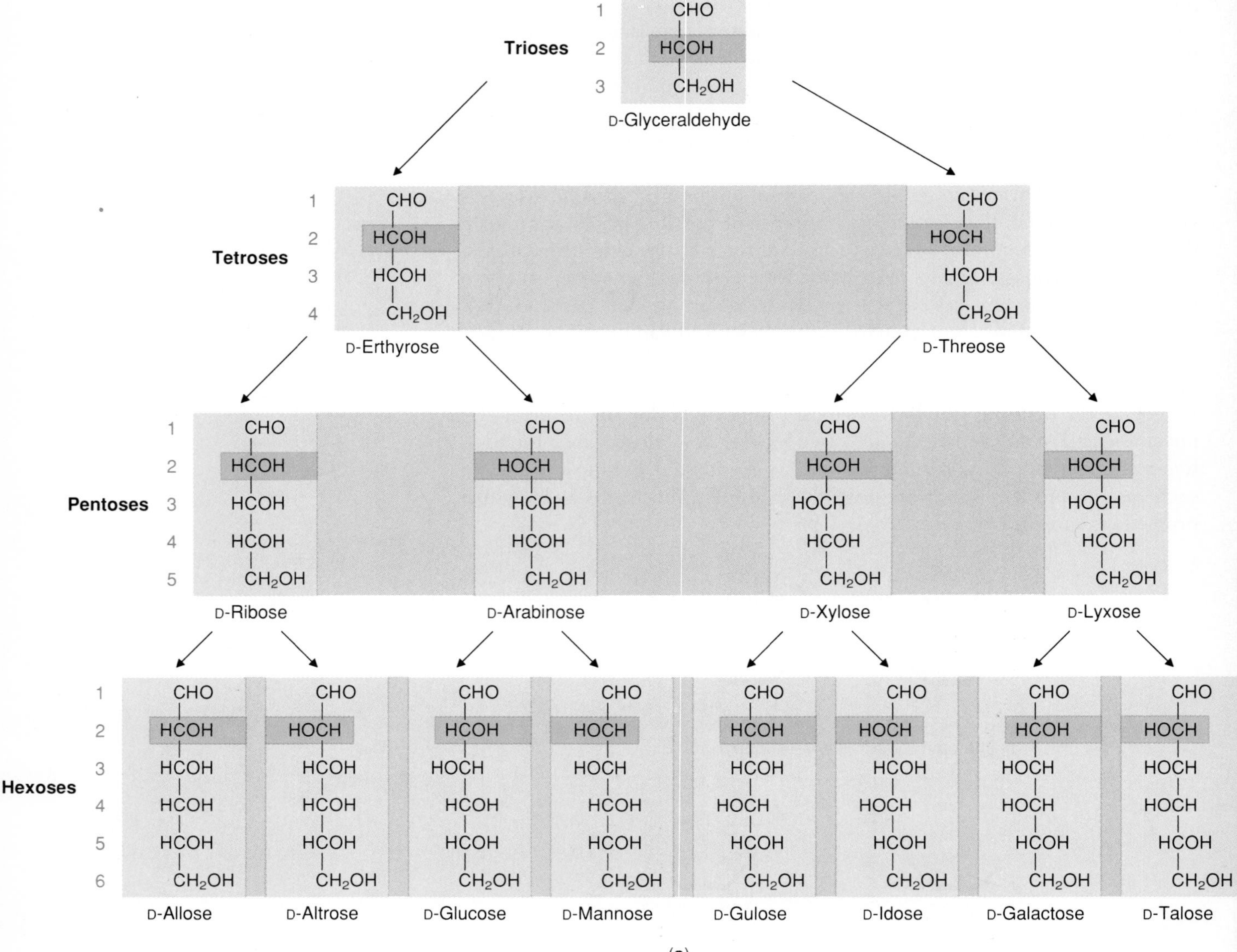

Figure 8.11
Stereochemical relationships between the D-aldoses (**a**) and between the D-ketoses (**b**). Relationships between pairs of diastereoisomers are indicated. Each series is generated by addition of one carbon (heavily shaded) just below the carbonyl carbon at each step. In each case, its two possible orientations generate a pair of diastereoisomers.

α-D-Sedoheptulopyranose

form and the less favored "boat" form. These two conformations are depicted as ball-and-stick models in Figure 8.13a. We will frequently depict them by skeletal diagrams in the ways shown in Figure 8.13b.

For both the boat and chair forms of pyranose rings, a molecular axis can be defined perpendicular to the central plane of the molecule. Bonds to substituents on ring carbons can then be classed as **axial** (**a**) or **equatorial** (**e**), depending on whether they are approximately parallel or perpendicular to the axis (Figure 8.13b). For most sugars, the chair form is more stable, because substituents on axial bonds tend to be more crowded in the boat form.

There exist naturally occurring monosaccharides with seven or eight carbons, termed **heptoses** and **octoses**, respectively (see Table 8.1). Most are of minor importance. However, one heptose, called sedoheptulose, plays a major role in the fixation of CO_2 in photosynthesis (see Chapter 19).

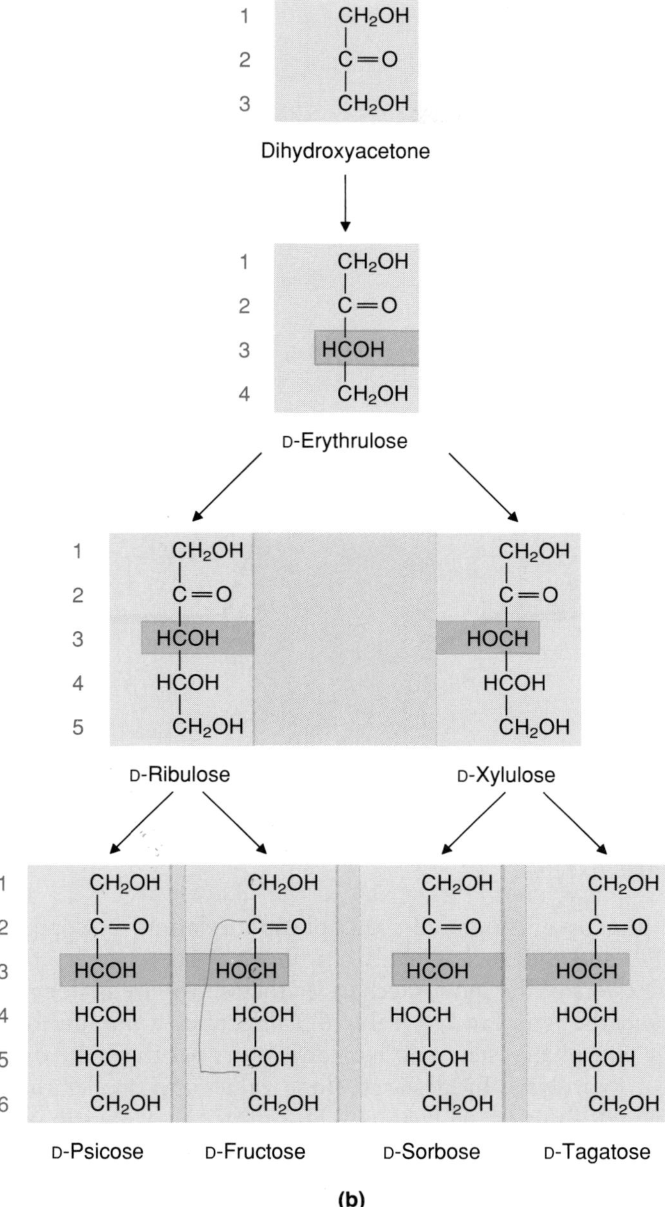

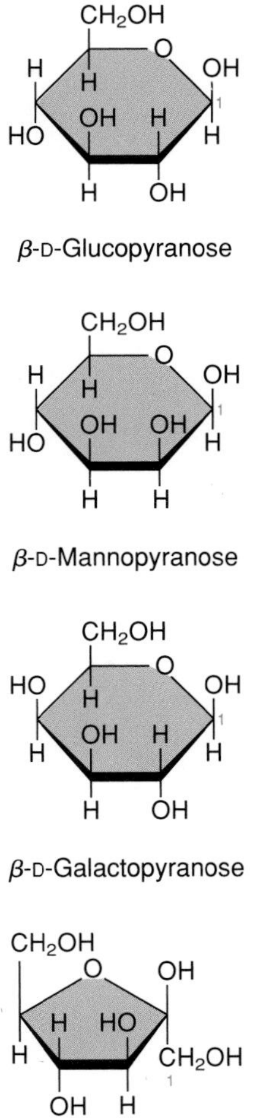

Figure 8.12
Haworth projections of the D enantiomers of the four most common hexoses. Only the β anomers are shown.

At this point, you may well have become confused by all of the terms used to describe the structures of sugar molecules—enantiomers, diastereoisomers, anomers, and ring conformations. For review, these are summarized in Figure 8.14.

Derivatives of the Monosaccharides

You will have noted from our discussion that the monosaccharides each carry a number of hydroxyl groups to which substituents might be attached or which could be replaced by other functional groups. An enormous number of sugars are modified in this way. We shall describe here only a small number of these—primarily those that play biologically important roles.

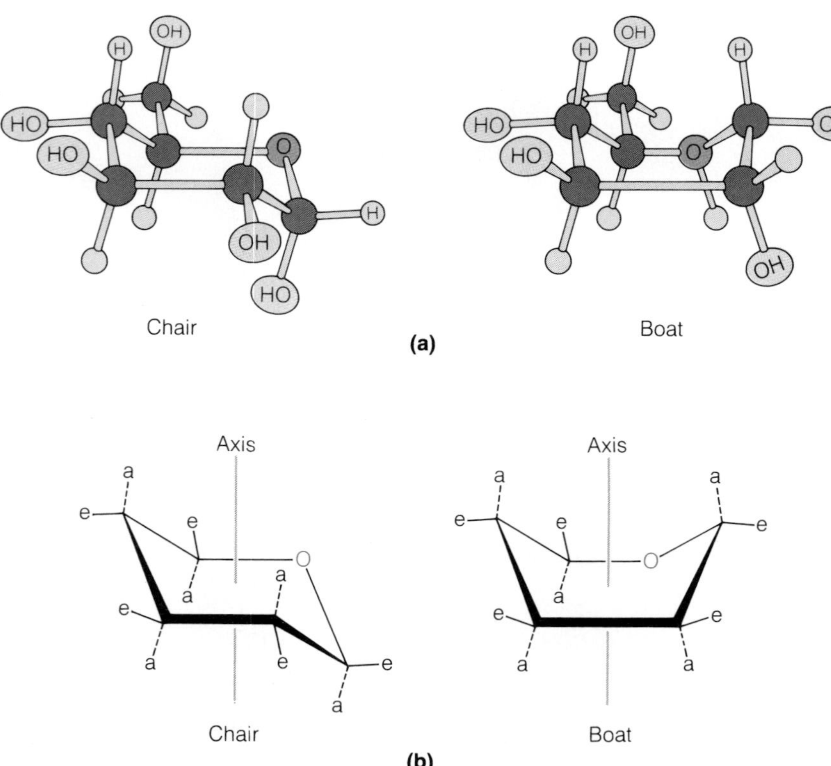

Figure 8.13
Three-dimensional representations of the pyranose ring in chair and boat conformations. (**a**) A ball-and-stick model of α-D-glucose; (**b**) a skeletal diagram of the bonding. Axial and equatorial bonds are indicated by *a* and *e*, respectively.

Phosphate Esters

We have already encountered sugar phosphorylation, in compounds like AMP, ATP, and the nucleic acids. As we shall see in later chapters, the phosphate esters of the monosaccharides themselves are major participants in many metabolic pathways. Table 8.3 illustrates a number of the more important phosphate esters and includes values for the standard state free energies of hydrolysis. In all cases, these values are smaller than the free energy of hydrolysis of ATP; thus ATP can act as a phosphate donor to monosaccharides. On the other hand, the fact that hydrolysis of the phosphate esters of sugars is thermodynamically favorable allows these derivatives to behave as "activated" compounds in many metabolic reactions.

Sugar phosphate esters are quite acidic, with pK_a values for the first two stages of ionization of about 1–2 and 6–7, respectively (see Table 8.3).

$$\mathrm{R{-}O{-}P(=O)(OH){-}OH} \underset{H^+}{\overset{H^+}{\rightleftharpoons}}_{pK_{a1}} \mathrm{R{-}O{-}P(=O)(O^-){-}OH} \underset{H^+}{\overset{H^+}{\rightleftharpoons}}_{pK_{a2}} \mathrm{R{-}O{-}P(=O)(O^-){-}O^-}$$

Consequently, these compounds exist under physiological conditions as a mixture of the mono- and dianions.

Acids and Lactones

Oxidation of monosaccharides can proceed in a variety of ways, depending on the oxidizing agent used. For example, mild oxidation of an aldose with alkaline Cu(II) (Fehling's solution) produces the **aldonic acids:**

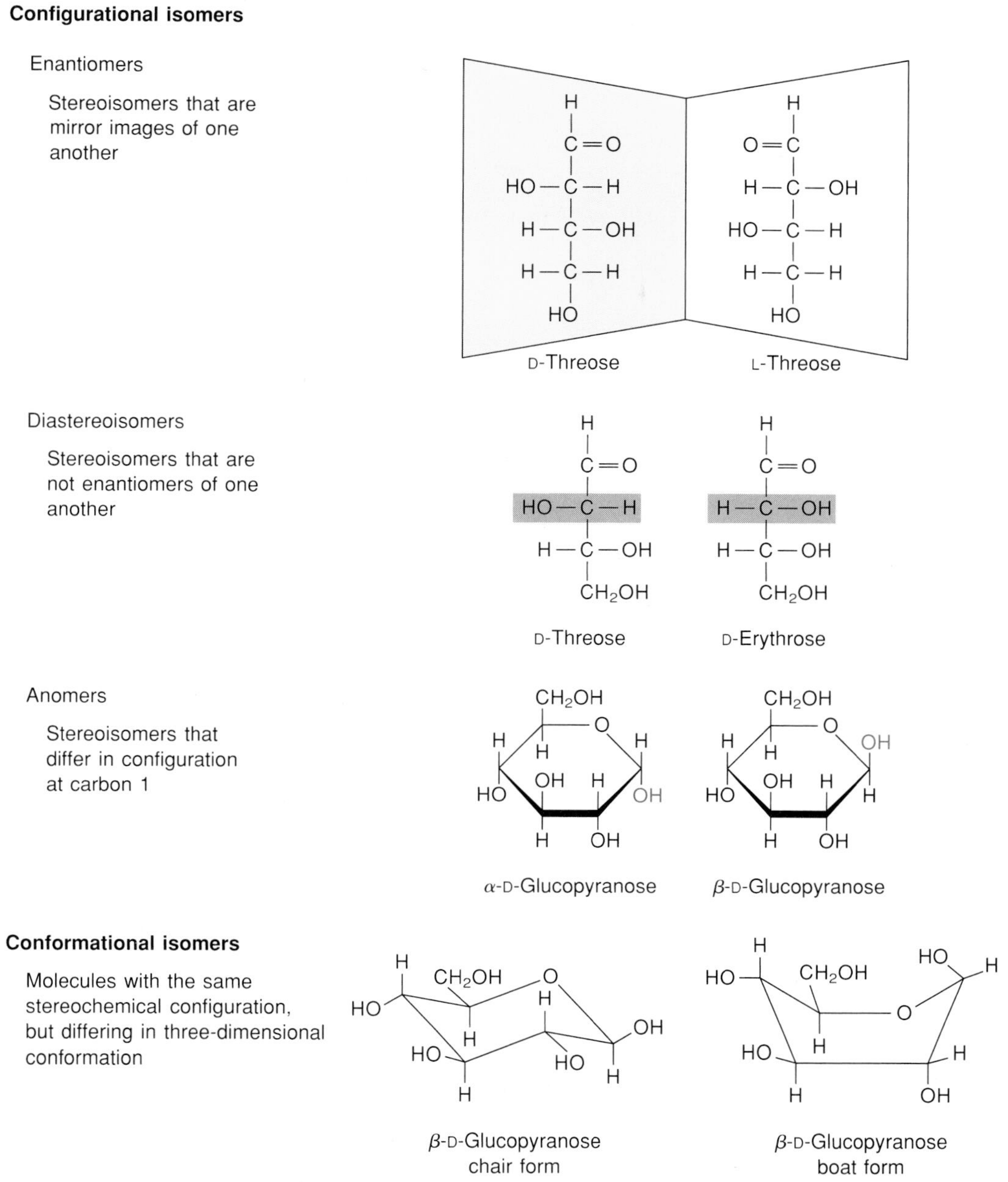
Configurational isomers
Enantiomers
Stereoisomers that are mirror images of one another
D-Threose
L-Threose
Diastereoisomers
Stereoisomers that are not enantiomers of one another
D-Threose
D-Erythrose
Anomers
Stereoisomers that differ in configuration at carbon 1
α-D-Glucopyranose
β-D-Glucopyranose
Conformational isomers
Molecules with the same stereochemical configuration, but differing in three-dimensional conformation
β-D-Glucopyranose chair form
β-D-Glucopyranose boat form

Figure 8.14
A review of nomenclature describing the configurations and conformations of sugar molecules.

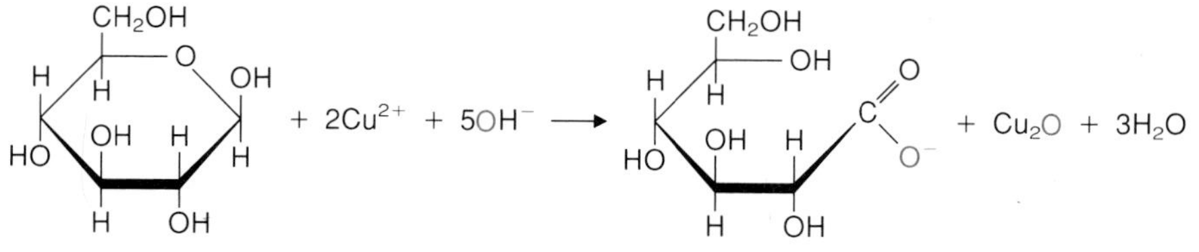
+ 2Cu2+ + 5OH−
+ Cu2O + 3H2O

β-D-Glucopyranose **D-Gluconic acid**

The production of a red precipitate of Cu_2O is a classic sugar test and was used formerly to test for excess sugar in the urine of persons thought to have diabetes. This method has now been replaced by more specific enzyme as-

Table 8.3
A number of biochemically important phosphate esters of monosaccharides

Name	Structure	$\Delta G^{\circ\prime}$ (kJ/mol)	pK_{a1}	pK_{a2}
D-Glyceraldehyde-3-phosphate	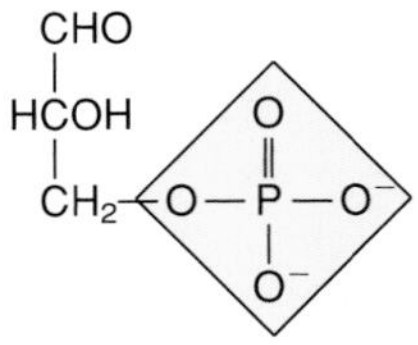 CHO–HCOH–CH$_2$–O–P(=O)(O$^-$)–O$^-$	~−12	2.10	6.75
β-D-Glucose-1-phosphate	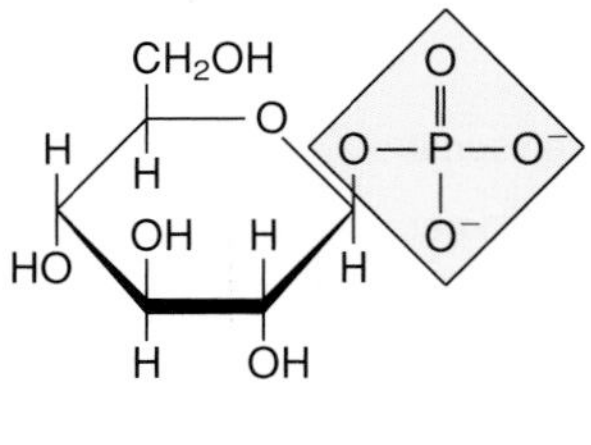	−20.9	1.10	6.13
β-D-Glucose-6-phosphate	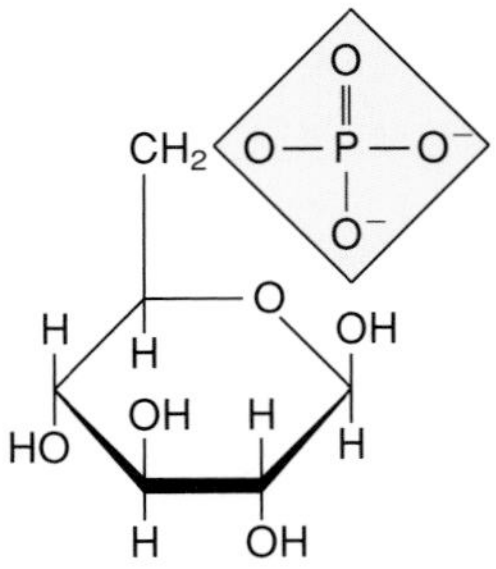	−13.8	0.94	6.11
α-D-Fructose-6-phosphate	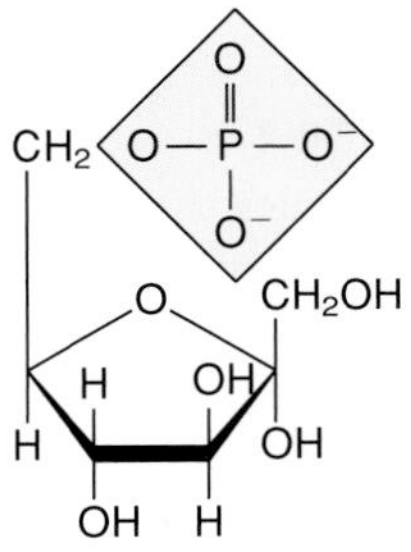	−13.8	0.97	6.11

says. Free aldonic acids, such as gluconic acid, are in equilibrium in solution with **lactones.**

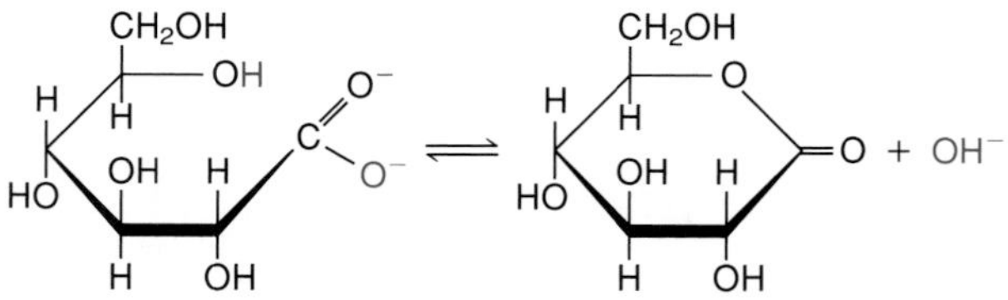

D-Gluconic acid ⇌ D-δ-Gluconolactone + OH^-

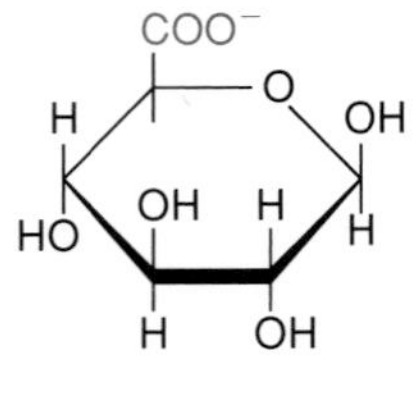

β-D-Glucuronic acid

Enzyme-catalyzed oxidation of monosaccharides gives a variety of specific products, including lactones and **uronic acids** such as **glucuronic acid,** in which oxidation has occurred on carbon 6. Uronic acids are, as we shall see later, important constituents of certain natural polysaccharides.

Alditols

Reduction of the carbonyl group on a sugar gives rise to the class of polyhydroxy compounds called **alditols.** Important naturally occurring ones are **erythritol, D-mannitol,** and **D-glucitol,** often called **sorbitol:**

Erythritol* **D-Mannitol** **D-Glucitol (sorbitol)**

Each is named from the corresponding monosaccharide. Sorbitol accumulates in the lens of the eye in people suffering from diabetes and leads to the formation of cataracts.

Amino Sugars

Two amino derivatives of simple sugars are widely distributed in natural polysaccharides. These are **glucosamine** and **galactosamine:**

β-D-Glucosamine **β-D-Galactosamine**

Further modifications of these amines are common. For example, from β-D-glucosamine are derived

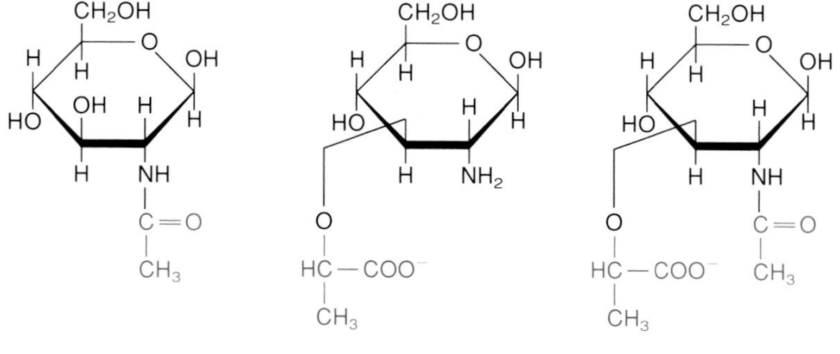

β-D-*N*-Acetylglucosamine **Muramic acid** ***N*-Acetylmuramic acid**

These sugar derivatives are important constituents of many natural polysaccharides (see below). Two others we shall encounter are

* Erythritol, although it contains chiral carbons, is not optically active, because it has a plane of symmetry, between C-2 and C-3.

β-D-N-Acetylgalactosamine

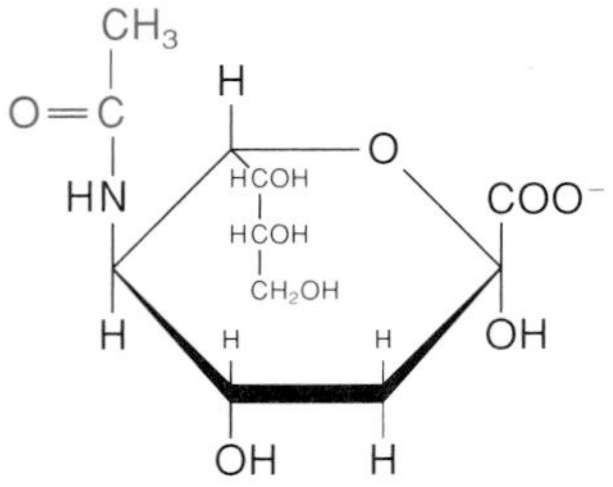

N-Acetylneuraminic acid (sialic acid)

We shall usually encounter the modified sugars—especially the amino sugars—as monomer residues in complex oligosaccharides and polysaccharides. To aid in writing the structures of such molecules, it is useful to have a shorthand notation, as is used in describing nucleic acid and protein structure. Therefore, a set of abbreviations has been defined for the simple sugars and their derivatives. A number of the most important ones are listed in Table 8.4.

Glycosides

Elimination of water between the anomeric hydroxyl of a cyclic monosaccharide and the hydroxyl group of another compound yields a **glycoside.** The reaction may be described as the transformation of a hemiacetal to an acetal. A simple example is the formation of methyl-α-D-glucopyranoside:

α-D-Glucopyranose + CH_3OH —(Acidic solution)→ Methyl-α-D-glucopyranoside + H_2O

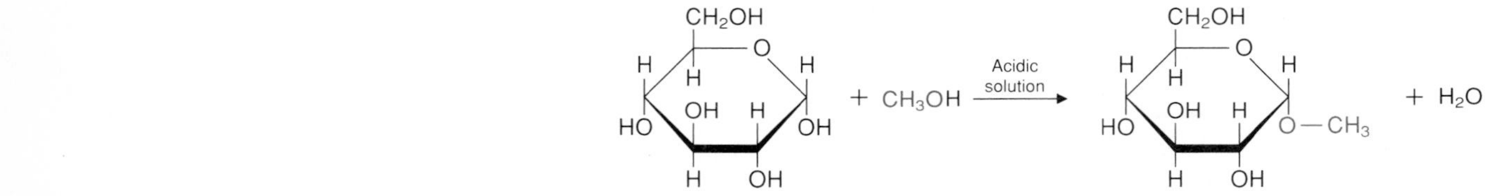

α-D-Glucopyranose **Methyl-α-D-glucopyranoside**

Unlike the anomers of the sugars themselves, the anomeric glycosides (i.e., methyl-α-D-glucopyranoside and methyl-β-D-glucopyranoside) do not interconvert by mutarotation in the absence of an acid catalyst.

Table 8.4
Abbreviations for some common monosaccharide residues

Monosaccharides		Monosaccharide Derivatives	
Arabinose	Ara	Gluconic acid	GlcA
Fructose	Fru	Glucuronic acid	GlcUA
Fucose	Fuc	Galactosamine	GalN
Galactose	Gal	Glucosamine	GlcN
Glucose	Glc	*N*-Acetylgalactosamine	GalNAc
Lyxose	Lyx	*N*-Acetylglucosamine	GlcNAc
Mannose	Man	Muramic acid	Mur
Ribose	Rib	*N*-Acetylmuramic acid	MurNAc
Xylose	Xyl	*N*-Acetylneuraminic acid = sialic acid	NeuNAc = Sia

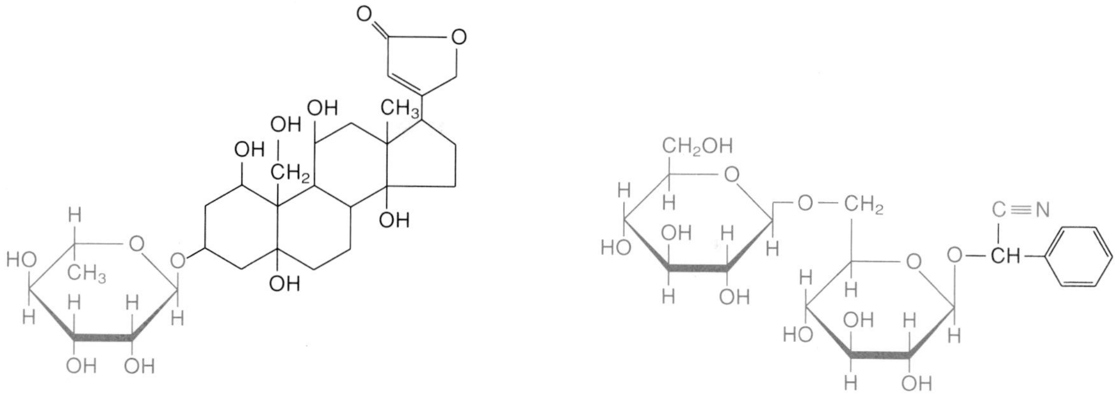

(a) Ouabain

(b) Amygdalin

Figure 8.15
Two naturally occurring glycosides.

Many glycosides are found in plant and animal tissues. Some are very toxic substances, primarily because they act as inhibitors of enzymes involved in ATP utilization. An example is **ouabain** (Figure 8.15a), which inhibits the action of the enzymes that pump Na^+ and K^+ ions across cell walls to maintain necessary electrolyte balance. Ouabain comes from an African shrub and was discovered when it was noted that Somali natives dipped arrowheads in an extract from the plant. It now finds use in treatment of some cardiac conditions. Another glycoside, **amygdalin** (Figure 8.15b), is toxic for a very different reason. Found in the seeds of bitter almonds, this substance yields HCN on hydrolysis.

Oligosaccharides

Just as monosaccharides can form glycosidic bonds with other kinds of hydroxyl-containing compounds, they can do so with one another. Such bonding gives rise to the *oligosaccharides* and *polysaccharides.*

Oligosaccharide Structures

The simplest, and biologically most important, oligosaccharides are the *disaccharides,* made up of two residues. As Table 8.5 shows, the disaccharides are encountered in many ways in living organisms. Some, like *sucrose, lactose,* and *trehalose,* are soluble energy stores in plants and animals. Others, like *maltose* and *cellobiose,* can be regarded primarily as intermediate products in the degradation of much longer polysaccharides. Still others, like *gentiobiose,* are found principally as constituents of more complex, naturally occurring substances. The structures of these important disaccharides are depicted in Figure 8.16, p. 280. Four major features distinguish disaccharides from one another:

1. *The two specific sugar monomers involved.* These may be of the same kind, as the two glucose residues in maltose, or they may be different, as the glucose and fructose residues in sucrose.

2. *The carbons involved in the linkage.* Although many possibilities exist, the most common are $1 \rightarrow 1$ (as in trehalose), $1 \rightarrow 2$ (as in sucrose), $1 \rightarrow 4$ (as in lactose), and $1 \rightarrow 6$ (as in gentiobiose).

Table 8.5
Occurrence and biochemical roles of some representative disaccharides

Disaccharide	Structure	Natural Occurrence	Physiological Role
Sucrose	Glc (α1→β2) Fru	Many fruits, seeds, roots, honey	A final product of photosynthesis; used as primary energy source in many organisms
Lactose	Gal β(1→4) Glc	Milk, some few plant sources	A major animal energy source
α,α-Trehalose	Glc (α1→α1) Glc	Yeast, other fungi, insect blood	A major circulatory sugar in insects; used for energy
Maltose	Glc α(1→4) Glc	Plants (starch) and animals (glycogen)	The dimer of the starch and glycogen polymers
Cellobiose	Glc β(1→4) Glc	Plants (cellulose)	The dimer of the cellulose polymer
Gentiobiose	Glc β(1→6) Glc	Some plants (e.g., gentians)	Constituent of plant glycosides and some polysaccharides

Note that all disaccharides involve the anomeric hydroxyl of at least one sugar as a participant in the bond.

3. *The order of the two monomer units*, if they are of different kinds. The glycosidic linkage involves the anomeric carbon on one sugar, but in most cases the other is free. Thus the two ends of the molecule can be distinguished by chemical reactivity. For example, in lactose the glucose residue, having a free anomeric carbon and thus a potential free aldehyde group, could be oxidized by Fehling's solution but the galactose residue could not be. Lactose is therefore a reducing sugar, and the glucose residue is at its *reducing end*. The other end is called the *nonreducing end*. In sucrose neither residue has a potential free aldehyde group. Therefore sucrose is a nonreducing sugar.

4. *The anomeric configuration of the hydroxyl group on carbon 1 of each residue*. This is especially important for the anomeric carbon(s) involved in the glycosidic bond. The configuration may be either α (as in the disaccharides shown in Figure 8.16a) or β (as in those in 8.16b). This may seem a small difference, but it has a major effect on the shape of the molecule that is recognized readily by enzymes. For example, different enzymes are needed to catalyze the hydrolysis of maltose and cellobiose, even though both are dimers of D-glucopyranose. Furthermore, we shall see that in polysaccharides the anomeric orientation plays a critical role in determining the secondary structures adopted by these polymers.

A convenient way to describe the structures of these and more complex oligosaccharides is illustrated in Table 8.5. The rules are:

1. The sequence is written starting with the nonreducing end at the left.

2. Anomeric and enantiomeric forms are designated by prefixes (e.g., -α, D-).

3. The ring configuration is indicated by a suffix (*p* for pyranose, *f* for furanose).

4. The atoms between which glycosidic bonds are formed are indicated by numbers in parentheses between residue designations (e.g., (1→4) means a bond from carbon 1 of the residue on the left to carbon 4 of the residue on the right).

As an example, we write the structure of sucrose as

α-D-Glc*p*(1→2)-β-D-Fru*f*

In many cases, the nomenclature is further shortened by omitting the D/L designation (except in the unusual cases where L enantiomers are encountered) and by omitting the *p*/*f* suffix when the monomers have their usual ring forms. Thus, we would more likely write sucrose in the way shown in Table 8.5. The system can be applied to oligosaccharides of any length and can include branched structures as well (see below). If only one carbon involved in the linkage between two residues is anomeric, the convention is even more condensed [e.g., maltose = Glc α(1→4) Glc].

The list of biologically important oligosaccharides is by no means restricted to dimeric structures. Many trimers, tetramers, and even larger, yet specifically constructed, molecules are known. An example is the naturally occurring trisaccharide **raffinose** (Figure 8.17). Others will be encountered when we examine, in later sections of this chapter, the oligosaccharides attached to certain proteins and to cell surfaces. Tools of Biochemistry 13 describes techniques used to sequence oligosaccharides.

Stability and Formation of the Glycosidic Bond

The glycosidic bond between two monomers in an oligosaccharide is created by the elimination of a molecule of water. Thus, we might expect the synthesis of lactose to proceed as follows:

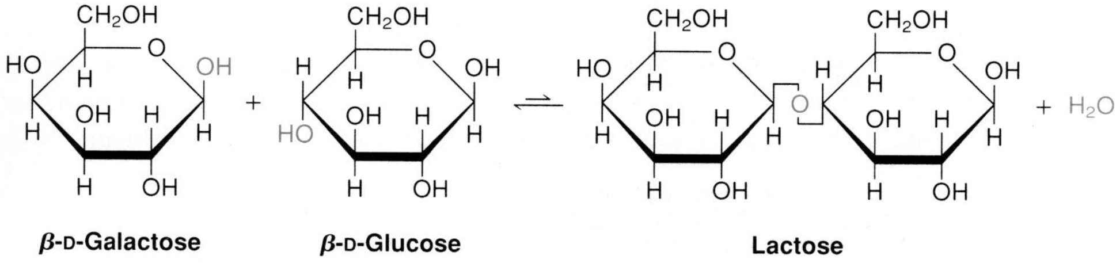

β-D-Galactose + **β-D-Glucose** ⇌ **Lactose** + H_2O

This reaction is analogous to the elimination of water between amino acids in the formation of polypeptides or between nucleotides in the formation of nucleic acids. As in those cases, the reaction as written is thermodynamically unfavored. The *hydrolysis* of oligosaccharides and polysaccharides is favored under physiological conditions by a standard state free energy change of about −15 kJ/mol, corresponding to an equilibrium constant of about 800. Nevertheless, like peptides and oligonucleotides, saccha-

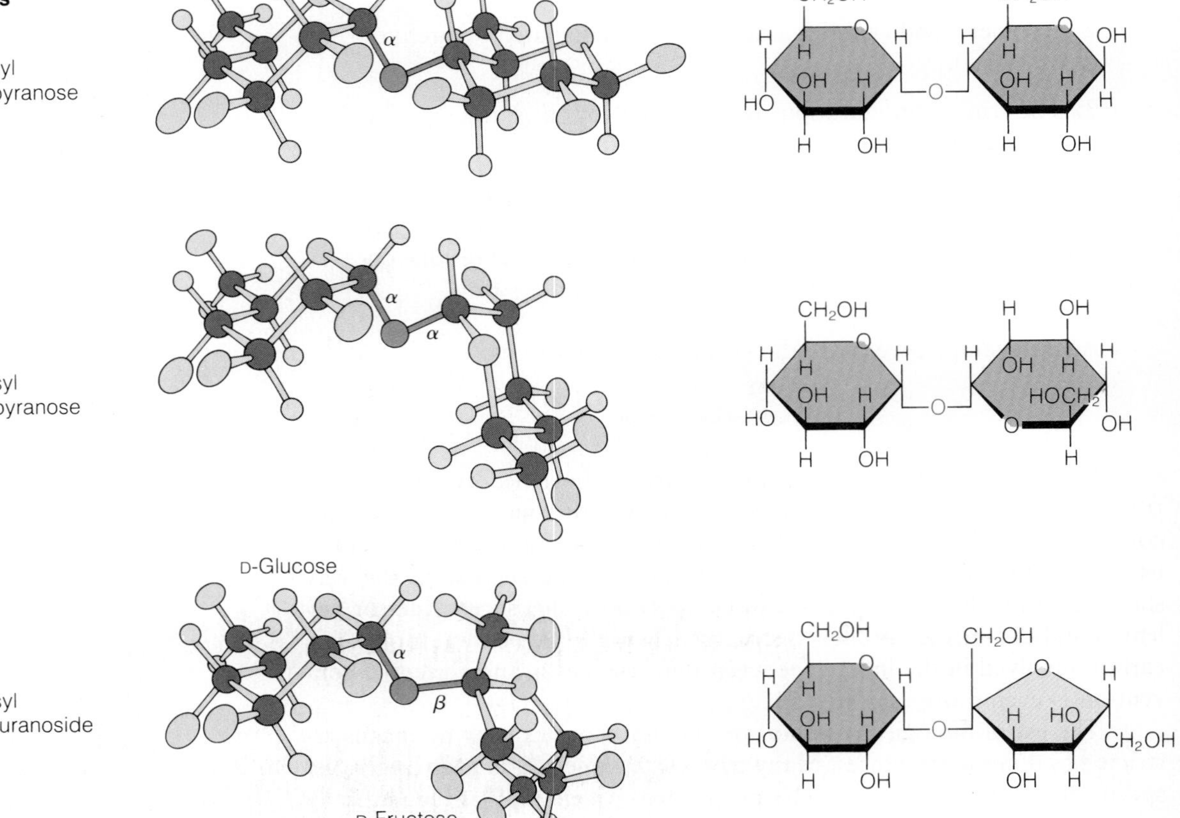

Figure 8.16
Structures of some important disaccharides. (**a**) Examples linked with the α anomer of carbon 1: maltose, trehalose, and sucrose. The anomeric oxygens are emphasized in bright red. (**b**) Examples which are β-linked: cellobiose, lactose, and gentiobiose. To the right are shown Haworth projections of the same molecules.

ride polymers are sufficiently metastable to persist for long periods unless their hydrolysis is catalyzed by enzymes or acid. So the situation is the same as we have encountered with the other important biopolymers: the breakdown of oligosaccharides and polysaccharides in vivo is controlled by the presence of specific enzymes. Furthermore, synthesis of these sugar polymers never proceeds in living organisms by reactions like the one we have just shown. As in the case of protein and nucleic acid synthesis, activated monomers are required. For example, the formation of lactose in mammary tissue proceeds through the reactions shown in Figure 8.18. A phosphate bond between UDP and galactose is hydrolyzed to drive the coupling. A specific enzyme, **lactose synthase**, is the catalyst. The reaction is further favored by the hydrolysis of the inorganic pyrophosphate produced.

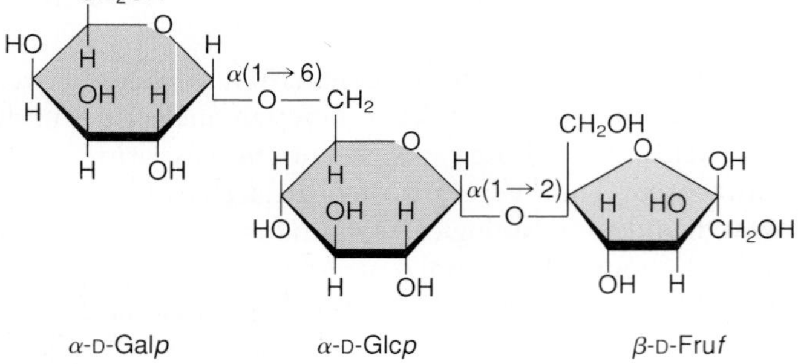

Figure 8.17
Structure of raffinose, a trisaccharide found in sugar beets.

α(1→6) α(1→2)

α-D-Gal*p* α-D-Glc*p* β-D-Fru*f*

(b) DISACCHARIDES with β connections

Cellobiose
β-D-glucopyranosyl
(1→4) β-D-glucopyranose

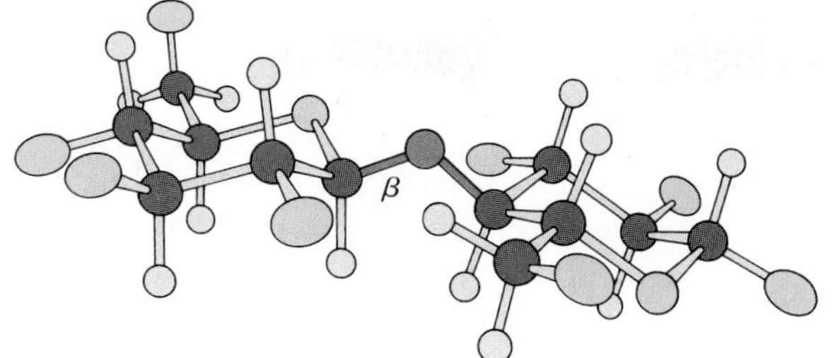

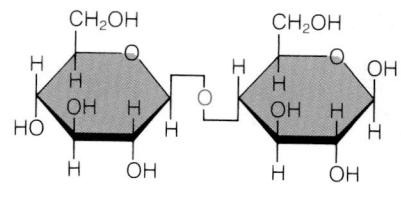

Lactose
β-D-galactopyranosyl
(1→4) β-D-glucopyranose

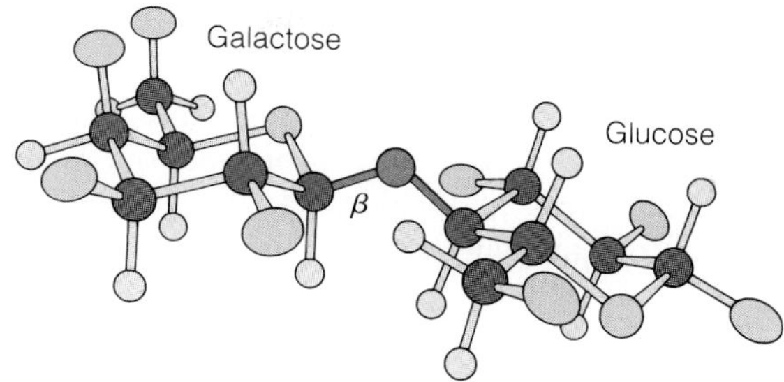

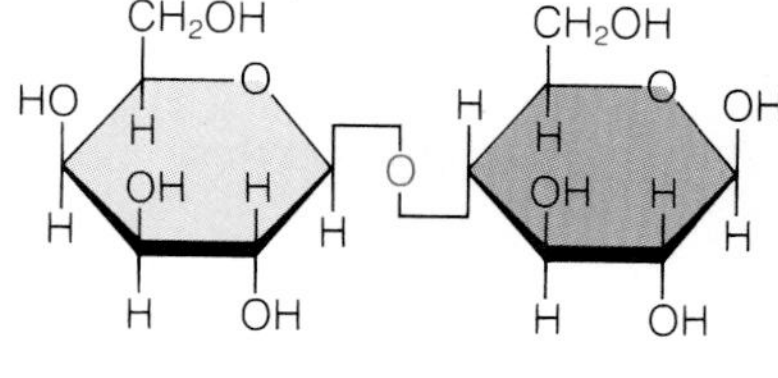

Gentiobiose
β-D-glucopyranosyl
(1→6) β-D-glucopyranose

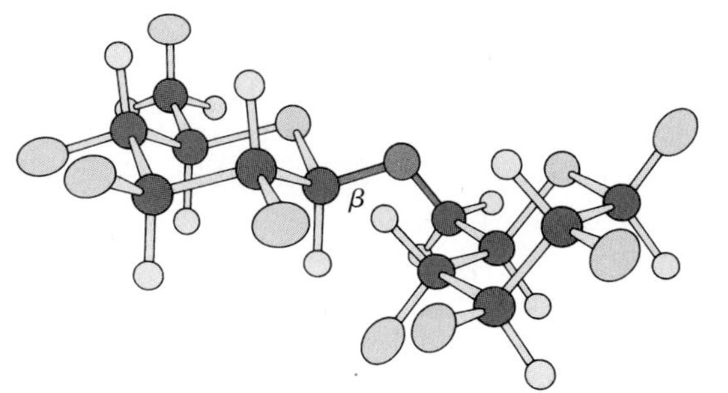

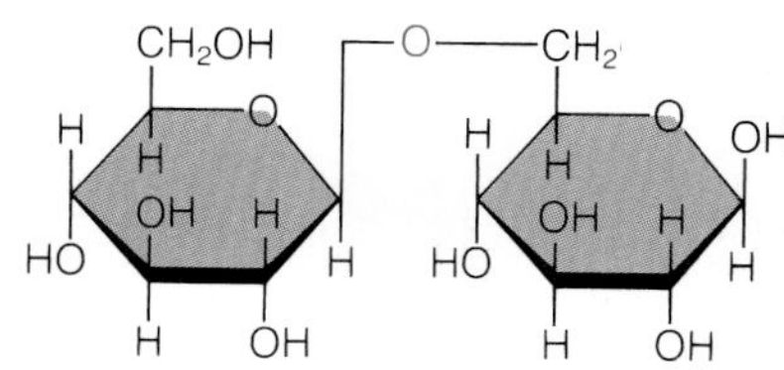

There is one important way in which oligo- and polysaccharide synthesis differs from synthesis of nucleic acids and proteins. These sugar polymers are never copied from template molecules. Instead, in the formation of oligo- or polysaccharides, a *different enzyme* is employed to catalyze the linkage of each kind of monomer unit. Details of the mechanisms of synthesis of oligo- and polysaccharides are given in Chapter 16.

Polysaccharides

Polysaccharides fulfill a wide variety of functions in living organisms. Some, like **starch** and **glycogen** (animal starch), serve mainly to store sugars in plants and animals. Others, like **cellulose** and **chitin,** are structural materials, analogous to the structural proteins. Still others, such as some **cell wall polysaccharides,** act as cell-specific recognition signals and thus play an informational role. It is simplest to consider these molecules in terms of these functional categories.

As with polypeptides and polynucleotides, the sequence of monomer residues in a polysaccharide defines its primary structure. Whereas proteins usually have complicated sequences, polysaccharides often have rather simple primary structures. In some cases (e.g., cellulose) the polymer is made from only one kind of monomer residue (here, β-D-glucose); these are referred to as **homopolysaccharides.** If two or more kinds of residues are involved, the polymer is called a **heteropolysaccharide.** In contrast to protein and nucleic acid molecules, which are almost always of defined length, polysaccharide chains often grow to indeterminate sizes.

The functional reasons for these differences are not hard to find. A storage material, such as starch, need convey no information nor adopt a very complicated three-dimensional form. It is simply a bin in which to put away glucose molecules for future use. Many structural polysaccharides, like structural proteins, form extended, regular secondary structures, well suited to the formation of fibers or sheets. Often a regular repetition of some simple mono- or disaccharide motif will serve this function. (Recall, for comparison, the simple and repetitive amino acid sequences of collagen and silk fibroin described in Chapter 6.) The only polysaccharides in which well-defined and complex sequences are found are some of those on cell surfaces or those attached to specific glycoproteins. Because these serve to identify, they must convey information. This function requires precisely defined "words" in the polysaccharide language, just as nucleic acid sequences spell out information in their own language.

Storage Polysaccharides

The principal storage polysaccharides are **amylose** and **amylopectin,** which constitute starch in plants, and **glycogen,** which is stored in animal and microbial cells. Starch is stored in **granules** in almost every kind of cell, but grain seeds, tubers, and unripe fruits are especially rich in this material. Glycogen is also stored in granules, in the liver, which acts as a central energy storage organ in many animals. Glycogen is also abundant in muscle tissue, where it is more immediately available for energy release. Some of these storage structures are depicted in Figure 8.19.

Amylose, amylopectin, and glycogen are all polymers of α-D-glucopyranose. These homopolysaccharides are members of the general class of **glucans,** the polymers of glucose. The three polymers differ only in the kinds of linkages between glucose residues. Amylose is a linear polymer, involving exclusively α(1→4) links between adjacent glucose residues. Amylopectin (Figure 8.20) and glycogen are both branched polymers, since they contain, in addition to the α(1→4) links, some α(1→6) links as well. The branches in glycogen are somewhat more frequent and shorter than those in amylopectin, but in most respects the structures of these two polysaccharides are very similar.

The very regular and simple primary structure of amylose allows a regular secondary structure for this molecule. As with polynucleotides and polypeptides, the details of this structure have come from x-ray diffraction studies. In fact, amylose was the first biopolymer whose structure was elucidated by this method. Because of the α(1→4) link, each residue is angled with respect to the next, favoring a regular helical conformation (Figure 8.21). However, this helix is not especially stable, and in the absence of interaction with other molecules, amylose is more likely to be found as a random coil structure. One molecule that does stabilize the amylose helix is

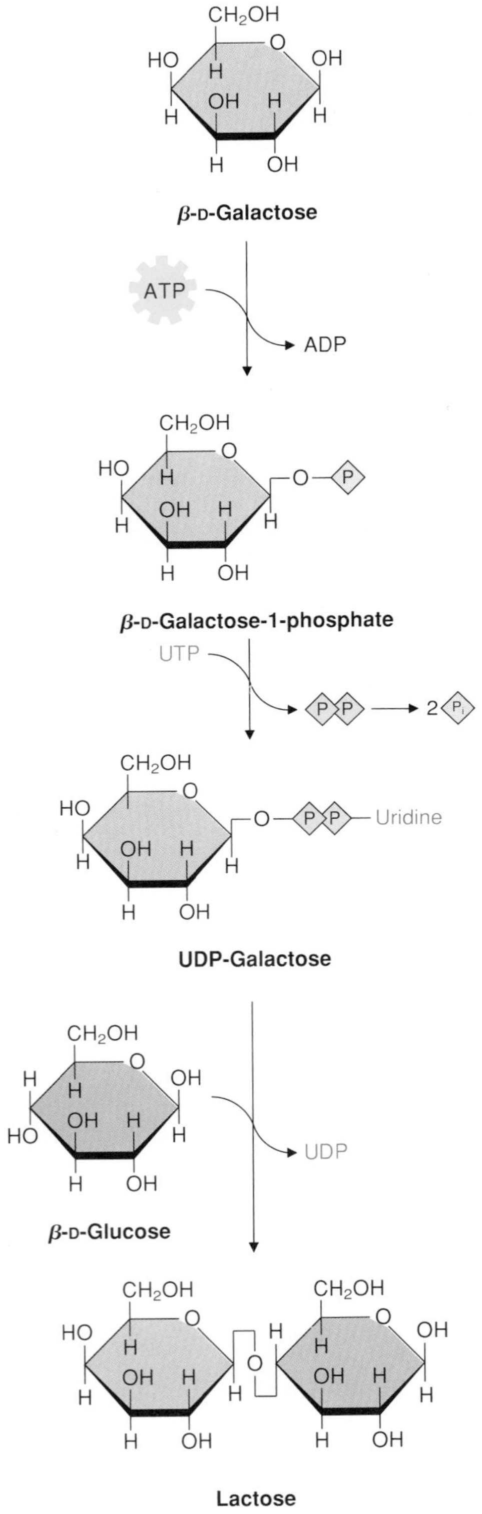

β-D-Galactose

ATP → ADP

β-D-Galactose-1-phosphate

UTP → P–P → 2 P_i

UDP-Galactose

β-D-Glucose

UDP

Lactose

Figure 8.18
Formation of lactose in vivo. The reaction shown occurs in the formation of milk in mammary tissue. Galactose is first phosphorylated by ATP and then transferred to uridine diphosphate (UDP). Formation of UDP-galactose is favored thermodynamically by the subsequent hydrolysis of the inorganic pyrophosphate. The UDP-galactose then transfers the galactose to glucose, with the accompanying cleavage of a phosphate bond. The reaction is catalyzed by the enzyme lactose synthetase.

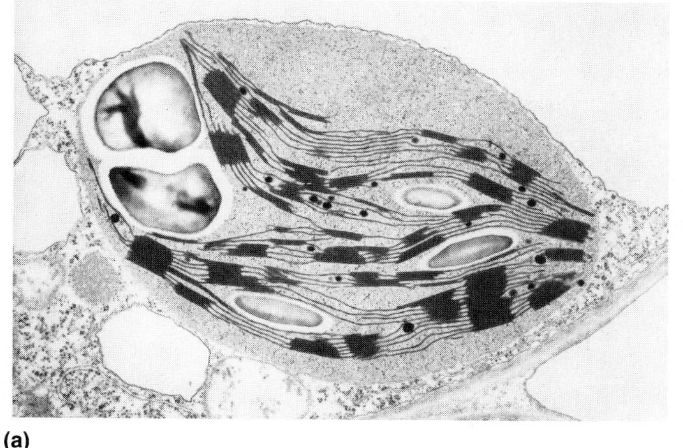

(a)

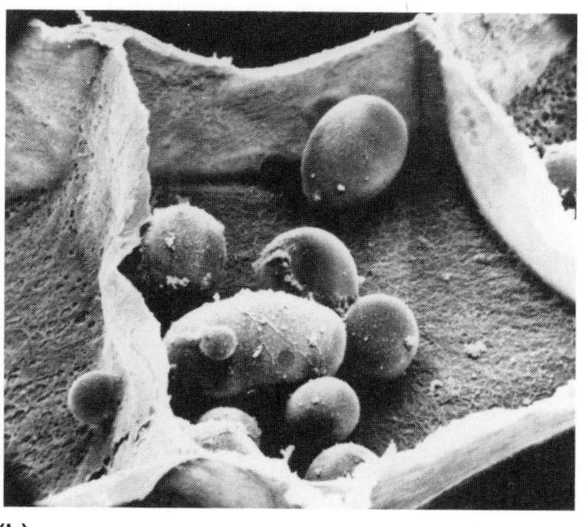

(b)

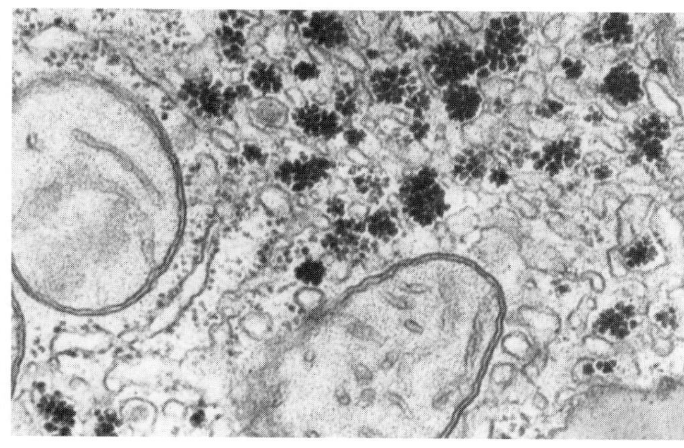

(c)

Figure 8.19
Storage of starch and glycogen in granules. (**a**) Starch granules in a plant leaf chloroplast. (**b**) Starch granules in potato tuber cells. (**c**) Glycogen granules in liver.

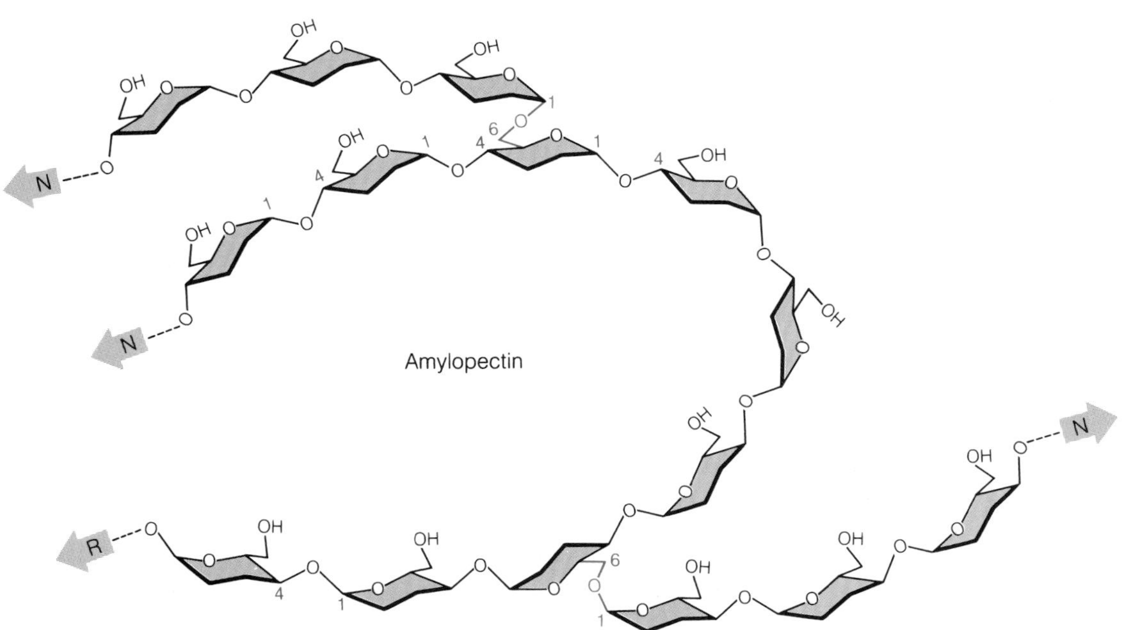

Figure 8.20
Primary structure of amylopectin. The nonreducing ends are indicated by N on the diagram. The reducing end is denoted by R.

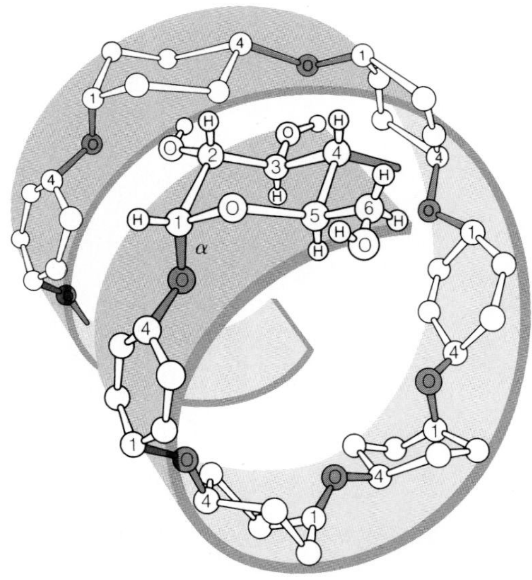

Figure 8.21
Secondary structure of amylose: the amylose helix. The orientation of successive glucose residues favors helix generation. Note the large "core." Glycosidic linkages are shown in red.

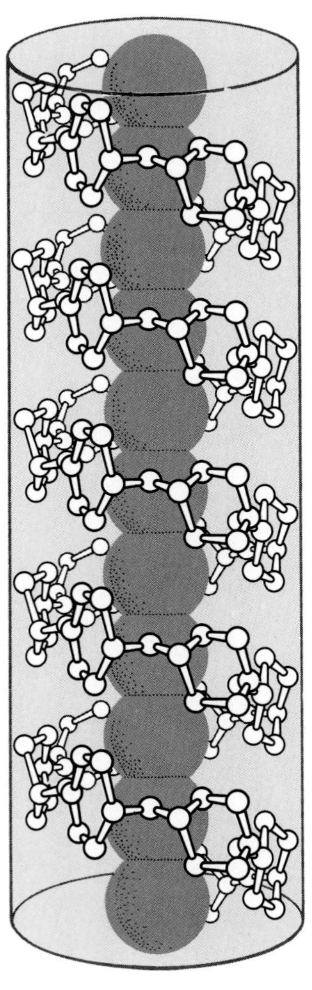

Figure 8.22
The amylose–iodine complex. Rows of iodine atoms fit neatly in the core of the amylose helix.

iodine, for strings of iodine atoms can fit nicely into the hollow nonpolar core of the structure (Figure 8.22). The formation of this complex, which is intensely blue in color, has long been used as a qualitative test for starches. In fact, the color can even be used to distinguish between kinds of starches. The shorter iodine polymers within the short amylopectin or glycogen helices absorb light at a lower wavelength than the long strings of iodine molecules that can form in an amylose helix. The branched nature of amylopectin and glycogen inhibits the formation of helices, for the helix requires 6 residues for each turn; the branches are of the order of 10–20 residues in length in amylopectin and 8 in glycogen.

The storage polysaccharides are admirably designed to serve their function. Glucose and even maltose are small, rapidly diffusing molecules, which are difficult to store. Were such small molecules present in large quantities in a cell, they would give rise to a very large cell osmotic pressure, which would in most cases be deleterious. Therefore, most cells build the glucose into long polymers, so that large quantities can be stored in a semi-insoluble state. Whenever glucose is needed, it can be obtained by selective degradation of the polymers by specific enzymes. These processes are discussed in detail in Chapter 13, but one aspect should be mentioned now. Most of the enzymes employed attack the chains at their nonreducing ends, releasing one glucose residue at a time. The branched structure of both amylopectin and glycogen is such that there are *many* nonreducing ends on each molecule (see Figure 8.20), which allows rapid mobilization of glucose when it is needed. The linear chain of amylose, on the other hand, is used mainly for long-term storage of glucose.

Structural Polysaccharides

Plants do not seem to synthesize or use fibrous structural proteins (like keratin and collagen) but rely entirely on special polysaccharides. Animals use both kinds of materials. Because each structural use requires different properties, a great variety of structural polysaccharides exists. We shall begin by considering those from plants.

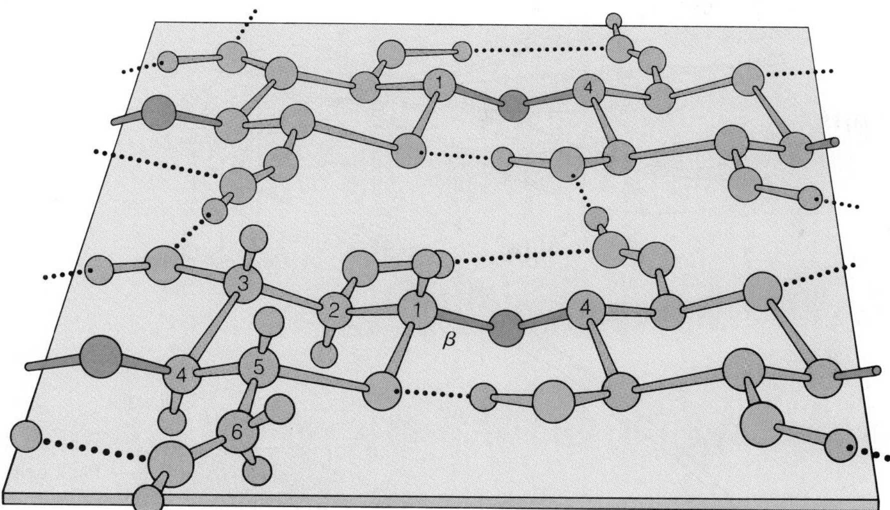

Figure 8.23
Cellulose structure. The $\beta(1\rightarrow4)$ linkages of cellulose generate a planar structure. The parallel cellulose chains are linked together by a network of hydrogen bonds.

Cellulose. The major polysaccharide in woody and fibrous plants (like trees and grasses), cellulose is the most abundant single polymer in the biosphere. Like amylose, cellulose is a linear polymer of D-glucose (and hence is also a glucan), but in this case the sugar residues are connected by $\beta(1\rightarrow4)$ linkages (Figure 8.23). This seemingly small difference from starch has remarkable structural consequences. Cellulose can exist as fully extended chains, with each glucose residue flipped by 180° with respect to its neighbor. In this extended form, the chains can form ribbons that pack side by side with a network of hydrogen bonds within and between them. This arrangement is reminiscent of the β-sheet structure in silk fibroin, and as in fibroin, the fibrils of cellulose have great mechanical strength but limited extensibility.

The small difference in linkage in cellulose and starch has another important consequence: animal enzymes that are able to catalyze the cleavage of the $\alpha(1\rightarrow4)$ link in starch cannot cleave cellulose. Ruminants such as cows can digest cellulose only because they contain in their digestive tracts symbiotic bacteria that produce the necessary **cellulases.** Termites manage to eat woody substances in a similar fashion, for they harbor in their guts protozoans capable of cellulose digestion. Many fungi also produce such enzymes, which is why mushrooms can live on wood as a carbon source.

It should not be presumed that the fibrous parts of plants are made up exclusively from cellulose. A variety of other polysaccharides are present in plant cell walls. These include the **xylans,** which are polymers with $\beta(1\rightarrow4)$-linked D-xylopyranose, often with substituent groups attached; the **glucomannans;** and many other polymers. Often these polysaccharides are grouped together under the term hemicellulose.

$\cdots\beta$-D-Xyl*p*$(1\rightarrow4)[\beta$-D-Xyl*p*$(1\rightarrow4)]_7$-β-D-Xyl*p*$(1\rightarrow4)$-β-D-Xyl*p*$(1\rightarrow4)\cdots$
Acetyl at C-2 or C-3 | 4-O-Meα-D-Glc*p*$(1\rightarrow2)$

A typical xylan structure

$\cdots\beta$-D-Glc*p*$(1\rightarrow4)$-β-D-Man*p*$(1\rightarrow4)$-β-D-Man*p*$(1\rightarrow4)$-β-D-Man*p*$(1\rightarrow4)\cdots$
β-D-Gal*p*$(1\rightarrow6)$ Acetyl at C-2 or C-3

A typical glucomannan structure

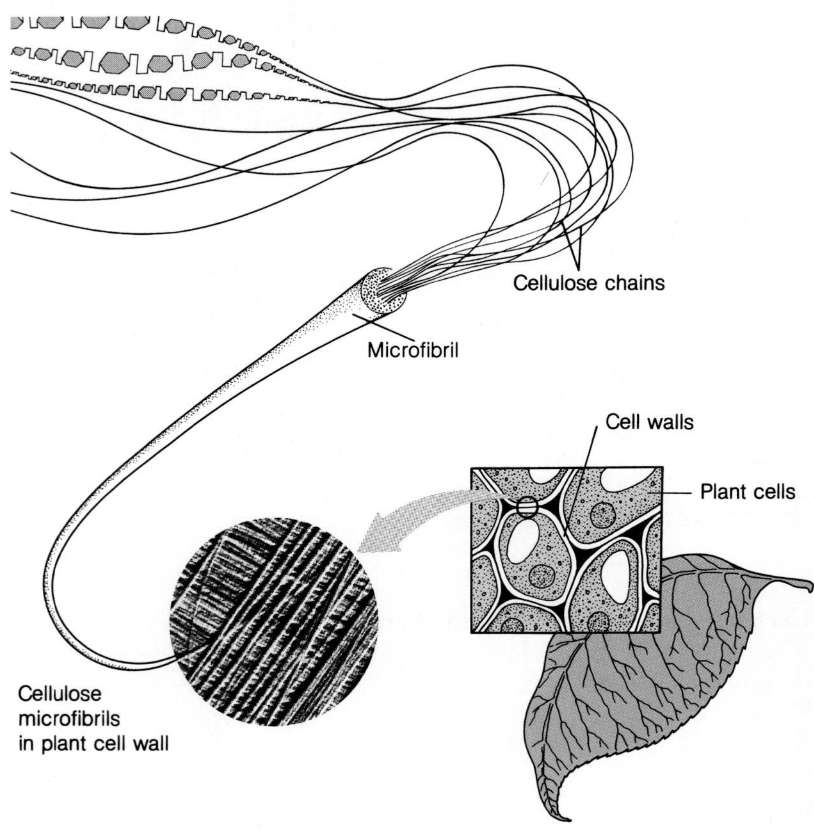

Figure 8.24
Organization of plant cell walls. Microfibrils of cellulose are embedded in a matrix of hemicellulose. Note that the fibers are laid down in a crosshatched pattern to give strength in all directions.

The cell wall of a plant is a complex structure, made up of several layers. Microfibrils of cellulose are laid down in a crosshatched pattern (Figure 8.24) and impregnated with a matrix of the other polysaccharides. The same principle is used when glass fibers are embedded in a tough resin to produce strong, durable sheets of fiberglass.

Cellulose is not confined exclusively to the plant kingdom. The marine invertebrates called tunicates contain considerable quantities of cellulose in their hard outer mantle. There are even reports of small amounts of cellulose in human connective tissue. However, as a structural material, cellulose seems to have been largely passed over in animal evolution. In the fungi, there is extensive use of other glucans, with $\beta(1\rightarrow3)$ or $\beta(1\rightarrow6)$ linkages between glucose residues, as structural polysaccharides.

CHITIN. A homopolymer of *N*-acetyl-β-D-glucosamine, chitin has a structure basically similar to that of cellulose, except that the hydroxyl on carbon 2 of each residue is replaced by an acetylated amino group:

Chitin

Chitin is widely distributed among the kingdoms of organisms. It is a minor constituent in most fungi and some algae, where it often substitutes for cellulose or other glucans. In dividing yeast cells, chitin is found in the septum that forms between the separating cells. The best known role of chitin, however, is in invertebrate animals; it constitutes a major structural material in the exoskeletons of many arthropods and mollusks. In many of these exoskeletons, chitin forms a matrix on which mineralization takes place, much as collagen acts as a matrix for mineral deposition in vertebrate bones. The evolutionary implications are interesting. As animals evolved to the size where rigid body parts became essential, quite different paths were taken. The vertebrates developed a mineral skeleton on a collagen matrix. Annelids such as earthworms also use collagen, but in a segmented exoskeleton. The arthropods and mollusks also developed exoskeletons, but these were built on chitin—a carbohydrate rather than a protein matrix.

Glycosaminoglycans

One group of polysaccharides is of major structural importance in vertebrate animals. These are the **glycosaminoglycans,** formerly called **mucopolysaccharides.** Important examples are the **chondroitin sulfates** and **keratan sulfates** of connective tissue, the **dermatan sulfates** of skin, and **hyaluronic acid.** All are polymers of repeating disaccharide units, in which one of the sugars is either *N*-acetylgalactosamine or *N*-acetylglucosamine. Some examples are shown in Figure 8.25.

A major function of the glycosaminoglycans is the formation of a matrix to hold together the protein components of skin and connective tissue. An example is given in Figure 8.26, which illustrates the protein–carbohydrate (or **proteoglycan**) complex in cartilage. The filamentous structure is built on a single long hyaluronic acid molecule, to which extended core proteins are attached noncovalently. These, in turn, have chondroitin sulfate and keratan sulfate chains covalently bound to them through serine side chains. In cartilage, this kind of structure binds collagen (Chapter 6) and helps hold the collagen fibers in a tight, strong network. The binding apparently involves electrostatic interactions between the sulfate and/or carboxylate groups and basic side chains in collagen.

The glycosaminoglycan complexes of connective tissue constitute one of the very few examples in which the element silicon enters into biology. Some of the carbohydrate chains are cross-linked by bridges of the type

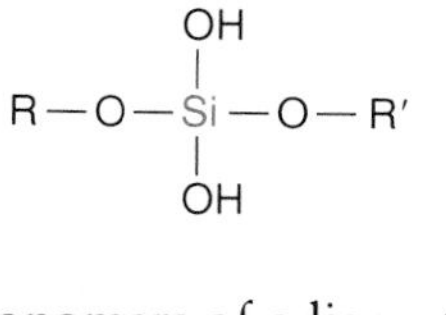

$$\mathrm{R{-}O{-}\overset{\overset{\displaystyle OH}{|}}{\underset{\underset{\displaystyle OH}{|}}{Si}}{-}O{-}R'}$$

where R and R′ are sugar monomers of adjacent chains. There is about one silicon atom for every hundred sugar monomers.

Hyaluronic acid has other functions in the body than as a structural component. The polymer itself is very soluble in water and is present in synovial fluid of joints and in the vitreous humor of the eye. It appears to act as a viscosity-increasing agent or lubricating agent in these fluids.

Another highly sulfated glycosaminoglycan is **heparin;** one fragment of its complex chain is shown in the margin. Heparin appears to be a natural anticoagulant and is found in many body tissues. It binds strongly to a blood protein, antiprothrombin III, and the complex inhibits enzymes of the blood-clotting process (see Chapter 11). Heparin is used medicinally to inhibit clotting in blood vessels.

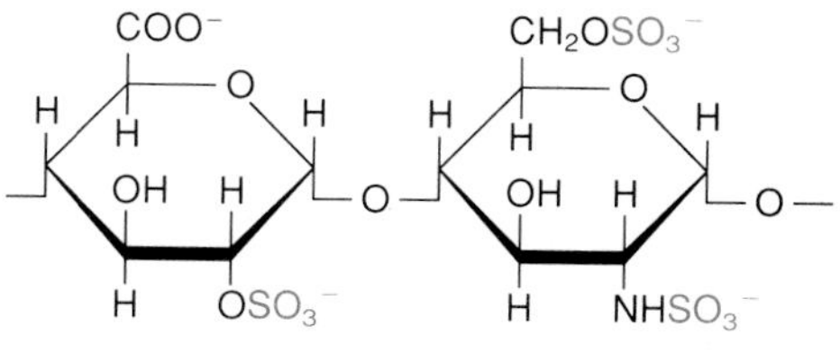

Heparin

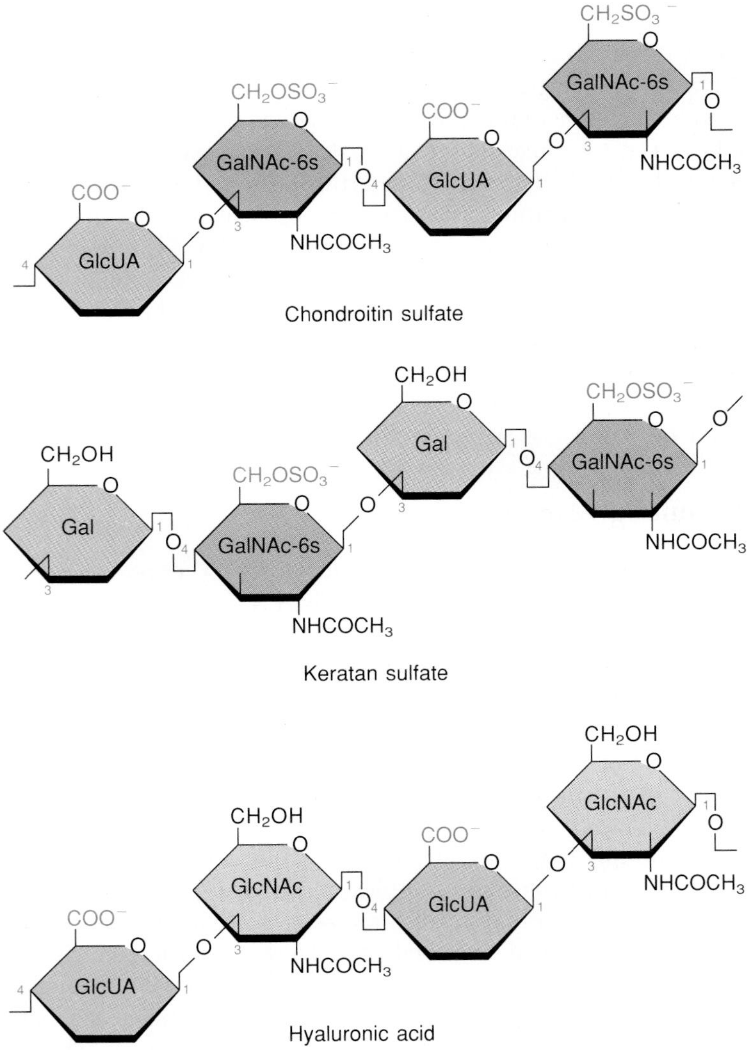

Figure 8.25
Repeating structures of some glycosaminoglycans. In each case, the repeating unit is a disaccharide, of which two are shown for each structure. Abbreviations are in Table 8.4; suffix 6s means a sulfate group on carbon 6. To simplify the figure, hydrogens and non-reacted hydroxyls are not shown.

Chondroitin sulfate

Keratan sulfate

Hyaluronic acid

The glycosaminoglycans are interesting examples of how sugar residues can be modified to provide polymers with a wide variety of properties and functions. In this respect, these polysaccharides approach polypeptides in complexity of structure and function.

Bacterial Cell Wall Polysaccharides

In Chapter 1 we noted that bacteria and most other unicellular organisms possess a *cell wall.* The nature of this cell wall is the basis for categorizing bacteria into two major classes: those that retain the Gram stain (a dye–iodine complex), which are called **Gram-positive** bacteria, and those that do not, which are termed **Gram-negative.** Gram-positive bacteria have a cell wall with a cross-linked, multilayered polysaccharide–peptide complex called **peptidoglycan** at the surface, outside the lipid cell membrane (Figure 8.27a). Gram-negative bacterial cell walls also contain peptidoglycan, but it is single-layered and covered by an outer lipid membrane layer (see Figure 8.27b). This difference allows the Gram stain to be washed from Gram-negative bacteria.

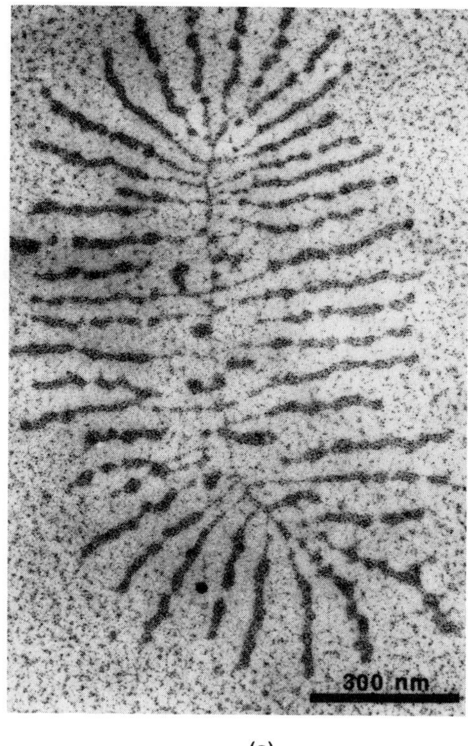

(a)

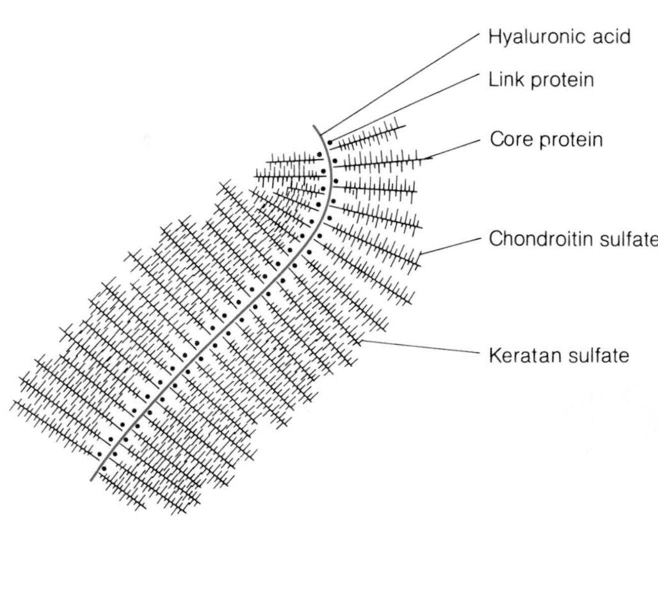

(b)

Figure 8.26
Proteoglycan structure in bovine cartilage. (**a**) An electron micrograph of a proteoglycan aggregate. (**b**) A schematic drawing of the structure shown in (**a**). Keratan sulfate and chondroitin sulfate are covalently linked to extended core protein molecules. The latter are noncovalently attached to a long hyaluronic acid molecule with the aid of a "link protein."

The chemical structure of the peptidoglycan of a Gram-positive bacterium is shown in Figure 8.28. Long polysaccharide chains, which are strictly alternating copolymers of *N*-acetylglucosamine and *N*-acetylmuramic acid, are cross-linked through short peptides. These peptides have unusual structures. Attached to the lactic acid moiety of the *N*-acetylmuramic acid is a tetrapeptide with the sequence

(L-Ala)–(D-Glu)–(L-Lys)–(D-Ala)

This peptide is unusual in two respects: it contains some D-amino acids, and the glutamic acid residue is linked into the chain through its γ-carboxyl instead of the usual α-carboxyl linkage. To the ϵ-amino group of each lysine residue is attached a glycine pentapeptide, which is bonded at its other end to the terminal D-Ala residue of an adjacent chain. The result is the formation of a covalently cross-linked structure that envelopes the bacterial cell. The entire cell wall can be regarded as a single enormous molecule made up of multiple layers of cross-linked peptidoglycan strands. In addition to the components mentioned above, elongated molecules of substances called **lipoteichoic acids** (see Chapter 9) protrude from the membrane through the peptidoglycan wall. The existence of the cell wall protects bacteria from plasmolysis when they are in the blood of host animals.

In the Gram-negative bacteria, the peptidoglycan layer is much thinner. Although the same basic polysaccharide structure is present, the peptide chains and their linkage are somewhat different (see Figure 8.27b). We shall return to the discussion of bacterial cell walls and membranes in Chapter 9, where we will be able to include the membrane and other lipid components to provide a more complete picture.

Clearly, assembly of a structure as complex as the bacterial cell wall requires a battery of enzymes and reactions. A number of antibiotics (for

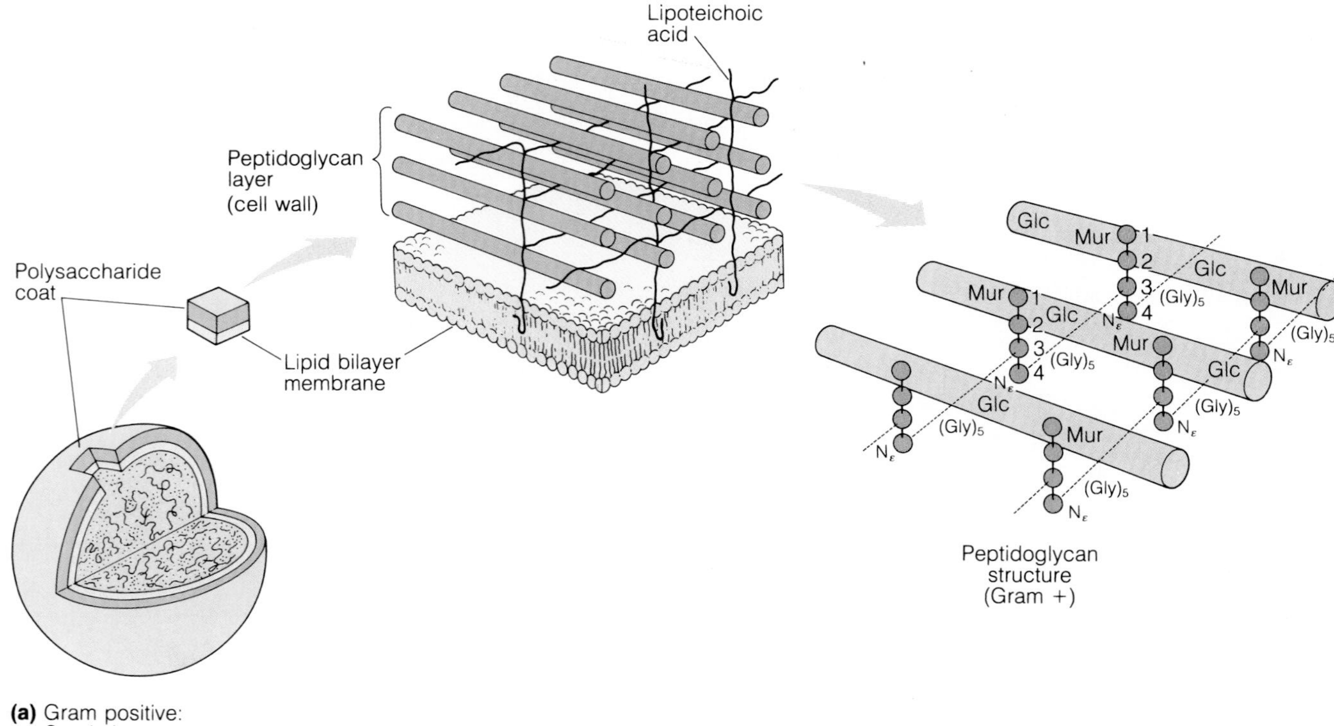

(a) Gram positive: *Staphylococcus aureus*

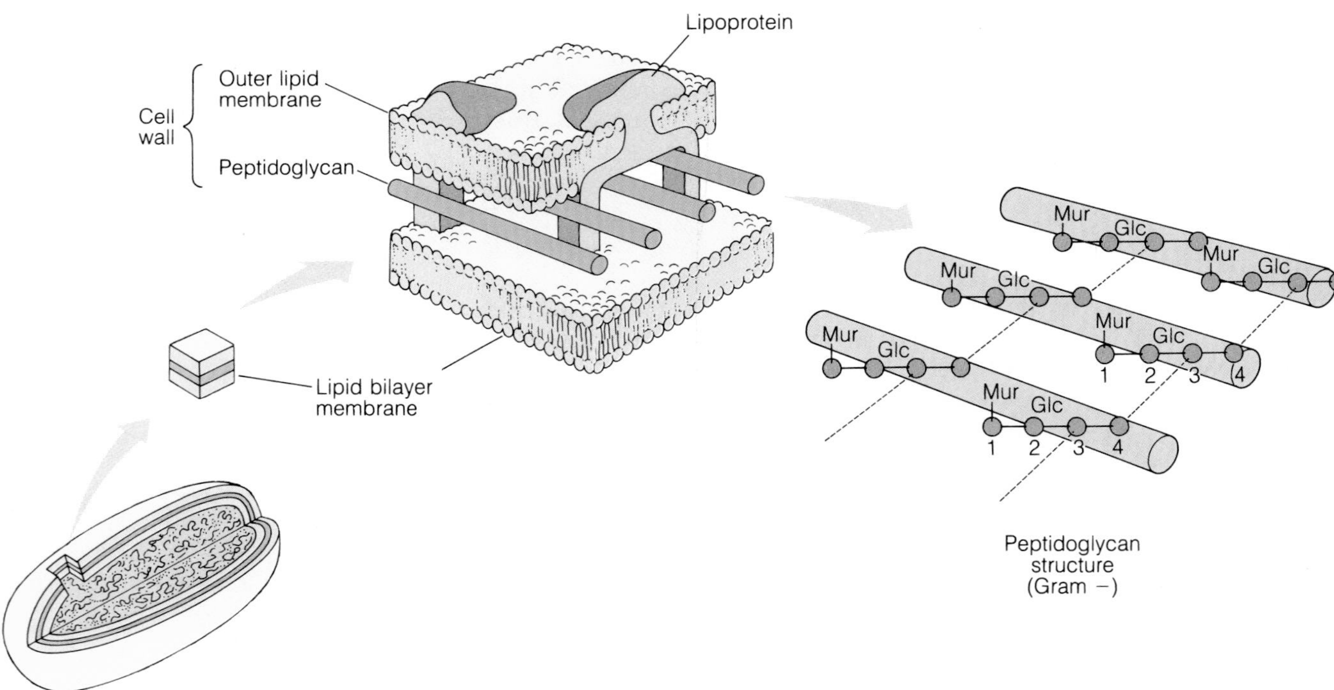

(b) Gram negative: *Escherichia coli*

Figure 8.27
Schematic drawings of bacterial cell walls. (**a**) A representative Gram-positive bacterium, *Staphylococcus aureus.* (**b**) A representative Gram-negative bacterium, *Escherichia coli.* Cylinders represent polysaccharide chains, and strings of circles represent peptide chains; dotted lines show links between them. Note the much greater thickness of the peptidoglycan layer in Gram-positive bacteria.

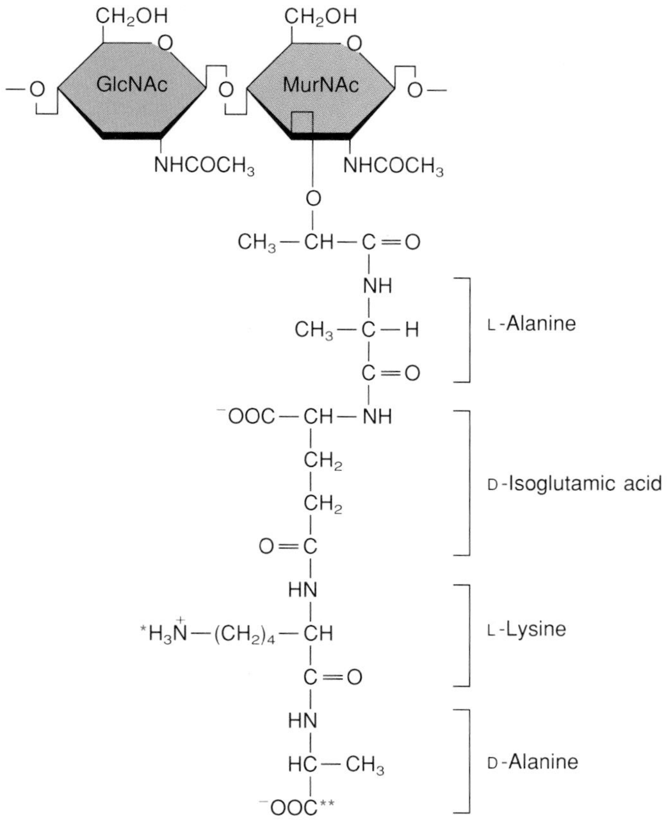

Figure 8.28
Chemical structure of the peptidoglycan of Gram-positive bacteria. Cross-links are formed by pentaglycine chains between the ε-amino group of the lysine on one chain (*) and the C-terminal carboxyl group of another (**).

example, penicillin) inhibit bacterial growth by interfering with the formation of the peptidoglycan layer. It is appropriate to defer discussion of these antibiotics until we consider the carbohydrate synthetic pathways in Chapter 16. However, we should note that one class of naturally occurring antibiotic substances acts, not by interfering with cell wall synthesis, but by attacking the peptidoglycan layer itself. These are the **lysozymes,** enzymes with wide distribution—they are found in bacteriophage, egg white, and human tears, for example. Lysozymes catalyze the hydrolysis of the glycosidic links between GlcNAc and MurNAc residues in the polysaccharide. Thus, they literally dissolve the cell wall, resulting in lysis and bacterial death.

Glycoproteins and Glycolipids

Many proteins carry covalently attached oligosaccharide or polysaccharide chains. There is an astonishing variety of these complexes, which are known as **glycoproteins,** and they serve many different functions; a few representative examples are listed in Table 8.6. Saccharide chains (**glycans**) can be linked to proteins in two major ways (Figure 8.29). The **O-linked glycans** are usually attached by a glycosidic bond between *N*-acetylgalactosamine and the hydroxyl group of a threonine or serine residue, although in a few cases—collagen, for example—hydroxylysine or hydroxyproline is employed. **N-linked glycans** are attached, usually through *N*-acetylglucosamine, to an asparagine residue. A common sequence surrounding the asparagine is –X–Asn–X–Thr–, where X may be any amino acid residue.

Table 8.6
Carbohydrate moieties of some glycoproteins

Glycoprotein	Oligosaccharide[a] and Attachment Site	No. of Chains in Proteins	Function of Protein
Fish antifreeze protein	Gal–GalNAc–Thr	From 4 to 50 in different proteins	Serves to lower body fluid freezing point
Sheep submaxillary mucin	Sia–GalNAc–Ser (or Thr)	Many	Lubrication
Ribonuclease B	$(Man)_6$–GlcNAc–GlcNAc–Asn	One	Enzyme
Hen ovalbumin	Man–Man \ Man–GlcNAc–GlcNAc–Asn / Man–Man (–Man) (only one of many variants)	One	Storage protein in egg white
Human IgG	Sia–Gal–GlcNAc–Man \ Man–GlcNAc–GlcNAc(–Fuc)–Asn / Sia–Gal–GlcNAc–Man (many other variants in other antibodies)	Two	Antibody molecule

[a] For compactness, the anomeric forms and linkages have been eliminated. For details, see R. C. Hughes, *Glycoproteins* (London: Chapman and Hall, 1983).

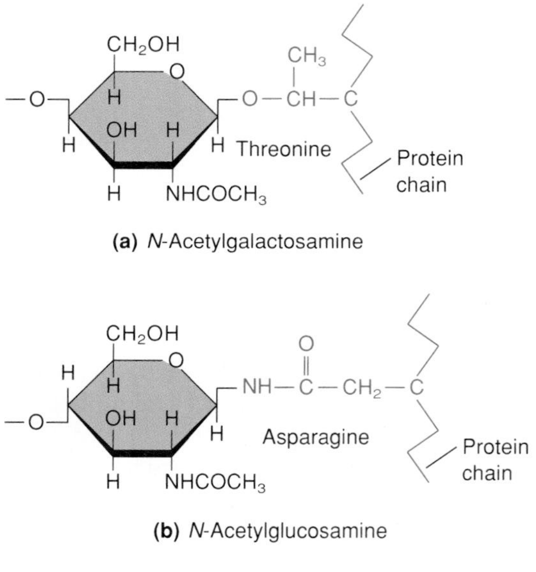

Figure 8.29
Two ways to link oligosaccharide chains to proteins. (**a**) O-linked chains are connected through C-1 of *N*-acetylgalactosamine to the hydroxyl of threonine or serine, via a glycosidic bond. (**b**) N-linked chains are bonded to the amide side chain of an asparagine residue. In this case *N*-acetylglucosamine is almost always the first sugar residue.

N-Linked Glycans

Careful study of many glycoproteins has revealed an enormous variety of N-linked oligosaccharide side chains, often exhibiting a complex branched structure. However, a common motif is often seen; the following structure serves as a foundation for further elaboration:

$$\begin{matrix}\text{Man } \alpha(1 \rightarrow 6) \\ \text{Man } \alpha(1 \rightarrow 3)\end{matrix} \text{Man } \beta(1 \rightarrow 4)\text{-GlcNAc } \beta(1 \rightarrow 4)\text{-GlcNAc } \beta(1 \rightarrow \text{N})\text{-Asn}$$

This structure can be seen, for example, in the glycan moieties of ovalbumin and the immunoglobulins. The structure of the oligosaccharide found attached to a human immunoglobulin G (IgG) is shown in Table 8.6. The residue denoted Fuc is α-L-fucose.

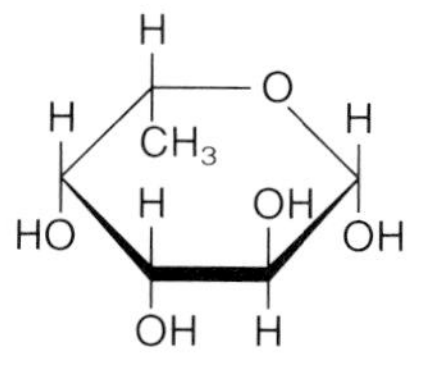

H, H, O, H, CH_3, H, OH, HO, OH, OH, H

α-L-Fucose

The immunoglobulins are an important example of the informational function of the glycan chains on glycoproteins. You will recall from Chapter 7 that every immunoglobulin has carbohydrate attached to the constant domain of each heavy chain. These constant domains allow the different types of immunoglobulin to be recognized, both for proper tissue distribution and for interaction with phagocytic cells, which will destroy the antigen–antibody complex. A part, at least, of this recognition is based on differences in the oligosaccharide chains.

The oligosaccharides of some glycoproteins circulating in the blood (immunoglobulins, for example) also serve as "time clocks" to signal old protein molecules for destruction. During circulation, the sialic acid residues at the termini of these oligosaccharides are slowly cleaved off. When they are gone, receptors in the surfaces of liver cells recognize this and bind the proteins, after which the proteins are engulfed by the cells and destroyed.

A further use of N-linked oligosaccharides is in intracellular targeting in eukaryotic organisms. In Chapters 16 and 28 we will see how proteins destined for certain organelles or for excretion are marked specifically by oligosaccharides during posttranslational processing and then sent to their proper destinations.

O-Linked Glycans

Many proteins carry O-linked oligosaccharides and they serve a variety of functions. Antarctic fish contain a glycoprotein that serves as an "antifreeze," preventing freezing of body fluids even in extremely cold water. The **mucins**, glycoproteins found extensively in salivary secretions, contain many short O-linked glycans. The highly extended and highly hydrated mucins increase the viscosity of the fluids in which they are dissolved. Some O-linked glycans also appear to function for intracellular targeting.

Blood Group Substances

A most important group of oligosaccharides are the **blood group antigens.** On some cells these are attached as O-linked glycans to membrane proteins. Alternatively, the oligosaccharide may be linked to a lipid molecule to form a **glycolipid** (see Chapter 9). The lipid portion of the molecule helps attach the antigen to the outside surface of erythrocyte membranes. It is these oligosaccharides that determine the blood group types in humans. Their presence in a blood sample is detected by blood typing—determining whether antibodies to a particular antigen cause the red cells of that blood sample to agglutinate. Although the system consisting of blood types A, B, AB, and O is probably most familiar to you, it is just one of 14 genetically characterized blood group systems, with more than 100 different blood group antigens. These substances are also present in many cells and tissues other than blood, but we often focus on blood because of the widespread use of typing in establishing familial relationships and preparing blood for transfusion.

For simplicity, we will concentrate on the ABO system as an example. Figure 8.30 depicts oligosaccharides corresponding to each of these blood types. Almost all humans can produce the type O saccharide, but addition of either galactose (to make type B) or *N*-acetylgalactosamine (to make type A) requires special enzymes (see Chapter 16 for details of this synthesis). Some individuals possess one of these enzymes, some possess the other, and

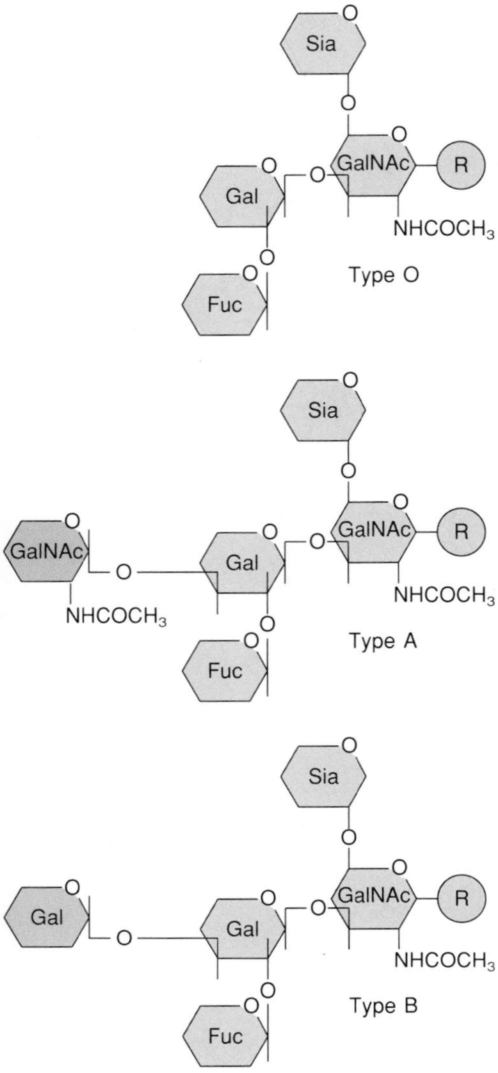

Figure 8.30
The blood group antigens. The O-oligosaccharide (top) does not elicit antibodies in most humans. The A and B antigens are formed by addition of GalNAc or Gal, respectively, to the O oligosaccharide. Each of the latter can elicit a specific antibody. In this figure R can represent either a protein molecule or a lipid molecule.

a few are heterozygous and can produce both. The latter individuals have type AB blood, with both A and B oligosaccharides present on cell surfaces.

Humans can produce antibodies against the A and B oligosaccharides, but the O type are nonantigenic. Normally, no one produces antibodies against his or her own antigen but does produce them against the other antigen type. Thus, an individual with type A blood carries antibodies directed against the B polysaccharide. If he or she accepts blood from a B-type donor, these antibodies will cause clumping and precipitation of the donated blood cells. Nor can a B individual safely accept blood from an A type. Persons with type O blood normally have antibodies against both A and B and thus can receive from neither. Those with AB type, since they carry both A and B antigens themselves, have antibodies against neither and thus can safely receive blood from any individual.

In donating blood, an inverse relationship holds. Those with type O blood, which carries no antigenic determinants, can safely donate to any—they are the "universal donors." Type AB individuals can donate *only* to other ABs; a person of any other type will carry antibodies to A, or B, or both. These relationships are summarized in Table 8.7.

Biologists are beginning to realize that such molecules as the blood group antigens represent only a special case of a much more general phenomenon—cell recognition. In a multicellular organism, it is essential that different kinds of cells be marked on their surfaces so that they can interact properly with other cells and molecules and so that an organism can recognize its own cells as immunologically distinct from foreign cells. In accord with this view is the growing appreciation that the surfaces of many cells are literally covered with saccharides, attached to either proteins or lipids in the cell membrane (Figure 8.31a). An extremely thick example is shown in Figure 8.31b, which depicts the **glycocalyx,** or polysaccharide coat, on the surface of an intestinal cell.

For oligosaccharides or polysaccharides to serve as recognition signals, there must be proteins that bind to them specifically. One such class is, of course, the immunoglobulins. Another, very diverse group of saccharide-binding proteins is the **lectins.** The lectins were first recognized in plant tissues, but it is now realized that they are widely distributed and play a great variety of roles. For example, lectins seem to be involved in interactions between cells and proteins of the intercellular matrix, such as collagen, and help to maintain tissue and organ structure. Lectins in the walls of intestinal bacteria help to bond these bacteria to the glycocalyx of the intestinal epithelium.

Why do oligosaccharides so often play the role of cellular markers? We do not know, but certain possibilities suggest themselves. First, oligosaccharides can present a remarkable variety of structures in relatively short chains. The multiple choices of monomers (including modified sugars), linkages, and branching patterns allow a vast but specific vocabulary. Second,

Table 8.7
Transfusion relationships between blood types

Person Has Blood Type:	Makes Antibodies Against:	Can Safely Receive Blood From:	Can Safely Donate Blood to:
O	A, B	O	O, A, B, AB
A	B	O, A	A, AB
B	A	O, B	B, AB
AB	None	O, A, B, AB	AB

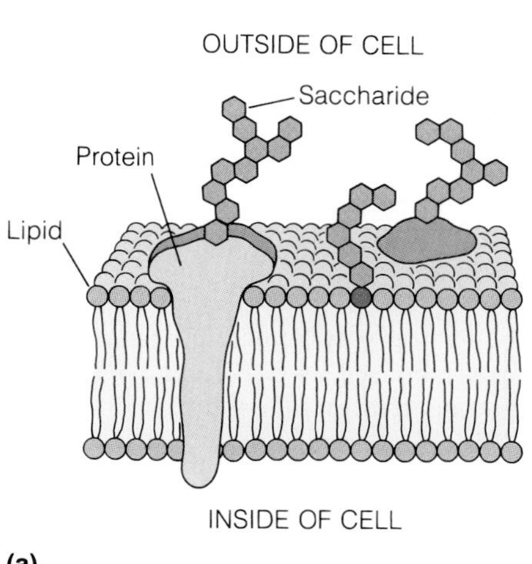

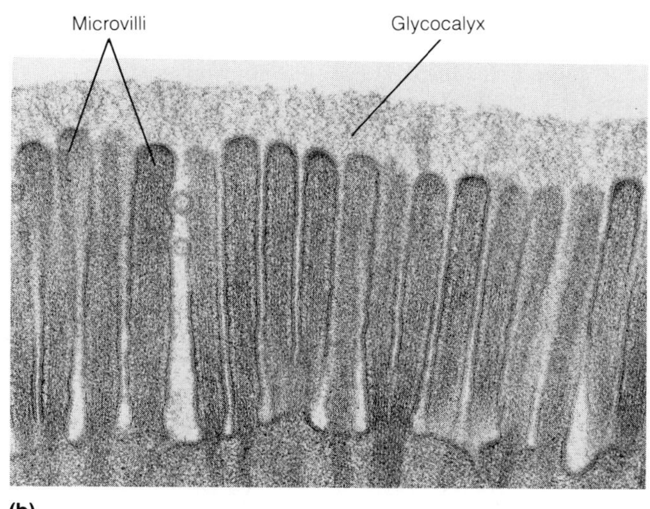

Figure 8.31
Cell surface recognition factors. (a) Schematic view of a lipid membrane. Oligosaccharides are attached to the outer surface, either through membrane-embedded proteins or special lipid molecules. (b) Electron micrograph of the surface of an intestinal epithelial cell. The projections, called microvilli, are covered on their outer surface by a layer of branched polysaccharide chains attached to proteins in the cell membrane. This carbohydrate layer, called the *glycocalyx,* is found on many animal cell surfaces.

oligosaccharides are especially potent antigens, which means that specific antibodies can be elicited swiftly against them (see Chapter 7). Whether this is the result of some intrinsic property of sugar molecules or of the antibody molecules is unclear. It is possible that antibodies evolved as a defense against bacteria, which have polysaccharide-rich walls, and thus favor saccharides as targets.

It seems likely that carbohydrate biochemistry, which has long been considered a rather prosaic part of the field, is about to enter an exciting new phase. As biochemists turn their attention more and more to the relationships between cells and tissues that make for an integrated organism, they find that many of these relationships are mediated by carbohydrates.

REFERENCES

General

Binkley, R. W. (1988) *Modern Carbohydrate Chemistry.* Marcel Dekker, New York. A comprehensive contemporary survey.

Ginsburg, V., and P. Robbins (eds.) (1984) *Biology of Carbohydrates.* Wiley, New York.

Lennarz, W. J. (ed.) (1980) *The Biochemistry of Glycoproteins and Proteoglycans.* Plenum, New York. Contains a number of useful review papers.

Pigman, W., and D. Horton (eds.) (1972) *The Carbohydrates,* 2nd ed. Academic Press, New York. A valuable source book for details of carbohydrate structure and chemistry.

Sugar Conformations

Angyal, S. J. (1969) The composition and conformation of sugars in solution. *Angew. Chem.* 8:157–226. A fine, clearly written review, drawing heavily on the earlier NMR data.

Barker, R., and A. S. Serianni (1986) Carbohydrates in solution: Studies with stable isotopes. *Acc. Chem. Res.* 19:307–313. A brief reveiw of recent ^{13}C NMR work.

Bock, K., and C. Pedersen (1983) Carbon-13 nuclear magnetic resonance spectroscopy of monosaccharides. *Adv. Carbohydrate Chem. Biochem.* 41:27–66.

Bacterial Cell Walls

Schockman, G. D., and J. F. Barnett (1983) Structure, function, and assembly of cell walls of Gram-positive bacteria. *Annu. Rev. Microbiol.* 37:501–527.

Tipper, D. J., and A. Wright (1979) The structure and biosynthesis of bacterial cell walls. In: *The Bacteria,* Vol. VII, edited by J. R. Sokatch and L. M. Denston, pp. 291–426. Academic Press, New York.

Glycoproteins

Hughes, R. C. (1983) *Glycoproteins.* Chapman and Hall, London. A small book, packed with information on glycoprotein structure and biosynthesis.

Vliegenhart, J. F. G., L. Dorland, and H. van Halbeek (1983) High resolution ^{1}H-nuclear magnetic resonance spectroscopy as a tool in the structural analysis of carbohydrates related to glycoproteins. *Adv. Carbohydrate Chem. Biochem.* 41:209–374.

Plant Saccharides

Goodwin, T. W., and E. I. Mercer (1983) *Introduction to Plant Biochemistry.* Pergamon, Oxford.

Oligosaccharides and Cell Recognition

Barondes, S. H. (1984) Soluble lectins: A new class of extracellular proteins. *Science* 223:1259–1264.

Labat-Robert, J., R. Timpl, and R. Ladiglas (eds.) (1986) *Structural Glycoproteins in Cell–Matrix Interaction.* Karger, New York.

PROBLEMS

1. Draw Haworth projections for the following:
   (a) 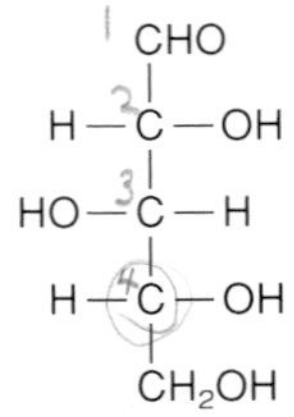

   in α-furanose form. Name the sugar.
   (b) The L isomer of the above
   (c) α-D-GlcNAc
   (d) α-D-fructofuranose

2. α-D-Galactopyranose rotates the plane of polarized light, but the product of its reduction with sodium borohydride (galacticol) does not. Explain the difference.

3. Using data in Table 8.2, calculate the standard state free energy change for the conversion of D-glucose from the α to the β anomer at 40°C.

4. The disaccharide α,β-trehalose differs from the α,α structure in Figure 8.16 by having an α(1→1)β linkage. Draw its structure.

5. A *reducing* sugar will undergo the Fehling reaction (p. 273), which requires a (potential) free aldehyde group. Which of the disaccharides shown in Figure 8.16 are reducing, and which are nonreducing?

6. *Dextrans* are polysaccharides produced by certain species of bacteria. They are glucans, with primarily α(1→6) linkages. There is frequent α(1→3) branching. Draw a Haworth projection of a portion of a dextran, including one branch point.

7. What is the natural polysaccharide whose repeating structure can be symbolized by GlcUAc β(1→3) GlcNAc, with these units connected by β(1–4) links?

8. Decide whether the structures shown are R or S in the absolute system.

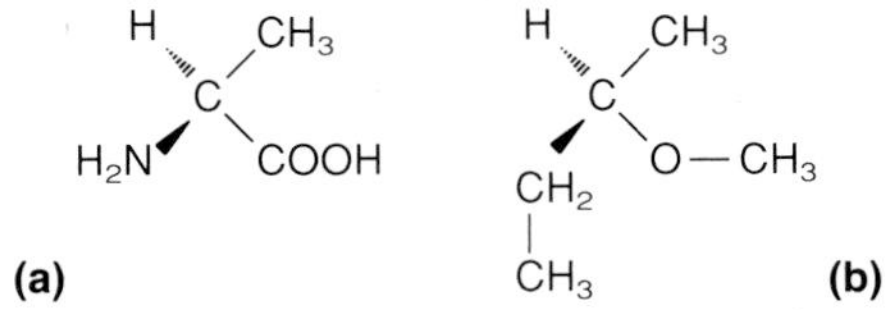

9. The reagent periodate (IO_4^-) oxidatively cleaves the carbon–carbon bonds between two carbons carrying hydroxyl groups. Explain how periodate oxidation might be used to distinguish between methyl glycosides of glucose in the pyranose and furanose forms.

10. Penicillin is very effective against certain living bacteria but has no effect on dead ones. Explain why.

TOOLS OF BIOCHEMISTRY 13

Sequencing Oligosaccharides

Determining oligosaccharide sequences presents problems similar to, but more difficult than, those encountered in protein sequencing. Because of the many types of monomers that may be encountered and the variety of linkages between them, no single method like the Edman degradation of polypeptides has been devised.

The first step in any sequence analysis is, as in polypeptide analysis, the determination of composition. The oligomer is hydrolyzed in acidic solution, yielding a mixture of monosaccharides. At present, these are almost always separated, identified, and quantitated by gas–liquid chromatography.

Table T13.1
Some specific glycosidases used in sequencing oligosaccharides

Enzyme Name	Source	Specificity
Exoglycosidases		
Neuraminidase	*Streptococcus pneumoniae*	Siaα(2→3 or 6)Gal or Siaα(2→6)GlcNAc
β-Galactosidase	*Streptococcus pneumoniae*	Galβ(1→4)GlcNAc
α-Fucosidase	*Clostridium perfringens*	Fucα(1→2)Gal
Endoglycosidases		
Endo-β-Galactosidase	*Escherichia freundii*	...GlcNAcβ(1→3)Galβ(1→4)Glc(GlcNAc)...
Almond emulsion	Bitter almond seeds	Cleaves bond to Asn in many N-linked oligosaccharides

Determination of the sequence itself is much more difficult. In the past, extensive use was made of chemical methods, but these have been largely supplanted by enzymatic cleavage of the oligomer followed by sophisticated methods for identification of the fragments. Researchers are now familiar with a large number of enzymes (**glycosidases**) that catalyze the cleavage of glycosidic bonds between sugar moieties. Some of these enzymes are very specific in their action. They may be divided into two groups: exoglycosidases, which remove the terminal residue from an oligosaccharide chain, and endoglycosidases, which catalyze cleavage within the chain. A few are listed in Table T13.1.

A simple example of the application of specific glycosidases to sequence determination is shown in Figure T13.1. The oligopeptide shown is a portion of one of several attached to the blood serum protein *orosomucoid.* The residue at the reducing end of the chain can be removed by neuraminidase. According to Table T13.1, the terminal residue must be sialic acid, attached to either Gal or GlcNAc. Sialic acid is found, and subsequent release of Gal by *Streptococcus* β-galactosidase indicates that the next residue is Gal, attached by a 1→4 linkage to GlcNAc. The latter residue is confirmed by its release by β-*N*-acetylglucosaminidase.

Although such procedures can often provide definitive information, they are laborious and not always effective. In recent years there has been remarkable development of high-resolution NMR and mass spectrometry as techniques for identification of complex oligosaccharides. The use of these methods is described in the reviews by Sweeley and Nunez and by Vliegenhardt et al. (see references) but is too technical to be described here.

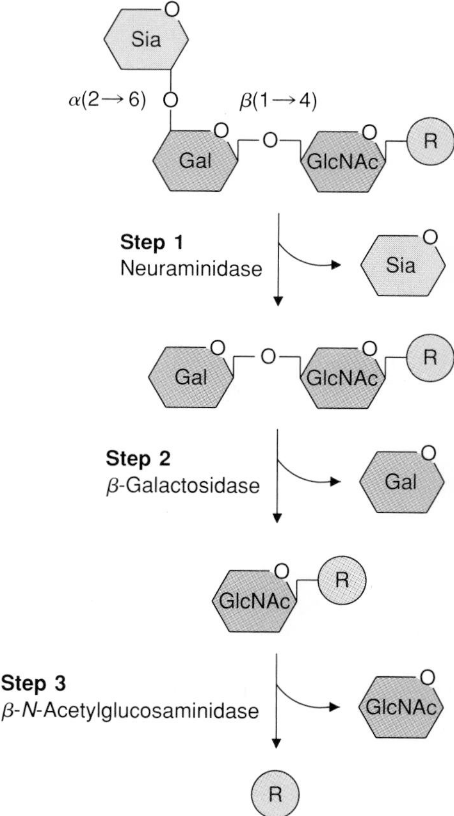

Figure T13.1
Successive cleavage of a portion of an oligosaccharide attached to the protein *orosomucoid.*

REFERENCES

Hughes, R. C. (1983) *Glycoproteins,* Chapter 2. Chapman & Hall, London. New York.

Sweeley, C. C., and H. A. Nunez (1986) Structural analysis of glycoconjugates by mass spectrometry and nuclear magnetic resonance. *Annu. Rev. Biochem.* 54:765–801.

Vliegenhardt, J. F. G., L. Dorland, and H. van Halbeek (1983) High resolution ^{1}H-nuclear magnetic resonance spectroscopy as a tool in the structural analysis of carbohydrates related to glycoproteins. *Adv. Carbohydrate Chem. Biochem.* 41:209–392.

CHAPTER 9

Lipids, Membranes, and Cellular Transport

The molecules we encounter in this chapter, the **lipids**, carry out multiple functions. Some—the fats—are used for energy storage, but a large fraction of cellular lipids are used to form the partitions that divide different compartments from one another and separate the cell from its surroundings. These partitions are the **lipid membranes** of the cell. The membranes are much more than passive walls, for they contain highly selective gates that promote the passage of certain materials in certain directions and block others altogether. It is this property of **selective membrane permeability** that allows each of the different parts of the cell to carry out its specific operations.

In this chapter we first examine the structure and behavior of lipid molecules and then describe the membranes they form and how selective transport through them is accomplished.

The Molecular Structure and Behavior of Lipids

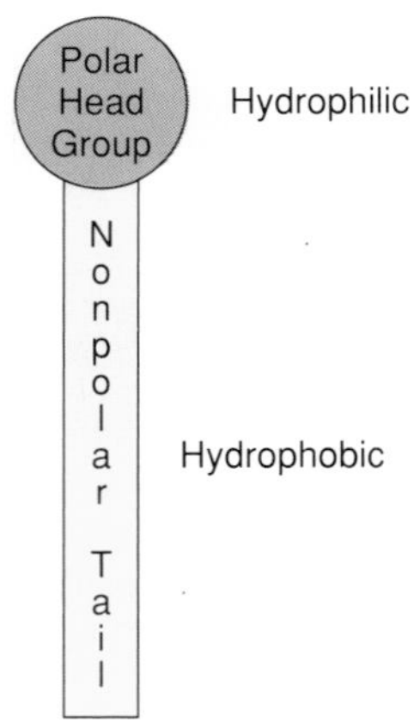

General lipid structure

Unlike the proteins, nucleic acids, and polysaccharides, lipids are not polymers. Rather, they are quite small molecules that have a strong tendency to associate together through noncovalent forces. Lipids are usually characterized by the kind of structure shown in the margin: a polar, hydrophilic "head" region connected to a very hydrophobic hydrocarbon "tail" portion. There are two fundamental reasons lipid molecules in an aqueous environment tend to clump together in noncovalent association. Just as hydrophobic groups in proteins exhibit an entropy-driven hydrophobic effect, so do the nonpolar tails of lipids. A second stabilizing force is the van der Waals interaction between the hydrocarbon regions of the molecules. On the other hand, the polar, hydrophilic head groups on lipid molecules tend to be associated with water. Lipids are, then, prime examples of the kind of amphipathic substance described in Chapter 2.

There are a number of possible consequences of this "molecular schizophrenia," as shown in Figure 2.16 (Chapter 2). From a biological point of view, the most important is the tendency of lipids to form micelles and membrane bilayers. Exactly what kind of structure is formed when a lipid is in contact with water depends on the specific molecular structure of the hydrophilic and hydrophobic parts of that lipid molecule. Thus, it is appropriate that we now examine some of the major types of lipids.

Fatty Acids

The simplest lipids are the **fatty acids,** which are also constituents of many more complex lipids. Their basic structure exemplifies the general lipid model described above: a hydrophilic carboxylate group is attached to one end of an (often long) hydrocarbon chain. An example is **stearic acid,** which is widely distributed in organisms. We show it here as the ionized form, the **stearate** ion:

$$^{-}O(O{=})CCH_2CH_2CH_2CH_2CH_2CH_2CH_2CH_2CH_2CH_2CH_2CH_2CH_2CH_2CH_2CH_2CH_3$$

Polar head group — Hydrocarbon tail

Stearate ion

Stearic acid is an example of a **saturated** fatty acid, one in which the carbons of the tail are all "saturated" with hydrogen atoms. A number of the more biologically important saturated fatty acids are listed in Table 9.1. Note that each has a common name (such as stearic acid) and a systematic name (in this case, *n*-octadecanoic acid).

Many important naturally occurring fatty acids are **unsaturated**—that is, they contain one or more double bonds (see Table 9.1). **Oleic acid,** which is found in many animal fats, yields the **oleate** ion:

$$^{-}O(O{=})CCH_2CH_2CH_2CH_2CH_2CH_2CH_2\overset{H}{C}{=}\overset{H}{C}CH_2CH_2CH_2CH_2CH_2CH_2CH_2CH_3$$

Oleate ion

In most of the naturally occurring unsaturated fatty acids the orientation about double bonds is cis rather than trans. This has an important effect on molecular structure, for each cis double bond inserts a bend into the hydrocarbon chain. Figure 9.1 presents a molecular comparison of the two 18-carbon acids described above. Although these are drawn as rigid structures, it must be remembered that there is freedom of rotation about each single bond in the tails; thus, many conformations are possible.

Although in most fatty acids the hydrocarbon chains are linear, there are examples (found primarily in bacteria) that contain branches or even cyclic structures (see Table 9.1).

To provide a more convenient and definitive way of referring to fatty acids, a system of abbreviations has been developed; it is illustrated in Table 9.1. The rules are simple: The number before the colon gives the total number of carbons, the number after the colon gives the count of double bonds, and the position of double bonds is indicated by a superscript. Thus, oleic acid is $18{:}1^{\Delta 9}$.

The fatty acids are weak acids, with pK_a values averaging about 4.5:

$$RCOOH \underset{pK_a \cong 4.5}{\rightleftharpoons} RCOO^- + H^+$$

Thus, they exist in the anionic form ($RCOO^-$) at physiological pH, and we should more properly speak of stearate and oleate, rather than stearic and oleic acids under these conditions. This dissociation has a most important consequence in fatty acid behavior. The charge on the carboxyl group

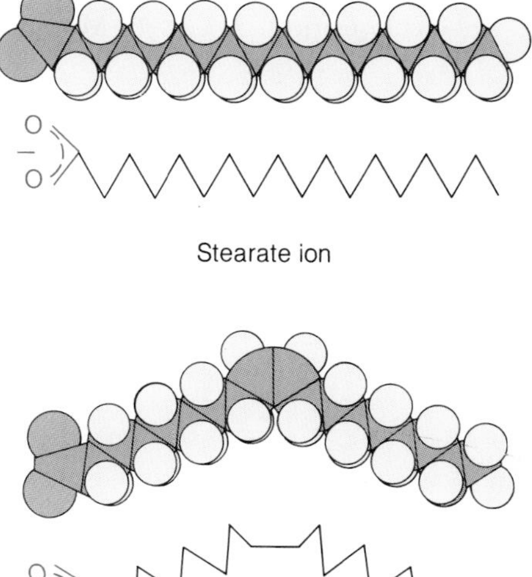

Figure 9.1
Molecular structures of stearate and oleate ions. Both molecules are depicted in their most extended form, with the skeletal structure shown below. Note the bend introduced in the oleate chain by the cis double bond.

Table 9.1
Some biologically important fatty acids

Common Name	Systematic Name	Abbreviation	Structure	Melting Point (°C)
A. Saturated fatty acids				
Capric	*n*-Decanoic	10:0	$CH_3(CH_2)_8COOH$	31.6
Lauric	*n*-Dodecanoic	12:0	$CH_3(CH_2)_{10}COOH$	44.2
Myristic	*n*-Tetradecanoic	14:0	$CH_3(CH_2)_{12}COOH$	53.9
Palmitic	*n*-Hexadecanoic	16:0	$CH_3(CH_2)_{14}COOH$	63.1
Stearic	*n*-Octadecanoic	18:0	$CH_3(CH_2)_{16}COOH$	69.6
Arachidic	*n*-Eicosanoic	20:0	$CH_3(CH_2)_{18}COOH$	76.5
Behenic	*n*-Docosanoic	22:0	$CH_3(CH_2)_{20}COOH$	81.5
Lignoceric	*n*-Tetracosanoic	24:0	$CH_3(CH_2)_{22}COOH$	86.0
Cerotic	*n*-Hexacosanoic	26:0	$CH_3(CH_2)_{24}COOH$	88.5
B. Unsaturated fatty acids				
Palmitoleic	*cis*-9-Hexadecenoic	$16{:}1^{\Delta 9}$	$CH_3(CH_2)_5CH{=}CH(CH_2)_7COOH$	0
Oleic	*cis*-9-Octadecenoic	$18{:}1^{\Delta 9}$	$CH_3(CH_2)_7CH{=}CH(CH_2)_7COOH$	16
Linoleic	*cis,cis*,9,12-Octadecadienoic	$18{:}2^{\Delta 9,12}$	$CH_3(CH_2)_4CH{=}CHCH_2CH{=}CH(CH_2)_7COOH$	5
Linolenic	all *cis*-9,12,15-Octadecatrienoic	$18{:}3^{\Delta 9,12,15}$	$CH_3CH_2CH{=}CHCH_2CH{=}CHCH_2CH{=}CH(CH_2)_7COOH$	−11
Arachidonic	all *cis*-5,8,11,14-Eicosatetraenoic	$20{:}4^{\Delta 5,8,11,14}$	$CH_3(CH_2)_4CH{=}CHCH_2CH{=}CHCH_2CH{=}CHCH_2CH{=}CH(CH_2)_3COOH$	−50
C. Branched and cyclic acids				
Tuberculostearic	*l*-D-10-Methyloctadecanoic		$CH_3(CH_2)_7CH(CH_3)(CH_2)_7COOH$	13.2
Lactobacillic	ω-(2-*n*-octylcyclopropyl)-octanoic		$CH_3(CH_2)_5CH{-}CH(CH_2)_9COOH$ (with CH_2 bridging the two CH carbons)	29

makes it extremely hydrophilic, promoting its solubility in water. At the same time, the long hydrocarbon tails are very hydrophobic. As a result, the fatty acids behave as typical amphipathic substances when we attempt to dissolve them in water. As shown in Figure 2.16 they tend to form **monolayers** at the air–water interface, with the hydrophilic carboxyl groups immersed and the hydrocarbon tails out of water. If they are shaken with water, fatty acids will make **micelles,** in which the hydrocarbon tails cluster together within the structure and the carboxylate heads are in contact with the surrounding water.

If fatty acids are mixed with water *and* an oily or greasy substance (e.g., a hydrocarbon) the micelles will form about droplets of the oil, emulsifying it. This behavior is the basis of the action of soaps and synthetic detergents. Soaps are obtained by hydrolyzing fats with alkalis such as NaOH or KOH (in earlier times, wood ashes were used) in a process called **saponification.** The fatty acids are released as either sodium or potassium salts, which are fully ionized. However, as cleansers, soaps have the disadvantage that the fatty acids are precipitated by the calcium or magnesium ions present in "hard" water, forming a scum and destroying the emulsifying action. Synthetic detergents have been devised that do not have this defect. One class is exemplified by **sodium dodecyl sulfate (SDS):**

$$^{-}O_3SO(CH_2)_{11}CH_3 + Na^+$$

We have already encountered this substance, for it is widely used in forming micelles about proteins for gel electrophoresis. There are also synthetic nonionic detergents, like **Triton X-100:**

$$H(OCH_2CH_2)_n{-}O{-}C_6H_4{-}C(CH_3)_2{-}CH_2{-}C(CH_3)_2{-}CH_3$$

The hydrophilic group here is the polyoxyethylene head group, which in the commercial product averages about 9.5 residues in length.

Although the fatty acids play important roles in metabolism, large quantities of the free acids or their anions are never found in living cells. Instead, they almost always occur as constituents of more complex lipids. We now turn to consideration of some of these classes of biologically important lipid molecules.

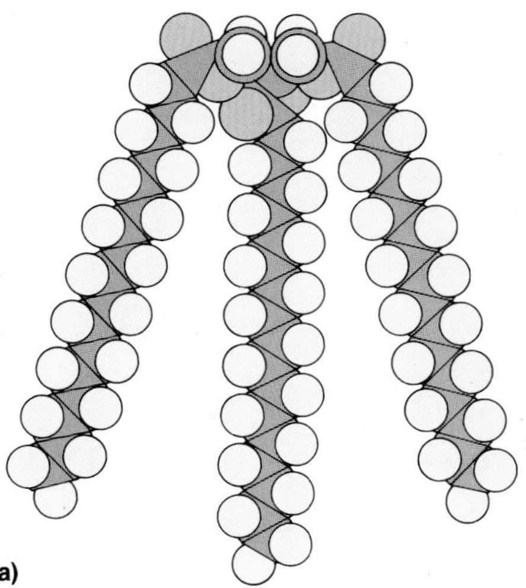

(a)

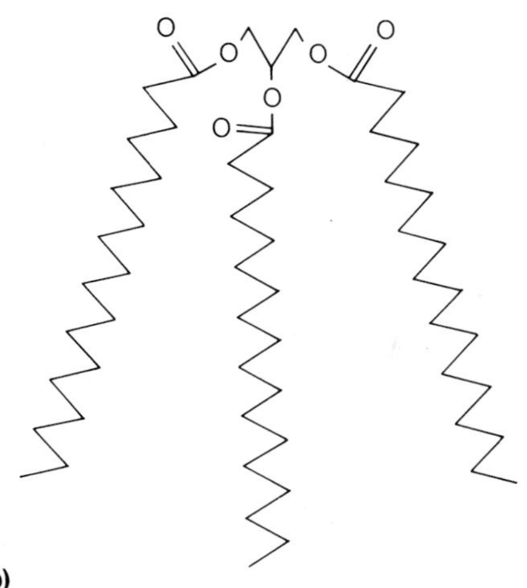

(b)

Figure 9.2
Tristearin, a fat. (**a**) Space-filling model; (**b**) skeletal structure.

Triacylglycerols: Fats

The long hydrocarbon chains of fatty acids are extraordinarily efficient for energy storage, because they contain carbon in a fully reduced form and will therefore yield a maximum amount of energy on oxidation. They are, in fact, much more efficient energy stores than are carbohydrates. (Explicit analysis of this difference is made in Chapter 17.) For this reason, lipids are used by many organisms, including humans, for energy storage.

Storage of fatty acids in organisms is largely in the form of **triacylglycerols, or fats.** These substances are *triesters* of fatty acids and **glycerol;** the general formula is

$$\begin{array}{l} H_2C{-}O{-}C({=}O){-}R_1 \\ \;\,| \\ HC{-}O{-}C({=}O){-}R_2 \\ \;\,| \\ H_2C{-}O{-}C({=}O){-}R_3 \end{array}$$

where R_1, R_2, and R_3 correspond to the hydrocarbon tails of various fatty acids. As a particular example, if $R_1 = R_2 = R_3 = (CH_2)_{16}CH_3$, the molecule **tristearin** (Figure 9.2) is obtained. Most triacylglycerols, however, contain a mixture of fatty acids, often including unsaturated ones. Table 9.2 lists the composition of some naturally occurring fats. Comparison of common experience with these fats and data in the table reveals an interesting correlation. Fats rich in unsaturated fatty acids (like olive oil) are liquid at room temperature, whereas those with a higher content of saturated fatty acids are more solid. Indeed, a wholly saturated fat is a quite firm solid, especially if the hydrocarbon chains are long. The reason is simple: long saturated chains can pack closely together, to form regular, semicrystalline

Table 9.2
Composition of some natural fats in percent of total fatty acids[a]

Number of C Atoms in Chain	Olive Oil	Butter	Beef Fat
Saturated			
4–12	2	11	2
14	2	10	2
16	13	26	29
18	3	11	21
Unsaturated			
16–18	80	40	46

[a]Numbers do not total 100% because each substance contains small amounts of other fatty acids.

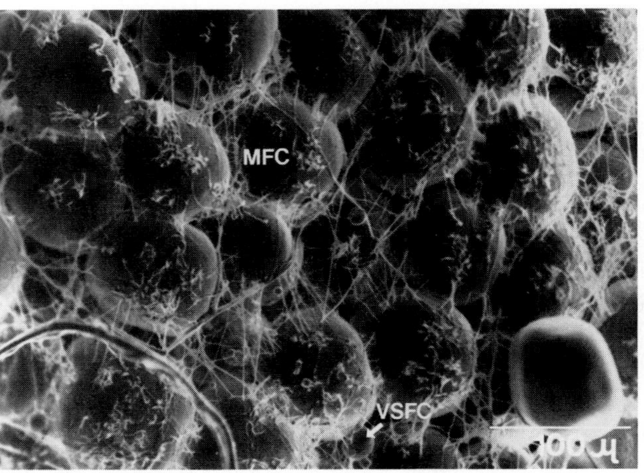

Figure 9.3
Adipocytes or animal fat storage cells. These are the kinds of cells that make up a large part of adipose tissue.

structures. In contrast, the kind of bend imposed by a cis double bond (see Figure 9.1) makes molecular packing more difficult. Indeed, partial **hydrogenation** of unsaturated fat oils (like corn oil) is used commercially to produce firmer fats, which can be used as butter substitutes such as oleomargarine.

Esterification with glycerol greatly diminishes the hydrophilic character of the head groups of the fatty acids. As a consequence, not only are triacylglycerols water-insoluble, they do not even form micelles very effectively. Fats are therefore stored in plant and animal cells as oily droplets in the cytoplasm. In **adipocytes,** animal cells specialized for fat storage, almost the entire volume of each cell is filled by a fat droplet (Figure 9.3). Such cells make up most of the adipose (fatty) tissue of animals.

Fat storage in animals serves three distinct functions:

1. Most fat in most animals is oxidized for the generation of ATP, to drive metabolic processes.

2. Some specialized cells (in "brown fat" of warm-blooded animals, for example) oxidize the triacylglycerols for heat production, rather than to make ATP.

3. In animals that must live in a cold environment, layers of fat cells under the skin serve as thermal insulation.

Waxes

Oleic acid	Oleyl alcohol
O=C—O (ester linkage)	
CH_2	CH_2
CH_2	CH_2
CH_2	CH_2
CH_2	CH_2
CH_2	CH_2
CH_2	CH_2
CH_2	CH_2
CH	CH_2
‖ CH	CH
CH_2	‖ CH
CH_2	CH_2
CH_2	CH_2
CH_2	CH_2
CH_2	CH_2
CH_2	CH_2
CH_2	CH_2
CH_3	CH_2
	CH_3

Figure 9.4
Structure of a typical wax. Waxes are formed by esterification of fatty acids and long-chain alcohols. The small head group (blue) can contribute little hydrophilicity, as compared to the hydrophobic contribution of the two long tails.

In the natural **waxes,** a long-chain fatty acid is esterified to a long-chain alcohol (Figure 9.4). This yields a head group that is only weakly hydrophilic, attached to two hydrocarbon chains. As a consequence, the waxes are completely water-insoluble. In fact, they are so hydrophobic that they often serve as water repellents, as in the feathers of some birds and the leaves of some plants. In some marine microorganisms, waxes are used instead of other lipids for energy storage. As with the triacylglycerols, the hardness of waxes is determined by chain length and degree of hydrocarbon saturation.

The Lipid Constituents of Biological Membranes

All biological membranes contain lipids as major constituents. The molecules that play the dominant roles in membrane formation all have highly polar head groups and, in most cases, *two* hydrocarbon tails. There is a molecular sense to this: If a large head group is attached to a single hydrocarbon chain, the molecule is wedge shaped and will tend to form spherical micelles (Figure 9.5a). A double tail yields a roughly cylindrical molecule, which can easily pack in parallel to form extended sheets of membranes. As indicated in Figure 9.5b, such membranes will be **bilayers,** with the hydrophilic head groups facing outward into the aqueous regions on either side. A number of classes of membrane-forming lipids share this type of structure; they differ principally in the nature of the head group. We shall describe a few examples of each.

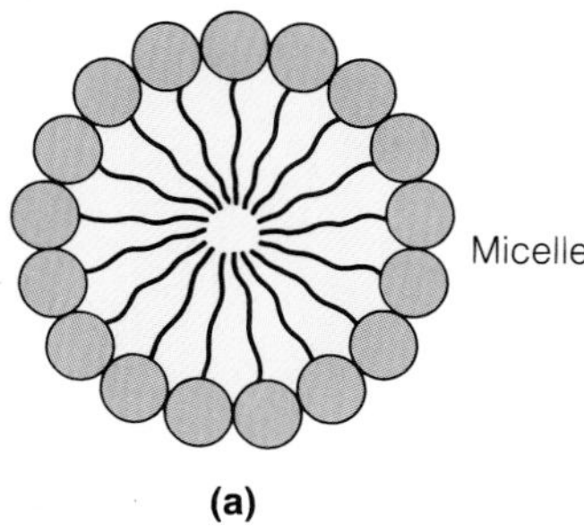

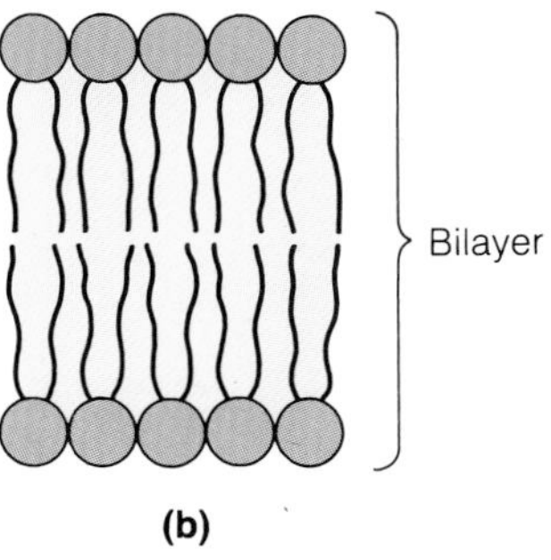

Figure 9.5
How molecular shape of lipids determines the structures they form. (**a**) The single tail of a fatty acid makes the molecule wedge-shaped, favoring micelle formation. (**b**) The multiple tails on membrane-forming lipids make the molecules more cylindrical, so that planar bilayer sheets can be formed.

Glycerophospholipids

Glycerophospholipids (also called **phosphoglycerides**) are the major class of naturally occurring **phospholipids,** lipids with phosphate-containing head groups. These compounds make up a significant fraction of the membrane lipids throughout the bacterial, plant, and animal kingdoms. All can be considered to be derivatives of glycerol-3-phosphate. Carbon 2 in glycerol-3-phosphate is a chiral center, and the naturally occurring glycerophospholipids are derivatives of the L enantiomer. The general structure of this group of compounds is shown in Figure 9.6. In panel (a) is depicted the stereochemical configuration. Panel (b) shows the molecule in the manner we will generally use to represent membrane lipids, with the hydrophobic tails drawn to the right and the hydrophilic head group to the left. Usually, R_1 and R_2 are acyl side chains derived from the fatty acids; often one is saturated, the other unsaturated. The R_3 group varies greatly, and it is this that confers the greatest variation in properties among the glycerophos-

Figure 9.6
Glycerophospholipid structure. (**a**) Stereochemical view of L-enantiomer. (**b**) A conventional visualization.

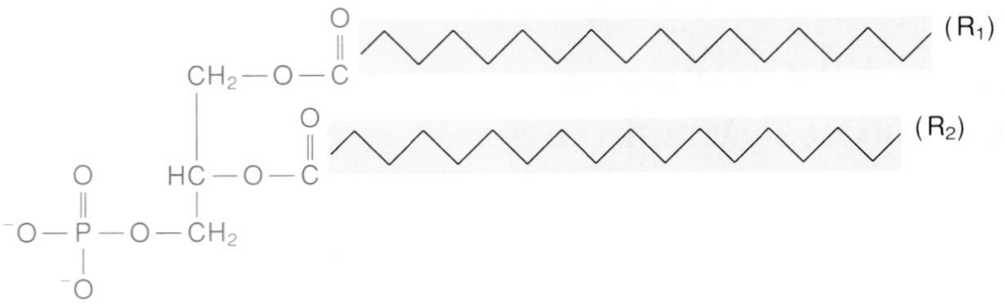

(a) Phosphatidic acid

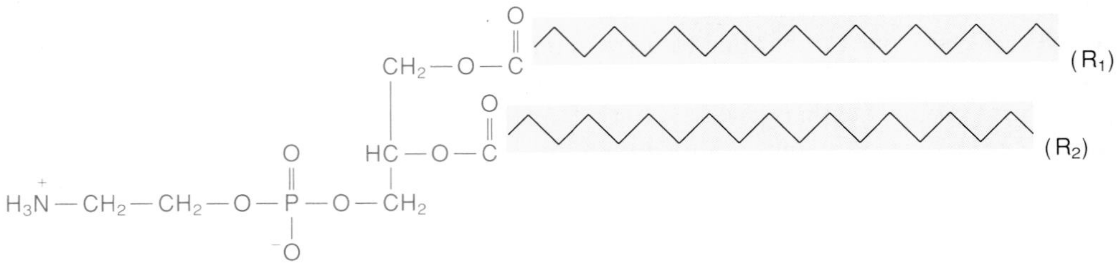

(b) Phosphatidylethanolamine

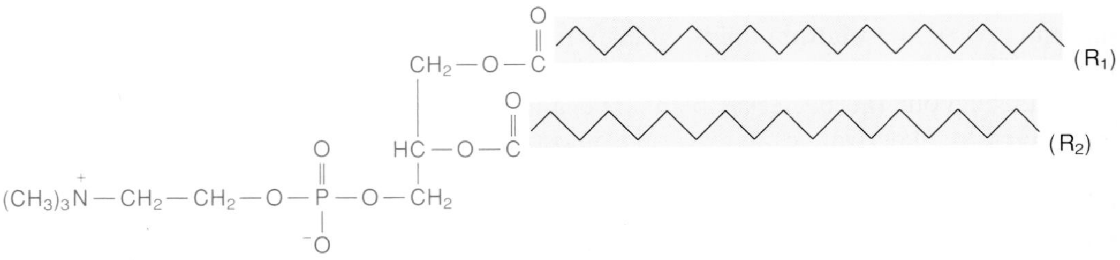

(c) Phosphatidylcholine

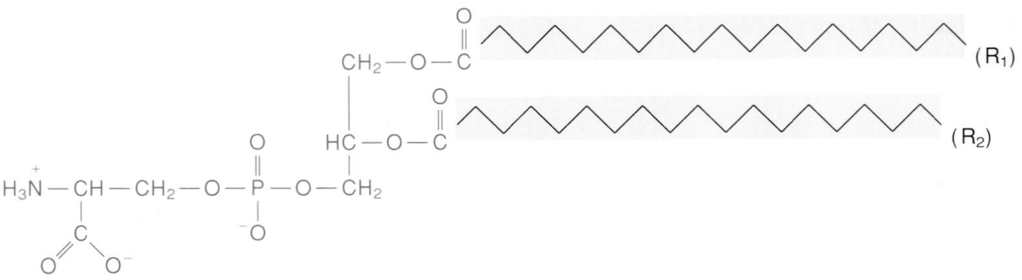

(d) Phosphatidylserine

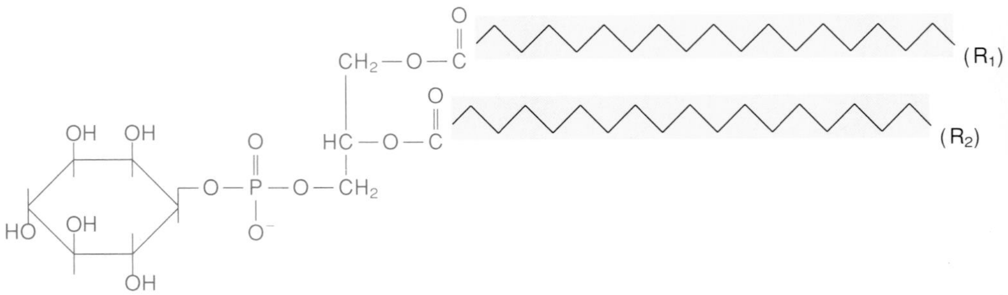

(e) Phosphatidylinositol

Figure 9.7
Examples of common glycerophospholipids. The hydrophobic R groups are indicated in yellow, the glyceryl moiety in black, and the very hydrophilic head groups in blue. All may be considered derivatives of phosphatidic acid (**a**).

pholipids. A gallery of the most common glycerophospholipids is shown in Figure 9.7, and their relative abundances in some membranes are given in Table 9.3. The simplest member of the group, **phosphatidic acid,** is only a minor membrane constituent; its principal role is as an intermediate in synthesis of other glycerophospholipids (described in Chapter 18). The names of glycerophospholipids are derived from phosphatidic acid: *phosphatidylcholine, phosphatidylethanolamine,* and so on. As Figure 9.7 shows, the glycerophospholipids have very polar head groups, all carrying some charge. Since the hydrocarbon tails are derived from the naturally occurring fatty acids in various combinations, an enormous variety of glycerophospholipids exists. For example, the erythrocyte membrane contains molecules with hydrocarbon chain lengths between 16 and 24 carbons, containing anywhere from 0 to 6 double bonds.

Sphingolipids and Glycosphingolipids

A second major class of membrane constituents is built on the long-chain amino alcohol **sphingosine,** rather than glycerol:

```
       H     H
       |     |
HO — C — C = C —(CH2)12 — CH3
       |   |
       |   H
       |
  H — C — NH2
       |
HO — CH2
```

Sphingosine = D-4-sphingenine

If a fatty acid is linked via an amide bond to the —NH_2 group, the class of **sphingolipids** referred to as **ceramides** is obtained:

```
       H          H
       |          |
HO — C ——— C = C —(CH2)12CH3
       |      |
       |      H   O
       |          ||
  H — C ——— N — C — R
       |      |
       |      H
       |
HO — CH2
```

General structure of a ceramide (R = hydrocarbon)

Further modification, by addition of groups to the hydroxyls, leads to a variety of other membrane lipids. An especially important example is **sphingomyelin,** in which one hydroxyl group is phosphorylated and a cationic group (**choline**) attached:

```
                                H          H
                                |          |
                         HO — C ——— C = C —(CH2)12CH3
                                |      |
                                |      H   O
                                |          ||
                   O         HC ——— N — C —(CH2)16CH3
         +         ||          |      |
(CH3)3NCH2CH2 — O — P — O — CH2      H
                   |
                   O⁻
```

Phosphocholine | Ceramide

Sphingomyelin

Table 9.3
Lipid composition of some biological membranes[a]

Lipid	Percent of Total Composition in			
	Human erythrocyte plasma membrane	Human myelin	Beef heart mitochondria	*E. coli* cell membrane
Phosphatidic acid	1.5	0.5	0	0
Phosphatidylcholine	19	10	39	0
Phosphatidylethanolamine	18	20	27	65
Phosphatidylglycerol	0	0	0	18
Phosphatidylinositol	1	1	7	0
Phosphatidylserine	8.0	8.0	0.5	0
Sphingomyelin	17.5	8.5	0	0
Glycolipids	10	26	0	0
Cholesterol	25	26	3	0
Others	0	0	23.5	17

[a]Data from C. Tanford, *The Hydrophobic Effect* (New York: Wiley, 1973).

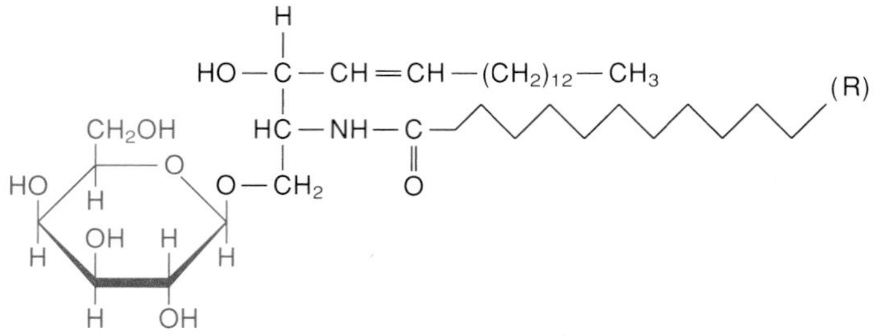

(a) Galactosylceramide

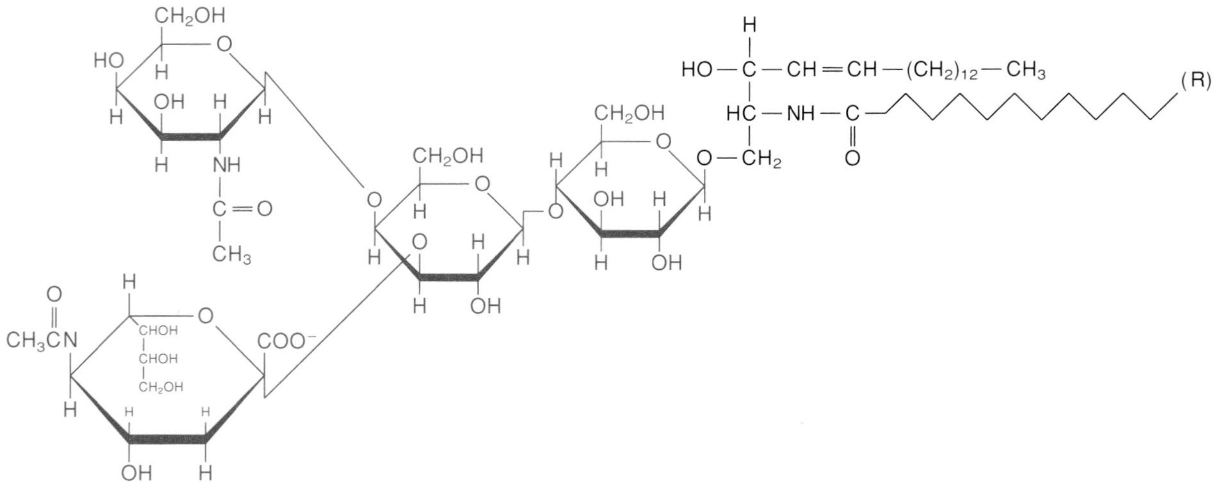

(b) GalNAcβ(1 $\longrightarrow$ 4)Galβ(1 $\longrightarrow$ 4)Glcβ(1 $\longrightarrow$ 1)ceramide
$\begin{pmatrix} 3 \\ \uparrow \\ \alpha 2 \end{pmatrix}$
Sia

Figure 9.8
Examples of glycosphingolipids. (**a**) A *cerebroside*, an important constituent of brain cell membranes. (**b**) A *ganglioside*. This particular ganglioside, GM_2, accumulates in neural tissue of infants who have the inherited defect called Tay–Sachs disease, hence it is also known as Tay–Sachs ganglioside. The defect lies in the lack of an enzyme that normally cleaves the terminal GalNAc.

(a)

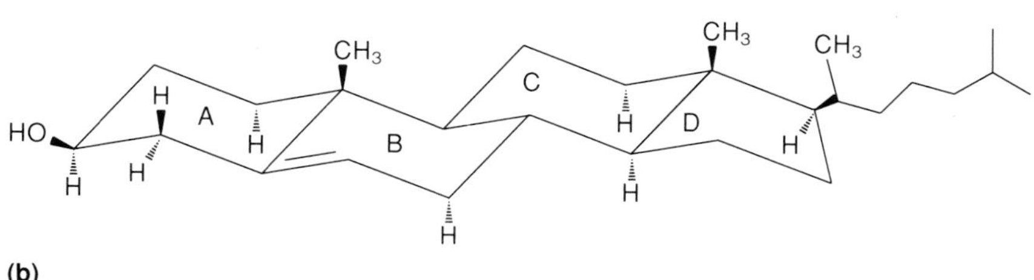

(b)

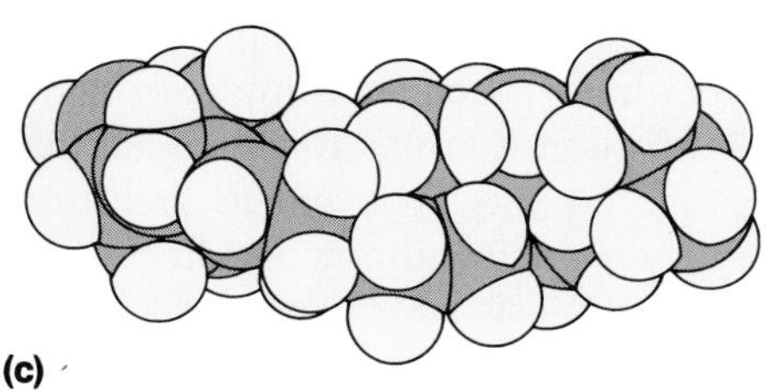

(c)

Figure 9.9
Cholesterol. (**a**) Structural formula; (**b**) skeletal model; (**c**) space-filling model.

In some of the membrane lipids built on sphingosine, the head group contains saccharides. Lipids containing saccharide groups go under the general name of **glycolipids.** The **glycosphingolipids** include such molecules as the **cerebrosides** and **gangliosides,** examples of which are shown in Figure 9.8. As the names of these compounds suggest, they are especially common in the membranes of brain and nerve cells.

Cholesterol

One important lipid constituent of many membranes bears little superficial resemblance to the compounds we have studied so far. This is **cholesterol,** the structure of which is shown in Figure 9.9. Cholesterol is a member of a large group of substances called **steroids,** which include a number of important hormones, among them the sex hormones of higher animals. In fact, cholesterol is the precursor for the synthesis of many of these substances (its role in these syntheses will be discussed in Chapter 18). Cholesterol is a weakly amphipathic substance, due to the hydroxyl group at one end of the molecule. The rest is readily soluble in the hydrophobic interiors of membranes. As the conformational structure in Figure 9.9b shows, the fused cyclohexane rings in cholesterol are all in the chair conformation. This makes cholesterol a bulky, rigid structure as compared with other hydrophobic membrane components such as the fatty acid tails. Thus, the cholesterol molecule fits awkwardly into membrane lipids and tends to disrupt regularity in membrane structure.

The Structure and Properties of Membranes

The membranes of living cells are remarkable bits of molecular architecture, with many and varied functions. To say that a membrane is essentially a phospholipid bilayer is a gross oversimplification. To be sure, the phospholipid bilayer forms the basic structure, but there is much more. A more realistic representation of a typical eukaryotic cell membrane is shown in Figure 9.10. A wide variety of specific proteins are contained in the membrane or are bound to its surface. Many of these proteins carry oligosaccharide groups that project into the surrounding aqueous medium. Other oligosaccharides are carried by glycolipids with the lipid portions inserted in the membrane. The two sides of the bilayer are usually different, both in lipid composition and in the placement and orientation of proteins and oligosaccharides. Let us examine this structure in more detail.

Motion in Membranes

A functioning biological membrane is not a rigid, frozen structure. In fact, the lipid and protein components are in constant motion. This can be demonstrated in a direct and dramatic way. If human and mouse cells, each carrying a distinctive fluorescent marker in its plasma membrane, are fused together, the two kinds of markers gradually become completely intermixed. This demonstrates that **lateral diffusion** (parallel to the membrane surface) can occur in the membrane. The rapidity with which such two-dimensional diffusion can occur depends on the membrane fluidity, which in turn depends on temperature and lipid composition. Under physiological conditions, the average time required for a phospholipid molecule to wan-

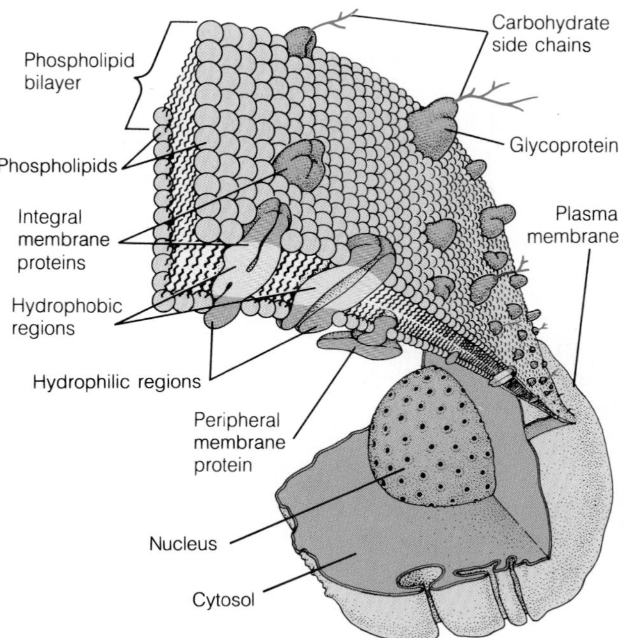

Figure 9.10
Structure of a typical cell membrane. In this schematic view, drawn according to the fluid mosaic model of membrane structure (p. 311), a strip of the plasma membrane of a eukaryotic cell has been peeled off. Proteins are embedded in and on the phospholipid bilayer; some of these are glycoproteins, carrying oligosaccharide chains. The membrane is about 7 nm thick.

der completely around a cell is on the order of seconds to minutes; membrane proteins also move, but more slowly.

The effects of temperature and composition on fluidity can be most simply studied using artificial membranes containing only one or a few kinds of lipids and no proteins (see Tools of Biochemistry 14, concerning the preparation of such membranes). Figure 9.11 depicts the behavior of a membrane made entirely from phosphatidylcholine carrying 16-carbon saturated chains (PC-16:0/16:0, in shorthand). At low temperatures the hydrocarbon tails pack together closely to form a nearly solid gel state. If the temperature is raised above 41°C, this regular order is lost, and the hydrocarbon tails begin to move about. The membrane "melts" to adopt a fluid **liquid crystalline** state. The temperature at which this happens is called the *transition temperature*. This abrupt change in membrane properties can be detected by a number of the techniques described in Tools of Biochemistry 14.

The transition temperature is very sensitive to the nature of the hydrocarbon tails. If a PC-14:0/14:0 membrane is used (with tails only two carbons shorter than those above) the transition temperature drops to 23°C. If a single cis double bond is incorporated into each 16-carbon tail (PC-16:1/16:1), melting occurs at −36°C! As explained earlier, such double bonds put bends in the chains, inhibiting their close packing; thus, they must be cooled to a lower temperature to produce the rigid gel. Changing the head group can also make a big difference: If phosphatidylethanolamine is substituted for phosphatidylcholine, the thermal transition in PE-16:0/16:0 is raised to 63°C. The sensitivity of the transition to lipid composition is shown dramatically by the fact that the small changes described above can change the transition temperature over a range of 100°C.

Biological membranes, which contain complex mixtures of lipid components plus protein, exhibit much broader phase transitions than those observed for synthetic bilayers of the kind described above. Because it is essential that the membranes in living cells be fluid, the membrane composition is regulated so as to keep the transition temperature below the body temperature of the organism. One example is found in bacteria, which will alter the saturated/unsaturated fatty acid ratio in their membranes in response to a change in the temperature at which they are grown. In the animal kingdom, there is the remarkable case of the reindeer's leg, the membranes of which increase in relative amount of unsaturated fatty acids near the hoof, which is usually cooler than the rest of the body.

Cholesterol has a specific and complex effect on membrane fluidity. As Figure 9.11b shows, it does not influence the transition temperature markedly, but it does broaden the transition. It has been hypothesized that this is because cholesterol can both stiffen the membrane above the transition temperature and inhibit regularity in structure formation below the transition temperature. Thus, it blurs the distinction between the gel and the fluid state. There is evidence that variations in cholesterol content are used to regulate membrane behavior in some organisms.

In contrast to the ease of lateral movement, the "flip-flop" of lipid molecules *across* synthetic lipid bilayers, from one side to the other, is much slower. The reason is not hard to see: If a phospholipid molecule were to turn from one face to the other, it would have to duck its very hydrophilic head into the inhospitable medium of the hydrocarbon tails and pass it through this region. Such an event is very unfavorable from an energetic point of view, and hence the process is slow.

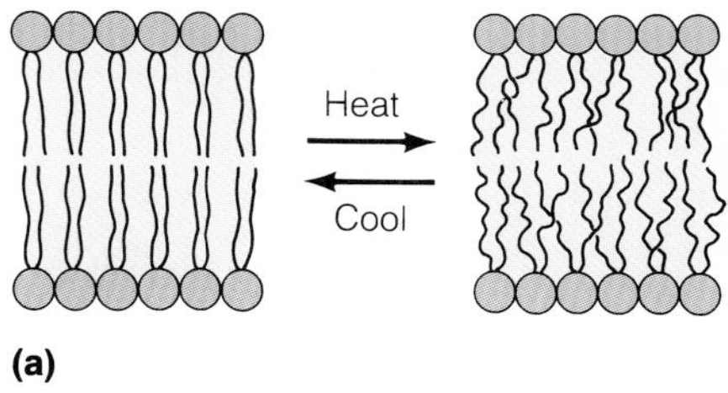

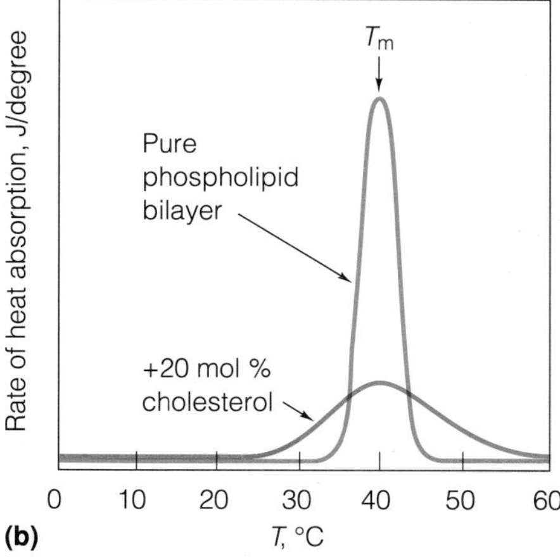

Figure 9.11
The gel–liquid crystalline transition in a lipid bilayer. (**a**) Schematic view of what happens at the transition temperature. Below this temperature the hydrocarbon tails are packed together in a nearly rigid, nearly crystalline gel state (left). Above the transition temperature, the chains become free to move about, and the interior of the membrane now resembles a liquid hydrocarbon (right). (**b**) Detection of the transition by calorimetry. Measurement of the heat absorbed as the temperature is raised shows a sharp "spike" at the transition temperature (T_m), where melting of the membrane takes place. The red curve, for a pure dipalmitoylphosphatidylcholine bilayer, shows a very sharp, well-defined transition. When 20 mol % cholesterol is mixed into the bilayer, the transition temperature is not changed, but the transition is broadened.

The Asymmetry of Membranes

Every biological membrane has two distinct faces, each facing a different environment. Examples can be seen in the illustrations of cell ultrastructure shown in Chapter 1: The plasma membrane of a cell faces the external environment on the outside and the cytosol on the inside, whereas the membrane around a chloroplast has the photosynthetic apparatus within and the cytosol without. Since they must deal with different surroundings, the two faces of a membrane are usually quite different in composition and structure.

This difference extends even to the level of phospholipid composition. Recall that all phospholipid membranes are bilayers; the individual layers are often called **leaflets.** The compositions of the two leaflets in the plasma membranes of several kinds of cells are shown in Figure 9.12. Not only are the individual lipids distributed very asymmetrically, but also the distribution varies considerably between cell types.

The consequences of such differences in phospholipid composition are numerous. Fluidity may be different on one side of the membrane or the other. The difference in charged groups on the two surfaces will contribute to the membrane potential (discussed later in this chapter). Glycoproteins and glycolipids carried in the outer leaflet of a plasma membrane contribute, via their oligosaccharide chains, to identification of cells (see Chapter 8).

It may seem paradoxical that such an asymmetric distribution can be established in cellular membranes. The existence of one particular kind of

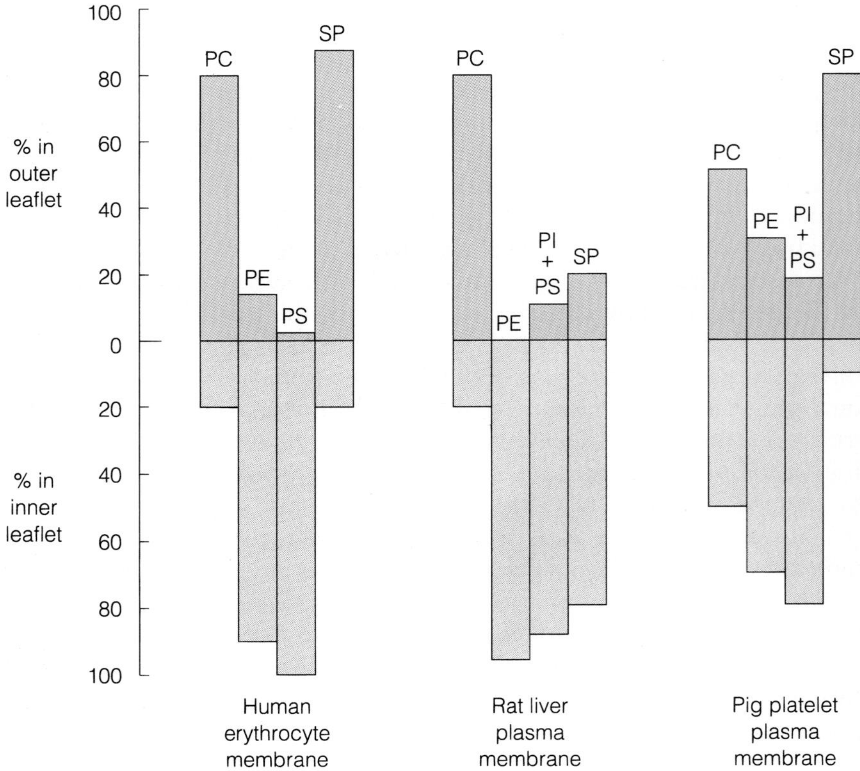

Figure 9.12
Phospholipid asymmetry in plasma membranes. Lipid composition in the two leaflets is graphed for human erythrocyte membrane, rat liver plasma membrane, and pig platelet plasma membrane. In each case, the fraction of each lipid in the *outer* leaflet is shown in green, the fraction in the *inner* leaflet in gray. Key to lipids: PC, phosphatidylcholine; PE, phosphatidylethanolamine; PS, phosphatidylserine; PI, phosphatidylinositol; SP, sphingomyelin.

lipid on the outside of the membrane would seem to require some method for preferentially transporting phospholipid molecules *across* membranes, since all lipid molecules are synthesized *inside,* in the cytosol. Yet we have seen that flip-flop transfer is a very slow process in synthetic bilayers. There is evidence, however, that transmembrane movement is much easier in natural membranes in the presence of proteins. A clear demonstration of this was provided by J. Rothman and E. Kennedy, using *Bacillus megaterium,* a Gram-positive bacterium. They used the reagent 2,4,6-trinitrobenzenesulfonic acid (TNBS), which reacts readily with the free amino groups on phosphatidylethanolamine (PE) (Figure 9.13). Reaction of the cell membrane of *B. megaterium* with TNBS showed that 33% of the PE was in the outer leaflet of the membrane and 67% in the inner. Rothman and Kennedy then investigated the distribution of *newly synthesized* PE, by giving the bacteria a short pulse labeling with [^{32}P]orthophosphate, which becomes incorporated in the new phosphatidylethanolamine. Immediately after labeling, *all* of the ^{32}P-containing PE was found in the inner leaflet, next to the cytosol. After only $\frac{1}{2}$ hour, however, the normal ratio had been reestablished. This transfer is enormously faster than that found for PE in artificial bilayers, indicating that some special process is at work. Whether membrane proteins in general simply facilitate the flip-flop of membrane lipids or there exist specific catalysts for the process is still unclear.

In any event, biological membranes are clearly dynamic structures. Not only must they continually expand as cells grow and divide, but also, even in resting cells, there appears to be a continual turnover and renewal of the membrane components. Indeed, this dynamic, nonequilibrium state is a necessity if asymmetry is to be maintained. From the kind of arguments presented in Chapter 3, we can see that the equilibrium state of a two-layer membrane would require an equal distribution of every component on either side. Like so many other biological systems, membranes exist as they do because they are *not* in equilibrium, but rather represent dynamic, steady-state structures.

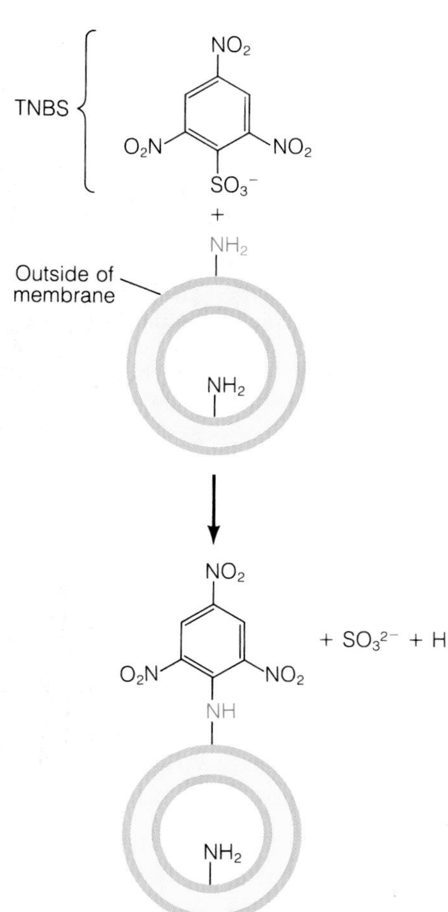

Figure 9.13
Demonstration of transmembrane movement. The reagent TNBS can be used to label selectively amino groups on the outside of membranes or vesicles, thereby providing evidence for asymmetry in membrane composition, and changes in that asymmetry resulting from transmembrane movement.

Proteins in Membranes: The Fluid Mosaic Model

Most of our current information concerning biological membranes is summarized by the **fluid mosaic model** proposed by S. J. Singer and G. L. Nicholson in 1972. This is the model depicted in Figures 9.10 and 9.15. The fluid, asymmetric lipid bilayer carries within it a host of proteins. Some of these, called **peripheral membrane proteins,** are only partially buried in the lipid matrix and are exposed at only one membrane face or the other. Thus, in the plasma membrane some proteins face into the cytosol, and others face the external environment. Other proteins, the **integral membrane proteins,** are largely buried within the membrane but are exposed on both faces. Integral proteins are frequently involved in transmitting either specific substances or chemical signals through the membrane. The whole membrane is a fluid mosaic of lipids and proteins, with the lipids moving readily in lateral directions but with greater difficulty across the membrane. Lateral motion of the proteins seems to be somewhat more limited, and these molecules almost never rotate in the membrane.

Membrane proteins possess special characteristics that distinguish them from other globular proteins. They often contain a high proportion of hydrophobic amino acids, particularly in the parts of the protein molecules that are embedded in the membrane (Figure 9.14). The segments of proteins

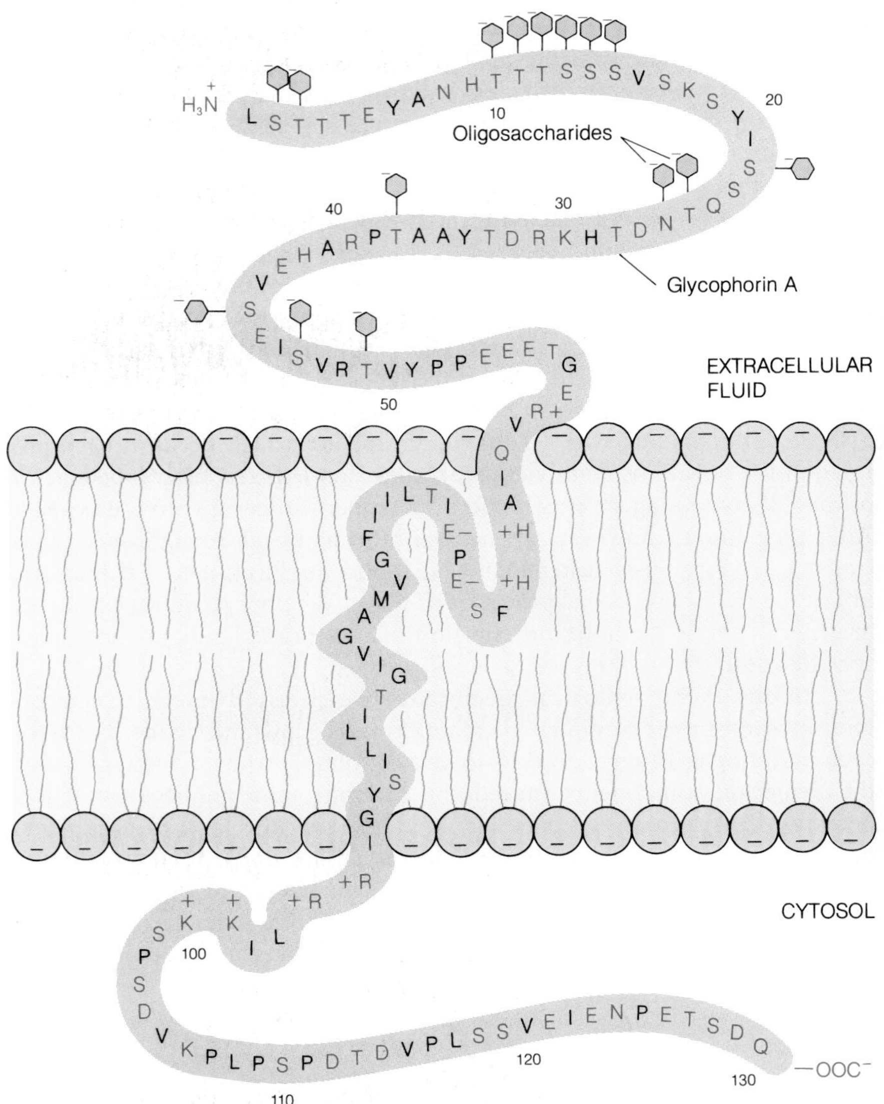

Figure 9.14
An integral membrane protein, glycophorin A. Positively charged amino acids are shown in red, other hydrophilics in blue. Hydrophobic amino acids are shown in black. This glycoprotein is an important constituent of erythrocyte membranes. According to present information, residues 1–61 are outside the cell membrane and carry numerous carbohydrate units (green). Residues 62–95 are largely hydrophobic and within the membrane. Of these, 74–95 are believed to form an α helix, and 62–73 may form a salt-bridged cluster. The high concentration of positively charged residues near the cytoplasmic face may interact with the negatively charged heads of phospholipid molecules.

that span membranes are often α-helical. Membrane proteins sometimes carry covalently attached lipid molecules; these are called **lipoproteins.** Because of their hydrophobic nature, lipoproteins are frequently found to be insoluble in water and must be solubilized with the aid of detergents to be studied. All integral membrane proteins are difficult to remove without disrupting the entire membrane structure.

The protein content varies greatly among different kinds of membranes (see Table 9.4) and appears to be directly related to the functions each must carry out. Mitochondrial inner membranes and bacterial cell wall membranes, which have multiple functions, are about 75% protein. The myelin of nerve fibers, which acts primarily as an electrical insulator, has a low protein content.

Our current understanding of membrane structure is a consequence of the development of sophisticated and elegant techniques. A number of these are described in Tools of Biochemistry 14, but two deserve special mention here. Of major importance is **freeze-fracture electron microscopy.** If a membrane sample is frozen quickly and then broken by a sharp blow from a

Table 9.4
Lipid, protein, and carbohydrate content of some membranes

Membrane	Percent by Weight		
	Protein	Lipid	Carbo-hydrate
Myelin	18	79	3
Human erythrocyte (plasma membrane)	49	43	8
Bovine retinal rod	51	49	0
Mitochondria (outer membrane)	52	48	0
Amoeba (plasma membrane)	54	42	4
Sarcoplasmic reticulum (muscle cells)	67	33	0
Chloroplast lamellae	70	30	0
Gram-positive bacteria	75	25	0
Mitochondria (inner membrane)	76	24	0

Adapted from G. Guidotti, *Annu. Rev. Biochem.* 41:731 (1972).

microtome knife, it frequently splits along the plane between the bilayer leaflets (Figure 9.15a). One layer is thus peeled back, revealing the internal structure of the membrane. The sample can then be metal-shadowed for electron microscopy (Figure 9.15b). In a variant called **freeze etching,** some of the ice is sublimed off before shadowing. This technique can uncover details of the membrane surface as well.

Much of our knowledge of membrane asymmetry comes from studies of **vesicles,** fragments of membrane that have resealed to form hollow shells, with an inside and an outside. Reagents can either be captured inside the vesicle or added only to the surrounding solution, so that they can react specifically with either outward-facing or inward-facing proteins or lipids.

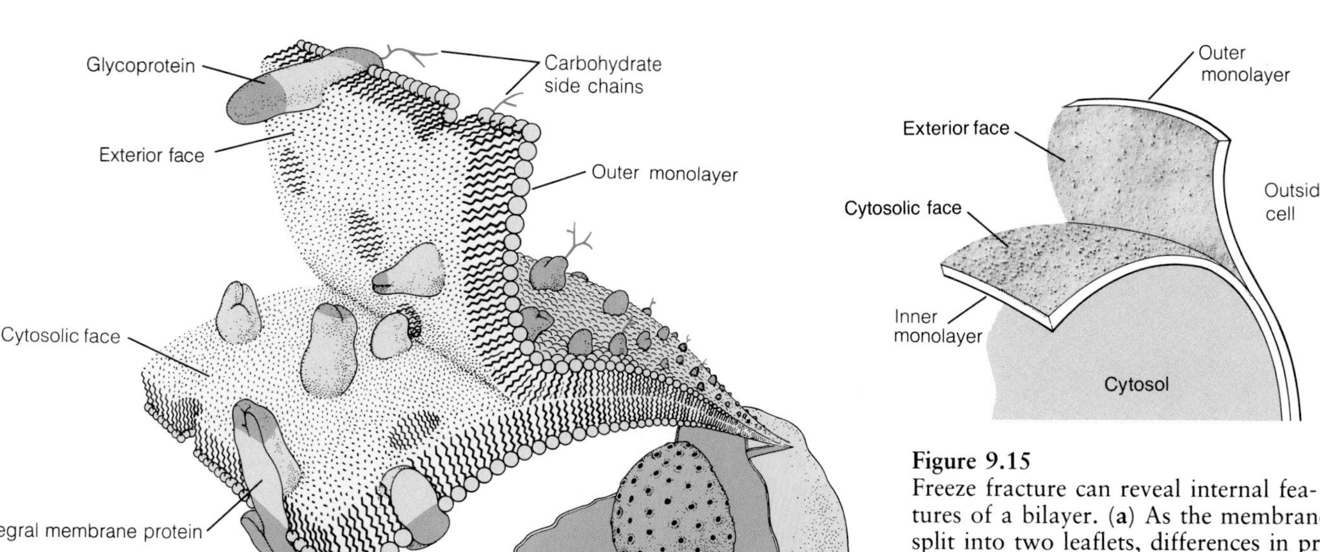

Figure 9.15
Freeze fracture can reveal internal features of a bilayer. (**a**) As the membrane is split into two leaflets, differences in protein composition are apparent. (**b**) Electron micrographs of a mouse kidney tubule cell show that the protoplasmic face is studded with intramembranous particles, whereas the exterior face is relatively smooth.

Figure 9.16
Electron micrograph of an erythrocyte ghost. In this negatively stained image, folds in the membrane appear as dark bands.

A membrane protein may be reacted covalently with a radioactively labeled reagent, isolated, and cleaved into peptides by proteases. Identification of which peptides are labeled can reveal which portions of the protein are on the inner and which are on the outer faces. In a similar way, lipids can be treated with enzymes or other reagents that cleave off or otherwise modify the head groups (see Figure 9.13). Experiments of this kind, performed from the inside or outside of vesicles, have provided much of the kind of information shown in Figure 9.12.

What emerges from all of these studies is that membranes are complex structures, with specific compositions for each of the two leaflets. To make the picture more concrete, let us consider in some detail the structure of one example, the plasma membrane of the erythrocyte (red blood cell).

The Erythrocyte Membrane: An Example of Membrane Structure

The erythrocytes of mammals are among the simplest of all cells. In their mature state in the circulating blood, they have lost nucleus, mitochondria, and internal membranes. They are essentially bags of hemoglobin, and they can easily be lysed to release their contents and produce membrane **ghosts** (Figure 9.16). These "ghosts" are really large vesicles that represent a nearly pure preparation of the plasma membrane of the cells. They have the lipid composition given in Table 9.3, distributed as shown in Figure 9.12. If the total protein content of the erythrocyte ghost is extracted with detergent and analyzed by SDS gel electrophoresis, the pattern shown in Figure 9.17a is obtained.

The erythrocyte plasma membrane contains far fewer different proteins than would be found in most other cell membranes, in keeping with the simple metabolism of these cells. What are these proteins, and what do they do? To answer, it is first necessary to distinguish between the peripheral and integral proteins. The distinction is based on the fact that peripheral proteins can be washed off the ghosts by simple changes in ionic strength and/or pH. In this way, we find that the 10 proteins listed at the top of Table 9.5 are peripheral. Furthermore, all of these turn out to be attached to the inside (cytosolic face) of the erythrocyte membrane. Freeze etching

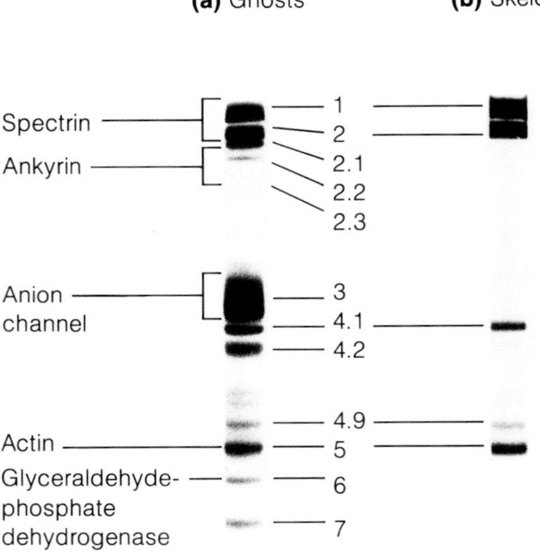

Figure 9.17
Gel electrophoretic analysis of erythyrocyte membrane proteins. (**a**) Proteins of the total ghost. Details concerning these proteins are given in Table 9.5. *Glycophorin* does not show up here because it does not stain with the dye that was used. (**b**) Proteins of the erythrocyte "skeleton," remaining after other proteins have been extracted with the detergent Triton X-100.

Table 9.5
The major proteins of the human red cell membrane[a]

Band No.[b]	Protein Name	Subunit Mol. Wt.	Probable State of Assembly	No. of Copies per Cell	Role
A. Peripheral proteins					
1	α-Spectrin	260,000	$\alpha_2\beta_2$ tetramers	10^5 tetramers	Membrane skeleton
2	β-Spectrin	225,000			
2.1	Ankyrin	215,000	Monomer	10^5	Links skeleton to band 3
4.1	—	78,000	?	2×10^5	Involved in spectrin junctions
4.2	—	72,000	?	2×10^5	?
4.9	—	45,000	?	5×10^4	?
5	Actin	43,000	Oligomers of 12–17 units	5×10^5	Involved in spectrin junctions
6	Glyceraldehyde-3-phosphate dehydrogenase	35,000	Tetramer	5×10^5	Glycolytic enzyme
7	—	29,000	?	5×10^5	?
8	—	23,000	?	10^5	?
B. Integral proteins					
3	—	89,000	Dimer	10^6	Ion channel
4.5		55,000	?	1.5×10^6	Glucose transport?
	Glycophorin A	31,000	Dimer	4×10^5	Cell recognition
	Glycophorin B	23,000	?	$\sim 10^5$	
	Glycophorin C	29,000	?	$\sim 10^5$	

[a]Most of the data are from V. Bennett, *Annu. Rev. Biochem.* 54:273–304 (1985).

[b]Band numbers in Figure 9.17. The glycophorins do not stain well with protein stains but can be detected by carbohydrate-specific stains.

reveals a nearly smooth outer surface, but the inside of the membrane and the inner surface are rich in protein particles.

The major integral proteins of this membrane are then the remainder of those shown in Table 9.5—glycophorin, band 3, and band 4.5 (we will identify these later). These proteins, plus much of the lipid material, can be extracted from membranes by use of the nonionic detergent Triton X-100.

Surprisingly, such treatment leaves intact a protein "skeleton," which retains the shape of the membrane ghost (see Figure 9.17b). This is a two-dimensional network of some of the peripheral proteins—mainly **spectrins, actin,** and bands 4.1 and 4.9. The 200-nm-long fibers in this network are made up of (α–β) pairs of spectrin molecules. These very elongated molecules contain a large fraction of α helix and appear to be linked at their ends through short chains of actin molecules, together with the band 4.1 protein. A schematic picture of this membrane skeleton is shown in Figure 9.18.

What purpose does this elaborate underpinning to the erythrocyte membrane serve? An obvious suggestion is that it helps to maintain the shape of the erythrocyte, despite the squeezings and buffetings a cell suffers in passing through the circulatory system. The erythrocyte is durable, typically surviving for about 120 days, or 10 million heartbeats. The discoid shape of the cell allows efficient exchange of O_2 and CO_2 to the hemoglobin inside, and even if that shape is momentarily deformed, the skeleton helps to regain it. Indeed, some forms of anemia, in which the symptoms are easy lysis of erythrocytes, can be traced to deficiencies in spectrin. Spectrin is not confined to erythrocytes; many other cell types contain similar membrane

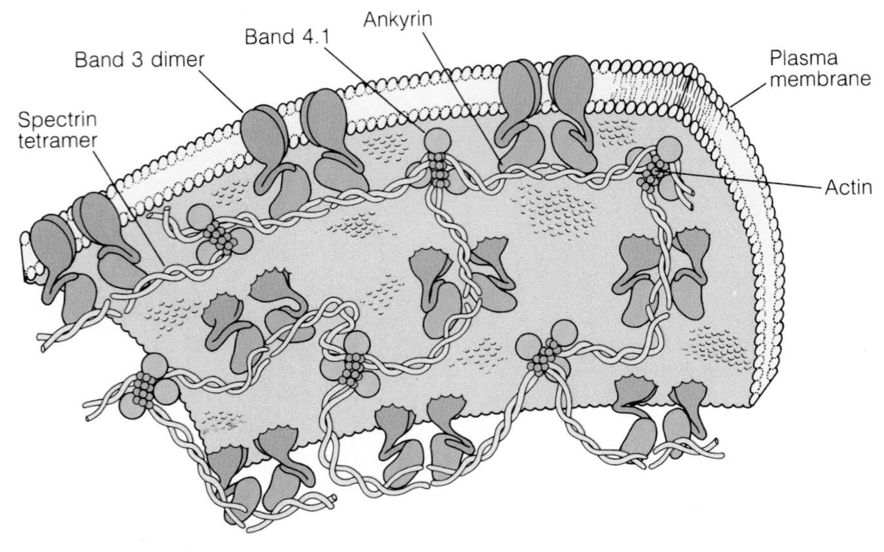

Figure 9.18
Model of the postulated structure of the erythrocyte membrane skeleton. The proteins are identified in Figure 9.17 and Table 9.5. Note that ankyrin "anchors" the membrane to the skeleton by interacting with both spectrin and the integral band 3 protein.

skeletons. It seems likely that there are connections between the membrane skeleton and the cellular cytoskeleton (Chapter 1) so that the membrane is linked to intracellular structure.

The major integral proteins of the red cell membrane play quite different roles. Band 3 is an 89-kilodalton protein that appears to exist in the membrane as a dimeric structure (Figure 9.19). This polypeptide chain passes back and forth through the membrane six times. Its major function is to serve as an anion channel, allowing the passage of HCO_3^- and Cl^- through the membrane, to facilitate CO_2 transport by the hemoglobin (as discussed in the next section). But band 3 protein does a number of additional interesting things. Its long N-terminal domain, which extends into the cytosol, is associated with a number of the enzymes of glycolysis, the major energy-producing pathway in the erythrocyte. These enzymes (including glyceraldehyde-phosphate dehydrogenase) are also involved in providing bisphosphoglycerate to regulate hemoglobin O_2 affinity (Chapter 7). Finally, the N-terminal domain of band 3 protein is associated with a peripheral protein, **ankyrin,** which acts to anchor the band 3 complex to the membrane skeleton (see Figure 9.18).

The other major integral red cell membrane proteins are the **glycophorins,** which apparently serve a variety of cell recognition functions. The primary structure of the best known of these, **glycophorin A,** is shown in Figure 9.14. The sequence includes a highly hydrophobic section, which is buried in the membrane. The N-terminal domain is outside the cell and carries a large number of oligosaccharide chains, most coupled to serines and threonines.

The asymmetry of the orientation of integral proteins should be emphasized. Just as the erythrocyte membrane is asymmetric in its distribution of lipid components, each kind of protein is oriented in a particular direction. Just how these remarkable membrane structures are assembled is only now beginning to be understood. As we shall see in Chapter 27, there appear to be special modes of protein synthesis that direct the placement of proteins in membranes, and assure their asymmetric orientation.

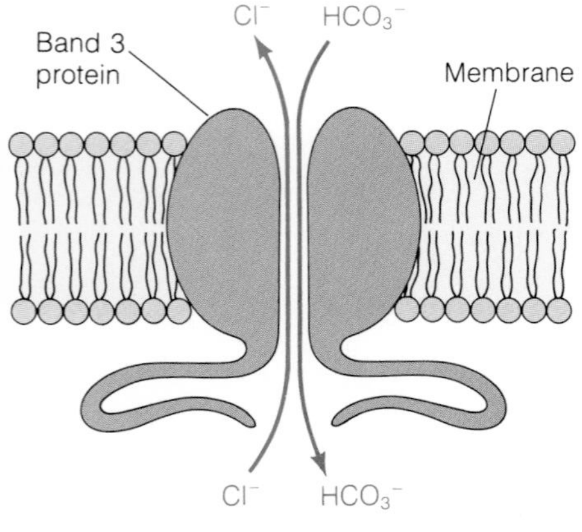

Figure 9.19
The ion-transport protein of erythrocyte membranes. This integral membrane protein (band 3 in Figure 9.17a) exists as a dimer in the membrane. It serves to exchange Cl^- and HCO_3^- across this membrane.

Transport Across Membranes

A cell or organelle cannot be either wholly open or wholly closed to its surroundings. Its interior must be protected from certain toxic compounds, and yet metabolites must be taken in and waste products removed. Since the cell must contend with thousands of substances, it is not surprising that membranes are complex structures.

In this section we consider the various ways in which molecules are transported across membranes. The three categories of transport—passive, facilitated, and active transport—are quite different and generally serve different purposes in the cell.

The Thermodynamics of Transport

Before considering the specific mechanisms of transport, it is useful to review some ideas that were presented in Chapter 3. There we discussed the general thermodynamic principles governing the transfer of substances across membranes or surfaces. It was shown that the free energy change, ΔG, for transporting 1 mol of a substance from a region in which its concentration is C_1 to a place where its concentration C_2 is given by

$$\Delta G = RT \ln \frac{C_2}{C_1} \tag{9.1}$$

According to this equation, if C_2 is less than C_1, ΔG is negative, and the process is thermodynamically favorable. As more and more substance is transferred (between two finite compartments) C_1 decreases and C_2 increases, until $C_2 = C_1$. At this point $\Delta G = 0$, and the system is at equilibrium. *Unless other factors are involved,* this is the ultimate state approached by transport across any membrane: a substance that can traverse the membrane will eventually come to the same concentration on each side. There is another, "kinetic" way to see this result. If the molecules are wandering into the membrane at random, the number entering from each side will be proportional to the concentration on that side. When the concentrations become equal, the rates of transport in the two directions will be the same, and no net transport will occur.

There are three circumstances under which this equalization can be circumvented, and each is important in the behavior of real membranes:

1. A substance may be preferentially bound by macromolecules confined to one side of the membrane. We may find that compound A is more concentrated inside a cell (in terms of total moles of A per unit volume) than outside. But much of A may be bound to some cellular macromolecules; that portion does not really count in equation (9.1), which simply states that the concentrations of *free* A on the two sides must be equal at equilibrium. An appropriate example is oxygen in erythrocytes. If we were to measure the *total* oxygen concentration in an erythrocyte, we would find it higher than the concentration of O_2 in the surrounding blood plasma. But the total concentration inside the cell includes that bound to hemoglobin. The *free* oxygen concentration in the fluids inside and outside an erythrocyte is the same at equilibrium.

2. A **membrane electrical potential** may be maintained which influences the distribution of ions. For example, if the inside of a cell is at an electrical potential negative with respect to the outside, cations will be favored on the inside. To put it most simply, the negative potential inside a cell tends to draw the positively charged cations in and drive negative anions out. This can be expressed mathematically in the following way: For an ion of charge Z, the free energy change for transport across a membrane now involves two contributions: the normal concentration term, as given in equation (9.1), plus a term describing the energy change in moving a mole of ions across the potential difference:

$$\Delta G = RT \ln \frac{C_2}{C_1} + ZF\,\Delta\psi \tag{9.2}$$

Here F is the Faraday constant (96.5 kJ volt^{-1} mol^{-1}) and $\Delta\psi$ is the membrane potential in volts. If $\Delta\psi$ is negative (in going from outside to inside) and Z is positive, the second term on the right of equation (9.2) makes a negative contribution to ΔG. That is, the transport *in* of cations is favored. For anions, of course, the opposite is true; they will be driven out. The equilibrium state ($\Delta G = 0$) will *not*, in this case, correspond to the same concentration of ions on the two sides of the membrane. Of course, the potential difference must be maintained, for if it were not, the transport of ions would obliterate it. So in a sense, maintaining an unequal concentration ratio by a membrane potential is a special case of active transport (discussed below), for energy must be expended continually to keep up the potential difference. Conversely, equation (9.2) may be interpreted to mean that if a difference in ionic concentration is maintained a potential will develop across the membranes (see Problem 7).

3. If some thermodynamically favored process is *coupled* to the transport, then the ΔG for this process must be included in the free energy equation. This is the general case of active transport, for which we can write

$$\Delta G = RT \ln \frac{C_2}{C_1} + \Delta G' \tag{9.3}$$

The quantity $\Delta G'$ could correspond to some thermodynamically favored reaction (like ATP hydrolysis) that was somehow coupled to the process of transport. This is clearly a generalization of equation (9.2), now allowing a variety of processes—not just those that maintain an electrical potential difference—to participate in the transport.

The important consequence of equations like (9.2) and (9.3) is that under certain circumstances, substances can be transported into or out of a cell or compartment even *against* an unfavorable concentration difference. We shall find, for example, that most cells manage to bring potassium ions in and export sodium ions, even though the intracellular potassium concentration is much higher than the concentration in the surrounding fluid, and the sodium ion concentration inside is much lower than that outside. In a similar way, many cells manage to scavange needed metabolites such as glucose from an environment very dilute in this compound.

With this background, we turn now to the mechanisms whereby substances are passed through membranes. We may introduce the problem by two questions: (1) Does the transport occur against unfavorable concentration gradients, or is it eventually limited by equation (9.1); that is, are equal free concentrations on both sides approached? (2) How fast does the transport occur? For example, some molecules that are not actively transported against a concentration gradient can still traverse some membranes very rapidly, whereas others do so so slowly as to be effectively excluded.

Passive Transport: Diffusion

Passive transport is that which is accomplished by the random wandering of molecules through membranes. The process is the same as the Brownian motion of molecules in any fluid, which is termed **molecular diffusion.** Passive transport will ultimately end up with the free concentration of the diffusing substance being the same on both sides of the membrane. The net rate of transport, J (in moles per square centimeter per second), is, as you might expect, proportional to the concentration difference $(C_2 - C_1)$ across the membrane:

$$J = -\frac{D_m(C_2 - C_1)}{l} \tag{9.4}$$

where l is the thickness of the membrane and D_m is the effective diffusion coefficient of the diffusing substance in the membrane. In agreement with equation (9.1), equation (9.4) says that net transport will stop when $C_2 = C_1$. If C_1 and C_2 are expressed in mol/cm^3 and l in cm, then D_m has the units cm^2/s. D_m is not the same as the diffusion coefficient (D) that the same molecule would have in aqueous solution, for it depends not only on the size and shape of the molecule and the viscosity of the membrane lipid but also on the solubility of the substance in the membrane. D_m can be related to the true diffusion coefficient, D_l, of the molecule in the lipid bilayer by $D_m = KD_l$, where K is the **partition coefficient** for the diffusing material between lipid and water. For ions and other hydrophilic substances, K will be a very small number. This has the effect that diffusion of such substances through lipid membranes is extremely slow. There is simply not enough of such a substance dissolved in the bilayer to provide rapid transport.

Since we usually do not know the thickness of membranes with exactness, studies of passive transport are often described in terms of a **permeability coefficient, *P*,** which can be measured experimentally:

$$J = -P(C_2 - C_1) \tag{9.5}$$

According to the equations above, P must be given by

$$P = \frac{KD_l}{l} \tag{9.6}$$

Table 9.6 lists permeability coefficients for a number of small molecules and ions in membranes. Although we expect ions to have low values of K, the relatively large permeability value for water is surprising. Biological mem-

Table 9.6
Permeability coefficients from some ions and molecules through membranes[a]

Membrane	Permeability Coefficient (cm/s) for				
	K^+	Na^+	Cl^-	Glucose	Water
Phosphatidylserine	$<9 \times 10^{-13}$	$<1.6 \times 10^{-13}$	1.5×10^{-11}	4×10^{-10}	5×10^{-3}
Human erythrocyte	2.4×10^{-10}	10^{-10}	1.4×10^{-4}[b]	2×10^{-5}[b]	5×10^{-3}
Squid axon—resting	5.6×10^{-8}	1.5×10^{-8}	1.0×10^{-8}	ND[c]	ND
Squid axon—excited	1.7×10^{-4}[b]	5×10^{-6}[b]	$\sim 10^{-8}$	ND	ND

[a] Data from M. K. Jain and R. C. Wagner, *Introduction to Biological Membranes* (New York: Wiley, 1980).

[b] Facilitated transport. Note that whenever facilitated transport is encountered, the permeability coefficient rises dramatically.

[c] ND = not determined.

branes are not, in fact, very good barriers against water. Although the reasons for this are not entirely clear, the fact is probably fortunate for life, for it allows cells to exchange water readily with their surroundings. When water loss is to be strenuously avoided, as in the leaves of desert plants, waxy substances provide a nearly impermeable barrier.

Facilitated Transport

For many substances, the slow transport provided by passive diffusion is simply insufficient for the functional and metabolic needs of cells and means must be found to increase transport rates. Examples are found in the mammalian erythrocyte. If we examine the permeability of erythrocyte membranes to chloride or bicarbonate ions, we find permeability coefficients of about 10^{-4} cm/s. This value is about 10 million times greater than the permeability coefficient for Cl^- in pure lipid bilayers (see Table 9.6). Clearly, some special mechanism is required to account for this difference. In fact, we have already encountered the transport apparatus, in the transmembrane protein called the band 3 protein of erythrocyte membranes (see Figures 9.17 and 9.18 and Table 9.5). This protein passes through the membrane, forming a channel or **ion pore** through which Cl^- and HCO_3^- can pass (Figure 9.19).

Exchange of Cl^- and HCO_3^- is essential to erythrocyte function. In the capillaries, HCO_3^- from CO_2 production in the tissues is taken up by erythrocytes. In order to maintain anion balance in the cell, chloride ions are released. The band 3 protein doesn't just form a hole in the membrane; the channel is very selective, exchanging HCO_3^- for Cl^- on a 1:1 basis. When the erythrocyte reaches the lungs, the hemoglobin is reoxygenated. Now the reverse happens; one Cl^- is taken in for every HCO_3^- released. Both the hemoglobin molecule and the erythrocyte membrane have evolved to facilitate this process. By contrast, such facilitated transport is not necessary for O_2; this tiny nonpolar molecule can move rapidly through the membrane by passive diffusion.

There is a second facilitated diffusion process in the erythrocyte membrane that is essential for the cell's survival. The small energy demands of an erythrocyte are met by glucose, readily available in the surrounding blood plasma. But as Table 9.6 shows, the passive transport of glucose through artificial phospholipid membranes is agonizingly slow: $P = 4 \times 10^{-10}$ cm/s. Therefore, erythrocytes need a mechanism for the facilitated diffusion of glucose. This, like the ion-exchange system, appears to be accomplished by a transmembrane protein, which increases the rate 50,000-

fold. The **transport protein,** whose exact identity is still a matter of some dispute (but it may be band 4.5; see Table 9.5), appears to be both a lipoprotein and a glycoprotein. It is quite selective in its facilitation; for example, D-glucose is transported orders of magnitude more rapidly than L-glucose. Facilitated transport of metabolites like glucose appears to be a common feature in cells.

A number of models have been hypothesized for facilitated diffusion, but most fall into one of the three classes shown in Figure 9.20. We have already encountered an example of 9.20a—the ion pore in the erythrocyte membrane. The existence of certain transmembrane carriers (Figure 9.20b) is also now well established; one of these is described on page 322. The flip-flop model (Figure 9.20c) is now considered unlikely, in view of the general difficulty of turning things over in membranes.

Some pores are gated; that is, they can be opened or closed in response to control mechanisms. A well-studied example is the **gap junction,** a kind of pore existing between adjacent cells in some animal tissues (Figure 9.21). Under normal circumstances these pores are open, allowing nonspecific exchange of small molecules between the cells. But certain signals can close the gate. For example, if Ca^{2+} levels in one cell increase above 5×10^{-5} M as a consequence of damage to that cell, the gap junctions to neighboring cells close. We shall encounter other examples of such gated pores in Chapter 29 when we consider the complex structures of nerve membranes. There, gated pores play a major role in the propagation of nerve impulses.

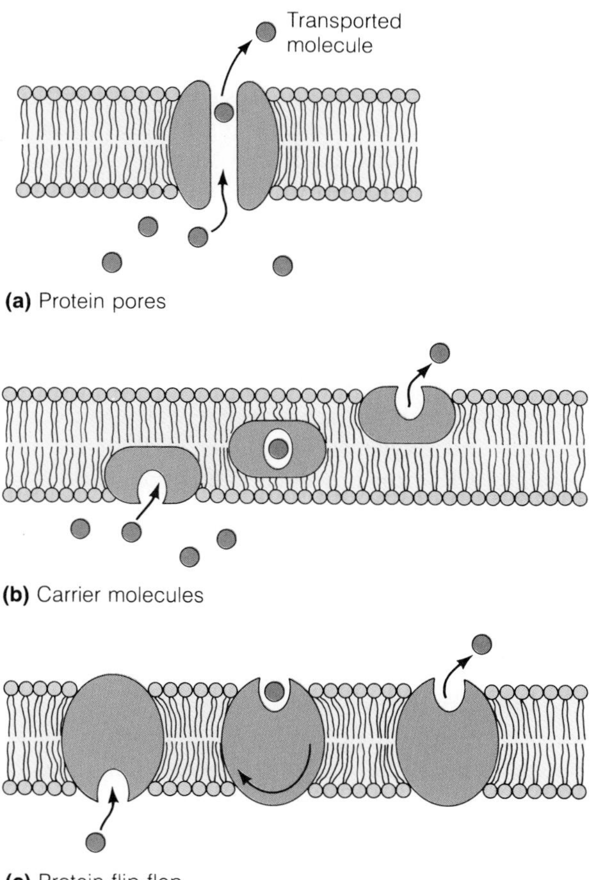

Figure 9.20
Possible models for facilitated transport. (**a**) Protein pores; (**b**) carrier molecules; (**c**) protein flip-flop. Both (**a**) and (**b**) are known to exist; (**c**) is now considered unlikely.

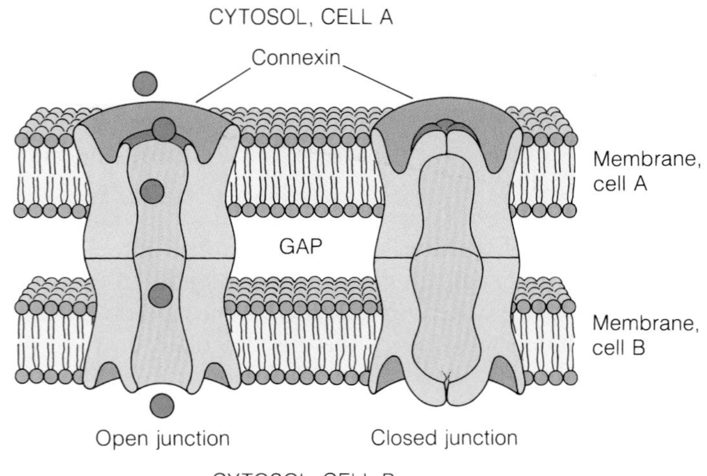

Figure 9.21
Gap junctions. A gap junction is a relatively nonspecific pore that connects two cells. The plasma membranes of adjacent cells contain hexagonal structures made from a protein called connexin. When these are matched up, a connection is made through which small molecules or ions can flow between cells. The gap junction is gated; it can close in response to certain stimuli. In this schematic illustration one is shown closed, another open.

A quite different kind of pore is provided by certain antibiotics. Many organisms have evolved the ability to synthesize antibiotic substances, and a considerable number of these are **ionophores,** facilitators of ion transport across membranes. For example, **gramicidin A,** produced by the bacterium *Bacillus brevis* acts as a cation-specific ion pore allowing a breakdown in the unequal balance between K^+ and Na^+ that is normally maintained in living cells. Gramicidin A is a 15-residue polypeptide, containing both L- and D-amino acids (Figure 9.22a). Gramicidin adopts an open helical conformation when dissolved in the membrane, but one molecule of the antibiotic is only long enough to traverse half the thickness of the membrane. An open pore forms only when two gramicidin molecules line up to form an end-to-end dimer (Figure 9.22b). Potassium ions can then pass through the channel. If the conductance of membranes is studied at very low gramicidin concentrations, a remarkable phenomenon is seen: the conductance changes up and down in quantized steps. Each step corresponds to the opening or closing of an individual channel, probably by the diffusing together or separating of two gramicidin molecules, one in each membrane leaflet.

Antibiotic ionophores also provide the best known examples of **carrier-facilitated transport** (see Figure 9.20b). For example, **valinomycin,** produced by a *Streptomyces,* has the structure shown in Figure 9.23. It is a cyclic polypeptide-like molecule, involving three repeats of the sequence (D-valine)–(L-lactate)–(L-valine)–(D-hydroxyisovalerate). Its folded conformation presents an outside surface rich in —CH_3 groups and an interior cluster of nitrogens and oxygens. The dimensions of the interior cavity nicely accommodate a K^+ ion but do not fit other cations as well. This structure is exactly what is needed for a cation carrier: the outer surface is hydrophobic, making the molecule soluble in the lipid bilayer, whereas the inside mimics in some ways the hydration shell that the cation would have in aqueous solution. A molecule like valinomycin can diffuse to one surface of a membrane, pick up an ion, and then diffuse to the other surface and release it. There is no *directed* flow, but the carrier in effect increases the solubility of the ion in the membrane.

A number of other ion-carrier antibiotics have the same kind of structure. These molecules are either cyclic or linear chains that can fold into cyclic structures, stabilized by hydrogen bonding. Their relative affinities for different ions vary greatly. For example, valinomycin has nearly a 20,000-fold preference for K^+ over Na^+, whereas the antibiotic *monensin* prefers Na^+ by tenfold.

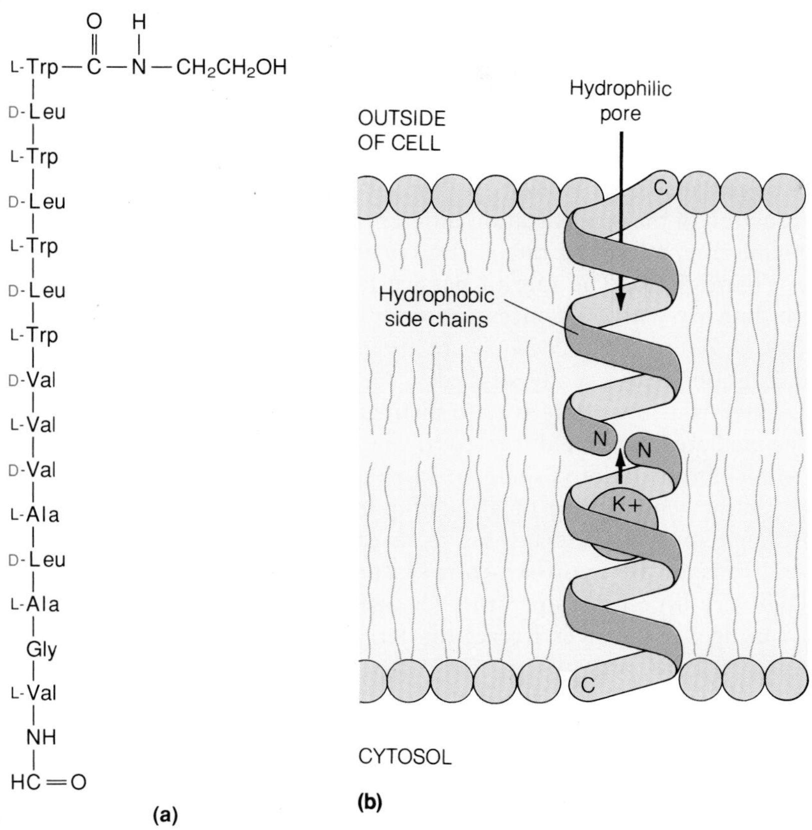

Figure 9.22
Gramicidin A, an antibiotic that acts as an ion pore. (**a**) Sequence of the polypeptide. Note the presence of several D-amino acids (red). (**b**) Two molecules of gramicidin A form a pore through the membrane by adopting a helical conformation, with hydrophobic side chains (brown) projecting into the lipid. The inside of the helix (blue) forms the hydrophilic pore.

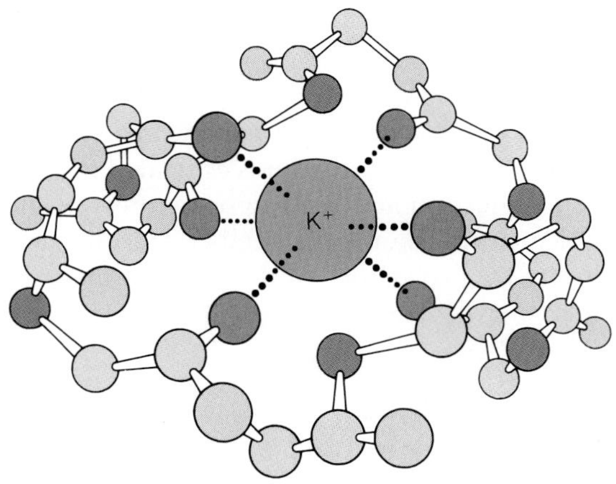

Figure 9.23
Valinomycin, an antibiotic that acts as an ion carrier. In this cyclic polymer, oxygens binding the K^+ ion are shown in red, nitrogens are blue, and carbons gray. The central cavity complexes a K^+ ion; the outside of the roughly spherical molecule is very hydrophobic.

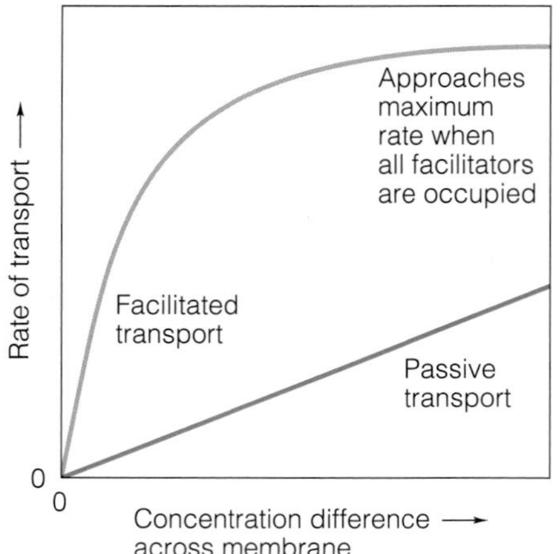

Figure 9.24
Distinguishing between facilitated transport and passive transport. If the rate of transport is graphed versus the concentration difference driving the transport, facilitated transport reaches a limiting rate at high concentrations, whereas passive transport gives a linear graph, as expressed by equation (9.5).

How can facilitated transport be distinguished from passive transport? Aside from the generally much higher transport rate, there is a simple test. Facilitated transport systems are *saturable*. Each carrier molecule can handle only one molecule or ion at a time. Each pore can accommodate only one or a few ions or molecules at any moment. There is a limited number of pores or carriers in any membrane. Thus, if we measure the rate of transport as a function of the concentration difference of substance transported across the membrane, a limiting rate is approached when all pores or carriers are busy (Figure 9.24). The rate of passive transport, on the other hand, increases linearly with the concentration difference, as expressed in equation (9.4) or (9.5) and Figure 9.24. There are no sites to saturate, at least at any concentration levels usually studied.

There is also an easy way to distinguish between pore-mediated and carrier-mediated facilitated transport. The latter should be extremely sensitive to membrane fluidity, since the carrier must actually move in the membrane. If the temperature of a membrane is lowered below its fluid–gel transition temperature, transport by a carrier like valinomycin virtually ceases. Transport by a pore structure like gramicidin A, on the other hand, is affected very little by temperature changes. The simple analogy is a ferry and a bridge: if the river freezes the ferry is stopped, but the bridge can continue to transport.

In conclusion, we must reiterate that even though facilitated transport is sometimes very fast and very selective, it is still only a special form of diffusion. The equilibrium state for a system exhibiting facilitated transport is the same as for passive transport—the substance transported will eventually end up at equal free concentrations on the two sides of the membrane.

Active Transport

Facilitated transport is useful in many biochemical processes, but in some situations it is imperative that cells or cellular compartments be able to transport substances against even very unfavorable concentration gradients. To take an extreme example, a calcium ion ratio of 30,000 must, under some circumstances, be established across membranes of the sarcoplasmic reticulum in muscle fibers (see below). According to equation (9.1), this ratio corresponds to $\Delta G = +26.6$ kJ/mol under physiological conditions, a formidable barrier. Transport against a concentration gradient is called **active transport.** Clearly, to pump ions against a gradient requires a free energy source of some kind. As you might expect, this energy usually comes from the hydrolysis of ATP. Altogether, it is estimated that some cells spend 30–50% of their ATP just on active transport. However, the hydrolysis of ATP can be coupled to transport in a number of different ways, some of them rather indirect. To give an idea of the range of these mechanisms, we shall now consider specific examples.

Ion Pumps: Direct Coupling of ATP Hydrolysis to Transport

The most common physiological example of active transport is the maintenance of sodium and potassium gradients across the plasma membranes of cells. The fluid surrounding cells in most animals is about 140 mM in Na^+ and 5 mM in K^+. Yet animal cells maintain, in their cytosol, an Na^+ concentration of about 10 mM and a K^+ concentration of about 100 mM. Thus, we have:

	Outside Cell	Inside Cell	Ratio (in/out)
Na^+	140 mM	10 mM	0.071
K^+	5 mM	100 mM	20

Even though Na^+ and K^+ pass very slowly through membranes by passive diffusion, such inequalities would ultimately vanish unless something were done to keep K^+ moving in and Na^+ out. This is accomplished by the action of the **sodium–potassium pump.** This molecular machine is a tetrameric protein, consisting of two large chains (α) of 95 kDa each and two smaller (β) subunits of 40 kDa each. The α subunits extend through the membrane, but the β subunits, which carry oligosaccharide groups, are exposed only on the outer surface (see Figure 9.25). The α subunits are enzymes that hydrolyze ATP, and the free energy change in that reaction is used to drive the transport.

Present estimates indicate that about two K^+ ions are pumped into the cell and three Na^+ ions are pumped out for every ATP hydrolyzed. Is this reasonable from a thermodynamic point of view? To answer this, we calculate the free energy required to take 3 mol of Na^+ from 10 mM to 140 mM and 2 mol of K^+ from 5 mM to 100 mM at 37°C. First let us calculate the free energy required to transport 3 mol of Na^+ from within the cell to outside. We must take into account the membrane potential of about 70 mV. The inside of the membrane is more negative than the outside, so this potential opposes the flow. Per mole of Na^+, we have

$$\Delta G = RT \ln \frac{C_{Na^+}(\text{out})}{C_{Na^+}(\text{in})} + Z_{Na^+} F \,\Delta\psi_{\text{in}\rightarrow\text{out}} \tag{9.7}$$

$$= 8.314 \frac{\text{J}}{{}^\circ\,\text{mol}} \times 310^\circ \times \ln \frac{140}{10} + 1 \times 96{,}480 \frac{\text{J}}{\text{volt} \cdot \text{mol}} \times 0.07 \text{ volt}$$

$$= 6800 \frac{\text{J}}{\text{mol}} + 6750 \frac{\text{J}}{\text{mol}} = 13{,}550 \frac{\text{J}}{\text{mol}} = 13.55 \frac{\text{kJ}}{\text{mol}}$$

For 3 mol, then, we have 3 × 13.55 kJ = 40.65 kJ. Now, in transporting K^+ in, the membrane potential is working for us. Per mole of K^+, we get

$$\Delta G = 8.314 \frac{\text{J}}{{}^\circ\,\text{mol}} \times 310^\circ \times \ln \frac{100}{5} + 1 \times 96{,}480 \frac{\text{J}}{\text{volt} \cdot \text{mol}} \times (-0.07 \text{ volt})$$

$$= 7719 \frac{\text{J}}{\text{mol}} - 6750 \frac{\text{J}}{\text{mol}} = 969 \frac{\text{J}}{\text{mol}} = 0.97 \frac{\text{kJ}}{\text{mol}}$$

or, for 2 mol, ΔG = 1.94 kJ. Note that the membrane potential is nearly sufficient, in itself, to maintain the K^+ gradient. The total free energy requirement for the outward transport of 3 mol of Na^+ and the inward transport of 2 mol of K^+ is then

$$\Delta G_{\text{total}} = 40.65 \text{ kJ} + 1.94 \text{ kJ} = 42.59 \text{ kJ}$$

At first glance, it would not appear that the hydrolysis of 1 mol of ATP would suffice to provide the necessary energy, for we have stated that $\Delta G^{\circ\prime}$ for ATP hydrolysis under physiological conditions is about −30 kJ/mol. However, in most cells, ATP is in much higher concentration than ADP or P_i. This means that the actual free energy change per mole is more like −45 to −50 kJ/mol. Thus, ATP hydrolysis is sufficient to maintain these concentration gradients under the observed stoichiometry, but could do little more.

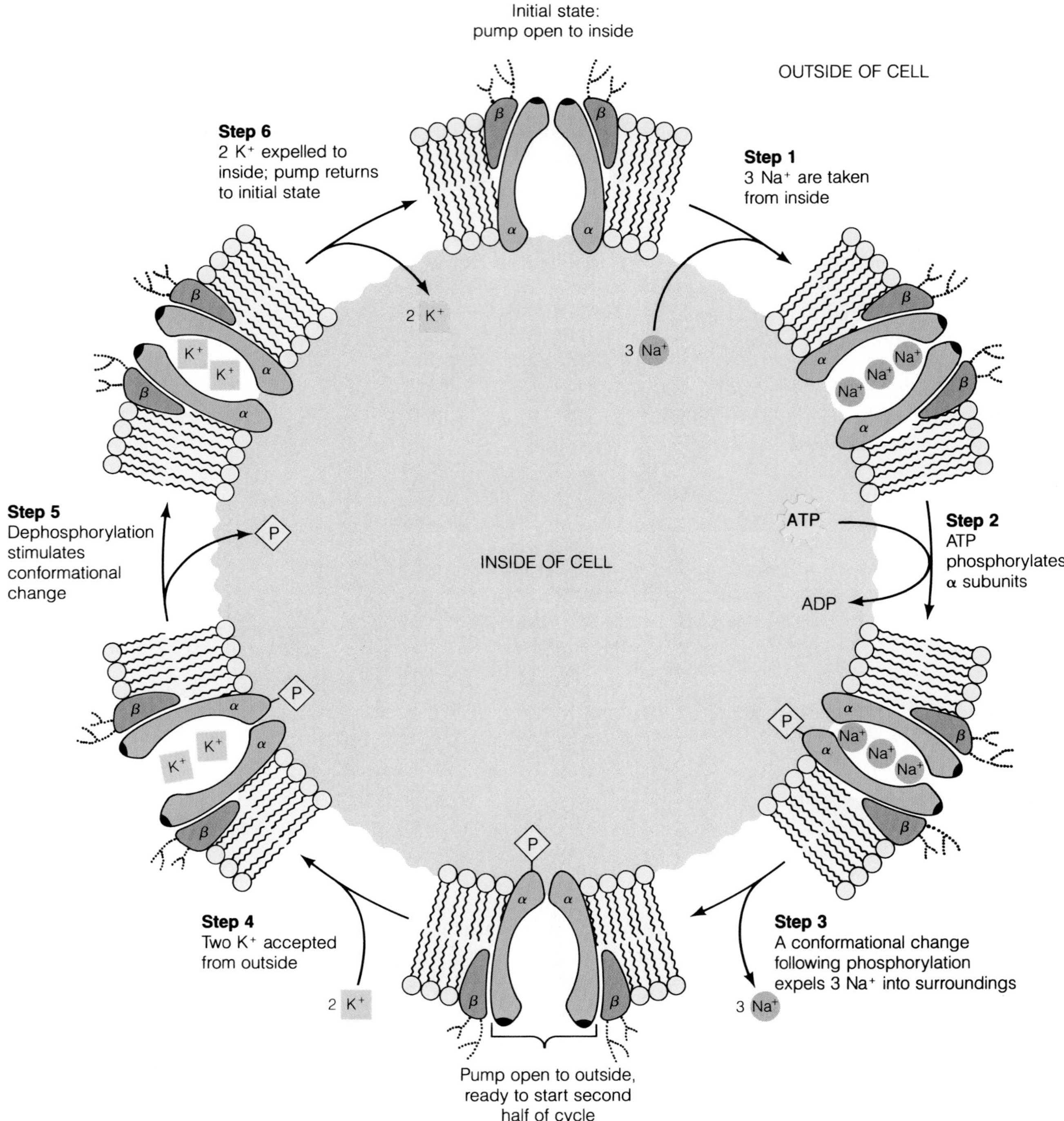
Initial state:
pump open to inside
OUTSIDE OF CELL
Step 6
2 K⁺ expelled to inside; pump returns to initial state
Step 1
3 Na⁺ are taken from inside
2 K⁺
3 Na⁺
Step 5
Dephosphorylation stimulates conformational change
ATP
Step 2
ATP phosphorylates α subunits
INSIDE OF CELL
ADP
Step 4
Two K⁺ accepted from outside
2 K⁺
Step 3
A conformational change following phosphorylation expels 3 Na⁺ into surroundings
3 Na⁺
Pump open to outside, ready to start second half of cycle

Figure 9.25
Schematic model of the sodium–potassium pump in operation. Steps of the pump's operation are shown in a cycle.

The sodium–potassium pump involves no violation of thermodynamic principles; the only requirement is that ATP hydrolysis and transport be *coupled*. This is apparently accomplished in a multistep process. A current model for the entire process is diagrammed in Figure 9.25. As the figure shows, key steps in the transport involve the phosphorylation of the enzyme by ATP and subsequent dephosphorylation to release phosphate. In this case, the coupling of the ATP hydrolysis to transport is direct.

Other kinds of ion pumps will be encountered in later chapters. Before we leave the subject, however, one point should be noted for future reference. *A pump driven backward can act as an energy generator.* In fact, the same kind of molecular mechanism described above, if faced with an opposite gradient, can often be used to *generate* ATP. Remarkably, this is probably the major way in which ATP is made in living organisms (see Chapter 15).

Cotransport Systems

There are other kinds of active transport that do not depend directly on ATP as an energy source, but employ ATP hydrolysis in an indirect way. You can imagine how this might occur, if you reflect on the fact that the kind of ATP-driven ion pumps described above can generate gross inequalities of ion concentrations across membranes. These ion gradients are far from equilibrium and hence represent in themselves a potential source of free energy. An example of how an ion gradient is used is provided by the **sodium–glucose cotransport** system of the small intestine (Figure 9.26). The transport of each glucose molecule from within the intestine into the cells of the intestine wall is accompanied by the simultaneous movement of one Na^+ ion in the same direction. Since a favorable Na^+ gradient is maintained by the ATP-driven Na^+–K^+ pump of these cells, glucose can be transported against an unfavorable gradient in glucose concentration. Glucose "piggybacks" on the thermodynamically favored Na^+ transport.

There is a large number of such cotransport systems, many of them utilized to move nutrients into cells. A few examples are listed in Table 9.7. Many use the Na^+ gradient as a driving force but some, like the **lactose permease** system in *E. coli,* depend on an H^+ gradient. As we shall see in later chapters, generation of H^+ gradients is a central step in energy production by most cells.

Table 9.7
Some cotransport systems

Molecule Transported	Ion Gradient Used	Organism or Tissue
Glucose	Na^+	Intestine, kidney of many animals
Amino acids	Na^+	Mouse tumor cells
Glycine	Na^+	Pigeon erythrocytes
Alanine	Na^+	Mouse intestine
Lactose	H^+	*E. coli*

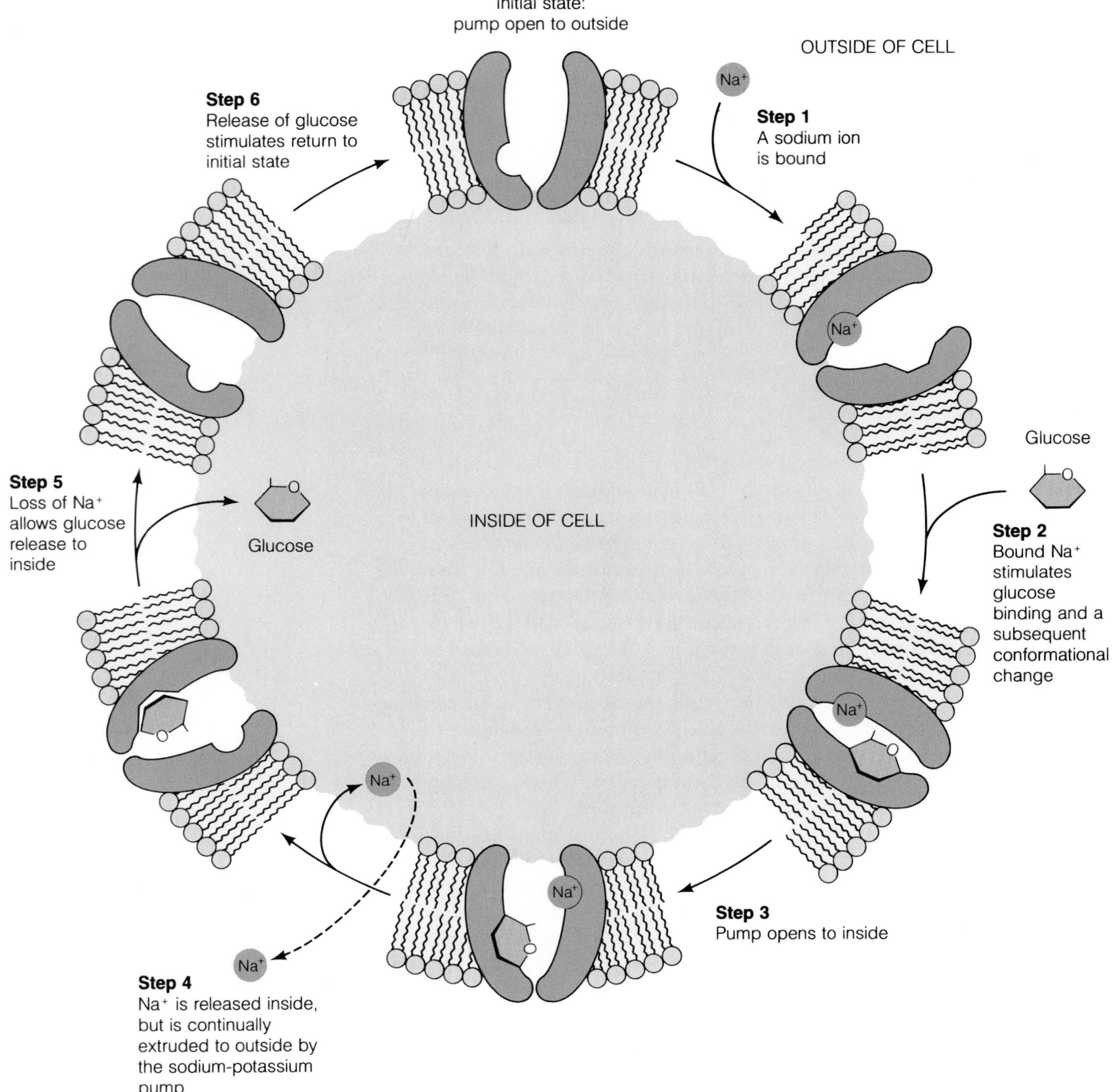

Figure 9.26
Sodium–glucose cotransport. Cotransport is used to pump many kinds of metabolites across cell membranes. In this example, transport of glucose is driven by the unequal concentration of Na^+ across the plasma membrane. The Na^+–K^+ pump (Figure 9.25) maintains this Na^+ concentration difference and thus keeps cotransport running.

Transport by Modification

One other method cells have of achieving transport against a gradient uses the following trick: Suppose a molecule, on moving into a cell by passive diffusion or facilitated transport, is chemically modified in such a way that it can no longer pass back through the membrane. The net result is that quantities of the modified molecule steadily accumulate within the cell. This method is used by many bacteria for the uptake of sugars. The sugars are phosphorylated, either during their diffusion through the membrane or as soon as they emerge into the cytosol. Membranes are impervious to the charged, phosphorylated monosaccharides, and thus they remain in the cell.

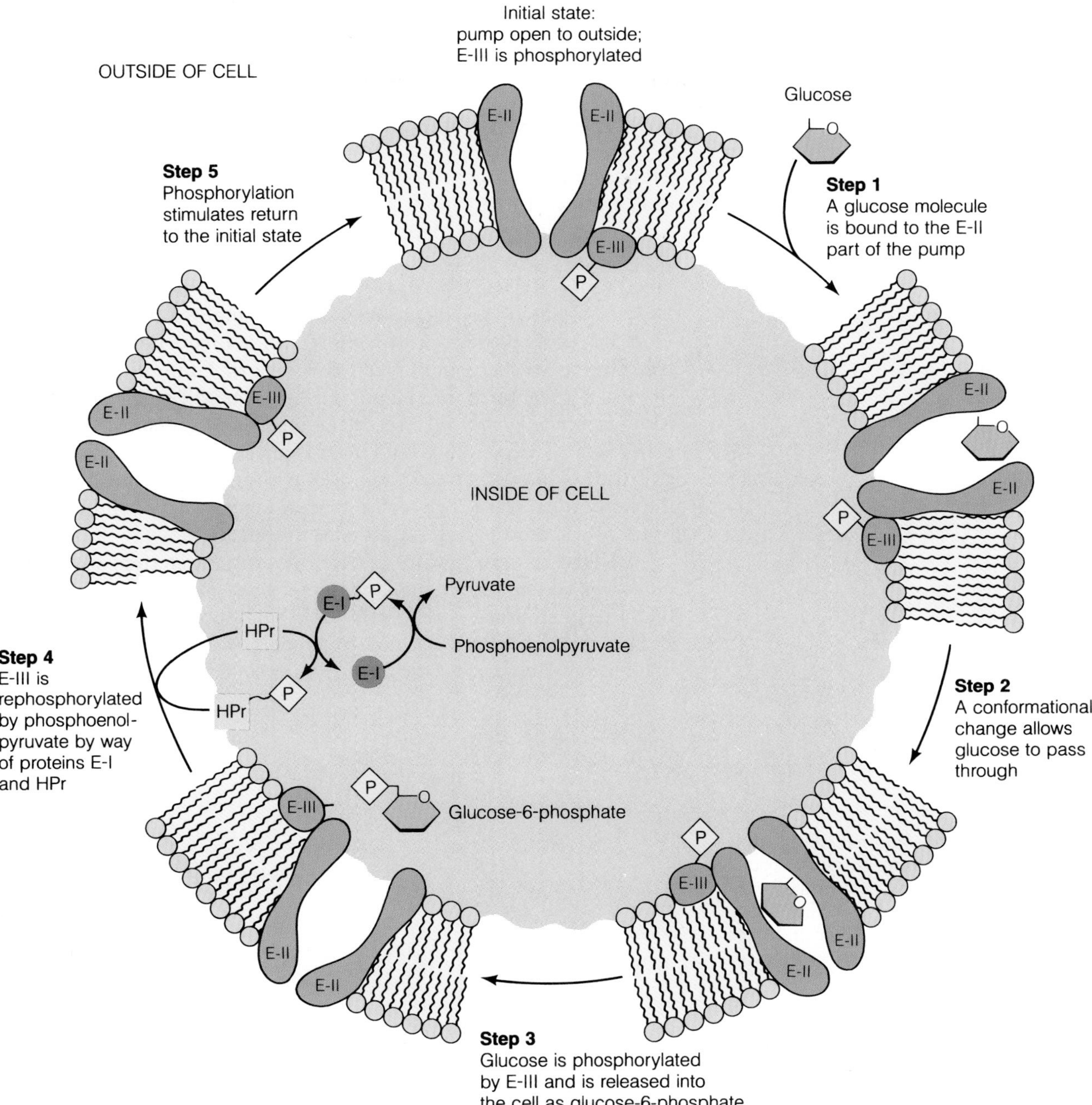
Initial state:
pump open to outside;
E-III is phosphorylated
OUTSIDE OF CELL
Glucose
E-II
E-III
P
Step 1
A glucose molecule
is bound to the E-II
part of the pump
Step 2
A conformational
change allows
glucose to pass
through
Step 3
Glucose is phosphorylated
by E-III and is released into
the cell as glucose-6-phosphate
Step 4
E-III is
rephosphorylated
by phosphoenol-
pyruvate by way
of proteins E-I
and HPr
Step 5
Phosphorylation
stimulates return
to the initial state
INSIDE OF CELL
Pyruvate
Phosphoenolpyruvate
E-I
HPr
Glucose-6-phosphate

Figure 9.27
Transport by modification. This example shows how glucose is effectively transported into some cells by coupling transport to phosphorylation. In addition to the transmembrane protein, several enzymes are involved. E-II acts as a specific sugar acceptor; E-III donates its phosphate to the sugar. To recharge E-III, two other intracellular enzymes (E-I and HPr) pass the phosphate from phosphoenolpyruvate, the ultimate high-energy phosphate donor.

In the best-studied example, the **phosphotransferase system** of *E. coli,* transport is facilitated by a transmembrane protein and the sugar molecule is apparently phosphorylated while in the pore. The ultimate phosphate donor is not ATP but **phosphoenolpyruvate,** which we will encounter as a potent phosphorylating agent in glycolysis. The phosphate group is actually passed via a series of enzymes before being attached to the sugar (Figure 9.27). This complicated mechanism allows selective control of the uptake of different sugars, depending on the metabolic needs of the cell. Furthermore, phosphorylation of monosaccharides is, as we shall see in Chapter 13, the first step in their metabolic utilization. The sugars are, as it were, already primed for metabolism.

Although such a transport mechanism looks very different from the direct coupling in ion pumps, it is basically the same—a high-energy phosphate compound has been hydrolyzed to accomplish the directed transport of a molecule across the membrane.

An Overview of Membrane Transport

This section has touched on only a few examples of cellular transport. We shall encounter many more in our later explorations of cellular metabolism. To gain an idea of how pervasive this process is, examine Figure 9.28, which shows many (but by no means all) of the substances transported across cell membranes.

Any functional membrane system employs a combination of the modes of transport described above. As a final integrating example, we consider again the complex cell envelope of Gram-negative bacteria, described in Chapter 8 (see Figure 8.27b). You will recall that such bacteria have an *inner membrane,* a rather thin *peptidoglycan layer,* and an *outer membrane.* The peptidoglycan layer is anchored to the outer membrane by covalently linked lipopolysaccharides with their lipid moieties embedded in the outer membrane. Between the peptidoglycan layer and the inner membrane is a region called the **periplasmic space.**

Nutrients passing into a Gram-negative bacterium like *E. coli* must cross this rather formidable-appearing series of barriers. Passage through the outer membrane is relatively easy. Small polar molecules and ions can diffuse through pores, formed by trimers of a protein called **porin.** For bulkier metabolites, specific channels are available for facilitated transport.

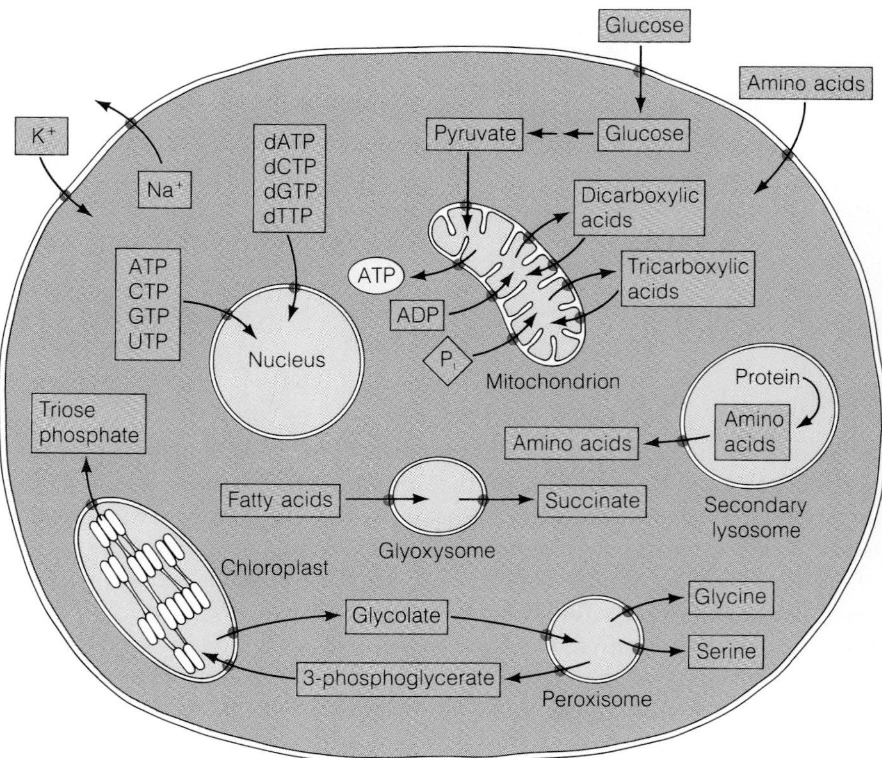

Figure 9.28
A "composite" plant–animal cell illustrating some of the common transport processes. All of the substances shown here, and many more, are transported in specific directions across cellular membranes. Red dots indicate specific transport sites.

The peptidoglycan layer itself, being a loose network of polysaccharide and polypeptide chains, offers little barrier to transport. However, the inner membrane is a different matter. Since it is the inner membrane that effectively seals off the cell, transport through it is highly regulated. The inner membrane is rich in specific pumps to promote the transport of desired nutrients and to maintain the necessary differences in ion concentration between the cytosol and the periplasmic space. It is not surprising that this membrane is about 75% protein.

There is a further elaboration to the transport of some nutrients through the bacterial envelope. Transport proteins have been identified that serve to carry particular molecules (e.g., histidine, lactose) across the periplasmic space and deliver them to the appropriate pumps for transport into the cell. Thus, the importation of nutrient materials through the envelopes of Gram-negative bacteria involves all of the kinds of transport we have discussed: passive diffusion, transport facilitated by both transmembrane proteins and carriers, and active transport.

REFERENCES

General

Gurr, A. I., and A. T. James (1976) *Lipid Biochemistry: An Introduction,* 2nd Ed. Halstead, New York. A valuable source for general information concerning lipids.

Finean, J. B., and R. Coleman (1984) *Membranes and Their Cellular Functions.* Blackwell Scientific, Oxford.

Vance, D. E., and J. E. Vance (eds.) (1985) *Biochemistry of Lipids and Membranes.* Benjamin/Cummings, Menlo Park, Calif. A collection of chapters, each written by an expert, providing up-to-date information on a wide variety of topics.

Membrane Asymmetry and Assembly

Lodish, H. F., and J. F. Rothmann (1979) The assembly of cell membranes. *Sci. Am.* 240:(1)48–63.

On den Kamp, J. A. F. (1979) Lipid asymmetry in membranes. *Annu. Rev. Biochem.* 48:47–71.

The Fluid Mosaic Model and Membrane Proteins

Eisenberg, D. (1984) Three dimensional structure of membrane and surface proteins. *Annu. Rev. Biochem.* 55:595–623.

Singer, S. J., and G. L. Nicholson (1972) The fluid mosaic model of the structure of membranes. *Science* 175:720–731. The classic paper presenting this model.

Unwin, N., and R. Henderson (1984) The structure of proteins in biological membranes. *Sci. Am.* 250(2):78–94.

The Erythrocyte Membrane

Bennett, V. (1985) The membrane skeleton of human erythrocytes and its implication for more complex cells. *Annu. Rev. Biochem.* 54:273–304.

Transport

Graves, J. S. (ed.) (1985) *Regulation and Development of Membrane Transport Processes.* Wiley, New York. A good collection of recent papers.

Jay, D., and C. Cautley (1986) Structural aspects of the red cell anion exchange protein. *Annu. Rev. Biochem.* 55:311–338.

Taniguchi, K., K. Suzuki, D. Kai, I. Matsuoka, K. Tomita, and S. Iida (1984) Conformational change of sodium and potassium-dependent adenosine triphosphatase. *J. Biol. Chem.* 259:15228–15233.

PROBLEMS

1. Give structures for the following, based on the data in Table 9.1.
   a. *cis*-9-dodecenoic acid
   b. $18:1^{\Delta 11}$
   c. A saturated fatty acid that would melt below 30°C.

2. Given the molecular components glycerol, fatty acid, phosphate, long-chain alcohol, and carbohydrate:
   a. Which *two* are present in both waxes and sphingomyelin?
   b. Which *two* are present in both fats and phosphatidylcholine?
   c. Which are present in a ganglioside but not in a fat?

3. The classic demonstration that cell plasma membranes are composed of *bilayers* depends on the following kind of data:

(i) The membrane lipids from 4.74×10^9 erythrocytes will form a *monolayer* of area $0.89\ m^2$ when spread on a water surface.

(ii) The surface of one erythrocyte is approximately $100\ (\mu m)^2$ in area.

Show that these data can be accounted for only if the erythrocyte membrane is a bilayer.

4. In the situations described below, what is the free energy change if 1 mol of Na^+ is transported across a membrane from a region where the concentration is 1 μm to a region where it is 100 mM? (Assume $T = 37°C$.)
   a. In the absence of a membrane potential.
   b. When the transport is opposed by a membrane potential of 70 mV.

   In each case, will hydrolysis of 1 mol of ATP suffice to drive the transport of 1 mol of ion? (Assume ΔG for ATP hydrolysis is about -50 kJ/mol under these conditions.)

5. Consider passive diffusion of ions across the erythrocyte membrane, as measured by the permeability coefficients in Table 9.6. Calculate the number of moles of K^+ that would diffuse across a single erythrocyte membrane in 1 minute, given the following data:

$$C_{K^+}\ (\text{inside}) = 100\ \text{mM}$$

$$C_{K^+}\ (\text{outside}) = 15\ \text{mM}$$

   Surface area of one erythrocyte = $100\ (\mu m)^2$. (Note: be careful of units.)

6. If the volume of the erythrocyte in Problem 5 is about $100\ (\mu m)^3$, what percent of the K^+ would escape by passive diffusion in 1 minute?

7. Suppose calcium ion is maintained within an organelle at a concentration 1000 times greater than outside (T = 37°C). What is the contribution of Ca^{2+} to the membrane potential? Which side of the organelle membrane is positive, which negative?

14

TOOLS OF BIOCHEMISTRY 14

Techniques for the Study of Membranes

There exists a battery of special techniques for the study of membrane structure and function. Examinations of membrane structure as it exists within cells depends heavily on electron microscopy, and virtually all of the variants of that method mentioned in Tools of Biochemistry 1 have been employed at one time or another, as have the special methods of freeze fracture and freeze etching described in this chapter. While electron microscopy has revealed much of the elaborate architecture of natural membranes, investigators often need to use simplified systems to study specific membrane properties. For these purposes, synthetic bilayers and vesicles are frequently used.

Preparation of Bilayers and Vesicles

Membranes from individual types of cells or purified organelles can usually be obtained by lysis of the cell or organelle, followed by differential centrifugation. In many cases, the membrane components will float to the top of the tube.

If, as shown in Figure T14.1, a membrane is extracted with organic solvents (a chloroform–ethanol mixture, for example), the soluble lipid constituents can be separated from insoluble protein and oligosaccharides. The lipid mixture can then, if desired, be fractionated by methods such as high-pressure liquid chromatography to yield pure lipid components and an analysis of the lipid content. Alternatively, the investigator may wish to use the entire lipid mixture from a membrane.

If the organic solvent in such a preparation is removed by evaporation, and the membrane lipids are dispersed in aqueous solution, they will form vesicles (also

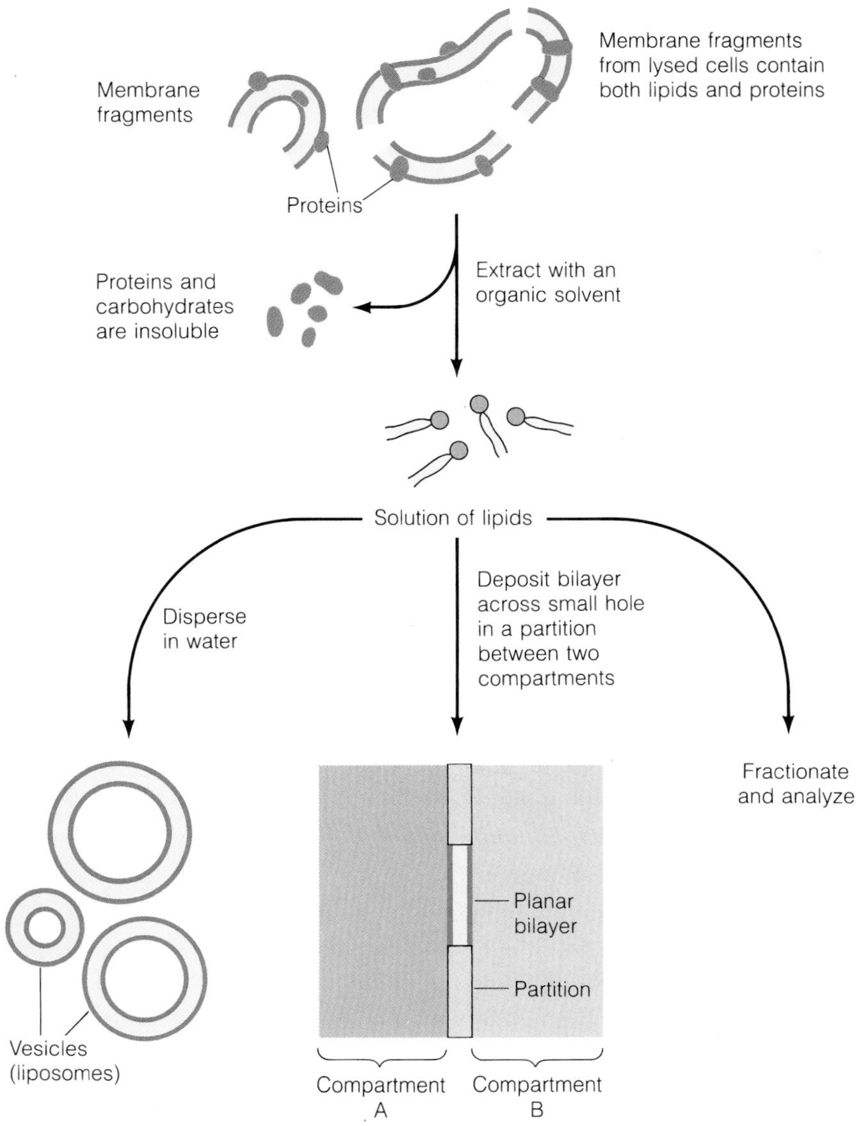

Figure T14.1
Preparation of vesicles and bilayers.

called *liposomes*)—small spherical bilayer structures. Alternatively, a bilayer may be spread across a small hole in a partition between two compartments. Such preparations are often used to study permeability across membranes.

The vesicles can be used for many kinds of studies. For example, it is possible to reconstitute specific transport systems by isolating the transport proteins through detergent solubilization and then adding them to a vesicle-forming system. In the presence of ATP, properly reconstituted systems will display active transport. Figure T14.2 demonstrates reconstitution of the Ca^{2+} pump of muscle cells. Vesicles made with specific mixtures of lipids and other components are also excellent objects for the study of such processes as the phase transition or diffusion in membrane.

In other kinds of experiments, it is desirable to retain all of the natural components in the membrane. This can often be done by carefully lysing cells or organelles, isolating the intact membranes, and then dispersing them in a solution in which they will reseal to form vesicles. In some cases, as shown with an erythrocyte in Figure T14.3, it is possible to adjust conditions so that the vesicles reseal preferentially either "right side out" or "inside out." Such preparations have provided much of our information concerning membrane asymmetry.

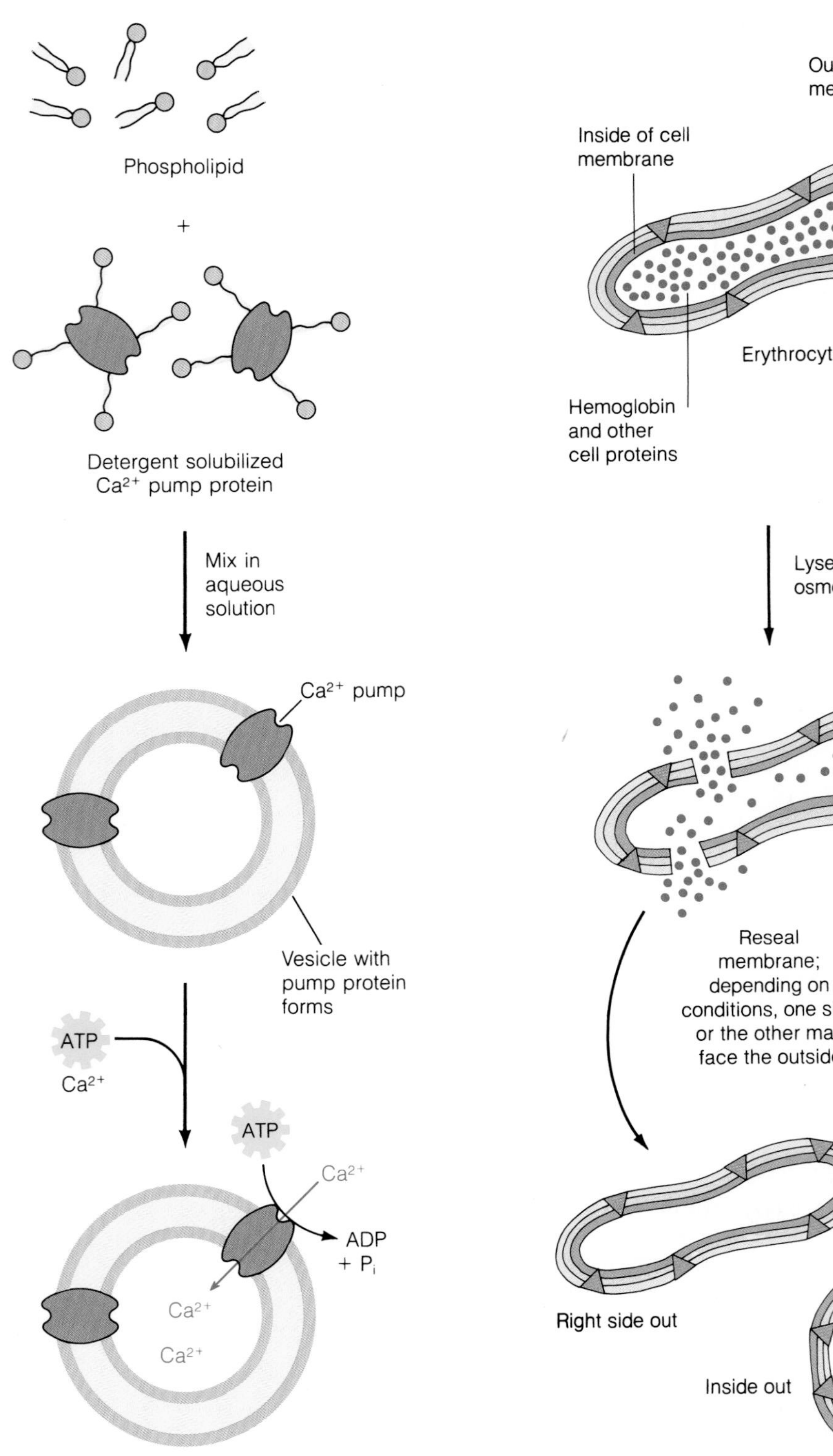

Figure T14.2
Reconstitution of the Ca^{2+} pump.

Figure T14.3
Preparation and resealing of erythrocyte ghosts.

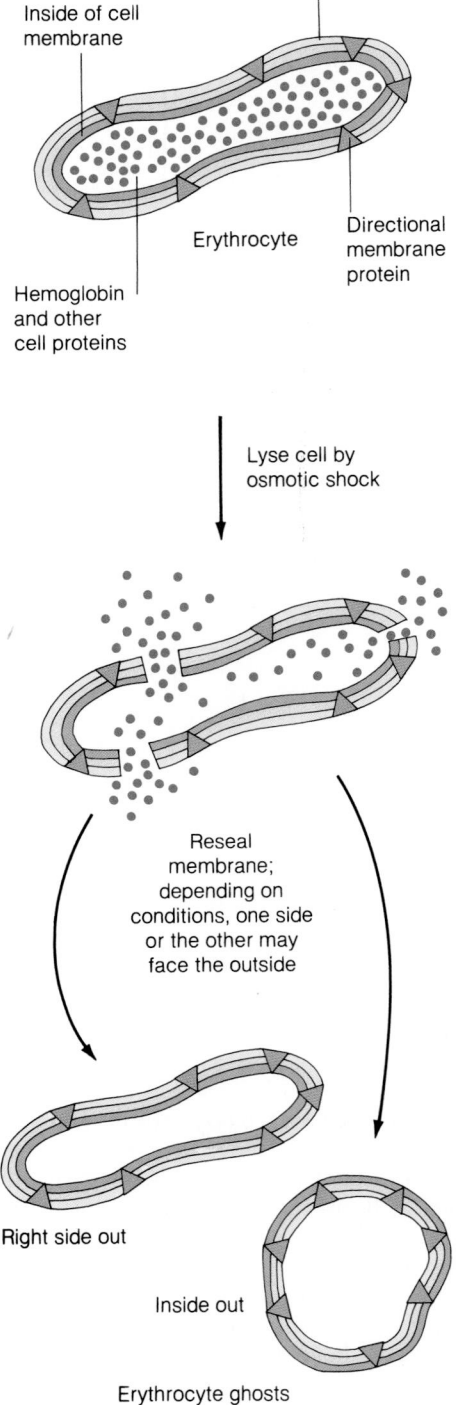

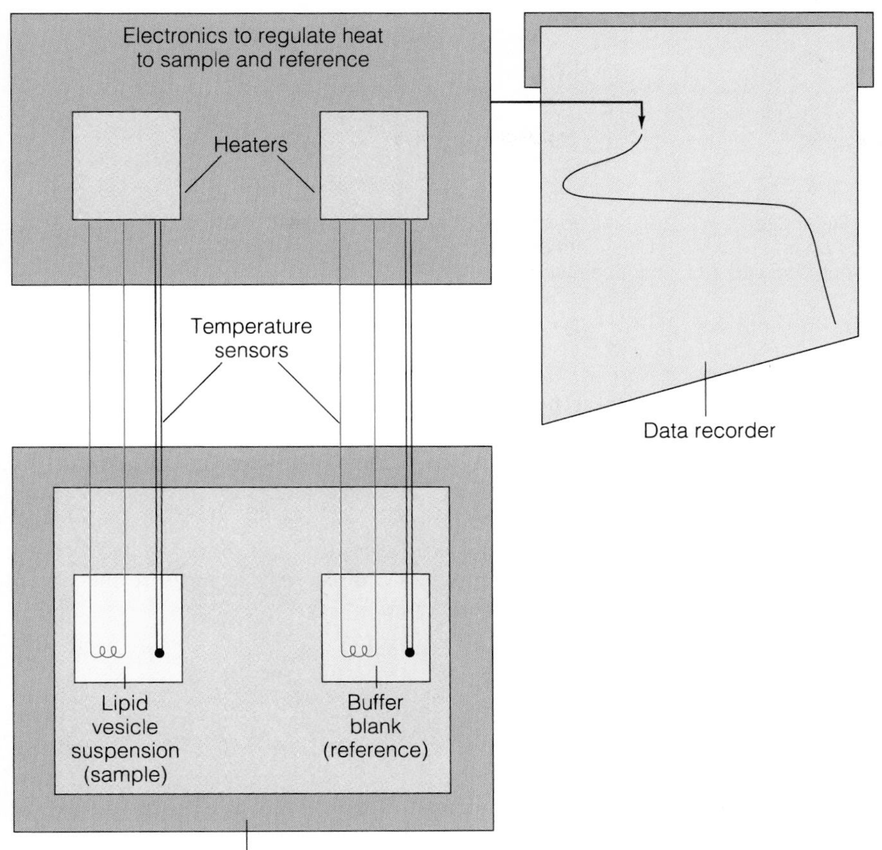

Figure T14.4
Differential scanning calorimetry.

Physical Techniques

Much of our information concerning structural transitions in membranes has come from **scanning calorimetry.** A very simplified schematic of a scanning calorimeter is shown in Figure T14.4. Samples of a lipid vesicle suspension and a buffer blank are heated in parallel, and the difference in energy input required to keep them at the same temperature is carefully monitored. As the transition temperature (T_m) is passed, more heat has to be put into the lipid sample to melt the membrane structure. This shows up as the kind of spike shown in Figure 9.11. The experiment reveals the transition temperature, the sharpness of the transition, and the total energy required.

To study membrane fluidity more directly, and to examine the motion of individual kinds of molecules in the membrane, a number of other techniques are employed. **Electron spin resonance** (ESR) has been of major importance. Electron spin resonance bears similarities to nuclear magnetic resonance but involves changes in the spin of unpaired electrons rather than nuclei. The resonance spectrum is sensitive, in both absorption line spacing and sharpness, to the environment of the unpaired electron. As in NMR, narrow spectral lines are characteristic of a fluid environment with rapid molecular motion, and broadened lines are observed when molecular motion is sluggish.

Most compounds do not have unpaired electrons, but certain nitroxide compounds, such as *tempocholine*, contain such an electron in an N—O bond. Tempocholine has been substituted for choline in phosphatidylcholine to act as a "reporter group," sensing the freedom of motion of the head groups in membranes. Other reporter groups or molecules can probe the fluidity of the inside of membranes, or the mobility of membrane proteins.

$HO-CH_2-CH_2-N^+(CH_3)_2-$ [ring with $N-O^{\cdot}$ and four CH_3]

Tempocholine

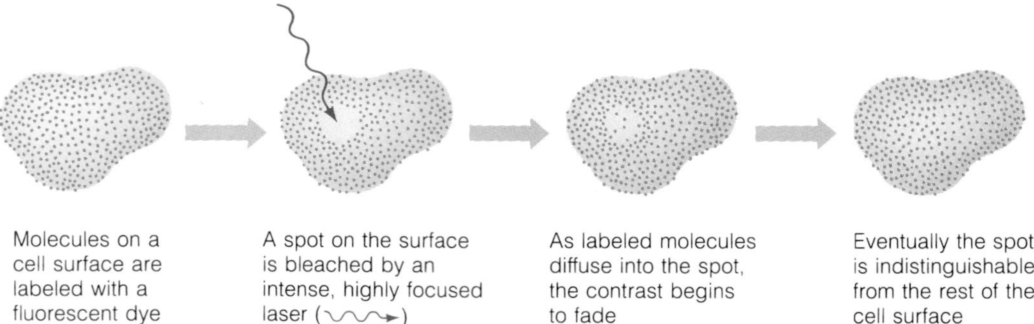

Molecules on a
cell surface are
labeled with a
fluorescent dye
A spot on the surface
is bleached by an
intense, highly focused
laser (⟿)
As labeled molecules
diffuse into the spot,
the contrast begins
to fade
Eventually the spot
is indistinguishable
from the rest of the
cell surface

Figure T14.5
Fluorescence photobleaching recovery.

A new and powerful technique is called **fluorescence bleaching recovery** (Figure T14.5). Selected molecules in cell membranes are given a fluorescent label. Then a tiny spot on the membrane surface is exposed to a high-intensity laser beam. This "bleaches" the fluorescent label, forming a bleached spot on the cell surface. The cell is then observed under a fluorescence microscope. As bleached molecules diffuse out of the spot, and unbleached molecules diffuse in, the spot gradually recovers its original fluorescence intensity. This provides a direct way to measure the lateral movement of selected molecules in membranes.

REFERENCES

Elson, E. L. (1986) Membrane dynamics studied by fluorescence correlation spectroscopy and photobleaching recovery. *Soc. Gen. Phys. Ser.* 40:367–383.

Fleischer, S., and L. Packer (eds.) (1974) *Methods in Enzymology,* Vol. 32. Academic Press, New York. This volume of a many-volume set focuses on techniques for membrane study.

Kornberg, R. D., and H. M. McConnell (1971) Inside–outside transitions of phospholipids in vesicle membranes. *Biochemistry* 10:1111–1120. Use of ESR methods to measure motion across membranes.

Madden, T. D. (1986) Current concepts in membrane protein reconstitution. *Chem. Phys. Lipids* 40:207–222.

PART III

Dynamics of Life: Catalysis and Control of Biochemical Reactions

CHAPTER 10

Enzymes: Biological Catalysts

Again and again we have alluded to the importance of the specific catalysts called **enzymes** in regulating the chemistry of cells and organisms. Catalysis is necessary to make many essential biochemical reactions proceed at a useful rate under physiological conditions. A reaction that takes many hours to approach completion cannot be metabolically useful to a bacterium that must divide in 20 minutes.

In fact, life takes advantage of the fact that most reactions must be catalyzed. In the complex milieu of the cell, there are countless *possible* reactions between the molecules. The cell seizes on specific catalysis to channel substances into *useful* pathways rather than into wasteful side reactions. Furthermore, the enzymes the cell employs are unusual catalysts in that the efficiency of their operation can in almost all cases be controlled so as to modulate the production of different substances in response to cellular and organismal needs.

Much of the remainder of this book is devoted to an elaboration of this theme. When we discuss the many aspects of metabolism in following chapters, we will encounter, at virtually every step of each pathway, a specific enzyme that has been tailored evolutionarily to catalyze the particular reaction. Thus it is essential that we begin our analysis of metabolism by examining what enzymes are and how they work.

The Role of Enzymes

What Enzymes Do: An Example

A **catalyst** is defined as a substance that increases the rate, or velocity, of a chemical reaction without itself being changed in the overall process. Most (but as we now know, not all) biological catalysts are protein enzymes. We have already encountered one example in the protein trypsin, which catalyzes hydrolysis of peptide bonds in proteins and polypeptides. The substance that is acted on by an enzyme is called the **substrate** of that enzyme. Thus, we can consider polypeptides to be appropriate substrates for trypsin.

It is useful, however, to begin our discussion of enzymes with a simpler reaction than hydrolysis of a polypeptide. Consider, for example, the isomerization between glyceraldehyde-3-phosphate and dihydroxyacetone phosphate. We have already considered this reaction in Chapter 8, and will find in Chapter 13 that it is an important step in the metabolic pathway called glycolysis. The reaction is

$$\underset{\textbf{Glyceraldehyde-3-phosphate (G3P)}}{\mathrm{H{-}C({=}O){-}CH(OH){-}CH_2OPO_3^{2-}}} \rightleftharpoons \underset{\textbf{Dihydroxyacetone phosphate (DHAP)}}{\mathrm{CH_2OH{-}C({=}O){-}CH_2OPO_3^{2-}}}$$

The protein that catalyzes this interconversion in cells is called **triose-phosphate isomerase.** Like many enzyme names, this one is descriptive of what the particular enzyme does: It catalyzes the isomerization of certain triose phosphates. As we have written the reaction, G3P is considered the substrate and DHAP the product. But we could equally well have written the reaction the other way around and called DHAP the substrate. Like all true catalysts, enzymes accelerate the rate of the reaction in *both* directions. *Enzymes do not affect the state of equilibrium in a reaction;* they just allow it to be attained much more quickly. The acceleration achieved by enzymatic catalysis is often enormous: in this case, for example, the reactions go more than 1 billion times faster in the presence of the enzyme.

The Diversity of Enzyme Function

Since metabolism is a very complex business, there are thousands of different kinds of enzymes. Many of these have been given common names. Some of the names, like *triose-phosphate isomerase,* are descriptive of the enzyme's function; others, like *trypsin,* are not. A rational naming and numbering system has been devised by the Enzyme Commission of the International Union of Biochemistry. Enzymes are divided into six major classes, with subgroups to define their functions more precisely. The major classes are:

1. *Oxidoreductases*—catalyze oxidation/reduction reactions.
2. *Transferases*—catalyze transfer of molecular groups from one molecule to another.
3. *Hydrolases*—catalyze hydrolytic cleavage.
4. *Lyases*—catalyze removal of a group from or addition of a group to a double bond, or other cleavages involving electronic rearrangement.
5. *Isomerases*—catalyze intramolecular rearrangement.
6. *Ligases*—catalyze reactions in which two molecules are joined.

The Enzyme Commission (EC) has given each enzyme a number like

EC 3.4.21.5. The first three numbers define major class, subclass, and sub-subclass, respectively. The last is a serial number in the sub-subclass, indicating the order in which each enzyme is added to the list, which is continually growing. For example, triose-phosphate isomerase is listed as EC 5.3.1.1. It is an isomerase and in the third subclass (enzymes that involve an oxidation in one part of the substrate molecule and reduction in another). It is in the first sub-subclass (those that interconvert aldoses and ketoses) and is the first entry (of 19) in this sub-subclass. Listings for almost all currently recognized enzymes, together with information on each and literature references, are given in the book *Enzyme Nomenclature* (Academic Press, 1984). The 2728 entries in this edition surely do not include all enzymes; more are being discovered all the time. Indeed, it has been estimated that the typical cell contains thousands of different kinds of enzymes.

Table 10.1 lists one example from each of the major classes of enzymes. We will discuss each of these reactions later in this book; the main point we wish to emphasize here is the enormous diversity of enzymatic functions.

Chemical Reaction Rates and the Effects of Catalysts

Before analyzing enzyme behavior in detail, we must review a bit of chemical kinetics to provide precise definitions of reaction rates and a clear understanding of what catalysts do.

Reaction Rates and Reaction Order

To understand what is meant by a reaction rate and how it might be measured, let us begin by considering the simplest possible reaction: the *irreversible* conversion of substance A to substance B:

$$A \longrightarrow B$$

The fact that we draw only one arrow here signifies that the reverse reaction ($B \rightarrow A$) proceeds to only an infinitesimal extent. The equilibrium state lies far to the right—essentially all B.

We can define the **reaction rate,** or **velocity** (V) at any instant as the time rate of formation of the product, in this case B:

$$V = \frac{d[B]}{dt} \tag{10.1}$$

The units of V are moles per liter per second, if [B] symbolizes molar concentration of B. If we note that for every B molecule formed an A molecule must disappear, it is clear that V could equally well be written as

$$V = -\frac{d[A]}{dt} \tag{10.2}$$

The change of each molecule of A into B is an independent event. Therefore, the rate decreases as the reaction proceeds and molecules of A are con-

Table 10.1
Examples of each of the major classes of enzymes

Class	Example (Reaction Type)	Reaction Catalyzed
1. Oxidoreductases	Alcohol dehydrogenase (EC 1.1.1.1) (oxidation with NAD^+)	CH_3CH_2OH (**Ethanol**) $\xrightarrow{NAD^+ \rightarrow NADH + H^+}$ CH_3-CHO (**Acetaldehyde**)
2. Transferases	Glucokinase (EC 2.7.1.2) (phosphorylation)	**D-Glucose** $\xrightarrow{ATP \rightarrow ADP}$ **D-Glucose-6-phosphate** ($CH_2OPO_3^{2-}$)
3. Hydrolases	Carboxypeptidase A (EC 3.4.17.1) (peptide bond cleavage)	$-NH-CHR_{n-1}-CO-NH-CHR_n-COO^-$ (**C-terminal of polypeptide**) $\xrightarrow{H_2O}$ $-NH-CHR_{n-1}-COO^-$ (**Shortened polypeptide**) + $H_3\overset{+}{N}-CHR_n-COO^-$ (**C-terminal residue**)
4. Lyases	Pyruvate decarboxylase (EC 4.1.1.1) (decarboxylation)	$^-OOC-CO-CH_3 + H^+$ (**Pyruvate**) $\longrightarrow CO_2 + H-CO-CH_3$ (**Acetaldehyde**)
5. Isomerases	Maleate isomerase (EC 5.2.1.1) (cis–trans isomerization)	**Maleate** ($^-OOC(H)C=C(H)COO^-$, cis) $\rightleftharpoons$ **Fumarate** ($^-OOC(H)C=C(H)COO^-$, trans)
6. Ligases	Pyruvate carboxylase (EC 6.4.1.1) (carboxylation)	$^-OOC-CO-CH_3$ (**Pyruvate**) $+ CO_2 \xrightarrow{ATP \rightarrow ADP + P_i}$ $^-OOC-CO-CH_2-COO^-$ (**Oxaloacetate**)

sumed. Mathematically, we state this by saying that the rate is proportional to [A]:

$$V = \frac{d[B]}{dt} = -\frac{d[A]}{dt} = k[A] \qquad (10.3)$$

The constant k is called the **rate constant** for the reaction. For this reaction, k will have units of $(\text{seconds})^{-1}$. This reaction is called **first order** because its rate depends on the first power of the concentration.

If we want to prove that a reaction is first order, or measure the rate constant, it is more convenient to have an equation that describes how the concentration of A changes with time during the reaction. This can be obtained by *integrating** equation (10.3), which gives

$$\ln \frac{[A]}{[A]_0} = -kt \tag{10.4a}$$

or

$$\frac{[A]}{[A]_0} = e^{-kt} \tag{10.4b}$$

where $[A]_0$ is the starting concentration, when $t = 0$. Equation (10.4b) tells us that the concentration of A decreases exponentially with time, as shown in Figure 10.1a. To test if a reaction is truly first order, we need only make a graph according to equation (10.4a), as shown in Figure 10.1b. An alternative procedure is shown in Figure 10.1c and d: One measures the *initial rate* of the reaction at different values of the initial concentration, $[A]_0$, and plots these initial rates against $[A]_0$. If the initial rate is proportional to $[A]_0$, the reaction is first order. Either method allows determination of the rate constant k from the slope of the appropriate graph.

Most biochemical processes cannot be described, over their full course, by equations as simple as (10.4). One reason is that many of the reactions and processes we encounter are *reversible*—that is, equilibrium does not lie far to one side—and as product accumulates, the reverse reaction becomes important. For example, we may have a reaction like this:

$$[A] \underset{k_{-1}}{\overset{k_1}{\rightleftharpoons}} [B]$$

Since A is being consumed in the reaction to the right and formed by the reaction to the left, the corresponding rate equation is

$$-V = \frac{d[A]}{dt} = -k_1[A] + k_{-1}[B] \tag{10.5}$$

Here, k_1 and k_{-1} are the rate constants for the first-order forward and reverse reactions. Such a reaction approaches a state of equilibrium, at which point the rates of the forward and reverse reactions become equal, and so the overall rate becomes zero. Then

$$0 = -k_1[A]_{eq} + k_{-1}[B]_{eq} \tag{10.6a}$$

or

$$\frac{[B]_{eq}}{[A]_{eq}} = \frac{k_1}{k_{-1}} = K \tag{10.6b}$$

where K is the equilibrium constant. For a reversible reaction that is first order in both directions, the equilibrium constant is always the ratio of the

---

* See any elementary calculus book for how this simple integration is carried out.

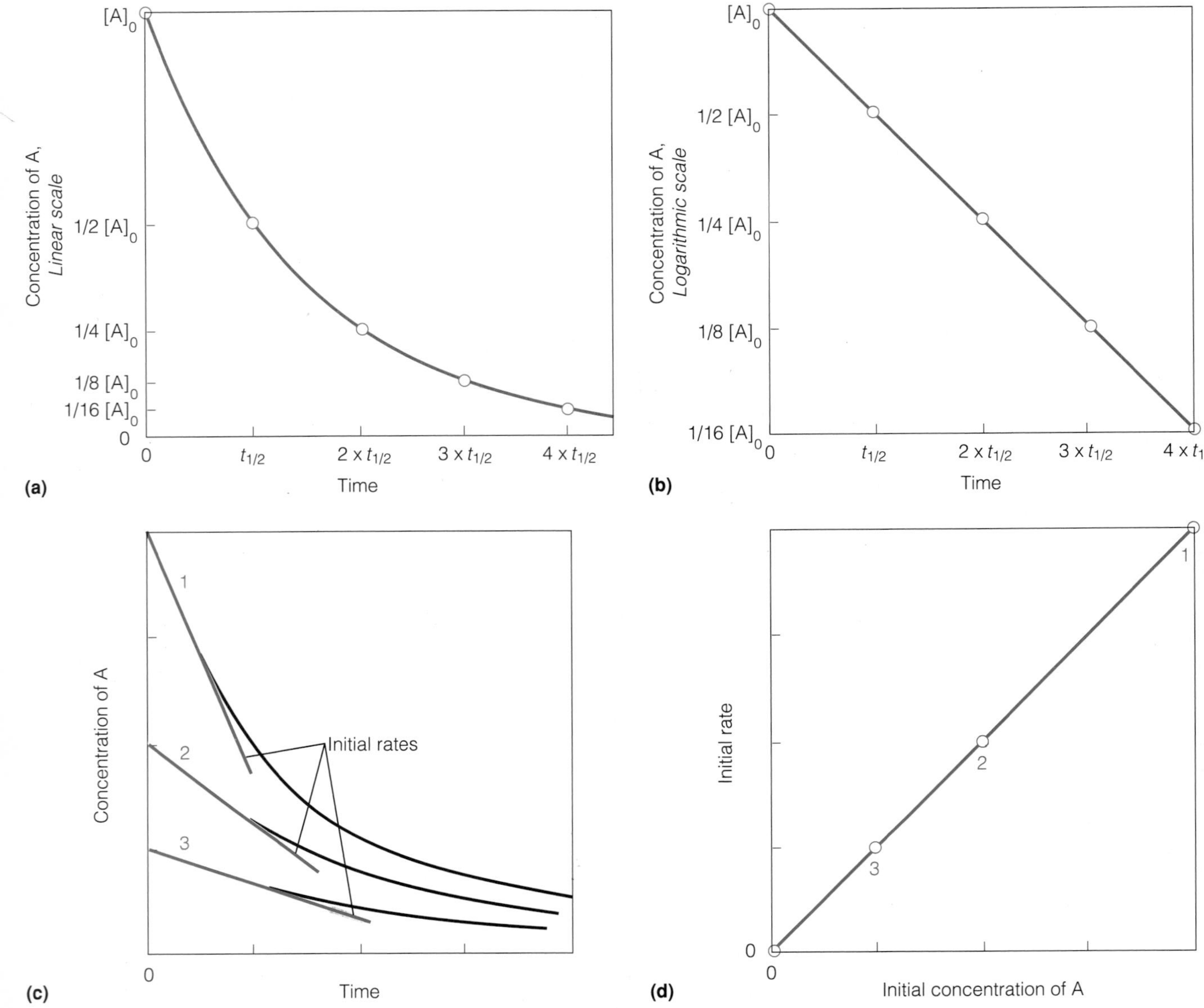

Figure 10.1
An irreversible first-order reaction. Tests for order and determination of k.
(a,b) Direct analysis of reaction time course. A graph of [A] versus t shows that the rate, defined as the slope of the line, decreases as the reaction continues. A graph of ln[A] versus t, when linear, indicates that the reaction follows equation (10.4a) and hence is first order. The slope of this line is equal to $-k$. In these graphs we have expressed time as multiples of the half-life ($t_{1/2}$) of the reactant. Note that for each interval of $t_{1/2}$, the reactant concentration is halved.
(c,d) Analysis of initial rates. First the initial rate is determined at different values of $[A]_0$. Then initial rates are plotted against $[A]_0$. If the initial rates are proportional to $[A]_0$, yielding a straight-line graph, the reaction is first order. The slope of the line is k.

forward and reverse rate constants. To describe the time course of the reaction during the approach to equilibrium, we must integrate equation (10.5). If the initial condition is pure A, ($[A]_0$) we obtain

$$[A] - [A]_{eq} = ([A]_0 - [A]_{eq})e^{-kt} \qquad (10.7)$$

where $[A]_{eq}$ is the equilibrium concentration of A, and $k = k_1 + k_{-1}$. The kind of graph obtained is shown in Figure 10.2. If we graph $\ln([A] - [A_{eq}])$ versus t for such a reaction, we obtain a straight line.

Many of the reactions encountered in biochemistry are not first order, but are of *second order* or greater complexity. A second-order reaction is encountered whenever two reactants must come together to form products. Simple examples are

$$2A \xrightarrow{k} C$$

$$A + B \xrightarrow{k'} D$$

The rate of such a reaction is proportional to the *second* power of concentration, for we would write for these two reactions

$$V = \frac{d[C]}{dt} = k[A]^2 \tag{10.8a}$$

and

$$V = \frac{d[D]}{dt} = k'[A][B] \tag{10.8b}$$

respectively. Obtaining the integrated expression for concentration as a function of time from equation (10.8a) is not difficult. If we note that two moles of A are consumed for every mole of C produced, equation (10.8a) may be rewritten

$$\frac{d[C]}{dt} = -\frac{1}{2}\frac{d[A]}{dt} = k[A]^2 \tag{10.9}$$

Integration yields

$$\frac{1}{[A]} = \frac{1}{[A]_0} + 2kt \tag{10.10}$$

which indicates that a graph of 1/[A] versus t should be linear for such a process. Equation (10.8b) leads to a more complicated expression, except in the case where $[B]_0 = [A]_0$, in which event equation (10.10) is followed.

Many more complicated kinds of reactions occur, including complex, multistep processes. We are not concerned with their kinetics at this point, although we shall see that enzyme-catalyzed reactions, when analyzed in detail, are generally more complicated than those described above.

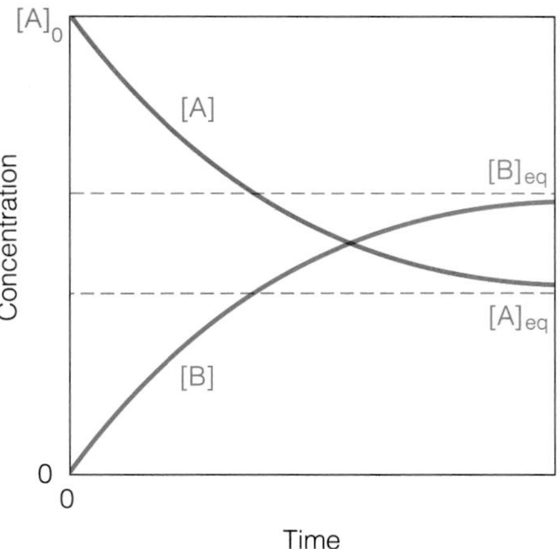

Figure 10.2
A reversible first-order reaction. Concentrations of A and B are graphed as they approach their equilibrium values.

Transition States and Reaction Rates

Our next concern is what determines the rate of a chemical reaction—that is, what makes a rate constant large or small? Thermodynamics gives no information concerning rates. If we draw a free energy diagram for a favorable reaction on the basis of thermodynamics alone it will look like Figure 10.3a. We can see that the free energy of the final state in this reaction is lower than that of the initial state. What may be of most importance in determining the rate is what happens in the *transition* from one state to the other, and equilibrium measurements do not reveal this.

There are many indications that molecules must be somehow *activated* before they can undergo reaction. It is clear that in a first-order reaction, a molecule is only occasionally in an energy state in which the process can occur. In a second-order reaction, not all encounters between reactants can be productive, for even in dilute solutions, reactable molecules collide with one another much more frequently than reaction occurs. Most encounters do not lead to reaction. Such considerations have given rise to the idea of an **energy barrier** to reaction and the concept of an **activated,** or **transition, state** (symbolized by ‡). The transition state is thought of as an intermediate stage in the reaction, often one in which the reacting molecule is strained or distorted or has an unfavorable electronic structure. The molecule must pass through the higher energy transition state if the reaction is to occur.

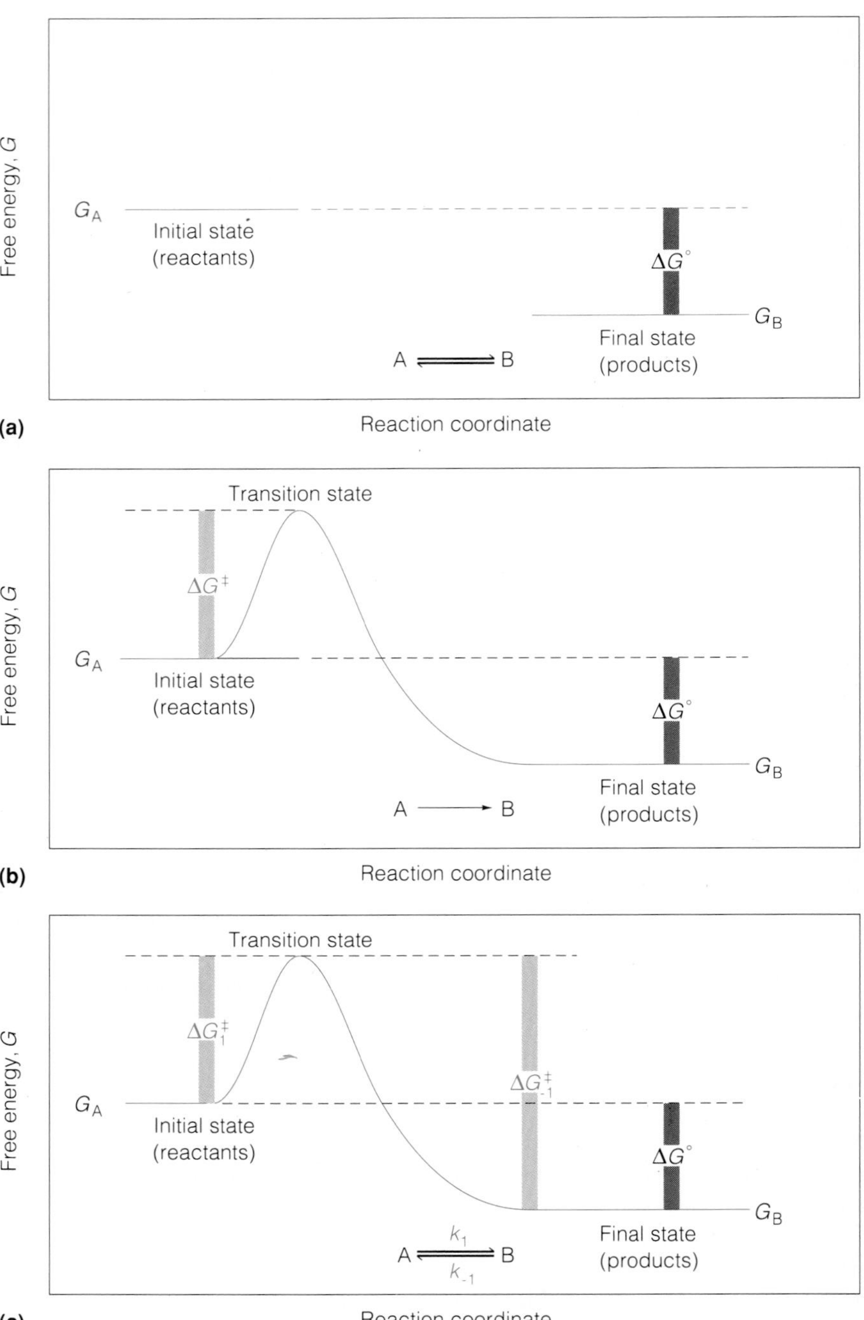

Figure 10.3
Free energy diagrams for a simple reaction: A ⇌ B. (**a**) Information provided by thermodynamic studies of the equilibrium: only the free energy difference between the initial state and the final state is revealed. ΔG° is the standard state free energy change for the reaction. G_A and G_B represent the *average* free energies per mole of A and B molecules. (**b**) Free energy diagram expanded to include the transition state (‡) through which the molecule must pass to go from A to B or vice versa. $\Delta G^\ddagger$ marks the energy of activation for the A → B transition. (**c**) The same diagram, now showing activation energies for both directions. Note that $\Delta G^\circ = \Delta G_1{}^\ddagger - \Delta G_{-1}{}^\ddagger$, as discussed in the text.

Such ideas about the transition state can be incorporated into the reaction diagram, as shown in Figure 10.3b. For the moment, we are concerned only with the reaction going in the direction A → B. For this process, the **energy of activation,** $\Delta G^\ddagger$, represents the additional free energy (above the average) that molecules must have to attain the transition state. We know that in any sample or solution, not all molecules have the same energy at any instant. If the barrier to reaction is high, only a small fraction of the molecules will have enough energy to surmount it, or only a small fraction of collisions will be energetic enough. If we assume that A molecules are in equilibrium between the initial state and the activated state, the fraction activated at any instant will be given by the familiar expression for a chemical equilibrium:

$$[A]^{\ddagger} = [A]^{\circ} e^{-\Delta G^{\ddagger}/RT} \qquad (10.11)$$

where $[A]^{\ddagger}$ represents the concentration of molecules having the activation energy and $[A]^{\circ}$ the total concentration. Once a molecule is activated to the transition state, it will hop to one side or the other—either back or over the barrier. Since only molecules that attain the transition state have the possibility of doing this, it is argued that the rate constant for a chemical reaction should be proportional to the population of the transition state, as expressed in equation (10.11):

$$k = Qe^{-\Delta G^{\ddagger}/RT} \qquad (10.12)$$

Since $\Delta G^{\ddagger} = \Delta H^{\ddagger} - T\,\Delta S^{\ddagger}$, this can be rewritten

$$k = Qe^{\Delta S^{\ddagger}/R} \cdot e^{-\Delta H^{\ddagger}/RT} = Q'e^{-\Delta H^{\ddagger}/RT} \qquad (10.13)$$

where Q' is a constant* called the *preexponential term.* Since we always expect $\Delta H^{\ddagger}$ to be positive, equation (10.13) (known as the Arrhenius equation) tells us that reactions should go faster at higher temperatures.

We can use the temperature dependence of reaction rates as a test to see if equation (10.13) is of the correct form. The equation can be rewritten, by taking natural logarithms of both sides, as:

$$\ln k = \ln Q' - \Delta H^{\ddagger}/RT \qquad (10.14)$$

Accordingly, a graph of ln k versus $1/T$ should be a straight line. Figure 10.4 is typical of the results for many reactions and argues that equation (10.13) is correct, at least in form. Such graphs also allow experimental determination of $\Delta H^{\ddagger}$.

It is particularly important to note that the energy barrier opposes reaction in *both* directions. In Figure 10.3c we depict the barriers $\Delta G_1^{\ddagger}$ and $\Delta G_{-1}^{\ddagger}$ for both the forward and reverse directions for a reversible reaction, first order in each direction. As we noted above, the equilibrium constant for such a reaction is given by $K = k_1/k_{-1}$. This is entirely consistent with equation (10.12), for according to that equation

$$K = \frac{Qe^{-\Delta G_1^{\ddagger}/RT}}{Qe^{-\Delta G_{-1}^{\ddagger}/RT}} = e^{-(\Delta G_1^{\ddagger} - \Delta G_{-1}^{\ddagger})/RT} \qquad (10.15)$$

The difference $(\Delta G_1^{\ddagger} - \Delta G_{-1}^{\ddagger})$ can be seen from Figure 10.3c to equal ΔG°, the overall free energy change for the process as written. Thus, equation (10.15) yields the relation between K and ΔG° given in Chapter 3 equation (3.31c):

$$K = e^{-\Delta G^{\circ}/RT} \qquad (10.16)$$

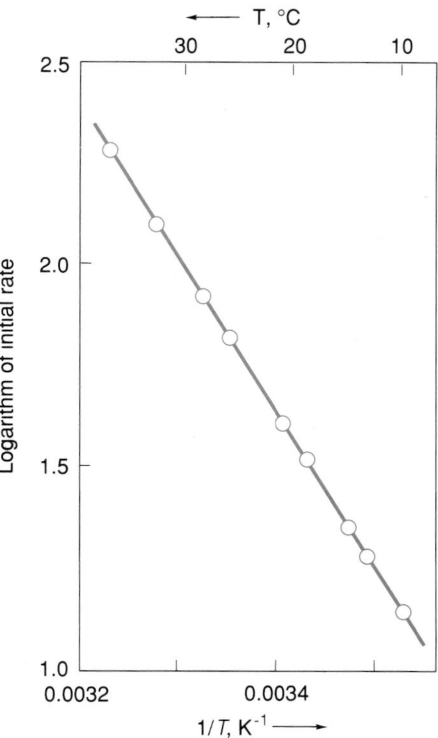

Figure 10.4
Test of the Arrhenius equation. The logarithm of the rate constant for the reaction L-malate → fumarate + H_2O is plotted versus the reciprocal of the absolute temperature. A straight line is obtained as predicted by equation (10.13). Note the strong dependence of rate on temperature: increasing the temperature by 10° produces almost a threefold increase in rate.

---

* According to some theories, Q is not exactly constant, but proportional to the absolute temperature. However, the temperature dependence introduced by this term will be small compared to the exponential term.

These results show directly why the ΔG° value cannot tell us how fast a reaction is: ΔG° itself provides *no* information concerning the height of the energy barrier in either direction.

What a Catalyst Does

A catalyst lowers the energy barrier to a reaction, thereby making the reaction go faster in either direction. However, the presence of a catalyst has no effect on the position of equilibrium. In Figure 10.5, the *difference* in barrier heights in the two directions is exactly the same whether a catalyst is present or not: ΔG° is unchanged, so is K, and so is k_1/k_{-1}, even though k_1 and k_{-1} may each be thousands or millions of times larger than they were for the uncatalyzed process.

We now ask *how* a catalyst lowers the energy barrier. To answer this question, let us examine $\Delta G^\ddagger$ a bit more closely.

Since $\Delta G^\ddagger = \Delta H^\ddagger - T\,\Delta S^\ddagger$, either a requirement for a great deal of energy (a large positive $\Delta H^\ddagger$) or a very improbable conformation in the transition state (a large negative $\Delta S^\ddagger$) will make $\Delta G^\ddagger$ large and positive and thus make a reaction slow. The energy contribution is easy to visualize. Very often, the reactive molecule or molecules must go through energy-demanding strained and distorted states in order for the reaction to occur. Sometimes the catalyst can reduce this energy requirement by binding the reactant molecule in an *intermediate* state that resembles the transition state but is of lower energy because of binding (Figure 10.6). As the figure shows, the result is that two very low transition energy barriers replace the single higher barrier. We distinguish such an intermediate state from a transition state by the fact that the former corresponds to an energy minimum, the latter to a maximum. An intermediate state is a semistable state of the molecule.

The entropy factor in the transition free energy reflects the fact that a particular *orientation* between reactants or parts of a molecule may be necessary to enter the transition state. For example, when collisions between molecules occur (as in a second-order reaction), most encounters are unproductive just because the molecules happen to be pointed the wrong way when they hit. A catalyst that can bind two reacting molecules in proper mutual orientation will increase their chances for reaction by lowering the entropy of activation (Figure 10.7). In a first-order reaction, which involves something happening *within* a single molecule, parts of the molecule must be oriented properly to allow the transition state to be reached. This is part

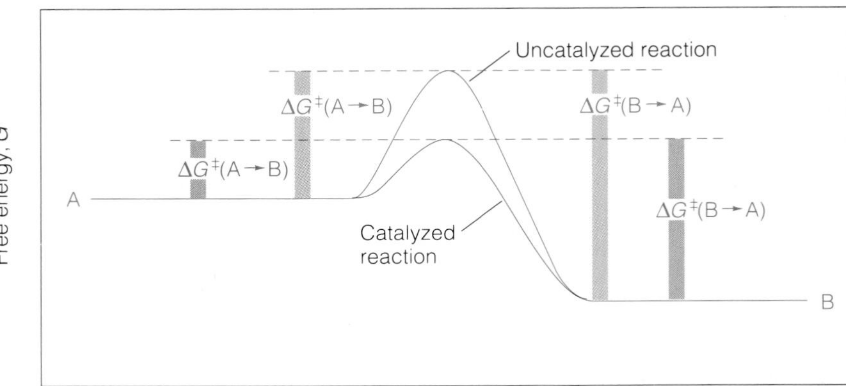

Figure 10.5
Effect of a catalyst. Here we have drawn the same free energy diagram as in Figure 10.3b but have provided an alternative *catalyzed* path (red) for the reaction. The catalyst lowers $\Delta G^\ddagger$, thus making it possible for a larger fraction of the reactant molecules to have the energy needed to reach this lowered transition state.

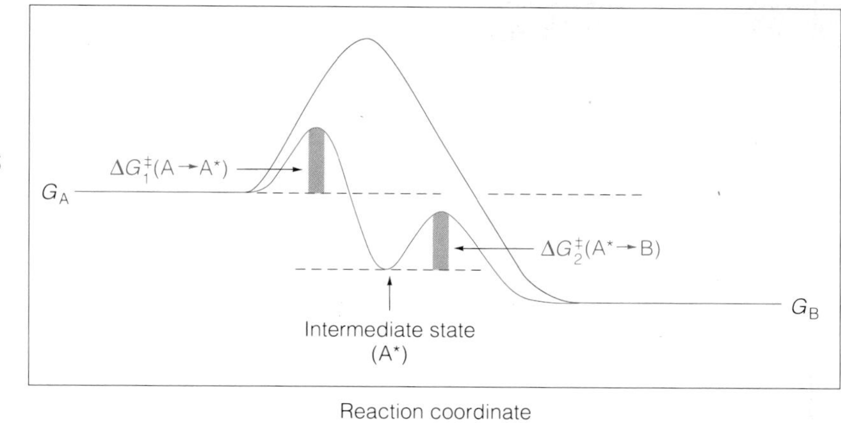

Figure 10.6
Importance of intermediate states. Often, a catalyst (such as an enzyme) will bind the substrate in a conformation that is a necessary intermediate in the reaction but that has a lower energy because of the favorable energetics of the binding to the catalyst. The activation energies for binding in the intermediate state, and for conversion of the intermediate to product, $\Delta G_1^{\ddagger}$ and $\Delta G_2^{\ddagger}$, respectively, are lower than the activation energy for the uncatalyzed reaction (Figure 10.3).

of what a catalyst does. As an external matrix, it can force molecules, or parts of molecules, into the conformation that favors reaction.

Some General Catalytic Mechanisms

Before we approach the question of how enzymes act as catalysts, it is worthwhile to consider some general molecular mechanisms of catalysis. As you will recall from organic chemistry, many reactions involve the attack by a *nucleophile* at a carbon atom. Common examples are found in the hydrolysis of esters or amides. We may regard an uncatalyzed hydrolysis, for example, as proceeding in the following manner:

$$R'-C(=O^{\delta-})^{\delta+}-N(H)-R + H_2O^{\delta-} \rightarrow [\text{transition state}] \rightarrow R'-C(=O)-OH + H-N(H)-R$$

Reactants **Transition state** **Products**

Here the nucleophile, H_2O, interacts with the partial positive charge on the carbonyl carbon. The uncatalyzed reactions are very slow because (1) there is only a small partial positive charge on the carbon and (2) water is a weak nucleophile because of the small negative charge at the oxygen. Consequently, formation of the transition state is difficult. The various general

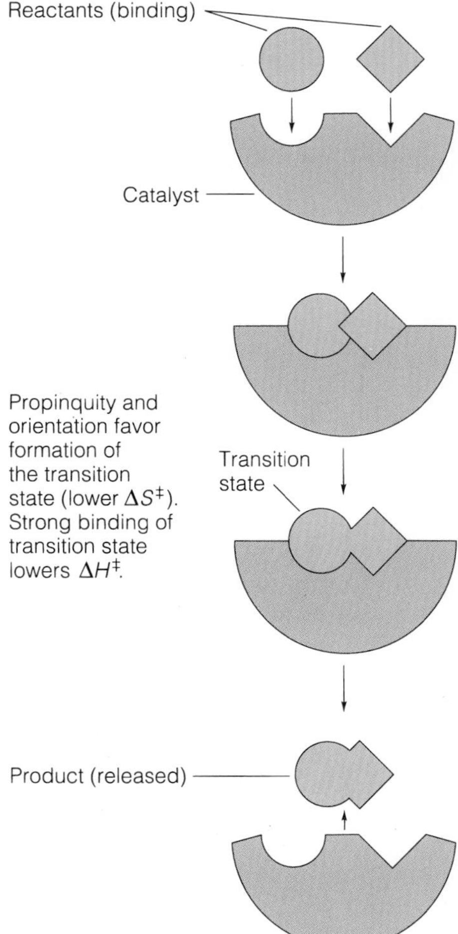

Figure 10.7
Entropic and enthalpic factors in catalysis. Two reactants are bound to an enzyme, which ensures their correct mutual orientation and propinquity and binds them most strongly in the transition state.

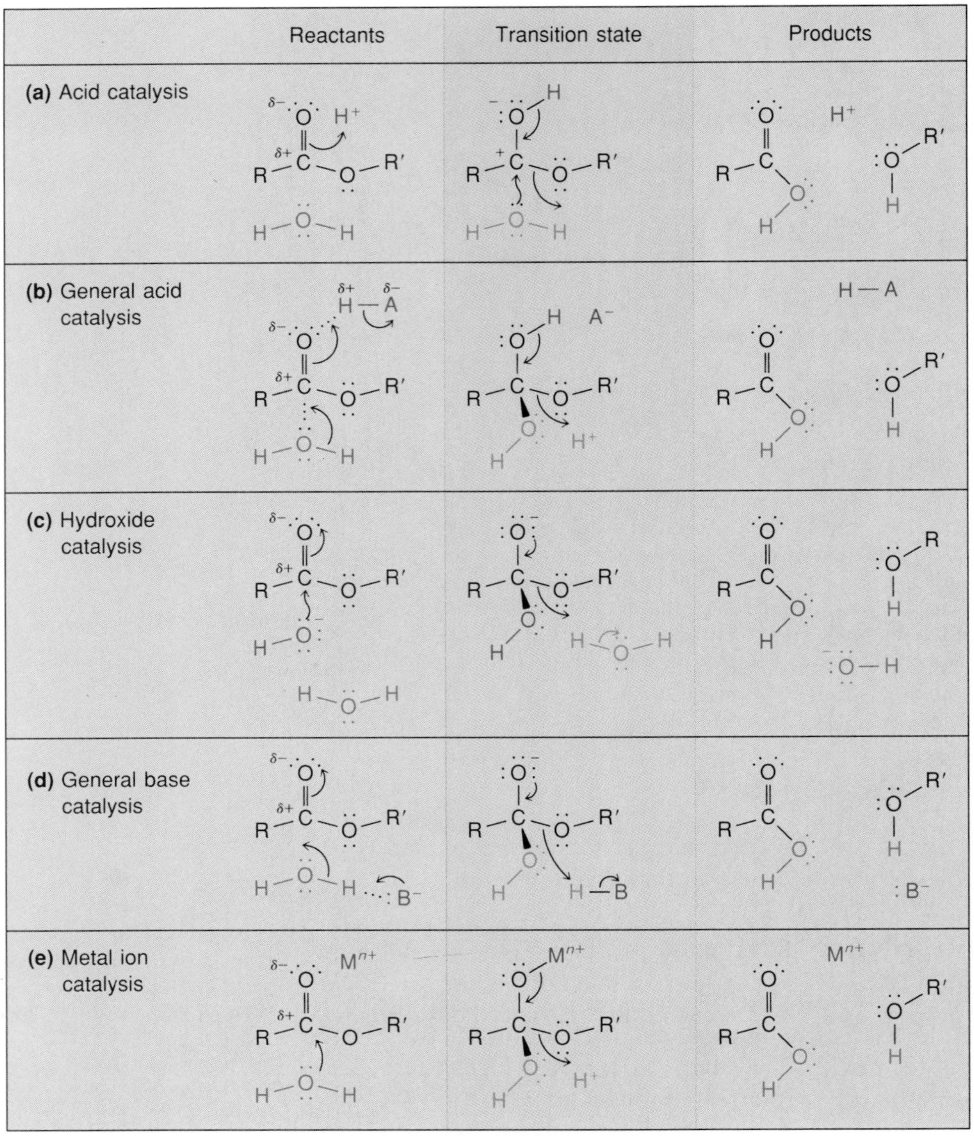

	Reactants	Transition state	Products
(a) Acid catalysis			
(b) General acid catalysis			
(c) Hydroxide catalysis			
(d) General base catalysis			
(e) Metal ion catalysis			

Figure 10.8
Transition state formation in hydrolysis reactions. The figure depicts ester hydrolysis, but the same principle would apply with amides.

classes of catalysis illustrated in Figure 10.8 all facilitate the formation of the transition state in one way or another. For example, in *acid catalysis* (a), a strong *electrophile* (H^+) first attacks the electron-rich carboxyl oxygen, shifting negative charges to that atom and leaving the carbon even more positively charged and susceptible to attack by water. In *general acid catalysis* (b), the proton is donated by a weakly acidic group. In *hydroxide catalysis* (c), a stronger nucleophile (OH^-) makes the initial attack on the carbonyl carbon. In *general base catalysis* (d), the hydroxyl is formed at the site of attack by extraction of a water proton. In *metal ion catalysis* (e), a positively charged cation interacts initially with the carbonyl oxygen to promote the necessary electron shift. In the following sections we shall find that these general mechanisms are frequently employed by enzymes, which may place an acidic group, a basic group, or a metal ion in just the right position to interact with the appropriate part of a substrate molecule. In addition, they usually bind the transition state strongly and thus stabilize it.

How Enzymes Act as Catalysts: Principles and Examples

General Principles: The Induced Fit Model

We have seen that the role of a catalyst is to decrease $\Delta G^{\ddagger}$, often by facilitating the formation of the transition state. An enzyme, which is a protein molecule, binds a molecule of substrate (or multiple substrates) into a region of the enzyme called the **active site** (Figure 10.9). The active site is often a pocket or cleft which is surrounded by amino acid side chains that help to bind the substrate and other side chains that play a role in the catalytic process. The complex tertiary structure of enzymes makes it possible for this pocket to fit the substrate quite closely, which explains the extraordinary specificity of enzyme catalysis. This possibility was realized as early as 1894 by the great German biochemist E. Fischer, who proposed a **lock-and-key** hypothesis for enzyme action: the enzyme accommodates the specific substrate as a lock does its specific key (see Figure 10.9a). Although the lock-and-key model explained enzyme specificity, it did not increase our understanding of the catalysis itself. This understanding came from an elaboration of Fischer's idea: what fits the enzyme active site best is a substrate molecule induced to take up a configuration approximating the transition state. This **induced fit** hypothesis, as proposed by D. Koshland in 1958, is still the dominant model for enzymatic catalysis.

As currently envisioned, induced fit implies distortion of the enzyme as well as the substrate. Thus, as depicted in Figure 10.9b, the stress produced in the distorted enzyme helps pull the substrate into the transition state conformation. A graphic visualization of this distortion of the enzyme can be seen with the enzyme *hexokinase*, which catalyzes the phosphorylation of glucose to glucose-6-phosphate. The structure of this enzyme has been determined by x-ray diffraction both in the absence of glucose and when glucose is bound. As Figure 10.10 shows, the binding of glucose causes two domains of the enzyme to fold toward each other, closing the binding site cleft about the substrate.

But enzymes do more than simply distort or position their substrates. Often, we find specific amino acid side chains poised in exactly the right places to aid in the catalytic process itself. In many cases, these are acidic or

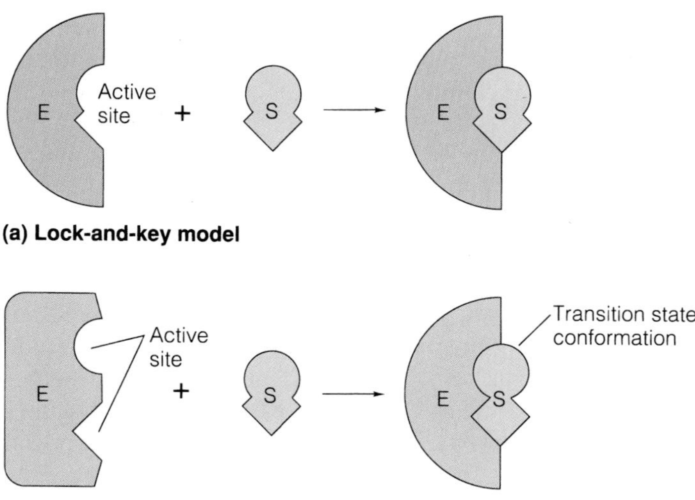

Figure 10.9
Two models for enzyme–substrate interaction. (**a**) The lock-and-key model. In this early model, the active site of the enzyme fits the substrate as a lock does a key. (**b**) The induced fit model. In this elaboration of the lock-and-key model, both enzyme and substrate are distorted on binding. The substrate is forced into a conformation approximating the transition state; the enzyme keeps the substrate under stress.

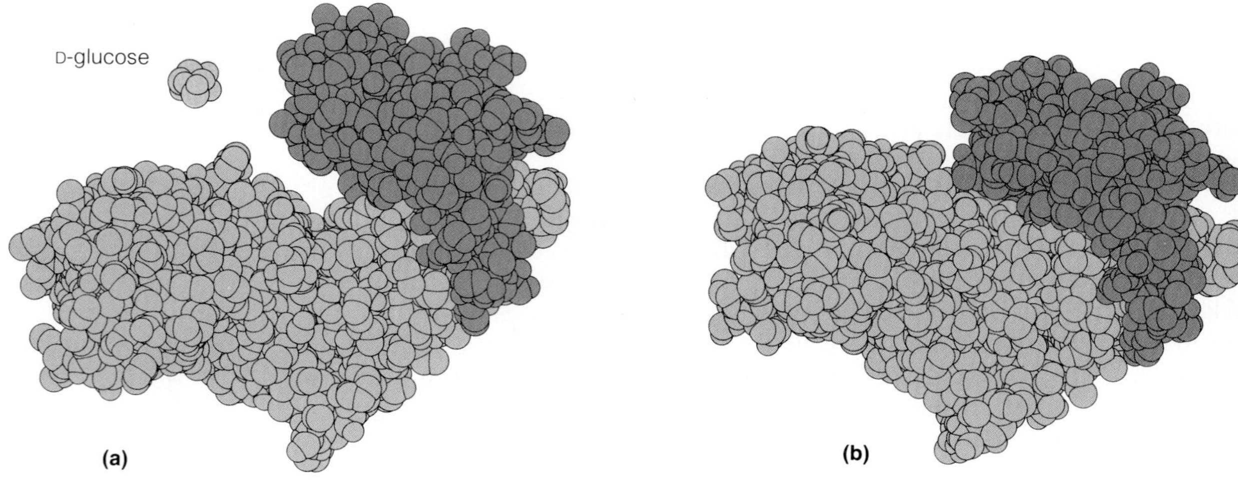

Figure 10.10
The conformational change in hexokinase induced by glucose binding. (**a**) Before glucose binding. (**b**) After glucose binding. The two major domains of the enzyme are shaded differently to distinguish them.

basic groups that can promote the kinds of general acid or base catalysis described above. In other instances, the enzyme holds a metal ion in exactly the right position to allow metal ion catalysis. Thus, an enzyme (1) binds the substrate or substrate(s), (2) lowers the energy of the transition state, and (3) directly promotes the catalytic event. When this process has been completed, the enzyme must be able to release the product or products and return to its original state, ready for another round of catalysis. For an enzyme (E) that catalyzes the conversion of a single substrate (S) into a single product (P) or vice versa, the simplest way to write the overall reaction is

$$S + E \rightleftharpoons ES$$

$$ES \longrightarrow E + P$$

Here ES represents the **enzyme–substrate complex,** in which we envision the substrate as being in or near the transition state. We shall see later that much more complex representations are possible, with one or more intermediate states, but the equations above give the essence of the process and emphasize the catalytic role of the enzyme.

To illustrate these general principles in a more concrete way, we turn to two examples, drawn from enzymes and reactions you have already encountered.

Triose-Phosphate Isomerase

We mentioned earlier that this enzyme catalyzes the interconversion of glyceraldehyde-3-phosphate (G3P) and dihydroxyacetone phosphate (DHAP). The structures of the enzyme and its complex with DHAP have been determined in detail by x-ray diffraction methods. Triose-phosphate isomerase is a dimer of two identical subunits. Each has the conformation shown in Figure 10.11a: a parallel β barrel with α helices in the interconnecting loops. The active site lies near the top of the barrel and can accommodate either G3P or DHAP (Figure 10.11b). Studies of the uncatalyzed reaction have shown that it proceeds through an intermediate state in which the molecule has been rearranged to an *enediol*:

$$\underset{\textbf{G3P}}{\mathrm{HC(=O)-CH(OH)-CH_2OPO_3^{2-}}} \rightleftharpoons \left[\underset{\textit{cis}\textbf{-enediol}}{\mathrm{HC(OH)=C(OH)-CH_2OPO_3^{2-}}}\right] \rightleftharpoons \underset{\textbf{DHAP}}{\mathrm{H_2C(OH)-C(=O)-CH_2OPO_3^{2-}}}$$

The enediol intermediate is unstable, in that it normally has a much more positive free energy than either G3P or DHAP. Thus, to catalyze the reaction, the enzyme must both promote the proton exchange to form the enediol and stabilize this intermediate.

The active site of triose-phosphate isomerase includes a glutamic acid residue (Glu 165), which has been found essential for enzymatic function. This glutamate is negatively charged at physiological pH and acts as a generalized base catalyst. Apparently, Glu 165, together with an acidic, proton-donating group (HA), serves to shuttle protons between the atoms of substrate in the following way:

$$\mathrm{E{-}B^-}\;\cdots\;\mathrm{HC(=O)-CH(OH)-CH_2OPO_3^{2-}}\;\cdots\;\mathrm{H{-}A{-}E} \rightleftharpoons \mathrm{E{-}BH}\;\cdots\;\mathrm{HC(OH)=C(OH)-CH_2OPO_3^{2-}}\;\cdots\;\mathrm{A^-{-}E} \rightleftharpoons \mathrm{E{-}B^-}\;\cdots\;\mathrm{H_2C(OH)-C(=O)-CH_2OPO_3^{2-}}\;\cdots\;\mathrm{H{-}A{-}E}$$

(1) G3P binds to the active site

(2) B^- (Glu 165) picks up the C_2 proton; HA donates a proton to make —OH on C_1

(3) A proton is returned from B—H to C_1; A^- accepts a proton from C_2—OH

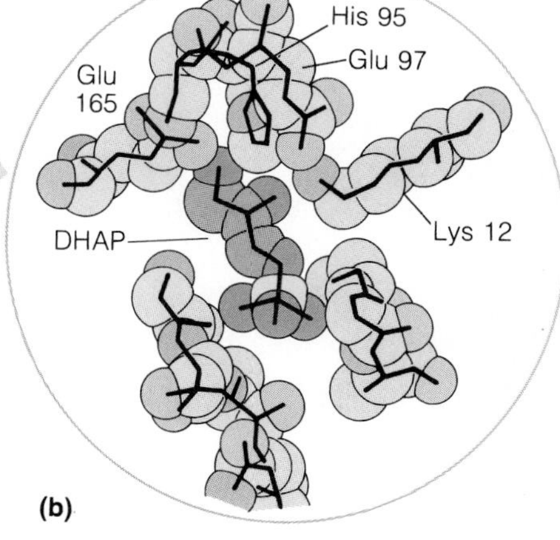

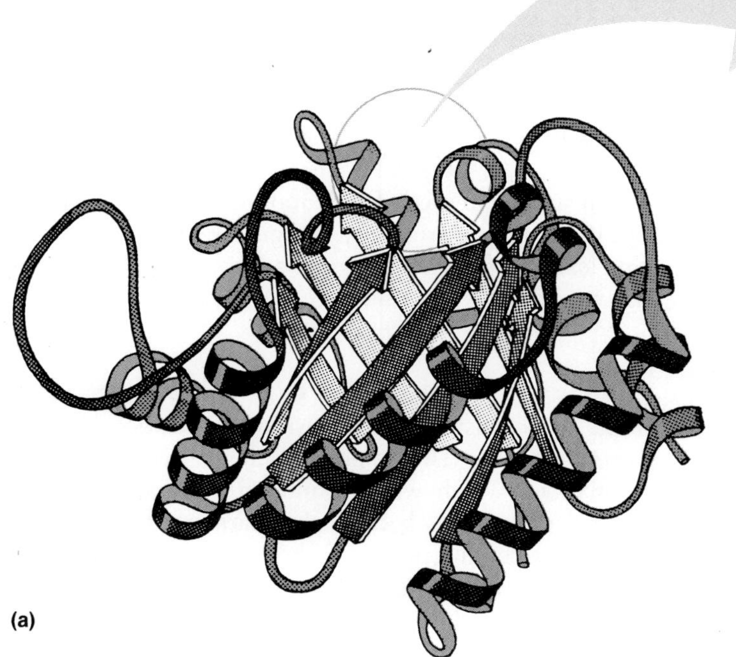

Figure 10.11
Triose-phosphate isomerase. (**a**) Structure of one of the two subunits in the enzymatically active dimer. The β-barrel structure is shown in gray and the active site region in green. (**b**) Detail of the active site in the yeast enzyme, with DHAP bound. Three residues critical to the reaction are Glu 165, His 95, and Lys 12.

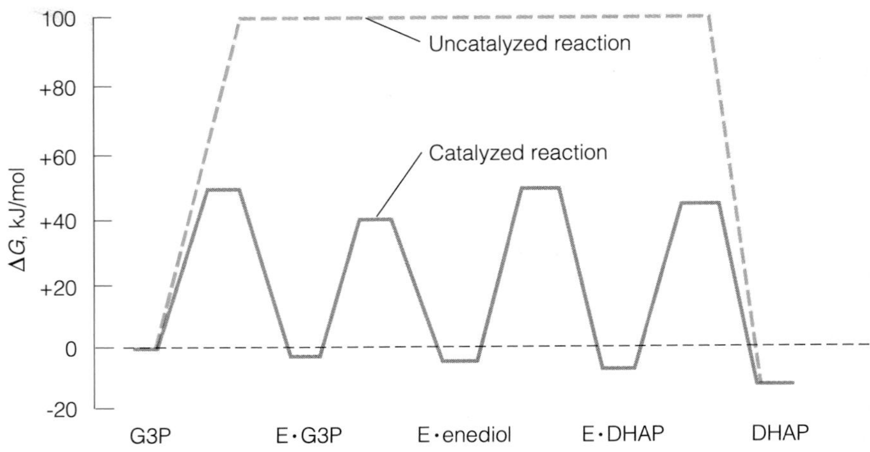

Figure 10.12
A free energy diagram for the reaction G3P $\rightleftharpoons$ DHAP. The reaction is catalyzed by triose-phosphate isomerase. The activation energy for each step is *much* lower than that for the uncatalyzed reaction. Steps are drawn rectangularly, since only heights of peaks and depths of valleys are known.

Other groups surrounding the substrate help to stabilize the enediol intermediate. These include Lys 12 and His 95, one of which (or possibly both) also serves as the general acid catalyst (HA).

Note what happens: First, enzyme provides a "cage," or binding site, into which the substrate can fit. Second, the presence of Glu 165 and Lys 12 or His 95 in exactly the right position favors the exchange of protons essential for the formation of the intermediate state. The importance of correct positioning has recently been demonstrated. Using site-directed mutagenesis, Glu 165 has been replaced by Asp, which moves the carboxyl only 0.1 nm farther away from the substrate. This change reduces the catalytic rate by a factor of 1000. Third, the environment is such as to stabilize the unfavorable enediol state. Both entropy (positioning) and enthalpy (stabilization) factors are involved in reducing the free energy barrier to reaction. Thus, we see that the catalysis of the G3P-to-DHAP conversion is a bit more complicated than the "simplest" scheme for enzyme catalysis described above. A minimal description of this reaction requires that we write

$$\begin{array}{rcll} E + G3P & \rightleftharpoons & E(G3P) & \text{binding of G3P} \\ E(G3P) & \rightleftharpoons & E(EDI) & \text{conversion to enediol} \\ E(EDI) & \rightleftharpoons & E(DHAP) & \text{conversion to DHAP} \\ E(DHAP) & \rightleftharpoons & E + DHAP & \text{release of DHAP} \end{array}$$

The reaction has been studied in great detail, and the free energy barriers to the individual steps have been determined. As Figure 10.12 shows, the presence of the enzyme leads to four low-energy barriers instead of the much higher barrier that would exist in its absence.

Serine Proteases

As a second example, let us consider the catalysis of peptide bond hydrolysis by one of the *serine proteases*. This important class of enzymes includes trypsin and chymotrypsin, which we encountered in Chapters 5 and 6. These enzymes are called proteases because they catalyze the hydrolysis of peptide bonds in polypeptides and proteins. There exist many kinds of proteases, involved in a variety of catalytic reactions. The *serine* proteases are distinguished by the fact that they all have a serine residue that plays a critical role in the catalytic process.

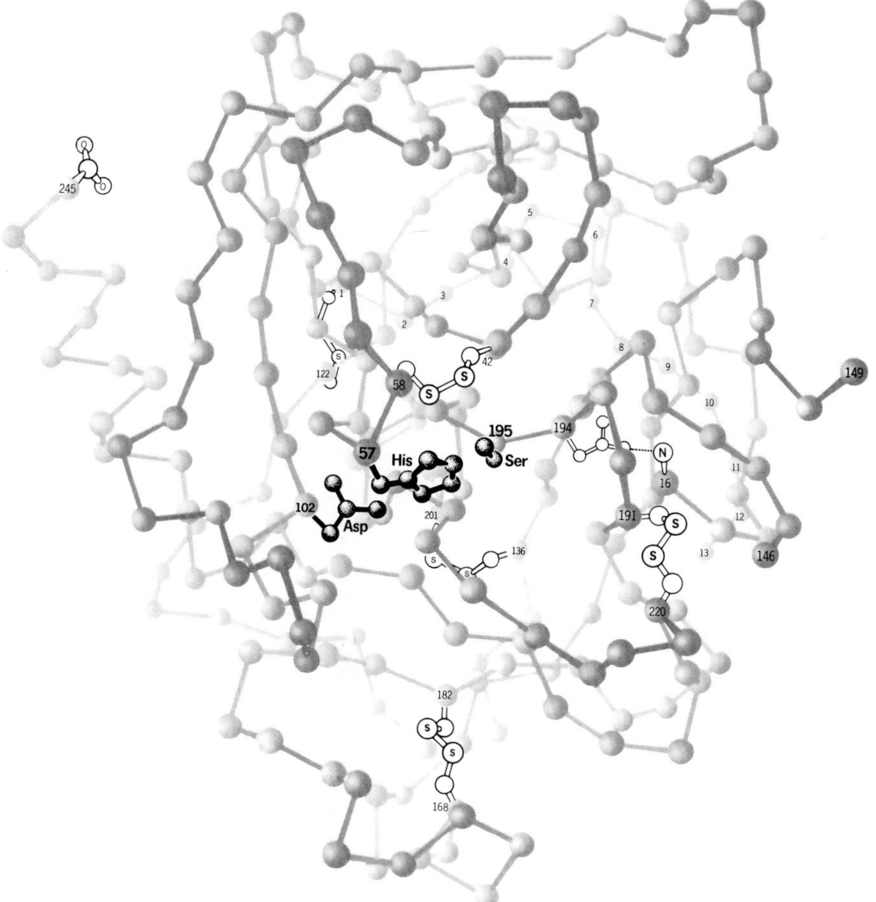

Figure 10.13
The structure of chymotrypsin. Critical residues in the active site are shown in black.

Each of the serine proteases preferentially cuts a polypeptide chain just to the carboxyl side of specific kinds of amino acids. Examples of sites of preferential cutting are shown in Chapter 5, Table 5.4. For example, trypsin cuts preferentially to the carboxylate side of basic amino acid residues like lysine or arginine, whereas chymotrypsin acts most strongly if there is a hydrophobic residue (like phenylalanine or leucine) in this position. The serine proteases also hydrolyze a wide variety of esters—a fact of little physiological importance, but one made use of by biochemists in kinetic studies.

Most of the serine proteases have similar three-dimensional structures and are obviously evolutionarily related. The active site regions of all of the serine proteases have a number of common factors. In particular, there are always an aspartic acid residue, a histidine residue, and a serine residue clustered about the active site depression. These are Asp 102, His 57, and Ser 195 in the structure of chymotrypsin shown in Figure 10.13. A fourth feature of the active site differs from one serine protease to another. This is a "pocket," always located close to the active site serine (Figure 10.14). In trypsin the pocket is deep and narrow, with a negatively charged carboxylate at its bottom—just the thing to catch and hold a long, positively charged side chain like lysine or arginine. In chymotrypsin it is wider and lined with hydrophobic residues, to accommodate a hydrophobic side chain. It is the nature of this pocket that gives each of the serine proteases its specificity.

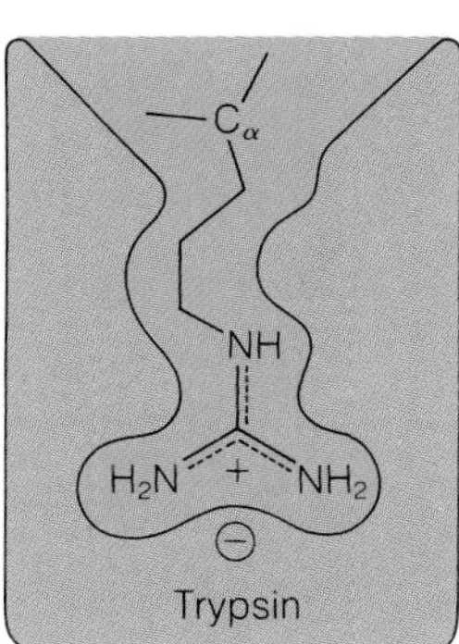

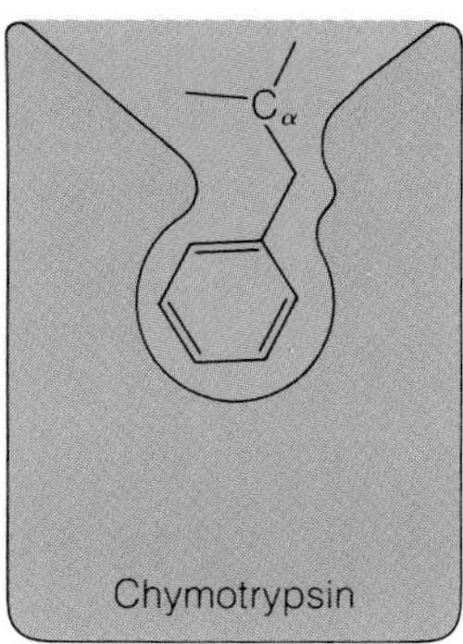

Figure 10.14
Different active site pockets account for the different specificities of serine proteases.

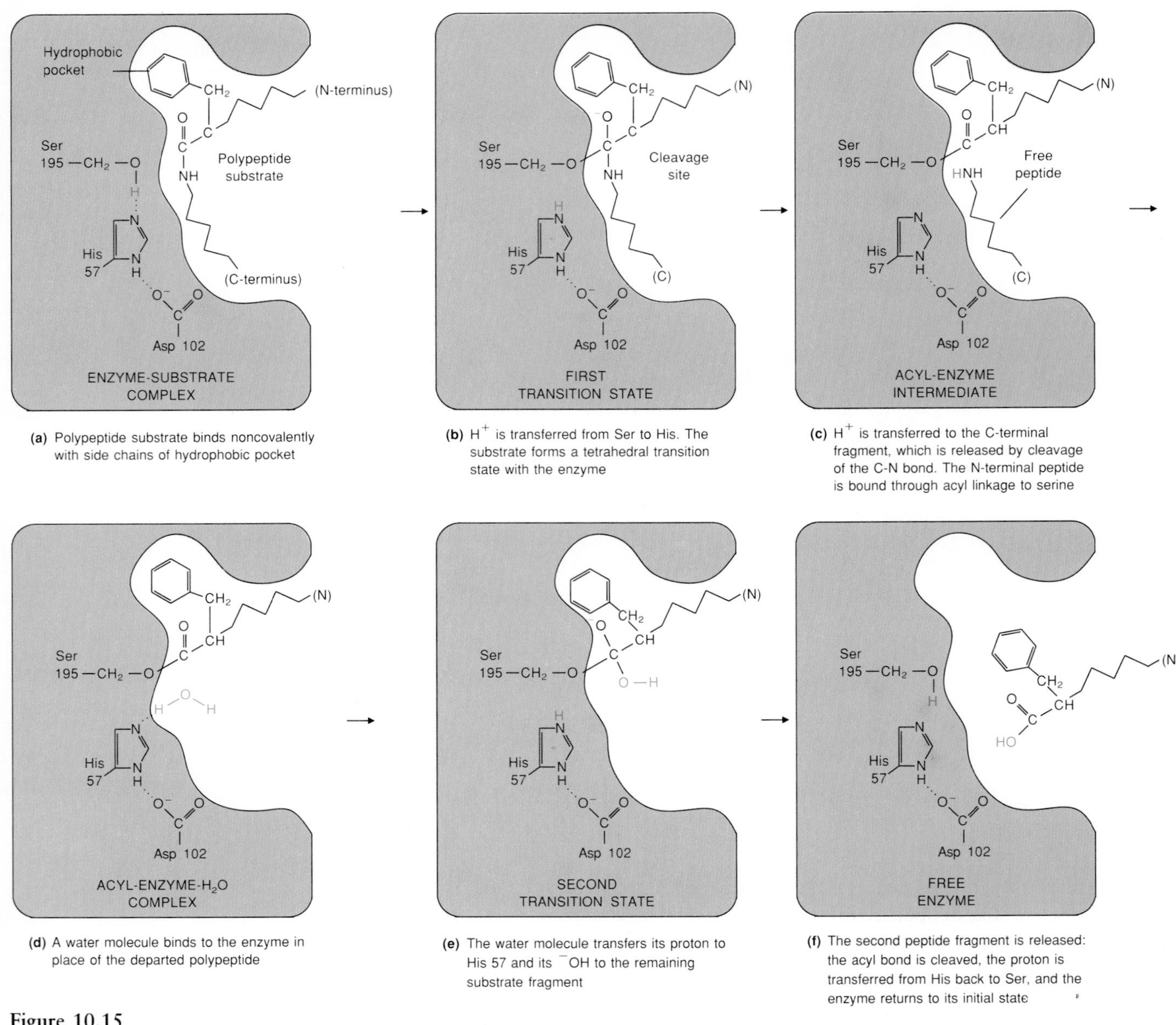

(a) Polypeptide substrate binds noncovalently with side chains of hydrophobic pocket

(b) H^+ is transferred from Ser to His. The substrate forms a tetrahedral transition state with the enzyme

(c) H^+ is transferred to the C-terminal fragment, which is released by cleavage of the C-N bond. The N-terminal peptide is bound through acyl linkage to serine

(d) A water molecule binds to the enzyme in place of the departed polypeptide

(e) The water molecule transfers its proton to His 57 and its ^-OH to the remaining substrate fragment

(f) The second peptide fragment is released: the acyl bond is cleaved, the proton is transferred from His back to Ser, and the enzyme returns to its initial state

Figure 10.15
Steps in the cleavage of a polypeptide chain, as catalyzed by chymotrypsin. The brown area represents the enzyme.

Catalysis of peptide bond hydrolysis by a serine protease (chymotrypsin, for example) proceeds as shown in Figure 10.15. First (a), the polypeptide chain that is to be cleaved is bound to the surface of the enzyme. Most of it binds nonspecifically, but the side chain of the residue to the N-terminal side of the peptide bond that is to be cleaved must fit in the active site pocket. This pocket defines not only the position of cutting but also the *stereospecificity* of these enzymes. If a D-amino acid residue were present at this position, the side chain would project in the wrong direction, *away* from the pocket. Serine proteases will therefore not cut next to D-amino acids.

This very specific binding of the substrate places the active site serine very close to the carbonyl group of the bond to be cleaved. Serine residues are not usually reactive, but this one is in an unusual environment—it also sits very close to His 57. The serine proton is transferred to the histidine ring (Figure 10.15b), leaving a negative charge. Normally, this transfer would also be unlikely, but it appears to be facilitated by Asp 102, which, by its negative charge, stabilizes the protonation of the histidine ring. Thus, a **charge transfer system** is involved. The activated serine can now attack the carbonyl of the substrate, forming the *tetrahedral transition state* shown in the figure. Cleavage of the peptide bond in this activated state occurs, yielding an **acyl enzyme intermediate** in which the N-terminal part of the polypeptide substrate is left covalently bound to the enzyme (Figure 10.15c). The C-terminal portion of the polypeptide extracts the proton (which was originally the serine proton) back from the histidine to form a new terminal amino group. This portion of the substrate chain then departs.

Until this point H_2O has not entered the picture. It comes in now (Figure 10.15d) to cleave the acyl intermediate. The water molecule positions itself between the acyl group and His 57, transfers one proton to the latter, and links to the acyl intermediate to form another tetrahedral transition state (Figure 10.15e). Note that what is happening now is essentially a reversal of the formation of the initial acyl intermediate, with the water molecule playing the role of the released portion of the polypeptide chain. Finally (Figure 10.15f), the proton is transferred from the histidine back to Ser 195, and the rest of the polypeptide chain is released. The enzyme is back in its original state, ready to catalyze the hydrolysis of another chain. The whole process has been a sophisticated version of general base catalysis (see Figure 10.8), with His 57 acting as the base and the active site serine playing an intermediate role.

Chymotrypsin is an excellent example of how enzymes operate. Note that both energetic stabilization of activated states and correct positioning of reactants are important in lowering the free energy of activation.

The Kinetics of Enzymatic Catalysis: An Introduction

We have described some enzyme-catalyzed reactions in considerable detail. From these examples it should be clear that the *minimal* equation to describe a simple, one-substrate, one-product reaction catalyzed by an enzyme will be

$$\mathrm{E} + \mathrm{S} \underset{k_{-1}}{\overset{k_1}{\rightleftharpoons}} \mathrm{ES} \xrightarrow{k_{\mathrm{cat}}} \mathrm{E} + \mathrm{P}$$

We have assumed that conditions are such that the reverse reaction between E and P is negligible. The catalytic formation of the product, with enzyme regeneration, will then be a simple first-order reaction. Therefore, we may express the reaction rate or velocity, defined as the rate of formation of products, just as we did in equation (10.3):

$$V = k_{\mathrm{cat}}[\mathrm{ES}] \qquad (10.17)$$

We will want to express this rate in terms of substrate and enzyme concentrations, so we need an expression for the concentration [ES] in these terms.

This can be arrived at in either of two ways, which we shall find give very similar results.

Assuming Equilibrium Between Enzyme and Substrate: The Michaelis–Menten Analysis

The first attempts to analyze enzyme kinetics were made in 1913 by Leonor Michaelis and Maude Menten. They assumed that the dissociation rate, as measured by k_{cat}, is *very* low compared to rates of formation and redissociation of the enzyme–substrate complex (k_1 and k_{-1}). If this is so, ES will always be nearly in equilibrium with E and S, so we may write

$$K_S = \frac{[E][S]}{[ES]} = \frac{k_{-1}}{k_1} \tag{10.18a}$$

or

$$[ES] = \frac{[E][S]}{K_S} \tag{10.18b}$$

where K_S is a *dissociation equilibrium constant.* Now the concentration of enzyme [E] in this equation is the concentration of *free* enzyme. What we will usually know in such a study is the *total* concentration of enzyme used in an experiment, $[E]_t$. Obviously,

$$[E] = [E]_t - [ES] \tag{10.19}$$

since every enzyme molecule is *either* free *or* complexed with substrate. If we substitute the latter expression for [E] in equation (10.18b), we find

$$[ES] = \frac{[E]_t[S]}{K_S} - \frac{[ES][S]}{K_S} \tag{10.20}$$

Rearranging gives

$$[ES]\left\{1 + \frac{[S]}{K_S}\right\} = \frac{[E]_t[S]}{K_S} \tag{10.21}$$

or, on dividing by $(1 + [S]/K_S)$ and simplifying:

$$[ES] = \frac{[E]_t[S]}{[S] + K_S} \tag{10.22}$$

If we insert this result in (10.17), we find

$$V = \frac{k_{cat}[E]_t[S]}{[S] + K_S} \tag{10.23}$$

Equation (10.23) is called the **Michaelis–Menten equation.** The graph of V versus [S] predicted by equation (10.23) is shown in Figure 10.16. Note that at high substrate concentrations a limiting velocity, V_{max}, is approached. This is because the enzyme molecules become *saturated;* every enzyme molecule is occupied by substrate and busy in carrying out the catalytic step.

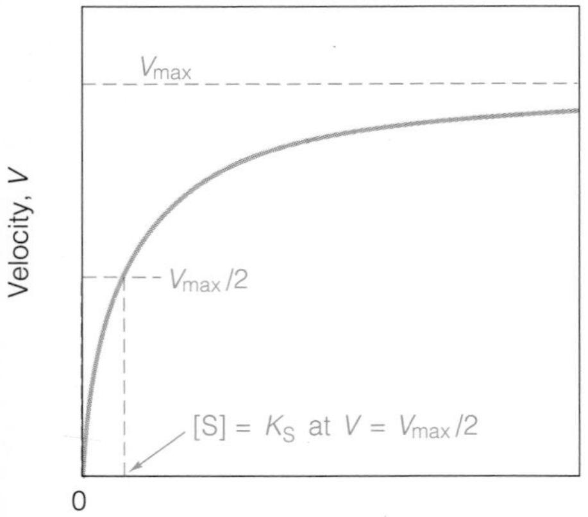

Figure 10.16
Variation of reaction velocity with substrate concentration according to Michaelis–Menten kinetics. Note that at the point where $[S] = K_S$, the reaction will have exactly half its maximum velocity. Exactly the same graph will describe the Briggs–Haldane equation (Equation 10.28) with K_M replacing K_S.

Mathematically, we express the saturation situation by noting that at very high substrate concentrations, $[S] \gg K_S$. Then

$$\lim_{\text{As}[S]\to\infty} V = \frac{k_{cat}[E]_t[S]}{[S]} = k_{cat}[E]_t = V_{max} \tag{10.24}$$

This means that an alternative way of writing equation (10.23) is

$$V = \frac{V_{max}[S]}{[S] + K_S} \tag{10.25}$$

Measurement of V_{max} allows us to determine k_{cat} from equation (10.24).

The Steady-State Assumption: The Briggs–Haldane Analysis

The derivation of equation (10.25) was based on the assumption that the rate of product formation was slow compared to complex formation and redissociation. Thus, enzyme, substrate, and complex were assumed to be in equilibrium. But this cannot ever be *exactly* true, since some of the enzyme–substrate complex is always going on to form *product*. Indeed, in some cases k_{cat} is large compared to k_{-1}. As we shall see, this usually makes little difference in the final result, for a more comprehensive model, presented by G. E. Briggs and J. B. S. Haldane in 1925, yields an equation identical in form to (10.25). The Briggs–Haldane model is based on the following argument: The more ES that is formed, the faster ES will dissociate; therefore, the ES concentration will first build up but then quickly reach a *steady state,* in which it remains almost constant. This steady state will persist until almost all of the substrate has been consumed (Figure 10.17). Mathematically, we treat the consequences as follows. The rate of change of [ES] with time must always be given by

$$\frac{d[ES]}{dt} = \underbrace{k_1[E][S]}_{\text{Formation of ES}} \underbrace{- k_{-1}[ES] - k_{cat}[ES]}_{\text{Breakdown of ES by two routes}} \tag{10.26}$$

So when $d[ES]/dt \cong 0$ (steady state) we have

$$[ES] \cong \left(\frac{k_1}{k_{-1} + k_{cat}}\right)[E][S] \tag{10.27}$$

This is *exactly* the same form as equation (10.18b), so we get the same kind of final result:

$$V = \frac{k_{cat}[E]_t[S]}{[S] + K_M} = \frac{V_{max}[S]}{[S] + K_M} \tag{10.28}$$

which is just like (10.25). However, instead of the equilibrium constant K_S, the equation now involves a ratio of rate constants expressed as K_M.

$$K_M = \frac{k_{-1} + k_{cat}}{k_1} \tag{10.29}$$

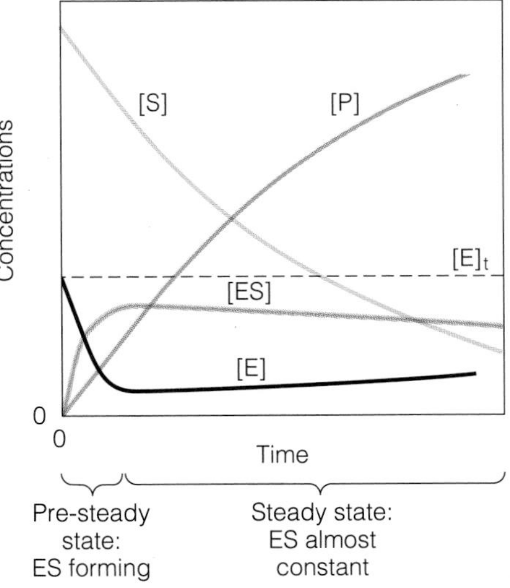

Figure 10.17
The steady state in enzyme kinetics. For a simple enzyme-catalyzed reaction ($S + E \rightleftharpoons ES \rightarrow E + P$) the graph shows how the concentrations of substrate [S], free enzyme [E], enzyme–substrate complex [ES], and product [P] vary with time. After a very brief initial period, the concentration [ES] reaches a steady state, in which ES is consumed as rapidly as it is formed, so $d[ES]/dt = 0$. The amounts of E and ES are greatly exaggerated for clarity. Note that $[E] + [ES] = [E]_t$ and that [ES] actually falls very slowly as substrate is consumed, and [E] accordingly rises.

Equation (10.28) should properly be called the Briggs–Haldane equation, but since it is of the same form as (10.25), the terms *Michaelis–Menten equation* and *Michaelis–Menten kinetics* are usually applied in both cases. The constant K_M is commonly referred to as the *Michaelis constant.* Everything said above concerning the shape of the V versus [S] curve and its behavior at low and high [S] applies equally well whether the original Michaelis–Menten or the Briggs–Haldane assumption is correct. We need only note that the quantity K_M is not a true equilibrium constant for dissociation of the enzyme–substrate complex, although it approaches that quantity when $k_{cat} \ll k_{-1}$. [Examine equation (10.29) to convince yourself that this is true.]

The Significance of K_M and k_{cat}

The Michaelis constant, K_M, is often associated with the strength of binding of substrate to enzyme. This is certainly true in the Michaelis–Menten limiting case, where k_{cat} is very small. Under these circumstances, a large K_M means that k_{-1} is much greater than k_1, and the enzyme binds substrate very weakly. But a very large value of k_{cat} can also lead to a large K_M, as equation (10.29) shows. Therefore the interpretation of K_M as an effective dissociation equilibrium constant for ES must be used with caution. Table 10.2 lists K_M values for a number of important enzymes.

The second constant, k_{cat}, is a direct measure of the catalytic production of product. The larger k_{cat} is, the more rapid the catalytic events on the enzyme surface must be. The units of k_{cat} are s^{-1}, so the reciprocal of k_{cat} can be thought of as a time—the time required by an enzyme molecule to "turn over" one substrate molecule. Alternatively, k_{cat} measures the number of substrate molecules turned over per enzyme molecule per second. Thus, k_{cat} is sometimes called the **turnover number.** Some typical turnover numbers are listed in Table 10.2.

Further insight into the meaning of these quantities can be gained by considering the situation at very low substrate concentrations. Under these circumstances, $[S] \ll K_M$, and most of the enzyme is free, so $[E]_t \cong [E]$. Then equation (10.28) becomes

$$V = \frac{k_{cat}}{K_M}[E][S] \qquad (10.30)$$

Table 10.2
Michaelis–Menten parameters for selected enzymes, arranged in order of increasing efficiency as measured by k_{cat}/K_M

Enzyme	Reaction Catalyzed	K_M (mol/L)	$k_{cat}(s^{-1})$	k_{cat}/K_M $(s^{-1})(mol/L)^{-1}$
Chymotrypsin	Ac–Phe–Ala $\xrightarrow{H_2O}$ Ac–Phe + Ala	1.5×10^{-2}	0.14	9.3
Pepsin	Phe–Gly $\xrightarrow{H_2O}$ Phe + Gly	3×10^{-4}	0.5	1.7×10^3
Tyrosyl-tRNA synthetase	Tyrosine + tRNA $\longrightarrow$ Tyrosyl-tRNA	9×10^{-4}	7.6	8.4×10^3
Ribonuclease	Cytidine 2′, 3′ cyclic phosphate $\xrightarrow{H_2O}$ Cytidine 3′-phosphate	7.9×10^{-3}	7.9×10^2	1×10^5
Carbonic anhydrase	$HCO_3^- + H^+ \longrightarrow H_2O + CO_2$	2.6×10^{-2}	4×10^5	1.5×10^7
Fumarase	Fumarate $\xrightarrow{H_2O}$ Malate	5×10^{-6}	8×10^2	1.6×10^8

Therefore, under these circumstances the ratio (k_{cat}/K_M) behaves as a second-order rate constant for the reaction between substrate and free enzyme. This ratio is important, for it provides a direct measure of enzyme efficiency and specificity—it shows what the enzyme and substrate can accomplish when abundant enzyme sites are available. Suppose an enzyme has a choice of two substrates, A or B, present at equal concentrations. Then under conditions where both substrates are dilute and are competing for the enzyme, we find

$$\frac{V_A}{V_B} = \frac{(k_{cat}/K_M)_A[E][A]}{(k_{cat}/K_M)_B[E][B]} = \frac{(k_{cat}/K_M)_A}{(k_{cat}/K_M)_B} \tag{10.31}$$

Table 10.3
Preference of chymotrypsin for different *N*-acetyl amino acid methyl esters, as measured by k_{cat}/K_M

Amino Acid in Ester	k_{cat}/K_M ((mol/L)$^{-1}$ s^{-1})
Glycine	0.13
Norvaline	3×10^2
Norleucine	3×10^3
Phenylalanine	1×10^5

Table 10.3 lists values of k_{cat}/K_M for chymotrypsin with various substrates. Within the group shown, k_{cat}/K_M varies 1 millionfold, showing the range of preference the enzyme has for even quite similar substrates. The preference to cleave next to hydrophobic residues is quite clear from these data.

A further significance of k_{cat}/K_M can be seen under extreme Briggs–Haldane conditions, where $k_{cat} \gg k_{-1}$. This means that the catalytic process is very fast and the enzyme efficiency depends *entirely* on the rate of loading the substrate onto the enzyme. Mathematically, this is expressed by noting that, since $K_M = (k_{-1} + k_{cat})/k_1$, we have

$$\frac{k_{cat}}{K_M} \cong \frac{k_{cat}}{k_{cat}/k_1} = k_1 \tag{10.32}$$

In other words, the overall rate of enzymatic reaction in this extreme case is determined solely by the true second-order rate constant (k_1) for combination of enzyme and substrate. A practical upper limit exists for such rates, determined by the frequency with which enzyme and substrate molecules can collide. If *every* collision results in formation of an enzyme–substrate complex, diffusion theory predicts that k_1 will have a value of about 10^8 to 10^9 (mol/L)$^{-1}$s^{-1}. Thus, an enzyme that approaches maximum possible efficiency in *every* respect will demonstrate this efficiency by having a value of k_{cat}/K_M in this range. As Table 10.2 shows, a number of enzymes actually approach this limit. Triose-phosphate isomerase, which we have already studied, is another example, having a value of $k_{cat}/K_M = 2.4 \times 10^8$ (mol/L)$^{-1}$ s^{-1}. In fact, it has been argued that this is an almost perfect enzyme, having evolved to nearly maximum efficiency. In support of this idea is the observation that triose-phosphate isomerases from organisms as evolutionarily distant as yeasts and vertebrates show very little change in structure. Apparently, this vital enzyme reached near perfection early in evolution and has changed little since that time.

Analysis of Kinetic Data: Testing the Michaelis–Menten Equation

Suppose we wish to see whether an enzyme-catalyzed reaction follows the Michaelis–Menten law [equation (10.28)] and determine the constants K_M and k_{cat}. How do we proceed? First, we will need some way to follow the formation of product, or the consumption of substrate, in order to measure reaction velocity.

A number of analytical methods for the measurement of rates are described in Tools of Biochemistry 15. One general point should be noted: In principle, we could simply mix enzyme and substrate and follow both the change in substrate concentration and reaction velocity with time, as shown

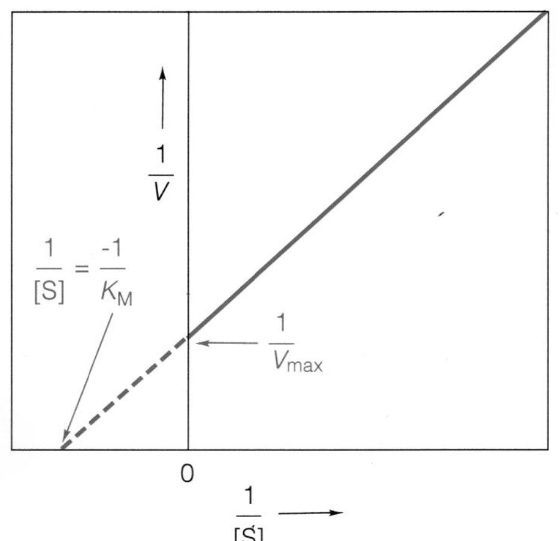

Figure 10.18
A Lineweaver–Burk plot, or double reciprocal plot. Here $1/V$ is graphed versus $1/[S]$ according to equation (10.33).

in Figure 10.1a. As substrate is consumed, the velocity decreases, until equilibrium is eventually reached. But measuring the instantaneous velocity at specific times during the reaction is difficult and usually inaccurate. It is usually easier to set up a series of experiments, all at the same enzyme concentration but different substrate concentrations, and measure the *initial* rate (Figure 10.1b). Since we know the initial [S] precisely, and the change in [S] versus t is almost linear in the initial stages, accurate data for V as a function of [S] can thus be obtained.

Given the data, how are they to be analyzed? The best way is to rearrange equation (10.28) in such a way that a *linear* graph can be obtained. Several kinds of graphs are possible, but it is most common to use a **double reciprocal plot,** also called a **Lineweaver–Burk plot** (Figure 10.18). If we simply invert both sides of equation (10.28), we easily arrive at the result:

$$\frac{1}{V} = \frac{K_M}{V_{max}[S]} + \frac{1}{V_{max}} \tag{10.33}$$

So, plotting $1/V$ versus $1/[S]$, we should get a straight line. From this, we find $1/V_{max}$ from the intercept of the line with the $1/V$ axis, that is, where $1/[S] = 0$ or [S] is infinitely large. Having V_{max}, and knowing $[E]_t$, we can calculate k_{cat} from equation (10.24). If we extrapolate the plot to where $1/V = 0$, we get

$$\left(\frac{K_M}{V_{max}[S]}\right)_{at\ 1/V=0} = -\frac{1}{V_{max}} \tag{10.34a}$$

or

$$[S]_{at\ 1/V=0} = -K_M \tag{10.34b}$$

Thus, the intercept of the line on the [S] axis gives K_M.

A Lineweaver–Burk plot provides a quick test for adherence to Michaelis–Menten kinetics and allows easy evaluation of the critical constants. As we shall see in the next chapter, it also allows discrimination between different kinds of enzyme inhibition and regulation. A disadvantage of a Lineweaver–Burk plot is that a long extrapolation is often required to determine K_M, with corresponding uncertainty in the result. Consequently, other ways of plotting the data are sometimes used. One alternative is to rearrange equation (10.28) into the form

$$V = V_{max} - \frac{K_M V}{[S]} \tag{10.35}$$

and graph V versus $V/[S]$. This yields what is called an **Eadie–Hofstee plot** (Figure 10.19).

V_max / K_M
Slope = −1/K_M
V_max
V/[S]
0
V

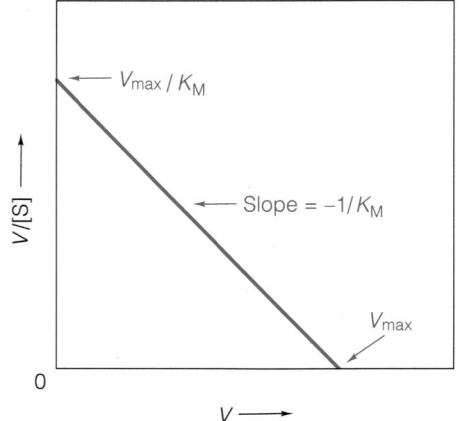

Figure 10.19
An Eadie–Hofstee plot. Graphing V versus $V/[S]$, we obtain V_{max} at $(V/[S]) = 0$ and K_M from the slope of the line.

Multisubstrate Reactions

Our discussions of enzyme kinetics have, to this point, centered on simple reactions in which one substrate molecule is bound to an enzyme and undergoes reaction there. In fact, such reactions are in the minority. Most biochemical reactions catalyzed by enzymes involve the reaction of two or more substrates, often with the production of multiple products. An example we have already discussed is proteolysis, in which there are two substrates (the polypeptide and water) and two products (the two fragments of the cleaved polypeptide chain). Phosphorylation of glucose, as catalyzed by hexokinase, is another such case: the two substrates are glucose and ATP and the products are glucose-6-phosphate and ADP.

When an enzyme binds two or more substrates and releases multiple products, the order of the steps becomes an important feature of the enzyme mechanism. Several major classes of mechanisms for multisubstrate reactions are recognized. We shall illustrate them with examples using two substrates, S1 and S2, and two products, P1 and P2.

RANDOM SUBSTRATE BINDING. In this case, either substrate can be bound first, although in many cases one substrate will be favored for initial binding, and its binding may promote the binding of the other. The general pathway is

$$\begin{array}{lcl} \text{either} & \mathrm{E} \xrightarrow{\mathrm{S1}} \mathrm{E \cdot S1} \xrightarrow{\mathrm{S2}} & \\ & & \mathrm{E \cdot S1 \cdot S2} \longrightarrow \mathrm{E + P1 + P2} \\ \text{or} & \mathrm{E} \xrightarrow{\mathrm{S2}} \mathrm{E \cdot S2} \xrightarrow{\mathrm{S1}} & \end{array}$$

The phosphorylation of glucose by ATP, with hexokinase as enzyme, appears to follow such a mechanism, although there is some tendency for glucose to bind first.

ORDERED SUBSTRATE BINDING. In some cases, one substrate *must* bind before a second substrate can bind significantly. We then have

$$\mathrm{E} \xrightarrow{\mathrm{S1}} \mathrm{E \cdot S1} \xrightarrow{\mathrm{S2}} \mathrm{E \cdot S1 \cdot S2} \longrightarrow \mathrm{E + P1 + P2}$$

This mechanism is often observed in oxidations of substrates by the coenzyme nicotinamide-adenine dinucleotide (NAD^+), which will be discussed in the next section.

THE "PING-PONG" MECHANISM. Sometimes the sequence of events in catalysis goes like this: One substrate is bound, one product is released, a second substrate comes in, and a second product is released. For obvious reasons, this is called a ping-pong reaction:

$$\mathrm{E} \xrightarrow{\mathrm{S1}} \mathrm{E \cdot S1} \xrightarrow{\mathrm{P1}\uparrow} \mathrm{E^*} \xrightarrow{\mathrm{S2}} \mathrm{E^* \cdot S2} \xrightarrow{\mathrm{P2}\uparrow} \mathrm{E}$$

Here E* is a modified form of the enzyme, often carrying a fragment of S1. A very nice example is the cleavage of a polypeptide chain by a serine protease. We describe the polypeptide as S = A · B, where A and B designate the C-terminal and N-terminal portions of the chain from the point of cleavage:

$$\mathrm{E} \xrightarrow{\mathrm{S}} \mathrm{E \cdot S} \xrightarrow{\mathrm{A}\uparrow} \mathrm{E^* \cdot B} \xrightarrow{\mathrm{H_2O}} \mathrm{E^* \cdot B \cdot H_2O} \xrightarrow{\mathrm{B}\uparrow} \mathrm{E}$$

Here E* · B and E* · B · H_2O indicate the acyl intermediates described in our earlier discussion.

Detailed Analysis of the Mechanisms of Complex Reactions

How does one actually analyze a complex enzyme-catalyzed reaction and determine the rate constants for different steps? As an example, let us consider the cleavage of a substrate by a serine protease such as chymotrypsin.

In the first place, it should be noted that we cannot study the step $E^* \cdot B + H_2O \rightarrow E^* \cdot B \cdot H_2O$. The concentration of water is essentially fixed in aqueous solution, and $[H_2O]$ is not a variable. Therefore, it will suffice to write the reaction as

$$E + S \underset{k_{-1}}{\overset{k_1}{\rightleftharpoons}} E \cdot S \xrightarrow[k_2]{\nearrow A} E^* \cdot B \xrightarrow[k_3]{} E + B$$

We have, then, a number of constants to determine. Steady-state measurements will not in themselves be sufficient. It can be shown that the steady-state velocity for the reaction written above is given by the equation

$$V = \frac{\left(\dfrac{k_2 k_3}{k_2 + k_3}\right)[E]_t[S]}{[S] + \left(\dfrac{K_S \cdot k_3}{k_2 + k_3}\right)} \tag{10.36}$$

In other words, the enzyme obeys Michaelis–Menten kinetics, with

$$k_{cat} = \frac{k_2 k_3}{k_2 + k_3} \tag{10.37a}$$

and

$$K_M = K_S \frac{k_3}{k_2 + k_3} \tag{10.37b}$$

$$K_S = \frac{k_{-1}}{k_1} \tag{10.37c}$$

To obtain the individual rate constants in such a case, measurements outside the steady-state range must be employed. One of the first indications that an intermediate might be involved in hydrolysis came from the observation of early stages in the kinetics of the hydrolysis of esters by chymotrypsin. If the formation of product A (see reaction above) is followed, it is found to rise quickly for a few minutes, until about 1 mol has been produced per mole of enzyme; after this, steady-state production is found (Figure 10.20).

This initial "burst" of production of the first product has been explained in the following way: For ester hydrolysis, k_3 is much smaller than k_2. Thus, the acyl intermediate forms quickly on each enzyme, with accompanying release of 1 mol of product A. But after this, more A can be formed only after each acyl intermediate breaks down and the enzyme becomes available again. The dissociation of the acyl intermediate has become the rate-limiting step.

Still faster measurements, using stopped-flow techniques (see Tools of Biochemistry 15), allow measurement of the rate of formation of the enzyme–substrate complex (ES), and measurements of the slow decay of the

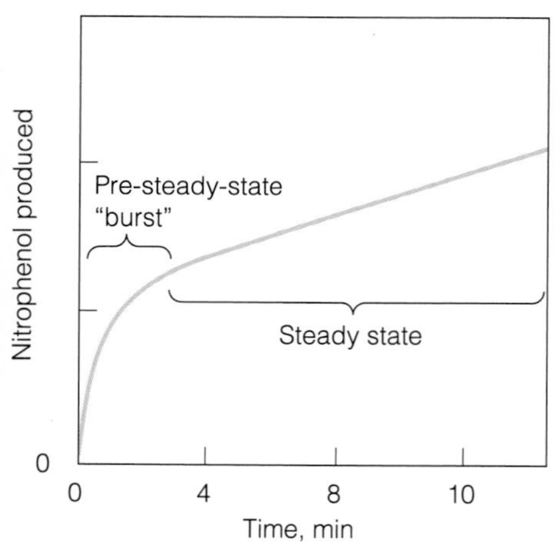

Figure 10.20
Pre-steady-state kinetics in the chymotrypsin-catalyzed hydrolysis of *p*-nitrophenylacetate. The production of the first product (*p*-nitrophenol) is followed spectrophotometrically after mixing enzyme and substrate. The initial burst of product formation ceases when enzyme is almost all in the acyl intermediate.

Table 10.4
Rate constants for the hydrolysis of two *N*-acyl amino acid esters by chymotrypsin

	Substrate	k_{cat} (s^{-1})	k_2 (s^{-1})	k_3 (s^{-1})	K_M (mM)	K_S (mM)
(1)	C_6H_5—C(=O)—N(H)—C(H)(CH_3)—C(=O)—O—C_2H_5	0.069	0.069	0.6	5.87	5.97
(2)	CH_3—C(=O)—N(H)—C(H)(CH_2—C_6H_4—OH)—C(=O)—O—C_2H_5	192	5000	200	0.663	17.2

acyl intermediates after substrate is exhausted provide k_3. By using a combination of such methods, together with steady-state studies, *all* of the constants in equation (10.36) can be obtained. Two examples of such detailed kinetic data are given in Table 10.4 for hydrolysis of *N*-acyl amino acid esters. In example (1) we find $k_{cat} \cong k_2$, and $K_M \cong K_S$. This is the result to be expected when the acylation reaction (k_2) is rate limiting, with $k_2 \ll k_3$. In the second instance, deacylation is rate limiting ($k_2 \gg k_3$) and $k_{cat} \cong k_3$. In this situation, as equation (10.37b) shows, $K_M = K_S(k_3/k_2)$.

From this example, it should be clear that a steady-state analysis is only a first step in the study of any enzyme and that a variety of techniques must be employed to unravel mechanisms.

Coenzymes, Vitamins, and Essential Metals

The complexity of globular protein structure and the variety of side chain residues available in a protein allow the formation of many kinds of catalytic sites. This, in turn, allows proteins themselves to act as efficient enzyme catalysts for many reactions. However, for some kinds of biological processes, the molecular vocabulary of protein side chains alone is not sufficient. A protein may require the help of some other small molecule or ion to carry out the reaction. The molecules that are bound to enzymes for this purpose are called **coenzymes.**

Coenzymes and What They Do: An Example

The structures of some of the most important coenzymes are shown in Figure 10.21. Each kind of coenzyme has a particular chemical function; some are oxidation–reduction agents, some facilitate group transfers, and so forth. There is a limited number of important coenzymes, but each of them may be associated with many different enzymes. We will encounter each of these in later chapters, often repeatedly. For the moment, we will consider as an example **nicotinamide-adenine dinucleotide** (**NAD^+**). As Figure 10.21a shows, this molecule contains two major parts: an adenosine diphosphate portion, linked through a ribose to **nicotinamide.** It is the latter

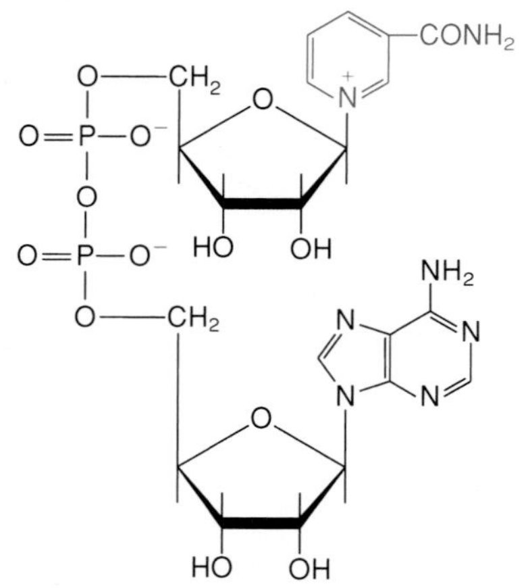

(a) Nicotinamide-adenine dinucleotide (NAD⁺):
oxidation–reduction

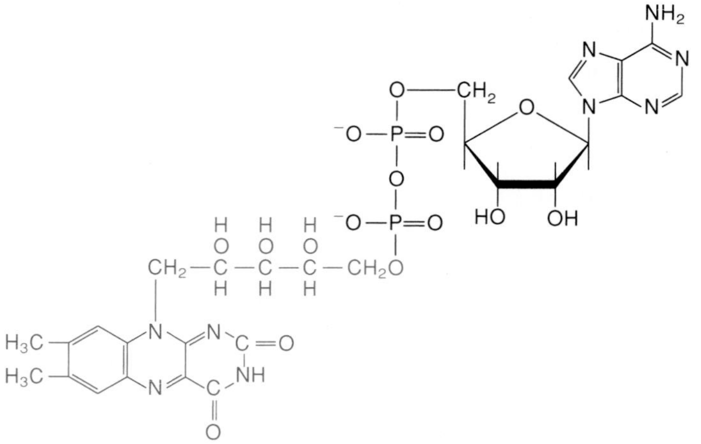

(b) Flavin-adenine dinucleotide (FAD):
oxidation–reduction

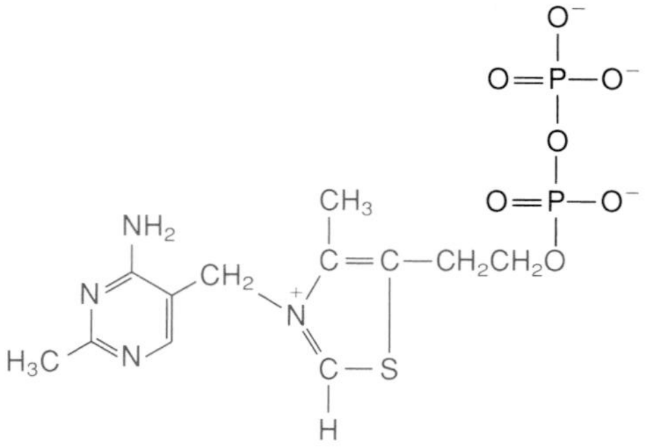

(c) Thiamine pyrophosphate:
group transfer

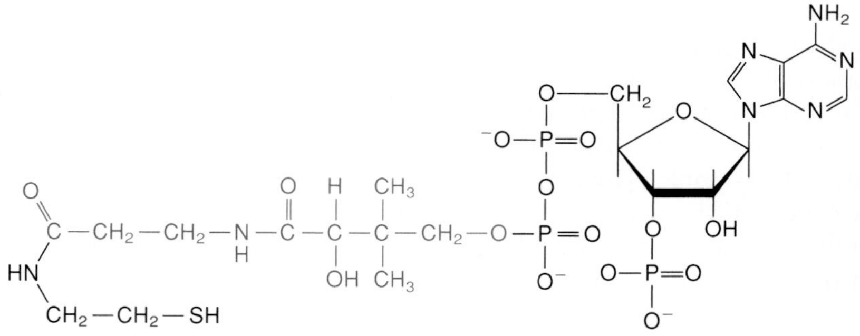

(d) Coenzyme A:
acyl transfer

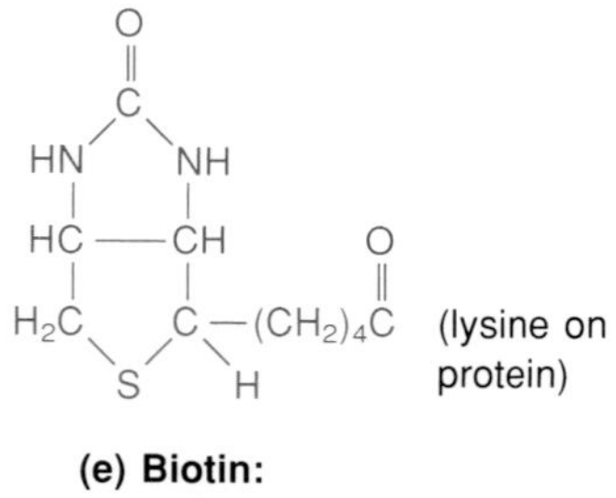

(lysine on protein)

(e) Biotin:
carboxylation

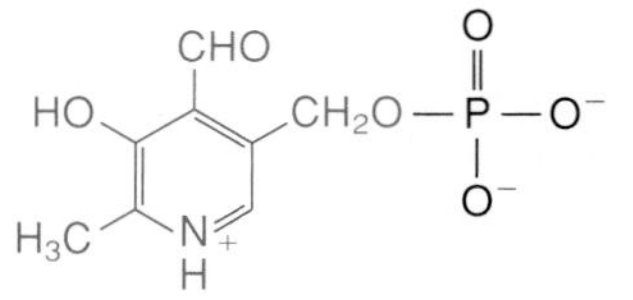

(f) Pyridoxal phosphate:
group transfer; transamination

Figure 10.21
Some important coenzymes. The kind of reaction each coenzyme participates in is indicated. The relationship of the coenzymes to vitamins is indicated by showing in blue the portion of the coenzyme derived from the corresponding vitamin (compare Table 10.5).

region that is the "business end" of the NAD^+ molecule, for the nicotinamide ring is able to be reduced readily and serve as an oxidizing agent:

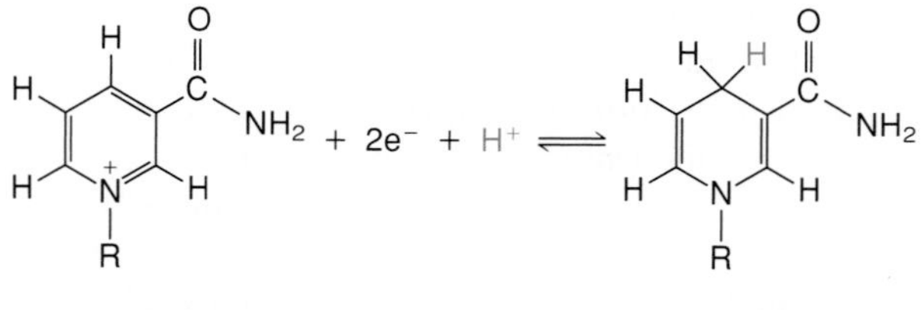

$$NAD^+ + 2e^- + H^+ \rightleftharpoons NADH$$

NAD⁺ **NADH**

Here R stands for the remainder of the molecule. In reduction, two electrons and a proton are added to the nicotinamide ring. The reaction is more correctly written as a **hydride ion** transfer: $NAD^+ + H^- \rightleftharpoons NADH$.

A typical reaction in which NAD^+ acts as an oxidizing agent is the conversion of alcohols to aldehydes or ketones (for example, by the *alcohol dehydrogenase* of liver).

$$CH_3{-}CH_2{-}OH \rightleftharpoons CH_3CHO + 2e^- + 2H^+$$

$$NAD^+ + 2e^- + H^+ \rightleftharpoons NADH$$

$$\underset{\textbf{Ethanol}}{CH_3CH_2OH} + NAD^+ \rightleftharpoons \underset{\textbf{Acetaldehyde}}{CH_3CHO} + H^+ + NADH$$

The fact that the C-linked H, and not the O-linked H, is transferred to NAD^+ can be demonstrated by studies using deuterated compounds. Furthermore, there is a strong stereospecificity in these reactions. Even when the hydroxyl carbon has *two* hydrogens attached (as with ethanol), a particular one of the hydrogens is transferred to NAD^+. This may seem surprising, since the hydroxyl carbon of ethanol is not a chiral center. How can a particular H be favored when the substrate molecule has a plane of symmetry? The answer lies in the fact that the enzyme surface, to which both NAD^+ and the alcohol are bound, is itself asymmetric. If even a symmetrical molecule like ethanol is bound by *three* points to an asymmetric object, we find that the two H atoms are no longer equivalent (Figure 10.22). Such considerations lie behind the high stereospecificity of many enzyme-catalyzed reactions, in contrast to nonenzymatic processes.

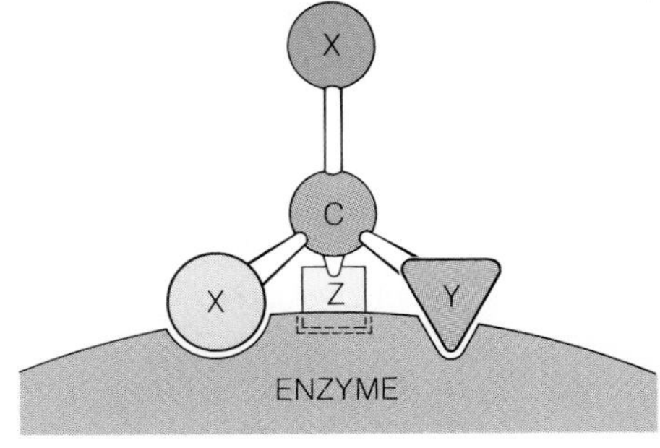

Figure 10.22
How the surface of an enzyme confers asymmetry in the reaction of a nonasymmetric substrate. If the molecule X_2CYZ makes three contacts with unique complementary groups on the asymmetric surface of an enzyme, the two X atoms can no longer be equivalent. Only a specific one of the two X atoms can contact the surface properly.

Coenzyme or Second Substrate?

Sometimes it is difficult to make a clear distinction between a true coenzyme and a second substrate in a reaction. Examples of the latter are provided by the reactions we have just discussed. The dehydrogenase enzymes, such as alcohol dehydrogenase, all have a strong binding site for the oxidized form, NAD^+. After oxidation of the substrate, the reduced form, NADH, leaves the enzyme and is reoxidized by other electron-acceptor systems in the cells. The NAD^+ so formed can now bind to another enzyme molecule and repeat the cycle. In such cases, NAD^+ is acting more like a second substrate than a true coenzyme. Yet NAD^+ and NADH differ from most substrates in that they are continually recycled in the cell and are used over and over again. In this behavior, they resemble coenzymes.

An example of NAD^+ behaving as a true coenzyme is found in the reaction involving **UDP-glucose 4-epimerase** shown in Figure 10.23. This enzyme facilitates synthesis of complex polysaccharides by interconverting UDP-glucose and UDP-galactose (see Chapter 8). The mechanism by which the hydroxyl at position 4 is changed in stereochemical orientation involves oxidation of the hydroxyl to a carbonyl as an intermediate state. In this case, NAD^+ and NADH never leave the enzyme; they are reduced and reoxidized in a cyclic fashion, providing a temporary resting place for electrons and the hydrogen from the substrate. This reaction provides a good example of what coenzymes do and why they are necessary. The carbonyl

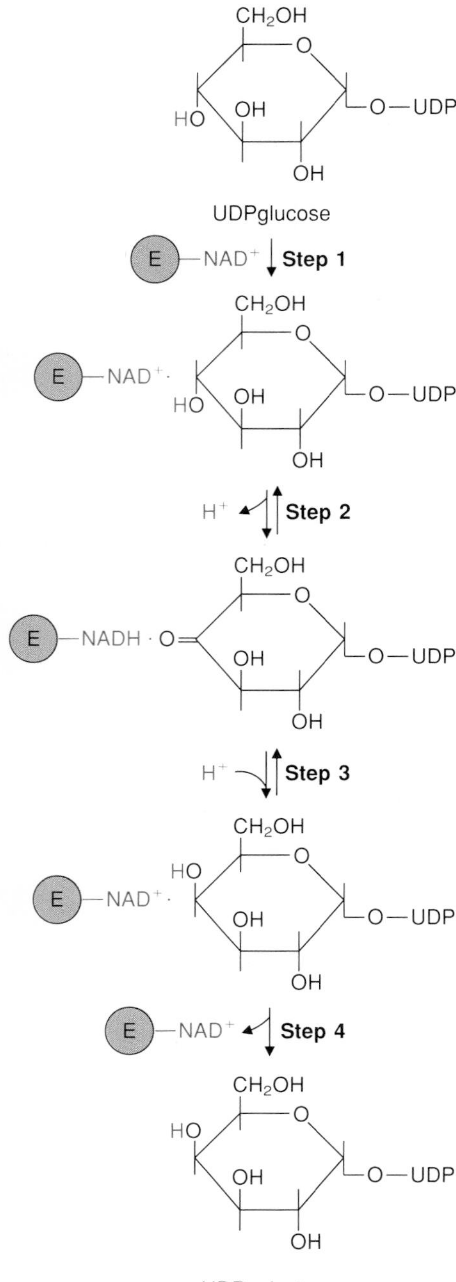

Figure 10.23
Proposed mechanism for UDP-glucose epimerase. UDP-glucose is bound to the enzyme (step 1). A proton is extracted to produce the carbonyl intermediate (step 2) and then reinserted with opposite enantiomeric orientation (step 3). The product, UDP-galactose, is then released (step 4). The reaction is, of course, reversible.

intermediate provides an excellent intermediate state for interconversion of the sugars, but none of the normal amino acid side chains of a protein are really well suited to promote this kind of oxidation and reduction. By binding NAD^+, the enzyme can carry out this function.

Coenzymes and Vitamins

Vitamins are organic molecules that are essential to the biological processes of higher organisms, but which cannot be synthesized by these organisms. Somewhere in the course of evolution the ability to synthesize these substances was lost, so we must now obtain them from our diet. The bacterium *E. coli* has no vitamin requirements; it can live on glucose plus a few inorganic salts. Humans, on the other hand, need over a dozen specific organic substances and suffer severe symptoms if they must subsist on a diet deficient in one or more of them.

Table 10.5 lists a number of the most important vitamins, together with their known functions and associated **deficiency diseases.** When we compare Figure 10.21 with Table 10.5 one point becomes immediately obvious: many of the major vitamins are closely related in molecular structure to the major coenzymes. In most cases, the vitamin contributes part of the total coenzyme structure; this part is shown in blue in Figure 10.21.

Other vitamins have equally important functions. **Vitamin C** (**ascorbic acid**) serves as a general reducing agent in many bodily processes. In addition, it is involved in hydroxylation reactions such as the formation of hydroxyproline in collagen. The vitamin C deficiency disease scurvy is characterized by degeneration of connective tissue (see Chapter 6). **Vitamin A** (**retinol**) is important in the visual system (see Chapter 29) but appears to have a number of other roles as well, for vitamin A deficiency causes many symptoms in addition to impaired vision.

You will note that the vitamins listed in Table 10.5 have been divided into two groups—water-soluble and fat-soluble vitamins. This distinction is important in the present day of "megavitamin" fads. Ingestion of moderately large amounts of the water-soluble vitamins is usually not harmful, since excesses are readily excreted. But the fat-soluble vitamins tend to accumulate in membranes and fat cells and can produce extremely toxic effects if high levels are reached. For example, excess ingestion of vitamin A or D over long periods results in brittleness of bones and other serious effects. Even six to ten times the recommended daily allowance of vitamin D can be toxic to children, if continued over long periods.

Metal Ions in Enzymes

Many enzymes contain metal ions, usually held by coordinate-covalent bonds from amino acid side chains, but sometimes bound by a prosthetic group like heme. These ions act in much the same way as coenzymes, conferring on the enzyme a property it would not possess in their absence. As Table 10.6 shows, the roles these ions play are diverse: some, like the zinc ion in **carboxypeptidase A** (Figure 10.24), act as metal catalysts for hydrolytic reactions such as those shown in Figure 10.8e. This protease is to be contrasted with the serine proteases, in which general base catalysis is employed for a similar purpose.

In other cases, the metal in an enzyme ion serves as a redox reagent. An example is the iron-containing enzyme **catalase,** which catalyzes the

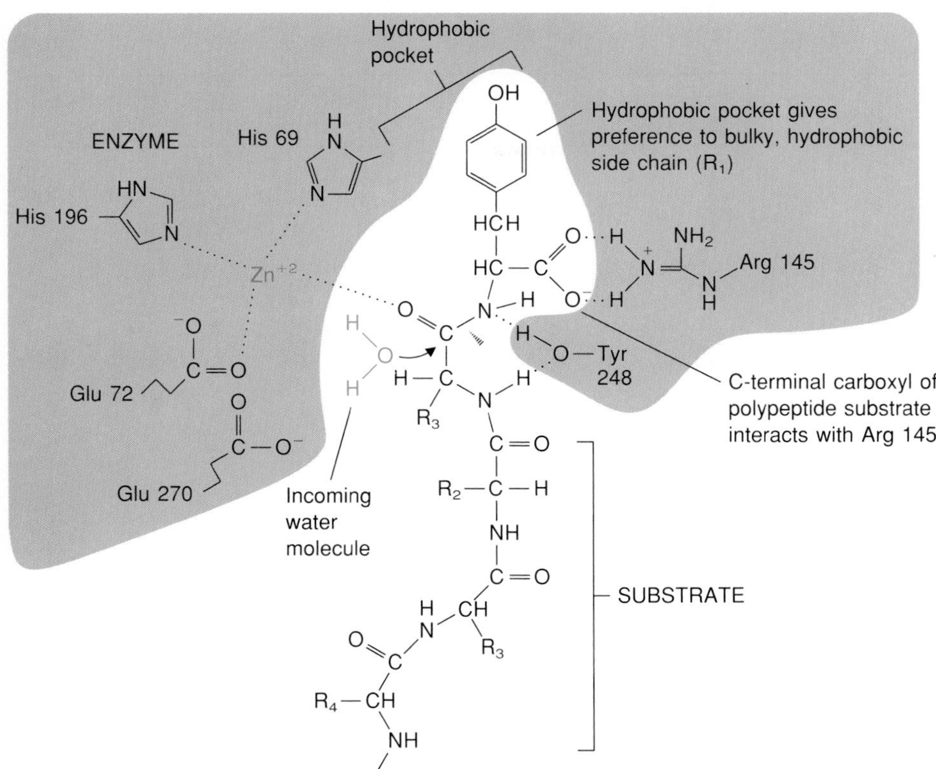

Figure 10.24
Active site of the protease carboxypeptidase A. The zinc atom serves as a metal ion catalyst to promote hydrolysis. The bond cleaved is indicated with a wedge.

breakdown of hydrogen peroxide, a potentially destructive agent in cells:

$$2H_2O_2 \xrightarrow[\text{Catalase}]{} 2H_2O + O_2$$

The reaction involves oxidation of one molecule of H_2O_2 and reduction of another. It can be catalyzed by free ferric ion in solution, with a rate about 3×10^4 times greater than that of the uncatalyzed reaction. But the enzyme catalase, which holds Fe^{3+} in protoporphyrin IX, accelerates the reaction 10^8 fold. In the reaction, Fe^{3+} is alternately reduced and reoxidized.

It is interesting to compare catalase with the heme oxygen-binding proteins, myoglobin and hemoglobin. As noted in Chapter 7, these also carry protoporphyrin IX, but in these the iron is maintained in the Fe^{2+} state. The wholly different role of the porphyrin–iron complex in catalase is a consequence of the very different three-dimensional structure of this enzyme.

In many other enzymatic reactions, certain ions are found to be necessary for catalytic efficiency, even though they may not remain permanently attached to the protein. For example, a number of enzymes that couple ATP hydrolysis to other processes require Mg^{2+} for efficient function. In at least some cases this is because the Mg–ATP complex (see Chapter 3) is a better substrate than ATP itself.

Text continues on p. 373.

Table 10.5
Major vitamins required in human nutrition

Name of Vitamin	Structure	Related Coenzyme	Deficiency Disease
Water-soluble vitamins			
Niacin (nicotinamide)	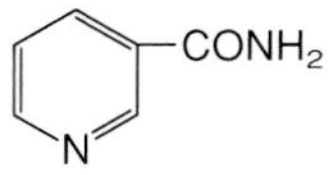 (pyridine ring, N) — $CONH_2$	NAD^+, $NADP^+$	Pellagra
Riboflavin (vitamin B_2)	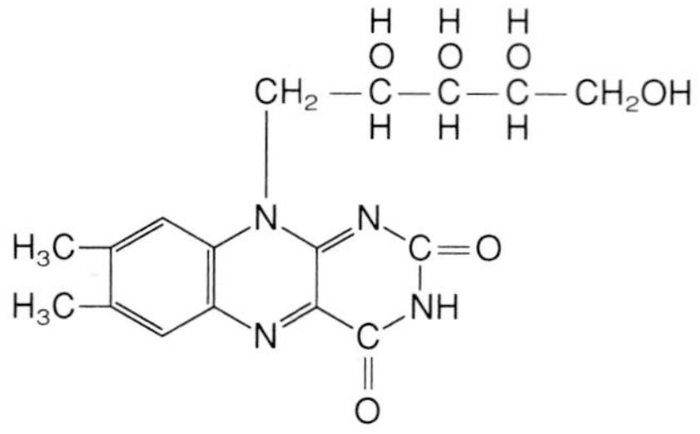 $CH_2-C(H)(OH)-C(H)(OH)-C(H)(OH)-CH_2OH$ attached to N of isoalloxazine ring bearing H_3C, H_3C, N, N, N, $C=O$, NH, $C=O$	FAD	Growth retardation
Thiamine (vitamin B_1)	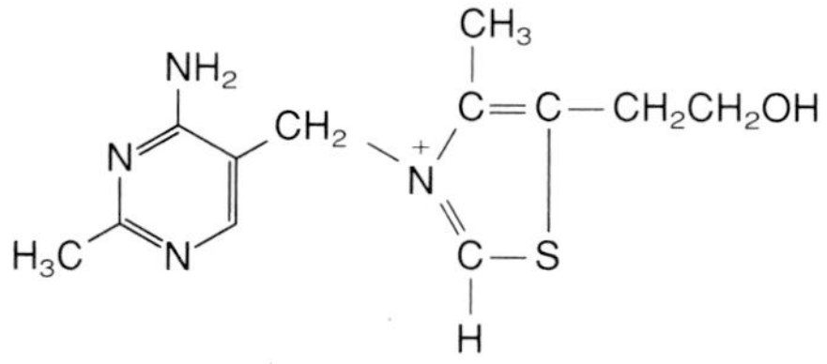 pyrimidine ring (NH_2, N, N, H_3C) — CH_2 — $\overset{+}{N}$ of thiazole ring (CH_3, $C=C-CH_2CH_2OH$, C—S, H)	Thiamine pyrophosphate	Beriberi
Pantothenic acid	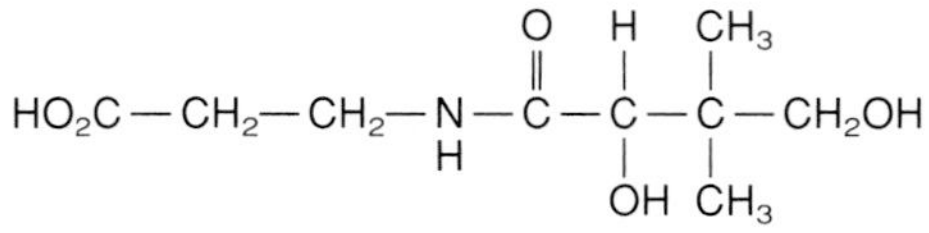 $HO_2C-CH_2-CH_2-NH-C(=O)-CH(OH)-C(CH_3)_2-CH_2OH$	Coenzyme A	Dermatitis (chickens)
Biotin	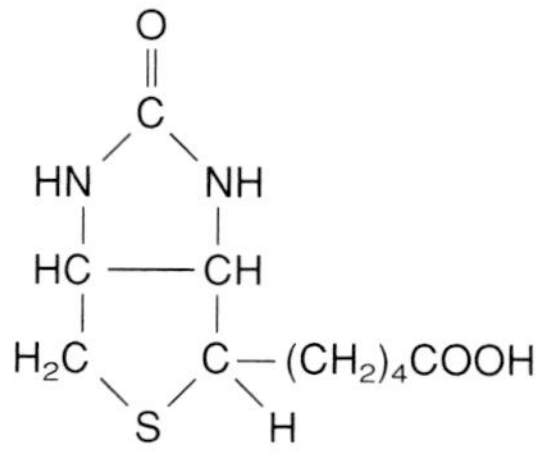 O=C(HN)(NH); HC—CH; H_2C, S, C(H)—$(CH_2)_4COOH$	Biotinylated enzymes	Dermatitis (humans)
Pyridoxal (vitamin B_6)	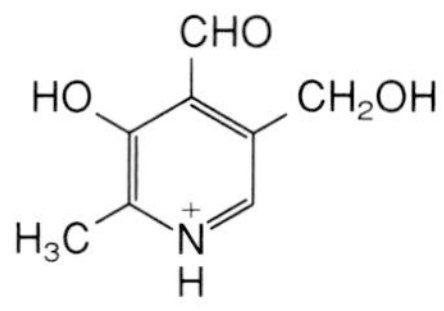 pyridinium ring with CHO, HO, CH_2OH, H_3C, $\overset{+}{N}$H	Pyridoxal phosphate	Dermatitis (rats); neurological symptoms
Folic acid	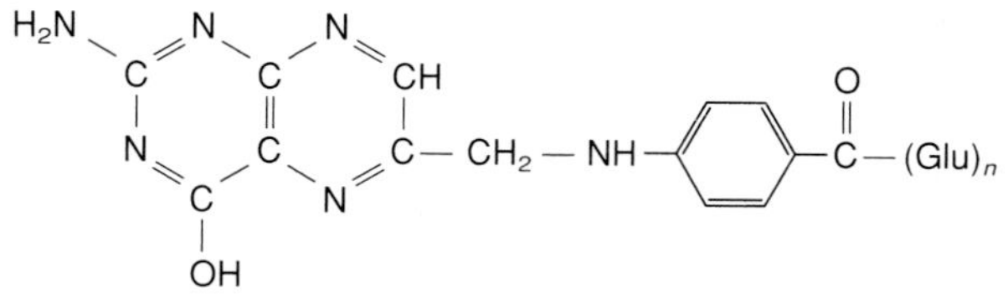 pteridine ring (H_2N, N, N, CH, N, N, OH) — $C-CH_2-NH-$(benzene ring)$-C(=O)-(Glu)_n$	Tetrahydrofolate	Anemias

Table 10.5 *(continued)*
Major vitamins required in human nutrition

Name of Vitamin	Structure	Related Coenzyme	Deficiency Disease
Lipoic acid	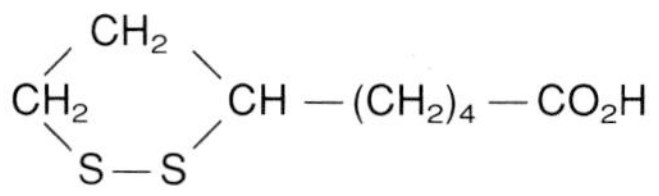	Attached to ε-lysine NH_2 in protein	Growth deficiencies
Cobalamin[a] (vitamin B_{12})	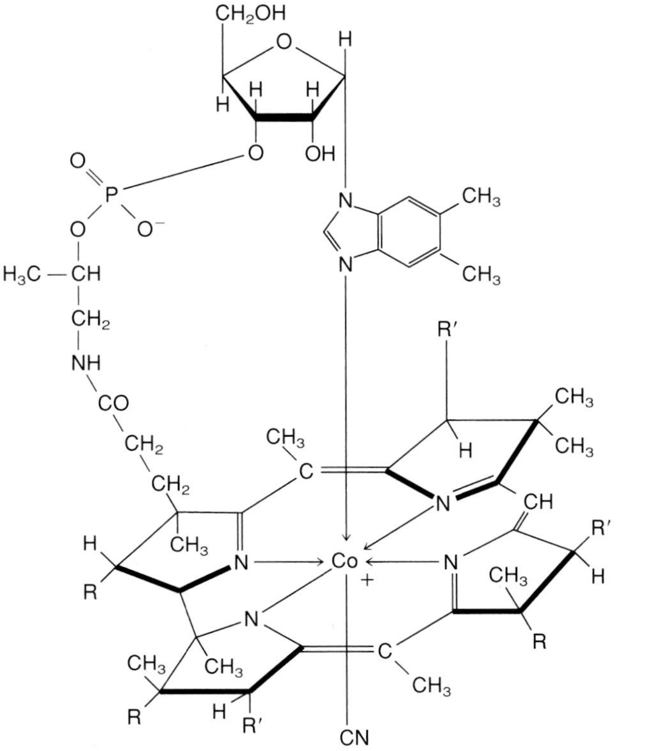	5′-Deoxyadenosyl cobalamin	Pernicious anemia
L-Ascorbic acid (vitamin C)	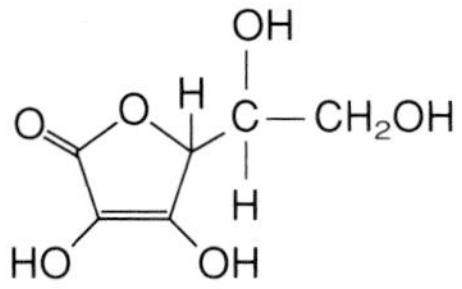	L-Ascorbic acid	Scurvy
Fat-soluble vitamins *trans*-Retinol (vitamin A)	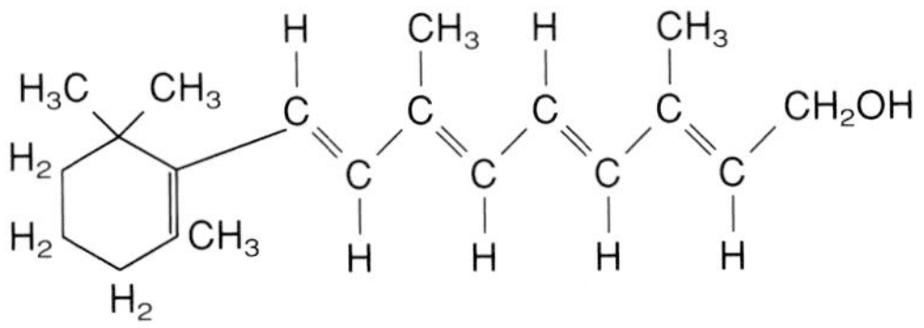	Associated with visual pigment	Night blindness, other effects

(Continues)

Table 10.5 *(continued)*
Major vitamins required in human nutrition

Name of Vitamin	Structure	Related Coenzyme	Deficiency Disease
Cholecalciferol (vitamin D)	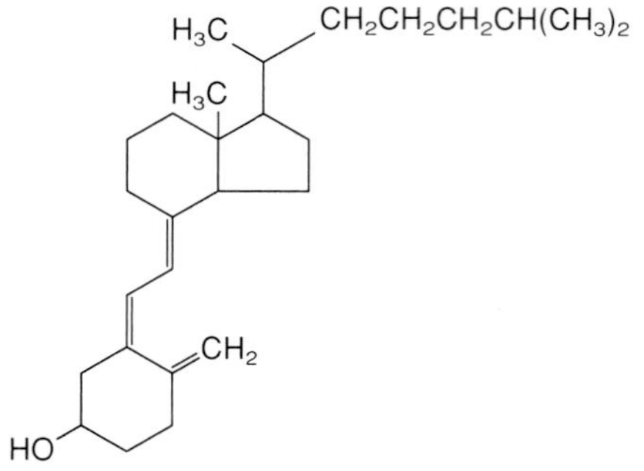 [structure: H_3C, $CH_2CH_2CH_2CH(CH_3)_2$, H_3C, CH_2, HO]	None	Rickets
Tocopherol (vitamin E)	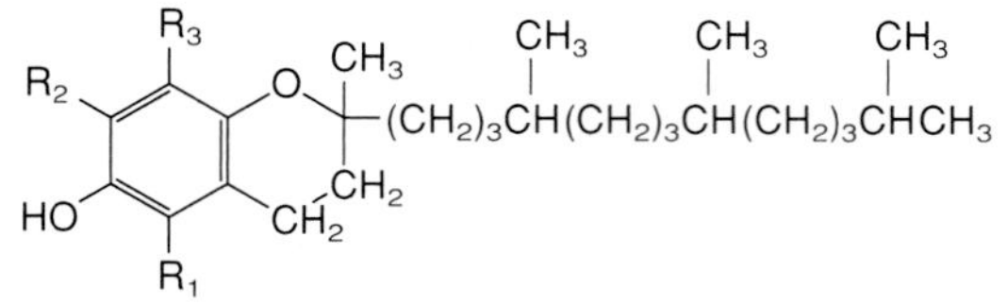 [structure: R_3, R_2, HO, R_1, O, CH_3, $(CH_2)_3CH(CH_3)(CH_2)_3CH(CH_3)(CH_2)_3CH(CH_3)CH_3$, CH_2, CH_2] (several variants, with R_1, R_2, R_3 = H or CH_3)	None	Reproductive and other problems in rats; uncertain in humans
Phylloquinone (vitamin K_1)	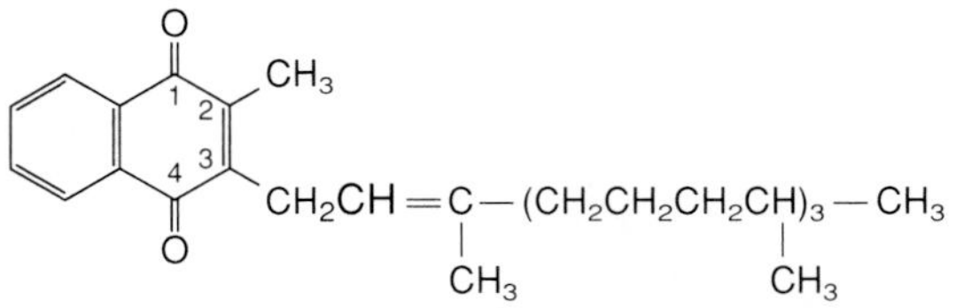 [structure: O, O, 1, 2, 3, 4, CH_3, $CH_2CH{=}C(CH_3){-}(CH_2CH_2CH_2CH(CH_3))_3{-}CH_3$]	None	Problems in blood clotting

[a] In the structure for cobalamin $R = CH_2CONH_2$; $R' = CH_2CH_2CONH_2$.

Table 10.6
Metals and trace elements important as enzymatic cofactors

Metal	Example of Enzyme	Role of Metal
Fe	Cytochrome oxidase	Oxidation/reduction
Cu	Ascorbic acid oxidase	Oxidation/reduction
Zn	Alcohol dehydrogenase	Helps bind NAD^+
Mn	Histidine ammonia-lyase	Aids in catalysis by electron withdrawal
Co	Glutamate mutase	Co is part of cobalamin coenzyme
Ni	Urease	?
Mo	Xanthine oxidase	Oxidation/reduction?
V	Nitrate reductase	Oxidation/reduction?
Se	Glutathione peroxidase	Replaces S in one cysteine in active site

Nonprotein Biocatalysts—Ribozymes

Throughout this chapter we have described how protein enzymes function as biocatalysts. Indeed, until very recently, it was assumed that *all* biochemical catalysis was carried out by proteins. But biochemistry is full of surprises, and the last few years have revealed something wholly unexpected: some RNA molecules can act like enzymes.

The first hint of this came from studies of **ribonuclease P,** an enzyme that cleaves the precursors of tRNAs to yield the functional tRNAs (Figure 10.25). It had been known for some time that active ribonuclease P contained both a protein portion and an RNA "cofactor," but it was generally assumed that the active site resided on the protein portion. However, careful studies of the isolated components by S. A. Altman and co-workers in 1983 revealed an astonishing fact: whereas the protein component alone was wholly inactive, the RNA by itself, if provided with either a sufficiently high concentration of magnesium or a small amount of magnesium plus the small basic molecule spermine, was capable of catalyzing the specific cleavage of pre-tRNAs. Furthermore, the RNA acted like a true enzyme, being unchanged in the process and obeying Michaelis–Menten kinetics. Addition of the protein portion of ribonuclease P does enhance the activity (k_{cat} is markedly increased) but is in no way essential for either substrate binding or cleavage.

At about the same time, another remarkable class of RNA-catalyzed reactions was discovered by T. Cech and his colleagues. Examining the removal of an intron (intervening sequence or IVS) from the pre-ribosomal RNAs of the protist *Tetrahymena,* they found that the reaction was autocatalytic—the rRNA *itself* catalyzed the excision of its 413-nucleotide intron and carried out the necessary resplicing as well (Figure 10.26a). Furthermore, the excised IVS itself went through a series of site-specific autocatalyzed reactions. The final product is a molecule called L-19 IVS—the intervening sequence with 19 more nucleotides removed. This molecule has further remarkable catalytic abilities; it is capable of either lengthening or shortening small oligonucleotides, in the manner shown in Figure 10.26b.

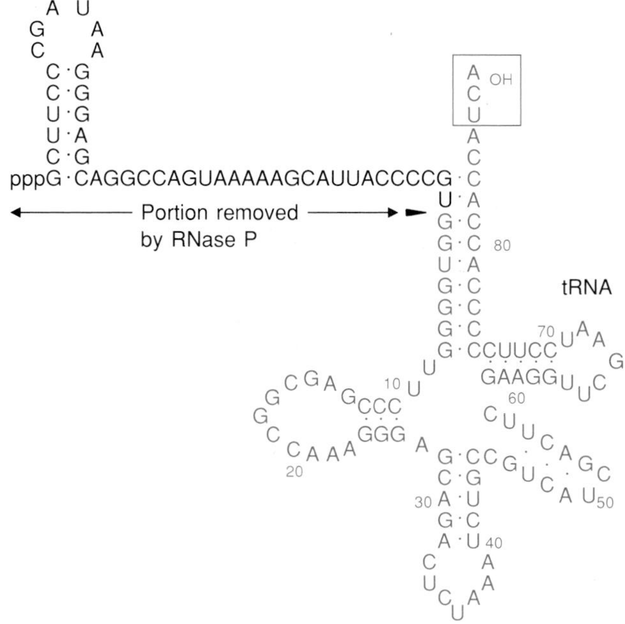

Figure 10.25
Production of tRNAs from pre-tRNAs catalyzed by ribonuclease P. A typical pre-tRNA is shown. The RNA portion of the RNA–protein complex called ribonuclease P is by itself capable of catalyzing the hydrolysis of the specific phosphodiester bond indicated by the wedge. The portion removed from the precursor is shown in black, and the resulting tRNA is in blue.

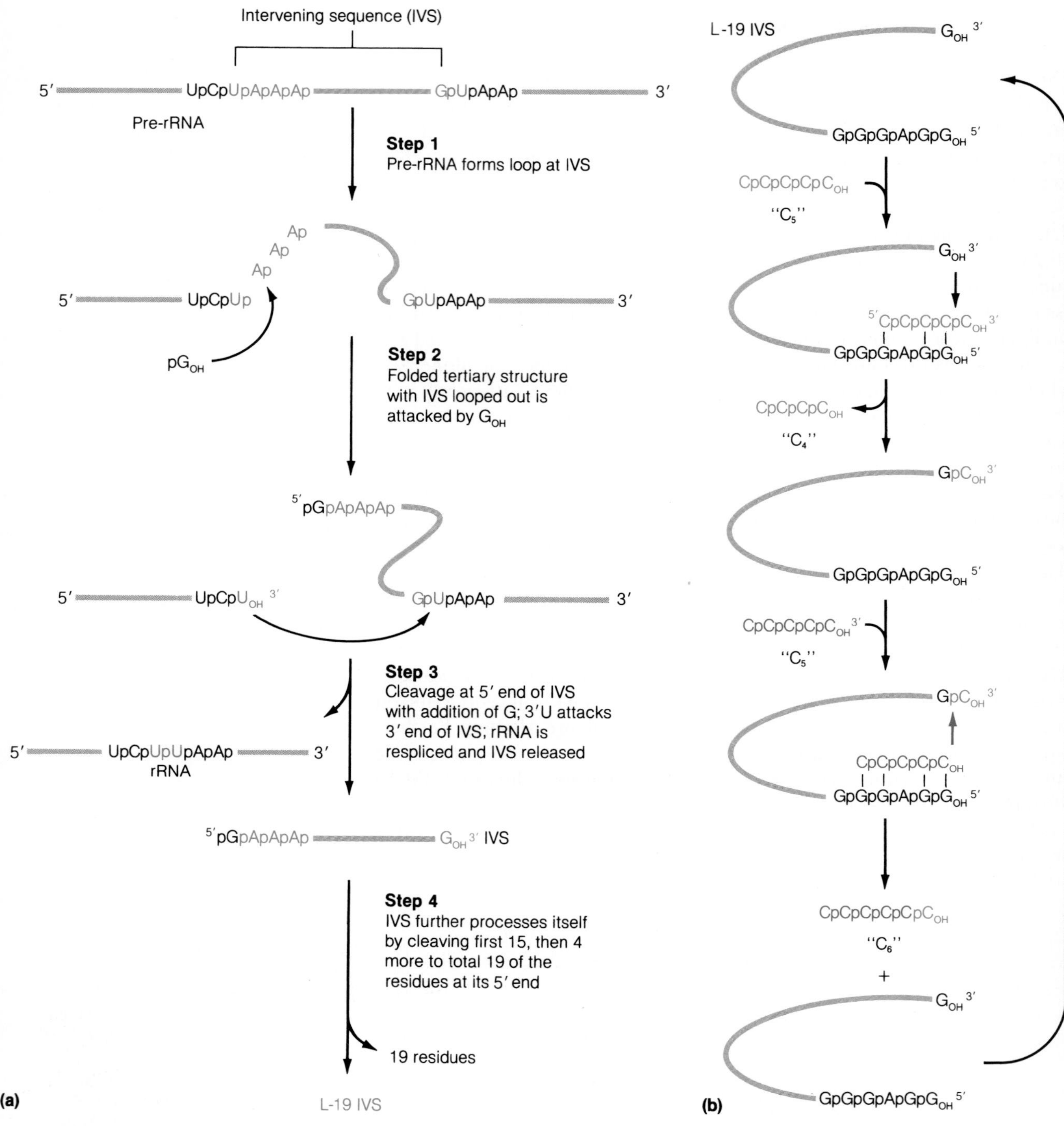

Figure 10.26
Catalysis by the intervening sequence in *Tetrahymena* pre-ribosomal RNA. (**a**) Self-excision and splicing of the intervening sequence (IVS). Note that a pG_{OH} is added in the reaction. A series of further autocatalyzed steps reduces the IVS to L-19 IVS. (**b**) $2C_5 \rightarrow C_4 + C_6$ conversion by L-19 IVS. This oligonucleotide can itself either shorten or elongate small oligonucleotides, acting here as a true ribozyme.

Although many biochemists were astounded by these discoveries, in retrospect we can see that there is no reason why RNA molecules should not have catalytic functions. As we saw in Chapter 4, RNA molecules can adopt a complex tertiary structure, just as proteins do, and this is the kind of structure that appears to be essential for effective biocatalysts.

Molecular Engineering: The Design of Modified and New Enzymes

Techniques now available to the biochemist and molecular biologist, together with our growing understanding of how enzymes function, allow an exciting possibility: the *creation,* in the laboratory, of new or modified enzymes. Several approaches are being taken in this quest, which has enormous potential in industry and medicine.

Site-Directed Mutagenesis

It is now possible to clone the genes for many enzymes and, using methods described in Tools of Biochemistry 22, make specific mutations at particular points. The method has proved extremely powerful in the study of enzyme mechanisms, as illustrated by the study of triose-phosphate isomerase described on page 354. But it is also being employed to change enzyme specificities. A study of the protease subtilisin by J. Wells and co-workers (see Oxander and Fox, 1987, in references) has focused on mutations at a specific site (residue 166) in the specificity pocket. This site is normally occupied by Gly, and the enzyme exhibits a preference to cleave polypeptide chains next to a bulky hydrophobic residue. Activity toward polypeptides containing glutamic acid in the same position is very low. Replacing Gly 166 by Lys increases the frequency of cutting next to glutamic acid by 500-fold.

Hybrid Enzymes

The rearrangement of genes that is possible with modern techniques allows the possibility of producing **fusion proteins**—proteins made from genes that have been spliced together from two or more sources. Thus, it may be possible to recombine binding sites and catalytic sites in novel ways. Another kind of hybrid enzyme is depicted in Figure 10.27. Here a synthetic oligonucleotide of defined sequence has been grafted onto the enzyme **staphylococ-**

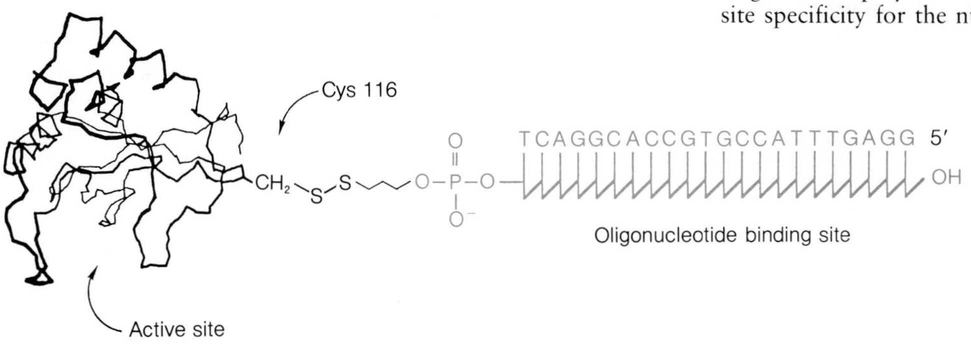

Figure 10.27
A hybrid enzyme. To the enzyme staphylococcal nuclease (black) has been attached an oligonucleotide of defined sequence (blue). The oligonucleotide base pairs with complementary sequences on single-strand polynucleotides, providing site specificity for the nuclease.

cal nuclease through a disulfide bond. Staphylococcal nuclease is efficient at cutting single-strand RNA or DNA, but does so with very little specificity. The oligonucleotide that has been attached to the enzyme in this experiment will bind to a specific complementary sequence in a single-strand nucleic acid. The enzyme then cuts next to that sequence. High specificity can thereby be designed into a normally nonspecific enzyme.

Catalytic Antibodies

As you will recall from Chapter 7, antibodies show high specificity in binding. Since enzymes supposedly function by binding favorably the transition state in a reaction, we may ask what will happen if one makes antibodies against molecules that are structurally analogous to the transition states of particular substrates? The answer is that these antibodies act like enzymes. Consider the following example. To prepare an enzyme with the ability to catalyze the hydrolysis of methyl *p*-nitrophenyl carbonate

$O_2N-C_6H_4-O-C(=O)-O-CH_3$

monoclonal antibodies were prepared against the compound

$O_2N-C_6H_4-O-P(=O)(O^-)-CH_2-CH_2-CH_2-CH_2-COO^-$

The tetrahedral phosphonate was chosen as an analog for the tetrahedral transition state in such hydrolyses. The antibody acted like an enzyme, obeying Michaelis–Menten kinetics, with $k_{cat} = 30\ \text{min}^{-1}$, $K_M = 350\ \mu\text{M}$.

It seems likely that biochemists have only begun to explore the possibilities of engineering enzymes for specific purposes.

REFERENCES

General

Boyer, P. D. (ed.) (1970) *The Enzymes*, 3rd ed. Academic Press, New York. This is a multivolume set. Volumes 1 and 2 cover general topics, and the many succeeding volumes deal with specific enzymes in great detail.

Creighton, T. E. (1983) *Proteins*. W. H. Freeman, New York. Chapter 9 of this excellent book contains a succinct summary of enzyme catalysis, together with many recent references to the appropriate literature.

Fersht, A. (1985) *Enzyme Structure and Mechanism*, 2nd ed. W. H. Freeman, New York. A very fine, up-to-date treatise on most all aspects of enzymology.

Webb, E. (1984) *Enzyme Nomenclature*. Academic Press, Orlando, Fla. This is the authoritative listing of enzymes prepared by the Nomenclature Committee of the International Union of Biochemistry.

Enzyme Mechanisms and Kinetics

Alber, T. C., R. C. Davenport, Jr., D. A. Giammona, E. Lolis, G. Petsko, and D. Ringe (1988) *Crystallography and Site Directed Mutagenesis of Yeast Triose Phosphate Isomerase: What We Can Learn from a "Simple" Enzyme*. Cold Spring Harbor Symp. Quant. Biol. 52:603–613.

Bennet, W. S., and T. A. Steitz (1978) Glucose-induced conformational change in yeast hexokinase. *Proc. Natl. Acad. Sci. USA* 75:4848–4852.

Cleland, W. W. (1977) Determining the chemical mechanisms of enzyme catalyzed reactions by kinetic studies. *Adv. Enzymol.* 45:273–387.

Koshland, D. E., Jr. (1973) Protein shape and biological control. *Sci. Am.* 229(4):52–64. A clearly written introduction.

Kraut, J. (1988) How do enzymes work? *Science* 242:533–540. A thoughtful reanalysis of transition state theory in terms of current information.

Walsh, C. T. (1977) *Enzymatic Reaction Mechanisms.* W. H. Freeman, New York. A comprehensive work with a very thorough treatment of coenzymes.

Proteases

Kraut, J. (1977) Serine proteases: Structure and mechanism of catalysis. *Annu. Rev. Biochem.* 46:331–358.

Steitz, T. A., and R. G. Shulman (1982) Crystallographic and NMR studies of the serine proteases. *Annu. Rev. Biochem. Biophys.* 11:419–444.

Stroud, R. M. (1974) A family of protein cutting proteins. *Sci. Am.* 231(1):24–88. A succinct overview.

Ribozymes

Cech, T. R. (1987) The chemistry of self-splicing RNA and RNA enzymes. *Science* 236:1532–1539.

McCorkle, G. M., and S. Altman (1987) RNA's as catalysts. *Concepts Biochem.* 64:221–226.

Molecular Engineering of Enzymes

Oxander, D. L., and C. F. Fox (eds.) (1987) *Protein Engineering.* Alan Liss, New York. A fascinating compilation of papers on techniques and results. Chapter 25 (by J. A. Wells and 12 collaborators) describes the work on subtilisin referred to in the text.

Schultz, P. G. (1988) The interplay between chemistry and biology in the design of enzymatic catalysts. *Science* 240:426–433. A short, very clearly written exposition of recent methods.

PROBLEMS

1. A substance A is consumed by a reaction of unknown order. The initial concentration is 1 mM, and concentrations at later times are as shown:

Time (min)	[A] (mM)
1	0.83
2	0.72
4	0.56
8	0.38
16	0.24

Test whether a first-order or second-order reaction best fits the data.

2. The enzyme urease catalyzes the hydrolysis of urea to ammonia plus carbon dioxide. At 21°C the uncatalyzed reaction has an activation energy of about 125 kJ/mol, whereas in the presence of urease this is lowered to about 46 kJ/mol. By what factor does urease increase the velocity of the reaction?

3. The maximum possible rate for a reaction between two molecules occurs when every collision results in reaction. In this case, the second-order rate constant is predicted to be

$$k = \frac{4\pi N}{1000}(D_1 + D_2)r_{12}$$

with units of $(\text{mol/L})^{-1}\ \text{s}^{-1}$. Here N is Avogradro's number (6.02×10^{23} molecules/mol), r_{12} is the critical distance for reaction, and D_1 and D_2 are the diffusion coefficients for the two participants. In a typical enzyme-catalyzed reaction, we might expect r_{12} to be about 10^{-7} cm, $D_{\text{enzyme}} = 5 \times 10^{-7}\ \text{cm}^2/\text{s}$, and $D_{\text{substrate}} = 5 \times 10^{-6}\ \text{cm}^2/\text{s}$. What does this predict for the value of k? Compare with values of $k_{\text{cat}}/K_{\text{M}}$ in Table 10.2.

4. The initial rate for an enzyme-catalyzed reaction has been determined at a number of substrate concentrations. Data are given below:

[S] (μmol/L)	V(μmol/L) min^{-1}
5	22
10	39
20	65
50	102
100	120
200	135
500	147

(a) Attempt to estimate V_{max} and K_{M} from a direct graph of V versus [S].

(b) Now use a Lineweaver–Burk plot to analyze the same data.

(c) Finally, try an Eadie–Hofstee plot of the same data.

5. (a) If the total enzyme concentration in Problem 4 was 1 nmol/L, what is k_{cat}?

(b) Calculate $k_{\text{cat}}/K_{\text{M}}$ for the enzyme reaction in Problem 4. Is this a fairly efficient enzyme? (See Table 10.2.)

6. (a) If we write the Michaelis–Menten equation as

$$\frac{d[\text{S}]}{dt} = -\frac{d[\text{P}]}{dt} = -\frac{V_{\text{max}}[\text{S}]}{K_{\text{M}} + [\text{S}]}$$

it should be possible to integrate it and obtain an expression for substrate concentration as a function of time. Do so, calling $[\text{S}]_0$ the initial substrate concentration at $t = 0$.

(b) Show from the result in (a) that under conditions where $[\text{S}]_0 \gg K_{\text{M}}$, the decrease in substrate concentration with time is approximately linear. Interpret this result.

7. Suggest a graphical method for using the equation obtained in Problem 6(a) (the integrated Michaelis–Menten equation) to determine V_{max} and K_{M}.

8. The catalytic efficiency of many enzymes depends on pH. Chymotrypsin shows a maximum value of k_{cat}/K_M at pH 8. Detailed analysis shows that k_{cat} increases rapidly between pH 6 and 7 and remains constant at higher pH. K_M increases rapidly between pH 8 and 10. Suggest explanations for these observations.

9. Suppose you had available a sample of the ribozyme IVS-19 from *Tetrahymena*. Describe the protocol for a simple experiment that would demonstrate the reaction shown in Figure 10.26b.

10. There has long been argument as to whether DNA, RNA, or proteins were the "primordial molecules" in the origin of life. Now most scientists believe RNA to be the best candidate. From what is now known about RNA, what two properties of this substance favor this hypothesis?

15

TOOLS OF BIOCHEMISTRY 15

How to Measure the Rates of Enzyme-Catalyzed Reactions

There are essentially two approaches to the study of enzyme kinetics. The first and simplest is to make measurements of rates under conditions where the steady-state approximation holds (see p. 359). Under these conditions, the Michaelis–Menten equation is often applicable, and determination of the reaction velocity as a function of substrate and enzyme concentrations will yield K_M and k_{cat}. Almost all enzymatic studies at least start in this way. But if the experimenter wishes to learn more of the details of the mechanism, it is often important to carry out studies before the steady state has been attained. Such *pre-steady-state* experiments require the use of special fast techniques. In Chapter 10 we have described how a combination of such approaches can be used to dissect a complex enzymatic process and to understand it in detail. Here we describe some of the experimental techniques that can be employed.

Analysis in the Steady State

The steady state in most enzymatic reactions is established within seconds or a few minutes and persists for many minutes or even hours thereafter. Therefore, extreme rapidity of measurement is not important, and many techniques are available to the experimenter wishing to follow the reaction. The most commonly used are described below.

Spectrophotometry

Spectrophotometric methods are simple and accurate (see Tools of Biochemistry 10). However, an obvious requirement is that either a substrate or a product of the reaction absorb light in a spectral region where other substrates or products do not. Classic examples are reactions that generate or consume NADH. NADH absorbs quite strongly at 340 nm, but NAD^+ does not absorb in this region. Thus we could, for example, follow the oxidation of ethanol to acetaldehyde, as catalyzed by alcohol dehydrogenase, by measuring the formation of NADH spectrophotometrically. Even if the reaction being studied does not involve a light-absorbing substance, it may be possible to *couple* this reaction to another, very rapid, reaction that does.

Fluorescence

The applications of fluorescence are very similar to those of spectrophotometry, and the problems are similar: a substrate or a product must have a distinctive fluores-

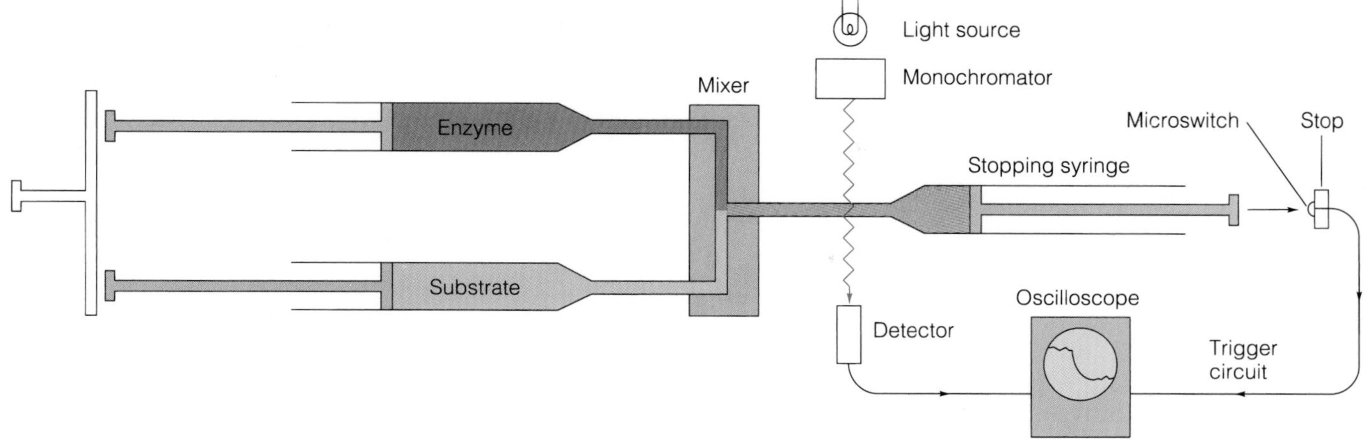

Figure T15.1
Typical stopped-flow apparatus.

cence emission spectrum (see Tools of Biochemistry 10). However, fluorescence often has the advantage of very high sensitivity, so extremely dilute solutions may be employed, enabling one to greatly extend the concentration range over which studies are practicable.

Automatic Titration

If the reaction produces or consumes acid or base, it can be followed by using a device called a *pH-stat.* A glass electrode senses the pH of the solution, and its signal is used to actuate a motor-driven syringe that titrates acid or base into the reaction vessel to keep the pH constant. The record of acid or base consumed is then a record of the progress of the enzymatically catalyzed reaction.

Radioactivity Assays

If a substrate is labeled with a radioactive isotope that will be lost or transferred during the reaction to be studied, measurement of changes in radioactivity can be an extremely sensitive kinetic method. This requires, of course, that the labeled compound can be separated very quickly at different, precisely defined times during the reaction. An example is a method often used with radioactive ATP. The ATP can be adsorbed on charcoal-impregnated filter disks by very fast filtration of aliquots from the reaction mixture. The radioactivity can then be measured in a scintillation counter (see Tools of Biochemistry 16).

Analysis of Very Fast Reactions

Reactions that are extremely rapid require special techniques in order to investigate the pre-steady-state processes. Three major methods are currently employed.

Stopped Flow

Figure T15.1 shows a stopped-flow apparatus. Enzyme and substrate are initially in separate syringes. The syringes are very rapidly driven, to deliver their contents through a mixing chamber and into a third, "stopping" syringe. This triggers a detector to begin observing (for example, by light absorption or fluorimetry) the solution in the tube connecting the mixer to the stopping syringe. Flow rates can easily be made as high as 1000 cm/s. If the observation point is 1 cm from the mixer, the detection system first sees a mixture that is 1 ms "old." The reaction can then be followed for as long as desired—often for a period of only a few seconds. The limitations of the method are imposed only by the initial "dead time" (in the example above, 1 ms) and the rapidity of the detection system (usually a fast oscilloscope is used to follow an absorbance or fluorescence change).

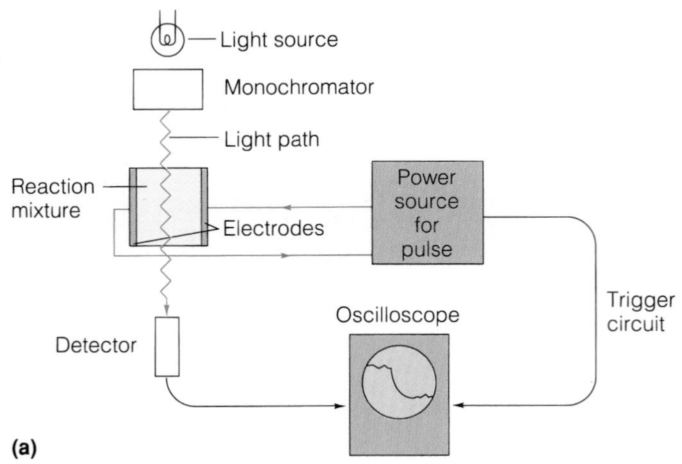

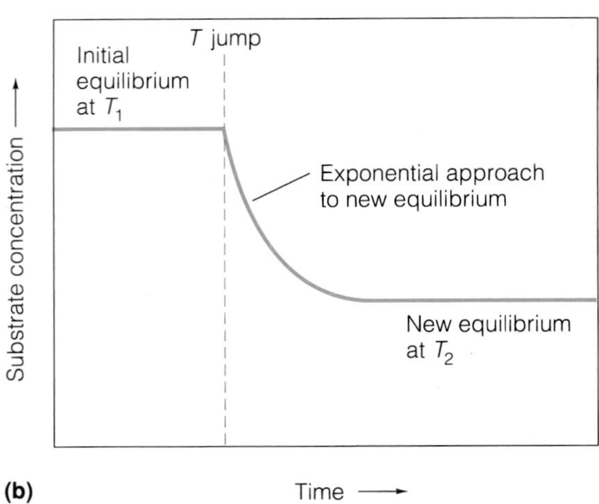

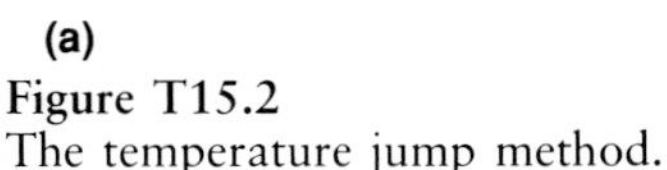

Figure T15.2
The temperature jump method.

Temperature Jump

Some processes are so fast that they are essentially completed in the dead time of a stopped-flow apparatus. The experimenter may then turn to temperature jump (*T*-jump) methods. The basic apparatus and principle of the method are shown in Figure T15.2a and b, respectively. A reaction mixture that is at equilibrium at a temperature T_1 is suddenly jumped to a temperature T_2. The position of equilibrium will shift, and reaction must occur to attain this new equilibrium. A rapid jump in temperature (5–10°C in 1 μs) can be obtained by passing a large burst of electrical current between electrodes immersed in the reaction mixture. Even more rapid jumps (10–100 ns) can be obtained if a pulsed infrared laser is used to heat the mixture. The relaxation to a new equilibrium, followed by absorption or fluorescence measurements, will be an exponential process. For a simple reaction the change in reactant concentration is given by

$$\Delta[\mathrm{A}] = (\Delta[\mathrm{A}])_{\mathrm{total}} e^{-t/\tau} \qquad \text{(T15.1)}$$

where τ is called the *relaxation time* and can be related to the rate constants for the reaction. For example, for the simple reversible isomerization ($\mathrm{A} \underset{k_{-1}}{\overset{k_1}{\rightleftharpoons}} \mathrm{B}$) we see by comparison with equation (10.7) that

$$\frac{1}{\tau} = k_1 + k_2 \qquad \text{(T15.2)}$$

More complex reactions involve multiple relaxation times and more complex curves than expressed by equation (T15.1).

Although a number of other techniques are employed for fast reactions, including some newly developed NMR methods, those described above are the ones most widely used. If we consider the variety of techniques available to the experimenter, we can see that they cover a very wide time range—altogether, times from nanoseconds to hours can be studied.

REFERENCES

Fersht, A. (1985) *Enzyme Structure and Mechanism*, 2nd ed. W. H. Freeman, New York. Chapters 4, 6, and 7 contain much information about techniques and up-to-date references.

Himori, K. (1979) *Kinetics of Fast Enzyme Reactions*, Halstead Press, New York.

CHAPTER 11

The Regulation of Enzyme Activity

The analogy between a living cell and a factory (Chapter 1) is especially appropriate when we consider enzyme regulation. We note that a cell has certain raw materials available to it and must produce specific products from them. The machines that facilitate these transformations in the cell are the enzymes. Often, as we shall see, they are arranged in "assembly lines" to carry out the necessary sequential steps in a metabolic pathway.

No factory will operate efficiently if every machine is operating at its maximum rate. The capabilities of machines vary greatly, and if all were running "all out," massive problems would soon arise. Intermediate products would pile up in some assembly lines, and certain parts of the finished product would be produced in vast excess. Different assembly lines might draw on the same raw material, and the faster ones could deplete the supplies so completely that other, equally important lines would have to shut down. Obviously, *coordination* and *regulation* are required to run a large factory efficiently.

The same kinds of problems could occur if the enzymatic machinery of the cell were not regulated exquisitely. The efficiencies with which individual enzymes operate must be controlled in a manner that reflects the availability of substrates, the utilization of products, and the overall needs of the cell. Regulation is accomplished in many ways—through the noncovalent or covalent binding of other molecules to the enzyme, through chemical modification of the enzyme, and through control of enzyme synthesis and degradation. We discuss each of these mechanisms of regulation in this chapter.

We begin by describing a mode of regulation of enzymatic activity that has been especially useful to the biochemist studying enzymes. This is **inhibition**—the decrease in enzymatic activity by binding of a molecule called an **inhibitor** to the enzyme. Studies of inhibition have been especially useful in revealing mechanisms of enzyme function.

Enzyme Inhibition

Many different kinds of molecules inhibit enzymes, and they act in a variety of ways. A major distinction must be made between **reversible** and **irreversible** inhibition. The former involves *noncovalent* binding of the inhibitor and can be reversed immediately by removal of the inhibitor. In irreversible

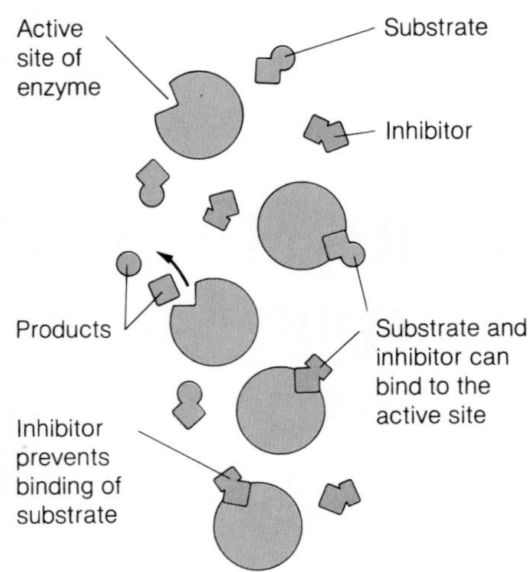
Active site of enzyme
Substrate
Inhibitor
Products
Substrate and inhibitor can bind to the active site
Inhibitor prevents binding of substrate

Figure 11.1
Competitive inhibition. Both substrate (green) and inhibitor (red) can fit the active site. Substrate can be processed by the enzyme, but inhibitor cannot.

inhibition, a molecule is *covalently* bound to the enzyme and incapacitates it. As you might expect, reversible inhibition is used primarily for rapid cellular response, whereas irreversible inhibition is encountered more frequently when slow or permanent changes occur. Toxic substances are often irreversible inhibitors of key enzymes.

Reversible Inhibition

The various modes of reversible inhibition all involve the noncovalent binding of an inhibitor to the enzyme, but they differ in the mechanisms by which they decrease the enzyme's activity and in how they affect the kinetics of the reaction.

COMPETITIVE INHIBITION. Suppose there exists a molecule that so closely resembles the substrate for an enzyme-catalyzed reaction that the enzyme will accept it in its binding site. If this molecule can also be processed by the enzyme, it is merely a competing alternative substrate. However, if the molecule binds to the active site but *cannot* undergo the catalytic step, it simply wastes the enzyme's time. Such a molecule is called a **competitive inhibitor** (Figure 11.1).

For whatever fraction of the time a competitive inhibitor molecule is occupying the active site, the enzyme is unavailable for catalysis. The overall effect is as if the enzyme cannot bind substrate as well when the inhibitor is present. Thus, we expect that the enzyme would act as if its K_M were increased by the presence of the inhibitor. Mathematically, we express these ideas by writing the reaction scheme as

$$\begin{array}{l} \mathrm{E} + \mathrm{S} \overset{K_M}{\rightleftharpoons} \mathrm{ES} \xrightarrow{k_{cat}} \mathrm{E} + \mathrm{P} \\ + \\ \mathrm{I} \\ K_I \updownarrow \\ \mathrm{EI} \end{array}$$

Here I stands for the inhibitory substance and K_I is a dissociation constant for inhibitor binding, defined as $K_I = [E][I]/[EI]$. We can solve the rate equations just as was done in Chapter 10, noting that now

$$\underset{\text{Total enzyme}}{[E]_t} = \underset{\text{Free enzyme}}{[E]} + \underset{\text{Enzyme bound to substrate}}{[ES]} + \underset{\text{Enzyme bound to inhibitor}}{[EI]} \tag{11.1}$$

where $[EI] = [E][I]/K_I$, [I] being the concentration of free inhibitor. The result is

$$V = \frac{k_{cat}[E]_t[S]}{[S] + K_M(1 + [I]/K_I)} \tag{11.2a}$$

which may be rewritten as

$$V = \frac{k_{cat}[E]_t[S]}{[S] + K_M^{app}} = \frac{V_{max}[S]}{[S] + K_M^{app}} \tag{11.2b}$$

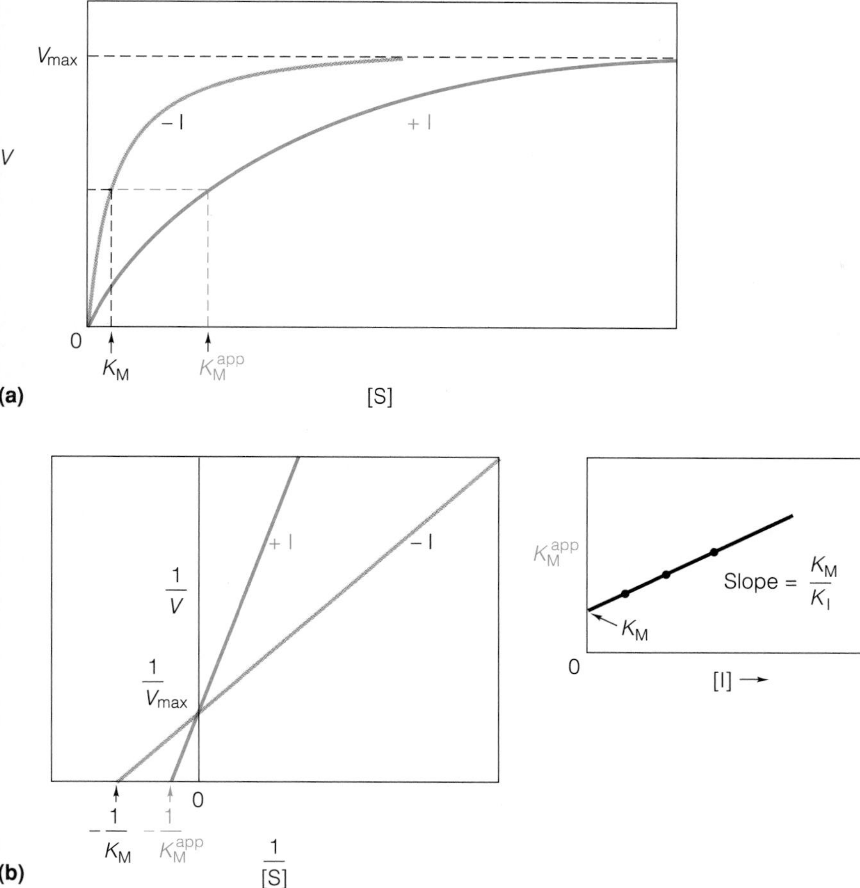

Figure 11.2
Effects of competitive inhibition on enzyme kinetics. (**a**) Reaction velocity (V) versus substrate concentration [S]. Addition of inhibitor (I) decreases rate, but V_{max} approached at high [S] is the same. The apparent K_M is higher in the presence of inhibitor. (**b**) Lineweaver–Burk plots. The fact that the lines cross at the same V_{max} proves that this is competitive inhibition. K_M^{app} values can be determined in the way described in Chapter 10. If the experiment is repeated at different I concentrations, K_I can be determined as shown in the insert.

This looks just like the Michaelis–Menten equation, with an "apparent" K_M given by $K_M^{app} = K_M(1 + [I]/K_I)$. As we predicted, increasing [I] causes an increase in the apparent K_M. Note that V_{max} is unchanged, for as [S] becomes very large, V approaches V_{max}, just as in the absence of inhibition. Physically, this simply means that if we make [S] very large at a given [I], the numerous substrate molecules will outcompete the inhibitor. The effect of competitive inhibition on a graph of V versus [S] is shown in Figure 11.2.

Since the system, at a given [I], still obeys an equation of the Michaelis–Menten *form,* we must expect that Lineweaver–Burk plots and Eadie–Hofstee plots will still be linear graphs, with K_M (but not V_{max}) changed by the presence of inhibitor. As Figure 11.2b shows, this is exactly what happens. If we graph K_M^{app} versus [I], we can determine both the true K_M and K_I as shown in the insert to the figure.

A clear example of what a competitive inhibitor is like is shown in Figure 11.3. The molecule UpA is an excellent substrate for the enzyme ribonuclease (p. 187). But if the oxygen atom at the cleavage site in the substrate UpA is replaced by a CH_2 group to form the phosphonate analog UpcA, a strong competitive inhibitor is formed. Ribonuclease binds the analog strongly enough in the active site to allow x-ray diffraction studies of the complex, but it cannot cleave the phosphonate bond.

A variant of competitive inhibition is **nonproductive binding.** Sometimes a substrate molecule has an extra way of fitting into the binding site, a way in which the normal catalytic event cannot occur:

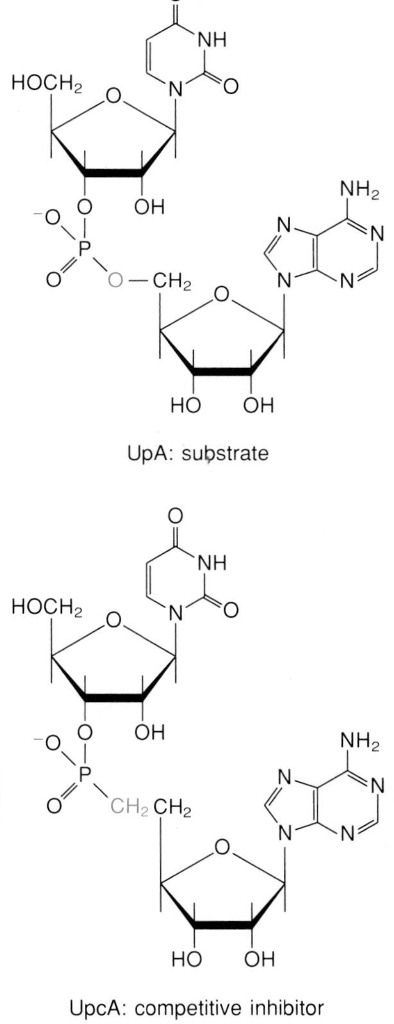

UpA: substrate

UpcA: competitive inhibitor

Figure 11.3
A substrate (UpA) and a competitive inhibitor (UpcA) for the enzyme ribonuclease. The differences between the two molecules are shown by the atoms in red.

$$\begin{array}{l} \mathrm{E} + \mathrm{S} \underset{}{\overset{K_M}{\rightleftharpoons}} \mathrm{ES} \xrightarrow{k_{cat}} \mathrm{E} + \mathrm{P} \\ + \\ \mathrm{S} \\ K'_S \updownarrow \\ \mathrm{ES}' \end{array}$$

In such cases both K_M and k_{cat} are modified, for even at saturating concentrations of substrates a fraction of the substrate molecules will be bound in the nonproductive mode. An example is found in the action of carboxypeptidase on dipeptides such as glycyl-L-tyrosine. Carboxypeptidase, you will recall, cleaves the C-terminal residues from polypeptide chains. Dipeptides are cleaved only slowly, presumably because they can be bound in a way in which the N-terminal amino group interferes with the catalytic site.

Noncompetitive Inhibition. This form of inhibition occurs when a molecule or ion can bind to a *second* site on an enzyme surface (not the active site) in such a way that it modifies k_{cat}. It might, for example, distort the enzyme so that the catalytic process is not as efficient (Figure 11.4). Such a **noncompetitive inhibitor** can be a molecule that does not resemble the substrate at all but has a strong affinity for a second binding site. The simplest case to consider is one in which the inhibitor molecule does not interfere in any way with substrate binding but completely prevents the catalytic step. In this case, inhibitor will bind equally well to both E and ES. We then diagram the reactions as follows:

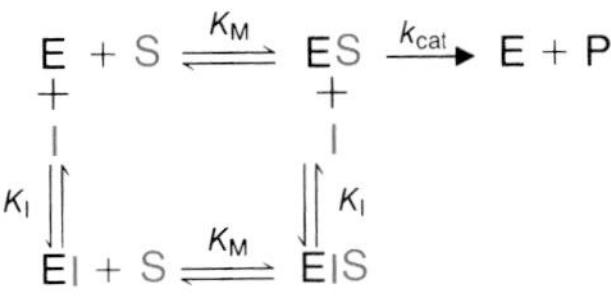

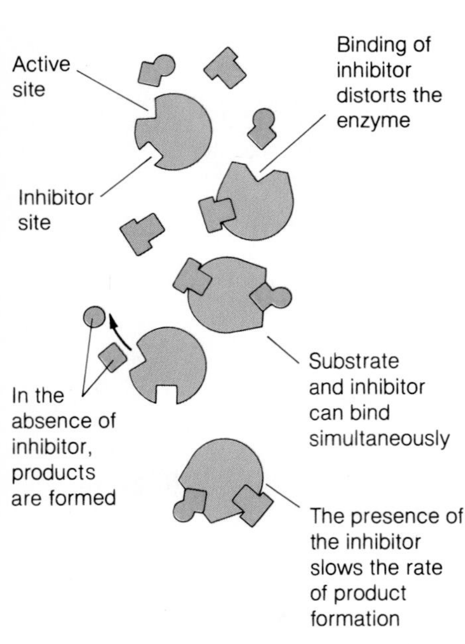

Figure 11.4
Noncompetitive inhibition. The inhibitor (red) binds at a different site on the enzyme surface than the substrate (green). In this simple example, the inhibitor does not interfere with substrate *binding*, but inhibits the catalytic event. Inhibitor and substrate can bind independently.

Mathematical analysis yields

$$V = \frac{\{k_{cat}/(1 + [\mathrm{I}]/K_\mathrm{I})\}[\mathrm{E}]_t[\mathrm{S}]}{[\mathrm{S}] + K_M} = \frac{k_{cat}^{app}[\mathrm{E}]_t[\mathrm{S}]}{[\mathrm{S}] + K_M} \tag{11.3}$$

The result is exactly what we would expect; the apparent K_M is uninfluenced by inhibitor, but the apparent k_{cat} (given by $k_{cat}/([1 + [\mathrm{I}]/K_\mathrm{I}])$ *decreases* with increasing [I]. Therefore V_{max} is changed in this case (Figure 11.5), for at high [S] we find

$$V \longrightarrow V_{max}^{app} = k_{cat}^{app}[\mathrm{E}]_t = \frac{k_{cat}[\mathrm{E}]_t}{1 + [\mathrm{I}]/K_\mathrm{I}} \tag{11.4}$$

The effect of noncompetitive inhibition on a Lineweaver–Burk plot is shown in Figure 11.5b. Both k_{cat} and K_I may be determined by graphing $1/V_{max}^{app}$ versus [I] (insert).

In reality, the situation is often more complex than the simple case we have described. For example, the complex ESI may be able to undergo the catalytic process *slowly*, or the binding of inhibitor may modify *both* k_{cat} and K_M. The latter case is called, for obvious reasons, **mixed inhibition.** The kinetic equations in such circumstances become more complicated and will not be considered here.

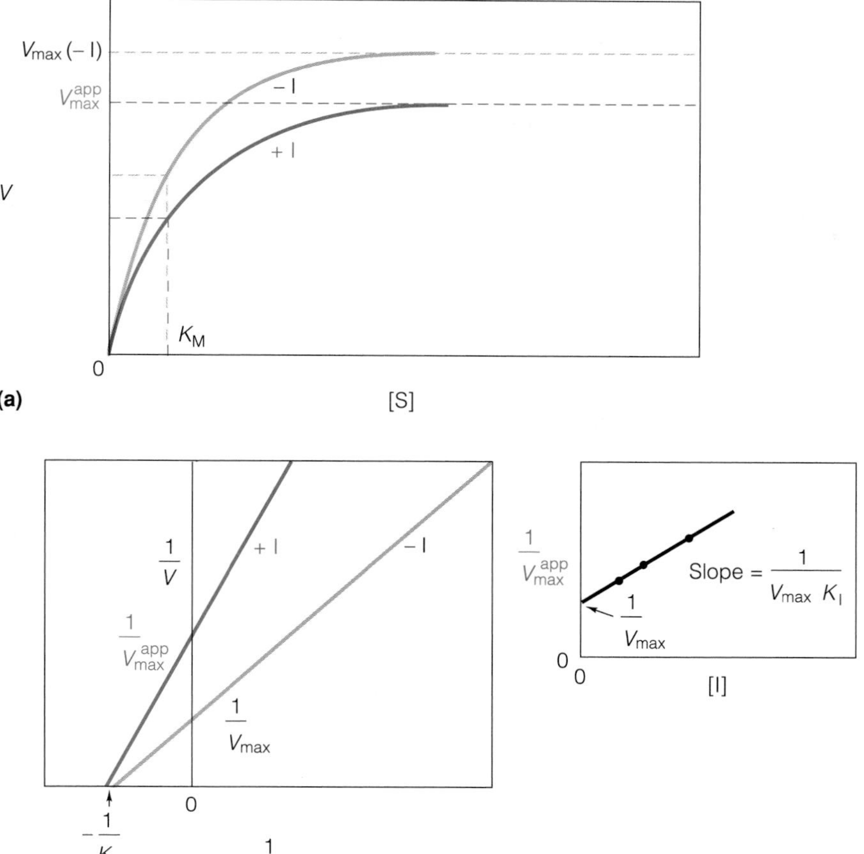

Figure 11.5
Effects of noncompetitive inhibition on enzyme kinetics. (**a**) Reaction velocity (V) versus [S]. In this simple example, K_M is not affected, but V_{max} is decreased because the enzyme is not as catalytically efficient in the presence of the inhibitor. (**b**) Lineweaver–Burk plots. Note how this can be clearly distinguished from competitive inhibition (Figure 11.2b). The insert shows how K_I can be determined.

Uncompetitive Inhibition. There is a third class of reversible inhibition, in which the inhibitor binds *only* to the enzyme–substrate complex:

$$\mathrm{E + S} \underset{}{\overset{K_M}{\rightleftharpoons}} \mathrm{ES} \xrightarrow{k_{cat}} \mathrm{E + P}$$
$$\mathrm{ES} + \mathrm{I} \underset{}{\overset{K_I}{\rightleftharpoons}} \mathrm{ESI}$$

If this binding blocks catalysis (Figure 11.6) we find the following rate equation:

$$V = \frac{k_{cat}[\mathrm{E}]_t[\mathrm{S}]}{[\mathrm{S}](1 + [\mathrm{I}]/K_I) + K_M} \qquad (11.5)$$

When data for such an **uncompetitive inhibition** are graphed on a Lineweaver–Burk plot (Figure 11.7b) we find that both the apparent k_{cat} and K_M are changed by the addition of inhibitor. A series of parallel lines is obtained.

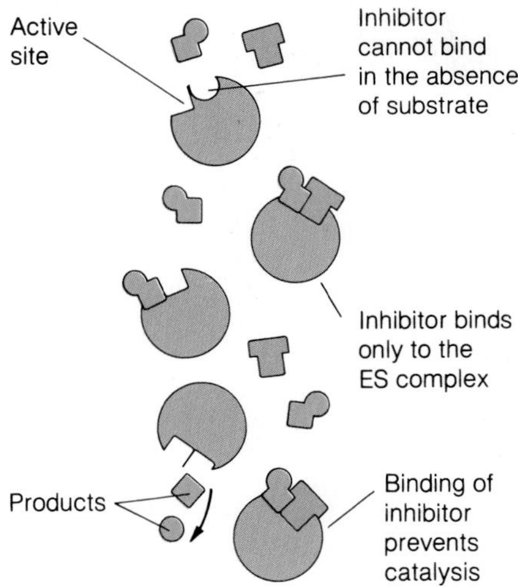

Figure 11.6
Uncompetitive inhibition. This form of inhibition differs from noncompetitive inhibition in that the inhibitor (red) can bind *only* if the ES complex is formed. Here we suggest this by a change in the structure of the inhibitor binding site on substrate binding.

Irreversible Inhibition

Some substances combine *covalently* with enzymes so as to inactivate them irreversibly. Almost all **irreversible enzyme inhibitors** are toxic substances,

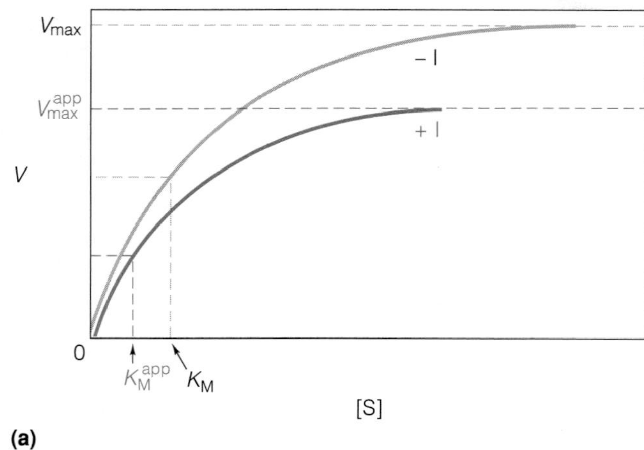

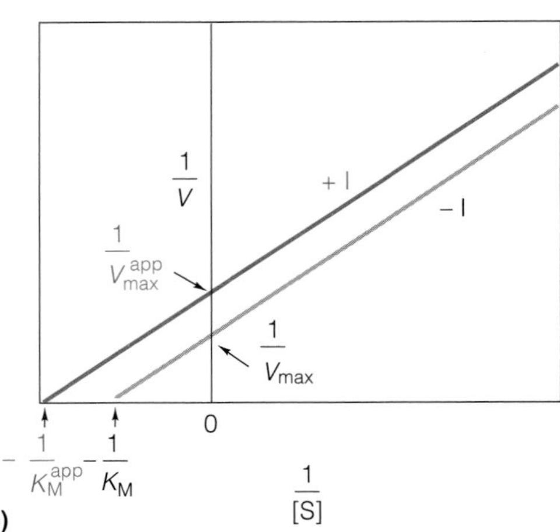

Figure 11.7
Effects of uncompetitive inhibition on enzyme kinetics. (**a**) Reaction velocity (V) versus [S]. Since the inhibitor effectively "siphons off" some of the enzyme substrate complex, K_M actually decreases in the presence of I. V_{max} decreases because the inhibited enzyme is less catalytically effective, as in noncompetitive inhibition. (**b**) Lineweaver–Burk plots.

either natural or man-made. Table 11.1 lists a number of these. In most cases, such substances react with some functional group in the active site to block the site from substrate or to leave it catalytically inactive.

A typical example of an irreversible competitive inhibitor is found in **diisopropyl fluorophosphate (DFP),**

$$(CH_3)_2CH{-}O{-}\underset{\underset{O}{\|}}{\overset{\overset{F}{|}}{P}}{-}O{-}CH(CH_3)_2$$

This compound reacts rapidly and irreversibly with serine hydroxyl groups to form the covalent *adduct* shown in Figure 11.8. Therefore, DFP acts as an irreversible inhibitor of enzymes that contain an essential serine in their active site. These include, among others, the serine proteases and the enzyme **acetylcholinesterase.** It is the inhibition of acetylcholinesterase that makes DFP such an exceedingly toxic substance to animals. The enzyme is essential for nerve conduction (see Chapter 29) and its inhibition causes rapid paralysis of vital functions. Many insecticides and nerve gases are potent acetylcholinesterase inhibitors. An example of each is given in Table 11.1.

Some of the most effective of the irreversible inhibitors bind strongly to the active site because they contain a group of atoms in a configuration resembling the transition state. Examples of such **transition state analogs** include DFP and sarin (Table 11.1), which have a tetrahedral structure surrounding the phosphorus atom very similar to the tetrahedral transition state in many hydrolytic enzymes.

In other cases, irreversible inhibitors may be extremely selective because they resemble the substrate sufficiently to be strongly bound in the active site, facilitating covalent adduct formation. An example is **tosyl-L-phenylalaninechloromethyl ketone (TPCK).** TPCK is an excellent inhibitor for chymotrypsin, because the phenyl group fits nicely into the active site pocket, positioning the chlorine to react with the imidazole ring of His 57. A large number of such specific irreversible inhibitors have been synthesized to aid in the analysis of enzyme mechanisms and to control enzyme activity. For example, a biochemist who is using chymotrypsin to hydrolyze a protein can stop the reaction instantly at any point by simply adding TPCK. Another use for such substances is to label active site residues of an enzyme specifically to aid in their identification. Such active site-directed irreversible

E — Active site serine
$$E{-}CH_2{-}O{-}\underset{\underset{O}{\|}}{P}(O{-}CH(CH_3)_2)_2$$

Figure 11.8
Adduct formed when diisopropyl fluorophosphate reacts with a serine group on a protein. The covalent bond renders the catalytically important serine ineffective in catalysis. The adduct also may block the site to substrate.

Table 11.1
Irreversible enzyme inhibitors

Name	Formula	Source	Mode of Action
Cyanide	CN^-	Bitter almonds	Reacts with enzyme metal ions (i.e., Fe, Zn, Cu)
Diisopropyl fluorophosphate (DFP)	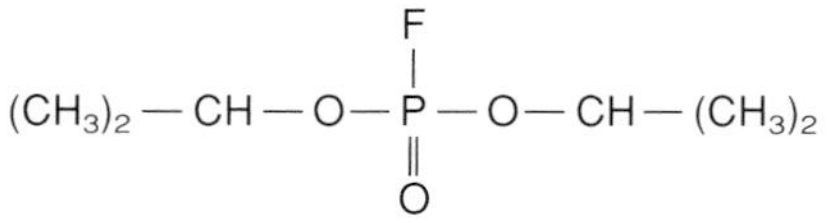 $(CH_3)_2-CH-O-P(F)(=O)-O-CH-(CH_3)_2$	Synthetic	Inhibits enzymes with active site serine
Sarin	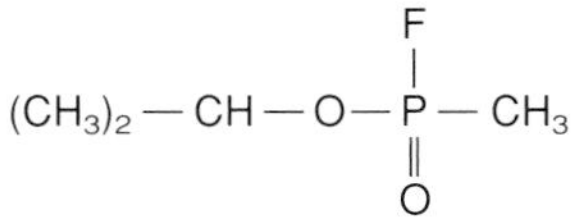 $(CH_3)_2-CH-O-P(F)(=O)-CH_3$	Synthetic (nerve gas)	Like DFP
N-tosyl-L-phenyl-alaninechloro-methyl ketone (TPCK)	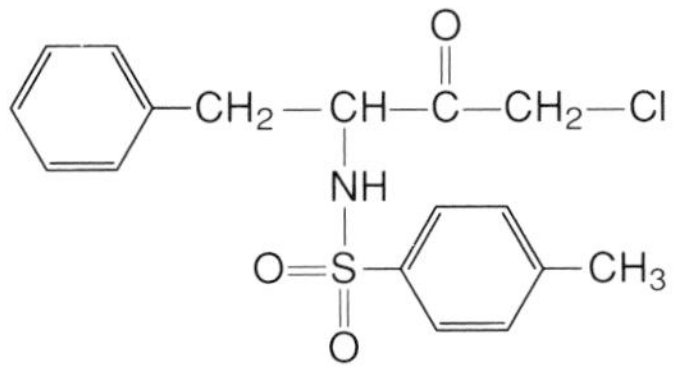 $C_6H_5-CH_2-CH(NH-SO_2-C_6H_4-CH_3)-C(=O)-CH_2-Cl$	Synthetic	Reacts with His 57 of chymotrypsin
Physostigmine	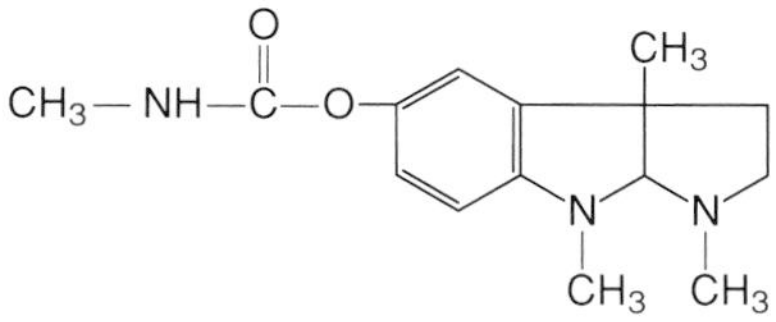 $CH_3-NH-C(=O)-O-$ (tricyclic ring system with CH_3, two $N-CH_3$)	Calabar beans	Forms acyl derivative with acetylcholinesterase, other enzymes
Parathion	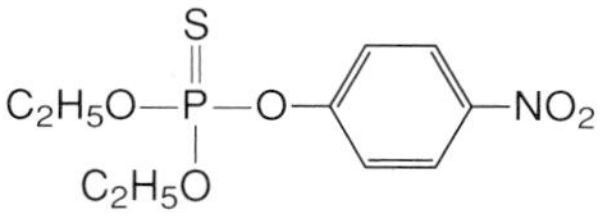 $C_2H_5O-P(=S)(OC_2H_5)-O-C_6H_4-NO_2$	Synthetic (insecticide)	Acetylcholinesterase inhibition
Penicillin	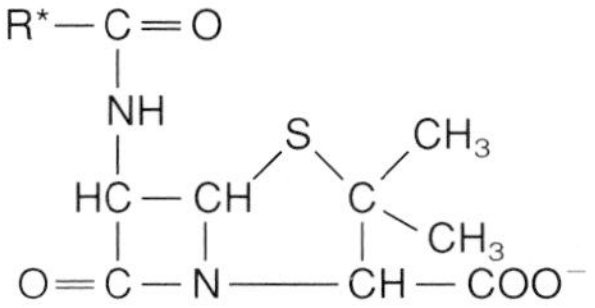 $R^*-C=O$; NH; $HC-CH$, S, $C(CH_3)_2$; $O=C-N-CH-COO^-$	From *Penicillium* fungus	Inhibits enzymes in bacterial cell wall synthesis

R^* = variable group; differs on different penicillins.

inhibitors are sometimes called **affinity labels.** In some cases an affinity label is unreactive until it is acted on by the enzyme, at which point it binds irreversibly. These are called **suicide inhibitors,** because the enzyme "kills" itself.

Many natural toxins are irreversible enzyme inhibitors. The alkaloid **physostigmine,** which is contained in calabar beans, is toxic because it is a potent inhibitor of acetylcholinesterase. The *penicillin* antibiotics also act as irreversible inhibitors of serine-containing enzymes used in bacterial cell wall synthesis (see Chapter 8).

Allosteric Regulation

Regulation of enzyme activity is, we have argued, essential to cellular function. Some of this occurs through direct interaction of the substrates and products of each enzyme-catalyzed reaction with the enzyme itself, and is referred to as **substrate level control.** As our analysis of kinetics has shown, the higher substrate levels are, the more rapidly reaction occurs. Conversely, high levels of product, which can also bind to the enzyme, tend to inhibit the conversion of substrate to product. As an example, consider the first step in glycolysis (Chapter 13)—the phosphorylation of glucose to yield glucose-6-phosphate. The enzyme hexokinase, which catalyzes this reaction, is inhibited by its product, glucose-6-phosphate. If glycolysis or glycogen formation is blocked for any reason, glucose-6-phosphate will accumulate; this will inhibit hexokinase and slow down further entry of glucose into the pathway.

However, substrate level control is not sufficient for the regulation of many metabolic pathways. In many instances, it is essential to have an enzyme regulated by some substance quite different from the substrate or immediate product. This requires quite different mechanisms.

Feedback Regulation

We have emphasized that most metabolic pathways resemble assembly lines. The simplest common pattern is like this:

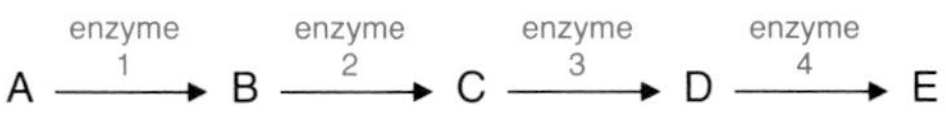

$$A \xrightarrow[1]{\text{enzyme}} B \xrightarrow[2]{\text{enzyme}} C \xrightarrow[3]{\text{enzyme}} D \xrightarrow[4]{\text{enzyme}} E$$

where A is the initial reactant or raw material; B, C, and D are intermediate products; and E is the final product.

The final product of this pathway will probably be used in some other pathway. Similarly, the "raw material" may also participate in some other set of processes. Suppose the utilization of E suddenly slows down. If everything kept going as it was, E would accumulate, and at the same time consumption of A would continue. But this is inefficient. A modern, automated assembly line would handle the problem by keeping an inventory on E and, as E accumulated, sending a signal back to slow the line. The cell can control generation of the final product through activation (⊕) or inhibition (⊘) of a step in the pathway. The most efficient step to slow would be the *first* step—the conversion of A to B. So the A → B "machine" should be regulated by the amount of E.

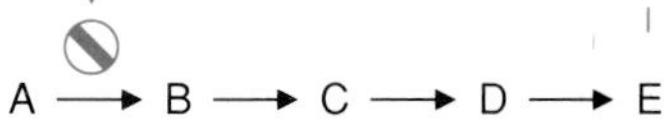

$$A \xrightarrow{\oslash} B \longrightarrow C \longrightarrow D \longrightarrow E$$

This process is called **feedback control** or, more precisely, **negative feedback control** because an *increase* in the concentration of E leads to a *decrease* in its rate of production. Note that by inhibiting the *first* step, we prevent both unwanted utilization of A and accumulation of E. Furthermore, since most biochemical processes are reversible to some extent, generation of a large quantity of E will tend to build up the concentration of intermediate products. A feedback control mechanism could prevent possible undesired effects on metabolism.

This is one way in which metabolic pathways work. Other situations require more complicated patterns. For example, consider a slightly more complex case, in which A is led into two pathways, which lead to two products needed in roughly equivalent amounts. Then a scheme like this emerges:

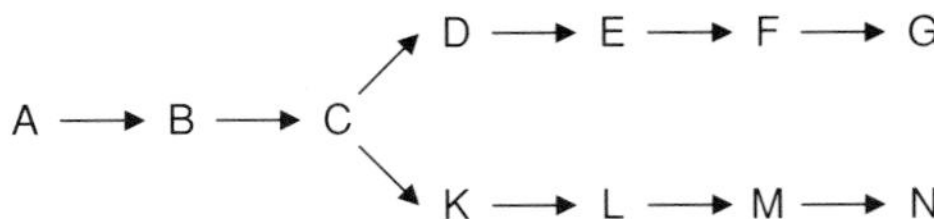

To control the pathways so that G and N keep in balance, high concentrations of G might *inhibit* the C → D enzyme and/or *activate* the C → K enzyme. Conversely, N might inhibit the C → K enzyme and/or activate the C → D enzyme. Finally, it might be useful to have both G and N *mutually* inhibit the A → B enzyme, to provide overall control. In a real situation, all or only some of these controls might operate. Such situations arise, for example, in the synthesis of the purine and pyrimidine monomers that go into making DNA, since approximately equal quantities of all four deoxyribonucleotides are required for DNA replication.

It is important to note that both inhibition *and* activation of enzymes are essential to regulated metabolism. Furthermore, control by the end products of pathways means that the necessary inhibitions and activations *must* be produced by molecules that come from far down the "assembly line" and therefore bear no resemblance to either the substrates or direct products of the enzymes to be regulated. To attain this kind of control, organisms have evolved a special class of enzymes, capable of **allosteric regulation.**

Allosteric Enzymes

Allosteric enzymes are invariably multisubunit proteins, with multiple active sites. They exhibit cooperativity in substrate binding (**homoallostery**) and regulation of their activity by other, effector molecules (**heteroallostery**).

We have already studied an example of allosteric control of protein function. Hemoglobin (Chapter 7) is a four-subunit protein that has four binding sites for its "substrate," oxygen. The binding of oxygen is cooperative and is influenced by other molecules and ions. The basic ideas that were presented in the analysis of hemoglobin function apply equally well to allosteric enzymes.

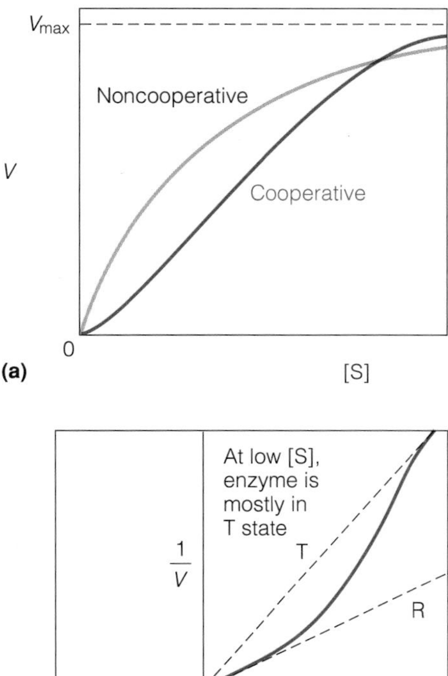

Figure 11.9
Effect of cooperative substrate binding on enzyme kinetics. (**a**) Comparison of V versus [S] curves for a noncooperative enzyme (green) and an allosteric enzyme with cooperative binding (red). Both are assumed to have the same V_{max}. Compare this with myoglobin and hemoglobin binding curves in Chapter 7. (**b**) Lineweaver–Burk plot corresponding to the cooperative binding curve shown above. The T state has a high K_M^T (binds S weakly). As more S is bound, T → R equilibrium shifts toward R, with stronger binding and a lower K_M^R.

HOMOALLOSTERY. Let us first consider the homoallosteric effects (cooperative substrate binding). In Chapter 7 we contrasted O_2 binding by the single-subunit protein myoglobin with that by the multisubunit hemoglobin. Myoglobin gave a hyperbolic binding curve; hemoglobin, with its cooperative binding, gave a sigmoidal curve. As Figure 11.9a shows, *exactly the same comparison* holds for the V versus [S] curves for single-site enzymes obeying Michaelis–Menten kinetics and multisite enzymes showing cooperative binding of substrate molecules. The same kind of reasoning applies: an enzyme that binds substrate cooperatively will behave, at low substrate concentration, as if it were poor at substrate binding (that is, as if it had a large K_M). But as the substrate levels are increased and more is bound, the enzyme becomes more and more effective, for it binds substrate more avidly in the last sites to be filled (see Figure 11.9b). We imagine this happening, as with hemoglobin, because the enzyme undergoes, as substrate is bound, a transition from a weak binding state (T state) to a strong binding state (R state). The kinds of models that have been used to describe O_2 binding by hemoglobin (i.e., the MWC model or KNF model, p. 226) can equally well account for the kinetics exhibited by enzymes that show cooperative substrate binding.

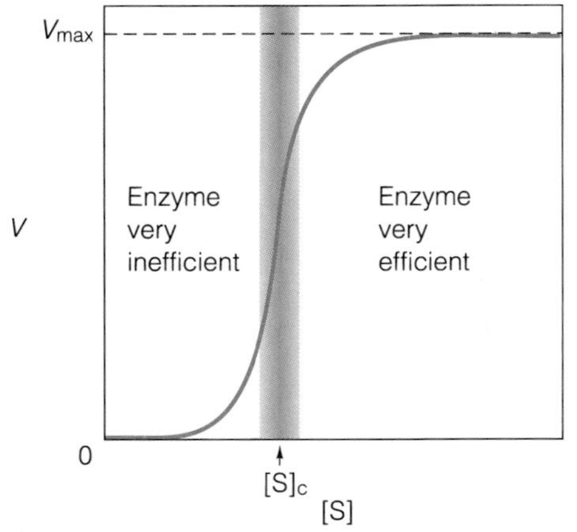

Figure 11.10
Effect of extreme homoallostery. The *V*–[S] curve is shown for a hypothetical enzyme with *very* high cooperativity in substrate binding. At concentrations below $[S]_c$, the enzyme is almost inactive; above this it is very active. Substrate can easily accumulate to the level $[S]_c$, but at higher concentrations it will be processed rapidly.

What physiological function do sigmoidal kinetics fulfill? In extreme cases, they can act so as to regulate substrate levels to quite constant values. Consider a substrate that is being supplied constantly by other reactions and is acted on by an enzyme that exhibits the extreme cooperativity shown in Figure 11.10. Substrate can easily accumulate up to the critical level $[S]_c$; the enzyme is essentially inactive at lower [S]. But any further increase leads to a greatly increased activity, so that the substrate level will be maintained near the value $[S]_c$. Although real allosteric enzymes rarely if ever exhibit curves as extremely sigmoidal as that in Figure 11.10, the principle remains: multisubunit enzymes may help to maintain the homeostasis of a dynamic system.

HETEROALLOSTERY. The major function of allosteric control is found in the role of **heteroallosteric effectors,** which may be either inhibitors or activators. These are the analogs, in enzyme kinetics, of the CO_2, DPG, and H^+ that so elegantly regulate O_2 binding by hemoglobin. The activation and inhibition of enzymes by allosteric effectors are the key to the kind of complex feedback control described at the beginning of this section. If an enzyme molecule can exist in two conformational states (T and R) that differ dramatically in the strength with which substrate is bound or in the catalytic rate, then its kinetics can be controlled by any other substance that, in binding to the protein, shifts the $T \rightleftharpoons R$ equilibrium. Allosteric *inhibitors* shift the equilibrium toward T, and *activators* shift it toward R (Figure 11.11). Some enzymes are regulated by multiple inhibitors and activators, allowing extremely subtle metabolic control.

Aspartate Carbamoyltransferase: Example of an Allosteric Enzyme

An excellent example of allosteric regulation is provided by the enzyme **aspartate carbamoyltransferase** (also known as aspartate transcarbamoylase, or ATCase), a key enzyme in pyrimidine synthesis. As can be seen from Figure 11.12, ATCase stands at a crossroads in biosynthetic pathways. Glutamine, glutamate, and aspartate are, of course, also used in protein synthesis; but once aspartate has been carbamoylated to form *N-carbamoyl-L-aspartate* (CAA) the molecule is committed to pyrimidine synthesis. In bacteria like *E. coli,* the activity of ATCase is regulated to respond to needs for pyrimidine for nucleotide synthesis. It is, as shown in Figure 11.13, inhibited by cytidine triphosphate (CTP) and activated by ATP. Both responses make physiological sense; when CTP levels are already high, more pyrimidines are not needed, whereas high ATP signals an energy-rich cell condition under which DNA and RNA synthesis will be active.

ATCase is, as you would expect, a multisubunit protein. Its quaternary structure has been examined in some detail and is depicted in Figure 11.14. There are six *catalytic* subunits, in two tiers of three, held together by six *regulatory* subunits. Pairs of regulatory subunits appear to connect catalytic subunits in the two tiers. The substrate binding sites are on the catalytic subunits. Each regulatory subunit has a site that can bind *either* CTP *or* ATP. These two molecules thus compete for the same sites, so that the activity of ATCase is regulated by the *ratio* of ATP to CTP in the cell.

As in the case of hemoglobin, the allosteric regulation of ATCase involves changes in the quaternary structure of the molecule. Conformations of the R and T states have been determined by x-ray diffraction. As Figures

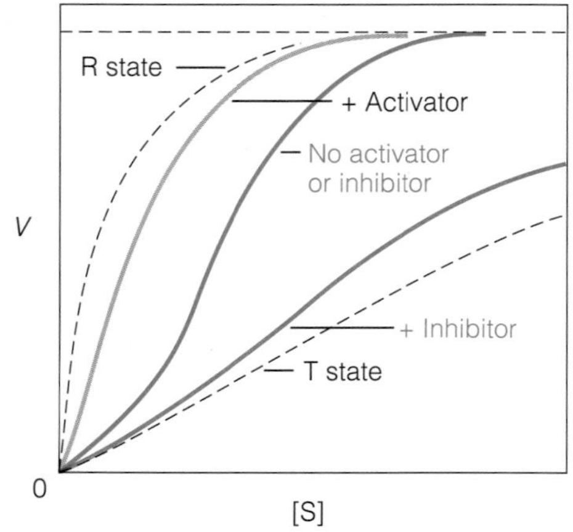

Figure 11.11
Heteroallosteric control of an enzyme. In the absence of activation or inhibitors, the *V* versus S curve has the sigmoidal shape shown in blue. Activators shift the system toward the R state; inhibitors stabilize the T state.

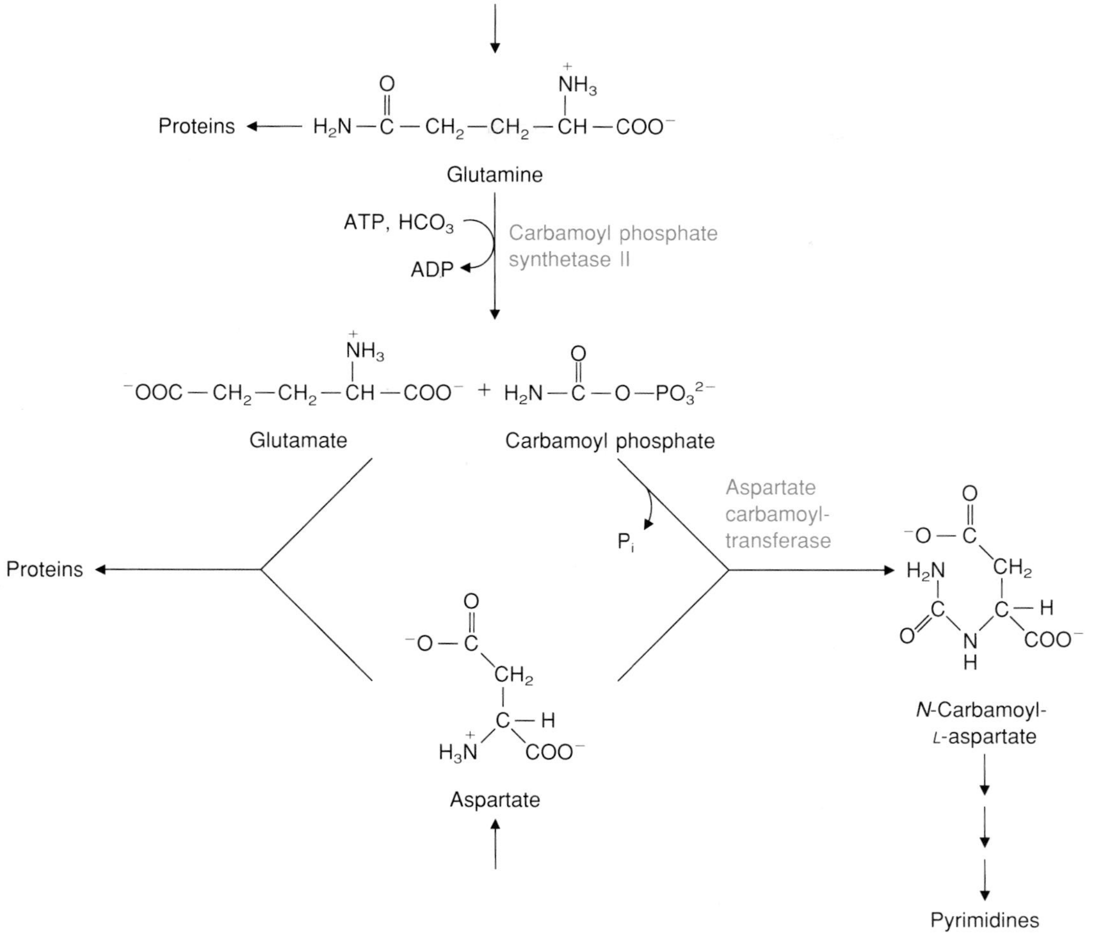

Proteins ← $H_2N{-}C(=O){-}CH_2{-}CH_2{-}CH(\overset{+}{N}H_3){-}COO^-$

Glutamine

ATP, HCO_3^- → ADP — Carbamoyl phosphate synthetase II

$^-OOC{-}CH_2{-}CH_2{-}CH(\overset{+}{N}H_3){-}COO^-$ + $H_2N{-}C(=O){-}O{-}PO_3^{2-}$

Glutamate — Carbamoyl phosphate

Proteins ←

Aspartate carbamoyltransferase, P_i

Aspartate → N-Carbamoyl-L-aspartate → → → Pyrimidines

Figure 11.12
Metabolic role of aspartate carbamoyltransferase. Formation of *N*-carbamoyl-L-aspartate from carbamoyl phosphate and aspartate is the first committed step in a series of reactions that leads to synthesis of pyrimidine nucleotides. Control at or near this point is essential. In prokaryotes, it is the aspartate carbamoyltransferase that is regulated; in most eukaryotes, regulation is on the carbamoyl-phosphate synthetase II.

11.15 and 11.16 show, a major rearrangement of subunit positions occurs in the T → R transition.

Virtually every metabolic pathway we shall encounter in the following chapters is subject to complex feedback control, and in almost all cases multisubunit, allosteric enzymes are employed. The pattern of control, even in a given pathway, is not the same in every organism. To take a relevant example, while ATCase is the major control point in the pyrimidine pathway in bacteria, eukaryotes regulate at the preceding step—the synthesis of carbamoyl phosphate (see Figure 11.12). In mammals, the **carbamoyl-phosphate synthetase II** is inhibited by UDP, UTP, CTP, dUDP, and UDP-glucose. These compounds all inhibit binding of the ATP substrate. In addition, glycine acts as a competitive inhibitor for glutamine.

It should be clear at this point that organisms can regulate metabolism in complex and subtle ways through allosteric enzymes. But this kind of regulation is not sufficient for all needs. We turn now to an entirely different kind of mechanism.

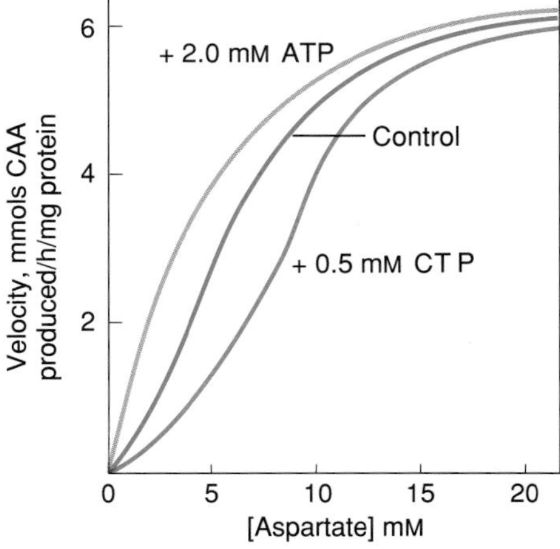

Figure 11.13
Regulation of aspartate carbamoyltransferase. ATP is an activator, CTP an inhibitor. The curve marked "control" shows the behavior in the absence of both. CAA, *N*-carbamoyl aspartate, is the product of the reaction.

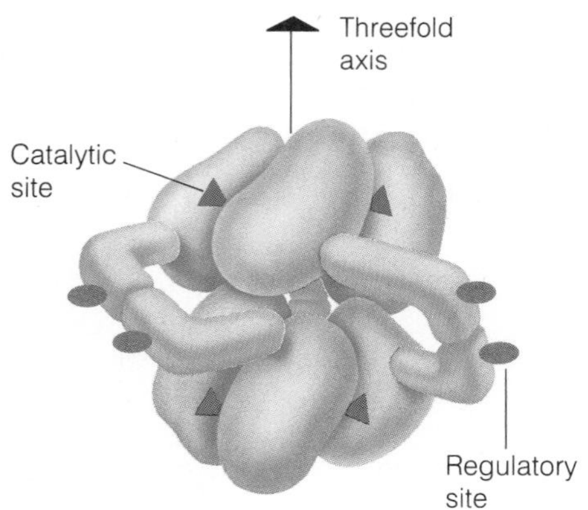

Figure 11.14
Structure of aspartate carbamoyltransferase. In this schematic view of the enzyme, the six catalytic subunits are shown in green, the six regulatory subunits in orange. Six catalytic sites (gray triangles) lie in or near the grooves between the catalytic subunits. Regulatory sites (red circles) lie on the outer surfaces of the regulatory subunits. The molecule has one threefold axis and three twofold axes (D_3 symmetry). This is a "side" view of the molecule with the threefold axis in the plane of the paper.

Covalent Modifications Used to Regulate Enzyme Activity

In the factory analogy, allosteric regulation can be thought of as the feedback control of continuously running machines. But any large factory also has machinery that is used only from time to time and is left on standby until needed. The same is true for the cell. In this section, we discuss enzymes that are wholly inactive until they are changed by a **covalent modification** and then begin to function. In some cases the modification acts in the opposite direction, to inactivate otherwise active enzymes. Some such modifications can be reversed; others cannot.

Glycogen Phosphorylase: Activation by Phosphorylation

An excellent example of regulation by covalent modification is found in the enzyme **glycogen phosphorylase** of skeletal muscle. It acts to liberate glucose residues from their storage in glycogen by cleaving and phosphorylating a residue from the nonreducing end of glycogen chains in the process of **glycogenolysis** (Figure 11.17; Chapter 13). The glucose-1-phosphate that is released is available as an energy source for the cell.

Clearly, it would not be advantageous for a muscle cell to have phosphorylase active at all times. A resting muscle cell can be energized adequately by the supply of glucose from the blood, so even a low level of phosphorylase activity would be wasting glycogen reserves. Yet the reaction shown in Figure 11.17 must be available on demand. The problem is solved by having muscle phosphorylase in two forms, called phosphorylase *a* and *b*. **Phosphorylase *b*** exists in resting cells as an inactive dimer. Another enzyme, **phosphorylase kinase,** can catalyze the addition of a phosphate to serine 14, which has the effect of "locking" the N-terminal portion of each

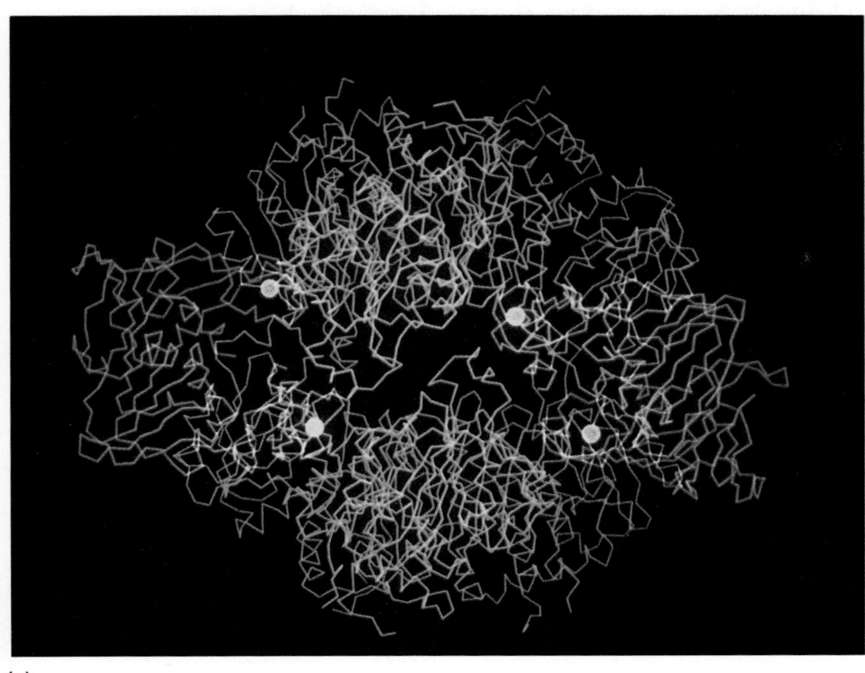

(a)

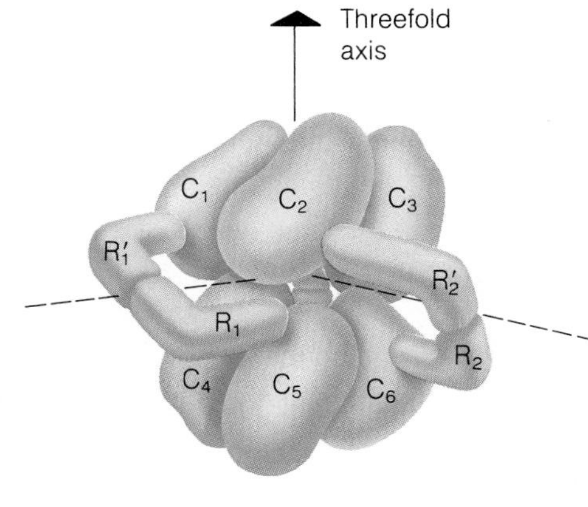

(a) ATCase: T state

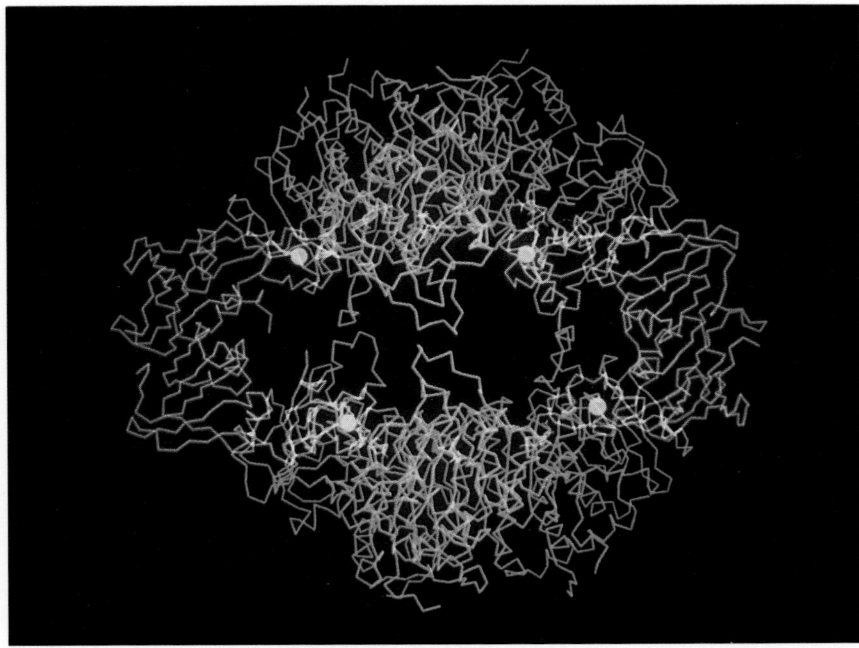

(b)

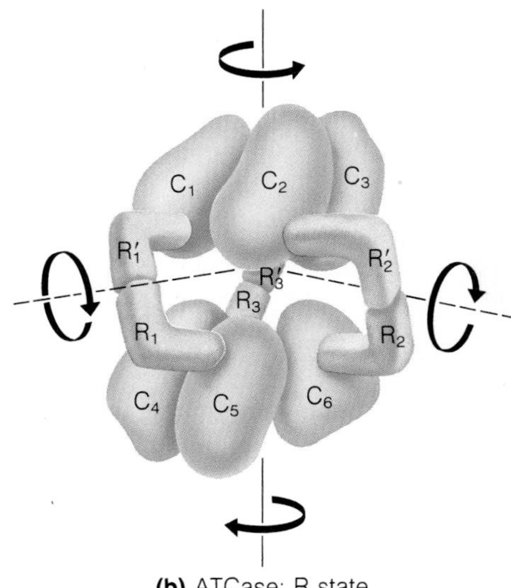

(b) ATCase: R state

Figure 11.15
The T → R transition in aspartate carbamoyltransferase. Structures in the T (**a**) and R (**b**) states as determined by x-ray diffraction. Color lines show the course of the peptide chain in the catalytic subunits (green) and the regulatory dimers (orange).

Figure 11.16
Schematic views of the ATCase T → R transition. The transition involves a rotation of the regulatory subunits, which pushes the two tiers of catalytic subunits apart and rotates them slightly about the threefold axis. (**a**) T state; (**b**) R state, showing direction of subunit rotation.

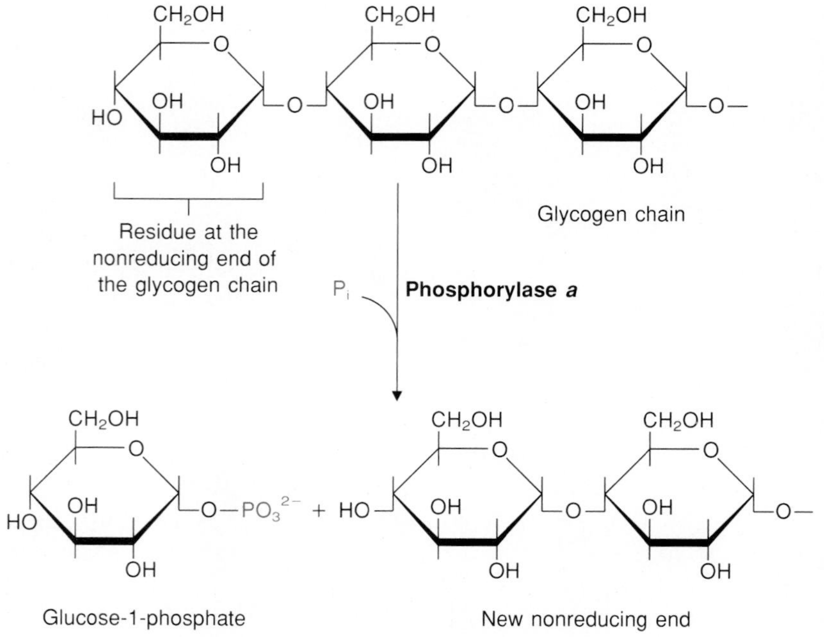

Figure 11.17
The reaction catalyzed by glycogen phosphorylase. Long glycogen chains are stored in muscle cells as energy reserves. When phosphorylase is activated, it cleaves and phosphorylates, releasing glucose-1-phosphate for energy production.

CH_2OH CH_2OH CH_2OH

Residue at the nonreducing end of the glycogen chain

Glycogen chain

P_i **Phosphorylase *a***

$-O-PO_3^{2-}$ + HO–

Glucose-1-phosphate

New nonreducing end

chain, which lacks defined structure in the *b* form. This change is communicated to the rest of the enzyme, resulting in the formation of enzymatically active **phosphorylase *a*** (Figure 11.18). When glycogen breakdown is no longer required, a **phosphatase** removes the phosphate, and phosphorylase reverts to its inactive *b* form.

The activation of phosphorylase *b* is actually the last step in a series of reactions that are initiated when a cell surface receives the hormone epinephrine or glucagon. In a series of reactions known as a **regulatory cascade** (to be described in Chapter 12) the signal from a single hormone molecule is

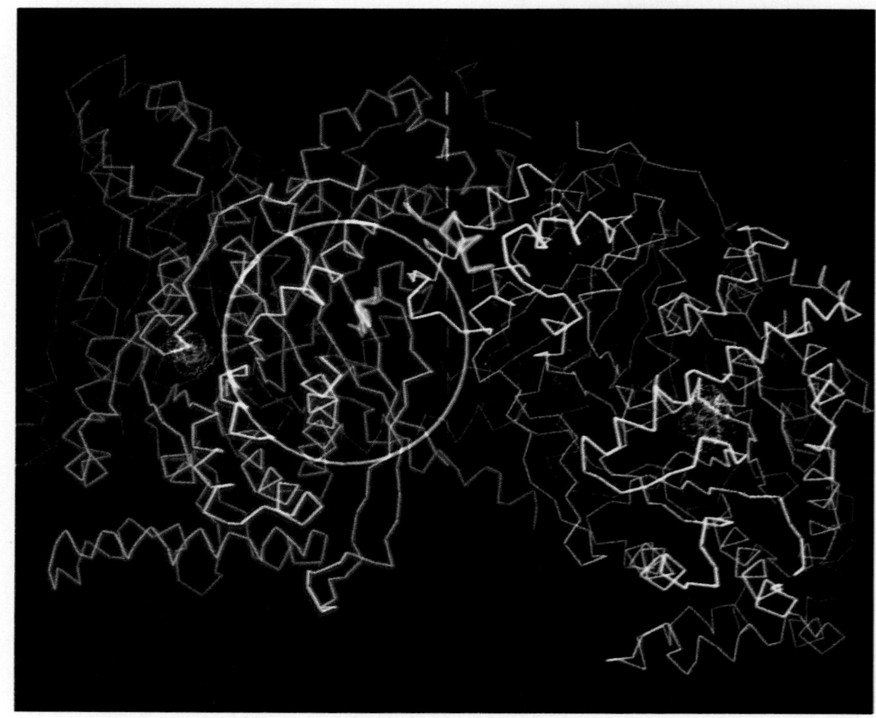

Figure 11.18
Three-dimensional structure of rabbit muscle glycogen phosphorylase *a*, as determined by x-ray crystallography. The red clouds depict van der Waals radii of two glucose molecules bound in catalytic sites of the dimeric protein. Two molecules of bound 5′-AMP are shown, one in red at the upper left region of the subunit depicted in green. The orange circle encloses the AMP-binding locus of the blue subunit, with the AMP shown also in orange. Binding of AMP causes a large conformational change, which is responsible for the activation of phosphorylase *b*. Amino acid residues that change their main-chain position by more than 0.4 Å on AMP binding are colored yellow in the green subunit and magenta in the blue subunit.

amplified to convert many molecules of phosphorylase *b* to phosphorylase *a*. Calcium ions, which are released in a muscle cell when it is stimulated to contract, augment this process. Thus, the muscle can respond when extra energy is needed.

But the regulation of phosphorylase activity is even more sophisticated than this. The dimeric phosphorylase is an allosteric enzyme, inhibited by glucose and ATP and activated by AMP. This, too, makes metabolic sense. If a signal arrives that activates the phosphorylase system in a cell that *already* has adequate glucose supplies, the enzyme is ready but is inhibited until those supplies are depleted. On the other hand, high AMP levels signify a low energy charge in the cell and the need for mobilization of energy reserves. Even phosphorylase *b* is potentially active, but is strongly inhibited by glucose. At low glucose levels it will initiate glycogen breakdown even in the absence of the hormonal signal necessary to convert it to the *a* form.

This behavior of phosphorylase illustrates another important aspect of enzyme regulation: very often two or more levels of control are "overlaid" on a particular enzyme, to allow appropriate response to complicated situations. In the case of phosphorylase, it works as follows: The hormones epinephrine and glucagon are secreted and circulated to cells when the organism as a *whole* needs a burst of energy. They effect a preparation for glycogen breakdown in all muscle cells. But cells well supplied with instantly available energy (high glucose or high ATP/AMP ratio) need not respond by breaking down glycogen at high rates until these energy reserves are depleted. Conversely, individual cells with low energy supplies can draw on glycogen by release of the inhibition of the *b* form.

The regulation of enzymes by phosphorylation is very common, and we shall encounter many more examples in later chapters.

Pancreatic Proteases: Activation by Cleavage

An entirely different kind of covalent enzyme activation is exemplified by the **pancreatic proteases.** These include a number of enzymes—for example, trypsin, chymotrypsin, elastase, and carboxypeptidase—some of which we have already discussed. All are synthesized in the pancreas. They are secreted through the pancreatic duct into the duodenum of the small intestine in response to a hormone signal generated when food passes from the stomach. They are not, however, synthesized in their final, active form, for a battery of potent proteases free in the cells of the pancreas would obviously present serious problems. Rather, they are made as slightly longer, catalytically inactive molecules, called **zymogens.** The names given to the zymogens of the enzymes mentioned above are **trypsinogen, chymotrypsinogen, proelastase,** and **procarboxypeptidase.** The zymogens must themselves be cleaved proteolytically to yield the active enzymes. This process is diagrammed in Figure 11.19.

The first step is the activation of trypsin in the duodenum. A hexapeptide is removed from the N-terminal end of trypsinogen by **enteropeptidase,** a protease secreted by duodenal cells. This action yields the active trypsin, which then activates the other zymogens by specific proteolytic cleavages. In fact, once some active trypsin is present, it will activate other trypsinogen molecules to make more trypsin; its activation is autocatalytic. This is an example of the kind of cascade process frequently observed when enzymes are activated by covalent modification. The production of just a few trypsin molecules leads quickly to many more, for each enzyme molecule, when activated, can process many more every minute. These can in turn activate

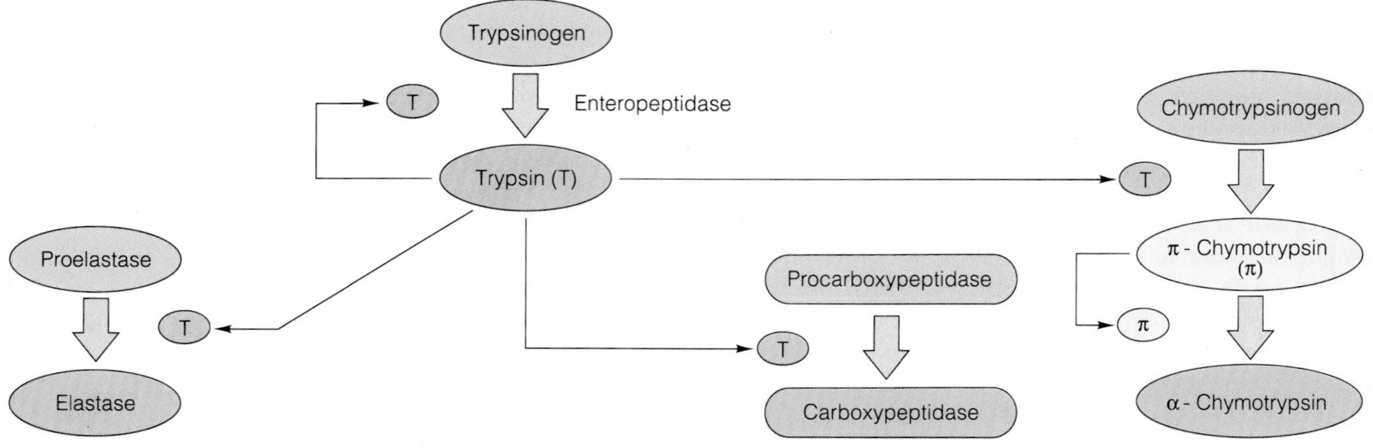

Figure 11.19
Schematic view of pancreatic zymogen activation. Zymogens are in red, active proteases in yellow or green.

the other zymogens. A most spectacular example of such a process occurs in blood clotting, which we describe below.

The activation of chymotrypsinogen to chymotrypsin is one of the most complex and best studied examples of proteolytic activation of an enzyme; it is illustrated in Figure 11.20. In the first step trypsin cleaves the bond between arginine 15 and isoleucine 16. The N-terminal peptide remains attached to the rest of the molecule because of the disulfide bond between residues 1 and 122. The product, called **π-chymotrypsin,** is a fully active enzyme.

Just how the cleavage of one peptide bond transforms a wholly inactive protein into a fully active one can now be understood as a result of detailed x-ray diffraction studies of the zymogen and the enzyme. When the peptide bond is cleaved, a new, positively charged N-terminal residue is created at Ile 16. This residue shifts its position and forms a salt bridge with Asp 194, the neighbor of the active site Ser 195 (see p. 357). This in turn triggers further conformational changes in the neighborhood of the active site. These changes include the correct modeling of the active site pocket and movement of main chain amino groups of residues 193 and 195 to where they can hydrogen-bond to the substrate oxyanion in the tetrahedral transition state. Thus, both the binding pocket and the catalytic site are formed correctly only after the peptide bond between Arg 15 and Ile 16 has been cleaved.

Chymotrypsin is not quite "finished" by this cleavage. There are more autocatalytic cleavages that remove residues 14–15 and 147–148 from the molecule, to produce the final **α-chymotrypsin** which is the principal form found in the digestive tract. The functional significance (if any) of these secondary cleavages is unknown.

This battery of enzymes, trypsin, chymotrypsin, elastase, and carboxypeptidase, together with the pepsin of the stomach and other proteases secreted by the intestinal wall cells, is capable of ultimately digesting most ingested proteins into free amino acids, which can be absorbed by the intestinal epithelium. The enzymes themselves are continually subjected to mutual and autodigestion, so that high levels of these enzymes never accumulate in the intestine.

The generation of such a potent group of proteases is a source of danger to the pancreas, even if they are made in their zymogen forms. Since trypsin activation can be autocatalytic, the presence of even a single active trypsin molecule could set the whole chain in motion prematurely. Therefore the pancreas protects itself further by synthesizing a small protein called **pancreatic trypsin inhibitor.** This competitive inhibitor binds so tightly to the active site of trypsin that it effectively inactivates it even at very low concentration. The bonding between trypsin and its inhibitor is among the strongest noncovalent associations known in biochemistry; the equilibrium constant is about 10^{13} $(\text{mol/L})^{-1}$, corresponding to $\Delta G° = -75$ kJ/mol. Only a tiny amount of trypsin inhibitor is present—far less than can inhibit all of the potential trypsin in the pancreas. Thus, only a fraction of the trypsin generated in the duodenum is inhibited, and the rest can be activated. Since protection is limited, zymogen activation can sometimes be triggered in the pancreas—for example, if the pancreatic duct is blocked. The active enzymes then begin to digest the pancreatic tissue itself. This condition, called *acute pancreatitis*, is extremely painful and sometimes fatal.

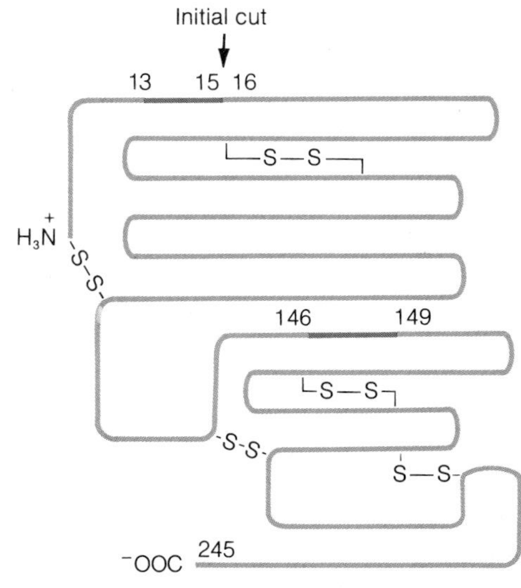

Figure 11.20
Activation of chymotrypsinogen. A schematic view of the molecule. Cleavage between amino acids 15 and 16 (arrow) results in the formation of π-chymotrypsin (green and gray regions). Subsequent removal of the segments shown in gray yields α-chymotrypsin. Note that disulfide bonds hold the structure together despite these cleavages.

Blood Clotting

Activation of zymogens is the key to another biologically important process—the clotting of vertebrate blood. If a blood clot is examined in the electron microscope, it is found to be composed of striated fibers of a protein called *fibrin* (Figure 11.21a). The fibrin monomers are elongated molecules, about 46 nm long, that stick together in an staggered array as shown in Figure 11.21b and c. The fibrin monomers are derived from a precursor, **fibrinogen,** by proteolytic cleavages that release small **fibrinopeptides.** Loss of these peptides uncovers positions at which the fibrin molecules can stick together. After the clot is formed, it is further stabilized by covalent cross-links between glutamine and lysine residues.

The proteolysis of fibrinogen to fibrin is catalyzed by the serine protease **thrombin.** Thrombin has sequence and structural similarities to trypsin, but, as a protease with a very specific function, it cleaves only a few types of bonds, mainly (Arg–Gly). Thrombin itself is produced from **prothrombin** by another specific protease; in fact, as Figure 11.22 shows, there is a whole cascade of proteolytic activation reactions that lead ultimately to the formation of a fibrin clot. Involved are a series of proteases referred to as *factors.* In damaged tissues, the proteins **kininogen** and **kallikrein** activate factor XII (also called **Hageman factor**), which in turn activates factor XI—and the cascade of reactions proceeds as shown. This is called the **intrinsic pathway.** Alternatively, damage to blood vessels leads to the release of **tissue factor** and activation of factor VII, starting the **extrinsic pathway.** The two pathways merge in the activation of factor X, which will proteolyze and thereby activate prothrombin.

Some of the activation steps require auxiliary proteins. For example, activation of factor X in the intrinsic pathway by factor IX (**Christmas factor**) requires a 330-kDa protein called **antihemophilic factor** (factor VIII). It is the partial or complete absence of factor VIII activity that is the cause of classic **hemophilia.** The gene for factor VIII is carried on the X chromosome, so women, who have two copies of this chromosome, can be heterozygous carriers of the trait but will exhibit the symptoms only if they are homozygous. However, a male descendent who receives on his single X chromosome the damaged copy of the factor VIII gene will experience more or less severe difficulty in blood clotting. The condition can now be treated by frequent transfusions of a blood serum fraction concentrated in factor VIII. The gene for this protein has recently been cloned and expressed in

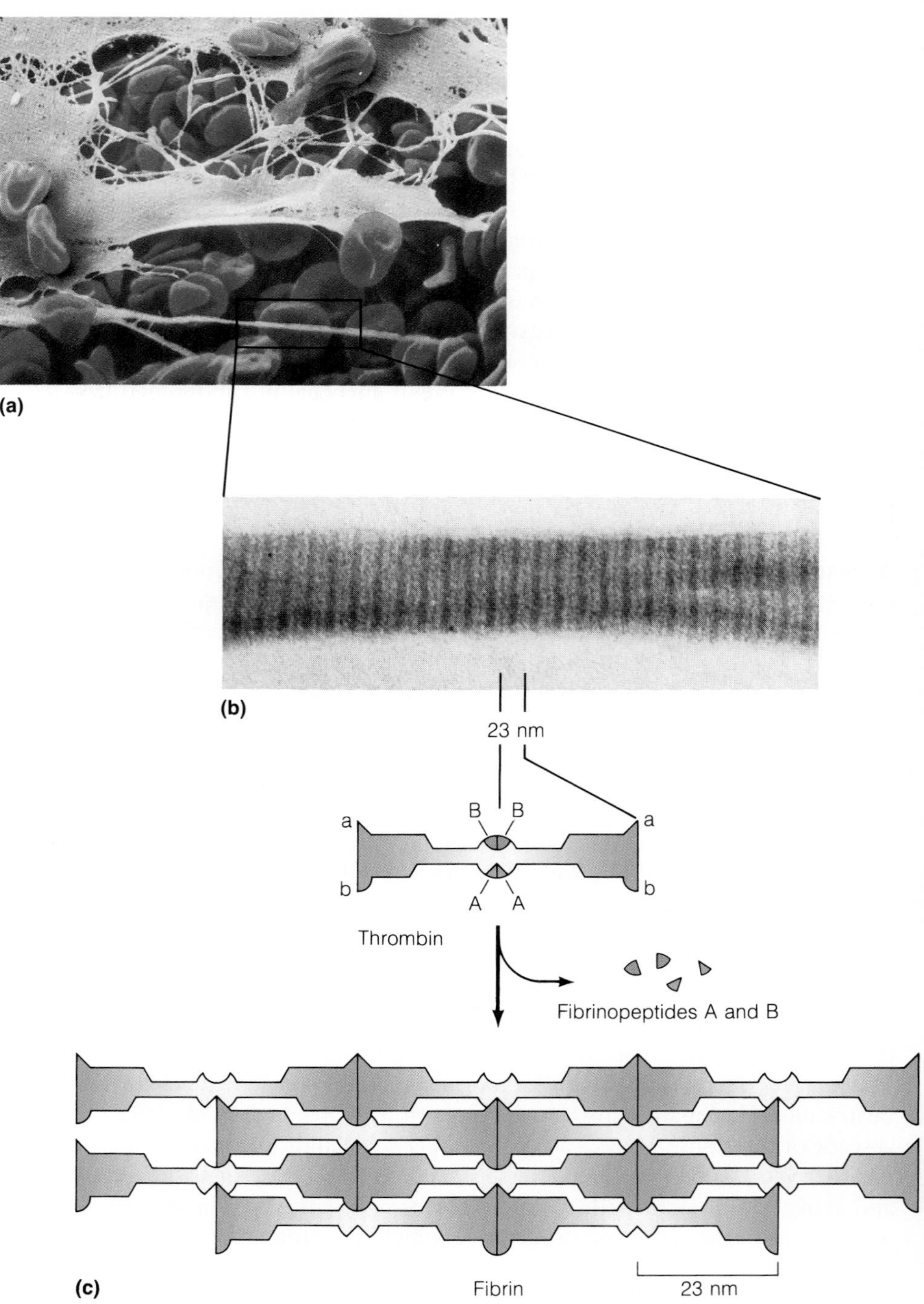

Figure 11.21
Formation of a blood clot. (**a**) Red blood cells enmeshed in the insoluble strands of a fibrin clot. (**b**) Electron micrograph of part of a fibrin fiber. (**c**) Schematic view of how fibrin monomers are thought to associate to form a clot fiber. Removal of fibrinopeptides A and B from fibrinogen by thrombin makes sites accessible for association with complementary sites a and b on adjacent monomers. The molecules are believed to overlap as shown, since the spacing seen in the fibers (see **b** above) is 23 nm—exactly half the length of the fibrinogen molecule.

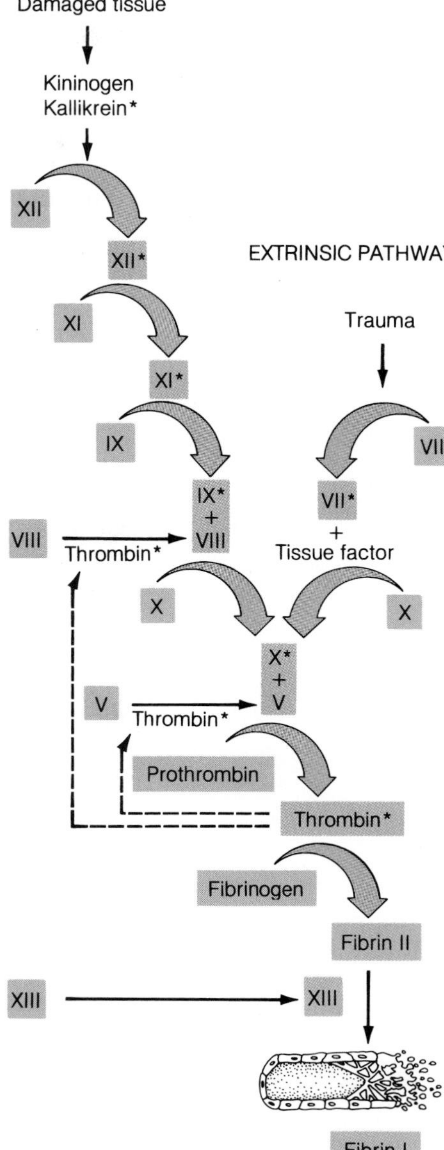

Factor Key

II	Prothrombin
V	Proaccelerin
VII	Proconvectin
VIII	Antihemophilic Factor
IX	Christmas Factor
X	Stuart Factor
XI	Thromboplastin Antecedent
XII	Hageman Factor
XIII	Fibrin Stabilizing Factor

Figure 11.22
The cascade process in blood clotting. Each factor in the pathway can exist in an inactive form (red) or an active form (green). The cascade of proteolytic activations can start from exposure of blood at damaged tissue surfaces (intrinsic pathway) or from internal trauma to blood vessels (extrinsic pathway). The common result is activation of fibrinogen to clotting fibrin. Auxiliary factors that aid some steps are also shown. Asterisk (*) denotes serine proteases.

bacteria, and the availability of this synthetic factor VIII may allow such patients to avoid the dangers of regular transfusions.

Regulation by Control of Enzyme Synthesis and Degradation

All the forms of enzyme regulation we have described so far have one feature in common: they modify the activity of enzymes that are already present in the cell or tissue. Organisms have still another way in which enzymatic activity can be regulated: they can control the *synthesis* and *degradation* of specific enzymes in response to changing needs.

Regulation of Enzyme Synthesis

Regulation of synthesis is usually employed only for enzymes that are needed in special circumstances or at particular stages of organismal development. In multicellular organisms, these are frequently enzymes needed only in one particular tissue or cell type. The enzymes that perform general "housekeeping" functions, essential to every cell at all times, are never regulated in this manner.

There are two levels at which enzyme synthesis might be controlled—transcription or translation—but the former is used much more. Two specific examples, one from a prokaryotic organism and one from a eukaryote, illustrate the general features of such regulation.

β-GALACTOSIDASE SYNTHESIS IN *E. COLI*. The bacterium *E. coli* usually does not have the opportunity to utilize the disaccharide lactose as an energy source. Thus, it does not have a continual requirement for enzymes needed for lactose metabolism. But if it finds itself in a lactose-rich medium, it is prepared. The *E. coli* genome carries the genes to code for special proteins required to deal with the metabolism of this sugar. These include a **galactose permease,** which is not an enzyme but a membrane transport protein specific for facilitating transport of lactose into the bacterium, and the enzyme **β-galactosidase,** which catalyzes the breakdown of lactose into glucose and galactose (see Chapter 8). Both of these monosaccharide products can readily be metabolized by *E. coli*.

The messenger RNAs for the permease and β-galactosidase genes, and the protein gene products themselves, are present in very low levels in bacteria growing in a lactose-free medium. There are, it is estimated, only a few β-galactosidase molecules per cell. But if lactose is introduced into the medium, the small amount that can enter the cell via the available permease triggers transcription of the gene. Just how this happens is a fascinating story, which we will discuss in Chapter 26. The point is that within a few minutes there are enough permease molecules to ensure a rapid influx of lactose and several thousand β-galactosidase molecules available to cleave that lactose. If lactose is removed, the permease and β-galactosidase genes are quickly shut down and return to their quiescent state. Many other examples of similar **enzyme induction** are found in microbiology.

EGG PROTEIN SYNTHESIS IN CHICKEN OVIDUCT. The oviduct of chickens is an excellent example of a specialized eukaryotic tissue. As the chick matures, certain cells in the lining of the oviduct become specially differentiated into what are known as **tubular cells.** These cells secrete most of the proteins that form the egg white. As eggs in sexually mature chickens pass through the oviduct, they are coated with a layer of these proteins. Four proteins—*ovalbumin, conalbumin, ovomucoid,* and *lysozyme*—make up about 90% of the egg white and are the major special protein products of the tubular cells. These proteins are not synthesized in other chicken cells, even though every cell in the animal carries the appropriate genes.

The tubular cells are abundant in the oviducts of chickens even at the age of a few weeks, but synthesis of egg white proteins is not detectable at this stage. Normally, transcription of these genes commences only at sexual maturity, in response to the sex hormone *estradiol*. However, synthesis can be stimulated even at a few weeks of age by hormone administration. The hormone stimulates transcription of the egg white protein genes *only* in the tubular oviduct cells, even though many cells in the chicken are exposed to this substance.

This kind of regulation of enzyme synthesis is especially important in eukaryotic organisms, where different cells, carrying out specialized functions, require special enzymes.

Regulation by Enzyme Degradation

An enzyme produced at a particular developmental stage or in response to a particular environmental situation may not be needed later. In many cases, its continued presence would be deleterious to the organism. If enzymes were "immortal," serious problems would be created by their persistent presence.

Individual enzyme molecules have very different lifetimes; some last in the cell for many days, others for only minutes or less. We now know that specific mechanisms exist for the proteolytic degradation of cellular proteins. The mechanisms of this process and the way in which some enzymes are targeted for rapid degradation are described in Chapters 20 and 28. However, two points are important to our present discussion.

First, the more rapidly degraded enzymes are most often those that occupy key control points in metabolism. This means that the level of these enzymes is under rapid control. If not needed, they rapidly disappear, but they can be resynthesized quickly when required. A long-lived protein, on the other hand, is much more sluggish in response.

Second, damaged or aberrant proteins are selected for swift degradation. A faulty enzyme in a major metabolic pathway is a malfunctioning machine. At best it is worthless, and it may be deleterious. So these are selectively destroyed.

A summary of the many ways in which enzyme activity can be regulated is depicted in Figure 11.23. As we have emphasized, different kinds of control are employed for different purposes, yet the activity of any particular enzyme may be regulated at several levels.

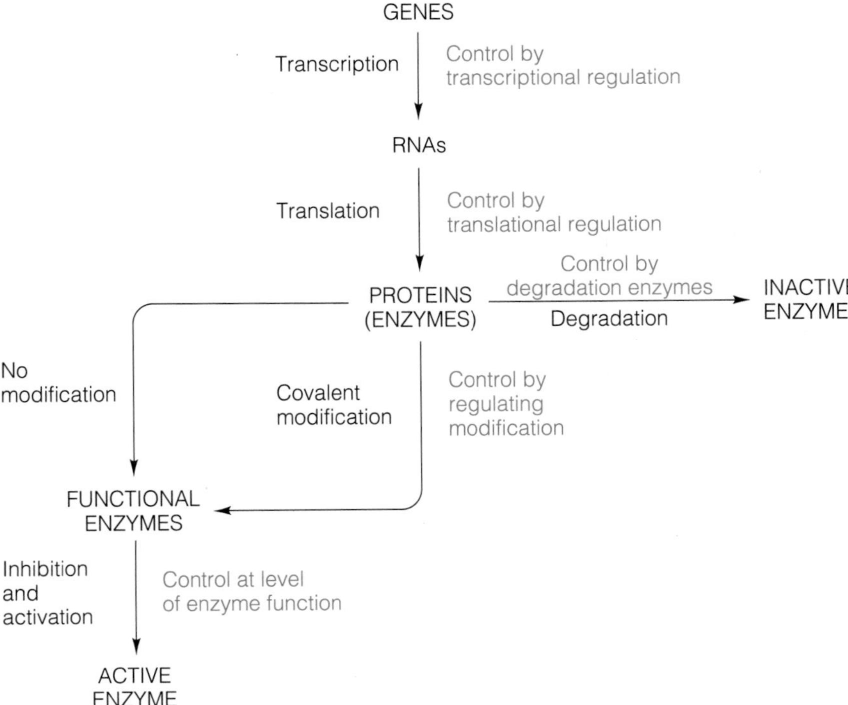

Figure 11.23
Summary of the different levels at which enzymatic activity can be controlled.

REFERENCES

General

Fersht, A. (1985) *Enzyme Structure and Mechanism,* 2nd ed. W. H. Freeman, New York. Contains excellent chapters on all the topics we have considered here.

Segel, I. H. (1975) *Enzyme Kinetics.* Wiley, New York.

Active Site-Directed Inhibitors

Abeles, R. H. (1983) Suicide enzyme inactivators. *Chem. Eng. News* 61(38):48–56. A brief, clear discussion.

Rando, R. R. (1975) Mechanisms of action of naturally occurring irreversible enzyme inhibitors. *Acc. Chem. Res.* 8:281–288.

Allosteric Regulation

Krause, K. L., K. W. Volz, and W. N. Lipscomb (1985) Structure at 2.9 Å resolution of aspartate carbamoyltransferase complexed with the bisubstrate analog *N*-(phosphonacetyl)-L-aspartate. *Proc. Natl. Acad. Sci. USA* 82:1643–1647. Detailed structure of this protein, which has become a model for allosteric enzymes.

Monod, J., J.-P. Changeux, and F. Jacob (1963) Allosteric proteins and cellular control systems. *J. Mol. Biol.* 6:306–329. This is the paper that introduced the concept of allosteric control.

Schachman, H. K. (1974) Anatomy and physiology of a regulatory enzyme—aspartate transcarbamoylase. *Harvey Lectures* 68:67–113. A clear exposition of ATCase regulation.

Regulation by Phosphorylation

Cohen, P. (1982) The role of protein phosphorylation in neural and hormonal control of cellular activity. *Nature* 296:613–620.

Sprang, S. R., K. R. Acharya, D. I. Goldsmith, et al. (1988) Structural changes in glycogen phosphorylase induced by phosphorylation. *Nature* 236:215–221.

Zymogen Activation

Bode, W., and R. Huber (1986) Crystal structure of pancreatic serine endopeptidases. In: *Molecular and Cellular Basis of Digestion,* edited by P. Desnuelle, H. Sjorstrom, and O. Noren, pp. 213–234. Elsevier, New York.

Neurath, H. (1986) The versatility of proteolytic enzymes. *J. Cell. Biochem.* 32:35–49. A contemporary survey of proteases.

Blood Clotting

Davie, E. W. (1986) Introduction to the blood coagulation cascade and the cloning of blood coagulation factors. *J. Protein Chem.* 5:247–253.

Doolittle, R. F. (1984) Fibrinogen and fibrin. *Annu. Rev. Biochem.* 53:195–229.

Lawn, R. M., and G. A. Vehar (1986) The molecular genetics of hemophilia. *Sci. Am.* 254(3):48–65. A readable contemporary account.

Regulation of Enzyme Synthesis

Gottesman, S. (1984) Bacterial regulation: Global regulatory networks. *Annu. Rev. Genet.* 18:415–442.

Jacob, F., and J. Monod (1961) Genetic regulatory mechanism in the synthesis of proteins. *J. Mol. Biol.* 3:318–356. The early classic paper on regulation of synthesis.

Palmiter, R. (1972) Regulation of protein synthesis in chick oviduct. *J. Biol. Chem.* 247:6450–6461.

Regulation of Protein Degradation

Creighton, J. E. (1984) *Proteins,* Chapter 10. W. H. Freeman, New York. A succinct discussion of the control of degradation, with further references.

Rogers, S., R. Wells, and M. Rechsteiner (1986) Amino acid sequences common to rapidly degraded protein: The PEST hypothesis. *Science* 234:364–368.

PROBLEMS

1. The steady-state kinetics of an enzyme are studied in the absence and presence of an inhibitor (inhibitor A). The initial rate is given as a function of substrate concentration in the table below.

	V (mmol L^{-1} min^{-1})	
[S] (mmol/L)	No inhibitor	Inhibitor A
1.25	1.72	0.98
1.67	2.04	1.17
2.50	2.63	1.47
5.00	3.33	1.96
10.00	4.17	2.38

(a) What kind of inhibition (competitive, noncompetitive, uncompetitive) is involved?

(b) Determine V_{max} and K_M in the absence and presence of inhibitor.

2. The same enzyme as in Problem 1 is studied in the presence of a different inhibitor (inhibitor B). In this case, two different concentrations of inhibitor are used. Data are:

	V (mmol L^{-1} min^{-1})		
[S] (mmol/L)	No inhibitor	3 mM inhibitor B	5 mM inhibitor B
1.25	1.72	1.25	1.01
1.67	2.04	1.54	1.26
2.50	2.63	2.00	1.72
5.00	3.33	2.86	2.56
10.00	4.17	3.70	3.49

(a) What kind of inhibitor is inhibitor B?

(b) Determine the apparent V_{max} and K_M at each inhibitor concentration.

(c) Estimate K_I from these data.

3. In Chapter 10 we discussed Eadie–Hofstee plots as an alternative to Lineweaver–Burk plots for expression of kinetic data. Sketch what Eadie–Hofstee plots would look like for a series of experiments at different concentrations of:

   (a) A competitive inhibitor.

   (b) A noncompetitive inhibitor.

4. We have discussed the role of TPCK as an irreversible inhibitor of chymotrypsin. Design a comparable inhibitor for trypsin.

5. The following data are found for the steady-state kinetics of a multisubunit enzyme:

[S] (mmol/L)	Initial rate (mmol L^{-1} s^{-1})
0.25	0.26
0.33	0.45
0.50	0.92
0.75	1.80
1.00	2.50
2.00	4.10
4.00	4.80

   (a) Demonstrate that this enzyme does not follow Michaelis–Menten kinetics.

   (b) Estimate V_{max}.

6. In Chapter 7 we used Hill plots to describe cooperative binding to allosteric proteins. If we assume, for an allosteric enzyme, that V_{max} is the same for T and R states, then the reaction velocity should be V_{max} times the fraction of sites occupied. On this basis, show how you would use Hill plots to analyze kinetic data for allosteric enzymes. Show in a sketch what the effect of a strong activator and a strong inhibitor would be.

7. In a few instances, multisubunit enzymes have been demonstrated to exhibit *negative cooperativity;* that is, the binding of the first substrate molecule markedly decreases the affinity for binding of subsequent molecules.

   (a) Show what a Hill plot would look like for an enzyme exhibiting negative cooperativity.

   (b) The MWC theory (see Chapter 7) cannot account for negative cooperativity, although the KNF theory can. Explain why this is so.

8. Although the MWC theory describing binding to a multisite protein is quite complicated, we can easily derive equations for a simple case. Assume a dimeric enzyme, with two sites, with states T and R. For simplicity, assume that the R state can bind one or two molecules of substrate, but the T state cannot bind. We then have the equations

$$\mathrm{T} \rightleftharpoons \mathrm{R} \qquad L = \frac{[\mathrm{T}]}{[\mathrm{R}]}$$

$$\mathrm{R} + \mathrm{S} \rightleftharpoons \mathrm{RS} \qquad K = \frac{2[\mathrm{R}][\mathrm{S}]}{[\mathrm{RS}]}$$

$$\mathrm{RS} + \mathrm{S} \rightleftharpoons \mathrm{RS_2} \qquad K = \frac{[\mathrm{RS}][\mathrm{S}]}{2[\mathrm{RS}]}$$

   The factor 2 appears because there are two equivalent sites on each R to bind S.

   (a) From this, show that the fraction of sites occupied at substrate concentration [S] is

$$\theta = \frac{[\mathrm{S}]}{K}\,\frac{1 + [\mathrm{S}]/K}{L + (1 + [\mathrm{S}]/K)^2} = \frac{V}{V_{max}}$$

   (b) Picking a large value of L (say 10^5) and $K = 10^{-6}$ M, show that a sigmoidal curve is obtained for V/V_{max} vs. [S].

9. Suggest the effects of each of the following mutations on the physiological role of chymotrypsinogen:

   (a) Arg 15 ⟶ Ser

   (b) Cys 1 ⟶ Ser

   (c) Thr 147 ⟶ Ser

10. It has been suggested that one "marker" determining the rate at which enzymes are degraded in vivo is the nature of the N-terminal residue. Using recombinant DNA technology, suggest a way to test this.

CHAPTER 12

Introduction to Metabolism

A chemist carrying out an organic synthesis rarely runs more than one reaction in a single reaction vessel. This is essential to prevent side reactions and to optimize the yield of the desired product. Yet a living cell carries out hundreds of reactions simultaneously, with each reaction sequence controlled in such a way that unwanted accumulations or deficiencies of intermediates and products do not occur. Reactions of great mechanistic complexity and stereochemical selectivity proceed smoothly under mild conditions—1 atmosphere pressure, moderate temperature, and a pH near neutrality.

A principal task of the biochemist is to understand the remarkable processes by which a cell regulates its myriad chemical reaction sequences and, in so doing, controls its internal environment. In Chapters 10 and 11 we discussed the properties of individual enzymes and control mechanisms that affect their activity. Here we begin to focus on specific reaction sequences, or **pathways;** on the relationship between each pathway and cellular architecture; on the biological importance of each pathway; and on control mechanisms that regulate **flux,** or the flow of material through each pathway. A primary focus of this and the next few chapters will be on the generation of metabolic energy. At the same time, we shall see that energy-yielding pathways also generate intermediates used in biosynthetic processes. Thus, although at the outset we will focus most on degradation of organic compounds for provision of energy, you should constantly be aware that metabolism is a continuum, with many of the same reactions playing roles in both degradative and biosynthetic processes.

Orientation to Metabolism

Metabolism: The Totality of Chemical Reactions in Living Matter

The living cell can be described thermodynamically as an isothermal open system which perpetuates and replicates itself by means of consecutive linked organic reactions that are promoted by catalysts which the cell itself

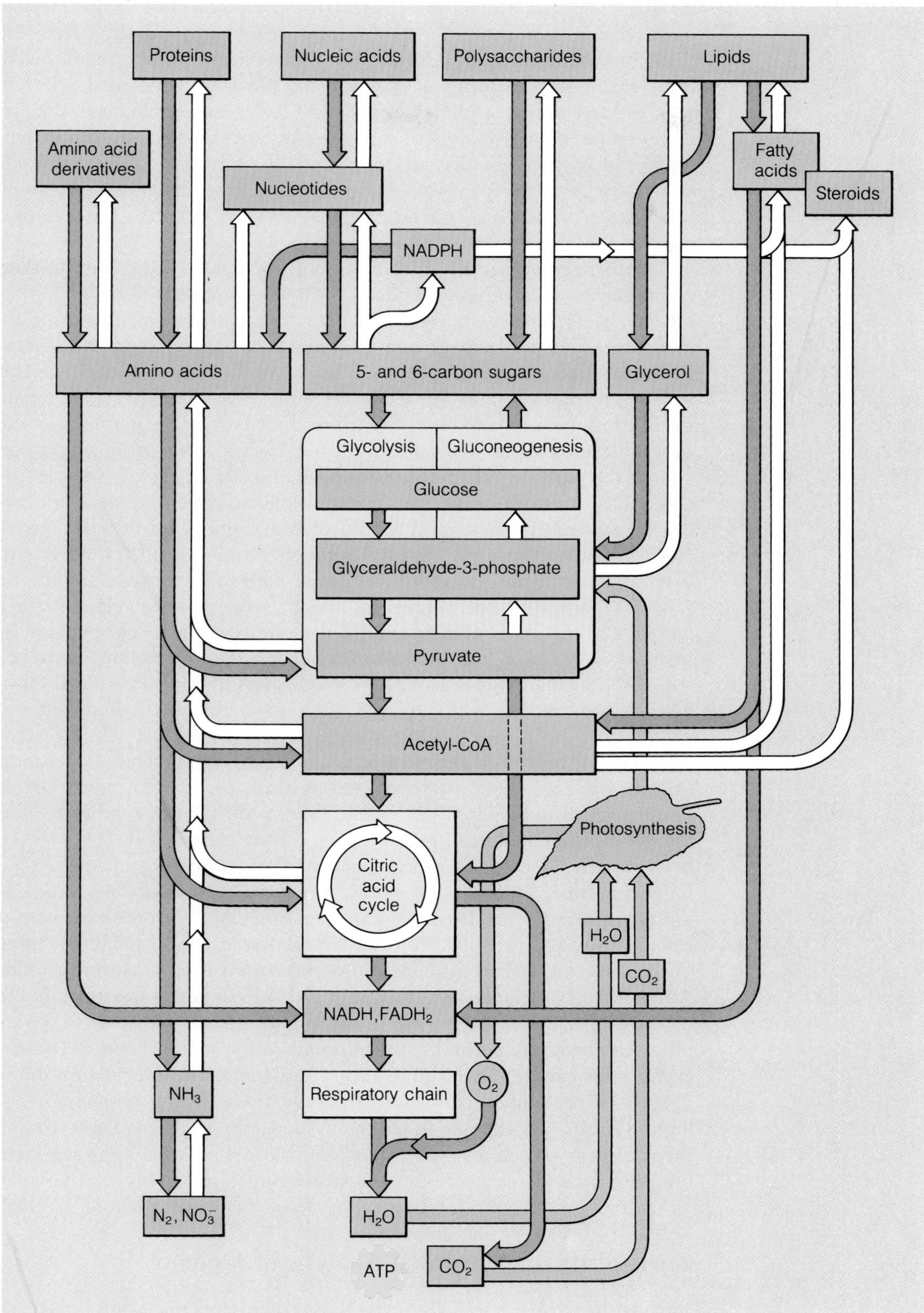

Figure 12.1
Overview of metabolism, showing the central pathways and the key intermediates. In general, catabolic pathways proceed downward (gray arrows) and anabolic pathways upward (white arrows). Pathways confined to autotrophs are shown in green. Key metabolites are shown in blue.

produces. The totality of these consecutive linked reactions is called **metabolism.** Any participant in a metabolic reaction, whether substrate, intermediate, or product, is called a **metabolite.** Metabolism itself can be subdivided into two major categories: **anabolism,** the processes concerned primarily with the assembly of complex organic molecules, and **catabolism,** the processes related to degradation of complex substances, with concomitant generation of energy. The distinction is partially blurred, because many substrates for anabolic pathways are formed as **intermediates** in catabolic processes. Nevertheless, even though it is somewhat arbitrary, the division is useful.

Other terms that we shall use in our discussion include **intermediary metabolism** and **energy metabolism.** Although somewhat ill defined, intermediary metabolism comprises all the reactions concerned with storing and generating metabolic energy and with using that energy in biosynthesis of low-molecular-weight compounds. The reactions of intermediary metabolism can be thought of as those that do not involve a nucleic acid template; the information needed to specify each reaction is provided within the structure of the enzyme catalyzing that reaction. Energy metabolism simply consists of the pathways that store or generate metabolic energy. Chapters 12 through 23 present intermediary metabolism, and the major focus in Chapters 13 through 19 is on its energetic aspects. Figure 12.1 depicts an overview of metabolism, with primary emphasis on the so-called central pathways of intermediary metabolism. These pathways are substantially the same in many different organisms, and they account for relatively large amounts of mass transfer and energy generation within a cell; they are the quantitatively major pathways. The schematic view presented in Figure 12.1 will reappear many times as we proceed through intermediary metabolism; in each case the pathways on which discussion focuses is highlighted in color.

Organisms vary widely in their ability to carry out metabolic reactions or pathways; a major distinction among them pertains to nutritional requirements. **Autotrophs** (from Greek, "self-feeding") can synthesize all of their organic compounds from inorganic carbon, supplied as CO_2, whereas **heterotrophs** ("feeding on others") can synthesize their organic metabolites only from other organic compounds, which they consume. This represents the primary difference between plants and animals. With the exception of rare insect-eating plants, such as the Venus flytrap, higher plants obtain all of their organic carbon through photosynthetic fixation of carbon dioxide. Animals feed on plants and other animals and synthesize their metabolites by transforming the molecules they consume.

Although the above discussion points to the sun as the sole source of energy, this concept is not quite accurate, given recent information on extremely **thermophilic** organisms. These organisms live at temperatures as high as 200°. Typically they are found in or near hydrothermal vents deep in the ocean, or geothermal vents in active volcanic regions. Although metabolic information on thermophiles is still accumulating, it is clear that much or most of their metabolic energy comes from the internal heat of the earth.

Goals of the Experimental Analysis of Metabolism

Starting from the premise that metabolism represents the totality of chemical reactions in living matter, how does a biochemist actually approach metabolism in the laboratory, breaking it into experimentally attainable objectives? For a particular metabolic process, the biochemist seeks to (1) identify reactants, products, and cofactors, plus the stoichiometry, for each reaction involved; (2) understand how the rate of each reaction is controlled in the tissue of origin; and (3) identify the physiological function of each reaction and control mechanism. For these goals to be met, the enzyme catalyzing each reaction in a pathway must be isolated and characterized. Later in this chapter we shall describe some experimental approaches used to attain the above goals.

The task of extrapolating from test tube biochemistry to the intact cell (objective 3 above) is especially challenging and important. For instance, what is the direction in which an enzymatic reaction proceeds in vivo? Over the history of biochemistry many reactions that were demonstrated in vitro were later shown to proceed in the opposite direction in vivo. The mitochondrial enzyme that synthesizes ATP from ADP, for example, was originally isolated as an ATPase—an enzyme that hydrolyzes ATP to ADP and P_i. Therefore, it is not enough to isolate an enzyme and to demonstrate that it catalyzes a particular reaction in the test tube. We must show also that the same enzyme is catalyzing the same reaction in intact tissue, usually a far more difficult task.

Life in an Oxidizing Environment

EXERGONIC NATURE OF MOST SUBSTRATE OXIDATIONS. Since we shall be concerned at first with catabolism, it is appropriate to look briefly at the kinds of reactions that generate energy. As we saw in Chapter 3, a thermodynamically unfavorable reaction will proceed smoothly in the nonfavored direction only if it can be coupled to an exergonic reaction. In principle, any exergonic reaction can serve this purpose, provided that it releases sufficient free energy. In living systems most of the needed energy release is derived from the *oxidation* of organic substrates. Oxygen is a strong oxidant; it has a marked tendency to attract electrons, becoming reduced in the process. Because of this tendency and because of the abundance of oxygen in our atmosphere, it is not surprising that living systems have evolved the ability to derive energy from the oxidation of organic substrates.

ESSENTIAL COUPLING OF EXERGONIC AND ENDERGONIC REACTIONS. In a thermodynamic sense the biological oxidation of organic substrates is comparable to nonbiological oxidations such as combustion of wood in a fire. Biological oxidations are far more complex, however. When wood is burned, all of the energy is released as heat; useful work cannot be performed, except through the action of a device such as a steam engine. In biology, by contrast, some of the free energy released is captured as chemical energy and used to drive a diversity of biochemical processes. As discussed in Chapter 3, this occurs largely through the synthesis of ATP, a molecule designed to do various kinds of biological work. It is important to realize, however, that the free energy release is identical whether we consider a wood fire involving complete oxidation of the glucose polymer cellulose, *or* combustion of glucose in a calorimeter, *or* the complete metabolic oxidation of glucose.

$$C_6H_{12}O_6 + 6O_2 \longrightarrow 6CO_2 + 6H_2O \qquad \Delta G^\circ = -2870 \text{ kJ/mol}$$

A major difference between combustion and catabolism is that in the latter about 40% of the released 2870 kJ is made available to drive the synthesis of ATP from ADP and P_i.

Relatively few metabolic reactions involve the direct transfer of electrons from a reduced substrate to oxygen. Rather, a series of coupled oxidation–reduction reactions occurs (the **electron transport chain,** also called the **respiratory chain**), with the electrons passed to intermediate electron carriers such as NAD^+ and finally transferred to oxygen, the so-called **terminal electron acceptor.** By this means the potential energy stored in the organic substrate is released in small increments, making it easier both to control oxidation and to capture some of the energy as it is released. This is somewhat analogous to the generation of hydroelectric power, where a series of small dams on a river generates more electricity than a single large dam, even though the total distance through which the water drops is the same in both cases.

ENERGY YIELDS, RESPIRATORY QUOTIENTS, AND REDUCING EQUIVALENTS. If metabolic energy comes primarily from oxidative reactions, it follows that *the more highly reduced a substrate, the higher its potential for generating biological energy.* We can use a calorimeter to measure the heat output from oxidation of fat or carbohydrate (the measurements give enthalpy of oxidation figures in calories, which, as noted in Chapter 3, are readily convertible to joules). When we carry out these measurements, we find that the combustion of fat provides more heat energy. In other words, fat has a higher **caloric content** than carbohydrate. This is because the carbons in fat are in general more highly reduced than those in carbohydrate. For example, compare the oxidation of glucose with that of a typical saturated fatty acid, palmitic acid (shown here as the acid, rather than the palmitate ion, so that we can write a balanced equation).

$$C_6H_{12}O_6 + 6O_2 \longrightarrow 6CO_2 + 6H_2O$$

$$C_{16}H_{32}O_2 + 23O_2 \longrightarrow 16CO_2 + 16H_2O$$

Glucose yields 15.64 kJ/g, while palmitic acid yields 38.90 kJ/g. The more highly reduced carbons of the fatty acid contain more protons and electrons to combine with oxygen on the path to CO_2 than do those of the sugar. This can be seen from the ratio of moles of CO_2 produced to moles of O_2 consumed during oxidation, a term called the **respiratory quotient,** or **RQ.** From the above equations you can see that RQ for glucose is 1.0 ($6\ CO_2/6\ O_2$), while that for palmitic acid is 0.70. In general, the lower the RQ for a substrate, the more oxygen consumed per carbon oxidized and the greater the potential per mole of substrate for generating energy.

Another way to express this relationship is to say that more reducing equivalents are derived from the oxidation of fat than from that of carbohydrate. A **reducing equivalent** can be defined roughly as 1 mol of hydrogen atoms (one proton and one electron per H atom). For example, two reducing equivalents are used in the reduction of one oxygen atom to water.

$$\tfrac{1}{2}O_2 + 2e^- + 2H^+ \longrightarrow H_2O$$

You should realize that not all metabolic energy comes from oxidation by oxygen. Substances other than oxygen can serve as terminal electron acceptors. For example, some bacteria (including some found in deep-sea hydrothermal vents) reduce sulfur to sulfide as the terminal electron transfer reaction, while others reduce nitrite to ammonia. Also, many microorganisms either can or must live **anaerobically** (in the absence of oxygen). These organisms derive their energy from **fermentations,** which are defined most simply as energy-yielding catabolic pathways that proceed with no net oxidation. A good example is the production of ethanol and CO_2 from glucose, which we will discuss in Chapter 13.

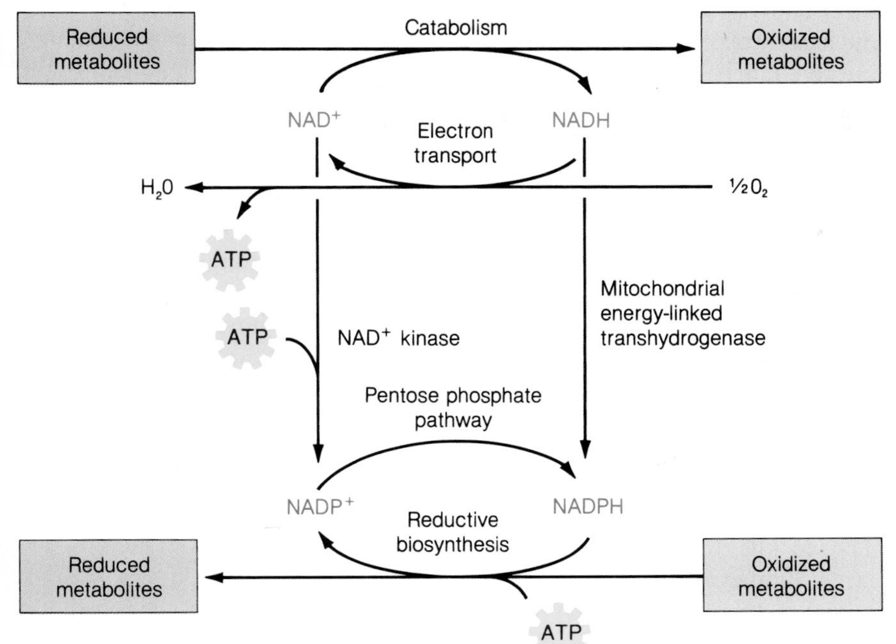

Figure 12.2
Oxidative electron flows in catabolism and reductive electron flows in biosynthesis, as seen in nicotinamide nucleotide metabolism. NAD^+ is the cofactor for most enzymes that act in the direction of substrate oxidation. NADPH, which usually functions as a reductive cofactor, is synthesized either from $NADP^+$ in the pentose phosphate pathway (Chapter 14) or through the action of mitochondrial energy-linked transhydrogenase (Chapter 15). $NADP^+$ is synthesized from NAD^+ by an ATP-dependent kinase reaction.

If catabolism involves primarily energy-yielding oxidative pathways, it follows that biosynthesis of complex organic compounds requires both energy and reducing equivalents. For example, we know that both carbons of acetate are used for fatty acid biosynthesis

$$8CH_3COO^- \rightarrow\rightarrow\rightarrow CH_3(CH_2)_{14}COO^- \quad \text{(palmitate)}$$

Since 14 of the 16 carbon atoms of palmitate are at the methylene level and 1 is at the methyl level, many reducing equivalents must be made available. The major source of electrons for reductive biosynthesis is NADPH. $NADP^+$ and NADPH are identical to NAD^+ and NADH, respectively, but they have an additional phosphate esterified at C-2′ on the adenylate moiety. NAD^+ and $NADP^+$ are equivalent in their thermodynamic tendency to accept electrons; they have equal **standard reduction potentials** (Chapter 15). However, nicotinamide nucleotide-linked enzymes that act primarily in a catabolic direction usually use the NAD^+-NADH pair, while those acting primarily in anabolic pathways use $NADP^+$ and NADPH. These relationships are summarized in Figure 12.2.

CONNH2
N+
O—CH2
⁻O—P=O
O
⁻O—P=O
O—CH2
adenine
OH OH
OH O(P)

$NADP^+$
Nicotinamide adenine dinucleotide phosphate

Energy Transductions

In Chapter 3 we pointed out that living cells capture free energy released in catabolism largely as ATP. The chemical energy stored in the anhydride bonds of ATP can be converted to other forms of energy, a process called **energy transduction.** What special properties of ATP equip it to function as the "energy currency" of the cell in these transductions? In principle, any of a dozen or more high-energy metabolites could play this role—deoxyATP, for example, or UTP, or even NAD^+. All of these compounds contain energy-rich bonds which, like ATP, yield about 30 kJ on hydrolysis (under standard conditions). However, evolution has generated an array of enzymes that preferentially bind ATP and use its free energy of hydrolysis to drive endergonic reactions. There is no obvious chemical reason why other nucleoside triphosphates couldn't play this role just as well; in fact, they do, but to a much more limited extent. For example, GTP provides much of the energy needed for protein synthesis.

ATP happens to be abundant in most cells, and this is related to its role as energy currency. Intracellular ATP concentrations range from 2 to 5 mM, considerably higher than the levels of other nucleoside triphosphates. Equally important, ATP is far more abundant than ADP or AMP, which means that under conditions in the cell, there is a strong thermodynamic tendency for the potential energy stored in ATP to be used in the synthesis of other high-energy compounds. Other nucleoside triphosphates are synthesized from the corresponding diphosphates and ATP through the action of **nucleoside diphosphokinase,** an enzyme of broad specificity for ribonucleotides and deoxyribonucleotides.

$$\text{ATP} + \text{CDP} \rightleftharpoons \text{ADP} + \text{CTP}$$

The enzyme transfers the γ-phosphate from many triphosphates to any of a large number of nucleoside diphosphates; thus, the reaction above is but one of many reactions catalyzed by this enzyme. The equilibrium constant is essentially unity, but since ATP levels are so high, synthesis of the other triphosphates is favored under intracellular conditions.

More interesting is the situation where ATP drives the synthesis of a compound more unstable than itself. Recall from Chapter 3 that several important metabolites have higher standard free energies of hydrolysis than ATP. One such compound is **creatine phosphate,** which is abundant in vertebrate skeletal muscle, where it participates in shuttling phosphate bond energy from mitochondria, where it is generated as ATP, to myofibrils (Chapter 29), where that bond energy is transduced to the mechanical energy of contraction.

Creatine phosphate is synthesized from creatine and ATP in a reaction catalyzed by **creatine phosphokinase.**

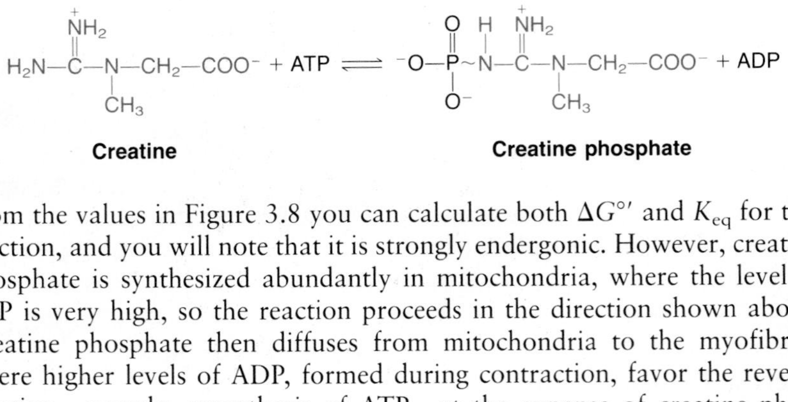

$$H_2N{-}\overset{\overset{+}{N}H_2}{\overset{\|}{C}}{-}\underset{CH_3}{\underset{|}{N}}{-}CH_2{-}COO^- + \text{ATP} \rightleftharpoons {}^-O{-}\underset{O^-}{\underset{|}{\overset{O}{\overset{\|}{P}}}}{\sim}\overset{H}{\overset{|}{N}}{-}\overset{\overset{+}{N}H_2}{\overset{\|}{C}}{-}\underset{CH_3}{\underset{|}{N}}{-}CH_2{-}COO^- + \text{ADP}$$

Creatine **Creatine phosphate**

From the values in Figure 3.8 you can calculate both $\Delta G^{\circ\prime}$ and K_{eq} for this reaction, and you will note that it is strongly endergonic. However, creatine phosphate is synthesized abundantly in mitochondria, where the level of ATP is very high, so the reaction proceeds in the direction shown above. Creatine phosphate then diffuses from mitochondria to the myofibrils, where higher levels of ADP, formed during contraction, favor the reverse reaction—namely, resynthesis of ATP—at the expense of creatine phosphate cleavage to creatine. This is an excellent example of the importance, in predicting the direction of a reaction in vivo, of considering both the free energy change *under standard conditions* and the *actual concentrations of all reactants and products.*

$$\begin{array}{l} H{-}N{\sim}\overset{\overset{O}{\|}}{P}{-}O^- \\ \quad\;\; | \qquad | \\ H_2\overset{+}{N}{=}C \quad O^- \\ \qquad | \\ \qquad NH \\ \qquad | \\ \qquad CH_2 \\ \qquad | \\ \qquad CH_2 \\ \qquad | \\ \qquad CH_2 \\ \qquad | \\ H{-}C{-}\overset{+}{N}H_3 \\ \qquad | \\ \qquad COO^- \end{array}$$

Arginine phosphate

Muscle metabolism in some invertebrate animals involves a different compound, **arginine phosphate,** which functions identically in shuttling high-energy phosphate from mitochondria to sites of muscle contraction.

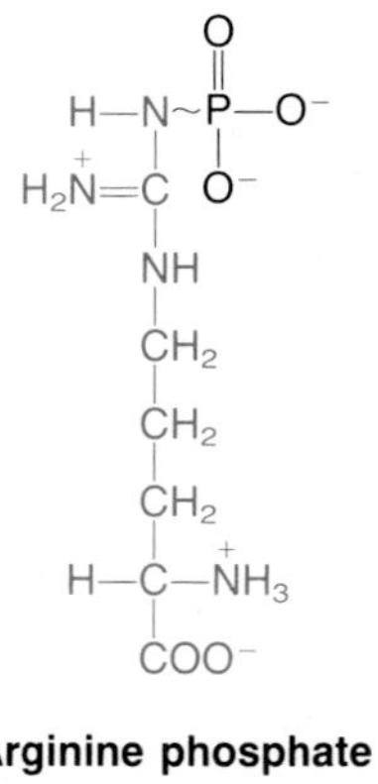

Two other important high-energy metabolites, 1,3-bisphosphoglycerate and phosphoenolpyruvate, also have values for $\Delta G^{\circ\prime}$ of hydrolysis higher than that of ATP. Each of these super-high-energy compounds is formed readily in cells because of the maintenance of intracellular concentrations, particularly of ATP, that are far removed from standard values.

As we have seen, the energy stored in the ATP molecule is utilized as chemical energy, to drive the synthesis of other high-energy compounds. It also can be transduced to other forms of energy, including the transduction of chemical to mechanical energy, as in muscle contraction or ciliary motion. Transduction to electrical energy occurs in depolarization of a membrane or the defensive reaction of an electric eel.

Freeways on the Metabolic Road Map

The Three Stages of Metabolism

Undoubtedly you have seen metabolic charts—those huge, maplike wall hangings that adorn biochemistry laboratories and offices. Figure 12.1 is a highly simplified metabolic chart. Faced with the thousands of individual reactions that constitute metabolism, how do we approach our presentation of this vast topic? Our first concern is with central pathways—those that play indispensable roles and that are widespread or universal in biology. Also, our initial focus will be on energy metabolism. Given these conditions, we shall first consider the degradative processes that are most important in energy generation—carbohydrate and lipid breakdown. We shall also consider how these substances are biosynthesized. These are the reactions sitting in the middle of the metabolic charts and illustrated with the biggest arrows—freeways, so to speak, on the metabolic road map.

It is convenient to think of metabolism in three stages, based on structural complexity of the metabolites. First are the reactions that interconvert polymeric metabolites, such as polysaccharides or proteins, with their monomeric constituents, monosaccharides or amino acids. In the catabolism of carbohydrates or lipids, this stage of metabolism generates monomeric substrates such as simple sugars, sugar phosphates, fatty acids, or glycerol. In the second stage these monomers are further degraded to simple intermediates, such as pyruvate, or other compounds not readily identifiable as carbohydrate, lipid, or protein. In the third stage are the pathways that carry out ultimate degradation to CO_2, H_2O, and NH_3. In aerobic organisms the principal stage 3 pathway is the **citric acid cycle** (Chapter 14). This cyclic pathway accepts simple carbon compounds derived from carbohydrate, lipid, or protein, such as pyruvate or succinate, and oxidizes them to CO_2. Using the freeway analogy again, we will see that numerous on-ramps from the highways and byways of stage 1 and stage 2 metabolism lead to the citric acid cycle; all catabolic pathways converge at this point. These relationships are indicated in Figure 12.1 and can be schematized more simply as in Figure 12.3.

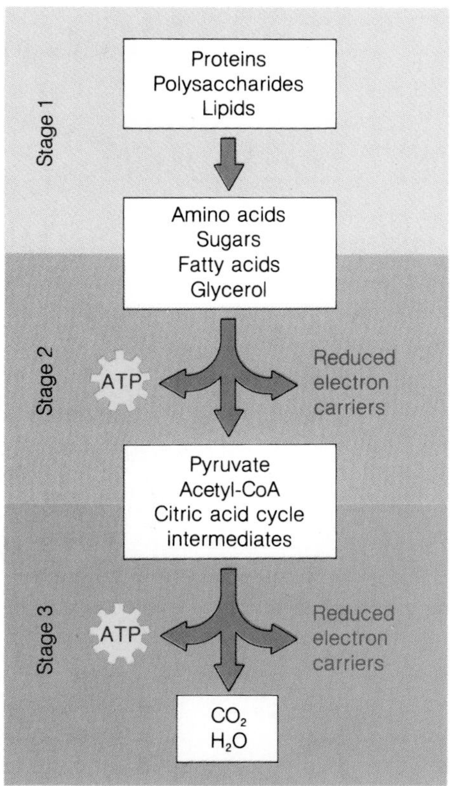

Figure 12.3
The three stages of metabolism, illustrated for the catabolism of proteins, carbohydrates, and lipids. Catabolic streams converge in the citric acid cycle.

The Major Metabolic Pathways

We will explore the major metabolic pathways in detail in the next several chapters; we name them here to indicate their part in the overall metabolic process. Oxidative reactions of the citric acid cycle drive ATP biosynthesis, primarily through **electron transport** and **oxidative phosphorylation.** Other pathways deliver fuel to the citric acid cycle. **Glycolysis** is a stage 2 pathway that catabolizes monosaccharide sugars to pyruvate, whose major fate in some tissues is oxidation in the citric acid cycle. Similarly, fatty acids in aerobic organisms are broken down by **β-oxidation** to yield an activated two-carbon fragment, **acetyl-coenzyme A.** These two carbons can be oxidized further in the citric acid cycle; in the anabolic direction, they provide substrates for synthesis of fatty acids and cholesterol.

Existence of Separate Biosynthetic and Degradative Pathways

As you look at Figure 12.1, note that some pathways seem to operate simply as the reversal of other pathways. For example, fatty acids are synthesized from acetyl-CoA, but they are also converted to acetyl-CoA by β-oxidation. Similarly, pyruvate is a precursor to glucose-6-phosphate via **gluconeogenesis,** which looks at first glance like a simple reversal of glycolysis. In fact, gluconeogenesis is the biosynthesis of carbohydrates from noncarbohydrate precursors. It is important to realize that in such cases the two opposed pathways are quite distinct from one another. They may share some common intermediates or enzymatic reactions, but they are different pathways, regulated by distinct mechanisms and with different enzymes catalyzing their regulated reactions. In the case of fatty acid synthesis from acetyl-CoA and oxidation to acetyl-CoA, the two processes take place in different cell compartments—fatty acid synthesis in the cytosol, β-oxidation in the mitochondria. Biosynthetic and degradative pathways are rarely if ever simple reversals of one another, even though they often begin and end with the same metabolites.

This distinction is important for two reasons. First, for a pathway to proceed in a particular direction, it must be exergonic in that direction. If a pathway (e.g., glycolysis) is strongly exergonic, it follows that the reversal of that pathway is just as strongly endergonic under the same conditions.

Equally important is the need to control the flow of metabolites in relation to the bioenergetic status of a cell. When the energy level is high, ATP levels are high, and there is less need for carbon to be oxidized in the citric acid cycle. At such times carbon can be stored, through biosynthesis of fats and carbohydrates. Thus, fatty acid synthesis, gluconeogenesis, and other pathways come into play. In contrast, when ATP levels are low, stored carbon must be mobilized to generate substrates for oxidation in the cycle. Using separate pathways for the biosynthetic and degradative processes is crucial for control, so conditions that activate one pathway tend to inhibit the opposed pathway and vice versa. Consider what would happen, for example, if fatty acid synthesis and oxidation took place in the same cell compartment and in uncontrolled fashion. Two-carbon fragments released by oxidation would be immediately used for resynthesis. Biochemists call this situation a **futile cycle.** No useful work is done, and the net result is simply consumption of the ATP used in the endergonic reactions of fatty acid synthesis.

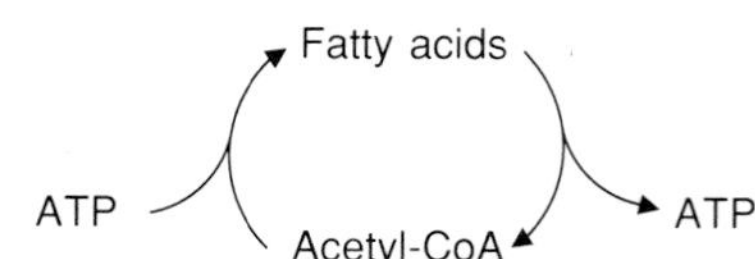

A futile cycle could result from the interconversion of fructose-6-phosphate with fructose-1,6-bisphosphate in carbohydrate metabolism. The two reactions involved follow.

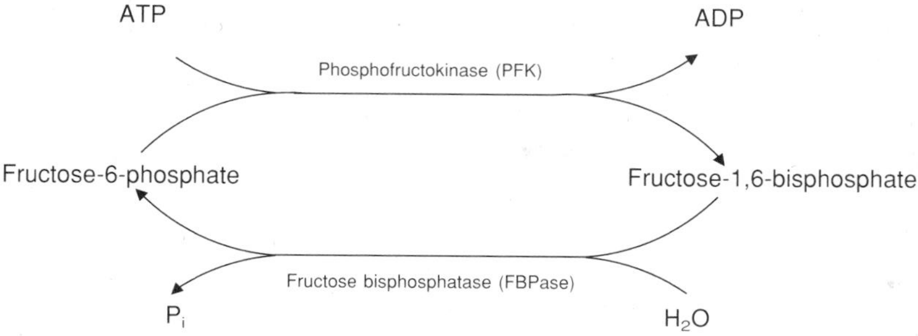

PFK catalyzes a reaction of glycolysis (sugar catabolism), while FBPase participates in a sugar biosynthetic pathway, gluconeogenesis. The net effect of the operation of both reactions is the hydrolysis of ATP to ADP and P_i. However, both of these enzymes respond to allosteric effectors, such that one enzyme is inhibited by conditions that activate the other. Thus, the futile cycle is effectively controlled, even though the two enzymes occupy the same cell compartment; for this reason the term **substrate cycle** is preferable to futile cycle.

Recent studies on metabolic control have suggested that a substrate cycle represents an efficient regulatory mechanism, because a small change in the activity of either or both enzymes can have a much larger effect on the flux of metabolites in one direction or the other. You can verify this in Problem 4 at the end of this chapter.

Major Metabolic Control Mechanisms

The substrate cycles just discussed represent one way to regulate metabolism. The living cell uses a marvelous array of additional regulatory devices to control its functions. The mechanisms that act primarily to control enzyme activity, such as substrate concentration and allosteric control, were discussed in Chapter 11. **Compartmentation** represents another control mechanism, with the flow of a metabolite into or out of a cell compartment or tissue controlling the fate of that metabolite. Metabolic pathways are also controlled through regulation of intracellular enzyme concentration at the level of either enzyme synthesis or degradation. Overlying all of these mechanisms are the actions of **hormones,** chemical messengers that act at all levels of regulation, as we shall see later in this chapter and in Chapter 23.

Control of Enzyme Levels

If you were to assay a cell-free extract of a particular tissue for several different enzymes and determine intracellular concentrations of those enzymes, you would find a tremendous variation in concentration. Enzymes of the central energy-generating pathways might be present at many thousands of molecules per cell, whereas enzymes that have limited or specialized functions might be present at fewer than a dozen molecules per cell. Two-dimensional gel electrophoresis of a cell extract (Tools of Biochemistry 16) gives an impression, from the varying spot intensities, of the wide variations in amounts of individual proteins in a particular cell.

Furthermore, the level of a single protein can vary widely under different environmental conditions. The mechanism of this variation is particularly well understood in bacteria. When a utilizable substrate is added to a bacterial growth milieu, the abundance of the enzymes needed to utilize the substrate may increase, through synthesis of new protein, from less than one molecule per cell to many thousand molecules. This is known as enzyme **induction.** Similarly, the presence of the end product of a pathway may turn off the synthesis of enzymes needed to synthesize that end product (**repression**).

For some time it was thought that understanding the control of the intracellular level of a protein was primarily a question of knowing what controlled the *synthesis* of that protein—in other words, a question of genetic regulation. More recently we have learned that intracellular protein *degradation* is also important in determining the enzyme level. Clearly, the intracellular concentration of an enzyme molecule is determined by both its rate of synthesis and its rate of degradation.

Control of Enzyme Activity

Basically, the *catalytic activity* of an enzyme molecule can be controlled in two ways: (1) by interaction with ligands, and (2) by covalent modification of the protein molecule.

Ligands that control enzyme activity can be polymeric; for example, protein–protein interactions can affect enzyme activity, and several enzymes of nucleic acid metabolism are activated by binding to DNA. Here, however, we are more concerned with interactions with low-molecular-weight ligands, principally substrates and allosteric effectors. With respect to substrates, measurements of intracellular levels, or **pool sizes,** of low-molecular-weight metabolites show that they usually have approximate molar concentrations lower than the K_M values for the enzymes that act on them but within an order of magnitude of these values. That is, substrate concentrations usually lie within the first-order ranges of substrate concentration–velocity curves for the enzymes that act on them, and hence enzyme activities are responsive to small changes in substrate concentration.

We saw in Chapter 11 that allosteric activation or inhibition usually acts on committed steps of a metabolic pathway, often initial reactions. The effectors function by binding at specific regulatory sites, thereby affecting subunit–subunit interactions in the enzyme protein. This in turn either facilitates or hinders the binding of substrates.

Such a mechanism operates in an obvious way to control product formation if a pathway is unidirectional and unbranched. However, some substrates are involved in numerous pathways, so many branch points exist. This presents a special metabolic problem, for if the end product of one branch feeds back and inhibits the first committed enzyme in its synthesis, then synthesis of the other product will be inhibited, even if that compound is not present in excess. The enzyme **deoxycytidylate deaminase** shows one effective mechanism for control at a branch point, in this case a branch point in pyrimidine nucleotide metabolism.

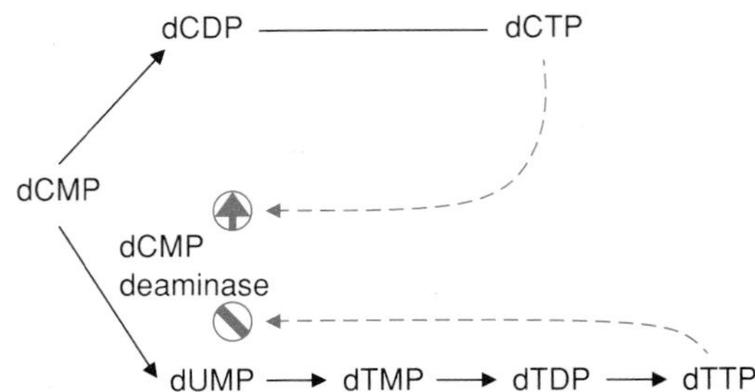

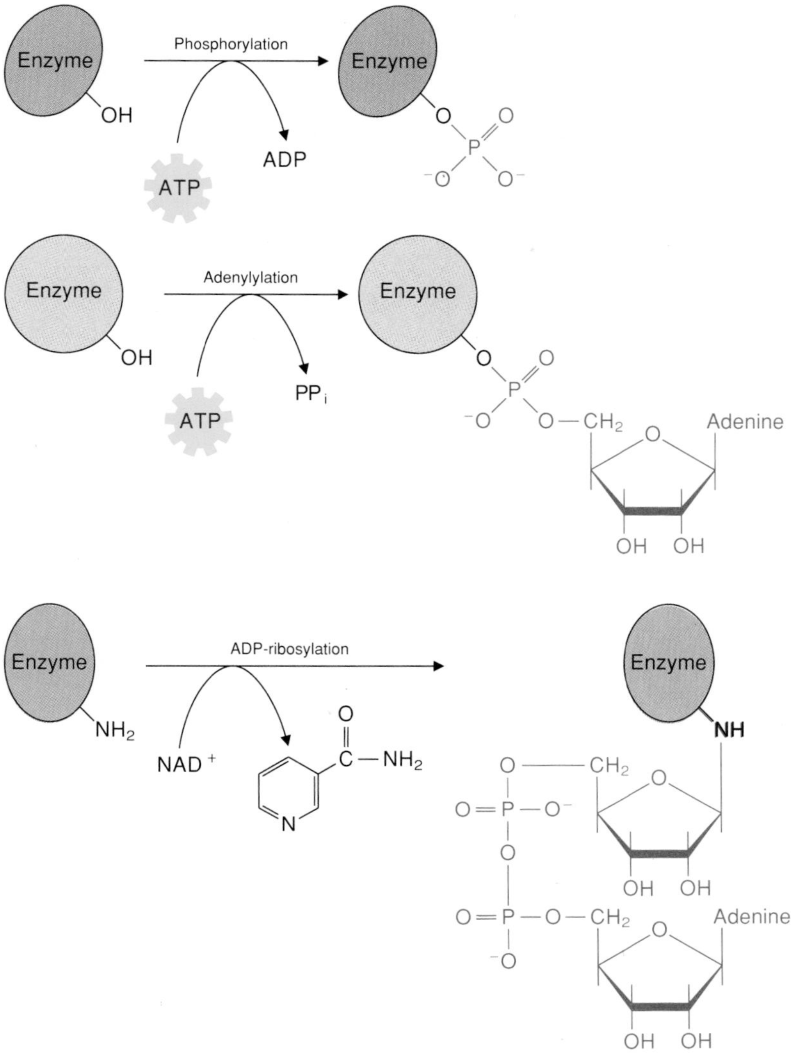

Deoxycytidine monophosphate (dCMP) is a precursor of dCTP and dTTP, both of which are used for DNA synthesis and are needed in roughly equal amounts. To regulate the flow of metabolites past this branch point, dCMP deaminase is allosterically activated by dCTP and inhibited by dTTP.

Covalent modification of enzyme structure represents another efficient way to control enzyme activity. In Chapter 11 we introduced the regulation of glycogen phosphorylase activity, effected through the ATP-dependent phosphorylation of a specific serine residue. Other proteins can undergo phosphorylation of threonine or tyrosine residues. More complex reactions involve **adenylylation,** the transfer of an adenylate moiety from ATP, and **ADP-ribosylation,** the transfer of an ADP-ribosyl moiety from NAD^+ (Figure 12.4).

A good example of covalent modification, discussed more fully in Chapter 20, is the control of **glutamine synthetase** activity in bacteria. This enzyme catalyzes the conversion of glutamate to glutamine, which in turn participates in many biosynthetic pathways. Activity of this enzyme, and its sensitivity to an array of feedback inhibitors, is controlled by adenylylation of one tyrosine residue in each of the enzyme's 12 identical subunits. Enzymes that catalyze both the addition and removal of the adenylate moiety

Figure 12.4
Three types of covalent modifications that control the activities of enzymes. The target residue for phosphorylation or adenylylation is usually serine, threonine, or tyrosine, while ADP-ribosylation can involve arginine or a modified histidine residue.

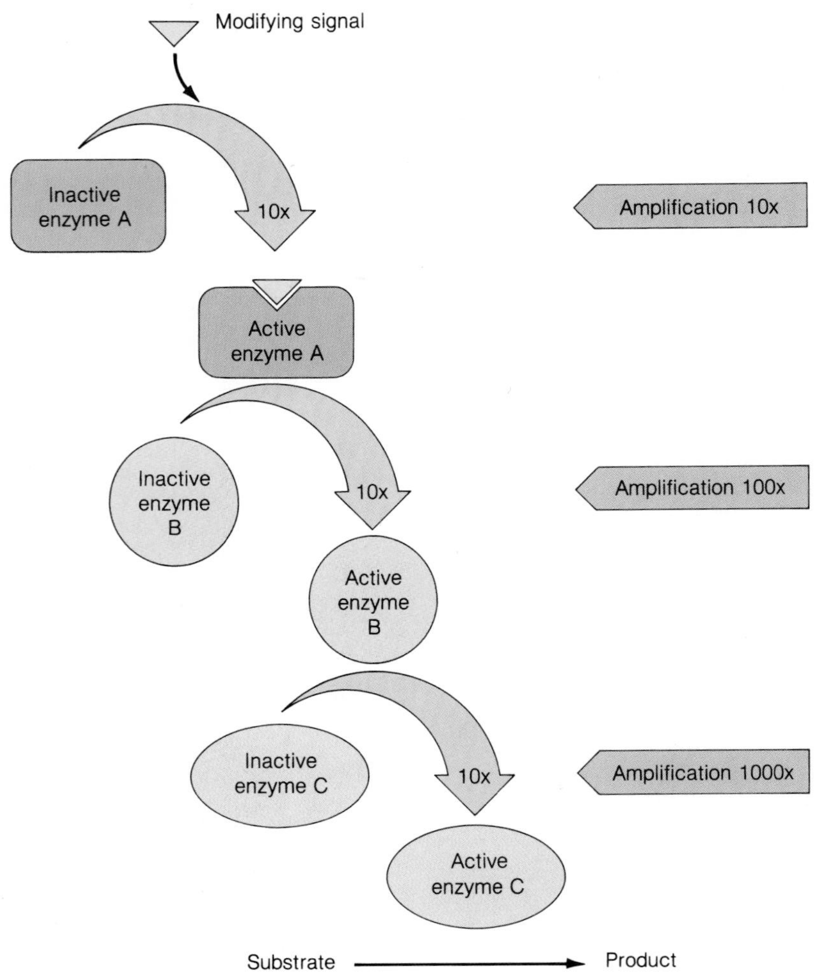

Figure 12.5
A generalized regulatory cascade.

are in turn controlled by covalent modification, with the ultimate signals coming from the levels of key metabolites.

Control through covalent modification is often associated with regulatory cascades, which we introduced in Chapter 11. Modification activates an enzyme, which in turn acts on a second enzyme, which may activate yet a third enzyme, which finally acts on the substrate, as shown in Figure 12.5. Since enzymes act catalytically, this provides an efficient way to *amplify* the original biological signal. Suppose that the original signal modifying enzyme A activates it 10-fold, that modified enzyme A then activates enzyme B 10-fold, and that enzyme C is finally activated 10-fold. Thus, with the involvement of relatively few molecules of enzyme, a pathway is activated by 1000-fold. By contrast, a single protein modification that activated one enzyme by that large a factor would be very hard to imagine.

The breakdown of glycogen in animal cells is a critical process, providing carbohydrate substrates for energy generation. The events leading to glycogen breakdown, which are controlled by enzyme phosphorylation and dephosphorylation, represent the first well-understood regulatory cascade (for details see Chapter 13). Another well-studied cascade is involved in the process of blood clotting (Chapter 11). The latter cascade is characterized by another type of covalent modification, namely proteolytic cleavage at specific sites, which converts an inactive enzyme precursor to an active catalyst. Unlike the modifications shown in Figure 12.4, however, modification by proteolysis is irreversible.

Intense interest in protein modification reactions has focused on the protein products of **oncogenes** (cancer-causing genes), many of which have been found to be protein kinases. Aberrant activities of these proteins, which phosphorylate other proteins within a cell and alter their activities, may be involved in the transformation of a normal cell to a cancer cell. This will be discussed further in Chapter 23.

Compartmentation

We have already described the physical division of labor that exists in a cell; enzymes participating in the same process are localized to a particular *compartment* within the cell. For example, RNA polymerases are found in the nucleus and nucleolus, where DNA transcription occurs, while the enzymes of the citric acid cycle are all found in mitochondria. Figure 12.6 presents the intracellular locations of a number of metabolic pathways.

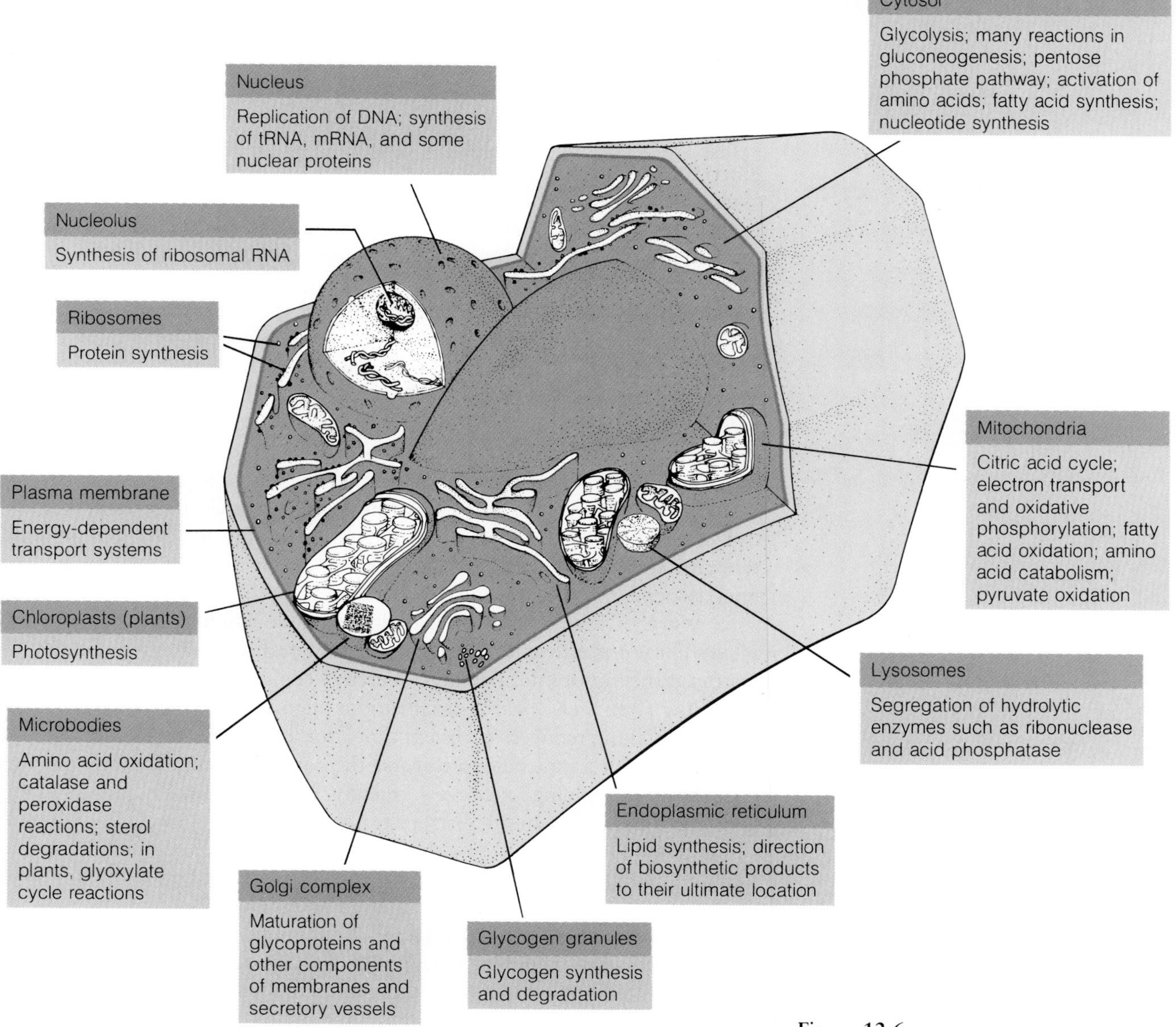

Figure 12.6
Intracellular locations of major metabolic pathways.

In addition to making the cell a more efficient machine, compartmentation has an important regulatory function. This function derives largely from the selective permeability of membranes to different metabolites, which controls the passage of intermediates from one compartment into another. Typically, intermediates remain trapped within an organelle, while specific carriers allow substrates to enter and products to exit. Flux through a pathway can be regulated by controlling the rate at which a substrate enters the compartment. For example, one of the ways in which insulin activates carbohydrate utilization is by increasing the number of glucose transporters in the plasma membrane, so that glucose is more readily taken into cells for catabolism or for synthesis of glycogen.

Compartmentation is more than a matter of tucking enzymes into the proper organelles. Juxtaposition of enzymes that catalyze sequential reactions acts to localize substrates by reducing opportunities for diffusion; the product of one reaction is released close to the active site of the next enzyme in the pathway. The enzymes may be juxtaposed morphogenetically, by being bound to one another in a membrane, as are the enzymes of mitochondrial electron transport, or they may be part of a highly organized multiprotein complex, such as the pyruvate dehydrogenase complex, a major entry point to the citric acid cycle.

Compartmentation can also result from weak interactions among enzymes that are readily isolated in soluble form. For example, conversion of glucose to pyruvate by glycolysis occurs in the cytosol, and all of the enzymes involved can readily be separated from one another. There is reason to believe, however, that they interact within the cytosol to facilitate the multistep glycolytic pathway. The concept of intracellular interactions among readily solubilized enzymes has developed from the realization that the cytosol is much more highly structured than was formerly thought. High-resolution electron micrographs of mammalian cell cytosol reveal the outlines of an organized structure that has been termed the **cytomatrix;** a model for this structure is shown in Figure 12.7. It is likely that such structures form as a result of the extremely high concentrations of proteins inside cells, which decrease the concentration of water and drive weakly interacting proteins to associate. Evidence of other such complexes exists, including associations of the enzymes that synthesize DNA precursors. However, because it is difficult to reproduce intracellular conditions in vitro, the existence of weakly interacting complexes has been difficult to demonstrate unequivocally.

Whether highly structured or loosely associated, multienzyme complexes allow for efficient control of reaction pathways. Enzyme complexes restrict diffusion of intermediates, thereby keeping their average concentrations low (but their local concentrations high, at enzyme catalytic sites). This complexing reduces "transient time" for a pathway, that is, the average time needed for a molecule to traverse the pathway. Thus, the flux through a pathway can change relatively quickly in response to a change in the concentration of the first substrate for that pathway.

Hormonal Regulation

Overlaid and interspersed with the regulatory mechanisms operating within a cell are messages dispatched from other tissues and organs. Such messages are transmitted by hormones. A **hormone** is a chemical messenger—a substance synthesized in specialized cells and carried via the circulation to remote target cells, where it interacts with specific receptors, resulting in spe-

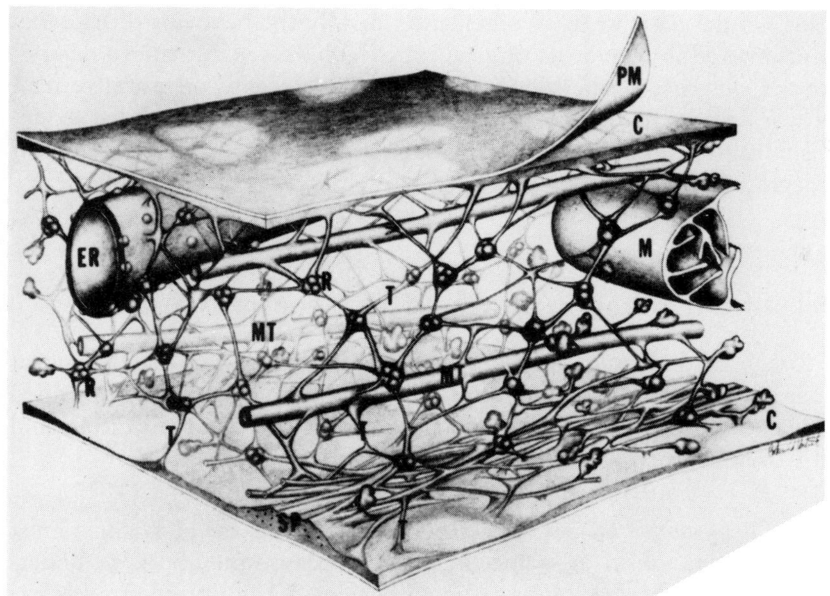

Figure 12.7
Model of the structural organization of the cytomatrix. The model is based on high-voltage electron microscopic examination of cultured mammalian fibroblast cells. PM = plasma membrane; C = cortex; ER = endoplasmic reticulum; MT = microtubules; R = free ribosomes; M = mitochondrion. Approximate magnification, ×150,000.

cific metabolic changes in the target cell. These changes are effected by all of the regulatory mechanisms we have discussed thus far: changes in the activity of particular enzymes, changes in enzyme concentration, and changes in membrane permeability to particular substrates. Insulin, a hormone that is released from the pancreas and affects target cells in liver, muscle, and adipose tissue, acts at all three levels to promote anabolic pathways and inhibit catabolism. Most other hormones have just one mechanism of action. Epinephrine, for example, functions through activation or inactivation of target enzymes, while the steroid hormones act primarily at the level of gene expression, by activating the synthesis of target proteins. Hormone action is discussed in more detail in Chapter 23.

Experimental Analysis of Metabolism

In Vitro and in Vivo Approaches

A goal of the biochemist is to understand, for a particular metabolic process, all of the chemical reactions that are part of that process. This necessitates the fractionation of cell-free preparations, so that the biochemist can isolate enzymes and other components and identify their actions. Cell-free, or in vitro, preparations can be manipulated in ways that intact cells cannot—for example, by addition of substrates and cofactors that will not pass through cell membranes. The researcher attempts to duplicate in vitro the process known to occur in vivo.

Preparation of cell-free components can be highly destructive of biological organization. Cells can be broken open by sonic oscillation, shear forces, or enzymatic digestion of cell walls. These harsh treatments inevitably lead to mixing of components that were in separate compartments in the intact cell. This mixing provides numerous opportunities for the biochemist to misinterpret data obtained from in vitro systems. A good example comes from studies on protein biosynthesis, a process that occurs on a messenger RNA template (Chapter 4). The existence of messenger RNA was predicted in 1961 on the basis of in vivo evidence, mainly the behavior of bacterial mutants altered in regulation of gene expression. It was difficult to demonstrate the existence of messenger RNA in vitro, because the putative template was present in very small amounts and was rapidly degraded by enzymes in cell-free extracts. Only when investigators learned how to prevent this degradation could they prove the existence of messenger RNA.

Levels of Organization at Which Metabolism Is Studied

The foregoing discussion reveals the necessity of studying metabolism at various levels of biological organization, from intact organism to purified chemical components. Here we discuss these levels of organization and what can be learned at each level.

WHOLE ORGANISM. Despite the difficulties involved, the biochemist must investigate metabolism in whole organisms, because the aim is to understand chemical processes in intact living systems. Radioisotopic tracers are widely used for studies at this level, as described in Tools of Biochemistry 16. A classical example is Konrad Bloch's studies of cholesterol synthesis in the 1940s. Bloch injected ^{14}C-labeled acetate into rats and followed the flow of label into intermediates by sacrificing rats at various times and analyzing radioactive compounds present in the liver. In designing experiments of this type, the investigator must worry about the efficiency of transport of the labeled precursor to the organs of interest, uptake into cells, and competition of the exogenous precursors and preexisting pools of unlabeled intermediates.

Sometimes these difficulties, particularly the problem of transport, can be circumvented by using a whole organism that is itself a single cell. In plant biochemistry, Melvin Calvin elucidated the path of carbon fixation in photosynthesis by using the green alga *Chlorella*. Radiolabeled bicarbonate was administered to cell suspensions for a few seconds, and then the labeled compounds were extracted and analyzed chromatographically.

Many diagnostic tests in clinical medicine are in vivo metabolic experiments. A good example is the **glucose tolerance test.** The subject consumes a large oral dose of glucose, and its level in the blood is then determined at intervals over several hours. The glucose tolerance test is used to diagnose diabetes and other disorders of carbohydrate metabolism.

In recent years an exciting new technique—nuclear magnetic resonance (NMR) spectroscopy—has been used for "noninvasive monitoring" of intact cells and organs. As explained in Tools of Biochemistry 10, compounds containing certain atomic nuclei can be identified from an NMR spectrum, which measures shifts in frequency of electromagnetic radiation absorbed. It is possible to determine an NMR spectrum of whole cells or of organs or tissues in an intact plant or animal. For the most part, macromolecular components do not contribute to the spectrum, nor do compounds that are present at less than about 0.5 mM. The nuclei most commonly used in this in vivo NMR technique are ^{1}H, ^{31}P, and ^{13}C. Figure 12.8 shows ^{31}P NMR spectra that represent components in the human forearm muscle. The five major peaks correspond to the phosphorus nuclei in orthophosphate, creatine phosphate, and the three phosphates of ATP. Since ADP and AMP can also be detected, and since peak area is proportional to concentration, one can determine the energy status of intact cells. For example, an energy-rich muscle has lots of creatine phosphate, whereas a fatigued muscle has none and also has decreased ATP and increased ADP and AMP levels. This approach to metabolic monitoring is finding wide applicability in monitoring recovery from heart attacks, where cellular ischemia (insufficient oxygenation) damages cells by reducing relative ATP content. The technique can also be used to study metabolite compartmentation, flux rates through major metabolic pathways, and intracellular pH.

ISOLATED OR PERFUSED ORGAN. Some of the difficulties in transporting a precursor or inhibitor to the desired organ can be circumvented by isolating the organ before carrying out the experiment. One usually **perfuses** the organ during the experimental manipulations; this involves pumping through the organ a buffered isotonic solution containing nutrients, drugs, or hormones. This solution partly takes the place of the circulation that now is absent, delivering nutrients and removing waste products. The researcher can also perfuse an organ within a living animal, following appropriate surgical procedures. Of course, perfusion is much less efficient than circulation; thus, experiments at this level must be of limited duration.

The circulation problem can be partly overcome by cutting the tissue into thin slices before the experimental manipulations begin. One loses the structural integrity of the organ, but most of the cells remain intact, and they are in much better contact with the fluid that bathes the tissue. The cells may be better oxygenated and supplied with substrates than in a whole organ. The elucidation of the citric acid cycle resulted largely from experiments with slices of liver and heart.

WHOLE CELLS. Tissue slices are not as widely used as they used to be, partly because methods are now available for disaggregating an organ or tissue into its component cells. Liver, kidney, and heart cells can be prepared by treatment of the organ with trypsin or with collagenase, an enzyme that degrades the collagen that forms much of the extracellular matrix and, hence, holds the organ together. For plant cells, enzymes such as cellulase or pectinase, which attack the cell wall, can be used to make comparable preparations. Whether from animals or plants, such cell preparations retain full metabolic activity for many hours.

Any organ from a plant or animal contains a complex mixture of different cell types. There are several means of fractionating the cells after disaggregation of an organ, to obtain preparations enriched in one cell type. One method is centrifugation to separate cells on the basis of size. In recent years the **fluorescence-activated cell sorter** has come into widespread use. In one application a cell suspension is treated with a fluorescent-tagged antibody to a cell surface component that binds differentially to different cell types. Cells pass in single file through a laser beam and are physically separated based on the amount of fluorescence recorded from each cell. Such machines can sort several thousand cells per second.

Uniformity in a cell population is often achieved by growth of cells in **tissue culture.** Disaggregated cells of an organ or tissue can, with special care, be induced to grow in a medium containing cell nutrients and protein growth factors. The cells grow and divide independently of one another,

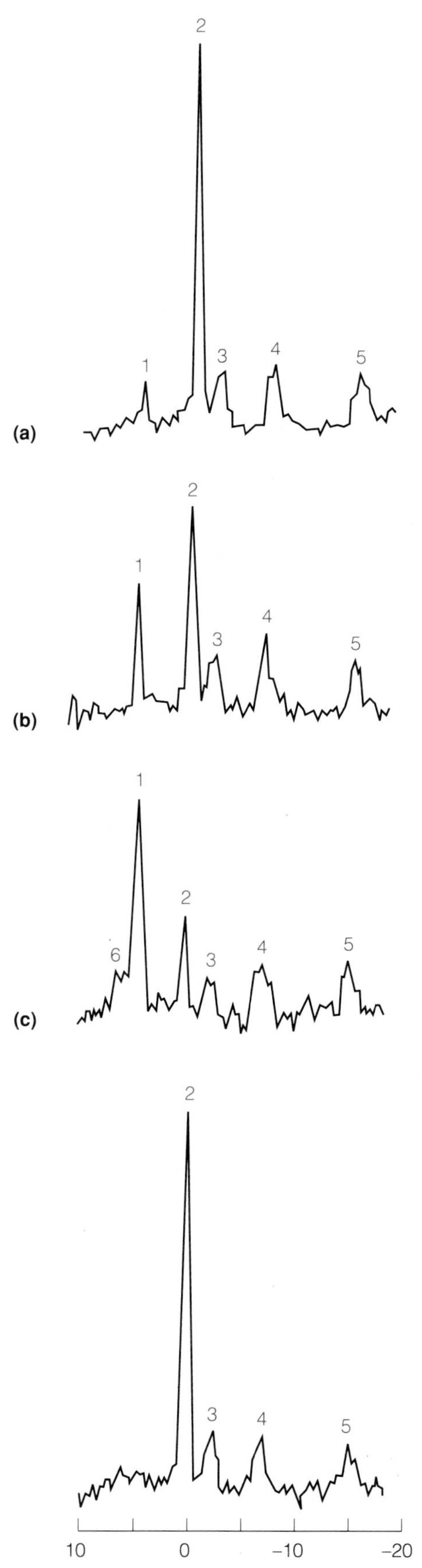

Figure 12.8
^{31}P NMR spectra of human forearm muscle (**a**) before, (**b** and **c**) during, and (**d**) after exercise. (**b**) One minute into a 19-minute exercise period; (**c**) the 19th minute. Peak heights are proportional to intracellular concentrations. Peak 1 = P_i; 2 = creatine phosphate; 3 = ATP γ-phosphate; 4 = ATP α-phosphate; 5 = ATP β-phosphate.

much like the cells in a bacterial culture. Although animal cells usually cease growth after a certain number of divisions in culture, variant lines arise that are capable of indefinite growth in culture, as long as they are adequately nourished. In such cultures, **clonal** cell lines can be generated in which all of the cells in a line are derived from a single cell, so that they are genetically similar and metabolically uniform. This uniformity is a boon for many biochemical investigations. For example, much of our understanding of virus replication comes from experiments where one can infect a large number of identical cells in culture simultaneously and then follow the metabolic changes by sampling the cell culture at various times after infection.

One problem with tissue culture is that cells adapted to long-term growth in culture take on characteristics different from those of their parent cells, which were originally embedded in plant or animal tissue. The maintenance of specialized cell characteristics in culture always presents a challenge.

CELL-FREE SYSTEMS. Problems of transport through membranes are obviated by working with broken cell preparations. Animal cells are easily ruptured by mild shear forces, suspension in hypotonic medium, or freezing and thawing. Bacterial cells have a rigid cell wall that requires vigorous treatment such as sonic oscillation or grinding in the presence of abrasive substances. Enzymatic digestion with lysozyme is often used to open bacterial cells under relatively mild conditions. Yeasts and plants have especially tough cell walls that usually require combinations of enzymatic and mechanical treatment to be broken.

Initial metabolic experiments in unfractionated cell-free extracts usually lead to attempts to localize a metabolic pathway within a particular cell compartment. Much of our understanding of DNA replication and transcription in eukaryotic cells comes from investigations with isolated nuclei, and isolated mitochondria have yielded much of our understanding of respiratory electron transport and oxidative phosphorylation. Cell organelles are usually isolated by **differential centrifugation** (Figure 12.9). Lysis is carried out in isotonic sucrose solutions and generally yields morphologically intact organelles. These can be separated by centrifugation at different speeds and for different lengths of time. Typically, nuclei, mitochondria, chloroplasts, lysosomes, and microsomes can all be at least partially separated from each other, with the contents of the cytosol remaining in the supernatant after the final centrifugation step.

PURIFIED COMPONENTS. Since the biochemist seeks to understand each biological process at the molecular level, an important task is to purify to homogeneity all of the factors involved in that process. Often, as in the case of the citric acid cycle, this is simply a matter of purifying the individual enzymes involved, determining the substrate and cofactor requirements of each, recombining the purified enzymes, and showing that the entire process can be catalyzed by purified components. This process is called **reconstitution.** Some pathways require cell constituents other than enzymes, such as the ribosomes and transfer RNAs needed for protein synthesis.

In purifying individual components the biochemist continually runs the risk of losing factors essential for normal control or for some other aspect of the process under study. Avoiding such pitfalls requires painstaking experiments in which the researcher establishes criteria for the biochemical integrity of a pathway and carries out reconstitution experiments at each stage of a fractionation. An excellent example of this approach is the

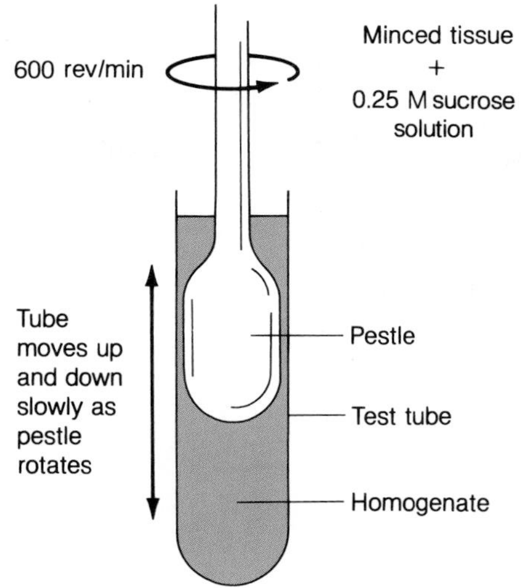

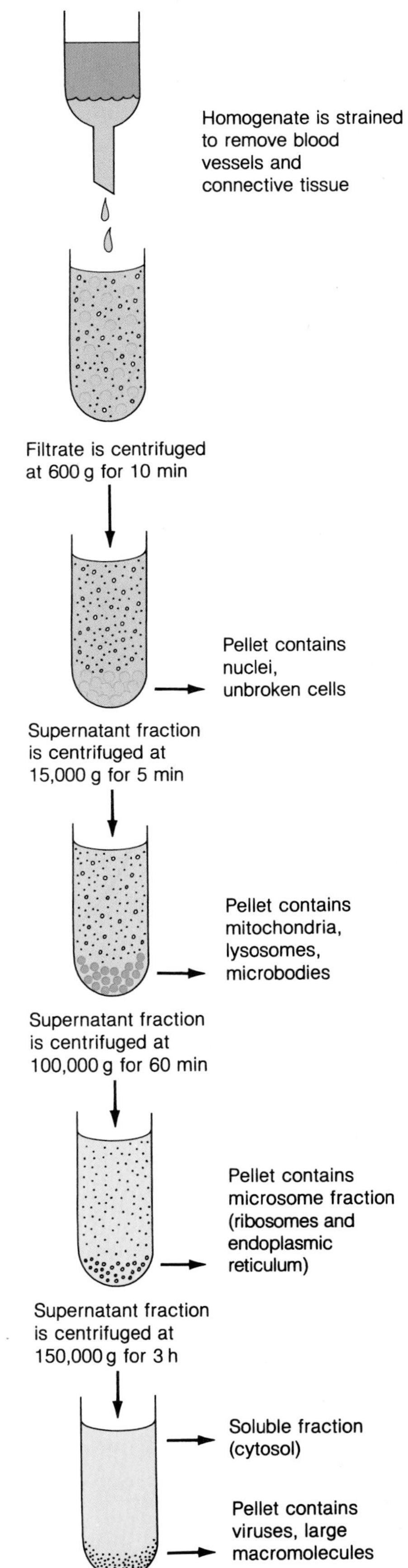

Figure 12.9
Subcellular fractionation of organelles by cell homogenization followed by differential centrifugation.

fractionation of enzymes and other proteins needed to catalyze the various steps in DNA replication; we discuss this technique in Chapter 24.

The Utility of Metabolic Probes

An invaluable aid to metabolic investigation is the ability to interfere specifically with one or a small number of reactions in a pathway; the consequences of such interference can be extremely informative. Two kinds of probes are most widely used—*metabolic inhibitors* and *mutations*. Both probes provide the opportunity to block a specific reaction in vivo and to determine the results of the blockade in order to help identify the metabolic role of a reaction. An excellent example of the use of metabolic inhibitors—identification of the order of electron carriers in the respiratory electron transport chain—is described in Chapter 15.

Inhibitors can present difficulties, such as poor transport into cells or multiple sites of action. More specific inhibition can be achieved by selection of mutant strains deficient in the enzyme of interest. The earliest example is the work of George Beadle and Edward Tatum in the 1940s with the bread mold *Neurospora crassa.* A number of x-ray-induced mutants required arginine for growth, in addition to the constituents of minimal medium. Beadle and Tatum found that their mutations affected different enzymes in the arginine biosynthetic pathway. Each mutation would accumulate the intermediate that was the substrate for the deficient enzyme. If a culture filtrate from mutant A allowed mutant B to grow without added arginine, one could conclude that mutation A blocked an enzymatic step *later* in the pathway than the step blocked by mutation B (Figure 12.10). In this way the mutations could be ordered in terms of the reactions controlled. Ultimately, the accumulating intermediates were identified and the pathway was elucidated.

In addition to identifying pathways, mutants have been used to elucidate genetic regulatory mechanisms. The earliest successes came from studies by François Jacob and Jacques Monod, of the Pasteur Institute in Paris, who isolated dozens of *Escherichia coli* mutants with defects in regulation

A —I→ B —II→ C —III→ D —IV→ E

Mutant defective in enzyme:	Accumulates metabolite:	Requires for growth metabolite:	Can feed mutant defective in enzyme:
I	A	B, C, D, or E	–
II	B	C, D, or E	I
III	C	D, or E	I or II
IV	D	E	I, II, or III

Figure 12.10
Identification of a hypothetical metabolic pathway by analysis of mutants defective in individual steps of the pathway. Metabolite C is identified as the substrate for enzyme III by the absence of enzyme III in mutants that accumulate C. Also, D and E are identified as metabolites that follow C in the pathway, because feeding either D or E to mutants defective in III bypasses the genetic block and allows the cells to grow.

of lactose catabolism or with abnormalities in virus–host relationships. These data led ultimately to the discovery of mRNA and the repressor–operator mechanism of genetic regulation (see Chapter 26).

Until recently, the genetic approach to metabolism was limited by the fact that relatively few biological systems, primarily viruses and single-celled organisms, were amenable to biochemical–genetic manipulation. Moreover, it was necessary to have either a workable plan to select desired mutations or the patience to analyze hundreds or thousands of different organisms, searching for the desired mutant. The ability to isolate genes from virtually any organism by recombinant DNA techniques has greatly extended our ability to isolate desired mutations. The cloned gene can be subjected to site-directed mutagenesis (see Tools of Biochemistry 22), so that almost any desired DNA structural change can be made at the will of the investigator. The current challenge is reintroducing this gene back into an intact organism.

Combining genetics with the use of metabolic inhibitors presents another powerful approach. As an example, consider DNA gyrase, one of the topoisomerases mentioned in Chapter 4. This enzyme is inhibited by a number of compounds, including **nalidixic acid.** When administered to whole bacteria, nalidixic acid inhibits DNA replication, which suggests that DNA gyrase plays an essential role in DNA replication. However, since nalidixic acid may inhibit DNA replication by blocking a different enzyme, stronger evidence is needed. Such evidence was obtained when mutants resistant to nalidixic acid were obtained and found to contain an altered form of DNA gyrase, also resistant to this inhibitor. Thus, a single mutation abolished both nalidixic acid sensitivity of the enzyme and the ability of cells to replicate their DNA, suggesting strongly that DNA gyrase plays an essential role in DNA replication.

At this stage we have described the general strategy of metabolism, identified the major pathways, described how pathways are regulated, and identified experimental approaches to understanding metabolism. The stage is now set for detailed descriptions of metabolic pathways, which we begin in the next chapter with carbohydrates.

REFERENCES

Experimental Techniques in the Study of Metabolism

Alberts, B., D. Bray, J. Lewis, M. Raff, K. Roberts, and J. D. Watson (1989) *Molecular Biology of the Cell*, ed 2. Garland Publishing Co., New York, pp. 135–198. Chapter 4 of this outstanding book, entitled "How Cells Are Studied," describes many of the experimental techniques mentioned in this chapter and has an exhaustive reference list.

Bottomley, P. A., R. J. Herfkens, L. S. Smith, S. Brazzamano, R. Blinder, L. W. Hedlund, J. L. Swain, and R. W. Redington (1985) Noninvasive detection and monitoring of regional myocardial ischemia in situ using depth-resolved ^{31}P NMR spectroscopy. *Proc. Natl. Acad. Sci. USA* 82:8747–8751. This article describes an approach to the use of whole-cell NMR as a clinical diagnostic tool.

Radda, G. K. (1986) The use of NMR spectroscopy for the understanding of disease. *Science* 233:640–645. This noninvasive technique for investigating the bioenergetics of tissues and organs is finding numerous applications in human medicine, some of which are described here.

Compartmentation and Intracellular Enzyme Organization

Bessman, S. P., and C. L. Carpenter (1985) The creatine–creatine phosphate energy shuttle. *Annu. Rev. Biochem.* 54:831–862. A contemporary account of the bioenergetic role of creatine phosphate, which points up the importance of compartmentation as a metabolic control phenomenon.

Srere, P. A. (1987) Complexes of sequential metabolic enzymes. *Annu. Rev. Biochem.* 56:89–124. A timely review of evidence supporting structural organization of sequential metabolic pathways, including both membranous complexes and those involving soluble enzymes.

Enzyme Control and Metabolic Regulation

Chock, P. B., S. G. Rhee, and E. R. Stadtman (1980) Interconvertible enzyme cascades in cellular regulation. *Annu. Rev. Biochem.* 49:813–845. Stadtman is chiefly responsible for elucidating the regulatory cascade affecting glutamine synthetase, so he and his colleagues write about this subject with special authority.

Koshland, D. E. (1984) Control of enzyme activity and metabolic pathways. *Trends Biochem. Sci.* 9:155–159. A brief but provocative essay, in the special 100th issue of TIBS.

Newsholme, E. A., R. A. J. Challiss, and B. Crabtree (1984) Substrate cycles: Their role in improving sensitivity in metabolic control. *Trends Biochem. Sci.* 9:277–280. A brief but lucid discussion of substrate cycle control, with several examples.

Nishizuka, Y. (1984) Protein kinases in signal transduction. *Trends Biochem. Sci.* 9:163–166. A short review about a topic that receives brief treatment in this chapter but will be encountered repeatedly throughout the book.

PROBLEMS

1. Write a balanced equation for the complete oxidation of each the following, and calculate the respiratory quotient for each substance.
   (a) Ethanol
   (b) Acetic acid
   (c) Stearic acid
   (d) Oleic acid
   (e) Linoleic acid

2. Given what you know about the involvement of nicotinamide nucleotides in oxidative and reductive metabolic reactions, predict whether the following intracellular concentration ratios should be (1) unity, (2) greater than unity, or (3) less than unity. Explain your answers.
   (a) $[NAD^+]/[NADH]$
   (b) $[NADP^+]/[NADPH]$

   Since NAD^+ and $NADP^+$ are essentially equivalent in their tendency to attract electrons, discuss how the two concentration ratios might be maintained inside cells at greatly differing values.

3. Free energy changes under intracellular conditions differ markedly from those determined under standard conditions. $\Delta G^{\circ\prime}$ for ATP hydrolysis to ADP and P_i is -30.5 kJ/mol. Calculate $\Delta G'$ for ATP hydrolysis in a cell at 37°, which contains ATP at 3 mM, ADP at 0.2 mM, and P_i at 50 mM.

4. Consider a substrate cycle operating with enzymes X and Y in the following hypothetical metabolic pathway.

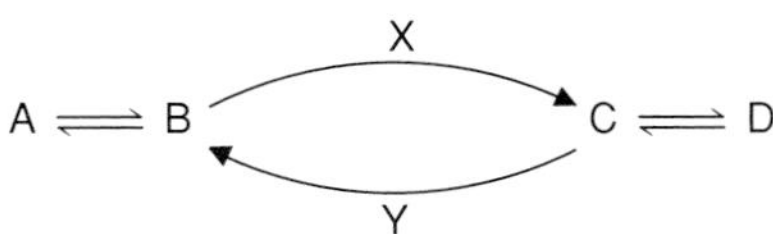

   (a) Under intracellular conditions the activity of enzyme X is 100 pmol/10^6 cells/s, and that of enzyme Y is 80. What is the direction and rate of metabolic flux between B and C?
   (b) Calculate the effect on metabolic flux rate and direction of (1) an inhibitor that reduces the activity of X by 10%; (2) an activator that increases the activity of X by 10%; (3) doubling the activity of enzyme Y.
   (c) Discuss possible experimental approaches to the question of whether this substrate cycle plays a significant role in regulating the pathway from A to D.

5. Two-dimensional gel electrophoresis of proteins in a cell extract provides a qualitative way to compare proteins with respect to intracellular abundance. Describe a quantitative approach to the determination of number of molecules of an enzyme per cell.

6. Mammalian cells growing in culture were labeled for a long time in $[^3H]$thymidine to estimate the rate of DNA synthesis. The thymidine administered had a specific activity of 3000 cpm/pmol. At intervals, samples of culture were taken and acidified to precipitate nucleic acids. The rate of incorporation of isotope into DNA was 1500 cpm/10^6 cells/min. A portion of culture was taken for determination of specific activity of the intracellular dTTP pool, which was found to be 600 cpm/pmol.
   (a) What fraction of the intracellular dTTP is synthesized from the exogenous precursor?

(b) What is the rate of DNA synthesis, in terms of molecules per minute per cell of thymine nucleotides incorporated into DNA?

(c) How might you determine the specific activity of the dTTP pool?

7. Suppliers of radioisotopically labeled compounds usually provide each product as a mixture of labeled and unlabeled material. Unlabeled material is added deliberately as a **carrier**, partly because the specific activity of the carrier-free product is too high to be useful and partly because the product is more stable at lower specific activities. Using the radioactive decay law, calculate the following.

(a) The specific activity of carrier-free [^{32}P]orthophosphate, in mCi/μmol.

(b) In a preparation of uniform-label [^{3}H]leucine, provided at 10 mCi/μmol, the fraction of H atoms that are radioactive.

TOOLS OF BIOCHEMISTRY 16

Radioisotopes and the Liquid Scintillation Counter

Radioisotopes revolutionized biochemistry when they became available to investigators shortly after World War II. Radioisotopes extend by several orders of magnitude the sensitivity with which chemical species can be detected. Traditional chemical analysis can detect and quantify molecules in the micromole or nanomole range (i.e., 10^{-6} to 10^{-9} mol). A compound which is "tagged" or "labeled," that is, which contains one or more atoms of a radioisotope, can be detected in picomole or even femtomole amounts (i.e., 10^{-12} or 10^{-15} mol). This is why radiolabeled compounds are called **tracers**; they allow one to follow specific chemical or biochemical transformations in the presence of a huge excess of nonradioactive material.

Introductory chemistry tells us that isotopes are different forms of the same element; they have different atomic weights but the same atomic number. Thus, the chemical properties of the different isotopes of a particular element are virtually identical. Isotopic forms of an element exist naturally, and substances enriched for rare isotopes can be isolated and purified from natural sources. Most of the isotopes used in biochemistry, however, are produced in nuclear reactors. Simple chemical compounds produced in such reactors are then converted to radiolabeled biochemicals by chemical and enzymatic synthesis.

Stable Isotopes

Although radioisotopes are widely used in biochemistry, stable isotopes are also used as tracers, particularly when convenient radioisotopes are not available. For example, the two rare isotopes of hydrogen include a stable isotope ($^{2}H_1$, or **deuterium**) and a radioactive isotope ($^{3}H_1$, or **tritium**). Among the many uses of stable isotopes in biochemical research, we mention three applications here. First, incorporation of a stable isotope often increases the density of a material, because the rare isotopes usually have higher atomic weights than their more abundant counterparts. This presents a way to separate labeled from nonlabeled compounds physically, for example, in equilibrium gradient centrifugation (Chapter 24). Second, compounds labeled with stable isotopes, particularly ^{13}C, are widely used in nuclear magnetic resonance studies of molecular structure and reaction mechanisms (see Tools of Biochemistry 10). Third, stable isotopes can be used as tracers in the absence of suitable radioisotopes, as noted above. For example, there are no radioisotopes of oxygen or nitrogen, so ^{18}O and ^{15}N are useful tracers of these respective elements. Stable isotopes are detected and quantified in the mass spectrometer, which is more laborious and expensive than methods used to detect radioisotopes.

Table T16.1 gives information about the isotopes, both stable and radioactive, that have found the greatest use in biochemistry.

Table T16.1
Some useful isotopes in biochemistry

Isotope	Stable or Radioactive	Emission	Half-Life	Maximum Energy (mev[a])
^{2}H	Stable			
^{3}H	Radioactive	β	12.1 years	0.018
^{13}C	Stable			
^{14}C	Radioactive	β	5568 years	0.155
^{15}N	Stable			
^{18}O	Stable			
^{24}Na	Radioactive	β (and γ)	15 hours	1.39
^{32}P	Radioactive	β	14.2 days	1.71
^{35}S	Radioactive	β	87 days	0.167
^{45}Ca	Radioactive	β	164 days	0.254
^{59}Fe	Radioactive	β (and γ)	45 days	0.46, 0.27
^{131}I	Radioactive	β (and γ)	8.1 days	0.335, 0.608

[a] mev, Million electron volts.

The Nature of Radioactive Decay

The atomic nucleus of an unstable element can decay, giving rise to one of three types of ionizing radiation: α-, β-, or γ-rays. Only β- and γ-emitting radioisotopes are useful in biochemical research, as indicated in Table T16.1. A β-ray is an emitted electron, and a γ-ray is a high-energy photon. γ-Ray detectors have found wide use in immunological research, because γ-emitting isotopes of iodine are available, and antibodies, like many proteins, can easily be iodinated without substantial changes in their biological properties. The great majority of biochemical uses of radioisotopes, however, involve β-emitters.

Radioactive decay is a first-order kinetic process. The probability that a given atomic nucleus will decay is affected neither by the number of preceding decay events that have occurred nor by interaction with other radioactive nuclei; it is an intrinsic property of that nucleus. Thus, the number of decay events occurring in a given time interval is related only to the number of radioactive atoms present. This gives rise to the **law of radioactive decay,** which can be stated mathematically as

$$N = N_0 e^{-\lambda t} \tag{T16.1}$$

where N_0 is the number of radioactive atoms at time 0, N is the number remaining at time t, and λ is a radioactive decay constant for a particular isotope, related to the intrinsic instability of that isotope. This equation states that the *fraction* of nuclei in a population that decays within a given time interval is constant. For this reason, a more convenient parameter than the decay constant λ is the **half-life,** $t_{1/2}$, the time required for half of the nuclei in a sample to decay. The half-life is equal to $-\ln 0.5/\lambda$, or $+0.693/\lambda$. The half-life, like λ, is an intrinsic property of a given radioisotope (see Table T16.1).

The basic unit of radioactive decay is the **curie** (Ci); this is defined as an amount of radioactivity equivalent to that in 1 g of radium, namely 2.22×10^{12} disintegrations per minute (dpm). Since biochemists usually work with much smaller amounts of radioactive material, it is more convenient to speak of the **millicurie** (mCi) and the **microcurie** (μCi), amounts of radioactivity corresponding, respectively, to 2.2×10^{9} and 2.2×10^{6} dpm. Because detectors of radioactivity rarely record every decay event in a sample—that is, they are not 100% efficient—we often speak of radioactivity in terms of the decay events actually recorded: counts per minute, or cpm. Counting efficiency is the percentage of decay events actually recorded, as determined, for example, by reference to standards. A counter that is 50% efficient for a given isotope would show a count rate of 1.1×10^{5} cpm for a 0.1-μCi sample.

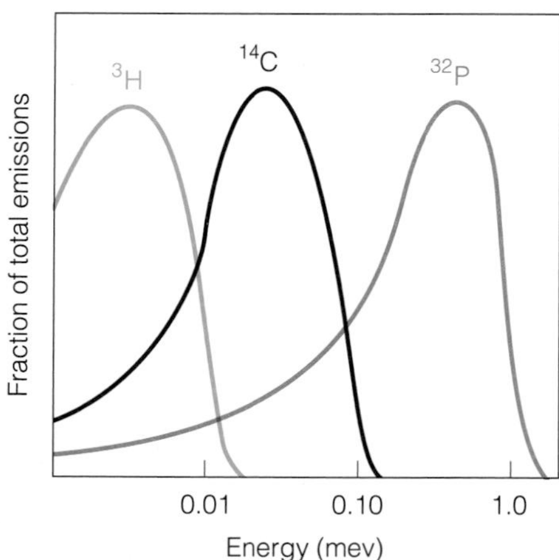

Figure T16.1
Energy spectra for the β-emitting isotopes most widely used in biochemistry—^{3}H, ^{14}C, and ^{32}P.

Detection of Radioactivity: The Liquid Scintillation Counter

Geiger counters were in widespread use two to three decades ago, but virtually all work with β-emitting isotopes is now done with the **liquid scintillation counter,** for two principal reasons. First, the scintillation counter has much higher efficiency than the Geiger counter, and second, one can simultaneously determine radioactivity from two different isotopes in the same sample.

In scintillation counting the sample is dissolved or suspended in an organic solvent, although aqueous mixtures are available as well. Also present are one or two fluorescent organic compounds, or **fluors.** A β-particle emitted from the sample has a high probability of hitting a molecule of the solvent; this excites the solvent molecule, driving an electron into a higher-energy orbital. When that electron returns to the ground state, a photon of light is emitted. The photon is absorbed by a molecule of the fluor, which in turn becomes excited. Fluorescence involves the absorption of light at a given energy, followed by emission of that light at lower energy, or longer wavelength. A photomultiplier detects that tiny flash of light and for each disintegration converts it to an electrical signal, which is recorded and counted.

Whereas γ-rays are emitted at distinct and characteristic energy values, a β-emitting isotope will display a range of energy values among its emissions. Each β-emitter shows a characteristic **energy spectrum**—that is, a plot of energy (in million electron volts, or mev) against the probability of an individual emission having that energy. Figure T16.1 shows the energy spectra for three widely used radioisotopes, ^{3}H, ^{14}C, and ^{32}P. Note that the maximum energy of the ^{14}C emissions is about 10-fold higher than that of ^{3}H and that the ^{32}P maximum is an additional 10-fold higher than that of ^{14}C. We call tritium a weak β emitter and ^{32}P a strong β emitter.

These differences in emission energies are exploited in the liquid scintillation counter so that one can simultaneously quantitate two isotopes in the same sample (a dual-label experiment). A strong β-emission excites more fluor molecules than a weak emission, so a brighter light flash is produced. This can be detected by setting electronic discriminators so that energy values falling only within a desired range are recorded; one sets a "window" that is optimized for detection of a particular isotope. This reduces counting efficiency, because only a fraction of the emissions from the higher-energy isotope are counted, but it does permit great selectivity. Most scintillation counters have three separate **counting channels,** or three different sets of discriminators. Thus, one can count simultaneously in three different windows.

Because radioactive decay is a random process, the more counts observed during the analysis of a sample, the more closely the measured radioactivity will approach the true decay rate. In practice, this means that one wants to count a sample for a long enough time to accumulate several thousand counts. Also, one must consider the signal-to-noise ratio and design an experiment so that the measured radioactivity of the samples of interest will be many times higher than **background radioactivity,** or the counts recorded in the absence of a radioactive sample.

Some Uses of Radioisotopes in Biochemistry

Radioisotopes have a great variety of uses in metabolic investigations; in this section we describe a few examples. Tracer experiments can be categorized in terms of the time of exposure of the biological system to the radioisotope and include (1) equilibrium labeling, (2) pulse labeling, and (3) pulse-chase labeling.

In **equilibrium labeling** the exposure to the tracer is relatively long, so that each labeled species reaches a constant **specific radioactivity,** or **specific activity.** The specific activity is a measure of the relative abundance of radioactive molecules in a labeled sample and is reported as radioactivity per unit mass—for example, cpm/μmol.

Experiments to identify metabolic precursors often involve equilibrium labeling conditions; the investigator administers a radiolabeled precursor so as to maximize the chance of detecting label in the product. A good example, mentioned earlier in this chapter, is Konrad Bloch's study of cholesterol biosynthesis, in which ^{14}C-labeled acetate was administered to rats. Bloch then isolated cholesterol from the liver and determined, by chemical degradation, the specific carbon atoms in cholesterol that had incorporated radioactivity.

Equilibrium labeling can be used as part of a procedure for purifying biological species of interest. An example is Mark Ptashne's purification of a genetic repressor, before a biological assay was available—indeed, before the existence of repressors had been fully proved. Ptashne had two bacterial strains, one designed to overproduce the putative repressor, the other identical except for inability to synthesize repressor. The first strain was grown for several generations in the presence of [^{14}C]leucine, and the second was grown similarly in [^{3}H]leucine. The cultures were mixed and proteins fractionated, with the expectation that repressor would be the only protein present that was labeled with ^{14}C but not ^{3}H. One protein with a high ^{14}C/^{3}H ratio was isolated and subsequently shown to be the repressor.

Another use of equilibrium labeling is to measure rates of biological processes from the rates of labeling by low-molecular-weight precursors. A good example is the widespread use of radiolabeled thymidine to follow rates of DNA synthesis, either in bacterial or cell cultures or in intact organisms. Incorporation of thymidine into macromolecules other than DNA is negligible, so one merely samples the population with time and observes the incorporation into acid-insoluble material. However, this measurement alone does not give the true rate of DNA replication. In any labeling process of this sort one must consider the metabolism of the labeled precursor en route to its ultimate destination. As shown in Figure T16.2, thymidine is converted first to thymidine monophosphate, then to the diphosphate, and finally to the triphosphate, which is the actual substrate for DNA synthesis. At each stage the labeled precursor mixes with intracellular pools of unlabeled precursors, which dilute its specific radioactivity. The labeling experiment gives only a rate of DNA labeling, in cpm incorporated per cell per unit time. To go from this to a true rate, in

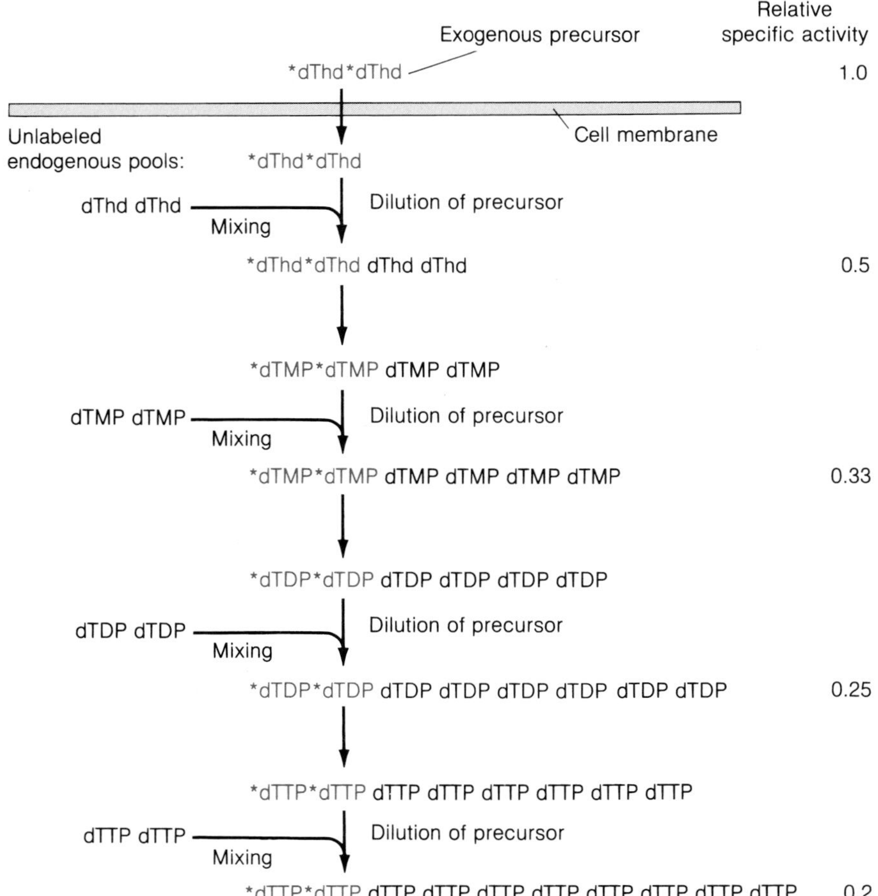

Figure T16.2
Dilution of an exogenous metabolic precursor (radiolabeled thymidine; *dThd) by unlabeled endogenous pools. In the example shown the specific activity of thymidine triphosphate is reduced to 20% of that of the exogenous thymidine; thus, the rate of incorporation of radioactivity into DNA would lead to underestimation of the true rate of DNA synthesis by fivefold.

molecules incorporated per cell per unit time, one must divide by the specific activity of the immediate precursor, in this case, dTTP. This requires isolation of dTTP in sufficient amount and purity to allow determination of both its mass and radioactivity.

Overlaid on these difficulties is the possibility of compartmentation; in this case, there may be two or more pools of dTTP in a cell, each of which becomes labeled at different rates but only one of which is used for DNA replication. There are various ways to deal with this situation, the best being to use a mutant strain unable to synthesize thymidine nucleotides. All thymidine compounds must then be supplied exogenously, and all intracellular dTTP pools will reach the same specific activity, namely that of the exogenous precursor.

Pulse labeling involves administration of an isotopic precursor for an interval that is short relative to the process under study. Radiolabel accumulates preferentially in the shortest-lived species, that is, the earliest intermediates in a metabolic pathway. Melvin Calvin identified the pathway of photosynthetic carbon fixation by labeling green algae with $^{14}CO_2$ for just a few seconds. If labeling was carried out for 10 seconds, a dozen or more radioactive compounds could be detected. After a 5-second pulse of radioactivity, only a single compound was labeled, namely 3-phosphoglycerate. This led ultimately to the discovery of ribulose-1,5-bisphosphate carboxylase as the first enzyme in the photosynthetic carbon fixation pathway (Chapter 19).

In a **pulse-chase** experiment one administers label for a short time and then rapidly reduces the specific activity of the isotopic precursor to prevent further incorporation. This can be done by adding unlabeled precursor at a molar excess of about 1000-fold, which greatly dilutes the radioisotope still present. One samples at various times afterward, to determine the metabolic fate of the material labeled during the pulse. Messenger RNA was originally detected by pulse-labeling bacterial cultures. Traditional analytical methods could not detect mRNA because of its low abundance and its metabolic instability in bacterial cells. When label incorporated into mRNA by a pulse of [^{32}P]orthophosphate or labeled uridine was chased out, the label was ultimately found to be distributed uniformly in all cellular RNA species. This showed that the metabolic instability of mRNA involves its degradation to nucleotides, which can then be used for synthesis of other RNA species. These experiments are described more fully in Chapter 26.

A final application of radioisotopes is **radioautography**, in which one incorporates an isotopic precursor into a biomolecule and prepares an image of the radiolabeled molecule on a sheet of photographic film. Chapter 24 shows several radioautograms of individual DNA molecules made radioactive by growth of *E. coli* in the presence of [^{3}H]thymidine. Of greater interest here is **two-dimensional gel electrophoresis,** a technique that allows visualization of virtually all the proteins in a cell or tissue. In the example shown in Figure T16.3, an *E. coli* culture was pulse-labeled with [^{35}S]methionine. An extract of the labeled proteins was resolved first by SDS polyacrylamide gel electrophoresis (downward from upper left corner) and then by isoelectric focusing, which separates proteins from left to right as shown. The gel was dried and used to expose a sheet of x-ray film. Each spot represents one protein, whose rate of synthesis is related to intensity of the spot. The molecular weight and isoelectric point given on the radioautogram, along with other information, allow identification of many of the protein spots.

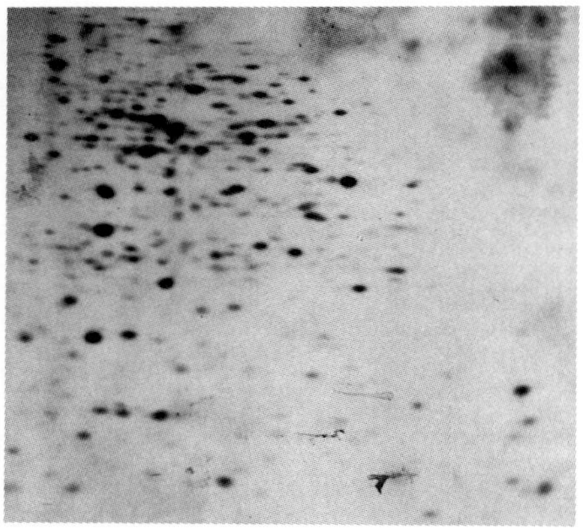

Figure T16.3
Two-dimensional electrophoresis of an extract of [^{35}S]methionine-labeled *E. coli* proteins.

REFERENCES

Freifelder, D. (1982) *Physical Biochemistry,* 2nd ed. W. H. Freeman, San Francisco. Chapter 5 of this book presents a clear description of techniques in radioactive labeling and counting.

PART IV

Dynamics of Life: Energy, Biosynthesis, and Utilization of Precursors

CHAPTER 13

Carbohydrate Metabolism I: Anaerobic Processes in Generating Metabolic Energy

Our detailed study of metabolism begins with the anaerobic phases of carbohydrate metabolism. Most of this chapter is devoted to **glycolysis,** the initial pathway in the catabolism of carbohydrates (Figure 13.1). The term glycolysis is derived from Greek words meaning "sweet" and "splitting." These are literally correct terms, for glycolysis is the pathway by which six-carbon sugars (which are sweet) are split, yielding a three-carbon compound, pyruvate. During this process some of the potential energy stored in the hexose structure is released and used to drive the synthesis of ATP from ADP. Glycolysis can proceed under anaerobic conditions, with no net oxidation of the sugar substrates taking place. **Anaerobes,** microorganisms that live in oxygen-free environments, can derive all of their metabolic energy from this process. However, aerobic cells use glycolysis as the first part of the complete oxidation of carbohydrates; in these cells glycolysis is the initial, anaerobic part of an overall degradation pathway that involves considerable oxygen consumption.

Glycolysis is an appropriate point to begin a detailed study of metabolism, for several reasons. First, it was the earliest metabolic pathway to be understood in detail. Second, the pathway is nearly universal in living cells. Third, the regulation of glycolysis is particularly well understood. Last but not least is the central metabolic role played by this pathway in generating both energy and metabolic intermediates for other pathways.

Although cells can metabolize a variety of hexose sugars via glycolysis, glucose is the major carbohydrate fuel for most cells. Indeed, some animal tissues, such as brain, normally use glucose as the sole energy source, and all energy generation in such cells begins with glycolysis. Most cells, however, can utilize other sugars, and we shall explore how they are converted to intermediates in glycolysis.

Although the molecular details of the conversion of glucose to pyruvate are identical in most organisms, pyruvate undergoes a variety of metabolic fates. These will concern us in this chapter, as will the processes by which stored carbohydrate in the form of polysaccharides is made available for utilization in glycolysis.

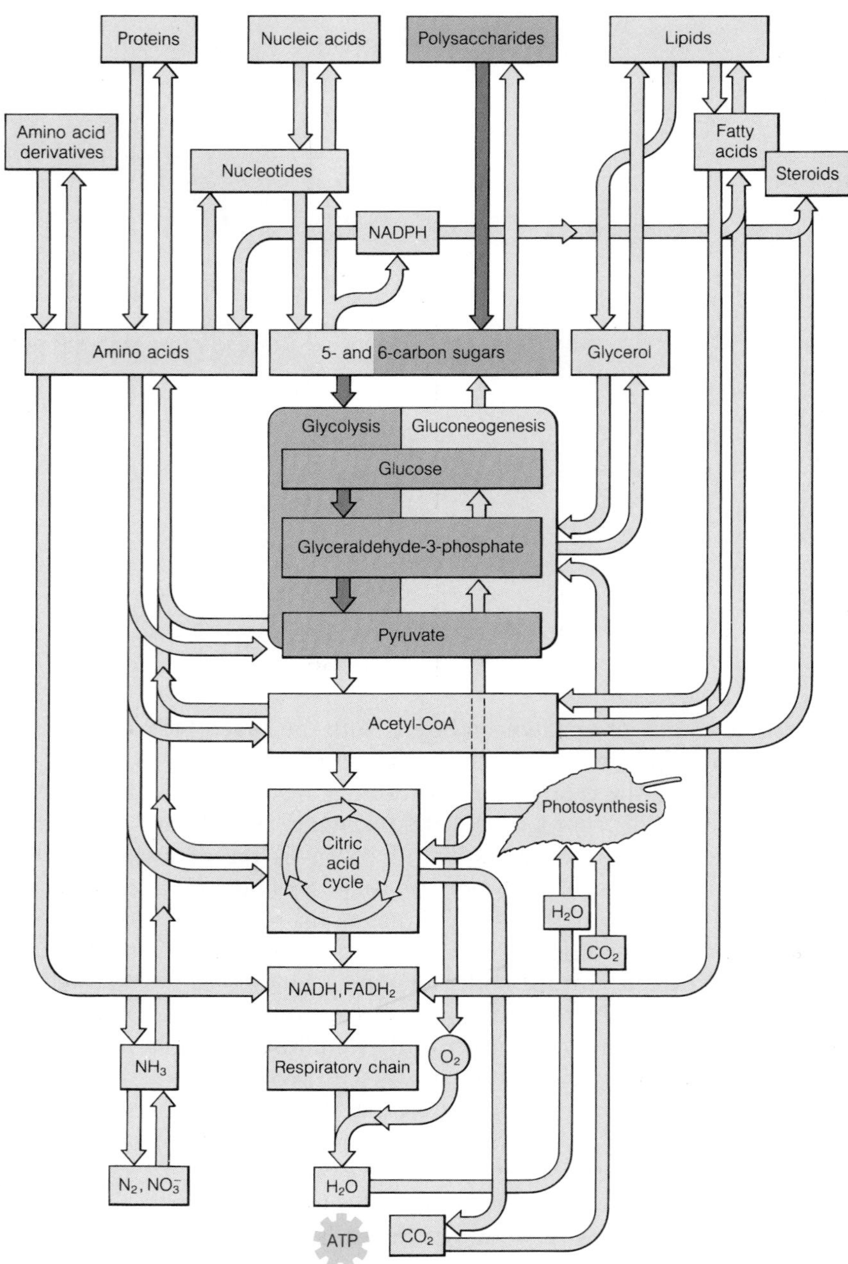

Figure 13.1
Glycolysis (purple) in the overall scheme of metabolism.

Glycolysis: An Overview

Relation of Glycolysis to Other Pathways

The generation of metabolic energy from carbohydrates begins with glycolysis, the 10-step pathway that converts one molecule of glucose to two molecules of pyruvate, with concomitant generation of two molecules of ATP. The breakdown of storage polysaccharides and the metabolism of oligosaccharides yield glucose, related hexose sugars, and sugar phosphates, all of which find their way into the glycolytic pathway. We focus initially on the pathway as it begins with glucose and then discuss the routes for entry of other carbohydrates.

The 10 reactions between glucose and pyruvate can be considered as two distinct phases, schematized in Figure 13.2. The first five reactions constitute an **energy investment phase,** in which sugar phosphates are synthesized at the expense of ATP conversion to ADP, and the six-carbon substrate is split to two three-carbon sugar phosphates. The last five reactions represent an **energy generation phase;** here the triose phosphates are converted to energy-rich compounds, which transfer phosphate to ADP, leading to ATP synthesis. The net yield, per mole of glucose metabolized, is 2 mol of ATP and 2 mol of pyruvate. Note that reducing equivalents are generated as well, in the form of NADH.

In aerobic organisms glycolysis is the first step in the complete combustion of glucose to CO_2 and water. The second step is oxidation of pyruvate to acetyl-CoA, and the final process is oxidation of the acetyl group carbons in the citric acid cycle. Chapter 14 presents the latter processes in detail. Also discussed in Chapter 14 is an alternative pathway for glucose utilization, the pentose phosphate pathway. While the latter pathway can operate to oxidize glucose carbons to CO_2, its primary functions are to generate reducing equivalents for biosynthetic processes and pentose phosphates for nucleotide synthesis. Glycolysis also provides biosynthetic intermediates. Thus, glycolysis is both an anabolic and a catabolic pathway, with an importance that extends beyond the synthesis of ATP and substrates for the citric acid cycle.

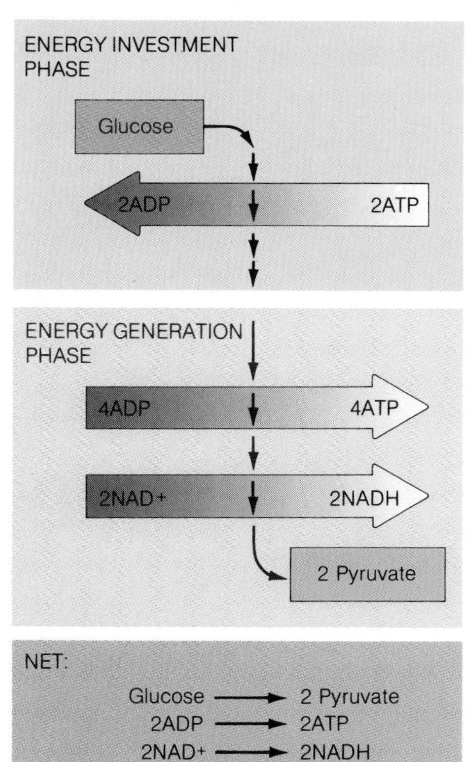

Figure 13.2
The two phases of glycolysis and the products of glycolysis.

Aerobic and Anaerobic Glycolysis; Fermentations

Glycolysis is an ancient metabolic pathway that was probably used by the earliest known bacteria, some 3.5 billion years ago. Since this was about 1 billion years before the earliest known photosynthetic organisms began contributing O_2 to the earth's atmosphere, glycolysis had to function under completely anaerobic conditions—with no net change in the oxidation state as substrates are converted to products. However, note in Figure 13.2 that the conversion of glucose to pyruvate involves the concomitant reduction of 2 mol of NAD^+ to NADH. For the pathway to operate anaerobically, NADH must be reoxidized to NAD^+ by transferring its electrons to an **electron acceptor** so that a *a steady state is maintained.* Microorganisms growing under anaerobic conditions can generate additional energy by transferring electrons to inorganic substances such as sulfate ion or nitrate ion, or they can reduce organic substrates. Most straightforward is the route used by lactic acid bacteria, which simply use NADH to reduce pyruvate to lactate, via the enzyme **lactate dehydrogenase.**

$$\text{Pyruvate} + \text{NADH} + \text{H}^+ \rightleftharpoons \text{L-lactate} + \text{NAD}^+ \qquad \Delta G^{\circ\prime} = -25.1 \text{ kJ/mol}$$

This makes glycolysis part of a **fermentation,** since there is no net change in oxidation state. The lactic acid fermentation is important in the manufacture of cheese. Another important fermentation involves cleavage of pyruvate to acetaldehyde and CO_2, with the acetaldehyde then reduced to ethanol by **alcohol dehydrogenase.**

$$\text{CH}_3\text{CHO} + \text{NADH} + \text{H}^+ \rightleftharpoons \text{CH}_3\text{CH}_2\text{OH} + \text{NAD}^+$$

As carried out by yeasts, this fermentation generates the alcohol in alcoholic beverages. Yeasts used in baking also carry out the alcoholic fermentation;

the CO_2 produced by pyruvate decarboxylation causes bread to rise, and the ethanol produced evaporates during baking. Some other fermentations of industrial importance lead to isopropanol, butyric acid, propionic acid, and acetic acid.

Animal cells are similar to lactic acid bacteria in that pyruvate, when produced faster than it can be oxidized through the citric acid cycle, is reduced to lactate. The production of lactate from glucose is called **anaerobic glycolysis.** During strenuous exertion skeletal muscle cells derive most of their energy from anaerobic glycolysis.

By contrast, consider a cell undergoing active **respiration,** which can be defined as the oxidative breakdown and release of energy from nutrient molecules by reaction with oxygen. In these cells pyruvate is oxidized to acetyl-CoA, which enters the citric acid cycle. The NADH produced during glycolysis is reoxidized through the mitochondrial electron transport chain (Chapter 15), with the electrons transferred ultimately to O_2, the terminal electron acceptor. The conversion of glucose to pyruvate in a respiring cell is called **aerobic glycolysis.**

The Crucial Early Experiments

For as long as humans have used yeasts in baking and in brewing, glycolysis has been exploited, even though it was not understood until this century. Louis Pasteur's demonstration, in 1856, that fermentations are carried out by microorganisms ranks as a milestone in the history of science. The dominant viewpoint of the time, however, was that a process such as the fermentation of glucose to ethanol was so complex that it could not be reproduced outside a living cell. In 1896, as we saw in Chapter 1, the German scientists Hans and Eduard Buchner showed that fermentation could occur under cell-free conditions.

In 1905 Arthur Harden and William Young found that inorganic phosphate, when added to yeast extract, stimulated and prolonged the fermentation of glucose. During fermentation the inorganic phosphate disappeared from the reaction medium, which led Harden and Young to suggest that fermentation was functioning via the formation of one or more sugar phosphate esters.

This opened the door to dissection of the individual chemical reactions involved in fermentation, a feat accomplished in Germany in the 1930s, largely by G. Embden, O. Meyerhof, and O. Warburg. In fact, glycolysis is often referred to as the **Embden–Meyerhof pathway.** These investigations identified 10 different reactions, virtually identical in a wide range of organisms, leading from glucose to pyruvate. Glycolysis is the first metabolic pathway to have been elucidated as a series of defined chemical reactions. By now, all of the enzymes involved have been isolated in crystalline form from diverse sources.

Strategy of Glycolysis

Glycolysis is such an important pathway that we shall examine each of its 10 reactions in some detail. Before doing so, let us look at the pathway as a whole. First, recall from Chapter 12 that in eukaryotic cells glycolysis occurs in the cytosol, while the further oxidation of pyruvate occurs in mitochondria. (Certain trypanosomes, the parasitic protozoans that cause African sleeping sickness, present an interesting exception, with glycolysis carried out in an organized cytoplasmic organelle, the **glycosome.**)

Figure 13.3 presents an abbreviated look at the conversion of glucose to pyruvate. In the "energy investment" phase of the first five reactions, the

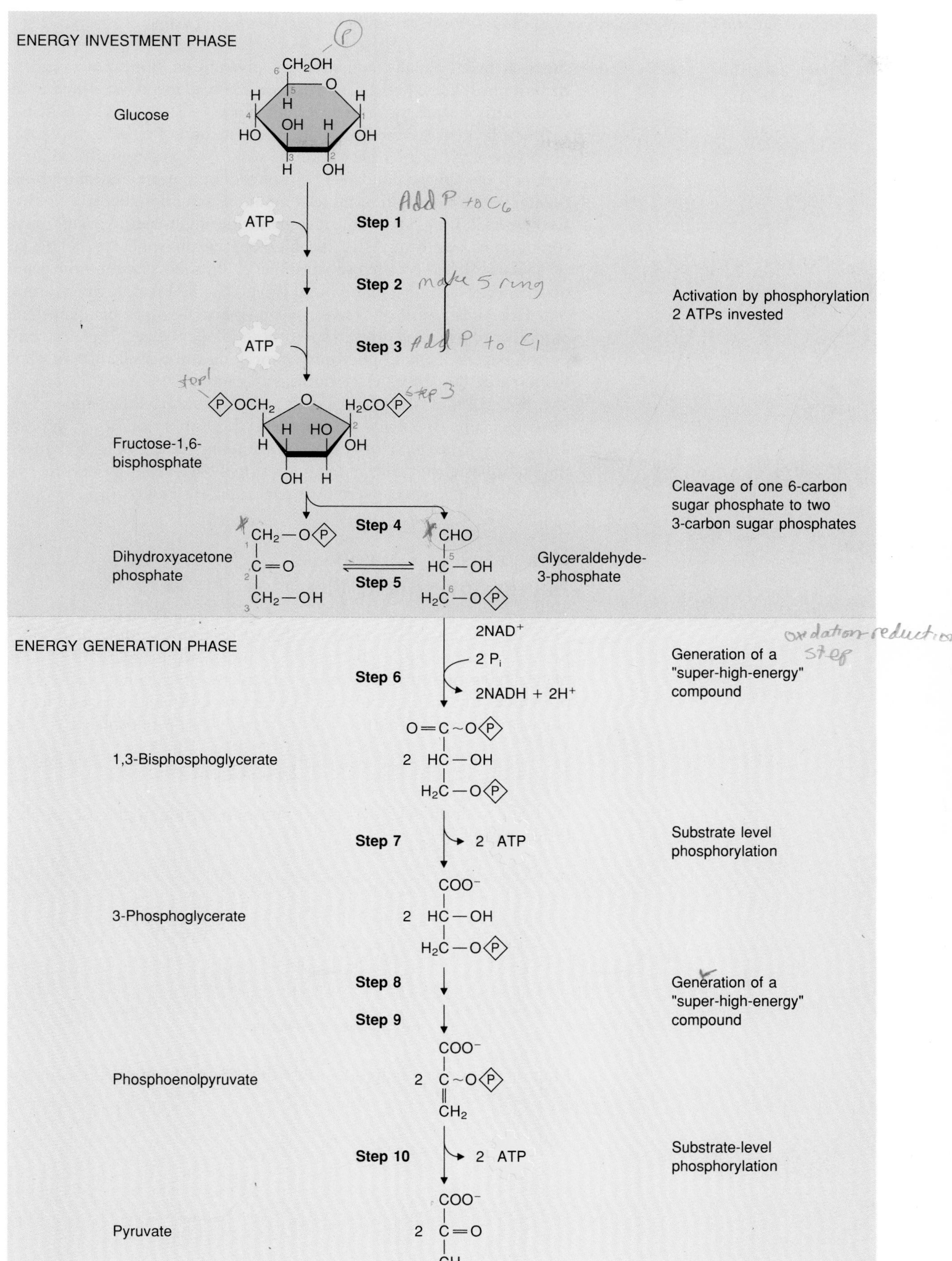

Figure 13.3
An overview of glycolysis, showing the key intermediates and reactions in each of the two major phases.

sugar is metabolically activated by phosphorylation. This yields a six-carbon phosphorylated sugar, **fructose 1,6-bisphosphate,** which undergoes cleavage to yield 2 mol of triose phosphate.

In the "energy generation" phase (reactions 6 to 10), the triose phosphates undergo further activation to yield two compounds containing energy-rich phosphate bonds—**1,3-bisphosphoglycerate** and **phosphoenolpyruvate.** Recall that both of these compounds have a higher $\Delta G^{\circ\prime}$ of hydrolysis than ATP; they can be considered as super-high-energy compounds. Both compounds readily transfer the high-energy phosphate to ADP, yielding ATP. This process is called **substrate-level phosphorylation**—the generation of an energy-rich phosphate bond driven by the breakdown of a more energy-rich substrate. Substrate-level phosphorylation is distinguished from **oxidative phosphorylation,** the synthesis of ATP driven by electron transport (Chapter 15), and **photophosphorylation,** the utilization of photosynthetic energy to drive ATP synthesis (Chapter 19).

Since 2 mol of triose phosphate are metabolized per mole of glucose, the yield from the two substrate-level phosphorylations of glycolysis is 4 mol of ATP per mole of glucose. Subtracting the 2 mol of ATP invested in the first phase, we realize a net gain of two ATP molecules synthesized per molecule of glucose converted to pyruvate (see Figure 13.2).

Reactions of Glycolysis: Energy Investment Phase

Now let us consider in sequence the 10 reactions leading from glucose to pyruvate, numbering each reaction as indicated in Figure 13.3. The complete names of substrates and products are given when each reaction is presented, but in the text these names are shortened for simplicity; thus, glucose-6-phosphate is the same as α-D-glucose-6-phosphate.

Reaction 1: The First ATP Investment

We begin with the ATP-dependent phosphorylation of glucose, catalyzed by **hexokinase.**

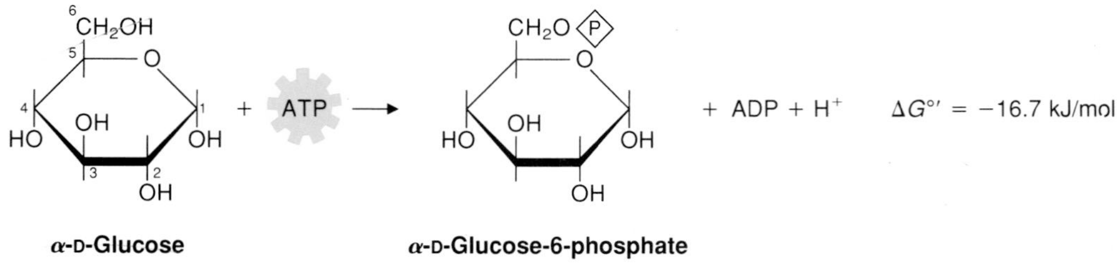

α-D-Glucose + ATP ⟶ α-D-Glucose-6-phosphate + ADP + H^+ $\quad \Delta G^{\circ\prime} = -16.7$ kJ/mol

α-D-Glucose **α-D-Glucose-6-phosphate**

Magnesium ion is required, because the reactive form of ATP is its chelated complex with Mg^{2+}.

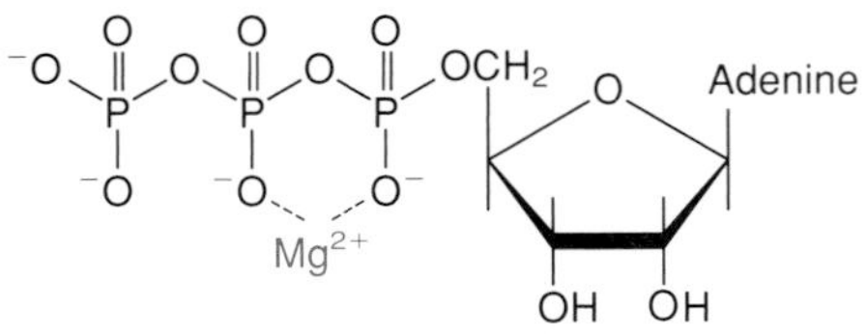

Magnesium-ATP complex

Hexokinase exists in various forms in different organisms but is generally characterized by low specificity for sugars and low K_M for the sugar substrate (about 0.1 mM). The low specificity allows phosphorylation of various hexose sugars, including fructose and mannose, and thereby permits their utilization via glycolysis. As noted in Chapter 11, hexokinase is feedback-inhibited by its product, glucose-6-phosphate, a mechanism that controls the influx of substrates into the glycolytic pathway. Recall also that the structure of hexokinase provides striking evidence for the induced fit model of enzyme catalysis.

Because intracellular glucose levels are usually far higher than the K_M value for hexokinase, the enzyme often functions in vivo at saturating substrate concentrations. In vertebrate liver another enzyme, **glucokinase,** allows assimilation of glucose when it is present at levels far exceeding the K_M for hexokinase. Like hexokinase, glucokinase phosphorylates glucose to glucose-6-phosphate. However, glucokinase has a much higher K_M for glucose, and it is highly specific for glucose. Unlike hexokinase, glucokinase is not feedback-inhibited by the product, glucose-6-phosphate. Since glucokinase comes into play when glucose levels in the liver exceed those needed for energy generation, it is likely that its role is to generate substrates for energy storage—as glycogen or fatty acids. These relationships are indicated in Figure 13.4.

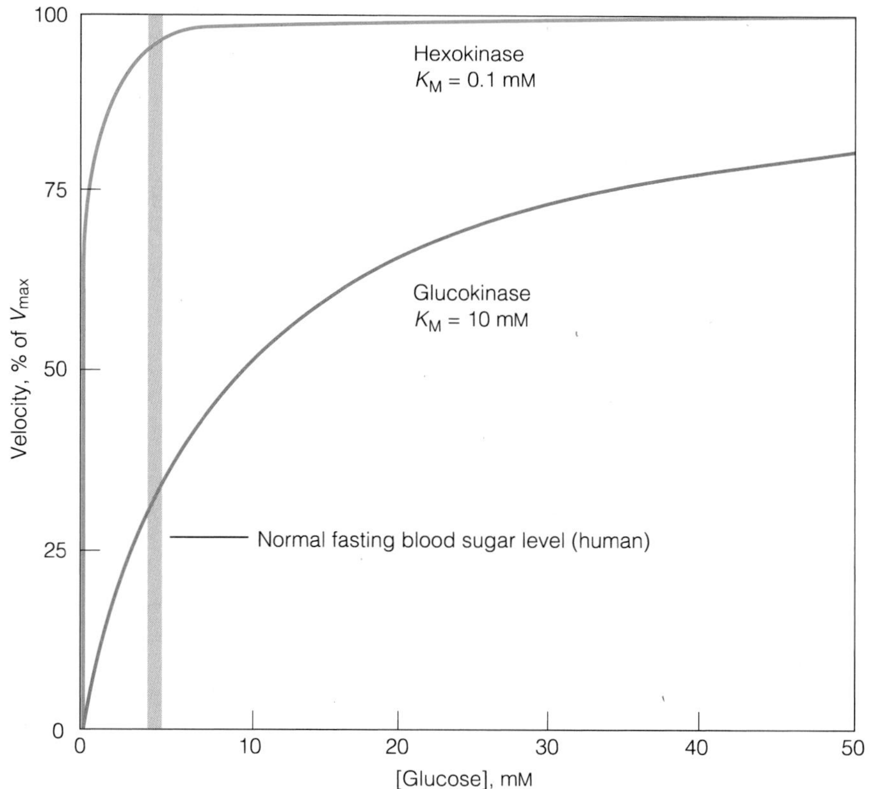

Figure 13.4
Relationship between hexokinase and glucokinase: substrate saturation curves for hexokinase and glucokinase. At a normal blood glucose level (4.4 mM) hexokinase is essentially saturated and cannot respond to small changes in blood glucose level; however, glucokinase can increase its activity when blood glucose levels rise, for example, after a meal.

Reaction 2: Isomerization of Glucose-6-phosphate

The next reaction, catalyzed by **phosphoglucoisomerase,** is the readily reversible isomerization of the aldose, glucose-6-phosphate, to the corresponding ketose, **fructose-6-phosphate.** Mg^{2+} is also required.

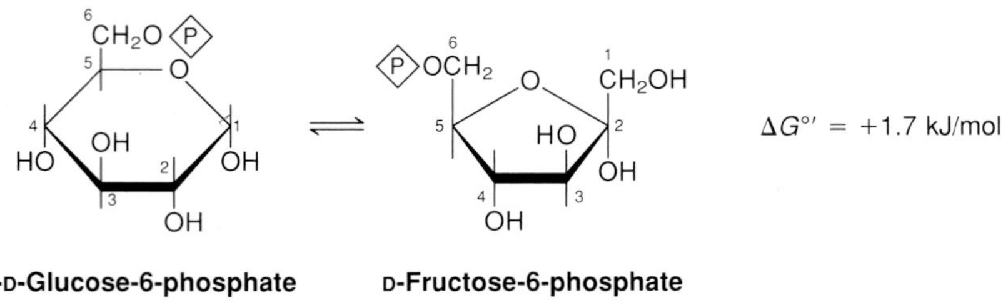

α-D-Glucose-6-phosphate ⇌ D-Fructose-6-phosphate $\Delta G^{\circ\prime} = +1.7$ kJ/mol

This reaction proceeds via an enediol intermediate, as shown below.

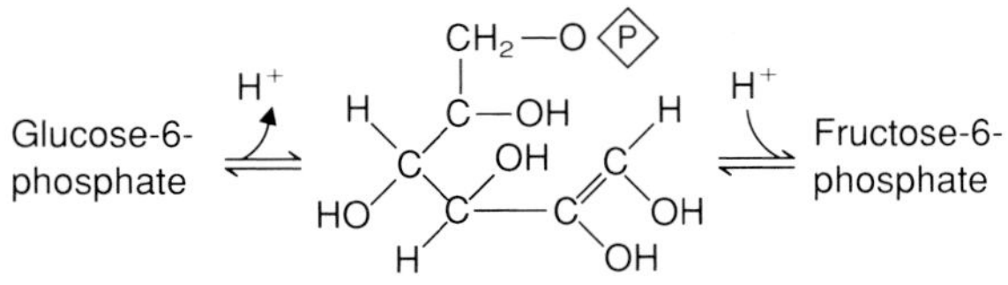

Glucose-6-phosphate ⇌ (enediol intermediate) ⇌ Fructose-6-phosphate

The effect of transferring the carbonyl oxygen from carbon 1 to carbon 2 is that the hydroxyl group generated at carbon 1 can be readily phosphorylated in the next reaction.

Reaction 3: The Second ATP Investment

The next enzyme, **phosphofructokinase,** carries out a second ATP-dependent phosphorylation, to give a hexose derivative phosphorylated at both carbons 1 and 6.

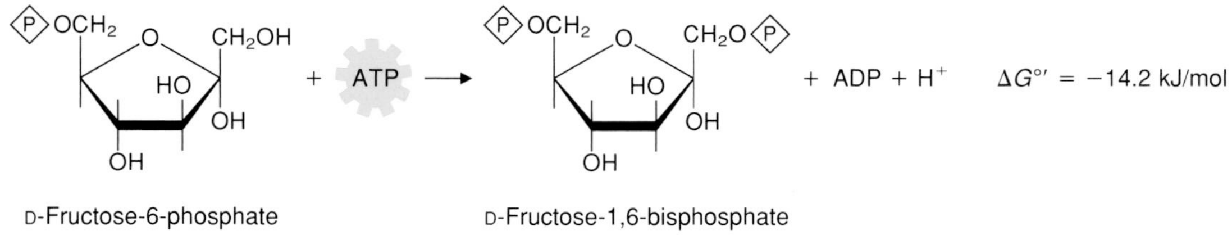

D-Fructose-6-phosphate + ATP → D-Fructose-1,6-bisphosphate + ADP + H^+ $\Delta G^{\circ\prime} = -14.2$ kJ/mol

The product, fructose-1,6-bisphosphate, was formerly called fructose-1,6-diphosphate; renaming was done to show that the two phosphates are separate, rather than linked as in ADP.

Like the phosphorylation of glucose, this reaction is sufficiently exergonic to be essentially irreversible in vivo. This is important, because phosphofructokinase (PFK) represents the primary site for regulation of the flow of carbon through glycolysis. PFK is an allosteric enzyme whose activity is acutely sensitive to the energy status of the cell, as well as to the levels of various other intermediates, particularly citrate and fatty acids. Interactions with allosteric effectors, which are discussed later in this chapter, activate PFK. Consequently, they increase carbon flux through glycolysis when there is a need to generate more ATP and inhibit it when there are ample stores of ATP or substrates for oxidation.

Higher plants contain a unique form of PFK, which couples the synthesis of fructose-1,6-bisphosphate from fructose-6-phosphate to the cleavage of pyrophosphate ion.

$$\text{Fructose-6-phosphate} + PP_i \longrightarrow \text{fructose-1,6-bisphosphate} + P_i$$

This enzyme, which is comparable in activity to ATP-dependent phosphofructokinase, seems to represent an alternative route to catalysis of the third step in glycolysis.

Reaction 4: Cleavage to Two Triose Phosphates

The enzyme **fructose-1,6-bisphosphate aldolase,** usually called **aldolase,** catalyzes a reaction that is the reverse of an aldol condensation (reaction of an alcohol with an aldehyde). Here occurs the "splitting of sugar" that is connoted by the term *glycolysis,* because here the six-carbon compound fructose-1,6-bisphosphate is cleaved to give two three-carbon intermediates.

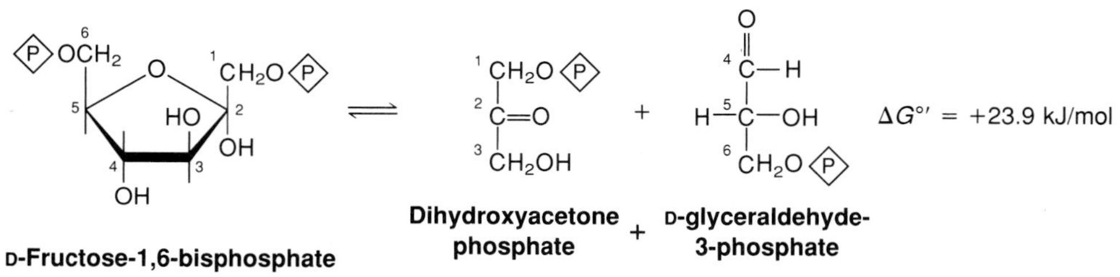

D-Fructose-1,6-bisphosphate ⇌ Dihydroxyacetone phosphate + D-glyceraldehyde-3-phosphate $\Delta G^{\circ\prime} = +23.9$ kJ/mol

This reaction illustrates an important metabolic principle. Note that the reaction is strongly endergonic under standard conditions, such that the leftward direction as written is favored. However, from the actual intracellular concentrations of the reactant and products, as determined in rabbit skeletal muscle, one can calculate a ΔG of -1.3 kJ/mol, consistent with the observation that the reaction runs rightward in vivo. This illustrates the importance of considering conditions *in the cell,* and *not* standard state conditions, in deciding which direction of a reaction is favored.

Aldolase from most vertebrate sources is a tetrameric protein. The enzyme activates the substrate for cleavage by condensing the keto carbon at position 2 with a lysine ϵ-amino group in the active site to give a **Schiff base** intermediate, as shown in Figure 13.5. Recall that a Schiff base is a condensation product between an amino group and a carbonyl group. The activated substrate undergoes abstraction of a proton from the hydroxyl group at carbon 4, followed by elimination of the resulting enolate ion.

Aldolase from bacteria and lower eukaryotes is a dimeric protein that requires Zn^{2+} for activity and does not proceed via a Schiff base intermediate.

Reaction 5: Isomerization of Dihydroxyacetone Phosphate

The aldolase reaction yields two three-carbon sugar phosphates, both of which are further catabolized in subsequent reactions of the pathway. The function of the next enzyme, **triose-phosphate isomerase,** is to convert one

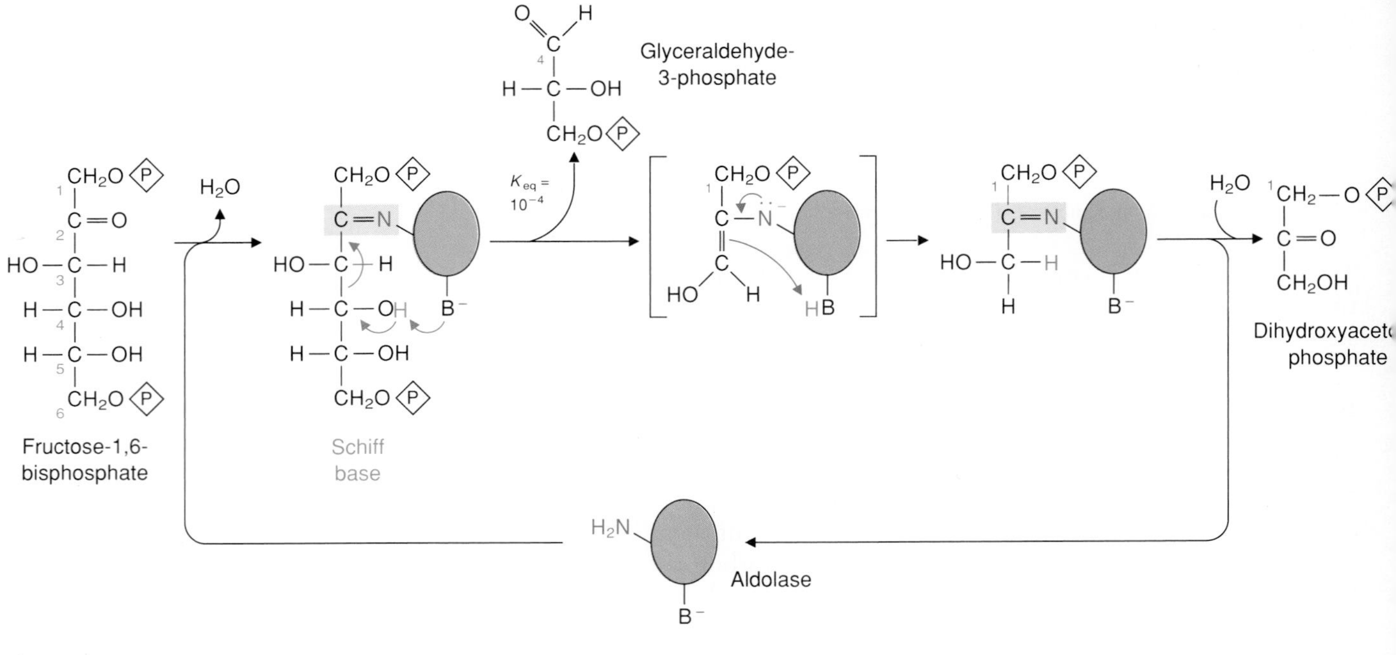

Figure 13.5
Reaction mechanism for fructose-1,6-bisphosphate aldolase. The figure shows the Schiff base intermediate between the substrate and the active site lysine residue. B is a basic residue on the enzyme, which accepts a proton from the hydroxyl on C-4 and returns it after cleavage between C-3 and C-4.

of these products, dihydroxyacetone phosphate, to glyceraldehyde-3-phosphate, the substrate for the next glycolytic reaction. We have encountered this enzyme before in Chapter 10.

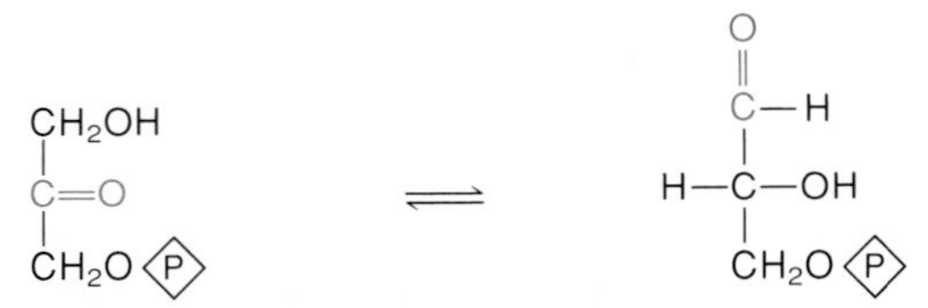

$$\text{Dihydroxyacetone phosphate} \rightleftharpoons \text{D-Glyceraldehyde-3-phosphate} \qquad \Delta G^{\circ\prime} = +7.6 \text{ kJ/mol}$$

Dihydroxyacetone phosphate **D-Glyceraldehyde-3-phosphate**

This reaction is also somewhat endergonic under standard conditions, but the intracellular concentration of glyceraldehyde-3-phosphate is low, drawing the reaction toward the right as written. Like the phosphoglucoisomerase reaction, the isomerization of dihydroxyacetone phosphate proceeds via an enediol intermediate.

At this point glycolysis has expended two ATP molecules and converted one hexose sugar to two triose phosphates, each of which is next metabolized to give high-energy compounds that can drive the synthesis of ATP. Thus, the energy investment phase of the cycle is complete, and the energy generation phase is about to begin.

Reactions of Glycolysis: Energy Generation Phase

Reaction 6: Generation of the First Energy-Rich Compound

Mechanistically, this reaction, catalyzed by **glyceraldehyde-3-phosphate dehydrogenase,** is among the most interesting in glycolysis. It is certainly one of the most important steps, partly because it generates the first high-energy intermediate and partly because it generates a pair of reducing equivalents. The overall reaction is as follows.

$$\begin{matrix} \text{O} \\ \| \\ \text{C}-\text{H} \\ | \\ \text{H}-\text{C}-\text{OH} \\ | \\ \text{CH}_2-\text{O}\text{(P)} \end{matrix} + \text{NAD}^+ + \text{P}_i \rightleftharpoons \begin{matrix} \text{O} \\ \| \\ \text{C}\sim\text{O}\text{(P)} \\ | \\ \text{H}-\text{C}-\text{OH} \\ | \\ \text{CH}_2-\text{O}\text{(P)} \end{matrix} + \text{NADH} \qquad \Delta G^{\circ\prime} = +6.3 \text{ kJ/mol}$$

D-Glyceraldehyde-3-phosphate **1,3-Bisphosphoglycerate**

This reaction involves a two-electron oxidation of the carbonyl carbon of glyceraldehyde-3-phosphate to the carboxyl level, a reaction that is normally quite exergonic. However, the overall reaction is slightly *endergonic* (under *standard* conditions), because the enzyme utilizes most of the energy released to drive the synthesis of a super-high-energy compound, 1,3-bisphosphoglycerate. This compound contains a carboxylic-phosphoric acid anhydride, or an **acyl-phosphate group,** at position 1, a functional group with a very high standard free energy of hydrolysis, −49.4 kJ/mol. This enzyme also requires a coenzyme, NAD^+, to accept electrons from the substrate being oxidized.

Because the acyl-phosphate group is much more energy rich than the phosphate anhydride of ATP, 1,3-bisphosphoglycerate can drive the synthesis of ATP from ADP. Indeed, it does so in the next reaction in the sequence, the first of two substrate-level phosphorylations in glycolysis. Because of the importance of understanding how ATP is synthesized, much attention has focused on understanding how the super-high-energy compounds in substrate-level phosphorylation are synthesized.

For glyceraldehyde-3-phosphate dehydrogenase, that understanding derived in large part from an old observation that glycolysis is inhibited by iodoacetate and by heavy metals such as mercury. Both compounds react with free sulfhydryl groups, as shown below for iodoacetate.

$$\text{RSH} + \text{ICH}_2\text{COO}^- \longrightarrow \text{RS}-\text{CH}_2\text{COO}^- + \text{HI}$$

The finding that these compounds inhibit glycolysis specifically by inhibition of glyceraldehyde-3-phosphate dehydrogenase strongly implied that the enzyme contains one or more essential thiol groups. We now know that the reaction proceeds as outlined in Figure 13.6, starting with formation of a **thiohemiacetal** group involving the substrate carbonyl group and a cysteine thiol group on the enzyme. The thiohemiacetal is next oxidized by NAD^+ to give an acyl-enzyme intermediate, or thioester. Thioesters are high-energy compounds; phosphorolysis of this thioester by P_i preserves much of the energy as the acyl phosphate, which is the product.

The overall stoichiometry of the reaction involves reduction of 1 mol of NAD^+ to $NADH + H^+$. This is the source of the NADH formed in glycolysis, which was first identified in Figure 13.2.

Reaction 7: The First Substrate-Level Phosphorylation

As noted above, 1,3-bisphosphoglycerate, because of its high group transfer potential, has a strong tendency to transfer its acyl phosphate group to ADP, with resultant formation of ATP. This substrate-level phosphorylation reaction is catalyzed by **phosphoglycerate kinase,** as shown on the next page.

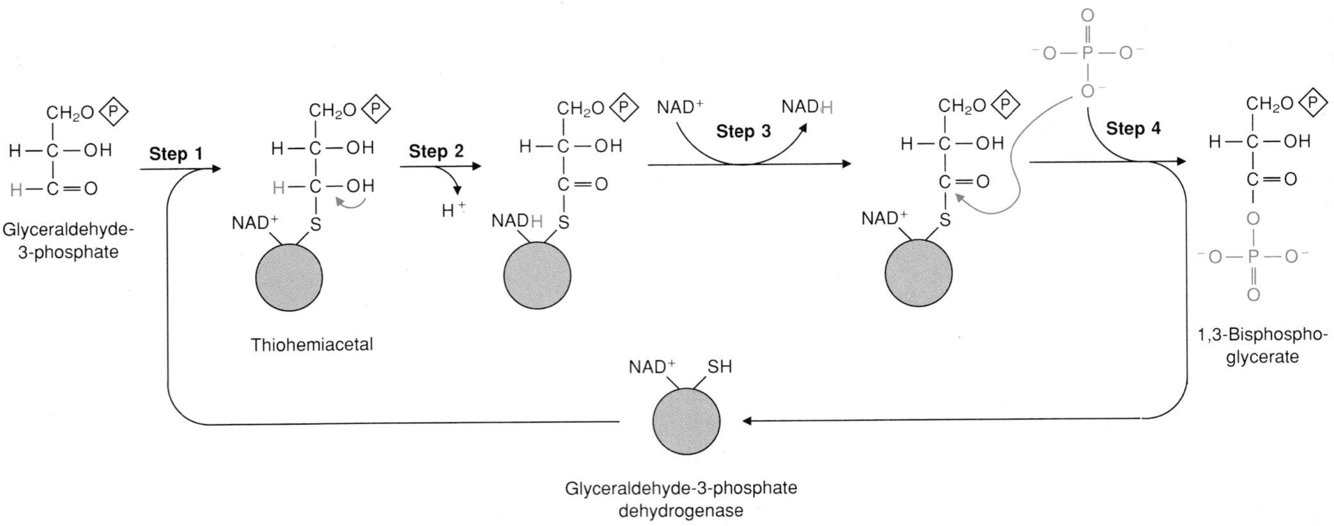

Figure 13.6
Reaction pathway for glyceraldehyde-3-phosphate dehydrogenase. Step 1, formation of the initial thiohemiacetal intermediate between glyceraldehyde-3-phosphate and the enzyme. Step 2, oxidation of this intermediate by bound NAD^+ to give an acyl enzyme intermediate. Step 3, reoxidation of the bound NADH. Step 4, phosphorolytic cleavage of the thioester bond in the acyl enzyme intermediate.

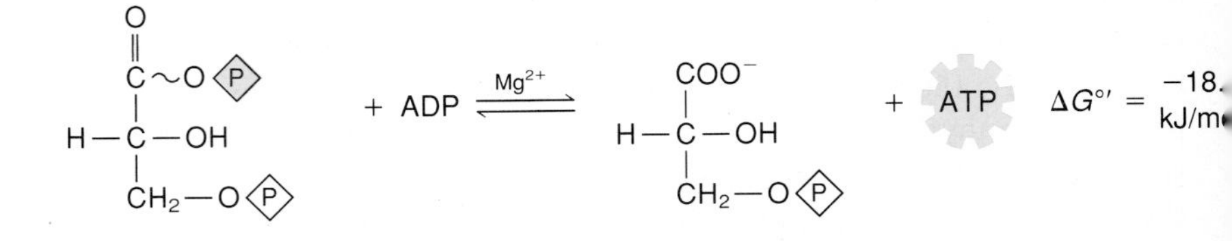

1,3-Bisphosphoglycerate + ADP $\xrightleftharpoons{Mg^{2+}}$ 3-Phosphoglycerate + ATP $\Delta G^{\circ\prime} = -18.$ kJ/m

At this stage the net ATP yield from the glycolytic pathway is zero. Recall that two ATPs per mole of glucose were invested to generate 2 mol of triose phosphate; the above reaction generates one ATP per mole of triose phosphate, or two per mole of glucose. The pathway as a whole becomes exergonic in the remaining three reactions. This involves activation of the remaining phosphate, which, in 3-phosphoglycerate, has a relatively low free energy of hydrolysis.

Reaction 8: Preparing for Synthesis of the Next High-Energy Compound

Activation of 3-phosphoglycerate begins with an isomerization catalyzed by **phosphoglycerate mutase**, a transfer of phosphate from position 3 to position 2 of the substrate.

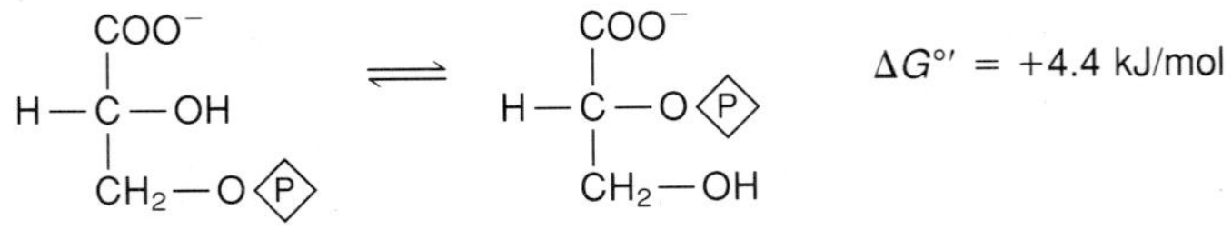

3-Phosphoglycerate $\rightleftharpoons$ 2-Phosphoglycerate $\Delta G^{\circ\prime} = +4.4$ kJ/mol

The reaction is slightly endergonic under standard conditions; again, the intracellular level of 3-phosphoglycerate is high relative to that of 2-phosphoglycerate, so that in vivo the reaction proceeds to the right without difficulty. The enzyme requires Mg^{2+} as a cofactor, and the reaction proceeds via the intermediate 2,3-bisphosphoglycerate.

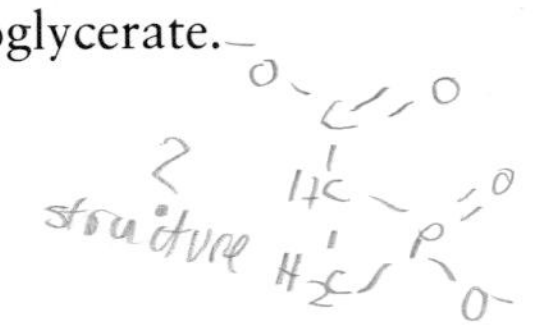

Reaction 9: Synthesis of the Second High-Energy Compound

Reaction 9, catalyzed by **enolase,** generates another super-high-energy compound, phosphoenolpyruvate, which participates in the second substrate-level phosphorylation of glycolysis. The enzyme requires either Mg^{2+} or Mn^{2+} and is inhibited by fluoride ion.

$$\underset{\textbf{2-Phosphoglycerate}}{\begin{array}{c} COO^- \\ | \\ H-C-O\text{Ⓟ} \\ | \\ CH_2OH \end{array}} \underset{}{\overset{Mg^{2+}}{\rightleftharpoons}} \underset{\textbf{Phosphoenolpyruvate}}{\begin{array}{c} COO^- \\ | \\ C\sim O\text{Ⓟ} \\ \| \\ CH_2 \end{array}} + H_2O \qquad \Delta G^{\circ\prime} = +1.7 \text{ kJ/mol}$$

The reaction involves a simple dehydration, or α,β-elimination, and the overall free energy change is small. However, the effect is to increase enormously the free energy of hydrolysis of the phosphate bond—from −15.6 kJ/mol for 2-phosphoglycerate to −61.9 kJ/mol for phosphoenolpyruvate. Carbon 2 of phosphoenolpyruvate is "locked into" the unfavored enol configuration, and, as discussed in Chapter 3, the great thermodynamic instability of the unstable enolpyruvate is chiefly responsible for the large negative free energy of hydrolysis of phosphoenolpyruvate.

Reaction 10: The Second Substrate-Level Phosphorylation

In the last reaction, catalyzed by **pyruvate kinase,** phosphoenolpyruvate transfers its phosphate to ADP in another substrate-level phosphorylation.

$$\underset{\textbf{Phosphoenolpyruvate}}{\begin{array}{c} COO^- \\ | \\ C\sim O\text{Ⓟ} \\ \| \\ CH_2 \end{array}} + H^+ + ADP \xrightarrow[K^+]{Mg^{2+}} \underset{\textbf{Pyruvate}}{\begin{array}{c} COO^- \\ | \\ C=O \\ | \\ CH_3 \end{array}} + ATP \qquad \Delta G^{\circ\prime} = -31.4 \text{ kJ/mol}$$

The enzyme requires K^+ and either Mg^{2+} or Mn^{2+}. Even though the reaction involves the endergonic synthesis of ATP, the overall reaction is strongly exergonic, because, as noted above, the spontaneous tautomerization of the product, enolpyruvate, to the highly favored keto form provides a strong thermodynamic drive in the forward direction.

The pyruvate kinase reaction is another site for metabolic regulation. In vertebrate liver the enzyme, a tetramer of M_r (molecular weight) about 250,000, is allosterically inhibited at high ATP concentrations and activated by fructose-1,6-bisphosphate. The synthesis of the liver enzyme is under dietary control; intracellular activity may increase as much as 10-fold from increased enzyme synthesis, or induction, as a result of high carbohydrate ingestion. This may contribute to the efficacy of "carbohydrate loading," the practice of eating a great deal of carbohydrate before an athletic event requiring great endurance, such as a marathon run. Augmented pyruvate kinase levels increase the rate at which energy can be generated by glycolysis.

Pyruvate kinase activity in the liver is also regulated by phosphorylation and dephosphorylation of the enzyme protein. The dephosphorylated form is far more active than the phosphorylated form. This process, which is under hormonal control, diverts phosphoenolpyruvate to gluconeogenesis when fatty acid oxidation and the citric acid cycle are already operating at rates sufficient to meet the energy needs of the cell. By contrast, virtually all of the phosphoenolpyruvate produced in muscle is converted to pyruvate.

Human genetic deficiencies of erythrocyte pyruvate kinase have been studied. Major clinical problems are related to oxygen transport through the bloodstream to the tissues. Accumulation of phosphoenolpyruvate caused by the enzyme deficiency leads to accumulation of other glycolytic intermediates, plus 2,3-bisphosphoglycerate. Because the latter compound regulates the oxygen affinity of hemoglobin (Chapter 7), the enzyme deficiency impairs the oxygen transport process.

The pyruvate kinase reaction converts the overall glycolytic pathway from an energy-neutral process to one that involves net synthesis of ATP. Two high-energy phosphates per mole of hexose are generated here, to go with the two generated by phosphoglycerate kinase. Subtracting the two invested at hexokinase and phosphofructokinase gives a net yield of two high-energy phosphates per mole of glucose—not a high yield, to be sure, but the process can meet the energy requirements of many anaerobes. Moreover, subsequent metabolism of pyruvate through aerobic pathways generates much additional high-energy phosphate.

Table 13.1 summarizes the reactions of glycolysis, showing free energy changes and ATP yields at each step.

Metabolic Fates of Pyruvate

Lactate Metabolism

Pyruvate represents an important metabolic branch point. Its fate depends crucially on the oxidation state of the cell, which is related to the reaction catalyzed by glyceraldehyde-3-phosphate dehydrogenase (reaction 6). Recall that this reaction converts 1 mol of NAD^+ per mole of triose phosphate to NADH. *This NADH must be reoxidized to NAD^+ if the cell is to maintain homeostasis.* During aerobic glycolysis this NADH can be oxidized by the mitochondrial electron transport chain, with the electrons transferred ultimately to oxygen. The oxidation of NADH itself yields additional energy, with 2 or 3 mol of ATP synthesized from ADP per mole of NADH oxidized (Chapter 15). Since 2 mol of NADH are produced per mole of glucose entering the pathway, aerobic glycolysis yields considerably more ATP than anaerobic glycolysis.

In anaerobic organisms or in aerobic cells that are undergoing very high rates of glycolysis, the NADH generated in glycolysis cannot be reoxidized in the mitochondrion. When this occurs, NADH must be used to drive the reduction of an organic substrate in order to maintain homeostasis. As noted earlier, that substrate is pyruvate itself, both in eukaryotic cells and in lactic acid bacteria. The enzyme catalyzing this reaction is lactate dehydrogenase (page 435). The equilibrium for this reaction lies far to the right. The relationship between glyceraldehyde-3-phosphate dehydrogenase and lactate dehydrogenase can be seen in Figure 13.7, which depicts the energy profile of anaerobic glycolysis. NADH produced by the former enzyme is stoichiometrically oxidized by the latter, so that during anaerobic glycolysis or lactic acid fermentation, an overall electron balance is maintained.

In vertebrates some tissues, such as red blood cells, derive most of their energy from anaerobic metabolism. Skeletal muscle, which derives most of its energy from respiration when at rest, relies heavily on glycolysis during exertion, when glycogen stores are rapidly broken down, or mobilized, to

Table 13.1
Summary of glycolysis

Reaction	Enzyme	ATP yield	$\Delta G^{\circ\prime}$ (kJ/mol)	ΔG (kJ/mol)
Glucose (G)				
1. ATP → ADP + H^+	Hexokinase (HK)	−1	−16.7	−33.5
Glucose-6-phosphate (G6P)				
2.	Phosphoglucoisomerase (PGI)		+1.7	−2.5
Fructose-6-phosphate (F6P)				
3. ATP → ADP + H^+	Phosphofructokinase (PFK)	−1	−14.2	−22.2
Fructose-1,6-bisphosphate (FBP)				
4.	Aldolase (ALD)		+23.8	−1.3
Glyceraldehyde-3-phosphate (G3P) + dihydroxyacetone phosphate (DHAP)				
5.	Triose-phosphate isomerase (TPI)		+7.5	+2.5
2 Glyceraldehyde-3-phosphate				
6. NAD^+ + P_i → NADH + H^+	Glyceraldehyde-3-phosphate dehydrogenase (G3PDH)		+6.3	−1.7
2 1,3-Bisphosphoglycerate (BPG)				
7. ADP + H^+ → ATP	Phosphoglycerate kinase (PGK)	+2	−18.8	+1.3
2 3-Phosphoglycerate (3PG)				
8.	Phosphoglyceromutase (PGM)		+4.6	+0.8
2 2-Phosphoglycerate (2PG)				
9. → H_2O	Enolase (ENO)		+1.7	−3.3
2 Phosphoenolpyruvate (PEP)				
10. ADP + H^+ → ATP	Pyruvate kinase (PK)	+2	−31.4	−16.7
2 Pyruvate (PYR)				
Net		+2	−35.5	−76.6

ΔG values are estimated from the approximate intracellular concentrations of glycolytic intermediates in rabbit skeletal muscle.

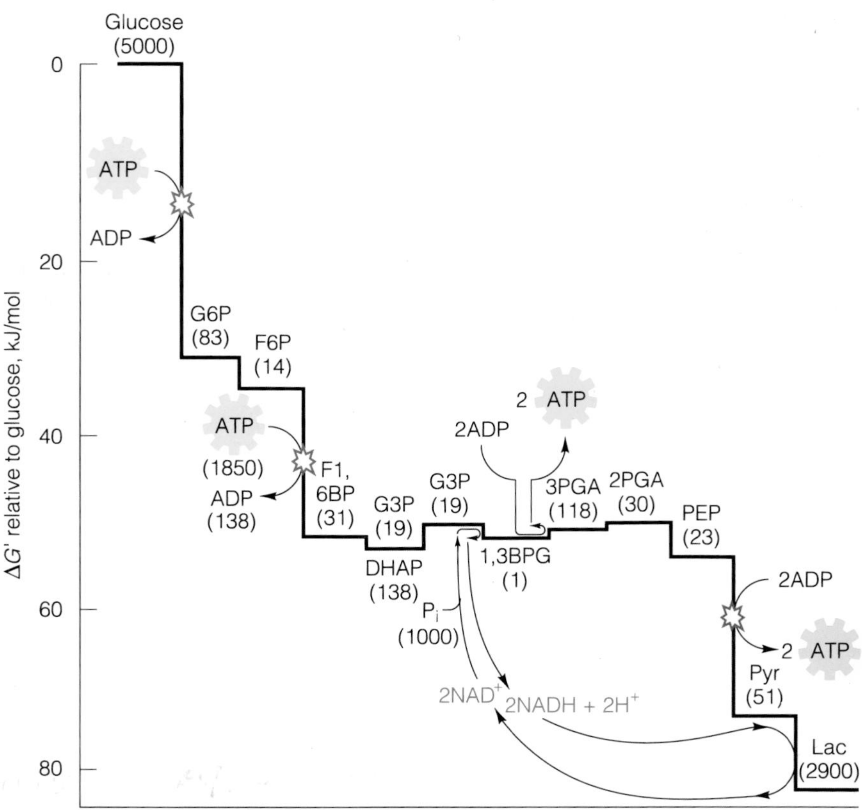

Figure 13.7
Energy and electron profile of anaerobic glycolysis. The graph shows the $\Delta G'$ for each reaction, calculated from $\Delta G^{\circ\prime}$ values and the estimated molar concentration of each intermediate in the human erythrocyte. The numbers in parentheses are approximate micromolar concentrations of the intermediates in the human erythrocyte. Note three points: (1) two of the four ATPs generated are used to repay the initial ATP investment; (2) the reducing equivalents generated by glyceraldehyde-3-phosphate dehydrogenase must be used to reduce an organic substrate in anaerobiosis; (3) enzymes subject to allosteric control are those catalyzing reactions that are so highly exergonic as to be virtually irreversible (red stars).

provide substrates for glycolysis. Normally the lactate produced diffuses from the tissue and is transported through the bloodstream to highly aerobic tissues, such as heart and liver. The aerobic tissue can catabolize lactate further, through respiration, or can convert it back to glucose, through gluconeogenesis. However, if lactate is produced in large quantities, it cannot be readily consumed. Then, as we discussed in Chapter 7, the blood pH falls and the Bohr effect functions to increase oxygen supplies to the tissues.

Until relatively recently it was thought that lactate accumulation in animals was largely a consequence of anaerobic metabolism, which occurs when the need for tissues to generate energy exceeds their capacity to oxidize the pyruvate produced in glycolysis. Recent metabolic studies, including ^{31}P NMR analyses of the levels of phosphorylated intermediates in living muscle cells during exercise, paint a somewhat different picture. These studies show that even in fully oxygenated muscle tissue, as much as 50% of the glucose metabolized is converted to lactate. The release of this lactate to the circulation, and its uptake in other tissues, is seen as part of the coordination in different tissues of all of the major energy-storing and energy-generating pathways. The mechanisms involved represent a current area of active investigation.

Isoenzymes of Lactate Dehydrogenase

Lactate dehydrogenase exists in animal tissues in multiple molecular forms. Different molecular forms of an enzyme catalyzing the same reaction are called **isoenzymes,** or **isozymes.** Lactate dehydrogenase was the first enzyme that established the physical basis for the existence of isoenzymes. Most tissues contain five isoenzymes of lactate dehydrogenase; these can be resolved electrophoretically, as shown in Figure 13.8.

Lactate dehydrogenase is a tetrameric protein consisting of two different types of subunits, called M and H. M subunits predominate in skeletal muscle and liver, while H subunits predominate in heart. M and H subunits combine randomly with each other, so that the five major isoenzymes have the compositions M_4, M_3H, M_2H_2, MH_3, and H_4. Because of random subunit reassortment, the isoenzymic composition of a tissue is determined primarily by the activities of the genes specifying the two subunits.

The physiological need for the existence of different forms of this enzyme is not yet clear. Some have tried to relate it to the fact that heart-type isozymes have relatively low K_M values for pyruvate and are sensitive to allosteric inhibition by pyruvate. In any case, the tissue specificity of isoenzyme patterns is useful in clinical medicine. Such pathological conditions as myocardial infarction, infectious hepatitis, or muscle diseases involve cell death of affected tissue, with release of cell contents to the blood. The pattern of LDH isoenzymes in the blood serum is representative of the tissue that released these isoenzymes. This information can be used to diagnose such conditions and to monitor the progress of treatment.

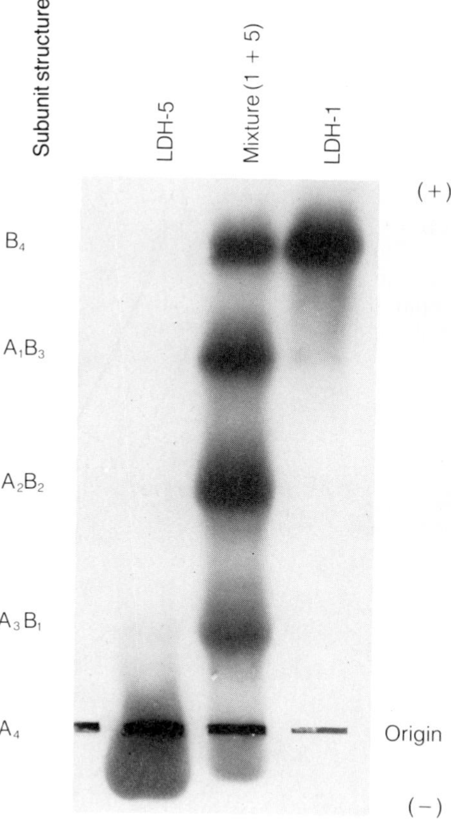

Figure 13.8
Structural basis for the existence of isoenzymes of lactate dehydrogenase. Preparations were subjected to electrophoresis in a starch gel, which was then treated to reveal bands containing enzymatically active protein. LDH-1 is a tetramer containing only the H subunit, while LDH-5 contains only M subunits. The middle lane depicts an experiment in which equal amounts of LDH-1 and LDH-5 were mixed; the subunits were dissociated and then allowed to reassociate. The presence of five different enzyme forms and their relative amounts show that individual M and H subunits can associate randomly to form tetramers of mixed subunit composition.

Ethanol Metabolism

Pyruvate has numerous alternative fates in anaerobic microorganisms. As noted earlier, lactic acid bacteria simply reduce it to lactate. The alcoholic fermentation in yeast generates ethanol, which starts with the nonoxidative decarboxylation of pyruvate to acetaldehyde, catalyzed by **pyruvate decarboxylase.** This reaction is followed by the NADH-dependent reduction of acetaldehyde to ethanol, catalyzed by alcohol dehydrogenase.

$$\underset{\textbf{Pyruvate}}{CH_3-\overset{O}{\overset{\|}{C}}-COO^-} \xrightarrow[\text{Pyruvate decarboxylase}]{H^+ \quad CO_2} \underset{\textbf{Acetaldehyde}}{CH_3-\overset{O}{\overset{\|}{C}}-H} \underset{\text{Alcohol dehydrogenase}}{\overset{NADH + H^+ \quad NAD^+}{\rightleftharpoons}} \underset{\textbf{Ethanol}}{CH_3-CH_2-OH}$$

The first reaction requires **thiamine pyrophosphate** as a coenzyme. This coenzyme, derived from vitamin B_1, participates in a number of group transfer reactions involving an activated aldehyde moiety (see Chapter 14).

Animal tissues also contain alcohol dehydrogenase, even though ethanol is not a major metabolic product in animal cells. Some of the major metabolic consequences of ethanol intoxication result from the action of alcohol dehydrogenase in the liver. First, there is massive reduction of NAD^+ to NADH, which decreases energy generation via the citric acid cycle because of reduced activity of NAD^+-dependent dehydrogenases. Second, acetaldehyde is quite toxic, and many of the unpleasant effects of hangovers result from actions of acetaldehyde and its subsequent metabolites.

Energy and Electron Balance Sheets

By writing a balanced chemical equation for glycolysis, we can compute the energy yield accompanying conversion of 1 mol of glucose. For anaerobic glycolysis or for a lactic acid fermentation we can write the following balanced equation.

$$\text{Glucose} + 2\text{ADP} + 2P_i \longrightarrow 2\text{ lactate} + 2\text{ATP} + 2H_2O$$

Similarly, we can write a balanced equation for alcoholic fermentation.

$$\text{Glucose} + 2\text{ADP} + 2P_i + 2H^+ \longrightarrow 2\text{ ethanol} + 2CO_2 + 2\text{ATP} + 2H_2O$$

Note first that both processes are electronically balanced; NAD^+ and NADH, both of which participate in the reaction pathways, do not appear in the overall reactions. This is also shown in Figure 13.7.

During aerobic glycolysis the nicotinamide nucleotides do appear in the overall equation, as shown.

$$\text{Glucose} + 2\text{ADP} + 2\text{P}_i + 2\text{NAD}^+ \longrightarrow 2\text{ pyruvate} + 2\text{ATP} + 2\text{H}_2\text{O} + 2\text{NADH} + 2\text{H}^+$$

The NADH is reoxidized by the mitochondrial electron transport chain, and we know empirically that the generation of 1 mol of NADH produced in the cytosol leads to formation of about 3 mol of ATP in the mitochondrion.

$$2\text{NADH} + 2\text{H}^+ + \text{O}_2 + 6\text{ADP} + 6\text{P}_i \longrightarrow 2\text{NAD}^+ + 8\text{H}_2\text{O} + 6\text{ATP}$$

Summing the above two equations, you can see why aerobic glycolysis yields 8 mol of ATP per mole of glucose.

$$\text{Glucose} + 8\text{ADP} + 8\text{P}_i + \text{O}_2 \longrightarrow 2\text{ pyruvate} + 8\text{ATP} + 10\text{H}_2\text{O}$$

Metabolism of glucose to either lactate or ethanol represents a nonoxidative process, as you can see by comparing the empirical formulas for glucose ($C_6H_{12}O_6$) and lactate ($C_3H_6O_3$). Clearly, there is no change in the overall oxidation state of the carbons, since the numbers of hydrogens and oxygens bound per carbon atom are identical for glucose and lactate. The same is true for ethanol plus CO_2, when one counts the atoms in both. By contrast, pyruvate is more highly oxidized than glucose, as seen from its empirical formula ($C_3H_4O_3$).

Note also that glycolysis, whether aerobic or anaerobic, releases but a small fraction of the potential energy stored in the glucose molecule. As noted earlier, the complete combustion of glucose to CO_2 and H_2O releases 2870 kJ/mol of free energy under standard conditions. As we shall see in the next chapter, 38 mol of ATP are synthesized from ADP per mole of glucose carried completely through glycolysis and the citric acid cycle. This represents about 40% of the energy released that is captured as ATP. Since catabolism of glucose to lactate or pyruvate yields but 2 or 8 mol of ATP, respectively, you can see that most of the potential energy originally present in glucose is still waiting to be released after glycolysis. For this reason, aerobic metabolism is much more efficient than anaerobic metabolism, and aerobic organisms in general are more successful and widespread than anaerobic organisms. The early evolution of aerobic metabolism made possible the large, active animals we see today. Nevertheless, some large animals still derive a large fraction of their metabolic energy from glycolysis, under certain physiological circumstances. A good example is the crocodile—torpid (and aerobic) for much of its life, yet capable of short bursts of intensely rapid movement. In the latter circumstance glycolysis, coupled with the breakdown of carbohydrate energy stores, represents a rapid, if inefficient, way to mobilize energy.

Regulation of Glycolysis

Glycolysis is closely coordinated with other major pathways of energy generation and utilization, notably synthesis and breakdown of glycogen (or starch), gluconeogenesis, the pentose phosphate pathway, and the citric acid cycle. Metabolic factors that control glycolysis tend to regulate other pathways in a coordinated fashion. Thus, it is difficult to consider regulation of glycolysis in isolation from these other pathways, and we shall return to this topic after we have presented the major pathways in energy metabolism (Chapter 23). However, it is important here to describe two key glycolytic enzymes that serve as regulatory targets—phosphofructokinase (the major target) and pyruvate kinase.

The Pasteur Effect

The recognition that glycolysis is controlled primarily by the activity of phosphofructokinase developed largely from a discovery made over a century ago by Louis Pasteur: when anaerobic yeast cultures metabolizing glucose were exposed to air, the rate of glucose utilization decreased dramatically. It became clear that this phenomenon, the **Pasteur effect,** involves the inhibition of glycolysis by oxygen. What is the mechanism of this effect if oxygen is not an active participant in glycolysis? The needed insight came from analyses of the intracellular contents of glycolytic intermediates in aerobic and anaerobic cells. In turn, this discovery required techniques for the rapid interruption of metabolism and extraction of metabolites. Such a technique is **freeze-clamping,** in which tissue is rapidly compressed between metal plates cooled to liquid nitrogen temperatures. The solid tissue can then be powdered and extracted for analysis.

The latter experiments revealed that when oxygen is introduced to anaerobic cells, the levels of all the glycolytic intermediates from fructose-1,6-bisphosphate onward *decrease,* while all of the *earlier* intermediates accumulate to higher levels. This finding is consistent with the idea that the metabolic flux through phosphofructokinase is specifically decreased in the presence of O_2.

Oscillations of Glycolytic Intermediates

Other important conclusions emerged from the discovery that the intracellular levels of glycolytic intermediates are not constant under many conditions but undergo periodic variations, or oscillations, as shown in Figure 13.9. These variations can be visualized most readily by irradiating a yeast cell suspension with near-ultraviolet light and following fluorescence of the whole cells at 450 nm. The major contributor to this fluorescence is NADH. Thus, this type of experiment monitors changes in the intracellular NADH pool with time. Oscillations are a common feature of feedback-controlled systems, and the cyclic variations in levels of glycolytic intermediates provide important clues to regulatory mechanisms affecting glycolysis. Moreover, there is reason to believe that oscillations reduce the dissipation of free energy released during glycolysis.

While the fluorescence of a yeast cell suspension is increasing, NADH is accumulating. Under these conditions glycolysis is turned on and NADH is being produced faster than it can be used to reduce pyruvate. Presumably, during this period one or more regulatory substances are also accumulating; once they have accumulated sufficiently to activate glycolysis, the NADH level falls—until the supply of regulators is depleted to the point that the pathway is inhibited again. This cycle occurs repeatedly.

Once investigators realized that the intracellular levels of NADH were varying periodically, they began sampling extracts of oscillating cells to

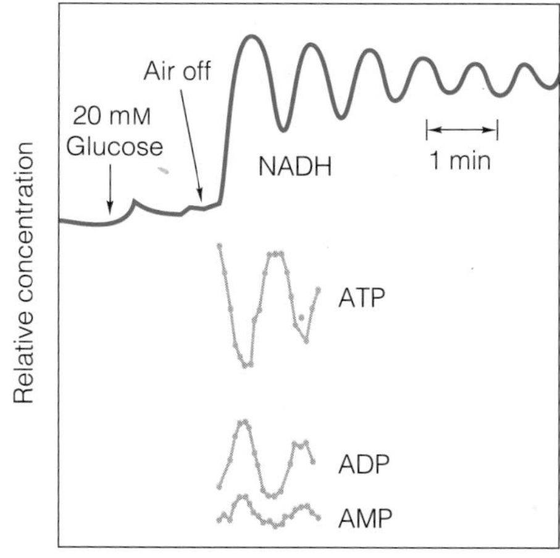

Figure 13.9
Periodic oscillations of the levels of glycolytic intermediates in yeast cells undergoing glycolysis. Upper tracing (blue) depicts continuous monitoring of fluorescence of the cell suspension, which is related to the intracellular concentration of NADH. The nucleotide levels (orange), determined in a parallel experiment in which samples of culture were removed at various times, extracted, and assayed for their contents of ATP, ADP, and AMP, show similar oscillations.

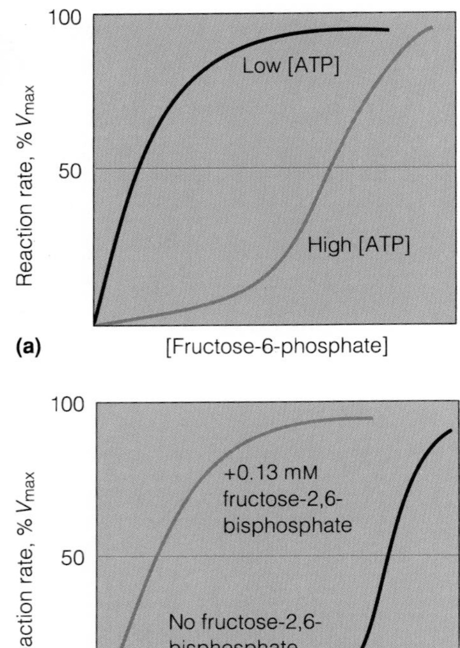

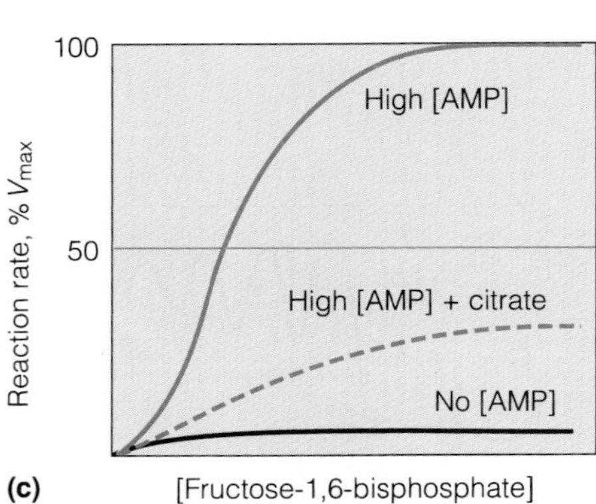

Figure 13.10
Allosteric control of phosphofructokinase. (a) How ATP increases the apparent K_M for substrate fructose-6-phosphate. (b) Activation by fructose-2,6-bisphosphate. (c) Activation by fructose-1,6-bisphosphate, strong dependence on allosteric activation by AMP, and inhibition by citrate.

analyze the intracellular levels of other glycolytic intermediates. These intermediates were found to rise and fall periodically as well. Note from Figure 13.9 that the levels of ADP and AMP rise and fall precisely in phase with NADH, while the level of ATP is 180° out of phase. This pattern suggests that the activity of glycolysis is in some way dependent on the adenylate energy charge (Chapter 2): when the charge is high, the pathway is turned off; when it is low, the pathway is activated. These and other observations pointed to phosphofructokinase as the major point of regulation.

Allosteric Regulation of Phosphofructokinase

Phosphofructokinase is a complex multisubunit enzyme; in muscle it exists as a homotetramer of M_r about 360,000, which undergoes reversible dissociation to a dimeric form. Some of the allosteric activators and inhibitors of the enzyme act by influencing the interconversion of the dimer and the tetramer.

As noted earlier, *activators* of phosphofructokinase include AMP, ADP, and the reaction product fructose-1,6-bisphosphate. A recently discovered compound, **fructose-2,6-bisphosphate,**

Fructose-2,6-bisphosphate

activates the enzyme at very low concentrations and is considered to be the major regulator controlling carbon flux through glycolysis and gluconeogenesis in the liver (Figure 13.10). The synthesis of this compound is catalyzed by another form of phosphofructokinase, called PFK-2 (where the glycolytic enzyme is called PFK-1). PFK-2 activity in turn is regulated by reversible phosphorylation and dephosphorylation, controlled ultimately by cyclic AMP. The details of these interactions will be presented when we discuss the control of gluconeogenesis in Chapter 16.

The most significant *inhibitors* of phosphofructokinase, from a biological standpoint, are ATP and citrate. The effect of ATP may seem anomalous, since ATP is a substrate and hence essential for the reaction. As an inhibitor, ATP binds to a separate site on the enzyme, with lower affinity. At low ATP concentrations the substrate saturation curve for fructose-6-phosphate is nearly hyperbolic. At high ATP levels the curve becomes sigmoidal and is shifted far to the right (see Figure 13.10a). Thus, inhibition is achieved because the apparent affinity for fructose-6-phosphate is greatly reduced.

The control of PFK by adenine nucleotides represents a way in which energy metabolism responds to the adenylate energy charge. At high energy charge, the relative abundance of ATP signals that the energy-yielding glycolytic pathway should diminish in activity; the signal involves inhibition of PFK. Conversely, a high AMP or ADP level signals that energy charge is low and that flux through glycolysis should increase. Inhibition by citrate represents another energy level sensor. At high energy charge flux through the citric acid cycle diminishes, through mechanisms that will be discussed in Chapter 14. Under these conditions citrate accumulates and is transported out of mitochondria. Interaction with PFK in the cytosol can signal that energy generation is adequate, and hence the production of citric acid cycle precursors via glycolysis can be diminished.

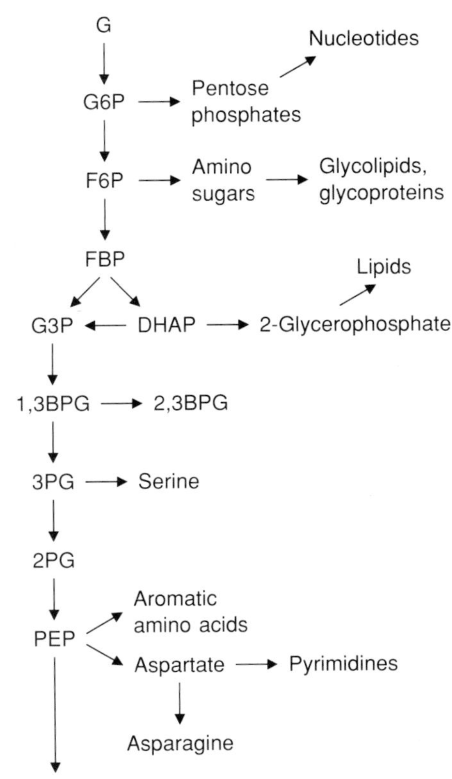

Figure 13.11
Major alternative fates of glycolytic intermediates in biosynthetic pathways.

Control of Pyruvate Kinase

Earlier we described mechanisms by which pyruvate kinase helps to regulate glycolysis. This multisubunit enzyme is also inhibited by ATP, in a fashion kinetically similar to the effect of ATP on PFK; high ATP levels reduce the apparent affinity of pyruvate kinase for its other substrate, phosphoenolpyruvate. A second allosteric effect is the **feedforward activation** of pyruvate kinase by fructose-1,6-bisphosphate. This effect, the converse of feedback inhibition, ensures that carbon passing the first regulated step in the pathway will be able to complete its passage through glycolysis and that undesirable accumulation of intermediates will not occur. A third feedback control effect is inhibition of pyruvate kinase by acetyl-CoA, which is derived largely from fatty acid oxidation. This allows the cell to reduce glycolytic flux when ample substrates are available from fat breakdown.

Phosphofructokinase and pyruvate kinase regulate the flow of intermediates from glucose-6-phosphate to pyruvate. Glycolysis is regulated also at the points of entry of carbon into the pathway. We have already mentioned hexokinase inhibition by glucose-6-phosphate as one such control site. The other major control point, at least in animal metabolism, is the breakdown of glycogen, catalyzed by **glycogen phosphorylase.** This extremely important process is discussed in detail later in this chapter.

Glycolysis as Both a Catabolic and an Anabolic Pathway

Why should the cell regulate glycolysis at more than one point? Any answer to this question should take into account that glycolysis not only generates ATP and provides pyruvate for oxidation via the citric acid cycle but also is a biosynthetic pathway. Intermediates in glycolysis are precursors to a number of compounds, particularly lipids and amino acids. These processes will be discussed at appropriate points throughout the book. At this point, Figure 13.11 identifies some of the major biosynthetic roles of glycolytic intermediates. This figure illustrates why glycolysis is considered a major metabolic thoroughfare. Many pathways lead into glycolysis, and many pathways diverge from it, creating a substantial flux through the pathway. Thus, even though the amount of fuel for oxidative degradation is ample, the regulatory demands placed on the cell are too complex to meet with a single rate-controlling reaction.

Figure 13.12 summarizes the regulatory relationships between glycolysis and other metabolic pathways.

Entry of Other Sugars into the Glycolytic Pathway

Thus far our discussion of glycolysis has focused on glucose as a source of carbon for this pathway. Many other sources of carbohydrate energy are available, whether through digestion of foodstuffs or utilization of endogenous metabolites. This section focuses on the utilization of monosaccharides other than glucose, of disaccharides, and of glycerol derived from fat metabolism. These pathways are summarized in Figure 13.13.

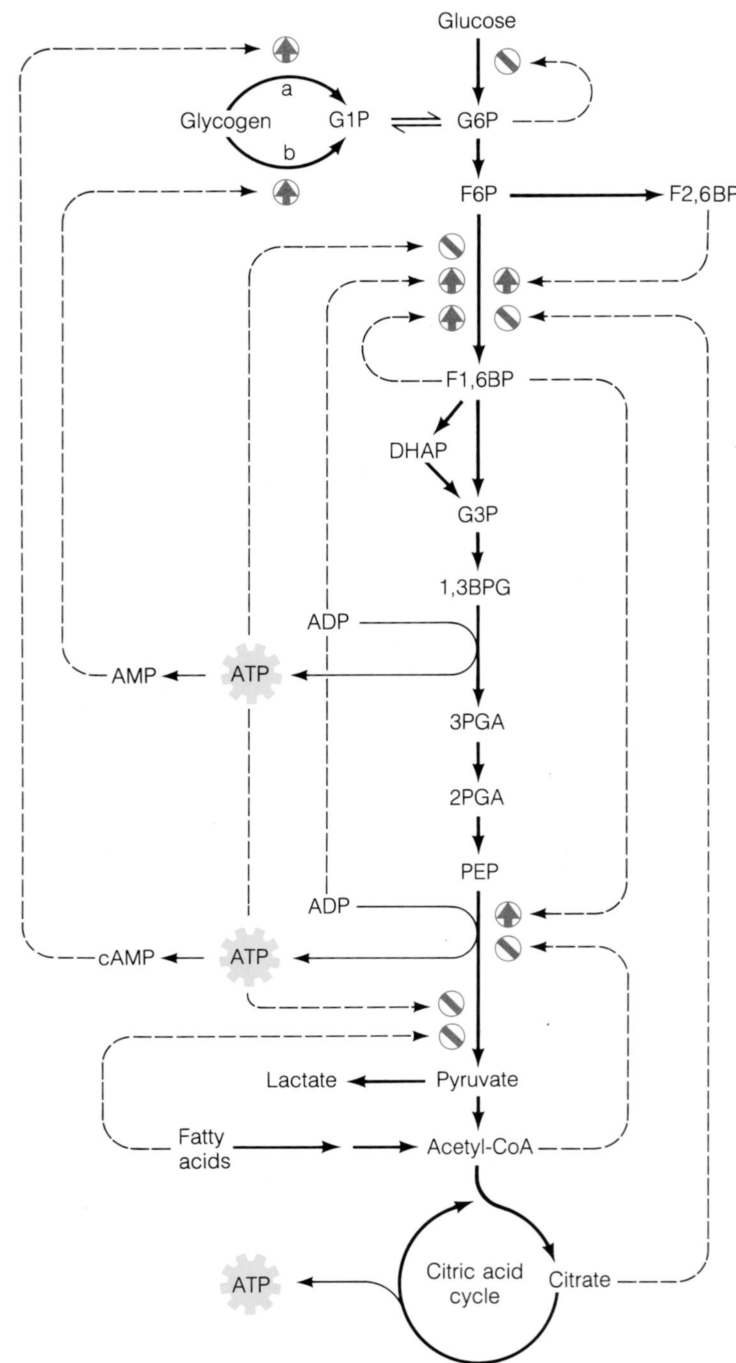

Figure 13.12
Regulation of glycolysis, showing points of coordination with other metabolic pathways. F2,6BP = fructose-2,6-bisphosphate; other abbreviations as in Table 13.1. The control of glycogen breakdown is presented later in this chapter.

Monosaccharide Metabolism

As stated earlier, hexokinase has a broad substrate specificity; thus, it can participate in utilization of the three major hexoses other than glucose: galactose, fructose, and mannose.

GALACTOSE UTILIZATION. D-Galactose is derived principally from hydrolysis of the disaccharide lactose, particularly abundant in milk. The main route for galactose utilization (Figure 13.14) involves its ATP-dependent

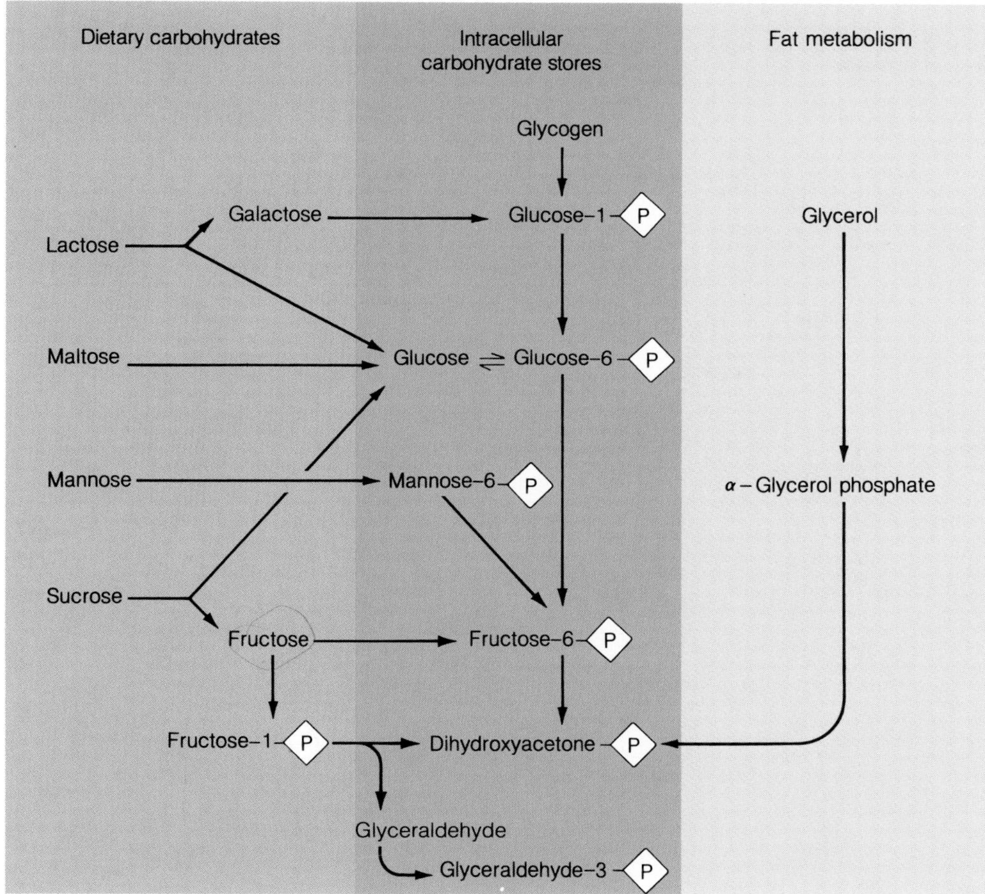

Figure 13.13
Routes for utilizing substrates other than glucose in glycolysis. In animals most of the carbohydrate other than glucose and glycogen comes from the diet, while most of the glycerol is derived from lipid catabolism.

conversion to galactose-1-phosphate, catalyzed by **galactokinase.** Transformation of galactose-1-phosphate to glucose-1-phosphate involves epimerization at carbon 4. However, galactose-1-phosphate must be metabolically activated before epimerization can occur, by a transferase reaction with a nucleotide-linked sugar, **uridine diphosphate glucose,** or UDP-Glc.

Uridine diphosphate glucose

This gives another nucleotide-linked sugar, **uridine diphosphate galactose,** abbreviated UDP-galactose or UDP-Gal. The NAD^+-linked enzyme **UDP-galactose 4-epimerase** converts this substrate to UDP-Glc, as shown in Figure 13.15. UDP-Glc is then cleaved by **UDP-Glc pyrophosphorylase** to give UTP and glucose-1-phosphate. This reaction, which is reversible, involves cleavage of the phosphoric acid anhydride bond in UDP-Glc by addition across that bond of the elements of pyrophosphoric acid. Glucose-1-phos-

Figure 13.14
Pathway for utilizing galactose by converting it to glucose-6-phosphate.

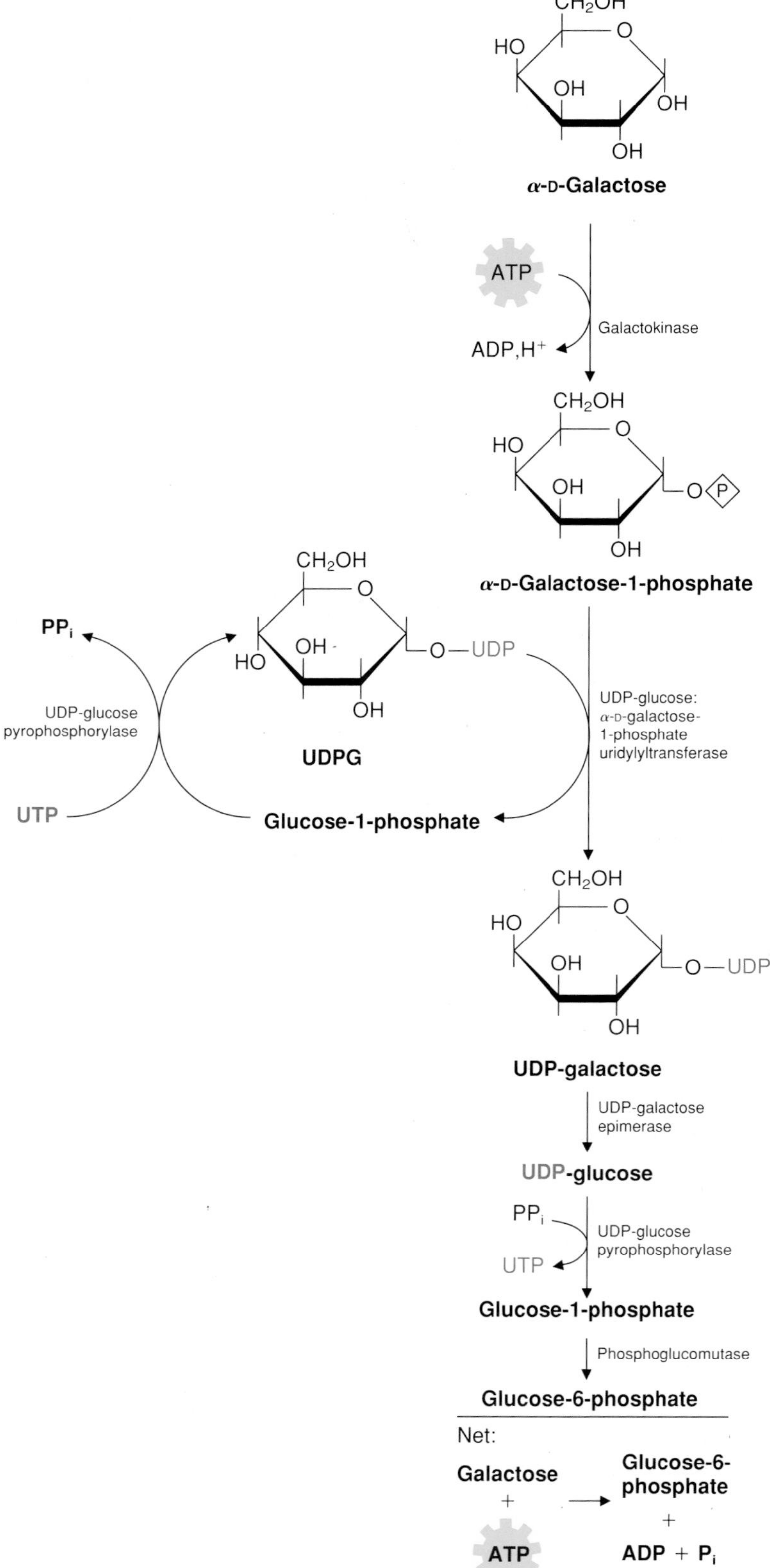

phate is then converted to glucose-6-phosphate by **phosphoglucomutase,** an enzyme whose action is discussed later in this chapter.

The pathway shown in Figure 13.14 is used in reverse in mammary gland for the biosynthesis of galactose, used ultimately to synthesize lactose in milk. The reversal in this tissue is driven by the high rate at which UDP-galactose is utilized in lactose synthesis.

A variety of genetic disorders in humans go by the generic name **galactosemia.** These all involve a failure to metabolize galactose, so that galactose and/or galactose-1-phosphate accumulate in the blood and tissues. Clinical consequences include mental retardation, visual cataracts, and enlargement of the liver and other organs. This disorder results from hereditary deficiency of any one of three enzymes involved in galactose utilization—UDP-glucose:α-D-galactose-1-phosphate uridylyltransferase (the most common form), galactokinase, or UDP-galactose 4-epimerase. Since the major dietary source of galactose is lactose in milk, the symptoms usually occur in infants. The condition can be alleviated by withholding from the diet milk and milk products.

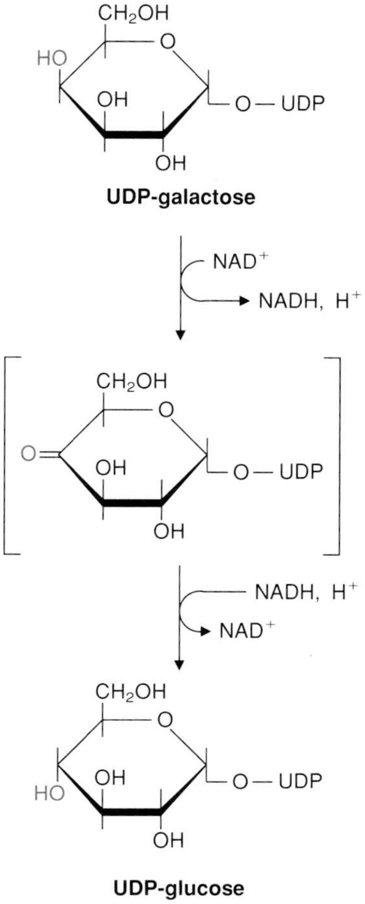

FRUCTOSE UTILIZATION. Fructose is present as the free sugar in many fruits, and it is also derived from hydrolysis of sucrose. Phosphorylation of fructose in most tissues yields fructose-6-phosphate, a glycolytic intermediate. A different pathway is involved in vertebrate liver, where the enzyme **fructokinase** phosphorylates fructose to **fructose-1-phosphate** (F1P). This is then cleaved in an aldolase reaction, catalyzed either by fructose-1,6-bisphosphate aldolase or by a specific enzyme. In either case, cleavage products are dihydroxyacetone phosphate, a glycolytic intermediate, and D-glyceraldehyde. The latter is then phosphorylated in an ATP-dependent reaction to give the glycolytic intermediate glyceraldehyde-3-phosphate. This pathway of utilization bypasses phosphofructokinase regulation and may account for the ease with which dietary sucrose is converted to fat.

MANNOSE UTILIZATION. Finally, among the major hexoses, mannose arises through digestion of foods containing certain polysaccharides or glycoproteins. The hexokinase-catalyzed phosphorylation of mannose to mannose-6-phosphate is followed by isomerization of the latter product to fructose-6-phosphate.

UDP-galactose

NAD^+ → $NADH, H^+$

$NADH, H^+$ → NAD^+

UDP-glucose

Figure 13.15
Reaction pathway for UDP-galactose 4-epimerase, showing the catalytic role of NAD^+ in oxidizing and then reducing the substrate.

Disaccharide Metabolism

The three disaccharides most abundant in foods are maltose, lactose, and sucrose. In animal metabolism these are hydrolyzed in cells lining the small intestine, to give the constituent hexose sugars.

$$\text{Maltose} + H_2O \xrightarrow{\text{Maltase}} 2\text{D-glucose}$$

$$\text{Lactose} + H_2O \xrightarrow{\text{Lactase}} \text{D-galactose} + \text{D-glucose}$$

$$\text{Sucrose} + H_2O \xrightarrow{\text{Sucrase}} \text{D-fructose} + \text{D-glucose}$$

The hexose sugars then find their way to tissues through the bloodstream and are catabolized as described in the previous section.

In many humans the enzyme lactase disappears from the intestinal mucosal cells after age 4 to 6, when milk drinking usually decreases. This causes **lactose intolerance,** a situation in which ingestion of milk or lactose-containing milk products causes intestinal distress, because of bacterial action on the lactose that accumulates.

Plants and microorganisms have different pathways for metabolizing disaccharides. Bacteria metabolize sucrose through the action of **sucrose phosphorylase.**

$$\text{Sucrose} + P_i \rightleftharpoons \text{D-glucose-1-phosphate} + \text{D-fructose}$$

Glycerol Metabolism

The digestion of triacylglycerols and most phospholipids generates glycerol as one product. In animals, glycerol enters the glycolytic pathway first by the action in liver of **glycerokinase.**

$$\underset{\textbf{Glycerol}}{\begin{array}{c} CH_2OH \\ | \\ H-C-OH \\ | \\ CH_2OH \end{array}} + ATP \longrightarrow \underset{\textbf{Glycerol-3-phosphate}}{\begin{array}{c} CH_2OH \\ | \\ H-C-OH \\ | \\ CH_2O\text{Ⓟ} \end{array}} + ADP + H^+$$

The product is then oxidized by **glycerol-3-phosphate dehydrogenase,** yielding the glycolytic intermediate dihydroxyacetone phosphate.

$$\underset{\textbf{Glycerol-3-phosphate}}{\begin{array}{c} CH_2OH \\ | \\ H-C-OH \\ | \\ CH_2-O\text{Ⓟ} \end{array}} + NAD^+ \longrightarrow \underset{\textbf{Dihydroxyacetone phosphate}}{\begin{array}{c} CH_2OH \\ | \\ C=O \\ | \\ CH_2-O\text{Ⓟ} \end{array}} + NADH + H^+$$

Catabolism of Polysaccharides

In animal metabolism there are two primary sources of glucose derived from polysaccharides: (1) digestion of dietary polysaccharides, chiefly starch from plant foodstuffs and glycogen from meat, and (2) mobilization of the animal's own glycogen reserves. Recall from Chapter 8 that starch, the major nutrient polysaccharide of plants, consists of the unbranched glucose polymer amylose and the branched polymer **amylopectin.** Glucose residues in both polymers are linked by $\alpha(1 \rightarrow 4)$ glycosidic bonds, while the branch points in amylopectin involve both $\alpha(1 \rightarrow 4)$ and $\alpha(1 \rightarrow 6)$ linkages. Glycogen is chemically similar to amylopectin, except that it is more highly branched and is of higher molecular weight. Many microorganisms store carbohydrate as glycogen also.

Hydrolytic and Phosphorolytic Cleavages

Polysaccharide digestion and glycogen mobilization both involve sequential cleavage of monosaccharide units from nonreducing ends of glucose polymers. The first of these processes occurs via *hydrolysis* and the other via *phosphorolysis.* These processes are chemically similar; hydrolysis involves

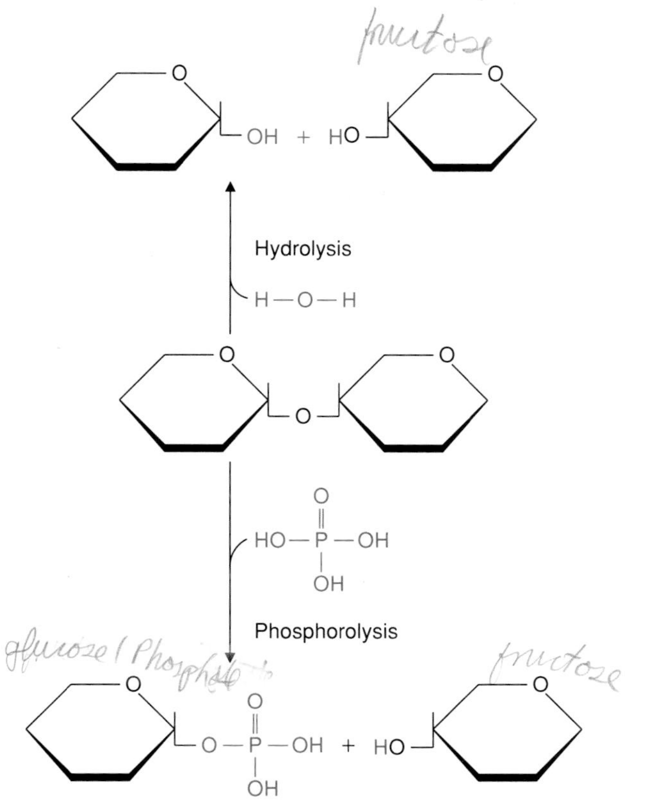

Figure 13.16
Cleavage of a glycosidic bond by hydrolysis or phosphorolysis. This formal diagram shows how the elements of water or phosphoric acid, respectively, are added across a glycosidic bond.

the cleavage of a bond by addition across that bond of the elements of water, while a phosphorolytic cleavage occurs by addition of the elements of phosphoric acid, as illustrated in Figure 13.16. An enzyme catalyzing a phosphorolysis is called a **phosphorylase,** to be distinguished from a *phosphatase* (or, more precisely, a *phosphohydrolase*), which catalyzes the hydrolytic cleavage of a phosphate ester bond.

Energetically speaking, the advantage of a phosphorolytic mechanism is that mobilization of glycogen yields most of its monosaccharide units in the form of sugar phosphates. These can be converted to glycolytic substrates directly, without the investment of additional ATP. By contrast, starch digestion yields glucose plus some maltose; ATP and the hexokinase reaction are necessary to initiate glycolytic breakdown of these sugars.

Unlike mobilization of stored carbohydrate, digestion of carbohydrate occurs largely in the intestine, and its products must be absorbed into the bloodstream and then transported into cells that utilize those products. Since sugar phosphates, like other charged compounds, are inefficiently transported across cell membranes, the digestion of polysaccharides to yield hexose sugars makes good metabolic sense.

In plant metabolism we are not concerned with digestion, because, except for carnivorous plants such as the Venus flytrap, plants synthesize both monosaccharides and energy storage polysaccharides via photosynthesis. However, the same enzymatic mechanisms are used to mobilize starch in plant metabolism as in animals—both hydrolysis and phosphorolysis, with the former predominating.

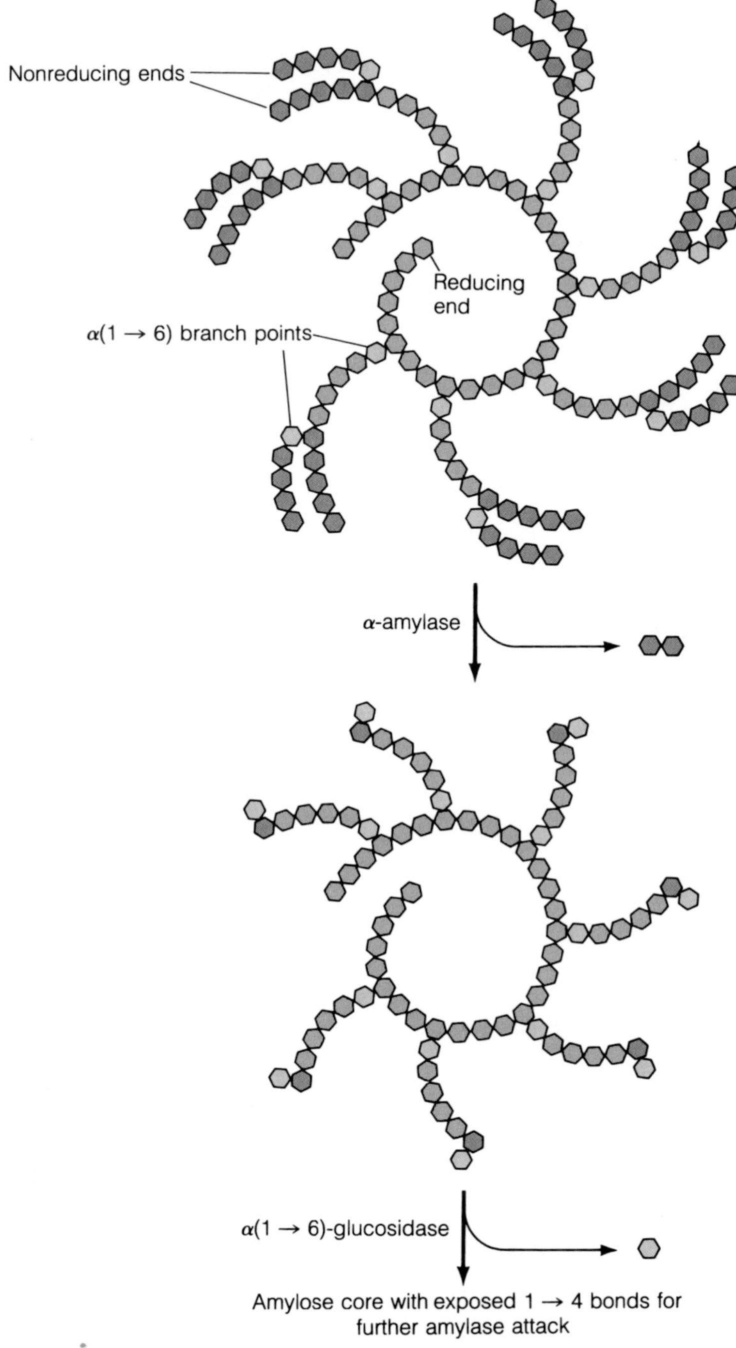

Figure 13.17
Actions of α-amylase and α(1→6)-glucosidase in digestion of amylopectin or glycogen. α-Amylase alone cannot cleave 1→6 glycosidic bonds in the branched polymer, and a limit dextrin (gray) accumulates unless both enzymes are present. α(1→6) branch points, green; maltose units, purple.

Starch and Glycogen Digestion

In animals the digestion of starch and glycogen begins in the mouth, with the action of **α-amylase** secreted in saliva. This enzyme cleaves the α(1→4) linkages of both polymers. In the intestine, digestion continues, aided by α-amylase secreted by the pancreas. This enzyme can completely degrade amylose to maltose and some glucose, but it only partially degrades amylopectin and glycogen, since α-amylase cannot cleave the α(1→6) linkages found at branch points. The product of exhaustive digestion of amylopectin or glycogen by α-amylase is called a **limit dextrin;** its continued degradation requires the action of a "debranching enzyme," *α(1→6)-glucosidase* (also called isomaltase). This action exposes a new group of $\alpha(1\rightarrow4)$-linked branches, as shown in Figure 13.17. These can be attacked by α-amylase until a new set of $\alpha(1\rightarrow6)$-linked branches is reached. The end result of the sequential action of these two enzymes is the complete breakdown of starch or glycogen to glucose plus some maltose. Glucose is then absorbed into the bloodstream and transported to various tissues for utilization.

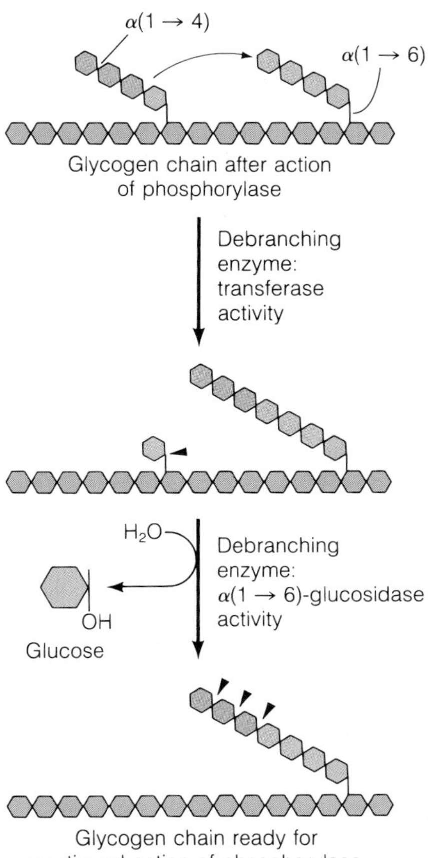

Figure 13.18
The debranching process in glycogen catabolism. Phosphorylase cleaves to within four residues of the branch point (green). Glucose residues (red) are then transferred to another branch (purple). The combined action of phosphorylase and the debranching enzyme creates new branches for attack by phosphorylase.

Glycogen Mobilization

The mobilization of glycogen involves sequential phosphorolytic cleavages of $\alpha(1\rightarrow4)$ bonds, catalyzed by **glycogen phosphorylase.** In plants, starch is similarly mobilized by the action of **starch phosphorylase.** Both reactions release glucose-1-phosphate from nonreducing ends of the glucose polymer.

$$\text{Glucose } \alpha(1\rightarrow4)\text{glucose}\alpha(1\rightarrow4)\text{glucose}\alpha(1\rightarrow4)\text{glucose} \cdots$$

$$\downarrow P_i \text{ Phosphorylase}$$

$$\alpha\text{-D-Glucose-1-}\text{Ⓟ} + \text{Glucose}\alpha(1\rightarrow4)\text{glucose}\alpha(1\rightarrow4)\text{glucose} \cdots$$

The reaction is thermodynamically reversible ($\Delta G^{\circ\prime} = +3.1$ kJ/mol), but the relatively high intracellular levels of inorganic phosphate cause this reaction to operate in vivo almost exclusively in the degradative, rather than the synthetic, direction. Note that the reaction proceeds with retention of configuration at carbon 1. The mechanism is probably comparable to that catalyzed by lysozyme.

Like α-amylase, phosphorylases cannot cleave past $\alpha(1\rightarrow6)$ branch points; in fact, cleavage stops four glucose residues from a branch point.

The debranching process involves the action of a second enzyme, **oligo(α1,4→α1,4)glucantransferase,** as shown in Figure 13.18. This "debranching enzyme" catalyzes two reactions. First is the transferase activity; the enzyme removes three of the remaining glucose residues and transfers this trisaccharide moiety intact to the end of some other outer branch. Next, the remaining glucose residue, which is still attached to the chain by an $\alpha(1\rightarrow6)$ bond, is cleaved by the $\alpha(1\rightarrow6)$-glucosidase activity of the same debranching enzyme. This yields one molecule of free glucose and a branch of three $\alpha(1\rightarrow4)$-linked glucose residues. This newly exposed branch is now available for further attack by phosphorylase.

At this point you might wonder why the glycogen breakdown scheme has evolved to include this complex debranching process. The importance of storing carbohydrate energy in the form of a highly branched polymer may well lie in an animal's need to generate energy very quickly following appropriate stimuli. Glycogen phosphorylase attacks *exoglycosidic* bonds; it cleaves sequentially from nonreducing ends. Obviously, the more of such ends that exist in such a polymer, the faster the polymer can be mobilized.

To be metabolized via glycolysis, the glucose-1-phosphate produced by phosphorylase action must be converted to glucose-6-phosphate. This isomerization is accomplished by phosphoglucomutase. The reaction is mechanistically interesting, because the enzyme is a phosphoprotein, and glucose-1,6-bisphosphate is involved as an intermediate. The enzyme-bound phosphate is transferred directly from a serine residue to the substrate to give the intermediate, glucose-1,6-bisphosphate.

$$\text{Enzyme-phosphate} + \text{glucose-1-phosphate} \rightleftharpoons \text{enzyme} + \text{glucose-1,6-bisphosphate}$$

$$\text{Glucose-1,6-bisphosphate} + \text{enzyme} \rightleftharpoons \text{enzyme-phosphate} + \text{glucose-6-phosphate}$$

$$\text{Net: Glucose-1-phosphate} \rightleftharpoons \text{glucose-6-phosphate} \qquad \Delta G^{\circ\prime} = -7.3 \text{ kJ/mol}$$

The serine residue that carries the phosphate group is unusually reactive, as shown by the fact that phosphoglucomutase, like chymotrypsin and other serine proteases, is irreversibly inhibited by diisopropylfluorophosphate. The inhibition, like that of chymotrypsin, involves acylation of only the active site serine.

Most of the glycogen in vertebrate animals is stored as granules in cells of liver and skeletal muscle. A major function of the liver is to provide glucose for metabolism by other tissues. This is accomplished both through glycogen mobilization and through gluconeogenesis. Both processes yield phosphorylated forms of glucose, which cannot exit from liver cells. Conversion to free glucose involves the action of **glucose-6-phosphatase,** which hydrolyzes glucose-6-phosphate to glucose and phosphate. This enzyme is also present in kidney and intestine. By contrast, muscle glycogen serves primarily as a source of glucose-6-phosphate for catabolism within muscle cells. Accordingly, glucose-6-phosphatase is absent from muscle, as it is from brain, which depends almost exclusively on glucose from the blood as its primary energy source. This ensures that glucose-6-phosphate formed from glycogen cannot diffuse out of these cells, since, as mentioned earlier, sugar phosphates do not readily traverse cell membranes.

Regulation of Glycogen Breakdown

In Chapter 11 we mentioned the control of glycogen breakdown, or glycogenolysis, as a particularly well-understood example of a regulatory cascade, a process in which the intensity of an initial regulatory signal is amplified manyfold through a series of enzyme activations. This is particularly important in the case of glycogenolysis, because fright, for example, or the need to catch prey, can trigger an instantaneous requirement for increased energy generation and utilization. Glycogen represents the most immediately available *large-scale* source of metabolic energy, and hence it is important that animals be able to activate glycogen mobilization very rapidly. Moreover, glycogen breakdown is the hormone-controlled process for which the molecular action of the hormone was first understood in detail. Therefore, although the process was introduced earlier, we present it here in more detail.

Structure of Glycogen Phosphorylase. In skeletal muscle glycogen phosphorylase is a dimer containing two identical polypeptide chains, each of 92,000 daltons (see Figure 11.18, Chapter 11). The enzyme exists in two interconvertible forms—the relatively *active* phosphorylase *a* and the relatively *inactive* phosphorylase *b*. A serine residue at position 14 of each polypeptide chain is phosphorylated in the active form, phosphorylase *a*, while the unmodified serine residues are present in the less active phosphorylase *b*. As shown in Figure 13.19, activation is catalyzed by a specific **phosphorylase *b* kinase,** which transfers phosphate from ATP to the two serine residues. Deactivation is brought about by a specific **phosphorylase phosphatase.**

Control of Phosphorylase Activity. Phosphorylase *b* kinase is also activated by phosphorylation of the enzyme from an inactive to an active form. The enzyme catalyzing the latter reaction is called **cyclic AMP-dependent protein kinase** because it catalyzes the phosphorylation of a number of different proteins. Its activity in turn is controlled by **adenosine-3′,5′-**

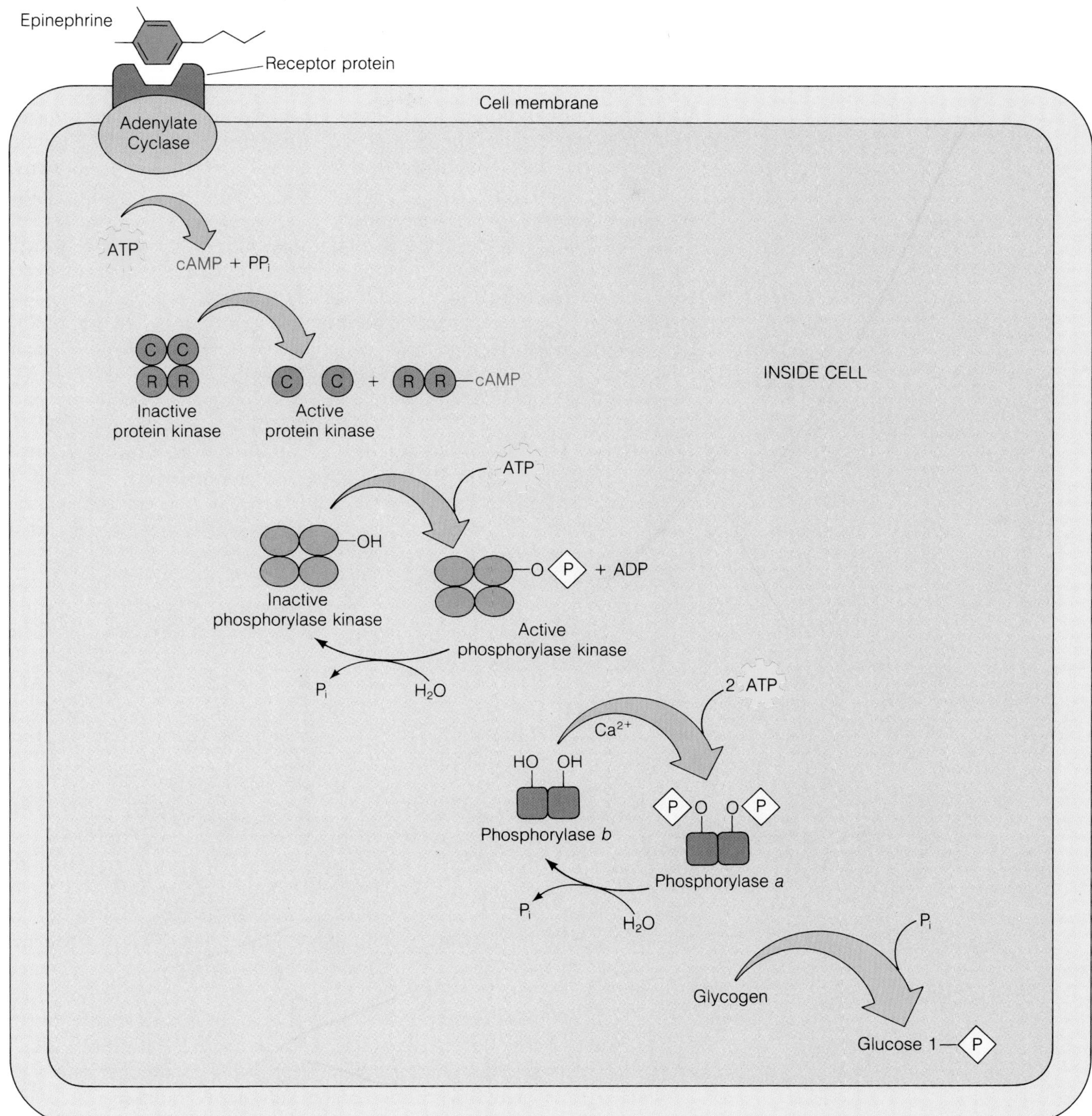

Figure 13.19
The regulatory cascade controlling glycogen breakdown, as it might occur in a muscle cell after epinephrine stimulation. Inactive protein kinase is a tetramer, with two catalytic subunits (C) and two regulatory subunits (R). The enzymatically active form is the C monomer. Phosphorylase kinase is a tetramer consisting of nonidentical subunits (one of which is calmodulin). The activation of phosphorylase *b* converts a nonphosphorylated dimer to a phosphorylated dimer consisting of identical subunits. Each reaction in the cascade series amplifies the metabolic signal, such that binding of very few epinephrine molecules to receptors triggers release of glucose-1-phosphate from intracellular glycogen stores. Inactivation of the pathway involves the actions of specific phosphatases, which remove the phosphates from phosphorylase kinase and phosphorylase *a*.

monophosphate, or cyclic AMP. Cyclic AMP plays numerous roles in regulating metabolism, both in eukaryotes and in prokaryotes.

Produced in response to hormonal stimulation, cyclic AMP is called a *second messenger* because it receives messages from outside the cell, in the form of hormonal stimuli, and transmits them within the cell. This transmission involves the activation of some metabolic processes and the inhibition of others. In glycogenolysis cyclic AMP exerts a rapid and efficient activation. At the same time it inhibits glycogen synthesis through a separate regulatory cascade, to be discussed in Chapter 16.

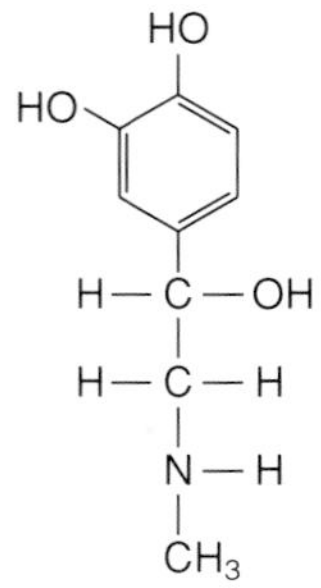

HO
HO
H—C—OH
H—C—H
N—H
CH_3

Epinephrine

The primary hormone promoting glycogenolysis in muscle is **epinephrine,** which is secreted from the adrenal medulla and binds to specific receptors on muscle cell membranes. Liver glycogen mobilization is stimulated largely by the pancreatic peptide hormone **glucagon.** In both cases, binding of the hormone at the membrane stimulates the membrane-bound enzyme **adenylate cyclase,** which catalyzes the synthesis of cyclic AMP from ATP. Cyclic AMP in turn activates protein kinase, which catalyzes the phosphorylation of phosphorylase *b* kinase. This in turn catalyzes the phosphorylation of phosphorylase *b* to *a* and, hence, the activation of glycogen breakdown, through the action of phosphorylase *a*. These events explain how the secretion of relatively few molecules of hormone, such as epinephrine, can, within just a few moments, trigger a massive conversion of glycogen to glucose-1-phosphate.

Epinephrine, also known as adrenaline, is the principal hormone governing the "fight or flight" response to various stimuli. In addition to stimulating glycogenolysis, the hormone triggers a variety of physiological events, such as increasing depth and frequency of heartbeats. These effects are also mediated via cyclic AMP, as discussed further in Chapter 23. Cyclic AMP also regulates other metabolic processes, including the stimulation of fat breakdown and the inhibition of glycogen synthesis. We shall return to these effects as we proceed through metabolism.

PROTEINS IN THE GLYCOGENOLYTIC CASCADE. Our presentation of the glycogenolytic cascade started with the phosphorylase reaction and then worked backward, to the initial hormonal signal. Now let us start with the hormone and work forward, with emphasis on the proteins involved (again, refer to Figure 13.19). The hormone binds to a specific receptor located on the outside of the cytoplasmic membrane. This leads to activation of adenylate cyclase, which is bound to the inside of the membrane; the mechanism of activation is discussed in Chapter 23.

Cyclic AMP-dependent protein kinase is a tetramer consisting of two *catalytic* subunits, C, and two *regulatory* subunits, R. The tetramer, C_2R_2, is catalytically inactive. Binding of cyclic AMP to the R subunits causes the tetramer to dissociate, giving the catalytically active monomer (C). This protein now catalyzes the phosphorylation of phosphorylase *b* kinase. The kinase is a complex multisubunit protein. One of its subunits is a protein called **calmodulin,** or *cal*cium-*modul*ating prote*in*. Calcium ion has long been known as an important physiological regulator, particularly of processes related to nerve conduction and muscle contraction (Chapter 29). Most of these effects are mediated through the binding of Ca^{2+} to calmodulin, which amplifies small changes in intracellular Ca^{2+} concentration.

Calmodulin is a small protein (M_r ~17,000) of highly conserved amino acid sequence. It contains four calcium ion-binding sites (Figure 13.20). Each site binds Ca^{2+} with a K_d of about 10^{-6} M, consistent with observations that calcium can effect intracellular metabolic changes in concentrations as low as 1 μM. Binding stimulates a major conformational change in the protein, leading to a more compact and more highly helical

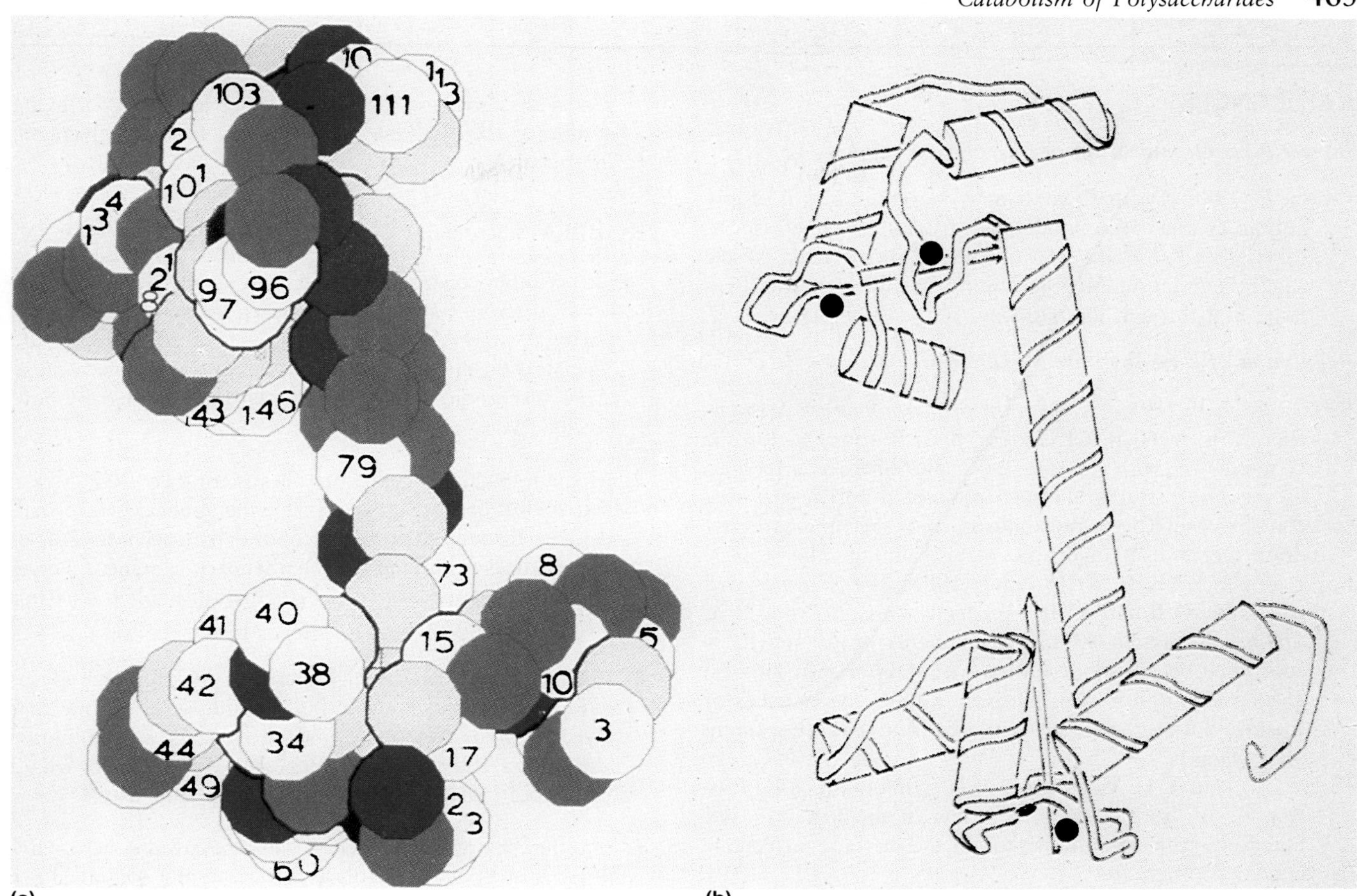

Figure 13.20
Three-dimensional structure of bovine brain calmodulin as determined by x-ray crystallography. (**a**) Space-filling model. Each colored ball represents one residue. (**b**) Each α-helical region is shown, with black circles representing the calcium-binding sites. Note the α-helical region connecting the two ends of the dumbbell-shaped molecule, each end of which binds two Ca^{2+} ions. It is likely that this helix undergoes length changes as a result of calcium binding and that these are responsible for changes in affinity for calcium-regulated targets.

structure, which augments the affinity of calmodulin for a number of regulatory target proteins. In the case of phosphorylase *b* kinase, the binding of calmodulin is essential for catalytic activity of the enzyme. Thus, the glycogenolysis cascade is dependent on intracellular calcium concentration as well as on cyclic AMP levels. This is particularly important in muscle, where contraction is stimulated by calcium release (Chapter 29). Thus, Ca^{2+} plays a dual role, in provision of the energy substrates needed to support muscle contraction and in contraction itself.

Nonhormonal Control of Glycogenolysis. Glycogen breakdown is under nonhormonal, as well as hormonal, control. Recall that phosphorylase *b* is relatively inactive; this form of the enzyme is activated allosterically by 5′-AMP (but not by cyclic AMP). Usually this activation does not occur in the cell, because ATP, which is far more abundant and which does not activate phosphorylase *b*, competes with AMP for binding to the enzyme. However, under energy-deprived conditions AMP may accumulate, at the expense of ATP breakdown, to the extent that phosphorylase *b* and hence glycogenolysis are activated. Thus, the mobilization of energy reserves from glycogen can be brought about either by hormonal stimulation, reflecting a physiological need for increased ATP production, or by an allosteric mechanism triggered when the energy level is deficient for maintenance of normal functions.

In this chapter we described the initial phases of carbohydrate catabolism. Most of the energy from carbohydrate catabolism comes from the further oxidation of pyruvate, which we pursue in the next two chapters.

REFERENCES

Intracellular Organization of Glycolytic Enzymes

Ovádi, J. (1988) Old pathway—new concept. Control of glycolysis by metabolite-modulated dynamic enzyme associations. *Trends Biochem. Sci.* 13:486–490. A recent review suggesting that interactions among glycolytic enzymes regulate flux through the pathway.

Regulation of Carbohydrate Metabolism

Beitner, R. (ed.) (1985) *Regulation of Carbohydrate Metabolism,* Vols. I and II. CRC Press, Boca Raton, Fla. Review chapters deal with fructose-2,6-bisphosphate as a biological regulator, structural organization of glycolytic enzymes, control of glycogen metabolism, and aberrant carbohydrate metabolism in disease states, among other topics.

Boyer, P. D., and E. G. Krebs (eds.) (1986) *The Enzymes,* 3rd ed., Vols. XVII and XVIII. Academic Press, Orlando, Fla. These volumes discuss enzymes systems whose activity is controlled by phosphorylation, including protein kinases, glycogen phosphorylase, pyruvate kinase, the enzymes of fructose-2,6-bisphosphate synthesis, and phosphoprotein phosphatases.

Klee, C. B., and T. C. Vanaman (1982) Calmodulin. *Adv. Protein Chem.* 35:213–321. A review of the structure and function of this ubiquitous protein and its roles in calcium-regulated processes.

Martin, B. R. (1987) *Metabolic Regulation. A Molecular Approach.* Blackwell Scientific Publications, Oxford. Most of the regulatory mechanisms emphasized in this slim volume are those involving hormonal systems and ions, particularly calcium.

Analysis of Carbohydrate Metabolism by in Vivo NMR

Chance, B., and K. Wasserman (1986) Anaerobiosis, lactate, and gas exchange during exercise, I and II. *Federation Proc.* 45:2904–2957. Reports from a symposium in which modern metabolic techniques, including ^{31}P NMR, were used to relate muscular exertion to the control of carbohydrate metabolism.

Shulman, R. G. (1988) High resolution NMR in vivo. *Trends Biochem. Sci.* 13:37–39. A minireview that summarizes how NMR studies have changed some of our concepts about energy metabolism and its regulation.

Sprang, S., et al. (1988) Structural changes in glycogen phosphorylase induced by phosphorylation. *Nature* 336:215–221. The structures of phosphorylase *a* and phosphorylase *b*, as determined by x-ray crystallography.

Oscillations of Glycolytic Intermediates

Hess, B., and A. Boiteux (1971) Oscillatory phenomena in biochemistry. *Annu. Rev. Biochem.* 40:237–258. Much has been learned about the control of glycolysis from the study of metabolite oscillations. This review focuses on glycolysis but also shows the general importance of oscillatory phenomena.

Richter, P. H., and J. Ross (1981) Concentration oscillations and efficiency: Glycolysis. *Science* 211:715–716. A theoretical discussion of the energetic advantages to a living system of the oscillations observed in levels of glycolytic intermediates.

PROBLEMS

1. Intracellular concentrations in resting muscle are as follows: fructose-6-phosphate, 2 mM; fructose-1,6-bisphosphate, 10 mM; AMP, 1 mM; ADP, 2 mM; ATP, 10 mM; P_i, 50 mM. Is the phosphofructokinase reaction in muscle *more* or *less* exergonic than under standard conditions? By how much?

2. Methanol is highly toxic, not because of its own biological activity but because it is converted metabolically to formaldehyde, through action of alcohol dehydrogenase. Part of the medical treatment for methanol poisoning involves administration of large doses of ethanol. Explain why this treatment is effective.

3. Refer to Figure 13.7, which indicates $\Delta G'$ for each glycolytic reaction under intracellular conditions. Assume that glyceraldehyde-phosphate dehydrogenase was inhibited with iodoacetic acid. Which glycolytic intermediate would you expect to accumulate most rapidly, and why?

4. In different organisms sucrose can be cleaved either by hydrolysis or by phosphorolysis. Calculate the ATP yield per mole of sucrose metabolized by anaerobic glycolysis starting with either (a) hydrolytic or (b) phosphorolytic cleavage.

5. Suppose it was possible to label glucose with ^{14}C at any position or combination of positions. For yeast fermenting glucose to ethanol, which form or forms of labeled glucose would give the *most* radioactivity in CO_2 and the *least* in ethanol?

6. Write balanced chemical equations for each of the following: (a) anaerobic glycolysis of 1 mol of sucrose, cleaved initially by sucrose phosphorylase; (b) aerobic glycolysis of 1 mol of maltose; (c) fermentation of one glucose residue in starch to ethanol, with the initial cleavage involving α-amylase.

7. Because of the position of arsenic in the periodic table, arsenate (AsO_4^{-3}) is chemically similar to inorganic phosphate and is used by phosphate-requiring enzymes as an alternative substrate. However, organic arsenates are quite unstable and spontaneously hydrolyze. Arsenate is known to inhibit glycolysis. Identify the target enzyme, and explain the mechanism of inhibition.

8. Suppose that you made some wine whose alcohol content was 10% w/v (i.e., 10 g of ethanol per 100 ml of wine). The initial fermentation mixture would have had to contain what molar concentration of glucose or its equivalent to generate this much ethanol? Is it likely that an initial fermentation mixture would contain that much glucose? In what other forms might the fermentable carbon appear?

CHAPTER 14

Oxidative Processes: Citric Acid Cycle and Pentose Phosphate Pathway

In Chapter 13 we explored the initial, anaerobic, phase of carbohydrate degradation. Next we follow the subsequent aerobic reactions by which carbohydrates are ultimately oxidized to carbon dioxide and water (Figure 14.1). This chapter is not primarily about carbohydrate metabolism, however. That is because the pathway involved, the citric acid cycle, is the central oxidative pathway in respiration, the process by which *all* metabolic fuels—carbohydrate, lipid, and protein—are catabolized in aerobic organisms and tissues.

Although respiration and glycolysis both conserve about 40% of energy released in the form of ATP, much less total energy is generated, per mole of substrate catabolized, in glycolysis than in respiration. Because lactate, like other end products of anaerobic carbohydrate catabolism, is nearly as complex a molecule as the starting material, glucose, relatively little of the potential energy stored in the glucose molecule is released by its conversion to lactate.

Far more energy is generated in the subsequent reactions by which lactate or pyruvate becomes completely oxidized to CO_2. That energy release, which in eukaryotic cells occurs in mitochondria, is in the form of exergonic dehydrogenation reactions that generate reduced electron carriers; these carriers are next reoxidized in the mitochondrial respiratory (electron transport) chain. The latter process involves a series of coupled oxidation–reduction reactions by which the electrons released from oxidizable substrates are passed from electron carrier to electron carrier until they reach the terminal electron acceptor, oxygen, which becomes reduced to water. The free energy released from some of these reactions drives the synthesis of ATP from ADP and orthophosphate, through oxidative phosphorylation. This chapter focuses on fates of the oxidizable substrates, and the next chapter focuses on the chain of electron carriers and the synthesis of ATP. In short, if we look at the end products of respiration—CO_2 and H_2O—our emphasis in this chapter is on the generation of CO_2 and in Chapter 15 on the generation of H_2O.

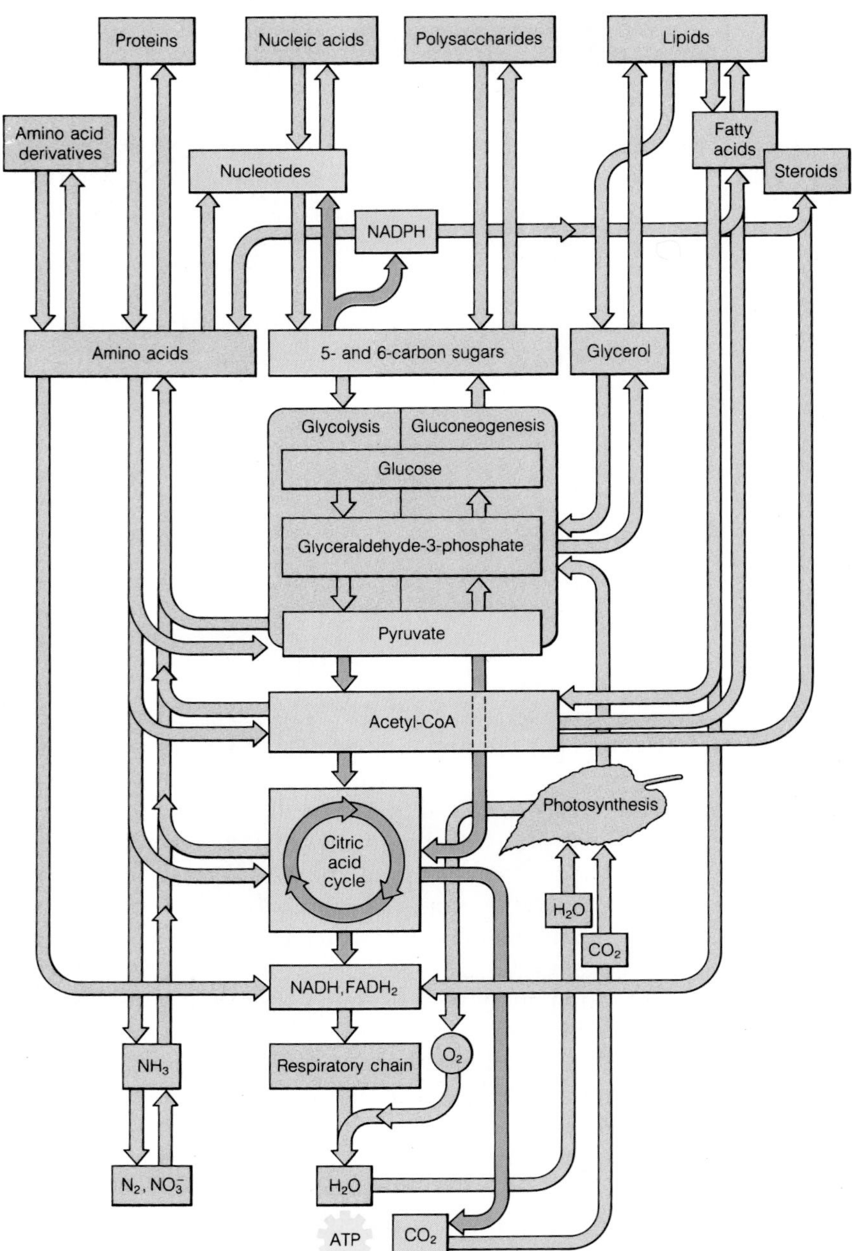

Figure 14.1
Overview of intermediary metabolism, highlighting the citric acid cycle and processes of carbohydrate oxidation (red).

Overview of Pyruvate Oxidation and the Citric Acid Cycle

The Three Stages of Respiration

It is convenient to think of the oxidation of organic substrates as a three-stage process, as schematized in Figure 14.2. Stage 1 is the generation of an activated two-carbon fragment—the acetyl group of **acetyl-coenzyme A,** or **acetyl-CoA.** Stage 2 is oxidation of those two carbon atoms in the citric acid cycle. Stage 3 is electron transport and oxidative phosphorylation, in which the reduced electron carriers generated in the citric acid cycle become reoxidized, with concomitant synthesis of ATP. Stage 1 is distinctive in that it

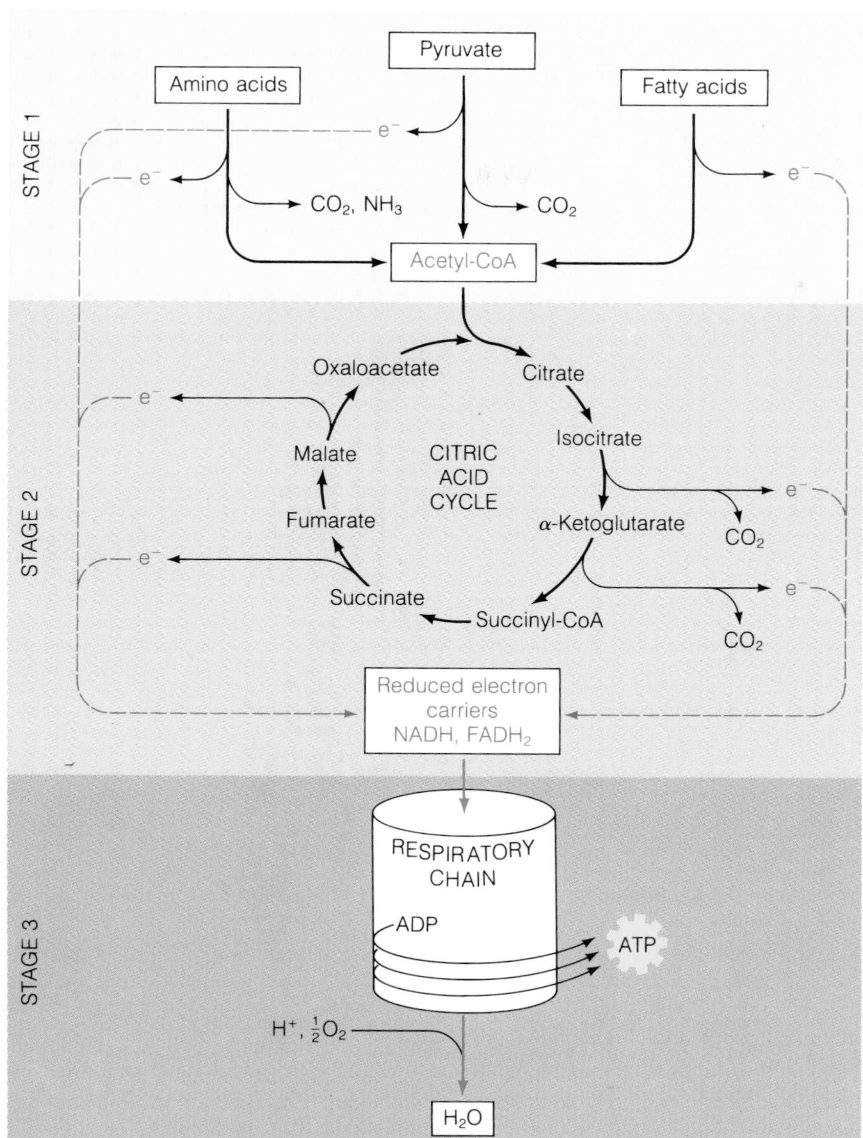

Figure 14.2
The three stages of respiration.

represents a family of pathways. Carbon from carbohydrate enters stage 1 as pyruvate. Fat breakdown generates acetyl-CoA primarily through β-oxidation of fatty acids (Chapter 17), while several different pathways generate acetyl-CoA and citric acid cycle intermediates from amino acid catabolism (Chapter 20).

Reactions of the first two stages of respiration occur within the interior, gel-like matrix of the mitochondrion, while electron transport and oxidative phosphorylation are catalyzed by membrane-bound enzymes in cristae—the threadlike projections of inner mitochondrial membrane into the matrix. These structural and biochemical relationships are discussed further when we explore the third stage of respiration in the next chapter.

Coenzyme A and the Activation of Carbon

Because carbon is introduced to the citric acid cycle as acetyl-CoA, this is a suitable point to discuss why, in chemical terms, coenzyme A is so well suited for its biological roles.

Coenzyme A was named originally for its role in activating acyl groups. The coenzyme is derived metabolically from ATP, from the water-soluble vitamin **pantothenic acid,** and from β-mercaptoethylamine.

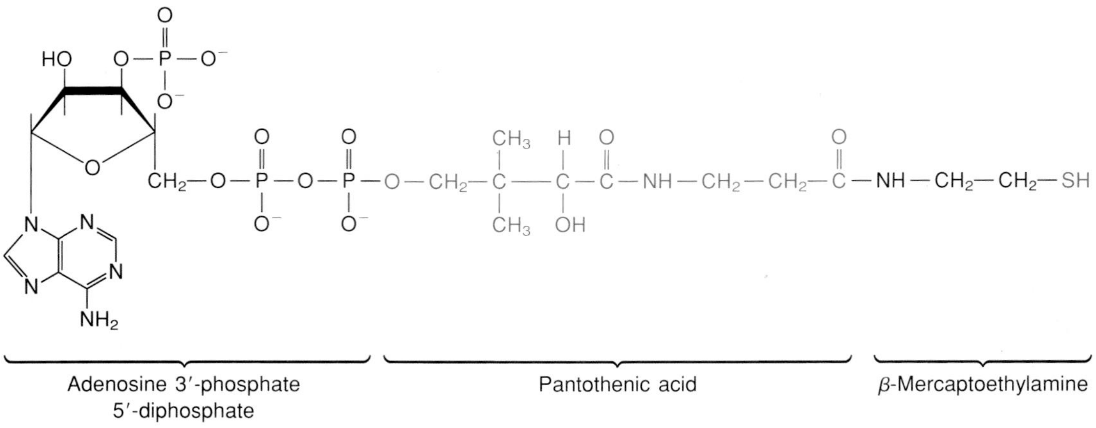

Adenosine 3′-phosphate 5′-diphosphate — Pantothenic acid — β-Mercaptoethylamine

Coenzyme A

A free thiol on the latter moiety is the functionally significant part of the coenzyme molecule; the rest of the molecule provides enzyme binding sites. In acyl-CoAs, such as acetyl-CoA, the acyl group is linked to the thiol group to form a thioester.

$$\underset{\textbf{Coenzyme A}}{\text{CoA—SH}} + \underset{\textbf{Acetyl unit}}{\text{—}\overset{\text{O}}{\overset{\|}{\text{C}}}\text{—CH}_3} \longrightarrow \underset{\textbf{Acetyl CoA}}{\text{CoA—S—}\overset{\text{O}}{\overset{\|}{\text{C}}}\text{—CH}_3}$$

We shall designate the unacylated form of coenzyme A as CoA-SH and the acylated form as $\text{R—}\overset{\text{O}}{\overset{\|}{\text{C}}}\text{—S—CoA}$.

The energy-rich nature of thioesters, as compared with ordinary esters, is related primarily to resonance stabilization. Esters can resonate between two forms (Figure 14.3). Stabilization involves π-electron overlap, giving partial double-bond character to the C—O link. By contrast, the larger atomic size of S as compared with O reduces the π-electron overlap between C and S, so that the C═S structure does not contribute significantly to resonance stabilization. Thus, the thioester is *destabilized* relative to an ester, so that its $\Delta G^{\circ\prime}$ of hydrolysis is increased.

The lack of double-bond character in the C—S bond of acyl-CoAs makes this bond weaker than the corresponding C—O bond in ordinary esters, in turn making the thioalkoxide ion a good leaving group in nucleophilic displacement reactions. Thus, the acyl group is readily transferred to other metabolites, as occurs, in fact, in the first reaction of the citric acid cycle.

Discovery of the Citric Acid Cycle

The essence of the citric acid cycle is the joining of the two activated carbons of acetyl-CoA to a four-carbon substrate, followed by reactions that oxidize two carbon atoms to CO_2 and regenerate the original four-carbon substrate. Hence, the pathway is cyclic, because the end product is identical to one of the original substrates, allowing the pathway to begin the cycle anew by reaction with another molecule of acetyl-CoA.

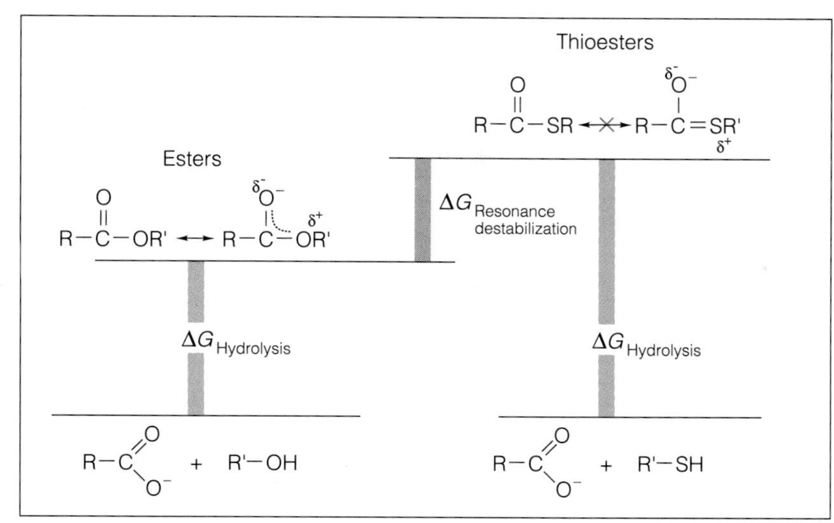

Figure 14.3
Lack of resonance stabilization (X) as the basis for the higher ΔG of hydrolysis for thioesters as compared with ordinary esters. The free energies of the hydrolysis products are identical for the two classes of compounds.

The idea that organic fuels are oxidized via a cyclic pathway was proposed in 1937 by Hans Krebs, on the basis of studies for which he later shared a Nobel Prize with Fritz Lipmann (the discoverer of coenzyme A). Krebs was studying oxygen consumption in minced pigeon breast muscle, a tissue with a very high rate of respiration, and made several fundamental observations. First, the anions of several tricarboxylic acids, including **citrate, isocitrate,** and ***cis*-aconitate,** stimulated both oxygen consumption and pyruvate oxidation out of proportion to the amounts added, suggesting catalytic roles for these ions.

CH_2-COO^- / $HO-C-COO^-$ / CH_2-COO^-

Citrate

CH_2-COO^- / $H-C-COO^-$ / $H-C-COO^-$ / OH

Isocitrate

CH_2-COO^- / $C-COO^-$ ‖ $H-C-COO^-$

***cis*-Aconitate**

Second, **malonate,** an analog of **succinate** and a known inhibitor of succinate dehydrogenase, blocked the oxidation of pyruvate, pointing to a role for succinate dehydrogenase in pyruvate oxidation.

COO^- / CH_2 / COO^-

Malonate

COO^- / CH_2 / CH_2 / COO^-

Succinate

Third, malonate-inhibited cells accumulated citrate, **α-ketoglutarate,** and succinate, suggesting that citrate and α-ketoglutarate are both normal precursors of succinate.

COO^- / CH_2 / CH_2 / $C=O$ / COO^-

α-Ketoglutarate

COO^- / CH_2 / $C=O$ / COO^-

Oxaloacetate

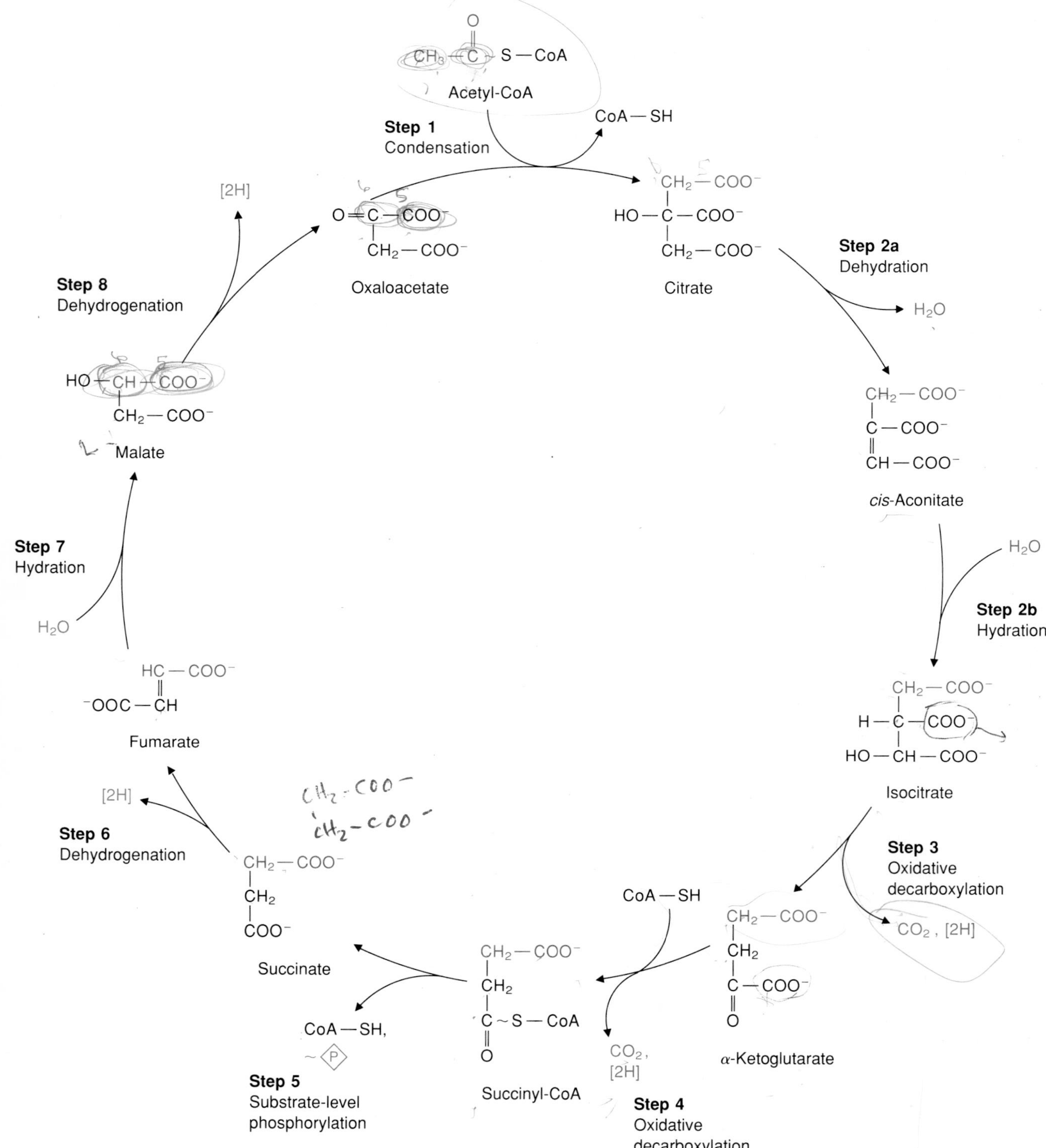

Figure 14.4
Reactions of the citric acid cycle. Carbon atoms entering as acetyl-CoA are shown in red, to depict the fates of these two atoms through one cycle as far as malate.

Finally, addition of pyruvate plus **oxaloacetate** to this system led to accumulation of citrate in the medium.

From these observations plus the structures of the organic acids that could stimulate respiration, Krebs proposed the sequence of reactions involved *and* the cyclic nature of the pathway. He postulated that a cyclic sequence could result from a reaction between pyruvate and oxaloacetate to give citrate plus CO_2. We now know that pyruvate must first be oxidized and that acetyl-CoA is the species that reacts with oxaloacetate. We know also of an activated intermediate form of succinate, **succinyl-CoA.** Except for these changes, the pathway as proposed by Krebs was correct; the reactions are outlined in Figure 14.4.

The citric acid cycle is also known by other names: the Krebs cycle, after its discoverer; and the tricarboxylic acid (TCA) cycle, because it was apparent from the outset that tricarboxylic acids are involved as intermediates, whereas it became clear only later that citric acid was one of those intermediates.

Strategy of the Citric Acid Cycle

Since the function of the citric acid cycle is to oxidize organic metabolites, it is appropriate to review the oxidation and reduction of organic compounds. Quantitative aspects of biological oxidations are presented in Chapter 15; the discussion here is largely qualitative.

Oxidation involves the loss of electrons from a substrate. The substrate that loses electrons is an electron donor, and the electrons are transferred to an electron acceptor, which thereby becomes reduced. Carbon atoms become oxidized either through loss of hydrogen or through combination with oxygen. The latter removes electrons from the shell about a carbon nucleus, because the electronegativity of the oxygen draws shared electrons toward its own nucleus. Similarly, when an organic compound loses hydrogen, it loses the electron associated with that hydrogen. Thus, either process involves a loss of electrons from the carbon atom undergoing oxidation. *Formally,* the two processes are equivalent.

A point of potential confusion arises in naming enzymes that catalyze oxidation reactions. Because of the conservation of matter, electrons released in an oxidation must be transferred to an electron acceptor. Since most metabolic oxidations involve loss of hydrogen from the electron donor, we call enzymes that catalyze those reactions **dehydrogenases.** The term **oxidase** is reserved for those enzymes in which molecular oxygen itself is the electron acceptor. Oxidases, therefore, catalyze only the small proportion of oxidation reactions in which molecular oxygen is a direct participant. The latter category of reactions is discussed further in Chapter 15.

Referring to Figure 14.4, let us preview the citric acid cycle in a bit more detail, focusing on the metabolic fates of the two carbons that enter the cycle as acetyl-CoA. Acetyl-CoA reacts with a four-carbon organic acid, oxaloacetate, to yield a six-carbon tricarboxylic acid, citrate. This enters into a series of seven reactions during which two carbons are released as CO_2 and the remaining four carbons are regenerated as oxaloacetate, which is ready to begin the process again. Hence the cyclic nature of the pathway: oxaloacetate is present at the beginning, to condense with an activated two-carbon fragment, and it is present at the end, after two carbons have been oxidized to CO_2.

Pyruvate Oxidation: A Major Route for Acetyl-CoA Formation

The Overall Reaction: Pyruvate to Acetyl-CoA

The primary fuel for direct oxidation in the citric acid cycle is acetyl-CoA. Later chapters discuss how this is formed in catabolism of fatty acids and amino acids. Here we focus on the oxidative decarboxylation of pyruvate, formed largely in glycolysis.

The overall reaction looks relatively simple, but in fact it is quite complicated. The enzyme system, called the **pyruvate dehydrogenase complex,** catalyzes the oxidative decarboxylation of pyruvate.

$$CH_3-\overset{O}{\overset{\|}{C}}-COO^- + NAD^+ + CoA-SH \longrightarrow CH_3-\overset{O}{\overset{\|}{C}}\sim SCoA + NADH + CO_2 \qquad \Delta G^{\circ\prime} = -33.5 \text{ kJ/mol}$$

pyr.

Three individual enzymes and five different coenzymes, including the two—coenzyme A and NAD^+—that appear in the overall equation, participate in this reaction. These enzymes and coenzymes constitute a highly organized multienzyme complex, found in prokaryotic cells and in eukaryotic cell mitochondria (Figure 14.5). The overall reaction catalyzed by the enzymes of the pyruvate dehydrogenase complex is highly exergonic and for all practical purposes is irreversible inside cells.

Biochemistry of Coenzymes in Pyruvate Oxidation: Flavins, Thiamine Pyrophosphate, and Lipoic Acid

To fully understand the reactions in pyruvate oxidation, it is essential to understand the biochemistry of the five coenzymes involved. All were introduced in Chapter 10, and the actions of coenzyme A and NAD^+ have been described. Here we must digress to present the remaining three—**flavin adenine dinucleotide, thiamine pyrophosphate,** and **lipoic acid.**

The story of these coenzymes is in large part the story of the vitamins from which they are derived. It began with observations in Asia, nearly a century ago, regarding the paralytic disease **beriberi.** Polished rice (with the husks removed) was a dietary staple in the region. It was found that the husks, or bran, from rice contained a substance that relieved the paralysis. Casimir Funk, a Polish biochemist, reasoned that beriberi and three other nutritional diseases—**pellagra, rickets,** and **scurvy**—all resulted from lack of a dietary trace substance, which he called vitamine (literally, amine of life). This led to adoption of the generic term, **vitamin.**

The idea that vitamins could prevent disease was not widely accepted, because of the dominance of Pasteur's idea that diseases were caused only by infection. Now, however, we recognize the existence of about a dozen vitamins. The first known example is thiamine, the component of rice bran that is nutritionally deficient in beriberi. Later the vitamin deficiencies associated with pellagra, rickets, and scurvy were related to vitamins B_6, D, and C, respectively.

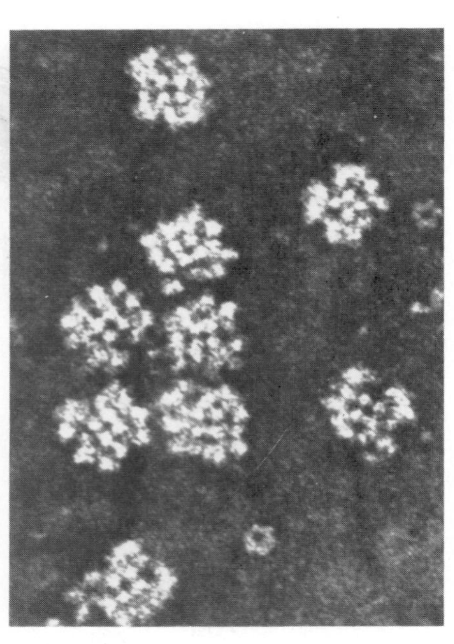

Figure 14.5
Electron micrograph of the pyruvate dehydrogenase complex from *E. coli.*

FLAVIN COENZYMES. **Flavin adenine dinucleotide,** or **FAD,** is one of two coenzymes derived from vitamin B_2, or **riboflavin** (Figure 14.6). The other is the simpler **flavin mononucleotide** (FMN), or riboflavin phosphate. The functional part of both coenzymes is the **isoalloxazine ring,** which serves as a two-electron acceptor. The ring is attached to **ribitol,** an open-chain version of ribose with the aldehyde carbon reduced to the hydroxymethyl level. The 5′ carbon of ribitol is linked to phosphate in FMN, while FAD is an adenylylated derivative of FMN. Thus, these compounds are somewhat analogous to nicotinamide mononucleotide and NAD^+, respectively.

Enzymes that use a flavin cofactor are called **flavoproteins.** FMN and FAD undergo virtually identical electron transfer reactions, and the use of one or the other coenzyme is largely a matter of structural moieties involved in binding the coenzyme to the apoenzyme. In a few cases that binding is covalent. However, in most cases the flavin is bound tightly but noncovalently, such that the coenzyme can be reversibly dissociated from the holoenzyme.

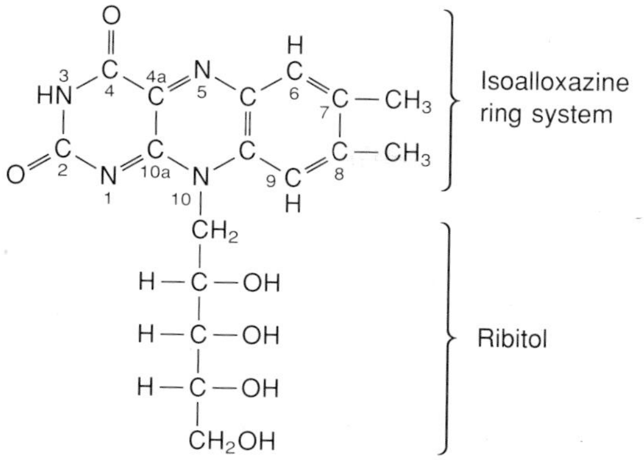

Isoalloxazine ring system

Ribitol

Riboflavin

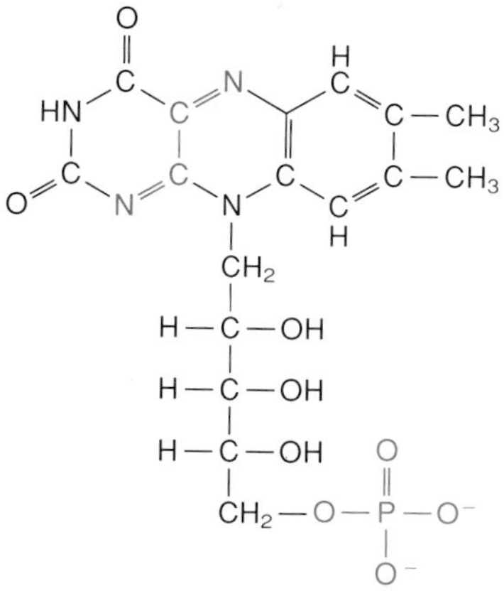

Flavin mononucleotide (FMN)
(also called riboflavin phosphate)

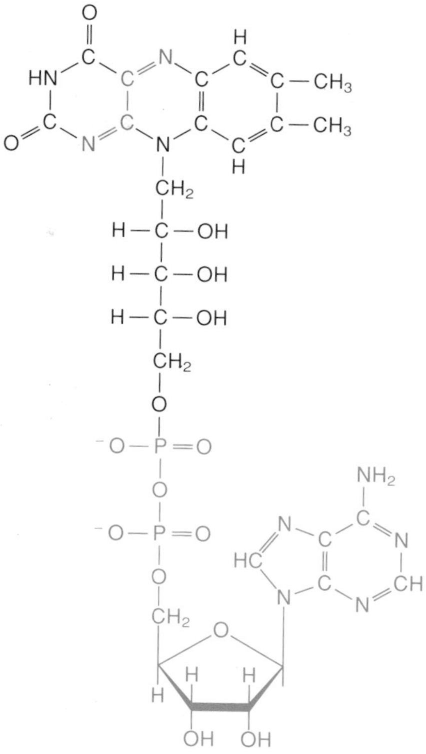

Flavin adenine dinucleotide (FAD)

Figure 14.6
Structure of riboflavin and its coenzymes. The figure identifies (in red) the cluster of two carbon and two nitrogen atoms in the isoalloxazine ring system that participates in flavin oxidation–reduction reactions.

Like the nicotinamide coenzymes, the flavins undergo two-electron oxidation and reduction reactions. The flavins, however, are distinctive in having a stable one-electron-reduced species, a **semiquinone** free radical, as shown in Figure 14.7. This can be detected spectrophotometrically; while oxidized FAD and FMN are bright yellow, and fully reduced flavins are colorless, the semiquinone intermediate is either red or blue, depending on pH. The stability of the intermediate gives flavins a catalytic versatility lacked by nicotinamide coenzymes, in that they can interact with either two-electron or one-electron donor–acceptor pairs. Flavoproteins can also interact directly with oxygen. Thus, some, but not all, flavoproteins are oxidases.

THIAMINE PYROPHOSPHATE. Because it was the first of the B vitamins to be identified, thiamine is also called vitamin B_1. The vitamin is structurally complex, but its conversion to the coenzyme form, thiamine pyrophosphate, or TPP, involves simply an ATP-dependent pyrophosphorylation.

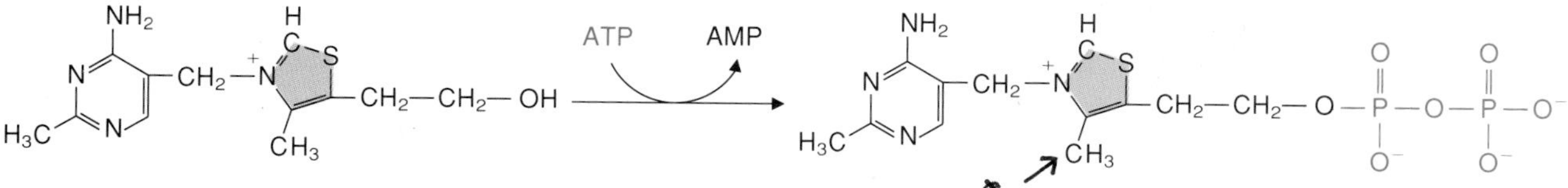

ATP → AMP

Thiamine

Thiamine pyrophosphate

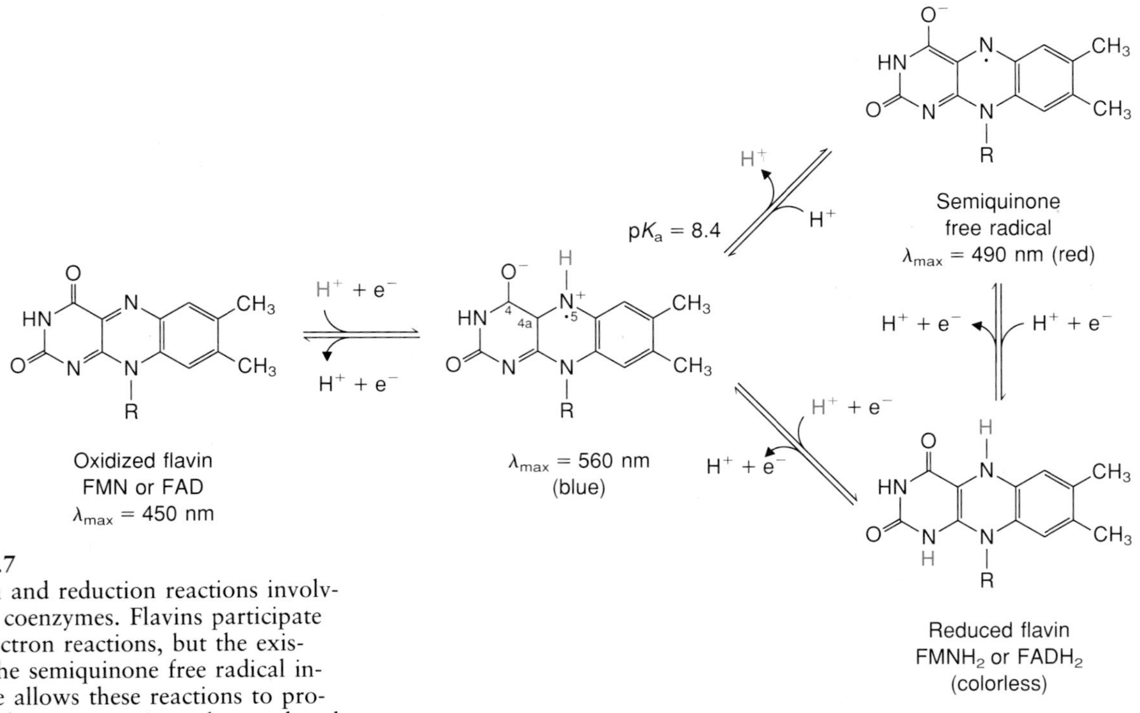

Oxidized flavin
FMN or FAD
$\lambda_{max} = 450$ nm

$\lambda_{max} = 560$ nm
(blue)

$pK_a = 8.4$

Semiquinone
free radical
$\lambda_{max} = 490$ nm (red)

Reduced flavin
$FMNH_2$ or $FADH_2$
(colorless)

Figure 14.7
Oxidation and reduction reactions involving flavin coenzymes. Flavins participate in two-electron reactions, but the existence of the semiquinone free radical intermediate allows these reactions to proceed one electron at a time; thus, reduced flavins can readily be oxidized by one-electron acceptors. The spectral maxima for oxidized flavin and the protonated and deprotonated forms of the semiquinone intermediate are indicated. In both forms the unpaired electron is delocalized between N-5 and C-4a.

Thiamine pyrophosphate contains two heterocyclic rings, a substituted pyrimidine and a thiazole. The latter is the reactive moiety—specifically, the rather acidic carbon between the sulfur and the nitrogen (Figure 14.8). This carbon forms a carbanion (step 1), which in turn can attack the carbonyl carbon of α-keto acids, such as pyruvate, as well as other carbonyl compounds (step 2). When an α-keto acid is a substrate, the resulting addition compound undergoes a nonoxidative decarboxylation (step 3). The positive charge on the thiazole ring nitrogen allows the ring to act as an electron sink, so that the decarboxylation proceeds without a change in the oxidation state of the bound carbonyl carbon. When the substrate is pyruvate, this intermediate is called **active acetaldehyde,** or more accurately, **hydroxyethyl-TPP** (step 4). The two-carbon fragment can be released hydrolytically as acetaldehyde, as in the pyruvate decarboxylase reaction of the yeast alcoholic fermentation. In the pyruvate dehydrogenase complex, the hydroxyethyl moiety becomes oxidized concomitantly with its transfer to another coenzyme. Thus, in general terms TPP functions in the generation of an activated aldehyde species, which may or may not undergo oxidation as it is transferred to an acceptor.

LIPOIC ACID. In the pyruvate dehydrogenase reaction, that next acceptor is lipoic acid, which is the internal disulfide of 6,8-dithiooctanoic acid. The coenzyme is joined to its apoenzyme via an amide bond linking the carboxyl group of lipoic acid to a lysine ε-amino group. Thus, the reactive species is an amide, called **lipoamide.**

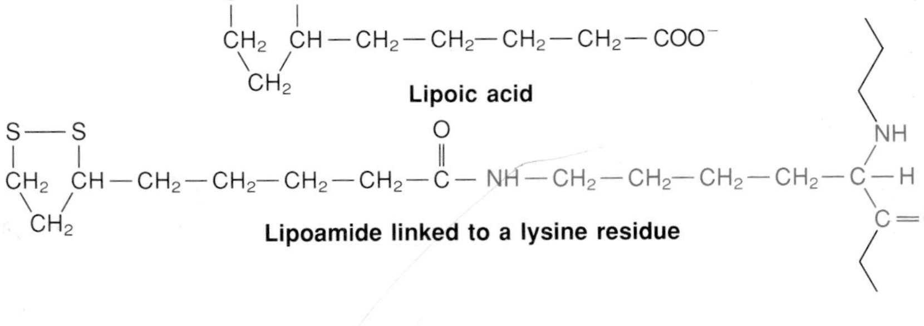

Lipoic acid

Lipoamide linked to a lysine residue

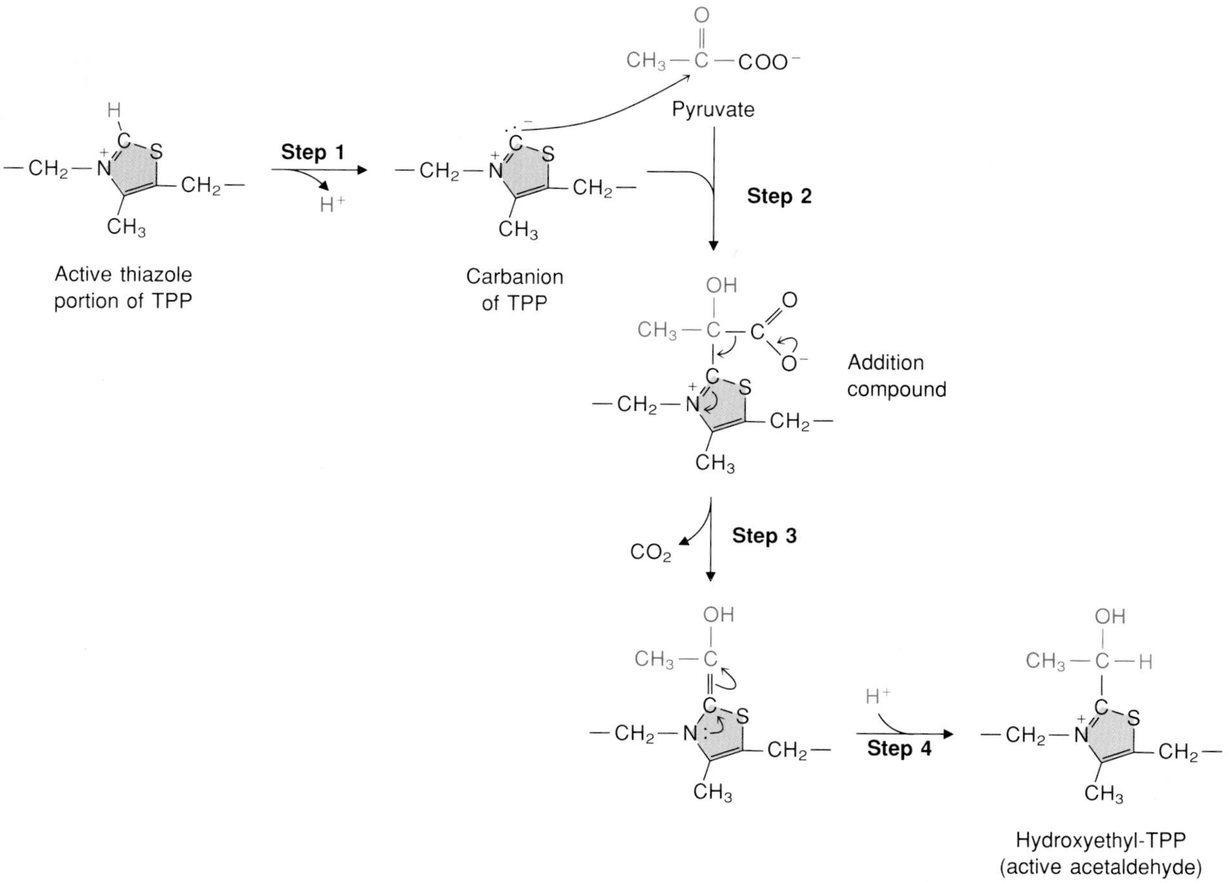

Figure 14.8
Mechanism of action of thiamine pyrophosphate in the pyruvate decarboxylase reaction. The key reaction (step 2) is attack by the carbanion of TPP on the carbonyl atom of pyruvate, followed by nonoxidative decarboxylation of the coenzyme-bound pyruvate. Thus, the two-carbon fragment (red) bound to TPP remains at the aldehyde oxidation level, and this compound is called active acetaldehyde.

Transfer of the active aldehyde moiety from TPP to the sulfur on carbon 6 of lipoamide involves simultaneous oxidation of that moiety, coupled to reduction of the disulfide. This generates an acyl group, which, in pyruvate dehydrogenase, is transferred next to coenzyme A.

Operation of the Pyruvate Dehydrogenase Complex

Now let us see how all of these components function together to effect the conversion of pyruvate to acetyl-CoA. As stated earlier, the overall reaction involves three enzymatic activities. In *E. coli* the complex has a mass of about 4.6 million daltons, slightly larger than a ribosome; the mammalian liver complex is about twice as large. The bacterial complex contains about 24 polypeptide chains of **pyruvate decarboxylase** (E_1), about 24 chains of **dihydrolipoyl transacetylase** (E_2), and about 12 chains of a flavoprotein **dihydrolipoyl dehydrogenase** (E_3). The complex in eukaryotic cells contains small amounts of two regulatory enzymes as well—a kinase that phosphorylates three serine residues in E_1 and a phosphatase that removes those phosphates. Regulation of the activities of this complex will concern us later in this chapter.

The function of each individual enzyme in the pyruvate dehydrogenase complex is shown in Figure 14.9. A central feature of this reaction sequence is the interaction of the lipoamide moiety first with hydroxyethyl-TPP bound to E_1, next with CoA-SH on E_2 to yield acetyl-CoA, and finally with bound FAD on E_3 to regenerate the disulfide species. Attachment of the eight-carbon lipoic acid chain with a six-carbon lysine residue gives a flexi-

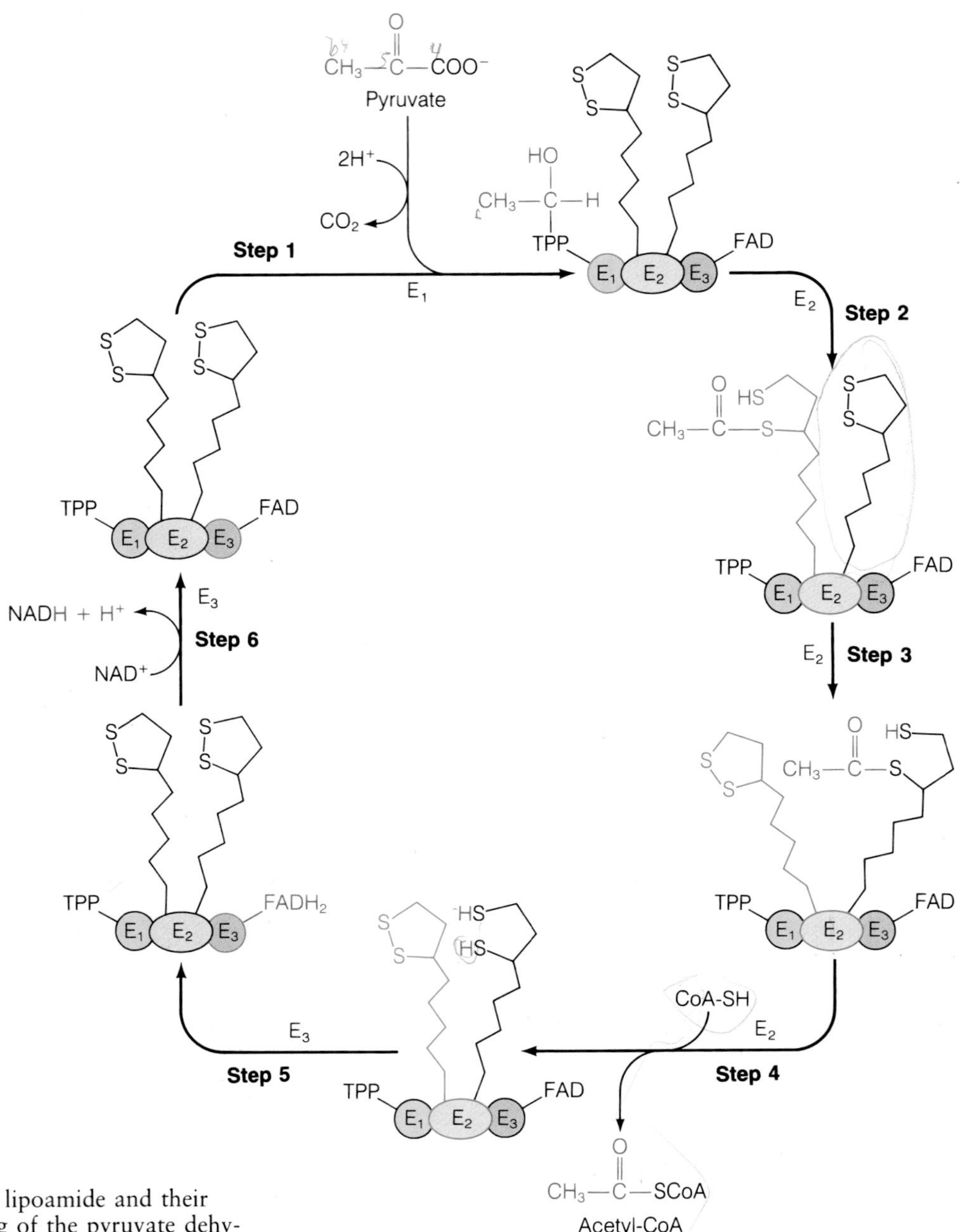

Figure 14.9
Swinging arms of lipoamide and their role in functioning of the pyruvate dehydrogenase complex. E_1 accepts a two-carbon aldehyde group from the substrate (step 1); this group is transferred to the first swinging arm on E_2 and simultaneously oxidized (step 2). The resultant acetyl group is then transferred to the second swinging arm (step 3), which positions it for transfer to CoA-SH (step 4). Finally, E_3 oxidizes the reduced lipoamide swinging arm (step 5), with oxidation of the reduced flavin by NAD^+ (step 6).

ble chain that, when fully extended, is some 14 Å in length. However, spectroscopic studies indicate that bound TPP on E_1 is some 45 to 60 Å away from bound FAD on E_3. Thus, a single lipoamide arm on E_2 cannot swing all the way between participating sites on E_1 and E_3. The problem is solved through the participation of at least two lipoamide arms on each E_2 chain, with transfer of the acetyl group from one arm to the next. Since each 14-Å arm can arc about twice that distance, two arms can connect functional groups up to 56 Å apart. Operation of these swinging arms is shown diagrammatically in Figure 14.9. The first arm accepts the two-carbon fragment from E_1 and then transfers it as an acetyl group to the second arm. From there it is transferred directly to coenzyme A. The reduced lipoamide arm, with two thiol groups, is then reoxidized to the cyclic disulfide by E_3.

Physical juxtaposition of the enzymes of the complex allows the overall reaction to proceed smoothly, without unwanted side reactions or diffusion of intermediates from catalytic sites. The pyruvate dehydrogenase complex represents one of the best-understood examples of how cells can achieve economy of function by juxtaposition of enzymes catalyzing sequential reactions.

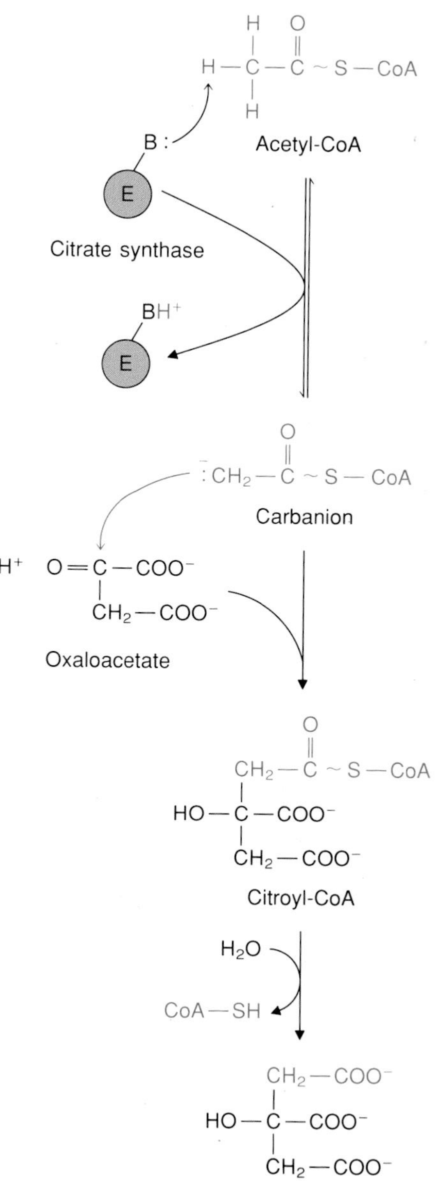

Figure 14.10
Mechanism of the citrate synthase reaction. B is a basic residue in the enzyme molecule.

The Citric Acid Cycle

Figure 14.4 presented the entire citric acid cycle, showing the structure of each intermediate. The importance of this pathway merits examination of each reaction. First, recall that the cycle begins with condensation of a two-carbon compound (acetyl-CoA) with a four-carbon compound (oxaloacetate) to give a six-carbon organic anion, citrate. Subsequent transformations of citrate oxidize two of its carbons to CO_2, with regeneration of oxaloacetate.

Reaction 1: Introduction of Two Carbon Atoms as Acetyl-CoA

The initial reaction, catalyzed by **citrate synthase,** is akin to an aldol condensation.

$$\underset{\textbf{Acetyl-S-CoA}}{CH_3-\overset{O}{\overset{\|}{C}}-S-CoA} + \underset{\textbf{Oxaloacetate}}{\overset{O}{\overset{\|}{C}}(-COO^-)-CH_2-COO^-} + H_2O \longrightarrow \underset{\textbf{Citrate}}{HO-C(-COO^-)(CH_2-COO^-)(CH_2-COO^-)} + CoA-SH + H^+ \qquad \Delta G^{\circ\prime} = -32.2\ \text{kJ/mol}$$

The methyl carbon of the activated acetyl moiety on acetyl-CoA loses a proton, with nucleophilic attack of the resultant carbanion on the carbonyl carbon of oxaloacetate (Figure 14.10). This generates the highly unstable **citroyl-CoA,** which spontaneously hydrolyzes while enzyme bound, to yield the products. As expected for the first committed step in a pathway, the reaction is highly exergonic and has been considered to be a site of regulation for the overall pathway. Structural analysis of citrate synthase (Figure 14.11) gives excellent evidence for the induced fit model of enzyme catalysis.

Reaction 2: Isomerization of Citrate

Citrate contains a tertiary alcohol functional group, which is difficult to oxidize. Isomerization, catalyzed by **aconitase,** generates the secondary alcoholic compound, isocitrate, which is much more readily oxidized. The reaction involves successive dehydration and hydration, through *cis*-aconitate as a dehydrated intermediate, which remains enzyme bound.

$$\underset{\textbf{Citrate}}{\begin{matrix} CH_2-COO^- \\ HO-C-COO^- \\ H-C-COO^- \\ H \end{matrix}} \underset{H_2O}{\overset{H_2O}{\rightleftharpoons}} \underset{\textbf{[\textit{cis}-Aconitate]}}{\begin{matrix} CH_2-COO^- \\ C-COO^- \\ \| \\ CH-COO^- \end{matrix}} \underset{H_2O}{\overset{H_2O}{\rightleftharpoons}} \underset{\textbf{Isocitrate}}{\begin{matrix} CH_2-COO^- \\ H-C-COO^- \\ HO-C-COO^- \\ H \end{matrix}} \qquad \Delta G^{\circ\prime} = +6.3\ \text{kJ/mol}$$

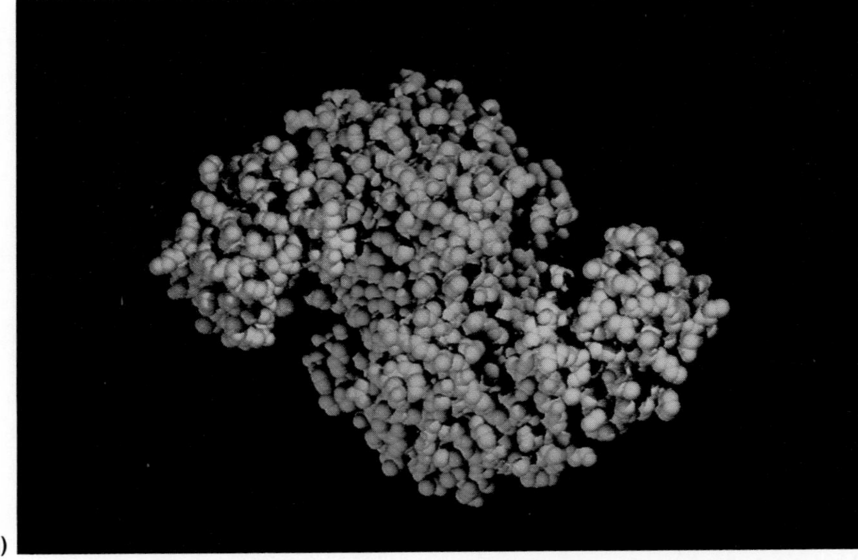

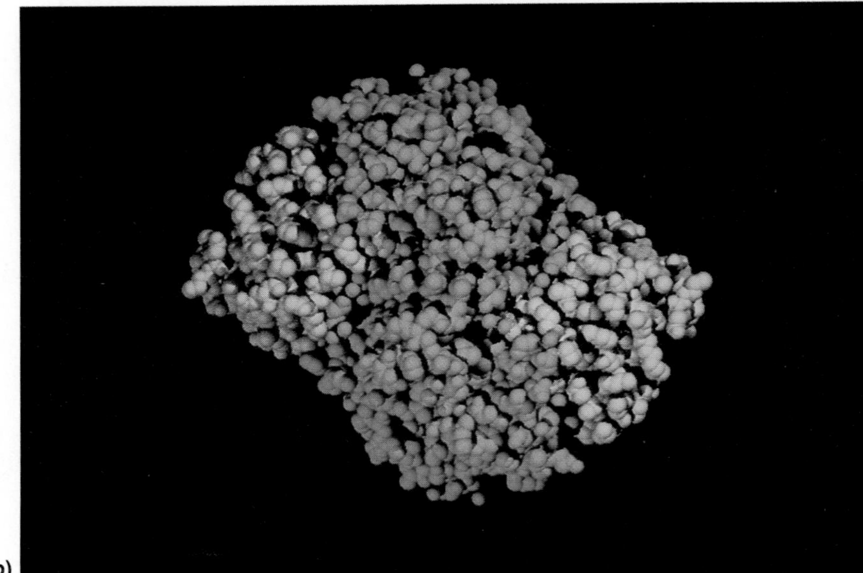

Figure 14.11
Three-dimensional structure of pig heart citrate synthase. This enzyme shows excellent evidence for the induced fit model of enzyme catalysis (Chapter 10). In the absence of substrates the enzyme crystallizes in an "open" form (**a**), with large clefts in both catalytic domains of the homodimeric protein. Binding of substrates or CoA-SH causes the enzyme to adopt a "closed" conformation, with the clefts essentially filled (**b**).

An equilibrium mixture of these three acids at 25° contains about 90% citrate, 4% *cis*-aconitate, and 6% isocitrate, but the exergonic nature of preceding and subsequent reactions draws the reaction to the right as written. The enzyme contains nonheme iron and acid-labile sulfur in a cluster called an **iron–sulfur center,** normally associated with oxidoreductases. In aconitase the iron forms a complex with citrate, through two carboxyl groups and one hydroxyl group on the citrate molecule.

Aconitase is the target site for the toxic action of **fluoroacetate,** a plant product that has been used as a rat poison. Fluoroacetate blocks the citric acid cycle by its metabolic conversion, via citrate synthase, to **fluorocitrate,** which is a potent inhibitor of aconitase. Fluoroacetate is an example of a **suicide substrate.** This is a compound which is not by itself toxic to cells but which resembles a normal metabolite closely enough that it undergoes metabolic transformation to a product that does inhibit a crucial enzyme; the cell "commits suicide" by transforming the analog to a toxic product. In this case fluoroacetate is converted to fluoroacetyl-CoA, which serves as a substrate for citrate synthase. However, the resultant fluorocitrate cannot be acted on by aconitase.

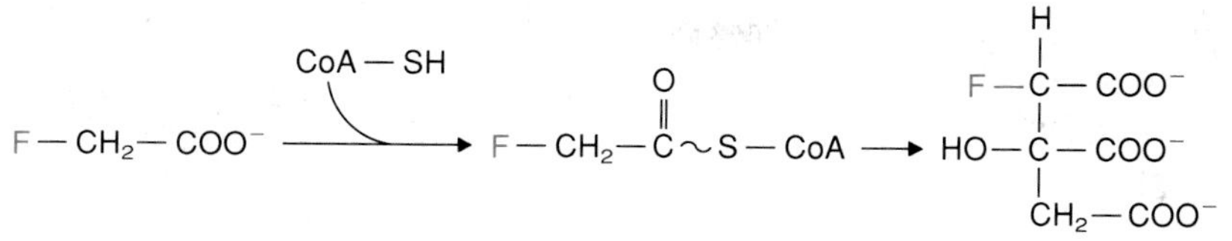

$$F-CH_2-COO^- \xrightarrow{CoA-SH} F-CH_2-\overset{O}{\overset{\|}{C}}\sim S-CoA \longrightarrow HO-\underset{CH_2-COO^-}{\underset{|}{\overset{F-\overset{H}{\overset{|}{C}}-COO^-}{\overset{|}{C}}}}-COO^-$$

Fluoroacetate **Fluoroacetyl-CoA** **Fluorocitrate**

The aconitase reaction is stereospecific; of the four possible diastereoisomers of isocitrate, only one is produced. Aconitase is one of the earliest enzymes for which stereospecificity was demonstrated and elucidated. To describe those developments, we must first present the next reaction in the citric acid cycle.

Reaction 3: Generation of CO_2 by an NAD^+-Linked Dehydrogenase

The first of two oxidative decarboxylations in the cycle is catalyzed by **isocitrate dehydrogenase.**

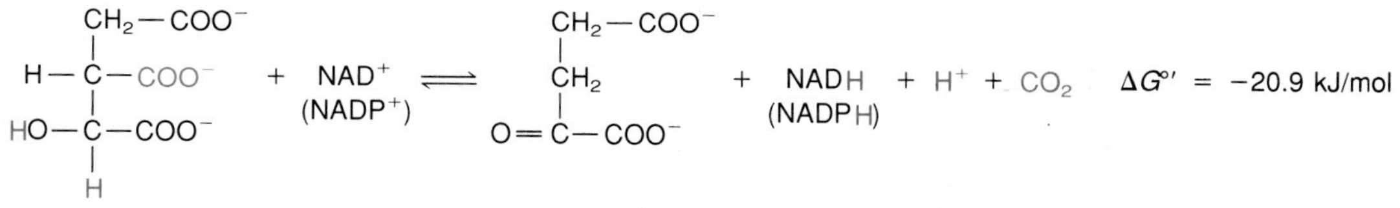

$$\text{Isocitrate} + NAD^+ \,(NADP^+) \rightleftharpoons \alpha\text{-Ketoglutarate} + NADH\,(NADPH) + H^+ + CO_2 \qquad \Delta G^{\circ\prime} = -20.9 \text{ kJ/mol}$$

Isocitrate: CH_2-COO^- / $H-C-COO^-$ / $HO-C-COO^-$ / H

α-Ketoglutarate: CH_2-COO^- / CH_2 / $O=C-COO^-$

Isocitrate **α-Ketoglutarate**

The reaction involves dehydrogenation to **oxalosuccinate,** an unstable enzyme-bound intermediate that spontaneously decarboxylates before release of the product.

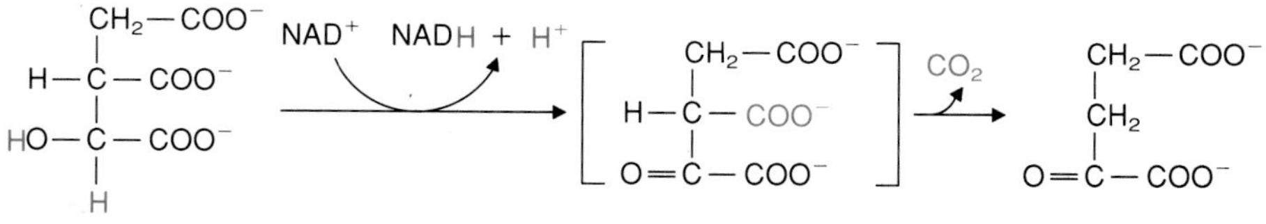

$$\text{Isocitrate} \xrightarrow{NAD^+ \;\; NADH + H^+} [\text{Oxalosuccinate}] \xrightarrow{CO_2} \alpha\text{-Ketoglutarate}$$

Oxalosuccinate: CH_2-COO^- / $H-C-COO^-$ / $O=C-COO^-$

Isocitrate **Oxalosuccinate** **α-Ketoglutarate**

A mitochondrial form of the enzyme that is specific for NAD^+ is probably the chief participant in the citric acid cycle, but an $NADP^+$-specific form of the enzyme is present also, in both cytosol and mitochondria.

As noted above, the stereospecificity of enzyme–substrate interactions was first shown in an experiment involving aconitase. That experiment, carried out in the late 1940s, also cast into question whether citrate was a participant in the overall cycle. Carboxyl-labeled acetate was administered to animal tissue preparations; this labels citrate via its conversion to acetyl-CoA. Since citrate is a symmetrical compound, it was expected that aconi-

$CH_3—\overset{*}{C}OO^-$
Acetate

↓ CoA—SH

$CH_3—\overset{*}{C}(=O)—S—CoA$

↓ Oxaloacetate

$CH_2—\overset{*}{C}OO^-$
$HO—C—COO^-$
$CH_2—COO^-$
Citrate

Aconitase

$CH_2—\overset{*}{C}OO^-$ / $H—C—COO^-$ / $H—C—COO^-$ / OH	$CH_2—COO^-$ / $H—C—COO^-$ / $H—C—\underset{*}{C}OO^-$ / OH
Isocitrate	Isocitrate
↓	↓
$CH_2—\overset{*}{C}OO^-$ / CH_2 / $C(=O)—COO^-$	$CH_2—COO^-$ / CH_2 / $C(=O)—\underset{*}{C}OO^-$
α-Ketoglutarate (form A)	α-Ketoglutarate (form B)

(a)

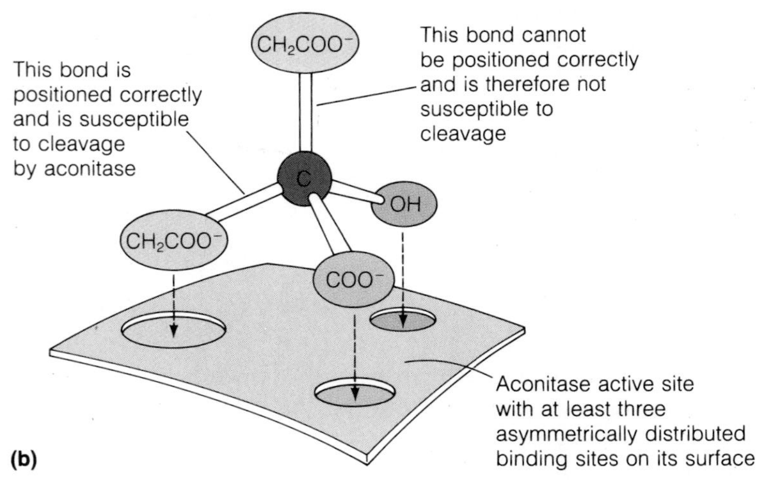

Figure 14.12
Asymmetric binding of the symmetrical citrate to the active site of aconitase. (a) The result expected if citrate were metabolized as a symmetrical compound; a ^{14}C label on the acetate carboxyl would appear in both carboxyls of α-ketoglutarate. (b) How binding of citrate at three sites on the enzyme positions one and only one of the two carboxymethyl groups for the dehydration and rehydration reactions catalyzed by the enzyme.

tase could transfer the alcoholic hydroxyl in either direction (Figure 14.12a). Thus, α-ketoglutarate was expected to be labeled equally in both carboxyl groups. However, only the γ-labeled form was detected, suggesting that only asymmetric tricarboxylic acids could be involved in these early steps of the cycle, which would rule out the symmetrical citrate. However, Alexander Ogston pointed out that if the enzyme bound the substrate at three points or more, then the binding site itself was asymmetric and could bind the substrate in only one way (Figure 14.12b). Thus, citrate was the first substance to be recognized as **prochiral,** revealing a general stereochemical principle of enzyme-substrate binding.

Reaction 4: Generation of a Second CO_2 by a Multienzyme Complex

The fourth reaction is entirely comparable to the pyruvate dehydrogenase reaction. A keto acid substrate undergoes oxidative decarboxylation, with concomitant formation of an acyl-CoA.

$$\underset{\alpha\text{-Ketoglutarate}}{CH_2COO^-—CH_2—C(=O)—COO^-} + NAD^+ + CoA—SH \longrightarrow \underset{\text{Succinyl-CoA}}{CH_2COO^-—CH_2—C(=O)—S—CoA} + CO_2 + NADH + H^+ \qquad \Delta G^{\circ\prime} = -33.5 \text{ kJ/mol}$$

This reaction is catalyzed by the **α-ketoglutarate dehydrogenase complex,** an enzyme cluster similar to the pyruvate dehydrogenase complex, with three analogous enzyme activities and the same five coenzymes—TPP, NAD^+, FAD, lipoic acid, and CoA-SH. The first reaction, catalyzed by α-ketoglutarate decarboxylase, is shown in Figure 14.13. Subsequent transfer of the four-carbon unit to lipoic acid and oxidation by FAD and NAD^+ are analogous to the reactions shown in Figure 14.9 for the pyruvate dehydrogenase complex. One important distinction between these two multienzyme complexes is that the regulatory activities associated with the pyruvate dehydrogenase complex are not present in the α-ketoglutarate dehydrogenase complex.

At this point two carbon atoms have been introduced as acetyl-CoA (by citrate synthase), and two have been lost. Because of the stereochemistry of the aconitase reaction, the two carbon atoms lost as CO_2 are not the same as the two atoms introduced as acetyl-CoA. In the remainder of the cycle the four-carbon intermediate, succinyl-CoA, is converted to the four-carbon product, oxaloacetate, with two of the four steps involving dehydrogenation reactions.

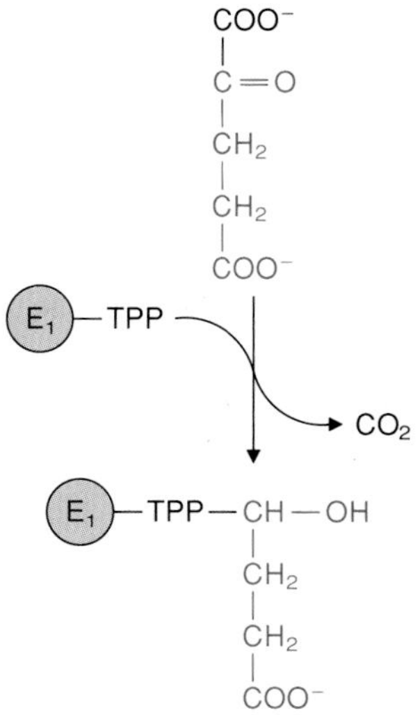

COO^-–C=O–CH_2–CH_2–COO^- + E$_1$–TPP → E$_1$–TPP–CH(OH)–CH_2–CH_2–COO^- + CO_2

Figure 14.13
The first reaction catalyzed by α-ketoglutarate decarboxylase (E_1).

Reaction 5: A Substrate-Level Phosphorylation

Succinyl-CoA is an energy-rich compound, and its potential energy is used to drive the formation of an energy-rich phosphate bond. This reaction, catalyzed by **succinyl-CoA synthetase,** is comparable to the two substrate-level phosphorylation reactions that we encountered in glycolysis, except that in animal cells the energy-rich nucleotide product is not ATP, but GTP.

$$\text{Succinyl-CoA} + P_i + \text{GDP} \rightleftharpoons \text{succinate} + \text{GTP} + \text{CoA-SH} \qquad \Delta G^{\circ\prime} = -2.9 \text{ kJ/mol}$$

Much of the GTP formed drives the ultimate synthesis of ATP, through the action of nucleoside diphosphokinase.

$$\text{GTP} + \text{ADP} \rightleftharpoons \text{GDP} + \text{ATP} \qquad \Delta G^{\circ\prime} = 0.0 \text{ kJ/mol}$$

In plants and bacteria, however, ATP is formed directly, through reaction of succinyl-CoA with ADP. In either case the reaction involves formation of the anhydride succinyl phosphate, which then activates the enzyme by phosphorylating a specific histidine residue. This *N*-phosphohistidine residue (Figure 14.14) then transfers its phosphate to the nucleoside diphosphate substrate.

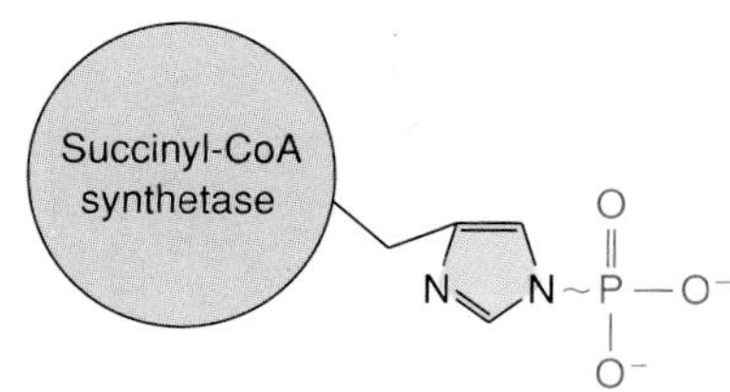

Figure 14.14
Structure of the *N*-phosphohistidine residue in the reaction catalyzed by succinyl-CoA synthetase.

Reaction 6: A Flavin-Dependent Dehydrogenation

At this stage we have lost two carbons from the six-carbon starting material; completion of the cycle involves conversion of the four-carbon succinate to the four-carbon oxaloacetate. The first of the three reactions involved, catalyzed by **succinate dehydrogenase,** is the FAD-dependent dehydrogenation of two saturated carbons to a double bond.

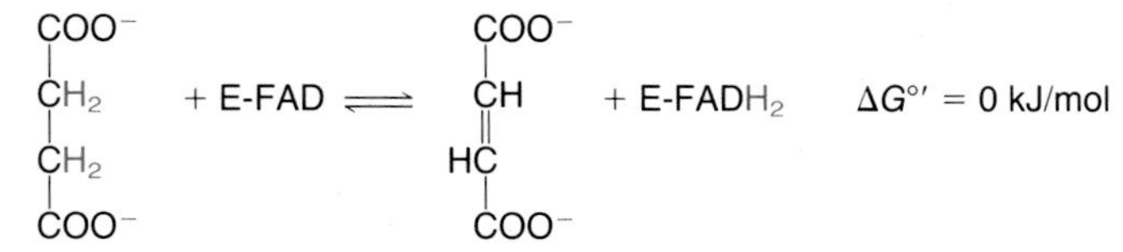

$$\underset{\textbf{Succinate}}{COO^-\text{–}CH_2\text{–}CH_2\text{–}COO^-} + \text{E-FAD} \rightleftharpoons \underset{\textbf{Fumarate}}{COO^-\text{–}CH{=}HC\text{–}COO^-} + \text{E-FADH}_2 \qquad \Delta G^{\circ\prime} = 0 \text{ kJ/mol}$$

This is followed by hydration of that double bond and, finally, dehydrogenation of an α-hydroxy acid to the α-keto acid oxaloacetate. Saturated carbons are more difficult to oxidize than a C—O bond. Therefore, the redox coenzyme for succinate dehydrogenase is not NAD^+ but the more powerful oxidant FAD, in conjunction with an iron–sulfur center in the enzyme molecule.

In succinate dehydrogenase the coenzyme is bound covalently to the enzyme protein, designated E, through a specific histidine residue.

FAD—CH_2—N(histidine ring)—CH_2—E

Histidine

The enzyme is tightly bound to the mitochondrial inner membrane (Chapter 15). The importance of this is that the reduced flavin, which must be reoxidized for the enzyme to act again, is reoxidized through interaction with the mitochondrial electron transport system, also bound to the membrane.

Note that action of succinate dehydrogenase is also stereospecific, with only the trans isomer, fumarate, formed, and not the cis isomer, maleate.

Reaction 7: Hydration of a Carbon–Carbon Double Bond

This stereospecific hydration is catalyzed by **fumarate hydratase,** more commonly called **fumarase.**

$$\underset{\textbf{Fumarate}}{^{-}OOC{-}CH{=}CH{-}COO^{-}} + H_2O \rightleftharpoons \underset{\textbf{L-Malate}}{^{-}OOC{-}CH(OH){-}CH_2{-}COO^{-}} \qquad \Delta G^{\circ\prime} = -3.8\ \text{kJ/mol}$$

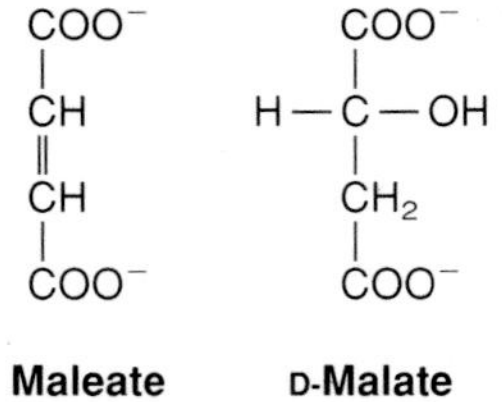

Maleate **D-Malate**

The cis isomer of fumarate, namely maleate, is not a substrate for the forward reaction, nor can the enzyme act on D-malate in the reverse direction.

Reaction 8: A Dehydrogenation That Generates a Substrate for the Next Turn of the Cycle

Finally, the cycle is completed with the NAD^+-dependent dehydrogenation of malate to oxaloacetate, catalyzed by **malate dehydrogenase.**

$$\underset{\textbf{L-Malate}}{^{-}OOC{-}CH(OH){-}CH_2{-}COO^{-}} + NAD^{+} \rightleftharpoons \underset{\textbf{Oxaloacetate}}{^{-}OOC{-}C({=}O){-}CH_2{-}COO^{-}} + NADH + H^{+} \qquad \Delta G^{\circ\prime} = +29.7\ \text{kJ/mol}$$

This highly endergonic reaction proceeds to the right as written because the highly exergonic citrate synthase reaction keeps intramitochondrial oxaloacetate levels exceedingly low (below 10^{-6} M).

Stoichiometry and Energetics of the Citric Acid Cycle

Now let us review what has been accomplished in one turn of the citric acid cycle; this is summarized in Table 14.1. First, a two-carbon fragment (acetyl-CoA) combined with a four-carbon acceptor (oxaloacetate). Next, two carbons were removed as CO_2 as the resultant citrate was further metabolized. Third, four oxidation reactions occurred, with NAD^+ serving as coenzyme for three and FAD as coenzyme for the fourth. Fourth, high-energy phosphate was generated directly at only one reaction (succinyl-CoA synthetase). Finally, oxaloacetate was regenerated; it is now ready to start the cycle again by condensation with another acetyl-CoA.

We can write a balanced chemical equation representing the sum of the eight reactions involved in one turn of the cycle.

$$\text{Acetyl-CoA} + 2H_2O + 3NAD^+ + FAD + GDP + P_i \longrightarrow 2CO_2 + 3NADH + 3H^+ + FADH_2 + \text{CoA-SH} + GTP$$

The GTP formed in the animal succinate synthetase reaction is energetically equivalent to ATP, because nucleoside diphosphokinase can convert the GTP that is formed to ATP. Therefore, we shall substitute ATP for GTP in the following discussion.

Now if we take into account the pyruvate dehydrogenase reaction, and if we recall that each molecule of glucose generates two molecules of pyruvate, we can write the following equation for catabolism of glucose through glycolysis and the citric acid cycle.

$$\text{Glucose} + 6H_2O + 10NAD^+ + 2FAD + 4ADP + 4P_i \longrightarrow 6CO_2 + 10NADH + 10H^+ + 2FADH_2 + 4ATP$$

Table 14.1
Reactions of the citric acid cycle

Reaction	Enzyme	$\Delta G^{\circ\prime}$ (kJ/mol)
1. Acetyl-CoA + oxaloacetate + H_2O $\rightarrow$ citrate + CoA-SH + H^+	Citrate synthase	−32.2
2a. Citrate $\rightleftharpoons$ *cis*-aconitate + H_2O	Aconitase	
2b. *cis*-Aconitate + H_2O $\rightleftharpoons$ isocitrate	Aconitase	+6.3
3. Isocitrate + NAD^+ $\rightleftharpoons$ α-ketoglutarate + CO_2 + NADH + H^+	Isocitrate dehydrogenase	−20.9
4. α-Ketoglutarate + NAD^+ + CoA-SH $\rightleftharpoons$ succinyl-CoA + CO_2 + NADH + H^+	α-Ketoglutarate dehydrogenase complex	−33.5
5. Succinyl-CoA + P_i + GDP $\rightleftharpoons$ succinate + GTP + CoA-SH	Succinyl-CoA synthetase	−2.9
6. Succinate + FAD (enzyme-bound) $\rightleftharpoons$ fumarate + $FADH_2$ (enzyme-bound)	Succinate dehydrogenase	0
7. Fumarate + H_2O $\rightleftharpoons$ L-malate	Fumarase	−3.8
8. L-Malate + NAD^+ $\rightleftharpoons$ oxaloacetate + NADH + H^+	Malate dehydrogenase	+29.7
	NET	−57.3

At this point the ATP yield per mole of glucose metabolized has not increased greatly over the yield from glycolysis: 2 mol of ATP per glucose in glycolysis alone, 4 mol in the above equation. Most of the ATP generated during glucose oxidation is not formed directly from reactions of glycolysis and the citric acid cycle, but from the reoxidation of reduced electron carriers in the respiratory chain. These compounds, NADH and $FADH_2$, are themselves energy rich, in the sense that oxidation of these reduced electron carriers is highly exergonic. As electrons are transferred from these reduced carriers stepwise toward molecular oxygen, the coupled synthesis of ATP from ADP takes place, with about 3 mol of ATP per NADH reoxidized and about 2 mol of ATP per $FADH_2$ reoxidized. As we will see in Chapter 15, this generates about 38 mol of ATP per mole of glucose oxidized to CO_2 and water.

Regulation of Pyruvate Dehydrogenase and the Citric Acid Cycle

Because the citric acid cycle is a source of biosynthetic intermediates, as well as a route for generating metabolic energy, regulation of the cycle is somewhat more complex than if it were only an energy-generating pathway. As in glycolysis, regulation occurs both at the level of entry of fuel into the cycle and at the level of control of key reactions within the cycle. Fuel enters the cycle primarily as acetyl-CoA, which arises from carbohydrate via pyruvate dehydrogenase and from lipid primarily from the β-oxidation of fatty acids. Since fatty acid oxidation is discussed in Chapter 17, let us concentrate here on control of pyruvate dehydrogenase.

Control of Pyruvate Oxidation

Activity of the pyruvate dehydrogenase complex is controlled by allosteric inhibition and, as noted earlier, by a covalent modification that is in turn controlled by the energy state of the cell. E_2, the transacetylase component, is inhibited by acetyl-CoA and activated by CoA-SH. E_3, the dihydrolipoyl dehydrogenase component, is inhibited by NADH and activated by NAD^+. ATP is an allosteric inhibitor of the complex, and AMP is an activator. Thus, the activity of this key reaction is coordinated with the energy charge, the NAD^+/NADH ratio, and the ratio of acetylated to free coenzyme A.

In the pyruvate dehydrogenase complex from mammalian sources, regulation occurs also by covalent modification of the pyruvate decarboxylase component of the complex (E_1 in Figure 14.9). As shown in Figure 14.15, this involves phosphorylation and dephosphorylation of serine residues in E_1. NADH and acetyl-CoA both activate a component of the complex, **pyruvate decarboxylase kinase,** which phosphorylates three specific serine residues, resulting in loss of activity of pyruvate dehydrogenase. A specific **pyruvate decarboxylase phosphatase** hydrolytically removes the bound phosphate and reactivates the complex. The phosphatase is activated by Ca^{2+} and Mg^{2+}. Because ATP and ADP differ in their affinities for Mg^{2+}, the concentration of free Mg^{2+} reflects the ATP/ADP ratio within the mitochondrion. Thus, pyruvate dehydrogenase responds to ATP levels by being turned off when ATP is abundant and further energy production is unneeded. In mammalian tissues at rest, much less than half of the total pyruvate dehydrogenase is in the active, nonphosphorylated form. The complex can be turned on when low ATP levels signal a need to generate more

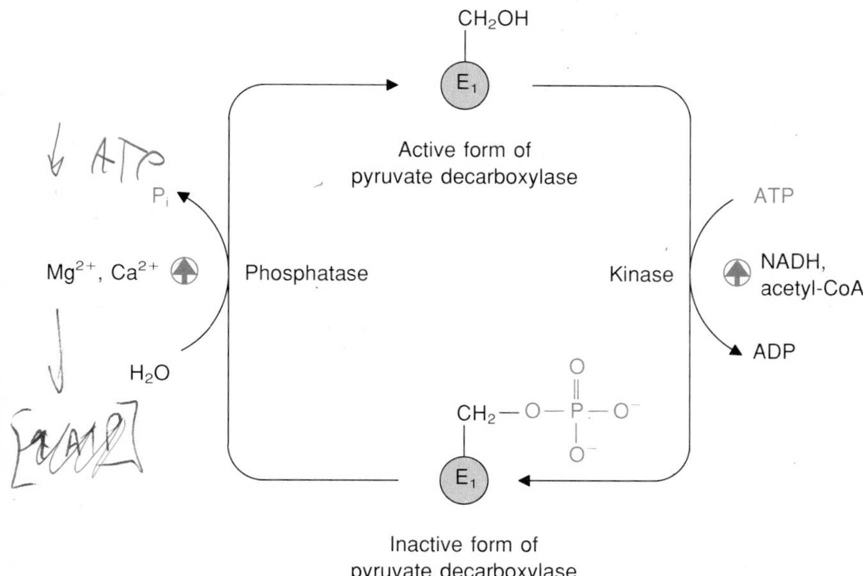

Figure 14.15
Regulation of the pyruvate dehydrogenase complex by phosphorylation and dephosphorylation of the complex. The respective kinase and phosphatase act at three particular serine residues of the pyruvate decarboxylase component (E_1).

ATP. The kinase protein is an integral part of the pyruvate dehydrogenase complex, while the phosphatase is but loosely bound.

Control of the Citric Acid Cycle

Flux through the citric acid cycle is controlled by allosteric interactions, but the concentrations of substrates also play a critical role. Though details of regulation vary among different cells and tissues, the major effects are as summarized in Figure 14.16.

Perhaps the most important factor controlling citric acid cycle activity is the intramitochondrial ratio of [NAD^+] to [NADH]. NAD^+ is a substrate for three cycle enzymes, as well as for pyruvate dehydrogenase. Under conditions that *decrease* the [NAD^+]/[NADH] ratio, such as inhibition of electron transport or substantial consumption of ethanol, the low concentration of NAD^+ can limit the activities of these dehydrogenases.

In some mammalian tissues, notably liver, the levels of citrate vary as much as tenfold, and at low levels flux through the citrate synthase reaction is limited by substrate availability. Recall that in some animal tissues citrate is also a prime regulator of flux through *glycolysis,* via allosteric regulation of phosphofructokinase. However, some tissues, such as heart, cannot transport citrate out of mitochondria, so interaction with cytosolic PFK probably does not occur to a significant extent.

Key sites for allosteric regulation are the reactions catalyzed by isocitrate dehydrogenase and α-ketoglutarate dehydrogenase. In many cells isocitrate dehydrogenase is activated by ADP and is inhibited directly by NADH (this is in addition to the indirect diminution of activity seen at low [NAD^+]/[NADH] ratios). α-Ketoglutarate dehydrogenase activity is inhibited by succinyl-CoA and by NADH. The mechanisms are comparable to the mechanisms by which pyruvate dehydrogenase activity is controlled by levels of acetyl-CoA and NADH.

For some time it was thought that citrate synthase represented another site for allosteric control. The enzyme is subject to inhibition by NADH, NADPH, or succinyl-CoA. However, measurements of intramitochondrial levels of oxaloacetate, acetyl-CoA, and citrate show that the enzyme is operating close to equilibrium conditions; that is, the ratio [citrate]/([acetyl-

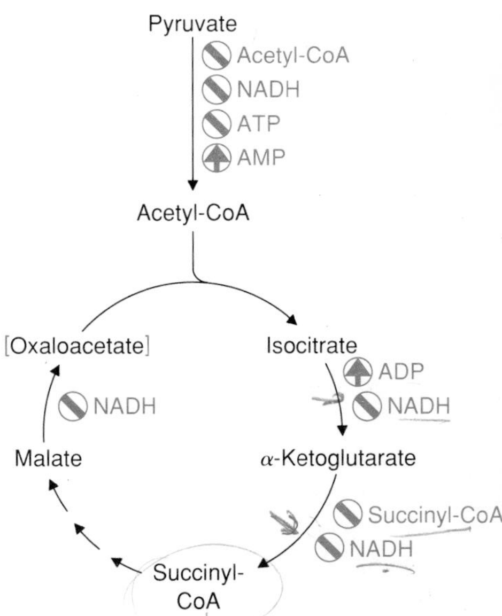

Figure 14.16
Major regulatory factors controlling the citric acid cycle. NADH can inhibit either through allosteric interactions or as a reflection of reduced NAD^+ availability, or both. [ ] indicates concentration dependence.

CoA] × [oxaloacetate]) is close to K_{eq} for the citrate synthase reaction. On the other hand, it is evident that intramitochondrial levels of oxaloacetate, which are quite low, can exercise substrate-level control over flux through citrate synthase.

To summarize, citric acid cycle flux is responsive to the energy state of the cell, through allosteric activation of isocitrate dehydrogenase by ADP; to the redox state of the cell, through flux rate limitation caused when intramitochondrial [NAD^+] decreases; and to the availability of energy-rich compounds, through inhibition of relevant enzymes by acetyl-CoA or succinyl-CoA.

Anaplerotic Sequences—The Need to Replace Cycle Intermediates Used for Biosynthesis

So far, our discussion of the citric acid cycle has focused on its role in catabolism and energy generation. The cycle also serves as an important source of biosynthetic intermediates. Figure 14.17 summarizes the most important anabolic pathways involved. These pathways tend to draw carbon from the cycle by utilizing intermediates in the cycle. Succinyl-CoA is used in the synthesis of heme and other porphyrins. Oxaloacetate and α-ketoglutarate are the keto acid analogs of the amino acids aspartate and glutamate, respectively, and are used in the synthesis of these and other amino acids by transamination (p. 490). In some tissues citrate is transported from mitochondria to the cytosol, where it is cleaved to provide acetyl-CoA for fatty acid biosynthesis. Since these and other reactions tend to deplete citric acid cycle intermediates by drawing carbon away, operation of the cycle would be impaired were it not for the existence of other processes that replenish the stores of citric acid cycle intermediates. These are called **anaplerotic** pathways, from a Greek word that means "filling up." In most cells the flow of carbon out of the cycle is balanced by these anaplerotic reactions, so that the intramitochondrial concentrations of citric

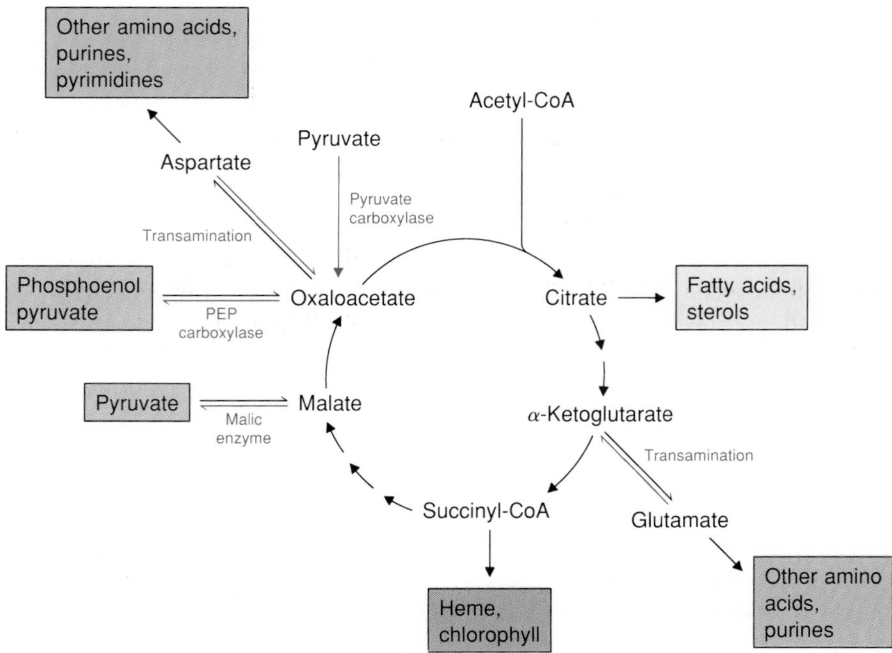

Figure 14.17
Major biosynthetic roles of some citric acid cycle intermediates. Anaplerotic pathways for replenishment of these intermediates are shown in red.

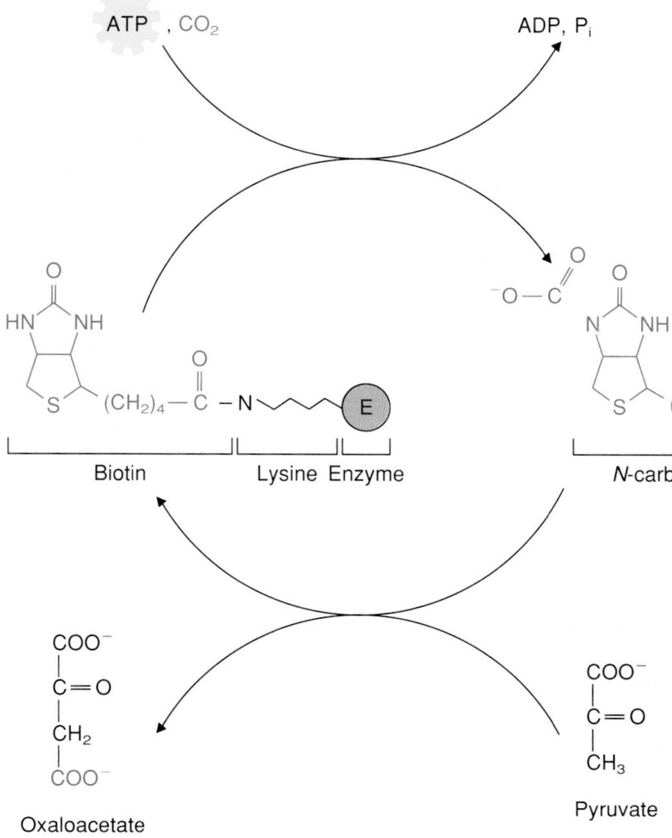

Figure 14.18
Enzyme-bound biotin as a carrier for CO_2 in the pyruvate carboxylase reaction. The CO_2 that is transferred is shown in red. Biotin (blue) linked to the ϵ-amino group of a lysine residue acts as a swinging arm, much like the lipoamide swinging arms of α-keto acid dehydrogenase complexes.

acid cycle intermediates remain constant with time. The anaplerotic processes are also summarized in Figure 14.17.

In animals the most important anaplerotic reaction, particularly in liver and kidney, is catalyzed by **pyruvate carboxylase,** which catalyzes the reversible, ATP-dependent carboxylation of pyruvate to give oxaloacetate.

$$\underset{\textbf{Pyruvate}}{\begin{array}{c}CH_3\\|\\C{=}O\\|\\COO^-\end{array}} + HCO_3^- + ATP \rightleftharpoons \underset{\textbf{Oxaloacetate}}{\begin{array}{c}COO^-\\|\\CH_2\\|\\C{=}O\\|\\COO^-\end{array}} + ADP + P_i + H^+ \qquad \Delta G^{\circ\prime} = -2.1\ \text{kJ/mol}$$

The enzyme is activated allosterically by acetyl-CoA; in fact, it is all but inactive in the absence of this effector. This represents a feedforward activation, because the effect of acetyl-CoA accumulation is to promote its own utilization by stimulating the synthesis of oxaloacetate. The latter compound in turn reacts with acetyl-CoA, via the citrate synthase reaction. Alternatively, oxaloacetate can be used for carbohydrate synthesis, via gluconeogenesis, and acetyl-CoA accumulation can be seen as a signal that adequate carbon is available for some of it to be stored as carbohydrate.

In Chapter 10 we identified biotin as a cofactor in most carboxylation reactions involving CO_2. Pyruvate carboxylase is a tetrameric protein containing four molecules of biotin, each bound covalently through an amide bond involving the ϵ-amino group of a lysine residue (Figure 14.18). The enzyme operates by a two-step mechanism, starting with an ATP-dependent carboxylation of the enzyme to give **N-carboxybiotin.** This activated derivative then transfers the carboxyl group directly to pyruvate.

In plants and bacteria an alternative route leads directly from phosphoenolpyruvate to oxaloacetate. Because phosphoenolpyruvate is such an energy-rich compound, this reaction, catalyzed by **phosphoenolpyruvate carboxylase,** requires neither an energy cofactor nor biotin.

$$\text{Phosphoenolpyruvate} + HCO_3^- \rightleftharpoons \text{oxaloacetate} + P_i$$

In addition to pyruvate carboxylase and phosphoenolpyruvate carboxylase, a third anaplerotic process is provided by an enzyme commonly known as **malic enzyme** but more officially as **malate dehydrogenase (decarboxylating; NADP$^+$).** This enzyme catalyzes the reductive carboxylation of pyruvate to give malate.

$$\underset{\textbf{Pyruvate}}{CH_3{-}C({=}O){-}COO^-} + HCO_3^- + NADPH + H^+ \rightleftharpoons \underset{\textbf{L-Malate}}{{}^-OOC{-}CH_2{-}CH(OH){-}COO^-} + NADP^+ + H_2O$$

This reaction is formally equivalent to the isocitrate dehydrogenase reaction operating in reverse, with pyruvate and malate being analogous, respectively, to α-ketoglutarate and isocitrate.

Transamination reactions can be regarded as anaplerotic also, because they are readily reversible. In transamination an amino acid transfers its amino group to a keto acid and is thereby converted itself to a keto acid; the mechanism is discussed in Chapter 20.

$$R_1{-}CH(\overset{+}{N}H_3){-}COO^- + R_2{-}C({=}O){-}COO^- \rightleftharpoons R_1{-}C({=}O){-}COO^- + R_2{-}CH(\overset{+}{N}H_3){-}COO^-$$

Glutamate and aspartate undergo transamination to generate the citric acid cycle intermediates α-ketoglutarate and oxaloacetate, respectively.

Glutamate	**α-Ketoglutarate**	**Aspartate**	**Oxaloacetate**
$^-OOC{-}CH_2{-}CH_2{-}CH(\overset{+}{N}H_3){-}COO^-$	$^-OOC{-}CH_2{-}CH_2{-}C({=}O){-}COO^-$	$^-OOC{-}CH_2{-}CH(\overset{+}{N}H_3){-}COO^-$	$^-OOC{-}CH_2{-}C({=}O){-}COO^-$

Hence, cells containing amino acids in abundance can convert them via transamination to citric acid cycle intermediates. Another enzyme, **glutamate dehydrogenase,** presents another route for synthesis of α-ketoglutarate from glutamate.

$$\text{Glutamate} + NAD(P)^+ + H_2O \rightleftharpoons \alpha\text{-ketoglutarate} + NAD(P)H + \overset{+}{N}H_4$$

This enzyme, which we discuss in more detail in Chapter 20, uses either NADH or NADPH as the reductive cofactor.

Finally, many plants and bacteria can convert two-carbon fragments to four-carbon citric acid cycle intermediates via the glyoxylate cycle, as described in the next section.

Glyoxylate Cycle: An Anabolic Variant of the Citric Acid Cycle

Metabolically, plant and animal cells differ in many important respects. Of particular concern to us here is the fact that plant cells, along with some microorganisms, can carry out the net synthesis of carbohydrate from fat. This conversion is crucial to the development of seeds, in which a great deal of energy is stored in the form of triacylglycerols. (In fact, most vegetable oils available in grocery stores are mixtures of triacylglycerols derived from seeds.) When the seeds germinate, triacylglycerol is broken down and converted to sugars, which provide energy and raw material needed for growth of the plant.

Plants synthesize sugars by using the **glyoxylate cycle,** which can be considered an anabolic variant form of the citric acid cycle. To understand the importance of this cycle, consider first the two primary fates of acetyl-CoA in animal metabolism—oxidation through the citric acid cycle, and the synthesis of fatty acids. Because of the irreversibility of the pyruvate dehydrogenase reaction, acetyl-CoA cannot be converted to pyruvate and hence cannot participate in the *net* synthesis of carbohydrate. To be sure, the two carbons of acetyl-CoA can be incorporated into oxaloacetate, which is an efficient gluconeogenic precursor. However, since two carbons are lost in this part of the citric acid cycle, there is no net accumulation of carbon in carbohydrate.

The glyoxylate cycle is a cyclic pathway that converts two acetyl units, as acetyl-CoA, to succinate. The pathway uses some of the same enzymes as the citric acid cycle, but it bypasses the two reactions where carbon is lost during the citric acid cycle. The second mole of acetyl-CoA is brought in during this bypass (Figure 14.19). Thus, each turn of the cycle brings in two two-carbon fragments and results in the net synthesis of one four-carbon molecule, namely succinate. This occurs in the **glyoxysome,** a specialized organ that carries out both β-oxidation of fatty acids to acetyl-CoA and utilization of that acetyl-CoA in the glyoxylate cycle (Figure 14.20). The succinate generated is transported from the glyoxysome to the mitochondrion, where it is converted, via reactions 6, 7, and 8 of the citric acid cycle, to oxaloacetate. The oxaloacetate is readily utilized for carbohydrate synthesis via gluconeogenesis.

The glyoxylate cycle also allows many microorganisms to metabolize two-carbon substrates, such as acetate. *Escherichia coli,* for example, can grow in a medium that provides acetate as the sole carbon source, as can many fungi, protozoans, and algae.

Now let us examine the individual reactions of the glyoxylate cycle. As noted, acetyl-CoA is provided from fatty acid oxidation. Alternatively, acetate itself is converted to acetyl-CoA by **acetate thiokinase.**

$$\text{Acetate} + \text{CoA-SH} + \text{ATP} \rightleftharpoons \text{acetyl-CoA} + \text{AMP} + \text{PP}_i$$

Next, acetyl-CoA condenses with oxaloacetate to give citrate, just as in the citric acid cycle, and citrate reacts with aconitase to give isocitrate. At this point the glyoxylate cycle diverges from the citric acid cycle. The next reac-

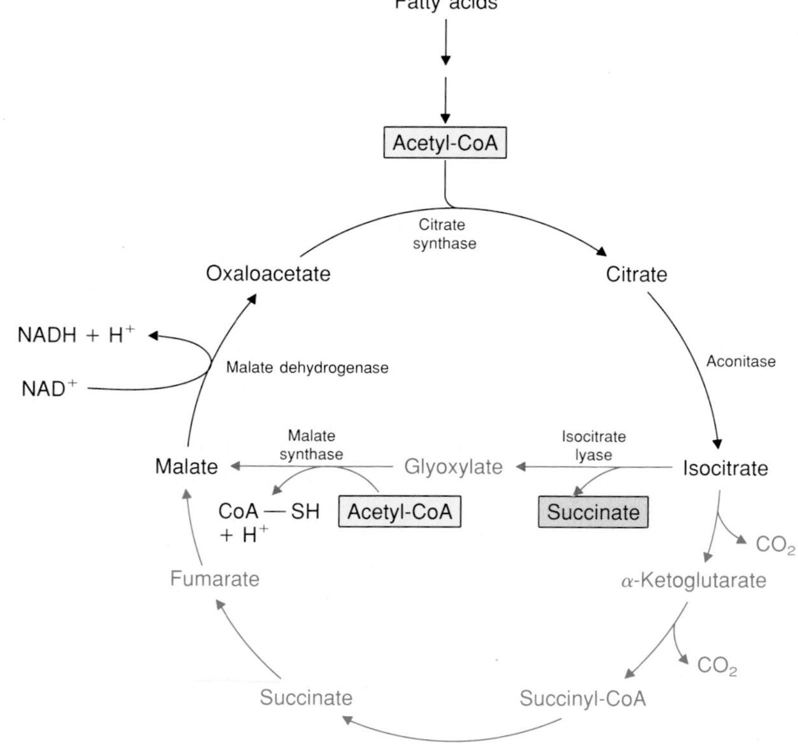

Figure 14.19
Reactions of the glyoxylate cycle. The reactions catalyzed by isocitrate lyase and malate synthase (red) bypass the five citric acid cycle steps between isocitrate and malate (shown in blue).

tion, catalyzed by **isocitrate lyase,** cleaves isocitrate to glyoxylate and succinate.

$$\begin{array}{c}\mathrm{CH_2{-}COO^-}\\ |\\ \mathrm{H{-}C{-}COO^-}\\ |\\ \mathrm{HO{-}C{-}COO^-}\\ |\\ \mathrm{H}\end{array} \longrightarrow \begin{array}{c}\mathrm{CH_2{-}COO^-}\\ |\\ \mathrm{CH_2{-}COO^-}\end{array} + \begin{array}{c}\mathrm{O{=}C{-}COO^-}\\ |\\ \mathrm{H}\end{array}$$

Isocitrate **Succinate** **Glyoxylate**

Glyoxylate then accepts acetate from another acetyl-CoA, in a reaction catalyzed by **malate synthase.**

$$\begin{array}{c}\mathrm{O{=}C{-}COO^-}\\ |\\ \mathrm{H}\end{array} + \mathrm{CH_3{-}\overset{\overset{\displaystyle O}{\|}}{C}{\sim}S{-}CoA} + \mathrm{H_2O} \longrightarrow \begin{array}{c}\mathrm{COO^-}\\ |\\ \mathrm{HO{-}C{-}H}\\ |\\ \mathrm{CH_2}\\ |\\ \mathrm{COO^-}\end{array} + \mathrm{CoA{-}SH} + \mathrm{H^+}$$

Glyoxylate **Acetyl-CoA** **Malate**

Mechanistically, this reaction is comparable to that of citrate synthase, involving nucleophilic attack of the carbanion form of acetyl-CoA on a carbonyl carbon, in this case the aldehyde carbon of glyoxylate. The malate is then dehydrogenated to regenerate oxaloacetate. Although this reaction, along with reactions of citrate synthase and aconitase, is identical to a citric acid cycle reaction, these three reactions are catalyzed by isozymic forms of each enzyme that are specialized for their role in the glyoxylate cycle. In plants these isoenzymes are compartmentalized in glyoxysomes.

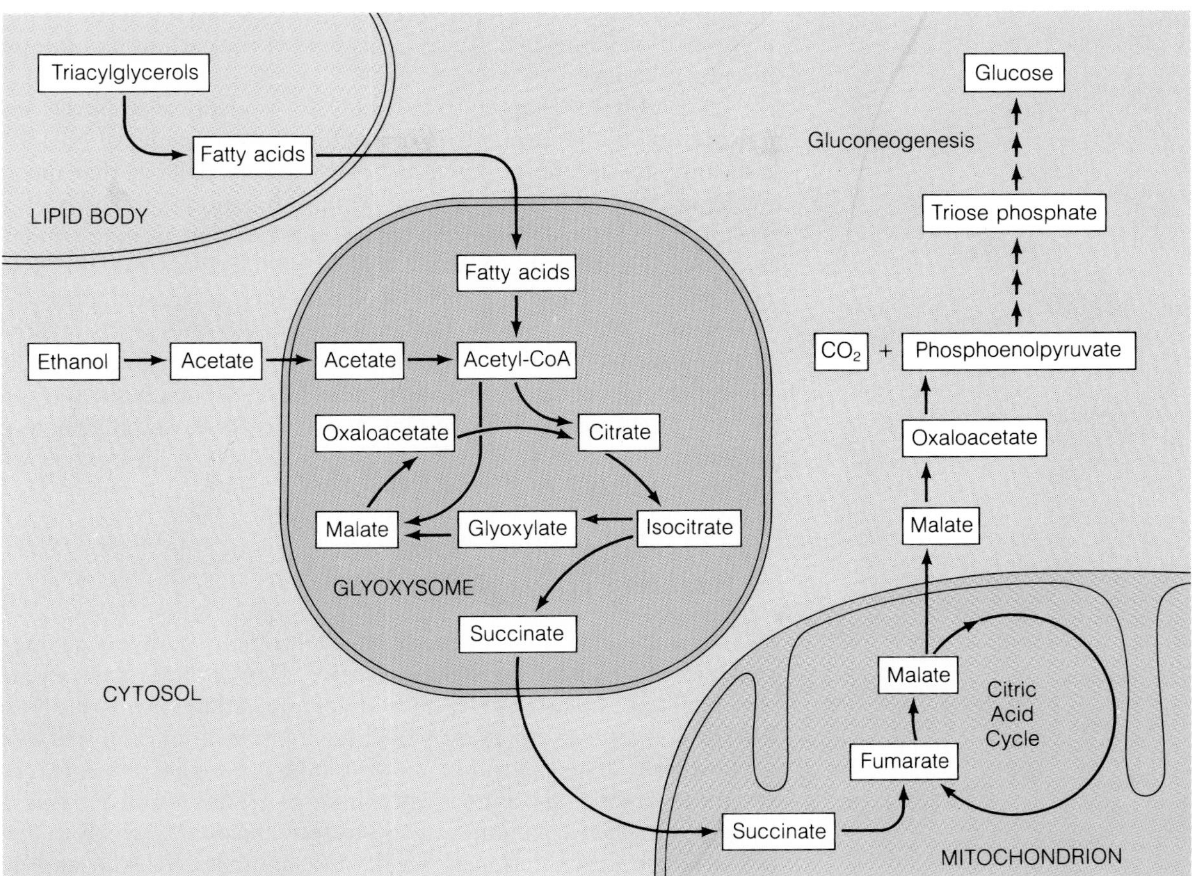

Figure 14.20
Intracellular relationships involving the glyoxylate cycle. Fatty acids released in lipid bodies are oxidized in glyoxysomes to acetyl-CoA, which can also come directly from acetate. Acetyl-CoA is then converted to succinate in the glyxoylate cycle. The succinate is metabolized, via the citric acid cycle in mitochondria, to oxaloacetate, which is readily converted to sugars by gluconeogenesis.

As noted earlier, the glyoxylate cycle results in the net conversion of two two-carbon fragments to a four-carbon compound, succinate, as shown by the following balanced equation.

$$2\text{ Acetyl-CoA} + NAD^+ \longrightarrow \text{succinate} + NADH + H^+ + 2\text{CoA-SH}$$

The primary fate of succinate is its entry into gluconeogenesis via its conversion to oxaloacetate.

An Alternative Pathway for Carbohydrate Oxidation: The Pentose Phosphate Pathway

The predominant pathway for glucose catabolism is glycolysis to yield pyruvate, followed by oxidation to CO_2 in the citric acid cycle. An alternative pathway, the **pentose phosphate pathway**, operates to varying extents in different cells and tissues. The role of this pathway is primarily anabolic rather than catabolic, but we present the pathway in this chapter because it does involve oxidation of glucose and in certain modes it can operate to oxidize glucose completely to CO_2 and water.

The pentose phosphate pathway has two primary functions: (1) to provide NADPH for reductive biosynthesis and (2) to provide ribose-5-phosphate for nucleotide and nucleic acid biosynthesis. In addition, the pathway operates to metabolize dietary pentose sugars, derived primarily from the digestion of nucleic acids. In plants, a variant of the pentose phosphate pathway operates in reverse as part of the carbon fixation process of photosynthesis.

Recall from Chapter 10 that $NADP^+$ is identical to NAD^+ except for the additional 2′ phosphate on one of the ribose moieties of $NADP^+$. Metabolically, the difference between NAD^+ and $NADP^+$ is that nicotinamide nucleotide-linked enzymes whose primary function is to *oxidize* substrates use the NAD^+/NADH pair, whereas enzymes functioning primarily in a *reductive* direction use $NADP^+$ and NADPH. Since NADPH is used for fatty acid and steroid biosynthesis, tissues such as adrenal gland, liver, adipose tissue, and mammary gland are rich in enzymes of the pentose phosphate pathway. NADPH is also the ultimate electron source for reduction of ribonucleotides to deoxyribonucleotides for DNA synthesis, so rapidly proliferating cells generally have high activity of pentose phosphate pathway enzymes. The pathway, which operates exclusively in the cytosol, is outlined in Figure 14.21.

The Oxidative Phase: Generation of Reducing Power as NADPH

It is convenient to think of the pentose phosphate pathway as operating in two phases—oxidative and nonoxidative. Two of the first three reactions in this pathway are oxidative, each involving reduction of one $NADP^+$ to NADPH. As shown in Figure 14.22, the first reaction, catalyzed by **glucose-6-phosphate dehydrogenase,** oxidizes glucose-6-phosphate to the corresponding **lactone** (an internal ester linking carbons 1 and 5). This is hydrolyzed by a specific **lactonase** to **6-phosphogluconate,** which undergoes an oxidative decarboxylation to yield CO_2, another NADPH, and a pentose phosphate, **ribulose-5-phosphate.** The net result of the oxidative phase is generation of 2 mol of NADPH, oxidation of one carbon to CO_2, and synthesis of 1 mol of pentose phosphate.

The Nonoxidative Phase: Fates of Pentose Phosphates Tailored to Metabolic Needs

Ribulose-5-phosphate is next converted to ribose-5-phosphate by **phosphopentose isomerase.**

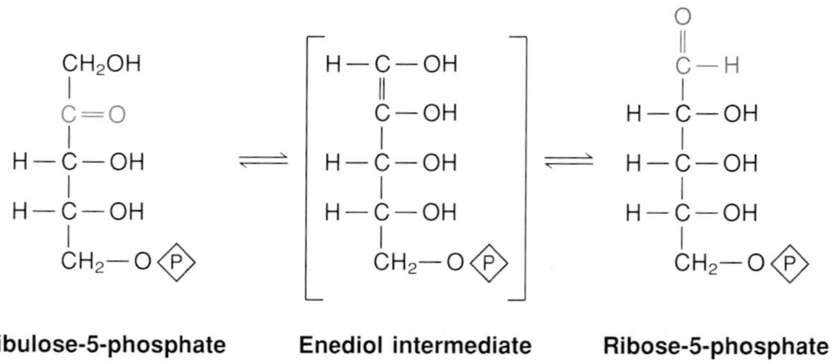

Ribulose-5-phosphate		Enediol intermediate		Ribose-5-phosphate
CH_2OH		H—C—OH		O=C—H
C=O		‖ C—OH		H—C—OH
H—C—OH	⇌	H—C—OH	⇌	H—C—OH
H—C—OH		H—C—OH		H—C—OH
CH_2—O—Ⓟ		CH_2—O—Ⓟ		CH_2—O—Ⓟ

The reaction proceeds via an enediol intermediate, just as in two different reactions of glycolysis—those catalyzed by triose-phosphate isomerase and phosphoglucoisomerase.

At this stage the primary functions of the pathway have been fulfilled, namely the generation of NADPH and ribose-5-phosphate. We can write a balanced equation for what has transpired thus far.

$$\text{Glucose-6-phosphate} + 2\text{NADP}^+ \longrightarrow \text{ribose-5-phosphate} + \text{CO}_2 + 2\text{NADPH} + 2\text{H}^+$$

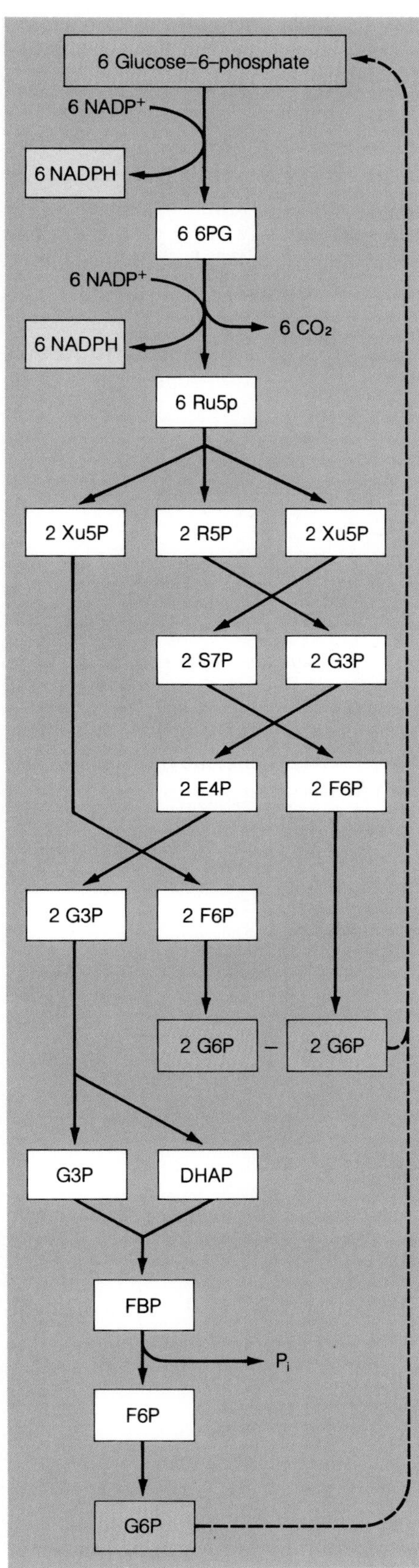

Figure 14.21
Overview of the pentose phosphate pathway. The figure shows recycling of glucose-6-phosphate to generate NADPH. Abbreviations: G6P = glucose-6-phosphate; 6PG = 6-phosphogluconate; Ru5P = ribulose-5-phosphate; Xu5P = xylulose-5-phosphate; R5P = ribose-5-phosphate; S7P = sedoheptulose-7-phosphate; G3P = glyceraldehyde-3-phosphate; E4P = erythrose-4-phosphate; DHAP = dihydroxyacetone phosphate; FBP = fructose-1,6-bisphosphate; F6P = fructose-6-phosphate.

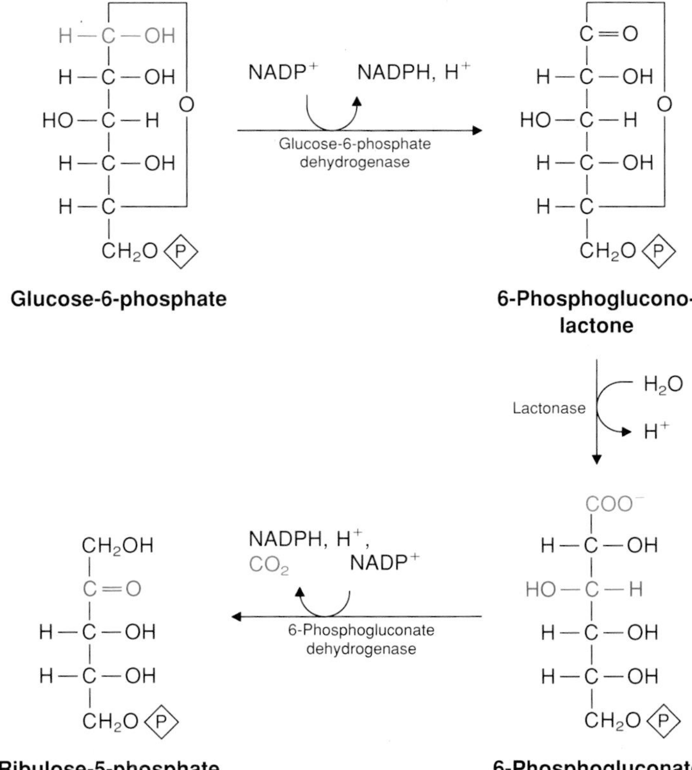

Figure 14.22
Oxidative phase of the pentose phosphate pathway.

Many cells need the NADPH for reductive biosynthesis but do not need the ribose-5-phosphate in such large quantities. We must consider, then, how this ribose-5-phosphate is catabolized. This process involves a series of sugar phosphate transformations that may look complicated but that have a simple result. *The reaction sequence converts three five-carbon sugar phosphates to two six-carbon sugar phosphates and one three-carbon sugar phosphate.* The hexose phosphates formed can be catabolized either by recycling through the pentose phosphate pathway or by glycolysis. The triose phosphate is glyceraldehyde-3-phosphate, a glycolytic intermediate. Three enzymes are involved: **phosphopentose epimerase, transketolase,** and **transaldolase.**

Phosphopentose epimerase converts ribulose-5-phosphate to its epimer, xylulose-5-phosphate.

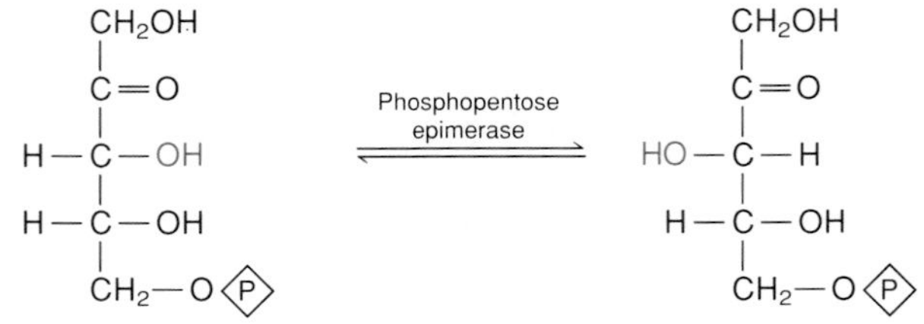

Phosphopentose epimerase

Ribulose-5-phosphate **Xylulose-5-phosphate**

One mole of xylulose-5-phosphate reacts with one mole of ribose-5-phosphate in a reaction catalyzed by transketolase. This enzyme transfers a two-carbon fragment from xylulose-5-phosphate to ribose-5-phosphate to give a triose phosphate, glyceraldehyde-3-phosphate, and a seven-carbon sugar, **sedoheptulose-7-phosphate.**

Transketolase

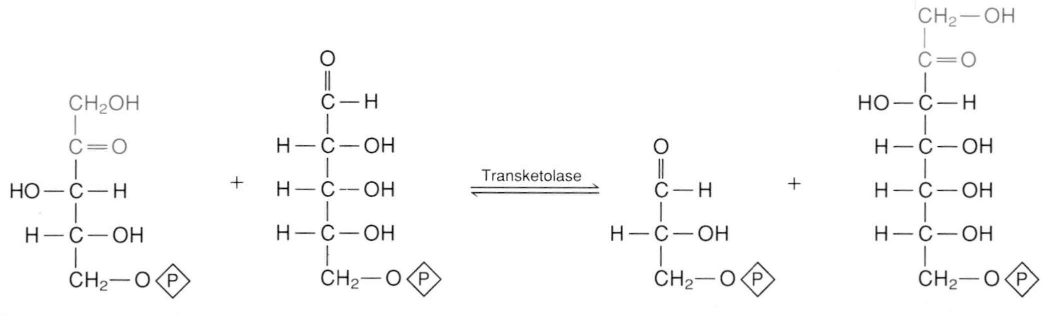

Xylulose-5-phosphate **Ribose-5-phosphate** **Glyceraldehyde-3-phosphate** **Sedoheptulose-7-phosphate**

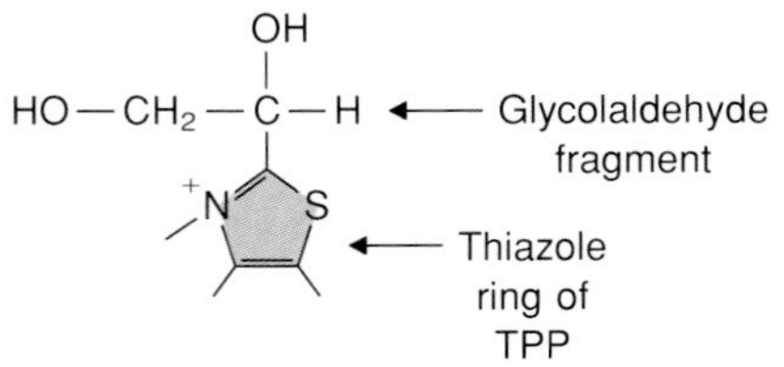

Glycolaldehyde fragment

Thiazole ring of TPP

The two-carbon fragment transferred is an activated **glycolaldehyde** ($HOCH_2$—CHO) fragment. Recall that pyruvate dehydrogenase transfers an active *acetaldehyde* fragment, with the aid of thiamine pyrophosphate (TPP). Transketolase also requires TPP as a cofactor, with the two-carbon fragment bound transiently to carbon 2 of the thiazole ring of TPP. The mechanism of activation and transfer of two-carbon fragments is very similar in the reactions catalyzed by these two enzymes.

Next, the two products of the transketolase reaction are acted on by transaldolase, with transfer of a three-carbon **dihydroxyacetone** unit from the seven-carbon substrate to the three-carbon substrate. The products are a four-carbon and a six-carbon sugar phosphate, **erythrose-4-phosphate** and fructose-6-phosphate, respectively.

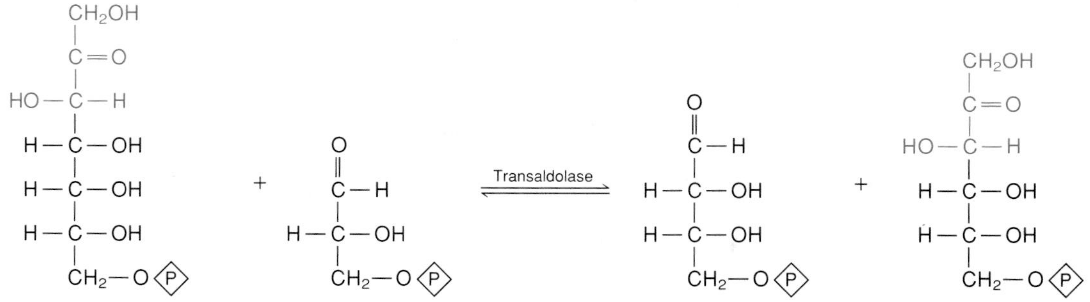

Sedoheptulose-7-phosphate + **Glyceraldehyde-3-phosphate** $\xrightleftharpoons{\text{Transaldolase}}$ **Erythrose-4-phosphate** + **Fructose-6-phosphate**

The net result of these two reactions is the conversion of two five-carbon sugar phosphates to a four-carbon and a six-carbon sugar phosphate. Transaldolase activates the ketose substrate by forming a Schiff base with a lysine residue on the enzyme (Figure 14.23). Protonation of the Schiff base leads to carbon–carbon bond cleavage, much as occurs in the fructose-6-phosphate aldolase reaction of glycolysis, with release of a four-carbon aldose phosphate. The dihydroxyacetone unit remains bound as a resonance-stabilized carbanion, which then attacks the carbonyl carbon of glyceraldehyde-3-phosphate, in an aldol condensation reaction. Hydrolysis of the Schiff base yields the six-carbon product, fructose-6-phosphate. Finally, transketolase acts on another molecule of xylulose-5-phosphate, transferring a glycolaldehyde fragment to erythrose-4-phosphate and generating a three-carbon and a six-carbon product—glyceraldehyde-3-phosphate and fructose-6-phosphate, respectively.

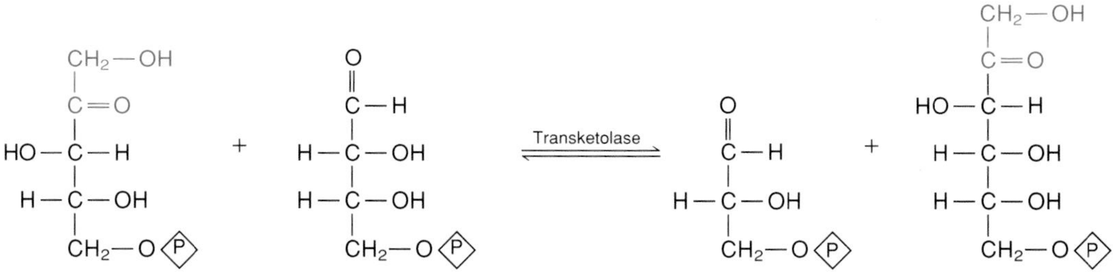

Xylulose-5-phosphate + **Erythrose-4-phosphate** $\xrightleftharpoons{\text{Transketolase}}$ **Glyceraldehyde-3-phosphate** + **Fructose-6-phosphate**

So far, the pathway has required input of three pentose phosphate molecules—two for the first transketolase reaction and one for the second. Thus, to summarize the pathway to this point, we must consider three molecules of glucose-6-phosphate passing through the oxidative phase.

$$\text{3 Glucose-6-phosphate} + 6\text{NADP}^+ \longrightarrow \text{3 pentose-5-phosphate} + 6\text{NADPH} + 6\text{H}^+ + 3\text{CO}_2$$

Now the rearrangements of the nonoxidative phase result in conversion of three pentose phosphates to two six-carbon and one three-carbon sugar phosphates.

$$\text{3 Pentose-5-phosphate} \longrightarrow \text{2 fructose-6-phosphate} + \text{glyceraldehyde-3-phosphate}$$

Thus, we can write a balanced equation for the pathway as described.

$$\text{3 Glucose-6-phosphate} + 6\text{NADP}^+ \longrightarrow \text{2 fructose-6-phosphate} + 3\text{CO}_2 + \text{glyceraldehyde-3-phosphate} + 6\text{NADPH} + 6\text{H}^+$$

Figure 14.23
Mechanism of the transaldolase reaction. The substrate, sedoheptulose-7-phosphate, becomes linked to the enzyme (E) via a Schiff base involving the keto carbon of the sugar and a lysine residue on the enzyme (step 1). Splitting of the activated substrate yields free erythrose-4-phosphate plus an enzyme-bound carbanion (step 2). The latter attacks the carbonyl carbon of glyceraldehyde-3-phosphate to form enzyme-bound fructose-6-phosphate (step 3), which is released by hydrolysis of the Schiff base (step 4).

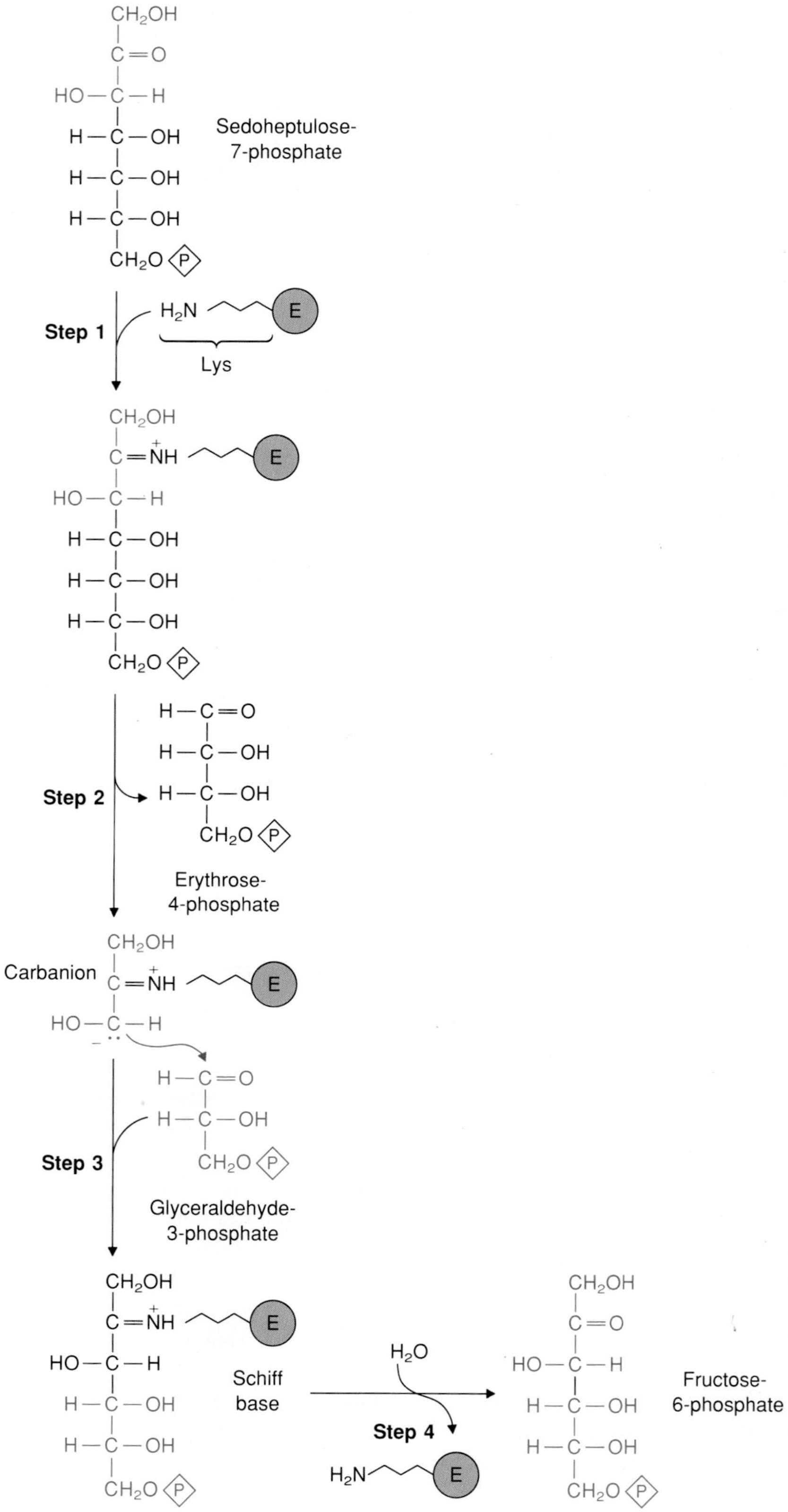

Alternatively, we can write a balanced equation that shows the complete oxidation of one mol of hexose phosphate to CO_2.

$$6\ \text{Hexose-6-phosphate} + 12\text{NADP}^+ \longrightarrow 5\ \text{hexose-6-phosphate} + 6CO_2 + 12\text{NADPH} + 12H^+$$

The actual fate of the sugar phosphates depends on the metabolic needs of the cell in which the pathway is occurring (Figure 14.24). If the primary need is for nucleotide synthesis, the major product is ribose-5-phosphate and most of the rearrangements of the nonoxidative phase do not take place. If the primary need is for NADPH generation, the nonoxidative phase generates compounds that can easily be reconverted to glucose-6-phosphate, for subsequent passage through the oxidative phase. In this

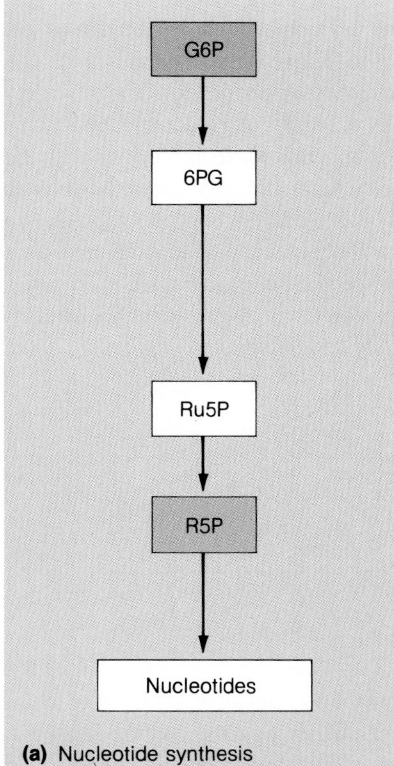

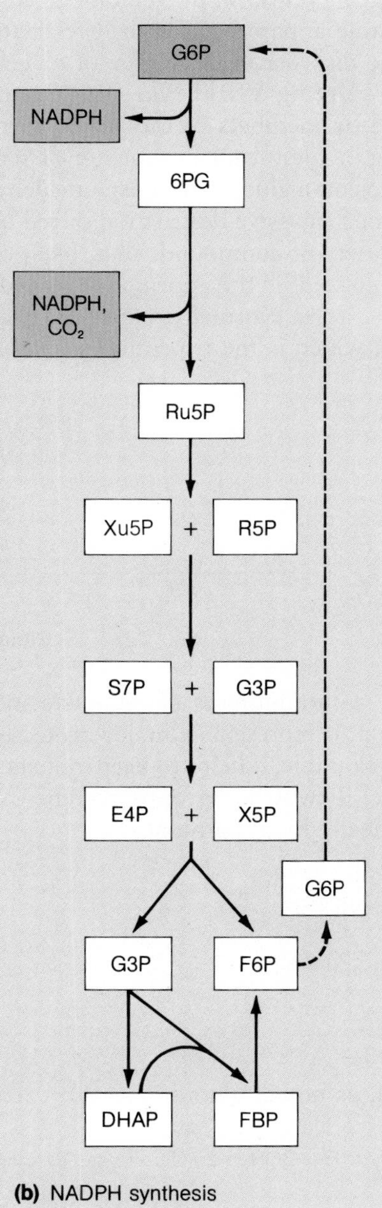

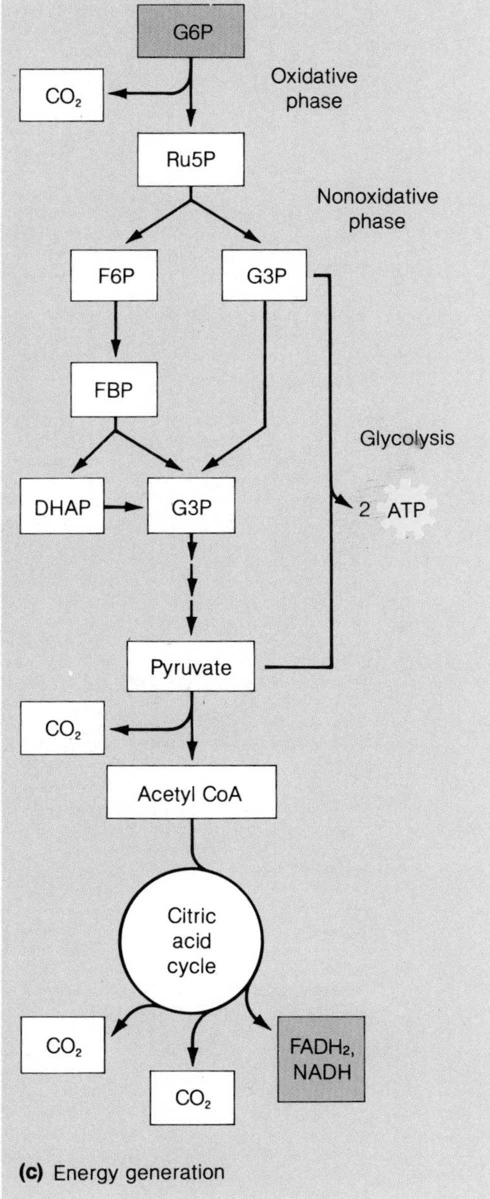

Figure 14.24
Different modes of operation of the pentose phosphate pathway to meet varying metabolic needs. (a) When the primary need is for biosynthesis, the primary product is ribose-5-phosphate. (b) The primary need is for reducing power. Fructose phosphates are reconverted to glucose-6-phosphate, for reoxidation in the oxidative phase. (c) The primary need is for energy, with the reaction products being oxidized through glycolysis and the citric acid cycle.

mode, repeated turns of the cycle result ultimately in the complete oxidation of glucose-6-phosphate to CO_2 and water.

Finally, in a cell with moderate needs for NADPH and for pentose phosphates, the fructose-6-phosphate and glyceraldehyde-3-phosphate produced in the nonoxidative phase can be further catabolized by glycolysis and the citric acid cycle. Because of the multiple metabolic needs of a cell for biosynthesis, it is unlikely that any one of these three modes operates exclusively in any one cell.

Human Genetic Disorders Involving Pentose Phosphate Pathway Enzymes

The pentose phosphate pathway is particularly active in the generation of reducing power in the red blood cells of vertebrates. The importance of this became apparent through investigation of a fairly widespread human genetic disorder, a deficiency of glucose-6-phosphate dehydrogenase.

During World War II the antimalarial drug primaquine was widely used by members of the armed forces. As a result of treatment with this drug, a significant proportion of servicemen, who were later found to be deficient in glucose-6-phosphate dehydrogenase, suffered a severe hemolytic anemia (massive destruction of red blood cells). They were also sensitive to a variety of compounds that, like primaquine, generate oxidative stress, as manifested by the appearance of hydrogen peroxide and organic peroxides in red cells. Normally, peroxides are inactivated via reduction by **glutathione,** which is the tripeptide γ-glutamylcysteinylglycine.

$$\underbrace{H_3\overset{+}{N}-\underset{\displaystyle COO^-}{\underset{|}{CH}}-CH_2-CH_2-\overset{\displaystyle O}{\overset{\|}{C}}}_{\gamma\text{-Glu}}-\underbrace{NH-\underset{\displaystyle \underset{\displaystyle SH}{\underset{|}{CH_2}}}{\underset{|}{CH}}-\overset{\displaystyle O}{\overset{\|}{C}}}_{\text{Cys}}-\underbrace{NH-CH_2-COO^-}_{\text{Gly}}$$

Glutathione

Glutathione is abundant in most cells and, because of its free thiol group, it represents a major protective mechanism against oxidative stress. For example, it helps to keep cysteine thiol groups in proteins in the reduced state. If two thiol groups become oxidized, they can be reduced nonenzymatically by glutathione.

$$\text{Protein}\begin{matrix}S\\ |\\ S\end{matrix} + 2\,\gamma\text{-Glu}-\underset{\displaystyle SH}{\underset{|}{\text{Cys}}}-\text{Gly} \rightleftharpoons \text{Protein}\begin{matrix}SH\\ \\ SH\end{matrix} + \begin{matrix}\gamma\text{-Glu}-\text{Cys}-\text{Gly}\\ |\\ S\\ |\\ S\\ |\\ \gamma\text{-Glu}-\text{Cys}-\text{Gly}\end{matrix}$$

And, as noted, glutathione also carries out the reduction of peroxides.

$$R-OOH + 2\,\gamma\text{-Glu}-\underset{\displaystyle SH}{\underset{|}{\text{Cys}}}-\text{Gly} \rightleftharpoons \begin{matrix}\gamma\text{-Glu}-\text{Cys}-\text{Gly}\\ |\\ S\\ |\\ S\\ |\\ \gamma\text{-Glu}-\text{Cys}-\text{Gly}\end{matrix} + H_2O + ROH$$

Oxidized glutathione (GSSG) is reduced by the NADPH-dependent enzyme **glutathione reductase.**

$$GSSG + NADPH + H^+ \rightleftharpoons 2GSH + NADP^+$$

This FAD-dependent enzyme acts essentially unidirectionally, so that the ratio of reduced glutathione (GSH) to oxidized glutathione in most cells is about 500 to 1.

In the erythrocyte, a particularly important role of glutathione is to maintain hemoglobin in the reduced (Fe^{2+}) state. Because of this, the erythrocyte is particularly sensitive to glutathione depletion. And, because the pentose phosphate pathway is virtually the only pathway in erythrocytes for generation of NADPH, the erythrocyte is especially vulnerable to conditions that impair flux through this pathway and thereby lower intracellular NADPH levels.

In most cases of glucose-6-phosphate dehydrogenase deficiency the enzyme in red cells is not totally inactive but instead is decreased in activity by about 10-fold. Individuals with this deficiency are asymptomatic until stressed, that is, until primaquine or a related agent generates enough peroxides that the available GSH becomes depleted. Reduction of the resultant GSSG back to GSH is impaired because NADPH levels are inadequate to allow glutathione reductase to function. This causes methemoglobin (Fe^{3+}) to accumulate at the expense of hemoglobin, which in turn changes the structure of the cell, weakening the membrane and rendering it sensitive to rupture, or hemolysis.

Interestingly, glucose-6-phosphate dehydrogenase deficiency, like sickle-cell trait, is associated with resistance to malaria caused by *Plasmodium falciparum* (see Chapter 7). Thus, the deficiency is a positive survival trait in tropical and subtropical regions of the world, where malaria is common. This explains the observation that glucose-6-phosphate dehydrogenase deficiency is seen most frequently among individuals of African or Mediterranean extraction.

Another genetic disorder related to the pentose phosphate pathway is the **Wernicke–Korsakoff syndrome.** This is a mental disorder, coupled with loss of memory and partial paralysis, that develops when affected individuals suffer a moderate thiamine deficiency (i.e., a deficiency not in itself severe enough to cause beriberi). The symptoms often become manifest in alcoholics, whose diets are apt to be vitamin deficient.

The basis for the Wernicke–Korsakoff syndrome is an alteration of transketolase that reduces its affinity for thiamine pyrophosphate by some 10-fold. Other TPP-dependent enzymes, notably pyruvate dehydrogenase and α-ketoglutarate dehydrogenase, are not affected. Symptoms of the disease become manifest when TPP levels drop below the values needed to saturate the abnormal transketolase. Normal individuals contain a transketolase that binds TPP strongly enough that no change in enzyme function occurs as a result of these slight to moderate thiamine deficiencies.

Both glucose-6-phosphate dehydrogenase deficiency and the Wernicke–Korsakoff syndrome illustrate the interdependence of genetic and environmental factors in the onset of clinical disease. Symptoms of the hereditary change become apparent only after some kind of moderate stress that does not affect normal individuals.

In this chapter we have seen how organic compounds are oxidized, yielding CO_2 and reduced electron carriers. Most of the energy that drives ATP synthesis comes from reoxidation of those reduced electron carriers in mitochondria, as we discuss in the next chapter.

REFERENCES

Regulation of the Citric Acid Cycle

Atkinson, D. E. (1977) *Cellular Energy Metabolism and Its Regulation.* Academic Press, New York. Provocative, even controversial, remarks by the person who originated the concept of adenylate energy charge.

McCormack, J. G., and R. M. Denton (1986) Ca^{2+} as a second messenger within mitochondria. *Trends Biochem. Sci.* 11:258–262. Brief review of the relationship between intramitochondrial calcium levels and the demand for energy generation.

Organization of Mitochondrial Enzymes

Collins, J. H., and L. J. Reed (1977) Acyl group and electron pair relay system: A network of interacting lipoyl moieties in the pyruvate and α-ketoglutarate dehydrogenase complexes from *Escherichia coli. Proc. Natl. Acad. Sci. USA* 74:4223–4227. Development of the concept of lipoamide swinging arms.

Reed, L. J. (1974) Multienzyme complexes. *Acc. Chem. Res.* 7:40–46. A review by the man primarily responsible for our current understanding of the pyruvate dehydrogenase and α-ketoglutarate dehydrogenase complexes.

Srere, P. A. (1985) Organization of proteins within the mitochondrion. *Organized Multienzyme Systems: Catalytic Properties,* edited by G. R. Welch, pp. 1–63. Academic Press, New York. Discusses the intramitochondrial location of enzymes of the citric acid cycle and oxidative phosphorylation, plus the evidence that enzymes catalyzing sequential reactions may be linked, even though located in separate compartments.

Experimental Background to the Citric Acid Cycle

Krebs, H. A. (1970) The history of the tricarboxylic acid cycle. *Perspect. Biol. Med.* 14:154–170. A historical account by the man responsible for most of the history.

Protection Against Intracellular Oxidation

Halliwell, B., and J. M. C. Gutteridge (1986) Iron and free radical reactions: Two aspects of antioxidant protection. *Trends Biochem. Sci.* 11:372–374. Provocative ideas about the role of iron in protection against reactive oxygen species.

Meister, A., and M. E. Anderson (1983) Glutathione. *Annu. Rev. Biochem.* 52:711–760. Reviews the chemistry and biochemistry of this important biological reductant.

Ortiz de Montellano, P. (ed.) (1986) *Cytochrome P-450: Structure, mechanism, and biochemistry.* Plenum, New York. A multiauthored treatise on this diverse and important group of enzymes.

The Glyoxylate Cycle

Tolbert, N. E. (1981) Metabolic pathways in peroxisomes and glyoxysomes. *Annu. Rev. Biochem.* 50:133–158. Reviews the cell biology and biochemistry of the glyoxylate cycle, primarily in plants.

PROBLEMS

1. Design a radiotracer experiment that would allow you to determine which proportion of glucose catabolism in a given tissue preparation occurs through the pentose phosphate pathway and which proportion through glycolysis and the citric acid cycle. Assume that you can synthesize glucose labeled with ^{14}C in any desired position or combination of positions. Assume also that you can trap CO_2 after administration of labeled glucose and determine its radioactivity.

2. Write a balanced chemical equation for the pentose phosphate pathway in each of the three modes depicted in Figure 14.24: (a) where ribose-5-phosphate synthesis is maximized; (b) where NADPH production is maximized, by conversion of the sugar phosphate products to glucose-6-phosphate for repeated operations of the pathway; (c) where the fructose-6-phosphate and glyceraldehyde-3-phosphate generated by each passage through the pathway are catabolized via glycolysis and the citric acid cycle.

3. Consider the fate of pyruvate labeled with ^{14}C in either carbon 1 (carboxyl), 2 (carbonyl), or 3 (methyl). Predict the fate of each labeled carbon during one turn of the citric acid cycle.

4. Suppose that aconitase did *not* bind its substrate asymmetrically. What fraction of the carbon atoms introduced as acetyl-CoA would be released in one turn of the cycle? In two turns?

5. [*methyl*-^{14}C]Pyruvate was administered to isolated liver cells in the presence of sufficient malonate to block succinate dehydrogenase completely. After a time, isocitrate was isolated and found to contain label in both carbon 2 and carbon 5, that is,

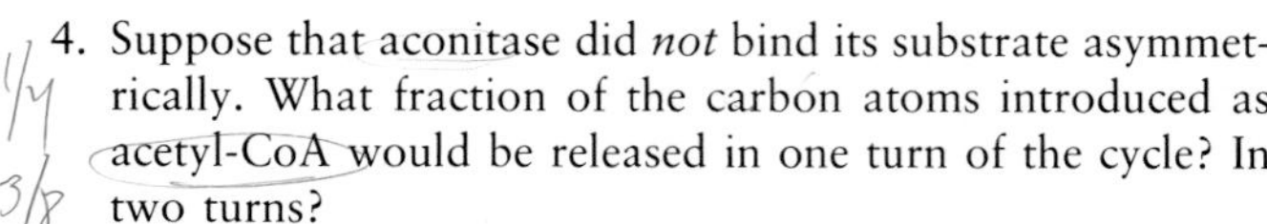

How do you explain this result?

6. Considering the evidence that led Krebs to propose a cyclic pathway for oxidation of pyruvate, discuss the type of experimental evidence that might have led to realization of the cyclic nature of the glyoxylate pathway.

7. Which carbon or carbons of glucose, if metabolized via glycolysis and the citric acid cycle, would be most rapidly lost as CO_2?

8. [1-^{14}C]Ribose-5-phosphate is incubated with a mixture of purified transketolase, transaldolase, phosphopentose epimerase, and glyceraldehyde-3-phosphate. Predict the distribution of radioactivity in the erythrose-4-phosphate and fructose-6-phosphate that are formed in this mixture.

9. Would you expect NAD^+ or CoA-SH to affect the activity of pyruvate dehydrogenase kinase? Briefly explain your answer.

10. In deciding which form of isocitrate dehydrogenase plays the more important role in the citric acid cycle—the NAD^+-dependent or the $NADP^+$-dependent form—what kinds of information would help you?

11. Referring to Figure 14.9, write a balanced equation for each of the six reactions catalyzed by the pyruvate dehydrogenase complex, and indicate whether each reaction is catalyzed by E_1, E_2, or E_3.

CHAPTER 15

Biological Oxidations, Electron Transport, and Oxidative Phosphorylation

The average adult human generates enough metabolic energy to synthesize his or her own weight in ATP *every day*. How is this massive amount of energy mobilized? Glycolysis and the citric acid cycle by themselves generate relatively little energy as ATP. However, six dehydrogenation steps—one in glycolysis, another in the pyruvate dehydrogenase reaction, and four more in the citric acid cycle—collectively reduce 5 mol of NAD^+ to NADH and 1 mol of FAD to $FADH_2$ per mole of glucose. Reoxidation of these reduced electron carriers generates most of the energy needed for ATP synthesis, the process that will concern us for much of this chapter. *Anaerobic* cells maintain an electronic steady state by transferring electrons from reduced carriers to some organic substrate, as in the reduction of pyruvate to lactate in anaerobic glycolysis. By contrast, *aerobic* cells and tissues transfer electrons in a stepwise process, from reduced carriers to molecular oxygen.

The aerobic metabolism of pyruvate generates many more reducing equivalents than does anaerobic catabolism. In respiration those reducing equivalents take the form of NADH and $FADH_2$, which in eukaryotic cells become reoxidized by electron transport proteins bound to the inner mitochondrial membrane. A series of coupled oxidation and reduction reactions occurs, with electrons being passed along a series of electron carriers—the electron transport chain, or respiratory chain (Figure 15.1). The final step is reduction of O_2 to water. The overall electron transport sequence is quite exergonic; one pair of reducing equivalents, generated from NADH, suffices to drive the coupled synthesis of 3 mol of ATP from ADP and P_i by oxidative phosphorylation. How is the energy released from the oxidative reactions of the respiratory chain harnessed, or **coupled,** to drive the synthesis of ATP? The mechanism of this coupling will concern us throughout this chapter. In addition, we will consider a number of other important metabolic roles oxygen plays in aerobic cells.

The Mitochondrion—Scene of the Action

Our comprehension of biological oxidations requires understanding both the chemistry of oxidation–reduction reactions and the cell biology of the mitochondrion. Before reviewing the chemistry, let us describe the intracellular sites where these reactions occur. Cellular metabolism generates reduced compounds in all of the major compartments of a eukaryotic cell. As noted earlier, glycolysis takes place in the cytosol of eukaryotic cells, while pyruvate oxidation, fatty acid β-oxidation, and the citric acid cycle occur within the mitochondrial matrix. Individual cells vary widely in the abundance and structure of their mitochondria. Most vertebrate cells contain several hundred mitochondria, but the number can be as low as one or as high as one hundred thousand.

The mitochondrion consists of four distinct subregions, shown in Figure 15.2—the outer membrane, the inner membrane, the intermembrane space, and the matrix, located within the inner membrane. The inner membrane is highly folded into **cristae** that project into, and often nearly through, the interior of the mitochondrion. Since respiratory proteins are bound to the inner membrane, the density of cristae is related to the respiratory activity of a cell. For example, heart muscle cells, which have very high rates of respiration, contain mitochondria with densely packed cristae. By contrast, liver cells have much lower respiration rates and mitochondria with more sparsely distributed cristae.

Whatever the compartment in which biological oxidations occur, all of these processes generate reduced electron carriers, primarily NADH. Most of this NADH is reoxidized, with concomitant ATP production, by the enzymes of the respiratory chain, firmly embedded in the inner membrane. The inner membrane itself consists of about 70% protein and 30% lipid. About half of the inner membrane protein in beef heart mitochondria consists of proteins directly involved in electron transport and oxidative phosphorylation. Most of the remaining proteins are involved in transport of substances into and out of mitochondria. By contrast, a completely different set of proteins is bound to the outer membrane, including enzymes of amino acid oxidation, fatty acid elongation, membrane phospholipid biosynthesis, and enzymatic hydroxylations.

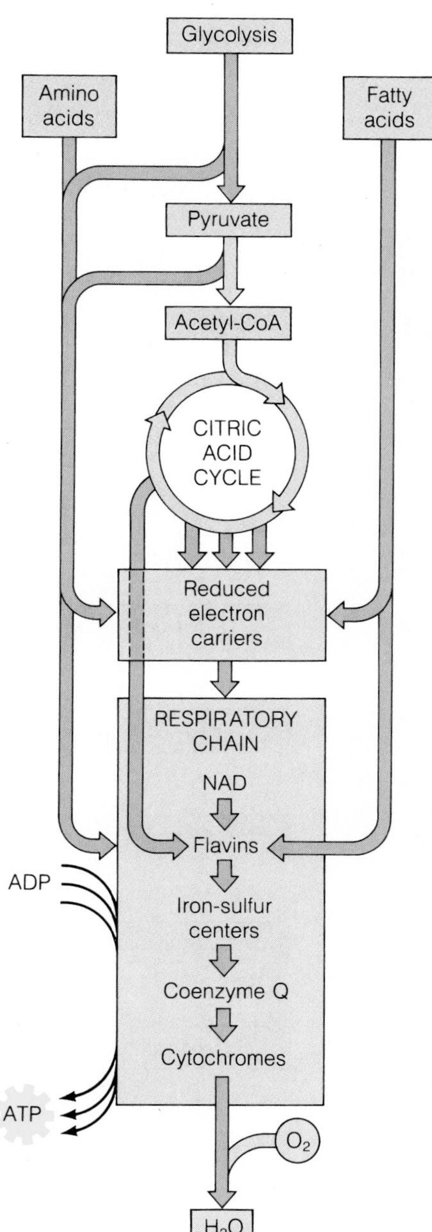

Figure 15.1
Overview of respiration, with processes discussed in this chapter in orange.

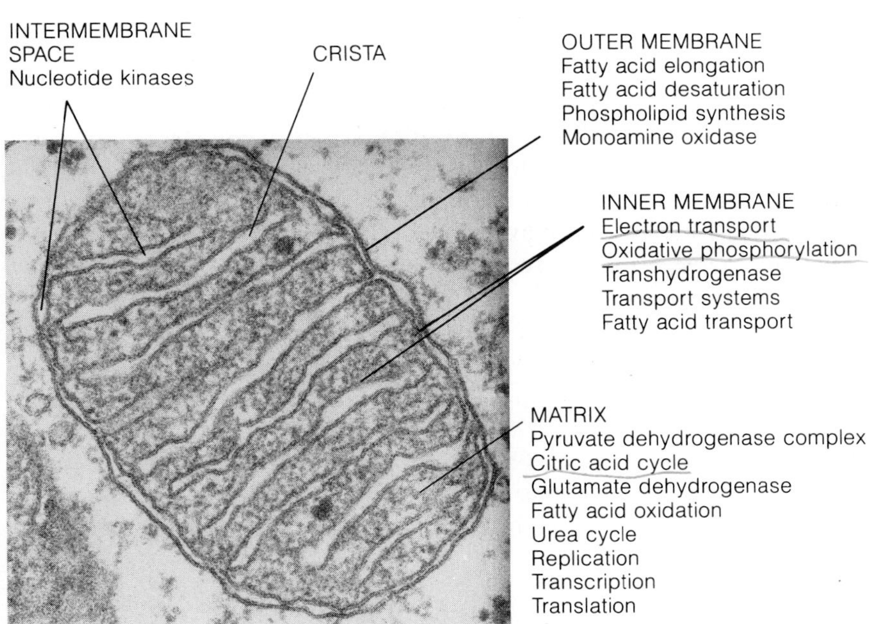

Figure 15.2
A mitochondrion from a pancreatic cell, shown as a thin section in the electron microscope. The major intramitochondrial compartments are shown, along with principal enzymes and pathways localized to each compartment. (Magnification ×155,000.)

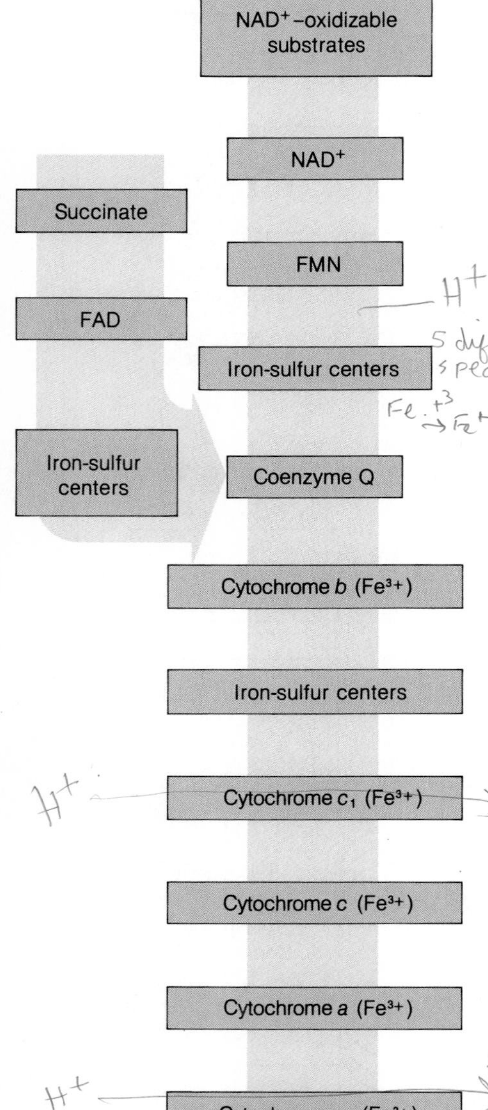

Figure 15.3
The sequence of respiratory electron carriers in mitochondria, for oxidation of succinate and of NAD$^+$-linked substrates.

Embedded within the inner membrane are the protein electron carriers, primarily cytochromes, that constitute the respiratory chain. On the surface of cristae lies an enzyme system that synthesizes ATP from ADP and orthosphosphate, within the matrix. Energy to drive this unfavored reaction is derived from the coupled exergonic reactions of electron transport, which generate an energy-rich proton gradient across the inner membrane.

Critical to comprehension of these processes was the isolation of physiologically intact mitochondria, using differential centrifugation of cell homogenates. This was accomplished in the late 1940s by Eugene Kennedy and Albert Lehninger, who demonstrated that isolated mitochondria could synthesize ATP from ADP and P_i in vitro, but only if an oxidizable substrate was present as well. Much of our understanding of the sequence of electron carriers, and of the localization of specific enzymes within mitochondria, has come from our ability to fractionate mitochondria into specific multienzyme complexes, each capable of catalyzing part of the overall reaction sequence of electron transport and oxidative phosphorylation.

The situation in prokaryotic cells is comparable, although different electron carriers are involved. However, since prokaryotic cells lack organelles, all of the electron carriers and enzymes of oxidative phosphorylation are bound to the inner membrane of the cell surface. Therefore, electron transport and oxidative phosphorylation occur at the cell periphery. As discussed in Chapter 19, there is reason to believe that mitochondria and chloroplasts, both of which contain rudimentary genetic machinery, are descended from free-living primitive prokaryotic cells.

Biological Oxidations

Biological electron transport consists of a series of linked oxidations and reductions, or redox reactions. To understand the logic behind the sequence of reactions in the respiratory chain, and the mechanisms by which metabolic energy is generated from these reactions, you must understand how to calculate the free energy available from a redox reaction. Note from Figure 15.3 the identities of the major respiratory electron carriers—nicotinamide nucleotides; flavin nucleotides; **iron–sulfur centers** (complexes of Fe^{2+} and sulfur); **coenzyme Q**, a low-molecular-weight, lipophilic carrier; and the heme proteins called cytochromes. Although the redox chemistry of these biological electron carriers may seem more complicated than that governing simple inorganic reactions, such as oxidation of ferrous to ferric ions, the basic principles are identical.

Quantitation of Reducing Power; Standard Reduction Potential

Redox chemistry is comparable in many ways to acid–base chemistry, which we discussed in Chapter 2. In a protonic equilibrium we have an acid and its conjugate base, which represent a proton donor and a proton acceptor, respectively.

$$\text{Proton donor} \rightleftharpoons \text{proton acceptor} + H^+$$

Similarly, in a redox reaction we have a donor and acceptor of *electrons*.

$$\text{Electron donor} \rightleftharpoons \text{electron acceptor} + e^-$$

As you know, free protons and free electrons exist at negligible concentrations in aqueous media, so the above are merely half-reactions in an overall acid–base or redox equilibrium. A complete redox reaction must show as one reactant an electron acceptor, which becomes reduced by gaining electrons. Of the two substrates the electron donor is the **reductant,** which becomes oxidized while transferring electrons to the other substrate, the **oxidant.**

$$\text{Reductant} + \text{oxidant} \rightleftharpoons \text{oxidized reductant} + \text{reduced oxidant}$$

A simple example is the oxidation of ferrous ion by cupric ion.

$$\underset{\textbf{Reductant}}{Fe^{2+}} + \underset{\textbf{Oxidant}}{Cu^{2+}} \rightleftharpoons \underset{\substack{\textbf{Oxidized}\\\textbf{reductant}}}{Fe^{3+}} + \underset{\substack{\textbf{Reduced}\\\textbf{oxidant}}}{Cu^{+}}$$

Ferrous ion is the reductant in this reaction, because it is the electron donor. The reductant is thus comparable to the acid in a protonic equilibrium.

Critical to our understanding of acid–base chemistry is the concept of pK_a, which represents a quantitative measure of the tendency of an acid to lose a proton. In the same sense, our understanding of biological oxidations demands a comparable measure of the tendency of a reductant to lose electrons. Such an index is provided by the **standard reduction potential,** or E_0'. In acid–base equilibria we arbitrarily define water, with a pK_a of 7.0, as neutral. Anything with a pK_a below 7, which tends to protonate water, is called an acid, and compounds that tend to be protonated by water are called bases. Redox chemistry also employs a reference standard, namely the standard hydrogen electrode in an electrochemical cell.

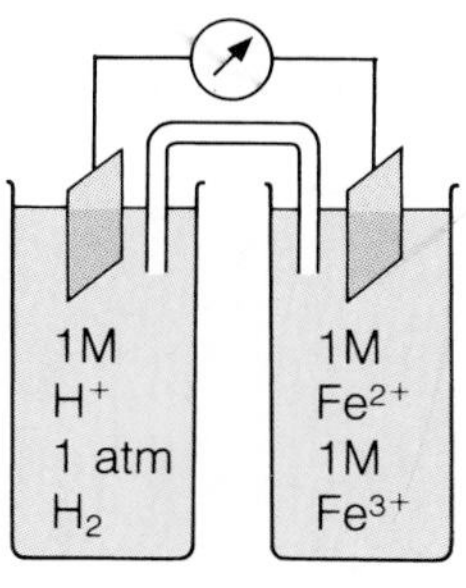

An electrochemical cell consists of two **half-cells,** each containing an electron donor and its conjugate acceptor (see margin). In the diagram the left-hand beaker constitutes the reference half-cell, a standard hydrogen electrode. The right-hand beaker contains the test half-cell, with the solution containing the test electron donor and its conjugate acceptor, each at 1 M concentration. In this example, the solution contains ferrous and ferric ions, each at 1 M. Electrical neutrality is maintained with an agar bridge. The galvanometer which links the two half-cells measures the **electromotive force,** or **emf,** in volts. This is a measure of the potential, or "pressure" for electrons to flow from one half-cell to the other. Depending on whether H_2 has a greater or lesser tendency than the test electron donor to lose electrons, those electrons may flow either toward or away from the reference half-cell. If H_2 loses electrons more readily than the test electron donor, then electrons will flow from the reference to the test half-cell, thereby oxidizing H_2 in the reference cell and reducing the test electron acceptor, ferric ion, and recording a positive emf. However, if the test electron donor loses electrons more readily, then electrons flow in the reverse direction, reducing H^+ to H_2 in the reference half-cell and causing a negative emf to be recorded. The stronger oxidant, whether H^+ or the test electron acceptor, will draw electrons away from the other half-cell and become reduced.

The tendency of an electron donor to reduce its conjugate acceptor is called the **reduction potential,** ***E.*** Under standard conditions (25°C, donor and acceptor each at 1 M) this term becomes the **standard reduction potential,** E_0. The emf recorded in an electrochemical cell as described above is the *difference* between the reduction potentials in the reference and test half-cells. If both half-cells are at standard conditions, we measure the difference in E_0 values.

By convention, E_0 for the hydrogen electrode is set at 0.00 volts. Any redox couple that tends to donate electrons to the standard hydrogen electrode has a negative value of E_0, which can be determined directly as the recorded emf in a half-cell linked to a standard hydrogen half-cell. A positive E_0 means that electrons from H_2 are flowing toward the test cell and reducing the electron acceptor, or, in other words, that the test acceptor is oxidizing H_2. *The higher the value of* E_0 *for a redox couple, the stronger an oxidant is the electron acceptor of that couple.* For example, we know from experience that ferric ion is a strong oxidant. This is borne out when we measure the reduction potential of a standard Fe^{3+}/Fe^{2+} half-cell with reference to a standard hydrogen half-cell. That value is +0.77 V. By contrast, a high negative value of E_0 connotes a ready tendency to lose electrons and, hence, strong reducing power.

Standard conditions for biochemists include a pH value of 7.0, a condition that does not obtain in the standard hydrogen electrode, which contains H^+ at 1.0 M. Therefore, biochemists use a modified term, $\boldsymbol{E_0'}$, which is the standard reduction potential measured at pH 7.0 (10^{-7} M H^+). These are the values used in this book and most other biochemical references. E_0' values for a number of biochemically important redox pairs are recorded in Table 15.1. Note that for the hydrogen ion itself E_0' is −0.42 V, significantly different from the standard value determined at pH 0 (i.e., at 1 M H^+). You should realize that all of the E_0' values in Table 15.1 were recorded with reference to a standard hydrogen electrode. Thus, the NADH + H^+/NAD^+ couple, with an E_0' of −0.32 V, will lose electrons to a *standard* hydrogen electrode but will tend to gain electrons from a hydrogen electrode at pH 7.0. This illustrates the general observation that *in a spontaneous reaction electrons flow from the half-cell of lower to that of higher potential.* Thus, the O_2/H_2O couple, with an E_0' of +0.82 V, has a high tendency to oxidize other substances. However, there is very little tendency for water to become oxidized to O_2, because none of the common biological oxidants has a higher E_0 than O_2/H_2O. (That is precisely what occurs in photosynthesis, as we shall see in Chapter 19.)

Free Energy Changes from Oxidation–Reduction Reactions

To recapitulate, the higher the value of E_0' for a redox couple, the greater is the tendency for that couple to participate in oxidation of another substrate. We can describe this tendency in quantitative terms, because free energy changes are directly related to differences in reduction potential.

$$\Delta G^{\circ\prime} = -nF\,\Delta E_0' = -nF[E_0'(\text{acceptor}) - E_0'(\text{donor})] \qquad (15.1)$$

where n is the number of electrons transferred in the half-reactions, F is Faraday's constant (96,500 J mol^{-1} V^{-1}), and $\Delta E_0'$ is the difference in standard reduction potentials between the two redox couples.

As an example, consider the oxidation of ethanol catalyzed by alcohol dehydrogenase.

$$\text{Ethanol} + \text{NAD}^+ \rightleftharpoons \text{acetaldehyde} + \text{NADH}^+ + \text{H}^+$$

The two half-reactions are as follows, both written in the direction of reduction.

Table 15.1
Standard reduction potentials of interest in biochemistry

Oxidant	Reductant	n	E_0', V
Acetate + CO_2	Pyruvate	2	−0.70
Succinate + CO_2	α-Ketoglutarate	2	−0.67
Acetate	Acetaldehyde	2	−0.60
O_2	O_2^-	1	−0.45
Ferredoxin (oxidized)	Ferredoxin (reduced)	1	−0.43
$2H^+$	H_2	2	−0.42
Acetoacetate	β-Hydroxybutyrate	2	−0.35
Pyruvate + CO_2	Malate	2	−0.33
NAD^+	NADH + H^+	2	−0.32
$NADP^+$	NADPH + H^+	2	−0.32
FMN (enzyme bound)	$FMNH_2$ (enzyme bound)	2	−0.30
Lipoate (oxidized)	Lipoate (reduced)	2	−0.29
1,3-Bisphosphoglycerate	Glyceraldehyde-3-phosphate + P_i	2	−0.29
Glutathione (oxidized)	Glutathione (reduced)	2	−0.23
FAD	$FADH_2$	2	−0.22
Acetaldehyde	Ethanol	2	−0.20
Pyruvate	Lactate	2	−0.19
Oxaloacetate	Malate	2	−0.17
α-Ketoglutarate + NH_4^+	Glutamate	2	−0.14
Methylene blue (oxidized)	Methylene blue (reduced)	2	0.01
Fumarate	Succinate	2	0.03
CoQ	$CoQH_2$	2	0.04
Cytochrome *b* (+3)	Cytochrome *b* (+2)	1	0.07
Dehydroascorbate	Ascorbate	2	0.08
Cytochrome c_1 (+3)	Cytochrome c_1 (+2)	1	0.23
Cytochrome *c* (+3)	Cytochrome *c* (+2)	1	0.25
Cytochrome *a* (+3)	Cytochrome *a* (+2)	1	0.29
$\frac{1}{2}O_2 + H_2O$	H_2O_2	2	0.30
Ferricyanide	Ferrocyanide	2	0.36
Nitrate	Nitrite	1	0.42
Cytochrome a_3 (+3)	Cytochrome a_3 (+2)	1	0.55
Fe (+3)	Fe (+2)	1	0.77
$\frac{1}{2}O_2 + 2H^+$	H_2O	2	0.82

E_0' is the standard reduction potential at pH 7 and 25°C, n is the number of electrons transferred, and each potential is for the partial reaction written as: oxidant $+ne^- \rightarrow$ reductant.

$$NAD^+ + 2H^+ + 2e^- \rightleftharpoons NADH \qquad E_0' = -0.32\ V$$

$$Acetaldehyde + 2H^+ + 2e^- \rightleftharpoons ethanol \qquad E_0' = -0.197\ V$$

Since in the reaction as written ethanol is the electron donor and NAD^+ the acceptor, we calculate $\Delta E_0'$ as follows.

$$\Delta E_0' = E_0'\ (\text{acceptor}) - E_0'\ (\text{donor}) = -0.32 - (-0.197) = -0.123\ V \quad (15.2)$$

The standard free energy change is given, then, by

$$\Delta G^{\circ\prime} = -nF\,\Delta E_0' = -2(96{,}500)(-0.123) = +23{,}739\ J/mol = +23.74\ kJ/mol \quad (15.3)$$

Note from this example that a negative $\Delta E_0'$ value gives a positive $\Delta G^{\circ\prime}$ and hence corresponds to a reaction that is *not* favored in the direction written. Note also that if we were to calculate $\Delta G^{\circ\prime}$ for the reverse reaction (reduction of acetaldehyde by NADH), the term E_0' (acceptor) − E_0' (donor) would be +0.123 V, and $\Delta G^{\circ\prime}$ would have the same numerical value but opposite sign.

The values given in Table 15.1 allow calculation of free energy changes only under standard conditions. To calculate reduction potentials under nonstandard conditions we use the **Nernst equation,**

$$E' = E_0' + \frac{2.303RT}{nF} \log \frac{[\text{electron acceptor}]}{[\text{electron donor}]} \tag{15.4}$$

where R is the gas constant (8.312 J deg^{-1} mol^{-1}), T is the absolute temperature, and 2.303 is the conversion factor from natural to common logarithms. At 25°C the term $2.303RT/nF$ has the value 0.059 for a one-electron transfer and 0.03 for a two-electron transfer (where $n = 2$). Thus, for a two-electron transfer we can use a simplified version of the above equation,

$$E' = E_0' + 0.03 \log \frac{[\text{electron acceptor}]}{[\text{electron donor}]} \tag{15.5}$$

Note that this equation is of the same form as the Henderson–Hasselbalch equation (pH = pK_a + log[proton acceptor]/[proton donor]). In the same sense, then, that pK_a is defined by the midpoint of an acid titration curve, E_0' is defined by the midpoint of an electrochemical titration, where electron acceptor and electron donor are present in equal concentrations.

Each of the coupled redox reactions in biological electron transport involves the transfer of electrons from one redox couple to another couple of higher reduction potential. Thus, each individual redox reaction in the sequence is exergonic under standard conditions. For electrons entering the respiratory chain as NADH, the overall reaction sequence is given by the following equation.

$$\text{NADH} + \text{H}^+ + \tfrac{1}{2}\text{O}_2 \rightleftharpoons \text{NAD}^+ + \text{H}_2\text{O}$$

This sequence is strongly exergonic.

$$\Delta G^{\circ\prime} = -nF\,\Delta E_0' = -2(96{,}500)[0.82 - (-0.32)] = -220 \text{ kJ/mol} \tag{15.6}$$

Experimentally, we know that the oxidation of 1 mol of NADH in the respiratory chain proceeds concomitantly with synthesis of 3 mol of ATP from ADP and P_i. Since $\Delta G^{\circ\prime}$ for ATP hydrolysis is −30.5 kJ/mol, synthesis of three ATPs requires 91.5 kJ under standard conditions, giving an efficiency for oxidative phosphorylation of about 40%, when determined under standard conditions. Since $\Delta G'$ for ATP under intracellular conditions is significantly higher (about 40 kJ/mol), the intracellular efficiency is probably somewhat higher than 40%.

Electron Transport

Electron Carriers in the Respiratory Chain

If you compare the sequence of respiratory electron carriers (Figure 15.3) with the standard reduction potentials of those carriers (Table 15.1), you will see that E_0' for each carrier increases in the same order as the sequence of their use in electron transport. This suggests that each individual oxidoreduction reaction in electron transport is exergonic under standard conditions and that electrons flow in continuous fashion from low-potential to high-potential carriers. Very neat, but is it real? After all, we have seen that glycolysis and the citric acid cycle both proceed smoothly despite the involvement of some highly endergonic reactions. We shall explore the several lines of evidence by which the currently accepted pathway of electron transport was determined. First, however, let us become better acquainted with the participants—the electron carriers and the enzymes involved.

NADH AND NADH DEHYDROGENASE. Numerous dehydrogenases in the cell generate NADH. The NADH becomes oxidized in the first step of electron transport by the enzyme **NADH dehydrogenase,** which contains flavin mononucleotide as a tightly bound prosthetic group and catalyzes the following reaction.

$$\mathrm{NADH + H^+ + FMN \rightleftharpoons NAD^+ + FMNH_2}$$

The enzyme is a large, multisubunit complex with about 25 separate polypeptide chains. The complex transfers electrons from reduced flavin to another respiratory carrier, coenzyme Q. The complex can also accept electrons from NADPH, although rather inefficiently. A more widely used route for NADPH oxidation, when that becomes necessary, is the transfer of two reducing equivalents from NADPH to NAD^+ via an energy-linked enzyme called **transhydrogenase.**

$$\mathrm{NADPH + NAD^+ \rightleftharpoons NADP^+ + NADH}$$

The NADH dehydrogenase complex also contains a number of iron–sulfur centers, which transfer electrons onward from $FMNH_2$. Iron–sulfur centers consist of nonheme iron complexed to sulfur in three known ways, as shown in Figure 15.4. The simplest form, designated **FeS,** involves one iron tetrahedrally complexed with the thiol sulfurs of four cysteine residues. The second form, denoted $\mathbf{Fe_2S_2}$, contains two irons, each complexed with two cysteine residues and two inorganic sulfides. The third and most complicated form ($\mathbf{Fe_4S_4}$) contains four irons, four sulfides, and four cysteine residues. NADH dehydrogenase contains both Fe_2S_2 and Fe_4S_4 centers. In all of these centers the iron can undergo cyclic oxidoreduction between ferrous and ferric states, as shown below for the NADH dehydrogenase complex. Since the electrons are then used to reduce coenzyme Q, a more descriptive name for this complex is **NADH-coenzyme Q reductase.**

$NADH + H^+$ → NAD^+; FMN → $FMNH_2$; $FMNH_2$ → FMN; $Fe^{3+}S$ → $Fe^{2+}S$; $Fe^{2+}S$ → $Fe^{3+}S$; CoQ → $CoQH_2$

Several important redox proteins contain iron–sulfur centers, including other proteins in the respiratory chain. This group goes by the collective term **iron–sulfur proteins,** or **nonheme iron proteins.**

COENZYME Q. The respiratory electron carrier coenzyme Q was discovered in the late 1950s, following the observation that treatment of isolated

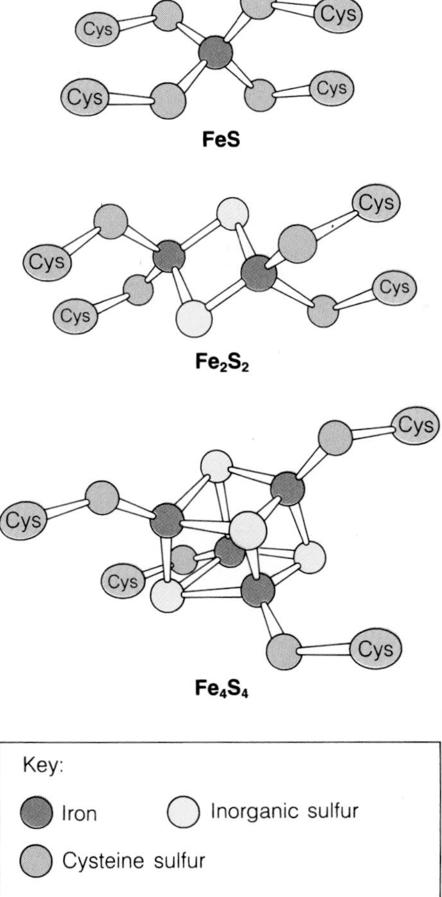

Figure 15.4
Structures of iron–sulfur centers.

Oxidized coenzyme Q_{10} (CoQ)

Reduced coenzyme Q_{10} ($CoQH_2$)

Semiquinone form of coenzyme Q

mitochondria with an organic solvent such as isooctane completely abolished the ability of the mitochondria to oxidize substrates. Addition of a concentrated isooctane extract completely restored the oxidative capacities of the mitochondria, suggesting the presence of an extremely lipophilic electron carrier that was but loosely linked to protein. This carrier was found to be a benzoquinone linked to a number of isoprene units, usually 10 in mammalian cells and 6 in bacteria. Because the substance is ubiquitous in living cells, one group of researchers named it **ubiquinone,** while another called it coenzyme Q, or **CoQ.** The term Q_{10} is used to specify the form of CoQ containing 10 isoprene units.

The isoprenoid tail gives the molecule its apolar character, which allows CoQ to diffuse rapidly through the inner mitochondrial membrane.

CoQ functions to draw electrons into the respiratory chain, not only from NADH but also from succinate and from intermediates in fatty acid oxidation. Succinate dehydrogenase uses an FAD coenzyme, as noted in Chapter 14. Unlike the other citric acid cycle enzymes, succinate dehydrogenase is an inner membrane protein. The enzyme can thus transfer electrons directly from its bound $FADH_2$ to the other membrane-bound respiratory carriers. Like NADH dehydrogenase, succinate dehydrogenase transfers electrons via iron–sulfur centers to coenzyme Q, and it is more completely named **succinate-coenzyme Q reductase.** Other flavoprotein dehydrogenases, such as the one involved in fatty acid oxidation (Chapter 17), also transport electrons to electron acceptors via CoQ.

Because CoQ is subsequently oxidized by cytochromes, it can be seen as a collection point, gathering electrons from several flavoprotein dehydrogenases and passing them along to cytochromes for ultimate transport to O_2. Further, since CoQ oxidoreduction proceeds through a **semiquinone** intermediate, CoQ is a suitable interface between the two-electron flavin carriers and the one-electron cytochromes.

CYTOCHROMES. Finally, we come to the cytochromes, a group of red or brown heme proteins having distinctive visible-light spectra. These proteins were first characterized and their role in respiration demonstrated by the Englishman David Keilin. Using a hand spectroscope, Keilin observed red-brown pigments in the flight muscles of insects. During muscular exertion (when an immobilized fly tried to free itself), the spectra of these pigments underwent marked changes. This led Keilin to postulate a role for these substances in carrying electrons from biological fuels to oxygen.

The major respiratory cytochromes are classified as *a, b,* or *c,* depending on the wavelengths of the spectral absorption peaks. Figure 15.5 shows the spectral characteristics of typical *a*-, *b*-, and *c*-type cytochromes. Cytochrome c_1 has a spectrum similar to that of cytochrome *c,* but the α and γ absorption peaks are shifted slightly toward the red. Cytochromes *b, c,* and c_1 all contain iron complexed with protoporphyrin IX, the heme found in hemoglobin and myoglobin. In cytochromes *c* and c_1, but not *b,* the heme is linked covalently to the protein, via thioether bonds formed between two of the vinyl side chains and two cysteine residues (Figure 15.6). Among the respiratory electron carriers are found three *b*-type cytochromes, distinguished by small spectral differences.

Cytochromes *a* and a_3 contain a modified form of heme, called heme A, in which two of the side chains are modified (see Figure 15.6). Cytochromes *a* and a_3 evidently represent identical heme A moieties, attached to the same polypeptide chain, but in different environments. Each of the hemes is associated with a copper ion, located close to the heme iron. Cytochromes undergo oxidoreduction through the complexed metal, which cycles between +2 and +3 states of heme iron and +1 and +2 states for the copper in cytochromes a and a_3. Thus, the cytochromes are one-electron carriers.

Cytochrome c is a small protein (M_r = 13,000), which is readily extracted in soluble form from mitochondrial preparations. Because it is small and relatively abundant, detailed structural studies have been carried out with this protein. Its three-dimensional structure was one of the earliest determined for any protein by x-ray crystallography (Figure 15.7). The amino acid sequence of the protein is highly conserved, with nearly 50% identity between residues at corresponding positions of cytochromes c in organisms as diverse as yeast and human.

Unlike cytochrome c, the other cytochromes are integral membrane proteins and are exceedingly difficult to dissociate from the membrane. Accordingly, we know less about the structures of these proteins. Cytochromes

Figure 15.5
Absorption spectra of cytochromes, in both oxidized (red) and reduced (blue) states. (**a**) Cytochrome b from *Neurospora;* (**b**) cytochrome c from horse heart; (**c**) beef heart cytochrome oxidase (which contains both cytochromes a and a_3).

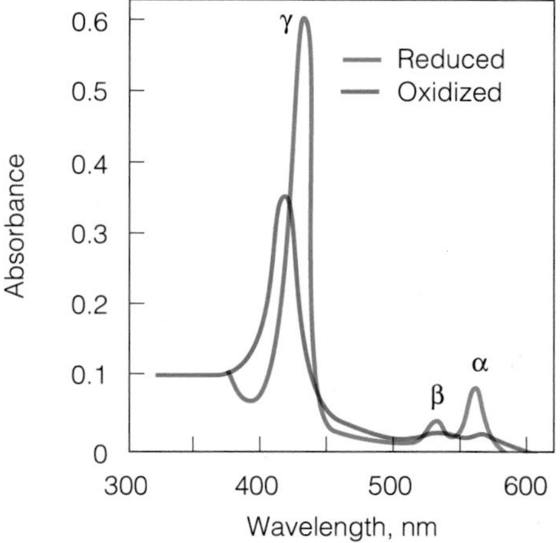

(**a**) Cytochrome b

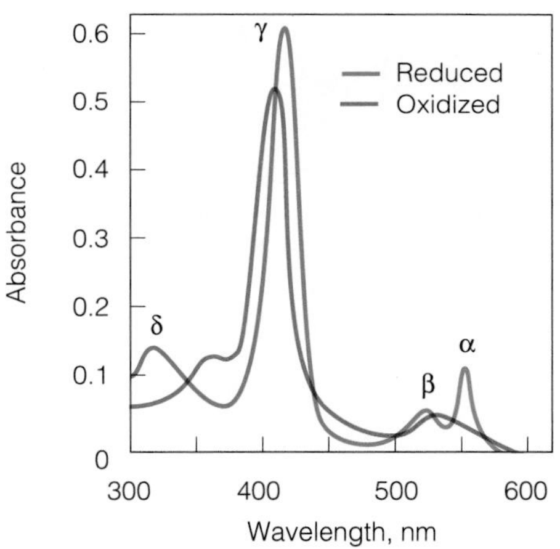

(**b**) Cytochrome c

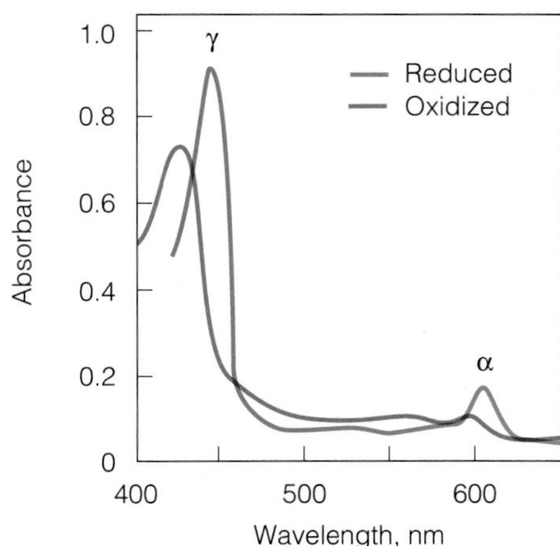

(**c**) Cytochromes a and a_3

Figure 15.6
The hemes found in cytochromes. (**a**) The covalent bond formed between heme and cytochromes c and c_1. The vinyl groups on heme are linked to the thiol groups of two cysteine residues (red). (**b**) Heme A, the form found in cytochromes a and a_3. Note the modified side chains—a formyl group (red) and a 17-carbon side chain (blue).

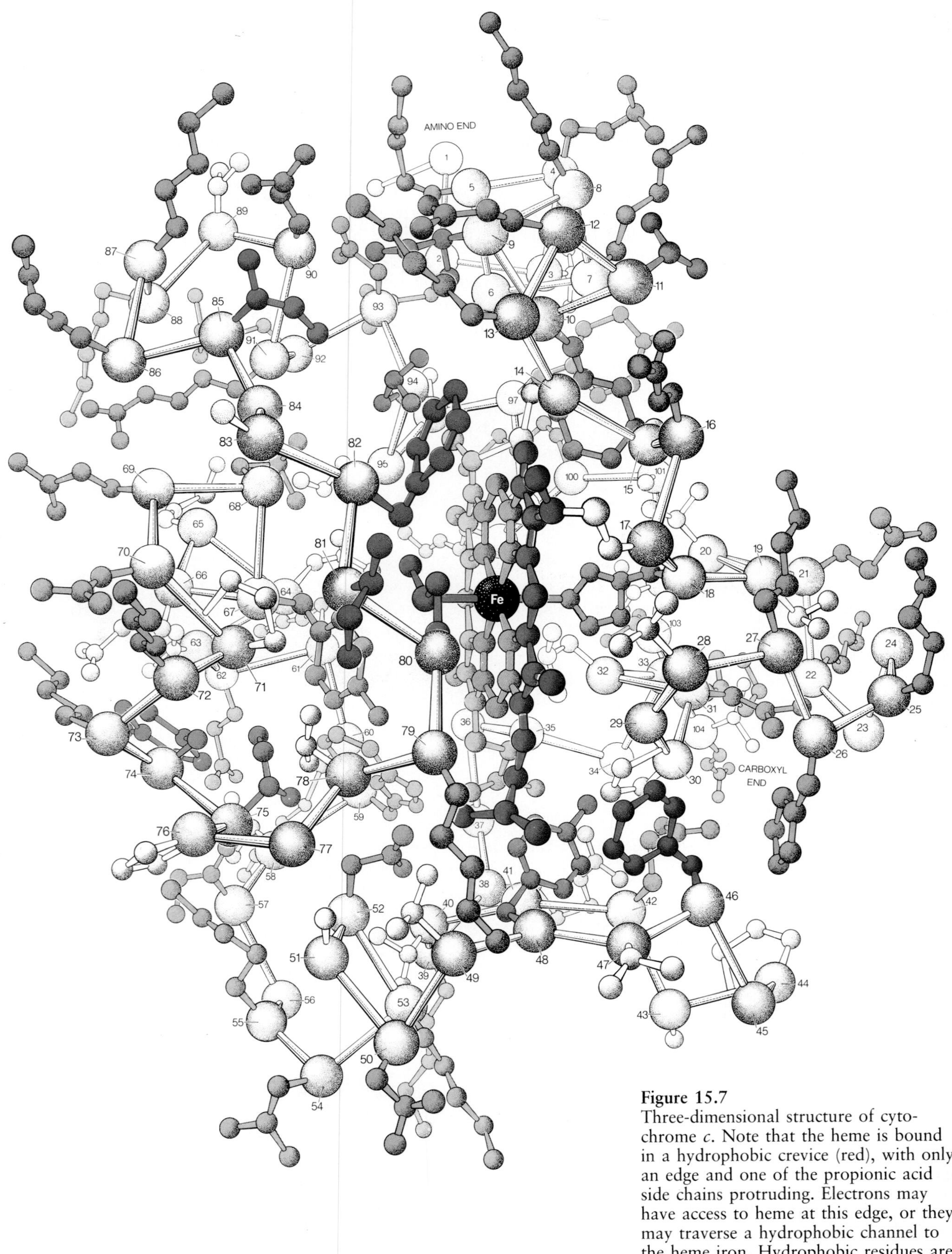

Figure 15.7
Three-dimensional structure of cytochrome *c*. Note that the heme is bound in a hydrophobic crevice (red), with only an edge and one of the propionic acid side chains protruding. Electrons may have access to heme at this edge, or they may traverse a hydrophobic channel to the heme iron. Hydrophobic residues are in red.

a and a_3 form part of a large multiprotein complex called **cytochrome oxidase.** This complex catalyzes the ultimate step in electron transport, namely the reduction of oxygen to water. In eukaryotes the cytochrome oxidase complex contains up to 13 polypeptide chains, three of which are known to be encoded by the mitochondrial genome and synthesized within the mitochondrion. The rest are encoded by nuclear genes.

Determining the Sequence of Respiratory Electron Carriers

Now let us return to the question of how we learned the specific sequence in which electrons are carried from reduced substrates to oxygen. Since E_0' values for the respiratory electron carriers increase in the same order as their sequence in the respiratory chain, the energy of biological oxidations is released in a series of exergonic reactions. By converting E_0' values to $\Delta G°'$ values, as is done in Figure 15.8, we see that three of these individual reactions release more free energy than 30.5 kJ. Since that is the $\Delta G°'$ for ATP hydrolysis, these three reactions represent sites at which the endergonic synthesis of ATP could be coupled to an exergonic reaction, with an overall free energy decrease.

However, all of these considerations involve standard conditions, and there is no doubt that conditions are much different in the mitochondrion. In particular, the hydrophobic nature of the membrane environment changes E values in ways that are difficult to predict or to measure. We shall outline here three experimental techniques that have been used to identify both the actual order of electron carriers and the specific sites at which respiration is linked to phosphorylation. We focus first on the order of electron carriers. Three major experimental approaches have been used: (1) spectrophotometric techniques to measure the redox status of electron carriers in intact mitochondria; (2) use of specific respiratory inhibitors and artificial electron acceptors; and (3) fractionation of mitochondria into respiratory subassemblies, each capable of catalyzing specific portions of the overall sequence.

Difference Spectra. For nicotinamide nucleotides, flavin nucleotides, and cytochromes, the absorption spectrum for the reduced carrier differs from that of its oxidized counterpart. We should, therefore, be able to scan the absorption spectrum of a mixture of these carriers and ascertain what proportion of each is in the oxidized state and what proportion in the reduced state. The sensitivity of the technique is increased if one obtains a

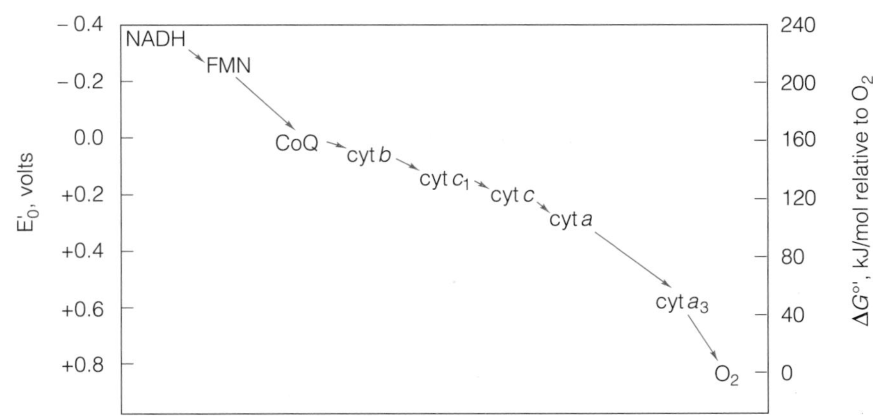

Figure 15.8
Standard reduction potentials of the major respiratory electron carriers, along with $\Delta G°'$ values corresponding to these potentials. Three reactions in the respiratory chain have $\Delta G°'$ values greater than 30.5 kJ/mol, the $\Delta G°'$ for ATP hydrolysis: FMN → CoQ, cyt b → cyt c_1, and cyt a → O_2.

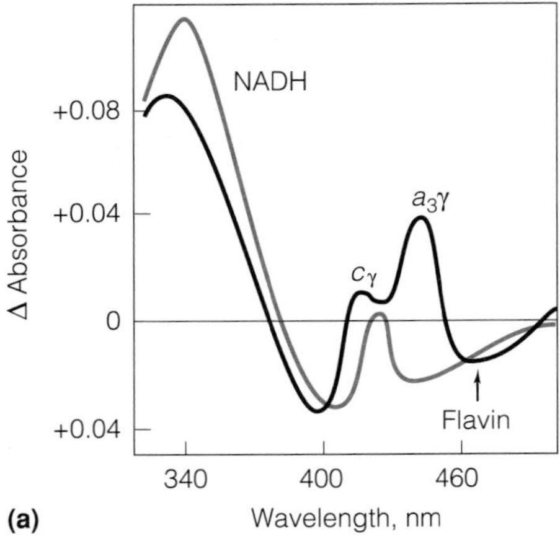

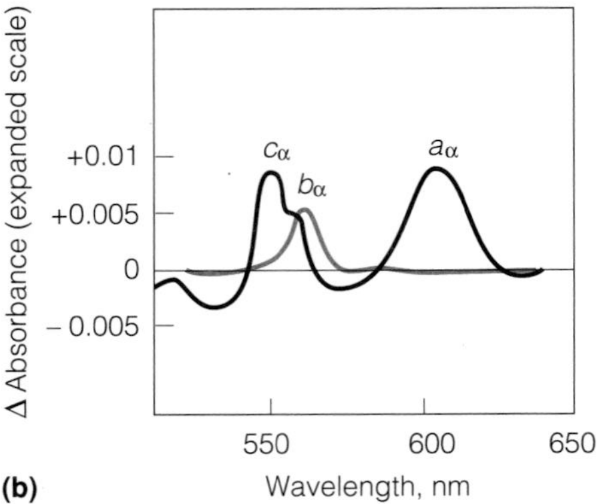

Figure 15.9
Difference spectra of rat liver mitochondria, as recorded in a double-beam spectrophotometer. The black line shows the difference spectrum of fully reduced versus fully oxidized mitochondria. Mitochondria reduced with substrate under anaerobic conditions were in the sample chamber and fully oxidized mitochondria in the reference chamber. The peaks and shoulders refer to NADH, flavin, and the α and γ absorption bands, as indicated, for cytochromes a, a_3, b, and c. The red line shows the effect of adding a respiratory inhibitor, antimycin A, which blocks electron flow from cytochrome b to c_1. The inhibitor causes all of the carriers beyond cytochrome b to become fully oxidized, while NADH, flavin, and cytochrome b are in the reduced state. Note the expanded absorbance scale beyond 500 nm.

difference spectrum. Here the sample cuvette contains the mixture of electron carriers under study, and the reference cuvette contains, not a blank, but an equimolar mixture of carriers in a known state, for instance, entirely oxidized. Thus, any small absorbance changes, either positive or negative, result from reduction of a portion of the carriers in the test sample.

When these carriers are embedded in the mitochondrion, however, the task becomes quite difficult. Mitochondrial suspensions are turbid, and the resultant light scattering makes it impossible to measure difference spectra with ordinary spectrophotometers. Britton Chance improved this technique by orders of magnitude when he developed a dual-wavelength, double-beam spectrophotometer that allowed him to obtain difference spectra with intact mitochondria. In the example shown in Figure 15.9 the reference beam contains mitochondria saturated with oxygen, so that all carriers are oxidized, while the sample beam contains anaerobic mitochondria plus an oxidizable substrate, so that all of the carriers are reduced. The difference spectrum identifies wavelengths of maximal and minimal absorbance, where readings on test suspensions can allow a determination of the proportion of an electron carrier that is in the oxidized state or the reduced state. For example, the higher the absorbance at 340 nm, the greater the proportion of the NAD^+/NADH couple that is present as NADH; and the lower the negative absorbance at 460 nm, the greater the proportion of flavin nucleotide that is in the oxidized form.

Two important observations were made soon after the introduction of this technique in the mid-1950s. First, in actively respiring mitochondria, NADH was found largely in the reduced state, while cytochrome a_3 was largely oxidized. For the intermediate carriers the proportion in the oxidized state increased in the same order as the order of their presumed function in respiration. Second, when oxygen was added to anaerobic mitochondria, and difference spectra were obtained at intervals after oxygen addition, it was possible to determine the order in which each carrier went from fully reduced to partially oxidized. That order was the same as the presumed order of function in respiration.

Inhibitors and Artificial Electron Acceptors. Further information was obtained from difference spectrophotometry in conjunction with exogenous compounds that functioned either as respiratory inhibitors or as artificial electron donors or acceptors. The sites of action of several important inhibitors, as we understand them now, are shown in Figure 15.10. These inhibitors include (1) **rotenone,** a plant product from South America that is used as an insecticide and that blocks electron flow from NADH to coenzyme Q; (2) **amytal,** a barbiturate drug that acts at the same site; (3) **antimycin A,** a *Streptomyces* antibiotic that blocks electron flow from cytochrome b to c_1; and (4) the cytochrome oxidase inhibitors cyanide, azide, and carbon monoxide. Cyanide and azide react with the oxidized form of the cytochrome target, while CO reacts with the reduced form. As we noted earlier, the chief toxic effect of CO on animals results from its strong affinity for the oxygen binding sites of hemoglobin.

To see the utility of respiratory inhibitors, consider what happens in actively respiring mitochondria to which antimycin A is added. Since electrons cannot flow from cytochrome b to c_1, all of the carriers before cytochrome b become reduced, while all of the subsequent carriers become fully oxidized, analogous to the accumulation of water upstream of a dam in a river. This is a crossover point, like that described for phosphofructokinase when glycolysis is inhibited at high ATP levels (Chapter 13). Crossover points in the respiratory chain can be detected by dual-beam spectropho-

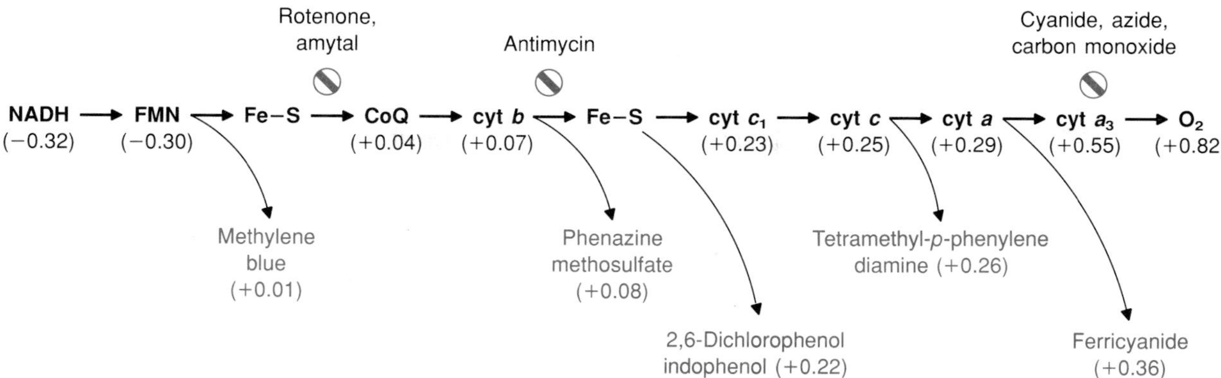

Figure 15.10
Some useful respiratory inhibitors and their sites of inhibition (red), and some artificial electron acceptors (blue). Each acceptor is positioned according to its E_0' value (in parentheses), identifying the most likely site at which an acceptor will withdraw electrons from the respiratory chain when added to mitochondria.

tometry, which allows identification of all the carriers lying *before* a site of inhibition (NAD, flavin, CoQ, and cytochrome *b*) and all those lying *beyond* that site (cytochromes *c*, *a*, and a_3).

Artificial electron donors and acceptors are compounds that can either feed in or draw off electrons from the respiratory chain in spontaneous nonenzymatic reactions. For example, 2,6-dichlorophenol-indophenol (DCIP) can spontaneously oxidize cytochrome *b*, but probably not c_1, because of the E_0' values involved (see Figure 15.10). This allowed a demonstration that cytochrome *b* lies beyond the entry point for electrons from succinate, as well as from NADH. The key observation was the finding that addition of DCIP to cyanide-inhibited mitochondria allows these mitochondria to oxidize both NAD^+-linked substrates and succinate. In this system electrons flow from substrate to cytochrome *b* and then to the exogenous electron carrier, DCIP.

RESPIRATORY COMPLEXES. Mitochondria can be disrupted by mechanical treatment, such as sonic oscillation, or by nonionic detergents such as **digitonin,** which preferentially solubilizes the outer membrane but tends not to denature enzymes. By combinations of these techniques, one can fractionate mitochondria into four separate enzyme complexes, each of which contains part of the entire respiratory sequence, plus a fifth complex that catalyzes ATP synthesis from ADP. Analysis of each complex for the presence of electron carriers, as well as for reactions catalyzed, has helped to establish the currently accepted sequence of carriers.

Figure 15.11 summarizes what we know about the structural components and functional activities of each of these subassemblies, which are denoted complexes I, II, III, IV, and V. Each of the oxidative complexes (i.e., I–IV) contains several integral membrane components, and each accepts electrons from a relatively mobile electron carrier (one that is not tightly membrane bound) and passes electrons along to another mobile carrier. These mobile carriers are NADH, coenzyme Q, cytochrome *c*, and oxygen. Complex I is the NADH dehydrogenase complex, or NADH-coenzyme Q reductase, and complex II is succinate-CoQ reductase. Electrons from NADH and succinate, respectively, are passed to coenzyme Q, which then passes electrons, via complex III, to cytochrome *c*. Cytochrome *c* is oxidized in turn by complex IV, which is the cytochrome oxidase complex. Energy released by the actions of complexes I, III, and IV drives the synthesis of ATP by complex V, by mechanisms discussed later in this chapter.

Analysis of the components of these complexes leaves some still unanswered questions, for example, how the *b*-type cytochrome functions in

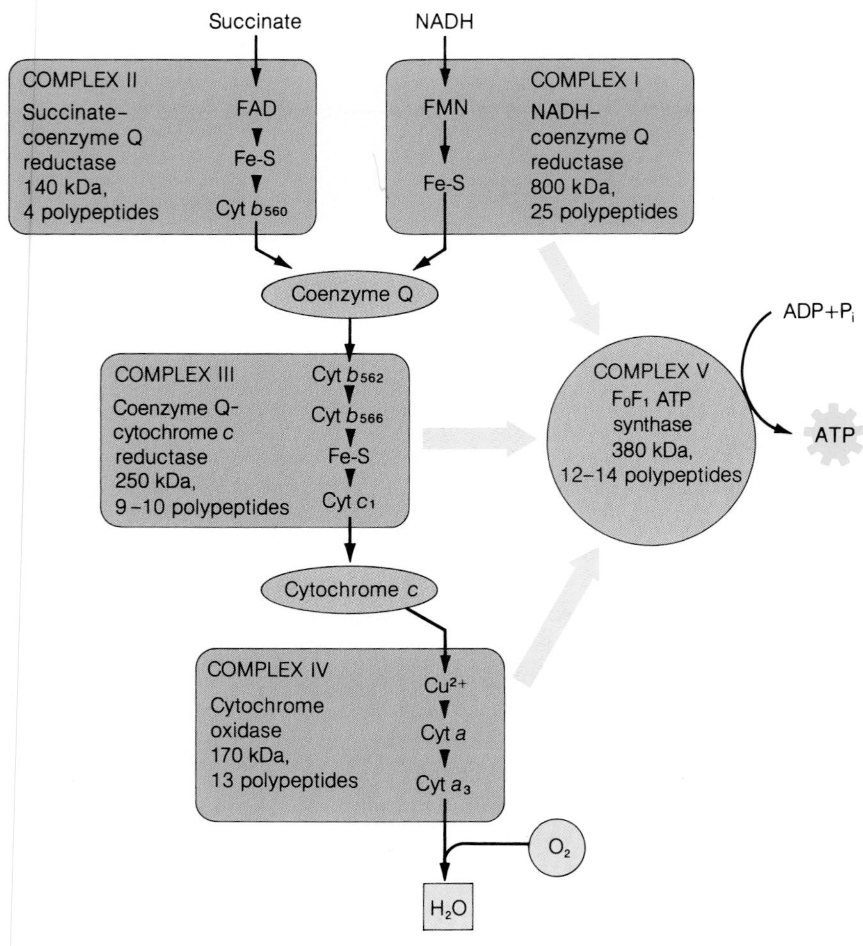

Figure 15.11
Multiprotein complexes in the respiratory assembly, showing the prosthetic groups associated with each complex. Complexes I–IV are involved in electron transport; complex V uses the energy derived from electron transport to synthesize ATP. The subscripts for the *b* cytochromes denote their spectral maxima. The two *b* hemes in complex III are bound to the same polypeptide chain.

complex II, or whether the coppers in complex IV (cytochrome oxidase) are obligatory electron carriers.

Shuttling Electron Carriers into Mitochondria

Reduced electron carriers generated in glycolysis and the citric acid cycle must transfer reducing equivalents into the mitochondrial matrix, because that is where the pyruvate dehydrogenase complex and the dehydrogenases of the citric acid cycle are located. Specific transport systems are required, since the NADH generated by a cytosolic enzyme, such as glyceraldehyde-3-phosphate dehydrogenase, cannot itself traverse the mitochondrial membrane to be oxidized by the respiratory chain. Therefore, the reducing equivalents must be *shuttled* to respiratory assemblies in the inner mitochondrial membrane, without physical movement of the coenzyme. This process involves the reduction of an organic molecule by NADH in the cytoplasm, passage of the reduced molecule into the mitochondrion via a specific transport system, reoxidation of that compound inside the mitochondrion, and passage of the oxidized substrate back to the cytoplasm, where it can undergo the same cycle all over again.

The earliest known shuttle system is the dihydroxyacetone phosphate/glycerol-3-phosphate shuttle, which is particularly active in brain (and in the flight muscle of insects). As shown in Figure 15.12a, dihydroxyacetone phosphate (DHAP) is reduced by NADH in the cytosol, followed by passage of the resultant glycerol-3-phosphate into the mitochondrion, where it is

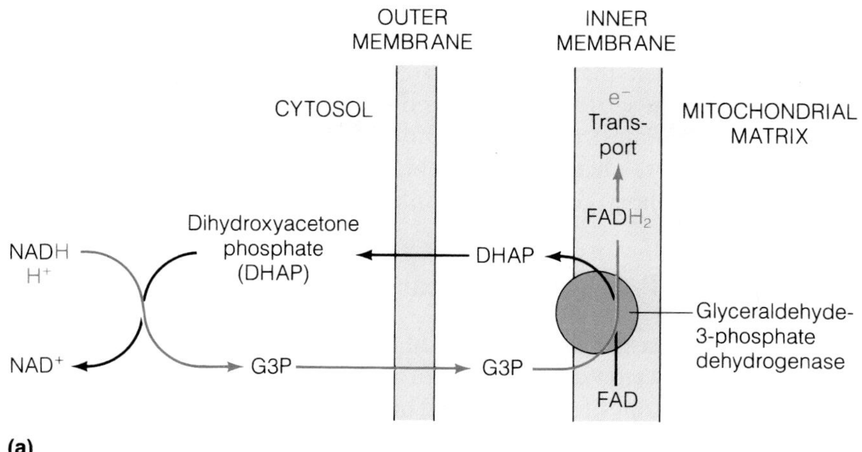

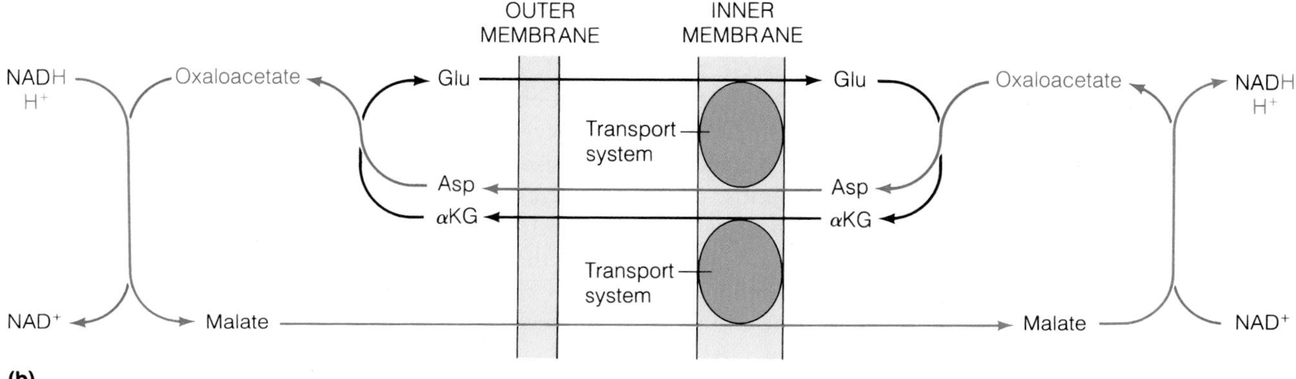

Figure 15.12
Shuttles for transfer of reducing equivalents from cytosol into mitochondria. (**a**) The glycerol-3-phosphate shuttle. (**b**) The malate–aspartate shuttle. Glu = glutamate, Asp = aspartate, αKG = α-ketoglutarate.

reoxidized by a flavoprotein, glycerol-3-phosphate dehydrogenase, bound at the outer face of the inner membrane. This involves reduction of FAD followed by transfer of an electron pair from $FADH_2$ to coenzyme Q, just as occurs in the oxidation of intramitochondrial NADH. Once dihydroxyacetone phosphate has returned to the cytosol, the net effect has been to transfer two reducing equivalents from cytosolic NADH to mitochondrial $FADH_2$ and hence on up the respiratory chain.

A different shuttle system, particularly active in liver and heart, is the malate/aspartate shuttle, shown in Figure 15.12b. Here a cytosolic isozyme of malate dehydrogenase, together with NADH, reduces oxaloacetate to malate, which passes into the matrix via a specific active transport system in the inner mitochondrial membrane. The malate is then reoxidized by the malate dehydrogenase of the citric acid cycle, which also uses NAD^+. Because oxaloacetate cannot cross the inner membrane, it is transaminated to aspartate, which is then transported out for reconversion to oxaloacetate, to begin the cycle anew. Because of the transamination involved, this process requires that α-ketoglutarate be continuously transported out of mitochondria and glutamate be continuously transported in.

Oxidative Phosphorylation

Having discussed how energy is generated from the mitochondrial oxidation of reduced substrates, we turn now to the question of how that energy is made available for ATP synthesis—in short, the mechanism of oxidative phosphorylation. Mechanistically, oxidative phosphorylation is far more complex than the substrate-level phosphorylation reactions in glycolysis and the citric acid cycle, and until recently this field of inquiry has been one of the most contentious biochemical research arenas. As stated by Efraim Racker, a leader in this field, "Anyone who is not confused about oxidative phosphorylation just doesn't understand the situation."

The P/O Ratio; Efficiency of Oxidative Phosphorylation

To measure the efficiency of oxidative phosphorylation, we must determine energy stored in the form of ATP as a fraction of total energy released in the oxidation of a substrate. This became possible once isolated mitochondria could be shown to carry out ATP synthesis coupled to electron transport, and one could measure the quantity of ATP synthesized per mole of substrate oxidized. What we usually measure is the **P/O ratio,** which is the number of molecules of ATP synthesized per pair of electrons carried through electron transport. ATP synthesis is quantitated as phosphate uptake—conversion of orthophosphate to organic phosphates—and electron pairs are quantitated as oxygen uptake, in atoms reduced to water. Oxygen uptake was formerly measured from the volume of oxygen consumed but is now almost universally determined with a recording oxygen electrode. Although there has been controversy about how to measure P/O ratios properly, it was widely agreed that the mitochondrial oxidation of NADH proceeds with a P/O ratio of 3 and, hence, that any substrate metabolized via a mitochondrial NAD^+-linked dehydrogenase should yield 3 mol of ATP per mole of NADH produced.

More recent work on the mechanism of oxidative phosphorylation, showing that the energy is derived from proton translocations across the inner membrane, has reopened the question of the true values of P/O ratios. When we discuss these mechanistic aspects later in this chapter, you will see that the stoichiometry of proton transport is still quite unsettled. Since the concept of the P/O ratio is useful for understanding the mechanism and control of oxidative phosphorylation, we shall continue using 3 as the P/O ratio for NAD^+-linked substrates, realizing that that value will probably require modification in the future.

With the above in mind, we can write a balanced equation for mitochondrial oxidation of NADH.

$$\text{NADH} + \text{H}^+ + \tfrac{1}{2}\text{O}_2 + 3\text{ADP} + 3\text{P}_i \rightleftharpoons \text{NAD}^+ + 4\text{H}_2\text{O} + 3\text{ATP}$$

As discussed earlier, the oxidation of NADH by O_2 has a $G^{\circ\prime}$ decrease of 220 kJ/mol. Reversal of ATP hydrolysis requires 30.5 kJ/mol under standard conditions. Therefore, coupling the synthesis of three ATPs to oxidation of one NADH traps 91.5 kJ (3 × 30.5), or about 42%, of the energy released (under standard conditions).

Oxidative Reactions That Drive ATP Synthesis

The most straightforward interpretation of the P/O ratio of 3 observed for NADH oxidation is that three of the individual reactions of the respiratory chain are sufficiently exergonic to drive the synthesis of one ATP molecule each. Three of these reactions have $\Delta G^{\circ\prime}$ values exceeding 30.5 kJ/mol, the minimum barrier that must be overcome to make the synthesis of each ATP

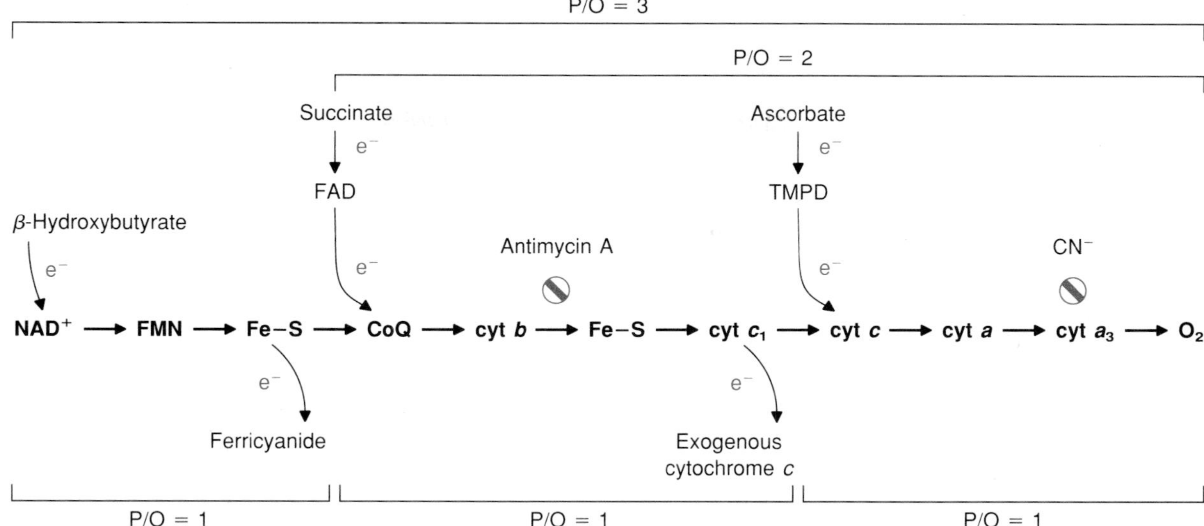

Figure 15.13
Experimental identification of "coupling sites" for oxidative phosphorylation. Electron transport was restricted to particular parts of the chain by use of selected electron donors, electron acceptors, and respiratory inhibitors, as indicated. For each segment of the chain that was thus isolated, P/O ratios were determined, as shown on the figure. Antimycin A was added in the experiment that identified the first coupling site, to block electron flow past cytochrome *b*.

exergonic (see Figure 15.8). Those three reactions are the oxidation of $FMNH_2$ by coenzyme Q, the oxidation of cytochrome *b* by cytochrome c_1, and the cytochrome oxidase reaction. An early interpretation of this observation was that each reaction represented a **coupling site** for ATP synthesis, that is, a reaction in which ATP synthesis is driven directly by the energy released from that reaction, in the same sense that ATP synthesis is driven by the hydrolysis of an energy-rich substrate in a substrate-level phosphorylation.

We now know that that concept is an oversimplification. As discussed later in this chapter, the coupling of energy release to energy utilization is indirect; energy generated in oxidation–reduction reactions forms a proton gradient across the inner membrane, and this in turn drives ATP synthesis. Nevertheless, the concept of coupling sites is useful, because it provided a framework for experiments identifying each of the above three reactions as individually capable of energizing the membrane for ATP synthesis, even when the entire electron transport chain is not operating. With that in mind, let us examine this evidence.

First, succinate is oxidized with a P/O ratio of 2, not 3, suggesting the existence of a coupling site before coenzyme Q in the respiratory chain. This was confirmed by blocking electron transport past cytochrome *b* with antimycin A. Ferricyanide was added as an artificial electron acceptor, so that electrons could continue to flow (Figure 15.13). Under these conditions NAD^+-linked substrates, such as β-hydroxybutyrate, were oxidized with a P/O ratio of 1, confirming the existence of one site before cytochrome *b*.

Another approach involved an artificial electron donor, ascorbate. In the presence of another electron acceptor, tetramethyl-*p*-phenylamine diamine (TMPD) (see Figure 15.10), electrons could be fed into the respiratory chain at cytochrome *c*. These electron carriers reduced cytochrome *c* nonenzymatically, and its oxidation via cytochrome oxidase proceeded with a P/O ratio of 1, thus localizing one coupling site to the right of cytochrome *c*. Finally, purified cytochrome *c*, when added to mitochondria, can withdraw electrons from the electron transport chain. The further addition of a cytochrome oxidase inhibitor, such as cyanide, forces electrons to exit from the respiratory chain at cytochrome *c*. Under these conditions succinate is oxidized with a P/O ratio of 1, which localizes a site between cytochromes *b* and *c*.

The Enzyme System for ATP Synthesis

Of the three major complexes involved in NADH oxidation—I, III, and IV—each can drive the synthesis of one ATP per electron pair. This was confirmed in elegant reconstitution studies involving the enzyme complex that catalyzes the synthesis of ATP (complex V). Let us review the discovery and nature of that complex and then discuss the reconstitution experiments.

When mitochondria are negatively stained with phosphotungstate, electron microscopy reveals that the cristae are covered with knoblike projections, on the matrix side, attached to the inner membrane by a short stalk (Figure 15.14); the knobs are known as **F_1 spheres.** Disruption of mitochondria by sonic oscillation generates fragments of inner membrane, which reseal in the form of closed vesicles. The membrane closes on itself inside out, so the knoblike projections are on the outside. These submitochondrial particles respire and synthesize ATP, just as do intact mitochondria. Racker and his colleagues showed that treatment of these vesicles with trypsin or urea caused the knobs to dissociate from the vesicles. After centrifugation to separate the "stripped" vesicles from the knobs, the vesicles could still oxidize substrates and reduce oxygen, but no ATP was synthesized. When knobs were added back to the vesicles, there was substantial reconstitution of particles that could now catalyze ATP synthesis dependent on the oxidation of exogenous substrates.

Because readdition of the knobs recoupled ATP synthesis to electron transport, they were originally called **coupling factors.** Purification of the knobs attached to the underlying stalks revealed a multiprotein aggregate of molecular weight about 380,000, consisting of about a dozen polypeptide chains—five or six proteins forming the F_1 knob, two proteins forming the stalk to which it is attached, and an attached complex, designated F_0, embedded in the inner membrane. The structure of the entire aggregate is schematized in Figure 15.15.

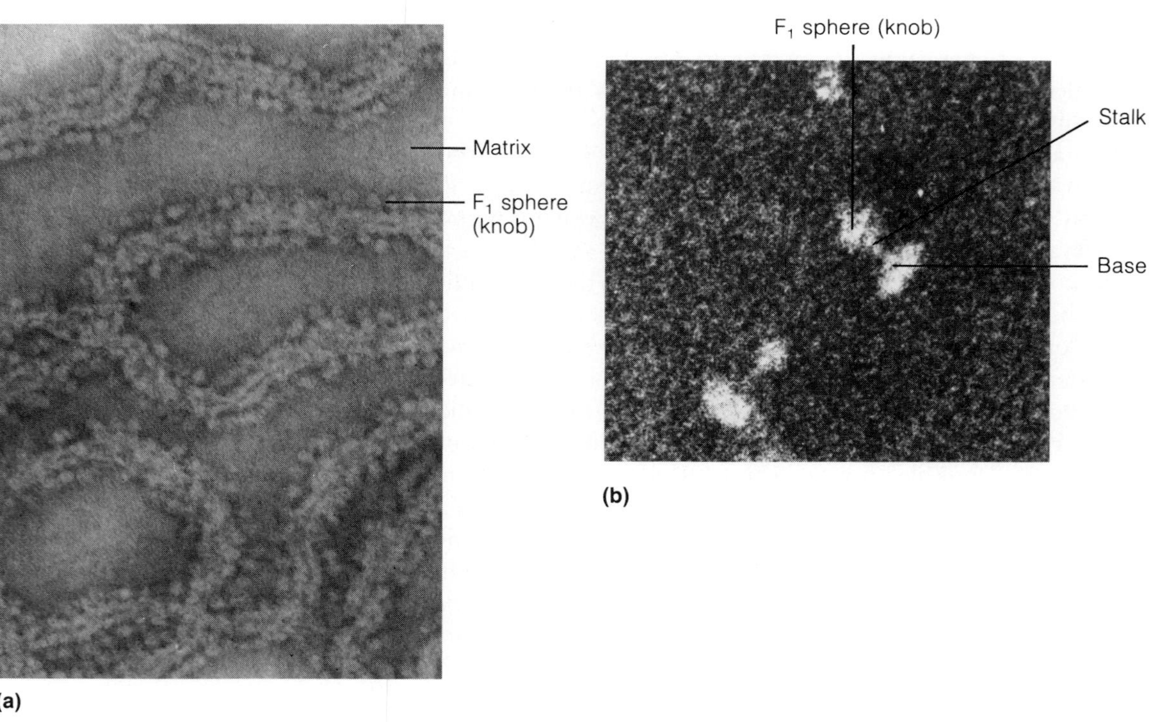

Figure 15.14
Fine structure of mitochondrial cristae, showing "knobs" in negatively stained preparations. (**a**) A portion of bovine heart mitochondrial inner membrane, showing knoblike projections along the matrix side of the membrane. (**b**) Purified preparations of knobs from rat liver mitochondria. The knob is attached by a short stalk to a base, which is embedded in the inner membrane of intact mitochondria.

One of the stalk proteins is the site of binding of the inhibitor **oligomycin.** The stalked knobs, which are called **F_0F_1 complexes,** have an ATPase activity in vitro, as does factor F_1 alone. It is virtually certain, from the reconstitution experiments mentioned above, that the true function of this activity in vivo is the reverse of ATPase, that is, synthesis of ATP from ADP and P_i. Therefore, either **ATP synthase** or **F_0F_1 coupling factor** is a suitable term to describe the stalked knobs. As we shall see in a later section, F_0 contains a specific channel, or pore, for passage of H^+ ions that have been pumped out of the mitochondrion by electron transport. Energy from this proton transport is used to drive ATP synthesis.

Now let us return to the identification of coupling sites for oxidative phosphorylation. Racker found that isolated submitochondrial respiratory complexes (complex I, complex II, etc.) could be reconstituted by sonic oscillation into artificial membranes (liposomes) containing purified phospholipids. When F_1 knobs were included in the sonication mixture, they were also incorporated into the vesicles. Each preparation had the electron transport properties of the original complex, plus a phosphorylation activity. For reconstituted complexes I, III, and IV, the P/O ratio was 1. For example, reconstituted complex III could transfer electrons from coenzyme Q to cytochrome *c*, with concomitant synthesis of 1 mol of ATP per pair of reducing equivalents. This type of evidence showed that respiratory complexes I, III, and IV each contains one coupling site. By contrast, complex II (succinate dehydrogenase) showed no ATP synthesis, confirming the absence of a coupling site in this complex.

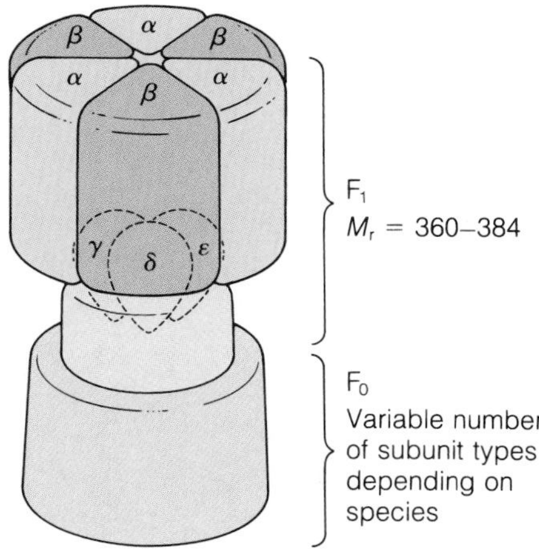

Figure 15.15
Structure of the F_0F_1 ATP synthase. The F_1 knob contains three $\alpha\beta$ subunit dimers and one copy each of subunits γ, δ, and σ. The subunit composition of the F_0 base is much more highly variable.

Respiratory States, Respiratory Control

Like any metabolic process, oxidative phosphorylation can occur only in the presence of adequate quantities of its substrates; it is controlled not by allosteric mechanisms, but simply by substrate availability. Those substrates include ADP, P_i, O_2, and an oxidizable metabolite that can generate reduced electron carriers—NADH and/or $FADH_2$. Under different metabolic conditions any one of these three substrates can limit the rate of oxidative phosphorylation.

The dependence of oxidative phosphorylation on ADP reveals an important general feature of this process: *respiration is tightly coupled to the synthesis of ATP.* Not only is ATP synthesis absolutely dependent on continued electron flow from substrates to oxygen, but in normal mitochondria electron flow occurs only when ATP is being synthesized as well. This regulatory phenomenon, called **respiratory control,** makes biological sense, because it ensures that substrates will not be oxidized wastefully; their utilization is controlled by the physiological need for ATP.

In most aerobic cells the level of ATP exceeds that of ADP by factors varying from 4 to 10. Thus, it is convenient to think of respiratory control as a dependence of respiration on ADP as a substrate for phosphorylation. If the energy demands on a cell cause ATP to be consumed at high rates, the resultant accumulation of ADP will stimulate respiration, with concomitant activation of ATP resynthesis. Conversely, in a relaxed and well-nourished cell, ATP accumulates at the expense of ADP, and the depletion of ADP limits the rate of both electron transport and its own phosphorylation to ATP. Thus, the energy-generating capacity of the cell is closely attuned to its energy demands. The energy state of the cell also has profound effects on the structure of the mitochondrion. In actively respiring mitochondria the inner membrane seems to fold on itself, leaving a greatly enlarged intermembrane

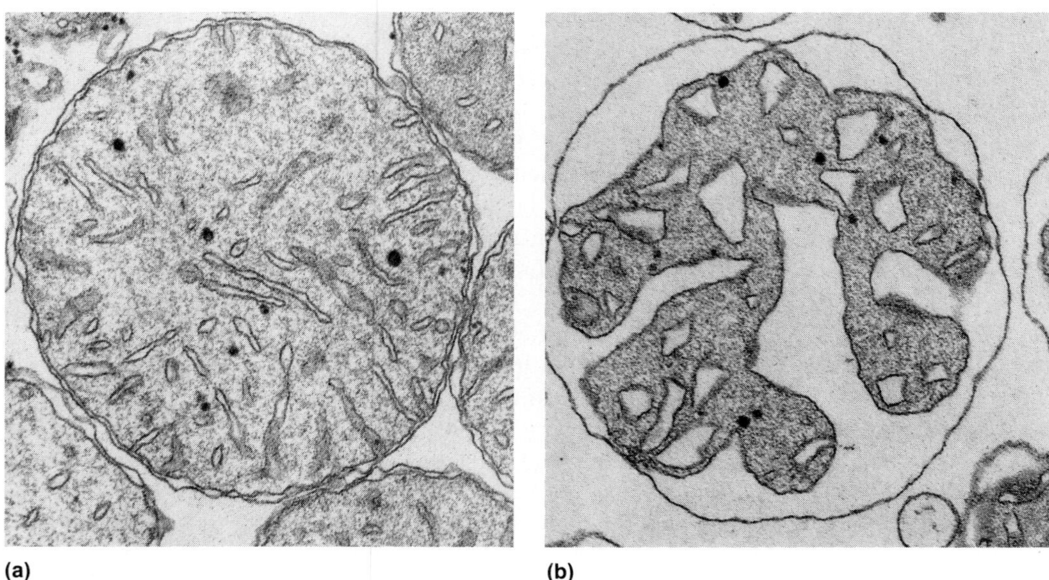

Figure 15.16
Morphological changes as rat liver mitochondria go from the resting state to a state of high respiration. (**a**) In resting mitochondria most of the internal space is occupied by matrix. (**b**) In the high-energy state the intermembrane space expands as the matrix volume decreases.

space (Figure 15.16). Experimentally, respiratory control is demonstrated by following oxygen utilization in isolated mitochondria (Figure 15.17). In the absence of added substrate or ADP, oxygen uptake, caused by oxidation of endogenous substrates, is slow. Addition of an oxidizable substrate, such as glutamate or malate, has but a small effect on the respiration rate. Subsequent addition of ADP stimulates respiration in a stoichiometric fashion; that is, oxygen uptake proceeds at an enhanced rate until all of the added ADP has been converted to ATP, and then returns to the basal rate. Addition of twice as much ADP causes twice the amount of oxygen uptake at the enhanced rate. If excess ADP is present instead of oxidizable substrate, the addition of substrate in limiting amounts will stimulate oxygen uptake until the substrate is exhausted.

Maintenance of respiratory control depends on the structural integrity of the mitochondrion. Disruption of the organelle causes electron transport to become uncoupled from ATP synthesis; under these conditions oxygen uptake proceeds at high rates even in the absence of added ADP. ATP synthesis is inhibited, even though electrons are being passed along the respiratory chain and used to reduce O_2 to water. Before carrying out experiments with freshly isolated mitochondria, biochemists usually ascertain that their mitochondria are tightly coupled, by determining rates of oxygen uptake in the presence and absence of added ADP. In carefully prepared mitochondria the ratio of these two O_2 uptake rates may be as high as 10. By contrast, aged or disrupted mitochondria may yield ratios as low as 1, showing an absence of coupling.

NO₂ / NO₂ / OH

2,4-Dinitrophenol (DNP)

F_3CO—C₆H₄—NH—N=C(CN)₂

Trifluorocarbonylcyanide phenylhydrazone (FCCP)

Uncoupling of respiration from phosphorylation can also be achieved chemically. A group of compounds, including **2,4-dinitrophenol** and **trifluorocarbonylcyanide phenylhydrazone,** are known as uncoupling agents. As shown in Figure 15.18, addition of an uncoupler to mitochondria stimulates oxygen utilization even in the absence of added ADP; obviously, no phosphorylation occurs under these conditions because there is no ADP to be phosphorylated. Another group of compounds, exemplified by the antibiotic oligomycin, act as inhibitors of oxidative phosphorylation. Addition of oligomycin to actively respiring, well-coupled mitochondria inhibits both oxygen uptake and ATP synthesis. However, there is no direct inhibition of oxygen transport, because the further addition of an uncoupler such as DNP greatly stimulates oxygen uptake. As mentioned in the previous section, oligomycin acts by direct inhibition of the F_0F_1 ATP synthase.

Under some natural conditions the ability to uncouple respiration from phosphorylation is highly desirable. Many mammals, particularly those born hairless, those that hibernate, and those that are cold-adapted, have special needs for maintenance of body temperature. Such animals have a special tissue, called brown fat, in the neck and upper back. Mitochondria in this tissue are specialized to generate heat from fat oxidation, uncoupled from phosphorylation. These mitochondria are especially rich in respiratory electron carriers, particularly cytochromes, which give this tissue its brown color. A comparable phenomenon is seen in the plant world among species that emerge in early spring, often when the ground is still covered with snow. The floral spike of the skunk cabbage is a particularly dramatic example; this tissue can maintain a temperature some 10 to 25° above ambient.

Figure 15.17
Experimental demonstration of respiratory control. Oxygen uptake is monitored in carefully prepared mitochondria. The addition of an exogenous oxidizable substrate, glutamate, stimulates respiration only slightly, unless ADP is added as well. Both ADP additions represent limiting amounts, with the second addition being twice the amount of the first, to show that the magnitude of oxygen uptake is stoichiometric. The slow oxygen uptake at the beginning results from endogenous substrates in the mitochondrion. ADP stimulates repiration only until it has all been converted to ATP. Oxygen uptake is recorded in μatoms (one half of a μmole of O_2), because one pair of electrons reduces one atom of O, not one molecule of O_2.

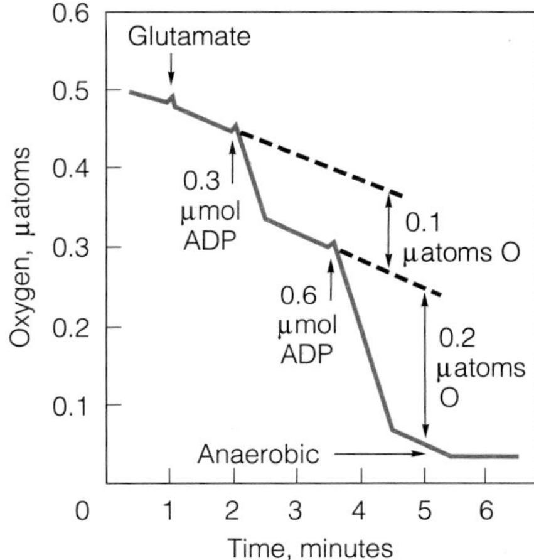

Mechanism of Oxidative Phosphorylation: Chemiosmotic Coupling

What is the mechanism by which energy released from respiration is harnessed to drive the synthesis of ATP? For the past four decades this question has been among the most controversial in all of biology. Even though most biochemists are agreed on the overall outlines of the answer, important unanswered questions remain.

Initial mechanistic interest in oxidative phosphorylation focused on a **chemical coupling** mechanism in which the energy released would be used directly for the synthesis of an energy-rich intermediate, as seen in substrate-level phosphorylations. However, such activated intermediates have never been demonstrated in oxidative phosphorylation.

An alternative model was the **conformational coupling** hypothesis. This theory explained the absence of chemically activated intermediates by proposing that energy from biological oxidations led one or more respiratory proteins to assume a thermodynamically unstable conformation. As the activated protein returned to its most stable conformation, the free energy release would be used to drive the synthesis of ATP. Again, the existence of conformationally activated respiratory proteins has not been demonstrated.

Today there is widespread acceptance of a third model, called **chemiosmotic coupling,** proposed in 1961 by British biochemist Peter Mitchell. In its most basic form, this model proposes that energy from electron transport drives an active transport system, which causes efflux of protons from the mitochondrial matrix; in other words, respiration "pumps" protons out of the matrix. This action generates an electrochemical gradient for protons, with a lower pH value outside the inner mitochondrial membrane than inside. The protons on the outside have a thermodynamic tendency to flow back in, so as to equalize pH on both sides of the membrane. Another way to state this is that free energy must be expended to maintain the proton

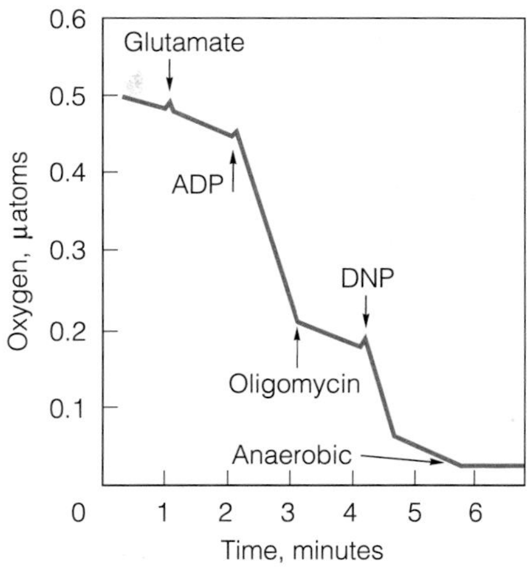

Figure 15.18
Effects of uncouplers and inhibitors of oxidative phosphorylation on oxygen uptake in isolated mitochondria. Excess ADP is added, but the addition of oligomycin inhibits phosphorylation and consequently slows respiration. Dinitrophenol (DNP) uncouples respiration from phosphorylation, so that O_2 uptake is stimulated even in the presence of oligomycin.

gradient. When protons do flow back into the matrix, that energy is dissipated, some of it being used to drive the synthesis of ATP. Although this model was fiercely resisted at first, overwhelming evidence has now accumulated in its support and Mitchell's achievements were recognized in 1978 with a Nobel Prize.

Let us now look at the chemiosmotic theory in more detail, as well as the experimental evidence supporting it. First, recall that some but not all of the reactions of electron transport transfer hydrogen ions (protons) as well as electrons; these reactions include the dehydrogenations of NADH, $FMNH_2$, and reduced coenzyme Q. Mitchell proposed that the enzymes catalyzing these dehydrogenations are asymmetrically arranged in the inner membrane, such that protons are always taken up from inside the matrix and released outside the inner membrane. Figure 15.19 shows how this might occur. This **proton pumping** by respiratory proteins results in conversion of the energy of respiration to osmotic energy, in the form of an electrochemical gradient, or a gradient of chemical concentration that establishes an electrical potential. The energy released from discharging this gradient can be coupled with the phosphorylation of ADP to ATP, with no isolatable intermediates being formed. This process involves the F_0F_1 ATP synthase. The F_0 portion of the complex spans the inner membrane and is thought to contain a specific channel for return of protons to the mitochondrial matrix. The free energy released as H^+ traverses this channel to return to the matrix is somehow harnessed to drive the synthesis of ATP, catalyzed by the F_1 component of the complex. This concept was introduced in Chapter 9.

Next we explore the experimental evidence for chemiosmotic coupling in some detail—partly because of its importance to an understanding of oxidative phosphorylation itself and partly because it provides insight into other important biological processes, including active transport and photosynthesis.

MEMBRANES CAN ESTABLISH PROTON GRADIENTS. When it became possible to measure changes in pH and electrical potential across mitochondrial membranes, it became clear that mitochondria can pump protons from the matrix to the intermembrane space. In fact, the pH value outside an actively respiring mitochondrion is about 1.4 units lower than in the matrix. The pH gradient also generates an electrical potential of about 0.14 V across the

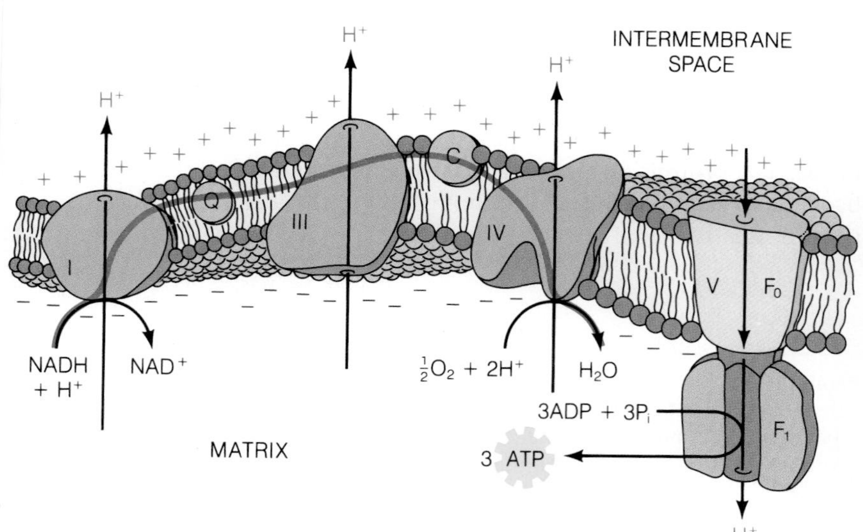

Figure 15.19
Vectorial transport of protons by complexes of the respiratory chain. It is likely that protons pumped by complexes I, III, and IV are not those removed from electron carriers that become oxidized. Proton reentry to the matrix, through the F_0 channel (complex V), provides the energy to drive ATP synthesis.

membrane, because of the net movement of positively charged protons outward across the inner membrane. The pH gradient and the membrane potential both contribute to an **electrochemical H^+ gradient,** or **proton motive force** as shown here and also in Chapter 9 (equation 9.2).

$$\underset{\text{Electrochemical } H^+ \text{ gradient}}{\Delta\mu_H} = \underset{\text{Membrane potential}}{\Delta\psi} - \underset{\text{pH gradient}}{2.3RT\,\Delta pH/F} \qquad (15.7)$$

$\Delta\mu_H$ is also called Δp, the proton motive force, or pmf. ΔpH has a negative value (−1.4) because it is measured from inside to outside; thus, the contribution of the pH gradient is positive: $-2.3RT(-1.4)/F$, which equals +0.084 V, and the $\Delta\mu_H$ is 0.14 + 0.084, or +0.224 V. From the relationship between standard reduction potential and $\Delta G°'$, we see that this value corresponds to a standard free energy change of about 21 kJ per mole of protons, which would generate about 42 kJ per pair of reducing equivalents. This is in the right range to drive the synthesis of 1 mol of ATP, although it is not universally accepted that only two protons are pumped at each coupling site. In fact, recent evidence suggests that the number of protons translocated per pair of reducing equivalents is as high as 12.

Comparable experiments have shown that both bacterial membranes and chloroplast thylakoid membranes pump protons, as does the mitochondrial inner membrane. This points up the fact that electrochemical proton gradients are used in energy transactions other than oxidative phosphorylation. Bacterial membranes transduce energy in this way both for oxidative phosphorylation and to drive flagellar motors that allow movement of the cell (Chapter 29). Proton pumping across the chloroplast thylakoid membrane drives ATP synthesis in photophosphorylation (Chapter 19). Proton gradients also drive active transport (Chapter 9), as well as two processes mentioned earlier in this chapter—heat production and the synthesis of NADPH by transhydrogenase.

AN INTACT INNER MEMBRANE IS REQUIRED FOR OXIDATIVE PHOSPHORYLATION. When the continuity of the membrane is interrupted, for example, by sonic oscillation, the resultant particles can carry out electron transport but not ATP synthesis. The necessity of an intact membrane for maintenance of a membrane potential is consistent with the idea that a proton gradient is essential for oxidative phosphorylation.

KEY ELECTRON TRANSPORT PROTEINS SPAN THE INNER MEMBRANE. The chemiosmotic theory states that protons are transported from the matrix across the inner membrane. If the respiratory proteins are to serve as proton pumps, then the electron carriers that carry protons, such as NADH dehydrogenase, should be in contact with both the inner and outer sides of the membrane. Moreover, these carriers should be asymmetrically located in the membrane, to account for transport in one direction—outward. Asymmetric location has been demonstrated by the use of agents that react with respiratory proteins but cannot themselves traverse the membrane, such as antibodies, proteolytic enzymes, or labeling reagents. Treatment of intact mitochondria with such reagents allows detection of proteins located at the outer surface of the inner membrane, while reaction with membrane vesicles allows access to the inner, or matrix, side. Such approaches have shown, for example, that the cytochrome oxidase complex binds to cytochrome *c* only on the cytosolic side. Moreover, 9 of the 13 subunits can be labeled from only one side or the other, indicating asymmetric placement of the complex in the membrane. Similar findings have now been made for the proteins of complexes I, II, and IV.

UNCOUPLERS ACT BY DISSIPATING THE PROTON GRADIENT. Recall that uncouplers of oxidative phosphorylation, such as dinitrophenol, allow respiration but block ATP synthesis. The pK_a of the phenolic hydroxyl group in DNP is such that it is normally dissociated at intracellular pH. However, a DNP molecule that approaches the inner membrane becomes protonated, because of the lower pH value in this vicinity. This increases the hydrophobicity of DNP, allowing it to diffuse into the membrane and by mass action to pass through. Once in the matrix, the higher pH causes the phenolic hydroxyl to deprotonate. Thus, the uncoupler has the effect of transporting H^+ back into the mitochondrion, bypassing the F_0 proton channel and thereby preventing ATP synthesis.

The phosphorylation inhibitor oligomycin acts quite differently. By binding to a specific protein of the F_0 complex, it blocks the flow of protons through the F_0 proton channel and hence inhibits oxidative phosphorylation directly. Proton pumping by the respiratory carriers is inhibited as well, but this is a secondary consequence of respiratory control; electron transport is diminished by the decreased rate of ATP synthesis.

To summarize these effects: an inhibitor such as DNP dissipates the proton gradient by transporting protons into mitochondria at sites other than the F_0 proton channel, whereas oligomycin blocks the F_0 channel so that the pH gradient cannot be dissipated at all.

Extensive data on the transport of other ions confirm that a functionally intact membrane is essential to oxidative phosphorylation. The antibiotic valinomycin (see Chapter 9) is an example of an **ionophore** ("ion carrier"). This lipid-soluble compound forms a specific complex with potassium ion. Since the complex is lipophilic and can diffuse into the membrane, just as protonated DNP does, valinomycin brings about the transport of K^+ through the inner membrane in much the same sense that DNP transports protons. Valinomycin acts by decreasing the $\Delta\psi$ component of the pmf, without a direct effect on the pH gradient. Another antibiotic, **nigericin,** acts as a K^+/H^+ **antiport;** it carries H^+ in one direction, coupled with the reverse transport of K^+. Thus, nigericin dissipates the ΔpH component of the pmf, with little effect on $\Delta\psi$. Neither compound alone is a particularly effective uncoupler of oxidative phosphorylation, but in combination both elements of the pmf are collapsed, and ATP synthesis is effectively inhibited.

GENERATION OF A PROTON GRADIENT PERMITS ATP SYNTHESIS WITHOUT ELECTRON TRANSPORT. Andre Jagendorf, while studying photosynthetic ATP production, provided important evidence for chemiosmotic coupling when he showed that ATP synthesis can proceed in chloroplast thylakoid membranes in the absence of electron transport, as long as a proton gradient is present. Chloroplasts were incubated at pH 4 for several hours and then quickly transferred to a buffer at pH 8. Thus, the inside of the thylakoid disk was at a lower pH than the outside (chloroplast membranes pump protons *inward,* not outward). Addition of ADP and P_i to these chloroplasts generated a simultaneous burst of ATP synthesis and dissipation of the pH gradient. Similar results have now been observed with mitochondria. These experiments show that the establishment of a proton gradient, even without a corresponding energy input, suffices to drive the synthesis of ATP.

A dramatic variation on Jagendorf's experiment involved a membrane protein, **bacteriorhodopsin,** from the photosynthetic bacterium *Halobacterium halobium.* Bacteriorhodopsin pumps protons when the bacteria are illuminated with light. Bacteriorhodopsin was isolated in a membrane-free form and then incorporated into synthetic vesicles, along with isolated liver mitochondrial F_0F_1 ATPase in the right orientation. When these vesicles were illuminated, ATP synthesis occurred, showing that phosphorylation can occur as a direct consequence of the formation of a proton gradient.

All of the above observations are consistent with the chemiosmotic theory of oxidative phosphorylation, and this model has now gained general acceptance. However, several important questions have yet to be answered. First, the mechanism of proton pumping has not been established. Mitchell proposed that the hydrogen atoms released from oxidized substrates became the effluxed protons. This would generate two protons per ATP molecule synthesized, and experiments suggest that three or four protons are transported per coupling site. Indeed, there is some question whether the pumping of two protons provides sufficient energy for ATP synthesis, given that $\Delta G'$ for ATP hydrolysis in the mitochondrion is higher than $\Delta G^{\circ\prime}$, the value determined under standard conditions. A closely related question is the source of the protons pumped by the cytochrome oxidase complex. None of the known electron carriers from cytochrome *b* onward transfers protons during its oxidation or reduction, except for O_2, which *consumes* protons rather than discharging them. Finally, what is the mechanism by which the potential energy of proton influx through the F_0 channel is harnessed to drive the synthesis of ATP by F_1? Recent studies show that ATP binds extremely tightly to isolated F_1 but that this affinity is greatly reduced when a proton gradient is generated. Since the reaction catalyzed by F_1 is known to be readily reversible, this suggests the release of newly synthesized ATP from the complex as the rate-limiting step in phosphorylation.

Mitochondrial Transport Systems

While the mitochondrial outer membrane is freely permeable to many substances, the permeability of the inner membrane is severely limited. The importance of this selective permeability should be evident from our discussions of electrochemical gradients and the shuttle systems used to transport reducing equivalents into the mitochondrion. We must also consider substrate transport, including the inward transport of intermediates for oxidation in the citric acid cycle, the export of intermediates used for biosynthesis in other cell compartments, and the exit of newly synthesized ATP. Properties of the principal transport systems are outlined in Figure 15.20.

First let us consider the substrates of oxidative phosphorylation, namely ATP, ADP, and orthophosphate. Two systems are involved, an **adenine nucleotide translocase** and a **phosphate translocase.**

The adenine nucleotide translocase spans the inner membrane, and it binds ADP to a specific site on the outer membrane surface. The protein couples the efflux of ATP from the matrix to the influx of an equivalent amount of ADP from the intermembrane space. Since this system exchanges ATP, with a charge of −4, for ADP, with a charge of −3, its action is driven by the proton gradient established through action of the F_0F_1 ATP synthase. It is generally true of the mitochondrial transport systems that at least one of the participants is moving down a concentration gradient, so that no further energy source is required. The adenine nucleotide translocase is inhibited by **atractyloside,** a naturally occurring toxic plant glycoside.

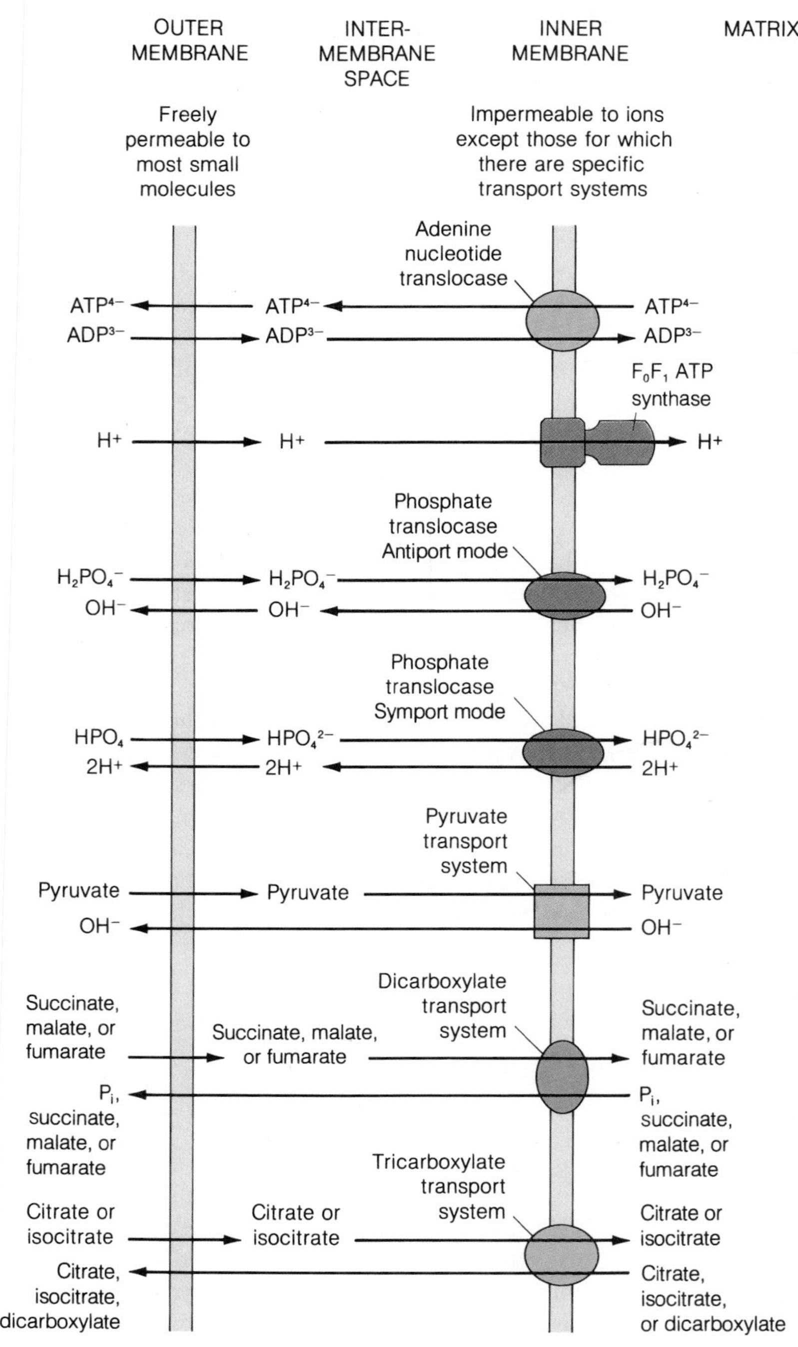

Figure 15.20
Major inner membrane transport systems for respiratory substrates and products. The phosphate translocase and the adenine nucleotide translocase move substrates for oxidative phosphorylation (ADP and P_i) into the mitochondrion and the product (ATP) out. The other transport systems move substrates and products for citric acid cycle oxidation into or out of the matrix, as dictated by the metabolic needs of the cell.

The phosphate translocase acts in either the *antiport* or the *symport* mode. As an antiporter it transports $H_2PO_4^-$ into the mitochondrion, coupled with the efflux of a hydroxide ion. In the alternative symport mode it transports HPO_4^{2-} into the matrix, along with two protons. Both modes of transport maintain electrical neutrality. The net effect of the adenine nucleotide and phosphate transport systems is to couple the inward transport of the substrates of oxidative phosphorylation, ADP and P_i, to the efflux of the product, ATP.

Next, let us consider the substrates for oxidation. The major substrate from carbohydrate catabolism is pyruvate, which, like phosphate, is exchanged for OH^-. Dicarboxylic acid substrates, namely succinate, fumarate, and malate, can be exchanged for each other or for orthophosphate in the dicarboxylate transport system. Similarly, the tricarboxylate transport system carries either citrate or isocitrate, coupled each with the other or with a dicarboxylic acid. The influx of fatty acids for β-oxidation involves an additional transport system, which is presented in Chapter 17.

In addition to the transport systems shown in Figure 15.20, there is an important transport system for calcium ion. Ca^{2+} regulates a number of metabolic processes in the cytosol, and its release from mitochondrial stores may represent one way to change the cytosolic concentration of Ca^{2+}. Calcium is transported inward by the membrane potential, which is more negative inside. It is effluxed by exchange with H^+ from the outside or, in heart mitochondria, for cytosolic H^+. Control of the efflux process helps to maintain cytosolic Ca^{2+} at very low levels.

Energy Yields from Oxidative Metabolism

Much of the past three chapters has been devoted to the pathways by which carbohydrates are oxidized to CO_2 and water. Finally, we are in a position to calculate the total energy yield and metabolic efficiency of these combined pathways. Let us review how much energy is recovered in the form of ATP from the entire oxidative catabolism of glucose. First we present a balanced equation for each of the three pathways involved, and then we estimate the amount of ATP that can be derived through oxidative phosphorylation from the reduced electron carriers.

Glycolysis:

$$\text{Glucose} + 2\text{ADP} + 2\text{P}_i + 2\text{NAD}^+ \longrightarrow 2\text{ pyruvate} + 2\text{ATP} + 2\text{NADH} + 2\text{H}_2\text{O} + 2\text{H}^+$$

Pyruvate dehydrogenase complex:

$$2\text{ Pyruvate} + 2\text{NAD}^+ + 2\text{CoA-SH} \longrightarrow 2\text{ acetyl-CoA} + 2\text{NADH} + 2\text{CO}_2$$

Citric acid cycle (including conversion of GTP to ATP):

$$2\text{ Acetyl-CoA} + 4\text{H}_2\text{O} + 6\text{NAD}^+ + 2\text{FAD} + 2\text{ADP} + 2\text{P}_i \longrightarrow 4\text{CO}_2 + 6\text{NADH} + 6\text{H}^+ + 2\text{FADH}_2 + 2\text{CoA-SH} + 2\text{ATP}$$

Net

$$\text{Glucose} + 10\text{NAD}^+ + 2\text{FAD} + 2\text{H}_2\text{O} + 4\text{ADP} + 4\text{P}_i \longrightarrow 6\text{CO}_2 + 10\text{NADH} + 8\text{H}^+ + 2\text{FADH}_2 + 4\text{ATP}$$

These three processes generate 4 mol of ATP directly, plus 10 mol of NADH and 2 mol of $FADH_2$. Although there is still some uncertainty about P/O ratios for oxidation of NADH and $FADH_2$, we shall use the values 3 and 2, respectively. Therefore, the total realizable ATP yield is 38 per mole of glucose oxidized. In prokaryotes and in cells using the glutamate–aspartate shuttle, these reducing equivalents are carried into the mitochondrion with no energy cost. However, cells using the glycerol phosphate shuttle must pay an energy cost. As described on p. 519, the electrons from cytosolic NADH enter the respiratory chain as $FADH_2$; therefore, the ATP yield from each of these two NADHs is 2, not 3. This decreases the overall ATP yield to 36. For this discussion we shall use 38 as the best estimate of total moles of ATP derived per mole of glucose oxidized. Recalling that $\Delta G^{\circ\prime}$ for glucose oxidation is -2870 kJ/mol and that for ATP hydrolysis is -30.5 kJ/mol, we can calculate the efficiency for operation of this biochemical machine: $(38 \times 30.5)/2870$, or 40%, under standard conditions. As noted earlier, the efficiency in vivo is probably higher.

Oxygen as a Substrate for Other Metabolic Reactions

In most cells at least 90% of the molecular oxygen consumed is utilized for oxidative phosphorylation. The remaining O_2 is used in a wide variety of specialized metabolic reactions. At least 200 known enzymes use O_2 as a substrate, and in this section we shall briefly categorize these enzymes. We also consider the metabolism of partially reduced forms of oxygen, which arise continually in all cells and are highly toxic because of their great reactivity.

Oxidases and Oxygenases

The term **oxidase** is applied to enzymes that catalyze the oxidation of a substrate by O_2 without incorporation of oxygen into the product. Since a two-electron oxidation is usually involved, the oxygen is converted to H_2O_2. Most oxidases utilize either a metal or a flavin coenzyme. D-Amino acid oxidases, for example, use FAD as a cofactor.

$$\mathrm{R{-}CH(\overset{+}{N}H_3){-}COO^-} + H_2O + \mathrm{FAD} \longrightarrow \mathrm{R{-}C({=}O){-}COO^-} + \overset{+}{N}H_4 + \mathrm{FADH_2}$$

$$\mathrm{FADH_2} + O_2 \longrightarrow \mathrm{FAD} + H_2O_2$$

Oxygenases are enzymes that incorporate oxygen from O_2 into the oxidized products. **Dioxygenases,** which incorporate both atoms of O_2 into one substrate, are of limited distribution. An example is tryptophan 2,3-dioxygenase which catalyzes the first reaction in tryptophan catabolism (below). This enzyme contains a heme cofactor.

$$\text{Tryptophan} \xrightarrow{O_2} N\text{-Formylkynurenine}$$

Tryptophan

N-Formylkynurenine

Far more widely distributed are **monooxygenases,** which incorporate one atom from O_2 into a product, the other atom being reduced to water. A monooxygenase has one substrate that accepts oxygen and another that furnishes two H atoms that reduce the other oxygen to water. Because two substrates are oxidized, this class of enzymes is also called **mixed-function oxidases.** The general reaction catalyzed by these enzymes is as follows.

$$\mathrm{AH} + \mathrm{BH_2} + \mathrm{O{-}O} \longrightarrow \mathrm{A{-}OH} + \mathrm{B} + H_2O$$

Since the substrate AH usually becomes hydroxylated by this class of enzymes, the term **hydroxylase** is also used. An example of this type of reaction is the hydroxylation of steroids. Here NADPH is the reductive cofactor, BH_2.

$$RH + NADPH + H^+ + O_2 \longrightarrow R{-}OH + NADP^+ + H_2O$$

Several compounds other than NADPH function as BH_2 in monooxygenase reactions, including the following examples, all of which we discuss in Chapter 21: α-ketoglutarate, in the hydroxylation of collagen proline residues to hydroxyproline; ascorbate, in the hydroxylation of dopamine in catecholamine biosynthesis; a methene bridge, in the degradation of heme; and a reduced heterocyclic compound, tetrahydrobiopterin, involved in hydroxylations of aromatic amino acids.

Cytochrome P-450

The most numerous hydroxylation reactions involve a family of heme proteins with the collective name **cytochrome P-450.** This family, which numbers in the hundreds, resembles mitochondrial cytochrome oxidase in being able to bind both O_2 and carbon monoxide. Cytochromes P-450 are distinctive, however, in that the reduced form of the heme, when complexed with carbon monoxide, absorbs light strongly at 450 nm. Cytochromes P-450 are usually found in the endoplasmic reticulum of eukaryotic cells, not in the mitochondria.

Cytochromes P-450 are involved in hydroxylating a large variety of compounds—hydrocarbon groups of fatty acids, steroid rings, and many xenobiotics (foreign compounds), including drugs such as phenobarbital and environmental carcinogens such as benzpyrene, a constituent of tobacco smoke. Hydroxylation of foreign substances usually increases their solubility and is a step in their detoxifications, or metabolism and excretion. However, cytochrome P-450-catalyzed hydroxylations are also involved in the activation of potentially carcinogenic substances to more reactive species. Typically, in one of these hydroxylations the cytochrome carries reducing equivalents from NADH or NADPH, which in the overall reaction functions as BH_2, as discussed above. A general mechanism for an NADPH-dependent hydroxylation reaction is shown in Figure 15.21. Cytochrome P-450 systems in liver participate in a wide variety of additional reactions, including epoxidations, dealkylations, deaminations, and dehalogenations.

A number of cytochromes P-450 are inducible in liver and other tissues; the synthesis of specific members of the family is stimulated by substrates that are metabolized by these enzymes. Inducers include drugs such as phenobarbital and other barbiturates.

Metabolism of Incompletely Reduced Oxygen

The complete reduction of one molecule of O_2 to water is a four-electron process. Oxidative metabolism continually generates partially reduced species of oxygen, which are far more reactive, and hence potentially more toxic, than O_2 itself. A one-electron reduction of O_2 yields **superoxide** ion, O_2^-; an additional electron yields hydrogen peroxide, H_2O_2; and a third electron yields a hydroxyl radical, OH·, and a hydroxide ion. Hydroxyl radicals in particular are extremely reactive and represent the most active mutagen derived from ionizing radiation. All of these species are generated intracellularly and must be converted to less reactive species if the organism is to survive.

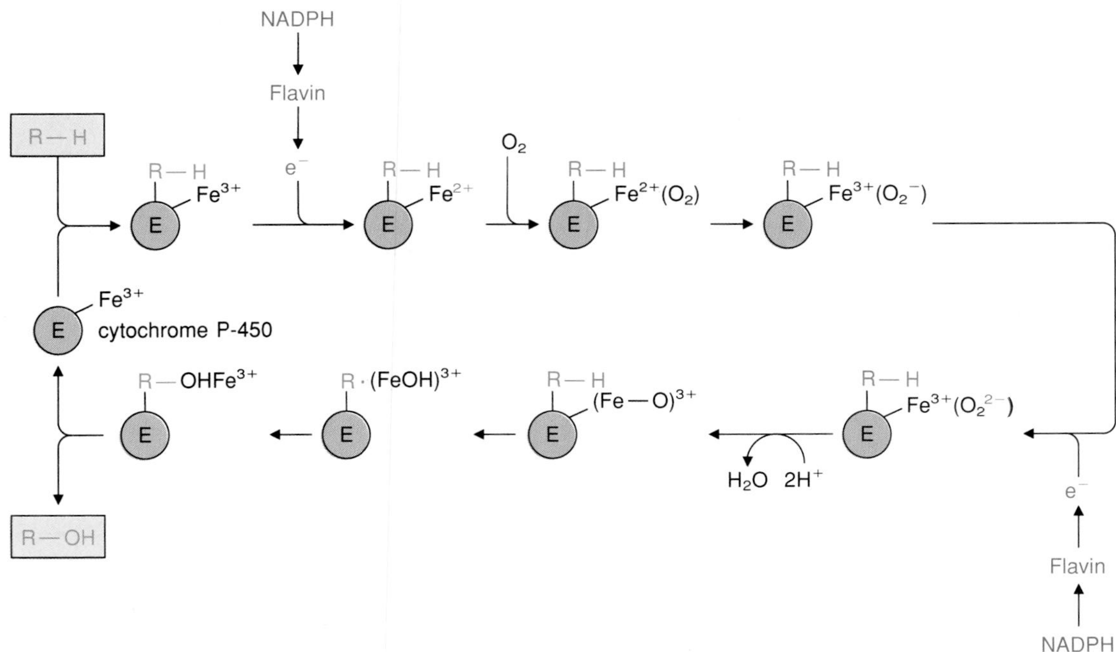

Figure 15.21
Scheme for an enzymatic hydroxylation involving NADPH and cytochrome P-450. A flavoprotein enzyme delivers electrons one at a time from NADPH for the two single-electron reductions depicted.

Thymine glycol

Increasing attention has focused on the toxicity of reactive oxygen species, the protective mechanisms employed by cells, and the consequences of incomplete protection. Bruce Ames, in particular, has championed the idea that oxidative damage is a key event in carcinogenesis and aging. Particularly intriguing are his measurements in urine of **thymine glycol,** an oxidation product of thymine residues in DNA. Ames has proposed the analysis of thymine glycol in urine as the basis for a simple test of the extent to which intracellular DNA has suffered oxidative damage.

Let us briefly consider the major enzyme systems used to protect cells against reactive oxygen species. The first line of defense is **superoxide dismutase,** a family of metalloenzymes that catalyze a **dismutation** (a reaction in which two identical substrate molecules have different fates). Here one molecule of superoxide is oxidized and one is reduced.

$$O_2^- + O_2^- + 2H^+ \longrightarrow H_2O_2 + O_2$$

A cytosolic form of this enzyme in eukaryotes contains copper and zinc; a manganese-containing form is found in both mitochondria and bacterial cells; and a related iron-containing form is found in bacteria, cyanobacteria, and some plants.

Hydrogen peroxide is metabolized either by **catalase,** a widely distributed enzyme, or by a more limited family of **peroxidases.** Catalase is a heme protein with an extremely high turnover rate (>40,000 molecules per second); it catalyzes the following reaction.

$$2H_2O_2 \longrightarrow 2H_2O + O_2$$

Peroxidases, which are widely distributed in plants, reduce H_2O_2 to water at the expense of oxidation of an organic substrate. An example of a peroxidase is found in erythrocytes, which are especially sensitive to peroxide accumulation. (Recall from Chapter 14 our discussion of the consequences of peroxide accumulation in glucose-6-phosphate dehydrogenase deficiency.) Within erythrocytes is found a selenium-containing enzyme, **glutathione peroxidase,** which reduces H_2O_2 to water, along with the oxidation of glutathione.

$$2GSH + H_2O_2 \longrightarrow GSSG + 2H_2O$$

Glutathione peroxidase is interesting in that it contains one residue per mole of an unusual amino acid, **selenocysteine,** an analog of cysteine that contains selenium in place of sulfur.

In some cases the production of reactive oxygen species is not a metabolic accident, but a normal part of the functioning of a cell. For example, certain white blood cells contribute to defense against infectious agents by phagocytosis. Such cells can literally engulf a bacterial cell. This event is followed by a **respiratory burst,** a rapid increase in oxygen uptake, stimulated by mechanisms still under study. Much of this oxygen is reduced to superoxide ion and to H_2O_2. These compounds contribute to killing of the engulfed bacterium. Thus, while most cells contain elaborate mechanisms to protect against toxicity of reactive oxygen species, the respiratory burst involves a deliberate and controlled production of these species.

The last three chapters have traced the path of carbon as it is catabolized from carbohydrate to CO_2. In the next chapter we consider biosynthetic processes, starting with those that lead to carbohydrate.

REFERENCES

Historical Background

Lehninger, A. L. (1965) *The Mitochondrion: Molecular Basis of Structure and Function.* Benjamin, New York. An account of the earlier work by one who contributed much to it.

Mitchell, P. (1979) Keilin's respiratory chain concept and its chemiosmotic consequences. *Science* 206:1148–1159. Mitchell's Nobel Prize address, in which he describes the genesis of the chemiosmotic theory of energy coupling.

General References

Ernster, L. (ed.) (1984) *Bioenergetics.* Elsevier, Amsterdam. A series of excellent review articles.

Tzagoloff, A. (1982) *Mitochondria.* Plenum, New York. A concise, well-illustrated book-length review of mitochondrial structure and function.

Yaffe, M., and G. Schatz (1984) The future of mitochondrial research. *Trends Biochem. Sci.* 9:179–182. The authors argue that researchers should concentrate on interrelationships between mitochondria and the rest of the cell, rather than considering mitochondria as isolated entities.

Mechanisms in Electron Transport

Hatefi, Y. (1985) The mitochondrial electron transport and oxidative phosphorylation system. *Annu. Rev. Biochem.* 54:1015–1069. An authoritative and up-to-date review.

Kowal, A. T., J. E. Morningstar, M. K. Johnson, R. R. Ramsay, and T. P. Singer (1986) Spectroscopic characterization of the number and type of iron–sulfur clusters in NADH:ubiquinone oxidoreductase. *J. Biol. Chem.* 261:9239–9245. A treatment of the use of electron paramagnetic resonance to describe the redox functions of iron–sulfur clusters.

Naqui, A., B. Chance, and E. Cadenas (1986) Reactive oxygen intermediates in biochemistry. *Annu. Rev. Biochem.* 55:137–166. A detailed presentation of the structure and mechanism of cytochrome oxidase.

Mechanisms in Oxidative Phosphorylation

Boyer, P. D. (1987) The unusual enzymology of ATP synthase. *Biochemistry* 26:8503–8507. This minireview thoughtfully relates the unusual subunit structure of the F_0F_1 complex to its function.

Chernyak, B. V., and I. A. Kozlov (1986) Regulation of H^+-ATPases in oxidative and photophosphorylation. *Trends Biochem. Sci.* 11:32–35. When the electrochemical proton gradient is low, what prevents F_1 ATPase from running in reverse and hydrolyzing ATP in a futile cycle? Discussed in this review.

Ferguson, S. J. (1986) The ups and downs of P/O ratios. *Trends Biochem. Sci.* 11:351–353. Discusses the complexities of predicting true P/O ratios in light of new mechanistic information about oxidative phosphorylation.

Pedersen, P. L., and E. Carafoli (1987) Ion-motive ATPases. I. Ubiquity, properties, and significance to cell function. *Trends Biochem. Sci.* 12:146–150. A review of the structure of the F_0F_1 ATPase.

Prince, R. C. (1988) The proton pump of cytochrome oxidase. *Trends Biochem. Sci.* 13:159–160. A minireview that asks the source and number of protons pumped in this last step of electron transport.

Oxygen Metabolism

Ames, B. N. (1983) Dietary carcinogens and anticarcinogens. *Science* 221:1256–1264. One of the earliest explicit proposals that reactive oxygen species are involved in mutation, cancer, and aging and that antioxidants in natural foods might counteract these effects.

Babior, B. M. (1987) The respiratory burst oxidase. *Trends Biochem. Sci.* 12:241–243. A minireview of superoxide production as an antibacterial defense mechanism.

Cerutti, P. (1985) Prooxidant states and tumor promotion. *Science* 227:375–381. Reviews the evidence that the reactivity of oxygen and its partially reduced derivatives is causally related to cancer.

Malmstrom, B. (1982) Enzymology of oxygen. *Annu. Rev. Biochem.* 51:21–59. A review of enzymes that use oxygen as a substrate.

White, R. E., and M. J. Coon (1980) Oxygen activation by cytochrome P450. *Annu. Rev. Biochem.* 49:315–356. An excellent review of mechanisms of action of cytochromes P-450.

PROBLEMS

1. Referring to Table 15.1 for E_0 values, calculate $\Delta G^{\circ\prime}$ for oxidation of malate by malate dehydrogenase.

2. When pure reduced cytochrome *c* is added to carefully prepared mitochondria along with ADP, P_i, antimycin A, and oxygen, the cytochrome *c* becomes oxidized, and ATP is formed, with a P/O ratio approaching 1.0.
   (a) Indicate the probable flow of electrons in this system.
   (b) Why was antimycin A added?
   (c) What does this experiment tell you about the location of coupling sites for oxidative phosphorylation?
   (d) Write a balanced equation for the overall reaction.
   (e) Calculate $\Delta G^{\circ\prime}$ for the above reaction, using E_0' values from Table 15.1 and a $\Delta G'$ value for ATP hydrolysis of about 40 kJ/mol under conditions in the mitochondrion.

3. Freshly prepared mitochondria were incubated with β-hydroxybutyrate, oxidized cytochrome *c*, ADP, P_i, and cyanide. β-Hydroxybutyrate is oxidized by an NAD^+-dependent dehydrogenase.

COO^- | CH_2 | H—C—OH | CH_3 → (with NAD^+ → NADH + H^+) → COO^- | CH_2 | C=O | CH_3

   The experimenter measured the rate of oxidation of β-hydroxybutyrate and the rate of formation of ATP.
   (a) Indicate the probable flow of electrons in this system.
   (b) How many moles of ATP would you expect to be formed per mole of β-hydroxybutyrate oxidized in this system?
   (c) Why is β-hydroxybutyrate added rather than NADH?
   (d) What is the function of the cyanide?
   (e) Write a balanced equation for the overall reaction occurring in this system.
   (f) Calculate the net free energy change ($\Delta G^{\circ\prime}$) in this system, using E_0' values from Table 15.1 and $\Delta G'$ for hydrolysis of ATP of about 40 kJ/mol in the mitochondrion.

4. If you were to determine the P/O ratio for oxidation of α-ketoglutarate, you would probably include some malonate in your reaction system. Why? Under these conditions, what P/O ratio would you expect to observe?

5. Of the various oxidation reactions in glycolysis and the citric acid cycle, the only one that does not involve NAD^+ is the succinate dehydrogenase reaction. What would $\Delta G^{\circ\prime}$ be for an enzyme that oxidizes succinate with NAD^+ instead of FAD? If the intramitochondrial concentration of succinate was 10-fold higher than that of fumarate, what minimum $[NAD^+]/[NADH]$ ratio in mitochondria would be needed to make this reaction exergonic?

6. Intramitochondrial ATP concentrations are about 5 mM, and phosphate concentration is about 10 mM. If ADP is fivefold more abundant than AMP, calculate the molar concentrations of ADP and AMP at an energy charge of 0.85. Calculate $\Delta G'$ for ATP hydrolysis under these conditions.

7. From E_0' values in Table 15.1, calculate the equilibrium constant for the glutathione peroxidase reaction.

8. In the early days of "mitochondriology," P/O ratios were determined from measurements of volume of O_2 taken up by respiring mitochondria and chemical assays for disappearance of inorganic phosphate. Now, however, it is possible to measure P/O ratios simply by measurements with a recording oxygen electrode. How might this be done?

9. Years ago there was interest in using uncouplers such as dinitrophenol as weight control agents. Presumably, fat could be oxidized without concomitant ATP synthesis for reformation of fat or carbohydrate. Why was this a bad idea?

	State 1	State 2	State 3	State 4	State 5
O_2 availability	Aerobic	Aerobic	Aerobic	Aerobic	Anaerobic
ADP level	Low	High	High	Low	High
Oxidizable substrate	Endogenous (low)	Near zero	High	High	High
Respiration rate	Slow	Slow	Fast	Slow	Zero
Rate-limiting component	ADP	Oxidizable substrate	Electron transport	ADP	O_2

10. Biochemists working with isolated mitochondria recognize five energy "states" of mitochondria, depending on the presence or absence of essential substrates for respiration—O_2, ADP, oxidizable substrates, and so forth. The characteristics of each state are given in the accompanying table.

    (a) On the graph, identify a state that might predominate under each set of conditions indicated with a letter.

    (b) To determine whether isolated mitochondria exhibit respiratory control, one determines the ratio of rates of oxygen uptake in two different states. Which states?

    (c) Which state probably predominates in vivo in skeletal muscle fatigued from a long and strenuous workout?

    (d) Which state probably predominates in resting skeletal muscle of a well-nourished animal?

    (e) Which state probably predominates in heart muscle most of the time?

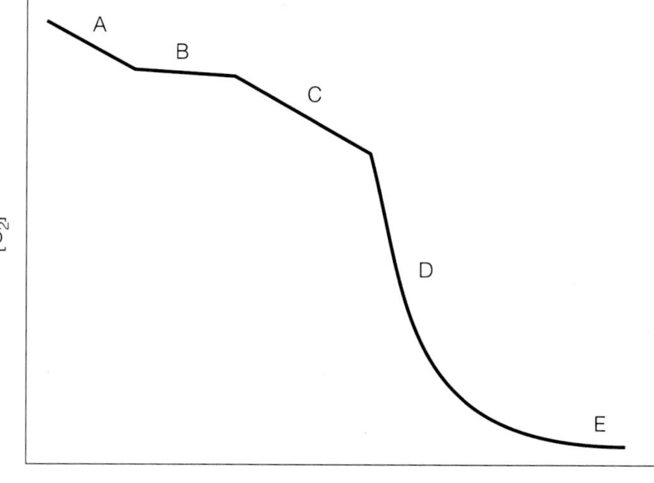

CHAPTER 16

Carbohydrate Metabolism II: Biosynthesis

In this chapter our focus turns from energy-yielding degradative pathways to energy-requiring biosynthetic pathways. The pathways of carbohydrate biosynthesis that receive emphasis in this chapter (Figure 16.1) include gluconeogenesis, the synthesis of glucose from noncarbohydrate precursors; the synthesis of polysaccharides, notably glycogen in animals; and the synthesis of **glycoconjugates,** molecules containing carbohydrate linked to a protein or lipid molecule. A supremely important carbohydrate biosynthetic process—photosynthesis in plants and some microorganisms—is presented separately in Chapter 19.

Here we encounter the first instances of a principle enunciated in Chapter 12: *biosynthetic processes are never simply the reversal of corresponding catabolic pathways.* Superficially, gluconeogenesis looks very much like glycolysis run in reverse, but different enzymatic reactions are used at crucial sites. These sites are points of strongly exergonic reactions that are controlled largely in reciprocal fashion, so that physiological conditions which activate glycolysis inhibit gluconeogenesis and vice versa. Very much the same picture will emerge from our discussion of glycogen synthesis as compared with glycogen mobilization.

Gluconeogenesis

Physiological Need for Glucose Synthesis in Animals

Most animal organs can metabolize a variety of carbon sources to generate their needed energy—triacylglycerols, various sugars, pyruvate, amino acids, and so forth. However, the brain and central nervous system require glucose as the sole or major carbon source; the same is true for some other organs, such as kidney medulla, testes, and erythrocytes. Consequently, animal tissues must be able to synthesize glucose from other precursors and also to maintain blood glucose levels within narrow limits—both for proper functioning of the brain and central nervous system and also to provide precursors for glycogen storage in other tissues (Figure 16.2). The glucose requirements of the human brain are relatively enormous—120 grams per day, compared with about 160 grams needed by the entire body. The amount of glucose that can be generated from the body's glycogen reserves at any time is about 190 grams, and the total amount of glucose in body

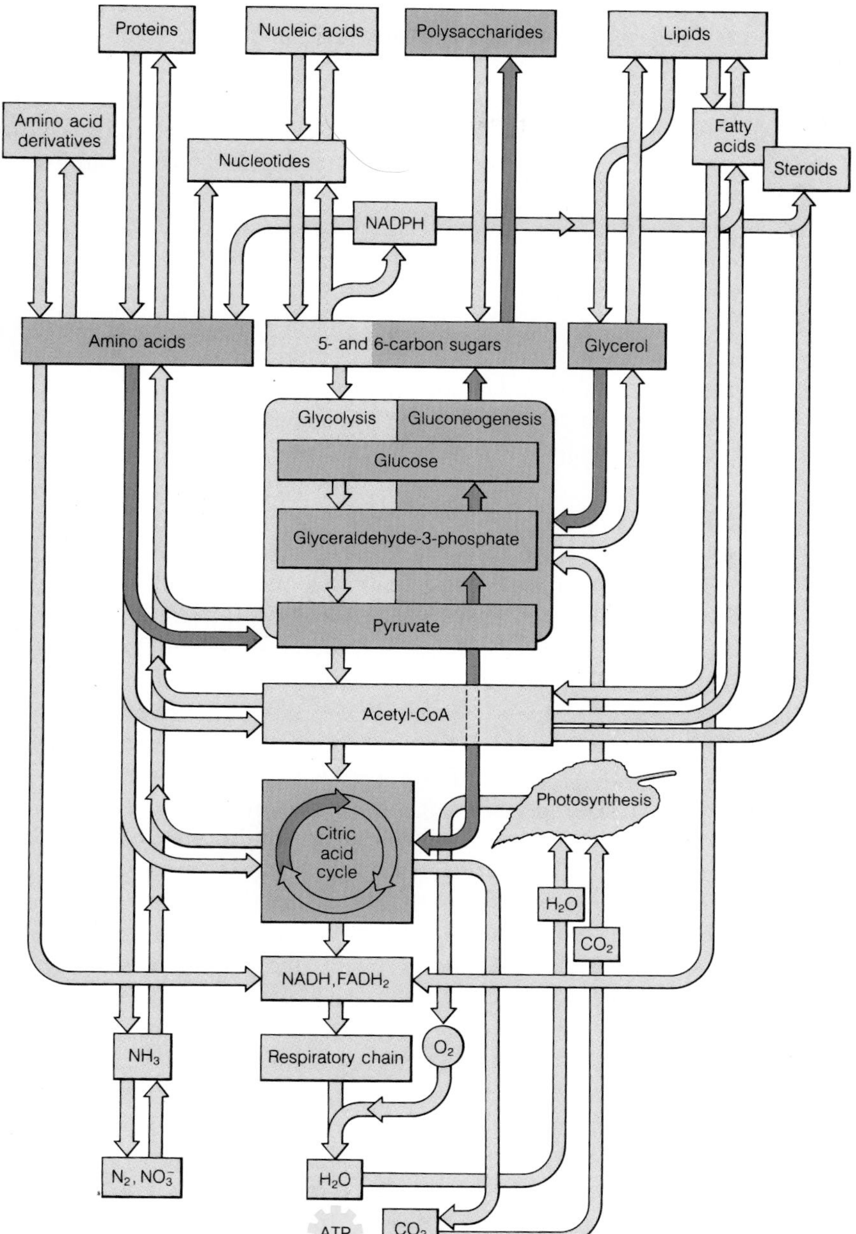

Figure 16.1
Intermediary metabolism, with the pathways presented in this chapter shown in blue.

fluids is about 20 grams. Thus, the readily available glucose reserves amount to about one day's supply. During periods of fasting for more than one day, glucose must be formed from other precursors. The same is true during intense exertion, where the glucose reserves are rapidly depleted.

The synthetic process is called **gluconeogenesis**—literally, the production of new glucose. Gluconeogenesis is defined as the biosynthesis of carbohydrate from three-carbon and four-carbon precursors, generally noncarbohydrate in nature. The principal substrates for gluconeogenesis are *lactate,* produced primarily from glycolysis in skeletal muscle and erythrocytes; *amino acids,* generated from dietary protein or from the breakdown of muscle protein during starvation; the specific amino acid *alanine,* produced in muscle through the pyruvate–alanine cycle (Chapter 20); and *glycerol,* derived from the catabolism of fats. The fatty acids released during lipid

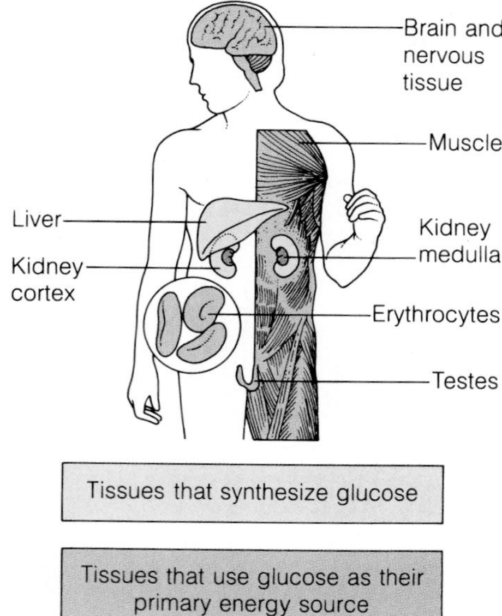

Figure 16.2
Synthesis and use of glucose in the human body. Liver and kidney cortex are the primary gluconeogenic tissues. Brain, muscle, kidney, medulla, and testes use glucose as the sole or primary energy source, but they lack the enzymatic machinery to synthesize it.

breakdown are primarily converted to acetyl-CoA and cannot be used for carbohydrate synthesis, except in organisms that have a functioning glyoxylate cycle.

Gluconeogenesis occurs primarily in the cytosol, although some precursors are generated in mitochondria and must be transported to the cytosol to be utilized. The major gluconeogenic organ in animals is the liver, with kidney cortex contributing in a minor way. The major fates of glucose formed by gluconeogenesis are catabolism by nervous tissue and utilization by skeletal muscle. In addition, glucose is the primary precursor for all other carbohydrates, including amino sugars, complex polysaccharides, and the carbohydrate components of glycoproteins and glycolipids. The need for glucose as a biosynthetic intermediate means that gluconeogenesis is an important pathway in plants and microorganisms, as well as in animals. The wealth of information on control of gluconeogenesis in animals leads us to concentrate on animal metabolism for the first part of this chapter.

Enzymatic Relationship of Gluconeogenesis to Glycolysis

Gluconeogenesis should be an easy pathway to learn, because it closely resembles glycolysis run in reverse. However, there are some important differences, which allow the pathway to run in the direction of glucose *synthesis* in the cell.

Earlier we explained that a metabolic pathway can proceed smoothly only if $\Delta G'$ is strongly negative for the overall pathway in the direction written. In Chapter 13 you learned that glycolysis from glucose to pyruvate is strongly exergonic; under typical intracellular conditions $\Delta G'$ is about -82 kJ/mol. How, then, can the conversion of pyruvate to glucose be made exergonic? The answer lies in the fact that three reactions of the glycolytic pathway are so strongly exergonic as to be irreversible—those catalyzed by hexokinase, phosphofructokinase, and pyruvate kinase. In gluconeogenesis, *different* enzymes are used at each of these steps, so that, for example, the conversion of fructose-1,6-bisphosphate to fructose-6-phosphate is not simply a reversal of the phosphofructokinase reaction. In essence, the three irreversible reactions of glycolysis are bypassed by enzymes specific to gluconeogenesis, reactions that run strongly in the direction of glucose synthesis (at an energy cost; see p. 543). The remaining seven reactions of gluconeogenesis are catalyzed by glycolytic enzymes, simply running in the reverse direction. The entire gluconeogenic pathway, from pyruvate to glucose, is summarized in Figure 16.3. We focus here on the three reactions that bypass the irreversible steps in glycolysis.

1. CONVERSION OF PYRUVATE TO PHOSPHOENOLPYRUVATE. This process involves two reactions, the first of which we encountered in Chapter 14. **Pyruvate carboxylase** catalyzes the biotin-dependent conversion of pyruvate to oxaloacetate. The enzyme absolutely requires acetyl-CoA for its activity.

$$\text{Pyruvate} + HCO_3^- + \text{ATP} \longrightarrow \text{oxaloacetate} + \text{ADP} + P_i + H^+$$

This is one of the *anaplerotic* reactions used to maintain levels of citric acid cycle intermediates in mitochondria. Pyruvate carboxylase generates oxaloacetate in the mitochondrial matrix. To be used for gluconeogenesis, that oxaloacetate must move from mitochondria to the cytosol, where the remainder of the pathway occurs. However, the mitochondrial membrane is

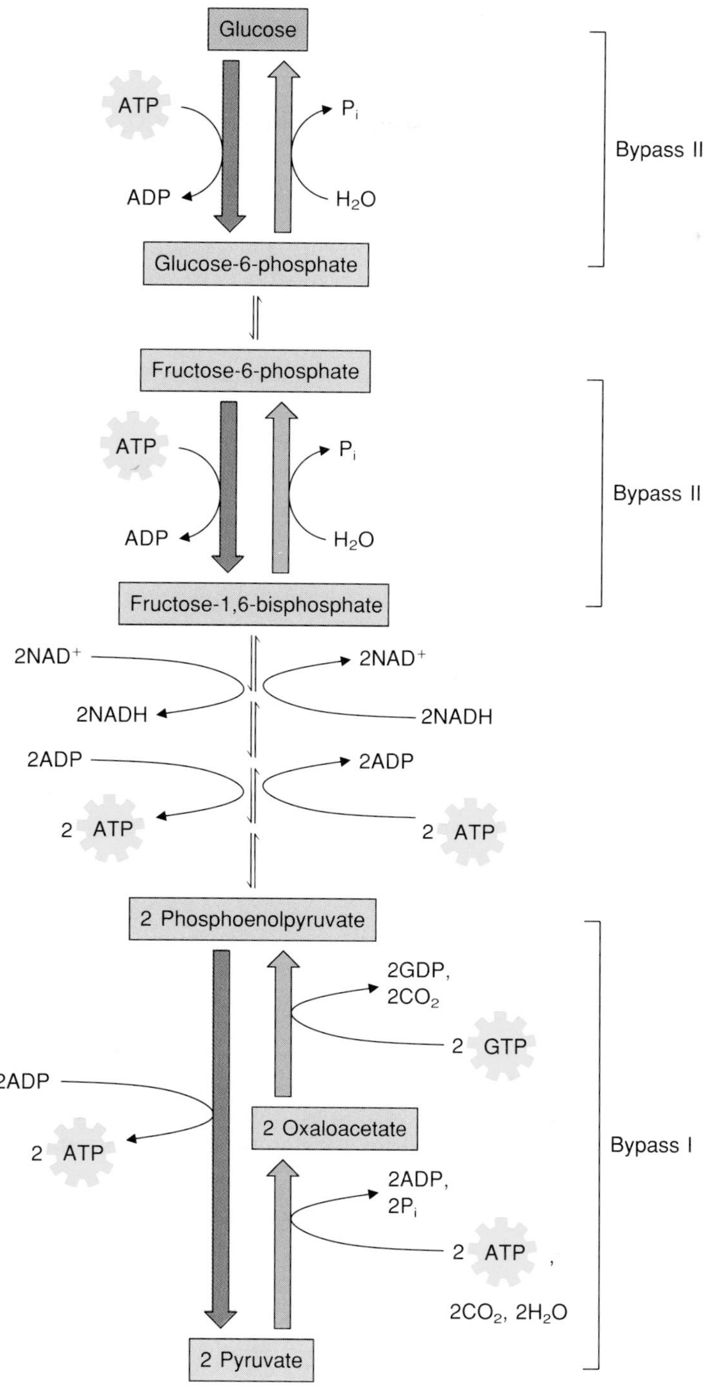

Figure 16.3
Reactions of glycolysis and gluconeogenesis. Irreversible reactions of glycolysis are shown in purple. The opposed reactions in gluconeogenesis, which bypass these steps, are shown in blue. Roman numerals identify bypass of pyruvate kinase (I), phosphofructokinase (II), and hexokinase or glucokinase (III).

relatively impermeable to oxaloacetate. Therefore, oxaloacetate is reduced by mitochondrial malate dehydrogenase to malate, which is transported into the cytosol and then reoxidized by cytosolic malate dehydrogenase. Alternatively, oxaloacetate can undergo transamination to give aspartate, which can exit the mitochondria and be reconverted to oxaloacetate in the cytosol. These shuttles act very similarly to those presented in Chapter 15 (see Figure 15.12b).

Once in the cytosol, oxaloacetate is acted on by **phosphoenolpyruvate carboxykinase** to give phosphoenolpyruvate.

$$\text{Oxaloacetate} + \text{GTP} \rightleftharpoons \text{phosphoenolpyruvate} + CO_2 + \text{GDP}$$

Note the use of GTP, rather than ATP, as an energy donor. Note also that the CO_2 which was fixed by pyruvate carboxylase is released in this reaction, so that no net fixation of CO_2 occurs. Phosphoenolpyruvate carboxykinase requires Mg^{2+} and is readily reversible. The enzyme is localized to the cytosol of most animal species, but in some species it is found in both cytosol and mitochondria. The level of the cytosolic form is regulated by hormonal conditions known to control gluconeogenesis. This coordinate regulation identifies the cytosolic enzyme as the primary gluconeogenic form of phosphoenolpyruvate carboxykinase.

The overall reaction for the bypass of pyruvate kinase is as follows.

$$\text{Pyruvate} + \text{ATP} + \text{GTP} + H_2O \longrightarrow \text{phosphoenolpyruvate} + \text{ADP} + \text{GDP} + P_i + 2H^+$$

$\Delta G^{\circ\prime}$ for the overall sequence is +0.8 kJ/mol; however, under intracellular conditions the sequence is quite exergonic, with $\Delta G'$ of about −25 kJ/mol. As seen from the summary reaction, two energy-rich phosphates must be invested for the synthesis of 1 mol of the super-energy-rich phosphoenolpyruvate.

2. CONVERSION OF FRUCTOSE-1,6-BISPHOSPHATE TO FRUCTOSE-6-PHOSPHATE. Phosphoenolpyruvate is converted to fructose-1,6-bisphosphate by glycolytic enzymes acting in reverse. Since the phosphofructokinase reaction of glycolysis is essentially irreversible, this reaction is bypassed in gluconeogenesis by a simple hydrolytic reaction, catalyzed by **fructose-1,6-bisphosphatase.**

$$\text{Fructose-1,6-bisphosphate} + H_2O \longrightarrow \text{fructose-6-phosphate} + P_i$$

$\Delta G^{\circ\prime}$ for this reaction is −16.3 kJ/mol, causing it to run to the right. The enzyme requires Mg^{2+} for activity. It is a multisubunit enzyme, which represents one of the major control sites regulating the overall pathway.

3. CONVERSION OF GLUCOSE-6-PHOSPHATE TO GLUCOSE. Fructose-6-phosphate, formed in the fructose-1,6-bisphosphatase reaction, undergoes isomerization by phosphoglucoisomerase to glucose-6-phosphate. This cannot be converted to glucose by reverse action of hexokinase or glucokinase, because of the high $\Delta G^{\circ\prime}$ of that reaction. Another enzyme specific to gluconeogenesis, **glucose-6-phosphatase,** comes into play instead. This bypass reaction also involves a simple hydrolysis.

$$\text{Glucose-6-phosphate} + H_2O \longrightarrow \text{glucose} + P_i \qquad \Delta G^{\circ\prime} = -12.1 \text{ kJ/mol}$$

This enzyme, which also requires Mg^{2+}, is found primarily in the endoplasmic reticulum of the liver. The significance of its intracellular location is not clear, because the product, glucose, is being formed for export to the bloodstream. Because most tissues, notably brain and skeletal muscle, lack this enzyme, they cannot complete the gluconeogenic pathway by releasing free glucose into the bloodstream—even though they might be capable of synthesizing glucose-6-phosphate from gluconeogenic precursors.

Stoichiometry and Energy Balance of Gluconeogenesis

We have emphasized that catabolic pathways generate energy, while anabolic pathways carry an energy cost. For gluconeogenesis we can estimate that cost. We can write a net equation for the conversion of 2 mol of pyruvate to glucose, as shown in the summary in Table 16.1. $\Delta G^{\circ\prime}$ for the overall process is about -37.6 kJ/mol. Clearly, the synthesis of glucose is costly to the cell in an energetic sense, because six energy-rich phosphate compounds are consumed (four ATP and two GTP), as well as 2 mol of NADH, which is the energetic equivalent of six more ATPs (recall that mitochondrial oxidation of 1 mol of NADH generates 3 mol of ATP).

By contrast, if glycolysis could operate in reverse, the net equation would show far less energy input—2 mol of NADH and 2 mol of high-energy phosphate.

$$\text{2 Pyruvate} + \text{2ATP} + \text{2NADH} + 2H^+ + 2H_2O \longrightarrow \text{glucose} + \text{2ADP} + 2P_i + 2NAD^+$$

However, this process is highly endergonic, with a $\Delta G^{\circ\prime}$ of $+83.7$ kJ/mol. Therefore, it is clear that the investment of four extra energy-rich phosphate bonds is essential if the net synthesis of glucose is to occur as an irreversible process.

Substrates for Gluconeogenesis

As indicated earlier, gluconeogenesis draws precursors from diverse sources, including lactate, amino acids, propionate, and glycerol. The pathways by which these substrates enter gluconeogenesis are shown in Figure 16.4, and discussed in this section.

LACTATE. In quantitative terms, lactate is the most significant gluconeogenic precursor. Recall from Chapter 13 that skeletal muscle derives much of its energy from glycolysis, particularly during intense exertion, when

Table 16.1
Sequential reactions in gluconeogenesis, from pyruvate to glucose

Reaction	
Pyruvate + CO_2 + ATP + H_2O $\longrightarrow$ oxaloacetate + ADP + P_i + $2H^+$	×2
Oxaloacetate + GTP $\rightleftharpoons$ phosphoenolpyruvate + CO_2 + GDP	×2
Phosphoenolpyruvate + H_2O $\rightleftharpoons$ 2-phosphoglycerate	×2
2-Phosphoglycerate $\rightleftharpoons$ 3-phosphoglycerate	×2
3-Phosphoglycerate + ATP $\rightleftharpoons$ 1,3-diphosphoglycerate + ADP	×2
1,3-Diphosphoglycerate + NADH + H^+ $\rightleftharpoons$ glyceraldehyde-3-phosphate + NAD^+ + P_i	×2
Glyceraldehyde-3-phosphate $\rightleftharpoons$ dihydroxyacetone phosphate	
Glyceraldehyde-3-phosphate + dihydroxyacetone phosphate $\rightleftharpoons$ fructose-1,6-bisphosphate	
Fructose-1,6-bisphosphate + H_2O $\longrightarrow$ fructose-6-phosphate + P_i	
Fructose-6-phosphate $\rightleftharpoons$ glucose-6-phosphate	
Glucose-6-phosphate + H_2O $\longrightarrow$ glucose + P_i	
Sum: 2 Pyruvate + 4ATP + 2GTP + 2NADH + $2H^+$ + $6H_2O$ $\longrightarrow$ glucose + $2NAD^+$ + 4ADP + 2GDP + $6P_i$ + $2H^+$	

The reactions in boldface type are those that bypass irreversible glycolytic reactions; the remaining reactions are reversible reactions of glycolysis. The first six reactions are to be multiplied by 2, because two three-carbon precursors are required to make one molecule of glucose.

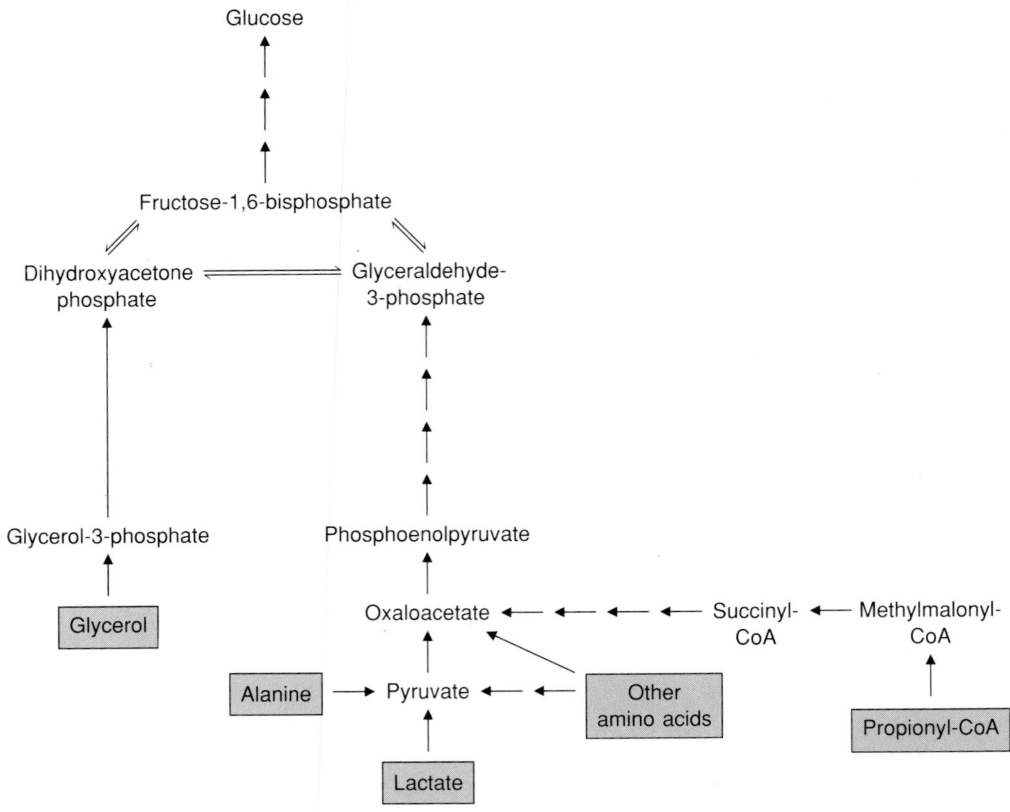

Figure 16.4
Outline of pathways by which the major gluconeogenic precursors are utilized for glucose synthesis. Note that both glucose and lactate are carried in the blood.

respiration cannot deliver sufficient oxygen to the tissues for complete oxidation of glucose. Under these conditions glycogen stores are mobilized, and glucose-6-phosphate is converted to pyruvate more rapidly than it can be further metabolized via the pyruvate dehydrogenase complex and the citric acid cycle. Lactate dehydrogenase is abundant in muscle, and the equilibrium strongly favors pyruvate reduction to lactate. Thus, lactate is released to the blood, whence it is readily taken up by gluconeogenic tissues, primarily liver; a substantial amount is also taken up by the heart and oxidized as fuel.

Lactate entering the liver is reoxidized to pyruvate; this pyruvate can then undergo gluconeogenesis to give glucose, which is returned to the bloodstream and taken up by muscle to regenerate the glycogen stores. This process, described originally by Carl and Gerti Cori and appropriately called the **Cori cycle,** is schematized in Figure 16.5. The pathway is particularly active during recovery from intense muscular exercise. During this time the breathing rate is elevated, and the increased oxidative metabolism generates more ATP, much of which is used to rebuild glycogen stores via gluconeogenesis. In a parallel process, the **glucose–alanine cycle,** pyruvate in peripheral tissues undergoes transamination to alanine, which is returned to the liver and used for gluconeogenesis. This pathway, which is presented in detail in Chapter 20, helps tissues to dispose of toxic ammonia formed during protein degradation.

Amino Acids. Like alanine, many other amino acids can readily be converted to glucose, primarily through degradative pathways that generate citric acid cycle intermediates, which can be converted to oxaloacetate. Such

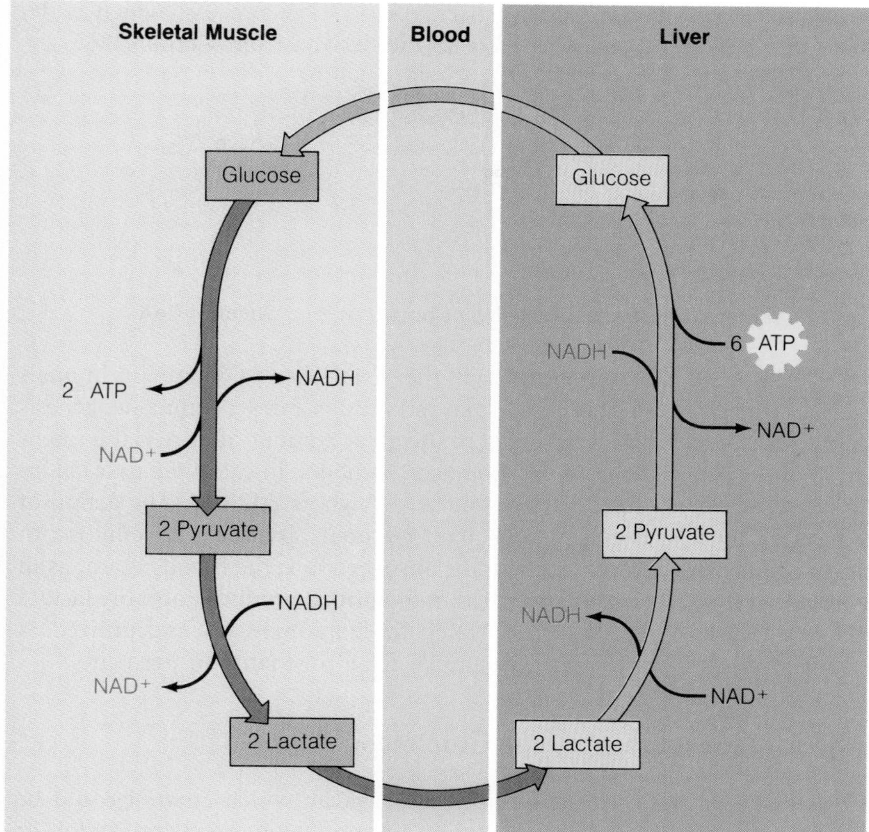

Figure 16.5
The Cori cycle. Lactate produced in muscle glycolysis is transported to the liver, for resynthesis by gluconeogenesis.

amino acids are called **glucogenic** (i.e., able to be converted to glucose), although *gluconeogenic* is probably a more descriptive term. Among the 20 amino acids found in proteins, only the catabolism of leucine and lysine does not generate gluconeogenic precursors. During fasting, when insufficient carbohydrate is ingested, the catabolism of muscle proteins is the major source for maintenance of normal blood glucose concentrations. The same is true in the disease diabetes mellitus, as discussed further in Chapter 23.

GLYCEROL. In general, lipids are poor gluconeogenic precursors. Catabolism of triacylglycerols yields fatty acids and glycerol. Fatty acids undergo β-oxidation to yield acetyl-CoA. In plants and bacteria acetyl-CoA can be incorporated into carbohydrate via the glyoxylate cycle. However, in animals acetyl-CoA cannot be converted to pyruvate or to any other gluconeogenic precursor. Hence, *fatty acids cannot undergo net conversion to carbohydrates*. Although it is true that two-carbon units from acetyl-CoA can proceed to oxaloacetate in the citric acid cycle, there is no net conversion, because two carbons are lost in each turn of the cycle. Therefore, aside from a minor contribution by odd-chain fatty acids (explained in the next section), the only fat breakdown product that can enter gluconeogenesis is glycerol. Its utilization involves phosphorylation, followed by dehydrogenation to dihydroxyacetone phosphate (see Figure 16.4).

PROPIONATE. In all organisms a three-carbon acyl-CoA, **propionyl-CoA,** is generated from the breakdown of some amino acids and the oxidation of fatty acids with odd numbers of carbon atoms. Propionyl-CoA enters gluconeogenesis via its conversion to succinyl-CoA. The process, which is detailed in Chapter 17, involves a coenzyme derived from vitamin B_{12}.

$$\underset{\text{Propionyl-CoA}}{CH_3-CH_2-\overset{O}{\overset{\|}{C}}-S-CoA} \xrightarrow{CO_2} \underset{\text{Methylmalonyl-CoA}}{CH_3-CH(COO^-)-\overset{O}{\overset{\|}{C}}-S-CoA} \longrightarrow \underset{\text{Succinyl-CoA}}{{}^-OOC-CH_2-CH_2-\overset{O}{\overset{\|}{C}}-S-CoA}$$

Propionate is very important in the metabolism of ruminant animals such as cattle. These animals display enormous rates of gluconeogenesis from a variety of substrates, owing to the great amount of bacterial fermentation occurring in their several stomach chambers. In cattle the four chambers of the stomach have a total volume as high as 70 liters. The action of various bacterial species degrades plant materials, particularly cellulose, to glucose. But before the glucose can be absorbed into the bloodstream, as in human digestion, it is fermented further to various products, notably lactate and propionate. Propionate is acylated to propionyl-CoA and utilized as described above, while lactate is simply dehydrogenated to pyruvate.

Ethanol Consumption and Gluconeogenesis

Although it is possible to visualize pathways by which ethanol could be converted to glucose, ethanol is actually a poor gluconeogenic precursor. In fact, ethanol strongly inhibits gluconeogenesis and can bring about **hypoglycemia,** a potentially dangerous decrease in blood glucose levels. This is because ethanol is metabolized primarily in the liver, by alcohol dehydrogenase.

$$\text{Ethanol} + NAD^+ \rightleftharpoons \text{acetaldehyde} + NADH + H^+$$

This reaction elevates the $[NADH]/[NAD^+]$ ratio in liver cytosol, which in turn shifts the equilibrium of the lactate dehydrogenase reaction toward lactate synthesis instead of pyruvate formation. The same mechanism shifts the equilibrium of cytosolic malate dehydrogenase so that oxaloacetate tends to be reduced to malate and hence becomes unavailable for gluconeogenesis. The resultant hypoglycemia can affect the parts of the brain concerned with temperature regulation. This, in turn, can lower the body temperature by as much as 2°C. Therefore, the time-honored practice of feeding brandy or whisky to those rescued from cold or wet conditions is counterproductive, although a small amount of alcohol can cause warming through vasodilation. Metabolically speaking, glucose would be far more effective in raising body temperature.

Regulation of Gluconeogenesis

Regulation of gluconeogenesis is crucial for many physiological functions, but particularly so for proper functioning of nervous tissue. While other organs can use a variety of energy sources, the well-being of the central nervous system demands maintenance of blood glucose levels within narrow limits. Gluconeogenic control is important also as an animal adjusts to muscular exertion or to cycles of feeding and fasting. Flux through the pathway must increase and decrease, based on the availability of lactate produced by the muscles, of glucose from the diet, or of other gluconeogenic precursors.

Much of the regulation of gluconeogensis occurs hormonally, through control of the synthesis of enzymes specific to gluconeogenesis, particularly phosphoenolpyruvate carboxykinase and glucose-6-phosphatase. These mechanisms will be presented, along with our general discussion of hormone action, in Chapter 23. Here we concentrate on mechanisms affecting *activities* of gluconeogenic enzymes. Some of these mechanisms involve hormone action also—in this case, hormones that act by controlling levels of cyclic AMP.

Reciprocal Regulation of Glycolysis and Gluconeogenesis

Gluconeogenesis and glycolysis both proceed largely in the cytosol. Since gluconeogenesis synthesizes glucose while glycolysis catabolizes glucose, it is evident that *gluconeogenesis and glycolysis must be controlled in reciprocal fashion*. In other words, intracellular conditions that activate one pathway tend to inhibit the other. If not for reciprocal control, glycolysis and gluconeogenesis would operate together as a giant futile cycle. Reciprocal regulation can be thought of in terms of the adenylate energy charge. Conditions of low energy charge tend to activate the rate-controlling steps in glycolysis while inhibiting carbon flux through gluconeogenesis. Conversely, gluconeogenesis is stimulated at high energy charge, under conditions where catabolic flux rates are adequate to maintain sufficient ATP levels.

Recall that glycolysis is controlled primarily by regulation of the three strongly exergonic reactions of the pathway—those catalyzed by hexokinase (or glucokinase in liver), phosphofructokinase, and pyruvate kinase. The opposed reactions in gluconeogenesis—those catalyzed by glucose-6-phosphatase, fructose-1,6-bisphosphatase, and the combination of pyruvate carboxylase and phosphoenolpyruvate carboxykinase—are also strongly exergonic and represent the chief targets for control of this pathway. Glucose-6-phosphatase is not known to be allosterically controlled, but its K_M for glucose-6-phosphate is far higher than intracellular concentrations of this metabolite; thus, intracellular activity is controlled in first-order fashion by the concentration of this substrate.

Figure 16.6 identifies the major allosteric activators and inhibitors of the key exergonic reactions in glycolysis and gluconeogenesis. For some time it was thought that the primary energy charge regulation of these pathways occurred at the interconversion of fructose-6-phosphate and fructose-1,6-bisphosphate, with phosphofructokinase being strongly activated by AMP and the opposed enzyme, fructose-1,6-bisphosphatase, being strongly inhibited by AMP. Thus, as energy charge decreased, glycolysis could be activated and gluconeogenesis inhibited by the opposed effects of AMP on phosphofructokinase and fructose-1,6-bisphosphatase, respectively. However, intracellular adenine nucleotide levels do not vary in close coordination with changes in flux through glycolysis and gluconeogenesis. This led investigators to seek other regulatory mechanisms, as described later.

Energy charge regulation is involved, however, in the control of one of the two isoenzyme forms of pyruvate kinase. In liver and in other gluconeogenic tissues the so-called L form of pyruvate kinase predominates, while the M form is found largely in muscle. The L form is inhibited both by ATP and by some amino acids, particularly alanine, the major gluconeogenic

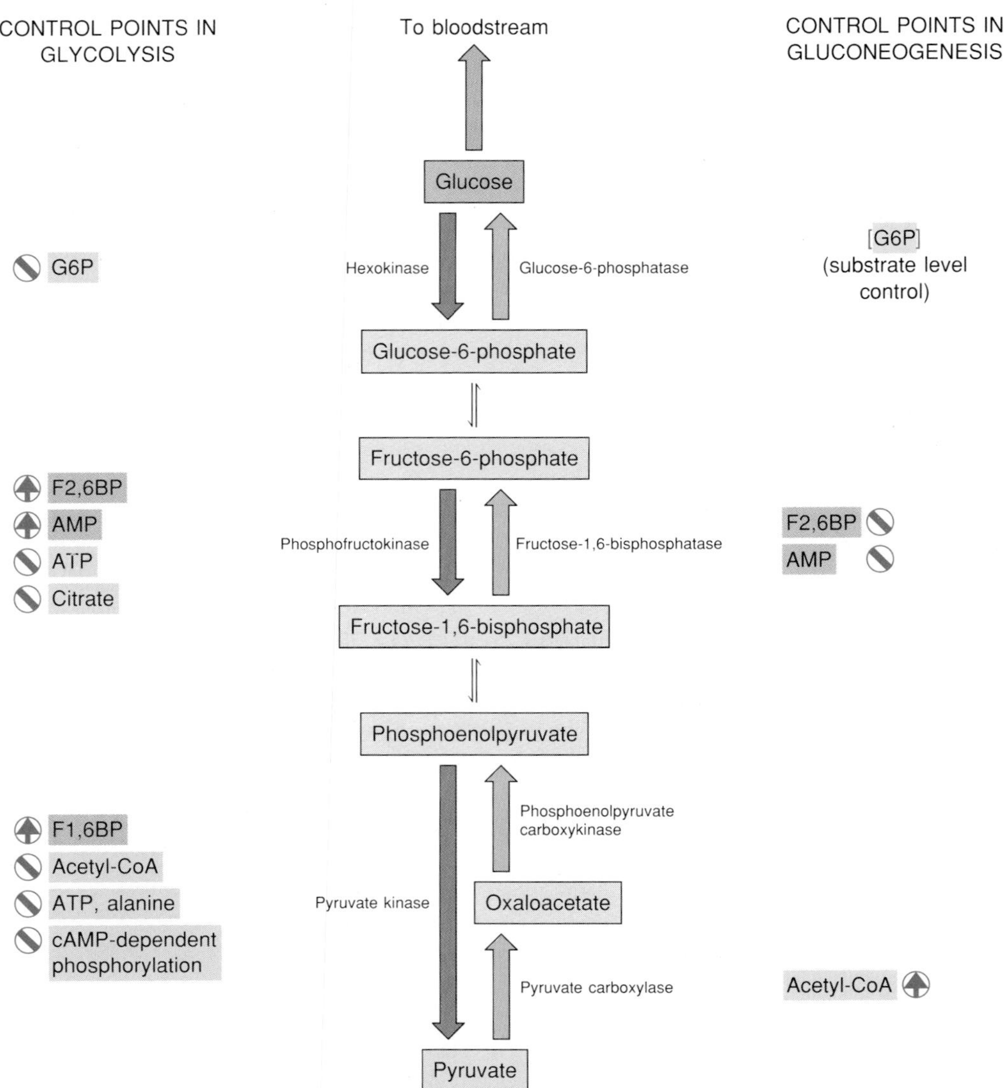

Figure 16.6
Major control mechanisms affecting glycolysis and gluconeogenesis. Strongly exergonic reactions are identified with blue (gluconeogenesis) or purple (glycolysis) arrows. Major regulators promoting gluconeogenesis are also shown in blue, while those promoting glycolysis are shown in purple.

precursor among the amino acids. This allows inhibition of glycolysis, with consequent activation of gluconeogenesis, specifically in gluconeogenic tissues, when ample energy and substrates are available. Control at the pyruvate kinase step allows conservation of high-energy phosphate in the phosphoenolpyruvate molecule.

Acetyl-CoA can also be seen as a reciprocal regulator of glycolysis and gluconeogenesis, acting on the enzymes that interconvert pyruvate and phosphoenolpyruvate. Acetyl-CoA is a required activator of pyruvate carboxylase and an inhibitor of pyruvate kinase and of pyruvate dehydrogenase complex. It can thus signal, when its levels rise, that adequate substrates are available to provide energy through the citric acid cycle and that more carbon can be shuttled into gluconeogenesis and ultimately stored as glycogen. There is some question whether the activation of pyruvate carboxylase by acetyl-CoA really represents a regulatory mechanism, because its intramitochondrial levels under most conditions are far higher than the concentration giving half-maximal stimulation. Thus, it is not clear that intramitochondrial pyruvate carboxylase activity can respond to changes in concentration of the activator.

Fructose-2,6-bisphosphate and the Control of Gluconeogenesis

Although the effects discussed above are all significant for controlling glycolysis and gluconeogenesis, developments since 1980 have identified fructose-2,6-bisphosphate as probably the most important regulator, through its effects on the interconversion of fructose-6-phosphate and fructose-1,6-bisphosphate. Fructose-2,6-bisphosphate is active at much lower concentrations than the other physiological regulators we have discussed. For example, fructose-2,6-bisphosphate inhibits fructose-1,6-bisphosphatase with a K_i of 2.5 μM, while AMP has a K_i for the same enzyme of 25 μM. Recall from Chapter 13 that fructose-2,6-bisphosphate is also an allosteric activator of phosphofructokinase. Thus, accumulation of this regulator has the effect of activating glycolysis and inhibiting gluconeogenesis. The level of fructose-2,6-bisphosphate itself is controlled ultimately by cyclic AMP, through the action of cAMP-dependent protein kinase.

As shown in Figure 16.7, fructose-2,6-bisphosphate is formed from fructose-6-phosphate by a newly discovered form of phosphofructokinase, called PFK-2 to distinguish it from the well-known PFK of glycolysis, which we can call PFK-1 for clarity. The activity of PFK-2 is in turn controlled by cyclic AMP. A 49,000-dalton subunit of the enzyme is subject to phosphorylation by cAMP-dependent protein kinase. This phosphorylation *decreases* the activity of PFK-2, by increasing its K_M for fructose-6-phosphate and by increasing the sensitivity of PFK-1 to allosteric inhibitors—citrate, ATP, and phosphoenolpyruvate.

Another activity, called **fructose-2,6-bisphosphatase,** cleaves fructose-2,6-bisphosphate back to fructose-6-phosphate. Control of this activity can contribute to regulation of the level of fructose-2,6-bisphosphate. Fructose-2,6-bisphosphatase is strongly inhibited by fructose-6-phosphate. More important, this enzyme is also regulated by cAMP-stimulated phosphorylation of a 49,000-dalton subunit, except that in this case phosphorylation *enhances* the activity of the enzyme.

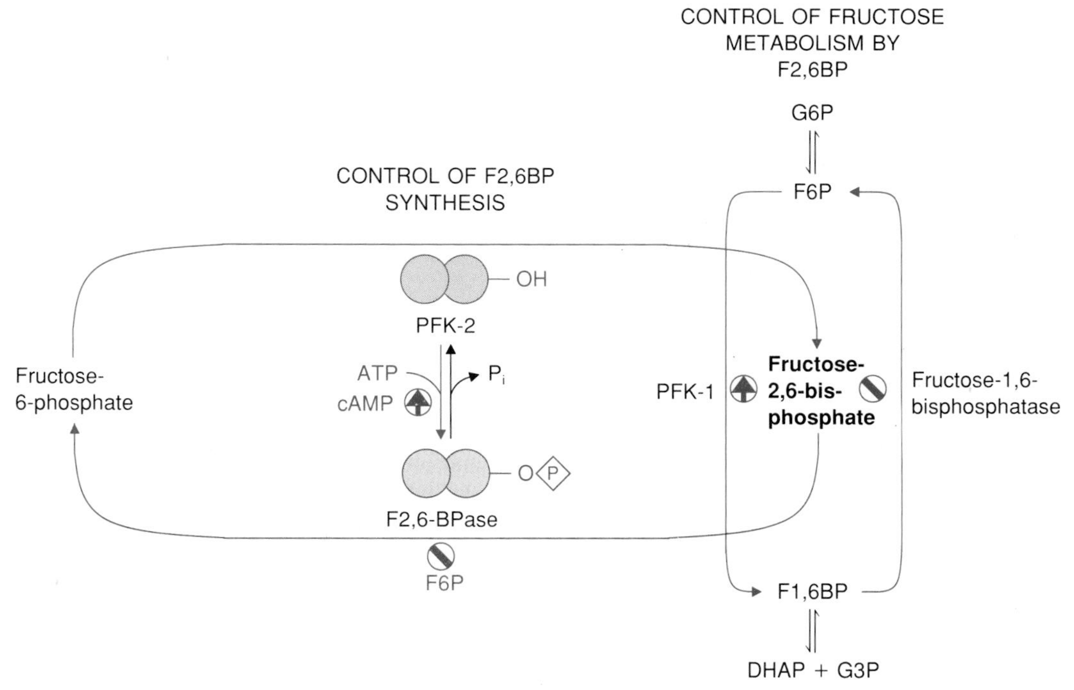

Figure 16.7
Biosynthesis and degradation of fructose-2,6-bisphosphate by action of a bifunctional enzyme. The phosphorylation of PFK-2 is carried out by cAMP-dependent protein kinase. Fructose-2,6-bisphosphate activates phosphofructokinase-1 and inhibits fructose-1,6-bisphosphatase. Reactions promoting glycolysis are shown in red, those favoring gluconeogenesis in blue.

The finding that both activities are regulated by phosphorylation of a 49,000-dalton protein led to the discovery that the same protein is involved in both effects. In fact, PFK-2 and fructose-2,6-bisphosphatase constitute a bifunctional enzyme. When the 49-kDa protein is in the *nonphosphorylated* form, the bifunctional enzyme catalyzes the synthesis of fructose-2,6-bisphosphate, and when the 49-kDa protein is phosphorylated, the phosphatase activity is activated. Thus, cAMP has a dual effect on fructose-2,6-bisphosphate levels, by inactivating PFK-2 and by stimulating fructose-2,6-bisphosphatase activities, and both effects tend to *decrease* the levels of fructose-2,6-bisphosphatase. This in turn reduces flux through glycolysis and *stimulates* gluconeogenesis by (1) reducing the stimulation of PFK-1 and (2) relieving the inhibition of fructose-1,6-bisphosphatase.

Recall from Chapter 13 that the primary hormone whose action raises cAMP levels in liver is the pancreatic hormone *glucagon*. Glucagon raises blood glucose levels by two distinct mechanisms: the cAMP stimulation of the regulatory cascade that causes glycogen breakdown and, as described here, the cAMP control of fructose-2,6-bisphosphate levels. Gluconeogenesis is also regulated by hormones such as insulin, as discussed in Chapter 23.

Fructose-2,6-bisphosphate plays a comparable regulatory role in plants, where, by inhibiting a cytosolic fructose-1,6-bisphosphatase, it controls the flow of three-carbon sugars, produced by photosynthesis, out of chloroplasts and into the pathway for sucrose synthesis in the cytosol.

Glycogen Biosynthesis

A major fate of glucose in animals is its use for glycogen synthesis. This is a good place to begin a discussion of polysaccharide synthesis, because glycogen synthesis illustrates general mechanisms used in synthesizing glycosidic bonds and because regulation of glycogen synthesis, in concert with control of glycogen breakdown, is of great metabolic importance.

For many years it was thought that reversal of the glycogen phosphorylase reaction described in Chapter 13 was the major route for glycogen synthesis. However, three observations could not be reconciled with this notion. First, epinephrine secretion activated glycogen metabolism only in the direction of breakdown; in fact, epinephrine inhibits glycogen biosynthesis. Second, intracellular levels of orthophosphate are relatively high, which would make it difficult on equilibrium grounds for phosphorylase to catalyze glycogen synthesis in vivo. Third, while phosphorylase can synthesize glycogen in vitro, the product is different from natural glycogen, particularly in being of much lower molecular weight.

Glycogen Synthase and the Branching Process

The true enzyme for glycogen biosynthesis was discovered shortly after the discovery in the late 1950s, by the Argentine biochemist Luis Leloir, of *uridine diphosphate glucose*, commonly called UDP-glucose or UDP-Glc. The first known role of UDP-Glc was as a substrate for glycogen biosynthesis, catalyzed by **glycogen synthase.** The latter is a tetrameric enzyme, bound tightly to intracellular glycogen granules.

Before discussing the reaction, let us consider how the substrate is synthesized from blood glucose. Glucose is phosphorylated by hexokinase or glucokinase to give glucose-6-phosphate, which is isomerized to glucose-1-phosphate by phosphoglucomutase. Glucose-1-phosphate reacts with uri-

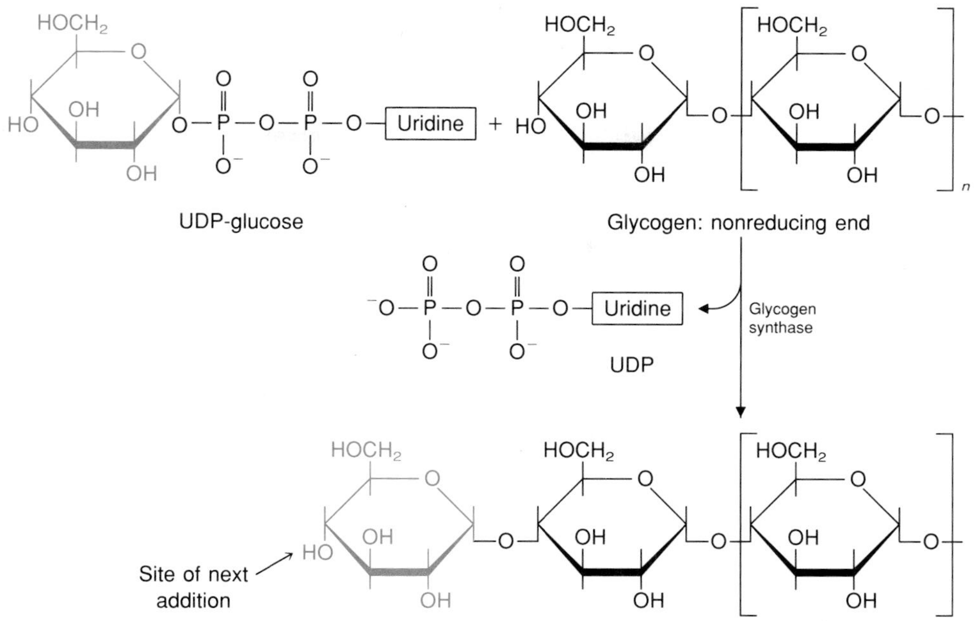

Figure 16.8
The glycogen synthase reaction.

dine triphosphate in a reversible reaction catalyzed by **UDP-Glc pyrophosphorylase.**

$$\text{UTP} + \text{glucose-1-phosphate} \rightleftharpoons \text{UDP-glucose} + \text{PP}_i \xrightarrow{H_2O} 2\text{P}_i$$

The reaction is drawn to the right by rapid enzymatic cleavage of pyrophosphate to orthophosphate, catalyzed by pyrophosphatase. Hydrolysis of pyrophosphate yields about 30 kJ/mol under standard conditions.

UDP-glucose is the immediate donor of a glucosyl residue to the nonreducing end of a glycogen branch, which must be at least four glucose residues in length. The reaction, depicted in Figure 16.8, is catalyzed by glycogen synthase and generates an $\alpha(1 \rightarrow 4)$ glycosidic linkage between carbon 1 of the incoming glucosyl moiety and carbon 4 of the glucose residue at the terminus of the glycogen chain. The enzyme continues to add glucose residues successively to the 4-hydroxyl group at the nonreducing end. Because UDP-Glc is an energy-rich compound, the glycogen synthase reaction is strongly exergonic, with a $\Delta G^{\circ\prime}$ of about -13.4 kJ/mol.

Glycogen synthesis involves both polymerization of glucose units and branching from $\alpha(1 \rightarrow 6)$ linkages. These branches are important because they increase the solubility of the polymer and also increase the number of nonreducing ends from which glucose-1-phosphate can be derived during glycogen mobilization. However, these branches cannot be introduced by glycogen synthase. Another enzyme, called **branching enzyme** but more accurately termed **amylo-(1,4→1,6)-transglycosylase,** comes into play (Figure 16.9). This enzyme transfers a terminal fragment, some 6 or 7 residues long, from a branch terminus at least 11 residues in length to a hydroxyl group at the 6-position of a glucose residue in the interior of the polymer. The reaction thus creates two termini for continued action by glycogen synthase, whereas just one existed before. The branching process does not involve a large free energy change, because of the chemical similarity of (1→4) and (1→6) linkages.

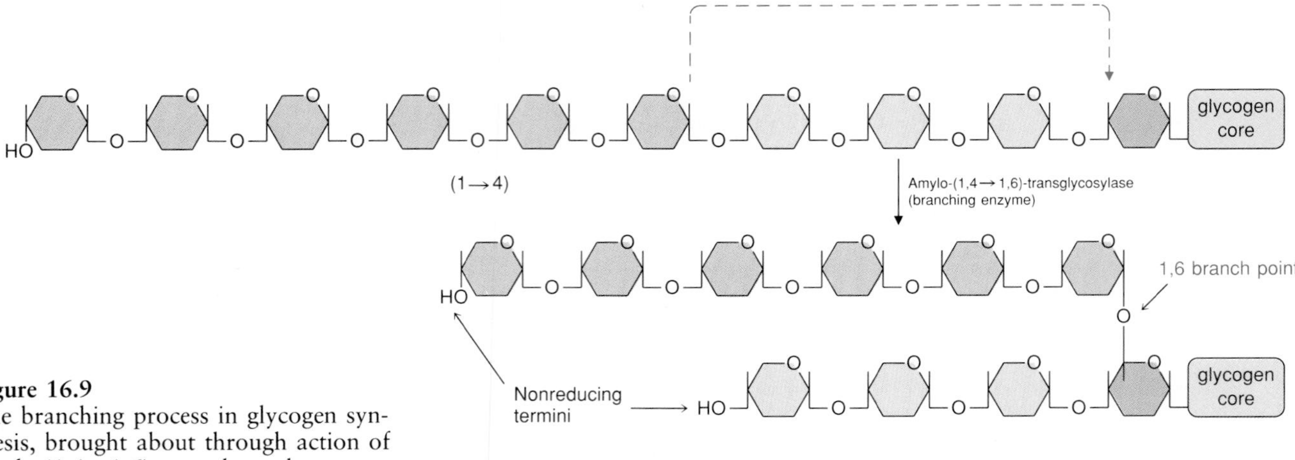

Figure 16.9
The branching process in glycogen synthesis, brought about through action of amylo-(1,4→1,6)-transglycosylase.

Reciprocal Relationship Between Glycogen Synthesis and Mobilization

Earlier we noted that epinephrine or glucagon secretion has the effect of inhibiting glycogen synthesis at the same time that it is promoting glycogenolysis. Both effects are mediated by the effects of cyclic AMP and cyclic AMP-dependent protein kinase. However, while the effect of the regulatory cascade presented in Chapter 13 (see Figure 13.19) is to *activate* glycogen phosphorylase, a comparable cascade has the end result of *inhibiting* glycogen synthase.

PHOSPHORYLATION OF GLYCOGEN SYNTHASE. Glycogen synthase from vertebrate tissues is a tetrameric protein comprising four identical subunits and having a total molecular weight of about 350,000. Like phosphorylase, glycogen synthase exists in phosphorylated and dephosphorylated states, with a phosphate reversibly bound to a serine residue on each subunit. Some of these reactions are catalyzed by the same protein kinases and phosphatase that act to regulate glycogenolysis. In fact, the enzyme we introduced as glycogen phosphorylase *b* kinase is also called **synthase-phosphorylase kinase (SPK)**, to emphasize its dual specificity.

The dephosphorylated form of glycogen synthase, known as **glycogen synthase I,** is the active form. Protein kinases phosphorylate serine residues and generate the less active **glycogen synthase D.** In kinetic terms, the phosphorylation of glycogen synthase from the I to the D form increases the K_M for UDP-glucose, *but only when the reactions are run in the absence of glucose-6-phosphate.* Glucose-6-phosphate is an allosteric effector, and when it is present at sufficient concentrations (>1 mM), there is no effect on the K_M for UDP-Glc. Practically, this means that the activity of glycogen synthase D is *dependent* on the presence of glucose-6-phosphate, which is the significance of the letter D to designate the less active form of glycogen synthase. By contrast, the dephosphorylated I form of the enzyme is *independent* of the presence of glucose-6-phosphate for its activity. In resting muscle the I form predominates, while during contraction the major species is the D form, which is active only at high glucose-6-phosphate concentrations.

REGULATORY SIGNIFICANCE OF GLUCOSE-6-PHOSPHATE. The relationship of glycogen synthase activity to glucose-6-phosphate concentration makes good metabolic sense. When glucose-6-phosphate levels are high, activation of glycogen synthase can serve as a signal to convert some of that glucose-6-phosphate into glycogen (via glucose-1-phosphate and UDP-Glc). This effect can come into play in the absence of a hormonal stimulus—comparable to the activation of glycogen phosphorylase *b* by 5′-AMP. These effects of glucose-6-phosphate and AMP represent allosteric mechanisms that regulate glycogen metabolism in understandable ways, by hormone-independent mechanisms.

CYCLIC AMP AND GLYCOGEN SYNTHASE REGULATION. Now let us examine the consequences of hormone release (Fig. 16.10). Just as depicted in Figure 13.19, the activation of adenylate cyclase by epinephrine or glucagon promotes the dissociation of cAMP-dependent protein kinase (C_2R_2) to give free catalytic monomers (C). C can phosphorylate glycogen synthase I directly. Alternatively, C phosphorylates SPK, which in turn activates this enzyme to phosphorylate glycogen synthase I to D. To complicate the picture somewhat, three additional protein kinases, less well understood, act on glycogen synthase I. Each of the five enzymes phosphorylates different serine residues, so that there are several different forms of glycogen synthase

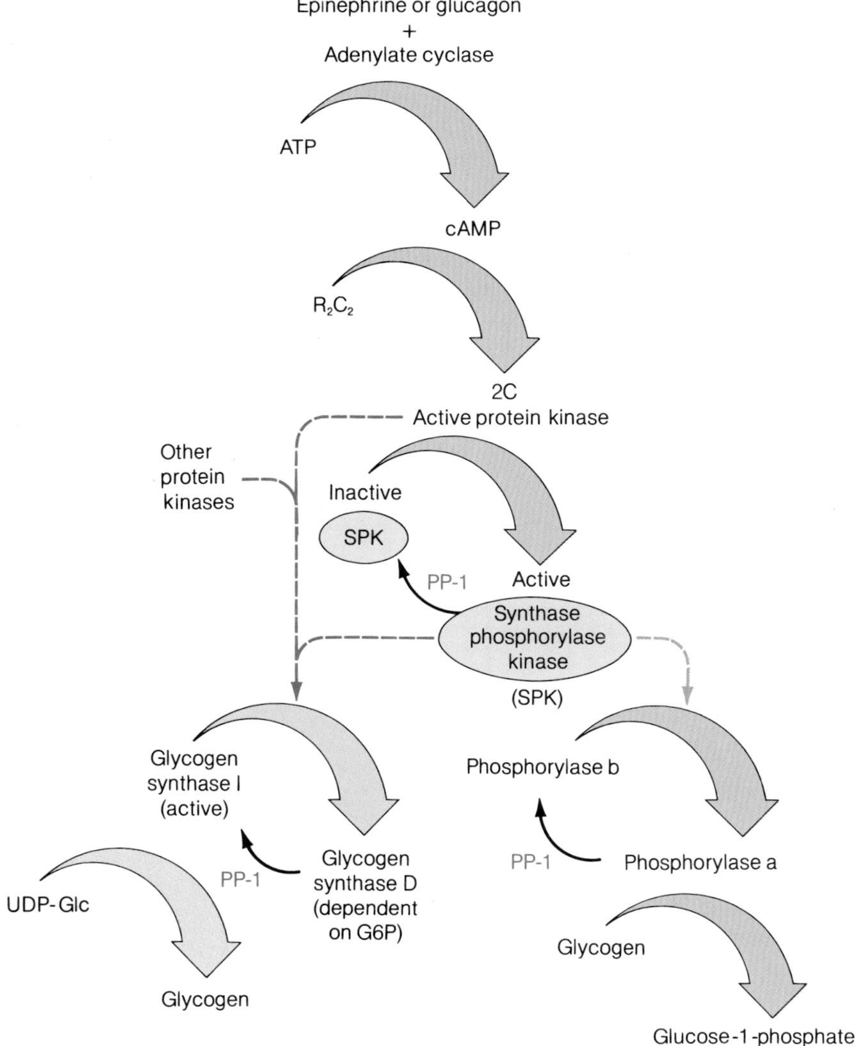

Figure 16.10
Regulatory cascades affecting glycogen synthesis and mobilization. Events following the activation of adenylate cyclase promote the simultaneous activation of glycogen breakdown and inhibition of glycogen synthesis. Four different protein kinases, including the phosphorylated form of SPK and the catalytic subunit of cAMP-dependent protein kinase, can inactivate glycogen synthase I by converting it to the less active D form. PP-1 = phosphoprotein phosphatase; R_2C_2 = cAMP-dependent protein kinase; SPK = synthase-phosphorylase kinase.

D, and it is an oversimplification to speak of just two forms. In general, as more sites are phosphorylated the activity of the enzyme progressively decreases because of the following progressive changes: (1) decreased affinity for the substrate, UDP-glucose; (2) decreased affinity for the allosteric activator, glucose-6-phosphate; and (3) increased affinity for ATP and P_i, both of which tend to antagonize the activation by glucose-6-phosphate. Thus, there is a graded series of responses to changing metabolic conditions, involving a series of different protein kinases. The control mechanisms affecting the three kinases other than cAMP-dependent protein kinase and SPK are not yet well understood.

Whichever kinase is used, the net effect of phosphorylating glycogen synthase is inhibition of the enzyme, with consequent inhibition of glycogen synthesis. Note that the glycogen *synthesis* cascade has one less cycle than the glycogen *breakdown* cascade, because C can phosphorylate glycogen synthase directly, whereas it can act on glycogen phosphorylase only through its action on SPK. The extra cycle allows for more sensitive regulation of glycogen breakdown than of its synthesis, which is consonant with needs of animals for exceedingly rapid energy generation in muscle. In fact, experimental observations show that the maximum rate of muscle glycogen breakdown is some 300-fold higher than that of glycogen synthesis.

DEPHOSPHORYLATION OF GLYCOGEN SYNTHASE D. A number of different phosphatases act to regenerate the dephosphorylated forms of glycogen synthase and SPK. Of primary physiological importance is **phosphoprotein phosphatase,** also called **PP-1.** The activity of this enzyme in turn is controlled by a protein called **phosphoprotein phosphatase inhibitor, PI-1.** A phosphorylated form of PI-1 is active as an inhibitor, while the dephosphorylated form is inactive. Phosphorylation is carried out by C, the cAMP-dependent protein kinase. Thus, as shown in Figure 16.11, cAMP exerts two effects in inhibiting glycogen synthesis: (1) inactivation of glycogen synthase and (2) inhibition of phosphoprotein phosphatase, whose activity would tend to restore activity of glycogen synthase.

Phosphoprotein phosphatase is also one of several sites of action of

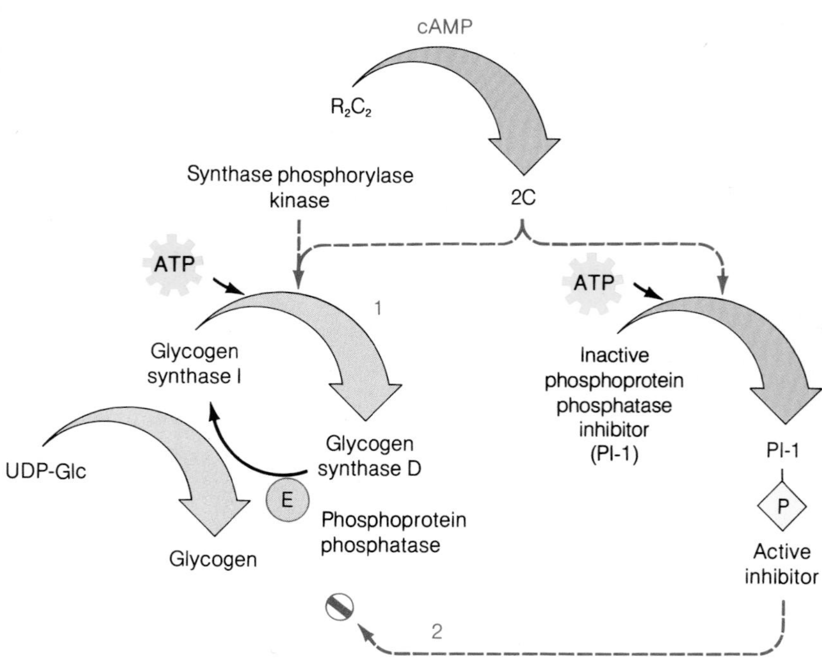

Figure 16.11
Regulation of glycogen synthase activity through cAMP-mediated control of phosphoprotein phosphatase activity.

insulin. Secretion of this hormone stimulates the phosphatase activity, which tends to oppose the effects of glucagon and epinephrine.

Functions of Glycogen Stores in Muscle and Liver

Glycogen is the major energy source for muscle contraction. Because liver derives most of its own metabolic energy from fatty acid oxidation, liver glycogen has quite a different function—as a source for blood glucose, to be transported to other tissues for catabolism. Liver serves primarily as a "glucostat," sensing the blood glucose levels and adjusting synthesis and breakdown of glycogen accordingly. As befits this role, the liver contains relatively large glycogen stores, from 2 to 8% of the weight of the organ. In liver the maximal rates of glycogen synthesis and degradation are about equal, while in muscle the maximal rate of glycogenolysis exceeds that of glycogen synthesis by some 300-fold. Although the enzymology of glycogen synthesis and breakdown is similar in liver and muscle, the endocrine control in liver is quite different, as we shall discuss in Chapter 23. The enzymes differ structurally as well.

Congenital Defects of Glycogen Metabolism in Humans

Several of the enzymes of glycogen metabolism can undergo inactivation by mutation, though not necessarily with lethal consequences. The clinical symptoms of these conditions, called **glycogen storage diseases,** can be quite severe and usually result from storage of abnormal quantities of glycogen or storage of glycogen with abnormal properties. This accumulation results from failure of an abnormal glycogen to be broken down.

Among the earliest glycogen storage diseases to be described was **von Gierke's disease,** named for a German physician who studied an 8-year-old girl with chronically enlarged liver. After her death from influenza, her liver was found to contain 40% glycogen. The glycogen appeared normal but could not be degraded by extracts of the girl's liver, only by extracts of other livers. Today it is recognized that these symptoms can result from deficiency of either glucose-6-phosphatase or the debranching enzyme; in the latter case phosphorylase can degrade glycogen only until branch points are reached and no farther.

Table 16.2 provides information on several of the glycogen storage diseases that have been characterized. Among the most serious clinically is the type I disease, resulting from functional lack of glucose-6-phosphatase. Individuals with this condition can break down glycogen normally, but since they are deficient in ability to cleave G6P to glucose for release from liver to the bloodstream, they are chronically hypoglycemic. In a less severe form of this disease blood glucose levels are normal, except after stress, when the normal hyperglycemic response is inhibited. One form of this disease (type Ia) results from deficiency of glucose-6-phosphatase itself; the type Ib disease involves deficiency of a specific **translocase,** a protein that helps transport glucose-6-phosphatase to its site in the lumen of the endoplasmic reticulum.

Other forms of glycogen storage diseases have abnormalities that can be easily understood in terms of the known enzymatic defect. In type III patients, with a defective debranching enzyme, glycogen with very short outer branches accumulates, which leads to enlargement of the liver. By contrast, type IV disease, which is associated with a defective branching enzyme, involves accumulation of glycogen with very long outer branches. Early death from liver failure is often observed. Type V, VI, VII, and VIII diseases have less severe symptoms. For instance, individuals with type V

Table 16.2
Human congenital defects of glycogen metabolism

Type	Common Name	Enzyme Deficiency	Glycogen Structure	Organ Affected
Ia	von Gierke's disease	Glucose-6-phosphatase	Normal	Liver, kidney, intestine
Ib		Glucose-6-phosphatase translocase	Normal	Liver
II	Pompe's disease	$\alpha(1\rightarrow 4)$glucosidase	Normal	Generalized
III	Cori's disease	Debranching enzyme	Short outer chains	Liver, heart, muscle
IV	Andersen's disease	Branching enzyme	Abnormally long unbranched chains	Liver and other organs
V	McArdle–Schmidt–Pearson disease	Muscle glycogen phosphorylase	Normal	Skeletal muscle
VI	Hers' disease	Liver glycogen phosphorylase	Normal	Liver, leukocytes
VII		Muscle phosphofructokinase	Normal	Muscle
VIII		Liver phosphorylase kinase	Normal	Liver

disease, with a deficiency of muscle glycogen phosphorylase, usually show no symptoms until about age 20. Once symptoms have appeared, the principal ones are severe muscle cramps on exercise and failure of lactate to accumulate in blood after exercise.

Biosynthesis of Other Polysaccharides

Synthesis of other polysaccharides involves the same mechanisms as we have just presented for glycogen, particularly the use of nucleotide-linked sugars as activated biosynthetic intermediates. In this section we consider briefly the synthesis of the most abundant and widely distributed polysaccharides—starch, cellulose, dextran, and several heteropolysaccharides.

UDP-glucose is used in some plant species for the synthesis of cellulose, a straight-chain glucose homopolymer with $\beta(1\rightarrow 4)$ linkages. The mechanism is identical to that of glycogen synthesis, except for the stereochemistry of glycosidic bond formation. Different nucleotide-linked sugars are also active in polysaccharide synthesis. *Adenosine* diphosphate glucose and *cytidine* diphosphate glucose are the substrates for cellulose biosynthesis in some plants. Moreover, ADP-glucose is the intermediate used in plant starch synthesis and in the synthesis of glycogen in bacterial cells. The structural polysaccharide chitin is synthesized in insects from UDP-*N*-acetylglucosamine, the product being a $\beta(1\rightarrow 4)$-linked polymer of the sugar. The synthesis of hyaluronic acid in animals involves two enzymes, because the product is a heteropolymer with a strictly alternating sequence of glucuronic acid and *N*-acetylglucosamine. The two enzymes use as substrates UDP-glucuronate and UDP-*N*-acetylglucosamine, respectively.

An intriguing exception to the use of nucleoside diphosphate sugars as activated intermediates is the biosynthesis in some bacteria of **dextran**, an $\alpha(1\rightarrow 6)$-linked polymer of glucose with $\alpha(1\rightarrow 2)$, $\alpha(1\rightarrow 3)$, or $\alpha(1\rightarrow 4)$ branch points. The polymerization, catalyzed by **dextran sucrase**, involves a transglycosylation reaction with sucrose as the substrate.

$$n \text{ Sucrose} \longrightarrow \text{glucose}_n \text{ (dextran)} + n \text{ fructose}$$

Several bacteria growing in the human oral cavity synthesize large quantities of dextran, which is a major component of dental plaque. Hence the concern of nutritionists with sucrose consumption.

The synthesis of sucrose itself in plants occurs via a nucleotide-linked sugar, as shown below.

UDP-glucose + fructose-6-phosphate ⟶ sucrose-6-phosphate + UDP

sucrose-6-phosphate + H_2O ⟶ sucrose + P_i

Biosynthesis of Amino Sugars

Amino sugars are major constituents of **glycoconjugates**—macromolecules containing covalently bound oligosaccharide chains. As described in Chapter 8, the glycoconjugates include glycoproteins and glycolipids. Here we consider the biosynthesis of some of the building blocks for glycoconjugates, the amino sugars themselves.

Glucose can serve as a metabolic precursor for all other sugars, although in most animals other sugars are available in the diet as well. The immediate precursor to the earliest-formed amino sugar is not glucose, but fructose-6-phosphate. The nitrogen comes from the amide group of glutamine, in an essentially irreversible reaction catalyzed by **glutamine: fructose-6-phosphate amidotransferase.**

Fructose-6-phosphate (CH_2OH–C=O–HO–C–H–H–C–OH–H–C–OH–CH_2OⓅ) + **Glutamine** (O=C–NH_2–CH_2–CH_2–H–C–$\overset{+}{N}H_3$–COO^-) ⟶ **Glucosamine-6-phosphate** (O=C–H–H–C–NH_2–HO–C–H–H–C–OH–H–C–OH–CH_2OⓅ) + **Glutamate** (COO^-–CH_2–CH_2–H–C–$\overset{+}{N}H_3$–COO^-)

Note that the enzyme catalyzes both amide group transfer and an internal oxidoreduction of the sugar, with oxidation of C-1 coupled to reduction of C-2.

Glucosamine-6-phosphate is then converted to UDP-*N*-acetylglucosamine by a three-step sequence, shown in Figure 16.12.

The biosynthesis of sialic acid proceeds via an epimerization of UDP-*N*-acetylglucosamine, comparable to the interchange of UDP-glucose and UDP-galactose (Figure 16.13). The product, UDP-*N*-acetylgalactosamine, is converted in a short sequence to ***N*-acetylmannosamine-6-phosphate.** The latter undergoes a reaction with phosphoenolpyruvate, somewhat akin to an aldol condensation. The product, ***N*-acetylneuraminic acid 9-phosphate,** is cleaved to give *N*-acetylneuraminic acid, or sialic acid. Metabolic activation of sialic acid for oligosaccharide biosynthesis involves not a nucleoside diphosphate sugar but a nucleoside *monophosphate* sugar, **cytidine monophosphate–sialic acid,** or CMP-sialic acid.

CTP + sialic acid ⟶ CMP-sialic acid + PP_i

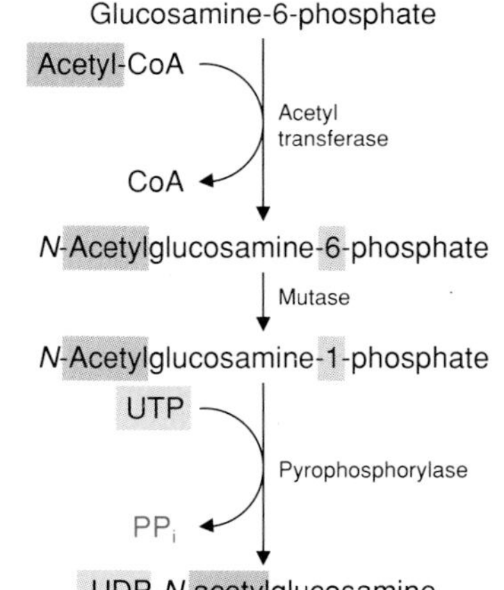

Figure 16.12
Pathway for biosynthesis of UDP-*N*-acetylglucosamine from glucosamine-6-phosphate.

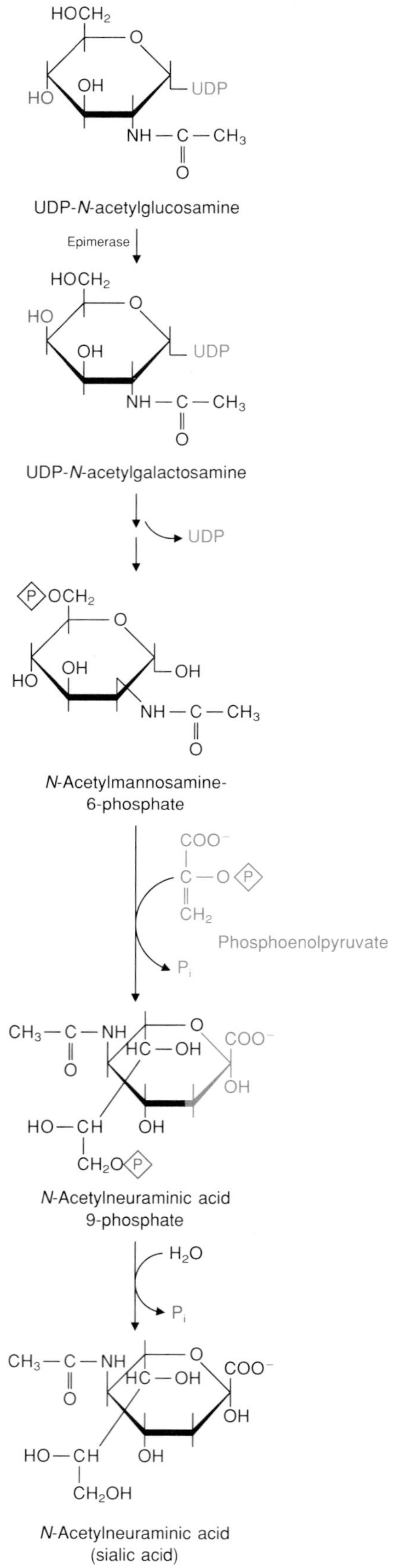

Interestingly, CMP-sialic acid is synthesized in the nucleus of animal cells, whereas all other known nucleotide-linked sugars are synthesized in the cytosol.

A fascinating aspect of amino sugar metabolism involves the biosynthesis of **N-acetyl-β-lactosamine,** a constituent of the carbohydrate portion of glycoproteins.

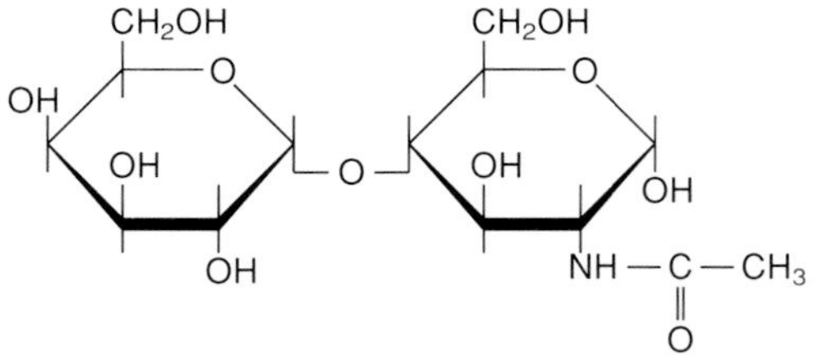

N-Acetyl-β-lactosamine

The synthetic reaction is catalyzed by **galactosyltransferase.**

UDP-galactose + N-acetylglucosamine ⟶ UDP + N-acetyllactosamine

The enzyme contains a single polypeptide subunit, but the presence of an additional subunit changes the specificity of the enzyme so that lactose is synthesized instead.

UDP-galactose + glucose ⟶ UDP + lactose

This modified enzyme, termed **lactose synthase,** is found in animals only in mammary gland, where it synthesizes the major sugar of milk. The polypeptide that modifies the specificity of the enzyme is the mammary gland protein **α-lactalbumin.** Synthesis of α-lactalbumin is activated hormonally in mothers shortly after giving birth. The protein combines with preexisting galactosyltransferase, changes its specificity, and activates the large amount of lactose synthesis needed for milk production.

Biosynthesis of Glycoconjugates

As indicated in Chapter 8, the covalent binding of carbohydrate to protein or lipid brings about large changes in the physical properties of these substances that allow them to serve specialized biochemical functions. In glycoproteins rich in carbohydrate (50% or more), also called **proteoglycans,** the presence of sulfated polysaccharides or sialic acid provides a net negative charge that make these substances effective biological lubricants. In glycolipids the carbohydrate portion provides a polar head group, allowing these components to be inserted into membranes, where they occur abundantly in the outer layer of the plasma membrane of most cells. The biosynthesis of glycolipids is presented in Chapter 18.

As noted in Chapter 8, the vast structural diversity possible among carbohydrates allows them to be used for molecular recognition in glycoproteins, even when present in very small amounts. The carbohydrate components of cell surface glycoproteins direct the interaction of these cells with

UDP-*N*-acetylglucosamine

Epimerase

UDP-*N*-acetylgalactosamine

UDP

N-Acetylmannosamine-6-phosphate

Phosphoenolpyruvate

P_i

N-Acetylneuraminic acid 9-phosphate

H_2O

P_i

N-Acetylneuraminic acid (sialic acid)

Figure 16.13
Biosynthesis of *N*-acetylneuraminic acid (sialic acid) from *N*-acetylglucosamine.

other cells and with their environment. Such interactions involve processes as diverse as cell movements in development, cell adhesion, cell motility, cell growth control, oncogenic transformation, and **endocytosis** (internalization of material from the extracellular milieu).

In this section we consider the biosynthesis of the carbohydrate components of some selected glycoconjugates—the oligosaccharide components of glycoproteins and the peptidoglycan and O-antigen portions of bacterial cell walls. The examples chosen have general biological interest, and they illustrate mechanisms used in synthesizing the vast known array of glycoconjugates. In general, those mechanisms involve the action of specific **glycosyltransferases** that, like glycogen synthase, transfer a monosaccharide unit from a nucleotide-linked sugar to a nonreducing end of an oligosaccharide chain, or to an appropriate functional group on the protein component. As stated earlier, the nucleotide-linked sugars are synthesized in the cytosol, with the exception of CMP-sialic acid, which is formed in the nucleus. The glycosyltransferases are bound to membranes of the smooth or rough endoplasmic reticulum or the Golgi apparatus. A particularly fascinating part of glycoprotein synthesis is the protein "sorting" or "traffic" that occurs as oligosaccharide chains are growing. In some cases the structure of the oligosaccharide chain or chains on a protein helps to direct that protein to the proper intracellular location for the next step in oligosaccharide synthesis and, ultimately, to the site where the mature protein will reside.

O-Linked Oligosaccharides: Blood Group Antigens

Mammalian glycoproteins are classified as **O-linked** or **N-linked.** As indicated in Chapter 8, N-linked glycoproteins contain an *N*-acetylglucosamine residue linked to the amide nitrogen of an asparagine residue in the protein, while the most common O linkage involves a terminal *N*-acetylgalactosamine residue in the oligosaccharide linked to a serine or threonine residue. The A, B, AB, and O blood group substances are the best known examples of the latter.

The sugars in these oligosaccharides are fucose (Fuc), galactose (Gal), *N*-acetylgalactosamine (GalNAc), and sialic acid (Sia). The synthetic pathway involves a series of membrane-bound glycosyltransferases, each of which uses a nucleotide-linked sugar substrate, as shown in Figure 16.14. The pathway leads to two different pentasaccharide products, the A antigen and the B antigen. The last enzyme in synthesis of the A antigen is a glycosyltransferase involving UDP-GalNAc, while in synthesis of the B antigen the last reaction involves UDP-Gal. Individuals with type A blood have the former glycosyltransferase and synthesize the A antigen, while type B individuals have the galactosyltransferase. AB individuals are heterozygotes; that is, they have both enzymes and carry both blood group substances on their erythrocytes. Type O individuals carry neither glycosyltransferase and hence carry the tetrasaccharide identified as O substance in Figure 16.14.

N-Linked Oligosaccharides: Glycoproteins

Quite a different process is involved in the synthesis of N-linked oligosaccharides. Here, oligosaccharide assembly occurs not on the polypeptide chain but on a *lipid-linked intermediate.* A precursor oligosaccharide is then transferred to a polypeptide chain, which is itself still in the midst of being

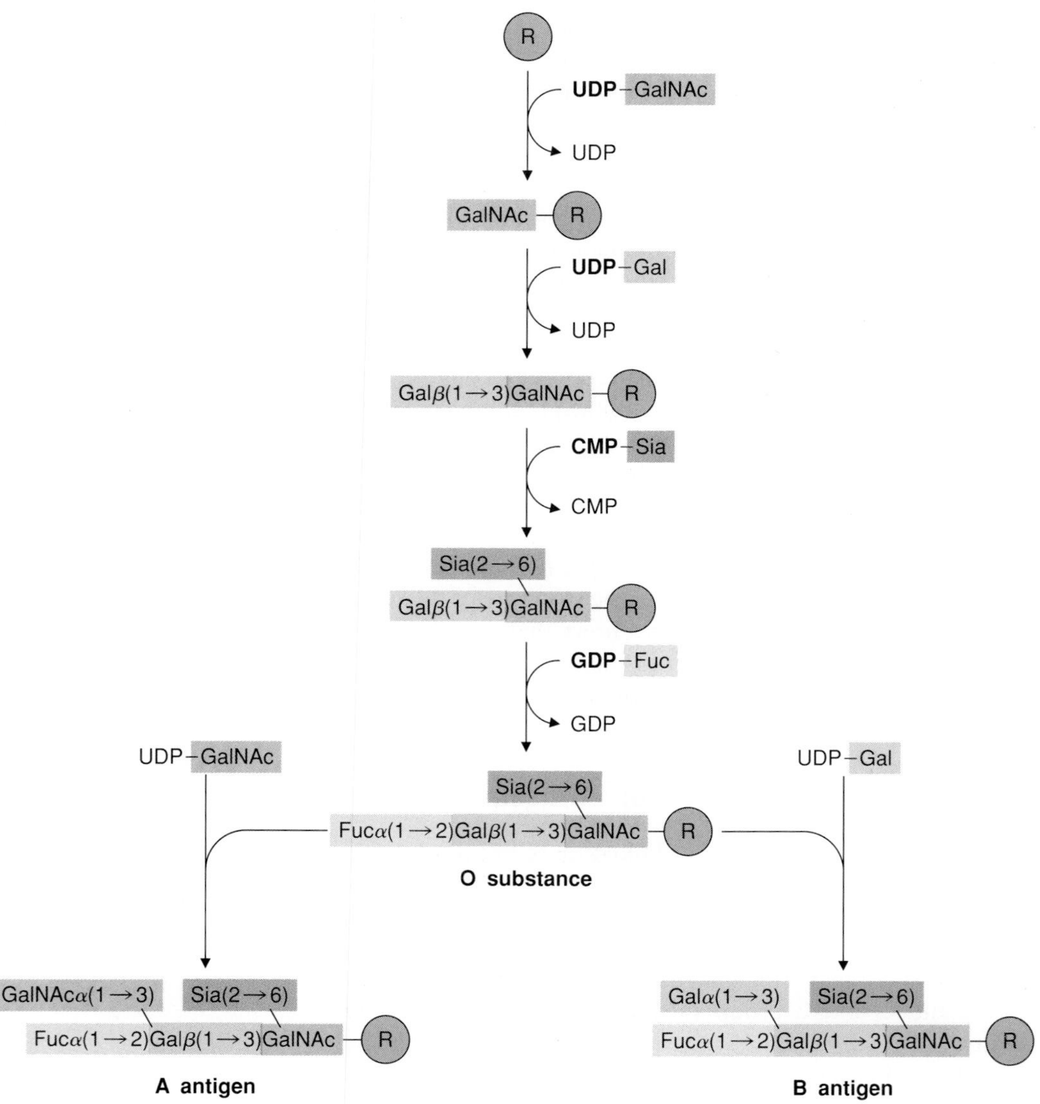

Figure 16.14
Biosynthesis of O-linked oligosaccharide units on glycoproteins of the O, A, and B group substances.

synthesized. This is called a **cotranslational** reaction. Finally, the transferred oligosaccharide is subject to further processing steps as it passes from the rough and smooth endoplasmic reticulum through the Golgi apparatus.

The many N-linked glycoproteins of known structure can be categorized in terms of three basic oligosaccharide structures, as summarized in Figure 16.15—**high mannose, complex,** and **hybrid.** All of the known N-linked oligosaccharides have a common core pentasaccharide structure: Manα1→3(Manα1→6)Manβ1→4GlcNAcβ1→4GlcNAc-Asn (the boxed areas in the figure). That core is assembled as part of a larger oligosaccharide intermediate linked to the isoprenoid lipid compound **dolichol phosphate.**

$$\mathrm{H(CH_2{-}\overset{\displaystyle CH_3}{\overset{|}{C}}{=}CH{-}CH_2)_{18\text{–}20}{-}CH_2{-}\overset{\displaystyle CH_3}{\overset{|}{C}H}{-}CH_2{-}CH_2{-}O{-}\underset{\underset{\displaystyle O^-}{|}}{\overset{\overset{\displaystyle O}{\|}}{P}}{-}O^-}$$

Dolichol phosphate

In vertebrate tissues dolichol contains 18 to 20 isoprenoid units with two **trans** double bonds, the remainder being **cis,** except for a saturated terminal isoprene unit. Dolichol is synthesized from the same pathway that yields cholesterol, other sterols, and other isoprenoid compounds (Chapter 18) and then converted to dolichol phosphate (Dol-P) by a CTP-requiring kinase.

The first step in glycoprotein synthesis is assembly of a lipid-linked oligosaccharide intermediate, which serves as precursor to all known N-linked oligosaccharides. That process takes place in the endoplasmic reticulum (ER). Figure 16.16 summarizes the biosynthetic pathway leading to this intermediate. The first seven sugars are transferred from nucleoside diphosphate sugars, UDP-GlcNAc and GDP-Man. Each reaction is catalyzed by a separate glycosyltransferase. The first of these enzymes is specifically inhibited by the antibiotic **tunicamycin,** which blocks the reaction of UDP-GlcNAc and Dol-P and, hence, inhibits the synthesis of all N-linked glycoproteins.

The next seven glycosyltransferases utilize as substrates not nucleotide-linked sugars but dolichol-linked intermediates, Dol-P-Man and Dol-P-Glc. These in turn are synthesized from Dol-P and GDP-Man or UDP-Glc, respectively.

The actual site of synthesis of the lipid-linked oligosaccharide has not been established, because nucleotide-linked sugars cannot penetrate to the lumen of the rough ER. Current data suggest that the $Man_5GlcNAc_2$ intermediate (the second lipid-linked sugar shown in Figure 16.16) is assembled on the cytosolic side of the ER membrane, followed by a translocation to the luminal side, where the dolichol phosphate-mediated steps occur. The next step, transfer of the oligosaccharide unit to a polypeptide acceptor, does occur in the lumen, catalyzed by a specific **oligosaccharyltransferase.** The acceptor site is an asparagine residue in the sequence Asn–X–Ser/Thr, where X is any amino acid. The acceptor site must be accessible in a loop or a bend in the polypeptide chain, and this is probably why the transfer occurs simultaneously with translation of the acceptor polypeptide. The three glucosyl residues on the transferred oligosaccharyl unit somehow facilitate the transfer but are not absolutely required. For virtually all glycoproteins these glucosyl residues are removed in subsequent processing steps, which begin in the lumen of the rough endoplasmic reticulum, and continue as the nascent glycoprotein moves into the smooth endoplasmic reticulum and ultimately through various cisternae of the Golgi apparatus.

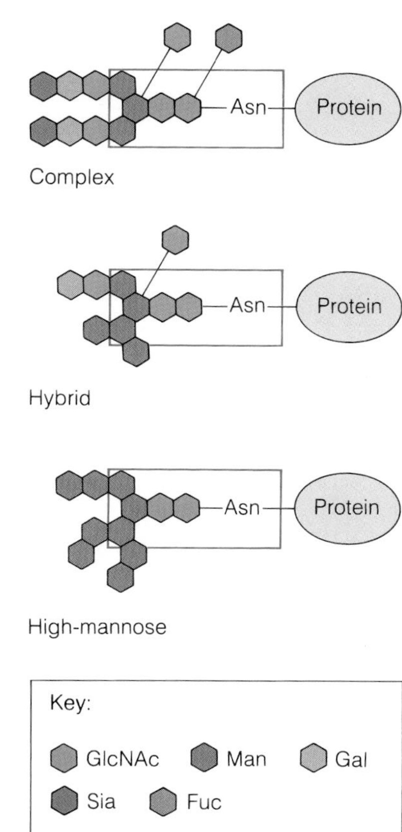

Figure 16.15
Structures of the major types of asparagine-linked oligosaccharides. The boxed areas contain the core common to all known N-linked structures.

A great variety of oligosaccharide structures is generated during this processing. Part of the diversity arises from differences in conformation of the protein moiety in the vicinity of a carbohydrate chain, which affect the accessibility of the chain to glycosidases and glycosyltransferases. Oligosaccharide chains probably generate recognition sites for targeting each processed compound to different sites within the Golgi, where specific membrane-bound processing enzymes are found. In all cases processing begins (after transfer to a polypeptide chain) with removal of the three glucosyl residues in the rough ER, followed by removal of some of the mannosyl residues in the Golgi apparatus, as schematized in Figure 16.17. Those glycoproteins destined to be of the "complex" type are further processed by addition of N-acetylglucosamine, followed by further trimming of mannosyl residues. Fucosyl, galactosyl, and sialyl residues are added from appropriate nucleotide-linked sugars by specific glycosyltransferases.

A graphic demonstration of the power of processing to direct intracellular protein traffic is provided by the generation of mannose-6-phosphate residues in one specific class of glycoproteins, the lysosomal acid hydrolases.

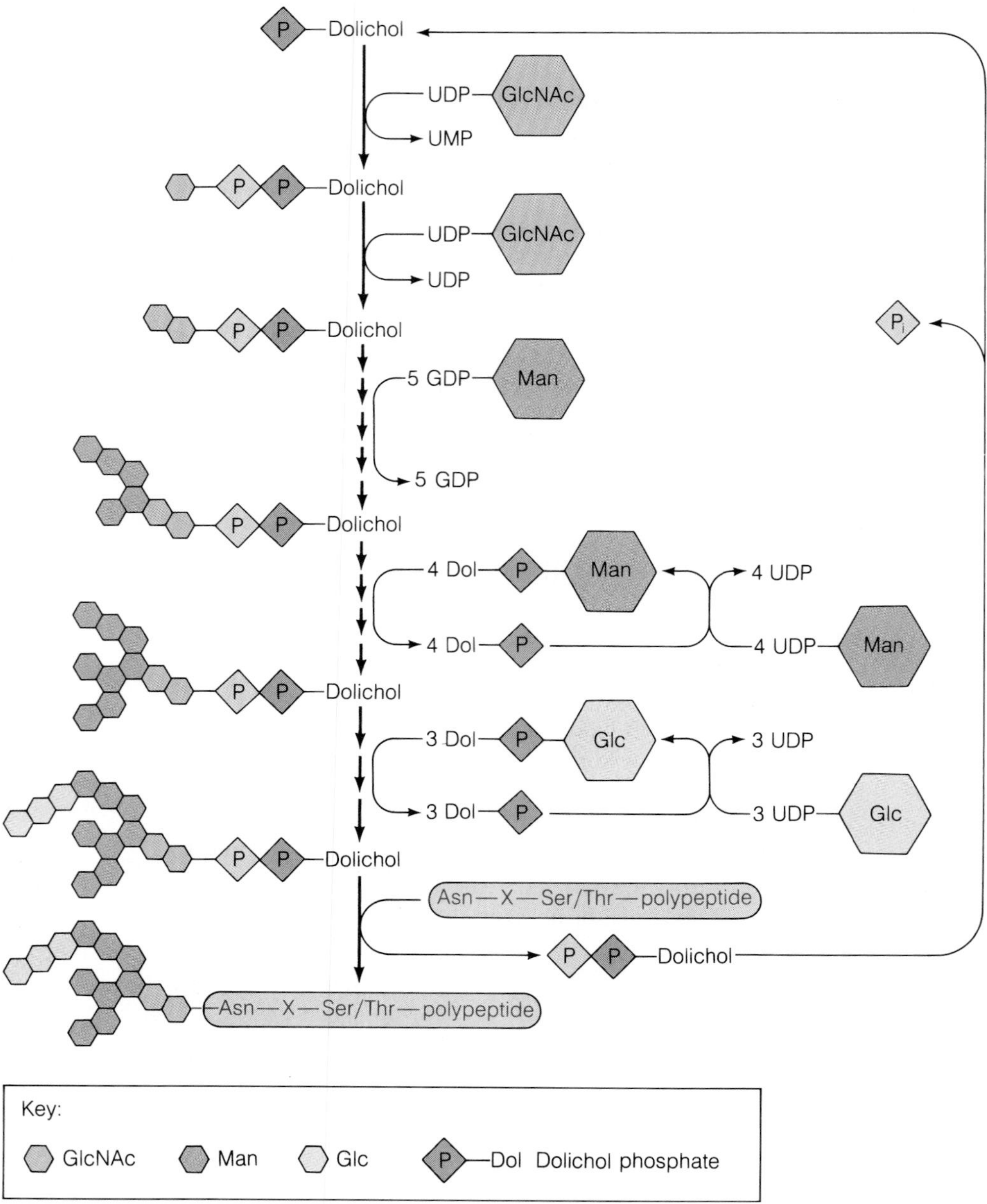

Figure 16.16
Biosynthesis of the lipid-linked oligosaccharide intermediate. The five sequential mannosyl transfer reactions from GDP-mannose are catalyzed by separate glycosyltransferases, as are the four mannosyl transfers from mannose phosphoryl-dolichol. The latter is synthesized in turn from GDP-mannose. The acceptor site on the polypeptide chain is an asparagine residue two positions to the N side of a serine or threonine. The whole process occurs in the endoplasmic reticulum.

All known enzymes of this type contain between one and five mannose-6-phosphate units, which evidently help to *target* proteins to, and through, the lysosomal membrane. This role of mannose-6-phosphate in directing protein traffic is confirmed by the existence of a rare and fatal congenital abnormality called **I-cell disease.** In this condition lysosomal function is defective, because the acid hydrolase content of lysosomes is very low as the result of a deficiency of the first glycosyltransferase that generates a mannose-6-phosphate residue. Hence, lysosomal enzymes cannot be targeted to their ultimate destinations but instead are secreted into the extracellular milieu. When purified and characterized, these enzymes are also found to lack mannose-6-phosphate in their oligosaccharide chains.

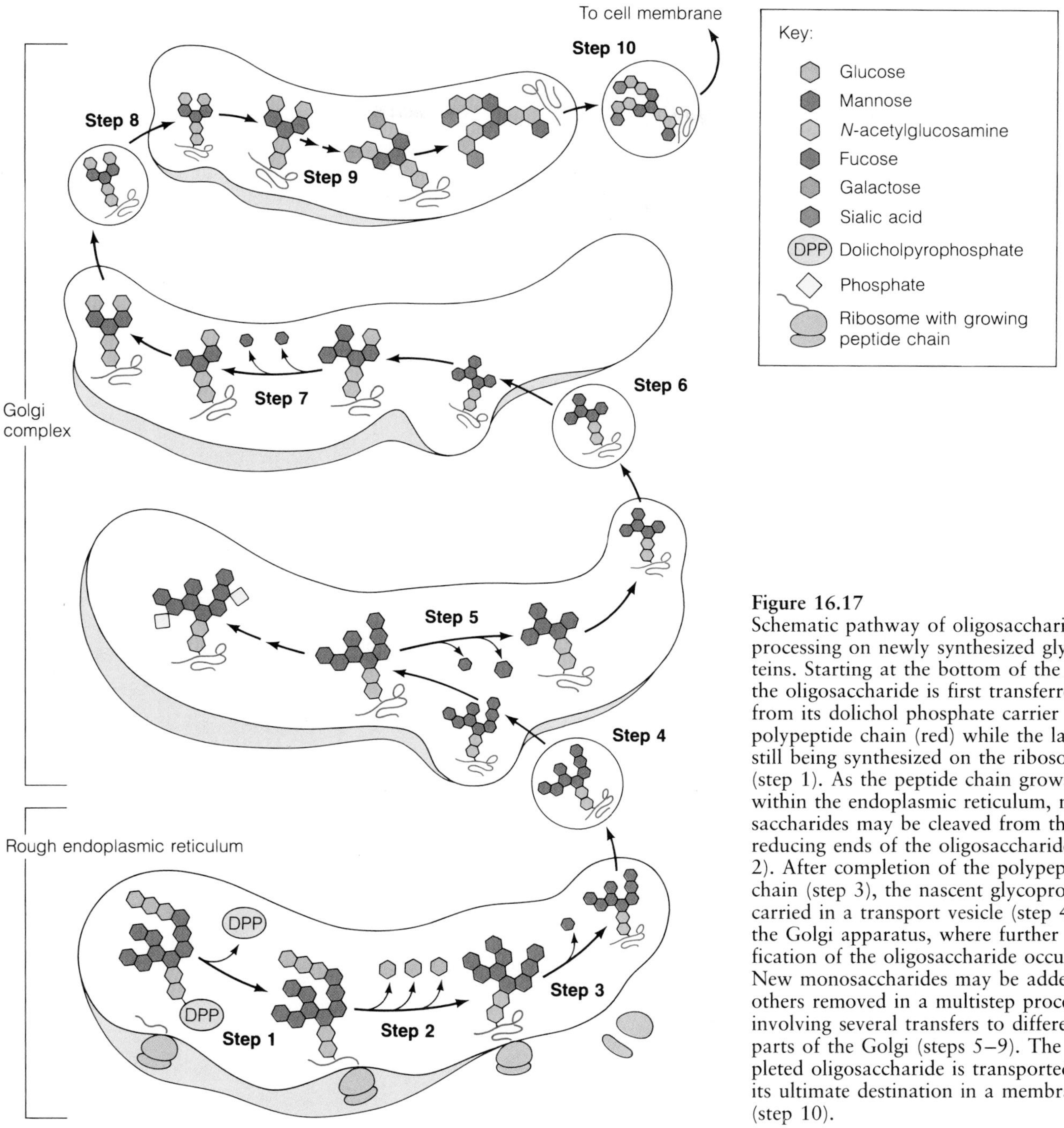

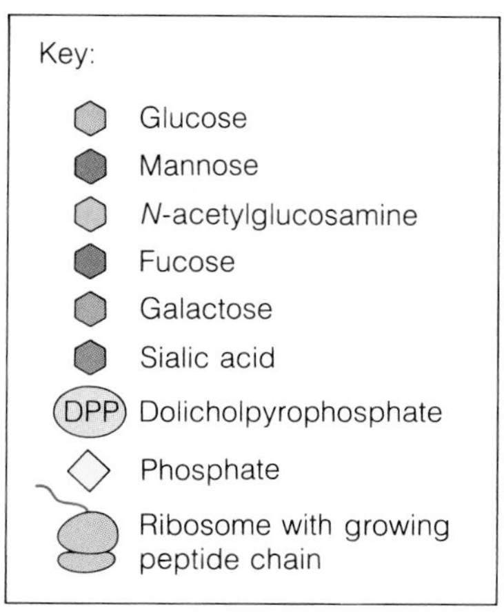

Figure 16.17
Schematic pathway of oligosaccharide processing on newly synthesized glycoproteins. Starting at the bottom of the figure, the oligosaccharide is first transferred from its dolichol phosphate carrier to a polypeptide chain (red) while the latter is still being synthesized on the ribosome (step 1). As the peptide chain grows within the endoplasmic reticulum, monosaccharides may be cleaved from the nonreducing ends of the oligosaccharide (step 2). After completion of the polypeptide chain (step 3), the nascent glycoprotein is carried in a transport vesicle (step 4) to the Golgi apparatus, where further modification of the oligosaccharide occurs. New monosaccharides may be added and others removed in a multistep process involving several transfers to different parts of the Golgi (steps 5–9). The completed oligosaccharide is transported to its ultimate destination in a membrane (step 10).

Microbial Cell Wall Polysaccharides: Peptidoglycan

Recall from Chapter 8 that Gram-positive bacteria contain a rigid peptidoglycan cell wall surrounding the cytoplasmic membrane. Gram-negative bacteria contain, in addition to the cytoplasmic membrane and peptidoglycan layer, a third layer, the outer membrane, a complex structure that contains lipoproteins and lipopolysaccharides (refer back to Figure 8.27). The structural diversity of these macromolecules among different bacterial species is enormous, and space permits introduction of only two of the most interesting pathways. The first pathway, biosynthesis of the peptidoglycan layer in *Staphylococcus aureus*, is of interest for two reasons: first, the pathway is the site of action of several important antibiotics, notably penicillins, and second, much of the biosynthetic process occurs on the outside of the cell, where there is no ready supply of energy in the form of ATP. Recall from Chapter 8 that bacterial peptidoglycans consist of polymeric chains of amino sugars cross-linked by oligopeptide chains, to form a huge three-dimensional network in which the entire peptidoglycan layer of a cell is one giant macromolecule (see Figure 8.27). The glycan, or polysaccharide, chain is an alternating polymer of *N*-acetylglucosamine and *N*-acetylmuramic acid, the latter a derivative of *N*-acetylglucosamine. In *S. aureus* the carboxyl groups of all the *N*-acetylmuramic acid residues are linked to the terminal amino group of the tetrapeptide L-alanyl-D-γ-isoglutaminyl-L-lysyl-D-alanine. Each cross-link takes the form of a pentaglycine chain that joins the carboxyl group of a D-alanine residue to the ϵ-amino group of a lysine residue in an adjacent oligopeptide.

The biosynthesis of peptidoglycan can be considered in three distinct stages: (1) synthesis of *N*-acetylmuramylpentapeptide, (2) formation of the polysaccharide chain by polymerization of *N*-acetylglucosamine and *N*-acetylmuramylpentapeptide, and (3) cross-linking of individual peptidoglycan strands. Much of this pathway was elucidated in studies of the action of penicillin, which kills bacterial cells by blocking synthesis of the cell wall through inhibition of peptidoglycan synthesis.

SYNTHESIS OF *N*-ACETYLMURAMYLPENTAPEPTIDE. The first step begins with synthesis of UDP-*N*-acetylmuramic acid from UDP-*N*-acetylglucosamine.

Phosphoenolpyruvate

P_i

H^+ + NADPH → NAD^+

UDP-*N*-acetylglucosamine

UDP-*N*-acetylmuramic acid

One mole of phosphoenolpyruvate is transferred, to give the 3-enolpyruvyl ether of UDP-*N*-acetylglucosamine. The pyruvyl group is reduced by NADPH to give the 3-*O*-D-lactyl ether of *N*-acetylglucosamine; this compound is *N*-acetylmuramic acid. Then the peptide is built up, one residue at a time, as shown in Figure 16.18. Unlike the synthesis of polypeptide chains in protein synthesis (Chapter 27), no messenger RNA template and no ribosomes are involved here. The specificity lies in the sequential actions of a series of ATP-dependent ligases, which add first L-alanine, followed by D-glutamate (later amidated to D-isoglutamine), then L-lysine (linked to the γ-carboxyl group of glutamate), and finally the dipeptide D-alanyl-D-alanine. One of these two D-alanyl residues will be removed at a later step (see Figure 16.20).

FORMATION OF THE POLYSACCHARIDE CHAIN. The next stage is polymerization of *N*-acetylglucosamine and *N*-acetylmuramylpentapeptide to give a linear peptidoglycan chain. This process involves a lipid carrier, **undecaprenol phosphate,** comparable to the isoprenoid compound dolichol encountered earlier.

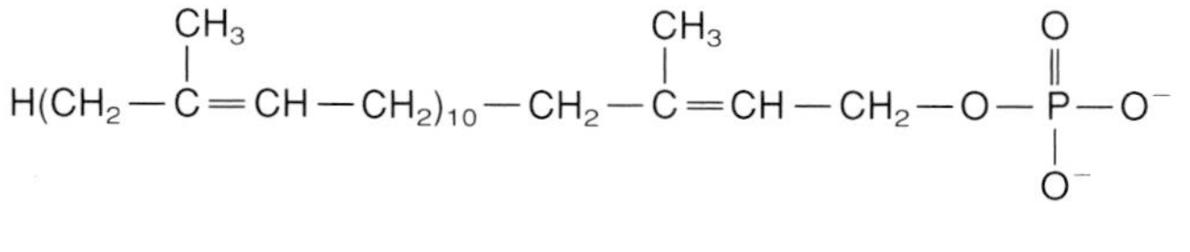

$$H(CH_2-\overset{CH_3}{\overset{|}{C}}=CH-CH_2)_{10}-CH_2-\overset{CH_3}{\overset{|}{C}}=CH-CH_2-O-\underset{O^-}{\underset{|}{\overset{O}{\overset{\|}{P}}}}-O^-$$

Undecaprenol phosphate

Undecaprenol phosphate is a 55-carbon compound containing 11 isoprenoid units, with phosphate linked at the terminus. To this phosphate is transferred the *N*-acetylmuramylpentapeptide moiety from UDP-*N*-acetylmuramylpentapeptide (Figure 16.19, step 1, p. 566). This compound then accepts *N*-acetylglucosamine from UDP-*N*-acetylglucosamine (step 2), followed by the sequential addition of five glycyl residues, from glycyl transfer RNA (step 3). It seems likely at this stage that the phospholipid carrier then serves to transport the peptidodisaccharide unit through the membrane (step 4), because polymerization—addition to the reducing end of a preexisting peptidoglycan chain—occurs outside the cell (step 5). The antibiotics **bacitracin** and **vancomycin** inhibit specific steps in this process at the sites shown in Figure 16.19.

CROSS-LINKING OF PEPTIDOGLYCAN STRANDS. Finally, the cross-linking occurs between adjacent chains, also outside the cell. This involves a **transpeptidation** reaction, with the cleavage of one peptide bond providing the energy needed to drive the formation of another peptide bond. This means that the free energy needed to drive cross-link formation was built into the structures while they were still accessible to ATP, inside the cell. As shown in Figure 16.20, the transpeptidation involves nucleophilic attack by the free terminal amino group of the pentaglycine chain on the amide carbon linking the terminal D-alanines in an adjacent chain.

Figure 16.18
Biosynthesis of UDP-*N*-acetylmuramylpentapeptide from UDP-*N*-acetylmuramic acid. The sugar structure is shown in outline form.

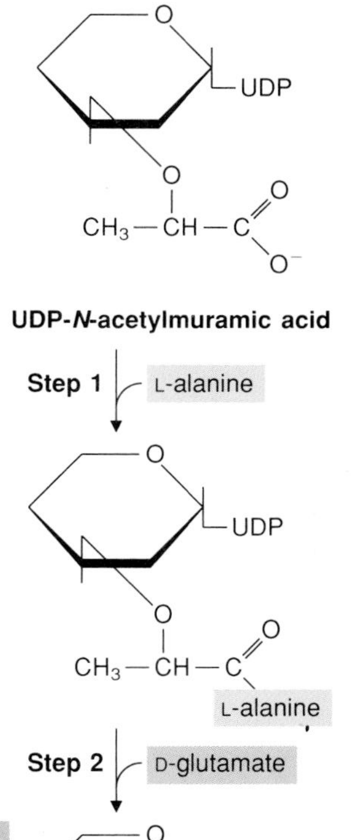

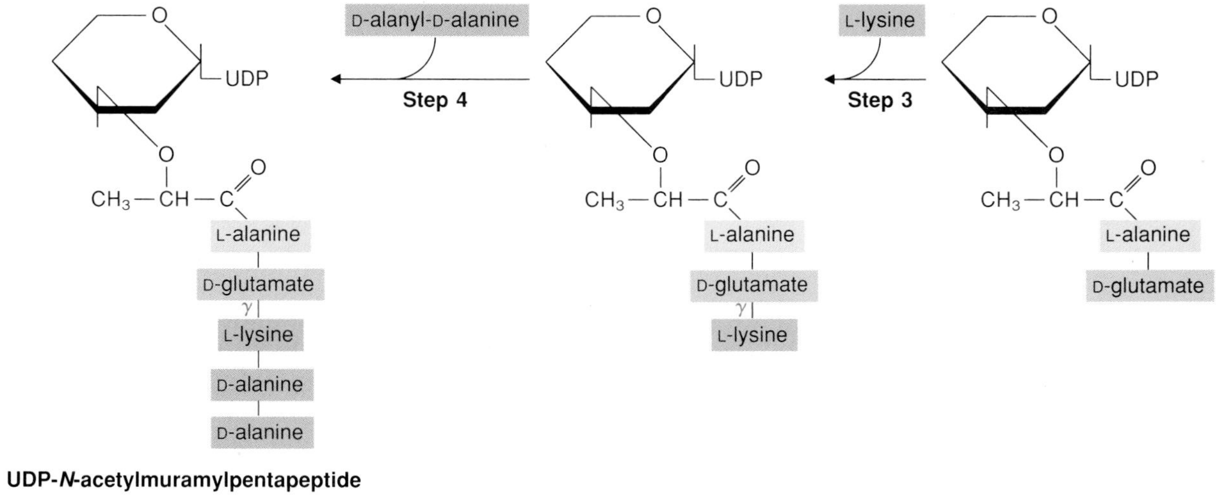

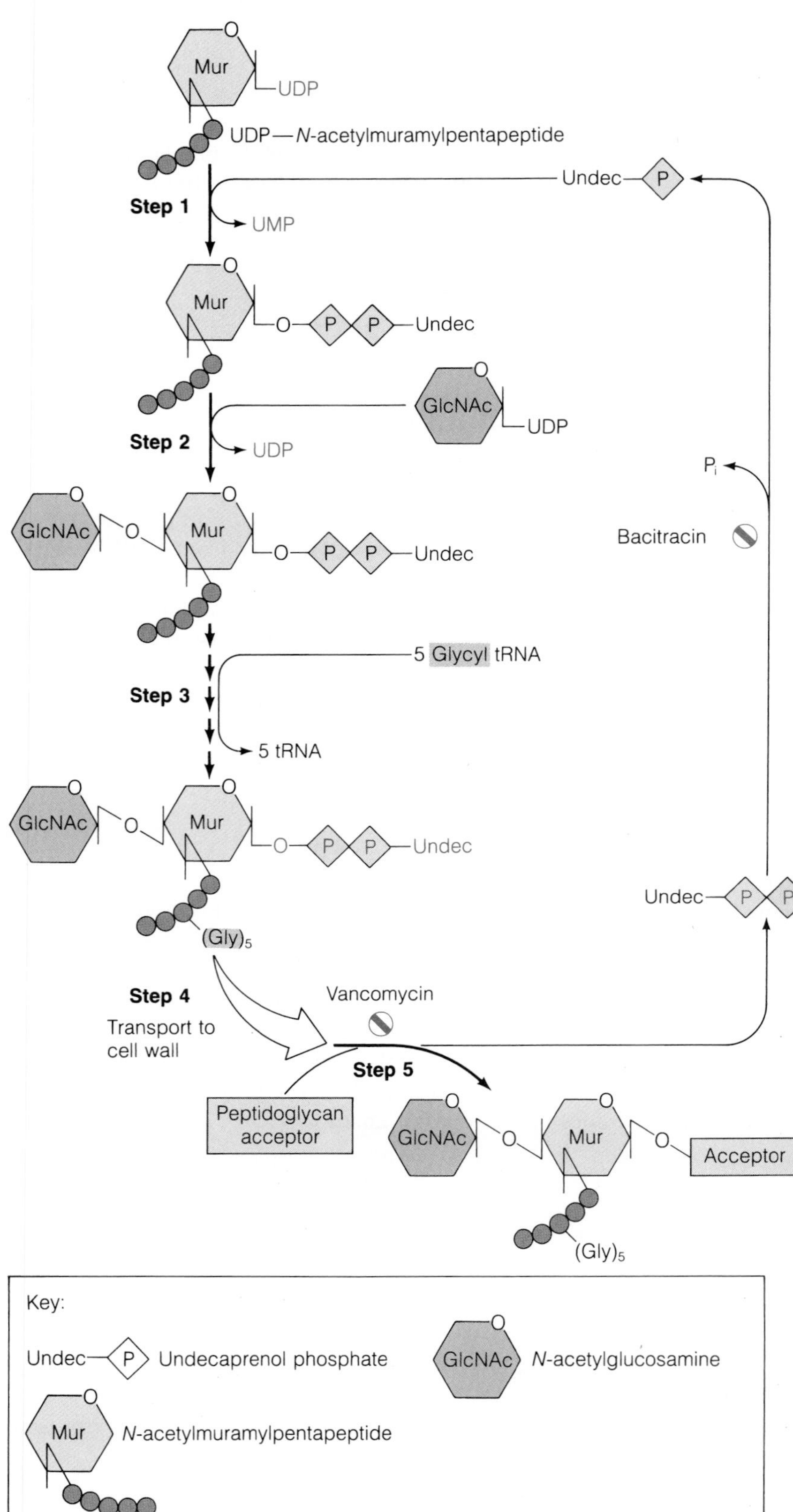

Figure 16.19
Synthesis of the linear peptidoglycan molecule of *S. aureus*. Not shown is an ATP-dependent amidation of the D-glutamate residue. Sites of inhibition by antibiotics are identified.

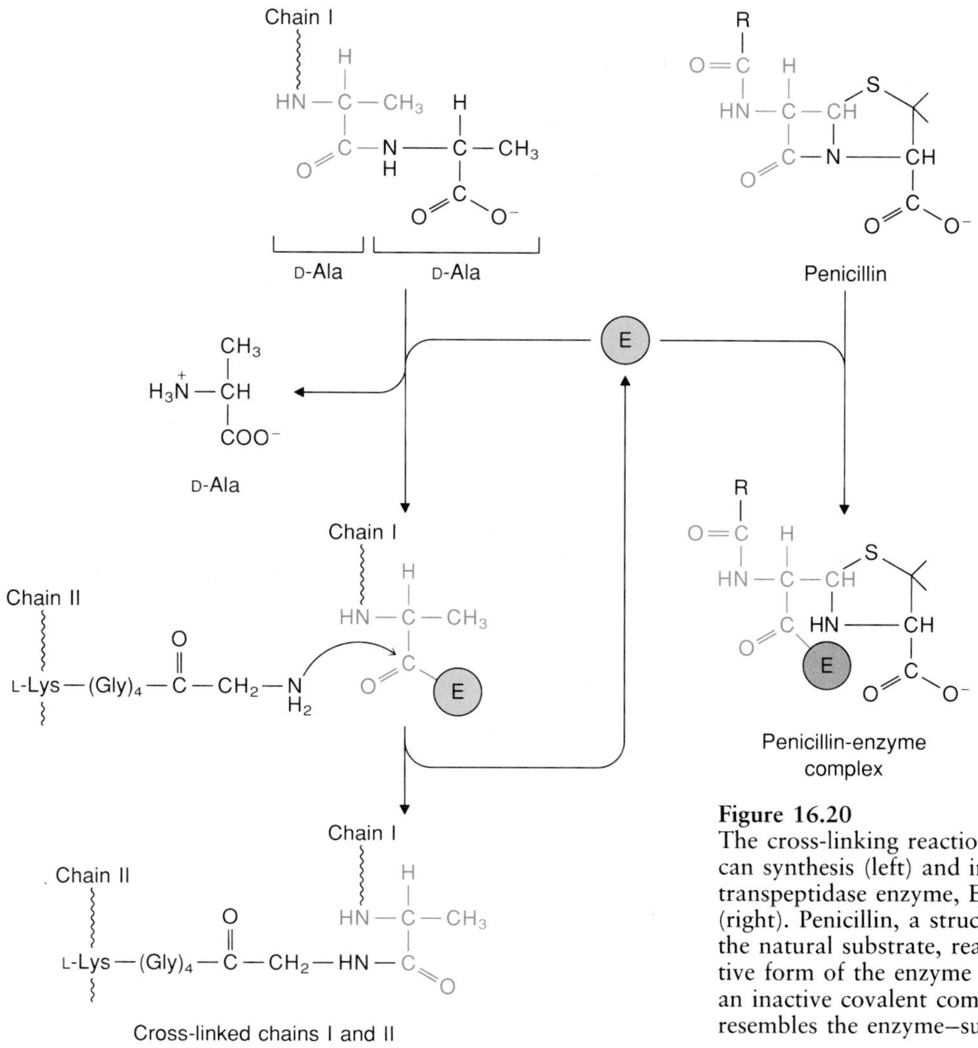

Figure 16.20
The cross-linking reaction in peptidoglycan synthesis (left) and inhibition of the transpeptidase enzyme, E, by penicillin (right). Penicillin, a structural analog of the natural substrate, reacts with the active form of the enzyme (green) to form an inactive covalent complex (red) that resembles the enzyme–substrate complex.

The cross-linking reaction is the target for the action of two important classes of antibiotics, the penicillins and the cephalosporins. Penicillin is thought to react irreversibly with the transpeptidase that catalyzes cross-linking. That enzyme normally forms an acyl–enzyme intermediate, via the penultimate D-alanine of the pentapeptide chain (Figure 16.20). Penicillin evidently resembles the terminal dipeptide of this structure to the point that it can also react with the transpeptidase. The reaction is driven in part by the strain built into the four-membered **lactam ring,** for that ring opens during the reaction. Penicillin has been widely studied as an "ideal" antibiotic, because the cross-linking reaction has no counterpart in animal metabolism. Since the bacterial cell must continue to synthesize cell wall in order to grow and divide, inhibition of a step in this process provides a completely specific way to interfere with the growth of bacterial pathogens. Unfortunately, resistance to penicillin can be acquired. This resistance usually involves the synthesis, directed by an extrachromosomal gene, of **lactamase,** an enzyme that hydrolyzes the lactam ring of penicillin and destroys its ability to interfere with peptidoglycan synthesis.

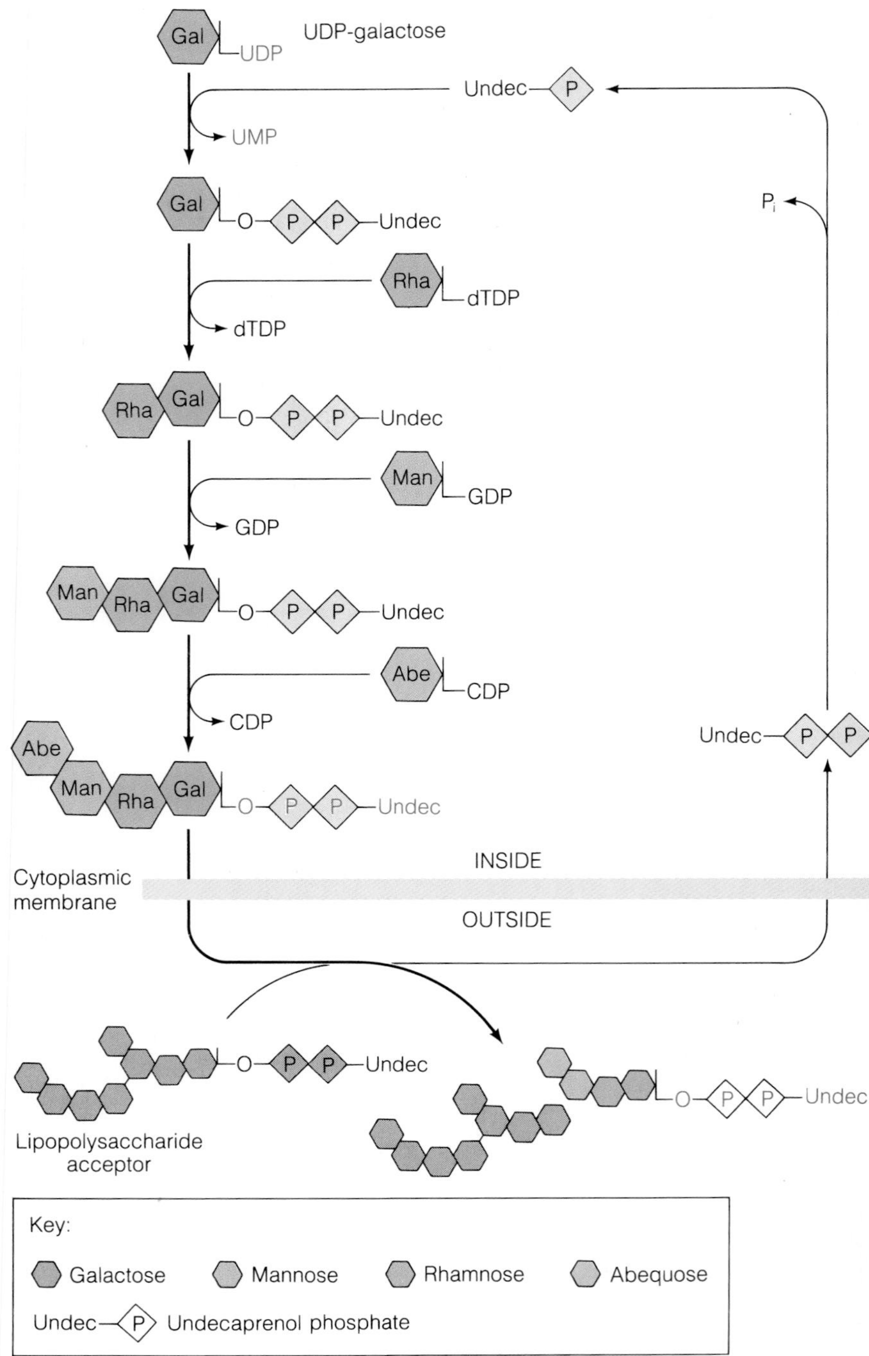

Figure 16.21
Biosynthesis of the repeating oligosaccharide unit of the O antigen of *Salmonella typhimurium*. The first four reactions occur in the cytoplasmic membrane. Transfer of the activated tetrasaccharide unit to the unactivated terminus of a growing polysaccharide unit occurs on the outside of the membrane.

Microbial Cell Wall Polysaccharides: O-Antigens

The second example we consider is the biosynthesis of the O antigen of *Salmonella typhimurium*. This is the major lipopolysaccharide component of the outer membrane. Lipopolysaccharides contain repeating oligosaccharide units that are attached to a basal core polysaccharide. The latter is, in turn, attached to a complex called lipid A. The repeating oligosaccharide units protrude like minute fibers from the outer membrane surface. Since they represent the outer surface of the cell and are composed of specific carbohydrate structures, these fibers are intensely immunogenic—hence the term O antigen that is applied to the fibers. Production of antibodies directed against O antigens represents a primary defense mechanism used by vertebrates against bacterial infection. Bacteria have responded evolutionarily by being able to change O-antigen structure by extremely rapid genetic change, the mechanism of which is not fully understood. Consequently, there exist hundreds of different serotypes of bacteria such as *S. typhimurium*, each with a different O-antigen repeating unit.

In wild-type *S. typhimurium* the O-antigen repeating unit has the structure abequose(α1→3)mannose(α1→4)rhamnose(β1→3)galactose. (The two new sugars are shown in the margin.) The oligosaccharide unit is assembled on a lipid carrier, undecaprenol phosphate, the same carrier we encountered in peptidoglycan synthesis. A lipid-linked tetrasaccharide is assembled as shown in Figure 16.21 and then passes outward through the outer membrane, where the activated galactose of a lipid-linked polymer attacks the mannose of the activated tetrasaccharide.

We will return to processes in carbohydrate biosynthesis in Chapter 19 when photosynthesis is presented. In the next two chapters we consider the metabolism of the second major class of biomolecules, the lipids.

REFERENCES

Gluconeogenesis

Benedetti, A. B., R. Fulceri, M. Ferro, and M. Comporti (1986) On a possible role for glucose 6-phosphatase in the regulation of liver cell cytosolic concentration. *Trends Biochem. Sci.* 11:284–285. Proposes a rationale for the localization of glucose-6-phosphatase to the endoplasmic reticulum.

Pilkis, S. J., M. R. El-Maghrabi, and T. H. Claus (1988) Hormonal regulation of hepatic gluconeogenesis and glycolysis. *Annu. Rev. Biochem.* 57:755–784. A contemporary review of fructose-2,6-bisphosphate and the hormonal control of levels of this gluconeogenic regulator.

Glycogen Metabolism

Chock, P. B., S. G. Rhee, and E. R. Stadtman (1980) Interconvertible enzyme cascades in cellular regulation. *Annu. Rev. Biochem.* 49:813–843. This review contains a concise but detailed summary of the interlocking control mechanisms that regulate glycogen biosynthesis and degradation.

Howell, R. R. and J. C. Williams (1983) The glycogen storage diseases. In: *The Metabolic Basis of Inherited Disease*, 4th ed., edited by J. B. Stanbury, J. B. Wyngaarden, and D. S. Fredrickson, pp. 141–166. McGraw-Hill, New York. A review of clinical and biochemical features of glycogen storage diseases in what is arguably the single most authoritative reference on human congenital disorders.

Leloir, L. F. (1983) Long ago and far away. *Annu. Rev. Biochem.* 52:1–16. A personal reminiscence, describing the author's Nobel Prize-winning role in the discovery of nucleotide-linked sugars and the mechanism of glycogen synthesis.

Glycoprotein Synthesis

Elbein, A. D. (1987) Inhibitors of the biosynthesis and processing of N-linked oligosaccharide chains. *Annu. Rev. Biochem.* 56:497–534. The complexity of the synthetic and processing steps has made inhibitors especially valuable in dissecting out the roles of individual reactions.

Hirschberg, C. B., and M. D. Snider (1987) Topography of glycosylation in the rough endoplasmic reticulum and Golgi apparatus. *Annu. Rev. Biochem.* 56:63–88. This article achieves a nice synthesis of the cell biology and biochemistry needed to understand glycoprotein processing.

Kornfeld, S. (1987) Trafficking of lysosomal enzymes. *FASEB J.* 1:462–468. A recent review of the system where intracellular targeting by N-linked oligosaccharides is best understood.

Lodish, H. F. (1989) Transport of secretory and membrane glycoproteins from the rough endoplasmic reticulum to the Golgi, a rate-limiting step in protein maturation and secretion. *J. Biol. Chem.* 263:2107–2110. A recent minireview of this important field that represents a blending of biochemistry and cell biology.

Sharon, N. (1984) Glycoproteins. *Trends Biochem. Sci.* 9:199–201. A brief review of the structures, biological roles, and biosynthetic pathways for these conjugated proteins.

Bacterial Cell Wall Polysaccharide Synthesis

Tipper, D. J., and A. Wright (1979) The structure and biosynthesis of bacterial cell walls. In: *The Bacteria*. Vol. VII, *Mechanisms of Adaptation*, edited by J. R. Sokatch and L. N. Ornston, pp. 291–426. Academic Press, New York.

A detailed review including pathways of complex carbohydrate synthetic processes.

Walsh C. T. (1989) Enzymes in the D-alanine branch of bacterial cell wall peptidoglycan assembly. *J. Biol. Chem.* 264:2393–2396. This minireview describes mechanisms in this pathway, which represents the site of action of penicillin and other antibiotics.

PROBLEMS

1. How many ATPs are consumed in the conversion of each of the following to a glucosyl residue in glycogen?
   (a) Dihydroxyacetone phosphate
   (b) Fructose-1,6-bisphosphate
   (c) Pyruvate
   (d) Glucose-6-phosphate

2. How many high-energy phosphates are generated in (a) converting 1 mol of glucose to lactate? (b) converting 2 mol of lactate to glucose?

3. Avidin is a protein that binds extremely tightly to biotin; hence, it is a potent inhibitor of biotin-requiring enzyme reactions. Consider glucose biosynthesis from each of the following substrates and predict which of these pathways would be inhibited by avidin.
   (a) Lactate
   (b) Oxaloacetate
   (c) Fumarate
   (d) Fructose-6-phosphate
   (e) Phosphoenolpyruvate

4. $^{14}CO_2$ was bubbled through a suspension of liver cells that was undergoing gluconeogenesis from lactate to glucose. Which carbons in the glucose molecule would become radioactive?

5. Propose synthetic pathways for nucleotide sugar intermediates involved in the biosynthesis of *S. typhimurium* O antigen: UDP-Gal, GDP-Man, CDP-abequose, and dTDP-L-rhamnose.

6. Write a balanced equation for each of the following reactions or reaction sequences.
   (a) The reaction catalyzed by PFK-2
   (b) The conversion of 2 mol of oxaloacetate to glucose
   (c) The conversion of glucose to UDP-Glc
   (d) The conversion of 2 mol of glycerol to glucose
   (e) The conversion of 2 mol of malate to glucose-6-phosphate

7. Sketch curves for reaction velocity versus [fructose-6-phosphate] for the phosphorylated *and* nonphosphorylated forms of PFK-2.

8. Based on information presented on p. 552, sketch curves relating glycogen synthase reaction velocity to [UDP-glucose], for both the I and D forms of the enzyme, in the presence and absence of glucose-6-phosphate.

9. Glycogen synthesis and breakdown are regulated primarily at the hormonal level. However, important *nonhormonal* mechanisms also control the rates of synthesis and mobilization. Describe these nonhormonal regulatory processes.

10. For a bacterial cell that contains enzymes of the glyoxylate cycle, calculate the number of ATPs consumed in the biosynthesis of 1 mol of glucose from acetyl-CoA. How many moles of acetyl-CoA are required?

CHAPTER 17

Lipid Metabolism I: Fatty Acids and Triacylglycerols

Like the carbohydrates we have discussed in previous chapters, lipids play roles both in energy metabolism and in aspects of biological structure and function. The latter include their roles in membrane structure and as steroid hormones, certain vitamins, and other biological regulators. This chapter focuses on bioenergetic aspects of lipid metabolism. We shall discuss the synthesis and breakdown of energy storage lipids, as well as fatty acid oxidation and biosynthesis (Figure 17.1), processes that are quite similar among plants, animals, and microorganisms. We shall also present some topics related strictly to animal metabolism—fat digestion, absorption, storage, and mobilization. Membrane lipids and lipids of more complex metabolic functions are covered in Chapter 18.

As we saw in Chapter 9, the great bulk of the lipid in most organisms is in the form of triacylglycerols (formerly called *triglycerides*). The term fat, or neutral fat, refers to this most abundant class of lipids. A mammal may contain 5 to 25%, or more, of its body weight as lipid, with as much as 90% of this in the form of triacylglycerols. Most of this fat, which is stored in adipose tissue, constitutes the primary energy reserve. In animal systems fat is stored in specialized cells, the adipocytes, where giant fat globules occupy most of the intracellular space (see Figure 9.3). Plant seeds store great quantities of fat to provide energy to the developing plant embryo (Figure 17.2). Because plant lipids contain mostly unsaturated fatty acids, the triacylglycerols of seeds are largely in the form of liquid oils.

Triacylglycerols play roles other than in energy storage. Fat serves to cushion organs against shock, and it provides an efficient thermal insulator, particularly in marine mammals, which must maintain a body temperature far higher than that of the seawater in which they live.

Lipids as Energy Reserves

Recall that most of the carbon in triacylglycerols is more highly reduced than the carbon in carbohydrates. Obviously, the carboxyl carbons of fatty acids are highly oxidized, but most of the fatty acid carbons are at the highly reduced methylene level. Thus, metabolic oxidation of fat consumes more oxygen, gram per gram, than oxidation of carbohydrate. A correspondingly larger amount of metabolic energy is released. The complete metabolic oxidation of triacylglycerols yields about 37 kJ/g or more, compared with

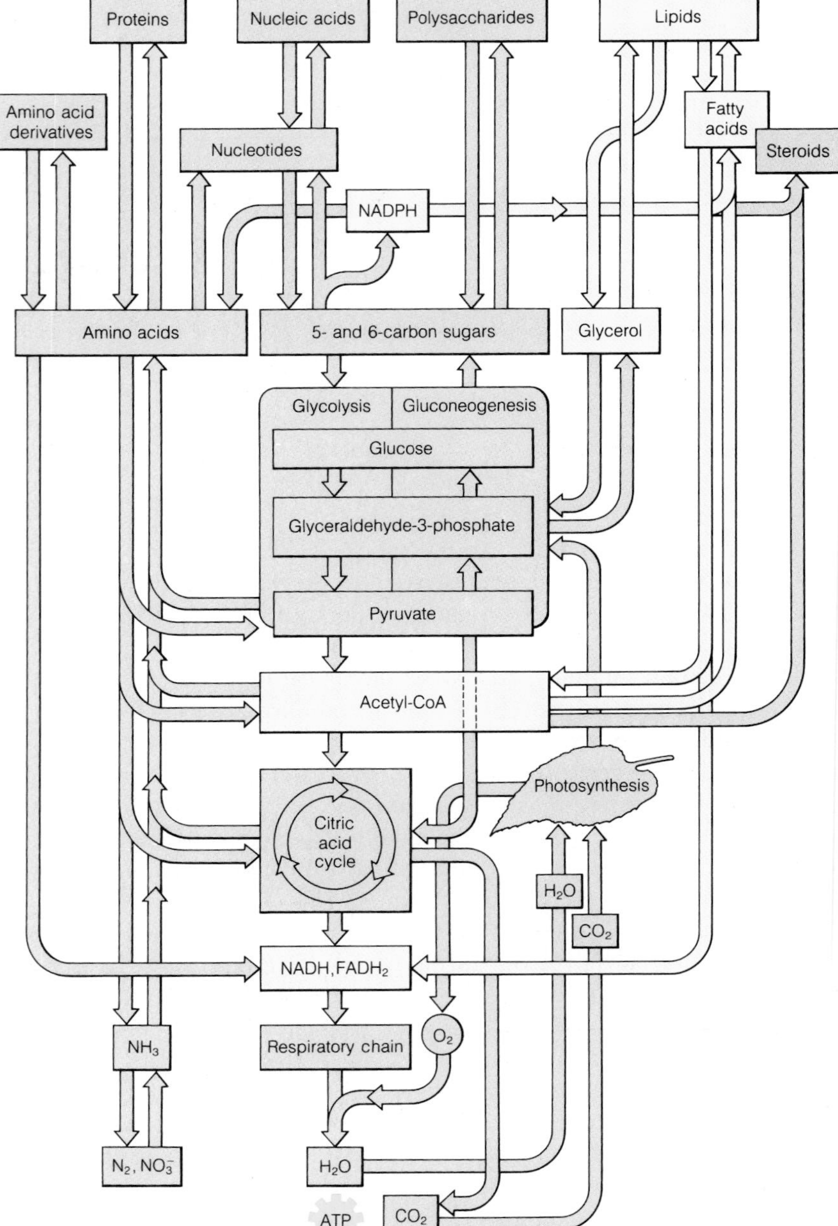

Figure 17.1
Overview of intermediary metabolism with fatty acid and triacylglycerol pathways highlighted in yellow.

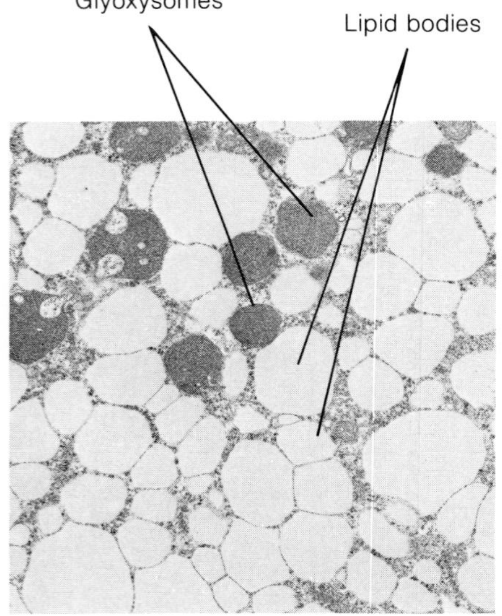

Figure 17.2
Fat storage in a plant seedling. The electron micrograph (×6500) shows a cell from a cucumber cotyledon (seed leaf) a few days after germination. Fat stored in lipid bodies is degraded, oxidized, and converted to carbohydrate in neighboring glyoxysomes (or microbodies) to support the growth of the plant.

about 17 kJ/g for carbohydrates and proteins. Adding to this difference between fat and carbohydrate is the hydrophilic nature of glucose polymers: glycogen binds about 2 grams of water per gram of carbohydrate. Fat, being extremely apolar, is anhydrous. Thus, since 1 gram of intracellular glycogen contains but $\frac{1}{3}$ gram of anhydrous glucose polymer, intracellular fat contains about six times as much potential metabolic energy, on a mass basis, as intracellular glycogen. This is to obvious advantage in some specialized situations, such as in hibernating animals, where several months' worth of food must be stored, or in the flight muscles of small birds, where weight is at a premium. In addition, the insolubility of fat allows it to be stored in cells without affecting intracellular osmotic pressure.

Little wonder, then, that fat is the major energy source in most cells. A typical 70-kg human may have fuel reserves of 400,000 kJ in total body fat and about 100,000 kJ in total protein (mostly muscle protein). By contrast, the glycogen stores amount to just 2500 kJ of available energy and the total glucose about 170 kJ. Fat stores are maintained from the diet; about 40% of the caloric value of Western diets comes from fat. In addition, we shall see that carbohydrate ingested in excess of its ability to be catabolized or stored as glycogen is readily converted to fat.

Most of the energy derived from fat breakdown comes from oxidation of the constituent fatty acids. Fatty acid oxidation provides the major energy source for many animal tissues. Brain is the only tissue that is unable to use fatty acids as a significant energy supply. However, under conditions of starvation, when blood glucose levels decrease, brain can adjust to use a class of lipid-related compounds called ketone bodies, which we discuss later in this chapter.

Utilization of Triacylglycerols in Animals

Triacylglycerols are derived from two primary sources: (1) the diet and (2) mobilization of fat stored in adipocytes. Processes by which both sources are utilized in animals are presented in this section and summarized in Figure 17.3.

Fat Digestion and Absorption

The major factor that animals must cope with in the digestion, absorption, and transport of dietary lipids is their insolubility in aqueous media. Essen-

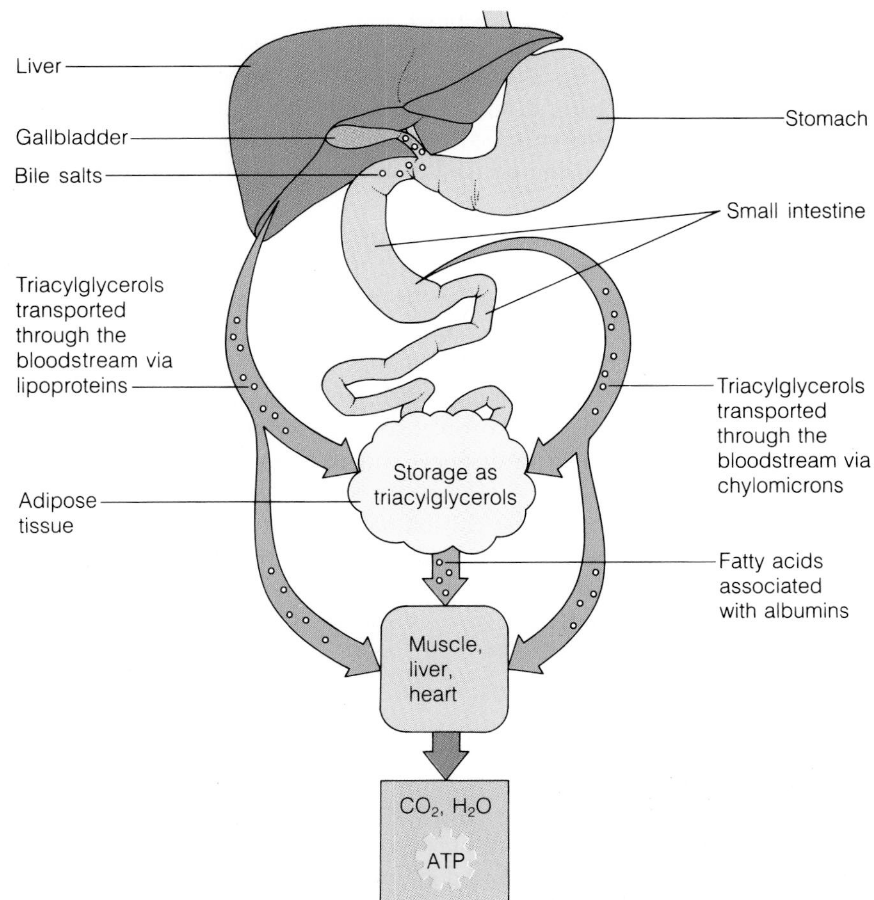

Figure 17.3
Overview of fat digestion, absorption, storage, and mobilization in the human.

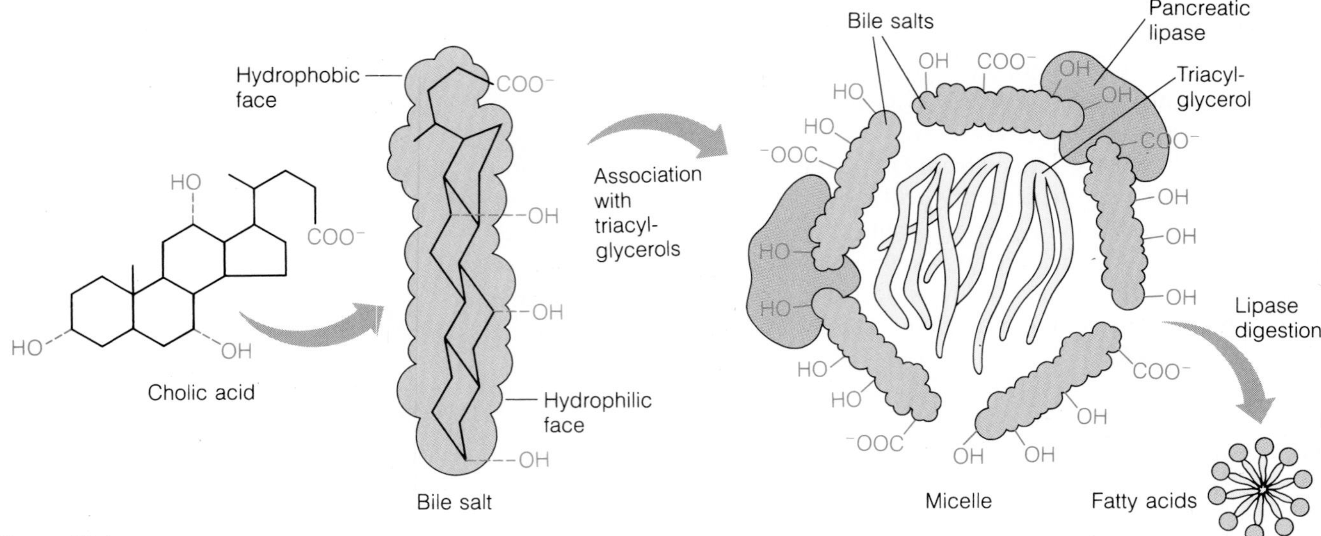

Figure 17.4
Action of bile salts in emulsifying fats in the intestine. The hydrophobic surface of the bile salt molecule associates with triacylglycerol, and a number of these complexes aggregate to form a micelle, with the polar surface of the bile salt facing outward. This allows association with pancreatic lipase, whose action liberates free fatty acids in a much smaller micelle, which can be absorbed through the intestinal mucosa.

tial to the normal digestion and intestinal absorption of lipids are **bile salts,** detergent substances secreted from the gallbladder (Figure 17.4; also discussed in Chapter 18). These are salts of **bile acids,** such as cholic acid, which are derived from cholesterol.

A bile salt molecule contains a hydrophobic surface and a hydrophilic surface. This allows bile salts to dissolve at an oil–water interface, with the hydrophobic surface in contact with the apolar phase and the hydrophilic surface in contact with the aqueous phase. This detergent action emulsifies lipids and yields micelles, which allows digestive attack by water-soluble enzymes and facilitates the absorption of lipid through the intestinal mucosa. Most of the digestion occurs through action of **pancreatic lipase,** an unusual calcium-requiring enzyme that catalyzes a reaction at an oil–water interface; the substrate being cleaved is in an apolar phase, and the other substrate, of course, is water.

The products of fat digestion comprise a mixture of free fatty acids and mono- and diacylglycerols. Less than 10% of the original triacylglycerol remains unhydrolyzed. During absorption through mucosal cells, some resynthesis of triacylglycerols must occur from the hydrolysis products. We know this because much of the absorbed lipid is in the form of triacylglycerols complexed with protein in a class of lipoproteins called **chylomicrons.** The chylomicron contains an oil droplet coated with more polar lipids and a skin of protein, which help to disperse and partially solubilize the fat for transport to tissues.

In general, fats containing substantial amounts of unsaturated fatty acids, which tend to be liquid at body temperature, are relatively easily absorbed, while lipids containing mostly saturated fatty acids are digested and absorbed more slowly.

Transport of Fat to Tissues: Lipoproteins

As noted above, after absorption lipids become associated with proteins to form **lipoproteins.** The latter aid in the transport of lipids to tissues, either for energy storage or for oxidation. Free lipids are all but undetectable in blood. The **apolipoprotein** components, or polypeptide chains (apoproteins), of lipoproteins are synthesized mainly in the liver, though about 20% are produced in intestinal mucosal cells.

Table 17.1
Properties of major human plasma lipoprotein classes

Lipoprotein	Density (g/ml)	Composition (wt %)				
		Protein	Free cholesterol	Cholesterol esters	Phospho-lipids	Triacyl-glycerols
Chylomicrons	0.92–0.96	1–2	1–3	2–4	3–8	90–95
VLDL	0.95–1.006	10.4	5.8	13.9	15.2	53.4
IDL	1.006–1.019	17.8	6.5	22.5	21.7	31.4
LDL	1.019–1.063	25.0	8.6	41.9	20.9	3.5
HDL_2	1.063–1.12	42.6	5.2	20.3	30.1	2.2
HDL_3	1.12–1.21	54.9	2.6	16.1	25.0	1.4
VHDL	1.21–1.25	62.4	0.3	3.2	28.0	4.6

NOTE: Some classifications include IDL and LDL as a single category, with a density range of 1.006–1.063. Data are from D. E. Vance and J. E. Vance (eds.), *Biochemistry of Lipids and Membranes* (Menlo Park, Calif.: Benjamin/Cummings, 1985).

As lipids are being absorbed, they combine in the blood plasma with the polypeptide chains to give distinct families of lipoproteins. The major lipoprotein families are classified in terms of their density, as determined by centrifugation (Table 17.1). Since lipids are of much lower density than proteins, the lipid content of a lipoprotein class is inversely related to its density; the higher the lipid abundance, the lower the density. The standard lipoprotein classification includes, in increasing order of density: chylomicrons, **very low-density lipoprotein** (VLDL), **low-density lipoprotein** (LDL), **high-density lipoprotein** (HDL), and **very high-density lipoprotein** (VHDL). Realizing that the differences among these general categories are more quantitative than qualitative, we now recognize three subcategories: intermediate density lipoprotein (IDL), which is part of the LDL class, and two classes of HDL—HDL_2 and HDL_3. Lipoproteins in each class contain characteristic apoproteins and have distinctive lipid compositions. A total of nine major apolipoproteins are recognized in human lipoproteins. Table 17.1 presents the protein and lipid composition of each lipoprotein class, and Table 17.2 gives information about the apolipoproteins. Despite the large differences among the lipoprotein classes, they share common structural features, notably a spherical shape that can be detected by electron microscopy. The hydrophobic parts, both lipid and apolar amino acid residues, form an inner core, and hydrophilic protein structures and polar head groups of phospholipids are on the outside.

Some apolipoproteins have specific biochemical activities other than their roles as passive carriers of lipid from one organ to another. For instance, apo C-2 is an activator of triacylglycerol hydrolysis by **lipoprotein**

Table 17.2
Apolipoproteins of the human plasma lipoproteins

Apolipoprotein	Molecular Weight	Lipoprotein Distribution
Apo A-1	28,331	HDL
Apo A-2	17,380	HDL
Apo B-48	240,000	Chylomicrons
Apo B-100	500,000	VLDL, LDL
Apo C-1	7,000	HDL, VLDL
Apo C-2	8,837	Chylomicrons, VLDL, HDL
Apo C-3	8,751	Chylomicrons, VLDL, HDL
Apo D	32,500	HDL
Apo E	34,145	Chylomicrons, VLDL, HDL

Data are from D. E. Vance and J. E. Vance (eds.), *Biochemistry of Lipids and Membranes* (Menlo Park, Calif.: Benjamin/Cummings, 1985).

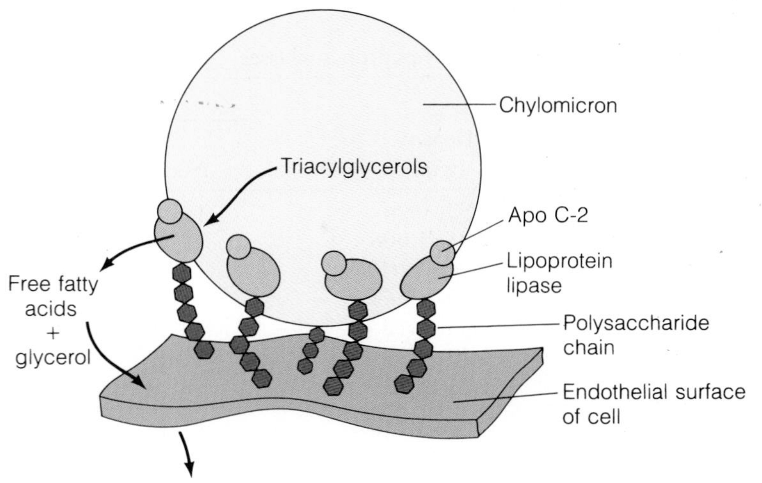

Figure 17.5
Schematic view of the binding of a chylomicron to lipoprotein lipase. The enzyme linked to the cell surface by a polysaccharide chain anchors the chylomicron, with additional help from the enzyme–apo C-2 interaction.

lipase, a cell surface enzyme that hydrolyzes triacylglycerols in lipoproteins. A human deficiency of apo C-2 is associated with elevated triacylglycerol levels in blood. Human variant forms of apo C-3 and apo E are associated with premature atherosclerosis, as is a low circulating HDL level (see Chapter 18).

Together, the lipoproteins help to maintain in solubilized form some 500 mg of total lipid per 100 ml of human blood in the postabsorptive state, after the contents of a meal have been digested and absorbed into the bloodstream. Of this 500 mg, typically about 120 mg is triacylglycerol, 220 mg is cholesterol (two-thirds esterified with fatty acids, one-third free), and 160 mg is phospholipids, principally phosphatidylcholine and phosphatidylethanolamine. As discussed in the next chapter, serum cholesterol levels in this range, which were formerly considered to be normal, are now considered to be unhealthy.

The association of lipids with proteins not only solubilizes lipids but also aids in their transport into cells. Triacylglycerols are transported to tissues either in chylomicrons or in VLDL. At the cell surface the triacylglycerols are cleaved by lipoprotein lipase to give glycerol and free fatty acids. Lipoprotein lipase lies at the end of a polysaccharide chain, which helps to anchor the chylomicron, as schematized in Figure 17.5. As noted above, interaction between the protein apo C-2, on the chylomicron surface, and lipoprotein lipase itself is essential for activation of this hydrolysis.

After absorption into the cell, the glycerol and fatty acids derived from lipoprotein lipase action can be either catabolized to generate energy or, in adipose cells, used to resynthesize triacylglycerols.

Mobilization of Stored Fat

In general, the transport of dietary fat to storage depots is not regulated; whatever appears in the body from the diet is absorbed, and most of it is transported to adipose tissue for storage. The lack of control of this process can be seen simply in the prevalence of obesity among humans, where fat is stored in excess of its need to supply energy. By contrast, the release of fat from storage depots in adipose tissue is controlled hormonally, to meet the needs of the organism for energy generation.

The catabolism of fat begins with the hydrolysis of triacylglycerol to yield glycerol plus 3 mol of free fatty acid (often abbreviated FFA). About 95% of the energy derived from subsequent oxidation of the fat comes from

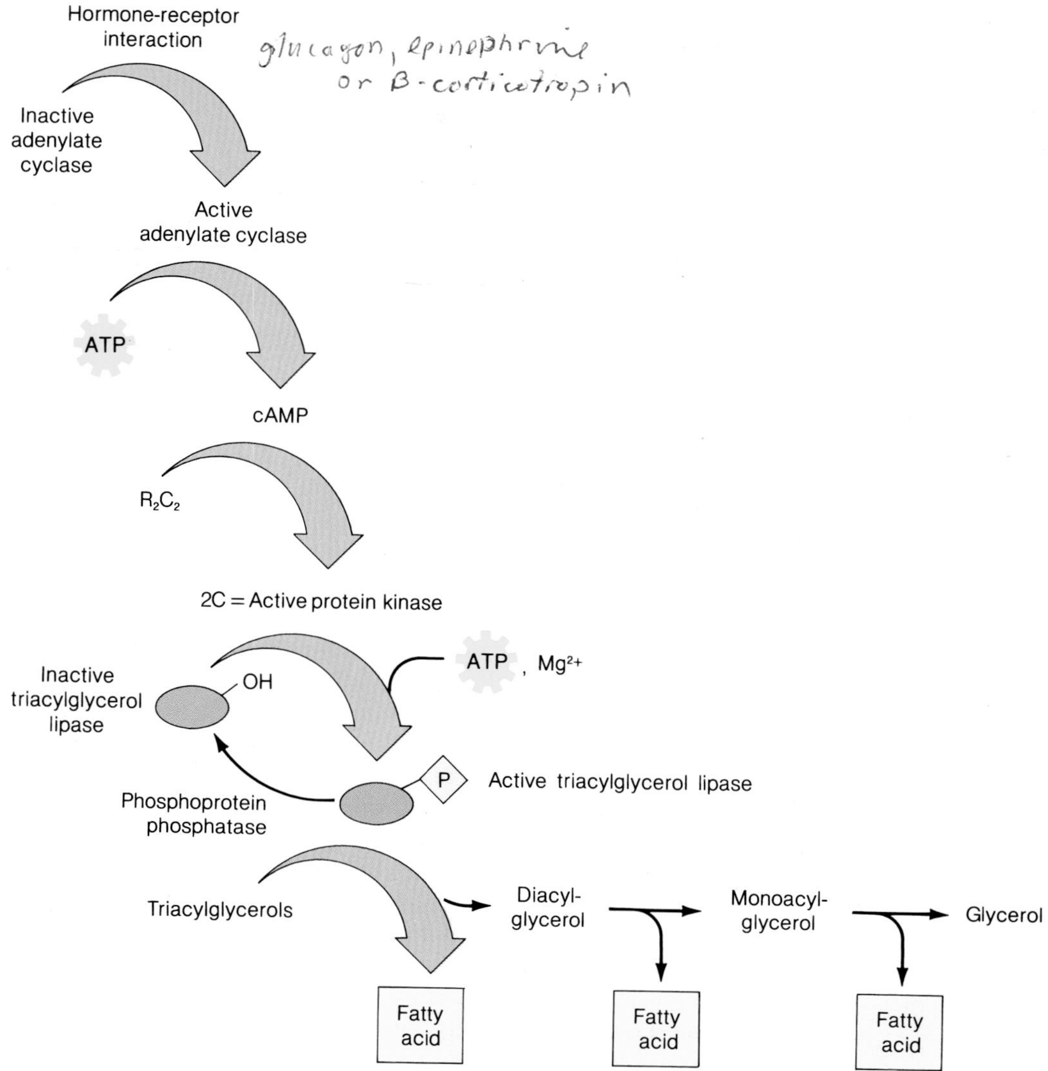

Figure 17.6
Control of triacylglycerol lipase activity by a cyclic AMP-mediated cascade system.

the fatty acids, with the remaining 5% coming from glycerol. All of the carbons from fatty acids are catabolized to two-carbon fragments, as acetyl-coenzyme A, except for the small proportion of fatty acids that contain odd-numbered chains.

The release of metabolic energy stored in triacylglycerols is comparable to the mobilization of carbohydrate energy stored in animal glycogen, in that the first step of fat breakdown—its hydrolysis to glycerol and fatty acids—is hormonally regulated. The first reaction is controlled by a cascade involving cyclic AMP (Figure 17.6). Depending on the physiological state, either glucagon or epinephrine or β-corticotropin activates adenylate cyclase, as was described in Chapter 13. This, in turn, activates protein kinase, which phosphorylates and thereby activates an enzyme called **triacylglycerol lipase,** also called **hormone-sensitive lipase.** This enzyme catalyzes the hydrolytic release of fatty acid from carbon 1 or 3 of the glycerol moiety. The fatty acid release is followed by the action of a di- and a monoacylglycerol lipase, in turn. Together, these three enzymes degrade the original molecule to glycerol and three unesterified fatty acids. In adipose tissue the primary hormonal effects are mediated by epinephrine in stress situations and by glucagon during fasting.

The triacylglycerol hydrolysis products exit the adipocyte by passive diffusion and find their way to the blood plasma, where the fatty acids become bound to **albumin.** This is the most abundant plasma protein, about 50% of the total plasma protein in humans. The protein, with an M_r of 66,200, contains 17 disulfide bridges. Each molecule of albumin can bind up to 10 molecules of free fatty acid, although the actual amount bound is usually far lower. Fatty acids are released from albumin and taken up by tissues largely by passive diffusion, so that fatty acid uptake into cells is driven primarily by concentration. Most of the glycerol released to the bloodstream is taken up by liver cells, where it serves as a gluconeogenic precursor to glucose.

Fatty Acid Oxidation

Early Experiments

The nature of the pathway by which fatty acids are oxidized was revealed as early as 1904, in a brilliant series of experiments by a German, Franz Knoop. The experiments were inspired because they involved the first known use of metabolic tracers, more than 40 years before radioactive tracers became available. Knoop fed dogs a series of fatty acids in which the terminal methyl group was substituted by a phenyl group. The expectation was that these analogs would follow metabolic pathways similar to those used for oxidizing normal fatty acids.

In fact, when the fed fatty acid had an even-numbered carbon chain, the final breakdown product, recovered from urine, was phenylacetic acid. When the fed fatty acid had an odd-numbered chain, the product was benzoic acid (Figure 17.7).

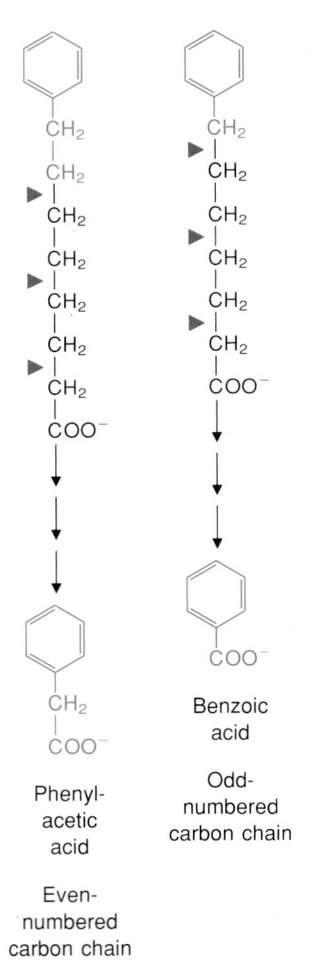

Figure 17.7
Oxidation of phenyl derivatives of fatty acids in Knoop's experiment. ▼-presumed sites of cleavage of these model fatty acids.

These results led Knoop to propose that fatty acids are oxidized in a stepwise fashion, with initial attack on carbon 3 (the β-carbon with respect to the carboxyl group). This attack would release the terminal two carbons, and the remainder of the fatty acid molecule could undergo another oxidation. Release of a two-carbon fragment would occur at each step in the oxidation. With the analogs, the process would be repeated until the remaining acid, either phenylacetic or benzoic acid, could not be further metabolized and would be excreted in the urine.

The next major development came in the 1940s, when Luis Leloir and Albert Lehninger independently demonstrated fatty acid oxidation in cell-free liver homogenates. Lehninger showed that ATP was essential for this process, and he proposed that ATP somehow activates the carboxyl group of the fatty acid. Working with Eugene Kennedy, Lehninger also showed that the process occurs in mitochondria and that it releases two-carbon fragments which are oxidized in the citric acid cycle. Feodor Lynen, in Munich, contributed greatly with the demonstration that the ATP-dependent activation esterifies the fatty acid carboxyl group with the thiol group of coenzyme A, and it was shown that all of the intermediates in the subsequent oxidative reactions are fatty acyl-CoAs. Thus, by the mid-1950s the basic outlines of the fatty acid oxidation pathway were clear. Now let us discuss the pathway as it is understood today.

Fatty Acid Activation and Transport into Mitochondria

For the most part, fatty acids arise in the cytosol, either through biosynthesis (as discussed later in this chapter) or through triacylglycerol transport from fat depots outside the cell. Since the inner mitochondrial membrane is impermeable to free fatty acids or acyl-CoAs, a specific transport system comes into play to move fatty acyl-CoAs into the mitochondrial matrix, where oxidation occurs. That transport system operates hand in hand with the metabolic activation needed to initiate the **β-oxidation pathway.**

A series of **fatty acyl-CoA ligases,** specific for short-chain, medium-chain, or long-chain fatty acids, catalyze formation of the fatty acyl thioester conjugate with coenzyme A.

$$\text{R—COO}^- + \text{ATP} + \text{CoA—SH} \rightleftharpoons \text{R—}\overset{\overset{\displaystyle\text{O}}{\|}}{\text{C}}\text{—S—CoA} + \text{AMP} + \text{PP}_i \qquad \Delta G^{\circ\prime} = -0.8 \text{ kJ/mol}$$

The ligase specific for long-chain acids is a membrane-bound enzyme, found in both the endoplasmic reticulum and the outer mitochondrial membrane; the short- and medium-chain enzymes are found primarily in the mitochondrial matrix. The long-chain enzyme, which plays the predominant role in initiating fatty acid oxidation, acts on fatty acids with chain lengths of 10 to 20 carbons, while the medium-chain enzyme acts on 4- to 12-carbon chains, and the short-chain enzyme prefers acetate and propionate.

Chemically, the energy-rich thioester link in long-chain fatty acyl-CoAs is identical to that of acetyl-CoA (see Chapter 14). In the latter case the energy to drive acetyl-CoA formation came from that released during the oxidation of pyruvate. The fatty acyl-CoA ligases, on the other hand, use a two-step mechanism involving cleavage of ATP to drive the endergonic thioester formation (Figure 17.8). First comes activation of the carboxyl group by ATP to give a **fatty acyl adenylate,** with concomitant release of pyrophosphate. Next, the activated carboxyl group is attacked by the thiol group of CoA, thereby displacing AMP and forming the fatty acyl-CoA derivative. When we discuss protein synthesis in Chapter 27, you will see that the carboxyl groups of amino acids are activated in very similar fashion.

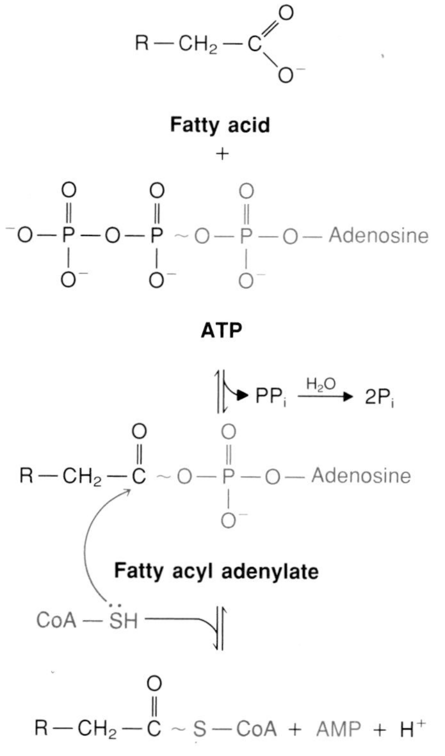

$$\text{R—CH}_2\text{—C}(=\text{O})\text{O}^-$$

Fatty acid

+

$$^-\text{O—}\overset{\overset{\displaystyle\text{O}}{\|}}{\underset{\underset{\displaystyle\text{O}^-}{|}}{\text{P}}}\text{—O—}\overset{\overset{\displaystyle\text{O}}{\|}}{\underset{\underset{\displaystyle\text{O}^-}{|}}{\text{P}}}\sim\text{O—}\overset{\overset{\displaystyle\text{O}}{\|}}{\underset{\underset{\displaystyle\text{O}^-}{|}}{\text{P}}}\text{—O— Adenosine}$$

ATP

$$\rightleftharpoons \; \to \text{PP}_i \xrightarrow{H_2O} 2\text{P}_i$$

$$\text{R—CH}_2\text{—}\overset{\overset{\displaystyle\text{O}}{\|}}{\text{C}}\sim\text{O—}\overset{\overset{\displaystyle\text{O}}{\|}}{\underset{\underset{\displaystyle\text{O}^-}{|}}{\text{P}}}\text{—O— Adenosine}$$

Fatty acyl adenylate

$$\text{CoA — }\ddot{\text{S}}\text{H} \rightleftharpoons$$

$$\text{R—CH}_2\text{—}\overset{\overset{\displaystyle\text{O}}{\|}}{\text{C}}\sim\text{S—CoA} + \text{AMP} + \text{H}^+$$

Fatty acyl-CoA

Figure 17.8
Mechanism of fatty acyl-CoA ligase reactions. The figure shows reversible formation of the activated fatty acyl adenylate, nucleophilic attack by the thiol sulfur of CoA—SH on the activated carboxyl group, and the quasi-irreversible pyrophosphatase reaction, which draws the overall reaction to the right.

Fatty acid activation is readily reversible, since each fatty acyl-CoA, like ATP itself, is an energy-rich compound. However, the direction of this reaction in cells is far to the right, because of the active pyrophosphatase present in most cells.

$$PP_i + H_2O \longrightarrow 2P_i \qquad \Delta G^{\circ\prime} = -28.8 \text{ kJ/mol}$$

Thus, the overall reaction (sum of the two previous reactions) proceeds far in the direction of completion.

$$R{-}COO^- + ATP + CoA{-}SH + H_2O \longrightarrow R{-}\overset{O}{\overset{\|}{C}}{-}S{-}CoA + AMP + 2P_i \qquad \Delta G^{\circ\prime} = -29.7 \text{ kJ/mol}$$

Fatty acyl-CoAs are formed outside the mitochondrion; hence, they must move through the inner mitochondrial membrane to become oxidized. This involves transfer of the fatty acyl moiety to a carrier called **carnitine.** The reaction, catalyzed by **carnitine acyltransferase I,** yields a derivative, **fatty acyl-carnitine,** which can traverse the inner membrane.

$$R{-}CH_2{-}\overset{O}{\overset{\|}{C}}{\sim}S{-}CoA + H{-}\underset{\underset{COO^-}{|}}{\underset{CH_2}{|}}{\overset{\overset{\overset{+}{N}(CH_3)_3}{|}}{\overset{CH_2}{|}}{C}}{-}OH \rightleftharpoons H{-}\underset{\underset{COO^-}{|}}{\underset{CH_2}{|}}{\overset{\overset{\overset{+}{N}(CH_3)_3}{|}}{\overset{CH_2}{|}}{C}}{-}O{\sim}\overset{O}{\overset{\|}{C}}{-}CH_2{-}R + CoA{-}SH$$

Fatty acyl-CoA **Carnitine** **Fatty acyl-carnitine** **CoA-SH**

The enzyme is located in the outer surface of the outer mitochondrial membrane.

Once formed, fatty acyl-carnitine is translocated through the inner membrane, being exchanged for free carnitine. Another enzyme, **carnitine acyltransferase II,** located on the matrix side of the inner membrane, completes the transfer process.

$$R{-}\overset{O}{\overset{\|}{C}}{-}O{-}\text{carnitine} + CoA{-}SH \rightleftharpoons R{-}\overset{O}{\overset{\|}{C}}{-}S{-}CoA + \text{carnitine}$$

Note that fatty acyl-carnitines are ordinary esters, in contrast to the energy-rich thioesters such as fatty acyl-CoAs. However, acyl-carnitines are, like thioesters, energy rich; otherwise, the acyltransferase reactions would not be freely reversible, as they are. Note also that the occurrence of this rather complex transport mechanism, which is analogous to mitochondrial shuttle mechanisms we discussed in Chapter 15, requires distinct cytosolic and intramitochondrial pools of coenzyme A. This separation helps to maintain separately regulated pools of substrates for fatty acid synthesis and degradation.

The β-Oxidation Pathway

Once inside the mitochondrial matrix, fatty acyl-CoAs are oxidized as predicted by Knoop, with initial oxidation of the β-carbon and a series of steps that each release a two-carbon fragment, in the form of acetyl-CoA, from the fatty acid undergoing oxidization. The pathway is cyclic in that each step (which involves four reactions) ends with formation of an acyl-CoA, shortened by two carbons, which undergoes the same process in the next step, or cycle. For example, palmitoyl-CoA, derived from a 16-carbon fatty acid, undergoes seven cycles of oxidation to give 8 mol of acetyl-CoA. Each cycle releases one two-carbon unit, concomitant with two two-electron oxidation reactions.

As shown in Figure 17.9, each cycle in the oxidation of a saturated fatty acyl-CoA involves the following reactions: (1) dehydrogenation to give an enoyl derivative; (2) hydration of the resultant double bond, with the β-carbon undergoing hydroxylation; (3) dehydrogenation of the hydroxyl

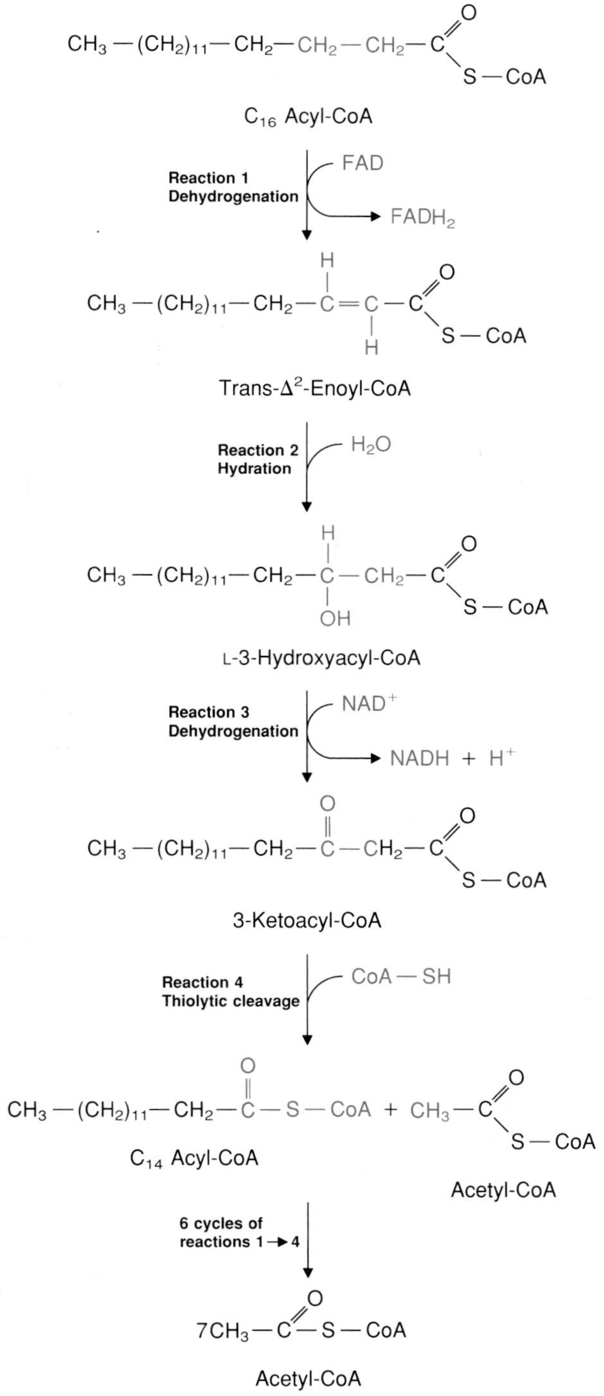

$$CH_3-(CH_2)_{11}-CH_2-CH_2-CH_2-C(=O)-S-CoA$$

C_{16} Acyl-CoA

Reaction 1
Dehydrogenation

FAD → $FADH_2$

$$CH_3-(CH_2)_{11}-CH_2-CH=CH-C(=O)-S-CoA$$

Trans-Δ^2-Enoyl-CoA

Reaction 2
Hydration

H_2O

$$CH_3-(CH_2)_{11}-CH_2-CH(OH)-CH_2-C(=O)-S-CoA$$

L-3-Hydroxyacyl-CoA

Reaction 3
Dehydrogenation

NAD^+ → $NADH + H^+$

$$CH_3-(CH_2)_{11}-CH_2-C(=O)-CH_2-C(=O)-S-CoA$$

3-Ketoacyl-CoA

Reaction 4
Thiolytic cleavage

CoA—SH

$$CH_3-(CH_2)_{11}-CH_2-C(=O)-S-CoA + CH_3-C(=O)-S-CoA$$

C_{14} Acyl-CoA

Acetyl-CoA

6 cycles of reactions 1→4

$$7CH_3-C(=O)-S-CoA$$

Acetyl-CoA

Figure 17.9
Outline of the β-oxidation of fatty acids. In the diagram a 16-carbon saturated fatty acyl-CoA (palmitoyl-CoA) undergoes seven cycles of oxidation to yield eight molecules of acetyl-CoA.

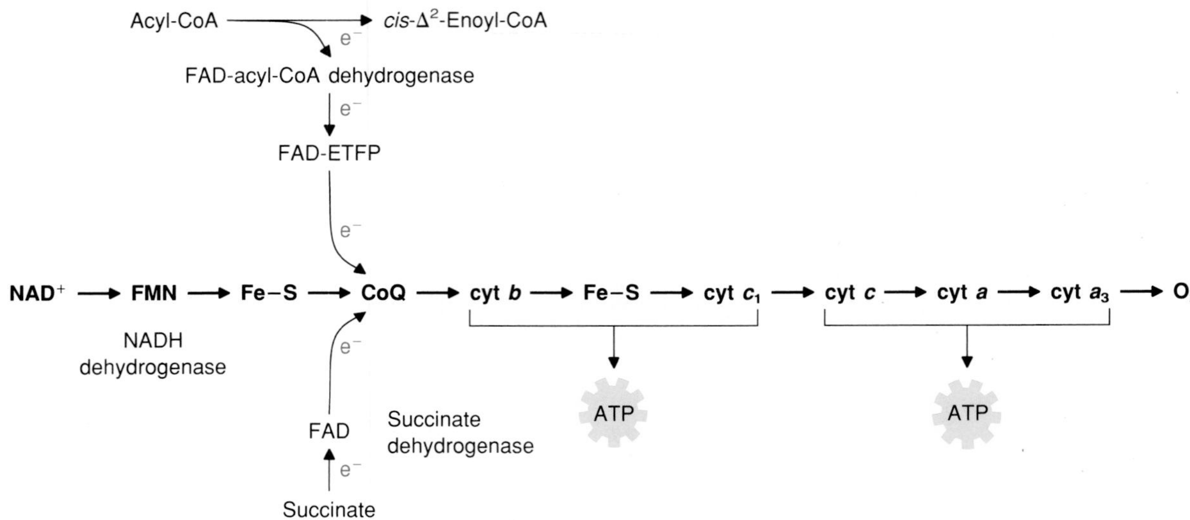

Figure 17.10
Fate of reducing equivalents derived from fatty acyl-CoA dehydrogenation. Enzyme-bound FAD becomes reduced and then transfers its electrons to ETFP, which in turn passes them to coenzyme Q, which is also a collection point for electrons from NADH dehydrogenase and succinate dehydrogenase.

group; (4) cleavage by attack of a second molecule of coenzyme A on the β-carbon, to release acetyl-CoA and a fatty acyl-CoA two carbons shorter than the original substrate. Oxidation of unsaturated fatty acyl-CoAs is slightly different, as discussed in a later section.

Acetyl-CoA from β-oxidation enters the citric acid cycle, where it is oxidized to CO_2 in the same fashion as the acetyl-CoA derived from the oxidation of pyruvate. Like the citric acid cycle, β-oxidation generates reduced electron carriers, whose reoxidation in the mitochondria generates ATP via oxidative phosphorylation from ADP. Now let us describe the individual reactions in detail.

THE INITIAL DEHYDROGENATION (REACTION 1). The first reaction is catalyzed by a **fatty acyl-CoA dehydrogenase,** which dehydrogenates between the α- and β-carbons to give a *trans*-Δ^2-enoyl-CoA as the product.

$$\text{Fatty acyl—S—CoA} + \text{E—FAD} \longrightarrow trans\text{-}\Delta^2\text{-enoyl—S—CoA} + \text{E—FADH}_2$$

Three forms of this enzyme exist and are specific for short-, medium-, or long-chain fatty acyl-CoAs, respectively. Each enzyme carries a tightly bound FAD prosthetic group. The reduced enzyme-bound flavin then contributes a pair of electrons to an inner membrane protein, the **electron-transferring flavoprotein (ETFP).** These electrons are passed in turn to coenzyme Q and are then shuttled along the respiratory chain, yielding two ATPs via oxidative phosphorylation (Figure 17.10). In this respect ETFP is comparable to NADH dehydrogenase and succinate dehydrogenase, which we discussed in Chapter 15; all three are flavoproteins that transfer electrons to the mobile electron carrier coenzyme Q.

HYDRATION AND DEHYDROGENATION (REACTIONS 2 AND 3). Mechanistically, fatty acid oxidation closely resembles the reactions of succinate oxidation in the citric acid cycle. Both proceed via an FAD-dependent desaturation, followed by hydration and an NAD^+-dependent dehydrogenation. In fatty acyl-CoA oxidation the latter two reactions are catalyzed by **enoyl-CoA hydratase** and **3-hydroxyacyl-CoA dehydrogenase,** respectively. Both reactions are stereospecific.

$$trans\text{-}\Delta^2\text{-Enoyl—S—CoA} + H_2O \longrightarrow \text{L-3-hydroxyacyl—S—CoA}$$

$$\text{L-3-Hydroxyacyl—S—CoA} + NAD^+ \longrightarrow \text{3-ketoacyl—S—CoA} + NADH + H^+$$

Because carbon 3 is β with respect to the carboxyl carbon, the products of these two reactions are sometimes called L-β-hydroxyacyl-CoA and β-ketoacyl-CoA, respectively. Hence the term β-oxidation, since the β-carbon is oxidized in each cycle.

THIOLYTIC CLEAVAGE (REACTION 4). The fourth and last reaction in each cycle of the β-oxidation pathway involves attack of the nucleophilic thiol sulfur of coenzyme A on the electron-poor keto carbon of 3-ketoacyl-CoA, with cleavage of the α–β bond and release of acetyl-CoA.

$$R-\overset{O}{\overset{\|}{C}}-CH_2-\overset{O}{\overset{\|}{C}}-S-CoA + CoA-SH \longrightarrow R-\overset{O}{\overset{\|}{C}}-S-CoA + CH_3-\overset{O}{\overset{\|}{C}}-S-CoA$$

The other product is a shortened fatty acyl-CoA, ready to begin a new cycle of oxidation.

Since the above reaction involves cleavage by a thiol, it is referred to as a **thiolytic cleavage,** by analogy with hydrolysis, which involves cleavage by water. The enzyme is commonly called **β-ketothiolase,** or simply thiolase. The enzyme catalyzes a two-step reaction in which an essential thiol group on the enzyme (E-SH) attacks the substrate, with formation of an acyl–enzyme intermediate and acetyl-CoA; free CoA—SH then attacks the intermediate.

As noted earlier, the overall oxidation pathway as just described is applicable to the most abundant fatty acids—those that contain even numbers of carbon atoms and are fully saturated; presently, we shall describe the variations in this pathway that permit oxidation of other fatty acids. For the saturated, even-chain fatty acyl-CoAs, oxidation simply proceeds stepwise, with two carbons lost after each cycle as acetyl-CoA. For the C_{16} palmitoyl-CoA, the example shown in Figure 17.6, the first cycle yields acetyl-CoA plus the C_{14} myristoyl-CoA; a second cycle, acting on the latter substrate, yields acetyl-CoA plus the C_{12} lauroyl-CoA. In the seventh and last cycle the 3-hydroxyacyl-CoA dehydrogenase reaction yields acetoacetyl-CoA. Thiolytic cleavage of this substrate yields 2 mol of acetyl-CoA. Thus, palmitic acid oxidation involves six successive cycles, each of which yields 1 mol of acetyl-CoA, and a seventh cycle, which yields two. Other saturated even-chain fatty acids are degraded identically; for example, stearic acid oxidation involves eight steps, with two acetyl-CoAs resulting from the last cycle.

Energy Yield from Fatty Acid Oxidation

We can now write a balanced equation for the overall degradation of palmitoyl-CoA to 8 mol of acetyl-CoA.

$$\text{Palmitoyl-CoA} + 7\text{CoA}-\text{SH} + \text{FAD} + 7\text{NAD}^+ + 7\text{H}_2\text{O} \longrightarrow 8\text{ acetyl-CoA} + 7\text{FADH}_2 + 7\text{NADH} + 7\text{H}^+$$

Each of the products is metabolized exactly as described earlier for oxidation of carbohydrates. Acetyl-CoA is catabolized via the citric acid cycle, while $FADH_2$ and NADH transfer electrons to the respiratory chain through ETFP and NADH dehydrogenase, respectively. Thus, we can easily compute the metabolic energy yield from fatty acid oxidation in terms of moles of ATP synthesized from ADP. Recall from Chapter 15 that oxidation of acetyl-CoA in one turn of the citric acid cycle yields 12 ATPs and that the P/O ratios for oxidation of flavoproteins and NADH are 2 and 3, respectively. The following summation gives the total energy yield.

Reaction	ATP Yield
Activation of palmitate to palmitoyl-CoA	-1
Oxidation of 8 acetyl-CoA	$8 \times 12 = 96$
Oxidation of 7 FAD	$7 \times 2 = 14$
Oxidation of 7 NADH	$7 \times 3 = 21$
Sum: Palmitate $\longrightarrow CO_2 + H_2O$	130

From this you can calculate the ATP yield per carbon oxidized to CO_2 as 130/16, or about 8.2. The same value for glucose is 6.3 (38 ATPs formed per 6 carbons oxidized). This quantitatively supports the statement made earlier, that the energy yield from fat oxidation is higher than that from carbohydrate.

Oxidation of Unsaturated Fatty Acids

We saw in Chapter 9 that many fatty acids in natural lipids are unsaturated; they contain one or more double bonds. Since these are in the cis configuration, they cannot be simply acted on by enoyl-CoA hydratase, which acts only on trans compounds. Two additional enzymes, **enoyl-CoA isomerase** and **2,4-dienoyl-CoA reductase,** must come into play for these fatty acids to be oxidized. To illustrate the action of the isomerase, let us consider the oxidation of the 18-carbon Δ^9 compound, oleic acid, which contains a cis double bond between carbons 9 and 10 (Figure 17.11). Oleic acid is activated and transported into mitochondria just as the saturated fatty acids are. The first three cycles of oxidation are identical to those involved in unsaturated fatty acid oxidation. The product of the third cycle is the CoA ester of a 12-carbon fatty acid with a cis double bond between carbons 3 and 4. Not only is the double bond in the wrong configuration to be hydrated, it is also in the wrong position. The enoyl-CoA isomerase enzyme converts this *cis*-Δ^3-enoyl-CoA to the corresponding *trans*-Δ^2-enoyl-CoA, which can now be acted on by enoyl-CoA hydratase. This hydration and all subsequent reactions are identical to those already described for saturated fatty acids.

The other auxiliary enzyme, 2,4-dienoyl-CoA reductase, comes into play during the oxidation of polyunsaturated fatty acids, such as linoleic acid (Figure 17.12). This 18-carbon fatty acid contains cis double bonds between carbons 9 and 10 and between carbons 12 and 13. Linoleyl-CoA undergoes three cycles of β-oxidation, just as does oleyl-CoA, to give a C_{12} acyl-CoA with cis double bonds between carbons 3 and 4 and between 6 and 7. The isomerase converts the Δ^3 double bond from cis to trans. There follow hydration, dehydrogenation, and thiolytic cleavage to give acetyl-CoA plus a 10-carbon enoyl-CoA, unsaturated between carbons 4 and 5. Action of acyl-CoA dehydrogenase yields a dienoyl-CoA, unsaturated at C-4–C-5 and at C-2–C-3. The NADPH-dependent 2,4-dienoyl-CoA reductase converts this to a C_{10} *cis*-Δ^3-enoyl-CoA. The isomerase comes into play once more, generating a *trans*-Δ^2-enoyl-CoA, which undergoes the remaining cycles of β-oxidation normally.

By these means both mono- and diunsaturated 18-carbon fatty acids can undergo degradation to 9 mol of acetyl-CoA. There is, of course, a slight reduction in the overall energy yield. With these two auxiliary en-

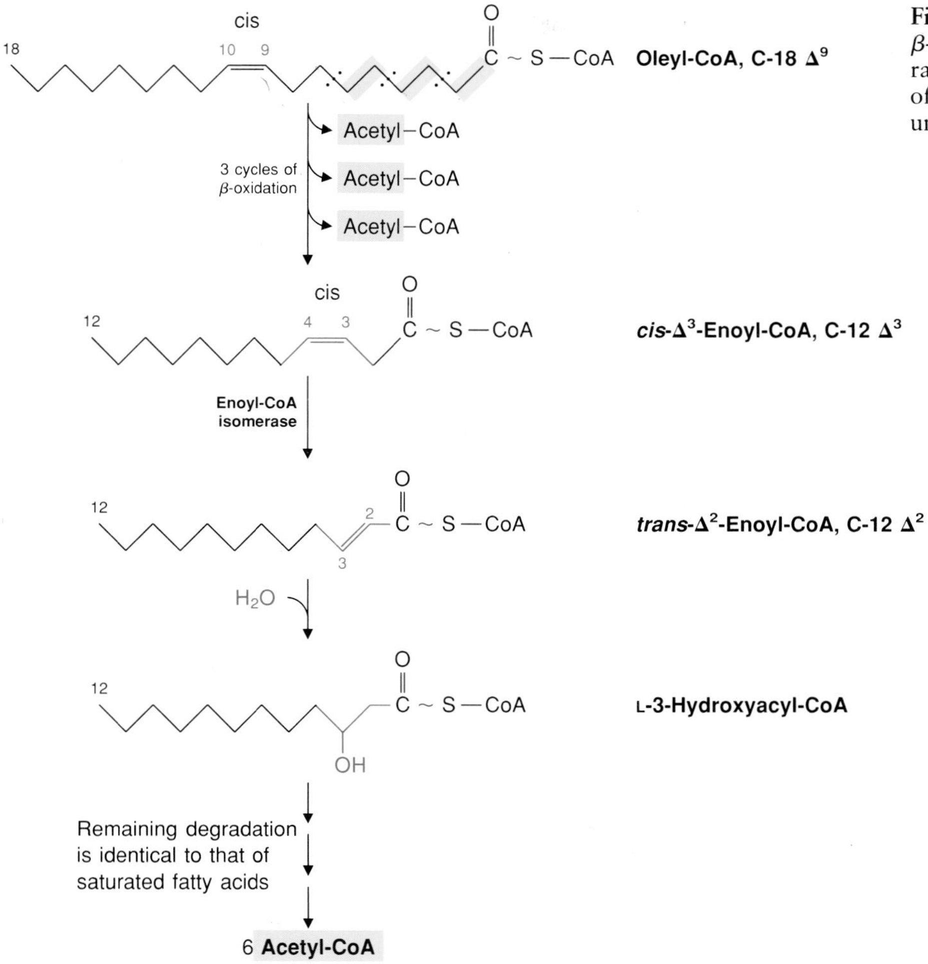

Figure 17.11
β-Oxidation pathway for a monounsaturated fatty acyl-CoA, showing the action of enoyl-CoA isomerase. Two-carbon units are shown in yellow.

zymes all of the even-chain polyunsaturated fatty acids can be similarly degraded, with the following exception. Recent studies have shown that a significant portion of dietary lipid contains unsaturated fatty acids with double bonds in the trans configuration. These fatty acids arise through microbial action in the digestive systems of ruminant mammals and also chemically, through partial hydrogenation of fats and oils. Thus, they are relatively abundant in dairy products and in margarine. Metabolic pathways for oxidizing trans unsaturated acids and nutritional implications of their long-term ingestion have not yet been resolved.

Oxidation of Fatty Acid with Odd-Numbered Carbon Chains

Though most of the fatty acids in natural lipids contain even-numbered carbon chains, a small proportion have odd-numbered carbon chains. The latter group presents a special metabolic problem, which is solved in a novel way. The substrate for the last cycle of β-oxidation of an odd-chain acyl-CoA is the five-carbon homolog of acetoacetyl-CoA; thiolytic cleavage of this yields 1 mol each of acetyl-CoA and **propionyl-CoA.**

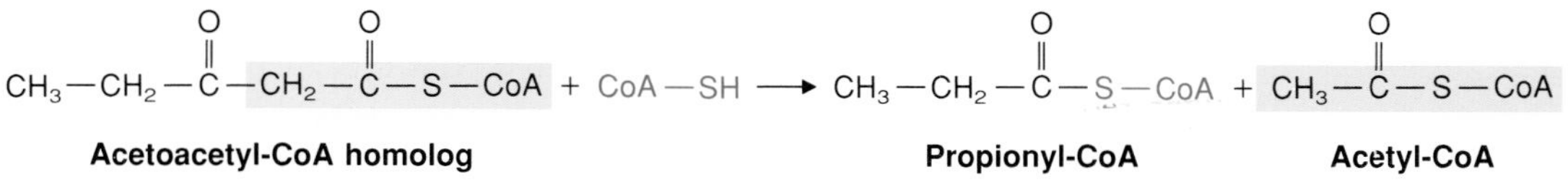

$$\underset{\text{Acetoacetyl-CoA homolog}}{CH_3-CH_2-\overset{O}{\overset{\|}{C}}-CH_2-\overset{O}{\overset{\|}{C}}-S-CoA} + CoA-SH \longrightarrow \underset{\text{Propionyl-CoA}}{CH_3-CH_2-\overset{O}{\overset{\|}{C}}-S-CoA} + \underset{\text{Acetyl-CoA}}{CH_3-\overset{O}{\overset{\|}{C}}-S-CoA}$$

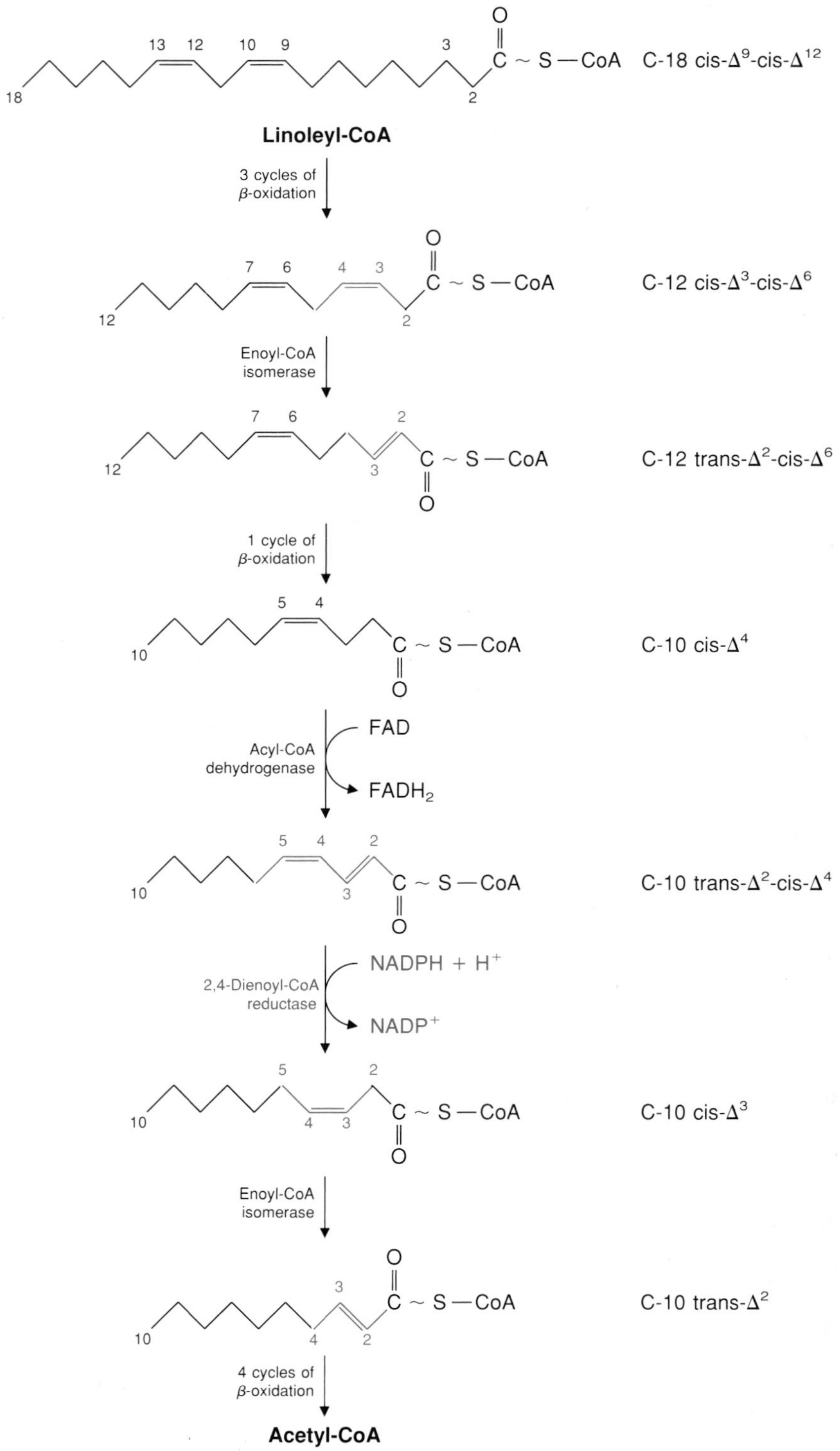

Figure 17.12
β-Oxidation pathway for polyunsaturated fatty acids. This example, linoleyl-CoA, shows points for the action of enoyl-CoA isomerase and 2,4-dienoyl-CoA reductase.

Unlike acetyl-CoA, which is catabolized via the citric acid cycle, propionyl-CoA must be further metabolized before its carbon atoms can enter the citric acid cycle for complete oxidation to CO_2. That further metabolism (Figure 17.13) involves first the ATP-dependent carboxylation of propionyl-CoA, catalyzed by the biotin-containing enzyme **propionyl-CoA carboxylase.** The product then undergoes epimerization to its L stereoisomer by action of **D-methylmalonyl-CoA epimerase.** Next, this branched-chain acyl-CoA derivative is converted to the corresponding straight-chain compound, which happens to be succinyl-CoA, by a most unusual reaction. The enzyme, **L-methylmalonyl-CoA mutase,** requires a cofactor derived from vitamin B_{12}; that cofactor is called **deoxyadenosylcobalamin.** Mechanistically, this reaction is quite interesting. However, since B_{12} coenzymes are also involved in amino acid metabolism, we shall examine this and other B_{12} cofactor-dependent reactions in Chapter 20.

Inability to catabolize propionyl-CoA properly has severe consequences in humans. If there is defective activity of L-methylmalonyl-CoA mutase or of the synthesis of the deoxyadenosylcobalamin coenzyme, L-methylmalonyl-CoA accumulates and exits from cells as methylmalonic acid. This causes a severe acidosis (lowering of blood pH) and also damages the central nervous system. This rare condition, called **methylmalonic acidemia,** is usually fatal in early life. In cases where synthesis of the cofactor from vitamin B_{12} is deficient, the disease can sometimes be successfully treated by administering large doses of vitamin B_{12}.

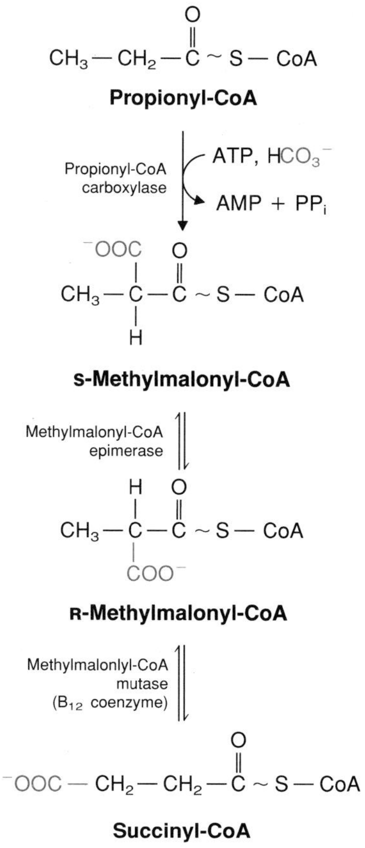

$CH_3—CH_2—C(=O)\sim S—CoA$

Propionyl-CoA

Propionyl-CoA carboxylase; ATP, HCO_3^- → AMP + PP_i

$^-OOC—C(CH_3)(H)—C(=O)\sim S—CoA$

S-Methylmalonyl-CoA

⇅ Methylmalonyl-CoA epimerase

$CH_3—C(H)(COO^-)—C(=O)\sim S—CoA$

R-Methylmalonyl-CoA

⇅ Methylmalonlyl-CoA mutase (B_{12} coenzyme)

$^-OOC—CH_2—CH_2—C(=O)\sim S—CoA$

Succinyl-CoA

Figure 17.13
Pathway for catabolism of odd-numbered fatty acid carbon chains.

Control of Fatty Acid Oxidation

In most cells fatty acid oxidation is controlled by availability of substrates for oxidation, the fatty acids themselves. In animals this availability is controlled in turn by the hormonal control of fat mobilization in adipocytes. Since the function of adipose tissue is to store fat for utilization in other cells, it makes good metabolic sense for breakdown and release of this stored fat to be regulated by hormones, which are extracellular messengers. Recall from page 577 that triacylglycerol lipase activity is regulated by hormonally initiated regulatory cascades involving cyclic AMP. The action of glucagon or epinephrine causes fat breakdown and release, which leads ultimately to fatty acid accumulation in other cells.

In liver the level of **malonyl-CoA** provides another regulatory mechanism, controlling the uptake of fatty acyl-CoAs from the cytosol into mitochondria. As we shall see shortly, malonyl-CoA is the first committed intermediate in fatty acid biosynthesis. In liver, malonyl-CoA is an inhibitor of carnitine acyltransferase I, the first enzyme in the acylcarnitine shuttle. Thus, malonyl-CoA has the dual effect of activating fatty acid biosynthesis (as the product of the rate-limiting reaction) and inhibiting fatty acid oxidation, by preventing its transport into mitochondria.

Peroxisomal β-Oxidation of Fatty Acids

A modified version of the β-oxidation pathway occurs in **peroxisomes,** organelles that are present in most eukaryotic cells. Peroxisomes are quite similar to plant cell glyoxysomes, except that peroxisomes lack the enzymes of the glyoxylate pathway. Both peroxisomes and glyoxysomes carry out a β-oxidation pathway in which the FAD-linked acyl-CoA dehydrogenase transfers electrons not to the respiratory electron transport chain but directly to oxygen. The latter is reduced to hydrogen peroxide, which in turn is acted on by catalase.

$$E—FAD + R—CH_2—CH_2—C(=O)—S—CoA \longrightarrow R—CH{=}CH—C(=O)—S—CoA + E—FADH_2$$

$$E—FADH_2 + O_2 \longrightarrow E—FAD + H_2O_2$$

$$H_2O_2 \longrightarrow H_2O + \tfrac{1}{2}O_2$$

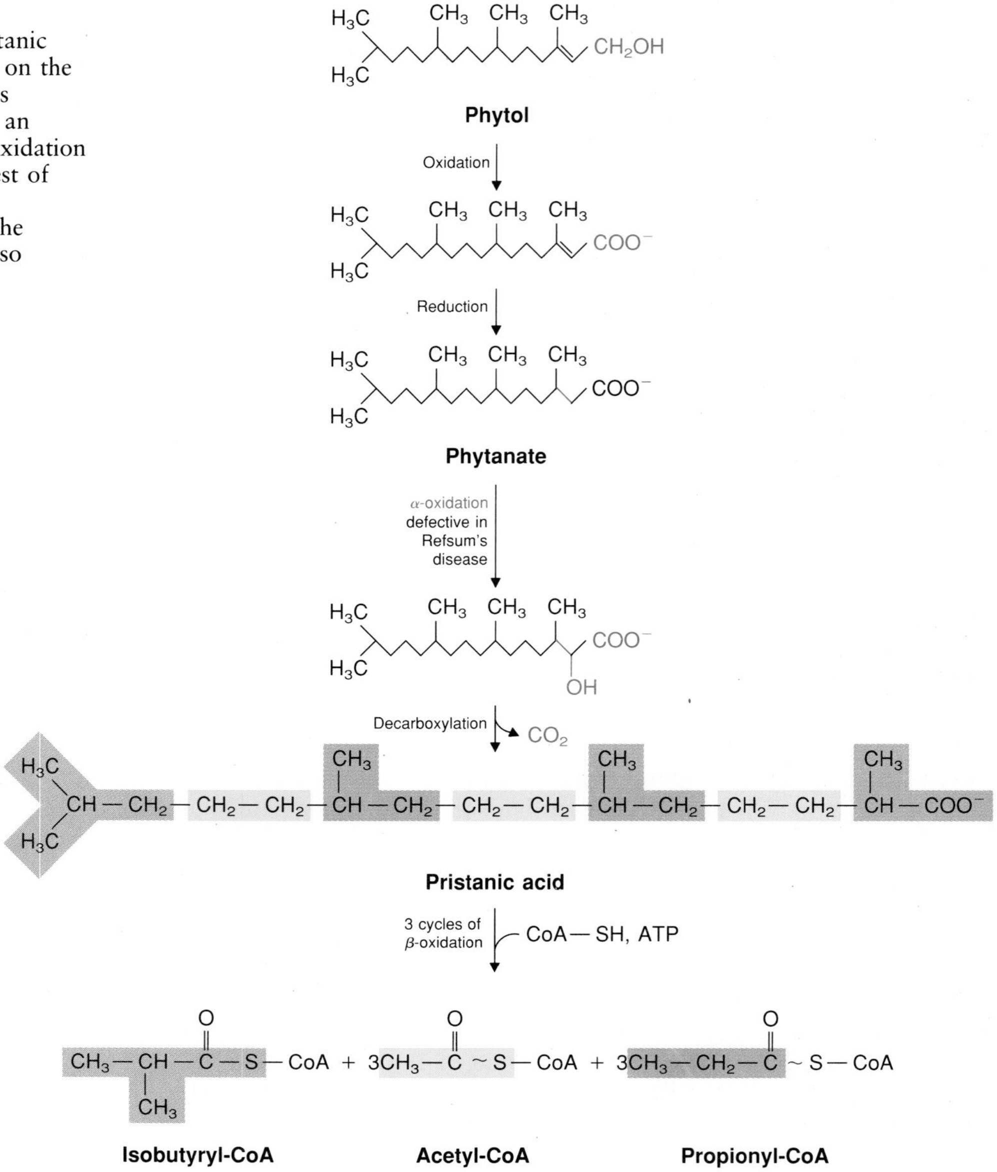

Figure 17.14
The α-oxidation pathway for phytanic acid oxidation. The methyl group on the β-carbon of phytanic acid prevents β-oxidation of this compound, so an additional pathway, involving α-oxidation comes into play. After that, the rest of the molecule can be degraded by β-oxidation. In Refsum's disease the α-oxidation pathway is defective, so β-oxidation cannot proceed.

Because electrons are not shuttled into the respiratory chain, the peroxisomal pathway is not coupled to energy production, but it does generate heat. In animal peroxisomes the pathway proceeds only as far as C_4 and C_6 acyl-CoAs; butyryl-CoA and hexanoyl-CoA are not substrates for the flavin-linked enzyme. However, the acyl groups can be transferred to carnitine for transport into mitochondria, where oxidation can be completed. By contrast, plant glyoxysomes carry the oxidation to acetyl-CoA, which is used for carbohydrate synthesis via the glyoxylate pathway. The function of the peroxisomal pathway is not yet clear, but current evidence points toward a role in the initial stages of oxidizing very long-chain fatty acids and other lipids.

α-Oxidation of Fatty Acids

Although β-oxidation is the major pathway for fatty acid degradation, a minor pathway oxidizes certain fatty acids with initial oxidation occurring

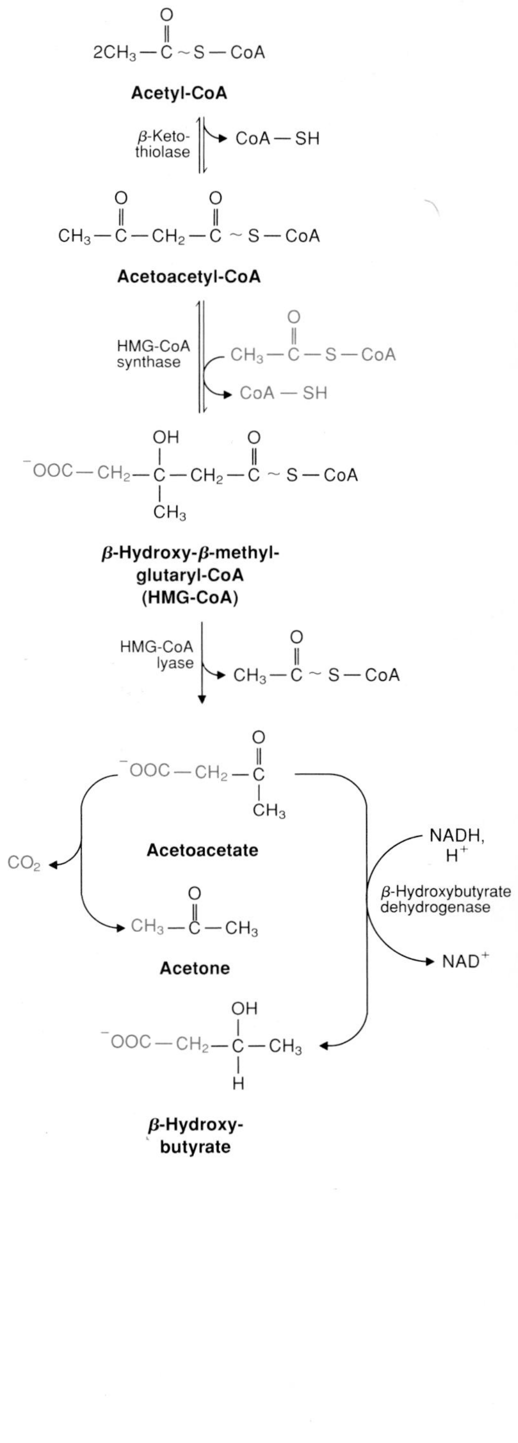

Figure 17.15
Biosynthesis of ketone bodies in the liver. The three compounds commonly called ketone bodies are boxed. Acetone is formed in very small quantities, possibly by nonenzymatic decarboxylation of acetoacetate.

$2CH_3-\overset{O}{\overset{\|}{C}}\sim S-CoA$

Acetyl-CoA

β-Ketothiolase → CoA—SH

$CH_3-\overset{O}{\overset{\|}{C}}-CH_2-\overset{O}{\overset{\|}{C}}\sim S-CoA$

Acetoacetyl-CoA

HMG-CoA synthase; $CH_3-\overset{O}{\overset{\|}{C}}-S-CoA$ → CoA — SH

$^{-}OOC-CH_2-\overset{OH}{\overset{|}{C}}(CH_3)-CH_2-\overset{O}{\overset{\|}{C}}\sim S-CoA$

β-Hydroxy-β-methylglutaryl-CoA (HMG-CoA)

HMG-CoA lyase → $CH_3-\overset{O}{\overset{\|}{C}}\sim S-CoA$

$^{-}OOC-CH_2-\overset{O}{\overset{\|}{C}}-CH_3$

Acetoacetate

CO_2

$CH_3-\overset{O}{\overset{\|}{C}}-CH_3$

Acetone

NADH, H^+; β-Hydroxybutyrate dehydrogenase; NAD^+

$^{-}OOC-CH_2-\overset{OH}{\overset{|}{C}}(H)-CH_3$

β-Hydroxybutyrate

on the α-carbon, rather than the β-carbon. This came to light through analysis of a rare and very severe congenital neurological disorder called **Refsum's disease.** Patients with this condition accumulate large amounts of an unusual fatty acid, **phytanic acid,** which is derived from **phytol,** a constituent of chlorophyll (Figure 17.14). The methyl group on carbon 3 of phytol prevents β-oxidation of this substrate. However, the α-carbon can evidently undergo oxidation, to give **pristanic acid,** a substrate that can be degraded by β-oxidation. In Refsum's disease the α-oxidation pathway is defective, and the phytanic acid is not converted to a compound that can be degraded. The only known treatment is to feed a diet containing little or no chlorophyll. This is difficult because it rules out both leafy green vegetables and meat that comes from herbivorous animals (such as beef).

Ketogenesis

Thus far we have written about acetyl-CoA as though it had only two major metabolic fates, either oxidation to CO_2 in the citric acid cycle, or biosynthesis of fatty acids. Another major pathway comes into play in mitochondria, when acetyl-CoA accumulates beyond its capacity to be oxidized or used for fatty acid synthesis. That pathway is **ketogenesis,** and it leads to a class of compounds called **ketone bodies.**

Particularly when levels of oxaloacetate are low, so that flux through citrate synthase is impaired, 2 mol of acetyl-CoA undergo a reversal of the thiolase reaction to give acetoacetyl-CoA (Figure 17.15). This can react in turn with a third mole of acetyl-CoA to give ***β*-hydroxy-*β*-methylglutaryl-CoA** (HMG-CoA), catalyzed by **HMG-CoA synthase.** When formed in cytosol, HMG-CoA is an early intermediate in cholesterol biosynthesis (Chapter 18). In mitochondria, however, HMG-CoA is acted on by **HMG-CoA lyase** to yield **acetoacetate** plus acetyl-CoA. Acetoacetate undergoes either NADH-dependent reduction to give ***β*-hydroxybutyrate** or, in very small amounts, spontaneous decarboxylation to acetone. Collectively, acetoacetate, acetone, and β-hydroxybutyrate are called ketone bodies, even though the latter compound does not contain a carbonyl group.

Ketogenesis can be considered an "overflow pathway." It is stimulated when acetyl-CoA accumulates due to deficient carbohydrate utilization, so that oxaloacetate levels are low; this reduces flux through citrate synthase and causes acetyl-CoA to accumulate. Ketogenesis occurs primarily in liver, because of the high levels of HMG-CoA synthase in that tissue. Ketone bodies are transported from liver to other tissues, where acetoacetate and β-hydroxybutyrate can be reconverted to acetyl-CoA for energy generation. The reconversion involves enzymatic transfer of a CoA moiety from succinyl-CoA to acetoacetate, yielding acetoacetyl-CoA and succinate.

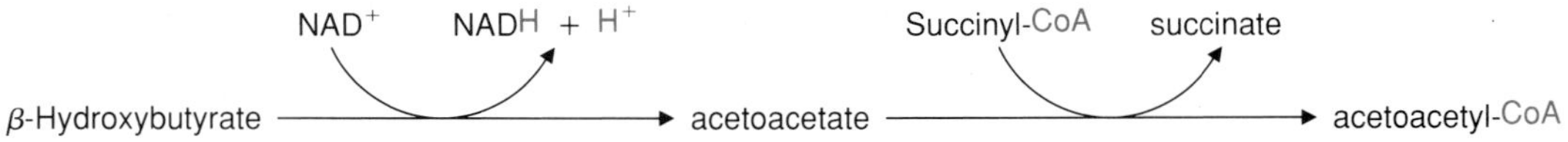

$$\beta\text{-Hydroxybutyrate} \xrightarrow{NAD^+ \;\to\; NADH + H^+} \text{acetoacetate} \xrightarrow{\text{Succinyl-CoA} \;\to\; \text{succinate}} \text{acetoacetyl-CoA}$$

Some tissues, notably heart, routinely derive much or most of their metabolic energy from the oxidation of ketone bodies supplied by the liver. Other tissues, particularly the brain, increase their utilization of ketone bodies when insufficient glucose is available (see Chapter 3).

Fatty Acid Biosynthesis

Relationship of Fatty Acid Synthesis to Carbohydrate Metabolism

We have noted that the vast majority of the stored fuel in most animal cells is in the form of fat. However, a large proportion of the caloric intake of many animal diets—certainly most human diets—is carbohydrate. Since carbohydrate storage reserves are strictly limited, there must be efficient mechanisms for conversion of carbohydrate to fat. In this section our primary focus is on fatty acid synthesis.

As schematized in Figure 17.16, a central metabolite is acetyl-CoA, which comes both from the pyruvate dehydrogenase reaction and from fatty acid β-oxidation. Acetyl-CoA is, in turn, converted in the cytosol to fatty acids. Thus, acetyl-CoA is derived from both fat breakdown and carbohydrate breakdown and is also the major fat precursor. However, *acetyl-CoA cannot undergo net conversion to carbohydrate*. This is because of the virtual irreversibility of the pyruvate dehydrogenase reaction. As noted in Chapter 14, the glyoxylate cycle in plants and some microorganisms permits a bypass of this step, with net conversion of acetyl-CoA to gluconeogenic precursors. However, *in animals the conversion of carbohydrate to fat is unidirectional*. Moreover, while the fatty acid synthetic pathway is subject to regulation, the total capacity for fat storage is not.

Early Studies of Fatty Acid Synthesis

Early in the twentieth century, when it became evident that most fatty acids in lipids contain even-numbered chains, it was reasonable to expect that the biosynthetic process would involve some stepwise addition of activated

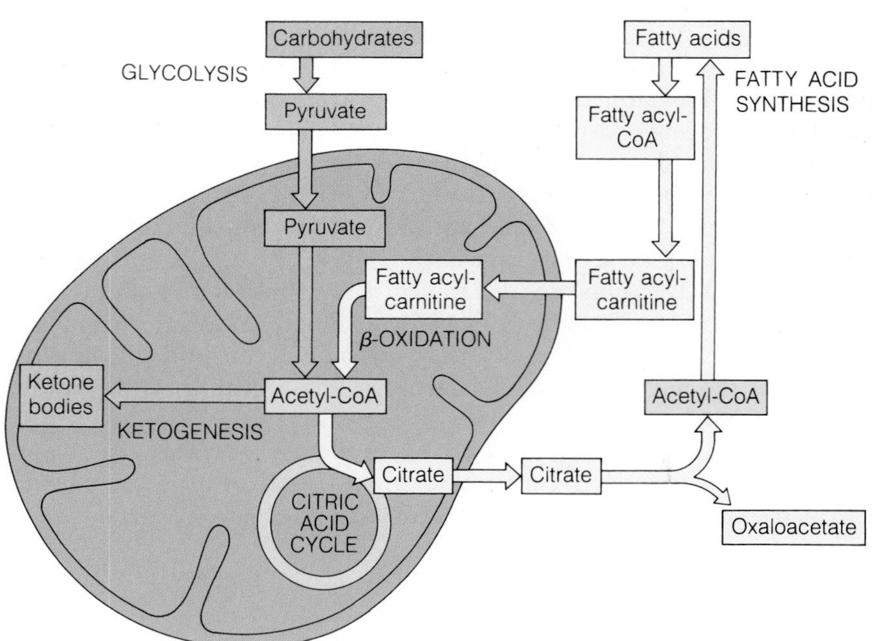

Figure 17.16
Acetyl-CoA as a key intermediate between fat and carbohydrate metabolism. Arrows identify major routes of formation or utilization of acetyl-CoA. Citrate serves as a carrier to transport acetyl units from mitochondrion to cytosol for fatty acid synthesis.

two-carbon fragments, in the same sense that oxidation proceeds two carbons at a time. Indeed, this was demonstrated experimentally in the 1940s, in one of the first metabolic experiments using isotopic tracers. David Rittenberg and Konrad Bloch fed to mice acetate labeled with the stable isotopes ^{13}C and deuterium ($C^2H_3{}^{13}COO^-$) and found both isotopes incorporated into fatty acids.

Once the β-oxidation pathway had been discovered, it was generally thought that fatty acid synthesis would proceed simply by a reversal of its degradation pathway. However, when biochemists began to fractionate enzyme systems capable of synthesizing fatty acids, they found that the activities of β-oxidation were lacking from their purified fractions. The key discovery that established fatty acid synthesis as an entirely different pathway was Salih Wakil's observation in the late 1950s that fatty acid synthesis has an absolute requirement for bicarbonate. The carbon from that bicarbonate was not incorporated in the final product; rather, the role of bicarbonate was catalytic. These observations led to the discovery of a three-carbon compound, malonyl-CoA, as the first committed intermediate in fatty acid biosynthesis. Today we know that, while the chemistries of fatty acid synthesis and degradation are similar, the pathways differ in the enzymes involved, acyl group carriers, stereochemistry of the intermediates, electron carriers, intracellular location, and regulation. Indeed, fatty acid metabolism is one of the best examples of the statement that biosynthetic pathways are never the simple reversal of degradative pathways.

$$^-OOC{-}CH_2{-}\overset{\displaystyle O}{\overset{\|}{C}}{-}S{-}CoA$$

Malonyl-CoA

The overall process of fatty acid synthesis is similar in all prokaryotic and eukaryotic systems analyzed to date. Three separate enzyme systems catalyze, respectively, (1) biosynthesis of palmitate from acetyl-CoA, (2) chain elongation starting from palmitate, and (3) desaturation. In eukaryotic cells the first pathway occurs in the cytosol, while chain elongation occurs both in mitochondria and in the endoplasmic reticulum, and desaturation occurs in the endoplasmic reticulum.

Biosynthesis of Palmitate from Acetyl-CoA

As noted above and illustrated in Figure 17.17, the chemistry of palmitate synthesis is remarkably similar to that of palmitate oxidation run in reverse. The major distinctions are the need for an activated intermediate, malonyl-CoA, at each two-carbon addition step and the use of NADPH-requiring enzymes in the reductive reactions. Details of these and the other reactions are presented below.

SYNTHESIS OF MALONYL-COA. The first committed step in fatty acid biosynthesis is the formation of malonyl-CoA from acetyl-CoA, catalyzed by **acetyl-CoA carboxylase.**

$$\underset{\textbf{Acetyl-CoA}}{CH_3{-}\overset{\displaystyle O}{\overset{\|}{C}}{-}S{-}CoA} + ATP + HCO_3^- \longrightarrow \underset{\textbf{Malonyl-CoA}}{^-OOC{-}CH_2{-}\overset{\displaystyle O}{\overset{\|}{C}}{-}S{-}CoA} + ADP + P_i + H^+$$

Like other committed steps in biosynthetic pathways, this reaction is so exergonic as to be virtually irreversible. Like other enzymes catalyzing carboxylation reactions, acetyl-CoA carboxylase has a biotin cofactor, covalently bound via a lysine ϵ-amino group. The reaction proceeds via a covalently bound *N*-carboxybiotin intermediate.

$$\text{E-biotin} + ATP + HCO_3^- \longrightarrow \text{E-}N\text{-carboxybiotin} + ADP + P_i$$

$$\text{E-}N\text{-carboxybiotin} + \text{acetyl-CoA} \longrightarrow \text{malonyl-CoA} + \text{E-biotin}$$

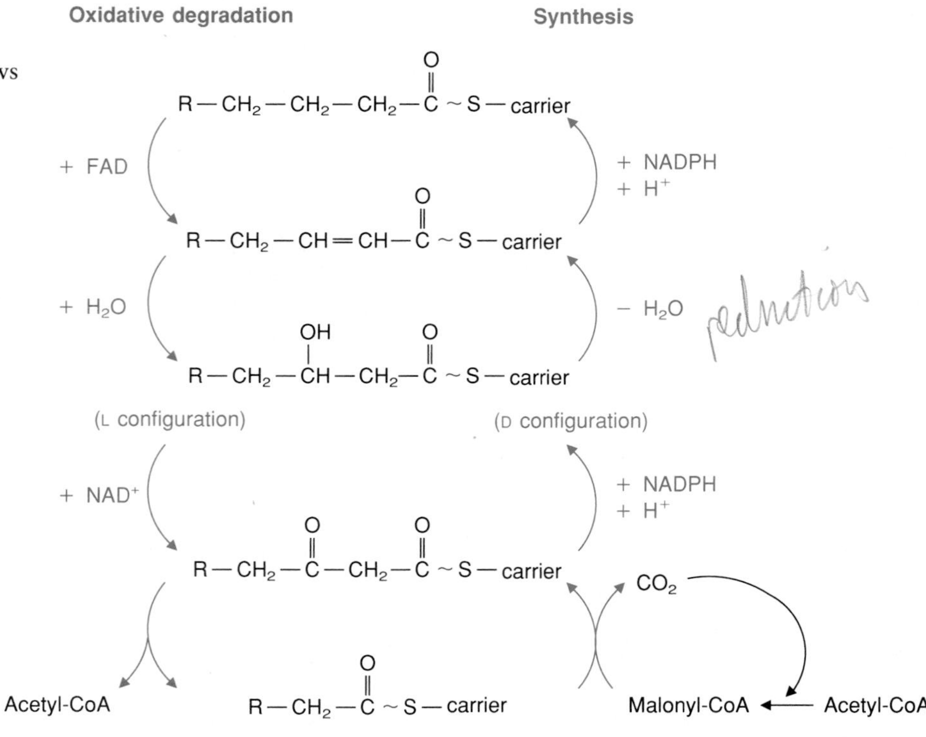

Figure 17.17
Chemical similarities between fatty acid oxidation and synthesis. The figure shows a single cycle of oxidation (down) or addition (up) of one two-carbon fragment.

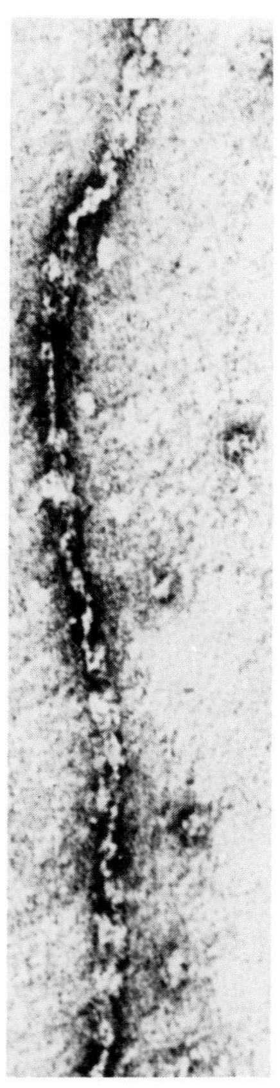

Figure 17.18
Electron micrograph of the enzymatically active filamentous form of acetyl-CoA carboxylase.

The prokaryotic form of this enzyme, exemplified by the enzyme purified from *E. coli,* consists of three separate proteins: (1) a small carrier protein that contains the bound biotin, (2) a **biotin carboxylase,** which catalyzes the first of the above reactions, and (3) a **transcarboxylase,** which transfers the activated carboxyl group from *N*-carboxybiotin to acetyl-CoA. The hydrocarbon chains in both biotin and its associated lysine residue act as a flexible swinging arm, which allows the biotin to interact with the catalytic sites of both catalytic subunits.

By contrast, acetyl-CoA carboxylase in eukaryotes consists of a single protein containing two identical polypeptide chains, each of M_r about 230,000. The dimeric protein itself is inactive, but in the presence of citrate it polymerizes to a filamentous form, of M_r 4–8 × 10^6, that can readily be visualized in the electron microscope (Figure 17.18). The equilibrium between inactive protein dimers and the active filamentous form, and its control by metabolic intermediates, probably represents a mechanism for regulating fatty acid biosynthesis. The primary physiological regulator appears to be not citrate, but long-chain fatty acyl-CoAs, which promote depolymerization of the active form.

ACYL CARRIER PROTEIN. All of the intermediates in fatty acid oxidation are activated via their linkage to coenzyme A. A similar activation is involved in fatty acid synthesis, but the carrier is different; it is a small protein (77 residues in *E. coli*) called **acyl carrier protein,** or ACP. The chemistry of activation is identical to that in acyl-CoAs. Recall from Chapter 14 that the reactive sulfhydryl group in CoA is part of a phosphopantetheine moiety, derived from pantothenic acid. Similarly, in ACP, a phosphopantetheine moiety is linked to a serine group in the polypeptide (Figure 17.19). It is not immediately clear why the synthetic pathway should use a protein carrier, but it is possible that a swinging arm mechanism is involved in the subse-

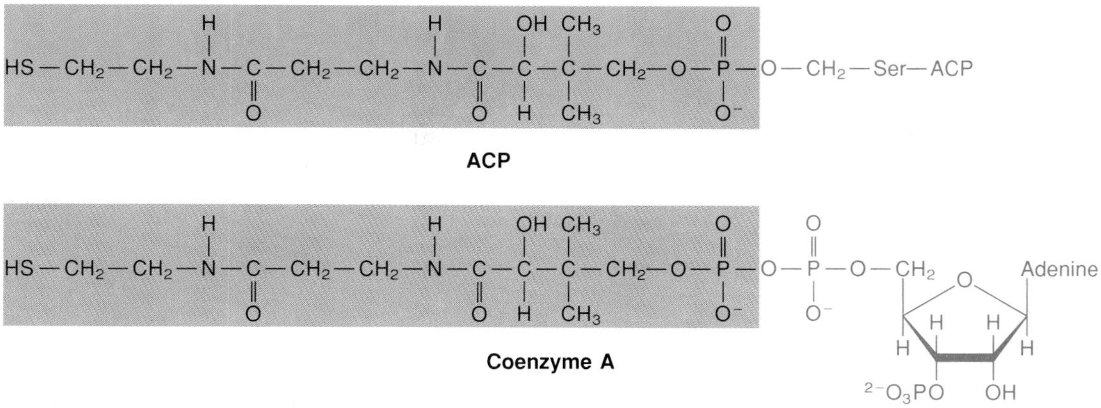

$$HS-CH_2-CH_2-\overset{H}{\underset{}{N}}-\underset{\|}{\overset{}{C}}(=O)-CH_2-CH_2-\overset{H}{N}-C(=O)-\overset{OH}{\underset{H}{C}}-\overset{CH_3}{\underset{CH_3}{C}}-CH_2-O-\overset{O}{\underset{O^-}{P}}-O-CH_2-Ser-ACP$$

ACP

$$HS-CH_2-CH_2-\overset{H}{N}-C(=O)-CH_2-CH_2-\overset{H}{N}-C(=O)-\overset{OH}{\underset{H}{C}}-\overset{CH_3}{\underset{CH_3}{C}}-CH_2-O-\overset{O}{\underset{O^-}{P}}-O-\overset{O}{\underset{O^-}{P}}-O-CH_2-\text{ribose}(\text{Adenine}; 3'\text{-}{}^{2-}O_3PO, 2'\text{-}OH)$$

Coenzyme A

Figure 17.19
Phosphopantetheine as the reactive unit in CoA and ACP. In each case the phosphopantetheine moiety is shown in purple.

quent steps of fatty acid synthesis, as well as in the formation of malonyl-CoA. ACP is the most abundant protein in *E. coli,* with about 1.5×10^6 molecules per cell.

ACP becomes involved in fatty acid synthesis through the actions of **malonyl-CoA-ACP transacylase** and **acetyl-CoA-ACP transacylase.** In each case the acyl group is transferred from acyl-CoA to ACP. Because the energy-rich bonds in acyl-CoAs and acyl-ACPs are identical, these reactions are readily reversible.

$$\text{Acetyl-CoA} + \text{ACP} \rightleftharpoons \text{acetyl-ACP} + \text{CoA—SH}$$

$$\text{Malonyl-CoA} + \text{ACP} \rightleftharpoons \text{malonyl-ACP} + \text{CoA—SH}$$

While the malonyl transacylase is highly specific, the acetyl transacylase can react to some extent with other acyl-CoA substrates. In fact, this is how the synthesis of odd-chain fatty acids begins, with propionyl-CoA as the substrate, instead of acetyl-CoA.

FROM MALONYL-COA TO PALMITATE. As noted earlier, the fatty acid chain is built up by successive additions of two-carbon units. Each cycle of addition consists of seven reactions, starting with acetyl-CoA carboxylase. The reaction pathway is identical in all known organisms, but the protein chemistry involved is startlingly variable. In *E. coli,* in other bacteria, and in plants the reactions are catalyzed by seven distinct enzymes, which can be separately purified. By contrast, in animals and in lower eukaryotes all of the activities are associated in a highly structured multienzyme complex called **fatty acid synthase.** Let us first focus on the reactions and then consider the nature of the fatty acid synthase complex.

The first three reactions are identical in each two-carbon addition cycle. They are reactions we have already presented—acetyl-CoA carboxylase, malonyl-CoA-ACP transacylase, and acetyl-CoA-ACP transacylase (Figure 17.20). Next, for the first cycle of synthesis, we start with 1 mol each of malonyl-ACP and acetyl-ACP, and in four reactions we generate 1 mol of butyryl-ACP (Figure 17.21). These four reactions remarkably resemble the four reactions (in reverse) of fatty acid oxidation (see Figure 17.17). The synthetic cycle proceeds via *condensation, reduction, dehydration, reduction,* while oxidation (in reverse) involves *thiolytic cleavage, dehydrogenation, hydration, dehydrogenation.* Some of the major differences are evident from the figure; ACP is the acyl carrier for synthesis, and NADPH is the electron carrier for both reductive steps. Also, the use of a malonyl group has no counterpart in fatty acid oxidation.

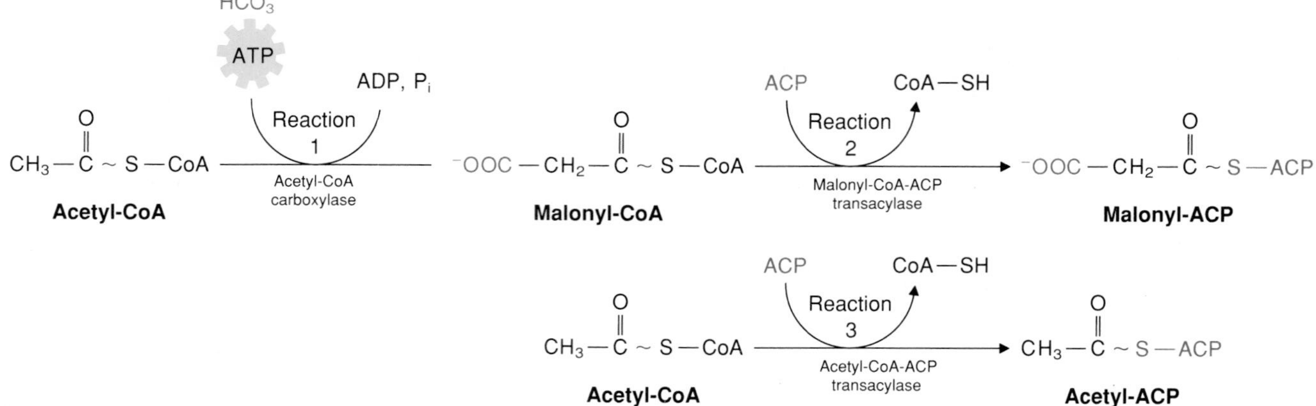

Figure 17.20
The first three reactions in fatty acid synthesis, leading to malonyl-ACP and acetyl-ACP.

What is the molecular logic of using malonyl-ACP as a donor of an acetyl unit? Probably it is that the condensation of two activated acetyl units is quite endergonic; a comparable reaction in reverse—namely the thiolytic cleavage of acetoacetyl-CoA—is strongly exergonic. However, the carboxyl group of *malonyl-ACP* is a good leaving group. Mechanistically, this carboxyl group activates the methylene carbon to act as a nucleophile and to attack the electrophilic keto carbon of acetyl-ACP. The involvement of ATP to drive this endergonic reaction is apparent but indirect, for ATP participated in the original synthesis of malonyl-CoA from acetyl-CoA.

This process explains the early observation that bicarbonate is not incorporated into the final product; rather, all of the carbons come from acetate.

The condensation product, a β-ketoacyl-ACP thioester, is now reduced to a D-3-hydroxyacyl-ACP (reaction 5 of Figure 17.21). By contrast, the 3-hydroxyacyl-CoAs produced in fatty acid oxidation have the L configuration. Dehydration of the 3-hydroxyacyl-ACP (reaction 6) yields a *trans*-Δ^2-enoyl-ACP, which undergoes a second reduction (reaction 7) to yield a fatty acyl-ACP—butyryl-ACP in the first cycle of synthesis. To start the second cycle butyryl-ACP reacts with another molecule of malonyl-ACP, and the product of the second cycle is hexanoyl-ACP. The same pattern continues until the product of the seventh cycle, palmitoyl-ACP, undergoes hydrolysis to yield palmitate and free ACP.

Like most biosynthetic pathways, this one requires both *energy* (as ATP) and *reducing equivalents* (as NADPH). The quantitative requirement can be seen from the stoichiometry of the complete seven-cycle process.

$$\text{Acetyl-CoA} + 7\ \text{malonyl-CoA} + 14\text{NADPH} + 7\text{H}^+ \longrightarrow \text{palmitate} + 7\text{CO}_2 + 14\text{NADP}^+ + 8\text{CoA} + 6\text{H}_2\text{O}$$

To see the ATP requirement one must consider the synthesis of the 7 mol of malonyl-CoA.

$$7\ \text{Acetyl-CoA} + 7\text{CO}_2 + 7\text{ATP} \longrightarrow 7\ \text{malonyl-CoA} + 7\text{ADP} + 7\text{P}_i + 7\text{H}^+$$

Hence, the following equation describes the overall process.

$$8\ \text{Acetyl-CoA} + 7\text{ATP} + 14\text{NADPH} \longrightarrow \text{palmitate} + 14\text{NADP}^+ + 8\text{CoA} + 6\text{H}_2\text{C} + 7\text{ADP} + 7\text{P}_i$$

Multifunctional Proteins in Fatty Acid Synthesis. As noted above, the enzymes of fatty acid synthesis in animals were originally found to con-

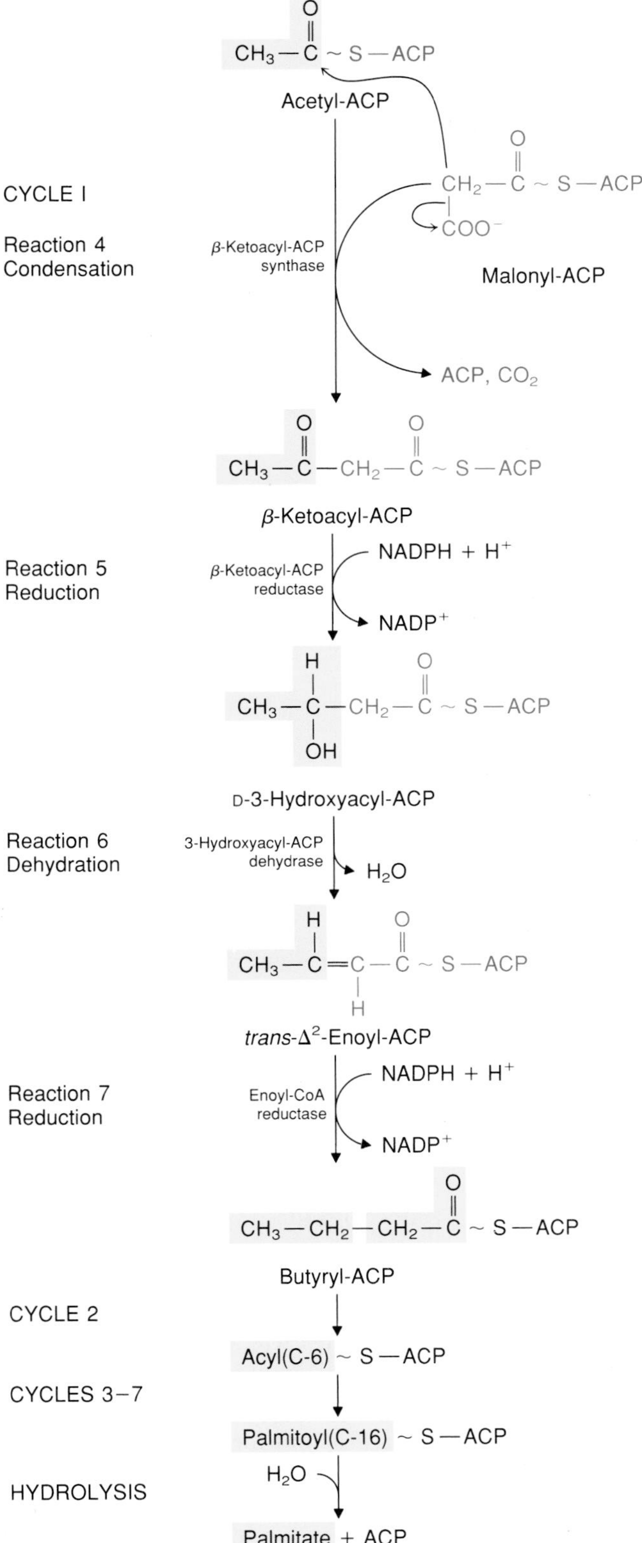

Figure 17.21
Synthesis of palmitate, starting with malonyl-ACP and acetyl-ACP. The first cycle of four reactions generates butyryl-ACP, which reacts with a second mole of malonyl-ACP, leading to a second cycle of two-carbon addition. A total of seven such cycles generates palmitoyl-ACP. Hydrolysis of this releases palmitate.

stitute a tightly coupled multienzyme complex. In the mid-1970s both biochemical and genetic studies in yeast revealed that the complex actually contained two multifunctional polypeptide chains. This complex has a mass of about 2.3 million daltons and can readily be visualized by electron microscopy (Figure 17.22). The complex contains six molecules each of two

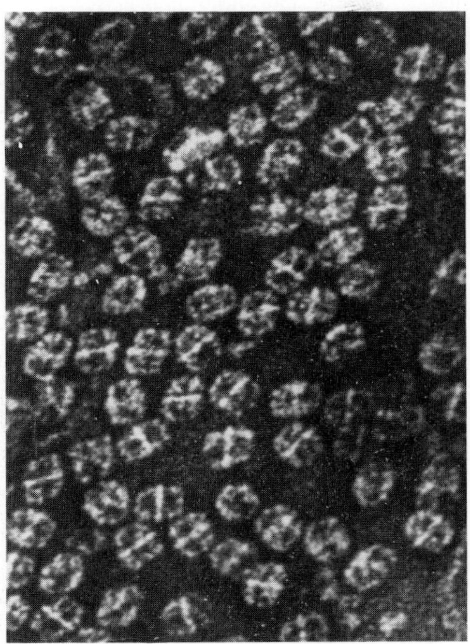

Figure 17.22
Electron micrograph of the fatty acid synthase complex from yeast. ×200,000.

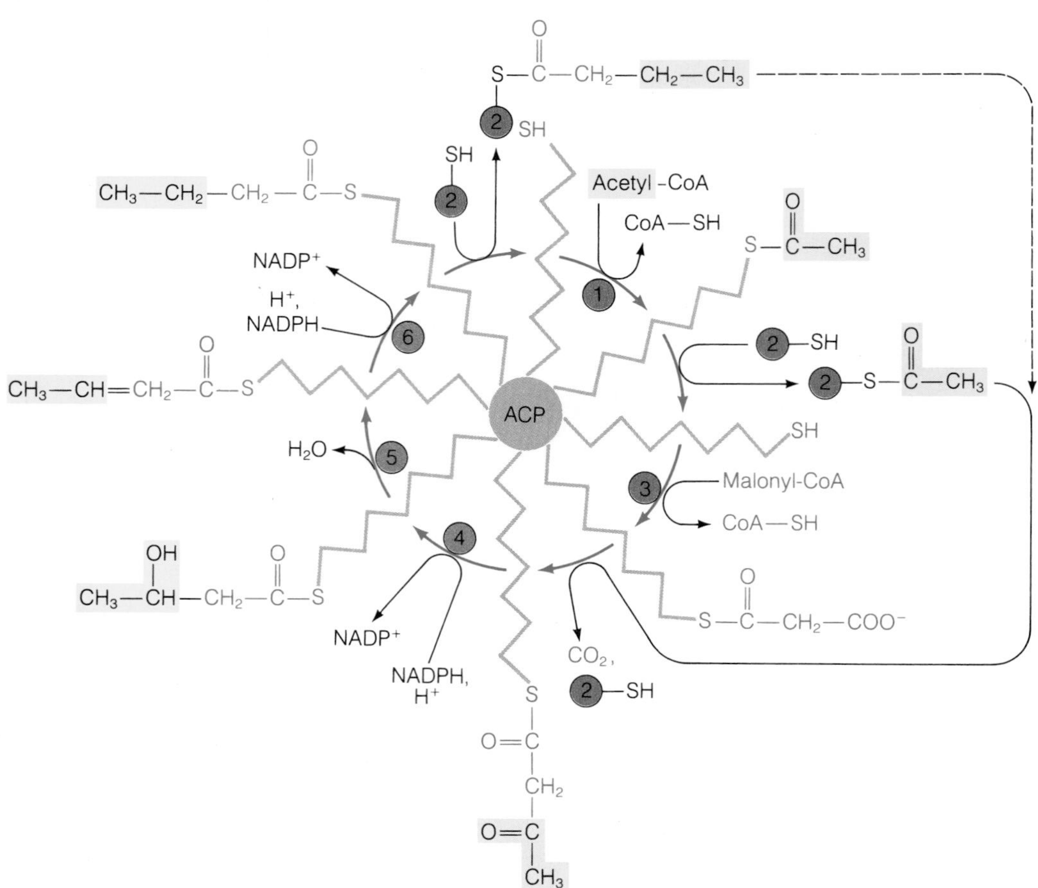

Figure 17.23
A phosphopantetheine swinging arm as a mechanism to bring acyl groups in contact with all of the active sites of fatty acid synthesis. The individual active sites (numbered circles) are (1) acetyl-CoA-ACP transacylase, (2) β-ketoacyl-ACP synthase, (3) malonyl-CoA-ACP transacylase, (4) β-ketoacyl-ACP reductase, (5) β-hydroxy-acyl-ACP dehydrase, and (6) enoyl-ACP reductase. The cycle begins with transfer of the acetyl group from acetyl-CoA to the phosphopantetheine swinging arm (site 1). The acetyl group is then transferred to a cysteine thiol group on β-ketoacyl-ACP synthase (site 2). At the end of the first cycle the resultant butyryl group is transferred to the same cysteine thiol group, so that another round (dashed line) can begin by transfer of a malonyl group from malonyl-CoA to the phosphopantetheine swinging arm.

polypeptide chains, which are called subunit A and subunit B. Subunit A (M_r = 185,000) contains the acyl carrier protein, the condensing enzyme, and the β-ketothioester reductase, while subunit B (M_r = 175,000) contains the remaining four activities.

In vertebrate tissues the complex is smaller, with two identical subunits of about 240,000 each. Evidently each subunit contains an ACP region plus all of the enzyme activities involved. An activity catalyzing the final release of palmitate is present as well. The yeast complex lacks the latter activity, because the final product of fatty acid synthase activity in yeast is not palmitate but palmitoyl-CoA.

It is interesting that in eukaryotic complexes the ACP is not a small protein, as it is in bacteria, but apparently a specific domain of a vastly larger polypeptide chain that contains several enzyme activities as well. It seems likely that the bound phosphopantetheine serves as a swinging arm to bring the acyl moiety into contact with the several active sites, just as we discussed for acetyl-CoA carboxylase and for the pyruvate dehydrogenase complex (Figure 17.23). There is already considerable information available on the architecture of this protein. Limited proteolysis of chicken liver fatty acid synthase yields products that represent globular domains of the multifunctional protein, each with specific catalytic activities. These plus more recent physical studies yield a model of the fatty acid synthase, as shown in Figure 17.24.

It has long been known that intermediates between acetyl-CoA and palmitate do not accumulate in cells that are synthesizing fatty acids. The basis for this is clear, if the intermediates are covalently bound to the ACP

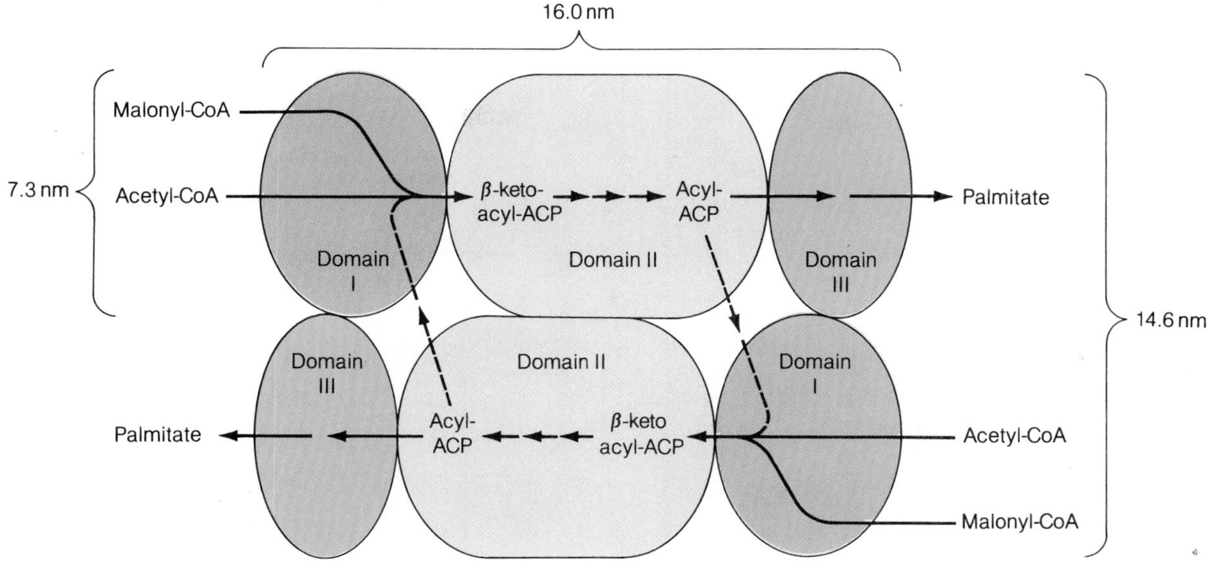

Figure 17.24
Model of chicken liver fatty acid synthase, based on limited proteolysis, high-resolution electron microscopy, and small-angle neutron scattering. Each of the two polypeptide chains in the dimer can be resolved into three separate domains by limited proteolysis. Domain I contains active sites 1, 2, and 3 (see Figure 17.23). Domain II contains active sites 4, 5, and 6. Domain III contains the site for hydrolysis of palmitoyl-ACP with release of palmitate. Individual polypeptide chains cannot carry out repeated cycles of two-carbon addition, indicating that domains on each of the two chains must interact with one another as the fatty acid chain is built up.

domain of a multifunctional protein, as they are in eukaryotic cells. This arrangement ensures that substrates need not seek catalytic sites by random diffusion. Moreover, since the genetic apparatus needs to regulate the synthesis of only one or two polypeptide chains instead of seven, this type of organization is attractive from the standpoint of coordinating the activities involved. Multifunctional proteins have now been described in purine and pyrimidine nucleotide synthesis, amino acid synthesis, coenzyme metabolism, and DNA replication as well.

Transport of Acetyl Units and Reducing Equivalents into the Cytosol. Since acetyl-CoA is generated in the mitochondrial matrix, it must be transported to the cytosol for use in fatty acid synthesis. Like longer-chain acyl-CoAs, acetyl-CoA cannot permeate the inner membrane. A shuttle system is used, which is interesting both because it provides a control mechanism for fatty acid synthesis and because it generates much of the NADPH needed for the process (Figure 17.25). The shuttle involves citrate, which is formed in mitochondria from acetyl-CoA and oxaloacetate in the first step of the citric acid cycle. When citrate is being generated in excess of the amount needed for oxidation in the cycle, it is transported through the mitochondrial membrane to the cytosol. There it is acted on by **citrate lyase** to regenerate acetyl-CoA and oxaloacetate

$$\text{Citrate} + \text{ATP} + \text{CoA—SH} \longrightarrow \text{acetyl-CoA} + \text{ADP} + \text{P}_i + \text{oxaloacetate}$$

at the expense of one ATP. Oxaloacetate cannot directly return to the mitochondrial matrix, because the inner membrane lacks a transporter for this compound. First it is reduced by a cytosolic malate dehydrogenase to malate, which can be oxidatively decarboxylated by the malic enzyme (p. 490).

$$\text{Oxaloacetate} + \text{NADH} + \text{H}^+ \longrightarrow \text{malate} + \text{NAD}^+$$

$$\text{Malate} + \text{NADP}^+ \longrightarrow \text{pyruvate} + \text{CO}_2 + \text{NADPH} + \text{H}^+$$

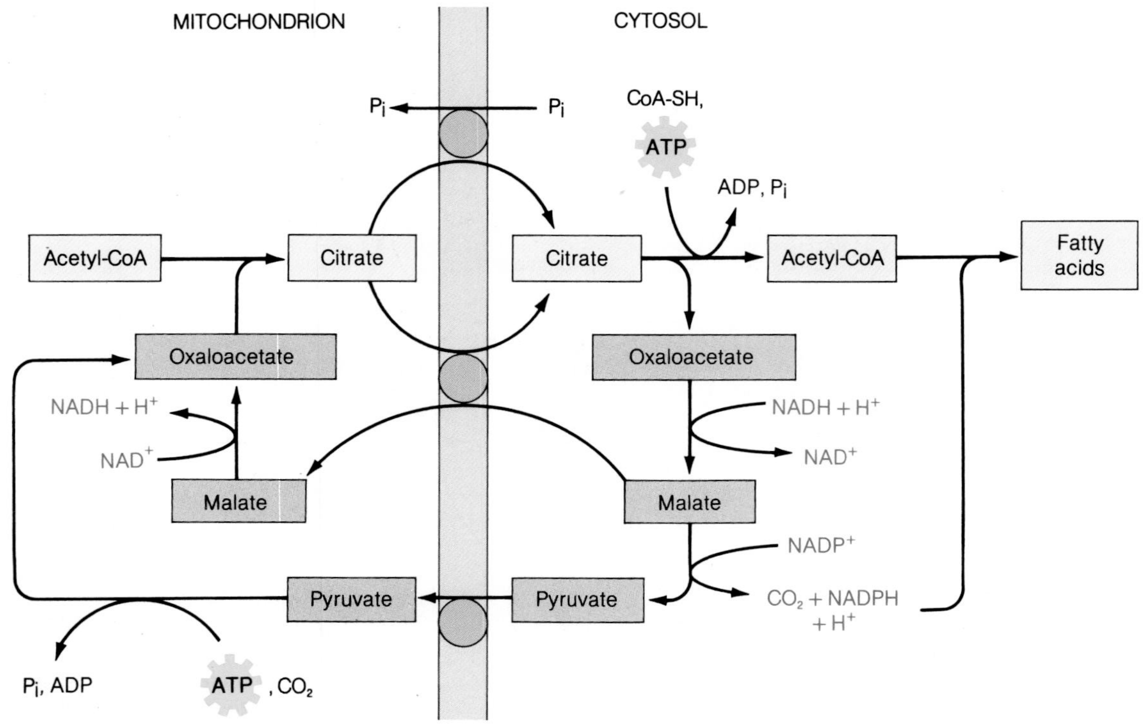

Figure 17.25
Shuttle mechanism for transfer of acetyl units and reducing equivalents from mitochondria to cytosol, for use in fatty acid synthesis. Citrate must be exchanged for a carrier as it moves out of the mitochondrion. Some is evidently exchanged for orthophosphate and some for malate. The malate that is not exchanged generates some of the NADPH for fatty acid synthesis, through action of the malic enzyme. Purple circles represent transport systems located in the mitochondrial membrane.

The resultant pyruvate diffuses back into mitochondria, where it is reconverted to oxaloacetate by pyruvate carboxylase.

$$\text{Pyruvate} + CO_2 + \text{ATP} + H_2O \longrightarrow \text{oxaloacetate} + \text{ADP} + P_i + H^+$$

The net reaction catalyzed by these three enzymes is

$$\text{NADP}^+ + \text{NADH} + \text{ATP} + H_2O \longrightarrow \text{NADPH} + \text{NAD}^+ + \text{ADP} + P_i + H^+$$

Some of the malate formed returns to the mitochondrion and is exchanged for citrate. However, for each mole of malate remaining in the cytosol, 1 mol of NADPH is generated. Most of the remainder of the 14 mol of NADPH required to synthesize 1 mol of palmitate is generated in the cytosol via the pentose phosphate pathway.

Elongation of Fatty Acid Chains

Since fatty acid synthase action leads primarily to palmitate, we must consider the processes that lead from palmitate to give the variations observed among fatty acids in both chain length and degree of unsaturation. In eukaryotic cells elongation occurs in both mitochondria and endoplasmic reticulum. The latter, so-called microsomal, system has far greater activity and is the one described here. It is similar to the fatty acid synthase sequence, but it involves acyl-CoA derivatives and separate enzymes. The first reaction is a condensation between malonyl-CoA and a long-chain fatty acyl-CoA substrate.

$$\underset{\textbf{Acyl-CoA}}{R-\overset{O}{\overset{\|}{C}}-S-CoA} + \underset{\textbf{Malonyl-CoA}}{{}^-OOC-CH_2-\overset{O}{\overset{\|}{C}}-S-CoA} \longrightarrow \underset{\boldsymbol{\beta}\textbf{-Ketoacyl-CoA}}{R-\overset{O}{\overset{\|}{C}}-CH_2-\overset{O}{\overset{\|}{C}}-S-CoA} + CoA-SH + CO_2$$

The resultant β-ketoacyl-CoA undergoes NADPH-dependent reduction, dehydration of the hydroxyacyl-CoA, and another NADPH-dependent reduction to give a saturated fatty acyl-CoA two carbons longer than the original substrate.

$$\underset{\beta\text{-Ketoacyl-CoA}}{R-\overset{O}{\overset{\|}{C}}-CH_2-\overset{O}{\overset{\|}{C}}\sim S-CoA} \xrightarrow[\quad]{NADPH + H^+ \rightarrow NADP^+} \underset{\beta\text{-Hydroxyacyl-CoA}}{R-\overset{OH}{\underset{H}{C}}-CH_2-\overset{O}{\overset{\|}{C}}\sim S-CoA} \xrightarrow{\rightarrow H_2O}$$

$$\underset{\text{Enoyl-CoA}}{R-CH=CH-\overset{O}{\overset{\|}{C}}\sim S-CoA} \xrightarrow{NADPH + H^+ \rightarrow NADP^+} \underset{\text{Acyl-CoA}}{R-CH_2-CH_2-\overset{O}{\overset{\|}{C}}\sim S-CoA}$$

At least two different condensing enzymes are present in endoplasmic reticulum, one of which acts on unsaturated fatty acyl-CoAs. Apparently, one common set of enzymes carries out the remaining three reactions.

Fatty Acid Desaturation

The most common monounsaturated fatty acids in animal lipids are oleic acid, an 18:1 *cis*-Δ^9 acid, and palmitoleic acid, a 16:1 *cis*-Δ^9 compound. These are synthesized from stearate and palmitate, respectively, by a microsomal system called **fatty acyl-CoA desaturase.** The overall reaction for stearoyl-CoA desaturation is as follows.

$$\text{Stearoyl-CoA} + NADH + H^+ + O_2 \longrightarrow \text{oleyl-CoA} + NAD^+ + 2H_2O$$

Note that both substrates undergo two-electron oxidations in this reaction. The overall electron transfer in this reaction involves another enzyme, the flavin-dependent **cytochrome b_5 reductase** (Figure 17.26).

In addition to the Δ^9 desaturating system described above, mammalian cells contain Δ^5 and Δ^6 desaturases. The activities of these enzymes are subject to complex hormonal control. All three activities are enhanced by

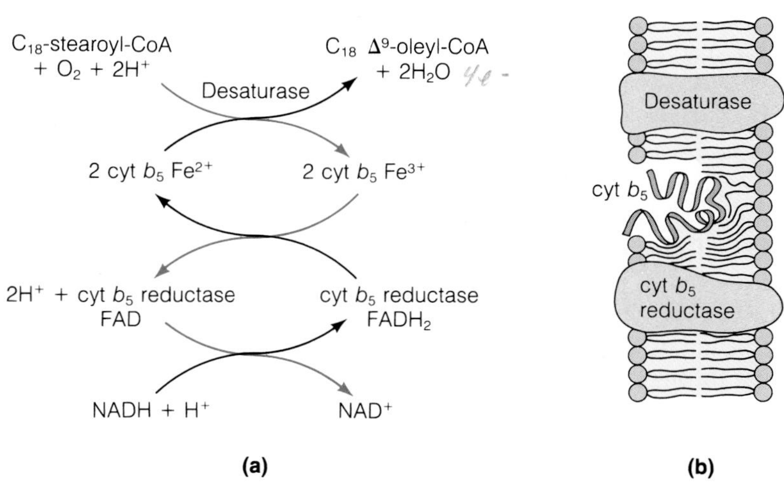

Figure 17.26
Fatty acid desaturation system. (**a**) The redox reactions involved; (**b**) a model for the locations of the three proteins involved in the membrane.

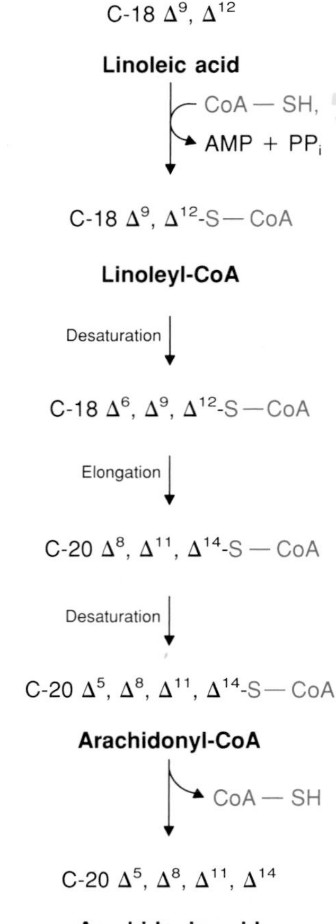

Figure 17.27
Pathway for conversion of linoleic acid to arachidonic acid in mammals. While arachidonyl-CoA can be cleaved as shown to give arachidonic acid, most of the arachidonyl-CoA is used for phospholipid synthesis, and arachidonic acid for eicosanoid biosynthesis is derived from hydrolysis of these phospholipids.

insulin, but other hormones have differential effects—activating one desaturase and inhibiting others. The significance of these effects remains a focus of active investigation.

Mammals are unable to introduce double bonds beyond Δ^9 in the fatty acid chain. Hence, they cannot synthesize either linoleic acid (18:2 *cis*-Δ^9, Δ^{12}) or linolenic acid (18:3 *cis*-Δ^9, Δ^{12}, Δ^{15}). These are called **essential fatty acids** because they must be provided in the diet. After ingestion in animals, they are, in turn, substrates for further desaturation and elongation reactions. Particularly important is the pathway summarized in Figure 17.27, which leads from linoleic acid to **arachidonic acid** (20:4 *cis*-Δ^5, Δ^8, Δ^{11}, Δ^{14}). Arachidonic acid is the precursor to a class of compounds called the **eicosanoids.** As discussed in Chapter 18, these include two important classes of metabolic regulators, the prostaglandins and the thromboxanes.

Control of Fatty Acid Synthesis

To a large extent fatty acid biosynthesis is controlled by hormonal mechanisms. Much of the fatty acid synthesis that occurs in animals takes place in adipose tissue, where fat is being stored for release and transport to other tissues on demand, to help meet their energy needs. As extracellular messengers, hormones are well suited to playing these interorgan regulatory roles.

Figure 17.28 summarizes the major effects in regulation of fatty acid synthesis in animal cells. Insulin acts to stimulate fatty acid synthesis. One of its effects is to stimulate glucose entry into cells. This increases flux through glycolysis and the pyruvate dehydrogenase reaction, which provides acetyl-CoA for fatty acid synthesis. Insulin also activates the pyruvate dehydrogenase complex, by stimulating its dephosphorylation to the active form. The mechanism of this effect is not clear, but it may involve increases in levels of Ca^{2+}, which stimulates pyruvate dehydrogenase phosphatase.

Another site for regulation (not shown in the figure) is the transfer of acetyl units from the mitochondrial matrix to the cytosol, where fatty acid synthesis occurs. There is evidence that citrate lyase activity is controlled by phosphorylation and dephosphorylation of the enzyme, but the physiological mechanisms involved are not yet known.

The first enzyme whose action is committed to fatty acid synthesis is acetyl-CoA carboxylase. As noted earlier, this enzyme must undergo a reversible polymerization in order to be active. Although the polymerization reaction can be controlled in vitro by either citrate or ATP, current evidence points away from physiological roles for these effectors. Long-chain fatty acyl-CoAs at low levels prevent depolymerization, with concomitant inactivation, of the enzyme, and this represents an apparent feedback inhibition of the pathway. The levels of fatty acyl-CoAs are lowered by insulin, presenting another mechanism by which insulin stimulates fatty acid synthesis.

Finally, there is evidence that fatty acid synthesis is controlled by the availability of reducing equivalents. Recall that NADPH comes from both the transport of citrate out of mitochondria and the pentose phosphate pathway. The latter pathway in turn is controlled through inhibition by NADPH of glucose-6-phosphate dehydrogenase and 6-phosphogluconate dehydrogenase. Typically, about 60% of the NADPH for fatty acid synthesis comes from the pentose phosphate pathway. When acetate is added to an experimental tissue preparation, that proportion can rise to 80%, indicating that flux through the pentose phosphate pathway can rise when the demand for NADPH increases.

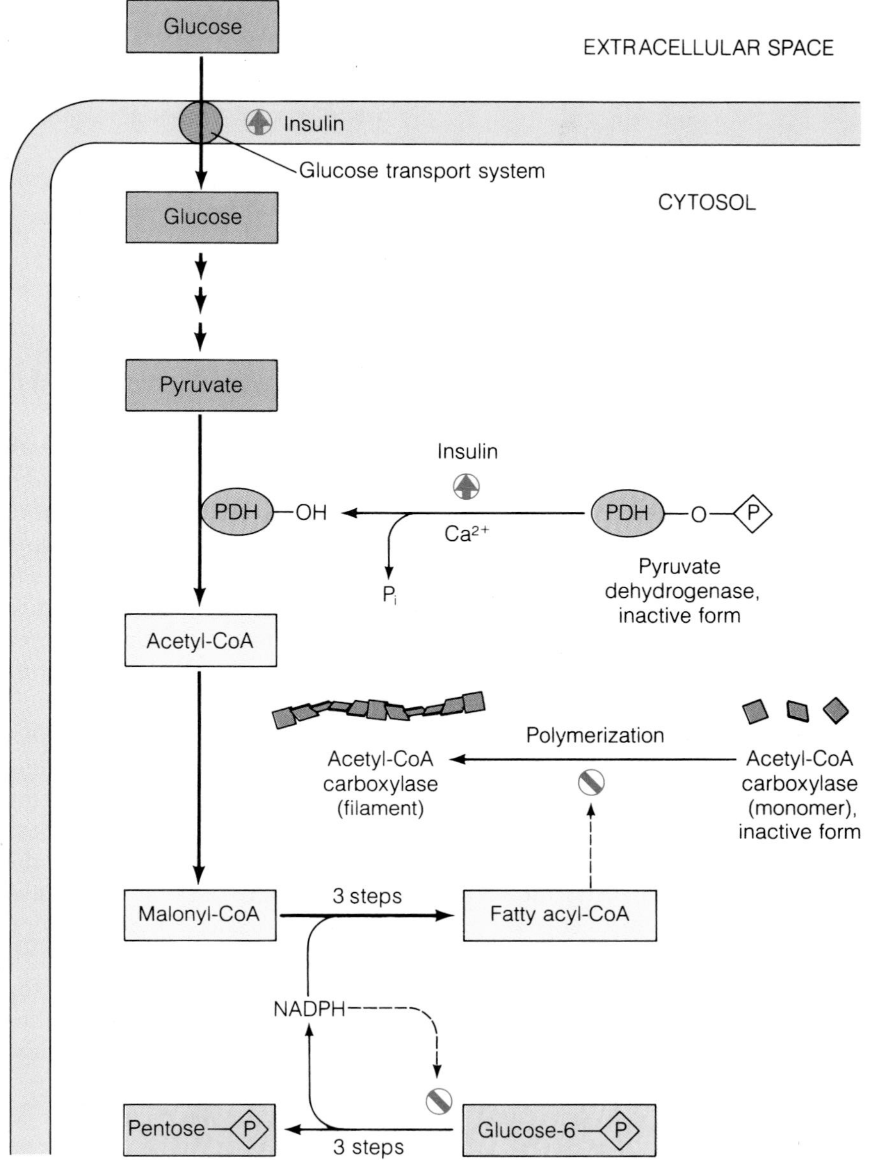

Figure 17.28
Regulation of fatty acid synthesis in eukaryotes.

Biosynthesis of Triacylglycerols

Fatty acyl-CoAs, along with glycerol-3-phosphate, serve as the major precursors to triacylglycerols. Glycerol-3-phosphate is derived either from the reduction of the glycolytic intermediate dihydroxyacetone phosphate, catalyzed by **glycerol phosphate dehydrogenase,** or by the ATP-dependent phosphorylation of glycerol.

$$\text{Dihydroxyacetone phosphate} + \text{NADH} + \text{H}^+ \longrightarrow \text{L-glycerol-3-phosphate} + \text{NAD}^+$$

$$\text{Glycerol} + \text{ATP} \longrightarrow \text{glycerol-3-phosphate} + \text{ADP}$$

Glycerol-3-phosphate undergoes two successive esterifications with fatty acyl-CoAs to yield **diacylglycerol-3-phosphate.**

Fatty acyl-S—CoA + glycerol-3-phosphate ⟶ monoacylglycerol-3-phosphate + CoA—SH

Monoacylglycerol-3-phosphate + fatty acyl-S—CoA ⟶ diacylglycerol-3-phosphate + CoA—SH

Diacylglycerol-3-phosphate, also called **phosphatidic acid,** is a precursor both to phospholipids and to triacylglycerols. The pathway to triacylglycerols involves hydrolytic removal of the phosphate, followed by transfer of another fatty acyl moiety from an acyl-CoA.

Phosphatidic acid + H_2O ⟶ 1,2-diacylglycerol + P_i

1,2-Diacylglycerol + fatty acyl-CoA ⟶ triacylglycerol + CoA—SH

As we have noted, triacylglycerols represent the major form in which energy can be stored. Normally in an adult animal synthesis and degradation are balanced, so that there is no net change in the total body amount of triacylglycerols. If dietary intake exceeds caloric needs, proteins, carbohydrate, or fat can all readily provide acetyl-CoA to drive the synthesis of fatty acids and triacylglycerols. On the other hand, fat reserves allow animals to go for rather long times without eating and still maintain adequate energy levels. This does generate some metabolic stresses, as we shall explore further in Chapter 23.

Hibernating animals have adapted remarkably well to cope with such stresses. For example, bears store huge amounts of fat just before beginning a hibernation that may last as long as 7 months. During this period all of the bear's energy comes from breakdown of this stored fat. Moreover, the bear excretes so little water that the water released from fat oxidation meets the animal's needs. Similarly, the glycerol released from triacylglycerols provides a source of gluconeogenic precursors.

As noted above, phosphatidic acid serves as precursor both to triacylglycerols and to phospholipids and other glycerol-containing lipids. In the next chapter we shall concern ourselves with the biosynthesis and metabolism of complex lipids, starting with the pathways that utilize phosphatidic acid.

REFERENCES

General References

Martin, B. R. (1987) *Metabolic Regulation. A Molecular Approach,* Blackwell Scientific Publications, London. A concise paperback book that clearly summarizes hormonal and allosteric regulation of the major metabolic pathways in animals.

Vance, D. E., and J. E. Vance (eds.) (1985) *Biochemistry of Lipids and Membranes.* Benjamin/Cummings, Menlo Park, Calif. This multiauthored text contains a number of timely and detailed reviews. Particularly relevant to this chapter are articles entitled Oxidation of fatty acids, pp. 116–142; Fatty acid synthesis in eucaryotes, pp. 143–180; Fatty acid desaturation and chain elongation in eucaryotes, pp. 181–212; Metabolism of triacylglycerols, pp. 213–241; and Metabolism of cholesterol and lipoproteins, pp. 404–474.

Lipid and Lipoprotein Metabolism in Animals

Breslow, J. (1985) Human apolipoprotein molecular biology and genetic variation. *Annu. Rev. Biochem.* 54:699–727. Reviews the molecular properties and genetic mapping data for these proteins, plus known human mutations in the apolipoprotein genes.

Hansen, H. S. (1986) The essential nature of linoleic acid in mammals. *Trends Biochem. Res.* 11:263–265. The discovery of linoleic acid-rich lipids in epidermal cells suggests a specific role for this essential fatty acid in water retention.

Stanbury, J. B., J. B. Wyngaarden, D. S. Fredrickson, J. L. Goldstein, and M. S. Brown (1989) *Metabolic Basis of Inherited Diseases,* 6th ed. McGraw-Hill, New York. Chapters 29–35 present human disorders of lipid and lipoprotein metabolism.

Fatty Acid Oxidation

Tolbert, N. E. (1981) Metabolic pathways in peroxisomes and glyoxysomes. *Annu. Rev. Biochem.* 50:133–157. A complete description of the enzymology and functions of peroxisomal β-oxidation.

Fatty Acid Synthesis

Wakil, S. J., J. K. Stoops, and V. C. Joshi (1983) Fatty acid synthesis and its regulation. *Annu. Rev. Biochem.* 52:537–579. A detailed discussion of the protein chemistry of the eukaryotic multifunctional proteins involved in fatty acid synthesis.

PROBLEMS

1. Calculate the metabolic energy yield from oxidation of palmitic acid, taking into account the energy needed to activate the fatty acid and transport it into mitochondria. Do the same for stearic acid, linoleic acid, and oleic acid.

2. If palmitic acid is subjected to complete combustion in a bomb calorimeter, one can calculate a standard free energy of combustion of 9788 kJ/mol. From the ATP yield of palmitate oxidation, what is the metabolic efficiency of this process, in terms of kilojoules saved as ATP per kilojoule released?

3. Calculate the number of ATPs generated by the complete metabolic oxidation of tripalmitin. Hydrolysis of the triacylglycerol occurs at the cell surface. Consider the energy yield from catabolism of glycerol, as well as from the fatty acids. Calculate the ATP yield per carbon atom oxidized, and compare it with the energy yield from glucose.

4. Write a balanced equation for the *complete* metabolic oxidation of each of the following (include O_2, ADP, and P_i as reactants and ATP, CO_2, and H_2O as products): (a) stearic acid; (b) oleic acid; (c) palmitic acid; (d) linoleic acid.

5. Calculate the number of ATPs generated from the metabolic oxidation of the four carbons of acetoacetyl-CoA to CO_2. Now consider the homolog derived from oxidation of an odd-numbered carbon chain, namely propionoacetyl-CoA. Calculate the net ATP yield from oxidation of the five carbons of this compound to CO_2.

6. 2-Bromopalmitoyl-CoA inhibits the oxidation of palmitoyl-CoA by isolated mitochondria but has no effect on the oxidation of palmitoylcarnitine. What is the most likely site of inhibition of 2-bromopalmitoylcarnitine?

7. When the identical subunits of chicken liver fatty acid synthase are dissociated in vitro, all of the activities can be detected in the separated subunits except for the β-ketoacyl synthase reaction and the overall synthesis of palmitate. Explain these observations.

8. Mammals cannot undergo *net* synthesis of carbohydrate from acetyl-CoA, but the carbons of acetyl-CoA can be incorporated into glucose and amino acids. Present pathways by which this could come about.

9. Describe a pathway whereby some of the carbon from a fatty acid with an odd-numbered carbon chain could undergo a net conversion to carbohydrate.

10. How many tritium atoms are incorporated into palmitate when fatty acid synthesis is carried out in vitro with the following labeled substrate?

$$^{-}OOC{-}C\,{}^{3}H_2{-}\overset{\overset{\large O}{\|}}{C}{-}S{-}CoA$$

11. What would be the effect on fatty acid synthesis of an increase in intramitochondrial oxaloacetate level? Briefly explain your answer.

CHAPTER 18

Lipid Metabolism II: Phospholipids, Steroids, Isoprenoids, and Eicosanoids

Chapter 17 was concerned primarily with energetic aspects of lipid metabolism—synthesis and oxidation of fatty acids and metabolism of triacylglycerols. Lipids play several additional roles, primarily as membrane components and as biological regulators. Our concerns in this chapter are with the roles played by these more complex lipids, as well as their pathways for synthesis and degradation (Figure 18.1). This chapter focuses on the following major classes: *glycerophospholipids* (also called *phosphoglycerides*), which are primarily membrane components but which play some specialized regulatory roles; *sphingolipids*, which in animals are found primarily in nervous tissue; *steroids* and other *isoprenoid* compounds, which function as hormones, vitamins, and membrane constituents; and the *eicosanoids*, a class of biological regulators. Two major topics involving lipids and regulation are discussed later in this book: the actions of steroid hormones, in Chapters 23 and 28, and the second-messenger regulatory role played by inositol phospholipids, in Chapter 23.

Metabolism of Phospholipids Containing Glycerol

The most abundant phospholipids are those derived from glycerol. These **glycerophospholipids** are found primarily as components of membranes. In animals they also participate in the transport of triacylglycerols and cholesterol, by coating the surface of lipoproteins. In addition, phospholipids play specific roles in processes as diverse as blood clotting and lung functioning. The major phospholipids other than glycerophospholipids are the sphingolipids.

Biosynthesis of Phospholipids in Bacteria

It is appropriate to begin our discussion of phospholipid metabolism in the prokaryotic kingdom, partly because of the relative simplicity of the biological systems involved. In bacteria, phospholipids may constitute 10% of the dry weight of the cell, yet their only known role is as components of membranes. In *E. coli,* the most widely studied microorganism, the membranes contain but three phospholipids in significant amounts—phosphatidylethanolamine (75–85%), phosphatidylglycerol (10–20%), and

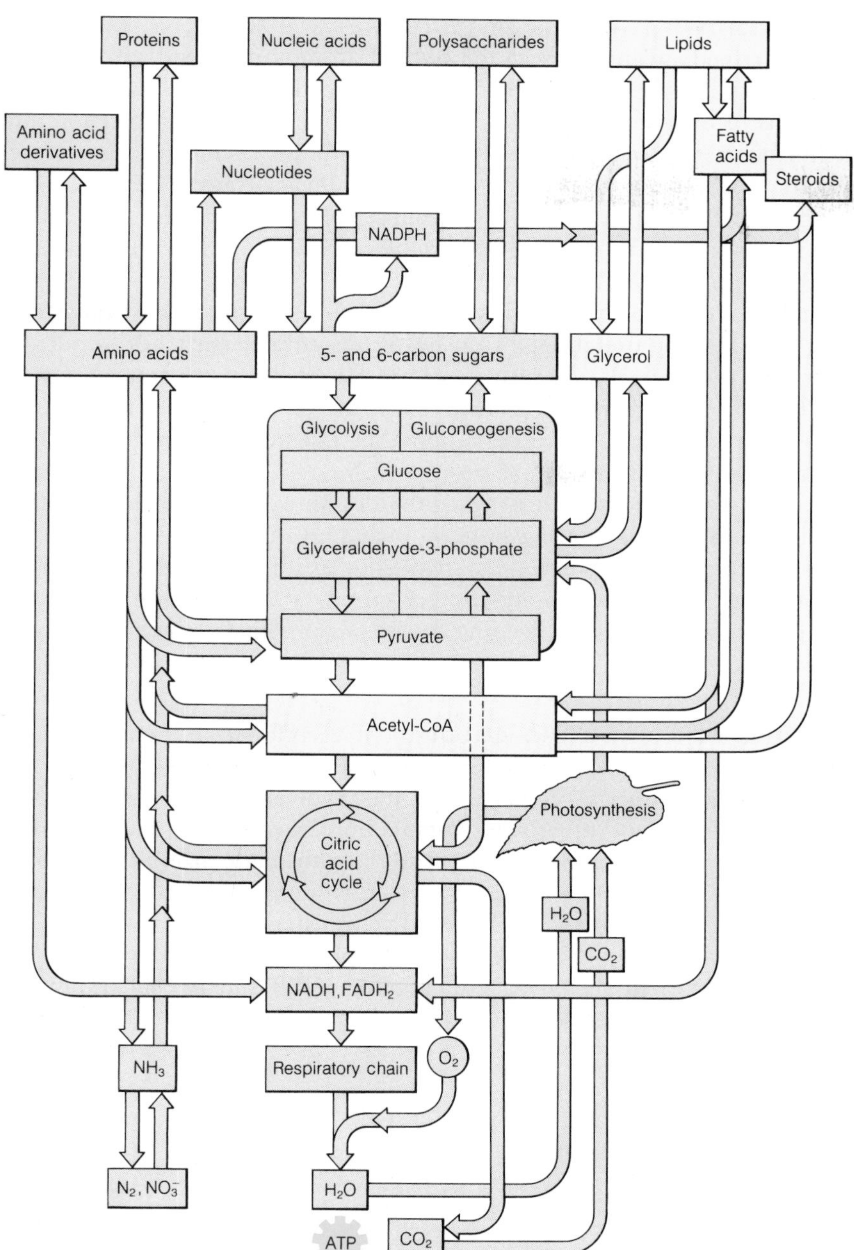

Figure 18.1
Metabolism of complex lipids and steroids (yellow) in relation to the major pathways in intermediary metabolism.

cardiolipin (5–15%). The fatty acid content of these lipids is also simple, with three dominating—palmitate (16:0), palmitoleate (16:1-Δ^9), and *cis*-vaccenate (18:1-Δ^{11}).

Because they are easy to grow in large quantities, bacteria provide abundant sources for large-scale isolation of the enzymes involved in lipid metabolism. Much of our earliest information on both phospholipid synthesis and fatty acid synthesis came from studies with *E. coli.* More recently, physiological studies on bacterial mutants have offered insight into control of membrane lipid synthesis and, in particular, the mechanisms used for temperature regulation of the fatty acid content of these lipids. Recall from Chapter 9 that, when grown at low temperature, bacteria increase the unsaturation of their membrane fatty acids to maintain optimal fluidity. One of the most intriguing areas of contemporary lipid biochemistry involves genetic analysis of mechanisms for maintaining the optimal pattern of unsaturation at a given temperature.

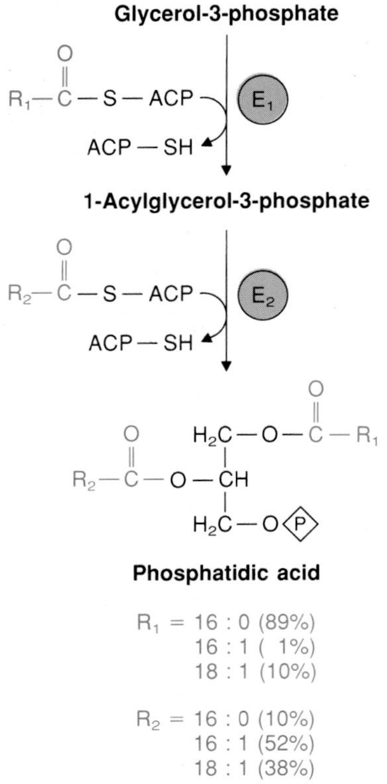

Figure 18.2
Synthesis of phosphatidic acid in bacteria. The percentages refer to proportion of fatty acids in *E. coli* inserted at each of the two indicated positions. E_1 = *sn*-glycerol-3-phosphate acyltransferase; E_2 = 1-acyl-*sn*-glycerol-3-phosphate acyltransferase. *Note:* In Chapter 9 the polar head groups were shown on the left for clarity. This chapter uses the stereochemically preferred designations.

Glycerol-3-phosphate
R_1—C(=O)—S—ACP → ACP—SH (E_1)
1-Acylglycerol-3-phosphate
R_2—C(=O)—S—ACP → ACP—SH (E_2)
Phosphatidic acid

R_1 = 16 : 0 (89%)
16 : 1 (1%)
18 : 1 (10%)

R_2 = 16 : 0 (10%)
16 : 1 (52%)
18 : 1 (38%)

BIOSYNTHESIS OF PHOSPHATIDIC ACID. We saw in Chapter 17 that a common intermediate in synthesis of both triacylglycerols and glycerophospholipids is diacylglycerol-3-phosphate, or phosphatidic acid. In *E. coli* the synthesis of phosphatidic acid occurs concomitantly with its insertion into the membrane, through the action of two membrane-bound acyltransferases (Figure 18.2). The first, ***sn*-glycerol-3-phosphate acyltransferase** (E_1), transfers primarily saturated fatty acyl groups from the respective acyl carrier protein derivatives to position 1 of *sn*-glycerol-3-phosphate (*sn* stands for stereospecifically numbered).

The second enzyme, **1-acyl-*sn*-glycerol-3-phosphate acyltransferase** (E_2), catalyzes the preferential transfer of unsaturated fatty acyl groups to position 2. Fatty acyl-CoA derivatives can also serve as substrates for these enzymes. The specificities of the two enzymes for saturated or unsaturated acyl moieties explain the observation that unsaturation in glycerophospholipids is found primarily at position 2.

Though the pattern in *E. coli* provides an excellent model for phospholipid biosynthesis in other systems, including animals, it is far from universal, even within prokaryotic systems. For example, the glycerolipids of archaebacteria contain phytanoyl ether groups at both positions 1 and 2. The biosynthetic pathways leading to these compounds have not yet been elucidated.

SYNTHESIS OF POLAR HEAD GROUPS. In carbohydrate metabolism we noted the use of nucleotide derivatives to activate relatively inert hydroxyl groups for biosynthesis—for example, the involvement of uridine nucleotides in glycogen biosynthesis. In phospholipid biosynthesis, nucleotides of either cytidine or deoxycytidine activate the phosphate group of phosphatidic acid for subsequent transfer of a potential polar head group. The reaction is catalyzed by **phosphatidate cytidylyltransferase,** which forms an anhydride bond between the phosphates of phosphatidic and cytidylic (or deoxycytidylic) acids. The reaction between CTP and phosphatidic acid yields **CDP-diacylglycerol.**

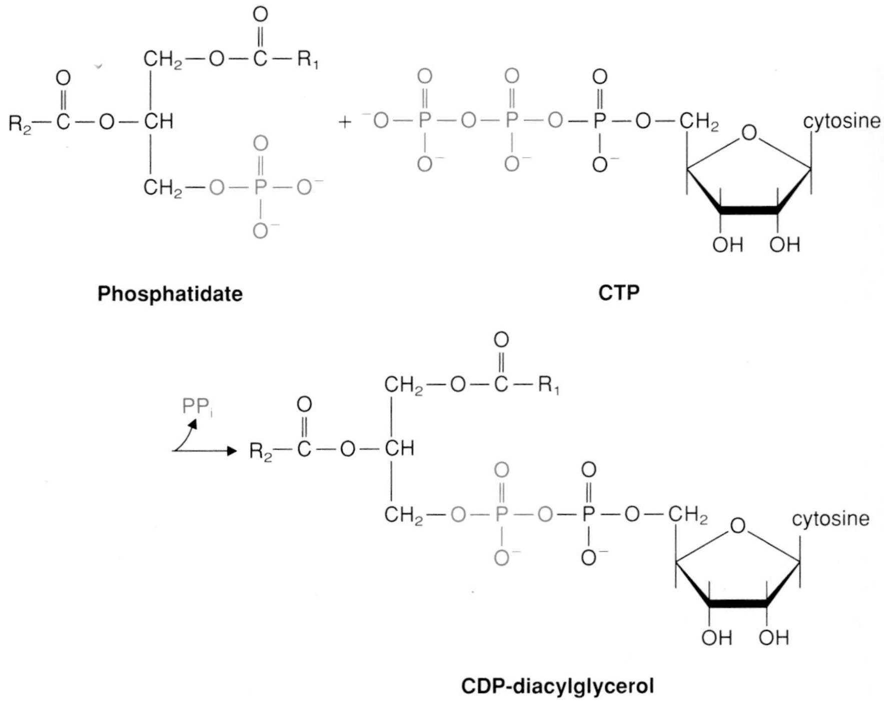

Phosphatidate + CTP → CDP-diacylglycerol + PP_i

Phosphatidate

CTP

cytosine

OH OH

CDP-diacylglycerol

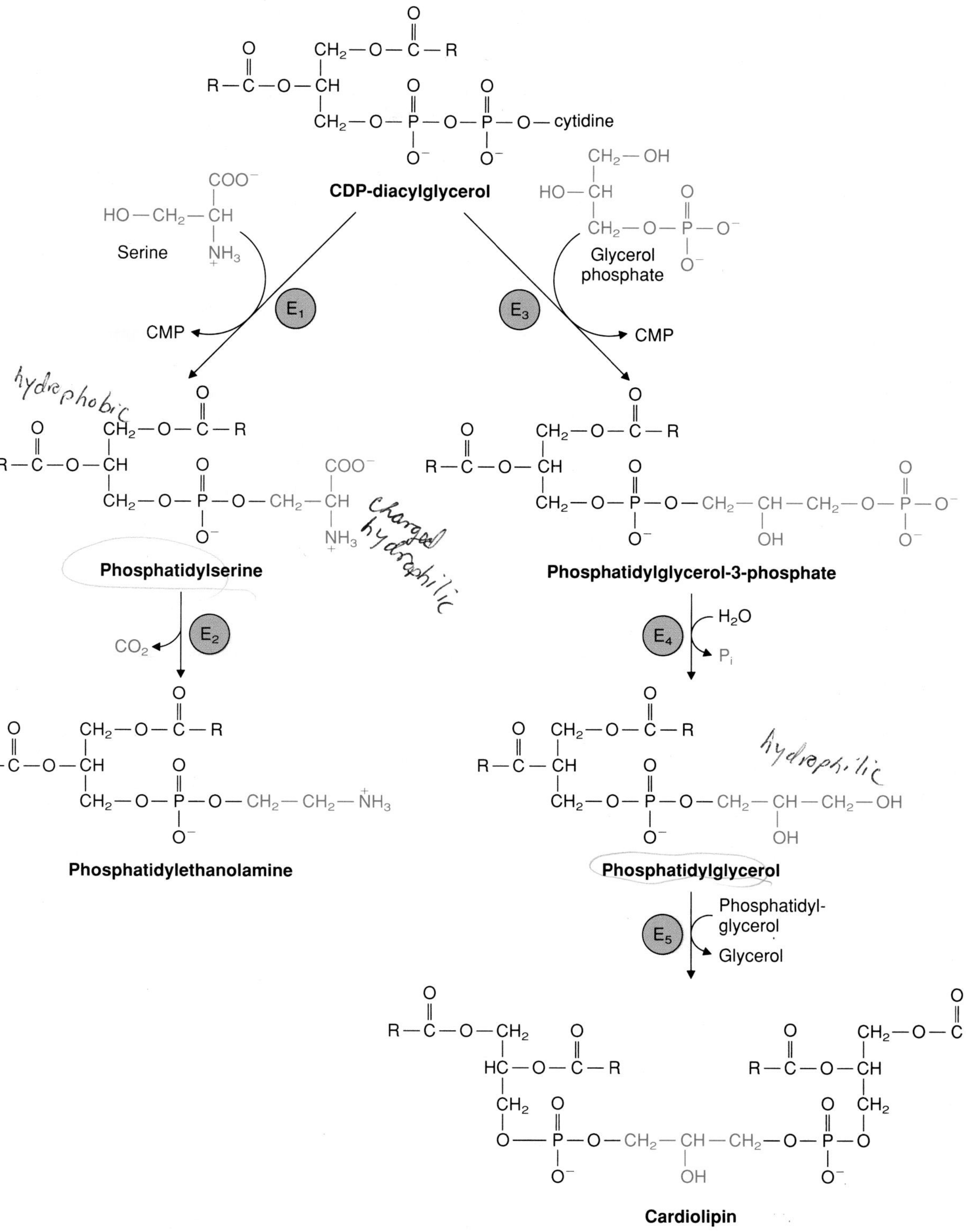

CDP-diacylglycerol

Serine

Glycerol phosphate

E_1 CMP

E_3 CMP

Phosphatidylserine

Phosphatidylglycerol-3-phosphate

E_2 CO_2

E_4 H_2O P_i

Phosphatidylethanolamine

Phosphatidylglycerol

E_5 Phosphatidylglycerol Glycerol

Cardiolipin

Figure 18.3
Synthesis of polar head groups of bacterial phospholipids. E_1 = phosphatidylserine synthase; E_2 = phosphatidylserine decarboxylase; E_3 = glycerophosphate phosphatidyltransferase; E_4 = phosphatidylglycerol phosphate phosphatase; E_5 = cardiolipin synthase.

CDP-diacylglycerol participates in two pathways, one leading to phosphatidylethanolamine and the other to phosphatidylglycerol and cardiolipin (Figure 18.3). In the first pathway, **phosphatidylserine synthase** (E_1 in the figure) exchanges serine for CMP, and this is followed by a decarboxylation that yields phosphatidylethanolamine. In the route to phosphatidylglycerol and cardiolipin, **glycerophosphate phosphatidyltransferase** (E_3) exchanges glycerol phosphate for CMP, and enzymatic cleavage of the phosphate ester yields phosphatidylglycerol. Finally, **cardiolipin synthase** (E_5) catalyzes the exchange of an additional phosphatidylglycerol molecule for glycerol itself to give cardiolipin, which is simply diphosphatidylglycerol.

CONTROL OF PHOSPHOLIPID SYNTHESIS IN PROKARYOTES. The genetic analysis of *E. coli* phospholipid metabolism is fairly advanced, in the sense that the structural genes for most of the enzymes involved have been identified and mapped, mutant phenotypes have been analyzed in detail, and several of the genes have been cloned and sequenced. However, we still know rather little about how phospholipid synthesis is regulated. We do know that phospholipid synthesis is somehow coordinated with the synthesis of other macromolecular precursors, so that the cell is normally in a state of balanced growth: proteins, nucleic acids, and membranes are produced at the same rate. Part of this control involves the "stringent response" directed by a regulatory nucleotide that we discuss in Chapter 22.

Glycerophospholipid Metabolism in Eukaryotes

Most eukaryotic cells contain six classes of glycerophospholipids—the same phosphatidylethanolamine (PE), phosphatidylglycerol, and cardiolipin found in bacteria, plus phosphatidylserine (PS), phosphatidylcholine (PC), and phosphatidylinositol. The increased number generates a greater diversity of biosynthetic pathways, but the metabolic strategies of eukaryotes are comparable to those we have just reviewed. As in bacterial metabolism, the major intermediates are phosphatidic acid and diacylglycerol.

Eukaryotic cells display three biosynthetic routes to phosphatidic acid. The major pathway, starting with glycerol-3-phosphate, is similar to that used by bacteria (see Figure 18.2), except that the acyltransferases use acyl-CoAs as substrates in preference to acyl-ACPs. A second pathway starts with dihydroxyacetone phosphate, which accepts a fatty acyl moiety at position 1 from an acyl-CoA, followed by reduction to 1-acylglycerol-3-phosphate and a second acylation.

$$\underset{\textbf{DHAP}}{CH_2OH{-}CO{-}CH_2O\text{(P)}} \xrightarrow{R_1{-}CO{-}S{-}CoA} \underset{\textbf{1-Acyl-DHAP}}{R_1COOCH_2{-}CO{-}CH_2O\text{(P)}} \xrightarrow{NADPH} R_1COOCH_2{-}CH(OH){-}CH_2O\text{(P)} \xrightarrow{R_2{-}CO{-}S{-}CoA} \underset{\textbf{Phosphatidic acid}}{R_1COOCH_2{-}CH(OOCR_2){-}CH_2O\text{(P)}}$$

The third pathway begins with diacylglycerol and represents a way to reutilize the carbon backbone of a phospholipid. Diacylglycerol, usually generated by dephosphorylation of phosphatidic acid, can be rephosphorylated by a specific **diacylglycerol kinase.** CDP-diacylglycerol is synthesized as previously described for bacteria. Both CDP-diacylglycerol and diacylglycerol itself are important intermediates in different pathways leading to phospholipids.

PATHWAYS TO PHOSPHATIDYLCHOLINE AND PHOSPHATIDYLETHANOLAMINE. The most abundant phospholipids in most eukaryotic cells are phosphatidylcholine and phosphatidylethanolamine. Both can be synthesized from phosphatidylserine, or through alternative pathways that start with free choline or ethanolamine, respectively. Since choline and ethanolamine arise largely through the turnover of preexisting phospholipids, the latter pathways can be considered routes for reutilization of these breakdown products. The significance of reutilizing choline lies in the fact that the three methyl groups of choline are derived from the amino acid methionine. As we discuss in Chapter 20, methionine is limiting in many animal diets, and it is essential for the reutilization of scarce metabolites. In fact, choline itself is an essential component of most animal diets.

The pathway for utilization of choline, which predominates in most animal cells, is summarized in Figure 18.4. Choline is phosphorylated, and the resultant **phosphocholine** undergoes a cytidylyltransferase reaction, similar to that described for bacterial diacylglycerol metabolism, to give **CDP-choline.** The phosphocholine moiety of this intermediate is transferred to diacylglycerol, yielding phosphatidylcholine. The first enzyme in the pathway, **choline kinase,** is cytosolic, while the second enzyme, **CTP:phosphocholine cytidylyltransferase,** is found in both cytosolic and microsomal fractions. The last enzyme, **CDP-choline:1,2-diacylglycerol**

$HO-CH_2-CH_2-\overset{+}{N}(CH_3)_3$

Choline

E_1: ATP → ADP

$^-O-P(=O)(O^-)-O-CH_2-CH_2-\overset{+}{N}(CH_3)_3$

Phosphocholine

E_2: CTP → PP_i

$\text{Cytidine}-O-P(=O)(O^-)-O-P(=O)(O^-)-O-CH_2-CH_2-\overset{+}{N}(CH_3)_3$

CDP-choline

E_3: Diacylglycerol ($CH_2-O-C(=O)-R_1$; $R_2-C(=O)-O-CH$; CH_2-OH) → CMP

$CH_2-O-C(=O)-R_1$; $R_2-C(=O)-O-CH$; $CH_2-O-P(=O)(O^-)-O-CH_2-CH_2-\overset{+}{N}(CH_3)_3$

Phosphatidylcholine

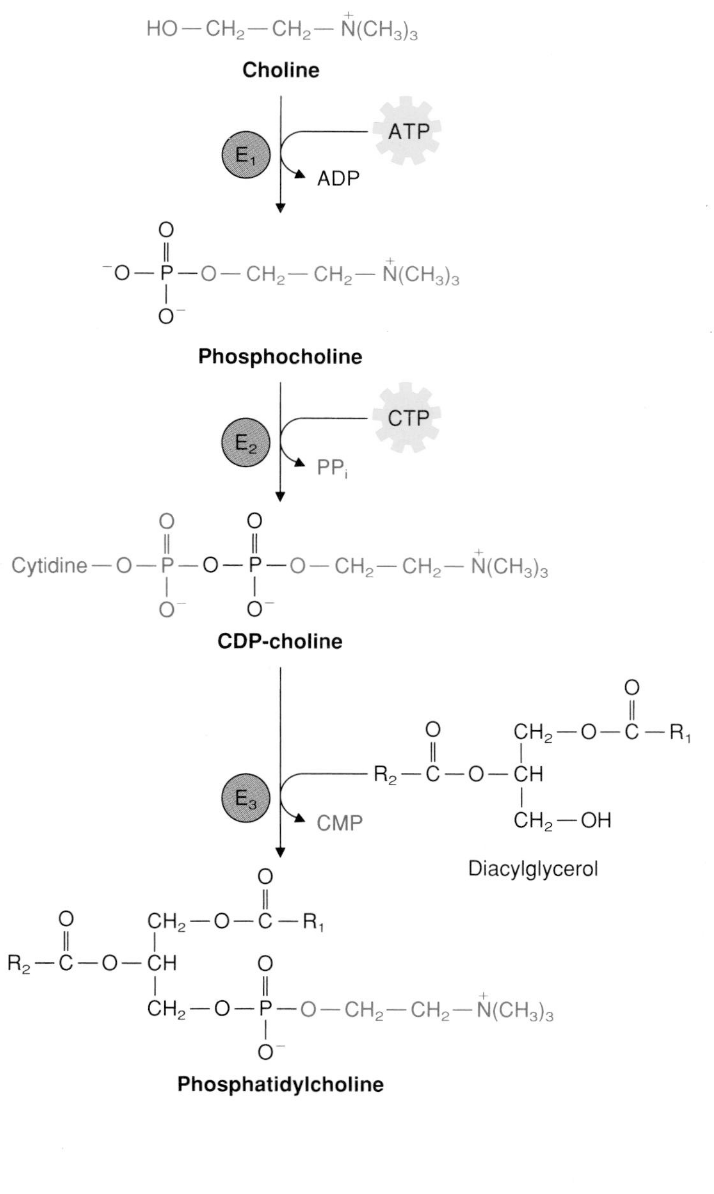

Figure 18.4
Reutilization of choline for synthesis of phosphatidylcholine. An identical pathway converts ethanolamine to phosphatidylethanolamine. E_1 = choline kinase; E_2 = CTP:phosphocholine cytidylyltransferase; E_3 = CDP-choline:1,2-diacylglycerol cholinephosphotransferase.

cholinephosphotransferase, is membrane bound in the endoplasmic reticulum. Recent evidence suggests that only the membrane-bound form of the enzyme is active and that the rate of phosphatidylcholine synthesis is controlled in part by translocation of this enzyme between cytosolic and membranous forms. Translocation is apparently controlled in turn by reversible phosphorylation and dephosphorylation of the enzyme, but the ultimate regulatory signals are not yet known.

The reutilization route to phosphatidylethanolamine involves the same reactions, with ethanolamine substituted for choline. The same enzyme carries out phosphorylation of both choline and ethanolamine, but the following reactions are carried out by different enzymes in the two pathways.

The phosphatidylserine pathway to PE and PC begins with the action of **phosphatidylserine decarboxylase,** a mitochondrial enzyme that converts phosphatidylserine to phosphatidylethanolamine, as seen also with bacteria.

H^+ CO_2

PS Decarboxylase

PS: $R_2{-}C(=O){-}O{-}CH(CH_2{-}O{-}C(=O){-}R_1){-}CH_2{-}O{-}P(=O)(O^-){-}O{-}CH_2{-}CH(COO^-){-}\overset{+}{N}H_3$

PE: $R_2{-}C(=O){-}O{-}CH(CH_2{-}O{-}C(=O){-}R_1){-}CH_2{-}O{-}P(=O)(O^-){-}O{-}CH_2{-}CH_2{-}\overset{+}{N}H_3$

PE serine-transferase

Ethanolamine Serine

AdoMet

AdoMet

AdoMet

PC: $R_2{-}C(=O){-}O{-}CH(CH_2{-}O{-}C(=O){-}O{-}R_1){-}CH_2{-}O{-}P(=O)(O^-){-}O{-}CH_2{-}CH_2{-}\overset{+}{N}(CH_3)_3$

There is also a calcium-activated transferase, **phosphatidylethanolamine serinetransferase,** which exchanges free ethanolamine for the serine moiety of phosphatidylserine, yielding phosphatidylethanolamine and serine. This enzyme is found in endoplasmic reticulum and the Golgi complex. This readily reversible enzyme can also function in the other direction, to synthesize phosphatidylserine from phosphatidylethanolamine.

Next, phosphatidylethanolamine undergoes three successive methylations, all catalyzed by the same enzyme, to give phosphatidylcholine. In animals this pathway occurs primarily in the liver. The methyl group donor for these reactions is an activated derivative of methionine, ***S*-adenosylmethionine.** The product of methyl group transfer is ***S*-adenosylhomocysteine.** As we shall see in Chapter 20, AdoMet is involved in many additional biological methylations.

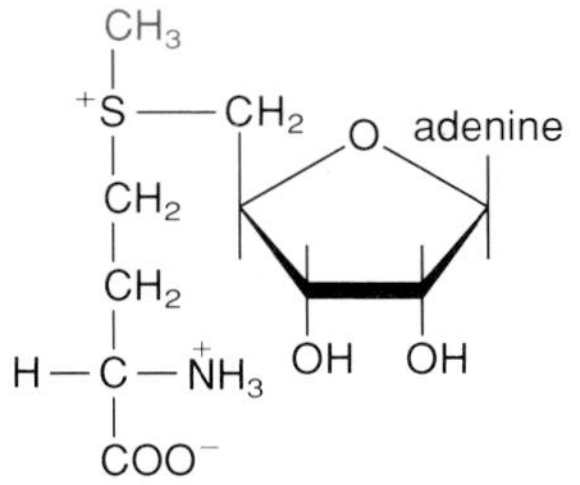

S-Adenosylmethionine (AdoMet)

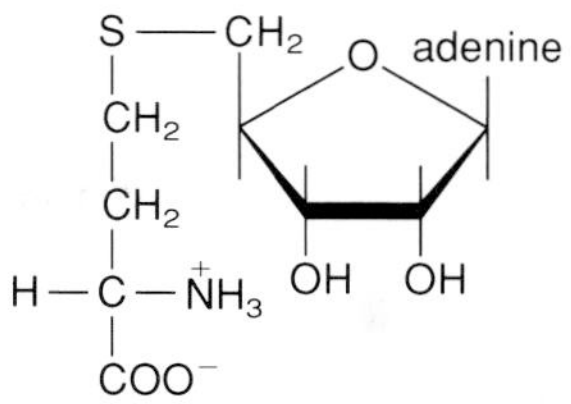

S-Adenosylhomocysteine (AdoHcy)

REDISTRIBUTION OF PHOSPHOLIPID FATTY ACIDS: LUNG SURFACTANT AND PHOSPHOLIPASES. Having discussed the biosynthesis of polar head groups, let us now focus on the fatty acid constituents of phospholipids. A variety of isotope labeling experiments show that phospholipids, even after insertion into membranes, are not metabolically inert. Specifically, the fatty acyl chains can change in response to varying environmental conditions or needs. Does this involve modification of resident fatty acid chains or substitution of new fatty acids? To answer this question, investigators have explored the biosynthesis of the phospholipid component of **lung surfactant,** a lipid-and-protein-containing substance which is secreted from lung and which, by maintaining high surface tension, prevents collapse of the alveoli when air is expelled. Lung surfactant contains some 50 to 60% **dipalmitoylphosphatidylcholine,** a form of phosphatidylcholine in which palmitoyl chains occupy both positions 1 and 2. Infants afflicted with **respiratory distress syndrome** show defects in the metabolism of lung surfactant—deficiencies, not yet identified, in either synthesis or secretion of this substance.

Since phospholipids are usually synthesized with an unsaturated chain at position 2, we can ask how the saturated chain arises in surfactant. Although the question is not yet settled, current evidence rules out the possibility that a fatty acid chain resident on position 2 becomes modified in situ to a palmitoyl chain. Two remaining possibilities are (1) transfer of palmitoyl chains from an acyl donor such as palmitoyl-CoA to glycerophosphorylcholine and (2) hydrolytic cleavage of acyl chains from phosphatidylcholine, followed by transfer of palmitoyl chains from a suitable acyl group donor.

Whatever the actual pathway for lung surfactant synthesis, it is apparent that phospholipid molecules in general can be "retailored" by cleavage and replacement of fatty acyl chains on carbons 1 and 2. Hydrolytic cleavage can occur through the action of **phospholipase A_2,** and reacylation would most probably occur from an acyl-CoA. Phospholipid remodeling by this mechanism is now known to be quite widespread.

Phospholipase A_2 is one of a class of four enzymes that hydrolyze specific bonds in phospholipids; the others are phospholipases A_1, C, and D (Figure 18.5). Phospholipases have been useful reagents in studies of both lipid and membrane structure. Phospholipase A_2 is an abundant component of some snake venoms, which have the ability to cause hemolysis, or rupture of red blood cells. The release of one fatty acyl chain from phosphatidylcholine yields 1-acylglycerophosphorylcholine, more commonly known as **lysolecithin** (phosphatidylcholine itself is commonly called lecithin). Lysolecithin derives its name from the fact that, as an excellent detergent, it solubilizes membranes and hence causes cells to lyse; erythrocytes are particularly susceptible to this action.

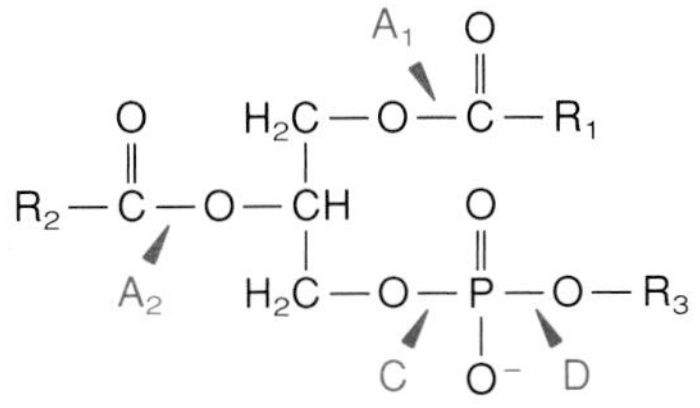

Figure 18.5
Specificities of phospholipases A_1, A_2, C, and D.

Other critical functions are emerging for phospholipases, particularly A_2. As discussed later in this chapter, several important regulators, including prostaglandins, are synthesized from arachidonic acid. Current evidence indicates that synthesis of these compounds is triggered by phospholipase-catalyzed release of arachidonic acid from membrane phospholipids. Another postulated role for phospholipase A_2 is in repair of damaged membrane phospholipids. Fatty acids are susceptible to nonenzymatic attack by oxygen or reactive oxygen species such as superoxide to give fatty acid peroxides, for instance,

O—O—H

COO^-

When a membrane phospholipid undergoes peroxidation of a fatty acyl chain, the structure of the membrane is distorted, and the function of the membrane can be affected. Current evidence suggests that phospholipase A_2 can remove these abnormal fatty acids from phospholipids still resident in a lipid bilayer, which leads to replacement of the damaged acyl chains by normal fatty acids.

SYNTHESIS OF OTHER GLYCEROPHOSPHOLIPIDS. Figure 18.6 summarizes the major pathways for synthesis of phosphatidylserine, phosphatidylglycerol, cardiolipin, and phosphatidylinositol. In yeast the CDP-diacylglycerol pathway to phosphatidylserine predominates, while in animals phosphatidylserine is synthesized primarily by the calcium-activated exchange between phosphatidylethanolamine and serine. However, the ultimate source of ethanolamine for this process is not yet known. The synthesis of phosphatidylglycerol, which is largely confined to mitochondria, is identical to the route used in bacteria. However, its conversion to cardiolipin involves CDP-diacylglycerol, rather than a second mole of phosphatidylglycerol, as the second substrate.

The biosynthesis of phosphatidylinositol, catalyzed by **phosphatidylinositol synthase,** involves CDP-diacylglycerol and L-*myo*-inositol (Figure 18.7). The latter is one of nine possible stereoisomers of hexahydroxyhexane; it is synthesized from D-glucose-6-phosphate. Phosphati-

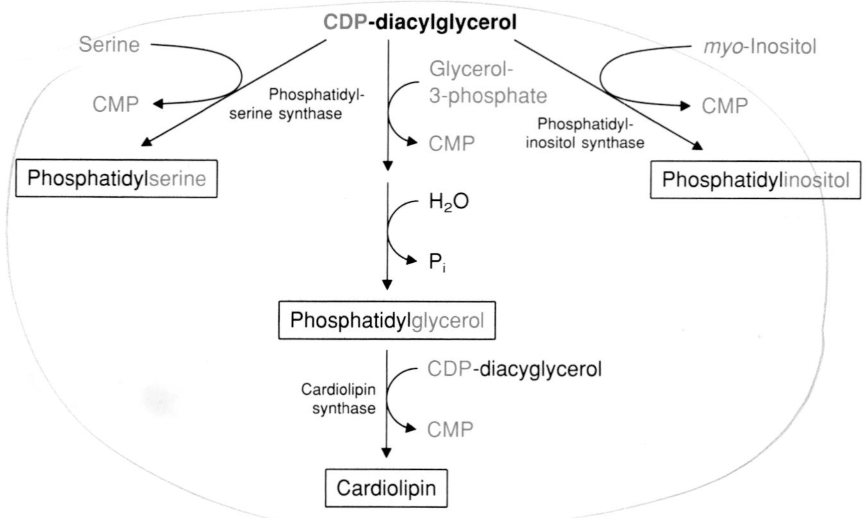

Figure 18.6
Biosynthetic routes to phosphatidylserine, phosphatidylglycerol, cardiolipin, and phosphatidylinositol in eukaryotic cells. CDP-diacylglycerol is converted to phosphatidylglycerol as shown in Figure 18.2.

dylinositol undergoes two successive phosphorylations to yield phosphatidylinositol-4-phosphate and phosphatidylinositol-4,5-bisphosphate, both of which are present in small but appreciable amounts. All three of these lipids, which are collectively termed **phosphoinositides,** are enriched in arachidonic acid at position 2; this evidently occurs via the deacylation–reacylation process we discussed earlier.

It has long been known from ^{32}P labeling studies that the phosphoinositides are in a state of active metabolic flux; they are synthesized and degraded rapidly, particularly in nervous tissue, and particularly in response to the binding of neurotransmitters. The phosphoinositides play important roles as second messengers in **transmembrane signaling,** the transmission of an extracellular signal to some element of the intracellular metabolic apparatus. Our current understanding of these events is presented in Chapter 23.

Quite recently, another metabolic role for phosphatidylinositol has come to light through studies on variant forms of a cell surface glycoprotein of the protozoal parasite *Trypanosoma brucei.* This organism escapes immunological detection by rapidly changing the structure of this surface glycoprotein. The glycoprotein has been found to be linked, through its terminal carboxylate group, to a glycosylated form of phosphatidylinositol. This linkage provides both an anchor, binding the protein to the membrane, and a site for cleavage, when the organism replaces one glycoprotein with another at its surface. There is evidence that this mechanism is generally used by eukaryotic cells to control the concentrations of particular proteins bound at the cell surface.

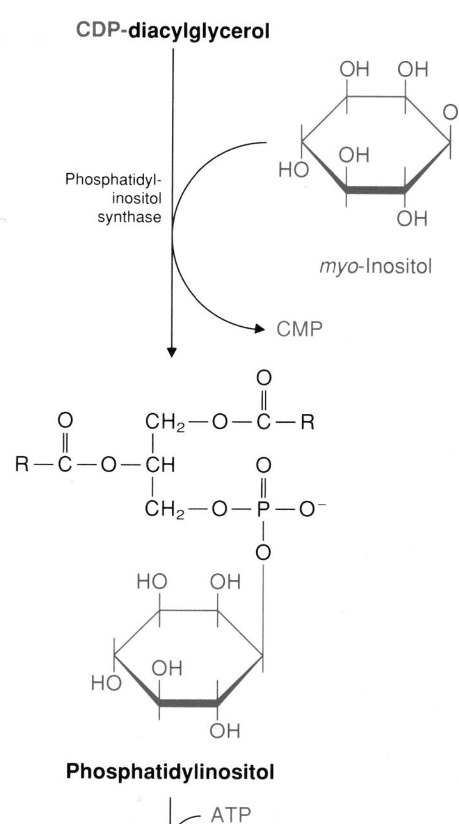

ATP
ADP

Phosphatidylinositol-4-phosphate

ATP
ADP

Phosphatidylinositol-4,5-bisphosphate

Figure 18.7
Biosynthesis of phosphoinositides.

Ether Phospholipids. Ether lipids are those containing an alkyl, rather than an acyl, group linked to one of the oxygen atoms of glycerol.

Phospholipid with an alkyl ether

Phospholipid with an alkenyl ether

Alkyl and alkenyl phospholipids are widely distributed, but their abundance in tissues varies greatly. For example, consider the **plasmalogens,** or **vinyl ethers,** phospholipids that contain an alkenyl ether at position *sn*-1 of glycerol. These constitute some 50% of all phospholipids found in heart tissue but are virtually undetectable in many other tissues. So far, little is known about the functional significance of the vast abundance of plasmalogens in the heart.

The biosynthesis of ether phospholipids (Figure 18.8) begins with 1-acyldihydroxyacetone phosphate (see p. 608). This undergoes exchange of an alkyl group for the acyl group (E_1); the saturated fatty alcohol used in this reaction is derived from NADPH-dependent reduction of the corresponding fatty acyl-CoA. Carbon 2 is then reduced from the keto to the hydroxyl level (E_2) and acylated (E_3). This gives the 1-alkyl analog of phosphatidic acid, which is converted to saturated ether phospholipids, or **glyceryl ethers,** by the pathways already presented for phospholipid biosynthesis. The primary route leads to the glyceryl ether of phosphatidylethanolamine; the serine and choline analogs arise by base exchange and methylation, respectively.

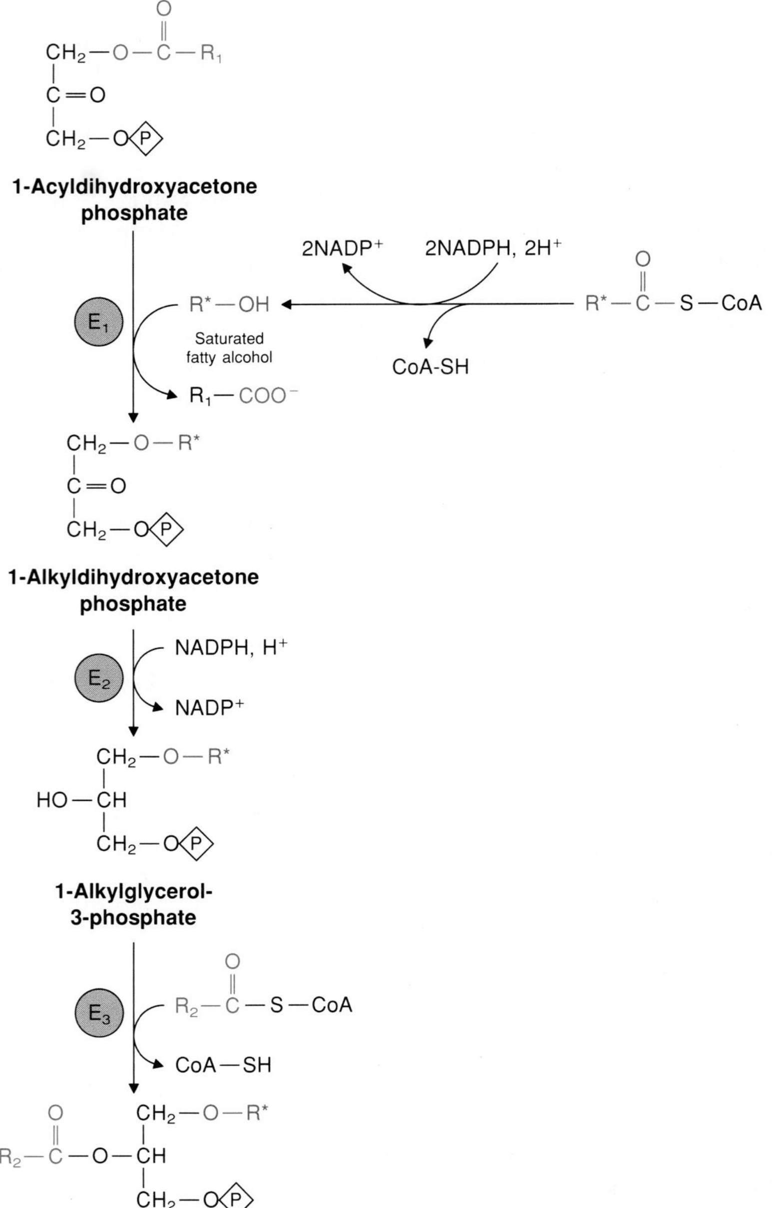

Figure 18.8
Biosynthetic routes to alkyl ether phospholipids.

The synthesis of plasmalogens then involves desaturation of the alkyl group at position *sn*-1 (Figure 18.9). The microsomal enzyme system involved, like that used for desaturation of stearoyl-CoA (Chapter 17), requires O_2, NADH, and cytochrome b_5.

A recently discovered ether lipid, called **platelet-activating factor,** has the structure **1-alkyl-2-acetylglycerophosphocholine.** Physiologically, this compound is perhaps the most potent compound known. At concentrations as low as 10 *picomolar* (10^{-11} M) it causes blood platelets to aggregate, and injection of as little as 60 ng will reduce blood pressure in hypertensive rats. This lipid is synthesized via acetylation of the corresponding 1-alkylglycerophosphocholine by acetyl-CoA.

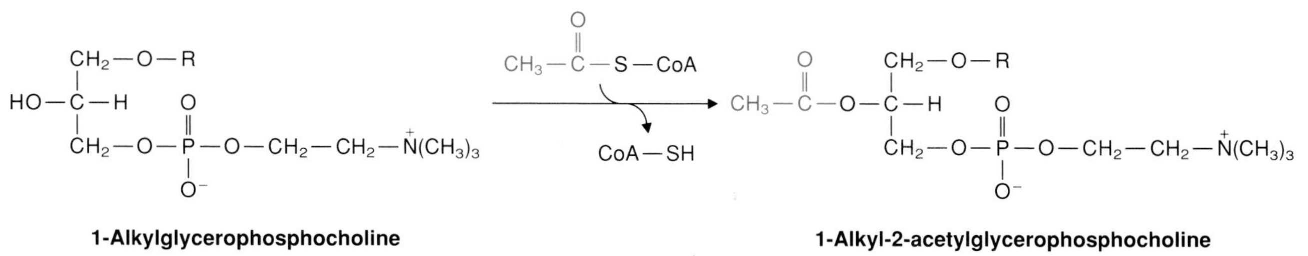

$CH_3{-}C(=O){-}S{-}CoA \rightarrow CoA{-}SH$

1-Alkylglycerophosphocholine → **1-Alkyl-2-acetylglycerophosphocholine**

The discovery of a phospholipid with such a striking biological activity was without precedent, and it has opened a fascinating new realm of biochemistry. The factor evidently acts through binding to a high-affinity receptor on the surface of susceptible cells, but the subsequent metabolic reactions have not yet been clarified.

Ether-containing lipids are quite abundant in the membranes of halophilic ("salt-loving") microorganisms. These bacteria and protozoans grow in media with NaCl concentrations as high as 4 M. While we don't know the relationship between ether lipids and the ability to grow in a high-salt environment, it may be related to the greater stability of alkyl ethers against hydrolysis, as compared with acyl esters.

Intracellular Transport of Membrane Phospholipids. Of the six major classes of glycerophospholipids, phosphatidylglycerol and cardiolipin are found primarily in mitochondrial membranes and are synthesized in mitochondria. The remaining four classes are synthesized simultaneously with their insertion into the cytosolic side of membranes of the endoplasmic reticulum. From there they undergo translocation to the luminal side of the membrane and ultimately transport to other membranes—the nuclear envelope, mitochondrial membranes, and the plasma membrane. Just how these events occur is the subject of one of the most active areas of contemporary cell biology. The three major questions are the following. (1) How do phospholipid molecules move from one side of a membrane to the other? (2) How do phospholipid molecules move from one site to another within the cell? (3) How is phospholipid transport directed to specific organelles to account for the differences in phospholipid composition of membranes within a single cell?

Investigations of transmembrane movement of phospholipids (question 1) use specific lipid probes that allow detection of a lipid on only one side of a bilayer. As mentioned in Tools of Biochemistry 14, one such approach involves the use of a **spin label,** a lipid analog that is detectable from its electron paramagnetic resonance spectrum. Such measurements show that transbilayer movement, or "flip-flop," does occur spontaneously but is quite slow. Since measurements in vivo show much faster transbilayer movement, proteins or other factors may promote flip-flop in living cells.

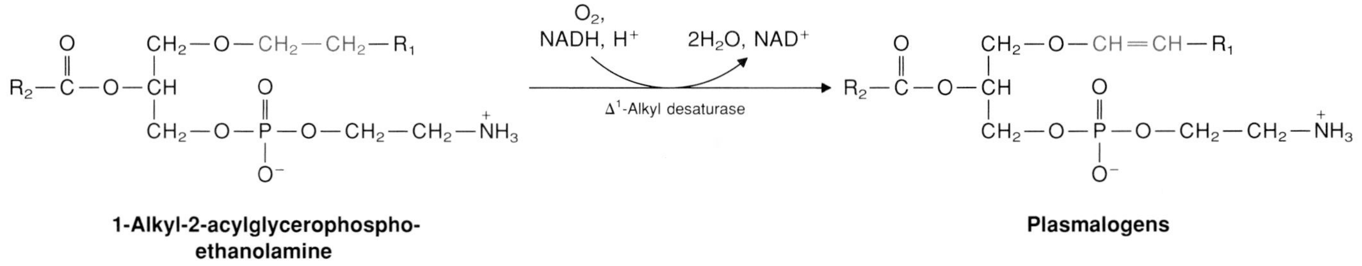

O_2, NADH, H^+ → $2H_2O$, NAD^+ (Δ^1-Alkyl desaturase)

1-Alkyl-2-acylglycerophospho-ethanolamine → **Plasmalogens**

Figure 18.9
Desaturation of 1-alkyl-2-acylglycerophosphoethanolamine (the alkyl analog of phosphatidylethanolamine) to the corresponding vinyl ether, yielding plasmalogens.

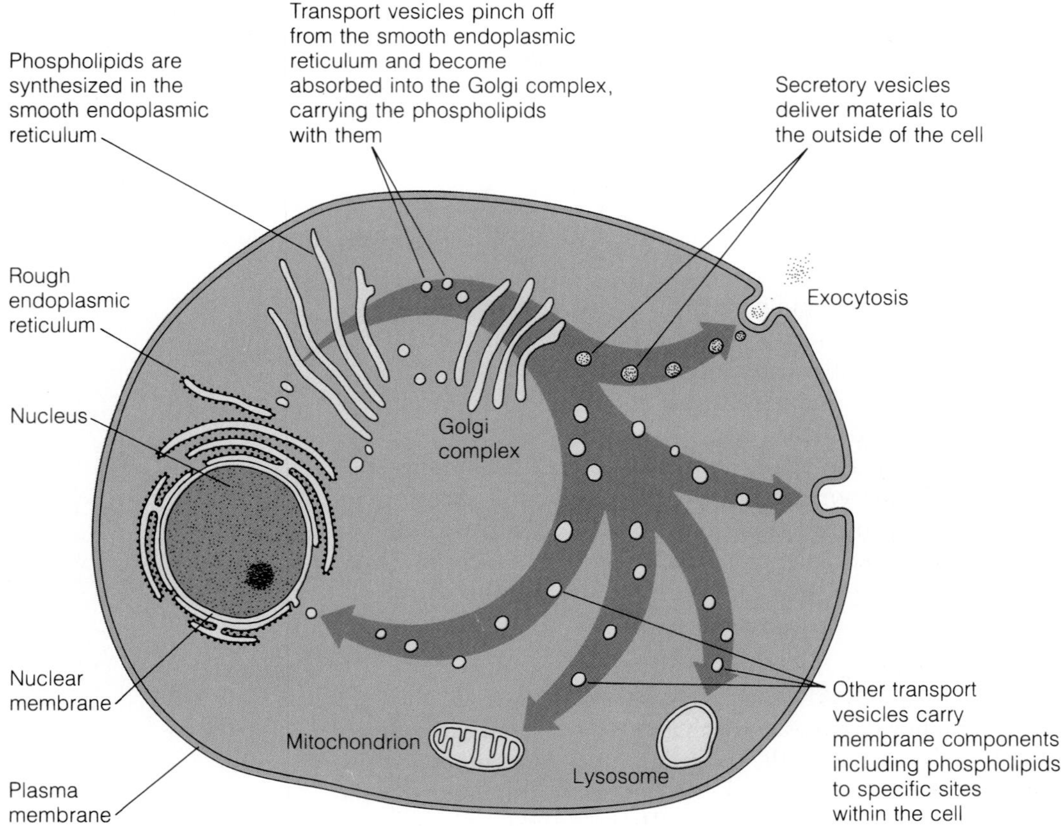

Transport vesicles pinch off from the smooth endoplasmic reticulum and become absorbed into the Golgi complex, carrying the phospholipids with them
Phospholipids are synthesized in the smooth endoplasmic reticulum
Secretory vesicles deliver materials to the outside of the cell
Rough endoplasmic reticulum
Exocytosis
Nucleus
Golgi complex
Nuclear membrane
Mitochondrion
Lysosome
Plasma membrane
Other transport vesicles carry membrane components including phospholipids to specific sites within the cell

Figure 18.10
Intracellular transport of membrane phospholipids. Phospholipid synthesis is completed on the cytosolic side of the endoplasmic reticulum bilayer; this bilayer fuses with cisternae of the Golgi complex, which buds off to yield vesicles that fuse with the plasma membrane and with membranes of other organelles.

Transport of phospholipids within the cell (question 2 above) involves largely the transfer of fragments of membranes of the ER into the Golgi complex (Figure 18.10). Membrane vesicles are constantly pinched off from the Golgi, and these vesicles, containing secretory products, fuse with the plasma membrane for secretion of their contents via **exocytosis** (transport out of the cell). It seems likely that this is a route not only for extracellular secretion but also for transport of membrane lipids to the plasma membrane. Probably comparable processes transport membrane lipids to mitochondria, plant chloroplasts, and nuclei, although these processes are not as well understood.

To explain the asymmetry of membrane lipid composition within a given cell (question 3 above), one can postulate the existence in Golgi membranes of specific targeting proteins—proteins that preferentially associate with certain lipids and that have an affinity for certain organelles.

Another mechanism involves the action of **phospholipid exchange proteins**—cytosolic proteins that bind a phospholipid and can catalyze its exchange with a corresponding membrane lipid. The protein-bound lipid ends up in the membrane, and the membrane lipid becomes bound to the cytosolic protein. This does not provide a mechanism for net transfer of lipid to a membrane, but it does allow for modulation of the lipid composition of a particular membrane.

Metabolism of Sphingolipids

Interest in sphingolipids focuses largely on their important role in nervous tissue and, related to this, a number of human genetic defects of sphingolipid metabolism. Sphingolipids are widely distributed also in the membranes of plant cells and in lower eukaryotes such as yeast.

Recall from Chapter 9 that sphingolipids are derivatives of the base *sphingosine*. Plant sphingolipids contain a slightly different form of this compound, called **phytosphingosine.** The sphingolipids include *ceramide* (*N*-acylsphingosine), *sphingomyelin* (*N*-acylsphingosine phosphorylcholine), and a family of carbohydrate-containing sphingolipids called neutral and acidic *glycosphingolipids;* the latter substances include *cerebrosides* and *gangliosides*. Ceramide serves as the precursor to both sphingomyelin and the glycosphingolipids. The pathway to ceramide starts with the synthesis of a sphingosine derivative, **sphinganine,** from palmitoyl-CoA and serine (Figure 18.11). After reduction of the resulting keto group, the amino group of sphinganine is acylated to give a ceramide. In animals the sphinganine unit of this compound is then desaturated to give a ceramide with a sphingosine base. Transfer of a phosphocholine unit from phosphatidylcholine yields sphingomyelin plus diacylglycerol.

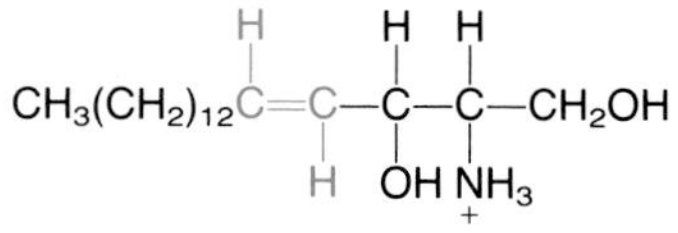

Sphingosine

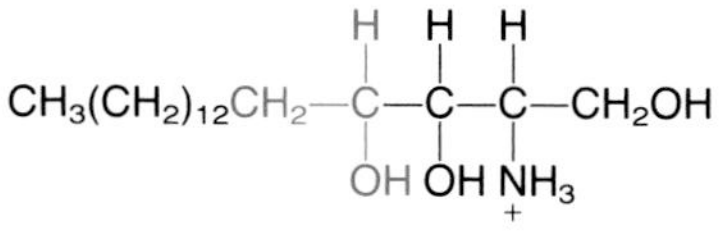

Phytosphingosine

The pathways leading to glycosphingolipids are more numerous, but the metabolic strategies are comparable to those we have encountered before in synthesis of the oligosaccharide chains of glycoproteins (Chapter 16). The pathways involve the stepwise addition of monosaccharide units, using nucleotide-linked sugars as the activated biosynthetic substrates and with ceramide as the initial monosaccharide acceptor. The sugar nucleotides involved in glycosphingolipid synthesis include UDP-glucose (UDP-Glc), UDP-galactose (UDP-Gal), UDP-*N*-acetylgalactosamine (UDP-GalNAc), and CMP-*N*-acetylneuraminic acid (CMP-Sia, or CMP-sialic acid). Figure 18.12 shows pathways leading to some of the most abundant glycosphingolipids.

Sphingolipids, especially sphingomyelin, are abundant components of the myelin sheath, a multilayered structure that protects and insulates cells of the central nervous system (Figure 18.13). In human myelin sphingolipids constitute some 25% of the total lipid. Sphingolipids are in a continuous state of metabolic turnover, both synthesis and degradation. Degradation occurs in the lysosomes, by a family of hydrolytic enzymes. These pathways are of great medical interest because of the existence of a group of congenital diseases called **sphingolipidoses** (also known as **lipid storage diseases**). Each condition is characterized by deficiency of one of the degradative enzymes, with concomitant accumulation within the lysosome of the substrate for the deficient enzyme (Table 18.1). In fact, structural analysis of the abnormal metabolites that accumulate helped to establish the degradative pathways, which are depicted in Figure 18.14. Most of these diseases are autosomal recessives, which means that two defective alleles of the gene encoding a particular enzyme must be present in an individual for disease symptoms to be manifest. Because of the large amounts of sphingolipids in nervous tissue, it is perhaps not surprising that most of the sphingolipidoses involve severely impaired central nervous system function.

The best known of the sphingolipidoses is **Tay–Sachs disease,** originally described in 1881, which is a deficiency of the lysosomal ***N*-acetylhexosaminidase A.** The enzyme deficiency causes accumulation of the ganglioside called GM_2, particularly in the brain. The disease is devastating, causing nervous system degeneration, mental retardation, blindness, and death, usually by the age of four.

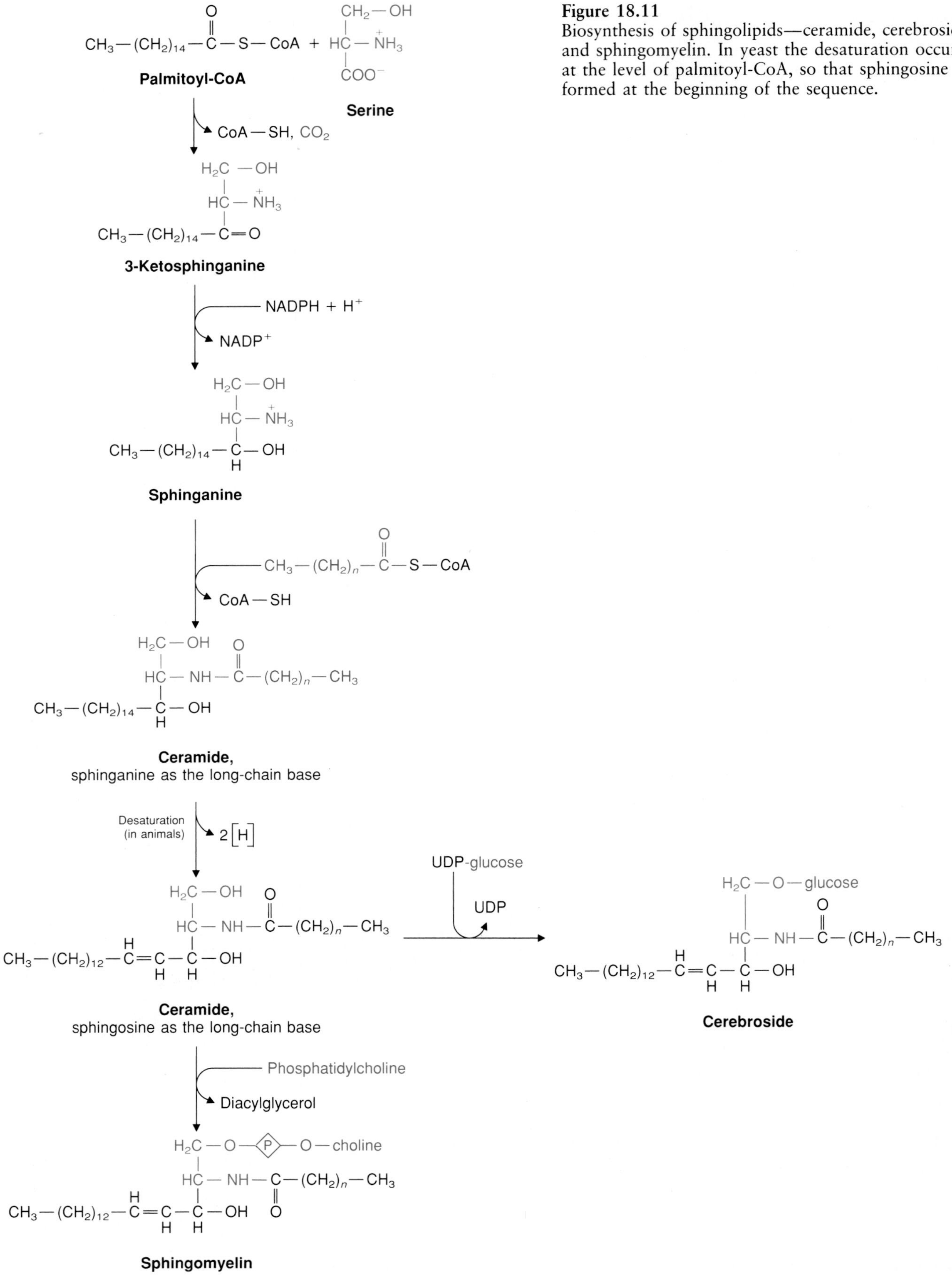

$CH_3-(CH_2)_{14}-C(=O)-S-CoA$ + $HC(CH_2-OH)(\overset{+}{N}H_3)(COO^-)$

Palmitoyl-CoA **Serine**

↓ → $CoA-SH$, CO_2

$CH_3-(CH_2)_{14}-C(=O)-CH(\overset{+}{N}H_3)-CH_2-OH$

3-Ketosphinganine

↓ $NADPH + H^+$ → $NADP^+$

$CH_3-(CH_2)_{14}-CH(OH)-CH(\overset{+}{N}H_3)-CH_2-OH$

Sphinganine

↓ $CH_3-(CH_2)_n-C(=O)-S-CoA$ → $CoA-SH$

$CH_3-(CH_2)_{14}-CH(OH)-CH(NH-C(=O)-(CH_2)_n-CH_3)-CH_2-OH$

Ceramide,
sphinganine as the long-chain base

↓ Desaturation (in animals) → 2[H]

$CH_3-(CH_2)_{12}-CH=CH-CH(OH)-CH(NH-C(=O)-(CH_2)_n-CH_3)-CH_2-OH$

Ceramide,
sphingosine as the long-chain base

→ UDP-glucose → UDP

$CH_3-(CH_2)_{12}-CH=CH-CH(OH)-CH(NH-C(=O)-(CH_2)_n-CH_3)-CH_2-O-glucose$

Cerebroside

↓ Phosphatidylcholine → Diacylglycerol

$CH_3-(CH_2)_{12}-CH=CH-CH(OH)-CH(NH-C(=O)-(CH_2)_n-CH_3)-CH_2-O-(P)-O-choline$

Sphingomyelin

Figure 18.11
Biosynthesis of sphingolipids—ceramide, cerebroside, and sphingomyelin. In yeast the desaturation occurs at the level of palmitoyl-CoA, so that sphingosine is formed at the beginning of the sequence.

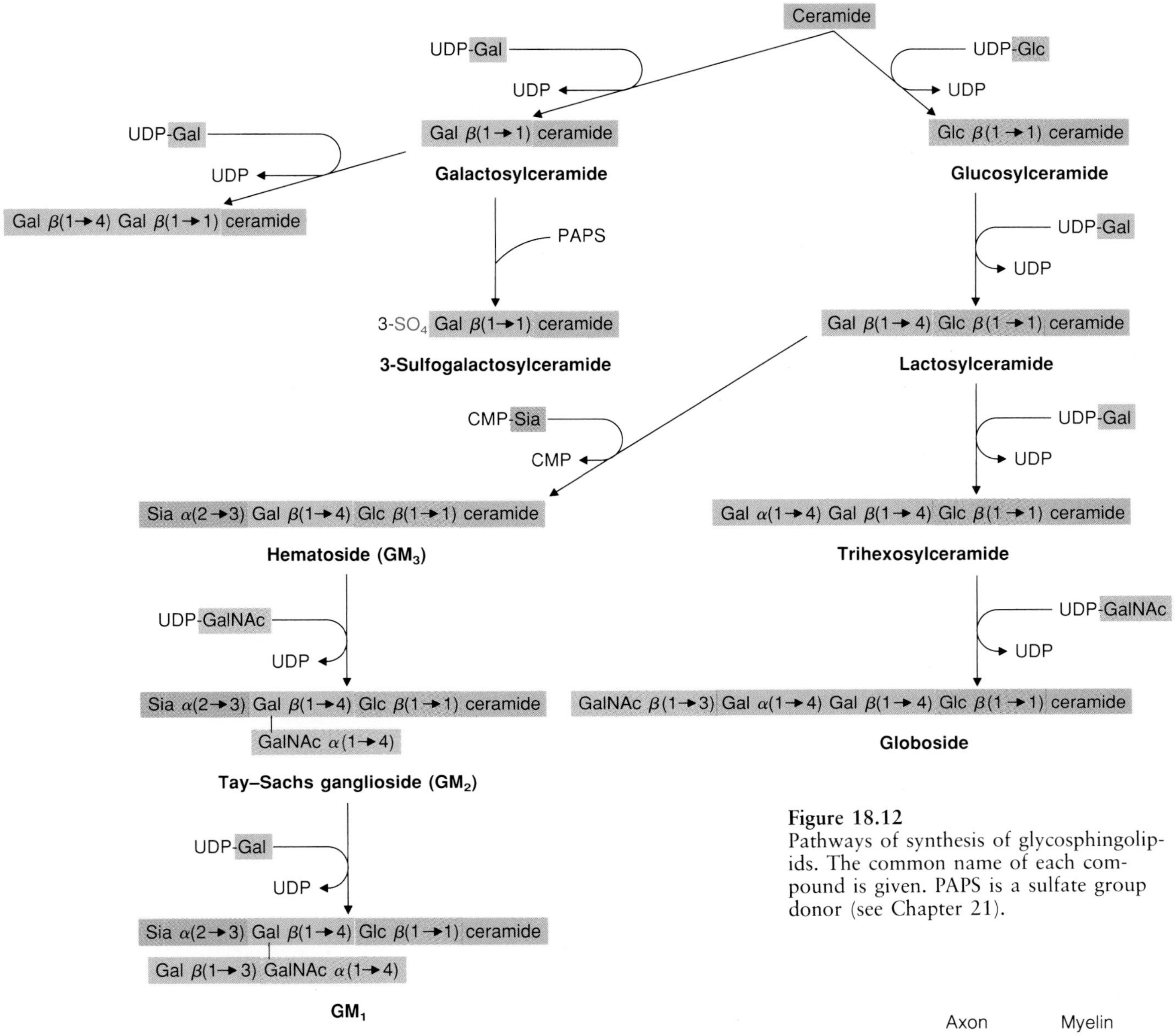

Ceramide

UDP-Gal → UDP

Gal $\beta(1 \to 1)$ ceramide

Galactosylceramide

UDP-Gal → UDP

Gal $\beta(1 \to 4)$ Gal $\beta(1 \to 1)$ ceramide

PAPS

3-SO_4 Gal $\beta(1 \to 1)$ ceramide

3-Sulfogalactosylceramide

UDP-Glc → UDP

Glc $\beta(1 \to 1)$ ceramide

Glucosylceramide

UDP-Gal → UDP

Gal $\beta(1 \to 4)$ Glc $\beta(1 \to 1)$ ceramide

Lactosylceramide

CMP-Sia → CMP

Sia $\alpha(2 \to 3)$ Gal $\beta(1 \to 4)$ Glc $\beta(1 \to 1)$ ceramide

Hematoside (GM_3)

UDP-GalNAc → UDP

Sia $\alpha(2 \to 3)$ Gal $\beta(1 \to 4)$ Glc $\beta(1 \to 1)$ ceramide
GalNAc $\alpha(1 \to 4)$

Tay–Sachs ganglioside (GM_2)

UDP-Gal → UDP

Sia $\alpha(2 \to 3)$ Gal $\beta(1 \to 4)$ Glc $\beta(1 \to 1)$ ceramide
Gal $\beta(1 \to 3)$ GalNAc $\alpha(1 \to 4)$

GM_1

UDP-Gal → UDP

Gal $\alpha(1 \to 4)$ Gal $\beta(1 \to 4)$ Glc $\beta(1 \to 1)$ ceramide

Trihexosylceramide

UDP-GalNAc → UDP

GalNAc $\beta(1 \to 3)$ Gal $\alpha(1 \to 4)$ Gal $\beta(1 \to 4)$ Glc $\beta(1 \to 1)$ ceramide

Globoside

Figure 18.12
Pathways of synthesis of glycosphingolipids. The common name of each compound is given. PAPS is a sulfate group donor (see Chapter 21).

While Tay–Sachs disease is rare in the general population, the defective gene is relatively common among Ashkenazic Jews (those of middle and northern European extraction). Among American Jews about 1 in 30 individuals carries the defective gene. Thus, two Jewish parents carry an appreciable risk of bearing a Tay–Sachs child. Because there is no known cure for the disease, attention has focused on prenatal detection; in fact, this was one of the first genetic diseases to be successfully diagnosed in conjunction with amniocentesis. At present the only known way to deal with Tay–Sachs disease, when it is detected in this way, is to terminate the pregnancy. In countries that carry out screening for the heterozygous state followed by abor-

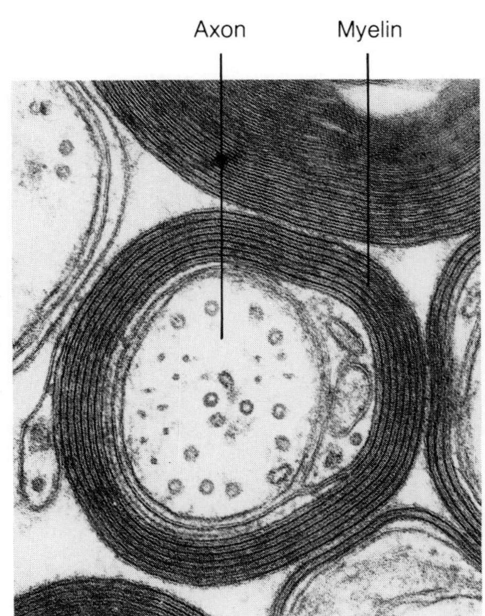

Figure 18.13
Electron micrograph of a myelinated axon from the spinal cord. Myelin, an insulating layer wrapping about the axon, is rich in sphingomyelin.

Table 18.1
Inherited diseases of sphingolipid catabolism

Disease	Defective Enzyme	Accumulated Intermediate
GM_1 gangliosidosis	① β-Galactosidase	GM_1 ganglioside
Tay–Sachs disease	② β-N-Acetylhexosaminidase A	GM_2 (Tay–Sachs) ganglioside
Fabry's disease	③ α-Galactosidase A	Trihexosylceramide
Gaucher's disease	④ β-Glucosidase	Glucosylceramide
Niemann–Pick disease	⑤ Sphingomyelinase	Sphingomyelin
Farber's lipogranulomatosis	⑥ Ceramidase	Ceramide
Globoid cell leukodystrophy	⑦ β-Galactosidase	Galactosylceramide
Metachromatic leukodystrophy	⑧ Arylsulfatase A	3-Sulfogalactosyl-ceramide
Sandhoff's disease	⑨ N-Acetylhexosaminidases A and B	GM_1 ganglioside and globoside

Numbers refer to enzymes shown in Figure 18.14.

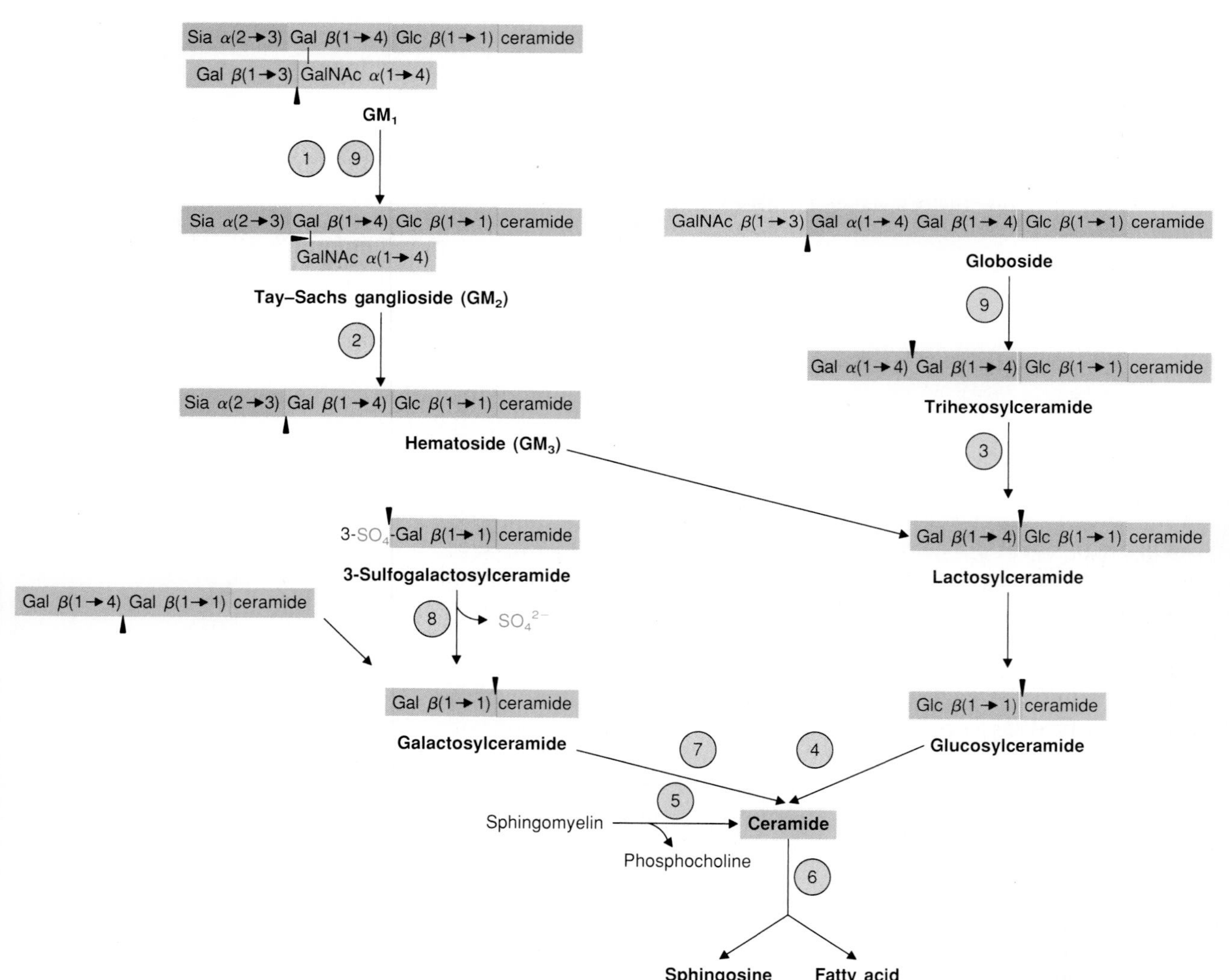

tion when necessary, the incidence of the disease has dropped virtually to zero.

Little is known about specific biochemical functions of sphingolipids, but their presence in the outer surface of plasma membranes of animal cells provides some tantalizing clues. Gangliosides are receptors for specific agents, such as cholera toxin, which binds to ganglioside GM_1, or influenza virus, which recognizes the sialic acid portion of certain gangliosides. The true roles of these gangliosides are unknown. Also, substantial changes in the glycolipid content of cell surfaces become evident after those cells undergo oncogenic transformation. Here, research attention focuses on both the mechanism of such changes and the potential for understanding the extent to which these changes are responsible for the altered properties of tumor cells, such as tumorigenesis, or the loss of growth control, adhesivity, or antigenic specificity.

Steroid Metabolism

We turn now to an extraordinarily large and diverse group of lipids, the **isoprenoids.** These are compounds built up from one or more five-carbon units, activated derivatives of **isoprene.**

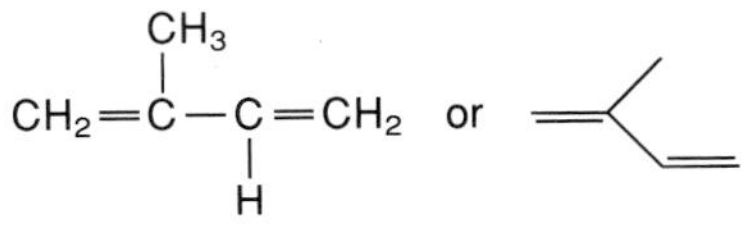

$$CH_2{=}C(CH_3){-}CH{=}CH_2$$

or (skeletal formula)

Isoprene

The family includes steroids and bile acids; the lipid-soluble vitamins; the dolichol and undecaprenol phosphates we encountered in glycoprotein synthesis; phytol, the long-chain alcohol in chlorophyll; gibberellins, the family of plant growth hormones; insect juvenile hormones; the major components of rubber; coenzyme Q; and many more.

Most of our discussion will focus on a single compound, **cholesterol.** As mentioned in Chapter 9, this lipid is a major component of eukaryotic cell membranes, where it participates in modulation of membrane fluidity. In animals it also serves as precursor to all of the steroid hormones and to the bile acids, which aid in fat digestion. And, of course, there is intense medical interest in cholesterol because of the relationships among diet, blood cholesterol levels, atherosclerosis, and heart disease. These biological relationships, coupled with the complex stereochemistry of its structure and the elegance of its biosynthetic pathway from a single low-molecular-weight precursor, have focused attention on this compound ever since its first isolation from gallstones, in 1784. Michael Brown and Joseph Goldstein have proclaimed cholesterol "the most highly decorated small molecule in biology," with 13 Nobel Prizes having been awarded to scientists, including themselves, who devoted major parts of their careers to cholesterol.

Figure 18.14
Lysosomal pathways for degradation of sphingolipids. The numbers refer to enzymes that are deficient in lipid storage diseases. The latter are identified in Table 18.1.

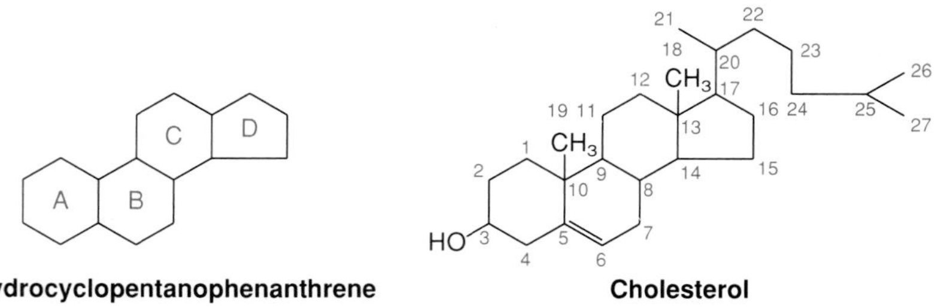

Figure 18.15
Ring identification system and carbon numbering system used for steroids.

Perhydrocyclopentanophenanthrene

Cholesterol

Some Structural Considerations

Steroids constitute a class of lipids that are derivatives of the saturated tetracyclic hydrocarbon **perhydrocyclopentanophenanthrene** (Figure 18.15). Note the letters used to denote the four rings—A, B, C, and D, with D being the five-membered ring—and the carbon numbering system. Cholesterol differs from the basic ring system in having an aliphatic chain at C-17, axial methyl groups at C-10 and C-13, a double bond in ring B, and a hydroxyl group in ring A. The alcoholic functional group makes cholesterol a **sterol,** which is the generic term used to identify steroid alcohols.

The cyclohexane rings of the steroids adopt puckered conformations, of which the more stable chair form predominates over the boat. This gives cholesterol a rigid molecular structure, with only the hydroxyl group generating a little polarity at one end (see Figure 9.9, Chapter 9). Much of the cholesterol is esterified at this position with a long-chain fatty acid, which makes the resultant cholesterol ester much more hydrophobic than cholesterol itself. The structure makes apparent how increasing concentrations of cholesterol in a membrane can reduce the fluidity of that membrane, by reducing the proportion of total lipid that can undergo a phase transition and by reducing the lateral mobility of polar lipids within the membrane.

During most of our treatment of steroid metabolism we shall use structural representations instead of the somewhat cumbersome three-dimensional configurational models. By convention, the methyl group at position 10 projects *above* the plane of the rings. This and all other substituents that project above the plane are denoted β and are drawn with a solid wedge. Substituents that project *below* the plane of the ring are called α and are denoted by a dashed wedge. Examples of α substituents can be seen in

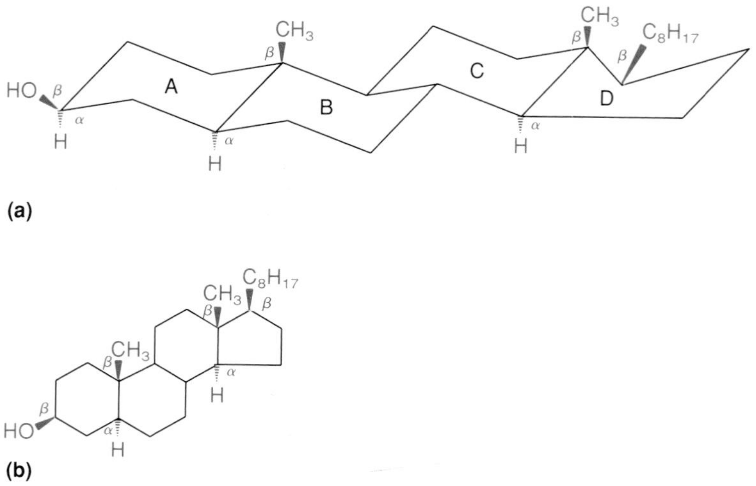

Figure 18.16
Structural conventions, with cholestanol as the example. α substituents project below the plane of the steroid ring system (dashed wedge), and β substituents project above that plane (solid wedge). The hydrogens at positions 5 and 14 have the α configuration, while the hydroxyl, the two methyl groups, and the aliphatic side chain at C-17 are all β substituents. **(a)** Three-dimensional representation; **(b)** two-dimensional representation.

the structure of **cholestanol,** one of the two fully saturated derivatives of cholesterol (Figure 18.16).

Biosynthesis of Cholesterol

The pathway by which cholesterol is synthesized is important because of the diversity of cholesterol metabolites and the elegance of the pathway itself. Isotopic tracer studies showed that all 27 carbons of cholesterol come from a two-carbon precursor—acetate. How could such a simple compound be built up to give a structure of the great complexity of cholesterol? That is what concerns us next.

EARLY STUDIES OF CHOLESTEROL BIOSYNTHESIS. Most of our early insights into cholesterol biosynthesis came from Konrad Bloch's laboratory, in the 1940s. Taking note of the fact that cholesterol biosynthesis in vertebrates is confined largely to the liver, Bloch fed to rats acetate with ^{14}C either in the methyl group position or in the carboxyl group. After each administration cholesterol was isolated from the liver and subjected to chemical degradation, with radioactive counting of the fragments. This established the pattern shown at right, in which each carbon of cholesterol was found to originate from either the methyl carbon (blue) or the carboxyl carbon (red) of acetate (actually, acetyl-CoA).

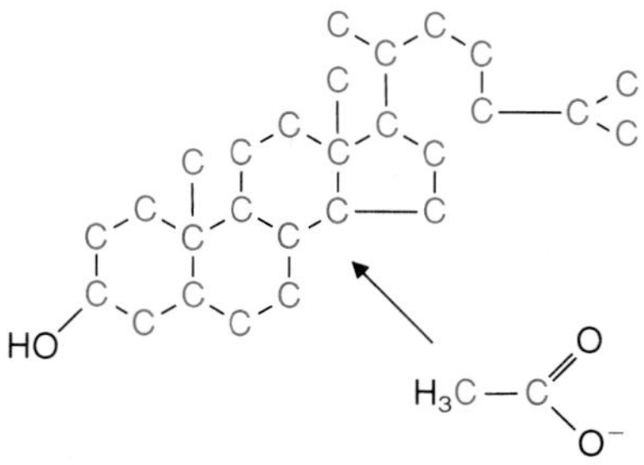

Other early insights came from the realization that the five carbons of isoprene could be derived metabolically from three molecules of acetate, and the prediction that cholesterol was a product of the cyclization of the linear C_{30} hydrocarbon **squalene.** Squalene contains six isoprene units, and its configuration makes it a plausible steroid precursor.

Squalene

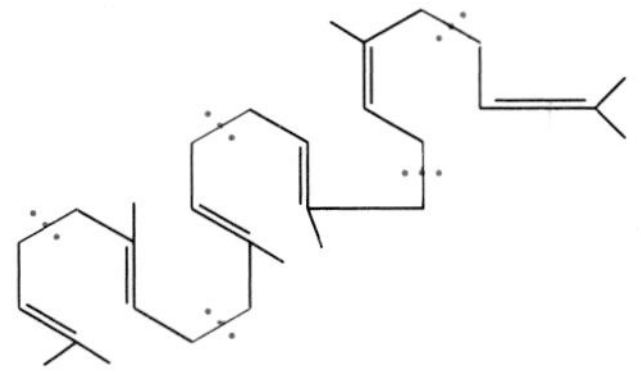

Postulated precyclization configuration of squalene (showing the six isoprene units)

In 1956 another important development occurred, when Karl Folkers discovered that a C_6 organic acid, **mevalonic acid,** could permit the growth of certain acetate-requiring strains of *Lactobacillus*. Folkers showed that mevalonic acid was readily converted to an activated C_5 isoprenoid compound, **isopentenyl pyrophosphate.** Interestingly, *Lactobacillus* does not synthesize steroids, but it uses the first several steps of the pathway to synthesize other isoprenoid compounds. In animals mevalonate is readily converted to squalene. Once this had been established, the stage was set for considering cholesterol biosynthesis as three distinct processes.

Stage 1. Conversion of a C_2 fragment to a C_6 isoprenoid (mevalonate).

Stage 2. Conversion of 6 mol of the C_6 mevalonate to the C_{30} squalene.

Stage 3. Cyclization of squalene and its transformation to the C_{27} cholesterol.

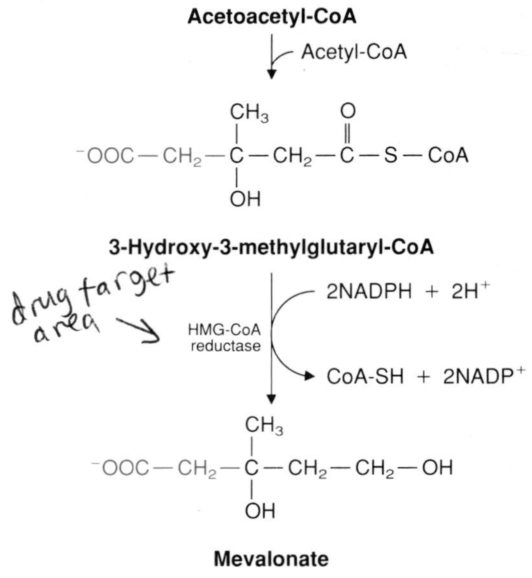

Figure 18.17
Biosynthesis of mevalonate.

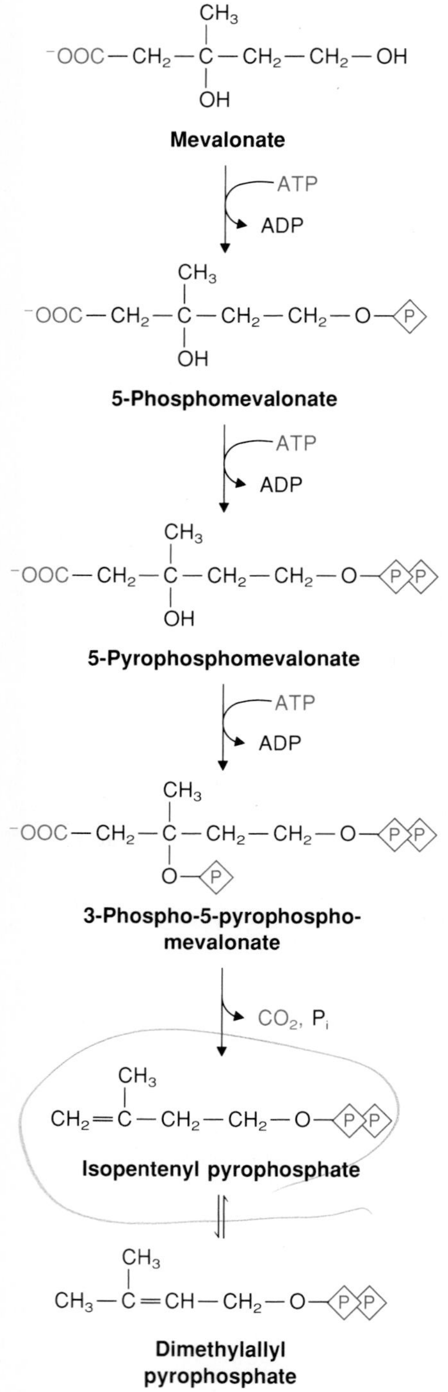

Figure 18.18
Conversion of mevalonate to isopentenyl pyrophosphate and to dimethylallyl pyrophosphate. The carboxyl carbon of mevalonate is lost as CO_2.

Now let us consider these three processes in detail.

STAGE 1. FORMATION OF MEVALONATE. The first part of the pathway is identical to reactions used in ketogenesis (Chapter 17), although it occurs in a different cell compartment. Ketogenesis occurs in mitochondria, while cholesterol biosynthesis occurs in the cytosol and the endoplasmic reticulum.

Two molecules of acetyl-CoA condense to give acetoacetyl-CoA, which reacts with a third molecule of acetyl-CoA to give 3-hydroxy-3-methylglutaryl-CoA (HMG-CoA). In mitochondrial matrix this cleaves to give acetoacetate plus acetyl-CoA. However, the HMG-CoA lyase that accomplishes this cleavage is missing from the endoplasmic reticulum, where cholesterol biosynthesis begins. Instead, **HMG-CoA reductase** catalyzes the NADPH-dependent reduction of the HMG-CoA to mevalonate (Figure 18.17). HMG-CoA reductase catalyzes the first committed reaction in cholesterol synthesis and is the major step that regulates the overall pathway.

STAGE 2. SYNTHESIS OF SQUALENE FROM MEVALONATE. The next several reactions occur in the cytosol. Mevalonate is activated by three successive phosphorylations (Figure 18.18). The third phosphorylation, probably at position 3, sets the stage for a decarboxylation via trans elimination, to give isopentenyl pyrophosphate. One molecule of this isomerizes to the C_5 **dimethylallyl pyrophosphate.** The latter reacts with a second molecule of isopentenyl pyrophosphate to give the C_{10} **geranyl pyrophosphate,** and still another molecule of isopentenyl pyrophosphate reacts with this to give the C_{15} **farnesyl pyrophosphate** (Figure 18.19).

Up to this point all of the reactions except that involving HMG-CoA reductase have occurred in the cytosol. The next enzyme, **farnesyl transferase** (also called squalene synthase), is bound to membranes of the endoplasmic reticulum. This NADPH-dependent enzyme joins 2 molecules of farnesyl pyrophosphate to give **presqualene pyrophosphate,** which then undergoes pyrophosphate elimination and rearrangement to yield squalene (Figure 18.20). All subsequent reactions occur in the endoplasmic reticulum.

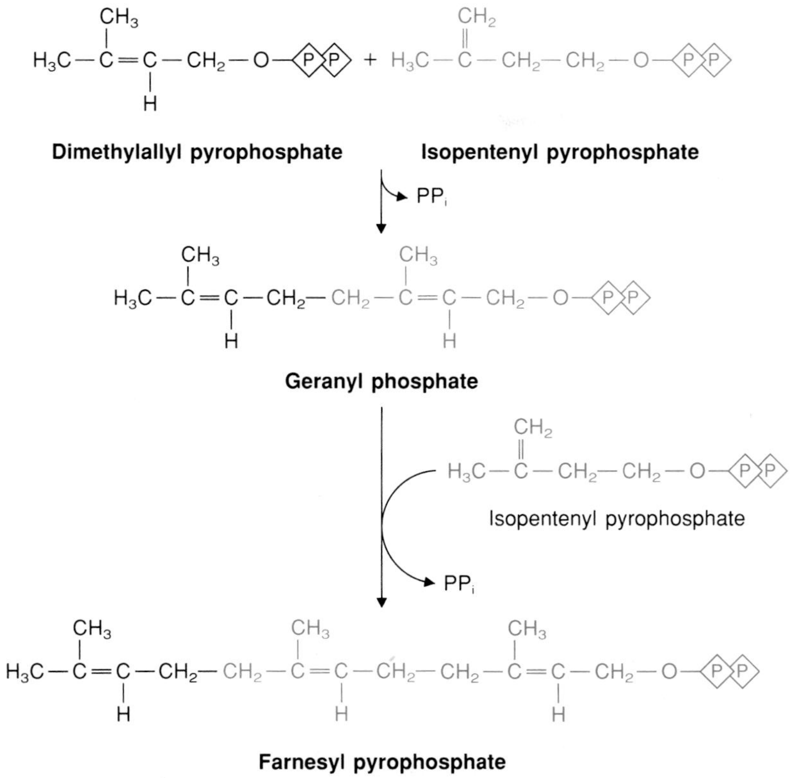

Figure 18.19
Conversion of isopentenyl pyrophosphate and dimethylallyl pyrophosphate to farnesyl pyrophosphate.

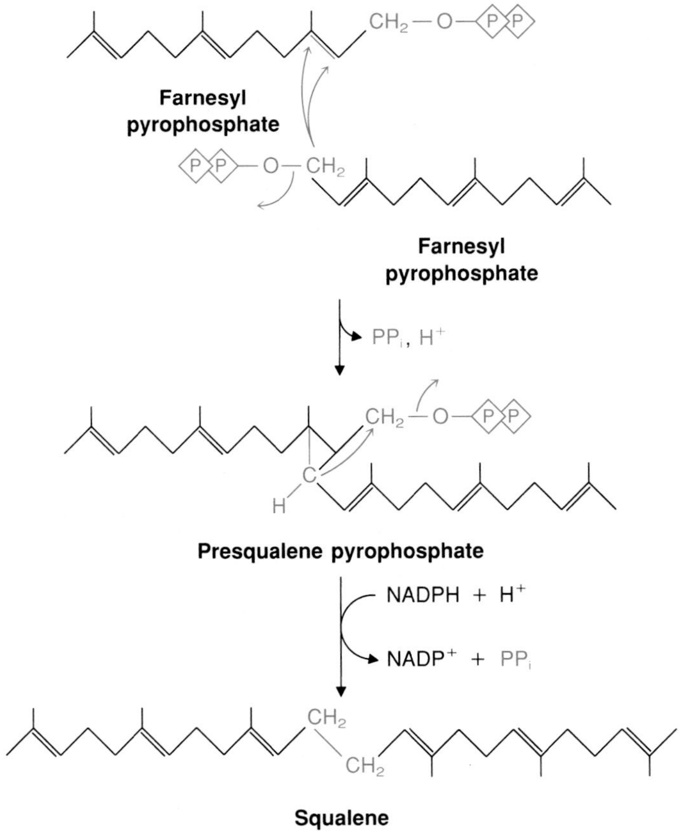

Figure 18.20
Conversion of farnesyl pyrophosphate to squalene. Note the head-to-head reaction between two molecules of farnesyl pyrophosphate, which involves the formation of an activated cyclopropane intermediate, coupled with loss of one pyrophosphate.

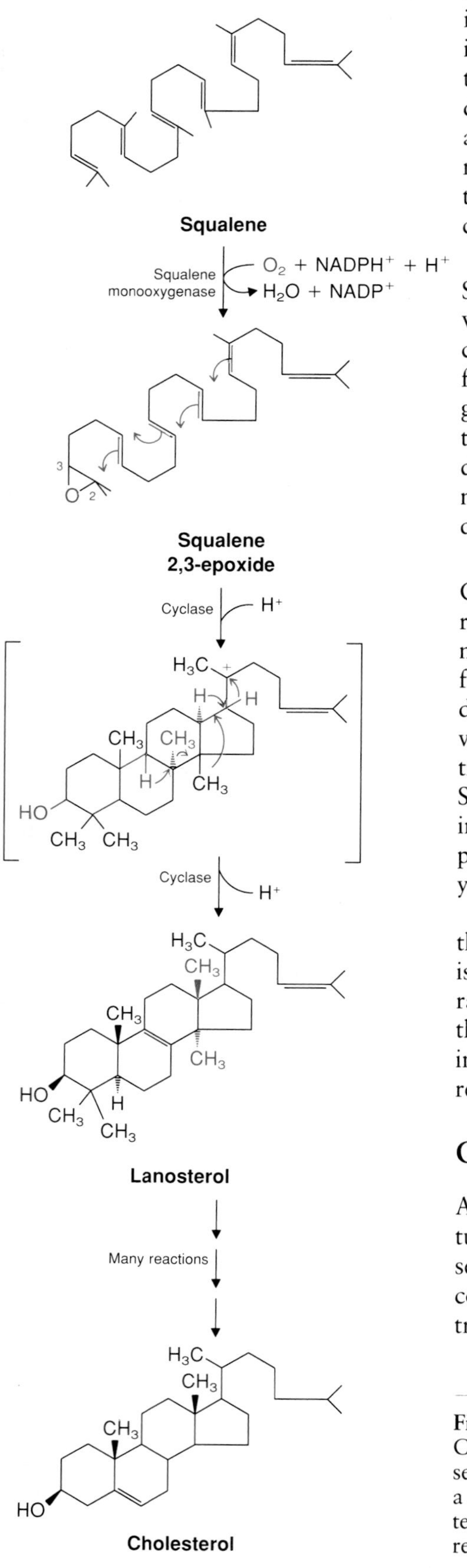

Squalene

Squalene monooxygenase (O_2 + NADPH$^+$ + H^+ → H_2O + NADP$^+$)

Squalene 2,3-epoxide

Cyclase (H^+)

Cyclase (H^+)

Lanosterol

Many reactions

Cholesterol

A remarkable feature of this part of the pathway is its stereochemistry. In the 1960s two British scientists, George Popjak and John Cornforth, identified 14 "stereochemical ambiguities" in this process, that is, 14 steps in the overall process that could go in either of two ways. For example, the triphosphorylated mevalonate derivative shown in Figure 18.18 could undergo decarboxylation by either a cis or trans elimination of the carboxyl and phosphoryl groups. Thus, 2^{14}, or 16,384, *different* stereochemical routes were possible for conversion of mevalonate to squalene. Remarkably, these scientists and their colleagues were able to determine the single stereochemical pathway that actually took place among these 16,384 possibilities.

STAGE 3. CYCLIZATION OF SQUALENE TO LANOSTEROL AND ITS CONVERSION TO CHOLESTEROL. The cyclization of squalene to lanosterol occurs in two steps. First, a mixed-function oxidase introduces an epoxide function at carbons 2 and 3 (Figure 18.21). Protonation of this functional group initiates a series of trans 1,2 shifts of methyl groups and hydride ions, to produce lanosterol. There follows a series of about 20 reactions involving double-bond reductions and three demethylations; one methyl group is removed from C-14 and two from C-4. The penultimate product is 7-dehydrocholesterol, which undergoes a final reduction to yield cholesterol.

CONTROL OF CHOLESTEROL BIOSYNTHESIS. As noted earlier, HMG-CoA reductase, the rate-limiting enzyme for cholesterol synthesis, represents the major target for regulation of the overall pathway. It has long been known from feeding studies that dietary cholesterol efficiently suppresses the endogenous synthesis of cholesterol. At least two distinct mechanisms are involved. First, *synthesis* of the enzyme is inhibited, probably at the transcriptional level, by cholesterol or a hydroxylated derivative of cholesterol. Second, the *activity* of the enzyme can be modulated through a mechanism involving cyclic phosphorylation and dephosphorylation of the reductase protein itself, although the importance of this control mechanism has not yet been established.

In vertebrates cholesterol synthesis is controlled beautifully, through the rate at which cholesterol enters cells from the bloodstream. Homeostasis is maintained by a mechanism that coordinates dietary intake of cholesterol, rate of endogenous cholesterol synthesis in the liver (and to a lesser extent in the intestine), and rate of cholesterol utilization by cells. The mechanism involves a specific receptor for low-density lipoproteins, the agents most responsible for transporting cholesterol in the bloodstream.

Cholesterol Transport and Utilization in Animals

As you know, one of the primary risk factors predisposing toward premature heart disease is an abnormally elevated level of cholesterol in the blood serum. Because of the medical importance of knowing both how cholesterol contributes to early heart attacks and how serum cholesterol levels are controlled, most of what we know about cholesterol metabolism comes from

Figure 18.21
Conversion of squalene to cholesterol. Formation of squalene epoxide leads to a series of double-bond electron shifts that close the four rings, and a migration of a carbon atom from C-14 to C-13 yields the first steroid intermediate, lanosterol. Many subsequent reactions lead to 7-dehydrocholesterol, which undergoes reduction to cholesterol.

studies on mammals, and that is what this section will emphasize. Crucial to our understanding is the research of Michael Brown and Joseph Goldstein, who showed in the mid-1970s that cholesterol uptake by cells is a receptor-mediated process and that the quantity of receptors themselves is subject to regulation.

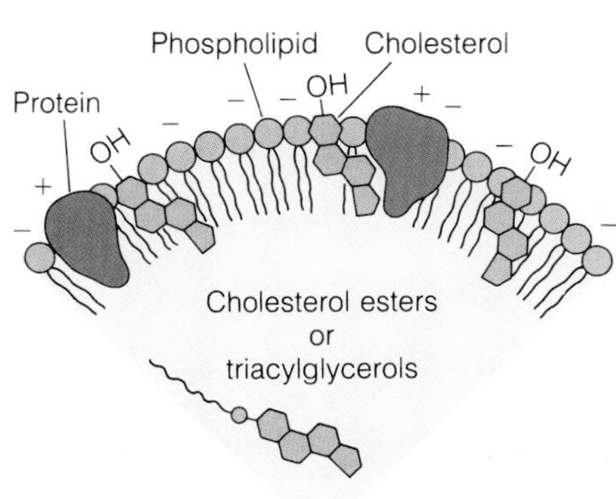

Figure 18.22
Generalized structure of a plasma lipoprotein.

Lipoproteins as Lipid Carriers in the Blood. Recall from Chapter 17 that triacylglycerols and cholesterol, being highly apolar, are transported through the bloodstream in the form of lipoproteins. Basically, a lipoprotein consists of an oil droplet—a droplet of triacylglycerol and/or cholesterol ester surrounded by a skin of polar lipids, primarily phospholipids, with the polar head group on the outside and with one or more apoproteins inserted into this outer shell (Figure 18.22). The cholesterol in the interior is esterified at its hydroxyl position with a long-chain fatty acid, usually unsaturated. The esters are synthesized in plasma from cholesterol and an acyl chain on phosphatidylcholine, through the action of **lecithin:cholesterol acyltransferase** (LCAT), an enzyme that is secreted from liver into the bloodstream.

$$\text{Phosphatidylcholine} + \text{cholesterol} \rightleftharpoons \text{lysolecithin} + \text{cholesterol ester}$$

Cholesterol esters are considerably more hydrophobic than cholesterol itself. The polar lipids on the exterior of lipoproteins allow the particles to interact with the surrounding milieu, so that lipoprotein particles remain dispersed throughout the blood, rather than aggregating in large drops that cannot be metabolized.

Chapter 17 listed seven classes of lipoproteins, but it is convenient for this discussion to consider just four major categories (see Table 17.1): (1) *chylomicrons,* which participate chiefly in transporting dietary triacylglycerols and cholesterol from the intestine to adipose tissue and liver; *very low-density lipoproteins* (VLDL), which deliver endogenously synthesized triacylglycerols from liver to adipose and other tissues; *low-density lipoproteins* (LDL), which transport cholesterol to peripheral tissues and regulate endogenous cholesterol levels in those tissues; and *high-density lipoproteins* (HDL), one of whose functions is to transport cholesterol from peripheral tissues to the liver.

Chylomicrons are by far the largest of the lipoproteins, and also the least dense because of their high triacylglycerol and low protein content. Chylomicrons are synthesized primarily in intestine, while the other lipoproteins are synthesized principally in the liver.

Of the four major lipoprotein classes, LDL is by far the richest in cholesterol. More than 40% of the weight of the LDL particle is cholesterol esters, while the total of esterified and free cholesterol amounts to well over half of the weight of the particle. The amounts of cholesterol and cholesterol esters associated with LDL are typically about two-thirds of the total plasma cholesterol (total plasma cholesterol ranges from 130 to 260 mg/100 ml, or 1.3 to 2.6 g/liter of human plasma). The particle contains a single molecule of apoprotein B-100 (M_r = 500,000) as its sole protein component. Since cholesterol biosynthesis is confined almost entirely to the liver with some occurring also in intestine, LDL plays an important role in delivering cholesterol to peripheral tissues.

The LDL Receptor and Cholesterol Homeostasis. The importance of understanding cholesterol homeostasis can be seen most directly by reviewing the consequences of prolonged **hypercholesterolemia,** or abnor-

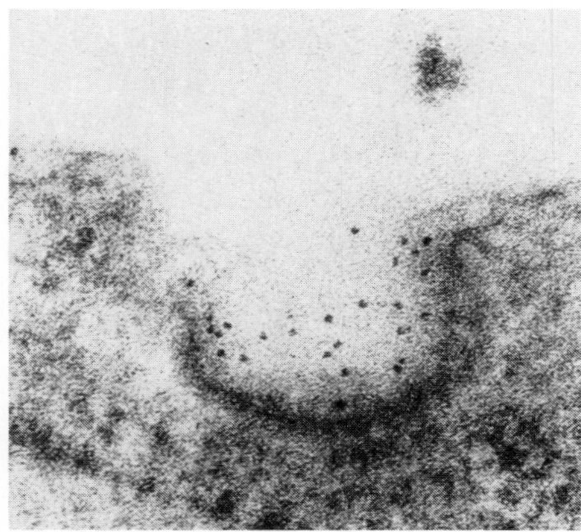

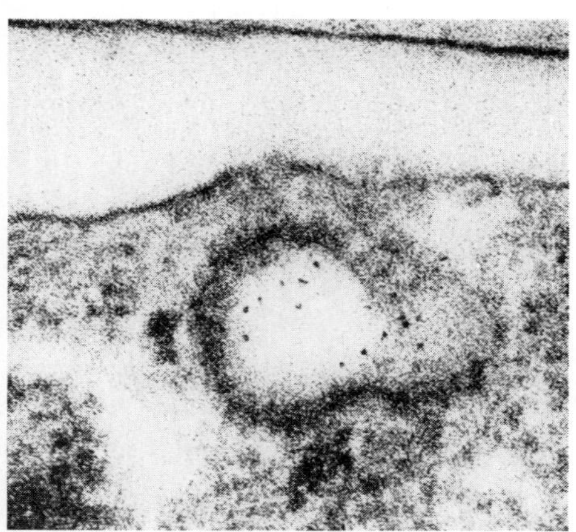

Figure 18.23
Visualization of LDL binding to receptors and receptor-mediated endocytosis. LDL was conjugated with ferritin to permit electron microscopic visualization. (**Top**) The LDL–ferritin (dark dots) binds to a coated pit on the surface of a cultured human fibroblast; (**bottom**) this region invaginates to form a coated vesicle.

mally high plasma cholesterol levels. Since cholesterol is highly insoluble, prolonged high levels in blood lead to cholesterol deposition on the inside walls of blood vessels. If not resolubilized over time, these fatty deposits eventually harden to atherosclerotic **plaque,** which ultimately blocks key blood vessels and causes myocardial infarctions, or heart attacks. Directly related is the question of how cholesterol is taken up from lipoproteins (primarily LDL) into cells, because cholesterol esters are too hydrophobic to traverse cell membranes.

In 1972 Brown and Goldstein began to study a hereditary condition called **familial hypercholesterolemia,** or FH. Individuals with the rare homozygous form of this disease (about 1 in 1 million) have grossly elevated levels of serum cholesterol, from 650 to 1000 mg/100 ml. They develop atherosclerosis (hardening of the arteries) early in life and usually die of heart disease before age 20. The more common heterozygous condition, characterized by one defective allele instead of two, affects about one individual in 500. These people have less severely elevated cholesterol levels, in the range of 250 to 500 mg/100 ml; they are at high risk to have heart attacks in their thirties and forties, although many enjoy a normal life span.

Crucial to the success of Brown and Goldstein was their ability to demonstrate the defective FH phenotype in cell culture. Cells from FH patients synthesized cholesterol at abnormally high rates in culture, while normal cells showed low rates of synthesis. When cultured in the presence of LDL, normal cells showed low activity of HMG-CoA reductase, while the absence of LDL in the culture medium generated activities some 50- to 100-fold higher. Their results suggested that normal fibroblasts were transporting cholesterol into the cell, where it regulates its own synthesis by suppressing the activity of the rate-limiting enzyme, HMG-CoA reductase. By contrast, fibroblasts from FH individuals showed high levels of reductase activity, whether cultured in the presence or absence of LDL, suggesting that they were deficient in ability to take up cholesterol from the medium.

The results suggested that cholesterol is taken into cells through the action of a specific receptor and that that receptor is deficient or defective in FH patients. In short order Brown and Goldstein and their colleagues demonstrated the existence of this receptor, the **LDL receptor,** and demonstrated a new mechanism by which cells can interact with their environment—**receptor-mediated endocytosis.** By conjugating LDL with an electron-dense material, the investigators were able to visualize the LDL receptor on cell surfaces, simply by asking where LDL became bound to the cell (Figure 18.23). These experiments showed that the receptors are clustered in a structure called a **coated pit,** an invagination whose most abundant protein is **clathrin,** a self-interacting protein that is capable of forming a cagelike structure (Figure 18.24).

Endocytosis is a process by which cells take up large molecules from the extracellular environment. Although LDL uptake involves a cell surface receptor, the interaction of LDL with its receptor is unlike the interaction of

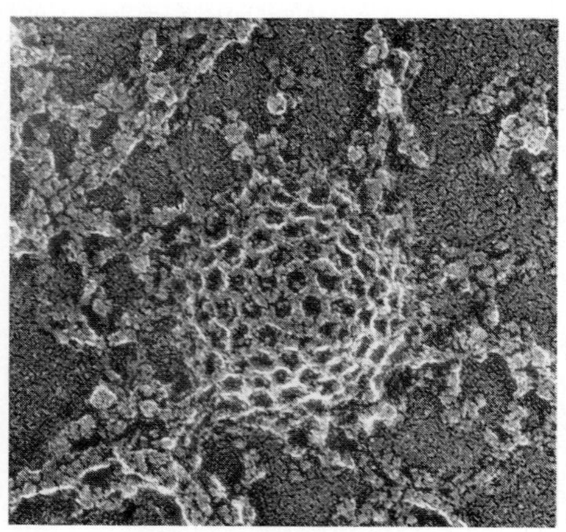

Figure 18.24
Coated pits and vesicles on the inner surface of the plasma membrane of cultured mammalian cells, visualized by freeze-fracture electron microscopy. The cagelike structure of the pits is due to the structure of clathrin, the major protein in these pits.

hormones such as epinephrine with their receptors. As discussed in Chapter 13, the binding of epinephrine at its receptor in the plasma membrane triggers intracellular metabolic changes, but the hormone itself does not enter the cell. By contrast, the binding of LDL to its receptor causes the entire LDL molecule to be taken up into the cell. The plasma membrane fuses in the vicinity of the LDL–receptor complex, and the coated pit becomes an endocytotic vesicle (Figure 18.25). Several of these vesicles fuse to form an **endosome.** The endosomes move to the lysosomes, where another membrane fusion takes place, putting the LDL–receptor complex in contact with the hydrolytic enzymes of the lysosome. The LDL apoprotein is hydrolyzed to amino acids, while the cholesterol esters are hydrolyzed to give free cholesterol. The receptor itself is recycled, moving back to the plasma membrane to pick up more LDL; about 10 minutes is required for each round trip.

Much of the cholesterol released moves to the endoplasmic reticulum, where it is used for membrane synthesis. The internalized cholesterol exerts three regulatory effects. (1) As mentioned earlier, it suppresses endogenous cholesterol synthesis, by inhibiting HMG-CoA reductase and also by sup-

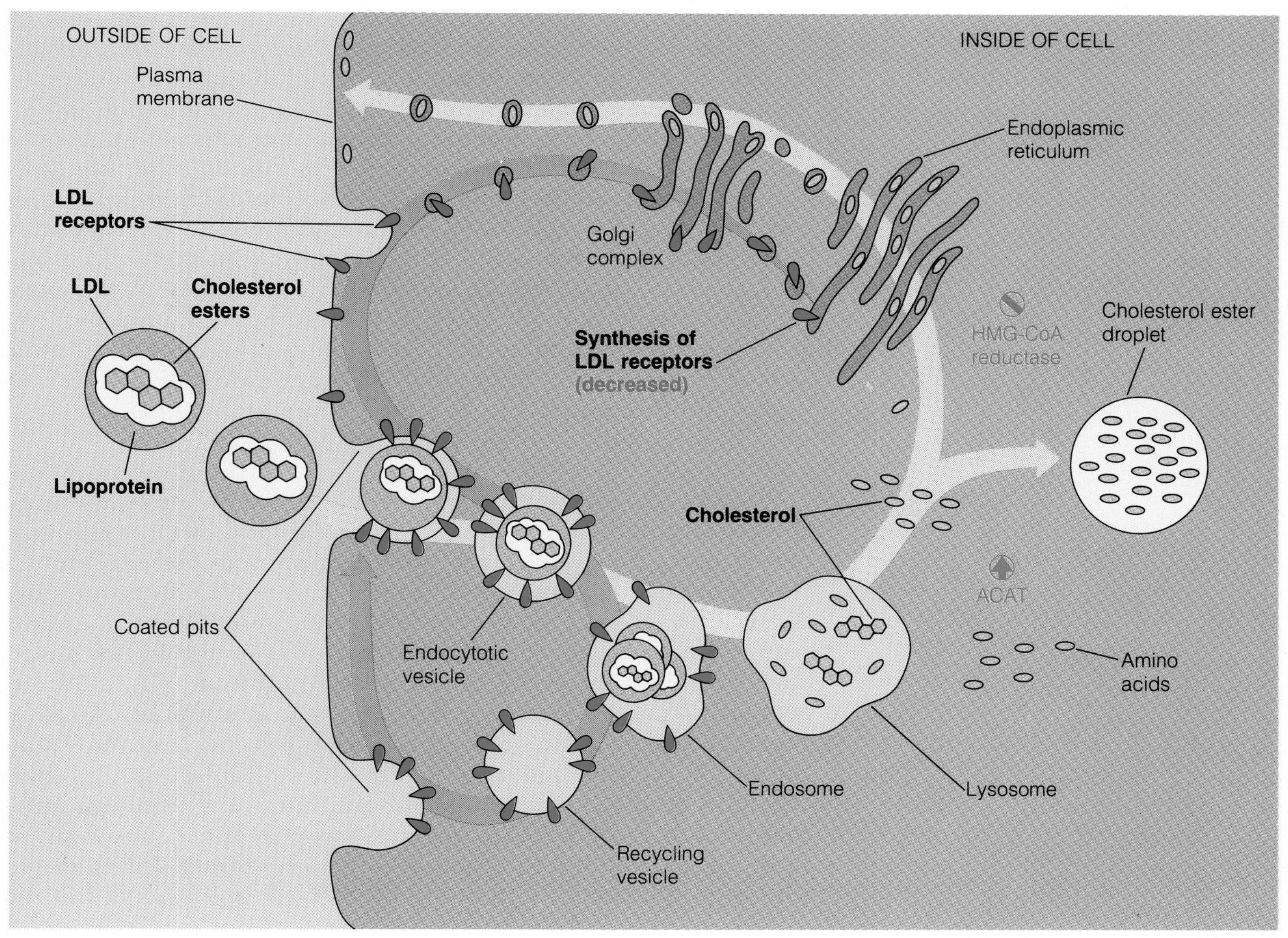

Figure 18.25
Involvement of the LDL receptor in cholesterol uptake and metabolism. After synthesis in the endoplasmic reticulum, the LDL receptor matures in the Golgi complex, then migrates to the cell surface, where it clusters in coated pits. After internalization of LDL, multiple vesicles fuse to form an organelle called the endosome. Proton pumping in the endosome membrane causes the pH to drop, which in turn causes LDL to dissociate from the receptor. The LDL apoprotein is degraded in lysosomes, and cholesterol has various fates. The LDL receptor remains in a vesicle, which returns to the plasma membrane to start the cycle anew. Regulatory actions of cholesterol are shown in red.

pressing transcription of the gene for this enzyme and accelerating degradation of the enzyme protein. (2) It activates **acyl-CoA:cholesterol acyltransferase** (ACAT), an intracellular enzyme that synthesizes cholesterol esters from cholesterol and a long-chain acyl-CoA. This promotes the storage of excess cholesterol in the form of droplets of cholesterol esters. (3) It regulates the synthesis of the LDL receptor itself, by somehow lowering the content of mRNA for the receptor. Decreased synthesis of the receptor ensures that cholesterol will not be taken into the cell in excess of the cell's needs, even when extracellular levels are very high. This regulatory mechanism explains why ingestion of excess cholesterol, as in many Western diets, leads directly to elevations of blood cholesterol levels: with intracellular cholesterol levels so well regulated, the extracellular cholesterol accumulates because it has nowhere else to go.

Current therapeutic approaches to lowering cholesterol levels are focused on developing inhibitors of HMG-CoA reductase, on the assumption that they will depress intracellular cholesterol levels and, hence, increase the production of LDL receptors for clearance of extracellular cholesterol from the blood. One such inhibitor, with the trade name **Lovastatin,** received approval in 1987 for treatment of patients with very high cholesterol levels.

Two additional points should be made about the life cycle of the LDL receptor. First, receptor-mediated endocytosis has been shown to be a widely used pathway for internalization of extracellular substances, including cell growth factors, insulin, and some viruses. Second, studies of many individuals with FH have revealed four different types of mutation that generate the FH phenotype, and these studies are giving a remarkably detailed picture of the receptor and its action. The receptor itself is a glycoprotein with an 839-residue polypeptide chain and 18 O-linked oligosaccharide chains. The receptor binds specifically the apolipoproteins B-100 and E. A cDNA, or DNA complementary to receptor messenger RNA, has been cloned, so the molecular nature of receptor mutations can be ascertained.

Receptor mutations are of several types. First, and most common, are those that lead to insufficient receptor synthesis. Second are mutations in which receptor is synthesized but fails to migrate from the endoplasmic reticulum to the Golgi complex, for transport to the cytoplasmic membrane. Third are mutations in which receptor is synthesized and processed normally and reaches the cell surface, but fails to bind LDL. Finally, there is a class of mutant receptors that reach the cell surface and bind LDL but fail to cluster in coated pits.

DIETARY FATTY ACIDS AND CHOLESTEROL METABOLISM. Investigations of the LDL receptor have enlarged our understanding of the relationships among dietary cholesterol, plasma cholesterol levels, and atherosclerosis. It is well established that prolonged hypercholesterolemia, whether caused by diet, heredity, or both, puts one at risk for heart disease. Further, the mechanisms involved are becoming clearer. What is still much less clear is the relationship between dietary fatty acids and cholesterol levels. It has long been known that a diet rich in saturated fatty acids is associated with high plasma cholesterol levels, whereas a diet rich in polyunsaturated fatty acids, as found in vegetable and fish oils, decreases cholesterol levels. We know very little about the mechanism by which saturation of dietary fat affects cholesterol and LDL levels. However, substitution of unsaturated fats for saturated fats in the diet tends to increase the levels of high-density lipoproteins while decreasing the levels of LDL. Though high levels of HDL are associated statistically with decreased risk of heart disease, the mechanism involved is not known. We do know that one function of HDL is to remove cholesterol from peripheral tissues and return it to the liver. A major metabolic fate of cholesterol in the liver is in the synthesis of bile acids, which are secreted into the intestine and ultimately excreted. Therefore, the beneficial effect of elevated HDL may derive simply from the fact that it promotes a process that removes cholesterol from the body.

Recent attention has focused on a class of polyunsaturated fatty acids particularly abundant in fish oils, the so-called n-3 or ω-3 fatty acids.* The addition of fish or fish oils, which are abundant in linolenic and other n-3 acids, to American and Western diets has dramatically lowered serum cholesterol and triacylglycerols. The mechanisms involved are not yet clear, but there seems to be a relationship with the biological regulators called prostaglandins and thromboxanes, which we discuss later in this chapter. Elongation and desaturation of *linoleic* acid, which is abundant in vegetable oils, lead to arachidonic acid, the precursor of prostaglandins and thromboxanes. *Linolenic* acid is elongated and desaturated to fatty acids with five and six double bonds (arachidonic acid has four). Current evidence suggests that n-3 fatty acids, such as linolenic acid and its metabolites, either decrease the production of arachidonic acid and its metabolites or displace arachidonic acid in phospholipid molecules, or both. This may prevent overproduction of prostaglandins or thromboxanes, although the relationships between these phenomena and the control of serum lipid levels are still quite obscure.

Bile Acids

Bile acids are steroid derivatives with detergent properties, which emulsify dietary lipids in the intestine and thereby promote fat digestion and absorption. They are secreted from the liver, stored in the gallbladder, and passed through the bile duct and into the intestine. Biosynthesis of bile acids represents the major metabolic fate of cholesterol, accounting for more than half of the 800 mg/day that is metabolized in the normal human adult. By contrast, steroid hormone synthesis amounts to only about 50 mg of cholesterol metabolized per day.

However, much more than 400 mg of bile acids is secreted into the intestine per day. Most of the bile acids secreted into the upper small intestine are absorbed in the lower small intestine and returned to the liver for reuse, through the portal blood. This process, which handles 20 to 30 g of bile acids per day, is called the **enterohepatic circulation.** Daily elimination of bile acids in the feces amounts to just 0.5 g/day or less.

The most abundant bile acids in humans are **cholic acid** and **chenodeoxycholic acid** (shown in Figure 18.26). These are usually conjugated with the amino acid **glycine** or **taurine.** The cholic acid conjugates with glycine and taurine are called **glycocholate** and **taurocholate,** respectively. All of the bile acids exist as salts at physiological pH, making the term *bile salts* somewhat more accurate than *bile acids*. Another bile salt, **deoxycholate,** is

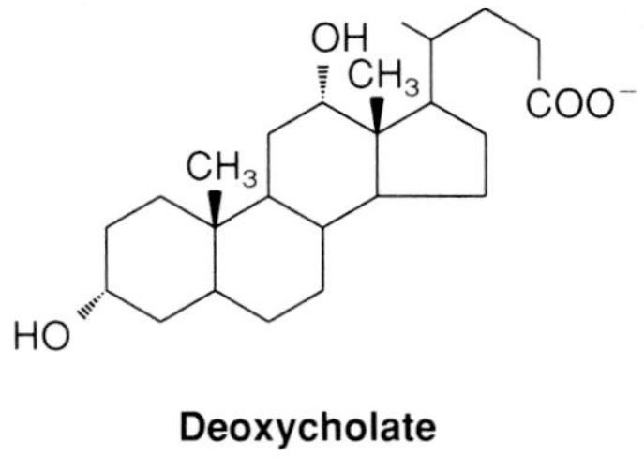

OH
CH_3
CH_3
COO^-
HO

Deoxycholate

* Biochemists ordinarily number the sites of fatty acid unsaturation from the carboxyl group, while nutritionists number unsaturation from the methyl group at the other end. Thus, linoleic acid, which we have called a C_{18}-Δ^9, Δ^{12} fatty acid, is n-6 in this terminology, because carbon 13 from the carboxyl group is carbon 6 from the methyl group. Similarly, linolenic acid, a C_{18}-Δ^9, Δ^{12}, Δ^{15} fatty acid, is n-3. In this section we shall use the nutritionists' terminology.

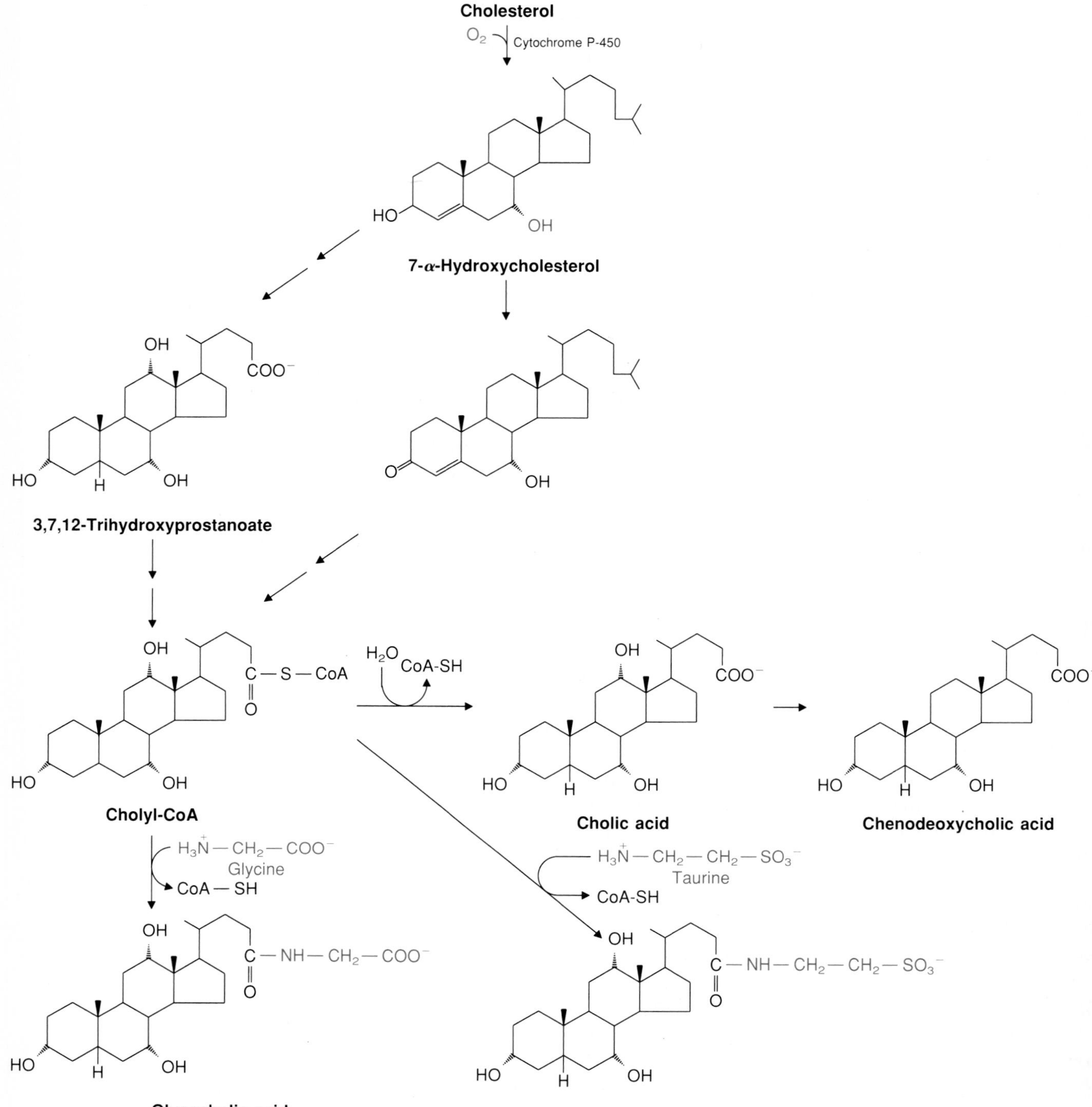

Figure 18.26
Biosynthesis of bile salts from cholesterol. The major pathway, shown at the right, begins with hydroxylation of cholesterol by a mixed-function oxidase.

abundant in the bile of some other mammals. It is widely used as a laboratory reagent, to solubilize membrane proteins.

In the biosynthetic routes to cholate, glycocholate, and taurocholate outlined in Figure 18.26, a series of hydroxylations occurs, catalyzed by microsomal mixed-function oxidases. The first of these, at C-7, is rate limiting and hence controls the overall pathway. Note also the dehydrogenation and reduction of the hydroxyl at carbon 3, which inverts its configuration and joins the A and B rings in a cis configuration.

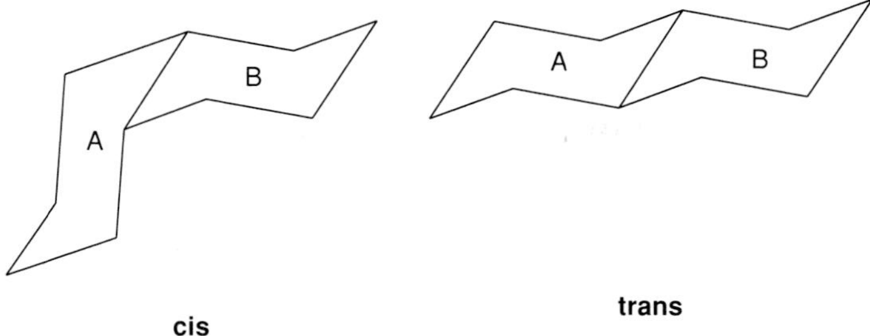

At present we know little about the mechanism of the side chain modification, which converts the eight-carbon saturated hydrocarbon to a five-carbon chain with a terminal carboxylate.

Steroid Hormones

Cholesterol is the biosynthetic source of all steroid hormones. The primary organs involved are the gonads and the adrenal cortex, plus the placenta in pregnant females. In this chapter we review the biosynthetic pathways to steroid hormones; their actions are discussed in Chapter 23. Five major classes of hormone will concern us: (1) **progesterone,** which regulates events during pregnancy; (2) the **glucocorticoids** (cortisol and corticosterone), which promote gluconeogenesis and suppress inflammation reactions; (3) the **mineralocorticoids** (aldosterone), which regulate ion balance by promoting reabsorption in the kidney of Na^+, Cl^-, and HCO_3^-; (4) the **androgens** (androstenedione and testosterone), which promote male sexual development and maintain male sex characteristics; and (5) the **estrogens** (estrone and estradiol), or female sex hormones, which support female characteristics. The structures of these hormones are shown in Figure 18.27. In each case the side chain in cholesterol is either greatly shortened or nonexistent.

A general feature of steroid hormones is that they are not stored for release after synthesis. Therefore, the level of a circulating hormone is controlled primarily by its rate of synthesis, which is often controlled ultimately by signals from the brain. These signals usually act through intermediary hormones. For example, the neurohormone **corticotropin-releasing hormone,** or CRH, is released from cells in the hypothalamus in response to central nervous system inputs. CRH stimulates release from the pituitary gland of **corticotropin,** or ACTH (**adrenocorticotropic hormone**), which in turn stimulates the synthesis of mineralocorticoids and glucocorticoids in adrenal cortex. Activation of steroid hormone synthesis involves stimulation of both hydrolysis of cholesterol esters and uptake of cholesterol into mitochondria of cells in the target organ. There an enzyme complex called **cholesterol desmolase** hydroxylates the side chain at C-20 and C-22 and cleaves it, to yield **pregnenolone.**

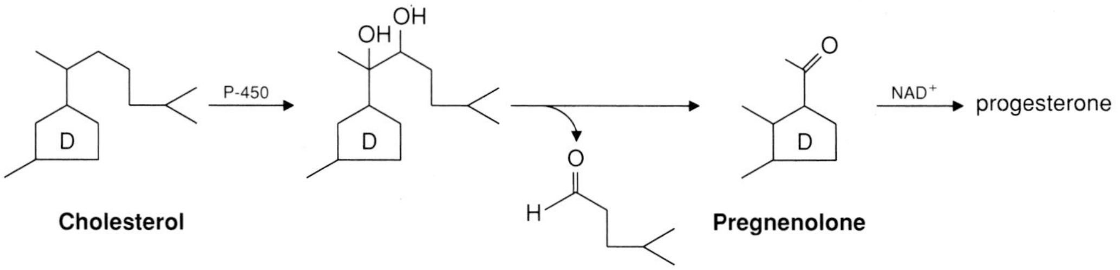

P-450 · NAD$^+$ → progesterone

Cholesterol · **Pregnenolone**

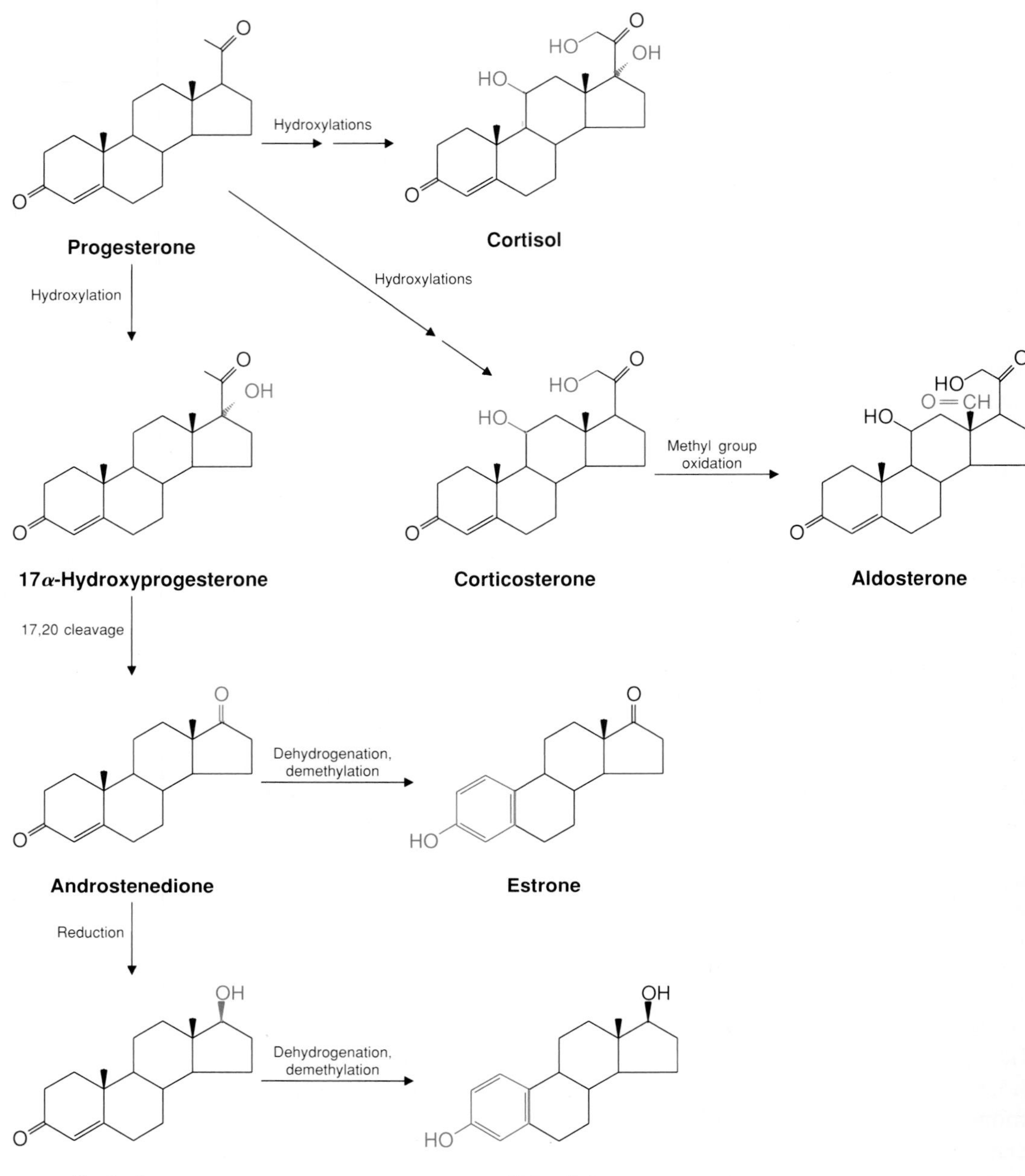

Figure 18.27
Biosynthetic routes from progesterone to other steroid hormones. Androstenedione is an androgen, and estrone is an estrogen.

These two hydroxylation reactions, as well as the subsequent hydroxylations in steroid hormone biosynthesis, involve mixed-function oxidases, which utilize O_2, NADPH, and a cytochrome P-450.

Dehydrogenation of pregnenolone yields progesterone, which has its own biological activity but is also the precursor to all other steroid hormones. The subsequent reactions, which chiefly involve hydroxylations and loss of the remaining two-carbon side chain, are summarized in Figure 18.27.

Other Isoprenoid Compounds

Lipid-Soluble Vitamins

The four lipid-soluble vitamins—A, D, E, and K—are all isoprenoid compounds, made up, like steroids, of activated five-carbon units. Thus, as a group they have a structural relatedness not seen among the water-soluble vitamins. On the other hand, while the water-soluble vitamins have a *functional* uniformity in that all are designed to carry mobile metabolic groups, the lipid-soluble vitamins are diverse in their functions.

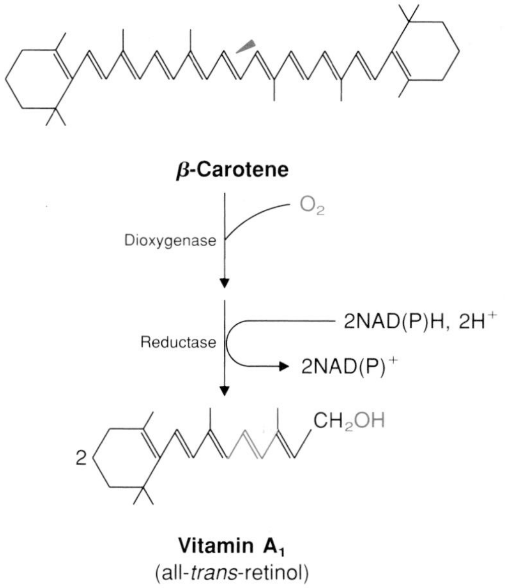

VITAMIN A. This isoprenoid alcohol, otherwise called ***trans*-retinol,** plays a key role in vision. It also plays an as yet undefined role in controlling animal growth. The vitamin can either be consumed in the diet or biosynthesized from **β-carotene,** an isoprenoid compound especially abundant in carrots (Figure 18.28). As discussed in Chapter 29, *trans*-retinol isomerizes to 11-*cis*-retinol, which then oxidizes to 11-*cis*-retinal. Combination of the retinal with the protein **opsin** in the eye gives the visual pigment **rhodopsin.** The visual process involves photochemical isomerization of 11-*cis*-retinal back to all-*trans*-retinal.

β-Carotene

Dioxygenase O_2

Reductase $2NAD(P)H, 2H^+$ → $2NAD(P)^+$

2 CH_2OH

Vitamin A_1
(all-*trans*-retinol)

Figure 18.28
Synthesis of vitamin A_1, all-*trans*-retinol. The 11-*cis* derivative is a component of the visual pigment, rhodopsin.

VITAMIN D. The most abundant form of vitamin D is vitamin D_3, or **cholecalciferol.** This is not truly a vitamin, because it is not required in the diet. Instead, it arises by synthesis from 7-dehydrocholesterol, an intermediate in cholesterol biosynthesis. It is more accurate to think of vitamin D_3 as a prohormone, because it is converted to a metabolite that acts analogously to a steroid hormone. Its action involves the regulation of calcium and phosphorus metabolism, particularly with respect to synthesis of the inorganic matrix of bone, which consists largely of calcium phosphate.

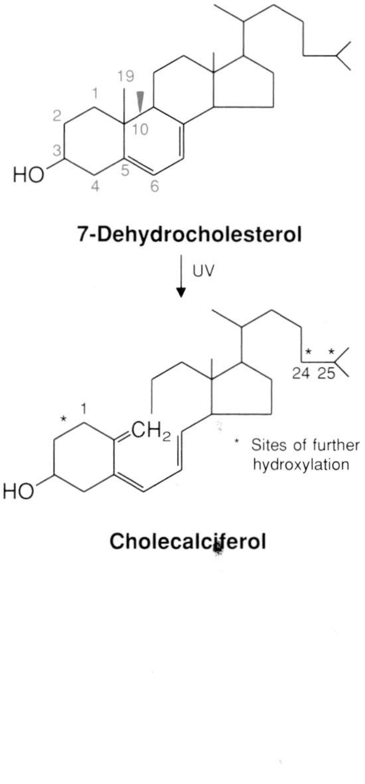

In skin cells 7-dehydrocholesterol undergoes ultraviolet photolysis to give cholecalciferol. Because the UV rays come from sunlight, insufficient sunlight exposure can cause a deficiency of vitamin D_3 and result in the bone malformation known as **rickets.** Vitamin D_3 is often added to dairy products, because sunlight exposure is limited in many regions for much of the year.

Cholecalciferol undergoes two successive hydroxylations, each catalyzed by a mixed-function oxidase. The first, at carbon 25, involves a microsomal enzyme system in liver. 25-Hydroxycholecalciferol is then transported to the kidney, where a mitochondrial enzyme hydroxylates it at carbon 1. This reaction is activated by **parathyroid hormone,** which is secreted from the parathyroid gland when calcium levels are low. When calcium levels are adequate, the second hydroxylation occurs at C-24, instead of C-1, to give an inactive metabolite.

1,25-Dihydroxycholecalciferol, or 1,25(OH)D_3, is the hormonally active form of vitamin D. This compound migrates to target cells in the intestine and in osteoblasts (bone cells), where it binds to protein receptors, which migrate to the cell nucleus. In intestine the hormone–receptor complex stimulates transcription, resulting in synthesis of a protein that stimulates calcium absorption into the bloodstream. In osteoblasts 1,25(OH)D_3 stimulates calcium uptake for deposition as calcium phosphate.

19 1 2 3 10 5 4 6 HO

7-Dehydrocholesterol

UV

24 25 * * 1 * CH_2 HO

* Sites of further hydroxylation

Cholecalciferol

VITAMIN E. This vitamin, also called **α-tocopherol,** was originally recognized in nutritional studies as an agent that prevented sterility in rats.

α-Tocopherol (vitamin E)

The vitamin appears to play an antioxidant role, particularly in preventing attack of peroxides on unsaturated fatty acids in membrane lipids. α-Tocopherol does prevent fatty acid peroxidation in vitro. However, vitamin E deficiency results in additional symptoms that are not relieved by other antioxidants. Therefore, additional biological roles seem likely for this vitamin.

VITAMIN K. This vitamin was originally discovered as a lipid-soluble substance involved in blood coagulation. Vitamin K_1, or **phylloquinone,** is found in plants; the quinone portion of this molecule has a largely saturated side chain. Another form of the vitamin, vitamin K_2, or **menaquinone,** is found largely in animals and bacteria.

Phylloquinone **Menaquinone**

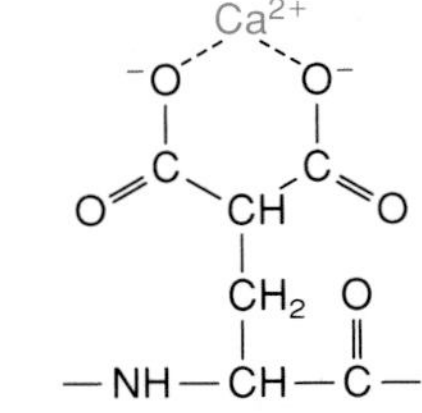

A γ-carboxyglutamate residue complexed with calcium

Menaquinone has a partly unsaturated side chain. In animals vitamin K_2 is essential for the carboxylation of glutamate residues in certain proteins, to give *γ-carboxyglutamate.* This modification allows the protein to bind calcium, which is essential to blood clotting (discussed in Chapter 11). Carboxylation of glutamate residues occurs in other proteins that are active in the mobilization or transport of calcium.

Terpenes

Terpene is a generic term for all compounds that are biosynthesized from isoprene precursors. Therefore, the compounds we have been discussing—cholesterol, bile acids, steroids, and lipid-soluble vitamins—are terpenes; they receive special attention because of their importance in animal metabolism. Here we take a brief glimpse at the vast range of other terpene compounds. They include insect hormones and plant growth hormones, as well as the lipid-linked sugar carriers we encountered in Chapter 16.

Terpenes are biosynthesized ultimately from isopentenyl pyrophosphate (C_5) and dimethylallyl pyrophosphate (C_5). When these combine to yield geranyl pyrophosphate (C_{10}), any terpene formed thereby is called a **monoterpene.** When a compound is formed from 1 mol of farnesyl pyrophosphate (C_{15}), the product is called a **sesquiterpene. Triterpenes** (C_{30}) are formed from 2 mol of farnesyl pyrophosphate. Geranylgeranyl pyrophosphate (C_{20}) yields either **diterpenes** (C_{20}) or **tetraterpenes** (C_{40}). The dolichols and undecaprenol, introduced in Chapter 16, are examples of **polyprenols** (polyisoprenoid alcohols); these have more than 50 carbons.

Structures of some common terpenes are given in Figure 18.29.

CLASS	EXAMPLE	FUNCTION
Monoterpenes	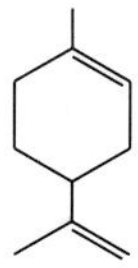 **Limonene**	Responsible for the characteristic odor of lemons
Sesquiterpenes	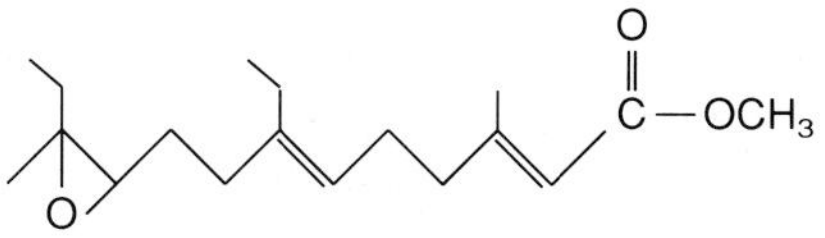 **Juvenile hormone I**	Controls metamorphosis in insects
Diterpenes	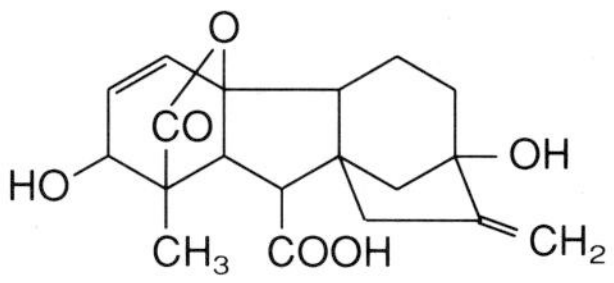 **Gibberellic acid**	Plant growth hormone
Triterpenes	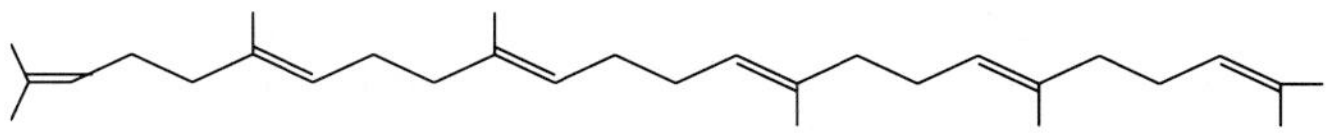 **Squalene**	Cholesterol precursor
Tetraterpenes	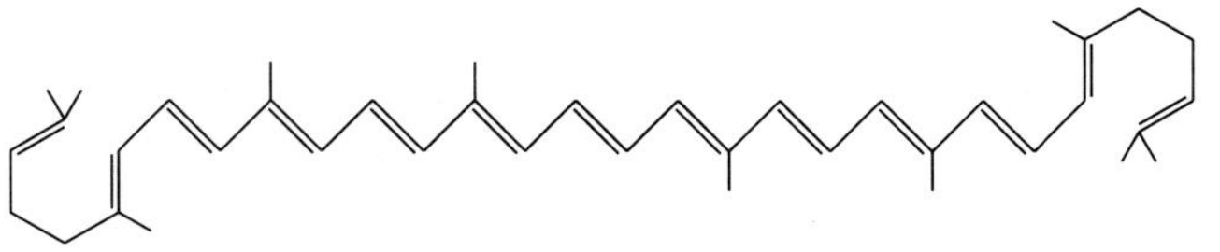 **Lycopene**	Tomato pigment
Polyprenols	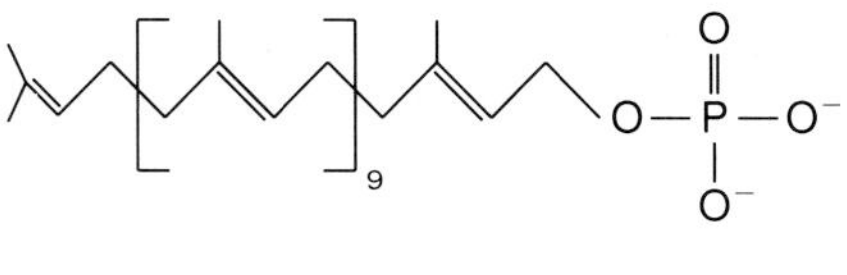 **Undecaprenol phosphate**	Sugar carrier for oligosaccharide synthesis
	 ***cis*-Polyisoprene**	Natural rubber

Figure 18.29
Some terpene compounds. These are representative of an enormous class of natural products.

Eicosanoids: Prostaglandins, Thromboxanes, and Leukotrienes

We turn finally to a class of lipids that are distinguished by their potent physiological properties, low levels in tissues, rapid metabolic turnover, and common metabolic origin. The most important of these compounds are the

prostaglandins; also included are the **thromboxanes** and **leukotrienes.** Collectively, these compounds are called **eicosanoids** because of their common origin from C_{20} polyunsaturated fatty acids, the eicosaenoic acids. The most abundant of these acids in most lipids, and the dominant precursor, is arachidonic acid, which is *cis*-Δ^5,*cis*-Δ^8,*cis*-Δ^{11},*cis*-Δ^{14}-eicosatetraenoic acid. Related C_{20} trienoic and pentaenoic acids serve as minor precursors to some prostaglandins and their relatives. In addition to the compounds discussed here, the eicosaenoic acids serve as precursors to another class of compounds, the hydroxyeicosaenoic acids and hydroperoxyeicosaenoic acids. The latter compounds are metabolic precursors to the leukotrienes.

The prostaglandins and the closely related thromboxanes are derived from a common pathway; a different pathway leads from arachidonic acid to the leukotrienes. Like hormones, the eicosanoids exert specific physiological effects on target cells. However, they are distinct from most hormones in that they act locally, near their sites of synthesis, and they are catabolized extremely rapidly. Moreover, the actions of a given prostaglandin seem to vary in different tissues. The biological properties of the eicosanoids have led to great interest in their medical use and in uses of their analogs.

Some Historical Aspects

The most important early chapters in prostaglandin research were written in Sweden. In the mid-1930s Ulf von Euler discovered that lipid extracts of human semen contained active compounds that, on injection into animals, stimulated smooth muscle contraction or relaxation and affected the blood pressure. Because of their presumed origin in the prostate gland, he named these compounds *prostaglandins.* Later it was realized that these compounds are widely distributed in animal tissues. The first structural elucidations were reported in the late 1950s, under the direction of Sune Bergstrom and Bengt Samuelsson, and biosynthetic pathways were described in the mid-1960s, in Sweden and in the Netherlands.

The biological properties of the prostaglandins attracted intense interest in the pharmaceutical industry, but initial progress was limited by the low availabilities of these compounds. Interest reached a peak in 1971 with the discovery that aspirin inhibits one of the enzymes in prostaglandin biosynthesis. This inhibition is now known as the major site of action of aspirin and other nonsteroidal anti-inflammatory drugs. Later in the 1970s the thromboxanes and leukotrienes were discovered.

Structure

The first two prostaglandins to be isolated were called prostaglandins E and F, respectively, because of their preferential solubility in ether (E) or in phosphate buffer (F for *fosfat,* the Swedish word). We now denote these compounds by PGE and PGF, respectively; all other prostaglandins are denoted by a letter, for instance, PGA and PGH. Each prostaglandin has a cyclopentane ring and two side chains, with a carboxyl group in one side chain. A subscript numeral denotes the number of double bonds in the two chains. The most abundant prostaglandins, those synthesized from arachidonic acid, contain two double bonds; thus, PGE_2 would be the prostaglandin E derived from arachidonic acid. Finally, a subscript α indicates that the hydroxyl group at C-9 is on the same side of the ring as the carboxyl group. Structures of the most common prostaglandins are shown in Figure 18.30, along with that of thromboxane A_2 (TxA_2). TxA_2 is the only known naturally occurring thromboxane; it was originally isolated from thrombocytes,

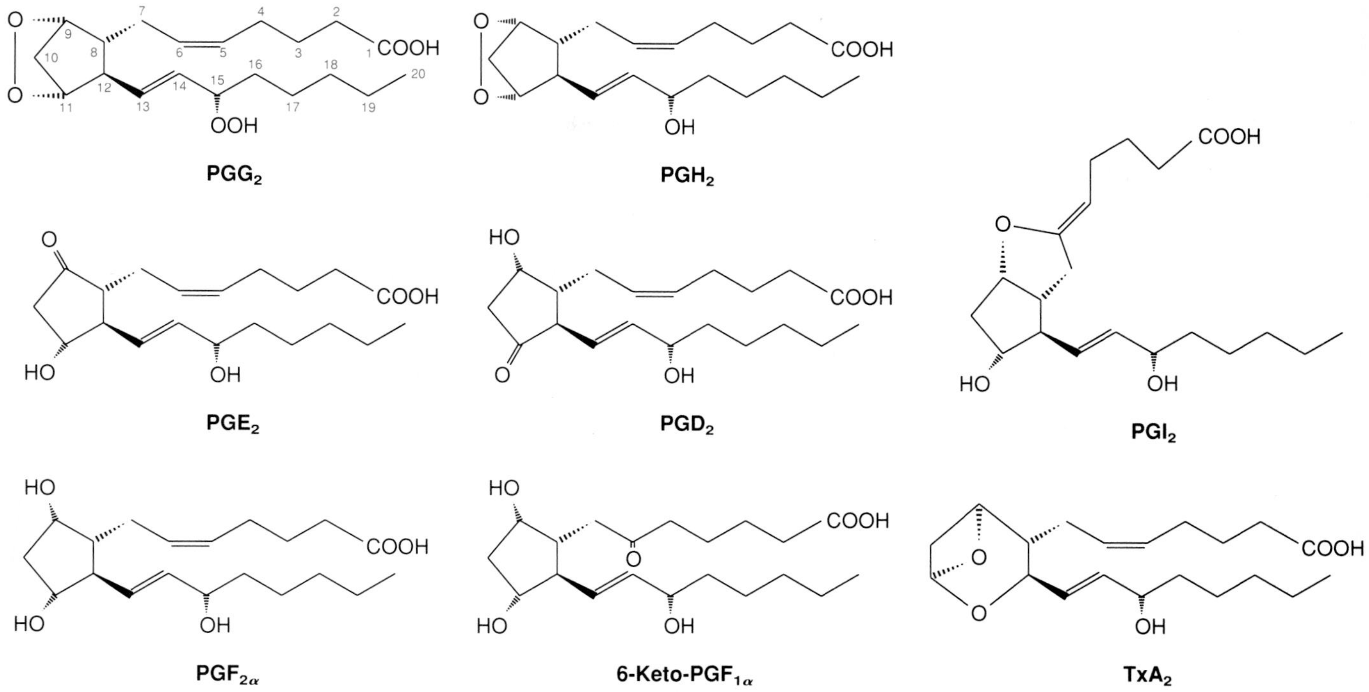

PGG_2 PGH_2 PGI_2 PGE_2 PGD_2 $PGF_{2\alpha}$ 6-Keto-$PGF_{1\alpha}$ TxA_2

Figure 18.30
Structures of the major prostaglandins and thromboxane derived from arachidonic acid. Numbering of carbons begins with the carboxyl group.

or blood platelets, as a compound that stimulated platelet aggregation, an early step in blood clotting. Note its structural resemblance to PGE_2, except for the cyclic ether ring. Another thromboxane, TxB_2, is a hydrolysis product of TxA_2.

Biosynthesis and Catabolism

Here we discuss only the biosynthesis of the 2-series of prostaglandins; the 1- and 3-prostaglandins are synthesized identically, from related C_{20} fatty acids. We can consider the biosynthetic pathways, which occur in endoplasmic reticulum, as three distinct stages: (1) release of arachidonic acid from membrane phospholipids; (2) oxygenation of arachidonate to yield PGH, a prostaglandin endoperoxide that serves as precursor to other prostaglandins; and (3) depending on the enzymes present in a cell, the conversion of PGH to other prostaglandins or to TxA_2. The biosynthetic pathways are summarized in Figure 18.31.

The release of arachidonic acid in stage 1 occurs as a result of tissue-specific stimuli by hormones such as bradykinin or epinephrine, or proteases such as thrombin. Pathological release can occur if membranes are perturbed; for example, the inflammation caused by bee stings is probably due to arachidonate release stimulated by the venom protein **melittin.** Release evidently involves the action of a specific phospholipase A_2 on phosphatidylcholine or phosphatidylethanolamine, yielding arachidonate, and the action of a phospholipase C on phosphatidylinositol, yielding a diacylglycerol, which in turn undergoes cleavage to give free arachidonate.

Free arachidonate is acted on, in stage 2, by **PGH synthase,** a bifunctional enzyme with two activities in a single heme-containing polypeptide chain. The first, a **cyclooxygenase,** introduces two molecules of O_2, one to form the ring and one to form a hydroperoxy group at C-15. The second

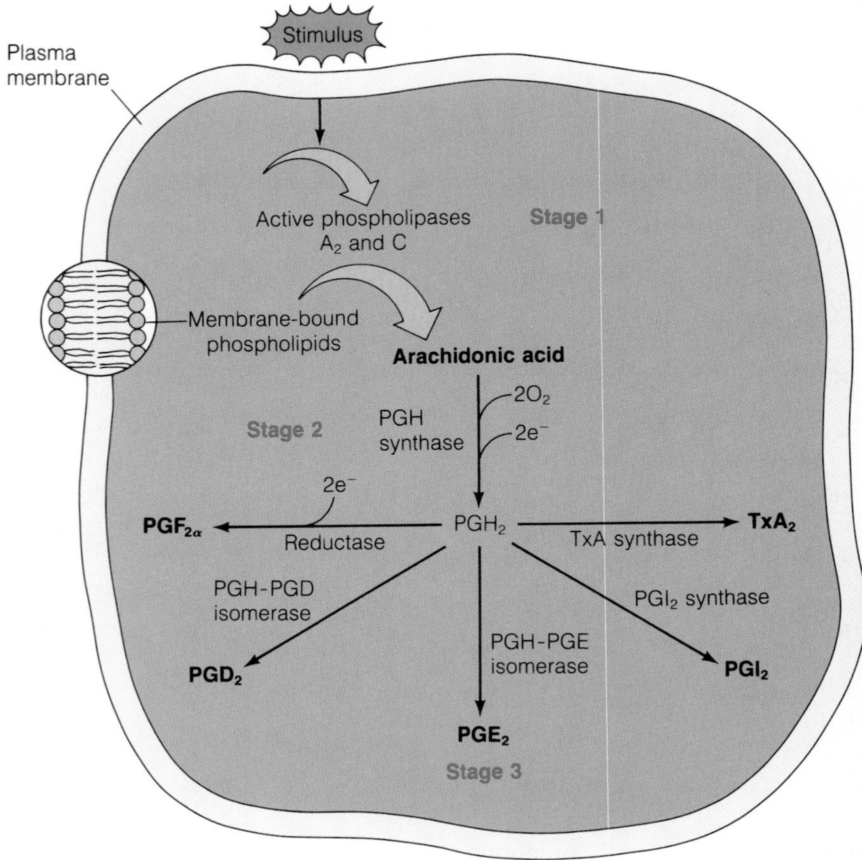

Figure 18.31
Summary of biosynthetic routes to the 2-series of prostaglandins (the most abundant group) and thromboxane A_2.

activity involves a two-electron reduction of the peroxide, to give PGH_2, with a hydroxyl group at C-15, as shown in Figure 18.32. The cyclooxygenase activity is the target for aspirin (acetylsalicylic acid), which acetylates a serine residue critical for activity and is in turn converted to salicylic acid.

In stage 3 a series of specific enzymes converts PGH_2 to other prostaglandins and to thromboxane A_2.

Another pathway leads from arachidonate to the leukotrienes. Leukotriene C was originally discovered in the class of white blood cells called polymorphonuclear leukocytes. It is a potent muscle contractant that is thought to be involved in the pathogenesis of asthma, through constriction of the small airways in the lung. Leukotrienes are formed from the initial attack on arachidonate of a **lipoxygenase,** which adds O_2 to C-5. A dehydration to give the epoxide coupled with isomerization of double bonds, followed by transfer of the thiol group of glutathione to one of the epoxide carbons, yields leukotriene C.

Leukotriene C

Subsequent modifications of the peptide chain yield related compounds, leukotrienes D and E.

All of the eicosanoids are metabolized extremely rapidly, with most failing to survive a single pass through the circulatory system. The lung is a major site of prostaglandin catabolism. The many catabolic pathways all seem to start with conversion to 15-keto-13,14-dihydro derivatives.

Arachidonic acid

Cyclooxygenase First oxygenation

Cyclooxygenase Second oxygenation

PGG_2

Reductase $2e^-$

PGH_2

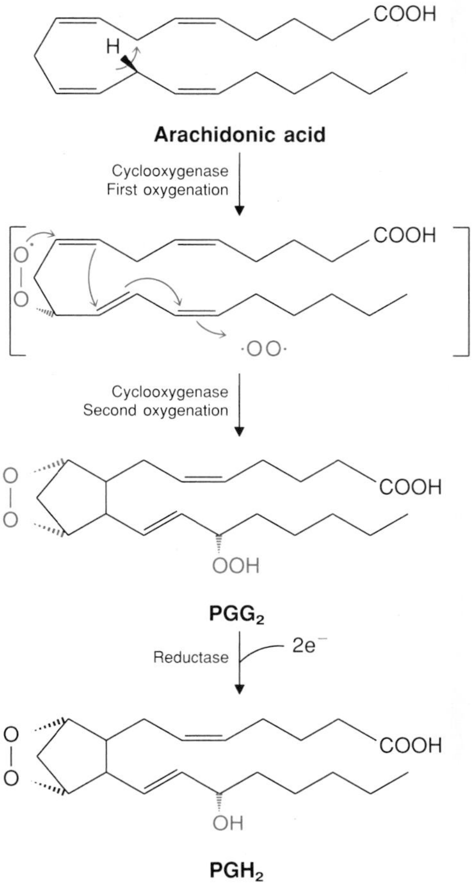

Figure 18.32
Probable mechanism for the cyclooxygenation of arachidonic acid by PGH synthase.

Biological Actions

As noted earlier, prostaglandins and their relatives can be considered as locally acting hormones. Evidently they act through binding to specific cellular receptors. We know relatively little about their subsequent effects at the molecular level, though there clearly are interactions with cyclic nucleotide metabolism. PGE stimulates adenylate cyclase in some cells, and PGF_2 has been reported to elevate levels of **cyclic guanosine monophosphate** in target cells.

Although little is known so far about molecular actions of the eicosanoids, knowledge of their physiology is being applied in useful ways. Since prostaglandin release is involved in the uterine muscle contraction that occurs in labor, PGF_2 is being used when it is necessary to induce labor in mothers at term. PGF_2 and PGE_2 are used as well to induce abortion in the second trimester, or to induce delivery in case of the death of a fetus. Prostaglandin derivatives are used in animal husbandry, to bring a group of female animals into heat simultaneously. PGI_2 is used to reduce the risk of blood clotting during cardiopulmonary bypass operations. PGE_1, a vasodilator, is being tested against various circulatory disorders, and forms of PGE, which also inhibit gastric secretion, are being used in treatment of stomach ulcers. Efforts in the pharmaceutical industry are devoted to developing longer-lived prostaglandin analogs.

REFERENCES

General References

Vance, D. E., and J. E. Vance (eds.) (1985) *Biochemistry of Lipids and Membranes,* Benjamin/Cummings, Menlo Park, Calif. This multiauthored textbook contains a number of timely and detailed reviews. Particularly relevant to this chapter are articles entitled Lipid metabolism in prokaryotes, pp. 73–115; Phospholipid metabolism in eukaryotes, pp. 242–270; Metabolism, regulation, and function of ether-linked glycerolipids, pp. 271–298; Sphingolipids, pp. 361–403; The eicosanoids, pp. 325–360; and Metabolism of cholesterol and lipoproteins, pp. 404–474.

Phospholipid Metabolism

Hanahan, D. J. (1986) Platelet activating factor: A biologically active phosphoglyceride. *Annu. Rev. Biochem.* 55:483–510. One of the most recent reviews of this highly potent phospholipid.

Low, M. G. (1989) Glycosyl-phosphatidylinositol: A versatile anchor for cell surface proteins. *FASEB* 3:1600–1608. This article summarizes new evidence for a role of phosphatidylinositol in anchoring cell surface proteins to membranes.

Stanbury, J. B., J. B. Wyngaarden, D. S. Fredrickson, J. L. Goldstein, and M. S. Brown (1989) *The Metabolic Basis of Inherited Disease,* 6th ed. McGraw-Hill, New York. Several chapters describe the biochemistry and clinical consequences of the lipid storage diseases.

Van Kujik, F. G. J. M., A. Sevanian, G. J. Handelman, and E. A. Dratz (1987) A new role for phospholipase A_2: Protection of membranes from lipid peroxide damage. *Trends Biochem. Sci.* 12:31–34. Summarizes recent evidence identifying phospholipase A_2 as a membrane repair enzyme.

Cholesterol Metabolism

Brown, M., and J. L. Goldstein (1986) A receptor-mediated pathway for cholesterol homeostasis. *Science* 232:34–47. Brown and Goldstein's Nobel Prize address, which chronicles the discovery and actions of the LDL receptor.

Kinsella, J. E. (1986) Dietary fish oils. *Nutrition Today* November/December: 7–14. A summary of the biochemical and nutritional aspects of n-3 fatty acids and their effects on lipid metabolism.

Eicosanoids

Needleman, P., J. Turk, B. A. Jakschik, A. R. Morrison, and J. B. Lefkowith (1986) Arachidonic acid metabolism. *Annu. Rev. Biochem.* 55:69–102. Several important regulatory molecules, including prostaglandins, thromboxanes, and leukotrienes, are derived metabolically from arachidonic acid; this review covers the field efficiently and completely.

Nelson, N. A., R. C. Kelly, and R. A. Johnson (1982) Prostaglandins and the arachidonic acid cascade. *Chem. Eng. News,* August 16:30–44. Written by three industrial scientists, this article focuses on the points of special interest to the pharmaceutical industry.

PROBLEMS

1. Would you expect the reaction catalyzed by cardiolipin synthase to be strongly exergonic or strongly endergonic? Explain your reasoning.

2. Phosphatidylserine (PS) is considered to be an intermediate in the biosynthesis of phosphatidylethanolamine (PE) in *E. coli,* yet PS is not found in appreciable amounts among *E. coli* membrane phospholipids. Since PS must be present in the membrane to serve as an intermediate, how might you explain its failure to accumulate to a significant extent? What kinds of experiments could test your proposed explanation?

3. Melittin is a protein in bee venom that activates phospholipase A_2. How might this effect contribute to the local inflammation that is caused by bee stings?

4. What would you expect to happen to levels of mevalonate in human plasma if one were to go from an ordinary to a vegetarian diet?

5. Write a balanced equation for the synthesis of *sn*-1-stearoyl-2-oleylglycerophosphorylserine, starting with glycerol, the fatty acids involved, and serine.

6. If mevalonate labeled with ^{14}C in the carboxyl carbon were administered to rats, which carbons of cholesterol would become labeled?

7. Which step in lipid metabolism would you expect to be affected by 3,4-dihydroxybutyl-1-phosphonic acid (shown below)? Explain your answer.

```
       CH2OH
        |
    HO—C—H        O
        |         ||
       CH2—CH2—P—O⁻
                  |
                  O⁻
```

8. Methyl-labeled [^{14}C]methionine at a specific activity of 2.0 millicuries per millimole was injected into rats. Six hours later the rats were sacrificed. Phosphatidylcholine was isolated from the liver and found to have a specific activity of 1.5 millicuries per millimole. Calculate the proportions of phosphatidylcholine synthesized by the phosphatidylserine pathway and by the pathway starting from free choline. What further information would you need for your calculated values to reflect the true rates of these processes?

9. Identify a pathway for utilization of the four carbons of acetoacetate in cholesterol biosynthesis. Carry your pathway as far as the rate-determining reaction in cholesterol biosynthesis.

CHAPTER 19

Photosynthesis

In earlier chapters, we have described in considerable detail the ways in which organisms extract, and store in ATP, a substantial portion of the energy available from the oxidation of carbohydrates. Using glucose as an example, we wrote for the overall reaction:

$$C_6H_{12}O_6 + 6O_2 \longrightarrow 6CO_2 + 6H_2O \qquad \Delta G^{\circ\prime} = -2870\,\frac{\text{kJ}}{\text{mol}}$$

and noted that as much as 40% of this energy could be recovered for useful biochemical work.

But life cannot depend on oxidative metabolism as its ultimate source of energy, and it cannot continue indefinitely returning organic carbon to the atmosphere as CO_2. The reaction above is only half of the great energy/carbon cycle of nature (Figure 19.1). The reverse of the carbohydrate oxidation reaction is accomplished by plants, algae, and some microorganisms, using the energy from sunlight to provide the enormous amount of free energy required:

$$6CO_2 + 6H_2O \xrightarrow{\text{Light energy}} C_6H_{12}O_6 + 6O_2$$

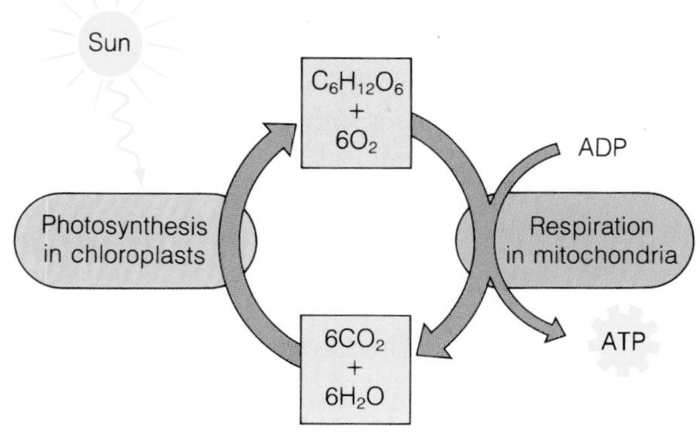

Figure 19.1
The carbon cycle in nature. Carbon dioxide and water are combined through photosynthesis in plants to form carbohydrates. In both plants and plant-eating animals, these carbohydrates are reoxidized to regenerate CO_2 and H_2O. Part of the energy is trapped in ATP.

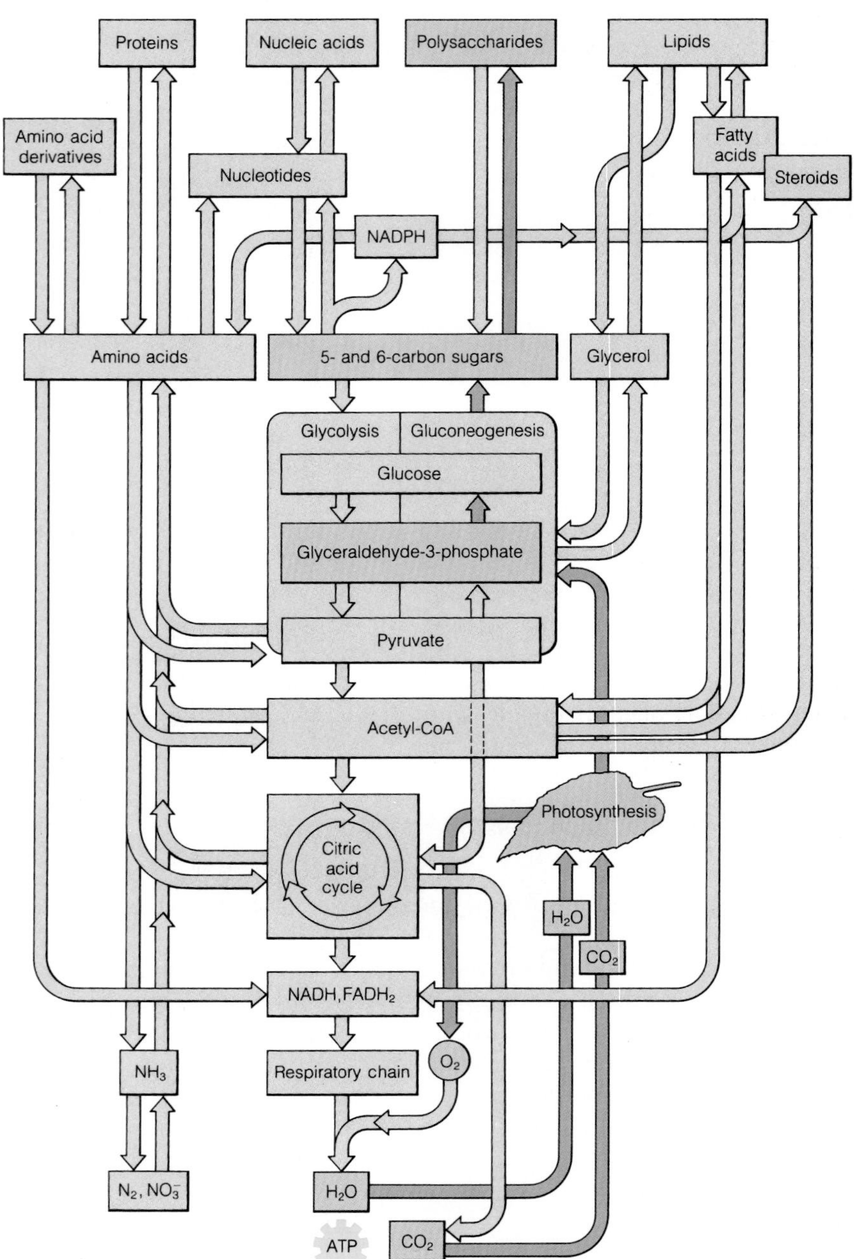

Figure 19.2
Role of photosynthesis (green) in metabolism.

This process is called **photosynthesis.** Not only does it provide carbohydrates for energy production in plants and animals; photosynthesis is also the major path through which carbon reenters the biosphere and is the principal source of oxygen in the earth's atmosphere. Indeed, it is believed that prior to the evolution of photosynthetic organisms the earth's atmosphere was devoid of oxygen (though rich in carbon dioxide). Prephotosynthetic organisms must have utilized abiotically synthesized energy-rich molecules for their metabolism. Without the advent of photosynthesis, these energy sources would have been wholly consumed, and life would have perished. The fossil record suggests that photosynthetic organisms first appeared about 3 billion years ago. Their gradual conversion of the primitive, nonoxidizing atmosphere of the earth to an oxidizing atmosphere paved the way for aerobic metabolism and the evolution of animals. Today, photosyn-

Table 19.1
Examples of some photosynthetic reactions

Organisms	Reductant	Reaction
Plants, algae, cyanobacteria	H_2O	$CO_2 + 2H_2O \longrightarrow [CH_2O] + H_2O + O_2$
Green sulfur bacteria	H_2S	$CO_2 + 2H_2S \longrightarrow [CH_2O] + H_2O + 2S$
Purple sulfur bacteria	$[HSO_3^-]$	$CO_2 + H_2O + 2[HSO_3^-] \longrightarrow [CH_2O] + 2[HSO_4^-]$
Nonsulfur photosynthetic bacteria	H_2 or many other reductants, such as lactate	$CO_2 + 2H_2 \longrightarrow [CH_2O] + H_2O$ $CO_2 + 2\left(\begin{array}{c} CH_3 \\ \vert \\ HC{-}OH \\ \vert \\ COO^- \end{array}\right) \longrightarrow [CH_2O] + H_2O + 2\left(\begin{array}{c} CH_3 \\ \vert \\ C{=}O \\ \vert \\ COO^- \end{array}\right)$ (lactate) (pyruvate)

thesis represents the ultimate source of energy for *almost** all life. It is used by plants, algae, and a wide variety of prokaryotes; all of these are food sources for other organisms. A comprehensive view of the relationship of photosynthesis to other pathways we have studied is shown in Figure 19.2.

The Basic Processes of Photosynthesis

The equation we have written above for the photosynthetic reaction is, of course, a great oversimplification. As you might expect, the actual process of photosynthesis involves many intermediate steps. Furthermore, a hexose itself is not the primary carbohydrate product, as we shall soon see. Therefore, the photosynthetic reaction is usually written in the more general form:

$$CO_2 + H_2O \xrightarrow{\text{Light energy}} [CH_2O] + O_2$$

where $[CH_2O]$ represents a general carbohydrate.

Since the burning of carbohydrates to form CO_2 is an oxidative process, converting CO_2 into carbohydrate must involve a *reduction* of the carbon. The ultimate reducing agent is usually H_2O, but there are photosynthetic processes in many bacteria that use other reductants. Thus, an even more general reaction can be written:

$$CO_2 + 2H_2A \xrightarrow{\text{Light energy}} [CH_2O] + H_2O + 2A$$

where H_2A is a general reductant and A is the oxidized product. Examples are given in Table 19.1. Comparison of these reactions suggests that the source of the oxygen released in photosynthesis by plants, algae, and cyanobacteria is H_2O, rather than CO_2. This was predicted in the early 1930s by

* Recent developments make it necessary to qualify the statement that *all* life derives its energy, directly or indirectly, from photosynthesis. It has been found that some bacteria, such as those associated with submarine biothermal vents, use the oxidation of substances like H_2S or H_2 as an alternate energy source, in the complete absence of light. This represents, however, only a small fraction of the energy flow in the biosphere.

C. B. van Niel, one of the pioneers in photosynthesis studies, and was confirmed by isotope labeling experiments using ^{18}O-labeled water and CO_2. It was found that neither of the oxygen atoms in O_2 comes from CO_2. Therefore, it is more correct to write the photosynthetic reactions in the fashion shown in Table 19.1, which makes it clear that one of the oxygens from CO_2 ends up in carbohydrate, the other in water:

$$CO_2 + 2H_2O \xrightarrow{\text{Light energy}} [CH_2O] + H_2O + O_2$$

However, there is no way in which light energy can be *directly* utilized to drive such a reaction, and H_2O does not *directly* reduce CO_2 under any known circumstances. What happens, in fact, is that the overall process is both chemically and physically separated into two subprocesses in photosynthetic organisms. A slightly more sophisticated version of what actually happens is shown in Figure 19.3. Energy from sunlight is used to carry out the photochemical oxidation of H_2O, in a series of steps called the **light reactions.** Two things are accomplished by this oxidation. First, the oxidizing agent $NADP^+$ is reduced to NADPH, producing reducing equivalents, and O_2 is released. Second, part of the energy from sunlight is captured by phosphorylating ADP to produce ATP. This process, which is called **photophosphorylation,** was one of the several major discoveries in this field by D. I. Arnon and his colleagues. Both NADPH and ATP are then utilized in the reductive synthesis of carbohydrate from CO_2 and water in the so-called **dark reactions** of photosynthesis. These reactions were originally named to emphasize the fact that they do not require the direct participation of light energy. Though this is true, the name carries the unfortunate implication that these synthetic reactions occur only in the dark. Nothing could be further from the truth—they occur at all times and are actually accelerated by light. Since the name "dark reactions" is well established, we shall retain it, but you should not be misled by it.

Before considering details of either the light reactions or dark reactions, it is appropriate to consider the sites of photosynthesis. Just as oxidative metabolism in eukaryotic cells is localized in mitochondria, there are organelles specialized for photosynthesis.

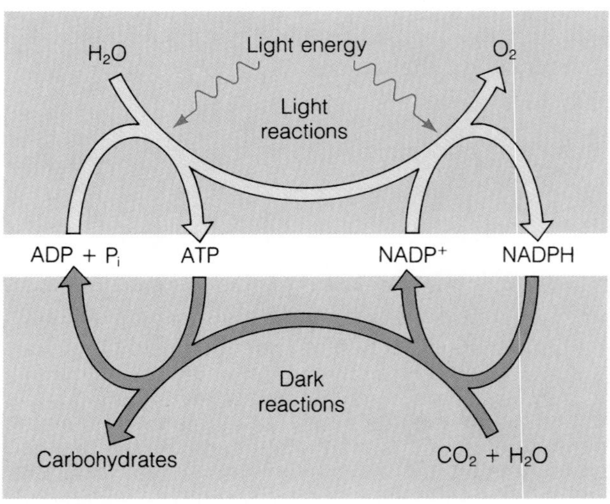

Figure 19.3
Overall process of photosynthesis, divided into "light" reactions and "dark" reactions. The light reactions require visible light as an energy source and produce reducing power (in the form of NADPH) and ATP to drive the synthetic dark reactions, which fix CO_2 into carbohydrates.

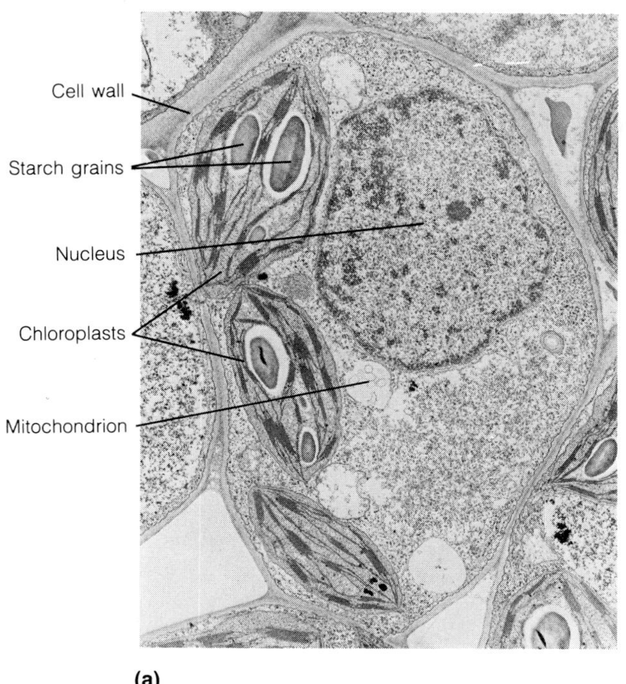

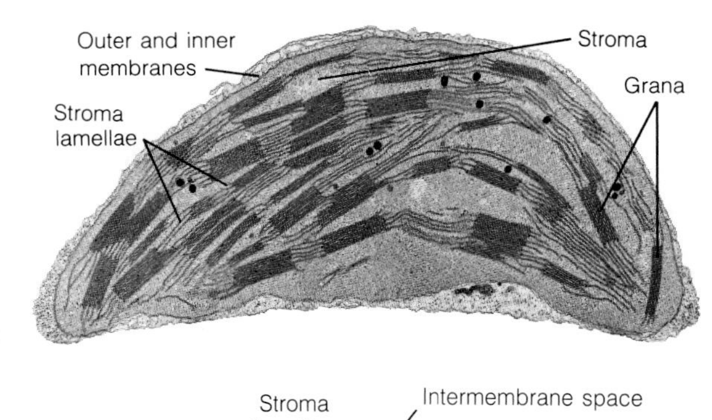

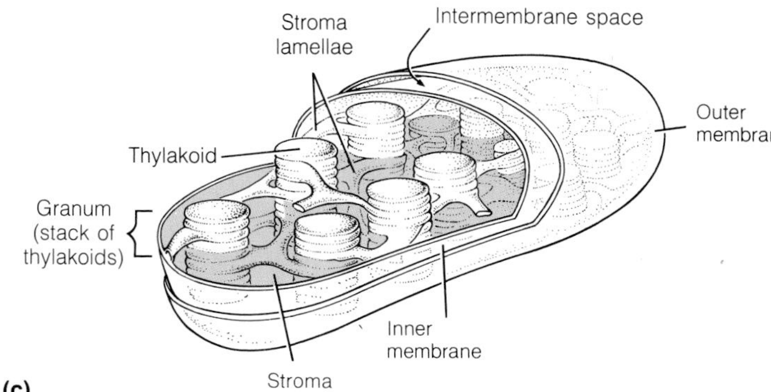

Figure 19.4
Chloroplasts, the photosynthetic organelles of green plants and algae. (**a**) Several chloroplasts are shown in a cross section of a parenchymal cell from a *Coleus* leaf. (**b**) Enlarged view of a single chloroplast from a leaf of timothy grass. (**c**) Schematic rendering of a chloroplast.

The Chloroplast

In all higher plants and algae, photosynthetic processes are localized in organelles called **chloroplasts.** In plants, most of the chloroplasts are found in **mesophyll** cells just under the leaf surface; each cell may contain 20 to 50 of these organelles (Figure 19.4a). The true algae also have chloroplasts, but often only one very large one is found in each cell.

Like mitochondria, chloroplasts are semiautonomous, carrying their own DNA to code for some of their proteins and the ribosomes necessary for translation of the appropriate messenger RNAs. There is now much evidence that chloroplasts evolved from unicellular organisms similar to cyanobacteria (blue-green algae). Such prokaryotic photosynthesizers do not contain chloroplasts, but have membrane structures that play the same roles as chloroplast membranes (Figure 19.5). To a certain extent, the cyanobacteria resemble free-living chloroplasts.

The internal structure of a chloroplast, as shown in Figure 19.4b and c, bears some resemblance to that of a mitochondrion. There are an outer, freely permeable membrane and an inner membrane that is selectively permeable. The inner membrane encloses a material analogous to the mitochondrial matrix—this is called the **stroma.** Immersed in the stroma are many flat, saclike membrane structures called **thylakoids,** which are often stacked like coins to form units called **grana** (see Figure 19.4c). Individual grana are irregularly interconnected by thylakoid extensions called **stroma lamellae.** The thylakoid membrane encloses an interior space, the lumen of the thylakoid.

The division of labor within a chloroplast is simple. Absorption of light and all of the light reactions occur within or on the thylakoid membranes. The ATP and $NADPH^+$ produced by these reactions are released into the surrounding stroma, where all of the synthetic dark reactions occur.

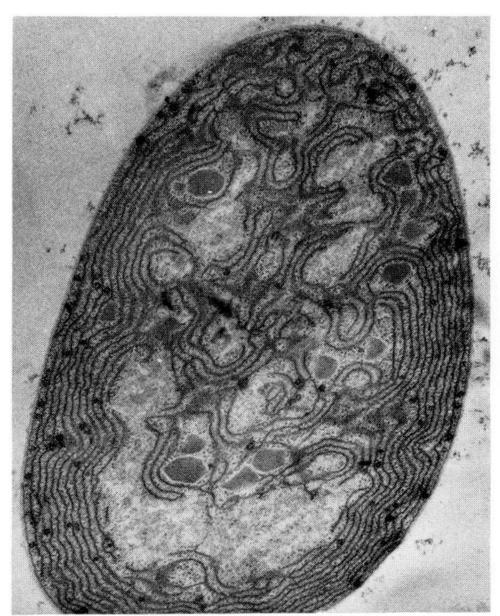

Figure 19.5
A cyanobacterium. This electron micrograph of a thin section of *Anabaena azollae* shows the folded membranes, which resemble the thylakoids of chloroplasts.

Obviously, there are analogies in structure and role between mitochondrial matrix and chloroplast stroma and between the inner membrane of the mitochondrion and the thylakoid membrane of the chloroplast. Indeed, we shall find that a very similar kind of chemiosmotic ATP generation is carried out across these membranes in both mitochondria and chloroplasts. To see how this occurs, we must first examine the light reactions in detail, beginning with the process of light absorption.

The Light Reactions

Absorption of Light: The Light-Harvesting System

To understand how the energy from sunlight can be captured and utilized, we must first examine the nature of electromagnetic radiation. The quantum mechanical theory of radiation states that light has two aspects: wavelike and particlelike. We can, for example, characterize a particular kind of radiation by its wavelength (λ) or frequency (ν). If waves with a length of λ are passing an observer at a velocity c, the number of waves passing per second is the frequency, ν. Thus

$$\nu = c/\lambda \tag{19.1}$$

where c is the velocity of light, 3×10^8 m/s.

The red light from a neon laser has a wavelength of 632.8 nm, or 6.328×10^{-7} m. Thus, its frequency will be $4.74 \times 10^{14}\ \mathrm{s}^{-1}$. These parameters—$\lambda$ and ν—characterize the *wave* aspects of the laser light. But to see how energy might be obtained from this light, it is necessary to consider the beam as a stream of light particles or **photons.** Each photon has a unit, or **quantum,** of energy associated with it, and the energy per photon is related to the frequency of the light by Planck's law:

$$E = h\nu \tag{19.2}$$

where h is Planck's constant, 6.626×10^{-34} J s. Thus, a neon laser can deliver light energy only in packets of 3.14×10^{-19} J. Since we are interested here in how radiation can promote chemical or biochemical processes, which are usually expressed on a molar basis, it is more appropriate to ask what energy a *mole* (6.02×10^{23}) of photons would have. For the neon laser light, this comes out to be 189 kJ. A mole of photons is called one **einstein.**

Figure 19.6 shows a graph of energy per mole of photons (einstein) as a function of wavelength, through the infrared, visible, and ultraviolet parts of the spectrum. For comparison, the energies associated with molecular vibrations and various covalent bonds are indicated. Photons of far-infrared radiation can do little except stimulate molecular vibrations, which appear as heat. Radiation in the far ultraviolet, on the other hand, comes in quanta with energies quite capable of breaking covalent bonds if absorbed by a molecule. Such radiation is chemically destructive, but fortunately most of it is screened from the earth's surface by the ozone layer. This is why threatened depletion of the ozone layer is of such serious concern.

Photosynthesis depends primarily on light in the visible and near-infrared regions of the spectrum, lying between the extremes described above. The ability to utilize radiation in this range has had clear evolutionary advantages for photosynthetic organisms. Most of the sun's energy that

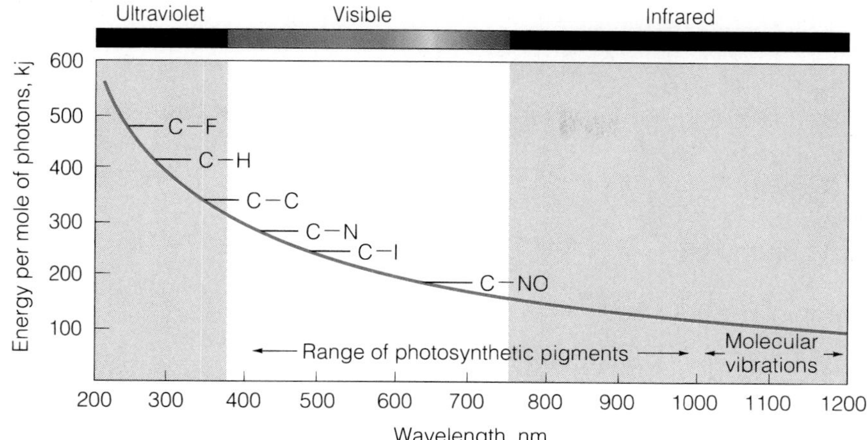

Figure 19.6
Energy per mole of photons (einstein) as a function of wavelength, compared with energies of several chemical bonds. Light in the UV and visible range has enough energy to directly break chemical bonds, whereas light in the long-wavelength portion of infrared region of the spectrum only causes heat-producing molecular vibrations.

reaches the earth's surface lies in these regions. The small amount of ultraviolet radiation that does get through can penetrate only a very short distance into water and thus would have been unavailable to primitive photosynthetic organisms living in the sea. The photons of far-infrared radiation have energies too low to be useful for any photochemical processes.

In response to this situation, photosynthetic organisms have evolved a set of pigments that efficiently absorb visible and near-infrared light. These are sometimes referred to as **chromophores**—entities that absorb light of specific wavelength. Structures of a number of the most important photosynthetic chromophores are shown in Figure 19.7. In Figure 19.8 the ab-

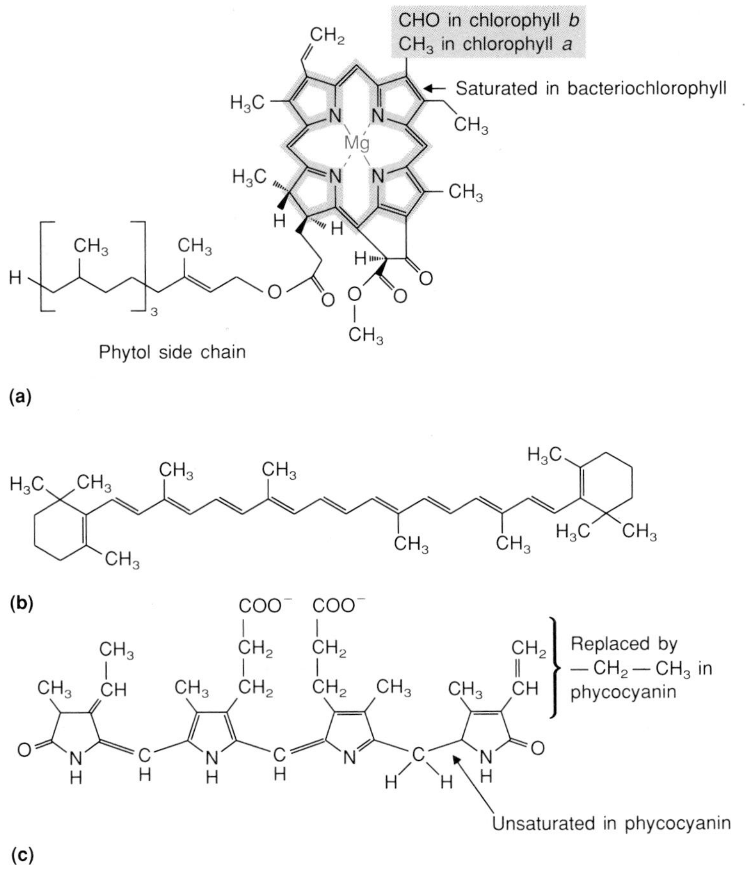

Figure 19.7
Photosynthetic pigments. (**a**) Chlorophylls; (**b**) β-carotene; (**c**) phycoerythrin and its analog, phycocyanin.

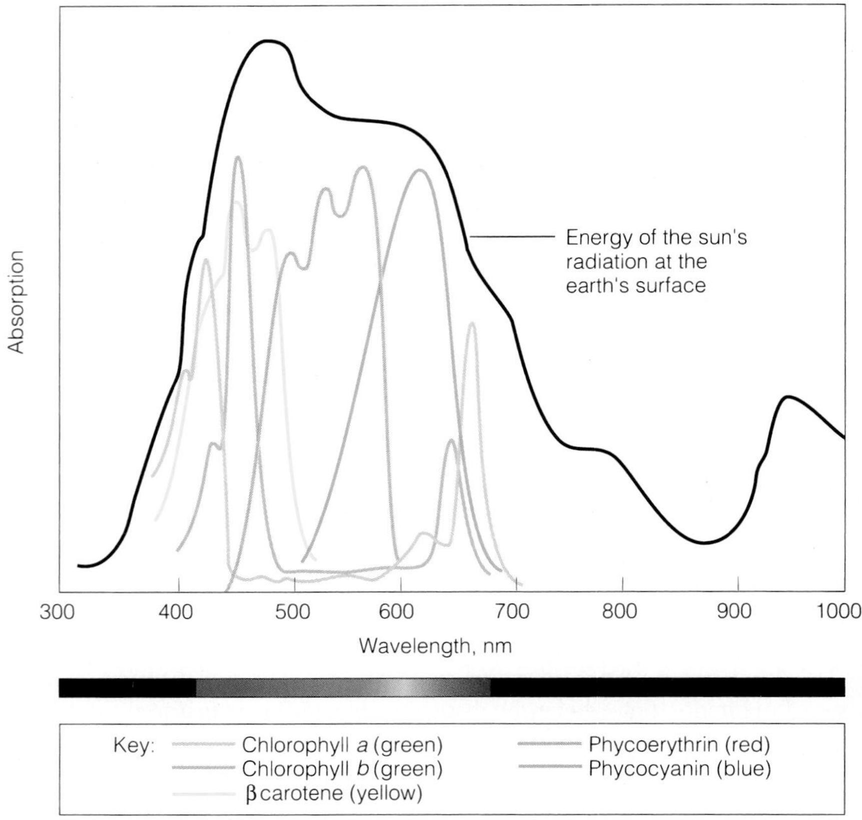

Figure 19.8
Absorption spectra of various plant pigments, compared with the spectral distribution of the light energy from the sun that reaches the earth's surface.

sorption spectra of these pigments are compared with the distribution of solar radiation in the spectrum. Together, they "blanket" the visible spectrum; scarcely a photon can come through that cannot be absorbed by one or the other. The most abundant pigments in higher plants are chlorophyll *a* and chlorophyll *b*. As Figure 19.7 shows, these molecules are related to the photoporphyrin IX found in hemoglobin and myoglobin (see Figure 7.1 in Chapter 7). However, the bound metal in the chlorophylls is Mg^{2+}, rather than Fe^{2+}. Molecules with large conjugated double-bond systems, like the chlorophylls or any of the accessory pigments, absorb light in the visible region of the spectrum. Since chlorophylls *a* and *b* absorb strongly in both the deep blue and red, the light that is *not* absorbed, but reflected, from leaves is green, the color we associate with most growing plants. The other colors exhibited, such as the red, brown, or purple of algae and photosynthetic bacteria, are accounted for by differing amounts of accessory pigments. Loss of chlorophylls in autumn leaves allows the colors of the accessory pigments to become evident. Some photosynthetic bacteria utilize pigments that absorb up to $\lambda = 1000$ nm, in the near infrared.

Chlorophyll and the accessory pigments are all contained in the **thylakoid membranes.** These membranes are somewhat unusual in composition: They contain only a small fraction of the usual phospholipids but are rich in glycolipids. They also contain much protein, and the photosynthetic pigments are bound by certain of these proteins. It appears likely that at least some of the pigments, such as chlorophylls *a* and *b*, also interact with the membrane lipids, for they carry long hydrophobic tails.

The assemblies of light-harvesting pigments in the thylakoid membrane, together with their associated proteins, are organized into well-defined **photosystems.** Each photosystem represents a structural unit dedicated to the task of absorbing a light photon and recovering some of its energy in a chemical form. To understand this, we must look a bit more closely into what can happen when a molecule absorbs a quantum of radiant energy. Recall from Tools of Biochemistry 10 that absorption in the visible region of the spectrum results in the molecule being excited from the ground state to a higher electronic state. In the case of the photosynthetic pigments, the electron so excited is one occupying a π orbital in the conjugated bond system. We described two ways in which the energy could be lost to return the molecule to its ground state: radiationless dissipation of the energy as heat, or reradiation as fluorescence. However, when similar absorbing molecules are packed tightly together, as in the photosystems, two other possibilities arise. First, the excitation energy may be passed from one molecule to an adjacent one—a process called **resonance transfer** (Figure 19.9a). Alternatively, the excited electron itself may be passed to a nearby molecule with a slightly lower excited state—an **electron transfer** reaction (Figure 19.9b). Both of these processes are important in photosynthesis.

The clue that eventually led to the recognition that resonance transfer played a role in photosynthesis came from measurements by Robert Emerson and William Arnold in the 1930s. They showed that when the photosynthetic system of the alga *Chlorella* was operating at maximum efficiency, only one O_2 molecule was evolved for every 2500 chlorophyll molecules. Most of the chlorophyll molecules are not directly engaged in the photochemical process itself but act, instead, as **antenna molecules,** absorbing photons and transferring the energy to a relatively few **reaction centers.** In other words, the energy of a photon absorbed by any antenna molecule in a photosystem wanders about the system in a random fashion (Figure 19.10). Eventually (meaning in about 10^{-10} s), the energy finds its way to a chlorophyll molecule in the reaction center. This molecule is like the other chlorophylls, but it is in a somewhat different environment so that its excited state energy level is a bit lower. Thus, it acts as a "trap" for quanta of energy absorbed by any of the other pigment molecules. It is the excitation of this reaction center that begins the actual photochemistry of the light reactions.

Figure 19.9
Two modes of energy transfer following photoexcitation. (**a**) Resonance energy transfer. Molecule I is excited by absorption of a photon of radiation. The energy transfers to adjacent molecule II, exciting this molecule as molecule I falls back to the ground state. (**b**) Electron transfer. An excited electron in molecule I is transferred to the slightly lower excited state of an adjacent molecule II. This leaves molecule I as a cation, molecule II as an anion.

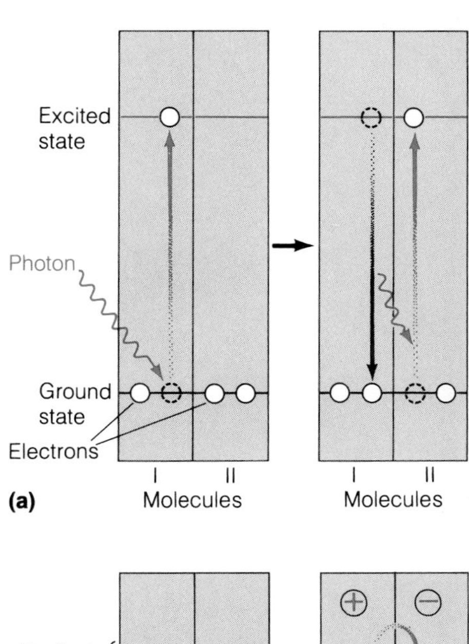

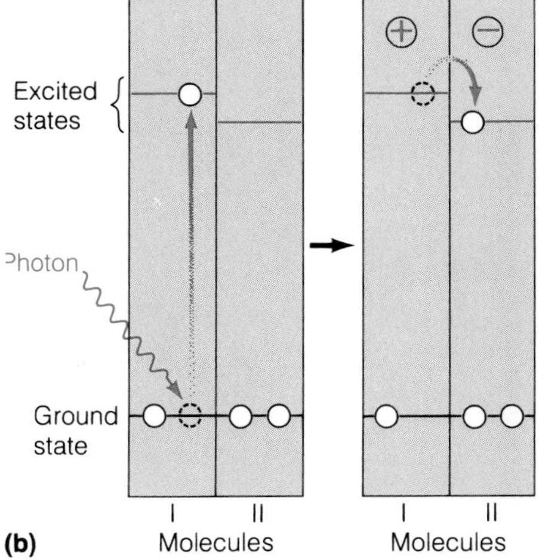

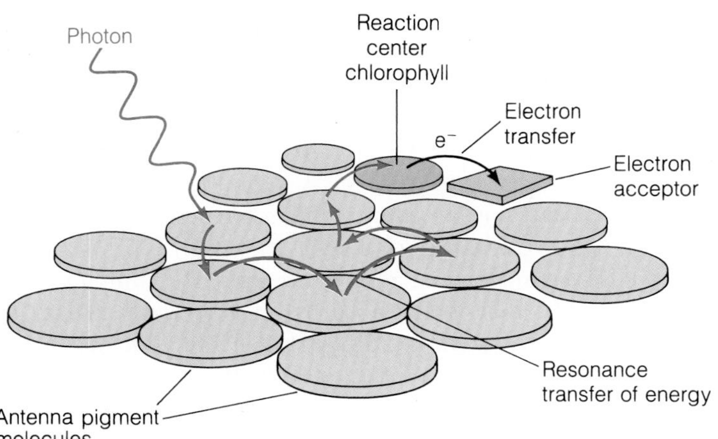

Figure 19.10
Resonance transfer of light energy among the antenna molecules of a photosystem. The excitation energy wanders from one molecule to another, until it reaches a reaction center. At the reaction center, electron transfer to an acceptor molecule occurs, and the energy is trapped.

The Photochemistry: Two Photosystems in Plants and Algae

Our understanding of the photochemical light reactions has developed from many elegant experiments in many different laboratories. A pioneering study in 1939, by Robert Hill at the University of Cambridge, showed that, when illuminated in the presence of any of a variety of electron acceptors, isolated chloroplasts could promote reduction. For example, when ferric ion is used, the reaction

$$4Fe^{3+} + 2H_2O \xrightarrow{\text{Light energy}} 4Fe^{2+} + 4H^+ + O_2$$

proceeded efficiently. A number of such reactions are known involving different inorganic oxidants; they are now referred to collectively as "Hill reactions." Such reactions, under normal circumstances, are very unfavorable; Fe^{3+}, for example, is a much weaker oxidant than O_2, and equilibrium should lie far to the left. Hill's discoveries showed that chloroplasts irradiated by light are capable of driving thermodynamically unfavorable reactions. The Hill reactions also demonstrated that the photosynthetic system can oxidize water to O_2 without any involvement of CO_2. This was the first clear indication that the light and dark reactions are separate processes, and it led ultimately to the discovery that the final electron acceptor in vivo is $NADP^+$, yielding NADPH.

Further studies revealed that *two* kinds of photosystems must be involved in photosynthesis in plants. The first hint came from experiments that measured the quantum efficiency of photosynthesis in algae. The **quantum efficiency** (*Q*) is the ratio of oxygen molecules evolved to photons absorbed. As the wavelength of the monochromatic light used was raised to about 700 nm (far red), an abrupt drop in *Q* was noted (Figure 19.11). This "red drop" was a strange observation, for chlorophylls in plants still show appreciable absorbance even at higher wavelengths. More remarkable was the observation that simultaneous illumination with yellow light (650 nm) produced a marked increase in the quantum yield at 700 nm. Even if the yellow light was switched off a few minutes before the measurement, the quantum yield still increased. The only reasonable explanation for these results is that two complementary photosystems exist, one absorbing most strongly at wavelengths around 700 nm and the other at shorter wavelengths. The action of *both* must be required for photosynthesis to proceed with maximal efficiency. These two photosystems have now been identified and characterized. Each photosystem is a multisubunit, transmembrane protein complex, carrying antenna pigments, reaction center chlorophyll molecules, and electron transport agents.

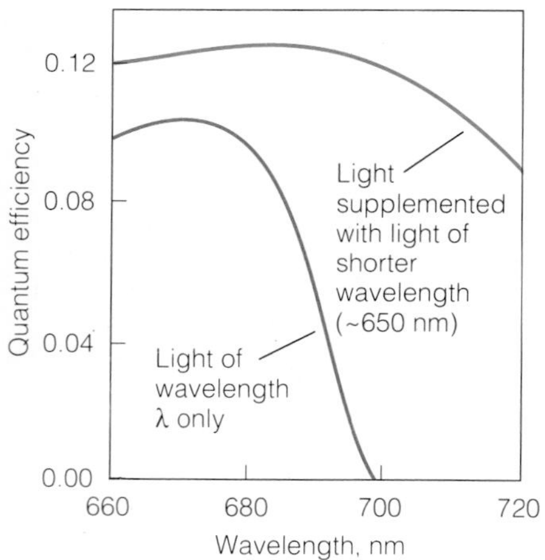

Figure 19.11
The red drop. Measurement of the quantum efficiency (*Q*) using monochromatic light shows a sharp drop in *Q* at about 700 nm, even though absorbance by chloroplasts continues above this wavelength. Supplementing the long-wavelength radiation with radiation at a shorter wavelength abolishes the red drop effect.

The photosystem showing absorbance up to 700 nm has been named **photosystem I (PSI)**; that which absorbs only to a wavelength of about 680 nm is called **photosystem II (PSII)**. In algae and all higher plants, these two photosystems are linked in series to carry out the complete sequence of the light reactions. The basic principles are illustrated in Figure 19.12, which depicts the path of electrons through the systems and places the major participants on a scale of reduction potential.

In both photosystems, the primary step is transfer of a light-excited electron from a reaction center (P680 or P700) into an electron transport chain. The ultimate source of electrons is the water molecule shown to the left in Figure 19.12; the final destination of the electrons is the molecule of

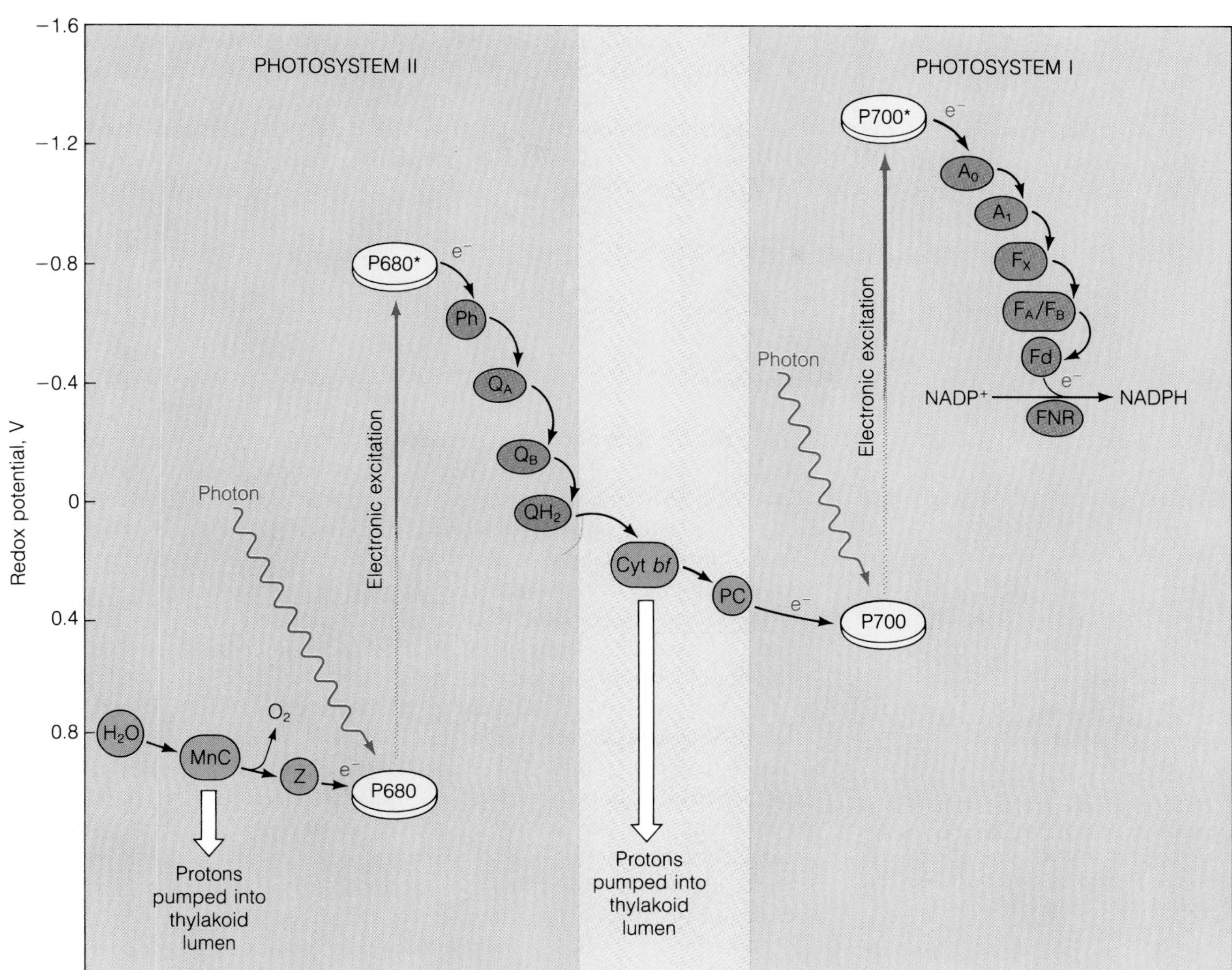

Figure 19.12
The two-system photosynthetic light reactions. Sometimes called the Z-scheme because of its pattern, this is the mode of photosynthesis used by higher algae and plants and by cyanobacteria. Basically, it is a mechanism for transferring electrons from water to $NADP^+$, using absorption of two photons to provide the energy. During the process, protons are pumped across the thylakoid membrane. MnC = manganese center; Z = donor to P680—probably a tyrosine radical; P680 = photosystem II reaction center chlorophyll; Ph = pheophytin acceptor; Q_A, Q_B = protein-bound plastoquinones; QH_2 = plastoquinol (reduced plastoquinone) in membrane; Cyt *bf* = cytochrome *bf* complex; PC = plastocyanin; P700 = photosystem I reaction center chlorophyll; A_0 = chlorophyll acceptor; A_1 = protein-bound phylloquinone; F_A, F_B, F_X = iron–sulfur clusters; Fd = ferredoxin; FNR = ferredoxin:$NADP^+$ oxidoreductase.

$NADP^+$ at the right, which is thereby reduced to NADPH. Along the way protons are pumped across the thylakoid membrane to produce a gradient. It should not surprise you to find that this proton gradient drives ATP production, since such a gradient produces ATP in mitochondria (Chapter 15). Thus, ATP and reducing power in the form of NADPH are the products of the light reactions. These are exactly what is needed to drive the syntheses carried out in the dark reactions. To examine these processes in detail, we begin with photosystem II, for this is where electrons enter the scheme.

Photosystem II

It is easiest to follow the events in a photosystem by starting with the absorption of a photon. A photon picked up by the light-harvesting system of photosystem II is funneled to a reaction center chlorophyll, designated P680 in Figure 19.12. Excitation of P680 raises the molecule from the ground state to an excited state at −0.8 volts. Thus, the excited P680 has become an excellent reducing agent, able to transfer quickly the excited electron to a lower-energy recipient—**pheophytin (Ph)**. The pheophytins are molecules identical to chlorophylls, except that two protons substitute for the centrally bound magnesium atom.

The electron is then transferred to a series of **plastoquinone** molecules (Q_A and Q_B) associated with PSII proteins. Ultimately, two electrons and two protons are picked up by the plastoquinone Q_B, and the reduced molecule is released into the lipid portion of the thylakoid membrane. The overall reduction of plastoquinone can be written:

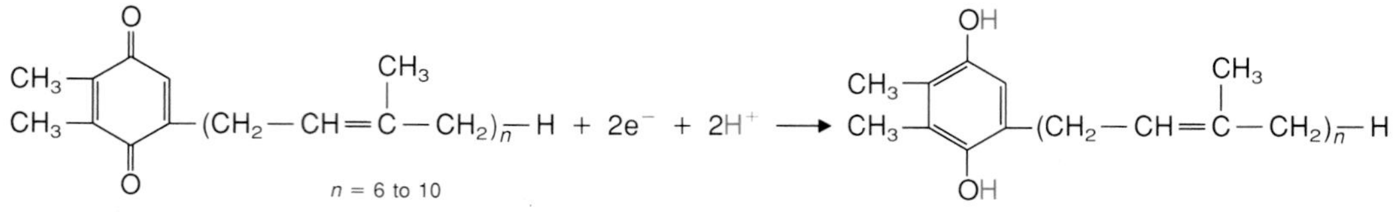

$$\text{Plastoquinone } (CH_3)_2C_6O_2-(CH_2-CH=C(CH_3)-CH_2)_n-H + 2e^- + 2H^+ \longrightarrow \text{Plastoquinol } (CH_3)_2C_6(OH)_2-(CH_2-CH=C(CH_3)-CH_2)_n-H$$

n = 6 to 10

Plastoquinone **Plastoquinol**

The reduced plastoquinone (**plastoquinol**) then interacts with a membrane-bound complex of cytochromes and iron–sulfur proteins, **cytochrome *bf*** (cyt *bf*). This complex catalyzes the transfer of the electrons to a copper protein, **plastocyanin** (Figure 19.13) and concomitantly pumps the two protons into the thylakoid lumen. In this function, cytochrome *bf* is playing a role analogous to that of the cytochrome reductase complex in mitochondria, which it resembles. Plastocyanin, a mobile protein in the thylakoid lumen, passes the electrons on to P700 reaction centers; we shall consider the importance of this process later.

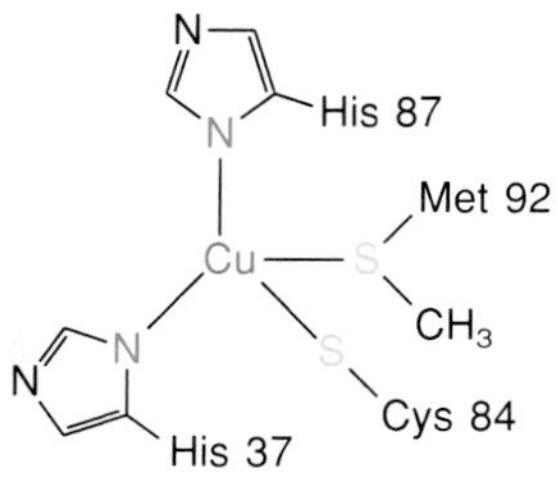

His 87
Met 92
Cu
CH_3
Cys 84
His 37

Figure 19.13
The electron-transfer copper center in the protein plastocyanin. The copper, which can alternate between +1 and +2 oxidation states, is held by two histidines and two sulfur-containing side chains.

You will have noted that the processes we have described so far have left P680 reaction centers deficient in electrons—in other words, oxidized to a strong oxidant, $P680^+$. These electrons are regained from water, as shown in the leftmost portion of Figure 19.12. The electron acceptor is a protein containing a cluster of four manganese atoms (MnC). This metalloprotein can exist in a series of overall oxidation states, which allows it to bind two water molecules, extract a total of four electrons from them, and pass the electrons on, through an intermediate called Z, to oxidized $P680^+$ reaction centers (Figure 19.14). In the process, the four protons are pumped from the stroma into the thylakoid lumen. Oxygen is released at this point and diffuses out of the chloroplast. We may summarize the reaction carried out by photosystem II as:

$$2H_2O \xrightarrow{\text{Light energy}} 4H^+ + 4e^- + O_2$$

with the understanding that the electrons have been delivered to plastocyanin, and protons have been pumped from the stroma into the thylakoid lumen.

Photosystem I

We have seen that in plants, which utilize two photosystems, photosystem II accomplishes the splitting of water, evolution of O_2, and generation of a proton gradient across the thylakoid membrane. However, the electrons from the water have not yet reached their final destination in NADPH. This is the task of photosystem I.

Photosystem I contains a reaction center chlorophyll, P700, and can absorb light up to 700 nm. Excitation by a photon absorbed by antenna chlorophylls raises P700 from a ground state to an excited state at about −1.3 V. As shown in Figure 19.12, electron transfer occurs; the excited electron is first passed through a chlorophyll acceptor (designated A_0), then through a molecule of phylloquinone (A_1, see p. 636), and finally through a series of three iron–sulfur proteins (F_X, F_B, and F_A). These proteins contain iron–sulfur clusters of the kinds depicted in Figure 15.4. Finally, the electron is transferred to **soluble ferredoxin** (Fd), which is present in the stroma. The enzyme (ferredoxin:$NADP^+$ oxidoreductase) catalyzes the transfer of electrons to $NADP^+$, after ferredoxin has been reduced by PSI.

$$2\text{Fd (red)} + H^+ + NADP^+ \xrightarrow[\text{reductase}]{} 2\text{Fd (ox)} + NADPH$$

In a sense ferredoxin, rather than $NADP^+$, can be considered the *direct* recipient of electrons from the pathway. Although much of the reduced ferredoxin is used to reduce $NADP^+$, some is used for other reductive reactions, which we discuss below.

The electrons that have been driven through photosystem I had their origin in electron transfer from P700 reaction centers. The oxidized reaction centers ($P700^+$) produced must be resupplied with electrons for photosynthesis to continue. In two-system photosynthesis, these are provided from photosystem II via plastocyanin.

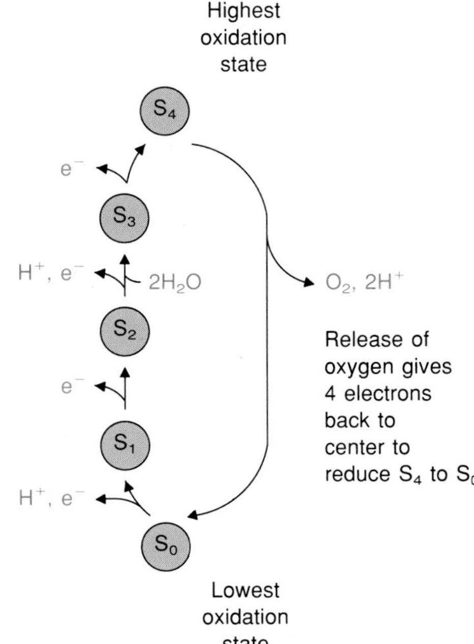

Figure 19.14
Cycle of oxidation states in the Mn center. This protein-bound cluster contains four Mn atoms. It is capable of four oxidation states, S_0 to S_4. It gives up four electrons sequentially, withdrawing them from the oxygens of bound water and releasing O_2. The center allows one electron at a time to pass to Z and then to P680.

Summation of the Two Systems: The Overall Reaction and ATP Generation

We can now summarize the entire scheme of noncyclic electron flow shown in Figure 19.12. Electrons are taken from water and end up in NADPH. We wrote for the reaction in photosystem II:

$$2H_2O \xrightarrow{\text{Light energy}} 4H^+ + 4e^- + O_2$$

The reactions of photosystem I, if written for four electrons and with all intermediates eliminated, are

$$4e^- + 2H^+ + 2NADP^+ \longrightarrow 2NADPH$$

Adding these two reactions, we find the following as a summation of the light reactions:

$$2H_2O + 2NADP^+ \xrightarrow{\text{Light energy}} 2H^+ + O_2 + 2NADPH$$

with the understanding that a number of protons have been pumped from the stroma into the thylakoid lumen during the passage of each electron through the system. Current estimates of the number are somewhat uncertain but are about two or three protons per electron. The net result is the generation of a proton gradient across the thylakoid membrane, with the lumen becoming more acidic than the stroma.

The pH difference produced across the thylakoid membrane can be very large—as much as 3.5 pH units in brightly illuminated chloroplasts. As in the case of ATP generation in mitochondria, these protons can pass back through the thylakoid membrane only through membrane-bound ATP synthase complexes. In chloroplasts these are called CF_1-CF_0 complexes, and they bear considerable resemblance to the F_1-F_0 complexes of mitochondria.

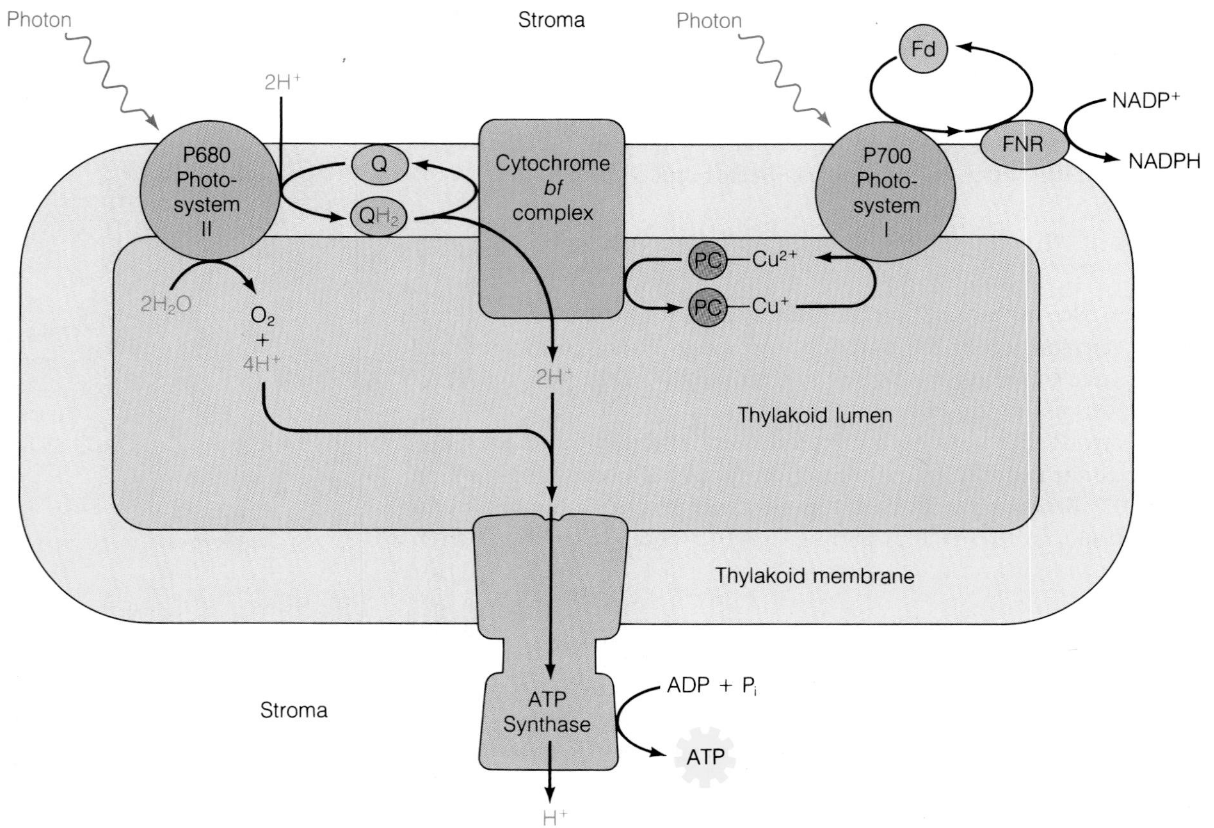

Figure 19.15
Summary view of the light reactions as they occur in the thylakoid. Photosystems I and II and the cytochrome *bf* complex are physically separate protein complexes embedded in the thylakoid membrane. Electron transfer from PSII to cytochrome *bf* is by diffusion of reduced plastoquinone in the membrane lipid, and transfer from *bf* to PSI is mediated by plastocyanin (PC), soluble in the lumen. The ATP synthase particles face into the stroma, so that ATP is generated in this compartment. Reduction of $NADP^+$ also occurs on or near the stromal surface of the membrane.

The pH gradient across the membrane corresponds to a ΔG of about -20 kJ/mol. It has been estimated that one ATP is produced for each three protons passing through the CF_1-CF_0 complex, a thermodynamically reasonable result. Since two or three H^+ are transported per electron, up to one ATP is generated for each electron that passes through the chain.

A summary view of the whole set of light reactions is shown in Figure 19.15. It should be noted that photosystems I and II, the cytochrome *bf* complex, and ATP synthase complexes are all individual entities, not necessarily contiguous in the thylakoid membrane. The components that link the photosystems and the *bf* complex are mobile—plastoquinone in the lipid phase of the membrane and plastocyanin in the thylakoid lumen. Thus, electrons can be moved over long spatial distances in this system. Indeed, careful analysis of the composition of grana indicates that the grana lamellae are rich in photosystem II; by contrast, the stroma lamellae are rich in photosystem I (Figure 19.16).

Cyclic Electron Flow

In the mechanisms we have just described, the electrons displaced from photosystem I by excitation are replaced by photosystem II, which receives them from water. An alternative to this process, called **cyclic electron flow,** is diagrammed in Figure 19.17. The use of this pathway depends on the levels of $NADP^+$. When $NADP^+$ is present in only small amounts, electrons excited in the P700 center are not transferred to $NADP^+$. Instead they are passed from ferredoxin to the cytochrome *bf* complex and from there returned via plastocyanin to the P700 ground state. The *bf* complex pumps

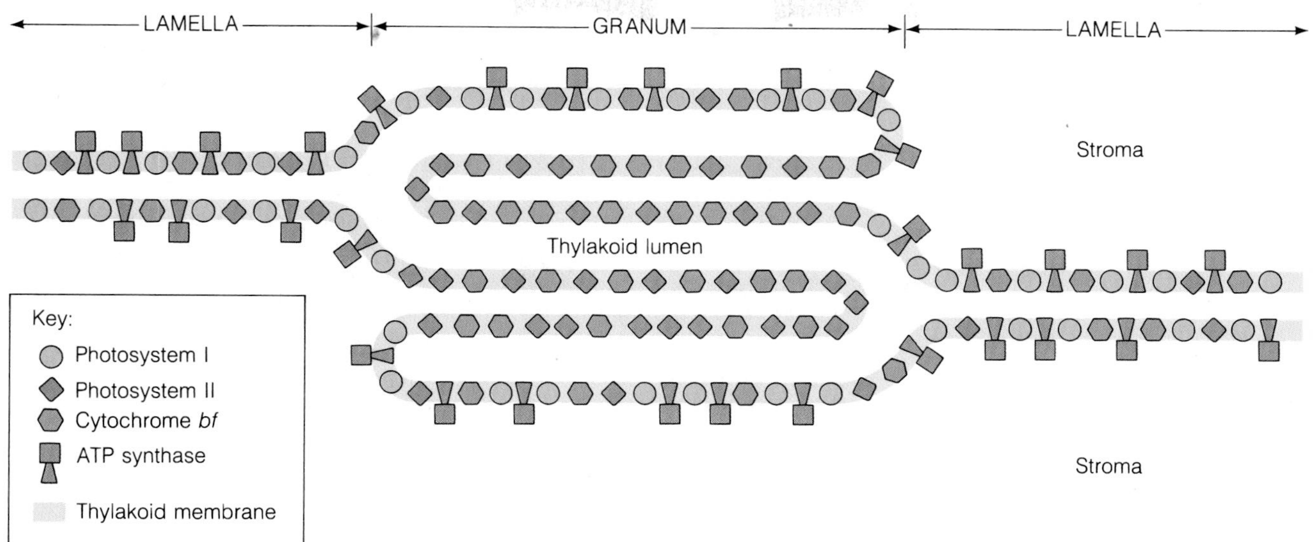

Figure 19.16
Model for the arrangement of components of the two photosystems on the thylakoid membrane. The stromal lamellae and the top and bottom surfaces of the grana, which are in contact with the stroma, are thought to be richer in photosystem I and ATP synthase particles. This allows $NADP^+$ reduction and ATP generation to occur in or near these stroma-facing surfaces. The interior layers of membrane in the grana are correspondingly richer in photosystem II.

protons across the thylakoid membrane during this process, thereby ensuring the generation of ATP. Approximately one ATP is generated for every two electrons that complete the cycle. However, no O_2 is released and no $NADP^+$ is reduced. This alternative process, sometimes called **cyclic photophosphorylation,** apparently serves to generate ATP under situations where the reductant NADPH is abundant and little $NADP^+$ is available as an electron acceptor. It may also play a more fundamental role. As we shall see, the requirements for ATP in the photosynthetic dark reactions are substan-

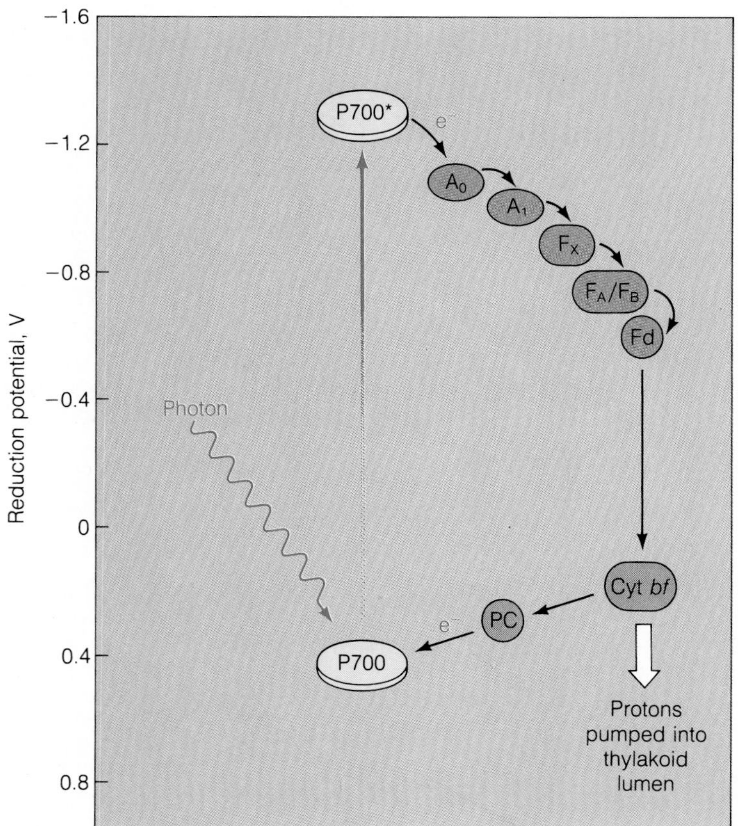

Figure 19.17
Cyclic electron flow. When levels of $NADP^+$ are low and levels of NADPH are high, electrons from the P700 center are passed to the cytochrome *bf* complex and hence back to P700. There is no $NADP^+$ reduction, but protons are pumped across the membrane and therefore ATP is generated. Symbols are as in Figure 19.12.

tial and may not always be fully met by noncyclic electron flow. Cyclic photophosphorylation, which produces ATP but no NADPH, may help maintain the necessary balance between ATP and NADPH production.

Reaction Centers in Photosynthetic Bacteria

Some of our most precise information concerning functioning of the light reactions in photosynthesis has come from studies of photosynthesizing bacteria. Pioneering studies by Roderick Clayton showed that reaction centers could be isolated in pure form. More recently, the entire molecular structure of a crystallized reaction center complex from the purple sulfur bacterium *Rhodopseudomonas viridis* was determined by Johann Deisenhofer, Hartmut Michel, and Robert Huber. This work received the Nobel Prize in chemistry in 1988.

A model of the reaction center structure is shown in Figure 19.18. It is a transmembrane protein, consisting of four polypeptides. The portion of the complex that lies outside the bacterial plasma membrane, in the peri-

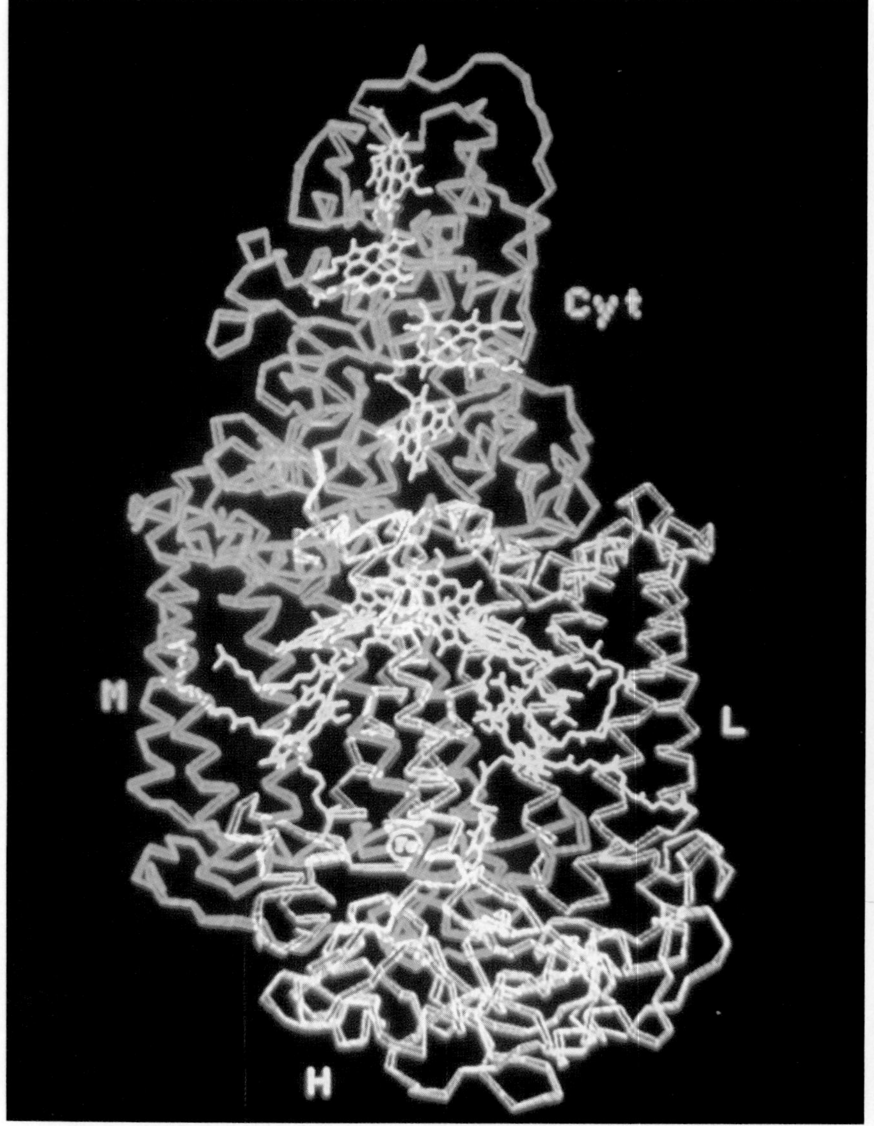

Figure 19.18
Reaction center from the purple photosynthetic bacterium *Rhodopseudomonas viridis*. This model of the whole enzyme complex was deduced from x-ray diffraction measurements. The cytochrome (green) with four heme groups (yellow) lies in the periplasmic space outside the bacterial plasma membrane. Subunits M and L span the membrane, each with five transmembrane α helices. It is these subunits that carry the four bacteriochlorophylls, two bacteriopheophytins, two quinones, and an iron atom, all involved in photon harvesting and electron transfer. Subunit H lies mostly on the cytosolic side of the membrane but has one membrane-spanning α helix.

plasmic space, is a cytochrome carrying four heme groups. Subunit H lies largely on the cytosolic face of the membrane. Buried within the membrane are two subunits (L and M) that are largely α-helical. They carry four bacteriochlorophyll *b* molecules, two bacteriopheophytins, two quinones, designated Q_A and Q_B, and a bound iron atom. Two of the chlorophylls lie very close together; these constitute the reaction center itself. Light absorbance is maximal in the near infrared, at about 870 nm, so the center is referred to as P870.

Chemically, the *Rhodopseudomonas* center complex resembles photosystem II in plants, as it contains pheophytins (bacteriopheophytin, BPh) and quinones. Studies of the kinetics of the reactions in isolated centers have clearly elucidated the electron pathway. Excitation of the reaction center leads very quickly (in about 10^{-12} s) to transfer of an electron to one of the two pheophytins. The electron is then passed on to Q_A and then to Q_B. These quinones are normally bound in the complex, but on receiving a second electron (and two protons) Q_B dissociates. It is believed to then move to a cytochrome bc_1 complex (rather like the *bf* complex), from which the electron is returned (via the cytochromes in the reaction center complex) to the reaction center. The entire process is depicted schematically in Figure 19.19.

Note that this is a cyclic electron flow, which causes protons to be pumped from the bacterial cytosol into the periplasmic space. The cytosols of such bacteria become quite alkaline as photosynthesis continues. Return of protons is through ATP synthase complexes, with generation of ATP.

The cyclic electron flow in these bacteria must be carefully distinguished from that which occurs through photosystem I in plants. The system employed here is, in its electron carriers, much more like photosystem

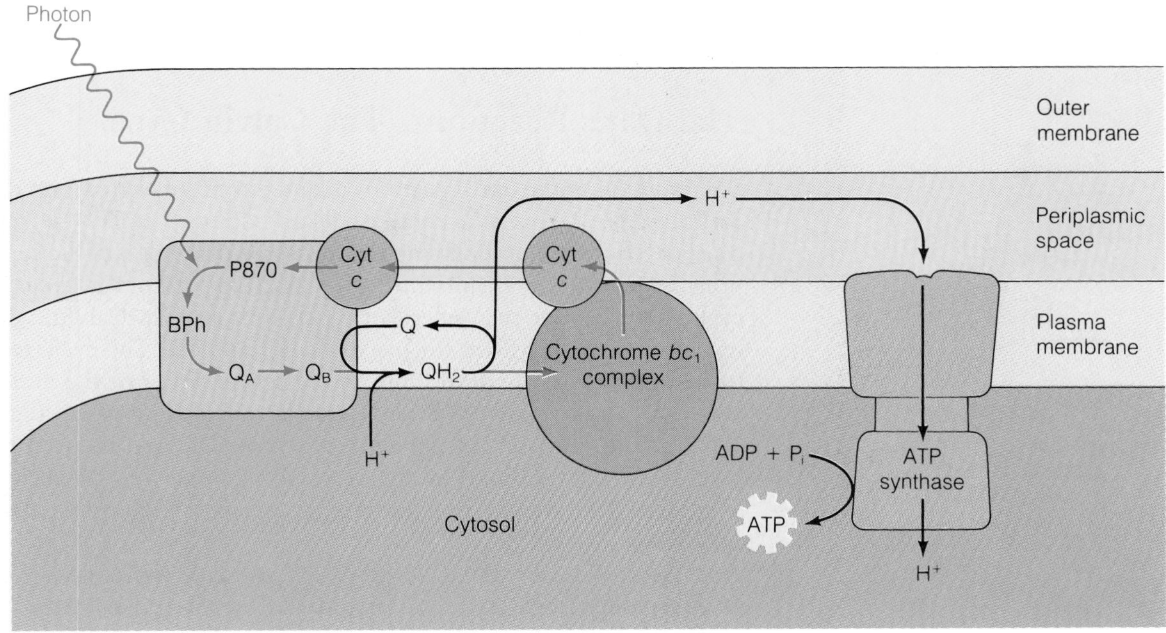

Figure 19.19
Postulated mechanism for bacterial photosynthesis. This process somewhat resembles Figure 19.15, with a reaction center and a membrane-bound cytochrome complex. However, there is only one kind of reaction center, and water is not split, nor is $NADP^+$ directly reduced.

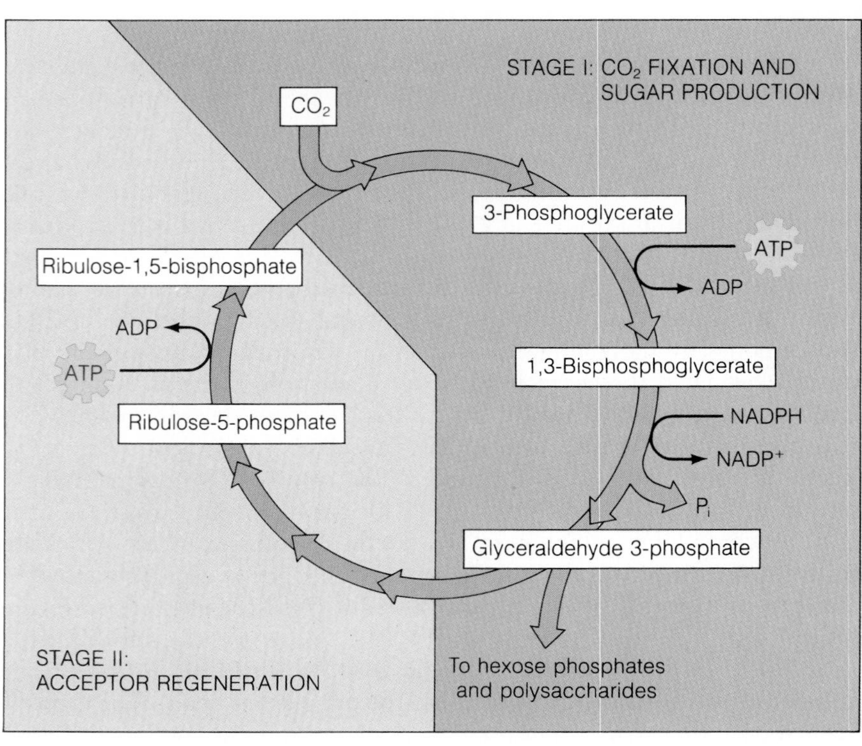

Figure 19.20
Schematic view of the Calvin cycle. The cycle may be divided into two stages. In stage I, CO_2 is fixed and glyceraldehyde-3-phosphate is produced. Part of this G3P is used to make hexose phosphates and, eventually, polysaccharides. Another fraction of the G3P is used in stage II, to regenerate the acceptor molecule, ribulose-1,5-bisphosphate.

II. Though no reducing power has been generated, these bacteria can carry out the dark reactions of photosynthesis by using ATP energy to transfer electrons from various substrates to $NADP^+$.

The Dark Reactions: The Calvin Cycle

The function of the dark reactions is to fix atmospheric carbon dioxide into carbohydrates. This is accomplished by adding one CO_2 at a time to an acceptor molecule and passing the molecule through a cyclic series of reactions, shown schematically in Figure 19.20. The series is called the **Calvin cycle,** named after the American biochemist Melvin Calvin, who in 1961 received the Nobel Prize for his work in this field. The cycle results in the formation of carbohydrates and in the regeneration of the acceptor molecule. The Calvin cycle can be envisioned as divided into two stages: In stage I, the carbon dioxide is trapped as a carboxylate and reduced to the aldehyde–ketone level found in sugars, resulting in net carbohydrate synthesis. Stage II is dedicated to regenerating the acceptor molecule. Let us examine each stage in turn.

As shown on the facing page, the acceptor molecule is **ribulose-1,5-bisphosphate (RuBP).** Carbon dioxide diffuses into the stroma of the chloroplast, where its addition at the carbonyl carbon of RuBP is catalyzed by the enzyme **ribulose-1,5-bisphosphate carboxylase/oxygenase** (also known as **Rubisco**). This is one of the most important enzymes in the biosphere and certainly the most abundant. It makes up about 15% of all chloroplast

Ribulose-1,5-bisphosphate → Enediol intermediate → (CO_2, H^+) → 2-Carboxy-3-keto-D-arabinitol-1,5-bisphosphate → (H_2O) → Hydrated intermediate → (H^+) → 2 Molecules of 3-phosphoglycerate

proteins, and it has been estimated that there are about 40 million tons of it in the world—about 20 pounds for every living person! As its name implies, the enzyme also has an alternative oxygenase activity. We shall see the significance of this later; for the moment will concentrate on its CO_2-fixing (carboxylase) function. The true substrate is the five-carbon enediol intermediate. This is carboxylated, and the second intermediate is hydrated and then cleaved to yield two molecules of 3-phosphoglycerate (3PG). The reaction is essentially irreversible, with $\Delta G° = -51.9$ kJ/mol.

Each molecule of 3PG is then phosphorylated by ATP, in a reaction catalyzed by **phosphoglycerate kinase.** The 1,3-bisphosphoglycerate so produced is then reduced to glyceraldehyde-3-phosphate (G3P) with accompanying loss of one phosphate. The reducing agent is NADPH, and the reaction is catalyzed by the enzyme **glyceraldehyde-3-phosphate dehydrogenase.**

3-Phosphoglycerate → (ATP → ADP; Phosphoglycerate kinase) → 1,3-Bisphosphoglycerate → (NADPH → $NADP^+ + P_i$; Glyceraldehyde-3-phosphate dehydrogenase) → Glyceraldehyde-3-phosphate

Both of these enzymes have been encountered before in connection with their roles in glycolysis (Chapter 13).

At this stage of the cycle CO_2 has already been fixed into a simple (three-carbon) monosaccharide. It is useful to note the requirements in ATP and NADPH up to this point. For each CO_2 molecule that has passed through these steps, two molecules of ATP have been hydrolyzed and two molecules of NADPH have been oxidized. However, it is more appropriate to keep accounts on a "per glucose" basis. Since 6 molecules of CO_2 will ultimately have to pass this route for every new molecule of hexose produced, the appropriate count at this point is *12* ATP and *12* NADPH.

At this point the pathway splits. Of the 12 molecules of G3P that have been produced, 2 are going to be used to make a molecule of a hexose. The remaining 10 will be utilized to regenerate the 6 molecules of ribulose bisphosphate necessary to maintain the cycle. That is, 10 three-carbon molecules will be converted to 6 five-carbon molecules.

Let us consider the hexose formation first. This is actually familiar ground, for it follows a portion of the gluconeogenic pathway described in

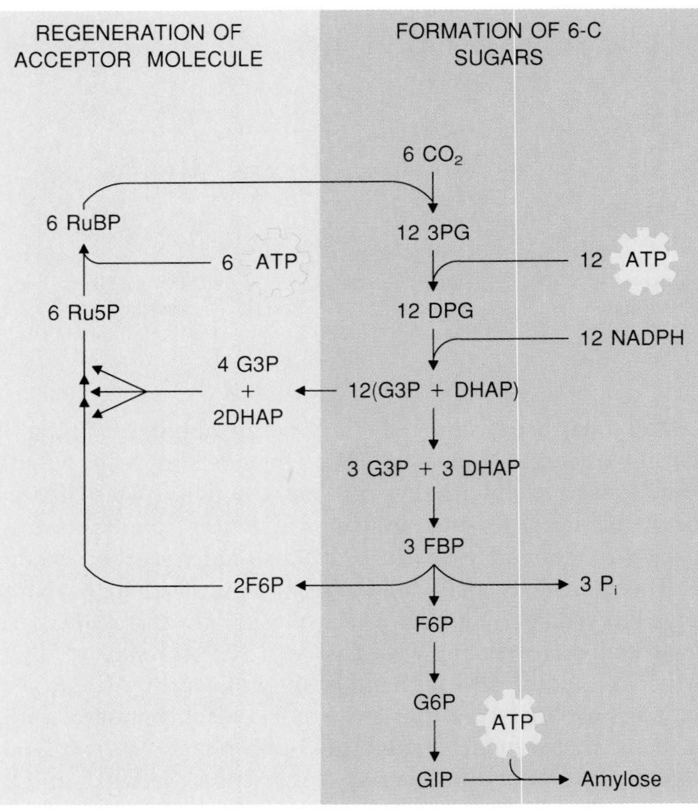

Figure 19.21
Stoichiometry of the Calvin cycle. In six turns of the cycle, 6 CO_2 molecules will have entered and bound to 6 molecules of ribulose-1,5-bisphosphate (RuBP) to yield 12 molecules of glyceraldehyde-3-phosphate (G3P). Since G3P is in isomeric equilibrium with dihydroxyacetone phosphate (DHAP), the 12 G3P may be considered to be an interconvertible stock of 12 molecules of (G3P + DHAP). Of these, six are used to make three molecules of fructose-1,6-bisphosphate (FBP), of which *one* is a net hexose product of the six turns. The other two FBP are used, together with the six remaining molecules of (G3P + DHAP), to form six molecules of ribulose-5-phosphate (Ru5P), which are then phosphorylated to regenerate the required six molecules of RuBP.

Chapter 16. The reactions are shown schematically in Figure 19.21. Recall that glyceraldehyde-3-phosphate can be isomerized to dihydroxyacetone phosphate (DHAP) by triose-phosphate isomerase. Thus, the 12 molecules of G3P produced can be considered to be an interconvertible pool of G3P and DHAP. A molecule of G3P and a molecule of DHAP can be combined, via the enzyme **fructose bisphosphate aldolase,** to yield fructose-1,6-bisphosphate (FBP). As Figure 19.21 shows, six of the G3P molecules follow this path, to yield three molecules of FBP. Of these, two will be employed in the regeneration pathway, but one is available as a net product of the Calvin cycle. This fructose-1,6-bisphosphate is dephosphorylated to yield fructose-6-phosphate (F6P) and then isomerized to glucose-6-phosphate (G6P) and finally to glucose-1-phosphate (G1P).

Glucose-1-phosphate is, in plants as in animals, the precursor to storage polysaccharide formation. However, instead of using UTP to activate the glucose monomer, as in glycogen formation, ATP is employed in the polymerization of plant starch, amylose:

$$\text{Glucose-1-phosphate} + \text{ATP} \longrightarrow \text{ADP-glucose} + \text{PP}_i$$

$$\text{ADP-glucose} + (\text{glucose})_n \longrightarrow (\text{glucose})_{n+1} + \text{ADP}$$

The reactions we have considered to this point can account for the net conversion of two molecules of G3P into one molecule of hexose. But to complete the Calvin cycle, it is necessary to regenerate six molecules of ribulose-1,5-bisphosphate.

This is accomplished by the set of reactions shown in Figure 19.22,

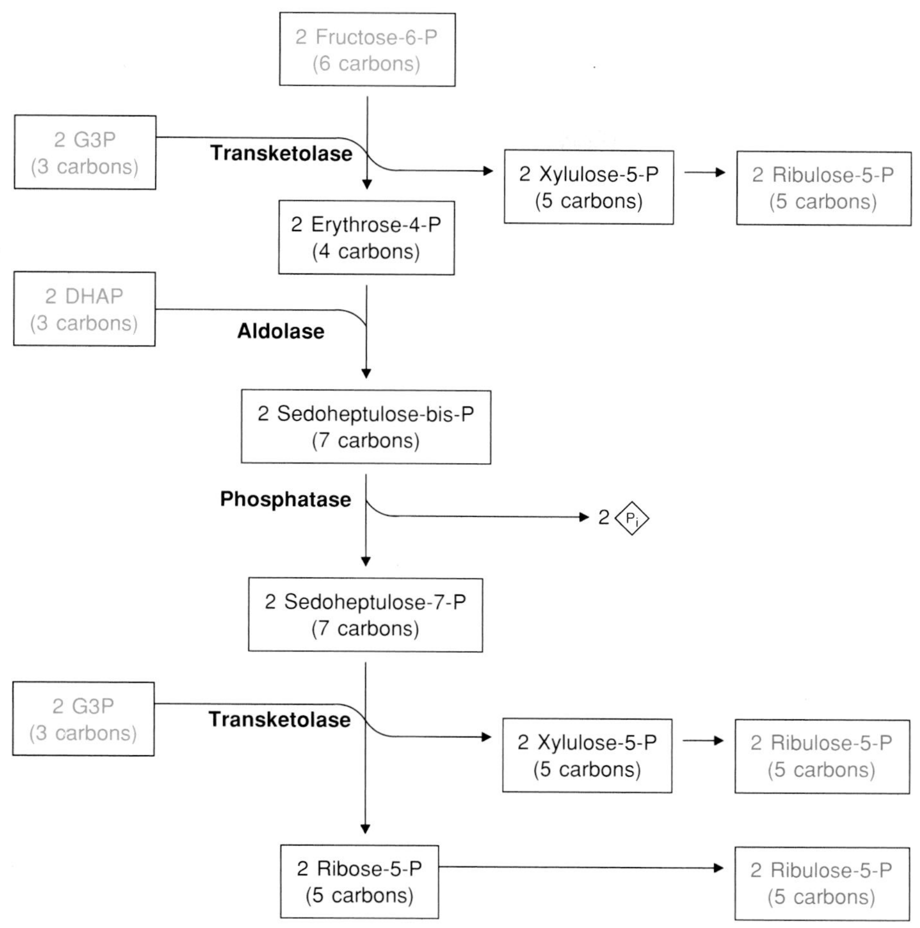

Figure 19.22
Regeneration phase of the Calvin cycle. The two molecules of fructose-6-phosphate entering at the top are combined with four molecules of G3P and two molecules of DHAP, to yield the required six molecules of ribulose-5-phosphate (Ru5P). Note the similarity of this pathway to parts of the pentose phosphate pathway running in reverse (Chapter 14).

which constitute the regenerative phase of the cycle schematized in Figure 19.21. Note that the *input* molecules in this somewhat complex reaction pathway are:

(a) Two molecules of DHAP and four molecules of G3P, from the six G3P that were diverted to the regeneration pathway in Figure 19.21.

(b) Two of the three molecules of fructose-6-phosphate that were produced from the remaining three G3P and three DHAP.

In order to make five-carbon molecules from six-carbon and three-carbon molecules, several rearrangements are required. These are accomplished by *transketolases* and *aldolases*. The structures of the sugars involved in these reactions are all given in Chapter 8; what is important here is the way in which two hexoses and six trioses have been rearranged and recombined to form six pentoses.

The final step in the regeneration of ribulose-1,5-bisphosphate is a phosphorylation, catalyzed by the enzyme **ribulose-5-phosphate kinase** and utilizing ATP. For six rounds of the cycle, this will require 6 ATPs in addition to the 12 already accounted for. Therefore, the requirements for synthesizing 1 mol of hexose from CO_2 are 12 mol of NADPH and 18 mol of ATP. The overall dark reaction may then be written as

$$6CO_2 + 18ATP + 12NADPH + 12H_2O \longrightarrow C_6H_{12}O_6 + 18ADP + 18P_i + 12NADP^+ + 6H^+$$

Summation of Two-System Photosynthesis

The Overall Reaction and Efficiency of Photosynthesis

The ATP and NADPH needed for the dark reactions are released into the stroma by the light reactions of photosynthesis. If we recall that two photons are required for every electron to pass through photosystems I and II, and that two electrons are required to reduce each $NADP^+$, then four photons are necessary for the production of each NADPH molecule. This corresponds to eight photons per O_2, in agreement with the quantum yields shown in Figure 19.11. For the 12 NADPH needed in the dark reaction, as written above, 48 photons must be absorbed. If we assume that these will also pump enough protons across the thylakoid membrane to yield the 18 ATP required, we may, as an approximation, write the light reactions as

$$6H_2O + 12NADP^+ + 18ADP + 18P_i + 6H^+ \xrightarrow{\text{48 photons}} 6O_2 + 12NADPH + 18ATP$$

Summing the light and dark reactions, we obtain

$$6H_2O + 6CO_2 \xrightarrow{\text{48 photons}} C_6H_{12}O_6 + 6O_2$$

This estimate of 48 photons assumes that noncyclic photophosphorylation provides enough ATP for the dark reactions. If, as many workers in the field believe, additional ATP from cyclic photophosphorylation is required, the number of photons needed will be greater.

We can, on the above basis, estimate the energy efficiency of photosynthesis. Forming a mole of hexose from CO_2 and water requires, as we have seen, 2870 kJ. The energy input per photon depends on the wavelength of light used. Assuming that light of 650 nm wavelength is used, 48 einsteins of such light correspond to about 8000 kJ (see Figure 19.6). From this, we estimate a theoretical efficiency of approximately 35%. Direct experimental measurements of the efficiency under optimal conditions give results in the same range or slightly lower. At high levels of illumination, where not all photons absorbed by the chloroplasts can be utilized for reaction center excitation, efficiency falls much lower.

Regulation of Photosynthesis

It should be evident that the so-called dark reactions of photosynthesis require careful regulation. They are dependent on the reductive power and ATP supplied by the light reactions and are therefore stimulated by the latter. There are two major ways in which this stimulation is accomplished. A central enzyme in the dark reactions, ribulose-1,5-bisphosphate carboxylase, is stimulated by high pH and by both CO_2 and Mg^{2+}. The pumping of protons from the stroma into the thylakoid lumen by the light reactions increases the stromal pH; at the same time Mg^{2+} ions enter the stroma to compensate for the positive charge of the H^+ ions that are lost. Recent experiments have revealed that there also exist directly light-dependent pathways for the stimulation of this enzyme.

Three other enzymes of the Calvin cycle are specifically activated by another light-dependent mechanism; these are sedoheptulose-1,7-

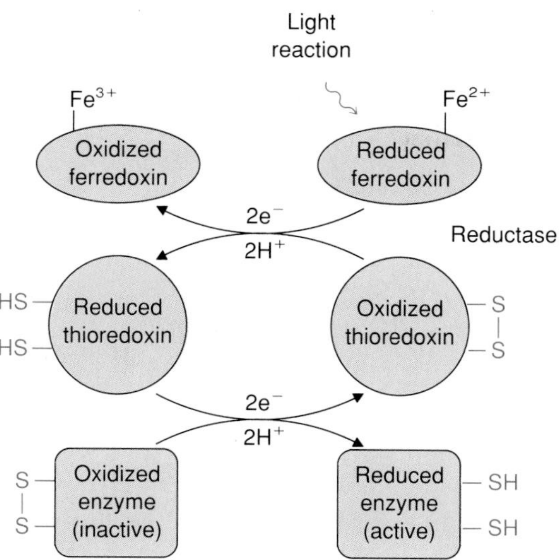

Figure 19.23
Thioredoxin-mediated activation of dark reaction enzymes by disulfide reduction. Additional functions of thioredoxin are presented in Chapter 22.

bisphosphatase, glyceraldehyde-3-phosphate dehydrogenase, and ribulose-5-phosphate kinase. This activation is dependent on reduction of disulfides to sulfhydryls in the enzymes, which is promoted by a disulfide exchange reaction with the protein thioredoxin (Figure 19.23). The reduction of thioredoxin, in turn, is promoted by the reduced form of ferredoxin via a reaction catalyzed by the enzyme **ferredoxin-thioredoxin reductase.** Since reduced ferredoxin will accumulate in strongly irradiated chloroplasts, this system provides another mechanism to stimulate the Calvin cycle reactions when the light reactions are very active. The same compound, reduced thioredoxin, also stimulates the CF_1-CF_0 complexes, ensuring a high rate of ATP generation when illumination is intense.

In the dark, the plant "turns into an animal" in terms of its biochemistry. Rather than photosynthetic synthesis and storage of carbohydrates, the plant begins to draw on its energy reserves, utilizing pathways familiar from our studies of animal catabolism: glycolysis, the citric acid cycle, and the pentose phosphate pathway. In general, these are inhibited in the presence of sunlight and become more active in the dark. The key light-inhibited enzymes are phosphofructokinase (glycolysis) and glucose-6-phosphate dehydrogenase (pentose phosphate pathway). The latter is turned off by the same reduced form of thioredoxin that activates Calvin cycle enzymes.

Photorespiration and the C_4 Cycle

Ribulose bisphosphate carboxylase is a peculiar enzyme. Under unusual atmospheric conditions, but especially when CO_2 is at low levels and O_2 is high, it can behave as an *oxygenase:*

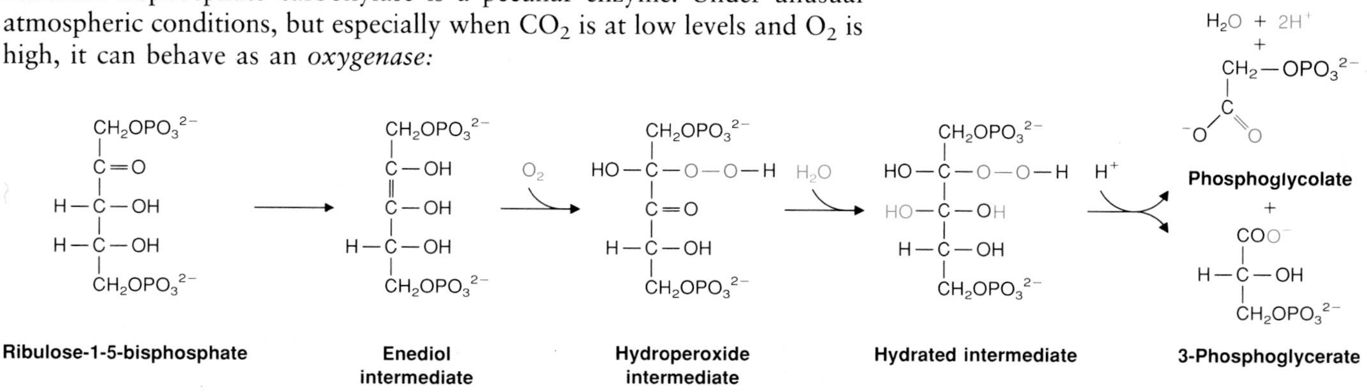

Ribulose-1-5-bisphosphate → **Enediol intermediate** $\xrightarrow{O_2}$ **Hydroperoxide intermediate** $\xrightarrow{H_2O}$ **Hydrated intermediate** $\xrightarrow{H^+}$ **Phosphoglycolate** + **3-Phosphoglycerate** ($H_2O + 2H^+$)

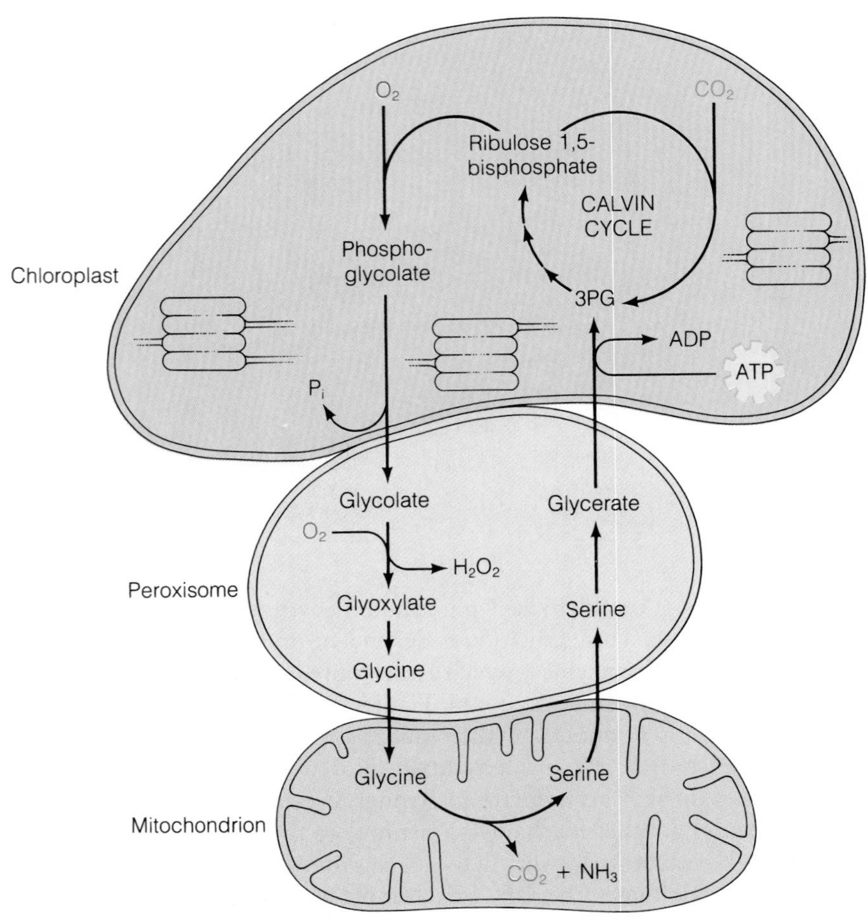

Figure 19.24
Photorespiration. Ribulose-1,5-bisphosphate can be diverted from the Calvin cycle, especially when the concentration of CO_2 is low. RuBP carboxylase/oxygenase catalyzes the oxidation of RuBP to form phosphoglycolate. In the reactions that follow, more O_2 is used, CO_2 is generated, and ATP is hydrolyzed, as metabolites pass from the chloroplast to nearby peroxisomes and mitochondria and then back into the chloroplast.

In this way it initiates a reaction pathway known as **photorespiration,** in which ribulose-1,5-bisphosphate is oxidized to 3-phosphoglycerate and **phosphoglycolate** in the chloroplast. As Figure 19.24 shows, the phosphoglycolate is then dephosphorylated and passed into organelles called *peroxisomes*. Here, it is further oxidized, yielding hydrogen peroxide and glyoxylate. The toxic H_2O_2 is broken down by catalase, and the glyoxylate is transamidated, producing glycine. The glycine enters mitochondria, where *two* molecules are converted into *one* molecule of serine, with accompanying decarboxylation and deamination. CO_2 and ammonia are released. The serine passes back into the peroxisome, where a series of reactions convert it to glycerate. Returning to the chloroplast, it is rephosphorylated (using ATP) to yield 3-phosphoglycerate.

However it is looked at, photorespiration appears to be a losing process. Note that:

1. Ribulose-1,5-bisphosphate is lost from the Calvin cycle.
2. The fixation of CO_2 is reversed: O_2 is consumed and CO_2 is released.
3. Only a part of the carbon is returned to the chloroplast.
4. ATP is needlessly expended.

It is very difficult to see any useful function for photorespiration. It essentially works against the entire photosynthetic process, undoing what

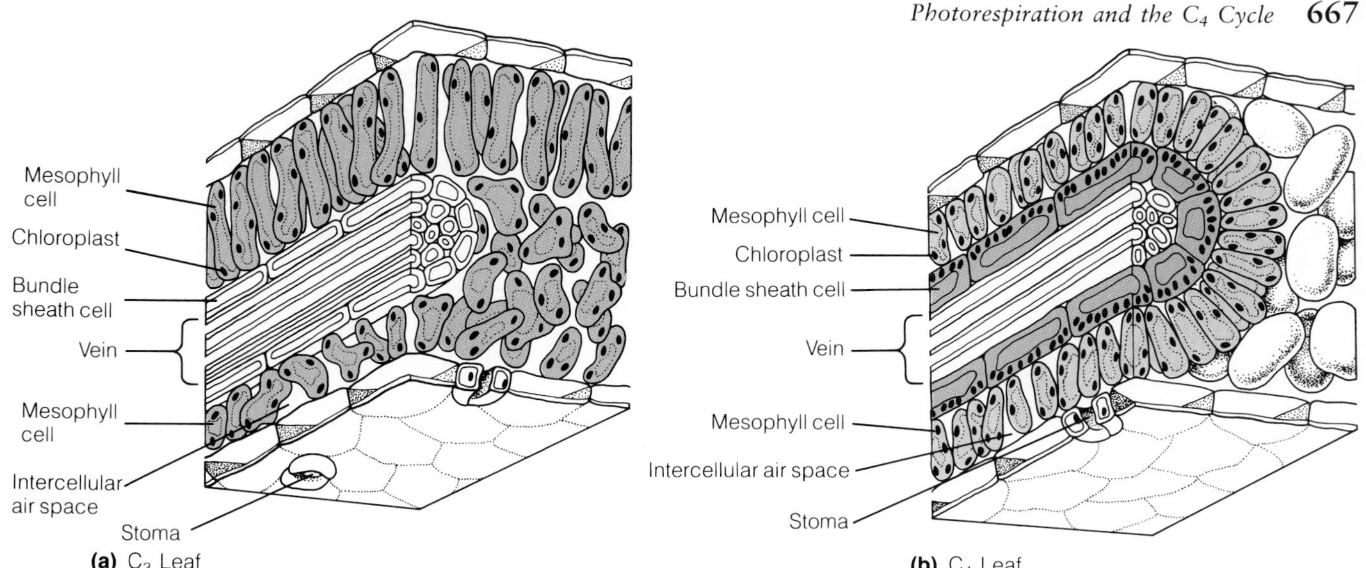

Figure 19.25
Structural differences between leaves of C_3 and C_4 plants. (**a**) Plants that utilize only the Calvin cycle (dark green) are called C_3 plants because they fix CO_2 directly into G3P, a three-carbon molecule. The mesophyll cells of these plants are the sites of most photosynthesis via the Calvin cycle. (**b**) C_4 plants utilize the C_4 cycle (light green) in their mesophyll cells to first fix CO_2 into C_4 compounds. These compounds are then delivered to the bundle sheath cells. Most of the Calvin cycle photosynthesis occurs in the bundle sheath cells of C_4 plants. The stoma are orifices through which the leaf exchanges gases and water vapor with the surroundings.

has been accomplished. For this reason, there are active research efforts to engineer plants genetically so that they lack this behavior.

Some plants have evolved their own way of dealing with this problem. Surprisingly, this is not by modification of the ribulose bisphosphate carboxylase/oxygenase enzyme to suppress the oxygenase function. Rather, such plants, which are called **C_4 plants,** have evolved an additional pathway to help conserve the CO_2. This is called the **C_4 cycle** or sometimes the **Hatch–Slack cycle,** after its discoverers, M. D. Hatch and C. R. Slack. This metabolic pathway is found in several crop species (maize and sugarcane, for example) and is important in tropical plants, which are exposed to intense sunlight and high temperatures. Although photorespiration occurs to some extent at all times in all plants, it is most active under conditions of high illumination, high temperature, and CO_2 depletion.

C_4 plants concentrate their photosynthesis in specialized **bundle sheath cells,** which lie below a layer of mesophyll cells (see Figure 19.25). The mesophyll cells, which are most directly exposed to external CO_2, contain the enzymes for the C_4 cycle. This pathway, as it operates in most C_4 plants, is shown in Figure 19.26. It is essentially a mechanism for trapping CO_2 in the four-carbon compound oxaloacetate (hence the term *C_4 cycle*) and passing it on to bundle sheath cells for fixation. The key to the efficiency of C_4 plants is that the CO_2-fixing enzyme used in this pathway, **phosphoenolpyruvate carboxylase,** is a much more efficient CO_2 scavenger than ribulose bisphosphate carboxylase. Thus, the mesophyll cells serve to pump CO_2 to the photosynthesizing bundle sheath cells. This helps to maintain high enough CO_2 levels in the bundle sheath cells that fixation, rather than photorespiration, is favored. Furthermore, if photorespiration *does* occur, the CO_2 that is released can be largely salvaged in the surrounding mesophyll cells and returned to the Calvin cycle.

As Figure 19.26 shows, the C_4 cycle costs the plant energy in the form of ATP. In fact, since ATP is hydrolyzed to AMP and inorganic phosphate in regenerating phosphoenolpyruvate, the expense is equivalent to *two* extra ATPs for every CO_2 molecule fixed. Nevertheless, the price appears to be worth paying under circumstances where photorespiration would dominate.

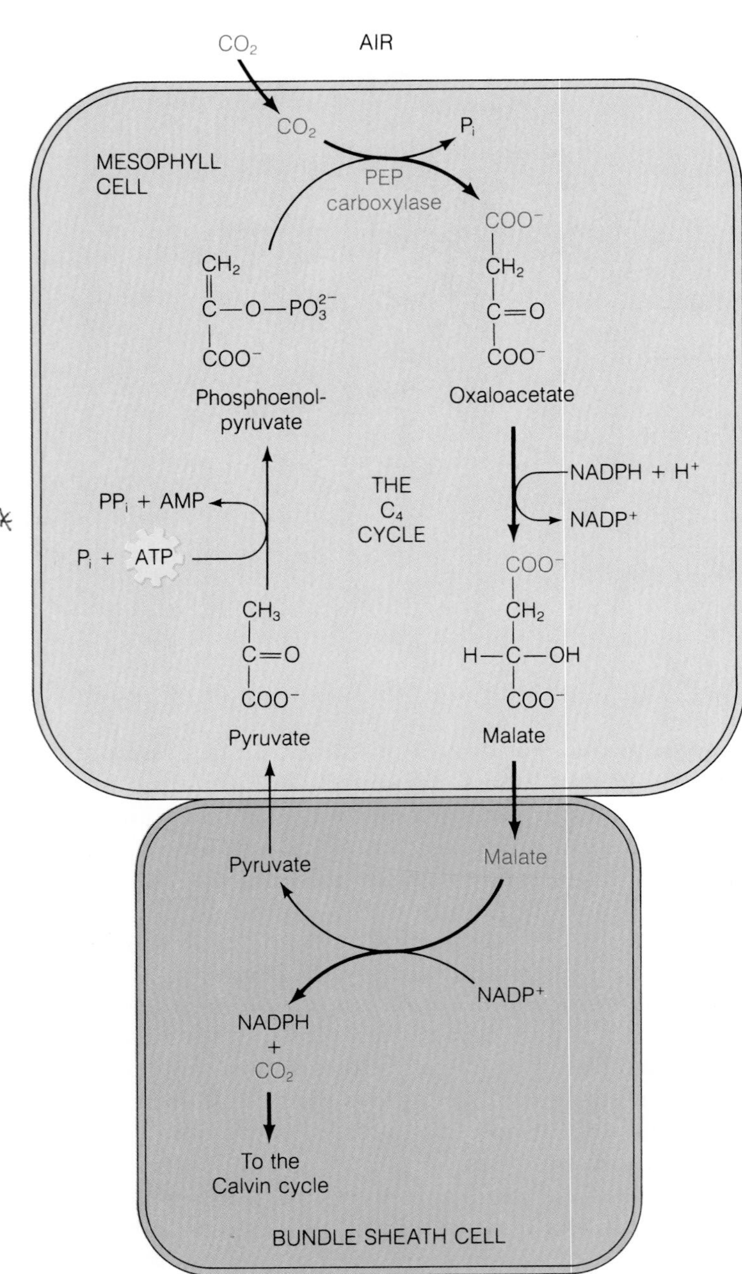

Figure 19.26
Reactions of the C_4 cycle. The CO_2 is transported from mesophyll cells to the bundle sheath cells by coupling it to phosphoenolpyruvate, forming oxaloacetate. This is reduced to malate, which is passed to the bundle sheath cells, where it is decarboxylated. The pyruvate product is returned to the mesophyll cells and phosphorylated to regenerate phosphoenolpyruvate.

* P_i + ATP → PP_i + AMP
equivent on
requiring 2 ATP

REFERENCES

General

Clayton, R. K. (1980) *Photosynthesis: Physical Mechanisms and Chemical Patterns.* Cambridge Univ. Press, Cambridge. An excellent summary of the more physical aspects of photosynthesis.

Cseke, C., and B. B. Buchanan (1986) Regulation of the formation and utilization of photosynthesis in leaves. *Biochim. Biophys. Acta* 853:43–64. An up-to-date, readable summary of the regulation of photosynthesis and associated processes.

Staehelin, L. A., and C. J. Arntzen (eds.) (1987) *Photosynthesis III. Photosynthetic Mechanisms and Light Harvesting Systems.* Springer-Verlag, Berlin. This book, volume 19 of the *Encyclopedia of Plant Physiology,* provides a number of up-to-date, although highly technical, articles on various aspects of the field.

Youvan, D. C., and B. L. Marrs (1987) Molecular mechanisms of photosynthesis. *Sci. Am.* 256:(6)42–48. A brief but lucid account of the current state of the field.

Chloroplasts

Barber, J. (1985) Organization and dynamics of protein complexes within the chloroplast thylakoid membrane. *Biochem. Soc. Trans.* 14:1–4. A succinct description of the functional structure.

Haliwell, B. (1982) *Chloroplast Metabolism—The Structure and Function of Chloroplasts in Green Leaf Cells.* Clarendon Press, Oxford.

Hoober, J. K. (1984) *Chloroplasts* Plenum Press, New York.

Light Reactions

Blankenship, R. E., and R. C. Prince (1985) Excited-state redox potentials and the Z-scheme of photosynthesis. *Trends Biochem. Sci.* 10:382–383.

Brudvig, G. W., and R. H. Crabtree (1986) Mechanism for photosynthetic O_2 evolution. *Proc. Natl. Acad. Sci. USA* 83:4586–4588.

Glazer, A. N., and A. Melis (1987) Photochemical reaction centers: Structure, organization, and function. *Annu. Rev. Plant Physiol.* 38:11–45.

Michel, H., O. Epp, and J. Deisenhofer (1986) Pigment–protein interactions in the photosynthetic reaction center from *Rhodopseudomonas viridis. EMBO J.* 5:2445–2451. An elegant study of the molecular mechanics of a bacterial reaction center.

Woodbury, N. W., M. Becker, D. Middendorf, and W. W. Parson (1985) Picosecond kinetics of the initial photochemical electron transfer reaction in bacterial photosynthetic reaction centers. *Biochemistry* 24:7516–7521. An example of the power of present-day fast kinetic methods.

Synthetic Reactions and Photorespiration

Chapman, M. S., S. W. Suh, P. Curmi, D. Cascio, W. W. Smith, and D. Eisenberg (1988) Tertiary structure of plant RuBisCo: Domains and their contacts. *Science* 241:71–74.

Ellis, R. J., and J. C. Gray (eds.) (1986) Ribulose bisphosphate carboxylase–oxygenase. *Philos. Trans. R. Soc. London Ser. B* 313:303–469. A volume containing a number of excellent papers on this important enzyme.

Heber, U., and G. H. Krause (1979) What is the physiological role of photorespiration? *Trends Biochem. Sci.* 4:32.

Zelich, I. (1975) Pathways of carbon fixation in green plants. *Annu. Rev. Biochem.* 44:123.

PROBLEMS

1. In cyclic photophosphorylation, it is estimated that two electrons must be passed through the cycle to pump enough protons to generate one ATP. Assuming that the $\Delta G^{\circ\prime}$ for hydrolysis of ATP under conditions existing in the chloroplast is about −50 kJ/mol, what is the corresponding percent efficiency of cyclic photophosphorylation, using light of 700 nm?

2. Assume a pH gradient of 4.0 units across a thylakoid membrane, with the lumen more acidic than the stroma. What is the *longest* wavelength of light that could provide enough energy per photon to pump one proton against this gradient, assuming a 20% efficiency in photosynthesis and $T =$ 25°C.

3. The reagent DCMU, 3-(3,4-dichlorophenyl)-1,1-dimethyl urea, inhibits electron transfer to plastoquinone in photosystem II. Predict effects of administering DCMU to (a) a suspension of illuminated plant chloroplasts and (b) a photosynthetic bacterium like *Rhodopseudomonas.*

4. Suppose a brief pulse of $^{14}CO_2$ is taken up by a green plant. (a) Trace the ^{14}C label through the steps leading to fructose-1,6-bisphosphate synthesis, showing which carbon atoms in each compound should carry the label during the first cycle. (b) Will all molecules of fructose-1,6-bisphosphate carry two ^{14}C atoms? Explain.

5. The flux of solar energy reaching the earth's surface is about 7 J/cm^2 s. Assume that *all* of this energy is utilized by a green leaf (10 cm^2 in area), with the maximal efficiency given on p. 664. How many moles of hexose could the leaf theoretically generate in an hour? You may use 600 nm for an average wavelength.

6. The substance dichlorophenyldimethylurea (DCMU) is an herbicide that inhibits photosynthesis by blocking electron transfer between plastoquinones in photosystem II.
   (a) Would you expect DCMU to interfere with cyclic photophosphorylation?
   (b) Normally, DCMU blocks O_2 evolution, but addition of ferricyanide to chloroplasts allows O_2 evolution in the presence of DCMU. Explain.

7. Suppose ribulose-5-phosphate, labeled with ^{14}C in carbon 1, is used as the substrate in dark reactions. In which carbon of 3PG will the label appear?

8. The following data, obtained by G. Bowes and W. L. Ogre (*J. Biol. Chem.* 247(1972):2171–2176), describe the kinetics of incorporation of CO_2 by Rubisco under N_2 and under pure O_2. Decide whether O_2 is a competitive or uncompetitive inhibitor.

$[CO_2]$ (mM)	Under N_2	Under O_2
0.20	16.7	10
0.10	12.5	5.6
0.067	8.3	4.2
0.050	7.1	3.2

9. It has been observed that Rubisco from tobacco leaves collected before dawn has a much lower specific activity than the enzyme collected at noon (J. C. Servaites, *Plant Physiol.* 78(1985):839–843). This difference persisted despite extensive dialysis, gel filtration, or heat treatment. However, precipitation of the predawn enzyme by 50% $(NH_4)_2SO_4$ restored the specific activity to the level of noon-collected enzyme. Suggest an explanation.

CHAPTER 20

Metabolism of Nitrogenous Compounds: Principles of Biosynthesis, Utilization, Turnover, and Excretion

Thus far our study of metabolism has been concerned primarily with compounds that can be degraded completely to carbon dioxide and water—in other words, compounds containing only carbon, hydrogen, and oxygen. We turn now to the metabolism of nitrogen-containing compounds—amino acids and their derivatives, nucleotides, and the polymeric nucleic acids and proteins (Figure 20.1). This chapter and the next two treat the metabolism of low-molecular-weight nitrogenous compounds, and Chapters 24–28 consider how these compounds are utilized for protein and nucleic acid biosynthesis. Because the biological availability of nitrogen is limited, and because the breakdown of nitrogenous compounds often yields toxic products, we will encounter some new metabolic principles as we tour this part of the biochemical scene.

The existence of 20 different amino acids in proteins implies the existence of 20 biosynthetic pathways and 20 degradative pathways. While this might seem intimidating to the student of metabolism, the numerous common features of anabolic and catabolic pathways for various amino acids should ease the task of learning. More important than many of the details of biosynthesis and degradation, however, are the numerous metabolic roles of amino acids other than as protein constituents, including functions as precursors to hormones, vitamins, coenzymes, porphyrins, pigments, and neurotransmitters.

These next three chapters will also indicate how much we have learned from naturally occurring human mutations, as well as from mutations generated in the laboratory, in cultured cells or in bacteria. Whereas a mutation that inactivates an enzyme in one of the central energy-generating or -storing pathways is likely to be lethal and, hence, not expressed in living individuals, mutations that affect amino acid metabolism are often not lethal and *are* found in living humans. While the clinical consequences of these mutations are often tragic, these **inborn errors of metabolism** have greatly enhanced our understanding of human biochemistry.

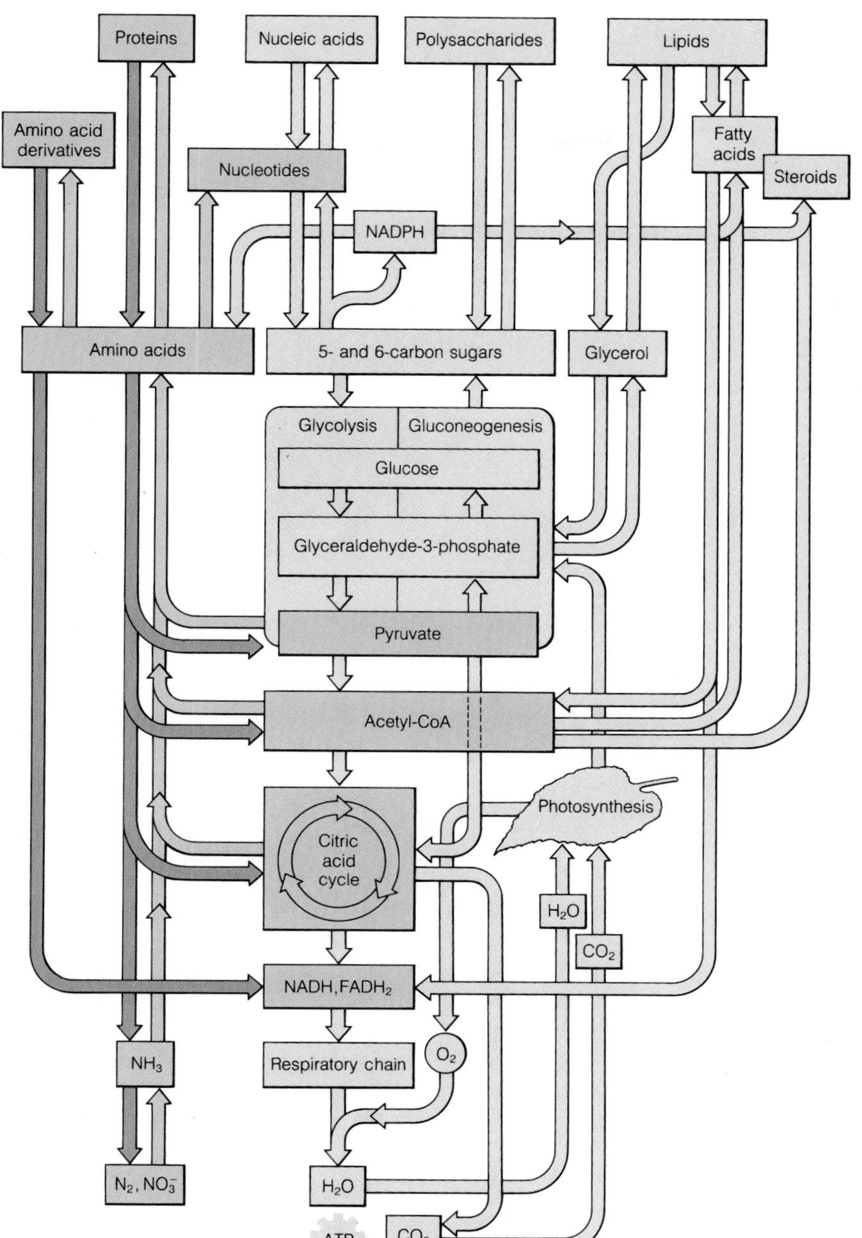

Figure 20.1
Pathways of nitrogen metabolism (purple) as they fit into the general pattern of intermediary metabolism.

Nitrogen Metabolism and the Biosphere

The Nitrogen Cycle; Inorganic Nitrogen Metabolism

For many organisms, both prokaryotic and eukaryotic, growth and reproduction are limited by the availability of utilizable nitrogen, which in turn is limited by the abilities of organisms to utilize different inorganic forms of nitrogen. All organisms can convert ammonia (NH_3) to organic nitrogen compounds, that is, substances containing C—N bonds. Fewer organisms can synthesize ammonia from the far more abundant forms of inorganic nitrogen—dinitrogen gas (N_2), the most abundant component of the earth's atmosphere, and nitrate ion (NO_3^-), a soil constituent essential for the

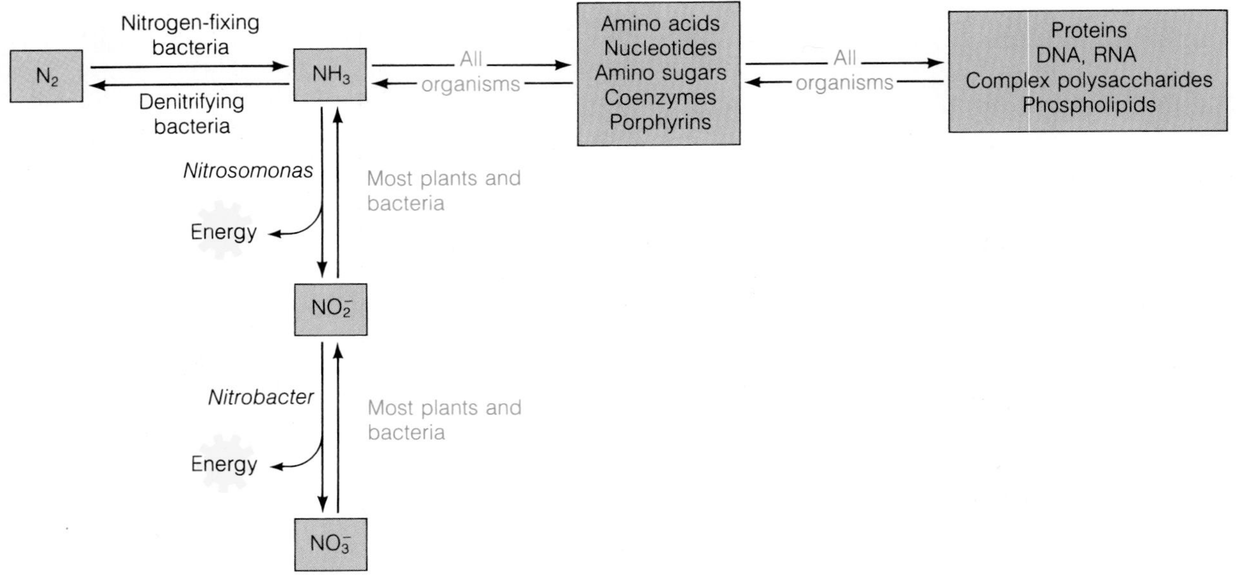

Figure 20.2
Relationships between inorganic (brown) and organic (purple) nitrogen metabolism. Interconversion of N_2, nitrate, and ammonia is limited in the biosphere, but steady-state levels of these species are maintained. All organisms can utilize ammonia for biosynthesis, and this is a major metabolic end product.

growth of most plants. The reduction of N_2 to NH_3, called **biological nitrogen fixation,** is carried out only by certain microorganisms, sometimes in symbiotic relationship with plants. The reduction of nitrate in the soil, by contrast, is widespread among both plants and microorganisms.

As in the consideration of any limited resource, it is useful to think in terms of a **nitrogen economy,** which focuses on questions of supply, demand, turnover, reutilization, growth, and maintenance of a steady state. Within the biosphere a balance is maintained between total inorganic and organic forms of nitrogen. The conversion of inorganic to organic nitrogen, which starts with nitrogen fixation and nitrate reduction, is counterbalanced by catabolism, **denitrification,** and decay (Figure 20.2). Catabolism yields ammonia and various organic nitrogenous end products, which can be metabolized by various bacteria: *Nitrosomonas* oxidizes ammonia to nitrite, and *Nitrobacter* oxidizes nitrite to nitrate. These oxidations are done to generate biological energy, just as other organisms derive energy from oxidation of carbohydrate or fat to CO_2. Other bacteria, the **denitrifying bacteria,** catabolize ammonia to N_2. Since our concern here is with the utilization of nitrogen for amino acid and nucleotide biosynthesis, our focus in the rest of this section will be on synthesis of ammonia from N_2 and from nitrate ion.

Biological Nitrogen Fixation

Although dinitrogen gas constitutes some 80% of the earth's atmosphere, its reduction to ammonia occurs in relatively few organisms—some free-living soil bacteria, such as *Klebsiella* and *Azotobacter;* cyanobacteria (blue-green algae); and symbiotic nodules on the roots of leguminous plants, such as beans or alfalfa, that have been infected with certain bacteria, notably of the genus *Rhizobium* (Figure 20.3). The infecting bacterium assumes a modified form, called a **bacteroid,** inside the cells of infected plants. Some trees, such as alder, also form nitrogen-fixing nodules and thus have the capacity to fix nitrogen.

Since nitrogen availability is the factor limiting the fertility of most soils, an understanding of biological nitrogen fixation is directly related to increasing the world's food supply. N_2 has a triply bonded structure ($N \equiv N$), with a bond energy of about 940 kJ/mol; this makes it extraordinarily difficult to reduce. Industrially, the reduction is done by a catalytic hydrogenation carried out at very high temperature and pressure. This process is used in the manufacture of ammonia-based fertilizers. Interest in the molecular details of biological nitrogen fixation has derived partly from hopes of supplanting this energy-intensive process with a means of ammonia production that can take place under milder conditions.

Formally, nitrogen fixation can be compared with photosynthesis. Both N_2 and CO_2 are stable inorganic compounds whose reduction requires both energy and low-potential electrons—electron carriers of very low E_0'. As we saw in Chapter 19, photosynthesis uses light to generate both energy (through photophosphorylation) and low-potential electrons (as ferredoxin). Comparable mechanisms used in nitrogen fixation are still not clear, because the enzymes involved are extremely sensitive to oxygen and can be studied only under anaerobic conditions. The major reason nitrogen can be fixed in root nodules of plants infected with *Rhizobium* is that the plant synthesizes a hemoglobinlike protein, called **leghemoglobin,** which is abundant in the nodule and which maintains an anaerobic environment by binding any O_2 that finds its way into the nodule.

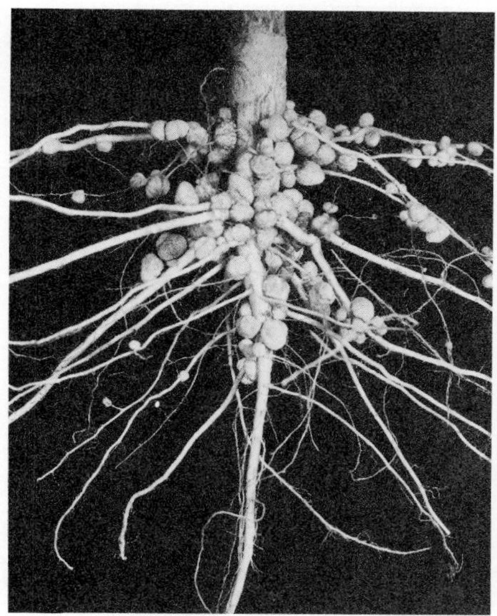

Figure 20.3
The site of nitrogen fixation in symbiotic root nodules of soybean plants infected by bacteria of the genus *Rhizobium*.

Mechanistic details of nitrogen fixation appear quite similar among the different species examined to date. The process is best understood in *Klebsiella,* and that is what we shall describe. The stoichiometry of the overall reaction is as follows.

$$N_2 + 8e^- + 8H^+ + 16ATP + 16H_2O \longrightarrow 2NH_3 + H_2 + 16ADP + 16P_i + 16H^+$$

Although we present a balanced equation, the precise number of ATP molecules in this process has not yet been established; in any case, a large amount is required. The ATP is generated through energy-yielding pathways of the organism, primarily carbohydrate catabolism. Electrons for N_2 reduction are derived from low-potential carriers, either ferredoxin or flavodoxin. Hydrogen is a by-product of nitrogen reduction. Some nitrogen-fixing species have the ability to "recycle" this hydrogen, to generate low-potential electrons for additional cycles of N_2 reduction.

The enzyme system responsible for N_2 reduction, called the **nitrogenase system,** consists of two separate proteins. As outlined in Figure 20.4, one protein, called **component I** or **nitrogenase,** catalyzes the reduction of N_2, while the other, called **component II** or **nitrogenase reductase,** transfers electrons from ferredoxin or flavodoxin to component I. Both proteins contain Fe_4S_4 iron–sulfur clusters, and component I contains, in addition, both molybdenum and a still-uncharacterized compound called molybdenum cofactor.

The genetics of nitrogen fixation is under intense study, because of the goal of transferring nitrogen-fixing capabilities to higher plants. In *Klebsiella* 18 genes known to be involved are linked within one region of DNA some 24,000 base pairs in length, the *nif* gene cluster. Products of these genes include the two polypeptide subunits of component I, the one subunit of component II, flavodoxin, and enzymes that synthesize molybdenum cofactor.

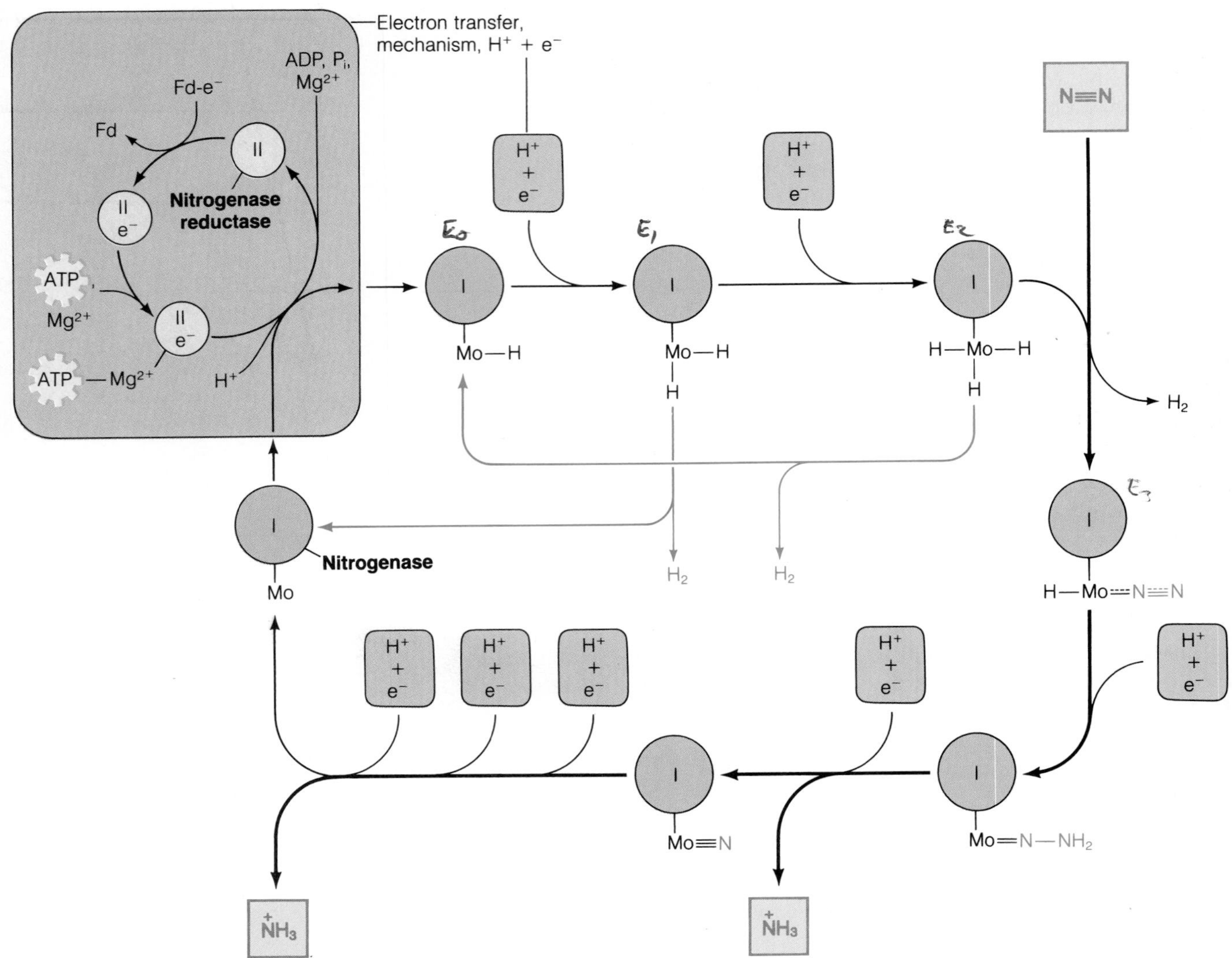

Figure 20.4
Schematic view of nitrogen fixation. Component I (brown) is nitrogenase (the molybdenum–iron protein), and component II (orange) is nitrogenase reductase (the iron protein). e^- is an electron. Fd-e^-, the electron donor that reduces component II, is either ferredoxin (in *Rhizobium*) or flavodoxin (in *Klebsiella*). Binding of ATP to reduced component II is thought to generate an altered conformation of component II, with a very low reduction potential. Interaction between reduced component II and component I transfers an electron, concomitant with the return of component II to a relaxed state and splitting of bound ATP. The same process occurs in subsequent electron transfers to component I (purple). These generate mono-, di-, and trihydride forms of Mo bound to component I. Binding of N_2 occurs concomitantly with release of two bound hydrogens as H_2. Hydrogen can also be released from the di- and trihydrides (blue arrows).

Nitrate Utilization

The ability to reduce nitrate to ammonia is common to virtually all plants, fungi, and bacteria. The first step, reduction of nitrate to nitrite (NO_2^-) is chemically difficult, and it involves a large and complex enzyme, **nitrate reductase.** The enzyme, with M_r about 800 kDa, contains bound FAD, molybdenum (bound to a molybdenum cofactor), and a cytochrome, called cytochrome 557 (which contains an Fe_4S_4 complex). The enzyme carries out the overall reaction

$$NO_3^- + NAD(P)H + H^+ \longrightarrow NO_2^- + NAD(P)^+ + H_2O$$

Plants use NADH as the electron donor, while fungi and bacteria use NADPH. The electrons are transferred to enzyme-bound FAD, thence to cytochrome 557, thence to molybdenum, and finally to the substrate.

Reduction of nitrite to ammonia is carried out in three steps ($NO_2^- \rightarrow NO^- \rightarrow NH_2OH \rightarrow NH_3$) by one enzyme, **nitrite reductase.** This enzyme contains one Fe_2S_2 center and one molecule of **siroheme,** a partially reduced iron porphyrin. The electron donor for each step is ferredoxin.

Utilization of Ammonia: Biogenesis of Organic Nitrogen

Although plants, animals, and bacteria derive their nitrogen from different sources, virtually all organisms share a few common routes for utilization of inorganic nitrogen in the form of ammonia. Ammonia in high concentrations is quite toxic, but at lower levels it is a central metabolite, serving as substrate for five reactions that convert it to organic nitrogen compounds. At physiological pH the dominant ionic species is ammonium ion, $\overset{+}{N}H_4$. However, the reactions to be presented involve the unshared electron pair of NH_3, and the latter, therefore, is the reactive species.

All organisms assimilate ammonia via reactions leading to glutamate, glutamine, and **carbamoyl phosphate.** Since carbamoyl phosphate is used only in the biosynthesis of arginine, urea, and the pyrimidine nucleotides, most of the nitrogen that finds its way from ammonia to amino acids and other nitrogenous compounds does so via the two amino acids glutamine and glutamate. The amino nitrogen of glutamate and the *amide* nitrogen of glutamine are both extremely active in biosynthesis.

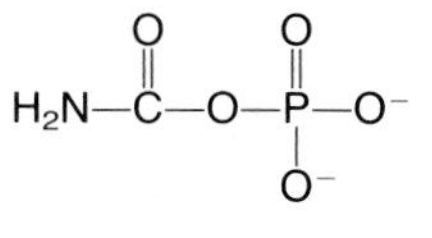

$$H_2N{-}\overset{O}{\overset{\|}{C}}{-}O{-}\underset{O^-}{\underset{|}{\overset{O}{\overset{\|}{P}}}}{-}O^-$$

Carbamoyl phosphate

Glutamate Dehydrogenase: Reductive Amination of α-Ketoglutarate

Glutamate dehydrogenase catalyzes the reductive amination of α-ketoglutarate.

$$\begin{array}{c}COO^-\\|\\CH_2\\|\\CH_2\\|\\C{=}O\\|\\COO^-\end{array} + NH_3 + NADH + 2H^+ \rightleftharpoons \begin{array}{c}COO^-\\|\\CH_2\\|\\CH_2\\|\\H{-}C{-}\overset{+}{N}H_3\\|\\COO^-\end{array} + H_2O + NAD^+$$

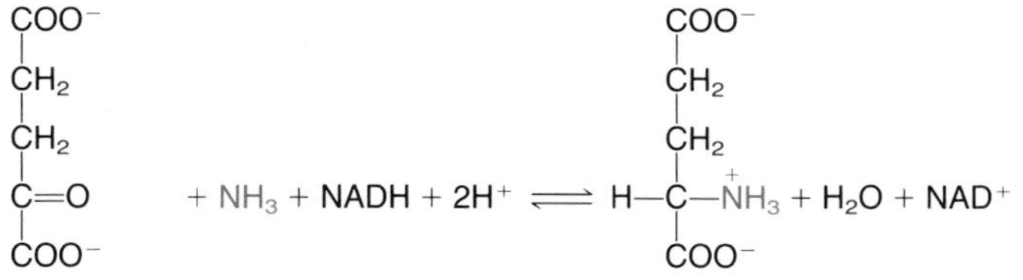

α-Ketoglutarate **Glutamate**

The reaction is reversible. Most bacteria and many plants contain an NADPH-specific form of the enzyme, which probably acts in the direction of glutamate formation. Consistent with this is the fact that bacteria growing with ammonia as their sole nitrogen source use this reaction as the primary route for nitrogen assimilation. On the other hand, in animal cells the enzyme probably functions in the catabolic direction, supplying α-ketoglutarate for oxidation in the citric acid cycle. The animal enzyme uses NAD^+ as its principal cofactor, but it can also use $NADP^+$. In animals glutamate dehydrogenase is a hexamer of identical subunits. It is located in mitochondria, which suggests a role in energy generation. Moreover, it is allosterically controlled, being inhibited by ATP or GTP, and stimulated by ADP or GDP. Thus, the enzyme is activated under conditions of low energy charge. Yeasts and fungi contain both types of glutamate dehydrogenase—each appropriately regulated, with one tailored for nitrogen assimilation and one functioning primarily in catabolism.

Another enzyme involved in glutamate synthesis is **glutamate synthase,** which catalyzes a reaction comparable to that of glutamate dehydrogenase, but which plays a much more widespread role in glutamate biosynthesis.

$$\alpha\text{-Ketoglutarate} + \text{glutamine} + NADPH + H^+ \longrightarrow 2\ \text{glutamate} + NADP^+$$

Because glutamate dehydrogenase has a higher K_M for ammonia than does glutamate synthase, the latter enzyme is more active in glutamate biosynthesis at the low ammonia concentrations seen in most cells. The enzyme as isolated from *E. coli* contains two types of subunits in an 800,000-dalton holoenzyme; one of the subunits contains FAD, FMN, and nonheme iron. The enzyme from plants uses NADPH and ferredoxin, while glutamate synthase from some fungi uses NADH instead of NADPH.

Glutamine Synthetase: Generation of Biologically Active Amide Nitrogen

Whether formed by glutamate dehydrogenase or by transamination (Chapter 14), glutamate can accept a second ammonia moiety to form glutamine in the reaction catalyzed by **glutamine synthetase.**

$$\text{Glutamate } ({}^{-}OOC{-}CH_2{-}CH_2{-}CH(\overset{+}{N}H_3){-}COO^{-}) + NH_3 + ATP + H^{+} \longrightarrow \text{Glutamine } (H_2N{-}C(=O){-}CH_2{-}CH_2{-}CH(\overset{+}{N}H_3){-}COO^{-}) + H_2O + ADP + P_i$$

Glutamate

Glutamine

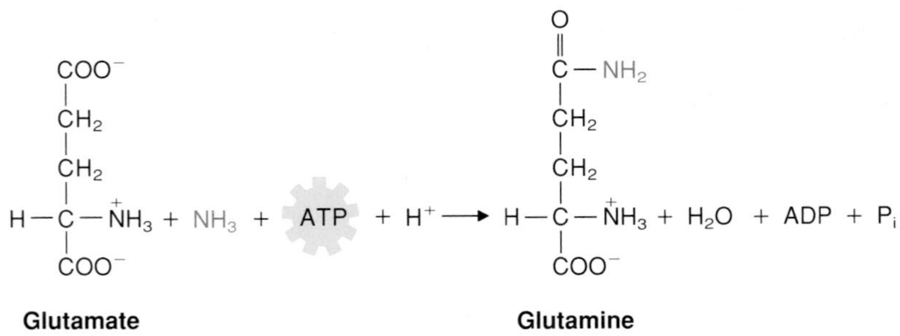

This enzyme is named a *synthetase,* rather than a *synthase,* because the reaction couples bond formation with the energy released from ATP hydrolysis. A synthase does not require ATP.

The glutamine synthetase reaction occurs via an acyl phosphate intermediate. ATP phosphorylates the δ-carbon of glutamate to give a carboxylic–phosphoric acid anhydride, which undergoes nucleophilic attack by the nitrogen of ammonia to give the amide product, glutamine.

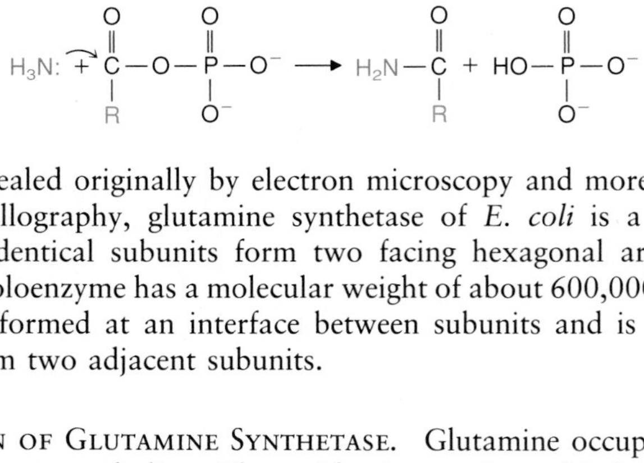

$$H_3N{:} + R{-}C(=O){-}O{-}P(=O)(O^{-}){-}O^{-} \longrightarrow H_2N{-}C(=O){-}R + HO{-}P(=O)(O^{-}){-}O^{-}$$

As revealed originally by electron microscopy and more recently by x-ray crystallography, glutamine synthetase of *E. coli* is a dodecamer, whose 12 identical subunits form two facing hexagonal arrays (Figure 20.5); the holoenzyme has a molecular weight of about 600,000. Each catalytic site is formed at an interface between subunits and is made up of residues from two adjacent subunits.

Regulation of Glutamine Synthetase. Glutamine occupies a central role in nitrogen metabolism. The amide nitrogen is used in biosynthesis of several amino acids (including glutamate, tryptophan, and histidine), purine and pyrimidine nucleotides, and amino sugars. In animals glutamine synthetase is a key participant in detoxifying ammonia formed from amino acid catabolism, particularly in brain; in fact, glutamate and glutamine are two of the most abundant free amino acids in brain cells. It is not surprising, therefore, that the glutamine synthetase reaction is tightly regulated.

As revealed primarily in *E. coli,* several remarkable control mechanisms interact with one another in complex ways. The activity of glutamine

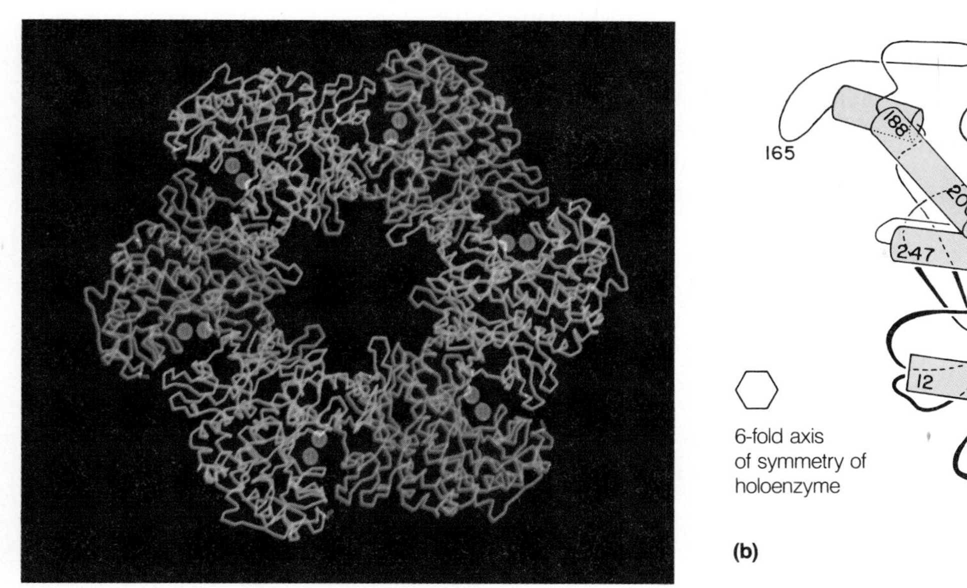
(a)

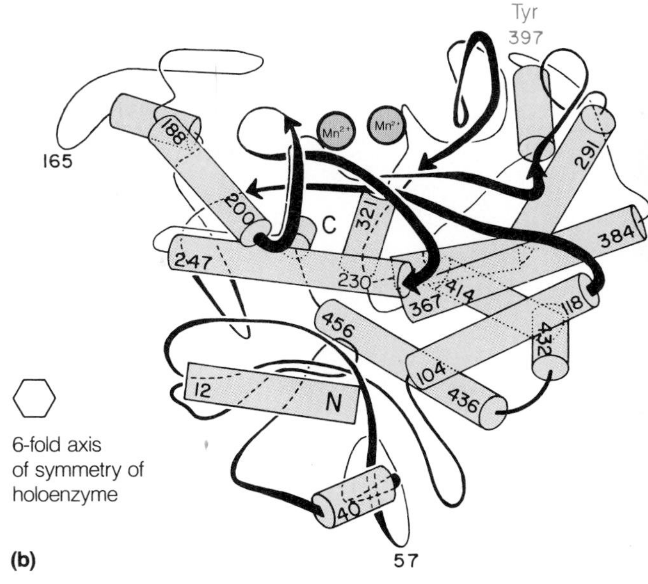

Figure 20.5
Structure of *E. coli* glutamine synthetase, based on x-ray crystallography. (**a**) Computer-generated top view of the holoenzyme. Only six subunits can be seen. (**b**) Schematic view of one of the 12 polypeptide chains. The adenylylation site (Tyr 397) and the two Mn^{2+} ions are shown in red.

synthetase is controlled by two distinct but interlocking mechanisms: (1) allosteric regulation by **cumulative feedback inhibition** and (2) covalent modification of the enzyme brought about by a regulatory cascade.

Cumulative feedback inhibition involves the action of eight specific feedback inhibitors. As shown in Figure 20.6, the eight inhibitors are either metabolic end products of glutamine or in some other way indicators of the general status of amino acid metabolism. Remarkably, each 50,000-dalton subunit of glutamine synthetase must contain binding sites for each of the eight inhibitors, as well as substrates and products. Each of the eight com-

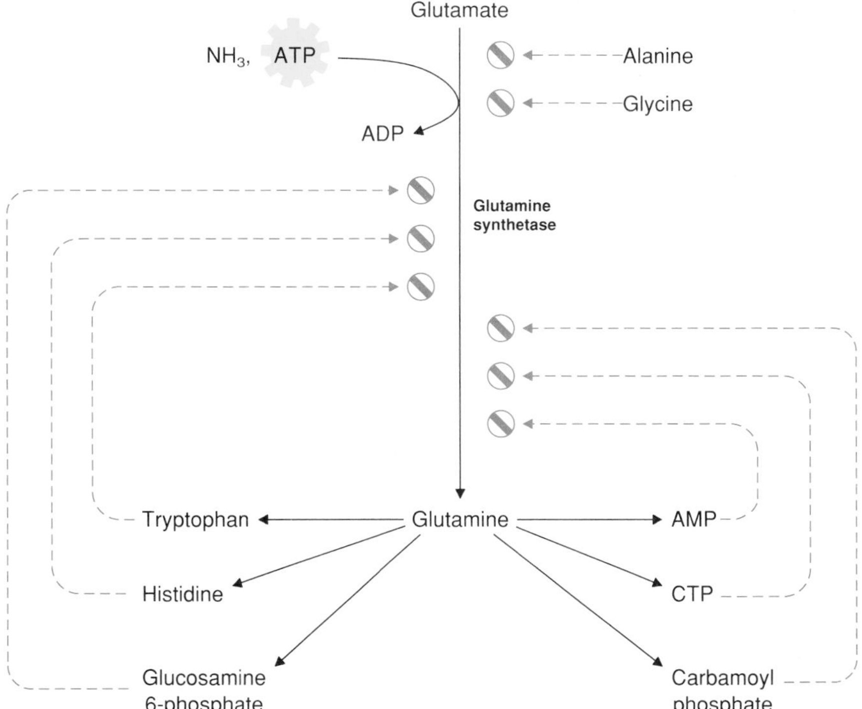

Figure 20.6
Cumulative feedback inhibition of bacterial glutamine synthetase. Six of the eight inhibitors are end products of glutamine metabolism. The other two, glycine and alanine, probably act as general indicators of the status of amino acid metabolism within the cell.

pounds alone gives only partial inhibition, but in combination the degree of inhibition is increased until a mixture of all eight gives virtually complete blockage. This makes good metabolic sense, because it ensures that an accumulation of the end product of one pathway does not shut off the supply of a substrate (glutamine) needed for other pathways.

Superimposed on the cumulative feedback inhibition is a mode of regulation involving covalent modification of the enzyme. Glutamine synthetase is regulated by **adenylylation:** a specific tyrosine residue in the enzyme reacts with ATP to form an ester between the phenolic hydroxyl group and the phosphate of the resultant AMP. That tyrosine residue lies very close to the catalytic site.

Enzyme —⟨phenyl⟩—O—P(=O)(O⁻)—O—CH₂—ribose(OH, OH)—Adenine

Adenylylation inactivates the catalytic site adjacent to the adenylylated tyrosine. Thus, an enzyme molecule with all 12 sites adenylylated is completely inactive, while partial adenylylation yields partial inactivation.

The processes of adenylylation and deadenylylation involve a complex series of regulatory cascades (Figure 20.7). Both reactions are catalyzed by

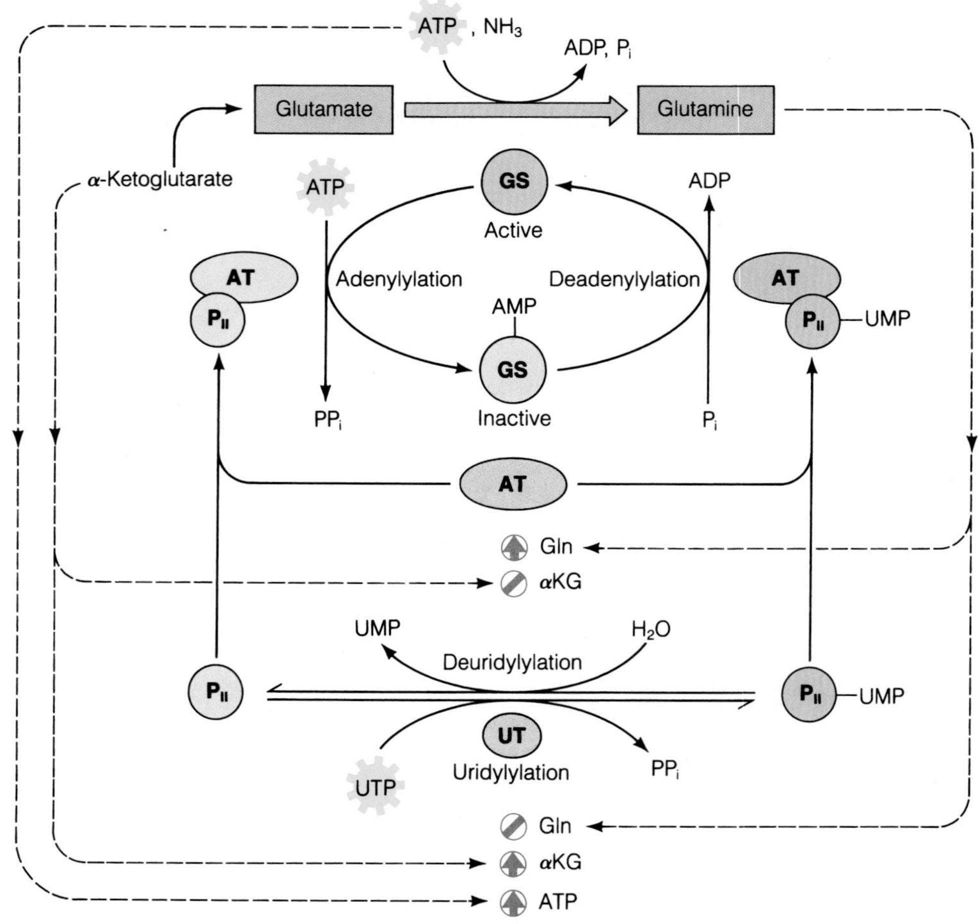

Figure 20.7
Regulation of the activity of *E. coli* glutamine synthetase. The complex of AT (adenylyl transferase) and P_{II} (a regulatory protein) catalyzes both the adenylylation and deadenylylation of glutamine synthetase (GS), depending on whether P_{II} is deuridylylated or uridylylated, respectively. Components identified in green tend to promote activity of glutamine synthetase, whereas red species are associated with enzyme inactivation. UT = uridylyl transferase.

the same enzyme: a complex of **adenylyl transferase** and a regulatory protein, $\mathbf{P_{II}}$. The molecular form of P_{II} determines whether the complex catalyzes adenylylation or deadenylylation. P_{II} is acted on by yet another enzyme, **uridylyl transferase,** which transfers a UMP residue to a specific site on the P_{II} molecule. The product, $\mathbf{P_{II}}$**-UMP,** reacts with adenylyl transferase to stimulate its *de*adenylylation of glutamine synthetase. The non-UMP-containing form of P_{II} converts adenylyl transferase to an adenylylating enzyme. Activity of the uridylyl transferase is in turn stimulated by ATP and α-ketoglutarate and inhibited by glutamine. The [α-ketoglutarate]/[glutamine] ratio is crucial in determining whether uridylylation is favored or inhibited. Deuridylylation of P_{II}-UMP is catalyzed by a different enzyme.

These regulatory cascades provide a responsive mechanism ensuring that, when the supply of activated nitrogen (glutamine) is high, its further biosynthesis is shut down; the non-UMP-containing form of P_{II} accumulates and activates the adenylylation activity of adenylyl transferase. This causes the AMP-containing, or less active, form of glutamine synthetase to accumulate. Conversely, when activated nitrogen supplies are low, α-ketoglutarate accumulates and, provided that ATP is also abundant, stimulates the activity of glutamine synthetase by the converse mechanism.

At present we know little about how glutamine synthetase is regulated in animal cells; we do know that these exquisite control mechanisms are not involved. Recent evidence suggests that interconversions between octameric and tetrameric forms of the enzyme may be involved.

Asparagine Synthetase: A Similar Amidation Reaction

An enzyme comparable to glutamine synthetase, namely **asparagine synthetase,** is also widespread, although the latter accounts for much less ammonia assimilation. Asparagine synthetase uses ammonia or glutamine in catalyzing the conversion of aspartate to asparagine.

$$\underset{\textbf{Aspartate}}{{}^-OOC-CH_2-CH(\overset{+}{N}H_3)-COO^-} + ATP + NH_3 \longrightarrow \underset{\textbf{Asparagine}}{H_2N-C(=O)-CH_2-CH(\overset{+}{N}H_3)-COO^-} + AMP + PP_i + H_2O$$

Note that this enzyme cleaves ATP differently than does glutamine synthetase. The enzyme also differs in that glutamine is strongly preferred as a substrate over ammonia. Probably the reactive species in bond formation is ammonia, which is generated at the active site by hydrolysis of the substrate glutamine.

Carbamoyl Phosphate Synthetase: Generation of an Intermediate for Arginine and Pyrimidine Synthesis

The final reaction of ammonia assimilation is catalyzed by **carbamoyl phosphate synthetase;** either ammonia or glutamine can serve as the nitrogen donor.

$$NH_4^+ + HCO_3^- + 2ATP \longrightarrow \text{carbamoyl phosphate} + 2ADP + P_i$$

$$\text{Glutamine} + HCO_3^- + 2ATP + H_2O \longrightarrow \text{carbamoyl phosphate} + 2ADP + P_i + \text{glutamate}$$

The bacterial enzyme can catalyze both reactions, although glutamine is the preferred substrate. Eukaryotic cells contain two forms of the enzyme. Form I, localized in mitochondria, has a preference for ammonia as substrate and is used in the arginine biosynthetic pathway. Form II, present in the cytosol, has a strong preference for glutamine. It is involved in pyrimidine nucleotide biosynthesis, as shown in part by its sensitivity to inhibition by UTP. The form II enzyme is part of a large protein with three distinct catalytic sites. As discussed in Chapter 22, this trifunctional enzyme catalyzes the first three reactions of pyrimidine nucleotide synthesis.

The Nitrogen Economy: Aspects of Amino Acid Synthesis and Degradation

Metabolic Consequences of the Absence of Nitrogen Storage Compounds

Protein and nucleic acid metabolism differs significantly from that of carbohydrates and lipids. Whereas carbohydrates and lipids can be stored for mobilization as needed by an organism for energy generation or for biosynthesis, there are no polymeric nitrogen compounds whose function is to be stored and released on demand. Although plants store some nitrogenous compounds, such as asparagine in asparagus, such compounds do not represent widely used nitrogen storage depots. The lack of such depots imposes special requirements on organisms, particularly because of the limited availability of utilizable nitrogen. Animals, for example, must continually replenish nitrogen supplies through the diet to replace nitrogen lost through catabolism. In much of the world, protein-rich foods cannot be produced in sufficient quantity to meet the nutritional needs of humans and domestic animals. When dietary protein is insufficient, proteins manufactured for other purposes, such as muscle contraction, are broken down and not replaced. Such consequences occur even when the diet contains an adequate caloric content of protein, if that protein does not contain the needed amino acids.

Just as we can think of a nitrogen economy for the biosphere, we can see it also in relation to individual organisms. Animals try to maintain nitrogen intake and excretion at equivalent rates. A well-nourished adult is said to be in **normal nitrogen balance** if the daily intake of nitrogen through the diet is equal to that lost through excretion and other processes, such as perspiration. Positive nitrogen balance, where normal nitrogen intake exceeds nitrogen loss, is seen during pregnancy, growth of a juvenile, or recovery from starvation. In senescence, starvation, and certain disease states, more nitrogen is lost than is taken in, and the individual is in a state of negative nitrogen balance. Plants and microorganisms commonly excrete very little nitrogen. Microorganisms often grow so rapidly that nitrogen released by catabolism is reassimilated, and in plants nitrogen is often available in such severely limited amounts that this factor itself limits cellular growth rates.

Biosynthetic Capacities of Organisms

Organisms vary widely in their ability to synthesize amino acids. Many bacteria and most plants can synthesize all of their nitrogenous metabolites

Table 20.1
Nutritional requirements for amino acids in mammals

Essential	Nonessential
Arginine, histidine, isoleucine, leucine, lysine, methionine, phenylalanine, threonine, tryptophan, valine	Alanine, asparagine, aspartate, cysteine, glutamate, glutamine, glycine, proline, serine, tyrosine

NOTE: In both humans and rats arginine and histidine are classified as essential amino acids, but nutritional studies show that they are required in the diet only during the growth of juveniles.

starting from a single nitrogen source, such as ammonia or nitrate. However, many microorganisms will use a preformed amino acid, when available, in preference to synthesizing that amino acid. For example, *Lactobacillus* has lost many biosynthetic capacities because it grows in milk, a very nutrient-rich environment; it must therefore be provided with all 20 amino acids if it is to be grown in the laboratory. Mammals are intermediate, being able to biosynthesize about half of the amino acids in quantities needed for growth and for maintenance of normal nitrogen balance. The amino acids that must be provided in the diet to meet an animal's metabolic needs are called **essential amino acids** (Table 20.1). Those that need not be provided because they can be biosynthesized in adequate amounts are called **nonessential amino acids.** In general, the essential amino acids include those with complex structures, including aromatic rings and hydrocarbon side chains. The nonessential amino acids include those that are readily synthesized from abundant metabolites, such as intermediates in glycolysis or the citric acid cycle.

Although most dietitians recommend for humans a protein intake of 100 grams per day or more, a human can do quite well on a diet containing as little as 20 grams per day, if that protein is of high nutritional quality—that is, if it contains abundant proportions of essential amino acids. In general, the more closely the amino acid composition of ingested protein resembles the amino acid composition of the animal eating the protein, the higher the nutritional quality of that protein. For humans, mammalian protein is of the highest nutritional quality, followed by fish, poultry, and plants. (In this context nutritional quality refers only to the single criterion of essential amino acid content.) Plant proteins in particular are often deficient in lysine, methionine, or tryptophan. A vegetarian diet can be nutritious, however, as long as it contains a variety of protein sources.

Transamination

In Chapter 14 we introduced transamination, a process whereby amino acids can replenish citric acid cycle intermediates. Transamination plays a somewhat broader role in amino acid metabolism, in that it provides a route for redistribution of amino acid nitrogen. In other words, if glutamate is the most abundant product of ammonia assimilation, transamination uses glutamate nitrogen to synthesize other amino acids. Transamination reactions are catalyzed by enzymes called **transaminases,** or more properly, **aminotransferases.** As mentioned in Chapter 14, transamination involves transfer of the amino group, usually of glutamate, to an α-keto acid, with formation of the corresponding amino acid plus the α-keto derivative of glutamate, which is α-ketoglutarate. The reaction is shown on p. 682.

$$\text{Glutamate} + \alpha\text{-Keto acid} \rightleftharpoons \alpha\text{-Ketoglutarate} + \alpha\text{-Amino acid}$$

$$H{-}C(CH_2CH_2COO^-)(COO^-){-}\overset{+}{N}H_3 + R{-}C(=O){-}COO^- \rightleftharpoons {}^-OOC{-}CH_2{-}CH_2{-}C(=O){-}COO^- + H{-}C(R)(COO^-){-}\overset{+}{N}H_3$$

Specific aminotransferases exist in animal cells for the synthesis of all of the amino acids found in proteins, except threonine and lysine, as long as the corresponding keto acids are available. Thus, the inability of animal cells to synthesize most of the essential amino acids results from an inability to synthesize the carbon skeletons in the form of α-keto acids; the amino acids could be synthesized if the carbon skeletons were available.

Aminotransferases utilize a coenzyme, **pyridoxal phosphate,** that is derived from vitamin B_6. We will discuss shortly the mechanisms of pyridoxal phosphate-dependent reactions, all of which involve formation of an intermediate Schiff base (the condensation product of an amino group with a carbonyl function). The transamination pathway is shown in Figure 20.8.

Because transamination reactions have equilibrium constants close to unity, the direction in which a particular transamination will proceed is controlled in large part by the intracellular concentrations of substrates and products. This means that transamination can be used not only for amino acid synthesis, but also for degradation of amino acids that accumulate in excess of need. In degradation the transaminase works in concert with glutamate dehydrogenase, as shown below for the degradation of alanine.

$$\text{Alanine} + \alpha\text{-ketoglutarate} \xrightarrow{\text{Amino-transferase}} \text{pyruvate} + \text{glutamate}$$

$$\text{Glutamate} + NAD^+ + H_2O \xrightarrow{\text{Glutamate DH}} \alpha\text{-ketoglutarate} + NADH + NH_4^+$$

$$\text{Net:}\quad \text{Alanine} + NAD^+ + H_2O \longrightarrow \text{pyruvate} + NADH + NH_4^+$$

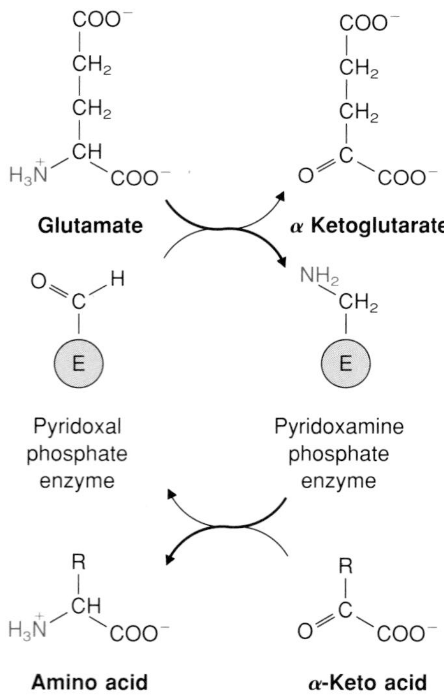

Figure 20.8
Reaction pathway for transamination. In the reaction shown glutamate reacts with enzyme-bound pyridoxal phosphate to give α-ketoglutarate plus pyridoxamine phosphate. The latter then reacts with an α-keto acid by the reverse of the pathway shown, to regenerate pyridoxal phosphate and give an amino acid product. The overall reaction is reversible.

Thus we see transamination as a mechanism for amino acid synthesis *or* degradation. Since the mixture of amino acids available to a cell is rarely present in the proportions needed to synthesize the specific proteins of that cell, transamination plays an important role in bringing the composition of such a mixture into line with the organism's needs. It also participates in funneling excess amino acids toward catabolism and energy generation.

Because of its key role as a product of ammonia assimilation, glutamate is a star player in transamination. Most, but not all, aminotransferases use glutamate/α-ketoglutarate as one of the two amino/keto acid pairs involved. Aminotransferases involving aspartate/oxaloacetate and alanine/pyruvate are also quite abundant. In fact, two such enzymes are important in the clinical diagnosis of human disease—serum glutamate-oxaloacetate transaminase (SGOT) and serum glutamate-pyruvate transaminase (SGPT). These enzymes, abundant in heart and in liver, are released from cells as part of the cell injury that occurs in myocardial infarction, infectious hepatitis, or other damage to either organ. Assays of these enzyme activities in blood serum can be used both in diagnosis and in monitoring the progress of a patient during treatment.

Protein Turnover

Proteins are like low-molecular-weight metabolic intermediates in that they are subject to continuous biosynthesis and degradation, a process called

protein turnover. For an intracellular protein whose total concentration does not change with time, the steady-state level is maintained by synthesis of the protein at a rate just sufficient to replenish protein lost by degradation. Many of the amino acids released during protein turnover are reutilized in the synthesis of new proteins.

Quantitative Features of Protein Turnover

The macroscopic dimensions of protein turnover can be appreciated by considering a day in the life of a 70-kilogram man. That person typically will consume 100 grams of protein during the day and, since he is in normal nitrogen balance, will excrete an equivalent amount of nitrogenous end products. Yet isotope labeling studies show that about 400 grams of protein are synthesized per day, and 400 grams broken down. About three-quarters of the released amino acids are reused in protein synthesis, with the remainder being degraded and the nitrogen excreted. Thus, the total amino acid pool consists of 500 g/day—100 ingested and 400 released via protein degradation. From this pool 400 grams are used in protein synthesis and 100 grams are catabolized and excreted.

An important point is that all of the proteins in the body are represented among the 400 grams broken down in a typical day. Extensive pulse-chase experiments in laboratory animals show that protein degradation follows first-order kinetics. For a particular protein, individual molecules are degraded at random, such that a semilogarithmic plot of isotope remaining in a protein versus time is linear. Thus, one can determine the metabolic half-life of a particular protein. Proteins exhibit tremendous variability in half-lives, from a few minutes to many months; in the rat the average protein has a half-life of 1 or 2 days. Table 20.2 gives information about the half-lives of specific proteins.

As you might expect, proteins that are secreted into an extracellular environment, such as digestive enzymes, polypeptide hormones, and anti-

Table 20.2
Protein turnover: Half-lives and intracellular sites of degradation

	Intracellular Location			
Half-life (hours)	Nucleus	Cytosol	Mitochondria	Endoplasmic reticulum and plasma membrane
<2	Oncogene products	Ornithine decarboxylase, tyrosine aminotransferase, protein kinase C	δ-Aminolevulinic acid synthetase	HMG-CoA reductase
2–8	—	Tryptophan oxygenase, cAMP-dependent protein kinase	—	γ-Glutamyltransferase
9–40	Ubiquitin	Calmodulin, glucokinase	Acetyl-CoA carboxylase, alanine aminotransferase	LDL receptor, cytochrome P-450
41–200	Histone H1	Lactate dehydrogenase, aldolase, dihydrofolate reductase, phytochrome P-670	Cytochrome oxidase, pyruvate carboxylase, cytochrome *c*	Cytochrome b_5, cyt b_5 reductase
>200	Histones H2A, H2B, H3, H4	Hemoglobin, glycogen phosphorylase	—	Acetylcholine receptor

This table represents just a few examples of the many proteins whose half-lives have been determined in different organisms. Data from M. Rechsteiner, S. Rogers, and K. Rote, *Trends Biochem. Sci.* 12:390–394 (1987).

bodies, turn over quite rapidly, whereas proteins that play a predominantly structural role, such as collagen of connective tissue, are much more stable metabolically. Enzymes catalyzing rate-determining steps in metabolic pathways are also short-lived. Indeed, for some enzymes the *rate of breakdown* is an important regulatory factor in controlling intracellular enzyme levels.

By contrast, proteins that do not represent metabolic control points turn over relatively slowly. In the rat, cytochrome *c* has a half-life of nearly a week, while in the human, hemoglobin is as long-lived as the erythrocyte in which it resides (about 120 days). But why should such proteins turn over at all, if their degradation does not represent a metabolic control mechanism? Isn't such turnover wasteful of energy? Let us consider that point.

Biological Importance of Protein Turnover

Like all other intracellular constituents, proteins are subjected to a barrage of environmental influences that can affect their structure, conformation, and biological activity. Though damaged DNA molecules can undergo enzyme-catalyzed repair, proteins lack this capability. Protein turnover could be seen as a quality control system in which the random nature of the process means that both normal and modified proteins are degraded and replaced. Recent work, however, suggests that the process is nonrandom; protein molecules that have become chemically altered are preferentially degraded. A certain chemical change may "mark" a protein molecule and target it for degradation by a proteolytic enzyme that specifically recognizes that mark.

Though much remains to be learned about intracellular protein degradation, a great deal has been learned within the past decade. We know that in bacteria, mutant proteins are degraded much more rapidly than their wild-type counterparts. Evidently evolution has generated proteins whose conformation renders maximum stability in the intracellular environment, and most structural changes decrease this stability.

Protein turnover also represents a route for cellular adaptation to altered environmental conditions. For example, in many bacteria extensive proteolysis is one of the metabolic events interlinked with **sporulation,** or spore formation. Spores represent a heat-stable form of the microorganism; they metabolize at negligible rates and can remain dormant for months or years. When metabolic conditions induce a growing cell to sporulate, extensive protein turnover occurs, with the amino acids released being used to synthesize proteins of the spore. This quasi-dormant state can be maintained indefinitely, with germination to vegetative cells occurring after improvement in the environmental conditions.

Intracellular Proteases and Sites of Turnover

Because most proteins are used intracellularly, most turn over within the cell. The earliest intracellular proteases to be characterized were in lysosomes. It seems likely, however, that proteins are degraded in all of the major cell compartments, inasmuch as proteolytic enzymes are found throughout the cell. In eukaryotic cells at least four major proteases have been found in the cytosol—two Ca^{2+}-activated proteases called **calpains,** a large (700 kDa) multisubunit neutral protease, and a still larger ATP-dependent protease. These enzymes are distinct from the lysosomal proteases, called **cathepsins,** which are designed to function in an acidic milieu. It has not yet been established which of these enzymes participate in protein turnover. However, it seems likely that extracellular proteins taken up by a cell, and long-lived cellular proteins, are degraded in lysosomes, while selective protein turnover related to metabolic regulation occurs in other compartments.

The idea that a proteolytic enzyme should require ATP for its action may seem anomalous, since the process of protein degradation is by itself quite exergonic. Although we don't know the reason in mechanistic terms, ATP hydrolysis may be involved in controlling the rate of proteolysis. In any event, the need to invest energy highlights the selectivity and biological importance of turnover. In bacteria it is clear that an ATP-dependent protease is involved in protein turnover. One such enzyme in *E. coli,* the product of the *lonA* gene, has been extensively characterized.

Chemical Signals for Turnover

The turnover rates for different proteins vary by as much as a thousandfold, while differences in protein stability, as measured by denaturation in vitro, may be much less. Four structural features are currently thought to be determinants of turnover rate: (1) **ubiquitination,** (2) oxidation of particular residues, (3) **PEST sequences,** and (4) particular N-terminal residues.

UBIQUITINATION. Ubiquitin is a small, heat-stable protein found in all eukaryotic cells; it derives its name from its widespread distribution. Ubiquitin undergoes an ATP-dependent reaction with proteins, which condenses the carboxyl terminus of ubiquitin with lysine amino groups. Such modified proteins are degraded soon afterward. We shall say more about ubiquitin in Chapter 28.

OXIDATION OF AMINO ACID RESIDUES. Earl Stadtman and co-workers have shown that many proteins undergo mixed-function oxidation of particular residues, particularly lysine amino groups, and that this modification marks these proteins for subsequent degradation. Consistent with this idea is the recent isolation of a protease, from both *E. coli* and rat liver, that cleaves oxidized glutamine synthetase in vitro but does not attack the native enzyme. Other forms of oxidative damage have also been correlated with increased turnover of proteins—specifically, the attack of oxygen radicals on tryptophan, tyrosine, histidine, and cysteine residues.

PEST SEQUENCES. Examination of the amino acid sequences of short-lived proteins ($t_{1/2} < 2$ hours) shows that virtually all contain one or more regions rich in proline, glutamate, serine, and threonine. From the one-letter designations for these amino acids (P, E, S, and T, respectively), these regions, some 12 to 60 residues long, have been called PEST sequences. Very few longer-lived proteins contain these regions. Although this is largely circumstantial evidence, which says nothing about the biochemical function of the PEST sequences, the pattern does result from inspection of several dozen amino acid sequences. Thus, it seems likely that the PEST region is part of a recognition scheme for the enzyme systems that degrade short-lived proteins.

N-TERMINAL AMINO ACID RESIDUE. Techniques of site-directed mutagenesis have recently revealed that the intracellular half-life of a particular yeast protein varies considerably depending on the identity of its N-terminal amino acid residue. An N-terminal residue of Phe, Leu, Asp, Lys, or Arg is correlated with great instability, while proteins with N-terminal Met, Ser, Ala, Thr, Val, or Gly are far more stable (see Table 28.6 in Chapter 28).

These and other observations indicate that specific structural features on proteins convey information about the metabolic stability of the proteins. The molecular nature of that information processing and the identities of the enzymes involved remain to be determined.

Amino Acid Degradation and Metabolism of Nitrogenous End Products

Common Features of Amino Acid Degradation Pathways

In animals whose dietary protein intake exceeds the need for protein synthesis and other biosyntheses, the excess nitrogen is largely degraded, with the carbon skeletons being metabolized in the citric acid cycle. Protein can thus be a significant contributor toward an animal's energetic requirements. In contrast, plants and bacteria generally synthesize most of their own amino acids and regulate the anabolic pathways so that excesses rarely develop. Generally, microorganisms utilize preformed amino acids in preference to synthesizing their own. Many bacteria can satisfy all of their requirements for nitrogen *and* carbon from a single amino acid. The degradative pathways active here are in general similar to those described in animals.

With a few exceptions, the first step in amino acid degradation involves removal of the α-amino group to give the corresponding α-keto acid. This is usually effected by transamination, with the concomitant synthesis of glutamate from α-ketoglutarate. Glutamate is then converted back to α-ketoglutarate by glutamate dehydrogenase. Thus, the net process is the deamination of amino acid to the corresponding keto acid plus ammonia. The same net conversion can also be catalyzed (although to a lesser extent) by **L-amino acid oxidase,** a flavoprotein enzyme found in liver and kidney.

$$\mathrm{R{-}\underset{\overset{|}{\overset{+}{N}H_3}}{CH}{-}COO^- + FMN + H_2O \longrightarrow R{-}\underset{\overset{\|}{O}}{C}{-}COO^- + FMNH_2 + NH_3}$$

$$\mathrm{FMNH_2 + O_2 \longrightarrow FMN + H_2O_2}$$

The peroxide formed is decomposed by catalase. Kidney and liver are also rich in an FAD-containing **D-amino acid oxidase,** whose function has yet to be described since the unnatural D isomers of amino acids are quite rare.

Once the nitrogen has been removed, the carbon skeleton can, depending on the physiological state of the organism, either proceed toward oxidation in the citric acid cycle or be used for biosynthesis of carbohydrate. Figure 20.9 shows the entry points into the citric acid cycle for the breakdown products of each of the amino acids. The individual pathways are presented in Chapter 21.

Amino acids whose skeletons generate pyruvate or oxaloacetate, such as alanine or aspartate, are efficiently converted to carbohydrates, via gluconeogenesis. Amino acids leading to acetyl-CoA or acetoacetyl-CoA, such as leucine, contribute heavily toward ketogenesis. The terms **glucogenic** and **ketogenic** have been used to classify amino acids as generators primarily of carbohydrate or ketone bodies, respectively.

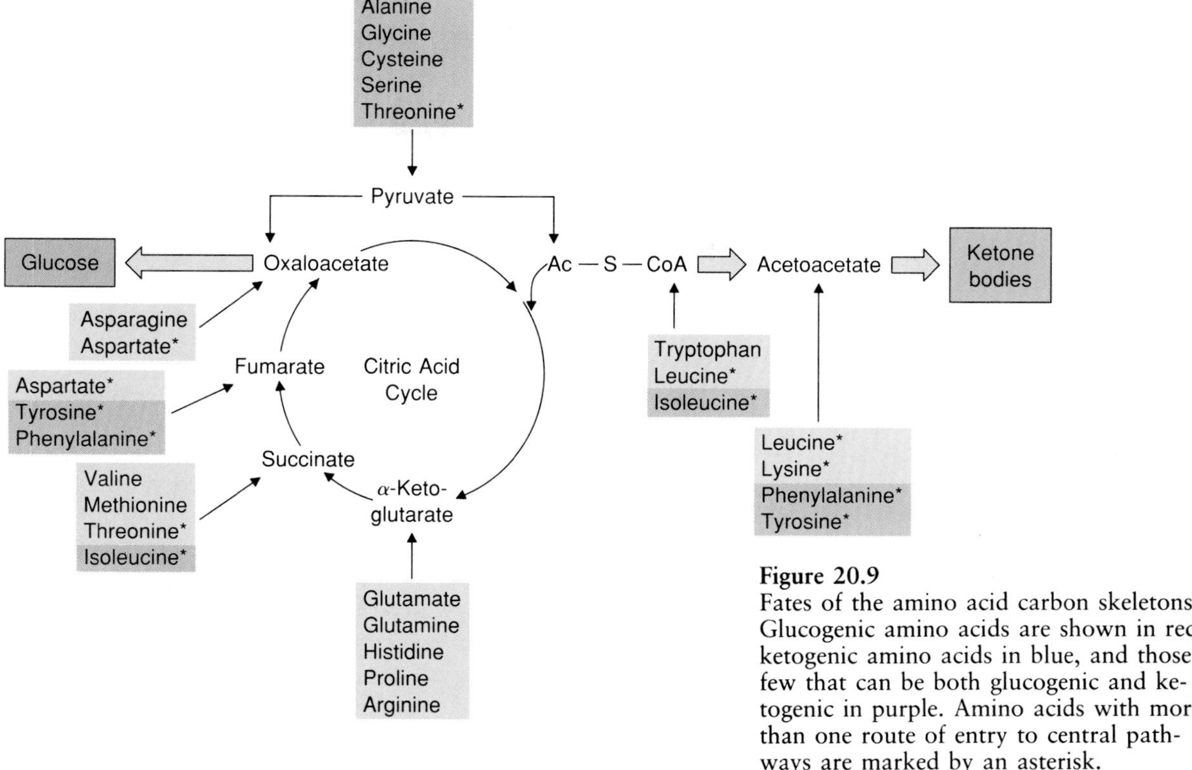

Figure 20.9
Fates of the amino acid carbon skeletons. Glucogenic amino acids are shown in red, ketogenic amino acids in blue, and those few that can be both glucogenic and ketogenic in purple. Amino acids with more than one route of entry to central pathways are marked by an asterisk.

Detoxification and Excretion of Ammonia; Krebs–Henseleit Urea Cycle

Although ammonia is a universal participant in amino acid synthesis and degradation, its accumulation in abnormal concentrations has toxic consequences. Therefore, cells undergoing active amino acid catabolism must be able to detoxify and/or excrete the ammonia as fast at it is generated. For most aquatic animals, which can take in and pass out unlimited quantities of water, ammonia simply dissolves in the water and diffuses away. Since terrestrial animals must conserve water, they convert ammonia to a form that can be excreted without large water losses. Birds, terrestrial reptiles, and insects convert most of their excess ammonia to **uric acid,** an oxidized purine. Because uric acid is quite insoluble, it precipitates and can be excreted without a large water loss and without building up osmotic pressure. This is particularly important during the part of each animal's lifetime that is spent in the egg. The biosynthesis of uric acid occurs by the route used to synthesize purine nucleotides, a pathway presented in Chapter 22. Most mammals excrete the bulk of their nitrogen in the form of **urea.** This compound is highly soluble, and, being electrically neutral, does not affect the pH when it accumulates, as does ammonia.

Uric acid

$$H_2N-\overset{\overset{\displaystyle O}{\|}}{C}-NH_2$$

Urea

Urea is synthesized almost exclusively in the liver. The pathway, which is cyclic, was discovered by Hans Krebs and Kurt Henseleit in 1932, five years before the other cycle for which Krebs's name is famous. Krebs and Henseleit were investigating the pathway of urea synthesis by adding possible precursors to liver slices and then measuring the amount of urea produced. When arginine was added, urea was produced in 30-fold molar excess over the amount of arginine administered. Similar results were seen if either of two structurally related amino acids, **ornithine** or **citrulline,** was substituted for arginine. Since these three amino acids seemed to function catalytically to promote urea synthesis, Krebs and Henseleit proposed the existence of a cyclic pathway:

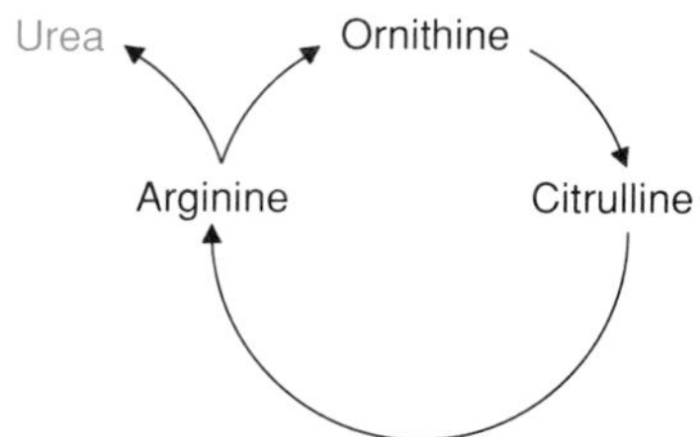

That proposal was correct, as was subsequently confirmed by isolation of the enzymes involved and identification of the biosynthetic route to ornithine, which begins the pathway. These details are shown in Figure 20.10. Ornithine serves as a "carrier," with the carbon and two nitrogen atoms that eventually constitute urea being added to the ornithine molecule. Ornithine itself is synthesized from glutamate, by a pathway that we present in Chapter 21. The source of the carbon and one nitrogen atom in urea is carbamoyl phosphate, which reacts with ornithine, via the enzyme **ornithine transcarbamoylase,** to give citrulline. The second nitrogen comes from aspartate, which reacts with citrulline to form **argininosuccinate,** through the action of **argininosuccinate synthetase.** Next, **argininosuccinase** cleaves argininosuccinate in a nonhydrolytic, nonoxidative reaction to give arginine and fumarate. Arginine is cleaved hydrolytically by **arginase,** to regenerate ornithine and yield one molecule of urea.

The enzyme arginase is responsible for the cyclic nature of the urea biosynthetic pathway. Virtually all organisms synthesize arginine from ornithine by the reactions shown in Figure 20.10. However, only **ureotelic** organisms (those excreting most of their nitrogen as urea) contain arginase and, hence, only those organisms carry out the cyclic pathway. Interestingly, the capacity to synthesize arginase develops in frogs at the same time that they undergo metamorphosis from the tadpole stage to the adult animal. Since the tadpole lives in water, it can excrete ammonia; the adult frog, being adapted to a terrestrial life-style, develops the ability to synthesize urea.

The nitrogen carried into the urea cycle via aspartate comes ultimately from ammonia, through a linkage between the urea cycle and the citric acid cycle. Fumarate is converted, via the citric acid cycle, to oxaloacetate, which undergoes transamination to aspartate; the nitrogen is derived ultimately from ammonia through glutamate dehydrogenase.

The net reaction for one turn of the urea cycle is as follows.

$$CO_2 + \overset{+}{N}H_4 + 3ATP + \text{aspartate} + 2H_2O \longrightarrow \text{urea} + 2ADP + 2P_i + AMP + PP_i + \text{fumarate}$$

Since two molecules of ATP are required to reconvert AMP to ATP, four high-energy phosphates are consumed in each turn of the cycle. Thus, the synthesis of this excretion product is energetically expensive. Extremely protein-rich diets are thus not necessarily beneficial. Since much energy is consumed both in **fixing** nitrogen and in **excreting** it, dietary intake of protein beyond its need for biosynthesis of nitrogenous metabolites is energetically wasteful.

Following its synthesis, urea is transported in the bloodstream to the kidneys, which filter it for excretion. Measurements of blood urea nitrogen

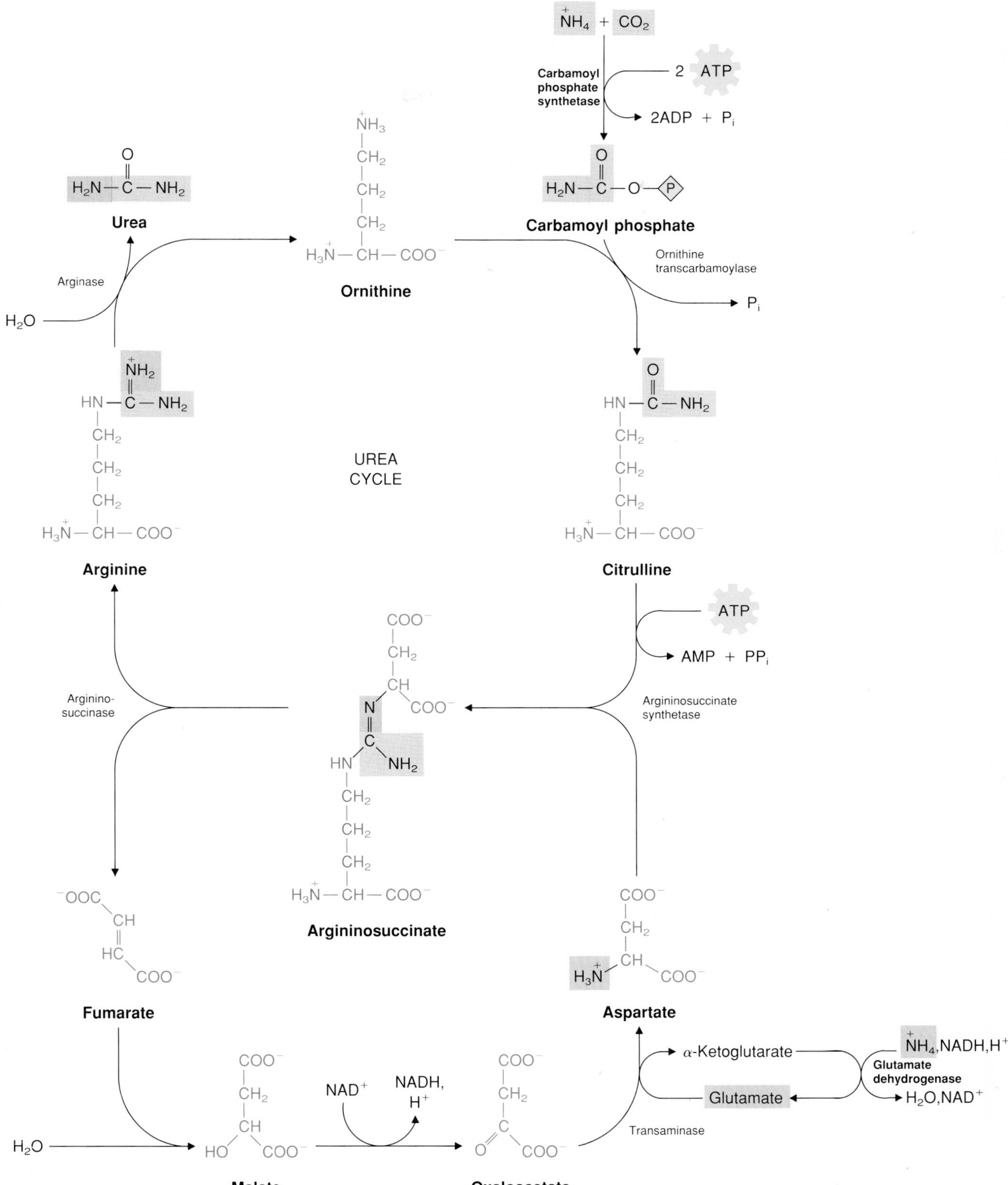

Figure 20.10
The Krebs–Henseleit urea cycle. Note the enzymes that incorporate the CO_2 and $\overset{+}{N}H_4$ that ultimately appear in urea—carbamoyl phosphate synthetase and glutamate dehydrogenase, respectively.

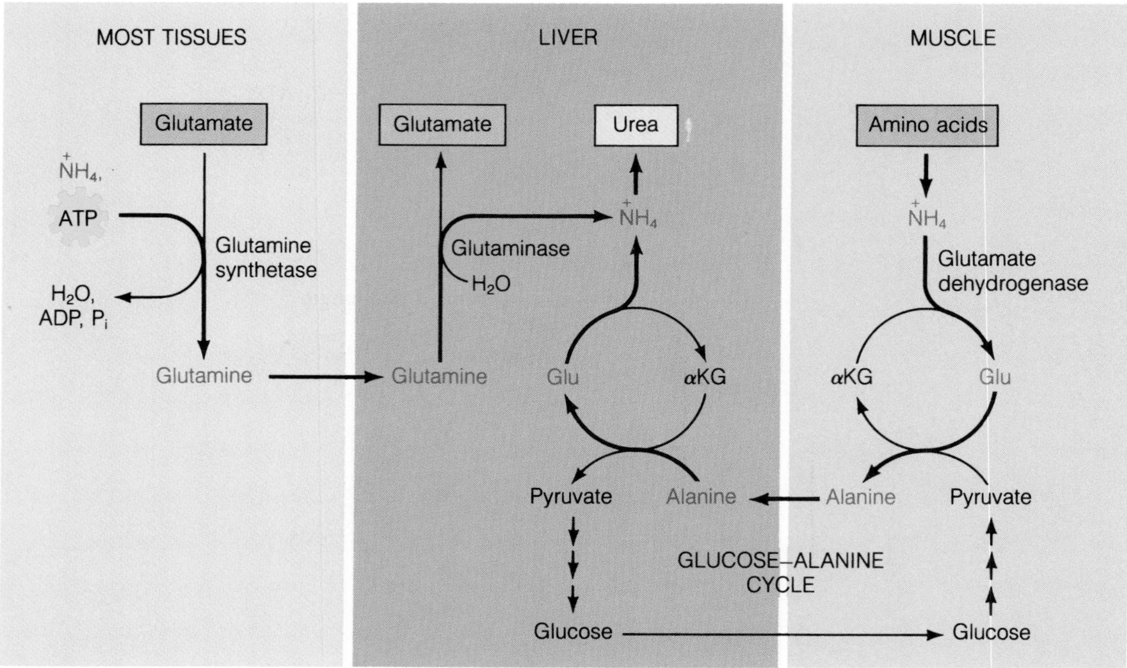

Figure 20.11
Transport of ammonia to the liver for urea synthesis, using as the carrier either glutamine (most tissues) or alanine (muscle).

(BUN) levels represent a sensitive clinical test of kidney function, since filtration and removal of urea are impaired in cases of malfunction. Analogously, blood ammonia measurements are a sensitive test of liver function. Liver damage, whether acute (hepatitis, poisoning) or chronic (alcoholic cirrhosis) reduces activity of the urea cycle. The accumulation of ammonia is toxic to the brain, and this is related to the comatose condition seen in advanced cases of chronic alcoholism.

The reactions of the urea cycle occur in both mitochondria and cytosol of liver cells. Glutamate dehydrogenase, the citric acid cycle enzymes, carbamoyl phosphate synthetase I, and ornithine transcarbamoylase are localized in the mitochondrion, while the rest of the cycle occurs in the cytosol.

Though the liver is the site of urea synthesis, all animal organs degrade amino acids and produce ammonia. Two mechanisms are involved in transporting ammonia from other tissues to liver for its eventual conversion to urea (Figure 20.11). Most tissues use glutamine synthetase to convert ammonia to the nontoxic, and electrically neutral, glutamine. The glutamine is transported in the blood to the liver, where it is cleaved hydrolytically by **glutaminase.**

$$\text{Glutamine} + H_2O \longrightarrow \text{glutamate} + \text{ammonia}$$

Muscle, which derives most of its energy from glycolysis, uses a different route, the glucose–alanine cycle. Glycolysis generates pyruvate, which undergoes transamination with glutamate to give alanine and α-ketoglutarate. The glutamate in turn has acquired its nitrogen from ammonia, via glutamate dehydrogenase. The resultant alanine is transported to the liver, where it loses its nitrogen by a reversal of the above processes. This yields ammonia for urea synthesis, plus pyruvate. The latter undergoes gluconeogenesis to give glucose, which is released to the blood for transport back to the muscle. This cyclic process helps the liver get rid of two unusable products—ammonia and pyruvate—with the carbon from the pyruvate being returned ultimately as the readily utilizable glucose.

Although urea is an end product, recent studies on hibernating animals have shown that urea can be reutilized for amino acid synthesis. During its six- or seven-month hibernation the black bear does not urinate. Urea that accumulates in the bladder is somehow reabsorbed and returned to the tissues for amino acid synthesis. The pathways of utilization are not known, but presumably involve hydrolytic cleavage of urea to ammonia. Current investigations are asking whether bears can synthesize any essential amino acids under these conditions.

Coenzymes Involved Primarily in Nitrogen Metabolism

Among the various coenzymes and water-soluble vitamins are three whose principal roles are played in amino acid and/or nucleotide metabolism. Although all have been mentioned before, we shall consider their actions in detail here. These cofactors include pyridoxal phosphate, the cofactor for transamination and other reactions of amino acid metabolism; the folic acid coenzymes, which transfer single-carbon functional groups in synthesizing nucleotides and certain amino acids; and the B_{12}, or cobalamin, coenzymes, which participate in the synthesis of methionine.

Pyridoxal Phosphate

Vitamin B_6 was discovered in the 1930s as the result of nutritional studies with rats fed vitamin-free diets. The vitamin as originally isolated is **pyridoxine,** named from its structural similarity to pyridine. Pyridoxine contains a hydroxymethyl group on the para position of the pyridine ring. However, in the coenzyme this group is oxidized to an aldehyde. In addition the hydroxymethyl group at position 5 is phosphorylated. Pyridoxal phosphate (which we shall abbreviate PyP) is the predominant coenzyme form, with pyridoxamine phosphate being an intermediate in transamination reactions.

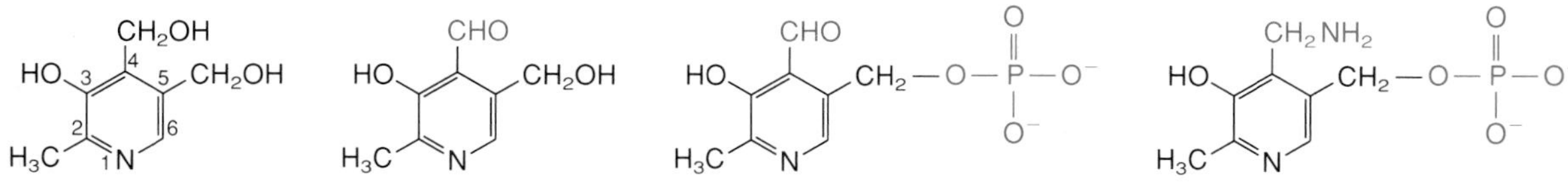

Pyridoxine **Pyridoxal** **Pyridoxal phosphate** **Pyridoxamine phosphate**

Nutritional requirements for vitamin B_6 are so low that dietary deficiency states are rarely observed in humans. However, many drugs and poisons *induce* deficiency states, usually by reacting with the aldehyde group and thereby sequestering the coenzyme. A well-understood situation is the induced B_6 deficiency arising during treatment of tuberculosis. The antimycobacterial agent **isoniazid** (isonicotinic acid hydrazide) reacts covalently with pyridoxal to make it unavailable for phosphorylation by pyridoxal kinase.

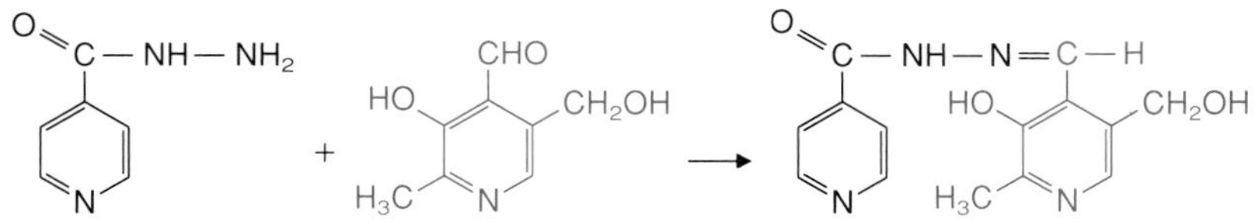

Isoniazid **Pyridoxal** **Hydrazone**

Because the mycobacterium contains low levels of the kinase, its growth is effectively blocked by the agent. Prolonged treatment with this drug can generate a B_6 deficiency in the patient by the same mechanism, unless the diet is supplemented with the vitamin.

Pyridoxal phosphate is a remarkably versatile coenzyme. In addition to its involvement in transamination reactions, it participates in amino acid decarboxylations, racemizations, and numerous modifications of amino acid side chains. A key clue to the mechanism by which this coenzyme functions was the finding in Esmond Snell's laboratory, in the 1940s, that all of the known PyP-requiring enzyme reactions can be catalyzed in the absence of any enzyme by pyridoxal itself. Certain metal ions are required as well, such as Al^{3+} or Cu^{2+}. Although the reaction rates were much lower than those catalyzed by enzymes, the model studies permitted a detailed analysis that led to formulation of a unified mechanism for the action of PyP-requiring enzymes. The metal ion was postulated to stabilize a Schiff base formed between pyridoxal and the amino acid substrate, shown at left. Normally this role would be played by an amino acid residue in the active site of the enzyme.

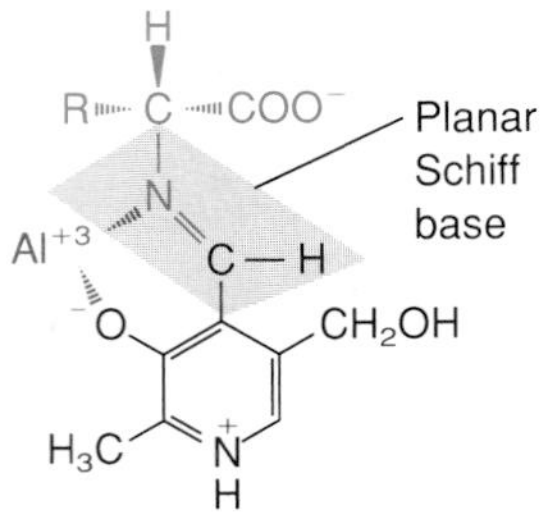

We now know that all pyridoxal phosphate-requiring enzymes act via the formation of a Schiff base between the amino acid and coenzyme. A cation, whether a metal as in the nonenzymatic model system or a proton, is essential to bridge the phenolate ion of the coenzyme and the imino nitrogen of the amino acid. This bridging maintains planarity of the structure, essential for catalysis. The most important catalytic feature of the coenzyme is the electrophilic nitrogen of the pyridine ring, which acts as an *electron sink,* drawing electrons away from the amino acid and stabilizing a carbanion intermediate. Interactions with the enzyme determine which bond is broken in the substrate and hence what specific reactions are catalyzed. *All of the known reactions of PyP enzymes can be described mechanistically in the same way:* formation of a planar Schiff base intermediate, followed by formation of a resonance-stabilized carbanion, generating a quinonoid intermediate, as shown in Figure 20.12. Depending on the bond labilized, this can lead to the first half of a transamination (as shown), to decarboxylation, to racemization, or to numerous side chain modification, such as β-elimination.

Although pyridoxal phosphate is the coenzyme for all of these reactions, we now know that the reactive species is not the aldehyde group, but rather an aldimine, formed between the coenzyme and an ϵ-amino group of a lysine residue in the active site. This bond can be reduced by sodium borohydride, to give irreversible bonding of the coenzyme to the active lysine residue.

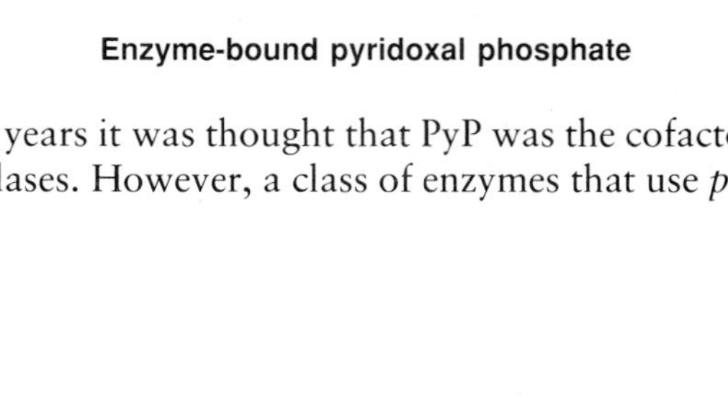

Enzyme-bound pyridoxal phosphate

For many years it was thought that PyP was the cofactor for all amino acid decarboxylases. However, a class of enzymes that use *pyruvic acid* as a cofactor is now known. Although structurally dissimilar to PyP, pyruvate evidently is mechanistically similar in this role; it is covalently bound to an amino group in the enzyme (although via an amide with its carboxyl group). Evidently during catalysis the electron-poor keto oxygen plays a role comparable to that of the PyP ring nitrogen, as shown below for the enzyme **histidine decarboxylase** of *Lactobacillus*.

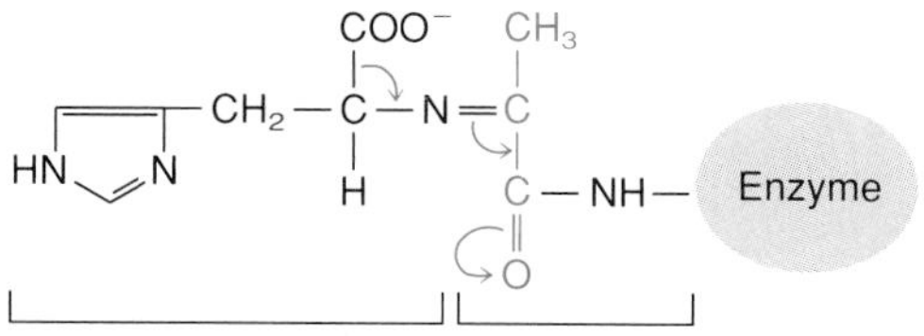

Histidine substrate Pyruvate

Tetrahydrofolate Coenzymes and One-Carbon Metabolism

Discovery and Chemistry of Folic Acid. The vitamin **folic acid** was discovered in the 1930s, when it was found that people with a certain type of **megaloblastic anemia** could be cured by treatment with yeast or liver extracts. The condition is characterized, like all anemias, by reduced levels of erythrocytes. The cells that remain are characteristically large and immature, suggesting a role for the vitamin in cell proliferation and/or maturation. The active component in the extracts was also shown to be essential for growth of chicks and to be required in the growth media for certain bacteria, notably *Lactobacillus casei* and *Streptococcus faecalis*. The latter findings allowed development of a rapid bioassay based on growth of these bacteria, and isolation and structural identification soon followed. The vitamin was found to be abundant in leafy green vegetables such as spinach, so it was named folic acid, from the same root as "foliage."

Chemically, folic acid is formed from three distinct moieties: (1) a bicyclic, heterocyclic **pteridine** ring, 6-methylpterin; (2) ***p*-aminobenzoic**

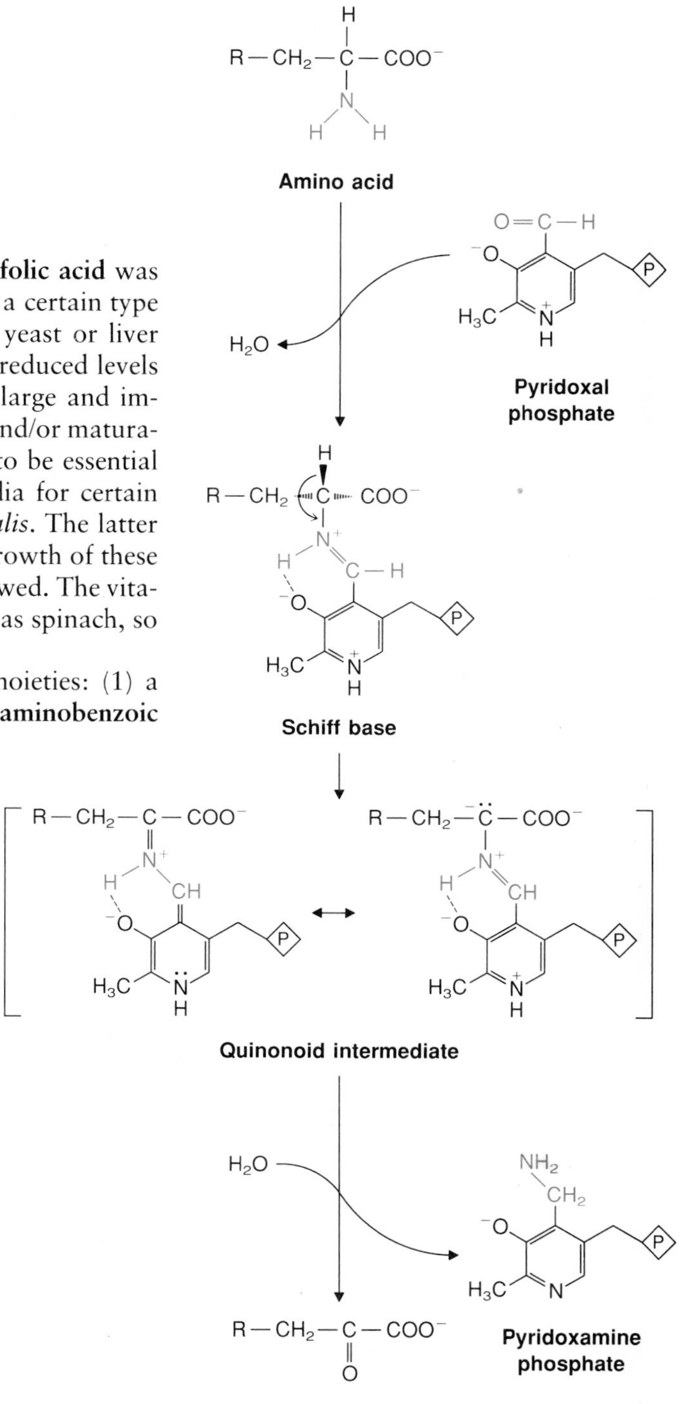

Amino acid

Pyridoxal phosphate

Schiff base

Quinonoid intermediate

Pyridoxamine phosphate

α-Keto acid

Figure 20.12
Involvement of pyridoxal phosphate in transamination. The figure shows the action of the positively charged pyridinium ion as an electron sink. Formation of a Schiff base intermediate leads to formation of a carbanion, which is resonance stabilized by interconversion with a quinonoid intermediate. This hydrolyzes to yield a keto acid product plus pyridoxamine phosphate. The stereochemistry of the initial Schiff base determines which bond is positioned to be broken and to form a carbanion intermediate.

acid (PABA), which is itself required for the growth of many bacteria; and (3) glutamic acid. These three moieties are shown in the overall structure given below.

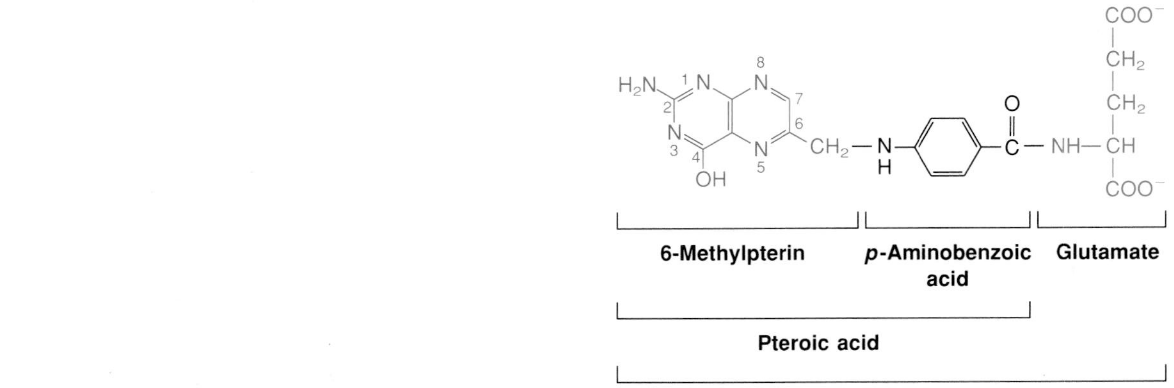

6-Methylpterin | *p*-Aminobenzoic acid | Glutamate

Pteroic acid

Pteroylglutamic acid (folic acid)

The pteridine ring was already known in nature, having been discovered in a large class of biological pigments. Insect wings and eyes contain pteridine pigments, as do the skins of amphibians and fish. Butterfly wings are particularly abundant in pteridines and were the first source from which any such compounds were identified structurally.

In the structure of folic acid, 6-methylpterin is linked through the amino group of PABA to form **pteroic acid,** which is linked in turn via an amide to glutamate, to form **pteroylmonoglutamate.** Naturally occurring folates may differ from this compound in the number of glutamate residues per molecule of vitamin, which ranges from three to eight or more. These residues are linked to one another, not by the familiar peptide bond, but rather by a modified peptide bond involving the α-amino group and the γ-carboxyl group. A typical representative of these folate polyglutamates is shown below.

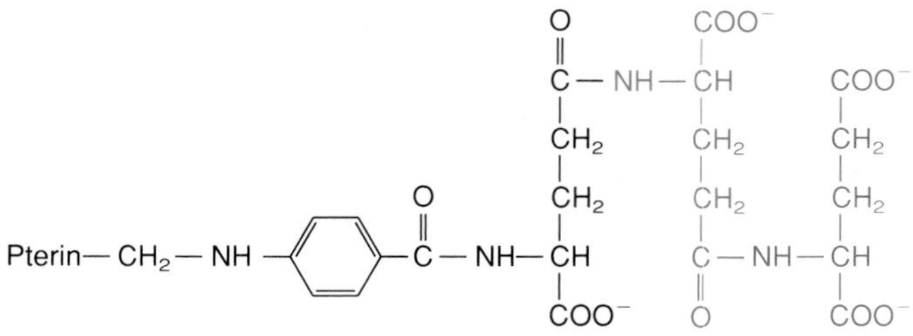

Pteroyl-γ-triglutamate

Most enzymes that use folate coenzymes bind more tightly to polyglutamated forms than to monoglutamates. The major need for the additional glutamate residues is probably for intracellular retention of folates. Most animal cells can transport only the monoglutamated form into cells. However, this form can also be transported out of cells, so conjugation with additional glutamate residues converts the folate to a form that cannot exit the cell.

Conversion of Folate to Tetrahydrofolate. Once inside a cell, folate is converted to active forms by two successive reductions of the pyrazine part of the pteridine ring. Both reactions are catalyzed by the NADPH-specific enzyme **dihydrofolate reductase.** The first reduction yields 7,8-

dihydrofolate (H_2folate), and the second reduction yields **5,6,7,8-tetrahydrofolate** (H_4folate).

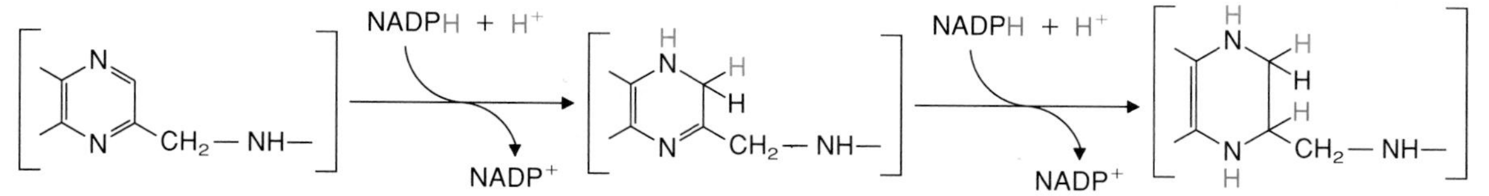

$NADPH + H^+$ → $NADP^+$

Folate → **Dihydrofolate** → **Tetrahydrofolate**

For reasons that will become clear later, dihydrofolate is the preferred substrate and, hence, the one whose name is lent to the enzyme.

Dihydrofolate reductase has been thoroughly studied because it is the target for action of a number of clinically useful **antimetabolites.** An antimetabolite is a synthetic compound, usually a structural analog of a normal metabolite, that interferes with the utilization of the metabolite to which it is related structurally. As early as 1948 two analogs of folate, **aminopterin** and **amethopterin** (also known as **methotrexate**), had been synthesized and found to induce remissions in acute leukemias.

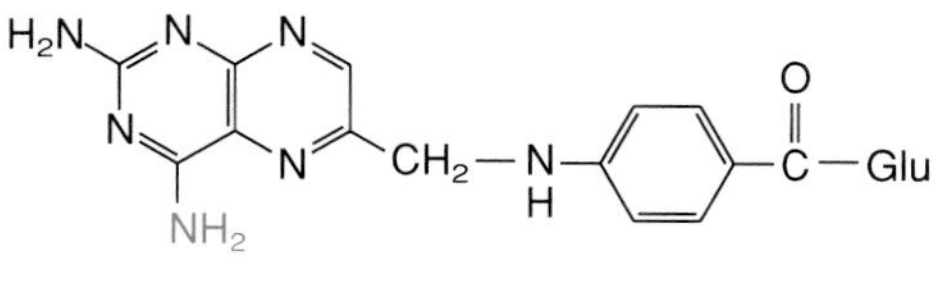

Aminopterin (4-aminofolate)

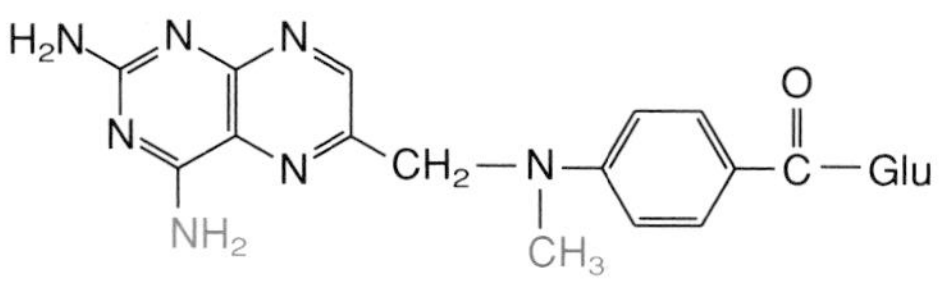

Amethopterin (4-amino-10-methylfolate)

A decade later it was found that these compounds are effective because they inhibit dihydrofolate reductase, binding to the enzyme at least a thousandfold more tightly than do the normal substrates. Thus, these analogs act to block the utilization of folate and dihydrofolate. Nearly two more decades passed before a detailed understanding of the mechanism of inhibition became available, through crystallization of enzyme–inhibitor complexes and determination of their three-dimensional structure. Figure 20.13 shows the residues that are in contact with the bound ligands of the methotrexate–dihydrofolate reductase complex.

Folate analogs such as methotrexate have been used in treating many different cancers in addition to leukemia. They are effective because of the involvement of dihydrofolate reductase in the biosynthesis of thymine nucleotides and, hence, of DNA. This relationship is discussed in more detail in Chapter 22.

The entire concept of antimetabolites as drugs arose from early work on folate metabolism. Before World War II one of the few effective antibacterial drugs available was **sulfanilamide,** one of the class of sulfonamide drugs. A British biochemist, D. D. Woods, noted a structural similarity between sulfanilamide and *p*-aminobenzoate, which was known to be essential for bacterial growth. Before anything was known about the relation between PABA and folic acid, Woods proposed that sulfanilamide acts by blocking the normal utilization of PABA, and he coined the term *antimetabolite*. PABA is not required for growth of animal cells, so the drug is not toxic to human cells. Years later, when the pathway of folate biosynthesis had been established, it was learned that Woods was right; the enzyme incorporating PABA is inhibited by sulfonamides. Since animals cells do not carry out the synthetic pathway, but instead take up fully formed folate from the diet, they are not harmed by the drug. The concept elucidated by Woods, of seeking a metabolic difference between normal tissue and a path-

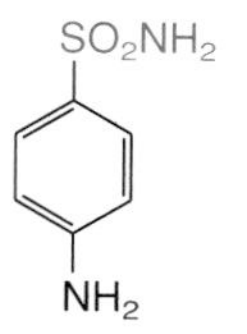

Sulfanilamide 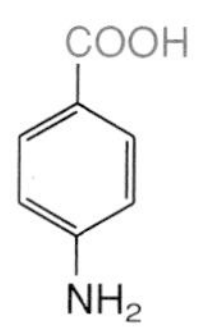 **p-Aminobenzoic acid**

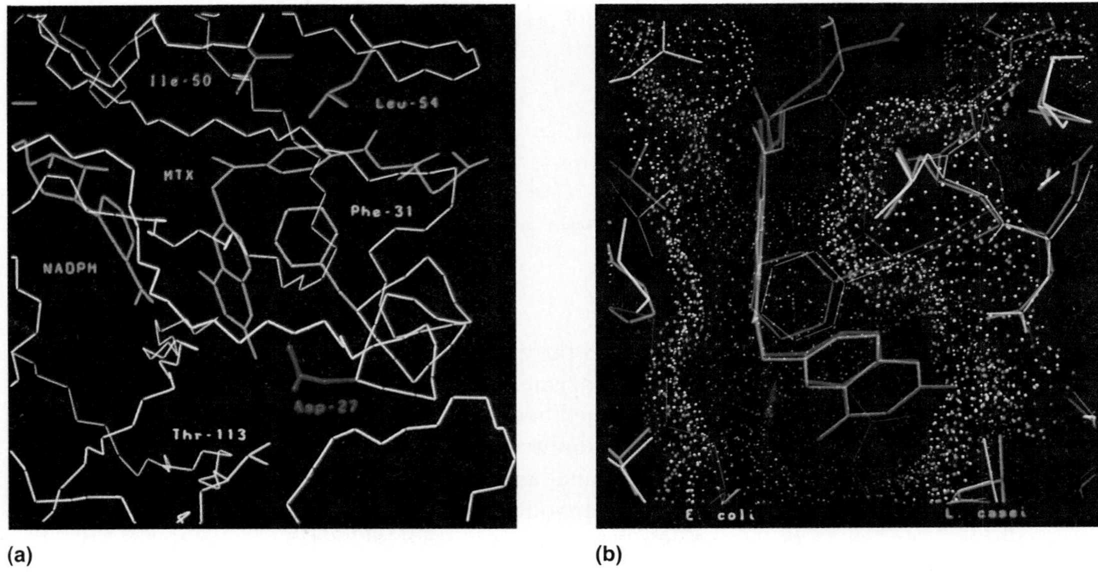

Figure 20.13
(**a**) A computer graphic representation of the active site of *E. coli* dihydrofolate reductase. The structure shows bound methotrexate (orange) and NADPH (turquoise), and amino acid residues important in binding ligands and in catalysis. (**b**) Superposition of the active sites of dihydrofolate reductases from *E. coli* (gold) and *Lactobacillus casei* (blue). All residues within 5 Å of the bound methotrexate were included. Even though there is just 55% homology between these two amino acid sequences, there is remarkable congruence between the three-dimensional structures.

ological process and of exploiting that difference chemically, has had an enormous impact on the field of pharmacology.

METABOLISM OF SINGLE-CARBON UNITS. The coenzymatic function of tetrahydrofolate is the mobilization and utilization of single-carbon functional groups. These reactions are involved in the metabolism of serine, glycine, methionine, and histidine, among the amino acids, and in the biosynthesis of purine nucleotides and the methyl group of thymine.

Tetrahydrofolate binds single-carbon units at the methyl, methylene, and formyl oxidation levels, equivalent in oxidation level to methanol, formaldehyde, and formic acid, respectively (Figure 20.14). Single-carbon groups on H_4folate can be carried on N^5 or N^{10}, or bridged between N^5 and N^{10}. Formation of a cyclic bridged adduct involves dehydration, so a methylene group (—CH_2—) is formally equivalent to a hydroxymethyl group in an unbridged compound, while a formyl group cyclizes to become a **methenyl** group (—CH═). In addition, there is a single-carbon adduct of tetrahydrofolate in which the single-carbon unit contains a nitrogen atom as well. In this **formimino** group (—C═NH), the carbon atom is at the same oxidation level as a formyl group.

Tetrahydrofolate can acquire single-carbon units from diverse sources. For example, bacteria that can grow on formate as the sole carbon source do so via the ATP-dependent activation of formate to N^{10}-formyl-H_4folate. The degradation of histidine, in both bacterial and animal cells, yields N^5-formimino-H_4folate, as does the bacterial fermentation of purines. However, most organisms derive most of their activated single-carbon units from the β-carbon of serine and the subsequent oxidation of glycine. The former reaction is catalyzed by **serine transhydroxymethylase.** This is a reversible reaction that, in the direction shown,

$$\underset{\textbf{Serine}}{H{-}\underset{COO^-}{\overset{CH_2OH}{C}}{-}\overset{+}{N}H_3} + \underset{\textbf{H}_4\textbf{folate}}{\text{N(H)},\ CH_2{-}HN{-}} \underset{}{\overset{PyP}{\rightleftharpoons}} \underset{\textbf{Glycine}}{H_2C{-}\overset{+}{N}H_3,\ COO^-} + \underset{\boldsymbol{N^5,N^{10}}\textbf{-Methylene-H}_4\textbf{folate}}{\text{N},\ CH_2,\ CH_2{-}N{-}} + H_2O$$

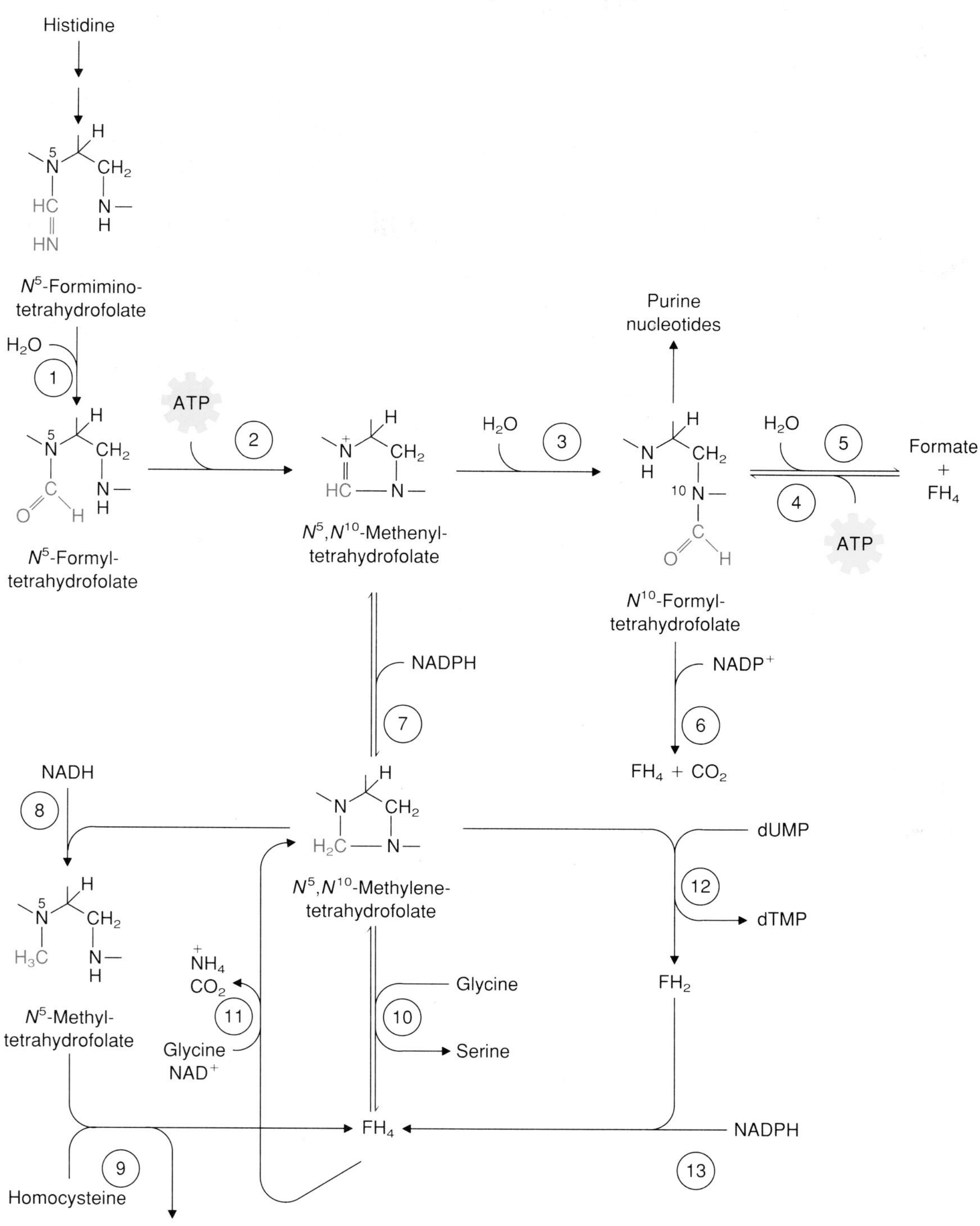

Figure 20.14
Metabolic reactions involving synthesis, interconversion, and utilization of single-carbon adducts of tetrahydrofolate. FH_2 and FH_4 are abbreviations for dihydrofolate and tetrahydrofolate, respectively. Enzymes involved are: (1), cyclodeaminase; (2), methenyltetrahydrofolate synthetase; (3), methenyltetrahydrofolate cyclohydrolase; (4), formyltetrahydrofolate synthetase; (5), formyltetrahydrofolate hydrolase; (6), formyltetrahydrofolate dehydrogenase; (7), methylenetetrahydrofolate dehydrogenase; (8), methylenetetrahydrofolate reductase; (9), homocysteine methyltransferase (also called methionine synthetase); (10), serine transhydroxymethylase; (11), glycine cleavage system; (12), thymidylate synthase; (13), dihydrofolate reductase.

yields glycine and N^5,N^{10}-methylene-H_4folate but that can also be used for serine biosynthesis as needed. The enzyme also requires pyridoxal phosphate, so the actual substrate is the Schiff base formed from serine and PyP.

Glycine can yield an additional molecule of N^5,N^{10}-methylene-H_4folate through action of the **glycine cleavage system,** a multienzyme complex located in mitochondria. This represents the chief catabolic route for glycine in most organisms. The overall reaction is as shown,

$$\text{Glycine} + \text{H}_4\text{folate} + \text{NAD}^+ \longrightarrow N^5,N^{10}\text{-methylene-H}_4\text{folate} + \text{CO}_2 + \text{NH}_3 + \text{NADH} + \text{H}^+$$

but the overall pathway is mechanistically similar to that catalyzed by the pyruvate dehydrogenase complex. Four proteins are involved, along with pyridoxal phosphate, FAD, and lipoic acid.

Once a single-carbon unit has been activated via its attachment to tetrahydrofolate, it can undergo interconversions such as change in oxidation state, or it can be used directly in a biosynthetic reaction. Figure 20.14 shows most of the known reactions involving tetrahydrofolate coenzymes. Note the reactions that involve change in the oxidation level of the bound single-carbon unit—the reversible oxidation of N^5,N^{10}-methylene-H_4folate to N^5,N^{10}-methenyl-H_4folate, catalyzed by **N^5,N^{10}-methylenetetrahydrofolate dehydrogenase** (reaction 7), and the irreversible reduction of N^5,N^{10}-methylene-H_4folate to the 5-methyl derivative, brought about by **N^5,N^{10}-methylenetetrahydrofolate reductase** (reaction 8). Many organisms contain multifunctional enzymes or complexes that help to channel these scarce and/or unstable intermediates; most eukaryotes combine activities 3, 4, and 7 into one trifunctional protein.

As shown in Figure 20.14, single-carbon units derived from tetrahydrofolate coenzymes are used in synthesis of purine nucleotides, thymine nucleotides, and methionine, in addition to the reactions we have discussed. Further, in prokaryotes N^{10}-formyl-H_4folate participates in synthesis of **N-formylmethionyl-tRNA,** which is involved in initiation of protein synthesis. In the synthesis of thymine nucleotides, catalyzed by **thymidylate synthase**

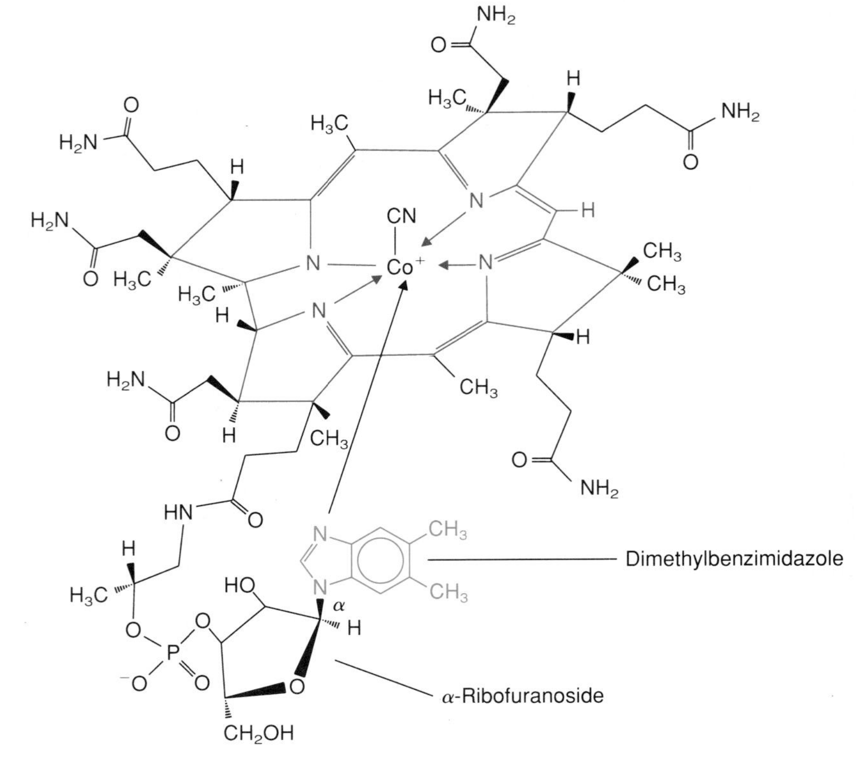

Figure 20.15
Structure of vitamin B_{12} as originally isolated (cyanocobalamin). The corrin ring is shown in red.

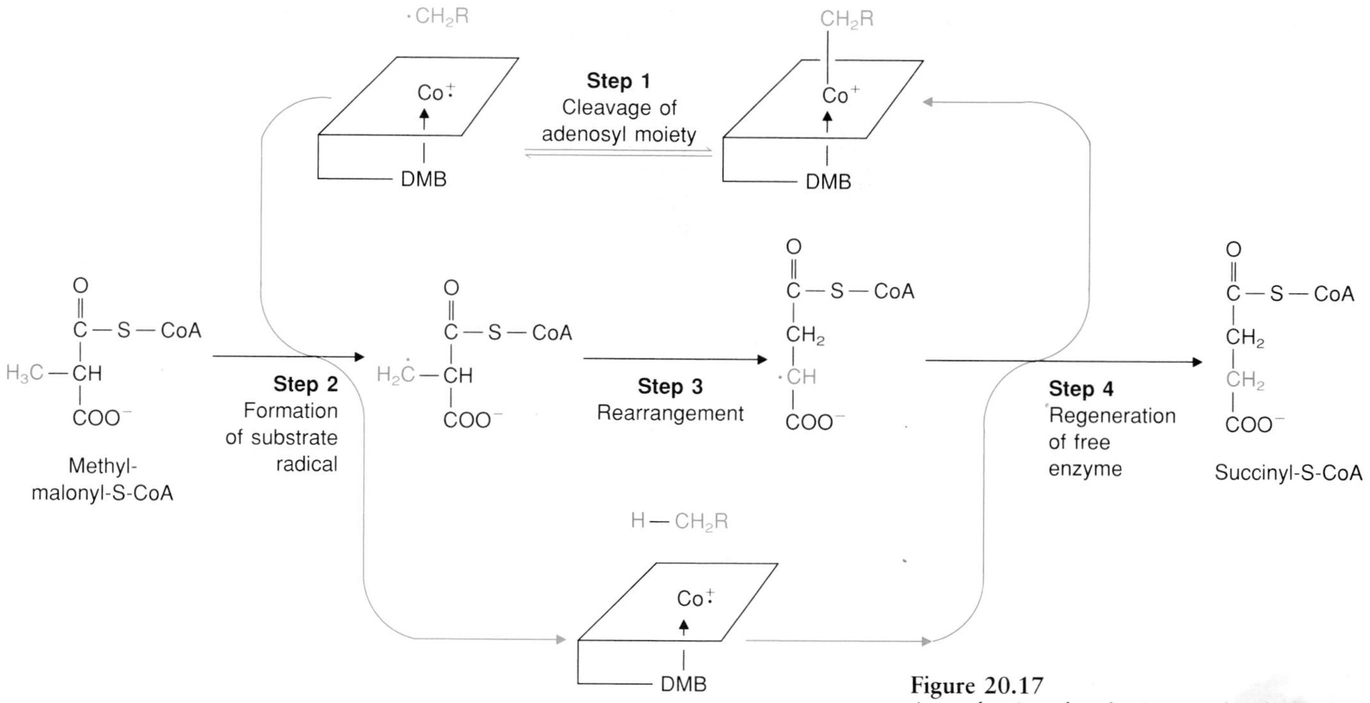

Figure 20.17
A mechanism for the intramolecular rearrangement catalyzed by methylmalonyl-CoA mutase. This mechanism is consistent with experimental observations and may account for other B_{12}-dependent rearrangments. CH_2R is the deoxyadenosyl moiety. The precise mechanism by which the free radical form of the substrate rearranges is not known.

ever, this treatment only hastens the onset of the far more severe neurological symptoms. Since these neurological problems are never seen in simple folate deficiency, what is the metabolic relationship between these two vitamins?

As outlined in Figure 20.18, the probable explanation is as follows. (1) When B_{12} levels are low, flux through the methionine synthetase reaction decreases; since adequate dietary methionine is probably available, there is no immediate disturbance of protein metabolism. (2) Reduction of N^5,N^{10}-methylene-H_4folate to N^5-methyl-H_4folate continues. (3) Since methionine synthetase is the only mammalian enzyme known to act on N^5-methyl-H_4folate, the decreased intracellular activity of this enzyme causes N^5-methyl-H_4folate to accumulate, at the expense of depleted pools of the other tetrahydrofolate coenzymes. Thus, even though total folate levels may seem ample, there is a *functional* folate deficiency, with insufficient levels of the formyl and methylene derivatives needed for synthesis of nucleic acid precursors.

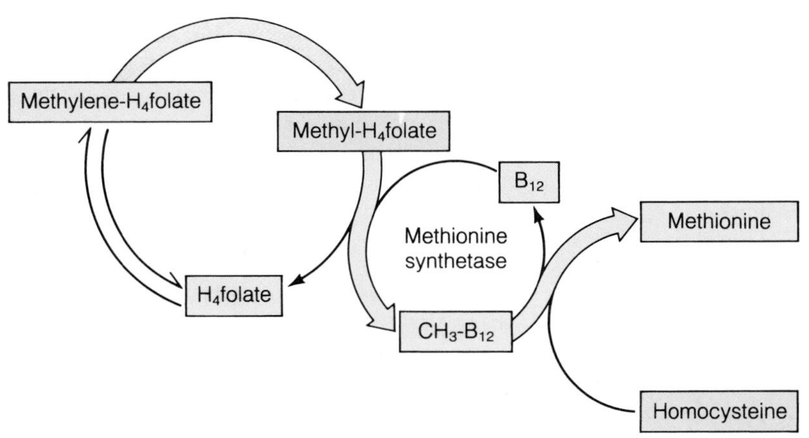

Figure 20.18
A relationship between folate and B_{12} metabolism, based on the apparent folate deficiency present in early stages of B_{12} deficiency. The diagram identifies intermediates that either accumulate (green) or are depleted (red) as a result of decreased flux through methionine synthetase.

This does not explain why untreated pernicious anemia progresses to a neurological disease, because simple folate deficiency anemias show no such complications. According to one hypothesis, abnormalities in fatty acid metabolism develop because of reduced flux through the methylmalonyl-CoA mutase reaction. This in turn could lead to abnormalities in the lipid components of nervous tissue.

At this stage we have considered general features of nitrogen utilization and excretion, general features of amino acid synthetic and degradative pathways, and all of the coenzymes known to function in amino acid and nucleotide metabolism. Thus, we are prepared to consider the synthesis, metabolic functions, and degradation of individual amino acids and nucleotides.

REFERENCES

Inorganic Nitrogen Metabolism

Evans, H. J., A. R. Harker, H. Papen, S. J. Russell, F. J. Hanus, and M. Zuber (1987) Physiology, biochemistry, and genetics of the uptake hydrogenase in rhizobia. *Annu. Rev. Microbiol.* 41:335–361. This review describes the mechanism of hydrogen evolution during nitrogen fixation, as well as routes to harness the loss of energy involved.

Gallon, J. R., and A. E. Chaplin (1987) *An Introduction to Nitrogen Fixation.* Cassell, London. A recent book-length review.

General Aspects of Nitrogen Metabolism

Guppy, M. (1986) The hibernating bear: Why is it so hot, and why does it cycle urea through the gut? *Trends Biochem. Sci.* 11:274–276. A discussion of the metabolic peculiarities of hibernating animals vis-à-vis protein metabolism.

Umbarger, H. E. (1978) Amino acid biosynthesis and its regulation. *Annu. Rev. Biochem.* 47:533–557. A discussion focused on regulation in plant, bacterial, and fungal systems, written by a leading contributor to the field.

Walsh, C. T. (1979) *Enzymatic Reaction Mechanisms.* Freeman, San Francisco. An excellent book, particularly valuable in the context of amino acid metabolism, one-carbon metabolism, cobalamin coenzymes, and oxygenases.

Wellner, D., and A. Meister (1981) A survey of inborn errors of metabolism and transport in man. *Annu. Rev. Biochem.* 50:911–968. Reveals how much can be learned about the metabolic roles of enzymes through the study of rare human mutations.

Protein Turnover

Davies, K. J. (1986) Intracellular proteolytic systems may function as secondary antioxidant defenses: An hypothesis. *J. Free Radicals Biol. Med.* 2:155–173. This article summarizes evidence that protein turnover occurs largely to remove from cells proteins that have suffered irreversible oxidative damage.

Hershko, A. (1988) Ubiquitin-mediated protein degradation. *J. Biol. Chem.* 263:15237–15240. A recent minireview of this process and its role in protein turnover.

Rechsteiner, M., S. Rogers, and K. Rote (1987) Protein structure and intracellular stability. *Trends Biochem. Sci.* 12:390–394. A minireview of the enzymology, mechanisms, and biological roles of protein turnover.

Folate and B_{12} Coenzymes

Benkovic, S. J., and R. L. Blakley (1984) *Folates and Pterins,* Vol. 1. Academic Press, New York. A book-length review, which covers one-carbon metabolism and mechanisms of action of folate conenzymes.

Benkovic, S. J., C. A. Fierke, and A. M. Naylor (1988) Insights into enzyme function from studies on mutants of dihydrofolate reductase. *Science* 239:1105–1110. Because of its importance as a target for chemotherapy, this enzyme has received much attention. This article summarizes mechanistic studies of altered enzymes produced by site-directed mutagenesis.

Halpern, J. (1985) Mechanisms of coenzyme B_{12}-dependent rearrangements. *Science* 227:869–875. Discussion of experimental evidence for the mechanism outlined in this chapter.

Metzler, D. E. (1977) *Biochemistry, the Chemical Reactions of Living Cells,* pp. 444–464. Academic Press, New York. Arguably the best available review of pyridoxal phosphate mechanisms, by one of the masters of the field.

PROBLEMS

1. Identify the most likely additional substrates, products, and coenzymes for each reaction in the following imaginary pathway.

$$^{-}OOC{-}CH(NH{-}CHO){-}CH_2{-}CH_2{-}CH_2{-}CH(\overset{+}{N}H_3){-}COO^{-} \longrightarrow {}^{-}OOC{-}CH(NH{-}CHO){-}CH_2{-}CH(CH_3){-}CH(\overset{+}{N}H_3){-}COO^{-} \longrightarrow {}^{-}OOC{-}CH(\overset{+}{N}H_3){-}CH_2{-}CH(CH_3){-}CH(\overset{+}{N}H_3){-}COO^{-} \longrightarrow$$

$$H_2N{-}C({=}O){-}CH(\overset{+}{N}H_3){-}CH_2{-}CH(CH_3){-}CH(\overset{+}{N}H_3){-}COO^{-} \longrightarrow H_2N{-}C({=}O){-}CH(\overset{+}{N}H_3){-}CH_2{-}CH(CH_3){-}C({=}O){-}COO^{-} \longrightarrow H_2N{-}C({=}O){-}CH(NH{-}CH_3){-}CH_2{-}CH(CH_3){-}C({=}O){-}COO^{-}$$

2. The diagram below shows the biosynthesis of B_{12} coenzymes starting with the vitamin. DMB is dimethylbenzimidazole.

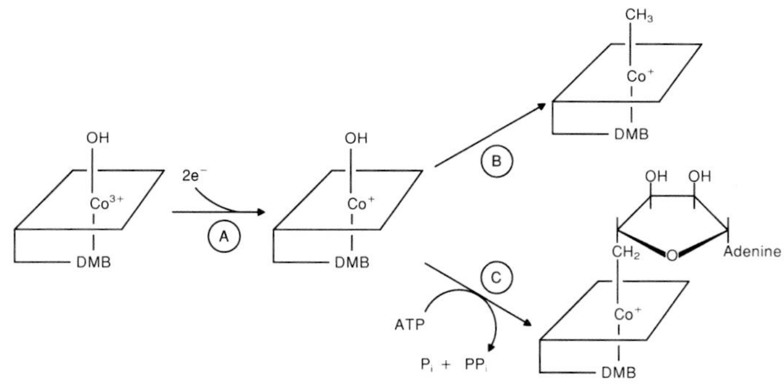

(a) What one additional substrate or cofactor is required by enzyme B?

(b) Genetic deficiency in animals of enzyme C would result in excessive urinary excretion of what compound?

(c) Some forms of the condition described above can be successfully treated by injection of rather massive doses of vitamin B_{12}. What kind of genetic alteration in the enzyme would be consistent with this result?

(d) Genetic deficiency in animals of enzyme B will result in excessive urinary excretion of what amino acid?

3. Using the principles described in the text regarding pyridoxal phosphate mechanisms, propose a mechanism for the reaction catalyzed by serine transhydroxymethylase.

4. A physician treating a patient with megaloblastic anemia might be tempted to treat the patient with folic acid, to see if a simple folate deficiency is involved. If the patient has pernicious anemia, this treatment hastens the onset of the far more serious neurological symptoms. Why?

5. Use numbers 1 to 5 to identify each carbon atom in the product of this reaction. What is the coenzyme?

$$^{-}\overset{5}{O}OC{-}\overset{4}{C}H_2{-}\overset{3}{C}H_2{-}\overset{2}{C}H(\overset{+}{N}H_3){-}\overset{1}{C}OO^{-} \longrightarrow {}^{-}OOC{-}CH(CH_3){-}CH(\overset{+}{N}H_3){-}COO^{-}$$

6. Based on the mechanism for methylmalonyl-CoA mutase shown in Figure 20.17, propose a mechanism for the diol dehydrase reaction.

$$HO{-}CH_2{-}CH_2{-}OH \longrightarrow CH_3CHO$$

7. The precise mechanism of ammonia toxicity to the brain is not known. Speculate on a possible mechanism, based on possible effects of ammonia on levels of key intermediates in energy generation.

8. Mutants of *Neurospora crassa* that lack carbamoyl-phosphate synthetase I (CPS I) require arginine in the medium in order to grow, while mutants that lack carbamoyl-phosphate synthetase II (CPS II) require a pyrimidine, such as uracil. A priori, one would expect the active CPS II in the arginine mutants to provide sufficient carbamoyl phosphate for arginine synthesis, and the active CPS I in the pyrimidine mutants to "feed" the pyrimidine pathway. Explain these observations.

9. In some forms of leukemia the proliferating white blood cells contain very low levels of asparagine synthetase. Some years ago there was interest in treating these leukemias by purifying the enzyme asparaginase from *E. coli* and injecting it into the bloodstream of leukemic patients. Asparaginase catalyzes the hydrolysis of asparagine to aspartate plus ammonia. What is the rationale behind this mode of therapy, and why might you expect it not to work?

10. Indicate whether each statement is true or false, and briefly explain your answer.

(a) In general, the metabolic oxidation of protein in mammals is less efficient, in terms of energy conserved, than the metabolic oxidation of carbohydrate or fat.

(b) Since the nitrogen of glutamate can be redistributed by transamination, glutamate should be a good supplement for nutritionally poor proteins.

(c) Arginine is a nonessential amino acid for mammals, because liver is abundant in the enzymes of arginine synthesis.

(d) Alanine is an essential amino acid because it is a constituent of every protein.

CHAPTER 21

Metabolism of Nitrogenous Compounds: Amino Acids

In considering the metabolism of individual amino acids our approach is to organize the 20 amino acids into families or groups that are related structurally or metabolically. Within each family we shall discuss biosynthesis and degradation of each amino acid and the major roles of each as an intermediate en route to metabolites other than protein. Because of the great diversity of amino acid functions, and because of the complexity of some of the biosynthetic and catabolic pathways, we shall devote most of our attention to the processes and pathways that are most widespread in biology, that interrelate with other metabolic processes, that illustrate important biological principles, or that are mechanistically interesting. In many cases descriptions of the pathways are presented largely through the illustrations.

The Citric Acid Cycle Superfamily

About half of the 20 amino acids are biosynthesized more or less directly from intermediates in the citric acid cycle or from pyruvate. We include here glutamate, aspartate, and alanine, which can be formed by transamination from α-ketoglutarate, oxaloacetate, and pyruvate, respectively. The family also includes glutamine and asparagine, which are formed directly from glutamate and aspartate, respectively; and proline and arginine, which are formed in short pathways from glutamate. Several other amino acids—threonine, methionine, and isoleucine—are derived from aspartate, but we have assigned them to other families and will consider their metabolism later in the chapter.

The metabolic relationships that will concern us in this citric acid cycle "superfamily" are outlined in Figure 21.1.

Synthesis and Catabolism of Glutamate, Aspartate, Alanine, Glutamine, and Asparagine

Transamination provides major routes for both synthesis and degradation of glutamate, aspartate, and alanine. Glutamate dehydrogenase and glutamate synthase, discussed in Chapter 20, present additional routes for glutamate synthesis. Glutamine is synthesized from glutamate by glutamine synthetase, and its major catabolic route is hydrolysis to glutamate by **glutaminase,** an important enzyme in ammonia metabolism in animals. Similarly, asparagine is synthesized from aspartate by asparagine synthetase, while hydrolytic cleavage back to aspartate, catalyzed by **asparaginase,** is

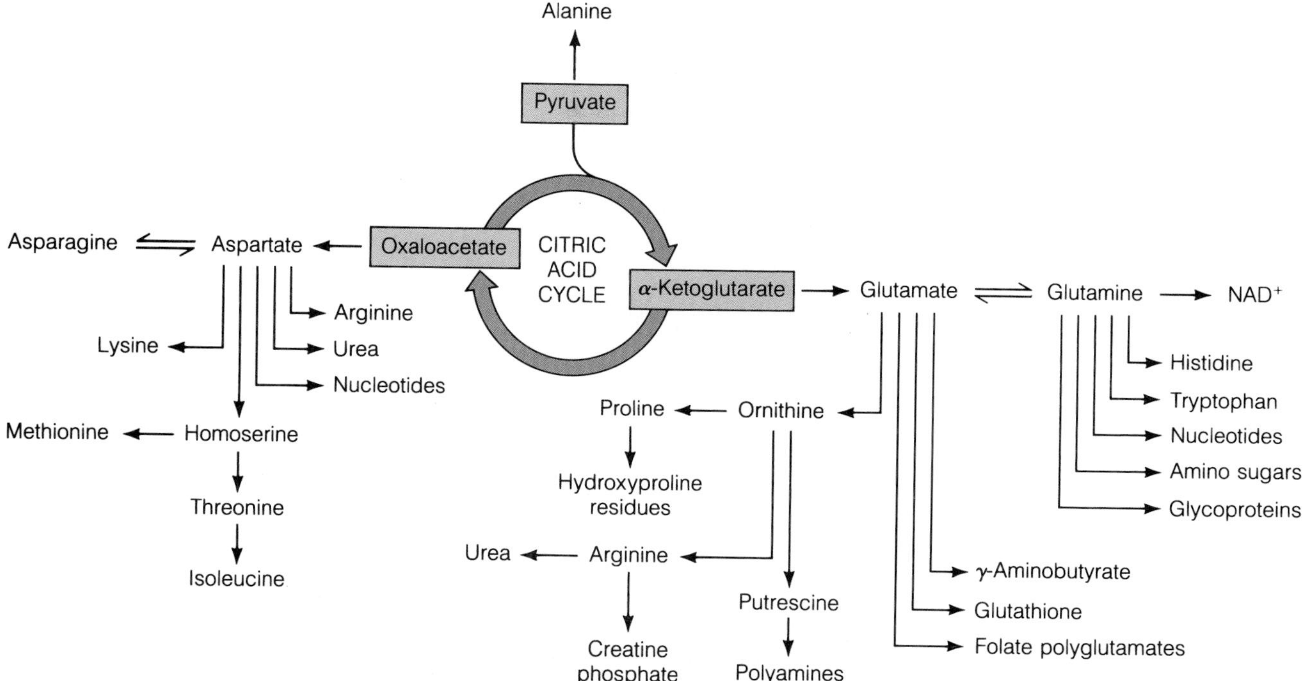

Figure 21.1
Metabolic relationships involving the family of amino acids derived from citric acid cycle intermediates.

the principal degradative pathway. Aspartate is catabolized either by transamination or by deamination by **aspartase,** to give fumarate and ammonia.

Alanine and asparagine both have quite limited metabolic functions in addition to serving as protein constituents. In animals, alanine participates in the glucose–alanine cycle as a carrier of carbon for gluconeogenesis from muscle to liver. Glutamine, aspartate, and glutamate, on the other hand, are all extremely active. Glutamine contributes its amide nitrogen in reactions leading to other amino acids, to nucleotides, and to amino sugars and glycoproteins. These reactions are catalyzed by a group of similar ATP-dependent ligases, the **amidotransferases.** One amidotransferase reaction was presented in Chapter 16—the biosynthesis of glucosamine-6-phosphate from fructose-6-phosphate, involved in the biosynthesis of amino sugars. Other amidotransferase reactions are presented later in this chapter and in Chapter 22.

The nitrogen of aspartate is used in the biosynthesis of arginine and urea, as noted in Chapter 20. Similar reactions are involved in purine nucleotide synthesis, and the entire aspartate molecule is used in pyrimidine nucleotide biosynthesis; both of these processes are discussed in Chapter 22. Finally, in plants and bacteria aspartate is a precursor to three other amino acids via its conversion to **homoserine,** as shown below. Separate pathways then lead from homoserine to methionine, threonine, and isoleucine.

$$\underset{\textbf{Aspartate}}{^{-}OOC{-}CH_2{-}CH(\overset{+}{N}H_3){-}COO^{-}} \xrightarrow[\text{Aspartokinase}]{ATP \rightarrow ADP} \underset{\boldsymbol{\beta}\textbf{-Aspartyl phosphate}}{O{=}C(O{-}PO_3^{2-}){-}CH_2{-}CH(\overset{+}{N}H_3){-}COO^{-}} \xrightarrow{NADPH + H^+ \rightarrow NADP^+ + P_i} \underset{\textbf{Aspartic }\boldsymbol{\beta}\textbf{-semialdehyde}}{CHO{-}CH_2{-}CH(\overset{+}{N}H_3){-}COO^{-}} \xrightarrow{NADPH + H^+ \rightarrow NADP^+} \underset{\textbf{Homoserine}}{CH_2OH{-}CH_2{-}CH(\overset{+}{N}H_3){-}COO^{-}}$$

Also, in bacteria and plants, aspartic β-semialdehyde is a precursor to lysine.

The first enzyme in the above pathway, **aspartokinase,** is a major site for regulation of each biosynthetic pathway. *Escherichia coli* and other bacteria contain three distinct forms of aspartokinase, each with its own mode of allosteric regulation. In *E. coli* one form is specifically inhibited by threonine and one by lysine. The third enzyme is not feedback-inhibited by methionine, but synthesis of this particular enzyme is repressed by methionine.

Intermediary Metabolism of Glutamate

BIOSYNTHESIS OF ORNITHINE, ARGININE, AND CREATINE PHOSPHATE. As you can see in Figure 21.1, glutamate is perhaps the most active of all the amino acids. An important reaction of glutamate is the energy-requiring reduction of the γ-carboxyl group, to give *glutamic γ-semialdehyde* (Figure 21.2). This is comparable to the reduction of aspartate to aspartic semialdehyde. In plants and bacteria glutamate is acetylated before reduction to **N-acetylglutamate,** with the acetyl group being removed a couple of steps later. Acetylation prevents cyclization of the molecule after reduction, by condensation between the resulting carbonyl carbon and the amino group.

Glutamic γ-semialdehyde is a precursor to both proline and ornithine, the latter via a transamination at the aldehyde group. You know already that ornithine is involved in synthesizing arginine and urea, the latter via the Krebs–Henseleit cycle. Arginine is involved in a cyclic pathway in muscle, leading to the energy storage compound creatine phosphate (Figure 21.3). The guanidino group is transferred to glycine, with the regeneration of ornithine. The resulting compound, **guanidinoacetic acid,** is methylated by *S*-adenosylmethionine to give creatine, followed by phosphorylation of the latter to creatine phosphate, catalyzed by creatine kinase.

Ornithine undergoes decarboxylation to give 1,4-diaminobutane, which has the trivial name **putrescine** because of its original isolation from rotting meat. Putrescine is the precursor to a class of important and ubiquitous compounds, the **polyamines.** Since their synthesis involves *S*-adenosylmethionine as well, we shall defer this discussion to the next section, when we present the metabolism of methionine.

BIOSYNTHESIS OF PROLINE, HYDROXYPROLINE, AND COLLAGEN. Glutamic γ-semialdehyde leads in a short pathway to proline. First the aldehyde and amino group undergo spontaneous cyclization to give **Δ^1-pyrroline carboxylic acid.** An NADPH-dependent reduction follows, to give proline.

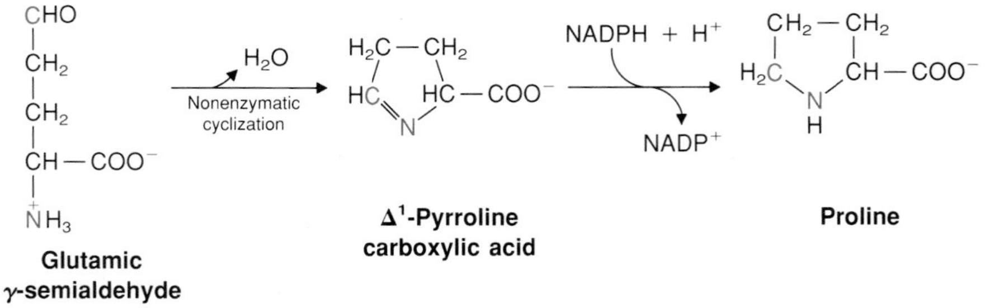

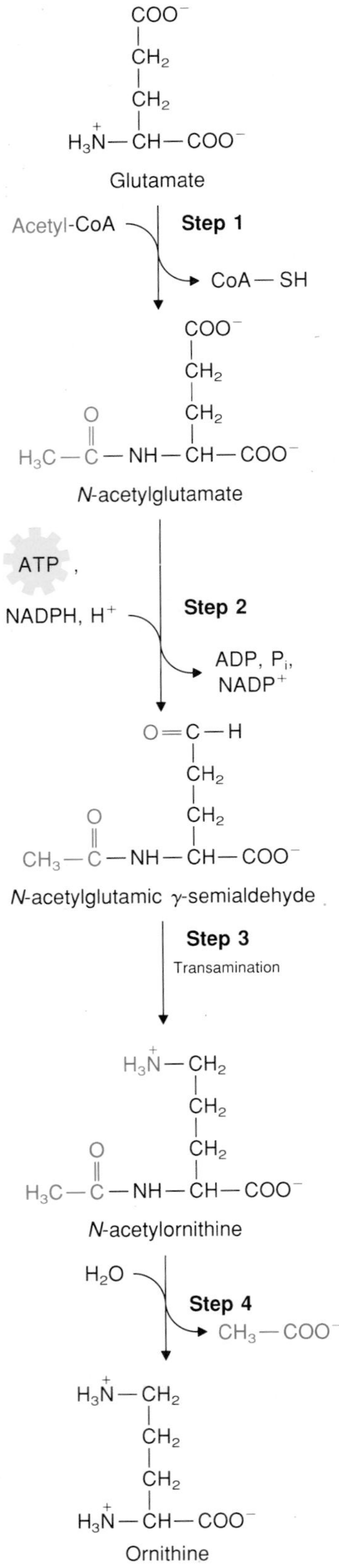

CHO
CH2
CH2
CH—COO⁻
NH3+
Glutamic γ-semialdehyde

H2O
Nonenzymatic cyclization

H2C—CH2
HC HC—COO⁻
N
Δ^1-Pyrroline carboxylic acid

NADPH + H+
NADP+

CH2—CH2
H2C CH—COO⁻
N
H
Proline

COO⁻
CH2
CH2
H3N+—CH—COO⁻
Glutamate

Acetyl-CoA
Step 1
CoA—SH

COO⁻
CH2
CH2
O
H3C—C—NH—CH—COO⁻
N-acetylglutamate

ATP,
NADPH, H+
Step 2
ADP, Pi,
NADP+

O=C—H
CH2
CH2
O
CH3—C—NH—CH—COO⁻
N-acetylglutamic γ-semialdehyde

Step 3
Transamination

H3N+—CH2
CH2
CH2
O
H3C—C—NH—CH—COO⁻
N-acetylornithine

H2O
Step 4
CH3—COO⁻

H3N+—CH2
CH2
CH2
H3N+—CH—COO⁻
Ornithine

Figure 21.2
Biosynthesis of ornithine from glutamate. The reduction of *N*-acetylglutamate (step 2) probably begins with phosphorylation of the carboxyl group by ATP, followed by NADPH-dependent reduction of the activated carboxyl group.

In plants glutamic γ-semialdehyde is synthesized from *N*-acetylglutamic γ-semialdehyde by transfer of the acetyl group on the latter to glutamate.

$$N\text{-Acetylglutamic } \gamma\text{-semialdehyde} + \text{glutamate} \rightleftharpoons N\text{-acetylglutamate} + \text{glutamic } \gamma\text{-semialdehyde}$$

Proline serves as a precursor to **hydroxyproline,** an amino acid found exclusively in structural proteins. These include collagen, the principal protein of connective tissue in animals, and cell wall proteins of still undetermined function in plants. Early labeling studies in animals showed that free hydroxyproline is poorly incorporated into collagen, whereas proline is readily incorporated, with label found in both proline and hydroxyproline residues. This led to the finding that hydroxyproline residues are generated by **posttranslational modification,** reactions that occur after completion of a polypeptide chain. The nonhydroxylated collagen precursor is called **procollagen.** In this polypeptide, a proline residue two positions to the carboxyl side of a glycine residue is the preferred substrate for the action of **procollagen-proline hydroxylase** (Figure 21.4). This unusual enzyme requires ferrous iron, ascorbic acid, molecular oxygen, and α-ketoglutarate. The latter is oxidized during the reaction to succinate and CO_2. One of the atoms from O_2 is incorporated into succinate, while the other ends up in the hydroxyproline hydroxyl group. While of great interest mechanistically and from the standpoint of control of collagen formation, the reaction is perhaps of greatest interest because it represents one of the few well-defined roles for ascorbic acid, or vitamin C. A vitamin C deficiency, or scurvy, involves defects in connective tissue function, and it seems likely that these derive from defective synthesis or maturation of collagen in connective tissue.

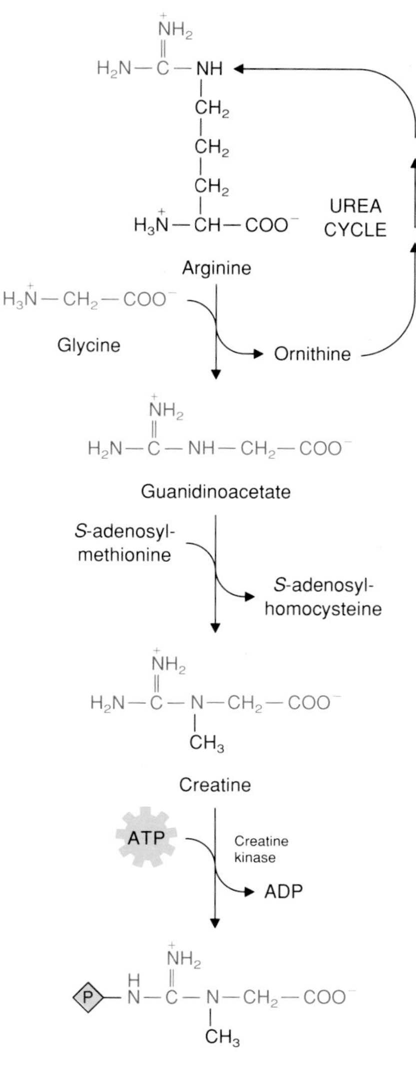

Figure 21.3
Biosynthesis of creatine phosphate from arginine.

γ-Aminobutyric Acid. Glutamate is one of several amino acids that serve as precursors to compounds that function in transmission of nerve impulses. Glutamate undergoes a pyridoxal phosphate-dependent decarboxylation to **γ-aminobutyric acid,** or GABA.

$$\underset{\textbf{Glutamate}}{H_3\overset{+}{N}-CH(COO^-)-CH_2-CH_2-COO^-} \xrightarrow{\nearrow CO_2} \underset{\gamma\textbf{-Aminobutyric acid}}{CH_2(\overset{+}{N}H_3)-CH_2-CH_2-COO^-}$$

This compound functions as an **inhibitory neurotransmitter,** a substance that inhibits transmission of an impulse from one cell to another in the central nervous system. We shall discuss neurotransmission later in this chapter and in Chapter 29.

Other Functions of Glutamate. Glutamate has at least two other major and ubiquitous metabolic fates, each involving specific enzyme systems: (1) the synthesis of glutathione, along with cysteine and glycine; and (2) the synthesis, via an ATP-dependent conjugating system, of polyglutamate forms of folic acid and its coenzymes.

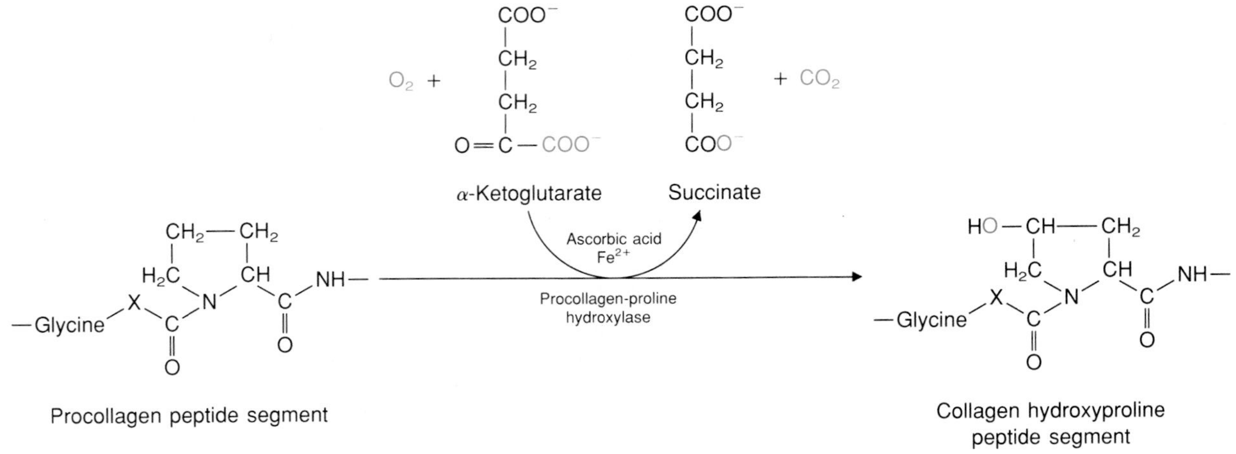

Figure 21.4
Enzymatic hydroxylation of procollagen proline residues in the synthesis of collagen. The fates of the two atoms of O_2 are indicated in red.

Metabolism of Sulfur-Containing Amino Acids

Fixation of Inorganic Sulfur

Like carbon and nitrogen, sulfur is made available to organisms largely in the form of inorganic compounds—principally sulfate, although some bacteria can synthesize organic compounds from elemental sulfur or from sulfite. Like the fixation of CO_2 and of N_2, the utilization of sulfate requires metabolic activation to a form that can readily undergo reduction. The process is largely confined to plants and bacteria. The end product of reduction is H_2S. The activated sulfate compound is **3′-phosphoadenosine-5′-phosphosulfate** (PAPS).

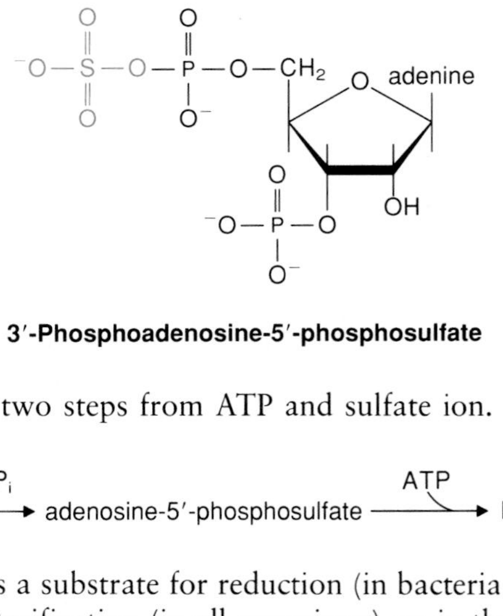

3′-Phosphoadenosine-5′-phosphosulfate

PAPS is formed in two steps from ATP and sulfate ion.

$$SO_4^{2-} + ATP \xrightarrow{\;PP_i\;} \text{adenosine-5'-phosphosulfate} \xrightarrow{\;ATP\;} PAPS + ADP + P_i$$

PAPS serves both as a substrate for reduction (in bacteria) and as an active agent for sulfate esterification (in all organisms), as in the synthesis of sulfated polysaccharides such as chondroitin sulfate. In plants adenosine 5′-phosphosulfate, rather than PAPS, is the substrate for reduction.

The sulfate in PAPS is reduced to sulfite (SO_3^{2-}) through the action of **thioredoxin,** a small protein (M_r ca. 12,000) that contains two reversibly oxidizable thiol groups. Thioredoxin is involved in several other intracellular redox reactions, as described in Chapters 19 and 22. The sulfite is subsequently reduced by **sulfite reductase,** a large and complex enzyme that catalyzes a six-electron transfer. The electrons are shuttled along a pathway involving NADPH, FAD, FMN, an iron–sulfur center, and the porphyrin siroheme. No intermediates accumulate, just the product H_2S.

Synthesis of Cysteine and Methionine in Plants and Bacteria

Out discussion of sulfur-containing amino acids deals only with cysteine and methionine. Although cystine is found in protein hydrolysates, this amino acid disulfide is synthesized at the polypeptide level, by posttranslational oxidation of cysteine residues. It normally does not exist in cells as the free amino acid; when it does accumulate, it is reduced nonenzymatically by glutathione.

CYSTEINE BIOSYNTHESIS. Plants and microorganisms utilize H_2S for the synthesis of cysteine, with serine providing the carbon skeleton. Some bacteria can condense these two substrates directly, via a pyridoxal phosphate-dependent enzyme.

$$\underset{\textbf{Serine}}{\mathrm{H{-}\overset{\displaystyle CH_2OH}{\underset{\displaystyle COO^-}{C}}{-}\overset{+}{N}H_3}} + H_2S \longrightarrow \underset{\textbf{Cysteine}}{\mathrm{H{-}\overset{\displaystyle CH_2SH}{\underset{\displaystyle COO^-}{C}}{-}\overset{+}{N}H_3}} + H_2O$$

Plants and most microorganisms use *O*-acetylserine as the substrate reacting with H_2S.

$$\text{Serine} + \text{acetyl-CoA} \longrightarrow O\text{-acetylserine} + \text{CoA}$$

$$O\text{-Acetylserine} + H_2S \longrightarrow \text{cysteine} + \text{acetate} + H_2O$$

METHIONINE BIOSYNTHESIS. In plants and bacteria cysteine provides the sulfur for methionine synthesis, with the carbon coming from homoserine. Homoserine first reacts with succinyl-CoA to form **O-succinylhomoserine.** This reaction evidently is a control point, because the enzyme catalyzing this reaction is feedback inhibited by methionine (Figure 21.5). O-Succinylhomoserine reacts with cysteine to give **cystathionine,** a thioether compound. Cleavage of cystathionine then occurs, with the sulfur thereby becoming linked to the four-carbon side chain that started as homoserine. The resulting amino acid, homocysteine, is converted to methionine through the action of methionine synthetase, by N^5-methyltetrahydrofolate and B_{12}, as presented in Chapter 20.

Methionine as the Source of Cysteine Sulfur in Animals

Methionine is classified as an essential amino acid for mammals, while cysteine is considered nonessential. Actually, cysteine sulfur is derived from methionine in animals, so cysteine is nonessential only as long as the diet contains adequate methionine.

The synthesis of cysteine in animals (Figure 21.6) resembles the reverse of the methionine biosynthetic pathway we just discussed. Homocysteine condenses with serine, in a pyridoxal phosphate-dependent reaction catalyzed by **cystathionine synthase,** to give cystathionine. This then cleaves, also in a PyP-dependent reaction, to give cysteine. The four-carbon side chain deaminates, the reaction products being ammonia and α-

Figure 21.5
Biosynthesis of methionine from homoserine, as it occurs in plants and bacteria.

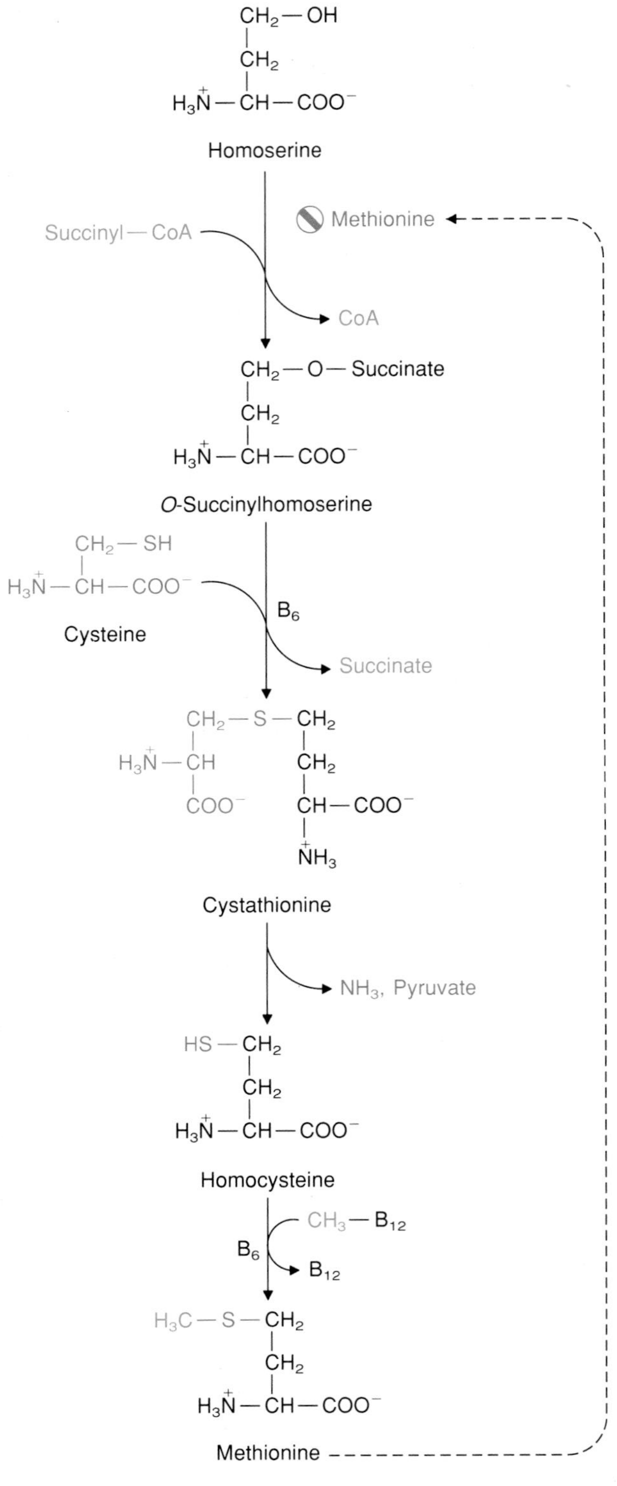

Figure 21.6
Cystathionine as an intermediate in cysteine biosynthesis.

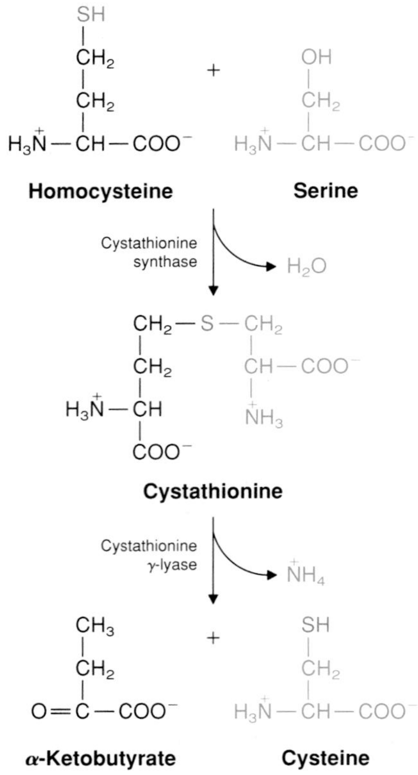

ketobutyrate. In plants and prokaryotes α-ketobutyrate can be used for isoleucine biosynthesis.

Cystathionine plays a central role in the pathways leading to both methionine and cysteine. In methionine synthesis from cysteine, cystathionine cleaves to bring about the transfer of sulfur from a three-carbon to a four-carbon side chain. Just the converse occurs during the synthesis of cysteine from methionine.

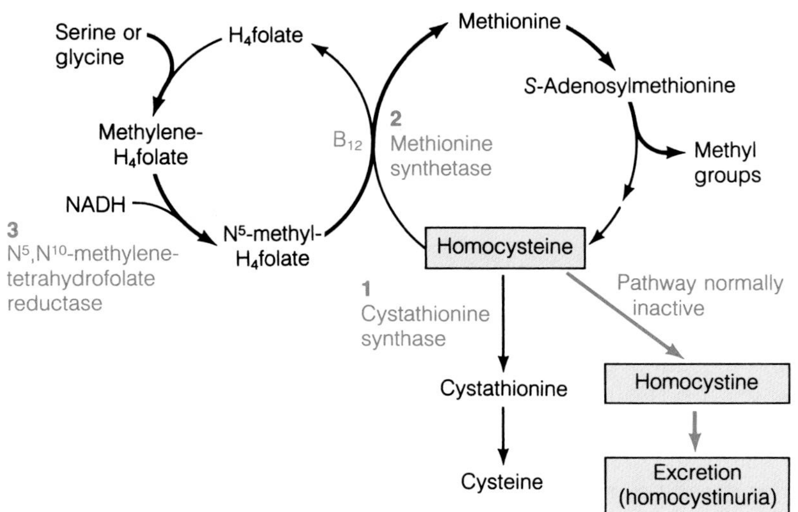

Figure 21.7
Enzyme deficiencies causing homocystinuria (blue). A defect in cystathionine synthetase (1) causes homocysteine to accumulate because conversion to cystathionine is impaired. A deficiency of methionine synthetase (2) or methylenetetrahydrofolate reductase (3) blocks the conversion of homocysteine to methionine. In either case the homocysteine that accumulates is oxidized to homocystine, the form in which it is excreted (orange).

In humans a genetic deficiency of cystathionine synthetase leads to a condition called **homocystinuria,** in which there is overaccumulation of homocysteine, as evidenced by excessive urinary excretion of homocystine (the oxidation product of homocysteine). The condition results in severe mental retardation and dislocation of the lens of the eye. After the cystathionine synthetase deficiency was described, it was found that similar symptoms resulted from deficiencies of either of two related enzymes—methionine synthetase or N^5,N^{10}-methylenetetrahydrofolate reductase. From Figure 21.7 you can see how a deficiency in any one of these three enzymes would cause homocysteine accumulation.

Metabolism of Glutathione

As described in Chapter 14, cysteine plays an important metabolic role as a constituent of **glutathione,** or γ-glutamylcysteinylglycine. This tripeptide, abundant in all cells, protects against two different kinds of metabolic stress. First, it can nonenzymatically reduce a number of substances, such as peroxides or free radicals, which might accumulate in cells under oxidizing conditions. Second, through the action of widely distributed animal enzymes, the **glutathione *S*-transferases,** it participates in detoxification of many substances, such as xenobiotics (foreign organic compounds not produced in metabolism) or electrophiles produced through the action of cytochrome P-450-linked oxidases. Glutathione reacts with such a compound (denoted RX) as shown below, followed by cleavage of the γ-glutamyl and glycyl residues and then acetylation by acetyl-CoA to give a **mercapturic acid.** This is a more soluble, less toxic derivative of the original compound, which can then be excreted in the urine.

RX + γ-Glu—Cys(—SH)—Gly (**Glutathione**) → (X^- released) → R—S—Cys(γ-Glu)(Gly) → (Glu + Gly released) → R—S—CH_2—CH(NH_3^+)—COO^- (**R-cysteine conjugate**) → (Ac—S—CoA → CoA—SH) → R—S—CH_2—CH(NH—Ac)—COO^- (**Mercapturic acid**)

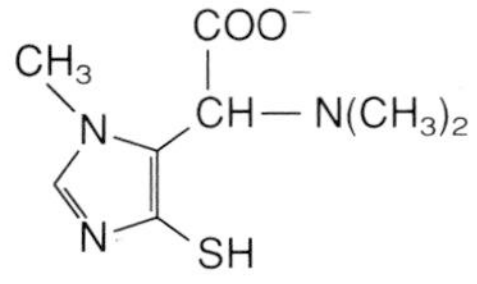

Ovothiol

A recently described sulfur amino acid, **ovothiol,** plays a role comparable to that of glutathione in fertilized eggs. Ovothiol, which is present at 5 mM concentration in sea urchin eggs, protects the egg against oxidative damage by peroxides produced at the egg surface early in fertilization. Ovothiol is in turn reduced by glutathione.

S-Adenosylmethionine and Biological Methylation

***S*-Adenosylmethionine** (**AdoMet**) is a metabolically activated form of methionine. It serves as an activated methyl group donor and also as the donor of propylamino groups in the biosynthesis of polyamines. One of the most important methyl group transfer reactions has already been mentioned, when we described the biosynthesis of phosphatidylcholine from phosphatidylethanolamine in Chapter 18.

Before discussing the roles of AdoMet, let us first see how this centrally important compound is synthesized. Methionine reacts with ATP, in a reaction that cleaves on the inner side of the α-phosphate of ATP. The adenosyl moiety is transferred directly to the sulfur of methionine. The other product, inorganic tripolyphosphate, is cleaved by the enzyme, to provide much of the thermodynamic drive for this reaction.

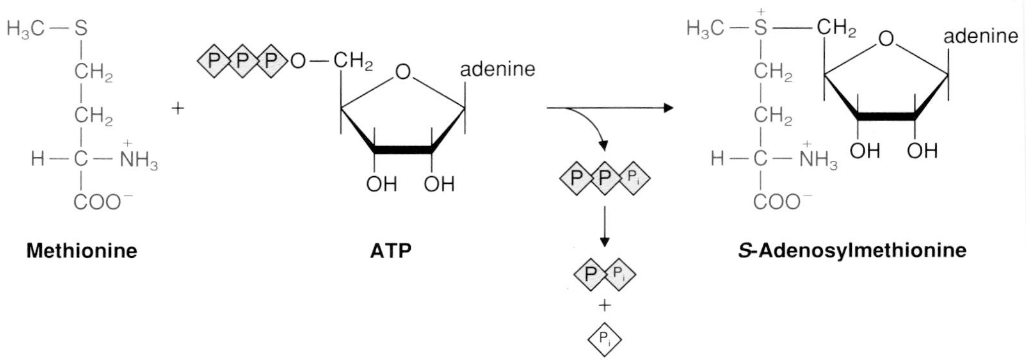

Methionine + ATP → *S*-Adenosylmethionine

The presence of three substituent groups on the sulfur generates a positively charged **sulfonium** compound. Since the sulfonium moiety is quite unstable, there is a strong thermodynamic tendency for one of the substituent groups to be lost; in other words, AdoMet has a very high group transfer potential. Most, though not all, of the group transfer reactions involving AdoMet are **transmethylations,** in which the methyl group is transferred to an acceptor, with the other product being *S*-adenosylhomocysteine (AdoHcy). The synthesis of creatine from guanidinoacetic acid (see Figure 21.3) is a good example.

Table 21.1 lists a number of biologically important AdoMet-dependent transmethylations. Note that the substrates can be polymeric proteins or nucleic acids. We discuss the functions of DNA and RNA methylation in later chapters. However, a brief discussion of protein methylation is in order here. Methylatable residues include, in different proteins, lysine, arginine, and residues containing free carboxyl groups. Although we know little about specific functions, we know that histones become methylated, with specific arginine and lysine residues becoming modified at particular times in the cell cycle. We also know that ϵ-*N*-trimethyllysine, derived specifically from the hydrolysis of methylated protein, serves as the precursor to carnitine, whose role in fatty acyl group transfer across membranes was presented in Chapter 17. Furthermore, we know that in bacteria protein meth-

Table 21.1
Some AdoMet-dependent transmethylations

Methyl Group Acceptor	Methylated Product
Norepinephrine	Epinephrine
Guanidinoacetic acid	Creatine
Phosphatidylethanolamine	Phosphatidylcholine
DNA-adenine or -cytosine	DNA-N-methyladenine or 5-methylcytosine
tRNA bases	Methylated tRNA bases
Nicotinamide	N^1-methylnicotinamide
Protein amino acid residues	Methylated amino acid residues

ylation plays an important role in **chemotaxis,** the process whereby bacteria sense a concentration gradient of a chemical substance in the medium and move either toward or away from it. Chemotaxis is being studied as a rudimentary model for sensory transduction; it involves the cyclic methylation and demethylation of a group of proteins called **MCPs,** or methylatable chemotactic proteins. (We discuss chemotaxis further in Chapter 29.) Finally, there are indications that protein methylation helps to control rates of protein turnover; by blocking sites of ubiquitination, methylation evidently helps to protect proteins from turnover.

The central metabolic role of AdoMet can be appreciated if you keep in mind that, except for a few reactions in bacterial metabolism, *the only known methyl group transfer that does not involve AdoMet is the synthesis of methionine itself.* As noted in Chapter 20, methyl groups are generated de novo through the reduction of N^5,N^{10}-methylenetetrahydrofolate to N^5-methyltetrahydrofolate. The methyl group is transferred to yield methionine via methyl-B_{12} and the action of methionine synthetase. There is, however, an additional route for methionine synthesis that involves a different kind of transmethylation. *S*-Adenosylhomocysteine, formed from AdoMet-dependent transmethylations, cleaves hydrolytically to yield adenosine and homocysteine. Homocysteine is remethylated to methionine, either via methionine synthetase or by transmethylation from **glycine betaine.** The latter is formed from the oxidation of choline, as shown below.

$$\underset{\textbf{Choline}}{\overset{\displaystyle CH_2OH}{\underset{\displaystyle CH_2-\overset{+}{N}(CH_3)_3}{|}}} \xrightarrow{\text{Oxidation}} \underset{\textbf{Glycine betaine}}{\overset{\displaystyle COO^-}{\underset{\displaystyle CH_2-\overset{+}{N}(CH_3)_3}{|}}} \xrightarrow[\displaystyle COO^- - CH_2-N(CH_3)_2]{\text{Homocysteine}} \underset{\textbf{Methionine}}{S-CH_3-CH_2-CH_2-H-C(\overset{+}{N}H_3)-COO^-}$$

The betaine, which is a quaternary amine derivative of an amino acid, has a positive charge and a high group transfer potential, as does AdoMet itself. Since the methyl groups on glycine betaine came originally from AdoMet, the existence of this transmethylation process is consistent with the statement that all methyl groups come from methionine except those used in the synthesis of methionine.

S-Adenosylmethionine plays an additional role in plant metabolism, as the precursor to the hormone **ethylene,** which promotes plant growth and development and induces the ripening of fruit. Although mechanistic details are not yet clear, we know that the main carbon skeleton of methionine, rather than the methyl group, is split off in this process, to yield **1-aminocyclopropane-1-carboxylic acid.** This is fragmented as shown to yield ethylene as one of the products.

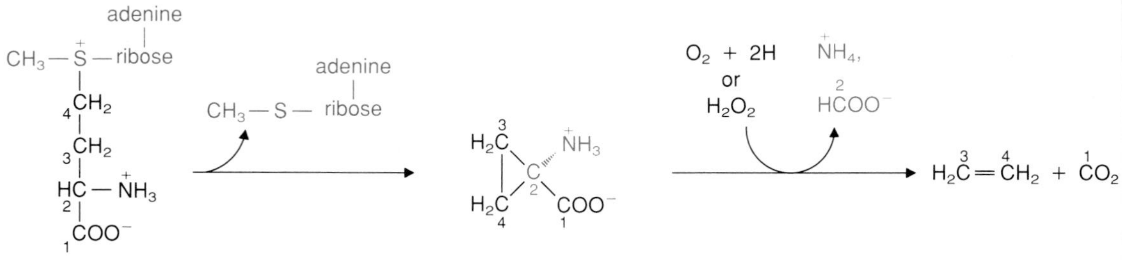

AdoMet

CH_3—S—ribose (adenine)

$O_2 + 2H$ or H_2O_2

$\overset{+}{N}H_4$, $HCOO^-$

$H_2C{=}CH_2 + CO_2$

1-Aminocyclopropane-1-carboxylate

Ethylene

Isotope labeling studies with methionine fed to plants have revealed the source of each carbon atom.

It is likely that a similar mechanism is involved in the synthesis of fatty acids that contain a cyclopropane ring in their hydrocarbon chains.

Polyamines

The synthesis of ethylene involves transfer, not of the methyl group, but of the four-carbon moiety of AdoMet-bound methionine. A comparable process occurs during synthesis of the widely distributed polyamines, **spermine** and **spermidine** (Figure 21.8). The names of these substances suggest the source of their original detection—human semen. These polyamines and the diamine putrescine are widely distributed cationic cell components, being especially abundant in rapidly proliferating cells. Although putrescine and its homolog **cadaverine** are classified among the polyamines, they are in fact diamines; they are synthesized by decarboxylation of ornithine and lysine, respectively. **Ornithine decarboxylase,** which synthesizes putrescine, is a highly regulated enzyme whose activity responds to many hormonal stimuli. It has an extremely short metabolic half-life (approximately 10 minutes).

As polycations, the polyamines play multiple roles in stabilizing intracellular conformations of negatively charged nucleic acids or membranes. In bacteriophage T4, for instance, about 40% of the negative charge on the viral DNA is neutralized by polyamines. Transfer RNA from some sources contains two molecules of tightly bound spermine or spermidine per molecule of transfer RNA, apparently helping to stabilize the cloverleaf structure of tRNA. Polyamines are often added to the medium when one is trying to isolate a cell organelle or a macromolecular assembly in the native conformation. Polyamines also may regulate the progression of cells through the cell cycle, because inhibition of ornithine decarboxylase by **difluoromethylornithine** arrests cells in the G_1 phase of the cycle.

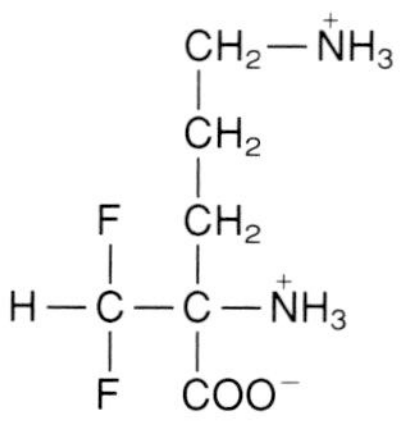

Difluoromethylornithine

Putrescine serves as the precursor to spermidine, then to spermine, through the AdoMet-mediated transfer of active propylamino groups (see Figure 21.8). First, AdoMet is decarboxylated by **AdoMet decarboxylase.** This reaction is an example of a decarboxylase that contains as its cofactor not pyridoxal phosphate, but a residue of bound pyruvate. The resulting propylamino group is then transferred to putrescine to give spermidine, and a second group is transferred to spermidine, by a different enzyme, to give spermine. The other product of these reactions is methylthioadenosine, which undergoes phosphorolytic cleavage to adenine and **5′-methylthioribose 1-phosphate.** The latter compound is used for resynthesis of methionine by another pathway, not shown here.

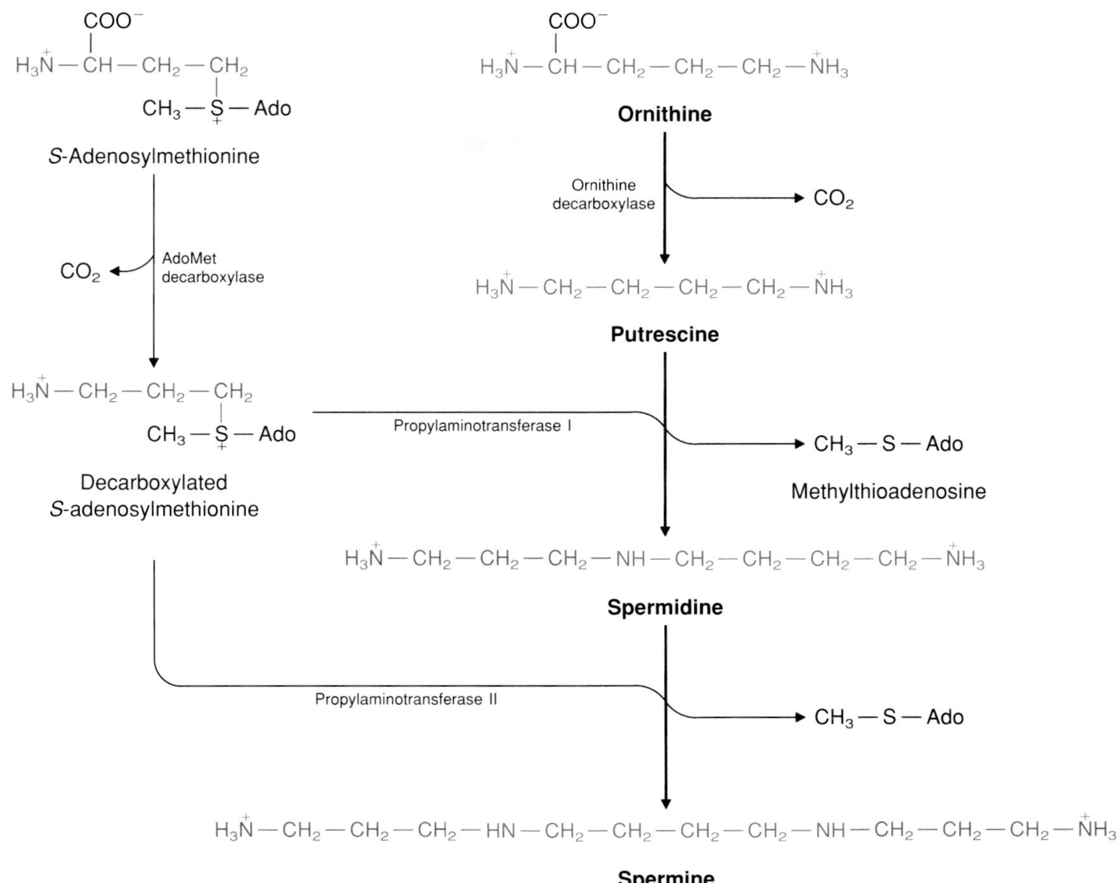

Figure 21.8
Biosynthesis of putrescine, spermidine, and spermine.

Catabolism of Cysteine and Methionine

The principal catabolic pathways for cysteine and methionine are outlined in Figure 21.9. β-Mercaptopyruvate, the transamination product of cysteine, can undergo desulfuration by several routes in different organisms, to give pyruvate and H_2S, sulfite (SO_3^{2-}), thiosulfate ($S_2O_3^{2-}$), or thiocyanate (SCN^-). A mitochondrial enzyme called **rhodanese** catalyzes several reactions, including the formation of thiocyanate from thiosulfate and cyanide ion. This reaction is involved in the detoxification of cyanide.

The catabolism of methionine proceeds largely through homocysteine, which is catabolized via cystathionine to give α-ketobutyrate. The latter reacts with CoA and ATP to yield propionyl-CoA, which is then converted to succinyl-CoA by the B_{12}-requiring pathway we have discussed before. In addition, α-ketobutyrate has a biosynthetic role, in the synthesis of isoleucine.

Aromatic Amino Acid Metabolism

In this section we consider the metabolism of phenylalanine, tyrosine, tryptophan, and histidine. Histidine is not usually classified among the aromatic amino acids, but because there are several similarities between its metabolism and those of the others, we shall consider it an "honorary aromatic" for purposes of our discussion. Synthesis of these heterocyclic and/or aromatic rings from noncyclic precursors involves complex chemistry. As with other lengthy biosynthetic pathways, such as the synthesis of vitamins, most

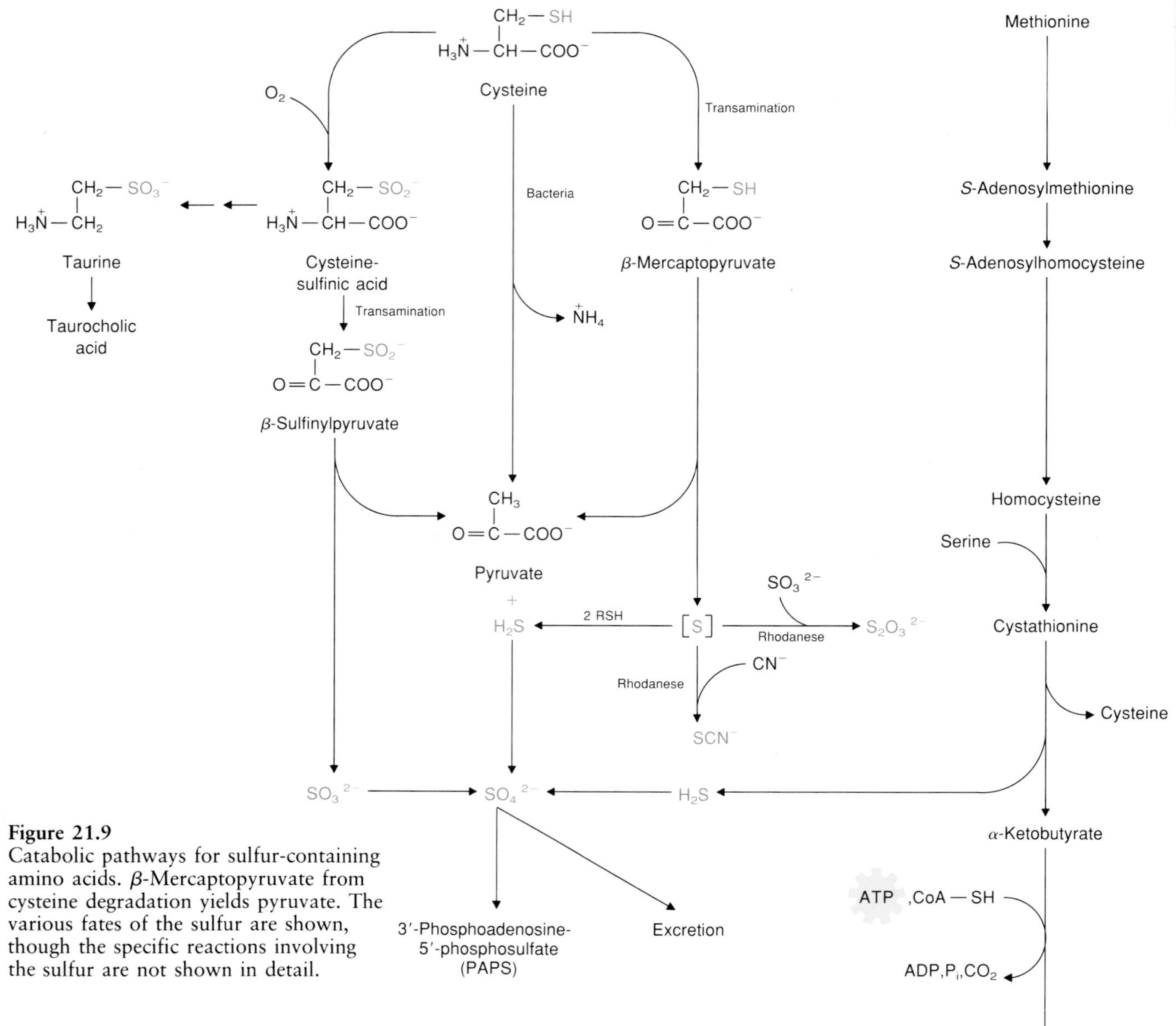

Figure 21.9
Catabolic pathways for sulfur-containing amino acids. β-Mercaptopyruvate from cysteine degradation yields pyruvate. The various fates of the sulfur are shown, though the specific reactions involving the sulfur are not shown in detail.

A complex polymer of C_9 aromatic units

Lignin

of the aromatic biosynthetic capabilities have been lost during animal evolution. The synthetic pathways we shall discuss are limited to plants and bacteria—with one exception. That exception is the hydroxylation of phenylalanine to tyrosine, an interesting and important reaction in animal metabolism.

Biosynthesis of Aromatic Rings: The Shikimic Acid Pathway

A single branched pathway in microorganisms and plants leads to synthesis of phenylalanine, tyrosine, and tryptophan, as well as other important substances. Since these latter substances include **lignin,** the major constituent of woody tissue, it is apparent that a tremendous amount of carbon in the biosphere passes through this pathway.

Individual reactions of the pathway were established in bacteria, where a large number of auxotrophic mutants could be isolated and characterized genetically (mapping), physiologically (identification of compounds that could satisfy growth requirements), and biochemically (identification of intermediates accumulating when a given step is blocked). So far as is known, the processes are quite similar in plants.

The key findings were first, that all of the carbon in phenylalanine and tyrosine was derived from erythrose 4-phosphate and phosphoenolpyruvate, and second, that a class of mutants requiring phenylalanine, tyrosine, tryptophan, *p*-aminobenzoic acid, and *p*-hydroxybenzoic acid for growth could have all five requirements met by a single compound—**shikimic acid.** We now know that these mutants are blocked in the formation of shikimate (the fourth reaction in Figure 21.10) and that when it is provided all of the subsequent steps in the pathway can occur. Note from Figure 21.10 that an unbranched pathway leads past shikimic acid to **chorismic acid.** Figure 21.11 shows that from chorismic acid one pathway leads to **prephenic acid,** with subsequent branches leading to phenylalanine and tyrosine. Another pathway leads through **anthranilic acid** to tryptophan; still another pathway leads to *p*-aminobenzoic acid; and a final pathway leads, via *p*-hydroxybenzoic acid, to coenzyme Q. Thus, the shikimic acid pathway is responsible for biosynthesis of virtually all aromatic compounds, since the products we have just mentioned serve in turn as precursors to other aromatic compounds.

Interesting control mechanisms are involved in regulating these pathways. Studies on the genetic control of tryptophan biosynthesis have provided some of our most important insights into transcriptional regulation. More recent studies on the reactions leading to chorismic acid have revealed the existence of several multifunctional enzymes—single polypeptide chains containing two or more active sites for catalysis of sequential reactions—obviously, an efficient way to control several reactions jointly.

Great attention has focused in recent years on the sixth reaction of the shikimic acid pathway, **5-enoylpyruvylshikimate-3-phosphate synthase (EPSP synthase),** in higher plants. This enzyme is specifically inhibited by a

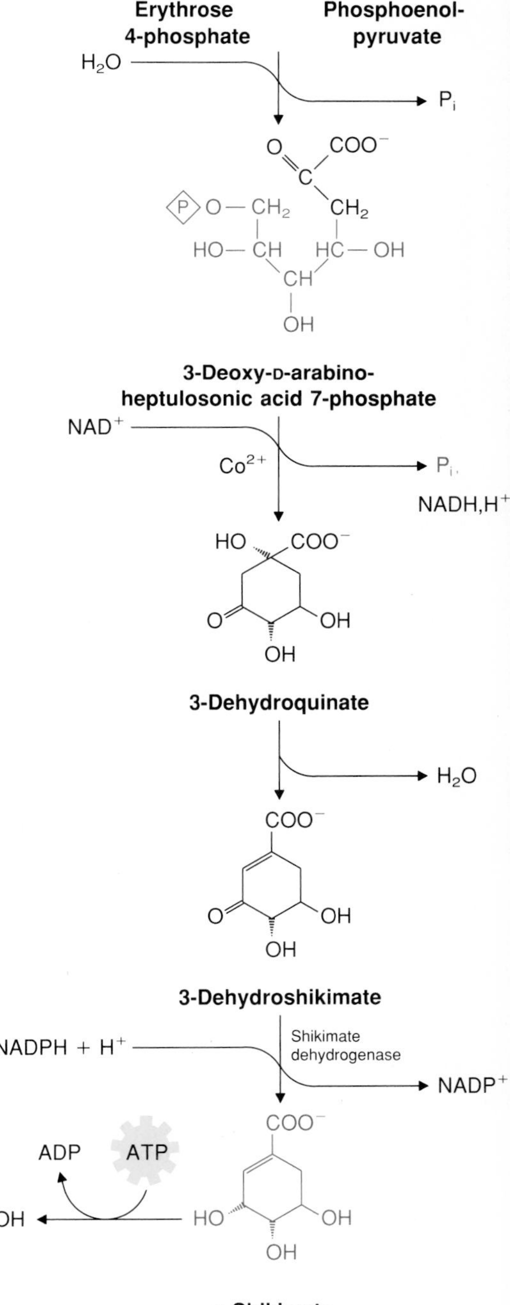

Figure 21.10
The shikimic acid pathway for biosynthesis of aromatic amino acids, I. This figure depicts in detail the unbranched pathway from erythrose 4-phosphate and phosphoenolpyruvate to chorismic acid. The first reaction is driven by loss of phosphate from phosphoenolpyruvate. In the second reaction an unusual cobalt-requiring enzyme effects ring closure with dehydrogenation and loss of the second phosphate. Dehydration in the third step yields dehydroshikimate, which is then reduced by NADPH to shikimate. A three-carbon side chain is then attached via phosphorylation of shikimate and reaction with a second molecule of phosphoenolpyruvate. Dephosphorylation of this intermediate gives chorismate, from which point the pathway branches.

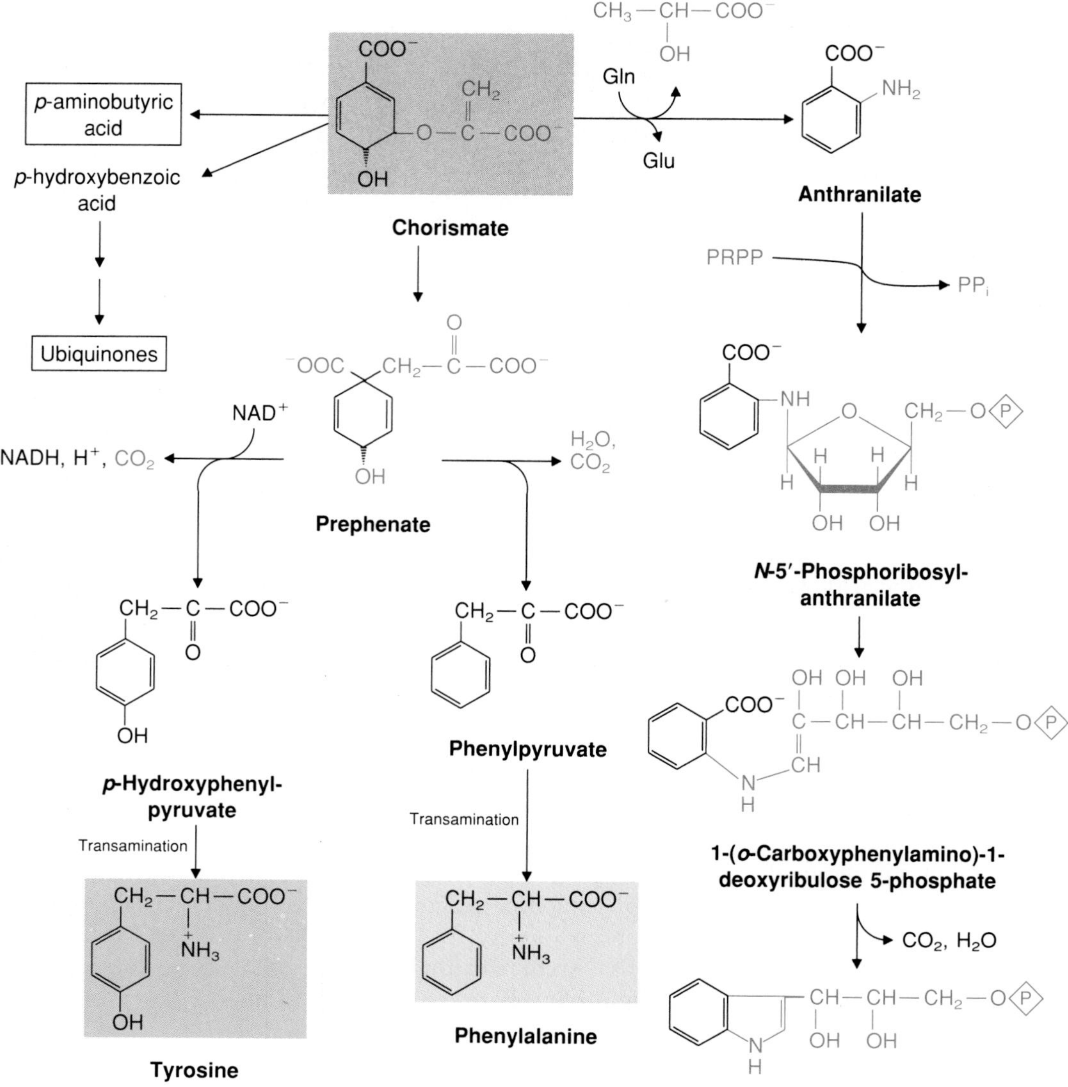

Figure 21.11
The shikimic acid pathway for biosynthesis of aromatic amino acids, II. Going down from chorismate, note its isomerization to prephenate, a branch point en route to phenylalanine and tyrosine. Decarboxylation and dehydration of prephenate yield phenylpyruvate, which leads directly to phenylalanine by transamination. Alternatively, oxidative decarboxylation gives *p*-hydroxyphenylpyruvate, the immediate precursor of tyrosine. The pathway to the right from chorismate involves exchange of the amide nitrogen of glutamine for the side chain of chorismate. The enzyme, **anthranilate synthetase,** catalyzes the first committed step en route to tryptophan. Carbon 1 of 5-phosphoribosyl-1-pyrophosphate (PRPP) is linked to the nitrogen of anthranilate, driven by loss of pyrophosphate from carbon 1. The sugar ring opens, followed by a decarboxylation and ring closure to give an indole compound, indole-3-glycerol phosphate. The three-carbon side chain is exchanged for that of serine in the final step.

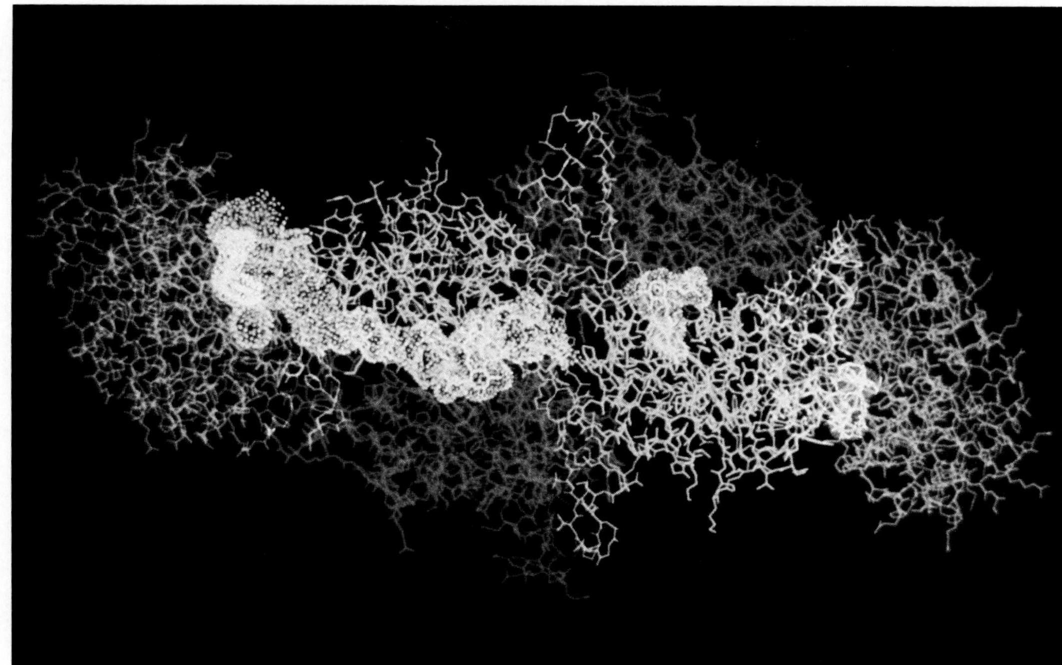

Figure 21.12
Structure of the $\alpha_2\beta_2$ tryptophan synthase complex from *Salmonella typhimurium.* The α subunits are shown in blue, the N-terminal domain of the β subunits in yellow, and the β C-terminal domains in red. Light blue shows an interior tunnel that connects the active sites on the left-most α and β subunits. The active sites on the rightmost α and β subunits are shown in red and light blue respectively.

compound called **glyphosate.** The growth of most crop and weed plants is inhibited by glyphosate, which is an effective broad-spectrum herbicide. A recent achievement of biotechnology is the transfer of genes conferring resistance to glyphosate into crop plants. This should allow simplified and effective weed control: spraying a field would eliminate all plants except the genetically engineered species.

The first reaction in the pathway from anthranilic acid to tryptophan involves an activated sugar derivative, **5-phosphoribosyl-1-pyrophosphate** (PRPP), which plays an important role in nucleotide synthesis (Chapter 22). In bacteria the genes encoding these enzymes are linked in a linear array, the **tryptophan operon.** As discussed in Chapter 26, an operon is a linked set of genes whose expression is regulated jointly at the level of transcription. The tryptophan operon has taught us a great deal about the organization and expression of genes, owing in particular to the work of Charles Yanofsky and his colleagues. In *E. coli* the final enzyme in the pathway, **tryptophan synthase,** is an $\alpha_2\beta_2$ dimer. Isolated α and β subunits catalyze partial reactions, as shown below, with the holoenzyme catalyzing a concerted reaction in which indole does not dissociate from the enzyme surface but immediately reacts with serine to give tryptophan.

$$\text{Indole-3-glycerol phosphate} \xrightarrow{\alpha\text{ subunit}} \text{indole} + \text{3-phosphoglyceraldehyde}$$

$$\text{Indole} + \text{serine} \xrightarrow{\beta\text{ subunit}} \text{tryptophan} + H_2O$$

Remarkably, x-ray crystallography shows that the intermediate, indole, is transported from the α subunit active site to the β active site, a distance of 2.5 nm, through a tunnel in the interior of the protein molecule (Figure 21.12).

Biosynthesis of Histidine

The biosynthesis of histidine presents several parallels with the shikimic acid pathway, with respect to methods used for its elucidation, complexity and nature of the reactions involved, elegance of genetic regulation of the pathway, and utility of much of the information gained from basic studies of the pathway. However, the histidine pathway is distinctive in being unbranched. As established largely in the laboratories of Bruce Ames and Philip Hartman, the pathway is shown in Figure 21.13. Ten individual reactions are involved, starting with an unusual reaction that joins ATP and PRPP; both compounds contribute carbon to the product.

The 10 structural genes for the enzymes of histidine synthesis in enteric bacteria are linked to one another in the same order as the order of the reactions of the pathway. This set of genes, the **histidine operon,** is coordinately regulated at the transcriptional level, and all 10 genes are transcribed to give one large messenger RNA, which is translated to give the 10 enzymes. Once the genes and gene products had been identified, Bruce Ames was able to use the mutant bacteria that had been generated in the course of these investigations to search for mutagens in the environment. In the **Ames test** researchers can count mutations simply by measuring the rate at which mutants that cannot synthesize histidine (histidine **auxotrophs**) mutate to a form that can synthesize the amino acid (**prototrophs**). One treats a culture of auxotrophs with a suspected mutagen, plates the bacteria on medium containing no histidine, and counts the colonies that appear as the result of reversion mutations. Using this system, Ames and colleagues reported a very high correlation between compounds known to be carcinogenic in animals and those found to be mutagens in this test. Thus, the Ames test provides a quick and inexpensive way to search for suspected carcinogens in the environment. Moreover, these findings provided support for the idea that cancer arises as the result of a series of somatic cell mutations, an idea we discuss further in Chapter 23.

Aromatic Amino Acid Utilization in Plants

Phenylalanine and tyrosine serve as precursors to an enormous number of plant substances, ranging from the polymeric lignin to tannins and pigments, to many of the flavor components of spices. In fact, their presence in such substances as cinnamon oil, wintergreen oil, bitter almonds, nutmeg, cayenne pepper, vanilla bean, cloves, and ginger gave this group of compounds the name *aromatic amino acids*. They are also precursors to many of the nearly 3000 **alkaloids,** nitrogenous substances that are synthesized by specific plants. Although their roles in plant metabolism are not well understood, many of the alkaloids—including morphine, a well-known narcotic, and colchicine, an inhibitor of microtubule assembly—have potent physiological effects in animals.

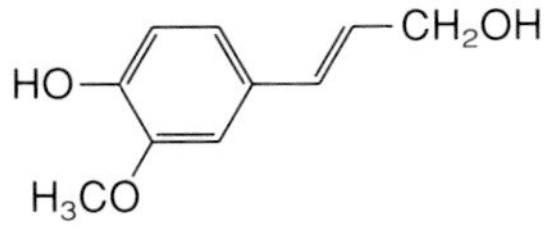

HO, H_3CO, CH_2OH

Coniferyl alcohol

Much of the utilization of phenylalanine and tyrosine by plants involves the action of specific lyases, which cleave out ammonia. The products lead to **coniferyl alcohol,** the precursor to both lignin and a number of flavor constituents.

Tryptophan is utilized for synthesis of a plant growth hormone. As shown below, the transamination product of tryptophan yields **indole-3-acetic acid,** or **auxin.**

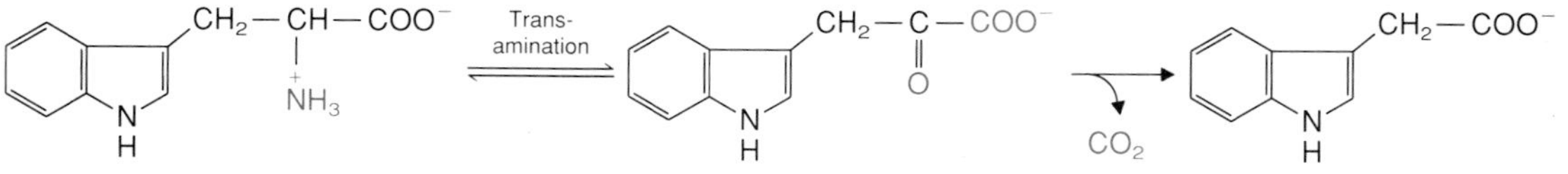

$CH_2—CH—COO^-$ (NH_3^+) —Trans-amination→ $CH_2—C(=O)—COO^-$ → (CO_2) $CH_2—COO^-$ (N–H indole rings)

Tryptophan **Indole-3-acetic acid**

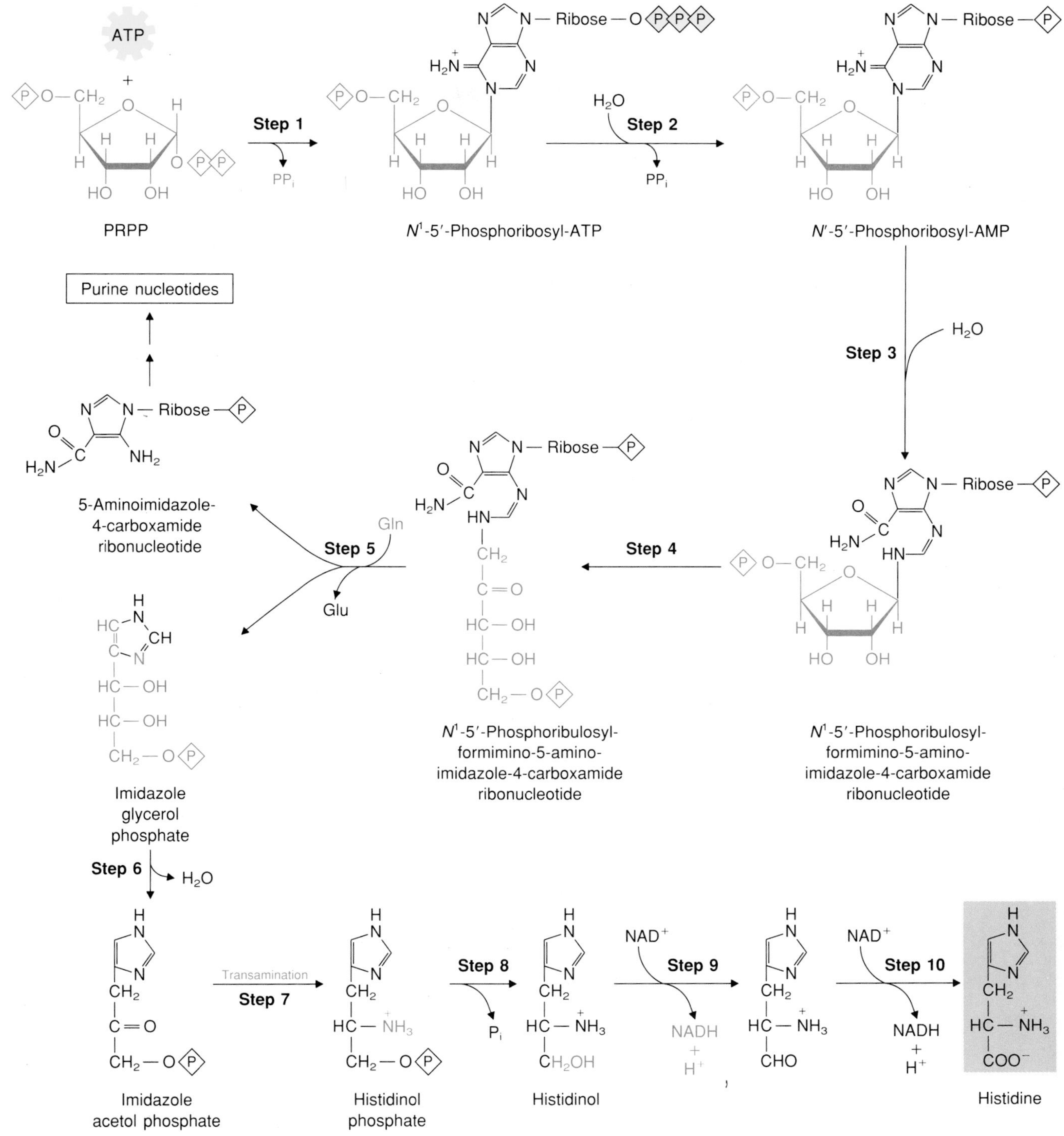

Figure 21.13
The pathway for histidine biosynthesis. After activation of a purine ring by reaction with PRPP (step 1), the purine ring opens to give the third intermediate (steps 2 and 3). The ribose ring derived from PRPP then opens (step 4). The transfer of amide nitrogen from glutamine (step 5) is accompanied by cleavage and ring closure, giving the first imidazole compound, imidazole glycerol phosphate. The other product is an intermediate in the synthesis of purine nucleotides. Imidazole glycerol phosphate is transformed to histidine by a straightforward sequence involving dehydration (step 6), transamination (step 7), dephosphorylation (step 8), and dehydrogenation (steps 9 and 10).

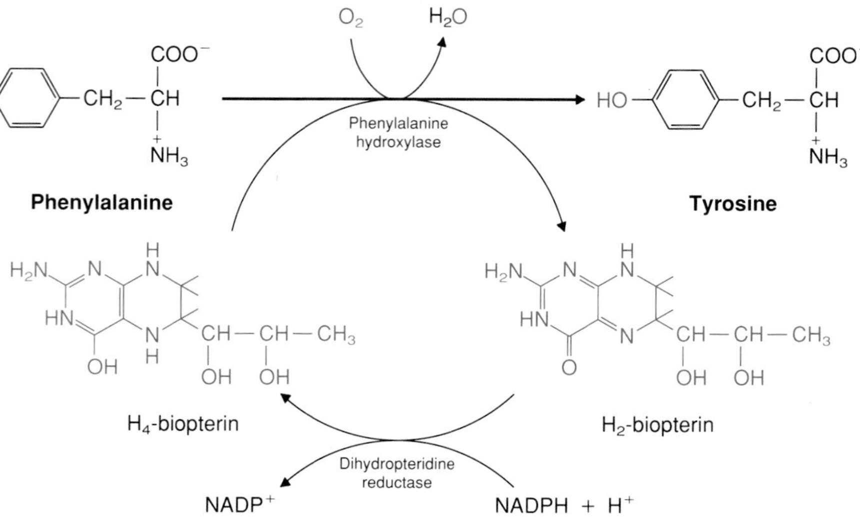

Figure 21.14
Conversion of phenylalanine to tyrosine, showing the phenylalanine hydroxylase and dihydropteridine reductase reactions.

Aromatic Amino Acid Metabolism in Animals

Although animal cells do not synthesize aromatic rings de novo, the intermediary metabolism of aromatic amino acids in animals is extensive, and it involves many biologically important processes. These include the biosynthesis of tyrosine from phenylalanine; the utilization of tyrosine in synthesis of pigments and hormones; and the utilization of tyrosine, tryptophan, and histidine in synthesis of **biogenic amines,** an important class of biological regulators. Aromatic amino acid metabolism also provides the backdrop for much of our early understanding of human biochemical genetics.

TYROSINE BIOSYNTHESIS. The only known reaction of aromatic amino acid biosynthesis in animals is the conversion of phenylalanine to tyrosine, catalyzed by **phenylalanine hydroxylase.** This interesting enzyme is a mixed-function oxygenase that uses a pterin cofactor, **tetrahydrobiopterin.** The reaction oxidizes tetrahydrobiopterin to the quinonoid isomer of dihydrobiopterin, as shown in Figure 21.14. The coenzyme is regenerated through the action of the NADPH-requiring **dihydropteridine reductase** (analogous, but not identical, to dihydrofolate reductase). This enzyme system occurs almost entirely in the liver.

A hereditary deficiency of phenylalanine hydroxylase is responsible for **phenylketonuria** (PKU), a condition that afflicts about one in 10,000 newborn infants in Western Europe and the United States. PKU is an autosomal recessive trait, which means that the heterozygous condition is rather common (about 2%), so two parents heterozygous for the trait have one chance in four of having a phenylketonuric child. In phenylketonuria phenylalanine accumulates to very high levels because of the block in conversion to tyrosine, and much of this phenylalanine is metabolized via pathways that are normally little used—particularly transamination to phenylpyruvate, and also subsequent conversion of phenylpyruvate to phenyllactate and phenylacetate. These compounds are excreted in urine in enormous quantities (1 to 2 grams per day).

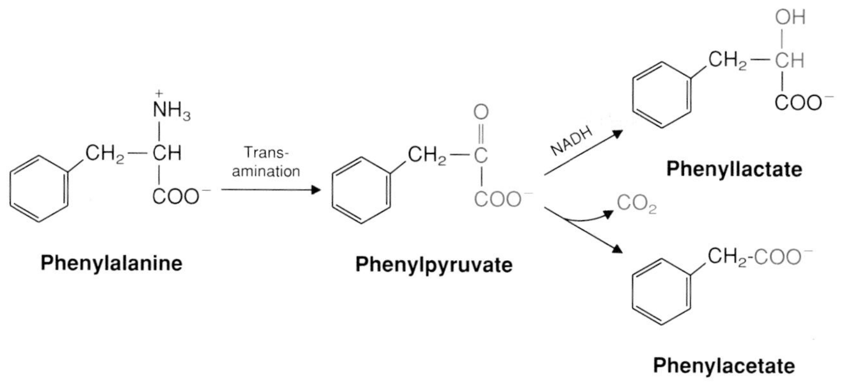

If undetected and untreated, PKU leads to profound mental retardation; the precise biochemical causation has not yet been identified. Fortunately, PKU can readily be detected at birth, and many hospitals carry out routine screening of newborns. If the condition is detected early, the onset of retardation can be prevented by feeding for several years a synthetic diet low in phenylalanine and rich in tyrosine. Since the use of this synthetic diet is quite expensive, there has been much interest in prenatal diagnosis of PKU. The human gene for phenylalanine hydroxylase has recently been cloned, providing the basis for a test, using nucleic acid hybridization, that can be applied in any cell.

In recent years a different form of PKU has been described which results from a hereditary deficiency of dihydropteridine reductase. This condition is much rarer than classical PKU and, because tetrahydrobiopterin is involved in other hydroxylations, much more severe in its symptoms.

In animals phenylalanine plays no significant role other than as a component of proteins and a precursor to tyrosine; its catabolism normally proceeds via tyrosine.

TYROSINE UTILIZATION AND CATABOLISM. Tyrosine plays several important roles in animal metabolism: it is a precursor to thyroid hormones, to the biological pigments called **melanins,** and to an important class of biological regulators, the **catecholamines** (discussed in the next section).

Thyroid hormones stimulate a number of metabolic processes, through activation of the transcription of particular genes (see Chapter 23). The synthesis of thyroid hormones, principally **thyroxine** and **triiodothyronine,** occurs at the level of tyrosine residues in a specific protein, **thyroglobulin** (Figure 21.15). This protein undergoes degradation to yield the free hormones, which are transported to their sites of action through the bloodstream. The synthesis of thyroid hormones involves iodination of the tyrosine ring. This process occurs in the thyroid gland, which concentrates iodide ion from the blood serum for this purpose. One result of iodine deficiency is **goiter,** hypertrophy of the thyroid gland. Before iodized salt came into widespread use, goiter was endemic in regions whose soil was deficient in iodine.

The synthesis of melanins occurs in pigment-producing cells, the **melanocytes.** Only one enzyme is involved, a copper-containing oxygenase called **tyrosinase.** This enzyme catalyzes first the hydroxylation of tyrosine to **3,4-dihydroxyphenylalanine,** usually called by the acronym **dopa** (Figure 21.16). Dopa, which also acts as a cofactor for this reaction, is the substrate for a subsequent oxidation, to **dopaquinone,** also catalyzed by tyrosinase. The subsequent reactions, leading to melanins, occur spontaneously in vitro. However, tyrosinase has been shown to catalyze another reaction, on the branch of the pathway leading to black melanin—the oxidation of 5,6-

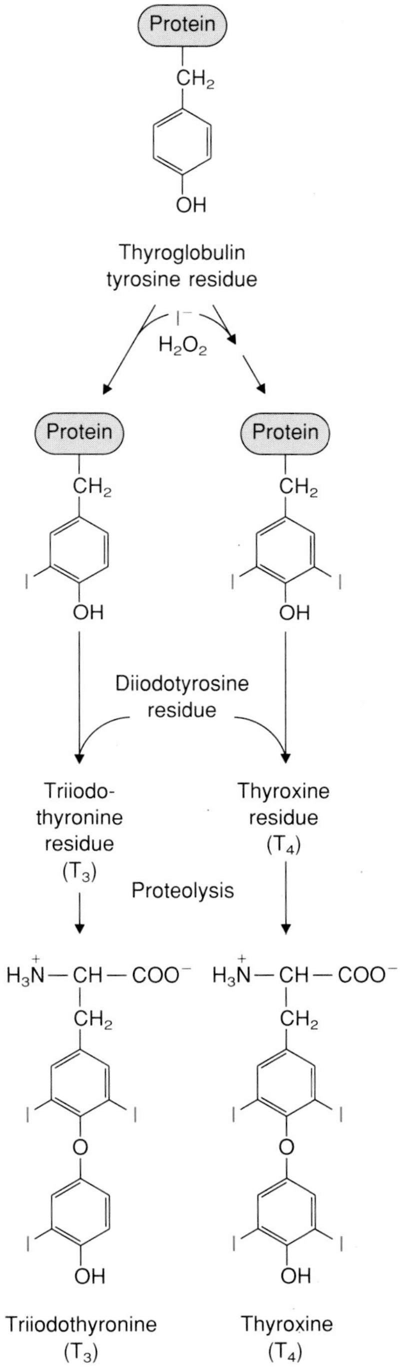

Figure 21.15
Biosynthesis of thyroid hormones as residues in the protein thyroglobulin. The iodinated forms of tyrosine—triiodothyronine (T_3) and thyroxine (T_4)—are released from these proteins by proteolytic degradation.

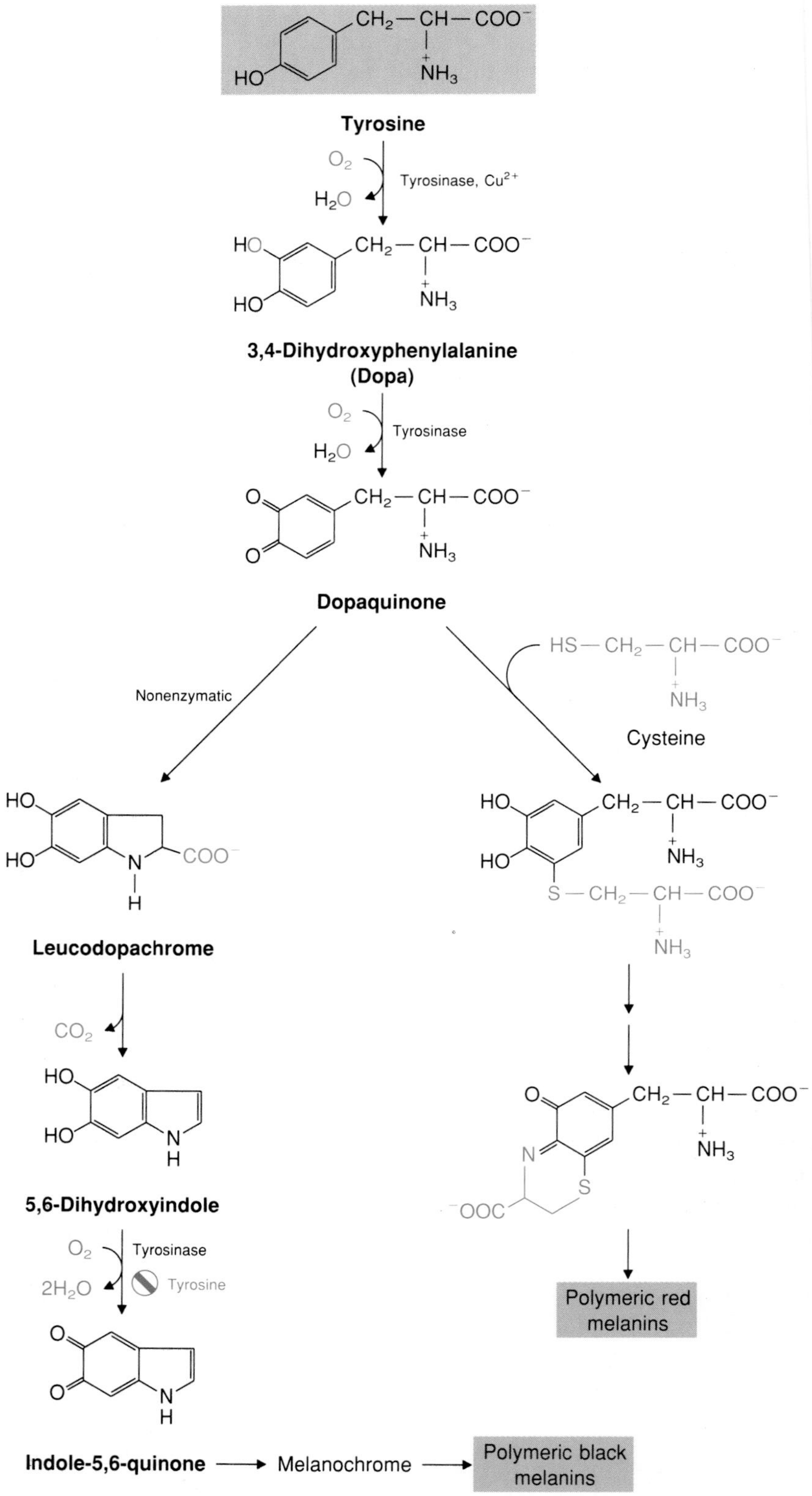

Figure 21.16
Biosynthetic pathways from tyrosine to melanins.

Tyrosine
O_2 → H_2O: Tyrosinase, Cu^{2+}
3,4-Dihydroxyphenylalanine (Dopa)
O_2 → H_2O: Tyrosinase
Dopaquinone
Nonenzymatic
Cysteine
Leucodopachrome
CO_2
5,6-Dihydroxyindole
O_2 → $2H_2O$: Tyrosinase; Tyrosine
Indole-5,6-quinone → Melanochrome → Polymeric black melanins
Polymeric red melanins

dihydroxyindole to indole 5,6-quinone. This reaction also requires dopa as a cofactor, and it is inhibited by tyrosine; this apparently provides a means for regulation of pigment formation.

In another branch of the pathway, dopaquinone reacts with cysteine en route to a related series of polymers, the red melanins. An individual's pigmentation is determined by the relative amounts of red and black melanins in the skin. These in turn result from the distribution and density of melanocytes in the basal layers of the skin, as well as the activities of the pathways leading to the different melanins. A genetic deficiency of tyrosinase causes an individual to lack pigmentation. Such individuals are called **albinos.**

The principal catabolic route for tyrosine involves first its transamination by **tyrosine aminotransferase,** an enzyme in the liver whose level is regulated hormonally (Figure 21.17). The product, *p*-hydroxyphenylpyruvate, is acted on by ***p*-hydroxyphenylpyruvate dioxygenase,** an unusual copper-containing enzyme, which catalyzes a ring hydroxylation, decarboxylation, and side chain migration. The product, **homogentisic acid,** is oxidized by an iron-containing enzyme, **homogentisic acid dioxygenase,** that cleaves the ring to yield a straight-chain eight-carbon compound that isomerizes to **fumarylacetoacetate.** The latter ultimately cleaves to yield fumarate and acetoacetate, both of which are catabolized by standard energy-yielding pathways.

A hereditary deficiency of the enzyme homogentisic acid dioxygenase causes a condition that was known for centuries as the "dark urine disease" but is now called **alkaptonuria.** Homogentisic acid accumulates and is excreted in large amounts in the urine. Homogentisic acid oxidizes on standing, especially if the urine is alkalinized, and causes the urine to become dark. Although the clinical symptoms of the disease are not severe, it is of considerable historical interest. Early in this century a physician named Archibald Garrod examined pedigrees of the families of afflicted individuals, and in 1908 he proposed correctly that the deficiency affected a gene controlling a normal reaction in the breakdown of aromatic compounds. In other words, he proposed that one gene encodes one enzyme, long before the chemical nature of either genes or enzymes was known.

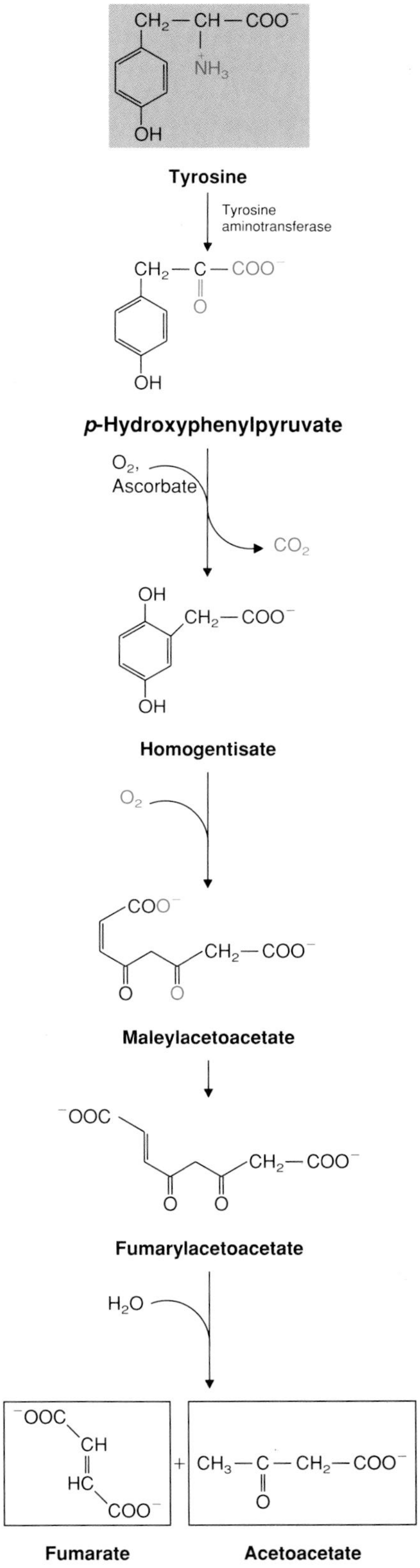

TRYPTOPHAN. Tryptophan is transformed by many pathways, of which just two concern us here—the major catabolic route, proceeding via **kynurenine** to glutaryl-CoA, and the synthesis of nicotinamide nucleotides (Figure 21.18). The first reaction in the degradative pathway is catalyzed by **tryptophan oxygenase,** an iron heme protein whose level is controlled by two interesting mechanisms—induction by certain hormones, and stabilization in vivo by its substrate, tryptophan, which increases enzyme levels by protecting the protein against intracellular degradation.

NAD^+ can be synthesized either from tryptophan or from the vitamin nicotinic acid (see Chapter 10). For both pathways the last reaction is catalyzed by a glutamine-requiring amidotransferase.

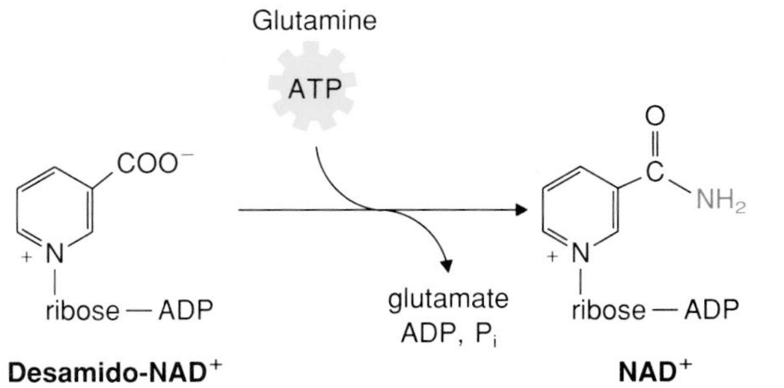

Figure 21.17
Catabolism of tyrosine to fumarate and acetoacetate.

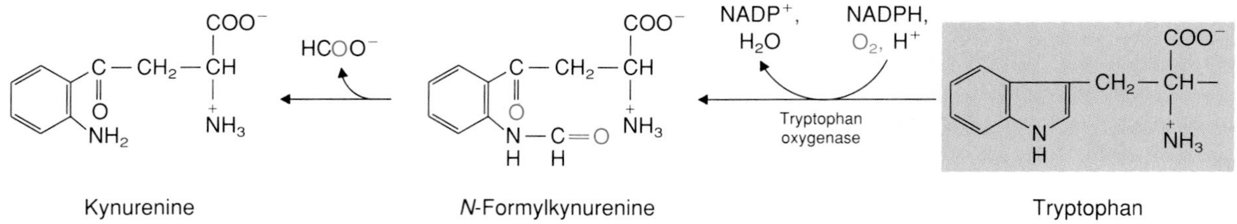

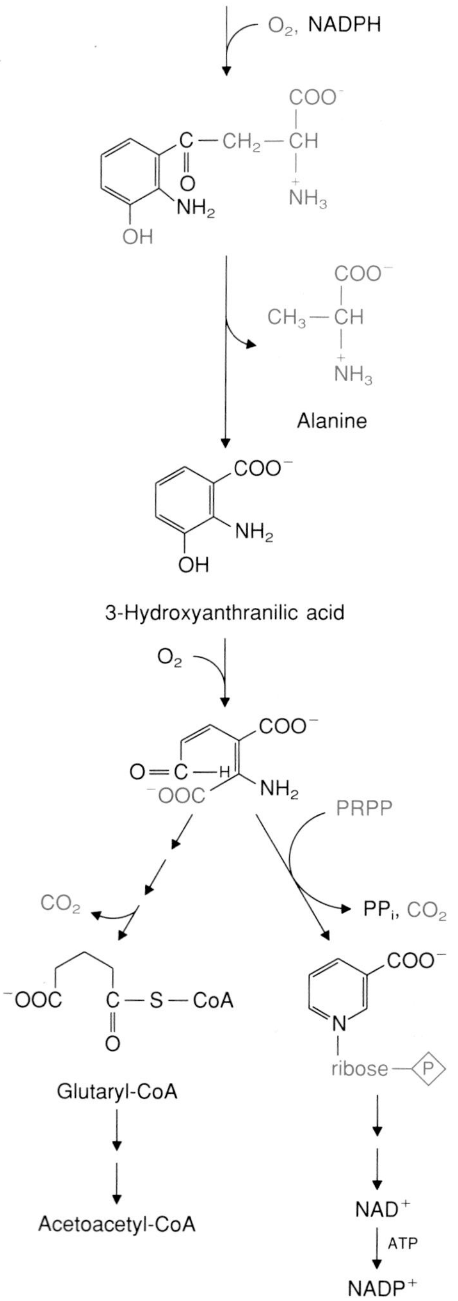

Figure 21.18
Metabolic fates of tryptophan, showing two major pathways—the synthetic pathway to the nicotinamide nucleotides and the major catabolic pathway that degrades the major part of the tryptophan molecule to acetoacetyl-CoA and CO_2.

Both pathways probably contribute significantly to NAD^+ biosynthesis, as can be inferred from studies of the nicotinamide deficiency disease **pellagra.** Pellagra was formerly endemic in regions, such as the southern United States, where corn is a dietary staple. Since corn proteins contain very little tryptophan, deficiencies of this amino acid were common. The symptoms of tryptophan deficiency are identical to those of nicotinamide, as expected if a major role of tryptophan is as a nicotinamide substitute.

HISTIDINE. Histidine metabolism generates an important biogenic amine, **histamine,** which we shall discuss in the next section. Figure 21.19 presents the major catabolic route for histidine. This amino acid, like tryptophan, does not undergo transamination at the start of its breakdown. Rather, a specific lyase cleaves out ammonia to give **urocanic acid.** Two subsequent steps bring about reduction and ring opening to yield **formiminoglutamic acid,** which is of interest because it serves as a donor of active one-carbon fragments. The formimino group is transferred to tetrahydrofolate, with the other product being glutamic acid. Formiminoglutamic acid accumulates when tetrahydrofolate is deficient, a fact that has been used as the basis of clinical tests for folate or B_{12} deficiency.

Amino Acid Decarboxylation and Biogenic Amines

Many basic compounds with amine functional groups play important roles in regulating mammalian metabolism. In Chapter 13 we considered epinephrine as a stimulator of cyclic AMP formation. In this chapter we have also mentioned creatine, the polyamines, and γ-aminobutyric acid. The term **biogenic amines** has been applied to this class of compounds—amines produced by living systems and important to life.

Several of the biogenic amines are formed largely by decarboxylation reactions involving aromatic amino acids or their derivatives. **Serotonin,** or 5-hydroxytryptamine, is derived from tryptophan, and **histamine** is derived from histidine.

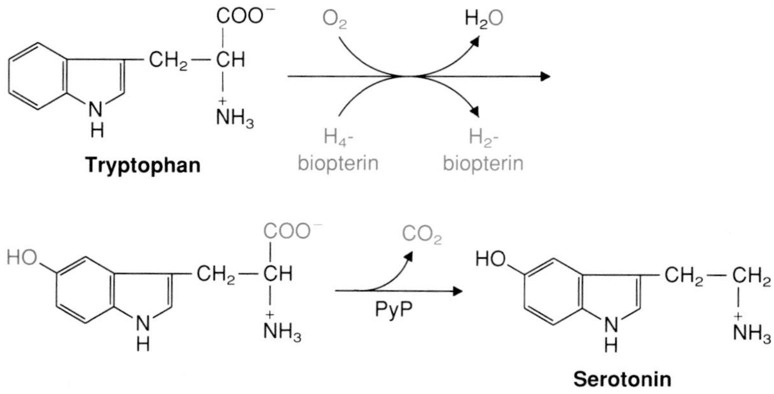

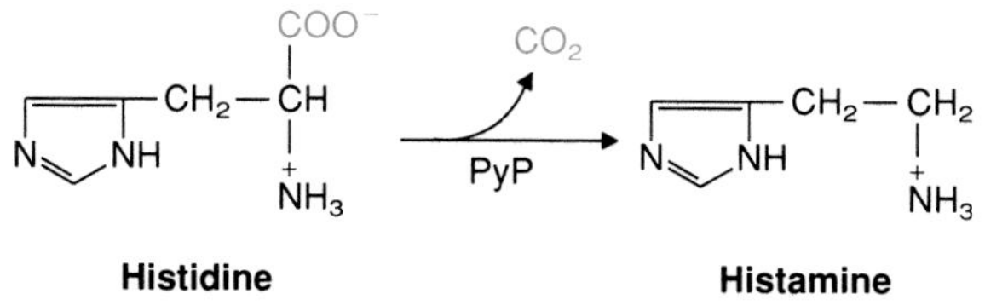

Histidine → (CO_2, PyP) → Histamine

The **catecholamines**—dopamine, norepinephrine, and epinephrine—are derived from tyrosine (Figure 21.20). Whereas histamine is formed simply by decarboxylation of histidine, both tryptophan and tyrosine undergo enzymatic hydroxylations before the decarboxylations. Both hydroxylations occur via tetrahydrobiopterin-dependent hydroxylases, comparable in mechanism to phenylalanine hydroxylase. Note that the product of tyrosine hydroxylation is dopa, which we described earlier as a product of tyrosinase action. However, tyrosinase is localized to melanocytes, while most catecholamine synthesis occurs in the adrenal medulla and in the central nervous system.

Once formed, dopa undergoes decarboxylation to give dopamine. The latter serves in turn as substrate for a copper-containing monooxygenase, *dopamine β-hydroxylase,* giving norepinephrine, which in turn is methylated by *S*-adenosylmethionine to give epinephrine. Although dopamine and norepinephrine are intermediates in epinephrine synthesis, each has its own roles to play as well.

Norepinephrine is a neurotransmitter in the central nervous system and also at nerve endings in the sympathetic nervous system. As discussed in Chapter 29, neurotransmission involves release of chemical transmitters from nerve endings and their uptake following binding to specific receptors in neighboring nerve cells. While moving from one cell to the other, norepinephrine is rapidly catabolized, starting with the action of either of two enzymes—**catecholamine O-methyltransferase** (COMT), which catalyzes

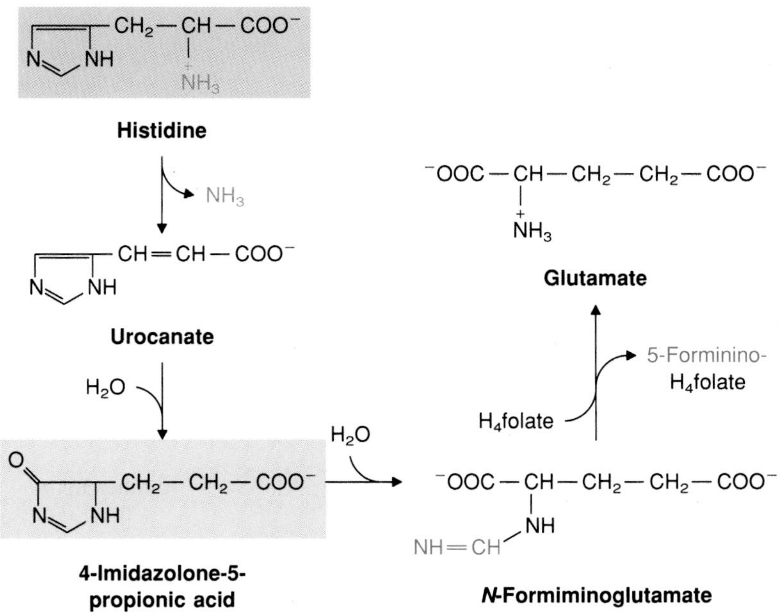

Histidine → (NH_3) → Urocanate → (H_2O) → 4-Imidazolone-5-propionic acid → (H_2O) → *N*-Formiminoglutamate → (H_4folate → 5-Formimino-H_4folate) → Glutamate

Figure 21.19
Catabolism of histidine, showing the conversion of the methene carbon of the imidazole ring to a one-carbon tetrahydrofolate adduct.

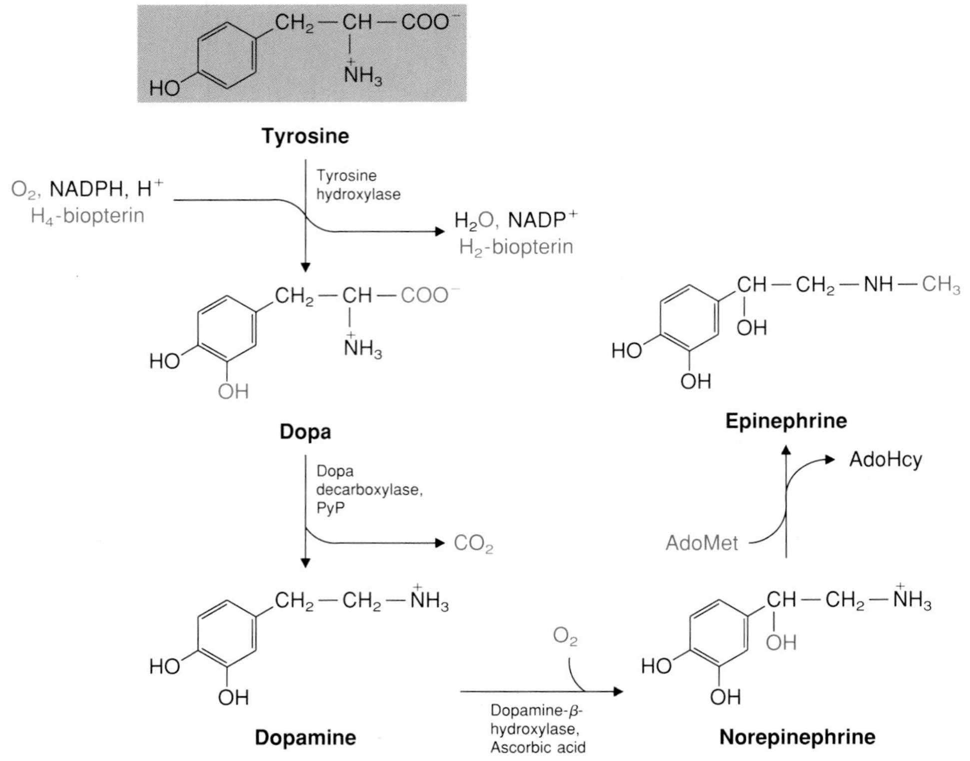

Tyrosine

Tyrosine hydroxylase

O_2, NADPH, H^+
H_4-biopterin

H_2O, $NADP^+$
H_2-biopterin

Dopa

Dopa decarboxylase, PyP

CO_2

Dopamine

O_2

Dopamine-β-hydroxylase, Ascorbic acid

Norepinephrine

AdoMet

AdoHcy

Epinephrine

Figure 21.20
Biosynthesis of the catecholamines—dopamine, norepinephrine, and epinephrine—from tyrosine.

an AdoMet-dependent transmethylation, or **monoamine oxidase (MAO)**, a flavoprotein that oxidizes primary amines to aldehydes.

Dopamine also plays a role in central nervous system transmission, possibly as an inhibitory agent. Such a role for dopamine was strongly implied when it was learned that dopamine levels are abnormally low in a particular region of the brain of patients with **Parkinsonism,** a severe neurological disorder. Attempts to treat such patients with dopamine were futile, because this substance, after injection, does not cross the blood–brain barrier and hence cannot remedy the deficiency. However, the dopamine precursor, dopa, does cross the blood–brain barrier, and in many patients with parkinsonism daily doses of dopa have provided dramatic clinical improvement.

Circumstantial evidence also links abnormal dopamine metabolism to schizophrenia, which may result partly from excess firing of **dopaminergic neurons**—neurons regulated by dopamine. Part of the circumstantial evidence is the close structural relationship between dopamine and **mescaline,** a product of the peyote cactus known to induce a quasi-schizophrenic state. **Amphetamine** is another catecholamine analog with potent psychopharmacological properties.

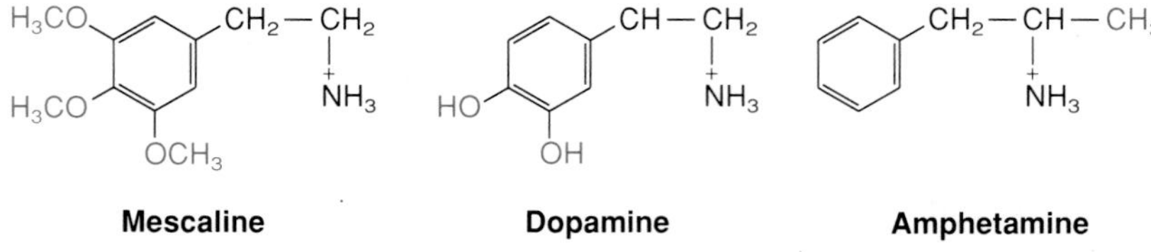

Mescaline

Dopamine

Amphetamine

Serotonin plays multiple regulatory roles in the nervous system, possibly including neurotransmission. It is produced in the pineal gland, where it serves as precursor to **melatonin** (*O*-methyl-*N*-acetylserotonin). The pineal is known to regulate light–dark cycles in animals, and the levels of serotonin and melatonin undergo cyclic variations in phase with these cycles. Thus, although the cycle-related actions of these compounds are not yet known, they point to serotonin and melatonin as regulators of sleep and wakefulness. Serotonin is also secreted by cells in the small intestine, where it regulates intestinal peristalsis. Finally, serotonin is a potent vasoconstrictor that helps regulate blood pressure.

Histamine is another substance with multiple biological actions. When secreted in the stomach, it promotes the secretion of hydrochloric acid and of pepsin, both of which aid digestion. It is a potent vasodilator, released locally in sites of trauma, inflammation, or allergic reaction. The local enlargement of blood capillaries is the basis for the reddening that occurs in inflamed tissues. Release of histamine in trauma contributes to the dangerous lowering of blood pressure that can lead to shock. A large number of **antihistamines** are in use to treat allergies and other inflammations.

Much of our understanding of biogenic amine function has come through study of compounds, either synthetic or natural, that either potentiate or antagonize actions of the amines. We have mentioned dopa and mescaline. Other examples include **lysergic acid diethylamide** (LSD), which evidently acts by mimicking the effects of serotonin at receptors in the central nervous system, and **chlorpromazine,** a tranquilizer that evidently blocks the action of dopamine, also by receptor binding.

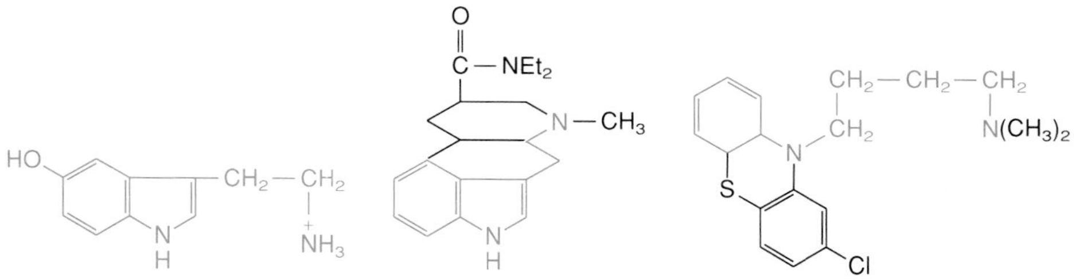

HO CH_2—CH_2 N H $\overset{+}{N}H_3$

Serotonin

O C—NEt_2 N—CH_3 N H

Lysergic acid diethylamide

CH_2—CH_2—CH_2 CH_2 N S $N(CH_3)_2$ Cl

Chlorpromazine

Serine, Glycine, and Threonine; Porphyrin Metabolism

Synthesis and Utilization of Serine, Glycine, and Threonine

Although serine, glycine, and threonine do not form a natural grouping, we consider them together partly because of the hydroxyl group shared by serine and threonine and partly because of the close interconnection between serine and glycine via the serine transhydroxymethylase reaction. Serine is quite active metabolically; we have already considered its roles in biosynthesis of phospholipids and cysteine, as well as its major contribution of activated one-carbon units to the pool of tetrahydrofolate coenzymes. Glycine also plays multiple roles, including contributions to the one-carbon pool and as a precursor to glutathione, to purine nucleotides (Chapter 22),

Figure 21.21
Metabolic interconversions and fates of serine and glycine.

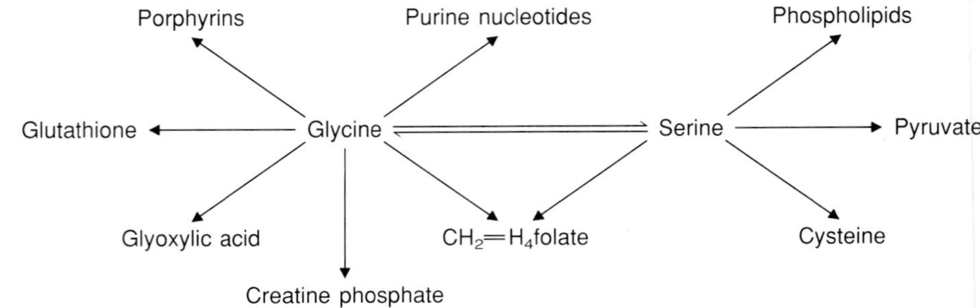

and to porphyrins. Figure 21.21 summarizes the metabolic fates of glycine and serine. By contrast, threonine plays but one significant role other than as a constituent of proteins: it is a precursor to isoleucine in plants and microorganisms.

All three of these amino acids are synthesized and degraded by rather simple pathways. Serine can be synthesized from glycine, via the serine transhydroxymethylase reaction. However, this reaction probably proceeds more often in the reverse direction, as the major biosynthetic route to glycine and to activated one-carbon compounds. Most serine biosynthesis occurs in a three-step sequence from the glycolytic intermediate 3-phosphoglycerate.

$CH_2{-}O{-}P$, $H{-}C{-}OH$, COO^- **3-Phosphoglycerate** $\xrightarrow{NAD^+ \;\; NADH,\, H^+}$ $CH_2{-}O{-}P$, $C{=}O$, COO^- **3-Phosphopyruvate**

$\xrightleftharpoons{\text{Transamination}}$ $CH_2{-}O{-}P$, $H{-}C{-}\overset{+}{N}H_3$, COO^- **3-Phosphoserine** $\xrightarrow{H_2O \;\; P_i}$ CH_2OH, $H{-}C{-}\overset{+}{N}H_3$, COO^- **Serine**

In bacteria the synthesis of serine is controlled through feedback inhibition of the first and third reactions by serine. Since threonine is an essential amino acid, its synthesis is limited to plants and prokaryotes. Threonine synthesis begins with homoserine, which is derived from aspartate (p. 705). Homoserine undergoes a phosphorylation, followed by a pyridoxal phosphate-dependent reaction that simultaneously cleaves out the phosphate and causes the hydroxyl group to migrate from the γ- to the β-carbon.

$CH_2{-}OH$, CH_2, $H{-}C{-}\overset{+}{N}H_3$, COO^- **Homoserine** $\xrightarrow{ATP \;\; ADP}$ $CH_2{-}O{-}P$, CH_2, $H{-}C{-}\overset{+}{N}H_3$, COO^- ***O*-Phosphohomoserine** $\xrightarrow{P_i}$ [CH_2, $\|$, CH, $H{-}C{-}\overset{+}{N}H_3$, COO^-] $\xrightarrow{H_2O}$ CH_3, $H{-}C{-}OH$, $H{-}C{-}\overset{+}{N}H_3$, COO^- **Threonine**

The major route for glycine degradation is via the mitochondrial glycine cleavage system (Chapter 20), to yield NH_3, CO_2, and N^5,N^{10}-methylenetetrahydrofolate. This reaction is particularly important in photorespiration in plants (Chapter 19). Glycolate produced in photorespiration is oxidized to glyoxylate, which undergoes transamination to glycine. One molecule of glycine undergoes the glycine cleavage reaction, and the resultant N^5,N^{10}-methylenetetrahydrofolate reacts with a second molecule of glycine, to give serine. Serine in turn is converted to 3-phosphoglycerate, which undergoes the normal photosynthetic carbon reduction cycle.

Serine is catabolized both by its conversion to glycine and by the action of **serine–threonine dehydratase.** This pyridoxal phosphate-dependent enzyme converts serine to pyruvate and threonine to α-ketobutyrate (Figure 21.22). The latter compound is converted to propionyl-CoA and CO_2; the propionyl-CoA is metabolized to succinyl-CoA by the same pathway used in odd-chain fatty acid oxidation and methionine catabolism. In plants and prokaryotes α-ketobutyrate has a biosynthetic role as well, as the first intermediate en route to isoleucine biosynthesis.

An alternative route for threonine catabolism is its reversible cleavage to glycine and acetaldehyde, catalyzed by **threonine aldolase.**

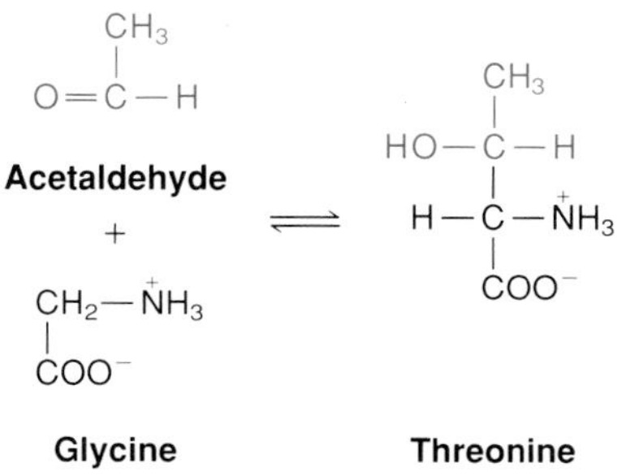

$$\underset{\textbf{Acetaldehyde}}{CH_3-CHO} + \underset{\textbf{Glycine}}{{}^{+}NH_3-CH_2-COO^-} \rightleftharpoons \underset{\textbf{Threonine}}{CH_3-CH(OH)-CH({}^{+}NH_3)-COO^-}$$

Though this reaction could in principle lead to threonine biosynthesis in animals, acetaldehyde is highly toxic and cannot accumulate to the point where it would be a biosynthetic intermediate.

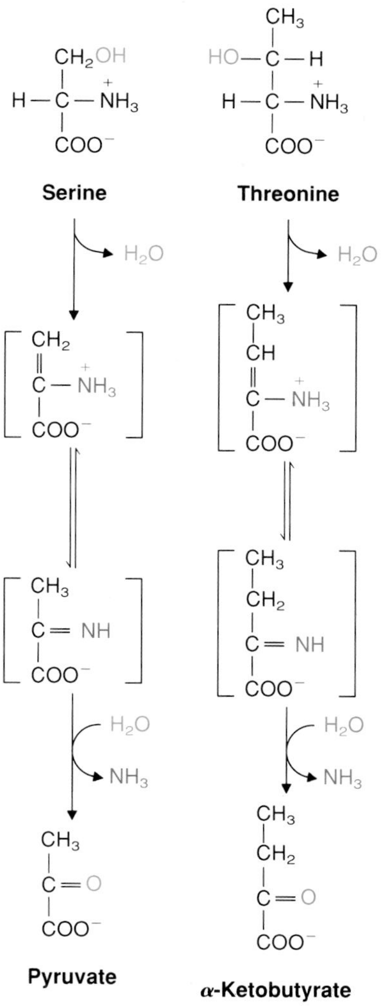

$$\underset{\textbf{Serine}}{HOCH_2-CH({}^{+}NH_3)-COO^-} \xrightarrow{-H_2O} [CH_2=C({}^{+}NH_3)-COO^-] \rightleftharpoons [CH_3-C(=NH)-COO^-] \xrightarrow[-NH_3]{+H_2O} \underset{\textbf{Pyruvate}}{CH_3-CO-COO^-}$$

$$\underset{\textbf{Threonine}}{CH_3-CH(OH)-CH({}^{+}NH_3)-COO^-} \xrightarrow{-H_2O} [CH_3-CH=C({}^{+}NH_3)-COO^-] \rightleftharpoons [CH_3-CH_2-C(=NH)-COO^-] \xrightarrow[-NH_3]{+H_2O} \underset{\alpha\textbf{-Ketobutyrate}}{CH_3-CH_2-CO-COO^-}$$

Figure 21.22
Reactions catalyzed by serine–threonine dehydratase.

Biosynthesis of Tetrapyrroles; the Succinate–Glycine Pathway to Porphyrins

A major metabolic fate of glycine is its utilization for tetrapyrrole biosynthesis. There are four classes of tetrapyrrole compounds in biology: the widely distributed **heme;** the **chlorophylls** of plants and photosynthetic bacteria; the **phycobilins,** photosynthetic pigments of algae; and the **cobalamins,** notably vitamin B_{12} and its derivatives. All of these compounds are synthesized from a common precursor, **δ-aminolevulinic acid.** Figure 21.23 illustrates the relationships among the various synthetic pathways.

Since many of the details of tetrapyrrole biosynthesis in plants and bacteria are not yet established, we shall concentrate here on the well-understood porphyrin synthetic pathway, which leads to heme. This pathway is widespread in animal tissues and, as far as is known, is the same in all animals. Seven reactions are involved, and they occur in two different cell compartments. The first reaction occurs in mitochondria, followed by three reactions in the cytosol, and finally, three more mitochondrial reactions. As we shall see, this compartmentation provides the opportunity for a novel control mechanism for the pathway.

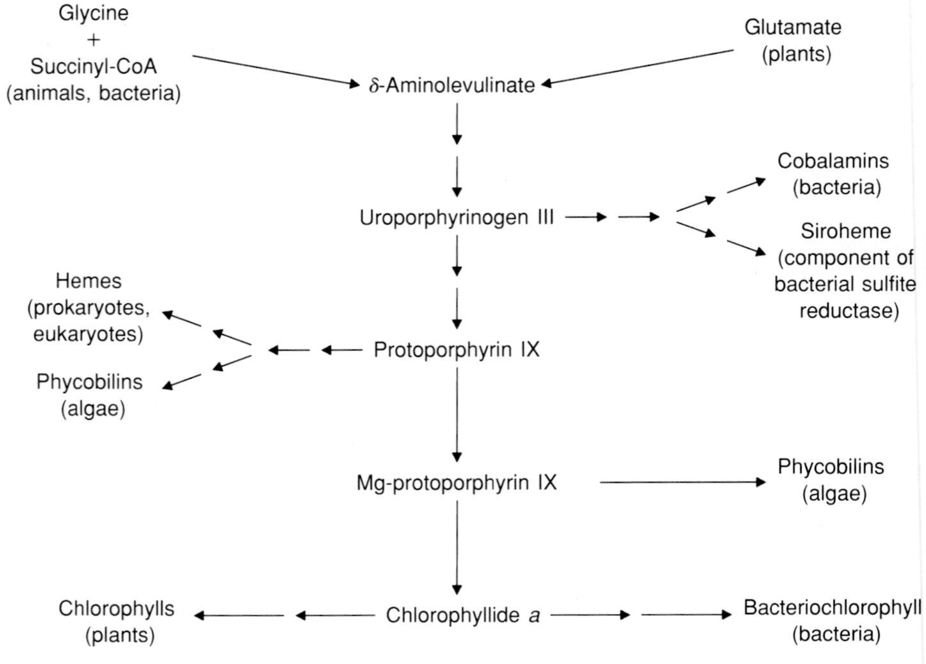

Figure 21.23
Biosynthetic pathways to tetrapyrroles: heme, chlorophylls, phycobilins, and cobalamins.

Early labeling studies revealed that *all of the nitrogen of heme is derived from glycine, and all of the carbon is derived from succinate and glycine*. Hence, this synthesis is often called the **succinate–glycine pathway.** The first reaction is catalyzed by a pyridoxal phosphate-dependent enzyme, **δ-aminolevulinic acid synthetase,** or ALA synthetase. As shown in Figure 21.24, the binding of glycine to PyP probably activates the α-carbon of glycine for an attack on the thioester carbon of succinyl-CoA. Decarboxylation follows, to give the product ALA.

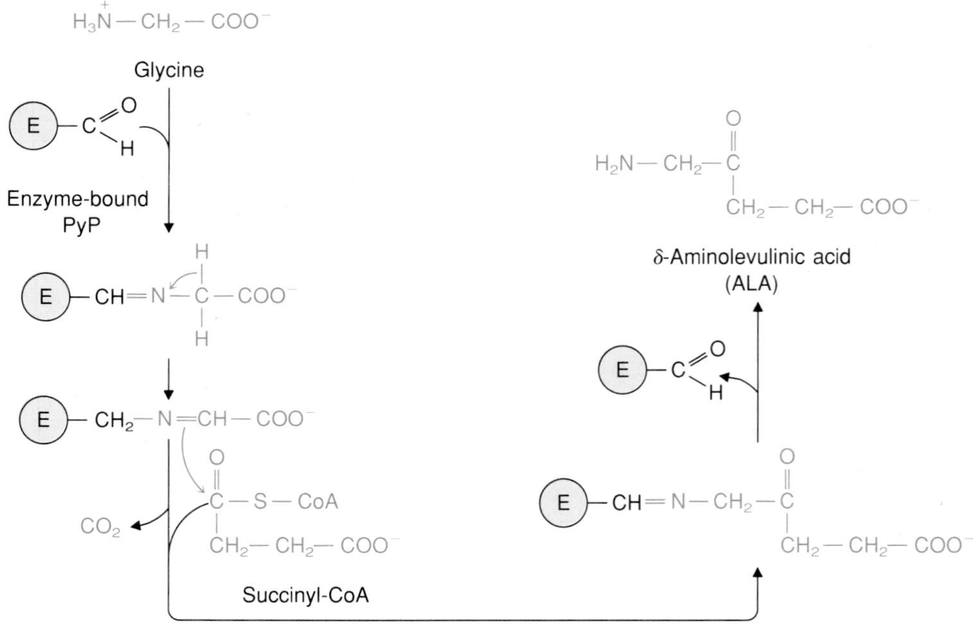

Figure 21.24
The δ-aminolevulinic acid synthetase reaction. E-CHO is the enzyme–pyridoxal phosphate complex.

In plants ALA is formed by a completely different pathway, a three-step sequence beginning with glutamate (Figure 21.25). The first reaction links glutamate, through its carboxyl group, to a specific transfer RNA, just as occurs in protein synthesis. The carboxyl group thus activated is then reduced by NADPH, giving glutamate 1-semialdehyde, which finally undergoes an internal transamination to give ALA. Because the major end product of the pathway in plants is chlorophyll, the synthesis of ALA is regulated by light; the identity of the specific light-regulated steps is now under active investigation.

Whether in plants, animals, or microorganisms, the remainder of the porphyrin synthetic pathway involves three distinct processes: (1) synthesis of a substituted pyrrole compound, **porphobilinogen;** (2) condensation of four porphobilinogen molecules to yield a partly reduced precursor called a **porphyrinogen;** and (3) modification of the side chains, dehydrogenation of the ring system, and introduction of iron, to give the porphyrin product, heme. Before going through these reactions, note the side chains found in porphyrins (see the legend for Figure 21.26).

Now let us review heme synthesis in some detail (Figure 21.26). Two molecules of ALA condense in the cytoplasm to form one molecule of porphobilinogen; the reaction is catalyzed by **ALA dehydratase.**

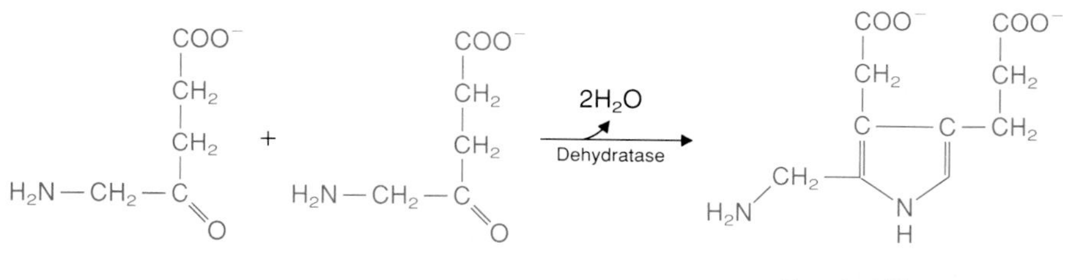

Next, four molecules of porphobilinogen combine in a PyP-requiring deaminase reaction to give the first tetrapyrrole compound. Two different proteins are involved in this reaction: **uroporphyrinogen I synthetase** and **uroporphyrinogen III cosynthetase.** The former protein is enzymatically active by itself, but when it acts alone, the product is the undesired symmetrical compound uroporphyrinogen I. Interaction between the synthetase and cosynthetase allows one of the rings to flip during the combining reaction, so that the product is the asymmetric **uroporphyrinogen III.** The symmetric compound and some metabolites derived from it are synthesized as side products, but in low amounts. They are nonfunctional.

In a hereditary condition called **congenital erythropoietic porphyria** the cosynthetase is defective, and the type I porphyrins accumulate beyond the capacity of the body to excrete them. Their accumulation causes the urine to turn red, the skin to become acutely photosensitive, the erythrocytes to be destroyed prematurely, and the teeth to become fluorescent. Because insufficient heme is synthesized, these individuals are quite anemic. It has been speculated that those labeled as vampires in medieval folktales suffered from this condition, which explains their preference for the dark, their bizarre appearance, and their propensity for drinking blood. In fact, individuals with congenital erythropoeitic porphyria can be treated by injections of heme.

Related to the above condition, though distinct from it, is **acute intermittent porphyria,** which results from deficiency in the synthetase; this causes ALA and porphobilinogen to accumulate in the liver. The condition

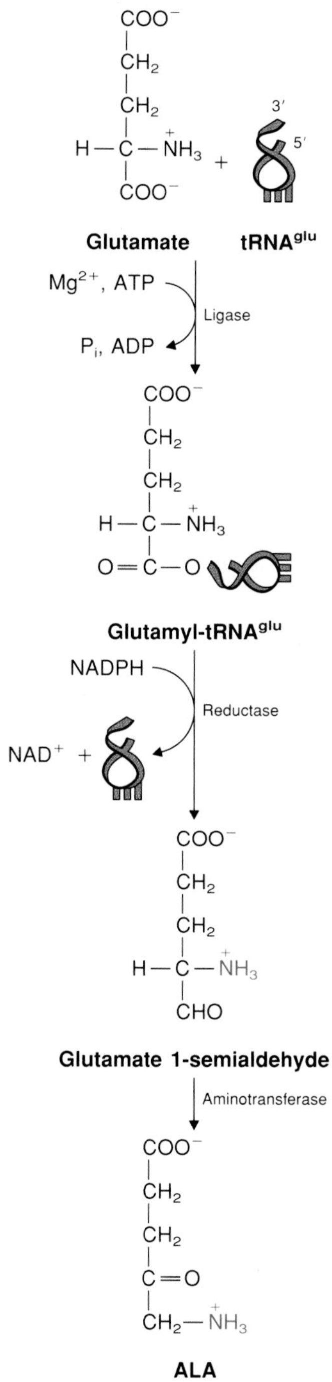

Figure 21.25
Synthesis of δ-aminolevulinic acid in plants.

is accompanied by episodes of acute abdominal pain and neurological disorders.

Uroporphyrinogen III undergoes decarboxylation of its acetic acid side chains, and the product reenters the mitochondrion for further side chain modifications, then ring oxidation to yield a fully conjugated system, and, finally, the insertion of iron. The last reaction can proceed spontaneously, but it is catalyzed by **ferrochelatase,** an enzyme on the inner mitochondrial membrane that also requires a reducing agent.

Being the first committed step in heme synthesis, the ALA synthetase reaction is the major control point. Heme and related compounds feedback-inhibit the enzyme. Heme also has two other important effects. At low concentrations heme inhibits the *synthesis* of ALA synthetase at the translational level. At higher levels heme somehow blocks the *translocation* of ALA synthetase from the cytosol, where it is synthesized on ribosomes, into the mitochondrion, where it acts. Heme also inhibits the ferrochelatase reaction. A number of drugs and poisons cause excessive heme synthesis. In some cases this results from stimulation of the synthesis of cytochrome P-450, which increases the demand for heme and hence activates ALA synthetase.

Porphyrin biosynthesis is being exploited as a target for the action of weedkillers. The idea is to spray weeds in the dark with ALA. The pathway to chlorophyll begins, and when it becomes light, the pathway is completed, and chlorophyll is produced in such massive amounts that the plant weakens and dies.

Degradation of Heme in Animals

Although considerable heme occurs in cytochromes, by far the most abundant porphyrin compound in vertebrates in the heme of hemoglobin. Therefore, the story of porphyrin degradation is largely the story of heme degradation (Figure 21.27). Lacking nuclei, the erythrocytes are incapable of renewal and self-destruct after characteristic intervals. In humans the average erythrocyte life span is 120 days, whereupon aged erythrocytes are transported to the spleen for degradation. Amino acids released from the

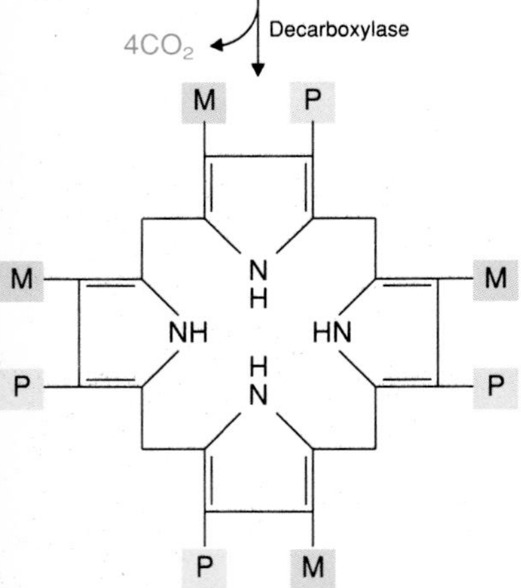

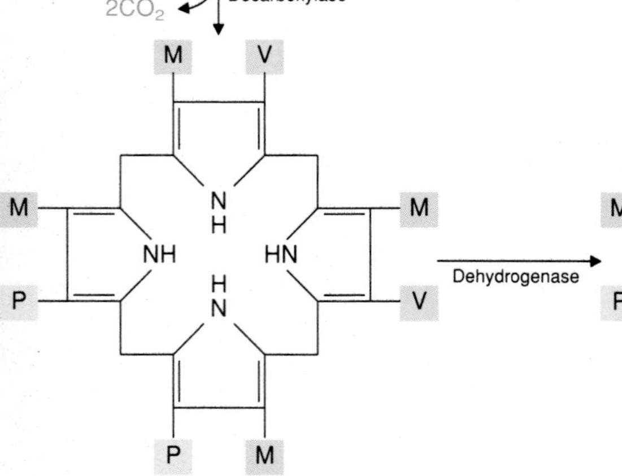

Figure 21.26
Biosynthesis of heme from porphobilinogen. Letter abbreviations indicate side chains: P (propionic acid), $—CH_2—CH_2—COO^-$; A (acetic acid), $—CH_2—COO^-$; V (vinyl), $—CH{=}CH_2$; M (methyl), $—CH_3$.

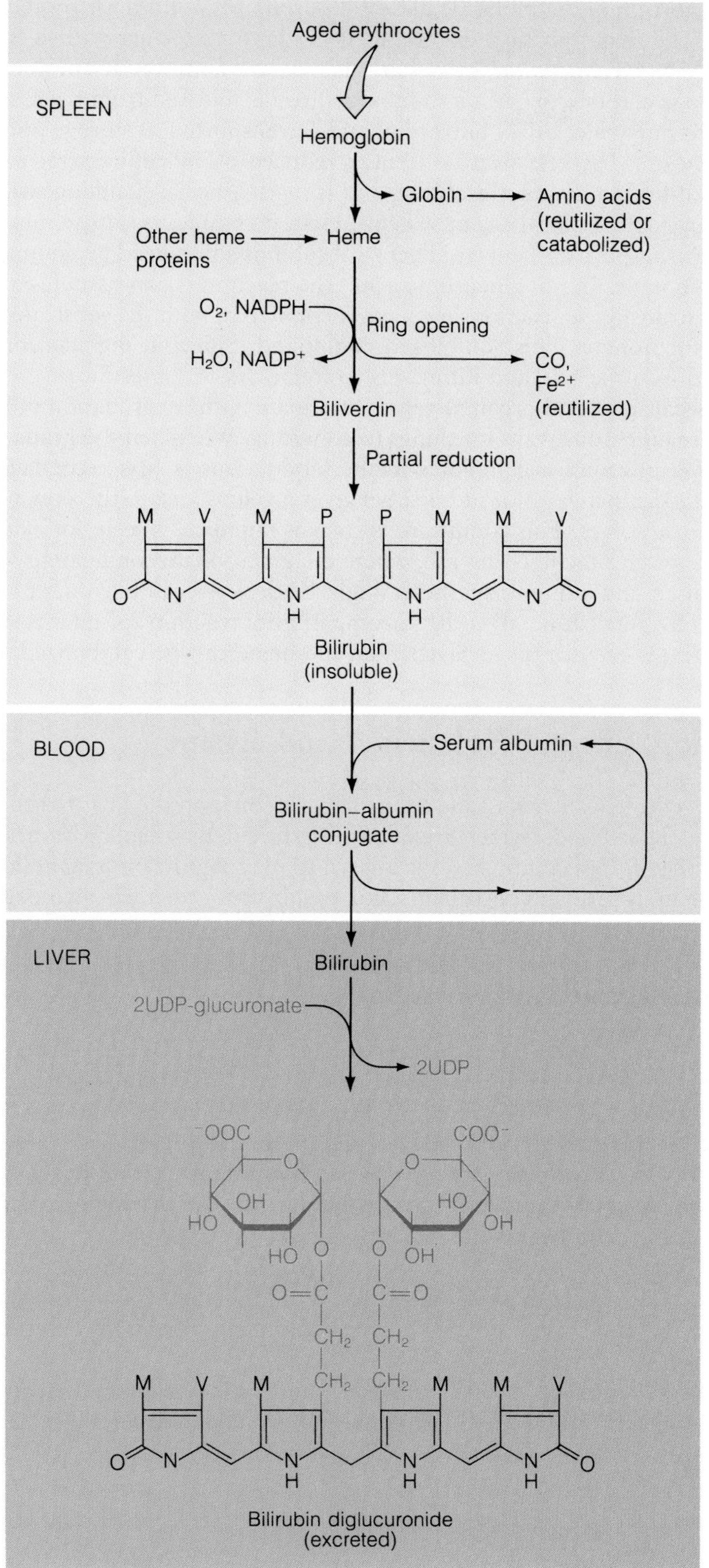

Figure 21.27
Catabolism of heme. Most of the heme comes from breakdown of aged erythrocytes, but some comes from cytochromes and other heme proteins.

globin portion of the molecule are catabolized or reutilized for protein synthesis. The heme portion undergoes degradation, starting with a mixed-function oxidase reaction that opens the ring and converts one of the methene bridge carbons to carbon monoxide. Iron is released from the resulting linear tetrapyrrole, called **biliverdin,** and is transported to storage pools for reutilization. The tetrapyrrole is next reduced to **bilirubin,** which is excreted. Bilirubin is quite insoluble, and it merits special handling, with the participation of several organ systems. First, it complexes with serum albumin for transport to the liver. There, bilirubin is solubilized by conjugation with two molecules of **glucuronic acid.** The reaction is comparable to other glycosyltransferase reactions we have encountered, with the substrate being **UDP-glucuronate.** This solubilized compound, **bilirubin diglucuronide,** is secreted into the bile and ultimately excreted via the intestine.

Because several organ systems participate in the degradation of heme, there are numerous ways for things to go wrong. When heme degradation is defective, bilirubin accumulates in the blood. This is first recognized because the distinctive color of bilirubin gives a yellow cast to the skin and the whites of the eyes. This condition, known as **jaundice,** is seen, for example, in acute or chronic liver disease, where the glucuronate conjugating system is impaired and albumin synthesis might be defective; in bile duct obstruction, as by a gallstone; or in Rh incompatibility reactions of infants, where erythrocytes are destroyed faster than the heme can be catabolized.

Valine, Leucine, Isoleucine, and Lysine

Valine, leucine, isoleucine, and lysine have in common the fact that they are essential amino acids, which are not synthesized in mammalian tissues. Further, none of these amino acids is known to play significant metabolic roles other than as protein constituents and as substrates for their own degradation.

Valine, Leucine, and Isoleucine

Valine, leucine, and isoleucine are structurally related and, as you might expect, they share certain reactions and enzymes in their biosynthetic pathways (Figure 21.28). The last four reactions in valine biosynthesis and in isoleucine biosynthesis are catalyzed by the same four enzymes. Valine biosynthesis begins with transfer of a two-carbon fragment from hydroxyethyl thiamine pyrophosphate to pyruvate. Similar transfer of a two-carbon unit to α-ketobutyrate begins the pathway to isoleucine.

The keto analog of valine, α-ketoisovalerate, is a precursor to leucine (Figure 21.29). Lengthening the carbon chain by one involves transfer of a two-carbon unit from acetyl-CoA, with subsequent loss of one carbon in an oxidative decarboxylation.

Degradation of leucine, isoleucine, and valine in animals starts with transamination followed by oxidative decarboxylation of the respective keto acids.

$$\underset{\text{Leu, Ile, or Val}}{R-\overset{\overset{+}{N}H_3}{\overset{|}{C}H}-COO^-} \xrightarrow{\text{Transamination}} R-\overset{\overset{O}{\|}}{C}-COO^- \xrightarrow[\text{Branched-chain } \alpha\text{-keto acid dehydrogenase}]{NAD^+,\ CoA-SH \ \rightarrow\ NADH + H^+,\ CO_2} R-\overset{\overset{O}{\|}}{C}-S-CoA$$

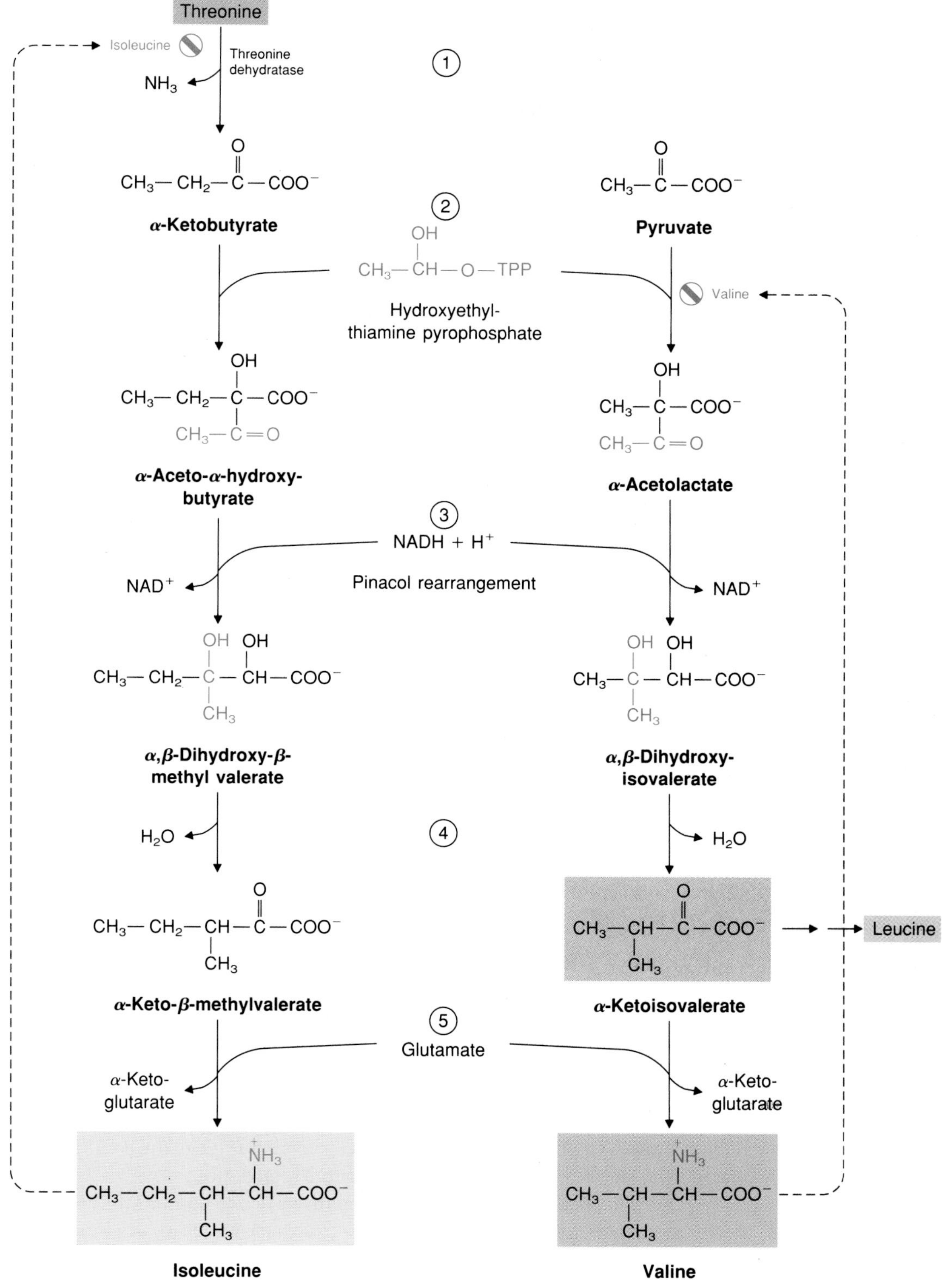

Figure 21.28
Biosynthesis of valine and isoleucine. After threonine dehydratase, one set of enzymes catalyzes the comparable reactions in valine and isoleucine synthesis: reaction 2, acetohydroxy acid synthase; reaction 3, acetohydroxy acid isomeroreductase; reaction 4, dihydroxy acid dehydratase; reaction 5, transaminase.

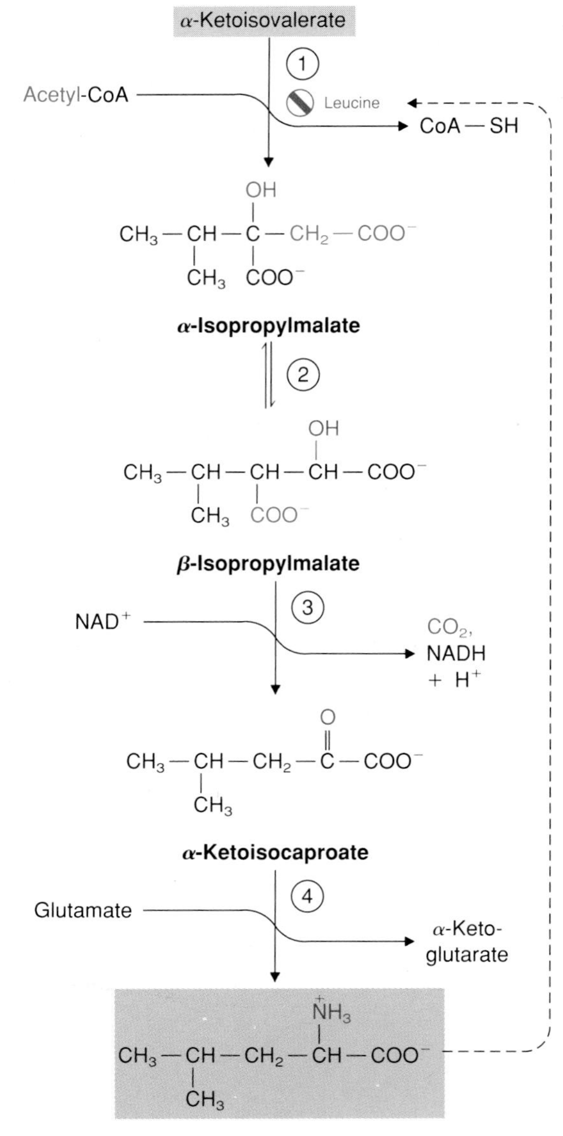

Figure 21.29
Biosynthesis of leucine, starting with an intermediate in valine synthesis. Reaction 1, isopropylmalate synthase; reaction 2, isomerase; reation 3, dehydrogenase; reaction 4, aminotransferase.

The latter reaction is carried out by a multienzyme complex, called the **branched-chain α-ketoacid dehydrogenase complex,** which is similar in structure and mechanism to the pyruvate dehydrogenase and α-ketoglutarate dehydrogenase complexes. In a rare human disorder called **maple syrup urine disease** this complex is defective. All three keto acids accumulate in the urine, and their characteristic odor gives the condition its name. The condition involves severe mental retardation.

The remainder of the degradative pathways are comparable to fatty acid oxidation, except that the intermediates must undergo a debranching not seen with β-oxidation.

Lysine

Lysine metabolism is distinguished partly by the complexity of both its synthetic and degradative pathways and partly by the fact that there are two distinct biosynthetic pathways. The **diaminopimelic acid pathway** operates in bacteria, some lower fungi, algae, and higher plants. This pathway, shown in Figure 21.30, begins with condensation of pyruvate with aspartate

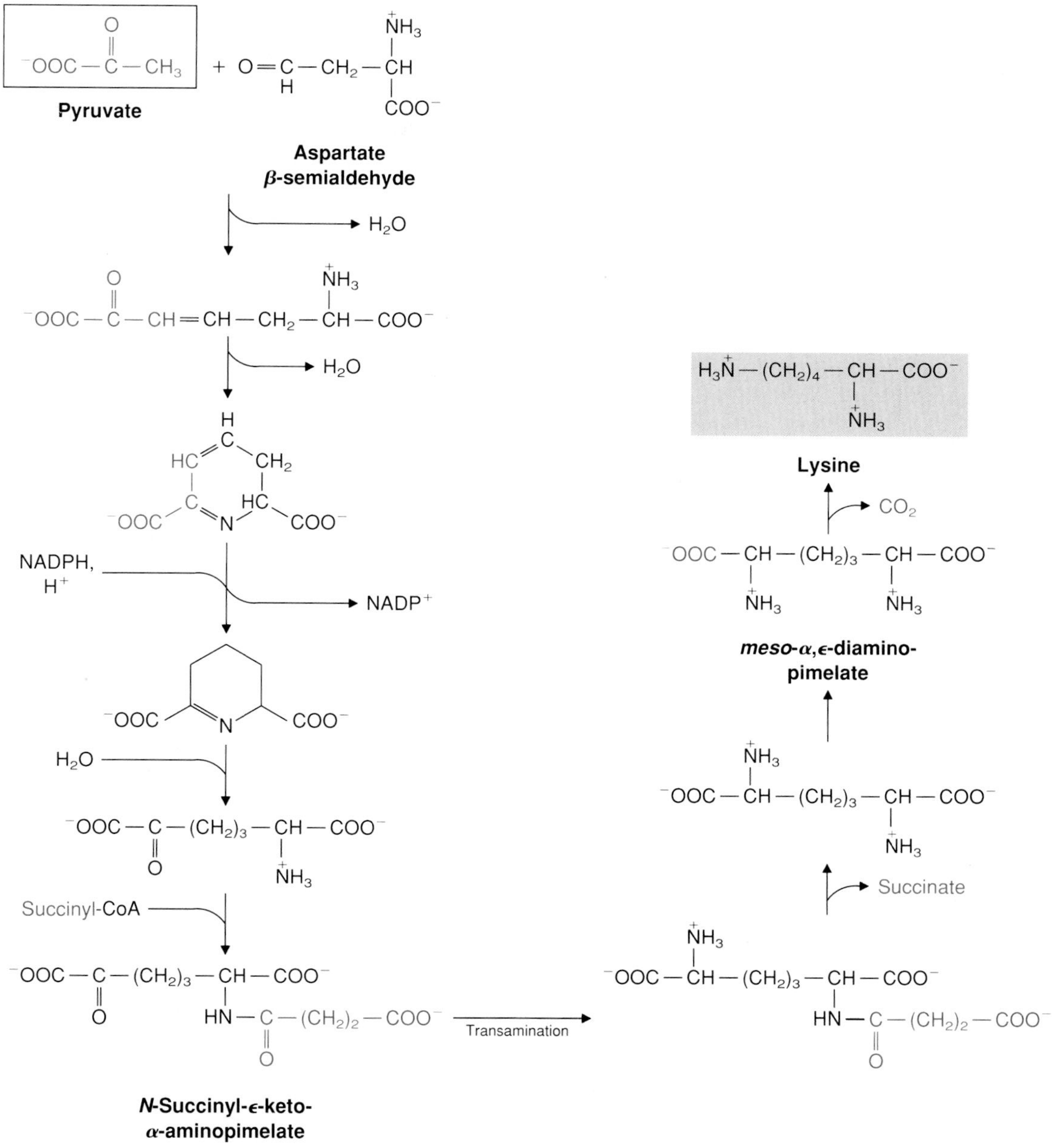

Figure 21.30
The diaminopimelate pathway to lysine, as it occurs in plants, bacteria, and some fungi.

β-semialdehyde. One intermediate in the pathway, diaminopimelic acid, also serves an important function as a constituent of bacterial cell walls.

The **α-aminoadipic acid pathway** is less widespread, functioning in other lower fungi, in higher fungi, and in the protist *Euglena*. Perhaps the most interesting feature of this pathway is the strong resemblance between the first four reactions, starting with acetyl-CoA and α-ketoglutarate, and four corresponding reactions in the citric acid cycle, starting with acetyl-CoA and oxaloacetate (Chapter 14).

The degradation of lysine does not begin with transamination. In animal tissues the major pathway is a series of five reactions in mitochondria, which lead to α-aminoadipic acid. Further degradation of this compound occurs through a multistep pathway in which the main carbon chain ultimately joins central metabolic pathways as acetoacetyl-CoA. The first several intermediates in the degradative pathway are identical to the last biosynthetic intermediates in the α-aminoadipic acid pathway.

REFERENCES

General Aspects of Amino Acid Synthesis and Its Regulation

Kishore, G. M., and D. M. Shah (1988) Amino acid biosynthesis inhibitors as herbicides. *Annu. Rev. Biochem.* 57:627–663. Summarizes current research that uses amino acid biosyntheses as specific targets for the action of weed control agents.

Walsh, C. T. (1979) *Enzymatic Reaction Mechanisms.* Freeman, San Francisco. An excellent book, particularly valuable in the context of amino acid metabolism, one-carbon metabolism, cobalamin coenzymes, and oxygenases.

Wellner, D., and A. Meister (1981) A survey of inborn errors of metabolism and transport in man. *Annu. Rev. Biochem.* 50:911–948. Reveals how much can be learned about the metabolic roles of enzymes through the study of rare human mutations.

Zalkin, H., and D. J. Ebbole (1988) Organization and regulation of genes encoding biosynthetic enzymes in *Bacillus subtilis. J. Biol. Chem.* 263:1595–1599. A recent minireview, with numerous references to the older literature.

The Citric Acid Cycle Superfamily

Cooper, J. B., J. A. Chen, G.-J. van Holst, and J. E. Varner (1987) Hydroxyproline-rich glycoproteins of plant cell walls. *Trends Biochem. Sci.* 12:24–27. Hydroxyproline was long thought to be present only in animal connective tissue proteins. This article summarizes its occurrence and roles in plant structural proteins.

Kivirikko, K. I., R. Myllylä, and T. Pihlajaniemi (1989) Protein hydroxylation: Prolyl 4-hydroxylase, an enzyme with four cosubstrates and a multifunctional subunit. *FASEB J.* 3:1609–1617. A recent review of the mechanism of proline hydroxylation and its role in collagen synthesis.

Sulfur-Containing Amino Acids

Pegg, A. E. (1988) Polyamine metabolism and its importance in neoplastic growth and as a target for chemotherapy. *Cancer Res.* 48:759–774. One of several excellent reviews of the metabolism and functions of these compounds and of the potential for chemotherapeutic manipulation of these reactions.

Aromatic Amino Acids

Bloom, F. E. (1988) Neurotransmitters: Past, present, and future directions. *FASEB J.* 1:32–41. A current review of neurotransmitters and their actions, by one of the leaders.

Fitzpatrick, P. F., and J. J. Villafranca (1987) Mechanism-based inhibitors of dopamine β-hydroxylase. *Arch. Biochem. Biophys.* 257:231–250. Neuropharmacologists are interested in this enzyme as a point at which to control catecholamine biosynthesis. This review describes progress in understanding the mechanism of the reaction and identifying useful inhibitors.

Porphyrin Metabolism

Granick, S., and S. I. Beale (1978) Hemes, chlorophylls, and related compounds: Biosynthesis and metabolic regulation. *Adv. Enzymol.* 40:33–203. An excellent and comprehensive review of tetrapyrrole metabolism.

Kannangara, C. G., S. P. Gough, P. Bruyant, J. K. Hoober, A. Kahn, and D. von Wettstein (1988) $tRNA^{glu}$ as a cofactor in δ-aminolevulinate biosynthesis. *Trends Biochem. Sci.* 13:139–143. This minireview describes the unexpected role of tRNA in the biosynthesis of δ-aminolevulinic acid in plants.

Valine, Leucine, and Isoleucine

Yeaman, S. J. (1986) The mammalian 2-oxoacid dehydrogenases: a complex family. *Trends Biochem. Sci.* 11:293–296. This minireview compares and contrasts the complexes involved in oxidizing pyruvate, α-ketoglutarate, and the branched-chain keto acids derived from catabolism of valine, leucine, and isoleucine.

PROBLEMS

1. A clinical test sometimes used to diagnose folate deficiency or B_{12} deficiency is a histidine tolerance test, where one injects a large dose of histidine into the bloodstream and then carries out a series of biochemical determinations. What histidine metabolite would you expect to accumulate in a folate- or B_{12}-deficient patient, and why?

2. In bacteria much of the putrescine is synthesized, not from ornithine, but from arginine, which decarboxylates to yield *agmatine*. Formulate a plausible pathway from arginine to putrescine, using this intermediate.

$$H_2N-\overset{\overset{+}{NH_2}}{\overset{\|}{C}}-NH-CH_2-CH_2-CH_2-CH_2-\overset{+}{N}H_3$$

Agmatine

3. The mitochondrial form of carbamoyl phosphate synthetase is allosterically activated by *N*-acetylglutamate. Briefly describe a rationale for this effect.

4. *Psilocybin* is a hallucinogenic compound found in some mushrooms. Present a straightforward pathway for its biosynthesis from one of the aromatic amino acids.

O—(P)
CH₂—CH₂—N(CH₃)₂
N
H

Psilocybin

5. One can identify phenylketonurics and PKU carriers (hetrozygotes) by means of a phenylalanine tolerance test. One injects a large dose of phenylalanine into the bloodstream and measures its clearance from the blood by meas-

uring serum phenylalanine levels at regular intervals. Sketch curves showing relative blood phenylalanine concentration versus time that you would expect to be displayed by (a) a PKU patient, (b) a heterozygote, and (c) a normal individual. What kind of tolerance test could you devise to distinguish between PKU resulting from either phenylalanine hydroxylase deficiency or dihydropteridine reductase deficiency?

6. Formaldehyde reacts nonenzymatically with tetrahydrofolate to generate N^5,N^{10}-methylenetetrahydrofolate. [^{14}C]Formaldehyde can be used to prepare serine labeled in the β-carbon. What else would be needed?

   [^{14}C]Serine, prepared as described above, is useful for many things, but you would probably not want to use it for studies on protein synthesis because it would label nucleic acids, carbohydrates, and lipids, as well as proteins. Indicate how each of these classes of compounds could become labeled by this precursor.

7. If oxidation of acetyl-CoA yields 12 ATPs per mole through the citric acid cycle, how many ATPs will be derived from the complete metabolic oxidation of 1 mol of alanine in a mammal? Would the corresponding energy yield in a fish be higher or lower? Why? How much energy would be derived from the metabolic oxidation of 1 mol of leucine to CO_2, H_2O, and NH_3? Of tyrosine?

8. Some bacteria contain three different forms of aspartokinase, each with its own mode of regulation. Based on the roles of aspartokinase, as discussed in the text, propose a regulatory scheme applicable to each form of aspartokinase.

9. Proline betaine is a putative osmoprotector in plants and bacteria, helping to prevent dehydration of cells.

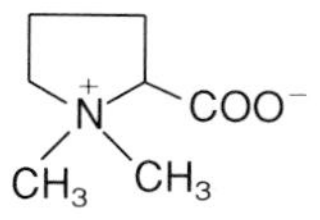

   $\overset{+}{N}$ COO^- CH_3 CH_3

   Propose a plausible pathway for biosynthesis of this compound.

10. Most bacterial mutants that require isoleucine for growth also require valine. Why? Which enzyme or reaction would be defective in a mutant requiring only isoleucine (not valine) for growth?

11. Describe a series of allosteric interactions that could adequately control the biosynthesis of valine, leucine, and isoleucine.

CHAPTER 22

Nucleotide Metabolism

We have encountered nucleotides repeatedly during our exploration of biochemistry. They serve as building blocks of nucleic acids, as critical elements in energy metabolism, as carriers of activated metabolites for biosynthesis (such as nucleoside diphosphate sugars), as structural moieties of coenzymes, and, finally, as metabolic regulators and signal molecules (for instance, cyclic AMP). In this chapter we discuss pathways of biosynthesis and degradation of purine and pyrimidine nucleotides, and we explore regulation of these processes—particularly critical in pathways leading to DNA replication. We discuss nucleotide biosynthetic enzymes as targets for the action of antimicrobial and anticancer drugs, and we describe the metabolic consequences of certain heritable alterations of nucleotide metabolism. We will encounter a motif that represents a widespread metabolic strategy, namely the use of multifunctional proteins and multienzyme complexes to channel and compartmentalize scarce or unstable intermediates.

Before starting this chapter you may find it useful to review the information on nucleotide structure in Chapter 4. You should be aware also of the distinction between nucleosides and nucleotides. On complete hydrolysis, a *nucleoside* yields at least 1 mol each of a sugar and a heterocyclic base, whereas a *nucleotide* yields at least 1 mol each of a sugar, a base, and inorganic phosphate. A *mononucleotide* contains only 1 mol each of base and sugar; it may contain more than one phosphate. On occasion we refer to the bases found in nucleic acids as **nucleobases,** to distinguish them from the much larger general class of organic bases.

Outlines of Pathways in Nucleotide Metabolism

Biosynthetic Routes: De Novo and Salvage Pathways

Unlike the other classes of metabolites we have encountered, nucleotides and their constituent bases and nucleosides are not required to meet nutritional needs. Most organisms can synthesize purine and pyrimidine nucleotides from low-molecular-weight precursors in amounts sufficient for their needs. These so-called **de novo** pathways are essentially identical throughout the biological world (Figure 22.1). Most organisms can also synthesize nucleotides from nucleosides or nucleobases that become available either in the diet or through enzymatic breakdown of nucleic acids. These processes are called **salvage** pathways, because they involve the utilization of preformed purine and pyrimidine compounds that would otherwise be lost to biodegradation. Salvage pathways represent important targets for therapy

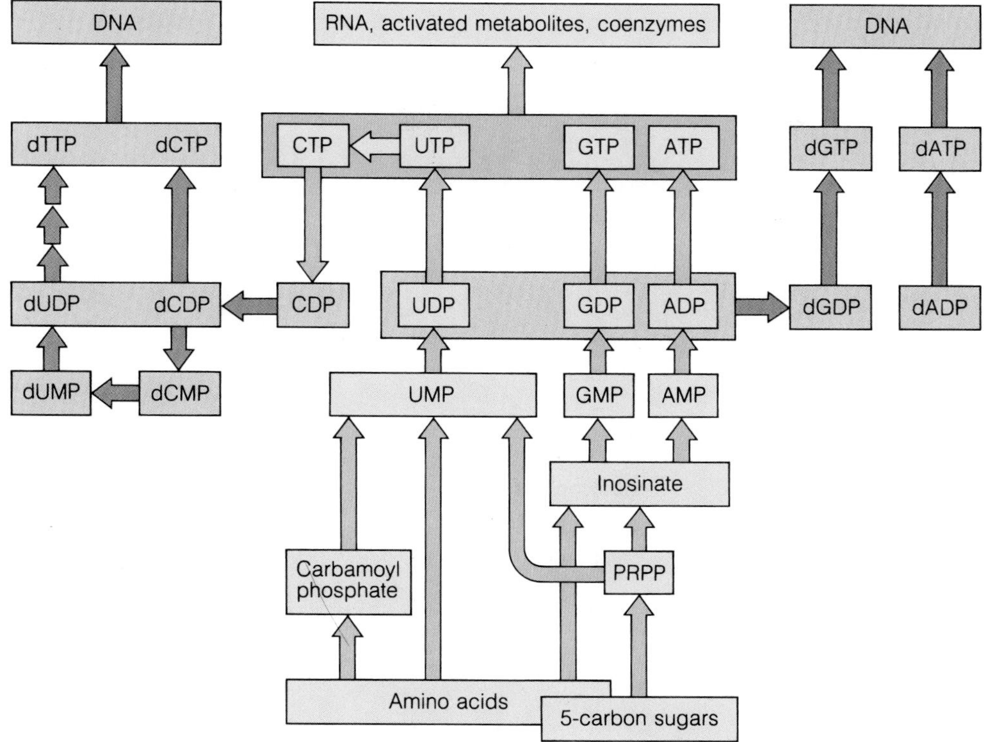

Figure 22.1
De novo pathways for nucleotide biosynthesis and involvement of nucleotides in other biosynthetic pathways.

of microbial or parasitic diseases, as sites for manipulation of biological systems (e.g., in studies of mutagenesis or preparation of monoclonal antibodies), and as biological processes in which genetic alterations have severe and far-reaching consequences.

Nucleic Acid Degradation and the Importance of Nucleotide Salvage

Since salvage, or reutilization, of purine and pyrimidine bases uses molecules released by nucleic acid degradation, let us begin by briefly considering these processes (Figure 22.2). Degradation can occur intracellularly (for example, through the turnover of unstable messenger RNA species), as a result of cell death, or, in animals, through digestion of nucleic acids ingested in the diet.

In animals the extracellular hydrolysis of nucleic acids represents the major route by which nucleobases and nucleosides become available. The breakdown processes are comparable to those involved in protein digestion. Cleavage processes begin at internal positions, in this case phosphodiester bonds. Catalysis occurs via **endonucleases,** such as pancreatic deoxyribonuclease or ribonuclease, which function to digest nucleic acids in the small intestine. Endonucleolytic cleavages yield oligonucleotides, which are then cleaved **exonucleolytically** (at linkages near the ends of molecules) by nonspecific enzymes called **phosphodiesterases.** The products are mononucleotides—nucleoside 5′- or 3′-monophosphates, depending on the specificities of the enzymes involved. Nucleotides can then be cleaved hydrolytically, by a group of phosphomonoesterases called **nucleotidases,** to yield orthophosphate plus the corresponding nucleoside. Although hydrolytic cleavage of the resultant nucleoside does occur, the most common route for cleavage to the base involves the action of a **nucleoside phosphorylase.** Like glycogen phosphorylase, nucleoside phosphorylases cleave a glycosidic bond by add-

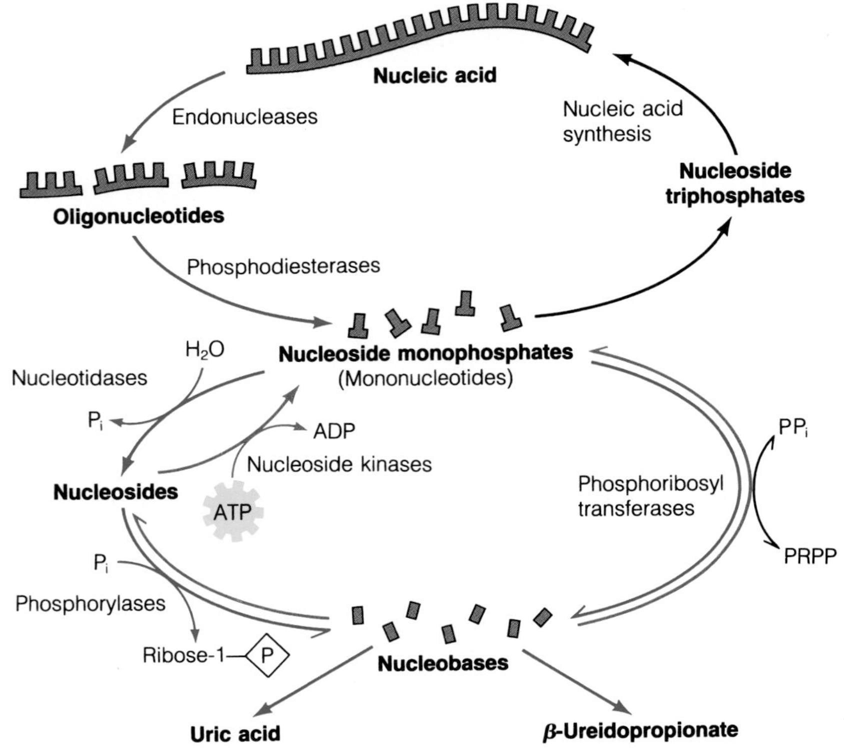

Figure 22.2
Relationships between nucleic acid catabolism (blue) and resynthesis of nucleotides by salvage pathways (red).

ing across it the elements of inorganic phosphate, to yield the corresponding base plus ribose-1-phosphate.

$$\text{Uridine} + {}^{-}O-\overset{O}{\overset{\|}{P}}(O^{-})-OH \rightleftharpoons \text{uracil} + \text{Ribose-1-phosphate}$$

These reactions are readily reversible, such that a nucleoside phosphorylase can also catalyze the first step in salvage synthesis of nucleotides from free nucleobases. When that occurs, the product nucleoside can be phosphorylated by ATP, through the action of a **nucleoside kinase.** These enzymes are not universal; for example, animal cells contain neither a guanosine kinase nor a uridine phosphorylase, although comparable enzymes are found in other organisms.

If bases or nucleosides are not reutilized for nucleic acid synthesis via salvage pathways, the purine and pyrimidine bases are further degraded, to uric acid or β-ureidopropionate, as indicated in Figure 22.2. We consider the pathways involved later in this chapter.

PRPP, a Central Metabolite in de Novo and Salvage Pathways

An alternative route for synthesis of nucleoside 5′-phosphates directly from free bases (also shown in Figure 22.2) involves a class of enzymes called **phosphoribosyltransferases** and an activated sugar phosphate, **5-phospho-**

α-D-ribosyl-1-pyrophosphate, or **PRPP.** This compound, which we introduced in Chapter 21 as an intermediate in histidine and tryptophan biosynthesis, is a key intermediate in the de novo synthesis of both purine and pyrimidine nucleotides. It is formed through the action of **PRPP synthetase,** which activates carbon 1 of ribose-5-phosphate by transferring to it the pyrophosphate moiety of ATP.

ATP + **Ribose-5-phosphate** ⇌ **5-Phospho-α-D-ribosyl-1-pyrophosphate** + AMP

A phosphoribosyltransferase reaction (shown below) is thermodynamically reversible.

PRPP + guanine ⇌ **GMP** + PP_i

In principle, therefore, such a reaction could participate in nucleotide breakdown. However, in vivo, pyrophosphate is rapidly cleaved by pyrophosphatase to give inorganic phosphate. This dictates that phosphoribosyltransferases most commonly act in the direction of nucleotide biosynthesis.

De Novo Biosynthesis of Purine Nucleotides

Early Studies on de Novo Purine Synthesis

The reactions of de novo purine nucleotide biosynthesis were identified some three decades ago, in the laboratories of John Buchanan and Robert Greenberg. This biochemical tour de force still merits description because of the combination of techniques applied to the problem.

Elucidation of purine synthesis began with the realization that birds excrete most of their excess nitrogen compounds in the form of *uric acid,* an oxidized purine. Thus, researchers were able to identify low-molecular-weight precursors to purines by administering isotopically labeled compounds to pigeons, crystallizing uric acid from the droppings, and, by selective chemical degradation, determining which positions were labeled by which precursors. This procedure yielded the pattern shown in Figure 22.3. At that time N^{10}-formyltetrahydrofolate was not known, but compounds such as formate or serine labeled in the hydroxymethyl carbon readily labeled C-2 and C-8 of uric acid.

Next, two related antibiotics, **azaserine** and **6-diazo-5-oxonorleucine,** were identified as potent inhibitors of purine nucleotide synthesis. Recognition that these compounds are structural analogs of glutamine led to the eventual realization that azaserine and 6-diazo-5-oxonorleucine

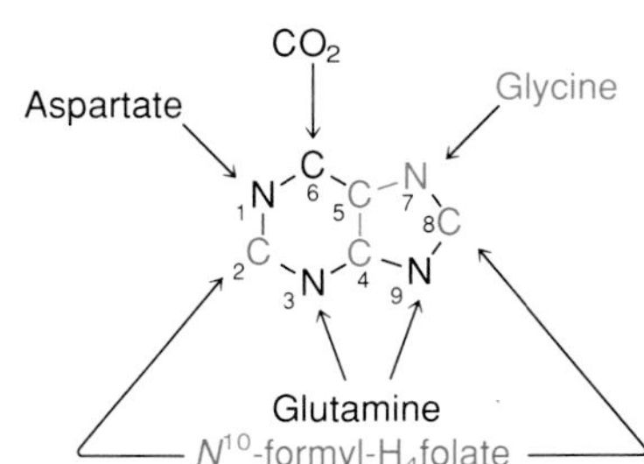

Figure 22.3
Low-molecular-weight precursors to the purine ring, as established with isotopic tracer studies of uric acid synthesis.

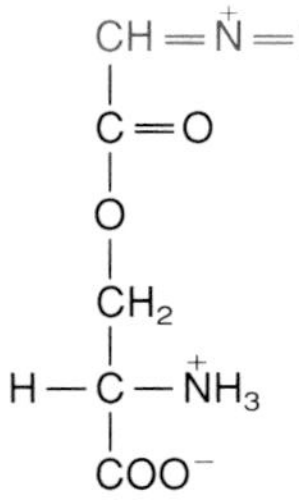

$$CH{=}\overset{+}{N}{=}\overset{-}{N}-C(=O)-O-CH_2-CH(\overset{+}{N}H_3)-COO^-$$

Azaserine

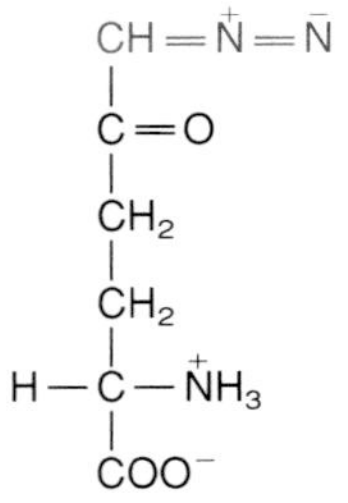

$$CH{=}\overset{+}{N}{=}\overset{-}{N}-C(=O)-CH_2-CH_2-CH(\overset{+}{N}H_3)-COO^-$$

6-Diazo-5-oxonorleucine (DON)

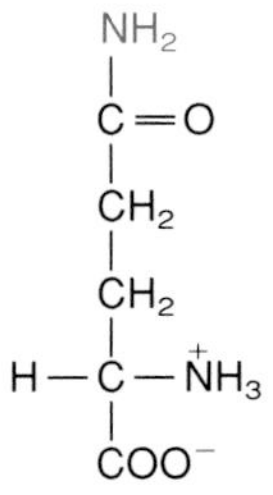

$$NH_2-C(=O)-CH_2-CH_2-CH(\overset{+}{N}H_3)-COO^-$$

Glutamine

are irreversible inhibitors of a class of enzymes, the **glutamine amidotransferases,** that catalyze the ATP-dependent transfer of the amido nitrogen of glutamine to an acceptor. Three such reactions occur in purine nucleotide synthesis.

In later experiments bacteria were treated with sulfonamide drugs, such as sulfanilamide. The bacteria excreted large quantities of a red compound identified as an oxidation product of **5-aminoimidazole-4-carboxamide ribonucleotide** (**AICAR**). Since sulfonamides block the synthesis of folate coenzymes, the accumulation of AICAR under these conditions suggested that a folate coenzyme participates in the next reaction. Moreover, since AICAR resembles an incomplete purine nucleotide, these observations suggested that the pathway proceeds at the nucleotide level—in other words, that the purine ring is assembled while already attached to the ribose-5-phosphate moiety.

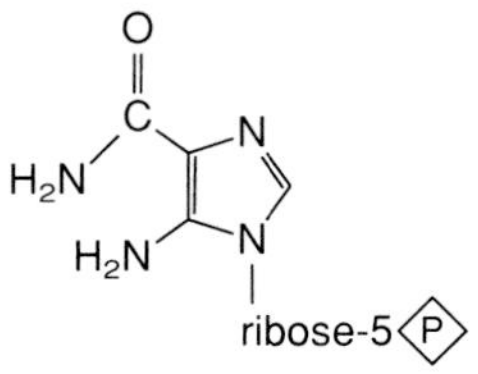

5-Aminoimidazole-4-carboxamide ribonucleotide (AICAR)

Purine Synthesis from PRPP to Inosinic Acid

Figure 22.4 summarizes the pathway leading from PRPP, the first intermediate, to the first fully formed purine nucleotide, **inosine 5′-monophosphate (IMP)**, also called **inosinic acid.** This is the 5′-ribonucleotide of the purine base **hypoxanthine.** Note that there are two glutamine amidotransferase reactions in this process, reactions 1 and 4. These differ mechanistically in that PRPP amidotransferase (reaction 1) does not require ATP, since the substrate has been activated by ATP in the previous step. An inversion of configuration occurs in reaction 1, as the amido nitrogen displaces the pyrophosphate moiety. The latter is an excellent leaving group, giving a simple nucleotide (5-phosphoribosylamine), which carries the β-configuration at carbon 1 of the sugar, as do all the common nucleotides.

In reaction 2 a glycine molecule is transferred, with the aid of ATP, to the nitrogen of phosphoribosylamine. This is followed by a **transformylase** reaction, in which a formyl group is transferred from N^{10}-formyltetrahydrofolate to the growing purine ring. As noted above, reaction 4 is catalyzed by an ATP-dependent amidotransferase. Reaction 5 is an ATP-dependent ring closure, giving the imidazole portion of the purine ring. Reaction 6 is a reversible carboxylation reaction, which is noteworthy in that it does *not* require biotin. Reactions 7 and 8 result in the transfer of a

Figure 22.4
De novo biosynthesis of the purine ring, from PRPP to inosinic acid. In this and subsequent figures ribose-Ⓟ refers to a ribose 5-phosphate moiety in a nucleotide. Enzyme $1(E_1)$ = PRPP amidotransferase; 2 = GAR synthetase; 3 = GAR transformylase; 4 = FGAR amidotransferase; 5 = FGAM cyclase; 6 = AIR carboxylase; 7 = SAICAR synthetase; 8 = SAICAR lyase; 9 = AICAR transformylase; 10 = IMP synthase.

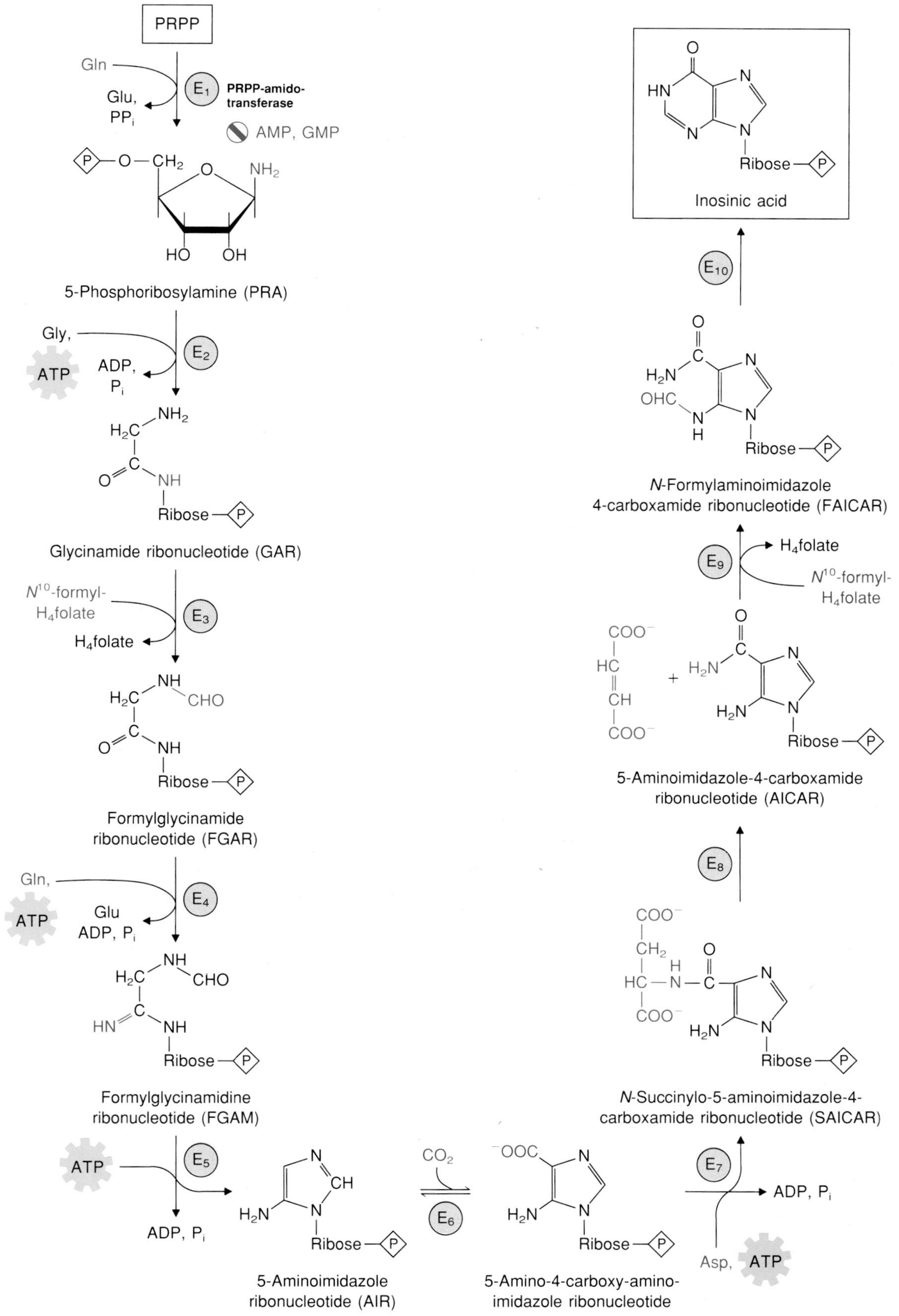

PRPP

Gln → Glu, PP_i (E_1) PRPP-amido-transferase (inhibited by AMP, GMP)

5-Phosphoribosylamine (PRA)

Gly, ATP → ADP, P_i (E_2)

Glycinamide ribonucleotide (GAR)

N^{10}-formyl-H_4folate → H_4folate (E_3)

Formylglycinamide ribonucleotide (FGAR)

Gln, ATP → Glu, ADP, P_i (E_4)

Formylglycinamidine ribonucleotide (FGAM)

ATP → ADP, P_i (E_5)

5-Aminoimidazole ribonucleotide (AIR)

CO_2 (E_6)

5-Amino-4-carboxy-amino-imidazole ribonucleotide

Asp, ATP → ADP, P_i (E_7)

N-Succinylo-5-aminoimidazole-4-carboxamide ribonucleotide (SAICAR)

(E_8)

5-Aminoimidazole-4-carboxamide ribonucleotide (AICAR)

N^{10}-formyl-H_4folate → H_4folate (E_9)

N-Formylaminoimidazole 4-carboxamide ribonucleotide (FAICAR)

(E_{10})

Inosinic acid

nitrogen from aspartate, by a mechanism identical to that used to convert citrulline to arginine in the urea cycle (Chapter 20). First the entire aspartate molecule is transferred to the carboxyl group of 4-carboxy-5-imidazole ribonucleotide (reaction 7), and this is followed by an α,β-elimination reaction. This yields AICAR, the intermediate shown earlier to accumulate in sulfonamide-treated bacteria.

Reaction 9 is another transformylase reaction, with a single-carbon group transferred from N^{10}-formyltetrahydrofolate. Finally, an internal condensation reaction (reaction 10) yields the first purine compound, inosinic acid.

Interestingly, the two transformylase enzymes, which catalyze reactions 3 and 9, exist as a multienzyme complex, in association with serine transhydroxymethylase and a trifunctional enzyme, formylmethenyl-methylene-tetrahydrofolate synthetase. The latter protein contains three activities of folate coenzyme synthesis. As shown in Figure 22.5, this juxtaposition of functionally related enzymes probably represents a route for generating the labile tetrahydrofolate cofactors at their sites of utilization. Recently it has been found in vertebrate liver that three of the activities, those catalyzing reactions 2, 3, and 5 in Figure 22.4, are associated with a single polypeptide chain. This could be either for channeling of intermediates or for joint regulation of these activities.

Figure 22.5
Reactions catalyzed by a multienzyme complex for transformylation reactions in purine nucleotide synthesis. The three activities that are associated with a single protein molecule are shown in green.

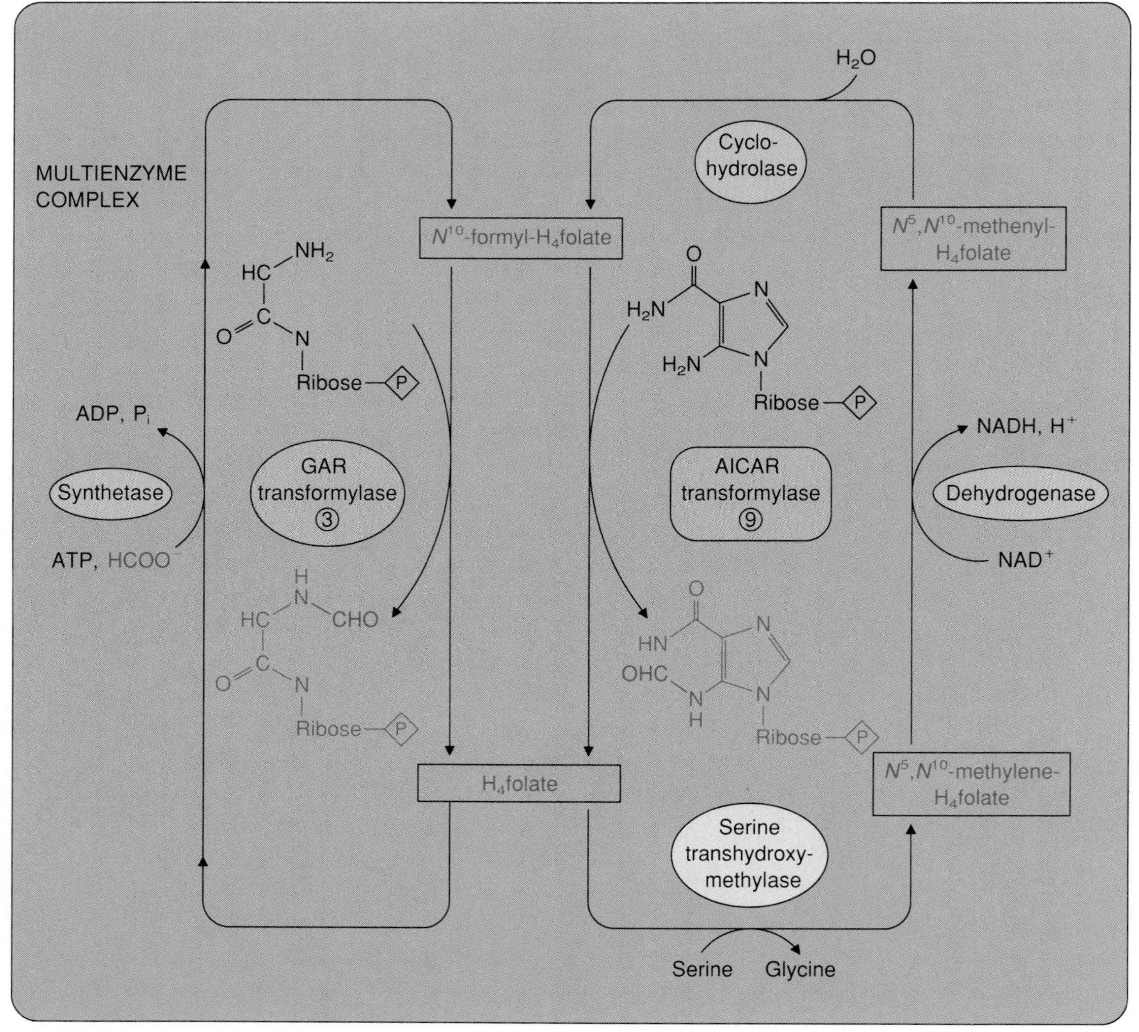

Control over the biosynthesis of inosinic acid is provided through feedback regulation of early steps in purine nucleotide synthesis; PRPP synthetase is inhibited by various purines, particularly AMP, ADP, and GDP, while PRPP amidotransferase (reaction 1 of Figure 22.4) is inhibited allosterically by AMP and GMP.

Synthesis of ATP and GTP from Inosinic Acid

Inosinic acid represents a branch point in purine nucleotide synthesis. Figure 22.6 depicts the conversion of this compound both to adenosine 5′-monophosphate and to guanosine 5′-monophosphate. The pathway to guanine nucleotides begins with an NAD^+-dependent hydroxylation of the purine ring, yielding **xanthosine monophosphate.** There follows a glutamine-dependent amidotransferase reaction, which yields GMP. The route to AMP involves the transfer of nitrogen from aspartate to IMP, by a mechanism similar to that of reactions 7 and 8 in the de novo synthesis of the purine ring: first a succinylonucleotide intermediate is formed, and then an α,β-elimination reaction yields AMP plus fumarate. An interesting difference is that the energy to drive the aspartate transfer reaction comes not from ATP but from GTP. This may represent a way to control the proportions of IMP that go to adenine and guanine nucleotide synthesis. GTP accumulation

Figure 22.6
Pathways from inosinic acid to GMP and AMP. Enzyme G-1 = IMP dehydrogenase; G-2 = XMP aminase; A-1 = adenylosuccinate synthetase; A-2 = adenylosuccinate lyase.

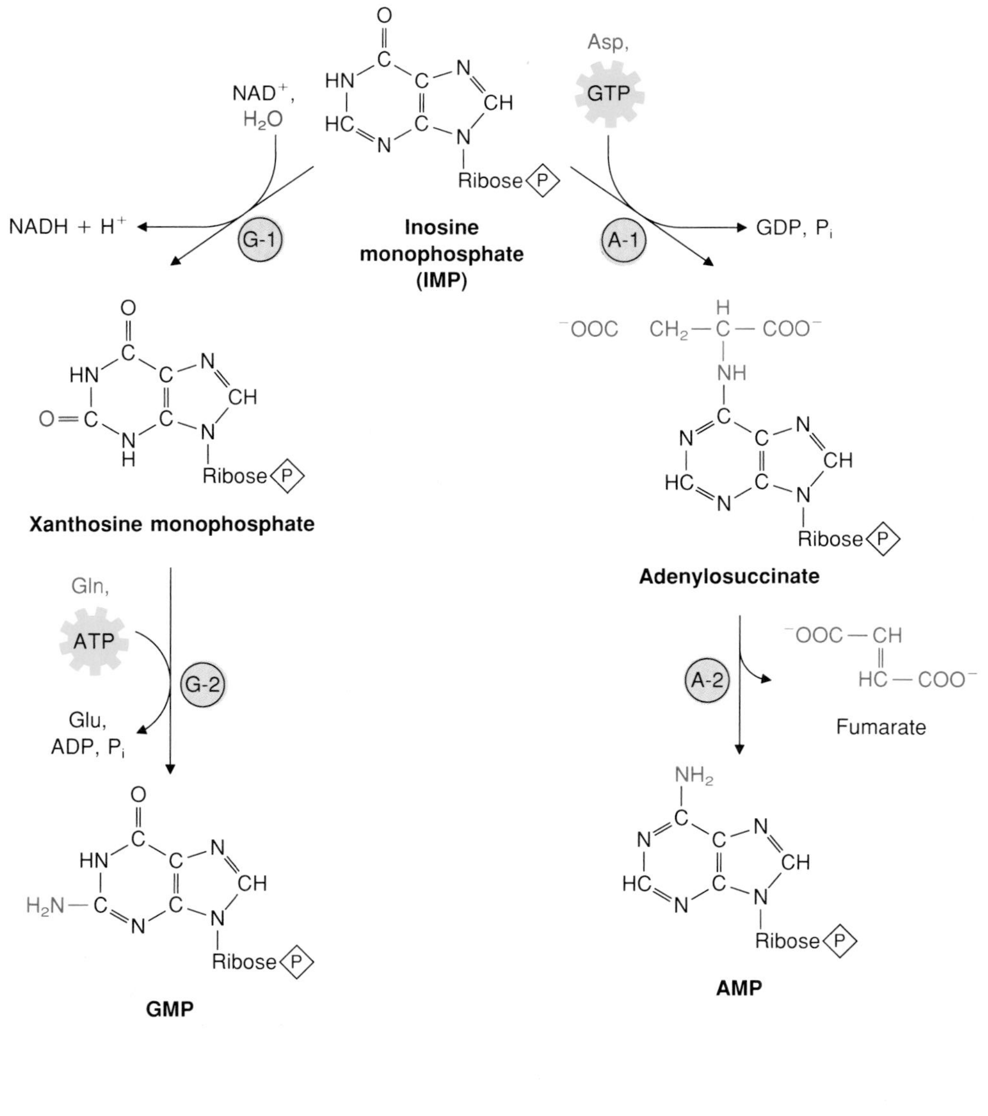

would tend to promote the pathway toward adenine nucleotides. Also, because the conversion of xanthosine monophosphate to GMP is ATP dependent, accumulation of ATP could promote guanine nucleotide synthesis.

Nucleotides are active in metabolism primarily as the nucleoside triphosphates. GMP and AMP are converted to their corresponding triphosphates through two successive phosphorylation reactions. Conversion to the diphosphates involves specific ATP-dependent kinases.

$$\text{GMP} + \text{ATP} \xrightleftharpoons{\text{Guanylate kinase}} \text{GDP} + \text{ADP}$$

$$\text{AMP} + \text{ATP} \xrightleftharpoons{\text{Adenylate kinase}} 2\text{ADP}$$

Phosphorylation of ADP to ATP occurs through energy metabolism—oxidative phosphorylation or substrate-level phosphorylations. ATP can also be formed from ADP through the action of adenylate kinase, acting in the reverse of the direction shown above.

ATP is the phosphate donor for conversion of GDP (and most other nucleoside diphosphates) to the triphosphate level through the action of **nucleoside diphosphokinase.** This enzyme is highly active but nonspecific, with regard to both phosphate donor and acceptor.

$$\text{GDP} + \text{ATP} \xrightleftharpoons{\text{NDP kinase}} \text{GTP} + \text{ADP}$$

Since ATP is by far the most abundant nucleoside triphosphate in most cells, equilibrium considerations dictate that it is used most readily as the donor of the γ (outer) phosphate in synthesis of other nucleoside triphosphates.

In most organisms the biosynthesis of deoxyribonucleotides for DNA synthesis begins at the ribonucleoside diphosphate level, with reduction of the ribose moiety to 2′-deoxyribose. This process is presented in detail later. In purine metabolism subsequent phosphorylation by nucleoside diphosphokinase yields the deoxyribonucleoside triphosphates, dATP and dGTP.

Let us briefly summarize points of feedback regulation in de novo purine biosynthesis. We have already described the control of IMP synthesis. Beyond that point GMP controls its own biosynthesis by inhibiting the conversion of IMP to XMP, and AMP controls its own formation by inhibiting the synthesis of adenylosuccinate. Additional control is exerted at the level of deoxyribonucleotide biosynthesis, as we discuss later.

Utilization of Adenine Nucleotides in Coenzyme Biosynthesis

An important metabolic role of purine nucleotides is the synthesis of coenzymes, primarily those containing an adenylate moiety. These include the flavin nucleotides, the nicotinamide nucleotides, and coenzyme A.

$$\text{Riboflavin} \xrightarrow{\text{ATP} \to \text{ADP}} \text{riboflavin 5}'\text{-phosphate (FMN)} \xrightarrow{\text{ATP} \to \text{PP}_i} \text{flavin adenine dinucleotide (FAD)}$$

$$\text{Nicotinic acid} \xrightarrow{\text{PRPP}} \text{(nicotinate)}\text{-Ribose-}(P) \xrightarrow{\text{ATP}} \text{(nicotinate)}\text{-Ribose-}(P)(P)\text{-ribose-adenine} \xrightarrow[\text{ATP}]{\text{gln}} \text{NAD}^+ \xrightarrow{\text{ATP}} \text{NADP}^+$$

$$\text{Pantothenic acid} \xrightarrow{\text{ATP}} \text{4-phosphopantothenate} \xrightarrow[+\text{ ATP}]{\text{cysteine}} \text{4-phosphopantothenylcysteine} \xrightarrow{-\text{CO}_2} \xrightarrow{\text{ATP}} \xrightarrow{\text{ATP}} \text{coenzyme A}$$

Purine Degradation and Clinical Disorders of Purine Metabolism

Formation of Uric Acid

Purine nucleotide catabolism yields uric acid, by routes shown in Figure 22.7. The specific pathways utilized vary among organisms and among tissues of the same organisms. For example, AMP is either deaminated to yield inosinic acid (IMP) or hydrolyzed to yield adenosine. Deamination is particularly active in muscle, while hydrolysis predominates in most other animal tissues. Hydrolysis of nucleoside monophosphates is not always involved in nucleotide catabolism. Nucleotides cannot be readily transported across cell membranes. However, nucleosides can be taken up into cells by specific transport mechanisms. It is likely that dephosphorylation of nucleotides at the cell surface is used primarily to permit their entry into cells as nucleosides, followed by salvage routes to nucleotide resynthesis.

In degradative pathways adenosine is deaminated by **adenosine deaminase** to give inosine. Both inosine and guanosine can be formed by hydrolysis of the respective nucleoside monophosphates, which are acted on by **purine-nucleoside phosphorylase** to give hypoxanthine and guanine, respectively. Guanine is deaminated to xanthine by **guanine deaminase,** an enzyme abundant in mammalian brain and liver. Hypoxanthine is oxidized to xanthine, and xanthine to uric acid, by **xanthine oxidase.** This enzyme, which oxidizes several other heterocyclic nitrogen compounds, contains bound FAD, molybdenum, and nonheme iron. Electrons derived from the oxidation of substrates are passed to each of these carriers, which ultimately reduce oxygen to H_2O_2. Xanthine oxidase is particularly abundant in milk and liver.

Purine catabolism in primates ends with uric acid, which is excreted. However, most animals further oxidize the purine ring, to **allantoin** and then to **allantoic acid,** which is either excreted (in some fishes) or further catabolized to urea (in most fishes, some mollusks, and amphibians) or ammonia (in some marine invertebrates). Figure 22.8 shows the pathway involved.

Excessive Accumulation of Uric Acid: Gout

Uric acid and its urate salts are quite insoluble. This is advantageous to egg-laying animals, because it provides a route for disposition of excess

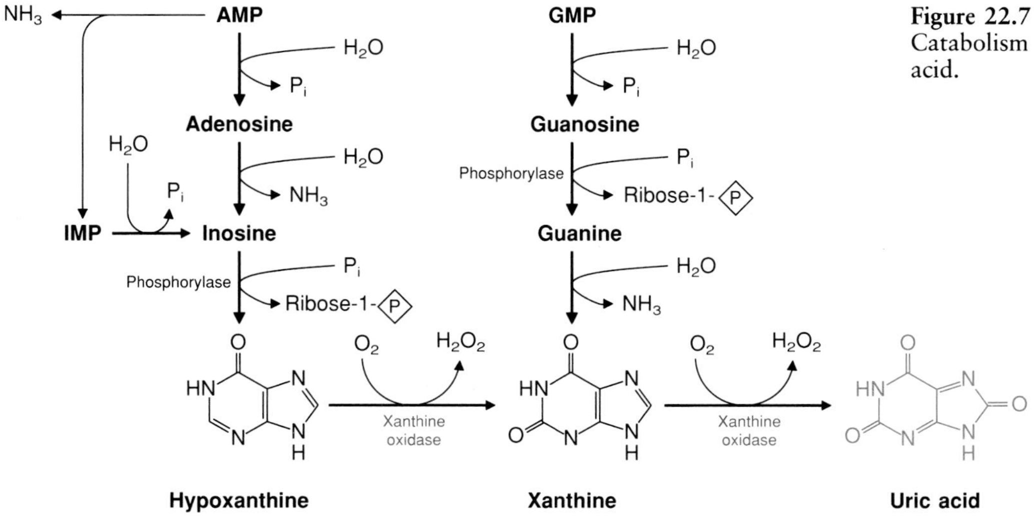

Figure 22.7
Catabolism of purine nucleotides to uric acid.

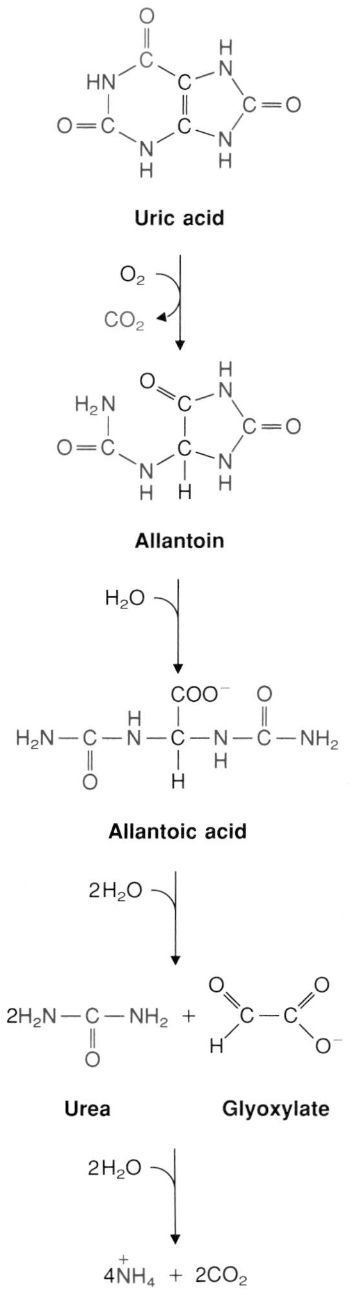

Uric acid

O_2 → CO_2

Allantoin

H_2O

Allantoic acid

$2H_2O$

Urea + Glyoxylate

$2H_2O$

$4NH_4^+ + 2CO_2$

Figure 22.8
Catabolism of uric acid to ammonia and CO_2, via the intermediates allantoin and allantoic acid.

nitrogen in a closed environment: the waste material simply precipitates in situ. However, the insolubility of urates can present difficulties in mammalian metabolism. In humans about three individuals out of 1000 suffer from **hyperuricemia**—chronic elevation of blood uric acid levels well beyond normal limits. Although the biochemical reasons for hyperuricemia vary, the condition goes by the single clinical name of **gout.** Prolonged or acute elevation of blood urate leads to its precipitation, as crystals of sodium urate, in the synovial fluid of joints. These precipitates cause inflammation, resulting in a painful arthritis which, if untreated, leads ultimately to severe degeneration of the joints. Eating and drinking purine-rich foods are apt to stimulate acute gouty attacks in susceptible individuals. Because such foods include "rich" items such as liver, sweetbreads, anchovies, and wine (yeast), gout is historically associated with an excess of high living.

Most forms of gout result from excessive biosynthesis of purine nucleotides, far beyond the body's needs or capacity for disposition. Several specific enzymatic defects have been described, as shown in Figure 22.9. One form (1) is characterized by abnormally elevated levels of PRPP synthetase, which may result from insensitivity to feedback inhibition by purine nucleotides. Since the intracellular activity of PRPP amidotransferase is probably controlled in part by the concentrations of substrates, an elevation of the steady-state pool of PRPP increases flux through the amidotransferase reaction, which represents the first, and rate-limiting, step in de novo purine biosynthesis (see Figure 22.4). Gout can also result from a change in the structure of PRPP amidotransferase that renders it less sensitive to feedback inhibition by purine nucleotides (2). This loss of control also increases flux through the rate-limiting step. Another form of gout results from deficiency of the salvage enzyme **hypoxanthine–guanine phosphoribosyltransferase,** or HGPRT (3). This is one of two phosphoribosyltransferases in animal purine metabolism; the other is specific for adenine.

We still don't know exactly why an HGPRT deficiency should increase the rate of purine nucleotide synthesis. A reasonable explanation is that the HGPRT reaction, when active, consumes PRPP. Decreased flux through this reaction, when the enzyme is deficient, can raise the steady-state level of PRPP, thereby increasing the intracellular activity of PRPP amidotransferase through mass action. However, other factors may be involved as well.

Many cases of gout can be successfully treated by the antimetabolite **allopurinol,** a structural analog of hypoxanthine that acts by strongly inhibiting xanthine oxidase. Since hypoxanthine and xanthine are more soluble than uric acid, and can thus be more readily excreted, their accumulation following allopurinol treatment is not deleterious.

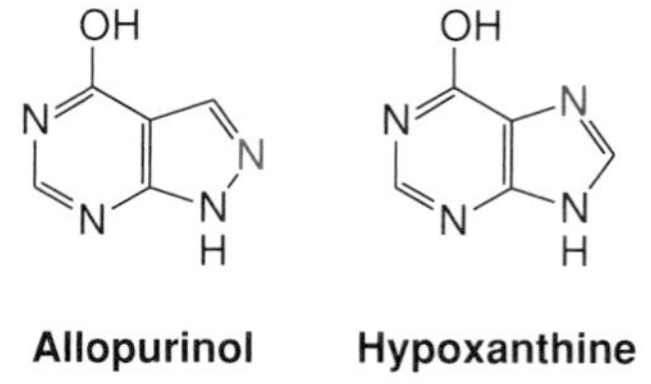

Allopurinol　　Hypoxanthine

Careful analysis of patients with simple gout resulting from HGPRT deficiency reveals low but significant residual levels of the affected enzyme; evidently the mutations involved alter the catalytic activity of the enzyme but do not abolish it completely. Far more serious consequences ensue when the enzyme is totally absent, as we shall see in the next section.

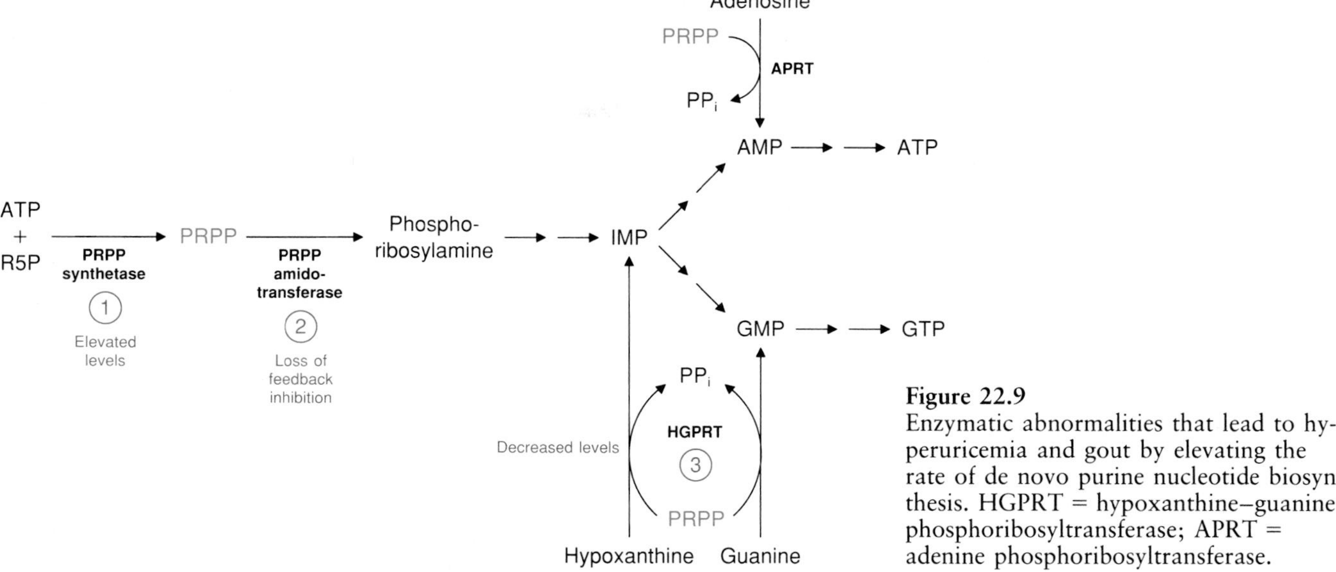

Figure 22.9
Enzymatic abnormalities that lead to hyperuricemia and gout by elevating the rate of de novo purine nucleotide biosynthesis. HGPRT = hypoxanthine–guanine phosphoribosyltransferase; APRT = adenine phosphoribosyltransferase.

Dramatic Consequences of a Severe HGPRT Deficiency: Lesch–Nyhan Syndrome

The condition resulting from a total HGPRT deficiency was first described in 1964 by Michael Lesch, who was then a medical student, and his faculty mentor, William Nyhan. Lesch–Nyhan syndrome is a sex-linked trait, because the structural gene for HGPRT is located on the X chromosome. Patients with this condition display a severe gouty arthritis, but they also have a dramatic malfunction of the nervous system, manifested as behavioral disorders, learning disability, and hostile or aggressive behavior, often self-directed. In the most extreme cases patients nibble at their fingertips, or, if restrained, their lips, causing severe self-mutilation. Nyhan has likened this behavior to "nailbiting, with the volume turned up." The biochemical reason for this bizarre behavioral pattern is unknown, but the condition, even though rare, is the object of much interest because it is clear that the aberrations derive ultimately from the single well-characterized enzyme deficiency affecting HGPRT levels. At the moment there is no successful treatment, and afflicted individuals have such severe gouty arthritis that they rarely live beyond 20 years. However, the condition can be diagnosed prenatally through amniocentesis, as cells from Lesch–Nyhan fetuses are unable to incorporate radiolabeled hypoxanthine into nucleic acids. Lesch–Nyhan syndrome is an attractive target for gene therapy, though the genetic engineering presents formidable obstacles.

Unexpected Consequences of Defective Purine Catabolism: Immunodeficiency

A surprising feature of human purine metabolism came to light in 1972, through studies on a hereditary condition called **severe combined immunodeficiency syndrome.** Patients with this condition are susceptible, often fatally so, to infectious diseases because of a total inability to mount an immune response to antigenic challenge. In this condition both B and T lymphocytes are affected; neither class of cells can proliferate as they must if antibodies are to be synthesized. In many such cases the immunodeficiency results from a heritable lack of the degradative enzyme **adenosine deaminase (ADA).** This baffling discovery led to intense research, which has now generated a reasonable explanation for the situation. First, adenosine deaminase can act on deoxyadenosine, which results from the degradation of DNA. Second, the white blood cells have abundant levels of salvage enzymes, including nucleoside kinases; thus, adenosine and deoxyadenosine that accumulate are readily converted in white cells to their respective nucleotides. These include dATP, which is known to be a potent inhibitor of DNA replication because it inhibits the synthesis of deoxyribonucleotides from ribonucleotides (discussed later in this chapter).

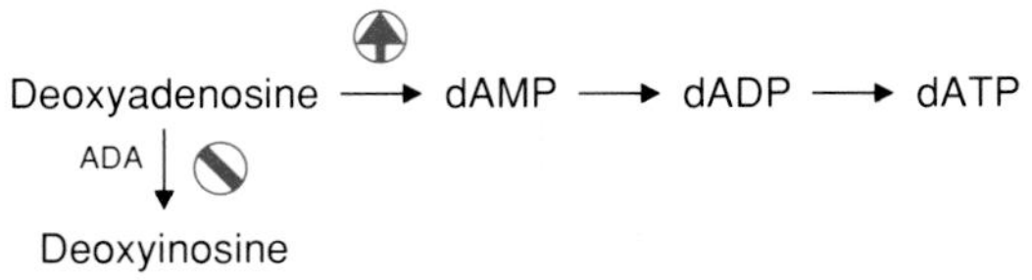

Deoxyadenosine → dAMP → dADP → dATP

ADA ↓

Deoxyinosine

White cells must proliferate for an immune response to occur. In turn, their ability to proliferate, and hence to mount an immune reaction, requires ample synthesis of DNA and its precursors. Additional mechanisms may be involved, because dATP has been found to kill white cells even when they are not proliferating.

A less severe immunodeficiency results from the lack of another purine degradative enzyme, **purine-nucleoside phosphorylase (PNP).** Decreased activity of this enzyme leads to accumulation primarily of dGTP. This also affects DNA replication, but less severely than does dATP. Interestingly, the phosphorylase deficiency destroys only the T class of lymphocytes and not the B cells.

Pyrimidine Nucleotide Metabolism

De Novo Biosynthesis of the Pyrimidine Ring

Now let us turn our attention to pyrimidine nucleotide biosynthesis, which is much simpler than formation of the structurally more complex purine nucleotides.

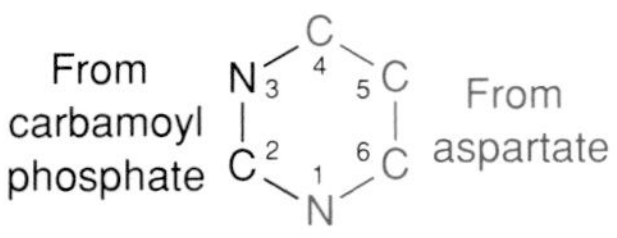

From carbamoyl phosphate: N3, C2

From aspartate: C4, C5, C6, N1

Like purine biosynthesis, the pathway is identical in virtually all organisms that have been studied. As summarized in Figure 22.10, however, there are two major distinctions from the purine pathway. First, the pyrimidine ring is assembled as a free base, with conversion to a nucleotide occurring later in the pathway, when the base **orotic acid** is converted to orotidine monophosphate, or OMP. Second, the pyrimidine pathway is unbranched. Uridine triphosphate, one of the two common ribonucleoside triphosphates, and hence an end product of the pathway, is also the substrate for formation of cytidine triphosphate, the other end product.

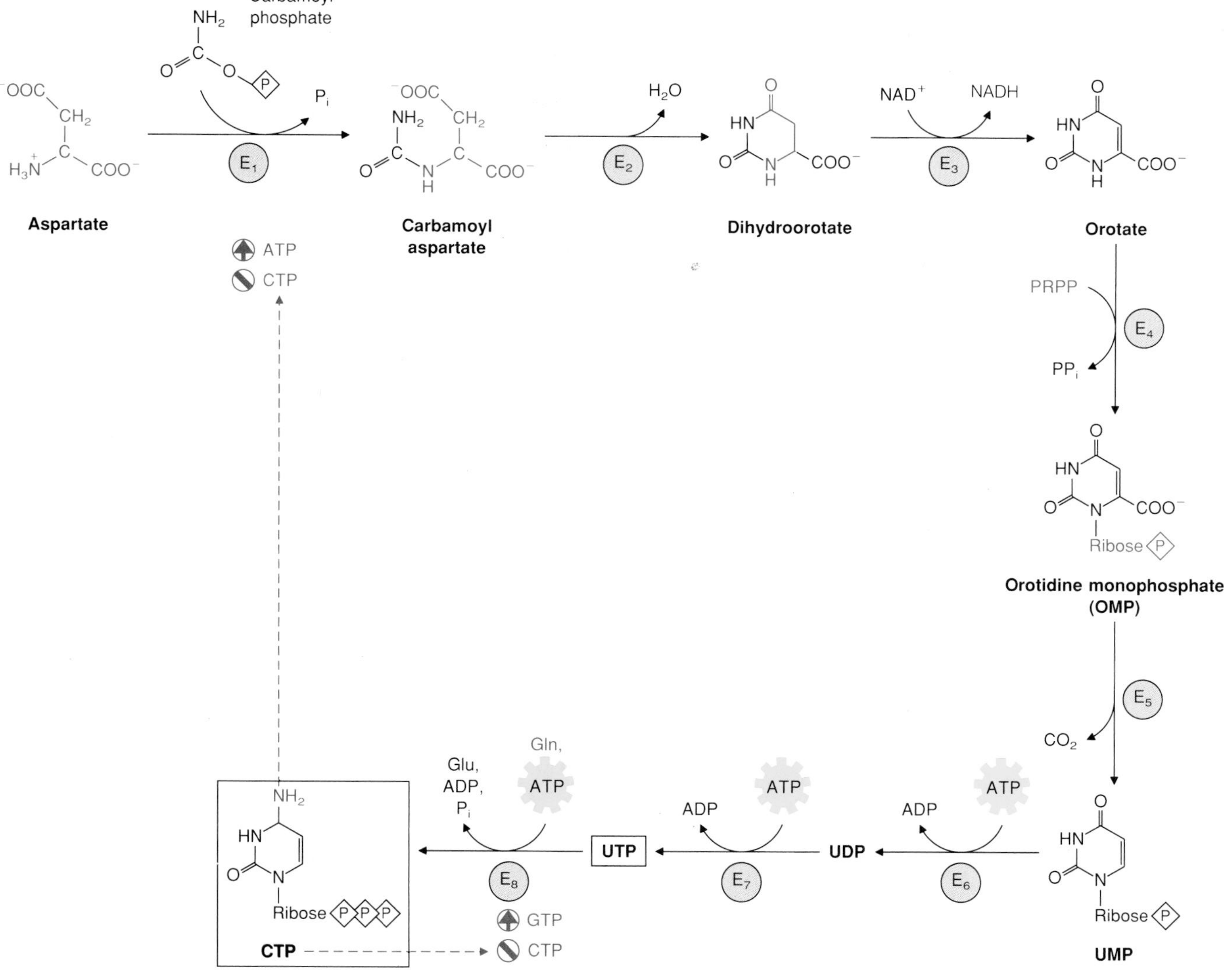

Figure 22.10
De novo synthesis of pyrimidine nucleotides. The pathway starts with aspartate and carbamoyl phosphate. Sites of allosteric control are indicated. E_1 = aspartate transcarbamoylase; E_2 = dihydroorotase; E_3 = dihydroorotate dehydrogenase; E_4 = orotate phosphoribosyltransferase; E_5 = orotidylate decarboxylase; E_6 = UMP kinase; E_7 = nucleoside diphosphokinase; E_8 = CTP synthetase.

Control of Pyrimidine Biosynthesis in Bacteria

Pyrimidine synthesis begins with formation of carbamoyl phosphate, a reaction presented in Chapter 20. However, the first reaction committed solely to pyrimidine synthesis is the formation of carbamoyl aspartate from carbamoyl phosphate and aspartate, catalyzed by aspartate transcarbamoylase (ATCase). In enteric bacteria this enzyme represents a marvelous example of feedback control, as was discussed at length in Chapter 11. Recall that the enzyme is inhibited by the end product CTP and activated by ATP, the latter possibly representing a mechanism to keep purine and pyrimidine biosyntheses in balance. Recall also that the enzyme contains six each of two types of subunits, arranged as two catalytic trimers and three regulatory dimers.

Bacteria also regulate pyrimidine metabolism through control of the *synthesis* of ATCase and the other enzymes. The rate of transcription of an operon encoding both of the ATCase subunits can vary by as much as 150-fold, depending on the intracellular level of UTP. The higher the UTP concentration, the lower the rate of transcription of these genes.

Multifunctional Enzymes in Eukaryotic Pyrimidine Synthesis

Aspartate transcarbamoylase in eukaryotes is strikingly different from the *E. coli* enzyme. This came to light through analysis of ATCase inhibition by **N-phosphonoacetyl-L-aspartate (PALA).**

PALA **Putative transition state complex**

This compound, synthesized as an analog of the putative transition state complex formed between the two substrates, inhibits pyrimidine synthesis in mammalian cells. However, cells eventually develop resistance to it, because levels of ATCase rise in these cells beyond the capacity of PALA to inhibit all of the activity. Surprisingly, these resistant cells contained similarly elevated levels of carbamoyl phosphate synthetase and dihydroorotase. The explanation for this observation came with the discovery of a single protein containing three identical polypeptide chains, each of M_r about 230,000, that catalyzes all three reactions.

George Stark has given this trifunctional enzyme the acronym CAD. He showed that the protein accumulates in PALA-resistant cells because the gene encoding the protein becomes amplified in response to the selective pressure exerted by PALA: resistant cells contain many more copies of the gene than the normal complement of two copies per diploid cell. This phenomenon of gene **amplification** has now been observed many times in eukaryotic cells exposed to prolonged and specific stresses.

Not only are the first three steps of pyrimidine synthesis catalyzed by one protein in mammalian cells, but also the last two reactions (orotate phosphoribosyltransferase and orotidylate decarboxylase) are catalyzed by another protein, which has been called **UMP synthase.** We don't know much about how this affects regulation of pyrimidine biosynthesis, but the juxtaposition of active sites may allow "channeling": efficient transfer of intermediates from one catalytic site to the next, such that they do not diffuse to sites where they might be consumed by other reactions. We still have much to learn about how pyrimidine biosynthetic intermediates are channeled, because dihydroorotate dehydrogenase (E_3 in Figure 22.10) is localized in mitochondria, whereas the bi- and trifunctional enzymes are both cytosolic. Since intermediates must travel into and then out of the mitochondria, the kinetic advantage provided by channeling of the first and last steps would seen to be negated.

Another site for control of pyrimidine nucleotide synthesis is the amidotransferase, **CTP synthetase,** which converts UTP to CTP. This enzyme is inhibited allosterically by its product CTP and is activated allosterically by GTP.

Salvage Synthesis and Pyrimidine Catabolism

Pyrimidine nucleotides are also synthesized by salvage pathways, in reactions comparable to those already discussed for purines: phosphorylases

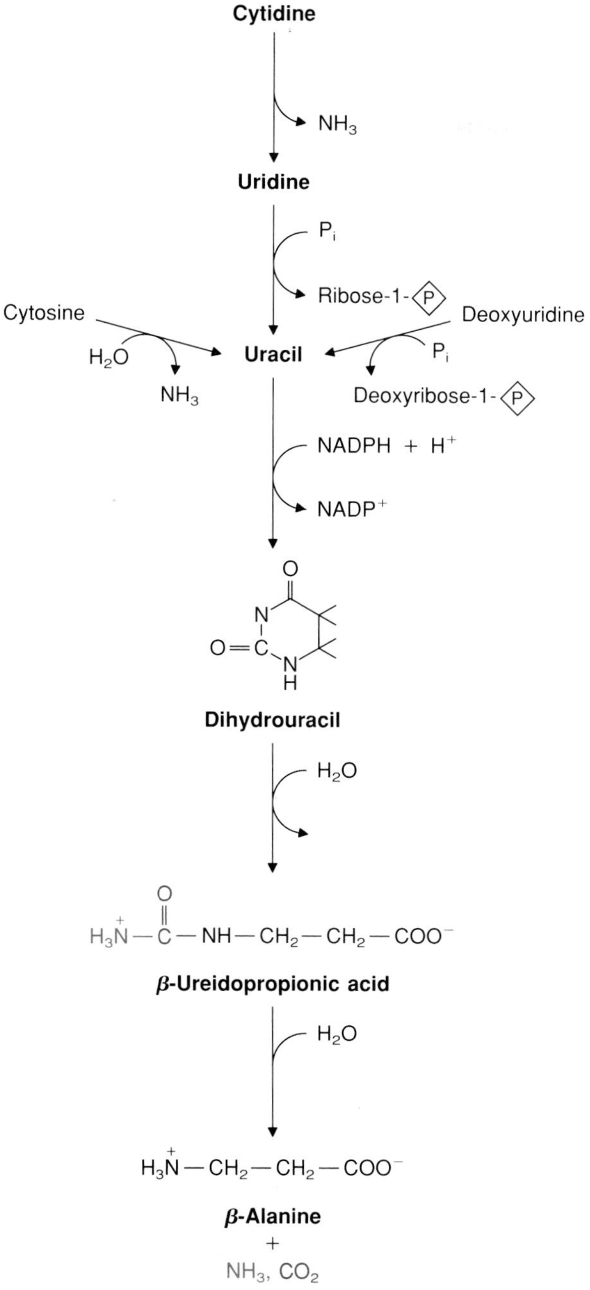

Figure 22.11
Catabolic pathways in pyrimidine nucleotide metabolism.

and kinases. The catabolic pathways for pyrimidines, summarized in Figure 22.11, are simpler than those for purines. Since the intermediates are relatively soluble, there are few known derangements of pyrimidine breakdown. One of the breakdown products, β-alanine, can be used in the biosynthesis of coenzyme A.

Deoxyribonucleotide Biosynthesis and Metabolism

Most cells contain 5 to 10 times as much RNA as DNA. Therefore, most of the carbon that flows through nucleotide synthetic pathways goes into ribonucleoside triphosphate (rNTP) pools. However, the relatively small frac-

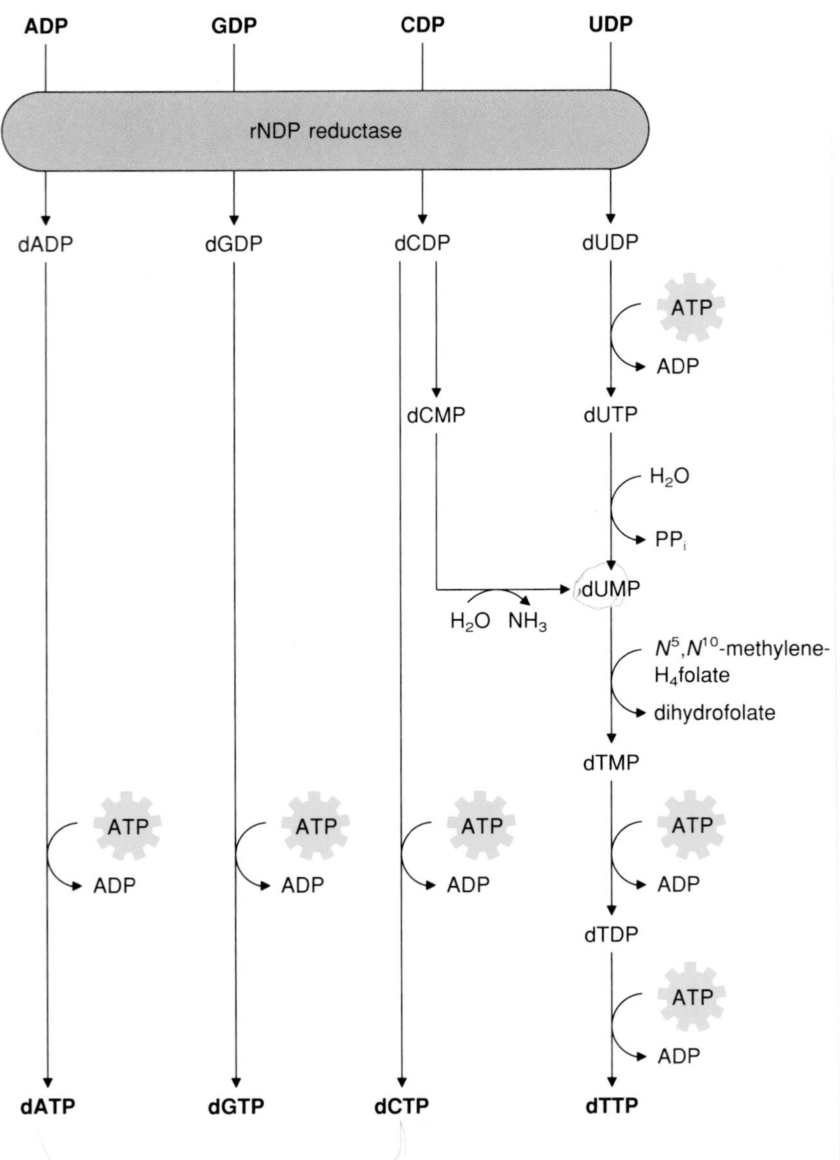

Figure 22.12
Overview of deoxyribonucleotide biosynthesis.

tion that is diverted to the synthesis of deoxyribonucleoside triphosphates (dNTPs) is of paramount importance in the life of the cell. dNTPs are used almost exclusively in the biosynthesis of DNA; consequently, there are especially close regulatory relationships between DNA synthesis and dNTP metabolism—more so than seen between other macromolecular biosyntheses and the pathways that provide their precursors. Overall pathways of dNTP biosynthesis are shown in Figure 22.12.

Recalling that DNA differs chemically from RNA in the nature of the sugar and in the identity of one of the pyrimidine bases, we can focus our discussion of deoxyribonucleotide biosynthesis on two specific processes—the conversion of ribose to deoxyribose, and the conversion of uracil to thymine. Both of these processes occur at the nucleotide level. Both processes are of great interest mechanistically, as target sites for chemotherapy for cancer or infectious diseases, and from the standpoint of regulation. Accordingly, we shall present both processes in some detail.

Reduction of Ribonucleotides to Deoxyribonucleotides

Mechanistically, the reduction of ribonucleotides to deoxyribonucleotides involves replacement of the 2′ hydroxyl moiety of the sugar by a hydride ion, with retention of configuration; an unusual free radical mechanism is involved. Biologically, the reaction represents the first committed step in the flow of precursors to DNA; because of this and because of the specialized roles of dNTPs as DNA precursors, the reaction is controlled by intricate regulatory mechanisms.

In cyanobacteria, some bacteria, and *Euglena* the substrates for reduction are the ribonucleoside triphosphates, and the deoxyadenosyl form of coenzyme B_{12} is involved. However, most organisms carry out the reduction at the nucleoside diphosphate level. All four common ribonucleoside diphosphates (rNDPs) are reduced by the same enzyme, **ribonucleoside-diphosphate reductase**, or **rNDP reductase.**

STRUCTURE OF rNDP REDUCTASE. As found in bacterial and mammalian cells, rNDP reductase is an $\alpha_2\beta_2$ tetramer. The *E. coli* enzyme contains two proteins—B1, with two identical α polypeptide chains of M_r 87,000 each, and B2, consisting of two 43,000-dalton β chains. The structure of the enzyme is shown schematically in Figure 22.13. The catalytic site contains residues from both subunits. Electrons for the reduction come from **redox-active** thiol groups located on the B1 protein, so called because they undergo cyclic oxidation and reduction during the reaction.

The B2 dimer contains an unusual tyrosine free radical that is involved in the reaction and an oxygen atom bridging two ferrous ions. This **iron center** probably functions to stabilize the free radical. The B1 protein contains two classes of regulatory sites, which we shall discuss shortly. Finally, recent evidence indicates that the B1 protein contains an additional pair of redox-active thiol groups. The mammalian enzyme is quite similar in structure, but the two protein components are called M1 and M2.

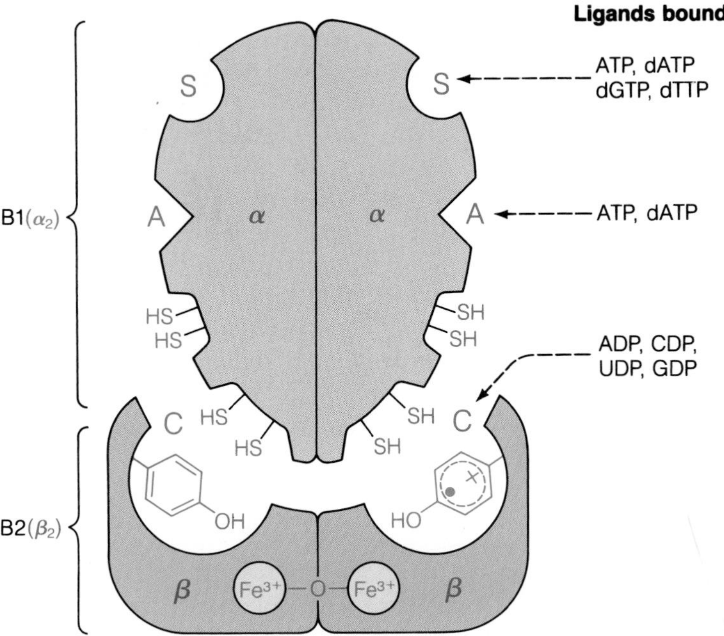

Figure 22.13
A model for *E. coli* ribonucleoside diphosphate reductase. The figure shows two separate proteins (B1 and B2), each a homodimer. Both β subunits in the B2 protein contain active site tyrosine residues, one of which is a free radical. An iron center with a bound oxygen atom stabilizes the radical. Each α subunit of the B1 protein contains a pair of redox-active thiols in the catalytic site (C), an additional pair of redox-active thiols at a nearby site, and two allosteric sites—activity sites (A) and specificity sites (S).

MECHANISM OF RIBONUCLEOTIDE REDUCTION. Although we do not yet know the complete mechanism of the rNDP reductase reaction, we can formulate a plausible mechanism based on the following observations. (1) Radiolabeling studies show that cleavage of the C—H bond at position 3′ of the sugar occurs during the reaction. (2) The reaction proceeds with retention of configuration at C-2′; this rules out displacement of the hydroxyl group by a hydride ion in an S_n2 reaction. (3) The thiol groups undergo oxidation during the reaction. (4) The tyrosine free radical participates in the reaction. This was shown first by the fact that **hydroxyurea,** an inhibitor of rNDP reductase, reversibly destroys the free radical. A more elegant demonstration is depicted in Figure 22.14. The free radical gives a characteristic electron paramagnetic resonance spectrum. Tyrosine-122, the residue thought to generate the radical in the *E. coli* enzyme, was changed to phenylalanine by site-directed mutagenesis of the cloned gene (this technique is described in Chapter 25). The modified protein was inactive and showed no evidence for the existence of a free radical.

HO—NH—C(=O)—NH_2

Hydroxyurea

A probable mechanism for rNDP reductase, based on the above observations, is shown in Figure 22.15. The enzyme abstracts a hydrogen atom from C-3′ of the substrate (step 1). This is followed by loss of a hydroxide ion from C-2′, giving a resonance-stabilized radical carbonium ion (step 2). The enzyme transfers an electron pair from the redox-active thiols and a hydrogen atom from tyrosine-122 (probably not in one step), to give the nucleotide product (step 3). Recent evidence indicates that the active site sulfurs are now reduced by disulfide exchange with another pair of redox-active thiols in the B1 subunit; this is not shown in the figure. The disulfide resulting from step 3 is reduced by an external cofactor (step 4), regenerating the active form of the enzyme.

SOURCE OF ELECTRONS FOR rNDP REDUCTION. Electrons for the reduction of ribonucleotides come ultimately from NADPH, but they are shuttled to rNDP reductase by a coenzyme that is unusual because it is itself a protein (Figure 22.16). The first known member of this class of redox-active proteins is **thioredoxin,** a small (M_r = 12,000) protein with two thiol groups in the sequence Cys–Gly–Pro–Cys. These thiols undergo reversible oxidation and reduction, and they can reduce the thiol groups in the active site of rNDP reductase. Thioredoxin is reduced by NADPH via the action of a flavoprotein enzyme, **thioredoxin reductase.**

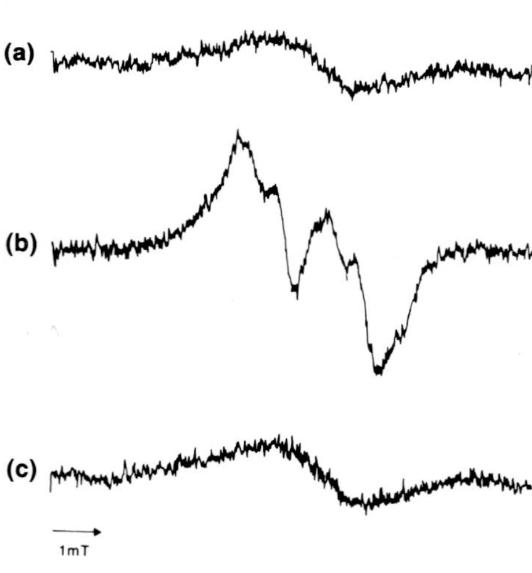

Figure 22.14
Electron spin resonance spectra of *E. coli* cells containing the cloned gene for the small subunit of ribonucleotide reductase. (**a**) Bacteria containing no clone; (**b**) bacteria containing the cloned wild-type gene; (**c**) bacteria containing a mutant cloned gene in which tyrosine-122 has been changed to phenylalanine.

Since its discovery, thioredoxin has been found to have many activities in vitro, suggesting an astonishing range of biological functions. Some of these are listed in Table 22.1. Whether thioredoxin is the true intracellular cofactor for ribonucleotide reduction was brought into question by the isolation of *E. coli* mutants lacking this protein. Since these mutants were capable of DNA replication, investigators looked in these cells for other redox proteins that could interact with rNDP reductase. Such a protein was found and named **glutaredoxin,** because of its ability to be reduced by glutathione. Mutants lacking both thioredoxin and glutaredoxin are not viable, suggesting that at least one of these proteins must function for ribonucleotide reduction to occur. Whichever of these proteins is the principal cofactor for rNDP reductase, the ultimate electron source is NADPH, because glutathione undergoes reduction via the NADPH-dependent glutathione reductase reaction (see Figure 22.16).

REGULATION OF RIBONUCLEOTIDE REDUCTASE ACTIVITY. Because deoxyribonucleotides are used only for DNA synthesis, and because one enzyme

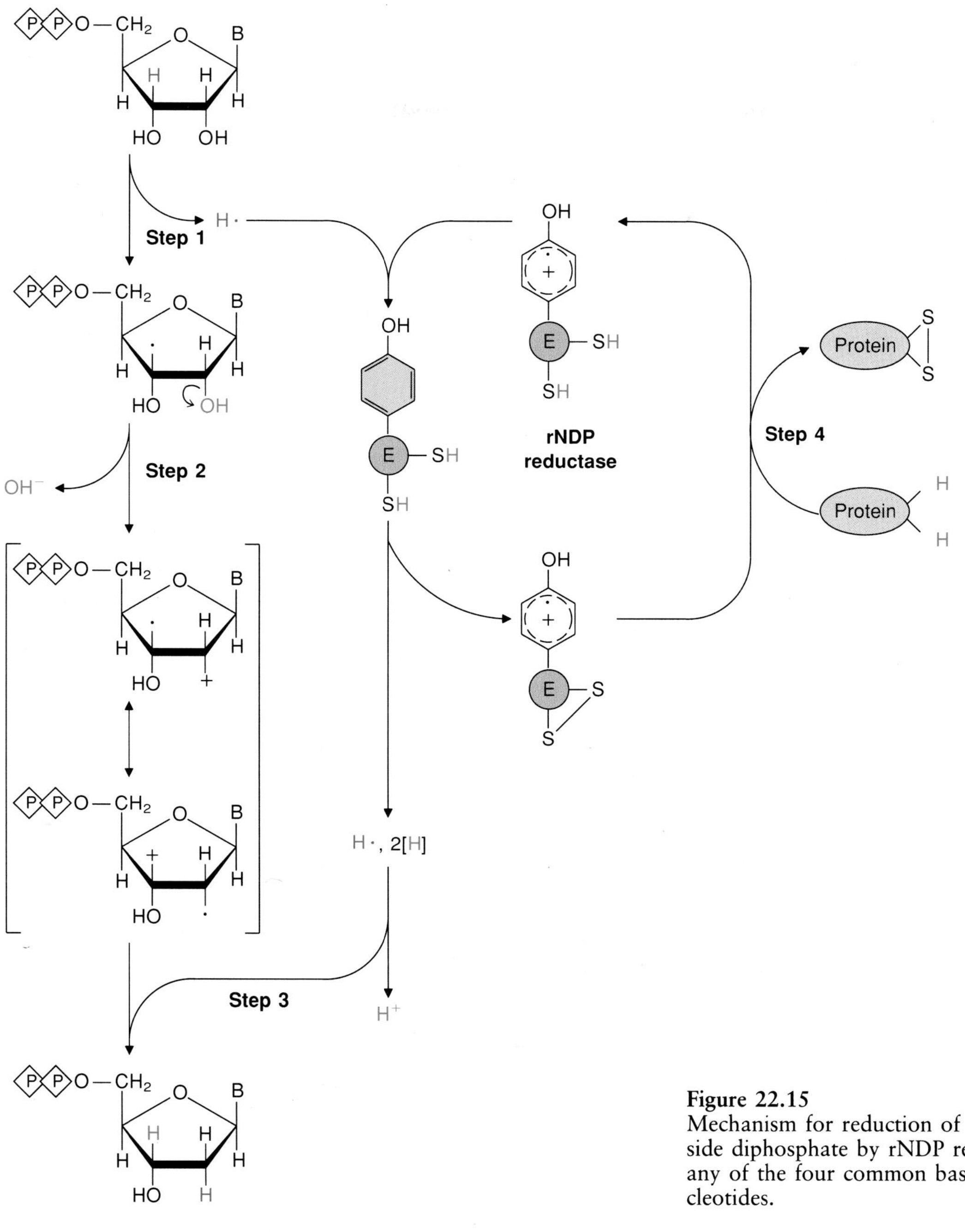

Figure 22.15
Mechanism for reduction of a ribonucleoside diphosphate by rNDP reductase. B is any of the four common bases in ribonucleotides.

system is used for reduction of all four ribonucleotide substrates, regulation of both the *activity* and the *specificity* of ribonucleotide reductase is essential to maintain balanced pools of DNA precursors. This is achieved through binding of nucleoside triphosphate effectors to *two classes* of regulatory sites on the B1 subunit (two of each site per molecule in the *E. coli* enzyme; see Figure 22.13). The **activity sites** bind either ATP or dATP, with relatively low affinity, while the **specificity sites** bind either ATP, dATP, dGTP, or dTTP, all with relatively high affinity. Binding of ATP at the activity sites tends to increase the catalytic efficiencies of ribonucleotide reductase, while dATP acts as a general inhibitor of all four reactions catalyzed by the enzyme. Binding of nucleotides at the specificity sites modulates the activities of the enzyme toward different substrates, so as to maintain

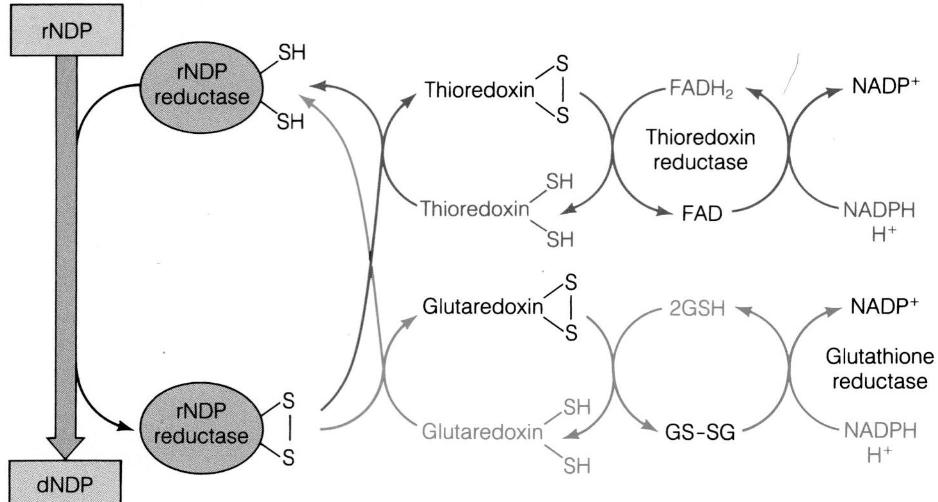

Figure 22.16
Reductive electron transport sequences in the action of ribonucleoside diphosphate reductase. Either thioredoxin or glutaredoxin can reduce the oxidized form of the reductase.

balanced rates of production of the four dNTPs. For example, binding of dTTP activates the enzyme for reduction of GDP but decreases its ability to reduce either UDP or CDP. Table 22.2 summarizes the principal regulatory effects.

These effects are seen in vitro with purified enzymes. There is ample reason to conclude that similar regulatory effects are also operating in intact cells. For example, either deoxyadenosine or thymidine will inhibit DNA synthesis when administered to intact cells. Measurements of intracellular pools of dNTPs show that in deoxyadenosine-treated cells the dATP pools expand (as expected from the effects of salvage pathways), while dTTP, dGTP, and dCTP pools shrink. This is probably the reason why white blood cells cannot proliferate as needed in immunodeficiency states associated with a lack of adenosine deaminase (p. 753): dATP accumulation in these cells blocks deoxyribonucleotide synthesis and, hence, DNA replication.

Another example comes from cell biology, where researchers often **synchronize** cell cultures, that is, manipulate the cells so that all are brought to the same phase of the cell cycle. Synchrony can be accomplished by

Table 22.1
Biological activities of thioredoxin

Activity	Source of Thioredoxin
Cofactor for ribonucleotide reduction	All organisms
Protein folding (thioredoxin promotes correct disulfide bond formation)	All organisms
Possible control of insulin levels, through control of insulin reduction	Animals
Control of melanin formation (people with high levels of thioredoxin reductase tan easily)	Animals
Regulation of photosynthetic carbon fixation	Plants
Sulfite reduction (see Chapter 21)	Plants, bacteria
Essential subunit of viral DNA polymerase	Bacteriophage T7
Maturation of filamentous phages by an unknown mechanism	Single-stranded DNA bacteriophages

Table 22.2
Regulation of the activities of mammalian ribonucleotide reductase

Nucleotide Bound in			
Activity site	Specificity site	Activates Reduction of	Inhibits Reduction of
ATP	ATP or dATP	CDP, UDP	
ATP	dTTP	GDP	CDP, UDP
ATP	dGTP	ADP	CDP, UDP[a]
dATP	Any effector		ADP, GDP, CDP, UDP

[a] dGTP inhibits the reduction of pyrimidine nucleotides by the mammalian enzyme but not by the *E. coli* enzyme.

thymidine block, in which thymidine is added to the cells to inhibit DNA synthesis. This prevents the passage of cells from the G1 to the S phase of the cell cycle. Transferring the cells to medium containing no thymidine relieves the inhibition, allowing the cells to progress synchronously from G1 into S. dNTP pool measurements show that in thymidine-blocked cells dTTP accumulates, as expected from salvage synthesis, while there is a specific depletion of dCTP, as expected from the effects of dTTP on ribonucleotide reductase activity. Indeed, addition of deoxycytidine restores normal dCTP pools (by salvage synthesis) and relieves the thymidine block.

Further support for the control of rNDP reductase activity in vivo comes from isolation of mammalian cell mutant lines whose growth is not inhibited by deoxyribonucleosides. rNDP reductase from these altered cells shows modifications in either activity or specificity sites, which render these enzymes less susceptible to inhibition by dNTP effectors. Interestingly, some of these cell lines display both dNTP pool abnormalities and **mutator phenotypes;** that is, they show increased rates of spontaneous mutation at a variety of genetic loci. Comparable observations have been made with mutant cells altered in other enzymes of nucleotide metabolism, including CTP synthetase and deoxycytidylate deaminase.

Biosynthesis of Thymine Nucleotides

The previous section explored the first metabolic reaction committed to DNA synthesis—the formation of deoxyribonucleoside diphosphates through the action of rNDP reductase. Once formed, three of the diphosphates—dADP, dGDP, and dCDP—are converted directly to the corresponding triphosphates by nucleoside diphosphokinase. Biosynthesis of thymidine triphosphate occurs partly from the dUDP produced via the reductase and partly from deoxycytidine nucleotides; the ratio varies in different cells and organisms. The pathways are summarized in Figure 22.17. Two pathways de novo lead to deoxyuridine monophosphate (dUMP), the substrate for synthesis of thymine nucleotides: (1) dUDP is phosphorylated to dUTP, which is then cleaved by a highly active diphosphohydrolase, **dUTPase;** and (2) dCDP is dephosphorylated to dCMP, which then undergoes deamination to dUMP by an aminohydrolase called **dCMP deaminase.** The latter enzyme represents a branch point for pyrimidine dNTP synthesis; it requires dCTP as an allosteric activator and is inhibited by dTTP.

E. coli and some other bacteria use a different route to dUMP. Deamination is carried out at the triphosphate level by **dCTP deaminase,** and the resultant dUTP is cleaved by dUTPase to dUMP and PP_i.

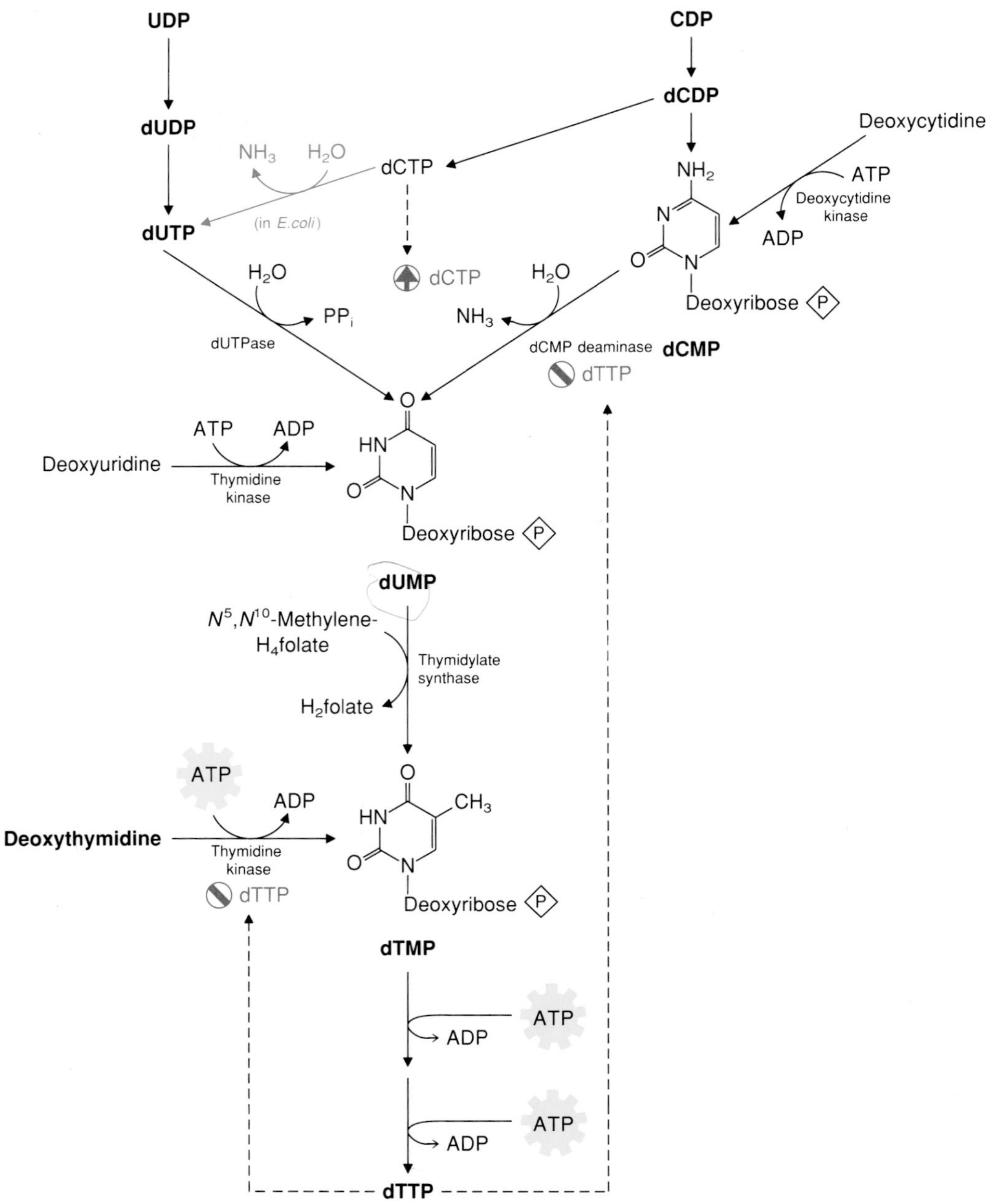

Figure 22.17
Salvage and de novo synthetic pathways to thymine nucleotides.

However it is formed, dUMP serves as substrate for formation of thymidine monophosphate (dTMP), catalyzed by **thymidylate synthase.** This enzyme transfers a one-carbon unit, at the methylene level of oxidation, and reduces it to the methyl level. The one-carbon donor is N^5,N^{10}-methylenetetrahydrofolate, which in this unusual reaction serves as a redox cofactor as well, to give dihydrofolate as the other reaction product (Figure 22.18). The cofactor must then be reduced, by dihydrofolate reductase, and it must acquire another methylene group, most commonly via serine transhydroxymethylase. Interruption of any of these steps of the cycle interferes with thymine nucleotide formation.

Deoxyuridine Nucleotide Metabolism

In addition to its biosynthetic function in forming dUMP for thymine nucleotide formation, dUTPase plays an important role in excluding uracil from

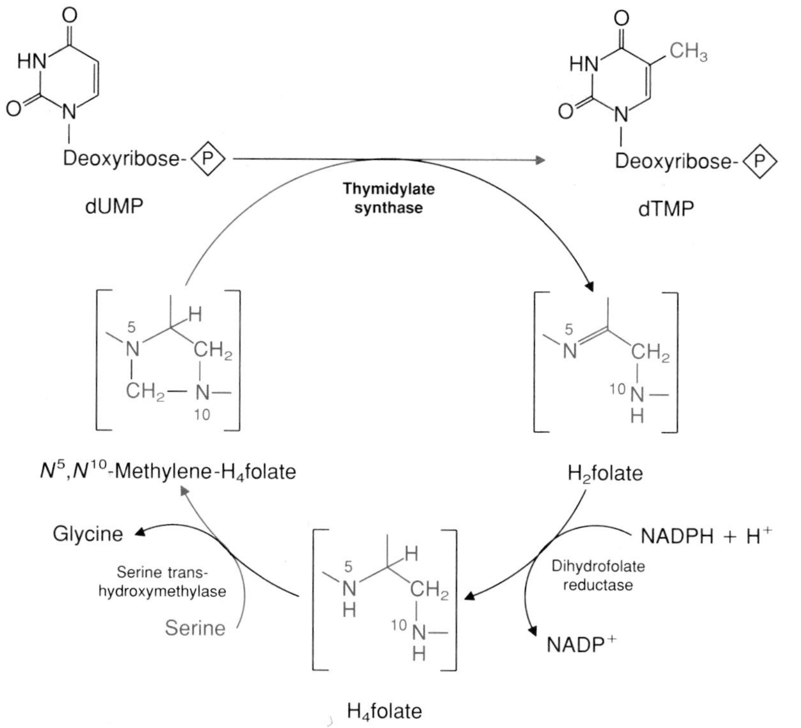

Figure 22.18
Relationship between thymidylate synthase and enzymes of tetrahydrofolate metabolism. Only the active portion of the folate coenzyme is shown.

DNA. Were dUTP not rapidly degraded, it could serve as a satisfactory substrate for DNA polymerases. In fact, as we shall see in Chapter 24, cells have a rather elaborate mechanism to ensure that any dUMP residues that do find their way into DNA are efficiently excised. Since uracil is virtually identical to thymine with respect to its base-pairing properties, why is it so important for the cell to allow only thymine to be stably incorporated into DNA? The answer probably lies in the fact that dUMP residues can arise in DNA not only by dUTP incorporation but also by the spontaneous deamination of dCMP residues. Since the latter, which occurs at an appreciable rate, would change the sense of a genetic message, it seems advantageous for the cell to maintain genetic stability by having a surveillance system that excises dUMP residues no matter how they arise.

Salvage Routes to Deoxyribonucleotide Synthesis

Of the various deoxyribonucleoside kinases, one that merits special mention is **thymidine kinase (TK)**. This enzyme is allosterically inhibited by dTTP (see Figure 22.17). Activity of thymidine kinase in a given cell is closely related to the proliferative state of that cell. During the cell cycle, activity of TK rises dramatically as cells enter S phase, and in general rapidly dividing cells have high levels of this enzyme.

For reasons not yet clear, the salvage pathway to dTTP, which involves thymidine kinase, competes very efficiently with the thymidylate synthase-mediated de novo pathway. Radiolabeled thymidine is widely used for isotopic labeling of DNA—for example, in radioautographic investigations or to estimate rates of intracellular DNA synthesis. In most cases exogenous thymidine is incorporated into DNA much more efficiently than would be expected from the activities of the enzymes involved. Although this apparent channeling process is extremely useful to scientists, its mechanism is not known. The channeling may be a manifestation of a more general tendency of the enzymes of deoxyribonucleotide biosynthesis to associate with one another. Multienzyme complexes that catalyze multiple reactions in dNTP synthesis have been isolated from microbial, plant, and animal cells. There is some reason to believe that such complexes help to regulate DNA replication by coordinating it with the synthesis of its precursors.

Also worthy of mention is **deoxycytidine kinase,** a salvage enzyme that is feedback inhibited by dCTP. This enzyme also functions as a deoxyadenosine kinase and a deoxyguanosine kinase. However, since the K_m values for these purine substrates are far higher than the K_m for deoxycytidine, it seems likely that deoxycytidine is the physiological substrate except when the purine deoxyribonucleosides are available at very high levels (as is the case in the inherited immunodeficiencies discussed earlier in this chapter). Unlike thymidine kinase, whose activity fluctuates over the course of the cell cycle, the activity of deoxycytidine kinase stays relatively constant.

Thymidylate Synthase, A Target Enzyme for Chemotherapy

A goal of chemotherapy—the treatment of diseases with chemical agents—is to exploit a biochemical difference between the disease process and the host tissue in order to interfere selectively with the disease process. Many chemotherapeutic agents were originally discovered by chance, through testing of analogs of normal metabolites. Most of these agents are limited in their effectiveness by unanticipated side effects, incomplete selectivity, and the development of drug resistance. One of the most exciting areas of modern biochemical pharmacology is drug architecture—the design of specific inhibitors based on knowledge of the molecular structure of the site to which the inhibitor will bind and the mechanism of action of the target molecule. For drugs whose target is an enzyme, it is necessary to know the three-dimensional structure of the enzyme and its mechanism of action. This requires a fusion of x-ray crystallography, classical bio-organic chemistry, site-directed mutagenesis, and computer-aided molecular graphics. Thymidylate synthase is an excellent example of the utility of these approaches.

Recall from our discussion of antimetabolites in Chapter 20 that the goal of treating diseases with chemicals is to attack selectively a metabolic process that is specific to the pathological condition. Because thymidylate synthase participates in the synthesis of a deoxyribonucleotide, any disease that involves uncontrolled cell proliferation can in principle be treated with inhibitors of thymidylate synthase—blocking the production of an essential DNA precursor should inhibit DNA replication with minimal effects on other processes. Cells that are not undergoing rapid proliferation should be relatively immune to such agents. Thus, cancer and a wide range of infectious diseases should be amenable to treatment by this approach.

None of this was recognized in the mid-1950s; in fact thymidylate synthase had not yet been discovered. It was known that certain tumor cells took up and metabolized uracil much more rapidly than normal cells. Without knowing the metabolic fates of uracil in detail, Charles Heidelberger hoped to kill tumor cells selectively by treatment with analogs that would block uracil metabolism in tumor cells. To that end, Heidelberger undertook the chemical synthesis of **5-fluorouracil** (**FUra**) and its deoxyribonucleoside, **5-fluorodeoxyuridine** (**FdUrd**). Both compounds were found to be potent inhibitors of DNA synthesis. Their action as inhibitors involves their intracellular conversion to **5-fluorodeoxyuridine monophosphate**

(**FdUMP**), a dUMP analog that acts as an irreversible inhibitor of thymidylate synthase.

Salvage pathways

Thymidine kinase

ADP ATP

5-Fluorouracil (FUra)

5-Fluorodeoxyuridine monophosphate (FdUMP)

5-Fluorodeoxyuridine (FdUrd)

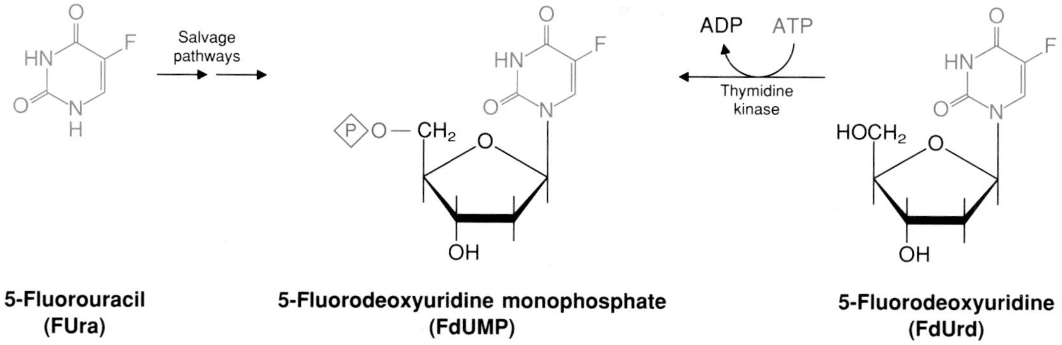

Both fluorouracil and fluorodeoxyuridine have found use in cancer treatment. However, the fluorinated pyrimidines are not completely selective in their effects. For example, fluorouracil can be incorporated into RNA by salvage routes normally used for uracil, thereby interfering with the function of messenger RNA in both cancer and normal cells. Clearly, a detailed understanding of the active site of thymidylate synthase would permit the design of completely specific enzyme inhibitors.

Analysis of the binding of 5-fluorodeoxyuridylate to thymidylate synthase has opened the door to understanding the mechanism of the reaction and the structure of the active site. FdUMP is a true **mechanism-based inhibitor,** in that irreversible binding occurs only in the presence of N^5, N^{10}-methylenetetrahydrofolate. Presumably, binding of the coenzyme induces a conformational change in the active site that duplicates early steps in the catalytic reaction and leads to irreversible FdUMP binding.

When Daniel Santi subjected the inhibited enzyme to proteolytic digestion, he isolated a peptide in which a covalent bond linked a cysteine sulfur with C-6 of the FdUMP pyrimidine ring. There is good evidence that this is a covalently bonded ternary complex, with FdUMP linked to the methylene carbon of the coenzyme.

Methylene-H_4folate (partial structure)

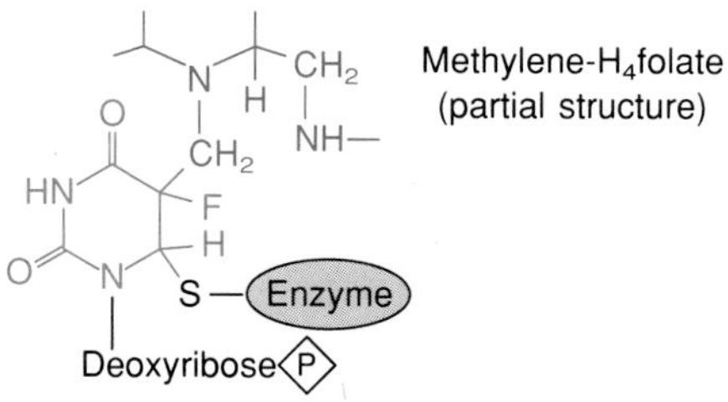

Complex between enzyme-bound FdUMP and N^5,N^{10}-methylene-H_4folate

The structure of this complex suggested the mechanism outlined in Figure 22.19. A cysteine thiolate ion on the enzyme initiates a nucleophilic attack on C-6 of the pyrimidine substrate (step 1). This generates a resonance-stabilized anion; C-5 now becomes a nucleophile, attacking the methylene carbon of the coenzyme (step 2). Loss of a proton from C-5 (step 3) initiates an electron shift that leads to transfer of the hydrogen at C-6 of the cofactor to the pyrimidine (step 4), consistent with observations that this hydrogen is incorporated quantitatively into the thymidylate methyl group. This oxidizes the cofactor to dihydrofolate, which dissociates in step 5, and

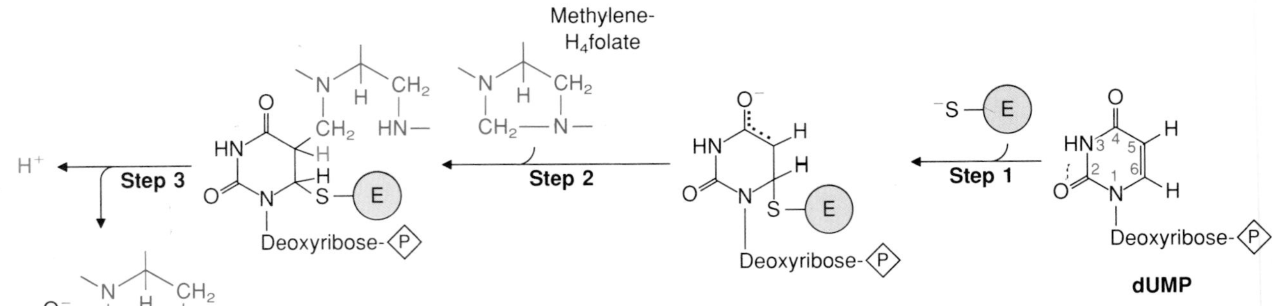

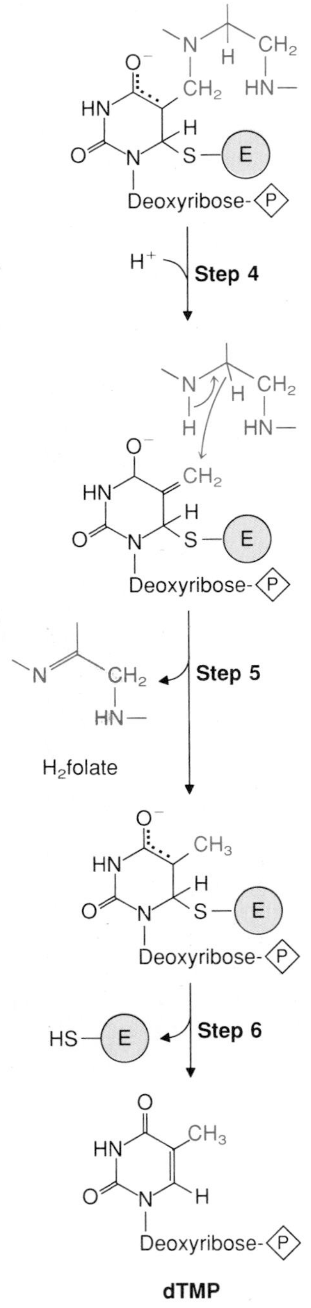

Figure 22.19
Mechanism for the reaction catalyzed by thymidylate synthase (E).

in step 6 the covalent enzyme–substrate bond is broken, with reformation of the double bond between C-5 and C-6 of the pyrimidine ring.

Inhibition by FdUMP results from the electronegativity of fluorine, which generates a C—F bond at C-5 that cannot be broken. Thus, the reaction pathway cannot proceed to step 3.

Confirmation of this mechanism has come about with evidence that the substrate, dUMP, forms a covalent bond with the coenzyme during normal catalysis. More important, a three-dimensional model of thymidylate synthase from *Lactobacillus casei* was presented in 1987 by Robert Stroud, Daniel Santi, and their colleagues (Figure 22.20). Thymidylate synthase is highly conserved evolutionarily, with about 20% of its residues being invariant among mammalian, bacterial, viral, fungal, and protozoal sequences. The crystal structure shows a high percentage of these conserved residues in the cleft thought to represent the active site. The protein is a homodimer, and conserved residues from both subunits line the cleft and, presumably, contribute to catalysis. Cysteine-198, the residue that becomes linked to FdUMP, lies in this cleft, close to another conserved residue, Arg-218, which may lower the pK_a of the cysteine thiol and generate the reactive thiolate ion. Several conserved lysine residues lie nearby and are thought to represent binding sites for the polyglutamate tail on the folate coenzyme (thymidylate synthase binds folate polyglutamates about 100-fold more tightly than the monoglutamate).

Because of the difficulty of crystallizing an enzyme–substrate or enzyme–coenzyme complex, computer graphics have been used to model the active site as it would appear during catalysis (Figure 22.21). These efforts, plus mechanistic studies of variant enzymes produced by site-directed mutagenesis and structural determination of thymidylate synthases from other species, should pave the way for rational development of highly specific and irreversible enzyme inhibitors. The same type of approach is under way in dozens of laboratories, focused on drug receptors that include membrane-bound or intracellular receptor proteins and nucleic acids, as well as enzymes.

Targets other than cancers are also susceptible to attack by inhibition of thymidylate synthase. For example, parasitic protozoans, such as those that cause malaria, synthesize an unusual form of thymidylate synthase—actually, a bifunctional enzyme, with both thymidylate synthase and dihydrofolate reductase activities. Knowing the active site structure of this enzyme should allow development of inhibitors that would block this specific enzyme but not thymidylate synthase of the animal or human host.

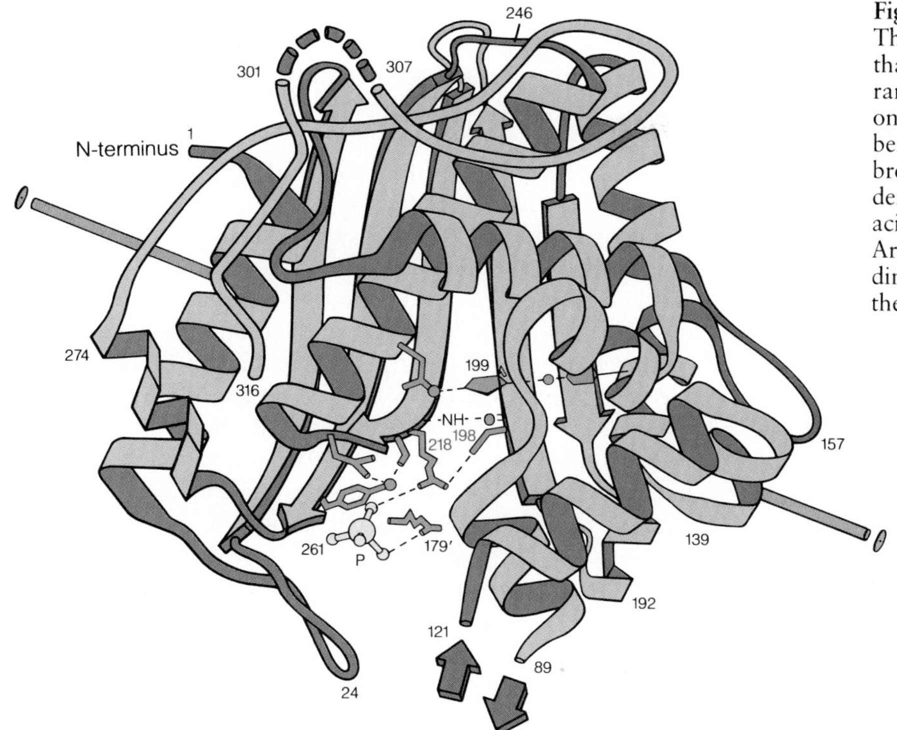

Figure 22.20
Three-dimensional model of thymidylate synthase from *Lactobacillus casei*, showing arrangement of helices and extended chain within one monomer. The amino terminus is at number 1. The broken chain at 301–307 and the break between 89 and 121 correspond to disordered regions in the structure. Conserved amino acid residues are shown in the active site cleft. Arg-179′ is from the adjacent monomer in the dimeric enzyme. Note the axis of symmetry of the homodimer.

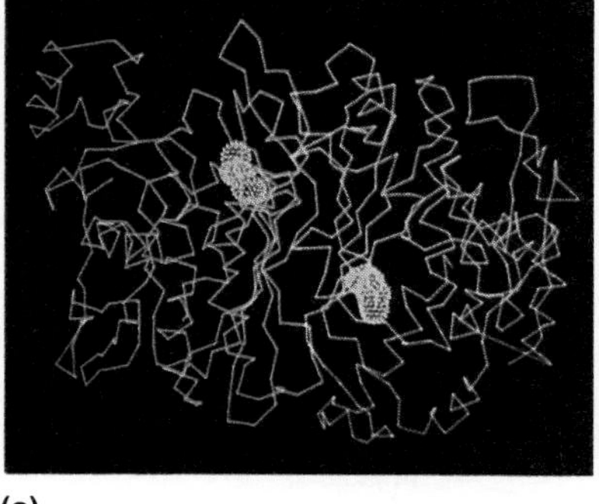

(a)

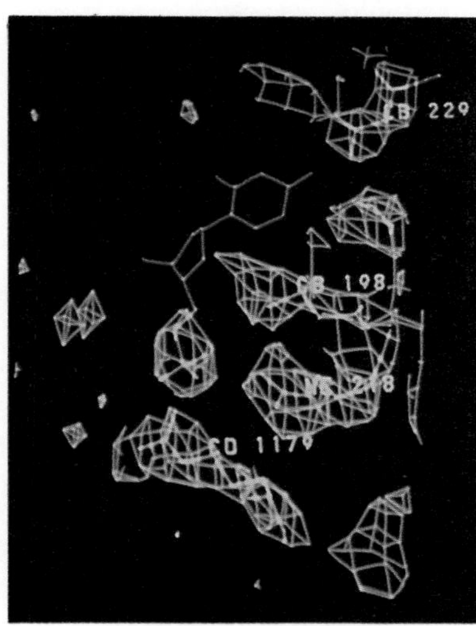

(b)

Figure 22.21
Thymidylate synthase. (**a**) α-carbon tracing of the thymidylate synthase dimer with each monomer colored differently. The enzyme is viewed approximately down the twofold axis (90° away from the line of view in Figure 22.20). The white blobs represent bound dUMP. (**b**) Computer-generated model of the active site of thymidylate synthase. dUMP and the side chains of amino acids in the active site are shown as red stick figures, while the electron density around each group is shown in blue. 229 = Asn; 198 = Cys; 218 = Arg; 179 = Arg.

Virus-Directed Alterations of Nucleotide Metabolism

That viruses can redirect the metabolism of cells that they infect first came to light in 1957 through studies of nucleotide biosynthesis in *E. coli* bacteria infected by the T-even bacteriophages—T2, T4, and T6. G. R. Wyatt and Seymour Cohen had shown in 1952 that DNA of these viruses contains no cytosine, but instead contains 5-hydroxymethylcytosine, with most of the hydroxymethyl groups further modified by being in glycosidic linkage with glucose moieties.

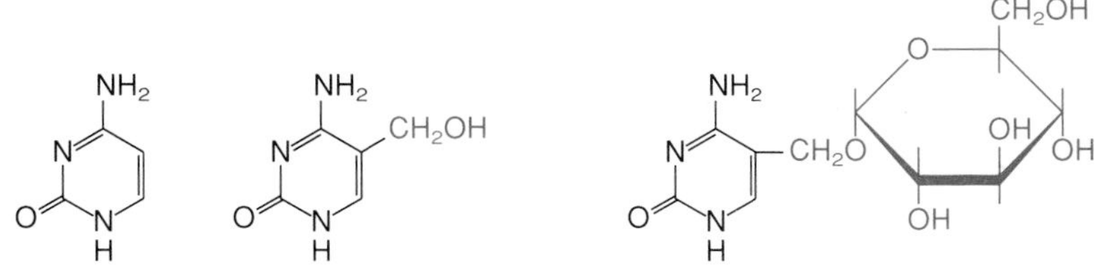

Continuing work showed that infection initiates the synthesis of virus-coded enzymes, which carry out these modifications (Figure 22.22). The principal enzymes in T-even phage-infected bacteria include a **dCTPase,** which cleaves dCTP to dCMP, a **dCMP hydroxymethylase,** which transfers a one-carbon group to dCMP at the hydroxymethyl oxidation level, and a **deoxyribonucleoside monophosphate kinase,** which can phosphorylate the

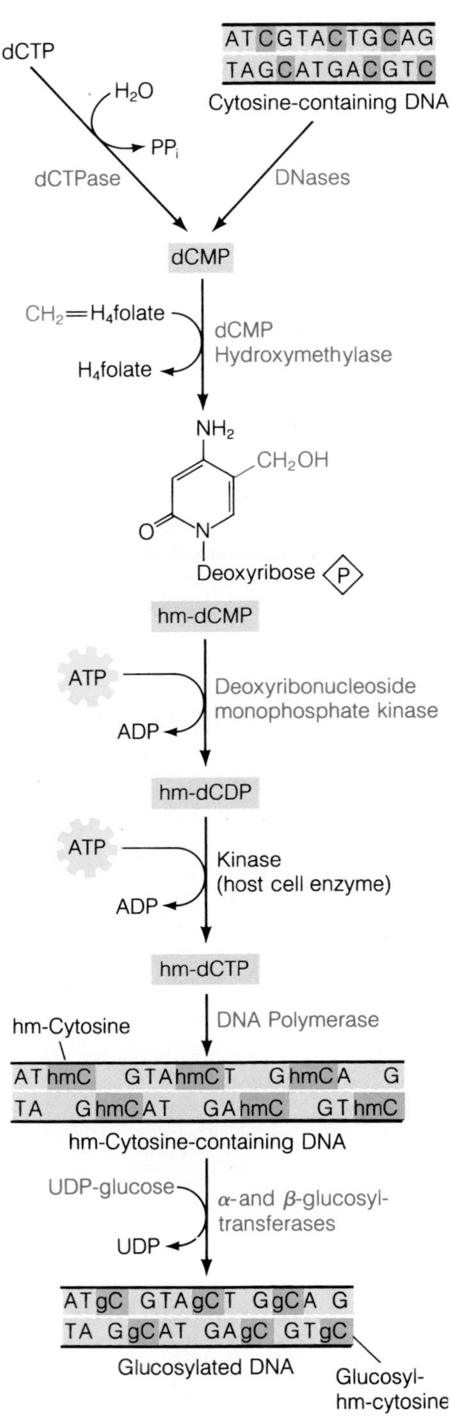

Figure 22.22
Metabolic pathways leading to nucleotide modifications in T-even phage-infected *E. coli*. Virus-coded enzymes are shown in blue. hm = hydroxymethyl.

resultant 5-hydroxymethyl-dCMP. Phosphorylation of 5-hydroxymethyl-dCDP to the triphosphate is catalyzed by nucleoside diphosphokinase of the host cell. The glucosylation reactions occur after the modified nucleotide has been incorporated into DNA. In T4 phage-infected bacteria there are two **glucosyltransferase** reactions, one of which transfers glucose in the α configuration and one in the β. In addition, the viral genome specifies several deoxyribonucleases, which specifically cleave cytosine-containing DNA. This helps the virus to abolish expression of host cell genes, and it also provides a source of precursors for viral DNA synthesis.

Although base substitutions in DNA are rather unusual, about a dozen instances have now come to light in which one of the four common deoxyribonucleotides is replaced, wholly or in part, by a chemically altered derivative that retains the base-pairing specificity of the replaced nucleotide. Some *Bacillus subtilis* phages, for example, substitute uracil for thymine in their DNA. Other *B. subtilis* phages contain 5-hydroxymethyluracil in place of thymine. A phage of *Xanthomonas oryzae* substitutes 5-methylcytosine for one of its DNA cytosine residues. In each case investigated, the virus directs the nucleotide modifications through synthesis of virus-encoded enzymes that create novel metabolic pathways in the cells they infect.

Plant and animal viruses do not contain extensive nucleic acid base modifications of the type found in bacteriophages. (Most organisms contain a small proportion of methylated nucleic acid bases, but these modifications occur after polymerization.) However, virus-coded enzymes are often produced to help the infected cell augment its synthesis of nucleic acid precursors. In some cases the virus-specified enzyme differs from its host cell counterpart to the extent that one can design selective enzyme inhibitors and thereby achieve a specific chemotherapy directed against the virus-infected cell. The best current example is the use of **acyclovir** in treating herpes simplex virus infections. The herpesviruses are large DNA-containing viruses whose genomes encode several enzymes, including **deoxypyrimidine kinase.** Originally discovered as a thymidine kinase, the enzyme also phosphorylates dTMP. More important, however, is the extremely broad substrate specificity of the nucleoside kinase activity of this enzyme. Two nucleosides that are readily phosphorylated by this enzyme are **5-iododeoxyuridine,** a thymidine analog, and **acycloguanosine** (acyclovir), an analog related to deoxyguanosine.

5-Iododeoxyuridine **Acycloguanosine**

In both cases the compound is converted ultimately to the 5′-triphosphate, which interferes with DNA replication. Since uninfected cells phosphorylate acycloguanosine not at all and iododeoxyuridine to only a limited extent, one has a means of blocking DNA replication and, hence, virus growth selectively in the infected cells.

Biological and Medical Importance of Nucleotide Analogs

The examples of the fluorinated pyrimidines and of acycloguanosine have made clear the utility of nucleoside and nucleotide analogs. Since nucleotides are very poorly transported into cells because of the negative phosphate charge, most of the compounds we shall discuss are actually nucleosides or nucleoside derivatives. They function as nucleotide antagonists, after uptake into the cell and phosphorylation. Therefore, we shall use the generic term *nucleotide analogs* in this section.

Nucleotide Antagonists as Chemotherapeutic Agents

The enzymes of nucleotide synthesis have been widely studied as target sites for the action of antiviral or antimicrobial drugs. As noted earlier, the aim is to identify a biochemical distinction between comparable processes of the host and the disease process.

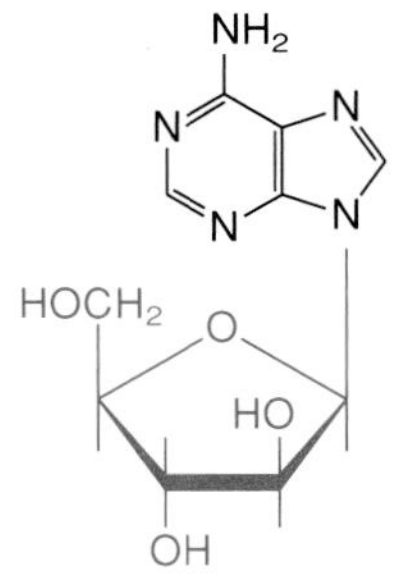

Arabinosyladenine (araA)

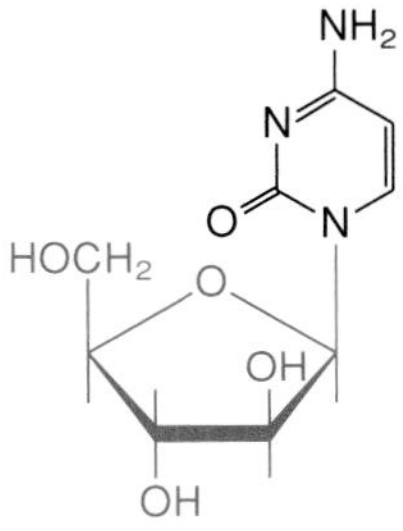

Arabinosylcytosine (araC)

ANTIVIRAL NUCLEOSIDE ANALOGS. The extremely broad substrate specificity of the viral deoxypyrimidine kinase, which forms the basis for the effectiveness of acycloguanosine, is also the basis for the antiviral effectiveness of **arabinosyladenine** (**araA**), now being used to treat viral encephalitis, a neurological disease caused by another member of the herpesvirus family. Because araA is susceptible to degradation by adenosine deaminase, it is often administered in combination with an inhibitor of the latter enzyme. The arabinose analog of deoxycytidine, **arabinosylcytosine** (**araC**), is being used in cancer chemotherapy. Both of the arabinose analogs interfere with DNA replication after conversion to the triphosphate.

Other analogs receiving considerable attention are those being used to combat acquired immune deficiency syndrome (AIDS) caused by the human immunodeficiency virus (HIV). One such analog, **3′-azido-3′-deoxythymidine** (**AZT**), is the first drug approved in the United States for the treatment of AIDS.

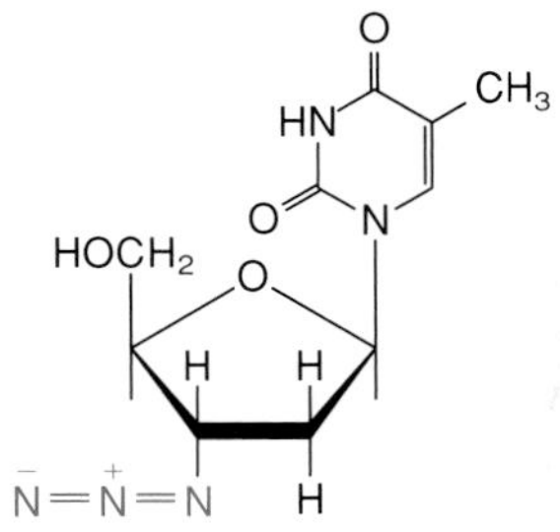

3′-Azido-3′-deoxythymidine 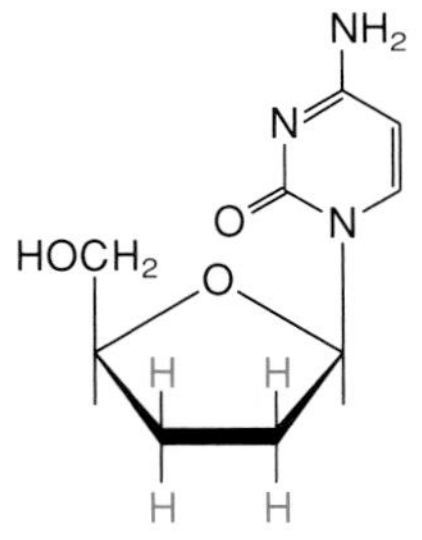 **2′,3′-Dideoxycytidine** 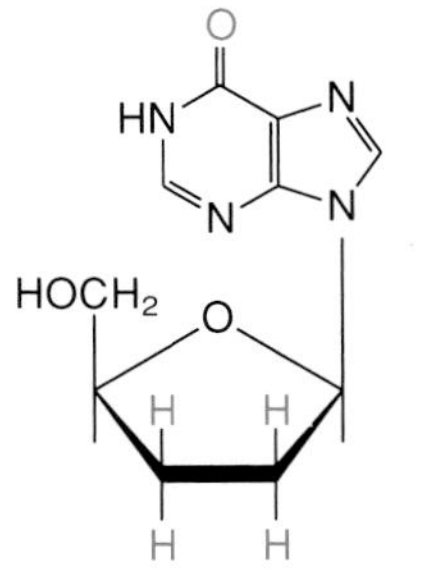 **2′,3′-Dideoxyinosine**

This nucleoside is anabolized to the corresponding 5′-triphosphate, which is an inhibitor of viral reverse transcriptase, the enzyme that makes a DNA copy of the viral RNA (see Chapter 24). Other nucleoside analogs, **2′,3′-dideoxycytidine** (DDC) and 2′,3′-dideoxyinosine (ddI) act by conversion to the corresponding triphosphate, which is incorporated into DNA but then blocks further replicative chain elongation because of the absence of a 3′-hydroxyl terminus.

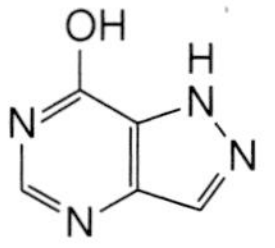

Formycin B

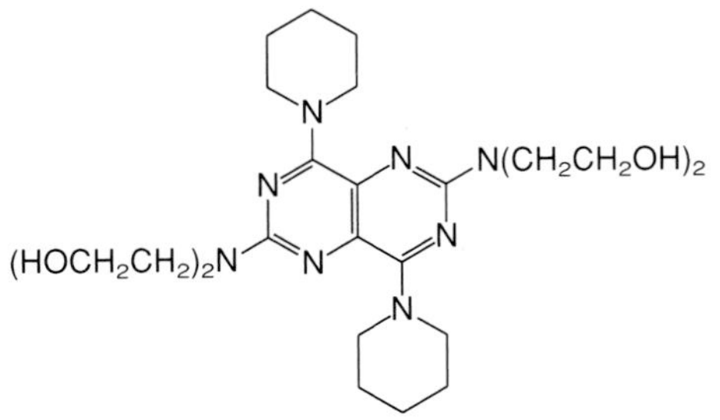

Dipyridamole

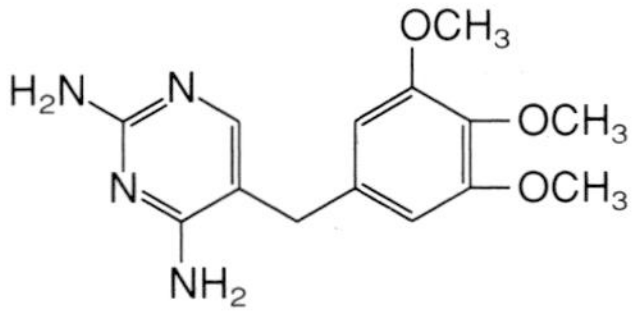

Trimethoprim

PURINE SALVAGE AS A TARGET. An important biochemical anomaly is found in parasitic protozoans such as *Plasmodium,* which is responsible for malaria, and *Leishmania,* which is a debilitating but usually nonfatal disease affecting the skin and visceral organs. Parasitic protozoans lack the capacity for de novo purine synthesis, and they depend entirely on salvage of nucleosides and bases provided by the host. Compounds such as allopurinol (p. 752) and **formycin B** inhibit the growth of these organisms in culture, partly through inhibition of salvage enzymes and partly through the ability of the salvage enzymes to anabolize the analog, while corresponding enzymes of the host do not. For example, allopurinol is converted to an analog of inosinic acid and then to an AMP analog and is finally incorporated into RNA, where it interferes with messenger RNA coding in protein synthesis.

Inhibition of nucleoside salvage may be more effective if the analog is administered along with an inhibitor of nucleoside transport. In animal cells a single protein, the nucleoside transporter, appears to be responsible for uptake of a wide variety of nucleosides. Functioning of this protein is inhibited by **dipyridamole,** which blocks the uptake of most nucleosides.

FOLATE ANTAGONISTS. Recall from Chapter 20 that the folic acid analog methotrexate was found long ago to induce remissions in certain acute leukemias. What is the basis for this selectivity? Since folate cofactors play essential roles in synthesizing precursors to DNA, RNA, protein, *and* phospholipids, one would expect inhibition of tetrahydrofolate synthesis to be toxic to all cells. However, there is a rationale for the selective toxicity of folate antagonists against proliferating cells. Recall that the thymidylate synthase reaction oxidizes methylenetetrahydrofolate to dihydrofolate; this is the only known tetrahydrofolate-requiring reaction that does not regenerate tetrahydrofolate. From the reactions shown in Figure 22.19, one can predict that inhibition of dihydrofolate reductase blocks the recycling of dihydrofolate back to tetrahydrofolate. Under these conditions the rate at which all of the intracellular reduced folates become oxidized is directly related to the intracellular activity of thymidylate synthase, which in turn is coordinated with the rate of DNA synthesis. Thus, proliferating cells, with rapid rates of DNA replication, will exhaust their tetrahydrofolate stores sooner than nonproliferating cells.

While antimetabolites such as fluorouracil or methotrexate do attack proliferating tissue selectively, they are also toxic to normal cells. Dangerous side effects are seen against tissues that must proliferate as part of their normal function; such tissues include intestinal mucosa and components of the immune system. Equally serious is the development of drug-resistant cell variants, in which levels of the target enzyme rise beyond the point where they can be controlled with inhibitors. Robert Schimke and his colleagues investigated the mechanism for the rise in dihydrofolate reductase activity, which can be several hundredfold in methotrexate-resistant cell lines. They found that prolonged incubation of cells in methotrexate leads to selective amplification of the gene encoding dihydrofolate reductase; the number of DNA copies of this gene per cell increases many times.

Another class of dihydrofolate reductase inhibitors is exemplified by **trimethoprim.** This compound is a specific inhibitor of dihydrofolate reductases of prokaryotic origin. Trimethoprim and its relatives are widely used to treat both bacterial infections and certain forms of malaria. The success of these drugs derives from the fact that they are extremely weak inhibitors of vertebrate dihydrofolate reductases. Trimethoprim is often administered in conjunction with a sulfonamide drug to inhibit the *synthesis* of folate and hence to block sequential steps in the same pathway.

Nucleotide Analogs and Mutagenesis

Some analogs are excellent mutagens, useful both in the isolation of mutants and in studies on the mechanism of mutagenesis. Two such analogs are **2-aminopurine** (2AP) and **5-bromodeoxyuridine** (BrdUrd).

N N H$_2$N N N H

2-Aminopurine

O Br HN O N deoxyribose

5-Bromodeoxyuridine

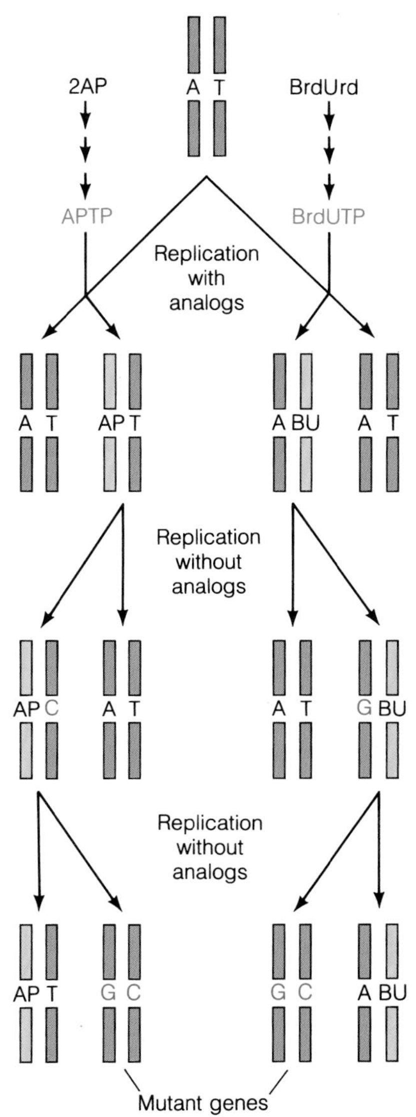

Figure 22.23
Mechanisms of mutagenesis by nucleotide analogs. 2-Aminopurine (2AP) is converted by salvage pathways to dAPTP, the dNTP analog of dATP, and BrdUrd is converted to BrdUTP, the dNTP analog of dTTP. The first round of replication occurs in the presence of the analog, and the second and third rounds occur in its absence. Replication of only the analog-containing duplex is shown in the second and third rounds. AP is a 2-aminopurine nucleotide residue in DNA, and BU is a bromodeoxyuridine nucleotide residue. In the pathways shown both analogs change an A-T base pair to a G-C base pair (red letters). Other pathways can occur as well.

2-Aminopurine is incorporated into DNA in place of adenine, but when a 2AP-containing template replicates, the analog occasionally base-pairs with cytosine rather than thymine. Thus, incorporation of 2AP changes an A-T pair in DNA to a G-C pair (Figure 22.23).

Bromodeoxyuridine functions similarly, but it has other uses as well. It is an excellent thymidine analog, because the van der Waals radius of the bromine atom is close to that of the methyl group. Hence, it is efficiently incorporated into DNA. Mispairing of BrdUrd with a deoxycytidine residue on replication of a BrdUrd-containing template can lead to mutagenesis by changing an A-T base pair to a G-C (see Figure 22.23). Alternatively, BrdUTP can compete with dCTP for incorporation opposite G in the template (not shown).

Because bromine is much heavier than a methyl group, it imparts an increased density to substituted DNA, thereby providing a physical basis for separating replicating from nonreplicating DNA (see Chapter 24). Finally, radiobiologists make use of BrdUrd because it is a radiosensitizing agent. Bromo-dUMP residues in DNA debrominate readily when that DNA is irradiated with UV or near-UV light. This generates free radicals, which cause various kinds of damage to the DNA structure.

Nucleotide-Metabolizing Enzymes as Selectable Genetic Markers

Because most cells can synthesize nucleotides de novo, the enzymes of salvage synthesis are usually nonessential for cell viability. Moreover, as we have seen, a large number of inhibitors of these enzymes are available. Consequently, nucleotide-metabolizing enzymes are being widely used as **selectable genetic markers.** This means that it is possible to devise selective growth conditions such that only cells lacking a particular enzyme, or containing a particular enzyme, will grow. For example, **6-thioguanine** is a purine analog that is metabolized to a toxic intermediate by HGPRT.

SH N N H$_2$N N N H

6-Thioguanine

Culture of cells in thioguanine-containing medium allows growth of only the cells lacking active HGPRT. Similarly, one can isolate cells lacking thymidine kinase by selecting for a bromodeoxyuridine-resistant phenotype, because TK must be active in order to anabolize BrdUrd to a toxic metabolite. This allows one to measure forward mutation rates in somatic cell genetic analysis by analyzing mutation to these drug-resistant phenotypes.

By the same token, one can select for mutation in the reverse direction by adjusting culture conditions so that the capacity for salvage synthesis is essential to cell viability. A common technique, both in somatic cell genetic analysis and in preparation of monoclonal antibodies, is **cell fusion.** Two cell lines of different origins are mixed under conditions where some of them can physically fuse, resulting in two heterologous nuclei in one cytoplasm. One can select for these cell hybrids by adjusting culture conditions so that only the hybrids will grow in **HAT medium** (normal cell culture medium augmented with hypoxanthine, aminopterin, and thymidine). Aminopterin inhibits dihydrofolate reductase and hence blocks de novo purine and thymidylate synthesis. Cells can survive, but only if they have active HGPRT, to utilize the hypoxanthine for purine synthesis, and TK, to utilize the thymidine for thymidylate synthesis.

Genes with selectable phenotypes represent an important adjunct to the use of recombinant DNA technology for introduction of novel genetic material into cells—animal, plant, or microbial. We will say more about this in Chapter 25, but for now just consider a recombinant DNA molecule in which a gene to be cloned is linked in tandem with a gene that encodes a methotrexate-resistant form of dihydrofolate reductase. After introduction of this DNA into recipient cells containing the normal methotrexate-sensitive dihydrofolate reductase, one simply grows the cells in the presence of methotrexate. Cells that have acquired the desired gene and its fusion partner have a selective growth advantage and can easily be isolated.

Nucleotides as Metabolic Regulators

Nucleotides play various roles, still being identified, in regulating cellular metabolism and, more specifically, in helping cells adjust to environmental changes.

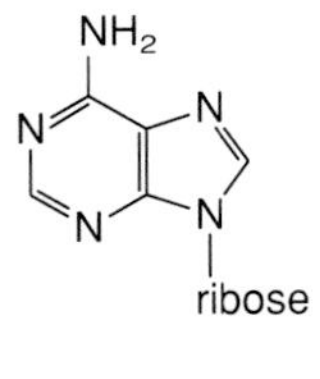

Adenosine

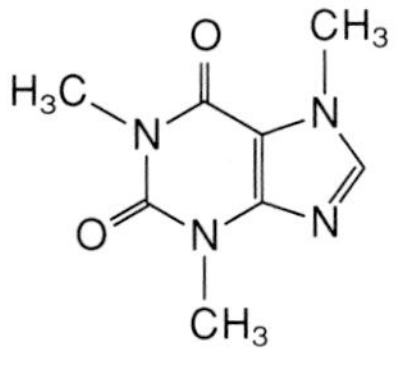

Caffeine

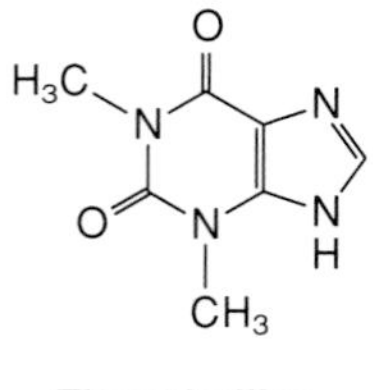

Theophylline

Cyclic AMP, Guanine Nucleotides, and Adenosine

The roles of cyclic AMP in hormone action have been discussed, and cyclic AMP's role in regulating bacterial transcription is presented in Chapter 26. Guanine nucleotides also participate in hormone action, as described in Chapter 23.

The nucleoside adenosine plays an important role in regulating activities of the central nervous system in animals. Binding of adenosine to membrane-associated cell receptors regulates the formation of cAMP via adenylate cyclase, just as we have seen with other hormones. Pharmacologically, adenosine has a depressant effect on central nervous system activities. Methyl xanthines that act as stimulants, such as **caffeine,** from coffee, and **theophylline,** from tea, antagonize the actions of adenosine at its receptors, and such action is thought to be responsible for the biological activities of these compounds.

"Magic Spot" Nucleotides—ppGpp and pppGpp

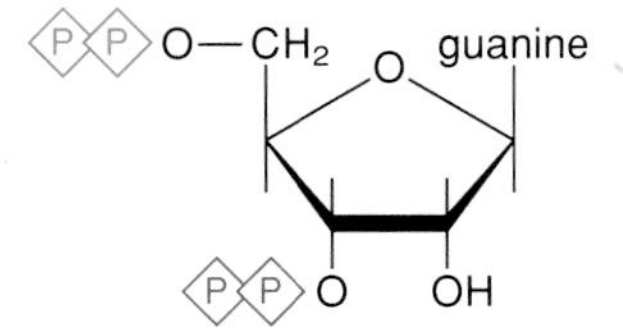

P P O—CH2
O
guanine
P P O
OH

ppGpp

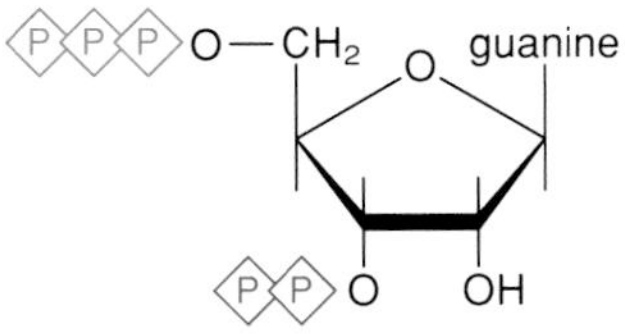

P P P O—CH2
O
guanine
P P O
OH

pppGpp

Bacterial metabolism is controlled by a system called the **stringent response.** A number of biosynthetic processes are inhibited when one or more amino acids become limiting, including phospholipid biosynthesis and the synthesis of ribosomal RNA and transfer RNA. The cell adjusts to blocked protein synthesis by turning off certain biosynthetic pathways. Two compounds accumulate during this stringent response; since they were originally detected as spots of unknown nature on thin-layer chromatograms, they were called "magic spots" 1 and 2 and later identified as guanosine 5′-diphosphate 3′-diphosphate (**ppGpp**) and guanosine 5′-triphosphate 3′-diphosphate (**pppGpp**).

GTP cleavage is essential for movement of messenger RNA along the ribosome during protein synthesis. When protein synthesis is inhibited, the ribosome still consumes GTP but does not cleave it to GDP and P_i. Instead, a ribosome-associated protein called *stringent factor* forms pppGpp from GTP and ATP, and ppGpp from GDP and ATP. The precise mechanisms by which these magic spot nucleotides control metabolism are still not known, but involve the regulation of ribosomal and transfer RNA synthesis.

Nucleotides as Alarmones

Bruce Ames has coined a term to describe compounds such as cAMP and ppGpp—**alarmones,** analogous to hormones in that they regulate cell metabolism, and to alarms in that they sense when all is not well and help the cell adjust to its new conditions.

In the mid-1980s Ames focused on a group of dinucleotides that are formed via the aberrant functioning of aminoacyl-tRNA synthetases, enzymes involved in protein synthesis. Under some circumstances a nucleotide can substitute for the normal tRNA substrate, with formation of adenylylated dinucleotides such as **diadenosine 5′,5‴-P^1,P^4-tetraphosphate** (Ap_4A) or its triphosphate analog, Ap_3A. Ames has proposed that Ap_3A represents, at least in bacteria, an alarmone that helps a cell respond to oxidative stress, imposed by agents such as hydrogen peroxide. There is evidence in animal cells that Ap_4A serves as a positive regulator for DNA replication. While the proposed roles for these compounds have not been established conclusively, the alarmone concept is appealing, and the discovery of new regulatory roles for nucleotides is likely to continue.

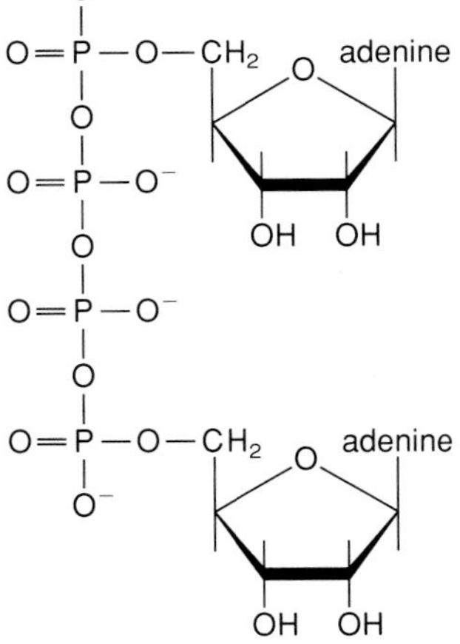

O⁻
O=P—O—CH2
O
adenine
O
O=P—O⁻
OH OH
O
O=P—O⁻
O
O=P—O—CH2
O
adenine
O⁻
OH OH

Ap_4A

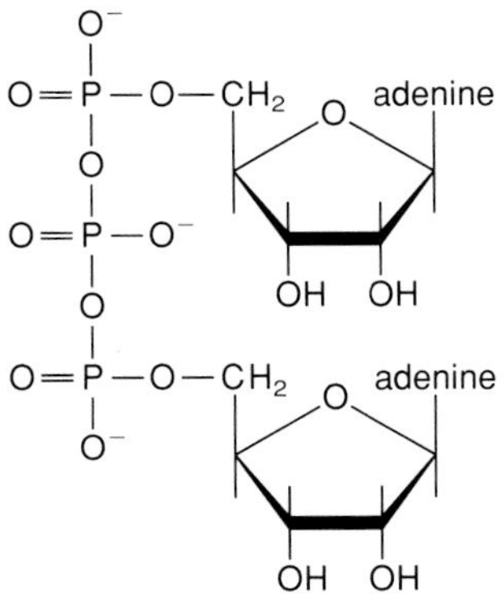

O⁻
O=P—O—CH2
O
adenine
O
O=P—O⁻
OH OH
O
O=P—O—CH2
O
adenine
O⁻
OH OH

Ap_3A

NAD$^+$ Metabolism, poly(ADP)-Ribosylation, and Programmed Cell Death

As we discuss in Chapter 25, when cells sustain DNA damage, a series of events is set in place to repair that damage. A number of additional events occur as well. The presence of gapped DNA structures, which contain lengthy single-stranded regions, somehow activates an enzyme called **poly(ADP)-ribosyltransferase.** This enzyme transfers single or multiple ADP-ribose moieties to carboxyl groups of nuclear proteins. The source of these moieties is NAD$^+$.

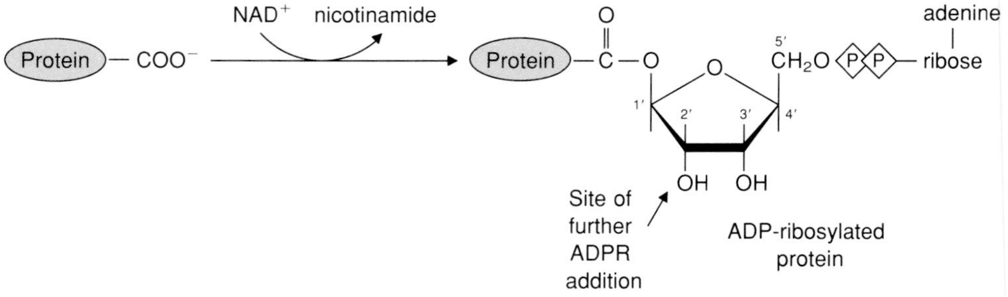

Protein—COO^- —(NAD$^+$ → nicotinamide)→ Protein—C(=O)—O—(1′)ribose(2′-OH, 3′-OH, 4′)—5′ CH_2O—P—P—ribose—adenine

Site of further ADPR addition (2′-OH)

ADP-ribosylated protein

This ADP-ribosylation can occur to such a great extent that pools of NAD$^+$ and its precursor, ATP, are depleted, and the cells die. Although this doesn't seem to make much biological sense, there is a likely rationale. Some forms of DNA repair are much less accurate than DNA replication. Thus, an extensively damaged cell may recover but sustain so many mutations that it is seriously impaired in function, or it becomes a cancer cell. Poly(ADP)-ribosylation has been proposed as an element of *programmed cell death*—a system ensuring that a cell will not survive if it is so badly damaged that its recovery would harm the organism. Like the alarmone concept, this model fits current data, but it awaits rigorous confirmation.

2′,5′-Oligoadenylate and the Action of Interferon

Interferons are a class of small animal glycoproteins ranging in molecular weight from 26,000 to 38,000. Their presence in cells induces an antiviral state in which the replication of many viruses is inhibited. Interferon synthesis is induced by double-stranded RNA, which is an intermediate in the replication of many RNA viruses.

Interferon acts in two ways: first, by direct inhibition of translation of viral messenger RNAs, and second, by induction of an endonuclease that specifically degrades viral messenger RNAs. Interferon stimulates the synthesis of an enzyme, **oligo-2′,5′-adenylate synthetase,** that is activated by double-stranded RNA. This enzyme catalyzes the synthesis of oligoadenylates (called 2–5A) that are linked not by 3′,5′ phosphodiester bonds, as in nucleic acids, but by 2′,5′ linkages. An example is the following trinucleotide. 2′,5′-Oligoadenylates activate an endogenous ribonuclease that degrades viral messenger RNA, thereby blocking the synthesis of viral proteins.

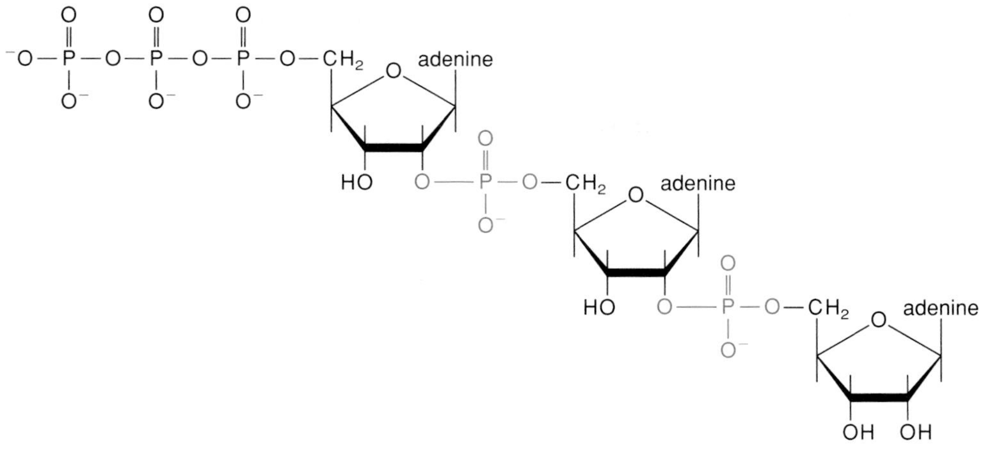

Tri-2′,5′-adenylate

REFERENCES

Enzymes of Nucleotide Metabolism

Blakley, R. L., and S. J. Benkovic (1984) *Folates and Pterins,* Vol. 1. This multiauthored book contains recent reviews on dihydrofolate reductase, purine metabolism, and pyrimidine biosynthesis.

Howell, E. E., J. E. Villafranca, M. S. Warren, S. J. Oatley, and J. Kraut (1986) Functional role of aspartic acid-27 in dihydrofolate reductase revealed by mutagenesis. *Science* 231:1123–1128. One of several penetrating analyses of the function of this enzyme using a combination of x-ray crystallography and site-directed mutagenesis.

Kornberg, A. (1989) *DNA Replication,* 2nd ed. Freeman, San Francisco. The first chapter of this book contains an excellent summary of nucleotide metabolism.

Nucleotides and Biological Regulation

Bochner, B. R., P. C. Lee, S. W. Wilson, C. W. Cutler, and B. N. Ames (1984) AppppA and related adenylylated nucleotides are synthesized as a consequence of oxidation stress. *Cell* 37:225–232. A paper from the Ames laboratory that describes ideas about alarmones and presents some data.

Daly, J. W. (1982) Adenosine receptors: Targets for future drugs. *J. Med. Chem.* 25:197–207. A nice review of the biochemistry and pharmacology of adenosine receptors.

de Serres, F. J. (ed.) (1985) *Genetic Consequences of Nucleotide Pool Imbalance.* Plenum, New York. Proceedings of a symposium, which discusses regulation of intracellular nucleotide levels and genetic and metabolic consequences of imbalanced pools.

Gaal, J. C., and C. K. Pearson (1986) Covalent modification of proteins by ADP-ribosylation. *Trends Biochem. Sci.* 11:171–175. Reviews several ADP-ribosylation processes in addition to its role in programmed cell death.

Zamecnik, P. C. (1983) Diadenosine $5',5'''$-P^1,P^4-tetraphosphate (Ap_4A): Its role in cellular metabolism. *Anal. Biochem.* 134:1–10. A review of the idea that this compound plays a regulatory role in DNA metabolism.

Nucleotide Analogs and Chemotherapy

Hardy, L. W., J. S. Finer-Moore, W. R. Montfort, M. O. Jones, D. V. Santi, and R. M. Stroud (1987) Atomic structure of thymidylate synthase: Target for rational drug design. *Science* 235:448–455. This article describes determination of the crystal structure of this enzyme and its implications.

Myers, C. E. (1981) Pharmacology of the fluoropyrimidines. *Pharmacol. Rev.* 33:1–15. Reviews the chemistry and biochemistry of these antimetabolites, as well as their use as drugs.

Robins, R. K. (1986) Synthetic antiviral agents. *Chem. Eng. News,* January 27, pp. 28–40. A comprehensive and timely review of the chemistry and biochemistry of antiviral agents, most of which act as nucleotide antimetabolites.

Stark, G., M. Debatisse, E. Giulotto, and G. M. Wahl (1989). Recent progress in understanding mechanisms of mammalian gene amplification. *Cell* 57:901–908. The most comprehensive recent treatment of this subject.

Deoxyribonucleotide Metabolism

Holmgren, A., C.-I. Bränden, H. Jörnvall, and B.-M. Sjöberg (1986) *Thioredoxin and Glutaredoxin Systems.* Raven, New York. Proceedings of a symposium that explored the roles of these proteins in ribonucleotide reductase, as well as their other roles.

Mathews, C. K., L. K. Moen, and R. G. Sargent (1988) Enzyme interactions in deoxyribonucleotide synthesis. *Trends Biochem. Sci.* 13:394–397. A discussion of the evidence for multienzyme complexes and substrate channeling in DNA precursor biosynthesis.

Reichard, P. (1988) Interactions between deoxyribonucleotide and DNA synthesis. *Annu. Rev. Biochem.* 57:349–375. A review discussing the extent to which DNA replication is coordinated with the synthesis of its precursors.

Reichard, P., and A. Ehrenberg (1983) Ribonucleotide reductase, a radical enzyme. *Science* 221:514–519. A review of the mechanism of action of this enzyme, with references to papers describing its regulation.

PROBLEMS

1. Identify each reaction catalyzed by (a) a nucleotidase; (b) a phosphorylase; (c) a phosphoribosyltransferase.

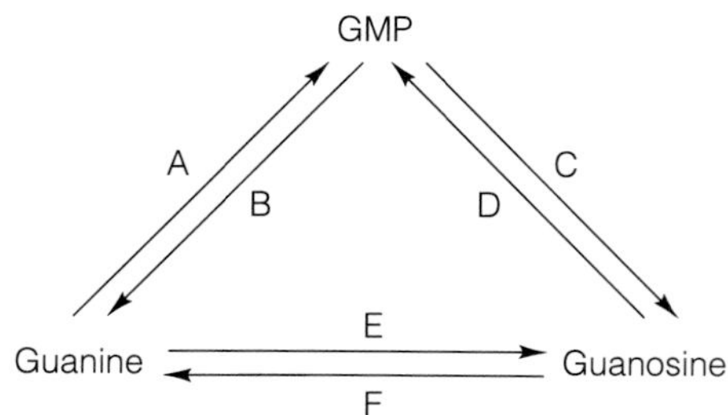

2. Describe the metabolic fate of 5-bromodeoxyuridine. Would you expect BrdUrd to *inhibit* DNA replication? Briefly explain your answer.

3. Predict the effects of the following compounds on intracellular nucleoside triphosphate levels. For each answer plot percent initial nucleotide level as a function of time after administration of the agent, for each of the four nucleotides (semiquantitatively). Consider not only the primary effect of the compound but also any indirect effects caused by nucleotide accumulation or depletion on allosteric enzymes.
   (a) Effect of thymidine on dNTP levels.
   (b) Effect of trimethoprim on bacterial rNTP levels.
   (c) Effect of fluorodeoxyuridine on dNTP levels.
   (d) Effect of hydroxyurea on dNTP levels.
   (e) Effect of azaserine on rNTP levels.

4. Radioactive uracil can be used to label all of the pyrimidine residues in DNA. Using either names or structures, present pathways for conversion of uracil to dTTP and to dCTP. For each reaction show the involvement of cofactors, and identify sites of allosteric regulation.

5. Similarly, hypoxanthine (HX) can be used to label purine residues. As in Problem 4, write reactions showing the conversion of hypoxanthine to dATP and dGTP.

6. Leukemia is a neoplastic (cancerous) proliferation of white blood cells. Clinicians are currently testing deoxycoformycin, an adenosine deaminase inhibitor, as a possible antileukemic agent. Why might one expect this to be effective?

7. Under what conditions might one expect a deficiency of hypoxanthine–guanine phosphoribosyltransferase to affect the rate of *pyrimidine* nucleotide biosynthesis? How might one estimate the rate of pyrimidine nucleotide biosynthesis in living animals or people?

8. A classic way to isolate thymidylate synthase-negative mutants of bacteria is to treat a growing culture with thymidine and trimethoprim. Most of the cells are killed, and the survivors are greatly enriched in thymidylate synthase-negative mutants.
   (a) What phenotype would allow you to identify these mutants?
   (b) What is the biochemical rationale for the selection? (That is, why are the mutants not killed under these conditions?)
   (c) How would the procedure need to be modified to select mammalian cell mutants defective in thymidylate synthase?

9. As stated in the text, mammalian cells can become resistant to the lethal action of methotrexate by amplifying the dihydrofolate reductase gene copy number, so that intracellular enzyme levels become very high. What other biochemical or genetic changes in cells could cause them to become resistant to methotrexate?

10. As stated in the text, bacteriophages have been discovered with the following base substitutions in their DNA: (a) dUMP completely substituting for dTMP; (b) 5-hydroxymethyl-dUMP completely substituting for dTMP; (c) 5-methyl-dCMP completely substituting for dCMP. For any one of these cases, formulate a set of virus-coded enzyme activities that could lead to the observed substitution. Write a balanced equation for each reaction you propose.

11. Radioisotope "suicide techniques" are often used in the selection of mutants. In one such technique cells are grown in the presence of [^{3}H]thymidine at very high specific activity. The cells are then stored frozen to allow for decay of some of the incorporated radioactivity. Decay of dTMP residues in DNA causes strand breakage and other potentially lethal events. Thus, cells that have incorporated thymidine into their DNA are very likely to be killed by this regimen.
    (a) Identify an enzyme deficiency that would allow a mutant to survive such a regimen.
    (b) Select any enzyme of nucleotide metabolism, and devise a suicide procedure that could select for mutants deficient in that enzyme.

12. The alarmone concept postulates that a small molecule, such as a nucleotide, accumulates in response to specific stress conditions and that it controls metabolic steps involved in response to that stress. If a given nucleotide is shown to accumulate during a given stress, discuss what must be done to show that it is acting as an alarmone.

CHAPTER 23

Integration and Control of Metabolic Processes

This chapter represents a transition in our study of biochemistry. Our emphasis in Chapters 12 through 22 has been on metabolism of small molecules and the aspects of metabolism most directly related to the generation, storage, and utilization of energy. Our emphasis in the last section of this book is on biological information and the metabolism of large molecules. However, metabolism is a continuum, with thousands of chemical reactions occurring simultaneously, all the time, in one living organism. Any distinctions made in organizing the presentation of biochemistry are artificial; they are made for the convenience of human beings who teach and study biochemistry.

In this chapter we focus on intermediary metabolism in terms of the functioning of a differentiated, multicellular organism. The primary emphasis is on vertebrate animals, because they have been the objects of the most intensive study. Where appropriate, we will describe findings from plant physiology, invertebrate zoology, and microbiology that have contributed to our current understanding.

In presenting the major pathways of energy metabolism in earlier chapters our major focus has been on the cell and its individual enzymes and other components. Here we review the metabolic profiles of the major organs: which fuels they use, which they generate, and how they interact under stress to maintain appropriate energy balance. The interactions among organs and tissues are controlled in large part by hormonal signals, and we have already discussed some of the mechanisms involved. Our goal here is to describe the broader area of chemical signals *between cells* and how they regulate biological functions in general. Where detailed information is available, we will describe molecular mechanisms for the chemical transmission of messages from cell to cell.

Keep in mind that metabolism is controlled to a very great extent by the availability of substrates for specific metabolic pathways. In general, substrate concentrations within cells fall below saturating levels for the enzymes that metabolize them. This means that reaction fluxes through particular enzymes vary as the concentrations of the substrate vary. A good example is the metabolic adaptation that occurs during a marathon run. Once the glycogen stores in liver and muscle are exhausted, flux through glycolysis decreases in muscle, not for any hormonal reason but simply because glucose phosphates are less available. Hormonal adjustments then occur to allow utilization of fatty acids in muscle, but the primary factor in the metabolic decision made by the cell—what substrate to catabolize for energy—is the concentration of each of the utilizable substrates.

Interdependence of the Major Organs in Vertebrate Fuel Metabolism

In this section we look at metabolism not in just one cell but as the totality of chemical reactions in a complex multicellular animal. We emphasize the specialized roles played by each of the major organs in fuel metabolism—brain, muscle, liver, adipose tissue, and heart—and we describe the varying relationships among these organs as the animal encounters different physiological conditions.

In a differentiated organism each tissue must be provided with fuels that it can utilize, in amounts sufficient to meet its own energy needs and to perform its specialized roles. The kidney, for example, must generate ATP for the osmotic work of transporting solutes against a concentration gradient for excretion. Muscle must generate ATP for the mechanical work of contraction, and in the heart that energy supply must be continuous. The liver generates ATP for biosynthetic purposes, whether for plasma protein synthesis, cholesterol generation, fatty acid synthesis, gluconeogenesis, or the production of urea for nitrogen excretion. Energy production must meet needs that vary widely, depending on level of exertion, composition of fuel molecules in the diet, time since last feeding, and so forth. For example, in humans the daily caloric intake may vary by fourfold, depending on the level of exertion—from 1500 to 6000 kcal/day in an average-sized human or, in the language of thermodynamics, from 6000 to 25,000 kJ/day.

The major organs involved in fuel metabolism vary in their levels of specific enzymes, so that each organ is specialized for the storage, utilization, and generation of different fuels. The major fuel depots are *triacylglycerols,* stored primarily in adipose tissue; *protein,* most of which exists in skeletal muscle; and *glycogen*, which is stored in both liver and muscle. Now let us review how the mobilization of each depot is controlled, and how the organs involved communicate with each other, to meet the energy needs of the animal. This information is summarized in Figure 23.1.

Metabolic Division of Labor Among the Major Organs

BRAIN. The brain is the most fastidious, and one of the most voracious, of organs. It must generate ATP in large quantities to maintain the membrane potentials essential for transmission of nervous impulses. Under normal conditions the brain utilizes only glucose to meet its prodigious energy requirement, which amounts to about 60% of the glucose consumption of a human at rest. The brain's need for about 120 grams of glucose per day is equivalent to 1760 kJ—about 15% of the total energy consumed. The quantitative requirement for glucose remains quite constant, even when an animal is at rest or asleep. The brain is a highly aerobic organ, and its metabolism demands some 20% of the total oxygen consumed by a human. Since the brain has no significant glycogen or other fuel reserves, the supply of both oxygen and glucose cannot be interrupted, even for a short time. However, the brain can adapt during fasting, to use ketone bodies instead of glucose as a major fuel.

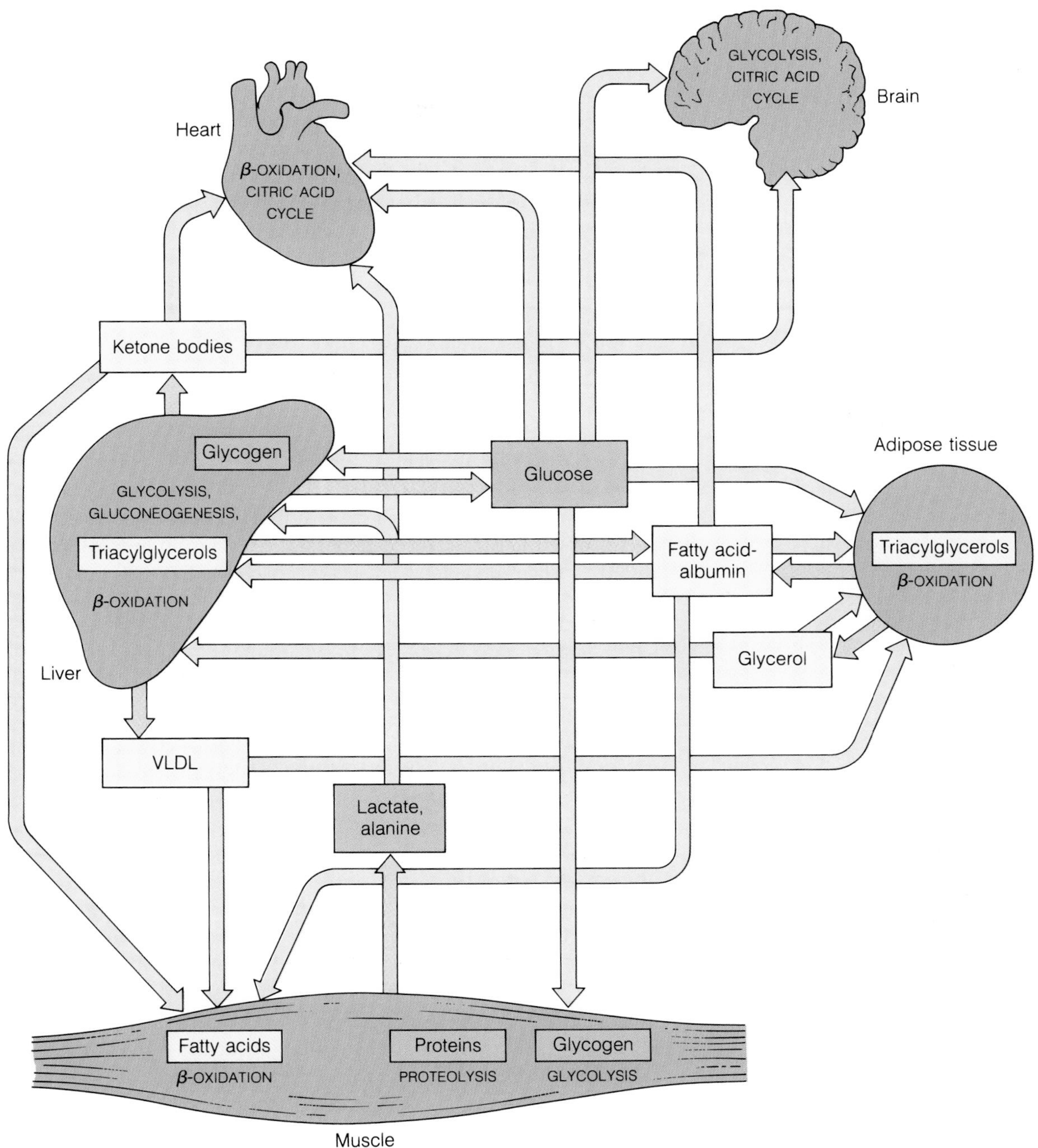

Figure 23.1
Metabolic interactions among the major fuel-metabolizing organs. The figure emphasizes the major fuel metabolites imported (blue arrows) and exported (gray arrows) by each organ and the major energy pathways in each organ.

MUSCLE. Muscle can utilize a variety of fuels—glucose, fatty acids, or ketone bodies. Skeletal muscle varies widely in its energy demands and the fuels it consumes, in line with its wide variations in activity. In resting muscle, fatty acids represent the major energy source; during exertion, glucose is the primary source. In the early stages of a period of exertion the glucose comes from mobilization of the muscle's glycogen reserves. Skeletal muscle stores about three-fourths of the total glycogen in humans, with most of the rest being stored in the liver. However, glucose released from muscle glycogen cannot be released from the cell for utilization by other tissues. This is because muscle lacks the enzyme glucose-6-phosphatase, so glucose phosphates derived from glycogen cannot be converted to glucose and released from the cell.

During exertion the flux rate through glycolysis exceeds that through the citric acid cycle, so lactate accumulates and is released. Another metabolic product is alanine, produced via transamination from pyruvate in the glucose–alanine cycle. Both lactate and alanine are transported through the bloodstream to the liver, where they are reconverted through gluconeogenesis to glucose, for return to the muscle and other tissues by the Cori cycle.

Muscle contains another readily mobilizable source of energy—its own protein. However, protein represents an inefficient form of metabolic energy. Moreover, the breakdown of muscle protein to meet energy needs is harmful to an animal, which must move about in order to survive.

Finally, recall that muscle has an additional energy reserve in creatine phosphate, which generates ATP without the need for metabolizing fuels. This reserve is exhausted early in a period of exertion and must be replenished, along with glycogen stores, as muscle rests after prolonged exertion.

Heart. The metabolism of heart muscle differs from that of skeletal muscle in three important respects. First, the variation in work output is far less than that seen in skeletal muscle. Second, the heart is a completely aerobic tissue, whereas skeletal muscle can function anaerobically for limited periods. Mitochondria are much more densely packed in heart than in other cells, constituting nearly half the volume of a heart cell. Third, the heart contains negligible energy reserves as glycogen or lipid, although there is a small amount of creatine phosphate. Therefore, the supply of both oxygen and fuels from the blood must be continuous to meet the unending energy demands of the heart. The heart uses a variety of fuels—glucose, fatty acids, lactate, and ketone bodies.

Adipose Tissue. Adipose tissue represents the major fuel depot for an animal; the total stored triacylglycerols amount to some 565,000 kJ in an average-sized human. This is enough fuel, metabolic complications aside, to sustain life for a couple of months in the absence of further caloric intake.

The adipose cell is designed for continuous synthesis and breakdown of triacylglycerols, although breakdown is stimulated by the activation of hormone-sensitive lipase. Because adipose cells lack the enzyme glycerol kinase, some glucose catabolism must occur for triacylglycerol synthesis to take place—specifically, the formation of dihydroxyacetone phosphate, for reduction to glycerol 3-phosphate. Glucose acts as a sensor in adipose tissue metabolism: when glucose levels are adequate, continuing production of dihydroxyacetone phosphate generates enough glycerol 3-phosphate for resynthesis of triacylglycerols from the released fatty acids. When intracellular glucose levels fall, the concentration of glycerol 3-phosphate falls also, and fatty acids are released from the adipocyte for export as the albumin complex to other tissues.

Liver. A major metabolic role of liver is the synthesis of fuel components for utilization by other organs. In fact, most of the low-molecular-weight metabolites that appear in the blood through digestion are taken up by the liver for this metabolic processing. The liver is a major site for fatty acid synthesis. It also produces glucose, both from its own glycogen stores and from gluconeogenesis, the latter using lactate from muscle, glycerol from adipose tissue, and the amino acids not needed for protein synthesis. Ketone

Table 23.1
Profiles of the major vertebrate organs in fuel metabolism

Tissue	Fuel Store	Preferred Fuel	Fuel Sources Exported
Brain	None	Glucose (ketone bodies during starvation)	None
Skeletal muscle (resting)	Glycogen	Fatty acids	None
Skeletal muscle (exertion)	None	Glucose	Lactate
Heart muscle	Glycogen	Fatty acids	None
Adipose tissue	Triacylglycerol	Fatty acids	Fatty acids, glycerol
Liver	Glycogen, triacylglycerol	Amino acids, glucose, fatty acids	Fatty acids, glucose, ketone bodies

bodies for use by the heart are also manufactured largely in the liver. In liver the level of malonyl-CoA, which is related to the energy status of the cell, is a determinant of the fate of fatty acyl-CoAs. When fuel is abundant, malonyl-CoA accumulates and inhibits carnitine acyltransferase I, preventing the transport of fatty acyl-CoAs into mitochondria for β-oxidation and ketogenesis. On the other hand, shrinking malonyl-CoA pools signal the cells to transport fatty acids into the mitochondrion, for generation of energy and fuels.

An important role of liver is to buffer the level of blood glucose. It does this through the action of glucokinase, an enzyme peculiar to liver, with a high K_M (about 10 mM) for glucose. Thus, liver alone is able to respond to high blood glucose levels by increasing the phosphorylation of glucose, which results eventually in its deposition as glycogen. Glucose-6-phosphate accumulation activates the D form of glycogen synthetase. In addition, glucose itself binds to glycogen phosphorylase *a*, increasing the susceptibility of phosphorylase *a* to dephosphorylation, with consequent inactivation. Thus, in addition to hormonal effects, described below, liver senses the fed state and acts to store fuel derived from glucose. Liver also senses the fasted state and increases the synthesis and export of glucose when blood glucose levels are low. (Other organs also sense the fed state, notably the pancreas, which adjusts its glucagon and insulin outputs accordingly.)

To meet its internal energy needs, the liver can use a variety of fuel sources, including glucose, fatty acids, and amino acids.

Fuel metabolism in the major vertebrate organs is summarized in Table 23.1.

Hormonal Regulation of Fuel Metabolism

In animals a supremely important regulatory function is the maintenance of blood glucose levels within rather narrow limits, particularly for proper functioning of the nervous system. Of course, blood glucose levels vary, depending on nutritional status. Several hours after a meal, the normal level in humans is about 80 mg % (80 mg per 100 ml of blood), or 4.4 mM. Shortly after a meal, that level might rise to 120 mg %. In response, homeostatic mechanisms come into play to promote uptake of glucose into cells and its utilization by tissues. Similarly, when glucose levels fall several hours after a meal, other mechanisms promote both glucose release, from intracellular glycogen stores, and gluconeogenesis, so that the normal level is maintained. Some of the homeostatic mechanisms were mentioned in the previous section; others involve hormonal regulation. Though we discuss hormone action in some detail later in this chapter, it is appropriate to look here at some of the hormones involved in fuel metabolism.

Actions of the Major Hormones

The most important hormone promoting glucose utilization is insulin, while both glucagon and epinephrine act to increase blood glucose levels (Figure 23.2). The major effects of these agents are summarized in Table 23.2.

INSULIN. Insulin is a 5.8-kDa protein that is synthesized in the pancreas. The pancreas has both **endocrine** cells, which secrete hormones directly into the bloodstream, and **exocrine** cells, which secrete zymogen precursors of digestive enzymes into the upper small intestine. The endocrine tissue, which takes the form of cell clusters known as islets of Langerhans, contains at least four different cell types, each capable of synthesizing one hormone. The A cells produce glucagon, the D cells **somatostatin,** and the P cells a recently discovered **pancreatic hormone.** Insulin is synthesized in the B cells.

The simplest way to describe the several actions of insulin is to say that *it signals the fed state* and thereby acts to promote the (1) uptake of fuel substrates into some cells, (2) storage of fuels—lipids and glycogen, and (3) biosynthesis of macromolecules—the production of nucleic acids and protein. Its various effects include an increase in the permeability of muscle and adipose tissue to glucose, with increased uptake of glucose; activation of glycolysis in liver; increased synthesis of fatty acids and triacylglycerols in liver and adipose tissue; inhibition of gluconeogenesis in liver; increased glycogen synthesis in liver and muscle; increased uptake of amino acids into muscle with consequent activation of muscle protein synthesis; and inhibition of protein degradation.

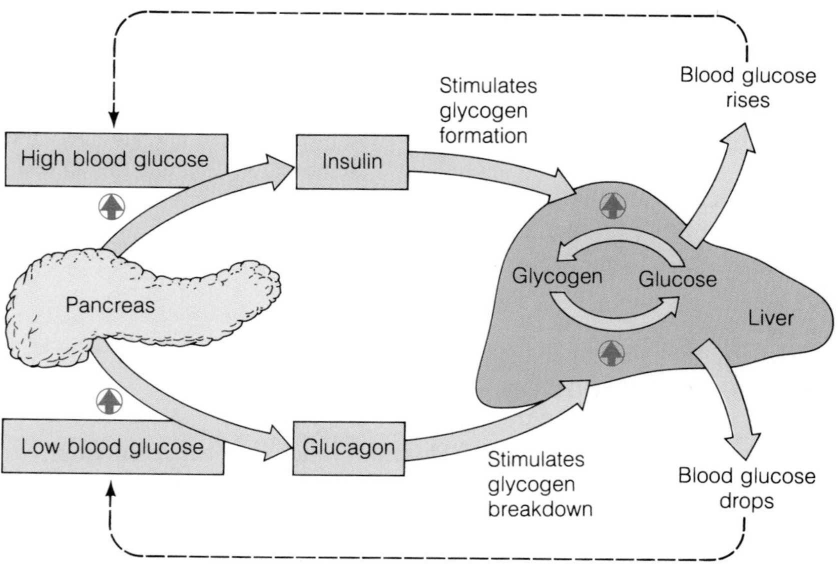

Figure 23.2
Control of blood glucose levels by pancreatic secretion of insulin and glucagon. Green: conditions resulting from high glucose levels; pink: conditions resulting from low glucose levels.

GLUCAGON. A 3.5-kDa polypeptide, glucagon is synthesized by A cells of the islets of Langerhans in the pancreas. These endocrine cells sense the blood glucose concentration and release the hormone in response to low levels (see Figure 23.2). The primary target of glucagon is the liver, and its principal effect is to increase cyclic AMP levels in liver cells. The resultant metabolic cascades, discussed in Chapters 13 and 16, promote glycogenolysis and inhibit glycogen synthesis. In addition, by activating the hydrolysis of fructose-2,6-bisphosphate, cAMP activates glycolysis and inhibits gluconeogenesis. Glucagon also brings about inhibition of pyruvate kinase in the liver, causing phosphoenolpyruvate to accumulate. The level of pyruvate decreases, both because its synthesis from PEP is blocked and because it continues to be converted to PEP, via the pyruvate carboxylase and phosphoenolpyruvate carboxykinase reactions. Accumulation of phosphoenolpyruvate promotes gluconeogenesis, while the inhibition of pyruvate kinase diminishes the glycolytic flux rate.

Glucagon also raises cAMP levels in adipose tissue, where its chief effect is to promote triacylglycerol mobilization.

CATECHOLAMINES. The catecholamines epinephrine and norepinephrine, when released from presynaptic nerve endings, function as neurotransmitters (Chapters 21 and 29). When released from adrenal medulla in response to low blood glucose levels, they interact with second messenger systems in many tissues, with varied effects. In muscle, catecholamines activate adenylate cyclase, with concomitant activation of glycogenolysis and inhibition of glycogen synthesis. Triacylglycerol breakdown is also stimulated, providing fuel for the muscle tissue. In consequence, glucose uptake into muscle is

Table 23.2
Major hormones controlling fuel metabolism in animals

Hormone	Biochemical Actions	Physiological Actions
Insulin	↑Cell permeability to glucose and amino acids	Signals fed state
	↑Glycolysis	↑Fuel storage
	↑Glycogen synthesis	↓[Blood glucose]
	↓Gluconeogenesis	↑Cell growth and differentiation
	↑Triacylglycerol synthesis	
	↓Lipolysis	
	↓Protein degradation	
	↑Protein, DNA, and RNA synthesis	
Glucagon	↑[cAMP] in liver and adipose tissue	↑Glucose release from liver
	↑Glycogenolysis	↑[Blood glucose]
	↓Glycogen synthesis	
	↑Triacylglycerol hydrolysis	
Catecholamines	↑[cAMP] in muscle	↑Glucose release from liver
	↑Triacylglycerol mobilization	
	↑Glycogenolysis	↓Glucose utilization by muscle
	↓Glycogen synthesis	↑[Blood glucose]

diminished, and in turn blood glucose levels rise. Epinephrine also inhibits insulin secretion and stimulates glucagon secretion; these effects tend to increase glucose production and release by the liver. The net result is to increase blood glucose levels. Unlike glucagon, the catecholamines have short-lived metabolic effects.

Responses to Metabolic Stress: Starvation, Diabetes

An excellent way to understand how the interorgan and hormonal relationships we have discussed actually integrate fuel metabolism is to examine the effects of metabolic stress. In this section we consider two examples—prolonged fasting, where the intake of fuel substrates is inadequate; and **diabetes mellitus,** where a functional insufficiency of insulin impairs the ability of the body to utilize glucose, even when it is present in abundance.

First let us review how glucose levels are maintained during normal feeding cycles (Figure 23.3). The elevated blood glucose levels shortly after a carbohydrate-containing meal stimulate the secretion of insulin and suppress the secretion of glucagon. Together these effects promote uptake of glucose into the liver, stimulate glycogen synthesis, and suppress glycogen breakdown. Glucokinase activity increases in response to elevated glucose levels, providing substrates for glycogen synthesis. In addition, activation of acetyl-CoA carboxylase stimulates fatty acid synthesis, with much that is synthesized in liver being transported to adipose tissue, as triacylglycerols in very low-density lipoproteins. There, the increased levels of glycolytic intermediates act to stimulate triacylglycerol synthesis. Finally, increased glucose uptake into muscle stimulates glycogen synthesis in that tissue as well.

Several hours later, when blood glucose levels begin to fall, the above events are reversed. Insulin secretion slows, and glucagon secretion increases. This promotes glycogen mobilization via the cAMP-dependent cascade mechanisms that activate glycogen phosphorylase and inactivate glycogen synthetase. Triacylglycerol breakdown is activated as well, via the action of hormone-sensitive lipase, generating fatty acids for use as fuel by liver and muscle. At the same time the decrease in insulin levels reduces glucose utilization by muscle, liver, and adipose tissue. Consequently, glucose produced in the liver is largely exported to the blood, allowing its continued utilization by the brain.

STARVATION. Suppose that food intake is denied not just for a few hours, as described above, but for many days. Since a 70-kg human can store at most the equivalent of 6700 kJ of energy as glycogen, this supplier of blood glucose will be exhausted in just a few hours. Since it is critical for brain function that blood glucose levels be maintained near 4.4 mM, how does the organism adapt in order to continue to provide that glucose? It does so by adjusting metabolically to increase the utilization of fuels other than carbohydrate.

Before we discuss those adjustments, recall the other major energy stores: about 565,000 kJ as triacylglycerol, largely in adipose tissue, and 100,000 kJ as mobilizable proteins, largely in muscle. These stores provide sufficient energy to permit survival for up to several months. However, utilization of both stores presents problems. Triacylglycerol mobilization generates energy largely as acetyl-CoA, whose further oxidation in the citric acid cycle requires oxaloacetate. Recall from Chapter 14 that oxaloacetate and other citric acid cycle intermediates are used in other metabolic reactions and must be replenished via anaplerotic pathways. The most important of these processes is the pyruvate carboxylase reaction, with most of the

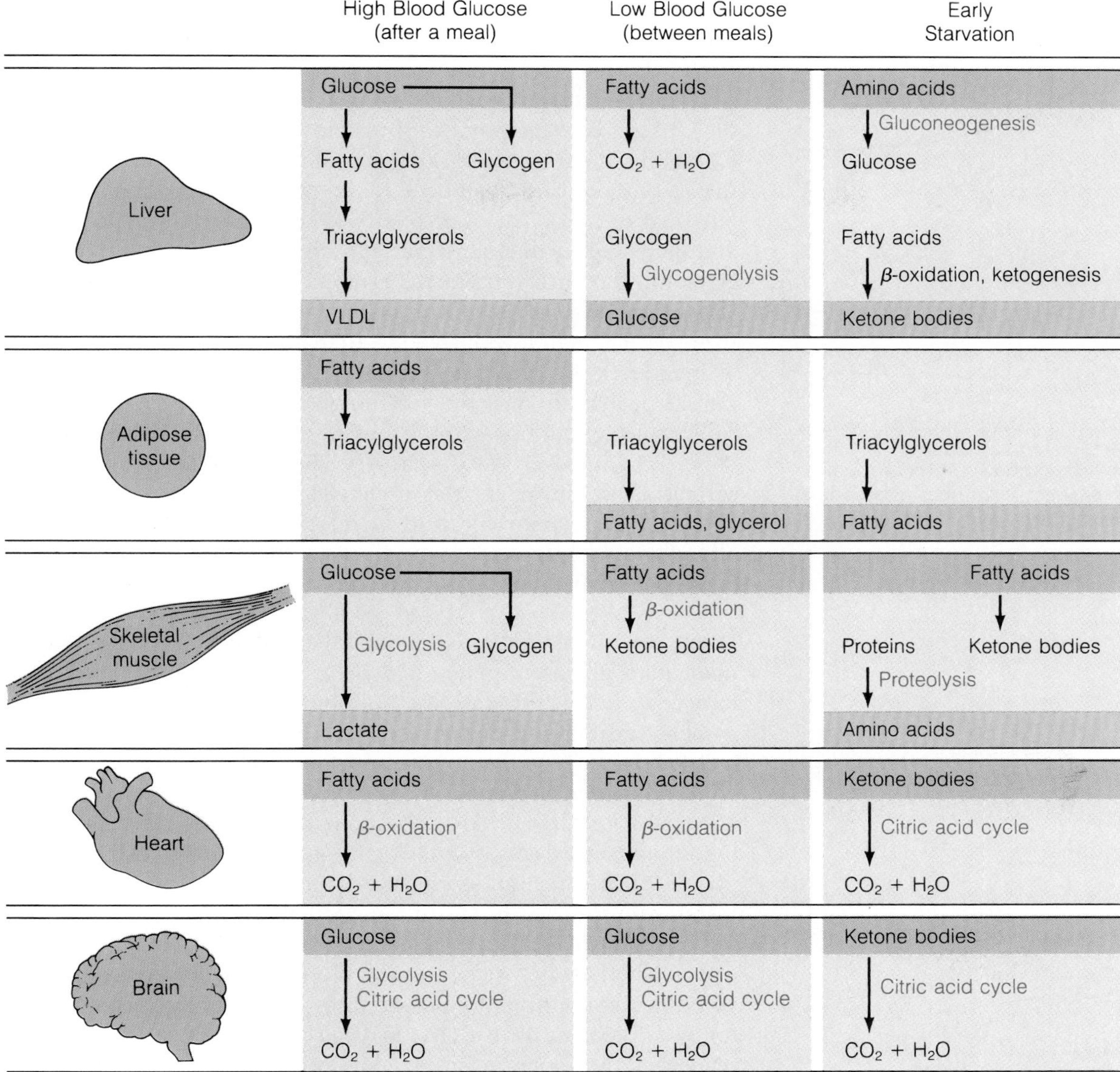

Figure 23.3
Major events in fuel storage, retrieval, and utilization in fed and unfed states. Purple indicates fuels imported into the tissue; green indicates fuels exported from the tissue.

pyruvate coming from carbohydrate catabolism. When carbohydrate availability is limited, the resupply of citric acid cycle intermediates is limited, and flux through the cycle may be reduced.

During carbohydrate limitation, citric acid cycle intermediates can be provided from other sources. For example, the glycerol released from lipolysis can be used but is not produced in amounts adequate to maintain levels of citric acid cycle intermediates. Alternatively, these intermediates can be produced from protein catabolism and transamination; however, this is energetically wasteful, and it has the undesirable effect of wasting the muscle and weakening the fasting subject. Nevertheless, during the first few days of starvation proteolysis is accelerated, because amino acids for protein synthesis are not present in sufficient amounts to counterbalance protein turnover, which continues at normal rates.

A major fate of the amino acids released is gluconeogenesis, as the body attempts to cope with the absence of glycogen stores by synthesizing

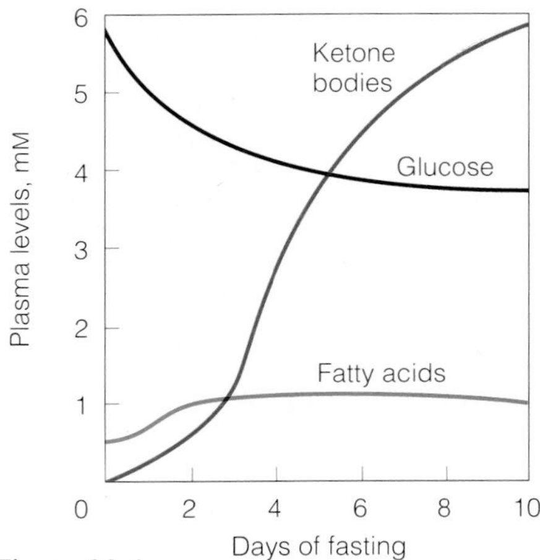

Figure 23.4
Metabolic adaptation to starvation, as shown by changes in human plasma levels of glucose, fatty acids, and ketone bodies.

its own glucose. During this time the liver and muscle are shifting to fatty acids as the dominant fuels for their own use.

Meanwhile, the increased use of carbon for gluconeogenesis diminishes the amount of oxaloacetate available to combine with acetyl-CoA in the citric acid cycle. Since fat breakdown has been activated, both acetyl-CoA and reduced electron carriers accumulate in the liver to the point where the acetyl-CoA cannot all be oxidized, and ketone bodies begin to accumulate (Figure 23.4). The brain adapts to the somewhat reduced glucose levels and increased ketone body concentrations by beginning to utilize acetoacetate and β-hydroxybutyrate. This trend continues for the duration of starvation; on the third day the brain derives about one-third of its energy needs from ketone bodies, while by day 40 that usage has increased to two-thirds. This reduces the need for gluconeogenesis and spares the mobilization of muscle protein. In fact, the loss of muscle protein *decreases* by about fourfold late in starvation—from about 75 grams consumed per day on day 3 to about 20 grams per day on day 40. The metabolic changes accompanying starvation compromise the organism's abilities to respond to further stresses, such as extreme cold or infection. However, the adaptations do allow life to continue for many weeks without food intake, the total period being determined largely by the size of the fat deposits.

Diabetes. In starvation, glucose utilization is abnormally low because of inadequate glucose supplies. In diabetes mellitus, glucose utilization is similarly low, but in this case the hormonal stimulus to glucose utilization—namely, insulin—is defective. Glucose is actually present in excessive amounts, and its spilling over into urine is what gives the disease its name, which comes from the Latin for "excessive sweetened urine." The consequences of insulin deficiency are comparable to those of starvation in revealing important aspects of interorgan metabolic relationships.

Diabetes mellitus is the third leading cause of death in the United States, affecting nearly 5% of the population. It is not a single disease, but rather a family of diseases. The major cause of juvenile, or insulin-dependent, diabetes is destruction of the B cells of the pancreas, probably due to viral infection or an autoimmune disorder or a combination of both. Some forms of diabetes have a genetic origin. Mutations in insulin structure can render the hormone inactive, while other mutations cause defects in the conversion of preproinsulin or proinsulin to the active hormone. Still other genetic forms of the disease involve defects in structure of the insulin receptor or in its intracellular activities that promote glucose utilization.

Whatever the cause of the functional insulin deficiency, diabetes mellitus can truly be called "starvation in the midst of plenty." The failure of insulin to act normally in promoting glucose utilization, with resultant glucose accumulation in the blood, starves the cells of nutrients, promoting metabolic responses similar to those of fasting (Figure 23.5). Liver cells attempt to generate more glucose by stimulating gluconeogenesis. Most of the substrates come from amino acids, which in turn come largely from degradation of tissue proteins. Glucose cannot be reutilized for resynthesis of amino acids or of fatty acids, so a diabetic may lose weight even while consuming what would normally be adequate calories in the diet.

As cells attempt to generate usable energy sources, triacylglycerol depots are mobilized. Fatty acid oxidation is elevated, with concomitant generation of acetyl-CoA. Flux through the citric acid cycle may decrease, because of the accumulation of reduced electron carriers. In liver both effects accelerate ketone body formation, generating increased levels of organic

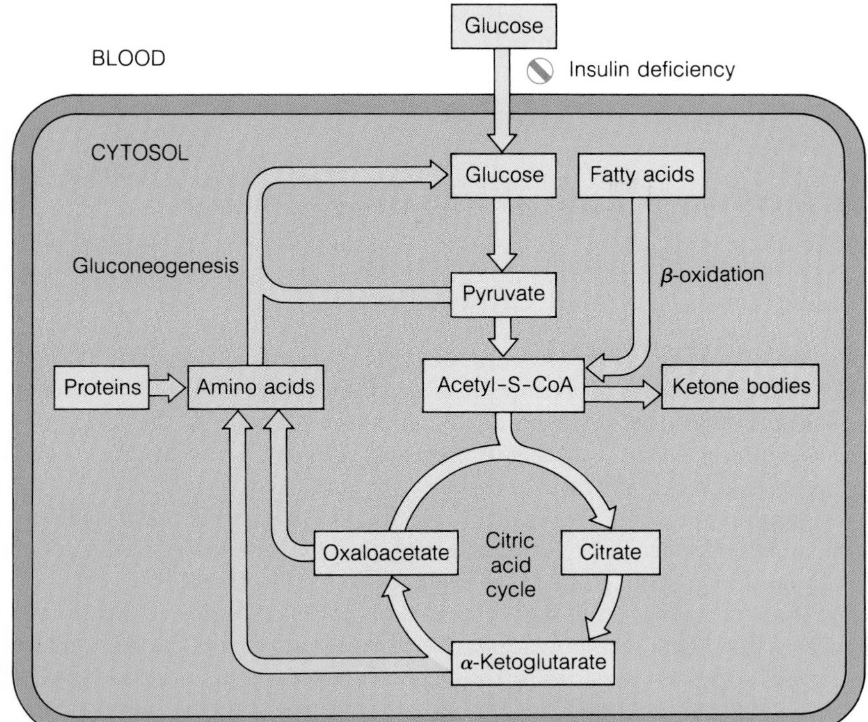

Figure 23.5
Metabolic aberrations in diabetes. Insulin deficiency reduces glucose catabolism. Gluconeogenesis from amino acids and citric acid cycle intermediates is then stimulated, as the cell attempts to remedy its perceived lack of utilizable glucose. Fatty acid oxidation and ketogenesis are increased. Red = pathways activated; green = pathways diminished in insulin deficiency.

acids in the blood. These can lower the blood pH from the normal value of 7.4 to levels as low as 6.8. Decarboxylation of acetoacetate, which is stimulated at low pH, generates acetone, which can be smelled on the breath of patients in severe uncontrolled diabetic situations. A special danger is that such people may lose consciousness and this, coupled with a sweet organic odor on the breath, may give the impression that they are drunk, when in fact their very lives are in jeopardy.

The excessive concentrations of glucose in body fluids generate other metabolic problems, quite different from anything seen in starvation. At blood glucose levels above 10 mM the kidney can no longer reabsorb all of the glucose out of the blood filtrate, and glucose is spilled into the urine, sometimes in amounts close to 100 grams per day. Large amounts of water are excreted along with the glucose, and under these conditions the kidney cannot reabsorb most of this water. In fact, the earliest indications that a person has diabetes are often frequent and excessive urination, coupled with excessive thirst. Long before biochemistry was a science, the loss of nutrients, excessive urination, and breakdown of fat and protein were recognized as hallmarks of diabetes; as early as the second century A.D. diabetes was described as a "melting-down of the flesh and limbs into urine."

When diabetes strikes in childhood (the juvenile, or insulin-dependent form of the disease, representing about 10% of all cases), the metabolic imbalance is usually more severe and difficult to control than in the milder and more common adult-onset form. The latter can often be controlled by diet, whereas treatment for juvenile diabetes usually involves daily self-injection of insulin. For many years this insulin was purified from bovine pancreas, and its high cost, coupled with occasional problems resulting from the minor structural differences between human and bovine insulin, led the fledgling biotechnology industry to attempt to produce human insulin through recombinant DNA techniques. In the late 1970s the gene for human insulin was cloned into *E. coli* in a form that allowed it to be expressed, and in 1982 cloned human insulin became the first recombinant DNA product to be approved for human use.

Extracellular Chemical Signals

Pheromones, Hormones, Prostanoids, Neurotransmitters, and Growth Factors

Coordination of the metabolic activities of various organs and tissues necessarily entails the action of extracellular signals—not only hormones but also neurotransmitters, growth factors, and pheromones. Each of these is a substance synthesized in one class of cells and released, to be transmitted to **target** cells and control their activities.

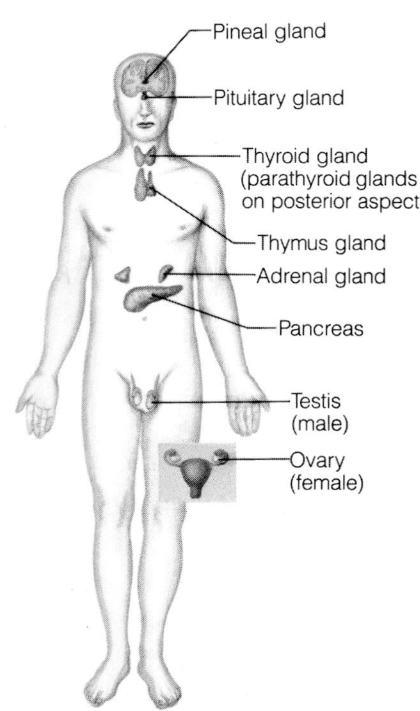

Figure 23.6
The major human endocrine glands and control centers.

Insulin, glucagon, and epinephrine are all *hormones,* a term that was coined in 1904 to describe **secretin,** a substance that is released in the upper small intestine and that acts in the stomach to stimulate the flow of gastric juice as an aid to digestion. Early research on hormones revealed little about how they act, but it did show fundamental similarities among a great many different hormones. First, they are secreted by specific tissues, now called *endocrine glands.* Second, they are secreted directly into the bloodstream, rather than being excreted through ducts or stored in bladders. Thus, the response to a hormonal signal comes as a direct result of its secretion. Figure 23.6 shows the locations of the major endocrine organs in the human body.

Hormones stimulate metabolic activities in tissues remote from the secretory organ. They are active at exceedingly low concentrations, in the micromolar to picomolar ranges. Further, most hormones are metabolized rapidly, with consequent inactivation. Hence, their effects are often short-lived.

The low concentrations of hormones and their metabolic lability have made it difficult to assay levels of a particular hormone, which is essential if we want to elucidate its mechanism of action. Until the 1960s it was usually necessary to use a bioassay. For example, **oxytocin,** which stimulates uterine contractions in labor, was assayed by adding hormone to strips of uterine muscle and measuring the length of the strips before and after administration of hormone. The introduction of **radioimmunoassay** revolutionized the field of hormone assay; this technique is described in Tools of Biochemistry 17.

A special class of hormones includes the prostanoids—prostaglandins, thromboxanes, and leukotrienes, which we discussed in Chapter 18. These mediators act like hormones but are distinctive in their extreme metabolic lability, their synthesis in many cell types, and their actions primarily on cells close to those that secreted them.

Another class of intercellular mediators, the **pheromones,** are distinctive in that they are transmitted between cells in different organisms—usually organisms of opposite sex, because one function of pheromones is as sex attractants, whose transmission stimulates reproductive behavior. Yet another class of mediators, the **neurotransmitters,** are released from nerve cells but act on adjacent rather than distant cells. Still other messengers are classified as extracellular **growth factors.** These differ from hormones in that their growth-stimulating activities are continuous, rather than being short-lived in response to a burst of secretion. Distinctions between these classes of regulators are somewhat indefinite. For example, epinephrine and norepinephrine, when secreted in the central nervous system, act as neurotransmitters, whereas epinephrine acts as a hormone when secreted from adrenal medulla.

An Outline of Hormone Action

Among the five major classes of chemical messengers, the hormones are most completely understood, and we shall discuss their actions in some detail for much of the rest of this chapter. The mechanism of hormone action, now quite well understood in some instances, should provide a basis for understanding other modes of **signal transduction,** the process whereby an external chemical signal evokes an intracellular metabolic charge.

Until the 1950s we knew practically nothing about molecular mechanisms of hormone action. A popular theory was that a hormone stimulates a metabolic pathway by binding directly to the rate-determining enzyme for that pathway and activating it. Our current understanding is based largely on research discussed in Chapter 13, namely studies of the effect of epinephrine in stimulating glycogen mobilization. These investigations, carried out largely in the laboratories of Earl Sutherland and Edwin Krebs, showed that epinephrine does not enter cells, as it must if a rate-limiting enzyme is to be activated directly. Instead, epinephrine binds to a macromolecular receptor at the cell surface, thereby setting in motion a series of intracellular events in which the key role is played by a **second messenger,** cyclic AMP (see Figure 13.19, Chapter 13). The hormone itself is the first messenger. Today we know that all hormones so far investigated act through binding to specific receptors, although those receptors may be located inside the target cell. Second messengers are often used to transmit the message to the target metabolic pathway, though not all hormone actions involve a second messenger.

Chemically, the hormones in vertebrate metabolism include peptides or polypeptides, such as insulin or glucagon; steroids, including glucocorticoids and the sex hormones; and amino acid derivatives, including the catecholamines and thyroxine. Hormone actions include enzyme activation or inhibition via second messengers, as noted for epinephrine and glucagon; stimulation of the synthesis of particular proteins, through activation of particular genes; and selective increases in the cellular uptake of certain metabolites, as mentioned for insulin. We know quite a lot about second messenger actions, and we are currently learning a great deal about genetic regulatory effects.

Hierarchical Nature of Hormonal Control

Hormonal regulation involves a hierarchy of cell types acting on each other to either stimulate or modulate the release and action of a hormone. The secretion of hormones from endocrine cells is stimulated by chemical signals from regulatory cells, which occupy a higher position in this hierarchy (Figure 23.7). Hormonal action is controlled ultimately by the central nervous system. The master coordinator in mammals is the **hypothalamus,** a specialized center of the brain. The hypothalamus receives and processes sensory inputs from the environment via the central nervous system. In response it produces a number of hypothalamic hormones, some of them called **releasing factors.** These factors act on the pituitary, which is located just beneath the hypothalamus. Releasing factors stimulate either the anterior or the posterior portion of the pituitary to release specific hormones. Other hypo-

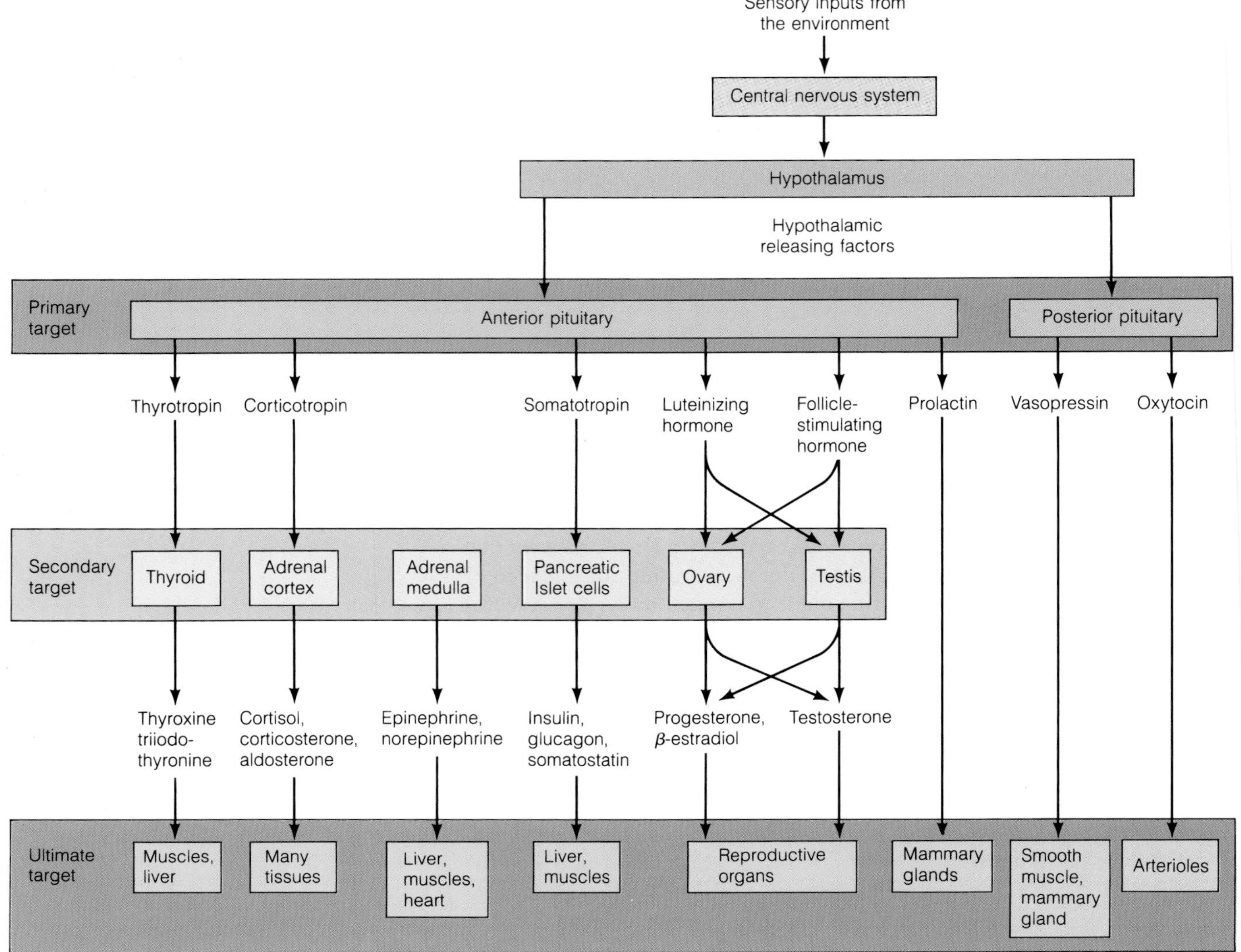

Figure 23.7
Hierarchical nature of hormone action in vertebrates. The pituitary regions represent the first target, for control by the hypothalamus. Pituitary hormones act on secondary targets, namely the other endocrine glands, and in most cases the outputs from these glands are the hormones that act on the other targets, essentially all of the other organs and tissues.

thalamic hormones act to inhibit the secretion of particular pituitary hormones. Some pituitary hormones stimulate target tissue directly. For example, **prolactin** stimulates mammary glands to produce milk. However, most pituitary hormones act on endocrine glands that occupy an intermediate, or secondary position in the hierarchy, stimulating them to produce hormones that exert the ultimate actions on target tissues. Pituitary hormones that act on other endocrine glands are called **tropic hormones** or **tropins.** An example is **adrenal corticotropic hormone** (ACTH), also called **β-corticotropin.** This peptide is secreted from the anterior pituitary, and it stimulates the adrenal cortex to produce glucocorticoids and mineralocorticoids, which in turn act on a number of tissues.

The action of a hormone is self-limiting because of the existence of feedback loops, in which secretion of a hormone sets in motion a series of events that leads to inhibition of that secretion. As you can see in Figure 23.8, for example, the secretion of β-corticotropin from the pituitary is stimulated by **corticotropin releasing factor (CRF)**, a hypothalamic hormone that is a 41-residue polypeptide. Hypothalamic cells contain glucocorticoid receptors, which sense the elevated levels of circulating glucocorticoids that result from stimulation of the adrenal cortex. Binding of glucocorticoids to these receptors has the effect of inhibiting the further release of CRF, thus completing a feedback loop.

To summarize the hierarchical nature of hormone action: the central nervous system transmits signals to the hypothalamus, which produces factors that either stimulate or inhibit the release of hormones from the pituitary. These hormones stimulate other endocrine glands, each of which releases a hormone that acts on a target tissue and elicits a specific metabolic response. Alternatively, a pituitary hormone may act directly on a target tissue. The action of a hormone sets in motion events that ultimately limit that action.

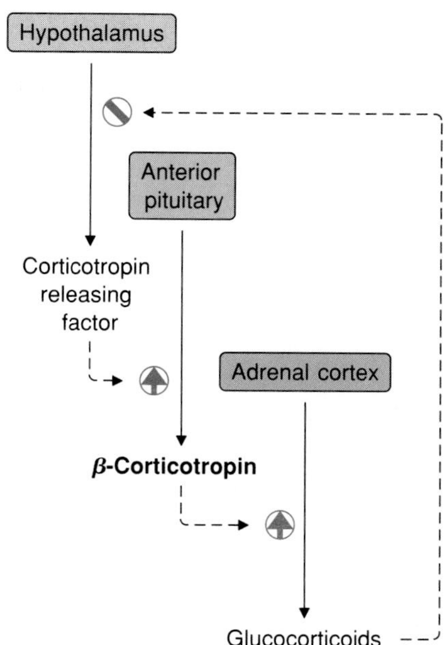

Figure 23.8
Regulation of the action of a hormone (β-corticotropin, or ACTH) by feedback loops.

Hormone Classification: Structures and Actions

In terms of their general modes of action, we recognize four hormonal mechanisms, two of them fairly well characterized. The peptide hormones and most of the amine hormones bind to receptors at the cell surface and act through second messengers. Synthesis of the second messenger inside the cell is stimulated by binding of the hormone at the cell periphery. Accumulation of the second messenger evokes metabolic changes inside the cell.

Some hormone receptors, notably that for insulin, are transmembrane proteins with a ligand binding site on the extracellular side and a catalytic activity in a domain on the cytosolic side. In the insulin receptor that activity is a protein tyrosine kinase. Still other receptors are transmembrane proteins that can act as specific ion channels. Binding of the hormone on the outer side of the membrane opens the channel and permits certain ions to flow across the membrane.

Hormones that can traverse the plasma membrane act through binding to intracellular receptors. These include the steroid hormones and the thyroid hormones. Inside the cell, the hormone–receptor complex binds to specific DNA sequences in chromatin in the cell nucleus and stimulates the transcription of specific genes. The best-studied example is the action of estrogen in chickens during development of eggs. Binding of the receptor–estrogen complex to DNA enormously stimulates the transcription of genes, including that for ovalbumin, the major protein of egg white. Although much remains to be learned about how hormones control gene expression, it is clear that a general effect of steroid hormones is to stimulate the synthesis of certain proteins in target tissues. Thus, the effects on metabolism result from changes in the *levels* of key proteins. The hormones that act through second messengers, by contrast, usually bring about changes in *activity* of key proteins and enzymes. Consequently, the effects of peptide hormones tend to be immediate and relatively short-lived, while steroids have delayed and longer-term actions.

Current evidence suggests that hormones that act through membrane-bound receptors exert their regulatory effects from the outside of the cell. However, some hormone–receptor complexes can undergo internalization via receptor-mediated endocytosis, the same process by which cholesterol becomes internalized through binding of LDL to its receptor (Chapter 18). Insulin and glucagon are taken into cells by this mechanism, but there is no evidence that the hormone continues to act after internalization. On the other hand, this process, known as **down-regulation,** is a way to control the action of the hormone, for internalization of the receptor decreases the number of receptors remaining at the cell surface to bind more hormone.

Synthesis of Hormones: Peptide Hormone Precursors

We have already presented the synthesis of steroid hormones (Chapter 18) and of catecholamines and thyroid hormones (Chapter 21), both of which

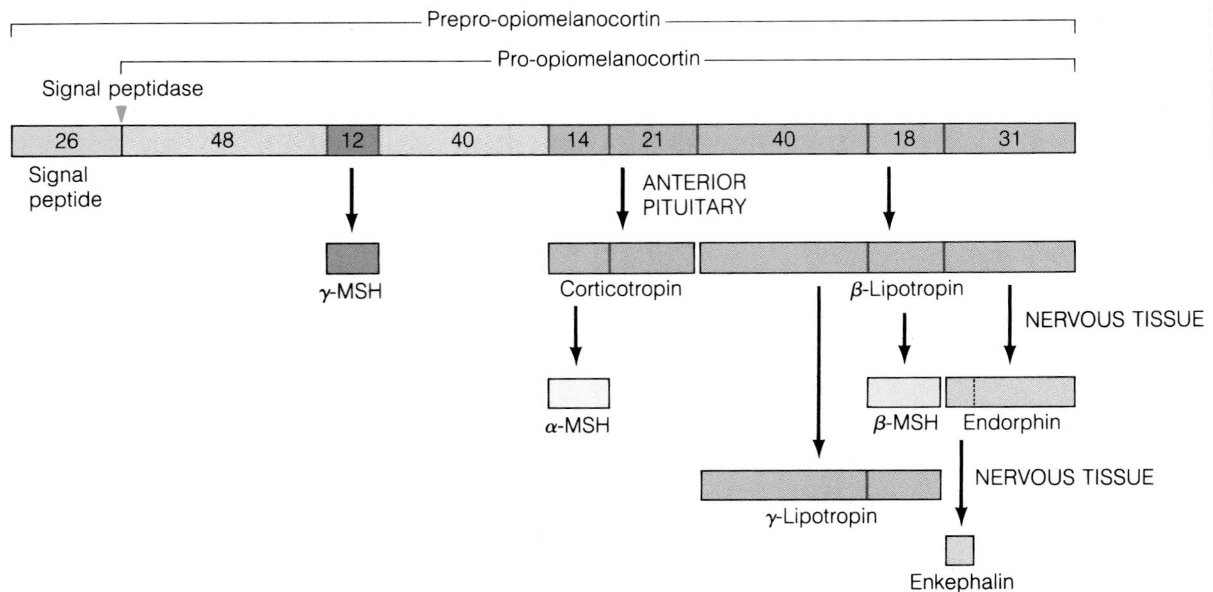

Figure 23.9
Structure of pro-opiomelanocortin, showing each of its peptide hormone domains. The structure shown is the primary translation product, *prepro-opiomelanocortin,* from which a signal peptide is cleaved to give pro-opiomelanocortin. Most cleavage sites are pairs of basic amino acids. The number of residues in each polypeptide is indicated. The cleavage pathway varies in different cell types, to yield different ensembles of peptide hormones.

occur via straightforward metabolic pathways. The synthesis of peptide hormones is interesting in that in nearly all cases studied so far they are synthesized as inactive precursors and converted to active hormones by proteolytic processing. Studies of the synthesis of insulin, which we presented in Chapter 5, provided the first evidence for this phenomenon. Recall that this hormone contains two polypeptide chains, of 21 and 30 residues, with two interchain disulfide bridges and one intrachain bridge. These chains are formed by cleavage from an 81-residue polypeptide, called **proinsulin.** The first product of translation of the insulin gene is the 105-residue *preproinsulin*. Cleavage from this of a 24-residue N-terminal segment gives proinsulin, which undergoes folding, disulfide bond formation, and cleavage to give the active hormone, insulin (see Figure 5.21, Chapter 5).

All known polypeptide hormones are synthesized in "prepro" form, with a signal sequence and additional sequence(s) that are cleaved out during maturation of the hormone. A particularly interesting case is that seen when a single polypeptide sequence contains two or more distinct hormones. Perhaps the most fascinating example is a pituitary multihormone precursor that contains sequences for β-lipotropin, melanocyte-stimulating hormone (MSH), endorphin, and enkephalin, as well as ACTH. This precursor, called **pro-opiomelanocortin,** derives its name from its role as precursor to endogenous **opi**ates, **melano**cyte-stimulating hormone, and **corti**cotropin. Even more interesting is the fact that pro-opiomelanocortin is cleaved at different sites in different cells, so that different cell types produce different ensembles of hormones derived from this precursor. In the pituitary, cleavage occurs to generate ACTH and β-lipotropin, while processing in the central nervous system yields endorphin and enkephalin, among other products. Figure 23.9 shows the cleavage sites and the hormone sequences carried within each region of the polyprotein.

Mechanisms of Hormone Action

In this section we describe mechanisms of action of representative classes of hormones, focusing on the most widely studied chemical messengers and on mechanisms that have general applicability to an understanding of hormone action.

Signal Transduction: Receptors

As noted earlier, all known hormones interact with target cells by initially binding to a macromolecular receptor, located in either the plasma membrane or the interior of the cell. Since the receptor participates in transduction of the signal from the external messenger to some component of the metabolic machinery, it must have at least one additional functional site. The activity of this site is altered by hormone binding just as the catalytic site of an allosteric enzyme is altered by binding effectors at remote sites.

EXPERIMENTAL STUDY OF RECEPTORS. Intermolecular interactions involving hormone receptors can be studied experimentally by methods comparable to those used in enzymology. Binding of a hormone to its receptor is saturable and resembles Michaelis–Menten kinetics. Most hormones bind quite tightly, with dissociation constants in the range of 0.1 micromolar to 1.0 picomolar. The ability of a tissue to respond to hormonal stimulation is a function of the receptor density of cells in that tissue.

Binding of radioactive hormone or an analog can be used to quantitate receptors, for use either in purifying receptors or in determining receptor density in a given cell type. The tight binding between hormones and their receptors can also be used to advantage in designing purification protocols involving affinity chromatography. In fact, one of the earliest successful applications of this technique was the purification of the insulin receptor, using columns of immobilized insulin. This eased a task that is normally very difficult, for two reasons. First, membrane-bound receptors must be solubilized prior to purification, without irreversible inactivation. Second, most hormone receptors are present in exceedingly small amounts; for example, adipose tissue contains the insulin receptor at only about 10^4 receptor molecules per cell.

AGONISTS AND ANTAGONISTS. Often the ligand used in receptor-binding assays or in affinity chromatographic purification is not the hormone itself but an analog that happens to bind to the receptor, sometimes more tightly than the natural hormone. The analog may show little or no structural resemblance to the natural ligand, but may show a stereochemical relationship when one constructs three-dimensional models. An example is the potent synthetic estrogen **diethylstilbestrol.** This compound is 3 times more potent than **17β-estradiol,** which in turn is 10 times more potent than other natural estrogens. Diethylstilbestrol has been used in purifying and characterizing estrogen receptors. The synthetic hormone has had its uses outside the laboratory as well, including its addition to feed to stimulate growth of beef cattle, until the fairly recent discovery that it is carcinogenic. At present we don't know whether the carcinogenicity of diethylstilbestrol is related to its estrogen activity or to some other undiscovered biological property of the molecule.

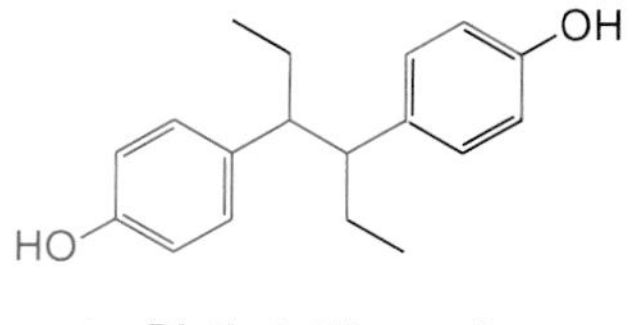

HO OH

Diethylstilbestrol

HO OH

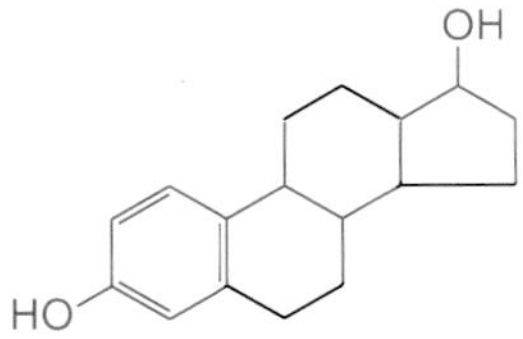

17-β Estradiol

Diethylstilbestrol is an example of a hormone **agonist**—an analog that binds productively to a receptor and mimics its biological activity. An agonist is comparable to an alternative substrate for an enzyme: its binding to a receptor is productive, in that it evokes a metabolic response comparable to that of binding the hormone. The other type of analog is the **antagonist,** which binds to receptors but does not provoke the normal biological response. An antagonist is to a receptor as a competitive inhibitor is to an enzyme, in that both antagonists and competitive inhibitors compete with a normal ligand (hormone or substrate, respectively) for binding to a specific site on a protein and by so binding inhibit a normal biological process.

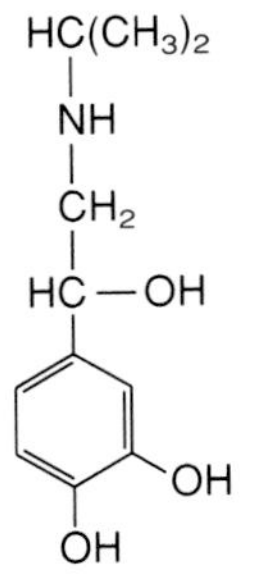

Agonists and antagonists have been very useful in studies of the stereochemistry of binding sites on receptors. In turn, these investigations are useful in drug design, where one wants to activate or inactivate specifically certain classes of receptors. Specific agonists and antagonists have great medicinal value. For example, the agonist **isoproterenol** is used to treat asthma, because it mimics the effects of catecholamines in relaxing bronchial muscles in the lung.

Isoproterenol

Classes of Catecholamine Receptors. Studies with a great many agonists and antagonists of the catecholamines have revealed the existence in vertebrates of four different catecholamine receptors, each of which has a distinctive pattern of response to these analogs. These are called the α_1-, α_2-, β_1-, and **β_2-adrenergic** receptors (named from **adrenaline,** the older name used for epinephrine). Receptors of the β type are those we encountered before, in our discussions of epinephrine-induced glycogenolysis and lipolysis (Chapters 13 and 17). Adrenergic receptors of different types, in different tissues, have various physiological effects, some of which are summarized in Table 23.3.

Receptors and Adenylate Cyclase as Distinct Components of Signal Transduction Systems. In the early stages of hormone research, when the epinephrine-stimulated mobilization of glycogen was the only defined biochemical response to hormone–receptor interaction, adenylate cyclase was found to be a membrane-bound enzyme. Since the receptor is also membrane-bound, it was thought for some time that the receptor *was* adenylate cyclase and that epinephrine binding activated the enzyme. However, two observations argued against that interpretation. First, other hormones were found to activate adenylate cyclase; more than a dozen are now

Table 23.3
Some biological actions associated with adrenergic receptors

Receptor Class	Target Tissue	Effect
α_1	Iris of the eye	Contraction
	Intestine	Decreased motility
	Salivary glands	Potassium and water secretion
α_2	Pancreatic β cells	Decreased secretion
	Blood platelets	Aggregation
	Fat cells	Decreased lipolysis
	Stomach	Decreased motility
α (subtype not identified)	Arterioles in skin, mucosa	Constriction
	Bladder sphincter	Contraction
	Male sex organs	Ejaculation
β_1	Heart	Increased rate, force, and depth of contraction
	Fat	Increased lipolysis
	Intestine	Decreased motility
β_2	Lung	Muscle relaxation
	Liver	Increased glycogenolysis
	Intestine	Decreased motility

Adapted from L. S. Goodman and A. C. Gilman (eds.), *Pharmacological Basis of Therapeutics,* 7th ed. (New York: Macmillan, 1985), p. 72.

known, including glucagon, ACTH, melanocyte-stimulating hormone, and luteinizing hormone. Adenylate cyclase seemed unlikely to have that many allosteric activation sites. Second, binding of catecholamines to the α_2 class of receptors was found to *inhibit* adenylate cyclase, rather than activating it.

Both of the above observations strongly indicated that the receptor and adenylate cyclase are distinct proteins. In fact, the resolution of β-adrenergic receptors from adenylate cyclase was observed experimentally in 1977. This was an important development, because it showed that this hormonal response system has far more flexibility and versatility than previously thought. A wide variety of hormones could exert a multitude of biological effects through a common mechanism, namely activation or inhibition of cyclic AMP synthesis. The diversity of signals and responses was built into both the diversity of receptors and the diversity of enzymes in target cells whose activities could be stimulated or inhibited by cyclic AMP-stimulated phosphorylation. It was soon learned that transduction of the hormonal signal to adenylate cyclase involved a *third* class of proteins. Before discussing these proteins, let us briefly review some common structural features of the receptors that have been analyzed.

Because receptors are embedded in the membrane and are present in very small amounts, their isolation in amounts sufficient for structural analysis is a Herculean task. Cloning the receptor genes, by techniques described in Chapter 25, has been indispensable in gaining complete amino acid sequence information. Among the several receptor proteins whose amino acid sequences were known by early 1988—including both α_2- and β_2-adrenergic receptors—there are some surprising structural similarities. The proteins are of comparable size, with 415 to 480 residues, including seven conserved regions that are rich in hydrophobic amino acids. It seems clear that these represent regions of α helix that are embedded in the membrane and linked by hydrophilic loops, projecting into both the extracellular environment and the cytosol, with the recognition site for hormone or neurotransmitter on the extracellular side. The structure of the β_2-adrenergic receptor is shown in Figure 23.10. Currently, researchers are identifying domains of function within the overall structure of this and other receptors. By manipulating cloned receptor genes it is possible to produce hybrid receptors, for example, with an N-terminal α_2 receptor sequence and a C-terminal β_2 sequence. Experiments with these hybrid receptors are permitting identification of regions of the protein that are in contact with the hormone and other components of the signal transduction machinery.

Signal Transduction: G Proteins

Signal transduction mechanisms in which cyclic AMP participates involve three separate proteins: (1) a hormone receptor, (2) adenylate cyclase, and (3) a member of a third protein family. The third class is designated **G proteins,** because of their ability to bind guanine nucleotides. In 1971 it was discovered that GTP is required for the activation of adenylate cyclase by β-adrenergic agonists, and late in the decade the basis for this requirement emerged. There are several known G proteins which interact with receptor systems that activate or inhibit adenylate cyclase. The two best characterized are G_s, a family of G proteins that are involved in *stimulation* of adenylate cyclase, and a closely related family of proteins, G_i, which are involved in responses that *inhibit* adenylate cyclase. Although both of these proteins interact with other receptors as well, it is useful to describe their functions in terms of the adrenergic receptors.

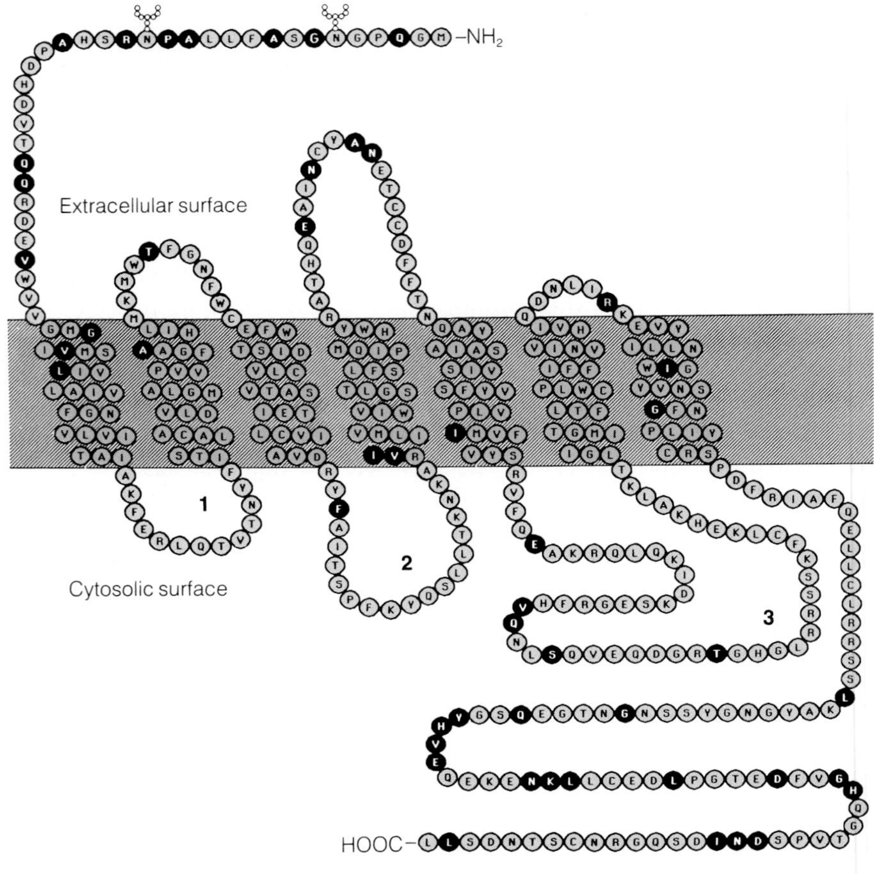

Figure 23.10
Amino acid sequence of the β_2-adrenergic receptor, showing the seven conserved transmembrane domains (orange), the three extracellular and three cytoplasmic loops (blue), and two oligosaccharide units on the extracellular side. Receptor interaction with G proteins is controlled by reversible phosphorylation of C-terminal serine and threonine residues.

Actions of G Proteins. The G proteins are membrane proteins that, in the unstimulated state, bind guanosine diphosphate, GDP. In a hormone response leading to stimulation of adenylate cyclase, the binding of extracellular hormone or agonist to a receptor, typically a β-adrenergic receptor, causes a conformational change that stimulates the receptor to interact with a nearby molecule of G_s. This in turn stimulates the dissociation of bound GDP from G_s, to be replaced by GTP. When GTP is bound, G_s is converted to a protein that interacts with adenylate cyclase and *activates* the target enzyme, producing cyclic AMP from ATP. This results in activation of cAMP-dependent protein kinase, with consequent phosphorylation of target proteins, such as phosphorylase b kinase in cells that activate glycogen phosphorolysis. To summarize, this signal transduction pathway involves the following causal chain of events, depicted in Figure 23.11: (1) hormone binding to receptor; (2) receptor interaction with G_s, stimulating release of GDP and association of GTP with G_s; (3) stimulation of adenylate cyclase by the GTP-bound G_s; (4) stimulation by cAMP of intracellular reactions.

Continued activation depends on the presence of bound GTP. The hormonal response is limited, and hence controlled, by the presence of a slow GTPase activity on the G protein. Thus, bound GTP is cleaved (ever so slowly) to GDP, with concomitant loss of ability to stimulate adenylate cyclase.

The G_i protein functions similarly, but in response to extracellular signals whose response is the *inhibition* of adenylate cyclase, typically α_2 agonists. Here the binding of GTP provokes an interaction of G_i with ade-

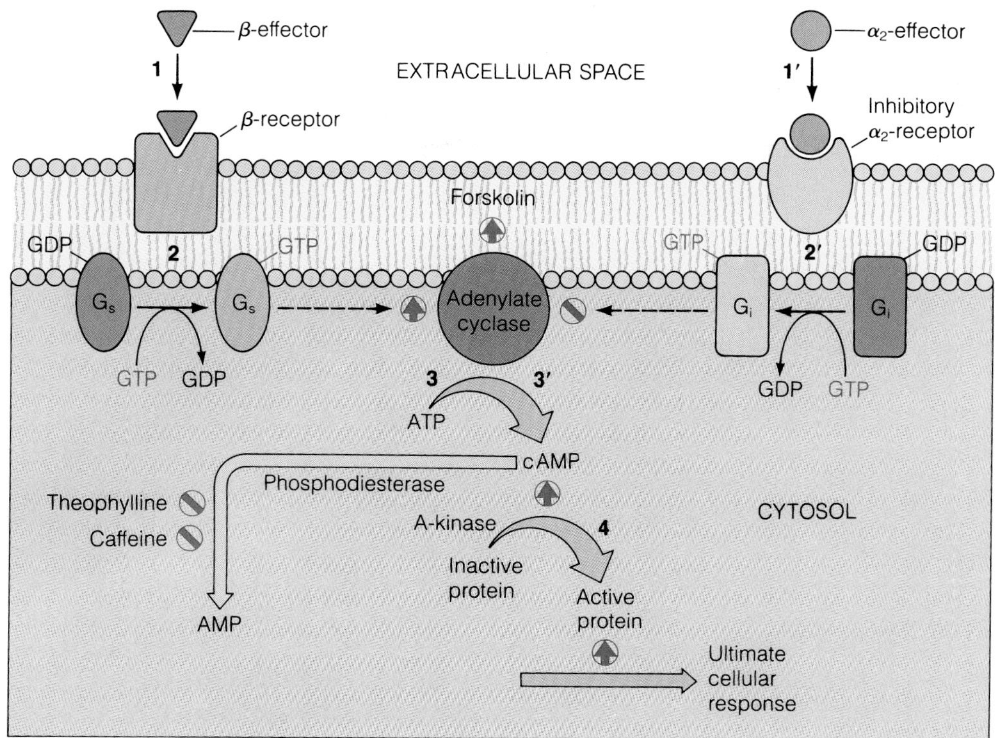

Figure 23.11
Signal transduction pathways involving G proteins and adenylate cyclase. cAMP binding activates A-kinase (a cAMP-dependent protein kinase) by releasing free catalytic subunits. Forskolin, a natural herbal product, functions as an adenylate cyclase activator. The hydrolytic catabolism of cAMP is inhibited by theophylline and caffeine. Steps 1, 2, and 3 are involved in a stimulatory response, where adenylate cyclase is activated, while steps 1′, 2′, and 3′ are involved in an inhibitory response. As noted elsewhere, the receptors span the entire membrane.

nylate cyclase, which inhibits the target enzyme and decreases intracellular cAMP levels (steps 1′ to 3′ in Figure 23.11).

Structure of G Proteins. Structural studies of G proteins reveal a trimeric structure for both G_s and G_i, as shown in Figure 23.12—a relatively large subunit (α), a slightly smaller subunit (β), and an 8-kDa subunit (γ). Current evidence suggests that β and γ are identical in both G_s and G_i. The guanine nucleotide-binding site and its associated GTPase activity are both located on α. Research indicates that a hormonal stimulus leads to dissociation of the G protein, with the α-GTP complex moving through the membrane until it encounters a molecule of adenylate cyclase. The slow GTPase activity eventually reconverts α-GTP to α-GDP, and the α-GDP complex dissociates from adenylate cyclase and rejoins the β-γ complex.

Consequences of Blocking GTPase. The importance of the GTPase activity as a control of the hormone response can be seen in the consequences of blocking the GTPase activity. Blocking can be achieved in vitro by substituting GTP with **GTPγS,** a nonhydrolyzable GTP analog in which a sulfur atom substitutes for the oxygen between the β and γ phosphates of GTP.

$^-$O—P(=O)(O$^-$)—S—P(=O)(O$^-$)—O—P(=O)(O$^-$)—O—CH$_2$ (ribose, OH OH) guanine

GTPγS

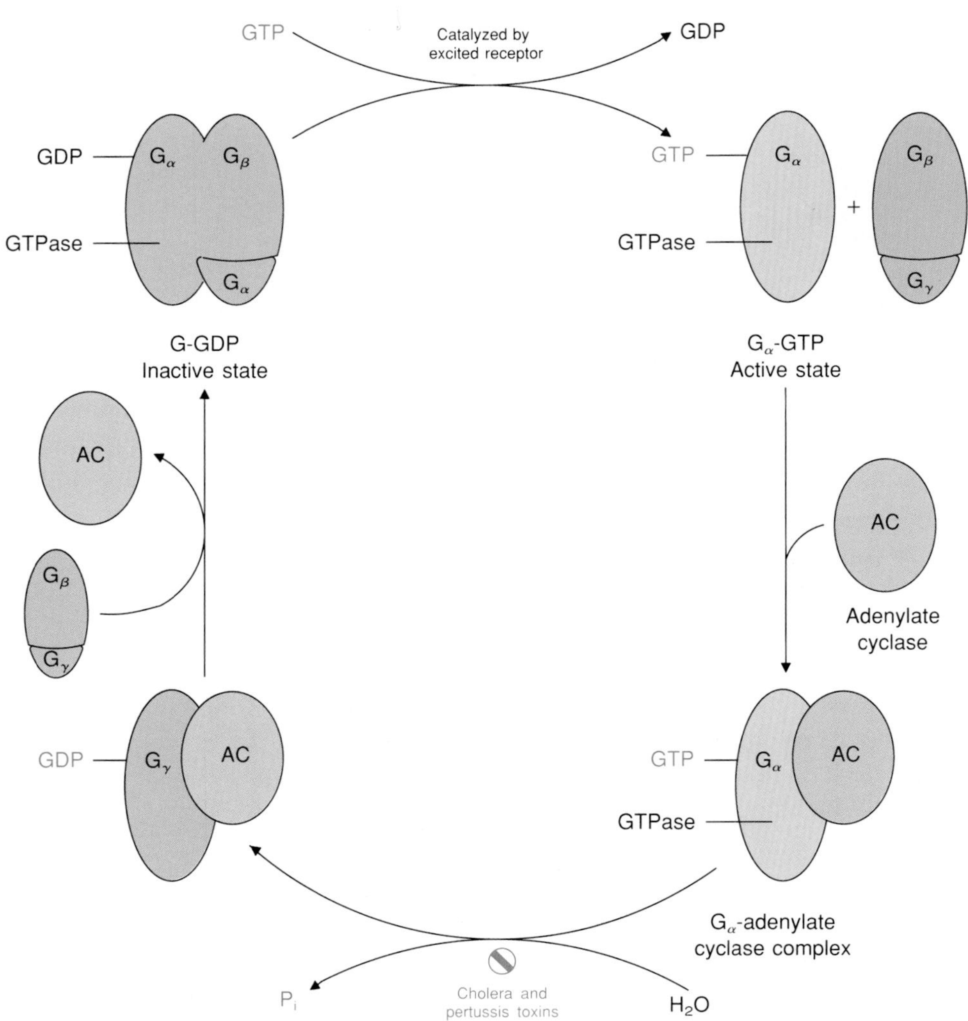

Figure 23.12
The G protein dissociation and reassociation cycle, showing the site of action of cholera and pertussis toxins.

In a G_s-dependent system the result is irreversible activation of the target adenylate kinase. More dramatic are the effects of bacterial toxins that have G proteins as their biological targets. The toxin of *Vibrio cholerae* is an enzymatic protein with the ability to cleave NAD^+ and transfer its ADP-ribose moiety to a specific site in the α subunit of G_s. This modification of G_s inhibits its GTPase and converts α to an irreversible activator of adenylate cyclase.

$$NAD^+ + \alpha_s \longrightarrow \text{nicotinamide} + \text{ADP-ribosyl-}\alpha_s$$

In the intestine this promotes uncontrollable secretion of water and Na^+ and is responsible for the severe dehydration and loss of salt that accompany cholera. A component of the toxin of *Bordetella pertussis*, which causes whooping cough, has a similar effect on the α subunit of the G_1 protein.

G Proteins in the Visual Process. There are remarkable similarities between the actions of G proteins in transmitting hormonal signals and the transmission of signals from light, in vision. Much of our understanding of G proteins in hormonal signal transduction came from studies of a G protein called **transducin** in the visual process. The extracellular stimulus and the biochemical end point are quite different in vision as compared with hormone action, but the transmembrane signaling processes are surprisingly similar.

As we shall see in Chapter 29, the extracellular signal in vision is a photon of light, and the membrane receptor is **rhodopsin,** an abundant membrane protein in the outer segment of rod cells in the retina. A photochemical change in the structure of rhodopsin causes it to activate transducin so that it binds GTP.

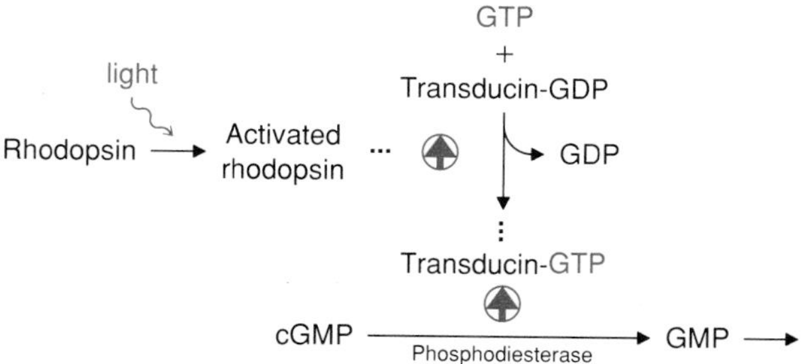

The transducin-GTP complex is an activator of a specific **phosphodiesterase,** which cleaves a cyclic nucleotide, **guanosine 3′, 5′-monophosphate** (cyclic GMP, or cGMP). Cleavage of cGMP, in turn, is the stimulus to intracellular reactions that generate a visual signal to the brain. Thus, the stimulated hydrolysis of cGMP is the visual analog of the stimulated *synthesis* of cAMP in β-adrenergic responses.

The G protein mechanism is used in other signal transduction pathways. At least ten different G proteins are known to exist. Table 23.4 summarizes the structural similarities among four well-characterized members of this family—G_s, G_i, transducin, and G_0, a G protein that is present specifically in brain and acts as an inhibitor of adenylate cyclase. As more G proteins are characterized, additional roles for these proteins are certain to be found. For example, recent evidence implicates a G protein mechanism in olfactory transductions (sense of smell).

Table 23.4
Properties of purified G proteins

Parameter	G_s	G_i	G_0	Transducin
Signal detector	β-Adrenergic, glucagon, other receptors	α_2-Adrenergic, muscarinic, opiate, other receptors	Unknown	Rhodopsin
Effector protein	Adenylate cyclase	Adenylate cyclase	Unknown	cGMP phosphodiesterase
Function	Stimulation	Inhibition	Unknown	Stimulation
Subunit M_r				
α	44, 44.5, 46	41	39	39
β	35 and 36	35 and 36	35 and 36	36
γ	8	8	8	8
Toxin susceptibility	Cholera	Pertussis	Pertussis	Pertussis and cholera
Location	Most cells	Most cells	Brain	Retinal rod outer segments

Adapted from L. Stryer and H. R. Bourne, *Annu. Rev. Cell Biol.* 2:396 (1986).

G Proteins and Oncogenes

One of the most exciting areas of G protein research comes from their structural similarity to a family of proteins called ***ras*** proteins. This class of proteins was originally discovered as the product of an oncogene in *ra*t *s*arcoma viruses, a class of tumor viruses. We shall discuss the genome structure of tumor viruses in Chapter 25. Suffice it to say here that a tumor virus genome acquires, during its evolution, one or more genes from cells of the infected host. During or after its transfer to the viral genome, this cellular gene undergoes one or more mutations. Expression of the altered gene (called an **oncogene**) in infected cells contributes to the virus-stimulated transformation of the infected cell to a cancer cell.

The *ras* gene is one of a family of genes, now numbering a few dozen, that have been found in tumor virus genomes and are responsible for the ability of the viruses to cause tumors in cells they infect. In each case analyzed to date the viral oncogene is closely related to a cellular gene, and it is likely that the virus acquired its oncogene from the genome of an infected cell as it evolved. The cellular counterpart to a viral oncogene is called a **proto-oncogene.** Because *ras* was the first oncogene for which these relationships were elucidated, it is worthwhile to analyze further the *ras* proto-oncogene and its conversion to an oncogene.

As noted above, *ras* genes exist in normal, noncancerous cells. Mutant forms of the *ras* gene can appear in cells, either through infection by a retrovirus bearing a mutant form of the *ras* gene (a *ras* oncogene), or through mutations affecting the expression or structure of the resident *ras*

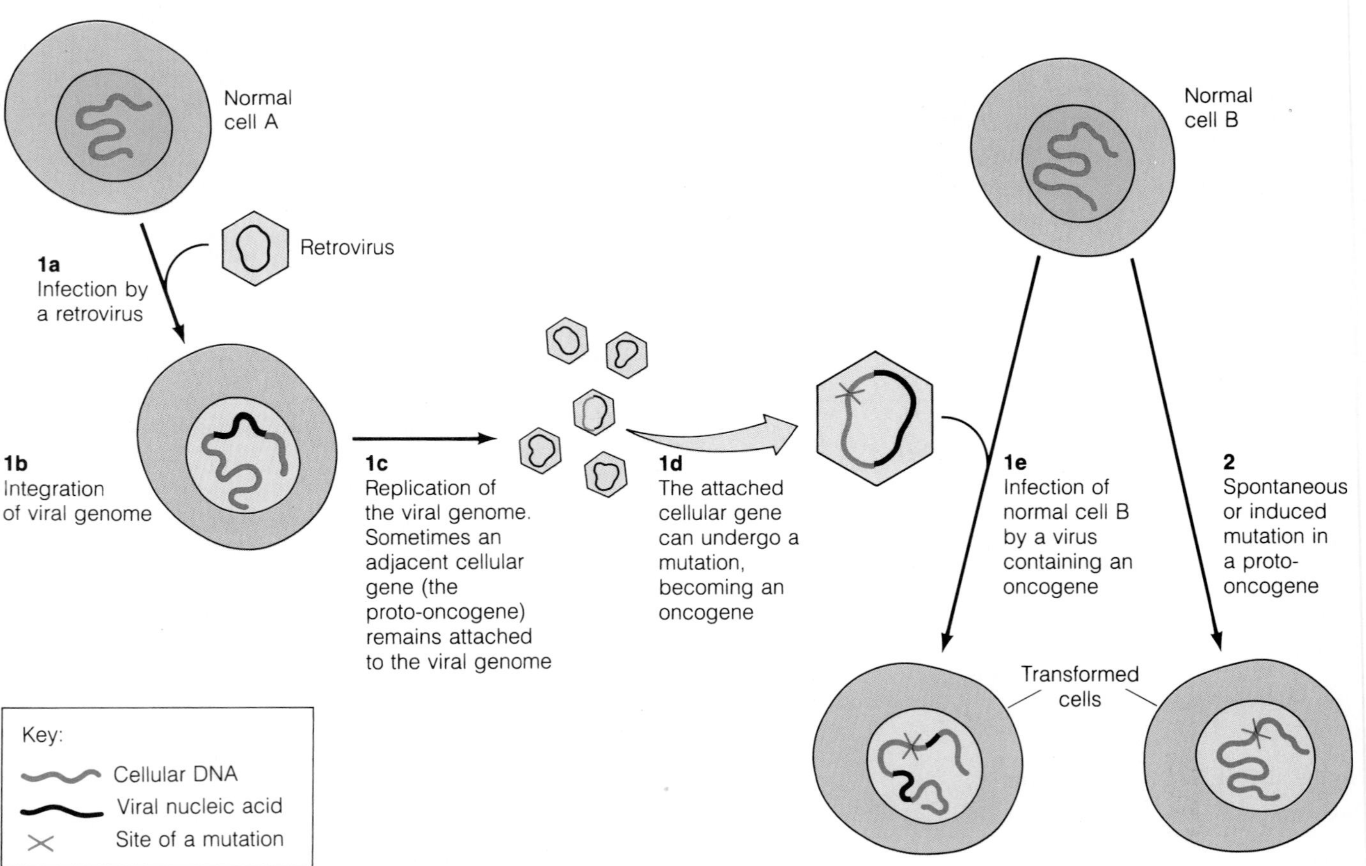

Figure 23.13
Relationships between oncogenes and proto-oncogenes. A proto-oncogene is a normal cellular gene that can be converted to an oncogene, and cause transformation to a cancer cell, by two possible mechanisms—(1) infection by a retrovirus, which integrates into a chromosomal site next to a proto-oncogene and carries that gene along in its own genome when the virus replicates; or (2) mutation of the cellular proto-oncogene. Mutation of the integrated cellular DNA converts the proto-oncogene to an oncogene, which can cause transformation when this virus infects another cell.

proto-oncogene (Figure 23.13). A number of *ras* genes have now been isolated by gene cloning, from normal tissues and from tumor viruses. Nucleotide sequences of *ras* oncogenes often differ from those of their normal cellular counterpart *ras* genes by as little as one nucleotide in the entire DNA sequence.

All of the known *ras* genes encode a 21-kDa protein, with regions homologous to sequences in the α subunit of G proteins. Like the α subunit, the *ras* proteins bind guanine nucleotides. Normal *ras* proteins possess a GTPase activity, as do G proteins, whereas most *ras* oncogene proteins lack this activity. This suggests that normal *ras* proteins function like G proteins in regulating aspects of cellular metabolism yet to be discovered. Mutant forms of the *ras* proteins cannot carry out normal regulation, perhaps in some cases because of their inability to hydrolyze GTP, and this loss of metabolic control is one of the events that converts a normal cell to a cancer cell.

Lending support to this model was the determination in early 1988 of the three-dimensional structure of a *ras* protein, crystallized as its complex with GDP (Figure 23.14). Of great interest in this structure is the fact that amino acid residues known to be changed in mutations that generate *ras* oncogenes are positioned close to the bound guanine nucleotide. This is consistent with the idea that the interactions between the proto-oncogene protein and guanine nucleotides are important to metabolic control and that this control is lost when a normal cell is transformed to a cancer cell. Much more work will be required to solidify this hypothesis.

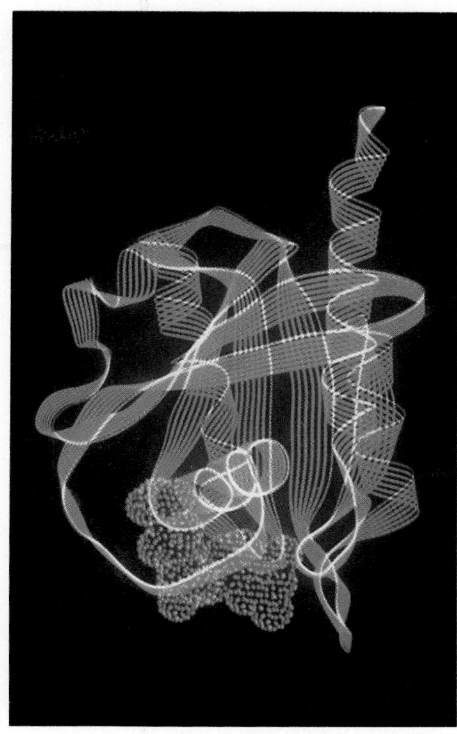

Figure 23.14
Structure of a GDP-*ras* protein complex. A polypeptide containing the N-terminal 171 residues of a 188-residue *ras* protein was produced by gene cloning and crystallized. The ribbon shows the polypeptide chain, and the GDP structure is shown in green.

Second Messenger Systems

We now have a fairly complete picture of the action of a large class of hormones. Transduction of a signal from a remote tissue and across a membrane involves the linked actions of receptor, G protein, and adenylate cyclase. These events either stimulate or inhibit the synthesis of a second messenger, cyclic AMP, inside the cell. A number of intracellular processes are controlled in turn by the level of the second messenger.

Cyclic AMP is not the only second messenger. Much interest has focused on cyclic GMP, although its roles as a second messenger are just now coming into focus. Many cells contain a cGMP-stimulated protein kinase that, like the cAMP-activated enzyme, contains both catalytic and regulatory subunits.

Calcium ion has also been considered a second messenger. Many cells respond to extracellular stimuli by altering their intracellular calcium concentration, which in turn exerts biochemical changes either by itself or through its interaction with calmodulin (Chapter 13). Calcium levels themselves are controlled in large part through the action of second messengers, including cAMP. In many nerve and muscle cells the activation of adenylate cyclase results in an influx of extracellular calcium. cAMP activates a voltage-dependent **calcium channel** in the presynaptic nerve membrane, allowing calcium ions to flow into the cell and triggering synaptic transmission (Chapter 29). That activation may involve phosphorylation by cAMP-dependent protein kinase of a protein component of the channel.

The calcium influx in muscle cells triggers muscle contraction and is responsible, for example, for the increased rate and force of heartbeats caused by β-adrenergic agonists. Because cAMP regulates calcium influx, it is probably more appropriate to call calcium a third messenger than a second messenger.

Phosphatidylinositol, a Source of Second Messengers

Calcium levels can be increased also by release from *intracellular* calcium stores. Access to these intracellular stores is controlled by a rather recently discovered set of messengers, the **phosphoinositide system.** Although similar in many respects to the adenylate cyclase system, the phosphoinositide system is distinctive in that the hormonal stimulus activates a reaction that generates *two* second messengers, not just one.

The earliest experimental observations regarding this signal transduction system occurred in 1953, when Mabel Hokin and Lowell Hokin noted that administration of acetylcholine to pancreatic secretory cells led to rapid synthesis and turnover of the phosphatidylinositol fraction of membrane phospholipids. Similar observations were made in other systems stimulated by hormones, neurotransmitters, or growth factors. However, more than two decades elapsed before a unifying concept emerged to explain these observations.

We now know that a specific lipid in the phosphoinositide family, namely, **phosphatidylinositol 4,5-bisphosphate** (PIP_2), is a membrane-associated storage form for *two* second messengers. As shown in Figure 23.15, binding of an agonist to an appropriate receptor (step 1) stimulates a G protein to bind GTP (step 2), just as occurs during the activation of adenylate cyclase. However, this G protein activates a *different* membrane-bound enzyme, **phospholipase C,** which in turn cleaves PIP_2 to yield two products (step 3)—***sn*-1,2-diacylglycerol** (**DAG**) and **inositol 1,4,5-trisphosphate** (**$InsP_3$**).

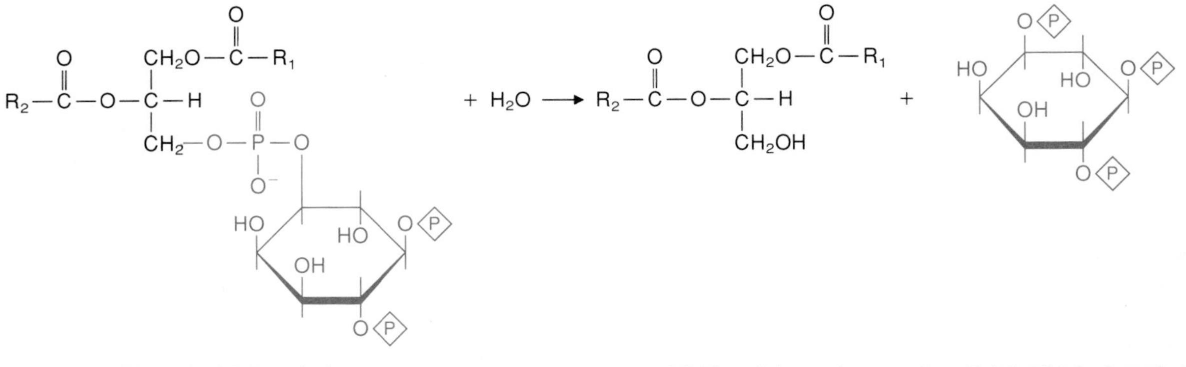

Phosphatidylinositol 4,5-bisphosphate (PIP_2) + H_2O ⟶ ***sn*-1,2-Diacylglycerol (DAG)** + **Inositol 1,4,5-trisphosphate ($InsP_3$)**

Both of these products act as second messengers. Therefore, the cleavage of PIP_2 by phospholipase C is the functional equivalent of the synthesis of cyclic AMP by adenylate cyclase.

The second messenger role of inositol trisphosphate is to stimulate the release of calcium from its intracellular stores in the endoplasmic reticulum (step 4). This has various effects on intracellular metabolism, as noted earlier, but it also contributes toward the second messenger role of diacylglycerol, which is the activation of membrane-bound protein kinase C (step 5). This enzyme requires for its activity *calcium* (hence the "C" designation) and a *phospholipid* (specifically, phosphatidylserine). Diacylglycerol stimulates the activity by greatly increasing the affinity of the enzyme for calcium ions. This requirement is specific for the *sn*-1,2-DAG; neither the 1,3- nor the 2,3-isomer is active. The enzyme phosphorylates specific serine and threonine residues in target proteins. As with cAMP-stimulated protein kinase,

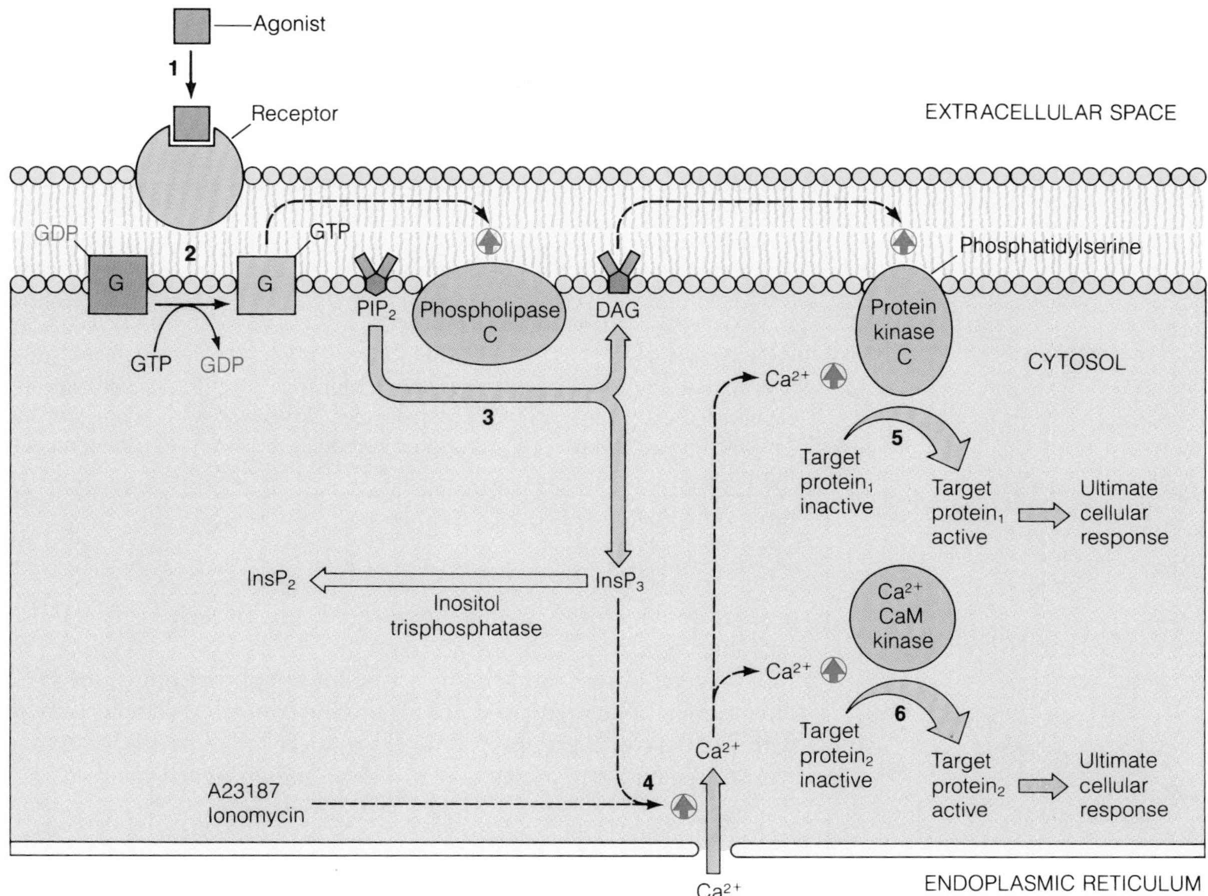

Figure 23.15
Signal transduction pathways involving phosphoinositide turnover. DAG = *sn*-1,2-diacylglycerol; PIP_2 = phosphatidylinositol 4,5-bisphosphate; $InsP_3$ = inositol 1,4,5-trisphosphate; $InsP_2$ = inositol 1,4-bisphosphate. Most of the effects of calcium result from its binding to calmodulin (CaM). A23187 is a calcium ionophore, which can be used experimentally to release calcium from intracellular stores. Probably most receptors are membrane-spanning proteins.

the specific cellular responses to protein kinase C activation (step 6) depend on the ensemble of target proteins that become phosphorylated in a given cell. Some of the known target proteins are the insulin receptor, β-adrenergic receptor, glucose transporter, HMG-CoA reductase, cytochrome P-450, and tyrosine hydroxylase.

Now let us briefly consider the metabolism of inositol trisphosphate after its release from PIP_2. Three sequential hydrolytic steps yield inositol, which is then reincorporated into phosphatidylinositol, as discussed in Chapter 18, to regenerate PIP and PIP_2. The last hydrolytic step, the hydrolysis of inositol monophosphate to inositol, is specifically inhibited by **lithium ion.** This inhibits the resynthesis of $InsP_3$ by depleting the cell of inositol.

$$\text{Inositol monophosphate} + H_2O \longrightarrow \text{inositol} + P_i$$

Since the phosphoinositide messenger system is widely used in nervous tissue, it has been proposed (but not proved) that this action of lithium explains its efficacy in the treatment of manic-depressive psychosis.

Because many metabolic processes are known to be controlled by calcium fluxes and by phosphorylation of specific proteins, the phosphoinositide system has great versatility as a control mechanism. The possibility of using either DAG or $InsP_3$, or both, mechanisms as the result of a single extracellular stimulus further increases this versatility, and exploration of phosphoinositide-related control systems is one of the most active research

Table 23.5
Some cellular processes controlled by the phosphoinositide second messenger system

Extracellular Signal	Target Tissue	Cellular Response
Acetylcholine	Pancreas	Amylase secretion
Acetylcholine	Pancreas (islet cells)	Insulin release
Acetylcholine	Smooth muscle	Contraction
Vasopressin	Liver	Glycogenolysis
Thrombin	Blood platelets	Platelet aggregation
Antigens	Lymphoblasts	DNA synthesis
Antigens	Mast cells	Histamine secretion
Growth factors	Fibroblasts	DNA synthesis
Spermatozoa	Eggs (sea urchin)	Fertilization
Light	Photoreceptors *(Limulus)*	Phototransduction
Thyrotropin releasing hormone	Pituitary anterior lobe	Prolactin secretion

Adapted from M. J. Berridge, *Sci. Am.* (Oct.) 127:147 (1985).

areas in contemporary biology. A partial listing of such systems is presented in Table 23.5.

Several observations imply a role for the phosphoinositide system not only in metabolic regulation but also in the control of cellular growth. First is the activity of a group of natural products called **phorbol esters,** part of whose structure resembles that of DAG (shown in red).

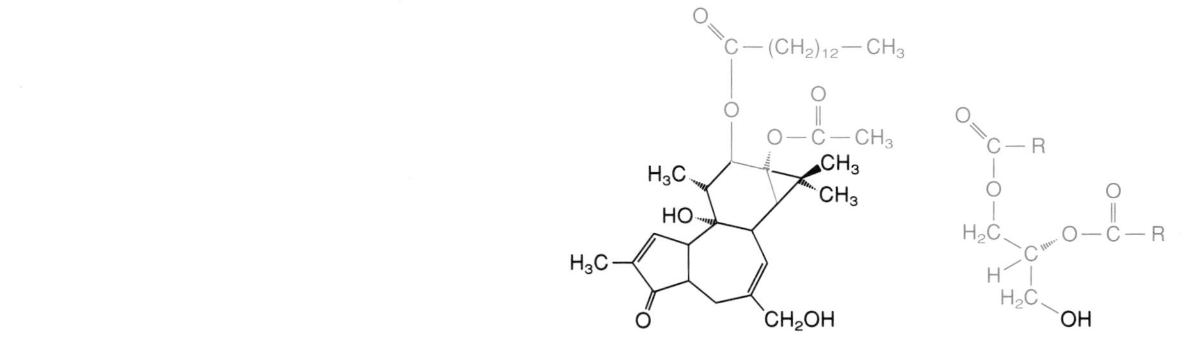

A phorbol ester, 1-*O*-tetradecanoylphorbol-13-acetate

***sn*-1,2-Diacylglycerol**

These compounds stimulate the formation of tumors when applied along with a carcinogen to experimental animals; compounds that act this way are called **tumor promoters.** Phorbol esters have been found to activate protein kinase C independently of diacylglycerol. This finding is consistent with the hypothesis that protein kinase C activation is part of the normal growth control process that becomes perturbed in tumorigenesis. Second, some cell growth factors, notably **platelet-derived growth factor** (**PDGF**), function by interacting with cell surface receptors and stimulating the hydrolysis of inositol trisphosphate. Finally, certain oncogenes are closely related to the functioning of this messenger system. An oncogene called ***sis*** is closely related in structure to the gene encoding PDGF, and two other proto-oncogenes, ***src*** and ***fos,*** are apparently involved in the synthesis of PIP_2 from phosphatidylinositol and its monophosphate. Clearly, further analysis of this system should tell us a great deal about normal growth control mechanisms and how genetic abnormalities contribute to the formation of cancer cells.

The Insulin Receptor and Clues to Insulin Action

The effects of insulin are numerous and complex as shown in Table 23.2. Insulin has a general growth-promoting property, seen both in short-term metabolic effects, which promote polysaccharide and fat synthesis, and in long-term effects, which stimulate nucleic acid and protein synthesis. Indeed, insulin can be considered a growth factor as well as a metabolic regulator, for it is a required component of defined tissue culture media for growth of mammalian cells. No other hormone has such a multiplicity of known effects. Moreover, insulin acts on almost all cells in the body.

Despite a vast research effort devoted to insulin, we know surprisingly little about its mechanisms of action. That action must involve a transmembrane signaling process, because, as noted earlier, insulin acts through a receptor located at the plasma membrane. No second messenger associated with this receptor has yet been identified, though insights have come from studies of the properties of purified insulin receptor. These investigations were aided greatly by the cloning in 1985 of a cDNA, or DNA molecule complementary to the messenger RNA for the human insulin receptor.

The insulin receptor, schematized in Figure 23.16, is a glycoprotein with an $\alpha_2\beta_2$ tetrameric structure. The tetramer is stabilized by disulfide bonds. Both the α chain, of 735 residues, and the β chain, of 620 residues, are translated from a single mRNA. The α chain, which is thought not to span the membrane, is believed to bind insulin near its C-terminus. The β chain has a transmembrane domain, with its C-terminus in the cell interior. That C-terminal region contains a protein kinase activity that is specific for phosphorylating tyrosine residues. The kinase activity is stimulated by binding of insulin to the extracellular part of the receptor. The intracellular portion of the insulin receptor is subject to autophosphorylation, by its own tyrosine kinase. It may also be phosphorylated by other enzymes, such as protein kinase C. The ensemble of proteins phosphorylated by the insulin receptor has not yet been identified, but there is good evidence that the tyrosine kinase activity is required for the major actions of insulin. One could, for example, envision that a membrane-linked glucose transport system becomes activated on phosphorylation.

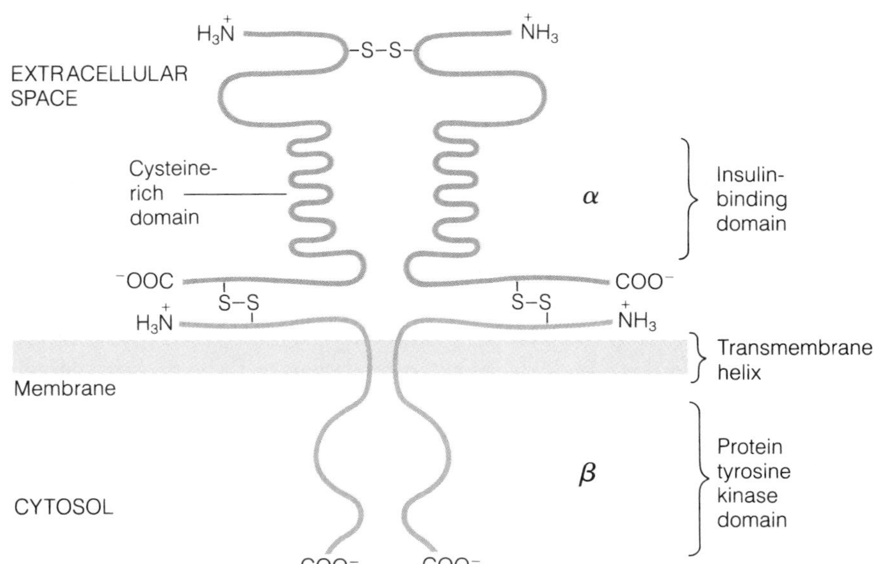

Figure 23.16
Model of the insulin receptor. The precise locations of disulfide bridges have not been established. The interior C-terminal portion of the β chains contains the tyrosine kinase activity.

Since insulin can be considered as a growth factor, it is of great interest that protein tyrosine kinase activity is associated with several other growth factor receptors, including those for platelet-derived growth factor, **epidermal growth factor** (EGF), and a peptide **insulin-like growth factor-1** (IGF-1). These receptors may represent a family of closely related proteins, for the tyrosine kinase domain of the insulin receptor shares amino acid sequence homology with the EGF receptor. Moreover, there is evidence that the action of insulin as a growth factor is mediated through its binding to the IGF-1 receptor.

Tantalizing clues to the growth control abnormalities in cancer cells have come from two related findings. First, protein tyrosine kinase activities are associated with certain oncogene and proto-oncogene products. Rous sarcoma virus, the first known cancer-causing virus, contains a tyrosine kinase activity associated with its oncogene product; the oncogene in this virus is called ***src.*** Mutational inactivation of the *src*-encoded tyrosine kinase activity causes loss of the ability of the virus to transform infected cells to the cancer phenotype. Second, domains of the receptors for insulin and various growth factors are structurally related to known oncogene products. The EGF receptor is highly homologous with an oncogene called *erb-B,* while the tyrosine kinase domain of the insulin receptor β chain shows sequence homology with the **v-*ros*** oncogene. The *erb-B* oncogene product is a truncated form of the EGF receptor, which contains a permanently activated tyrosine kinase activity. These findings suggest that normal growth control involves metabolic cascades that include specific tyrosine phosphorylations. Altered genes for these tyrosine kinases, whether carried in viral genomes or created by mutations of cellular genes, perturb these cascades and abolish normal control of cell growth. Obviously, much work must be done to complete this picture.

Steroid and Thyroid Hormones: Intracellular Receptors

Steroids have been detected in virtually all eukaryotes and some prokaryotes. While their involvement in biological regulation has not been confirmed in many lower eukaryotes, their small size and ability to traverse membranes suit them particularly well for regulatory roles. In vertebrates steroid hormones function as genetic regulators, controlling the rate of synthesis of particular proteins (Table 23.6). It has long been known that steroid and thyroid hormone action promotes the *accumulation* of specific proteins; later this was shown to result from increased synthesis of those proteins. In some cases the synthesis of a target protein is *decreased.*

These regulatory effects occur at the level of transcription of steroid-responsive genes. Steroid and thyroid hormones both act by binding to specific *intracellular* protein receptors. These receptors exist at concentrations of only about 10^4 molecules per cell, which has made their purification difficult. However, because they bind to hormones very tightly, it has been possible to purify the proteins by affinity chromatography. The receptor proteins that have been characterized are highly asymmetric in shape. Binding of hormone causes a major conformational change in the protein, which is detected as a decrease in sedimentation coefficient. Amino acid sequences for several steroid and thyroid receptors have now been determined from base sequencing of the complementary DNAs (Figure 23.17). All of these sequences show a region of homology that represents a DNA-binding domain. A region of great sequence diversity is found near the C-terminus; this almost certainly represents the hormone-binding domain.

Table 23.6
Target organs for steroid and thyroid hormones and major proteins whose synthesis is affected

Hormone Class	Target Organ	Protein[a]
Glucocorticoids	Liver	Tyrosine aminotransferase
		Tryptophan oxygenase
		α-Fetoprotein (↓)
		Metallothionein
	Liver, retina	Glutamine synthetase
	Kidney	Phosphoenolpyruvate carboxykinase
	Oviduct	Ovalbumin
	Pituitary	Pro-opiomelanocortin
Estrogens	Oviduct	Ovalbumin
		Lysozyme
	Liver	Vitellogenin
		apo-VLDL
Progesterone	Oviduct	Ovalbumin
		Avidin
	Uterus	Uteroglobin
Androgens	Prostate	Aldolase
	Kidney	β-Glucuronidase
	Oviduct	Albumin
1,25-Dihydroxyvitamin D_3	Intestine	Calcium-binding protein
Thyroid hormones	Liver	Carbamoyl phosphate synthetase
		Malic enzyme
	Pituitary	Growth hormone
		Prolactin (↓)
Ecdysone (in insects)	Epidermis	Dopa decarboxylase
	Fat body[b]	Vitellogenin

[a] Synthesis of each indicated protein is increased by the hormone, except for those two identified by (↓).

[b] The fat body is an organ in insects that plays some of the same roles as liver and adipose tissue.

Controversy still exists over whether steroid and thyroid receptors predominate in the cytosol or in the nucleus. In either case it is evident that after hormone binding the hormone–receptor complex becomes concentrated in the nucleus, driven by its very high affinity for certain sites on the genome—the sites involved in regulating the expression of target proteins. Chapter 28 discusses the effects of this binding on expression of the target genes.

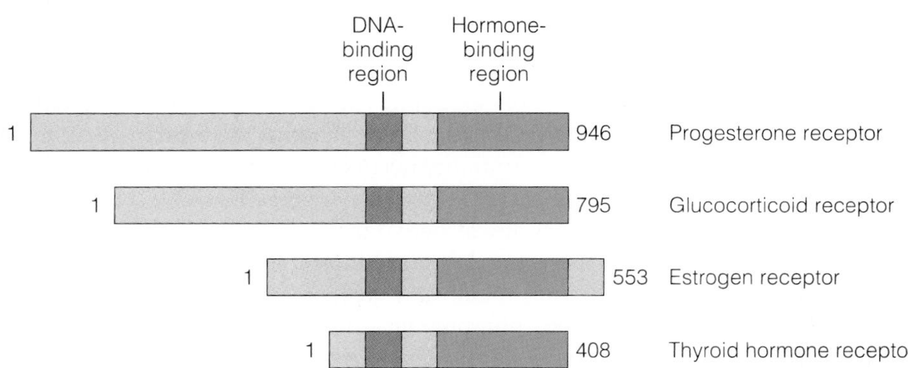

Figure 23.17
Structural similarities among receptors for steroid and thyroid hormones. All four share a common DNA-binding domain. The hormone-binding domain is at the C-terminus. Functions of the N-terminal domains are not yet known.

The utility of a set of long-term-acting regulators is evident from a couple of examples. Estrogens and progesterone regulate the female reproductive cycle. In humans these hormones interact over a 4-week cycle to prepare the uterus for implantation of a fertilized ovum. Muscle tissue builds up in the wall of the uterus and blood flow to the uterus increases; clearly these events require new protein synthesis. These processes stop when a pituitary signal triggers decreased release of the hormones, causing sloughing off of cells in the uterine lining and the beginning of menstrual bleeding.

The actions of glucocorticoids are comparable, in that control of the synthesis of particular proteins allows for long-term metabolic adaptation. For estrogens, however, control of reproductive metabolism is exercised over a several-week period, while the secretion of glucocorticoids is a means of adaptation to longer-term stress. This includes stimulation of gluconeogenesis and synthesis of a variety of proteins, including some that counteract the effects of inflammation. Unlike estrogens, which act only in reproductive tissues, the glucocorticoids influence cells in a wide variety of target tissues.

Just as the receptors for peptide hormones have been shown to resemble certain oncogenes, similar relationships are being revealed for the intracellular receptors. For example, the **c-*erb-A*** gene, a cellular counterpart of a viral oncogene, **v-*erb-A,*** encodes a thyroid hormone receptor. Thus, a pattern is emerging in which mutational alteration affecting any of the major signaling systems—steroid and thyroid hormones, peptide hormones, amine hormones, G proteins, or insulin—is associated with loss of metabolic control that accompanies the transformation of a normal cell to a cancer cell.

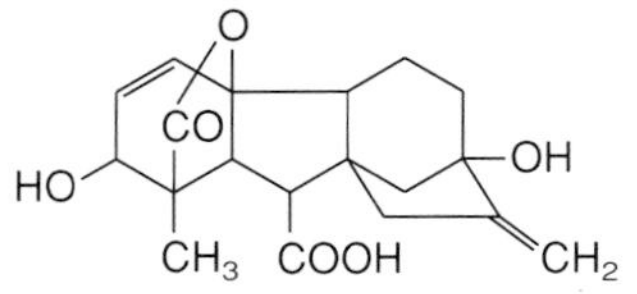

Plant Hormones

Our understanding of the molecular actions of plant hormones is far less advanced than that of the vertebrate animal hormones. In part this is because many of the plant hormones are growth factors, so that growth is the only readily measurable parameter. Furthermore, plant membranes are more difficult than animal membranes to isolate and study.

We encountered the five major classes of plant hormones in earlier metabolic chapters. To recapitulate, the five classes, illustrated in Figure 23.18, are (1) the diterpene *gibberellins,* derived from isopentenyl pyrophosphate; (2) the sesquiterpene *abscisic acid,* also derived from isopentenyl pyrophosphate; (3) the *cytokinins,* purine bases with a terpenoid side chain; (4) *auxins,* with the tryptophan metabolite *indole-3-acetic acid* being the most active; and (5) *ethylene,* which comes from the methionyl moiety of *S*-adenosylmethionine. Thus, these compounds are chemically quite distinct from animal hormones, although polypeptide factors are known to control sexual reproduction in fungi.

Additional differences from animal hormones lie in the diversity of effects of a given hormone and in the hormonal activity of a large number of structurally related species. For example, at least a dozen different cytokinins have been characterized structurally. Though all contain adenine with a side chain on N-6, great variability occurs in the structure of that side chain.

Auxins are synthesized in apical buds (at the tip) of growing shoots. They stimulate growth of the main shoot and inhibit lateral shoot development. A class of auxin-binding membrane proteins may represent auxin receptors. Auxin action involves pumping protons out of the cell, possibly in conjunction with a membrane ATPase. This raises the intracellular pH and

Gibberellic acid (GA3) [a gibberellin]

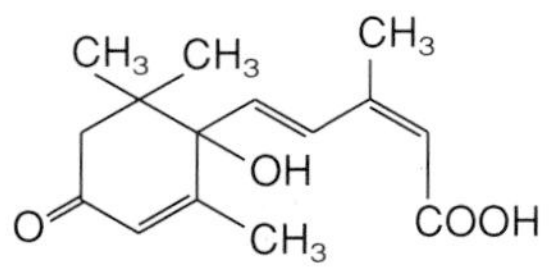

Abscisic acid (ABA)

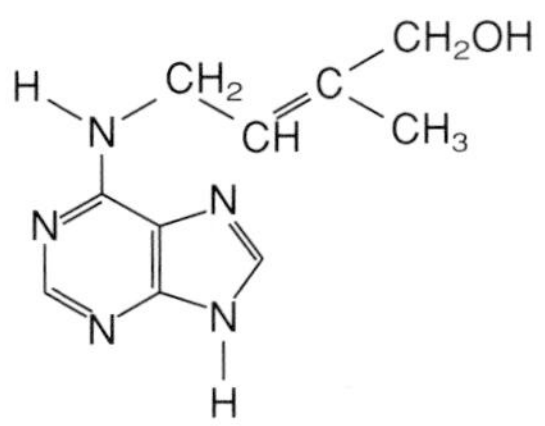

Zeatin [a cytokinin]

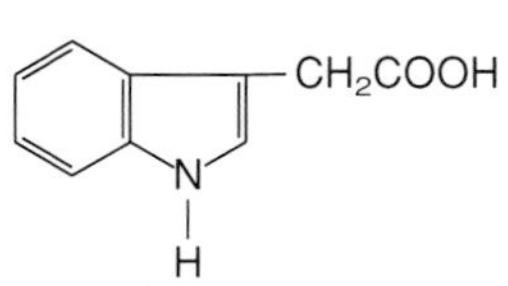

Indole-3-acetic acid (IAA) [an auxin]

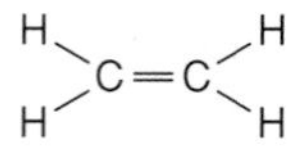

Ethylene

Figure 23.18
Representatives of the five major classes of plant hormones.

may be responsible for subsequent events—partial cell wall degradation, which is necessary for growth to occur, and increased RNA and protein synthesis.

The other classes of plant hormone are harder to describe in biochemical terms. Cytokinins are produced in the roots, and they promote growth and differentiation in many tissues. Cytokinins and auxins work together, and the ratio of cytokinin to auxin is often crucial in determining whether a plant will grow or differentiate. In many cases single plant cells in tissue culture can regenerate whole plants. Here, the investigator must experiment with different proportions of cytokinins and auxins, searching empirically for the proportions that allow the best growth of normal plants.

Gibberellins are also growth-promoting hormones and may act to stimulate expression of certain genes. Particular messenger RNAs are present in increased amounts as a result of gibberellin administration, supporting the idea that gibberellins act at the gene level.

Ethylene is considered a hormone of senescence. It stimulates fruit ripening and the aging of flowers, and it inhibits seedling growth. It also redirects auxin transport to promote transverse, rather than longitudinal, growth of plants.

Abscisic acid counteracts the effects of most other plant hormones. It inhibits germination, growth, budding, and leaf senescence. Abscisic acid synthesis in leaves is stimulated by wilting. Physiological evidence suggests a role for abscisic acid in ion and water balance.

REFERENCES

General

Hanks, S. K., A. M. Quinn, and T. Hunter (1988) The protein kinase family: Conserved features and deduced phylogeny of the catalytic domains. *Science* 241:42–51. The number of known protein kinases is approaching 100, and they are involved in every metabolic regulatory process. This review admirably summarizes the field.

Norman, A. W., and G. Litwack (1987) *Hormones*. Academic Press, New York. An 806-page textbook written by two leading researchers.

Second Messenger Systems

Berridge, M. B. (1987) Inositol trisphosphate and diacylglycerol: Two interacting second messengers. *Annu. Rev. Biochem.* 56:159–194. One of the most recent and comprehensive of many reviews of this field.

Freissmuth, M., P. J. Casey, and A. G. Gilman (1989) G proteins control diverse pathways of transmembrane signaling. *FASEB* 3:2125–2131. The most comprehensive recent review of G protein function.

Majerus, P. W., T. M. Connolly, V. S. Bansal, et al (1988) Inositol phosphates: Synthesis and degradation. *J. Biol. Chem.* 263:3051–3055. A recent minireview of inositol trisphosphate metabolism.

Nishizuka, Y. (1986) Studies and perspectives of protein kinase C. *Science* 233:305–312. A review describing one of the targets of the phosphoinositide messenger system.

Oncogenes and Growth Factors

de Vos, A. M., L. Tong, M. V. Milburn, P. M. Matias, J. Jancarik, S. Noguchi, S. Nishimura, K. Miura, E. Ohtsuka, and S.-H. Kim (1988) Three-dimensional structure of an oncogene protein: Catalytic domain of human c-*H-ras* p21. *Science* 239:888–895. The first crystallographic structure determination for a protein related to an oncogene.

Santos, E., and A. R. Nebrada (1989) Structural and functional properties of *ras* proteins. *FASEB* 3:2151–2163. A summary of the similarities between G proteins and the *ras* family of oncogene and proto-oncogene products.

Temin, H. M. (1988) Evolution of cancer genes as a mutation-driven process. *Cancer Res.* 48:1697–1701. A Nobel Prize winner reviews the relationship between proto-oncogenes and oncogenes and the actions of the latter.

Yarden, Y., and A. Ullrich (1988) Growth factor receptor tyrosine kinases. *Annu. Rev. Biochem.* 57:443–478. Summarizes the functional and evolutionary relationships among genes and oncogenes related to the insulin and PDGF receptors.

Receptors

Czech, M. P., J. K. Klarlund, K. A. Yagaloff, et al (1988) Insulin receptor signaling. Activation of multiple serine kinases. *J. Biol. Chem.* 263:11017–11020. A recent description of the action of the insulin receptor.

Kobilka, B. K., T. S. Kobilka, K. Daniel, J. W. Regan, M. G. Caron, and R. J. Lefkowitz (1988) Chimeric α_2-,β_2-adrenergic receptors. *Science* 240:1310–1316. Describes the manipulation of cloned receptor genes for structure–function analysis of adrenergic receptors.

Lefkowitz, R. J., and M. G. Caron (1988) Adrenergic receptors. Models for the study of receptors coupled to guanine nucleotide regulatory proteins. *J. Biol. Chem.* 263:4993–4996. This minireview summarizes the structural similarities among receptor proteins that interact with G proteins.

Williams, L. T. (1989) Signal transduction by the platelet-derived growth factor receptor. *Science* 243:1564–1570. A contemporary review of the action of an important membrane-spanning growth factor receptor.

G Proteins

Allende, J. E. (1988) GTP-mediated macromolecular interactions: The common features of different systems. *FASEB J.* 2:2356–2367. This article describes the similarities among different families of GTP-binding proteins, including G proteins, transducin, and elongation factors in protein synthesis.

Gilman, A. G. (1987) G proteins: Transducers of receptor-generated signals. *Annu. Rev. Biochem.* 56:615–650. A review by one of the pioneers in this field.

Snyder, S. H., P. B. Sklar, and J. Pevsner (1988) Molecular mechanisms of olfaction. *J. Biol. Chem.* 263:13971–13975. This minireview mentions evidence for G protein involvement in the sense of smell.

Synthesis of Peptide Hormones

Fisher, J. M., and R. H. Scheller (1988) Prohormone processing and the secretory pathway. *J. Biol. Chem.* 263:16515–16518. A recent review that describes how peptide hormones are formed by cleavage from high-molecular-weight precursors.

Mechanisms of Hormone Action

Prestwich, G. D. (1987) Chemistry of pheromone and hormone metabolism in insects. *Science* 237:999–1006. A review of an important class of biological regulators that received little attention in this chapter.

Ringold, G. M. (1985) Steroid hormone regulation of gene expression. *Annu. Rev. Pharmacol. Toxicol.* 25:529–566. A relatively recent review by one of the leaders in the field.

Rosen, O. (1987) After insulin binds. *Science* 237:1452–1458. This article focuses less on the insulin receptor and more on the metabolic consequences of insulin action.

Yamamoto, K. R. (1985) Steroid receptor regulated transcription of specific genes and gene networks. *Annu. Rev. Genet.* 19:209–252. Although this review covers some of the same ground as Ringold's (see above), the emphasis here is more highly focused on the properties of steroid receptors and their interactions with the genome.

PROBLEMS

1. Marathon runners preparing for a race engage in "carbo loading" to maximize their carbohydrate reserves. This involves eating large quantities of starchy foods. Why is starch preferable to candy or sugar-rich foods?

2. Supposing that an average human consumes energy at the rate of 1500 kcal/day at rest and that long-distance running consumes energy at 10 times that rate, how long would the glycogen reserves last during a marathon run? Recall that 1 kcal is equivalent to 4.183 kJ.

3. What proportion of the total energy consumption supports brain function in an average resting human? What proportion in a human running in a marathon?

4. Proteolysis increases during the early phases of fasting, but later it decreases as the body adapts to using alternative energy sources. Since feedback control mechanisms have not been described for intracellular proteases, how might you explain these apparent changes in protease activity?

5. Glucose has been found to react nonenzymatically with hemoglobin, through Schiff base formation between C-1 of glucose and the amino termini of the β chains. How might this finding be applied in monitoring diabetic patients?

6. Ketone bodies are exported from liver for use by other tissues. Since many tissues can synthesize ketone bodies, what enzymatic property of liver might contribute to its special ability to export these compounds?

7. G proteins show some amino acid sequence homology with certain bacterial proteins that are involved in protein synthesis; these proteins, called elongation factors, also bind and hydrolyze GTP. How might one determine whether plant tissues contain G proteins or similar entities?

8. List a couple of factors that make it advantageous for peptide hormones to be synthesized as inactive prohormones that are activated by proteolytic cleavage.

9. Describe how you might approach the problem of identifying a receptor for a prostaglandin.

10. Name a hormone whose concentration would probably not be amenable to analysis by radioimmunoassay, and explain your answer.

11. Describe a mechanism (hypothetical) by which a steroid hormone might act to increase intracellular levels of cyclic AMP.

TOOLS OF BIOCHEMISTRY 17

Radioimmunoassay

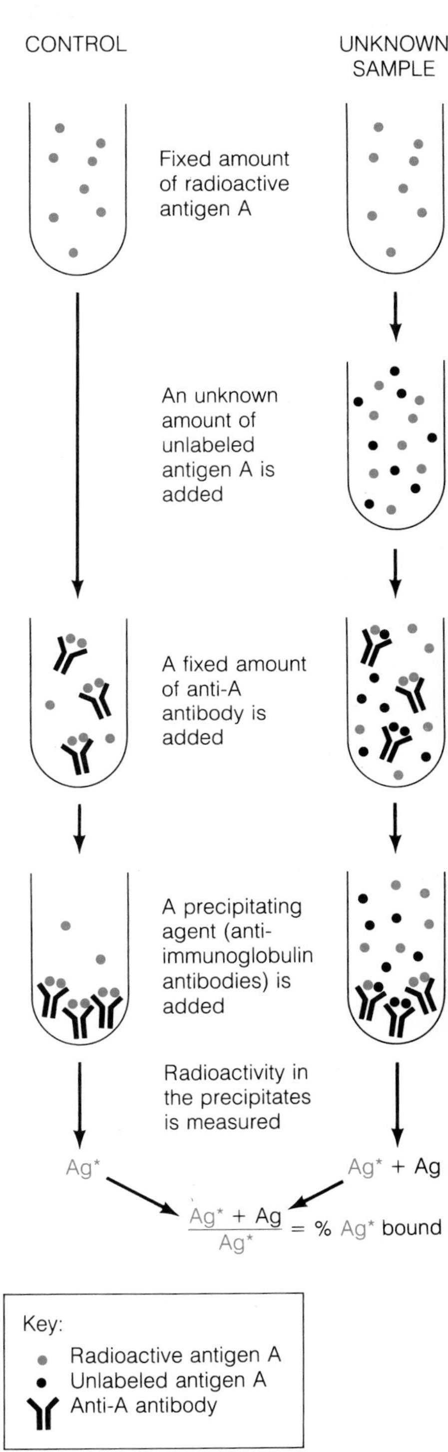

Figure T17.1
Principle of radioimmunoassay. In the example shown the amounts of labeled and unlabeled antigens are equal, which yields a fifty percent reduction in radioactive hormone bound.

Because most hormones are present at exceedingly low concentrations in biological materials (ca. 10^{-10} M), they cannot be quantitated by standard methods, such as fluorescence measurements or HPLC. Moreover, a hormone is usually surrounded by excessive amounts of chemically related substances, which would interfere with standard analytical techniques.

Radioimmunoassay (**RIA**) is a powerful technique developed in the 1960s by Solomon Berson and Rosalyn Yalow. RIA combines the specificity of antigen–antibody reactions with the sensitivity of radioisotopic detection. Radioimmunoassay allows the detection of any substance to which an antibody can be prepared by injection of an antigen into an experimental animal. The antigen can then be detected in minute amounts in the presence of a huge excess of chemically similar substances, provided those substances do not interfere with the antigen–antibody reaction.

The principle of radioimmunoassay is quite simple, as shown in Figure T17.1. One first determines the radioactivity bound to a small amount of antibody when that antibody is incubated with a large excess of radiolabeled antigen at a known specific activity. In this case the antigen is the hormone being assayed. To generate a standard curve, which relates reduction in bound radioactivity to amount of nonradioactive hormone, one then incubates the same quantities of antibody and radiolabeled antigen with an aliquot of nonradioactive antigen. The unlabeled antigen competes with labeled antigen for binding sites on the antibody molecules. Thus, the bound radioactivity decreases. This operation is carried out with increasing amounts of nonradioactive antigen. The data are shown in Figure T17.2. To determine the percentage of antigen bound, one then incubates antibody and radiolabeled antigen in the same proportions, along with an aliquot of the biological material being assayed. The unlabeled antigen in this aliquot dilutes the bound radioactivity by an amount proportional to the nonradioactive antigen present. One uses the standard curve to determine that amount quantitatively.

It is necessary to separate bound from unbound antigen, so that the radioactivity counted comes from only bound antigen. This can be done by precipitating the antigen–antibody complexes, either with antiserum raised against the antibody being used, or with "Staph A protein" (a cell surface protein from the bacterium *Staphylococcus aureus,* which efficiently binds antigen–antibody complexes).

Radioimmunoassay is used to analyze many biological substances that are present in minute amounts, including steroids, cyclic nucleotides, peptide hormones, prostaglandins, and drugs.

Figure T17.2
A standard curve for RIA, determined by varying the amount of unlabeled hormone in the presence of constant amounts of radiolabeled hormone and antibody.

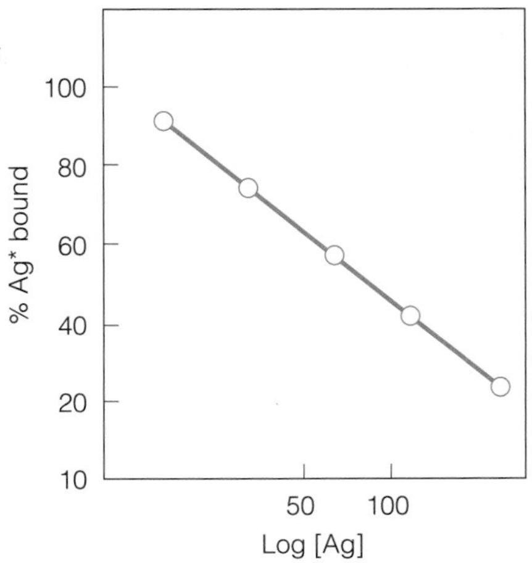

REFERENCES

Berson, S. A., and R. S. Yalow (1971) Radioimmunoassay: A status report. In: *Immunobiology,* edited by R. A. Good and O. W. Fisher. Sinauer Associates, New York. A description of the technique written by its inventors.

Van Vunakis, H., and J. J. Langone (eds.) (1980) Immunological methods. *Methods in Enzymology,* Vol. 70. Academic Press, New York. A multiauthored manual with articles about all the major immunological techniques.

PART V

Information

CHAPTER 24

Information Copying: Replication

The last major part of our study of biochemistry concerns the storage, retrieval, processing, and transmission of biological information. The processes involved, which we can call **informational metabolism,** are distinguished from intermediary metabolism in the following way. In virtually every case we have encountered so far, all the information that specifies the nature of a reaction lies within the structure of the enzyme involved. The three-dimensional structure of the protein determines which substrates are bound and which reaction is catalyzed. Of course, all metabolic reactions are controlled ultimately by genetic information, which specifies the structures and properties of enzymes. However, the reactions we shall encounter from here on are distinguished by the direct involvement of genetic information—specifically, the requirement for a *template,* which functions along with the enzyme to specify the nature of the reaction catalyzed. The biological templates, as we discussed in Chapter 4, are nucleic acids—either DNA or RNA. In general, the template plays a passive role, determining which specific substrates will be bound, while the enzyme continues to specify the nature of the reaction catalyzed once substrates have been bound. A surprising and important development is the recent finding that RNA can function as an active catalyst, as well as a passive template. This unexpected property of RNA was discussed in Chapter 10.

Major Processes in Informational Metabolism

Recall from Chapter 4 that the major processes in informational metabolism are *replication* (Figure 24.1), where DNA serves as the template for its own synthesis; *transcription,* where the information encoded in DNA specifies the structure of an RNA product; and *translation,* where RNA serves as the template for synthesis of a particular polypeptide chain. We can compare these processes to the expression of verbal information encoded in a language. Replication and transcription use only the four-letter nucleotide language of nucleic acids, while translation converts a message from the nucleotide language to the twenty-letter amino acid language of proteins.

In addition to their use of a template, the processes of replication, transcription, and translation are similar in that each can be considered to

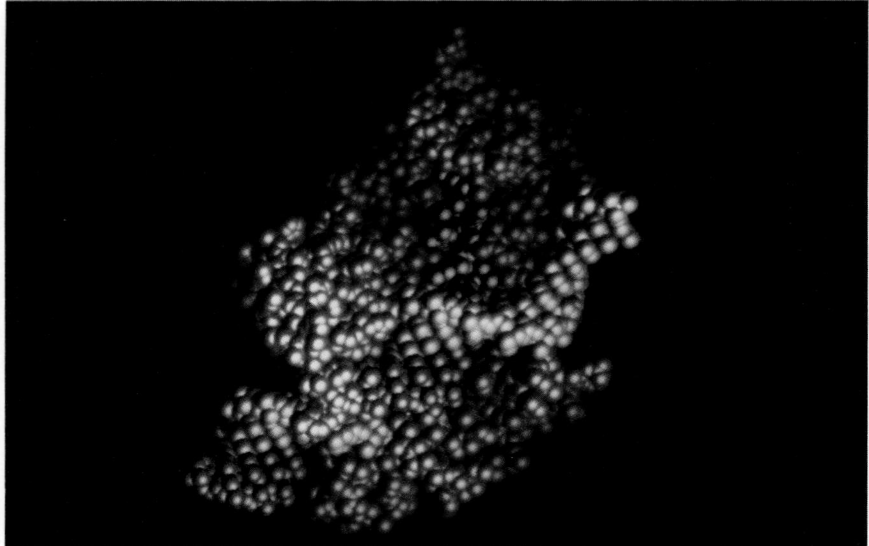

Figure 24.1
Template-directed copying of information encoded in DNA, the basic process in DNA replication. The figure shows the three-dimensional structure of the Klenow fragment of DNA polymerase I of *E. coli,* based on x-ray crystallography. Orange: enzyme (Klenow fragment of DNA polymerase I); gold: template DNA strand; blue and green: primer DNA strand; green: 3′ end of primer DNA strand, bound in 3′ exonuclease site. For additional information see p. 828.

consist of three distinct subprocesses: initiation, chain elongation, and termination. In principle, the initiation phase presents the most efficient target for regulation of each overall process and, indeed, most regulatory processes occur at this level.

In this chapter we concern ourselves with replication, of both DNA and RNA genomes. Chapter 25 deals with several processes in which DNA is a substrate as well as a template—genetic recombination, repair, modification and restriction, transposition, and gene amplification. Chapters 26 and 27 focus on transcription and translation, respectively. The experimental background for our initial understanding of these processes came largely from research with prokaryotic organisms—bacteria and their viruses—and thus much of the content of this and the following two chapters focuses on prokaryotes. In Chapter 28 we discuss some distinctive features of eukaryotic gene expression. Finally, in Chapter 29 we discuss the processing of information at the organismic level—for example, the transmission of signals through the nervous system. There, our focus will be on the whole organism rather than on the individual cell.

A Brief Review of Genetic Terminology

To a greater extent than in intermediary metabolism, our current picture of informational metabolism has come from genetics, as well as from biochemistry. Though you may be familiar with terms and concepts in genetics, it is useful to review several key terms at this point. The totality of genetic information in an organism, the **genome,** can be thought of most directly as the base sequences of all of the DNA molecules in a cell. These sequences are laid out as individual genes, each on a chromosome or on a small piece of extrachromosomal DNA. A **genetic map** identifies the relative positions of particular genes on chromosomes. As we discuss in Chapter 25, one can generate a **linkage map** from purely genetic analysis, or a **physical map,** which locates the actual positions of genes on a chromosome. Any characteristic of an organism that can be detected in terms of appearance, structure, or some measurable property is called the **phenotype.** The genetic composition of an individual, which specifies phenotypic properties, is the

genotype. Individuals with different genotypes may have the same phenotype; in Chapter 18, for example, we learned that humans with different mutations affecting the LDL receptor can have a common phenotype—the elevated serum cholesterol levels that are associated with familial hypercholesterolemia.

A diploid individual in a natural environment usually has at least one functional copy of each gene and displays a **wild-type** (genetically normal) phenotype. A gene in such an organism can undergo a mutation, thereby generating a mutant genotype, which may be detected as a mutant phenotype. For example, albinism represents a readily observable phenotype, resulting from a mutation that inactivates tyrosinase (Chapter 21). A mutant site within a gene can undergo a second mutation, which restores the wild-type genotype. This type of mutation is called a **reversion.** Alternatively, a wild-type phenotype can sometimes be restored by a mutation at a different site; this occurrence is called **second-site reversion,** or **suppression.**

DNA Replication as Revealed from Early Experiments

How does a molecule copy itself? As with most metabolic processes, our early ideas about the mechanism of DNA replication were shaped largely by in vivo experiments—those carried out with intact cells—and the complementary approach—isolation of enzymes and proteins involved in DNA replication—began later with the discovery of DNA polymerase. We begin our discussion of replication with the early experiments that revealed the general nature of the process.

Semiconservative Nature of DNA Replication

In 1953, when Watson and Crick proposed the double-helical, base-paired model for DNA structure, much of the ensuing excitement came not simply from the aesthetic beauty of their model but from predictions engendered by the model about how DNA copies itself during cell division. For the first time the *structure* of a biological molecule gave specific clues to *function.*

As we described in Chapter 4, the DNA copying mechanism involves unwinding the two strands of a parental DNA duplex, with each strand serving as template for the synthesis of a new strand, bound to and wound about that parental strand. Complete replication of a DNA molecule would yield two "daughter" duplexes, each consisting of one-half parental DNA (one strand of the original duplex) and one-half new material. This mode of replication is called **semiconservative,** because half of the original material is conserved in each of the two copies (Figure 24.2). It is distinguished from two other possible modes: **conservative,** in which one of the two daughter duplexes is the conserved parental duplex while the other is synthesized de novo, and **dispersive,** where parental material is scattered through the structures of the daughter duplexes.

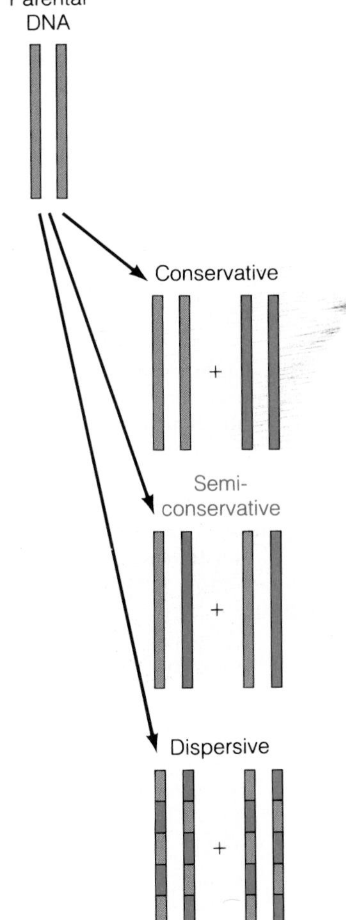

Figure 24.2
Distinguishing features of conservative, semiconservative, and dispersive models of DNA replication. Brown = parental DNA; blue = daughter DNA.

The first experimental test of this model came in 1958, when Matthew Meselson and Franklin Stahl realized that molecules that differ in density by very small amounts could be separated from each other in equilibrium density gradients. This technique allowed them to follow the fate of density-labeled DNA through several rounds of replication.

The density label was applied by growing *Escherichia coli* in medium containing the heavy isotope of nitrogen, ^{15}N (as ammonium chloride), for

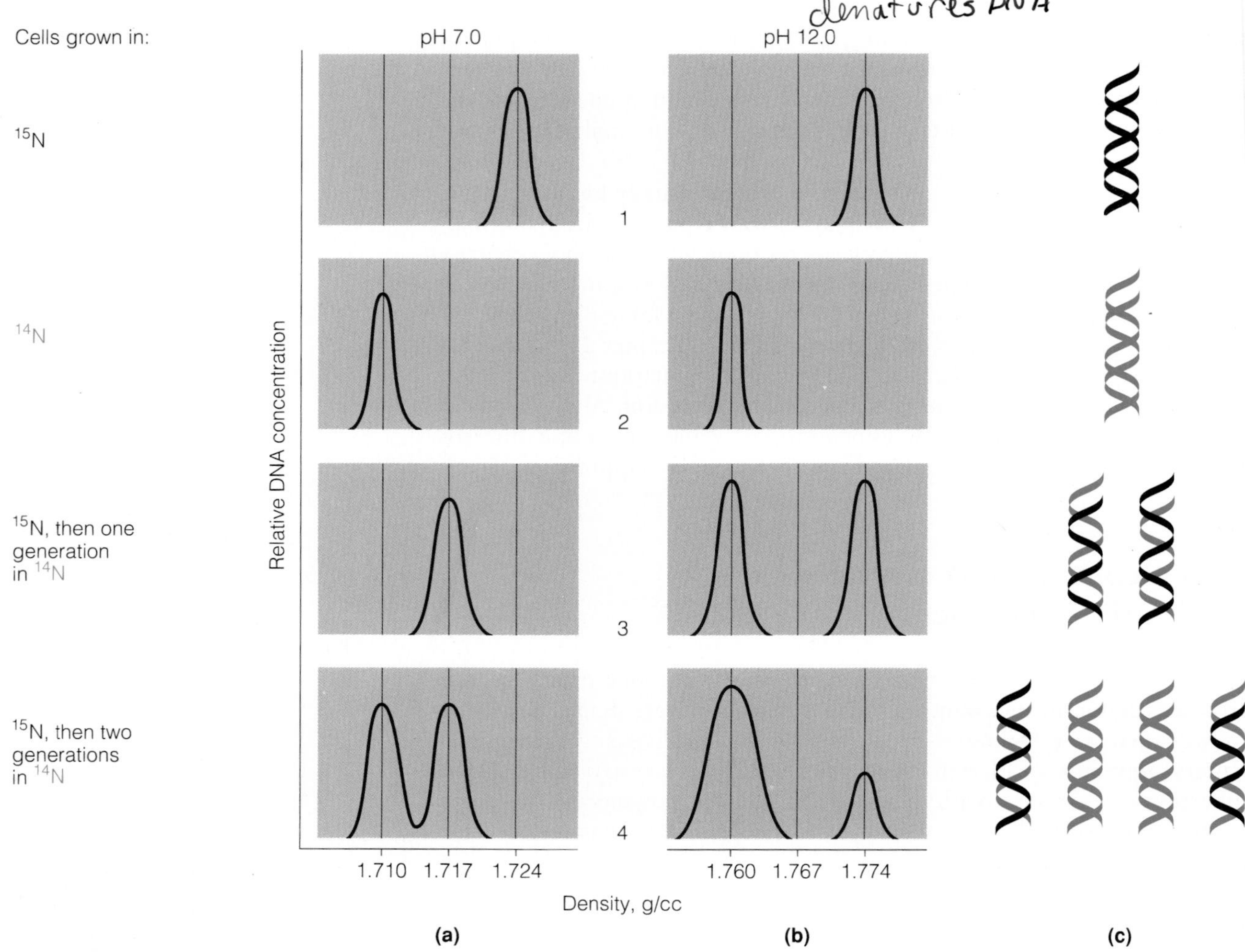

Figure 24.3
The Meselson–Stahl experiment demonstrating semiconservative DNA replication. The CsCl gradient profiles show density distributions of DNAs isolated from *E. coli* after growth in heavy (^{15}N) or light (^{14}N) medium. Gradients were run in (**a**) neutral or (**b**) alkaline CsCl solutions. Column (**c**) shows the DNA species assumed to be present under each set of conditions.

many generations, so that DNA achieved a higher density through extensive substitution of ^{15}N for ^{14}N in its purine and pyrimidine bases. When isolated and centrifuged to equilibrium at pH 7.0, this DNA formed a single band in a region of the gradient corresponding to a buoyant density of 1.724 g/ml (Figure 24.3, graph a1). By contrast, when DNA from bacteria grown in light medium (containing ^{14}N) was similarly analyzed, it banded at a density of 1.710 g/ml (graph a2).

Now when density-labeled bacteria grown in heavy medium were transferred to *light* medium, containing ^{14}N, the DNA isolated after one generation of growth banded exclusively at an intermediate density, 1.717 g/ml (graph a3). This result is expected if the newly replicated DNA is a *hybrid* molecular species, consisting of one-half parental material and one-half new DNA (synthesized in light medium). If the bacteria were cultured for an additional generation in light medium, two equal-sized bands were seen, one light and one of hybrid density (graph a4), as expected if the hybrid-density DNA underwent a second round of semiconservative replication.

These results were consistent with the idea that each replicated chromosome contains one parental strand and one daughter strand, but the data did not preclude alternative forms of semiconservative replication, which might involve DNA strand breakage. These models were ruled out by centrifugal analysis of the density-labeled DNAs at pH 12, where DNA denatures and strands separate (Figure 24.3, column b). When centrifuged to equilibrium after one generation in light medium, the DNA formed two bands, one light (equal to that seen on analysis of ^{14}N-grown bacteria) and one heavy (graph b3). DNA analyzed after a second round of replication in light medium showed three-fourths light and one-fourth heavy material (graph b4). The inescapable conclusion was that the replicative hybrid contains one complete strand of parental DNA and one complete strand of newly synthesized DNA (Figure 24.3, column c).

As discussed in the next chapter, analysis of density-labeled DNAs has been widely used for studies of DNA repair and recombination, as well as replication. However, stable isotopes are rarely used for these experiments, partly because the small density differences that can be achieved limit the resolution of this technique. What is often used as an alternative density label is 5-bromodeoxyuridine, which, as noted in Chapter 22, is efficiently incorporated into DNA in place of thymidine. This gives a much larger density increment in labeled DNA.

Sequential Nature of Replication

The Meselson–Stahl experiment showed that DNA is partitioned at replication, with parental DNA yielding two daughter replicas, each containing one old and one new strand. However, the experiment said nothing about the replication mechanism. For example, one could visualize complete strand unwinding, followed by the synthesis of a daughter strand on each of the separated parental templates. Because single-stranded DNA might be unstable inside the cell, it was more attractive to think of strand unwinding occurring simultaneously with the synthesis of daughter strands—in other words, to think of replication as *sequential* and *ordered*. This means that replication would start at one or a few fixed points on a chromosome, with **replication forks** emanating from these fixed points, or **replication origins.** Each fork would consist of two daughter strands being elongated on parental templates simultaneously with unwinding of the parental duplex (see Figure 4.13).

The sequential nature of replication was shown originally by genetic experiments and by direct visualization of replicating molecules. The genetic experiments were based on the premise that a chromosome caught in the midst of replication has two copies of every gene that has replicated and one copy of every nonreplicated gene. Thus, the closer a gene to a replication origin, the higher its **copy number,** or number of copies per cell. Genetic analysis can determine the copy number of genetic markers distributed about the chromosome. (A **genetic marker** is simply an allele of a particular gene whose relative abundance can be quantitated experimentally.) Plotting marker frequency as a function of genetic map position can in turn establish the approximate location of an origin and predict the direction of replication.

This approach was originally applied to the replication of *E. coli* DNA. The earliest experiments indicated that the bacterial chromosome contains a single origin of replication and that the fork moves in one direction from this origin. A similar interpretation was made from the earliest visualization of replicating *E. coli* DNA. John Cairns accomplished the exceedingly difficult task of lysing *E. coli* cells under conditions in which some of the fragile DNA molecules disentangled themselves without being broken by shear forces. When these molecules were made radioactive, by prior growth of the bacteria in medium containing [^3H]thymidine, the disentangled DNA could be visualized by radioautography (Figure 24.4). Although 2 months' exposure time was needed, the results were worth the wait. Ex-

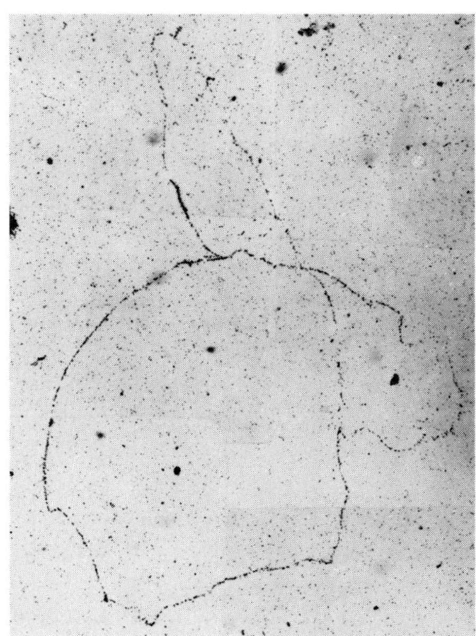

Figure 24.4
Radioautogram of a replicating *E. coli* chromosome after two generations of growth in [^3H]thymidine.

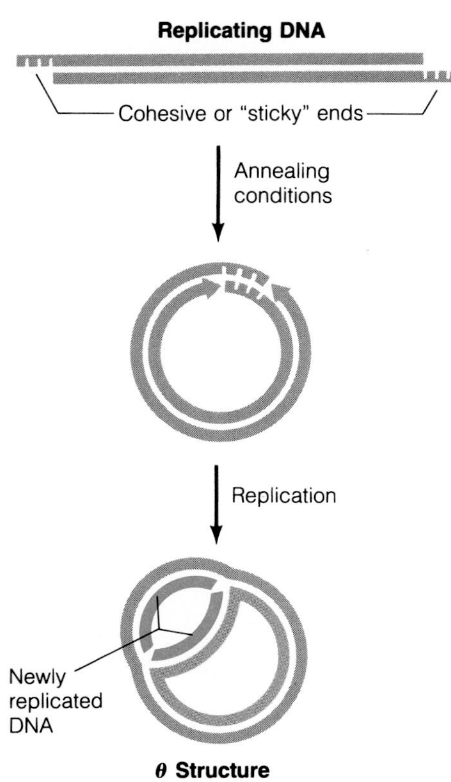

(b)

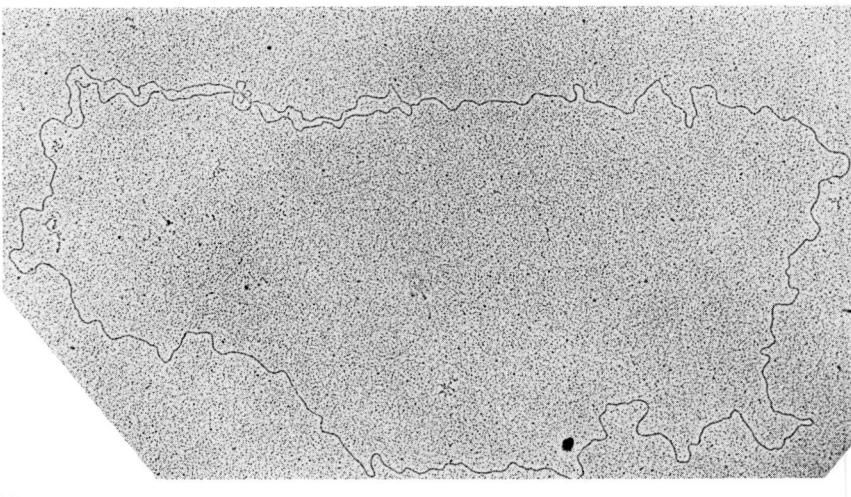

Figure 24.5
Bidirectional replication of lambda phage DNA. (**a**) Linear DNA followed by circularization involving base pairing of the cohesive ends. After covalent closure, replication can begin, yielding a theta structure. (**b**) Electron micrograph of a theta structure, or circular DNA molecule in the midst of replication.

amination of individual molecules, such as that shown in the figure, revealed that virtually all of the intrabacterial DNA was present as one giant, circular molecule, more than 1 *millimeter* long. (Recall from Chapter 4 that these molecules must be folded and supercoiled in order to fit inside bacterial cells that are just a few *micrometers* in length.)

Some of the molecules, such as that shown in Figure 24.4, contained two branch points and were identified as molecules caught in the midst of a round of replication. In the experiment described, DNA had been labeled by two generations of bacterial growth in [^{3}H]thymidine-containing medium. Determination of the **grain density,** or number of dark spots on the film per unit distance, allowed Cairns to identify regions of the structure that had replicated either once or twice; the latter have twice the grain density of the former. Cairns suggested that one of the two branch points represented an origin of replication and the other represented a replication fork, moving in one direction away from this origin. However, the results were also consistent with the idea of **bidirectional replication**—one fixed origin and two replication forks, each moving away from the origin until they join halfway around the circular template.

Origin and Direction of Replication

By the early 1970s it became possible to isolate replicating DNA in sufficient quantity for its analysis at higher resolution. Such experiments showed that most (though not all) DNA molecules replicate bidirectionally. The earliest definitive experiments were carried out by electron microscopic analysis of replicating DNA from the bacterial virus, or bacteriophage, λ. DNA isolated from the virus particle is a linear duplex, 17.5 μm long, containing 48,502 base pairs. It has an additional interesting feature: each of the two 5′ DNA termini is extended by 12 nucleotides, to give a short, single-stranded tail at each end. The two tails are complementary in base sequence, so that under annealing conditions, which promote DNA renaturation, the linear DNA spontaneously circularizes by formation of Watson–Crick base pairs between these short tails. The tails are called **cohesive ends** or **sticky ends** (Figure 24.5a). The resultant single-strand interruptions, or **nicks,** become covalently closed in vivo, as can be shown by the isolation of circular phage DNA molecules from infected cells. These circles can be visualized directly by coating DNA with a basic protein, spreading the coated DNA on an electron microscope grid, and shadowing the preparation with a heavy metal, which allows visualization of the DNA–protein complex (Figure 24.5b; see also Tools of Biochemistry 1).

When partially replicated λ DNA molecules are isolated, one sees a population of circles, each with two branch points, comparable to the structures seen by Cairns in replicating *E. coli* DNA, but much smaller. These are called **theta structures** because of their resemblance to the Greek letter θ (Figure 24.5a). A technique called **denaturation mapping** permits alignment and mapping of replication forks within a population of circular replicating molecules. This technique is based on the fact that λ DNA has several regions rich in adenine and thymine, which melt preferentially when DNA molecules are examined under partially denaturing conditions. These single-stranded regions, which can be visualized electron microscopically, provide a marker, allowing one to align circular replicating molecules with respect to the cohesive ends of the linear duplex. When a population of partially replicated molecules was analyzed, the data could be interpreted only in terms of bidirectional replication, with one specific replication origin, located toward the right end of the linear DNA molecule, and with replication forks proceeding in both directions from that origin.

In the early 1970s, genetic experiments at higher resolution indicated that the *E. coli* chromosome also replicates bidirectionally from a single fixed origin. This conclusion was reinforced by a variation of Cairns's radioautographic experiment, which led to an unambiguous conclusion. The experiment involved specific labeling of DNA synthesized during termination of one round of replication and initiation of the next. If replication is *bidirectional,* DNA labeled during termination would be labeled 180° away from the replication origin, where the next round would commence. Conversely, if replication is *unidirectional,* termination and reinitiation would occur in adjacent regions of the chromosome. The chromosome was labeled by growth of an *E. coli* culture in medium containing [^{3}H]thymidine. When replication was nearly complete, the specific activity was increased severalfold, so that any DNA synthesized during termination and subsequent reinitiation would be recognized by its higher grain density on a radioautogram. The experiment clearly showed that termination and reinitiation occur on opposite sides of the chromosome (Figure 24.6); thus, replication is bidirectional.

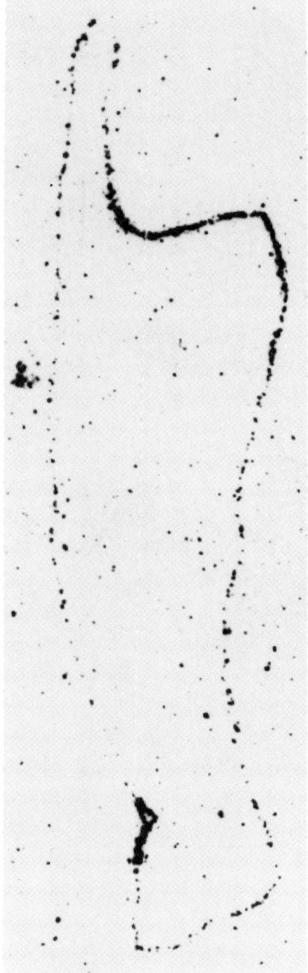

Figure 24.6
Bidirectional replication. Radioautography demonstrates that replication termination and initiation (heavy grain density) occur on opposite sides of the *E. coli* chromosome. The chromosome was labeled by long-term growth in [^{3}H]thymidine, and the specific activity was increased fourfold just before a synchronized culture was about to complete a cycle of replication.

Carefully timed radioautographic experiments also allow estimation of the rate of replication fork movement. The number of base pairs of DNA replicated in a given time is measured simply from the length of a labeled segment in micrometers and the number of base pairs in 1 μm (about 3000 if DNA is in the B form, with 0.34 nm per base pair). An optimally nourished *E. coli* cell at 37° shows a rate of fork movement of about 850 base pairs per second.

A simpler method for aligning replicating circular molecules came with the discovery of **restriction endonucleases.** As discussed in the next chapter, these are enzymes that recognize specific oligonucleotide sequences and catalyze double-strand breaks at or near these sites. For example, the enzyme ***Eco*RI** cleaves DNA within the following hexanucleotide sequence.

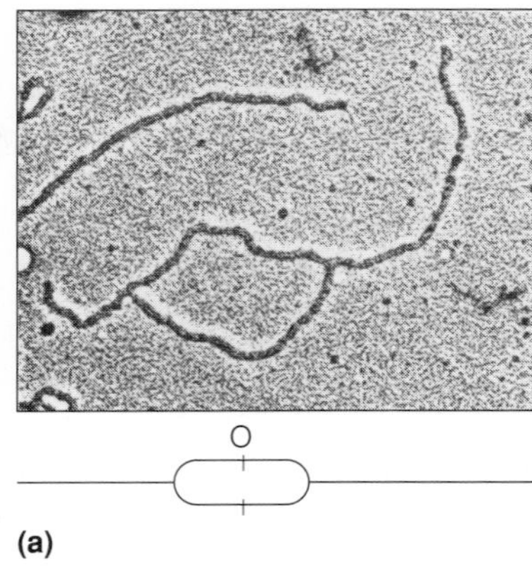

(a)

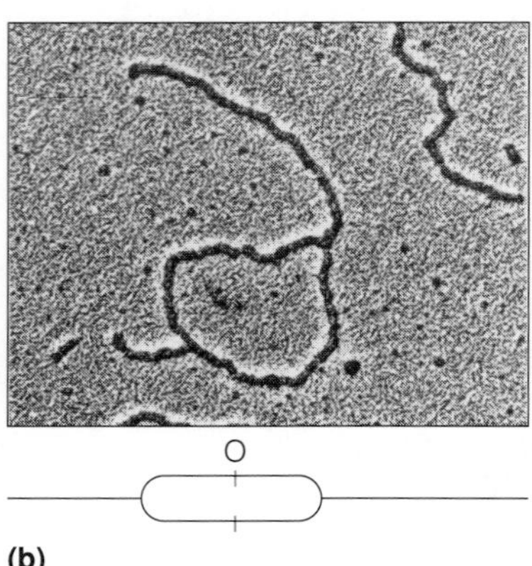

(b)

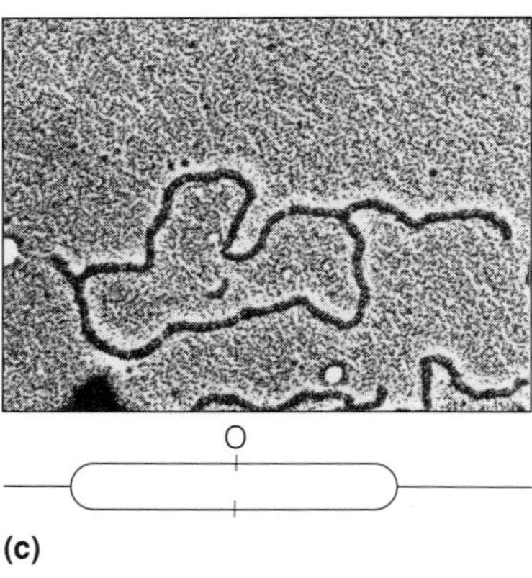

(c)

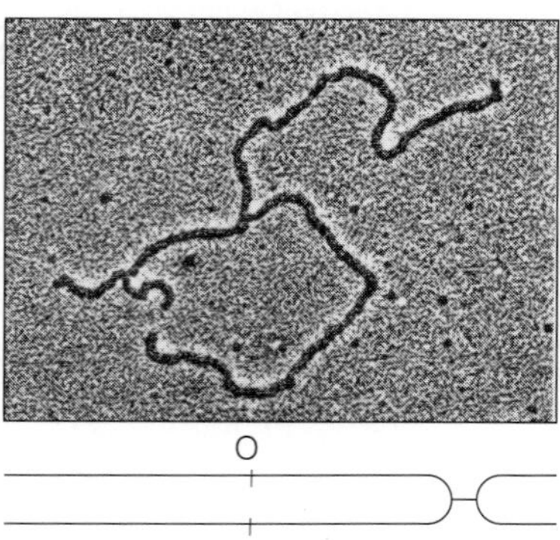

(d)

5′ ------ GAATTC ------

3′ ------ CTTAAG ------

To see how this enzyme can be used to analyze replication, consider the DNA genome of the small animal virus SV40. This DNA duplex, some 5224 nucleotide pairs long, contains just one site for *Eco*RI cleavage. Thus, treatment with this enzyme "linearizes" DNA, that is, converts the circles to linear molecules of equivalent length. Partially replicated SV40 DNA consists of a population of theta structures, just as seen with phage λ. Treatment of such a population with *Eco*RI allows one to align these molecules with respect to the cleavage site, in the same sense that denaturation mapping allows alignment with respect to AT-rich regions. Figure 24.7 shows the data expected for bidirectional replication. Like λ and *E. coli*, SV40 replicates bidirectionally from one fixed origin.

Do these experiments mean that all DNA molecules are circular, with bidirectional replication and one fixed origin? Not at all; we shall encounter examples of unidirectional replication as we go along, as well as the use of more than one origin per chromosome. In fact, there seem to be numerous variations on the theme we have developed thus far. Recent experiments suggest that some eukaryotic DNAs in early development, when cell division is rapid, replicate by a mechanism that does not involve a replication fork, perhaps with strand unwinding uncoupled from the synthesis of daughter strands.

Units of Replication: The Replicon

Early work on **extrachromosomal genetic elements** contributed importantly to our understanding of the control of DNA replication. In bacteria these elements consist of circular DNA molecules, called **plasmids,** which are much smaller than the chromosome and which replicate independently of the mechanisms that control the replication of chromosomal DNA. Some plasmids, called **sex factors,** specify gene products that allow bacteria to mate with one another and undergo sexual reproduction. Others, called **bacteriocinogenic factors,** provide a defense mechanism. They encode proteins that are released from the cell and that kill other bacteria; these proteins are called **bacteriocins.** Of great medical importance are **drug resistance factors**—plasmids carrying genes that specify resistance to antibiotics; for example, penicillin-resistant strains of bacteria often harbor a plasmid that specifies **β-lactamase,** an enzyme that destroys β-lactam antibiotics such as penicillin. Because plasmids are readily transferred from cell to cell, drug-resistant bacterial populations are arising with alarming frequency as the result of indiscriminate use of antibiotics to combat infectious diseases.

A few plasmids, called **episomes,** can undergo **integration**—insertion of the plasmid DNA into the chromosomal DNA sequence. When this happens, the integrated episome DNA no longer replicates autonomously but instead replicates in synchrony with the chromosome into which it has been

Figure 24.7
Analysis of SV40 DNA replication by linearization of branched circles. The electron micrographs show individual molecules that had been treated with *Eco*RI nuclease. Beneath each is an interpretation of the molecule, derived from the length of each branch of the micrograph and alignment of the ends. O denotes the bidirectional origin.

inserted (Figure 24.8). Thus, whereas a nonintegrated plasmid may be present at several dozen copies per chromosome and replicate independently of the chromosome, the integrated episome is present at one copy per chromosome and can replicate only when the chromosome does so.

Now consider the relationship between an episome and a bacterium carrying a temperature-sensitive *(ts)* mutation in a chromosomal gene whose product is essential to initiating DNA replication. Such a bacterium cannot initiate a round of DNA replication at a moderately elevated temperature (42°C). Integration of an episome reverses the temperature-sensitive phenotype, allowing DNA replication to be initiated at the higher temperature. However, replication is initiated not from the ordinary origin on the chromosome but from a site within the episome (see Figure 24.8c).

These observations led to the concept that the initiation of DNA replication is controlled by a small cluster of genes—a sequence of nucleotides at which replication begins (the origin) and structural genes for proteins whose action on the origin sequence initiates replication at that point. These genes, plus all of the DNA that is replicated under their control, are collectively termed a **replicon.** The bacterial chromosome constitutes a single replicon, because its entire replication commences from one fixed origin. The same is true for an extrachromosomal plasmid. However, when a plasmid integrates as an episome, the chromosome containing the episome constitutes one replicon, with replication usually initiated at the chromosomal origin.

Although a chromosome is one large molecule of DNA, a single chromosome is not always a single replicon, with replication of the chromosome driven from a single origin. Radioautographic analysis of eukaryotic DNA replication shows that mammalian cells have between 10^3 and 10^4 replication origins per complete genome. Replication is bidirectional from these origins, with each fork progressing until it merges with the fork started from initiation of the adjacent replicon. Since the human genome contains 46 chromosomes, this means that a typical chromosome consists of several dozen replicons.

Multiple replication origins are important for eukaryotic DNA replication because eukaryotic cells contain a great deal of DNA that must be replicated within a limited period of each cell division cycle (see Figure 28.1a). During mitosis, this **cell cycle** consists of four distinct phases: *M*, mitosis; G_1, pre-DNA synthetic phase; *S*, DNA synthetic phase; and G_2, post-DNA synthetic phase. The letter G stands for gap, a period when neither DNA nor the cell is actively dividing.

Radioautographic analysis shows that eukaryotic DNA replication rates are about 10-fold lower than prokaryotic rates. As you can estimate from the data in Table 24.1, it could take weeks for a human cell to replicate all of its DNA once, if each chromosome contained but a single origin and two replication forks. As it is, DNA replication occupies only a fraction of the cell division cycle; in a 24-hour cell cycle the S phase is typically 6 to 8 hours.

An apparently anomalous situation occurs in bacteria: more time is needed to replicate the single chromosome than is required for the complete cell division cycle. A bacterial culture can double in cell number every 20 minutes, but 40 minutes is required for each round of chromosome replication. How can cells divide in *less* time than needed for replication of the chromosome? The answer is that in rapidly dividing cells a replicating DNA molecule can reinitiate a new round of replication from the origin before completion of the round in progress. In other words, initiation of replication occurs more frequently than once every 40 minutes. This can be demonstrated by radioautography, as shown in Figure 24.9. Thus, when a cell

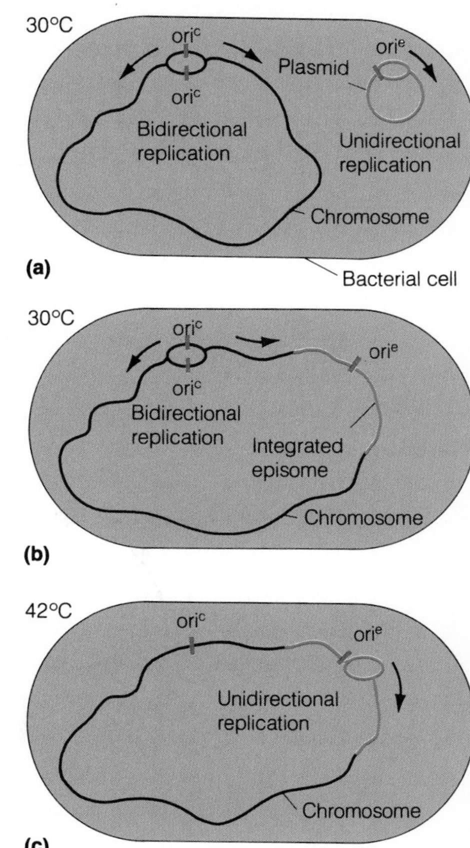

Figure 24.8
Replication driven by an integrated episome. (**a**) An *E. coli* cell with a *ts* defect in replication initiation, carrying a nonintegrated plasmid with a unidirectional origin (ori^e). Chromosome replication initiates bidirectionally from ori^c. (**b**) When the plasmid integrates as an episome, it replicates only when the chromosome replicates—bidirectionally from ori^c. (**c**) When chromosomal replication is inhibited by shift to a nonpermissive temperature (42°C), replication of the entire chromosome can be driven from the unidirectional origin (ori^e) on the integrated episome.

Table 24.1
Quantitative parameters of DNA replication in different cells[a]

	Replication Process	
	E. coli	Human
DNA content, number of nucleotide pairs per cell	3.9×10^6	ca. 10^9
Rate of replication fork progression, μm/min	30	3
DNA replication rate, nucleotides/sec per replication fork	850	60–90
Number of replication origins per cell	1	10^3–10^4
Hours required for complete genome replication	0.67	8
Hours required for one complete cell division	0.33	24

[a]The data are for an *E. coli* cell optimally nourished and cultured at 37°C. The values for human cells represent data from HeLa cells, which were originally derived from a tumor and have been maintained in culture for many years.

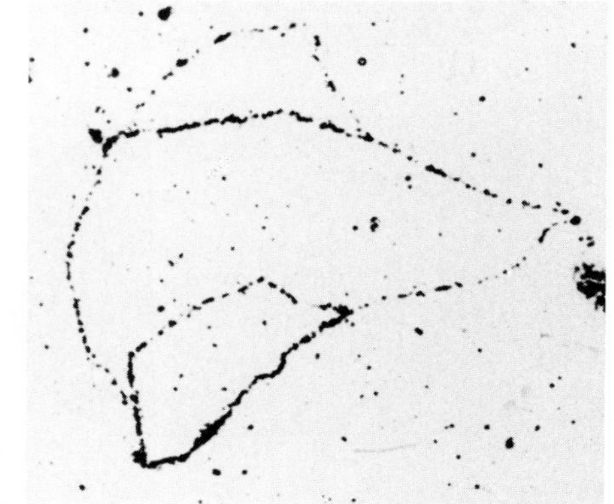

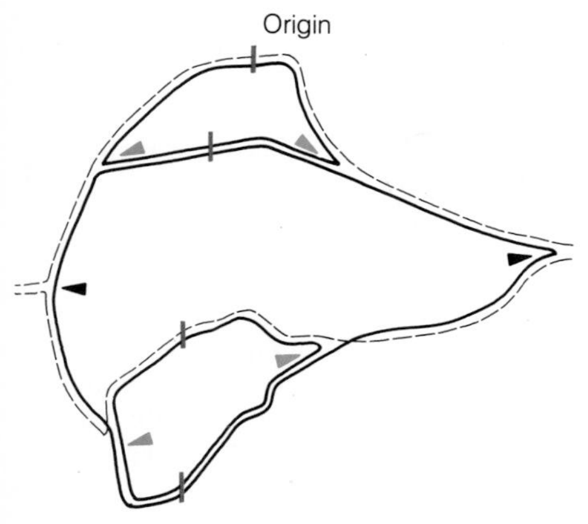

Figure 24.9
Reinitiation of a round of replication before termination of the previously initiated round. Labeling with [^{3}H]thymidine was started simultaneously with initiation of chromosome replication, and reinitiation occurred later. Radioactivity shows that one of the reinitiated branches contains two labeled strands, while the other contains only one. In the lower diagram solid lines denote radiolabeled DNA, and dashed lines represent unlabeled DNA. Black wedge = first-generation replication fork; blue wedge = second-generation replication fork.

divides, each daughter cell receives a chromosome that is far along into its next round of replication. The experiments that established this constitute some of the strongest evidence that DNA replication is controlled primarily at the level of initiation.

DNA Polymerases: Enzymes Catalyzing Polynucleotide Chain Elongation

Discovery of DNA Polymerase

Most of what we have presented so far about DNA replication was determined by in vivo approaches. The complementary in vitro investigations began in the mid-1950s, with Arthur Kornberg's search for an enzyme capable of catalyzing nucleotide incorporation into DNA. From his previous achievements in bioenergetics, Kornberg realized that the substrates for DNA replication should be activated derivatives of the DNA nucleotide residues. Thus, the endergonic synthesis of phosphodiester bonds between nucleotide residues could be coupled to the exergonic breakdown of activated substrates, just as occurs in synthesis of other macromolecules, such as glycogen. Kornberg correctly predicted that these activated nucleotides would be the 2′-deoxyribonucleoside 5′-triphosphates (dNTPs), and he synthesized them chemically. When radioactively labeled dNTPs were incubated with an extract of soluble proteins of *E. coli,* a small amount of radioactivity was incorporated into high-molecular-weight material. The enzyme required added DNA, plus Mg^{2+}. The incorporation of radioactivity into acid-insoluble material provided an assay for purification of the enzyme, which was named **DNA polymerase** (official name **deoxynucleotidyltransferase**). The enzyme catalyzes the following reaction, where dNMP is any deoxyribonucleoside monophosphate and dNTP is any triphosphate.

$$\underset{}{\text{dNTP}} + \underset{\text{DNA}}{(\text{dNMP})_n} \xrightleftharpoons{Mg^{2+}} \underset{\text{elongated DNA}}{(\text{dNMP})_{n+1}} + \text{PP}_i$$

The product is covalently attached to the DNA that was added to start the reaction. Note that the reaction is readily reversible; in cells and in crude preparations the reaction is drawn to the right by pyrophosphatase action on the other product of the reaction. Thus, two energy-rich phosphates are expended per nucleotide incorporated. This is important because of the mechanism of the reaction. The 3′ hydroxyl group of the primer terminus carries out nucleophilic attack on the inner (α) phosphate of the deoxyribonucleotide substrate, which leads to covalent bond formation. Since the 3′ hydroxyl is a very weak nucleophile, the triphosphate moiety on the dNTP substrate is essential to provide a good leaving group, which drives the reaction in an energetic sense.

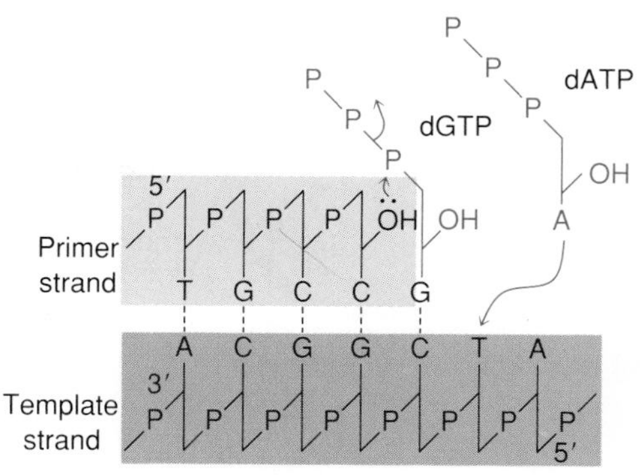

Figure 24.10
The DNA polymerase reaction. dGTP, which has been fitted to a C in the template, is about to be incorporated into the growing chain by nucleophilic attack by the 3′ hydroxyl group on the α phosphate of dGTP. The next nucleotide, dATP, is about to be fitted to the template.

Structure and Activities of DNA Polymerase I

The DNA polymerase discovered by Kornberg was later shown to be one of three different DNA polymerases in bacterial cells. The Kornberg enzyme is now called DNA polymerase I; we shall use this term henceforth and describe the other two polymerases later in this chapter.

ROLES OF DNA IN THE POLYMERASE REACTION. DNA plays two distinct roles in the DNA polymerase reaction. One role is as a **primer,** providing a 3′ hydroxyl terminus, to which polymerase action adds deoxyribonucleotide residues. Second, DNA provides a **template,** specifying which nucleotides are to be incorporated into a growing polynucleotide chain (Figure 24.10). Each incoming nucleotide is selected by fitting to the opposite base in the template strand, by Watson–Crick base-pairing interactions. In the example shown in Figure 24.10, dATP has been selected as the next nucleotide to be incorporated, because the opposite residue in the template strand is dTMP. Thus, the product of the reaction is covalently bound to the 3′ terminus of the primer, and the nucleotide sequence is complementary to that of the template strand. The product and primer strands have opposite polarities, as in natural duplex DNAs. The need for a 3′ primer terminus and a single-stranded DNA template limits the kinds of reactions that the enzyme can catalyze in vitro.

As shown in Figure 24.11a, the enzyme can copy around a circular single-stranded template, such as the DNA extracted from small bacteriophages, like ϕX174 or M13, as long as a primer is present, but it cannot join the ends. When the template is linear, polymerase copies only to the 5′ end of the template and then it dissociates (Figure 24.11b). If a single strand of DNA contains internally complementary sequences, it can form a "hairpin" structure and serve as both template and primer (Figure 24.11c). In similar fashion the enzyme can fill in a **gap,** or a single-strand interruption in a duplex that has one or more nucleotides missing (Figure 24.11d). When polymerase action has closed the gap to a nick (a single-strand interruption with no missing nucleotides), the enzyme dissociates. The enzyme can also initiate synthesis at a nick, provided that the 3′ terminus is a hydroxyl group. Typically, when this occurs the 5′ end of the pre-existing DNA is displaced in advance of the nick. This is called **strand displacement** synthesis (Figure 24.11d). Under some conditions the displaced strand is degraded hydrolytically. This process, called **nick translation,** is due to an additional enzymatic activity of the DNA polymerase molecule.

MULTIPLE ACTIVITIES IN A SINGLE POLYPEPTIDE CHAIN. When DNA polymerase I was purified from *E. coli,* it was found to consist of a single polypeptide chain (M_r = 109,000). The purified enzyme contains two nuclease activities, one of which degrades single-stranded DNA from the 3′ end (called the **3′-exonuclease**), and one of which degrades base-paired

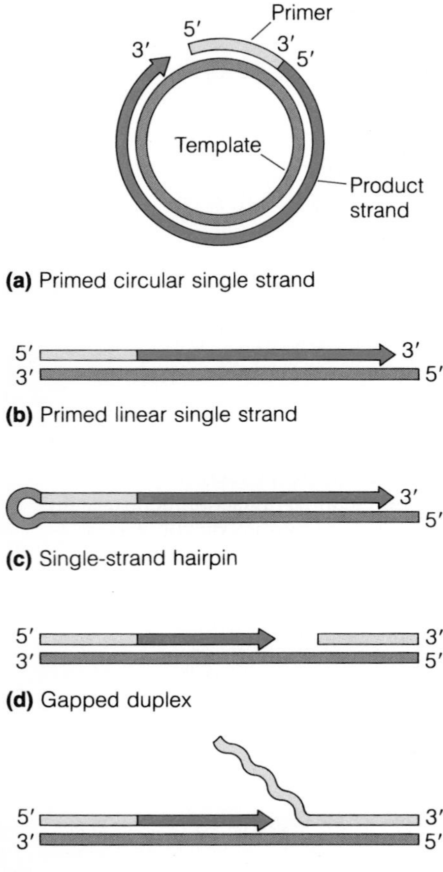

Figure 24.11
DNA substrates that can be acted on by purified DNA polymerase. Each blue arrow identifies a 3′ hydroxyl terminus at which chain extension is occurring.

DNA from the 5′ terminus (the **5′-exonuclease**). We now know that the 3′-exonuclease serves a "proofreading" function, to improve the accuracy with which a DNA template is copied. The activity will remove any improperly base-paired nucleotides from the growing 3′ end of a polydeoxynucleotide chain. Thus, if the polymerase makes an error and inserts an incorrect nucleotide, the 3′-exonuclease activity can recognize and excise the incorrect nucleotide, giving the polymerase activity a second chance to insert the nucleotide specified by the template. We shall say more about the 3′-exonuclease when we discuss the general problem of the fidelity of DNA replication later in this chapter.

The 5′-exonuclease activity plays two known roles. These came to light through studies of its role in nick translation. Under certain conditions, when DNA polymerase I binds to nicked duplex DNA at the site of a nick, it does not catalyze strand displacement synthesis. Instead, the 5′-exonuclease activity cleaves nucleotides, in advance of the polymerase activity, as the entire enzyme moves in a 5′ → 3′ direction along the chain being extended (Figure 24.12). Thus, synthesis of new DNA is coordinated precisely with degradation of preexisting DNA, with one old nucleotide being excised for each new nucleotide incorporated into a growing chain. No *net* synthesis of DNA occurs, and the only result is movement of the site of the nick in a 5′ → 3′ direction; hence, the term *nick translation*. Conditions that favor either nick translation or strand displacement are not yet known. Nick translation is used for radioactive labeling of DNA fragments in the laboratory (Chapter 25).

What sort of biological sense does this make, since no net synthesis of DNA occurs? First, nick translation is a way to remove DNA that has been damaged (for example, by radiation or chemicals), with the synthesis of correct DNA being coordinated with excision of damaged DNA (see Chapter 25). Second, the 5′-exonuclease can act on RNA, when that RNA is base-paired to a DNA of complementary nucleotide sequence. As we discuss later in this chapter, RNA plays a special role as a primer for biological DNA replication.

STRUCTURE OF DNA POLYMERASE I. The three catalytic activities of DNA polymerase I have been partially localized to regions of the long polypeptide chain, through limited proteolysis of the enzyme by subtilisin or trypsin. As shown by Hans Klenow, this proteolysis splits the 109-kDa polypeptide into a small N-terminal fragment (M_r = 34,000) and a large C-terminal fragment (M_r = 76,000). The large fragment (also called the Klenow fragment) contains the polymerase and 3′-exonuclease activities; the small fragment contains the 5′-exonuclease activity. How the three catalytic sites are arranged in space is very important to an understanding of how each of the two nuclease activities functions in concert with the polymerase. Crystallographic study of the large fragment has revealed a striking feature of the structure—a deep crevice, just large enough to accommodate B-form DNA, with a flexible subdomain that might allow bound DNA to be completely surrounded (Figure 24.13). Cocrystallization of Klenow fragment with a short duplex DNA shows that this is indeed the DNA-binding site and that bound DNA is almost completely surrounded by protein, as if the DNA helix were "threaded" into the protein. An interesting feature of the structure, revealed by study of mutant forms of the enzyme, is that the 3′-exonuclease active site is quite far—about 3 nm—from the polymerase active site. This suggests that DNA must rotate through the protein by a length of about eight nucleotides to move the primer 3′ terminus from one active site to the other.

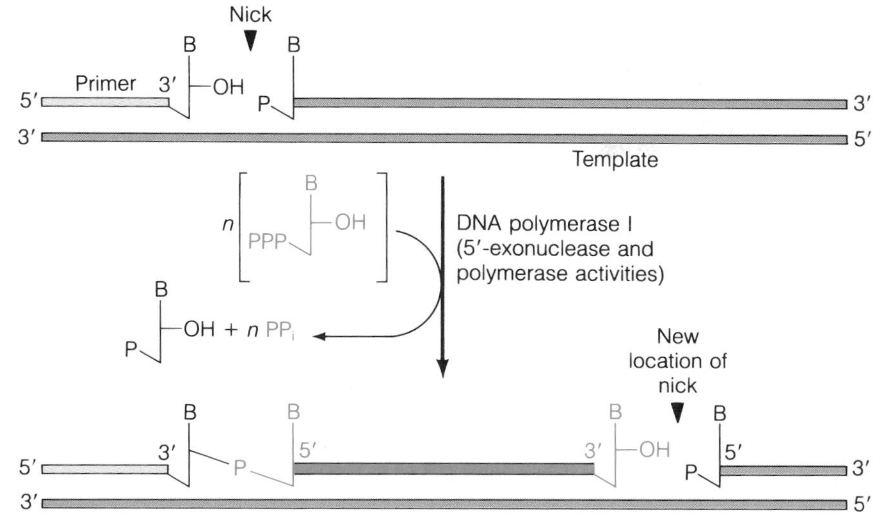

Figure 24.12
Nick translation. Newly incorporated nucleotides are shown in blue. B is any base (A, C, G, or T).

Since the 5′-exonuclease activity is absent in the Klenow fragment, this unnatural enzyme is a useful laboratory reagent for synthesis of DNA in vitro when one wishes specifically to avoid DNA degradation.

DNA Polymerases II and III

For his discovery of DNA polymerase, Arthur Kornberg was awarded the Nobel Prize in 1959. Although this was indeed a monumental discovery, several properties of the enzyme as isolated from *E. coli* were difficult to reconcile with the idea that this enzyme catalyzes the major nucleotide incorporation reactions in biological DNA replication. First, the enzyme in vitro is too slow, with a V_{max} of about 20 nucleotides per second, in comparison with replicative chain growth in vivo of about 500 nucleotides per second. Second, with about 400 molecules of enzyme per cell, the enzyme is present in vast excess over the small number (less than 10) of replication

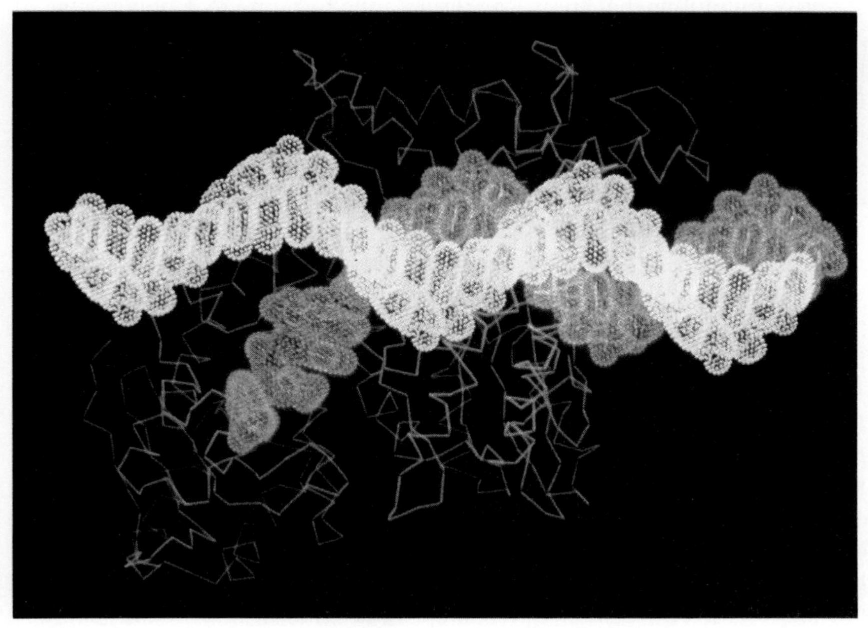

Figure 24.13
Structure of the complex between Klenow fragment of DNA polymerase I and DNA, as shown by co-crystallization of enzyme and DNA. Note the distance between polymerase and 3′ exonuclease active sites.

Table 24.2
DNA polymerases of *E. coli*

Characteristic	Polymerase I	Polymerase II	Polymerase III
Structural gene	*polA*	*polB*	*polC*
M_r	109,000	120,000	140,000
Number of molecules per cell	400	100	10
V_{max}, nucleotides/sec	16–20	2–5	250–1000
3′-Exonuclease	+	+	+[a]
5′-Exonuclease	+	−	+
Mutant phenotype[b]	UV^sMMS^s	None	DNA^{ts}

[a]This activity is carried on the ϵ subunit of pol III holoenzyme.

[b]UV^sMMS^s, sensitive to ultraviolet light, sensitive to methyl methanesulfonate; DNA^{ts}, temperature sensitive in DNA replication.

forks per cell. Third, DNA polymerase can extend chains only in a 5′ → 3′ direction (from the 5′ end). A complete understanding of replication requires knowing how both of the antiparallel chains of a DNA duplex are replicated within the same fork. Fourth, polymerase cannot initiate the synthesis of new DNA chains but can only extend from preexisting 3′ hydroxyl termini. Finally, genetic evidence suggested the existence of other polymerases, as well as additional enzymes and proteins essential to DNA replication, as described below.

An important development was the isolation by John Cairns, in 1969, of an *E. coli* mutant that was deficient in DNA polymerase I activity. With this very abundant enzyme absent from the cell, it became possible to detect the existence of two additional DNA polymerase activities in *E. coli* cells. These were named DNA polymerases II and III, with the original Kornberg polymerase now being named DNA polymerase I.

The properties of these three DNA polymerases are summarized in Table 24.2. Polymerase III was assigned the major role in nucleotide incorporation during replication, on the basis of its high V_{max}; note also that there are few molecules of the enzyme per cell, as expected if it functions only at replication forks. More important is the existence of temperature-sensitive mutations that specify a thermolabile form of DNA polymerase III, and in which DNA replication in vivo is blocked at high temperature. Since both temperature-dependent phenotypes can be traced to a single mutation, this presents powerful evidence that polymerase III plays an essential role in replication. The gene encoding the catalytic subunit of polymerase III was named *polC*. By convention, bacterial genes are given an italicized lowercase three-letter designation related to the function of the gene (*pol* for polymerase) and a capital letter denoting the order of discovery of the gene. *polA* and *polB*, the structural genes for polymerases I and II, respectively, were discovered earlier than *polC*. Gene *products* are given the corresponding capitalized and nonitalicized three-letter designations; thus, PolC is the product of the *polC* gene.

Mutants in polymerase II display no known phenotype, and the role of that enzyme is still not known. But what about polymerase I? The original *polA* mutant described by Cairns had no recognizable phenotype affecting replication. The bacteria were abnormally sensitive to ultraviolet irradiation and to certain alkylating agents that react with DNA. This discovery led to elucidation of the role of polymerase I in DNA repair. However, it was later found that the Cairns mutant, while lacking polymerase activity, did make a partial *polA* product, namely the small N-terminal fragment that contains

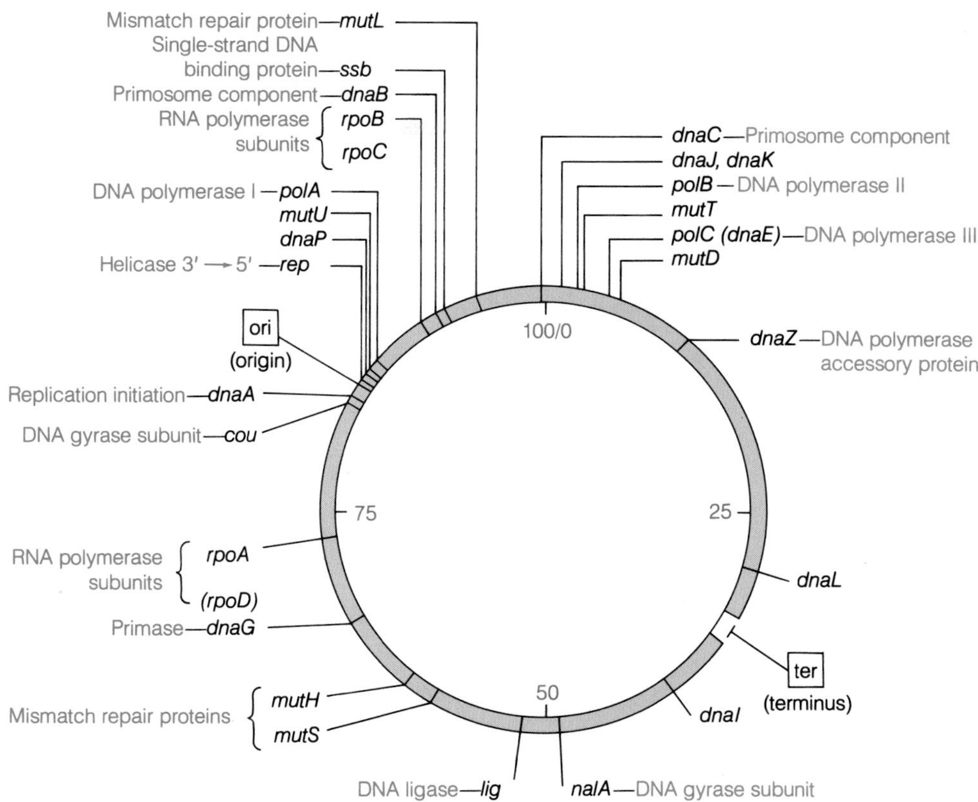

Figure 24.14
Partial genetic map of *E. coli,* showing some genes whose products are involved in DNA replication or repair. Some of the gene products are identified (blue). *dna* genes are those that play essential roles in replication. The *mut* genes specify proteins that, when mutant, cause elevated spontaneous mutation rates.

the 5′-exonuclease activity. Still later it was found that mutants defective in the 5′ exonuclease activity are also defective in DNA replication. Thus, it was established that two polymerases, I and III, play essential roles in DNA replication.

Clearly, the existence of mutants defective in a specific polymerase has been essential to identification of the true roles played by each polymerase. As we shall see, many other proteins are involved in DNA replication. Bacterial genetics has been indispensable for identifying these other proteins and their roles. Figure 24.14 shows part of the genetic map of *E. coli,* identifying the positions of genes known to be involved in replication. Some of the genes were originally discovered without knowledge of the nature of the gene product, simply by the existence of temperature-sensitive mutants. Typically, *ts* mutants grow at 30°C but not at a moderately elevated temperature such as 42°C. Temperature-sensitive mutations can affect any function essential to growth. When the mutation affects a protein indispensable for DNA replication, the cells display thermolabile replication. After mapping, these replication-specific genes were assigned names: *dnaA, dnaB, dnaC,* and so forth. When the gene originally called *dnaE* was found to encode DNA polymerase III, that gene was given the slightly more descriptive name *polC.*

The DNA Polymerase III Holoenzyme

The *polC* gene encodes a single polypeptide chain of M_r about 140,000. Although this protein has an intrinsic polymerase activity, in bacterial cells it functions as part of a multiprotein aggregate called the **DNA polymerase**

Table 24.3
Subunit composition of *E. coli* DNA polymerase III species

Subunit	Gene (where known)	M_r	Function
α	*polC*	130,000	Polymerase III
ϵ	*dnaQ (mutD)*	27,500	3′-Exonuclease
θ	Unknown	10,000	
τ	*dnaZX*	71,000	Core enzyme dimerization
γ	*dnaZ*[a]	52,000	Processivity
δ	Unknown	32,000	Processivity
χ[b]		14,000	
ψ[b]		12,000	
β	*dnaN*	40,600	Initiation complex formation

Polymerase species	Subunit composition
Pol III catalytic subunit	α
Pol III core enzyme	α-ϵ-θ
Pol III′ core enzyme dimer	$(\alpha\text{-}\epsilon\text{-}\theta\text{-}\tau)_2$
Pol III*	$(\alpha\text{-}\epsilon\text{-}\theta\text{-}\tau\text{-}\gamma\text{-}\delta)_2$
Pol III holoenzyme	$(\alpha\text{-}\epsilon\text{-}\theta\text{-}\tau\text{-}\gamma\text{-}\delta\text{-}\beta)_2$

[a] The γ subunit is encoded in the N-terminal portion of the structural gene for τ and in the same reading frame.

[b] These have not been established as functional subunits of the holoenzyme.

SOURCE: C. S. McHenry, *Annu. Rev. Biochem.* 57:519–550 (1988).

III holoenzyme. As noted in Table 24.3, the holoenzyme contains seven different polypeptide chains. Some of these subunits are essential for normal DNA replication in vivo, as shown by the existence of *ts* mutations in genes encoding these polypeptides. Three of these subunits—α, ϵ, and θ—form the **polymerase III core enzyme.** In the presence of two molecules of an additional subunit, τ, the core enzyme dimerizes. The binding of two additional polypeptides, γ and δ, converts the aggregate to a species called **pol III***. This binds to a large polypeptide, called β, to give the holoenzyme. Binding of these additional subunits increases the rate of the reaction and its **processivity,** that is, the ability of the enzyme to remain associated with a template–primer once it has become bound. Processivity is defined as the number of nucleotides incorporated per binding event between a polymerase molecule and a 3′ primer terminus. Clearly, in a cell with very few molecules of replicative polymerase and very few replication forks, it is advantageous for a molecule, once bound, to remain bound. While polymerase I has a processivity of about 20, the polymerase III holoenzyme incorporates many thousand nucleotides per binding event.

Eukaryotic DNA Polymerases

Eukaryotic cells contain four distinct DNA polymerases—α, β, γ, and a recently described enzyme called δ. The eukaryotic enzymes are distinguished from each other in large part by their intracellular locations, kinetic properties, and responses to inhibitors. Even though the genetic analyses are not nearly as advanced as those for bacteria, the biological roles of these polymerases have been established with some confidence.

Table 24.4 summarizes the properties of DNA polymerases α, β, and γ. We shall say more about polymerase δ below. DNA polymerase γ is

Table 24.4
Properties of eukaryotic DNA polymerases

	α	β	γ
Cell compartment	Nucleus	Nucleus	Mitochondrion
Biological role	Nuclear DNA replication	DNA repair	Mitochondrial DNA replication
Number of subunits	4–8	1	4 (identical)
M_r of catalytic subunit, kDa	120–180	30–50	50
K_M for dNTPs, μM	10	10	0.5
Processivity	Moderate	Low	High
Associated exonuclease	No[a]	No	No
Heat sensitivity	Low	High	High
Sensitivity to sulfhydryl reagents	High	Low	High
Sensitivity to 2′,3′-dideoxynucleoside triphosphates	Low	High	High
Sensitivity to arabinosylCTP	High	Low	Low
Sensitivity to aphidicolin	Yes	No	No

[a] A cryptic, or masked, 3′-exonuclease has recently been found associated with polymerase α from *Drosophila*.

localized to mitochondria. It is a highly processive enzyme, with high affinity for its deoxyribonucleotide substrates. It participates in the replication of mitochondrial DNA, and it seems also to replicate the DNAs of certain small viruses that do not encode their own DNA polymerase. Polymerase β is found in the nucleus. The enzyme has low processivity; it dissociates from the template after each nucleotide incorporation step. The enzyme, which has an alkaline pH optimum, seems to be involved in DNA repair.

Most interest has focused on DNA polymerase α, which is thought to be the replicative polymerase—the eukaryotic counterpart of polymerase III in prokaryotes. This function was established primarily through the use of an interesting inhibitor called **aphidicolin,** a fungal product with a steroid-like structure, which inhibits replicative DNA synthesis specifically in eukaryotic cells. Polymerase α is distinct from polymerases β and γ in being sensitive to inhibition by aphidicolin. Some aphidicolin-resistant eukaryotic cell mutants contain an aphidicolin-resistant form of DNA polymerase α and show aphidicolin-resistant DNA replication in vivo. This provides strong evidence that polymerase α participates in DNA replication, because a single mutational event changes the effects of the inhibitor both on the enzyme and on the DNA replication process.

HO CH_2OH H H_3C H HO HOH_2C H CH_3

Aphidicolin

The mechanism by which aphidicolin inhibits DNA polymerase remains mysterious. Kinetically, the inhibition appears to be competitive with dNTP substrates, particularly with dCTP. Since there is no discernible structural similarity between dCTP and the rather apolar aphidicolin, it is not clear how both compounds could compete for binding to the same site on the enzyme.

Like prokaryotic pol III, DNA polymerase α is a multisubunit enzyme. However, the subunit structure is not clear, partly because of variation in the different systems studied, but mostly because the subunits are extraordinarily sensitive to proteolytic degradation during isolation and purification. Molecular weights from 120 to 180 kDa have been reported for the catalytic subunit, and the total number of reported subunits varies between four and eight.

DNA polymerase δ has been characterized relatively recently and has not yet been shown to exist in all eukaryotic cells. Polymerase δ is a multisubunit enzyme that is distinct from the other three in that it does contain an associated 3′-exonuclease activity and thus can proofread newly replicated DNA. Polymerase δ, like α, is inhibited by aphidicolin, which suggests that it might also be a replicative polymerase. Some consider polymerase δ to be a modified form of polymerase α, since some monoclonal antibodies raised against α also react with polymerase δ. An attractive hypothesis, is that α and δ are involved in lagging and leading strand synthesis, respectively.

Viral DNA Polymerases

Small viruses, like bacteriophage ϕX174 and M13, replicate their DNA largely by using enzymatic machinery of the host cell. By contrast, larger viruses encode most of their own replication enzymes and proteins. Among the largest DNA viruses are the T-even bacteriophages, T2, T4, and T6, which infect *E. coli;* the herpesviruses, including herpes simplex, which cause ocular and genital infections in mammals; and vaccinia virus, which is used to vaccinate against smallpox.

The DNA polymerase specified by bacteriophage T4 is of special interest because it was the first DNA polymerase from any organism shown to participate directly in DNA replication. The demonstration took place in 1965, six years before DNA polymerase III of *E. coli* had even been discovered. Phage mutants that specified a thermolabile species of DNA polymerase were unable to replicate their DNA at moderately elevated temperatures (42°C). Moreover, phage that reverted to the wild-type phenotype simultaneously regained a temperature-stable DNA polymerase and the ability to replicate their DNA at 42°. Thus, the same gene was shown to control both the structure of the DNA polymerase and the ability of the virus to replicate its DNA.

DNA polymerases specified by herpesviruses are of interest because they can serve as target sites for the action of antiviral drugs such as acyclovir for herpes simplex (see Chapter 22).

Interest in vaccinia viral DNA replication derives partly from the fact that it occurs in the cytosol. Since cellular DNA polymerase α resides within the nucleus, there must be a cytosolic form of DNA polymerase for replication of the viral genome. This form could result from movement of cellular polymerase α out of the nucleus or from vaccinia encoding its own DNA polymerase; the latter was found to be the case. Both the herpes and vaccinia polymerases are inhibited by aphidicolin, which seems to make them resemble replicative polymerases of the eukaryotic cells they infect. However, unlike DNA polymerase α, both viral enzymes contain but one subunit, like prokaryotic polymerases. Also, both are distinctive in that each contains an associated 3′-exonuclease activity.

Bacteriophage Systems for Studying Replication Mechanisms

From our discussion of DNA polymerases, it should be clear that several other enzymes and proteins must participate in the total replication of a biologically complete DNA molecule. The properties of polymerases alone do not tell us how DNA replication is initiated, how both antiparallel daughter strands are extended within a single replication fork, how strands of the parental duplex are unwound to expose single-stranded templates for new DNA synthesis, or how newly replicated DNA assumes its final supercoiled structure. For the most definitive answers to these questions, investigators have turned to easily manipulated bacterial viruses, or bacteriophages.

Throughout the recent history of biochemistry, bacterial viruses have provided instructive systems for analysis of the structure, organization, and expression of genetic information—they can be thought of as windows on normal cellular processes. Part of the attraction of phage systems is the ease of growth of the host cell; *E. coli* multiplies rapidly in a simple buffered solution of a few inorganic salts plus a single carbon source, such as glucose. Furthermore, the ease with which mutations can be selected and mapped, in either the viral or the host cell genome, allows genetic analysis to complement biochemical approaches. The development of mammalian tissue culture has allowed, in comparable fashion, the use of animal viruses to illuminate aspects of eukaryotic DNA replication. We shall explore these later, but for now let us concentrate on the bacterial viruses. The phages that have been most widely used include T4, T7, and the small circular single-stranded DNA phages (φX174, G4, M13, and fd). All of these phages infect *E. coli,* and all have special properties that have allowed illumination of different aspects of DNA metabolism.

Bacteriophage T4

Bacteriophage T4 was the first virus for which any biochemical genetic information was developed, that is, the identification of the specific protein products of particular genes. As shown in Figure 24.15, T4 is a large virus with an icosahedral head and a long and complex tail. Infection begins with attachment of the tail fibers to a specific receptor in the outer membrane of the host bacterium (Figure 24.16). Contraction of the tail forces the tail core through the cell surface, providing a tube for transit of the viral DNA from the inside of the phage head to the interior of the cell. Shortly thereafter, the

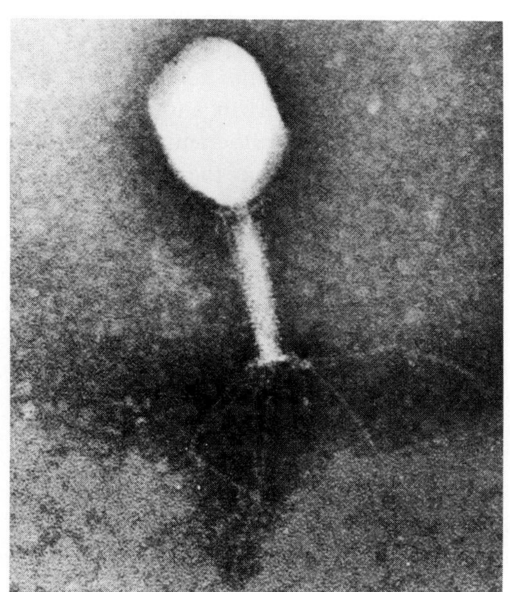

(a)

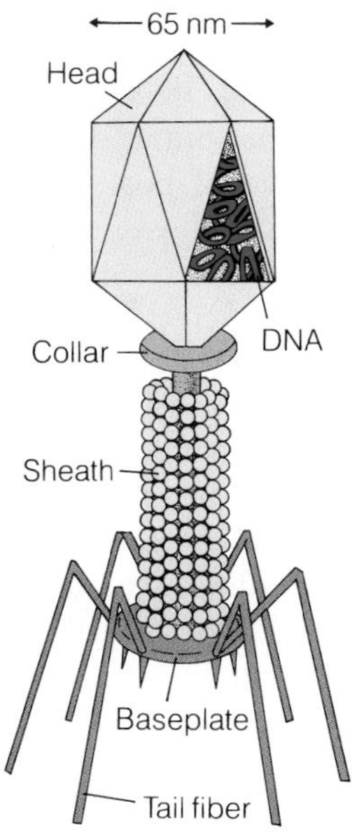

(b)

Figure 24.15
T4 bacteriophage. (**a**) Electron micrograph; (**b**) schematic visualization.

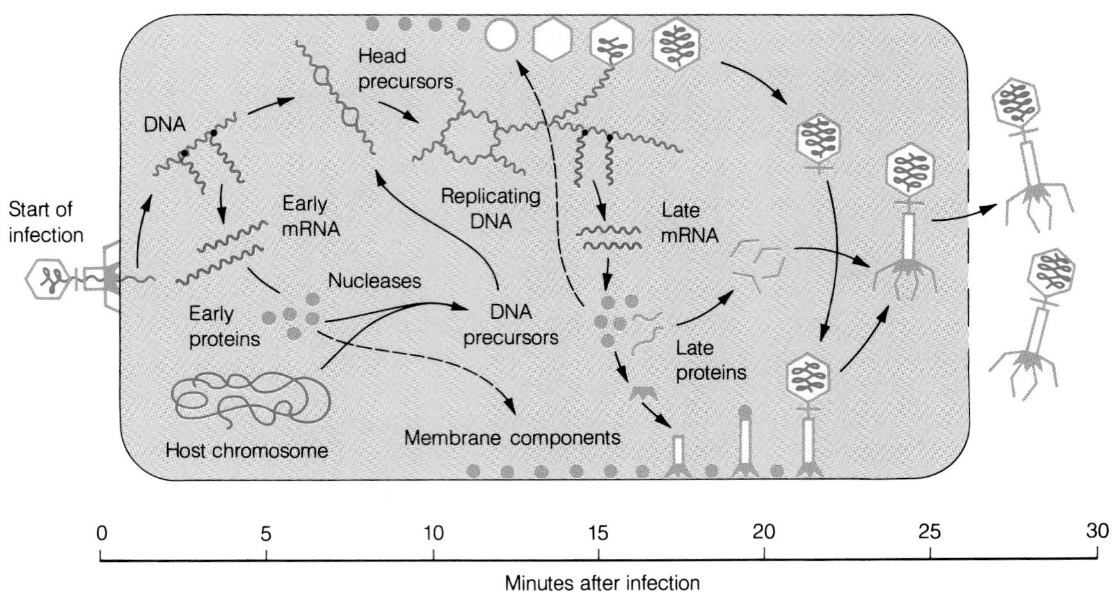

Figure 24.16
The T4 infective cycle. Immediately after injection of DNA into the host cell, early genes are transcribed, yielding enzymes involved in DNA precursor biosynthesis and DNA replication. Replication of the linear DNA molecule is initiated bidirectionally from multiple origins. Recombination among newly replicated molecules yields giant circular replicative intermediates. Replicating DNA serves as the template for late gene transcription, yielding structural proteins. Independent subassembly pathways generate heads, tails, and tail fibers. Packaging of DNA into heads occurs concomitantly with head maturation and DNA replication. Newly formed virus particles are released by lysis of the cell.

expression of host cell genes is turned off, and a coordinated expression of viral genes results in the production of about 200 new T4 particles per cell by about 30 minutes after infection.

The T4 particle contains a linear duplex DNA some 170 kilobase pairs (kb) in length. However, because some information is repeated at both ends of the molecule (**terminal redundancy**), the genome consists of just 166 kb. The DNA contains 5-hydroxymethylcytosine instead of cytosine, and the hydroxymethyl groups are glucosylated. In Chapter 22 we described the metabolic pathways that generate these modifications.

The biochemical genetics of T4 was derived from the isolation of hundreds of **conditional lethal** mutants—mutants whose growth is blocked under *restrictive* conditions, but which can be propagated under *permissive* conditions. The most useful conditional lethal mutations in T4 are (1) *ts* mutations, in which an essential protein has become thermolabile, and (2) **amber** mutations, in which expression of an essential gene is blocked in one kind of host cell (a **restrictive host**) but allowed in another strain (a **permissive host**). As discussed in Chapter 27, an *amber* mutation causes premature termination of a genetic message, except in the presence of an altered bacterial transfer RNA gene, called a **suppressor.** For this reason *amber* mutations are more accurately called **suppressor-sensitive** mutations.

When T4 *amber* and *ts* mutations were mapped genetically, they were found to define a *circular* linkage map (Figure 24.17). The map is circular, although the DNA is linear, because individual molecules have different end points. If the genome is represented by the alphabet, some molecules begin at A, some at E, some at K, and so forth. The sequences are said to be **circularly permuted.** About two dozen genes on the map encode enzymes and proteins essential for DNA replication, as shown by defective DNA replication when cells are infected under restrictive conditions with a phage containing a mutation in one of these genes. A few genes encode enzymes essential for DNA precursor biosynthesis; for example, gene 42 specifies the

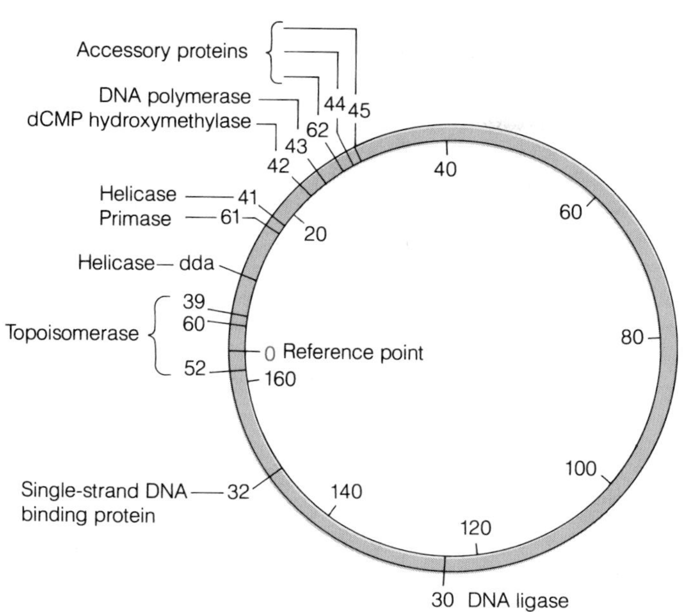

Figure 24.17
The T4 genomic map, showing the positions of genes whose products participate in DNA metabolism. The inner numbers represent distances in kilobase pairs. The reference point (0) represents the divide between two particular genes, *rIIA* and *rIIB*.

enzyme **deoxycytidylate hydroxymethylase** (discussed in Chapter 22), which synthesizes 5-hydroxymethyldeoxycytidine monophosphate. This still leaves a lot of gene products, which apparently participate in the replication of DNA itself. Gene 43, for example, codes for the T4 DNA polymerase. What these gene products are and what they do will concern us for much of our remaining discussion of DNA replication.

Bacteriophage T7

This virus is smaller and simpler than T4. Like T4, it has a linear duplex genome with a small terminal redundancy (160 base pairs). Unlike T4, it has unmodified DNA bases, and mapping conditional lethal mutations generates a linear genetic map. As shown in Figure 24.18, genes that are expressed *early* in infection control DNA replication and transcription. The late genes encode virus structural proteins. The entire nucleotide sequence of T7 DNA is known. T7 has been informative regarding mechanisms by which a duplex DNA initiates replication.

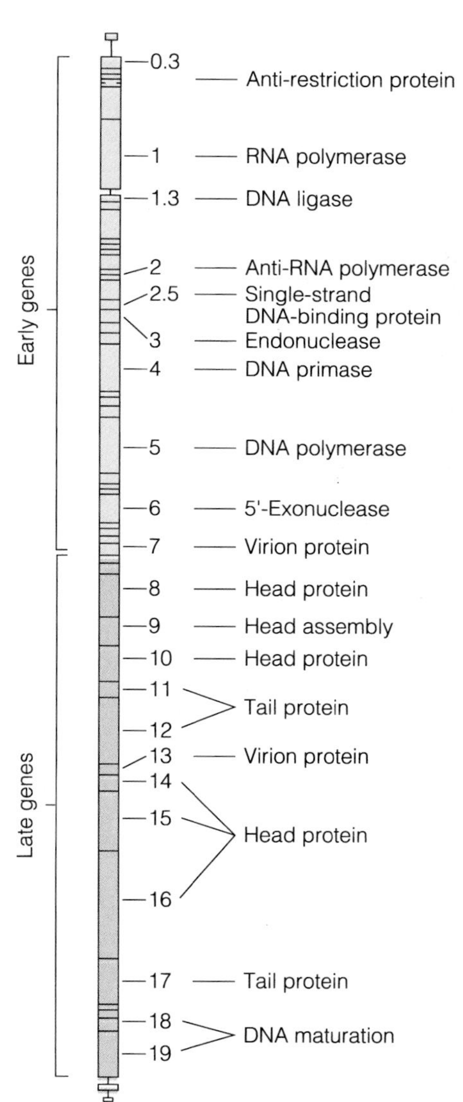

Figure 24.18
The T7 genomic map. The early genes (0.3 to 6) participate in DNA replication.

Small Circular DNA Phages: ϕX174, G4, M13, fd

Two classes of very small phages have been indispensable for defining the mechanism by which the chromosome of the host *E. coli* cell is replicated. This is because the genomes of these viruses are so small that they depend almost entirely on proteins of the host cell to replicate their own DNA. All of these viruses have as their genome one molecule of circular *single-stranded* DNA, long enough to encode about 10 proteins. The **spherical phages,** such as ϕX174 and G4 (which actually have an icosahedral shape), have a DNA about 5.7 kb in length; the **filamentous phages,** fd and M13, have a slightly longer DNA, about 6.4 kb in length.

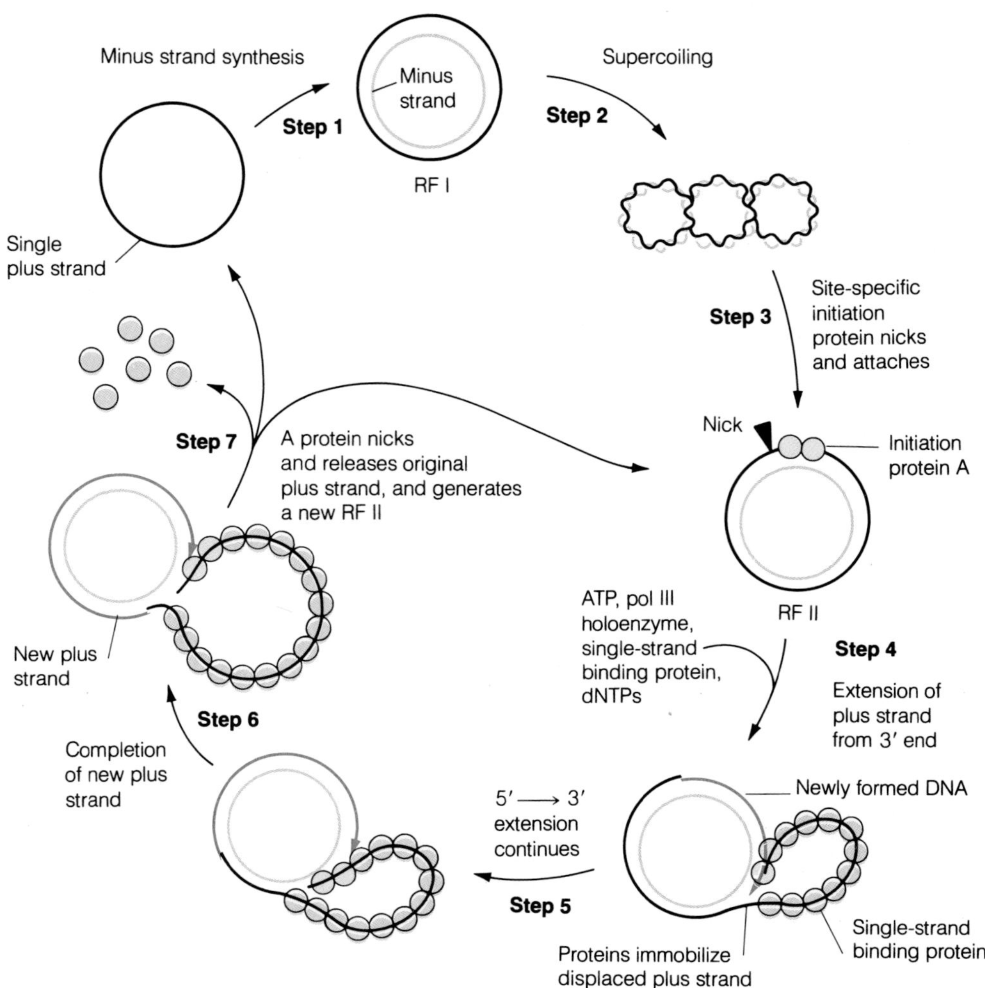

Figure 24.19
Replication scheme for single-stranded phage DNAs.

While molecular details vary, all of these phages replicate their DNA by the pathway schematized in Figure 24.19. The single-stranded genome, after passage into the host cell, serves as template (step 1) for production of a circular duplex DNA, called a **replicative form,** or **RF** (a covalently closed circular duplex is called **RFI**). In this species the DNA strand derived from the viral particle is called the **plus strand,** and the newly synthesized complement is called the **minus strand.** The duplex RF becomes supercoiled through the action of DNA gyrase (step 2; see p. 844) and then undergoes replication by a **rolling circle** mechanism to give single-stranded progeny DNA. First, a phage-coded, site-specific endonuclease forms a nick in the plus strand (step 3), yielding **RFII,** a circular molecule with at least one single-strand interruption. Polymerase extends from the 3′ hydroxyl terminus of the nick and proceeds around the circular minus strand template (steps 4, 5, and 6). The 5′ end of the plus strand is displaced from the template and is ultimately packaged into virus particles. When polymerase has proceeded once around the circle, the site-specific endonuclease acts again (step 7), releasing the displaced plus strand and a new RFII. This circular duplex contains within its one single-strand interruption a 3′ hydroxyl terminus to which nucleotides can be added in a subsequent round of plus strand synthesis.

Replication Proteins Other Than DNA Polymerases

Now that we have described some of the biological systems used to analyze DNA replication, we can discuss some of the proteins involved in addition to DNA polymerases. As you can see from the genetic maps of *E. coli* and T4 (Figures 24.14 and 24.17), about a dozen additional proteins are required for replication of duplex DNAs. These proteins are involved in such functions as unwinding strands of the parental duplex, extending the 5′-terminated daughter strand in a replication fork, and separating newly synthesized circular DNAs near the completion of a round of replication.

DNA Ligase: Discontinuous Replication of DNA

DNA ligase, originally discovered as an activity thought to be involved in genetic recombination, is an enzyme that covalently closes nicks in double-stranded DNA. The nick must contain 3′ hydroxyl and 5′ phosphoryl termini, and the nucleotides being linked must be juxtaposed in the duplex structure. As shown in Figure 24.20, DNA ligase is activated by adenylylation of a lysine residue in the active site. The enzyme in turn adenylylates the 5′ terminal phosphate of the DNA substrate, thereby activating it for nucleophilic attack by the 3′ hydroxyl group with phosphodiester bond formation and displacement of AMP. The T4 phage enzyme uses ATP to adenylylate the enzyme, as do eukaryotic DNA ligases. However, the enzyme from *E. coli* and other bacteria uses *NAD*$^+$ instead. Instead of being a redox cofactor, the dinucleotide in this enzyme is cleaved, to yield adenylylated enzyme plus nicotinamide mononucleotide (NMN).

The discovery of DNA ligase in 1967 led to the resolution of an important problem in DNA replication: How can daughter strands of opposite polarity be elongated in one replication fork if DNA polymerases can extend chains only in a 5′ → 3′ direction? In 1968 Reiji Okazaki proposed that ligase could participate in synthesis of the 5′-terminated daughter strand, with synthesis proceeding overall in the same direction as movement of the replication fork. Okazaki proposed that the 5′-terminated strand could be synthesized in small pieces *backward from the direction of fork*

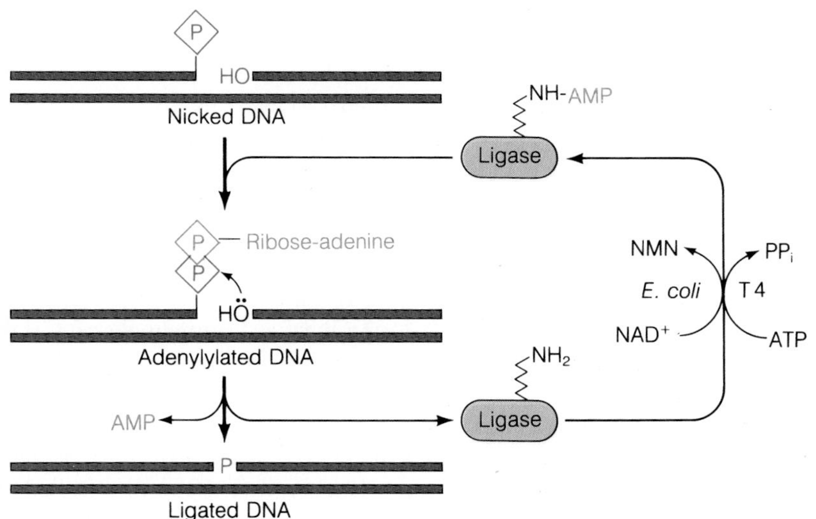

Figure 24.20
The reaction catalyzed by DNA ligase.

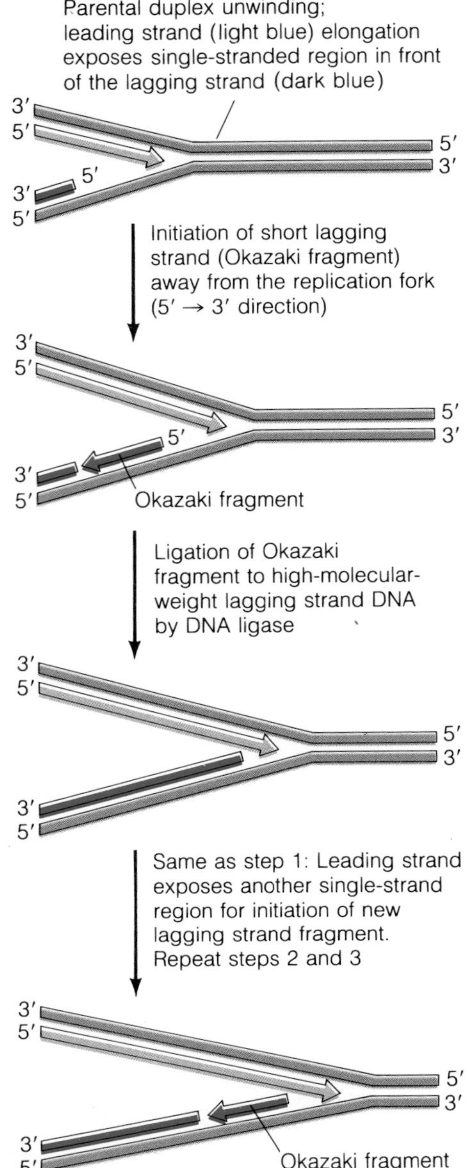

Parental duplex unwinding; leading strand (light blue) elongation exposes single-stranded region in front of the lagging strand (dark blue)
3′
5′
5′
3′
5′
3′
5′
Initiation of short lagging strand (Okazaki fragment) away from the replication fork (5′ → 3′ direction)
3′
5′
5′
3′
5′
3′
5′
Okazaki fragment
Ligation of Okazaki fragment to high-molecular-weight lagging strand DNA by DNA ligase
3′
5′
5′
3′
3′
5′
Same as step 1: Leading strand exposes another single-strand region for initiation of new lagging strand fragment. Repeat steps 2 and 3
3′
5′
5′
3′
3′
5′
Okazaki fragment

Figure 24.21
The Okazaki model of discontinuous chain growth in DNA replication.

movement (Figure 24.21). Each piece could then be attached covalently to the growing daughter strand through the action of DNA ligase. In principle, the 3′-terminated daughter chain (the **leading strand**) could be replicated continuously, with chain elongation proceeding in the same direction as the replication fork. However, the 5′-terminated chain, or **lagging strand,** was seen as being synthesized discontinuously, in small segments, which were inevitably named **Okazaki fragments.**

Supporting evidence for this model came from two sources. First, *E. coli* cultures were pulse labeled with [^{3}H]thymidine, to label preferentially the most recently synthesized DNA. Sedimentation analysis in sucrose gradients showed that most of this newly synthesized DNA is in the form of short fragments, roughly 1000 nucleotides in length. When the pulse was "chased" by incubation of the pulse-labeled culture in nonradioactive medium, the label was found in high-molecular-weight DNA, as expected if DNA is synthesized as short fragments that subsequently become covalently attached to high-molecular-weight DNA. The action of DNA ligase in the latter process was demonstrated after the discovery that gene 30 in the phage T4 genome is the structural gene for DNA ligase. When cells were infected with gene *ts* 30 mutants at 42°C, all of the nascent DNA accumulated as short fragments. Only when the culture was transferred to 30°C, so that ligase could act, was it possible to demonstrate the conversion of Okazaki fragments to high-molecular-weight DNA.

Primase: Synthesis of RNA Leader Sequences

The discovery of discontinuous DNA replication answered one question, but it posed another: Since DNA polymerases cannot initiate the synthesis of new DNA chains, but only extend chains from preexisting 3′ hydroxyl termini, what provides those termini, both during the initiation of chromosome replication and for initiation of each Okazaki fragment? When it was found that replication of phage M13 was sensitive to an inhibitor of transcription and that DNA polymerase could extend from 3′ hydroxyl termini of *RNA* base paired to DNA, it became evident that oligoribonucleotides might provide primers for biological DNA replication.

Some of the earliest evidence for RNA priming in DNA replication came from an important experiment carried out by Reiji and Tuneko Okazaki. *E. coli* cells were made permeable to exogenous nucleotides by brief treatment with a buffer containing toluene. DNA synthesis occurred when these permeabilized bacteria were incubated with a mixture of α-[^{32}P]deoxyribonucleoside triphosphates and unlabeled ribonucleoside triphosphates (Figure 24.22). DNA was then isolated and treated with mild alkali to hydrolyze RNA. Now, if DNA synthesis starts from 3′ RNA termini, *there should exist for each Okazaki fragment one RNA–DNA joint,* or deoxyribonucleotide residue covalently linked via a radioactive phosphate group to a ribonucleotide residue. Alkaline hydrolysis would now *transfer* that phosphate from the 5′ position of the original deoxyribonucleotide substrate to the 3′ position of the ribonucleotide at the joint. Sure enough, the Okazakis found that for each Okazaki fragment formed in this system one radiolabeled phosphate was transferred to a ribonucleotide.

In *E. coli* the product of the *dnaG* gene is the enzyme that synthesizes RNA primers, while the comparable reaction in phage T4 is carried out by the product of gene 61. In both cases the enzyme, called **primase,** is active only in the presence of other proteins, which create a complex called the **primosome.** This complex also participates in unwinding parental DNA strands ahead of the replication fork. **Priming** involves insertion of a ribonucleoside 5′-triphosphate opposite a base-paired deoxyribonucleotide residue in the template DNA, followed by sequential ribonucleotide additions to the 3′ hydroxyl terminus, just as occurs with DNA polymerases. At some point the primosome dissociates and DNA polymerase begins to extend from the 3′ hydroxyl terminus of the RNA primer.

In T4 the RNA primers are five residues long, with the general sequence pppApCpXpYpZ, where X, Y, and Z can be any nucleotide. Primer synthesis can start opposite the second nucleotide of any GTT sequence in DNA. Since such sequences would occur on a random basis at intervals of about 256 nucleotides (1 in 4^3), and since Okazaki fragments in T4 are about 1500 nucleotides long, it is apparent that not every GTT sequence directs synthesis of a primer.

For Okazaki fragments to be ligated to high-molecular-weight DNA, the RNA primers must be excised and replaced with deoxyribonucleotides. In *E. coli,* DNA polymerase I seems to be involved in this process, through its nick translation activity; the removal of ribonucleotides from the 5′ end of the primer can be coordinated with their replacement by deoxyribonucleotides (Figure 24.23). In agreement with this idea, the *polA* mutant isolated

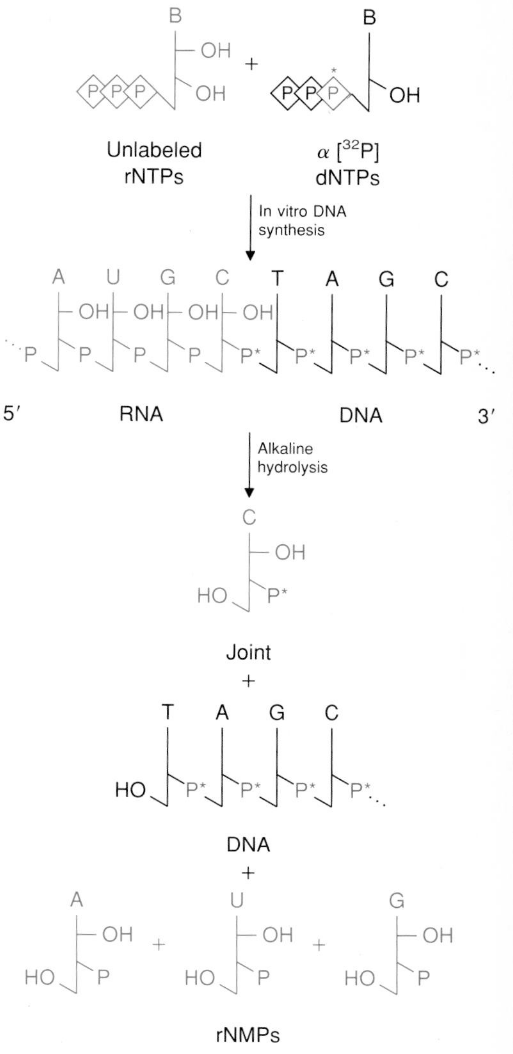

Figure 24.22
The transfer experiment that demonstrated the existence of RNA primers in DNA replication. Each Okazaki fragment generated one radiolabeled ribonucleotide from its RNA–DNA joint after alkaline hydrolysis.

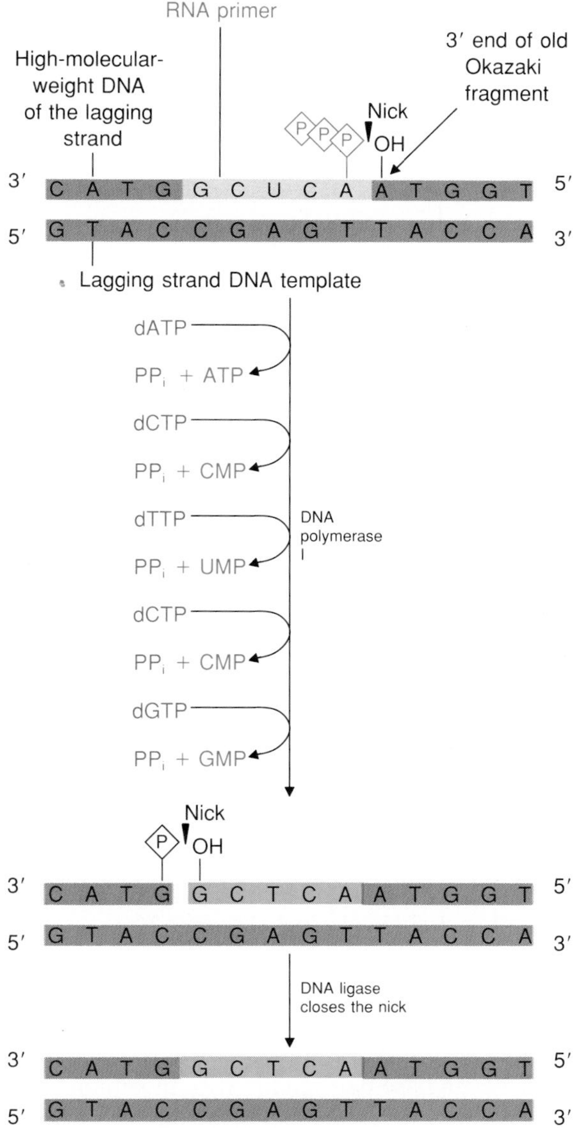

Figure 24.23
Nick translation activities of DNA polymerase I as a means of excising RNA primers.

by Cairns (p. 830) shows a delay in conversion of Okazaki fragments to high-molecular-weight DNA. That it does so at all is probably due to the 5′-exonuclease activity remaining in this mutant, plus the activity of another polymerase (perhaps polymerase II). *PolA* mutants lacking the 5′-exonuclease as well as the polymerase are totally defective in fragment sealing and display a conditional lethal phenotype.

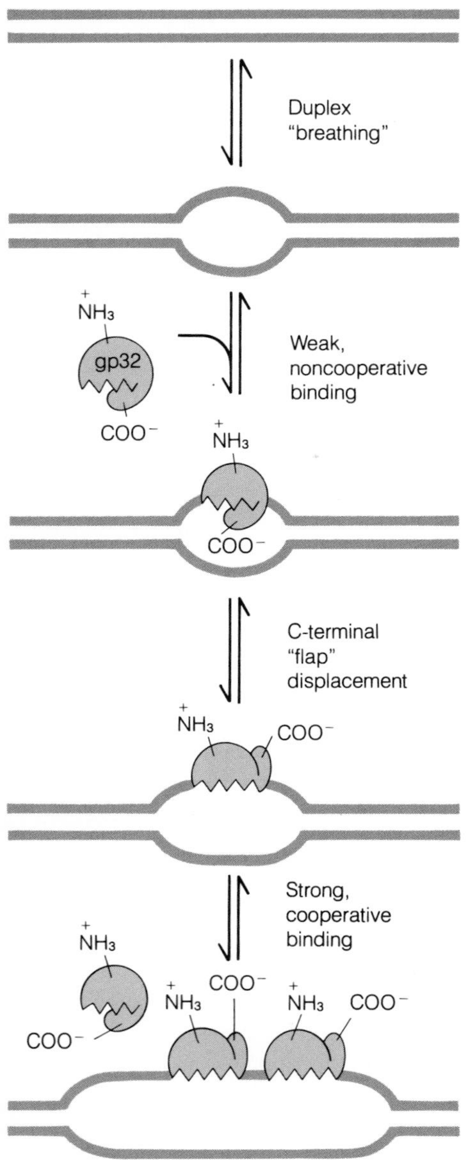

Figure 24.24
Model for gp32 action in facilitating DNA denaturation or renaturation. The conformation of the C-terminal domain prevents cooperative binding and, hence, strand unwinding, when only a short region of single-stranded DNA is exposed. Exposure of a longer region, after binding one molecule of gp32, allows displacement of C-terminal domain, permitting cooperative binding and further extension of the denatured region. For simplicity, binding is shown on only one of the two denatured DNA strands.

Polymerase-Accessory Proteins: Processivity

DNA replication must be an exceedingly efficient process. As we have seen, *E. coli* replicates its entire chromosome in 40 minutes with only two replication forks, each of which involves the action of two molecules of DNA polymerase III. Since there are only 10 molecules of this enzyme per cell, it is clearly advantageous for DNA polymerase, once bound to its template, to remain bound, in other words, to act processively. In T4 three proteins, called **polymerase-accessory proteins,** act to increase the processivity of the gene 43 DNA polymerase, as well as its catalytic activity. As noted earlier, several of the proteins of the *E. coli* polymerase III holoenzyme also act to increase processivity. In T4 two proteins, the products of genes 44 and 62, form a tight complex that has a DNA-dependent ATPase activity. It is likely that ATP cleavage during replication facilitates the movement of DNA polymerase along its template.

In *E. coli* the polymerase-accessory proteins play roles in addition to improving processivity. One such protein, τ, causes the polymerase complex to dimerize. It is likely that both polymerase molecules in a replication fork are associated with one another. This means that the lagging strand polymerase can remain bound at the replication fork, even after completion of synthesis of an Okazaki fragment.

Single-Strand DNA-Binding Proteins: Maintaining Optimal Template Conformation

One of the earliest replication proteins to be identified, other than DNA polymerase itself, was a T4 protein called either **single-strand DNA-binding protein** or **helix-destabilizing protein.** In one of the earliest applications of affinity chromatography, Bruce Alberts immobilized DNA by binding it to cellulose and analyzed the T4 proteins that were retained by a column of this material. One particularly abundant protein was shown to be the product of gene 32, because the protein isolated from a *ts* gene 32 mutant was unable to bind to DNA at a restrictive temperature. Since gene 32 mutants were known to be defective in DNA repair and genetic recombination, as well as DNA replication, it was clear that the protein played multiple roles in DNA metabolism.

Analysis of the purified gp32 (an abbreviation for gene product 32) showed that it binds specifically to single-stranded DNA. Moreover, binding is strongly *cooperative,* meaning that the protein is far more likely to bind to DNA adjacent to a site already occupied than to an isolated site. In other words, binding of one gp32 molecule facilitates the binding of others, and the protein tends to bind in clusters. Thus, gp32 promotes the denaturation of DNA; its presence lowers the melting temperature of DNA by as much as 40°C (Figure 24.24).

Action of this protein is important to keep the template in an extended, single-stranded conformation, with the purine and pyrimidine bases exposed so that they can base-pair readily with incoming nucleotides. Since replication also involves the re-formation of duplex structures, with one strand each of parental and daughter DNA, it is of interest that gp32 facilitates the renaturation of single-stranded DNA, as well as the denaturation of duplexes.

How can a single protein promote both duplex formation and duplex unwinding? The answer seems to lie in special design features of the gp32 molecule. Note that the protein does not itself unwind DNA strands; rather, it stabilizes single-stranded DNA by binding *after* a region of DNA is unwound. Limited proteolysis of gp32 removes a C-terminal fragment. This renders the protein a stronger DNA denaturant in vitro, with a small increase in its equilibrium binding affinity for DNA. As shown in Figure 24.24, the C-terminal domain provides a "flap" that partly covers the DNA-binding domain of the protein. When a short region of single-stranded DNA is exposed, through partial and reversible unwinding of the duplex (sometimes called "breathing"), gp32 can bind, but to a small site (3 or 4 nucleotides), with the flap down. Occupancy of a complete site (7 to 10 nucleotides) is thereby inhibited, rendering binding both weak and noncooperative. When a longer single-stranded site is made available, through continued "breathing" of the helix, the flap can move up, permitting the complete site to be occupied, and promoting further denaturation.

The partially proteolyzed gp32 lacks the ability to renature DNA in vitro, and it seems likely that the conformation of the native gp32 is a critical determinant of whether the protein functions primarily as a denaturant ("flap up" position) or renaturant ("flap-down" position). The position of the flap is determined in part through interaction with other proteins at the replication fork. Thus, the protein can act both to stabilize a single-strand template as it pairs with incoming nucleotides, and to facilitate reformation of a duplex after DNA polymerase has passed by.

The amino acid sequence of gp32 shows a cluster of tyrosine residues, spaced on the average 7.7 residues apart, in a region of the protein involved in DNA binding. Although the three-dimensional structure of this protein is not yet known, it is likely that these residues are spaced so as to bind to DNA by **intercalation,** or fitting, of the aromatic rings between pairs of bases.

Single-strand DNA-binding proteins have now been found in many organisms. The protein of *E. coli* (specified by the *ssb* gene) also binds cooperatively to single-stranded DNA. However, the mechanism of binding appears to be quite different; DNA is wrapped about the outer surface of this protein. Moreover, under some conditions the SSB protein binding shows *negative* cooperativity. It has been suggested that the protein binds in different modes depending on whether it is participating in replication, DNA repair, or recombination.

Helicases: Unwinding DNA Ahead of the Fork

Single-strand DNA-binding proteins are not DNA denaturants; they stabilize single-stranded DNA but cannot actively unwind duplex DNA strands. Such unwinding must occur if single-stranded templates are to be exposed for polymerase action. The **helicases** are a class of proteins that have this ability; they catalyze the ATP-dependent unwinding of double-stranded DNA.

Generally, a region of single-stranded DNA is necessary for binding a helicase to a DNA substrate. The enzyme then moves in a specific direction, coupling its movement with ATP hydrolysis and with strand unwinding (Figure 24.25). The T4 phage genome specifies two known helicases, the products of genes 41 and *dda*. Gp41 binds to gp61 (the T4 primase) to form

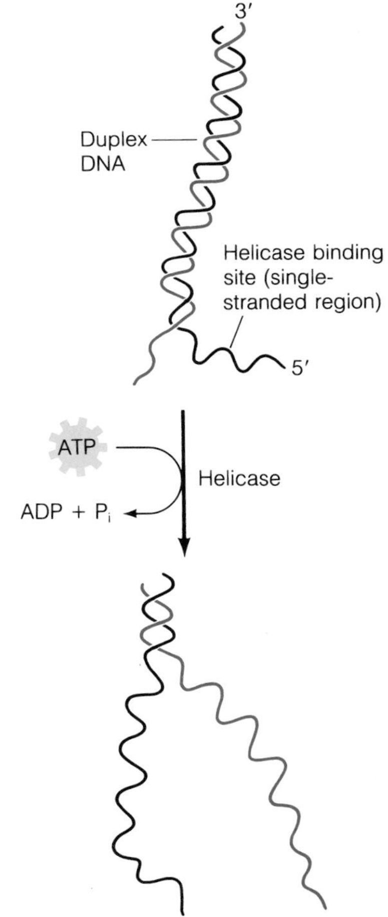

Figure 24.25
Action of DNA helicases in catalyzing the ATP-dependent unwinding of duplex DNA.

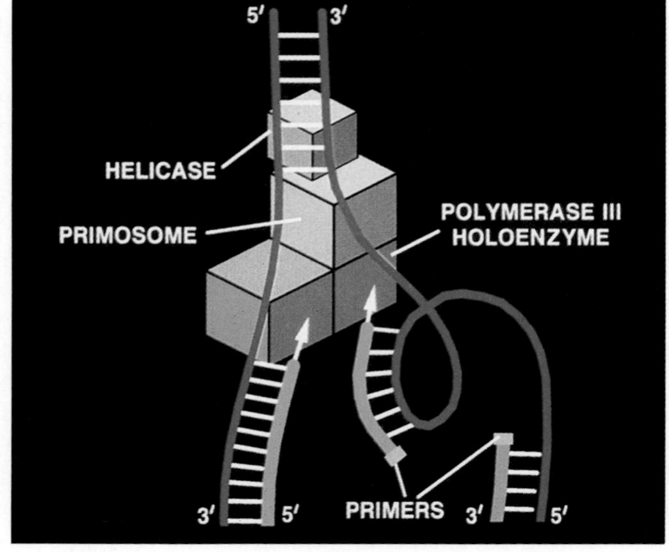

Figure 24.26
Simultaneous extension of leading and lagging DNA strands by a dimeric DNA polymerase complex. Helicase is unwinding parental DNA strands, the primosome is synthesizing RNA primers, and a polymerase holoenzyme is extending leading (lower) and lagging (upper) daughter strands, respectively.

a *primosome* for phage DNA replication (Figure 24.26). This complex moves in a 5′ → 3′ direction along the template for lagging strand synthesis, facilitating the unwinding of the parental duplex and at intervals stopping to synthesize an RNA primer. The other helicase, gp*dda*, moves along the other parental strand. The latter helicase is not absolutely required for DNA replication, whereas gp41 is essential. However, gp*dda* increases the rate of replication, and it plays the additional role of displacing bound transcription proteins, so that a genome can be replicated even though some of its genes are serving as templates for transcription.

E. coli contains four known helicases, called I, II, III, and Rep. Not all of these proteins are essential for DNA replication. Helicase I participates in the replication that occurs during bacterial mating, or **conjugation** (Figure 24.27). In "male" strains of *E. coli* the chromosome replicates during conjugation by a rolling circle mechanism, with the displaced single strand being transferred from the male, or donor, strain to the female, or recipient. Helicases II and III have the correct polarity to move along the template for lagging strand synthesis, while Rep moves along the template for the leading strand. Rep is required for the replication of small circular phage genomes in *E. coli,* but the protein is not absolutely required for replication of the bacterial chromosome itself.

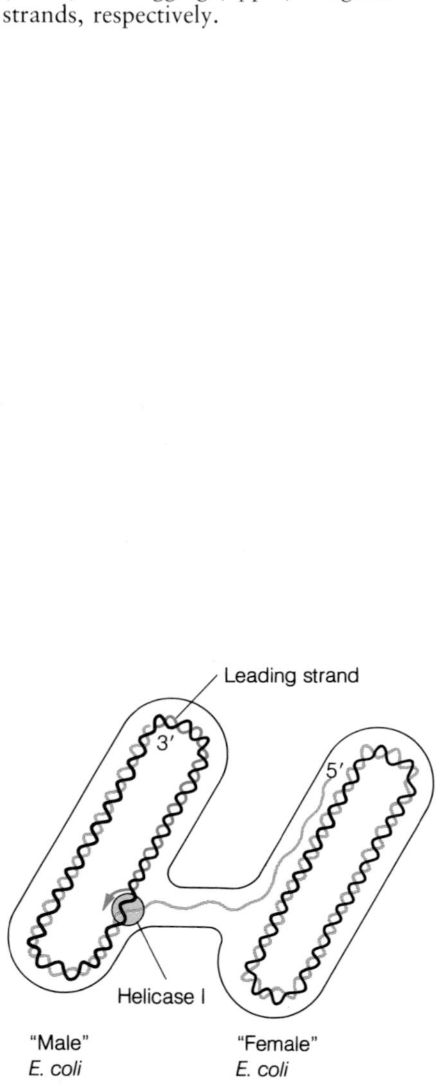

Leading strand
3′
5′
Helicase I
"Male"
E. coli
"Female"
E. coli

Figure 24.27
Helicase involvement in bacterial conjugation. The single strand that is transferred to the "female" is created by rolling circle replication. After a nick is made in the "male" chromosome, helicase unwinds the two strands. The nicked strand (gray) is elongated (blue) at the 3′ end, displacing the 5′ end and transferring it into the female cell. Later, the transferred DNA strand can undergo recombination with the recipient chromosome, thereby transferring genetic markers to the female.

Topoisomerases: Relieving Torsional Stress

Bidirectional replication of the circular *E. coli* chromosome unwinds about 100,000 base pairs per minute. Were there not some mechanism for relieving this torsional stress, DNA ahead of the fork would become overwound as DNA at the fork became unwound, and replication could not be sustained. A "swivel" mechanism for relieving this stress can be provided by **topoisomerases,** a group of enzymes that can interconvert different topological isomers of DNA (see Chapter 4). This process is seen most simply in vitro by the relaxation of supercoiled DNA. One can incubate negatively supercoiled DNA with a purified topoisomerase and observe, by gel electrophoresis, the intermediate stages in conversion of the supercoiled substrate to relaxed circular DNA containing no superhelical turns (Figure 24.28).

This analysis reveals the existence of two general classes of topoisomerases—type I enzymes, which change the linking number in units of one, and type II enzymes, which change the linking number in units of two.

A type I topoisomerase breaks just one strand of the duplex (Figure 24.29a). The enzyme remains covalently attached to the 5′ end of the broken strand, with the 3′ end moving far enough for the unbroken strand to pass through the single-strand interruption. The hydroxyl group on the 3′ end then attacks the activated, covalently bound 5′ phosphate, closing the nick; in fact, type I topoisomerase was originally called "nicking-closing enzyme." By contrast, a type II topoisomerase catalyzes a double-strand break, with both strands of the unbroken duplex passing *through* the gap that is created (Figure 24.29b). A plausible model for this reaction posits the wrapping of DNA about the enzyme molecule, so that DNA is immobilized and the ends can be resealed. Indeed, electron microscopic analysis indicates that DNA is wrapped about a type II topoisomerase molecule (Figure 24.30). Some type II topoisomerases require ATP hydrolysis for relaxation of a supercoiled duplex to occur, even though relaxation is an exergonic reaction.

E. coli contains both type I and type II topoisomerases. The type II enzyme is called **DNA gyrase,** because it has an activity in addition to the relaxation of superhelical DNA—the ability to introduce negative superhelical turns into closed circular DNA. ATP is required for this reaction. DNA gyrase plays an essential role in DNA replication, as shown by the properties of gyrase inhibitors. The enzyme contains two copies of each of two subunits, A and B. The A subunit is the target for binding of **nalidixic acid,** a compound long known to inhibit DNA replication. Another replication inhibitor, **novobiocin,** binds to the B subunit and inhibits an ATPase activity associated with that subunit. Mutant bacteria resistant to nalidixic acid or novobiocin show structural alterations in subunit A or B, respectively.

An archaebacterium, *Sulfolobus,* has recently been found to contain a **reverse gyrase,** an enzyme that introduces *positive* supercoils into relaxed DNA. At this writing the significance of this reaction is unknown, since intracellular circular DNAs are found only in the negatively supercoiled state.

Type II topoisomerases catalyze interconversions in addition to those between relaxed and supercoiled DNA, including knotting and unknotting of DNA and **catenation** and **decatenation** of circular DNAs (Figure 24.31). Since a circular DNA nearing the end of a round of replication will generate two interlinked circles, it is likely that type II topoisomerase action is necessary for termination of replication of a circular DNA. As two forks approach each other at the replication terminus, steric barriers will ultimately interfere with the unwinding activities of topoisomerases ahead of the two forks (Figure 24.32). At this stage the two incompletely replicated chromosomes are still interlinked, or **catenated,** and decatenation may precede the completion of replication. Molecular details of the termination event are not yet known, but we do know that in *E. coli* termination normally occurs in a specific region of the chromosome and that nucleotide sequences on either side of this region tell replication forks to slow down. Therefore, if one fork arrives too early, it is still in the vicinity when the other one arrives; this coordination of fork movement may be necessary for the action of DNA gyrase in termination.

Since all circular DNAs in a bacterial cell are negatively supercoiled, it is likely that DNA gyrase introduces superhelical turns in these molecules concomitantly with, or shortly after, the termination of replication. In

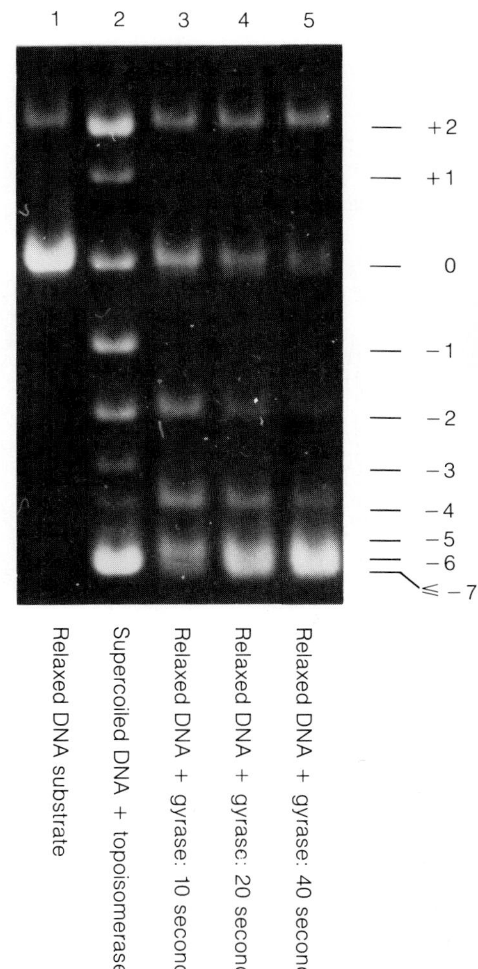

Figure 24.28
Action of type I and type II topoisomerases, as shown by gel electrophoretic analysis. Lane 2 shows the pattern from treatment of supercoiled DNA with type I topoisomerase. Lanes 3, 4, and 5 show relaxed circles treated with DNA gyrase for different times. Note that more different topoisomers can be seen in topoisomerase I reaction mixtures, as expected if changes in the linking number (ΔL) occur in units of one, while gyrase changes L in units of two.

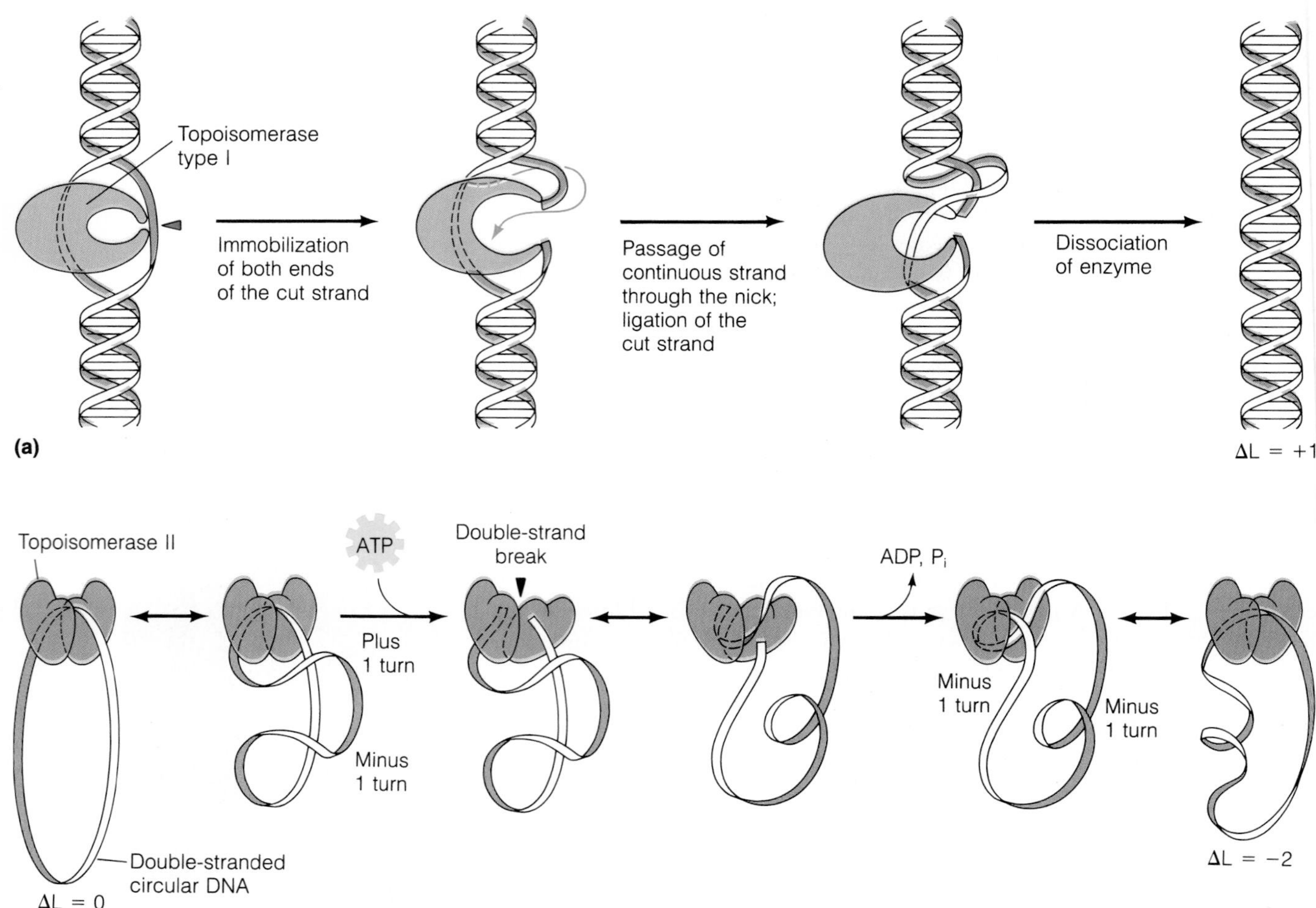

Figure 24.29
Mechanisms of action of type I and II topoisomerases. (**a**) Type I enzyme breaks one strand and immobilizes the 5′ end, allowing unbroken strand to pass through the nick before resealing. (**b**) Type II enzyme acting as a DNA gyrase can bind and induce a positive superhelical turn, compensated by a negative turn elsewhere on the DNA. The enzyme catalyzes a double-strand break, with the ends immobilized while the unbroken duplex passes through the gap. Resealing converts the positive supertwist to a negative one, giving the overall molecule a ΔLk of −2.

Figure 24.30
Computer-generated model of a type II topoisomerase–DNA complex, based on electron microscopic analysis. The topoisomerase is DNA gyrase of *E. coli.*

higher cells mitochondrial and chloroplast DNAs are negatively supercoiled, as are circular viral DNAs. Since DNA gyrase has not yet been observed in eukaryotes, it is not yet clear how this supercoiling occurs.

T4 specifies a type II topoisomerase, the product of genes 39, 52, and 60. The enzyme can relax supercoiled DNAs but, unlike its *E. coli* counterpart, it cannot introduce such turns in vitro. In this respect it resembles type II topoisomerases from a number of animal species. T4 topoisomerase mutants are defective in initiation of chromosome replication, but the role of topoisomerase in this process remains unknown.

Uracil-DNA *N*-glycosylase: Removal of Incorporated Uracil

Uracil can base pair with adenine in a DNA duplex, and DNA polymerases readily accept deoxyuridine triphosphate as a substrate in place of deoxythymidine triphosphate. Yet cells possess a rather elaborate two-stage mechanism that prevents deoxyuridylate residues from accumulating in DNA.

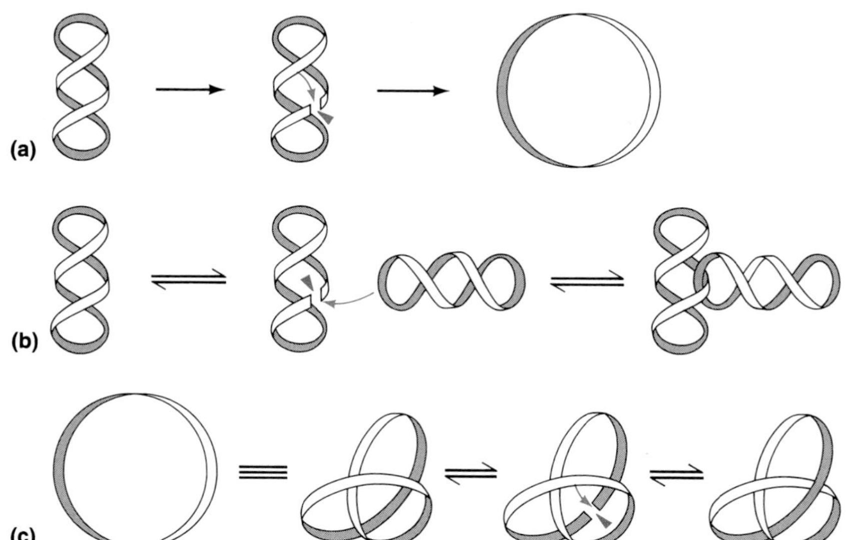

Figure 24.31
The types of topological interconversions catalyzed by type II topoisomerases. (**a**) Relaxation; (**b**) catenation and decatenation; (**c**) knotting and unknotting.

The first of these mechanisms was described in Chapter 22 and involves an active deoxyuridine triphosphatase that cleaves dUTP to dUMP and PP_i. The second involves an enzyme, **uracil-DNA *N*-glycosylase,** that removes any dUMP residues that might have arisen through occasional misincorporation of a dUTP that escaped the action of dUTPase. As shown in Figure 24.33, the enzyme hydrolytically cleaves the glycosidic bond between N-1 of uracil and C-1 of deoxyribose, to yield free uracil and DNA containing an **apyrimidinic site,** that is, a sugar with no base attached. Another enzyme, **apyrimidinic endonuclease,** recognizes this site and cleaves the phosphodiester bond on the 5′ side of the deoxyribose moiety. This is followed in bacteria by the nick translation activity of DNA polymerase I, which excises the deoxyribose phosphate at this site, along with several more nucleotide residues downstream from this site. The excised nucleotides are replaced with fresh deoxyribonucleotides, including dTTP at the original site occupied by dUMP. The process is completed with DNA ligase action.

Why go to all this trouble just to replace a nucleotide that does not affect the information encoded in DNA? The likely answer is that uracil misincorporated for thymine is probably not the true target of this DNA repair system. Uracil residues in DNA can also arise through spontaneous deamination of cytosine residues. The latter alteration does change the genetic sense, because it converts a G-C base pair to a G-U pair, and in subsequent rounds of replication the U-containing strand would give rise to an A-T base pair. Probably the uracil repair system is in place to prevent this, and the system does not discriminate between uracils paired with either adenines or guanines. Consistent with this model is the **hypermutable** phenotype displayed by mutants lacking an active dUTPase; such strains display elevated rates of spontaneous mutagenesis, resulting from the accumulation of dUMP residues in their DNAs.

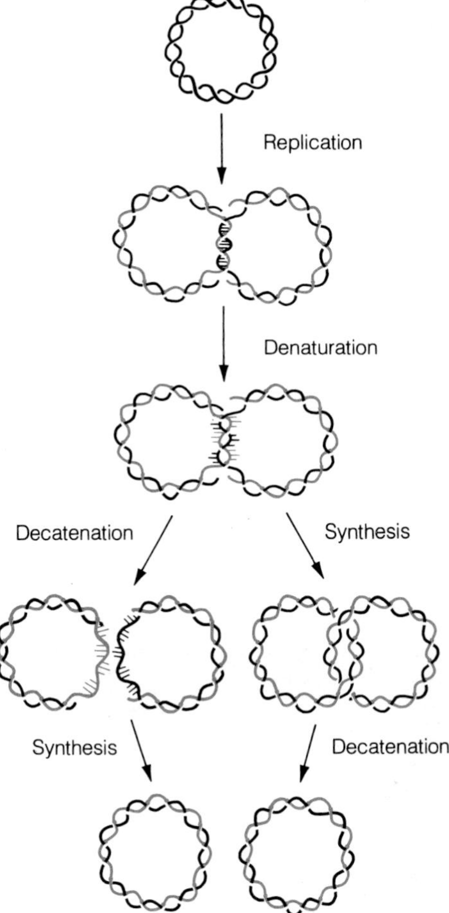

Figure 24.32
Topoisomerase action in termination of replication. Proximity of approaching replication forks renders gyrase ineffective. The still-interlinked circles can denature, then decatenate, followed by completion of synthesis. Alternatively, synthesis can be completed, followed by decatenation.

Reconstruction of Replication Machines

We have seen that many proteins must function at or near the replication fork, in a coordinated fashion, for rapid and accurate replication of both strands of double-helical DNA. Much of our understanding of these events has come from successful attempts to reconstruct part or all of a multiprotein replication apparatus, or *replisome,* from purified components. The most informative systems are those using mostly T4 phage proteins, those using mostly T7 phage proteins, and those using mostly *E. coli* proteins to replicate the genomes of small single-stranded DNA phages. Comparable approaches are now being used successfully with eukaryotes, but for now let us concentrate on the three prokaryotic systems.

The similarities among these three systems add up to a unified picture of events occurring at the replication fork. All three systems use comparable components to carry out the same kinds of subprocesses—parental strand unwinding, increasing DNA polymerase processivity, and so forth. The individual components of each system are listed in Table 24.5. Note that T4 and T7, with relatively large genomes, use few if any proteins encoded by the host, whereas the small phages depend almost entirely on bacterial proteins to replicate their own DNAs.

Each of these three systems has made distinct contributions to our understanding of DNA replication. The T4 system has been especially useful because it catalyzes DNA chain extension in vitro at a rate and accuracy approaching those of replication in intact cells. This system has taught us about processivity and how it is achieved, about the functions of single-strand DNA-binding proteins, and about RNA priming. However, the T4 system does not carry out the initiation of chromosome replication, nor does it replicate complete T4 DNA molecules. The T7 system has been useful because it was the first in vitro system to initiate replication of a duplex DNA at or very close to the origin.

The small phage systems are most useful for the insight they provide into replication of a cellular DNA, namely the chromosome of the host, *E. coli.* Particularly fruitful have been studies on initiation of replication of parental plus strands to yield RF. As shown in Figure 24.34, phages M13 and G4 initiate from a "hairpin loop" created by internal base pairing in the single-stranded template, while ϕX174 uses the mechanism by which lagging strand RNA primers are initiated during chromosomal DNA replication. Analysis of ϕX174 initiation has generated insight into the assembly and action of the *E. coli* primosome, which, with seven components, is clearly more complex than the T7 and T4 primosomes (one and two proteins, respectively).

Taken together, studies on these three systems point to the general picture of the replication fork schematized in Figure 24.35. In this picture both parental strands are being replicated by a dimeric polymerase, with unreplicated DNA ahead of each polymerase being coated with SSB. Because the lagging strand polymerase replicates backward from the direction of fork movement, the section of lagging strand DNA that is being replicated forms a loop, both ends of which are linked to the holoenzyme complex. At the moment depicted in the figure the loop is being elongated at both ends—to the left by polymerase action, which is extending the 3′ end of an Okazaki fragment toward the 5′ end of previously replicated DNA, and to the right by helicase action ahead of the fork, which exposes unreplicated lagging strand DNA, which is then coated by SSB. The loop reaches its

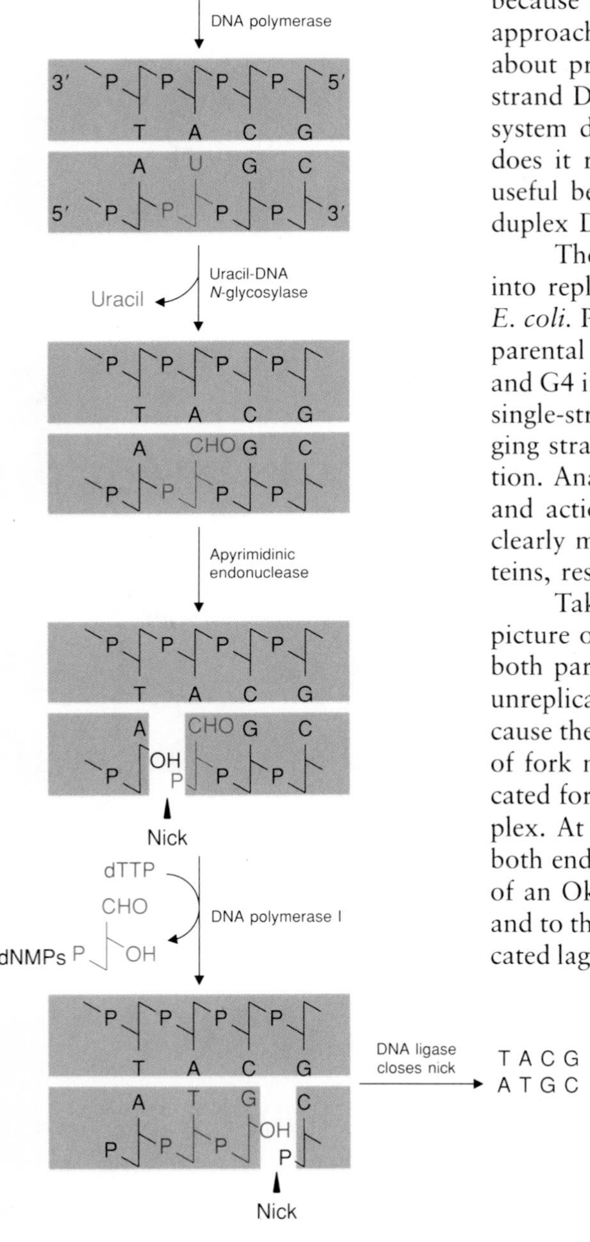

Figure 24.33
Action of the DNA-uracil repair system. Uracil-DNA *N*-glycosylase removes uracil, leaving an apyrimidinic site. A specific endonuclease recognizes this site and cleaves on the 5′ side. Nick translation by DNA polymerase I and DNA ligase action remove the apyrimidinic nucleotide and several adjacent nucleotides and seal the nick.

Table 24.5
Components of purified replication systems

	System		
Activity	T4	T7	Single-strand DNA phages
DNA polymerase	gp43	gp5	pol III core α-ϵ-θ
Accessory proteins	gp45, gp44/62	*E. coli* thioredoxin[a]	pol III holoenzyme subunits τ-γ-δ-β
Single-strand DNA binding	gp32	gp2.5 or *E. coli* SSB	*E. coli* SSB
Helicase	gp41, gp*dda*	gp4	*E. coli* Rep
Primase	gp61	gp4	Different for different phages (see Figure 24.34)
Primer excision	*E. coli* RNase H[b]?	gp6	*E. coli* DNA pol I
Okazaki fragment sealing	gp30	gp1.7	*E. coli* DNA ligase
Initiation	gp39, gp52, gp60, *E. coli* RNA pol	gp1, T7 DNA polymerase	gpA (ϕX174, G4), RNA pol + gp2 (M13)

[a] *E. coli* thioredoxin is absolutely required for DNA polymerase activity of the T7 gene 5 product.
[b] Ribonuclease H is an enzyme that specifically cleaves RNA out of RNA–DNA hybrids.

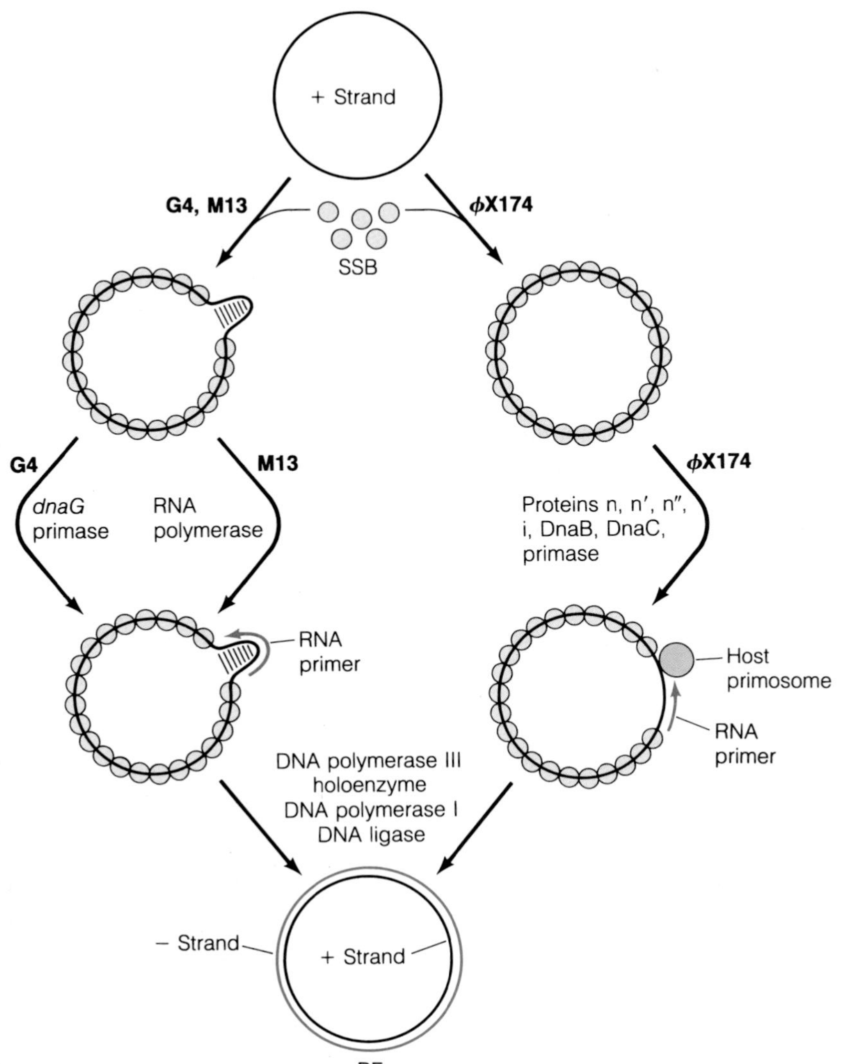

Figure 24.34
Conversion of single-stranded DNA phage plus strands to RF, by G4, M13, and ϕX174. Unlike M13 and G4, the replicating ϕX174 chromosome assembles a primosome identical to that used in replication of the *E. coli* chromosome. SSB, single-strand DNA-binding protein.

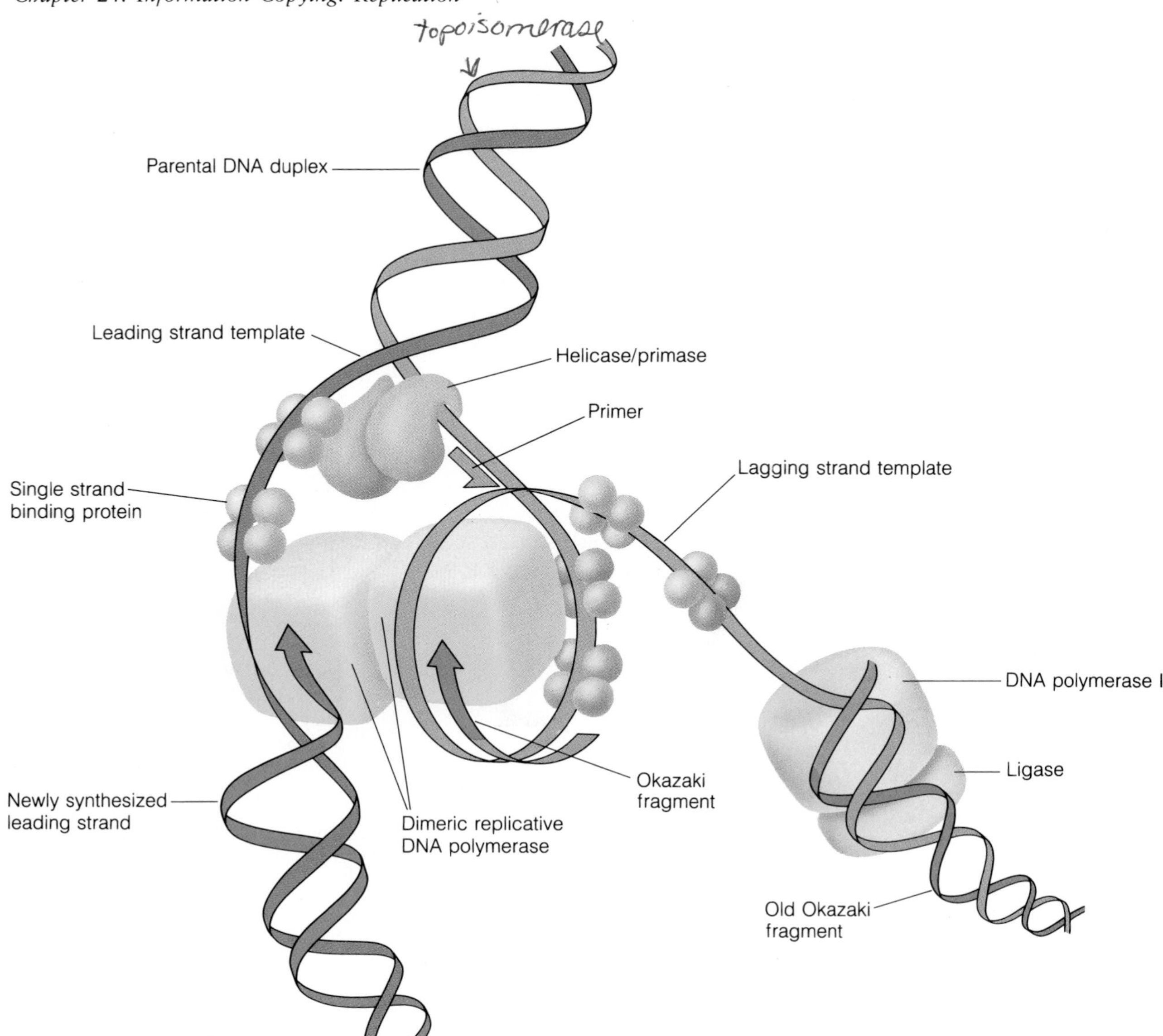

Figure 24.35
Action of different components at a replication fork. Each polymerase is a holoenzyme, complete with accessory proteins that bind the enzyme to its template.

maximal length soon after the Okazaki fragment has elongated as far as it can. Then the next primer initiation signal to be exposed by strand unwinding serves as the template for synthesis of another Okazaki fragment, with the lagging strand polymerase now acting on the stretch of DNA that became coated with SSB during synthesis of the previous Okazaki fragment.

Initiation of DNA Replication

When we ask how DNA replication is initiated from an origin, we are really asking three distinct but interrelated questions. (1) What signals are involved in terms of site-specific interactions of replication proteins with origin sequences? (2) How do these proteins act? (3) How is the process controlled? Most available evidence implicates initiation as the major target for control of replication. In general, however, we know much less about the control of replication than we know about the control of transcription or translation. It is evident that intracellular contacts, of a still undefined type, link the replication apparatus to other cellular structures, so that there is the potential for coordination between DNA replication and the cell cycle. In prokaryotes it is likely that replication occurs at a site attached to the cell membrane, while eukaryotic DNA replication may occur at a DNA- and protein-containing structure called the nuclear matrix (Chapter 28). The nature and significance of these physical linkages have remained elusive. Most of the events in replication have been duplicated in soluble cell-free systems, without components of the membrane or nuclear matrix. Therefore, it has been difficult to define a biochemical role for the membrane or matrix in replication. In this section we describe the systems for which our understanding of initiation mechanisms is most advanced.

Requirements for Initiation of Replication

Since replication proceeds from fixed origins, we can recognize two requirements for initiation: (1) a nucleotide sequence that specifically binds initiation proteins and (2) a mechanism that generates a primer terminus for the action of DNA polymerases. A number of phage, bacterial, plasmid, and organelle replication origins have been isolated by gene cloning and their nucleotide sequences have been determined. In general, these origins include repeated sequences of either identical or opposite polarity (**direct repeats** or **inverted repeats,** respectively). This raises the possibility that initiation proteins bind in multiple copies or that the DNA assumes an unusual configuration, such as a hairpin structure like that involved in the initiation of replication of the plus strand of phages M13 and G4 (see Figure 24.34).

The two most straightforward ways to generate a primer terminus at the origin are (1) nicking a strand of the parental duplex to expose a 3′ hydroxyl terminus and (2) synthesizing an RNA primer, as occurs in lagging strand chain elongation, to expose a 3′ hydroxyl *ribonucleotide* terminus. Phages ϕX174 and G4 initiate the conversion of RF DNA to single-stranded progeny DNA by a site-specific endonuclease, the viral *cisA* gene product. By contrast, duplex DNA replication, where studied, seems to occur without nicking of the parental duplex and with the synthesis of RNA primers.

RNA Primers and Initiation of Replication

As we discuss in Chapter 26, transcription of double-helical DNA involves partial unwinding of the duplex, with the RNA that is synthesized being complementary to one of the two template strands. How is DNA unwound at the origin? The discovery that a type II topoisomerase is required to initiate T4 phage DNA replication suggested that such an enzyme might bind to DNA in a site-specific manner and introduce negative superhelical turns. This unwinding at the origin could expose a short single-stranded region for synthesis of an RNA primer. Indeed, *E. coli* RNA polymerase is required for initiation of T4 DNA replication. Initiation of *E. coli* replication probably involves partial unwinding at the origin as well, beginning with the action of the *dnaA* gene product, which binds to several sites in the 245-base-pair sequence called *oriC*. The *dnaA* protein is activated by binding to specific phospholipids. This finding may illuminate the role of the membrane in initiation of replication.

In neither T4 nor *E. coli* has the mechanism of duplex unwinding at the origin yet been revealed. However, recent studies of phage λ indicate that a complex of phage and host proteins acts to underwind DNA at the origin. As shown in Figure 24.36, the λ gene O protein binds in four dimeric copies to the origin, which contains four direct repeats, each 18 nucleotide

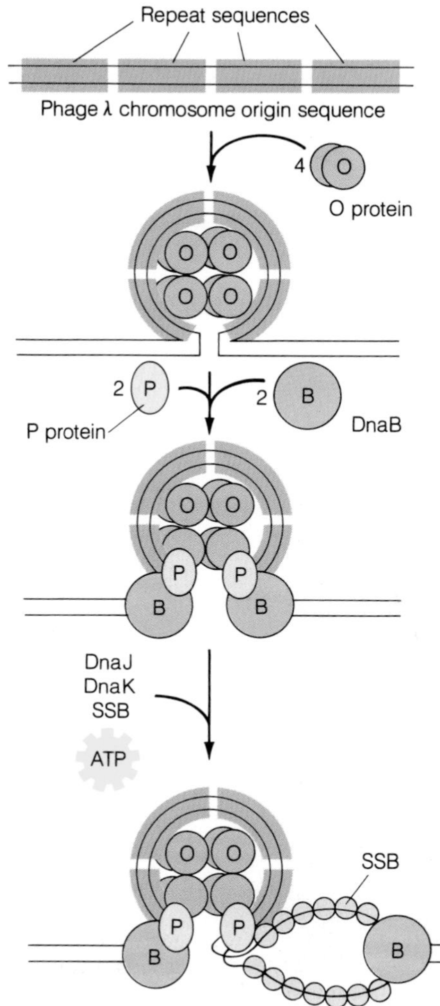

Figure 24.36
Unwinding origin-sequence DNA in the initiation of λ phage DNA replication. Four molecules of phage O protein bind to repeat sequences in the origin. The λ P protein and host DnaB protein then bind, possibly immobilizing DNA at the boundaries of the origin. Addition of host proteins DnaJ, DnaK, and SSB, along with ATP, allows DnaB to act as a helicase, unwinding the origin. The resultant DNA strands are coated with SSB. This allows priming with DnaG (not shown), followed by DNA chain extension with DNA polymerase III holoenzyme.

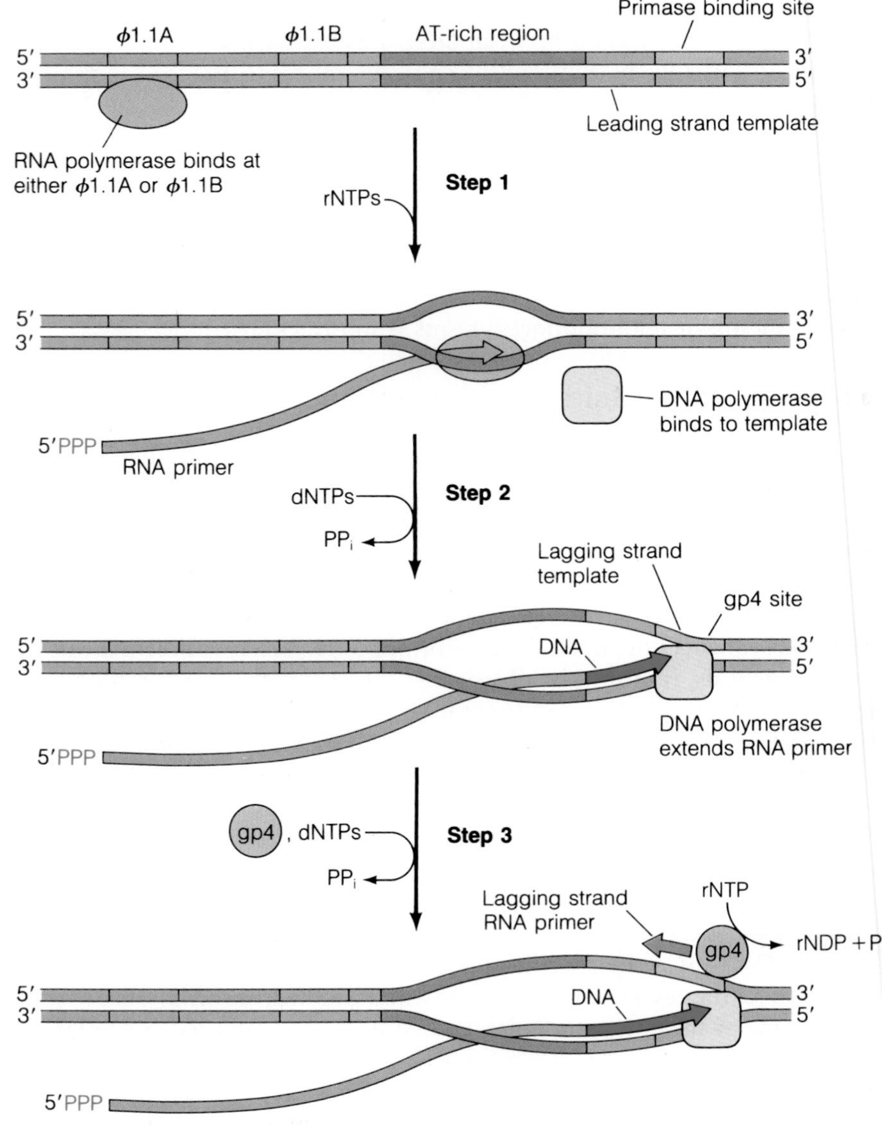

Figure 24.37
Initiation of T7 DNA replication catalyzed by purified proteins. Synthesis of an RNA primer by T7 RNA polymerase can initiate either at ϕ1.1A or ϕ1.1B, both of which are promoters for T7 RNA polymerase. Extension of the primer by T7 DNA polymerase exposes a site for binding of the helicase–primase (gp4) on the lagging strand, initiating a chain in the leftward direction.

pairs long. These segments have clusters of deoxyadenylate residues, which permits DNA to bend, so that it can form the small-diameter circle needed to wrap around the O protein.

Next, with the participation of the phage gene P product, the host *dnaB* gene product comes into play; the latter protein has the helicase activity associated with the *E. coli* primosome. Additional *E. coli* proteins, DnaJ and DnaK, are required to release the *dnaB* protein from P, so that its helicase activity can proceed to unwind DNA to the right of the origin, coat it with SSB, and leave it in a state to be transcribed, generating a rightward-moving replication fork. How a bidirectional fork gets started is not yet clear. This is an interesting question, partly because λ phage is known to switch from bidirectional to unidirectional replication late in infection.

Studies of T7 phage DNA replication in vitro indicate that RNA polymerase itself can unwind enough DNA to transcribe the origin. We know that RNA polymerase is capable of both strand unwinding, to expose the template strand, and rewinding, after it has transcribed a particular region. T7 codes for its own RNA polymerase, the product of gene 1. As discussed in Chapter 26, this enzyme responds to transcription initiation signals, or **promoters,** different from those used by the RNA polymerase of the host, *E. coli.*

In the presence of purified T7 DNA polymerase and T7 RNA polymerase, plus nucleoside triphosphates, T7 DNA initiates replication in vitro from a site at or very near the biological origin. As shown in Figure 24.37, RNA polymerase synthesizes a short primer (step 1), which is extended rightward by DNA polymerase (step 2). Eventually a recognition site for the gene 4 helicase–primase is exposed on the other strand. Addition of purified gp4 is necessary at this point for initiation of lagging strand synthesis (step 3). Though not shown in the figure, purified SSB (of either phage or host origin) also must be added for initiation of a leftward-moving fork. Probably the same proteins and functions occur during initiation of T7 DNA replication in vivo, but this has not yet been established with certainty.

Plasmid DNA Replication: Control of Initiation by RNA

The replication of several small circular DNAs, including plasmids and organelle DNA, is being investigated in cell-free systems. An interesting feature of plasmid DNA replication is the control of initiation exercised by "antisense RNA." In the *E. coli* plasmid ColE1, RNA can be synthesized at the replication origin from the DNA strand opposite that used to synthesize the initiation primer. This inhibits the initiation of replication, because the primer RNA becomes tied up as part of an RNA–RNA duplex. Further analysis of this system, particularly the control of antisense RNA synthesis, may reveal how the initiation of DNA replication itself is controlled.

Mitochondrial DNA: Two Unidirectional Replication Forks

A particularly interesting mode of replication initiation is seen with animal cell mitochondrial DNA. As shown in Figure 24.38, unidirectional replication of this 16-kb circular duplex starts from a fixed origin on one strand (the H, or heavy, strand). As replication commences, the parental L (light) strand is displaced by the nascent L strand, forming a structure called a displacement loop, or D loop, which can readily be visualized in the electron microscope. When this unidirectional replication apparatus has progressed two-thirds of the way around the circle, an L strand origin is exposed on the displaced strand, and unidirectional replication from this origin starts backward from the direction of H strand replication. The entire process requires about 1 hour, consistent with the much slower replication of eukaryotic than of prokaryotic DNAs. RNA priming is involved, as shown first by the fact that mature mitochondrial DNA contains a few ribonucleotides, apparently remnants of an inefficient primer excision system.

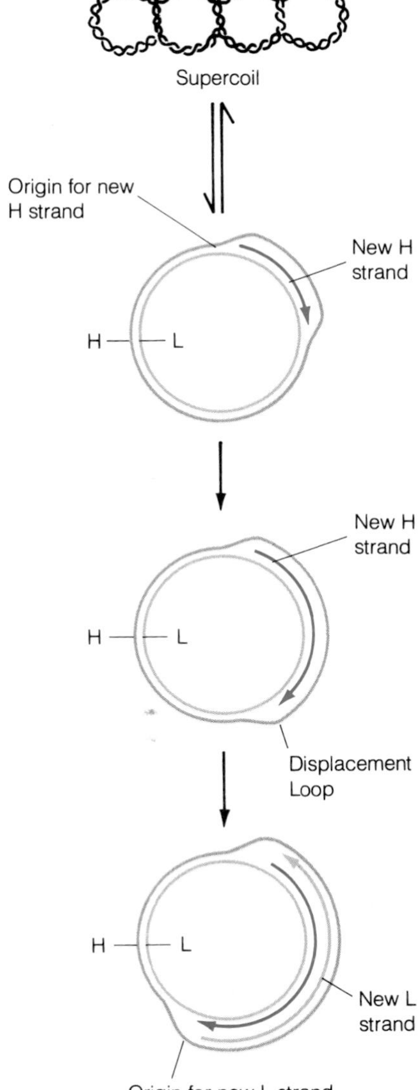

Figure 24.38
Replication of mitochondrial DNA. Parental heavy and light chains are shown in brown. New strands are in blue and red.

Replication of Linear Genomes

Thus far we have discussed in detail the replication of only circular DNA genomes. *Linear* genomes, including those of several viruses as well as the chromosomes of eukaryotic cells, face a special problem—how to complete replication of the lagging strand. Excision of an RNA primer from the 5′

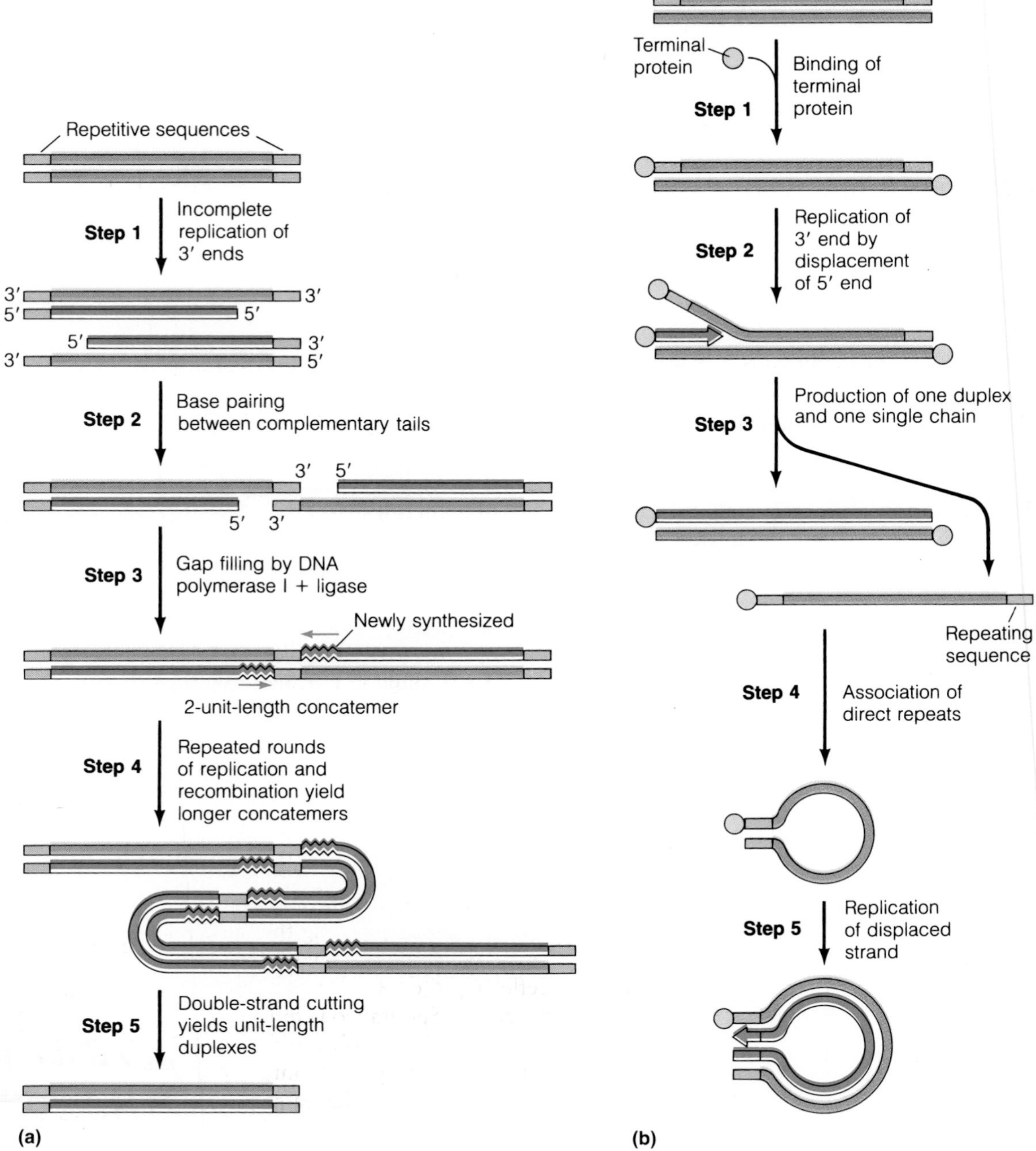

Figure 24.39
Mechanisms for replication of linear DNAs. (**a**) End-to-end recombination at terminal redundancies, as seen with T4 and T7; (**b**) use of 5′ genome-linked protein, as seen with adenovirus and ϕ29.

end of a linear molecule would leave a gap that cannot be filled by DNA polymerase action, because of the absence of a primer terminus to extend (Figure 24.39a, step 1). If this DNA could not be replicated, the chromosome would shorten a bit with each round of replication. Viruses reveal at least three strategies for dealing with this problem. Phages T4 and T7 exhibit terminal redundancy, so recombination can occur at the incompletely replicated ends of two nascent DNA molecules without loss of genetic information (Figure 24.39a, steps 2 and 3). This process is repeated in subsequent rounds of replication (step 4) until the end-to-end linear aggregate (called a **concatamer**) is more than 20 times the length of a single phage chromosome. A virally specified nuclease then cuts this giant DNA into genome-length pieces when it is packaged into phage heads (step 5).

Bacteriophage ϕ29 and the adenoviruses have evolved a quite different strategy (Figure 24.39b). The genomes of these viruses contain inverted repeat sequences at the ends. Replication begins at one end of the linear duplex, with a *protein*, called **terminal protein,** serving as the primer (step 1). This protein in adenovirus reacts with dCTP to form a dCMP residue covalently linked through its phosphate to a serine residue. The dCMP serves as primer for replication of the 3′-terminated strand, with the 5′ end being displaced as a single strand (step 2). Once this strand is completely displaced (step 3) it can form a short duplex region by base pairing with the repeat sequence at the other end of the molecule (step 4), which can serve as substrate for initiation of a second round of displacement synthesis (step 5), yielding complete replication of both strands. ϕ29 replicates similarly, except that the protein-bound primer nucleotide is dAMP, rather than dCMP.

Still another mechanism is involved in the replication of poxviruses, such as vaccinia (Figure 24.39c). The two strands at each end of this linear genome are covalently linked together. Such a structure, when approached by a replication fork, can support movement of the leading strand *around* the link between the strands, generating a primer for synthesis of the final lagging strand fragment.

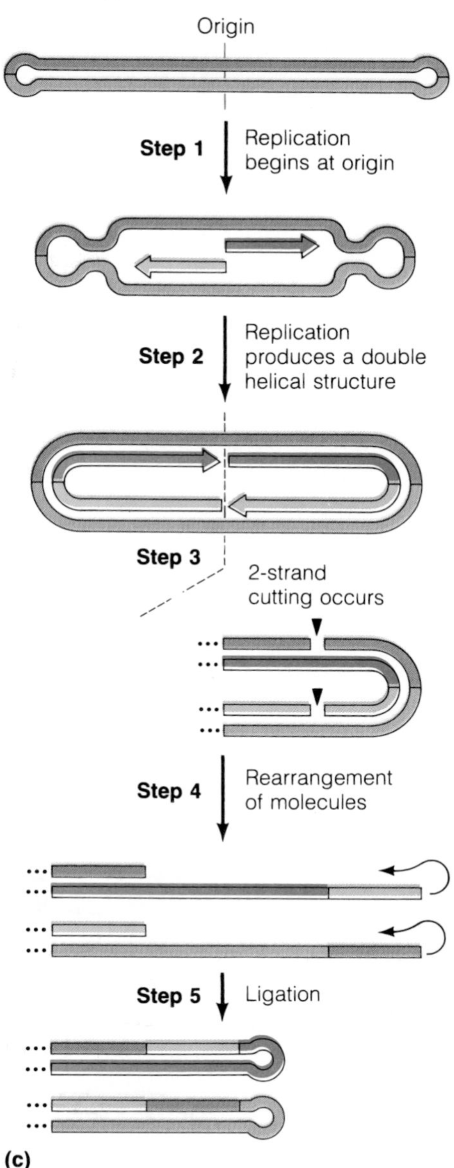

Figure 24.39 Continued
(c) Replication around hairpin loops at the termini, followed by two-strand cutting, as seen with poxviruses.

Fidelity of DNA Replication

DNA replication is by far the most accurate of known enzyme-catalyzed processes. From spontaneous mutation frequencies, which for all positions in a particular gene amount to about 10^{-6} per generation, we can estimate the chance that a particular nucleotide residue will be copied incorrectly as about 10^{-9} to 10^{-10} per base pair per round of replication. This very high accuracy cannot be explained simply from the energetics of DNA structure. The free energy of formation of a Watson–Crick base pair (A-T or G-C) differs from that of an incorrect pair (A-C or G-T) by only 4 to 13 kJ/mol (although the difference may be greater in the environment of a replication fork). If DNA polymerase were completely passive, only incorporating the nucleotide that spontaneously associates with a template base, errors would be made approximately 1 to 10% of the time, reflecting the relative abundance of correctly and incorrectly base-paired structures. Spontaneous mutation rates would be many orders of magnitude higher than is observed.

How might other components of the replication machinery contribute to the extremely low error rate of DNA replication? The earliest clue came from phage T4, when it was found that mutations that changed the viral DNA polymerase affected the accuracy with which other genes were replicated. Some polymerase mutations had a **mutator** phenotype, meaning that mutation rates at other loci were uniformly increased. Other polymerase mutations were **antimutators,** which seemed to improve the accuracy of DNA replication. Most of the *mutator* mutations were found to result from *decreased* activity of the polymerase-associated exonuclease. Since this activity has the ability to cleave incorrectly base-paired nucleotides from a 3′ terminus, a decreased activity of exonuclease would *decrease* the chance that a misincorporated nucleotide would be excised before the polymerase could add the next nucleotide. Similarly, *increased* 3′-exonuclease activity relative to polymerase would *increase* the chance that a mispaired nucleotide would be excised and hence would reduce the error rate. Thus, the accuracy of DNA replication could be ascribed in part to a proofreading activity that checks newly replicated DNA and corrects most errors.

Subsequent research revealed that DNA polymerase plays an additional role in determining accuracy. A number of polymerase mutator mutations affect not the proofreading activity but the polymerase activity itself, by increasing the probability that polymerase will cause mutation by *misinsertion,* that is, insertion of an incorrect base. The mechanism by which the polymerase activity itself controls the insertion of correct or incorrect nucleotides remains an active area of investigation.

Another error correction mechanism, *mismatch repair,* makes an additional contribution to the high accuracy of genetic information copying. As we describe in the next chapter, the mismatch repair system scans newly replicated DNA, excising residues that are not properly base paired and replacing them with the correct nucleotides.

The extent to which these fidelity-enhancing processes contribute to the accuracy of eukaryotic DNA replication is slowly being revealed. DNA polymerase α does not contain an associated 3′-exonuclease activity, and this has led to skepticism about the existence of exonucleolytic proofreading in eukaryotes. However, the proofreading activity could lie on another protein; indeed, for *E. coli* the proofreading nuclease activity *does* lie on a protein other than the polymerase III subunit, namely the *dnaQ* gene product. Moreover, recent investigation has revealed a "cryptic" 3′-exonuclease activity in a replicative polymerase from *Drosophila,* an activity that is undetectable except under certain conditions. Moreover, as noted earlier, polymerase δ does contain an associated 3′-exonuclease activity.

RNA Viruses: The Replication of RNA Genomes

We conclude this chapter with a few words about the replication of viral genomes consisting of RNA, not DNA. Virtually all known plant viruses contain RNA instead of DNA, as do several bacteriophages and many important animal viruses, including polio and influenza viruses. The **retroviruses,** which are responsible for many tumors as well as acquired immune deficiency syndrome, also contain RNA genomes.

RNA-Dependent RNA Replicases

Most, though not all, RNA viruses contain a genome consisting of a single molecule of single-stranded RNA. Usually that RNA is the "sense" strand for expression of genetic information at the level of translation. In other words, the RNA molecule that passes from the viral particle into the infected cell can serve directly as a messenger RNA, without having first to direct the synthesis of complementary-strand RNA. One of the early products of translation of this input genome is the enzyme **replicase,** or RNA-dependent RNA polymerase. This enzyme, sometimes after complexing with required polypeptide subunits of host cell origin, replicates the input RNA ("plus" strands), starting from the 3′ end, so that the new "minus" strand is laid down from its 5′ end to the 3′ end, the same as the direction in which DNA polymerases work. The newly replicated minus strands then serve as templates for synthesis of plus strands, which are packaged into progeny **virions,** or virus particles. More complex mechanisms come into play when the RNA genome is double stranded or segmented (three or four separate RNA molecules), or when the virion RNA is itself the minus strand.

Since the known RNA replicases lack proofreading activity, it is no surprise that viral RNA replication is far more error prone than DNA replication, and RNA viruses undergo mutation and evolution far more rapidly than the organisms they infect. This is clearly related to viral pathogenesis, in part because a virus population infecting a plant or animal can undergo change so rapidly that it can evade or counteract the host's defense mechanisms.

Replication of Retroviral Genomes

Different strategies for genome replication are involved in the action of retroviruses, so named because of the presence of a special enzyme, **reverse transcriptase.** In this class of viruses the single-stranded RNA genome achieves latency—the ability to persist in a host cell for a long period without pathological effects—by making a DNA copy of itself and inserting that copy into the host cell genome. The DNA copy is made by reverse transcriptase, which is packaged in virions and enters the infected cell along with the viral genome. As shown in Figure 24.40, reverse transcriptase first uses viral RNA as a template for synthesizing a complementary DNA strand (step 1); a specific transfer RNA molecule serves as the primer. A ribonuclease H activity of the enzyme then partially digests the RNA (step 2), and the structure circularizes by DNA–RNA base pairing (step 3). The nascent DNA chain is extended around the circle, and the tRNA primer is removed by RNase H activity (step 4). Strand displacement synthesis then occurs, with the RNA strand being displaced (step 5). The resultant double-stranded DNA circle then recombines with a site on a chromosomal DNA and is linearly inserted into that DNA in the process. Under these conditions the viral genome can persist in a noninfectious state for many years, with most of its own genes turned off. Environmental stresses, still undetermined, can trigger the excision of the integrated viral genome and return of the virus to an infectious state.

Studies of viral RNA replication have revealed two important general principles: (1) the chemical mechanism of nucleic acid replication seems to be universal, with a template strand providing the information and the 5′-triphosphate moieties of nucleotides providing the energy; and (2) the flow of information is not unidirectional from DNA to RNA, since RNA can serve as template for the synthesis of DNA.

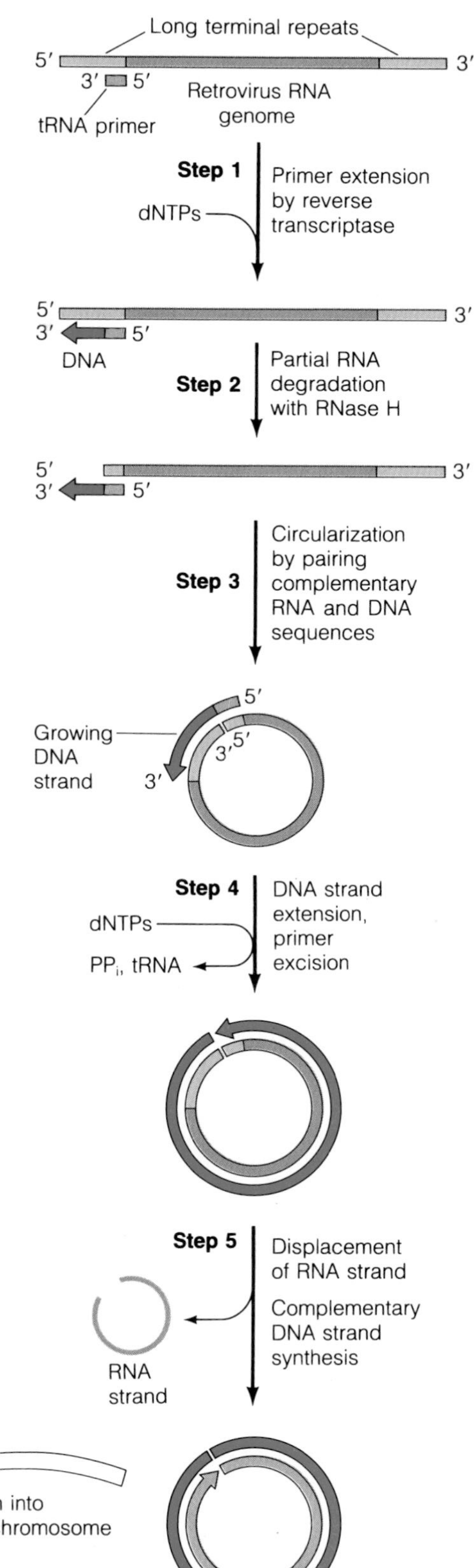

Figure 24.40
Simplified view of retrovirus life cycle. The RNA genome contains long terminal repeats (LTRs), to one of which a tRNA molecule binds. Primer extension, partial RNA digestion, and circularization generate a substrate for extensive DNA synthesis. Ultimately a circular duplex DNA molecule is formed, the likely substrate for integration into a host chromosome. Recent evidence suggests that two viral RNA molecules must interact at step 2 or 3.

REFERENCES

General References

Adams, R. L. P., J. T. Knowles, and D. P. Leader (1986) *The Biochemistry of the Nucleic Acids,* 10th ed. Chapman & Hall, New York. An up-to-date book-length review, with voluminous references, covering topics that will concern us for the remainder of this book.

Kornberg, A. (1988) DNA replication. *J. Biol. Chem.* 263:1–4. The first in a series of minireviews in JBC, conciselv worded, well referenced, and beautifully illustrated.

Kornberg, A. (1990) *DNA Replication,* 3rd ed. Freeman, New York. *The* definitive work on this subject.

Watson, J. D., N. H. Hopkins, J. W. Roberts, J. A. Steitz, and A. M. Weiner (1986) *Molecular Biology of the Gene,* 4th ed. Benjamin/Cummings, Menlo Park, Calif. Current molecular biology described with exceptional clarity, well illustrated, and exhaustively referenced.

DNA Polymerases

Joyce, C. M., and T. A. Steitz (1987) DNA polymerase I: From crystal structure to function via genetics. *Trends Biochem. Sci.* 12:288–293. A clearly written review of structure–function relationships in DNA polymerase I of *E. coli,* based on x-ray crystallography and site-directed mutagenesis.

McHenry, C. S. (1988) DNA polymerase holoenzyme III of *Escherichia coli. Annu. Rev. Biochem.* 57:519–550. This review covers the same material as Kornberg's JBC minireview, but in much greater detail.

So, A. G., and K. M. Downey (1988) Mammalian DNA polymerases α and δ: Current status in DNA replication. *Biochemistry* 27:4591–4595. This minireview advances the idea that these two polymerases participate in leading strand and lagging strand synthesis.

Replication Proteins Other than DNA Polymerase

Chase, J. W., and K. R. Williams (1986) Single-stranded DNA binding proteins required for DNA replication. *Annu. Rev. Biochem.* 55:103–136. A current review of both prokaryotic and eukaryotic helix-destabilizing proteins.

Drlica, K., and R. J. Franco (1988) Inhibitors of DNA topoisomerases. *Biochemistry* 27:2253–2259. This minireview reveals how inhibitors have provided clues to mechanisms of topoisomerase action.

Lohman, T. M., W. Bujalowski, and L. B. Overman (1988) *E. coli* single strand binding protein: A new look at helix-destabilizing proteins. *Trends Biochem. Sci.* 13:250–255. This article questions whether the T4 gene 32 protein is the best model for the involvement of single-strand DNA-binding proteins in replication.

Wasserman, S. A., and N. R. Cozzarelli (1986) Biochemical topology: Applications to DNA replication and recombination. *Science* 232:951–960. A well-illustrated article describing topoisomerase actions and the biological processes in which they participate.

Mechanism of DNA Replication

Allewell, N. (1988) Why does DNA bend? *Trends Biochem. Sci.* 13:193–195. Recent studies indicate that DNA bending at origin sequences is essential for initiation. This minireview discusses base sequence features associated with bending

Echols, H. (1986) Multiple DNA–protein interactions governing high-precision DNA transactions. *Science* 233:1050–1056. This review article touches on several interesting topics, specifically the evidence that site-specific DNA unwinding is involved in the initiation of phage λ DNA replication.

Fidelity of Nucleic Acid Synthesis

Holland, J., K. Spindler, F. Horodyski, E. Grabau, S. Nichol, and S. VandePol (1982) Rapid evolution of RNA genomes. *Science* 215:1577–1585. This review deals with the biological consequences of the absence of error-correcting mechanisms in viral RNA replicases.

Kunkel, T. A. (1988) Exonucleolytic proofreading. *Cell* 53:837–840. A timely minireview on one of the most important fidelity-enhancing mechanisms.

Loeb, L. A., and M. E. Reyland (1987) Fidelity of DNA synthesis. In: *Nucleic Acids and Molecular Biology,* edited by F. Eckstein and D. M. J. Lilley, pp. 157–173. Springer-Verlag, Berlin. A current review of this most accurate of enzymatic processes.

PROBLEMS

1. Describe an experimental approach to determining the *processivity* of a DNA polymerase, that is, the number of nucleotides incorporated per chain per polymerase binding event.

2. After Okazaki's proposal of discontinuous DNA chain growth, there was much controversy about whether both DNA strands are synthesized discontinuously, or only the lagging strand. Clearly, there is no need for the leading strand to be synthesized in fragments, but many workers found that pulse-labeled DNA fragments hybridized to both parental DNA strands. This indicated that both parental strands served as the template for synthesis of short DNA fragments. Propose an alternative mechanism by which leading strand replication could generate short fragments and propose an experimental test of your suggestion.

3. The buoyant density of DNA can be increased by incorporation of either heavy stable isotopes, such as ^{15}N, or base analogs, such as 5-bromouracil. This problem asks you to calculate the density increment generated by each technique. In all calculations the density of "light" *E. coli* DNA is 1.710 g/ml. The G + C content is 51 mol % (i.e., 51 mol of G + C per 100 mol of DNA nucleotides). Residue molecular weights of cesium salts of nucleotides are as follows: dAMP, 445; dTMP, 436; dGMP, 461; dCMP, 421; 5-bromo-dUMP, 501.

(a) Calculate the buoyant density of *E. coli* DNA when all of the ^{14}N atoms are replaced with ^{15}N atoms.

(b) Calculate the buoyant density of *E. coli* DNA when all thymine residues are replaced with 5-bromouracil.

4. A mixture of four $\alpha[^{32}P]$-labeled ribonucleoside triphosphates was added to permeabilized bacterial cells that were undergoing DNA replication, and incorporation into DNA was followed over time, as shown in the accompanying graph. After 10 minutes of incubation a 1000-fold excess of unlabeled ribonucleoside triphosphates was added, with the results shown in the graph.

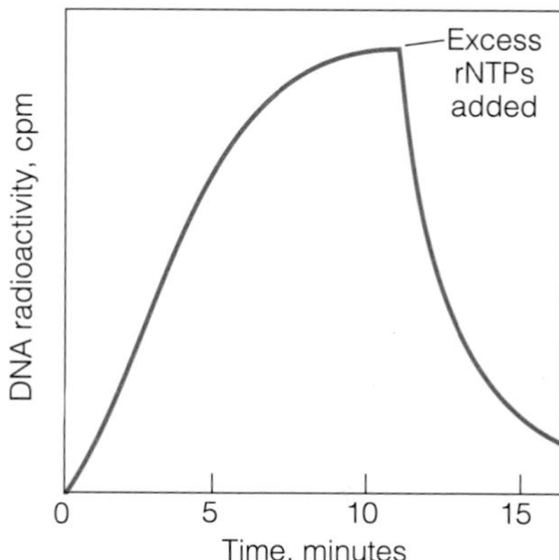

(a) Why was the excess of unlabeled rNTPs added?

(b) How could you tell that radioactivity is being incorporated as ribonucleotides, rather than an alternative such as reduction to deoxyribonucleotides, followed by incorporation?

(c) What does this experiment tell you about the process of DNA replication?

5. Deoxyadenylate residues in DNA undergo deamination fairly readily, as do deoxycytidylate residues.

(a) What is the product of dAMP deamination?

(b) The deamination product is known to base pair with A, C, or T. What would be the genetic consequences if this deaminated site in DNA were not repaired and if it paired with C on the next round of replication?

6. The *E. coli* chromosome is 1.28 mm long. Under optimal conditions the chromosome is replicated in 40 minutes.

(a) What is the distance traversed by one replication fork in 1 minute?

(b) If replicating DNA is in the B form (10.4 base pairs per turn), how many nucleotides are incorporated in 1 minute in one replication fork?

(c) If cultured human cells (such as HeLa cells) replicate 1.2 m of DNA during a 5-hour S phase and at a rate of fork movement one-tenth of that seen in *E. coli*, how many origins of replication must the cells contain?

(d) What is the average distance, in kilobase pairs, between these origins?

7. DNA ligase has the ability to relax supercoiled circular DNA in the presence of AMP but not in its absence.

(a) What is the mechanism of this reaction, and why is it dependent on AMP?

(b) How might one determine that supercoiled DNA had in fact been relaxed?

8. A recent paper reports that in mammalian cells, genes that are expressed in a particular cell are replicated during the first half of S phase, while genes not expressed in that cell are replicated in the latter half of S phase. Briefly describe an experiment that could lead to this conclusion.

9. Although DNA polymerases require both a template and a primer, the following single-stranded polynucleotide was found to serve as a substrate for DNA polymerase in the absence of any additional DNA.

3′ OH-TGAGCCCATAGCCGGGGCTCTAA-
CCGTAGACCACGAATAGCATTAGG-p 5′

Give the structure of the product of this reaction.

10. The 3′ exonuclease activity of *E. coli* DNA polymerase I was recently found to show no discrimination between correctly and incorrectly base-paired nucleotides at the 3′ terminus; properly and improperly base-paired nucleotides are cleaved at equal rates. How can this observation be reconciled with the fact that the 3′ exonuclease activity increases the accuracy with which template DNA is copied?

11. 2′,3′-Dideoxyinosine is currently receiving attention as an anti-AIDS drug. Propose a mechanism by which it might act to block the growth of the AIDS virus.

12. Aphidicolin-resistant mutants of mammalian cells often have alterations in the structure of DNA polymerase-alpha that render the enzyme insensitive to inhibition. However, some mutants show alterations, not in DNA polymerase, but in ribonucleotide reductase. How might a change in the latter enzyme cause an aphidicolin-resistant phenotype?

CHAPTER 25

Information Restructuring: Restriction, Repair, Recombination, Rearrangement, Amplification

Our focus in this chapter shifts from DNA as a *template* for its own replication to DNA as a *substrate,* in a number of processes which we categorize as **information restructuring.** Some of these processes preserve the information content of DNA, some represent mechanisms used for defense by organisms (Figure 25.1), some are involved in controlling normal gene expression or differentiation, some introduce genetic diversification, and some represent responses to environmental stresses. Some processes are related to several of these functions. For example, genetic recombination introduces diversity to a species, but it is also a response to environmental stress, a response that promotes survival of an individual organism.

From the biochemist's standpoint, information restructuring is distinguished from DNA replication in a significant regard. Replication is a central pathway that is a major metabolic event in at least part of the lifetime of all cells, and the activities of the enzymes and proteins involved are relatively high. As a result, their discovery and characterization, while elegant, were fairly straightforward. By contrast, information restructuring involves quantitatively minor pathways, with far less mass conversion per cell. Detection of the enzymes involved is correspondingly more difficult. For example, although DNA methylation may involve only one or a few sites per gene, the presence or absence of methyl groups can have profound effects on gene inactivation by enzymatic cleavage, gene expression at the transcriptional level, or the correction of replication errors. However, assay of DNA methylases and identification of methylated sites require experimental techniques much more sensitive than those used to analyze DNA replication, because of the low activities involved.

The processes we shall consider in this chapter include (1) restriction and modification, which serve as protective mechanisms in prokaryotic cells but are also important because they provide reagents for gene isolation by in vitro generation of recombinant DNA molecules; (2) metabolic responses to DNA structural damage, principally mutagenesis and repair; (3) recombination, whereby the contents of a genome are redistributed during sexual reproduction; (4) gene rearrangements, including transpositions of DNA segments from one chromosomal integration site into another and the joining of DNA segments from distant parts of a genome; and (5) gene amplifica-

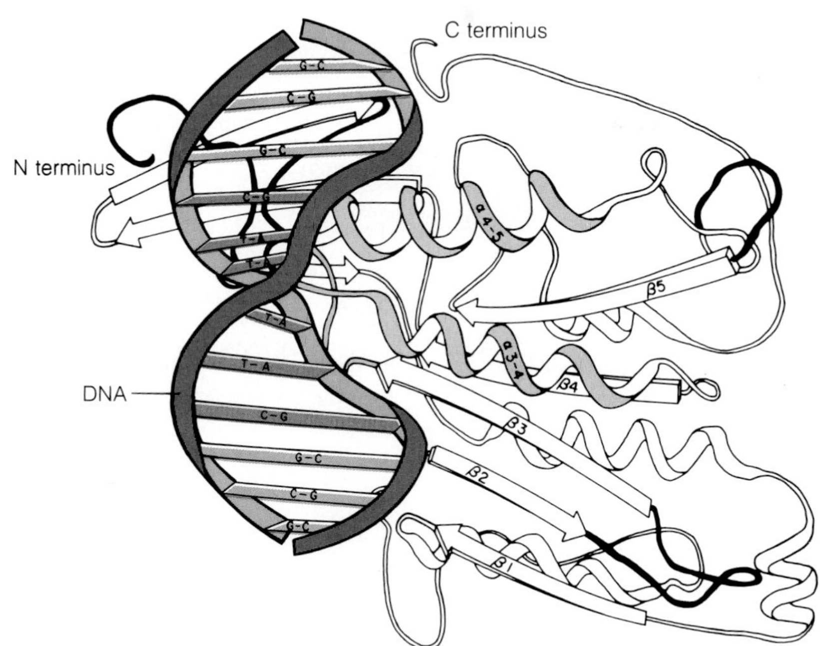

Figure 25.1
Enzymatic basis for DNA restriction, an important process in information restructuring. Schematic backbone drawing of one of the two subunits of *Eco*RI nuclease in a complex with its DNA substrate. The arrows represent β strands, the coils represent α helices, and the ribbons represent the DNA backbone. The two helices in the foreground form the central interface with the other subunit. The N-terminal "arm" wraps about the DNA substrate.

tion, an increase in the copy number of individual segments of DNA, which occurs both as a normal developmental process and as a response to environmental stresses. Collectively, these processes are essential to the survival of a species. They are also crucial to evolution, for they ensure the genetic diversification of individuals.

DNA Methylation

Before we examine the processes of information restructuring, let us first review the occurrence and synthesis of methylated bases in DNA. This discussion is relevant because DNA methylation underlies several important biological processes—not only restriction and modification, but also mismatch error correction (a DNA repair process) and the control of eukaryotic gene expression. Recall from Chapter 21 that *S*-adenosylmethionine (AdoMet) is the substrate for methylation of both RNA and DNA. Methylation occurs at the polynucleotide level, with transfer of a methyl group from AdoMet to an RNA or DNA nucleotide residue.

The sole methylated base found in eukaryotic DNA is 5-methylcytosine (mC), which is 3 to 5% as abundant as cytosine in animal DNA and much more abundant in some plant DNAs, but virtually absent from other DNAs, such as those of insects. Mechanistically, the synthesis of an mC residue in DNA is similar to the thymidylate synthase reaction (Chapter 22); a nucleophilic residue on the enzyme (probably a thiol group) attacks C-6 of the pyrimidine ring of cytosine, which in turn activates C-5 for formation of a carbon–carbon bond.

5-Methylcytosine (mC)

In prokaryotic DNA the major methylated bases are N^6-methyladenine and to a lesser extent N^4-methylcytosine. Fewer than 1% of the adenines and cytosines are methylated in prokaryotic DNAs.

***N*^{6}-Methyladenine (mA)**

Methylation in bacteria occurs at specific sites. In *E. coli* methylation of A residues in the sequence 5′...GATC...3′ is involved in mismatch error

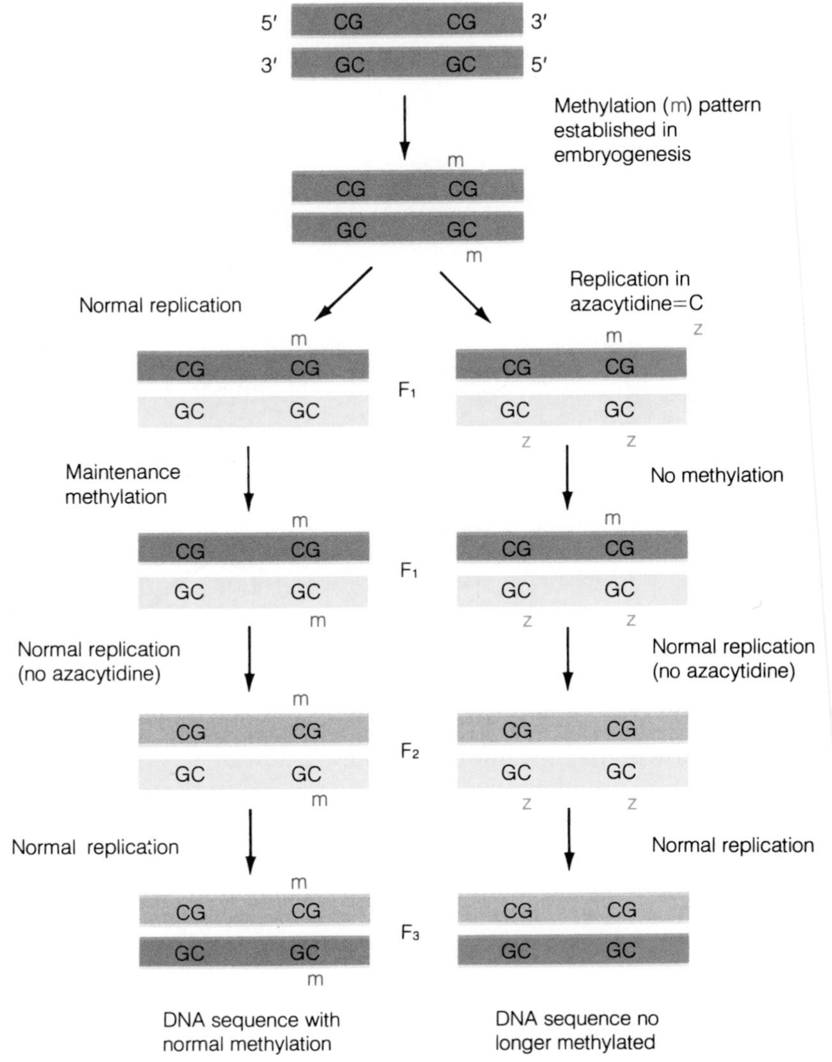

Figure 25.2
Maintenance methylation of eukaryotic DNA. 5-Azacytidine is an analog that can be incorporated into DNA but not methylated. Therefore, growth in this analog leads to DNA demethylation. $\underset{m}{C}$ = 5-methylcytosine; $\underset{z}{C}$ = 5-azacytidine.

correction (see p. 874), and methylation at other sites protects DNA against cleavage by restriction endonucleases (described in the next section). In animals methylation is found only in C residues that are immediately 5′ to G residues, that is, in a sequence ...CpG.... When such a C is methylated, so is the corresponding C in the complementary strand. In plant DNAs the methylated sequence is ...CpNpGp..., where N can be any base.

The biological significance of DNA methylation in prokaryotes is now fairly clear, but its importance in eukaryotes has not yet been defined. We do know that methylation at a particular site is a heritable phenomenon: when eukaryotic DNA replicates, a **maintenance methylase** ensures that all of the sites that were methylated in parental DNA are methylated in daughter DNA, as shown in Figure 25.2. The unmethylated sites remain unmethylated. How the original sites are selected for methylation during embryonic differentiation is not yet clear. Several lines of evidence suggest that methylation is responsible for the tissue-specific inactivation of genes during development. For example, gene inactivation can be reversed by treatment with **5-azacytidine,** a cytidine analog that is metabolized like cytidine to the analog of dCTP, which is then incorporated into DNA. Because of the N substi-

NH_2 ... ribose

5-Azacytidine

tution, position 5 cannot be methylated, and when that DNA subsequently replicates, the C that is incorporated is not methylated (see Figure 25.2). Azacytidine treatment can cause adult bone marrow cells to reactivate the synthesis of fetal hemoglobin, which is normally turned off during development. This analog is being tested as a treatment for β-thalassemia, where the β-like fetal globin chains could replace the missing β chains.

A great many studies support the concept that DNA methylation participates in the arrest of expression of developmentally related genes, but puzzling questions remain. How do insects regulate their developmental patterns of gene expression if their DNA is unmethylated? On the other hand, extensive substitution of mC for C has been described in specific DNA segments in *Neurospora*. Does this imply the existence of additional roles for eukaryotic DNA methylation?

Restriction and Modification

One of the most important developments in the recent history of biochemistry is the discovery of **restriction endonucleases,** enzymes that catalyze the double-strand cleavage of DNA at specific base sequences. The discovery of the enzymes arose through research on an interesting biological process—**host-induced restriction and modification.**

Biology of Restriction and Modification

Although restriction and modification were first described in 1952, it was not until Werner Arber's work in Switzerland, in the mid-1960s, that the biochemical basis for the phenomena became clear. The basic observation was as follows. Bacteriophage λ, when grown on the K12 strain of *E. coli* (which we shall call K), grows well in subsequent infections of the K strain, with every phage particle giving rise to a **plaque.** (A plaque is a cleared area on a Petri plate seeded with excess bacteria and a limited number of phages, resulting from the multiplication of one virus particle and the localized lysis of bacteria; see Figure 25.3). However, the phages would grow very poorly on infection of *E. coli* strain B, with only 0.01% of the infecting phages giving rise to a plaque. Most of the infecting phage DNA was broken down in these nonproductive infections.

Phages purified from the few plaques that formed could infect strain B with high efficiency. The same phenomenon was seen in reverse; phages grown on *E. coli* B infected strain K with low efficiency, but the few plaques that formed yielded phages that could infect strain K with high efficiency. Biologically, it appeared that phages are *modified* in a way peculiar to a specific bacterial host, and when those phages infect a host with a different modification system, they are *restricted,* meaning that their growth on that host is blocked. Since restriction was accompanied by DNA breakdown, it was reasonable to expect that restriction and modification involved structural changes in DNA.

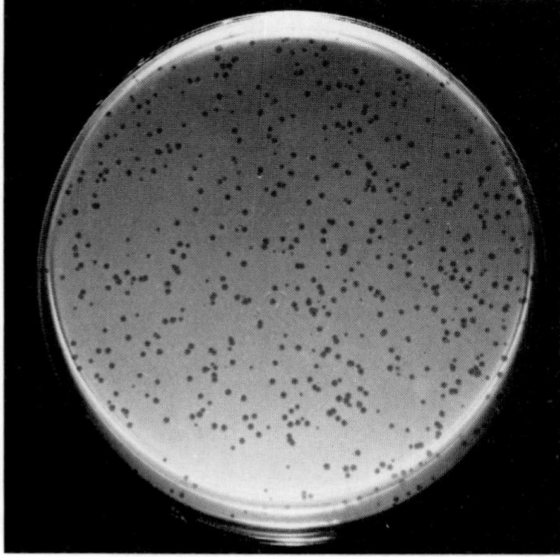

Figure 25.3
Bacteriophage plaques. A Petri plate was seeded with many bacteria (about 10^8) and a small number of bacteriophage particles (about 100). Each phage particle generates a cleared area, or plaque, by infecting and lysing all of the bacteria growing on that part of the plate.

Arber and his colleagues showed that the basis for modification is a host-specified pattern of DNA methylation. Phages grown in strain K are methylated in the same pattern as the chromosomal DNA of strain K itself and thereby become resistant to the restriction system of strain K. When these phages infect strain B, their DNA is susceptible to the B restriction system, and most of the DNA is broken down. However, a few of these

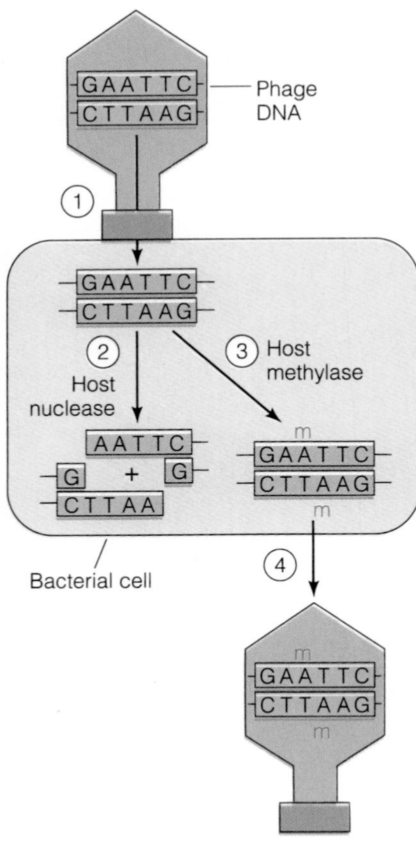

Figure 25.4
Host-induced restriction and modification. (1) A phage whose DNA is unmodified infects a bacterium with a restriction system that recognizes the DNA sequence –GAATTC–. Most phage DNA molecules are cleaved by the restriction nuclease (2), but the few that become methylated first are protected from attack (3). The phages that emerge contain modified (methylated) DNA (4). They are thus able to reinfect the same bacterial strain because they are not vulnerable to restriction by the host nuclease.

DNA molecules become methylated by the host modification system and escape restriction. The host-induced restriction-modification systems represent a defense mechanism that allows a bacterium to recognize a DNA as foreign and thereby to escape death by virus infection.

Restriction-modification systems are widespread among bacteria. Some are encoded by chromosomal genes and some by plasmids. In 1970 Hamilton Smith made the remarkable observation that a restriction nuclease he was studying catalyzed double-strand cleavage of DNA within a specific short nucleotide sequence. Hundreds of restriction endonucleases are now known to catalyze similar sequence-specific cleavages. Figure 25.4 schematizes restriction and modification phenomena, showing the operation of sequence-specific enzymes.

The importance of Smith's discovery was that, for the first time, one could isolate homogeneous DNA fragments of precisely defined length by treating DNA with a restriction nuclease in vitro and then resolving the fragments in the digest on an electrophoretic gel. Recall the utility of proteases such as trypsin, which cleave at specific sites on proteins, and their

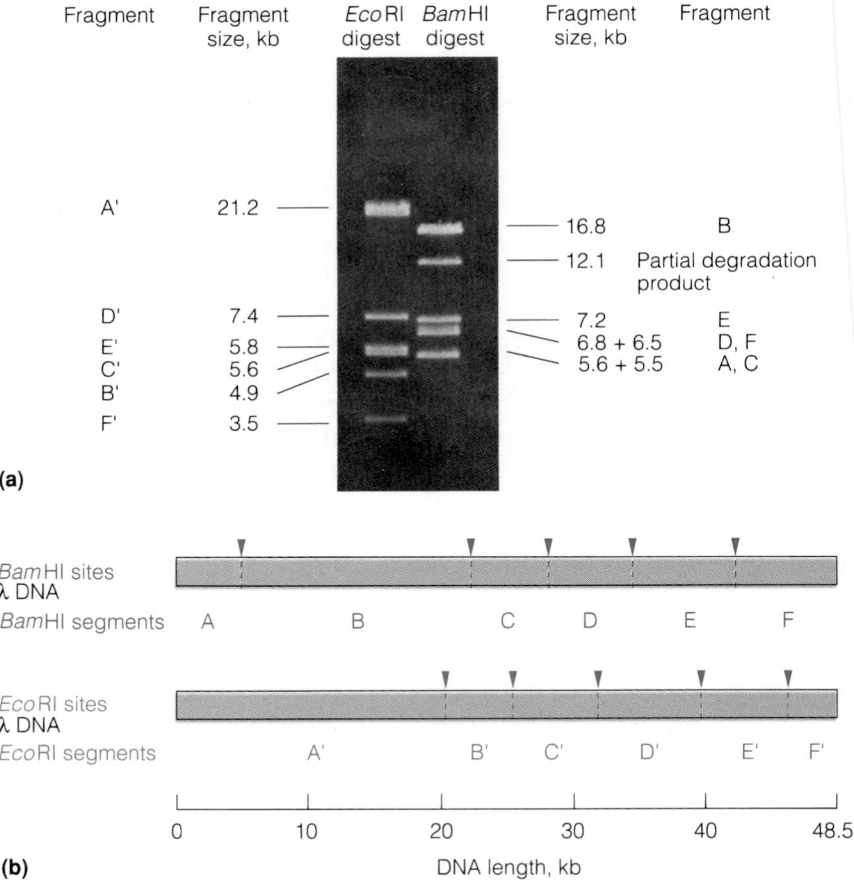

Figure 25.5
Fragmentation of phage λ DNA with *Eco*RI or *Bam*HI. (**a**) Experimental determination of fragmentation patterns. Restriction digests are subjected to agarose gel electrophoresis, and the fragments are visualized by staining the gel with ethidium bromide. Note that fragments with very similar sizes form but one band on a gel. (**b**) Maps of cleavage sites for each enzyme on the 48.5-kb λ DNA molecule. By convention, the fragments are assigned letters as shown. The "partial degradation product" in the *Bam*HI digest resulted from incomplete cleavage between fragments C and D in this experiment.

usefulness in fragmenting protein molecules for sequencing. Similarly, DNA fragments resolved by electrophoresis (Figure 25.5) can be arranged in order, so as to give physical maps of DNA molecules. Isolation of these homogeneous fragments makes possible their amplification by gene cloning.

Properties of Restriction and Modification Enzymes

We now recognize three different types of restriction-modification systems—types I, II, and III. Each system consists of two distinct enzyme activities, a DNA methylase and an endonuclease that catalyzes a double-strand DNA break. The sequence-specific endonucleases, those most widely used in molecular biology, are type II enzymes. Regardless of type, the enzymes are named as follows. The first three letters denote the bacterial species of origin, and a fourth letter may denote an individual strain. For example, the restriction system from *E. coli* K is called ***Eco*K.** If more that one enzyme system is found in a given strain, the different enzymes are designated by Roman numerals. For instance, ***Eco*RI** is one of two known restriction systems in *E. coli* strain R, and ***Hin*dIII** is one of three enzymes from the d strain of *Haemophilus influenzae* R.

Properties of each of the three types of restriction systems are discussed below and summarized in Table 25.1.

Type I. These enzymes contain both methylase and nuclease activities in one protein molecule, which contains three subunits; one subunit contains the nuclease activity, one the methylase, and one a sequence recognition determinant. The recognition site is not symmetrical, and cleavage occurs some distance (up to 10 kb) away from the recognition site (although methylation occurs within the recognition site). The enzyme remains bound to the recognition site, and DNA is looped out around it, with concomitant supercoiling. About 10^5 ATP molecules are hydrolyzed per cleavage event; energy is probably needed for both translocation of the enzyme and supercoiling of the DNA. For reasons still not clear, both ATP and AdoMet are required for the cleavage activity; AdoMet may be an allosteric activator, because it is not broken down during the reaction.

Table 25.1
Properties of Restriction Endonucleases

	Type I	Type II	Type III
Example	*Eco*B	*Eco*RI	*Eco*PI
Recognition site	TGAN$_8$TGCT	GAATTC	AGACC
Cleavage site	About 1 kb away from recognition site	Between G and A (both strands)	24–26 base pairs 3′ to recognition site
Methylation site	m (on A) TGAN$_8$TGCT ACTN$_8$ACGA m (on A)	m (on A) GAATTC CTTAAG m (on A)	m (on A) AGACC (only one strand methylated)
Nuclease and methylase in one enzyme?	Yes	No	Yes
Requirements for cleavage	ATP, Mg^{2+}, *S*-AdoMet	Mg^{2+} or Mn^{2+}	Mg^{2+}, *S*-AdoMet
Requirements for methylation	ATP, Mg^{2+}, *S*-AdoMet	*S*-AdoMet	Mg^{2+}, *S*-AdoMet

Table 25.2
Specificities of some type II restriction systems

Enzyme	Bacterial Source	Restriction and Modification Site*
*Bam*H1	*Bacillus amyloliquefaciens* H	G↓GATCC
*Bgl*II	*Bacillus globiggi*	A↓GATCT
*Eco*RI	*E. coli* RY13	G↓A$\overset{m}{A}$TTC
*Eco*RII	*E. coli* R245	C$\overset{m}{C}$↓GG
*Hae*III	*Haemophilus aegyptius*	GG↓$\overset{m}{C}$C
*Hga*I	*Haemophilus gallinarum*	GACGCNNNNN↓
*Hha*I	*Haemophilus haemolyticus*	G$\overset{m}{C}$G↓C
*Hind*II	*Haemophilus influenzae* Rd	GTPy↓Pu$\overset{m}{A}$C
*Hind*III	*Haemophilus influenzae* Rd	$\overset{m}{A}$↓AGCTT
*Hinf*I	*Haemophilus influenzae* Rf	G↓ANTC
*Hpa*I	*Haemophilus parainfluenzae*	GTT↓AAC
*Hpa*II	*Haemophilus parainfluenzae*	C↓$\overset{m}{C}$GG
*Msp*I	*Moraxella species*	C↓CGG
*Not*I	*Nocardia rubra*	GC↓GGCCGC
*Ple*I	*Pseudomonas lemoignei*	GAGTCNNNN↓
*Pst*I	*Providentia stuartii*	CTGCA↓G
*Sal*I	*Streptomyces albus* G	G↓TCGAC
*Sma*I	*Serratia marcescens* Sb	CC$\overset{m}{C}$↓GGG
*Xba*I	*Xanthomonas badrii*	T↓CTAGA

*The methylated base in each site, where known, is identified with the letter m. All sequences read 5′ to 3′, left to right. The cleavage on the opposite strand in each case can be inferred from the symmetry of the site (except for *Hga*I, which has an asymmetric site). Pu = purine; Py = pyramidine; N = any base.

Type II. Restriction nucleases of this type have been of great value to researchers, because most cut within the recognition sequence, making cleavage absolutely sequence specific. Most of the type II enzymes are homodimers, with subunit molecular weights in the range 30 to 40 kDa. A divalent cation is required for cleavage, but ATP is not required. Each type II nuclease has a counterpart methylase, which binds to the same recognition sequence and methylates one nucleotide within that sequence. A hemimethylated DNA (methyl group on one strand only) is a preferred substrate for the methylase, but not for the nuclease, which generally cleaves only when the recognition site is unmethylated. Cleavage generates 3′ hydroxyl and 5′ phosphate termini. Table 25.2 shows the recognition sites for several widely used type II nucleases; several hundred enzymes of this type have now been isolated. Most recognition sites are four, five, or six nucleotides in length, although a few type II enzymes recognize an eight-nucleotide sequence. Most show twofold rotational sequence symmetry, suggesting that the two enzyme subunits are also arranged symmetrically. Cleavage sites on the two strands may be offset by as much as four nucleotides (as in *Eco*RI), giving cuts with short, self-complementary, single-stranded termini.

Some enzymes cleave to give a 5′-terminated single-strand end ("overhang"), while others generate a 3′ overhang. Other type II nucleases, including *Sma*I and *Hin*dII, generate blunt-ended fragments, where the cutting sites are not offset. Not all type II nucleases are absolutely sequence specific. For example, *Hin*dII recognizes four different hexanucleotide sequences, and some enzymes (such as *Hga*I) cleave a few nucleotides outside the recognition sequence.

The year 1986 saw the first crystallographic structural determination of a restriction nuclease (*Eco*RI) complexed with a double-stranded oligonucleotide containing its DNA recognition sequence. Figure 25.1 shows one polypeptide subunit of the dimeric enzyme in contact with its DNA recognition sequence. The DNA is bound in a cleft, and the protein has an N-terminal "arm" that wraps about the DNA. Sequence specificity is maintained by 12 hydrogen bonds, which link the purine residues in the site to a glutamate and two arginine residues (not shown in the figure). Binding of the DNA alters its structure to generate "kinks"; the sequences immediately flanking the six-nucleotide cutting site adopt the A duplex conformation, while within the cutting site the B structure is retained by

```
GAATTC
||||||
CTTAAG
```

The other subunit, not shown in the figure, contacts the substrate identically, accounting for the ability of the enzyme to catalyze symmetrical cleavages within the cutting site.

Type III. These enzymes resemble more closely the type I systems than the type II systems. They contain both nuclease and methylase activities in a two-subunit enzyme. They differ from type I enzymes in that they do not require ATP, they modify just one DNA strand, and the cleavage site is fairly close to the recognition site.

The ability to cut DNA molecules at precisely known short sequences opened the door to a host of powerful research techniques. Several of these are discussed in Tools of Biochemistry 18–22.

DNA Repair

Types and Consequences of DNA Damage

Chemical changes occur in all biological macromolecules, whether due to environmental damage or to errors in synthesis. For most biopolymers, including RNA, protein, and membrane phospholipids, the effects of these changes are minimized by turnover and replacement of altered molecules. However, DNA is distinctive in that its information content must be transmitted virtually intact from one cell to another during cell division or reproduction of an organism. Thus, DNA has a special need for metabolic stability. This stability is maintained in two ways: (1) by a replication process of very high accuracy and (2) by mechanisms for correcting genetic information when DNA suffers damage. In Chapter 24 we described mechanisms used to ensure high replication accuracy, specifically 3′-exonucleolytic proofreading, which corrects errors made by DNA polymerases, and the uracil-DNA *N*-glycosylase pathway, which prevents mutations that might result from deamination of cytosine to uracil in DNA. Here we consider several processes for repairing DNA that is altered either by replicative errors that are not corrected or by environmental damage. The latter can result from chemical modification of DNA nucleotides or from photochemical changes ensuing from absorption of high-energy radiation.

Most studies of DNA repair have been carried out on samples altered by chemical alkylation or cross-linking of DNA strands, by deamination of DNA bases, or by ultraviolet irradiation. Much current attention is being focused on oxidative damage. The intracellular generation of reactive oxygen species (Chapter 15) causes the formation of DNA bases such as 8-hydroxyguanine or thymine glycol. At the moment we know less about the biological consequences of DNA oxidation than about the consequences of DNA irradiation or alkylation. Recent work indicates that oxidative damage is repaired by mechanisms similar to those of other repair processes.

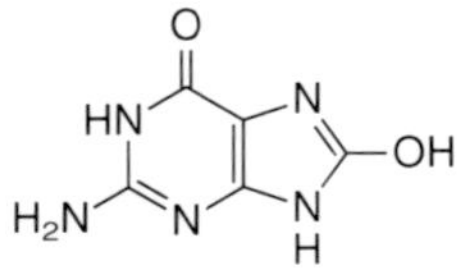

8-Hydroxyguanine

The consequences of irradiation or alkylation damage include mutagenesis, resulting from erroneous replication of a damaged template base, and cell death, resulting from inability of the replication apparatus to copy past a damaged site. Since it now appears that cancer results from an accumulation of somatic cell mutations, the mechanisms of DNA repair are under intense study as determinants of an animal's susceptibility to cancer.

Biologically Significant DNA Photoproducts: Pyrimidine Dimers

In studying mechanisms of DNA repair, one first needs to identify the chemical forms of DNA alteration that are subject to repair. Most of the early discoveries about DNA repair were made in studies with ultraviolet light-irradiated organisms. DNA was identified as the principal biological target for UV irradiation, partly through determination of **action spectra:** irradiation of bacteria or phage with UV light and determination of the wavelengths most effective in stimulating mutagenesis or death. Those wavelengths lie near 260 nm, where DNA light absorption is maximal.

When one examines either UV-irradiated DNA or the DNA extracted from a UV-irradiated organism, one detects small amounts of many different altered DNA constituents, called **photoproducts.** Prominent among these are intrastrand dimers consisting of two pyrimidine bases joined by a cyclobutane ring structure involving carbons 5 and 6 (Figure 25.6a). Such **thymine dimers**, formed from two adjacent DNA thymine residues, were identified quite early as a biologically significant photoproduct, because the relative abundance of thymine dimers in irradiated DNA correlated most closely with survival of irradiated phage or bacteria. Thus, the ability of an organism to survive ultraviolet irradiation was related directly to its ability to remove thymine dimers from its DNA. Dimerization draws the adjacent thymine residues together, which distorts the helix in such a way that replicative polymerization past this site is blocked.

For many years it was thought that ultraviolet light-induced *mutagenesis,* as well as cell death, was also caused primarily by cyclobutane thymine dimers. However, more recent data suggest that a different pyrimidine–pyrimidine dimer, called **6–4 photoproduct,** is the principal cause of UV-induced mutations. As shown in Figure 25.6b, these products are also dimers, linked via C-6 of the 5′ pyrimidine (either thymine or cytosine) and C-4 of the 3′ pyrimidine (usually cytosine). Supporting the idea that 6–4 photoproducts are responsible for mutations in UV-irradiated DNA is recent work showing that cyclobutane thymine dimers can be completely removed from UV-irradiated DNA by **photoreactivation** (see next section) without affecting the DNA's ability to cause mutations after it replicates.

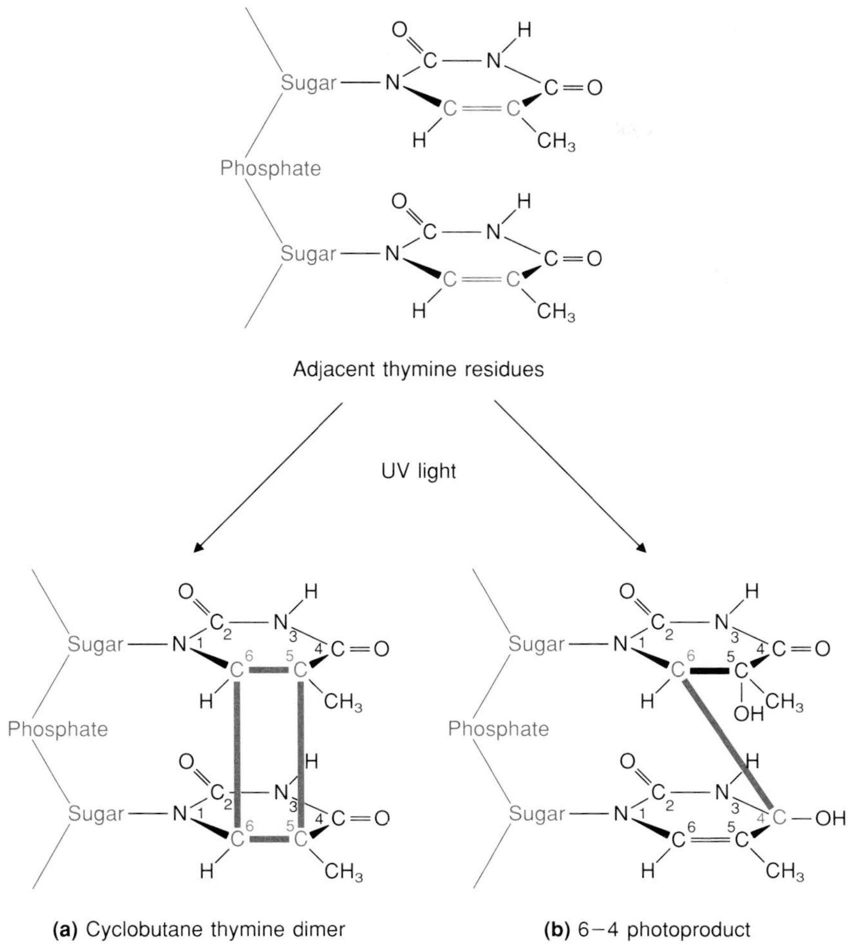

Figure 25.6
Structures of pyrimidine dimer photoproducts.

Direct Repair of Damaged DNA Bases: Photoreactivation and Alkyltransferases

Of the half-dozen known DNA repair processes, most involve removal of the damaged nucleotides, along with several adjacent residues, followed by replacement of the excised region using information encoded in the complementary (undamaged) strand. However, at least two processes involve reactions that directly change the damaged bases, rather than removing them.

PHOTOREACTIVATION. A widely distributed enzyme, called **photoreactivating enzyme,** or **DNA photolyase,** repairs pyrimidine dimers in the presence of visible light; a wavelength of 370 nm is most effective. The enzyme binds to DNA in the dark, specifically at the site of cyclobutane thymine dimers. In the presence of visible-wavelength light, the bonds linking the pyrimidine rings are broken, after which the enzyme can dissociate in the dark. Clues to the mechanism have come with the finding that the enzyme contains bound flavin-adenine dinucleotide in the reduced state; there is also a second chromophore, either a pterin or a modified flavin, depending on the source of the enzyme. (A chromophore is a structural moiety that absorbs light of characteristic wavelengths.) Current mechanistic studies suggest a process

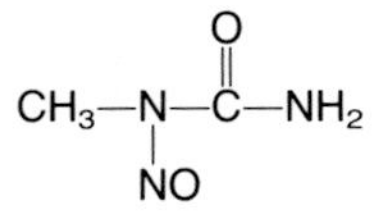

$$CH_3-N(NO)-C(=O)-NH_2$$

Methylnitrosourea (MNU)

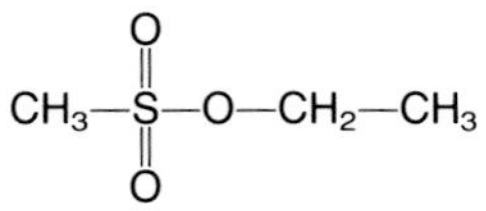

$$CH_3-S(=O)_2-O-CH_2-CH_3$$

Ethylmethanesulfonate (EMS)

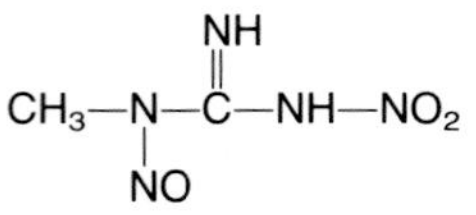

$$CH_3-N(NO)-C(=NH)-NH-NO_2$$

***N*-Methyl-*N*-nitro-*N′*-nitrosoguanidine (MNNG)**

akin to photosynthesis, with the second chromophore functioning as a light-harvesting factor, and $FADH_2$ functioning like the photochemical reaction center, translating light energy to generate an electron which is transferred to the dimer and contributes to breaking the pyrimidine–pyrimidine bonds.

O^6-ALKYLGUANINE ALKYLTRANSFERASE. Treatment of DNA with a methylating or ethylating reagent is comparable to ultraviolet irradiation in that various modified DNA bases are formed, some of which are lethal if not repaired and some of which are mutagenic. Some alkylating agents are used in cancer chemotherapy because of their ability to block DNA replication and, hence, cell proliferation. Some are widely used as mutagens in the laboratory. Several methylating or ethylating reagents are shown at left.

Whereas the spectrum of products formed varies with the reagent used, the target bases are primarily purines (phosphate oxygens are also targets). The most highly mutagenic of these products, **O^6-alkylguanine,** is mutagenic because the modified base has a very high probability of pairing with thymine when the modified strand replicates (Figure 25.7). Thus, alkylation of a DNA-guanine stimulates a particular kind of substitution mutation, namely, a **GC-to-AT transition** (Figure 25.8).

Repair of this type of damage involves an unusual enzyme, **O^6-alkylguanine alkyltransferase,** which transfers a methyl or ethyl group from an O^6-methyl- or ethylguanine residue to a cysteine residue in the active site of the protein. Remarkably, this enzyme can function only once; having become alkylated, it cannot remove the alkyl group, and the protein molecule turns over. Thus, the term "enzyme" is probably a misnomer. In bacteria the protein regulates its own synthesis, as well as the synthesis of a DNA-*N*-glycosylase, which participates in the removal of other types of alkylated purine residues. The regulation involves activation of the transcription of genes encoding these two proteins. There is evidence that the *alkylated* form of the alkyltransferase is the specific form of the transcriptional activator. If so, the cell could adapt to alkylation damage by using the alkylated protein as a specific signal to produce more of the proteins needed to repair the damage.

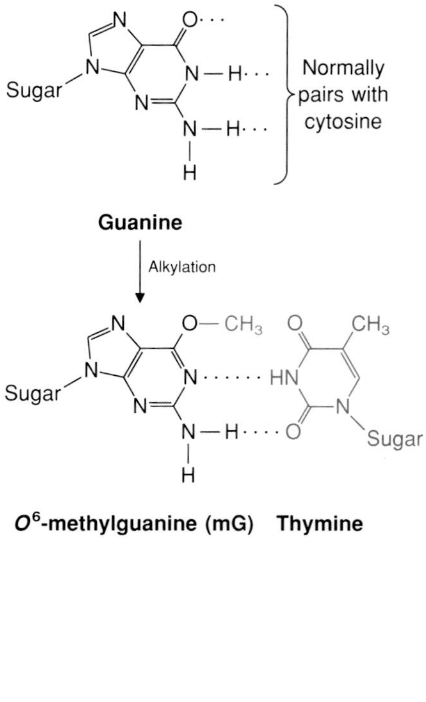

Figure 25.7
Mispairing of O^6-methylguanine with thymine in a DNA duplex.

Excision Repair: Excinucleases and DNA-*N*-Glycosylases

Excision repair is a process whereby a damaged section of a DNA chain is cut out, or excised, followed by the action of DNA polymerase and then DNA ligase to regenerate a covalently closed duplex at the site of the original damage. The process can start in either of two ways, as described below.

E. coli uses a three-subunit enzyme, encoded by the *uvrA, uvrB,* and *uvrC* genes, which recognizes the structural distortion created by a thymine dimer or by a bulky substituent group. As shown in Figure 25.9, the enzyme cleaves the damaged strand at two sites—eight nucleotides to the 5′ side of the damaged site (shown in the figure as a thymine dimer) and four or five nucleotides to the 3′ side—leaving a 12- or 13-nucleotide gap, with a 3′ hydroxyl group and a 5′ phosphate. The gap can then be filled by polymerase and ligase action. Helicase II, the product of the *uvrD* gene, is also required, presumably to unwind and remove the excised oligonucleotide, which is ultimately broken down by other enzymes. The UvrABC enzyme is not a classical endonuclease, since it cuts at two distinct sites, and the term **excinuclease** has been proposed, denoting its role in excision repair. This system is also involved in the repair of DNA damage that involves covalent cross-linking of the two strands to each other. Here, in order to preserve an intact template strand, the two strands are repaired sequentially, one after the other.

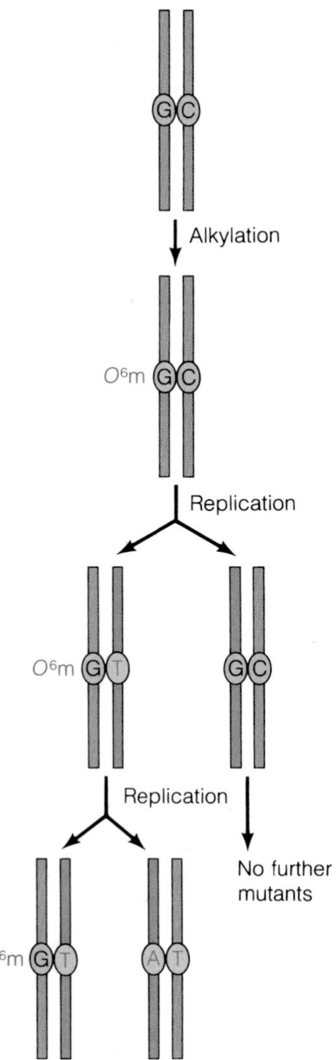

Figure 25.8
Pathway for mutagenesis caused by the presence of an O^6-methylguanine residue in DNA in place of guanine.

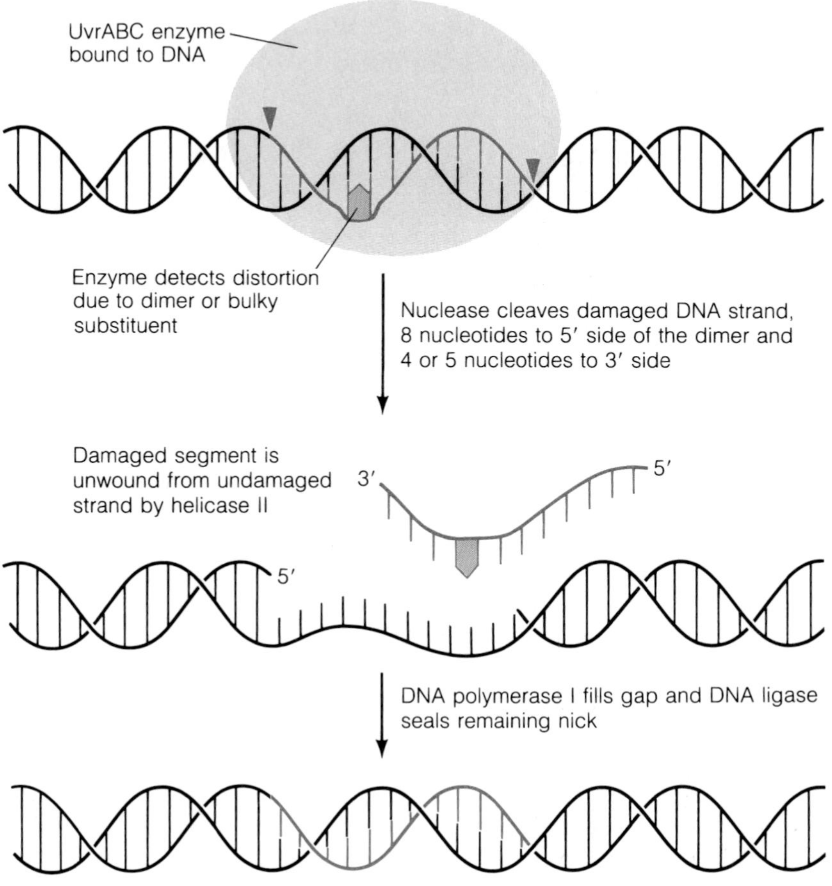

Figure 25.9
Excision repair of thymine dimers by the UvrABC excinuclease of *E. coli*. The same system is involved in repairing DNA with bulky adducts that distort the helix.

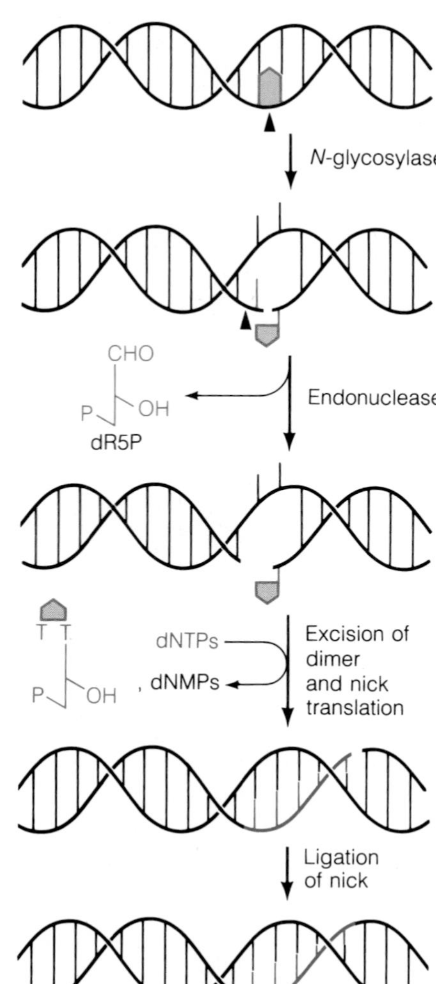

Figure 25.10
Excision repair of thymine dimers by the DNA-*N*-glycosylase mechanism, as carried out by the bifunctional repair enzyme of phage T4 or *M. luteus*.

The second excision repair system, studied in bacteriophage T4 and in *Micrococcus luteus*, involves a DNA-*N*-glycosylase mechanism of the type discussed in Chapter 24 for repair of uracil-containing DNA (see Figure 24.33). As shown in Figure 25.10, this enzyme recognizes the 5′ thymine in a dimer and cleaves the glycosidic bond, creating an apyrimidinic site. This leads to a nick translation process, whereby the damaged DNA is excised and replaced. Probably the first 3′ hydroxyl terminus for chain extension by the repair polymerase is created by enzymatic removal of deoxyribose 5-phosphate. A similar mechanism, also involving a DNA-*N*-glycosylase, is used to excise alkylated DNA purines other than O^6-methylguanine, including *N*-methyladenine, 3-methyladenine, and 7-methylguanine.

Excision repair also occurs in mammalian cells. This fact came to light in part through studies of a rare genetic disease called **xeroderma pigmentosum** (XP). XP is actually a family of diseases, in which one or more enzymes of the excision pathway are deficient. In cells cultured from an XP patient, ultraviolet DNA damage, which otherwise remains unrepaired, can be repaired by introduction into the cells of the bifunctional glycosylase–endonuclease specified by phage T4. Thus, there is a fundamental uniformity about the excision repair system. In affected humans, however, there is at present no known way to treat the condition. The biological consequences of XP include extreme sensitivity to sunlight and a high incidence of skin cancers. Although overexposure to the ultraviolet rays in sunlight increases the risk of skin cancer for normal humans, the greatly increased skin cancer frequency suffered by XP patients highlights the biological importance to mammals of UV repair pathways.

Postreplication Repair: Recombinational Repair and the SOS Response

Photoreactivation and excision repair are short-term metabolic responses; both processes can occur within a few minutes after DNA suffers chemical or ultraviolet damage. However, if the photoreactivation and excision repair systems are defective, or become saturated by DNA damage too extensive for their capacities, at least two longer-term systems can come into play in bacteria. Discussion of these systems will focus on thymine dimer repair, but it is likely that other types of damage can be repaired as well.

When a replicative polymerase encounters a thymine dimer, it cannot replicate past this site. Deoxyadenylate is incorporated opposite the first thymine base in the template. The double helix is distorted because of the thymine dimer, causing the structure to be recognized as a mismatch, and the 3′-exonucleolytic activity of DNA polymerase III cleaves out the newly incorporated dAMP. Thus, polymerase "idles" at the damage site, converting dATP to dAMP by a continual process of insertion and exonucleolytic cleavage. Synthesis of an Okazaki fragment can commence opposite the damaged site, leaving a gap opposite the thymine dimer. Of course, this gap would be lethal if unrepaired, because it would generate a double-strand break in the next round of replication.

Two distinct processes can repair the gap: (1) **recombinational repair,** or **daughter-strand gap repair**; and (2) **SOS repair**, or **error-prone repair.** Both processes depend critically on a protein called RecA. Bacteria carrying mutations in *recA* were originally characterized as defective in general recombination and DNA repair. We now know that *recA*$^-$ bacteria, which specify a defective RecA protein, have a complex phenotype, including defective DNA repair. We shall describe the protein later in this chapter, but for our purposes here two properties are important. First, RecA catalyzes **strand pairing,** or **strand assimilation,** a process whereby two different DNAs undergo homologous base pairing with each other. Second, RecA is a genetic regulator, activating the synthesis of many proteins that help a bacterium adapt to a variety of metabolic stresses. This adaptation, called the **SOS response,** is more fully described below, and the transcriptional activations that lead to the SOS response are presented in Chapter 26.

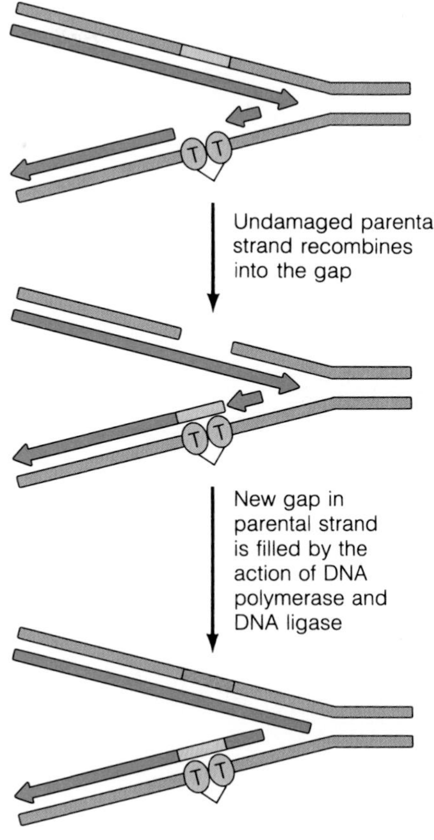
Replication stops at the thymine dimer, but resumes after the dimer is passed, creating a gap. The parental strand complementary to the dimer is shown in orange
T T
Undamaged parental strand recombines into the gap
T T
New gap in parental strand is filled by the action of DNA polymerase and DNA ligase
T T

Figure 25.11
Recombinational repair: transfer of undamaged parental DNA strand to daughter-strand gap.

Recombinational Repair. Since it is created by faulty replication, the gap that is generated opposite a thymine dimer is close to the replication fork. Therefore, it is also close to the corresponding region on the other daughter duplex (Figure 25.11). If that region has itself not sustained damage, the two duplexes can undergo recombination (promoted by the RecA protein). The uninvolved parental strand, which is complementary to the damaged parental strand, recombines into the gap, opposite the damaged site. A gap now exists in the previously undamaged arm, but since it lies opposite an undamaged template, it can be filled by action of DNA polymerase and DNA ligase. The thymine dimer itself is not repaired in this process, but the process allows time for the excision system to come in later and repair this damage. RecA protein is required for daughter-strand gap repair, particularly in the first reaction, where the undamaged parental strand undergoes pairing with the parental strand opposite the gap.

Error-prone Repair and the SOS Response. The bacterial *SOS response* is a metabolic alarm system that helps the cell to save itself in the presence of potentially lethal stresses. Inducers of the SOS response include ultraviolet irradiation, thymine starvation, treatment with certain DNA-modifying reagents such as the cross-linker mitomycin C, and inactivation of mutant genes essential to DNA replication. Responses include mutagenesis, filamentation (where cells elongate by growth but don't divide), activated excision repair, and activation of latent bacteriophage genomes. Mutagenesis occurs because under SOS conditions, the gaps formed opposite thymine dimers can be filled by replication rather than by daughter-strand transfer. In fact, this is the principal pathway by which mutagenesis is stimulated by ultraviolet light.

We do not yet know the mechanism by which gaps are filled during the SOS response, but we do know that replication of the dimer-containing template is extremely inaccurate. One possible explanation for the inaccuracy is that a normal polymerase is involved (recent evidence points to DNA polymerase II) but that its proofreading exonuclease is inhibited. The proteins involved include RecA and the products of two other RecA-inducible genes, *umuC* and *umuD* (*umu* stands for *u*ltraviolet *mu*tagenesis). Bacteria defective in either of these genes are more sensitive to ultraviolet light than are wild-type bacteria, but *umu*$^-$ mutants do not undergo UV-induced mutagenesis.

Since most mutations are probably deleterious, what is the advantage to the cell of accumulating mutations during DNA repair? Probably none, except that the alternative would be for the cell to die. In other words, extensive mutagenesis can be seen as a worthwhile price for the cell to pay, if a massive UV dose has overwhelmed all other repair pathways and if error-prone replication past a dimer is necessary for the cell to stay alive.

Mismatch Repair

Mismatches, or non-Watson–Crick base pairs in a DNA duplex, can arise through replication errors, through deamination of 5-methylcytosine in DNA to yield thymine, or through recombination between DNA segments that are not completely homologous. We best understand the correction of replication errors, so that is what we present here.

If DNA polymerase introduces an incorrect nucleotide, creating a non-Watson–Crick base pair, the error is normally corrected by 3′ exonucleolytic proofreading. If the error is not corrected immediately, the fully replicated DNA will contain a mismatch at that site. This can be corrected by another process, called **mismatch repair.** In *E. coli* the proteins that participate include the products of genes *mutH, mutL,* and *mutS.* (Mutations in any of these genes generate a **mutator phenotype,** or enhance spontaneous mutagenesis at many loci; hence, the name *mut*). Another required gene product, originally called MutU, has now been identified as DNA helicase II.

The mismatch correction system scans newly replicated DNA, looking for either mismatched bases or single-base insertions or deletions. When it finds such a mismatch, part of one strand containing the mismatched region is cut out and replaced (Figure 25.12). How does the mismatch repair system recognize the right strand to repair? After all, if it chose either strand randomly, it would choose incorrectly half the time and there would be little gain in replication accuracy. The answer is that the mismatch repair enzymes identify the newly replicated strand, because for a short period that DNA is unmethylated. In *E. coli* the sequence –GATC– is crucial, because it is the site that becomes methylated (by creation of a 6-methyladenine residue) fairly soon after replication. The mismatch repair enzymes look for –GATC– sequences that are not methylated. Recognition of an unmethylated GATC can target that strand for mismatch correction at a site as far as 1 kb or more away from the GATC site. Once the methylation system has acted on all GATC sites in the daughter strand, it is too late for the mismatch repair system to recognize the more recently synthesized DNA strand, and any advantage in total DNA replication fidelity is lost. When the system functions properly, it has the effect of increasing overall replication fidelity from about one error in 10^8 base pairs replicated to about one in 10^{10}.

The protein products of genes *mutH, mutL,* and *mutS* have been isolated by cloning and overexpression of the genes (see Tools of Biochemistry 19). Repair of mismatches in vitro occurs in the presence of these purified proteins, plus helicase II and the *E. coli* single-strand DNA-binding protein (SSB). The MutH protein binds specifically to DNA at the GATC sequence, and it has a nuclease activity that cleaves just 5′ to the G in an unmethylated GATC sequence. The MutS protein binds to DNA specifically at the site of the mismatch. (At this writing the function of MutL has not been described.) Current attention is focused on the mechanism of signal transduction, as well as the repair process itself. How do the proteins communicate with each other over the several hundred base pairs that may lie between the GATC sequence and the site of the mismatch? A mechanism akin to type I restriction endonucleases has been suggested. Binding of one protein or the other would stimulate a DNA-looping process, with the protein remaining bound at the first site until the protein bound at the other site was reached.

Recent evidence indicates that mammalian cells possess one or more mismatch repair systems. The nature of the strand selection mechanism remains to be clarified; clearly, it is not a methyladenine residue, since this is not found in eukaryotic DNA.

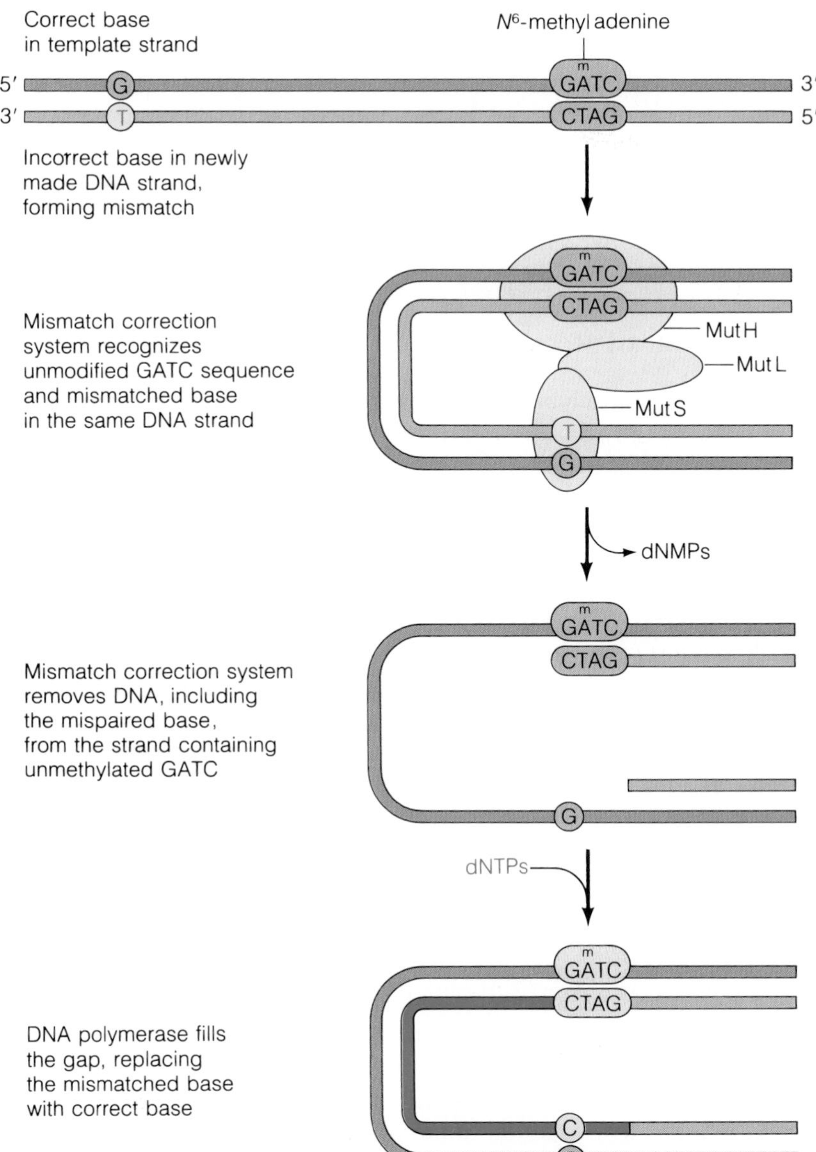
Correct base
in template strand
N^6-methyl adenine
5′
3′
G
T
m
GATC
CTAG
3′
5′
Incorrect base in newly
made DNA strand,
forming mismatch
Mismatch correction
system recognizes
unmodified GATC sequence
and mismatched base
in the same DNA strand
MutH
MutL
MutS
dNMPs
Mismatch correction system
removes DNA, including
the mispaired base,
from the strand containing
unmethylated GATC
dNTPs
DNA polymerase fills
the gap, replacing
the mismatched base
with correct base
C

Figure 25.12
Methyl-directed mismatch repair, acting to correct replication errors before the daughter DNA strand has become methylated. The actual length of the excised region has not yet been determined, nor has the sequence of protein–DNA or protein–protein interactions. MutH binds at the –GATC– sequence, and MutS binds at the site of the mismatch. Presumably, all three proteins (MutH, MutL, and MutS) interact, linking the mismatch to the –GATC– site before excision of the mismatched nucleotide occurs.

Recombination

Population genetics teaches us that the survival of a species depends on its ability to maintain genetic diversity, so that individuals can vary in their ability to respond to unforeseen environmental pressures. Diversity is maintained through both mutation, which alters single genes or small groups of genes in an individual, and **recombination,** which redistributes the contents of a genome among various individuals during reproduction. The term recombination usually conjures up the image of crossing over between paired sister chromosomes during meiosis in eukaryotes, and in fact our earliest information about recombination came from cytological and cytogenetic observations in *Drosophila*. However, recombination encompasses more processes and biological functions than those involved in sexual reproduction. As noted in the previous section, recombination is involved in the repair of certain types of DNA damage. Recombination also participates in the integration of certain bacteriophage or plasmid genomes into the chromosomal DNA of the infected bacterium; many viral genomes integrate into animal host cells as well. Recombination is also involved in **transposition,** a process that involves the movement of DNA from one chromosomal integration site to another. In some cases this type of recombination is a gene regulatory mechanism. The transposed gene may move from a site where it is inactive to another site where its transcription can be activated; alternatively, the integration of a transposable element may turn on the expression of adjacent genes. We discuss these processes later, in the context of gene rearrangements.

Classification of Recombination Processes

Though different mechanisms are involved, one feature is common to all recombinational processes—the formation of new DNA from two distinct parental molecules, such that genetic information from each parent is present in the new molecules. Nevertheless, different recombination processes have quite different requirements, both for nucleotide sequence homology among the recombining partners and for proteins and enzymes to catalyze the process. Meiotic recombination in diploid organisms requires extensive sequence homology between the recombining partners and accordingly is called **homologous recombination.** This term also applies to certain recombinational events between bacterial chromosomes. New DNA can be introduced into a bacterial cell by various processes: (1) conjugation during bacterial mating; (2) transformation, when DNA is taken up by cells; or (3) **transduction,** when bacterial DNA that was packaged into a phage particle is introduced by infection. In transduction the assembly of phage particles in an infected cell occasionally goes awry, with bacterial DNA assembled into a phage head. Reinfection of another bacterial cell now introduces that packaged bacterial DNA into the new cell.

If the introduced DNA contains a replication origin, as in a plasmid, it can replicate autonomously once inside a new bacterial cell. More often that DNA does not contain an origin, and its information can be expressed and maintained only if the DNA is taken up into the resident chromosome by recombination. Biochemical analysis of homologous recombination has focused on prokaryotic systems, which in turn have shed light on meiotic recombination. Most bacterial homologous recombination processes share a common requirement for the RecA protein or its counterpart.

Site-specific recombination was earliest studied in bacteriophage λ. After circularization, the λ chromosome can undergo either multiple rounds of replication leading to virus production or integration into a specific site on the host chromosome. In the latter condition, called **lysogeny,** most of the viral genes are inactivated, and the virus can maintain a long-term, nonlethal relationship with its host (Figure 25.13). Bacteriophages that can establish lysogeny are called **temperate phages,** in contrast to **virulent**

Table 25.3
Characteristics of different types of genetic recombination

Type	Requirement: Sequence homology	RecA protein or counterpart	Sequence-specific enzyme
Homologous	Yes	Yes	Yes
Site-specific	Yes (ca. 15 bases)	No	Yes
Transposition	No	No	Yes
Illegitimate	No	No	Unknown

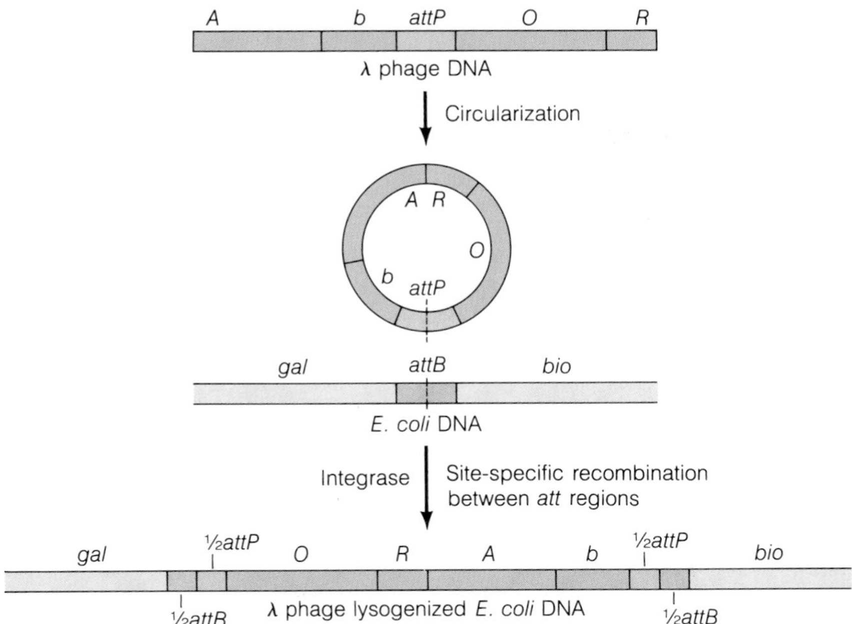

Figure 25.13
Site-specific recombination, establishing lysogeny in bacteriophage λ. The phage chromosome circularizes between genes *A* and *R* and undergoes recombination between the *attP* site and a corresponding region on the *E. coli* chromosome, *attB*, which lies between the *gal* and *bio* markers. The enzyme integrase carries out the site-specific recombinational event. *O* and *b* are additional genetic markers.

phages, such as phage T4, which always lyse their host cells after infection. In most (not all) cases of lysogeny, integration occurs at a specific site. Integrative recombination in phage λ was studied as a model for understanding homologous recombination between chromosomes. However, the processes were found to be quite different when it was learned that the phage and bacterial DNA sequences in the regions undergoing recombination show very limited homology (15 base pairs, to be exact). Moreover, the RecA protein is not required for this process. Rather, the virus specifies a site-specific enzyme, called **integrase,** in which specific DNA–protein interactions between the enzyme and the recombining partners, rather than extensive DNA–DNA sequence homology, determine the site where recombination will occur.

We recognize two other forms of recombination: transposition involves neither sequence homology nor the RecA protein but does require a special sequence on the donor DNA. **Illegitimate recombination,** an extremely rare event, possibly occurring by chance, involves neither sequence homology nor the action of any known protein. Table 25.3 summarizes the main distinctions between the four major types of recombination.

Homologous Recombination

Breaking and Joining of Chromosomes. Early thinking about recombination focused on mechanisms that involve breaking and rejoining of chromosomes. However, if recombination occurs by breakage and reunion, the sites of breakage must be precisely the same on both recombining chromosomes for intact genes to be regenerated. Some researchers favored alternative mechanisms, but in 1961 Matthew Meselson and Jean Weigle showed that recombination in fact occurs via breakage and rejoining of chromosomes. The demonstration, diagramed in Figure 25.14, involved a Meselson–Stahl type of experiment (see Figure 24.3, Chapter 24). *E. coli* was infected with two genetically marked λ phage populations, one of which had been density labeled by growth in ^{13}C–^{15}N medium. The phage particles resulting from this cross were centrifuged to equilibrium in a cesium chloride gradient. Phages with recombinant genotypes were recovered

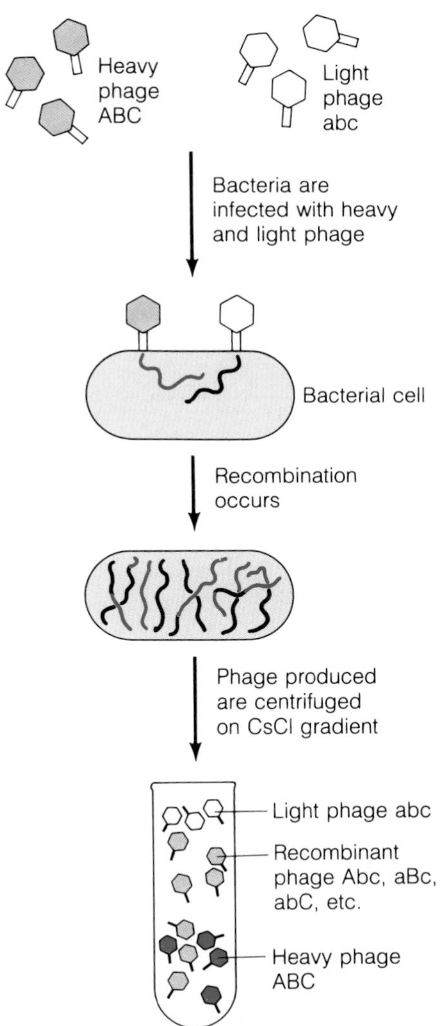

Figure 25.14
The Meselson–Weigle experiment, which established that genetic recombination occurs by breakage and reunion of DNA strands.

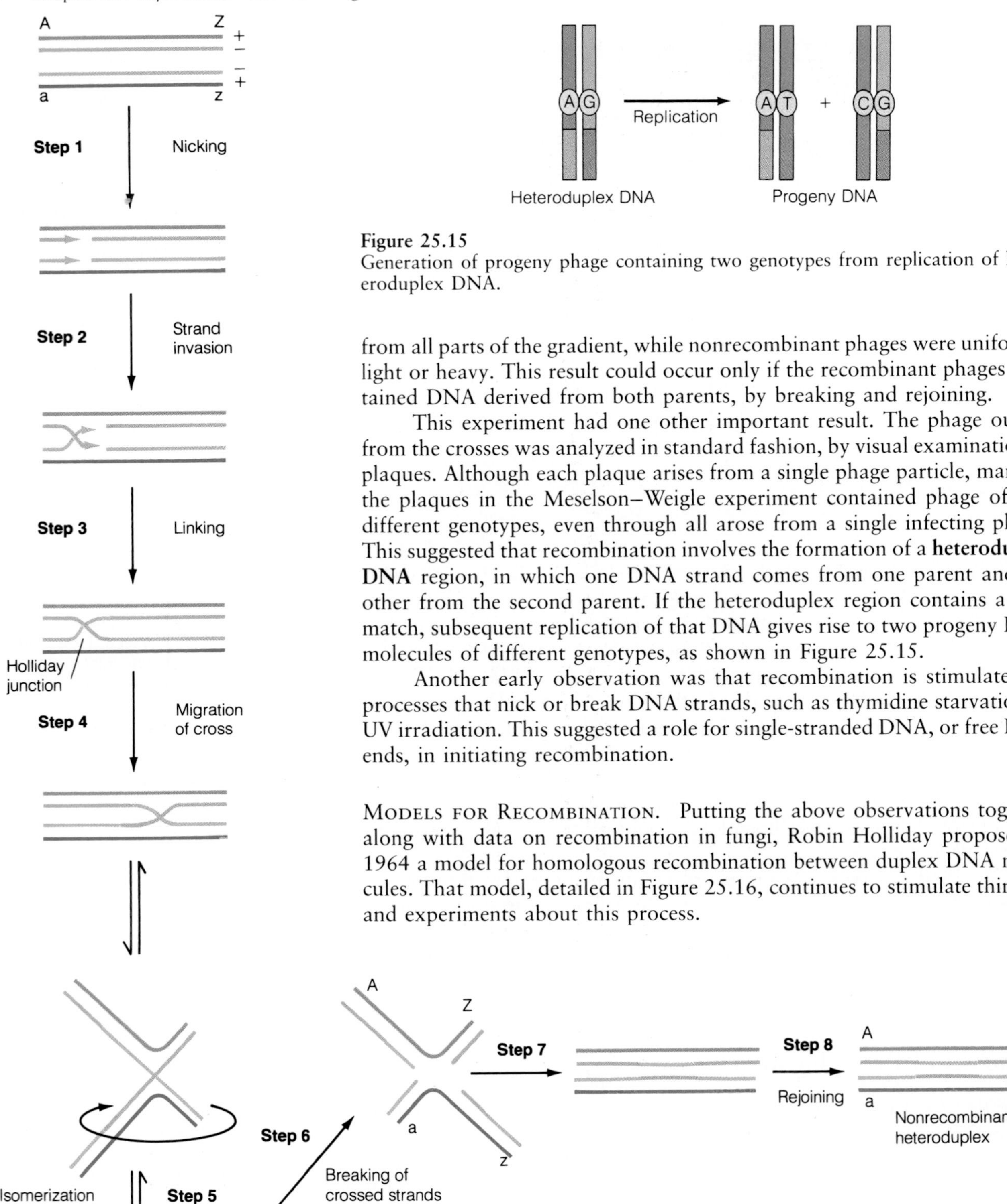

Figure 25.15
Generation of progeny phage containing two genotypes from replication of heteroduplex DNA.

from all parts of the gradient, while nonrecombinant phages were uniformly light or heavy. This result could occur only if the recombinant phages contained DNA derived from both parents, by breaking and rejoining.

This experiment had one other important result. The phage output from the crosses was analyzed in standard fashion, by visual examination of plaques. Although each plaque arises from a single phage particle, many of the plaques in the Meselson–Weigle experiment contained phage of two different genotypes, even through all arose from a single infecting phage. This suggested that recombination involves the formation of a **heteroduplex DNA** region, in which one DNA strand comes from one parent and the other from the second parent. If the heteroduplex region contains a mismatch, subsequent replication of that DNA gives rise to two progeny DNA molecules of different genotypes, as shown in Figure 25.15.

Another early observation was that recombination is stimulated by processes that nick or break DNA strands, such as thymidine starvation or UV irradiation. This suggested a role for single-stranded DNA, or free DNA ends, in initiating recombination.

Models for Recombination. Putting the above observations together along with data on recombination in fungi, Robin Holliday proposed in 1964 a model for homologous recombination between duplex DNA molecules. That model, detailed in Figure 25.16, continues to stimulate thinking and experiments about this process.

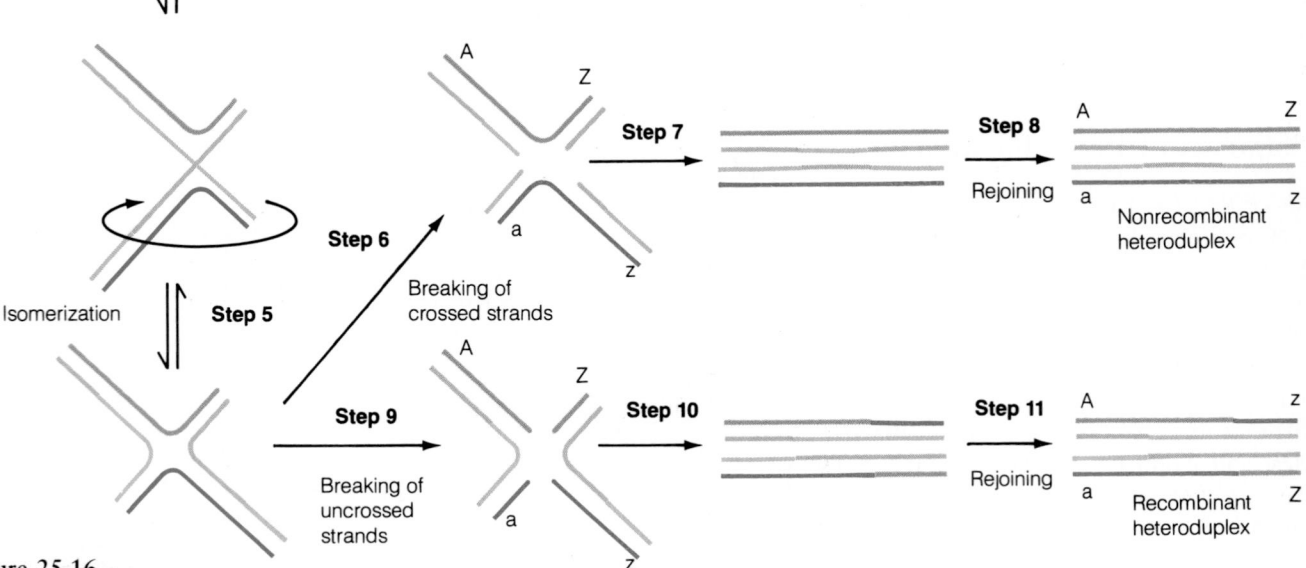

Figure 25.16
The Holliday model for homologous recombination.

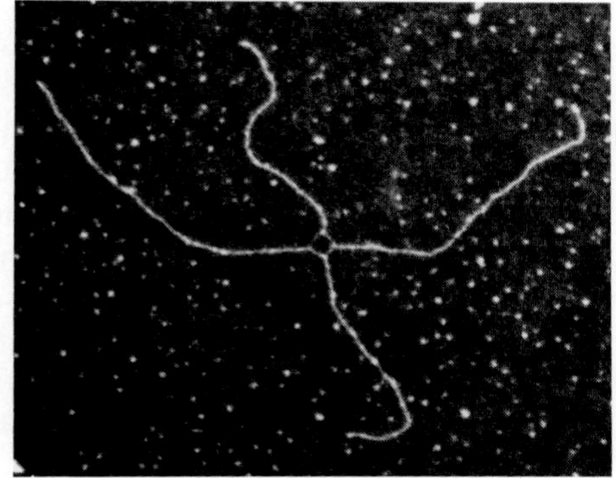

Figure 25.17
Electron microscopic visualization of a Holliday junction, created during recombination between two plasmid DNA molecules.

Holliday proposed that recombination begins with nicking at the same site on two paired chromosomes (step 1). Partial unwinding of the duplexes is followed by **strand invasion,** in which a free single-strand end from duplex I pairs with its unbroken complementary strand in duplex II, and vice versa (step 2). Enzymatic ligation generates a crossed-strand intermediate, called a **Holliday junction** (step 3). The crossed-strand structure can move in either direction by duplex unwinding and rewinding (step 4). The Holliday junction now "resolves" itself into two unbroken duplexes, by a process of strand breaking and rejoining. The process leading to recombination begins with isomerization of the Holliday structure (step 5), so that the crossed strands are those that were *not* broken in step 1. If the crossed strands break and rejoin without isomerization (steps 6–8), the products are nonrecombinant duplexes, each containing a heteroduplex region. However, when isomerization does occur, resolution of the resulting structure (steps 9–11) generates two chromosomes recombinant for DNA flanking the region and each containing a heteroduplex region.

Considerable evidence now supports the central tenets of the Holliday model, particularly the electron microscopic visualization of Holliday junctions (Figure 25.17). However, the model has been modified as new data have emerged. Matthew Meselson and Charles Radding proposed that recombination could start with a single nick (which eliminates the nagging question of how two duplexes could be nicked at precisely the same point). As shown in Figure 25.18, strand displacement synthesis occurs (step 1), with the displaced 5′ end of the nicked duplex, A, invading the homologous region of the unbroken duplex, B (step 2). Eventually the displaced loop on duplex B is cleaved and partially degraded (step 3), and the displaced 5′ end from duplex A is ligated to B (step 4). Isomerization then occurs, as in the Holliday model, with the originally unbroken strands crossed (step 5). An additional feature of the Meselson–Radding model is **branch migration,** a process of simultaneous strand unwinding and rewinding of both duplexes, which allows the site at which the strands cross to move. As a result, the final cuts of these strands may occur some distance from the site of either strand invasion or the original nick (steps 6 and 7).

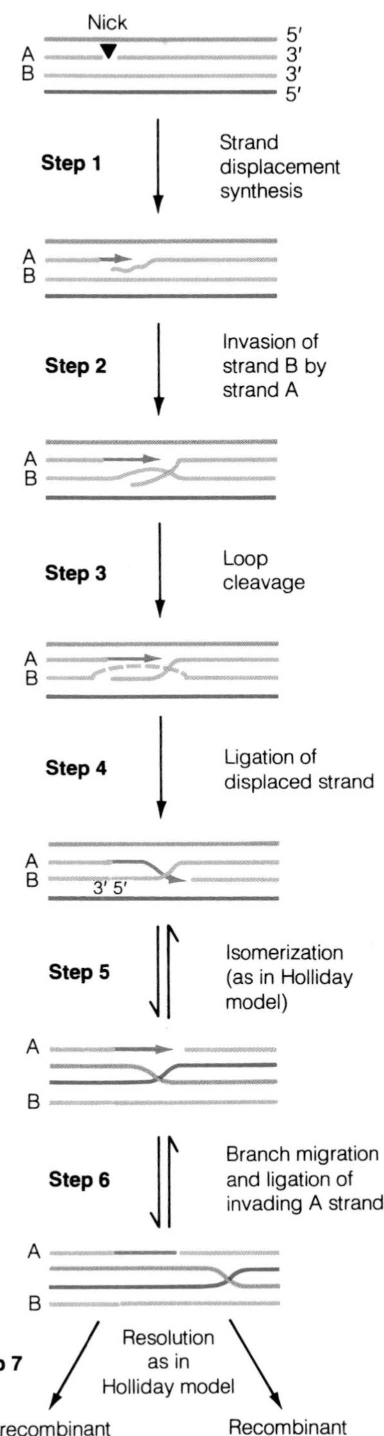

Figure 25.18
The Meselson–Radding model for homologous recombination.

PROTEINS INVOLVED IN HOMOLOGOUS RECOMBINATION. The above models explained most of the existing data on homologous recombination between paired chromosomes, particularly as studied in lower eukaryotes such as yeast. Moreover, the models could easily be adapted to explain daughter-strand gap repair or the recombination that occurs in bacteria after transformation or conjugation. Some of the proteins thought to participate, notably DNA polymerase, DNA ligase, and single-stranded DNA-binding protein, had been characterized and shown to participate in recombination. What about other proteins that must function if the above models are largely correct? For this we turn back to *E. coli* and its phages and the characteristics of bacterial mutants defective in recombination. Mutations conferring a recombination-defective (*rec*$^-$) phenotype map in several loci, and two important proteins have been characterized as the products of these genes. One of these, the RecA protein, was mentioned earlier. The other protein has been called exonuclease V or the RecBCD nuclease.

RecA is a multifunctional protein of M_r about 38,000. In recombination it acts to promote the pairing of homologous strands, as described earlier in connection with recombinational repair. Several strand-pairing reactions can be demonstrated in vitro, as summarized in Figure 25.19. For example, a circular, single-stranded DNA can invade a homologous linear duplex (such as the linearized replicative form of phage ϕX174 DNA). Strand invasion and assimilation reactions of this type have three requirements in order to be promoted by RecA in vitro: (1) sequence homology (at least 40–60 base pairs) between the reacting DNAs, (2) a free end on one or both of the reactants, and (3) a single-stranded region on one or both of the partners.

In vitro studies of these reactions reveal five distinct steps in RecA-promoted strand assimilation (Figure 25.20). In step 1, RecA coats the single-stranded DNA to form a nucleoprotein filament of regular structure, in which the DNA length is extended by about 50% relative to its length in a duplex. In step 2, known as **synapsis**, the coated single-stranded DNA reacts with a duplex, not necessarily at a homologous sequence. In step 3, **homologous alignment**, the two DNAs move with respect to each other until homologous sequences come into contact. This process requires ATP,

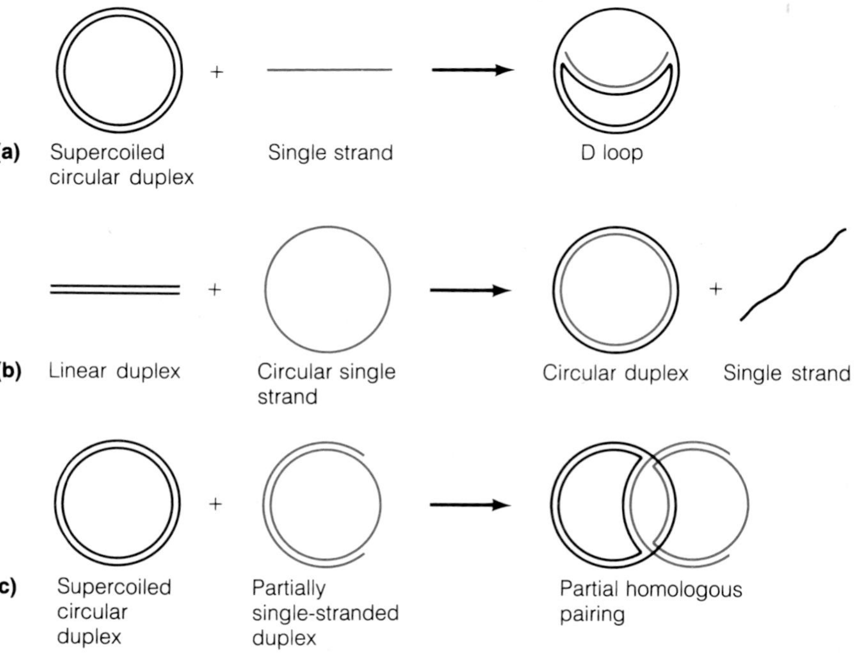

Figure 25.19
DNA strand-pairing reactions promoted by RecA protein in vitro. (**a**) Single strand invades a circular duplex, forming a D loop. Supercoiling in the circular molecule facilitates pairing. (**b**) Circular single strand invades a linear duplex, releasing one strand of the duplex. (**c**) Partial homologous pairing of a circular duplex (preferably supercoiled) and a partially single-stranded duplex.

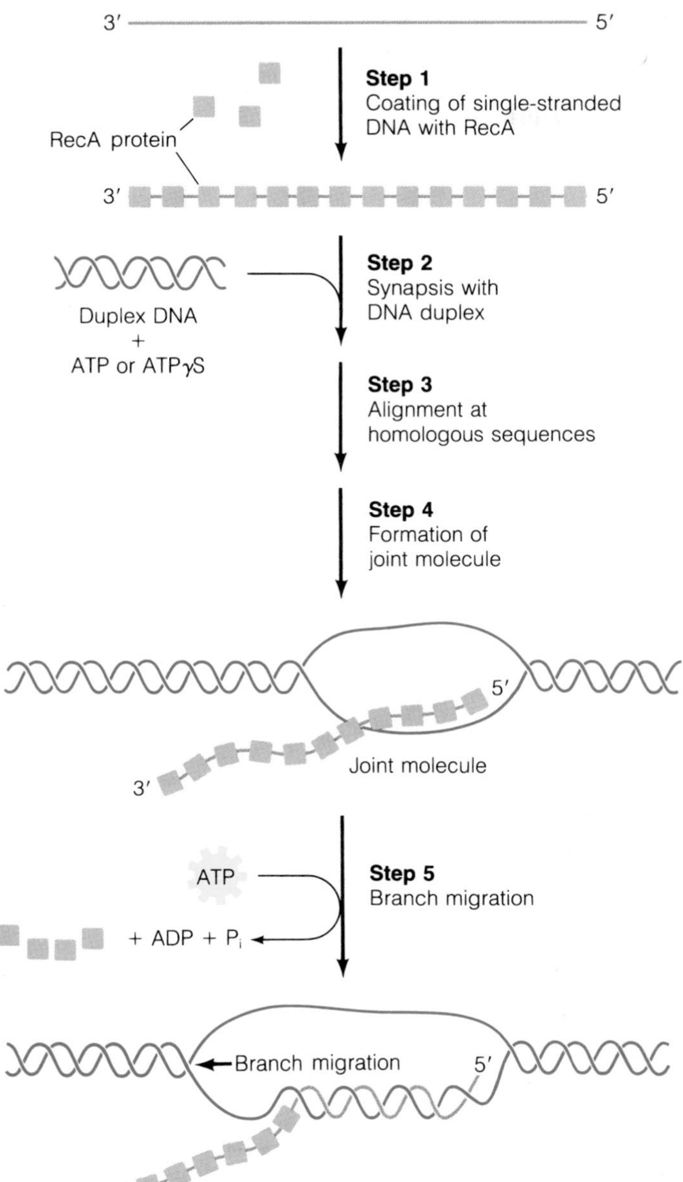

Figure 25.20
Action of RecA protein in promoting DNA strand pairing.

but that ATP need not be cleaved, because ATPγS, a noncleavable analog of ATP, can substitute. In step 4, a **joint molecule** is formed by base pairing between the two reacting molecules; RecA catalyzes local denaturation of the duplex partner and strand exchange with the single-stranded partner. Finally (step 5), **branch migration** occurs, essentially a continuation of joint molecule formation. The incoming strand pairs with its homolog, using the energy of ATP cleavage to advance (in a $5' \rightarrow 3'$ direction relative to the single-stranded DNA) and displacing the other strand as it moves. In this respect RecA acts as a helicase.

The process outlined above is supported by electron micrographs of joint molecules (Figure 25.21) and of DNA–RecA nucleoprotein filaments (Figure 25.22). The best pictures of joint molecules were obtained not with the *E. coli* RecA protein, but with a T4 phage-encoded counterpart, the product of gene *uvsX*.

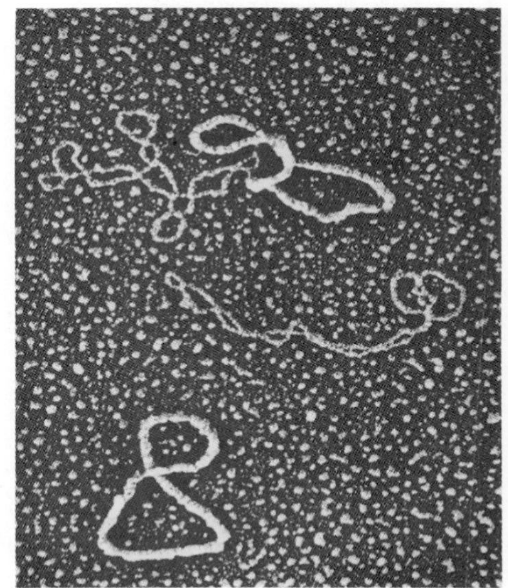

Figure 25.21
Electron microscopic visualization of joint molecules. For technical reasons this was obtained with the UvsX protein of T4, which is similar in its action to *E. coli* RecA. In the micrograph single-stranded DNA appears much thicker, because it is coated with UvsX protein. (Bottom) a circular duplex, (middle) a circular single-stranded DNA, and (top) a joint molecule between the two. Joint molecules can form between homologous sequences on circular molecules, but the strands are not interwound.

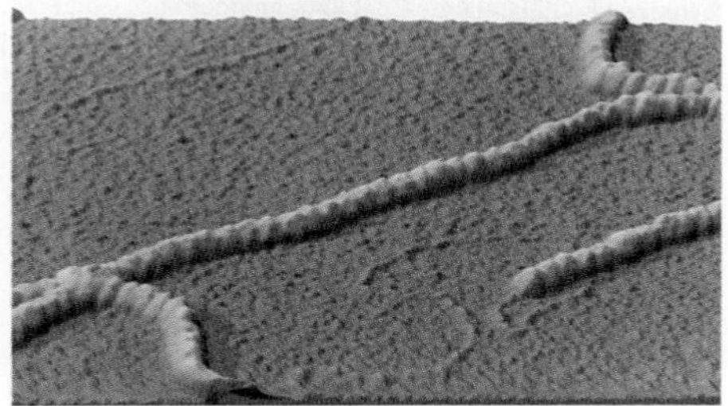

Figure 25.22
A RecA–DNA complex as visualized by scanning tunneling microscopy (thick structure). A helical structure to the complex can be seen. The thin structure represents uncoated DNA.

The properties of the RecBCD enzyme help to answer other important questions about recombination, notably the mechanism by which DNA becomes nicked in vivo to generate a free DNA end and a single-stranded region on which RecA can act. RecBCD, with an M_r of about 300,000, is, like RecA, a multifunctional enzyme, with a helicase activity, a strand renaturation activity, and ATP-dependent single-strand endo- and exonuclease activities. The helicase, of course, also requires ATP. The nuclease activities are inhibited in the presence of Ca^{2+}, so that the DNA unwinding and rewinding activities are easily observed electron microscopically. The protein binds to the end of one strand and unwinds at about 300 nucleotides per second, faster than it rewinds (200 nucleotides per second), generating two single-stranded loops, whose bases are held together by the enzyme. Such structures are called "rabbit ears" (Figure 25.23). Cleavage in one rabbit ear can generate both a free DNA end and a single-stranded region on which RecA can act. Figure 25.24 summarizes this process.

How does RecBCD decide when to cleave within a rabbit ear? Recent studies show that the endonuclease activity is site specific. It has long been known that recombination is favored near certain sites in many prokaryotes and lower eukaryotes. One such recombination-enhancing site, termed *Chi,* has been identified in *E. coli* as the octanucleotide sequence 5′-GCTGGTGG-3′. This is a recognition site for RecBCD enzyme, which cuts this DNA several nucleotides to the 3′ side of Chi. Three related octanucleotide sequences also stimulate DNA cutting and recombination, although to a lesser extent than Chi itself.

Additional proteins in *E. coli,* the products of genes *recE* and *recF*, also participate in homologous recombination, but their biochemical roles are not yet known.

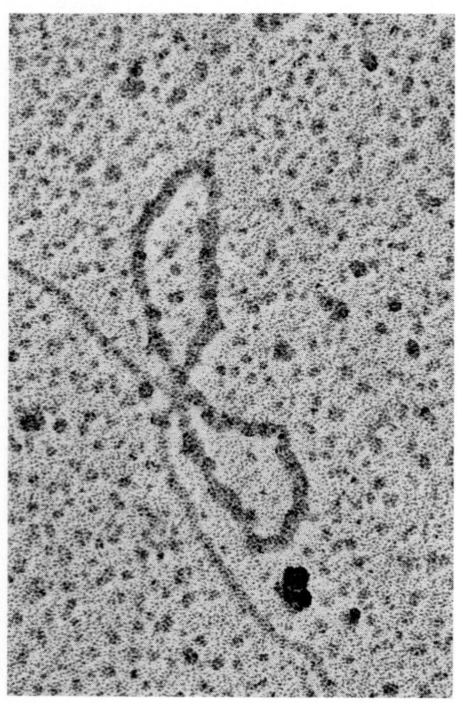

Figure 25.23
"Rabbit ears" created by the action of RecBCD enzyme on T7 phage DNA, with Ca^{2+} added to inhibit nuclease. The single-stranded loops appear thicker than duplex DNA because they are coated with single-strand DNA-binding protein.

Site-Specific Recombination

As discussed above, alignment of sites for homologous recombination occurs via DNA–DNA (base pairing) interactions. Another important class of recombination reactions is directed by highly specific DNA–protein interactions, although a short stretch of DNA homology occurs at the actual site of cutting and resealing. Our biochemical understanding of this site-specific recombination is most advanced for the mechanism by which temperate phages such as λ become integrated at specific sites on the chromosome of the infected bacterium. This process provides an important model for studying counterpart reactions in higher organisms, such as the DNA arrangements involved in the maturation of antibody-forming genes (p. 884).

The circularized λ chromosome integrates into a specific site on the *E. coli* chromosome, *attB*, which maps between genes involved in galactose utilization and biotin synthesis (the *gal* and *bio* markers), as was schematized in Figure 25.13. Integration occurs at a specific site on the phage chromosome, called *attP*. Recombination at this site can be duplicated in vitro, by using plasmids that contain cloned *attB* and *attP* sites (Figure 25.25). Required in addition are the phage protein called integrase, or Int (product of the *int* gene), and an *E. coli* protein called IHF (integration host factor). The Int protein functions like a topoisomerase, underwinding DNA in the region of *attP*. A specialized nucleoprotein structure forms, with the 230-base-pair *attP* region becoming wrapped about seven molecules of Int, each bound at a specific site. This structure becomes aligned with *attB*, which is only about 20 nucleotides long, and binds two molecules of Int. *attB* need not be supercoiled. In the core of both sites is a 15-base-pair region of complete homology. In each of these sequences Int

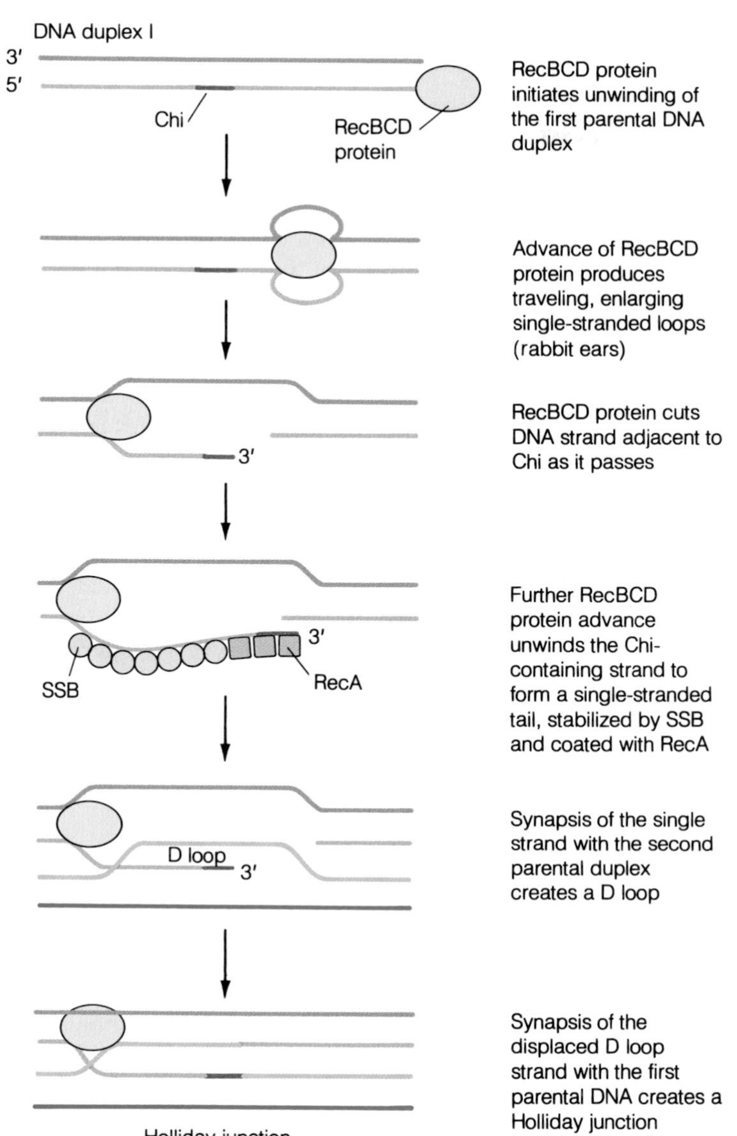

Figure 25.24
Model for the action of RecBCD, Chi sites (red), and RecA in promoting homologous recombination. RecA and SSB both participate in strand pairing, but the relationship between these proteins is not yet known.

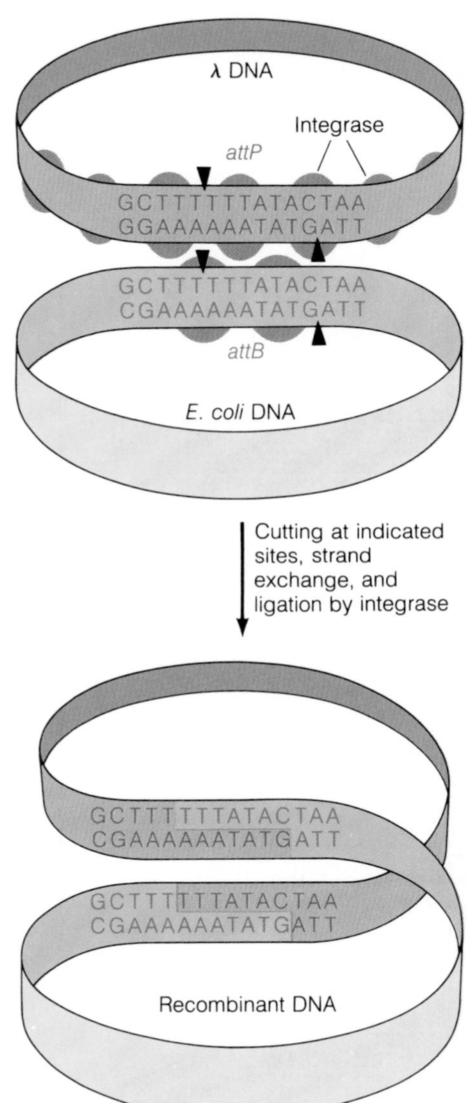

Figure 25.25
Integrative recombination in phage λ, as revealed by in vitro studies with phage and bacterial attachment sites cloned into plasmids. Integrase catalyzes cleavage within the region of homology at sites indicated by wedges, and then rejoins strands to create hybrid attachment sites. The action of IHF protein in this process is currently under study.

generates a staggered cut, with a seven-nucleotide overlap. The ends then exchange to form a Holliday junction intermediate. The bacterial and phage core sequences yield two hybrid sequences, each of which contains both phage and bacterial DNA. The multifunctional Int protein, which has already catalyzed site-specific cutting and strand exchanges, now completes the process with a DNA ligase reaction to join the ends covalently.

When a λ phage chromosome becomes integrated, it is essentially dormant. Almost all of its genes are turned off, and it replicates only as a part of the bacterial chromosome in which it resides. However, changes in gene expression (Chapter 26) can activate the integrated chromosome, or **prophage.** When this occurs, the above sequence of steps is reversed, with excision of a circular phage chromosome. An additional protein, called Xis, is required in addition to Int and IHF. Apparently, Xis and Int form a complex that binds much more tightly to the hybrid phage–bacterial attachment sites than does Int alone, and the complex helps to initiate the excisive recombination reaction.

Gene Rearrangements

Until the mid-1970s the genetic information content of an organism or population was considered to be static. All cells of a differentiated organism were thought to have identical DNA contents, with variations among different cells arising at the level of gene expression. Supporting this idea was the fact that in some plants, such as carrot, a single differentiated cell can be manipulated in culture so as to give rise to a complete and normal plant. However, more recent developments have shown a plasticity to DNA that had not been expected. In the course of normal development, segments of DNA can be deleted from the genome, can move from one site to another within a genome, or can duplicate themselves manyfold. In addition, mobile genetic elements have been described in many prokaryotes and eukaryotes. These are segments of DNA that can move from one chromosomal integration site to another.

Actually, the plasticity of DNA was predicted, but by very few scientists. Barbara McClintock's work on maize genetics, starting in the 1940s, led her to postulate genetic regulatory mechanisms effected through the action of mobile genetic elements. However, some three decades passed before the physical demonstration of such elements in bacteria focused attention on McClintock's pioneering work. For the remainder of this chapter we shall discuss three widely studied aspects of genome plasticity: the genetic basis for antibody variability in vertebrates, gene transposition, and gene amplification.

Immunoglobulin Synthesis: Generating Antibody Diversity

Recall from Chapter 7 that antibodies are proteins manufactured by the immune systems of vertebrates, which aid in defense against infectious agents and other substances foreign to the animal. The immune response, resulting from introduction of an antigen, elicits the formation of several highly specific antibodies. It is estimated that a human is capable of synthesizing more than 10 million distinct antibodies. Such great diversity is generated through the action of precisely controlled gene rearrangements, involving but a small fraction of the coding capacity of the genome. These rearrangements occur during differentiation of many individual clones of cells, each clone specialized for the synthesis of one and only one antibody. Other large protein families are diversified by similar mechanisms.

To see how this diversity is generated, let us look at one type of antibody, the immunoglobulin G, or **IgG**, class. Recall from Chapter 7 that these proteins consist of two heavy chains and two light chains. Each chain comprises two distinct segments—a variable domain of polypeptide sequence and a constant domain, which is virtually invariant among different IgG light or heavy chains. We will focus on the light chains and in particular the κ class of light chains. (Another class, λ, has somewhat different sequences in its constant region, but its development involves similar mechanisms.)

Figure 25.26 shows the organization of the precursor genes to κ chains and the rearrangements leading to one such gene in a differentiated antibody-producing cell. Each light chain is encoded by DNA sequences that are noncontiguous in the genome of undifferentiated cells but are all in the same chromosome. These sequences are called V (variable), C (constant), and J (joining). The human genome contains about 300 different V sequences,

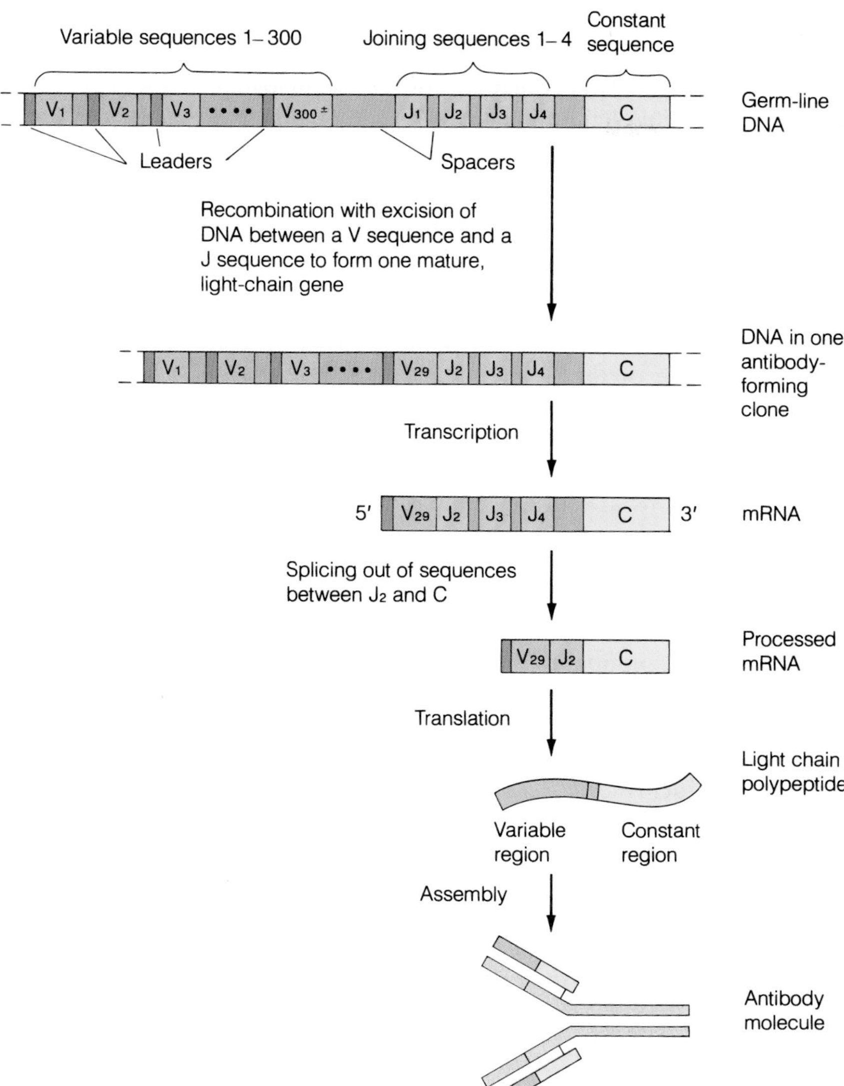

Figure 25.26
Gene rearrangements in antibody gene maturation. Arrangement of C, V, and J sequences encoding elements of antibody κ light chains, and transcription and processing of one light chain gene.

each of which encodes the first 95 amino acids of the variable region; 4 different J sequences, which encode the last 12 residues of the variable region and join it to the constant region; and 1 C sequence, which encodes the constant region. In an embryonic cell the V sequences, each preceded by a leader sequence containing a transcriptional activator that is not expressed, form a tight cluster; the J sequences form another cluster some distance away; and the C sequence follows shortly after the J cluster. Each J sequence is flanked by nonexpressed spacer sequences.

In the differentiation of one antibody-forming clone of cells, a gene rearrangement links one of the approximately 300 V sequences with one of the 4 J sequences. All of the DNA that lies between these two spliced sequences is deleted in this rearrangement and disappears from all progeny of this cell line. Any upstream V sequences (on the 5′ side; leftward in Figure 25.26) and downstream J sequences (on the 3′ side; to the right) remain in these cells but are not used in antibody synthesis.

Additional diversity is provided by the way in which the V and J sequences recombine. The cutting and splicing can occur within the terminal trinucleotide sequences of V and J in any way that yields one trinucleotide

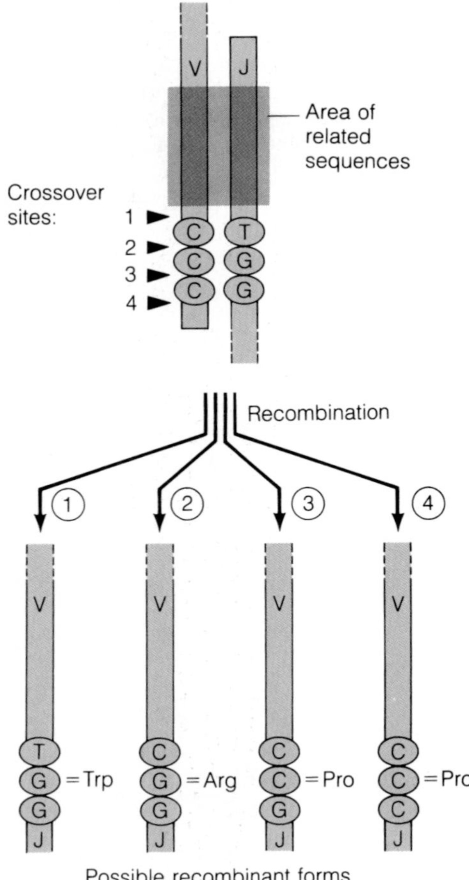

Figure 25.27
Generation of diversity by variable V–J joining mechanisms. Four crossover events are possible at a V–J junction, giving rise to three possible amino acids encoded at that site. Only one DNA strand is shown.

sequence in the spliced product (Figure 25.27). This increases the total number of different light chain sequences by about 2.5 (the average number of different amino acids encoded by four random triplets). Thus, the total number of possible light chain sequences that can be formed from 300 V sequences and 4 J sequences is about 3000 ($300 \times 4 \times 2.5$).

The final step in producing a light chain polypeptide involves joining of the C and J segments (see Figure 25.26). This occurs not at the DNA level, but at the level of messenger RNA. As discussed in Chapters 7 and 28, eukaryotic gene expression usually involves cutting and splicing of the messenger RNA, with excision of sequences that are not represented in the final gene product. In this case transcription yields an RNA molecule extending from the 5′ side of the V gene that is spliced to J to the 3′ side of C. Depending on which J region has been spliced to V in this cell, the RNA excised during splicing may contain sequences corresponding to other J regions.

Heavy chains are formed similarly—from V sequences, J sequences, and a class of sequences called D. In addition, there are eight different C sequences, which are also involved in the synthesis of other antibody classes. The total number of possible IgG heavy chains is about 5000. Since any light chain can combine with any heavy chain to form a complete IgG, the total possible number of IgG molecules is 3000×5000, or 1.5×10^7. In this way, enormous diversity can be generated from a very small fraction of the total DNA in germ-line cells. Even further diversity arises from the fact that V sequences undergo mutation at a high rate during development of the antibody-producing cell. Thus, cells that undergo the same V–J joining event may still differentiate to produce different IgGs.

Although the enzymes involved in the V–J joining reactions have not yet been isolated, it is presumed that some type of site-specific recombination is involved. Related DNA sequences are found to the 3′ side of each V sequence and to the 5′ side of each J sequence, and these presumably represent recognition sites for the enzymes involved in the joining reaction. Those sequences are shown below.

5′···V···CACAGTG···11 bases···ACATAAAC···3′
3′···J···GTGTCAC···22 bases···TGTTTTTG···5′

Note the presence of nearly identical seven-base palindromic sequences and nearly complementary eight-base AT-rich regions in these segments. These features could allow alignment of distant regions of the chromosome, with a process akin to that in phage λ integration recombining the sequences and excising the intervening DNA.

It is not clear whether both of the homologous chromosomes in a diploid antibody-forming cell undergo identical rearrangements. However, since each cell produces only one type of antibody, either that must occur, or else one chromosome is silenced after the other has completed its rearrangement.

Transposable Genetic Elements

In this section we discuss **transposable genetic elements**—genes that do not have a fixed location in a genome but can move from place to place within the genome, albeit with low frequency. Transposition occurs without benefit of DNA sequence homology. Although the existence of gene transposition had been predicted by Barbara McClintock's work on maize genetics, the first physical characterization of transposable elements arose from studies of antibiotic-resistant strains of bacteria. By the early 1970s it was known that genes conferring resistance to drugs such as tetracycline or penicillin were usually carried on plasmids, whose DNA sequences bore no detectable homology with chromosomal DNA sequences of the host. Nevertheless, the genes for antibiotic resistance would appear, with low frequency, in the chromosome of the bacterium or in the DNA of a phage that had infected that cell. The presence of new DNA inserted into the host or phage chromosome could be confirmed by restriction cleavage analysis of the DNA or by heteroduplex analysis in the electron microscope. The existence of these "jumping genes," which move from one chromosome to another in seemingly random fashion, greatly altered our views on gene organization and evolution. The new concepts were of much more than academic interest because they relate to the use of antibiotics to treat bacterial infections—specifically, to the ease with which populations of antibiotic-resistant bacteria can arise.

Transposable elements have been demonstrated in many eukaryotes, including maize, *Drosophila,* and yeast. However, we shall concentrate on bacteria, where the physical structures and transposition mechanisms are best understood. Let us first point out several distinctions between bacterial transposition and other recombinational mechanisms we have discussed. First, transposition does not require extensive DNA sequence homology. Furthermore, the fact that transposition occurs normally in a *recA*$^-$ host suggests that homologous recombination events are not involved. Second, DNA synthesis is involved in bacterial transposition. Transposition always involves duplication of the target site, the short sequence (3–12 base pairs) at which the transposable element is inserted. In many instances the transposable element is itself replicated, with one copy being deposited in the new sequence and one remaining in the donor sequence. Finally, transposable elements can restructure a host chromosome. A transposable element can move from one site to another within the same chromosome, producing two homologous sequences resident in the same chromosome. Depending on whether these sequences are oriented identically or in reverse, homologous recombination between them can yield a deletion or an inversion, as shown in Figure 25.28. Transposable elements also have other effects on the chromosomes they move to—either inactivation of any gene into which they move (where insertion interrupts the coding sequence) or activation of adjacent genes (where a promoter, or transcriptional activator, might be created next to the gene). Abortive transpositional events can cause deletions or inversions in the chromosome. Insertional inactivation of genes is useful for isolating mutants defective in specific functions and for mapping genes.

We recognize three different classes of transposable elements in bacteria, with general structures as shown in Figure 25.29. In classifying these elements we consider the involvement of two enzymes, **transposase** and **resolvase,** whose functions will be discussed later. Class I elements, which encode a transposase but not a resolvase, are of two types. The simplest transposable element, called an **insertion sequence** (or **IS**), consists simply of a gene for transposase, flanked by two short inverted repeat sequences of about 15 to 25 base pairs. A less simple structure called a **composite transposon** consists of a protein-encoding gene, such as a gene conferring antibiotic resistance, flanked by two insertion sequences, or IS-like elements. These elements may be in either identical or inverted orientations. Class II transposons contain but one set of short flanking direct repeat sequences. In addition to a protein-encoding gene (usually conferring antibiotic resistance) and a transposase gene, these elements encode a gene for

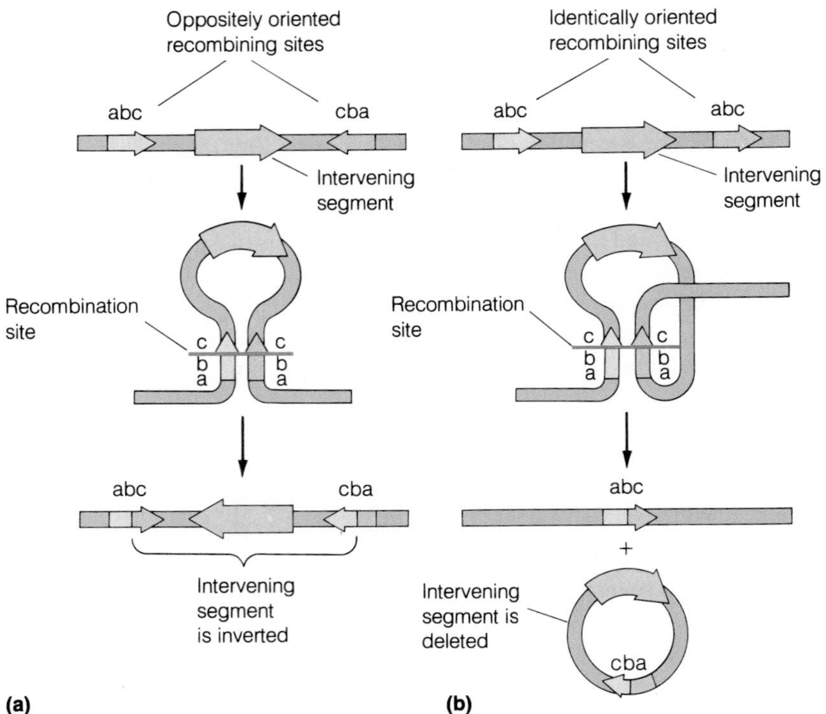

Figure 25.28
Genome rearrangements promoted by homologous recombination between two copies of the same transposable element. Depending on orientation of the two copies, either inversion (**a**) or deletion (**b**) can result.

resolvase. Finally, class III elements consist of a small group of bacteriophages, of which the best-known is phage Mu. This phage chromosome is known to insert at random in the host chromosome, by a transpositional mechanism, and it replicates its genome by a transpositional mechanism. One gene of this phage, A, encodes transposase. Another gene, B, encodes a protein with DNA-dependent ATPase activity. While class I and class II transposable elements synthesize transposase at such low levels that transposition occurs at frequencies of only 10^{-7} to 10^{-5} per generation, phage Mu integrates about 100 times per lytic infection. The B gene product is partially responsible for this far greater efficiency of transposition. Other genes encode structural and other proteins of the virus.

Table 25.4 summarizes the properties of a number of transposons and insertion sequences. Note that each transposon (conventionally referred to with the abbreviation *Tn*) and IS inserts at a specific target sequence of five or nine base pairs in the examples shown. Insertion involves a duplication of that site, and it results in two copies of the target sequence, one on each side of the integrated element (Figure 25.30). It seems likely that this results from the action of transposase, which generates a staggered cut that brackets the target sequence. Attachment of the mobile element to each end results in gaps, which are then filled and ligated to generate the flanking direct repeats.

Since the transposable element never exists as free linear DNA, how are ends of the element generated, to join with the ends of the staggered cut? The currently favored model, shown in Figure 25.31, involves transposase introducing both the staggered cuts in the target site and a nick at each of the 3′ ends of the element—precisely between the transposon sequence and the flanking direct repeat. Next, the free 5′ ends in the recipient DNA target sequence are joined to the 3′ ends of the element. Two possible outcomes

Class I

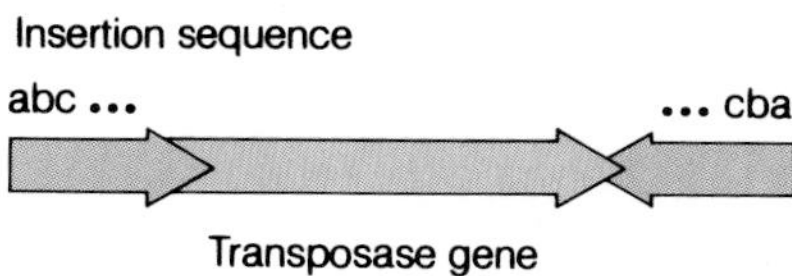

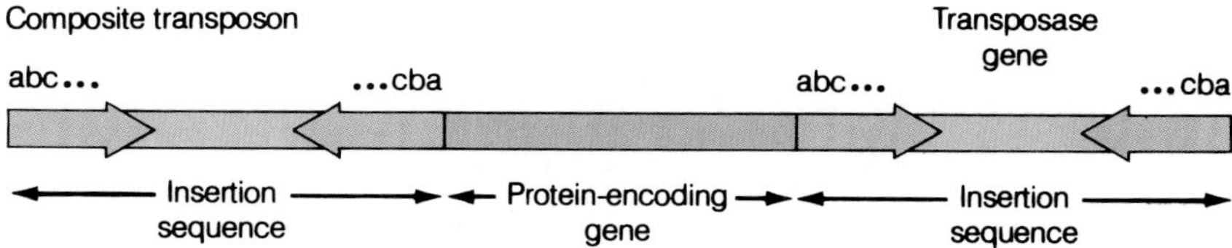

Class II

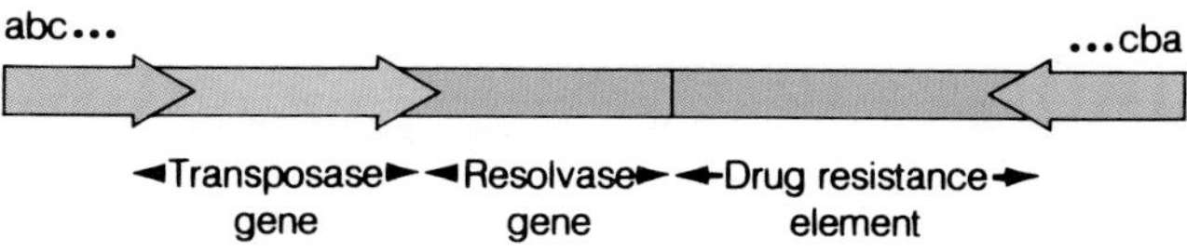

Class III

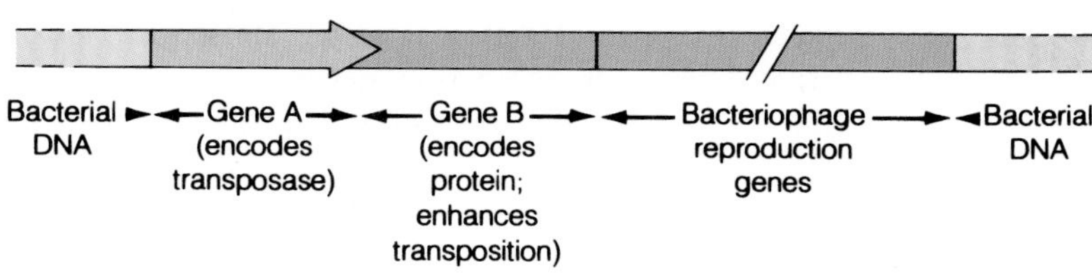

Figure 25.29
Structures of class I, class II, and class III mobile genetic elements. Inverted repeats are shown in purple.

Table 25.4
Structures of some transposable elements of *E. coli*

Element	Size (bp)	Target DNA (bp)	Resistance Conferred
Insertion sequences			
IS1	768	9	None
IS2	1327	5	None
IS10-R	1329	9	None
Composite transposons			
Tn5	5700	9	Kanamycin
Tn10	9300	9	Tetracycline
Tn2571	23,000	9	Chloramphenicol, fusidic acid, streptomycin, sulfonamides, and mercury
Class II transposons			
Tn3	4957	5	Ampicillin
Class III transposons			
Phage Mu	38,000	5	None

SOURCE: Excerpted from N. Kleckner, *Annu. Rev. Genet.* 15:354 (1981).

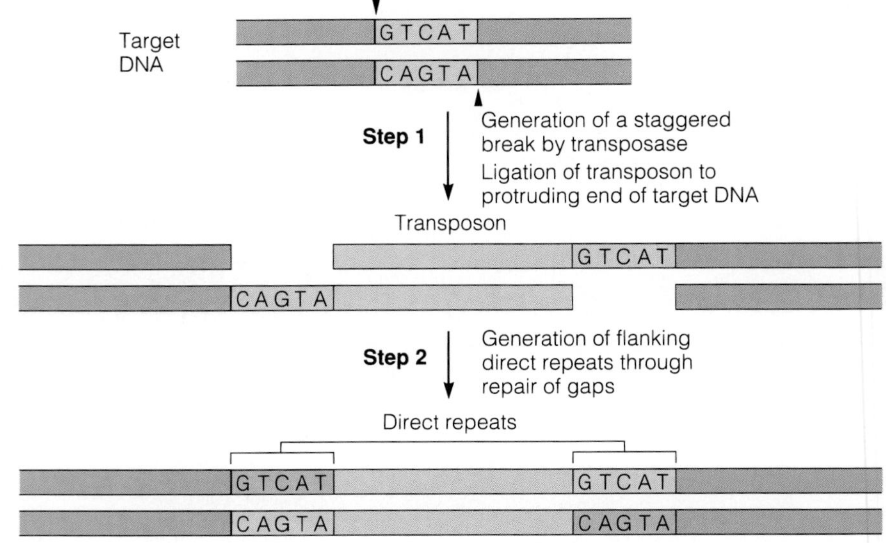

Figure 25.30
Model to explain the generation of direct repeats accompanying insertion of a transposon or insertion sequence.

are then possible. In **simple transposition** the joining is followed by cutting of the 5′ ends of the transposon, also immediately adjacent to the flanking sequences. This gives a gapped structure like that shown in Figure 25.30, which can be filled and closed by DNA polymerase and ligase. In this form of transposition only the target sequence is copied; the donor chromosome suffers a lethal double-strand break. Tn10 transposes by a conservative mechanism, with both original strands somehow transferred to the new location.

The other process, **replicative transposition,** requires the enzyme resolvase, so it occurs only with class II and class III elements. The 3′ ends of the target chromosome, after the first cutting and splicing, serve as replicative primers for copying both the gaps, as shown in Figure 25.31, and the two strands of the transposable element itself. Ligase action generates a **cointegrate,** a large circular structure containing both donor and target chromosomes with two freshly replicated copies of the transposable element. The other enzyme, resolvase, now catalyzes site-specific recombination between the two elements, resulting in one copy of the transposable element inserted into each of the two chromosomes. Studies of transposition of phage Mu, which occurs in cell-free systems, should soon reveal whether this model is correct and how transposition is regulated. These studies should also illuminate mechanisms of transposition in eukaryotic systems, notably yeast and *Drosophila.*

Retroviruses

Gene transposition in eukaryotic systems presents some strong similarities to and some distinct differences from transposition in bacteria. The first major distinction is the fact that in eukaryotes integration and excision are distinct processes; thus, the transposable element can be isolated in free form, often as a double-stranded circular DNA. Second, replication of that DNA often involves the synthesis of an RNA intermediate. Both of these properties are seen in the vertebrate retroviruses, perhaps the most widely studied class of eukaryotic transposable elements. As we noted in Chapter 24, these RNA viruses use reverse transcriptase to synthesize a circular duplex DNA, which can integrate into many sites of the host cell chromosome.

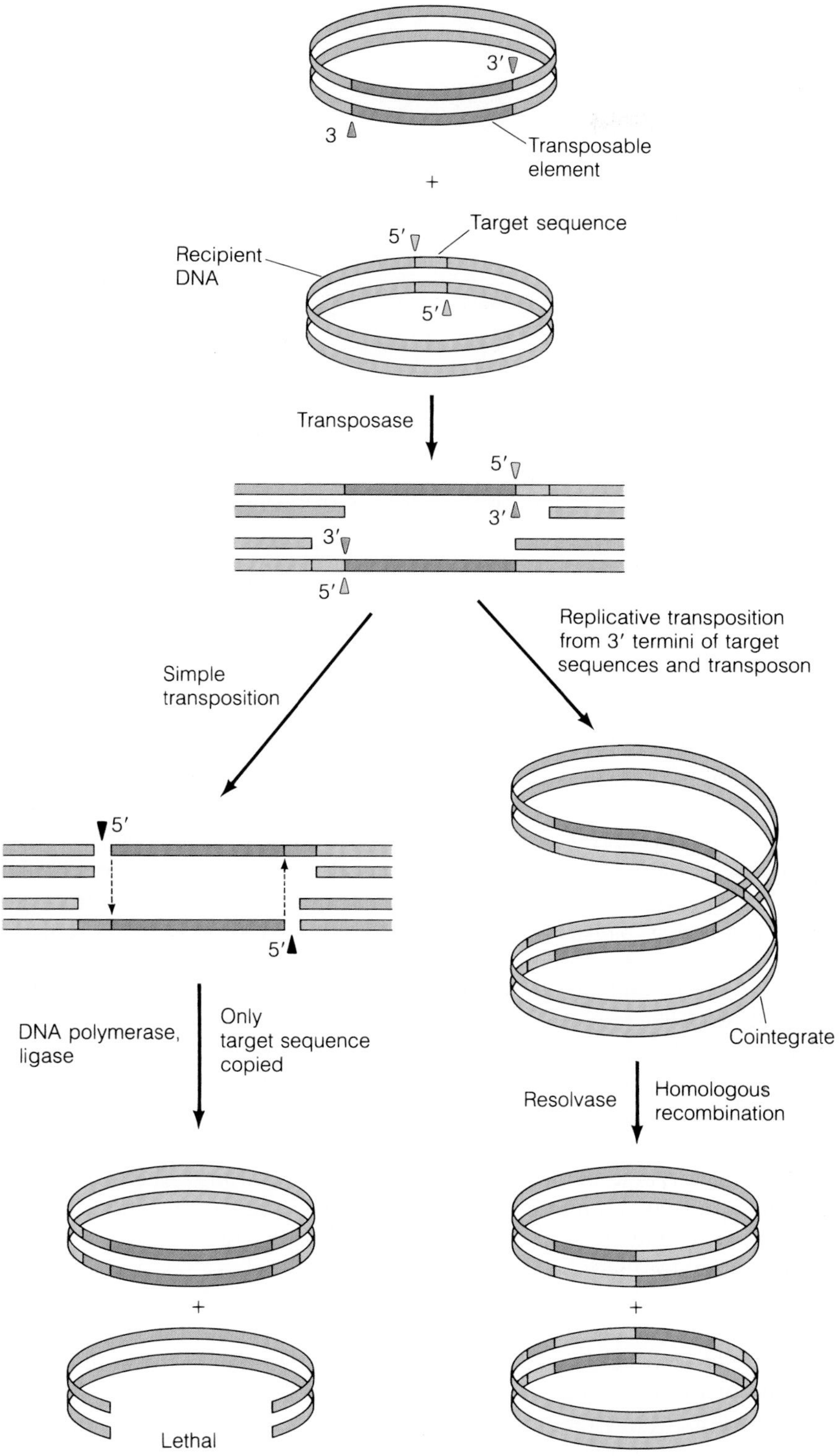

Figure 25.31
Models of simple transposition and replicative transposition.

The integrated retroviral genome bears remarkable resemblance to a bacterial composite transposon, as you can see by comparing Figure 25.32 with Figure 25.29. The prototypical retroviral genome has three structural genes—*gag*, which encodes a glycosaminoglycan that functions as a virion core protein; *pol*, which encodes the viral polymerase, or reverse transcriptase;

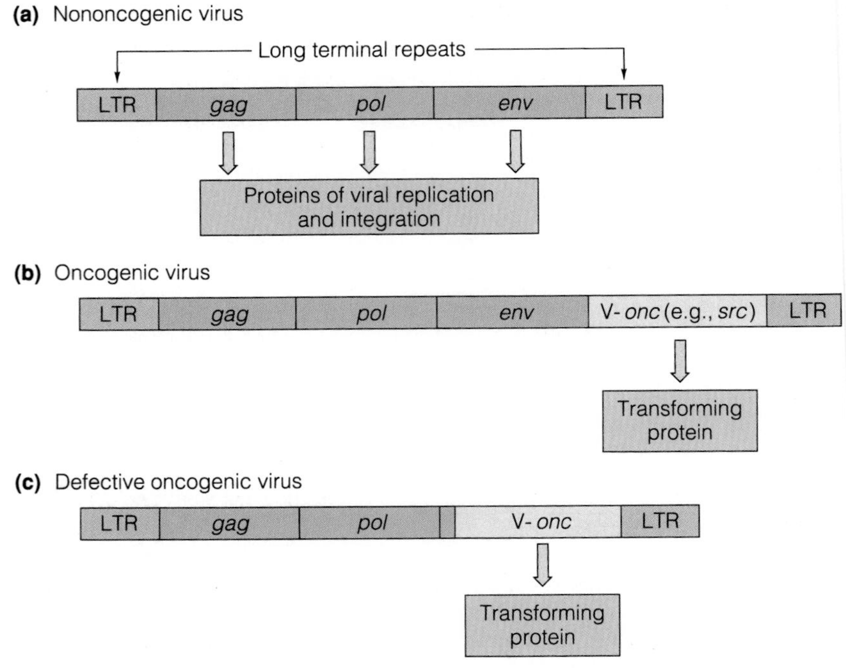

Figure 25.32
Structure of retroviral genomes in the integrated state. (**a**) A nononcogenic virus; (**b**) an oncogenic virus such as Rous sarcoma virus, showing the viral oncogene downstream (rightward) from the viral replication genes; (**c**) a defective oncogenic virus, such as Moloney murine sarcoma virus, with the viral oncogene replacing part or all of a gene (*env*) essential to virus replication.

and *env,* the major glycoprotein of the viral envelope. Flanking these structural genes are two direct repeats, the **long terminal repeats,** or LTRs, some 250 to 1400 base pairs each. Each LTR is flanked in turn by short inverted repeat sequences, 5 to 13 base pairs in length. Integration occurs by a mechanism that duplicates the target site, so that the integrated viral gene, called a **provirus,** is flanked by cellular direct repeats of 5 to 13 base pairs each.

Just as bacterial transposons carry passenger genes, so also do retroviruses. The earliest known retrovirus, Rous sarcoma virus, is also the first virus shown to contain an oncogene (Chapter 23). Rous sarcoma virus was isolated in 1911 and shown to cause tumors in chickens. Not until 1978, however, was the *src* gene identified and shown to be responsible for tumorigenesis. The *src* gene product is a 60-kDa protein with a protein kinase activity. A related but distinct sequence can be detected in the host genome. In Rous sarcoma virus the *src* oncogene lies to the 3′ side of the *env* gene. Other tumorigenic viruses contain the oncogene either inserted with, or substituting for, one of the genes *gag, pol,* and *env.* Since the loss of an essential gene makes it impossible for the virus to replicate, the latter class of virus can grow only in a cell coinfected with a **helper virus,** a related retrovirus that provides the missing function(s).

By various means one can show that the action of an oncogene is essential for the virus to effect **oncogenic transformation,** the change of a normal cell to a cancerous cell. For example, mutational inactivation of the *src* gene of Rous sarcoma virus does not affect the ability of the virus to replicate, but it does render the virus unable to cause tumors in infected animals. Because each viral oncogene is related in sequence to a cellular counterpart, it is presumed that the viral oncogene originated in a cell many generations ago and underwent independent mutations after being picked up by a viral genome. Moreover, because many of the oncogene products, like the Src protein, have protein kinase activity, they may be aberrant forms of normal cellular regulatory elements, and their expression may be involved in the abnormal growth control that characterizes tumor cells. An additional mechanism may be related to activation of cellular genes by insertion of proviral DNA. The leftmost LTR in an integrated provirus contains the transcriptional activator, or **promoter,** for the adjacent *gag* gene and the downstream *pol* and *env* genes. Because the LTRs are direct repeats, the rightmost LTR can activate transcription of cellular genes downstream from the integration site. If these genes include those involved in metabolic regulation, their overexpression may unbalance metabolism in some still undefined way and, hence, contribute to oncogenesis.

At present we know little about the mechanism of retroviral insertion and integration. However, viral insertion has recently been demonstrated in a cell-free system, so it is likely that future progress will be rapid.

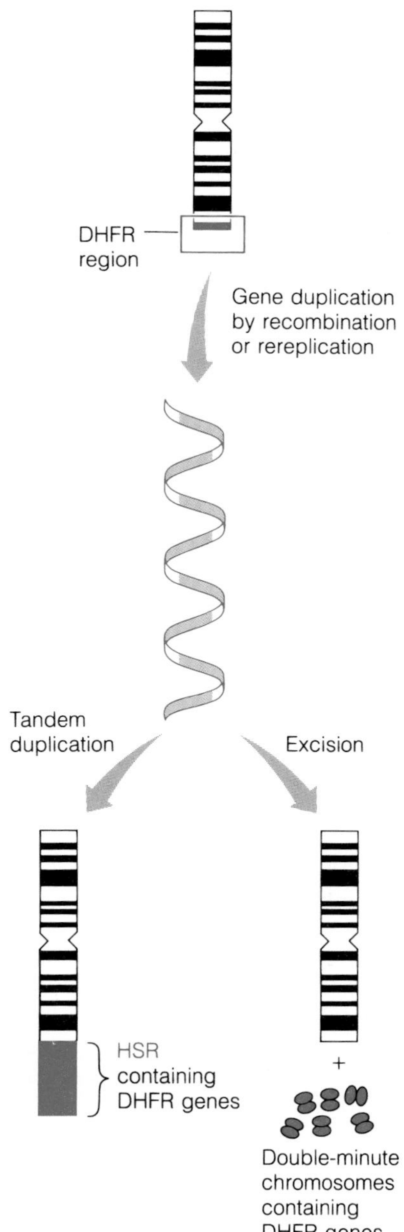

Figure 25.33
Dihydrofolate reductase (DHFR) gene amplification. The amplified sequences occur either on double-minute chromosomes or in tandemly repeated sequences in the homogeneously staining region (HSR) of a giant chromosome.

Gene Amplification

The final process we discuss in information restructuring is the selective amplification of specific regions of the genome. This occurs in normal developmental processes and as a consequence of particular metabolic stress situations.

It has long been known that during oogenesis in certain amphibians the genes encoding ribosomal RNAs increase in copy number by some 2000-fold, in preparation for the large amount of protein synthesis that must occur in early development. The amplified DNA is in the form of extrachromosomal circles, each of which contains several copies of the ribosomal DNA repeat and a replication origin. A similar situation has been analyzed in *Drosophila,* where genes encoding egg proteins are amplified at a particular developmental stage. In this case, however, the amplification results from repeated rounds of replication initiation within the amplified region, and the amplified sequences remain within the chromosome of origin.

Both types of mechanism apparently occur during development of certain drug-resistant mammalian cell lines in culture. This process has been studied most widely in cells that become resistant to the dihydrofolate reductase inhibitor, methotrexate. As discussed in Chapter 22, treatment of leukemia with methotrexate often leads to the emergence of drug-resistant leukemic cell populations, which contain vastly elevated levels of the target enzyme, dihydrofolate reductase (DHFR). In cultured cells overproduction of the enzyme usually results from specific amplification of a large DNA segment that includes the DHFR gene. In one process, tandem duplication of the DNA segment generates a giant chromosome with multiple gene copies, in what is called a **homogeneously staining region,** or **HSR** (Figure 25.33). Alternatively, a DNA segment containing the DHFR gene can be excised, apparently by a recombinational process, to form minichromosomes called **double-minute** chromosomes. Some resistant cells contain both types of amplified genes. Double-minute chromosomes are maintained within a cell only as long as selective pressure is maintained by growth of the cell in methotrexate. However, the chromosomally amplified phenotype is stable through many generations of cell growth.

Amplification of genes under selective conditions has been widely observed—for example, in development of pesticide-resistant forms of insects. The mechanism of amplification is not yet clear. However, it appears that any region of the genome has a low but finite probability of replicating or segregating irregularly as a cell proceeds through the cell cycle. One possibility is that homologous sequences on a chromosome undergo recombination. If these sequences are oriented identically, the region between the sequences can be duplicated; conversely, recombination between oppositely oriented sequences can lead to excision of the intervening region. It has also been suggested that amplification occurs via abnormal reinitiation of replication from a replicon origin sequence.

The presence of selective pressure, such as the continuous presence of methotrexate, promotes specifically the survival of cells that can respond to that pressure, for instance, by overproducing DHFR. Once two or more copies of the gene are present on a chromosome, additional copies can be generated by further recombinational events or by abnormalities of replication. Resistance is thus developed in stepwise fashion and occurs over many generations of growth. These findings have immense practical significance, because cancer chemotherapy often involves long-term treatment with low doses of an antimetabolite—precisely the conditions most likely to nullify the effect of treatment by generating drug-resistant cells. Findings on gene amplification not only have changed the way in which anticancer drugs are administered but also have been extended to the tumorigenic process itself. Investigators have recently found that specific oncogenes become amplified during the clinical progression of certain human tumors. Thus, gene amplification is seen as a mechanism in normal development, in cellular adaptation to stress, and in abnormal developmental processes. Also, the gene duplications that have occurred during evolution probably take place by similar pathways.

REFERENCES

Nucleic Acid Methylation

Doerfler, W. (1983) DNA methylation and gene activity. *Annu. Rev. Biochem.* 52:93–124. Reviews the biochemistry of methylation, its possible role in regulating eukaryotic gene expression, and its role in other processes, including recombination.

Holliday, R. (1987) The inheritance of epigenetic defects. *Science* 238:163–170. A review of the genetics of DNA methylation and the effects of defective methylation on gene expression.

Restriction and Modification

McClarin, J. A., C. A. Frederick, B.-C. Wang, P. Greene, H. W. Boyer, and J. M. Rosenberg (1986) Structure of the DNA–*Eco*RI endonuclease recognition complex at 3 Å resolution. *Science* 234:1526–1541. The first structure determination for a restriction enzyme.

DNA Repair

Bohr, V. A., and K. Wasserman (1988) DNA repair at the level of the gene. *Trends Biochem. Sci.* 13:429–432. Molecular genetic techniques are allowing analysis of repair of individual genes, for example, showing that actively expressed genes are more rapidly repaired than inactive genes.

Friedberg, E. C. (1985) *DNA Repair*. Freeman, New York. A book-length review that can be considered a companion volume to Kornberg's *DNA Replication*.

Grossman, L., P. R. Caron, S. J. Mazur, and E. Y. Oh (1988) Repair of DNA-containing pyrimidine dimers. *FASEB J* 2:2696–2701. One of the most recent reviews of the subject.

Imlay, J. A., and S. Linn (1988) DNA damage and oxygen radical toxicity. *Science* 240:1302–1642. Oxidative damage has been recognized rather recently as a major determinant of DNA damage, and this article describes the current status of the field.

Lahue, R. S., K. G. Au, and P. Modrich (1989) DNA mismatch correction in a defined system. *Science* 245:160–164. Modrich and colleagues describe elegant experiments focusing on mismatch repair through in vitro reconstitution of the process.

Sancar, A., and G. B. Sancar (1988) DNA repair enzymes. *Annu. Rev. Biochem.* 57:29–68. A review of the properties of photoreactivating enzymes and other repair systems by two of the leading contributors in the field.

Walker, G. C. (1985) Inducible DNA repair systems. *Annu. Rev. Biochem.* 54:425–457. Primarily a discussion of RecA-promoted processes, including the *uvr* system, recombinational repair, and the SOS response, with coverage also of the alkylguanine repair pathway.

Recombination

Kowalczykowski, S. C. (1987) Mechanistic aspects of the DNA strand exchange activity of *E. coli* RecA protein. *Trends Biochem. Sci.* 12:141–146. Describes the strand assimilation activity of this versatile protein.

Landy, A. (1989) Dynamic, structural, and regulatory aspects of λ site-specific recombination. *Annu. Rev. Biochem.* 58:913–950. This brief review addresses the generation of complex DNA–protein structures as intermediates in site-specific processes, including phage integration and excision.

Smith, G. R. (1988) Homologous recombination in procaryotes. *Microbiol. Rev.* 52:1–28. This review describes the actions of RecA and RecBCD in promoting recombination by a Meselson–Radding type of mechanism.

Gene Rearrangements

Berg D. E., and M. M. Howe (1989) *Mobile DNA*. American Society for Microbiology. Washington, D.C. A multi-authored volume, with chapters by virtually all contributors to the field of gene transposition.

Kingsman, A. J., and S. M. Kingsman (1988) Ty: A retroelement moving forward. *Cell* 53:333–335. Ty is a transposable element in yeast. The recent discovery that Ty is an RNA molecule at a stage in its life cycle makes it an excellent model for retroviruses.

McClintock, B. (1984) The significance of responses of the genome to challenge. *Science* 226:792–801. Barbara McClintock's Nobel Prize address, giving the history of the first description of mobile genetic elements.

Milstein, C. (1986) From antibody structure to immunological diversification of the immune response. *Science* 231:1261–1268. Milstein was awarded the Nobel Prize largely for his role in discovering monoclonal antibodies, but in this Nobel Prize lecture he discusses the general problem of generating antibody diversity.

Retroviruses

Bishop, J. M. (1987) The molecular genetics of cancer. *Science* 235:305–311. A brief summary of the relationships between retroviruses, viral oncogenes, their cellular counterparts, and genetic changes in carcinogenesis.

Varmus, H. (1988) Retroviruses. *Science* 240:1427–1435. This excellent review is part of a special issue of *Science* devoted to "Frontiers in Biotechnology: Biological Systems."

Gene Amplification

Painter R. B., B. R. Young, and L. N. Kapp (1987) Absence of DNA overreplication in Chinese hamster cells incubated with inhibitors of DNA synthesis. *Cancer Res.* 47:5595–5599. Experimental evidence challenging the notion that gene amplification results from aberrant initiation of DNA replication.

Stark, G. R., M. Debatisse, E. Giulotto, and G. M. Wahl (1989) Recent progress in understanding mechanisms of mammalian DNA amplification. *Cell* 57:901–908. A comprehensive contemporary review.

PROBLEMS

1. Of the restriction enzymes listed in Table 25.2, which generate flush, or blunt-ended, fragments? Of those that recognize offset sites and generate staggered cuts, which of these cuts cannot be converted to flush ends by the action of DNA polymerase? Why?

2. Homologous recombination in *E. coli* forms heteroduplex regions of DNA containing mismatched bases. Why are these mismatches not eliminated by the mismatch repair system?

3. Deficiencies in the activity of either dUTPase or DNA ligase stimulate recombination. Why?

4. pBR322 DNA (4.32 kb) was cleaved with *Hin*dIII nuclease and ligated to a *Hin*dIII digest of human mitochondrial DNA. One recombinant plasmid DNA was analyzed by gel electrophoresis of restriction cleavage fragments, with the following results: lane A: *Eco*RI-treated recombinant; lane B: *Hin*dIII-treated vector; lane C: *Hin*dIII-treated recombinant. (a) Why was the recombinant plasmid treated with *Eco*RI for determination of its size? (b) How might you explain the discrepancy between the size of the recombinant molecule and the sum of the sizes of the *Hin*dIII cleavage fragments? (c) Draw a diagram of the recombinant showing the locations of the *Hin*dIII cleavage sites.

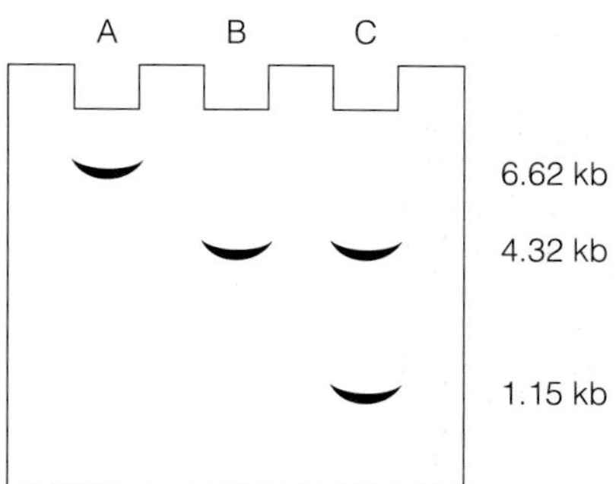

5. A small DNA molecule was cleaved with several different restriction nucleases, and the size of each fragment was determined by gel electrophoresis. The following data were obtained.

Enzyme	*Fragment size (kb)*
*Eco*RI	Two 1.3-kb fragments
*Hpa*II	One 2.6-kb fragment
*Hin*dIII	One 2.6-kb fragment
*Eco*RI + *Hpa*II	1.3, 0.8, 0.5
*Eco*RI + *Hin*dIII	0.6, 0.7, 1.3

(a) Is the original molecule linear or circular?

(b) Draw a map of restriction sites, showing distances between sites, that is consistent with the data presented.

(c) How many additional maps are compatible with the data?

(d) What would have to be done to locate the cleavage sites unambiguously with respect to each other?

6. *A.* pApGpApTpCpT *B.* pGpGpApTpCpC
C. pGpTpCpGpApC *D.* pCpTpGpCpApG
E. pGpTpTpApApC *F.* pGpApGpTpCpNpNpNpN

(a) Show the complete structure of any one of the foregoing restriction cleavage sites (both strands and cutting sites).

(b) Which two of the cleavages shown will yield fragments that *cannot* be rejoined by *E. coli* DNA ligase? Why not? (Note: In vitro DNA ligase shows strong preference for sealing a staggered, rather than blunt-ended, cut.)

(c) Cleavage products from two of these reactions can readily be joined to one another by DNA ligase, even though the two enzymes recognize different sites. Which two?

(d) If you wished to linearize a newly isolated plasmid DNA, which one of the sites shown would be *least* likely to exist once and only once in that DNA molecule? Assume that the DNA has equal contents of the four nucleotides.

7. A cloned 8.0-kb fragment of bacterial DNA contains a gene that you wish to map, with respect to its transcriptional end points. You have in hand a highly purified, highly radioactive mRNA transcribed from that gene, to use as a hybridization probe.

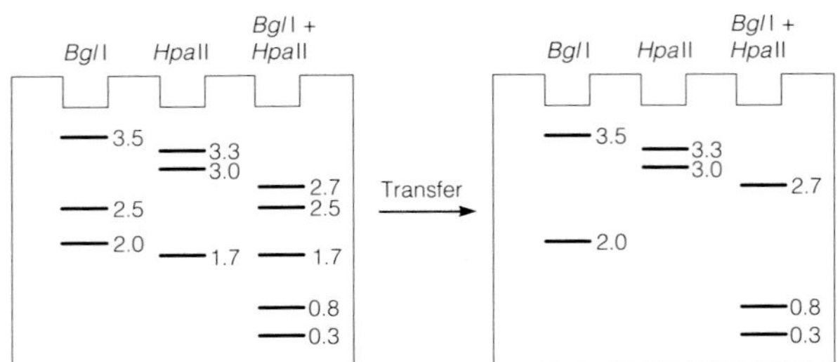

On the left are depicted the restriction cleavage patterns of the 8-kb fragment, visualized with ethidium bromide. To the right is shown a radioautogram that resulted when the fragments were transferred to nitrocellulose and probed with the radioactive RNA. On the following diagram, show the locations of all of the *Bgl*I and *Hpa*II restriction sites, and show the approximate end points of the transcript. How might you tells which end represents the 5′ end and which the 3′ end of the transcript?

*Bgl*I sites

0 kb———————8 kb

*Hpa*II sites

8. Suppose that you wanted to study retrovirus integration mechanisms by determining the nucleotide sequence at the integration site—several dozen nucleotides on each side of the viral–cellular DNA junction. Describe how to isolate DNA containing a junction site in amounts sufficient for sequence analysis.

9. In site-directed mutagenesis one often designs the mutagenic oligonucleotide not only to force a desired amino acid substitution but also to create a new restriction site in the mutant gene or eliminate a preexisting restriction site. Why is this desirable?

10. We wish to determine whether the serine residue of an enzyme is essential for its catalytic activity, by changing it to a glycine and testing the activity of the mutant protein. We also wish to remove one restriction site and introduce one site, without introducing additional amino acid changes. A maximum of two single-base changes can be tolerated in an 18-nucleotide primer and still have it hybridize to the template. Using the genetic codon assignments shown in Chapter 27 and the restriction site information in Table 25.2, design an oligonucleotide of 18 residues that can accomplish these objectives, starting with the indicated sequence in the cloned gene in the accompanying diagram. Show where cleavage sites are added and lost. Additional cleavage sites that might be useful are as follows: *Ava*I, C↓PyCGPuG; *Bal*I, TGG↓CCA; *Xma*I, C↓CCGGG; *Hae*II, PuGCGC↓Py; *Hinc*III, GTPy↓PuAC (Py and Pu are pyrimidine and purine, respectively).

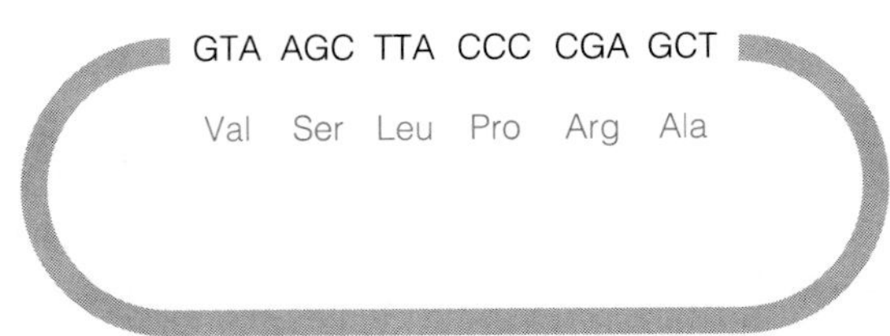

11. In site-directed mutagenesis using oligonucleotide primers, the proportion of mutant clones recovered usually falls well below the 50% that is expected. Why might this be so?

TOOLS OF BIOCHEMISTRY 18

Genome Mapping

One of the earliest applications of restriction enzymes, in the early 1970s, was in the generation of physical maps of DNA molecules. An aim of molecular biology is to determine the size and location of each gene in the genome of an organism. Until about 1970 all mapping involved determination of genetic linkage between markers, as assessed by measurements of recombination frequency following conjugation, transduction, or transformation. The first attempts to correlate genetic map position with physical location on a chromosome involved rather cumbersome electron microscopic techniques.

Restriction fragmentation is a far easier way to generate a physical map—in this case a map of cleavage sites for a given type II restriction nuclease. Referring back to Figure 25.5, you can see that cleavage of λ DNA with one enzyme, such as *Eco*RI, gives the lengths of all six fragments in the digest but gives no information about the order of these fragments in the intact phage genome. Comparable information comes from analysis of the *Bam*HI sites. However, if DNA is digested with both *Eco*RI and *Bam*HI, this double-digest pattern gives much more information about the placement of sites for both enzymes. One can often obtain from analysis of three DNA gels maps for two restriction sites, as you can see if you work Problems 5 and 7 of Chapter 25. If the data are difficult to interpret—for example, because a digest contains two or more different fragments that have the same size—the needed information can be obtained from a partial digest, that is, by running the restriction cleavage reactions only partly to completion. Another way to simplify the interpretation of mapping data is to label the 5′ ends with $\gamma[^{32}P]$ATP and polynucleotide kinase and then carry out partial digestion, followed by electrophoresis and radioautography of the labeled fragments (Figure T18.1). The radioautogram gives a ladder of fragments, comparable to that seen in a Maxam–Gilbert sequencing gel (Chapter 4), which allows ordering of sites from the 5′ end.

By methods such as these, the restriction maps of many small DNAs—from viruses, organelles, and plasmids—were generated in the late 1970s and early 1980s. Genomes up to about 2×10^5 nucleotide pairs in size could be mapped. However, the smallest cellular genomes are about 20 times that size, and the difficulty of mapping a genome increases disproportionately to increases in the size of the genome, because of the gigantic number of restriction fragments that must be analyzed. Clearly, new technologies are needed to handle the enormous amounts of data generated by the mapping effort.

In 1987 two laboratories described physical maps of the *E. coli* genome, which contains 4.7 million base pairs. One laboratory cloned and separately mapped *3400* restriction fragments resulting from cleavage of *E. coli* DNA by restriction nucleases that recognize six-base cutting sites. The other laboratory used a newly discovered restriction nuclease, ***Not*I**, which recognizes an eight-base cutting sequence and, hence, generates fewer (but larger) fragments. These fragments were separated by **pulsed field gel electrophoresis,** a technique that allows resolution of DNA fragments up to 1000 kb in length.

Now consider the effort that must be expended if the entire human genome is to be mapped—a task which has captured the imagination of many in the scientific community. The human genome contains about *3 billion* base pairs of DNA, making that genome several hundred times larger and more complex than the *E. coli* genome. In early 1988 it was estimated that the task of mapping and determining the

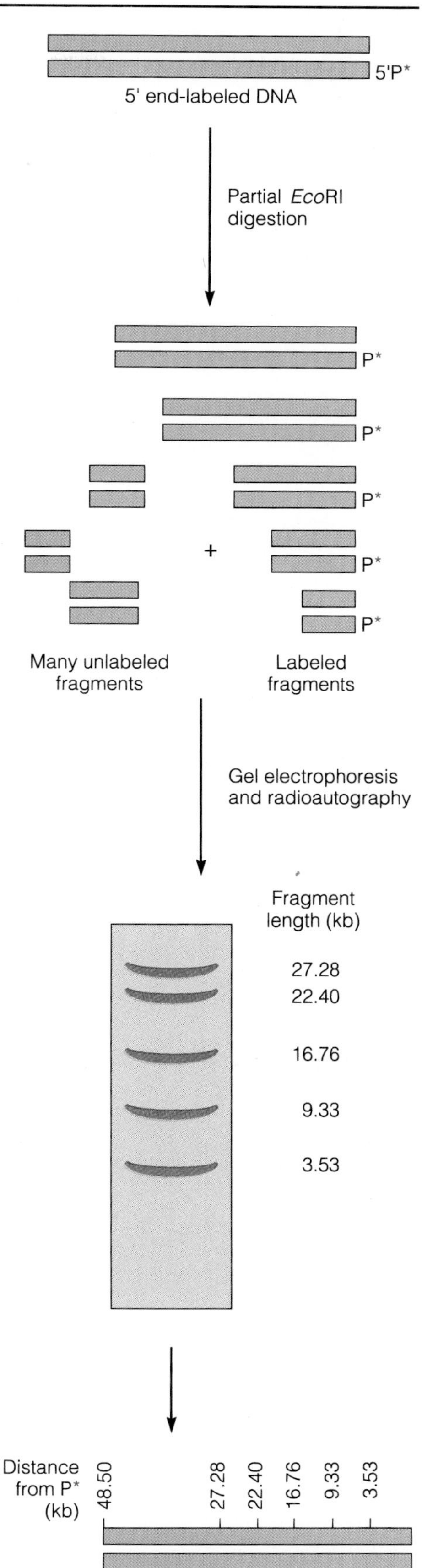

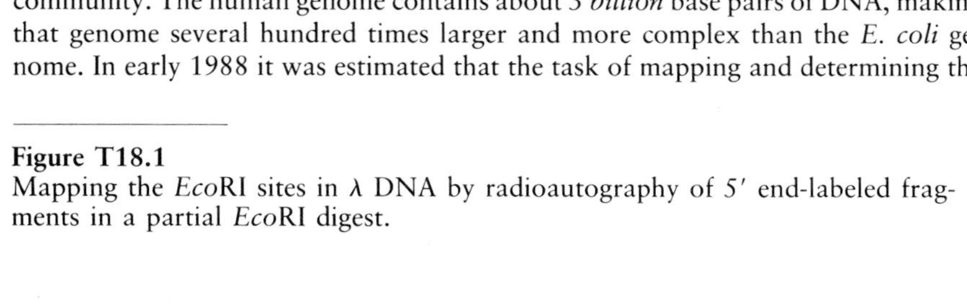

Figure T18.1
Mapping the *Eco*RI sites in λ DNA by radioautography of 5′ end-labeled fragments in a partial *Eco*RI digest.

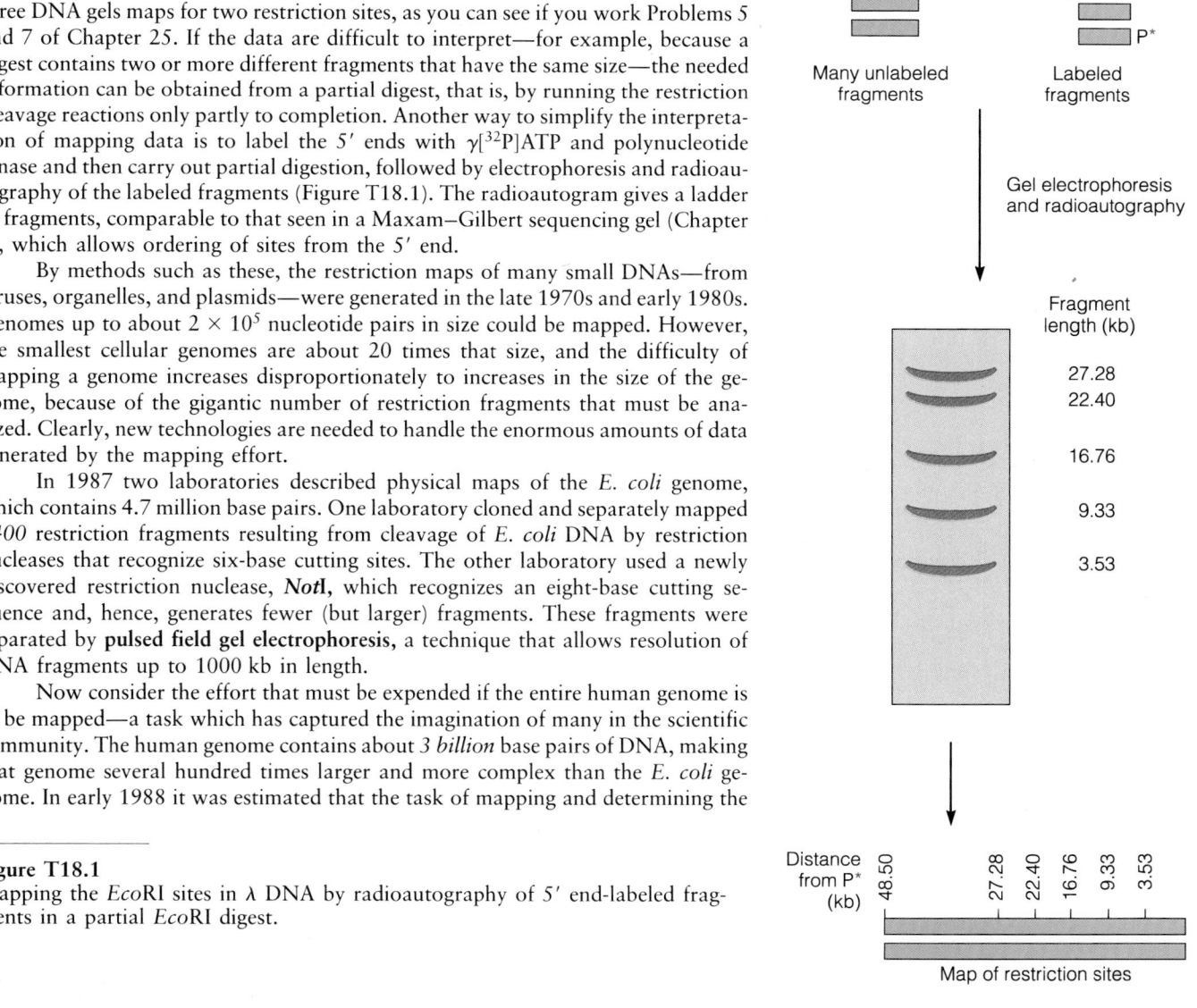

entire nucleotide sequence of the human genome would cost $3 billion and require 15 years of effort. That effort is now under way.

While mapping the human genome sounds very much like the "big science" embodied in the Manhattan Project or space exploration, it is reassuring to note that important discoveries are being made in human molecular genetics by individual laboratories, approaching manageable objectives. As described in Tools of Biochemistry 21, defective genes responsible for several hereditary disorders have been located on human chromosomes. Another example comes from mapping of human mitochondrial DNA. Mitochondrial genes are maternally inherited, because mitochondria from the sperm cell do not enter the fertilized egg. By comparing mitochondrial DNA restriction maps of 147 individuals of varying origins, Allan Wilson concluded that all mitochondrial lineages are derived from one woman, who lived in Africa about 200,000 years ago. These data have been overinterpreted by the lay press to mean that we are all descended from one woman. In fact, the data say nothing about the origins of the larger and less accessible nuclear genome. Nevertheless, the fact that relatively simple data can lead to any generalizations about human origins indicates the power of restriction mapping techniques.

REFERENCES

Cann, R. L., M. Stoneking, and A. C. Wilson (1987) Mitochondrial DNA and human evolution. *Nature* 325:31–36. Restriction mapping of human mitochondrial DNAs leads to conclusions about human origins.

Cold Spring Harbor Laboratory (1986) *Molecular Biology of* Homo sapiens. Cold Spring Harbor Laboratory, Cold Spring Harbor, N.Y. This symposium volume contains numerous articles about the use of molecular genetic technologies to map the human genome, to diagnose human genetic diseases, and to produce desired human gene products by molecular cloning.

Maniatis, T., E. F. Fritsch, and J. Sambrook (1989) *Molecular Cloning. A Laboratory Manual,* 2nd ed. Cold Spring Harbor Laboratory, Cold Spring Harbor, N.Y. The most recent edition of the definitive laboratory handbook of molecular biological methods.

19

TOOLS OF BIOCHEMISTRY 19

Gene Cloning

In classical biology a **clone** is a population of organisms that are genetically homogeneous because they were derived from a single ancestor. For example, all of the bacterial cells in one colony represent a clone, because they were derived from a single cell that was deposited at that location on a Petri plate. In 1973 Herbert Boyer and Stanley Cohen realized that they could use restriction enzymes to "clone" a gene in comparable fashion. What was actually done was to construct in vitro a novel DNA molecule consisting of a restriction fragment containing a gene of interest, linked to another DNA molecule capable of directing its own replication. This molecule could then be propagated biologically by introduction into a bacterial cell. As the bacterium grows, the novel DNA molecule replicates concomitantly. The novel DNA is called a **recombinant DNA molecule,** because it is formed by end-to-end joining of two different DNAs.

Cohen and Boyer realized that, because of the symmetry in type II restriction sites, enzymes giving staggered or offset cuts generated cohesive ends that could be joined covalently by DNA ligase, just as occurs in λ phage DNA replication (Chapter

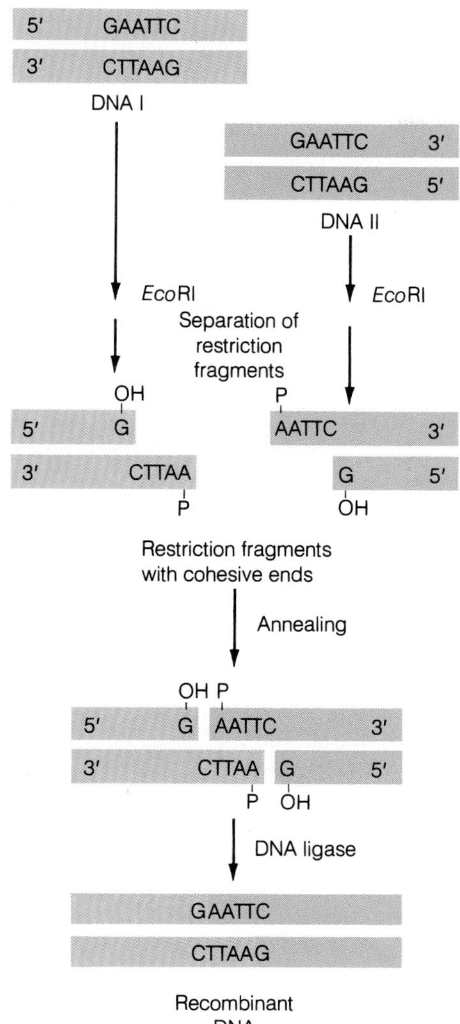

Figure T19.1
Creation of recombinant DNA molecules in vitro.

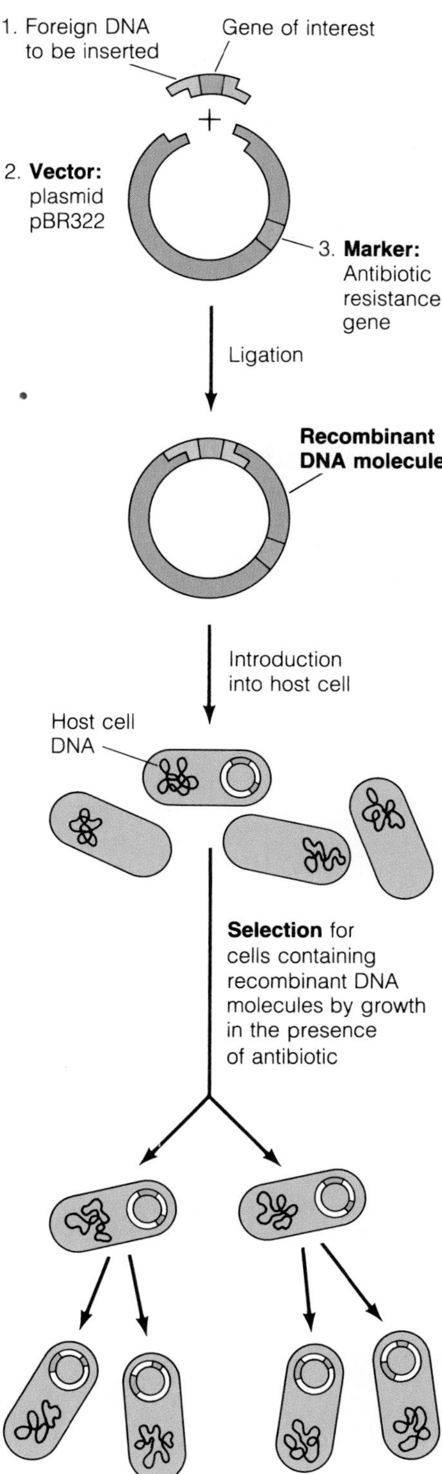

Figure T19.2
Cloning a fragment of DNA into a plasmid vector, followed by its introduction into bacteria.

24). DNA molecules from any sources are capable of being joined, as long as they have the same cohesive ends (Figure T19.1). After the DNA ligase reaction in vitro, the mixture of DNAs is introduced into bacteria that have been rendered permeable by treatment with calcium chloride, followed by heat shock. About 0.01% of the bacteria can undergo genetic transformation—the uptake of DNA and its replication in the recipient cell. This process is illustrated in Figure T19.2.

Successful cloning of a gene requires several elements. First, one needs a DNA fragment thought to contain the gene of interest. Usually this is a restriction fragment, but other means of fragmenting DNA, such as sonic oscillation, are often used when there is a possibility that a restriction enzyme would cut within the target gene. Cohesive ends can also be generated by synthesis. An enzyme called **terminal deoxynucleotidyltransferase** adds nucleotides to 3′ DNA termini, without a need for a template. Thus, one can put a poly(dC) "tail" on one DNA and a poly(dG) "tail" on the other, and these tails are complementary to each other, just like the cohesive ends created by restriction nucleases. Finally, techniques are now available for the end-to-end joining of blunt-ended DNA fragments.

Second, one needs a **vector,** or DNA molecule to which the target gene will become linked. A vector can be any DNA that contains an origin of replication and

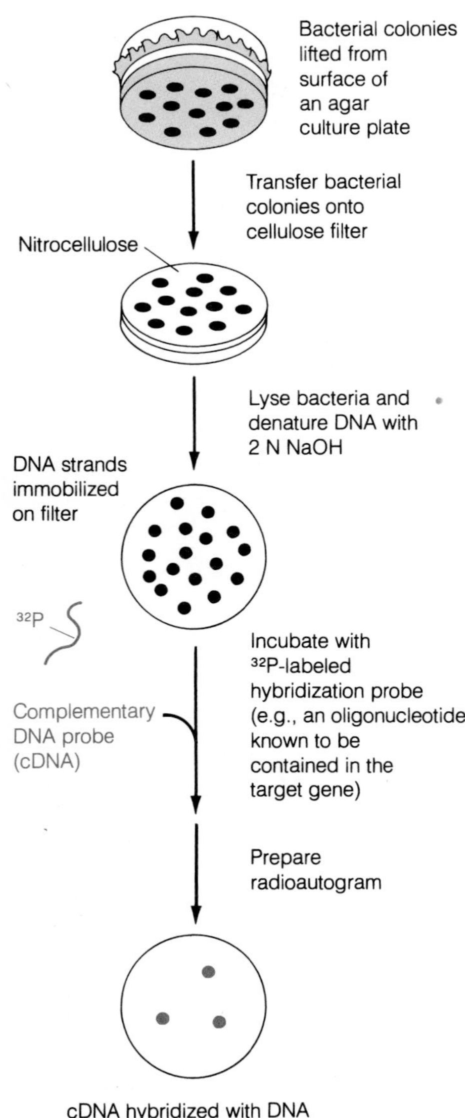
Bacterial colonies lifted from surface of an agar culture plate
Transfer bacterial colonies onto cellulose filter
Nitrocellulose
Lyse bacteria and denature DNA with 2 N NaOH
DNA strands immobilized on filter
32P
Complementary DNA probe (cDNA)
Incubate with 32P-labeled hybridization probe (e.g., an oligonucleotide known to be contained in the target gene)
Prepare radioautogram
cDNA hybridized with DNA from a particular colony makes that colony radioactive

Figure T19.3
Colony hybridization, using a radioactive oligonucleotide probe to identify bacteria containing cloned sequences homologous to that probe.

can replicate after its entry into a suitable cell (usually a bacterium, although genes can also be cloned into eukaryotic cells). Plasmids and bacteriophage chromosomes are most often used as vectors, because they can replicate independently of the host cell into which they are introduced. This allows one to amplify the recombinant DNA, that is, to replicate it to a much higher copy number than that of the chromosomal DNA.

Third, because ligation and transformation both occur with low efficiency, one needs a suitable screening technique—a method for identifying bacteria that contain the cloned gene in the presence of a huge excess of cells that do not. If there is reason to expect that the cloned gene will be expressed, one can select conditions where the cloned gene confers a selectable phenotype on the recipient bacterium, such as resistance to a particular drug or loss of a nutritional requirement. Reaction with an antibody against the product of the cloned gene is another way to screen for desired clones among a large number of transformants. However, if one doesn't know whether the cloned gene will be expressed, one can search for the desired *DNA sequences* instead of the product of the cloned gene. In one technique, called **colony hybridization,** one searches among individual bacterial colonies for DNA sequences homologous to a labeled nucleic acid probe (Figure T19.3). The hybridization probe can be either a restriction fragment of natural DNA or a chemically synthesized oligonucleotide known to contain a sequence represented in the desired clone. Alternatively, if the DNA has been cloned into a phage vector instead of a plasmid, one can screen individual phage plaques in much the same way (this technique is called **plaque hybridization**).

Screening for desired recombinants is aided greatly by the design of vectors with suitable restriction sites and selectable markers. pBR322 is a popular plasmid vector which is itself a recombinant DNA molecule produced in vitro. As shown in Figure T19.4, pBR322 is a circular DNA molecule containing a replication origin from a naturally occurring plasmid, ColE1, plus genes conferring resistance to the antibiotics ampicillin and tetracycline. Some restriction sites lie within either the amp^R or the tet^R genes, the genes conferring resistance to ampicillin and tetracycline, respectively. If one "clones into" the *Hin*dIII site, for example, by opening pBR322 with *Hin*dIII and ligating DNA to the resulting DNA ends, one will create a recombinant in which the cloned gene forms an insert that splits, and hence inactivates, the tet^R gene. This aids greatly in screening for clones; all bacteria that are transformed become ampicillin resistant. However, only the bacteria that contain a recombinant plasmid are also tetracycline sensitive. Transformation with the unaltered plasmid confers resistance to both antibiotics.

In 1987 a novel technique was introduced that allows amplification of a nucleic acid sequence without the need to clone it. This technique, called **polymerase chain reaction** or **PCR,** requires knowledge of sequences that flank the region to be amplified. Oligonucleotides complementary to these sequences are produced by automated chemical synthesis and used as primers in a special series of DNA polymerase-catalyzed reactions (Figure T19.5). First, DNA containing the sequences to be amplified is heat denatured and then annealed to the primers, which are present in excess (steps 1 and 2). Next, polymerase chain extension is carried out from the primer termini (step 3). Then a second cycle of heat denaturation, annealing, and primer extension is carried out. By using a thermostable form of DNA polymerase from a bacterial species that lives at high temperatures, one need not add more polymerase, because the enzyme is not inactivated at DNA-denaturing temperature. This cycle is repeated up to 30 times, and, as shown in the figure, each cycle increases the abundance of duplex DNA species bounded by the oligonucleotide primers. This technique has dozens of potential applications, including the analysis of minute amounts of DNA obtained as evidence in criminal investigations (see Tools of Biochemistry 21).

The variations on cloning and gene amplification techniques are virtually limitless, and several excellent books have been written on the subject. We shall see numerous examples throughout the rest of this book of the utility of gene cloning as a research tool. However, the excitement generated by recombinant DNA technology derived in large part from its potential for attacking practical problems. The earliest such example was the cloning and expression in *E. coli* of a gene encoding

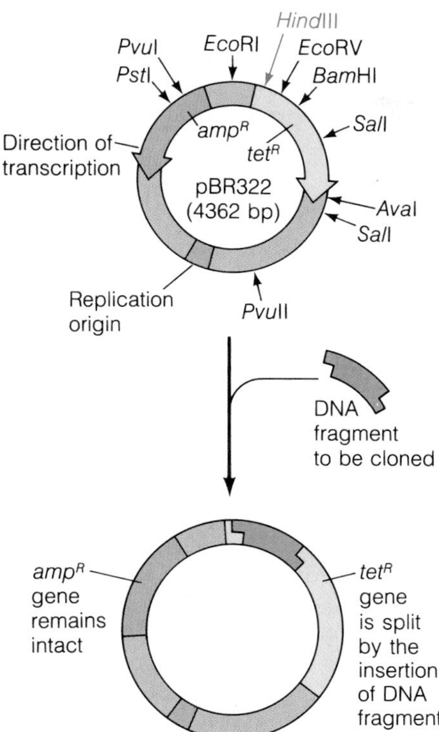

Figure T19.4
pBR322, a widely used plasmid cloning vector, showing some of the restriction sites, the direction of transcription of the ampicillin and tetracycline resistance genes, and the effect of cloning a novel sequence into the Hind III site.

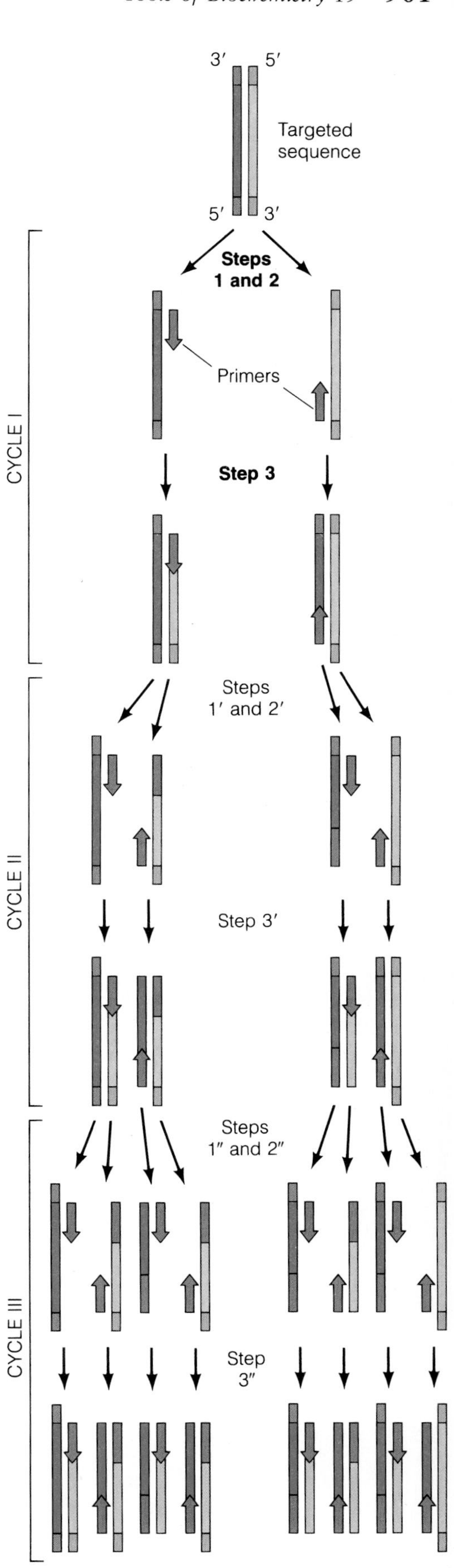

human insulin. This was accomplished in 1977, and by 1982 the purified recombinant product was available for use in treating diabetes. Other recombinant products include blood clotting factors, other polypeptide hormones (including pituitary growth hormone), and interferons.

Current interest is focused on genetic engineering of plants to endow them with desirable properties, such as resistance to cold, to dehydration, to viruses, or to insect pests. Nature has provided an excellent vehicle for introducing novel genes into plants. *Agrobacterium tumefaciens* is a pathogenic bacterium that causes a disease called crown gall, through transfer of some of its own DNA into infected plants. The disease-causing genes are carried on a plasmid, called the **Ti plasmid.** Foreign genes can be introduced into the plasmid by recombinant DNA methodology, and infection of plants by the resultant modified *Agrobacterium* can transfer these novel genes into plants. Current efforts focus on controlling the expression of these genes after infection and directing their integration into the cellular genome, so that the desirable characteristics become permanently incorporated into the genome.

Comparable gene transfer techniques exist for animals as well. In 1982, Ralph Brinster and Richard Palmiter introduced, by microinjection, a cloned gene for rat growth hormone into the male pronucleus of fertilized mouse eggs. The gene was cloned downstream from an inducible promoter to increase its chances of being expressed. After insertion of the microinjected eggs into the uterus of a foster-mother mouse, a significant number of offspring expressed the gene, produced high levels of

Figure T19.5
The polymerase chain reaction, for amplifying a segment of DNA without cloning it.

rat growth hormone, and grew to about twice the size of normal mice. While the early success of this experiment with **transgenic** animals was perhaps fortuitous, more recent developments have greatly increased our ability to achieve homologous recombination of DNA after transfer into mammalian cells. Thus, the technology is on the horizon for directing gene insertions or substitutions at desired intrachromosomal locations in living animals.

REFERENCES

Berger, S. L., and A. R. Kimmel (1987) *Methods in Enzymology,* Vol. 152, *Guide to Molecular Cloning Techniques.* Academic Press, San Diego, Calif. One of several excellent methods books describing recombinant DNA technologies.

Capecchi, M. R. (1989) The new mouse genetics: Altering the genome by gene targeting. *Trends Genetics* 5:70–76. A review of studies on homologous recombination of DNA after transfer into mammalian cells.

Koshland, D. E., ed (1989) The new harvest: Genetically engineered species. *Science* 24A:1275–1317. A special issue of *Science* featuring seven articles reviewing practical applications of gene cloning and genetic engineering.

Saiki, R. K., D. H. Gelford, S. Stoffel, et al. (1988) Primer-directed enzymatic amplification of DNA with a thermostable DNA polymerase. *Science* 239:487–494. A complete description of the polymerase chain reaction method for in vitro amplification of DNA sequences. A news article in *Science* (240:1408–1410, 1988) describes potential applications of the technique.

Westphal, H. (1989) Transgenic mammals and biotechnology. *FASEB J* 3:117–120. A recent review of gene transfer techniques in animals and their potential applications.

20

TOOLS OF BIOCHEMISTRY 20

Gene Sequencing with Dideoxynucleotides

In Chapter 4 we described the Maxam–Gilbert DNA sequencing technique, which involves base-specific cleavages of a 5′ end-labeled DNA molecule to give a series of fragments of varying lengths, which can be analyzed by gel electrophoresis. In 1976 Fred Sanger introduced an alternative sequencing methodology, which is now in more widespread use because it can be applied to longer fragments.

Sanger sequencing is similar in principle to Maxam–Gilbert sequencing in that it generates a set of fragments with a common 5′ origin and base-specific 3′ termini. However, those 3′ termini are created not by base-specific cleavage after synthesis of a full-length molecule, but by base-specific *interruption* of enzymatic synthesis of the molecule in vitro, by incorporation of nucleotide analogs that serve as chain terminators. The principle of the method is shown in Figure T20.1. First, one must clone the fragment to be sequenced into a vector that allows it to be isolated as single-stranded DNA (although methods are now available for sequencing from double-stranded DNAs). Bacteriophage M13 is most commonly used as a vector. One can isolate the double-stranded replicative form of this virus and clone the target sequence into it, just as one clones into a plasmid such as pBR322. After introduction of the double-stranded DNA into a bacterium, the cell produces virus particles, which contain single-stranded DNA. The M13 genome has been genetically engineered for this type of cloning—first, to contain suitable restriction sites for cloning,

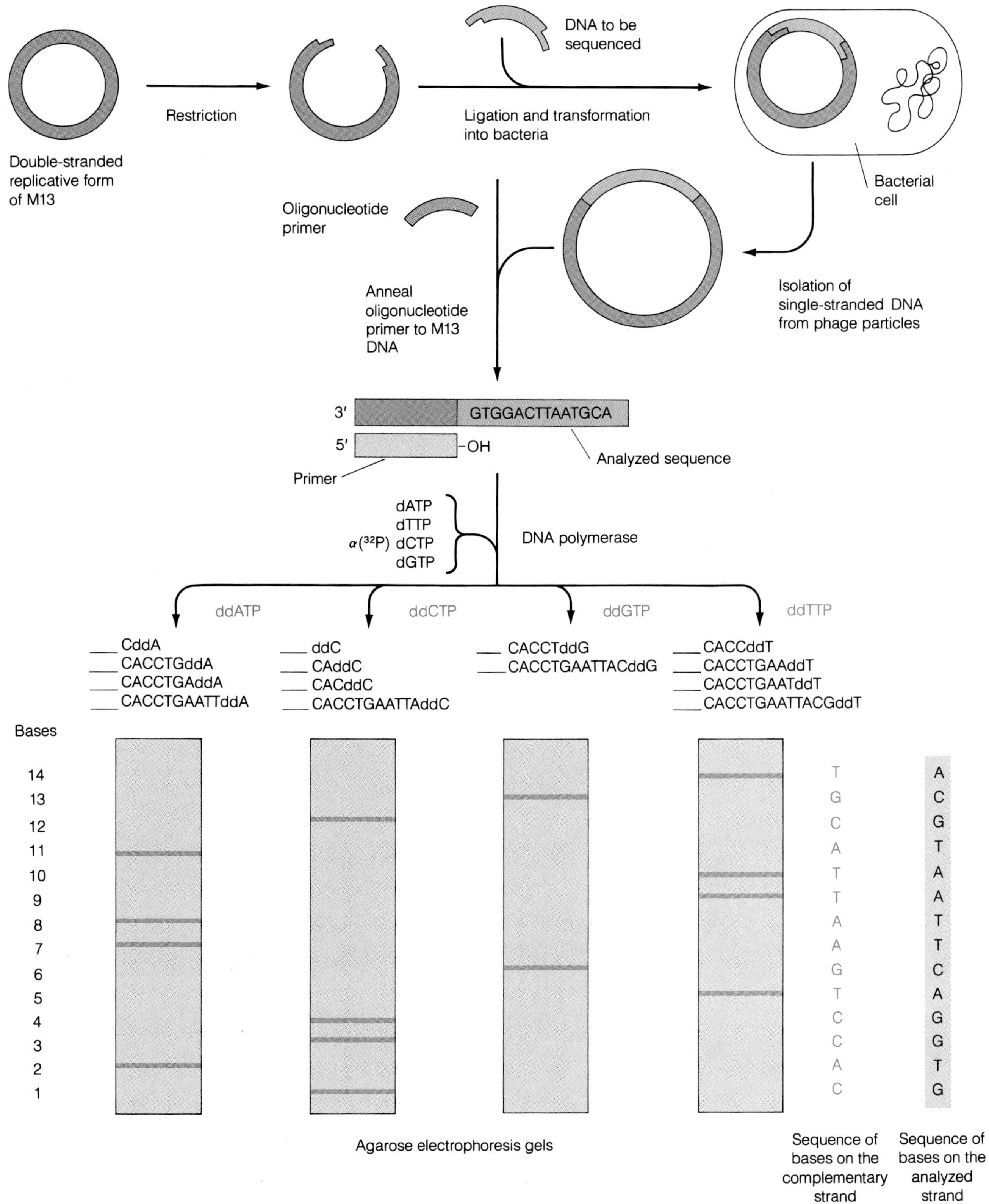

Figure T20.1
Cloning into M13 and sequencing by the Sanger method.

and second, to allow fast identification of plaques containing phage that have acquired DNA insert in the cloning process.

After isolation of single-stranded circular DNA from phage containing the desired insert, one uses this as the template for four DNA polymerase-catalyzed reactions. The primer is an oligonucleotide that is complementary to an M13 sequence lying just 3′ to the insert. Extension of this primer by DNA polymerase will copy the insert. These polymerase reactions are run in the presence of deoxyribonucleoside triphosphate analogs, the **2′,3′-dideoxyribonucleoside triphosphates,** which serve as terminators of chain extension because they lack 3′ hydroxyl termini. The dideoxy analog (ddATP) of adenosine triphosphate is shown below.

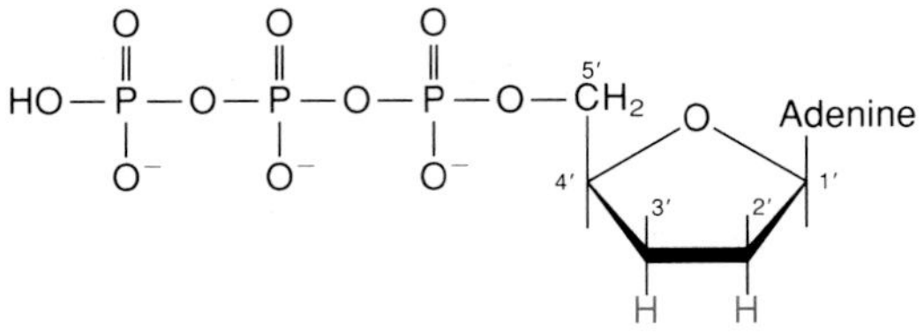

2′,3′-Dideoxyadenosine triphosphate

To generate a series of A-terminated fragments one runs the DNA polymerase reaction in the presence of equivalent concentrations of dATP, dCTP, dGTP, and dTTP, plus one-tenth equivalent of ddATP. When T is in the template strand, DNA polymerase occasionally inserts ddAMP instead of dAMP. When that happens, chain elongation stops and the fragment is released from the enzyme. Thus, a series of fragments of varying lengths accumulates, with each fragment identifying a T in the nucleotide sequence of the insert. Similarly, one identifies sites terminated by C, G, or T simply by running comparable polymerase reactions with the other three dideoxy analogs. Inclusion of a radioactive label in the polymerization mixture and gel electrophoresis followed by radioautography generate four sequencing ladders, comparable to those seen in Maxam–Gilbert sequencing, except that each ladder is absolutely base specific.

Recently introduced technology allows automated sequence determination by the Sanger method. The sequencing primer is derivatized at its 5′ end with a fluorescent dye that is red, blue, green, or yellow. Each of the four sequencing reactions is run with a differently colored primer, which imparts a characteristic fluorescence to all fragments terminated by A, T, G, or C, respectively. This allows for both sequence determination without radioisotopes and computer-assisted reading of the sequence gel and processing of the data.

REFERENCES

Church, G. M., and S. Kieffer-Higgins (1988) Multiplex DNA sequencing. *Science* 240:185–188. This article describes a variation of DNA sequencing methodology that enormously increases the amount of sequence data that can be obtained from one set of operations.

von Heinje, G. (1987) *Sequence Analysis in Molecular Biology: Treasure Trove or Trivial Pursuit.* Academic Press, New York. Uses and abuses of sequence data, with descriptions of how to access and use the sequence data banks.

TOOLS OF BIOCHEMISTRY 21

Southern Blotting

An important application of restriction nucleases is the detection and mapping of specific sequences in complex genomes. For example, you learned in Chapter 24 that λ phage DNA becomes integrated into the chromosome of the host bacterium during a lysogenic infection; tumor viruses also insert their genomes into chromosomes of infected animal cells. How do we know this? One way is by looking specifically for viral sequences in restriction digests of chromosomal DNA from infected cells.

Earlier we described the physical mapping of viral or plasmid DNAs by restriction enzyme cleavage, followed by electrophoretic separation of the fragments and their visualization with ethidium bromide. Any cellular genome, even from a bacterium, is so complex that such analysis would yield a "smear" consisting of a very large number of fragments, each present in minute amounts. However, because of the power of radiolabeling and nucleic acid reannealing techniques, one can detect specific DNA sequences within such a digest, even though they may be present at the level of one copy per genome, or as little as one-millionth of the total DNA present on a gel.

What is done is to radiolabel a sequence of interest to high specific activity and then use this labeled "probe" to search for complementary sequences among the fragments in a restriction digest. This is done by denaturing both the probe and the target restriction digest, reannealing, and then searching for radioactive duplex molecules. This is difficult to accomplish with DNA fragments embedded in an agarose gel, because of the fragility of the gel. However, in 1975 E. M. Southern devised a way to *transfer* the fragments from agarose to a sheet of nitrocellulose, following alkali denaturation of the fragments. As shown in Figure T21.1, the transfer is achieved by capillary action: the liquid associated with the agarose gel, in which the DNA fragments are dissolved, is literally blotted out, and the single-stranded DNA fragments in the denatured digest become irreversibly bound to the nitrocellulose sheet. For this reason, the technique is called **Southern blotting** or, more descriptively, **Southern transfer.** Alternatively, the transfer can be achieved electrophoretically.

The nitrocellulose sheet, now containing a replica of the original DNA gel, is incubated under annealing conditions with the radiolabeled probe, and DNA heteroduplexes are detected by radioautography after thorough washing of the nitrocellulose. A "prewash" treatment is also carried out to reduce nonspecific binding of the hybridization probe. In the example shown (Figure T21.2), radiolabeled phage λ DNA was used to probe an *Eco*RI digest of *E. coli* DNA lysogenic for λ. The fragments detected in the digest by this technique are identical to those seen by ethidium bromide staining of DNA from phage particles, except at the junctions between phage and bacterial DNA, where the *Eco*RI sites do not correspond with phage DNA ends. The increased molecular weights of these junction fragments can be understood only if the fragments contain *E. coli* sequences that flank a linearly inserted viral chromosome, the **prophage.** For illustrative purposes the hybridization pattern in Figure T21.2 is what would be seen if integration occurred at the ends of the linear phage DNA molecule (this does no occur; see Figure 25.13). Note that the terminal fragments are longer than their counterparts in DNA from phage particles, because the nearest *Eco*RI sites (X and Y) lie in neighboring bacterial DNA.

A variation of this technique, inevitably called **Northern blotting,** allows the detection of rare RNA molecules in a cell with similar sensitivity. Here one prepares total cell RNA and resolves different size classes electrophoretically, with subsequent transfer and probing with radiolabeled DNA, followed by radioautographic detection of DNA–RNA hybrid duplexes. The transfer technology is slightly different,

Figure T21.1
The principle of Southern blotting.

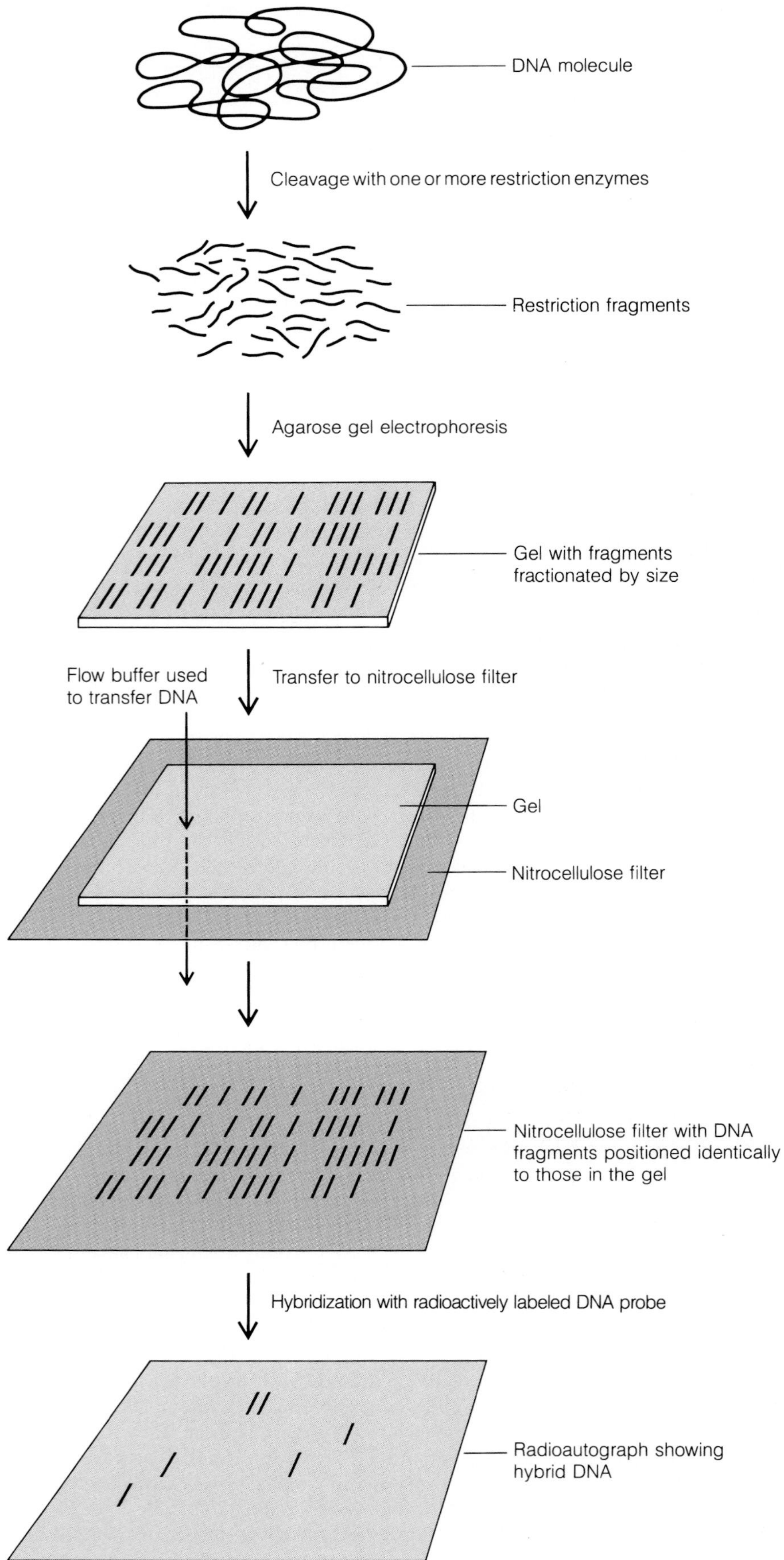

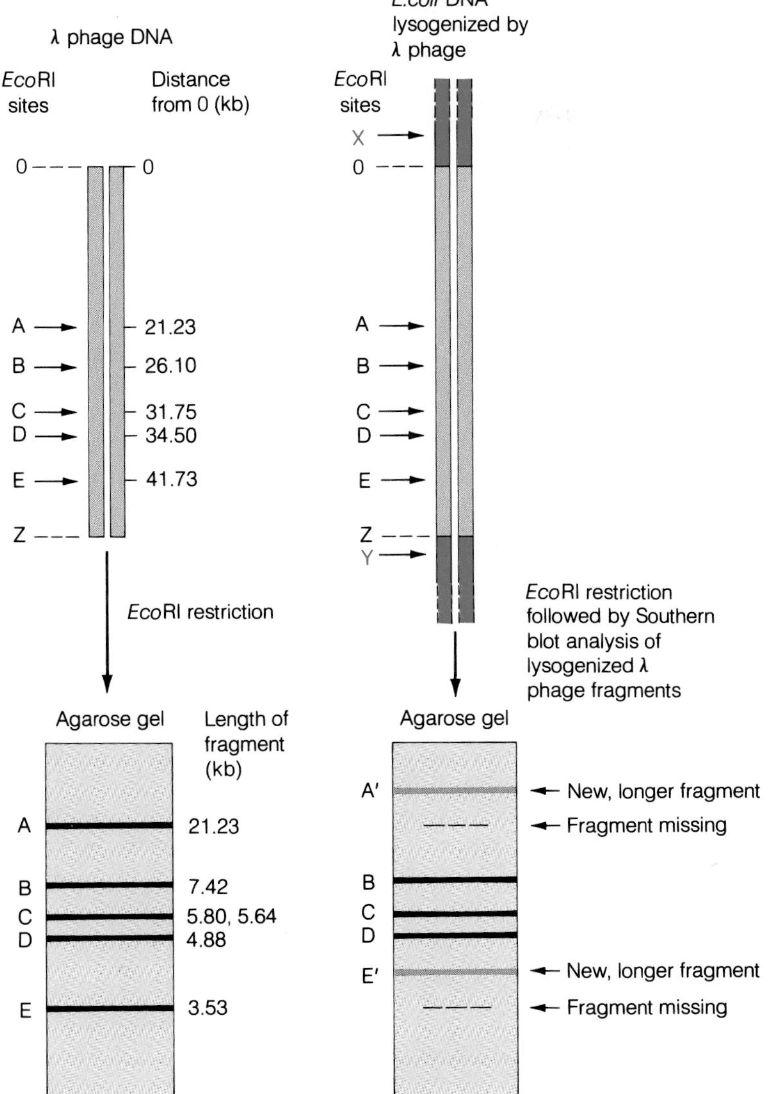

Figure T21.2
Application of Southern blotting to a demonstration that *E. coli* lysogenized with phage λ contains phage DNA integrated into the chromosome.

involving either covalent linkage of the RNA to diazobenzyloxymethyl (DBM) cellulose or transfer of formaldehyde-treated RNA to nitrocellulose.

Of the many applications of Southern transfer and hybridization, two will be mentioned here. First, Southern transfer is an invaluable adjunct to cloning, when one knows the amino acid sequence of at least part of the protein product of the target gene. From the sequence, one designs oligonucleotides that are expected to be homologous to part of the sequence. These oligonucleotides can then be chemically synthesized, radiolabeled, and used to screen a **library,** or large random collection of cloned DNA fragments from the organism of interest. The screening involves colony or plaque hybridization, as described in Tools of Biochemistry 19. Cloned segments homologous to the hybridization probe are then analyzed to see which ones contain a full-length copy of the target gene. Because this technique starts with a protein and works "backward" to isolate the structural gene for that protein, it is called **reverse genetics.**

A second application of Southern transfer is the use of **restriction fragment length polymorphisms (RFLPs)** to map the genes responsible for inherited diseases in humans. Many mutations either create a new restriction site in a genome or else destroy a previously existing site. If such a mutation occurs within or near a particular gene, the restriction fragmentation pattern will change, and this can be determined by Southern blotting and probing with a sequence that lies in the same region of the genome. For example, some cases of sickle-cell disease can be detected by an altered restriction pattern when human DNA is probed with part of the gene for the β-globin subunit. The sensitivity of the technique allows prenatal diagnosis of sickle-cell disease, using cells cultured from amniotic fluid surrounding the fetus. More recently, the technique is being used to map and isolate genes for hereditary illnesses where the responsible gene has not yet been identified, including cystic fibrosis, Huntington's chorea, susceptibility to some forms of cancer, and manic-depression.

Detection methods using nonradioactive probes are expected to find extensive applications in clinical diagnosis of infectious diseases, such as acquired immune deficiency syndrome (AIDS), which is caused by a retrovirus. The earliest developed AIDS test involves screening for antibody against a viral protein. However, it may take 4 months after infection for such an antibody to accumulate to detectable levels in the blood. Therefore, current efforts are focused on the development of rapid hybridization tests for the genome of the infecting virus, which should be detectable much earlier.

Finally, these techniques are finding use in criminology. By use of the polymerase chain reaction, it is now possible to isolate and analyze the DNA in a tissue sample as small as one human hair. Thus, a suspect leaving a hair or similarly small amount of tissue at the scene of a crime can be positively identified by an individual pattern of RFLPs—in short, a DNA "fingerprint."

REFERENCES

Martin, J. B. (1987) Molecular genetics: Applications to the clinical neurosciences. *Science* 238:765–772. This review describes RFLP analysis as a means of identifying human genes associated with specific diseases.

Marx, J. L. (1988) DNA fingerprinting takes the witness stand. *Science* 240:1616–1618. This news article describes the use of RFLP technology in forensic science.

22

TOOLS OF BIOCHEMISTRY 22

Site-Directed Mutagenesis

Analysis of the function of a protein involves altering the structure of the protein and then determining whether, and how, the biological functions of the protein are altered. Two methods have been used classically for generating modified proteins. One is to alter certain residues chemically by treatment with protein-modifying reagents. This approach lacks specificity, since all residues of a given amino acid may be modified, not just the one or two of special interest. Another approach is to mutagenize an organism with ultraviolet light, ionizing radiation, or chemical mutagens and then select for surviving organisms containing mutations that affect the protein of interest. The mutations can be identified by sequence analysis of the mutant gene or its protein product. The problem with this approach is the inability to target mutations to a given region of the gene, typically the part specifying the catalytic site of an enzyme molecule, or a regulatory region involved in DNA–protein interactions.

Once it became possible to clone the gene encoding a protein of interest, it was realized that one could systematically alter the gene at specific sites to generate virtually any desired mutation, a technique known as **site-directed mutagenesis.** Introduction of the mutant gene into a host cell, followed by its expression, could then yield the altered protein for study of the altered function. The earliest experiments involved the generation of deletions. For example, partial digestion of a cloned DNA with a restriction nuclease, followed by ligation, could yield mutant genes in which the region between two cleavage sites was deleted. Shorter deletions could be generated by opening at one cleavage site, followed by treatment with an exonuclease that would digest back from the two ends created by the cleavage. Such an enzyme is *Bal*31 exonuclease, which digests both strands starting from a cut site. The ends can be trimmed with S1 nuclease (an enzyme that cleaves only single-stranded DNA), followed by ligation under conditions that permit blunt DNA ends to join; often an oligonucleotide that contains a restriction site (e.g., *Bam*HI "linkers") is introduced at this stage. This approach can generate deletions of various lengths simply by varying the period of exonuclease treatment.

The most powerful and widely used method for site-directed mutagenesis allows the introduction of practically any mutation at any site, including single-base substitutions, short deletions, or insertions. The approach, illustrated in Figure T22.1, requires that the gene first be cloned into a single-stranded vector, such as phage M13. Next, one chemically synthesizes an oligonucleotide, about 20 nucleotides long, which is complementary in sequence to the cloned gene at the site of the desired mutation, *except in the center of the sequence.* Here the sequence contains one or two deliberate mistakes—either single nucleotides that do not base pair with the template or insertions or gaps of a few nucleotides. These alterations generate, on annealing to the cloned gene, either non-Watson–Crick base pairs or bases that have no partners, and create a "looping out" on annealing. The correctly matched bases on both sides of the mismatch cause it to remain annealed, despite the mismatch. One then uses DNA polymerase to synthesize around the circular vector from this primer, followed by ligation to created a closed circular duplex. After introduction of this duplex into bacteria, both strands replicate and yield phage. In principle, 50% of the phage should contain the desired mutation. In practice, that fraction is considerably less, but it can be increased by various techniques. One method utilizes a template containing extensive substitution of uracil for thymine; this is done by growing the M13 phage clone in an *E. coli* strain lacking dUTPase and uracil-DNA *N*-glycosylase (Chapter 24). Then, after DNA synthesis in vitro, the DNA is introduced into wild-type bacteria, where the action of uracil-DNA *N*-glycosylase selectively cleaves the template strand and leaves the desired mutant strand intact. If the investigator wants to produce a number of mutants with different amino acid substitutions at the same site, he or she can use a mixture of oligonucleotide primers, differing from each other only within the codon to be altered.

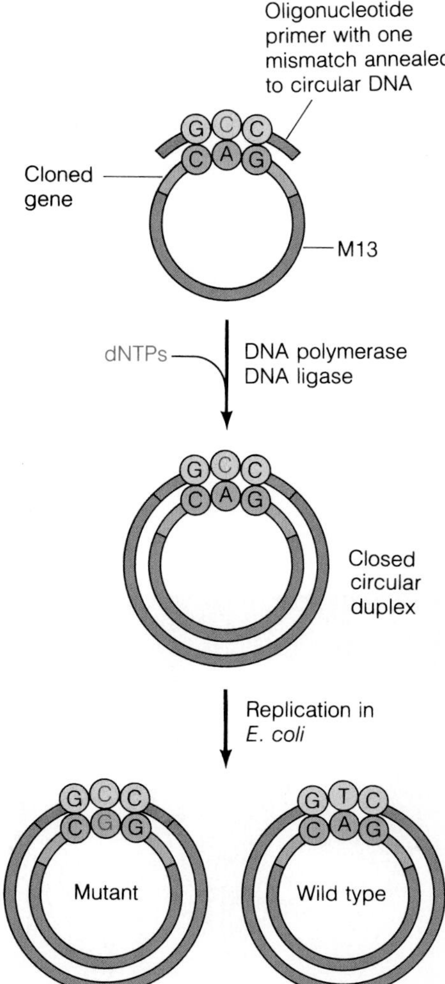

Figure T22.1
Use of a mismatched synthetic oligonucleotide primer to introduce mutations into a gene cloned into a single-stranded vector.

REFERENCES

Oxender, D. L., and C. F. Fox, eds. (1987) *Protein Engineering.* Alan R. Liss, New York. This multiauthored book describes several approaches to site-directed mutagenesis, as well as studies of deliberately altered proteins that help one to predict the types of mutations to use to generate proteins of desired properties.

CHAPTER 26

Information Transfer: Transcription

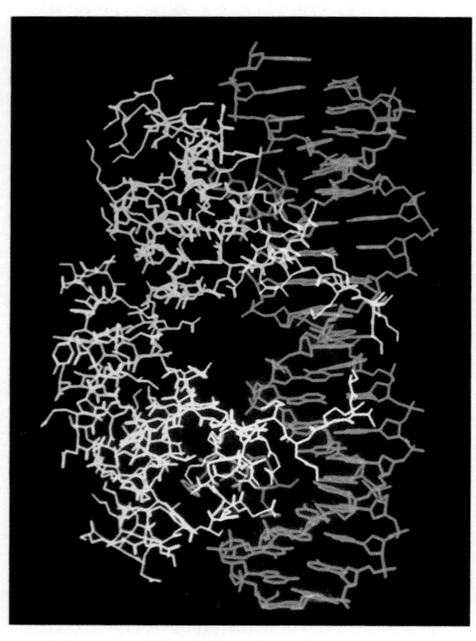

Figure 26.1
A DNA–protein complex involved in transcription. Transcription involves interactions of proteins with specific sites on DNA molecules. Shown here is a transcriptional regulatory protein, the *c*I repressor of bacteriophage λ, bound at an operator site. DNA is blue, and the two subunits of the dimeric repressor are white and gold, respectively. The "recognition helix" (p. 937) on each repressor monomer is shown in red.

We turn now to transcription, in which information stored in the nucleotide sequence of DNA is read out, by the template-dependent synthesis of a polyribonucleotide. From a mechanistic standpoint transcription is quite similar to DNA replication, particularly in the use of nucleoside triphosphate substrates and the template-directed growth of nucleic acid chains in a 5′→3′ direction. Two major differences are (1) because of the constraints of the genetic code, with few known exceptions only one DNA template strand is transcribed; and (2) only a small fraction of the entire genetic potential of an organism is realized in one cell. In a differentiated eukaryotic cell as little as 1% of the total DNA may be transcribed. Even in single-celled organisms, where virtually all of the DNA sequences can be transcribed, far fewer than half of all genes may be transcribed at any time. Therefore, much of what concerns us as we study transcription involves the mechanisms used to select particular genes and template strands for transcription, because this in large part governs the metabolic capabilities of a cell. Those mechanisms operate largely at the levels of initiation and termination of transcription, in large part through the operation of proteins that interact with DNA in a highly site-specific manner (Figure 26.1).

Our earliest insights into transcription developed from experiments with bacteria and their viruses. Our knowledge of prokaryotic transcription is, therefore, much deeper than our comprehension of the much more complex multicellular organisms. It has been estimated that our understanding of prokaryotic gene expression is about 30% complete, and that of eukaryotes is less than 0.01% complete. Therefore, this chapter focuses on well-understood prokaryotic systems, emphasizing where they have served as models for understanding eukaryotic transcription. The latter topic is covered more thoroughly in Chapter 28.

DNA as the Template for RNA Synthesis

The concept that RNAs are generated by template-directed copying of DNA base sequences is so firmly ingrained that we tend not to think about the critical experiments that led to our current understanding. This topic merits discussion, because it involves a fascinating intellectual history, with many participants and some brilliant deductive reasoning. Even some of the false starts yielded important contributions. Also, the experimental systems used in the early work are some of those, like the lactose operon and bacteriophages T4 and λ, which continue to be useful.

The involvement of RNA in information transfer was suspected from the time DNA was first identified as the genetic storehouse—both from the chemical similarity of RNA to DNA and from the knowledge that proteins are synthesized on ribosomes. The latter fact meant that in eukaryotic cells information must somehow be transferred from the nucleus, where information is stored in DNA, to the cytosol, where most of the ribosomes reside. However, the nature of that information transfer was not revealed until the existence of messenger RNA had been predicted and demonstrated. This demonstration was far from straightforward, because messenger RNA constitutes such a small proportion of total cellular RNA (1 to 3% in bacteria) that in ordinary fractionations its presence is masked by the much more abundant ribosomal and transfer RNAs.

The Predicted Existence of Messenger RNA

Until about 1960 it was thought that ribosomal RNA (rRNA) represented the set of templates for protein synthesis. Jacques Monod and François Jacob, at France's Pasteur Institute, questioned this idea, in part because rRNAs are homogeneous in size (5S, 16S, and 23S in bacteria, as we learned in Chapter 4), while the molecular weights of proteins vary over at least two orders of magnitude. Jacob and Monod predicted the existence of messenger RNA (mRNA) from their analyses of *E. coli* mutants that are altered in the control of lactose metabolism. We shall discuss these mutants later in this chapter, but the following information is relevant at this point.

Lactose utilization is controlled by three enzymes, whose genes are adjacent on the chromosome. One of the enzymes is **β-galactosidase,** which hydrolyzes lactose and other β-galactosides. When bacteria are grown in a medium containing simple salts and glucose as the sole carbon source, the levels of the lactose-utilizing enzymes are very low, with less than one molecule of β-galactosidase per cell. However, substitution of lactose or a related β-galactoside for glucose in the medium leads to rapid enzyme induction, or synthesis of the three enzymes (discussed also in Chapter 11). The β-galactosidase ultimately represents as much as 6% of the total soluble protein of the cell. Removal of lactose from the culture slows the further synthesis of enzyme molecules. The rapid synthesis and decay of β-galactosidase-forming capacity suggested that the template for synthesizing this enzyme is metabolically unstable—synthesized very fast on demand and degraded when a continued stimulus to induction is absent. Since rRNAs are stable, these species were unlikely to be intermediates in information transfer.

Jacob and Monod analyzed numerous *E. coli* mutants that displayed faulty control over induction of the lactose-utilizing enzymes. Some expressed all three genes at high levels even when lactose or a similar inducer was absent, and others could not induce any of the enzymes, even after addition of lactose. These experiments led to the notion of a repressor protein that acted to regulate the level of a messenger RNA and also responded to the presence of lactose.

The existence of similar repressor and mRNA molecules was also predicted by Monod's colleague, Andre Lwoff, who was analyzing mutations of bacteriophage λ that affect the ability of the phage to lysogenize the host (integration into the chromosome) or to become **induced** (excision from the chromosome followed by replication; see Chapter 25). Some mutants could

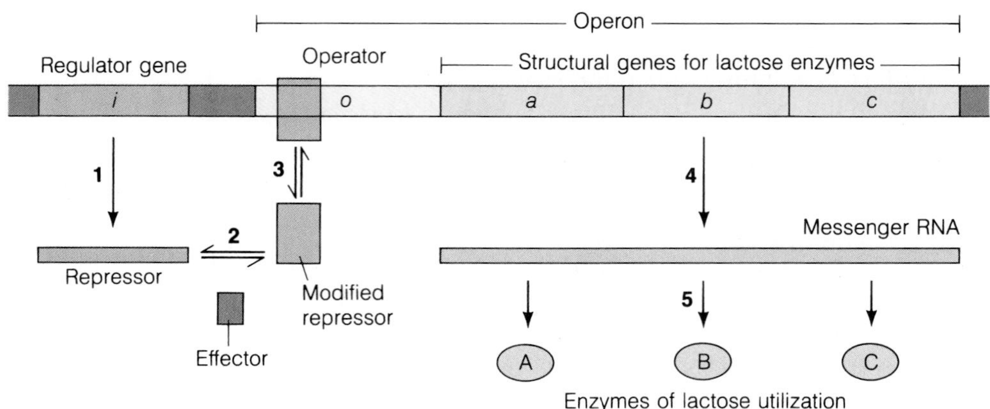

Figure 26.2
An early formulation of the operon model, as proposed by Jacob and Monod in 1961. (1) Regulator gene encodes a repressor molecule. (2) Small molecule effector alters equilibrium between conformational states of repressor. (3) Modified repressor binds to the operator, blocking transcription of the structural genes. (4) Enzyme induction yields messenger RNA, an RNA copy of the structural genes. (5) The mRNA sequence is translated into protein sequences—in this case the enzymes involved in lactose utilization.

not lysogenize, while others could not be induced. Generalizing from the phenotypes of the lactose utilization mutants and the phage λ mutants, Jacob and Monod in 1961 proposed a unifying hypothesis of gene regulation. They proposed that transcription involved copying of a DNA strand to give an RNA of complementary sequence. They proposed further that this process is controlled at the initiation stage. Hypothetical regulatory elements called **repressors** and **operators** controlled the synthesis of other hypothetical entities called **messenger RNAs** (a term they coined). In the lactose system messenger RNA was postulated to be a complementary copy of the DNA that encompassed the genes encoding the three enzymes of lactose utilization, as schematized in Figure 26.2. A set of contiguous genes whose expression is controlled by the same set of regulatory elements was termed an **operon,** and the Jacob–Monod hypothesis thus came to be known as the **operon model.**

Jacob and Monod postulated several properties for the predicted messenger RNA. First, a high rate of synthesis followed by rapid degradation would explain the rapid turn-on of the genes after induction and turn-off after removal of the inducer. Further, because of these two properties, mRNA was expected to accumulate rapidly but not to high steady-state levels. Since the messenger was seen as a copy of two or more contiguous genes, it was postulated to be fairly large and part of a heterogeneous size class of RNA. Finally, since the messenger RNA was seen as a complementary copy of DNA, its nucleotide sequence should be identical to that of one of the template DNA strands.

T2 Bacteriophage and the Demonstration of Messenger RNA

The first physical demonstration of mRNA came from work with T2 and T4 bacteriophages. Infection by these large viruses arrests all expression of host cell genes, and no accumulation of RNA can be detected after infection. However, in 1956 the use of radioisotopes led to detection of a "DNA-like RNA" in T2-infected *E. coli.* When infected cultures were pulse-labeled with [^{32}P]orthophosphate for 3 or 4 minutes, about 2% of the total RNA became radioactive. This T2 phage RNA had two properties that led to its eventual identification as viral messenger RNA. First, it was metabolically labile; in pulse-chase experiments, radioactivity was rapidly lost from T2 RNA. Second, it seemed to be a product of viral DNA metabolism, because its nucleotide composition was close to that of T2 DNA—rich in adenine and uracil and low in guanine and cytosine (T2 DNA contains about 67% A + T).

Sol Spiegelman obtained important additional evidence. First, sucrose gradient centrifugation of pulse-labeled RNA showed that the labeled material sediments heterogeneously and distinctly from any of the known rRNA or tRNA species (Figure 26.3). Second, Spiegelman established that this RNA is a viral gene product when he carried out the first DNA–RNA hybridization experiment, which showed that the labeled RNA was complementary in sequence to phage DNA. The earliest experiments were based on the fact that RNA is of higher density than DNA. Thus, a DNA–RNA hybrid could be detected in equilibrium gradient centrifugation as a species of intermediate density containing label derived from both DNA and RNA. Such a hybrid was formed when T2 RNA was heated and slowly cooled along with T2 DNA, but not when the DNA came from *E. coli.*

Additional support for the existence of phage messenger RNA came from a density-shift experiment carried out by François Jacob, Matthew Meselson, and Sydney Brenner. *E. coli* was grown in dense medium (^{13}C–^{15}N) and infected with T2 in light medium. After pulse labeling with radioactive amino acids, density analysis on CsCl gradients showed that phage proteins were synthesized on dense ribosomes, that is, ribosomes synthesized before infection. The fact that only phage proteins were being made ruled out any possibility that ribosomal RNA provided a template. Instead, this experiment pictured the ribosome as a nonspecific workbench, a particle on which any protein could be assembled, depending on which messenger template became associated with that workbench.

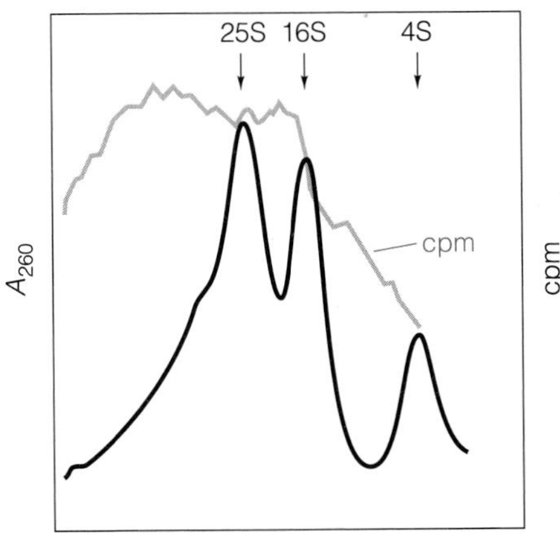

Figure 26.3
Sedimentation profile of total and pulse-labeled RNAs in T2 phage-infected *E. coli.* Total concentration of RNA in each fraction was determined by ultraviolet absorbance (A_{260}). The radioactivity profile (orange) shows the distribution of species synthesized during the pulse. Data are from K. Asano, *J. Mol. Biol.* 14:71–84 (1965).

RNA Dynamics in Uninfected Cells

What about uninfected bacteria? Spiegelman showed that pulse-labeled RNA from uninfected *E. coli* hybridized to *E. coli* DNA. At very short labeling intervals the sedimentation pattern showed incorporation into both rRNA and tRNA species *and* a heterogeneously sedimenting species. After a chase (Figure 26.4), the radioactivity profile followed the absorbance profile, showing that all RNA species were labeled to equivalent specific activities. This is consistent with the postulated short lifetime of messenger RNA ($t_{1/2}$ of 2 or 3 minutes). mRNA would reach its maximal radioactivity within just a few minutes, but during the chase mRNA turnover would release nucleotides that could flow into stable RNA species. Since these species do not turn over, label accumulates, and the fraction of total label in the stable RNA species continues to increase. Consistent with this idea, Spiegelman also showed that highly labeled ribosomal and transfer RNAs

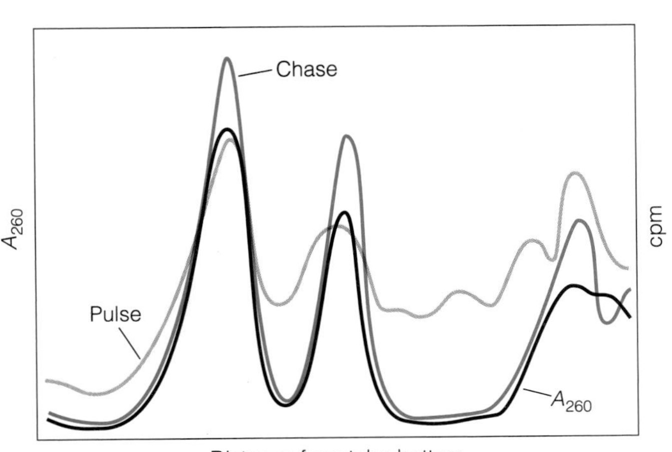

Figure 26.4
Pulse-labeled RNA species in uninfected bacteria and their fate in a chase. Data from M. Hayashi and S. Spiegelman, *Proc. Natl. Acad. Sci. U.S.A.* 47:1569 (1961). Orange line: cells labeled for 3 minutes; blue line: an identical culture, with the label chased by 0.7 generations of growth in nonradioactive medium after the pulse. Black line: A_{260} profile.

hybridize to *E. coli* DNA, demonstrating that all three major classes of RNA are synthesized from template DNA strands.

As noted earlier, the earliest DNA–RNA hybridization experiments involved equilibrium gradient centrifugation, a time-consuming and expensive technique. Spiegelman and his colleagues made the important discovery that single-stranded DNA binds irreversibly to membrane filters, such as nitrocellulose. This technique allowed rapid hybridization analysis of a large number of samples, because a radiolabeled RNA could hybridize to denatured DNA immobilized on a filter. After suitable treatment and washing of the filter, the amount of hybridization could be determined simply by placing the filter in a liquid scintillation counter and counting its radioactivity. The same principle—immobilization of nucleic acid on nitrocellulose followed by analysis of bound radioactivity—underlies Southern blotting and Northern blotting (described in Tools of Biochemistry 21), which are now much more widely used to analyze gene organization and expression.

The final chapter in the messenger RNA saga was the demonstration that isolated mRNAs have template activity in vitro; that is, they can program the synthesis of specific protein molecules in the presence of 20 amino acids and other factors. This will be presented in Chapters 27 and 28.

Enzymology of RNA Synthesis: RNA Polymerase

The earliest description of an enzyme capable of synthesizing RNA in vitro occurred in the late 1950s, at about the same time as the discovery of DNA polymerase I. This RNA-synthesizing enzyme, called **polynucleotide phosphorylase,** was quite different from DNA polymerase. The enzyme required no template, and it used ribonucleoside *diphosphates* (rNDPs) as substrates to produce a polynucleotide whose base composition matched the nucleotide composition of the reaction medium.

$$n\ \text{rNDP} \rightleftharpoons (\text{rNMP})_n + nP_i$$

It was thought that polynucleotide phosphorylase might be the major RNA-synthesizing enzyme, but the lack of a template requirement was troubling, as was the apparent absence of the enzyme from eukaryotic cells. Ultimately, polynucleotide phosphorylase turned out not to play a role in RNA synthesis but instead to participate in the *degradation* of bacterial messenger RNAs. However, the enzyme was of great value in the synthesis of polynucleotides used as templates for in vitro protein synthesis, when the genetic code was being elucidated (Chapter 27).

Investigators continued to search for an enzyme that would copy a DNA template in vitro. Such an enzyme was discovered almost simultaneously in four different laboratories. The enzyme, **DNA-directed RNA polymerase,** resembled DNA polymerase in the nature of the reaction catalyzed.

$$n\left\{\begin{matrix}\text{ATP}\\+\\\text{CTP}\\+\\\text{GTP}\\+\\\text{UTP}\end{matrix}\right. \xrightleftharpoons{Mg^{2+},\ \text{DNA}} \begin{bmatrix}\text{AMP}\\ \\\text{CMP}\\ \\\text{GMP}\\ \\\text{UMP}\end{bmatrix}_n + nPP_i$$

Biological Role of RNA Polymerase

RNA polymerase catalyzes the synthesis of all three *E. coli* RNA classes—mRNA, rRNA, and tRNA. This was shown in the following way. An antibiotic called **rifampicin** (Figure 26.5a) inhibits the synthesis of mRNA, rRNA, and tRNA in vivo. Rifampicin is an inhibitor of isolated RNA polymerase. When rifampicin-resistant mutants of *E. coli* were prepared, they were found both to contain a rifampicin-resistant form of RNA polymerase and to be capable of synthesizing all three RNA classes in vivo in the presence of rifampicin. Since a single mutation affected both the RNA polymerase and the rifampicin sensitivity of RNA synthesis in vivo, RNA polymerase had to be the one enzyme catalyzing all forms of transcription.

The situation is different in eukaryotes, which contain three distinct RNA polymerases, one each for synthesis of rRNA, mRNA, and small RNAs (tRNA plus the 5S species of rRNA). The enzymes are called RNA polymerases I, II, and III, respectively. The existence of separate enzymes was revealed partly because they differ in their sensitivity to inhibition by **α-amanitin** (Figure 26.5b), a toxin from the poisonous *Amanita* mushroom. RNA polymerase II is inhibited at low concentrations, RNA polymerase III is inhibited at high concentrations, and RNA polymerase I is quite resistant. It is thus possible to assay the three polymerases separately in a crude extract, without separating them first.

Figure 26.5 shows the structures of two additional inhibitors. **Cordycepin,** which is 3′-deoxyadenosine, is a transcription chain terminator because it lacks a 3′ hydroxyl group from which to extend. The nucleotide of cordycepin is incorporated into growing chains, confirming that transcriptional chain growth occurs in a 5′→3′ direction. Another important inhibitor is **actinomycin D,** which acts by binding to DNA. The tricyclic ring system (phenoxazone) intercalates between adjacent GC base pairs, and the cyclic polypeptide arms fill the nearby narrow groove.

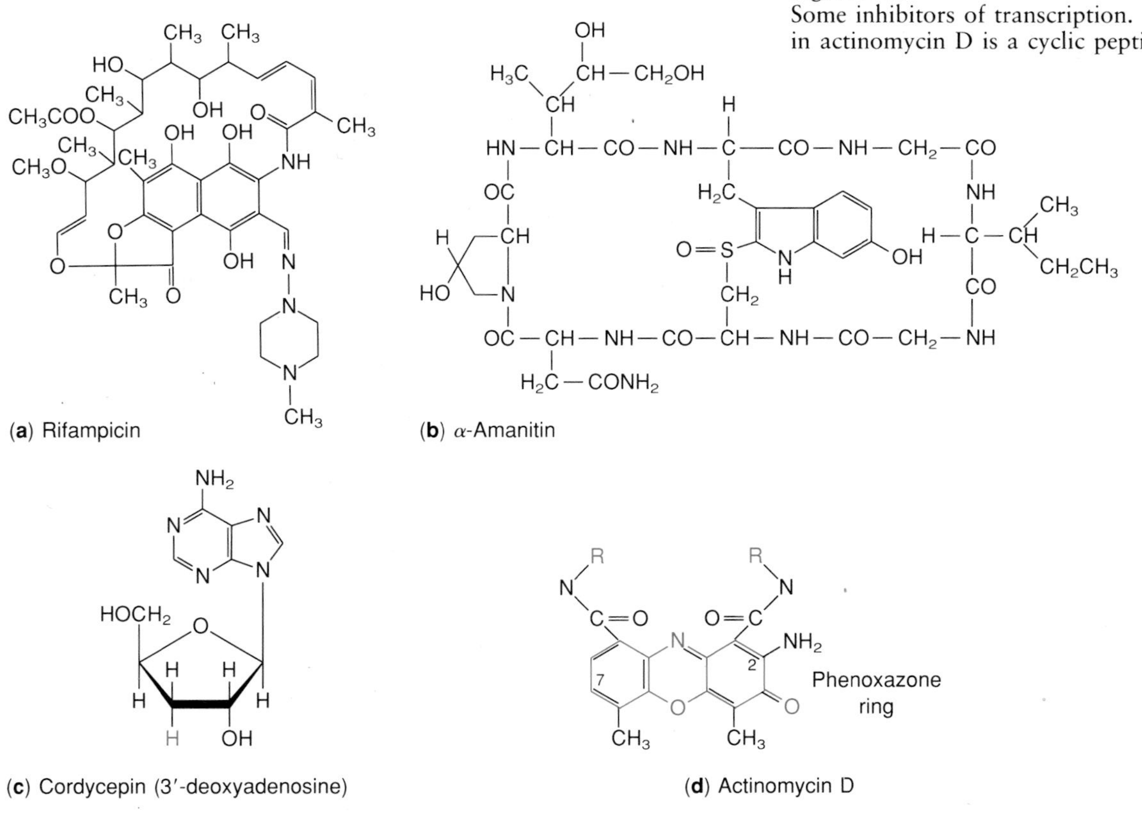

Figure 26.5
Some inhibitors of transcription. Each R in actinomycin D is a cyclic peptide.

(**a**) Rifampicin

(**b**) α-Amanitin

(**c**) Cordycepin (3′-deoxyadenosine)

(**d**) Actinomycin D

Since DNA polymerases and RNA polymerases catalyze reactions that are nearly identical chemically, it is interesting to compare some of their kinetic features. V_{max} for DNA polymerase III holoenzyme, at about 500–1000 nucleotides per second, is much higher than the chain growth rate for bacterial transcription—50 nucleotides per second, which is the same as V_{max} for purified RNA polymerase. While there are only about 10 molecules of DNA polymerase III per *E. coli* cell, there are some 3000 molecules of RNA polymerase, of which half might be involved in transcription at any instant. This fits in with observations that replicative DNA chain growth is rapid but occurs at few sites, while transcription is much slower but occurs at many sites. The result is that far more RNA than DNA accumulates in the cell. Like the DNA polymerase III holoenzyme, the action of RNA polymerase is highly processive. Once transcription of a gene has been initiated, RNA polymerase rarely, if ever, dissociates from the template until the specific signal to terminate has been reached.

Another important difference between DNA and RNA polymerases is the accuracy with which a template is copied. With an error rate of about 10^{-5}, RNA polymerase is far less accurate than replicative DNA polymerase holoenzymes. This is largely due to absence of exonucleolytic proofreading in RNA polymerase. Since RNA does not carry information from one cell generation to the next, an ultrahigh-fidelity template-copying mechanism is evidently not needed.

Structure of RNA Polymerase

When highly purified *E. coli* RNA polymerase is analyzed in denaturing electrophoretic gels, one observes five distinct polypeptide subunits, whose properties are summarized in Table 26.1. Two copies of the α subunit are present, along with one each of β, β', σ, and ω, giving M_r for the holoenzyme of about 450,000. The role of ω is not known, nor is it yet clear that it is an integral part of the holoenzyme. However, much has been learned about the functions of the other subunits, partly from reconstitution studies, in which the dissociated subunits are renatured and allowed to reassociate, with formation of active enzyme. Since ω is not required for reconstitution of active enzyme, we shall not include it in our subsequent discussions.

A useful approach to analyzing the function of each RNA polymerase subunit is **mixed reconstitution.** Here one recombines subunits isolated from two or more different RNA polymerases. For example, when β from a rifampicin-resistant form of RNA polymerase is recombined with α, β', and σ from wild-type cells, the reconstituted enzyme is rifampicin-resistant. This establishes β as the target for rifampicin inhibition, and since rifampicin is known to inhibit the initiation of transcription, β also must play a role in initiation. The fact that β is also the target for inhibition by an inhibitor of elongation called **streptolydigin** points to β as the subunit with the catalytic site for chain elongation.

Table 26.1
Subunit composition of *E. coli* RNA polymerase

Subunit	M_r	Number per Enzyme Molecule	Function
α	36,500	2	Chain initiation
β	151,000	1	Chain initiation and elongation
β'	155,000	1	DNA binding
σ	70,000	1	Promoter recognition
ω	11,000	1	Unknown

The σ subunit is easily dissociated from RNA polymerase—for example, by passing the purified enzyme through a carboxymethylcellulose column. The σ-free enzyme, called **core polymerase,** is still catalytically active. However, it binds to DNA at far more sites than does the RNA polymerase holoenzyme, and it shows no strand specificity. σ plays an important role in directing RNA polymerase to bind to template at the proper site for initiation—the **promoter** site—and to select the correct strand for transcription. The addition of σ to core polymerase reduces the affinity of the enzyme for nonpromoter sites by about 10^4, thereby increasing its specificity for binding to promoters.

These discoveries about σ, which were made in the early 1970s, suggested that gene expression might be regulated by having core polymerase interact with different forms of σ, which would in turn direct the holoenzyme to different promoters. In some instances this does occur. For example, when *Bacillus subtilis* is induced to sporulate (that is, to generate metabolically inert cells capable of later outgrowth), a new form of σ is produced. This form combines with core polymerase to redirect cellular metabolism for transcription of the genes involved in sporulation. Another example is apparent when an *E. coli* culture is stressed by a sudden temperature increase. In this case a new form of σ appears and directs the modified RNA polymerase to a different set of promoters, thereby activating transcription of a block of genes called heat-shock genes.

RNA polymerases from different prokaryotic sources are remarkably similar in subunit size and composition. In many instances mixed reconstitution can be observed, with active enzyme formed from subunits of the RNA polymerases of widely divergent prokaryotic species.

Although the multisubunit motif for RNA polymerases is the dominant structural theme, it is not universal. The best-known exception is RNA polymerase specified by bacteriophage T7. The genome of this virus was presented in Chapter 24 (Figure 24.18). The left-hand 20% of the T7 genome (through gene 1.3; about 7 kb) is transcribed early in infection by *E. coli* RNA polymerase. One of these early gene products (the product of gene 1) is a virus-specified RNA polymerase. This is a single-subunit enzyme (M_r = 98,000) that responds to different DNA control sequences and is responsible for all T7 transcription late in infection. These late transcriptional events initiate at promoters on the right-hand 80% of the DNA molecule.

Mechanism of Transcription

Like DNA replication and protein synthesis, transcription occurs in three distinct phases—initiation, elongation, and termination. Initiation and termination signals in the DNA sequence punctuate the genetic message by directing RNA polymerase to specific genes and by specifying where transcription will start, where it will stop, and which strand to transcribe. The signals involve both instructions encoded in DNA base sequences and interactions between DNA and proteins other than RNA polymerase. Most of our discussion will focus on prokaryotic RNA polymerases, exemplified by the widely studied *E. coli* enzyme, but the basic mechanics of transcription are similar in all organisms.

Initiation of Transcription: Interactions with Promoters

The overall process of initiation and elongation is summarized in Figure 26.6. The first step in transcription is binding of RNA polymerase to an

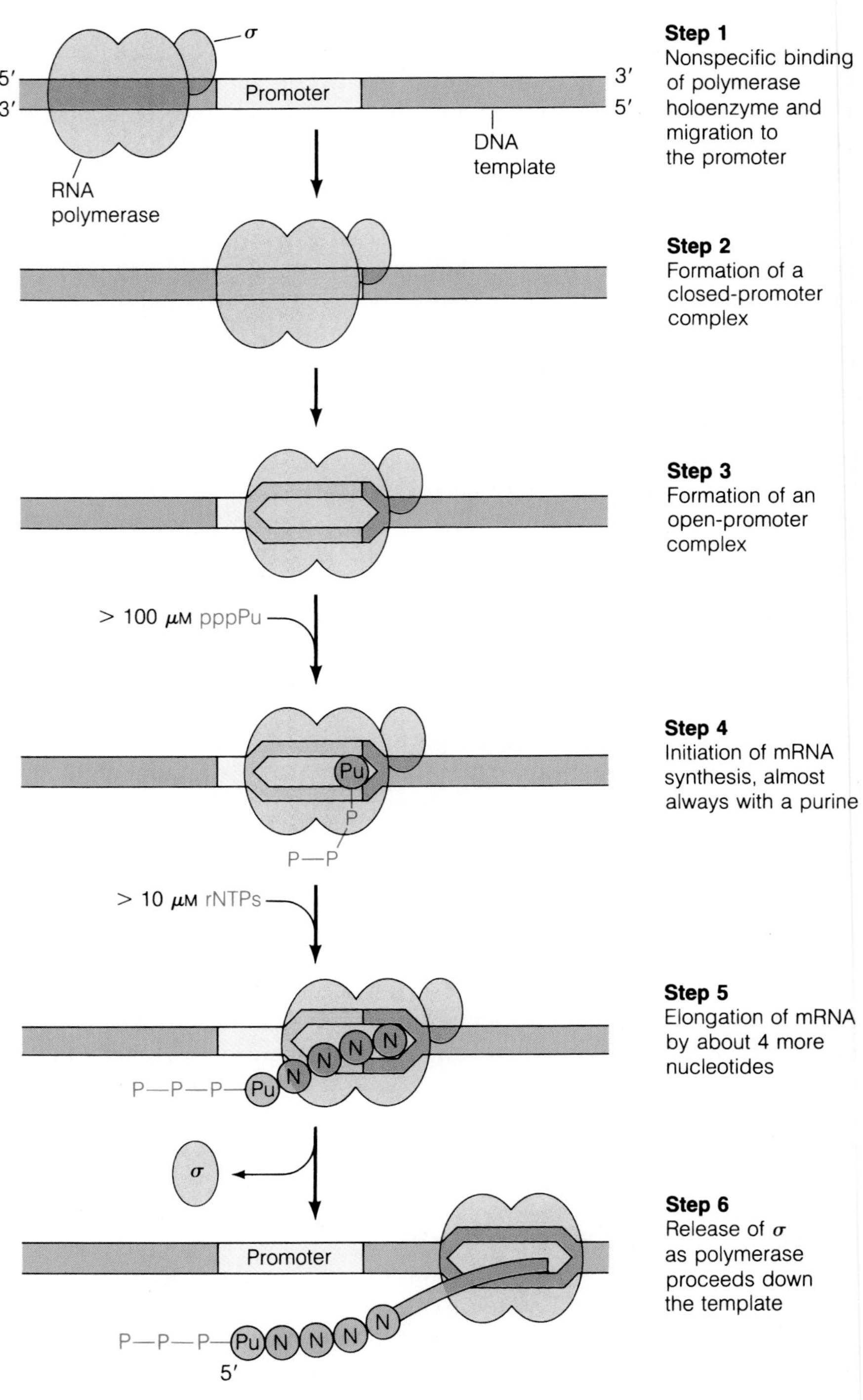

Figure 26.6
Steps in initiation and elongation of transcription by bacterial RNA polymerase.

initiation DNA site, the promoter. In Tools of Biochemistry 23 we describe **footprinting,** a technique that is used to determine promoter structure. For now let us concentrate on the mechanistic details of synthesizing polyribonucleotide chains.

RNA polymerase finds promoters by a search process, in which the holoenzyme binds nonspecifically to DNA, with low affinity, and then in some fashion migrates to another site, "searching" for a promoter, to which it binds with much higher affinity (Figure 26.6, step 1). σ factor is essential for this search, because, as noted earlier, the core enzyme does not bind to promoters more tightly than to nonpromoter sites. The search may involve movement between distant sites that are joined by a loop. Kinetic analysis suggests that a random search process, with multiple enzyme–DNA encounters, cannot account for the speed with which RNA polymerase finds promoters. By contrast, binding to DNA and then somehow moving along it reduces the complexity of the search from three dimensions to one.

The initial encounter between RNA polymerase holoenzyme and a promoter generates a **closed-promoter complex** (step 2). Whereas DNA strands unwind later in transcription, no unwinding is detectable in a closed-promoter complex. This complex forms with a K_a between 10^6 and 10^9 M^{-1} at 0.1 M NaCl. Binding is primarily electrostatic, for K_a is dependent on ionic strength. The complex is relatively labile, for it dissociates with a half-life of about 10 seconds.

Next, in a temperature-dependent reaction with a highly variable half-life (about 15 seconds to 20 minutes, depending on the promoter), RNA polymerase unwinds about 12 base pairs of DNA (from -9 to $+3$, where $+1$ represents the first nucleotide in the RNA molecule to be synthesized). A very stable complex forms at this site of unwinding. Because of the strand unwinding, this is called an **open-promoter complex** (step 3). This complex is much less susceptible to disruption by salt than is the closed-promoter complex. K_a can be as high as 10^{14}, and the complex has a half-life of 20 to 40 hours. This complex must form before transcription can begin.

Initiation and Elongation: Incorporation of Ribonucleotides

Having located a promoter and formed an open-promoter complex, the enzyme is ready to begin the synthesis of an RNA chain. RNA polymerase contains two binding sites for ribonucleoside triphosphates. One, which is used during elongation, binds any of the four common ribonucleoside triphosphates (rNTPs), with a half-saturating concentration of about 10 μM. The second site, used for initiation, binds ATP and GTP preferentially, with half-saturating concentrations of about 100 μM. Thus, most mRNAs have a purine at the 5′ end. Chain growth begins with binding of the template-specified rNTP at this site (step 4), followed by binding of the next nucleotide at the other site. Next, nucleophilic attack by the 3′ hydroxyl of the first nucleotide on the inner phosphorus of the second nucleotide generates the first phosphodiester bond and leaves an intact triphosphate moiety at the 5′ position of the first nucleotide. The σ subunit dissociates from the core enzyme soon after these events, and the core enzyme catalyzes virtually all of the phosphodiester bond formations. At this point transcription, as studied in vitro, can no longer be inhibited by adding rifampicin.

During *elongation* (steps 5 and 6) the core enzyme moves along the duplex DNA template, simultaneously unwinding the template to expose a single-stranded template for base pairing with the nascent transcript (the most recently synthesized RNA) and rewinding the template behind the 3′ end of the growing RNA chain. The amount of unwinding remains constant, at about 17 nucleotide pairs unwound per growing RNA chain. Unwinding occurs slightly ahead of the 3′ end of the growing RNA chain, for the nascent transcript is base paired to the template strand over about 12 base pairs, and rewinding of the parental duplex dissociates the transcript from its template.

Eventually RNA polymerase reaches a termination signal encoded in DNA. Let us postpone a consideration of the mechanisms involved until we have discussed promoter recognition in more detail.

Punctuation of Transcription: Promoter Recognition

Promoter recognition is a crucial step in transcription, from the standpoint of regulation as well as mechanism. In *E. coli* it is estimated that the most frequently transcribed genes initiate transcription about once every 10 seconds, while some genes are transcribed as infrequently as once per generation (30 to 60 minutes). Promoter recognition is a rate-limiting step for transcription. Since all genes in bacteria are transcribed by the same enzyme, variations in promoter structure must be largely responsible for variations in the frequency of initiation, which, as just observed, extend over several orders of magnitude.

What structural features in DNA direct RNA polymerase to bind at a promoter site and to form an open-promoter complex? The first hint of an answer came in 1975, when David Pribnow and Heinz Schaller independently examined the limited DNA sequence data available and revealed that each gene transcribed in *E. coli* shared a short adenine- and thymine-rich sequence, centered about 10 nucleotides to the 5′ side of the transcriptional start site (Figure 26.7). (Transcriptional start points are identified as described in Tools of Biochemistry 24.) There was some variation among the promoters analyzed, but a **consensus sequence** emerged within this conserved region. A consensus sequence comprises the bases that appear most frequently at each position when one compares a series of sequences thought to have a common function. Among the different initiation sequences analyzed in *E. coli,* that consensus sequence was TATAAT on the **antisense strand.** The antisense strand is the nontranscribed DNA strand; since it is complementary in sequence to the sense, or template, strand, it is identical in base sequence to the RNA product, within the region transcribed, but with T instead of U. Later, another region of conserved nucleotide sequence was found centered at nucleotide −35, with a consensus sequence of TTGACA.

The two conserved sequences are called the −35 region and the −10 region (the latter is also called the Pribnow box). No known natural pro-

Figure 26.7
Conserved sequences in promoters recognized by *E. coli* RNA polymerase and lengths of spacer sequences.

Promoter		−35 Region	Spacer	−10 Region	Spacer	RNA start
trp operon	G	TTGACA	N_{17}	TTAACT	N_7	A
$tRNA^{tyr}$	C	TTTACA	N_{16}	TATGAT	N_7	A
λP2	G	TTGACA	N_{17}	GATACT	N_6	G
lac operon	C	TTTACA	N_{17}	TATGTT	N_6	A
recA	C	TTGATA	N_{16}	TATAAT	N_7	A
lexA	G	TTCCAA	N_{17}	TATACT	N_6	A
T7A3	G	TTGACA	N_{17}	TACGAT	N_7	A
CONSENSUS		TTGACA		TATAAT		

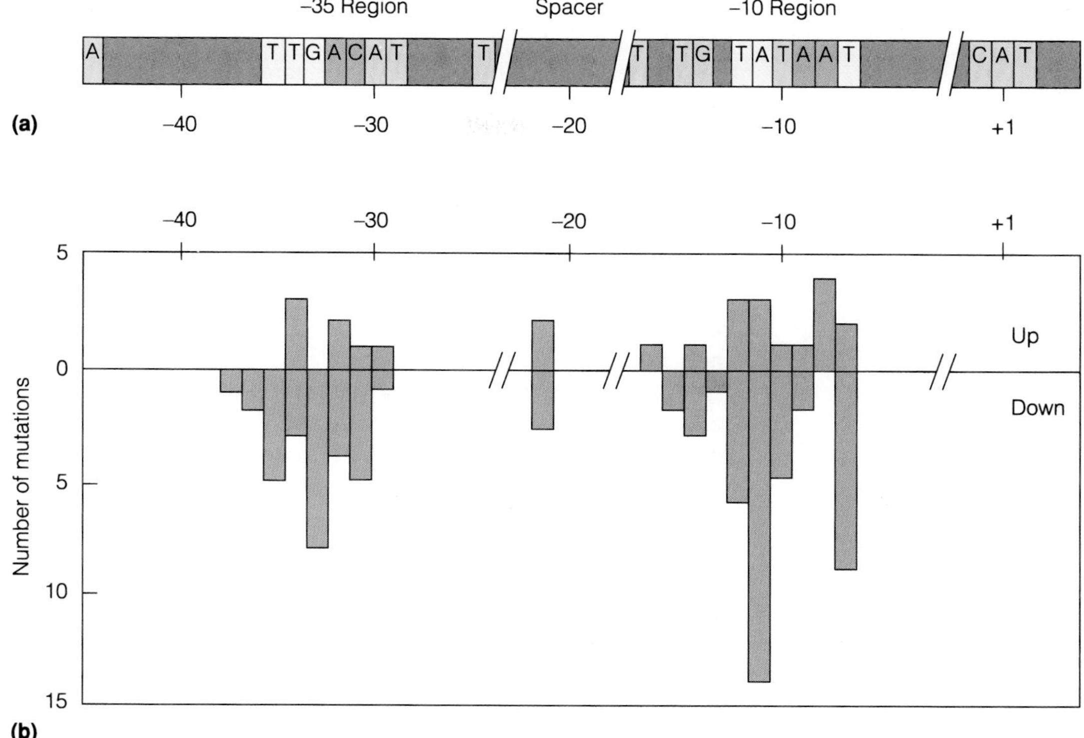

Figure 26.8
Survey of conserved nucleotides in *E. coli* promoters. Nucleotide sequences were compared among 114 known *E. coli* promoters. (**a**) Nucleotide positions that were invariant in at least 75% of the promoters are shown in yellow, moderately conserved nucleotides (50–75%) are shown in purple, and weakly conserved nucleotides (40–50%) are in blue. (**b**) The number of known up-promoter and down-promoter mutations affecting each site, among the entire set.

moter has −35 and −10 regions that are identical to the consensus sequences, but, in general, the more closely a promoter resembles the consensus sequences, the more efficient that promoter is in initiating transcription. Variations in promoter structure represent a simple way for the cell to vary rates of transcription from different genes. Figure 26.8 indicates the extent to which different nucleotides are conserved. Among 114 sequenced promoters for *E. coli* RNA polymerase, 6 of the 12 nucleotides in the two boxes are found in more than 75% of the promoters.

What evidence points to a functional role for these conserved sequences in binding RNA polymerase and initiating transcription? First, a variety of mutations map in promoter regions and affect transcription efficiency in vivo. As shown in Figure 26.8, most of the promoter mutations that have been sequenced change the structure of either the −35 region or the −10 region, pointing directly to those sequences as having the greatest effect on transcriptional initiation efficiency. In general, the mutations that increase promoter strength (**up-promoter** mutations) change either the −35 or the −10 region to more closely resemble the consensus sequences. **Down-promoter** mutations, which decrease promoter strength, change the sequence away from the consensus. Similar conclusions have been drawn from site-directed mutagenesis of promoters and analysis of their efficiency in vitro. The latter studies have confirmed the importance of spacing between the two regions. Although most natural promoters have a 17-nucleotide spacer between the −35 and −10 regions, many have 16 or 18. In vitro studies show that a 17-nucleotide spacer yields the most efficient promoter structure.

The other evidence for the importance of the −35 and −10 regions is that most of the DNA nucleotides in close contact with RNA polymerase are those in or near these two conserved sequences. This can be ascertained

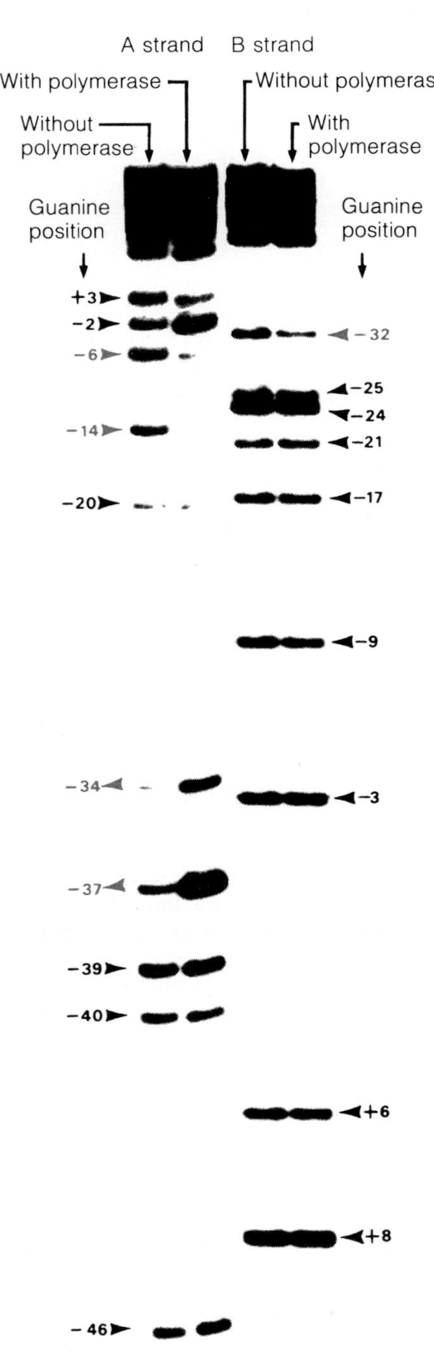

Figure 26.9
Identification of G residues that contact RNA polymerase in the *E. coli* tryptophan promoter. Enhancement of cleavage by binding of protein is shown by ◀; inhibition of cleavage is shown by ▶. Red numbers identify guanines within the −35 and −10 regions.

by determining the susceptibility of particular nucleotides to chemical modification, in the presence or absence of RNA polymerase. For example, one can determine the susceptibility of guanine residues to methylation by dimethyl sulfate. This involves methylating a 5′ end-labeled fragment, followed by cleavage, as in a Maxam–Gilbert sequencing reaction, and displaying the G-terminated fragments on an electrophoretic gel (Figure 26.9). In some cases band intensity is increased by the presence of RNA polymerase, showing that binding of protein enhances reactivity; in other cases reactivity is decreased, showing that binding of the protein protects that particular guanine from the methylating reagent. In either case the result indicates that the protein is in contact with the guanines in those regions and changes their reactivity.

In the promoter analyzed in Figure 26.9, the affected guanines—those at positions −37, −34, −32, −14, and −6—lie in or next to the −35 and −10 regions. By modifying reaction conditions one can also identify adenines that lie close to RNA polymerase. Also, by examining the reaction with ethylnitrosourea, one can identify phosphate oxygens of enhanced or decreased reactivity. Finally, one can identify thymines that are in contact with the protein, by substituting bromodeoxyuridine for thymine in DNA and then determining the extent to which each bromo-dUMP residue can become photocross-linked to protein. Note that these approaches, like footprinting, can identify critical nucleotides in any DNA site that binds a protein specifically, not just in promoters.

All of these techniques point to the −35 region and the −10 region, plus a few nucleotides upstream of −10, as the major contact points in an open-promoter complex. Figure 26.10 summarizes the results of these approaches as applied to one of the promoters for *E. coli* RNA polymerase on T7 DNA. Note that, because there are approximately two turns of the helix between the −35 and −10 regions, RNA polymerase is postulated to bind to DNA primarily on one side of the duplex, and the data on nucleotide reactivity support this conclusion.

Another factor affecting transcriptional efficiency, in addition to the base sequence of the promoter, is the superhelical tension on the DNA template. The relation between DNA topology and transcriptional efficiency is now receiving considerable attention. The relation is not clear, because transcription of some genes is activated in vivo when the template is highly supercoiled, for example, by inactivating topoisomerase I. Transcription of other genes, by contrast, is inhibited under these conditions. Recent evidence indicates that the act of transcription itself introduces superhelical turns into DNA. Since duplex DNA must unwind to be transcribed, movement of RNA polymerase along a template overwinds DNA ahead of the transcription complex (RNA polymerase, nascent transcript, and associated regulatory proteins) and underwinds behind the complex, as shown in Figure 26.11a. In a circular template the overwinding and underwinding cancel each other out, so that no torsional stress is created in the duplex DNA. However, if the template contains two or more genes being transcribed in opposite directions (Figure 26.11b), torsional stress is created, and topoisomerase action may be necessary to relieve that stress. Recent work indicates a similar pattern in eukaryotic transcription.

Interestingly, the promoter for transcription of DNA gyrase subunits becomes activated when the gene is in a relaxed state. Since gyrase functions to introduce superhelical turns, this seems to represent a feedback mechanism in which the cell responds appropriately to a signal that intracellular DNA is becoming too relaxed.

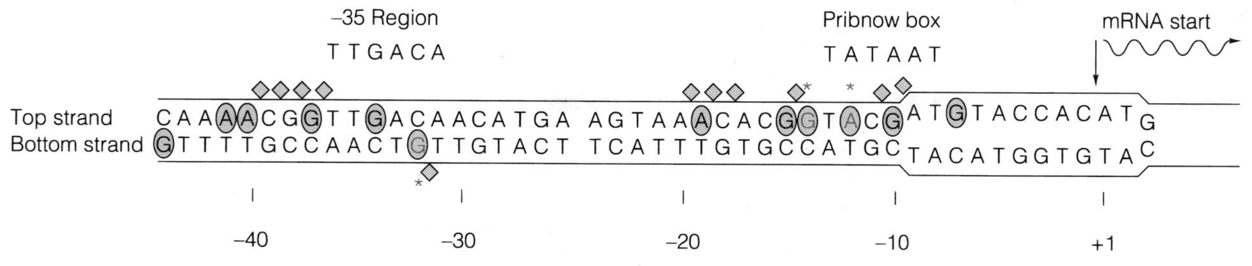

Figure 26.10
Structure of the T7 A3 promoter, one of the three *E. coli* promoters in T7 DNA. Nucleotides whose reactivities are changed by RNA polymerase binding are indicated as follows: orange, those protected from methylation by polymerase; purple, those whose susceptibility to methylation is enhanced by polymerase. * indicates methylated purines that interfere with polymerase binding. Diamonds indicate phosphate contacts. The two conserved regions of the promoter are precisely two helix turns apart.

Punctuation of Transcription: Termination

In bacteria we recognize two distinct types of termination events—those that depend on the action of a protein **termination factor,** called ρ, and those that are factor independent. Sequencing the 3′ ends of genes that terminate in a factor-independent manner reveals two structural features shared by many such genes: first, two symmetrical GC-rich segments that in the transcript have the potential to form a stem–loop structure, and second, a downstream run of four to eight A residues (Figure 26.12). How can these features help us understand the mechanism of factor-independent termination? The following explanation seems likely. First, RNA polymerase slows down, or pauses, when it reaches the first GC-rich segment, because the stability of G-C base pairs makes the template hard to unwind. In fact, RNA polymerase can be shown to pause for several minutes at a GC-rich segment. Second, the pausing gives time for the complementary GC-rich parts

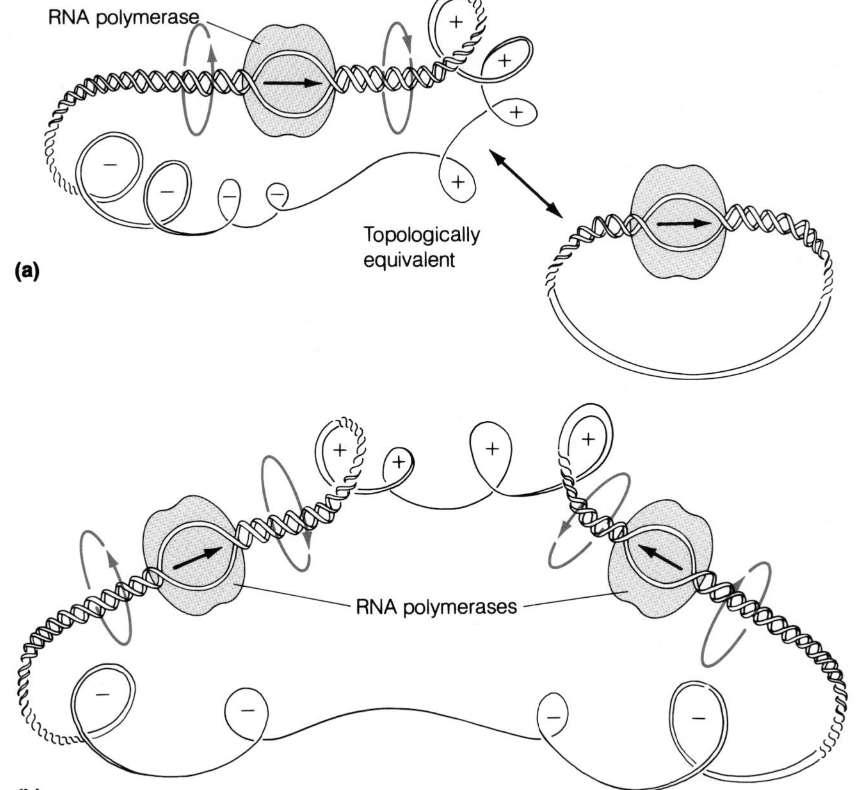

Figure 26.11
Supercoiling of DNA by transcription. RNA polymerase is moving from left to right. (**a**) Movement of the transcription complex produces overwinding ahead of the complex and underwinding behind it. In a circular template no torsional stress is created unless (**b**) two or more genes in the template are transcribed in opposite directions.

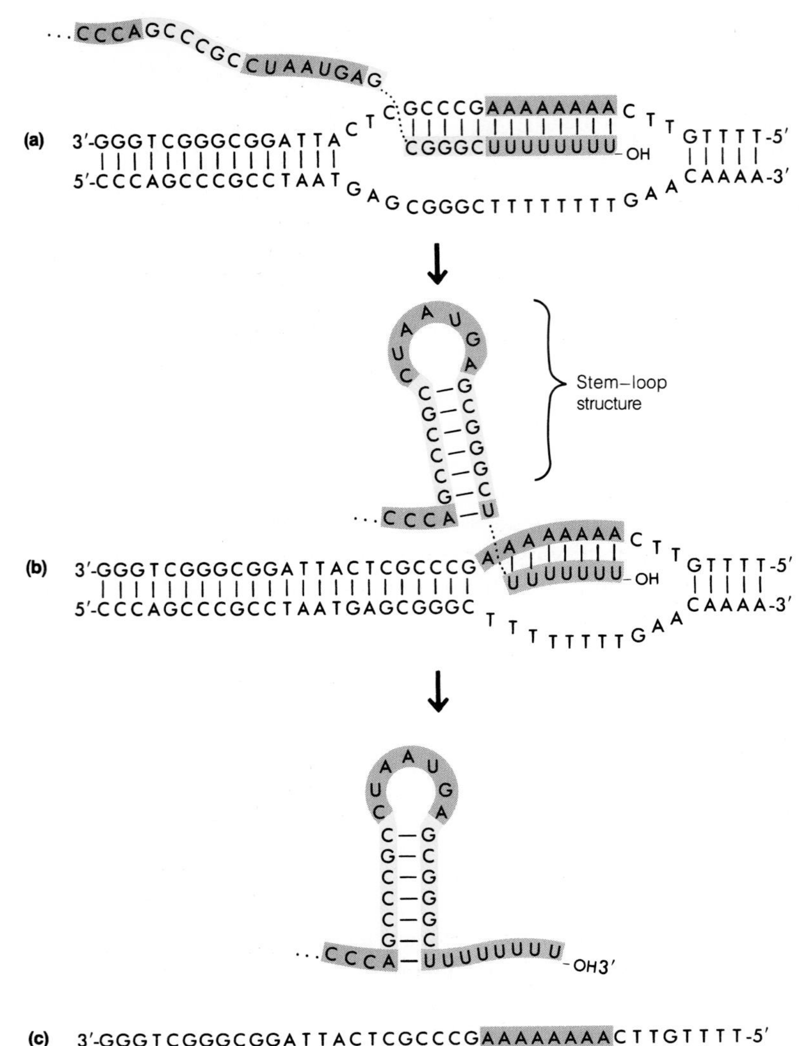

Figure 26.12
A model for factor-independent termination of transcription. (**a**) An A-rich segment of the template (tan) has just been transcribed into a U-rich mRNA segment. (**b**) RNA–RNA duplex, stabilized by G-C base pairs (yellow), eliminates some of the base pairing between template and transcript. (**c**) The unstable A-U bonds linking transcript to template hybrid dissociate, releasing the transcript.

of the nascent transcript to base pair with one another. In the process the downstream GC-rich segment is displaced from its template and, hence, the interaction between transcript and template is weakened. Further weakening, leading to dissociation, occurs when the A-rich segment is transcribed to give a series of A-U bonds (very weak), linking transcript to template.

Factor-dependent termination sites are less frequent, and the mechanism of this type of termination is more complex. The ρ protein, which is a hexamer composed of identical subunits, was originally discovered in studies on termination of λ phage DNA transcription in vitro. The protein, which has been characterized as a DNA–RNA helicase, contains a nucleoside triphosphatase activity, which is activated by binding to polynucleotides. Apparently ρ acts by binding to the nascent transcript at a specific site near the 3′ end, when RNA polymerase has paused (Figure 26.13). Then ρ moves along the transcript toward the 3′ end, with the helicase activity unwinding the 3′ end of the transcript from the template and causing its release.

It is not clear what causes RNA polymerase to pause at ρ-dependent termination sites. The action of another protein, NusA, is somehow involved. NusA was discovered during studies of **antitermination** in phage λ. Antitermination occurs early in the phage transcription program, when two ρ-dependent termination sites are inactivated, so that RNA polymerase can move past these sites and transcribe genes essential to phage development (p. 935). The action of a viral protein (the product of gene N) is essential for this to occur. The N protein interacts with NusA (*N utilization substance*) by a mechanism as yet unknown to accomplish this. Mutations in either N or the *E. coli nusA* gene interfere with antitermination and block phage development. The NusA protein evidently associates with RNA polymerase, and there is reason to believe that it binds at some point in transcription after the ρ factor has dissociated, because the two purified proteins compete with each other for binding to core RNA polymerase. Pausing of the core polymerase–NusA complex requires interaction with a DNA sequence called the *nut* site (for *N utilization*), but we don't yet know the mechanism of this pausing.

Further insight into termination mechanisms has come from an extensively studied regulatory mechanism called **attenuation.** Attenuation controls the rate of transcription of certain operons by terminating the synthesis of a nascent transcript before RNA polymerase has reached the structural genes. Attenuation, which has been investigated as a model for end-of-transcript termination, is discussed on page 941.

Regulation of Transcription

Bacteria respond rapidly to changes in their environment. As we have seen, the genes for lactose utilization are activated once *E. coli* cells sense the presence of lactose or a similar compound, and the lactose-utilizing enzymes are induced. Many other genes are similarly kept in a turned-off state until their products are needed. Still other genes, concerned principally with anabolism, are kept in a turned-on state unless the product of the anabolic sequence is present, and then the genes are **repressed,** or turned off. The metabolic logic involved is obvious: many genes are expressed only when the gene products are needed, either to utilize an available substrate, to synthesize a complex metabolite that is absent from the medium, or in some other way to respond to changing environmental conditions.

Both enzyme induction and repression act through control of transcription, primarily at the initiation step. Similarly, bacteriophages use primarily the transcription machinery of the host cell to express their own genes; host transcriptional control mechanisms are either used or modified to meet the needs of the virus. Consequently, studies of enzyme induction and repression and of bacteriophage development have provided our most penetrating insights into how transcription is controlled in prokaryotes. In turn, this work has provided paradigms for understanding the far more complex questions of how transcriptional regulation in multicellular organisms can account for differentiation and development. For these reasons, we shall focus our discussion of transcriptional regulation on the prokaryotic systems that have been most informative, with extension of these concepts to eukaryotes in Chapter 28.

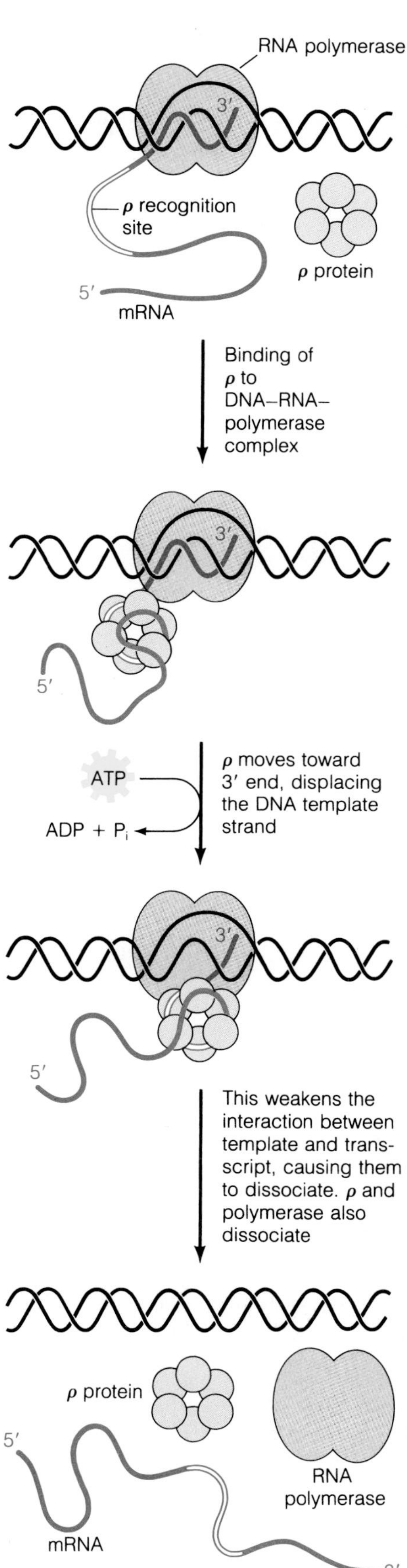

Figure 26.13
ρ factor-dependent termination. ρ binds to a site on the nascent transcript and unwinds the RNA–DNA duplex. Once ρ reaches RNA polymerase, interaction with bound NusA protein (not shown) leads to termination.

The Lactose Operon: Early Evidence for Transcriptional Control of Gene Expression

Allolactose

Isopropyl β-thiogalactoside

The lactose operon consists of three linked structural genes that encode enzymes of lactose utilization, plus adjacent regulatory genes and sites. The three structural genes—*z*, *y*, and *a*—encode respectively β-galactosidase, **β-galactoside permease** (a transport protein), and **thiogalactoside transacetylase,** a protein of still unknown metabolic function. In the presence of an inducer all three enzymes accumulate simultaneously, but to different levels. Lactose itself will induce the lactose operon, but the true intracellular inducer is **allolactose,** a minor product of β-galactosidase action. In the laboratory one usually uses a synthetic inducer such as **isopropyl thiogalactoside (IPTG),** which induces the lactose operon but is not cleaved by β-galactosidase; hence, its concentration does not change during an experiment.

Regulation in the Lactose Operon. A mutation in a structural gene, *z* for example, can inactivate its product (β-galactosidase) without affecting control of the other two genes. However, mutations in the regulatory regions mapping *outside* genes *z*, *y*, and *a* can affect expression of all three structural genes. In their early work, Jacob and Monod recognized two distinct phenotypes—**constitutive,** in which all three gene products are synthesized at high levels even when inducer is absent, and **noninducible,** meaning that all three enzyme activities remain low even after an addition of inducer. These mutations mapped in two sites, termed *o* and *i*. (Additional regulatory mutations mapped in other sites; we shall describe them presently.)

The original Jacob–Monod view of this system was presented in Figure 26.2; a more complete description of the *lac* operon is shown in Figure 26.14. Jacob and Monod proposed originally that the *i* gene product is a macromolecular **repressor,** which in the active form binds to DNA at a specific site, the **operator.** Since the operator lies adjacent to the structural genes of the operon, binding a macromolecule at that site blocks transcription (Figure 26.15a). Jacob and Monod proposed that transcription of the three genes is initiated from within or near the operator, to give a single **polycistronic messenger RNA,** or an RNA copy of all three genes. (The term **cistron** has genetic significance; for our purposes it is a region of a genome that encodes one polypeptide chain.)

The repressor also has a binding site for inducer; binding of IPTG, allolactose, or some other inducer at this site inactivates repressor by vastly decreasing its affinity for DNA (Figure 26.15b). This has the effect of activating transcription of *z*, *y*, and *a*, because dissociation of the repressor–inducer complex from the operator removes a steric block to binding of RNA polymerase at the initiation site. Thus, the introduction of lactose or a similar inducer activates synthesis of the gene products involved in its catabolism by removing a barrier to their transcription. This mode of regulation

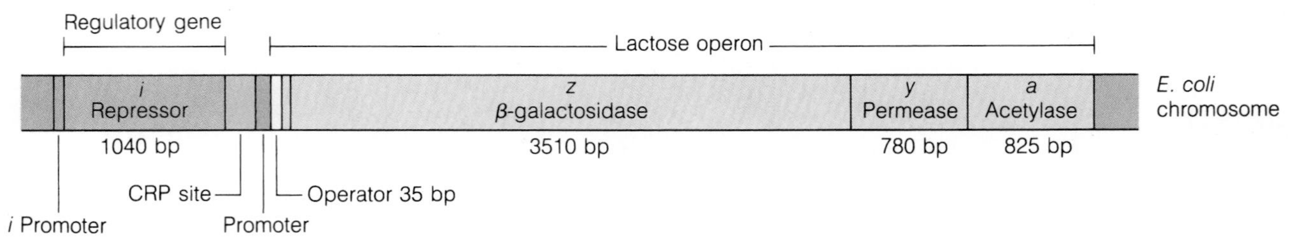

Figure 26.14
A current map of the lactose operon. The CRP site is the binding site for a regulatory factor, cAMP receptor protein.

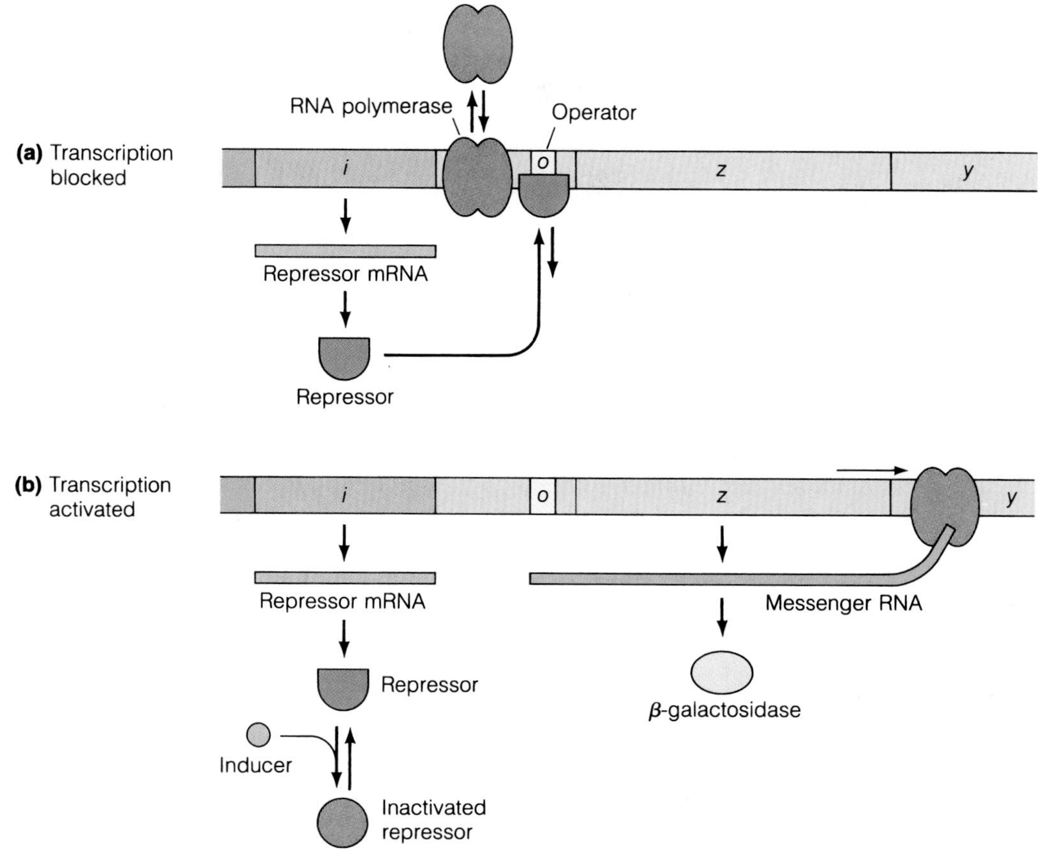

Figure 26.15
Configurations of the lactose operon. (**a**) Repressor binds to operator, inhibiting transcription of structural genes. (**b**) Inducer binds repressor, decreasing its affinity for operator. Diffusion of inactive repressor–inducer complex from operator permits transcription of structural genes.

is seen as essentially negative, since the active regulatory element (the repressor) is an inhibitor of transcription. Positive control was later discovered, involving the CRP site shown in Figure 26.14; we shall discuss this shortly.

Central to the success of Jacob and Monod in formulating such a detailed model was their ability to analyze regulatory *lac* mutations in partial diploids—cells containing one complete chromosome plus part of another, transferred in by conjugation. One copy of the operon resides in the chromosome, while an incomplete chromosome could be introduced into a cell, as part of the bacterial mating process. Noninducible mutations that mapped in *i* had a dominant phenotype, meaning that in the presence of both wild-type and mutant alleles of *i*, expression of the structural genes was low. Jacob and Monod predicted that these mutant repressors would be unable to bind inducer, so the mutant repressor would remain bound to DNA (at operator sites on both chromosomes) even when inducer was present.

Another class of *i* mutations had a constitutive phenotype. These generated repressors that were defective in operator binding and thus could not turn off gene expression. These mutations were recessive, because a normal repressor in the same cytosol could bind to all operators and inhibit transcription. On the other hand, the constitutive mutations mapping in *o* had a **cis-dominant** effect. That is, in a cell with one wild-type and one mutant operator, only the genes on the same chromosome as the mutant operator were expressed constitutively. This suggested that the operator does not encode a gene product, which would be capable of diffusing through the cytosol and acting on other chromosomes. Repressor mutations, on the other hand, are **trans-dominant,** meaning that the *i* gene product encoded by one operon can affect gene expression from other operons. Thus, the repressor was seen as a diffusible product, capable of acting on any DNA site in the cell to which it could bind.

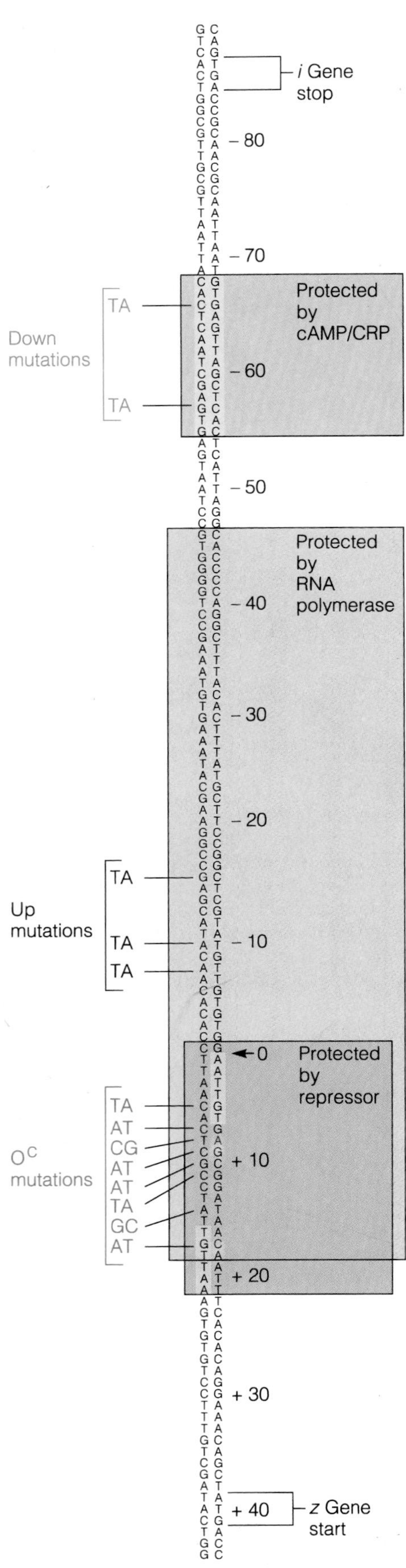

For a scheme proposed largely from indirect evidence, the operon model as advanced by Jacob and Monod has stood the test of time remarkably well. Three major modifications occurred as the system was subjected to further analysis. First, the promoter was discovered as an element distinct from the operator. Second, while the repressor was first thought to be *i*-gene *RNA*, its isolation proved that it is protein. Third, Jacob and Monod proposed that all transcriptional regulation was negative, that is, that binding a regulatory protein always inhibited transcription. The lactose operon, like many other regulated genes, also exhibits *positive* control of transcription, that is, *activation* of transcription by binding of a protein. The latter protein, which we describe on page 929, responds to the intracellular level of cyclic AMP, and it facilitates binding of RNA polymerase to the *lac* promoter.

Isolation and Properties of Repressor. The *lac* repressor was isolated in 1966 by Walter Gilbert and Benno Müller-Hill. Since this repressor constitutes only 0.001% of the total cell protein, Gilbert and Müller-Hill used mutants designed to overproduce the putative repressor, so as to maximize its synthesis (to about 2% of total protein). They then purified the protein on the basis of its ability to bind the synthetic inducer IPTG. The purified *lac* repressor is a tetramer, formed from four identical subunits, each with 360 amino acids (M_r 38,350). The protein binds IPTG with a K_a of about 10^6, and it binds nonspecifically to duplex DNA with a K_a of about 3×10^6. However, its specific binding at the *lac* operator is much tighter, with a K_a of 10^{13}. Like RNA polymerase, repressor seeks its operator site by first binding to DNA at any site and then moving in one dimension along the DNA, either by sliding or by transfer from one site to another, when the two sites are brought next to each other on adjacent loops of DNA.

Control by the *lac* repressor is exceedingly efficient, particularly in view of the minute amount of repressor present in an *E. coli* cell. The *i* gene is expressed at a very low rate, to give about 10 molecules of repressor tetramer per cell. Although this corresponds to a concentration of only about 10^{-8}M, this value is several orders of magnitude higher than the *dissociation* constant, meaning that in an uninduced cell the operator is bound by repressor more than 99.9% of the time. Hence, the very low levels of *lac* operon proteins in uninduced cells (less than one molecule per cell). However, binding of inducer reduces the affinity of the repressor–inducer complex for operator by many orders of magnitude, so that in induced cells the operator is bound less than 5% of the time.

The DNA site bound by *lac* repressor has been analyzed by footprinting and methylation protection experiments of the type described earlier for RNA polymerase (see Figure 26.9). As shown in Figure 26.16, the operator constitutes 35 base pairs, including 28 base pairs of symmetrical sequence, that is, sequence that is identical in both directions. This is not a perfect palindrome, because of the seven base pairs that do not show this symmetry. Base pairs that are included in this partial palindrome are shaded in the diagram at the top of p. 929.

Figure 26.16
The 122-base *lac* regulatory region. The inverted repeat sequence in the CRP binding site is shaded in blue, and the hyphenated palindrome of the operator is shaded in red. The binding sites for all three proteins (CRP, repressor, and RNA polymerase), as determined by DNase I footprinting, are boxed. The sequence changes encountered in promoter and operator mutations are shown, as well as the start codon for the *lacZ* protein and the stop codon for repressor protein (*i* gene). Nucleotides are numbered from 0 at the *lac* transcriptional start point.

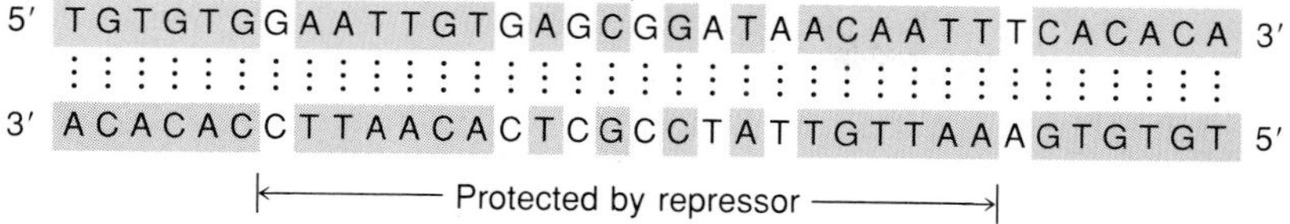

The transcriptional start point is included within this sequence, and the sequence partially overlaps the promoter. Twenty-four of the 35 base pairs are protected from DNase attack by repressor binding. Operator-constitutive (O^c) mutations involve changes in the central portion of this sequence of nucleotides, as shown in Figure 26.16. Note that the operator and promoter overlap, as determined by the regions of DNA protected by binding either repressor or RNA polymerase, respectively.

Regulation of the *lac* Operon by Glucose: A Positive Control System. The *lac* repressor–operator system acts to keep the operon turned *off* in the absence of utilizable β-galactosides. An overlapping regulatory system acts to turn the operon on only when alternative energy sources are unavailable. *E. coli* has long been known to utilize glucose in preference to most other energy substrates. When grown in a medium containing both glucose and lactose, the cells metabolize glucose exclusively until the supply is exhausted. Then growth slows, and the lactose operon becomes activated in preparation for continued growth using lactose. This phenomenon, originally called "glucose repression" or "catabolite repression," is now known to involve a transcriptional *activation* mechanism, which comes into play when glucose levels are low and exerts control through intracellular levels of cyclic AMP. This is summarized in Figure 26.17.

cAMP levels respond to the state of glucose metabolism in bacteria. Recall that in animal cells a rise in cAMP levels stimulates catabolic enzymes, which increase the levels of energy substrates. Those effects are mediated *metabolically*, through hormonal signals and triggering of metabolic cascades. In bacteria the mechanism of activation is *genetic*, but the end results are similar. In *E. coli* cAMP levels are low when intracellular glucose

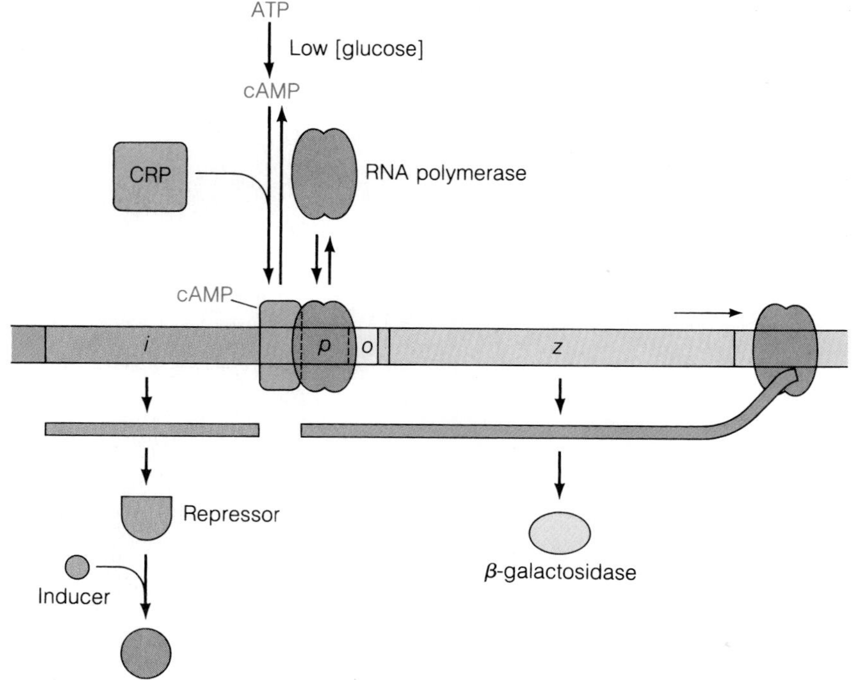

Figure 26.17
Activation of the *lac* operon. Repressor is inactivated by binding inducer, and CRP is activated by binding cyclic AMP. Binding of the CRP–cAMP complex to DNA facilitates initiation of transcription by RNA polymerase.

levels are high. The actual regulatory mechanism is not yet known. Adenylate cyclase apparently senses the intracellular level of an unidentified intermediate in glucose catabolism; hence, the current name for the regulatory process, **catabolite activation.** When glucose levels drop, cAMP levels rise, triggering activation of the lactose operon by virtue of its interaction with a protein formerly called **catabolite activator protein (CAP),** but now called **cAMP receptor protein (CRP).** This protein is a dimer, each of whose identical polypeptide chains contains 210 amino acid residues. When it binds cAMP, the protein undergoes a conformational change. The change greatly increases its affinity for certain DNA sites, including a site in the *lac* operon, adjacent to the RNA polymerase binding site. Binding at this site protects a DNA sequence from −72 to −52, as shown in Figure 26.16. Binding facilitates transcription of the *lac* operon by stimulating the binding of RNA polymerase to form a closed-promoter complex.

One reason why our understanding of CRP action is still incomplete is that the cAMP–CRP complex activates several different gene systems in *E. coli,* all of them involved with energy generation. These include operons for utilization of other sugars, including galactose, maltose, arabinose, and sorbitol, and several amino acids. Among the operons that have been analyzed, the DNA binding site of the cAMP-activated dimer varies considerably with respect to the transcriptional startpoint. Thus, it is difficult to propose a common mechanism of action. The structure of the CRP crystal (Figure 26.18) suggests a mechanism by which the protein binds to DNA. The protein contains a characteristic pair of α helices (designated E and F in the figure), which are joined by a turn. This structural motif, which was observed at about the same time in the structure of the λ phage Cro repressor (p. 936), is found in several DNA-binding regulatory proteins, suggesting common evolutionary origins for this family of proteins. We shall return to this motif and its regulatory significance when we discuss the repressors of phage λ.

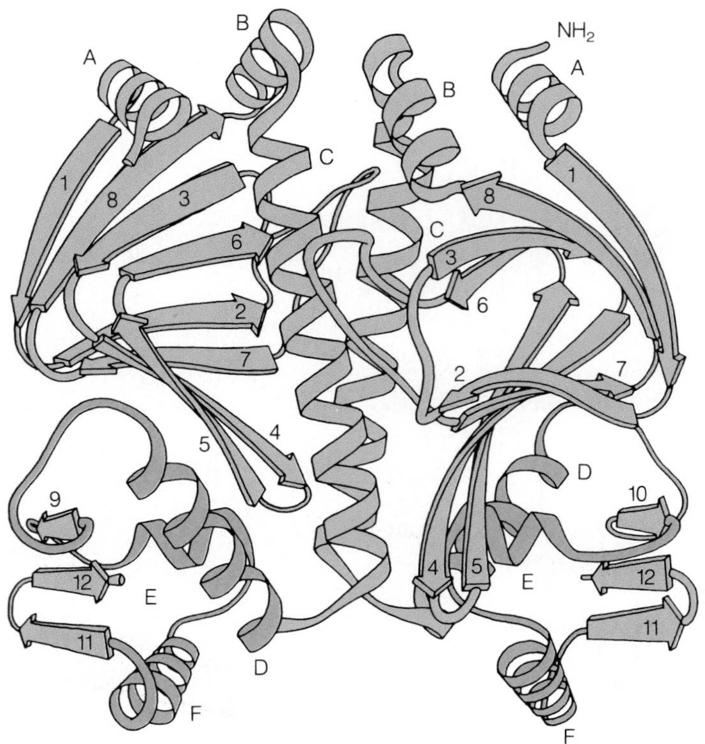

Figure 26.18
Structure of the CRP dimer. The C-terminal segment, consisting of α helices D, E, and F, constitutes the DNA-binding region. The F helices contain the DNA sequence binding determinants.

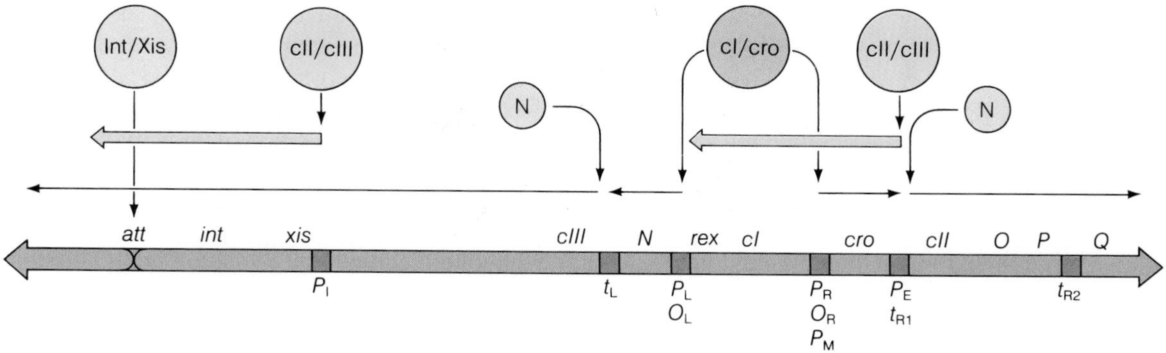

Figure 26.19
The early regulatory region of phage λ. Sites of action of the following regulatory proteins are shown: *cI* and cro repressors, *cII* activator, *cIII* (which acts to stabilize *cII*), the antiterminator N, and the integration and excision proteins (Int/Xis).

Bacteriophage λ: Multiple Operators, Dual Repressors, Models for DNA Binding Specificity

We now turn back to phage λ, a much larger and more complex genetic system than the lactose operon, but one that is regulated by much the same principles. In the lysogenic state most of the genes in λ are turned off, and a sequential program of gene activations is involved either in breaking lysogeny or in sustaining a lytic infection. Figure 26.19 shows the regulatory genes and sites that will concern us. (The genes shown on that figure are identified on page 932).

Genes and Mutations in the λ System. Phage mutants defective in establishing or maintaining lysogeny have phenotypes highly comparable to those defective in *lac* regulation, and these similarities helped Jacob and Monod to make the generalizations embodied in the operon model. The major λ phenotypes, and the comparable *lac* mutations, are summarized in Table 26.2. Recall that when λ infects a bacterial cell, two outcomes can result—either a cycle of lytic phage growth or establishment of lysogeny. The lysogenic response involves processes of gene inactivation comparable to those that keep the *lac* operon turned off when inducer is absent, namely binding of a repressor to an operator (actually, two operators in λ). When a defective repressor cannot bind to operators, the mutant phages give clear plaques when they are plated. Normally a λ plaque is turbid, because it contains not only phage but also lysogenized bacteria, which continue to grow. Clear-plaque mutants in λ map in three different genes—*cI*, *cII*, and *cIII*. *cI* is the structural gene for a repressor, while the other two genes control the synthesis of the *cI* protein.

Table 26.2
Phenotypes of comparable *lac* and λ mutations

lac Phenotype	Corresponding λ Phenotype	Regulatory Abnormality
Inducer-constitutive, recessive	Clear-plaque; cannot establish lysogeny	Repressor defective in operator binding
Operator-constitutive, cis-dominant	Virulent; can replicate in a superinfected immune lysogen	Operator unable to bind repressor
Noninducible, trans-dominant	Noninducible (cannot be induced by UV or other treatments)	Repressor cannot bind inducer or be inactivated

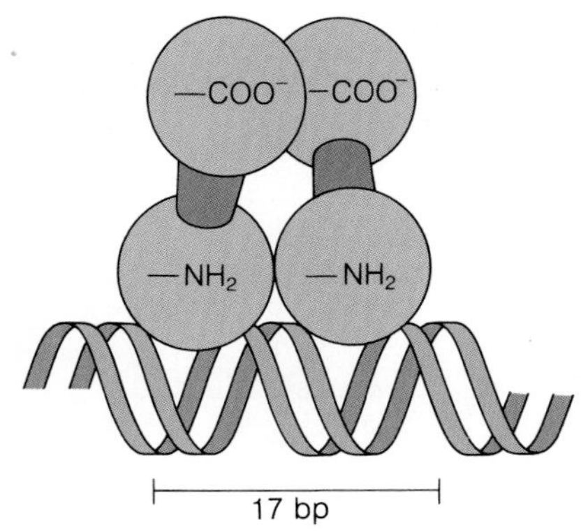

Figure 26.20
Dimeric structure of λ *cI* repressor and its binding to a 17-base pair region in a λ operator.

Virulent mutations, which map in operators, also convey a clear-plaque phenotype, but there is an important distinction. Bacteria lysogenic for λ are immune to infection by a second λ phage, because repressor in the cell binds to the operators of any input phage. This is true for *cI* mutants because they contain normal operators, which can bind repressors. However, virulent mutants can produce progeny phage in infection of an immune lysogen, because their operators cannot bind repressor after entry into the cell. These observations are similar in principle to those resulting from the experiments with partial diploids in the *lac* operon.

THE λ REPRESSOR AND ITS OPERATORS. In 1967 Mark Ptashne isolated the λ repressor encoded by *cI*, at about the same time as the *lac* repressor was isolated. The *cI* repressor is a dimeric protein, with a subunit M_r of 27,000, which binds through its N-terminal sequences to operator sites with a K_a of about 3×10^{13} M (Figure 26.20). Repressor–DNA interactions have been used to map and characterize what turned out to be two operators, one on each side of *cI*. The two operators control divergent transcriptional events from a central regulatory region—leftward (O_L) and rightward (O_R). As shown by footprinting, each operator contains three separate repressor-binding sites, each site some 17 base pairs in length; Figure 26.21 illustrates this for the O_R region. The sites are homologous, but not completely so (Figure 26.22), and the sites are separated by spacer regions of three to seven base pairs. Mutations that confer virulence map within the repressor-binding regions; a fully virulent mutant has at least two mutations—one in O_L and one in O_R.

The λ operators are remarkable in several respects other than their multiple repressor-binding sites. First, mutations that affect promoter activity lie between the repressor-binding sites. Thus, operators and promoters are *interspersed,* and the regulatory regions are more properly called O_LP_L and O_RP_R. Second, O_RP_R controls transcription from two distinct promoters—one rightward (P_R), one leftward (P_{RM}). Third, transcription from O_LP_L and O_RP_R is controlled by two different repressors—*cI* and a protein called Cro (the *cro* gene is an acronym, *c*I *r*epressor *o*ff). Fourth, under certain conditions the *cI* repressor is a transcriptional *activator,* not an inhibitor. Another novel feature of this regulatory system is that *cI* transcription is initiated from different promoters under different physiological conditions. All of these complexities are related to the need for orderly and efficient phage gene control in several quite different "life-styles"—(1) lytic infection; (2) *establishment* of lysogeny, when some genes for replication are being expressed; (3) *maintenance* of lysogeny, when replication genes are turned off; and (4) excision leading to lytic growth, after repressor inactivation.

EARLY GENES IN PHAGE λ. To understand the significance of *cI* gene regulation, we must describe several λ genes that are expressed early in infection (see Figure 26.19). Most of these genes we have encountered before. *cI* and *cro* code for repressors, as we have noted, and *cII* and *cIII* both act to stimulate *cI* synthesis. *rex* is a gene of still unknown function, the only gene aside from *cI* known to be expressed during lysogeny. We encountered *int, xis,* and *att* in Chapter 25, when we discussed site-specific recombination, and we encountered O and P in connection with the initiation of λ DNA replication. The N gene product interacts with NusA to prevent termination. The Q product activates late gene transcription.

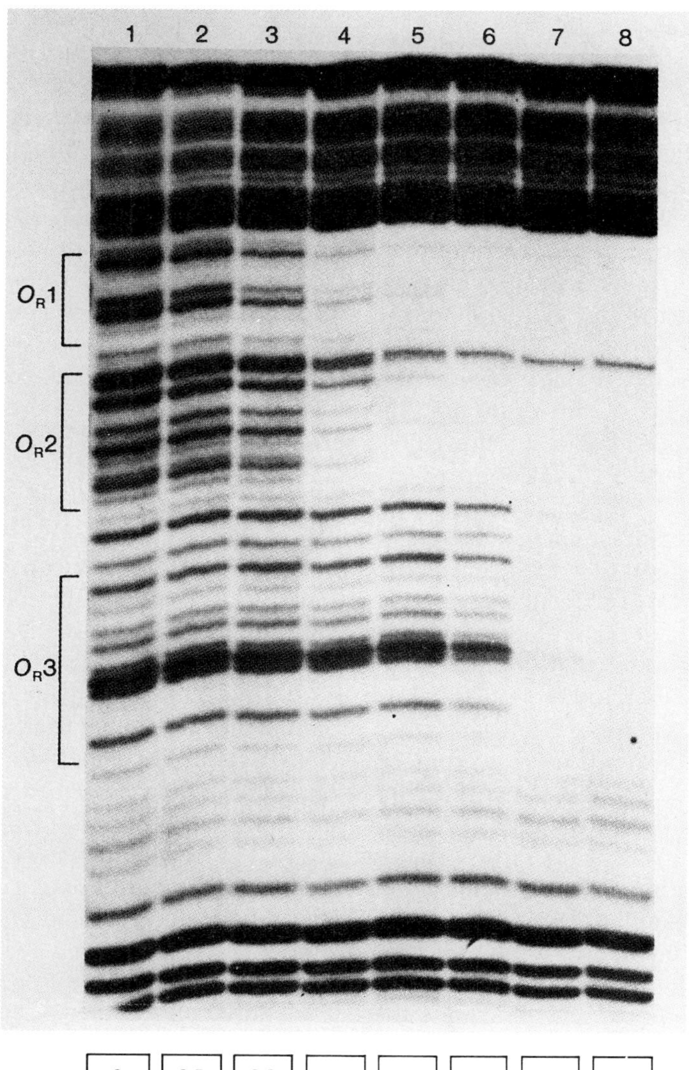

0	3.5	8.8	18.0	35	88	180	350

Repressor concentration, nM

Figure 26.21
Structures of the λ operators, determined from quantitative DNase I footprinting of a restriction fragment containing the three sites in O_R. Repressor concentration was increased from 0 (slot 1) to 350 nM (slot 8).

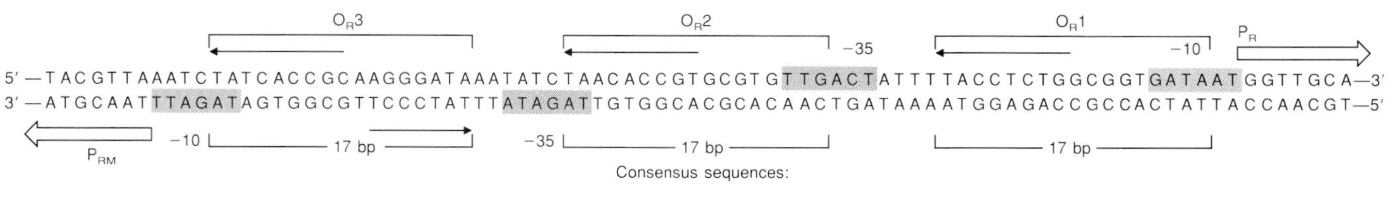

Operator (half-site):	5′—T A T C A C C G—3′
Promoter −35 region:	5′—T T G A C A—3′
Promoter −10 region:	5′—T A T A A T—3′

Figure 26.22
Nucleotide sequence of O_RP_R, showing each repressor-binding site (O_R1, O_R2, and O_R3), the leftward promoter P_{RM}, and the −35 and −10 boxes for each of the two promoters.

INTERACTIONS BETWEEN THE TWO λ REPRESSORS. The interspersed O_LP_L controls transcription of N, through interaction of its repressor-binding sites with the *cI* protein. However, most of the regulatory action occurs at O_RP_R, and it is here that the decision is made between lytic and lysogenic infection. Quantitative footprinting experiments (see Figure 26.21) show that of the three repressor-binding sites in O_RP_R, *cI* binds most tightly to site O_R1, less tightly to O_R2, and still less tightly to O_R3. Comparable experiments show that *cI* binding is *cooperative,* so that when one repressor dimer is bound at O_R1, affinity for a second molecule is increased at O_R2. Cro protein is a dimer of identical 66-residue subunits. It binds considerably less tightly to any of the sites than does *cI* and in the reverse order; that is, site O_R3 is favored, followed by approximately equal binding at O_R2 and O_R1. Binding is noncooperative.

Although Cro is a repressor, it can also be considered an **antirepressor** because it antagonizes the action of *cI* in a very specific way. To understand how this works, consider the transcriptional events occurring in the presence of varying levels of *cI* (Figure 26.23). Because of the cooperative binding of *cI* to its operators, both sites O_R1 and O_R2 are usually occupied in the lysogenic state, even though the intracellular concentration of *cI* is quite low (about 200 molecules per cell, or 10^{-7} M). This inhibits the rightward transcription of *cro* from its own promoter and *activates* the leftward transcription of *cI* from a promoter called P_{RM} (the M stands for maintenance, because this is the promoter from which *cI* is transcribed during maintenance of lysogeny). The −10 and −35 regions for the P_{RM} promoter lie within the operators. The evidence that *cI* really does activate its own transcription under these circumstances lies in the existence of a special class of *cI* mutants whose repressor can bind tightly at O_R1 and O_R2, but cannot stimulate *cI* transcription from P_{RM}.

During *establishment* of lysogeny, when lytic and lysogenic genes are competing to determine the fate of the viral genome, there is a need for larger amounts of *cI* repressor than can be transcribed from P_{RM}. At this time a different *cI* promoter, called P_{RE}, is activated (E stands for establishment). In this activation the *cII* protein binds specifically at the −35 region of P_{RE} and stimulates RNA polymerase binding at that site. Footprinting experiments establish that *cII* protein and RNA polymerase do bind at the same DNA site, on opposite sides of the helix. This transcriptional event yields a longer *cI* messenger RNA, which is more efficiently translated than the message synthesized from P_{RM}. The result is sufficient *cI* repressor to bind all three sites in O_R and, hence, to block both transcriptional initiation events. *cII* protein promotes lysogenization by an additional mechanism. By binding to the *int* gene promoter, it increases transcription of *int* and synthesis of integrase, resulting in enhanced integration of repressed λ prophage.

Clearly, the *cII* protein plays a central role in whether a λ-infected bacterium undergoes lytic growth or lysogeny. By itself P_{RE} is a very poor promoter for *E. coli* RNA polymerase, with only three of the six positions in the −10 region identical to those of the consensus Pribnow box, and no identities in the −35 region. For lysogeny to occur, sufficient *cII* protein must be present to stimulate transcription from P_{RE} and, hence, to synthesize *cI* repressor in abundance. *cII* protein is highly susceptible to proteolytic degradation. Proteases tend to be more abundant and active in bacteria grown in rich medium than in starved cells. Hence, a well-nourished cell tends to undergo lytic growth, because *cII* protein is inactivated. In a starved cell, it is to the advantage of the virus to lysogenize, preserving its genome until the cell is in a better condition to replicate the virus. Another regulatory protein, the product of the *cIII* gene, appears to act by inhibiting a cellular protease that is involved in degrading the *cII* protein. Thus, *cIII*

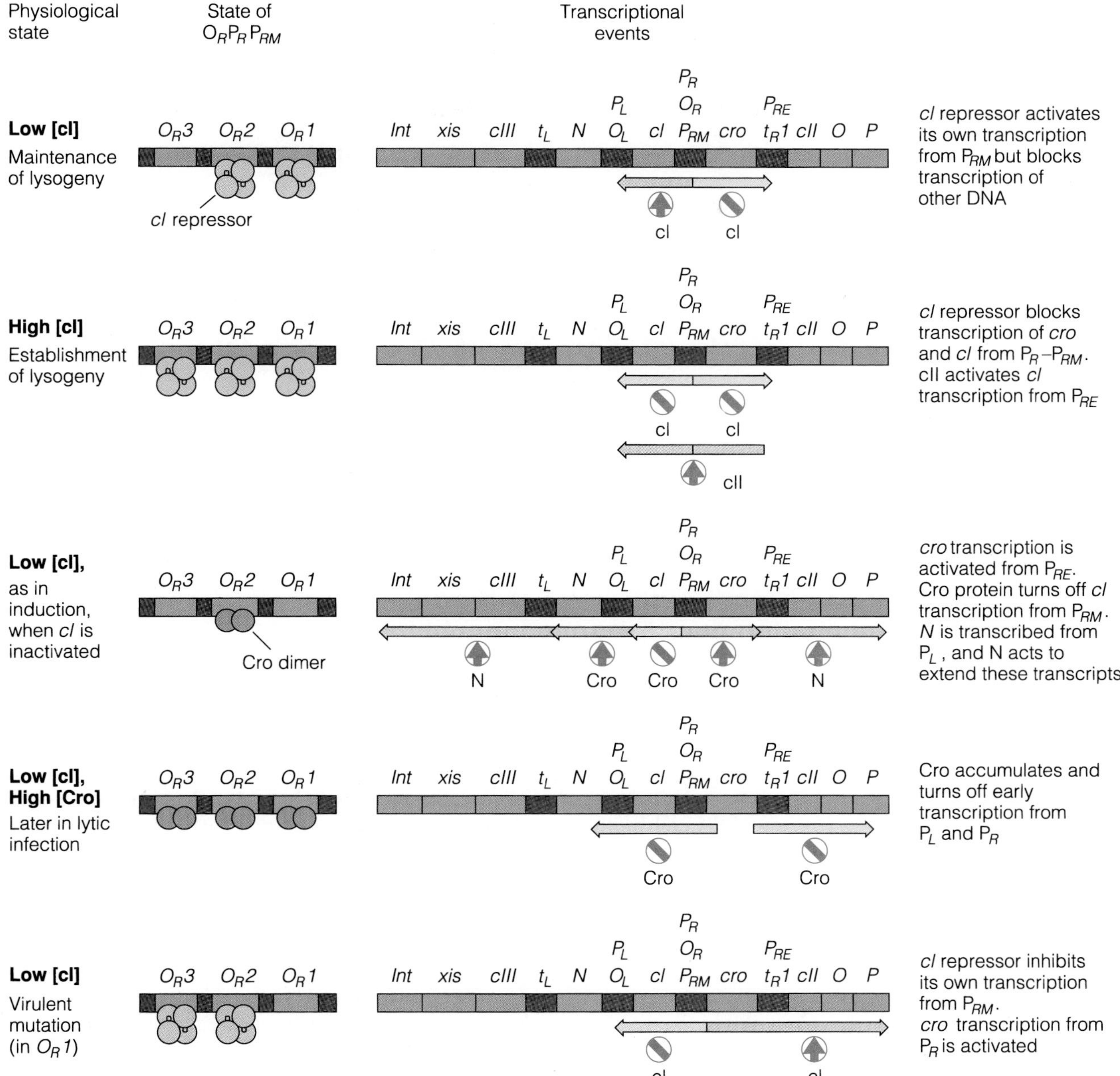

Figure 26.23
cI–Cro interactions in the $O_RP_RP_{RM}$ region under various physiological conditions, and the effects on transcription of the rest of the λ genome.

mutants have a clear-plaque phenotype, because they cannot prevent degradation of *cII* protein and hence cannot activate transcription from P_{RE}.

Now consider the events in prophage induction, when lysogeny is broken, leading to a lytic infection. First, *cI* repressor is inactivated (we shall see shortly how this occurs), and the O_R sites become unoccupied. This permits transcription of *cro* from P_R, and the Cro protein blocks further transcription of *cI* from P_{RM}. At the same time, leftward transcription from P_L generates the N protein, blocking transcriptional termination at the sites indicated in Figure 26.23 as t_R1 and t_L. Thus, the two early transcripts for Cro and N are extended to activate new genes. Leftward transcription generates Int and Xis proteins, necessary for prophage excision. Rightward transcription generates O and P, necessary for DNA replication.

Subsequent regulatory events, including the action of the gene Q protein, activate transcription of late-acting genes, which encode structural pro-

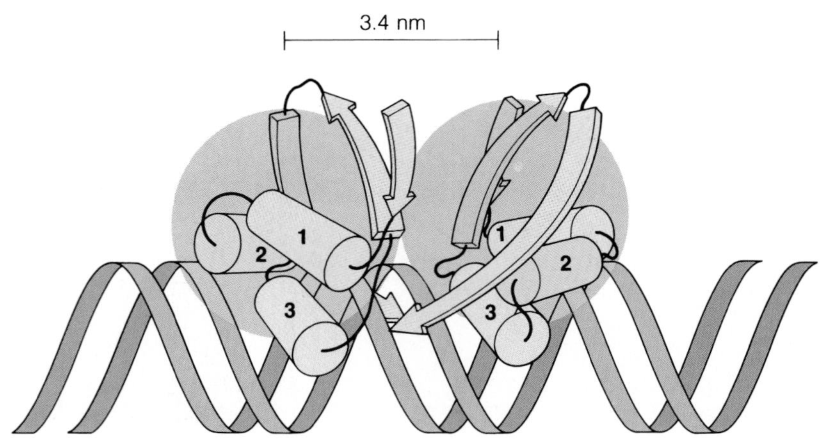

Figure 26.24
Proposed structure of the Cro dimer–operator complex.

teins of the virus. At this time it is desirable to suppress early gene transcription, so that the late proteins can be made at maximal rates. This involves further action of the Cro protein, which by this time has accumulated to the point where it can bind to both O_R1 and O_L1, blocking transcription from P_R and O_R, respectively.

STRUCTURE OF CRO, *cI* REPRESSOR, AND RELATED DNA-BINDING PROTEINS. Studies of the three-dimensional structures of Cro and *cI* repressors have yielded remarkable insight into mechanisms by which proteins recognize specific DNA sequences. This in turn has greatly enhanced our understanding of how transcription is regulated through specific DNA–protein interactions. In 1981 the crystal structure of Cro was determined, revealing 66 residues folded into three α-helical regions and three β sheets. Model-building studies showed that two of the helices, numbered 2 and 3 in Figure 26.24, could fit within the major groove of the DNA double helix. These two helices are separated by a short β turn. In the Cro dimer the two number 3 helices are 3.4 nm apart, the length of one turn of the DNA double helix. This distance suggested that the two subunits bound on the same side of the helix, in adjacent major groove sites, with the number 3 helices lying lengthwise in the grooves. This model was strongly supported by methylation and ethylation protection experiments, which identified the DNA functional groups in close contact with the protein.

The amino acid sequence within and between the 2 and 3 helices showed remarkable homology with corresponding sequences in a large family of sequence-specific DNA-binding proteins but *not* with DNA-binding proteins that showed no sequence preference. This suggested that the helix–turn–helix motif is a commonly evolved structural element in transcriptional regulatory proteins, at least in prokaryotes. A different motif, the "zinc finger," is present in a large number of eukaryotic DNA-binding proteins (and a few known prokaryotic proteins), as we will see in Chapter 28.

The cAMP receptor protein (CRP) has the helix–turn–helix motif in its helices E and F, as mentioned earlier. However, B-form DNA must bend in order to bind to the protein as postulated, and it was not immediately clear that this motif is used for DNA binding in CRP. Once the three-dimensional structure of λ *cI* repressor was shown also to have the helix–turn–helix structure, it seemed likely that the helix–turn–helix structure is involved in DNA binding in all of these proteins. The relevant three-dimensional similarities for Cro, CRP, and *cI* repressor are shown in Figure 26.25. Amino acid sequence homologies between these regions are shown for Cro

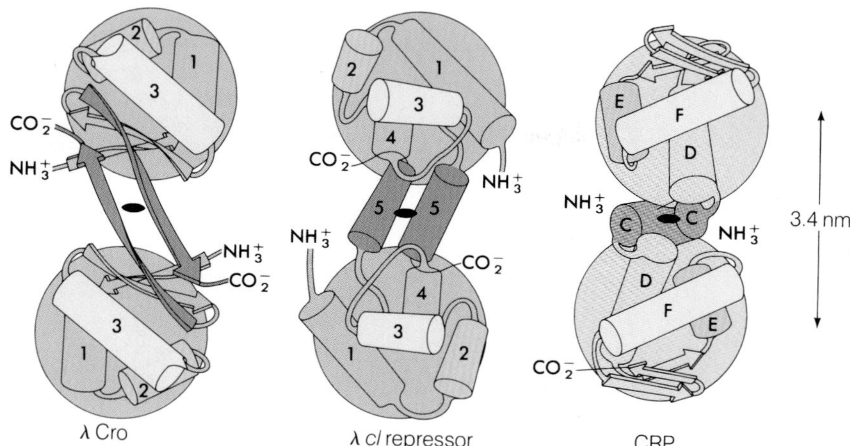

Figure 26.25
DNA binding faces of λ Cro, λ *cI* repressor, and CRP, showing the 2–3 helix–turn–helix motif (helices 2 and 3 in Cro and *cI*, E, and F in CRP). Each black ellipse represents an axis of symmetry.

and *cI* in Figure 26.26. Note that the sequences, while similar, are not identical. If they were identical, one would not be able to explain how Cro and *cI* repressors differ in their relative affinities for different operators. *cI* repressor contains an additional binding determinant—a pair of "arms," or short polypeptide segments that extend from helix 1 and are seen extending around the helix and establishing contacts on the other side (Figure 26.1). This probably explains why *cI* binds more tightly to its operators than Cro.

The α-3 helix is called the **recognition helix,** because its position deep within the major groove allows it to contact specific DNA bases and hence to determine sequence specificity of binding. The α-2 helix is in contact primarily with DNA phosphates. These electrostatic contacts strengthen binding but do not contribute to specificity. Supporting the concept of α-3 as a recognition helix is the fact that most *cI* mutations that reduce repressor binding to DNA alter the amino acid sequence in this region of the protein.

Model-building studies help to explain how Cro and *cI* can bind to the same operator sites with different binding affinities. As shown in Figure 26.27, the residues common to both proteins are seen as being in contact with DNA sequence elements common to all of the operators. In both proteins a glutamine residue interacts with one AT base pair, as shown. *cI* repressor establishes specificity through a contact in O_R1 with a unique alanine residue, while Cro can be in contact with three specific base pairs in O_R3 with unique asparagine and lysine residues. A recent experiment involved a "helix swap," with the α-3 helix recognition-determining amino

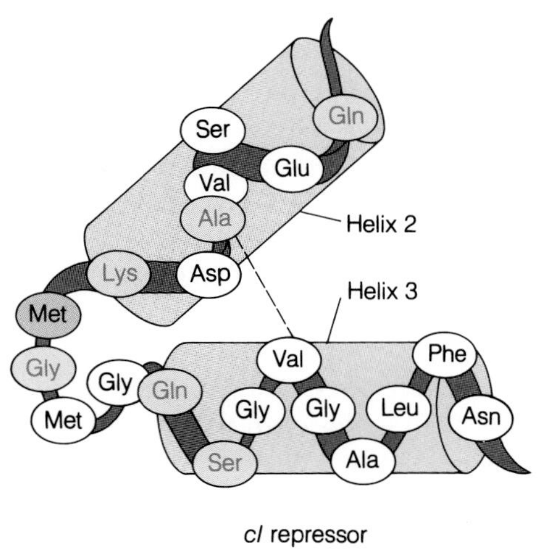

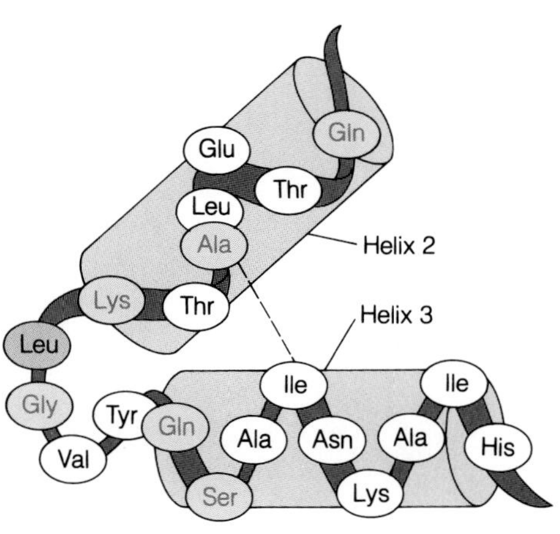

Figure 26.26
Conserved residues in the DNA-binding helices of λ *cI* repressor and Cro. Conservative substitutions are shown in purple and identities in red. The alanine in helix 2 contacts a residue in helix 3, which helps to position the helices with respect to each other in both proteins.

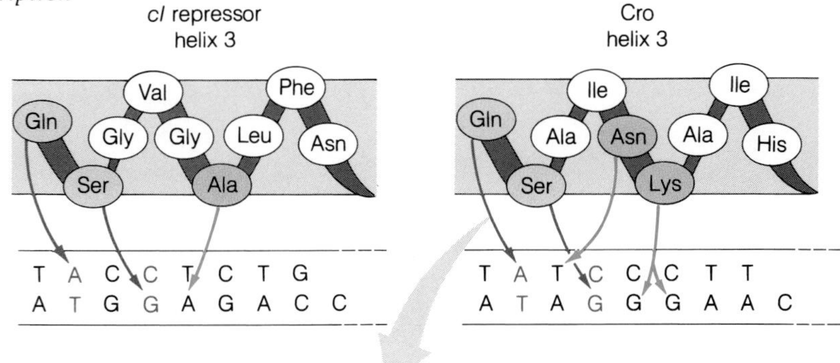

Figure 26.27
Specific amino acid–nucleotide contacts involving *cI* and Cro repressors. The conserved residues (red) bind to nucleotides common to all of the operators, and the unique residues (purple) bind to nonconserved nucleotides in the operators. Also shown is the structure of a glutamine residue in contact with an AT base pair.

acids from phage 434 repressor substituted in phage P22 repressor; the hybrid repressor assumed the binding specificity of phage 434. These experiments established that the helix–turn–helix motif is a common structural feature that allows regulatory proteins to control gene expression by binding tightly to DNA in a sequence-specific fashion.

The SOS Regulon: Activation of Multiple Operons by a Common Set of Environmental Signals

How is the λ *cI* repressor inactivated when the prophage is excised and begins a cycle of lytic growth? Various DNA-damaging treatments are known to induce λ prophages, including ultraviolet irradiation, inhibition of DNA replication, and chemical damage to DNA. Because of the similarity in genetic control between the λ and *lac* systems, investigators sought a small molecule, perhaps a nucleotide, that would accumulate after these treatments and that might be the ligand which binds to *cI* and inactivates it. Surprisingly, the λ repressor was found to be inactivated by a quite different mechanism—proteolytic cleavage. Analysis of this cleavage reaction revealed the SOS system described in Chapter 25 as one of the elements in error-prone DNA repair, in which the genes are controlled by a single repressor–operator system. Such a set of unlinked genes, regulated by a common mechanism, is called a **regulon.** The heat-shock genes, all activated by a transient temperature rise, constitute another regulon.

The control elements in the SOS regulon are the products of genes *lexA* and *recA*. We encountered RecA protein before, in its role of stimulating DNA strand pairing during recombination. Remarkably, this small protein has an enzymatic activity in addition to those involved in recombination; when bound to single-stranded DNA, it can stimulate proteolytic cleavage of the proteins encoded by *cI, lexA,* and *umuD*. LexA is a repressor that binds to at least 15 different operators scattered about the *E. coli* genome (Figure 26.28). Each operator controls the transcription of one or

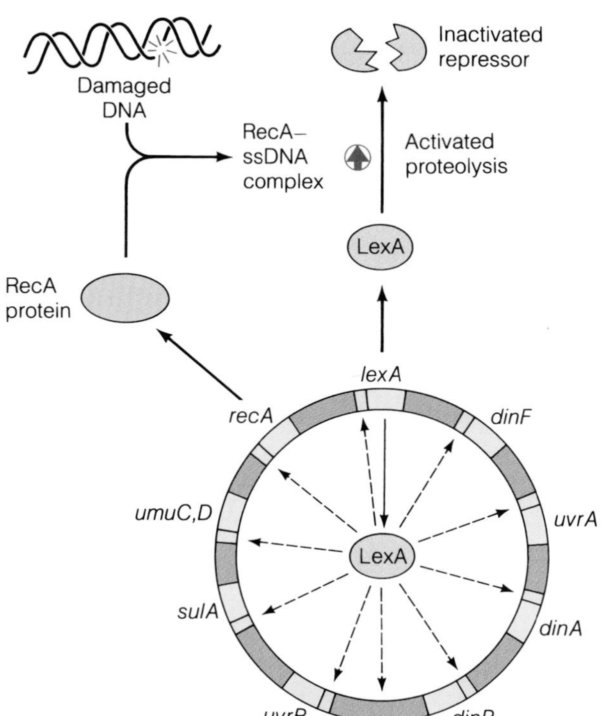

Figure 26.28
The SOS regulon. The figure shows locations on the *E. coli* chromosome of some of the genes controlled by the LexA repressor. *dinA*, *B*, and *F* are *d*amage-*in*ducible genes of unknown function. LexA repressor (pink) is inactivated by proteolysis, which is somehow enhanced by a complex of RecA protein (green) and single-stranded DNA.

more proteins that help the cell to respond after environmental damage that might harm the genetic apparatus. These proteins include the gene products of *uvrA* and *uvrB*, involved in excision repair; *umuC,D*, involved in error-prone mutagenesis; *sulA*, involved in cell division control; *recA* itself; *lexA* itself; and several genes of unknown function, including *dinA*, *dinB*, and *dinF*.

In a healthy cell *lexA* and *recA* are expressed at low levels, with sufficient LexA protein to turn off completely the synthesis of the other SOS genes. LexA protein does not completely abolish either *lexA* transcription or that of *recA*. The trigger that activates the SOS system after damage is thought to be single-stranded DNA. As we have seen, UV irradiation generates gapped DNA structures, and so do other conditions that induce the SOS system. RecA binding within a gap activates proteolysis by a mechanism not yet clear. Intracellular levels of LexA decrease, removing the LexA barrier to *recA* transcription. RecA protein accumulates in large amounts. Cleavage of the LexA protein activates transcription of all genes under *lexA* control. In a λ lysogen cleavage of λ *cI* repressor is stimulated as well, activating prophage excision and replication, as discussed in the previous section.

Sequencing of LexA operators has yielded a consensus sequence, with 7 highly conserved bases in a 20-base-pair region. As shown in Figure 26.29, different LexA-sensitive genes have this sequence located quite differently with respect to the transcriptional start site. Therefore, it appears that the exact location of bound repressor is not critical to ensure that transcription will be inhibited.

Biosynthetic Operons: Ligand-activated Repressors and Attenuation

The lactose operon is involved with catabolism of a substrate. Therefore, the gene products are not needed unless the substrate is also present to be consumed. A different situation is encountered with genes concerned with

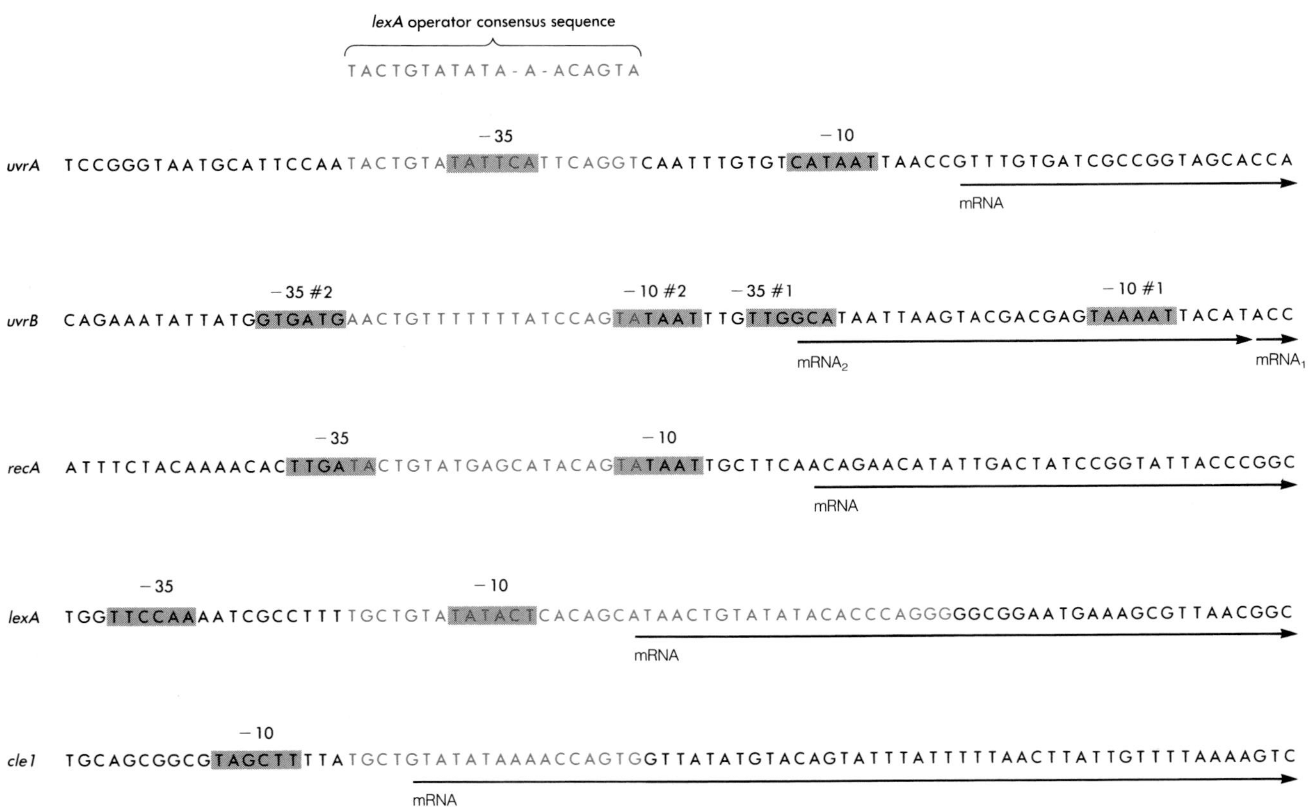

Figure 26.29
Sequences of the promoter–operator regions for several genes in the SOS regulon. The figure shows the *lexA* consensus binding sequence and the binding site, or sites, for LexA repressor (in red). The −10 and −35 regions are shaded in gray.

biosynthesis—of an amino acid, for example. Since biosynthesis consumes energy, it is to the cell's advantage to use the preformed amino acid, if it is available. Therefore, the regulatory goal is to repress gene activity, by turning *off* the synthesis of enzymes in the pathway when the end product is available. This is accomplished by a repressor design in which binding of a small-molecule ligand *activates* the repressor, rather than inactivating it. We can see this in the case of the *trp* operon, which controls the four reactions from chorismic acid to tryptophan (see Figure 21.11, Chapter 21). The operon consists of five linked structural genes whose transcription is controlled from a common promoter–operator regulatory region (Figure 26.30). The *trp* repressor is a 58-kDa protein, encoded by the unlinked *trpR* gene. The protein binds a low-molecular-weight ligand, namely tryptophan. However, in this case the protein–ligand complex is the *active* form of the protein, which binds to the operator and blocks transcription; when intracellular tryptophan levels decrease, the ligand–protein complex dissociates and the free protein ("aporepressor") leaves the operator, so that transcription is activated. If we call lactose an inducer in a catabolic system, it is appropriate to call tryptophan a **corepressor** in this anabolic system.

The crystal structure of the *trp* repressor–DNA complex shows a helix-turn-helix motif, comparable to that seen with the λcI, Cro, and *lac* repressors (Figure 26.31). Remarkably, this model shows no direct contacts between residues in the recognition helices and specific DNA bases. It is not yet clear how this protein specifically recognizes its DNA binding site.

The *trp* operon has an additional regulatory feature, now known to be involved in controlling numerous biosynthetic operons. Charles Yanofsky found that the activities of the *trp* enzymes varied over a 600-fold range under different physiological conditions, more than could be accounted for

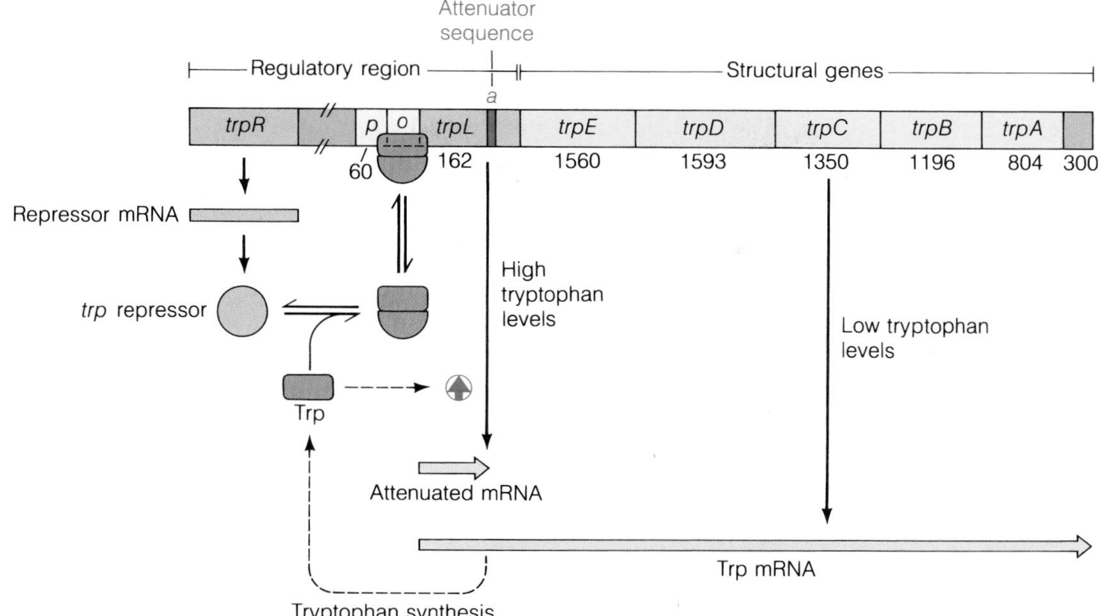

Figure 26.30
The *trp* operon. The figure shows regulation by *trp* repressor and by attenuation. *trpa,* the attenuator site, is shown in red.

by a repressor–operator mechanism. Analysis revealed a second mechanism, called **attenuation,** that involves abortive termination of *trp* operon transcription under conditions of tryptophan abundance. Note from Figure 26.30 a 162-nucleotide sequence called *trpL,* the *trp* leader region. 131 nucleotides from the 5′ end of this sequence is a site called *a,* the attenuator. When tryptophan levels are high, transcription terminates here, to give a 131-nucleotide transcript rather than the 7000-nucleotide *trp* mRNA.

Critical to understanding the mechanism of attenuation is the presence in the *trp* leader region of four oligonucleotide sequences capable of base pairing to form stem–loop structures (Figure 26.32). In the most stable conformation (Figure 26.33a) region 1 pairs with 2, and 3 pairs with 4, to give two stem–loops. The 3–4 structure, being followed by eight U's, is an efficient transcription terminator, since it resembles the factor-independent terminator structure we showed in Figure 26.12. When tryptophan levels are low (Figure 26.33b), formation of the 3–4 stem–loop is inhibited, and termination does not occur at the attenuation site. Note that region 1 contains two tryptophan codons (Figure 26.32). In prokaryotes translation is coupled to transcription, so a ribosome can begin translating a message from its 5′ end while the message is still being synthesized at its 3′ end. In this case, the ribosome stalls when it reaches the two tryptophan codons, because there is insufficient tryptophanyl-tRNA to translate them. The presence of the bulky ribosome prevents region 1 from base pairing with 2, leaving 2 free to base pair with 3. Since the 2–3 pairing creates a more stable structure than the 3–4 stem–loop, the transcriptional terminator cannot form, and the entire message is synthesized. Conversely, when tryptophan is abundant (Figure 26.33c), the ribosome does not stall, and the 3–4 stem–loop structure forms, leading to transcription termination on the 3′ side of 3–4.

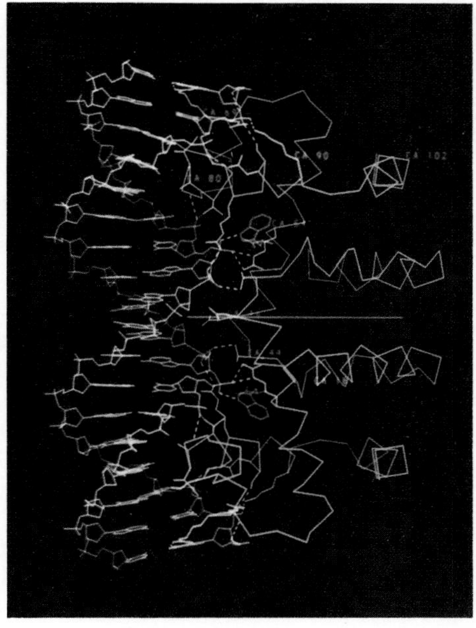

Figure 26.31
A model of the *trp* repressor-operator complex. DNA is in gold, the α-carbon trace of the repressor backbone is in blue, and the side chains that contact the operator are in green. The bound tryptophan is in red.

Neither the *trpR* system nor the attenuator is simply an on–off system; both respond in graded fashion to the intracellular tryptophan level. Even though both systems are controlled by the same signal, the action of two distinct control systems greatly extends the possible range of rates of transcription of the *trp* operon, giving maximum efficiency to regulation of these genes. At low tryptophan concentration the repressor–operator inter-

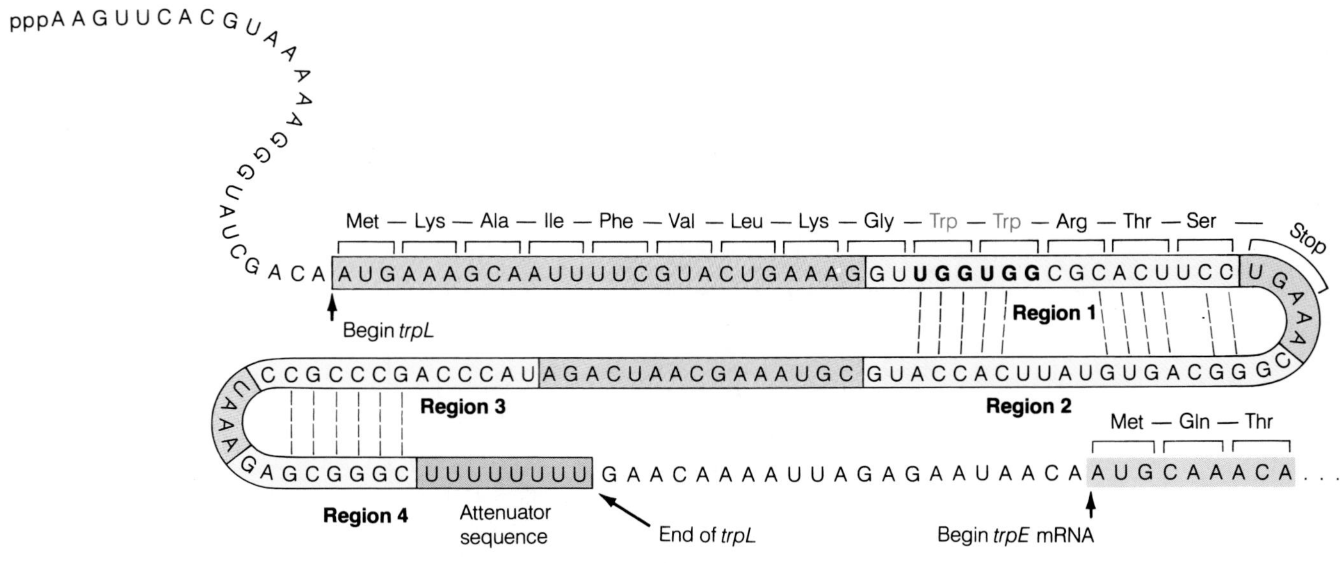

Figure 26.32
RNA base sequence of the *trp* leader region, showing (in yellow) the internally complementary sequences that participate in attenuation and the two *trp* codons in region 1.

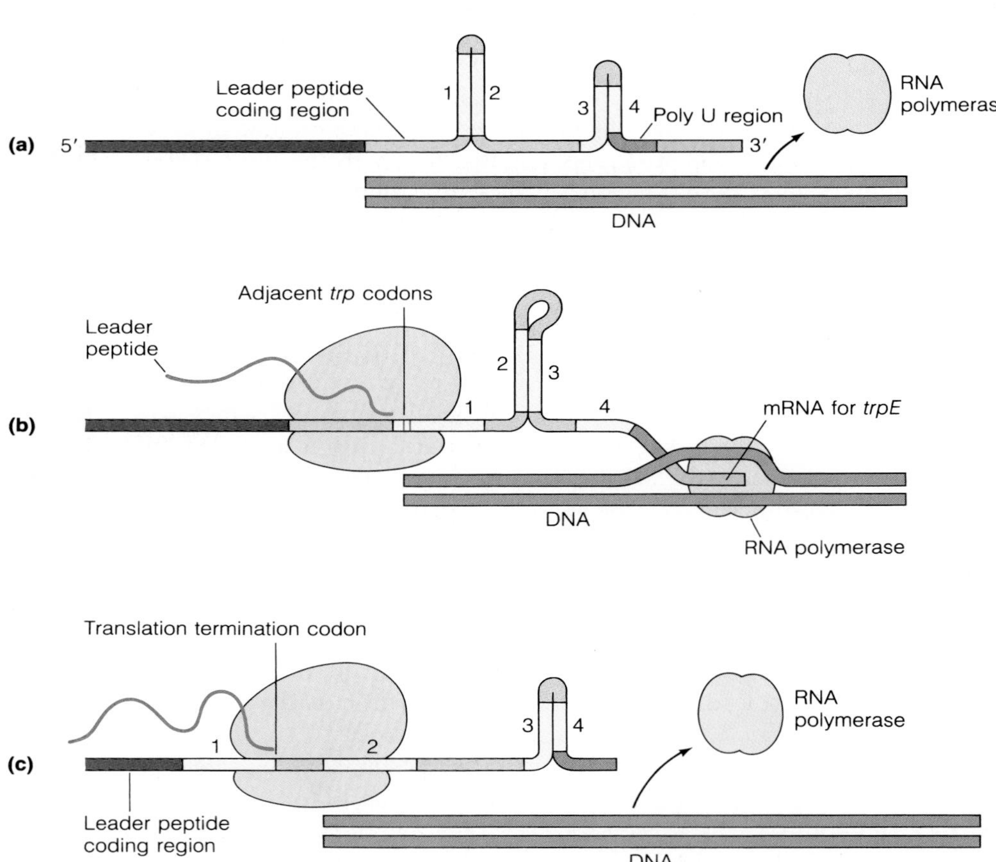

Figure 26.33
Mechanism of attenuation in the *trp* operon. (**a**) Most stable conformation for leader mRNA; (**b**) conformation for leader mRNA at low tryptophan levels; (**c**) conformation for leader mRNA at high tryptophan levels.

action is the principal regulatory mechanism, while attenuation comes into play primarily at moderate to high tryptophan levels.

Although the foregoing model was originally proposed simply by inspection of the *trp* leader sequence, it is now supported by several lines of evidence. Significant confirmation comes from the existence in other attenuation-controlled operons of "stalling sequences"—sequences at which movement of a ribosome is inhibited at a low concentration of the product of the operon. These include operons for synthesis of leucine, with four adjacent leucine codons in the leader sequence, and histidine, with seven.

Other Forms of Regulation

Biochemical analyses of the *lac*, λ phage, *trp*, and SOS regulatory systems have confirmed the central tenets espoused by Jacob and Monod—that gene expression is regulated at the level of transcription and that specific DNA–protein interactions control the rate of transcription primarily by regulating transcriptional initiation. These analyses have also revealed several important variations on the simple theme—including positive control of initiation, interspersed operators and repressors, dual proteins binding to the same site, multiple operons controlled by the same repressor, and abortive termination as a regulatory mechanism. Thus, it is clear that the repressor–operator mode of regulation is far from monolithic. Here we briefly describe two other well-studied operons that reveal additional themes.

THE GALACTOSE OPERON. The **galactose operon** (Figure 26.34) controls the utilization of galactose, one of the products of lactose cleavage by *lac* operon enzymes. The *gal* operon is regulated negatively by a repressor in a manner comparable to *lac* regulation, except that the repressor gene *(galR)* is unlinked to the structural genes. The novelty of *gal* comes from the existence of overlapping promoters, leading to transcripts that are initiated just five nucleotides apart. Transcription from the start point called S1 is catabolite activated; in vivo, S1 is silent when glucose is present, and in vitro, the cAMP–CRP complex must be present for transcription to start from this promoter. The other promoter, S2, is used when glucose is present. The details of this dual regulation are not clear, but it is significant that galactose has a biosynthetic fate in addition to its role as an energy sub-

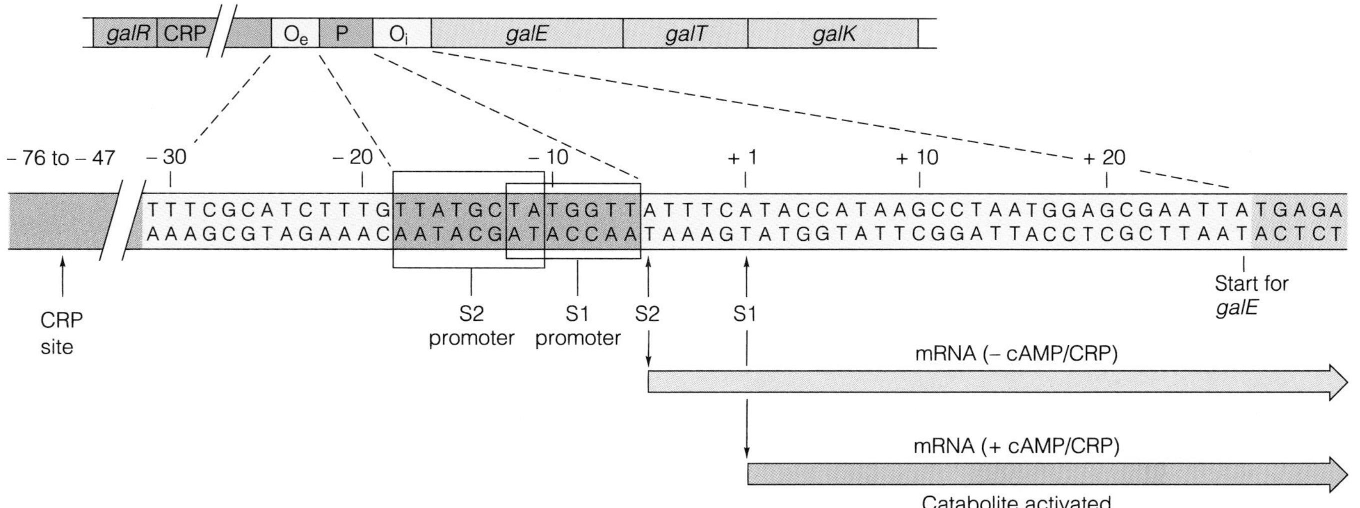

Figure 26.34
Map of the *E. coli gal* operon and nucleotide sequence of the regulatory region. The −10 regions of the two overlapping promoters are shown in brown. The transcriptional start site for each promoter is shown with an arrow. *galR*, repressor gene; *galE*, epimerase gene; *galT*, transferase gene; *galK*, kinase gene.

strate. UDP-galactose is used in synthesis of cell wall lipopolysaccharide. The second promoter may exist to ensure that UDP-galactose is made available even when the cell is using glucose as its prime energy source.

The Arabinose Operon. The **arabinose operon** presents another unusual regulatory feature, namely enzyme induction acting through an operon-specific *positive* transcriptional regulator (Figure 26.35). The three structural genes in this operon—*araB, A,* and *D*—encode enzymes that convert arabinose to xylulose-5-phosphate, a pentose phosphate pathway intermediate. *araC* encodes a regulatory protein that binds to the inducer of the operon, namely arabinose. Binding of the *araC*–arabinose complex at a site called *araI* activates transcription of *araBAD*, but only when the cAMP–CRP complex is bound at an adjacent site. Thus, while the *lac* operon requires one protein to be bound and one to be dissociated for maximal transcription, the *ara* operon requires two proteins to be bound at adjacent sites.

Another novel feature of *ara* regulation is that in the absence of arabinose, the *araC* protein acts as a *repressor* and binds to two operator sites within the operon, $araO_1$ and $araO_2$. The use of the same protein as both positive and negative control elements is unusual but not unprecedented. Recall that the *cI* repressor can either activate or inhibit its own synthesis, depending on the operator sites to which it binds.

Control by Antisense RNA. Lest you conclude that all of prokaryotic transcriptional regulation is simply a matter of repressors, operators, and transcriptional activators, let us mention two other regulatory mechanisms. While not strictly a regulator of transcription, **antisense RNA** has been shown to control the expression of several prokaryotic and eukaryotic genes. Antisense RNA is the transcription product of the DNA antisense strand—the strand that does not encode a protein; presumably it can pair with an mRNA species, thereby inhibiting translation of that species.

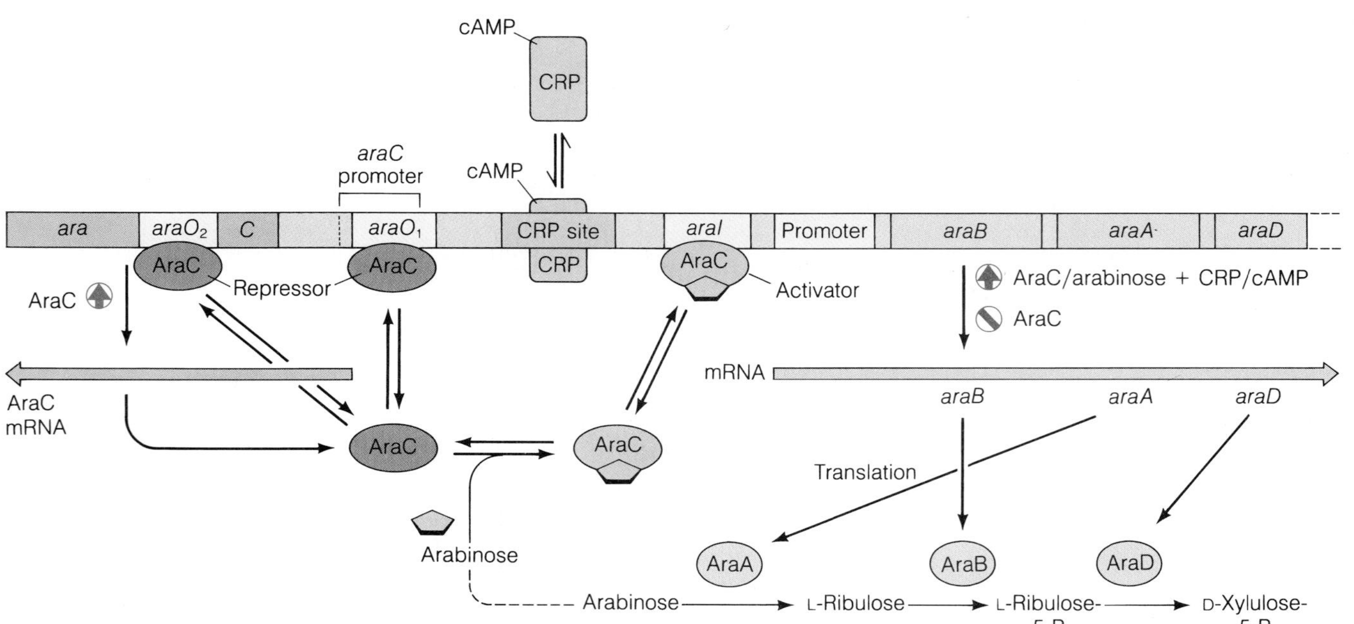

Figure 26.35
Structure of the *E. coli ara* operon. The AraC protein (yellow) acts as both a positive (green) and a negative (purple) transcriptional regulator. The AraC–arabinose complex binds at the *araI* site and, when the CRP–cAMP complex is bound nearby, activates transcription of the *araBAD* operon, thus promoting arabinose utilization through action of the AraB, AraA, and AraD enzymes. When intracellular arabinose levels are low, the uncomplexed AraC protein binds to two operators, $araO_1$ and $araO_2$. This binding activates transcription of the *araC* gene. Binding at $araO_2$ also leads to repression of the *araBAD* operon, by a mechanism not presented here.

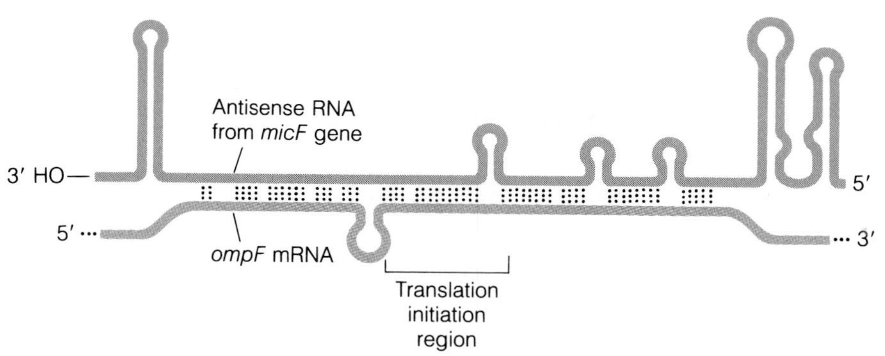

Figure 26.36
Inactivation of *ompF* mRNA by pairing with antisense RNA from the *micF* gene. A change in osmolarity stimulates transcription of the *micF* gene. The transcript is largely complementary to a region in *ompF* RNA that includes the translational start site. Hairpin loops within the sequences allow base pairing between complementary regions on the two mRNAs and in this way both messengers are prevented from serving as templates for protein synthesis.

A good example of antisense regulation involves *ompC* and *ompF,* two genes that encode outer membrane proteins in *E. coli.* These genes are osmoregulated: cells respond to growth in a medium of high osmolarity by shutting down the synthesis of OmpF protein and activating the synthesis of OmpC, so that the total amount of protein is constant. The *ompF* shutoff results from synthesis of an antisense RNA, the product of the *micF* gene, which is partly complementary to sequences in the 5′ end of *ompF* mRNA. The *micF* RNA inactivates the *ompF* message, presumably by forming a duplex RNA in vivo (Figure 26.36). The translational initiation sequences of OmpF mRNA are included in this duplex, which probably is responsible for blocking translation of this message. High osmolarity triggers the synthesis of this antisense RNA, by mechanisms not yet clear.

Another gene regulated by antisense RNA is *crp,* the structural gene for cAMP receptor protein. Also, the transposon Tn10 uses an antisense mechanism to regulate the level of its transposase. A promising approach in chemotherapy is the development of antisense oligonucleotides or chemically modified derivatives, which can be introduced into human cells and inactivate selected target genes. This approach is receiving intense scrutiny for possible treatment of viral diseases and cancer.

THE STRINGENT RESPONSE. The final control mechanism is one we introduced in Chapter 22—the **stringent response,** in which ribosomal RNA synthesis is inhibited when guanosine 3′,5′-tetraphosphate accumulates as a result of amino acid starvation. The regulatory action of ppGpp itself is not yet known, which points up our lack of detailed information about the control of ribosomal gene transcription. On the other hand, the events between ribosomal transcription and rRNA maturation are reasonably well understood, as we shall see in the next section.

Posttranscriptional Processing

mRNA Turnover

A major aspect of messenger RNA metabolism in eukaryotes is the events occurring *after* transcription, events that are necessary for messages to move from the nucleus to their sites of utilization in the cytosol. We discuss these events in Chapter 28. In prokaryotes, by contrast, mRNAs are used in protein synthesis directly. In fact, as we noted in our discussion of attenuation, a nascent mRNA serves as a template for translation while still in the process of its own synthesis.

The major posttranscriptional event in metabolism of prokaryotic mRNA is its own degradation, which in most cases is quite rapid. A few bacterial mRNAs, notably those encoding outer membrane proteins, are long-lived; however, most bacterial messages are degraded with half-lives of 2 or 3 minutes. This means that genes which are expressed must be transcribed continuously, and that most mRNA molecules are translated only a few times. Although this might seem energetically wasteful, it is consistent with prokaryotic life-styles, which necessitate rapid adaptation to environmental changes. Earlier we noted the selective advantage to bacteria of expressing the genes for lactose utilization only when an inducer is present. By the same token, it would be wasteful for the cell to continue producing these proteins after lactose was exhausted from the milieu. Rapid degradation of *lac* mRNA ensures that the energetically wasteful synthesis of these proteins will cease soon after the need for the proteins is gone.

Although we have known about the instability of bacterial mRNA for nearly three decades, we still know surprisingly little about the pathway of degradation. There are probably overlapping mechanisms, involving hydrolysis by nucleases and phosphorolysis by polynucleotide phosphorylase. We do know that degradation starts from the 5′ end, which is important because translation also starts from the 5′ end. If degradation were to start from the 3′ end, a ribosome starting from a 5′ end might never reach an intact 3′ end. There is reason to think that mRNA degradation sometimes starts with the action of ribonuclease III, an enzyme specific for duplex RNA, which could cleave in stem–loop structures and create sites for exonucleolytic attack. RNase III is actually involved in the maturation of certain phage mRNAs as they undergo posttranscriptional processing, but this is not known to occur with bacterial mRNAs.

Posttranscriptional Processing in the Synthesis of rRNA and tRNA

Both ribosomal RNAs and transfer RNAs are synthesized in the form of larger transcripts (pre-rRNA and pre-tRNA, respectively), which undergo cleavage at both ends of the transcript, en route to becoming mature RNAs. The total amount of DNA encoding these RNAs amounts to less than 1% of the *E. coli* genome, but because of the instability of mRNA (which is encoded by the remaining 99%), rRNA and tRNA constitute about 98% of the total RNA in a bacterial cell.

rRNA PROCESSING. The *E. coli* genome contains seven different operons for rRNA species. Each one encodes in a single transcript sequences for one copy each of 16S, 23S, and 5S rRNAs. Since the three species are used in equal amounts, the logic of this organization is apparent. Less easy to explain is the fact that each transcript also includes sequences for one to four tRNA molecules (Figure 26.37). Since rRNAs and tRNAs are all used in protein synthesis, the interspersion of rRNA and tRNA sequences may represent a means of coordinating the rates of synthesis of these RNAs, but specific mechanisms have not yet been revealed.

The initial transcript from each rRNA operon is a short-lived RNA molecule of 30S. The abnormal accumulation of this species in bacterial strains defective in RNase III first suggested a role for this enzyme in rRNA processing. In fact, one double-strand cut in each of two giant stem–loop regions releases precursors to 16S and 23S rRNAs, and the same probably occurs for 5S rRNA. Further maturation steps require the presence of par-

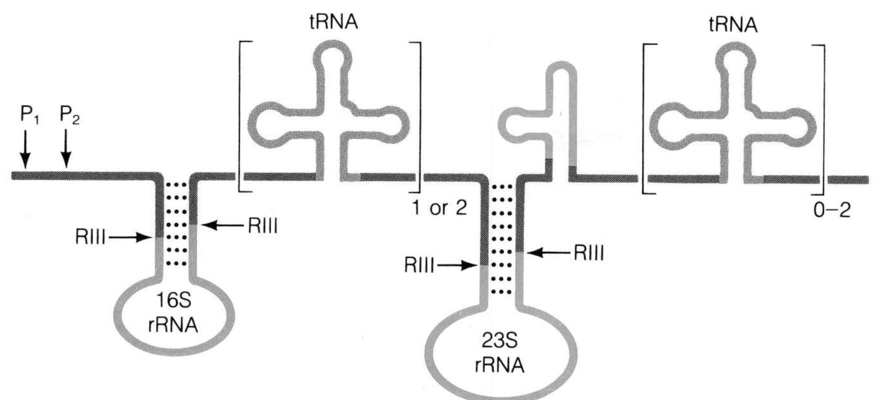

Figure 26.37
Structure of *E. coli* pre-rRNA. Two promoter sites (P_1 and P_2), RNase III cleavage sites (RIII) that release 16S and 23S species, and the locations of tRNA sequences embedded within the transcript are shown.

ticular ribosomal proteins, which begin to assemble on the precursor RNAs while transcription is still in progress. The tRNA sequences are processed to give mature tRNAs, along the same routes used for other tRNA species.

tRNA PROCESSING. Aside from the tRNAs embedded in pre-rRNA transcripts, the other tRNAs are synthesized in transcripts that contain one to seven tRNAs each, all surrounded by lengthy flanking sequences. The maturation steps are summarized in Figure 26.38, using as an example the well-studied case of the *E. coli* tyrosine tRNA species ($tRNA^{tyr}$). In this case maturation starts (step 1) with an endonuclease that cleaves at a stem–loop structure on the 3′ side of the tRNA sequence. This is followed by action of **ribonuclease D** (step 2), which carries out exonucleolytic cleavage to a point two nucleotides removed from the CCA sequence at the 3′ end. Next (step 3), the 5′ end is created by **ribonuclease P,** which cleaves to leave a phosphate on the 5′ terminal G. This enzyme creates the 5′ terminus of all tRNA

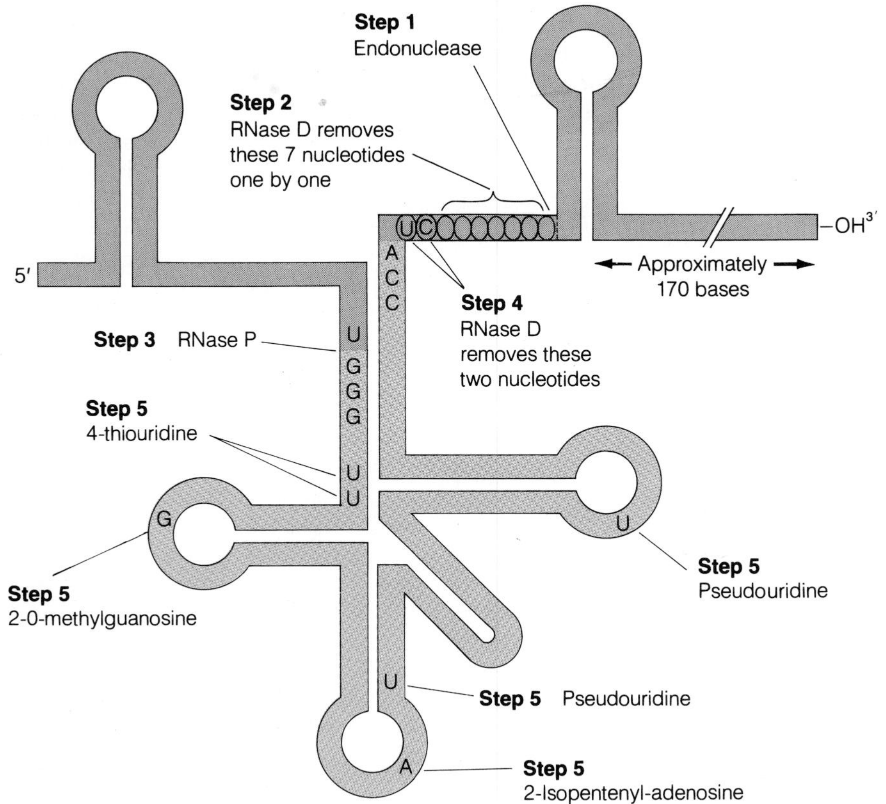

Figure 26.38
Steps in maturation of *E. coli* $tRNA^{tyr}$ from its transcript. The tRNA sequence is shown in purple.

molecules. It is not clear what structural features are recognized by RNase P, for different sequences are contained in the cleavage sites. Ribonuclease P is an amazing enzyme that consists of one RNA molecule of 377 nucleotides and one protein molecule of M_r about 20,000. Both components are necessary for full catalytic activity, but under nonphysiological conditions the RNA molecule alone can catalyze accurate cleavage. Thus, ribonuclease P is a ribozyme, a member of the small class of RNA enzymes that we introduced in Chapter 10.

Once the proper 5′ terminus has been created, ribonuclease D removes the remaining two nucleotides from the 3′ end (step 4). Should excessive "nibbling" occur through faulty control of RNase D activity, there is an enzyme that will restore the CCA end to any tRNA in a nontranscriptive fashion. This enzyme specifically recognizes the 3′ terminus of tRNAs that lack the CCA end and catalyzes sequential reactions with CTP, another CTP, and ATP.

Creation of the modified bases occurs at the final stage, including methylations, thiolations, reduction of uracil to dihydrouracil, and so forth. In the specific example shown, the modifications include formation of two pseudouridines, one 2-isopentenyladenosine, one O^2-methylguanosine, and one 4-thiouridine (step 5).

Having discussed the transcriptional reactions that generate all RNA species and the posttranscriptional processing that generates ribosomal and transfer RNAs, we are ready to see how these species work together in the synthesis of proteins. That is our concern in the next chapter.

REFERENCES

Mechanism of Transcription

Platt, T. (1986) Transcription termination and the regulation of gene expression. *Annu. Rev. Biochem.* 55:339–372. A detailed review of transcriptional termination.

Watson, J. D., N. H. Hopkins, J. A. Steitz, and A. M. Weiner (1986) *Molecular Biology of the Gene,* 4th ed. Benjamin/Cummings, Menlo Park, Calif. This excellent book discusses transcriptional regulation in more detail than is possible here and is a fine general reference as well.

Yager, T. D., and P. H. von Hippel (1987) "Transcript elongation and termination in *Escherichia coli.*" In: Escherichia coli *and* Salmonella typhimurium, edited by F. C. Neidhardt, pp. 1241–1275. American Society for Microbiology, Washington, D.C. This two-volume set presents a wealth of information about the biochemist's favorite microorganism, including this exhaustive review of transcriptional mechanisms.

DNA-Protein Interactions and Transcriptional Regulation

Berg, O. G., and P. H. von Hippel (1988) Selection of DNA binding sites by regulatory proteins. *Trends Biochem. Sci.* 13:207–211. This minireview summarizes, with key references, the factors that generate specificity in DNA–protein interactions.

Brennan, R. G., and B. W. Matthews (1989) The Helix-Turn-Helix DNA binding motif. *J. Biol. Chem.* 264:1903–1906. A recent minireview that discusses gene regulation in the context of surprising features of the *trp* repressor-operator complex.

Gartenberg, M. R., and D. M. Crothers (1988) DNA sequence determinants of CAP-induced bending and protein binding affinity. *Nature* 333:824–829. This paper describes the mechanism by which catabolite repressor protein binds to its site in catabolite-sensitive promoters, including the discovery that DNA bends at this site to bind the protein.

Pruss, G. J., and K. Drlica (1989) DNA supercoiling and prokaryotic transcription. *Cell* 56:521–523. Summarizes and integrates information pointing to DNA topology as a transcriptional control mechanism.

Ptashne, M. (1986) *A Genetic Switch: Gene Control and Phage* λ. Cell Press, Cambridge, Mass., and Blackwell Scientific, Palo Alto, Calif., 128 pages. A short, lucidly written, and beautifully illustrated book describing one of the most fruitful systems for analyzing transcriptional regulation and DNA–protein interactions.

Reznikoff, W. S., and L. Gold, eds. (1987). *Maximizing Gene Expression.* Butterworths, Boston. This multiauthored book is intended for those using recombinant DNA technology to optimize the production of cloned gene products. Particularly relevant to this chapter are articles by W. S. Reznikoff and W. R. McClure, "*E. coli* Promoters" and by D. E. Kennell, "The Instability of Messenger RNA in Bacteria."

Wang, J. C., and G. N. Giaever (1988) Action at a distance along a DNA. *Science* 240:300–304. This article summarizes three mechanisms by which binding a protein to one DNA site can affect events at a remote site, as occurs in transcriptional regulation—tracking, DNA looping, and topoisomerization.

Yanofsky, C. (1988) Transcription attenuation. *J. Biol. Chem.* 263:609–612. A concise and informative minireview by the discoverer of this phenomenon.

Posttranscriptional Processing

Cech, T. R., and B. L. Bass (1986) Biological catalysis by RNA. *Annu. Rev. Biochem.* 55:599–630. Both RNase P and the "ribozymes" involved in eukaryotic mRNA splicing are discussed in this review.

Green, P. J., O. Pines, and M. Inouye (1986) The role of antisense RNA in gene regulation. *Annu. Rev. Biochem.* 55:569–597. Summarizes knowledge of this newly discovered regulatory phenomenon in both prokaryotes and eukaryotes.

PROBLEMS

1. Outline an experimental approach to determining the average chain growth rate for transcription in vivo. Chain growth rate is the number of nucleotides polymerized per minute per RNA chain.

2. Outline an experimental approach to determining the average RNA chain growth rate during transcription of a cloned gene in vitro.

3. Measurements of RNA chain growth rates are often led astray by the phenomenon of pausing, in which an RNA polymerase molecule stops transcription when it reaches certain sites, for intervals that may be as long as several minutes. How might pausing be detected?

4. The active form of lactose repressor binds to the operator with a dissociation constant of 10^{-13} for the reaction $R + O \rightleftharpoons RO$. About 10 molecules per *E. coli* cell suffice to keep the operon turned off in the absence of inducer.
   (a) If the average *E. coli* cell has an intracellular volume of 0.3×10^{-12} ml, calculate the approximate intracellular concentration of repressor.
   (b) If the average cell contains two copies of the *lac* operon, calculate the approximate intracellular concentration of operators.
   (c) Calculate the average intracellular concentration of *free* operators under these conditions.
   (d) Explain how a cell with a haploid chromosome could contain an average of two copies of the *lac* operon.

5. Is attenuation likely to be involved in eukaryotic gene regulation? Briefly explain your answer.

6. RNA polymerase has a higher K_M for ribonucleoside triphosphates in the chain initiation step than in the elongation steps. How might this be important in regulating transcription?

7. Suppose you want to study the transcription in vitro of one particular gene in a DNA molecule that contains several genes and promoters. Without adding specific regulatory proteins, how might you stimulate transcription from the gene of interest relative to the transcription of the other genes on your DNA template?

8. For some time it was not clear whether *lac* repressor inhibits *lac* operon transcription by inhibiting the binding of RNA polymerase to its promoter or allowing transcription initiation but blocking elongation past the site of bound repressor. How might you distinguish between these possibilities?

9. The *tac* promoter, an artificial promoter made from a chemically synthesized oligonucleotide, has been introduced into a plasmid. It is a hybrid of the *lac* and *trp* promoters, containing the -35 region of one and the -10 region of the other. This promoter directs transcription initiation more efficiently than either the *trp* or *lac* promoters. Why?

10. A *lac* operon containing one mutation was cloned into a plasmid, which was introduced by transformation into a bacterium containing a wild-type *lac* operon. The three genes of the chromosomal operon were rendered noninducible in the presence of the plasmid.
    (a) What kind of mutation in the plasmid operon could have this effect?
    (b) Suppose the result of transformation was to cause the three plasmid *lac* genes to be expressed constitutively, at a high level. What type of plasmid gene mutation could have this result?

11. Several new genes in the SOS regulon were identified by an ingenious use of "Mud" phages. These are derivatives of phage Mu that have a promoterless β-galactosidase gene inserted at a particular point in this phage genome. How might these phages be used to identify genes whose expression is turned on after ultraviolet irradiation of bacteria?

12. Would you expect actinomycin D to be a competitive inhibitor of RNA polymerase? What about cordycepin? Briefly explain your answers.

TOOLS OF BIOCHEMISTRY 23

Footprinting: Identifying Protein-binding Sites on DNA

Transcription is controlled in large part through interactions of proteins with specific sites on DNA molecules, including operators and promoters. Such sites were initially identified through genetic analysis in the lactose system and phage λ. Biochemical analysis requires the identification and nucleotide sequence determination of a site, along with structural determination of the DNA–protein complexes that form. A technique called **footprinting,** usually involving DNase I protection, is widely used to identify such sites.

Footprinting can be used to identify any DNA site that binds a protein specifically, as long as the protein binds sufficiently tightly. For example, the method has received widespread use to identify promoters, by analysis of binding to RNA polymerase.

The principle of the method, outlined in Figure T23.1, is that binding of RNA polymerase to a specific DNA sequence should protect that DNA from attack by DNase I (pancreatic deoxyribonuclease). One uses γ[^{32}P]ATP and T4 polynucleotide kinase to prepare a 5′ end-labeled fragment of DNA containing a protein-binding site. One aliquot of the end-labeled DNA is mixed with the protein under study (step 1) and then incubated with DNase I under conditions where most chains are cleaved only once (step 2). Another aliquot is incubated with DNase I under identical conditions, except that the protein is absent (step 2′). Next, the two incubation mixtures are analyzed in adjacent lanes of a sequencing gel (step 3). One sees a ladder of fragments identical to that seen in a Maxam–Gilbert sequencing gel, except that there is little sequence specificity; the bands on the gel are spaced uniformly at one-nucleotide intervals. Any sites protected from DNase attack because of interaction with RNA polymerase yield either no band or a low-intensity band in the ladder from the DNA–protein complex, indicating that little or no cleavage occurred at that site. The blank region on the gel pattern from the DNA–protein complex (the "footprint") identifies the location and the size of the fragment in contact with RNA polymerase (or with any other site-specific DNA-binding protein).

Recent improvements in footprinting technology involve cleavage with chemical agents such as **methidiumpropyl-EDTA-Fe^{2+}** (MPE-Fe^{2+}).

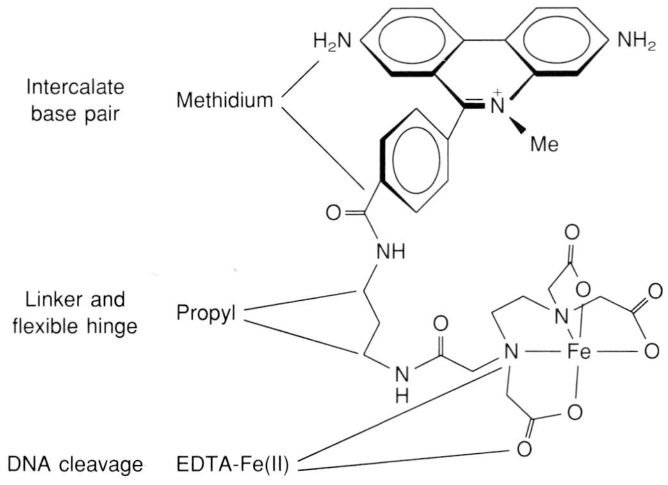

Intercalate base pair — Methidium

Linker and flexible hinge — Propyl

DNA cleavage — EDTA-Fe(II)

This compound intercalates between DNA bases, as does ethidium bromide, and catalyzes oxidation leading to cleavage at a nearby site. Because there is some sequence selectivity in DNase I attack and none with MPE-Fe^{2+}, the latter technique gives cleaner footprints.

Footprinting shows that RNA polymerase binds to a region about 60 base pairs in length, extending from about 40 nucleotides upstream, or on the 5′ side, of the transcriptional start site to about 20 nucleotides past that site. In other words, the binding site extends from nucleotide −40 to +20, where the template for the first nucleotide in the transcript is +1. That first nucleotide, or the 5′ end of the transcript, can be identified by various methods; the most widely used is **S1 nuclease mapping,** which is described in Tools of Biochemistry 24.

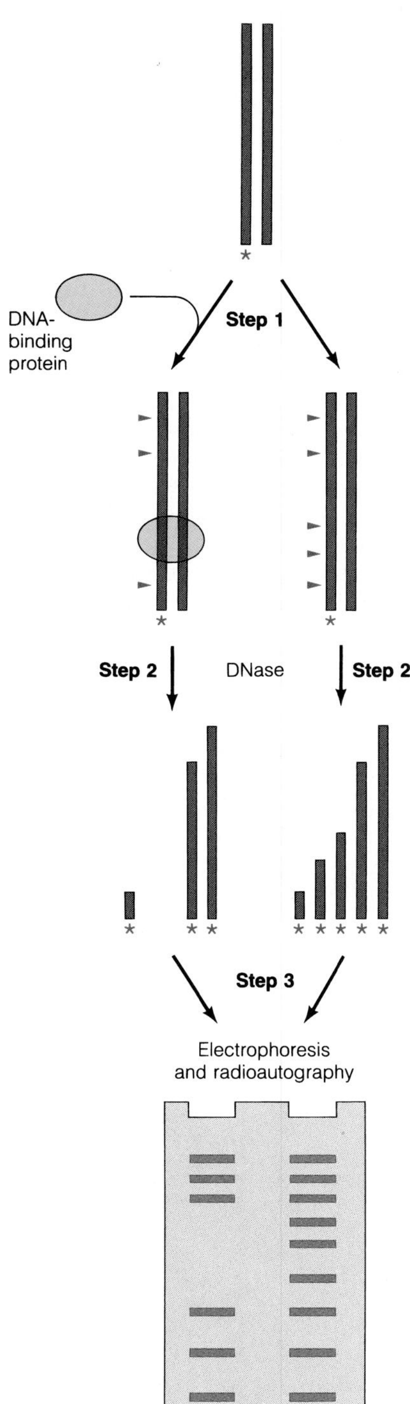

Figure T23.1
DNase footprinting as a tool to identify DNA sites that bind specific proteins.

24

TOOLS OF BIOCHEMISTRY 24

Mapping Transcriptional Start-Points

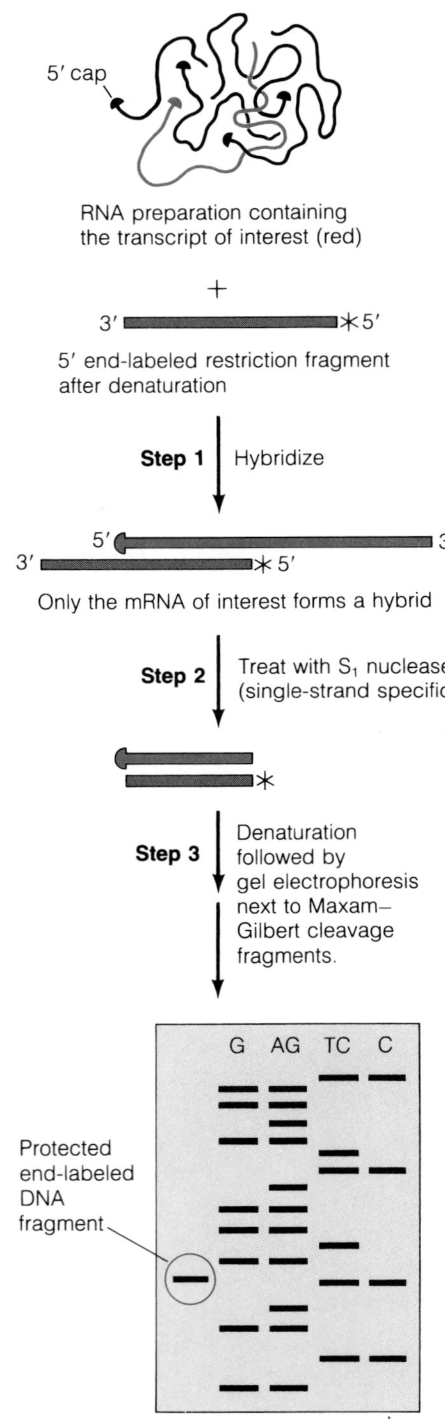
5′ cap
RNA preparation containing the transcript of interest (red)
+
3′
5′
5′ end-labeled restriction fragment after denaturation
Step 1
Hybridize
5′
3′
3′
5′
Only the mRNA of interest forms a hybrid
Step 2
Treat with S_1 nuclease (single-strand specific)
Step 3
Denaturation followed by gel electrophoresis next to Maxam–Gilbert cleavage fragments.
G
AG
TC
C
Protected end-labeled DNA fragment
Deduce where sequence of protected (complementary) DNA strand ends

Studies of transcriptional initiation and its control require methods for the accurate identification of transcriptional start points, specifically the DNA template nucleotide that encodes the 5′ nucleotide of the transcript. The low abundance of specific mRNAs and high turnover of nearly all bacterial mRNAs make this a challenging task.

Prokaryotic transcripts all have a 5′ triphosphate terminus on the first nucleotide, which can provide a handle for their identification. Since this identification requires purification of the transcript, less laborious methods are usually preferable. One such method, **S_1 nuclease mapping,** uses the fungal enzyme S_1 nuclease, which specifically and quantitatively cleaves single-stranded DNA and RNA (Figure T24.1). One needs the cloned gene and a restriction fragment thought to contain the 5′ end of the transcript. The fragment is 5′ end-labeled, as in the Maxam–Gilbert sequencing procedure, and cleaved asymmetrically with another restriction enzyme, so that only the template DNA strand is labeled. This fragment is denatured and hybridized to mRNA (Figure T24.1, step 1). The only double-stranded nucleic acid, then, should be a DNA–RNA hybrid with 3′ single-stranded extensions from the 5′ end of the transcript and the labeled 5′ end of the restriction fragment. Treatment with S_1 nuclease (step 2) then yields a fully duplex structure, with a labeled DNA strand whose length precisely measures the distance from the 5′ end of the transcript to the relevant restriction site. This distance can be identified by denaturing the DNA–RNA hybrid and running it on a sequencing gel, alongside a set of Maxam-Gilbert cleavage fragments (step 3).

An alternative technique, called **primer extension,** requires a 5′ end-labeled restriction fragment whose 3′ end lies downstream of the promoter. This is hybridized to an RNA preparation containing the message of interest (Figure T24.2). Then the mixture is incubated with reverse transcriptase and nonradioactive deoxyribonucleoside triphosphates. Extension from the 3′ end of the fragment continues until the 5′ end of the transcript is reached. The length of this labeled fragment, minus the length of the original primer, gives the distance from the 5′ end of the transcript to the 3′ end of the restriction fragment used. This method may not have sufficient resolution to identify the precise 5′ terminal nucleotide, but it often works when S_1 nuclease mapping does not.

Figure T24.1
S_1 nuclease mapping method for identifying the 5′ end of an RNA molecule.

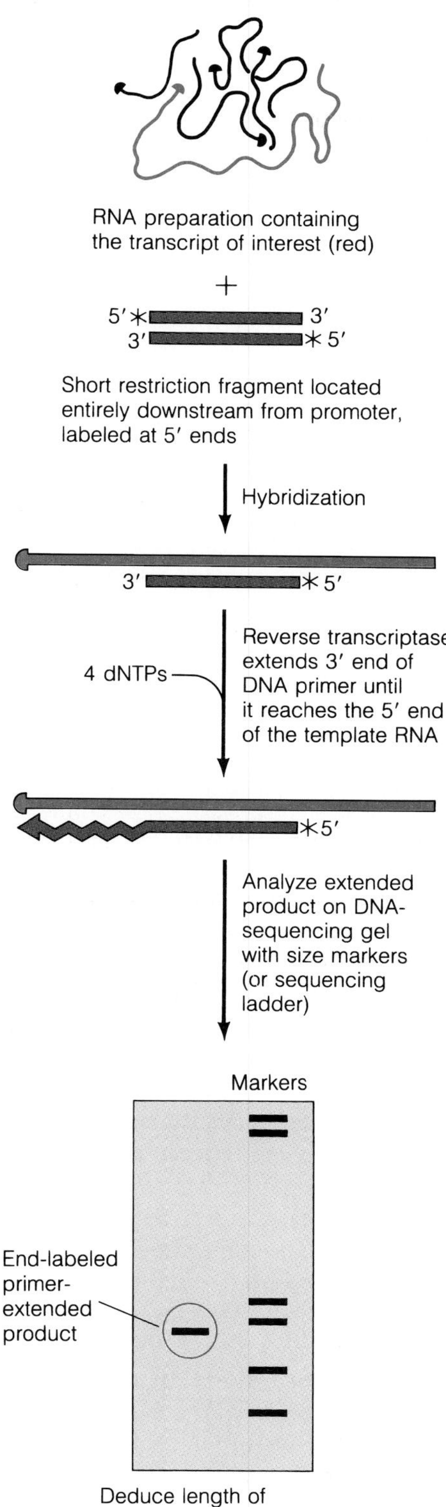
RNA preparation containing
the transcript of interest (red)
+
5′
3′
3′
5′
Short restriction fragment located
entirely downstream from promoter,
labeled at 5′ ends
Hybridization
3′
5′
4 dNTPs
Reverse transcriptase
extends 3′ end of
DNA primer until
it reaches the 5′ end
of the template RNA
5′
Analyze extended
product on DNA-
sequencing gel
with size markers
(or sequencing
ladder)
Markers
End-labeled
primer-
extended
product
Deduce length of
primer-extended product

Figure T24.2
Primer extension method for locating the 5′ end of a transcript.

CHAPTER 27

Information Decoding: Translation

The *translation* of mRNA into protein sequences is the critical step in the expression of genetic information. In Chapter 5 we described translation in a very elementary way. Before getting into the full complexities of the reactions, it is worthwhile to review the process briefly. Reference to Figures 5.19 and 5.20 will be helpful at this point.

An Overview of Translation

You will recall that the genetic code specifies a nucleotide triplet corresponding to each amino acid residue. In order to make the correspondence between individual amino acids and their corresponding codon triplets, a set of adaptor molecules is needed, each of which can be coupled to a specific amino acid and will recognize the corresponding codon in the messenger RNA. These adaptor molecules are the *transfer RNAs* (or tRNAs), which have been described in Chapter 4. Figure 27.1 illustrates the general structure of a tRNA. A specific enzyme called an **aminoacyl-tRNA synthetase** is required to make the attachment between each tRNA and the corresponding amino acid. Each tRNA contains a nucleotide sequence, called the *anticodon*, which is complementary to a codon for the particular amino acid. Thus, the whole set of tRNAs contained in a cell constitute a kind of molecular dictionary for the translation—they define the correspondences between words in the nucleic acid language and words in the polypeptide language.

In order to translate an mRNA, there must exist a machinery that brings the amino acid-charged tRNAs and the mRNA together, matching anticodon and codon triplets and joining amino acid residues together in the correct sequence. The cellular apparatus that accomplishes this feat is the *ribosome*, a particle composed of both RNA and proteins. A ribosome can bind mRNA and "read" it, accepting the charged tRNAs in the order dictated by the message and transferring their amino acid residues one by one to the growing polypeptide chain in the proper order.

The message is always read $5' \rightarrow 3'$, that is, starting near its $5'$ end and proceeding in the $3'$ direction. The polypeptide chain is translated starting with its N-terminal residue. This latter fact was established by Howard Dintzis in 1961, using reticulocytes, cells that synthesize hemoglobin as

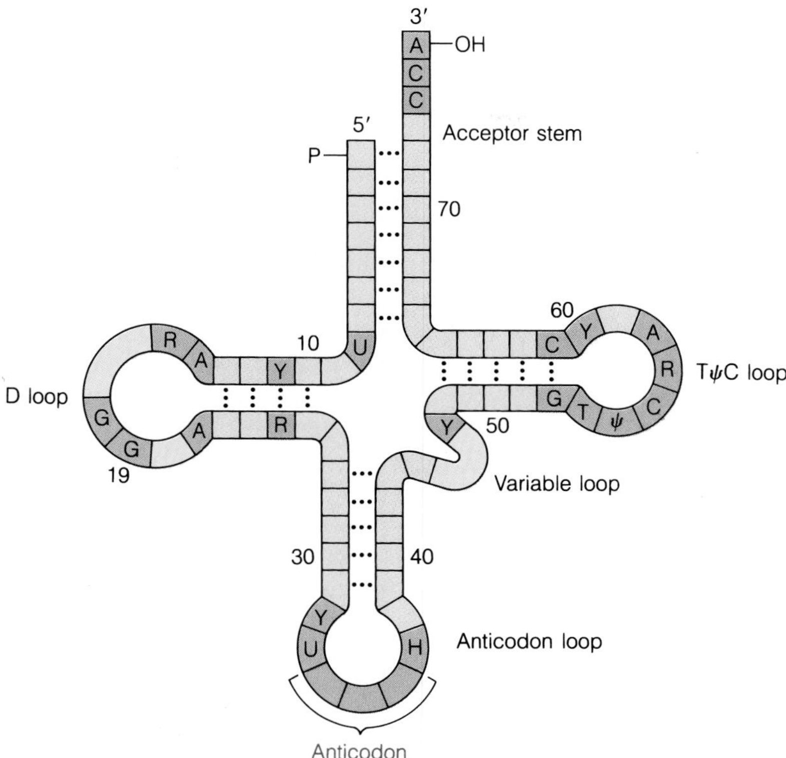

Figure 27.1
Generalized structure of a tRNA. The positions of invariant and rarely varied bases are shown in purple. Regions in the D loop and the variable loop that can contain different numbers of nucleotides are shown in blue. The anticodon is shown in orange. Code for bases: Y = pyrimidine; R = purine; H = hypermodified purine; ψ = pseudouridine; T = ribothymidine (see Figure 27.8).

their major product. Such cells were given a short "pulse" of ^{3}H-labeled leucine, and the completed hemoglobin molecules were isolated at various times after the pulse. These were cleaved into peptides by trypsin, and the radioactivity of peptides from various points in the chain was compared. As Figure 27.2 shows, immediately after the pulse label was added, it appeared only in the C-terminal region; it was being put into chains that were just being completed. At longer times after the pulse, radioactivity was found closer and closer to the N-terminus.

The simple picture of translation we have presented so far leaves a host of questions unanswered. How are tRNA and amino acid matched? How does the ribosome attach to the mRNA and move along it? How does it start and stop translation correctly? How does it avoid making mistakes? Where does the energy for all of this come from? To answer such questions, we must dissect the whole process of translation, with careful examination of each of its parts.

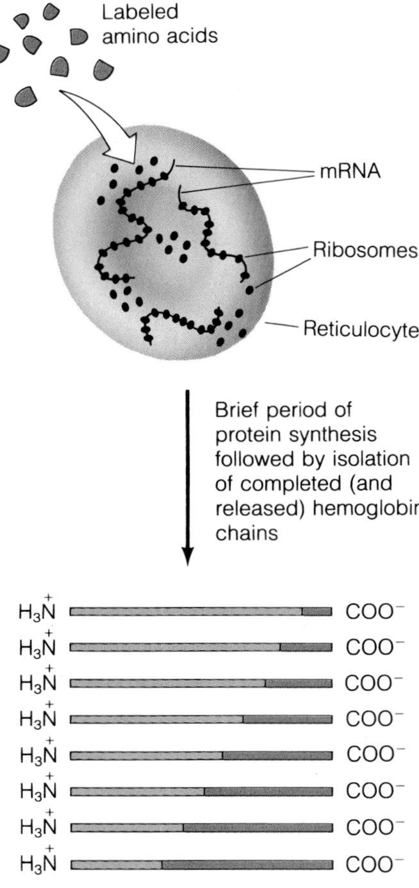

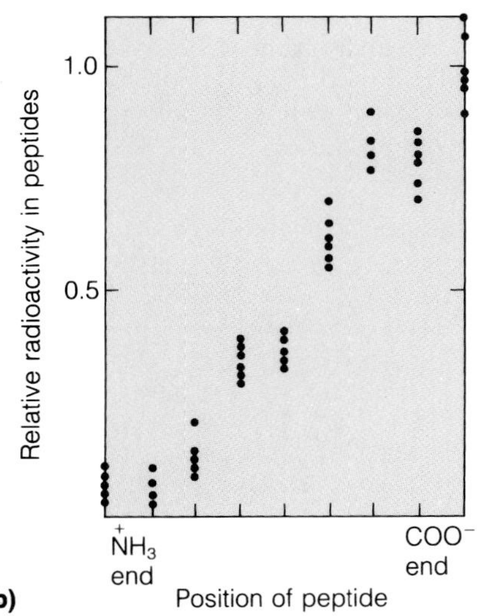

Figure 27.2
How the direction of translation was established. (**a**) Hemoglobin-synthesizing cells (reticulocytes) were given a brief pulse of radioactive leucine and the newly synthesized proteins were isolated. At first, the radioactivity appeared only in C-terminal portions of nearly finished chains (**b**), showing that translation starts at the N-terminus.

The Genetic Code

The genetic code was introduced in Chapter 5. Here, we will describe some of the crucial experiments used to deduce it and analyze a number of the code's more important features.

How the Code Was Deciphered

Following the discovery of mRNA (see Chapter 26), the immediate question was, "How is it read?" It was clear that a triplet code was required to account for all 20 amino acids, but many kinds of triplet codes are possible. Some conceivable ones are shown in Figure 27.3. An **overlapping code** like that shown in Figure 27.3a would be space-saving. This possibility was eliminated, however, by the observation that most mutations result in the change of a single amino acid; if the codons overlapped there should be a significant number of cases in which two adjacent residues are modified by mutation of the "overlap" residue. Furthermore, an overlapping code would lead to statistical regularities between neighboring amino acid residues in proteins—and this has never been observed. Another possibility was that the code was **punctuated.** That is, as shown in Figure 27.3b, some base or bases might serve as "spacing" between code words.

That the code was neither overlapping nor punctuated was elegantly demonstrated by Francis Crick and Sidney Brenner in 1961. Using conditions that encouraged deletion and insertion mutations in the bacteriophage T4 DNA, they showed that while insertions or deletions of one or two nucleotides led to nonsense proteins (because the reading frame had been shifted), adding or removing *three* always led to the insertion or deletion of one amino acid residue. Thus the code *must* be triplet, and the message must be read in units of three, without punctuation, from some starting point.

But which codons corresponded to which amino acids? The first answers were provided by Marshall Nirenberg and his colleagues, using a cell-free translation system. They found that extracts from bacterial cells contained all of the necessary ingredients (tRNAs, amino acids, aminoacyl synthetases, and ribosomes) to carry out polypeptide synthesis directed by a synthetic polynucleotide template added to the system. Thus, in 1961, Nirenberg and J. H. Matthaei showed that using polyuridylic acid, poly(U), as a

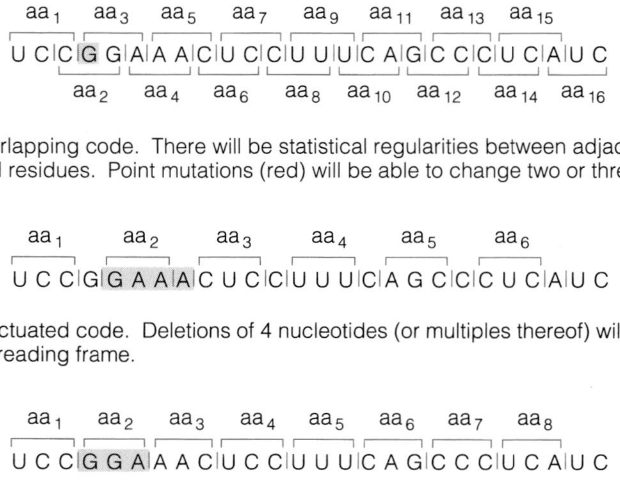

Figure 27.3
Three conceivable kinds of genetic codes. Early research on the nature of the code quickly showed that a nonoverlapping, unpunctuated code (c) fit all experimental observations.

aa$_1$ aa$_3$ aa$_5$ aa$_7$ aa$_9$ aa$_{11}$ aa$_{13}$ aa$_{15}$
U C C G G A A A C U C C U U U C A G C C C U C A U C
aa$_2$ aa$_4$ aa$_6$ aa$_8$ aa$_{10}$ aa$_{12}$ aa$_{14}$ aa$_{16}$

(a) Overlapping code. There will be statistical regularities between adjacent amino acid residues. Point mutations (red) will be able to change two or three residues.

aa$_1$ aa$_2$ aa$_3$ aa$_4$ aa$_5$ aa$_6$
U C C G G A A A C U C C U U U C A G C C C U C A U C

(b) Punctuated code. Deletions of 4 nucleotides (or multiples thereof) will restore the reading frame.

aa$_1$ aa$_2$ aa$_3$ aa$_4$ aa$_5$ aa$_6$ aa$_7$ aa$_8$
U C C G G A A A C U C C U U U C A G C C C U C A U C

(c) Unpunctuated code. Deletions of 3 nucleotides (or multiples thereof) will restore the reading frame. This is the actual form of the code.

template led to the production of polyphenylalanine. Thus, UUU must be a phenylalanine codon. In the same way, it was later shown that CCC codes for proline and AAA for lysine.

Using *homopolynucleotides* (all bases the same) as templates could yield only a few code words. So the development, by the Indian biochemist H. G. Khorana, of methods for the synthesis of polyribonucleotides of a complex but repeating structure provided many more opportunities. An example is shown in Figure 27.4. The repeating $(AAG)_n$ was found to give three different homopolymers: polylysine, polyarginine, and polyglutamic acid. This not only confirmed the importance of the reading frame but also showed that AAG, AGA, and GAA must be codons for these amino acids. The experiment did not, however, reveal which codon corresponds to which amino acid. Further experiments were needed to discriminate between the possible matches. When the repeating unit is a dimer or tetramer, rather than a triplet, more complicated polypeptides containing several amino acid residues in repeating sequences are formed. These experiments also confirmed that the code was triplet and unpunctuated.

Experiments of this kind provided many code words, but the method that allowed completion of the deciphering was much more direct. Transfer RNAs had been discovered in 1958, by M. B. Hoagland and his colleagues, and by the early 1960s many specific ones had been identified. Ribosomes had been discovered even earlier, and by 1959 it was clear that these were the sites of protein synthesis. Thus it was possible, in 1964, for Nirenberg and P. Leder to develop a new and rapid method for codon assignment. Synthetic triplets would bind to ribosomes and specify the binding of specific tRNAs. For example, UUU and UUC permitted binding of phenylalanine tRNAs to ribosomes, and CCC and CCU permitted the binding of prolyl-tRNA. Such experiments provided unequivocal evidence for the **redundancy** of the code, since several different codons were found to correspond to a single amino acid. By the combined use of these various techniques, the entire genetic code was established within a few years of intensive research.

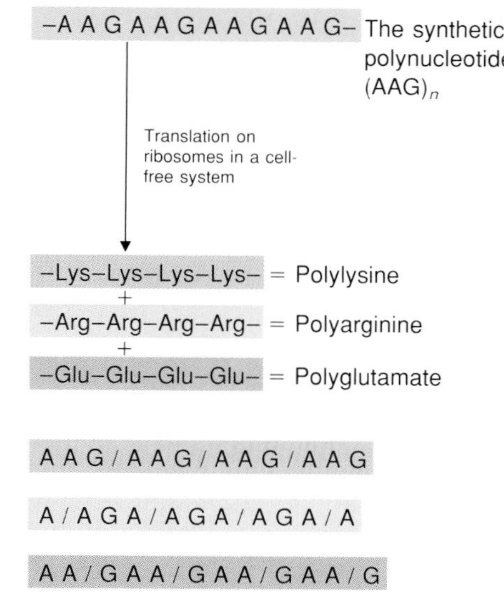

Figure 27.4
Use of synthetic polynucleotide with repeating sequences to decipher the code. This example shows how polypeptides derived from the $(AAG)_n$ polymer were used to confirm the triplet code. The polymer $(AAG)_n$ can yield three different polypeptides, depending on which reading frame is employed.

Features of the Code

The genetic code, as given in Figure 27.5, is *almost,* but not quite, universal. Throughout the whole of the prokaryotic, plant, and animal kingdoms, the same codons are used for the same amino acids, with very few exceptions. The exceptions are confined, as far as we know, to two cases. In the first, ciliated protozoans read AGA and AGG as "stop" signals rather than as *arginine*. Second, mitochondria make the set of changes shown in Table 27.1. Note that each of these changes decreases the information content of the third code letter in the mitochondrial code—it no longer makes a distinction between amino acids. This increased simplicity reduces the number of kinds of tRNA necessary in mitochondria and may be an adaptive increase in efficiency in these compact organelles.

The virtual universality of the code is illustrated by the fact that bacteria readily and correctly translate the mRNAs produced from human DNA inserted into them by recombinant DNA techniques. Why has the code remained virtually unchanged over so vast an evolutionary span? Perhaps it is because codon changes could be devastating. A single codon change could alter the sequence of virtually every protein made by the organism. Some of these changes would almost certainly have lethal effects. Therefore, codon changes are opposed by the most intense selective pressure during evolution. They represent changes in the most basic rules of the game.

Table 27.1
Differences between the common genetic code and the mitochondrial code

Codon	Common Code	Mitochondrial Code
AUA	Ile	Met
AGA	Arg	Stop
AGG	Arg	Stop
UGA	Stop	Trp

Figure 27.5
The genetic code. Chain termination, or "stop," codons are shown in red, and the usual "start" codon AUG is green. The codons GUG or UUG (pale green) are *occasionally* used as starts. When AUG, UUG, and GUG are used as start codons, they incorporate f-Met; otherwise they code for Met, Leu, and Val, respectively.

Second position

First position (5′ end)	U		C		A		G		Third position (3′ end)
U	UUU	Phe	UCU	Ser	UAU	Tyr	UGU	Cys	U
U	UUC	Phe	UCC	Ser	UAC	Tyr	UGC	Cys	C
U	UUA	Leu	UCA	Ser	UAA	Stop	UGA	Stop	A
U	UUG	Leu	UCG	Ser	UAG	Stop	UGG	Trp	G
C	CUU	Leu	CCU	Pro	CAU	His	CGU	Arg	U
C	CUC	Leu	CCC	Pro	CAC	His	CGC	Arg	C
C	CUA	Leu	CCA	Pro	CAA	Gln	CGA	Arg	A
C	CUG	Leu	CCG	Pro	CAG	Gln	CGG	Arg	G
A	AUU	Ile	ACU	Thr	AAU	Asn	AGU	Ser	U
A	AUC	Ile	ACC	Thr	AAC	Asn	AGC	Ser	C
A	AUA	Ile	ACA	Thr	AAA	Lys	AGA	Arg	A
A	AUG	Met	ACG	Thr	AAG	Lys	AGG	Arg	G
G	GUU	Val	GCU	Ala	GAU	Asp	GGU	Gly	U
G	GUC	Val	GCC	Ala	GAC	Asp	GGC	Gly	C
G	GUA	Val	GCA	Ala	GAA	Glu	GGA	Gly	A
G	GUG	Val	GCG	Ala	GAG	Glu	GGG	Gly	G

If you examine the code table shown in Figure 27.5 carefully, you will note that, in general, each amino acid is associated with the first two codon letters. For example, all Pro codons start with CC; all Val codons start with GU. As a rule, redundancy is most expressed in the third letter (i.e., ACU, ACC, ACA, and ACG all code for threonine). We will see later that there is a simple physical explanation for this common occurrence.

Because the messenger RNA is almost invariably longer than the part of the message to be translated, specific start and stop signals are required to properly begin and end translation. In almost all organisms, UAA, UAG, and UGA are used for stop signals. They do not correspond to any amino acid, but indicate that translation is to terminate, and both the mRNA and the polypeptide product are to be released by the ribosome. Curiously, the start signal commonly used in translation is AUG, which also serves as the single methionine codon. How does the ribosome know how to interpret this triplet? The answer is that the 5′ end of any message contains specific sequences that ensure that it is correctly attached to the ribosome. As the message begins to be read, the *first* AUG encountered is interpreted as a start signal, and translation begins. Although prokaryotic and eukaryotic cells handle this situation somewhat differently, the consequence is that *N*-formylmethionine (in prokaryotes) or methionine (in eukaryotes) is always the first amino acid incorporated into a polypeptide chain.* Any subsequent AUG encountered is treated as a signal to incorporate methionine within the sequence at that point. Very occasionally, GUG (normally valine) or UUG (normally leucine) will serve as a prokaryotic start codon when located near the 5′ end of a message. When they do, however, they invariably place *N*-formylmethionine in the first position. Elsewhere, they are read as normal codons.

* You may wonder whether this means *all* proteins start with *N*-fMet or Met. They do, as synthesized, but in many cases this residue is removed after translation (see p. 983).

The Major Participants in Translation: mRNA, tRNA, and Ribosomes

Structure of Prokaryotic mRNAs

As we described in Chapter 26, mRNAs are single-strand products of the transcription of genomic DNA, with special features that destine them to become attached to ribosomes and function properly in translation. In this chapter, we shall consider the properties of the prokaryotic type of mRNAs—those produced by bacteria, the viruses that infect them, and the mitochondria and chloroplasts of eukaryotic cells. The mRNAs transcribed from the *nuclear* genomes of eukaryotic cells and some viruses of eukaryotes have a somewhat different structure and are handled quite differently. We shall describe these in Chapter 28.

As an example of a prokaryotic messenger RNA, consider the RNA of the bacteriophage MS2. This small bacterial virus contains a single RNA molecule. This ribonucleic acid serves as a message, translated by the host translational apparatus, to make the only four polypeptide chains the virus requires: (1) a *coat protein,* (2) a *replicase* subunit that combines with host proteins to produce an enzyme complex that will replicate the viral RNA, (3) a protein involved in new phage assembly (the *A protein*), and (4) a small polypeptide that aids in cell lysis (the *lysis protein*). As shown in Figure 27.6, a copy of one of each of these genes is carried on the single message. Such multigene messages, which are called polycistronic messages, are common in prokaryotes. Each gene has its own start and stop signals. There is extra, untranslated RNA between the genes and at the ends. The regions 5′ to each gene contain sequences rich in A and G, which help to align the mRNA on the ribosome so that translation can begin at the proper points and in the correct reading frame. Such sequences, found on all prokaryotic mRNAs, are called *Shine-Dalgarno sequences,* after J. Shine and L. Dalgarno, who discovered them.

The MS2 RNA can be considered a typical prokaryotic messenger, except for two somewhat unusual features: (1) The start signal for the A protein is GUG instead of the usual AUG, and (2) The lysis protein is coded in a region that *overlaps* the 3′ end of the coat gene and the 5′ end of the replicase gene. The reading of the lysis protein RNA sequence is frameshifted by one residue from the reading frame used for translation of the other three proteins. Thus, the AUG start codon of the lysis protein is a

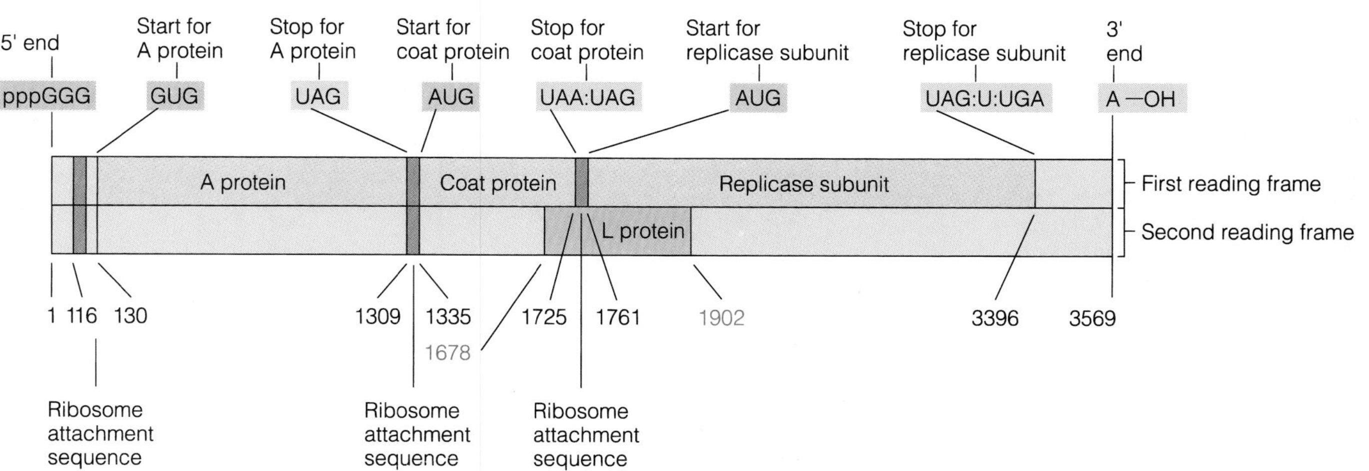

Figure 27.6
The RNA of bacteriophage MS2. This RNA molecule serves as a message for all four proteins required by this virus. Coding sequences for the A protein, coat protein, and replicase subunit are shown in gray. The message for the lysis protein (the L protein, shown in blue) actually *overlaps* the coat protein and replicase messages; it is translated in a different reading frame.

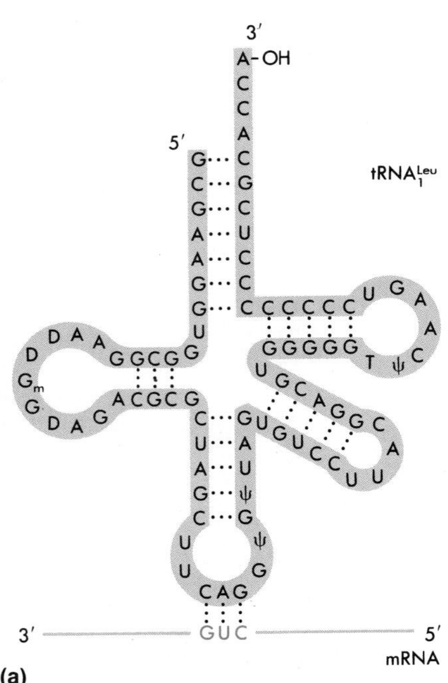

(a)

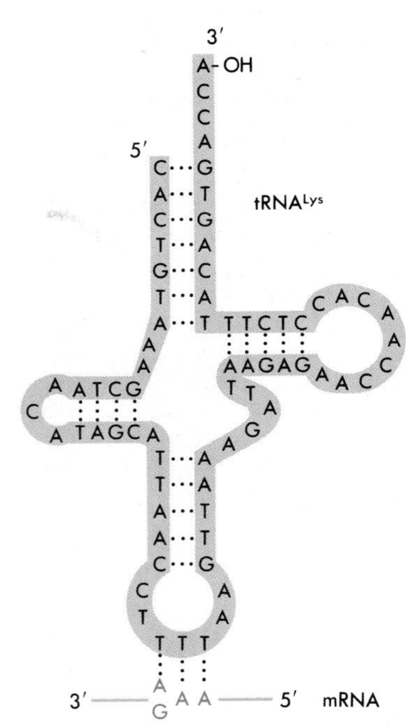

(b)

Figure 27.7
Some representative tRNAs. (**a**) A leucine tRNA from *E. coli.* (**b**) A human mitochondrial tRNA for lysine.

part of two codons in the coat protein frame, as is its stop codon (UAA), which lies within the replicase message. Such overlapping messages are not common, but they are found in a number of viral mRNAs. Although they impose obvious constraints on the protein coding, the economy of genome size that this affords appears to be important to some viruses.

Thus, the message of MS2 has the basic elements necessary for its function: sequences to align it properly on the ribosome and to start and stop translation at the proper points. As we shall see, the sequence also holds the possibilities for forming three-dimensional secondary and tertiary structure, which plays a role in regulation of the relative translations of the various protein products.

Transfer RNA Structure

Any cell, prokaryotic or eukaryotic, contains a battery of different types of tRNA molecules sufficient to incorporate all 20 amino acids into protein. This does not mean that there need be as many tRNA types as there are codons, for some tRNAs can recognize more than one codon, when the difference is in the third (3′) position. *Escherichia coli,* for example, has about 40 different tRNAs—plenty to code for all amino acids, but not as many as the 61 amino acid codons.

As far as we know, all tRNA molecules have the general structure shown schematically in Figure 27.1 and have similar sequences of about 70 to 80 nucleotides. There is, however, some variation in detail, as the examples shown in Figure 27.7 indicate. Furthermore, the tRNAs are unique among RNA molecules in their high content of unusual and modified bases, some of which are shown in Figure 27.8.

The **cloverleaf** models, of the kind shown in Figures 27.1 and 27.7, are useful in that they indicate the general pattern of hydrogen bonding and discriminate functional parts of the tRNA. The **anticodon loop** is shown at the bottom; it is complementary to the codon and will make base pairs with it. Since the codon and anticodon, when paired, constitute a short stretch of double-strand RNA, their directions must be antiparallel. In Figures 27.1 and 27.7 we have written the tRNA molecules with their 5′ ends to the left;

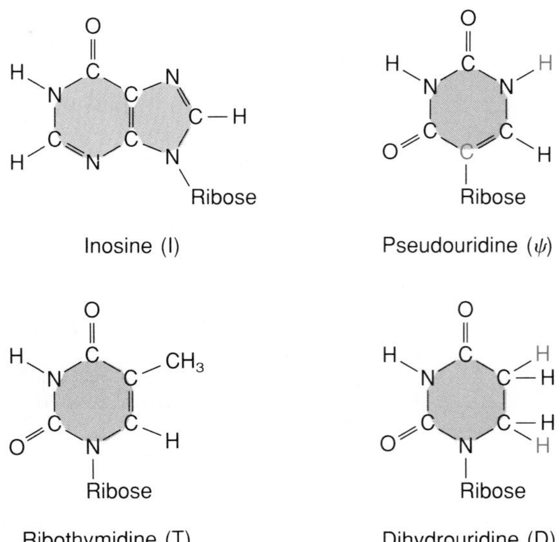

Figure 27.8
A sampling of the modified and unusual bases found in tRNAs.

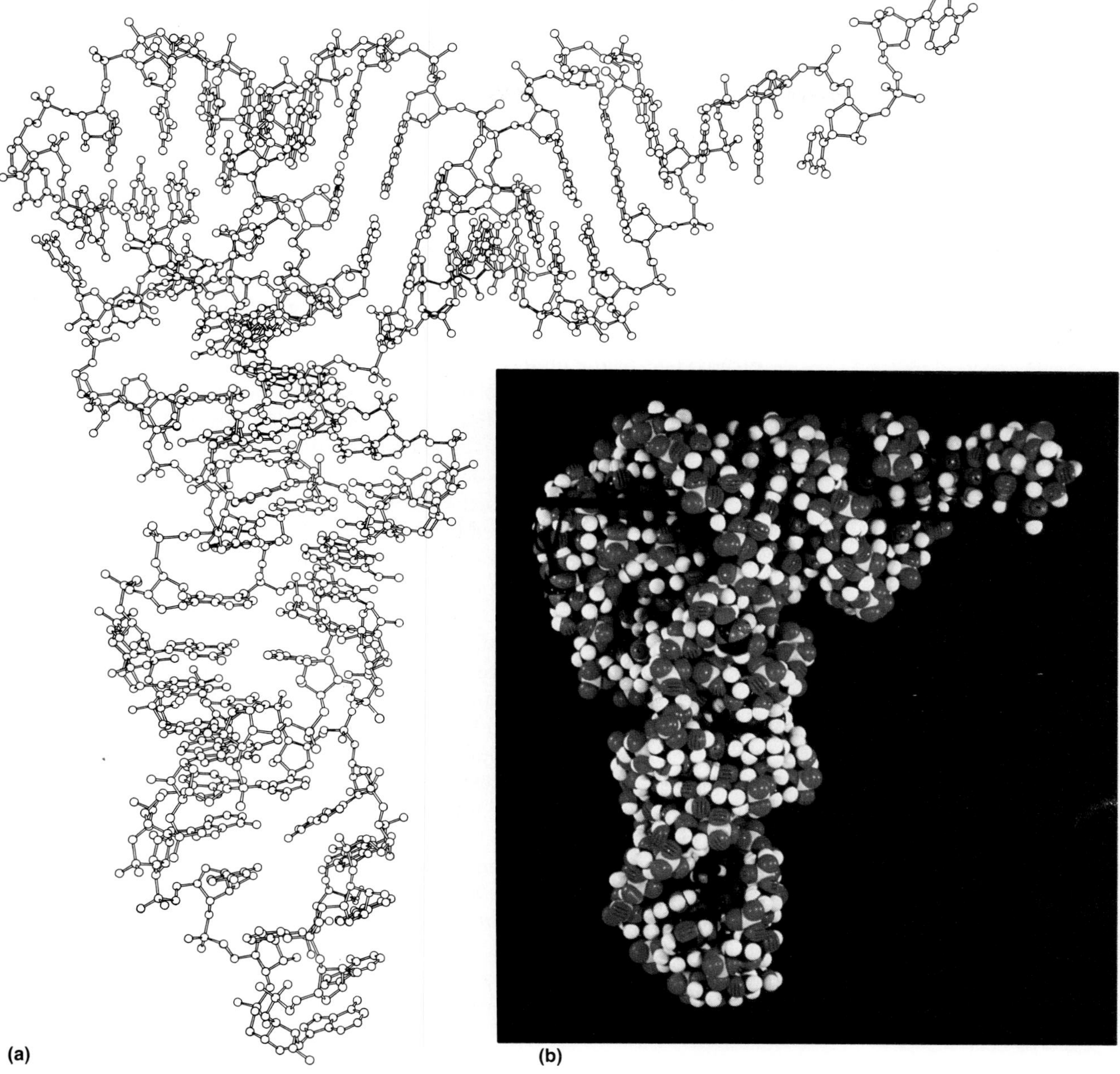

Figure 27.9
Model of yeast phenylalanine tRNA derived from x-ray diffraction studies. (**a**) Depicts all atomic positions; (**b**) is a space-filling model. The anticodon is at the bottom, the 3′ acceptor stem to the upper right.

therefore the messenger RNA, when shown in such figures, is written with its 5′ end to the right. In some later figures we will reverse this convention.

The **acceptor stem** at the top of the cloverleaf figure is where the amino acid will be attached, at the 3′ terminus of the tRNA. This stem always has the sequence 5′ . . . CCA-OH 3′. Other common features of tRNA molecules are the **D loop** and the **TψC loop,** regions that contain a substantial fraction of invariant positions. The so-called **variable loop** is indeed variable, both in nucleotide composition and in length, as Figure 27.7 demonstrates.

Although the cloverleaf models are convenient for depicting the primary structure and some elements of secondary structure, they are not good three-dimensional representations of tRNA molecules. X-ray diffraction studies of tRNA molecules have revealed that the real molecular shape is quite complex, as you can see in Figure 27.9 and Figure 4.22 (Chapter 4). As these figures show, a tRNA molecule looks rather like a hand-held drill or soldering gun. The anticodon loop is at the bottom of the grip, and the acceptor stem is at the working tip. The D loop and the TψC loop are folded inward in a complex fashion near the top of the grip, to provide a maximum of hydrogen bonding and base stacking. Some of the hydrogen-bonding patterns required to produce this folding are rather unusual (Figure 27.10). The three-dimensional shape of the tRNAs seems to be conserved even more than is the primary structure. A likely explanation is that such conservation is necessary so that each tRNA can fit onto the ribosome and carry out its function.

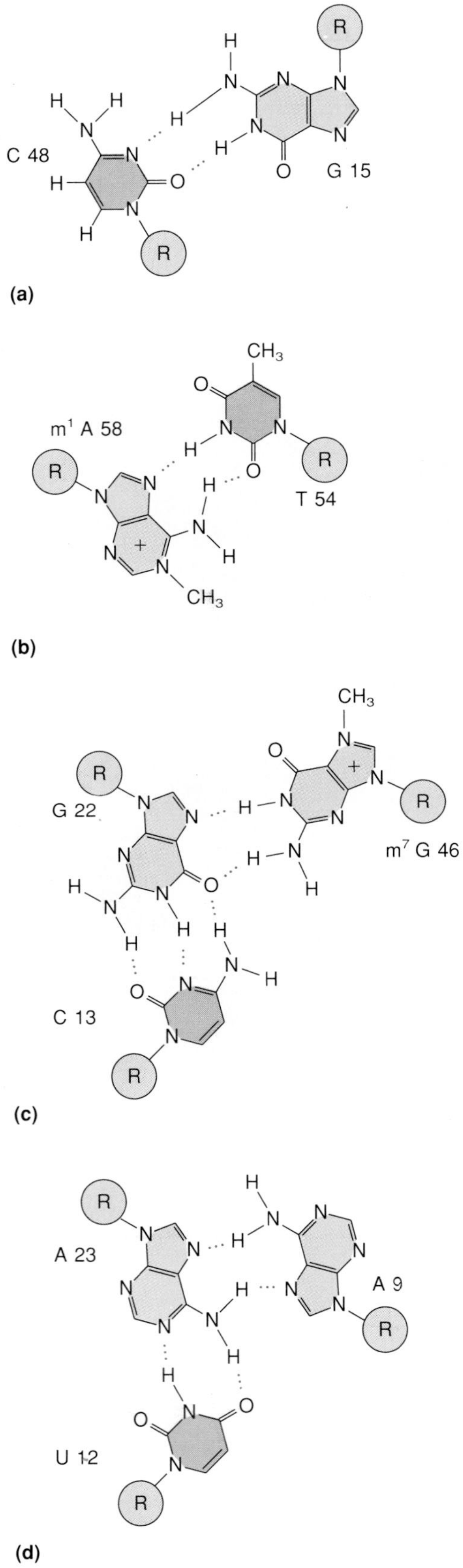

Coupling of tRNAs to Amino Acids

Amino acids are attached to tRNAs by a covalent bond between a ribose hydroxyl group of the invariant 3′ adenosine residue on the tRNA and the carboxylate of the amino acid. Pairing of the correct amino acid residues and the tRNAs is accomplished by a set of enzymes called **aminoacyl-tRNA synthetases.** In the bacterium *E. coli* there are 20 synthetases, each of which recognizes one amino acid. The reaction shown in Figure 27.11 proceeds in two steps: First, the amino acid is activated by ATP to form an **aminoacyl adenylate.** While bound to the enzyme, this intermediate reacts with the correct tRNA (again selected by the synthetase) to form the covalent bond and release AMP. Reaction can be with either the 2′ or 3′ hydroxyl group, and the amino acid can apparently shuttle back and forth between these sites.

You might expect that the synthetase would choose the tRNA on the basis of its anticodon, but recent studies indicate that this is not the case. In 1988, Ya-Ming Hou and Paul Schimmel showed that changing a single base pair (between residues 3 and 70 in the acceptor stem) of either tRNA^{Cys} or tRNA^{Phe} to the G-U pair found in tRNA^{Ala} caused the alanine synthetase to accept the cysteine or phenylalanine tRNAs and couple them to alanine. Other tRNAs appear to be recognized by their synthetases at still different locations.

It is essential that this matching of amino acid and tRNA be very accurate. Once the amino acid is attached to a tRNA molecule and the charged tRNA is released from the enzyme, there is no further chance for it to be "checked." If the wrong amino acid has been coupled to a tRNA, it will be incorporated into protein according to the tRNA anticodon. The following very clever experiment has demonstrated this. Cysteine residues were attached correctly to cysteinyl tRNA and then reduced to alanine (Figure 27.12). When these aminoacyl tRNAs were then used with a hemoglobin messenger in cell-free protein synthesis, alanine residues were incorporated in the positions normally reserved for cysteines.

The accuracy of amino acid/tRNA matching is guaranteed in several ways. As we mentioned above, each tRNA has a distinctive sequence, which allows a number of critical points for recognition by the enzyme. In addition, each synthetase is designed to make appropriate and discriminating interactions with the amino acid side chain. For example, tyrosyl and phenylalanyl tRNA synthetases both have a hydrophobic pocket that accommodates the aromatic ring. Further discrimination between these amino acids is made by the fact that the tyrosyl synthetase has groups placed so as to form hydrogen bonds with the tyrosine hydroxyl group (Figure 27.13).

Figure 27.10
Unusual base pairings in tRNA. All are from the yeast tRNA^{Phe} shown in Figure 27.9. (**a, b**) Some unusual pair matches. (**c, d**) Some examples of triple interactions. R represents the ribosyl residue of the RNA chain.

Even so, a single recognition event is apparently not always sufficient to discriminate with the needed accuracy. In such cases, a second "proofreading" is done at the time the tRNA is brought in. If the match is incorrect, the aminoacyl adenylate is hydrolyzed and the amino acid and AMP are released from the enzyme, which is then ready for another try. This combination of specific recognition and proofreading guarantees a final error frequency of only about one in 10,000, which ensures that most protein molecules will not incorporate errors of translation. (See Problems 7 and 8 at the end of this chapter.)

The Ribosome

We have now described the two kinds of participants that must be brought together to carry out protein biosynthesis—the mRNA and the set of tRNAs charged with the appropriate amino acids. The actors are, so to speak, in the wings, and all that is needed is a proper stage on which the events can unfold. This stage is provided by the ribosome. Because all pro-

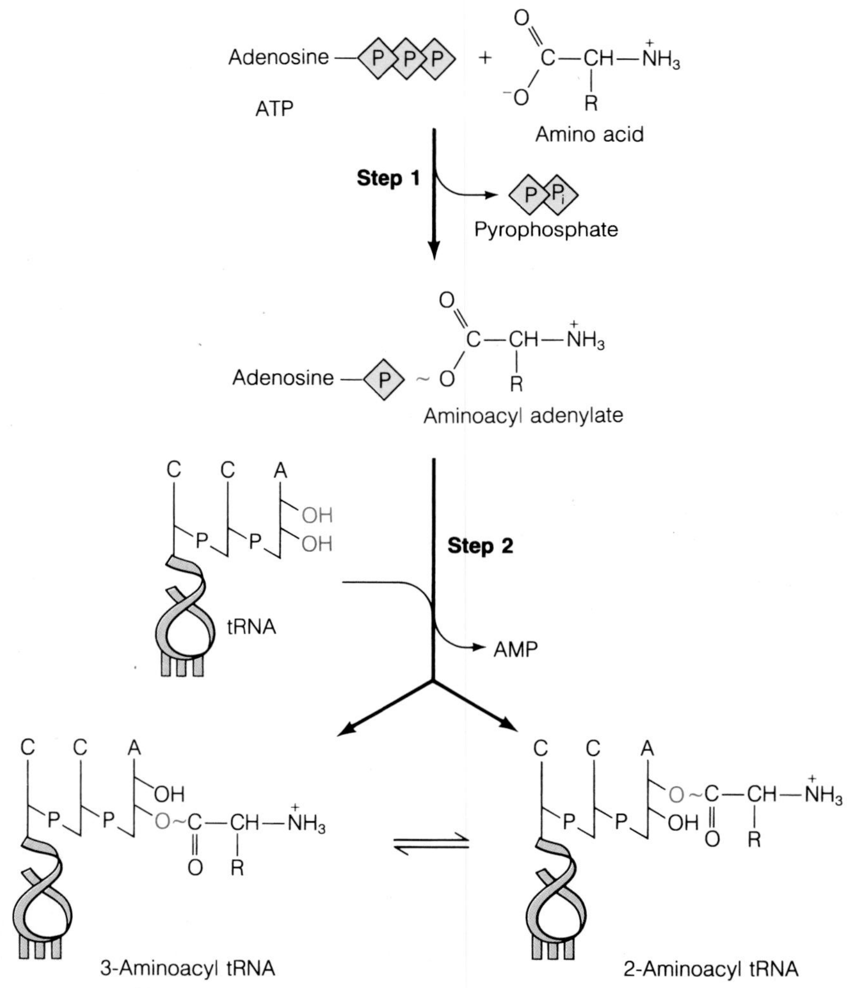

Figure 27.11
Formation of aminoacyl tRNAs. In step 1 the amino acid is adenylylated. In step 2 the proper tRNA is accepted by the synthetase, and the amino acid residue is transferred to the 2′ or 3′ OH of the 3′-terminal residue of the tRNA. All reactions occur on the synthetase.

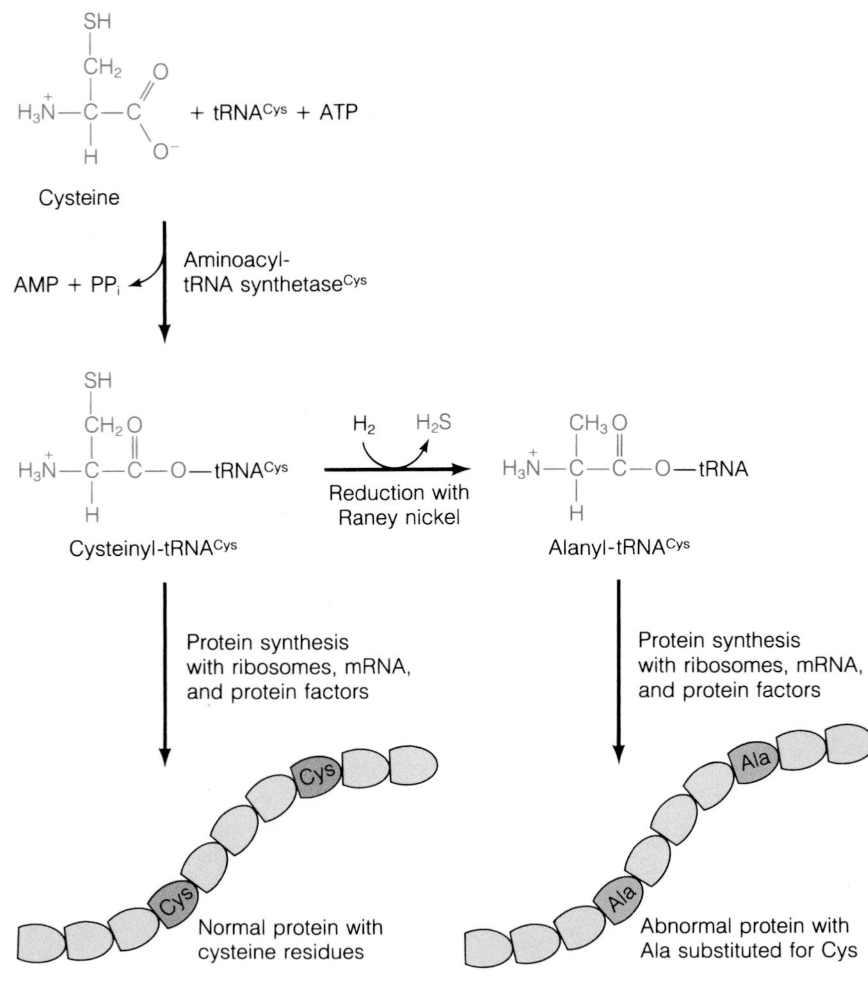

Figure 27.12
Demonstration that the incorporation of residues into proteins is determined by the tRNA coupled to the amino acid residue. Reduction of cysteine to alanine *after* the aminoacyl tRNA is formed leads to misincorporation of Ala into what should be Cys positions in the protein.

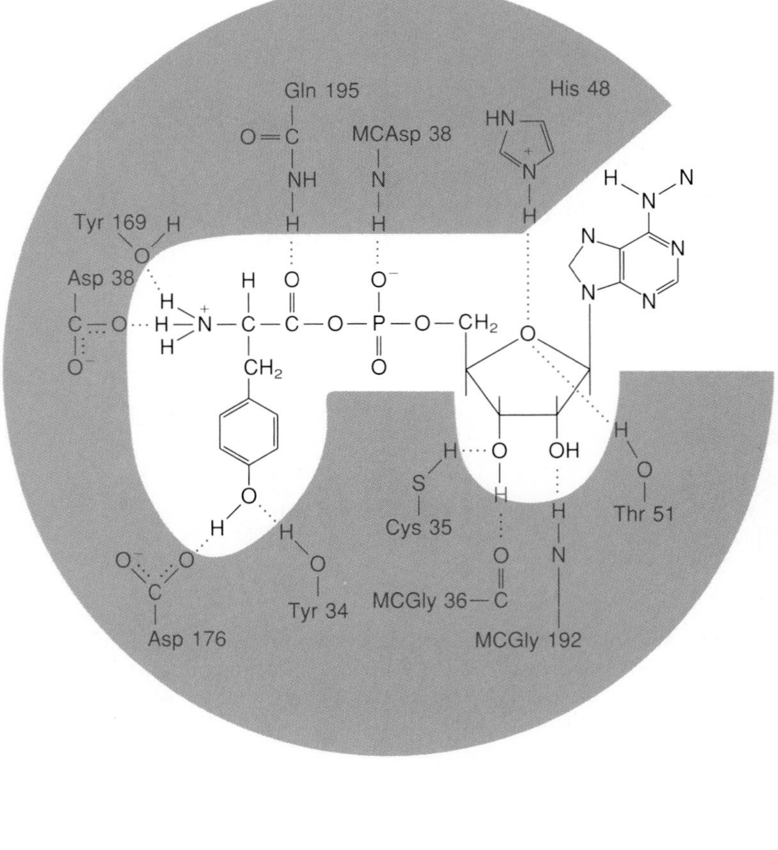

Figure 27.13
Recognition of an adenylylated amino acid by the proper synthetase. Shown is adenyl tyrosine bound to the tyrosyl synthetase (brown). A network of H bonds not only stabilizes the reaction but also serves to discriminate between different amino acid residues. MC designates main chain (backbone) carbonyl or amino groups participating in hydrogen bonding.

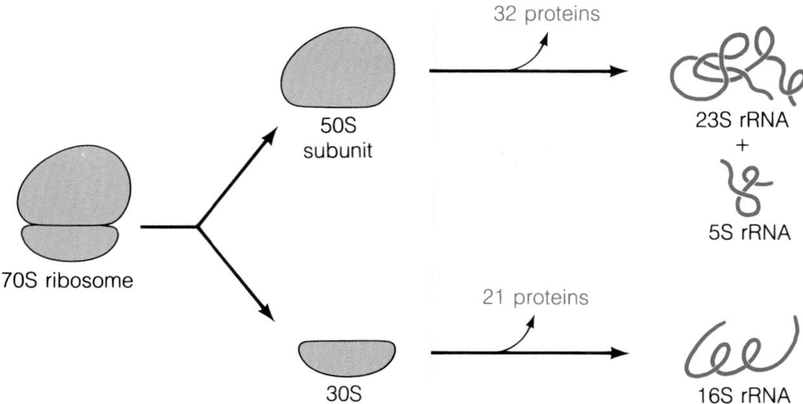

Figure 27.14
Components of a 70S prokaryotic ribosome.

tein synthesis depends on them, ribosomes are major cell constituents. An *E. coli* cell contains about 15,000 ribosomes, accounting for about 25% of the dried cell mass. A cell devotes a good deal of its energy to producing ribosomes and to using them in protein synthesis.

Functional prokaryotic ribosomes are particles with a sedimentation coefficient of 70S. They are composed of two subunits, a 30S particle and a 50S particle, each of which is a complex of specific ribosomal RNAs (called rRNAs) and proteins (Figure 27.14). The 30S subunit of *E. coli* ribosomes contains a single 16S rRNA molecule (1542 residues in length) plus 21 different proteins, at one copy of each. The 50S subunit contains two rRNA molecules—a 23S rRNA (2900 residues in *E. coli*) and a 5S rRNA (120 residues in *E. coli*). In addition, there are 32 different 50S proteins. Each of these, except for one termed L7/L12, is present in one copy. There are four copies of L7/L12. One protein (S20 = L26) is present in both the small and large subunits.

Ribosomal proteins can be cleanly separated by two-dimensional gel electrophoresis, as shown in Figure 27.15. All have now been sequenced, and the probable functions of some are known. A list of the *E. coli* ribosomal proteins is given in Table 27.2. The sequencing studies reveal no homologies between the different proteins in a ribosome, but comparison of sequences between corresponding proteins in the ribosomes of different organisms reveals considerable evolutionary conservatism. Thus, the ribosome is a complex object that evolved early in the history of life and has remained relatively unchanged. The sequences of many ribosomal RNAs (especially those from the small subunit) have also been determined, and they tell the same story. Indeed, because of their relatively slow evolutionary rates, rRNAs are useful as evolutionary yardsticks over vast phylogenetic distances. From them, we have learned that the eukaryotes, eubacteria, and archaebacteria represent three major kingdoms of organisms, sharing a common ancestor in the most distant past.

When the sequence of a 16S RNA is examined carefully, it is found to contain many regions of self-complementarity, which are capable of forming double-helical segments. A pattern like that shown in Figure 27.16 may seem so complex as to appear almost arbitrary, but comparison with other, even distantly related 16S RNA sequences shows that the double-strand regions are highly conserved. Indeed, the secondary structure seems more highly conserved than is the primary structure, for it is often found that there are compensatory mutations in double-helical regions so as to maintain base pairing. A schematic illustration like that in Figure 27.16 is analogous to the cloverleaf visualization of a tRNA. The actual rRNA must be

(*Text continues on p. 968*)

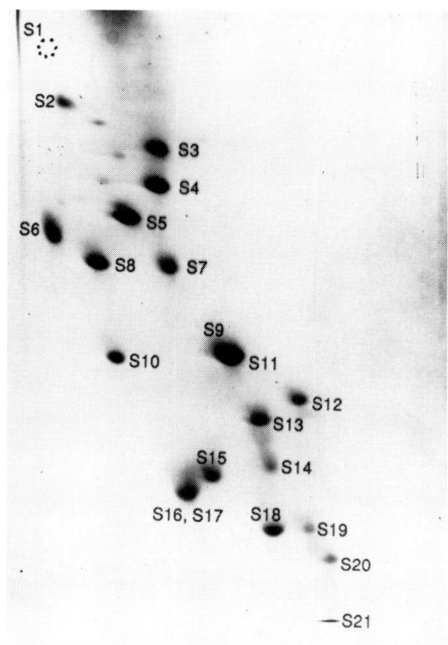

(a)

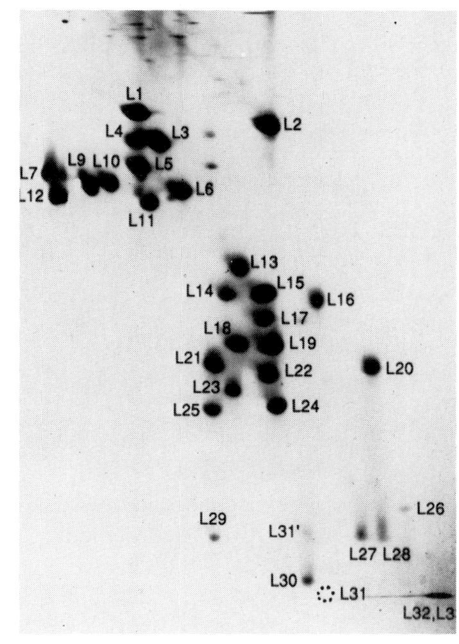

(b)

Figure 27.15
Two-dimensional gel electrophoresis of ribosomal proteins. (a) 30S proteins; (b) 50S proteins.

Table 27.2
Properties and functions of ribosomal proteins

Protein	Number of Amino Acid Residues	Presumed Function*
S1	557	mRNA binding
S2	240	mRNA binding
S3	232	mRNA binding; tRNA binding?
S4	203	Essential for 30S folding
S5	166	mRNA binding, tRNA binding
S6	131	———
S7	177	Essential for 30S folding
S8	129	Essential for 30S folding
S9	128	mRNA binding, tRNA binding?
S10	103	tRNA binding; mRNA binding?
S11	128	———
S12	123	Streptomycin site; probably involved in tRNA binding
S13	117	———
S14	98	tRNA binding; mRNA binding?
S15	87	Essential for 30S folding
S16	82	Essential for 30S folding
S17	83	Essential for 30S folding
S18	74	———
S19	91	tRNA binding
S20 = L26	86	———
S21	70	———
L1	233	———
L2	272	Peptidyl transfer?
L3	209	Essential for 50S folding
L4	201	Essential for 50S folding
L5	178	Peptidyl transfer?
L6	176	Peptidyl transfer
L7/L12	120	Binding of protein factors, translocation
L9	148	———
L10	163	———
L11	141	Peptidyl transfer
L13	142	———
L14	120	———
L15	144	Peptidyl transfer?
L16	136	Peptidyl transfer
L17	127	———
L18	117	Peptidyl transfer
L19	114	———
L20	117	Essential for 50S folding
L21	103	———
L22	110	———
L23	99	———
L24	103	Essential for 50S folding
L25	94	Peptidyl transfer?
L26 = S20	86	———
L27	84	Probably part of P-site; peptidyl transfer?
L28	77	———
L29	63	———
L30	58	———
L31	62	———
L32	56	———
L33	54	———
L34	46	———

* Where no specific function can be presently assigned, a line (———) is shown.

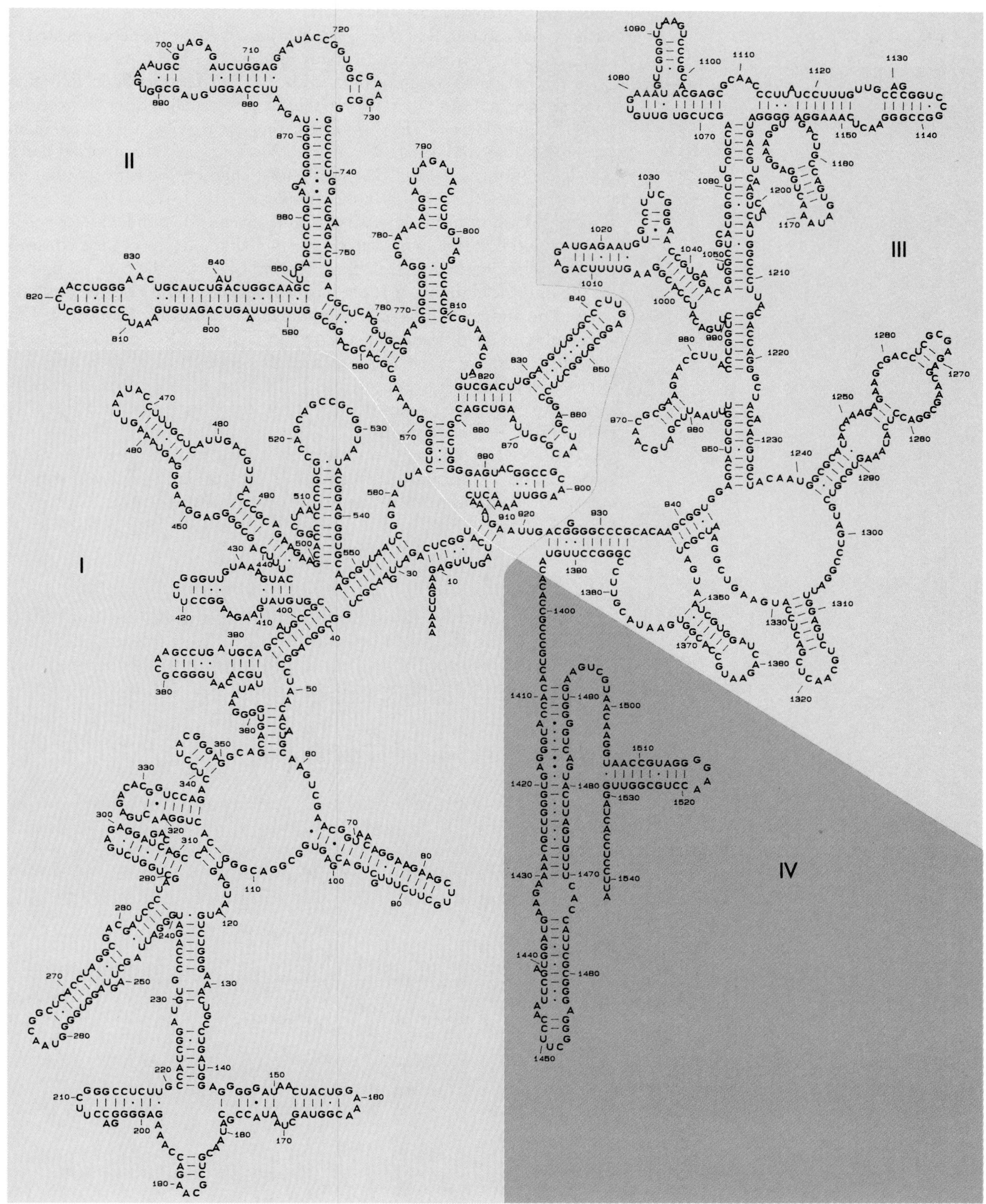

Figure 27.16
Postulated secondary structure of *E. coli* 16S rRNA. The sequence has been aligned to produce maximum base pairing between complementary segments. The molecule appears to have four major domains of folding (I–IV).

folded into a three-dimensional structure, just as the tRNA is. In the case of the ribosomal subunit, however, the structure will be further complicated by the presence of the ribosomal proteins binding to the RNA.

The ribosomal subunits (30S and 50S) are large enough that their overall shapes and how they fit together can be visualized by electron microscopy (Figure 27.17). The 70S particle they form is about 30 nm in diameter, and its total molecular weight (RNA plus protein) is approximately 1.5×10^6 g/mol. As we shall see in the next chapter, the ribosomes of eukaryotes are even larger and more complex.

Considering their compositional and structural complexity, it is surprising to find that ribosomes can be assembled in vitro from their constituent parts. This impressive feat, first accomplished for the 30S subunit by P. Traub and M. Nomura in 1968, has proved very useful in elucidating the roles and positions of ribosomal components. A schematic "flowchart" for the assembly of the 30S subunit is shown in Figure 27.18. Successful assembly seems to require that the proteins go on in a particular order: the first proteins to bind to RNA appear to help fold the RNA to provide sites for the rest. A complicated series of steps at different temperatures is required for the entire in vitro assembly.

The 30S and 50S subunits formed in such in vitro experiments are functional and can form 70S ribosomes on increasing divalent ion concentration. The equilibrium between 30S, 50S, and 70S particles,

$$30S + 50S \rightleftharpoons 70S$$

is in general shifted to the right by increasing concentrations of Ca^{2+} or Mg^{2+} and to the left by high concentrations of a monovalent cation such as Na^+ or K^+. Under the conditions existing in the cell, a substantial fraction of ribosomes are dissociated into free subunits. It is these subunits that enter into the series of translation reactions we shall consider next.

The Mechanisms of Translation

We now have described all of the major participants in the translation process: a messenger RNA, charged tRNAs, and the ribosome, where the actual translation events occur. Translation of a message can be divided into three stages: *initiation, elongation,* and *termination.* We shall describe these steps as they occur in prokaryotes. As we shall see in the following chapter, there

Table 27.3
Protein factors involved in translation

Factor	Approximate M	Binds GTP?	Approximate Number per Ribosome in Cell
Initiation			
IF1	9,000	No	1/7
IF2	120,000	Yes	1/7
IF3	22,000	No	1/7
Elongation			
EF-Tu	45,000	Yes	About 10 (varies)
EF-Ts	30,000	Yes	1
EF-G	80,000	Yes	1
Termination			
RF1	36,000	No	1/20
RF2	38,000	No	1/20
RF3	46,000	Yes	?

are significant, though not fundamental, differences in eukaryotic cells.

Each of these steps requires, in addition to the major participants listed above, a number of specific proteins. These are referred to as **initiation factors** (**IF**), **elongation factors** (**EF**), and **release factors** (**RF**). These proteins interact with the ribosome and the other major participants in order that each step may proceed. A listing of these factors, together with some of their properties and functions, is given in Table 27.3.

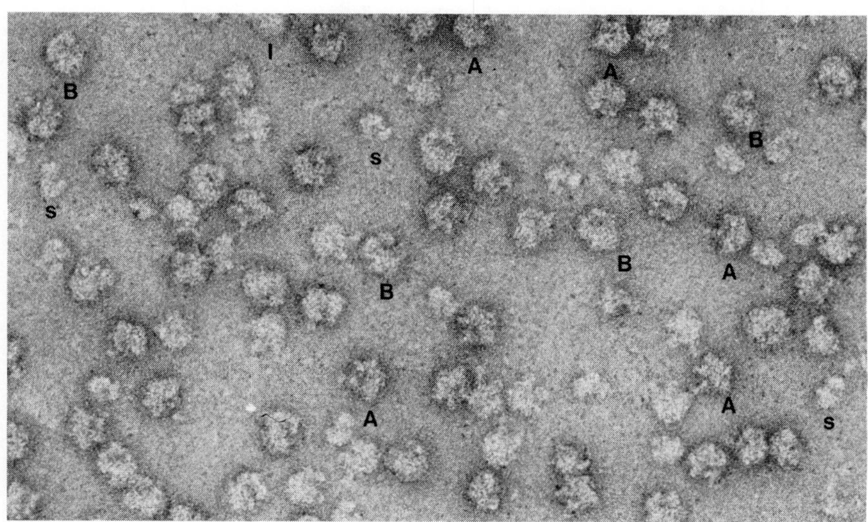

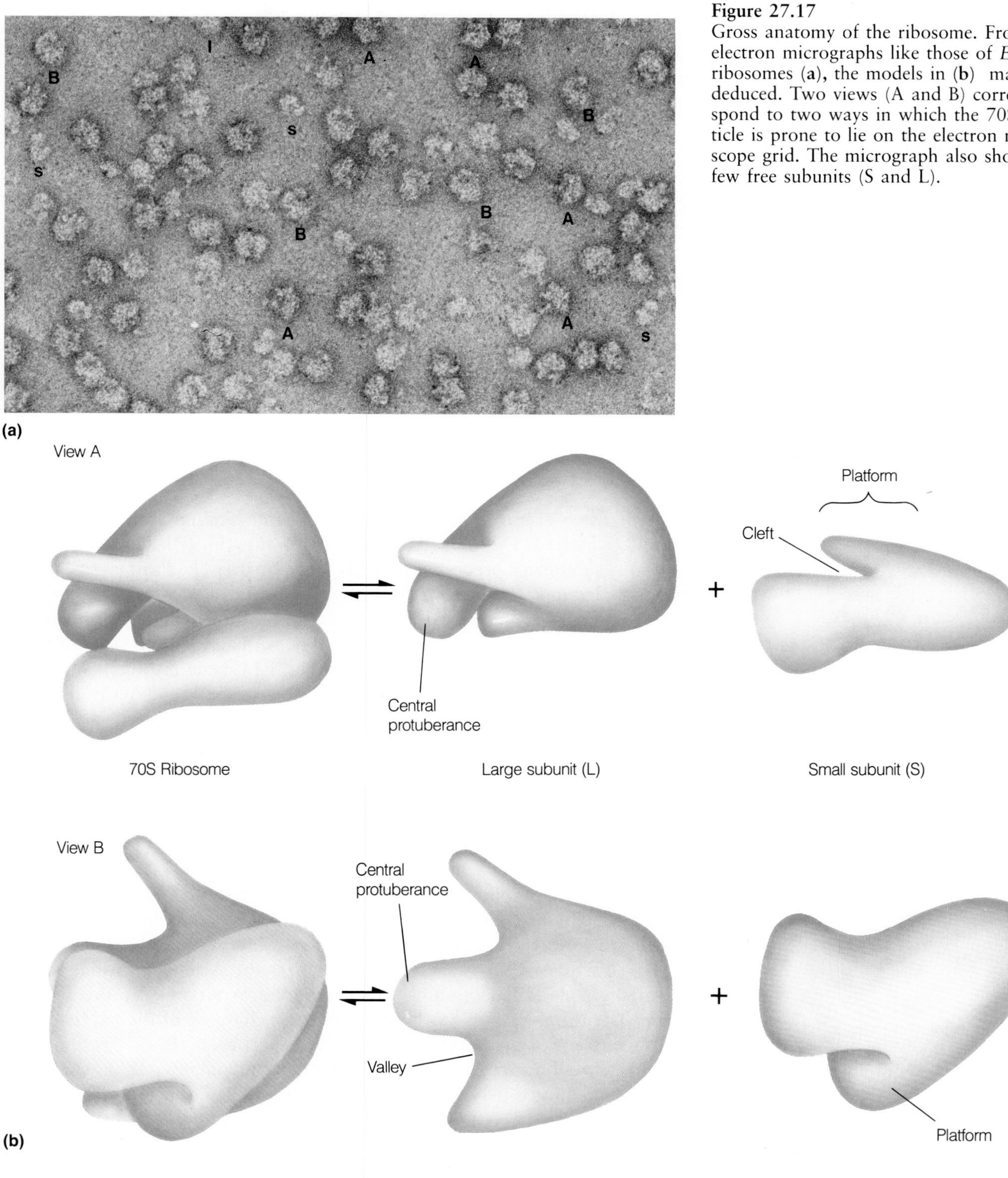

(a)

View A

Platform

Cleft

Central protuberance

70S Ribosome

Large subunit (L)

Small subunit (S)

View B

Central protuberance

Valley

Platform

(b)

Figure 27.17
Gross anatomy of the ribosome. From electron micrographs like those of *E. coli* ribosomes (**a**), the models in (**b**) may be deduced. Two views (A and B) correspond to two ways in which the 70S particle is prone to lie on the electron microscope grid. The micrograph also shows a few free subunits (S and L).

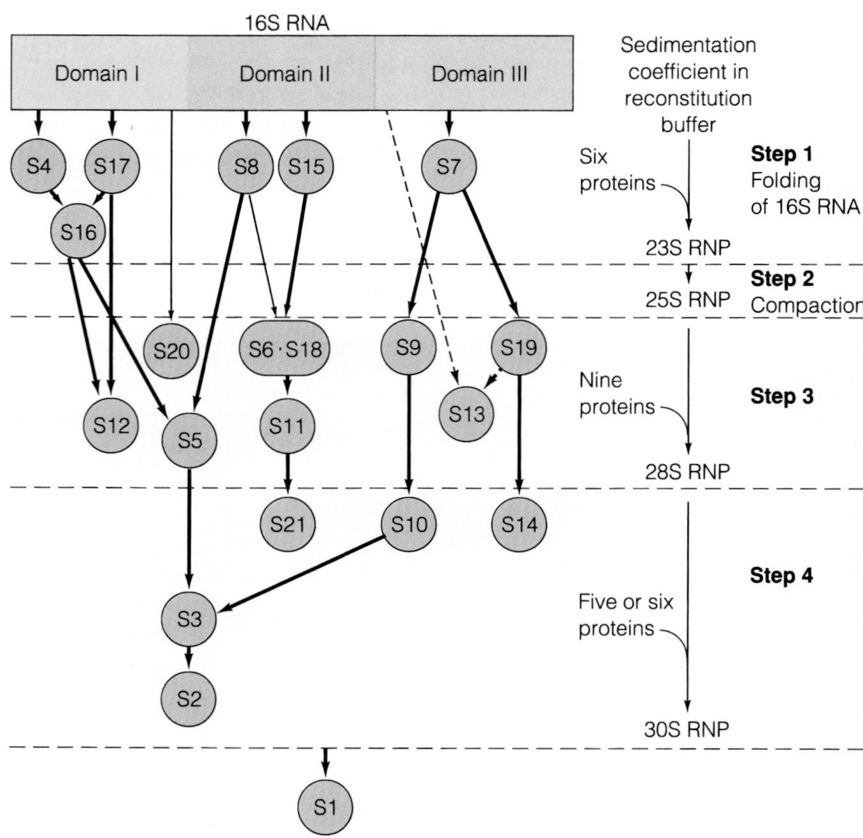

Figure 27.18
Assembly map for the 30S subunit. The map shows how a 30S ribosomal subunit of *E. coli* can be reassembled in the test tube from RNA and proteins (green). The order of addition shown here is important—some proteins must be added before others will fit in properly. Certain proteins are associated with each of the RNA domains I–III shown in Figure 27.16. Arrows between proteins indicate dependence in binding; the thicker the arrow, the more essential the preceding proteins. For example, binding of S3 in step 4 requires that S5 and S10 are already bound. The binding of S1 requires a nearly complete ribosomal structure, but it is difficult to tell just which proteins are essential. In any event, its binding completes the assembly, and a functional 30S subunit has been produced.

Initiation

The first step in translation (Figure 27.19) is the binding of the three initiation factors (IF1, IF2, and IF3) to a free 30S subunit. IF3 will bind only to *free* 30S subunits, a process apparently aided by IF1. The third factor, IF2, is bound carrying a molecule of GTP.

The 30S subunit, with these protein factors attached to it, is now ready to accept an mRNA and the first charged tRNA (step 2). The order of these additions is still uncertain, but it is clear that IF2 GTP is absolutely required.

The mRNA is attached to the ribosome near the 5′ end of the message, which is appropriate since all messages are translated in the 5′ → 3′ direction. In describing the structure of typical prokaryotic messages, we mentioned the presence of a specific binding sequence (the Shine-Dalgarno sequence) near the 5′ end of the mRNA. There exists, near the 3′ end of the 16S ribosomal RNA, a sequence that can base pair with these mRNA sequences. The sequence is 3′ . . . UCCUCC . . . 5′ and it will pair, for example, with the binding sequence of the MS2 A-protein message as shown in Figure 27.20. This pairing aligns the message correctly for the start of translation. The first tRNA can also be accepted at this point, along with release of IF3 to form the complete *30S initiation complex*.

This first tRNA is special. It recognizes the AUG codon that would normally code for methionine, but it actually carries an *N*-formylmethionine. The formyl group has been added *after* charging of the tRNA by an enzyme (**transformylase**) that recognizes the particular $tRNA^{fMet}$. Only the $tRNA^{fMet}$ will be accepted to form the initiation complex. As we shall see, all further charged tRNAs require the fully assembled ribosome. This initiation step therefore requires that all prokaryotic proteins are synthesized with the same N-terminal residue (*N*-formylmethionine). In almost all cases, the formyl group is removed during chain elongation; for many proteins the methionine itself is also cleaved off later.

Once IF3 has been released, the initiation complex gains a high affinity for a 50S subunit and will bind one from the available pool (see step 3,

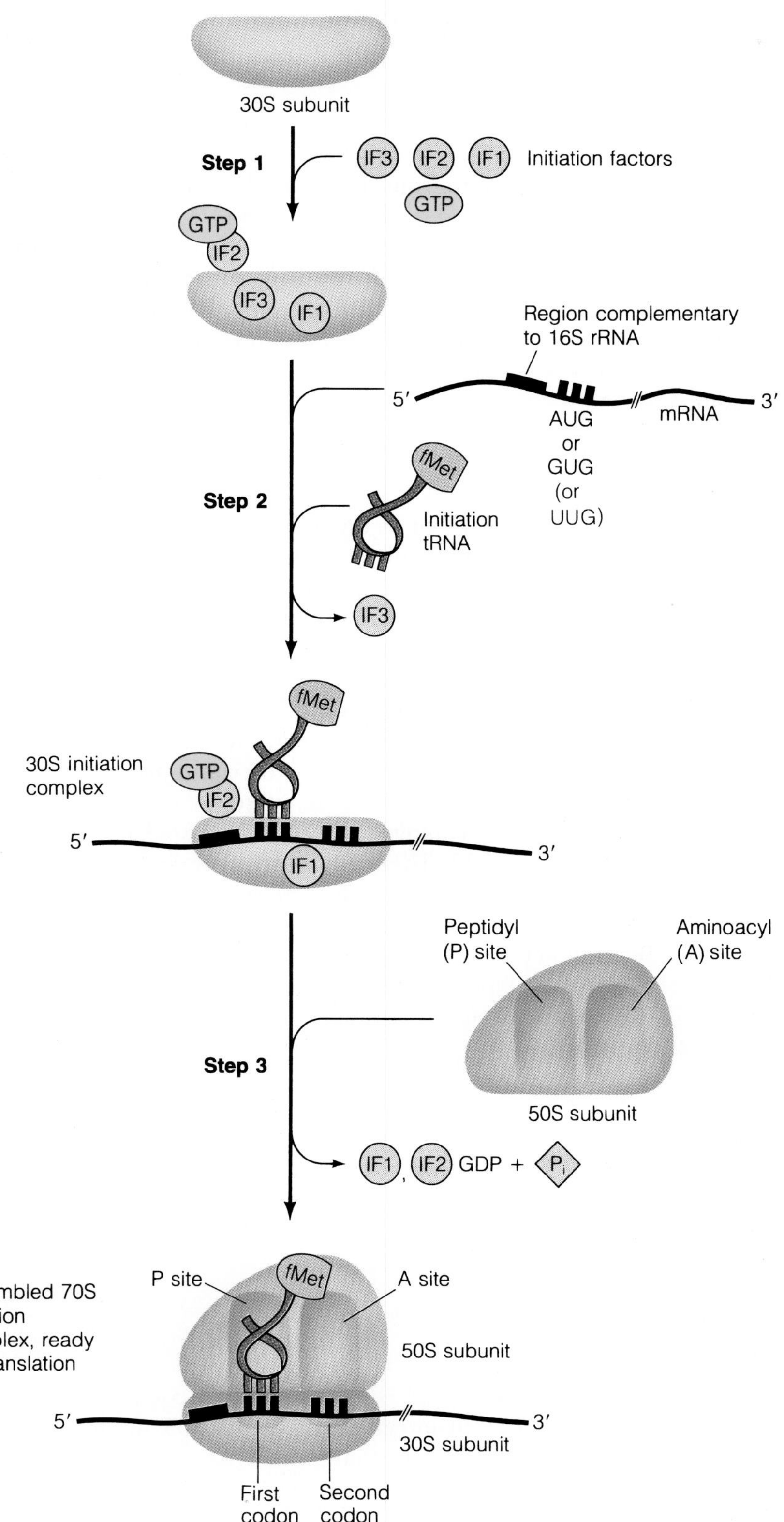

Figure 27.19
Initiation of protein biosynthesis in prokaryotes. Certain details of the process described in the text remain uncertain. For example, the exact order of binding of factors IF1, IF2, and IF3, tRNA, and mRNA is unclear. IF2, when carrying GTP, may actually transport the $tRNA^{fMet}$ to the ribosome. We show the first tRNA going into the P site. It is possible that it goes into the A site first and then is moved.

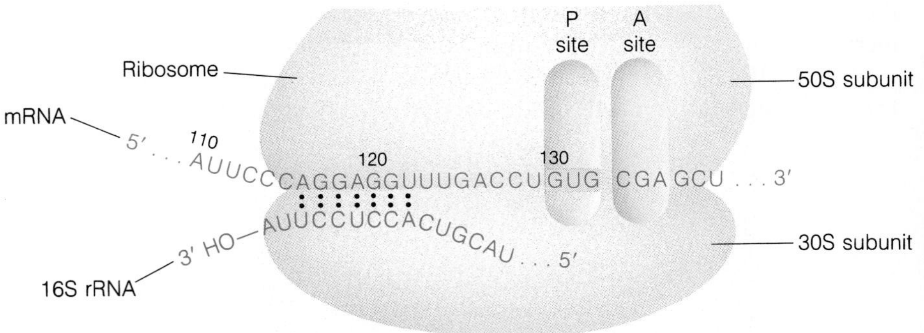

Figure 27.20
Pairing of the Shine-Dalgarno sequence of the A-protein message in MS2 RNA with the complementary sequence near the 3′ end of the 16S rRNA.

Figure 27.19). You will note from the figure that the 50S subunit has two sites for tRNA binding, called the P (peptidyl) and A (aminoacyl) sites. The mRNA has been bound to the ribosome in such a fashion that the AUG initiator codon with its bound $tRNA^{fMet}$ aligns with the P site. At this point, the GTP molecule carried by IF2 is hydrolyzed, and GDP, P_i, IF2, and IF1 are all released. The *70S initiation complex* so formed is ready to accept a second charged tRNA and begin elongation of the protein chain. Initiation is complete.

Elongation

Growth of the polypeptide chain on the ribosome occurs by a cyclic process. Figure 27.21 illustrates a single round in this cycle. Imagine it to be the first round following formation of the 70S initiation complex, or any subsequent round; they are all the same until a termination signal is reached.

Note that at the beginning of each cycle, the A site is empty. Aligned with this site is the mRNA codon corresponding to the *next* amino acid to be incorporated. The charged (aminoacylated) tRNA is escorted to this aminoacyl (A) site in a complex with a protein, the *elongation factor* EF-Tu, which also carries a molecule of GTP. When the appropriate charged tRNA is deposited into the A site, the GTP is hydrolyzed, and EF-Tu GDP is released. The EF-Tu GTP complex is then regenerated by the subsidiary cycle shown in Figure 27.22. There is proofreading at this step. After the charged tRNA is in place, it is checked both before and after the GTP hydrolysis and rejected if incorrect.

The next, and crucial, step is peptide bond formation (step 2, Figure 27.21). The nascent polypeptide chain, which was attached to the tRNA in the P site, is now transferred to the amino group of the amino acid carried by the A-site tRNA.

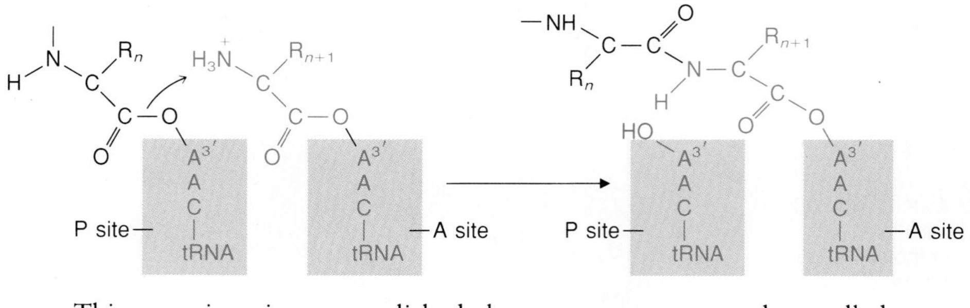

This reaction is accomplished by an enzyme complex called **peptidyltransferase,** which is an integral part of the 50S subunit. Ribosome reconstitution experiments indicate that proteins L6, L11, and L16 are critical parts of this enzymatic complex, and several others may be involved as well (see Table 27.2, Figure 27.28).

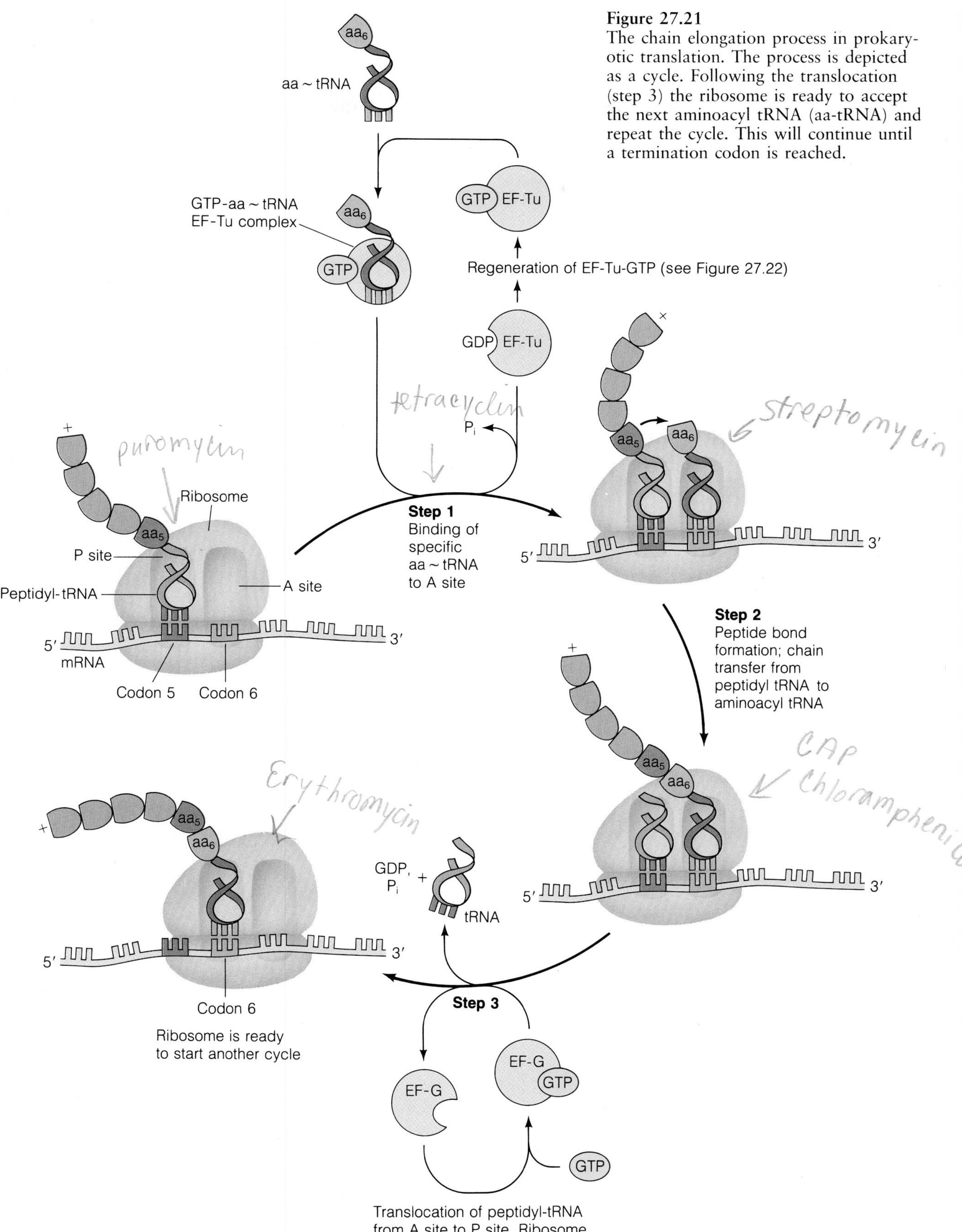

Figure 27.21
The chain elongation process in prokaryotic translation. The process is depicted as a cycle. Following the translocation (step 3) the ribosome is ready to accept the next aminoacyl tRNA (aa-tRNA) and repeat the cycle. This will continue until a termination codon is reached.

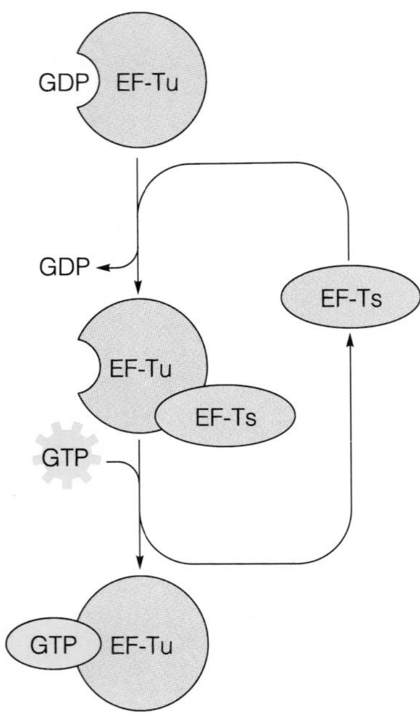

Figure 27.22
Regeneration of EF-Tu GTP by Tu-Ts exchange. This figure gives details of the "delivery cycle" shown at the top of Figure 27.21. Binding of the factor EF-Ts allows the release of GDP and binding of a new GTP to prepare EF-Tu for another cycle.

The next step (Figure 27.21, step 3) is called **translocation.** The tRNA in the A site (which now has the nascent polypeptide chain attached to it) is moved to the P site, and the now uncharged tRNA which had occupied the P site is expelled. In the process, the mRNA is pulled along, placing a new codon adjacent to the now empty A site. Like peptidyl transfer, this step also requires a protein factor (EF-G) with GTP bound to it and requires GTP hydrolysis.

A cycle is now complete. All is as it was at the start, except that now:

1. The polypeptide chain has grown by one residue.
2. The ribosome has moved along the mRNA by three residues, or one codon.

The whole process will be repeated again and again until a termination signal is reached.

Termination

The completion of the synthesis of a polypeptide on a ribosome is signaled by the translocation of one of the **terminator codons** (UAA, UAG, or UGA) into the A site. Under normal circumstances, there is no tRNA that recognizes these codons. Instead, there are **release factors,** proteins that participate in the termination process. In prokaryotes there are three such proteins (see Table 27.3). Two of these (RF1 and RF2) are capable of recognizing terminator codons when in the A site and will bind where a tRNA would normally be expected to bind. RF1 recognizes UAA and UAG; RF2 recognizes UAA and UGA. The third factor, RF3, plays a different role. It is a GTPase that appears to stimulate, via GTP binding and hydrolysis, the release process.*

The sequence of termination events is shown in Figure 27.23. After RF1 or lRF2 has bound to the ribosome, the peptidyltransferase transfers the C-terminal residue of the polypeptide chain from the P-site tRNA to a water molecule; this releases the peptide chain from the ribosome. The RF factors and GDP are then released, followed by the tRNA from the P site. The 70S ribosome is now unstable and dissociates to 50S and 30S subunits. The messenger RNA may or may not dissociate at this point. In some cases where polycistronic messages are being translated, the 30S subunit may simply slide along the mRNA until the next initiation codon is encountered and then begin a new round of translation. If the 30S subunit does dissociate, it will soon reattach to another message.

Understanding the process of termination helped clarify some peculiar observations concerning nonsense mutations. Recall from Chapter 7 that a *nonsense mutation* is one in which a codon for some amino acid has been mutated into a stop codon, so that the polypeptide chain terminates prematurely. Soon after such mutations were discovered, a mysterious class of **revertants** was found. These were not simply back-mutations in the mutated gene, but mutations in *other* genes that somehow *suppressed* the nonsense mutations. The phenomenon is called **intergenic suppression.** On examination, these suppressor mutations were found to lie in tRNA genes.

---

*Note that RF2 is the third GTP-binding protein we have encountered in translation. Recent studies indicate that IF2, EF-Tu, and RF2 all belong to a widespread class of G proteins, which serve as celluiar signals and transfer factors. The *ras* oncogene product (Chapter 23) appears to be a member of this family.

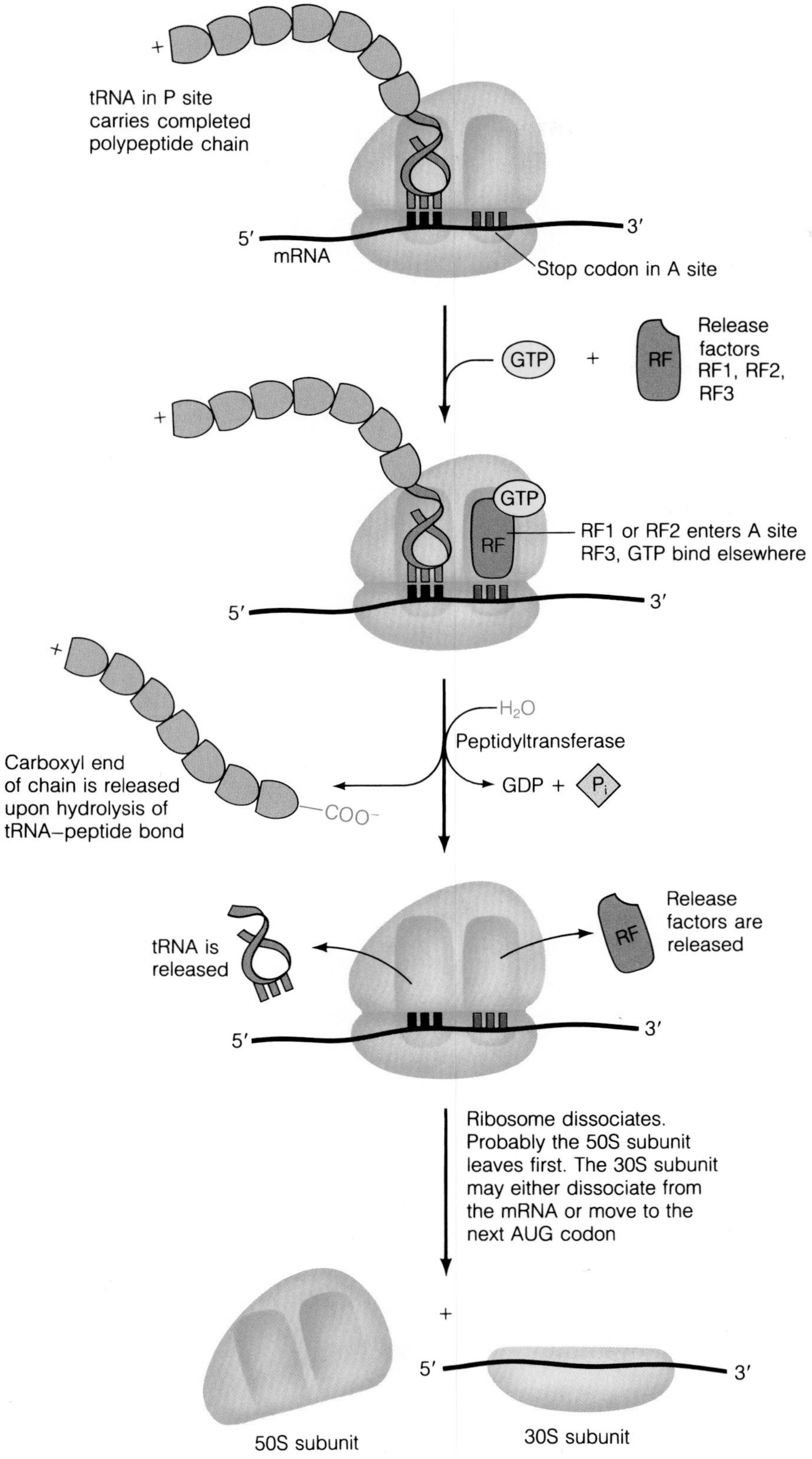

Figure 27.23
Sequence of events in the termination of translation in prokaryotes.

Consider the example shown in Figure 27.24. A nonsense mutation has changed a codon that normally specifies the amino acid tyrosine to a stop codon, causing abortive termination of the polypeptide chain. If, however, one of the several tyrosine tRNAs mutates in its anticodon region so as to recognize the stop codon, translation can sometimes proceed in the normal fashion. Thus, a mutation that might otherwise be lethal can be *suppressed* by such a change, and the microorganisms can survive. Clearly, they will still have problems, for the presence of such a mutated tRNA will interfere with the normal termination of other proteins. That they can survive at all depends on the fact that the suppressor mutation usually involves a minor tRNA species, little used in normal translation. Furthermore, such effects may be minimized by the fact that mRNAs frequently have two or more different stop signals in tandem (for example, see Figure 27.6). Even if the first is suppressed, the "emergency brake" still holds.

Suppressor mutations are by no means confined to correction of nonsense mutations. There are also mutated tRNAs that correct missense and even some that contain two or four bases in the anticodon loop, the so-called **frameshift suppressors.**

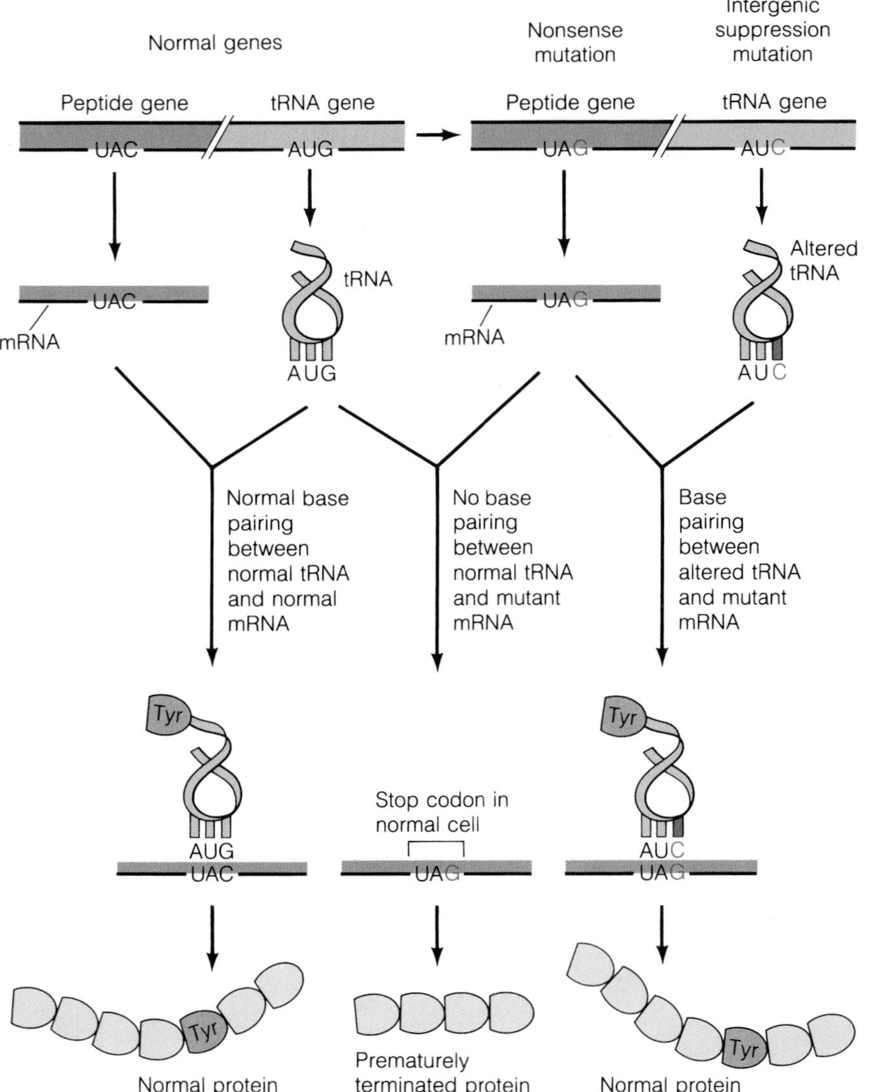

Figure 27.24
An intergenic suppression mutation can overcome a nonsense mutation. A nonsense mutation in a peptide gene produces a mutant mRNA which, instead of base pairing with the normal tRNA, codes for *stop* and consequently prematurely terminates the growing peptide chain. Another mutation (in the tRNA gene) can circumvent the first mutation by altering the tRNA anticodon so that it will base pair with the mutant mRNA. A normal protein is produced in this situation.

Rates and Energetics of Translation

Translation is a rapid process in prokaryotes. An *E. coli* ribosome, at 37°C, can synthesize a 300-residue polypeptide chain in about 20 seconds. Since chain elongation is the rate-limiting process, this means that the ribosome must pass through about 15 codons, or 45 nucleotides per second. But this by no means accounts for the total rate of protein synthesis, for many ribosomes may be simultaneously translating a given message. In fact, if one carefully lyses *E. coli* cells, **polyribosomes** like those shown in Figure 27.25 are observed. Apparently, as soon as one ribosome has moved clear of the 5′ region of the mRNA, another attaches. Under some conditions, there may be as many as 50 ribosomes packed onto an mRNA, with one finishing translation every few seconds. Since there are about 5000 mRNA molecules in an *E. coli* cell, this means that the microorganism is capable of producing about 1000 protein molecules per second.

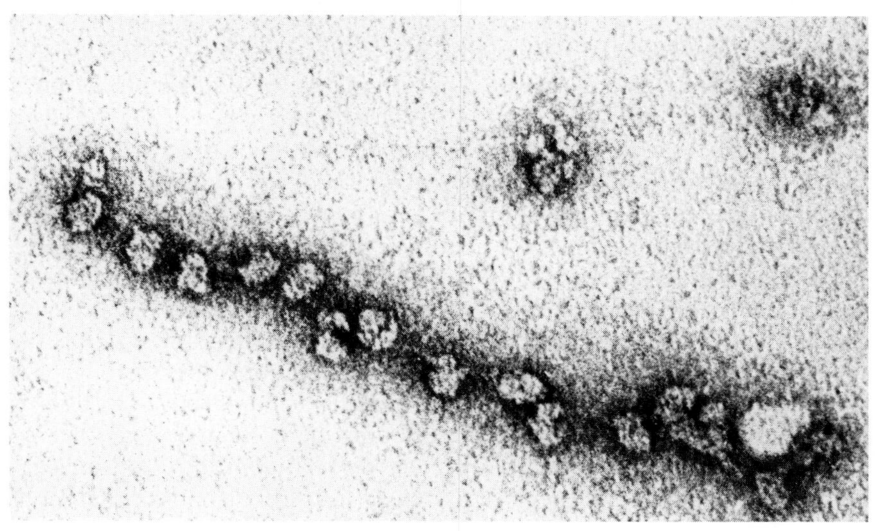

(a)

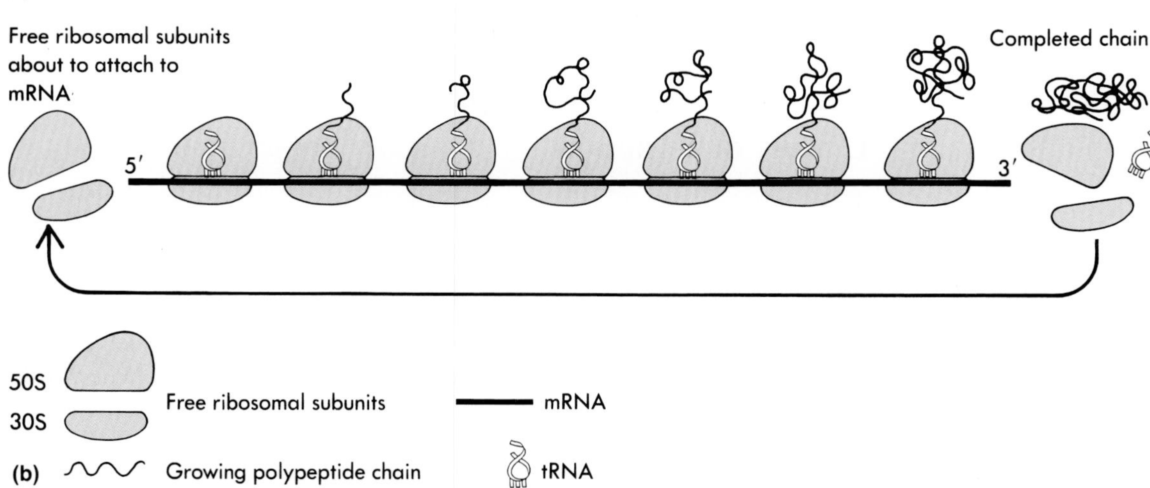

(b)

Figure 27.25
Polyribosomes. (a) Electron micrograph showing *E. coli* polyribosomes. Note that the ribosomes are closely clustered on an mRNA molecule that is still in the process of being transcribed. Translation does not wait for transcription to be finished. (b) Schematic picture of a polyribosome like that shown in (a). Each ribosome is to be imagined as moving from left to right.

The energy cost for this process is very large. If we examine the individual steps in protein synthesis described above, we can make the following estimate of the total energy budget for synthesizing a protein of N residues:

N	ATPs are required to charge the tRNAs
1	GTP is needed for initiation
$N - 1$	GTPs are required to form the N − 1 peptide bonds
$N - 1$	GTPs are necessary for the $N - 1$ translocation steps
1	GTP is required in termination
Sum = $3N$	

Altogether, then, about $3N$ high-energy phosphate molecules must be hydrolyzed to complete a chain of N units. Thus, a typical protein of 300 residues costs the cell about 40,000 kJ of free energy per mole, if we assume ATP or GTP hydrolysis yields about 40 kJ/mol under cellular conditions. Proteins are expensive!

If we express the same data in terms of the energy requirement for synthesis *per peptide* bond, we obtain a cost of about 120 kJ. Since the free energy of hydrolysis of a peptide bond is only about −2 kJ/mol, the price seems exorbitant. Why does not the cell have a mechanism to make every peptide bond for a few kilojoules? Certainly, an input of even 20 kJ/mol would be enough to make the synthesis process very favorable.

At this point a subtle and interesting complication emerges. The cell is making polypeptides of *defined* sequence. If it were simply throwing together amino acids at random, the free energy price could be much cheaper. But a chain of 300 residues, made from 20 different amino acids, can be put together in 20^{300} ($\cong 10^{390}$) different ways, whereas the cell needs *one* specific sequence. There is, in other words, a very large entropy price to be paid in making specific sequences. What this means at the mechanistic level is that every step in the assembly not only must be done with a free energy excess but also must, at critical points, be checked by a proofreading mechanism, which in turn costs energy. It is expensive to get a good translation of a book, for not only must the translators be expert and careful but also their work must be rechecked with great care.

The Fine Structure of the Ribosome: Details of Function

With its several RNA components and many proteins, the ribosome is a complicated piece of molecular machinery. Finding out just how it is put together and how it functions has been the goal of much research. So far, it has not been possible to apply x-ray diffraction successfully to ribosomes. It has proved difficult to obtain satisfactory crystals, and even if such crystals were available, the x-ray analysis of so complex a structure would be a truly formidable task. Consequently, researchers have been forced to use other, less direct methods to map protein and RNA components in the ribosome. Some of these methods are quite powerful techniques and are applicable to studies of other kinds of complex macromolecular structures. In Tools of Biochemistry 25 we briefly describe three of these methods: **cross-linking, immunoelectron microscopy,** and **low-angle neutron scattering.** Figure 27.26 depicts current ideas concerning the arrangement of proteins and RNA within the 30S subunit, as deduced from combined results of these three methods. Figure 27.27 shows how antibodies can be used to find

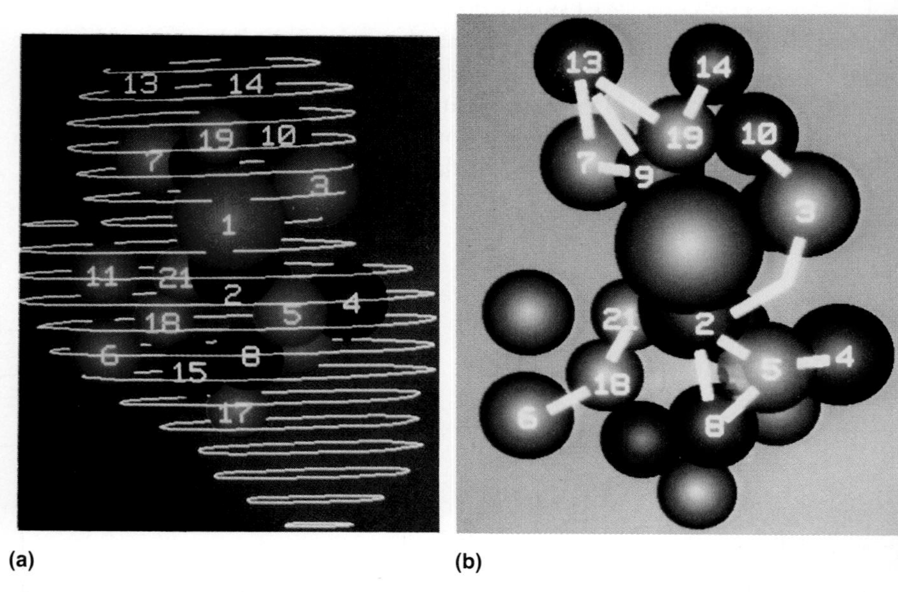

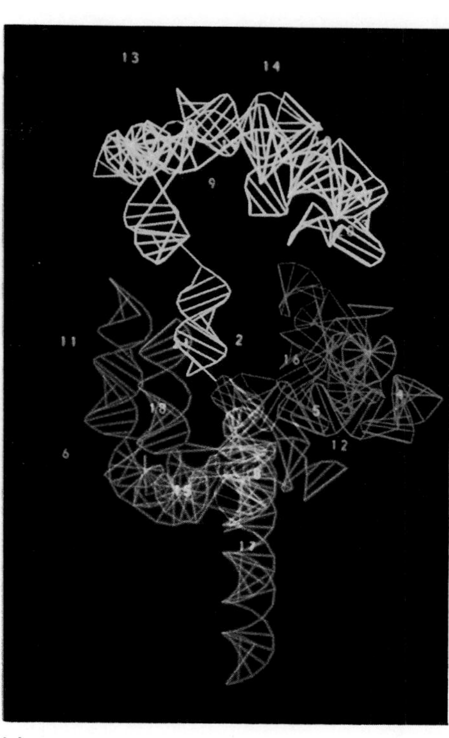

Figure 27.26
Three-dimensional arrangement of components of the 30S subunit. (**a**) Protein arrangement as deduced from neutron scattering experiments. The contour lines show the overall shape of the particle. Regions not occupied by proteins are filled by RNA. (**b**) Cross-linking pattern. The protein arrangement shown in (**a**) is reproduced, with bars between all protein pairs reported by two or more independent research groups to be cross-linked. Note that the cross-linking pattern agrees well with the neutron scattering data. (**c**) Proposed location of the 16S rRNA, as deduced from RNA-protein interaction studies.

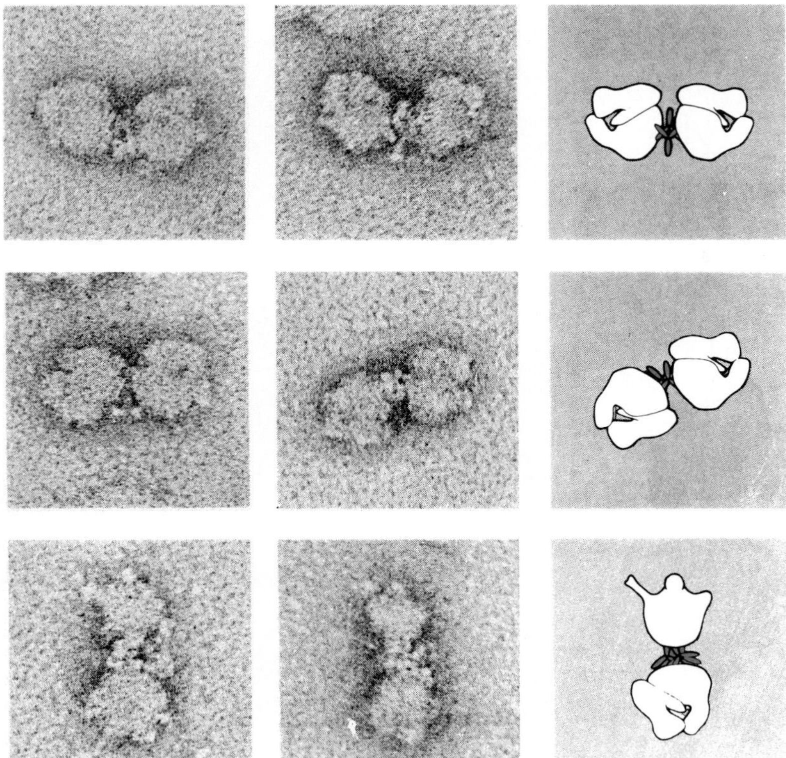

Figure 27.27
Use of antibodies to detect where the polypeptide chain emerges from the ribosome. Antibodies to β-galactosidase were used to interact with β-galactosidase chains at the point where they emerge from the ribosomes. The 70S ribosomes were found to be tied together by such antibodies at sites near the back of the 50S subunit. The technique is described in Tools of Biochemistry 25.

where the growing polypeptide chain emerges from the ribosome.

As a consequence of such studies, a quite complete three-dimensional picture of the prokaryotic ribosome has been developed. In Figure 27.28 we show those details that are of major importance in the translational function. Part (a) shows important features of the two subunits and part (b) shows two views of the whole ribosome. The "working region" of the 70S particle appears to lie largely in the cleft between the 30S and 50S subunits.

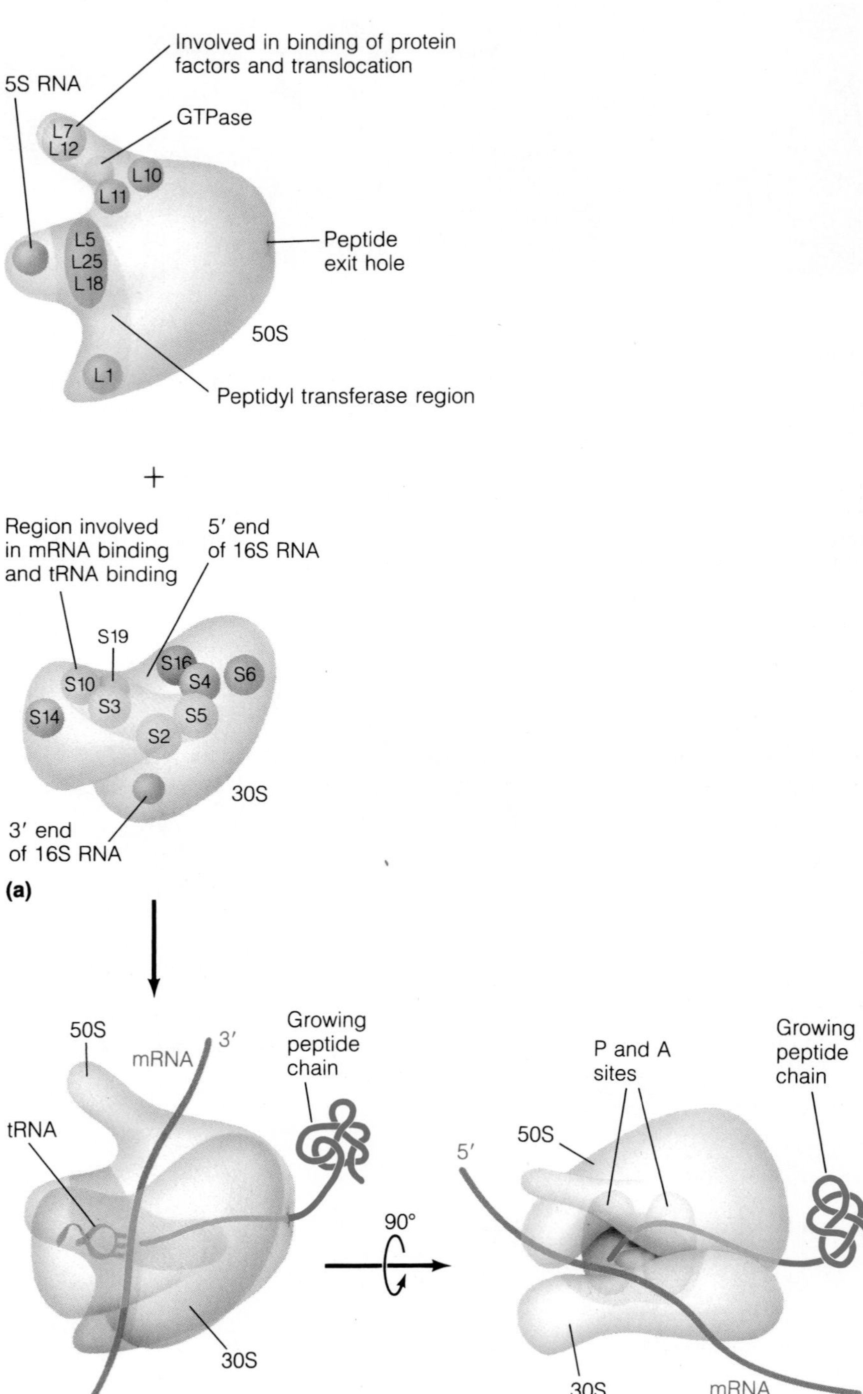

Figure 27.28
Some functional regions on the ribosome. Data from various techniques are combined here to show where things are and where things happen on prokaryotic ribosomes. Purple shading shows the location of the peptidyltransferase complex; yellow shading shows the region where mRNA and tRNAs bind. Positions of certain proteins are illustrated.

Table 27.4
Base-pairing capabilities in wobble pairs

Base at 5′ Position in Anticodon		Base at 3′ Position in Codon
G	pairs with	C or U
C	pairs with	G
A	pairs with	U
U	pairs with	A or G
I	pairs with	A, U, or C

Here are located the 3′ end of the 16S RNA with the sequence complementary to the Shine-Dalgarno sequence, the P and A sites for tRNA binding, and the peptidyltransferase complex. The messenger RNA seems to pass through the cleft, whereas the growing polypeptide chain extends perpendicular to it and emerges from the ribosome near the "back" of the structure. Because they are "hidden" within the ribosomal structure, about 50 nucleotides of the mRNA and 30 residues of the nascent polypeptide chain are protected from the action of hydrolytic enzymes during translation.

It is clear that the tRNA molecules bound in the P and A sites must have their anticodon regions close together, for they are interacting with adjacent codons on the mRNA. Furthermore, at least in the peptidyl transfer step, their 3′ stems must approach one another very closely to allow transfer of the growing polypeptide from one to the other.

It seems probable that there is a certain amount of flexibility in the binding of tRNAs to the ribosome–mRNA structure, even in the codon–anticodon interaction. One strong piece of evidence for this is that in some cases a single tRNA can recognize several different codons. The multiple recognition always involves the 3′ residue of the codon and therefore the 5′ residue of the anticodon.

To explain these observations, Francis Crick proposed in 1966 that the 5′ base of the anticodon was capable of "wobble" in its position during translation, allowing it to make alternative (non-Watson–Crick) hydrogen-bonding arrangements with several different codon bases (Figure 27.29a). Supporting this so-called wobble hypothesis is the observation that the 5′ anticodon position is often occupied by *inosine* (I), which is capable of several different hydrogen-bonding interactions. An I in the 5′ anticodon site can bond with A, U, or C in the 3′ codon site, depending on how it is oriented (Figure 27.29b). Considering both base-pairing possibilities and the observed selectivity of tRNAs, Crick proposed the set of "wobble rules" given in Table 27.4. The wobble hypothesis explains a feature of the genetic code that may seem unusual at first glance—multiple codons for a given amino acid most often involve differences in the third, or 3′, codon, which pairs with the 5′ base of the anticodon.

The Final Stages in Protein Synthesis: Folding and Covalent Modification

The polypeptide chain, as it emerges from the ribosome, is not a completed, functional protein. It must fold into its tertiary structure and may have to associate with other subunits. In some cases disulfide bonds must be formed, and other covalent modifications, such as hydroxylation of specific

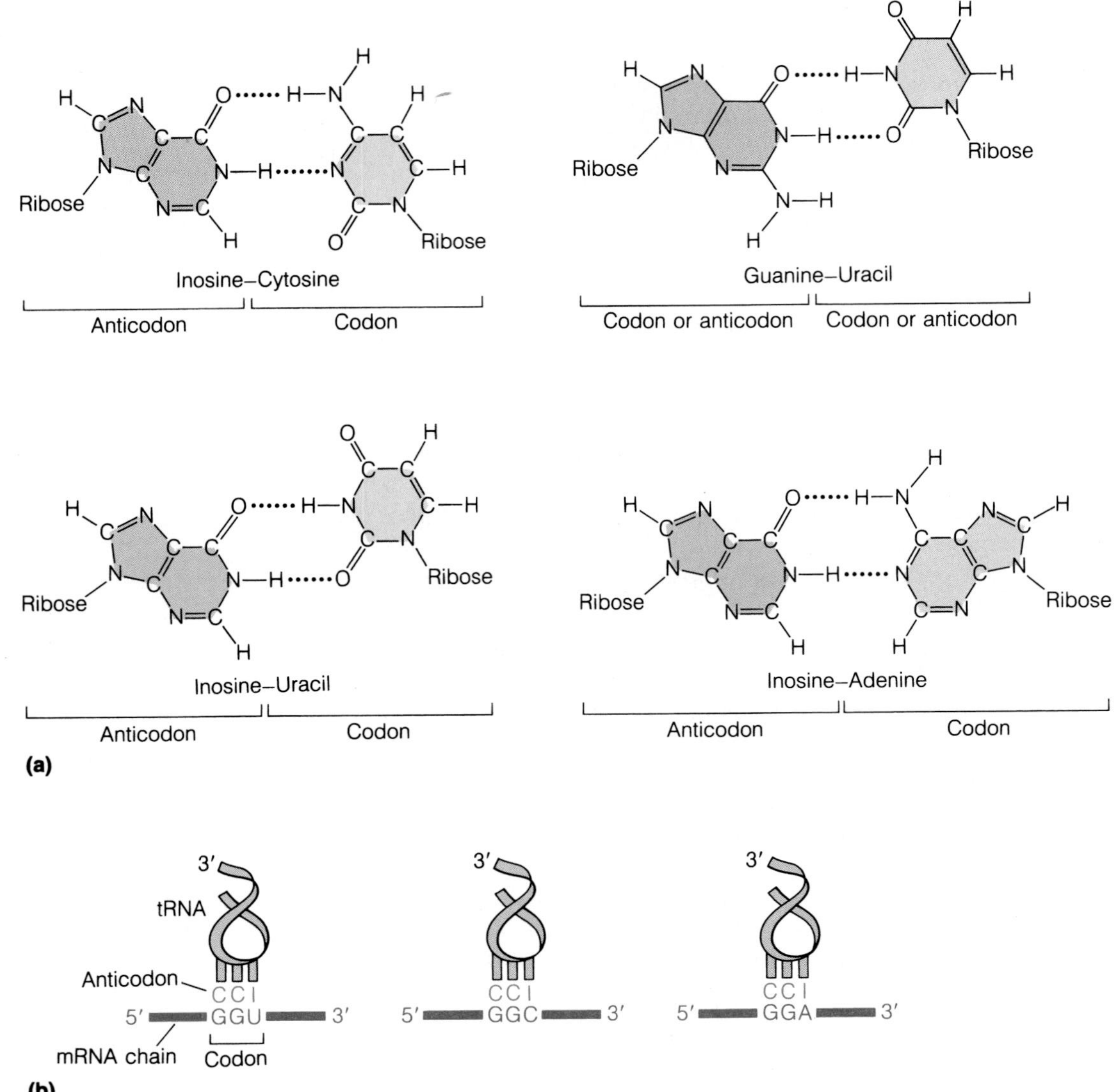

Figure 27.29
The wobble hypothesis. (**a**) Some of the hydrogen-bonding schemes involved in wobble pairing. (**b**) An example of wobble pairing: one tRNA recognizes three different glycine codons, making use of the versatility of inosine.

prolines and lysines or any of the other amino acid modifications we have discussed, must take place. In addition, many proteins are subjected to specific proteolytic cleavage to remove portions of the nascent chain.

Chain Folding

The cell need not wait until the entire chain is released from the ribosome to commence its finishing touches. The first portion of the nascent chain (about 30 to 40 residues) is protected within the ribosome itself, but almost as soon as the N-terminal end emerges, changes begin. There is good evidence, for example, that folding into the tertiary structure starts during translation and is nearly complete by the time the chain is released. For example, antibodies to *E. coli* β-galactosidase, which recognize the tertiary folding of the molecule, will attach to polyribosomes synthesizing this protein. This enzyme displays catalytic activity only as a tetramer. It has been demonstrated that nascent β-galactosidase chains, still attached to ribosomes, can associate with free subunits to form a functional enzyme. Thus, even quaternary structure can be partially established before synthesis is complete.

This behavior should not be surprising, if we recall (Chapters 6 and 7) that formation of the secondary, tertiary, and quaternary levels of protein structure is spontaneous. However, the results do tell us something more about the in vivo pathway for protein folding: it must start from the N-terminal end.

Covalent Modification

Some of the covalent modifications of polypeptide chains also occur during translation. We mentioned above that the *N*-formyl group is removed from most prokaryotic proteins. A special **deformylase** is employed in this reaction. In many cases, deformylation seems to happen almost as soon as the N-terminus emerges from the ribosome. Removal of the N-terminal methionine itself can also be an early event, but whether it happens or not apparently depends on the cotranslational folding of the chain. Presumably, in some cases this residue is "tucked away" and protected.

Some prokaryotic, and many eukaryotic, proteins experience much more severe proteolytic modifications. These are almost invariably proteins that are going to be exported from the cell or are destined for membrane locations. We will reserve discussion of the more complicated eukaryotic protein processing for the next chapter and concentrate here on what happens in prokaryotes.

Consider a bacterial protein that is to be exported from the cell or, in a gram-negative bacterium, delivered into the periplasmic space. Most prokaryotic proteins are synthesized on polyribosomes that are free in the cytosol, and it would be difficult to ensure their proper delivery through the bacterial cell membrane if they were simply released into the cytosol. Those destined for export, therefore, are synthesized on polyribosomes in contact with the membrane. There appears to be no *intrinsic* difference between free and membrane-bound ribosomes. Rather, the membrane attachment is a consequence of a special feature of those polypeptide chains that are destined for extracytosolic sites. Such proteins are characterized by highly hydrophobic **signal sequences** in their N-terminal regions. Representatives are listed in Table 27.5.

Table 27.5
N-terminal signal sequences of representative proteins

Protein					−20					−15					−10					−5				−1	↓ +1		
A. Prokaryotic																											
Prehemagglutinin (influenza virus)										M	A	I	I	Y	L	I	L	L	F	T	A	V	R	G	D	Q	...
Prealkaline phosphatase (*E. coli*)				M	K	Q	S	T	I	A	L	A	L	L	P	L	L	F	T	P	V	T	K	A	R	T	...
Prelipoprotein (*E. coli*)					M	K	A	T	K	L	V	L	G	A	V	I	L	G	S	T	L	L	A	G	C	S	...
B. Eukaryotic																											
Preinvertase (yeast)						M	L	L	Q	A	F	L	F	L	L	A	G	F	A	A	K	I	S	A	S	M	...
Prelysozyme (chicken)						M	R	S	L	L	I	L	V	L	C	F	L	P	K	L	A	A	L	G	K	V	...
Preproinsulin (rat)	M	A	L	W	M	R	F	L	P	L	L	A	L	L	V	L	W	E	P	K	P	A	Q	A	F	V	...

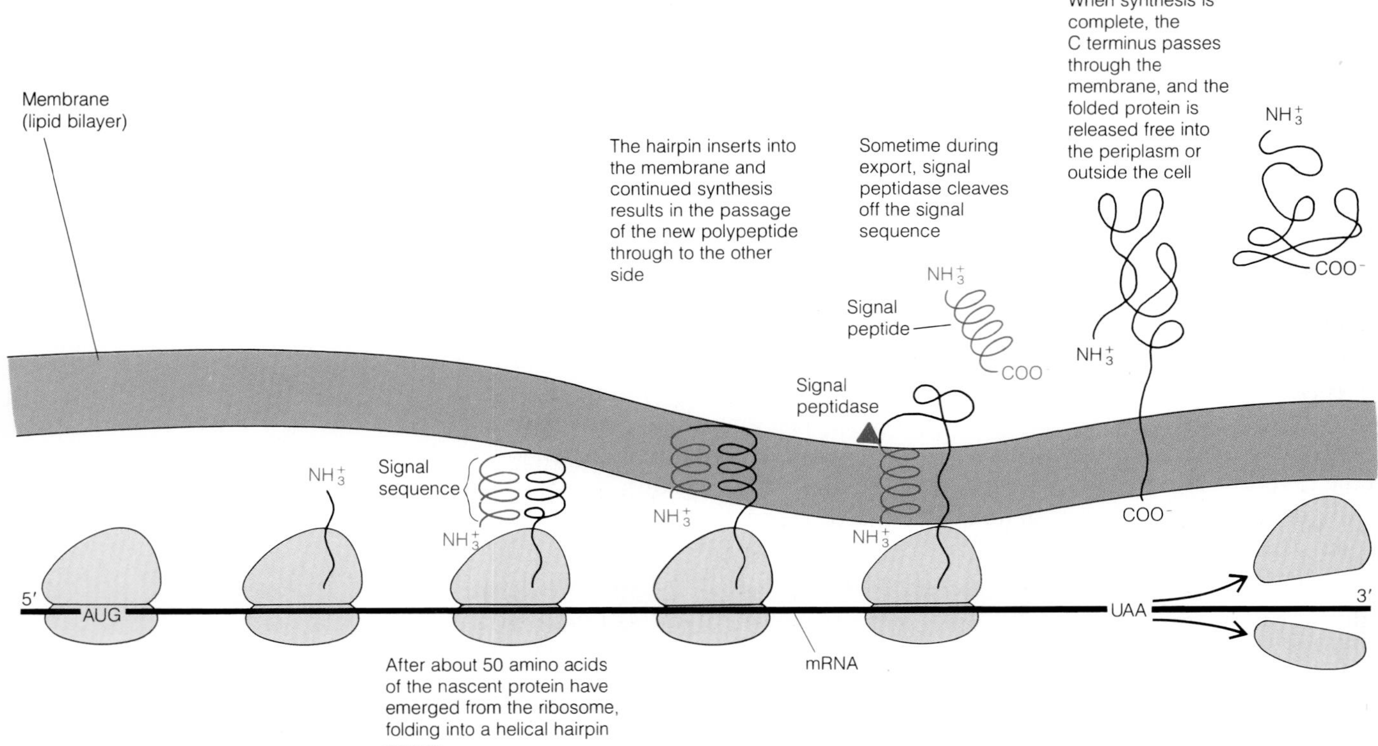

Figure 27.30
Translation accompanied by protein export through membranes. Polyribosomes synthesizing a protein destined for extracellular transport become attached to the cell membrane, through N-terminal signal sequences. The completed polypeptide is then passed through the membrane.

These sequences emerge first from the ribosome, and they are bound to and inserted into the membrane, as shown in Figure 27.30. This "locks" the ribosome to the membrane, and if it is a member of a polyribosomal chain, its neighbors, with their leader peptides, become attached as well. As synthesis proceeds, the polypeptide chain is extended through the membrane. At some point in the process, the signal sequence is cleaved off by a membrane-bound protease, which recognizes a particular site. As Table 27.5 shows, this cleavage is almost always to the carboxyl side of a glycine or alanine residue; the following residue is usually an acidic or basic one.

It seems likely that the fundamental mechanism for protein export is very similar in prokaryotic and eukaryotic cells, although it is used much more and in more complex ways in eukaryotes. The similarity in mechanism is of practical importance in genetic engineering. It is an aid, for example, in the production of cloned eukaryotic proteins (like insulin) in bacteria, for it allows their efficient export from the cell. Preproinsulin, carrying a signal sequence, can be passed from *E. coli* into the external environment.

In the preceding sections, we have followed the translation of proteins from initiation to the finished product. Like almost all biochemical processes, translation is regulated in the cell. We turn now to a discussion of that regulation.

The Regulation of Protein Synthesis

In most organisms, the overall control of protein biosynthesis occurs primarily at the transcriptional level. If need for a protein decreases, it is usually more energy efficient to simply shut off the transcription of the gene and not make an unnecessary mRNA. But there are some cases in which regulation is imposed at the translational level, and we shall consider a few prokaryotic examples in this section.

The major step at which control is imposed on translation appears to be initiation. This is not surprising, for two reasons. First, it is most efficient to block translation at its start, before ribosomes have been committed or significant energy expended. Second, the special requirements for fitting the mRNA properly into a ribosome allow numerous ways to block the process at this point. There are now known to be at least three mechanisms by which initiation of translation of selected messages can be controlled.

1. *The tertiary structure of the mRNA itself can be such as to prevent its attachment to the 30S ribosomal subunit.* This kind of control appears to explain the regulation of translation of the polycistronic message of the MS2 phage, described on p. 959 and in Figure 27.6. Consider the different requirements that translation of this message must satisfy: To make new viruses, *many* copies of the coat protein and a significant number of replicase subunits are needed. But only *one* copy of the A protein and only a small amount of the lysis protein are needed per virus. To simply translate the whole polycistronic message equally in all its parts would be woefully inefficient. What happens is this: The ribosome binding site at the 5′ end of the mRNA (which would start translation of the A-protein message) is normally blocked by tertiary folding of the mRNA molecule. Therefore, the ribosome normally begins with the coat protein, translating its message efficiently and proceeding on to the synthetase message. As synthetase is made, it stimulates replication of the viral mRNA itself. Copies of new plus strands that are still being transcribed have not yet folded into their final conformation. Therefore their 5′ sites are still open for initiation and can attach to ribosomes to translate A protein. But this happens only once in the life of each mRNA—at the time it is being synthesized.

The message for the lysis protein, which overlaps, in a different reading frame, with the coat protein message (see Figure 27.6), is apparently translated only from time to time, as a consequence of frameshift slip during translation of the coat protein.

2. *Proteins may bind to the mRNA, blocking initiation.* A most elegant example of this kind of control is found in the synthesis of the ribosomal proteins themselves. As Figure 27.31 shows, prokaryotic ribosomal proteins are coded by polycistronic messages. In each of these groups, there is *one* protein that is capable of binding at or near the 5′ end of the message

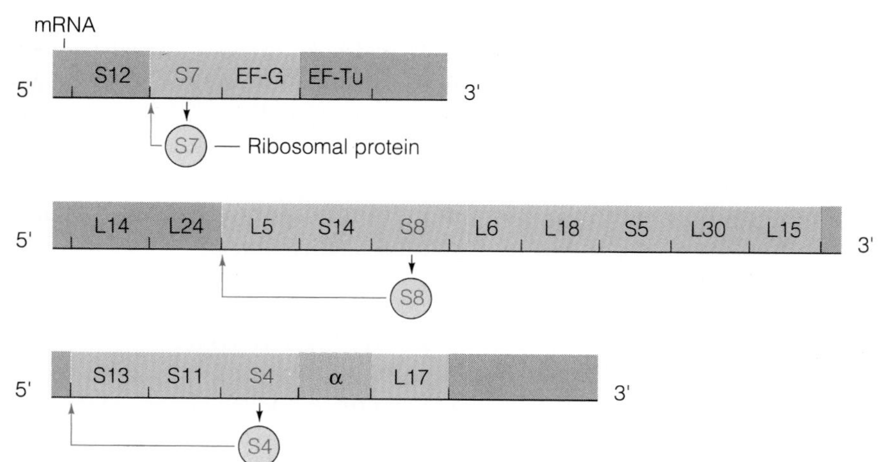

Figure 27.31
Regulation of synthesis of ribosomal proteins. Shown are three examples of polycistronic messages coding for ribosomal proteins. Note that these mRNAs also contain messages for EF-G, EF-Tu, and the α subunit (purple) of RNA polymerase. Each polycistronic message is controlled by the ribosomal protein (red). At each point the start of control is indicated by the head of the arrow. Control extends *only* over the cistrons shown in green; in other words, the RNA polymerase and EF-Tu messages are excluded from this control.

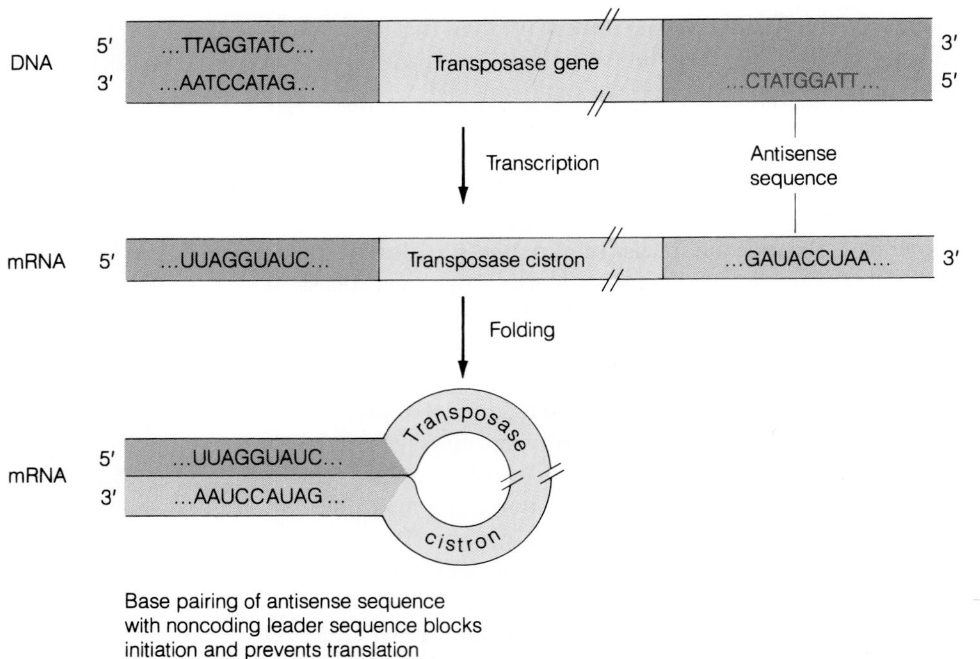

Figure 27.32
Blocking of translation initiation by antisense RNA. The transposase gene is used as an example.

and blocking its translation. This seems to occur because the mRNA has a tertiary structure similar to the normal binding site for the protein on rRNA.

The beauty of the system can be seen by comparing Figure 27.31 with Figure 27.18: The proteins that control ribosomal synthesis are also among the first to bind onto the rRNA in ribosome assembly. These proteins are the keys to ribosome construction, and they have a very high affinity for the appropriate rRNAs. If ribosomal RNA is abundant in the cell, they are snapped up, and synthesis of ribosomal proteins can proceed. But if rRNA is in short supply, the unutilized proteins will bind to the appropriate mRNAs, shutting down the synthesis of unneeded ribosomal proteins.

Although it is not a matter of translational control, it is appropriate to note that rRNA synthesis is itself under metabolic control. In prokaryotes, starvation for amino acids results in the accumulation of the unusual nucleotides ppGpp and pppGpp. These have a pronounced inhibitory effect on the transcription of rRNA, which inhibition in turn inhibits ribosomal protein synthesis. Thus, starvation conditions induce a general "belt tightening" in the biosynthetic apparatus.

3. *Blocking of initiation by antisense RNA.* A number of situations are now known where control of translation is effected by the synthesis, at another genomic locus, of RNA fragments corresponding to the noncoding strand of DNA complementary to the 5′ end of an mRNA. By formation of a double-strand structure, the initiation site is blocked. Such a mechanism regulates, for example, the translation of transposase mRNA (Figure 27.32). Another example was presented in Chapter 26.

The Inhibition of Translation by Antibiotics

The control of translation that we have described above is imposed by the cell itself, to regulate its own metabolism. There are, on the other hand, many instances in which some organisms produce substances that interfere with protein synthesis in other organisms. When the target organisms are bacteria, these substances are called *antibiotics*.

We have already encountered a number of other kinds of antibiotics. In Chapter 8, the action of the penicillins in inhibiting bacterial cell wall synthesis was described, and Chapter 9 discussed antibiotics such as gramicidin and valinomycin, which interfere with the ionic balance across membranes. The *sulfa drugs* (Chapter 20) are antimetabolites, and other antibiotics, such as *rifampicin* and *streptolydigin* (Chapter 26) block transcription in prokaryotes.

Since translation is such a complex and vital process, it perhaps is not surprising that it is a favorite target for antibiotics. A host of naturally occurring substances interfere with various stages in protein synthesis. We describe here the action of only a few of the most important ones, in the order of the steps in synthesis they inhibit.

The tetracyclines (Figure 27.33a) inhibit the binding of aminoacyl tRNAs to the ribosome. Although eukaryotic protein synthesis is also inhibited by tetracyclines, the drugs are not dangerous to eukaryotic cells, for the cell membranes of higher organisms are impermeable to them. Some bacteria develop tetracycline resistance by changing their membrane permeability to these drugs; others develop enzymes to inactivate them.

Streptomycin (Figure 27.33b), which is representative of the class of **aminoglycoside** antibiotics, also interferes with the normal pairing between aa-tRNAs and message codons. But streptomycin has its major effect by allowing incorrect pairing. Thus, it causes misreading of the message, to an extent that normal proteins are hardly synthesized at all. Bacteria may mutate to develop resistance to streptomycin. These mutations seem to occur mainly in protein S12, which lies near the cleft in the small subunit. This has been taken as evidence that S12 and this region of the ribosome are involved in tRNA binding.

Chloramphenicol (Figure 27.33c), also called chloromycetin, is one of a number of antibiotics that interfere with chain elongation. This molecule seems to act as a competitive inhibitor of the peptidyltransferase complex.

Erythromycin (Figure 27.33d) blocks the next step in elongation—the translocation process. Its binding site includes a specific region of the 23S RNA, and its action can be inhibited by an enzyme that methylates a specific A residue in this region. As with a number of other antibiotics, resistance to erythromycin can be conferred by the insertion of a "resistance" gene on a bacterial plasmid. In this case, the methylase gene is used. The use of such conferred antibiotic resistance provides a common method for screening bacterial clones for those containing particular plasmids (see Chapter 25). In this example, bacteria containing a plasmid carrying the methylase gene will grow in an erythromycin-containing medium, whereas those lacking the plasmid will be killed.

Puromycin (Figure 27.33e) causes premature chain termination. The molecule of puromycin bears structural similarity to the 3′ end of the tRNA, with its attached amino acid residue. Thus, it can enter the A site, be transferred to the growing chain, and cause its release.

There are sufficient differences between the translation machinery in prokaryotic and eukaryotic cells that most of the antibiotics that interfere with protein synthesis in bacteria are harmless to higher organisms. Despite these differences, we find in the next chapter that the main features of translation have been preserved over the whole of evolution. This process constitutes the fundamental connection between genetic information and its functional expression, and as such, has been subject to strong evolutionary constraints.

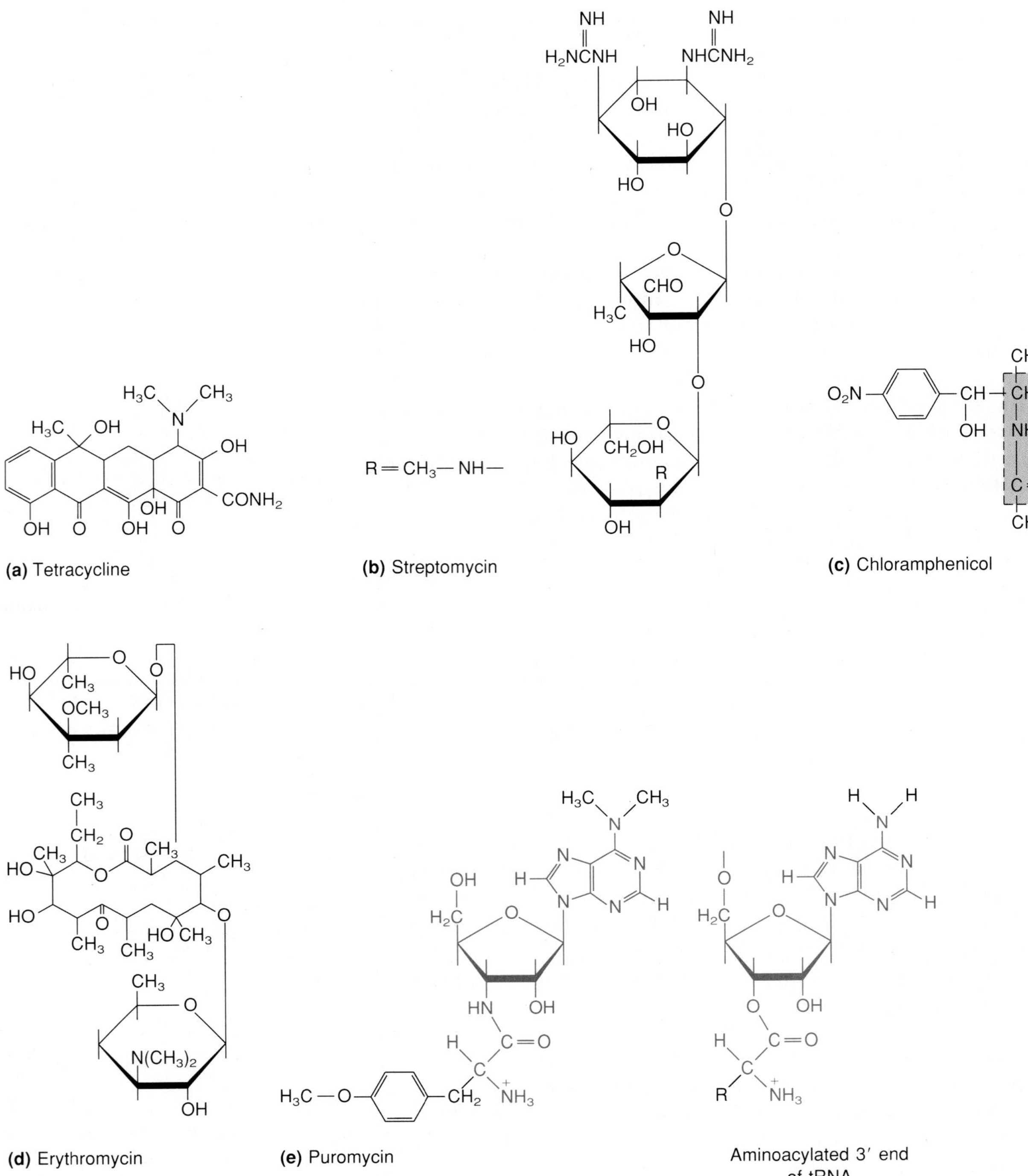

Figure 27.33
Antibiotics that act by interfering with protein biosynthesis. Note that chloramphenicol contains an amide link resembling a peptide bond (blue) and that a portion of puromycin (colored) resembles the 3′ end of an aminoacylated tRNA.

REFERENCES

General

Bermek, E. (ed.) (1985) *Mechanisms of Protein Synthesis.* Springer-Verlag, Berlin, New York.

Clark, B. F. C. and H. F. Petersen (eds.) (1984) *Gene Expression. The Translational Step and Its Control.* Munksgaard, Copenhagen.

Spirin, A. S. (1986) *Ribosome Structure and Protein Synthesis.* Benjamin/Cummings, Menlo Park, Calif. Concentrates on the ribosomes but also provides an excellent overview of the whole field.

The Genetic Code

Crick, F. H. C. (1958) On protein synthesis. *Symp. Soc. Exp. Biol.* 12:138–162. With great prescience, Crick foresees the nature of the translation mechanism.

Crick, F. H. C. (1966) Codon-anticodon pairing: The wobble hypothesis. *J. Mol. Biol.* 19:548–555.

Grosjean, F., and W. Fiers (1982) Preferential codon usage in prokaryotic genes. *Gene* 18:199–209.

Khorana, H. G. (1968) Nucleic Acid Synthesis in the Study of the Genetic Code. In: *Nobel Lectures, Physiology and Medicine (1963–1970),* pp. 341–343. American Elsevier, New York. A Nobel Prize winner's account of the deciphering of the code.

Woese, C. R. (1967) *The Genetic Code.* Harper & Row, New York.

Messenger RNA

Brenner, S., F. Jacob, and M. Meselson (1961) An unstable intermediate carrying information from genes to ribosomes for protein synthesis. *Nature* 190:576–581. Early evidence for mRNA.

Bronson, M. J., C. Squires, and C. Yanovsky (1973) Nucleotide sequence from tryptophan messenger RNA of *Escherichia coli.* The sequences corresponding to the amino terminal region of the first polypeptide specified by the operon. *Proc. Natl. Acad. Sci. U.S.A.* 70:2335–3339.

Shine, J., and L. Dalgarno (1974) The 3′-terminal sequence of *E. coli* 16S rRNA: Complementarity to nonsense triplets and ribosome binding sites. *Proc. Natl. Acad. Sci. U.S.A.* 71:1342–1346.

Transfer RNAs

Kim, S.-H. (1978) Three-dimensional structure of transfer RNA and its functional implications. *Adv. Enzymol.* 46:279–315.

Schimmel, P., D. Söll, and J. Abelson (eds.) (1979) *Transfer RNA.* Cold Spring Harbor Laboratory Press, Cold Spring Harbor, N.Y. A two-part treatise.

Aminoacyl-tRNA Synthetases and aa-tRNA Coupling

Bedoulte, H., and G. Winter (1986) A Model of Synthetase Transfer RNA Interaction as Deduced by Protein Engineering. *Nature* 320:371–373.

Fersht, A. R., R. J. Leatherborrow, and T. N. C. Welk (1986) Structure and reactivity of the tyrosyl-tRNA synthetase: The hydrogen bond in catalysis and specificity. *Philos. Trans. R. Soc. London Ser. A* 317:305–320.

Hou, Y.-M. and P. Schimmel (1988) A simple structural feature is a major determinant of the identity of a transfer RNA. *Nature* 333:140–145. How synthetase recognizes the tRNA.

Schimmel, P. (1987) Aminoacyl-tRNA synthetases: General scheme of structure–function relationships in the polypeptides and recognition of transfer RNAs. *Annu. Rev. Biochem.* 56:125–158.

Ribosomes

Capel, M. S., D. M. Engelman, B. R. Freeborn, et al. (1987) A complete mapping of the proteins in the small ribosomal subunit of *Escherichia coli. Science* 238:1403–1406.

Hardesty, B. and G. Kramer (eds.) (1986) *Structure, Function, and Genetics of Ribosomes.* Springer-Verlag, New York.

Lake, J. A. (1985) Evolving ribosome structure: Domains in archaebacteria, eubacteria, eocytes, and eukaryots. *Annu. Rev. Biochem.* 54:507–530. A "molecular anatomy" approach to evolution.

Stern S., T. Powers, L.-M. Chang chien, H. Noller (1989) RNA-protein interactions in 30S ribosomal subunits: Folding and function of 16S rRNA. *Science* 244:783–790.

Wittmann, H. G. (1983) Architecture of prokaryotic ribosomes. *Annu. Rev. Biochem.* 52:35–65.

The Translation Process

Caskey, C. T., W. C. Forrester, and W. Tate (1984) Peptide chain termination. In: *Alfred Benzor Symposium,* edited by B. Clark and H. Petersen, Vol. 19, pp. 457–466. Munksgaard, Copenhagen.

Gold, L., et al. (1981) Translational initiation in prokaryotes. *Annu. Rev. Microbiol.* 35:365–403.

Kurland, C. G. (1982) Translational accuracy *in vitro. Cell* 28:201–202.

Maitra, U, E. A. Stryer, and A. Chandhuri (1982) Initiation factors in protein synthesis. *Annu. Rev. Biochem.* 51:869–900.

Yarus, M., and R. Thompson (1983) Precision of protein biosynthesis. In: *Gene Function in Prokaryotes,* edited by J. Beckwith, J. Davies, and J. A. Gallant, pp. 23–63. Cold Spring Harbor Press, Cold Spring Harbor, N.Y.

Regulation

Campbell, K., G. Stormo, and L. Gold (1983) Protein-mediated translational repression. In: *Gene Function in Prokaryotes,* edited by J. Beckwith, J. Davies, and J. A. Gallant, pp. 185–210. Cold Spring Harbor Press, Cold Spring Harbor, N.Y.

Nomura, M., J. Yates, D. Dean, and L. Post (1980) Feedback regulation of ribosomal gene expression in *Escherichia coli. Proc. Natl. Acad. Sci. U.S.A.* 77:7084–7088.

Antibiotics

Cundliff, E. (1980) Antibiotics and prokaryotic ribosomes: Action, interaction and resistance. In: *Ribosomes: Structure, Function, and Genetics,* edited by G. Chambliss et al., pp. 377–412. University Park Press, Baltimore.

Jiminez, A. (1976) Inhibitors of translation. *Trends Biochem. Sci.* 1:28–29.

Posttranslational Modification

Griffiths, G. and K. Simons (1986) The *trans* Golgi network: Sorting at the exit site of the Golgi complex. *Science* 234:438–441.

Randall, L. L. and S. J. S. Hardy (1984) Export of protein in bacteria. *Microbiol. Rev.* 48:290–298.

PROBLEMS

1. The following synthetic polynucleotide is synthesized and used as a template for peptide synthesis in a cell-free system from *E. coli.*

   ...AUAUAUAUAUAU...

   What polypeptide would you expect to be produced?

2. If the same polynucleotide described in Problem 1 is used with a *mitochondria-derived* cell-free protein-synthesizing system, the product is

   ...Met–Tyr–Met–Tyr–Met–Tyr...

   What does this say about differences between the mitochondrial and bacterial codes?

3. When polynucleotides are synthesized with repeating triplets of nucleotide residues, two or three kinds of polypeptide chains will be produced in cell-free synthesis.
   (a) Explain why there are two different possible results.
   (b) Predict polypeptides produced when the following are used with an *E. coli* system:

   $(GUA)_n$
   $(UUA)_n$

4. What kind of repeating polynucleotide would yield a single polypeptide with a tetrapeptide repeating unit?

5. Suppose that the MS2 A-protein message (p. 972) is modified by insertion of an extra U to
   (a) the 3′ side of the ...AGGAGG... sequence.
   (b) the 5′ side of the ...AGGAGG... sequence.
   Predict the effect on translation in each case.

6. According to wobble rules, what codons should be recognized by the following anticodons? What amino acid residues do these correspond to?
   (a) 5′...ICC...3′
   (b) 5′...GCU...3′

7. Suppose that the probability of making a mistake in translation at each step is a small number, δ. Show that the probability that a given protein molecule, containing n residues, will be completely error free is then $p = (1 - \delta)^n$.

8. If we assume that the translational error frequency is $\delta = 1 \times 10^{-4}$, calculate the probability of making a perfect protein with
   (a) 100 residues
   (b) 1000 residues

9. Assuming that glucose is burned to CO_2 as an energy source, how many amino acid residues can be incorporated into a protein molecule for each glucose consumed by a cell?

10. We have shown that one way in which proteins can regulate their own production is by binding to mRNA so as to prevent translation. Suggest *other* ways by which feedback regulation of protein production might occur.

11. Why might you expect a small dose of puromycin to be less effective in repressing bacterial growth than an equivalent number of molecules of erythromycin?

12. The protein colicin E3 is a very effective inhibitor of protein synthesis in prokaryotes. This protein is a nuclease, specifically attacking a phosphodiester bond near the 3′ end of the 16S RNA. Suggest a mechanism for the effect of colicin E3 on translation.

TOOLS OF BIOCHEMISTRY 25

Ways to Map Complex Macromolecular Structures

As we continue to probe the structure of the cell, using ever gentler and more discriminating techniques, it becomes apparent that much of the cellular machinery is organized in complicated structures that are *assemblies* of macromolecules. The ribosome, with its several kinds of RNA and many kinds of proteins, is an excellent example. Direct analysis of such immense, complex structures by x-ray diffraction is a task still at the limits of our capabilities. But there are other ways in which we can learn something about the spatial arrangement of the various parts. We describe several of them here.

Chemical Cross-Linking

One way to learn about the arrangement of components in a particle is to see what can easily be attached to what. Consider an idealized particle (like that shown in Figure T25.1) that contains three protein molecules. If we have a bifunctional reagent (like one of those shown in Table T25.1) that can react with side chain residues

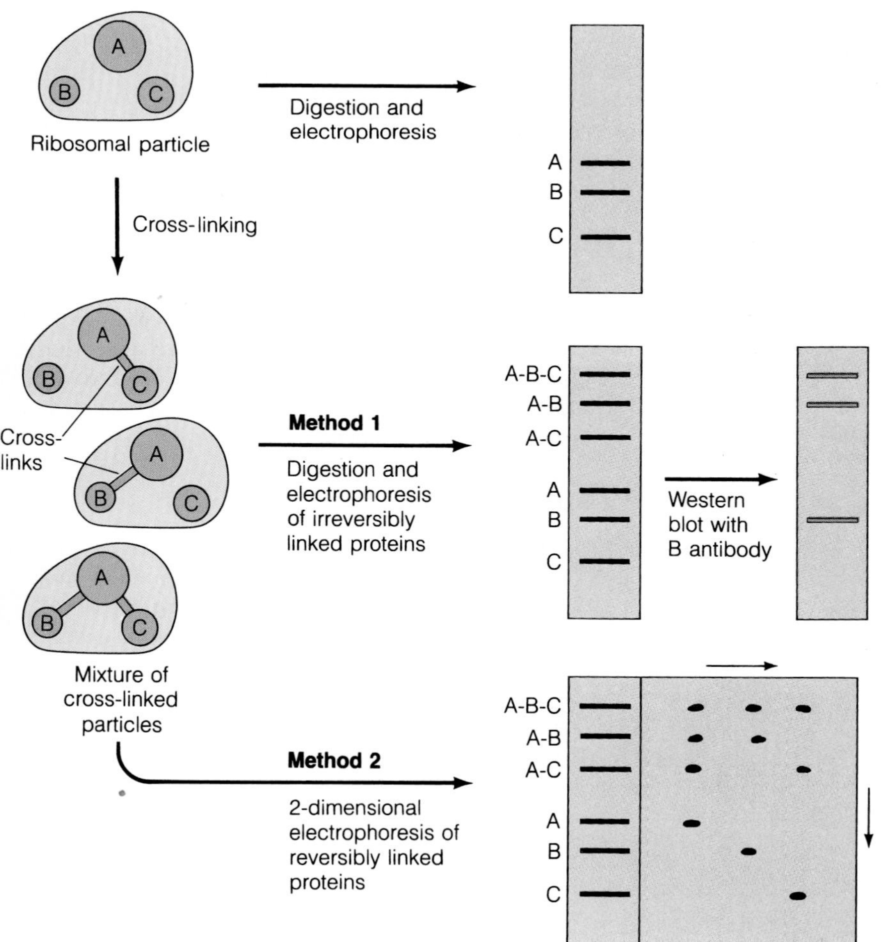

Figure T25.1
Use of chemical cross-linkers to determine propinquity of proteins in a complex particle.

Table T25.1
Some protein cross-linking reagents

Reagent	Formula	Reacts Primarily with	Cleavable?
Bis(*N*-maleimidomethyl) ether	(maleimide)N—CH_2—O—CH_2—N(maleimide)	Sulfhydryl group	No
2,2′-Dicarboxy-4,4′-azophenyldiisocyanate	O=C=N—C_6H_3(COOH)—N=N—C_6H_3(HOOC)—N=C=O	Amino groups	Yes, by reduction of —N=N—
Dimethyl suberimidate	$CH_3O—C(=NH)—(CH_2)_6—C(=NH)—OCH_3$	Amino group	Yes, by ammonia
Tetranitromethane	$C(NO_2)_4$	Phenolic groups (tyrosine)	No
Methyl-4-azidobenzoimidate[a]	$N_3—C_6H_4—C(=NH)—OCH_3$	Amino groups	No
Methyl [3-(*p*-azidophenyl) dithio]propionimidate[a]	$N_3—C_6H_4—S—S—(CH_2)_2—C(=NH)—OCH_3$	Amino groups	Yes, by reduction of —S—S—

[a]These reagents are photoactivatable. They may be first reacted, in the dark, through the imidate group to the right and then coupled to another group by activating the azide (N_3) with a flash of light.

to form cross-links between protein molecules, we can react lightly and then extract the protein as a mixture of cross-linked particles. Partners that have formed can be identified in various ways. If we have antibodies to the different proteins, we can identify dimers that have formed by a "Western blotting" technique (method 1, Figure T25.1; also see Tools of Biochemistry 12). Alternatively, we might use one of the "cleavable" cross-linkers shown in Table T25.1, together with two-dimensional gel electrophoresis (method 2, Figure T25.1). In any event, in the simple example shown, it is clear that protein A must lie between B and C, since it can be linked to either, but B and C do not form cross-linked dimers.

Immunoelectron Microscopy

Components that lie on the surface of a particle can be localized in a very direct manner by using antibodies prepared against them. The Y-shaped antibody molecules form bridges between two particles, connecting points at which the particular component lies near the surface. In Figure T25.2 we show the same particle used in Figure T25.1. The fact that protein B lies near the pointed end of the particle is made clear by the way in which anti-B antibodies tie two particles together. When appropriate hapten groups are attached at specific points on RNA molecules, the same method can be used to show when these RNA sequences are near the ribosomal surface. The ends of the ribosomal RNAs have been located in this way, as has the position at which the nascent peptide emerges from the ribosome (Figure 27.27).

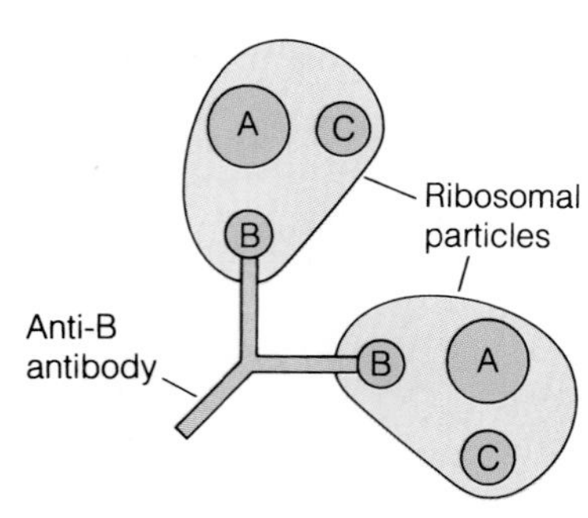

Figure T25.2
Locating proteins on a particle surface by antibody binding.

Low-Angle Scattering of X-Rays and Neutrons

Although crystal diffraction studies of particles like ribosomes may still be impracticable, we can learn much from studying the *scattering* of radiation from solutions of such particles.

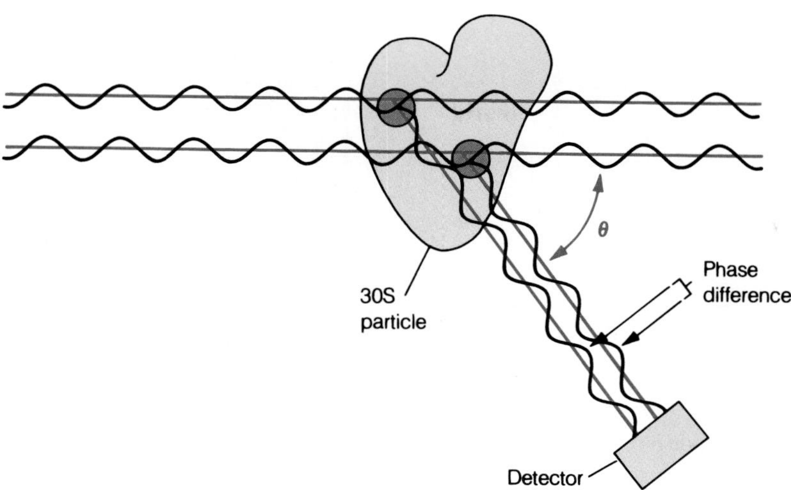

Figure T25.3
Scattering of x-rays or neutrons from different volume elements within a large particle produces mutual interference in the scattered waves.

As diagramed in Figure T25.3, electromagnetic waves scattered from a particle with dimensions much greater than the wavelength of the radiation will exhibit a dependence of scattering intensity on the angle of observation. This is because different regions within the particle are scattering out of phase at all angles from the incident beam other than 0°. At large angles this phase difference increases, causing partial cancellation of the scattered waves. This effect may be used to measure average dimensions of the particle. The intensity of scattering at angle θ (I_θ) compared with the scattering at angle 0 (I_0) is given (for small angles) by

$$\frac{I_\theta}{I_0} = e^{-(16\pi^4 R_G^2/3\lambda^4)\sin^2(\theta/2)} \qquad \text{(T25.1)}$$

Here λ is the wavelength of the radiation, and R_G is a quantity called the **radius of gyration,** a kind of average dimension of the particle. According to equation (T25.1), a graph of $\ln(I_\theta/I_0)$ versus $\sin^2(\theta/2)$ should be a straight line at low angles, with initial slope of $(-16\pi^4 R_G^2/3\lambda^4)$. Thus, measurement of the scattering at very low angles gives a measure of average particle size. At higher angles, the I_θ/I_0 curves have a more complex shape, with maxima and minima. These can be used to give additional information about the shape of the particle and its internal distribution of matter.

Although low-angle scattering of x-rays has been very useful in studying particles in solution, a much more powerful technique has recently been developed—**low-angle neutron scattering.** It may seem strange at first to think of neutrons as radiation, but one must remember that according to quantum mechanics any elementary particle has wavelike properties as well. The wavelength of a particle with mass m moving at velocity v is given by $\lambda = h/mv$, where h is Planck's constant. It turns out that "thermal neutrons" emerging from a nuclear reactor have a wavelength of a few tenths of a nanometer. Thus, they are of the right length for examination of details of macromolecular structure. More important is a peculiarity of neutron scattering: since neutrons interact primarily with the nuclei of atoms, the scattering power of different atoms varies greatly. Thus, nucleic acids and proteins scatter differently, and even hydrogen and deuterium have very different scattering powers. Since H_2O and D_2O differ in scattering, it is possible to use as solvents H_2O/D_2O mixtures that *match* the neutron scattering power of either the nucleic acid or protein portion of a nucleoprotein particle. Then, as shown in Figure T25.4, we can make either nucleic acid or protein "disappear" into the background and measure the radius of gyration of either component. In the example shown, the greater R_G observed for nucleic acid than for protein tells us that the nucleic acid is concentrated on the outside of the particle.

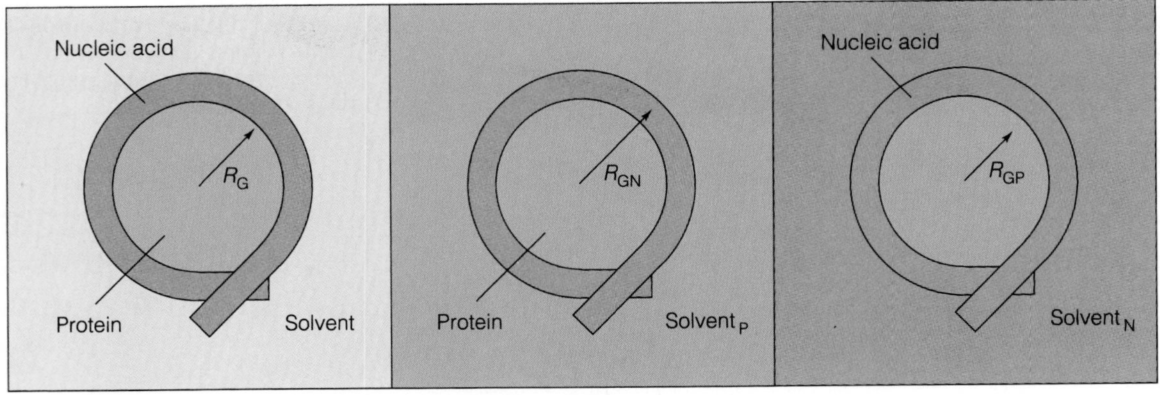

$Solvent_P$ = solvent that matches the scattering properties of the protein
$Solvent_N$ = solvent that matches the scattering properties of the nucleic acid
R_G = radius of gyration of the nucleic acid–protein complex
R_{GN} = radius of gyration of the nucleic acid
R_{GP} = radius of gyration of the protein

Figure T25.4
Use of selective solvent matching to show, by neutron scattering, that the nucleic acid is on the outside of a nucleoprotein particle.

An even more powerful variant of the same technique has been used to "map" the distances between particular pairs of proteins in complex particles. Suppose, as in Figure T25.5 we have reconstituted particles containing just two proteins (among the many in the particle) that have been prepared from deuterium-fed bacteria. These two proteins will be heavily deuterated and will have a neutron scattering power much different from that of the rest of the particle. If the H_2O/D_2O solvent is now mixed so as to match the average background in the nondeuterated portion of the particle, the two deuterated proteins will stand out in contrast. The neutron scattering pattern obtained will be dominated by the interference in scattering between these two proteins and can be used to measure their separation. In addition, the method allows one to measure the radius of gyration of a particular protein in situ.

Although neutron scattering can give much useful information, it is not a technique the average biochemist can use in the laboratory. Rather, one must arrange to go to one of a few locations in the world where large research reactors are fitted to do neutron scattering studies.

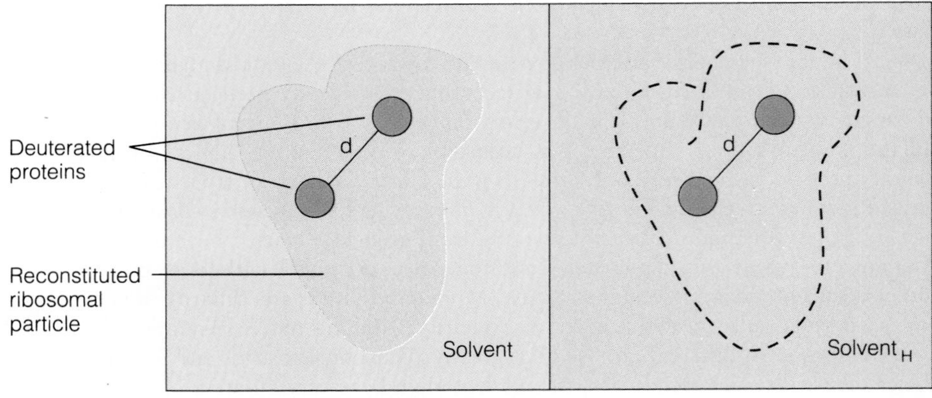

$Solvent_H$ = solvent matching the nondeuterated part of particle

Figure T25.5
Use of solvent matching to determine the distance between two selectively deuterated proteins in an undeuterated particle.

REFERENCES

Das, M., and C. F. Fox (1979) Chemical cross-linking in biology. *Annu. Rev. Biophys. Bioeng.* 8:165–193.

Engelman, D. M., and P. B. Moore (1976) Neutron scattering studies of the ribosomes. *Sci. Am.* 235(4):44–54.

Serdyuk, I. N., Grenader, A. K. and Zaccai, G. (1979) Study of the internal structure of *E. coli* ribosomes by neutron and x-ray scattering. *J. Mol. Biol.* 135:691–707.

Stoffler, G., and M. Stoffler-Meilicke (1983). The ultrastructure of ribosomes: An immunological approach. In: *Modern Methods in Protein Chemistry*, edited by H. Tschesche, pp. 409–457. Walter Gruyter Verlag, Berlin, New York.

Wold, F. (1972) Bifunctional reagents. *Methods Enzymol.* 25:623–651.

CHAPTER 28

Encoding and Expression of Genetic Information in Eukaryotes

So far, most of our discussion of the encoding, copying, and expression of genetic information has used examples from prokaryotic organisms. There are two reasons for this choice. First, most of our fundamental understanding of the processes of replication, transcription, and translation was first gained from studies of bacteria and viruses. Second, as we shall see, the corresponding processes in eukaryotes are almost invariably more complicated and not as completely understood. The more complex life-styles of eukaryotic organisms require that their genomes be much larger. Bacteria and viruses do not require enormous amounts of genetic information. A virus needs only a minimal genome to ensure its own replication, for it depends in large part on genetic information carried by the specific host. A bacterium such as *E. coli* is autonomous, but it carries only the genetic information needed for it and its descendants to function as unicellular organisms in very limited environments.

Eukaryotic organisms face very different problems. The higher eukaryotes, such as plants or animals, reproduce sexually, which means, for one thing, that the entire genetic program for the development of the whole *multicellular* organism must already be present in the fertilized egg. In such organisms, not all cells develop alike. Rather, the first cells of the developing embryo must **differentiate** to produce all of the varied tissues of the adult, and the information to direct this development has to be carried in the genome. The specialized tissues must be correctly interrelated and positioned and must function together in concert. This, in turn, means that another kind of information processing is required—communication between cells and tissues. Furthermore, eukaryotes interact with their surroundings in much more complex ways than do prokaryotes. The processing of all of these kinds of nongenetic information requires further biochemical complexity.

Such topics will constitute the subject matter of these last two chapters. We shall consider first how eukaryotes process the vast amount of *genetic* information they must carry to produce and maintain their complex structures. In the final chapter, we shall consider briefly how they process the continual stream of *nongenetic* information they received from the environment and how they react to it.

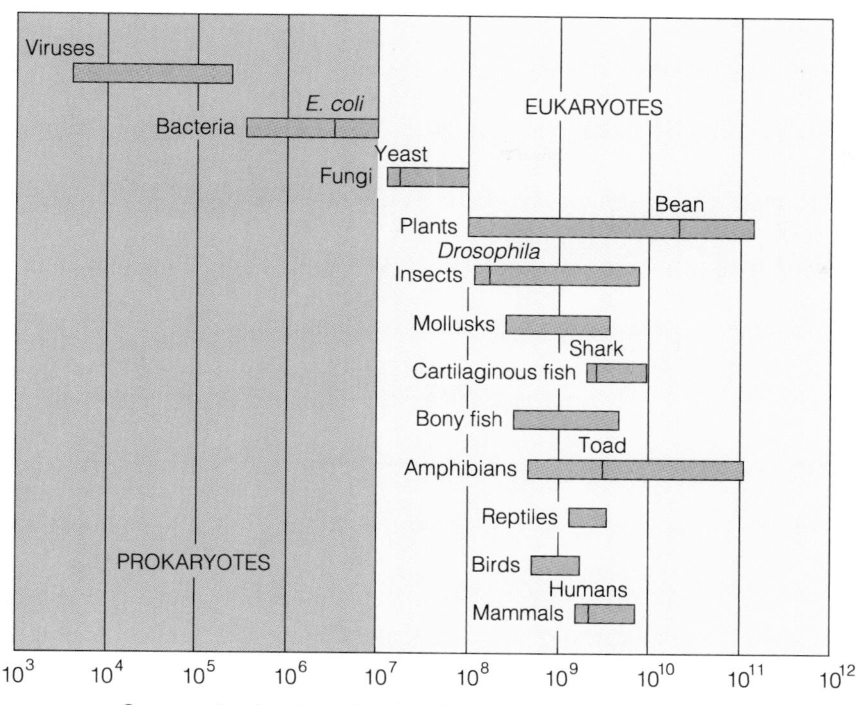

Figure 28.1
Genome size. The bars show the ranges of reported haploid genome sizes for different kinds of organisms. Note that many organisms have larger genomes than humans. The genome size scale is logarithmic.

The Eukaryotic Genome

Size of the Genome

Since the DNA of higher eukaryotes must code for all of the specialized proteins found in different tissues, we might expect such organisms to have a considerably larger amount of DNA than is found in a typical prokaryote such as *E. coli.* What is surprising is how much more there actually is. As Figure 28.1 shows, most animal or plant cells contain 100 to 10,000 times as much DNA as does *E. coli.* In some cases the factor approaches 100,000. The only exceptions to this difference between eukaryotes and prokaryotes are found among the lower fungi, such as yeast and *Neurospora,* which do not exhibit the complex tissue differentiation found in higher eukaryotes.

The enormous quantity of DNA in a plant or animal is not, however, simply correlated with greater organismal complexity. A lungfish, for example, is certainly no more complex a being than a human, yet it has a genome over 10 times larger. The fact that the human genome must be able to code for many more proteins than a bacterial cell (maybe 50 times as many) cannot alone explain its thousandfold greater size. Obviously, much DNA is present in eukaryotes that does not code for protein. What is this "other" DNA? There are several answers, but we begin with the observation that part of this DNA is distributed in two categories not found to any significant extent in prokaryotes: **repetitive sequences** and **introns.**

Repetitive Sequence DNA

In 1970 R. Britten and D. E. Kohne applied a new technique for analyzing the kinetics of DNA reassociation. In this method, which is described in Chapter 4, the total DNA from an organism is cut into pieces about 300 bp

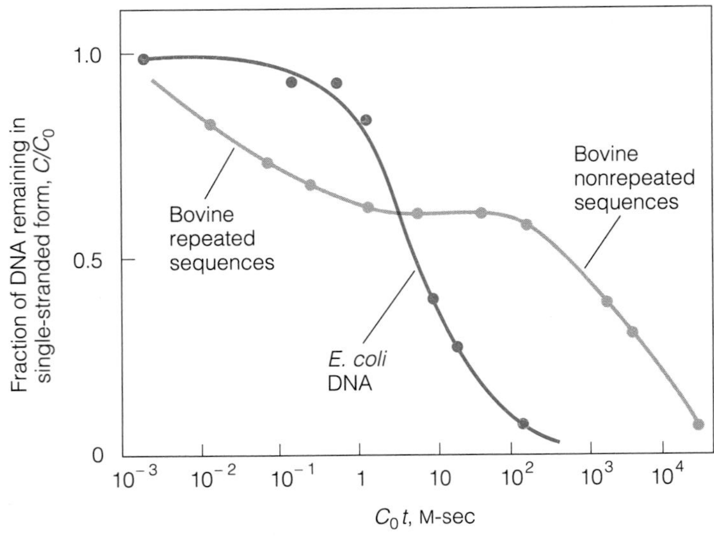

Figure 28.2
Comparison of the kinetics of reassociation of *E. coli* and bovine DNA. The C_0t scale can be thought of as a measure of time to reassociate. The curve for *E. coli* corresponds to that expected for a collection of single-copy genes in a genome of the *E. coli* size—about 4×10^6 bp. The curve for bovine DNA exhibits two steps in reassociation. One, in which C_0t is much greater than that for *E. coli* (more slowly reassociating since the genome is about 1000 times larger than the *E. coli* genome), corresponds to single-copy DNA (nonrepeated sequences). The other, in which C_0t is much less than for *E. coli* (rapidly reassociating), corresponds to DNA with repeated sequences. Many classes of repeated DNA are represented in this phase of the reassociation.

long, heated to cause strand separation, and then cooled to allow the DNA to reassociate. Any sequence that is present in multiple copies will reassociate quickly. When they used this method to study bovine DNA, Britten and Kohne were amazed to find that while a portion of the DNA reassociated at the rate expected for "single-copy" segments in the large genome, almost half of it reassociated *very* much more rapidly (Figure 28.2). To account for the rapidity with which some portions reassociated, we must assume that some of these sequences are reiterated as many as 10^5 to 10^6 times in each cell.

Further analysis has shown that these reiterated sequences can be divided into several categories. One type involves multiple tandem repetitions, over long stretches of DNA, of very short, simple sequences like $(ATAAACT)_n$. Such DNA can often be separated from the major portion of the DNA by sedimentation to equilibrium in density gradients (see Chapter 24). Repetitive DNA sequences that are A/T-rich (like that above) have a lower density than average-composition DNA, while G/C-rich sequences are more dense. Thus, in a density-gradient experiment, the reiterated, simple-sequence fragments band as "satellites" about the "main-band" DNA; they are sometimes referred to as **satellite DNAs.** In higher eukaryotes, satellite DNA usually makes up 10 to 20% of the total genome.

What function can such highly reiterated DNA sequences serve? They do not code for proteins, and most are not even transcribed into RNA. Some, at least, appear to play a structural role. Certain reiterated sequences have, for example, been found to be highly concentrated near the **centromeres** of chromosomes, the regions where sister chromatids are attached. They may serve as binding sites for proteins that attach the spindle fibers in mitosis. (See pp. 1001–1009 for discussion of the physical structure of chromosomes and the process of mitosis.)

There are many other classes of DNA sequences with varying degrees of repetition. Some of these are functional genes, and in many cases the repetitiveness seems to play a useful role, by allowing high levels of production of much-needed transcripts. Examples include the genes for ribosomal rRNAs, of which as many as several thousand copies may be present, and tRNA genes, with hundreds of copies of each type often found. The cell's continual need for large quantities of ribosomes and tRNAs for translation is met by having multiple copies of these genes. The same is true for the genes for some much-used proteins, such as those found in ribosomes and the histones that bind to eukaryotic DNA to form the chromatin structure. As we pointed out in Chapter 25, even single-copy genes are sometimes amplified to yield multiple copies, either in response to environmental stress or as part of normal development.

The significance of some other kinds of repeated DNA sequences remains mysterious. There exist large families of closely related or identical sequences whose members are scattered throughout the genome, rather than being clustered like the satellite DNAs. Some of these may represent control elements of kinds and with purposes we still do not understand. One of the most common such classes of repeating sequences in mammals consists of the so-called ***Alu* elements.** These are sequences about 300 bp long, of which there are nearly a million copies in the human genome. Their name reflects the common existence of a single site for the restriction endonuclease *Alu*I in most members of this class. The *Alu* sequences can be (inefficiently) transcribed into RNA, although they are not translated.

The function of the large number of *Alu* sequences remains uncertain, although some of them may contain origins for DNA replication. But it is also conceivable that many repetitive sequences such as these serve *no* useful function. They may simply exist in the genome as "molecular parasites." A way in which such sequences could spread through the genome has been proposed, on the basis of the observation that *Alu* sequences are flanked by short, repeated oligonucleotides resembling those of transposons (Chapter 25). In this view, *Alu* sequences, like other mobile genetic elements, may be inserted in the genome as reverse-transcriptase copies of the RNA that is transcribed from them. Recent studies suggest that the *Alu* sequences may have been derived from a small RNA (7SL RNA) involved in protein transport across membranes. We shall discuss this RNA on p. 1027.

A similar mechanism may also explain the abundance of many **pseudogenes,** or nonfunctional genes, in eukaryotic DNA. Pseudogenes can be recognized because they bear strong sequence similarity to functioning genes, from which they probably evolved. They are no longer transcribed, however, for some element required for transcription (often a flanking control region or promoter) is missing. Since their sequences are not expressed, pseudogenes are no longer under strong selective control in evolution. In a sense, it does not matter what happens to them, and as a consequence they can accumulate mutations that would be selected against in functional genes. Examples of pseudogenes can be seen in Figure 7.26, which shows the arrangement of α- and β-globin gene variants in human DNA.

Introns and Exons

A second reason for the large size of eukaryotic genomes is that most eukaryotic genes are interrupted by introns. In Chapter 7 we introduced the concept of introns and exons and pointed out that the β-globin gene consists of three exons interrupted by two noncoding intron regions. This kind of structure is common in eukaryotes and is often even more extreme than in the hemoglobin example. Consider the ovalbumin gene shown in Figure 28.3a. This codes for a protein 386 amino acid residue in length, corresponding to a message 1158 nucleotides long. Yet the total ovalbumin gene is about 7700 base pairs long, containing eight exons interspersed by seven introns. The difference between the ovalbumin gene and its mRNA are dramatically displayed when hybrid DNA-RNA molecules are examined in the

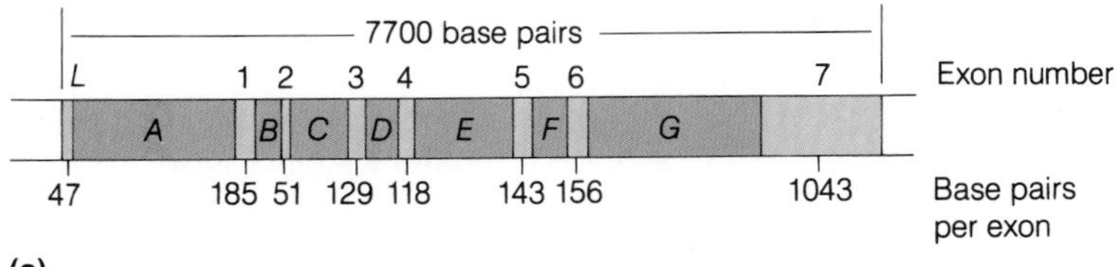

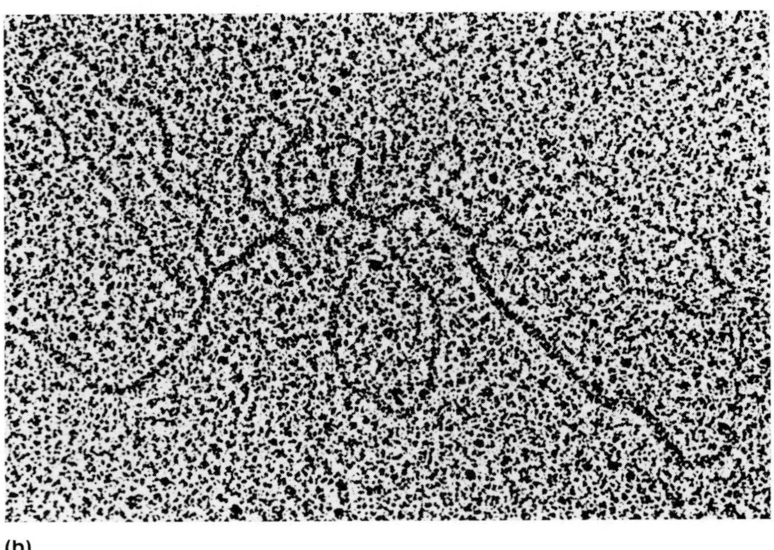

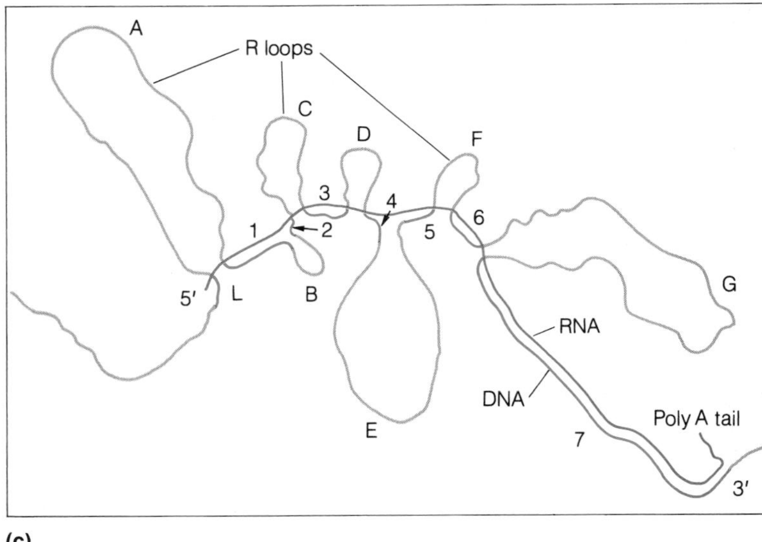

Figure 28.3
Exon–intron structure of the ovalbumin gene in chickens. (**a**) Map of the 7700-bp gene, showing exons 1–7 plus an untranslated leader sequence (blue) and introns A–G (brown). (**b**) Electron micrograph of a hybrid between the genomic DNA and ovalbumin mRNA. (**c**) Diagram showing how the intron regions loop out in R loops in such a hybrid. The RNA is shown in red.

electron microscope (Figure 28.3b). The genomic DNA pairs to the mRNA along the exons, but the introns are looped out, to form what are called **R-loops.**

Recent sequencing studies have shown that introns are present in most eukaryotic structural genes and frequently exceed exons in total length. The presence of introns provides an explanation for at least a part of the very large amount of DNA in eukaryotic cells. Interestingly, some of the "lower" eukaryotes like yeast have many fewer introns, and their genome size is correspondingly smaller.

Gene Families—Multiple Variants of a Gene

One characteristic of complex eukaryotic organisms is the expression of different *variants* of the same type of protein at different stages in development. We already encountered this in Chapter 7, where the embryonic (ζ and ϵ), fetal (α and γ), and adult (α and β) globins of mammals were described. For each such protein there must exist a corresponding gene, a feature that again acts to increase genome size. Figure 7.26 depicts the clusters of genes for the α and β classes of hemoglobins in humans. Each of these genes has the kind of exon–intron structure shown in Figure 7.23; in addition, the genes themselves are separated by long stretches of nontranscribed DNA. Some portions of these intervening regions must contain control signals, for the expression of the globin genes is under complex and subtle regulation. In the first place, although the globin gene clusters are found in many different kinds of cells, in the human body they are expressed *only* in the **erythropoietic,** or red blood cell-producing, cells. Furthermore, as we saw in Chapter 7, expression of each variant is strictly constrained to certain developmental stages. For example, in the early embryo, only the ζ and ϵ genes are being transcribed; all others are turned off. As development proceeds, transcription switches first to the fetal α and γ genes, and at about the time of birth the adult β variant begins to dominate and transcription of γ ceases (see Figure 7.25). This kind of developmental regulation is peculiar to eukaryotes—no prokaryote does this. This use of multiple variant genes is expensive in genome size: the human genome devotes about 100,000 bp of DNA—corresponding, for example, to about 1/50 of the *entire E. coli* genome—just to hemoglobin.

Many other gene families exist. Some, like the groups of genes encoding variants of the histones, seem to play developmental roles rather like those of the globin gene family. Others, like the immunoglobulin genes or genes for the interferons (p. 776), appear to exist in multiple forms in order to satisfy a multiplicity of similar but distinct needs. In each case, it seems likely that the members of a particular gene family have evolved by successive duplications of an original, ancestral gene.

In summary, we find that a number of quite different reasons account for the very large size of the genomes of eukaryotic organisms. At the same time, we still find it hard to rationalize the extreme variations in the amount of DNA that are sometimes observed even between closely related organisms. To take one example, among the amphibians alone there is a hundredfold range of genome sizes. The function, if any, of such enormous variation is still obscure, which suggests that there are some fundamental things about eukaryotes that we still do not understand.

The Physical Organization of Eukaryotic DNA—Chromosomes and Chromatin

Chromosomes

Almost all of the DNA of a typical eukaryote like *E. coli* is contained in a single, large circular DNA molecule, with a minor fraction present in small plasmids. This "prokaryotic chromosome" is supercoiled and complexed with proteins and exists free in the cytosol—although attached at one or

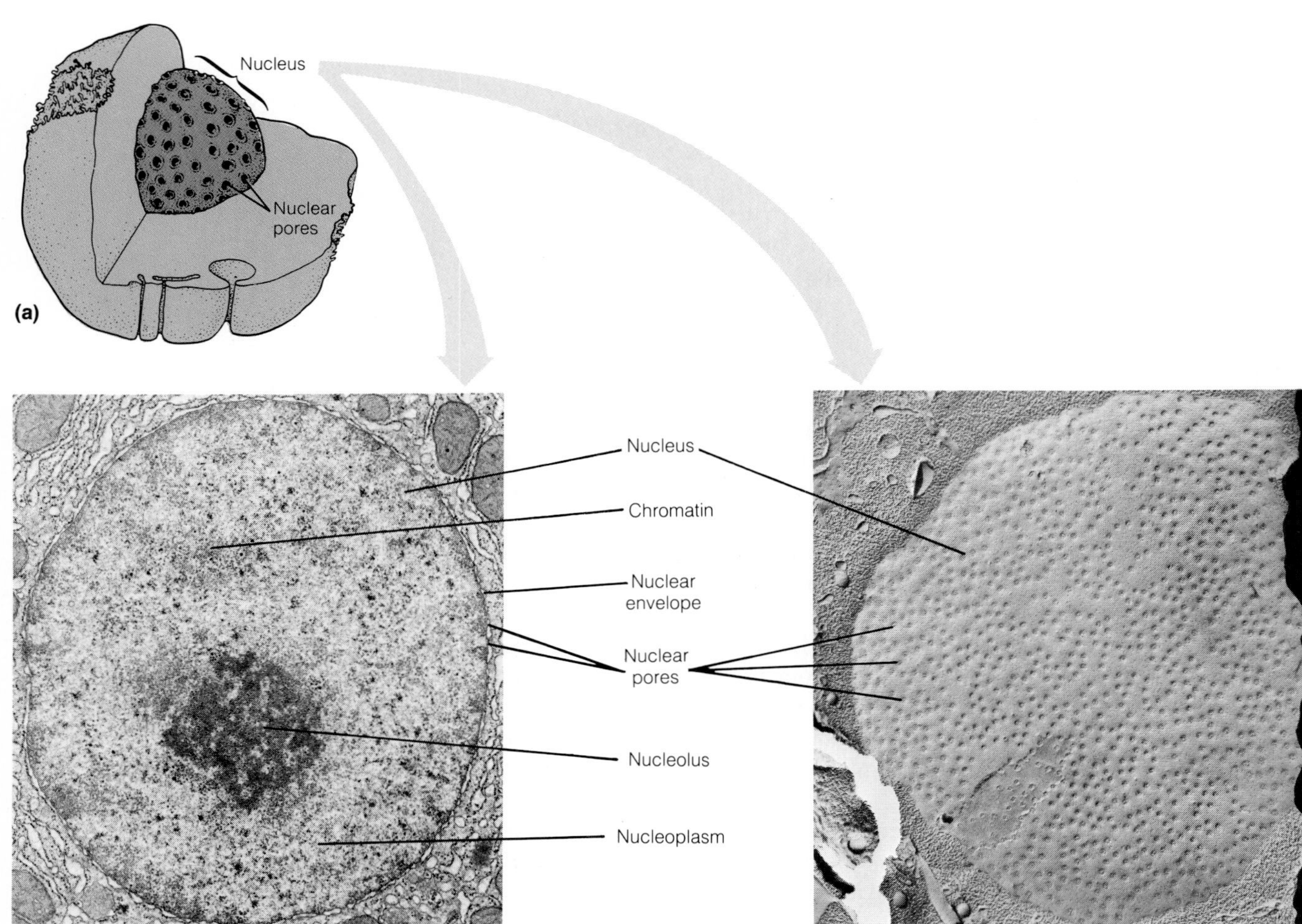

Figure 28.4
The nucleus. (**a**) Schematic, cutaway view of a typical animal cell, showing the position and relative size of the nucleus. (**b**) Electron micrograph of a section of a rat liver nucleus. (**c**) Freeze-fracture image of the surface of a *Drosophila* nucleus.

more points to the cell membrane. This structure is usually called the bacterial nucleoid.

The situation in eukaryotes is quite different. The typical cell's genome is divided into several, or many, eukaryotic chromosomes, each of which contains a single very large, linear DNA molecule. Although their size varies greatly between organisms and even between different chromosomes in a given species, these DNA molecules are commonly of the order of 10^7 to 10^9 bp in length.

Different eukaryotic species contain widely varying numbers of distinguishable chromosomes—from one (in an Australian ant) to 190 in one butterfly species. Whereas most prokaryotes are **haploid** (containing only one copy of their chromosome), most eukaryotic cells are **diploid;** that is, they carry two copies of each chromosome. For example, the human genome is made up of 23 different chromosomes, so that normal, diploid human cells have a total chromosome number of 46. Some eukaryotic cells are highly **polyploid,** carrying many copies of each chromosome.

Unlike the prokaryotic chromosome (nucleoid), the chromosomes of eukaryotes are not normally found free in the cytoplasm. In nondividing cells, the chromosomes are segregated within the nucleus (Figure 28.4) as a mass of fibers of a DNA–protein complex called chromatin (discussed below). The **nuclear envelope** is pierced by **nuclear pores,** which allow quite large molecules (even some proteins) to pass in and out of the nucleus. The protein translational processes are excluded from the nucleus, however, so transcription and translation cannot be directly coupled, as they are in prokaryotes, but occur in separate cellular compartments. This may be related to the fact that eukaryotic organisms must extensively splice and process the initial transcripts before they are ready to be exported from the nucleus to serve as messengers for translation.

During **mitosis**—the separation of chromosomes following DNA replication—the nuclear envelope breaks down and chromosomes condense into the compact structures shown in Figure 28.5.

The enormous amount of DNA in eukaryotic cells poses some very serious problems. First, there is the question of *compaction*. The diploid DNA content of a human cell is about 8×10^9 bp, corresponding to a total length of nearly 3 meters. Somehow, all this DNA must be packed into a nucleus about 10 μm (10^{-5} m) in diameter. Second, there is the problem of *selective transcription*. In a typical differentiated eukaryotic cell, only a small fraction of the DNA (5–10%) is ever read. Much, as we have seen, is nontranscribable. Many genes that do undergo transcription do so only in certain cell lines in particular tissues, and then often only under special circumstances. To maintain and regulate such complex programs of selective transcription, the accessibility of the DNA to RNA polymerases must be under strict control.

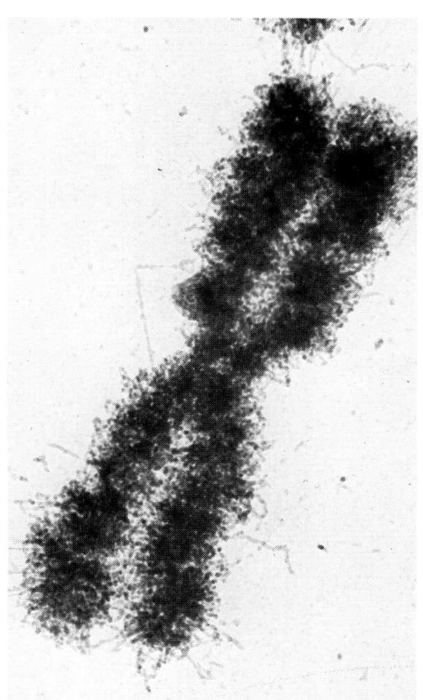

Figure 28.5
A mitotic chromosome. An electron microscope image of a human chromosome during the metaphase stage of mitosis. The constriction at the centromere and the lengthwise division into sister chromatids are clearly visible. The hairy surface is made of loops of highly coiled chromatin.

Chromatin

Both compaction and the control of gene expression in eukaryotes are achieved by having the DNA complexed with a set of special proteins. The resultant protein–DNA complex is called **chromatin.**

The DNA-binding proteins of chromatin fall into two classes. The major class consists of **histones.** There are five types of histones; their properties are outlined in Table 28.1. All are small, very basic proteins rich in lysine and arginine. Evolutionarily, they have been remarkably well conserved in amino acid sequence; histone H4, for example, shows only two substitutions between humans and peas and only eight substitutions between humans and yeast. The histones are the basic building blocks of chromatin structure. The nucleoids of prokaryotic cells also have proteins associated with the DNA, but these are unrelated to the histones and do not seem to form a comparable chromatin structure. Thus, a histone-containing chromatin structure is a uniquely eukaryotic feature.

In all kinds of eukaryotic nuclei, the histones are present in an amount of about 1 gram per gram of DNA. They are accompanied by a much more

Table 28.1
Properties of the major histone types

Histone Type	Molecular Weight	Number of Amino Acid Residues	Mol % Lys	Mol % Arg	Role
H1	22,500	244	29.5	1.3	Associated with linker DNA; helps form higher-order structure
H2A	13,960	129	10.9	9.3	Nucleosomal structure: two of each
H2B	13,774	125	16.0	6.4	
H3	15,273	135	9.6	13.3	
H4	11,236	102	10.8	13.7	

All data are for calf thymus histones, except for H1, which is from rabbit.

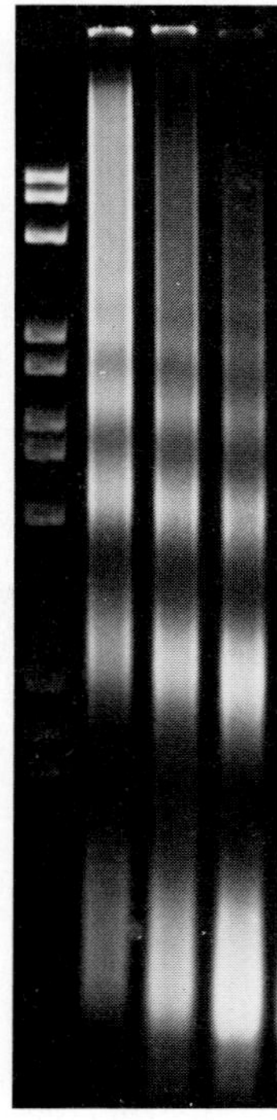

Figure 28.6
The kind of evidence that first suggested a repetitive structure in chromatin. In this gel, the three columns to the right show DNA fragments obtained after three successively longer digestions of chicken erythrocyte chromatin by micrococcal nuclease. The column to the left contains DNA restriction fragments as size markers.

diverse group of DNA-binding proteins, rather unimaginatively named **nonhistone chromosomal proteins.** The total amount of the latter varies greatly from one cell type to another, ranging from about 0.05 to 1 g/g DNA. They include a bewildering variety of proteins, such as polymerases and other nuclear enzymes, hormone receptor proteins, and regulatory proteins of many kinds. It is possible to count, on two-dimensional gels, approximately a thousand different nonhistone chromosomal proteins in a typical eukaryotic cell.

The association of proteins with eukaryotic DNA has long been recognized. Indeed, as early as 1888 the German chemist Albrecht Kossel isolated histones from nuclei and recognized them as basic substances that would bind to the nucleic acid. However, the precise role of histones was not understood until about 1974. Then, research in a number of laboratories showed that these proteins combined in a specific way to form a repeating element of chromatin structure, the nucleosome.

The Nucleosome

If naked (that is, not protein complexed) DNA is partially digested with a nonspecific endonuclease like micrococcal nuclease, which cuts double strands almost randomly, a broad smear of polynucleotide fragments is produced. But in the early 1970s, researchers in a number of laboratories found that if the same experiment is conducted with chromatin, or even with whole nuclei (into which the nuclease can easily penetrate through the nuclear pores), the DNA is cleaved in a quite specific, nonrandom way. On a polyacrylamide gel, the DNA from nuclease-digested chromatin yields a series of bands that are multiples of approximately 200 base pairs (Figure 28.6). This indicates that the nuclease can find easy access to the DNA only at regularly spaced points. At about the same time that these observations were made, other laboratories obtained electron micrographs of extended chromatin fibers, revealing a regular "beaded" pattern in the chromatin structure with beads about every 200 bp (Figure 28.7). If nuclease digestion of chromatin is continued, it slows down and nearly stops when about 30% of the DNA has been consumed. The remaining protected DNA is present in particles that correspond to the "beads" seen in the electron micrographs. These particles, called **nucleosomes** (or more precisely, *nucleosomal core particles*), have a definite composition that is practically invariant over the whole eukaryotic kingdom. They always contain 146 bp of DNA, wrapped about an octamer of histone molecules—two each of H2A, H2B, H3, and H4. The nucleosome has been crystallized, and x-ray diffraction studies reveal the structure shown in Figure 28.8. The DNA lies on the surface of the octamer and makes about 1.75 left-handed superhelical turns about it.

Although the nucleosome itself is a nearly invariant structure in eukaryotes, the way in which nucleosomes are spaced along the DNA varies considerably among organisms and even between tissues. The length of DNA between nucleosomes may vary from about 20 bp to over 100 bp. This internucleosomal DNA is occupied by the H1-type (very lysine-rich) histones and nonhistone proteins. Figure 28.9 illustrates the elements of chromatin structure.

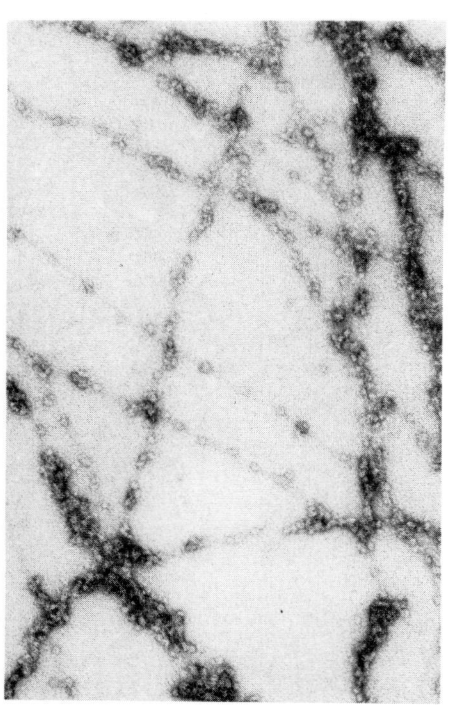

Figure 28.7
Beaded-fiber structure of chromatin. An electron micrograph of chromatin spread on a grid at low ionic strength and negatively stained. The spreading under these conditions unravels and stretches the condensed chromatin fibers to show the regularly spaced nucleosomes.

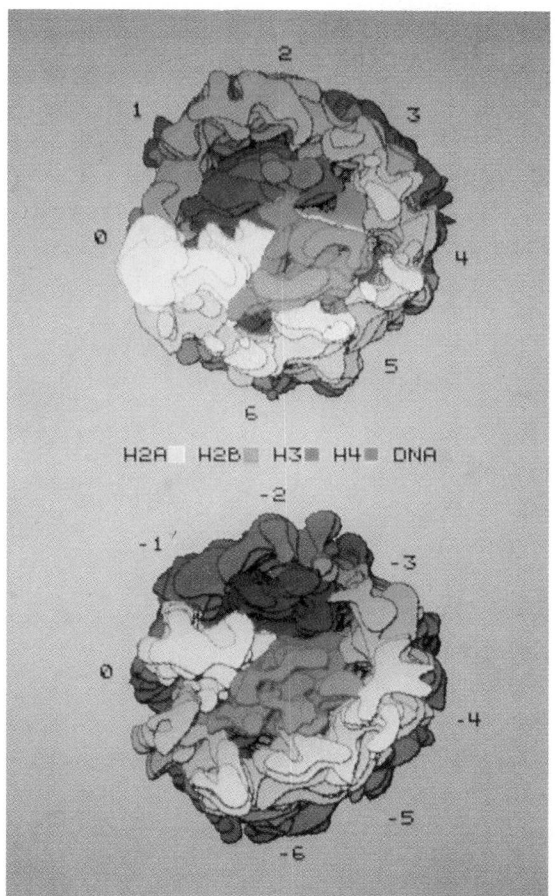

Figure 28.8
Structure of the nucleosome core particle, as revealed by x-ray diffraction. Views from the top and bottom of the core particle. The color code shows putative positions of the DNA and the four histones. The numbers correspond to turns of the DNA helix as measured from the two-fold axis of the particle (0).

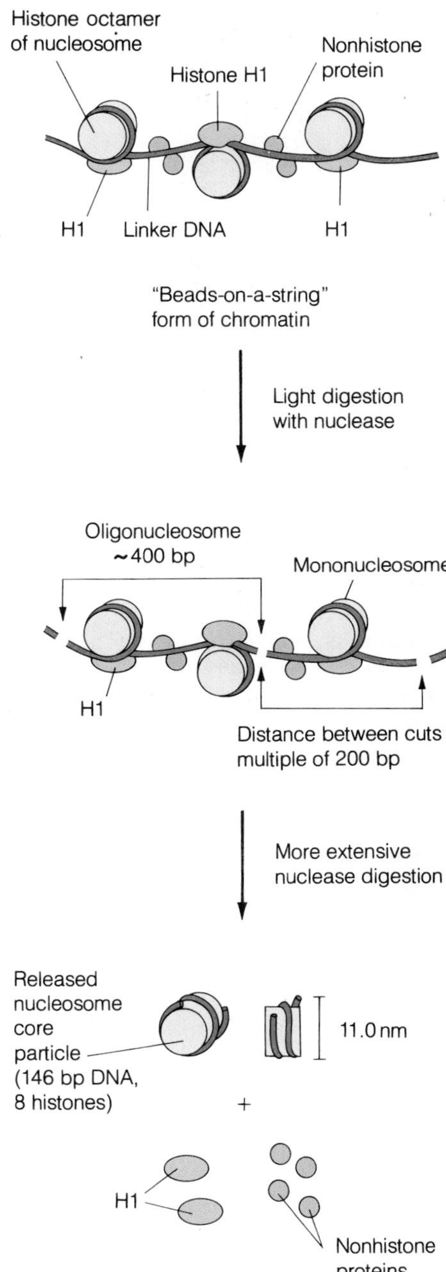

Figure 28.9
The elements of chromatin structure. At the top, the extended beaded string structure shown in Figure 28.10 is diagrammed. Light digestion with nuclease first releases mono- and oligonucleosomes; then, as linker DNA is further digested, nonhistone proteins and H1 are released, to yield the core particle whose structure is shown in Figure 28.8.

Higher-Order Chromatin Structure in the Nucleus

Wrapping DNA about histone cores to form nucleosomes accomplishes part of the compaction necessary to fit the eukaryotic DNA into the nucleus, for the strand is thereby shortened severalfold. Since wrapping about the core is left-handed, the formation of nucleosomes introduces negative supercoils in the DNA. As long as the histone cores are in place, this DNA is not under superhelical tension.

Although nucleosome formation shortens the chromatin strand approximately fivefold, it is clear that much of the chromatin in the nucleus is even more highly compacted. The next stage in compaction involves wrapping the beaded fiber into a helical *solenoidal* structure like that shown in Figure 28.10. These fibers are about 30 nm in diameter and may be further folded on themselves to make the thicker chromatin fibers visible in both metaphase chromosomes (see Figure 28.5) and in the nuclei of nondividing (interphase) cells (see Figure 28.4).

Evidence is emerging concerning still higher orders of folding in both metaphase and interphase chromatin. It has long been obvious that the metaphase chromosome is a highly organized structure. Dye staining of the chromosomes from a particular organism gives a very reproducible banding pattern, and **in situ hybridization** methods, in which specific sequences are located on chromosomes by hybridization with a complementary radioac-

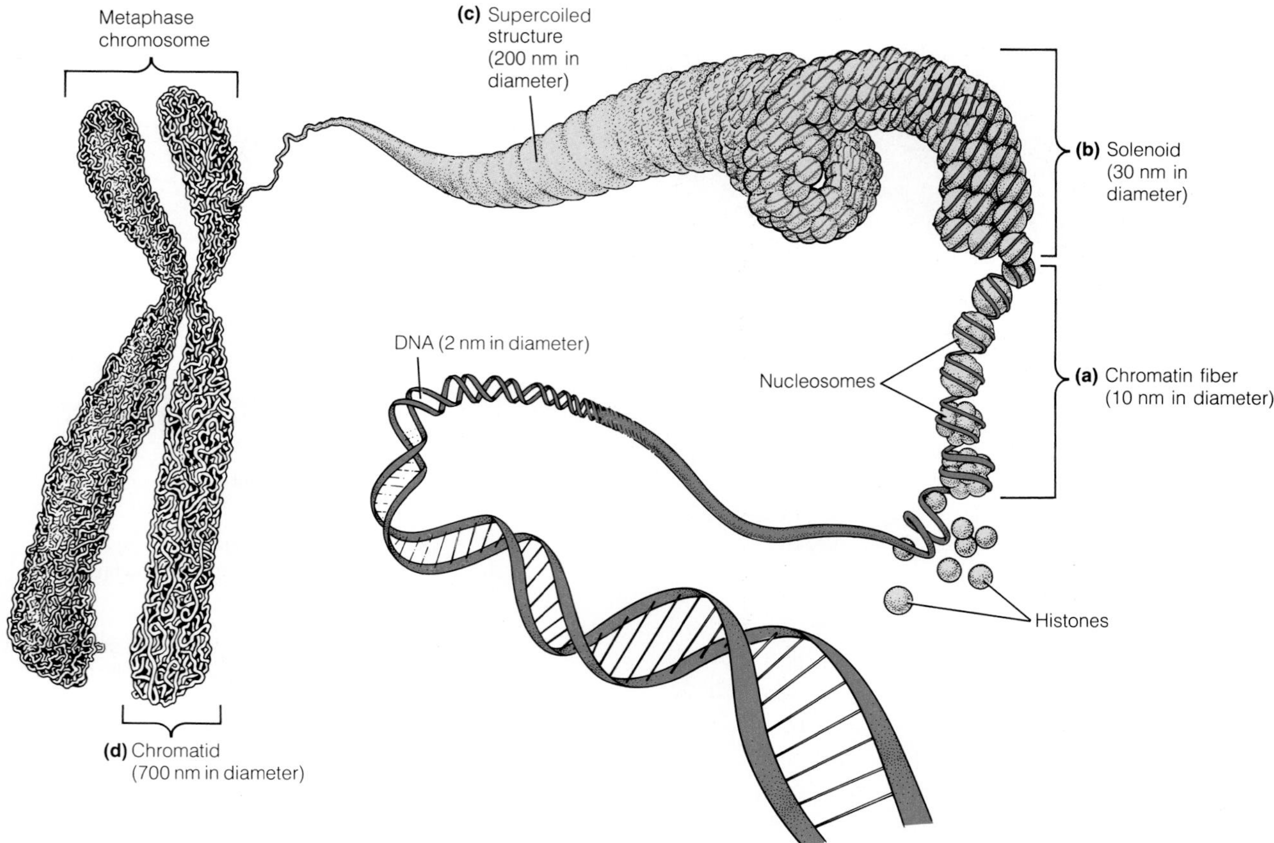

Figure 28.10
Levels of chromatin structure. The beaded string structure is a 10-nm fiber, which folds into a "solenoidal" 30-nm fiber with about six nucleosomes per turn. This can further fold to form thick 200-nm fibers that can be observed in electron micrographs of chromosomes or nuclei.

tive nucleic acid and then visualized by radioautography, show that particular DNA sequences are always located at the same places in specific chromosomes. Since the DNA in a eukaryotic chromosome is one long continuous strand, this implies that some kind of structure must be present to preserve this order. Recent evidence indicates what this may be. If metaphase chromosomes are treated with polyanions, like dextran sulfate, which strip off the histones and loosely bound nonhistone proteins, the DNA strands are seen to emerge as enormous loops from a scaffold of tightly bound protein. An electron micrograph of this structure is shown in Figure 1.5. Individual loops vary in size but may range up to 100,000 bp in length—about the size of the β-globin gene cluster, for example. Approximately 1000 such loops exist in the average chromosome. The proteins that form the scaffold from which the loops extend include some very interesting members, including topoisomerases. It has been hypothesized that topoisomerase molecules at the base of a loop would exert control on the supercoiling level of that particular loop; such changes in supercoiling would be in addition to that imposed by the nucleosomes. Changes in supercoiling may be involved in chromosome condensation and seem to be essential during replication and transcription. It seems likely that the structure of chromatin is dynamic, changing locally as the DNA is replicated or transcribed.

Evidence also exists for a similar but more diffuse scaffold in the interphase nucleus. Removal of histones and weakly bound nonhistone proteins from intact nuclei by high salt concentrations or detergents, together with digestion of most of the DNA by nucleases, leaves a structure that has been called the **nuclear scaffold** (or **nuclear matrix**) (Figure 28.11). This includes the laminar shell that lines the inside of the nuclear membrane, plus a network of fine fibers that seem to extend throughout the nucleus. When the chemical dissection is done very carefully, using the specific detergent lithium diiodosalicylate to remove the histones and most other proteins plus

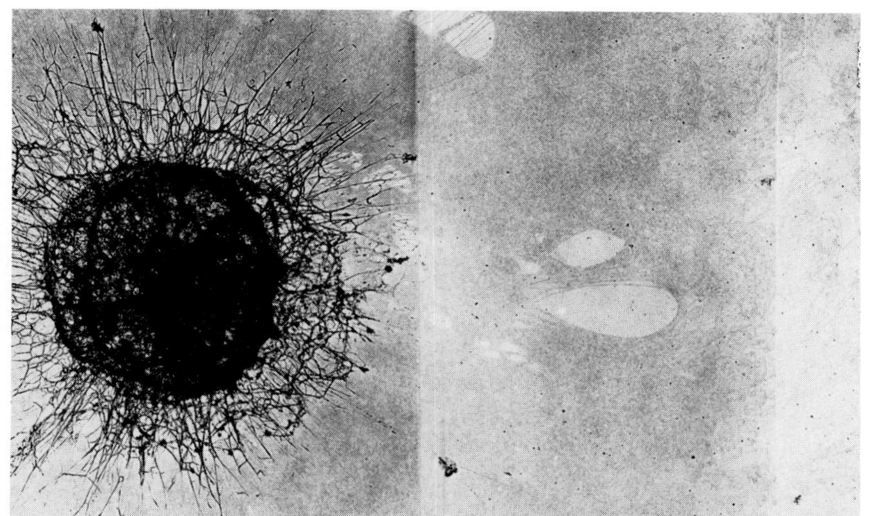

Figure 28.11
The nuclear scaffold. The nucleus of a HeLa cell at interphase has been treated with salt to remove most of the proteins. The remaining nuclear scaffold, or nuclear matrix, consists of portions of the nuclear envelope plus a fibrous protein structure within the nucleus. From these, very large DNA loops extend.

restriction endonucleases to cut the DNA, the remaining, attached fragments of DNA turn out to contain characteristic "attachment" sequences and to be spaced at rather long intervals along the genome. In fact, it appears that groups of concurrently expressed genes lie between adjacent attachment sites. Figure 28.12 illustrates this for the histone gene clusters in *Drosophila*. Although it has not yet been established with certainty, it is possible that at least part of the loop-and-scaffold structures are identical in metaphase and interphase chromatin. If so, they would provide part of the mechanism necessary for "keeping track" of chromatin structure through successive cell divisions. Furthermore, it now seems likely that this loop structure may play a role in the control of expression of groups of functionally linked genes. We shall have more to say about this later in this chapter.

With this background on the organization and architecture of the eukaryotic genome, we now turn to its major functions—replication and transcription.

Cell Division and DNA Replication in Eukaryotes

The Cell Cycle

The circumstances under which cells divide and DNA is replicated are somewhat more complicated in eukaryotes than in prokaryotes. As we mentioned in Chapter 24, DNA replication in bacteria is an almost continuous process, at least during exponential growth. The somatic cells of eukaryotes, on the other hand, typically divide much less frequently (and some, in mature organisms, not at all). Those that are dividing in growing tissues exhibit a well-defined cell cycle, which is almost always separated into several dis-

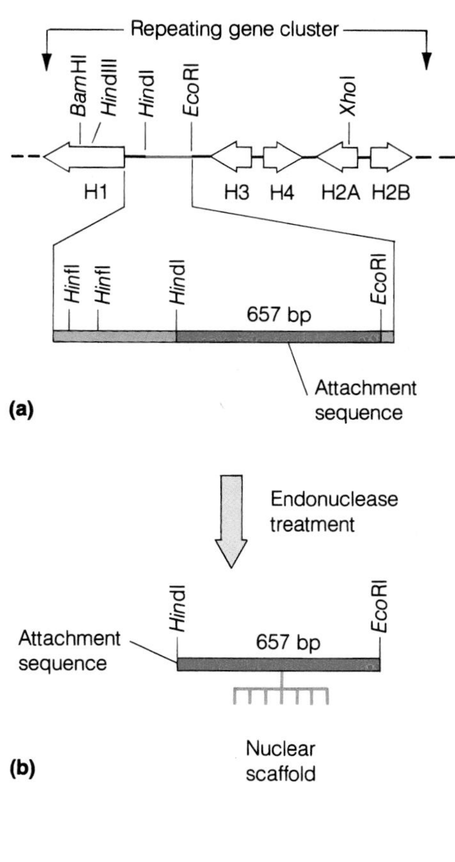

Figure 28.12
Attachment of gene clusters to the nuclear scaffold. (a) A map of the repeating histone gene clusters (white arrows) in *Drosophila*, as well as a number of restriction sites. If *Drosophila* nuclei are extracted with lithium diiodosalicylate to remove proteins gently and then digested with a collection of the restriction endonucleases shown, *only* the 657-bp *Hin*dI/*Eco*RI DNA fragments are left attached to the scaffold (**b**). The interpretation (**c**) is that the gene clusters exist in individual loops, the bases of which are tied to the scaffold.

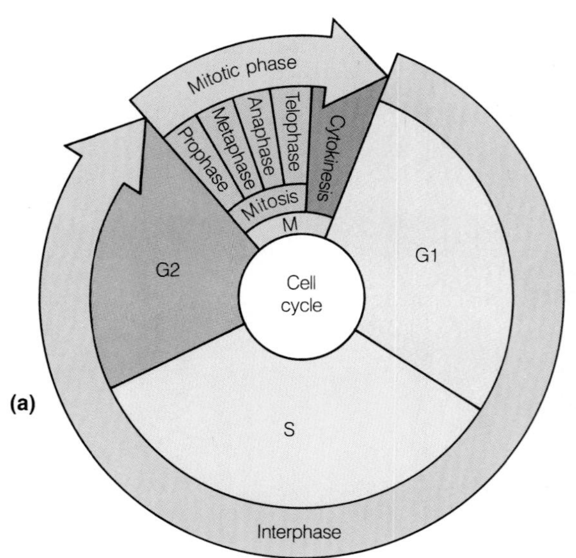

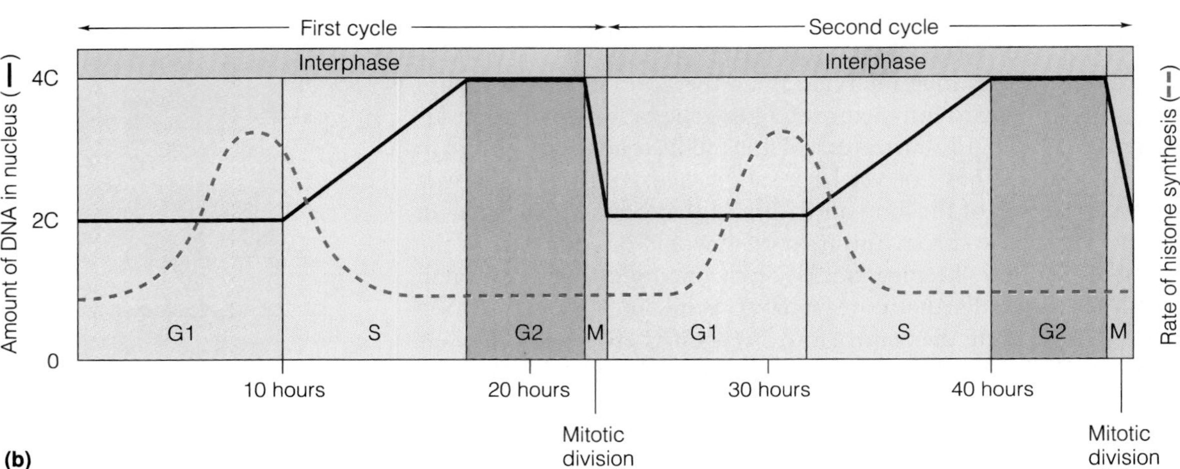

Figure 28.13
The eukaryotic cell cycle. (**a**) Schematic view. In a dividing cell the mitotic (M) phase alternates with an interphase, or growth, period. Interphase is divided into two gap phases (G1 and G2), separated by a synthetic (S) phase. During mitosis, the nucleus and chromosomes within it are divided; at cytokinesis, the cytosol is divided to make two daughter cells. (**b**) Changes in the amount of DNA (solid line) and rate of histone synthesis (red line) during two cell cycles. The DNA content is measured in units of the haploid genome (C). The time scale is typical of many eukaryotic cells.

tinct phases, as shown in Figure 28.13. Let us follow a typical eukaryotic cell through one cycle. We can arbitrarily begin the cycle in what is called the **G1 phase** (or first "gap") phase. At this point, it will contain two copies of each chromosome, the normal diploid state of a eukaryotic cell. This is indicated in Figure 28.13b by a DNA content of 2C—twice the haploid amount. Sometime in G1 phase the commitment to divide is triggered in an as yet unknown fashion. Since division will first require doubling of the DNA content, and the new chromatin will need new histones, synthesis of histones is one of the first indications of incipient DNA replication (see Figure 28.13).

The cell then enters synthesis, or **S phase,** during which the DNA is replicated and histones and nonhistone proteins are deposited on the daughter DNA molecules to reproduce the chromatin structures. We will describe details of this process later, but for the moment let us continue through the cycle. When chromatin replication is complete, the cell enters what is called the second gap, or **G2 phase;** it now has a DNA content four times the haploid amount (4C). Even in "rapidly" dividing eukaryotic cells, the total time required for G1, S, and G2 phases will be many hours. During this whole period, which is termed **interphase,** the chromatin is dispersed throughout the nucleus and is actively engaged in transcription.

Mitosis

At the end of G2, the cell is ready to enter the process called **mitosis,** during which it divides. Mitosis is a multi-stage process, and has been divided by biologists into the "phases" depicted in Figure 28.14. In **prophase,** the replicated chromosomes condense, to form the typical **metaphase** chromosome structures so often pictured (see Figure 28.5). The nuclear membrane disintegrates, and the **mitotic spindle** forms, contractile fibers pulling pairs of chromatids apart so that the daughter cells will receive identical sets of chromosomes. (The mechanism of this contractile process is described in Chapter 29.) In **telophase** the nuclear membrane then re-forms about each daughter nucleus, and the cell itself divides. After division the chromosomes of the daughter cells decondense, and a new G1 phase begins.

In many tissues of higher organisms, the G1 phase may become very prolonged after growth and tissue differentiation are complete. The most extreme examples are fully differentiated nerve cells, most of which never divide again in mature organisms. Such nondividing cells are in a permanently arrested G1 phase, which is often called G_0. On the other hand, some specialized stem cells, such as those found in the bone marrow and intestinal epithelium, undergo continuous division throughout the life of the organism. These stem cells provide new cells to replace ones that have been lost or damaged.

Meiosis

A quite different series of events occurs in the **germ cells** of organisms that reproduce sexually. These are the cells that give rise to the sperm and ova, and they undergo **meiotic division,** or **meiosis** (Figure 28.14b). This requires two successive divisions, ultimately yielding four *haploid* cells. The DNA content of the germ-line cell is first doubled, in a nearly normal cell cycle. However, in a phase known as prophase I before the first meiotic division, the pairs of homologous chromosomes (each of which comprises two identical sister chromatids, joined at the centromere) combine to form structures called **bivalents** (Figure 28.15). At this point recombination between homologous chromosomes can occur, so that the gene contents of parental chromosomes can become intermixed. In the first meiotic division the recombined parental diploid chromosomes are randomly segregated into the daughter cells. Note how the genetic information present in the initial germline cell has been scrambled in two ways: (1) there has been a recombination of segments of chromosomes, and (2) the two parental copies of each chromosome have been randomly sorted into the two daughter cells.

The first meiotic division is followed not by a normal second cycle, but by a short interphase (interphase II) during which *no* DNA replication occurs (Figure 28.16). The second meiotic division follows, in which each of the pairs of sister chromatids is divided, reducing the DNA content of the resulting cells to the haploid level. In the male, each of the four resulting cells matures into a sperm. In most higher eukaryotes, this maturation is accompanied by replacement of the histone and nonhistone proteins of the chromatin by small, basic polypeptides called **protamines,** which highly compact the DNA in the sperm head. In the female, the meiotic divisions are highly asymmetric, and only one of the four daughter cells from the meiotic divisions matures into an egg cell. The three small remaining cells, called **polar bodies,** do not develop further. In the egg cell the normal chromatin structure appears to be retained. If the egg is fertilized, the protamines are removed from the sperm DNA and replaced by histones. The chromosomes

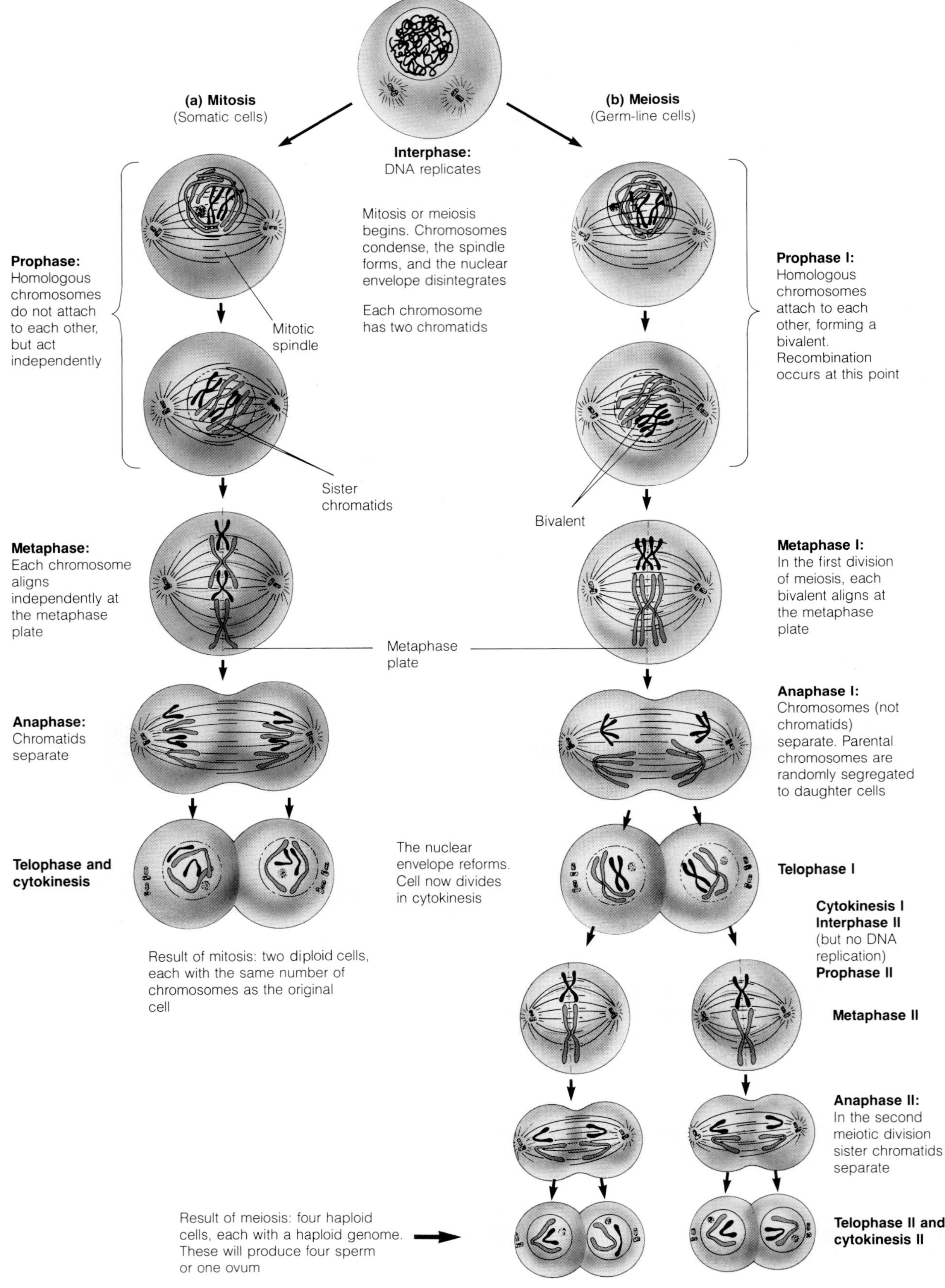
(a) Mitosis
(Somatic cells)
(b) Meiosis
(Germ-line cells)
Interphase:
DNA replicates
Mitosis or meiosis begins. Chromosomes condense, the spindle forms, and the nuclear envelope disintegrates
Each chromosome has two chromatids
Prophase:
Homologous chromosomes do not attach to each other, but act independently
Mitotic spindle
Sister chromatids
Prophase I:
Homologous chromosomes attach to each other, forming a bivalent. Recombination occurs at this point
Bivalent
Metaphase:
Each chromosome aligns independently at the metaphase plate
Metaphase plate
Metaphase I:
In the first division of meiosis, each bivalent aligns at the metaphase plate
Anaphase:
Chromatids separate
Anaphase I:
Chromosomes (not chromatids) separate. Parental chromosomes are randomly segregated to daughter cells
Telophase and cytokinesis
The nuclear envelope reforms. Cell now divides in cytokinesis
Telophase I
Cytokinesis I
Interphase II
(but no DNA replication)
Prophase II
Result of mitosis: two diploid cells, each with the same number of chromosomes as the original cell
Metaphase II
Anaphase II:
In the second meiotic division sister chromatids separate
Result of meiosis: four haploid cells, each with a haploid genome. These will produce four sperm or one ovum
Telophase II and cytokinesis II

Figure 28.14
Mitosis and meiosis. The cell entering either of these pathways was originally diploid. It has undergone DNA replication and is now in G2, with a DNA content of 4C. Mitosis (**a**) occurs in somatic cell division; meiosis (**b**) is confined to the germ cells producing sperm or ova.

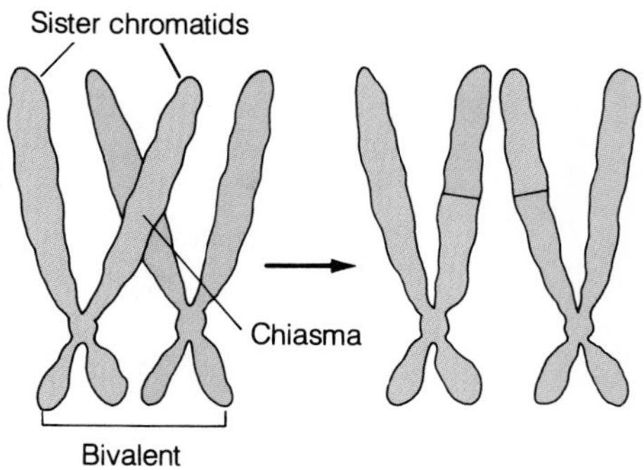

Figure 28.15
Genetic recombination by crossing over of fragments of chromatids in bivalents during meiosis. The point of attachment between chromatids is called a *chiasma*.

of the fertilized egg are diploid, each containing one DNA molecule from each parent.

Molecular Events in DNA Replication

We consider now in more detail the events that occur during the S phase of the cell cycle. The *basic* mechanisms of DNA replication are quite similar in eukaryotes and prokaryotes. In both cases replication is semiconservative and is continuous on one strand and discontinuous on the other. As in prokaryotes, eukaryotic replication begins with assembly of short RNA primer molecules, is followed by elongation by a DNA polymerase, and (on the discontinuous strand) is concluded by ligation of Okazaki fragments. A significant difference between eukaryotic and prokaryotic DNA replication is found in the size of the Okazaki fragments; they are much smaller in eukaryotes—only about 135 bases long. As was pointed out in Chapter 24, eukaryotic cells contain *four* DNA polymerases. Two of these—polymerases α and δ—are used during S-phase replication.

Replication of the nuclear genome in eukaryotes involves some special problems. The first results from the nucleosomal structure of chromatin. The polymerase must proceed through this complex structure, which must apparently be dismantled and then reconstructed on the daughter DNA molecules. As we shall see, it now appears that the arrangement of nucleosomes in chromatin and the placement of nonhistone proteins are far from random. Thus, not only must the DNA be faithfully copied, but also a whole highly organized chromatin structure must be regenerated.

Though there is much that we still do not understand about chromatin replication, we do have some information concerning the assembly of nucleosomes on the daughter strands. Both preexisting and newly synthesized

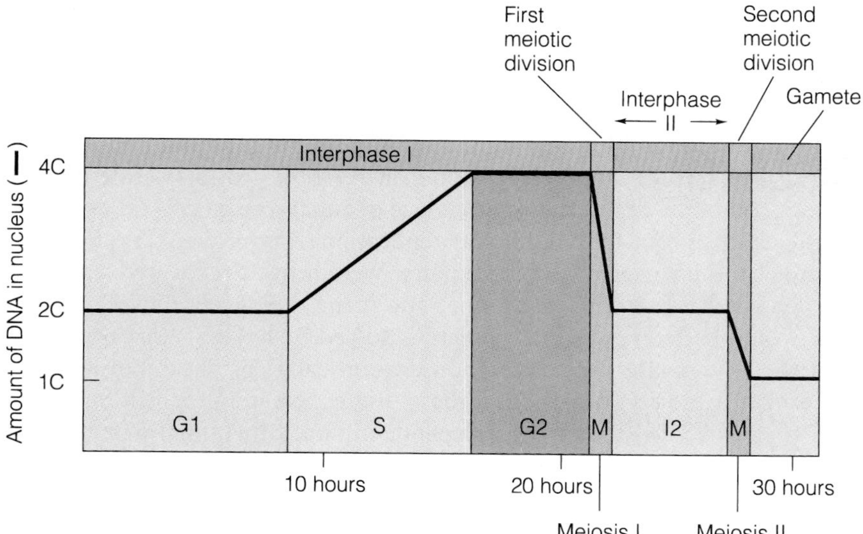

Figure 28.16
Nuclear DNA content during meiosis. Interphase II is much shorter than interphase I (sometimes it is undetectable) and does not involve an S phase in which DNA replication occurs.

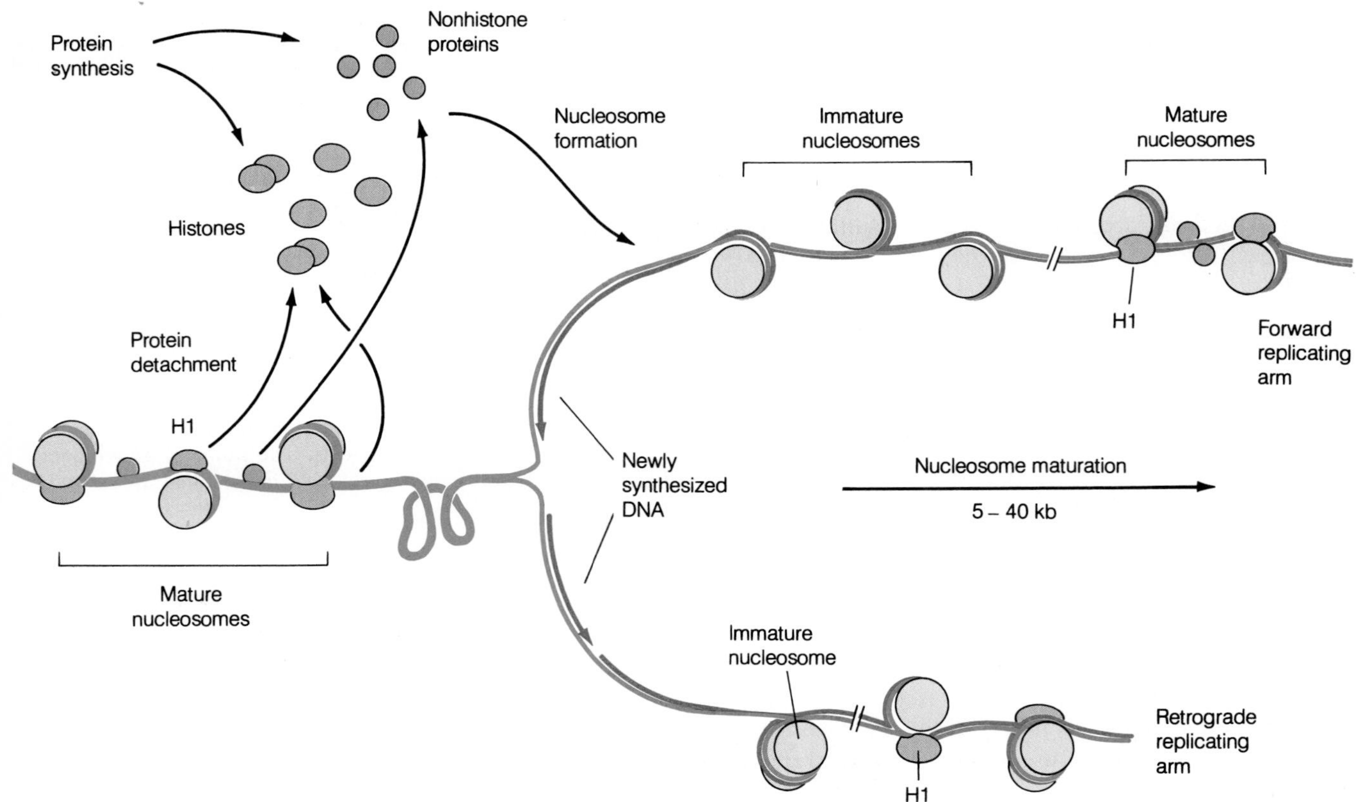

Figure 28.17
Model for chromatin replication. Nucleosomes on the parental DNA are dissociated as the replication fork approaches and are re-formed on the newly synthesized daughter molecules. Very little is known about this process, except that both old and new histones are utilized. The chromatin structure only slowly "matures"; the full organization does not seem to be established until many kilobases behind the moving fork. The figure does not include the polymerase and accessory proteins that must be present at the replication fork.

histones are employed, and present evidence indicates that they segregate randomly to new daughter strands. As Figure 28.17 indicates, the process requires that histone octamers and other proteins be detached from the parental chromatin strand ahead of the replication fork and then be reattached on one or the other of the daughter strands, along with new histones and nonhistone proteins. Just how the precise arrangement of nucleosomes and nonhistone proteins is reestablished after replication is still not understood. In fact, there *may* be changes in some instances, for it is generally observed that differentiation of cells in embryonic development occurs at the point of cell division. Such changes must be *programmed* ones, so their existence does not simplify the problem, but makes it still more complex.

Perhaps these complications account for the fact that the rate of motion of a replication fork is much slower in eukaryotes (about 1–5 kb/min) than in prokaryotes (about 100 kb/min in *E. coli*). This slow rate of fork motion, combined with the enormous size of eukaryotic genomes, raises the second serious problem with nuclear replication. A typical eukaryotic chromosome contains roughly 10^8 base pairs. Were it to be replicated from *one* origin, as is the *E. coli* chromosome, but at the rate given above, nearly a week would be required. The problem is solved by having *many* origins of replication—as many as several thousand on each chromosome. Replication proceeds bidirectionally from these origins, creating replication "bubbles" (Figure 28.18) that grow independently until they finally merge and the whole chromosome has been copied. Not all regions of the DNA begin replication at the same time in S phase. In fact, there is a tendency for

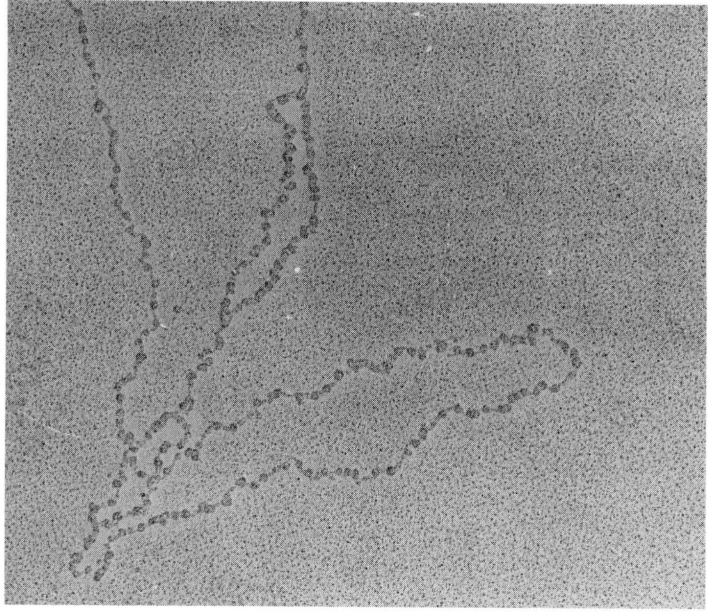

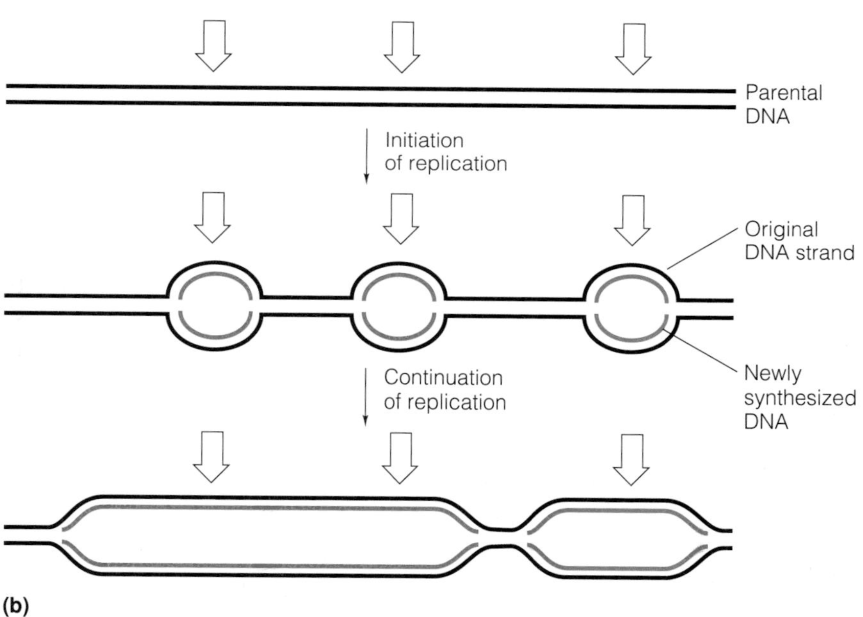

Figure 28.18
DNA replication in eukaryotes. (**a**) An electron micrograph of replicating *Drosophila* chromatin showing multiple replication bubbles. The dots on the fibers are nucleosomes. (**b**) A schematic view of eukaryotic replication.

transcriptionally active regions to replicate early, whereas inactive regions replicate later.

The precise nature of the origins of replication is still somewhat obscure, but useful hints have come from studies of yeast. Sequences have been recognized that are essential for the replication of plasmids in yeast cells; these are called **autonomously replicating sequences,** or **ARS.** Sequences with a similar function have been found to be essential for the replication of DNA viruses in animal cells. Most interestingly, very similar sequences are found within the *Alu* class of repeats.

Transcription and Its Control in Eukaryotic Cells

Transcription in eukaryotes is a much more complex process than it is in prokaryotes. Not only is there much more discrimination in what is to be transcribed and what is not, but this transcription is precisely programmed during development and tissue differentiation. Furthermore, the transcription machinery must somehow deal with the complicated levels of structure existing in eukaryotic chromatin.

Reflecting this complexity is the fact that eukaryotic cells have several different RNA polymerases, each with a specialized function. These have already been mentioned in Chapter 26 and are reviewed in Table 28.2. Aside from the special RNA polymerases functioning in mitochondria and chloroplasts, three enzymes are used to transcribe various portions of the nuclear genome. We shall consider each of these in turn.

Transcription of the Major Ribosomal RNA Genes by Polymerase I

The eukaryotic ribosome contains four rRNA molecules. The small subunit has an 18S rRNA, whereas the large subunit contains 28S, 5.8S, and 5S rRNA molecules. Of these, the 28S, 18S, and 5.8S are all produced from an initial 45S transcript (Figure 28.19a), and it is the special function of RNA polymerase I (Pol I) to carry out this transcription.

The gene for the 45S rRNA is present in multiple, tandemly arranged copies within the nucleolus (see Figure 28.4). This is, in fact, the site of ribosome assembly in eukaryotes; 5S rRNA from other regions of the nucleus and ribosomal proteins synthesized in the cytosol are transported into the nucleolus, where they combine with the 18S, 5.8S, and 28S RNAs to form the ribosomal subunits. These are then exported back into the cytosol.

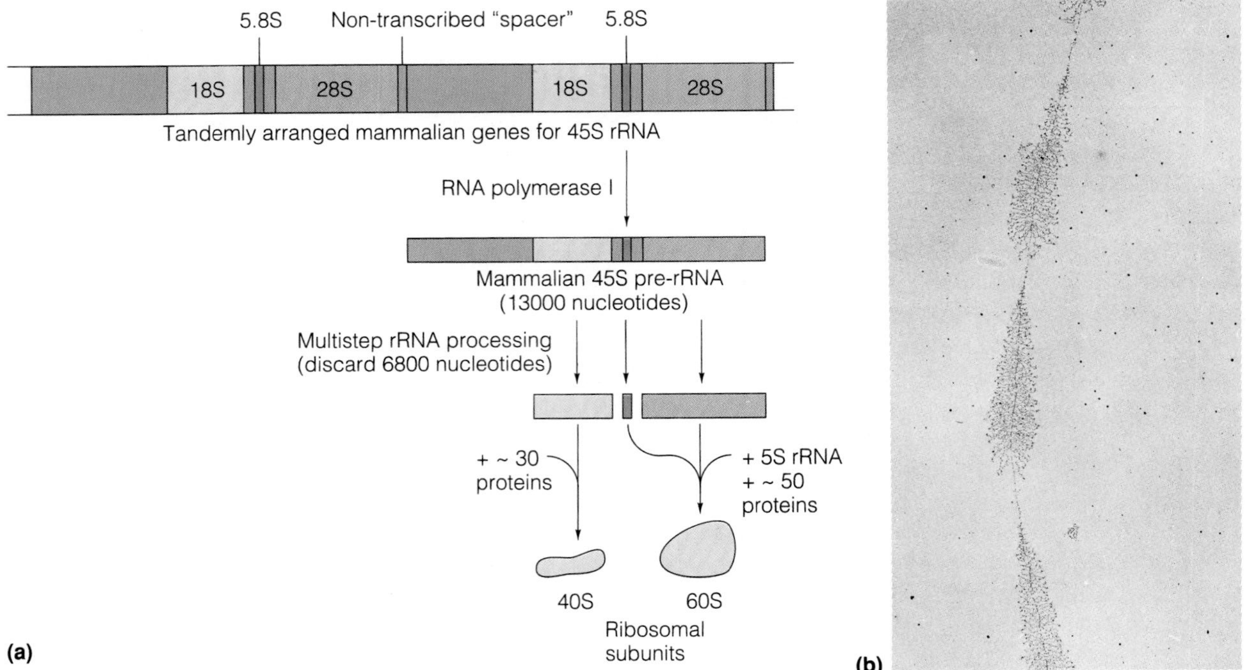

Figure 28.19
Transcription and processing of the major ribosomal RNAs in eukaryotes. (**a**) The 45S transcripts first produced are processed by removal of portions shown in tan, to yield the 18S, 5.8S, and 28S products. These are then assembled into ribosomal subunits, through addition of proteins. (**b**) Electron micrograph of spread nucleolar rRNA undergoing transcription. Tandemly arranged genes are being transcribed from top to bottom.

Table 28.2
Comparative properties of eukaryotic RNA polymerases

Polymerase	Location	RNAs Synthesized
I	Nucleus (nucleolus)	Pre-rRNA (except 5S)
II	Nucleus	Pre-mRNA, some small nuclear RNAs
III	Nucleus	Pre-tRNA, 5S rRNA, other small RNAs
Mitochondrial[a]	Mitochondrion	Mitochondrial
Chloroplast[a]	Chloroplast	Chloroplast

[a]These are quite similar to the prokaryotic RNA polymerase.

Transcription of the tandem copies of 45S rRNA can be beautifully visualized in the electron microscope (Figure 28.19b). The structure of the nucleolar chromatin has been a subject of some controversy, but it now appears that nucleosomes are not present, at least in the transcribed regions. This may be a specific chromatin modification to allow rapid and continuous transcription of these genes. After transcription, the 45S rRNA is processed to yield the components shown in Figure 28.19a; about 6800 nucleotides are discarded in this process. In the higher eukaryotes, little is yet known about the mechanism of this processing, the assembly of ribosomes, or the coordination of synthesis of ribosomal proteins and ribosomal RNA.

In some lower eukaryotes, like the protozoan *Tetrahymena,* the large-subunit RNA contains an intron near its 3′ end. Excision of this intron and splicing of the RNA are carried out by a remarkable and unexpected process. The RNA *itself* acts as the catalyst for the process, via the series of reactions described in Chapter 10 (see in particular Figure 10.26).

Polymerase III and the Transcription of Small RNA Genes

The genes transcribed by eukaryotic RNA polymerase III (Pol III) all share certain features. They are small, they are not translated, and they are unique in that the transcription is regulated by certain sequences that lie *within* the transcribed region. The major targets for Pol III are the genes for all the tRNAs and for the 5S ribosomal RNA. Like the major ribosomal genes described above, these small genes are present in multiple copies, but they are usually not grouped together in tandem arrays, nor are they localized in one region of the nucleus. Rather, they are scattered over the genome and throughout the nucleus.

Of all the genes transcribed by Pol III, the most thoroughly studied are those for 5S ribosomal RNA in the toad *Xenopus*. There are two sets of 5S rRNA genes in the *Xenopus* genome—those expressed specifically in the oocyte, which accumulate stores of 5S RNA for subsequent embryonic development, and another set that takes over transcription in somatic cells.

In vitro experiments have revealed that at least three protein factors in addition to polymerase III are needed for expression of the oocyte-specific genes (Figure 28.20). Two of these **transcription factors** (TFIIIB and TFIIIC)

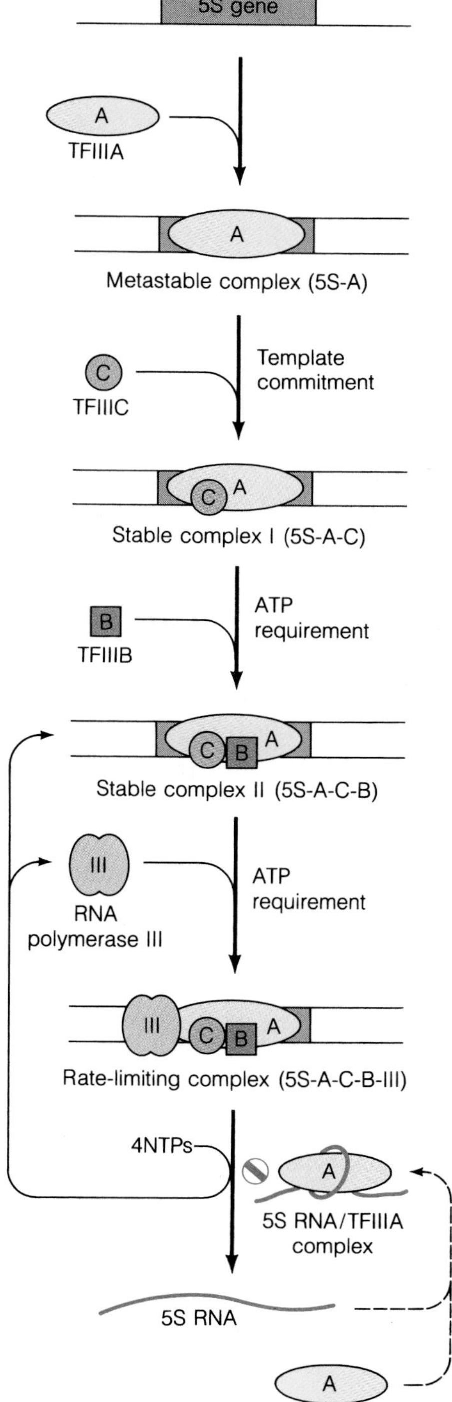

Figure 28.20
Binding of transcription factor TFIIIA prepares 5S rRNA genes for transcription. TFIIIA binding is essential for the subsequent binding of factors TFIIIC and TFIIIB, as well as of the RNA polymerase. (Precise binding sites for TFIIIB and C are not known.) Once the stable complex II has been formed, it will recycle with Pol III to produce many RNA copies. High concentrations of 5S RNA will compete with the 5S gene for TFIIIA.

appear to be more general factors that participate in the transcription of tRNA genes as well, but one, called TFIIIA, is specific for the 5S genes. The molecule of TFIIIA makes contact with DNA in at least two control regions *within* the gene (Figure 28.21a), and this somehow makes the gene accessible to TFIIIB, TFIIIC, and polymerase III. TFIIIA is an example of a recently discovered kind of DNA-binding protein, in which metal-binding "fingers" make contact with and identify DNA sequences (Figure 28.22). Most interestingly, TFIIIA *also* binds to the 5S RNA product; this provides autoregulation of 5S RNA production. As the gene product accumulates in the oocyte, more and more of the TFIIIA becomes associated with it, and less is available to stimulate transcription of the gene.

Many of the proteins involved in DNA binding and regulation seem to involve a few motifs common to both prokaryotes and eukaryotes. The metal finger motif is one, the "helix-turn-helix" (p. 936) is another. A third common motif is the "leucine zipper," a row of regularly spaced leucines on one side of an α-helix that can interdigitate with a set on another helix to dimerize some of these proteins.

The genes for eukaryotic tRNAs, also transcribed by Pol III, share certain features with 5S rRNA genes. Again, two regions *within* the gene are recognized by transcription factors. The tRNA transcripts are longer than the final tRNAs, for they contain introns that must be removed by a splicing process. In addition, tRNAs undergo extensive chemical modification to produce the unusual and modified bases so common in them.

Figure 28.21
Regions in the 5S rRNA genes and 5S rRNA that bind TFIIIA. (**a**) The central 60 bp of the gene. Tan indicates regions in the gene that are protected from DNase action by TFIIIA binding. Phosphates that make contact with the factor are noted by triangles, and G's whose modification blocks binding are underlined. (**b**) The whole 5S rRNA. Here, color indicates protection against ribonuclease by TFIIIA. In both (**a**) and (**b**) the purple sequences represent repetition of a CCUGG sequence that is important for folding the rRNA.

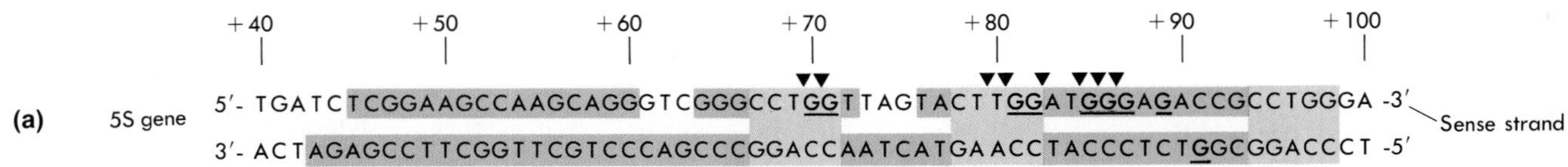

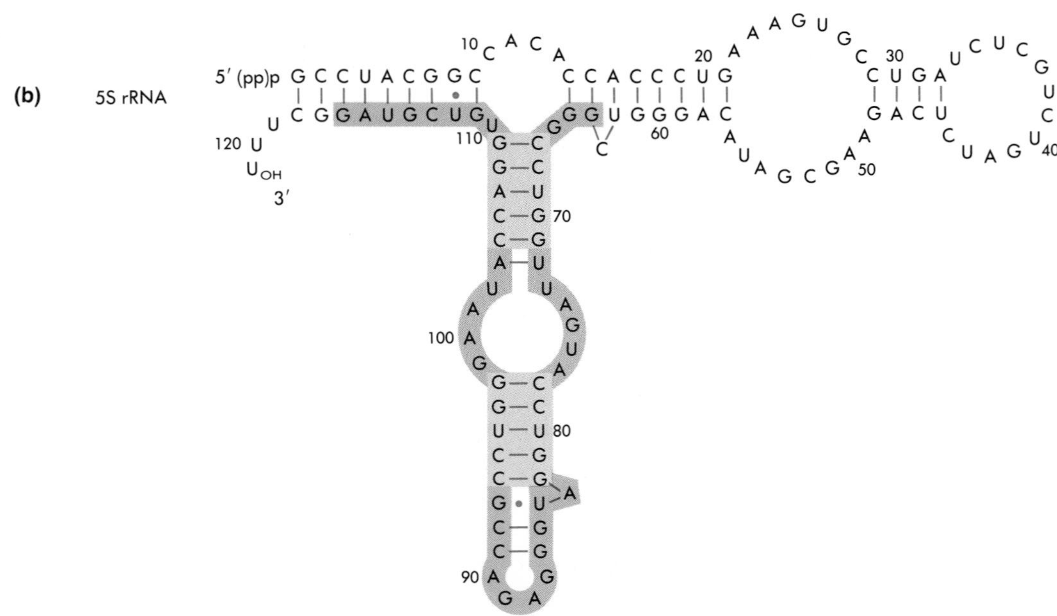

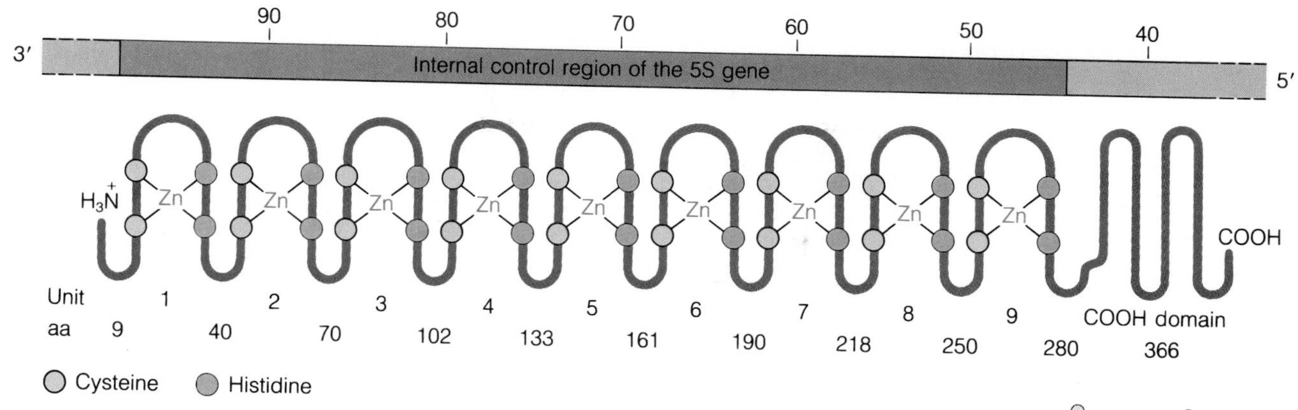

Polymerase II: Multiple Ways of Regulating Structural Gene Transcription

All of the structural genes (those coding for protein products) in the eukaryotic cell are transcribed by polymerase II. Different proteins are required in different tissues and at different developmental stages. Thus, precise control of polymerse II transcription is the rule, rather than the exception. The only genes whose transcription appears *not* to be tissue- or stage-specific are the "housekeeping" genes—those that code for proteins involved in processes found in every cell—the glycolytic pathway, for example.

Changes in Chromatin Structure Accompanying Transcription. To take a well-studied example of regulation, consider the β-globin genes in the chicken. These are expressed *only* in erythroid cells and then in a fixed developmental sequence. If one examines the chromatin structure in this gene cluster in embryonic cells that have not yet begun synthesis of *any* globin, it appears much the same as in any other cell in the chicken. But with the onset of globin synthesis, the whole β-globin domain in erythroid cells undergoes changes in chromatin structure. One way in which these changes are expressed is in a greater sensitivity to digestion by nucleases. At early stages in developing chick embryos, nuclease sensitivity is especially pronounced in the vicinity of the embryonic genes, which are the first to be transcribed. **Hypersensitive sites** appear in their 5′-flanking regions. With the shift from embryonic to adult genes, hypersensitive sites shift from the embryonic genes to the 5′ flanks of the adult genes.

In whatever gene is being actively transcribed, the regular chromatin structure appears to be perturbed. Some of the hypersensitive sites have been identified as nucleosome-free regions; they are able to bind specific proteins that facilitate polymerase attachment and transcription. In fact, it now appears that polymerase II always requires several accessory protein factors in order to initiate transcription. Some of these transcription factors may be required for most genes and probably bind to certain common sequences found in the 5′-flanking promoter region. These include the TATA and CAAT sequences, usually placed at about 25 bp and about 75 bp upstream from the start site (Figure 28.23). Most genes have a TATA sequence, but in some cases a CG-rich sequence is found instead of CAAT. As shown in Table 28.3, specific transcription factors appear to be associated with each sequence. The special nuclease-hypersensitive region for the β^A gene, which has been shown to bind proteins specific for globin gene synthesis, lies still farther upstream, centered at about −150 bp. The integrity of each

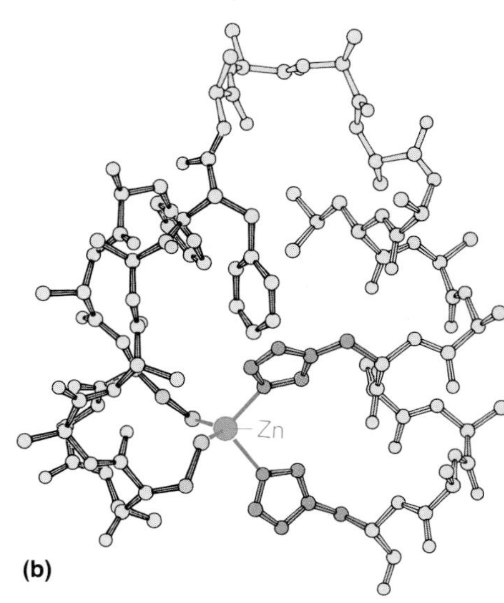

Figure 28.22
The zinc finger in TFIIIA. (**a**) Sequence of TFIIIA showing nine zinc fingers and presumed interaction with the control region of the 5S rRNA gene. The zinc finger domains are nearly, but not exactly, identical. (**b**) Model of a zinc finger. This structural motif is found in multiple copies in TFIIIA and a number of other DNA-binding proteins. It is thought to interact with the major groove in DNA. Cys (yellow) and His (blue) coordinate with Zn (red). The structure shown here is a model, based on common sequences in many zinc fingers, and on reasonable conformations.

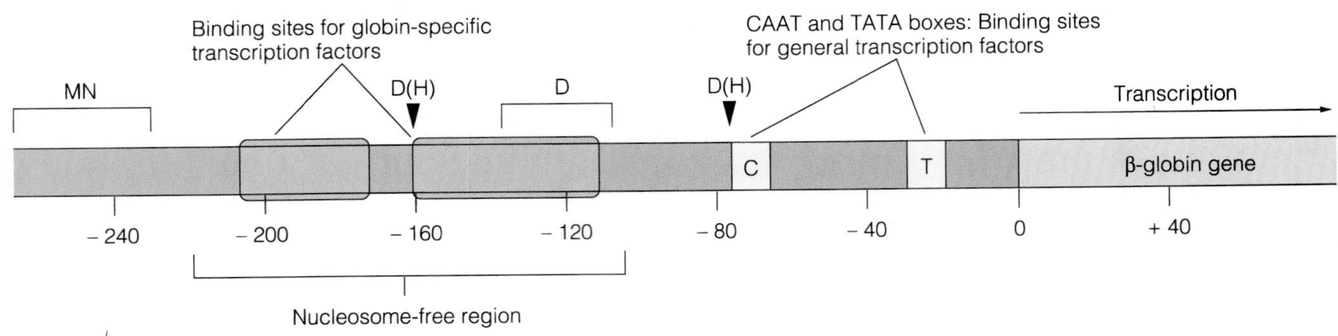

Figure 28.23
The 5′-flanking region of the chicken β-globin gene. The region shown is "upstream" from the transcription start site (O). The CAAT (C) and TATA (T) boxes, common to many eukaryotic genes, are shown in yellow. Regions of pronounced micrococcal nuclease (MN) or DNase I (D) sensitivity are indicated. The arrows D(H) are DNase-1 *hyper*sensitive sites. The region between −110 and −240 appears to be nucleosome free. Two protein-binding domains are shown in purple; these bind transcription factors.

of these control sequences appears to be essential—deletion or modification of any one will abolish or inhibit transcription of the gene.

Thus, several levels of control are superimposed in regulating the transcription of specialized genes like those in the globin clusters. First, in cells that are destined to transcribe a gene group, there are changes that occur over large domains of chromatin structure containing that group. These domains may correspond to the chromatin loops described earlier, and the general changes in nuclease sensitivity may result from changes in supercoiling, perhaps augmented by covalent modification of the histones. Next, the *specific* genes within the cluster that have been chosen for transcription at a particular developmental stage are marked by the binding of specific protein factors (e.g., Figure 28.23). This may in turn modify the chromatin structure within the gene so as to facilitate transcription. Remarkably, these modifications of chromatin structure appear to be transmitted to daughter cells in developing tissues. Details of these fascinating changes, so important to eukaryotic development and tissue specialization, are only now emerging.

HORMONAL REGULATION OF TRANSCRIPTION. For some eukaryotic genes, transcription is also dependent on hormone action. Chapter 23 has described the role certain hormones play in metabolic integration. Other hormones are found to be important in the regulation of gene transcription during development. A well-studied example includes the genes that code for the major chicken egg-white proteins—ovalbumin, ovomucoid, and lysozyme. Transcription of these genes occurs only in the tubular cells of the hen oviduct. Even in immature chicks the genomic domain containing the ovalbumin gene appears to have a somewhat different chromatin structure in these oviduct cells than in other tissues. But only on stimulation by the sex hormone estrogen (either on sexual maturation of the chick or following hormone administration) does transcription of ovalbumin genes commence. Withdrawal of administered hormone from an immature chick leads to an immediate cessation of transcription.

There are many examples of such hormonal control of transcription. In each case it appears that "target" cells contain specific proteins that are *hormone receptors*. These proteins bind the hormone and then become ca-

Table 28.3
Some general Pol II transcription factors

Factor	Approximate Molecular Weight	Sequence Bound
B	?	TATA (TATA box)
CTF	~60,000	GGNCAATCT (CAAT box)
Sp1	~100,000	GGGCGG (GC box)

pable of interacting either with specific DNA sites or with nonhistone regulatory proteins bound to such sites. Thus, both positive and negative regulation is possible. The hormone-binding receptor could act as a positive regulatory factor, directly facilitating polymerase II attachment and initiation of transcription. Alternatively, the hormone-binding receptor might interact with a *repressor* protein, augmenting or relieving the repression. Recent evidence suggests that the latter model may describe the response of chicken oviduct cells to estrogens.

ENHANCERS. Regulation of eukaryotic gene expression may occur over long distances in the genome. Some eukaryotic genes are regulated by **enhancer** sequences, which may lie hundreds or even thousands of base pairs distant from the gene. Effective enhancers may lie to either the 5′ or the 3′ side of the gene or in an intron within it. How such action at a distance occurs is not yet clear, but it may well have to do with changes in DNA torsion or supercoiling resulting from specific proteins binding to the enhancer sequences. This interpretation is supported by the observation that the effect of enhancers is generally unchanged even if they are excised and replaced in the opposite orientation or moved to the opposite strand. Some enhancer sites appear to be binding sites for hormone receptors; others may bind as yet unrecognized transcription factors.

To summarize, it is becoming clear that the control of polymerase II transcription is a complex, multistage process. The "program" for activating transcription of a particular structural gene in a particular tissue at a specific stage of development may look something like this:

If A, *and* B, *and* C (but *not* D) then: *GO*

Considering the great variety of proteins that must be expressed differently in the particular tissues of a eukaryote, it is not surprising that the controls should prove to be complicated.

Structural Gene Transcription and mRNA Processing

There are significant differences in the ways that messenger RNAs for protein-coding genes are produced and utilized in prokaryotic and eukaryotic cells. As you will recall from previous chapters, prokaryotic mRNAs are synthesized on the bacterial nucleoid in direct contact with the cytosol and are *immediately* available for translation. A specific nucleotide sequence at the 5′ end recognizes a ribosome site, allowing attachment of the ribosome and initiation of translation, often even before transcription of the message is completed.

In eukaryotes, the mRNA is produced in the nucleus and must be exported into the cytosol for translation. Furthermore, the initial product of transcription (**pre-mRNA**) includes all of the introns and substantial flanking regions. The introns must be removed before correct translation can occur. Therefore, eukaryotic mRNA requires extensive processing before it can be used as a protein template.

Even the termination of mRNA transcription is different in eukaryotes. Whereas the prokaryotic RNA polymerase recognizes terminator signals, which sometimes function with the aid of the ρ protein, the eukaryotic polymerase II routinely continues to transcribe well past the end of the gene (Figure 28.24). In doing so, it passes through one or more AATAAA signals, which lie beyond the 3′ end of the coding region. The pre-mRNA, carrying this signal as AAUAAA, is then cleaved by a special endonuclease that rec-

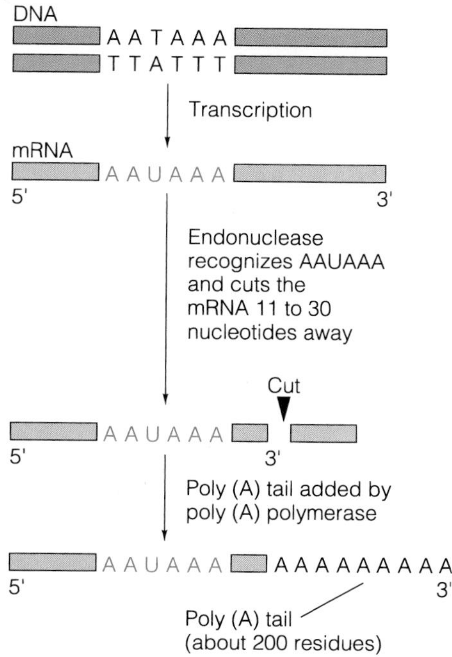

Figure 28.24
Termination of transcription in eukaryotes: addition of poly(A) tails. There is an AATAAA sequence near the 3′ end of most eukaryotic genes. When this is transcribed to AAUAAA, it provides a signal for endonuclease cleavage and poly(A) tailing.

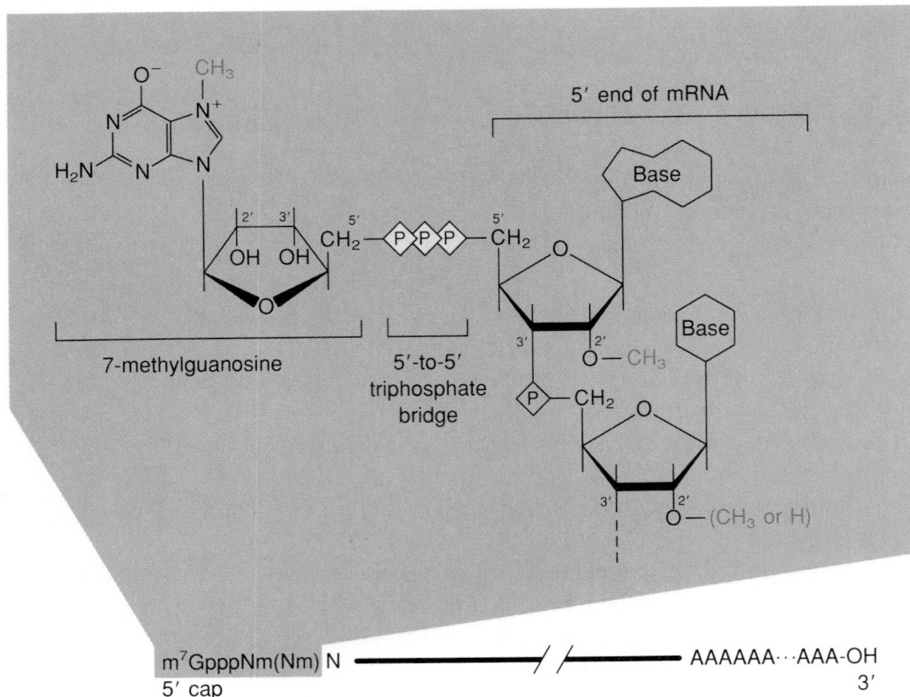

Figure 28.25
Overall structure of a fully processed eukaryotic message, including the cap site. Details of the 5′ cap region are shown. Methyl groups that are added are in red.

ognizes the signal and cuts at a site 11 to 30 residues 3′ to it. At this point, a "tail" of polyriboadenylic acid, poly(A), as much as 200 bases long, is added by a special non-template-directed polymerase. The function of the poly(A) tails of eukaryotic mRNAs is unknown. They cannot be essential for all messages, for some mRNAs (for example, those for histones in higher eukaryotes) do not even have them.

A second modification occurs at the 5′ end of the pre-mRNA. A GTP residue is added in *reverse* orientation to form what is known as a *cap*. The cap is further decorated by addition of methyl groups to the N-7 position of the guanine, and to one or two OH groups of nucleotides at the 5′ end of the chain (Figure 28.25). This cap structure will serve initially to position the mRNA on the ribosome for translation.

At this point, the pre-mRNA is still in the nucleus and has already been complexed with a number of nuclear proteins to form what are called **hnRNP particles**—heterogeneous nuclear ribonucleoprotein particles (Figure 28.26). It seems likely that tailing, capping, and other processing occurs in these structures, or in particles derived from them by the addition of further factors.

The most elegant part of the processing is the cutting and splicing that is necessary to excise introns from the pre-mRNA. We still do not know all of the details of splicing, but it seems to involve, in addition to special enzymes, certain **small nuclear RNAs,** called **snRNAs**. At the junctions between exons and introns there exist consensus sequences in the pre-mRNA (see Table 28.4). One of the snRNAs, called U1, contains a sequence that is complementary to the 5′ consensus sequences. It seems likely that it is interaction with U1 (and perhaps other snRNAs, in particular U2) that aligns the ends of the introns for correct excision and splicing. The initial step in splicing is the formation of an unusual loop to a 2′ OH group within the intron (Figure 28.27). At the same time, a G residue at the 3′ end of one exon is spliced to the 5′ end of the next. The looped intron is released and degraded. Much of the splicing is now thought to occur in a particular class

Table 28.4
Representative sequences at splice junctions

Protein, Intron	5′ Exon	Intron	3′ Exon
Ovalbumin, intron 3	. . . UCAG	GUACAG . . . UGUAUUCAG	UGUG
β-Globin, human, intron 1	. . . CGAG	GUUGGU . . . CACCCUUAG	GCUG
β-Globin, human, intron 2	. . . CAGG	GUGAGU . . . CCUCCACAG	CUCC
Immunoglobin I, L-VI	. . . UCAG	GUCAGC . . . UGUUUCGAG	GGGC
Rat pre-proinsulin	. . . CAAG	GUAAGC . . . CCCUGGCAG	UGGC
Consensus sequences[a]	—AG	GURAGY . . . YYYYY—AG	——

[a] Here R stands for purine, Y for pyrimidine. Residues listed for the consensus sequence are those found in two-thirds or more of over 100 cases analyzed. The residues shown in red are *invariant* in all cases analyzed.

Figure 28.26
The upper panel is an electron micrograph showing 40S hnRNP particles isolated from human cells. The particles have a diameter near 20 nm. The lower panel is a polyacrylamide slab gel showing the protein composition of fractions from a 15–30% linear sucrose density gradient sedimentation experiment. Sedimentation is from left to right. 40S hnRNP particles package 700 nucleotides of pre-mRNA into a repeating array of regular particles that are each composed of three copies of three different protein tetramers $(A1)_3B2$, $(A2)_3B1$, and $(C1)_3C2$. This fundamental stoichiometry is apparent in the gradient fractions near 40S that contain the monoparticles. The A, B, and C proteins have molecular weights between 32,000 and 42,000 daltons.

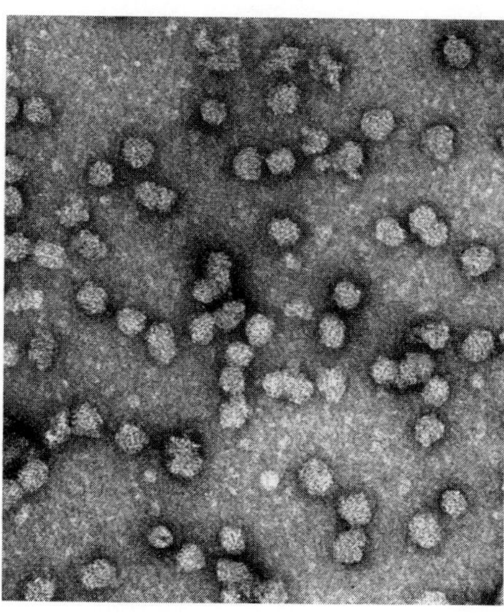

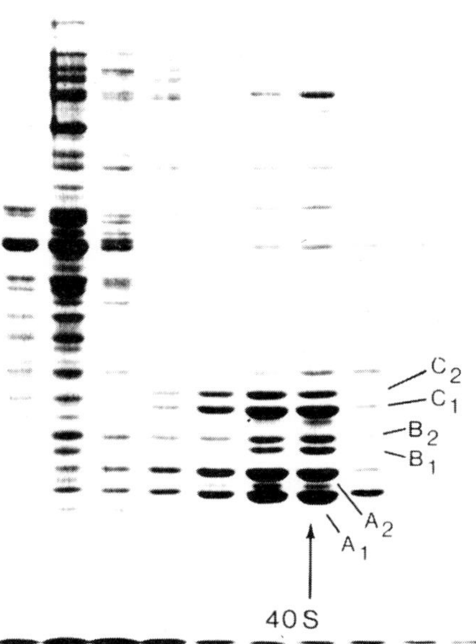

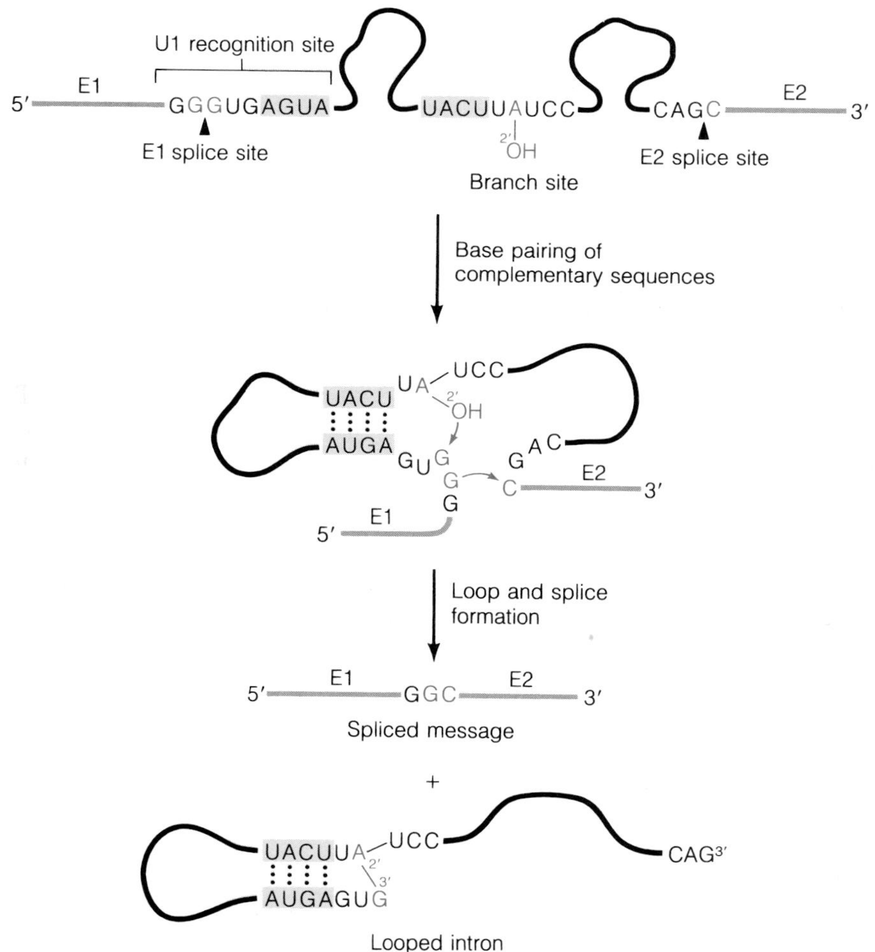

Figure 28.27
Proposed mechanism for mRNA splicing. The splice site, presumably with the aid of the small RNA U1, pairs with a sequence at the branch site to form a loop. The 2′ hydroxyl on the branch site A reacts with a G residue (blue) at the 3′ end of the exon 1 (E1). This frees the adjacent G (red) to transfer to the 5′ end of the following exon (E2). The products are a spliced message and a looped intron, which is then degraded.

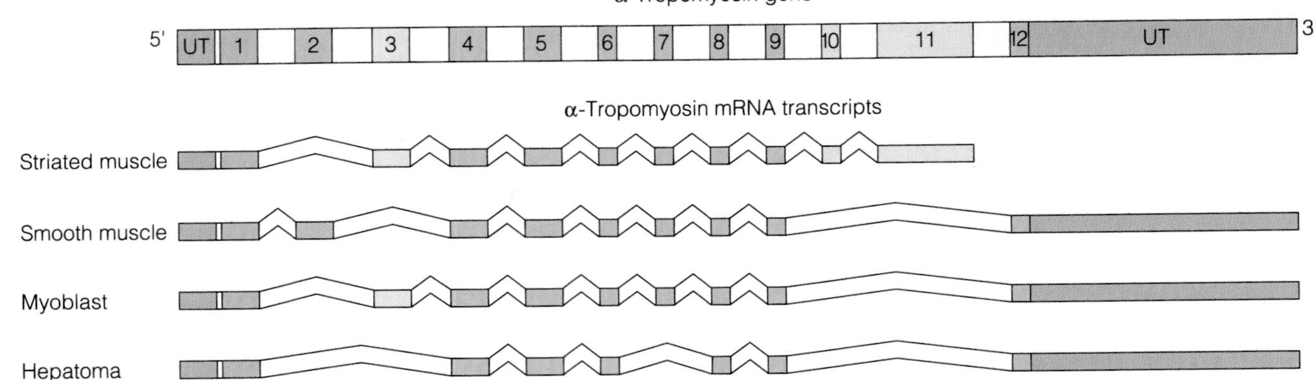

Figure 28.28
Some alternative splicings of transcripts of the α-tropomyosin gene. Exons are distinguished as follows: blue = in all tropomyosins; orange = striated muscle only; purple = smooth muscle; green = variable. UT (tan) denotes untranslated region. Introns are in white.

of hnRNP particles called **spliceosomes,** which contain proteins, small nuclear RNAs, and the mRNA being processed.

One of the more remarkable and significant discoveries in recent years is the phenomenon of *alternative* splicing. Some gene transcripts may be spliced in different ways, to include or exclude certain exons, in different tissues of an organism or at different developmental stages. We mentioned one example of this in Chapter 7, where it was pointed out that the heavy chains of immunoglobulins may or may not carry an N-terminal hydrophobic membrane-binding domain. A more dramatic example is shown in Figure 28.28, which depicts some of the alternative splicing modes for the muscle protein *α-tropomyosin.* Tropomyosin is involved in a number of different kinds of contractile systems in various cell types. Apparently, the need for functional domains coded for by different exons differs from one use of α-tropomyosin to another. Rather than having different genes expressed in different tissues, a single gene is employed, but the specific splicing patterns in different tissues provide a variety of α-tropomyosins. The economy of this phenomenon, in terms of genome size, should be obvious. Over 50 examples of eukaryotic genes exhibiting alternative splicing are known at the present, and the list will surely grow.

After capping, poly(A) tailing, and splicing are complete, the newly formed mRNA is exported from the nucleus, almost certainly through the nuclear pores. It is then attached to ribosomes for translation.

Translation in Eukaryotes

Mechanisms: Comparison with Prokaryotes

The mechanism for translating messenger RNA into protein in eukaryotic cells is basically the same as in prokaryotes: messenger RNA is read by ribosomes. There are, however, significant differences in both the ribosomes and the details of the translational mechanism. Since we have considered the prokaryotic translational system quite thoroughly in Chapter 27, we shall concentrate here on the differences.

RIBOSOMES. The eukaryotic ribosomal subunits are larger, 40S and 60S, combining to form a functional 80S ribosome. These correspond to the prokaryotic 30S, 50S, and 70S particles. The large subunit contains 28S (26S in yeast), 5S, and 5.8S RNAs, the last having no counterpart in prokaryotes. The small subunit (40S) has an 18S RNA (17S in yeast). The subunits contain more proteins than do the corresponding prokaryotic particles.

INITIATION FACTORS. The formation of the eukaryotic initiation complex requires many more protein factors than are needed in prokaryotes (see Table 28.5 and Figure 28.29; compare Figure 27.19). At least seven proteins are needed, and some of these are complex, multisubunit structures. The initiation factor eIF2 is recycled via a cyclic GDP–GTP exchange (see Figure 28.29).

INITIATION MECHANISM. While eukaryotic proteins initiate like their prokaryotic counterparts, with a special Met-tRNA reading an AUG codon, the methionine is not formylated. As shown in Figure 28.29, the eukaryotic mRNA is aligned correctly on the 40S subunit by the 5′ cap, rather than by the Shine-Dalgarno sequence used by prokaryotes. The ribosomal subunit then scans along the mRNA (an ATP-dependent process) until the first AUG is found. At this point initiation factors are released and the 60S subunit is attached to begin translation.

ELONGATION AND TERMINATION. The mechanism of chain elongation appears to be very similar in eukaryotes and prokaryotes. Eukaryotic chain termination, in contrast with prokaryotic termination, requires only one protein factor—eRF (see Table 28.5). This factor can recognize all three stop codons (UAA, UAG, UGA).

Table 28.5
Comparison of translational protein factors in prokaryotes and eukaryotes

Prokaryotic Factor	Eukaryotic Factor	Function
Initiation factors		
IF1		Involved in forming initiation complex
IF2	eIF2	
IF3	eIF3, eIF4C	
	CAP BPI	Involved in cap binding (eukaryotes)
	eIF4 A, B, F	Involved in search for first AUG
	eIF5	Helps dissociate eIF2, eIF3, eIF4C
	eIF6	Helps dissociate 60S subunit from inactive ribosomes
Elongation factors		
EF-Tu	eEF1α	Delivery of aminoacyl tRNA to ribosomes
EF-Ts	eEF1$\beta\gamma$	Aids in recycling factor above
EF-G	eEF2	Translocation factor
Release factors		
RF1	eFR	Release of complete polypeptide chain
RF2		

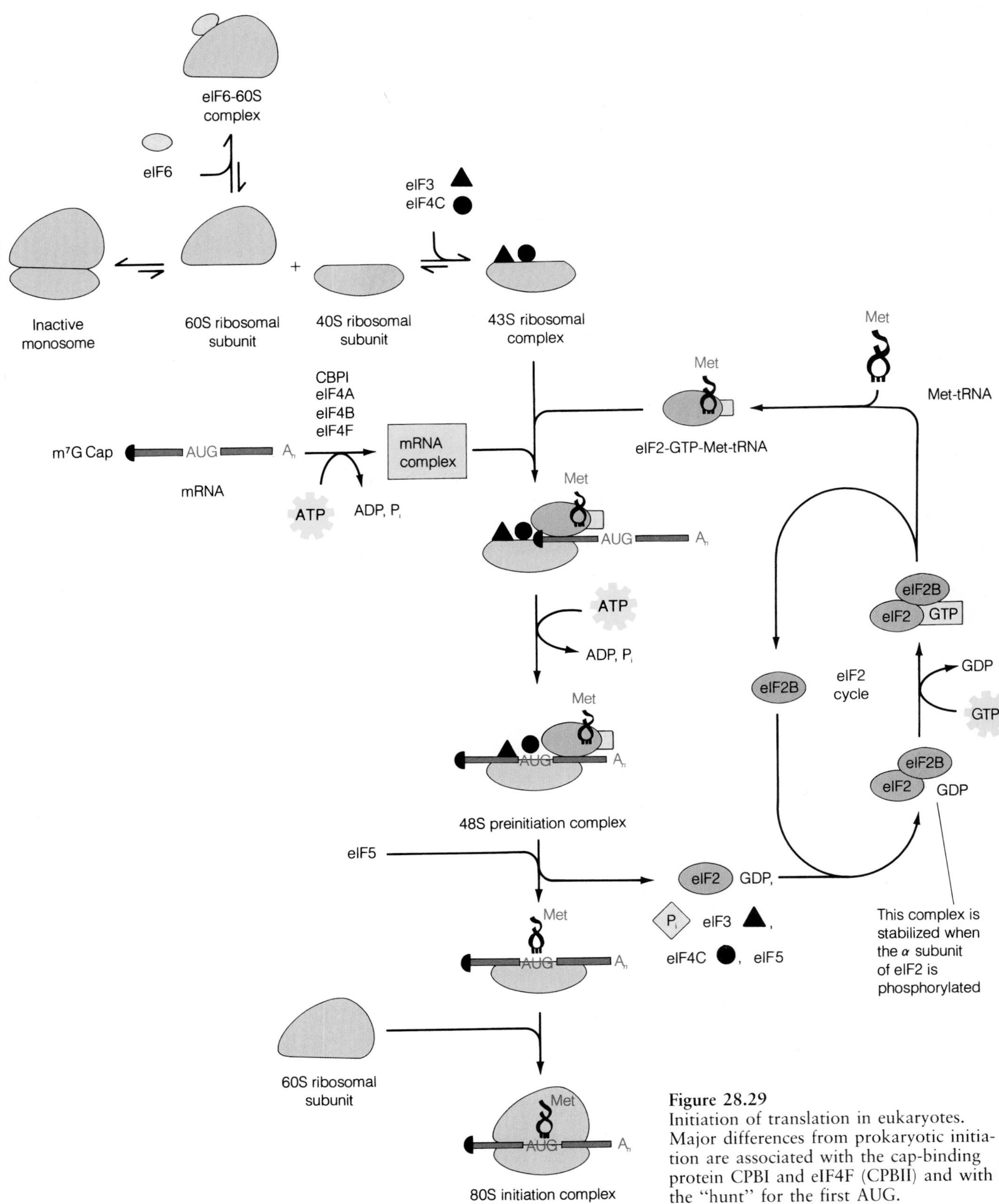

Figure 28.29
Initiation of translation in eukaryotes. Major differences from prokaryotic initiation are associated with the cap-binding protein CPBI and eIF4F (CPBII) and with the "hunt" for the first AUG.

Inhibitors of Translation

A number of the common inhibitors of prokaryotic translation are also effective in eukaryotic cells; these include interferon, pactamycin, tetracycline, and puromycin. There are also inhibitors that are effective *only* in eukaryotes. Two important ones are **cycloheximide** and **diphtheria toxin.** Cycloheximide blocks translocation and is often used in biochemical studies when processes must be studied in the absence of protein synthesis. Diphtheria toxin is an enzyme, coded for by a bacteriophage lysogenic in the bacterium *Corynebacterium diphtheriae.* It catalyzes a reaction in which NAD^+ adds an **ADP ribose** group to a specially modified histidine in the translocation factor eIF2 (Figure 28.30). Since the toxin is a catalyst, minute amounts can irreversibly block a cell's protein synthetic machinery; pure diphtheria toxin is one of the most deadly substances known.

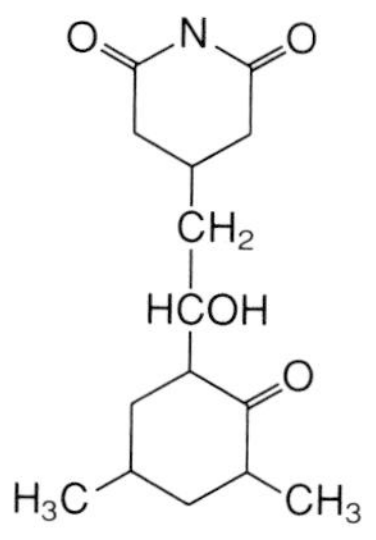

Cycloheximide

Control of Translation

Except in the case of the ribosomal proteins (p. 985), prokaryotes do not seem to make extensive use of control at the translational level. Eukaryotes, on the other hand, appear to use translational control much more widely. In part, this control occurs at the mRNA level. Specific mRNAs may be sequestered by combination with specific mRNA-binding proteins until needed. Some other mRNAs are rapidly degraded so that they do not persist in inappropriate phases of the cell cycle.

We still know little about the detailed mechanisms of translational regulation, but one example has been rather thoroughly studied—that concerned with hemoglobin synthesis in the reticulocytes of mammals. Al-

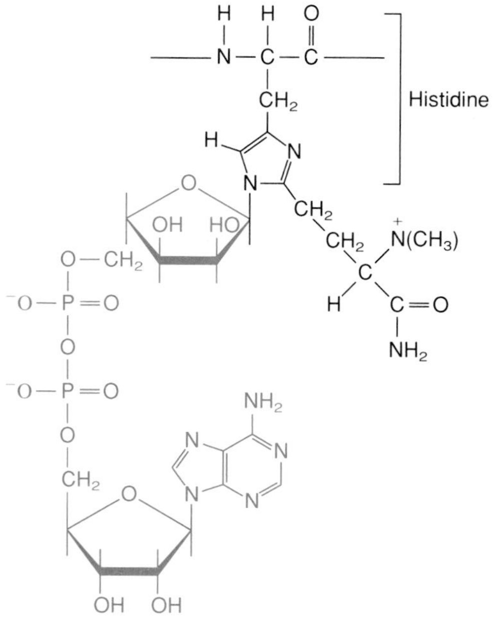

Figure 28.30
ADP-ribosylated diphthamide in eIF2. Synthesis of this derivative of histidine in eIF2 from NAD^+ is catalyzed by diphtheria toxin. eIF2 is inactivated and protein synthesis is therefore blocked. ADP ribose from NAD^+ is shown in blue. Diphthamide is in black.

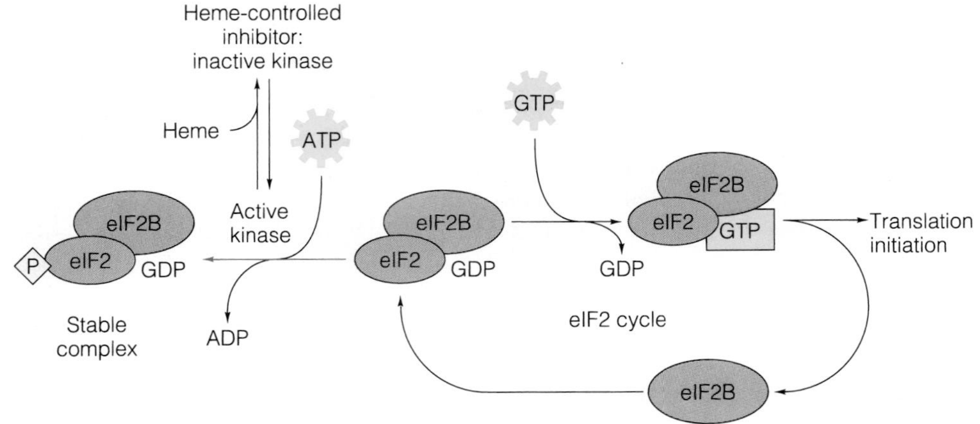

Figure 28.31
Regulation of translation in erythropoietic cells by heme levels. If heme levels fall, the heme-controlled kinase becomes active and phosphorylates eIF2 (red arrow). This blocks further translation by tying up this factor in a stable complex. When heme levels are adequate, eIF2 is available for translation initiation.

though immature red cells have lost their nuclei, they still have adequate reserves of mRNAs available for extensive globin synthesis, which is at this point their major function. Such synthesis would be inappropriate, however, if there were not an adequate supply of heme to produce complete hemoglobin molecules. The reticulocyte cells contain a protein kinase called **heme-controlled inhibitor** (HCI) (Figure 28.31). In the presence of adequate heme levels the kinase is inactive, but if heme levels fall, HCI becomes activated and specifically phosphorylates the initiation factor eIF2. When this occurs, the complex between eIF2 and eIF2B in the eIF2 cycle becomes unusually stable. The result is that all of the eIF2B and eIF2 are tied up and can no longer be recycled for new initiation. Hence, protein synthesis is halted, and no more globin is made until heme supplies are again adequate.

Protein Targeting in Eukaryotes

The eukaryotic cell is a multicompartmental structure. Its several organelles each require different proteins, only a few of which are synthesized by the organelles themselves. Most mitochondrial and chloroplast proteins, for example, are coded for in the nuclear genome, synthesized in the cytosol, and must be carefully distinguished from other newly synthesized proteins and selectively transported into the organelles. Some other new proteins will be destined for export out of the cell or into vesicles like lysosomes. The diversity of destinations for different proteins implies that there must be a sophisticated system for *labeling and sorting* newly synthesized proteins and ensuring that they end up in their proper places.

Proteins destined for the cytosol and those to be incorporated into mitochondria, chloroplasts, or nuclei are synthesized on polyribosomes free in the cytosol. The mitochondrial and chloroplast proteins, as initially synthesized, contain specific signal sequences at the N-terminal ends of the polypeptide chains. These sequences apparently recognize the appropriate organelle membranes and are taken in; the signal sequence is then removed. The situation with nuclear proteins remains unclear. Early theories held that these simply diffused into the nucleus through the nuclear pores and were then bound to chromatin. But strong evidence is emerging for "nuclear targeting" sequences, which may help some proteins—perhaps with others in tow—select the nucleus as their objective.

Proteins destined for cellular membranes, lysosomes, or extracellular transport utilize a special "distribution system." The key structures in this

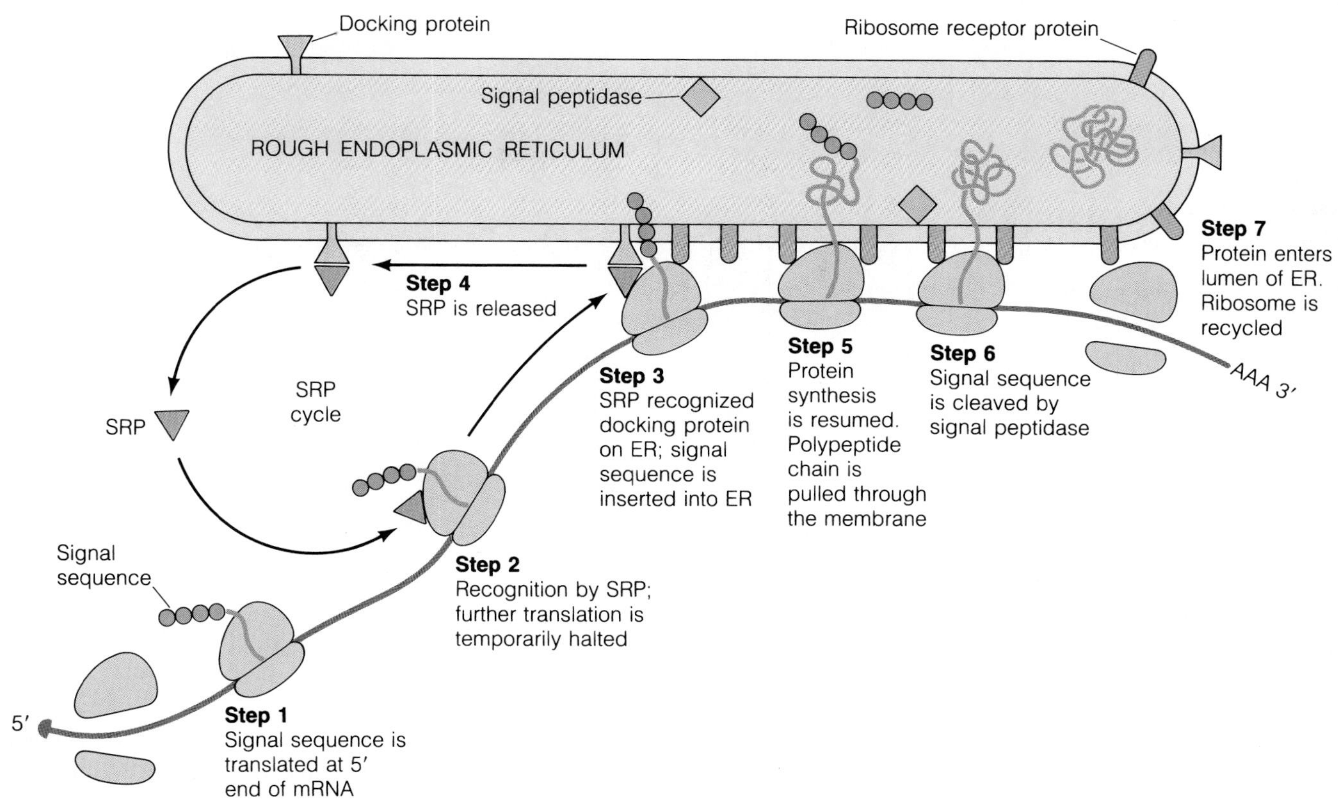
Docking protein
Ribosome receptor protein
Signal peptidase
ROUGH ENDOPLASMIC RETICULUM
Step 7
Protein enters lumen of ER. Ribosome is recycled
Step 4
SRP is released
Step 5
Protein synthesis is resumed. Polypeptide chain is pulled through the membrane
Step 6
Signal sequence is cleaved by signal peptidase
AAA 3′
SRP cycle
SRP
Step 3
SRP recognized docking protein on ER; signal sequence is inserted into ER
Signal sequence
Step 2
Recognition by SRP; further translation is temporarily halted
5′
Step 1
Signal sequence is translated at 5′ end of mRNA

Figure 28.32
The sequence of events in synthesis of proteins on the RER.

system are the rough endoplasmic reticulum (RER) and the Golgi complex (see also Chapter 16). The rough endoplasmic reticulum is a network of membrane-enclosed spaces within the cytosol.* The RER membrane is heavily coated on the outer, cytosolic surface with polyribosomes. The Golgi complex resembles the RER in that it is a stack of thin, membrane-bound sacs. However, the Golgi sacs are not interconnected, nor do they carry polysomes on their surfaces. The role of the Golgi complex is to act as a "switching center" for proteins with various destinations.

Those proteins to be directed to their destinations via the Golgi complex are synthesized by polyribosomes associated with the RER. Their synthesis begins with an N-terminal hydrophobic signal sequence (Figure 28.32, step 1). A special mechanism ensures that the signal peptide is inserted into the RER before translation has proceeded too far into the protein. This is accomplished by **signal recognition particles** (**SRP**s), containing several proteins and a small (7SL) RNA (see p. 999). These particles recognize the signal sequences of the appropriate nascent proteins and bind to them as they are being extruded from the ribosomes (step 2). The SRP appears to have two functions. First, its binding temporarily halts translation, so that no more than the N-terminal signal sequence extends from the ribosome. This prevents completion of the wrong protein (e.g., a protease destined for the lysosome) in the wrong place (i.e., the cytosol) and also prevents premature folding of the peptide chain. The second function of the SRP is to recognize a "docking protein," which binds the ribosome to the RER (step 3). The signal sequence is then inserted into the membrane of the

---

*The rough endoplasmic reticulum should be distinguished from the smooth endoplasmic reticulum, which has completely different functions, being primarily involved in lipid metabolism.

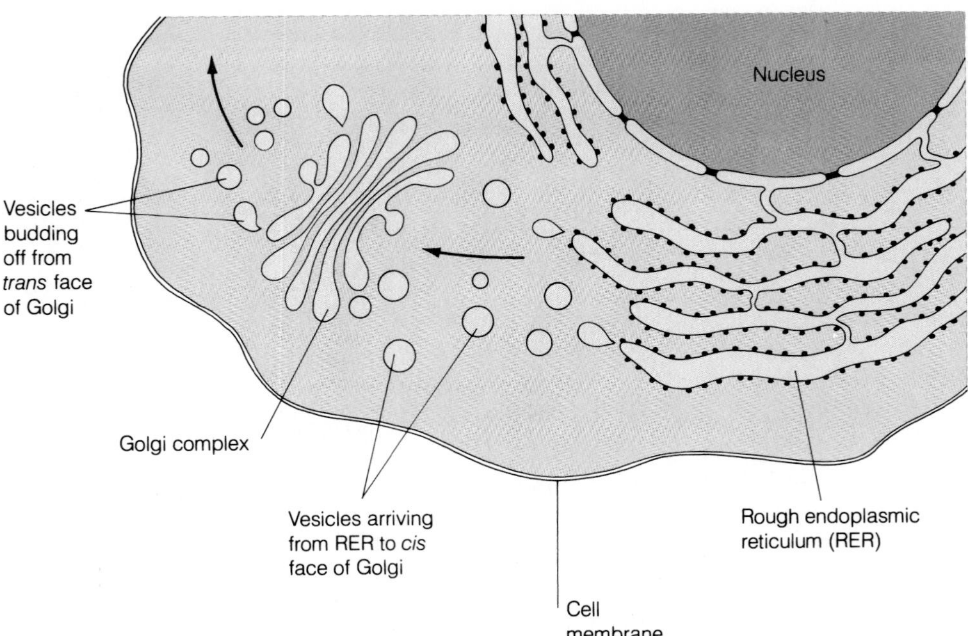

Figure 28.33
The rough endoplasmic reticulum (RER) and the Golgi complex. Note that vesicles bud off the RER and move to the *cis* face of the Golgi. Lysosomal vesicles bud from the *trans* portion of the Golgi.

RER, and the SRP is released (step 4), allowing translation to resume (step 5). The protein being synthesized is actually *pulled* through the membrane by an ATP-dependent process. After translation is complete, the next step depends on the intended destination of the protein. Proteins destined for export or for incorporation into the plasma membrane of the cell, for example, have signal sequences that can be cleaved off by an RER-associated protease (step 6). They are released into the lumen of the RER and further transported (step 7). Proteins that will remain in the endoplasmic reticulum have resistant signal peptides and thereby remain anchored to the membrane.

The proteins that enter the lumen of the RER undergo the first stages of glycosylation at this point. Vesicles carrying these proteins bud off the RER and move to the Golgi complex (Figure 28.33). Here the carbohydrate moieties of glycoproteins are completed (see Chapter 16 for details), and a final sorting occurs. The multiple membrane sacs that constitute the Golgi complex represent a multilayer arena for these processes. Vesicles from the ER enter at the cis face of the Golgi (that closest to the RER) and fuse with the Golgi membrane. Proteins are then passed, again via vesicles, to the intermediate layers. Finally, vesicles bud off from the trans face of the Golgi complex to form lysosomes, peroxisomes, or glyoxysomes or to travel to the plasma membrane. Most details of this sorting are still obscure. Carbohydrate moieties clearly play some part, for it is known that hydrolytic enzymes destined to end up in lysosomes carry a mannose-6-phosphate marker, which is recognized by receptors on the inner surface of the lysosome membrane. But this marker itself must surely be added as a consequence of some peculiarity in the amino acid sequences of these enzymes. It seems likely that recognition at this stage (when proteins are already folded) may involve features of the surface of three-dimensional protein structure.

Some vesicles travel to the plasma membrane and fuse with it, in the manner shown in Figure 28.34. In doing so, they both release their internal contents from the cell and add their membrane components to the plasma membrane itself. Proteins that have been incorporated into the vesicular membranes will remain as plasma membrane proteins. Note from Figure 28.34 that proteins on the *inner* surface of the vesicular membrane will end up on the *outer* surface of the plasma membrane. Since they were glycosylated while facing *into* the Golgi complex, the carbohydrate moieties will protrude on the *outer* surface of the cell. These mechanisms are capable, therefore, of explaining the pronounced asymmetry in plasma membrane protein composition that was noted in Chapter 9.

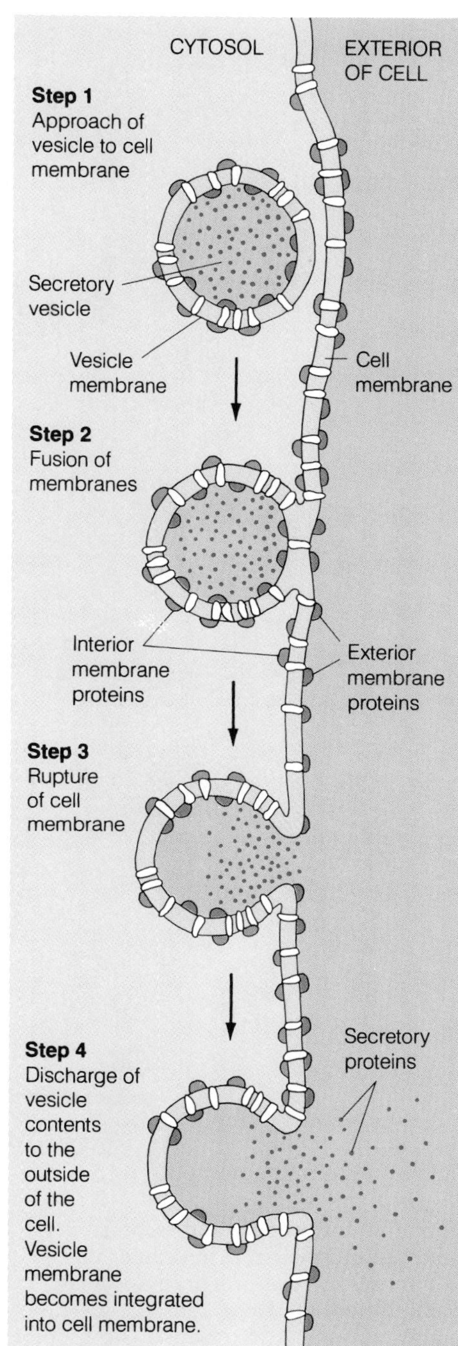

Figure 28.34
Exocytosis by fusion of vesicles with the plasma membrane. Note that the process not only exports materials but also adds to the plasma membrane, with internal and external membrane constituents correctly positioned.

The Fate of Proteins: Programmed Destruction

In Chapter 11 we pointed out that one mechanism for the control of enzymatic function was the selective degradation of certain enzymes. However, not only enzymes need to be destroyed in a programmed way. For example, some regulatory proteins may be essential in certain parts of the cell cycle and deleterious in other phases. Proteins that have become damaged must be removed. In some developmental processes it is necessary to remove whole organelles, or even entire cells and tissues.

Eukaryotic cells have two distinct methods for protein degradation. The lysosomes contain among their hydrolases proteolytic enzymes that will degrade any protein the lysosome entraps. Parallel to this is a cytosolic degradation system, which is, of necessity, highly selective. The danger inherent in having general, nonspecific proteases loose in the cytosol should be evident.

The Lysosomal System

The lysosomal particles budded from the Golgi complex, known as **primary lysosomes,** are bags of degradative enzymes. Over 50 different hydrolytic enzymes are contained in lysosomes, including proteases, nucleases, lipases, and carbohydrate-cleaving enzymes. The lysosomes play a number of different and important roles in cellular metabolism, as schematically depicted in Figure 28.35.

In some cells, such as those in the pancreas that secrete degradative enzymes, lysosomes migrate to the cell surface and release their contents into the exterior medium (path A).

Primary lysosomes may fuse with **autophagic vesicles,** which have engulfed organelles destined for destruction (path B), thereby adding their digestive enzymes to such vesicles.

In some kinds of cells—mainly certain white blood cells—primary lysosomes may fuse with **phagocytic vacuoles** that have engulfed nutrient materials at the cell surface (path C). The nutrients are digested and their amino acids, nucleotides, lipids, and other low-molecular-weight constituents released into the cytosol. Residual, undigested material is excreted when the heterophagic lysosomes and autophagic lysosomes find their way to the plasma membrane.

In certain special circumstances, lysosomes may open within the cell, spewing their degradative enzymes into the cytosol (path D). Such lysosomal **autolysis** occurs when cells themselves are destined for destruction, as in metamorphic changes during development. Such events are by no means rare—each of us has experienced them. For example, the webbing that is present between the toes and fingers of the early fetus is destroyed by just this kind of programmed cell death.

Cytosolic Protein Degradation—The Role of Ubiquitin

In contrast to the lysosomal enzymes, which are usually safely sequestered in their vesicles, any protease activity free in normal cytosol must be under

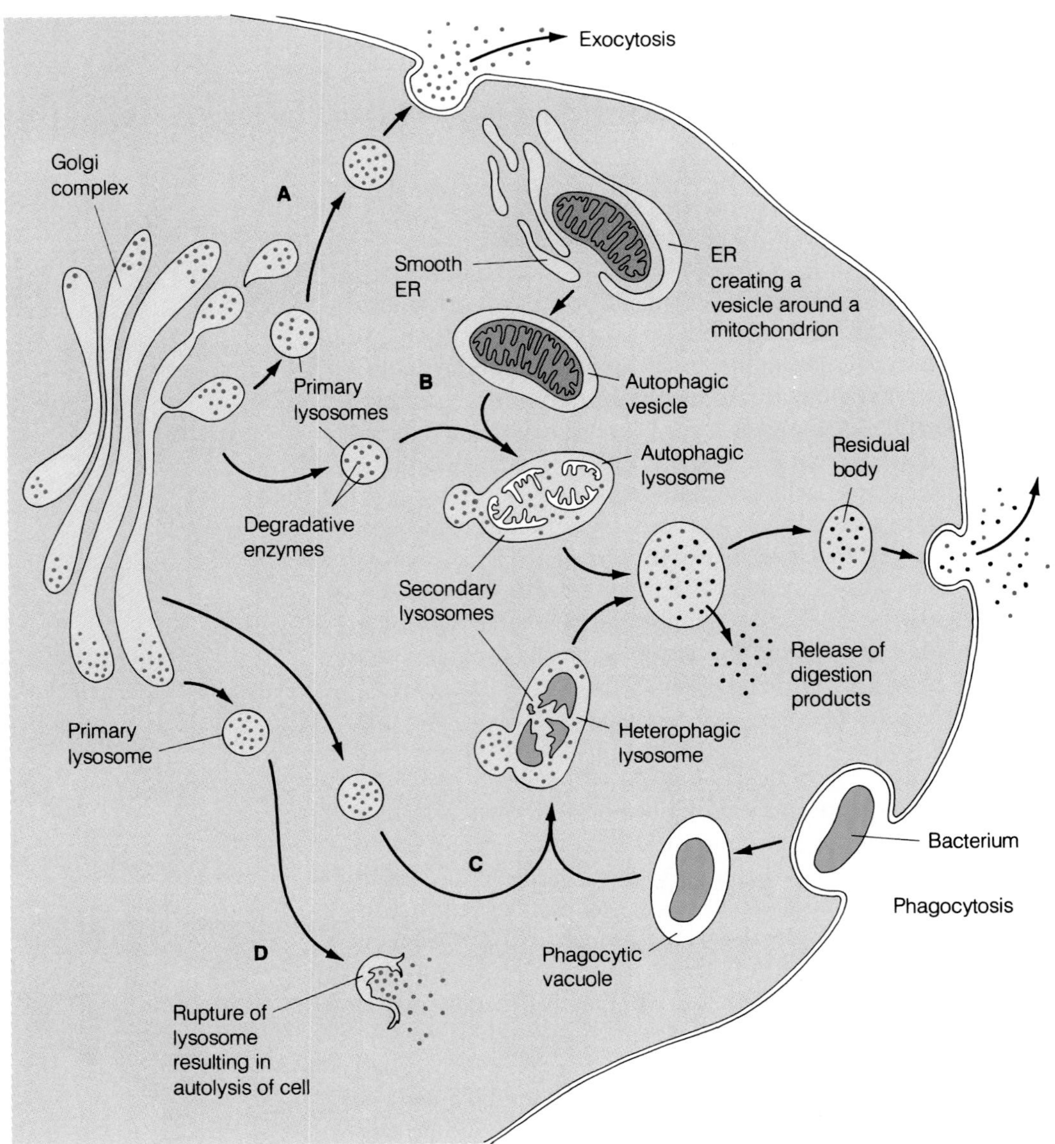

Figure 28.35
Formation of primary and secondary lysosomes and their role in cellular digestive processes. The primary lysosomes budded from the Golgi can take several pathways. Path A: Exocytosis—transports enzymes to outside of cell. Paths B and C: Phagocytosis—formation of phagic lysosomes for digesting organelles (autophagocytosis) or ingested matter (heterophagocytosis). Path D: Autolysis—destruction of the cell itself.

rigid control. It must attack only the proteins whose destruction is needed. These may include damaged proteins, incorrectly synthesized proteins, or proteins no longer required at a particular stage in the cell cycle. If we recall that protein hydrolysis is a thermodynamically favored reaction, it becomes clear that the enzymes participating in such cytosolic degradation must be more than simple catalysts for the hydrolytic process—otherwise destruction would be wholesale. Furthermore, there must be some means of "marking" the proteins to be attacked.

There appear to be a number of such systems in mammalian cytosol, but the best known utilizes a protein called **ubiquitin** as a marker. This is a 76-residue polypeptide, found in virtually every cell of every eukaryote. Its carboxyl group is activated by thiol coupling to an activating enzyme (Figure 28.36, step 1) in an ATP-dependent process. The ubiquitin moieties are then transferred to a second enzyme (step 2), which attaches them to ϵ-amino groups of lysine residues on the fated protein (step 3). Proteins so marked are then digested by specific cytosolic enzymes (step 4), which recognize the ubiquitin markers. Remarkably, ATP is required for both the

REFERENCES

General

Darnell, J., H. Lodish, and D. Baltimore (1986) *Molecular Cell Biology.* Scientific American Books, New York. An up-to-date and very well written account of the molecular biology of cells.

Watson, J. D., N. H. Hopkins, J. W. Roberts, J. A. Steitz, and A. M. Weiner (1987) *Molecular Biology of the Gene,* 4th ed. Benjamin/Cummings, Menlo Park, Calif. The latest edition, greatly expanded and revised, of Watson's classic book.

Repetitive DNA

Britten, R. J., and E. H. Davidson (1971) Repetitive and nonrepetitive DNA sequences and a speculation on the origins of evolutionary novelty. *Quart. Rev. Biol.* 46:111–138. An early paper on repetitive sequences.

Jelineck, W. R., and C. W. Schmid (1982) Repetitive sequences in eukaryotic DNA and their expression. *Annu. Rev. Biochem.* 51:813–844.

Pardue, M. L., and J. G. Gall (1970) Chromosomal localization of mouse satellite DNA. *Science* 168:1356–1358. A classic early paper using in situ hybridization.

Ullu, E., and C. Tschudi (1984) Alu sequences are processed 7SL RNA pseudogenes. *Nature* 312:171–172.

Exons, Introns, and Splicing

Bränden, C.-I., H. Eklund, C. Cambillau, and A. J. Pryor (1984) Correlation of Exons with Structural Domains in Alcohol Dehydrogenase. EMBO J. 3:1307–1310.

Breathnach, R., J. L. Mandel, and P. Chambon (1977) Ovalbumin gene is split in chicken DNA. *Nature* 270:314–319. A very early recognition of exon–intron structure.

Breitbart, R. E., A. Andreadis, and B. Nadal-Ginard (1987) Alternative splicing: A ubiquitous mechanism for the generation of multiple protein iso forms from single genes. *Annu. Rev. Biochem.* 56:467–495.

Padgett, R. A., P. J. Grabowski, M. M. Konarska, S. Seiler, and P. A. Sharp (1986) Splicing of messenger RNA precursors. *Annu. Rev. Biochem.* 55:1119–1150.

Sharp, P. A. (1985) On the origin of RNA splicing and introns. *Cell* 42:397–400.

Gene Families

Karlsson, S., and A. W. Nienhuis (1985) Developmental regulation of human globin genes. *Annu. Rev. Biochem.* 54:1071–1108.

Chromosomes and the Nucleus

Kavenoff, R., L. C. Klotz, and B. H. Zimm (1974) On the nature of chromosome-size DNA molecules. *Cold Spring Harbor Symp. Quant. Biol.* 38:1–8. The first measure of chromosomal DNA size.

Newport, J. W., and D. J. Forbes (1987) The nucleus: Structure, function, and dynamics. *Annu. Rev. Biochem.* 56:535–566.

Chromatin and Nucleosomes

Kornberg, R. (1974) Chromatin structure: A repeating unit of histones and DNA. *Science* 184:868–871. The classic paper introducing the subunit model for chromatin.

Olins, A. L., and D. E. Olins (1974) Spheroid chromatin units (nu bodies). *Science* 183:330–332. The first electron microscope evidence for subunit structure in chromatin.

Pedersen, D. S., F. Thoma, and R. T. Simpson (1986) Core particle, fiber, and transcriptionally active chromatin. *Annu. Rev. Cell Biol.* 2:117–147.

Richmond, T. J., J. T. Finch, B. Rushton, D. Rhodes, and A. Klug (1984) Structure of the nucleosomal core particle at 7Å resolution. *Nature* 311:532–537.

van Holde, K. E. (1988) *Chromatin.* Springer-Verlag, New York.

Cell Division and DNA Replication

DePamphilis, M. L., and P. M. Wassarman (1980) Replication of eukaryotic chromosomes: A close-up of the replication fork. *Annu. Rev. Biochem.* 49:627–666.

Goldman, M. A., G. P. Holmquist, M. C. Gray, L. A. Caston, and A. Nag (1984) Replication timing of genes and middle repetitive sequences. *Science* 224:686–692.

Prescott, D. M. (1976) *Reproduction of Eukaryotic Cells.* Academic Press, New York.

Transcription in Eukaryotes

Darnell, J. E., Jr. (1983) The processing of RNA. *Sci. Am.* 249:90–100.

Dynan, W. S., and R. Tjian (1985) Control of eukaryotic messenger RNA synthesis by sequence-specific DNA-binding proteins. *Nature* 316:774–778.

Gluzman, Y. (ed.) (1985) *Eukaryotic Transcription: The Role of Cis- and Trans-Acting Elements in Initiation.* Cold Spring Harbor Laboratory, Cold Spring Harbor, New York.

Miller, O. L. (1981) The nucleolus chromosomes, and visualization of genetic activity. *J. Cell Biol.* 91:15S–27S. The first EM pictures of transcription in eukaryotes.

Struhl, K. (1989) Helix-turn-helix, zinc finger, and leucine-zipper motifs for eukaryotic transcriptional regulatory proteins. *Trends Biochem. Sci.* 14:137–144.

Translation in Eukaryotes

Moldave, K. (1985) Eukaryotic protein synthesis. *Annu. Rev. Biochem.* 54:1109–1149.

Perez-Bereoff, R. (ed.) (1982) *Protein Biosynthesis in Eucaryotes.* Plenum, New York.

Protein Targeting

Dingwall, C. (1985) The accumulation of proteins in the nucleus. *Trends Biochem. Sci.* 10:64–66.

Garoff, H. (1985) Using recombinant DNA techniques to study protein targeting in the eucaryotic cell. *Annu. Rev. Cell Biol.* 1:403–445.

Meyer, D. F., E. Krause, and B. Dobberstein (1982) Secretory protein translocation across membranes: The role of the "docking" proteins. *Nature* 297:647–650.

Pfeffer, S. R., and J. A. Rothman (1987) Transport and sorting by the endoplasmic reticulum and Golgi. *Annu. Rev. Biochem.* 56:829–853.

Rothman, J. E. (1985) The compartmental organization of the Golgi apparatus. *Sci. Am.* 253(3):74–89.

Schatz, G., and R. A. Butow (1983) How are proteins imported into mitochondria? *Cell* 32:316–318.

Protein Degradation

Bachmair, A., D. Finley, and A. Varshavsky (1986) In vivo half-life of a protein is a function of its amino-terminal residue. *Science* 234:179–186.

Ciechanover, A. (1987) Regulation of the ubiquitin-mediated proteolytic pathway: Role of the substrate α-NH_2 group and transfer RNA. *J. Cell. Biochem.* 34:81–100.

Homeotic Genes and the Homeo Box

Gehring, W. J. (1985) The molecular basis of development. *Sci. Am.* 253(4):152–162.

Manley, J. L., and M. S. Levine (1985) The homeo box and mammalian development. *Cell* 43:1–2.

McGinnis, W., R. L. Garber, J. Wirz, A. Kuriowa, and W. J. Gehring (1984) A homologous protein-coding sequence in *Drosophila* homeotic genes and its conservation in other metazoans. *Cell* 37:403–408.

PROBLEMS

1. The average human chromosome contains about 1×10^8 bp of DNA.
   (a) If each base pair has a mass of about 660 daltons, and there are about 2 g of protein (histones plus nonhistones) per gram of DNA, how much does such a chromosome weigh, in grams?
   (b) If the DNA were extended, how long would it be?
   (c) An actual chromosome is about 5 μm in length. What is the approximate compaction ratio?

2. Formation of nucleosomes and wrapping them into a 30-nm solenoid provide part of the compaction of DNA in chromatin. If the solenoid contains six nucleosomes per turn and has a pitch of 100 nm, what is the approximate compaction ratio achieved? Comment on the comparison of this answer with that of Problem 1c.

3. (a) About how long would be required to replicate the average human chromosome described in Problem 1 from a single origin of replication, if a replication fork moves at about 6 kb/min?
   (b) Evidence suggests that origins are spaced about 100 kb apart in many eukaryotic DNAs. How long should replication then require? Compare this with the fact that S phase in most cells requires several hours, and suggest an explanation for the discrepancy.

4. Although protein hydrolysis is a thermodynamically favorable process, the selective hydrolysis of ubiquitin-marked proteins requires ATP. Suggest why this should be necessary.

5. Histone genes are unusual among eukaryotic genes in that they do not have introns, and histone mRNAs do not have poly(A) tails. Furthermore, in almost all eukaryotes, histone genes are arranged in multiple tandem domains, each domain carrying one copy of each of the five histone genes. Suggest an explanation for these features in terms of the special requirements for histone synthesis.

6. It has been suggested that the function of the polyA tail on a eukaryotic message may be to "ticket" the message—that is, each time the message is used one or more residues is removed, and the message is degraded after the tail is shortened below a critical length. Suggest an experiment to test this hypothesis.

7. In degradation of proteins via the ubiquitin pathway (Figure 28.36) only steps 1 and 4 are ATP–dependent. Why these two steps and no others?

8. A sample of chromatin was partially digested by the enzyme staphylococcal nuclease. The DNA fragments from this digestion were purified and run on a polyacrylamide gel, in comparison with a set of DNA restriction fragments used as markers. Distances of migration are given below. From these data, estimate the nucleosome repeat distance in the chromatin.

Marker DNA Fragment		Chromatin DNA Fragment
Size (bp)	*d* (cm)	*d* (cm)
94	40	30.5
145	34.2	19.2
263	25.2	14.4
498	16.7	11.5
794	11.5	—

9. Pancreatic deoxyribonuclease I (DNase I) is a nuclease that makes single-strand nicks on double-strand DNA. It has been observed that treatment of nucleosomal core particles with DNase I yields a peculiar result. When DNA from such a digestion is electrophorosed under denaturing conditions, single-strand fragments are observed with a very regular periodicity of about 10 bases. Explain this result in terms of the structure of the nucleosome.

10. Loss of the functional ability to cleave N-terminal methionines from proteins might be expected to have a major influence on the distribution of proteins in growing cells. Explain why.

CHAPTER 29

Information Processing and Expression in Multicellular Organisms

Genetic information, such as we have been discussing over the last four chapters, programs the embryonic development of multicellular organisms and largely dictates the structures of cells and tissues, down to the very macromolecules that each contains. But this kind of information alone is not sufficient to allow an organism, even a unicellular one, to exist in the real world. If it and its genetic lineage are to survive, every organism must be able to sense its continually changing environment, interpret these sensations, and produce appropriate actions. In other words, the life of every organism is conditioned by two kinds of information: the genetic information that it receives from its forebears, and the **environmental information** it receives from the world about it.

Examples are found even in the simplest of organisms. A bacterium, like *E. coli,* that has been feeding on glucose may be transferred to a medium in which lactose is its sole energy source. By transporting some lactose through the cell wall, the bacterium senses the information, and it responds appropriately, turning on the *lac* operon, which is part of its genetic heritage. Slightly more complex are the **chemotactic** responses of some bacteria: they sense and swim toward food sources, and away from noxious conditions. Although these responses need not be learned, they have this in common with the most complex types of animal behavior: information from the world outside the organism is evaluated and properly responded to, using genetically dictated mechanisms.

In multicellular organisms, proper response to external stimuli involves, of necessity, both specialized sense organs and communication of the information from these sensors to other parts of the organism. Such communication is of two kinds. The first is one we have already encountered: the transmission of chemical substances (for example, hormones) from one tissue to another. But the major way of handling such information, at least in higher animals, is through the generation and transmission of impulses through the nervous system. In vertebrates, this system becomes enormously complex—the human brain alone contains over 10^{12} cells. Systems of such complexity are difficult to study by current techniques, and consequently much attention has been given to the simpler nervous systems of lower animals.

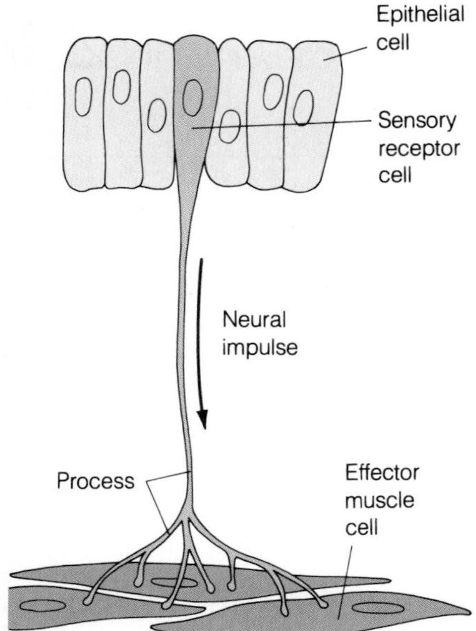

Figure 29.1
A very simple neural system. In the tentacles of the sea anemone sensory cells on the tentacle surface are connected directly by an axon to a muscle cell. A stimulus, such as touch, generates a neural impulse in the sensory cell that travels to the muscle, stimulating contraction.

Neural Systems and Neural Transmission

Structure of Neural Systems

The simplest prototypes of nerve networks are found in primitive creatures like the sea anenome. The anemone's tentacles contract on being touched. This contraction is stimulated by sensory receptor cells in the tentacle epithelium that have extended **processes** reaching to the muscle cells within the tentacle (Figure 29.1). When a sensory cell contacts any object, a neural impulse is transmitted to the muscles. Such a system can only make a direct and simple connection between stimulus and response and is incapable of learning.

In higher organisms, the nervous system is constituted mainly of special cells, called **neurons,** which serve only to transmit impulses. The structure of a typical mammalian neuron is shown in Figure 29.2. The cell body is rather like that of many other somatic cells, but the surface is extended into a variety of projections, or *processes.* Those called **dendrites** receive neural signals, which are passed over the cell surface to the **axon,** whence they are transmitted either to other neurons or to receptor cells in muscles or glands. Such a cell has multiple dendrites and an axon that is usually branched near its terminus, allowing it to make the complex interconnections required for complicated patterns of behavior.

There are three major types of neurons in higher organisms: **sensory neurons, interneurons** and **motor neurons.** Figure 29.3 shows a highly simplified scheme. Most of the sensory and motor neurons are part of the **peripheral nervous system (PNS),** whereas the interneurons are concentrated in the brain and spinal cord, the **central nervous system** (CNS). The CNS acts as an information processing center, where sensory inputs can be integrated and interpreted in a multitude of ways to regulate complex behavior.

Because there is not necessarily a direct connection between stimulus and response in the nervous system of a higher organism, information received in the past can influence responses of the moment, and memory and learning become possible. Marvelous as this system may be, our concern here is neither with neuroanatomy nor animal behavior. Rather, we shall center on the biochemical events accompanying the generation and transmission of neural impulses and how they produce effects in receptor cells like those in muscles and glands. We consider first the basic processes of neural transmission.

Excitable Membranes and Electrical Potentials

The axons and dendrites of nerve cells are remarkable organic structures for the transmission of electrical impulses. The motor neurons that control the muscles in a giraffe's feet, for example, have their cell bodies in the spinal cord, with axons extending for several meters. Yet these must be able to conduct impulses without significant signal loss, rapidly enough to allow the animal to run with agility! Nerve conduction is accomplished not by electron flow, as in wires, but by waves in the membrane electrical potential on the surface of the axon. When such a wave passes a point on an axon or dendrite, the **resting potential** of the membrane is momentarily changed to a moving **action potential** that constitutes the impulse. To understand how this occurs, we must examine how membrane potentials are generated and how they can be changed.

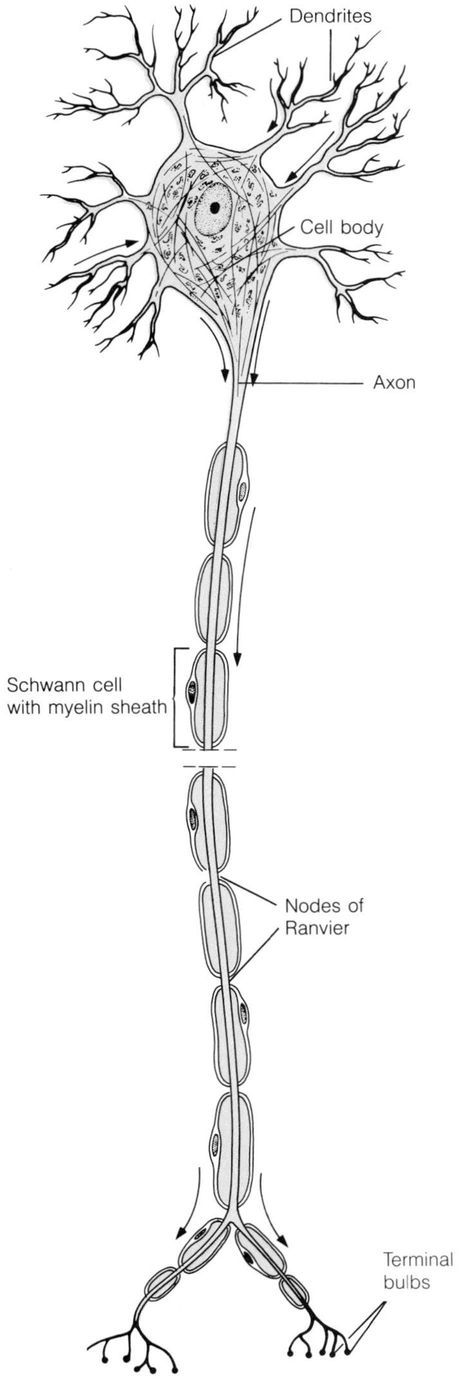

Figure 29.2
Structure of a typical mammalian motor neuron. This is the kind of nerve cell that transmits impulses to muscles or other nerve cells in higher animals. The cell body contains the nucleus and most of the other organelles. From the body project one long axon and many shorter dendrites. The dendrites serve as receptors of signals from other neurons and conduct them into the cell body. The axon transmits signals from this cell to others, via the terminal bulbs, which connect with other dendrites or with cells. Along the axon are Schwann cells, which envelop the axon in layers of an insulating myelin membrane. The Schwann cells are separated by nonmyelinated regions called the nodes of Ranvier.

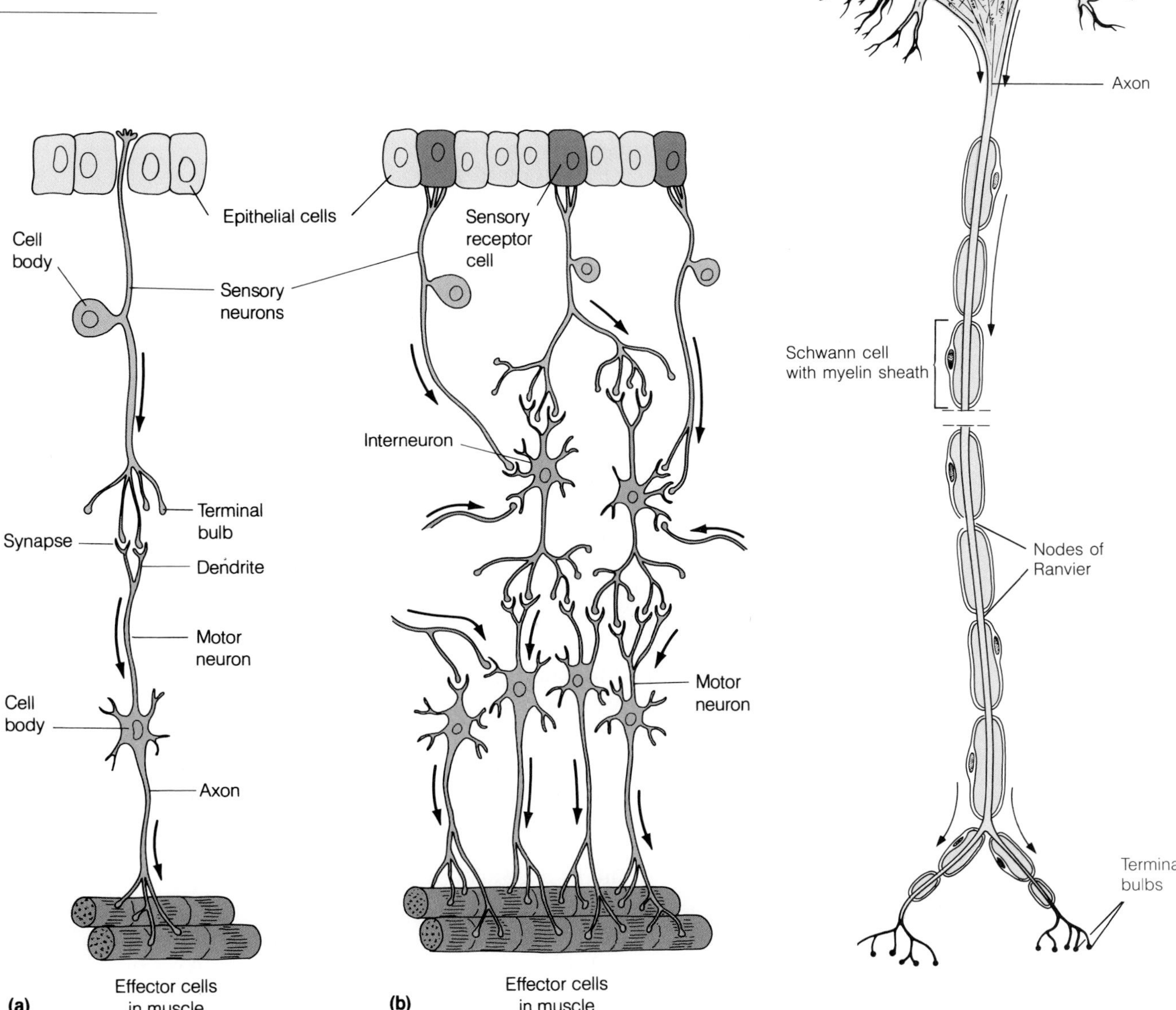

Figure 29.3
Simple and complex neural networks. (**a**) In some simple organisms, a sensory neuron, with receptors on the body surface, will connect directly (via a synapse) to a motor neuron, which transmits the signal to a muscle fiber. (**b**) A more common and complex arrangement is to have multiple connections, from many sensory receptor cells, transmitted via interneurons to effector cells. The interneurons receive impulses from other neurons in the net, which may stimulate or inhibit the signal to the effector.

First, we must understand the source and nature of the resting potential. We begin with an oversimplified model, which draws on our earlier discussion of the electrochemical potential difference across a semipermeable membrane (Chapter 9). Suppose we have an ion (M^Z) of charge Z that is present outside the membrane at concentration $[M^Z]_{out}$ and inside at concentration $[M^Z]_{in}$. As described in Chapter 9, the electrochemical potential on each side is given by

$$\mu_{out} = \mu^0 + RT\ln[M^Z]_{out} + FZ\psi_{out} \tag{29.1a}$$

$$\mu_{in} = \mu^0 + RT\ln[M^Z]_{in} + FZ\psi_{in} \tag{29.1b}$$

where F is the Faraday constant and ψ_{out} and ψ_{in} are the electrical potentials outside and inside the membrane. For the system to be at equilibrium we must have $\mu_{in} = \mu_{out}$, so

$$RT\ln[M^Z]_{out} + FZ\psi_{out} = RT\ln[M^Z]_{in} + FZ\psi_{in} \tag{29.2a}$$

or

$$\frac{RT}{ZF}\ln\frac{[M^Z]_{out}}{[M^Z]_{in}} = \psi_{in} - \psi_{out} = \Delta\psi \tag{29.2b}$$

For monovalent ions ($Z = \pm 1$) at room temperature (20°C), the equation reduces to

$$\Delta\psi = \pm 58\log_{10}\frac{[M]_{out}}{[M]_{in}} \tag{29.2c}$$

when $\Delta\psi$ is expressed in millivolts.

This result can be thought of in two ways. First, if we establish a potential difference across a membrane, any ion that can pass through the membrane can be present at different concentrations on the two sides at equilibrium. Mechanistically, the potential difference will "pull" ions across the membrane. Alternatively, equation (29.2b) says that if we somehow maintain a concentration difference across a membrane, an electrical potential difference will be produced if the membrane is permeable to the ion. For example, if an ion such as K^+ ($Z = +1$) was kept 10 times as concentrated inside as outside, $\Delta\psi$ would be -58 mV. The membrane would be polarized, with a potential 58 mV lower on the inside. If the ion was chloride, the potential would be $+58$ mV. Equation (29.2) is called the **Nernst equation.**

The Action Potential and Transmission on Axons

Two mechanisms tend to create ionic imbalance across almost any cellular membrane. The first is simply the fact that many molecules too large to pass through the membrane carry a net charge. Their counterions will tend to concentrate on the same side of the membrane with them. Second, specific ion pumps, as described in Chapter 9, are continually acting to concentrate certain ions on one side or the other.

This is exactly the situation that gives rise to the resting potential across the membrane of a nerve axon. A much-studied example is the giant axon of the squid. This is a favorite experimental tool, since squids are unusual in having axons as large as 1 mm in diameter. It is possible, as

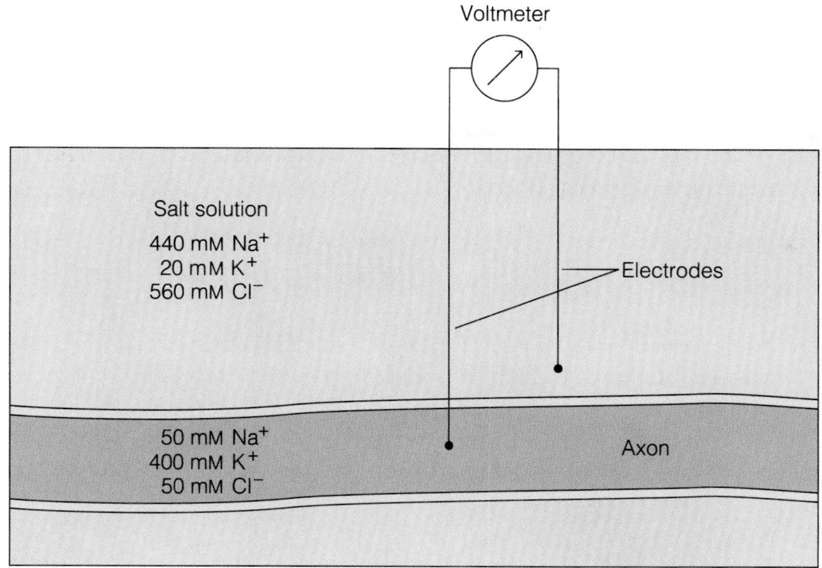

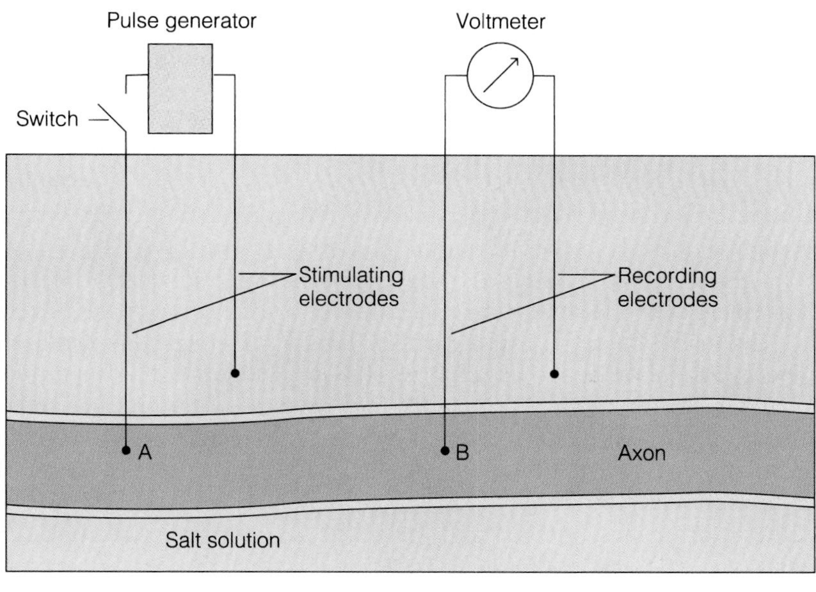

Figure 29.4
Use of squid giant axons for studies of neural transmission. (**a**) Measurement of the resting potential. Electrodes on either side of the axonal membrane record the potential differences. At the resting axon ion concentrations shown here, the voltmeter would read about −60 mV. (**b**) Arrangement for recording an action potential. If the axon is stimulated at point A by a depolarizing pulse, the traveling action potential will shortly pass point B, where it can be recorded.

shown in Figure 29.4, to insert an electrode into such an axon and measure the potential difference across the membrane. As with most cells, the Na^+/K^+ pump maintains large differences in cation concentrations, with K^+ being concentrated and Na^+ depleted inside the membrane. Concentrations in the resting axon are shown in Figure 29.4a.

We now have a situation more complex than can be described by the Nernst equation. Several ions are involved, each of which can pass, at least to some degree, through the membrane and which are maintained in unequal concentrations on the two sides. If we applied the Nernst equation to the distribution of K^+ alone, we would predict a potential of −75 mV (which we will call $\Delta\psi_{K^+}$). On the other hand, using the Nernst equation with the Na^+ concentration, we would find $\Delta\psi_{NA^+} = +55$ mV. When we measure the potential across a resting squid axonal membrane, we find a value of about −60 mV. What determines the actual potential in this case?

The key to the answer can be seen by noting that the *permeability* of the membrane to different ions is important. The various ions are not at true equilibrium across the membrane but are in a steady state, the position of which is determined in part by the individual permeabilities. This steady state can be described quantitatively by the **Goldman equation,** which may be written in the general form:

$$\Delta\psi = \frac{RT}{F} \ln \left\{ \frac{\Sigma_+ P_i[M_i^+]^{out} + \Sigma_- P_j[X_j^-]^{in}}{\Sigma_+ P_i[M_i^+]^{in} + \Sigma_- P_j[X_j^-]^{out}} \right\} \tag{29.3}$$

Here, the sums (denoted by Σ) are taken over all cations (Σ_+) and anions (Σ_-) with significant permeability, and the P's are the relative membrane permeabilities (see Chapter 9) for these ions. Note that if any one ion were to have a *much* greater permeability than any of the others, it would dominate in the Goldman equation, and (29.3) would reduce to the Nernst equation.

Now let us apply the Goldman equation to the situation in the squid giant axon. The only ions that contribute appreciably to the membrane potential in this case are K^+, Na^+, and Cl^-, and their relative permeabilities are $P_K = 1.0$, $P_{Na} = 0.04$, and $P_{Cl} = 0.45$. If these values are inserted into equation (29.3), together with the ion concentrations given in Figure 29.4a, we find $\Delta\psi = -61$ mV, in agreement with experimental observation.

The value of -61 mV has a vital significance. Since -61 mV lies much closer to -75 mV than to $+55$ mV, it means that at the potential existing across the squid axon membrane, K^+ is *much* closer to its equilibrium distribution than is Na^+. If the membrane were to become fully permeable to ions, the major event would be a massive influx of sodium ions, with an accompanying shift in the membrane potentials toward ψ_{Na^+}.

This is just what happens when an action potential is transmitted along a nerve. Axonal fibers have specific, voltage-sensitive "gates" in protein "channels" for the transport of Na^+ and K^+ through the membrane. Suppose we elaborate the experiment shown in Figure 29.4a in the way depicted in Figure 29.4b. At some distance from the recording electrode in the squid axon, we place a stimulating electrode, connected to a voltage source. If we apply at this electrode a pulse sufficient to depolarize the membrane locally by about 20 mV (i.e., to a potential of -40 mV, the threshold for opening the Na^+ channels), the Na^+ channel activation gates are opened. A flood of sodium ions rushes in, bringing the membrane potential up to about $+40$ mV within less than a millisecond. In terms of the Goldman equation (29.3) the high permeability to Na^+ ions makes these dominate, and the membrane potential *approaches* $\Delta\psi_{Na^+}$ ($+55$ mV). It does not reach this value, however, for further changes now occur. The stimulus also causes K^+ channel gates to open, but more slowly, and these pour forth potassium ions into the surrounding medium. This reverses the potential again, overshooting to about -70 V, closing the Na^+ channel inactivation gate (separate from the activation gate), and temporarily making the Na^+ channels refractory toward opening. The potential and permeability changes that occur in this brief period of a few milliseconds are shown in Figure 29.5.

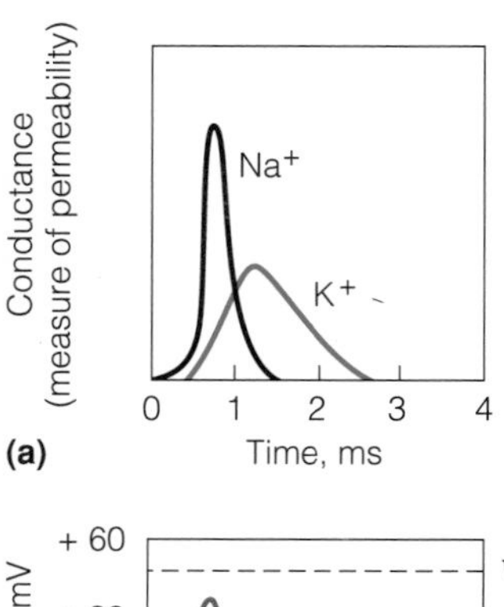

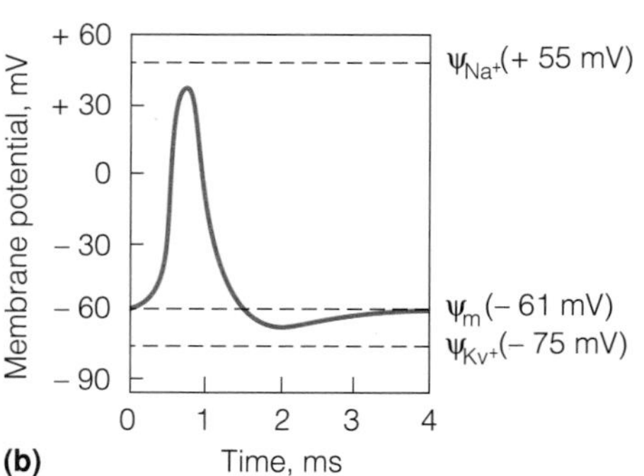

Figure 29.5
The action potential. (**a**) Membrane conductance changes at a point on an axon as a neural impulse passes. The membrane first becomes permeable to sodium ions, allowing a large inward rush of Na^+. This is followed by a decrease in the Na^+ permeability, followed by an outward flow of K^+. (**b**) Changes in the membrane potential accompanying the permeability changes shown in (**a**). As sodium ions rush in, the potential becomes positive. With the potassium flow, the potential decreases and overshoots, before leveling off at the resting potential (ψ_m). The potentials ψ_{Na^+} and ψ_{K^+} are the theoretical values for a membrane potential determined solely by either Na^+ or K^+, respectively.

All of this would be a dramatic, but localized, effect were it not for the phenomenon shown in Figure 29.6. As sodium ions pour in, they diffuse *ahead* of the region of stimulus and trigger the same round of depolarization in an adjacent section of the fiber. Thus a *wave* of depolarization proceeds down the axon. Following the wave front is the reverse polarization due to K^+ efflux, the refractory period when K^+ channels are open and the Na^+

(a) Time = 0

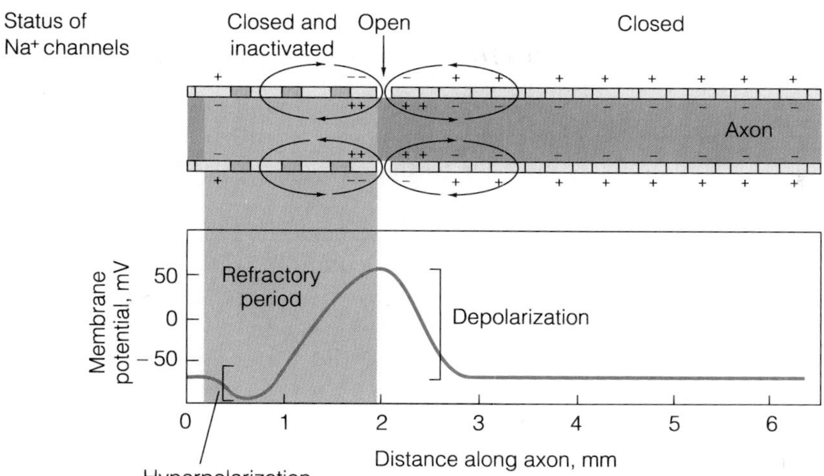

(b) Time = 1 ms

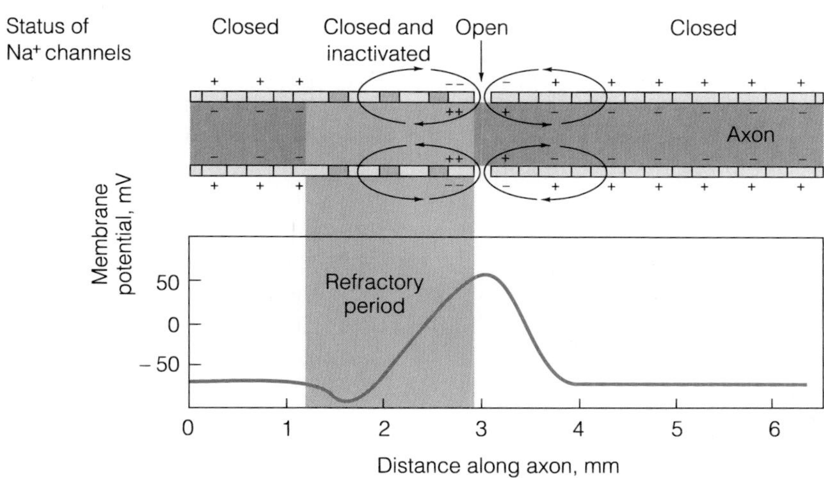

Figure 29.6
Transmission of the action potential. Shown are two "snapshots," taken at an interval of 1 ms, of potential along the axon. Red indicates inactivated channels. Arrows show Na^+ influx and K^+ efflux. (**a**) At time = 0, an action potential is occurring at the 2-mm position. The depolarization spreads down the axon, triggering development of the action potential downstream. (**b**) At time = 1 ms, the action potential peak has moved to the 3-mm position. The potential can move in only one direction, since after it has passed, the region behind the potential becomes refractory for a few milliseconds.

channels are inactivated, and the final return to the resting condition. Thus, the recording electrode in Figure 29.4b will see traveling past it, at some time after the stimulating pulse, exactly the same pattern of depolarization and reversal depicted in Figure 29.6. It is this traveling pulse that is called the *action potential.* The time required for the impulse to pass from stimulating electrode to recording electrode is proportional to the distance between the two and inversely proportional to the velocity of propagation of the pulse. Typical values for propagation of the action potential range from 1 to 100 meters per second.

Several features of the action potential arise directly from the fact that it represents a wave of membrane depolarization; these distinguish it clearly from the kind of conductance by electron flow seen in common electrical circuits.

1. The action potential does not appreciably decrease with distance transmitted. This is because it is continually renewed at each point along the axon.

2. The action potential is an all-or-none phenomenon. If the stimulus is sufficient to activate it, it occurs, and its magnitude is independent of the stimulus voltage as long as the stimulus is above the threshold value required for activation.

3. After an impulse has passed, the region of axon immediately behind it is unable to transmit another impulse for a time period of some milliseconds due to the refractory period.

These three features account for some of the peculiar properties of the neural systems of animals. Because of feature 2, neural networks are more like digital than analog circuits. A neuron fires, or it does not. Furthermore, intense stimuli do not give larger action potentials—they simply result in more frequent impulses. But because of the refractory period (feature 3) a neural system can become "saturated"; it can handle only so many impulses per second.

The properties of the axon itself have a major effect on how rapidly a single pulse can pass along it. Basically, the velocity depends on how rapidly the region of the axon ahead of the pulse can be depolarized to the threshold value necessary to open the Na^+ channel gates. This turns out to depend on two factors—the electrical *resistance* of the axon core and the electrical *capacitance* of the membrane. If either the resistance or the capacitance is low, Na^+ ions can spread rapidly ahead of their point of entry, and the pulse will move quickly. The resistance to Na^+ movement can be decreased by using a big axon; the capacitance can be decreased by insulating the axon.

Both approaches to increasing the speed of neural transmission have been utilized in different animals. Most invertebrates, and particularly mollusks, are sluggish creatures, but the squid is an exception. To put on a burst of speed, it must be able to expel water quickly by contraction of its mantle cavity. The axons that control the appropriate muscles are very large, allowing a fast response. Vertebrates have taken the insulation approach. Many vertebrate axons, particularly in the peripheral nervous system, are covered over most of their length by a discontinuous sheath of *myelin*, produced by special cells called *Schwann cells*. The myelin sheath is a wrapping of many layers of membrane about the axon, which insulates it and decreases its capacitance (Figure 29.7). The insulation is interrupted periodically at the **nodes of Ranvier** (see Figure 29.2). These nodes are the only points in a myelinated axon at which exchange of ions with the surroundings can occur; the Na^+ and K^+ channels are all concentrated here. In such an axon, the action potential jumps rapidly from one node to another, in a process known as **saltatory** (i.e., jumping) conduction. The efficiency of myelination

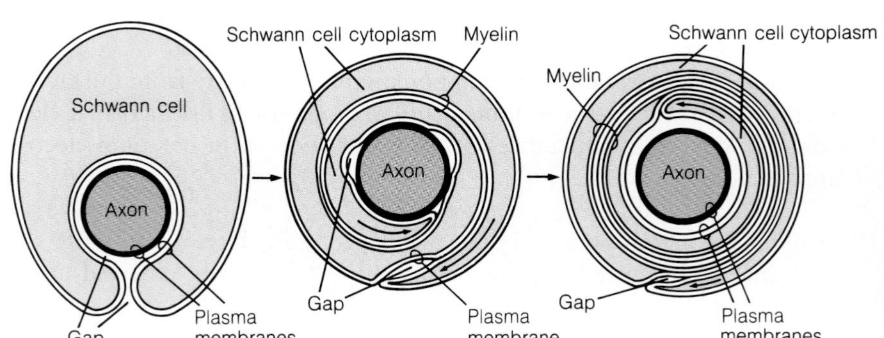

Figure 29.7
How an axon is myelinated. During neural development, Schwann cells envelop the axon. These cells then grow in a spiral fashion, wrapping many layers of the myelin membrane about the axon. As pointed out in Chapter 9, these membranes are unusually low in protein content and correspondingly high in lipids; therefore they make excellent insulators. The Schwann cells are spaced regularly along the axon, separated by the nodes of Ranvier, as shown in Figure 29.2. For an electron micrographic view of a myelinated axon see Figure 18.13 in Chapter 18.

can be judged from the fact that a 12-μm-diameter myelinated axon conducts as rapidly as a 600-μm-diameter, nonmyelinated squid axon.

Certain neurological diseases are associated with loss of myelin, the best known being **multiple sclerosis.** In this condition there is a gradual and still unexplained destruction of myelin from certain areas of the brain and spinal cord. Conduction of action potentials is slowed, with an accompanying loss of motor coordination.

Many extremely toxic substances have their effect by blocking the action of the specific ion gates necessary for development of the action potential. **Tetrodotoxin** is found in some organs of the puffer fish. This fish is considered a delicacy in Japan, where special chefs are trained and certified for their ability to remove the toxin-containing organs. Tetrodotoxin binds specifically to the Na^+ channel, blocking all ion movement. The same effect is produced by **saxitoxin,** contained in those marine dinoflagellates responsible for the "red tide." These microscopic algae are ingested by shellfish and can in turn be consumed by humans. These two toxins, which attack a fundamental process of the nervous system, are among the most poisonous substances known, and their accidental ingestion leads to many deaths every year. A third very poisonous substance, **veratridine,** is found in the seeds of a plant of the lily family, *Schoenocaulon officinalis.* This toxin also binds to the Na^+ channels, but blocks them in the "open" configuration.

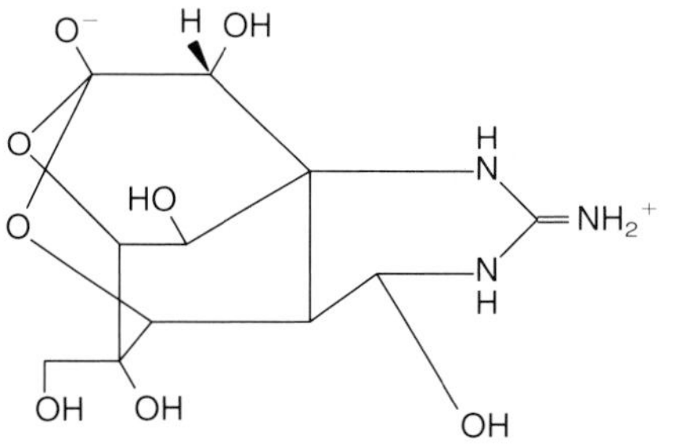

Tetrodotoxin

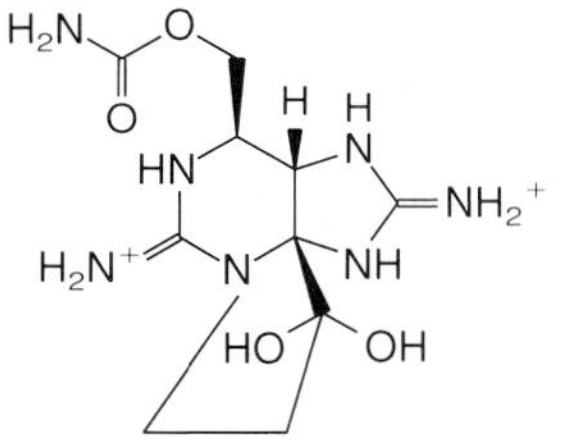

Saxitoxin

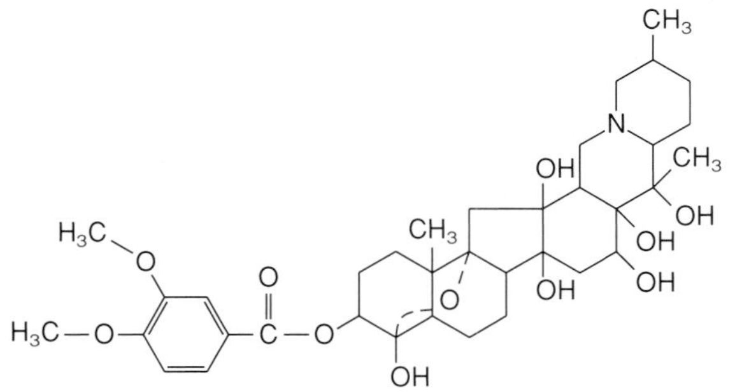

Veratridine

These toxins have proved to be very useful in studies of axonal structure and conduction, for their tight binding makes them excellent affinity labels for the Na^+ channel. Their use has enabled researchers to determine that nonmyelinated axons contain about 100 channels per square micrometer, whereas the nodes of Ranvier in myelinated fibers have a density about 200 times greater.

Such studies have revealed that the Na^+ channels are typically composed of one large (α) subunit, of mass about 260–300 kDa, plus one or more smaller (β) subunits of mass 30–40 kDa. In some cases the β subunits are missing, and it is clear that most of the functional properties of the channel are associated with the α subunit. Recently, cDNA clones for the α subunit have been obtained, and the amino acid sequence has been deduced. The protein is found to contain four hydrophobic domains, which probably form membrane-spanning walls about a central pore. Some segments of these domains contain an unusual repeating sequence in which every third residue is lysine or arginine, separated by two hydrophobic residues. Such sequences are suspected to be part of the gating mechanism, susceptible to conformational change when the membrane potential is altered.

Transmission Across Synapses

So far we have described the basic mechanism whereby a signal, in the form of an action potential, is transmitted through a single nerve cell. But how do cells in the neural network pass these signals from one to another or to other cells such as muscle cells, in which a response is to be evoked? Figure 29.3 shows axonal processes of one cell making connections with dendrites of other cells, or with other neural cell bodies, and with muscle cells. In this section we will examine these connections, which are called **synapses,** in detail.

Figure 29.8 depicts one common kind of synapse. The **terminal bulb** (also called the **synaptic knob**) of the **presynaptic axon** does not make direct contact with the **postsynaptic dendrite.** Rather, the terminal bulb of the axon and the dendrite are separated by a **synaptic cleft** about 20 nm wide. The electrical impulse is not transmitted directly across the synaptic cleft, but is carried by a chemical **neurotransmitter.** In the synapse depicted, the neurotransmitter is **acetylcholine;** therefore, this is termed a **cholinergic synapse.** The nerve impulse (action potential) moves down the presynaptic axon to the terminal bulb; the change in membrane potential in the bulb causes the opening of voltage-gated calcium channels, allowing Ca^{2+} ions to pass from the surrounding space into the axonal bulb. Within the bulb are **synaptic vesicles,** each containing about 10^3 to 10^4 acetylcholine molecules. The increase in Ca^{2+} concentration causes these vesicles to fuse with the axonal membrane, open, and expel their contents into the synaptic cleft by exocytosis. The postsynaptic membrane of the receptor dendrite carries specific acetylcholine receptors, to which the neurotransmitter diffuses. Binding of acetylcholine to the type of receptors shown in Figure 29.8 triggers the opening of channels in the postsynaptic membrane, initiating an action potential that can be passed on to the next axon.

The receptors involved in this kind of transmission of a neural impulse are referred to as **nicotinic** acetylcholine receptors, because they can bind

Figure 29.8
Transmission of a neural impulse across a cholinergic synapse. (**a**) Movement of the incoming action potential into the bulb of the presynaptic axon opens Ca^{2+} channels in the membrane, allowing Ca^{2+} to flow in. (**b**) The increased Ca^{2+} causes acetylcholine-containing vesicles to fuse with the presynaptic membrane, spilling the neurotransmitter, acetylcholine, into the synaptic cleft. (**c**) Acetylcholine diffuses across the cleft and is picked up by acetylcholine receptors in the postsynaptic membrane of the dendrite. (**d**) This produces local depolarization of the membrane, which may or may not be sufficient to trigger an action potential in the postsynaptic neuron. The action potential triggers only *if* the summation of excitory and inhibitory impulses reaching the postsynaptic neuron is above the threshold potential. (**e**) The action potential then propagates to the cell body, other dendrites, and down the axon.

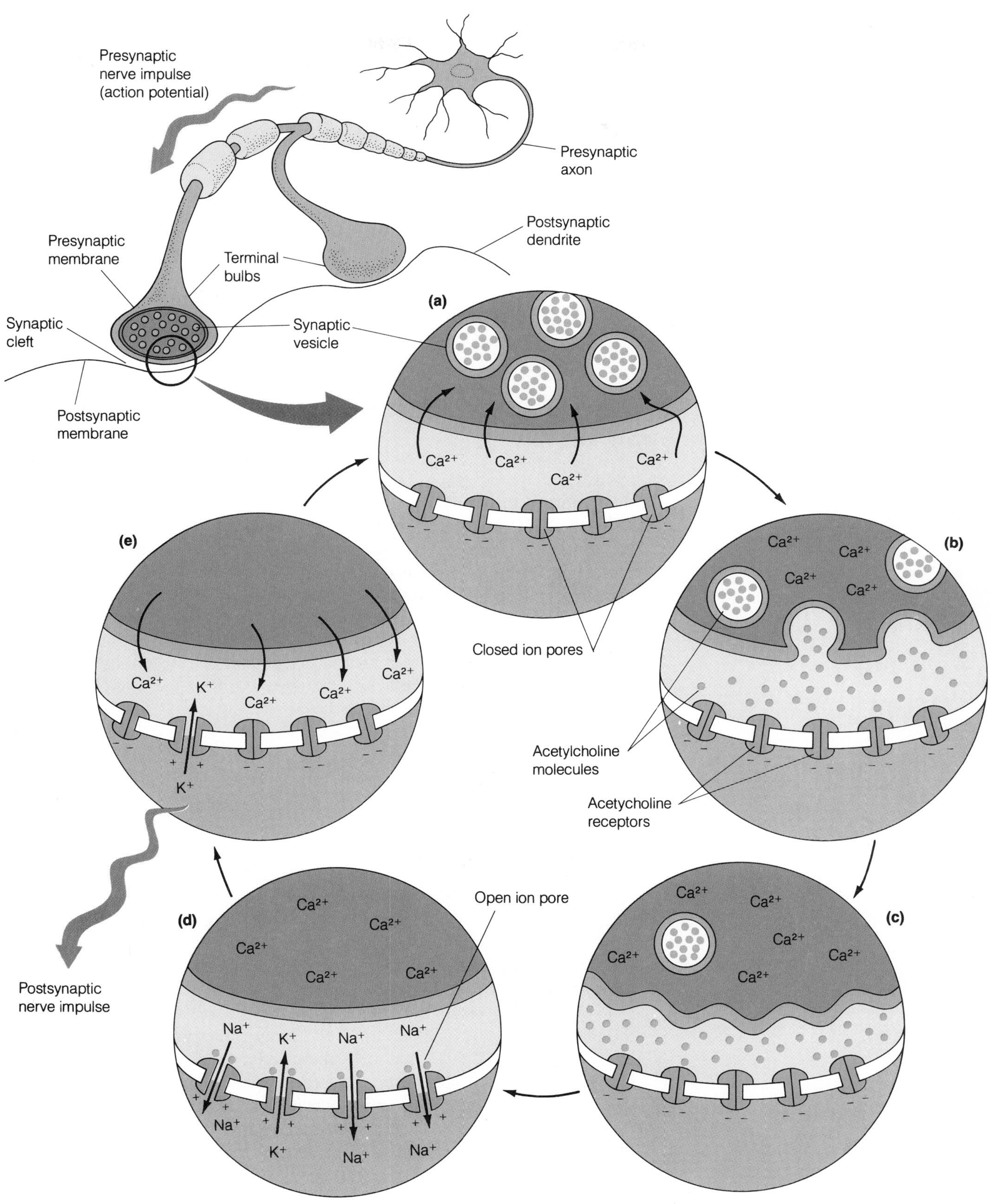
Presynaptic
nerve impulse
(action potential)
Presynaptic
axon
Postsynaptic
dendrite
Presynaptic
membrane
Terminal
bulbs
Synaptic
cleft
Synaptic
vesicle
Postsynaptic
membrane
(a)
Ca2+
Ca2+
Ca2+
Ca2+
Closed ion pores
(b)
Ca2+
Ca2+
Ca2+
Ca2+
Acetylcholine
molecules
Acetycholine
receptors
(c)
Ca2+
Ca2+
Ca2+
Ca2+
Ca2+
Ca2+
(d)
Open ion pore
Ca2+
Ca2+
Ca2+
Ca2+
Ca2+
Na+
K+
Na+
Na+
Na+
K+
Na+
Na+
(e)
Ca2+
K+
Ca2+
Ca2+
Ca2+
K+
Postsynaptic
nerve impulse

nicotine. As we shall see, there are other kinds of acetylcholine receptors, which convey quite different properties to their synapses.

Acetylcholine is synthesized from choline and acetyl-CoA in the axonal terminal bulbs, a reaction catalyzed by the enzyme **choline acetyltransferase:**

$$\underset{\textbf{Acetyl-CoA}}{H_3C-\overset{\overset{O}{\|}}{C}-S-CoA} + \underset{\textbf{Choline}}{HO-CH_2CH_2\overset{+}{N}(CH_3)_3} \xrightarrow[\text{Choline acetyltransferase}]{\text{HS-CoA}} \underset{\textbf{Acetylcholine}}{H_3C-\overset{\overset{O}{\|}}{C}-O-CH_2CH_2\overset{+}{N}(CH_3)_3}$$

After it has been released from vesicles and bound to the receptors, the neurotransmitter is rapidly hydrolysed by an enzyme present in the synaptic cleft, **acetylcholinesterase:**

$$\underset{\textbf{Acetylcholine}}{H_3C-\overset{\overset{O}{\|}}{C}-O-CH_2CH_2\overset{+}{N}(CH_3)_3} + H_2O \xrightleftharpoons[\text{Acetylcholinesterase}]{} \underset{\textbf{Acetate}}{H_3C-\overset{\overset{O}{\|}}{C}-O^-} + \underset{\textbf{Choline}}{HO-CH_2CH_2\overset{+}{N}(CH_3)_3} + H^+$$

As the product of hydrolysis, choline is a very poor neurotransmitter. Degradation of acetylcholine restores the resting potential in the postsynaptic membrane.

In order to ready the synapse for another impulse, the empty synaptic vesicles, which are returned to the axonal terminal bulb by endocytosis, must be refilled with acetylcholine. This is accomplished by an **acetylcholine transporter** protein, which brings newly synthesized acetylcholine into the vesicles by exchanging it for protons. The protons are pumped into the vesicles by a **vacuolar ATPase.** As they return to the cytosol, acetylcholine is transported in the opposite direction.

Viewed directly from the surface, the nicotinic acetylcholine receptors appear in the electron microscope to be doughnutlike structures extending through the membrane and containing a central pore (Figure 29.9). Isolation of the receptors by affinity chromatography shows them to be pentamers made of four kinds of glycoprotein subunits (molecular masses 54, 56, 58, and 60 kDa, in molar ratio 2:1:1:1). Reconstitution of receptors into lipid vesicles provides a system that can conduct ions when stimulated by acetylcholine. *Thus, it appears that the receptor and ion channel are a single unit.* The central pore presumably functions as the gated ion channel.

Genes for many of the components of the synaptic system have now been cloned, including choline acetyltransferase, acetylcholinesterase, and components of the nicotinic receptors.

Although postsynaptic membranes are typically packed densely with receptors (of the order of 20,000 per square micrometer), these regions constitute only a small fraction of the cellular surface in a typical neural tissue. Therefore, biochemists often turn to more specialized tissues for study of synaptic processes and biochemistry. Favorite experimental subjects for such studies are the electric organs of the electric ray (*Torpedo*) and the electric eel (*Electrophorus*). These organs contain stacks of cells called **electroplaques,** which have a high density ($10^5/(\mu m)^2$) of nicotinic acetylcholine receptors extending over one whole face of the cell. Depolarization of the membrane on this face, while the other face remains at the resting

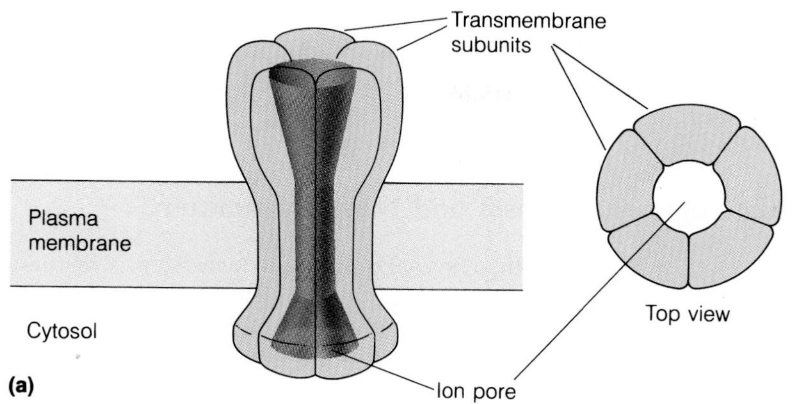

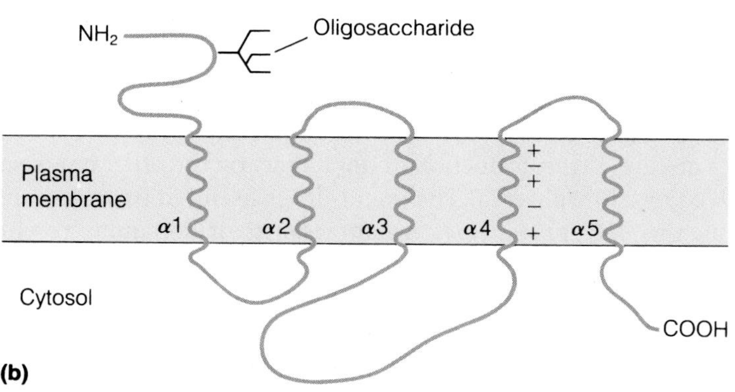

Figure 29.9
The nicotinic acetylcholine receptor. (**a**) Schematic model of the receptor. Five subunits combine to form a transmembrane structure, with an ion pore in the center. (**b**) There are four different kinds of subunits, but their sequences are all similar, and each has the kind of structure depicted here. Five α-helixes (α-1 to α-5) in each subunit traverse the membrane. The charged residues on helix α-4, which tend to be on one surface, probably line the wall of the pore.

potential, gives a $\Delta\psi$ of 130 mV across the cell. With thousands of cells stacked in series, potentials of several hundred volts are generated.

Inhibition of the Cholinergic Synaptic Transmission

Acetylcholinesterase is a serine esterase and consequently can be irreversibly inhibited by reagents that react with the active site serine. We encountered a number of these in Chapter 11—diisopropyl fluorophosphate, sarin, physostigmine, and parathion, for examples (see Table 11.1). As you might expect, acetylcholinesterase inhibitors are extremely toxic substances.

Another class of toxins acts on the acetylcholine receptor itself either by blocking it (*d*-tubocurarine, from curare, and the small protein toxins in some snake venoms, like cobratoxin) or by locking the ion channels open (*nicotine*). The former class are called **antagonists,** the latter **agonists.**

Tubocurarine, an antagonist

Nicotine, an agonist

Inhibitors of synaptic transmission have been very useful for biologists and biochemists seeking to understand synaptic function. Some are also of considerable utility in medicine—for example, as muscle relaxants, when used in carefully regulated doses.

Other Kinds of Synapses and Neurotransmitters

We have concentrated on the cholinergic synapse because it is the best understood at the present time. But there are *many* other substances known to be or suspected to be neurotransmitters, which function in different kinds of synapses. A partial listing of confirmed or probable transmitters is given in Table 29.1.

One major class of neurotransmitters consists of the *catecholamines,* tyrosine derivatives which include *dopamine, norepinephrine,* and *epinephrine* (see Chapter 21). Since epinephrine is also an adrenal hormone, the nerves whose synapses use catecholamines are called *adrenergic.* The importance of dopamine in neural transmission is emphasized by the fact that a number of major neurological diseases seem to be associated with improper dopamine regulation. Underproduction of dopamine in the brain leads to Parkinson's disease. Overproduction of dopamine, on the other hand, seems to be involved in schizophrenia. The major drugs useful in treatment of this disorder, such as *chlorpromazine,* are antagonists of dopamine receptors. The synthesis of these biogenic amines has been described in detail in Chapter 21.

Other major neurotransmitters are themselves amino acids or are derived from amino acids; these include glutamate, glycine, *serotonin* (from tryptophan), *histamine* (from histidine) and *γ-aminobutyric acid* (*GABA,* from glutamic acid). The biochemistry of these compounds also has been fully described in Chapter 21.

Table 29.1
Major neurotransmitters

Name	Function
Acetylcholine	Utilized for fast responses
Catecholamines Epinephrine Norepinephrine Dopamine L-Dopa Octopamine	Utilized in adrenergic synapses of brain, smooth muscle
Amino acids and their derivatives Glutamate Glycine γ-Aminobutyric acid (GABA) Histamine Serotonin	Some of these (i.e., GABA and glycine) are used primarily in inhibitory signals
Peptides Neurotensin Somatostatin Enkephalins	Neurotensin and somatostatin function only as neurotransmitters; the enkephalins act as neurohormones as well

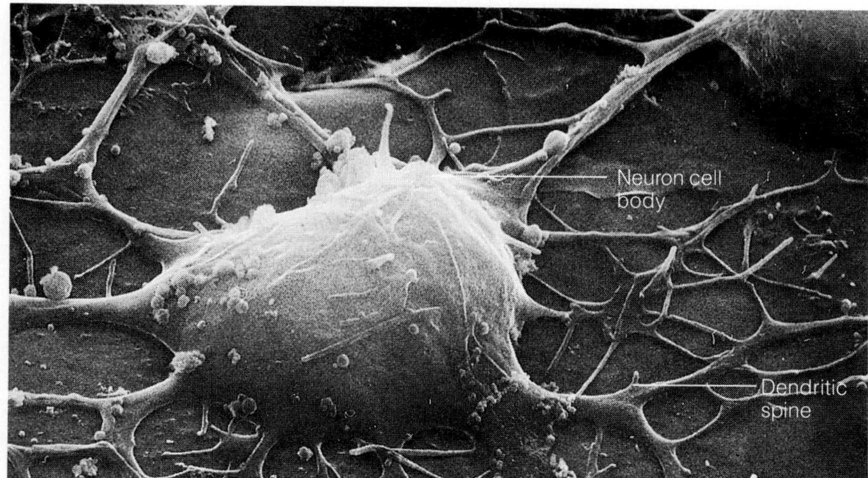

Figure 29.10
Multiple synapses on the body of a single neuron. This scanning electron micrograph gives an idea of the complexity of interconnection in the nervous system. Some of these synapses will be stimulatory, others inhibitory.

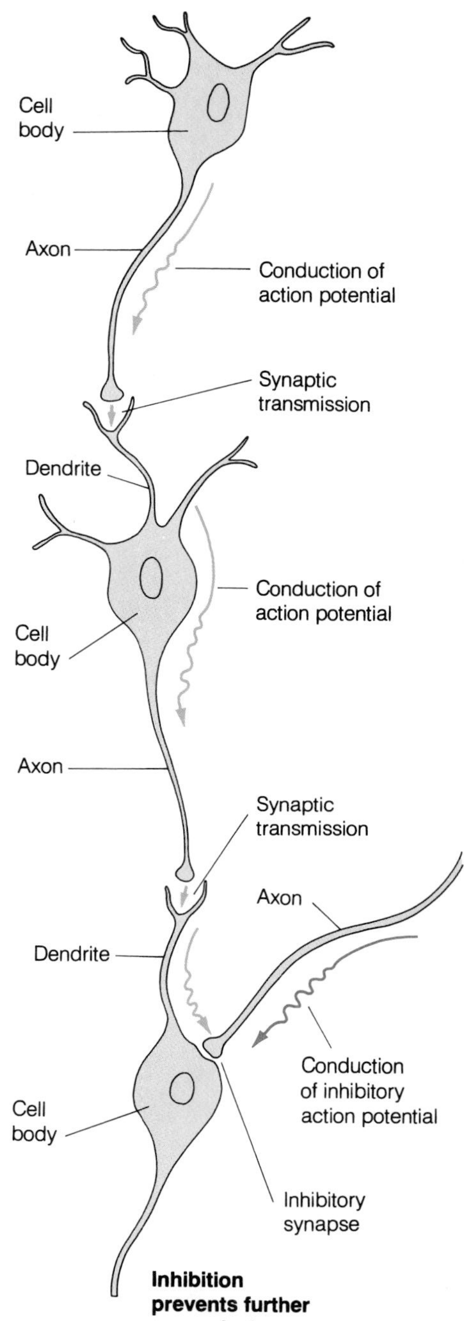

Figure 29.11
Transmission and inhibition in neural networks. In the hypothetical example shown, an action potential passes from the axon at the top, through the neuron in the center, to reach the neuron at the bottom. However, another signal has generated an inhibitory response at a synapse on the lower cell body. This blocks further transmission of the action potential.

The necessity for so many types of neurotransmitters and corresponding kinds of synapses probably derives from the complexity of vertebrate neural systems. Different synapses and different neurotransmitters have quite different properties. Some are rapid in their action, some are slow. Some, such as the nicotinic cholinergic synapses, are stimulatory and urge the neuron to transmit the impulse. Others (often those using GABA as a transmitter) are inhibitory; an impulse received at these synapses will *discourage* the transmission of another signal through the recipient neuron. This can occur, for example, by the opening of chloride channels. Whether the neuron fires will depend on the net summation of stimulatory and inhibitory inputs. A second class of receptor for acetylcholine, called the *muscarinic acetylcholine receptor,* can be inhibitory (though not by the Cl^- channel mechanism). Thus, whether acetylcholine is stimulatory or inhibitory depends on the type of receptor to which it binds. Inhibitory synapses play a major role in the regulation of neural transmission. A typical nerve cell, such as the motor neuron shown in Figure 29.10, will receive input from the axons from many different neurons. Figure 29.11 is a schematic illustration of transmission and inhibition in a neural network.

Certain small peptides, such as *somatostatin, neurotensin,* and the *enkephalins* also act as neurotransmitters (see Tables 29.1 and 29.2). In some cases, these compounds exhibit a second function as neurohormones, which will be described in the following section.

Finally, there are even specialized synapses that do not use neurotransmitter substances at all. Although transmission through a synapse via neurotransmitters can be quite rapid (about a millisecond), some responses must be even quicker than this will allow. In such cases, there is direct electrical–ionic conduction between neural cells, using gap junctions (see Chapter 9). Such **electrical synapses** are frequently found in animals that live in cold environments yet need to make quick motions. The low body temperatures of such creatures would slow the chemical and diffusion processes involved in chemical synapses; direct conduction provides a solution to this problem.

Table 29.2
Some peptides that act as neurohormones (H) or neurotransmitters (T)

Name	H/T	Sequence
β-Endorphin	H	YGGFMTSFKSQTPLVTLFKNAIIKNAYKKGE[a]
Met-enkephalin	H, T	YGGFM
Leu-enkephalin	H, T	YGGFL
Neurotensin	T	pELYENKPRRPYIL[b]
Somatostatin	T	AGCKNFFWKTFTSC

[a] The subsequence YGGF, common to β-endorphin and the enkephalins, appears to be essential for their narcotic effects.

[b] The p at the N-terminal end of neurotensin signifies that the glutamate has been cyclized to the "pyro" form.

CH_3 N HO O OH

Morphine

Neurohormones

Whereas transmitter substances carry out the actual passage of a signal across a synapse, other substances in nervous systems modify the way in which nerve cells respond to transmitters. These substances include peptides called *neurohormones.* The discovery of one class of neurohormones has an interesting history. Researchers studying opium addiction had detected the presence of so-called opiate receptors in brain tissue. Since it seemed unlikely that vertebrates would contain in their brains specific receptors directed toward a product of the poppy plant, a search was made for compounds present in the brain itself that recognized these sites. The search led, in the 1970s, to the finding of several small peptides, called *enkephalins* and *endorphins,* that are natural analgesics (see Table 29.2). The modification of neural signals by these substances appears to be responsible for the insensitivity to pain that is experienced under conditions of great stress or shock. The effectiveness of opiate analgesics such as **morphine** is an accidental consequence of the fact that these substances are also recognized by neurohormone receptors, despite their very different structure.

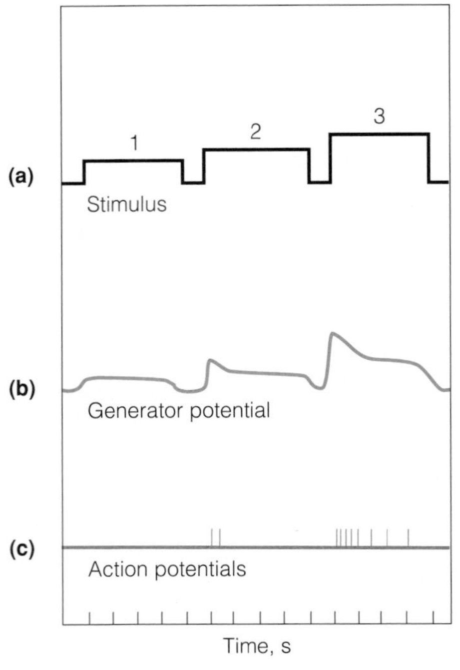

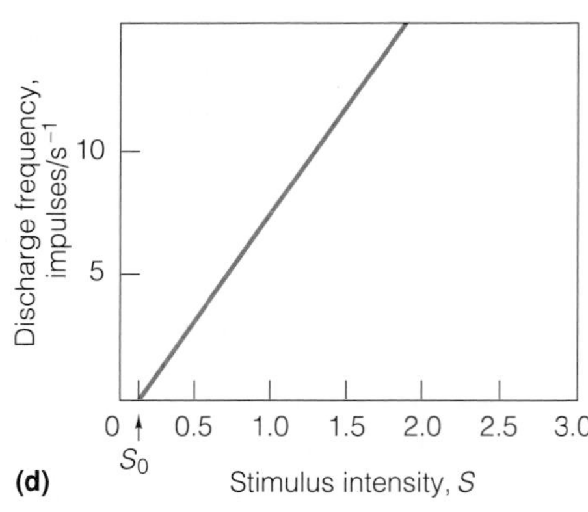

Figure 29.12
Stimulus and action potential. In (**a**) are shown three stimuli of increasing magnitude. In a sensor cell, these will evoke the generator potentials shown in (**b**). If the generator potential is above a threshold value, it will produce action potential pulses in the sensory neurons. In (**c**) it is seen that stimulus 1, which is below threshold, produces no pulses, whereas 2 and 3 produce a small number and a large number of pulses, respectively. There is, as shown in (**d**), a linear relationship between pulse frequency and stimulus intensity S above the threshold S_0.

The endorphins and enkephalins are synthesized as part of the much longer hormone percursor, prepro-opiomelanocortin. As described in Chapter 23, this precursor is cleaved to release both the neurohormones and a number of other hormones with entirely different functions (see Figure 23.9).

Sensory Transduction

Many of the signals transmitted through an animal's nervous system have their origins in stimuli received by the organism from its environment. In order to translate external stimuli into neural impulses, **sensory transducers,** or **sensors,** are required. Most animals have specialized kinds of sensors for a wide variety of stimuli—heat, cold, pain, touch, taste, smell, sound, and light, for example. Other, internal, sensors provide information about the positioning of the body and the state of internal organs.

All of these sensors share certain common features in their transduction of stimuli into nerve impulses. A stimulus (or sometimes a change in stimulus) produces a **generator potential** in the sensory cells by permitting membrane depolarization through altered ion flow. If this depolarization voltage is above a threshold value, it generates action potentials in the connecting sensory neurons. Stimulus intensity is reflected in the number and frequency of action potentials generated (Figure 29.12). In this way, the neural system of the organism is informed that there is a specific kind of stimulus worth noting and what its intensity is.

The Visual System

The sensory transduction system about which we know the most is that used for vision in vertebrate animals; consequently, we shall use it as an example. In the eye structure common to most higher animals, the **lens** focuses an image of the surroundings on the **retina.** The retina is a layer of cells, tightly packed with phototransducers. These are of two kinds: **rod cells** and **cone cells.** The rod cells are specialized for detection of low light levels, whereas the cone cells operate in bright light and in addition are specialized for color vision. Creatures like owls, for which acute night vision is critical and color distinction is less important, possess only rod cells.

A schematic drawing of a rod cell is shown in Figure 29.13. The **inner segment** of the cell has the structures found in other somatic cells—a nucleus, mitochondria, etc. But in addition, since it is a neural cell, the basal portion is a synaptic body, connecting to neurons in the retina. The **outer segment** is a specialized photosensitive system consisting of lamellar membrane disks. These membranes are rich in the protein **opsin,** which carries, as a covalently bound prosthetic group, the pigment **11-*cis*-retinal.** The complex of 11-*cis*-retinal and opsin is called **rhodopsin.** The retinal is bound to a lysine residue in opsin by a Schiff base linkage (Figure 29.14, step 1). When bound in this fashion, the retinal moiety exhibits very strong absorption in the region of the spectrum from 400 to 600 nm. This is, of course, why we humans refer to this as the "visible" region of the spectrum.

Absorption of a photon of radiation by 11-*cis*-retinal triggers a chain reaction of events. The excited retinal molecule undergoes conversion to an all-*trans* form (Figure 29.14, step 2). The retinal moiety then passes through a series of conformational changes, ultimately resulting in deprotonation of the Schiff base and release of the all-*trans*-retinal (steps 3 and 4). The retinal moiety, after release from the opsin, is isomerized in the dark to 11-*cis*-reti-

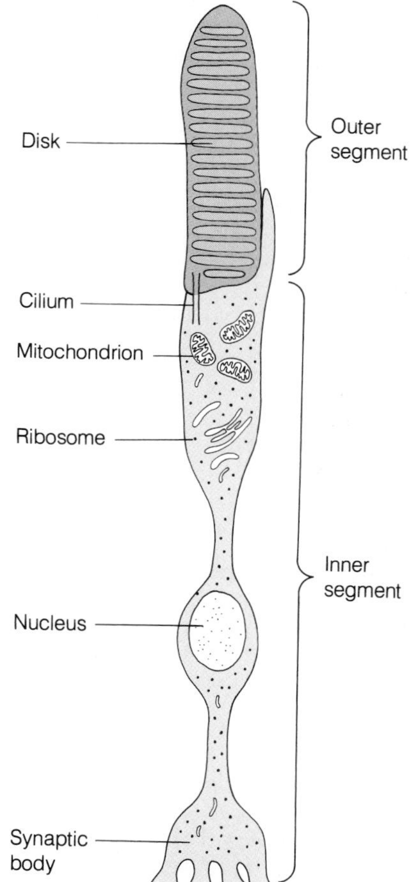

Figure 29.13
Schematic drawing of a rod cell. The outer segment is a stack of membranous disks, which contain the photoreceptive pigments. This segment is connected, via a thin cilium, to the inner segment, which contains the cell nucleus, cytosol, and synaptic body. The potential change produced in the outer segment travels to the synaptic body and is transmitted to one or more of the neurons of the retina (see Figure 29.17).

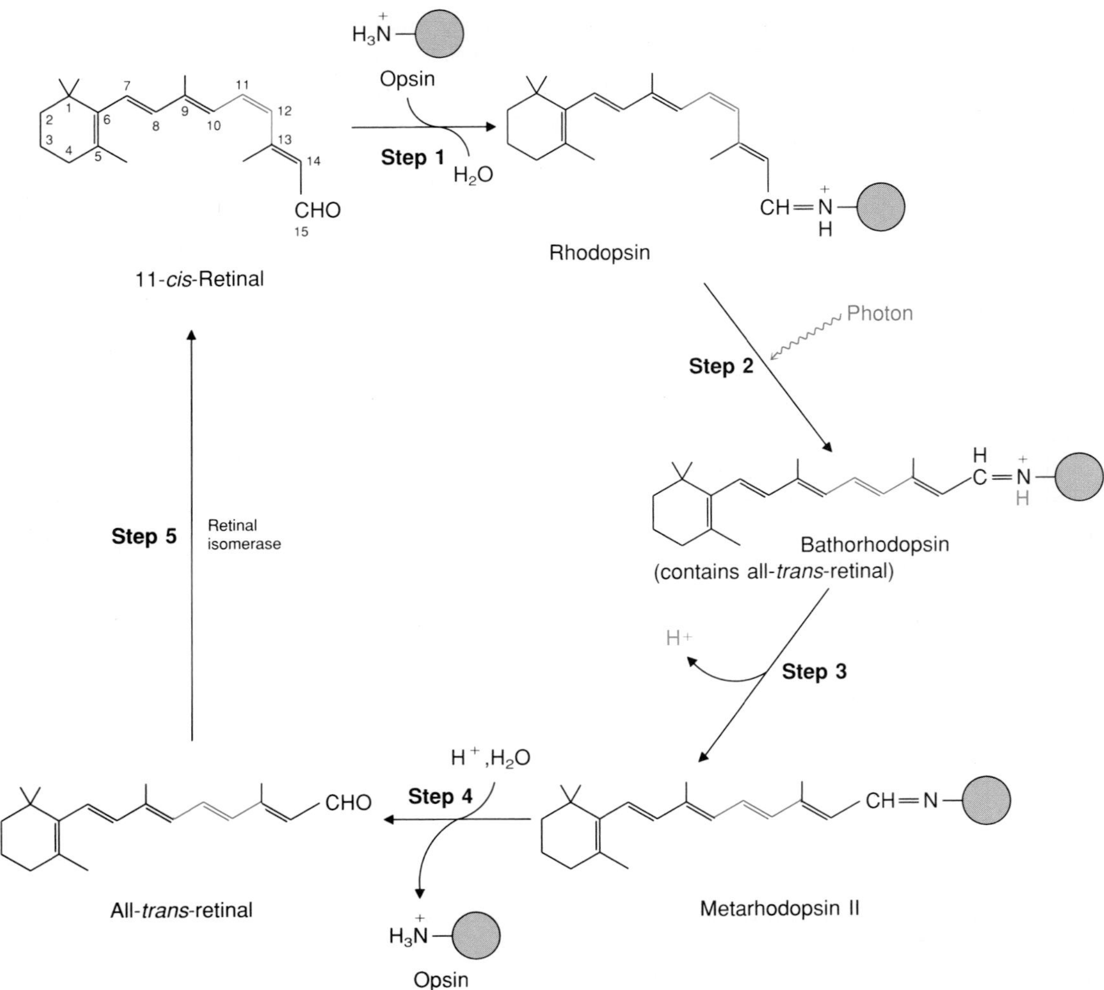

Figure 29.14
The chemical changes in photoreception. The compound 11-*cis*-retinal is present in the unstimulated photoreceptor as a Schiff base conjugate of the protein opsin, forming the complex called rhodopsin. On absorbing a photon, the 11-*cis*-retinal moiety is isomerized to all-*trans*-retinal. This goes through a series of intermediates, culminating in metarhodopsin II, in which the Schiff base is deprotonated. Metarhodopsin II is the form that activates transducin to initiate the cascade process described in the text. After about 1 second, metarhodopsin II dissociates into all-*trans*-retinal plus opsin. The all-*trans*-retinal is then isomerized back to 11-*cis*-retinal, which recombines with an opsin molecule to re-form rhodopsin.

nal by the enzyme **retinal isomerase** (step 5). The *cis* isomer can then be recaptured by another opsin molecule to reconstitute rhodopsin. The basic mechanism of rhodopsin excitation was first described by George Wald and co-workers in studies for which Wald received the Nobel Prize in 1968.

It is the activation of rhodopsin to the form metarhodopsin II (Figure 29.14, step 3) that is the key to the transduction of photon reception to a neural action potential. Only recently have the details become clear. The process depends on a peculiar property of the photoreceptor cell membrane. Recall that most neural membranes have resting potentials in the neighborhood of −60 mV, dominated by the K^+ distribution. But the membranes of rod and cone cells are appreciably "leaky" toward sodium, so their resting potential is only about −30 mV. Absorption of photons by these cells, with the accompanying activation of opsin, has the effect of *closing* Na^+ channels. The membrane potential thereby becomes more negative, changing in the direction of that dictated by the K^+ distribution (−75 mV; see above). Therefore, absorption of a pulse of light will produce the kind of hyperpolarization shown in Figure 29.15.

The mechanism that accomplishes this utilizes cyclic GMP. This compound acts to keep the Na^+ channels normally open, so that a decrease in the level of cyclic GMP will close channels and make the potential more negative. Activated opsin is capable of activating a protein called

transducin. Transducin is a G protein (see Chapter 23), and its activation by metarhodopsin II involves exchange of GTP for the GDP on inactive transducin (Figure 29.16). Activated transducin in turn activates a cGMP phosphodiesterase, which causes the level of cyclic GMP in the rod cell to decrease dramatically and close the Na^+ channel. This cascade process is so effective that a single photon can block the influx of more than 1 million Na^+ ions.

The result of this excitation is a pulse of hyperpolarization (a negative signal) in the light receptor cell. This appears to pose a paradox, for we have considered in preceding sections only the transmission of *positive* signals. To understand how excitation is passed to the ocular nerves, we must examine the neuroanatomy of the retina in a bit more detail. As Figure 29.17 shows, there are three major layers of neural cells in the retina: the rods and cones, the bipolar neurons, and the ganglial neurons, which connect to the cptic nerve. The hyperpolarized pulses from the rods and cones decrease the release of inhibitory transmitter to the bipolar cells; thus they evoke an opposite, excitatory, response in these cells. This is called **impulse inversion.** The excitatory response is transmitted to the ganglial neurons, resulting in

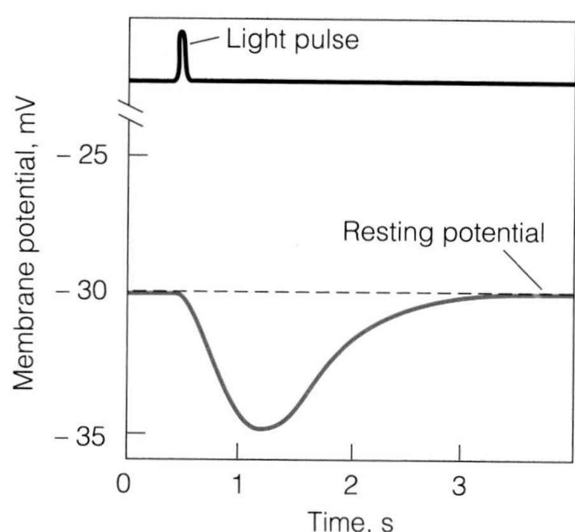

Figure 29.15
Response of a rod cell to a light pulse. A very short pulse of light (top) produces a hyperpolarization of the rod cell membrane that persists 1 to 2 seconds.

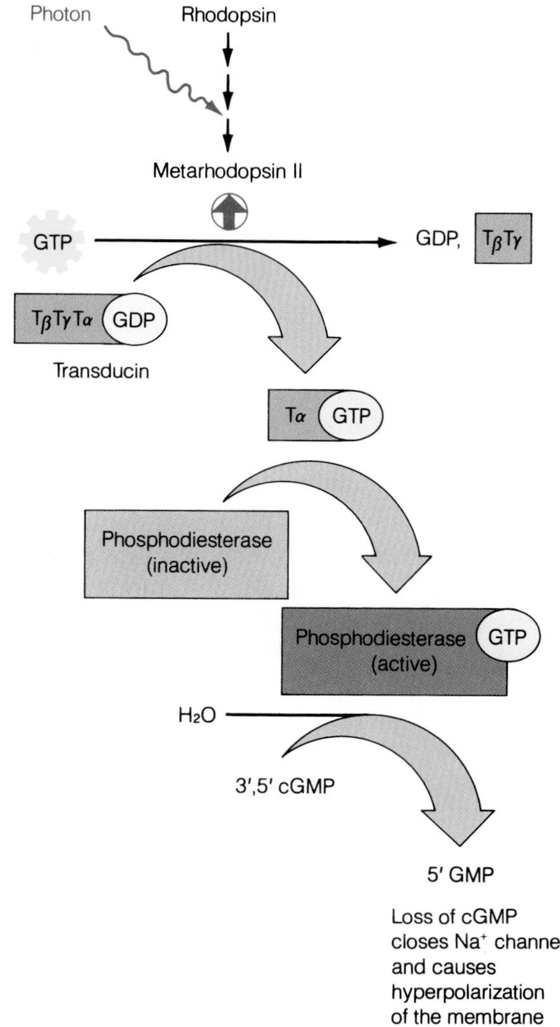

Figure 29.16
The visual cascade. Absorption of a photon by a rhodopsin molecule produces metarhodopsin II, as shown in Figure 29.14. Metarhodopsin II catalyzes the exchange of GTP for GDP on transducin. In this process, two of the transducin subunits (T_β and T_γ) are released, leaving GTP bound to the α subunit (T_α). This molecule binds to inactive phosphodiesterase, activating it to hydrolyze cyclic GMP (3′,5′-cGMP). As a consequence of the cascade, a rapid fall in cyclic GMP level occurs, which closes Na^+ channels in the membrane of the visual receptor.

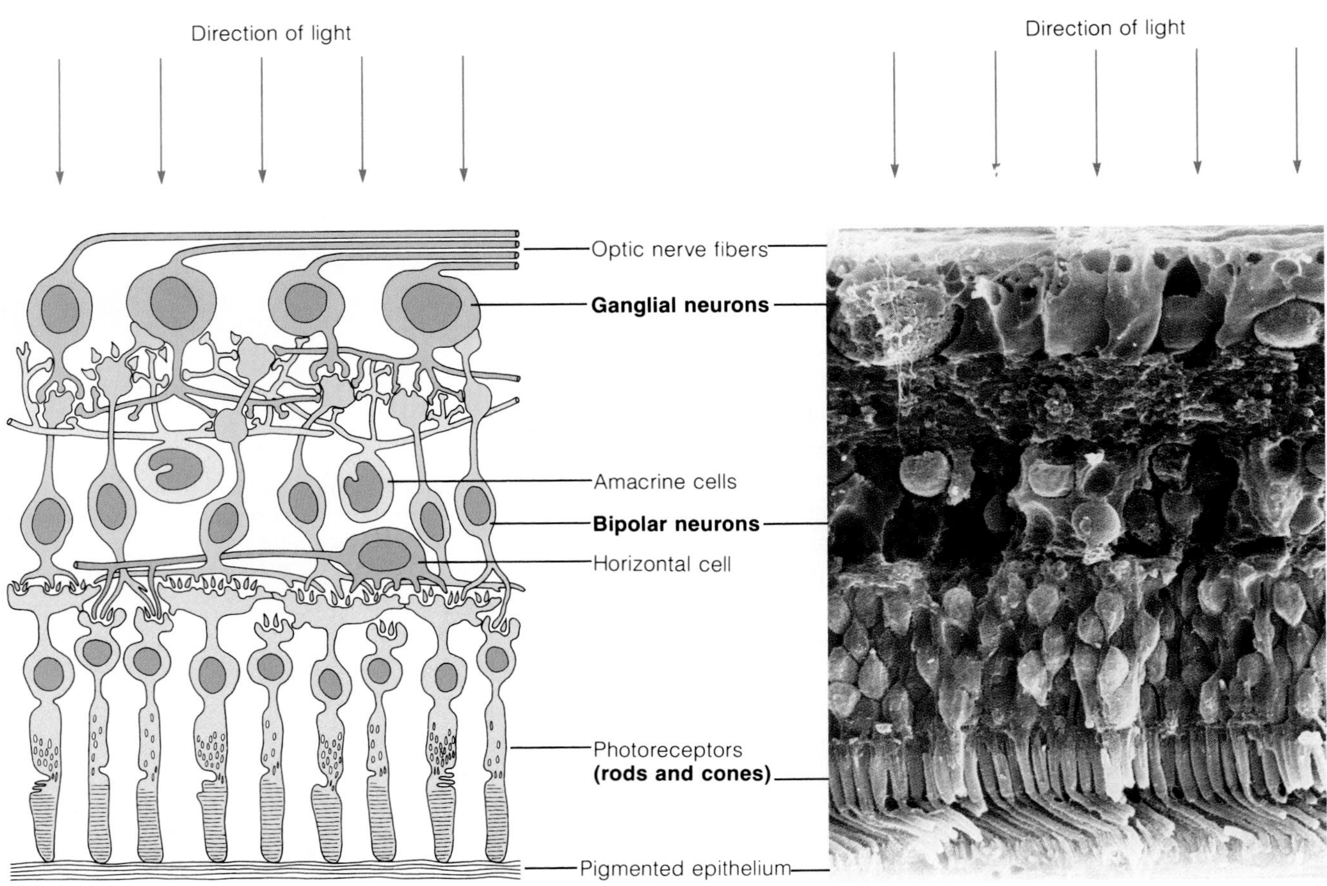

Figure 29.17
Schematic cross-section of the human retina. The photoreceptors (rods and cones) are actually at the *back* of the retina; light must pass through the intervening cell layer to reach the receptors. Impulses from the receptors are transmitted to the bipolar neurons, where they are passed to the ganglial neurons and the optic nerve (see Figure 29.18). The horizontal cells and amacrine cells integrate and correlate signals from multiple receptors and bipolar cells.

action potentials sent to the optic nerve (Figure 29.18). Of course, the overall organization of the retinal neurons is much more complex than the simple illustration would imply. For example, there are frequently multiple connections between several rod cells and one bipolar cell, and between several bipolar cells and one ganglial neuron, to allow for summation of signals from multiple rod cells to detect weak light. In addition, other cells in the retina (e.g., the horizontal cells) aid further in integrating responses.

We have said little about the cone cells so far. Structurally and functionally, they are very similar to the rod cells, with one important difference: there are three classes, each containing a different visual pigment. In humans, one absorbs maximally at 570 nm (yellow-red), the second at 530 nm (green), and the third at 440 nm (blue). Thus, the differential response from cone cells allows color vision. Since there is no summation of signals from cone cells (as there is for rod cells) they do not produce nearly so strong a response to weak light. This explains why our vision at night appears mainly in shades of gray and lacks clarity; we are using primarily the summed signals from rod cells spread over the retina.

The critical substance in vision is **retinal,** and mammals cannot synthesize it de novo. Thus, to maintain effective vision, we need a source such as β-carotene, or vitamin A (all-*trans*-retinol), which can be converted into all-*trans*-retinal by NAD^+ oxidation and cleavage, followed by isomerization to 11-*cis*-retinal (see Figure 29.14). Plants such as carrots are rich in β-carotene (top of p. 1055).

H_3C CH_3 CH_3 CH_3 H_3C CH_3 CH_3 CH_3 CH_3 CH_3

β-Carotene

Muscles: Effectors of Motility in Higher Animals

Once they receive sensory information about their environment, animals must respond. If we exclude such exceptional actions as bioluminescence and the release of chemicals like pheromones, the major ways animals act on the environment involve motions of either the whole or parts of the organism. Even the emission of sound is a muscular action, as is the injection of venom by an insect or a snake. In addition, equally important motions maintain the animal's internal world—the beating of its heart, the breathing of lungs or gills, and the peristaltic motions in the digestive system, for example.

For every such motion, there must be *effectors,* structural elements capable of exerting force in contraction or flexion. To exert a force against a resistance, energy must be expended. We shall find that the source of this energy for every kind of effector is the controlled hydrolysis of ATP.

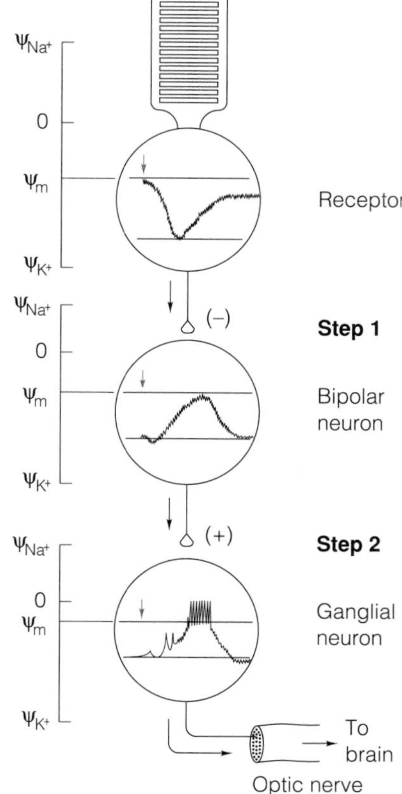

Figure 29.18
How a negative pulse from a visual receptor is converted into a positive action potential. Light (red arrow) induces a hyperpolarized pulse from the receptor, which decreases the release of an inhibitory neurotransmitter at receptor–bipolar neuron synapse (step 1). This disinhibition creates a positive pulse in the bipolar neuron, which is transmitted to the ganglial neuron (step 2), causing a burst of rapid action potential spikes to be sent to the optic nerve.

The Composition of Muscle

Vertebrates like ourselves possess, as physical effectors, three morphologically distinct kinds of muscle: **striated muscle, smooth muscle,** and **cardiac muscle.** It is the striated muscles that we most often associate with the term muscle, for these are the ones in arms, legs, eyelids, etc. that make possible voluntary motions. Smooth muscle surrounds internal organs like the blood vessels, intestines, and gallbladder, which are not under voluntary control and are capable of slow, sustained contractions. Cardiac muscle can be considered as a specialized form of striated muscle, adapted for the repetitive, involuntary beating of the heart.

All the kinds of muscle, as well as a number of other kinds of contractile systems we will encounter, have in common two major proteins: **actin** and **myosin.** As these proteins constitute the major elements in the contractile apparatus, we begin with a discussion of their properties.

Under physiological conditions, actin exists as a long, helical polymer (fibrous actin, or F-actin) of a globular protein monomer (G-actin). The G-actin monomer, shown in Figure 29.19, is a two-domain molecule of mass 42,000 daltons. Each G-actin monomer can bind one ADP or ATP. The helical stacking of the dumbbell-shaped monomers gives an actin fiber the appearance of two strands of beads wrapped about one another (see Figure 6.27, Chapter 6). Because of the asymmetry of the subunits, the

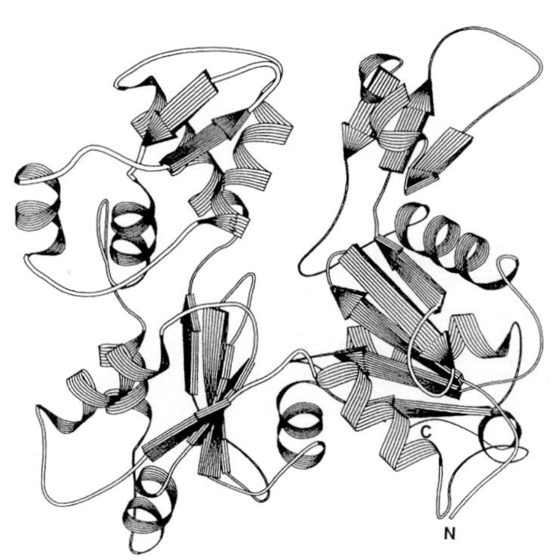

Figure 29.19
G-actin. This structure of the actin monomer, as deduced by x-ray diffraction, shows the two-domain conformation clearly.

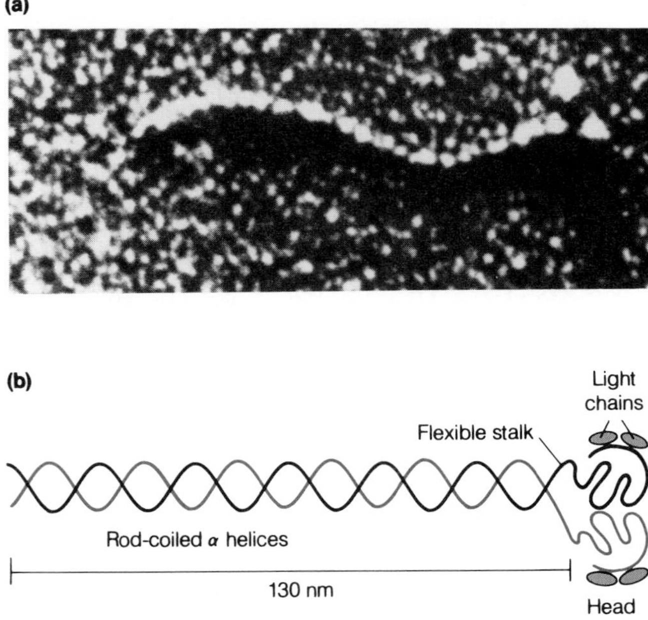

Figure 29.20
The myosin molecule. (**a**) An electron micrograph of a single myosin molecule. (**b**) The model depicts the six polypeptide chains. The two large subunits are connected by the intertwining of two α-helices in the rodlike tail of the molecule. The globular headpieces each carry two noncovalently bound light chains.

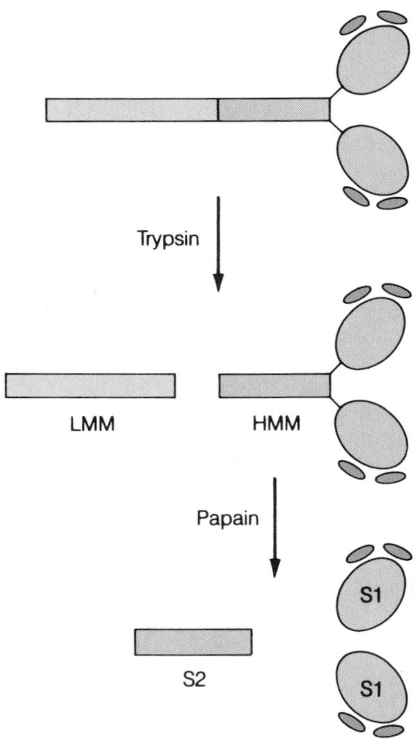

Figure 29.21
Dissection of myosin by proteases. Trypsin cleavage cuts in the tail to produce light meromyosin (LMM) and heavy meromyosin (HMM). Treatment of heavy meromyosin with the protease papain digests part of the stalk structure, allowing separation of the two headpieces, with their bound light chains, as S1 fragments, and releases the remainder of the tail as an S2 fragment.

F-actin fiber has a defined directionality. Polymerization of G-actin to F-actin is favored by adding Mg^{2+} or higher concentrations of monovalent ions. This polymerization also exhibits a preferred direction, especially when ATP is present.

The other major muscle protein is myosin (Figure 29.20). The functional myosin molecule is composed of six polypeptide chains: two identical heavy chains (M_r = 230,000) and two pairs of two kinds of light chains (M_r = 20,000) to form a functional complex of molecular weight 540,000 (see Figure 29.20b). The heavy chains have very long α-helical tails, which are interwound into a two-strand coiled coil, and globular head regions to which the light chains are bound. Between each head domain and tail domain is a flexible stalk. The tail domain can be cleaved at a specific point by trypsin to yield fragments called **light meromyosin** and **heavy meromyosin** (Figure 29.21). Further cleavage by papain cuts in the stalks to yield what are called S1 fragments, consisting of headpieces carrying the light chains. The ability to cleave the myosin molecule in these specific ways has helped researchers understand the functions of its several parts.

The functional domains of myosin play quite different roles. The tail regions have a pronounced tendency to aggregate, causing myosin molecules to form the kind of bipolar thick filaments shown in Figure 29.22. The headpieces, with their attached light chains, have a strong tendency to bind to actin. If this reaction is carried out with isolated S1 fragments, an actin filament will become "decorated" with these headpieces to produce an asymmetric "arrowhead" pattern which demonstrates the directionality of the actin fiber. In the presence of actin, whole myosin molecules or individual S1 fragments have ATPase activity, and the hydrolysis of ATP disrupts the binding. As we shall see, these observations have strong implications with respect to the mechanisms of muscle contraction. To understand this we must see how actin and myosin interact in the muscle structure.

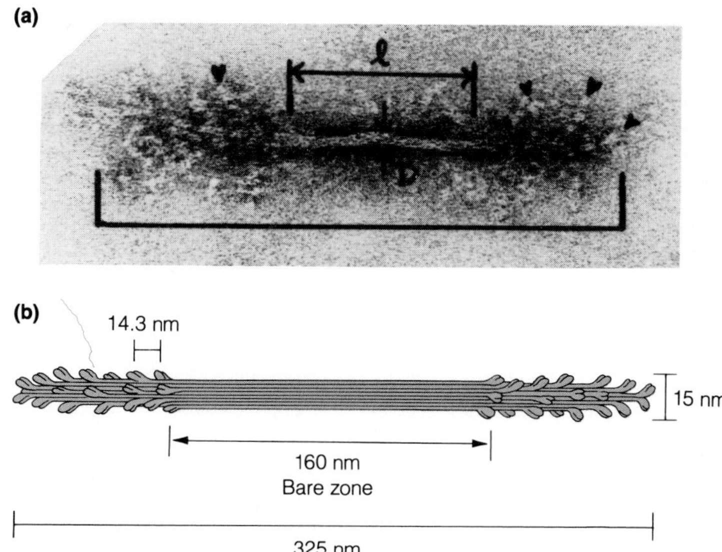

Figure 29.22
A thick filament of myosin molecules. (**a**) An electron micrograph; (**b**) a drawing of the structure, showing dimensions in nanometers. The projections are the pairs of headpieces on each myosin molecule.

The Structure of Muscle

We turn now to the fine structure of muscle. Figure 29.23 shows a typical vertebrate striated muscle at successive levels of organization. The individual muscle fibers, or **myofibers,** are actually very long (1–40 mm) multinucleate cells, formed by the fusion of muscle precursor cells called **myoblasts.** Each myofiber contains a bundle of **myofibrils,** which exhibit a very regular

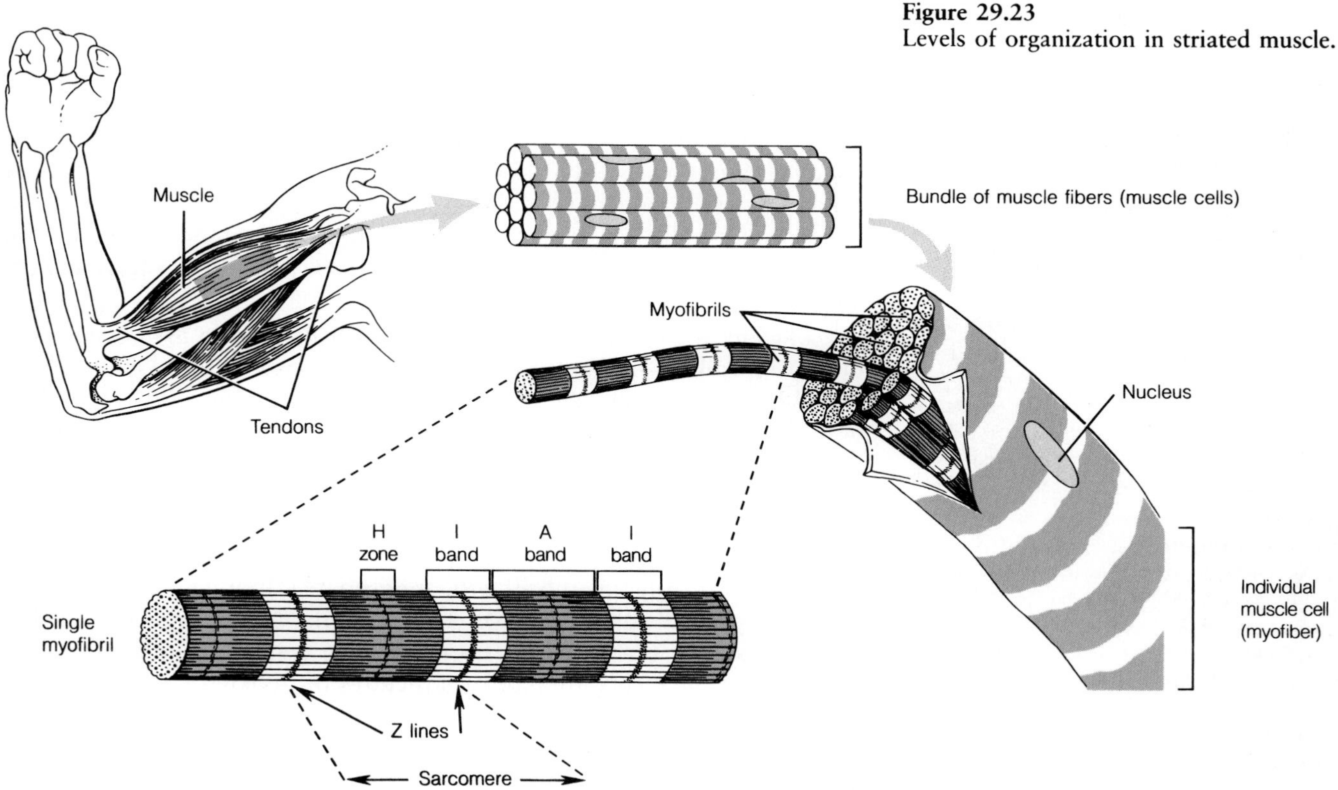

Figure 29.23
Levels of organization in striated muscle.

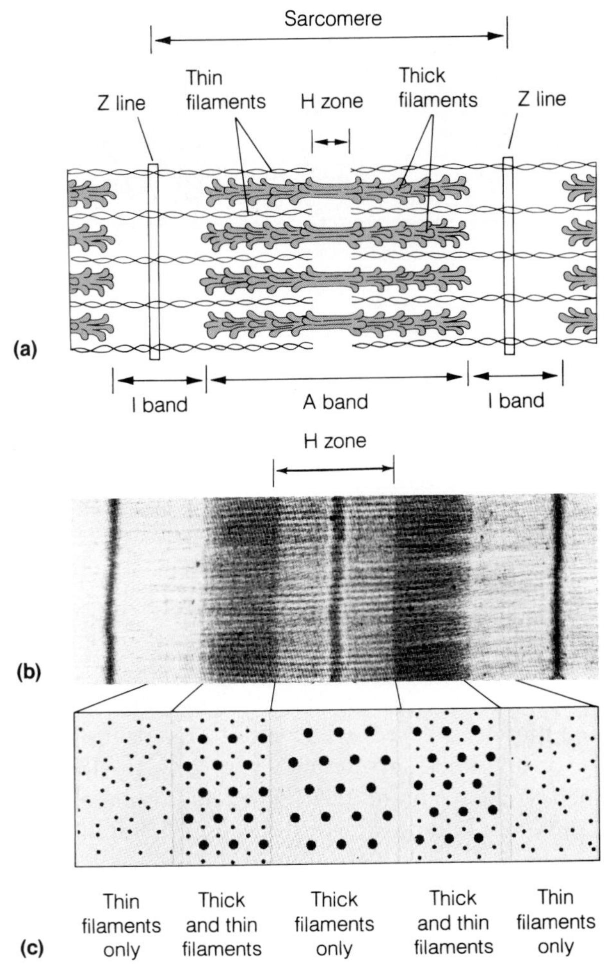

Figure 29.24
Muscle structure seen at the EM level. (a) A model of the sarcomere, the repeating unit in striated muscle. The I bands, A bands, and Z lines shown in Figure 29.23 are identified and structural elements of the sarcomere are indicated. (b) An electron micrograph, showing the same features. (c) A schematic drawing of cross sections of a sarcomere, in the various regions shown in parts (a) and (b). Thick fibers are indicated by heavy brown dots, thin fibers by small black dots.

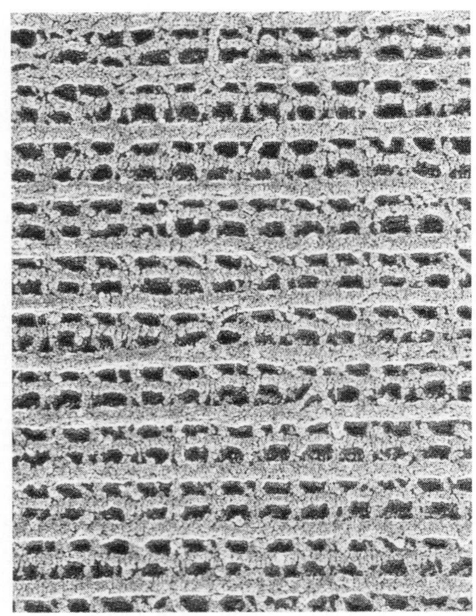

Figure 29.25
Cross-bridges between actin and myosin filaments, as revealed by high-resolution electron microscopy.

periodic structure when examined under the light microscope. Dark **A bands** alternate with lighter **I bands;** the latter are divided by thin lines called **Z lines** (or sometimes **Z disks**). At the center of the A band is found a lighter region called the H zone. The repeating unit of muscle structure can be taken as extending from one Z line to the next; it is called the **sarcomere** and is about 2.3 μm long in relaxed muscle.

The molecular basis for this periodic structure can be seen by electron microscopic studies of thin sections of muscle, as shown in Figure 29.24. **Thin filaments** of actin extend in both directions from the Z lines, interdigitating with myosin **thick filaments.** The regions in which the thick and thin filaments overlap form the dark areas of the A band. The I bands contain only thin filaments, which extend to the edges of the H zones. Within the H zones, only thick filaments are found.

The composition of the thick and thin filaments has been demonstrated by extracting myofibrils with appropriate salt or detergent solutions to remove virtually all of the myosin. Because the A bands are abolished in this process, the thick A-band fibers must be composed of myosin. The myosin thick filaments are bipolar structures of the kind shown in Figure 29.22, in which the helical tails of the myosin molecules join together, with the head groups projecting with a regular 14.3-nm spacing at either end. This spacing appears to be generated by a 14.3-nm periodicity in the amino acid sequence in the myosin tails.

That the thin filaments contain actin can be shown in the following way. If myofibrils from which myosin has been removed are perfused with a solution of S1 fragments, the thin filaments are decorated in the arrowhead pattern; thus, the thin filaments contain actin. Furthermore, the arrowhead patterns always point outward from the Z disks, demonstrating the polarity of these thin filaments. However, the thin filaments are not *entirely* composed of F-actin. As we shall see later, they contain other important proteins as well.

Finally, if we look very closely at electron micrographs of myofibrils, we can see small projections extending from the thick fibers, often contacting the actin filaments. These correspond to the headpieces of the myosin molecules. These *cross-bridges* between myosin and actin filaments (Figure 29.25) are the key to muscle contraction.

The Mechanism of Contraction: The Sliding Filament Model

Understanding the mechanism of muscle contraction has come both from the fine details of structure and from careful observations of changes in the banding pattern in the sarcomere during contraction. In a full contraction, each sarcomere shortens to a length of about 1.0 μm. During this process, the I bands and H zones disappear, and the Z disks move right up against the A band. Such observations led two independent groups of investigators, one led by Hugh Huxley, the other by Andrew Huxley, to propose in the 1950s the **sliding-filament model** of muscle contraction. This is depicted in Figure 29.26. According to this model, the myosin headpieces are presumed to "walk" along the interdigitated actin filaments, pulling them past, and thereby shortening the sarcomere.

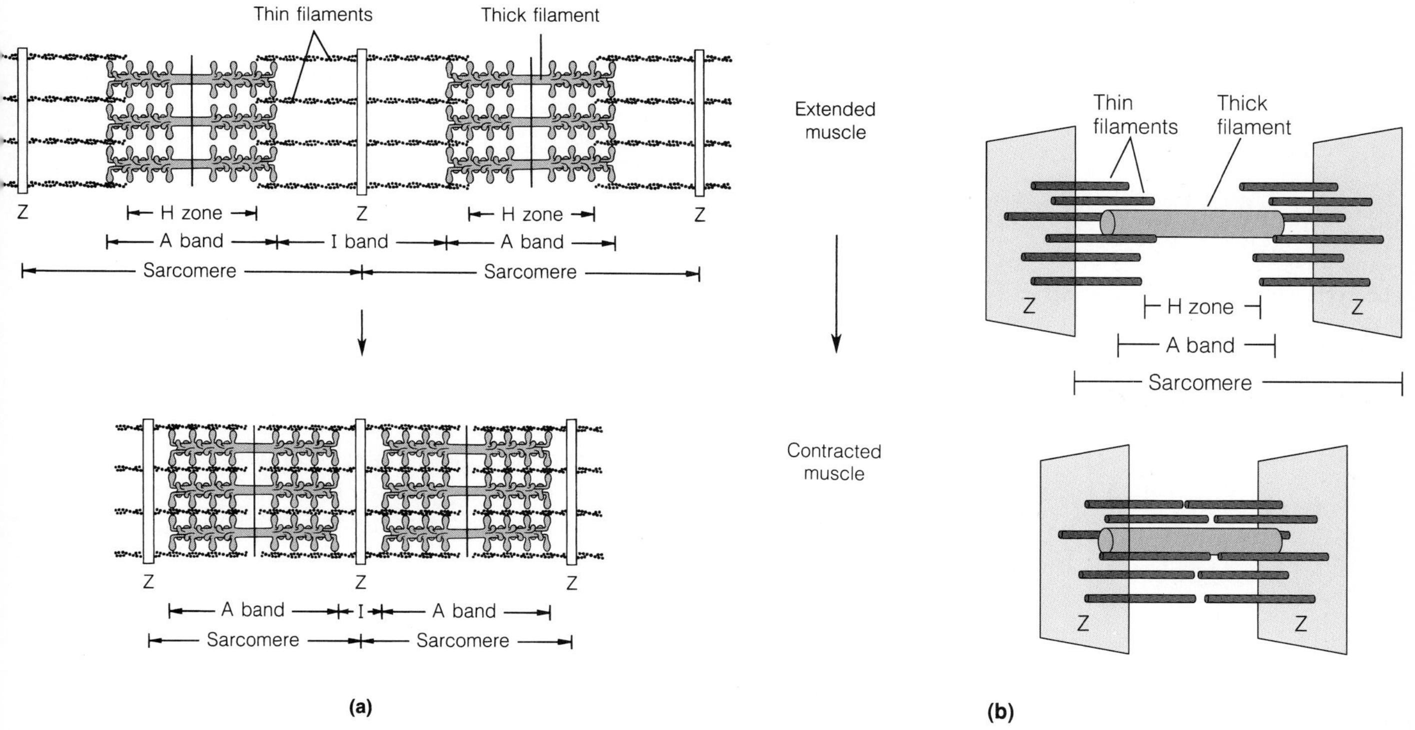

Figure 29.26
The sliding-filament model of muscle contraction. (**a**) Two sarcomeres, schematically shown extended and contracted. (**b**) A three-dimensional model of the same. Thick fibers are brown, thin fibers gray.

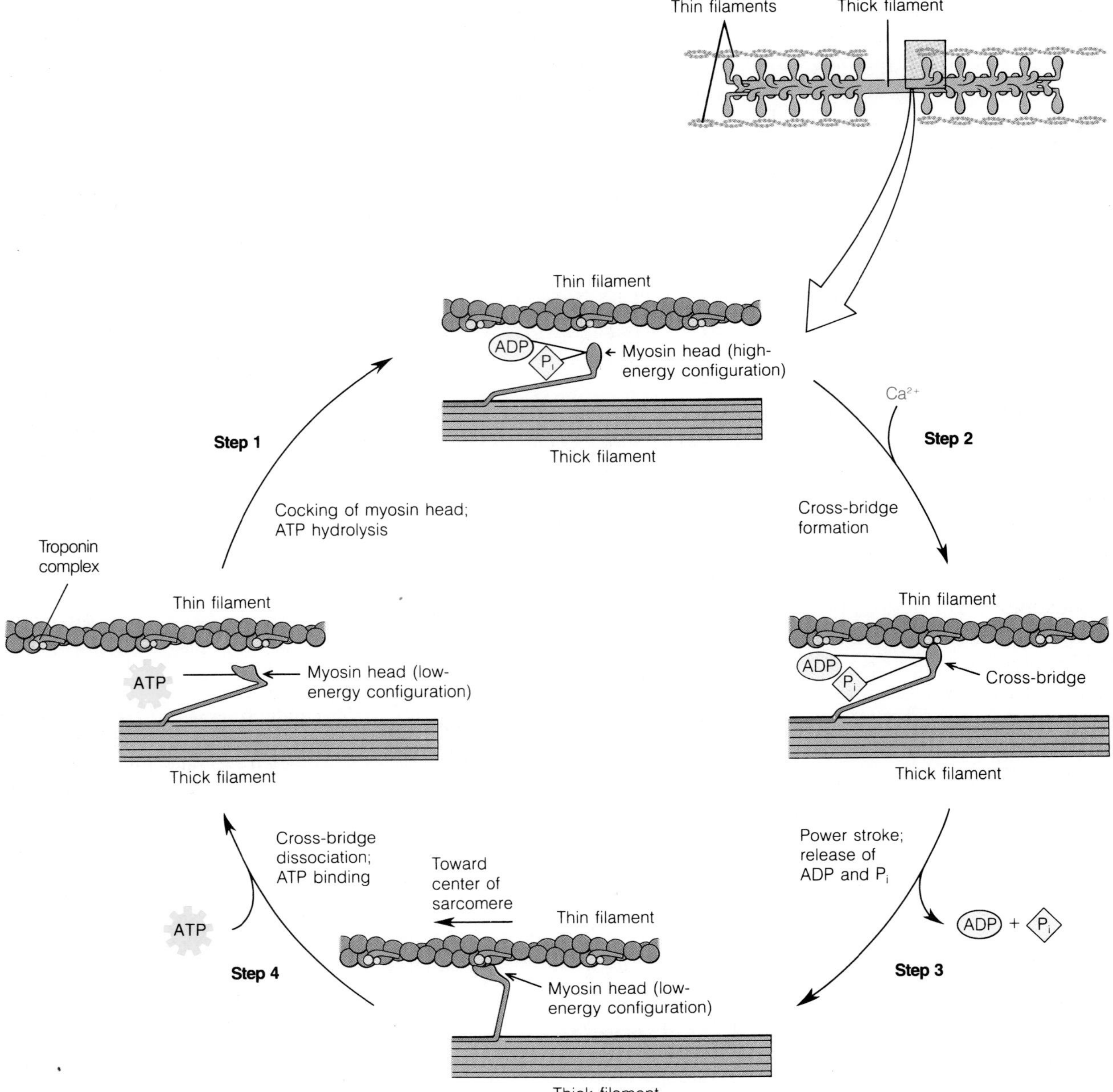

Figure 29.27
The cyclic process of muscle contraction. Stages in the interaction of one myosin headpiece with a thin filament are shown.

To produce such directed motion against an opposing force on the muscle, *energy* must be expended. You may expect that energy to be derived in some way from ATP hydrolysis, and our previous mention of the ATPase activity of the actin–myosin complex gives a hint of how this might occur. Recall that ATP binding and hydrolysis causes the *release* of the actin–myosin interaction. According to the presently accepted mechanism, as shown in Figure 29.27, each myosin headpiece is thought to take part in a repetitive cycle of making and breaking cross-bridges to an adjacent thin filament. The critical step is the hydrolysis of ATP on the myosin head (step 1). The energy released puts the myosin into a high-energy conformation. Then, on stimulation by Ca^{2+} (step 2), the myosin binds to the actin, with the cross-bridge still in the high-energy conformation. But on release of the

ADP and P_i, the myosin headpiece must revert to its low-energy conformation. This apparently involves a change in the relative orientation of the head and stalk of the myosin molecule with respect to the tail, which is buried in the thick filament. This motion pulls the thin filament toward the center of the sarcomere (step 3). Binding of a new ATP (step 4) is accompanied by release of the headpiece–actin interaction, completing the cycle and readying the headpiece for another stroke. Note that the actin filament has been moved with respect to the myosin, so that each headpiece makes successive steps along the thin filament. The "walking" is rather like that of a millipede—some legs are always in contact with the thin filament, so that it cannot slip back during contraction.

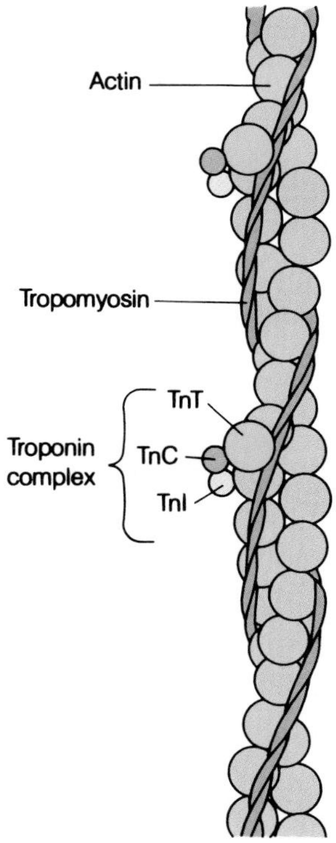

Stimulation of Contraction: The Role of Calcium

The critical substance in stimulating contraction is not ATP, which is generally available in the myofibril, but the Ca^{2+} that enters at step 2 of Figure 29.27. To understand how calcium regulates muscle contraction, we must examine the chemical structure of the thin filament in a bit more detail.

A thin fiber, as found in striated muscle, is more than just an F-actin polymer. Four other proteins are essential to its contractile function. One of these is **tropomyosin,** a fibrous protein that exists as elongated dimers lying along, or close to, the groove in the F-actin helix (Figure 29.28). Bound to each tropomyosin molecule are three small proteins called **troponins T, I,** and **C.** The presence of tropomyosin and the troponins act to inhibit the ATPase activity of the actin–myosin complex, *unless calcium is present at a concentration of about* 10^{-5} M. In resting muscle Ca^{2+} concentrations are in the neighborhood of 10^{-7} M, so that new cross-bridges cannot be formed, and ATP is not actively hydrolyzed. An influx of Ca^{2+} stimulates contraction, because the ion is bound by troponin C, causing a rearrangement of the troponin–tropomyosin complex. This makes sites on actin available for binding by the myosin headpieces. The postulated mechanism shown in Figure 29.29 permits step 2 and the subsequent steps in the cycle of Figure 29.27 to take place.

Figure 29.28
Schematic drawing of F-actin, together with the associated proteins that are present in the thin filaments of striated muscle. TnI, TnC, and TnT are troponins I, C, and T, respectively.

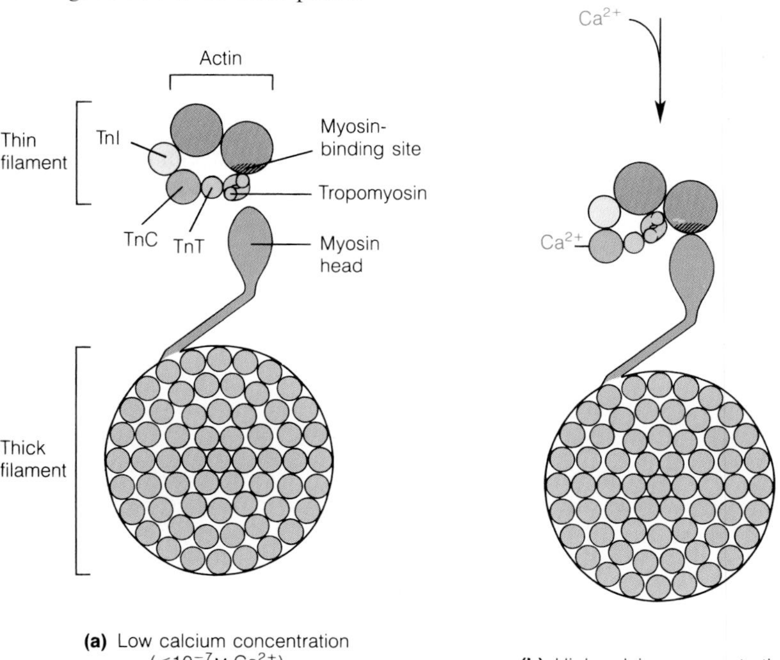

(a) Low calcium concentration ($<10^{-7}$ M Ca^{2+})

(b) High calcium concentration ($>10^{-6}$ M Ca^{2+})

Figure 29.29
Model for the regulation of contraction in striated muscle. A single myosin headpiece is shown next to a thin filament, in cross-sectional view. (**a**) Relaxed muscle. At low Ca^{2+}, the configuration of actin, tropomyosin and troponin in the thin filament is such that most myosin headpieces are blocked from contact with the thin fiber. (**b**) Calcium binding to troponin C causes a rearrangement in the thin-fiber components so that myosin-binding sites on the actin are made available. The cycle shown in Figure 29.27 can then occur, and the muscle contracts.

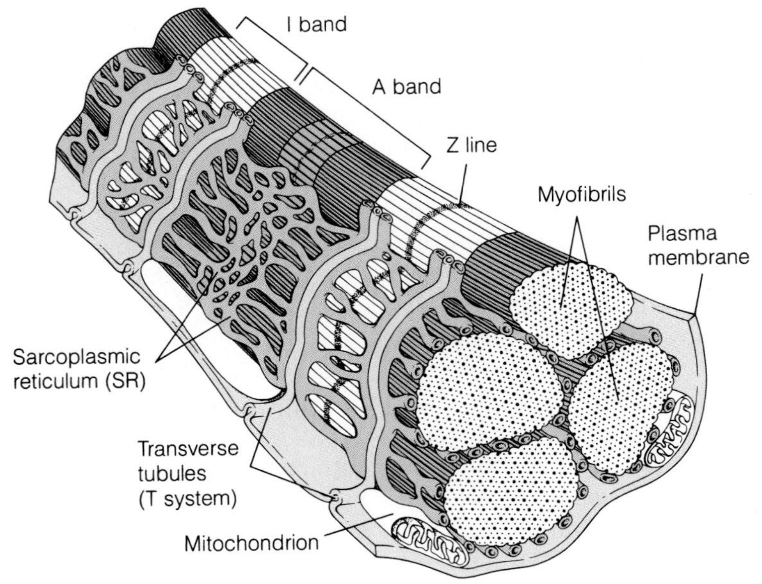

Figure 29.30
Structure of a myofiber, or muscle cell. The sarcoplasmic reticulum (SR) is a network of specialized endoplasmic reticulum tubules that surrounds the myofibrils. The SR accumulates Ca^{2+} in the resting muscle and discharges it into the myofibrils when a neural signal reaches the plasma membrane. The transverse tubules (or T system) are invaginations of the plasma membrane that make contact with the SR at many points, ensuring uniform response to the signal.

We have now traced the activation of muscle contraction to the influx of calcium into the myofibrils. But how does this influx occur? In particular, how can it be brought about by the nerve impulses that excite muscles to contract? The answer can be found from a closer examination of the myofiber (Figure 29.30). Each myofibril is surrounded by a membranous structure, the **sarcoplasmic reticulum.** The membrane of the sarcoplasmic reticulum is enriched in the Ca^{2+}-ATPase pumps described in Chapter 9, which transport Ca^{2+} from the myofibrils into the lumen of the reticulum. This maintains the level of Ca^{2+} in the myofibrils of resting muscle at a level of about 10^{-7} M, whereas the Ca^{2+} level within the lumen of the sarcoplasmic reticulum may be 10,000-fold higher. Impulses from motor nerves depolarize the membrane of the sarcoplasmic reticulum, opening gated Ca^{2+} channels that allow the high concentration of Ca^{2+} within the lumen to pour into the myofibrils, stimulating contraction. The signal is rapidly transmitted to the entire sarcoplasmic reticulum of a myofiber via **transverse tubules,** invaginations of the plasma membrane that connect at periodic intervals with the reticulum.

Although abrupt change in Ca^{2+} level is the common *signal* for muscle contraction, it obviously cannot provide the necessary energy. We turn now to the question of where the energy required for muscular work may come from.

Energetics and Energy Supplies

Basically, muscle is a mechanism for converting the chemical free energy released in ATP hydrolysis into mechanical work. The conversion can be remarkably efficient, approaching values up to 80% under optimal circumstances. This is a much higher efficiency than can be attained by synthetic chemical engines.

How is the ATP generated? Even in striated muscle the answer can vary, depending on the particular *kind* of muscle and its function. Striated muscles can be divided into two categories—"red" muscle, designed for relatively continuous use, and "white" muscle, employed for brief, often rapid motions. Red muscle owes its color to three factors: it is well supplied with blood vessels, has many mitochondria, and has large stores of myoglobin. Clearly, it is a kind of muscle that depends heavily on aerobic metabo-

Table 29.3
Comparison of red and white striated muscle

	Red	White
Relative fiber size	Small	Large
Mode of contraction	Slow twitch	Fast twitch (about 5 times faster)
Vascularization	Heavy	Lighter
Mitochondria	Many	Few
Myoglobin	Much	Little
Major stored fuel	Fat	Glycogen
Main source of ATP	Fatty acid oxidation	Glycolysis

lism in mitochondria; the primary energy source in red muscle is the oxidation of fat. White muscle, on the other hand, relies heavily on glycolysis, with glycogen as a primary energy source. Glycogen is excellent for quick energy production but cannot sustain activity for long periods. (See Table 29.3 for a more detailed comparison of red and white muscle and Chapter 23 for a more extensive discussion of fuel metabolism.)

The functional differences between the two types of striated muscle are clearly revealed in birds. In the domestic chicken, the pectoral flight muscles, used only for brief fluttering or short flights, are white, whereas the leg muscles are red. Wild flying birds, which make sustained flights but rarely walk, have the opposite distribution of "white" and "dark" meat.

Careful observation of ATP levels in red striated muscle has shown that the provision of energy is more complicated than might appear at first. The amount of ATP used in a single contraction may be greater than all the ATP immediately available to a sarcomere. Yet even after relatively long exercise, ATP levels in the sarcomeres remain essentially constant. Only after extreme exhaustion do they begin to fall. This suggests that ATP may be an intermediary, and not the ultimate energy storage compound in these muscles. Indeed, it has been known for many years that the high-energy compound steadily depleted during muscular activity is creatine phosphate. This compound is capable of phosphorylating ADP very efficiently, in a reaction catalyzed by the enzyme creatine kinase.

$$^{-}OOC-CH_2-N(CH_3)-C(=NH)-NH-\text{Ⓟ} + ADP \xrightleftharpoons[\text{kinase}]{\text{Creatine}} {}^{-}OOC-CH_2-N(CH_3)-C(=NH)-NH_3^{+} + ATP \qquad \Delta G^{\circ\prime} = -8\ \text{kJ/mol}$$

Creatine phosphate **Creatine**

Because the equilibrium lies well to the right, virtually all of the muscle adenylate is maintained in the ATP form, rather than as ADP or AMP, as long as creatine phosphate is available. The consumption of creatine phosphate during exercise, while an almost constant level of ATP is maintained, is clearly demonstrated by the NMR studies of whale muscle shown in Figure 12.8 (Chapter 12).

Creatine phosphate is synthesized in the muscle mitochondria and transported to the myofibrils; creatine is returned, in a process called the **creatine phosphate shuttle.** Among vertebrate animal tissues only muscle and, to a lesser extent, brain are rich in creatine phosphate. In some invertebrates arginine phosphate plays a similar role (see Chapters 12 and 21).

When exercise of striated muscle is particularly intense and sustained, oxygen supplies may not be sufficient to maintain the high level of aerobic metabolism demanded by the rapid consumption of creatine phosphate and ATP. Under such circumstances, anaerobic glycolysis, with accompanying lactate production, becomes more and more significant. There is, as you will recall from Chapter 7, a compensatory feedback: as lactate levels rise and blood pH falls, oxygen delivery to the tissues is enhanced because of the Bohr effect. Glycolysis in muscle draws on both blood glucose and muscle glycogen. As these are consumed and the ADP/ATP ratio becomes high, a second mechanism will provide a bit more ATP. The enzyme **myokinase** catalyses the disproportionation reaction

$$2\text{ADP} \underset{\text{Myokinase}}{\rightleftharpoons} \text{ATP} + \text{AMP}$$

Although the reaction has an equilibrium constant near unity, it is driven to the right by ATP consumption. The reaction can be shifted even farther by an enzyme-catalyzed deamidation of AMP to inosine monophosphate (IMP). Such "reserve" mechanisms are vital, for the very survival of an animal may depend on a last burst of muscular energy.

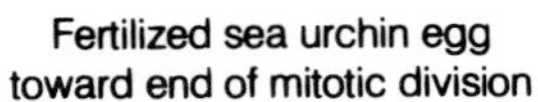

Fertilized sea urchin egg
toward end of mitotic division

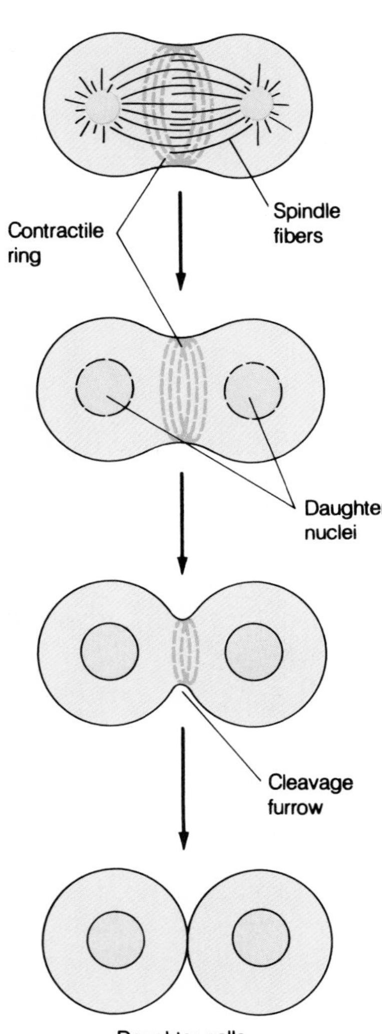

Figure 29.31
Cytokinesis in a fertilized sea urchin egg. The contractile ring that cleaves the cell in two is an actin–myosin complex, located just beneath the plasma membrane.

Nonmuscle Contractile and Motile Systems

The importance of muscle-generated motility for higher organisms should not overshadow the fact that a number of other systems convert chemical energy to mechanical work. Many of these operate at the cellular level, and they are vital to the physiological well-being of organisms ranging from bacteria to humans. We shall begin with some examples that use myosin and actin, but in rather different ways than in striated muscle.

Nonmuscle Actin and Myosin

Actin and, to a smaller extent, myosin are found in most eukaryotic cells, even those that are in no way involved in muscular tissues. They appear to play roles in cell motility and changes of cell shape. Actin seems to be a major component of the **cytoskeleton**—the fibrous array that pervades almost every kind of cell and gives it a specific shape (see Figure 12.7). In many instances, such as amoeboid motion and the penetration of eggs by sperm, polymerization and depolymerization of actin filaments seem to be the important processes. We shall consider here only one example in which a true actin–myosin contractile complex seems to be involved in an intracellular process. This is **cytokinesis,** the division of cells in the last stages of mitosis (see Chapter 28). The process can be followed with clarity in the sea urchin egg, a favorite model system for such studies. As mitosis approaches conclusion, and the daughter nuclei are clearly separated at the two poles of the cell, a ring of indentation is observed in the cell surface, defining a plane perpendicular to the mitotic spindle (Figure 29.31). This ring contracts, forming the **cleavage furrow** and eventually cutting the cell in two. Electron microscopy shows that the ring consists of fibers, and staining with fluorescent antibodies shows that these contain both actin and myosin.

The involvement of myosin in cytokinesis has been most elegantly demonstrated by S. Inoue and his colleagues in the experiment depicted in

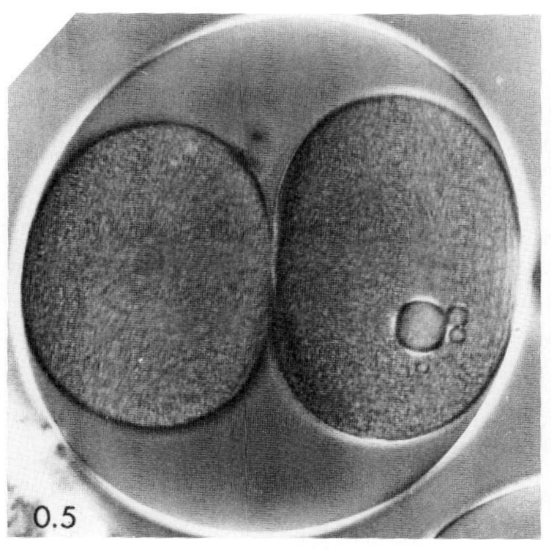

(a)

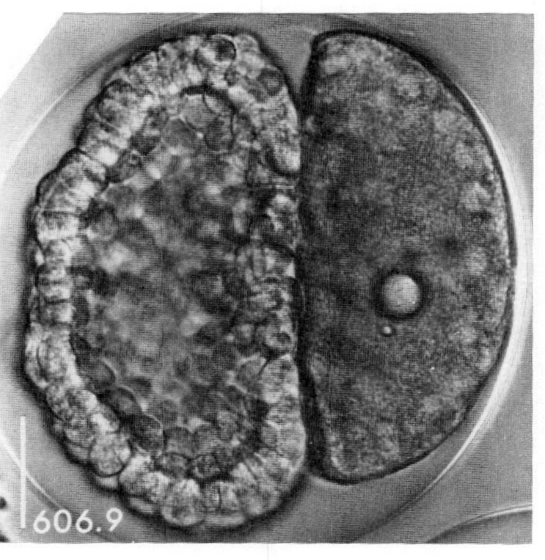

(b)

Figure 29.32
Demonstration that myosin is essential in cytokinesis, but not in mitosis. (**a**) A sea urchin embryo at the two-cell stage. The cell to the right has been injected with antimyosin antibodies (the oil droplet shows the site of injection). (**b**) After 10 hours, the cell to the left has undergone many divisions, to form half of a normal embryo. In the cell to the right, cytokinesis has been completely blocked. Mitosis has continued, however, as can be judged by the many new nuclei that have been formed in the cell at the right.

Figure 29.32. After the sea urchin egg had undergone one division, the daughter cell on the right was microinjected with antimyosin antibodies. After 10 hours, the control cell (left) had undergone many divisions, to construct half of an embryo. In the treated cell, mitosis continued, as evidenced by the many nuclei present, but *cytokinesis had been completely blocked.* Thus, myosin is essential for cytokinesis.

The experiment demonstrates one other important fact. Since mitosis continued even in the cell treated with antimyosin antibodies, the contractile process in the mitotic spindle apparently does *not* require the participation of myosin. There have been many reports of the presence of actin and myosin in the spindle, and it was long believed that they were essential for chromosome separation. But these and other experiments show that some other contractile systems must be involved. It is to this second general class of motility-generating systems that we now turn.

Motility from Microtubules

A class of motile systems completely different from and unrelated to the actin–myosin contractile systems is used in places as diverse as the mitotic spindle, protozoan and sperm flagella, and transport of molecules and organelles along axons, to name only a few. These systems are constructed from **microtubules,** very long, tubular structures built from a helical wrapping of the protein **tubulin.** There are two kinds of tubulin subunits, α and β, each of molecular weight 55,000. They are present in equimolar quantities in the microtubule, which can be considered a helical array of $\alpha\beta$ dimers (Figure 29.33). Seen from the side, the microtubule is found to consist of 13 rows of alternating α and β subunits. The association between the $\alpha\beta$ dimers to form a microtubule appears to be largely hydrophobic, for cooling a preparation of microtubules to 4°C causes complete dissociation into dimers. On reheating to a physiological temperature, the tubules spontaneously re-form. However, the tubular fiber in vivo is not as simple in composition as this might imply. A number of **microtubule-associated proteins** (**MAPs**) exist in the native structures. As we shall see, these minor constituents are very important in the biological functions of microtubules. We shall now consider some of the diverse roles microtubules play.

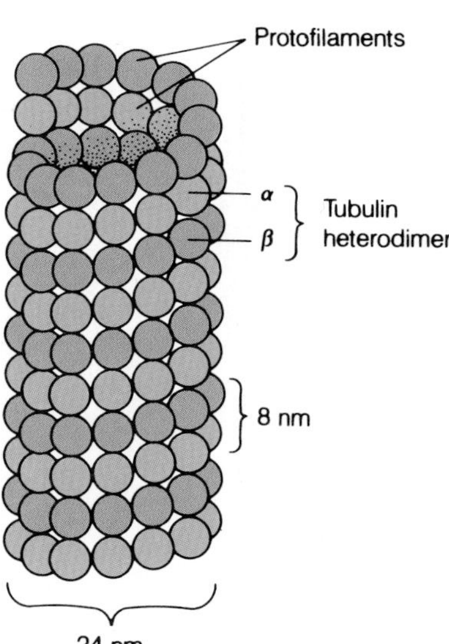

Figure 29.33
Model of a portion of a microtubule. The α and β tubulin subunits are shown in green and brown, respectively. The microtubule can be thought of either as a helical array of $\alpha\beta$ dimers or as 13 parallel rows of $\alpha\beta$ dimers. These rows are referred to as protofilaments.

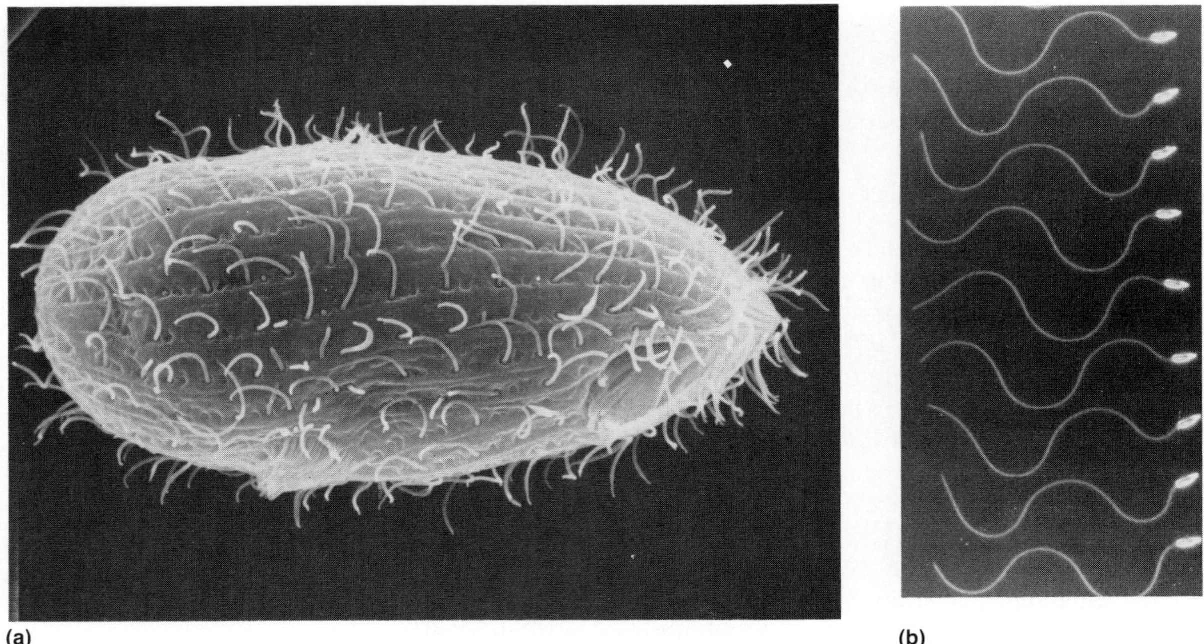

Figure 29.34
Cilia and flagella. (**a**) The protozoan *Tetrahymena* is covered with rows of cilia. (**b**) A sperm of the tunicate *Ciona*. A series of time-lapse photographs shows how undulations of the tail (flagellum) propel the sperm.

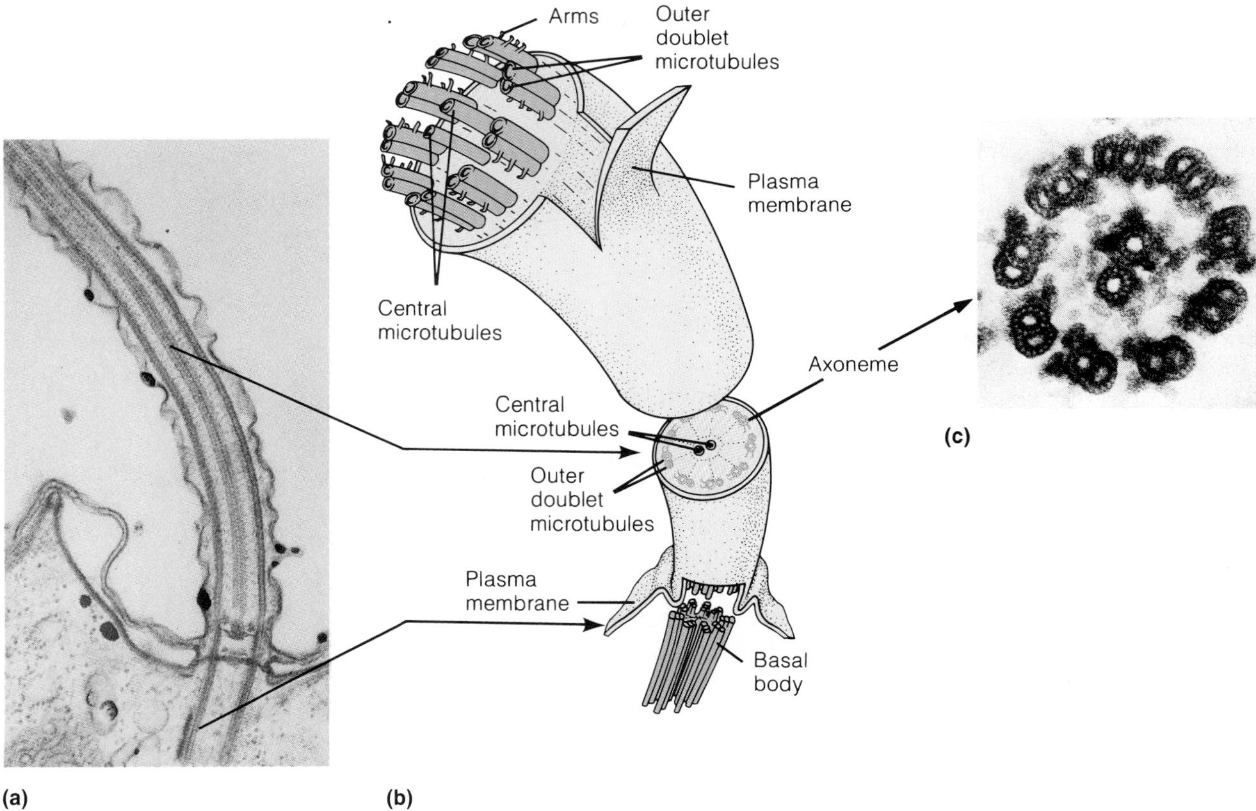

Figure 29.35
Ultrastructure of a cilium. (**a**) Electron micrograph of a longitudinal section, showing microtubules running the length of the appendage. (**b**) A schematic drawing of the structure. (**c**) An electron micrograph of a cross section of the axoneme, showing the 9 + 2 arrangement of outer doublets and inner tubules.

CILIA AND FLAGELLA. Many kinds of eukaryotic cells are propelled by the beating of cilia or flagella. Cilia are shorter and exert a coordinated "rowing" motion to move an organism through solution (Figure 29.34a). Eukaryotic flagella, like those in sperm tails, are longer and propel by an undulatory motion (Figure 29.34b). The structures of both types of motile appendages have many elements in common. In each case there is a central **axoneme,** which is a highly organized bundle of microtubules, connected to a **basal body,** an anchoring structure within the cell. The axoneme is enveloped by an extension of the plasma membrane (Figure 29.35).

The internal structure of the axoneme is truly remarkable. As the cross section in Figure 29.35 shows, the most obvious feature is the arrangement of microtubules known as a 9 + 2 array: two central microtubules ringed by nine microtubule doublets. The single microtubules in the center are complete, each having 13 rows of $\alpha\beta$ tubulin dimers. By contrast, each of the nine surrounding doublets is composed of one complete microtubule (the **A fiber**) to which is fused an incomplete microtubule, carrying only 10 or 11 rows of subunits (the **B fiber**). Closer inspection of electron micrographs reveals even greater complexity, as diagrammed in Figure 29.36. The outer doublets are periodically interconnected by a protein called **nexin** and carry at regular intervals "arms" composed of the protein **dynein.** In addition, **radial spokes** project from the outer doublets to connect with the central pair.

The full complexity of the structure is revealed only by gel electrophoresis studies of isolated axonemes. About 200 polypeptides can be resolved. Analysis indicates that there are at least 6 proteins in the spoke heads and 11 others in the arms of the spokes.

All of this apparatus seems to be involved in the motions of cilia and flagella. If ATP is added to isolated axonemes, adjacent doublets can be seen to slide past one another. The best current model holds that this occurs by "walking" of the dynein arms along the adjacent doublet (Figure 29.37a).

It is thought that the sliding of adjacent filaments past one another is translated into the beating of cilia and flagella in the manner depicted in Figure 29.37b. In this process, the central spokes and nexin connectors transform the sliding of doublets into bending of the whole cilium or flagellum. If connections within the axoneme are removed by careful proteolysis, ATP simply causes axonemes to extend and thin, as the outer doublets slide past one another with no stopping point.

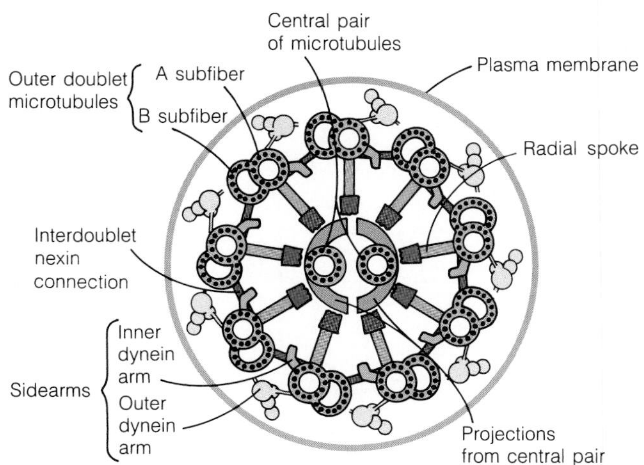

Figure 29.36
Diagram of the cross section of an axoneme. The dynein arms have ATPase activity and can cause adjacent doublets to slide with respect to each other. The nexin connections between doublets and the radial spoke system give stability to the whole assembly.

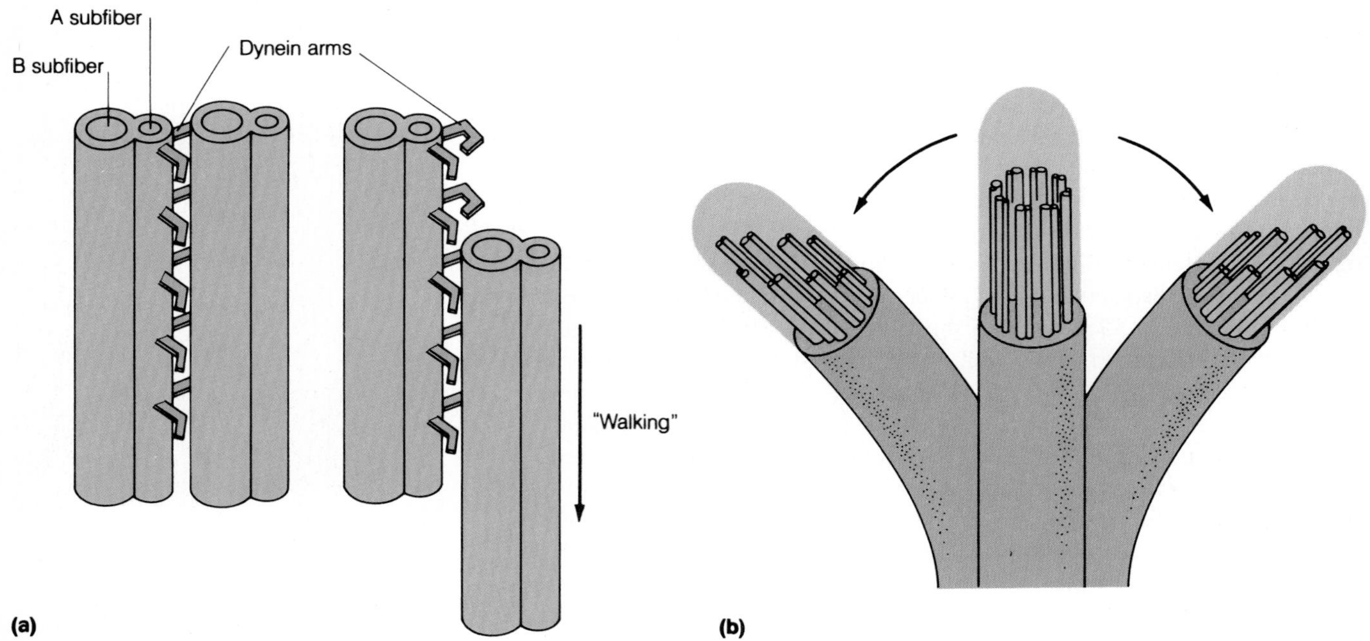

Figure 29.37
Model for the bending of cilia and flagella. (**a**) Isolated microtubule doublets can "walk" past one another through the making and breaking of dynein arm contacts, if ATP is present. (**b**) In a cilium or flagellum, long walks are prevented by the cross-connections between microtubules, but short motions, if synchronized properly on two sides of the cilium, could produce a back-and-forth bending motion.

It has been demonstrated that dynein has ATPase activity, with binding of ATP associated with the *breaking* of dynein cross-bridges. Comparison of the cilia and flagella mechanisms with the ATP-driven walking of myosin heads along the actin fiber in muscles is obvious, but there appears to be no relationship between the two systems at the protein structure level. Rather, we seem to be seeing here an example of convergent evolution at the level of molecular function.

MICROTUBULES AND MITOSIS. A number of studies have indicated that the mitotic spindle is composed mainly of microtubules. Specifically, various conditions that block the formation of microtubules (certain plant alkaloids, low temperature, high pressure) also prevent the formation of the spindle and the completion of mitosis. The mitotic spindle contains microtubules that have a variety of functions (Figure 29.38). Some (**polar microtu-**

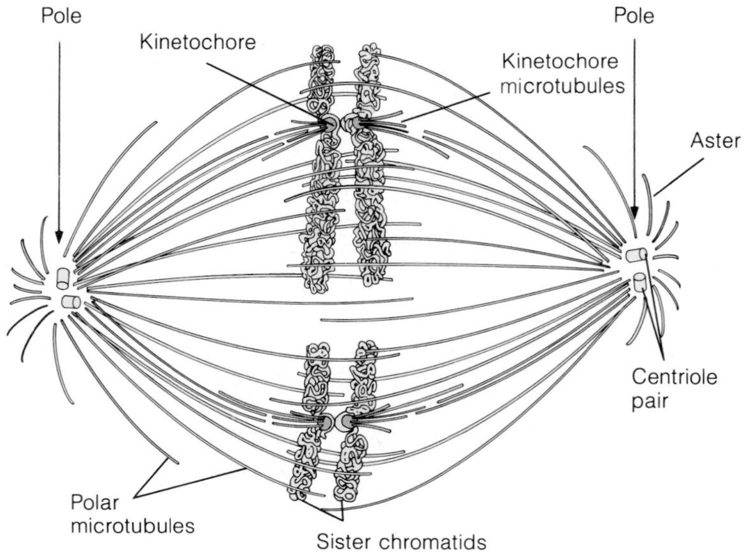

Figure 29.38
Microtubules in the mitotic spindle. Both the separation of the centrioles (poles) and the movement of the chromatids toward them appear to be accomplished by microtubules. Sets of microtubules (polar microtubules) radiate from the centrioles. Others (kinetochore microtubules) appear to emerge from points near the center of each chromosome (the kinetochores). Exactly how the motion of these elements is driven in mitosis is still unclear.

bules) extend between the poles, apparently pushing them apart. Others (**kinetochore microtubules**) extend from the poles to the kinetechores of the individual chromosomes and seem to pull the latter to the poles in telophase.

At the present time, controversy remains as to just how the microtubules accomplish the motions of poles and chromosomes that occur in mitosis. According to one model, all motion is a consequence of sliding of microtubules along one another, presumably by "walking" of dynein arms, as in ciliar motion. The polar microtubules are thought to slide on one another, pushing the poles apart, while the kinetochore microtubules slide on these, pulling the chromosomes toward the poles. In another model, the motion of kinetochore tubules is a consequence of their selective depolymerization at the polar ends. Active research is continuing on this problem.

INTRACELLULAR TRANSPORT. At one time, it was thought that all transport of materials within cells was either through passive diffusion or by active transport between compartments. There is now evidence that some proteins and organelles are transported over long distances along microtubules, which serve as tracks for the motion.

The clearest evidence comes from studies of transport in nerve axons. As you will recall, nerve axons can be very long, and rapid movement of materials between the cell body and the synaptic bulb presents a formidable task. The problem can be studied in a very direct way, using the mammalian sciatic nerve, which has very long axons extending from the cell body in the spinal cord. If radiolabeled amino acids are injected into the cell body, they will be incorporated into proteins by the ribosomes contained therein. After some time, the axon can be sectioned and the location of newly synthesized proteins determined. This method reveals that although transport rates vary greatly, some proteins, especially those associated with lipid vesicles, move as rapidly as 40 cm/day, much faster than could be accounted for by diffusion.

The squid giant axon, because of its size, has provided an excellent subject for studies of intracellular transport. As shown in Figure 29.39, small vesicles can actually be seen moving along microtubule bundles in such axons. A protein ATPase, **kinesin,** plays a major role in such motion. Vesicles to which kinesin is attached move along microtubules when ATP is supplied. Thus, it seems that the same general mechanism that accounts for the relative motion of microtubules is involved here, but major puzzles remain. For instance, what determines the *direction* of motion, especially in cases (as in Figure 29.39) where two objects speed along the same track in opposite directions? Clearly, the fine structure of the cell is capable of processes much more sophisticated than we might have expected.

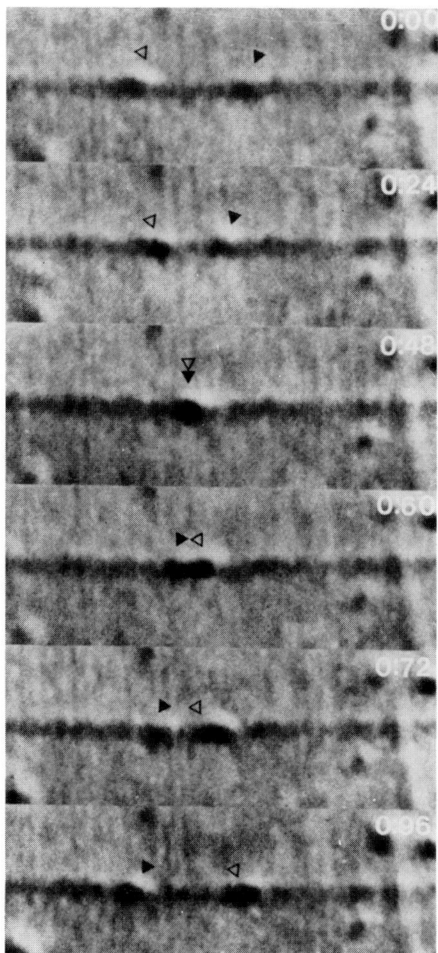

Figure 29.39
Movement on microtubule tracks. A series of time-lapse video micrographs showing bidirectional motion of organelle vesicles on a single microtubule filament from squid giant axon. The two organelles (open and solid triangles) move in opposite directions and pass each other. Elapsed time (seconds) is given in the upper right corner of each frame.

Bacterial Motility and Chemotaxis

It is appropriate to close this chapter by returning to the prokaryotic world, examining a system that is almost without rival in its elegance and simplicity. In bacterial chemotaxis we find the most rudimentary kind of a sensory system, and in bacterial motility a mechanism that would never have been believed until the evidence became clear. The general phenomenon of **taxis,** widespread throughout the animal and plant worlds, comprises responses to light, magnetic fields, and virtually any kind of chemical substance. The

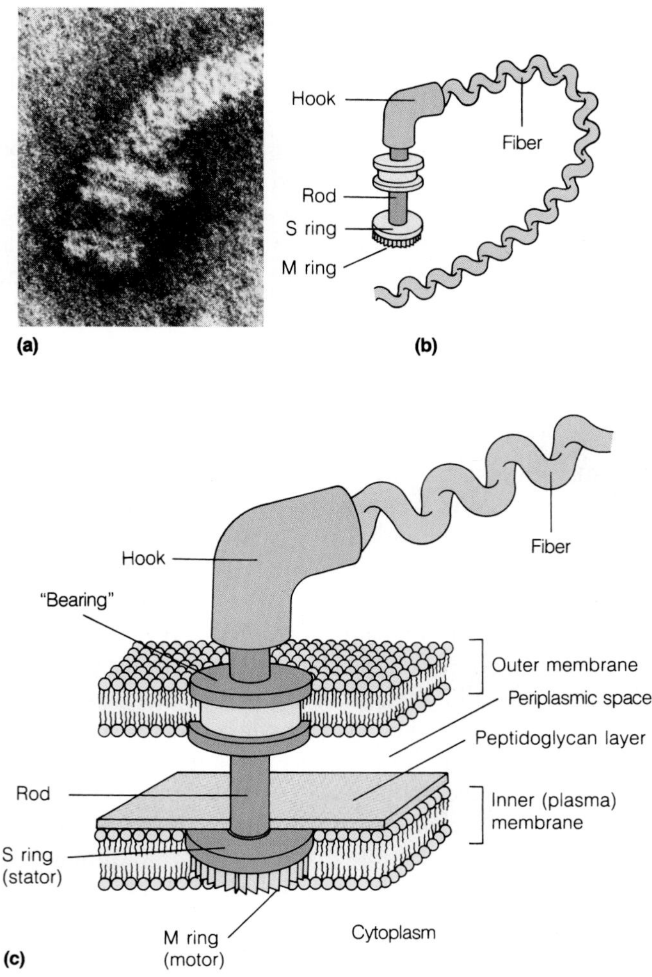

Figure 29.40
The bacterial flagellum and its motor. (**a**) Electron micrograph of a bacterial flagellum assembly. (**b**) The flagellum itself is a helical fiber. It is attached to a minute motor buried in the inner membrane of the bacterium. (**c**) A larger-scale drawing shows that the rod rotates within the S ring of the motor, driven by protons passing through the M ring from the periplasmic space into the cytosol.

latter response is referred to as **chemotaxis** and is demonstrated by a number of species of bacteria (like *E. coli*) that possess flagella for propulsion.

To understand how flagellated bacteria move preferentially toward attractants (often potential nutrients) and away from repellents (such as noxious chemicals), we must first understand their method of locomotion. The bacterial flagellum is a right-handed helical fiber, composed almost entirely of one fibrous protein, **flagellin.** It does not contain microtubules, actin, myosin, or any contractile system. For many years it was assumed that the bacterial flagellum underwent in-plane bending motions, like those of sperm tails. Thus, researchers were surprised to learn that, in fact, it *rotates*. This was demonstrated most simply when the flagellum of a bacterium was stuck to a glass plate by antiflagellin antibodies. Since the flagellum could no longer rotate, the bacterium did!

The remarkable structure that attaches the flagellum to the bacterium and generates the rotation is shown in Figure 29.40. The helical fiber of the flagellum is attached through a "hook" structure to a rod that passes through a "bearing" in the outer bacterial membrane and into the inner membrane. There it passes through a fixed protein "stator" (the S ring) and terminates in a multisubunit rotor. In other words, the flagellum is made to rotate by an ultramicroscopic motor. In a sense this is an electric motor, for

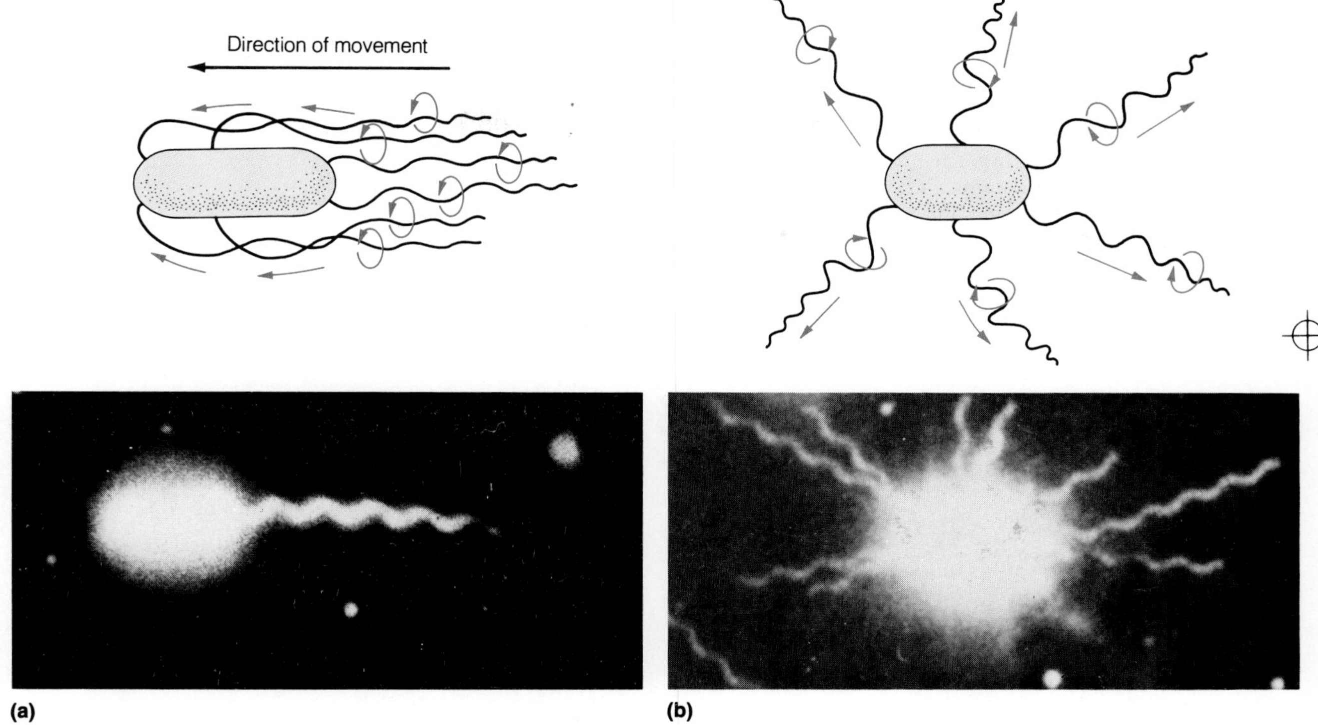

Figure 29.41
Effect of direction of flagellar rotation on cell behavior. (**a**) If the several flagella rotate counterclockwise, they are pulled together into a bundle and drive the bacterium in a straight line (swimming movement). (**b**) When rotation is clockwise, the flagella fly out in all directions, and the bacterium tumbles randomly.

the driving force comes from the proton gradient across the bacterial inner membrane. The motor runs at about 60 revolutions per second and requires the passage of about 1000 protons per revolution.

The structure has still one more remarkable property—it can be reversed. That is, it can rotate the flagellum in either a clockwise or counterclockwise direction. This is very important to the bacterium, for it allows for both steady, rectilinear motion and changes in direction. As Figure 29.41 shows, if the multiple flagella are all rotating counterclockwise, their right-hand helical sense makes them *push* together. They tend to be drawn together in a bundle, and propel the bacterium in a straight line, a movement known as **running.** But if they rotate clockwise, the flagella all *pull.* They fly out from the surface and attempt to pull in all directions. The result is that the bacterium **tumbles.**

In a neutral and uniform environment, periods of running lasting a few seconds alternate with brief periods of tumbling, and the bacterium wanders about randomly (Figure 29.42). The presence of a gradient of either a nutrient or a noxious repellent biases this distribution of running and tumbling. If a bacterium is moving up a gradient of nutrient, tumbling is delayed. So there is net motion toward the source of nutrition. Conversely, a bacterium moving away from a repellent continues to do so for longer than usual before tumbling, resulting in an avoidance reaction.

All of this implies that flagellated bacteria must have some mechanism to sense the existence of gradients of attractants or repellents and in some fashion relay this information to the motors of their flagella. This is, perhaps, the simplest example of how organisms employ information about their environment to modify their behavior. Here, no differentiation of tissues into sensor, neural, and effector cells is involved; everything takes place in one cell. How does it work?

A bacterium is too small to easily detect a gradient by sensing concentration differences along the length of the cell. Rather, it appears to respond

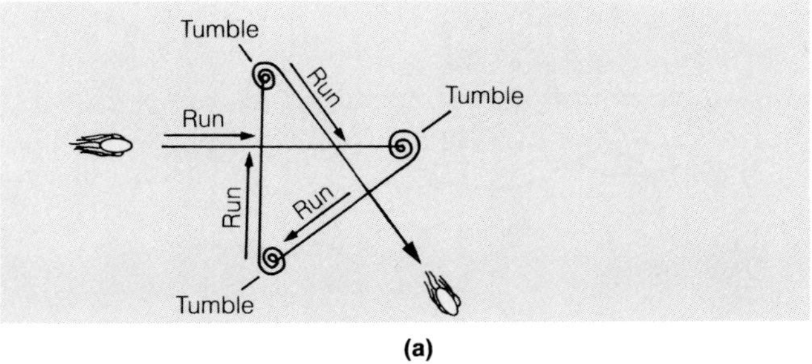

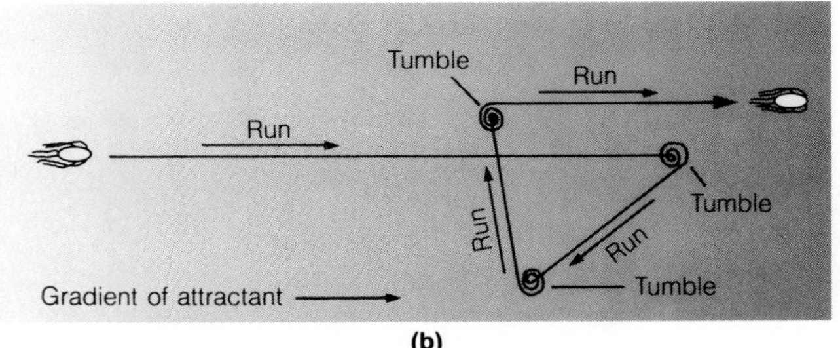

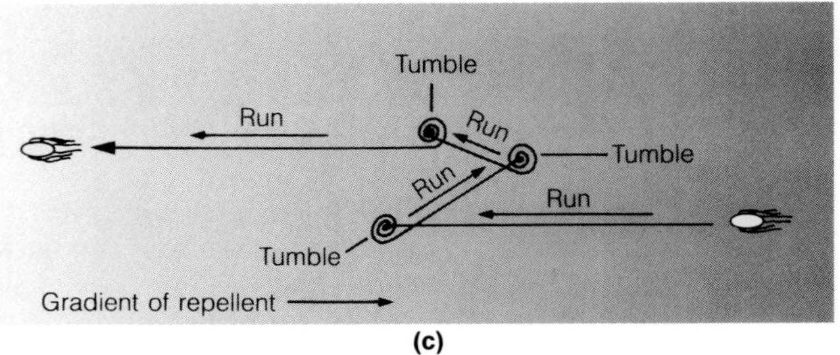

Figure 29.42
Chemotactic motion of bacteria. (**a**) In the absence of either attractants or repellents, the bacterium stops and tumbles frequently, each time starting out in a new, random direction. (**b**) When a gradient of an attractant is present, a bacterium heading toward the attractant tends to run for longer periods without tumbling. This biases its overall motion toward the direction in which the attractant is most concentrated. (**c**) A gradient of a repellent has the opposite effect, favoring long runs away from the repellent source.

to differences in the concentration of attractants or repellents at successive points in time as it moves through the medium. The sensors are transmembrane proteins, embedded in the inner membrane of the bacterium. There are a number of these proteins, each of which can bind, on its outer surface, certain kinds of attractant or repellent molecules. The cytosolic portion of the sensor proteins contains glutamate sites susceptible to methylation. This is why these sensors are called **methyl-accepting chemotaxis proteins (MCPs)**. Binding of attractants stimulates methylation, whereas binding of repellents promotes demethylation. Thus, the methylation state of each kind of MCP keeps a running tally of attractant or repellent concentration.

A cytosolic protein called **CheY** acts as an intermediate between MCPs and the bacterial motor. Somehow, low levels of methylation of MCPs (caused by either repellent binding or the absence of attractants) trigger an ATP-dependent modification of CheY, which in turn is transmitted to the motor, switching it to clockwise rotation and causing tumbling. This allows the bacterium to seek a new and possibly more profitable direction. On the other hand, if the level of an attractant is increasing, tumbling is inhibited by the high methylation, and swimming in the previous direction is maintained. Note that the varying levels of MCP methylation serve as a crude "memory" to aid the bacterium in its "choices."

Chemotaxis and the bacterial motor are fitting subjects with which to close this book. Here is the behavior of an organism expressed most directly at the molecular level, utilizing remarkable structures that have evolved to give these creatures an advantage in survival. In this example we see clearly how the chemical energy gained from consumption of foodstuffs is used to propel the organism toward better food sources.

There is much we still do not understand about even this "simplest" system. Yet it seems that with techniques now in hand and insights already gained, the potential for the understanding of life is almost without limit. So also, perhaps, are the surprises that life has in store for us.

REFERENCES

Neurotransmission

Catterall, W. A. (1986) Molecular properties of voltage-sensitive sodium channels. *Annu. Rev. Biochem.* 55:953–985.

Hille, B. (1984) *Ionic Channels of Excitable Membranes.* Sinauer Associates, Sunderland, Mass. The definitive book on this subject.

Kuffler, S. W., J. G. Nicholls, and A. R. Martin (1984) *From Neuron to Brain.* Sinauer Associates, Sunderland, Mass. The new edition of a classic.

Schofield, P. R., M. J. Darlinson, N. Fujita, et al. (1987) Sequence and functional expression of the GABA receptor shows a ligand-gated receptor super-family. *Nature* 328:221–227.

Stroud, R. M., and J. Finer-Moore (1985) Acetylcholine receptor structure, function, and evolution. *Annu. Rev. Cell Biol.* 1:317–351.

Vision

Chabre, M. (1985) Trigger and amplification mechanisms in visual phototransduction. *Annu. Rev. Biophys. Biophys. Chem.* 14:331–360. A broad, current review of the biochemistry of visual response.

Schnapf, J. L., and D. A. Baylor (1987) How photoreceptor cells respond to light. *Sci. Am.* 256(4):40–47. A less technical but excellent review.

Stryer, L. (1986) Cyclic GMP cascade of vision. *Annu. Rev. Neurosci.* 9:87–119.

Wald, G. (1968) The molecular basis of visual excitation. *Nature* 219:800–807. The Nobel address describing Wald's pioneering work in this field.

Muscle

Bessman, S. P., and C. L. Carpenter (1985) The creatine–creatine phosphate energy shuttle. *Annu. Rev. Biochem.* 54:831–862.

Cooke, R. (1986) The mechanism of muscle contraction. *CRC Crit. Rev. Biochem.* 21:53–118. A good review of the field.

Squire, J. M. (1986) *Muscle: Design, Diversity, and Disease.* Benjamin/Cummings, Menlo Park, Calif. The most current book in the field.

Warrick, H. M., and J. A. Spudich (1987) Myosin: Structure and function in cell motility. *Annu. Rev. Cell Biol.* 3:379–422.

Microtubules

Amos, L. W., R. W. Linck, and A. Klug (1976) Molecular structure of flagellar microtubules. In: *Cell Motility,* edited by R. Goldman, T. Pollard, and J. Rosenbaum, pp. 847–868. Cold Spring Harbor Press, Cold Spring Harbor, New York. A classic early paper on microtubule structure.

Dustin, P. (1978) *Microtubules.* Springer-Verlag, Berlin, Heidelberg, New York. A very complete synopsis and source for the earlier work in this field.

Goodenough, U., and J. E. Heuser (1985) Substructure of inner dynein arms, radial spokes, and the central pair/projection complex of cilia and flagella. *J. Cell Biol.* 100:2008–2018.

Schnapp, B. J., R. D. Vale, M. P. Sheetz, and T. S. Reese (1985) Single microtubules from squid axoplasm support bidirectional movement of organelles. *Cell* 40:455–462.

Summers, K. E., and I. R. Gibbons (1971) ATP-induced sliding of tubules in trypsin-treated flagella of sea urchin sperm. *Proc. Natl. Acad. Sci. U.S.A.* 68:3092–3096.

Vale, R. D., J. M. Scholey, and M. P. Sheetz (1986) Kinesin: Possible roles for a new microtubule motor. *Trends Biochem. Sci.* 11:464–468.

Chemotaxis and the Bacterial Motor

Boyd, A., and M. I. Simon (1982) Bacterial chemotaxis. *Annu. Rev. Physiol.* 44:501–517.

McNab, R. M. (1984) The bacterial flagellar motor. *Trends Biochem. Sci.* 9:185–190.

Meister, M., G. Lowe, and H. C. Berg (1987) The proton flux through the bacterial flagellar motor. *Cell* 49:643–650.

Segall, J. E., S. M. Block, and H. C. Berg (1986) Temporal comparisons in bacterial chemotaxis. *Proc. Natl. Acad. Sci. U.S.A.* 83:8987–8991.

Wolfe, A. J., M. P. Conley, T. J. Kramer, and H. C. Berg (1987) Reconstitution of signaling in bacterial chemotaxis. *J. Bacteriol.* 169:1878–1885.

PROBLEMS

1. Calculate the equilibrium membrane potentials to be expected across a membrane at 37°C, with an NaCl concentration of 0.1 M on the right and 0.01 M on the left, given the following conditions. In each case state which side is +, which −.
   (a) Membrane permeable only to Na^+.
   (b) Membrane permeable only to Cl^-.
   (c) Membrane permeable to both ions.

2. In each of a, b, and c of Problem 1, will any appreciable transport of material take place in establishing the potential?

3. Suppose the following concentrations exist inside and outside a retinal rod cell:

	Concentration (mM)	
	Inside	Outside
K^+	100	5
Na^+	10	140
Cl^-	10	100

   (a) If the relative permeabilities of K^+ and Cl^- are 1.0 and 0.45, respectively, what must be the relative permeability for Na^+ to give a potential of -30 mV, the value found across the resting rod cell membrane? Assume T = 37°C. [Hint: You must rearrange and solve the Goldman equation.]

   (b) If Na^+ gates were to close completely on stimulation by a photon, what value of the membrane potential would be reached?

4. A typical relaxed sarcomere is about 3 μm in length and contracts to about 2 μm in length. Within this, the thin filaments are about 1 μm long and the thick filaments are about 1.5 μm long.

   (a) Describe the overlap of thick and thin filaments in the relaxed and contracted sarcomere.

   (b) An individual "step" by a myosin head in one cycle pulls the thin filament about 7.5 nm. How many steps must each actin fiber make in one contraction?

5. Each gram of mammalian skeletal muscle consumes ATP at a rate of about 1×10^{-3} mol per minute during contraction. Concentrations of ATP and creatine phosphate in muscle are about 4 mM and 25 mM, and the density of muscle tissue can be taken to be about 1.2 g/cm^3.

   (a) How long could contraction continue on ATP alone?

   (b) If all creatine phosphate were converted into ATP and utilized as well, how long could contraction continue?

   (c) Finally, suppose all muscle glycogen (about 1% by weight of muscle) was used for glycolysis. How much additional contraction time could be sustained?

6. Within the myofibrils of resting muscle cells, the concentration of Ca^{2+} is about 10^{-7} M, whereas in the sarcoplasmic reticulum the value is about 10^{-2}. The myofibrillar side of the membrane is at a potential of about -75 mV with respect to the outside. Calculate ΔG for the transfer of 1 mol of Ca^{2+} from the myofibril to the sarcoplasmic reticulum at 37°C.

7. Tubulin binds GTP, and the polymerization of a microtubule is markedly stimulated when a tubulin molecule at the growing end carries GTP. Tubulin also has a GTPase activity, with low turnover number. Microtubules carrying GDP at their ends are more likely to lose tubulin monomers. Show how these facts can explain the remarkable observation that at certain GTP levels some microtubules in a mixture will grow while others spontaneously shrink.

8. Predict the effects on the behavior of a chemotactic bacterium of the following mutations:

   (a) A mutant deficient in CheY protein production.

   (b) An overproducer of CheY.

Answers to Problems

CHAPTER 2

1. (a) 3.54 kJ/mol.
   (b) 151 kJ/mol.

2. Since O=C=O is symmetric, the dipole moments associated with the O=C bonds will be equal and opposed vectors, so they will cancel. In C=O this does not happen.

3. The graph will be a reflection of the graph given, across the $E = 0$ line.

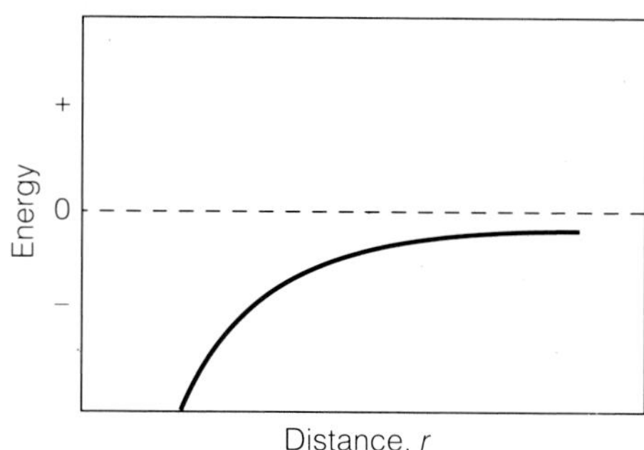

4. (a) 0.456.
   (b) 2.608 to first approximation; 2.609 to second approximation; 2.609 exact.
   (c) 3.108 to first approximation; 3.113 to second approximation; 3.113 exact.

5. (a) pH = 4.46.
   (b) pH = 2.57.

6. (a) 0.138 M.
   (b) $[KH_2PO_4] = 0.126$ M; $[Na_2HPO_4] = 0.174$ M.

7. pI = 11.1. Form I will be present in only insignificant amounts above pH 7.

8. The ionic strength of 1 M $(NH_4)_2SO_4$ is 3.0.

CHAPTER 3

1. $\Delta H_f - \Delta E_f = -0.165$ J/mol, $\left(\dfrac{\Delta H_f - \Delta E_f}{\Delta E_f}\right) \times 100 =$ 0.003%

2. (a) +926 kJ/mol.
   (b) 463 kJ/mol.

3. (a) $\Delta S° = +470$ J/° mol.
   (b) A solid material is being transformed entirely into gases. The greater freedom of motion of gas molecules results in an entropy increase.
   (c) $\Delta E° = 103.4$ kJ/mol.
   (d) Because the system does work on the surroundings, in expanding from a solid to a gas.

4. (a) $\Delta G° = -2872$ kJ/mol.
   (b) $\Delta G° = -1713$ kJ/mol.
   (c) 40%.

5. (a) 1.1×10^{-4} mM.
   (b) $\Delta G°' = -16.5$ kJ/mol.
   (c) 9.1 M. It is never reached because glucose-6-phosphate is continually consumed in other reactions. The system never reaches equilibrium.

6. 0.12 mM.

7. (a) $K = 5.44 \times 10^{-2}$; $(f_{G3P})_{eq} = 0.052$.
   (b) $\Delta G = -4.37$ kJ/mol.

8. (a) ΔS must be positive, because the increase in available states corresponds to an increase in entropy.
   (b) Since $\Delta G = \Delta H - T\,\Delta S$, a positive ΔS yields a negative contribution to ΔG (T is always a positive number). Thus, for proteins to be stable, which requires ΔG for the above to be positive, denaturation must involve a large positive ΔH and/or an additional negative contribution to ΔS. As we shall see in Chapter 6, both occur.

9. Given $\Delta G° = \Delta H° - T\,\Delta S°$ and $\Delta G° = -RT \ln K$, combination and rearrangement gives the desired result, which is called the van't Hoff equation. According to the equation, if $\Delta H°$ and $\Delta S°$ are independent of temperature, a graph of $\ln K$ vs. $1/T$ should have a slope of $-\Delta H°/R$.

10. (a) $\Delta H° = +59.0$ kJ/mol.
    (b) $\Delta S° = -103$ J/° mol. To obtain the true K, so as to get $\Delta G°$, one must divide K_w by 55.5, the molar concentration of pure water.

11. (a) $K = 16.8$.
    (b) Toward glucose-l-phosphate.

CHAPTER 4

1. It must be a single-strand DNA, since Chargaff's rules are not obeyed.

2. 3′ TGGCATTCCGAAATC 5′.

3.

```
          A—T
          |  |
          C···G
          |  |
          C···G
          |  |
          T···A
          |  |
          G···C
          |  |
          A···T
          |  |
···T—C—A···T—G—G— ···
   ⋮  ⋮        ⋮  ⋮
 A—G—T···A—C—C— ···
          |  |
          T···A
          |  |
          C···G
          |  |
          A···T
          |  |
          G···C
          |  |
          G···C
          |  |
          T—A
```

4. 5′ GTTACTGGACG 3′.

5. (a) 28 base pairs and 3 turns.
   (b) More tightly. You can see this most easily by noting that an infinite number of base pairs per turn corresponds to a wholly untwisted DNA.

6. (a) $L_0 = 200$.
   (b) $\sigma = -0.06$.

7. (a) $\Delta T = -2$; for one turn of DNA goes from a twist of $+1$ to -1.
   (b) $\Delta L = 0$; L cannot change unless the DNA is cut and resealed.
   (c) $\Delta W = +2$ to compensate for the change in ΔT. If the DNA were initially negatively supercoiled, its negative writhe would decrease.

8. 0.0586 mm.

9. 10^6 copies.

10. A = 20%, G = 30%, T = 20%, C = 30%

CHAPTER 5

1. (a) SYSMEHFRWGKPV.
   (b) 1624.

2. (a) The curve will exhibit inflections corresponding to two groups titrating near 4 (carboxyl terminus and glutamic side chain), one group near 7 (histidine), two near 9 (N-terminus and lysine), one near 10 (tyrosine), and one near 12 (arginine).
   (b) Approximately -2, $+2$, $+4$.
   (c) pI = 8.

3. (a) SYSMEHFR, WGK, PV.
   (b) SYSM*, EHFRWGKPV; M* = homoserine lactone.
   (c) SYSMEHF, RWGKPV.

4. (a) DSGPYKMEHFRWGSPPKD.
   (b) Cyanogen bromide.

5. (a) Two.
   (b) The bonds may be 1–3 and 11–15, or 1–11 and 3–15, or 1–15 and 3–11. Cleavage with trypsin will produce,

in the first case, two peptides plus arginine, and in the second and third cases, one peptide plus arginine. The latter two can be distinguished by *partial* oxidation of disulfide bonds; the mix of peptides produced will be distinguishable.

6. (a) One possibility, of many:

···UGUAAUUGUAAAGCGCCCGAGACCGCGCUUU
GUGCUCGACGAUGUCAACAACAU···

(b) Since it does not have an N-terminal methionine, at least some proteolytic cleavage must be involved in its synthesis.

7. Met–Phe–Pro–Ser–Tyr–Pro–Lys–Asp–Lys–Lys–Glu–

CHAPTER 6

1. (a) Left-handed. (b) 3.0.

2. They are spaced about three to four residues apart. Therefore, they will all lie on the same side of the α-helix. This suggests that this side of the helix may face the interior of the protein.

3. In the unfolded form of the protein, most of its hydrogen bond donors and acceptors can make H bonds to water. Therefore it is only the *difference* in hydrogen bonding energy that contributes to protein stability.

4. (a) $w = 3^{200} = 2.7 \times 10^{95}$.

   (b) Not all of these conformations will be sterically possible. But even if only 0.1% of these are allowed, there are still 2.7×10^{92}, a very large number.

5. (a) $$\begin{aligned}\Delta S_{\text{folding}} &= S_{\text{folded}} - S_{\text{unfolded}}\\ &= R \ln W_{\text{folded}} - R \ln W_{\text{unfolded}}\\ &= 8.314\ \text{J/° mol} \times [\ln 1 - \ln 2.7 \times 10^{92}]\\ &= -1769\ \text{J/° mol} = -1.77\ \text{kJ/° mol}\end{aligned}$$

   (b) $\Delta H_{\text{folding}} = 100 \times (-5\ \text{kJ/mol}) = -500\ \text{kJ/mol}$. Note: We have neglected the fact that some H bonds cannot be formed at helix ends. This will make only a small difference.

   (c) $$\begin{aligned}\Delta G_{\text{folding}} &= \Delta H_{\text{folding}} - T\,\Delta S_{\text{folding}}\\ &= -500\ \text{kJ/mol} - 298°(-1.77\ \text{kJ/° mol})\\ &= +27\ \text{kJ/mol}\end{aligned}$$

   Since $\Delta G_{\text{folding}}$ is >0 at 25°, the protein would not be stable. It would be stable at 0°C. This points out the importance of sources of stabilization other than backbone H bonds.

6. An α-helix from residues 4–11, β sheet between 14–19 and 24–30. There is very probably a β turn involving residues 20–23.

7. (a) C_4, held together by heterologous interaction, or D_2, held together by isologous interaction.

   (b) C_2, since each $\alpha\beta$ dimer forms an asymmetric unit.

8. (a) Using data in Table 5.1, we calculate the molecular weight to be 1072 g/mol. A 1 mg/cm^3 solution will be 9.33×10^{-4} M. From Figure 5.4a, we see that the molar extinction coefficient of Tyr is ~1000 l/cm mol. Phe does not contribute appreciably. Therefore ϵ is 0.93 cm^2/mg.

   (b) 2.8 mg/cm^3.

   (c) 5%.

CHAPTER 7

1. (a) 2.6 mm. (b) About 0.94.

2. (a) (1) 98%; (2) 56%.

   (b) About 42%.

   (c) About 70%.

3. Note that the chloride ion (Figure 7.16) interacts with positively charged groups in the $\alpha_1\beta_2$ interface, in the deoxy state. Thus, its presence stabilizes deoxyhemoglobin, and the higher the Cl^- concentration, the lower will be the O_2 affinity of hemoglobin.

4. (a) $P_{50} = 4.47$ mm. (Note: This very low value is a consequence of stripping, plus the fact that T is lower than physiological (37°.)

   (b) $n_H \cong 3.5$.

   (c) $P_{50T} = 28$ mm, $P_{50R} = 0.022$ mm. These are obtained by extrapolating the limiting lines on the Hill plot.

5. Hemoglobin concentration can influence affinity only if dissociation of the tetramer is occurring. The likely dissociation is to $\alpha\beta$ dimers, which will have lost some of the interactions that stabilize the T state. Thus, affinity should rise and P_{50} decrease.

6. (a) $P_{50} = 125$ mm.
   (b) $n_H = 3.1$.
   (c) Since $n_H = 3.1$, there must be *at least* 4 sites, because $n_H \leq n$. (Actually, this is an immense molecule with 24 binding sites.)

7. (a) This should break a salt bridge stabilizing the T state and hence increase affinity. This mutation is known; it is *hemoglobin Hiroshima*.
   (b) Since F8 is involved in heme binding, this mutation should be unstable; leucine will not ligate to the heme iron.
   (c) Since Arg 141 is involved in binding Asn 126 in the T state, this mutation should destabilize the T state and increase affinity. It is known as *hemoglobin Legano*.
   (d) This histidine is involved in binding of BPG (Figure 7.23). Its replacement by the negatively charged Asp should greatly reduce BPG binding and hence increase O_2 affinity.

8. (a) If one multiplies both sides of the first equation by $1 + Kc$, divides through by c, and rearranges, the second equation results.
   (b) According to the second equation, the slope of a graph of r/c versus r should be $-K$, and its intercept at $r/c = 0$ should give n. Using the data in this way, we get $K = 2.2 \times 10^4$ M^{-1} and $n = 2.1$. Since the true number of sites must be integral, we would choose $n = 2$.

CHAPTER 8

1. (a)

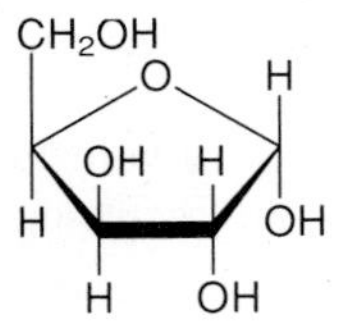

α-D-Xylofuranose

(b)

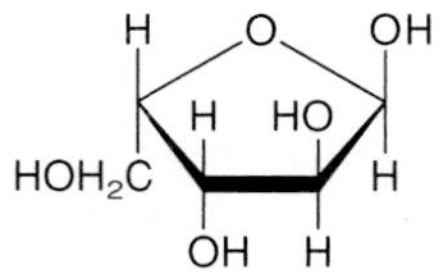

(c)

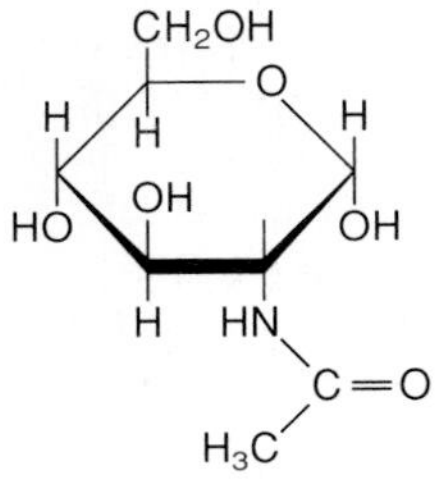

(d)

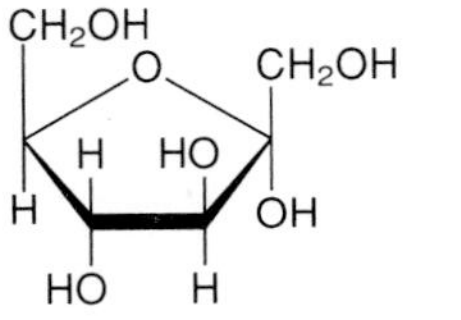

2. Galactitol has the structure shown below:

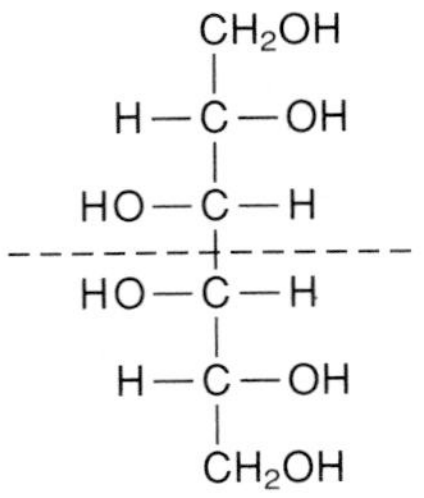

Since it has a plane of symmetry between C-3 and C-4, it is optically inactive. Such compounds, which contain asymmetric carbons but have no net optical activity, are called *meso* forms.

3. For the reaction

$$\alpha\text{-D-glucopyranose} \rightleftharpoons \beta\text{-D-glucopyranose}$$

we have $K = 64/36 = 1.78$. Therefore

$$\Delta G° = -RT \ln K = -8.314 \text{ J/° mol} \times 313° \times 0.577$$
$$= -1.50 \text{ kJ/mol}$$

4.

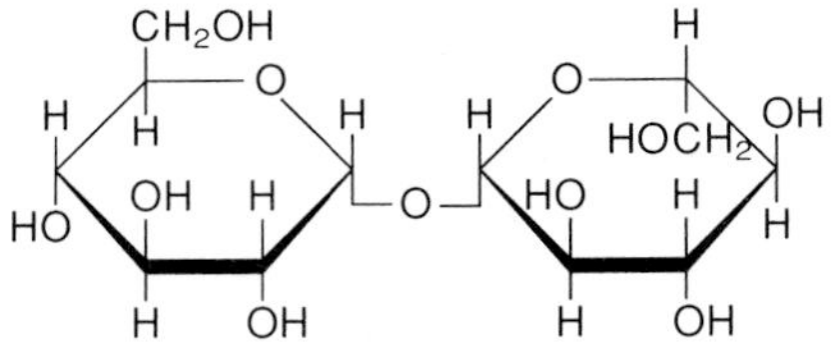

5. Reducing: maltose, cellobiose, lactose, gentiobiose. Nonreducing: sucrose, trehalose.

6.

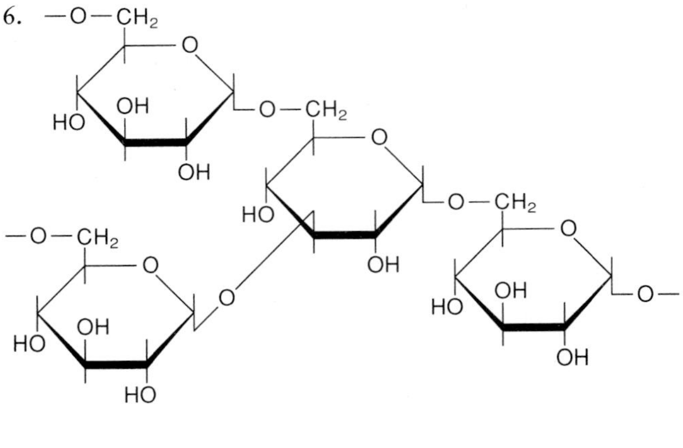

Note: For simplicity, H atoms are represented by vertical bars.

7. Hyaluronic acid.

8. (a) S; (b) R.

9. The pyranose form will contain three adjacent hydroxyl groups and thus two bonds will be cleaved, whereas the furanose form, with only a pair of adjacent hydroxyls available, will suffer cleavage of only one bond.

10. Penicillin interferes with a function of growing bacteria, in inhibiting proper cell wall formation. Dead bacteria, which are not forming new cell wall, are unaffected.

CHAPTER 9

1. (a) $CH_3CH_2CH{=}CH(CH_2)_7COOH$.

   (b) $CH_3(CH_2)_5CH{=}CH(CH_2)_9COOH$. (This is called vaccenic acid.)

   (c) According to Table 9.1, any with fewer than 10 carbons will melt below 30°C. An example would be $CH_3(CH_2)_6COOH$, *n*-octanoic acid.

2. (a) Fatty acid, long-chain alcohol.

   (b) Glycerol, fatty acid.

   (c) Carbohydrate, long-chain alcohol.

3. From the data, 4.74×10^9 cells would have a total surface area of 4.74×10^{11} $(\mu m)^2 = 0.474$ m^2. The ratio of monolayer area to cell surface is 0.89 $m^2/0.474$ $m^2 = 1.89$, very close to 2.00.

4. The concentration ratio is 10^5. Therefore

   (a) $\Delta G = 29.67$ kJ/mol.

   (b) $\Delta G = 36.4$ kJ/mol.

   In either case, hydrolysis of 1 mol of ATP would suffice to transport 1 mol of the ion.

5. $J = -P(C_2 - C_1)$; from Table 9.6, we have $P = 2.4 \times 10^{-10}$ cm/s. If we express $C_2 - C_1$ in mol/cm^3, J will have dimensions of mol/cm^2 s. Calculation yields $J = 2.04 \times 10^{-14}$ mol/cm^2 s. Therefore, the amount transferred in 1 minute across 100 $(\mu m)^2$ $(= 100 \times 10^{-8}$ cm$^2)$ will be

$$M = -1.224 \times 10^{-18} \text{ mol.}$$

6. The initial concentration inside = 100 mM = 0.1×10^{-3} mol/cm^3. The cell volume is 100 $(\mu m)^3$, or 100×10^{-12} cm^3. The amount initially present is

$$M = 0.1 \times 10^{-3} \text{ mol/cm}^3 \times 1 \times 10^{-10} \text{ cm}^3 = 1 \times 10^{-14} \text{ mol.}$$

   Therefore, the percent escaping in 1 min will be

$$\frac{1.224 \times 10^{-18} \text{ mol}}{1 \times 10^{-14} \text{ mol}} \times 100 = 1.224 \times 10^{-2}\%.$$

   Thus, unless facilitated transport is available, "leakage" of K^+ from cells is very slow.

7. If the calcium ion is *maintained* at this concentration difference, we must have, from equation (9.2), $V_{in-out} = -92$ mV. Since V_{in-out} is defined as the potential inside minus the potential outside, the inside of the organelle is negative with respect to the outside.

CHAPTER 10

1. Graphing the data according to first-order kinetics (ln ([A]/[A]$_0$) vs. t) gives a very curved line, but graphing according to second-order kinetics (1/[A] vs. t) gives a linear plot. Therefore a second-order reaction is indicated.

2. About 1.1×10^{14}.

3. $k = 4 \times 10^9$ (mol/L)$^{-1}$ s^{-1}. Some values of k_{cat}/K_m approach this limit.

4. (a) It is quite difficult to estimate V_{max} from these data. A value of about 150 $(\mu$mol/L$)^{-1}$ min^{-1} might be guessed at. Then, by taking K_M as the substrate concentration at $V_{max}/2$, we would estimate $K_M = 20$ μmol/L.

   (b) $V_{max} = 162$ $(\mu$mol/L) min^{-1}; $K_M = 32$ μmol/L. These values are considerably more reliable than those found in (a).

   (c) Best fit gives $V_{max} = 158$ $(\mu$mol/L) min^{-1}; $K_M = 30$ μmol/L.

5. (a) Using $V_{max} = 160$ $(\mu$mol/L) min^{-1}, $k_{cat} = 1.60 \times 10^5$ min^{-1} = 2667 s^{-1}

   (b) Using $K_M = 31$ μmol/L, $k_{cat}/K_M = 8.6 \times 10^7$ (mol/L)$^{-1}$ s^{-1}. This compares favorably with the largest values of k_{cat}/K_M.

6. (a) It is necessary only to rearrange the equation to

$$\left(\frac{K_M}{[S]} + 1\right) d[S] = -V_{max}\, dt$$

   and then integrate from $t = 0$ ([S] = [S]$_0$) to a finite time t. We obtain

$$[S]_0 - [S] - K_M \ln \frac{[S]}{[S]_0} = V_{max} t$$

   (b) When $[S]_0 \gg K_M$, the first term on the left will be much larger than the second, so

$$[S_0] - [S] \cong V_{max} t$$

   or

$$[S] = [S]_0 - V_{max} t$$

   This corresponds to a situation in which the enzyme is saturated with substrate, so substrate molecules are simply being consumed at the maximum rate.

7. One way is to rearrange the answer to 6(a) to

$$\frac{1}{t} \ln\left(\frac{[S]}{[S]_0}\right) = \frac{1}{K_M}\left(\frac{[S]_0 - [S]}{t}\right) - \frac{V_{max}}{K_M}$$

   and graph the left-hand side vs. ([S$_0$] − [S])/t. The slope will give K_M, and the intercept at $(1/t) \ln [S]/[S]_0 = 0$ will give V_{max}.

8. The change in k_{cat} between pH 6 and pH 7 must involve loss of a proton in the active site. The best candidate is His 57. The increase in K_M at higher pH must involve a change in the binding site. Although you could not decide from data we have given, the group involved is probably the N-terminal on Ile 16, created by the cleavage that activates chymotrypsin (see Chapter 11).

9. The simplest technique would be to incubate the ribozyme with a radiolabeled pentanucleotide and collect samples from time to time for electrophoresis. Autoradiography should initially show only the pentanucleotide, but with time both longer and shorter oligomers should appear on the gel.

10. RNA molecules possess the two criteria—the possibility of self-replication and catalytic capabilities—that would seem essential for the first "living" matter. No protein has the first, and no DNA has been shown to have the second.

CHAPTER 11

1. (a) Noncompetitive.
   (b) No inhibitor: $K_M = 2.5$ (mmol/L), $V_{max} = 5.0$ (mmol L^{-1} min^{-1}). With inhibitor: $K_M = 2.5$ (mmol/L), $V_{max} = 3.0$ (mmol L^{-1} min^{-1}).

2. (a) Competitive.
   (b) $V_{max} = 5.0$ (mmol L^{-1} min^{-1}) at all inhibitor concentrations.

[I] (mmol/L)	K_M(app) (mmol/L)
0	2.5
3	4.0
5	5.0

   (c) $K_I = 0.2$ (mmol/L).

3. According to equation (10.34), we can write the Michaelis–Menten equation as $V = V_{max} - K_M V/[S]$, so we plot V vs. $V/[S]$. The intercept at $V/[S] = 0$ will give V_{max}; the slope will be K_M(app). Therefore the graphs will look like the following sketches:

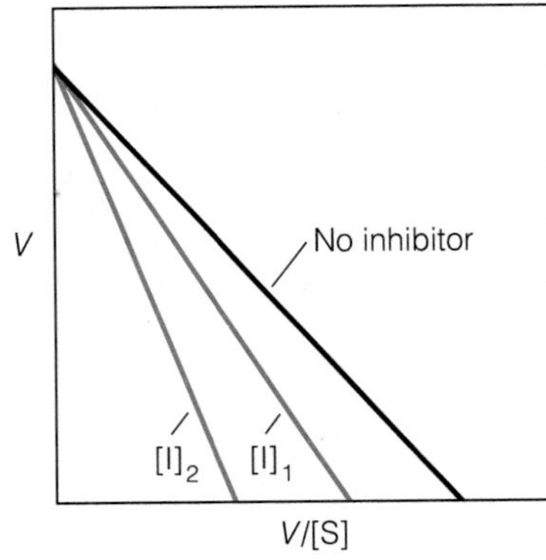

(a) Competitive inhibition
$[I]_1 < [I]_2$

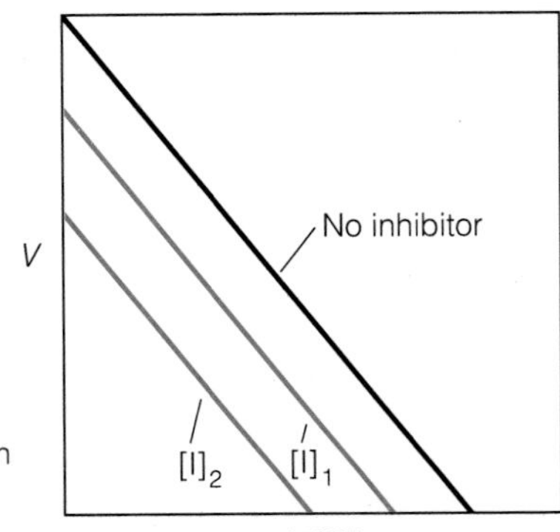

(b) Noncompetitive inhibition
$[I]_1 < [I]_2$

4. One should replace the hydrophobic group in TPCK with a positively charged group. A good candidate would be

$$H_3\overset{+}{N}-(CH_2)_4-\underset{\underset{\underset{\underset{\displaystyle C_6H_4-CH_3}{|}}{O=S=O}}{|}}{\underset{NH}{|}}{\overset{H}{\overset{|}{C}}}-\overset{O}{\overset{\|}{C}}-CH_2Cl$$

5. (a) If we make a V vs. [S] graph, the data show a slightly sigmoidal curve. A more convincing demonstration is to use a Lineweaver–Burk plot or an Eadie–Hofstee plot. Neither is linear.
   (b) From the Lineweaver–Burk plot, we can estimate V_{max} to be approximately 5 (mmol/L^{-1} min^{-1}).

6. If $V = \theta V_{max}$, then a graph of $\log[\theta/(1-\theta)]$ vs. log[S], as suggested in Chapter 7, translates into a graph of $\log[(V/V_{max})/(1 - V/V_{max})]$ vs. log[S] (see sketch). The linear limiting lines at low and high [S] will correspond approximately to the T and R states, respectively. An activator should shift the curve toward R, and an inhibitor toward T (see graph).

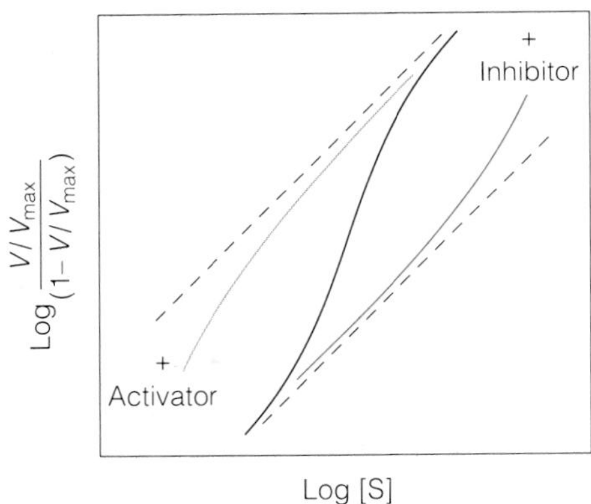

7. (a)

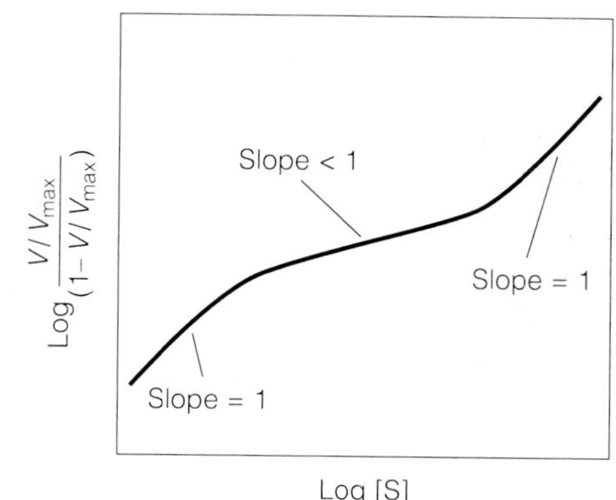

(b) In the MWC theory, the more favorable binding to the R state makes the equilibrium shift as [S] is increased; the shift can only be toward R. In the KNF theory, binding to one site can have *any* kind of effect on another, including inhibition.

8. (a) The key to the derivation is to note that

$$\theta = \frac{\text{number of sites occupied}}{\text{total number of sites}} = \frac{[RS] + 2[RS_2]}{2([T] + [R] + [RS] + [RS_2])}$$

Expressions given for [T], [RS], and $[RS_2]$ are inserted and the equation simplified.

(b) This must be done graphically. A large L and small K ensure a sigmoidal curve.

9. (a) Activation of the molecule by trypsin will be blocked.

(b) If activation occurs, the N-terminal peptide will no longer be constrained by an —S—S— bond and may be released.

(c) Automodification of π-chymotrypsin to α-chymotrypsin would be blocked.

10. The technique of site-directed mutagenesis allows us to change the amino acid sequences at will. However, the *first* residue in a newly synthesized protein will always be either methionine or *N*-formylmethionine. One would have to choose a protein in which the N-terminal Met is removed (a common occurrence) and make modifications in the *second* codon.

CHAPTER 12

1.

	Equation	*RQ*
(a)	$C_2H_5OH + 3O_2 \longrightarrow 2CO_2 + 3H_2O$	0.67*
(b)	$CH_3COOH + 2O_2 \longrightarrow 2CO_2 + 2H_2O$	1.0
(c)	$CH_3(CH_2)_{16}COOH + 26O_2 \longrightarrow 18CO_2 + 18H_2O$	0.69
(d)	$CH_3(CH_2)_{14}(CH)_2COOH + 25.5O_2 \longrightarrow 18CO_2 + 17H_2O$	0.71
(e)	$CH_3(CH_2)_{12}(CH)_4COOH + 25O_2 \longrightarrow 18CO_2 + 16H_2O$	0.71

*Example: RQ = $2CO_2/3O_2$.

2. (a) >1; (b) <1. Since NAD^+-dependent enzymes usually act to dehydrogenate substrates, an $[NAD^+]/[NADH]$ ratio greater than unity tends to drive reactions in that direction. Similarly, $[NADP^+]/[NADPH]$ ratios less than unity provide concentrations that tend to drive these reactions in the direction of substrate reduction.

3. −45.2 kJ/mol ($\Delta G' = \Delta G^{\circ\prime} + RT \ln[ADP][P_i]/[ATP]$).

4. (a) 20, in the direction B to C (i.e., 100 − 80).

(b) (1) The rightward flux decreases twofold. (2) The rightward flux increases by 50%. (3) Flux would change to leftward, with a value of 60.

(c) Specific inhibitors of enzyme X or Y would be useful. One could partially inhibit either activity and observe the effect on flux rate. If a substrate cycle is operating, the change in flux rate should be greater than the percent change in activity of either enzyme.

5. The most direct way is to purify the enzyme to homogeneity and determine both its molecular weight and turnover number. Then, from the activity observed in a crude extract, one can calculate the number of active enzyme molecules needed to achieve that activity (assuming that the extract does not contain inhibitors of the activity). Another approach is to treat the extract with a specific antibody to the enzyme of interest and quantitate the protein immunoprecipitated. One still needs to know the molecular weight to convert this value to number of molecules of enzyme.

6. (a) 0.20 (600 cpm/pmol ÷ 3000 cpm/pmol).

(b) 7500 cpm dTTP incorporated per minute per 10^6 cells, which gives 2.5 pmol per cell per minute per 10^6 cells, or 1.5×10^6 molecules per cell per minute.

(c) Prepare an acid extract of the cells (e.g., 5% trichloroacetic acid), separate the nucleotides by ion-exchange HPLC, and determine in the dTTP fraction its radioactivity and its mass, the latter from UV absorbance.

7. (a) 9131. (b) 2.6%. This problem is solved by using the differential form of the radioactive decay equation, $dN/dt = \lambda N$, where dN/dt is the radioactive dacay rate in dps, λ is $0.693/t_{1/2}$, and N is the number of radioactive atoms.

CHAPTER 13

1. Neither; the same.
$\Delta G' = \Delta G^{\circ\prime} + RT \ln([\text{FBP}][\text{ADP}]/[\text{F6P}][\text{ATP}])$

2. Ethanol will generate NADH, through action of alcohol dehydrogenase, and this will reduce formaldehyde back to methanol, also through alcohol dehydrogenase. The acetaldehyde formed can be metabolized further to acetate. Also, ethanol may simply compete with formaldehyde for binding to alcohol dehydrogenase.

3. Probably F1,6BP. As triose phosphate began to accumulate, the unfavorable equilibrium for the forward reaction might drive both DHAP and G3P back to F1,6BP. Of course, G3P would accumulate first, but probably not to significantly increased levels.

4. (a) 4 (2 each from glucose and fructose).
(b) 5 (3 from G6P + 2 from fructose).

5. C-3 or C-4; both become C-1 of pyruvate, which is lost as CO_2 in the pyruvate decarboxylase reaction.

6. (a) Sucrose + $5P_i$ + 5ADP $\longrightarrow$ 4 lactate + 5ATP + $4H_2O$ + $4H^+$.
(b) Maltose + $4P_i$ + 4ADP + $4NAD^+$ $\longrightarrow$ 4 pyruvate + 4ATP + $3H_2O$ + 4NADH + $4H^+$.
(c) (Glucose) residue + $2P_i$ + 2ATP $\longrightarrow$ 2 ethanol + 2ATP + $2CO_2$ + H_2O + $2H^+$.

7. Glyceraldehyde-3-phosphate dehydrogenase. The acyl arsenate analog of 1,3-bisphosphoglycerate spontaneously hydrolyzes.

8. 1.09 M (10% EtOH is 2.18 M, and 1 mol of glucose generates 2 mol of ethanol). No, because glucose is not an abundant free component of most fermentable plant sources. Starch, sucrose, or some other oligosaccharide or polysaccharide.

CHAPTER 14

1. Administer separately 1-[^{14}C]glucose and 6-[^{14}C]glucose, and measure initial rates of $^{14}CO_2$ formation. The ratio gives relative flux rates through the two pathways. For example, if the flux rates are equal, the ratio from C1-labeled glucose to C6-labeled glucose is 2:1.

2. (a) G6P + $2NADP^+$ + H_2O $\longrightarrow$ R5P + CO_2 + 2NADPH + $2H^+$
(b) G6P + $12NADP^+$ + $7H_2O$ $\longrightarrow$ $6CO_2$ + 12NADPH + $12H^+$ + P_i
(c) 3G6P + $6NADP^+$ + $25NAD^+$ + 5FAD + 13ADP + $10P_i$ + $5H_2O$ $\longrightarrow$ $18CO_2$ + 6NADPH + 25NADH + 13ATP + $5FADH_2$ + $12H^+$

3. C-1: all released as CO_2. C-2, C-3: all retained in oxaloacetate.

4. One turn: one-quarter. Two turns: three-eighths.

5. The action of pyruvate carboxylase on the labeled pyruvate would yield oxaloacetate labeled so that on a turn through the citric acid cycle it could label C-5 of isocitrate.

6. Addition to isolated glyoxysomes of citrate, isocitrate, glyoxylate, malate, or oxaloacetate would stimulate succinate formation out of proportion to the amount added.

7. C-3 and C-4, since these become the carboxyl group of pyruvate, which is lost in the pyruvate dehydrogenase reaction.

8. C-1 and C-3 of fructose-6-phosphate should be labeled. Erythrose-4-phosphate should be unlabeled.

9. Because NADH and acetyl-CoA activate the enzyme, it makes metabolic sense to expect that NAD^+ and CoA-SH would be inhibitory, and these inhibitions are observed.

10. First, the cytosolic location of much of the $NADP^+$-dependent enzyme raises suspicions about its role in the citric acid cycle. More important, cells that have high or low flux rates through the cycle have high or low activities, respectively, of the NAD^+-dependent enzyme. Activity of the $NADP^+$-dependent enzyme does not vary in coordination with activities of other cycle enzymes.

11. Pyruvate + TPP + H^+ $\xrightarrow[E_1]{\text{step 1}}$ hydroxyethyl-TPP + CO_2

$$\text{Hydroxyethyl-TPP} + \text{lip}\langle^{\text{S}}_{\text{S}} \xrightarrow[E_2]{\text{step 2}} \text{TPP} + \text{lip}\langle^{\text{S-Ac}}_{\text{SH}}$$

$$\text{lip}\langle^{\text{S-Ac}}_{\text{SH}} + \text{lip}\langle^{\text{S}}_{\text{S}} \xrightarrow[E_2]{\text{step 3}} \text{lip}\langle^{\text{S}}_{\text{S}} + \text{lip}\langle^{\text{S-Ac}}_{\text{SH}}$$

$$\text{lip}\langle^{\text{S-Ac}}_{\text{SH}} + \text{CoA-SH} \xrightarrow[E_2]{\text{step 4}} \text{lip}\langle^{\text{SH}}_{\text{SH}} + \text{CoA-S-Ac}$$

$$\text{lip}\langle^{\text{SH}}_{\text{SH}} + \text{FAD} \xrightarrow[E_3]{\text{step 5}} \text{lip}\langle^{\text{S}}_{\text{S}} + \text{FADH}_2$$

$$\text{FADH}_2 + \text{NAD}^+ \xrightarrow[E_3]{\text{step 6}} \text{FAD} + \text{NADH} + \text{H}^+$$

Sum: Pyruvate + CoA-SH + NAD^+ $\longrightarrow$ CoA-S-Ac + CO_2 + NADH

CHAPTER 15

1. +28.95 kJ/mol (from the equation $\Delta G^{\circ\prime} = -nF\,\Delta E_0'$; $E_0' =$ [−0.17 − (−0.32)].

2. (a) cyt c ⟶ cyt a ⟶ cyt a_3 ⟶ O_2.
   (b) To block oxidation of endogenous substrates.
   (c) One site is associated with cytochrome oxidase.
   (d) 2 cyt c-Fe^{2+} + $\frac{1}{2}O_2$ + $4H^+$ + ADP + P_i ⟶ 2 cyt c-Fe^{3+} + ATP + $2H_2O$
   (e) −70 kJ/mol (calculated as in Problem 1).

3. (a) β-Hydroxybutyrate ⟶ NAD^+ ⟶ FMN ⟶ Fe-S ⟶ CoQ ⟶ cyt b ⟶ Fe-S ⟶ cyt c_1 ⟶ cyt c.
   (b) 2, because cytochrome oxidase is bypassed.
   (c) Because NADH cannot freely enter the mitochondrion.
   (d) To block cytochrome oxidase, so that electrons exit the chain at cytochrome c.
   (e) β-Hydroxybutyrate + 2 cyt c-Fe^{3+} + 2ADP + $2P_i$ + $4H^+$ ⟶ acetoacetate + 2 cyt c-Fe^{2+} + 2ATP + $2H_2O$.
   (f) −35.8 kJ/mol (calculated as in Problem 1).

4. To block succinate dehydrogenase and measure phosphorylation resulting only from the α-ketoglutarate dehydrogenase reaction. P/O ratio = 4: 3 from NADH and one from the succinyl-CoA synthetase reaction.

5. +67.6 kJ/mol (solved as in Problem 1); 0.103, calculated from the expression $\Delta G' = \Delta G^{\circ\prime} + 2.3RT \log$ [fumarate][NADH]/[succinate][NAD^+], where $\Delta G' < 0$.

6. [ADP] = 1.44 mM, [AMP] = 0.29 mM (two simultaneous equations: the defining equation for energy charge, and [ADP] = 5 × [AMP]). $\Delta G' = -41.2$ kJ/mol (from $\Delta G' = \Delta G^{\circ\prime} + 2.3RT \log[\text{ADP}][\text{P}_i]/[\text{ATP}]$).

7. 2.2×10^5 (calculate $\Delta G^{\circ\prime}$ as in Problem 1, and then apply $\Delta G^{\circ\prime} = -2.3RT \log K_{eq}$).

8. Add ADP in limiting amount and measure O_2 uptake. The ratio of μmol ADP consumed to μatom oxygen taken up is identical to the P/O ratio (see Figure 15.18).

9. Because the energy not used for ATP was dissipated as heat, and the subjects developed uncontrollable fevers.

10. (a) A, 4; B, 2; C, 1; D, 3; E, 5.
    (b) 3 divided by 4 (ratio of uptakes in presence and absence of ADP, with substrate present for both measurements).
    (c) 5; substrates depleted.
    (d) 4; ADP level low because ATP level is high.
    (e) 3; rapid ATP production and turnover demands rapid and continuous O_2 uptake.

CHAPTER 16

1. (a) 4 ATPs:

   $$\text{2DHAP} \longrightarrow \text{F1,6BP} \xrightarrow{\text{2ATP}} \text{F6P} \longrightarrow \text{G6P} \longrightarrow \text{G1P} \xrightarrow{\text{2ATP}} \text{UDPG} \longrightarrow \text{glucosyl residue}$$

   (b) 3 (same as above, except 1 P per mole of F1,6BP).
   (c) 12 (see text).
   (d) 2.

2. (a) 2 (glycolysis).
   (b) 6 (gluconeogenesis).

3. (a)(because it is the only substrate that must go through the pyruvate carboxylase reaction).

4. None; the CO_2 that is fixed comes off in the PEP carboxykinase reaction.

5\.

Glucose-1-P + UTP ⟶ UDP-Glc + PP_i; UDP-Glc —(C-4 epimerization)⟶ UDP-Gal

Fructose-6-P ⟶ mannose-6-P ⟶ mannose-1-P + GTP ⟶ GDP-mannose + PP_i

Gal-1-P + CTP ⟶ CDP-Gal + PP_i; CDP-Gal —(reduction of C-2)⟶ —(reduction of C-6)⟶ CDP-D-abequose

Gal-1-P + dTTP ⟶ dTTP-Gal + PP_i; dTTP-Gal —(epimerization of C-3)⟶ —(epimerization of C-5)⟶ —(reduction of C-6)⟶ dTDP-L-rhamnose

6\. (a) Fructose-6-phosphate + ATP ⟶ fructose-2,6-bisphosphate + ADP

(b) 2 Oxaloacetate + 2ATP + 2GTP + 2NADH + $2H^+$ + $4H_2O$ ⟶ glucose + $2CO_2$ + $2NAD^+$ + 2ADP + 2GDP + $4P_i$

(c) Glucose + ATP + UTP ⟶ UDP-Glc + PP_i + ADP

(d) 2 Glycerol +2ATP + $2NAD^+$ + $2H_2O$ ⟶ glucose + 2ADP + 2NADH + $2H^+$ + $2P_i$

(e) 2 Malate + 3ATP + 2GTP + $3H_2O$ ⟶ glucose-6-phosphate + $2CO_2$ + 3ADP + 2GDP + $4P_i$

7\.

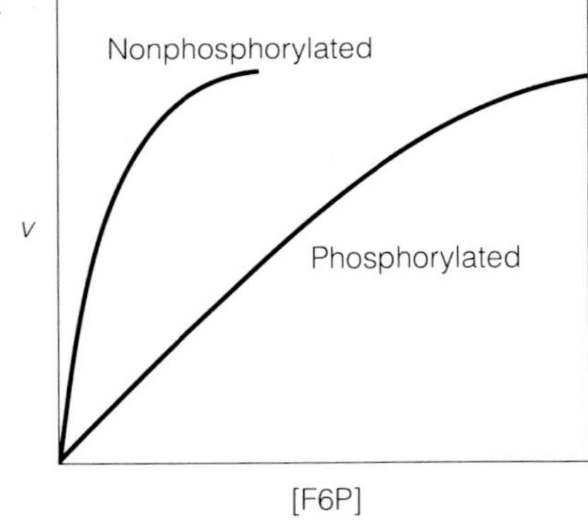

8\.

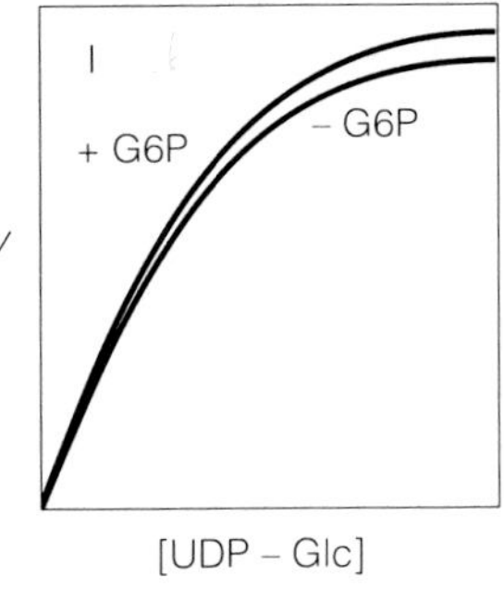

D
+ G6P
– G6P
v
[UDP – Glc]

9\. AMP activation of glycogen phosphorylase *b*; glucose-6-P activation of glycogen synthase D.

10\. 8 ATPs and 4 acetyl-CoAs:

Ac-CoA ⟶ citrate ⟶ isocitrate ⟶ glyoxylate ⟶ malate ⟶ oxaloacetate —(2ATP)⟶ PEP ⟶ 2PG ⟶ 3PG —(2ATP)⟶ 1,3BPG ⟶ G3P ⟶ F1,6BP —(2ATP)⟶ F6P ⟶ G6P ⟶ G1P —(2ATP)⟶ UDPG ⟶ glycosyl residue

CHAPTER 17

1. Palmitic acid, 130; linoleic acid, 142; stearic acid, 147; oleic acid, 145.

2. 40.5% [(130 × 30.5/9788) × 100%].

3. 412 (130 from each palmitate, 22 from glycerol). 412 ATP/51 carbons = 8.1 ATPs per carbon atom (6.3 for glucose).

4. Example: palmitic acid.

   CoA-SH + palmitate + ATP ⟶ palmitoyl-CoA + AMP + PP_i
   PP_i + H_2O ⟶ $2P_i$
   Palmitoyl-CoA + 7CoA-SH + 7FAD + $7NAD^+$ + $7H_2O$ ⟶
   8 acetyl-CoA + $7FADH_2$ + 7NADH + $7H^+$
   8 Acetyl-CoA + $16H_2O$ + $24NAD^+$ + 8FAD + 8ADP + $8P_i$ ⟶
   $16CO_2$ + 24NADH + $24H^+$ + 8ATP + $8FADH_2$ + 8CoA-SH

   Sum: Palmitate + $24H_2O$ + 15FAD + $31NAD^+$ + 8ADP + $6P_i$ ⟶ $16CO_2$ + AMP + $15FADH_2$ + 31NADH + $31H^+$ + 7ATP

   Now add the equations for the metabolic oxidation of NADH and $FADH_2$.

   31NADH + $31H^+$ + $15.5O_2$ + 93ADP + $93P_i$ ⟶
   $31NAD^+$ + 93ATP + $124H_2O$
   $15FADH_2$ + $7.5O_2$ + 30ADP + $30P_i$ ⟶
   15FAD + $45H_2O$ + 30ATP

   Sum: palmitate + $23O_2$ + 131ADP + $129P_i$ ⟶
   $16CO_2$ + AMP + 130ATP + $145H_2O$

5.

Acetoacetyl-CoA ⟶ 2 acetyl-CoA	0ATP
2Acetyl-CoA ⟶ $4CO_2$	24ATP (two turns of citric acid cycle)
Sum:	24ATP
Propionoacetyl-CoA ⟶ propionyl-CoA + acetyl-CoA	5ATP (one cycle of β-oxidation)
Propionyl-CoA ⟶ succinyl-CoA	−2ATP
Succinyl-CoA ⟶ $4CO_2$	24ATP (two turns of citric acid cycle)
Acetyl-CoA ⟶ $2CO_2$	12ATP (one turn of citric acid cycle)
Sum:	39ATP

6. Carnitine acyltransferase I; if inhibition occurred at a later step, then palmitoyl-carnitine oxidation would be inhibited, as well as that of palmitoyl-CoA.

7. The acyl-ACP produced by one subunit undergoes the next round of reductive two-carbon addition on the other subunit.

8. Acetyl-CoA $\xrightarrow{\text{citric acid cycle}}$ oxaloacetate $\xrightarrow{\text{PEP carboxykinase}}$ PEP $\xrightarrow{\text{gluconeogenesis}}$ glucose

   The main point is that the carbons lost in one turn of the citric acid cycle are not the ones that entered as acetyl-CoA in that cycle.

9. Propionyl-CoA ⟶ methylmalonyl-CoA ⟶ succinyl-CoA ⟶ oxaloacetate ⟶ phosphoenolpyruvate ⟶ glucose.

10. Fourteen; two from each molecule of labeled malonyl-CoA.

11. Increase in citrate levels would increase generation of acetyl-CoA in cytosol, hence stimulating fatty acid synthesis.

CHAPTER 18

1. Probably K_{eq} is close to unity, because the bond broken is identical to the bond created.

2. If phosphatidylserine synthase and phosphatidylserine decarboxylase (E_1 and E_2 in Figure 18.3) were juxtaposed, PS could never accumulate because, once formed by E_1, it would immediately react with E_2. If the two enzymes are tightly coupled in the membrane, then addition of radiolabeled PS to an enzyme system would not label PE, because E_2 would act only on PS generated by E_1.

3. By stimulating release of arachidonic acid from membrane phospholipids; these in turn would be converted to prostaglandins, which contribute to inflammation.

4. Substitution of vegetable for animal fats could decrease cholesterol levels; this would ultimately decrease inhibition of HMG-CoA reductase levels by cholesterol, which could result in increased mevalonate levels.

5. Stearic acid + oleic acid + 3ATP + CTP + serine ⟶ *sn*-1-stearoyl-2-oleyl-glycerophosphorylserine + 2AMP + $3PP_i$ + ADP + CMP

6. None.

7. Since this is a nonhydrolyzable analog of glycerol-3-phosphate, you might expect it to be acylated without difficulty to give the phosphonate analog of diacylglycerol. By acting as an analog of diacylglycerol, this could competitively inhibit the synthesis of CDP-diacylglycerol from phosphatidic acid.

8. 25% from the choline utilization pathway, 75% from the phosphatidylserine pathway. To know true intracellular rates of synthesis, you must also know the specific radioactivity of the final intermediate in each pathway, whether pools of intermediates are compartmentalized, and the rate of degradation of the product (see Tools of Biochemistry 16).

9. Acetoacetate + succinyl-CoA ⟶ acetoacetyl-CoA + succinate
   Acetoacetyl-CoA + acetyl-CoA ⟶ HMG-CoA + CoA-SH
   HMG-CoA + 2NADPH + $2H^+$ ⟶ mevalonate + $2NADP^+$ + CoA-SH

CHAPTER 19

1. 14.7%.

2. 1050 nm.

3. (a) Since photosystem II will be blocked, the Z-scheme will not be active. However, cyclic electron flow can continue.

   (b) The cyclic photophosphorylation, which is the sole photosynthesis mechanism in such organisms, utilizes electron transfer to quinones. Hence, the bacteria will be unable to undergo photosynthesis.

4. (a) After being attached to C-2 of ribulose-1,5-bisphosphate, the ^{14}C will become the carboxylate (C-1) of *one* of the two molecules of 3-phosphoglycerate. This becomes the carbonyl carbon (C-1) in G3P, or C-1 in DHAP. Upon condensation to form fructose-1,6-bisphosphate the ^{14}C will show up in carbons 3 and/or 4. (b) No; they may carry 0, 1, or 2 labeled carbons, depending on what combination of labeled and unlabeled trioses has been used.

5. Assuming 48 photons per mole of hexose, the leaf could theoretically produce 0.0263 mol, or 4.73 g of hexose in 1 hour. It will, in fact, produce only a small fraction of this, for not all photons are absorbed, nor do all absorbed serve to pass electrons through the photosynthetic pathway.

6. (a) No.

   (b) Addition of ferricyanide as an electron donor allows a Hill reaction.

7. Carbon 3.

8. Competitive.

9. The data suggest the strong binding of an inhibitor. It is apparently heat stable, so probably not a protein, but is released on ammonium sulfate precipitation. Other data indicate that the inhibitor is a low molecular weight phosphorylated compound.

CHAPTER 20

1. Deoxyadenosyl-B_{12}; H_4folate; ATP + glutamine; α-ketoglutarate + pyridoxal phosphate; *S*-adenosylmethionine.

2. (a) 5-Methyl-H_4folate.

   (b) Methylmalonate.

   (c) Decreased affinity of enzyme C for B_{12} coenzyme.

   (d) Homocystine (see Chapter 21).

3. Example: serine transhydroxymethylase

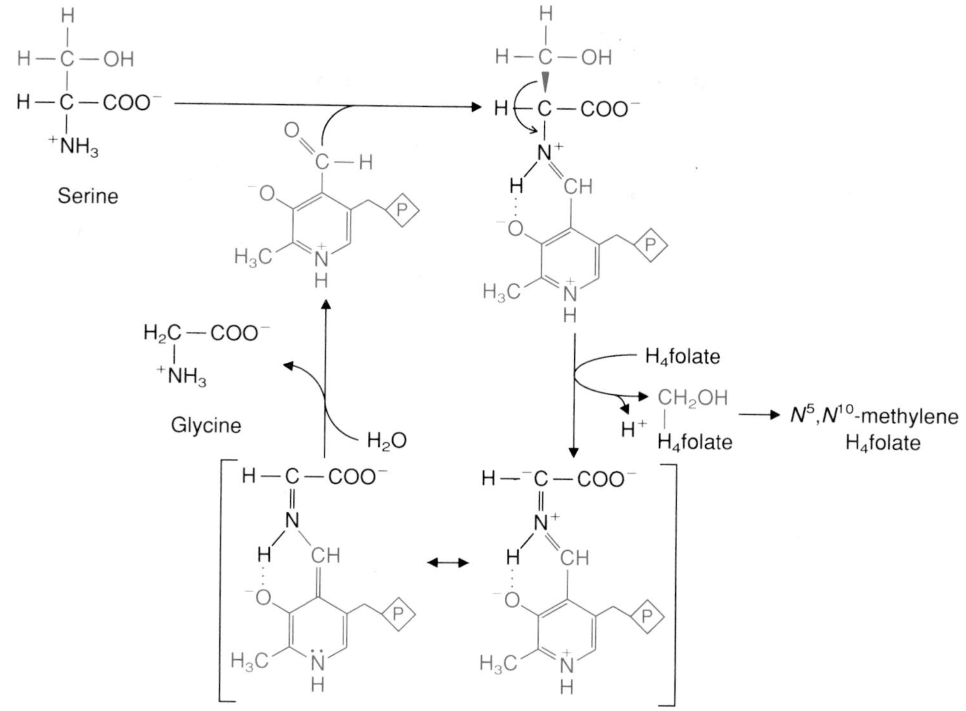

Serine

Glycine

H_2O

H_4folate

N^5,N^{10}-methylene H_4folate

4. Feeding folate will expand the intracellular pools of 5-methyl-H_4folate. This will increase flux through methionine synthetase and, hence, divert the already shrunk store of B_{12}, reducing its availability for other functions.

5. $^{-}OOC^5—C^4H—C^2H—C^1OO^{-}$ with C^3H_3 on C^4 and $\overset{+}{N}H_3$ on C^2

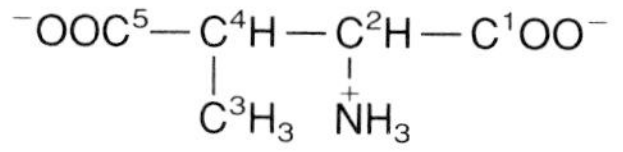

6.

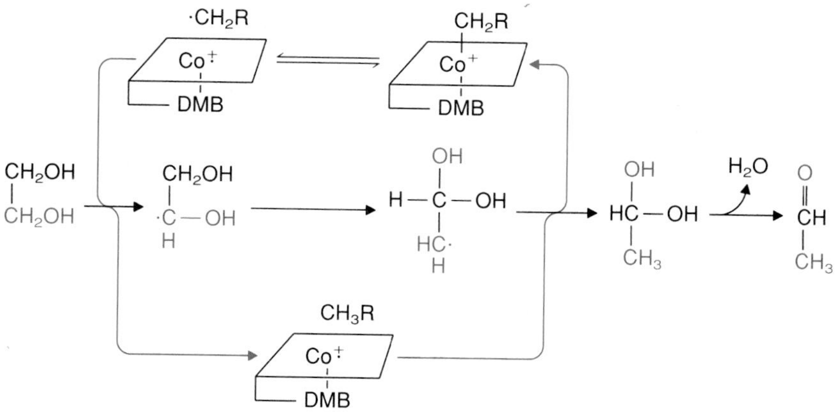

7. One plausible theory proposes that ammonia depletes pools of α-ketoglutarate through the glutamate dehydrogenase and glutamine synthetase reactions, which convert αKG to glutamate and glutamine, and that this diminishes ATP production by reducing flux through the citric acid cycle.

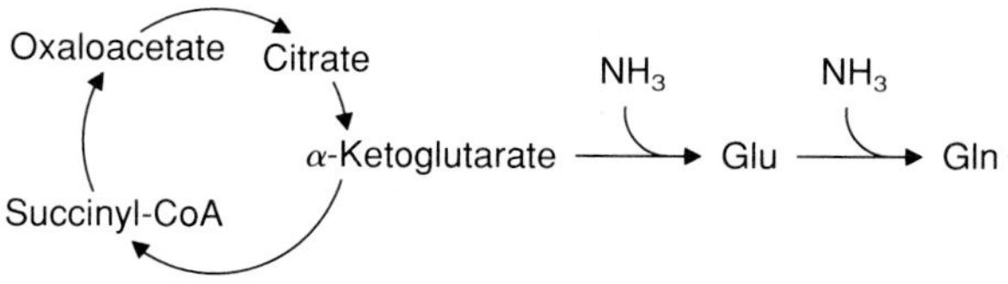

Oxaloacetate → Citrate → α-Ketoglutarate → Succinyl-CoA → Oxaloacetate

α-Ketoglutarate —(NH_3)→ Glu —(NH_3)→ Gln

8. CPS I is in mitochondria, and CPS II is in cytosol. Apparently, carbamoyl phosphate cannot cross the mitochondrial membrane, so that which is formed in mitochondria can be used only for arginine synthesis, and that formed in cytosol is used only for pyrimidine synthesis.

9. The injected enzyme should convert all circulating asparagine to aspartate. Normal cells would take up aspartate and resynthesize asparagine. However, leukemic cells would be unable to synthesize asparagine and would starve. On the other hand, an enzyme injected into the bloodstream would probably have too short a half-life to be effective. Moreover, being a foreign protein, it might cause an immunological reaction.

10. (a) True. The complete catabolism of amino acids yields CO_2, H_2O, and ammonia. However, the ammonia must be converted to urea for detoxification and excretion, and this requires ATP, which decreases the net ATP yield.

 (b) False. Glutamate can be used to synthesize essential amino acids only if the carbon skeletons are available as keto acids.

 (c) False. Arginine biosynthesis in liver is part of the urea cycle, and most of the arginine that is formed is cleaved to urea and ornithine. Little arginine is left over to meet the needs of other tissues.

 (d) False. While it is present in all proteins, alanine can be synthesized by mammalian cells and is not required in the diet.

CHAPTER 21

1. Formiminoglutamate, because the next reaction in its catabolism requires tetrahydrofolate.

2. Arginine —(PyP; CO_2 released)→ agmatine —(H_2O in; urea out)→ putrescine

3. *N*-Acetylglutamate is an intermediate in ornithine biosynthesis. Activity of the urea cycle requires both ornithine and carbamoyl phosphate. If insufficient carbamoyl phosphate is available, ornithine will accumulate, and this could cause accumulation of the precursor, *N*-acetylglutamate. This acts as a signal to stimulate carbamoyl phosphate synthesis to increase urea cycle flux.

4.

Tryptophan —(PyP; CO_2 released)→ indole–CH_2–CH_2–$\overset{+}{N}H_3$ —(O_2, H_4biopterin in; H_2O, H_2biopterin out)→ hydroxylated indole–CH_2–CH_2–$\overset{+}{N}H_3$ —(2AdoMet in; 2AdoHcy out)→ (OH)indole–CH_2–CH_2–$\overset{+}{N}H(CH_3)_2$ —(ATP in; ADP out)→ (Ⓟ–O)indole–CH_2–CH_2–$\overset{+}{N}H(CH_3)_2$

5.

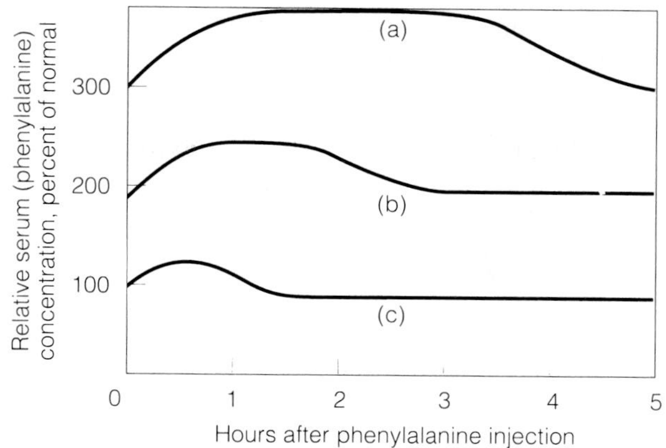

6.

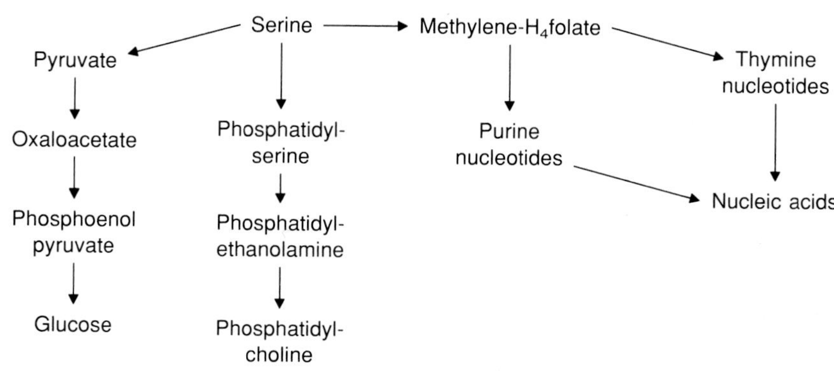

7. Sixteen

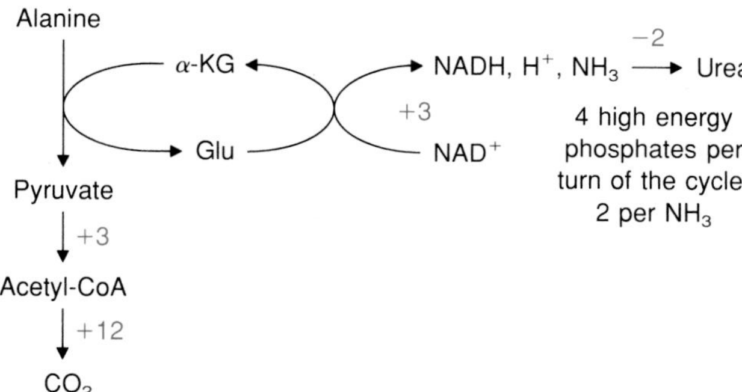

Leucine: 5 ATPs net, leucine ⟶ HMG-CoA.
Tyrosine: 1 ATP net, tyrosine ⟶ fumarate + acetoacetate.
You might expect all of these pathways to generate a little more energy in a fish, because it is not necessary to consume ATP in converting ammonia to urea for excretion.

8. One should be feedback inhibited or have its synthesis repressed by threonine, a second by methionine, and a third by lysine, because the aspartokinase reaction is involved in separate biosynthetic pathways leading to each of these three amino acids.

9.

AdoMet, AdoHcy; AdoMet, AdoHcy

Proline (N–H, COO^-) ⟶ N-methylproline (N–CH_3, COO^-) ⟶ N,N-dimethylproline (N^+, CH_3, CH_3, COO^-)

10. Because the same enzymes are involved in comparable steps of both isoleucine and valine biosynthesis. Threonine dehydratase.

11. Threonine → α-ketobutyrate: inhibited by isoleucine
α-Ketoisovalerate + acetyl-CoA → β-isopropyl malate: inhibited by leucine

Control of valine synthesis is more complicated, because three of the enzymes are involved in synthesis of all three amino acids. One could look for cumulative feedback inhibition—by valine, isoleucine, and leucine—of the first committed reaction: pyruvate + hydroxyethyl-TPP → α-acetolactate.

CHAPTER 22

1. (a) C. (b) E, F. (c) A, B.

2. BrdUrd $\xrightarrow[\text{ATP}]{\text{TK}}$ Br-dUMP $\xrightarrow[\text{ATP}]{}$ Br-dUDP $\xrightarrow[\text{diphosphokinase, ATP}]{\text{Nucleoside}}$ Br-dUTP

Br-dUTP can be incorporated into DNA in place of dTTP. However, it might also inhibit DNA synthesis by acting as a false feedback inhibitor of CDP reduction. Since dTTP is an allosteric inhibitor of CDP reduction by ribonucleotide reductase, Br-dUTP might have a similar effect. This could inhibit DNA replication by causing a dCTP deficiency.

3.

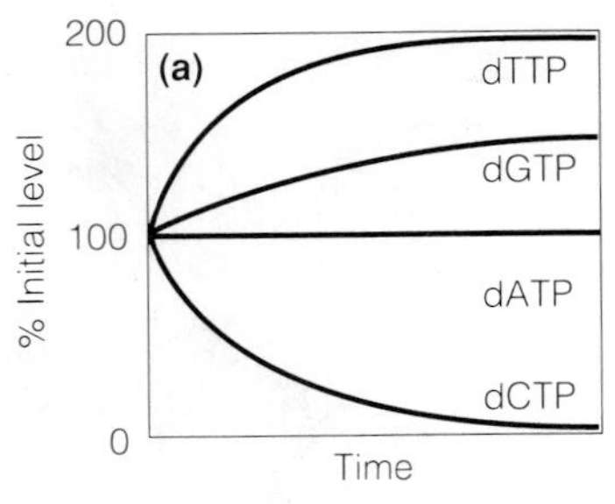

200
100
0
(b)
UTP
CTP
GTP, ATP
% Initial level
Time

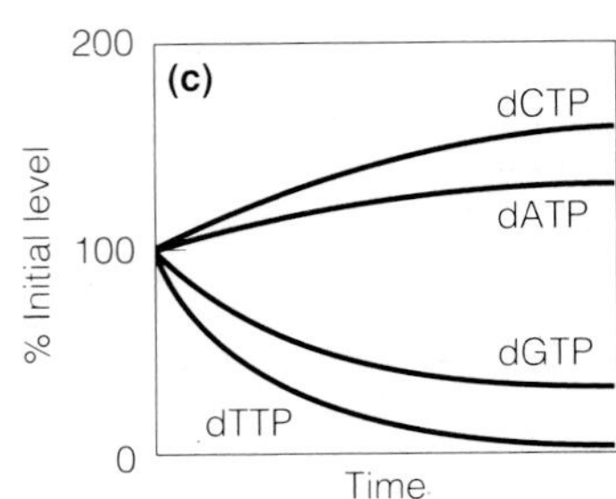

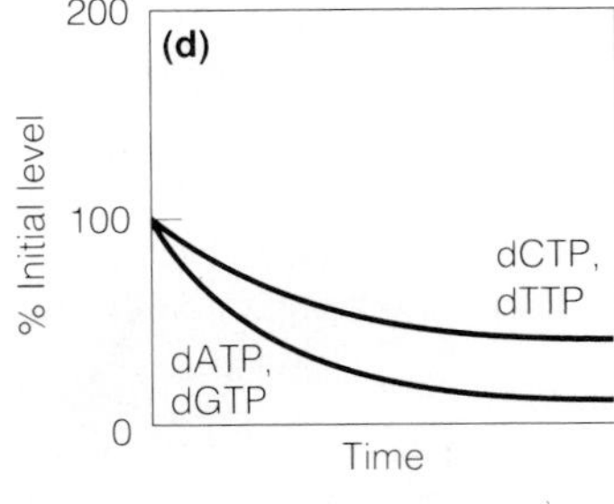

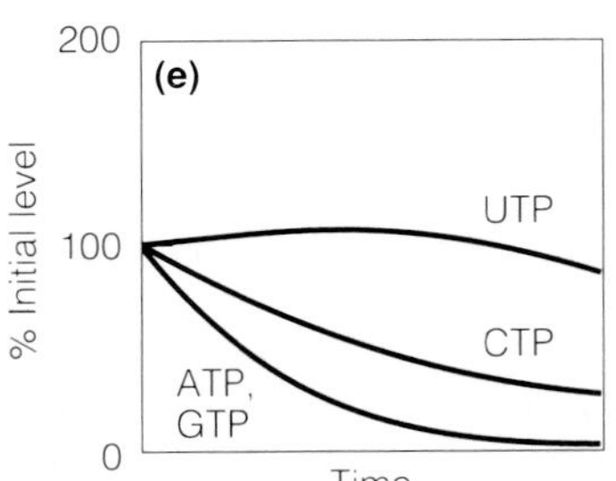

4. (Outline answer); *, sites of allosteric regulation.

Uracil → uridine → UMP → UDP → UTP →* CTP → CDP →* dCDP → dCTP

UDP ↓* dUDP → dUTP → dUMP → dTMP → dTDP

dTDP ↓ dTTP

5. (Outline answer); *, sites of allosteric regulation.

HX → IMP →* XMP → GMP → GDP →* dGDP → dGTP

IMP ↓* adenylosuccinate → AMP → ADP →* dADP → dATP

6. It might lead to adenosine and deoxyadenosine accumulation, as in adenosine deaminase deficiency. This would cause dATP to accumulate and shut off all activities of ribonucleoside diphosphate reductase.

7. Decreased flux through HGPRT could cause the substrates to accumulate, including PRPP. This could increase pyrimidine synthesis at the orotate phosphoribosyltransferase reaction, if PRPP levels are normally subsaturating for that enzyme. Orotate incorporation into nucleotide pools or nucleic acids would give a reasonable estimate of the rate of de novo pyrimidine nucleotide synthesis.

8. (a) Either a deficiency of thymidylate synthase activity, measured in a cell-free extract, or a growth requirement of cells for thymidine would confirm that the mutants are deficient in this enzyme.
   (b) Since the mutants have negligible flux through the thymidylate synthase reaction, they do not deplete their intracellular tetrahydrofolate. Thymidine satisfies the need for synthesis of thymine nucleotides.
   (c) Use methotrexate instead of trimethoprim, because trimethoprim is not an effective inhibitor of mammalian dihydrofolate reductase.

9. Mutation in DHFR gene that makes the enzyme resistant; mutation in transcriptional or translational control mechanism that causes overproduction of normal DHFR; transport defect that causes failure of methotrexate to be taken up by cell.

10. (a) Inhibitors of thymidylate synthase and dUTPase specified by the viral genome.
    (b) Virus-specified dUMP hydroxymethylase and hm-dUMP kinase:

$$\text{dUMP} \xrightarrow{CH_2{=}FH_4 \;\to\; FH_4} \text{hm-dUMP} \xrightarrow{ATP \;\to\; ADP} \text{hm-dUDP} \xrightarrow{ATP \;\to\; ADP} \text{hm-dUTP}$$

    (c) Virus-specified dCMP methylase, e.g.,

$$\text{dCMP} \xrightarrow{CH_2{=}FH_4 \;\to\; FH_2} CH_3\text{-dCMP} \xrightarrow{ATP \;\to\; ADP} CH_3\text{-dCDP} \xrightarrow{ATP \;\to\; ADP} CH_3\text{-dCTP}$$

11. (a) Mutants with a defect in DNA synthesis (no thymidine incorporated into DNA).
    (b) Mutants defective in thymidine kinase, thymidylate synthase, dihydrofolate reductase, or serine transhydroxymethylase can be selected for by their resistance to decay of incorporated [^{3}H]deoxyuridine.

12. Show that the putative alarmone *does* affect the rate of a target metabolic process, that mutants which lose the ability to respond to the stress also lose their ability to respond to the putative alarmone, that both abilities are regained in a single metabolic event, and that intracellular concentrations of the putative alarmone are in the range known to control the target process in vitro.

CHAPTER 23

1. It is presumed that starch increases blood glucose levels less than do simple sugars. Thus, there is less stimulation of insulin secretion. Insulin would tend to retard energy mobilization from intracellular stores—something not desirable during a marathon.

2. About 2.5 hours.

3. About 28% at rest, about one-tenth that value during a marathon.

4. The measured proteolysis represents the sum of rates of protein synthesis and breakdown. The actual rate of protein breakdown does not rise, but in the early stage of a fast, the utilization of amino acids in catabolic pathways reduces the concentrations needed to support protein synthesis at rates needed to counterbalance breakdown. Later, as fatty acids and ketones are used more for energy, amino acids are spared for this purpose and are more readily available to be used for protein synthesis.

5. Glucosylated hemoglobin accumulates in the blood of diabetics and can easily be measured. Because the glucosylation reaction is covalent, the level of glucosylated hemoglobin reflects the level of blood glucose over a period of time, while a simple determination of glucose reflects only the value at the time of blood sampling. Thus, a diabetic patient can be monitored with many fewer blood drawings and with greater accuracy if glucosylated hemoglobin is measured.

6. Liver contains low levels of the enzyme that synthesizes acetoacetyl-CoA from acetoacetate, ATP, and CoA-SH. Therefore, when liver synthesizes ketone bodies, they cannot readily be activated for catabolism within the hepatocyte; instead, they are released and ultimately utilized by other tissues.

7. Using recombinant DNA techniques (Chapter 25), one could prepare nucleic acid sequences that encode either an entire α subunit of a G protein or a GTP-binding domain and ask whether that "probe" can hybridize with DNA from a plant species of interest. Alternatively, one could ask whether antibodies against mammalian G proteins cross-react with any proteins in a plant extract. If so, the protein could be isolated and its properties analyzed (including its ability to bind guanine nucleotides). Another approach is to look for proteins that can be ADP-ribosylated by cholera toxin or pertussis toxin.

8. Some possible reasons: storage, control of activity, proper folding of the polypeptide chain, signal sequences for direction to the proper part of the cell.

9. Because prostaglandins are lipid derivatives and, hence, quite hydrophobic, it is likely that they would traverse the membrane and bind to intracellular receptors. Therefore, you might examine cytosolic extracts for prostaglandin binding activity. On the other hand, since some prostaglandins act through second messengers, the actions of their receptors may be different from those of steroid hormones.

10. Ethylene; as a gas, it would probably be impossible to prepare an antibody against it.

11. Since steroid receptor–hormone complexes activate the transcription of specific genes, it is conceivable that a steroid hormone could activate the transcription of the gene for adenylate cyclase and, hence, increase the steady-state level of this enzyme.

CHAPTER 24

1.

Use a 5′ end-labeled primer, and run a polymerase reaction in DNA excess (so that any polymerase that dissociates will rebind to a new chain). Stop reactions after a brief incubation and determine the molecular weight of radioactive material by gel electrophoresis. M_r tells how many nucleotides were incorporated per chain, which gives the processivity (number of nucleotides incorporated per unit time per chain).

2.

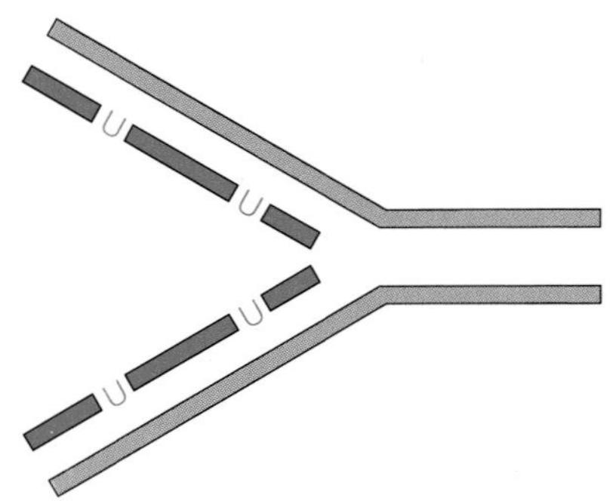

Incorporation of dUMP generates sites for chain breakage, as the uracil replacement process begins. If the single-strand breaks had not been fully repaired when DNA was isolated, short fragments would be seen. Fewer of these would be seen if the experiment were done with a mutant defective in uracil-DNA-*N*-glycosylase—and more if the experiment were done with a dUTPase-negative mutant.

3. (a) 1.725 g/cm^3.
(b) 1.772 g/cm^3.

4. (a) To eliminate virtually all further incorporation of radioisotope.
   (b) If incorporated as ribonucleotides, the labeled product will be hydrolyzed by mild alkali; if incorporated as deoxyribonucleotides, the product is alkali stable.
   (c) RNA primers are metabolically unstable; they are rapidly degraded, as they are replaced by deoxyribonucleotides in DNA.

5. (a) Inosinic acid.
   (b) Conversion of an A-T to a G-C base pair.

6. (a) 0.016 mm.
   (b) 9.79×10^4 nucleotides per fork per minute.
   (c) 1250 origins (2500 replication forks).
   (d) 2937 kilobase pairs.

7. (a) It's the reversal of the DNA ligase reaction; single-strand interruption relaxes supercoiled DNA.

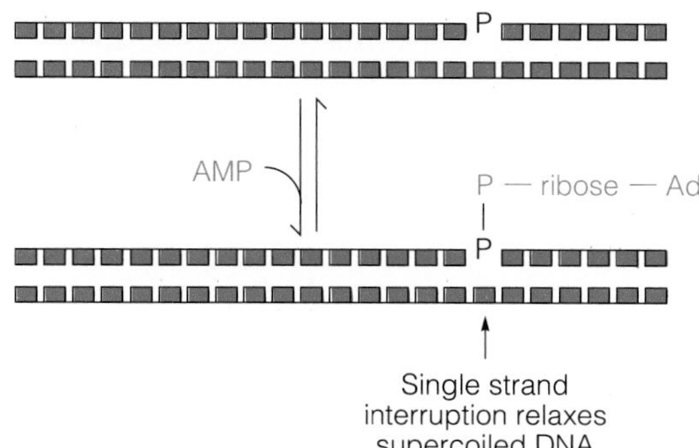

   (b) Analyze the DNA by gel electrophoresis.

8. Synchronize cells and label with 5-bromodeoxyuridine either early or late in S phase. Separate replicated DNA by CsCl equilibrium centrifugation. This can be analyzed by hybridization with cloned DNAs from genes known to be expressed or not expressed in those cells.

9.

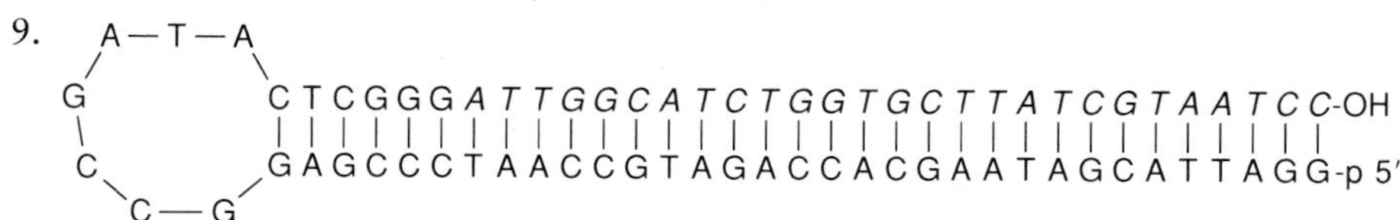

(Note. The 3′ exonuclease removes the mismatched A and T before polymerase action begins.)

10. The polymerase activity was found to extend more slowly from an improperly paired terminus. Thus, a mismatched base has a longer residence time at the 3′ terminus than a matched base, increasing the likelihood of cleavage by the exonuclease activity.

11. Dideoxyinosine is probably converted to the corresponding dideoxyribonucleoside triphosphate, ddITP. Incorporation of this nucleotide in place of dGTP would block further chain elongation, because of the absence of a 3′-OH terminus. If the AIDS reverse transcriptase incorporates ddITP more readily than do the cellular DNA polymerases, then replication of the viral genome would be selectively inhibited.

12. If the mutation affected one of the feedback control sites of ribonucleotide reductase, the cell could overproduce dNTPs to the extent that competitive inhibition in vivo of replicative DNA polymerases by aphidicolin could be overcome.

CHAPTER 25

1. Flush ends: *Eco*RII, *Hae*III, *Hind*II, *Hpa*I, *Sma*I. Not made flush by DNA polymerase (because the recessed end lacks a 3′ hydroxyl): *Hha*I, *Hga*I, *Pst*I, *Ple*I.

2. Because both strands are methylated.

3. dUTPase deficiency increases dUMP incorporation into DNA and increases subsequent excision repair. Ligase deficiency increases the mean lifetime of Okazaki fragments. Both conditions increase the number of single-strand interruptions, the structure that initiates recombination.

4. (a) To linearize circular DNA, so that its electrophoretic mobility can be compared with those of other linear DNA fragments.
   (b) The recombinant DNA contains two tandem 1.15-kb inserts.
   (c)

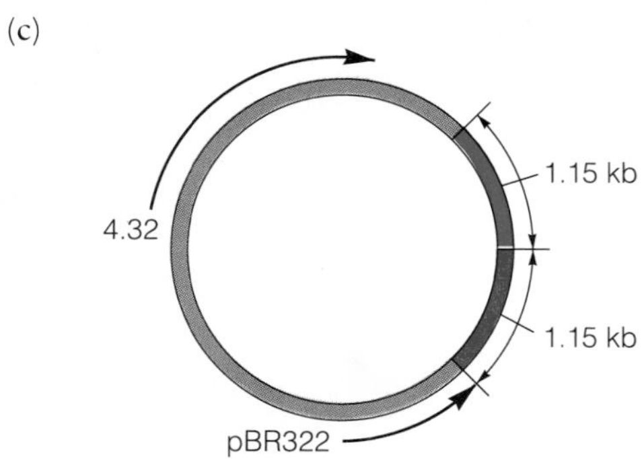

5. (a) Circular.
   (b)

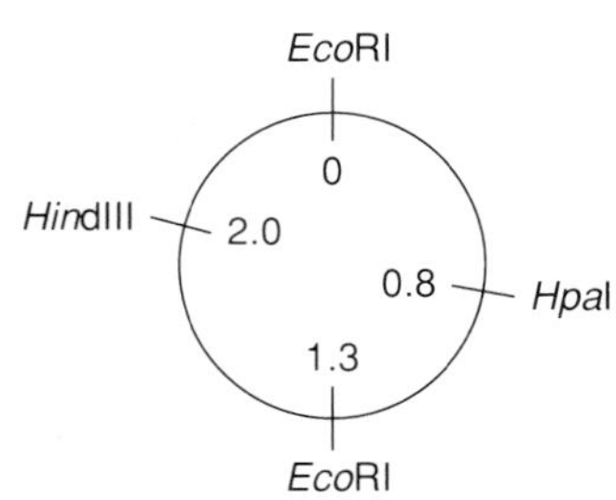

   (c) 7.
   (d) Cleave with *Hpa*II plus *Hind*III.

6. (a)

P P P P P P OH
A G A T C T
T C T A G A
HO P P P P P P

   (b) A and B.
   (c) E and F.
   (d) F.

7.

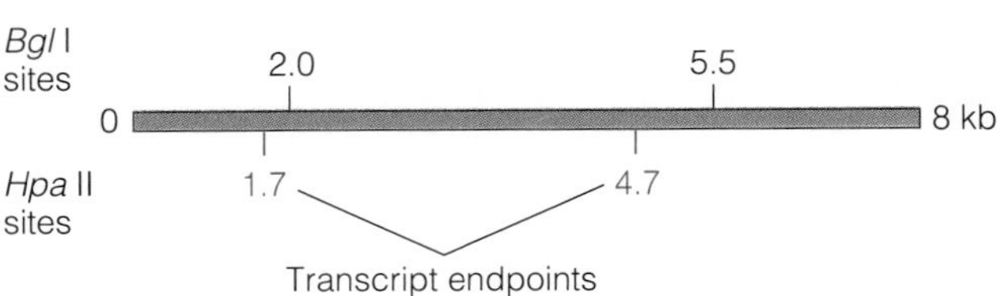

It would be laborious, but you could separate the DNA strands and determine which strand hybridizes to the RNA; this would determine the orientation of transcription, since you know the DNA 3′ and 5′ ends.

8. First, clone a large number of DNA fragments from cells containing the integrated virus. Next, carry out colony hybridization on these clones, using labeled viral DNA as the hybridization probe. From Southern blot analysis of clones containing viral DNA, identify viral DNA fragments that are larger than corresponding fragments from the nonintegrated double-strand viral DNA. These should represent viral sequences fused to adjacent cellular sequences. These can now be subcloned and sequenced.

9. To quickly identify the desired clones.

10.

Gly
GTA GGC TTA CCT CGA GCT
*Hind*III *Ava*I
Site lost Site gained

11. The methyl-directed mismatch repair system corrects many of the deliberately engineered mismatches. Also, many of the in vitro chain extension reactions do not proceed entirely around the circular template, as they must to undergo ligation.

CHAPTER 26

1. Pulse label cells with a radioactive RNA precursor, isolate RNA, and hydrolyze with mild alkali. This will yield one nucleoside per 3′ terminus. Total pmoles of labeled nucleotide in the hydrolysate divided by the number of pmoles of nucleosides gives the number of nucleotides incorporated per chain in the labeling interval.

2. First, devise conditions under which only the cloned gene is synthesized. You could initiate transcription by adding three ribonucleotides. At time zero add the fourth nucleotide and stop incubation after a fixed time. Use a small, measured amount of template so that the number of transcribing genes is roughly equal to the number of template molecules. Determine the average length of transcripts synthesized by gel electrophoresis.

3. By gel electrophoresis of radiolabeled RNA products. Any paused species will accumulate and can be detected as a heavier-than-expected band on a radioautograph of the gel.

4. (a) $\dfrac{(10 \text{ molecules/cell})/6.02 \times 10^{23} \text{ molecules/mol})}{0.3 \times 10^{-15} \text{ liters/cell}} =$ 5.5×10^{-8} M.

   (b) $\dfrac{10 \text{ molecules/cell}}{5.5 \times 10^{-8} \text{ M}} = \dfrac{2 \text{ molecules/cell}}{1.1 \times 10^{-8} \text{ M}}$

   (c) Approximately 2.5×10^{-14} M.

   (d) A rapidly growing cell could have two copies, on average, because each single chromosome is partly replicated.

5. It's unlikely, because in eukaryotes transcription occurs in the nucleus and translation in the cytosol. Thus, direct coupling between transcription and translation, essential for attenuation, is absent.

6. This can be a way to control total RNA synthesis when ribonucleotide pools are depleted, by controlling initiation and preventing the wasteful incorporation of nucleotides into RNA molecules that might not be completed.

7. A dinucleotide, complementary to the first two template nucleotides, can bypass the first nucleotide incorporation step and permit efficient transcription at low nucleotide concentrations.

8. Using footprinting, carry out competitive binding experiments to ask whether repressor inhibits RNA polymerase binding and vice versa. Also, examine transcription in vitro to determine whether transcription extends as far as the repressor binding site.

9. Because it uses the −35 region of *trp* and the −10 region of *lac*, the hybrid *tac* promoter more closely matches the consensus −35 and −10 sequences than does either *lac* or *trp*.

	Number of identities with consensus sequence
lac	9/12
trp	8/12
tac	10/12

10. (a) The plasmid *i* gene encodes a repressor that cannot bind inducer.

    (b) The plasmid operator cannot bind repressor.

11. Mu inserts randomly in the genome (Chapter 25). If it is inserted downstream from a damage-inducible promoter, UV irradiation of that cell will activate β-galactosidase synthesis from the integrated phage genome.

12. Actinomycin is a noncompetitive inhibitor because it binds to DNA, not to a substrate-binding site on the enzyme. Cordycepin acts as the nucleotide, 3′-deoxyATP. This should compete with ATP for utilization by RNA polymerase.

CHAPTER 27

1. . . .Ile–Tyr–Ile–Tyr–Ile–Tyr–. . . .

2. The mitochondrial code uses AUA to code for *Met*, not *Ile*.

3. (a) Some reading frames for some polynucleotides will be read as repeats to stop codons.

   (b) $(GUA)_n$ — Val_n or Ser_n (third reading frame is UAG = stop).
   $(UUA)_n$ — Leu_n or Tyr_n or Ile_n.

4. Any polynucleotide built from a tetranucleotide repeat, unless a stop codon is involved.

5. (a) Should block translation, since start signal will no longer be properly aligned.

   (b) Should have no effect, from what we know at present.

6. (a) $^{5'}$GGU$^{3'}$, $^{5'}$GGC$^{3'}$, or $^{5'}$GGA$^{3'}$; all are Gly codons.

   (b) $^{5'}$AGU$^{3'}$ or $^{5'}$AGC$^{3'}$; both are Ser codons.

7. If δ is the probability that an error is committed at each step, then $1 - \delta$ is the probability that an error has *not* been made in any one step. The probability that no error has been made in *any* of the n steps is then $(1 - \delta)^n$. This is the probability that the protein is entirely error-free.

8. (a) 0.990.

   (b) 0.904.

9. 36/3 = 12.

10. Possible mechanisms might include (a) the protein acting as a repressor of its own gene and (b) the protein, on binding to its own message, stimulating nuclease digestion of that message.

11. Each puromycin molecule becomes bound to the polypeptide chain it has aborted and is hence used only once. An erythromycin molecule can block an entire polyribosome and can be used repeatedly.

12. Cutting a phosphodiester bond near the 3′ end of 16S RNA might (and in fact does) remove the sequence that binds the Shine-Dalgarno sequence on the messenger RNA.

CHAPTER 28

1. (a) 3.3×10^{-13} grams = 0.33 picograms.
   (b) 3.4 cm.
   (c) 7×10^3.

2. If we take the average nucleosome (including linker) to contain about 200 bp, we get a compaction ratio of 40 for the 30-nm solenoidal fiber. Obviously, this must be greatly folded on itself to yield the overall ratio of over about 7000 found in the metaphase chromosome.

3. (a) If the origin is at the center, we note that each fork of the enlarging bubble need only replicate half the DNA, so 5.79 days would be required. Any other position of the origin would increase the time.
   (b) About 8.3 minutes. This assumes, however, that all replicating regions do so simultaneously. It is known that some regions replicate before others, which will increase the total time required for S phase.

4. A very specific recognition is required and perhaps a proofreading step as well. As in any biochemical process where error rates must be very low, accuracy surely can be purchased by free energy (compare translation).

5. Histones are required in very large amounts during only one brief period (the beginning of S phase) in the cell cycle. Furthermore, they are always required in equivalent amounts, to make nucleosomes. The presence of multiple copies aids in rapid transcription, and an equal number in each domain should help maintain balance in production. The absence of introns and poly(A) tails means that two of the major steps involved in processing most eukaryotic genes are avoided, so that histone mRNAs can be delivered rapidly to the cytosol.

6. One way would be to measure the distribution of poly A tail lengths in bulk mRNA. If a distribution was found that cut off at a particular minimal value, the hypothesis would be supported.

7. Step 1 forms a high-energy thioester bond. Step 4 is a reaction requiring specific recognition, and perhaps a proofreading step.

8. Using a graph of log (bp) vs. d for the markers, one can interpolate the chromatin fragments to obtain sizes of 185, 380, 578, and 772 bp. A repeat of about 195 bp is indicated, with the monosome being slightly degraded.

9. The simplest explanation comes from the hypothesis (since proved) that the DNA lies on the surface of the particle. Thus, each strand is maximally exposed approximately every 10 residues and is most susceptible to nicking at these periodically spaced points.

10. According to Table 28.6, proteins with N-terminal methionines are exceptionally stable. If *all* retained them, certain proteins that normally turn over rapidly would become long-lived.

CHAPTER 29

1. (a) -60 mV (right $-$).
   (b) $+60$ mV (right $+$).
   (c) 0.

2. No in (a) and (b). Yes in (c). Concentrations will equalize.

3. (a) 0.28.
   (b) -72.9 mV.

4. (a) In relaxed state, there is about 0.25 μm overlap on each side; in contracted state, about 0.75 μm.
   (b) ~67 steps.

5. (a) 0.0033 min $\cong$ 0.2 s.
   (b) 0.021 min = 1.2 s.
   (c) An additional 0.12 min.

6. 44.1 kJ/mol.

7. If the GTPase activity is slow, microtubules that happen to have GTP ends will be able to pick up GTP-tubulin and grow for a considerable time, until the GTP at the end is hydrolyzed. These will then tend to shrink as long as release of GDP-tubulin continues to reveal new GDP ends.

8. (a) Preference for counterclockwise rotation.
   (b) Preference for clockwise rotation.

Index

Page numbers in boldface refer to a definition and major discussion of the entry. F after a page number indicates a figure, T a table, and n a note.